Environmental Encyclopedia

Third Edition

Environmental Encyclopedia

Third Edition

Volume 1
A-M

Marci Bortman, Peter Brimblecombe,
Mary Ann Cunningham, William P. Cunningham,
and Bill Freedman, *Editors*

GALE®

THOMSON
™
GALE

Detroit • New York • San Diego • San Francisco • Cleveland • New Haven, Conn. • Waterville, Maine • London • Munich

THOMSON
GALE

Environmental Encyclopedia 3
Marci Bortman, Peter Brimblecombe, Bill Freedman, Mary Ann Cunningham, William P. Cunningham

Project Coordinator
Jacqueline L. Longe

Editorial
Deirdre S. Blanchfield, Madeline Harris, Chris Jeryan, Kate Kretschmann, Mark Springer, Ryan Thomason

Editorial Systems Support
Andrea Lopeman

Permissions
Shalice Shah-Caldwell

Imaging and Multimedia
Robert Duncan, Mary Grimes, Lezlie Light, Dan Newell, David Oblender, Christine O'Bryan, Kelly A. Quin

Product Design
Michelle DiMercurio, Tracey Rowens, Jennifer Wahi

Manufacturing
Evi Seoud, Rita Wimberley

ISBN 0-7876-5486-8 (set), ISBN 0-7876-5487-6 (Vol. 1),
ISBN 0-7876-5488-4 (Vol. 2), ISSN 1072-5083

Printed in the United States of America
10 9 8 7 6 5 4

CONTENTS

A

Abbey, Edward
Absorption
Acclimation
Accounting for nature
Accuracy
Acetone
Acid and base
Acid deposition
Acid mine drainage
Acid rain
Acidification
Activated sludge
Acute effects
Adams, Ansel
Adaptation
Adaptive management
Adirondack Mountains
Adsorption
Aeration
Aerobic
Aerobic sludge digestion
Aerobic/anaerobic systems
Aerosol
Aflatoxin
African Wildlife Foundation
Africanized bees
Agency for Toxic Substances and Disease Registry
Agent Orange
Agglomeration
Agricultural chemicals
Agricultural environmental management

Agricultural pollution
Agricultural Research Service
Agricultural revolution
Agricultural Stabilization and Conservation Service
Agroecology
Agroforestry
AIDS
Air and Waste Management Association
Air pollution
Air pollution control
Air pollution index
Air quality
Air quality control region
Air quality criteria
Airshed
Alar
Alaska Highway
Alaska National Interest Lands Conservation Act (1980)
Albedo
Algal bloom
Algicide
Allelopathy
Allergen
Alligator, American
Alpha particle
Alternative energy sources
Aluminum
Amazon basin
Ambient air
Amenity value
American box turtle
American Cetacean Society
American Committee for International Conservation
American Farmland Trust
American Forests
American Indian Environmental Office
American Oceans Campaign
American Wildlands
Ames test
Amoco Cadiz
Amory, Cleveland

ADVISORY BOARD

A number of recognized experts in the library and environmental communities provided invaluable assistance in the formulation of this encyclopedia. Our panel of advisors helped us shape this publication into its final form, and we would like to express our sincere appreciation to them:

Dean Abrahamson: Hubert H. Humphrey Institute of Public Affairs, University of Minnesota, Minneapolis, Minnesota

Maria Jankowska: Library, University of Idaho, Moscow, Idaho

Terry Link: Library, Michigan State University, East Lansing, Michigan

Holmes Rolston: Department of Philosophy, Colorado State University, Fort Collins, Colorado

Frederick W. Stoss: Science and Engineering Library, State University of New York—Buffalo, Buffalo, New York

Hubert J. Thompson: Conrad Sulzer Regional Library, Chicago, Illinois

CONTRIBUTORS

Margaret Alic, Ph.D.: Freelance Writer, Eastsound, Washington

William G. Ambrose Jr., Ph.D.: Department of Biology, East Carolina University, Greenville, North Carolina

James L. Anderson, Ph.D.: Soil Science Department, University of Minnesota, St. Paul, Minnesota

Monica Anderson: Freelance Writer, Hoffman Estates, Illinois

Bill Asenjo M.S., CRC: Science Writer, Iowa City, Iowa

Terence Ball, Ph.D.: Department of Political Science, University of Minnesota, Minneapolis, Minnesota

Brian R. Barthel, Ph.D.: Department of Health, Leisure and Sports, The University of West Florida, Pensacola, Florida

Stuart Batterman, Ph.D.: School of Public Health, University of Michigan, Ann Arbor, Michigan

Eugene C. Beckham, Ph.D.: Department of Mathematics and Science, Northwood Institute, Midland, Michigan

Milovan S. Beljin, Ph.D.: Department of Civil Engineering, University of Cincinnati, Cincinnati, Ohio

Heather Bienvenue: Freelance Writer, Fremont, California

Lawrence J. Biskowski, Ph.D.: Department of Political Science, University of Georgia, Athens, Georgia

E. K. Black: University of Alberta, Edmonton, Alberta, Canada

Paul R. Bloom, Ph.D.: Soil Science Department, University of Minnesota, St. Paul, Minnesota

Gregory D. Boardman, Ph.D.: Department of Civil Engineering, Virginia Polytechnic Institute and State University, Blacksburg, Virginia

Marci L. Bortman, Ph.D.: The Nature Conservancy, Huntington, New York

Pat Bounds: Freelance Writer,

Peter Brimblecombe, Ph.D.: School of Environmental Sciences, University of East Anglia, Norwich, United Kingdom

Kenneth N. Brooks, Ph.D.: College of Natural Resources, University of Minnesota, St. Paul, Minnesota

Peggy Browning: Freelance Writer,

Marie Bundy: Freelance Writer, Port Republic, Maryland

Ted T. Cable, Ph.D.: Department of Horticulture, Forestry and Recreation Resources, Kansas State University, Manhattan, Kansas

John Cairns Jr., Ph.D.: University Center for Environmental and Hazardous Materials Studies, Virginia Polytechnic Institute and State University, Blacksburg, Virginia

Liane Clorfene Casten: Freelance Journalist, Evanston, Illinois

Ann S. Causey: Prescott College, Prescott, Arizona

Ann N. Clarke: Eckenfelder Inc., Nashville, Tennessee

David Clarke: Freelance Journalist, Bethesda, Maryland

Sally Cole-Misch: Freelance Writer, Bloomfield Hills, Michigan

Edward J. Cooney: Patterson Associates, Inc., Chicago, Illinois

Terence H. Cooper, Ph.D.: Soil Science Department, University of Minnesota, St. Paul, Minnesota

Gloria Cooksey, C.N.E.: Freelance Writer, Sacramento, California

Mark Crawford: Freelance Writer, Toronto, Ontario, Canada

Neil Cumberlidge, Ph.D.: Department of Biology, Northern Michigan University, Marquette, Michigan

John Cunningham: Freelance Writer, St. Paul, Minnesota

Mary Ann Cunningham, Ph.D.: Department of Geology and Geography, Vassar College, Poughkeepsie, New York

William P. Cunningham, Ph.D.: Department of Genetics and Cell Biology, University of Minnesota, St. Paul, Minnesota

Richard K. Dagger, Ph.D.: Department of Political Science, Arizona State University, Tempe, Arizona

Tish Davidson, A.M.: Freelance Writer, Fremont, California

Stephanie Dionne: Freelance Journalist, Ann Arbor, Michigan

Frank M. D'Itri, Ph.D.: Institute of Water Research, Michigan State University, East Lansing, Michigan

Teresa C. Donkin: Freelance Writer, Minneapolis, Minnesota

David A. Duffus, Ph.D.: Department of Geography, University of Victoria, Victoria, British Columbia, Canada

Douglas Dupler, M.A.: Freelance Writer, Boulder, Colorado

Cathy M. Falk: Freelance Writer, Portland, Oregon

L. Fleming Fallon Jr., M.D., Dr.P.H.: Associate Professor, Public Health, Bowling Green State University , Bowling Green, Ohio

George M. Fell: Freelance Writer, Inver Grove Heights, Minnesota

Gordon R. Finch, Ph.D.: Department of Civil Engineering, University of Alberta, Edmonton, Alberta, Canada

Paula Anne Ford-Martin, M.A.: Wordcrafts, Warwick, Rhode Island

Janie Franz: Freelance Writer, Grand Forks, North Dakota

Bill Freedman, Ph.D.: School for Resource and Environmental Studies, Dalhousie University, Halifax, Nova Scotia, Canada

Rebecca J. Frey, Ph.D.: Writer, Editor, and Editorial Consultant, New Haven, Connecticut

Cynthia Fridgen, Ph.D.: Department of Resource Development, Michigan State University, East Lansing, Michigan

Andrea Gacki: Freelance Writer, Bay City, Michigan

Brian Geraghty: Ford Motor Company, Dearborn, Michigan

Robert B. Giorgis, Jr.: Air Resources Board, Sacramento, California

Debra Glidden: Freelance American Indian Investigative Journalist, Syracuse, New York

Eville Gorham, Ph.D.: Department of Ecology, Evolution and Behavior, University of Minnesota, St. Paul, Minnesota

Darrin Gunkel: Freelance Writer, Seattle, Washington

Malcolm T. Hepworth, Ph.D.: Department of Civil and Mineral Engineering, University of Minnesota, Minneapolis, Minnesota

Katherine Hauswirth: Freelance Writer, Roanoke, Virginia

Richard A. Jeryan: Ford Motor Company, Dearborn, Michigan

Barbara J. Kanninen, Ph.D.: Hubert H. Humphrey Institute of Public Affairs, University of Minnesota, Minneapolis, Minnesota

Christopher McGrory Klyza, Ph.D.: Department of Political Science, Middlebury College, Middlebury, Vermont

John Korstad, Ph.D.: Department of Natural Science, Oral Roberts University, Tulsa, Oklahoma

Monique LaBerge, Ph.D.: Research Associate, Department of Biochemistry and Biophysics, University of Pennsylvania, Philadelphia, Pennsylvania

Royce Lambert, Ph.D.: Soil Science Department, California Polytechnic State University, San Luis Obispo, California

William E. Larson, Ph.D.: Soil Science Department, University of Minnesota, St. Paul, Minnesota

Ellen E. Link: Freelance Writer, Laingsburg, Michigan

Sarah Lloyd: Freelance Writer, Cambria, Wisconsin

James P. Lodge Jr.: Consultant in Atmospheric Chemistry, Boulder, Colorado

William S. Lynn, Ph.D.: Department of Geography, University of Minnesota, Minneapolis, Minnesota

Alair MacLean: Environmental Editor, OMB Watch, Washington, DC

Alfred A. Marcus, Ph.D.: Carlson School of Management, University of Minnesota, Minneapolis, Minnesota

Gregory McCann: Freelance Writer, Freeland, Michigan

Cathryn McCue: Freelance Journalist, Roanoke, Virginia

Mary McNulty: Freelance Writer, Illinois

Jennifer L. McGrath: Freelance Writer, South Bend, Indiana

Robert G. McKinnell, Ph.D.: Department of Genetics and Cell Biology, University of Minnesota, St. Paul, Minnesota

Nathan H. Meleen, Ph.D.: Engineering and Physics Department, Oral Roberts University, Tulsa, Oklahoma

Liz Meszaros: Freelance Writer, Lakewood, Ohio

Muthena Naseri: Moorpark College, Moorpark, California

B. R. Niederlehner, Ph.D.: University Center for Environmental and Hazardous Materials Studies, Virginia Polytechnic Institute and State University, Blacksburg, Virginia

David E. Newton: Instructional Horizons, Inc., San Francisco, California

Robert D. Norris: Eckenfelder Inc., Nashville, Tennessee

Teresa G. Norris, R.N.: Medical Writer, Ute Park, New Mexico

Karen Oberhauser, Ph.D.: University of Minnesota, St. Paul, Minnesota

Stephanie Ocko: Freelance Journalist, Brookline, Massachusetts

Kristin Palm: Freelance Writer, Royal Oak, Michigan

James W. Patterson: Patterson Associates, Inc., Chicago, Illinois

Paul Phifer, Ph.D.: Freelance Writer, Portland, Oregon

Jeffrey L. Pintenich: Eckenfelder Inc., Nashville, Tennessee

Douglas C. Pratt, Ph.D.: University of Minnesota: Department of Plant Biology, Scandia, Minnesota

Jeremy Pratt: Institute for Human Ecology, Santa Rosa, California

Klaus Puettman: University of Minnesota, St. Paul, Minnesota

Stephen J. Randtke: Department of Civil Engineering, University of Kansas, Lawrence, Kansas

Lewis G. Regenstein: Author and Environmental Writer, Atlanta, Georgia

Linda Rehkopf: Freelance Writer, Marietta, Georgia

Paul E. Renaud, Ph.D.: Department of Biology, East Carolina University, Greenville, North Carolina

Marike Rijsberman: Freelance Writer, Chicago, Illinois

L. Carol Ritchie: Environmental Journalist, Arlington, Virginia

Linda M. Ross: Freelance Writer, Ferndale, Michigan

Joan Schonbeck: Medical Writer, Nursing, Massachusetts Department of Mental Health, Marlborough, Massachusetts

Mark W. Seeley: Department of Soil Science, University of Minnesota, St. Paul, Minnesota

Kim Sharp, M.Ln.: Freelance Writer, Richmond, Texas

James H. Shaw, Ph.D.: Department of Zoology, Oklahoma State University, Stillwater, Oklahoma

Laurel Sheppard: Freelance Writer, Columbus, Ohio

Judith Sims, M.S.: Utah Water Research Laboratory, Utah State University, Logan, Utah

Genevieve Slomski, Ph.D.: Freelance Writer, New Britain, Connecticut

Douglas Smith: Freelance Writer, Dorchester, Massachusetts

Lawrence H. Smith, Ph.D.: Department of Agronomy and Plant Genetics, University of Minnesota, St. Paul, Minnesota

Jane E. Spear: Freelance Writer, Canton, Ohio

Carol Steinfeld: Freelance Writer, Concord, Massachusetts

Paulette L. Stenzel, Ph.D.: Eli Broad College of Business, Michigan State University, East Lansing, Michigan

Les Stone: Freelance Writer, Ann Arbor, Michigan

Max Strieb: Freelance Writer, Huntington, New York

Amy Strumolo: Freelance Writer, Beverly Hills, Michigan

Edward Sucoff, Ph.D.: Department of Forestry Resources, University of Minnesota, St. Paul, Minnesota

Deborah L. Swackhammer, Ph.D.: School of Public Health, University of MinnesotaMinneapolis, Minnesota

Liz Swain: Freelance Writer, San Diego, California

Ronald D. Taskey, Ph.D.: Soil Science Department, California Polytechnic State University, San Luis Obispo, California

Mary Jane Tenerelli, M.S.: Freelance Writer, East Northport, New York

Usha Vedagiri: IT Corporation, Edison, New Jersey

Donald A. Villeneuve, Ph.D.: Ventura College, Ventura, California

Nikola Vrtis: Freelance Writer, Kentwood, Michigan

Eugene R. Wahl: Freelance Writer, Coon Rapids, Minnesota

Terry Watkins: Indianapolis, Indiana

Ken R. Wells: Freelance Writer, Laguna Hills, California

Roderick T. White Jr.: Freelance Writer, Atlanta, Georgia

T. Anderson White, Ph.D.: University of Minnesota, St. Paul, Minnesota

Kevin Wolf: Freelance Writer, Minneapolis, Minnesota

Angela Woodward: Freelance Writer, Madison, Wisconsin

Gerald L. Young, Ph.D.: Program in Environmental Science and Regional Planning, Washington State University, Pullman, Washington

HOW TO USE THIS BOOK

- The third edition of *Environmental Encyclopedia* has been designed with ready reference in mind.
- Straight **alphabetical arrangement** of topics allows users to locate information quickly.
- **Bold-faced terms** within entries direct the reader to related articles.
- **Contact information** is given for each organization profiled in the book.
- **Cross-references** at the end of entries alert readers to related entries not specifically mentioned in the body of the text.

- The **Resources** sections direct readers to additional sources of information on a topic.
- **Three appendices** provide the reader with a chronology of environmental events, a summary of environmental legislation, and a succinct alphabetical list of environmental organizations.
- A comprehensive **general index** guides readers to all topics mentioned in the text.

INTRODUCTION

Welcome to the third edition of the Gale *Environmental Encyclopedia*! Those of us involved in writing and production of this book hope you will find the material here interesting and useful. As you might imagine, choosing what to include and what to exclude from this collection has been challenging. Almost everything has some environmental significance, so our task has been to select a limited number of topics we think are of greatest importance in understanding our environment and our relation to it. Undoubtedly, we have neglected some topics that interest you and included some you may consider irrelevant, but we hope that overall you will find this new edition helpful and worthwhile.

The word environment is derived from the French *environ*, which means to "encircle" or "surround." Thus, our environment can be defined as the physical, chemical, and biological world that envelops us, as well as the complex of social and cultural conditions affecting an individual or community. This broad definition includes both the natural world and the "built" or technological environment, as well as the cultural and social contexts that shape human lives. You will see that we have used this comprehensive meaning in choosing the articles and definitions contained in this volume.

Among some central concerns of environmental science are:

- how did the natural world on which we depend come to be as it is, and how does it work?

- what have we done and what are we now doing to our environment—both for good and ill?

- what can we do to ensure a sustainable future for ourselves, future generations, and the other species of organisms on which—although we may not be aware of it—our lives depend?

The articles in this volume attempt to answer those questions from a variety of different perspectives.

Historically, environmentalism is rooted in natural history, a search for beauty and meaning in nature. Modern environmental science expands this concern, drawing on almost every area of human knowledge including social sciences, humanities, and the physical sciences. Its strongest roots, however, are in ecology, the study of interrelationships among and between organisms and their physical or nonliving environment. A particular strength of the ecological approach is that it studies systems holistically; that is, it looks at interconnections that make the whole greater than the mere sum of its parts. You will find many of those interconnections reflected in this book. Although the entries are presented individually so that you can find topics easily, you will notice that many refer to other topics that, in turn, can lead you on through the book if you have time to follow their trail. This series of linkages reflects the multilevel associations in environmental issues.

As our world becomes increasingly interrelated economically, socially, and technologically, we find evermore evidence that our global environment is also highly interconnected. In 2002, the world population reached about 6.2 billion people, more than triple what it had been a century earlier. Although the rate of population growth is slowing—having dropped from 2.0% per year in 1970 to 1.2% in 2002—we are still adding about 200,000 people per day, or about 75 million per year. Demographers predict that the world population will reach 8 or 9 billion before stabilizing sometime around the middle of this century. Whether natural resources can support so many humans is a question of great concern.

In preparation for the third global summit in South Africa, the United Nations released several reports in 2002 outlining the current state of our environment. Perhaps the greatest environmental concern as we move into the twenty-first century is the growing evidence that human activities are causing global climate change. Burning of fossil fuels in power plants, vehicles, factories, and homes release carbon dioxide into the atmosphere. Burning forests and crop residues, increasing cultivation of paddy rice, raising billions of ruminant animals, and other human activities also add to the rapidly growing atmospheric concentrations of heat trapping gases in the atmosphere. Global temperatures have begun to rise, having increased by about 1°F (0.6°C) in the second half of the twentieth century. Meteorologists predict that over the next 50 years, the average world temperature is likely to increase somewhere between 2.7–11°F (1.5–6.1°C). That may not seem like a very large change, but the difference between current average temperatures and the last ice age, when glaciers covered much of North America, was only about 10°F (5°C).

Abundant evidence is already available that our climate is changing. The twentieth century was the warmest in the last 1,000 years; the 1990s were the warmest decade, and 2002 was the single warmest year of the past millennium. Glaciers are disappearing on every continent. More than half the world's population depends on rivers fed by alpine glaciers for their drinking water. Loss of those glaciers could exacerbate water supply problems in areas where water is already scarce. The United Nations estimates that 1.1 billion people—one-sixth of the world population—now lack access to clean water. In 25 years, about two-thirds of all humans will live in water-stressed countries where supplies are inadequate to meet demand.

Spring is now occurring about a week earlier and fall is coming about a week later over much of the northern hemisphere. This helps some species, but is changing migration patterns and home territories for others. In 2002, early melting of ice floes in Canada's Gulf of St. Lawrence apparently drowned nearly all of the 200,000 to 300,000 harp seal pups normally born there. Lack of sea ice is also preventing polar bears from hunting seals. Environment Canada reports that polar bears around Hudson's Bay are losing weight and decreasing in number because of poor hunting conditions. In 2002, a chunk of ice about the size of Rhode Island broke off the Larsen B ice shelf on the Antarctic Peninsula. As glacial ice melts, ocean levels are rising, threatening coastal ecosystems and cities around the world.

After global climate change, perhaps the next greatest environmental concern for most biologists is the worldwide loss of biological diversity. Taxonomists warn that one-fourth of the world's species could face extinction in the next 30 years. Habitat destruction, pollution, introduction of exotic species, and excessive harvesting of commercially important species all contribute to species losses. Millions of species—most of which have never even been named by science, let alone examined for potential usefulness in medicine, agriculture, science, or industry—may disappear in the next century as a result of our actions. We know little about the biological roles of these organisms in the ecosystems and their loss could result in an ecological tragedy.

Ecological economists have tried to put a price on the goods and services provided by natural ecosystems. Although many ecological processes aren't traded in the market place,

we depend on the natural world to do many things for us like purifying water, cleansing air, and detoxifying our wastes. How much would it cost if we had to do all this ourselves? The estimated annual value of all ecological goods and services provided by nature are calculated to be worth at least $33 trillion, or about twice the annual GNPs of all national economies in the world. The most valuable ecosystems in terms of biological processes are wetlands and coastal estuaries because of their high level of biodiversity and their central role in many biogeochemical cycles.

Already there are signs that we are exhausting our supplies of fertile soil, clean water, energy, and biodiversity that are essential for life. Furthermore, pollutants released into the air and water, along with increasing amounts of toxic and hazardous wastes created by our industrial society, threaten to damage the ecological life support systems on which all organisms—including humans—depend. Even without additional population growth, we may need to drastically rethink our patterns of production and disposal of materials if we are to maintain a habitable environment for ourselves and our descendants.

An important lesson to be learned from many environmental crises is that solving one problem often creates another. Chlorofluorocarbons, for instance, were once lauded as a wonderful discovery because they replaced toxic or explosive chemicals then in use as refrigerants and solvents. No one anticipated that CFCs might damage stratospheric ozone that protects us from dangerous ultraviolet radiation. Similarly, the building of tall smokestacks on power plants and smelters lessened local air pollution, but spread acid rain over broad areas of the countryside. Because of our lack of scientific understanding of complex systems, we are continually subjected to surprises. How to plan for "unknown unknowns" is an increasing challenge as our world becomes more tightly interconnected and our ability to adjust to mistakes decreases.

Not all is discouraging, however, in the field of environmental science. Although many problems beset us, there are also encouraging signs of progress. Some dramatic successes have occurred in wildlife restoration and habitat protection programs, for instance. The United Nations reports that protected areas have increased five-fold over the past 30 years to nearly 5 million square miles. World forest losses have slowed, especially in Asia, where deforestation rates slowed from 8% in the 1980s to less than 1% in the 1990s. Forested areas have actually increased in many developed countries, providing wildlife habitat, removal of excess carbon dioxide, and sustainable yields of forest products.

In spite of dire warnings in the 1960s that growing human populations would soon overshoot the earth's carrying capacity and result in massive famines, food supplies have more than kept up with population growth. There is more than enough food to provide a healthy diet for everyone now living, although inequitable distribution leaves about 800 million with an inadequate diet. Improved health care, sanitation, and nutrition have extended life expectancies around the world from 40 years, on average, a century ago, to 65 years now. Public health campaigns have eradicated smallpox and nearly eliminated polio. Other terrible diseases have emerged, however, most notably acquired immunodeficiency syndrome (AIDS), which is now the fourth most common cause of death worldwide. Forty million people are now infected with HIV—70% percent of them in sub-Saharan Africa—and health experts warn that unsanitary blood donation practices and spreading drug use in Asia may result in tens of millions more AIDS deaths in the next few decades.

In developed countries, air and pollution have decreased significantly over the past 30 years. In 2002, the Environmental Protection Agency declared that Denver—which once was infamous as one of the most polluted cities in the United States—is the first major city to meet all the agency's standards for eliminating air pollution. At about the same time, the EPA announced that 91% of all monitored river miles in the United States met the water quality goals set in the 1985 clean water act. Pollution-sensitive species like mayflies have returned to the upper Mississippi River, and in Britian, salmon are being caught in the Thames River after being absent for more than two centuries.

Conditions aren't as good, however, in many other countries. In most of Latin America, Africa, and Asia, less than two % of municipal sewage is given even primary treatment before being dumped into rivers, lakes, or the ocean. In South Asia, a 2-mile (3-km) thick layer of smog covers the entire Indian sub-continent for much of the year. This cloud blocks sunlight and appears to be changing the climate, bringing drought to Pakistan and Central Asia, and shifting monsoon winds that caused disastrous floods in 2002 in Nepal, Bangladesh, and eastern India that forced 25 million people from their homes and killed at least 1,000 people. Nobel laureate Paul Crutzen estimates that two million deaths each year in India alone can be attributed to air pollution effects.

After several decades of struggle, a world-wide ban on the "dirty dozen" most dangerous persistent organic pollutants (POPs) was ratified in 2000. Elimination of compounds such as DDT, Aldrin, Dieldrin, Mirex, Toxaphene, polychlorinated biphenyls, and dioxins has allowed recovery of several wildlife species including bald eagles, perigrine falcons, and brown pelicans. Still, other toxic synthetic chemicals such as polybrominated diphenyl ethers, chromated copper arsenate, perflurooctane sulfonate, and atrazine are now being found accumulating in food chains far from anyplace where they have been used.

Solutions for many of our pollution problems can be found in either improved technology, more personal responsibility, or better environmental management. The question is often whether we have the political will to enforce pollution control programs and whether we are willing to sacrifice short-term convenience and affluence for long-term ecological stability. We in the richer countries of the world have become accustomed to a highly consumptive lifestyle. Ecologists estimate that humans either use directly, destroy, co-opt, or alter almost 40% of terrestrial plant productivity, with unknown consequences for the biosphere. Whether we will be willing to leave some resources for other species and future generations is a central question of environmental policy.

One way to extend resources is to increase efficiency and recycling of the items we use. Automobiles have already been designed, for example, that get more than 100 mi/gal (42 km/l) of diesel fuel and are completely recyclable when they reach the end of their designed life. Although recycling rates in the United States have increased in recent years, we could probably double our current rate with very little sacrifice in economics or convenience. Renewable energy sources such as solar or wind power are making encouraging progress. Wind already is cheaper than any other power source except coal in many localities. Solar energy is making it possible for many of the two billion people in the world who don't have access to electricity to enjoy some of the benefits of modern technology. Worldwide, the amount of installed wind energy capacity more than doubled between 1998 and 2002. Germany is on course to obtain 20% of its energy from renewables by 2010. Together, wind, solar, biomass and other forms of renewable energy have the potential to provide thousands of times as much energy as all humans use now. There is no reason for us to continue to depend on fossil fuels for the majority of our energy supply.

One of the widely advocated ways to reduce poverty and make resources available to all is sustainable development. A commonly used definition of this term is given in *Our Common Future*, the report of the World Commission on Environment and Development (generally called the Brundtland Commission after the prime minister of Norway, who chaired it), described sustainable development as: "meeting the needs of the present without compromising the ability of future generations to meet their own needs." This implies improving health, education, and equality of opportunity, as well as ensuring political and civil rights through jobs and programs based on sustaining the ecological base, living on renewable resources rather than nonrenewable ones, and living within the carrying capacity of supporting ecological systems.

Several important ethical considerations are embedded in environmental questions. One of these is intergenerational justice: what responsibilities do we have to leave resources and a habitable planet for future generations? Is our profligate use of fossil fuels, for example, justified by the fact that we have technology to extract fossil fuels and enjoy their benefits? Will human lives in the future be impoverished by the fact that we have used up most of the easily available oil, gas, and coal? Author and social critic Wendell Berry suggests that our consumption of these resources constitutes a theft of the birthright and livelihood of posterity. Philosopher John Rawls advocates a "just savings principle" in which members of each generation may consume no more than their fair share of scarce resources.

How many generations are we obliged to plan for and what is our "fair share?" It is possible that our use of resources now—inefficient and wasteful as it may be—represents an investment that will benefit future generations. The first computers, for instance, were huge clumsy instruments that filled rooms full of expensive vacuum tubes and consumed inordinate amounts of electricity. Critics complained that it was a waste of time and resources to build these enormous machines to do a few simple calculations. And yet if this technology had been suppressed in its infancy, the world would be much poorer today. Now nanotechnology promises to make machines and tools in infinitesimal sizes that use minuscule amounts of materials and energy to carry out valuable functions. The question remains whether future generations will be glad that we embarked on the current scientific and technological revolution or whether they will wish that we had maintained a simple agrarian, Arcadian way of life.

Another ethical consideration inherent in many environmental issues is whether we have obligations or responsibilities to other species or to Earth as a whole. An anthropocentric (human-centered) view holds that humans have rightful dominion over the earth and that our interests and well-being take precedence over all other considerations. Many environmentalists criticize this perspective, considering it arrogant and destructive. Biocentric (life-centered) philosophies argue that all living organisms have inherent values and rights by virtue of mere existence, whether or not

they are of any use to us. In this view, we have a responsibility to leave space and resources to enable other species to survive and to live as naturally as possible. This duty extends to making reparations or special efforts to encourage the recovery of endangered species that are threatened with extinction due to human activities.Some environmentalists claim that we should adopt an ecocentric (ecologically centered) outlook that respects and values nonliving entities such as rocks, rivers, mountains—even whole ecosystems—as well as other living organisms. In this view, we have no right to break up a rock, dam a free-flowing river, or reshape a landscape simply because it benefits us. More importantly, we should conserve and maintain the major ecological processes that sustain life and make our world habitable.

Others argue that our existing institutions and understandings, while they may need improvement and reform, have provided us with many advantages and amenities. Our lives are considerably better in many ways than those of our ancient ancestors, whose lives were, in the words of British philosopher Thomas Hobbes: "nasty, brutish, and short." Although science and technology have introduced many problems, they also have provided answers and possible alternatives as well.

It may be that we are at a major turning point in human history. Current generations are in a unique position to address the environmental issues described in this encyclopedia. For the first time, we now have the resources, motivation, and knowledge to protect our environment and to build a sustainable future for ourselves and our children. Until recently, we didn't have these opportunities, or there was not enough clear evidence to inspire people to change their behavior and invest in environmental protection; now the need is obvious to nearly everyone. Unfortunately, this also may be the last opportunity to act before our problems become irreversible.

We hope that an interest in preserving and protecting our common environment is one reason that you are reading this encyclopedia and that you will find information here to help you in that quest.

[*William P. Cunningham, Managing Editor*]

A

Edward Paul Abbey (1927 – 1989)
American environmentalist and writer

Novelist, essayist, white-water rafter, and self-described "desert rat," Abbey wrote of the wonders and beauty of the American West that was fast disappearing in the name of "development" and "progress." Often angry, frequently funny, and sometimes lyrical, Abbey recreated for his readers a region that was unique in the world. The American West was perhaps the last place where solitary selves could discover and reflect on their connections with wild things and with their fellow human beings.

Abbey was born in Home, Pennsylvania, in 1927. He received his B.A. from the University of New Mexico in 1951. After earning his master's degree in 1956, he joined the **National Park Service**, where he served as park ranger and fire fighter. He later taught writing at the University of Arizona.

Abbey's books and essays, such as *Desert Solitaire* (1968) and *Down the River* (1982), had their angrier fictional counterparts—most notably, *The Monkey Wrench Gang* (1975) and *Hayduke Lives!* (1990)—in which he gave voice to his outrage over the destruction of deserts and rivers by dam-builders and developers of all sorts. In *The Monkey Wrench Gang* Abbey weaves a tale of three "ecoteurs" who defend the wild west by destroying the means and machines of development—dams, bulldozers, **logging** trucks—which would otherwise reduce forests to lumber and raging rivers to **irrigation** channels.

This aspect of Abbey's work inspired some radical environmentalists, including **Dave Foreman** and other members of **Earth First!**, to practice "monkey-wrenching" or "ecotage" to slow or stop such environmentally destructive practices as **strip mining**, the **clear-cutting** of old-growth forests on **public land**, and the damming of wild rivers for flood control, hydroelectric power, and what Abbey termed "industrial tourism." Although Abbey's description and defense of such tactics has been widely condemned by many mainstream environmental groups, he remains a revered figure among many who believe that gradualist tactics have not succeeded in slowing, much less stopping, the destruction of North American **wilderness**. Abbey is unique among environmental writers in having an oceangoing ship named after him. One of the vessels in the fleet of the militant **Sea Shepherd Conservation Society**, the *Edward Abbey*, rams and disables **whaling** and drift-net fishing vessels operating illegally in international waters. Abbey would have welcomed the tribute and, as a white-water rafter and canoeist, would no doubt have enjoyed the irony.

Abbey died on March 14, 1989. He is buried in a **desert** in the southwestern United States.

[*Terence Ball*]

RESOURCES
BOOKS

Abbey, E. *Desert Solitaire*. New York: McGraw-Hill, 1968.
———. *Down the River*. Boston: Little, Brown, 1982.
———. *Hayduke Lives!* Boston: Little, Brown, 1990.
———. *The Monkey Wrench Gang*. Philadelphia: Lippincott, 1975.
Berry, W. "A Few Words in Favor of Edward Abbey." In *What Are People For?* San Francisco: North Point Press, 1991.
Bowden, C. "Goodbye, Old Desert Rat." In *The Sonoran Desert*. New York: Abrams, 1992.
Manes, C. *Green Rage: Radical Environmentalism and the Unmaking of Civilization*. Boston: Little, Brown, 1990.

Absorption

Absorption, or more generally "sorption," is the process by which one material (the sorbent) takes up and retains another (the sorbate) to form a homogenous concentration at equilibrium.

The general term is "sorption," which is defined as adhesion of gas molecules, dissolved substances, or liquids to the surface of solids with which they are in contact. In soils, three types of mechanisms, often working together, constitute **sorption**. They can be grouped into physical sorption, chemiosorption, and penetration into the solid mineral phase. Physical sorption (also known as **adsorption**) involves the attachment of the sorbent and sorbate through weak atomic and molecular forces. Chemiosorption involves chemical bonds similar to holding atoms in a molecule. Electrostatic forces operate to bond minerals via **ion exchange**, such as the replacement of sodium, magnesium, potassium, and **aluminum** cations (+) as exchangeable bases with **acid** (-) soils. While cation (positive **ion**) exchange is the dominant exchange process occurring in soils, some soils have the ability to retain anions (negative ions) such as **nitrates**, **chlorine** and, to a larger extent, oxides of sulfur.

Absorption and Wastewater Treatment

In on-site **wastewater** treatment, the **soil** absorption field is the land area where the wastewater from the **septic tank** is spread into the soil. One of the most common types of soil absorption field has porous plastic pipe extending away from the distribution box in a series of two or more parallel trenches, usually 1.5–2 ft (30.5–61 cm) wide. In conventional, below-ground systems, the trenches are 1.5–2 ft deep. Some absorption fields must be placed at a shallower depth than this to compensate for some limiting soil condition, such as a hardpan or high **water table**. In some cases they may even be placed partially or entirely in fill material that has been brought to the lot from elsewhere.

The porous pipe that carries wastewater from the distribution box into the absorption field is surrounded by gravel that fills the trench to within a foot or so of the ground surface. The gravel is covered by fabric material or building paper to prevent plugging. Another type of drainfield consists of pipes that extend away from the distribution box, not in trenches but in a single, gravel-filled bed that has several such porous pipes in it. As with trenches, the gravel in a bed is covered by fabric or other porous material.

Usually the wastewater flows gradually downward into the gravel-filled trenches or bed. In some instances, such as when the septic tank is lower than the drainfield, the wastewater must be pumped into the drainfield. Whether gravity flow or pumping is used, wastewater must be evenly distributed throughout the drainfield. It is important to ensure that the drainfield is installed with care to keep the porous pipe level, or at a very gradual downward slope away from the distribution box or pump chamber, according to specifications stipulated by public health officials. Soil beneath the gravel-filled trenches or bed must be **permeable** so that wastewater and air can move through it and come in contact with each other. Good **aeration** is necessary to ensure that the proper chemical and microbiological processes will be occurring in the soil to cleanse the percolating wastewater of contaminants. A well-aerated soil also ensures slow travel and good contact between wastewater and soil.

How Common Are Septic Systems with Soil Absorption Systems?

According to the 1990 U.S. Census, there are about 24.7 million households in the United States that use septic tank systems or cesspools (holes or pits for receiving sewage) for wastewater treatment. This figure represents roughly 24% of the total households included in the census.

According to a review of local health department information by the National Small Flows Clearinghouse, 94% of participating health departments allow or permit the use of septic tank and soil absorption systems. Those that do not allow septic systems have sewer lines available to all residents. The total volume of waste disposed of through septic systems is more than one trillion gallons (3.8 trillion l) per year, according to a study conducted by the U.S. Environmental Protection Agency's Office of Technology Assessment, and virtually all of that waste is discharged directly to the subsurface, which affects **groundwater** quality.

[*Carol Steinfeld*]

RESOURCES
BOOKS

Elliott, L. F., and F. J. Stevenson, *Soils for the Management of Wastes and Waste Waters.* Madison, WI: Soil Science Society of America, 1977.

OTHER

Fact Sheet SL-59, a series of the Soil and Water Science Department, Florida Cooperative Extension Service, Institute of Food and Agricultural Sciences, University of Florida. February 1993.

Acaricide
see Pesticide

Acceptable risk
see Risk analysis

Acclimation

Acclimation is the process by which an organism adjusts to a change in its **environment**. It generally refers to the ability of living things to adjust to changes in **climate**, and usually occurs in a short time of the change.

Scientists distinguish between acclimation and acclimatization because the latter adjustment is made under natural conditions when the organism is subject to the full range of changing environmental factors. Acclimation, however, refers to a change in only one environmental factor under laboratory conditions.

In an acclimation experiment, adult **frogs** (*Rana temporaria*) maintained in the laboratory at a temperature of either 50°F (10°C) or 86°F (30°C) were tested in an environment of 32°F (0°C). It was found that the group maintained at the higher temperature was inactive at freezing. The group maintained at 50°F (10°C), however, was active at the lower temperature; it had acclimated to the lower temperature.

Acclimation and acclimatization can have profound effects upon behavior, inducing shifts in preferences and in mode of life. The golden hamster (*Mesocricetus auratus*) prepares for hibernation when the environmental temperature drops below 59°F (15°C). Temperature preference tests in the laboratory show that the hamsters develop a marked preference for cold environmental temperatures during the pre-hibernation period. Following arousal from a simulated period of hibernation, the situation is reversed, and the hamsters actively prefer the warmer environments.

An acclimated microorganism is any microorganism that is able to adapt to environmental changes such as a change in temperature or a change in the quantity of oxygen or other gases. Many organisms that live in environments with seasonal changes in temperature make physiological adjustments that permit them to continue to function normally, even though their environmental temperature goes through a definite annual temperature cycle.

Acclimatization usually involves a number of interacting physiological processes. For example, in acclimatizing to high altitudes, the first response of human beings is to increase their breathing rate. After about 40 hours, changes have occurred in the oxygen-carrying capacity of the blood, which makes it more efficient in extracting oxygen at high altitudes. As this occurs, the breathing rate returns to normal.

[*Linda Rehkopf*]

RESOURCES
BOOKS

Ford, M. J. *The Changing Climate: Responses of the Natural Fauna and Flora.* Boston: G. Allen and Unwin, 1982.
McFarland, D., ed. *The Oxford Companion to Animal Behavior.* Oxford, England: Oxford University Press, 1981.
Stress Responses in Plants: Adaptation and Acclimation Mechanisms. New York: Wiley-Liss, 1990.

Accounting for nature

A new approach to national income accounting in which the degradation and depletion of natural resource stocks and environmental amenities are explicitly included in the calculation of net national product (NNP). NNP is equal to gross national product (GNP) minus capital depreciation, and GNP is equal to the value of all final goods and services produced in a nation in a particular year. It is recognized that **natural resources** are economic assets that generate income, and that just as the depreciation of buildings and capital equipment are treated as economic costs and subtracted from GNP to get NNP, depreciation of *natural capital* should also be subtracted when calculating NNP. In addition, expenditures on environmental protection, which at present are included in GNP and NNP, are considered defensive expenditures in accounting for **nature** which should not be included in either GNP or NNP.

Accuracy

Accuracy is the closeness of an experimental measurement to the "true value" (i.e., actual or specified) of a measured quantity. A "true value" can determined by an experienced analytical scientist who performs repeated analyses of a sample of known purity and/or concentration using reliable, well-tested methods.

Measurement is inexact, and the magnitude of that exactness is referred to as the error. Error is inherent in measurement and is a result of such factors as the **precision** of the measuring tools, their proper adjustment, the method, and competency of the analytical scientist.

Statistical methods are used to evaluate accuracy by predicting the likelihood that a result varies from the "true value." The analysis of probable error is also used to examine the suitability of methods or equipment used to obtain, portray, and utilize an acceptable result. Highly accurate data can be difficult to obtain and costly to produce. However, different applications can require lower levels of accuracy that are adequate for a particular study.

[*Judith L. Sims*]

RESOURCES
BOOKS

Jaisingh, Lloyd R. *Statistics for the Utterly Confused.* New York, NY: McGraw-Hill Professional, 2000.
Salkind, Neil J. *Statistics for People Who (Think They) Hate Statistics.* Thousand Oaks, CA: Sage Publications, Inc., 2000.

Acetone

Acetone (C_3H_6O) is a colorless liquid that is used as a solvent in products, such as in nail polish and paint, and in the manufacture of other **chemicals** such as **plastics** and fibers. It is a naturally occurring compound that is found in plants and is released during the **metabolism** of fat in the body. It is also found in volcanic gases, and is manufactured by the chemical industry. Acetone is also found in the **atmo-**

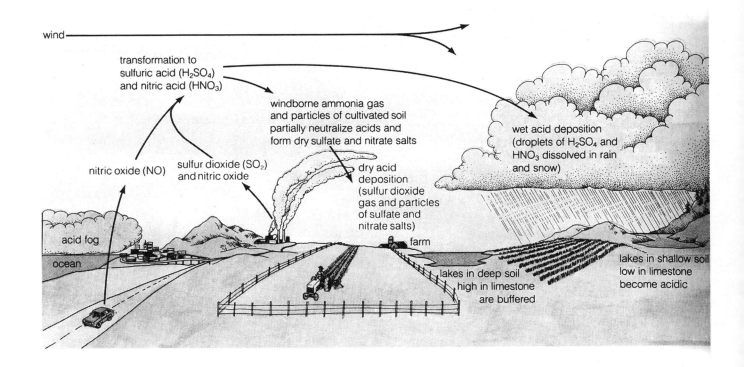

The basic mechanisms of acid deposition. (Illustration by Wadsworth Inc. Reproduced by permission.)

sphere as an oxidation product of both natural and **anthropogenic** volatile organic compounds (VOCs). It has a strong smell and taste, and is soluble in water. The evaporation point of acetone is quite low compared to water, and the chemical is highly flammable. Because it is so volatile, the acetone manufacturing process results in a large percentage of the compound entering the atmosphere. Ingesting acetone can cause damage to the tissues in the mouth and can lead to unconsciousness. Breathing acetone can cause irritation of the eyes, nose, and throat; headaches; dizziness; nausea; unconsciousness; and possible coma and death. Women may experience menstrual irregularity. There has been concern about the carcinogenic nature of acetone, but laboratory studies, and studies of humans who have been exposed to acetone in the course of their occupational activities show no evidence that acetone causes **cancer**.

[*Marie H. Bundy*]

Acid and base

According to the definition used by environmental chemists, an acid is a substance that increases the **hydrogen ion** (H⁺)

concentration in a solution and a base is a substance that removes hydrogen ions (H⁺) from a solution. In water, removal of hydrogen ions results in an increase in the hydroxide ion (OH⁻) concentration. Water with a **pH** of 7.0 is neutral, while lower pH values are acidic and higher pH values are basic.

Acid deposition

Acid precipitation from the **atmosphere**, whether in the form of dryfall (finely divided acidic salts), rain, or snow. Naturally occurring carbonic acid normally makes rain and snow mildly acidic (approximately 5.6 **pH**). Human activities often introduce much stronger and more damaging acids. Sulfuric acids formed from sulfur oxides released in **coal** or oil **combustion** or smelting of sulfide ores predominate as the major atmospheric acid in industrialized areas. Nitric acid created from **nitrogen oxides**, formed by oxidizing atmospheric **nitrogen** when any fuel is burned in an oxygen-rich **environment**, constitutes the major source of acid precipitation in such cities as Los Angeles with little industry, but large numbers of trucks and automobiles. The damage caused to building materials, human health, crops, and natural ecosystems by atmospheric acids amounts to billions of dollars per year in the United States.

Acid mine drainage

The process of mining the earth for **coal** and metal ores has a long history of rich economic rewards—and a high level of environmental impact to the surrounding aquatic and terrestrial ecosystems. Acid mine **drainage** is the highly acidic, sediment-laden **discharge** from exposed mines that is released into the ambient aquatic **environment**. In large areas of Pennsylvania, West Virginia, and Kentucky, the bright orange seeps of acid mine drainage have almost completely eliminated aquatic life in streams and ponds that receive the discharge. In the Appalachian coal mining region, almost 7,500 mi (12,000 km) of streams and almost 30,000 acres (12,000 ha) of land are estimated to be seriously affected by the discharge of uncontrolled acid mine drainage.

In the United States, coal-bearing geological strata occur near the surface in large portions of the Appalachian mountain region. The relative ease with which coal could be extracted from these strata led to a type of mining known as **strip mining** that was practiced heavily in the nineteenth and early twentieth centuries. In this process, large amounts of earth, called the **overburden**, were physically removed from the surface to expose the coal-bearing layer beneath. The coal was then extracted from the rock as quickly and cheaply as possible. Once the bulk of the coal had been mined, and no more could be extracted without a huge additional cost, the sites were usually abandoned. The remnants of the exhausted coal-bearing rock and **soil** are called the **mine spoil waste**.

Acid mine drainage is not generated by strip mining itself but by the nature of the rock where it takes place. Three conditions are necessary to form acid mine drainage: pyrite-bearing rock, oxygen, and iron-oxidizing bacteria. In the Appalachians, the coal-bearing rocks usually contain significant quantities of pyrite (iron). This compound is normally not exposed to the **atmosphere** because it is buried underground within the rock; it is also insoluble in water. The iron and the sulfide are said to be in a reduced state, i.e., the iron atom has not released all the electrons that it is capable of releasing. When the rock is mined, the pyrite is exposed to air. It then reacts with oxygen to form ferrous iron and sulfate ions, both of which are highly soluble in water. This leads to the formation of sulfuric acid and is responsible for the acidic nature of the drainage. But the oxidation can only occur if the bacteria *Thiobacillus ferrooxidans* are present. These activate the iron-and-sulfur oxidizing

reactions, and use the energy released during the reactions for their own growth. They must have oxygen to carry these reactions through. Once the maximum oxidation is reached, these bacteria can derive no more energy from the compounds and all reactions stop.

The acidified water may be formed in several ways. It may be generated by rain falling on exposed mine spoils waste or when rain and surface water (carrying **dissolved oxygen**) flow down and seep into rock fractures and mine shafts, coming into contact with pyrite-bearing rock. Once the acidified water has been formed, it leaves the mine area as seeps or small streams.

Characteristically bright orange to rusty red in color due to the iron, the liquid may be at a pH of 2–4. These are extremely low pH values and signify a very high degree of acidity. Vinegar, for example, has a pH of about 4.7 and the pH associated with **acid rain** is in the range of 4–6. Thus, acid mine drainage with a pH of 2 is more acidic than almost any other naturally occurring liquid release in the environment (with the exception of some volcanic lakes that are pure acid). Usually, the drainage is also very high in dissolved iron, manganese, **aluminum**, and suspended solids.

The acidic drainage released from the mine **spoil** wastes usually follows the natural **topography** of its area and flows into the nearest streams or **wetlands** where its effect on the **water quality** and **biotic community** is unmistakable. The iron coats the stream bed and its vegetation as a thick orange coating that prevents sunlight from penetrating leaves and plant surfaces. **Photosynthesis** stops and the vegetation (both vascular plants and algae) dies. The acid drainage eventually also makes the receiving water acid. As the pH drops, the fish, the invertebrates, and algae die when their **metabolism** can no longer adapt. Eventually, there is no life left in the stream with the possible exception of some bacteria that may be able to tolerate these conditions. Depending on the number and volume of seeps entering a stream and the volume of the stream itself, the area of impact may be limited and improved conditions may exist downstream, as the acid drainage is diluted. Abandoned mine spoil areas also tend to remain barren, even after decades. The colonization of the acidic mineral soil by plant **species** is a slow and difficult process, with a few **lichens** and aspens being the most hardy species to establish.

While many methods have been tried to control or mitigate the effects of acid mine drainage, very few have been successful. Federal mining regulations (**Surface Mining Control and Reclamation Act** of 1978) now require that when mining activity ceases, the mine spoil waste should be buried and covered with the overburden and vegetated **topsoil**. The intent is to restore the area to premining condition and to prevent the generation of acid mine drainage by limiting the exposure of pyrite to oxygen and water. Although some minor seeps may still occur, this is the single-most effective way to minimize the potential scale of the problem. Mining companies are also required to monitor the effectiveness of their restoration programs and must post bonds to guarantee the execution of abatement efforts, should any become necessary in the future.

There are, however, numerous abandoned sites exposing pyrite-bearing spoils. Cleanup efforts for these sites have focused on controlling one or more of the three conditions necessary for the creation of the acidity: pyrite, bacteria, and oxygen. Attempts to remove bulk quantities of the pyrite-bearing mineral and store it somewhere else are extremely expensive and difficult to execute. Inhibiting the bacteria by using **detergents**, solvents, and other bactericidal agents are temporarily effective, but usually require repeated application. Attempts to seal out air or water are difficult to implement on a large scale or in a comprehensive manner.

Since it is difficult to reduce the formation of acid mine drainage at abandoned sites, one of the most promising new methods of mitigation treats the acid mine drainage after it exits the mine spoil wastes. The technique channels the acid seeps through artificially created wetlands, planted with cattails or other wetland plants in a bed of gravel, limestone, or compost. The limestone neutralizes the acid and raises the pH of the drainage while the mixture of oxygen-rich and oxygen-poor areas within the wetland promote the removal of iron and other metals from the drainage. Currently, many agencies, universities, and private firms are working to improve the design and performance of these artificial wetlands. A number of additional treatment techniques may be strung together in an interconnected system of anoxic limestone trenches, settling ponds, and planted wetlands. This provides a variety of physical and chemical microenvironments so that each undesirable characteristic of the acid drainage can be individually addressed and treated, e.g., acidity is neutralized in the trenches, suspended solids are settled in the ponds, and metals are precipitated in the wetlands. In the United States, the research and treatment of acid mine drainage continues to be an active field of study in the Appalachians and in the metal-mining areas of the Rocky Mountains.

[*Usha Vedagiri*]

RESOURCES
PERIODICALS

Clay, S. "A Solution to Mine Drainage?" *American Forests* 98 (July-August 1992): 42-43.
Hammer, D. A. *Constructed Wetlands for Wastewater Treatment: Municipal, Industrial, Agricultural.* Chelsea, MI: Lewis, 1990.
Schwartz, S. E. "Acid Deposition: Unraveling a Regional Phenomenon." *Science* 243 (February 1989): 753-763.

Welter, T. R. "An 'All Natural' Treatment: Companies Construct Wetlands to Reduce Metals in Acid Mine Drainage." *Industry Week* 240 (August 5, 1991): 42-43.

Acid rain

Acid rain is the term used in the popular press that is equivalent to acidic deposition as used in the scientific literature. **Acid deposition** results from the deposition of airborne acidic pollutants on land and in bodies of water. These pollutants can cause damage to forests as well as to lakes and streams.

The major pollutants that cause acidic deposition are **sulfur dioxide** (SO_2) and **nitrogen oxides** (NO_x) produced during the **combustion** of **fossil fuels**. In the **atmosphere** these gases oxidize to sulfuric acid (H_2SO_4) and nitric acid (HNO_3) that can be transported long distances before being returned to the earth dissolved in rain drops (wet deposition), deposited on the surfaces of plants as cloud droplets, or directly on plant surfaces (**dry deposition**). Electrical utilities contribute 70% of the 20 million tons (21 million metric tons) of SO_2 that are annually added to the atmosphere. Most of this is from the combustion of **coal. Electric utilities** also contribute 30% of the 19 million tons of NO_x added to the atmosphere, and internal combustion engines used in automobiles, trucks, and buses contribute more than 40%. Natural sources such as forest fires, swamp gases, and volcanoes only contribute 1-5% of atmospheric SO_2. Forest fires, lightning, and microbial processes in soils contribute about 11% to atmospheric NO_x. In response to **air quality** regulations, electrical utilities have switched to coal with lower sulfur content and installed scrubbing systems to remove SO_2. This has resulted in a steady decrease in SO_2 emissions in the United States since 1970, with a 18-20% decrease between 1975 and 1988. Emissions of NO_x have also decreased from the peak in 1975, with a 9-15% decrease from 1975 to 1988.

A commonly used indicator of the intensity of acid rain is the **pH** of this rainfall. The pH of non-polluted rainfall in forested regions is in the range 5.0-5.6. The upper limit is 5.6, not neutral (7.0), because of carbonic acid that results from the dissolution of atmospheric **carbon dioxide**. The contribution of naturally occurring nitric and sulfuric acid, as well as organic acids, reduces the pH somewhat to less than 5.6. In **arid** and semi-arid regions, rainfall pH values can be greater than 5.6 due the effect of alkaline **soil** dust in the air. Nitric and sulfuric acids in acidic rainfall (wet deposition) can result in pH values for individual rainfall events of less than 4.0.

In North America, the lowest acid rainfall is in the northeastern United States and southeastern Canada. The lowest mean pH in this region is 4.15. Even lower pH values are observed in central and northern Europe. Generally, the greater the population density and density of industrialization the lower the rainfall pH. Long distance transport, however, can result in low pH rainfall even in areas with low population and low density of industries, as in parts of New England, eastern Canada, and in Scandinavia.

A very significant portion of acid deposition occurs in the dry form. In the United States, it is estimated that 30–60% of acidic deposition occurs as dry fall. This material is deposited as sulfur dioxide gas and very finely divided particles (aerosols) directly on the surfaces of plants (needles and leaves). The rate of deposition depends not only on the concentration of acid materials suspended in the air, but on the nature and density of plant surfaces exposed to the atmosphere and the atmospheric conditions(e.g., wind speed and humidity).

Direct deposition of acid cloud droplets can be very important especially in some high altitude forests. Acid cloud droplets can have acid concentrations of five to 20 times that in wet deposition. In some high elevation sites that are frequently shrouded in clouds, direct droplet deposition is three times that of wet deposition from rainfall.

Acid deposition has the potential to adversely affect sensitive forests as well as lakes and streams. Agriculture is generally not included in the assessment of the effects of acidic deposition because experimental evidence indicates that even the most severe episodes of acid deposition do not adversely affect the growth of agricultural crops, and any long-term soil **acidification** can readily be managed by addition of agricultural lime. In fact, the acidifying potential of the fertilizers normally added to cropland is much greater than that of acidic deposition. In forests, however, long-term acidic deposition on sensitive soils can result in the depletion of important **nutrient** elements (e.g., calcium, magnesium, and potassium) and in soil acidification. Also, acidic pollutants can interact with other pollutants (e.g., **ozone**) to cause more immediate problems for tree growth. Acid deposition can also result in the acidification of sensitive lakes and with the loss of biological productivity.

Long-term exposure of acid sensitive materials used in building construction and in monuments (e.g., zinc, marble, limestone, and some sandstone) can result in surface corrosion and deterioration. Monuments tend to be the most vulnerable because they are usually not as protected from rainfall as most building materials. Good data on the impact of acidic deposition on monuments and building material is lacking.

Nutrient depletion due to acid deposition on sensitive soils is a long-term (decades to centuries) consequence of acidic deposition. Acidic deposition greatly accelerates the very slow depletion of soil nutrients due to natural **weathering** processes. Soils that contain less plant-available calcium,

magnesium and potassium are less buffered with respect to degradation due to acidic deposition. The most sensitive soils are shallow sandy soils over hard bedrock. The least vulnerable soils are the deep clay soils that are highly buffered against changes due to acidic deposition.

The more immediate possible threat to forests is the **forest decline** phenomenon that has been observed in forests in northern Europe and North America. Acidic deposition in combination with other stress factors such as ozone, disease and adverse weather conditions can lead to decline in forest productivity and, in certain cases, to **dieback**. Acid deposition alone cannot account for the observed forest decline, and acid deposition probably plays a minor role in the areas where forest decline has occurred. Ozone is a much more serious threat to forests, and it is a key factor in the decline of forests in the Sierra Nevada and San Bernardino mountains in California.

The greatest concern for adverse effects of acidic deposition is the decline in biological productivity in lakes. When a lake has a pH less than 6.0, several **species** of minnows, as well as other species that are part of the food chain for many fish, cannot survive. At pH values less than about 5.3, lake trout, walleye, and smallmouth bass cannot survive. At pH less than about 4.5, most fish cannot survive (largemouth bass are an exception).

Many small lakes are naturally acidic due to organic acids produced in acid soils and acid bogs. These lakes have chemistries dominated by organic acids, and many have brown colored waters due to the organic acid content. These lakes can be distinguished from lakes acidified by acidic deposition, because lakes strongly affected by acidic deposition are dominated by sulfate.

Lakes that are adversely affected by acidic deposition tend to be in steep terrain with thin soils. In these settings the path of rainwater movement into a lake is not influenced greatly by soil materials. This contrasts to most lakes where much of the water that collects in a lake flows first into the **groundwater** before entering the lake via subsurface flow. Due to the contact with soil materials, acidity is neutralized and the capacity to neutralize acidity is added to the water in the form of bicarbonate ions (bicarbonate alkalinity). If more than 5% of the water that reaches a lake is in the form of groundwater, a lake is not sensitive to acid deposition.

An estimated 24% of the lakes in the Adirondack region of New York are devoid of fish. In one third to one half of these lakes this is due to acidic deposition. Approximately 16% of the lakes in this region may have lost one or more species of fish due to acidification. In Ontario, Canada, 115 lakes are estimated to have lost populations of lake trout. Acidification of lakes, by acidic deposition, extends as far west as Upper Michigan and northeastern Wisconsin, where many sensitive lakes occur and there is some evidence for

Acid rain in Chicago, Illinois, erodes the structures of historical buildings. (Photograph by Richard P. Jacobs. JLM Visuals. Reproduced by permission.)

acidification. However, the extent of acidification is quite limited.

[*Paul R. Bloom*]

RESOURCES
BOOKS

Bresser, A. H., ed. *Acid Precipitation*. New York: Springer-Verlag, 1990.
Mellanby, K., ed. *Air Pollution, Acid Rain and the Environment*. New York: Elsevier, 1989.
Turck, M. *Acid Rain*. New York: Macmillan, 1990.
Wellburn, A. *Air Pollution and Acid Rain: The Biological Impact*. New York: Wiley, 1988.
Young, P. M. *Acidic Deposition: State of Science and Technology. Summary Report of the U. S. National Acid Precipitation Program*. Washington, DC: U. S. Government Printing Office, 1991.

Acidification

The process of becoming more acidic due to inputs of an acidic substance. The common measure of acidification is a decrease in **pH**. Acidification of soils and natural waters by **acid rain** or **acid** wastes can result in reduced biological productivity if the pH is sufficiently reduced.

Acidity
see **pH**

Acoustics
see **Noise pollution**

Acquired immune deficiency syndrome
see **AIDS**

Activated sludge

The activated **sludge** process is an **aerobic** (oxygen-rich), continuous-flow biological method for the treatment of domestic and **biodegradable** industrial **wastewater**, in which organic matter is utilized by **microorganisms** for life-sustaining processes, that is, for energy for reproduction, digestion, movement, etc. and as a food source to produce cell growth and more microorganisms. During these activities of utilization and degradation of organic materials, degradation products of **carbon dioxide** and water are also formed. The activated sludge process is characterized by the suspension of microorganisms in the wastewater, a mixture referred to as the mixed liquor. Activated sludge is used as part of an overall treatment system, which includes primary treatment of the wastewater for the removal of **particulate** solids before the use of activated sludge as a secondary treatment process to remove suspended and dissolved organic solids.

The conventional activated sludge process consists of an **aeration** basin, with air as the oxygen source, where treatment is accomplished. Soluble (dissolved) organic materials are absorbed through the cell walls of the microorganisms and into the cells, where they are broken down and converted to more microorganisms, **carbon** dioxide, water, and energy. Insoluble (solid) particles are adsorbed on the cell walls, transformed to a soluble form by enzymes (biological catalysts) secreted by the microorganisms, and absorbed through the cell wall, where they are also digested and used by the microorganisms in their life-sustaining processes.

The microorganisms that are responsible for the degradation of the organic materials are maintained in suspension by mixing induced by the aeration system. As the microorganisms are mixed, they collide with other microorganisms and stick together to form larger particles called *floc*. The large flocs that are formed settle more readily than individual cells. These flocs also collide with suspended and colloidal materials (insoluble organic materials), which stick to the flocs and cause the flocs to grow even larger. The microorganisms digest these adsorbed materials, thereby re-opening sites for more materials to stick.

The aeration basin is followed by a secondary clarifier (settling tank), where the flocs of microorganisms with their adsorbed organic materials settle out. A portion of the settled microorganisms, referred to as sludge, are recycled to the aeration basin to maintain an active population of microorganisms and an adequate supply of biological solids for the **adsorption** of organic materials. Excess sludge is wasted by being piped to separate sludge-handling processes. The liquids from the clarifier are transported to facilities for disinfection and final **discharge** to receiving waters, or to tertiary treatment units for further treatment.

Activated sludge processes are designed based on the mixed liquor suspended solids (MLSS) and the organic **loading** of the wastewater, as represented by the **biochemical oxygen demand** (BOD) or **chemical oxygen demand** (COD). The MLSS represents the quantity of microorganisms involved in the treatment of the organic materials in the aeration basin, while the organic loading determines the requirements for the design of the aeration system.

Modifications to the conventional activated sludge process include:

• *Extended aeration.* The mixed liquor is retained in the aeration basin until the production rate of new cells is the same as the decay rate of existing cells, with no excess sludge production. In practice, excess sludge is produced, but the quantity is less than that of other activated sludge processes. This process is often used for the treatment of industrial wastewater that contains complex organic materials requiring long detention times for degradation.

• *Contact stabilization.* A process based on the premise that as wastewater enters the aeration basin (referred to as the contact basin), colloidal and insoluble organic biodegradable materials are removed rapidly by biological **sorption**, synthesis, and flocculation during a relatively short contact time. This method uses a reaeration (stabilization) basin before the settled sludge from the clarifier is returned to the contact basin. The concentrated flocculated and adsorbed organic materials are oxidized in the reaeration basin, which does not receive any addition of raw wastewater.

• *Plug flow.* Wastewater is routed through a series of channels constructed in the aeration basin; wastewater flows through and is treated as a plug as it winds its way through the basin. As the "plug" passes through the tank, the concentrations of organic materials are gradually reduced, with a corresponding decrease in oxygen requirements and microorganism numbers.

• *Step aeration.* Influent wastewater enters the aeration basin along the length of the basin, while the return sludge enters at the head of the basin. This process results in a more

uniform oxygen demand in the basin and a more stable **environment** for the microorganisms; it also results in a lower solids loading on the clarifier for a given mass of microorganisms.

• *Oxidation ditch.* A circular aeration basin (racetrack-shaped) is used, with rotary brush aerators that extend across the width of the ditch. Brush aerators aerate the wastewater, keep the microorganisms in suspension, and drive the wastewater around the circular channel.

[*Judith Sims*]

RESOURCES
BOOKS

Corbitt, R. A. "Wastewater Disposal." In *Standard Handbook of Environmental Engineering*, edited by R. A. Corbitt. New York: McGraw-Hill, 1990.
Junkins, R., K. Deeny, and T. Eckhoff. *The Activated Sludge Process: Fundamentals of Operation.* Boston: Butterworth Publishers, 1983.

Acute effects

Effects that persist in a biologic system for only a short time, generally less than a week. The effects might range from behavioral or color changes to death. Tests for acute effects are performed with humans, animals, plants, insects, and **microorganisms**. Intoxication and a hangover resulting from the consumption of too much alcohol, the common cold, and parathion **poisoning** are examples of acute effects. Generally, little tissue damage occurs as a result of acute effects. The term acute effects should not be confused with acute toxicity studies or acute dosages, which respectively refer to short-term studies (generally less than a week) and short-term dosages (often a single dose). Both chronic and acute exposures can initiate acute effects.

Ansel Easton Adams (1902 – 1984)
American photographer and conservationist

Ansel Adams is best known for his stark black-and-white photographs of **nature** and the American landscape. He was born and raised in San Francisco. Schooled at home by his parents, he received little formal training except as a pianist. A trip to Yosemite Valley as a teenager had a profound influence on him, and **Yosemite National Park** and the Sierra "range of light" attracted him back many times and inspired two great careers: photographer and conservationist. As he observed, "Everybody needs something to believe in [and] my point of focus is conservation." He used his photographs to make that point more vivid and turned it into an enduring legacy.

Adams was a painstaking artist, and some critics have chided him for an overemphasis on technique and for creating in his work "a mood that is relentlessly optimistic." Adams *was* a careful technician, making all of his own prints (reportedly hand-producing over 13,000 in his lifetime), sometimes spending a whole day on one print. He explained: "I have made thousands of photographs of the natural scene, but only those images that were most intensely felt at the moment of exposure have survived the inevitable winnowing of time."

He did winnow, ruthlessly, and the result was a collection of work that introduced millions of people to the majesty and diversity of the American landscape. Not all of Adams's pictures were "uplifting" or "optimistic" images of scenic wonders; he also documented scenes of **overgrazing** in the **arid** Southwest and of incarcerated Japanese-Americans in the Manzanar internment camp.

From the beginning, Adams used his photographs in the cause of **conservation**. His pictures played a major role in the late 1930s in establishing Kings Canyon National Park. Throughout his life, he remained an active, involved conservationist; for many years he was on the Board of the **Sierra Club** and strongly influenced the Club's activities and philosophy.

Ansel Adams's greatest bequest to the world will remain his photographs and advocacy of **wilderness** and the **national park** ideals. Through his work he not only generated interest in environmental conservation, he also captured the beauty and majesty of nature for all generations to enjoy.

[*Gerald L. Young*]

RESOURCES
BOOKS
Adams, Ansel. *Ansel Adams: An Autobiography.* New York: New York Graphic Society, 1984.

PERIODICALS
Cahn, R. "Ansel Adams, Environmentalist." *Sierra* 64 (May–June 1979): 31–49.
Grundberg, A. "Ansel Adams: The Politics of Natural Space." *The New Criterion* 3 (1984): 48–52.

Adaptation

All members of a population share many characteristics in common. For example, all finches in a particular forest are alike in many ways. But if many hard-to-shell seeds are found in the forest, those finches with stronger, more conical bills will have better rates of reproduction and survival than finches with thin bills. Therefore, a conical, stout bill can be considered an adaptation to that forest **environment**. Any specialized characteristic that permits an individual to survive and reproduce is called an adaptation. Adaptations may result either from an individual's genetic heritage or from its ability to learn. Since successful genetic adaptations are more likely to be passed from generation to generation through the survival of better adapted organisms, adaptation can be viewed as the force that drives biological **evolution**.

Adaptive management

Adaptive management is taking an idea, implementing it, and then documenting and learning from any mistakes or benefits of the experiment. The basic idea behind adaptive management is that natural systems are too complex, too non-linear, and too multi-scale to be predictable. Management policies and procedures must therefore become more adaptive and capable of change to cope with unpredictable systems.

Advocates suggest treating management policies as experiments, which are then designed to maximize learning rather than focusing on immediate resource yields. If the environmental and resource systems on which human beings depend are constantly changing, then societies who utilize that learning cannot rely on those systems to sustain continued use. Adaptive management mandates a continual experimental process, an on-going process of reevaluation and reassessment of planning methods and human actions, and a constant long-term monitoring of environmental impacts and change. This would keep up with the constant change in the environmental systems to which the policies or ideas are to be applied.

The Grand Canyon Protection Act of 1992 is one example of adaptive management at work. It entails the study and monitoring of the **Glen Canyon Dam** and the operational effects on the surrounding **environment**, both ecological and biological.

Haney and Power suggest that "uncertainty and complexity frustrate both science and management, and it is only by combining the best of both that we use all available tools to manage ecosystems sustainably." However, Fikret Berkes and colleagues claim that adaptive management can be attained by approaching it as a rediscovery of traditional ecological knowledge among **indigenous peoples**: "These traditional systems had certain similarities to adaptive management with its emphasis on feedback learning, and its treatment of uncertainty and unpredictability intrinsic to all ecosystems."

An editorial in the journal *Environment* offered the rather inane statement that adaptive management "has not realized its promise." The promise is in the idea, but implementation begins with people. Adaptive management, like **Smart Growth** and other seemingly innovative approaches to **land use** and environmental management, is plagued by the problem of how to get people to actually put into practice what is proposed. Even for practical ideas the problem remains the same: not science, not technology, but human willfulness and human behavior. For policies or plans to be truly adaptive, the people themselves must be willing to adapt.

Haney and Power provide the conclusion: "When properly integrated, the [adaptive management] process is continuous and cyclic; components of the adaptive management model evolve as information is gained and social and ecological systems change. Unless management is flexible and innovative, outcomes become less sustainable and less accepted by stakeholders. Management will be successful in the face of complexity and uncertainty only with holistic approaches, good science, and critical evaluation of each step. Adaptive management is where it all comes together."

[*Gerald L. Young*]

RESOURCES
BOOKS

Holling, C. S., ed. *Adaptive Environmental Assessment and Management.* NY: John Wiley & Sons, 1978.

PERIODICALS

Haney, Alan, and Rebecca L. Power. "Adaptive Management for Sound Ecosystem Management." *Environmental Management* 20, no. 6 (November/December 1996): 879–886.

McLain, Rebecca J., and Robert G. Lee. "Adaptive management: Promises and Pitfalls." *Environmental Management* 20, no. 4 (July/August, 1996): 437–448.

Shindler, Bruce, Brent Steel, and Peter List. "Public Judgments of Adaptive Management: A Response from Forest Communities." *Journal of Forestry* 96, no. 6 (June, 1996): 4–12.

Walters, Carl. *Adaptive Management of Renewable Resources.* NY: Macmillan, 1986.

Walters, Carl J. "Ecological Optimization and Adaptive Management." *Annual Review of Ecology and Systematics* 9 (1978): 157–188.

Adirondack Mountains

A range of mountains in northeastern New York, containing Mt. Marcy (5,344 ft; 1,644 m), the state's highest point. Bounded by the Mohawk Valley on the south, the St. Lawrence Valley on the northeast, and by the **Hudson River** and Lake Champlain on the east, the Adirondack Mountains form the core of Adirondack Park. This park is one of the earliest and most comprehensive examples of regional planning in the United States. The regional plan attempts to balance conflicting interests of many users at the same time as it controls environmentally destructive development. Although the plan remains controversial, it has succeeded

in largely preserving one of the last and greatest **wilderness** areas in the East.

The Adirondacks serve a number of important purposes for surrounding populations. Vacationers, hikers, canoeists, and anglers use the area's 2,300 wilderness lakes and extensive river systems. The state's greatest remaining forests stand in the Adirondacks, providing animal **habitat** and serving recreational visitors. Timber and mining companies, employing much of the area's resident population, also rely on the forests, some of which contain the East's most ancient old-growth groves. Containing the headwaters of numerous rivers, including the Hudson, Adirondack Park is an essential source of clean water for farms and cities at lower elevations.

Adirondack Park was established by the New York State Constitution of 1892, which mandates that the region shall remain "forever wild." Encompassing six million acres (2.4 million ha), this park is the largest wilderness area in the eastern United States—nearly three times the size of **Yellowstone National Park**. Only a third of the land within park boundaries, however, is owned by the state of New York. Private mining and timber concerns, public agencies, several towns, thousands of private cabins, and 107 units of local government occupy the remaining property.

Because the development interests of various user groups and visitors conflict with the state constitution, a comprehensive regional **land use** plan was developed in 1972–1973. The novelty of the plan lay in the large area it covered and in its jurisdiction over land uses on private land as well as **public land**. According to the regional plan, all major development within park boundaries must meet an extensive set of environmental safeguards drawn up by the state's Adirondack Park Agency. Stringent rules and extensive regulation frustrate local residents and commercial interests, who complain about the plan's complexity and resent "outsiders" ruling on what Adirondackers are allowed to do. Nevertheless, this plan has been a milestone for other regions trying to balance the interests of multiple users. By controlling extensive development, the park agency has preserved a wilderness resource that has become extremely rare in the eastern United States. The survival of this century-old park, surrounded by extensive development, demonstrates the value of preserving wilderness in spite of ongoing controversy.

In recent years forestry and **recreation** interests in the Adirondacks have encountered a new environmental problem in **acid** precipitation. Evidence of deleterious effects of **acid rain** and snow on aquatic and terrestrial vegetation began to accumulate in the early 1970s. Studies revealed that about half of the Adirondack lakes situated above 3,300 ft (1,000 m) have **pH** levels so low that all fish have disappeared. Prevailing winds put these mountains directly downstream of urban and industrial regions of western New York and southern Ontario. Because they form an elevated obstacle to weather patterns, these mountains capture a great deal of precipitation carrying acidic sulfur and **nitrogen oxides** from upwind industrial cities.

[*Mary Ann Cunningham*]

RESOURCES
BOOKS

Ciroff, R. A., and G. Davis. *Protecting Open Space: Land Use Control in the Adirondack Park.* Cambridge, MA: Ballinger, 1981.

Davis, G., and T. Duffus. *Developing a Land Conservation Strategy.* Elizabethtown, NY: Adirondack Land Trust, 1987.

Graham, F. J. *The Adirondack Park: A Political History.* New York: Knopf, 1978.

Popper, F. J. *The Politics of Land Use Reform.* Madison, WI: University of Wisconsin Press, 1981.

Adsorption

The removal of ions or molecules from solutions by binding to solid surfaces. **Phosphorus** is removed from water flowing through soils by adsorption on **soil** particles. Some pesticides adsorb strongly on soil particles. Adsorption by suspended solids is also an important process in natural waters.

AEC

see **Atomic Energy Commission**

AEM

see **Agricultural Environmental Management**

Aeration

In discussions of plant growth, aeration refers to an exchange that takes place in **soil** or another medium allowing oxygen to enter and **carbon dioxide** to escape into the **atmosphere**. Crop growth is often reduced when aeration is poor. In geology, particularly with reference to **groundwater**, aeration is the portion of the earth's crust where the pores are only partially filled with water. In relation to **water treatment**, aeration is the process of exposing water to air in order to remove such undesirable substances in drinking water as iron and manganese.

Aerobic

Refers to either an **environment** that contains molecular oxygen gas (O_2); an organism or tissue that requires oxygen for its **metabolism**; or a chemical or biological process that requires oxygen. Aerobic organisms use molecular oxygen in **respiration**, releasing **carbon dioxide** (CO_2) in return. These organisms include mammals, fish, birds, and green plants, as well as many of the lower life forms such as **fungi**, algae, and sundry bacteria and actinomycetes. Many, but not all, organic **decomposition** processes are aerobic; a lack of oxygen greatly slows these processes.

Aerobic/anaerobic systems

Most living organisms require oxygen to function normally, but a few forms of life exist exclusively in the absence of oxygen and some can function both in the presence of oxygen (aerobically) and in its absence (anaerobically). Examples of **anaerobic** organisms are found in bacteria of the genus *Clostridium*, in parasitic protozoans from the gastrointestinal tract of humans and other vertebrates, and in ciliates associated with sulfide-containing sediments. Organisms capable of switching between **aerobic** and anaerobic existence are found in forms of **fungi** known as yeasts. The ability of an organism to function both aerobically and anaerobically increases the variety of sites in which it is able to exist and conveys some advantages over organisms with less adaptive potential.

Microbial decay activity in **nature** can occur either aerobically or anaerobically. Aerobic **decomposers** of compost and other organic substrates are generally preferable because they act more quickly and release fewer noxious odors. Large **sewage treatment** plants use a two-stage digestion system in which the first stage is **anaerobic digestion** of **sludge** that produces flammable **methane** gas that may be used as fuel to help operate the plant. Sludge digestion continues in the aerobic second stage, a process which is easier to control but more costly because of the power needed to provide **aeration**. Although most fungi are generally aerobic organisms, yeasts used in bread making and in the production of fermented beverages such as wine and beer can metabolize anaerobically. In the process, they release ethyl alcohol and the **carbon dioxide** that causes bread to rise.

Tissues of higher organisms may have limited capability for anaerobic **metabolism**, but they need elaborate compensating mechanisms to survive even brief periods without oxygen. For example, human muscle tissue is able to metabolize anaerobically when blood cannot supply the large amounts of oxygen needed for vigorous activity. Muscle contraction requires an energy-rich compound called adenosine triphosphate (ATP). Muscle tissue normally contains enough ATP for 20–30 seconds of intense activity. ATP must then be metabolically regenerated from glycogen, the muscle's primary energy source. Muscle tissue has both aerobic and anaerobic metabolic systems for regenerating ATP from glycogen. Although the aerobic system is much more efficient, the anaerobic system is the major energy source for the first minute or two of exercise. The **carbon** dioxide released in this process causes the heart rate to increase. As the heart beats faster and more oxygen is delivered to the muscle tissue, the more efficient aerobic system for generating ATP takes over. A person's physical condition is important in determining how well the aerobic system is able to meet the needs of continued activity. In fit individuals who exercise regularly, heart function is optimized, and the heart is able to pump blood rapidly enough to maintain aerobic metabolism. If the oxygen level in muscle tissue drops, anaerobic metabolism will resume. Toxic products of anaerobic metabolism, including lactic **acid**, accumulate in the tissue, and muscle fatigue results.

Other interesting examples of limited anaerobic capability are found in the animal kingdom. Some diving ducks have an **adaptation** that allows them to draw oxygen from stored oxyhemoglobin and oxymyoglobin in blood and muscles. This adaptation permits them to remain submerged in water for extended periods. To prevent desiccation, mussels and clams close their shells when out of the water at low tide, and their metabolism shifts from aerobic to anaerobic. When once again in the water, the animals rapidly return to aerobic metabolism and purge themselves of the acid products of anaerobiosis accumulated while they were dry.

[*Douglas C. Pratt*]

RESOURCES
BOOKS

Lea, A.G.H., and Piggott, J. R. *Fermented beverage production.* New York: Blackie, 1995.

McArdle, W. D. *Exercise Physiology: Energy, Nutrition, and Human Performance.* 4th ed. Baltimore: Williams & Wilkins, 1996.

Stanbury, P. F., Whitaker, A., and Hall, S. J. *Principles of Fermentation Technology.* 2nd ed. Tarrytown, N.Y.: Pergamon, 1995.

PERIODICALS

Klass, D. L. "Methane from Anaerobic Fermentation." *Science* 223 (1984): 1021.

Aerobic sludge digestion

Wastewater treatment plants produce organic **sludge** as wastewater is treated; this sludge must be further treated before ultimate disposal. Sludges are generated from primary settling tanks, which are used to remove settable, **particulate**

solids, and from secondary clarifiers (settling basins), which are used to remove excess **biomass** production generated in secondary biological treatment units.

Disposal of sludges from wastewater treatment processes is a costly and difficult problem. The processes used in sludge disposal include: (1) reduction in sludge volume, primarily by removal of water, which constitutes 97–98% of the sludge; (2) reduction of the volatile (organic) content of the sludge, which eliminates nuisance conditions by reducing putrescibility and reduces threats to human health by reducing levels of **microorganisms**; and (3) ultimate disposal of the residues.

Aerobic sludge digestion is one process that may be used to reduce both the organic content and the volume of the sludge. Under aerobic conditions, a large portion of the organic matter in sludge may be oxidized biologically by microorganisms to **carbon dioxide** and water. The process results in approximately 50% reduction in solids content. Aerobic sludge digestion facilities may be designed for batch or continuous flow operations. In batch operations, sludge is added to a reaction tank while the contents are continuously aerated. Once the tank is filled, the sludges are aerated for two to three weeks, depending on the types of sludge. After **aeration** is discontinued, the solids and liquids are separated. Solids at concentrations of 2–45 are removed, and the clarified liquid supernatant is decanted and recycled to the wastewater treatment plant. In a continuous flow system, an aeration tank is utilized, followed by a settling tank.

Aerobic sludge digestion is usually used only for biological sludges from secondary treatment units, in the absence of sludges from primary treatment units. The most commonly used application is for the treatment of sludges wasted from extended aeration systems (which is a modification of the **activated sludge** system). Since there is no addition of an external food source, the microorganisms must utilize their own cell contents for metabolic purposes in a process called endogenous **respiration**. The remaining sludge is a mineralized sludge, with remaining organic materials comprised of cell walls and other cell fragments that are not readily **biodegradable**.

The advantages of using aerobic digestion, as compared to the use of **anaerobic digestion** include: (1) simplicity of operation and maintenance; (2) lower capital costs; (3) lower levels of **biochemical oxygen demand** (BOD) and **phosphorus** in the supernatant; (4) fewer effects from upsets such as the presence of toxic interferences or changes in **loading** and **pH**; (5) less odor; (6) nonexplosive; (7) greater reduction in grease and hexane solubles; (8) greater sludge **fertilizer** value; (9) shorter retention periods; and (10) an effective alternative for small wastewater treatment plants.

Disadvantages include: (1) higher operating costs, especially energy costs; (2) highly sensitive to ambient temperature (operation at temperatures below 59°F [15°C]) may require excessive retention times to achieve stabilization; if heating is required, aerobic digestion may not be cost-effective); (3) no useful byproduct such as **methane** gas that is produced in **anaerobic** digestion; (4) variability in the ability to dewater to reduce sludge volume; (5) less reduction in volatile solids; and (6) unfavorable economics for larger wastewater treatment plants.

[*Judith Sims*]

RESOURCES
BOOKS

Corbitt, R. A. "Wastewater Disposal." In *Standard Handbook of Environmental Engineering*, edited by R. A. Corbitt. New York: McGraw-Hill, 1990.
Gaudy Jr., A. F., and E. T. Gaudy. *Microbiology for Environmental Scientists and Engineers*. New York: McGraw-Hill, 1980.
Peavy, H. S., D. R. Rowe, and G. Tchobanoglous. *Environmental Engineering*. New York: McGraw-Hill, 1985.

Aerosol

A suspension of particles, liquid or solid, in a gas. The term implies a degree of permanence in the suspension, which puts a rough upper limit on particle size at a few tens of micrometers at most (1 micrometer = 0.00004 in). Thus in proper use the term connotes the ensemble of the particles and the suspending gas.

The atmospheric aerosol has two major components, generally referred to as coarse and fine particles, with different sources and different composition. Coarse particles result from mechanical processes, such as grinding. The smaller particles are ground, the more surface they have per unit of mass. Creating new surface requires energy, so the smallest average size that can be created by such processes is limited by the available energy. It is rare for such mechanically generated particles to be less than 1 μm (0.00004 in.) in diameter. Fine particles, on the other hand, are formed by condensation from the vapor phase. For most substances, condensation is difficult from a uniform gaseous state; it requires the presence of pre-existing particles on which the vapors can deposit. Alternatively, very high concentrations of the vapor are required, compared with the concentration in equilibrium with the condensed material.

Hence, fine particles form readily in **combustion** processes when substances are vaporized. The gas is then quickly cooled. These can then serve as nuclei for the formation of larger particles, still in the fine particle size range, in the presence of condensable vapors. However, in the **atmosphere** such particles become rapidly more scarce with increasing size, and are relatively rare in sizes much larger than a few micrometers. At about 2 μm (0.00008 in.), coarse and fine particles are about equally abundant.

Using the term strictly, one rarely samples the atmospheric aerosol, but rather the particles out of the aerosol. The presence of aerosols is generally detected by their effect on light. Aerosols of a uniform particle size in the vicinity of the wavelengths of visible light can produce rather spectacular optical effects. In the laboratory, such aerosols can be produced by condensation of the heated vapors of certain oils on nuclei made by evaporating salts from heated filaments. If the suspending gas is cooled quickly, particle size is governed by the supply of vapor compared with the supply of nuclei, and the time available for condensation to occur. Since these can all be made nearly constant throughout the gas, the resulting particles are quite uniform. It is also possible to produce uniform particles by spraying a dilute solution of a soluble material, then evaporating the solvent. If the spray head is vibrated in an appropriate frequency range, the drops will be uniform in size, with the size controlled by the frequency of vibration and the rate of flow of the spray. Obviously, the final particle size is also a function of the concentration of the sprayed solution.

[*James P. Lodge Jr.*]

RESOURCES
BOOKS

Jennings, S. G., ed. *Aerosol Effects on Climate*. Tucson, AZ: University of Arizona Press, 1993.
Reist, P. *Aerosol Science and Technology*. New York: McGraw-Hill, 1992.

PERIODICALS

Monastersky, R. "Aerosols: Critical Questions for Climate." *Science News* 138 (25 August 1990): 118.
Sun, M. "Acid Aerosols Called Health Hazard." *Science* 240 (24 June 1988): 1727.

Aflatoxin

Toxic compounds produced by some **fungi** and among the most potent naturally occurring carcinogens for humans and animals. Aflatoxin intake is positively related to high incidence of liver **cancer** in humans in many developing countries. In many farm animals aflatoxin can cause acute or chronic diseases. Aflatoxin is a metabolic by-product produced by the fungi *Aspergillus flavus* and the closely related **species** *Aspergillus parasiticus* growing on grains and decaying organic compounds. There are four naturally occurring aflatoxins: B$_1$, B$_2$, G$_1$, and G$_2$. All of these compounds will fluoresce under a UV (black) light around 425–450 nm providing a qualitative test for the presence of aflatoxins. In general, starch grains, such as corn, are infected in storage when the moisture content of the grain reaches 17–18% and the temperature is 79–99°F (26–37°C). However, the fungus may also infect grain in the field under hot, dry conditions.

African Wildlife Foundation

The African Wildlife Foundation (AWF), headquartered in Washington, DC, was established in 1961 to promote the protection of the animals native to Africa. The group maintains offices in both Washington, DC, and Nairobi, Kenya. The African headquarters promotes the idea that Africans themselves are best able to protect the wildlife of their continent. AWF also established two colleges of **wildlife management** in Africa (Tanzania and Cameroon), so that rangers and park and reserve wardens can be professionally trained. **Conservation** education, especially as it relates to African wildlife, has always been a major AWF goal—in fact, it has been the association's primary focus since its inception.

AWF carries out its mandate to protect Africa's wildlife through a wide range of projects and activities. Since 1961, AWF has provided a radio communication network in Africa, as well as several airplanes and jeeps for anti-poaching patrols. These were instrumental in facilitating the work of Dr. **Richard Leakey** in the Tsavo National Park, Kenya. In 1999, the African Hearlands project was set up and to try to connect large areas of wild land which is home to wild animals. They also attempt to involve people who live adjacent to protected wildlife areas by asking them to take joint responsibility for **natural resources**. The program demonstrates that land conservation and the needs of neighboring people and their livestock can be balanced, and the benefits shared. Currently there are four heartland areas: Maasai Steppe, Kilimanjaro, Virunga, and Samburu.

Another highly successful AWF program is the Elephant Awareness Campaign. Its slogan, "Only **Elephants** Should Wear Ivory," has become extremely popular, both in Africa and in the United States, and is largely responsible for bringing the plight of the African elephant (*Loxodonta africana*) to public awareness.

Although AWF is concerned with all the wildlife of Africa, in recent years the group has focused on saving African elephants, black **rhinoceroses** (*Diceros bicornis*), and mountain gorillas (*Gorilla gorilla berengei*). These **species** are seriously endangered, and are benefiting from AWF's Critical Habitats and Species Program, which works to aid these and other animals in critical danger.

From its inception, AWF has supported education centers, wildlife clubs, **national parks**, and reserves. There is even a course at the College of African Wildlife Management in Tanzania that allows students to learn community conservation activities and helps park officials learn to work with residents living adjacent to protected areas. AWF also involves teachers in its endeavors with a series of publications, *Let's Conserve Our Wildlife*. Written in Swahili, the series includes teacher's guides and has been used in both elementary schools and adult literacy classes in African villages. AWF also publishes the quarterly magazine *Wildlife News*.

[*Cathy M. Falk*]

RESOURCES

ORGANIZATIONS

African Wildlife Foundation., 1400 16th Street, NW, Washington, DC USA 20036 (202) 939-3333, Fax: (202) 939-3332, Email: africanwildlife@awf.org, <http://www.awf.org>

Africanized bees

The Africanized bee (*Apis mellifera scutellata*), or "killer bee," is an extremely aggressive honeybee. This bee developed when African honeybees were brought to Brazil to mate with other bees to increase honey production. The imported bees were accidentally released and they have since spread northward, traveling at a rate of 300 mi (483 km) per year. The bees first appeared in the United States at the Texas-Mexico border in late 1990.

The bees get their "killer" title because of their vigorous defense of colonies or hives when disturbed. Aside from temperament, they are much like their counterparts now in the United States, which are European in lineage. Africanized bees are slightly smaller than their more passive cousins.

Honeybees are social insects and live and work together in colonies. When bees fly from plant to plant, they help pollinate flowers and crops. Africanized bees, however, seem to be more interested in reproducing than in honey production or **pollination**. For this reason they are constantly swarming and moving around, while domestic bees tend to stay in local, managed colonies. Because Africanized bees are also much more aggressive than domestic honey bees when their colonies are disturbed, they can be harmful to people who are allergic to bee stings.

More problematic than the threat to humans, however, is the impact the bees will have on fruit and vegetable industries in the southern parts of the United States. Many fruit and vegetable growers depend on honey bees for pollination, and in places where the Africanized bees have appeared, honey production has fallen by as much as 80%. Beekeepers in this country are experimenting with "re-queening" their

An Africanized bee collecting grass pollen in Brazil. (Photograph by Scott Camazine. Photo Researchers Inc. Reproduced by permission.)

colonies regularly to ensure that the colonies reproduce gentle offspring.

Another danger is the propensity of the Africanized bee to mate with honey bees of European lineage, a kind of "infiltration" of the **gene pool** of more domestic bees. Researchers from the **U.S. Department of Agriculture** (USDA) are watching for the results of this interbreeding, particularly for those bees that display European-style physiques and African behaviors, or vice versa.

When Africanized bees first appeared in southern Texas, researchers from the USDA's Honeybee Research Laboratory in Weslaco, Texas, destroyed the colony, estimated at 5,000 bees. Some of the members of the 3-lb (1.4 kg) colony were preserved in alcohol and others in freezers for future analysis. Researchers are also developing management techniques, including the annual introduction of young mated European queens into domestic hives, in an attempt to maintain gentle production stock and ensure honey production and pollination.

As of 2002, there were 140 counties in Texas, nine in New Mexico, nine in California, three in Nevada, and all of the 15 counties in Arizona in which Africanized bee colonies had been located. There have also been reported colonies in Puerto Rico and the Virgin Islands. Southern

Nevada bees were almost 90% Africanized in June of 2001. Most of Texas has been labeled as a quarantine zone, and beekeepers are not able to move hives out of these boundaries. The largest colony found to date was in southern Phoenix, Arizona. The hive was almost 6 ft (1.8 m) long and held about 50,000 Africanized bees.

[*Linda Rehkopf*]

RESOURCES
PERIODICALS

"African Bees Make U.S. Debut." *Science News* 138 (October 27, 1990): 261.

Barinaga, M. "How African Are 'Killer' Bees?" *Science* 250 (November 2, 1990): 628–629.

Hubbell, S. "Maybe the 'Killer' Bee Should Be Called the 'Bravo' Instead." *Smithsonian* 22 (September 1991): 116–124.

White, W. "The Bees From Rio Claro." *The New Yorker* 67 (September 16, 1991): 36–53.

Winston, M. *Killer Bees: The Africanized Honey Bee in the Americas*. Cambridge: Harvard University Press, 1992.

OTHER

"Africanized Bees in the Americas." *Sting Shield.com Page*. April 25, 2002 [cited May 2002]. <http://www.stingshield.com/!ahbtitl.htm>.

Agency for Toxic Substances and Disease Registry

The Agency for Toxic Substances and Disease Registry (ATSDR) studies the health effects of hazardous substances in general and at specific locations. As indicated by its title, the Agency maintains a registry of people exposed to toxic **chemicals**. Along with the **Environmental Protection Agency** (EPA), ATSDR prepares and updates profiles of toxic substances. In addition, ATSDR assesses the potential dangers posed to human health by exposure to hazardous substances at Superfund sites. The Agency will also perform health assessments when petitioned by a community. Though ATSDR's early health assessments have been criticized, the Agency's later assessments and other products are considered more useful.

ATSDR was created in 1980 by the **Comprehensive Environmental Response, Compensation, and Liability Act** (CERCLA), also known as the Superfund, as part of the **U.S. Department of Health and Human Services**. As originally conceived, ATSDR's role was limited to performing health studies and examining the relationship between toxic substances and disease. The Superfund Amendments and Reauthorization Act (SARA) of 1986 codified ATSDR's responsibility for assessing health threats at Superfund sites. ATSDR, along with the national Centers for Disease Control and state health departments, conducts health surveys in communities near locations that have been placed on the Superfund's **National Priorities List** for clean up. ATSDR has preformed 951 health assessments in the two years after the law was passed. Approximately one quarter of these assessments were memos or reports that had been completed prior to 1986 and were simply re-labeled as health assessments.

These first assessments have been harshly criticized. The General Accounting Office (GAO), a congressional agency that reviews the actions of the federal administration, charged that most of these assessments were inadequate. Some argued that the agency was underfunded and poorly organized. Recently, ATSDR received less than 5% of the $1.6 billion appropriated for the Superfund project.

Subsequent health assessments, more than 200 of them, have generally been more complete, but they still may not be adequate in informing the community and the EPA of the dangers at specific sites. In general, ATSDR identifies a local agency to help prepare the health surveys. Unlike many of the first assessments, more recent surveys now include site visits and face-to-face interviews. However, other data on environmental effects are limited. ATSDR only considers environmental information provided by the companies that created the hazard or data collected by the EPA. In addition, ATSDR only assesses health risks from illegal emissions, not from "permitted" emissions. Some scientists contend that not enough is known about the health effects of exposure to hazardous substances to make conclusive health assessments.

Reaction to the performance of ATSDR's other functions has been generally more positive. As mandated by SARA, ATSDR and the EPA have prepared hundreds of toxicological profiles of hazardous substances. These profiles have been judged generally helpful, and the GAO praised ATSDR's registry of people who have been exposed to toxic substances.

[*Alair MacLean*]

RESOURCES

BOOKS

Environmental Epidemiology: Public Health and Hazardous Wastes. National Research Council. Committee on Environmental Epidemiology. Washington, DC: National Academy Press, 1991.

Lewis, S., B. Keating, and D. Russell. *Inconclusive by Design: Waste, Fraud and Abuse in Federal Environmental Health Research.* Boston: National Toxics Campaign Fund; and Harvey, LA: Environmental Health Network, 1992.

OTHER

Superfund: Public Health Assessments Incomplete and of Questionable Value. Washington, DC: General Accounting Office, 1991.

ORGANIZATIONS

The ATSDR Information Center, , (404) 498-0110, Fax: (404) 498-0057, Toll Free: (888) 422-8737, Email: ATSDRIC@cdc.gov, <http://www.atsdr.cdc.gov>

Agent Orange

Agent Orange is a **herbicide** recognized for its use during the Vietnam War. It is composed of equal parts of two **chemicals**: **2,4-D** and **2,4,5-T**. A less potent form of the herbicide has also been used for clearing heavy growth on a commercial basis for a number of years. However, it does not contain 2,4-D. On a commercial level, the herbicide was used in forestry control as early as the 1930s. In the 1950s through the 1960s, Agent Orange was also exported. For example, New Brunswick, Canada, was the scene of major Agent Orange spraying to control forests for industrial development. In Malaysia in the 1950s, the British used compounds with the chemical mixture 2,4,5-T to clear communication routes.

In the United States, herbicides were considered for military use towards the end of World War II, during the action in the Pacific. However, the first American military field tests were actually conducted in Puerto Rico, Texas, and Fort Drum, New York, in 1959.

That same year—1959—the Crops Division at Fort Detrick, Maryland initiated the first large-scale military **defoliation** effort. The project involved the aerial application of Agent Orange to about 4 mi^2 (10.4 km^2) of vegetation. The experiment proved highly successful; the military had found an effective tool. By 1960, the South Vietnamese government, aware of these early experiments, had requested that the United States conduct trials of these herbicides for use against guerrilla forces. Spraying of Agent Orange in Southeast Asia began in 1961. South Vietnam President Diem stated that he wanted this "powder" in order to destroy the rice and the food crops that would be used by the Viet Cong. Thus began the use of herbicides as a weapon of war.

The United States military became involved, recognizing the limitations of fighting in foreign territory with troops that were not accustomed to jungle conditions. The military wanted to clear communication lines and open up areas of **visibility** in order to enhance their opportunities for success. Eventually, the United States military took complete control of the spray missions. Initially, there were to be restrictions: the spraying was to be limited to clearing power lines and roadsides, railroads and other lines of communications and areas adjacent to depots. Eventually, the spraying was used to defoliate the thick jungle brush, thereby obliterating enemy hiding places.

Once under the authority of the military, and with no checks or restraints, the spraying continued to increase in

Deforestation of the Viet Cong jungle in South Vietnam. (AP/Wide World Photos. Reproduced by permission.)

intensity and abandon, escalating in scope because of military pressure. It was eventually used to destroy crops, mainly rice, in an effort to deprive the enemy of food. Unfortunately, the civilian population—Vietnamese men, women, and children—was also affected. The United States military sprayed 3.6 million acres (1.5 million ha) with 19 million gal (720 million l) of Agent Orange over nine years.

The spraying also became useful in clearing military base perimeters, cache sites, and waterways. Base perimeters were often sprayed more than once. In the case of dense jungle growth, one application of spray was made for the upper and another for the lower layers of vegetation. Inland forests, mangrove forests, and cultivated lands were all targets. Through Project Ranch Hand—the Air Force team assigned to the spray missions—Agent Orange became the most widely produced and dispensed defoliant in Vietnam.

Military requirements for herbicide use were developed by the Army's Chemical Operations Division, J-3, Military Assistance Command, Vietnam, (MACV). With Project Ranch Hand underway, the spray missions increased monthly after 1962. This increase was made possible by the continued military promises to stay away from the civilians

or to re-settle those civilians and re-supply the food in any areas where herbicides destroyed the food of the innocent. These promises were never kept. The use of herbicides for crop destruction peaked in 1965 when 45% of the total spraying was designed to destroy crops.

Initially, the aerial spraying took place near Saigon. Eventually the geographical base was widened. During the 1967 expansion period of herbicide procurement, when requirements had become greater than the industries' ability to produce, the Air Force and Joint Chiefs of Staff become actively involved in the herbicide program. All production for commercial use was diverted to the military, and the Department of Defense (DOD) was appointed to deal with problems of procurement and production. Commercial producers were encouraged to expand their facilities and build new plants, and the DOD made attractive offers to companies that might be induced to manufacture herbicides. A number of companies were awarded contracts. Working closely with the military, certain chemical companies sent technical advisors to Vietnam to instruct personnel on the methods and techniques necessary for effective use of the herbicides.

During the peak of the spraying, approximately 129 sorties were flown per aircraft. Twenty-four UC-123B aircraft were used, averaging 39 sorties per day. In addition, there were trucks and helicopters that went on spraying missions, backed up by such countries as **Australia**. C-123 cargo planes and helicopters were also used. Helicopters flew without cargo doors so that frequent ground fire could be returned. But the rotary blades would kick up gusts of spray, thereby delivering a powerful dose onto the faces and bodies of the men inside the plane.

The dense Vietnamese jungle growth required two applications to defoliate both upper and lower layers of vegetation. On the ground, both enemy troops and Vietnamese civilians came in contact with the defoliant. American troops were also exposed. They could inhale the fine misty spray or be splashed in the sudden and unexpected deluge of an emergency dumping. Readily absorbing the chemicals through their skin and lungs, hundreds of thousands of United States military troops were exposed as they lived on the sprayed bases, slept near empty drums, and drank and washed in water in areas where defoliation had occurred. They ate food that had been brushed with spray. Empty herbicide drums were indiscriminately used and improperly stored. Volatile fumes from these drums caused damage to shade trees and to anyone near the fumes. Those handling the herbicides in support of a particular project goal had the unfortunate opportunity of becoming directly exposed on a consistent basis. Nearly three million veterans served in Southeast Asia. There is growing speculation that nearly everyone who was in Vietnam was eventually exposed to some degree—far less a possibility for those stationed in urban centers or on the waters.

According to official sources, in addition to the Ranch Hand group at least three groups were exposed:

• A group considered secondary support personnel. This included Army pilots who may have been involved in helicopter spraying, along with the Navy and Marine pilots.

• Those who transported the herbicide to Saigon, and from there to Bien Hoa and Da Nang. Such personnel transported the herbicide in the omnipresent 55-gallon (208-l) containers.

• Specialized mechanics, electricians, and technical personnel assigned to work on various aircraft. Many of this group were not specifically assigned to Ranch Hand but had to work in aircraft that were repeatedly contaminated.

Agent Orange was used in Vietnam in undiluted form at the rate of 3–4 gal (11.4-15.2 l) per acre. 13.8 lb (6.27 kg) of the chemical 2,4,5-T were added to 12 lb (5.5 kg) of 2,4-D per acre, a nearly 50-50 ratio. This intensity is 13.3 lb (6.06 kg) per acre more than was recommended by the military's own manual. Computer tapes (HERBS TAPES) now available show that some areas were sprayed

as much as 25 times in just a few short months, thereby dramatically increasing the exposure to anyone within those sprayed areas. Between 1962 and 1971 an estimated 11.2 million gal (42.4 million l) of Agent Orange were dumped over South Vietnam.

Evaluations show that the chemical had killed and defoliated 90–95% of the treated vegetation. Thirty-six percent of all mangrove forest areas in South Vietnam were destroyed. Viet Cong tunnel openings, caves, and above ground shelters revealed to the aircraft after the herbicides were shipped in drums identified by an orange stripe and a contract identification number that enabled the government to identify the specific manufacturer. The drums were sent to a number of central **transportation** points for shipment to Vietnam.

Agent Orange is contaminated by the chemical **dioxin**, specifically TCDD. In Vietnam, the dioxin concentration in Agent Orange varied from **parts per billion** (ppb) to **parts per million** (ppm), depending on each manufacturer's production methods. The highest reported concentration in Agent Orange was 45 ppm. The **Environmental Protection Agency** (EPA) evacuated **Times Beach**, Missouri, when tests revealed **soil** samples there with two parts per billion of dioxin. The EPA has stated that one ppb is dangerous to humans.

Ten years after the spraying ended, the agricultural areas remained barren. Damaging amounts of dioxin stayed in the soil thus infecting the food chain and exposing the Vietnamese people. As a result there is some concern that the high levels of TCDD are responsible for infant **mortality**, **birth defects**, and spontaneous abortions that occur in higher numbers in the once sprayed areas of Vietnam. Another report indicates that thirty years after Agent Orange contaminated the area, there is 100 times as much dioxin found in the bloodstream of people living in the area than those living in non-contaminated areas of Vietnam. This is a result of the dioxin found in the soil of the once heavily sprayed land. The chemical is then passed on to humans through the food they eat. Consequently, dioxin is also spread to infants through the mother's breast milk, which will undoubtedly affect the child's development.

In 1991 Congress passed the Agent Orange Act(-Public Law 102-4), which funded the extensive scientific study of the long-term health effects of Agent Orange and other herbicides used in Vietnam. As of early 2002, Agent Orange had been linked to the development of peripheral neuropathy, type II diabetes, prostate **cancer**, multiple myeloma, lymphomas, soft tissue sarcomas, and respiratory cancers. Researchers have also found a possible correlation between dioxin and the development of spinal bifida, a birth defect, and childhood **leukemia** in offspring of exposed vets. It is important to acknowledge the **statistics** do not necessarily show a strong link between exposure to Agent Orange or TCDD and some of the conditions listed above. However, Vietnam veterans who were honorably discharged and have any of these "presumptive" conditions (i.e., conditions presumed caused by wartime exposure) are entitled to Veterans Administration (VA) health care benefits and disability compensation under federal law. Unfortunately many Vietnamese civilians will not receive any benefits despite the evidence that they continue to suffer from the affects of Agent Orange.

[*Liane Clorfene Casten and Paula Anne Ford-Martin*]

RESOURCES
BOOKS

Committee to Review the Health Effects in Vietnam Veterans of Exposure to Herbicides, Division of Health Promotion and Disease Prevention, Institute of Medicine. *Veterans and Agent Orange: Update 2000.* Washington, DC: National Academy Press, 2001.

PERIODICALS

"Agent Orange Exposure Linked to Type 2 Diabetes." *Nation's Health* 30, no. 11 (December 2000/January 2001): 11.
"Agent Orange Victims." *Earth Island Journal* 17, no. 1 (Spring 2002): 15.
Dreyfus, Robert. "Apocolypse Still." *Mother Jones* (January/February 2000).
Korn, Peter. "The Persisting Poison; Agent Orange in Vietnam." *The Nation* 252, no.13 (April 8, 1991): 440.
Young, Emma "Foul Fare." *New Scientist* 170, no. 2292 (May 26, 2001): 13.

OTHER

U.S. Veterans Affairs (VA). *Agent Orange* [June 2002]. <http://www.va.gov/agentorange>.

Agglomeration

Any process by which a group of individual particles is clumped together into a single mass. The term has a number of specialized uses. Some types of rocks are formed by the agglomeration of particles of sand, clay, or some other material. In geology, an agglomerate is a rock composed of volcanic fragments. One technique for dealing with **air pollution** is ultrasonic agglomeration. A source of very high frequency sound is attached to a smokestack, and the ultrasound produced by this source causes tiny **particulate** matter in waste gases to agglomerate into particles large enough to be collected.

Agricultural chemicals

The term agricultural chemical refers to any substance involved in the growth or utilization of any plant or animal of economic importance to humans. An agricultural chemical may be a natural product, such as urea, or a synthetic chemical, such as DDT. The agricultural **chemicals** now in use include fertilizers, pesticides, growth regulators, animal feed supplements, and raw materials for use in chemical processes.

In the broadest sense, agricultural chemicals can be divided into two large categories, those that promote the growth of a plant or animal and those that protect plants or animals. To the first group belong plant fertilizers and animal food supplements, and to the latter group belong pesticides, herbicides, animal vaccines, and antibiotics.

In order to stay healthy and grow normally, crops require a number of nutrients, some in relatively large quantities called macronutrients, and others in relatively small quantities called micronutrients. **Nitrogen**, **phosphorus**, and potassium are considered macronutrients, and boron, calcium, **chlorine**, **copper**, iron, magnesium, manganese among others are micronutrients.

Farmers have long understood the importance of replenishing the **soil**, and they have traditionally done so by natural means, using such materials as manure, dead fish, or compost. Synthetic fertilizers were first available in the early twentieth century, but they became widely used only after World War II. By 1990 farmers in the United States were using about 20 million tons (20.4 million metric tons) of these fertilizers a year.

Synthetic fertilizers are designed to provide either a single **nutrient** or some combination of nutrients. Examples of single-component or "straight" fertilizers are urea (NH_2CONH_2), which supplies nitrogen, or potassium chloride (KCl), which supplies potassium. The composition of "mixed" fertilizers, those containing more than one nutrient, is indicated by the analysis printed on their container. An 8-10-12 **fertilizer**, for example, contains 8% nitrogen by weight, 10% phosphorus, and 12% potassium.

Synthetic fertilizers can be designed to release nutrients almost immediately ("quick-acting") or over longer periods of time ("time-release"). They may also contain specific amounts of one or more trace nutrients needed for particular types of crops or soil. Controlling micronutrients is one of the most important problems in fertilizer compounding and use; the presence of low concentrations of some elements can be critical to a plant's health, while higher levels can be toxic to the same plants or to animals that ingest the micronutrient.

Plant growth patterns can also be influenced by direct application of certain chemicals. For example, the gibberellins are a class of compounds that can dramatically affect the rate at which plants grow and fruits and vegetables ripen. They have been used for a variety of purposes ranging from the hastening of root development to the delay of fruit ripening. Delaying ripening is most important for marketing agricultural products because it extends the time a crop can be transported and stored on grocery shelves. Other kinds of chemicals used in the processing, transporting, and storage of fruits and vegetables include those that slow down or speed up ripening (maleic hydrazide, ethylene oxide, potassium permanganate, ethylene, and acetylene are examples), that reduce weight loss (chlorophenoxyacetic **acid**, for example), retain green color (cycloheximide), and control firmness (ethylene oxide).

The term agricultural chemical is most likely to bring to mind the range of chemicals used to protect plants against competing organisms: pesticides and herbicides. These chemicals disable or kill bacteria, **fungi**, rodents, worms, snails and slugs, insects, mites, algae, termites, or any other **species** of plant or animal that feeds upon, competes with, or otherwise interferes with the growth of crops. Such chemicals are named according to the organism against which they are designed to act. Some examples are fungicides (designed to kill fungi), insecticides (used against insects), nematicides (to kill round worms), avicides (to control birds), and herbicides (to combat plants). In 1990, 393 million tons of herbicides, 64 million tons of insecticides, and 8 million tons of other pesticides were used on American farmlands.

The introduction of synthetic pesticides in the years following World War II produced spectacular benefits for farmers. More than 50 major new products appeared between 1947 and 1967, resulting in yield increases in the United States ranging from 400% for corn to 150% for sorghum and 100% for wheat and soybeans. Similar increases in **less developed countries**, resulting from the use of both synthetic fertilizers and pesticides, eventually became known as the Green Revolution.

By the 1970s, however, the environmental consequences of using synthetic pesticides became obvious. Chemicals were becoming less effective as pests developed resistances to them, and their toxic effects on other organisms had grown more apparent. Farmers were also discovering drawbacks to chemical fertilizers as they found that they had to use larger and larger quantities each year in order to maintain crop yields. One solution to the environmental hazards posed by synthetic pesticides is the use of natural chemicals such as juvenile hormones, sex attractants, and anti-feedant compounds. The development of such natural pest-control materials has, however, been relatively modest; the vast majority of agricultural companies and individual farmers continue to use synthetic chemicals that have served them so well for over a half century.

Chemicals are also used to maintain and protect livestock. At one time, farm animals were fed almost exclusively on readily available natural foods. They grazed on **rangelands** or were fed hay or other grasses. Today, carefully blended chemical supplements are commonly added to the diet of most farm animals. These supplements have been determined on the basis of extensive studies of the nutrients that contribute to the growth or milk production of cows,

sheep, goats, and other types of livestock. A typical animal supplement diet consists of various vitamins, minerals, amino acids, and nonprotein (simple) nitrogen compounds. The precise formulation depends primarily on the species; a vitamin supplement for cattle, for example, tends to include A, D, and E, while swine and poultry diets would also contain Vitamin K, riboflavin, niacin, pantothenic acid, and choline.

A number of chemicals added to animal feed serve no nutritional purpose but provide other benefits. For example, the addition of certain hormones to the feed of dairy cows can significantly increase their output of milk. **Genetic engineering** is also becoming increasingly important in the modification of crops and livestock. Cows injected with a genetically modified chemical, bovine somatotropin, produce a significantly larger quantity of milk.

It is estimated that infectious diseases cause the death of 15–20 of all farm animals each year. Just as plants are protected from pests by pesticides, so livestock are protected from disease organisms by immunization, antibiotics, and other techniques. Animals are vaccinated against species-specific diseases, and farmers administer antibiotics, sulfonamides, nitrofurans, arsenicals, and other chemicals that protect against disease-causing organisms.

The use of chemicals with livestock can have deleterious effects, just as crop chemicals have. In the 1960s, for example, the hormone diethylstilbestrol (DES) was widely used to stimulate the growth of cattle, but scientists found that detectable residues of the hormone remained in meat sold from the slaughtered animals. DES is now considered a **carcinogen**, and the U.S. **Food and Drug Administration** has banned its use in cattle feed since 1979.

[*David E. Newton*]

RESOURCES
BOOKS

Benning, L. E. *Beneath the Bottom Line: Agricultural Approaches to Reduce Agrichemical Contamination of Groundwater.* Washington, DC: Office of Technology Assessment, 1990.

———, and J. H. Montgomery. *Agrochemicals Desk Reference: Environmental Data.* Boca Raton, FL: Lewis, 1993.

———, and T. E. Waddell. *Managing Agricultural Chemicals in the Environment: The Case for a Multimedia Approach.* Washington, DC: Conservation Foundation, 1988.

Chemistry and the Food System, A Study by the Committee on Chemistry and Public Affairs of the American Chemical Society. Washington, DC: American Chemical Society, 1980.

Agricultural environmental management

The complex interaction of agriculture and **environment** has been an issue since the beginning of man. Humans grow food to eat and also hunt animals that depend on natural resources for healthy ongoing habitats. Therefore, the world's human population must balance farming activities with maintaining natural resources. The term agriculture originally meant the act of cultivating fields or growing crops. However, it has expanded to include raising livestock as well.

When early settlers began farming and ranching in the United States, they faced pristine wilderness and open prairies. There was little cause for concern about protecting the environment or population and for two centuries, the country's land and water were aggressively used to create a healthy supply of ample food for Americans. In fact, many American families settled in rural areas and made a living as farmers and ranchers, passing the family business down through generations. By the 1930s, the federal government began requiring farmers to idle certain acres of land to prevent oversupply of food and to protect exhausted **soil**.

Since that time, agriculture has become a complex science, as farmers must carefully manage soil and water to lessen risk of degrading the soil and its surrounding environment or depleting water tables beneath the land's surface. In fact, farming and ranching present several environmental challenges that require careful management by farmers and local and federal regulatory agencies that guide their activities. The science of applying principles of ecology to agriculture is called **agroecology**. Those involved in agroecology develop farming methods that use fewer synthetic (man-made) pesticides and fertilizers and encourage organic farming. They also work to conserve energy and water.

Soil **erosion**, converting land to agricultural use, introduction of **fertilizer** and pesticides, animal wastes, and irrigation are parts of farming that can lead to changes in quality or availability of water. An expanding human population has lead to increased farming and accelerated soil erosion. When soil has a low capacity to retain water, farmers must pump **groundwater** up and spray it over crops. After years of doing so, the local water table will eventually fall. This can impact native vegetation in the area.

The industry calls the balance of environment and lessening of agricultural effects sustainability or **sustainable development**. In some parts of the world, like in the High Plains of the United States or parts of Saudi Arabia, populations and agriculture are depleting water aquifers faster than the natural environment can replenish them. Sustainable development involves dedicated, scientifically based plans to ensure that agricultural activity is managed in such a way that aquifers are not prematurely depleted.

Agroforestry is a method of cultivating both crops and teres on the same land. Between rows of trees, farmers plant agricultural crops that generate income during the time it takes the trees to grow mature enough to produce earnings from nuts or lumber.

Increased modernization of agriculture also impacts the environment. Traditional farming practice, which continues in underdeveloped countries today, consists of subsistence agriculture. In subsistence farming, just enough crops and livestock are raised to meet the needs of a particular family. However, today large farms produce food for huge populations. More than half of the world's working population is employed by some agricultural or agriculturally associated industry. Almost 40% of the world's land area is devoted to agriculture (including permanent pasture). The growing use of machines, pesticides and man-made fertilizers have all seriously impacted the environment.

For example, the use of pesticides like DDT in the 1960s were identified as leading to the deaths of certain species of birds. Most western countries banned use of the pesticides and the bird populations soon recovered. Today, use of pesticides is strictly regulated in the United States.

Many more subtle effects of farming occur on the environment. When grasslands and wetlands or forests are converted to crops, and when crops are not rotated, eventually, the land changes to the point that entire species of plants and animals can become threatened. Urbanization also imposes onto farmland and cuts the amount of land available for farming.

Throughout the world, countries and organizations develop strategies to protect the environment, natural habitats and resources while still supplying the food our populations require. In 1992, The **United Nations Conference on Environment and Development** in Rio de Janeiro focused on how to sustain the world's natural resources but balance good policies on environment and community vitality. In the United States, the Department of Agriculture has published its own policy on sustainable development, which works toward balancing economics, environment and social needs concerning agriculture. In 1993, an Executive Order formed the President's Council on Sustainable Development (PCSD) to develop new approaches to achieve economic and environmental goals for public policy in agriculture. Guiding principles include sections on agriculture, forestry and rural community development.

According to the United States **Environmental Protection Agency** (EPA), Agricultural Environmental Management (AEM) is one of the most innovative programs in New York State. The program was begun in June 2000 when Governor George Pataki introduced legislation to the state's Senate and Assembly proposing a partnership to promote farming's good stewardship of land and to provide the funding and support of farmers' efforts. The bill was passed and signed into law by the governor on August 24, 2000.

The purpose of the law is to help farmers develop agricultural environmental management plans that control agricultural pollution and comply with federal, state and local regulations on use of land, water quality, and other environmental concerns. New York's AEM program brings together agencies from state, local, and federal governments, conservation representatives, businesses from the private sector, and farmers. The program is voluntary and offers education, technical assistance, and financial incentives to farmers to participate.

An example of a successful AEM project occurred at a dairy farm in central New York. The farm composted animals' solid wastes, which reduced the amount of waste spread on the fields. This in turn reduced pollution in the local watershed. The New York State Department of Agriculture and Markets oversees the program. It begins when a farmer expresses interest in AEM. Next, the farmer completes a series of five tiers of the program.

In Tier I, the farmer completes a short questionnaire that surveys current farming activities and future plans to identify potential environmental concerns. Tier II involves worksheets that document current activities that promote stewardship of the environment and help prioritize any environmental concerns. In Tier III, a conservation plan is developed that is tailored specifically for the individual farm. The farmer works together with an AEM coordinator and several members of the cooperating agency staff.

Under Tier IV of the AEM program, agricultural agencies and consultants provide the farmer with educational, technical, and financial assistance to implement best management practices for preventing pollution to water bodies in the farm's area. The plans use Natural Resources Conservation Service standards and guidance from cooperating professional engineers. Finally, farmers in the AEM program receive ongoing evaluations to ensure that the plan they have devised helps protect the environment and also ensures viability of the farm business.

Funding for the AEM program comes from a variety of sources, including New York's Clean Water/Clean Air Bond Act and the State Environmental Protection Fund. Local Soil and Water Conservation Districts (SWCDs) also partner in the effort, and farmers can access funds through these districts. The EPA says involvement of the SWCDs has likely been a positive factor in farmers' acceptance of the program.

Though New York is perceived as mostly urban, agriculture is a huge business in the state. The AEM program serves important environmental functions and helps keep New York State's farms economically viable. More than 7,000 farms participate in the program.

[*Teresa G. Norris*]

RESOURCES
BOOKS

Calow, Peter. *The Encyclopedia of Ecology and Environmental Management.* Malden, MA: Blackwell Science, Inc., 1998.

PERIODICALS
Ervin, DE, et al. "Agriculture and Environment: A New Strategic Vision." *Environment* 40, no. 6 (July-August, 1998):8.

ORGANIZATIONS
New York State Department of Agriculture and Markets, 1 Winners Circle, Albany, NY USA 12235 (518) 457-3738, Fax: (518)457-3412, Email: lauren.hoeffner@agmkt.state.ny.us, http://www.agmkt.state.ny.us

Sustainable Development, USA, United States Department of Agriculture, 14th and Independence SW, Washington, DC USA 20250 (202) 720-5447, Email: rbridge@oce.usda.gov, http://www.usda.gov

Agricultural pollution

The development of modern agricultural practices is one of the great success stories of applied sciences. Improved plowing techniques, new pesticides and fertilizers, and better strains of crops are among the factors that have resulted in significant increases in agricultural productivity.

Yet these improvements have not come without cost to the **environment** and sometimes to human health. Modern agricultural practices have contributed to the **pollution** of air, water, and land. **Air pollution** may be the most memorable, if not the most significant, of these consequences. During the 1920s and 1930s, huge amounts of fertile **topsoil** were blown away across vast stretches of the Great Plains, an area that eventually became known as the **Dust Bowl**. The problem occurred because farmers either did not know about or chose not to use techniques for protecting and conserving their **soil**. The soil then blew away during droughts, resulting not only in the loss of valuable farmland, but also in the pollution of the surrounding **atmosphere**.

Soil conservation techniques developed rapidly in the 1930s, including **contour plowing**, strip cropping, crop rotation, windbreaks, and minimum- or no-tillage farming, and thereby greatly reduced the possibility of **erosion** on such a scale. However, such events, though less dramatic, have continued to occur, and in recent decades they have presented new problems. When top soils are blown away by winds today, they can carry with them the pesticides, herbicides, and other crop **chemicals** now so widely used. In the worst cases, these chemicals have contributed to the collection of air pollutants that endanger the health of plants and animals, including humans. Ammonia, released from the decay of fertilizers, is one example of a compound that may cause minor irritation to the human respiratory system and more serious damage to the health of other animals and plants.

A more serious type of agricultural pollution are the **solid waste** problems resulting from farming and livestock practices. Authorities estimate that slightly over half of all the solid wastes produced in the United States each year—a total of about 2 billion tons (2 billion metric tons)—come from a variety of agricultural activities. Some of these wastes pose little or no threat to the environment. Crop residue left on cultivated fields and animal manure produced on **rangelands**, for example, eventually decay, returning valuable nutrients to the soil.

Some modern methods of livestock management, however, tend to increase the risks posed by animal wastes. Farmers are raising a larger variety of animals, as well as larger numbers of them, in smaller and smaller areas such as **feedlots** or huge barns. In such cases, large volumes of wastes are generated in these areas. Many livestock managers attempt to sell these waste products or dispose of them in a way that poses no threat to the environment. Yet in many cases the wastes are allowed to accumulate in massive dumps where soluble materials are leached out by rain. Some of these materials then find their way into **groundwater** or surface water, such as lakes and rivers. Some are harmless to the health of animals, though they may contribute to the eutrophication of lakes and ponds. Other materials, however, may have toxic, carcinogenic, or genetic effects on humans and other animals.

The **leaching** of hazardous materials from **animal waste** dumps contributes to perhaps the most serious form of agricultural pollution: the contamination of water supplies. Many of the chemicals used in agriculture today can be harmful to plants and animals. Pesticides and herbicides are the most obvious of these; used by farmers to disable or kill plant and animal pests, they may also cause problems for beneficial plants and animals as well as humans.

Runoff from agricultural land is another serious environmental problem posed by modern agricultural practices. Runoff constitutes a **nonpoint source** of pollution. Rainfall leaches out and washes away pesticides, fertilizers, and other **agricultural chemicals** from a widespread area, not a single source such as a sewer pipe. Maintaining control over nonpoint sources of pollution is an especially difficult challenge. In addition, agricultural land is more easily leached out than is non-agricultural land. When lands are plowed, the earth is broken up into smaller pieces, and the finer the soil particles, the more easily they are carried away by rain. Studies have shown that the **nitrogen** and **phosphorus** in chemical fertilizers are leached out of croplands at a rate about five times higher than from forest woodlands or idle lands.

The accumulation of nitrogen and phosphorus in waterways from chemical fertilizers has contributed to the acceleration of eutrophication of lakes and ponds. Scientists believe that the addition of human-made chemicals such as those in chemical fertilizers can increase the rate of eutrophication by a factor of at least 10. A more deadly effect is the **poisoning** of plants and animals by toxic chemicals leached off of farmlands. The biological effects of such chemicals are commonly magnified many times as they move up a **food chain/web**. The best known example of this phenomenon involved a host of biological problems—from reduced rates of reproduction to malformed animals to increased rates of death—attributed to the use of DDT in the 1950s and 1960s.

Sedimentation also results from the high rate of erosion on cultivated land, and increased sedimentation of waterways poses its own set of environmental problems. Some of these are little more than cosmetic annoyances. For example, lakes and rivers may become murky and less attractive, losing potential as **recreation** sites. However, sedimentation can block navigation channels, and other problems may have fatal results for organisms. Aquatic plants may become covered with sediments and die; marine animals may take in sediments and be killed; and cloudiness from sediments may reduce the amount of sunlight received by aquatic plants so extensively that they can no longer survive.

Environmental scientists are especially concerned about the effects of agricultural pollution on groundwater. Groundwater is polluted by much the same mechanisms as is surface water, and evidence for that pollution has accumulated rapidly in the past decade. **Groundwater pollution** tends to persist for long periods of time. Water flows through an **aquifer** much more slowly than it does through a river, and agricultural chemicals are not flushed out quickly.

Many solutions are available for the problems posed by agricultural pollution, but many of them are not easily implemented. Chemicals that are found to have serious toxic effects on plants and animals can be banned from use, such as DDT in the 1970s, but this kind of decision is seldom easy. Regulators must always assess the relative benefit of using a chemical, such as increased crop yields, against its environmental risks. Such as a risk-benefit analysis means that some chemicals known to have certain deleterious environmental effects remain in use because of the harm that would be done to agriculture if they were banned.

Another way of reducing agricultural pollution is to implement better farming techniques. In the practices of minimum- or no-tillage farming, for example, plowing is reduced or eliminated entirely. Ground is left essentially intact, reducing the rate at which soil and the chemicals it contains are eroded away.

[*David E. Newton*]

RESOURCES

BOOKS

Benning, L. E. *Agriculture and Water Quality: International Perspectives.* Boulder, CO: L. Rienner, 1990.

——, and L. W. Canter. *Environmental Impacts of Agricultural Production Activities.* Chelsea, MI: Lewis, 1986.

————, and M. W. Fox. *Agricide: The Hidden Crisis That Affects Us All.* New York: Shocken Books, 1986.

Crosson, P. R. *Implementation Policies and Strategies for Agricultural Non-Point Pollution.* Washington, DC: Resources for the Future, 1985.

Agricultural Research Service

A branch of the **U.S. Department of Agriculture** charged with the responsibility of agricultural research on a regional or national basis. The Agricultural Research Service (ARS) has a mission to develop new knowledge and technology needed to solve agricultural problems of broad scope and high national priority in order to ensure adequate production of high quality food and agricultural products for the United States. The national research center of the ARS is located at Beltsville, Maryland, consisting of laboratories, land, and other facilities. In addition, there are many other research centers located throughout the United States, such as the U.S. Dairy/Forage Research Center at Madison, Wisconsin. Scientists of the ARS are also located at Land Grant Universities throughout the country where they conduct cooperative research with state scientists.

RESOURCES

ORGANIZATIONS

Beltsville Agricultural Research Center, Rm. 223, Bldg. 003, BARC-West, 10300 Baltimore Avenue , Beltsville , MD USA 20705 , <http://www.ars.usda.gov>

Agricultural revolution

The development of agriculture has been a fundamental part of the march of civilization. It is an ongoing challenge, for as long as **population growth** continues, mankind will need to improve agricultural production.

The agricultural revolution is actually a series of four major advances, closely linked with other key historical periods. The first, the *Neolithic* or New Stone Age, marks the beginning of sedentary (settled) farming. Much of this history is lost in antiquity, dating back perhaps 10,000 years or more. Still, humans owe an enormous debt to those early pioneers who so painstakingly nourished the best of each year's crop. Archaeologists have found corn cobs a mere 2 in (5.1 cm) long, so different from today's giant ears.

The second major advance came as a result of Christopher Columbus' voyages to the New World. Isolation had fostered the development of two completely independent agricultural systems in the New and Old Worlds. A short list of interchanged crops and animals clearly illustrates the global magnitude of this event; furthermore, the current population explosion began its upswing during this period. From the New World came maize, beans, the "Irish" potato,

squash, peanuts, tomatoes, and **tobacco**. From the Old World came wheat, rice, coffee, cattle, horses, sheep, and goats. Maize is now a staple food in Africa. Several Indian tribes in America adopted new lifestyles, notably the Navajo as sheepherders, and the Cheyenne as nomads using the horse to hunt buffalo.

The Industrial Revolution both contributed to and was nourished by agriculture. The greatest agricultural advances came in **transportation**, where first canals, then railroads and steamships made possible the shipment of food from areas of surplus. This in turn allowed more specialization and productivity, but most importantly, it reduced the threat of starvation. The steamship ultimately brought refrigerated meat to Europe from distant Argentina and **Australia**. Without these massive increases in food shipments the exploding populations and greatly increased demand for labor by newly emerging industries could not have been sustained.

In turn the Industrial Revolution introduced major advances in farm technology, such as the cotton gin, mechanical reaper, improved plows, and, in this century, tractors and trucks. These advances enabled fewer and fewer farmers to feed larger and larger populations, freeing workers to fill demands for factory labor and the growing service industries.

Finally, agriculture has fully participated in the scientific advances of the twentieth century. Key developments include hybrid corn, the high responders in tropical lands, described as the "Green Revolution," and current genetic research. Agriculture has benefited enormously from scientific advances in biology, and the future here is bright for applied research, especially involving genetics. Great potential exists for the development of crop strains with greatly improved dietary characteristics, such as higher protein or reduced fat.

Growing populations, made possible by these food surpluses, have forced agricultural expansion onto less and less desirable lands. Because agriculture radically simplifies ecosystems and greatly amplifies **soil erosion**, many areas such as the Mediterranean Basin and tropical forest lands have suffered severe degradation.

Major developments in civilization are directly linked to the agricultural revolution. A sedentary lifestyle, essential to technological development, was both mandated and made possible by farming. Urbanization flourished, which encouraged specialization and division of labor. Large populations provided the energy for massive projects, such as the Egyptian pyramids and the colossal engineering efforts of the Romans.

The plow represented the first lever, both lifting and overturning the soil. The draft animal provided the first in a long line of nonhuman energy sources. Plant and animal selectivity are likely the first application of science and technology toward specific goals. A number of important crops bear little resemblance to the ancestors from which they were derived. Animals such as the fat-tailed sheep represent thoughtful cultural control of their lineage.

Climate dominates agriculture, second only to **irrigation**. Farmers are especially vulnerable to variations, such as late or early frosts, heavy rains, or **drought**. Rice, wheat, and maize have become the dominant crops globally because of their high caloric yield, versatility within their climate range, and their cultural status as the "staff of life." Many would not consider a meal complete without rice, bread, or tortillas. This cultural influence is so strong that even starving peoples have rejected unfamiliar food. China provides a good example of such cultural differences, with a rice culture in the south and a wheat culture (noodles) in the north.

These crops all need a wet season for germination and growth, followed by a dry season to allow spoilage-free storage. Rice was domesticated in the monsoonal lands of Southeast Asia, while wheat originated in the Fertile Crescent of the Middle East. Historically, wheat was planted in the fall, and harvested in late spring, coinciding with the cycle of wet and dry seasons in the Mediterranean region. Maize needs the heavy summer rains provided by the Mexican highland climate.

Other crops predominate in areas with less suitable climates. These include barley in semiarid lands; oats and potatoes in cool, moist lands; rye in colder climates with short growing seasons; and dry rice on hillsides and drier lands where paddy rice is impractical.

Although food production is the main emphasis in agriculture, more and more industrial applications have evolved. Cloth fibers have been a mainstay, but paper products and many **chemicals** now come from cultivated plants.

The agricultural revolution is also associated with some of mankind's darker moments. In the tropical and subtropical climates of the New World, slave labor was extensive. Close, unsanitary living conditions have fostered plagues of biblical proportions. And the desperate dependence on agriculture is all too vividly evident in the records of historic and contemporary **famine**. As a world, people are never more than one harvest away from global starvation, a fact amplified by the growing understanding of cosmic catastrophes.

Some argue that the agricultural revolution masks the growing hazards of an overpopulated, increasingly contaminated earth. Since the agricultural revolution has been so productive it has more than compensated for the population explosion of the last two centuries. Some appropriately labeled "cornucopians" believe there is yet much potential for increased food production, especially through scientific agriculture and **genetic engineering**. There is much room for optimism, and also for a sobering assessment of the environmental costs of agricultural progress. We must continually strive for answers to the challenges associated with the agricultural revolution.

[*Nathan H. Meleen*]

RESOURCES
BOOKS

Anderson, E. "Man as a Maker of New Plants and New Plant Communities." In *Man's Role in Changing the Face of the Earth,* edited by W. L. Thomas Jr. Chicago: The University of Chicago Press, 1956.

Doyle, J. *Altered Harvest: Agriculture, Genetics, and the Fate of the World's Food Supply.* New York: Penguin, 1985.

Gliessman, S. R., ed. *Agroecology: Researching the Ecological Basis for Sustainable Agriculture.* New York: Springer-Verlag, 1990.

Jackson, R. H., and L. E. Hudman. *Cultural Geography: People, Places, and Environment.* St. Paul, MN: West, 1990.

Narr, K. J. "Early Food-Producing Populations." In *Man's Role in Changing the Face of the Earth,* edited by W. L. Thomas, Jr. Chicago: The University of Chicago Press, 1956.

Simpson, L. B. "The Tyrant: Maize." In *The Cultural Landscape,* edited by C. Salter. Belmont, CA: Wadsworth, 1971.

PERIODICALS

Crosson, P. R., and N. J. Rosenberg. "Strategies for Agriculture." *Scientific American* 261 (September 1989): 128–32+.

Agricultural Stabilization and Conservation Service

For the past half century, agriculture in the United States has faced the somewhat unusual and enviable problem of overproduction. Farmers have produced more food than United States citizens can consume, and, as a result, per capita farm income has decreased as the volume of crops has increased. To help solve this problem, the Secretary of Agriculture established the Agricultural Stabilization and Conservation Service on June 5, 1961. The purpose of the service is to administer commodity and land-use programs designed to control production and to stabilize market prices and farm income. The service operates through state committees of three to five members each and committees consisting of three farmers in approximately 3,080 agricultural counties in the nation.

RESOURCES
ORGANIZATIONS

Agricultural Stabilization and Conservation Service, 10500 Buena Vista Court, Urbandale, IA USA 50322-3782 (515) 254-1540, Fax: (515) 254-1573.

Agriculture and energy conservation
see **Environmental engineering**

Agriculture, drainage
see **Runoff**

Agriculture, sustainable
see **Sustainable agriculture**

Agroecology

Agroecology is an interdisciplinary field of study that applies ecological principles to the design and management of agricultural systems. Agroecology concentrates on the relationship of agriculture to the biological, economic, political, and social systems of the world.

The combination of agriculture with ecological principles such as biogeochemical cycles, **energy conservation**, and **biodiversity** has led to practical applications that benefit the whole **ecosystem** rather than just an individual crop. For instance, research into **integrated pest management** has developed ways to reduce reliance on pesticides. Such methods include biological or biotechnological controls such as **genetic engineering**, cultural controls such as changes in planting patterns, physical controls such as quarantines to prevent entry of new pests, and mechanical controls such as physically removing weeds or pests.

Sustainable agriculture is another goal of agroecological research. Sustainable agriculture views farming as a total system and stresses the long-term **conservation** of resources. It balances the human need for food with concerns for the **environment** and maintains that agriculture can be carried on without reliance on pesticides and fertilizers.

Agroecology advocates the use of biological controls rather than pesticides to minimize agricultural damage from insects and weeds. Biological controls use natural enemies to control weeds and pests, such as ladybugs that kill aphids. Biological controls include the disruption of the reproductive cycles of pests and the introduction of more biologically diverse organisms to inhibit overpopulation of different agricultural pests.

Agroecological principals shift the focus of agriculture from food production alone to wider concerns, such as environmental quality, food safety, the quality of rural life, humane treatment of livestock, and conservation of air, **soil**, and water. Agroecology also studies how agricultural processes and technologies will be impacted by wider environmental problems such as global warming, **desertification**, or **salinization**.

The entire world population depends on agriculture, and as the number of people continues to grow agroecology is becoming more important, particularly in developing countries. Agriculture is the largest economic activity in the world, and in areas such as sub-Saharan Africa about 75% of the population is involved in some form of it. As population pressures on the world food supply increase, the application of agroecological principles is expected to stem the ecological consequences of traditional agricultural practices such as **pesticide poisoning** and **erosion**.

[*Linda Rehkopf*]

RESOURCES
BOOKS

Altieri, M. A. *Agroecology: The Scientific Basis of Alternative Agriculture.* Boulder, CO: Westview Press, 1987.
Carroll, D. R. *Agroecology.* New York: McGraw-Hill, 1990.
Gliessman, S. R., ed. *Agroecology.* New York: Springer-Verlag, 1991.

PERIODICALS
Norse, D. "A New Strategy for Feeding a Drowned Planet." *Environment* 34 (June 1992): 6–19.

Agroforestry

Agroforestry is a **land use** system in which woody perennials (trees, shrubs, vines, palms, bamboo, etc.) are intentionally combined on the same land management unit with crops and sometimes animals, either in a spatial arrangement or a temporal sequence. It is based on the premise that woody perennials in the landscape can enhance the productivity and sustainability of agricultural practice. The approach is especially pertinent in tropical and subtropical areas where improper land management and intensive, continuous cropping of land have led to widespread devastation. Agroforestry recognizes the need for an alternative agricultural system that will preserve and sustain productivity. The need for both food and forest products has led to an interest in techniques that combine production of both in a manner that can halt and may even reverse the ruin caused by existing practices.

Although the term agroforestry has come into widespread use only in the last 20–25 years, environmentally sound farming methods similar to those now proposed have been known and practiced in some tropical and subtropical areas for many years. As an example, one type of intercropping found on small **rubber** plantations (less than 25 acres/ 10 ha), in Malaysia, Thailand, Nigeria, India, and Sri Lanka involves rubber plants intermixed with fruit trees, pepper, coconuts, and arable crops such as soybeans, corn, banana, and groundnut. Poultry may also be included. Unfortunately, in other areas the pressures caused by expanding human and animal populations have led to increased use of destructive farming practices. In the process, inhabitants have further reduced their ability to provide basic food, fiber, fuel, and timber needs and contributed to even more **environmental degradation** and loss of **soil** fertility.

The successful introduction of agroforestry practices in problem areas requires the cooperative efforts of experts from a variety of disciplines. Along with specialists in forestry, agriculture, **meteorology**, **ecology**, and related fields, it is often necessary to enlist the help of those familiar with local culture and heritage to explain new methods and their advantages. Usually, techniques must be adapted to local circumstances, and research and testing are required to develop viable systems for a particular setting. Intercropping combinations that work well in one location may not be appropriate for sites only a short distance away because of important meteorological or ecological differences. Despite apparent difficulties, agroforestry has great appeal as a means of arresting problems with **deforestation** and declining agricultural yields in warmer climates. The practice is expected to grow significantly in the next several decades. Some areas of special interest include intercropping with coconuts as the woody component, and mixing tree legumes with annual crops.

Agroforestry does not seem to lend itself to mechanization as easily as the large scale grain, soybean and vegetable cropping systems used in industrialized nations because practices for each site are individualized and usually labor-intensive. For these reasons they have had less appeal in areas like the United States and Europe. Nevertheless, temperate zone applications have been developed or are under development. Examples include small scale **organic gardening and farming**, mining wasteland **reclamation**, and **biomass** energy crop production on marginal land.

[*Douglas C. Pratt*]

RESOURCES
BOOKS

Huxley, P. A., ed. *Plant Research and Agroforestry.* Edinburgh, Scotland: Pillans & Wilson, 1983.
Reifsnyder, W. S., and T. O. Darnhofer, eds. *Meteorology and Agroforestry.* Nairobi, Kenya: International Council for Research in Agroforestry, 1989.
Zulberti, E., ed. *Professional Education in Agroforestry.* Nairobi, Kenya: International Council for Research in Agroforestry, 1987.

AIDS

AIDS (acquired immune deficiency syndrome) is an infectious and fatal disease of apparently recent origin. AIDS is *pandemic*, which means that it is worldwide in distribution. A sufficient understanding of AIDS can be gained only by examining its causation (etiology), symptoms, treatments, and the risk factors for transmitting and contracting the disease.

AIDS occurs as a result of infection with the HIV (human immunodeficiency **virus**). HIV is a **ribonucleic acid** (RNA) virus that targets and kills special blood cells, known as helper T-lymphocytes, which are important in immune protection. Depletion of helper T-lymphocytes leaves the AIDS victim with a disabled immune system and at risk for infection by organisms that ordinarily pose no special hazard to the individual. Infection by these organisms is thus opportunistic and is frequently fatal.

The initial infection with HIV may entail no symptoms at all or relatively benign symptoms of short duration that may mimic infectious mononucleosis. This initial period is followed by a longer period (from a few to as many as 10 years) when the infected person is in apparent good health. The HIV infected person, despite the outward image of good health, is in fact contagious, and appropriate care must be exercised to prevent spread of the virus at this time. Eventually the effects of the depletion of helper T cells become manifest. Symptoms include weight loss, persistent cough, persistent colds, diarrhea, periodic fever, weakness, fatigue, enlarged lymph nodes, and malaise. Following this, the AIDS patient becomes vulnerable to chronic infections by opportunistic pathogens. These include, but are not limited to oral yeast infection (thrush), pneumonia caused by the fungus *Pneumocystis carinii*, and infection by several kinds of herpes viruses. The AIDS patient is vulnerable to Kaposi's sarcoma, which is a **cancer** seldom seen except in those individuals with depressed immune systems. Death of the AIDS patient may be accompanied by confusion, dementia, and coma.

There is no cure for AIDS. Opportunistic infections are treated with antibiotics, and drugs such as AZT (azidothymidine), which slow the progress of the HIV infection, are available. But viral diseases in general, including AIDS, do not respond well to antibiotics. Vaccines, however, can provide protection against viral diseases. Research to find a vaccine for AIDS has not yet yielded satisfactory results, but scientists have been encouraged by the development of a vaccine for feline leukemia—a viral disease that has similarities to AIDS. Unfortunately, this does not provide hope of a cure for those already infected with the HIV virus.

Prevention is crucial for a lethal disease with no cure. Thus, modes of transmission must be identified and avoided. Everyone is at risk, males constitute 52% and females 48% of the infected population. In 2002 there are about 40 million people infected with HIV or AIDS, and it is thought that this number will grow to 62 million by 2005. In the United States alone, 40,000 new cases are diagnosed each year. AIDS cases in heterosexual males and women are on the increase, and no sexually active person can be considered "safe" from AIDS any longer. Therefore, everyone who is sexually active should be aware of the principal modes of transmission of the HIV virus—infected blood, semen from the male and genital tract secretions of the female—and use appropriate means to prevent exposure. While the virus has been identified in tears, saliva, and breast milk, contagions by exposure to those substances seems to be significantly less.

[*Robert G. McKinnell*]

RESOURCES
BOOKS

Alcamo, I. E. *AIDS, the Biological Basis*. Dubuque, Iowa: William C. Brown, 1993.
Fan, H., R. F. Connor, and L. P. Villarreal. *The Biology of AIDS*. 2nd edition. Boston: Jones and Bartlett, 1991.
Stine, Gerald J. *AIDS Update 2002*. Prentice Hall, 2001.

Ailuropoda melanoleuca
see **Giant panda**

Air and Waste Management Association

Founded in 1907 as the International Association for the Prevention of **Smoke**, this group changed its name several times as the interests of its members changed, becoming the Air and **Waste Management** Association (A&WMA) in the late 1980s. Although an international organization for **environment** professionals in more than 50 countries, the association is most active in North America and most concerned with North American environmental issues. Among its main concerns are **air pollution control**, environmental management, and waste processing and control.

A nonprofit organization that promotes the basic need for a clean environment, A&WMA seeks to educate the public and private sectors of the world by conducting seminars, holding workshops and conferences, and offering continuing education programs for environmental professionals in the areas of **pollution control** and waste management. One of its main goals is to provide "a neutral forum where all viewpoints of an environmental management issue (technical, scientific, economic, social, political and public health) receive equal consideration." Approximately 10–12 specialty conferences are held annually, as well as five or six workshops. The topics continuously revolve and change as new issues arise.

Education is so important to A&WMA that it funds a scholarship for graduate students pursuing careers in fields related to waste management and **pollution** control. Although A&WMA members are all professionals, they seek to educate even the very young by sponsoring essay contests, science fairs, and community activities, and by volunteering

to speak to elementary, middle school, and high school audiences on environmental management topics.

The association's 12,000 members, all of whom are volunteers, are involved in virtually every aspect of every A&WMA project. There are 21 association sections across the United States, facilitating meetings at regional and even local levels to discuss important issues. Training seminars are an important part of A&WMA membership, and members are taught the skills necessary to run public outreach programs designed for students of all ages and the general public.

A&WMA's publications deal primarily with **air pollution** and waste management, and include the *Journal of the Air & Waste Management Association*, a scientific monthly; a bimonthly newsletter; a wide variety of technical books; and numerous training manuals and educational videotapes.

[*Cathy M. Falk*]

RESOURCES
ORGANIZATIONS

Air & Waste Management Association, 420 Fort Duquesne Blvd, One Gateway Center , Pittsburgh, PA USA 15222 (412) 232-3444, Fax: (412) 232-3450, Email: info@awma.org, <http://www.awma.org>

Air pollution

Air **pollution** is a general term that covers a broad range of contaminants in the **atmosphere**. Pollution can occur from natural causes or from human activities. Discussions about the effects of air pollution have focused mainly on human health but attention is being directed to environmental quality and amenity as well. Air pollutants are found as gases or particles, and on a restricted scale they can be trapped inside buildings as indoor air pollutants. Urban air pollution has long been an important concern for civic administrators, but increasingly, air pollution has become an international problem.

The most characteristic sources of air pollution have always been **combustion** processes. Here the most obvious pollutant is **smoke**. However, the widespread use of **fossil fuels** have made sulfur and **nitrogen oxides** pollutants of great concern. With increasing use of petroleum-based fuels, a range of organic compounds have become widespread in the atmosphere.

In urban areas, air pollution has been a matter of concern since historical times. Indeed, there were complaints about smoke in ancient Rome. The use of **coal** throughout the centuries has caused cities to be very smoky places. Along with smoke, large concentrations of **sulfur dioxide** were produced. It was this mixture of smoke and sulfur dioxide that typified the foggy streets of Victorian London, paced by such figures as Sherlock Holmes and Jack the Ripper, whose images remain linked with smoke and fog. Such situations are far less common in the cities of North America and Europe today. However, until recently, they have been evident in other cities, such as Ankara, Turkey, and Shanghai, China, that rely heavily on coal.

Coal is still burnt in large quantities to produce electricity or to refine metals, but these processes are frequently undertaken outside cities. Within urban areas, fuel use has shifted towards liquid and gaseous **hydrocarbons** (petrol and **natural gas**). These fuels typically have a lower concentration of sulfur, so the presence of sulfur dioxide has declined in many urban areas. However, the widespread use of liquid fuels in automobiles has meant increased production of **carbon monoxide, nitrogen** oxides, and volatile organic compounds (VOCs).

Primary pollutants such as sulfur dioxide or smoke are the direct **emission** products of the combustion process. Today, many of the key pollutants in the urban atmospheres are secondary pollutants, produced by processes initiated through photochemical reactions. The Los Angeles, California, type **photochemical smog** is now characteristic of urban atmospheres dominated by secondary pollutants.

Although the **automobile** is the main source of air pollution in contemporary cities, there are other equally significant sources. Stationary sources are still important and the oil-burning furnaces that have replaced the older coal-burning ones are still responsible for a range of gaseous emissions and **fly ash**. **Incineration** is also an important source of complex combustion products, especially where this incineration burns a wide range of refuse. These emissions can include **chlorinated hydrocarbons** such as **dioxin**. When **plastics**, which often contain **chlorine**, are incinerated, hydrochloric **acid** results in the waste gas stream. Metals, especially where they are volatile at high temperatures, can migrate to smaller, respirable particles. The accumulation of toxic metals, such as **cadmium**, on fly ash gives rise to concern over harmful effects from incinerator emissions. In specialized incinerators designed to destroy toxic compounds such as PCBs, many questions have been raised about the completeness of this destruction process. Even under optimum conditions where the furnace operation has been properly maintained, great care needs to be taken to control leaks and losses during transfer operations (**fugitive emissions**).

The enormous range of compounds used in modern manufacturing processes have also meant that there has been an ever-widening range of emissions from both from the industrial processes and the combustion of their wastes. Although the amounts of these exotic compounds are often rather small, they add to the complex range of compounds found in the urban atmosphere. Again, it is not only the deliberate loss of effluents through **discharge** from pipes and chimneys that needs attention. Fugitive emissions of volatile substances that leak from valves and **seals** often warrant careful control.

Air pollution control procedures are increasingly an important part of civic administration, although their goals are far from easy to achieve. It is also noticeable that although many urban concentrations of primary pollutants, for example, smoke and sulfur dioxide, are on the decline in developed countries, this is not always true in the developing countries. Here the desire for rapid industrial growth has often lowered urban **air quality**. Secondary air pollutants are generally proving a more difficult problem to eliminate than primary pollutants like smoke.

Urban air pollutants have a wide range of effects, with health problems being the most enduring concern. In the classical polluted atmospheres filled with smoke and sulfur dioxide, a range of bronchial diseases were enhanced. While **respiratory diseases** are still the principal problem, the issues are somewhat more subtle in atmospheres where the air pollutants are not so obvious. In photochemical **smog**, eye irritation from the secondary pollutant **peroxyacetyl nitrate** (PAN) is one on the most characteristic direct effects of the smog. High concentrations of **carbon** monoxide in cities where automobiles operate at high density means that the human heart has to work harder to make up for the oxygen displaced from the blood's hemoglobin by carbon monoxide. This extra stress appears to reveal itself by increased incidence of complaints among people with heart problems. There is a widespread belief that contemporary air pollutants are involved in the increases in **asthma**, but the links between asthma and air pollution are probably rather complex and related to a whole range of factors. **Lead**, from automotive exhausts, is thought by many to be a factor in lowering the IQs of urban children.

Air pollution also affects materials in the urban **environment**. Soiling has long been regarded as a problem, originally the result of the smoke from wood or coal fires, but now increasingly the result of fine black soot from diesel exhausts. The acid gases, particularly sulfur dioxide, increase the rate of destruction of building materials. This is most noticeable with calcareous stones, which are the predominant building material of many important historic structures. Metals also suffer from atmospheric acidity. In the modern photochemical smog, natural rubbers crack and deteriorate rapidly.

Health problems relating to indoor air pollution are extremely ancient. Anthracosis, or **black lung disease**, has been found in mummified lung tissue. Recent decades have witnessed a shift from the predominance of concern about outdoor air pollution into a widening interest in **indoor air quality**.

The production of energy from combustion and the release of solvents is so large in the contemporary world that it causes air pollution problems of a regional and global **nature. Acid rain** is now widely observed throughout the world. The sheer quantity of **carbon dioxide** emitted in combustion process is increasing the concentration of carbon dioxide in the atmosphere and enhancing the **greenhouse effect**. Solvents, such as carbon tetrachloride and **aerosol propellants** (such as **chlorofluorocarbons** are now detectable all over the globe and responsible for such problems as **ozone layer depletion**.

At the other end of the scale, it needs to be remembered that gases leak indoors from the polluted outdoor environment, but more often the serious pollutants arise from processes that take place indoors. Here there has been particular concern with indoor air quality as regards to the generation of nitrogen oxides by sources such as gas stoves. Similarly formaldehyde from insulating foams causes illnesses and adds to concerns about our exposure to a substance that may induce **cancer** in the long run. In the last decade it has become clear that **radon** leaks from the ground can expose some members of the public to high levels of this radioactive gas within their own homes. Cancers may also result from the emanation of solvents from consumer products, glues, paints, and mineral fibers (**asbestos**). More generally these compounds and a range of biological materials, animal hair, skin and pollen spores, and dusts can cause allergic reactions in some people. At one end of the spectrum these simply cause annoyance, but in extreme cases, such as found with the bacterium *Legionella*, a large number of deaths can occur.

There are also important issues surrounding the effects of indoor air pollutants on materials. Many industries, especially the electronics industry, must take great care over the purity of indoor air where a speck of dust can destroy a microchip or low concentrations of air pollutants change the composition of surface films in component design. Museums must care for objects over long periods of time, so precautions must be taken to protect delicate dyes from the effects of photochemical smog, paper and books from sulfur dioxide, and metals from sulfide gases.

[*Peter Brimblecombe*]

RESOURCES
BOOKS

Bridgman, H. *Global Air Pollution: Problems for the 1990s.* New York: Columbia University Press, 1991.

Elsom, D. M. *Atmospheric Pollution.* Oxford: Blackwell, 1992.

Kennedy, D., and R. R. Bates, eds. *Air Pollution, the Automobile, and Public Health.* Washington, DC: National Academy Press, 1988.

MacKenzie, J. J. *Breathing Easier: Taking Action on Climate Change, Air Pollution, and Energy Efficiency.* Washington, DC: World Resources Institute, 1989.

Smith, W. H. *Air Pollution and Forests.* 2nd ed. New York: Springer-Verlag, 1989.

Air pollution control

The need to control **air pollution** was recognized in the earliest cities. In the Mediterranean at the time of Christ, laws were developed to place objectionable sources of odor and **smoke** downwind or outside city walls. The adoption of **fossil fuels** in thirteenth century England focused particular concern on the effect of **coal** smoke on health, with a number of attempts at regulation with regard to fuel type, chimney heights, and time of use. Given the complexity of the air **pollution** problem it is not surprising that these early attempts at control met with only limited success.

The nineteenth century was typified by a growing interest in urban public health. This developed against a background of continuing industrialization, which saw smoke abatement clauses incorporated into the growing body of sanitary legislation in both Europe and North America. However, a lack of both technology and political will doomed these early efforts to failure, except in the most blatantly destructive situations (for example, industrial settings such as those around Alkali Works in England).

The rise of environmental awareness has reminded people that air pollution ought not to be seen as a necessary product of industrialization. This has redirected responsibility for air pollution towards those who create it. The notion of "making the polluter pay" is seen as a central feature of air **pollution control**. History has also seen the development of a range of broad air pollution control strategies, among them: (1) **Air quality** management strategies that set **ambient air** quality standards so that emissions from various sources can be monitored and controlled; (2) **Emission standards** strategy that sets limits for the amount of pollutant that can be emitted from a given source. These may be set to meet air quality standards, but the strategy is optimally seen as one of adopting best available techniques not entailing excessive costs (BATNEEC); (3) Economic strategies that involve charging the party responsible for the pollution. If the level of charge is set correctly, some polluters will find it more economical to install air pollution control equipment than continue to pollute. Other methods utilize a system of tradable pollution rights; (4) **Cost-benefit analysis**, which attempts to balance economic benefits with environmental costs. This is an appealing strategy but difficult to implement because of its controversial and imprecise nature.

In general air pollution strategies have either been air-quality or emission-based. In the United Kingdom, **emis-**

An industrial complex releases smoke from multiple chimneys. (Photograph by Josef Polleross. The Stock Market. Reproduced by permission.)

sion strategy is frequently used; for example the Alkali and Works Act of 1863 specifies permissible emissions of hydrochloric **acid**. By contrast, the United States has aimed to achieve air quality standards, as evidenced by the **Clean Air Act**. One criticism of using air quality strategy has been that while it improves air in poor areas it leads to degradation in areas with high air quality. Although the emission standards approach is relatively simple, it is criticized for failing to make explicit judgments about air quality and assumes that good practice will lead to an acceptable **atmosphere**.

Until the mid-twentieth century, legislation was primarily directed towards industrial sources, but the passage of the United Kingdom Clean Air Act (1956), which followed the disastrous **smog** of December 1952, directed attention towards domestic sources of smoke. While this particular act may have reinforced the improvements already under way, rather than initiating improvements, it has served as a catalyst for much subsequent legislative thinking. Its mode of operation was to initiate a change in fuel, perhaps one of the oldest methods of control. The other well-tried

aspects were the creation of smokeless zones and an emphasis on tall chimneys to disperse the pollutants.

As simplistic as such passive control measures seem, they remain at the heart of much contemporary thinking. Changes from coal and oil to the less polluting gas or electricity have contributed to the reduction in smoke and **sulfur dioxide** concentrations in cities all around the world. Industrial zoning has often kept power and large manufacturing plants away from centers of human population, and "superstacks," chimneys of enormous height are now quite common. Successive changes in automotive fuels—lead-free **gasoline**, low volatility gas, **methanol**, or even the interest in the electric automobile—are further indications of continued use of these methods of control.

There are more active forms of air pollution control that seek to clean up the exhaust gases. The earliest of these were smoke and grit arrestors that came into increasing use in large electrical stations during the twentieth century. Notable here were the **cyclone** collectors that removed large particles by driving the exhaust through a tight spiral that threw the grit outward where it could be collected. Finer particles could be removed by **electrostatic precipitation**. These methods were an important part of the development of the modern pulverized fuel power station. However they failed to address the problem of gaseous emissions. Here it has been necessary to look at burning fuel in ways that reduce the production of **nitrogen oxides**. Control of sulfur dioxide emissions from large industrial plants can be achieved by desulfurization of the flue gases. This can be quite successful by passing the gas through towers of solid absorbers or spraying solutions through the exhaust gas stream. However, these are not necessarily cheap options.

Catalytic converters are also an important element of active attempts to control air pollutants. Although these can considerably reduce emissions, they have to be offset against the increasing use of the **automobile**. There is much talk of the development of zero pollution vehicles that do not emit any pollutants.

Legislation and control methods are often associated with monitoring networks that assess the effectiveness of the strategies and inform the general public about air quality where they live. A balanced approach to the control of air pollution in the future may have to look far more broadly than simply at technological controls. It will become necessary to examine the way people structure their lives in order to find more effective solutions to air pollution.

[*Peter Brimblecombe*]

RESOURCES
BOOKS

Elsom, D. M. *Atmospheric Pollution*. Oxford: Blackwell, 1992.

Luoma, J. R. *The Air Around Us: An Air Pollution Primer*. Raleigh, NC: The Acid Rain Foundation, 1989.

Wark, K., and C. F. Warner. *Air Pollution: Its Origin and Control*. 3rd ed. New York: Harper & Row, 1986.

Air pollution index

The **air pollution** index is a value derived from an **air quality** scale which uses the measured or predicted concentrations of several criteria pollutants and other air quality indicators, such as coefficient of **haze** (COH) or **visibility**. The best known index of air **pollution** is the pollutant standard index (PSI).

The PSI has a scale that spans from 0 to 500. The index represents the highest value of several subindices; there is a subindex for each pollutant, or in some cases, for a product of pollutant concentrations and a product of pollutant concentrations and COH. If a pollutant is not monitored, its subindex is not used in deriving the PSI. In general, the subindex for each pollutant can be interpreted as follows

Air Pollution Stages	
Index Value	**Interpretations**
0	No concentration
100	National Ambient Air Quality Standard
200	Alert
300	Warning
400	Emergency
500	Significant harm

The subindex of each pollutant or pollutant product is derived from a PSI nomogram which matches concentrations with subindex values. The highest subindex value becomes the PSI. The PSI has five health-related categories:

PSI Range	Category
0 to 50	Good
50 to 100	Moderate
100 to 200	Unhealthful
200 to 300	Very unhealthful
300 to 500	Hazardous

Air quality

Air quality is determined with respect to the total **air pollution** in a given area as it interacts with meteorological conditions such as humidity, temperature and wind to produce an overall atmospheric condition. Poor air quality can manifest itself aesthetically (as a displeasing odor, for example), and can also result in harm to plants, animals, people, and even damage to objects.

As early as 1881, cities such as Chicago, Illinois, and Cincinnati, Ohio, had passed laws to control some types of **pollution**, but it wasn't until several air pollution catastrophes occurred in the twentieth century that governments began to give more attention to air quality problems. For instance, in 1930, **smog** trapped in the Meuse River Valley in Belgium caused 60 deaths. Similarly, in 1948, smog was blamed for 20 deaths in Donora, Pennsylvania. Most dramatically, in 1952 a sulfur-laden fog enshrouded London for five days and caused as many as 4,000 deaths over two weeks.

Disasters such as these prompted governments in a number of industrial countries to initiate programs to protect air quality. The year of the London tragedy, the United States passed the **Air Pollution Control** Act granting funds to assist the states in controlling airborne pollutants. In 1963, the **Clean Air Act**, which began to place authority for air quality into the hands of the federal government, was established. Today the Clean Air Act, with its 1970 and 1990 amendments, remains the principal air quality law in the United States.

The Act established a **National Ambient Air Quality Standard** under which federal, state, and local monitoring stations at thousands of locations, together with temporary stations set up by the **Environmental Protection Agency** (EPA) and other federal agencies, directly measure pollutant concentrations in the air and compare those concentrations with national standards for six major pollutants: **ozone**, **carbon monoxide**, **nitrogen oxides**, **lead**, particulates, and **sulfur dioxide**. When the air we breathe contains amounts of these pollutants in excess of EPA standards, it is deemed unhealthy, and regulatory action is taken to reduce the pollution levels.

In addition, urban and industrial areas maintain an **air pollution index**. This scale, a composite of several pollutant levels recorded from a particular monitoring site or sites, yields an overall air quality value. If the index exceeds certain values public warnings are given; in severe instances residents might be asked to stay indoors and factories might even be closed down.

While such air quality emergencies seem increasingly rare in the United States, developing countries, as well as Eastern European nations, continue to suffer poor air quality, especially in urban areas such as Bangkok, Thailand and **Mexico City, Mexico**. In Mexico City, for example, seven out of 10 newborns have higher lead levels in their blood than the World Health Organization considers acceptable. At present, many **Third World** countries place national economic development ahead of pollution control—and in many countries with rapid industrialization, high **population growth**, or increasing per capita income, the best efforts of governments to maintain air quality are outstripped by rapid proliferation of automobiles, escalating factory emissions, and runaway urbanization.

For all the progress the United States has made in reducing **ambient air** pollution, *indoor* air pollution may pose even greater risks than all of the pollutants we breathe outdoors. The **Radon** Gas and **Indoor Air Quality** Act of 1986 directed the EPA to research and implement a public information and technical assistance program on indoor air quality. From this program has come monitoring equipment to measure an individual's "total exposure" to pollutants both in indoor and outdoor air. Studies done using this equipment have shown indoor exposures to toxic air pollutants far exceed outdoor exposures for the simple reason that most people spend 90% of their time in office buildings, homes, and other enclosed spaces. Moreover, nationwide **energy conservation** efforts following the oil crisis of the 1970s led to building designs that trap pollutants indoors, thereby exacerbating the problem.

[*David Clarke and Jeffrey Muhr*]

RESOURCES
BOOKS

Brown, Lester, ed. *The World Watch Reader On Global Environmental Issues.* Washington, DC: Worldwatch Institute, 1991.
Council on Environmental Quality. *Environmental Trends.* Washington, DC: U. S. Government Printing Office, 1989.
Environmental Progress and Challenges: EPA's Update. Washington, DC: U. S. Environmental Protection Agency, 1988.

Air quality control region

The **Clean Air Act** defines an **air quality** control region (AQCR) as a contiguous area where air quality, and thus **air pollution**, is relatively uniform. In those cases where **topography** is a factor in air movement, AQCRs often correspond with airsheds. AQCRs may consist of two or more cities, counties or other governmental entities, and each region is required to adopt consistent **pollution control** measures across the political jurisdictions involved. AQCRs may even cross state lines and, in these instances, the states must cooperate in developing **pollution** control strategies. Each AQCR is treated as a unit for the purposes of pollution reduction and achieving National Ambient Air Quality Standards. As of 1993, most AQCRs had achieved national air quality standards; however the remaining AQCRs where standards had not been achieved were a significant group, where a large percentage of the United States population dwelled. AQCRs involving major metro areas like Los Angeles, New York, Houston, Denver, and Philadelphia were not achieving air quality standards because of **smog**, motor vehicle emissions, and other pollutants.

Air quality criteria

The relationship between the level of exposure to air pollutant concentrations and the adverse effects on health or public welfare associated with such exposure. **Air quality** criteria are critical in the development of **ambient air** quality standards which define levels of acceptably safe exposure to an air pollutant.

Air-pollutant transport

Air-pollutant transport is the advection or horizontal convection of air pollutants from an area where **emission** occurs to a downwind receptor area by local or regional winds. It is sometimes referred to as atmospheric transport of air pollutants. This movement of **air pollution** is often simulated with computer models for point sources as well as for large diffuse sources such as urban regions.

In some cases, strong regional winds or low-level nocturnal jets can carry pollutants hundreds of miles from source areas of high emissions. The possibility of transport over such distances can be increased through topographic channeling of winds through valleys. Air-pollutant transport over such distances is often referred to as long-range transport.

Air-pollutant transport is an important consideration in **air quality** planning. Where such impact occurs, the success of an air quality program may depend on the ability of **air pollution control** agencies to control upwind sources.

Airshed

A geographical region, usually a topographical basin, that tends to have uniform **air quality**. The air quality within an airshed is influenced predominantly by **emission** activities native to that airshed, since the elevated **topography** around the basin constrains horizontal air movement. Pollutants move from one part of an airshed to other parts fairly quickly, but are not readily transferred to adjacent airsheds. An airshed tends to have a relatively uniform **climate** and relatively uniform meteorological features at any given point in time.

Alar

Alar is the trade name for the chemical compound daminozide, manufactured by the Uniroyal Chemical Company. The compound has been used since 1968 to keep apples from falling off trees before they are ripe and to keep them red and firm during storage. As late as the early 1980s, up to 40% of all red apples produced in the United States were treated with Alar.

In 1985, the **Environmental Protection Agency** (EPA) found that UDMH (N,N-dimethylhydrazine), a compound produced during the breakdown of daminozide, was a **carcinogen**. UDMH was routinely produced during the processing of apples, as in the production of apple juice and apple sauce, and the EPA suggested a ban on the use of Alar by apple growers. An outside review of the EPA studies, however, suggested that they were flawed, and the ban was not instituted. Instead, the agency recommended

that Uniroyal conduct further studies on possible health risks from daminozide and UDMH.

Even without a ban, Uniroyal felt the impact of the EPA's research well before its own studies were concluded. Apple growers, fruit processors, legislators, and the general public were all frightened by the possibility that such a widely used chemical might be carcinogenic. Many growers, processors, and store owners pledged not to use the compound nor to buy or sell apples on which it had been used. By 1987, sales of Alar had dropped by 75%.

In 1989, two new studies again brought the subject of Alar to the public's attention. The consumer research organization Consumers' Union found that, using a very sensitive test for the chemical, 11 of 20 red apples they tested contained Alar. In addition, 23 of 44 samples of apple juice tested contained detectable amounts of the compound. The **Natural Resources Defense Council** (NRDC) announced their findings on the compound at about the same time. The NRDC concluded that Alar and certain other **agricultural chemicals** pose a threat to children about 240 times higher than the one-in-a-million risk traditionally used by the EPA to determine the acceptability of a product used in human foods.

The studies by the NRDC and the Consumers' Union created a panic among consumers, apple growers, and apple processors. Many stores removed all apple products from their shelves, and some growers destroyed their whole crop of apples. The industry suffered millions of dollars in damage. Representatives of the apple industry continued to question how much of a threat Alar truly posed to consumers, claiming that the carcinogenic risks identified by the EPA, NRDC, and Consumers' Union were greatly exaggerated. But in May of that same year, the EPA announced interim data from its most recent study, which showed that UDMH caused blood-vessel tumors in mice. The agency once more declared its intention to ban Alar, and within a month, Uniroyal announced it would end sales of the compound in the United States.

[*David E. Newton*]

RESOURCES
PERIODICALS

"Alar: Not Gone, Not Forgotten." *Consumer Reports* 52 (May 1989): 288–292.

Roberts, L. "Alar: The Numbers Game." *Science* 243 (17 March 1989): 1430.

———. "Pesticides and Kids." *Science* 243 (10 March 1989): 1280–1281.

Alaska Highway

The Alaska Highway, sometimes referred to as the Alcan (*Al*aska-*Can*ada) Highway, is the final link of a binational

transportation corridor that provides an overland route between the lower United States and Alaska. The first, all-weather, 1,522-mi (2,451 km) Alcan Military Highway was hurriedly constructed during 1942–1943 to provide land access between Dawson Creek, a Canadian village in northeastern British Columbia, and Fairbanks, a town on the Yukon River in central Alaska. Construction of the road was motivated by perception of a strategic, but ultimately unrealized, Japanese threat to maritime supply routes to Alaska during World War II.

The route of the Alaska Highway extended through what was then a **wilderness**. An aggressive technical vision was supplied by the United States **Army Corps of Engineers** and the civilian U.S. Public Roads Administration and labor by approximately 11,000 American soldiers and 16,000 American and Canadian civilians. In spite of the extraordinary difficulties of working in unfamiliar and inhospitable terrain, the route was opened for military passage in less than two years. Among the formidable challenges faced by the workers was a need to construct 133 bridges and thousands of smaller culverts across energetic watercourses, the infilling of alignments through a boggy muskeg capable of literally swallowing bulldozers, and working in winter temperatures that were so cold that vehicles were not turned off for fear they would not restart (steel dozer-blades became so brittle that they cracked upon impact with rock or frozen ground).

In hindsight, the planning and construction of the Alaska Highway could be considered an unmitigated environmental debacle. The enthusiastic engineers were almost totally inexperienced in the specialized techniques of arctic construction, especially about methods dealing with **permafrost**, or permanently frozen ground. If the integrity of permafrost is not maintained during construction, then this underground, ice-rich matrix will thaw and become unstable, and its water content will run off. An unstable morass could be produced by the resulting **erosion**, mudflow, slumping, and thermokarst-collapse of the land into subsurface voids left by the loss of water. Repairs were very difficult, and reconstruction was often unsuccessful, requiring abandonment of some original alignments. Physical and biological disturbances caused terrestrial landscape scars that persist to this day and will continue to be visible (especially from the air) for centuries. Extensive reaches of aquatic **habitat** were secondarily degraded by erosion and/or **sedimentation**. The much more careful, intensively scrutinized, and ecologically sensitive approaches used in the Arctic today, for example during the planning and construction of the trans Alaska pipeline, are in marked contrast with the unfettered and free-wheeling engineering associated with the initial construction of the Alaska Highway.

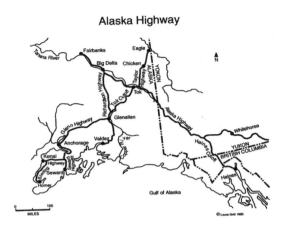

Alaska Highway

Map of the Alaska (Alcan) Highway. (Line drawing by Laura Gritt Lawson. Reproduced by permission.)

The Alaska Highway has been more-or-less continuously upgraded since its initial completion and was opened to unrestricted traffic in 1947. Non-military benefits of the Alaska Highway include provision of access to a great region of the interior of northwestern North America. This access fostered economic development through mining, forestry, trucking, and tourism, as well as helping to diminish the perception of isolation felt by many northern residents living along the route.

Compared with the real dangers of vehicular passage along the Alaska Highway during its earlier years, today the route safely provides one of North America's most spectacular **ecotourism** opportunities. Landscapes range from alpine **tundra** to expansive boreal forest, replete with abundantly cold and vigorous streams and rivers. There are abundant opportunities to view large mammals such as moose (*Alces alces*), caribou (*Rangifer tarandus*), and bighorn sheep (*Ovis canadensis*), as well as charismatic smaller mammals and birds and a wealth of interesting arctic, boreal, and alpine **species** of plants.

[*Bill Freedman Ph.D.*]

RESOURCES
BOOKS

Christy, J. *Rough Road to the North.* Markham, ON: Paperjacks, 1981.

PERIODICALS

Alexandra, V., and K. Van Cleve. "The Alaska Pipeline: A Success Story." *Annual Review of Ecological Systems* 14 (1983): 443–63.

Alaska National Interest Lands Conservation Act (1980)

Commonly known as the Alaska Lands Act, The Alaska National Interest Lands Conservation Act (ANILCA) law protected 104 million acres (42 million ha), or 28%, of the state's 375 million acres (152 million ha) of land. The law added 44 million acres (18 million ha) to the **national park** system, 55 million acres (22.3 million ha) to the fish and **wildlife refuge** system, 3 million acres (1.2 million ha) to the **national forest** system, and made 26 additions to the national wild and scenic rivers system. The law also designated 56.7 million acres (23 million ha) of land as **wilderness**, with the stipulation that 70 million acres (28.4 million ha) of additional land be reviewed for possible wilderness designation.

The genesis of this act can be traced to 1959, when Alaska became the forty-ninth state. As part of the statehood act, Alaska could choose 104 million acres (42.1 million ha) of federal land to be transferred to the state. This selection process was halted in 1966 to clarify land claims made by Alaskan **indigenous peoples**. In 1971, the Alaska Native Claims Settlement Act (ANSCA) was passed to satisfy the native land claims and allow the state selection process to continue. This act stipulated that the Secretary of the Interior could withdraw 80 million acres (32.4 million ha) of land for protection as national parks and monuments, fish and **wildlife** refuges, and national forests, and that these lands would not be available for state or native selection. Congress would have to approve these designations by 1978. If Congress failed to act, the state and the natives could select any lands not already protected. These lands were referred to as national interest or d-2 lands.

Secretary of the Interior Rogers Morton recommended 83 million acres (33.6 million ha) for protection in 1973, but this did not satisfy environmentalists. The ensuing conflict over how much and which lands should be protected, and how these lands should be protected, was intense. The environmental community formed the Alaska Coalition, which by 1980 included over 1,500 national, regional, and local organizations with a total membership of 10 million people. Meanwhile, the state of Alaska and development-oriented interests launched a fierce and well-financed campaign to reduce the area of protected land.

In 1978, the House passed a bill protecting 124 million acres (50.2 million ha). The Senate passed a bill protecting far less land, and House-Senate negotiations over a compromise broke down in October. Thus, Congress would not act before the December 1978 deadline. In response, the executive branch acted. Department of the Interior Secretary Cecil Andrus withdrew 110 million acres (44.6 million ha) from state selection and mineral entry. President Jimmy Carter then designated 56 million acres (22.7 million ha) of these lands as national monuments under the authority of the Antiquities Act. Forty million additional acres (16.2 million ha) were withdrawn as fish and wildlife refuges, and 11 million acres (4.5 million ha) of existing national forests were withdrawn from state selection and mineral entry. Carter indicated that he would rescind these actions once Congress had acted.

In 1979, the House passed a bill protecting 127 million acres (51.4 million ha). The Senate passed a bill designating 104 million acres (42.1 million ha) as national interest lands in 1980. Environmentalists and the House were unwilling to reduce the amount of land to be protected. In November, however, Ronald Reagan was elected President, and the environmentalists and the House decided to accept the Senate bill rather than face the potential for much less land under a President who would side with development interests. President Carter signed ANILCA into law on December 2, 1980.

ANILCA also mandated that the U.S. Geological Service (USGS) conduct biological and **petroleum** assessments of the coastal plain section of the **Arctic National Wildlife Refuge**, 19.8 million acres (8 million ha) known as area 1002. While the USGS did determine a significant quantity of oil reserves in the area, they also reported that petroleum development would adversely impact many native **species**, including caribou (*Rangifer tarandus*), snow geese (*Chen caerulescens*), and muskoxen (*Ovibos moschatus*). In 2001, the Bush administration unveiled a new **energy policy** that would open up this area to oil and **natural gas** exploration. In June 2002, a House version of the energy bill (H.R.4) that favors opening ANWR to drilling and a Senate version (S.517) that does not were headed into conference to reconcile the differences between the two bills.

[*Christopher McGrory Klyza and Paula Anne Ford-Martin*]

RESOURCES

BOOKS

Lentfer, Hank and C. Servid, eds. *Arctic Refuge: A Circle of Testimony.* Minneapolis, MN: Milkweed Editions, 2001.

OTHER

Alaska National Interest Lands Conservation Act. 16 USC 3101-3223; Public Law 96-487. [June 2002]. <http://www.access.gpo.gov/uscode/title16/chapter51_.html>.

Douglas, D. C., et al., eds. *Arctic Refuge Coastal Plain Terrestrial Wildlife Research Summaries.* Biological Science Report USGS/BRD/BSR-2002-0001. [June 2002]. <http://www.absc.usgs.gov/1002>.

ORGANIZATIONS

The Alaska Coalition, 419 6th St, #328 , Juneau, AK USA 99801 (907) 586-6667, Fax: (907) 463-3312, Email: info@alaskacoalition.org, <http://www.alaskacoalition.org>

Alaska National Wildlife Refuge
see **Arctic National Wildlife Refuge**

Alaska pipeline
see **Trans-Alaska pipeline**

Albedo

The reflecting power of a surface, expressed as a ratio of reflected radiation to incident or incoming radiation; it is sometimes expressed as a percentage. Albedo is also called the "reflection coefficient" and derives from the Latin root word *albus*, which means whiteness. Sometimes expressed as a percentage, albedo is more commonly measured as a fraction on a scale from zero to one, with a value of one denoting a completely reflective, white surface, while a value of zero would describe an absolutely black surface that reflects no light rays.

Albedo varies with surface characteristics such as color and composition, as well as with the angle of the sun. The albedo of natural earth surface features such as oceans, forests, deserts, and crop canopies varies widely. Some measured values of albedo for various surfaces are shown below:

Types of Surface	Albedo
Fresh, dry snow cover	0.80–0.95
Aged or decaying snow cover	0.40–0.70
Oceans	0.07–0.23
Dense clouds	0.70–0.80
Thin clouds	0.25–0.50
Tundra	0.15–0.20
Desert	0.25–0.29
Coniferous forest	0.10–0.15
Deciduous forest	0.15–0.20
Field crops	0.20–0.30
Bare dark soils	0.05–0.15

The albedo of clouds in the **atmosphere** is important to life on Earth because extreme levels of radiation absorbed by the earth would make the planet uninhabitable; at any moment in time about 50% of the planet's surface is covered by clouds. The mean albedo for the earth, called the planetary albedo, is about 30–35%.

[*Mark W. Seeley*]

Algal bloom

Algae are simple, single-celled, filamentous aquatic plants; they grow in colonies and are commonly found floating in ponds, lakes, and oceans. Populations of algae fluctuate with the availability of nutrients, and a sudden increase in nutrients often results in a profusion of algae known as algal bloom.

The growth of a particular algal **species** can be both sudden and massive. Algal cells can increase to very high densities in the water, often thousands of cells per milliliter, and the water itself can be colored brown, red, or green. Algal blooms occur in freshwater systems and in marine environments, and they usually disappear in a few days to a few weeks. These blooms consume oxygen, increase turbidity, and clog lakes and streams. Some algal species release water-soluble compounds that may be toxic to fish and shellfish, resulting in **fish kills** and **poisoning** episodes.

Algal groups are generally classified on the basis of the pigments that color their cells. The most common algal groups are blue-green algae, green algae, red algae, and brown algae. Algal blooms in freshwater lakes and ponds tend to be caused by blue-green and green algae. The excessive amounts of nutrients that cause these blooms are often the result of human activities. For example, **nitrates** and **phosphates** introduced into a lake from **fertilizer runoff** during a storm can cause rapid algal growth. Some common blue-green algae known to cause blooms as well as release nerve **toxins** are *Microcystis, Nostoc,* and *Anabaena.*

Red tides in coastal areas are a type of algal bloom. They are common in many parts of the world, including the **New York Bight**, the Gulf of California, and the Red Sea. The causes of algal blooms are not as well understood in marine environments as they are in freshwater systems. Although human activities may well have an effect on these events, weather conditions probably play a more important role: turbulent storms that follow long, hot, dry spells have often been associated with algal blooms at sea. Toxic red tides most often consist of genera from the dinoflagellate algal group such as *Gonyaulax* and *Gymnodinium.* The potency of the toxins has been estimated to be 10 to 50 times higher than cyanide or curare, and people who eat exposed shellfish may suffer from paralytic shellfish poisoning within 30 minutes of consumption. A fish kill of 500 million fish was reported from a **red tide** in Florida in 1947. A number of blue-green algal genera such as *Oscillatoria* and *Trichodesmium* have also been associated with red blooms, but they are not necessarily toxic in their effects. Some believe that the blooms caused by these genera gave the Red Sea its name.

The economic and health consequences of algal blooms can be sudden and severe, but the effects are generally not long lasting. There is little evidence that algal blooms have long-term effects on **water quality** or **ecosystem** structure.

[*Usha Vedagiri and Douglas Smith*]

RESOURCES
BOOKS

Lerman, M. *Marine Biology: Environment, Diversity and Ecology*. Menlo Park, CA: Benjamin/Cummings, 1986.

PERIODICALS

Culotta, E. "Red Menace in the World's Oceans." *Science* 257 (11 September 1992): 1476–77.

Mlot, C. "White Water Bounty: Enormous Ocean Blooms of White-Plated Phytoplankton Are Attracting the Interest of Scientists." *Bioscience* 39 (April 1989): 222–24.

Algicide

The presence of nuisance algae can cause unsightly appearance, odors, slime, and coating problems in aquatic media. Algicides are chemical agents used to control or eradicate the growth of algae in aquatic media such as industrial tanks, swimming pools, and lakes. These agents used may vary from simple inorganic compounds such as **copper** sulphate which are broad-spectrum in effect and control a variety of algal groups to complex organic compounds that are targeted to be species-specific in their effects. Algicides usually require repeated application or continuous application at low doses in order to maintain effective control.

Aline, Tundra
see **Tundra**

Allelopathy

Derived from the Greek words *allelo* (other) and *pathy* (causing injury to), allelopathy is a form of **competition** among plants. One plant produces and releases a chemical into the surrounding **soil** that inhibits the germination or growth of other **species** in the immediate area. These chemical substances are both acids and bases and are called secondary compounds. For example, black walnut (*Jugans nigra*) trees release a chemical called juglone that prevents other plants such as tomatoes from growing in the immediate area around

each tree. In this way, plants such as black walnut reduce competition for space, nutrients, water, and sunlight.

Allergen

Any substance that can bring about an allergic response in an organism. Hay fever and **asthma** are two common allergic responses. The allergens that evoke these responses include pollen, **fungi**, and dust. Allergens can be described as host-specific agents in that a particular allergen may affect some individuals, but not others. A number of air pollutants are known to be allergens. Formaldehyde, thiocyanates, and epoxy resins are examples. People who are allergic to natural allergens, such as pollen, are more inclined to be sensitive also to synthetic allergens, such as formaldehyde.

Alligator, American

The American alligator (*Alligator mississippiensis*) is a member of the reptilian family Crocodylidae, which consists of 21 **species** found in tropical and subtropical regions throughout the world. It is a species that has been reclaimed from the brink of **extinction**.

Historically, the American alligator ranged in the Gulf and Atlantic coast states from Texas to the Carolinas, with rather large populations concentrated in the swamps and river bottomlands of Florida and Louisiana. From the late nineteenth century into the middle of the twentieth century, the population of this species decreased dramatically. With no restrictions on their activities, hunters killed alligators as pests or to harvest their skin, which was highly valued in the leather trade. The American alligator was killed in such great numbers that biologists predicted its probable extinction. It has been estimated that about 3.5 million of these reptiles were slaughtered in Louisiana between 1880 and 1930. The population was also impacted by the fad of selling young alligators as pets, principally in the 1950s.

States began to take action in the early 1960s to save the alligator from extinction. In 1963 Louisiana banned all legalized **trapping**, closed the alligator **hunting** season, and stepped up enforcement of game laws against poachers. By the time the **Endangered Species Act** was passed in 1973, the species was already experiencing a rapid recovery. Because of the successful re-establishment of alligator populations, its endangered classification was downgraded in several southeastern states, and there are now strictly regulated seasons that allow alligator trapping. Due to the persistent demand for its hide for leather goods and an increasing market for the reptile's meat, alligator farms are now both legal and profitable.

An American alligator (*Alligator mississippiensis*). (Photograph by B. Arroyo. U. S. Fish & Wildlife Service. Reproduced by permission.)

Human fascination with large, dangerous animals, along with the American alligator's near extinction, have made it one of North America's best studied reptile species. Population pressures, primarily resulting from being hunted so ruthlessly for decades, have resulted in a decrease in the maximum size attained by this species. The growth of a reptile is indeterminate, and they continue to grow as long as they are alive, but old adults from a century ago attained larger sizes than their counterparts do today. The largest recorded American alligator was an old male killed in January 1890, in Vermilion Parish, Louisiana, which measured 19.2 ft (6 m) long. The largest female ever taken was only about half that size. ·

Alligators do not reach sexual maturity until they are about 6 ft (1.3 m) long and nearly 10 years old. Females construct a nest mound in which they lay about 35–50 eggs. The nest is usually 5–7 ft (1.5–2.1 m) in diameter and 2–3 ft (0.6–0.9 m) high, and decaying vegetation produces heat which keeps the eggs at a fairly constant temperature during incubation. The young stay with their mother through their first winter, striking out on their own when they are about 1.5 ft (0.5 m) in length.

[*Eugene C. Beckham*]

RESOURCES
BOOKS

Crocodiles. Proceedings of the 9th Working Meeting of the IUCN/SSC Crocodile Specialist Group, Lae, Papua New Guinea. Vol. 2. Gland, Switzerland: IUCN-The World Conservation Union, 1990.

Dundee, H. A., and D. A. Rossman. *The Amphibians and Reptiles of Louisiana.* Baton Rouge: LSU Press, 1989.

Webb, G. J. W., S. C. Manolis, and P. J. Whitehead, eds. *Wildlife Management: Crocodiles and Alligators.* Chipping Norton, Australia: Surrey Beatty and Sons, 1987.

OTHER

"Alligator mississippiensis in the Crocodilians, Natural History and Conservation." *Florida Museum of Natural History.* [cited May 2002] <http://www.flmnh.ufl.edu/cnhc/csp_amis.htm>.

"The American Alligator." *University of Florida, Gainesville.* [cited May 2002]. <http://agrigator.ifas.ufl.edu/gators>.

Alligator mississippiensis

see **Alligator, American**

All-terrain vehicle

see **Off-road vehicles**

Alpha particle

A particle emitted by certain kinds of radioactive materials. An alpha particle is identical to the nucleus of a helium atom, consisting of two protons and two neutrons. Some common alpha-particle emitters are uranium-235, uranium-238, radium-226, and radon-222. Alpha particles have relatively low penetrating power. They can be stopped by a thin sheet of paper or by human skin. They constitute a health problem, therefore, only when they are taken into the body. The inhalation of alpha-emitting **radon** gas escaping from bedrock into houses in some areas is thought to constitute a health hazard.

Alternative energy sources

Coal, oil, and **natural gas** provide over 85% of the total primary energy used around the world. Although figures differ in various countries, nuclear reactors and hydroelectric power together produce less than 10% of the total world energy. Wind power, active and passive solar systems, and **geothermal energy** are examples of alternative energy sources. Collectively, these make up the final small fraction of total energy production.

The exact contribution alternative energy sources make to the total primary energy used around the world is not known. Conservative estimates place their share at 3–4%, but some energy experts dispute these figures. **Amory Lovins** has argued that the **statistics** collected are based primarily on large **electric utilities** and the regions they serve. They fail to account for areas remote from major power grids, which are more likely to use **solar energy**, **wind energy**, or other sources. When these areas are taken into consideration, Lovins claims, alternative energy sources contribute as much as 11% to the total primary energy used in the United States. Animal manure, furthermore, is widely used as an energy source in India, parts of China, and many African nations, and when this is taken into account the percentage of the worldwide contribution alternative sources make to energy production could rise as high as 10–15%. Now an alternative energy source, wind power is one of the earliest forms of energy used by humankind. Wind is caused by the uneven heating of the earth's surface, and its energy is equal to about

2% of the solar energy that reaches the earth. In quantitative terms, the amount of kinetic energy within the earth's **atmosphere** is equal to about 10,000 trillion kilowatt hours.

The kinetic energy of wind is proportional to the wind velocity, and the ideal location for a windmill generator is an area with constant and relatively fast winds and no obstacles such as buildings or trees. An efficient windmill can produce 175 watts per square meter of propeller blade area at a height of 75 ft (25 m). The estimated cost of generating one kilowatt hour by wind power is about eight cents, as compared to five cents for hydropower and 15 cents for **nuclear power**. The largest two utilities in California purchase wind-generated electricity, and though this state leads the country in the utilization of wind power, Denmark leads the world. The Scandinavian nation has refused to use nuclear power, and it expects to obtain 10% of its energy needs from windmills.

Solar energy can be utilized either directly as heat or indirectly by converting it to electrical power using photovoltaic cells. Greenhouses and solariums are the most common examples of the direct use of solar energy, with glass windows concentrating the visible light from the sun but restricting the heat from escaping. Flatplate collectors are another direct method, and mounted on rooftops they can provide one third of the energy required for space heating. Windows and collectors alone are considered passive systems; an active solar system uses a fan, pump, or other machinery to transport the heat generated from the sun.

Photovoltaic cells are made of semiconductor materials such as silicon. These cells are capable of absorbing part of the solar flux to produce a direct electric current with about 14% efficiency. The current cost of producing photovoltaic current is about four dollars a watt. However, a thin-film technology is being perfected for the production of these cells, and the cost per watt will eventually be reduced because less materials will be required. Photovoltaics are now being used economically in lighthouses, boats, rural villages, and other remote areas. Large solar systems have been most effective using trackers that follow the sun or mirror reflectors that concentrate its rays.

Geothermal energy is the natural heat generated in the interior of the earth, and like solar energy it can also be used directly as heat or indirectly to generate electricity. Steam is classified as either dry (no water droplets), or wet (mixed with water). When it is generated in certain areas containing corrosive sulfur compounds, it is known as "sour steam," and when generated in areas that are free of sulfur it is known as "sweet steam." Geothermal energy can be used to generate electricity by the flashed steam method, in which high temperature geothermal brine is used as a heat exchanger to convert injected water into steam. The produced steam is used to turn a turbine. When geothermal

wells are not hot enough to create steam, a fluid which evaporates at a much lower temperature than water, such as isobutane or ammonia, can be placed in a closed system where the geothermal heat provides the energy to evaporate the fluid and run the turbine.

There are 20 countries worldwide that utilize this energy source, and they include the United States, Mexico, Italy, Iceland, Japan, and the former Soviet Union. Unlike solar energy and wind power, geothermal energy is not free of environmental impact. It contributes to **air pollution**, it can emit dissolved salts and, in some cases, toxic **heavy metals** such as **mercury** and **arsenic**.

Though there are several ways of utilizing energy from the ocean, the most promising are the harnessing of **tidal power** and **ocean thermal energy conversion**. The power of ocean tides is based on the difference between high and low water. In order for tidal power to be effective the differences in height need to be very great, more than 15 ft (3 m), and there are only a few places in the world where such differences exist. These include the **Bay of Fundy** and a few sites in China. Ocean thermal energy conversion utilizes temperature changes rather than tides. Ocean temperature is stratified, especially near the tropics, and the process takes advantage of this fact by using a fluid with a low boiling point, such as ammonia. The vapor from the fluid drives a turbine, and cold water from lower depths is pumped up to condense the vapor back into liquid. The electrical power generated by this method can be shipped to shore or used to operate a floating plant such as a cannery.

Other sources of alternative energy are currently being explored, some of which are still experimental. These include harnessing the energy in **biomass** through the production of wood from trees or the production of **ethanol** from crops such as sugar cane or corn. **Methane** gas can be generated from the **anaerobic** breakdown of **organic waste** in sanitary landfills and from **wastewater** treatment plants. With the cost of **garbage** disposal rapidly increasing, the burning of garbage is becoming a viable option as an energy source. Adequate air **pollution** controls are necessary, but trash can be burned to heat buildings, and municipal garbage is currently being used to generate electricity in Hamburg, Germany. In an experimental method known as *magnetohydrodynamics*, hot gas is ionized (potassium and sulfur) and passed through a strong magnetic field where it produces an electrical current. This process contains no moving parts and has an efficiency of 20–30%.

Ethanol and **methanol** can be produced from biomass and used in **transportation**; in fact, methanol currently powers Indianapolis race cars. **Hydrogen** could be valuable if problems of supply and storage can be solved. It is very clean-burning, forming water, and may be combined with oxygen in **fuel cells** to generate electricity. Also, it is not nearly as explosive as **gasoline**.

Of all the alternative sources, **energy conservation** is perhaps the most important, and improving **energy efficiency** is the best way to meet energy demands without adding to air and **water pollution**. One reason the United States survived the energy crises of the 1970s was that they were able to curtail some of their immense waste. Relatively easy lifestyle alterations, vehicle improvements, building insulation, and more efficient machinery and appliances have significantly reduced their potential energy demand. Experts have estimated that it is possible to double the efficiency of electric motors, triple the intensity of light bulbs, quadruple the efficiency of refrigerators and air conditioners, and quintuple the gasoline mileage of automobiles. Several **automobile** manufacturers in Europe and Japan have already produced prototype vehicles with very high gasoline mileage. Volvo has developed the LCP 2000, a passenger sedan that holds four to five people, meets all United States safety standards, accelerates from 0–50 MPH (0–80.5 km/hr) in 11 seconds, and has a high fuel efficiency rating. Alternative fuels will be required to meet future energy needs. Enormous investments in new technology and equipment will be needed, and potential supplies are uncertain, but there is clearly hope for an energy-abundant future.

[*Muthena Naseri and Douglas Smith*]

RESOURCES
BOOKS

Alternative Energy Handbook. Englewood Cliffs, NJ: Prentice Hall, 1993.

Brower, M. *Cool Energy: Renewable Solutions to Environmental Problems.* Cambridge: MIT Press, 1992.

Brown, Lester R., ed. *The World Watch Reader on Global Environmental Issues.* New York: W. W. Norton, 1991.

Goldemberg, J. *Energy for a Sustainable World.* New York: Wiley, 1988.

Schaeffer, J. *Alternative Energy Sourcebook: A Comprehensive Guide to Energy Sensible Technologies.* Ukiah, CA: Real Goods Trading Corp., 1992.

Shea, C. P. *Renewable Energy: Today's Contribution, Tomorrow's Promise.* Washington, DC: Worldwatch Institute, 1988.

PERIODICALS

Stein, J. "Hydrogen: Clean, Safe, and Inexhaustible." *Amicus Journal* 12 (Spring 1990): 33-36.

Alternative fuels
see **Renewable energy**

Aluminum

Aluminum, a light metal, comprises about 8% of the earth's crust, ranking as the third-most abundant element after oxygen (47%) and silicon (28%). Virtually all environmental aluminum is present in mineral forms that are almost insoluble in water, and therefore not available for uptake by organisms. Most common among these forms of aluminum are various aluminosilicate minerals, aluminum clays and sesquioxides, and aluminum **phosphates**.

However, aluminum can also occur as chemical **species** that are available for biological uptake, sometimes causing toxicity. In general, bio-available aluminum is present in various water-soluble, ionic or organically complexed chemical species. Water-soluble concentrations of aluminum are largest in acidic environments, where toxicity to non-adapted plants and animals can be caused by exposure to Al^{3+} and $Al(OH)^{2+}$ ions, and in alkaline environments, where $Al(OH)_4^-$ is most prominent. Organically bound, water-soluble forms of aluminum, such as complexes with fulvic or humic acids, are much less toxic than ionic species. Aluminum is often considered to be the most toxic chemical factor in acidic soils and aquatic habitats.

Amazon basin

The Amazon basin, the region of South America drained by the Amazon River, represents the largest area of **tropical rain forest** in the world. Extending across nine different countries and covering an area of 2.3 million square mi (6 million sq. km), the Amazon basin contains the greatest abundance and diversity of life anywhere on the earth. Tremendous numbers of plant and animal **species** that occur there have yet to be discovered or properly named by scientists, as this area has only begun to be explored by competent researchers.

It is estimated that the Amazon basin contains over 20% of all higher plant species on Earth, as well as about 20% of all birdlife and 10% of all mammals. More than 2,000 known species of freshwater fishes live in the Amazon river and represent about 8% of all fishes on the planet, both freshwater and marine. This number of species is about three times the entire ichthyofauna of North America and almost ten times that of Europe. The most astonishing numbers, however, come from the river basin's insects. Every expedition to the Amazon basin yields countless new species of insects, with some individual trees in the tropical forest providing scientists with hundreds of undescribed forms. Insects represent about three-fourths of all animal life on Earth, yet biologists believe the 750,000 species that have already been scientifically named account for less than 10% of all insect life that exists.

However incredible these examples of **biodiversity** are, they may soon be destroyed as the rampant **deforestation** in the Amazon basin continues. Much of this destruction is directly attributable to human **population growth**.

The number of people who have settled in the Amazonian uplands of Colombia and Ecuador has increased by 600% over the past 40 years, and this has led to the clearing of over 65% of the region's forests for agriculture.

In Brazil, up to 70% of the deforestation is tied to cattle ranching. In the past large governmental subsidies and tax incentives have encouraged this practice, which had little or no financial success and caused widespread environmental damage. Tropical soils rapidly lose their fertility, and this allows only limited annual meat production. It is often only 300 lb (136 kg) per acre, compared to over 3,000 lb (1,360 kg) per acre in North America.

Further damage to the tropical forests of the Amazon basin is linked to commercial **logging**. Although only five of the approximately 1,500 tree species of the region are extensively logged, tremendous damage is done to the surrounding forest as these are selectively removed. When loggers build roads move in heavy equipment, they may damage or destroy half of the trees in a given area.

The deforestation taking place in the Amazon basin has a wide range of environmental effects. The clearing and burning of vegetation produces **smoke** or **air pollution**, which at times has been so abundant that it is clearly visible from space. Clearing also leads to increased **soil erosion** after heavy rains, and can result in **water pollution** through **siltation** as well as increased water temperatures from increased exposure. Yet the most alarming, and definitely the most irreversible, environmental problem facing the Amazon basin is the loss of biodiversity. Through the irrevocable process of **extinction**, this may cost humanity more than the loss of species. It may cost us the loss of potential discoveries of medicines and other beneficial products derived from these species.

[*Eugene C. Beckham*]

RESOURCES
BOOKS

Caufield, C. *In the Rainforest: Report From a Strange, Beautiful, Imperiled World.* Chicago: University of Chicago Press, 1986.

Cockburn, A., and S. Hecht. *The Fate of the Forest: Developers, Destroyers, and Defenders of the Amazon.* New York: Harper/Perennial, 1990.

Collins, M. *The Last Rain Forests: A World Conservation Atlas.* London: Oxford University Press, 1990.

Cowell, A. *Decade of Destruction: The Crusade to Save the Amazon Rain Forest.* New York: Doubleday, 1991.

Margolis, M. *The Last New World: The Conquest of the Amazon Frontier.* New York: Norton, 1992.

Wilson, E. O. *The Diversity of Life.* Cambridge, MA: Belknap Press, 1992.

PERIODICALS

Holloway, M. "Sustaining the Amazon." *Scientific American* 269 (July 1993): 90–96+.

Ambient air

The air, external to buildings and other enclosures, found in the lower **atmosphere** over a given area, usually near the surface. **Air pollution** standards normally refer to ambient air.

Amenity value

The idea that something has worth because of the pleasant feelings it generates to those who use or view it. This value is often used in cost-benefit analysis, particularly in **shadow pricing**, to determine the worth of **natural resources** that will not be harvested for economic gain. A virgin forest will have amenity value, but its value will decrease if the forest is harvested, thus the amenity value is compared to the value of the harvested timber.

American alligator
see **Alligator, American**

American Box Turtle

Box turtles are in the Order Chelonia, Family Emydidae, and genus *Terrapene*. There are two major **species** in the United States: *carolina* (Eastern box turtle) and *ornata* (Western or ornate box turtle).

Box turtles are easily recognized by their dome-shaped upper shell (carapace) and by their lower shell (plastron) which is hinged near the front. This hinging allows them to close up tightly into the "box" when in danger (hence their name).

Box turtles are fairly small, having an adult maximum length of 4–7 in (10–18 cm). Their range is restricted to North America, with the Eastern species located over most of the eastern United States and the Western species located in the Central and Southwestern United States and into Mexico, but not as far west as California. Both species are highly variable in coloration and pattern, ranging from a uniform tan to dark brown or black, with yellow spots or streaks. They prefer a dry **habitat** such as woodlands, open brush lands, or **prairie**. They typically inhabit sandy **soil**, but are sometimes found in springs or ponds during hot weather. During the winter, they hibernate in the soil below the frost line, often as deep as 2 ft (60 cm). Their home range is usually fairly small, and they often live within areas less than 300 yd² (300 m²).

Eastern box turtle. (Photograph by Robert Huffman. Fieldmark Publications. Reproduced by permission.)

Box turtles are omnivorous, feeding on living and dead insects, earthworms, slugs, fruits, berries (particularly blackberries and strawberries), leaves, and mushrooms. They have been known to ingest some mushrooms which are poisonous to humans, and there have been reports of people eating box turtles and getting sick. Other than this, box turtles are harmless to humans and are commonly collected and sold as pets (although this should be discouraged because they are now a threatened species). They can be fed raw hamburger, canned pet food, or leafy vegetables.

Box turtles normally live as long as 30–40 years. Some have been reported with a longevity of more than one hundred years, and this makes them the longest-lived land turtle. They are active from March until November and are diurnal, usually being more active in the early morning. During the afternoons they typically seek shaded areas. They breed during the spring and autumn, and the females build nests from May until July, typically in sandy soil where they dig a hole with their hind feet. The females can store sperm for several years. They typically hatch three to eight eggs that are elliptically-shaped and about 1.5 in (4 cm) in diameter. Male box turtles have a slight concavity in their plastron that aids in mounting females during copulation. All four toes on the male's hind feet are curved, which aids in holding down the posterior portion of the female's plastron during copulation. Females have flat plastrons, shorter tails, and yellow or brown eyes. Most males have bright red or pink eyes. The upper jaw of both sexes ends in a down-turned beak.

Predators of box turtles include skunks, raccoons, foxes, snakes, and other animals. Native American Indians used to eat box turtles and incorporated their shells into their ceremonies as rattles.

[*John Korstad*]

RESOURCES
BOOKS

Conant, R. *A Field Guide to Reptiles and Amphibians of Eastern and Central North America.* Boston: Houghton Mifflin, 1998.

Tyning, T. F. *A Guide to Amphibians and Reptiles.* Boston: Little, Brown and Co., 1990.

OTHER

"Conservation and Preservation of American Box Turtles in the Wild." *The American Box Turtle Page.* Fall 2000 [cited May 2002]. <http://www.americanboxturtle.freeservers.com>.

American Cetacean Society

The American Cetacean Society (ACS), located in San Pedro, California, is dedicated to the protection of **whales** and other cetaceans, including **dolphins** and porpoises. Principally an organization of scientists and teachers (though its membership does include students and laypeople) the ACS was founded in 1967 and claims to be the oldest whale **conservation** group in the world.

The ACS believes the best protection for whales, dolphins, and porpoises is better public awareness about "these remarkable animals and the problems they face in their increasingly threatened habitat." The organization is committed to political action through education, and much of its work has been in improving communication between marine scientists and the general public.

The ACS has developed several educational resource materials on cetaceans, making such products as the "Gray Whale Teaching Kit," "Whale Fact Pack," and "Dolphin Fact Pack," which are widely available for use in classrooms. There is a cetacean research library at the national headquarters in San Pedro, California, and the organization responds to thousands of inquiries every year. The ACS supports marine mammal research and sponsors a biennial conference on whales. It also assists in conducting whale-watching tours.

The organization also engages in more traditional and direct forms of political action. A representative in Washington, DC, monitors legislation that might affect cetaceans, attends hearings at government agencies, and participates as a member of the International **Whaling** Commission. The ACS also networks with other conservation groups. In addition, the ACS directs letter-writing campaigns, sending out "Action Alerts" to citizens and politicians. The organization is currently emphasizing the threats to marine life posed by **oil spills**, toxic wastes from industry and agriculture, and particular fishing practices (including commercial whaling).

The ACS publishes a quarterly newsletter on whale research, conservation, and education, called *WhaleNews,*

and a quarterly journal of scientific articles on the same subjects, called *Whalewatcher.*

[*Douglas Smith*]

RESOURCES
ORGANIZATIONS

American Cetacean Society, P.O. Box 1391, San Pedro, CA USA 90733-1391 (310) 548-6279, Fax: (310) 548-6950, Email: acs@pobox.com, <http://www.acsonline.org>

American Committee for International Conservation

The American Committee for International Conservation (ACIC), located in Washington, DC, is an association of nongovernmental organizations (NGOs) that is concerned about international conservation issues. The ACIC, founded in 1930, includes 21 member organizations. It represents conservation groups and individuals in 40 countries. While ACIC does not fund conservation research, it does promote national and international conservation research activities. Specifically, ACIC promotes conservation and preservation of **wildlife** and other **natural resources**, and encourages international research on the **ecology** of **endangered species**.

Formerly called the American Committee for International Wildlife Protection, ACIC assists **IUCN—The World Conservation Union**, an independent organization of nations, states, and NGOs, in promoting natural resource conservation. ACIC also coordinates its members' overseas research activities.

Member organizations of the ACIC include the African Wildlife Leadership Foundation, **National Wildlife Federation**, **World Wildlife Fund** (US)/RARE, Caribbean Conservation Corporation, **National Audubon Society**, **Natural Resources Defense Council**, **Nature** Conservancy, International Association of Fish and Wildlife Agencies, and **National Parks and Conservation Association**. Members also include The Conservation Foundation, International Institute for **Environment** and Development; Massachusetts Audubon Society; Chicago Zoological Society; Wildlife Preservation Trust; Wildfowl Trust; School of Natural Resources, University of Michigan; **World Resources Institute**; Global Tomorrow Coalition; and The Wildlife Society, Inc.

ACIC holds no formal meetings or conventions, nor does it publish magazines, books, or newsletters. Contact: American Committee for International Conservation, c/o

Center for Marine Conservation, 1725 DeSales Street, NW, Suite 500, Washington, DC 20036.

[*Linda Rehkopf*]

American Farmland Trust

Headquartered in Washington, DC, the American Farmland Trust (AFT) is an advocacy group for farmers and farmland. It was founded in 1980 to help reverse or at least slow the rapid decline in the number of productive acres nationwide, and it is particularly concerned with protecting land held by private farmers. The principles that motivate the AFT are perhaps best summarized in a line from William Jennings Bryan that the organization has often quoted: "Destroy our farms, and the grass will grow in the streets of every city in the country."

Over one million acres (404,700 ha) of farmland in the United States is lost each year to development, according to the AFT, and in Illinois one and a half bushels of **topsoil** are lost for every bushel of corn produced. The AFT argues that such a decline poses a serious threat to the future of the American economy. As farmers are forced to cultivate increasingly marginal land, food will become more expensive, and the United States could become a net importer of agricultural products, damaging its international economic position. The organization believes that a declining farm industry would also affect American culture, depriving the country of traditional products such as cherries, cranberries, and oranges and imperiling a sense of national identity that is still in many ways agricultural.

The AFT works closely with farmers, business people, legislators, and environmentalists "to encourage sound farming practices and wise use of land." The group directs lobbying efforts in Washington, working with legislators and policymakers and frequently testifying at congressional and public hearings on issues related to farming. In addition to mediating between farmers and state and federal government, the trust is also involved in political organizing at the grassroots level, conducting public opinion polls, contesting proposals for incinerators and toxic waste sites, and drafting model **conservation easements**. They conduct workshops and seminars across the country to discuss farming methods and **soil conservation** programs, and they worked with the State of Illinois to establish the Illinois **Sustainable Agriculture** Society. The group is currently developing kits for distribution to schoolchildren in both rural and urban areas called "Seed for the Future," which teach the benefits of agriculture and help each child grow a plant.

The AFT has a reputation for innovative and determined efforts to realize its goals, and former Secretary of Agriculture John R. Block has said that "this organization has probably done more than any other to preserve the American farm." Since its founding the trust has been instrumental in protecting nearly 30,000 acres (12,140 ha) of farmland in 19 states. In 1989, the group protected a 507-acre (205-ha) cherry farm known as the Murray Farm in Michigan, and it has helped preserve 300 acres (121 ha) of farm and **wetlands** in Virginia's Tidewater region. The AFT continues to battle **urban sprawl** in areas such as California's Central Valley and Berks County, Pennsylvania, as well as working to support farms in states such as Vermont, which are threatened not so much by development but by a poor agricultural economy. The AFT promotes a wetland policy that is fair to farmers while meeting **environment** standards, and it recently won a national award from the **Soil and Water Conservation Society** for its publication *Does Farmland Protection Pay?*

The AFT has 20,000 members and an annual budget of $3,850,000. The trust publishes a quarterly magazine called *American Farmland*, a newsletter called *Farmland Update*, and a variety of brochures and pamphlets which offer practical information on **soil erosion**, the cost of community services, and estate planning. They also distribute videos, including *The Future of America's Farmland*, which explains the sale and purchase of development rights.

[*Douglas Smith*]

RESOURCES

ORGANIZATIONS

The American Farmland Trust (AFT), 1200 18th Street, NW, Suite 800, Washington, D.C. USA 20036 (202) 331-7300, Fax: (202) 659-8339, Email: info@farmland.org, <http://www.farmland.org>

American Forests

Located in Washington, DC, American Forests was founded in 1875, during the early days of the American **conservation** movement, to encourage **forest management**. Originally called the American Forestry Association, the organization was renamed in the later part of the twentieth century. The group is dedicated to promoting the wise and careful use of all **natural resources**, including **soil**, water, and **wildlife**, and it emphasizes the social and cultural importance of these resources as well as their economic value.

Although benefiting from increasing national and international concern about the **environment**, American Forests takes a balanced view on preservation, and it has worked to set a standard for the responsible harvesting and marketing of forest products. American Forests sponsors the Trees for People program, which is designed to help meet the national demand for wood and paper products by increasing the productivity of private woodlands. It provides educational and technical information to individual forest owners, as well as making recommendations to legislators and policymakers in Washington.

To draw attention to the **greenhouse effect**, American Forests inaugurated their **Global ReLeaf** program in October 1988. Global ReLeaf is what American Forests calls "a tree-planting crusade." The message is, "Plant a tree, cool the globe," and Global ReLeaf has organized a national campaign, challenging Americans to plant millions of trees. American Forests has gained the support of government agencies and local conservation groups for this program, as well as many businesses, including such Fortune-500 companies as Texaco, McDonald's, and Ralston-Purina. The goal of the project is to plant 20 million trees by 2002. In August of 2001, there had been 19 million trees planted. Global ReLeaf also launched a cooperative effort with the **American Farmland Trust** called Farm ReLeaf, and it has also participated in the campaign to preserve Walden Woods in Massachusetts. In 1991 American Forests brought Global ReLeaf to Eastern Europe, running a workshop in Budapest, Hungary, for environmental activists from many former communist countries.

American Forests has been extensively involved in the controversy over the preservation of old-growth forests in the American Northwest. They have been working with environmentalists and representatives of the timber industry, and consistent with the history of the organization, American Forests is committed to a compromise that both sides can accept: "If we have to choose between preservation and destruction of old-growth forests as our only options, neither choice will work." American Forests supports an approach to forestry known as *New Forestry*, where the priority is no longer the quantity of wood or the number of board feet that can be removed from a site, but the vitality of the **ecosystem** the timber industry leaves behind. The organization advocates the establishment of an Old Growth Reserve in the Pacific Northwest, which would be managed by the principles of New Forestry under the supervision of a Scientific Advisory Committee.

American Forests publishes the *National Registry of Big Trees*, which celebrated its sixtieth anniversary in 2000. The registry is designed to encourage the appreciation of trees, and it includes such trees as the recently fallen Dyerville Giant, a redwood tree in California; the General Sherman, a giant sequoia in Texas; and the Wye Oak in Maryland. The group also publishes *American Forests*, a bimonthly magazine, and *Resource Hotline*, a biweekly newsletter, as well as *Urban Forests: The Magazine of Community Trees*. It presents the Annual Distinguished Service Award, the John Aston Warder Medal, and the William B. Greeley Award, among others. American Forests has over 35,000 members, a staff of 21, and a budget of $2,725,000.

[*Douglas Smith*]

RESOURCES

ORGANIZATIONS

American Forests, P.O. Box 2000, Washington, DC USA 20013 (202) 955-4500, Fax: (202) 955-4588, Email: info@amfor.org, <http://www.americanforests.org>

American Indian Environmental Office

The American Indian Environmental Office (AIEO) was created to increase the quality of public health and environmental protection on Native American land and to expand tribal involvement in running environmental programs.

Native Americans are the second-largest landholders besides the government. Their land is often threatened by **environmental degradation** such as **strip mining, clear-cutting**, and toxic storage. The AIEO, with the help of the President's Federal Indian Policy (January 24, 1983), works closely with the U.S. **Environmental Protection Agency** (EPA) to prevent further degradation of the land. The AIEO has received grants from the EPA for environmental clean-up and obtained a written policy that requires the EPA to continue with the trust responsibility, a clause expressed in certain treaties that requires the EPA to notify the Tribe when performing any activities that may affect reservation lands or resources. This involves consulting with tribal governments, providing technical support, and negotiating EPA regulations to ensure that tribal facilities eventually comply.

The **pollution** of Dine Reservation land is an example of an environmental injustice that the AIEO wants to prevent in the future. The reservation has over 1,000 abandoned **uranium** mines that leak radioactive contaminants and is also home to the largest **coal** strip mine in the world. The **cancer** rate for the Dine people is 17 times the national average. To help tribes with pollution problems similar to the Dine, several offices now exist that handle specific environmental projects. They include the Office of Water, Air, Environmental Justice, Pesticides and Toxic Substances; Performance Partnership Grants; **Solid Waste** and Emergency Response; and the Tribal **Watershed** Project. Each of these offices reports to the National Indian Headquarters in Washington, DC.

At the Rio Earth Summit in 1992, the **Biodiversity** Convention was drawn up to protect the diversity of life on the planet. Many Native American groups believe that the convention also covered the protection of indigenous communities, including Native American land. In addition, the groups demand that prospecting by large companies for rare forms of life and materials on their land must stop.

Tribal Environmental Concerns

Tribal governments face both economic and social problems dealing with the demand for jobs, education, health care, and housing for tribal members. Often the reservations' largest employer is the government, which owns the stores, gaming operations, timber mills, and manufacturing facilities. Therefore, the government must deal with the conflicting interests of protecting both economic and environmental concerns. Many tribes are becoming self-governing and manage their own **natural resources** along with claiming the reserved right to use natural resources on portions of **public land** that border their reservation. As a product of the reserved treaty rights, Native Americans can use water, fish, and hunt anytime on nearby federal land.

Robert Belcourt, Chippewa-Cree tribal member and director of the Natural Resources Department in Montana stated:

"We have to protect **nature** for our **future generations**. More of our Indian people need to get involved in natural resource management on each of our reservations. In the long run, natural resources will be our bread and butter by our developing them through tourism and **recreation** and just by the opportunity they provide for us to enjoy the outdoor world."

Belcourt has fought to destroy the negative stereotypes of **conservation** organizations that exist among Native Americans who believe, for example, that conservationists are extreme tree-huggers and insensitive to Native American culture. These stereotypes are a result of cultural differences in philosophy, perspective, and communication. To work together effectively, tribes and conservation groups need to learn about one another's cultures, and this means they must listen both at meetings and in one-on-one exchanges.

The AIEO also addresses the organizational differences that exist in tribal governments and conservation organizations. They differ greatly in terms of style, motivation, and the pressures they face. Pressures on the **Wilderness Society**, for example, include fending off attempts in Washington, D.C. to weaken key environmental laws or securing members and raising funds. Pressures on tribal governments more often are economic and social in nature and have to do with the need to provide jobs, health care, education, and housing for tribal members. Because tribal governments are often the reservations' largest employers and may own businesses like gaming operations, timber mills, manufacturing facilities, and stores, they function as both governors and leaders in economic development.

Native Americans currently occupy and control over 52 million acres (21.3 million ha) in the continental United States and 45 million more acres (18.5 million ha) in Alaska, yet this is only a small fraction of their original territories.

In the nineteenth century, many tribes were confined to reservations that were perceived to have little economic value, although valuable natural resources have subsequently been found on some of these land. Pointing to their treaties and other agreements with the federal government, many tribes assert that they have reserved rights to use natural resources on portions of public land.

In previous decades these natural resources on tribal lands were managed by the Bureau of Indian Affairs (BIA). Now many tribes are becoming self-governing and are taking control of management responsibilities within their own reservation boundaries. In addition, some tribes are pushing to take back management over some federally managed lands that were part of their original territories. For example, the Confederated Salish and Kootenai tribes of the Flathead Reservation are taking steps to assume management of the National **Bison** Range, which lies within the reservation's boundaries and is currently managed by the U.S. **Fish and Wildlife Service**.

Another issue concerns Native American rights to water. There are legal precedents that support the practice of reserved rights to water that is within or bordering a reservation. In areas where tribes fish for food, mining pollution has been a continues threat to maintaining clean water. Mining pollution is monitored, but the amount of fish that Native Americans consume is higher than the government acknowledges when setting health guidelines for their consumption. This is why the AIEO is asking that stricter regulations be imposed on mining companies. As tribes increasingly exercise their rights to use and consume water and fish, their roles in natural resource debates will increase.

Many tribes are establishing their own natural resource management and environmental quality protection programs with the help of the AIEO. Tribes have established fisheries, **wildlife**, forestry, **water quality**, **waste management**, and planning departments. Some tribes have prepared comprehensive resource management plans for their reservations while others have become active in the protection of particular **species**. The AIEO is uniting tribes in their strategy and involvement level with improving environmental protection on Native American land.

[*Nicole Beatty*]

RESOURCES

ORGANIZATIONS

American Indian Environmental Office, 1200 Pennsylvania Avenue, NW, Washington, D.C. USA 20460 (202) 564-0303, Fax: (202) 564-0298, <http://www.epa.gov/indian>

American Oceans Campaign

Located in Los Angeles, California, the American Oceans Campaign (AOC) was founded in 1987 as a political interest group dedicated primarily to the restoration, protection, and preservation of the health and vitality of coastal waters, estuaries, bays, **wetlands**, and oceans. More national and conservationist (rather than international and preservationist) in its focus than other groups with similar concerns, the AOC tends to view the oceans as a valuable resource whose use should be managed carefully. As current president Ted Danson puts it, the oceans must be regarded as far more than a natural preserve by environmentalists; rather, healthy oceans "sustain biological diversity, provide us with leisure and **recreation**, and contribute significantly to our nation's GNP."

The AOC's main political efforts reflect this focus. Central to the AOC's lobbying strategy is a desire to build cooperative relations and consensus among the general public, public interest groups, private sector corporations and trade groups, and public/governmental authorities around responsible management of ocean resources. The AOC is also active in grassroots public awareness campaigns through mass media and community outreach programs. This high-profile media campaign has included the production of a series of informational bulletins (Public Service Announcements) for use by local groups, as well as active involvement in the production of several documentary television series that have been broadcast on both network and public television. The AOC also has developed extensive connections with both the news and entertainment industries, frequently scheduling appearances by various celebrity supporters such as Jamie Lee Curtis, Whoopi Goldberg, Leonard Nimoy, Patrick Swayze, and Beau Bridges.

As a lobbying organization, the AOC has developed contacts with government leaders at all levels from local to national, attempting to shape and promote a variety of legislation related to clean water and oceans. It has been particularly active in lobbying for strengthening various aspects of the **Clean Water Act**, the **Safe Drinking Water Act**, the Oil **Pollution** Act, and the **Ocean Dumping Ban Act**. The AOC regularly provides consultation services, assistance in drafting legislation, and occasional expert testimony on matters concerning ocean **ecology**. Recently this has included AOC Political Director Barbara Polo's testimony before the U.S. House of Representatives Subcommittee on Fisheries, Conservation, Wildlife, and Oceans on the substance and effect of legislation concerning the protection of **coral reef** ecosystems.

Also very active at the grassroots level, AOC has organized numerous cleanup operations which both draw attention to the problems caused by **ocean dumping** and make a practical contribution to reversing the situation. Concentrating its efforts in California and the Pacific Northwest, the AOC launched its "Dive for Trash" program in 1991. As many as 1,000 divers may team up at AOC-sponsored events to recover **garbage** from the coastal waters. In cooperation with the U.S. Department of Commerce's National

Maritime Sanctuary Program, the AOC is planning to add a marine environmental assessment component to this diving program, and to expand the program into Gulf and Atlantic coastal waters.

Organizationally, the AOC divides its political and lobbying activity into three separate substantive policy areas: "Critical Oceans and Coastal Habitats," which includes issues concerning estuaries, watersheds, and wetlands; "Coastal Water Pollution," which focuses on beach **water quality** and the effects of storm water **runoff**, among other issues; and "Living Resources of the Sea," which include coral reefs, fisheries, and marine mammals (especially **dolphins**). Activities in all these areas have run the gamut from public and legislative information campaigns to litigation.

The AOC has been particularly active along the California coastline and has played a central role in various programs aimed at protecting coastal wetland ecosystems from development and pollution. It has also been active in the Santa Monica Bay Restoration Project, which seeks to restore environmental balance to Santa Monica Bay. Typical of the AOC's multi-level approach, this project combines a program of public education and citizen (and celebrity) involvement with the monitoring and reduction of private-sector pollution and with the conducting of scientific studies on the impact of a various activities in the surrounding area. These activities are also combined with an attempt to raise alternative revenues to replace funds recently lost due to the reduction of both federal (**National Estuary Program**) and state government support for the conservation of coastal and marine ecosystems. In addition, the AOC has been involved in litigation against the County of Los Angeles over a plan to build flood control barriers along a section of the Los Angeles River. AOC's major concern is that these barriers will increase the amount of polluted storm water runoff being channeled into coastal waters. The AOC contends that prudent management of this storm water would be better used in recharging Southern California's scant **water resources** via storage or redirection into underground aquifers before this runoff becomes polluted.

In February of 2002, AOC teamed up with a new nonprofit ocean advocacy organization called Oceana. The focus of this partnership is the Oceans at Risk program that concentrates on the impact that wasteful fisheries have on the marine **environment**.

[*Lawrence J. Biskowski*]

RESOURCES

ORGANIZATIONS

American Oceans Campaign, 6030 Wilshire Blvd Suite 400, Los Angeles, CA USA 90036 (323) 936-8242, Fax: (323) 936-2320, Email: info@americanoceans.org, <http://www.americanoceans.org>

American Wildlands

American Wildlands (AWL) is a nonprofit wildland resource **conservation** and education organization founded in 1977. AWL is dedicated to protecting and promoting proper management of America's publicly owned wild areas and to securing **wilderness** designation for **public land** areas. The organization has played a key role in gaining legal protection for many wilderness and river areas in the U.S. interior west and in Alaska.

Founded as the American Wilderness Alliance, AWL is involved in a wide range of wilderness resource issues and programs including timber management policy reform, **habitat** corridors, rangeland management policy reform, riparian and **wetlands** restoration, and public land management policy reform. AWL promotes ecologically sustainable uses of public wildlands resources including forests, wilderness, **wildlife**, fisheries, and rivers. It pursues this mission through grassroots activism, technical support, public education, litigation, and political advocacy.

AWL maintains three offices: the central Rockies office in Lakewood, Colorado; the northern Rockies office in Bozeman, Montana; and the Sierra-Nevada office in Reno, Nevada. The organization's annual budget of $350,000 has been stable for many years, but with programs that are now being considered for addition to its agenda, that figure is expected to increase over the next few years.

The Central Rockies office in Bozeman considers its main concern timber management reform. It has launched the Timber Management Reform Policy Program, which monitors the U.S. **Forest Service** and works toward a better management of public forests. Since initiation of the program in 1986, the program includes resource specialists, a wildlife biologist, forester, water specialist, and an aquatic biologist who all report to an advisory council. A major victory of this program was stopping the sale of 4.2 million board feet (1.3 million m) of timber near the Electric Peak Wilderness Area.

Other programs coordinated by the Central Rockies office include: 1) *Corridors of Life Program* which identifies and maps wildlife corridors, land areas essential to the genetic interchange of wildlife that connect roadless lands or other wildlife habitat areas. Areas targeted are in the interior West, such as Montana, North and South Dakota, Wyoming, and Idaho; 2) The *Rangeland Management Policy Reform Program* monitors grazing allotments and files appeals as warranted. An education component teaches citizens to monitor grazing allotments and to use the appeals process within the U.S. Forest Service and **Bureau of Land Management**; 3) The *Recreation-Conservation Connection*, through newsletters and travel-adventure programs, teaches the public how to enjoy the outdoors without destroying **nature**. Six hundred travelers have participated in **ecotourism** trips through AWL.

AWL is also active internationally. The AWL/Leakey Fund has aided Dr. Richard Leakey's wildlife habitat conservation and elephant **poaching** elimination efforts in Kenya. A partnership with the Island Foundation has helped fund wildlands and river protection efforts in Patagonia, Argentina. AWL also is an active member of Canada's Tatshenshini International Coalition to protect that river and its 2.3 million acres (930,780 ha) of wilderness.

[*Linda Rehkopf*]

RESOURCES

ORGANIZATIONS

American Wildlands, 40 East Main #2, Bozeman, MT USA 59715 (406) 586-8175, Fax: (406) 586-8242, Email: info@wildlands.org, <http://www.wildlands.org>

Ames test

A laboratory test developed by biochemist Bruce N. Ames to determine the possible carcinogenic nature of a substance. The Ames test involves using a particular strain of the bacteria *Salmonella typhimurium* that lacks the ability to synthesize histidine and is therefore very sensitive to **mutation**. The bacteria are inoculated into a medium deficient in histidine but containing the test compound. If the compound results in DNA damage with subsequent mutations, some of the bacteria will regain the ability to synthesize histidine and will proliferate to form colonies. The culture is evaluated on the basis of the number of mutated bacterial colonies it produced. The ability to replicate mutated colonies leads to the classification of a substance as probably carcinogenic.

The Ames test is a test for mutagenicity not carcinogenicity. However, approximately nine out of 10 mutagens are indeed carcinogenic. Therefore, a substance that can be shown to be mutagenic by being subjected to the Ames test can be reliably classified as a suspected **carcinogen** and thus recommended for further study.

[*Brian R. Barthel*]

RESOURCES

BOOKS

Taber, C. W. *Taber's Cyclopedic Medical Dictionary*. Philadelphia: F. A. Davis, 1990.

Turk J., and A. Turk. *Environmental Science*. Philadelphia: W. B. Saunders, 1988.

Amoco Cadiz

This shipwreck in March 1978 off the Brittany coast was the first major supertanker accident since the **Torrey Canyon** 11 years earlier. Ironically, this spill, more than twice the size of the *Torrey Canyon*, blackened some of the same shores and was one of four substantial oil spills there since 1967. It received great scientific attention because it occurred near several renowned marine laboratories.

The cause of the wreck was a steering failure as the ship entered the English Channel off the northwest Brittany coast, and failure to act swiftly enough to correct it. During the next 12 hours, the *Amoco Cadiz* could not be extricated from the site. In fact, three separate lines from a powerful tug broke trying to remove the tanker before it drifted onto rocky shoals. Eight days later the *Amoco Cadiz* split in two.

Seabirds seemed to suffer the most from the spill, although the oil devastated invertebrates within the extensive, 20–30 ft (6-9 m) high intertidal zone. Thousands of birds died in a bird hospital described by one oil spill expert as a bird morgue. Thirty percent of France's seafood production was threatened, as well as an extensive kelp crop, harvested for **fertilizer**, **mulch**, and livestock feed. However, except on oyster farms located in inlets, most of the impact was restricted to the few months following the spill.

In an extensive journal article, Erich Grundlach and others reported studies on where the oil went and summarized the findings of biologists. Of the 223,000 metric tons released, 13.5% was incorporated within the water column, 8% went into subtidal sediments, 28% washed into the intertidal zone, 20–40% evaporated, and 4% was altered while at sea. Much research was done on chemical changes in the hydrocarbon fractions over time, including that taken up within organisms. Researchers found that during early phases, biodegradation was occurring as rapidly as evaporation.

The cleanup efforts of thousands of workers were helped by storm and wave action that removed much of the stranded oil. High energy waves maintained an adequate supply of nutrients and oxygenated water, which provided optimal conditions for biodegradation. This is important because most of the biodegradation was done by **aerobic** organisms. Except for protected inlets, much of the impact was gone three years later, but some effects were expected to last a decade.

[*Nathan H. Meleen*]

RESOURCES
PERIODICALS

Grove, N. "Black Day for Brittany: *Amoco Cadiz* Wreck." *National Geographic* 154 (July 1978): 124–135.

The *Amoco Cadiz* **oil spill in the midst of being contained.** (Photograph by Leonard Freed. Magnum Photos, Inc. Reproduced by permission.)

Grundlach, E. R., et al. "The Fate of *Amoco Cadiz* Oil." *Science* 221 (8 July 1983): 122–129.
Schneider, E. D. "Aftermath of the *Amoco Cadiz*: Shorline Impact of the Oil Spill." *Oceans* 11 (July 1978): 56–9.
Spooner, M. F., ed. *Amoco Cadiz Oil Spill*. New York: Pergamon, 1979. (Reprint of *Marine Pollution Bulletin*, v. 9, no. 11, 1978)

Cleveland Amory (1917 – 1998)
American Activist and writer

Amory is known both for his series of classic social history books and his work with the **Fund for Animals**. Born in Nahant, Massachusetts, to an old Boston family, Amory attended Harvard University, where he became editor of *The Harvard Crimson*. This prompted his well-known remark, "If you have been editor of *The Harvard Crimson* in your senior year at Harvard, there is very little, in after life, for you."

Amory was hired by *The Saturday Evening Post* after graduation, becoming the youngest editor ever to join that publication. He worked as an intelligence officer in the United States Army during World War II, and in the years after the war, wrote a trilogy of social commentary books, now considered to be classics. *The Proper Bostonians* was published to critical acclaim in 1947, followed by *The Last Resorts* (1948), and *Who Killed Society?* (1960), all of which became best sellers.

Beginning in 1952, Amory served for 11 years as social commentator on NBC's "The Today Show." The network fired him after he spoke out against cruelty to animals used in biomedical research. From 1963 to 1976, Amory served as a senior editor and columnist for *Saturday Review* magazine, while doing a daily radio commentary, entitled "Curmudgeon-at-Large." He was also chief television critic for *TV Guide*, where his biting attacks on sport **hunting** angered hunters and generated bitter but unsuccessful campaigns to have him fired.

In 1967, Amory founded The Fund for Animals "to speak for those who can't," and served as its unpaid president. Animal protection became his passion and his life's work, and he was considered one of the most outspoken and provocative advocates of animal welfare. Under his leadership, the Fund became a highly activist and controversial group, engaging in such activities as confronting hunters of **whales** and **seals**, and rescuing wild horses, burros, and goats. The Fund, and Amory in particular, are well known for their campaigns against sport **hunting and trapping**, the fur industry, abusive research on animals, and other activities and industries that engage in or encourage what they consider cruel treatment of animals.

In 1975, Amory published *ManKind? Our Incredible War on Wildlife*, using humor, sarcasm, and graphic rhetoric to attack hunters, trappers, and other exploiters of wild animals. The book was praised by *The New York Times* in a rare editorial. His next book, *AniMail*, (1976) discussed animal issues in a question-and-answer format. In 1987, he wrote *The Cat Who Came for Christmas*, a book about a stray cat he rescued from the streets of New York, which became a national best seller. This was followed in 1990 by its sequel, also a best seller, *The Cat and the Curmudgeon*. Amory had been a senior contributing editor of *Parade* magazine since 1980, where he often profiled famous personalities.

Amory died of an aneurysm at the age of 81 on October 14, 1998. He remained active right up until the end, spending the day in his office at the Fund for Animals and then passing away in his sleep later that evening. Staffers at both the Fund for Animals have vowed that Amory's work will continue, "just the way Cleveland would have wanted it."

[*Lewis G. Regenstein*]

RESOURCES
BOOKS

Amory, C. *The Cat and the Curmudgeon*. New York: G. K. Hall, 1991.
————. *The Cat Who Came for Christmas*. New York: Little Brown, 1987.

Cleveland Amory. (The Fund for Animals. Reproduced by permission.)

PERIODICALS

Pantridge, M. "The Improper Bostonian." *Boston Magazine* 83 (June 1991): 68–72.

Anaerobic

This term refers to an **environment** lacking in molecular oxygen (O_2), or to an organism, tissue, chemical reaction, or biological process that does not require oxygen. Anaerobic organisms can use a molecule other than O_2 as the terminal electron acceptor in **respiration**. These organisms can be either obligate, meaning that they cannot use O_2, or facultative, meaning that they do not require oxygen but can use it if it is available.

Organic matter **decomposition** in poorly aerated environments, including water-logged soils, septic tanks, and anaerobically-operated waste treatment facilities, produces large amounts of **methane** gas. The methane can become an atmospheric pollutant, or it may be captured and used for fuel, as in "biogas"-powered electrical generators. Anaerobic decomposition produces the notorious "swamp gases" that have been reported as unidentified flying objects (UFOs).

Anaerobic digestion

Refers to the biological degradation of either sludges or **solid waste** under **anaerobic** conditions, meaning that no oxygen is present. In the digestive process, solids are converted to noncellular end products.

In the anaerobic digestion of sludges, the goals are to reduce **sludge** volume, insure the remaining solids are chemically stable, reduce disease-causing pathogens, and enhance the effectiveness of subsequent dewatering methods, sometimes recovering **methane** as a source of energy. Anaerobic digestion is commonly used to treat sludges that contain primary sludges, such as that from the first settling basins in a **wastewater** treatment plant, because the process is capable of stabilizing the sludge with little **biomass** production, a significant benefit over **aerobic sludge digestion**, which would yield more biomass in digesting the relatively large amount of **biodegradable** matter in primary sludge.

The **microorganisms** responsible for digesting the sludges anaerobically are often classified in two groups, the **acid** formers and the methane formers. The acid formers are **microbes** that create, among others, acetic and propionic acids from the sludge. These **chemicals** generally make up about a third of the by-products initially formed based on a **chemical oxygen demand** (COD) mass balance, and some of the propionic and other acids are converted to acetic acid.

The methane formers convert the acids and by-products resulting from prior metabolic steps (e.g., alcohols, **hydrogen**, **carbon dioxide**) to methane. Often, approximately 70% of the methane formed is derived from acetic acid, about 10–15% from propionic acid.

Anaerobic digesters are designed as either standard- or high-rate units. The standard-rate digester has a solids **retention time** of 30–90 days, as opposed to 10–20 days for the high-rate systems. The volatile solids loadings of the standard- and high-rate systems are in the area of 0.5–1.6 and 1.6–6.4 Kg/m^3/d, respectively. The amount of sludge introduced into the standard-rate is therefore generally much less than the high-rate system. Standard-rate digestion is accomplished in single-stage units, meaning that sludge is fed into a single tank and allowed to digest and settle. High-rate units are often designed as two-stage systems in which sludge enters into a completely-mixed first stage that is mixed and heated to approximately 98°F (35°C) to speed digestion. The second-stage digester, which separates digested sludge from the overlying liquid and scum, is not heated or mixed.

With the anaerobic digestion of solid waste, the primary goal is generally to produce methane, a valuable source of fuel that can be burned to provide heat or used to power motors. There are basically three steps in the process. The first involves preparing the waste for digestion by sorting the waste and reducing its size. The second consists of constantly mixing the sludge, adding moisture, nutrients, and **pH** neutralizers while heating it to about 143°F (60°C) and digesting the waste for a week or longer. In the third step, the generated gas is collected and sometimes purified, and digested solids are disposed of. For each pound of undigested solid, about 8–12 ft^3 of gas is formed, of which about 60% is methane.

[*Gregory D. Boardman*]

RESOURCES
BOOKS

Corbitt, R. A. *Standard Handbook of Environmental Engineering.* New York: McGraw-Hill, 1990.

Davis, M. L., and D. A. Cornwell. *Introduction to Environmental Engineering.* New York: McGraw-Hill, 1991.

Viessman, W., Jr., and M. J. Hammer. *Water Supply and Pollution Control.* 5th ed. New York: Harper Collins, 1993.

Anemia

Anemia is a medical condition in which the red cells of the blood are reduced in number or volume or are deficient in hemoglobin, their oxygen-carrying pigment. Almost 100 different varieties of anemia are known. Iron deficiency is the most common cause of anemia worldwide. Other causes of anemia include **ionizing radiation**, **lead poisoning**, vitamin B$_{12}$ deficiency, folic **acid** deficiency, certain infections, and **pesticide** exposure. Some 350 million people worldwide—mostly women of child-bearing age—suffer from anemia.

The most noticeable symptom is pallor of the skin, mucous membranes, and nail beds. Symptoms of tissue oxygen deficiency include pulsating noises in the ear, dizziness, fainting, and shortness of breath. The treatment varies greatly depending on the cause and diagnosis, but may include supplying missing nutrients, removing toxic factors from the **environment**, improving the underlying disorder, or restoring blood volume with transfusion.

Aplastic anemia is a disease in which the bone marrow fails to produce an adequate number of blood cells. It is usually acquired by exposure to certain drugs, to **toxins** such as **benzene**, or to ionizing radiation. Aplastic anemia from **radiation exposure** is well-documented from the Chernobyl experience. Bone marrow changes typical of aplastic anemia can occur several years after the exposure to the offending agent has ceased.

Aplastic anemia can manifest itself abruptly and progress rapidly; more commonly it is insidious and chronic for several years. Symptoms include weakness and fatigue in the early stages, followed by headaches, shortness of breath, fever and a pounding heart. Usually a waxy pallor and hemorrhages occur in the mucous membranes and skin. **Resistance** to infection is lowered and becomes the major cause of death. While spontaneous recovery occurs occasionally, the treatment of choice for severe cases is bone marrow transplantation.

Marie Curie, who discovered the element radium and did early research into **radioactivity**, died in 1934 of aplastic anemia, most likely caused by her exposure to ionizing radiation.

While lead poisoning, which leads to anemia, is usually associated with occupational exposure, toxic amounts of lead can leach from imported ceramic dishes. Other environmental sources of lead exposure include old paint or paint dust, and drinking water pumped through lead pipes or lead-soldered pipes.

Cigarette smoke is known to cause an increase in the level of hemoglobin in smokers, which leads to an underestimation of anemia in smokers. Studies suggest that **carbon monoxide** (a by-product of smoking) chemically binds to hemoglobin, causing a significant elevation of hemoglobin values. Compensation values developed for smokers can now detect possible anemia.

[*Linda Rehkopf*]

RESOURCES
BOOKS

Harte, J., et. al. *Toxics A to Z.* Berkeley: University of California Press, 1991.
Nordenberg, D., et al. "The Effect of Cigarette Smoking on Hemoglobin Levels and Anemia Screening." *Journal of the American Medical Association* (26 September 1990): 1556.
Stuart-Macadam, P., ed. *Diet, Demography and Disease: Changing Perspectives on Anemia.* Hawthrone: Aldine de Gruyter, 1992.

Animal cancer tests

Cancer causes more loss of life-years than any other disease in the United States. At first reading, this statement seems to be in error. Does not cardiovascular disease cause more deaths? The answer to that rhetorical question is "yes." However, many deaths from heart attack and stroke occur in the elderly. The loss of life-years of an 85 year old person (whose life expectancy at the time of his/her birth was between 55 and 60) is, of course, zero. However, the loss of life-years of a child of 10 who dies of a pediatric **leukemia** is between 65 to 70 years. This comparison of youth with the elderly is not meant in any way to demean the *value* that reasonable

people place on the lives of the elderly. Rather, the comparison is made to emphasize the great loss of life due to malignant tumors.

The chemical causation of cancer is not a simple process. Many, perhaps most, chemical carcinogens do not in their usual condition have the potency to cause cancer. The non-cancer causing form of the chemical is called a "procarcinogen." Procarcinogens are frequently complex organic compounds that the human body attempts to dispose of when ingested. Hepatic enzymes chemically change the procarcinogen in several steps to yield a chemical that is more easily excreted. The chemical changes result in modification of the procarcinogen (with no cancer forming ability) to the ultimate **carcinogen** (with cancer causing competence). Ultimate carcinogens have been shown to have a great affinity for DNA, RNA, and cellular proteins, and it is the interaction of the ultimate carcinogen with the cell macromolecules that causes cancer. It is unfortunate indeed that one cannot look at the chemical structure of a potential carcinogen and predict whether or not it will cause cancer. There is no computer program that will predict what hepatic enzymes will do to procarcinogens and how the metabolized end product(s) will interact with cells.

Great strides have been made in the development of chemotherapeutic agents designed to cure cancer. The drugs have significant efficacy with certain cancers (these include but are not limited to pediatric acute lymphocytic leukemia, choriocarcinoma, Hodgkin's disease, and testicular cancer), and some treated patients attain a normal life span. While this development is heartening, the cancers listed are, for the most part, relatively infrequent. More common cancers such as colorectal carcinoma, lung cancer, breast cancer, and ovarian cancer remain intractable with regard to treatment.

These several reasons are why animal testing is used in cancer research. The majority of Americans support the effort of the biomedical community to use animals to identify potential carcinogens with the hope that such knowledge will lead to a reduction of cancer prevalence. Similarly, they support efforts to develop more effective chemotherapy. Animals are used under terms of the Animal Welfare Act of 1966 and its several amendments. The act designates that the U. S. Department of Agriculture is responsible for the humane care and handling of warm-blooded and other animals used for biomedical research. The act also calls for inspection of research facilities to insure that adequate food, housing, and care are provided. It is the belief of many that the constraints of the current law have enhanced the quality of biomedical research. Poorly maintained animals do not provide quality research. The law also has enhanced the care of animals used in cancer research.

[*Robert G. McKinnell*]

RESOURCES
PERIODICALS

Abelson, P. H. "Tesing for Carcinogens With Rodents." *Science* 249 (21 September 1990): 1357.

Donnelly, S., and K. Nolan. "Animals, Science, and Ethics." *Hastings Center Report* 20 (May-June 1990): suppl 1–132.

Marx, J. "Animal Carcinogen Testing Challenged: Bruce Ames Has Stirred Up the Cancer Research Community." *Science* 250 (9 November 1990): 743–5.

Animal Legal Defense Fund

Originally established in 1979 as Attorneys for Animal Rights, this organization changed its name to Animal Legal Defense Fund (ALDF) in 1984, and is known as "the law firm of the **animal rights** movement." Their motto is "we may be the only lawyers on earth whose clients are all innocent." ALDF contends that animals have a fundamental right to legal protection against abuse and exploitation. Over 350 attorneys work for ALDF, and the organization has more than 50,000 supporting members who help the cause of animal rights by writing letters and signing petitions for legislative action. The members are also strongly encouraged to work for animal rights at the local level.

ALDF's work is carried out in many places including research laboratories, large cities, small towns, and the wild. ALDF attorneys try to stop the use of animals in research experiments, and continue to fight for expanded enforcement of the Animal Welfare Act. ALDF also offers legal assistance to humane societies and city prosecutors to help in the enforcement of anti-cruelty laws and the exposure of veterinary malpractice. The organization attempts to protect wild animals from exploitation by working to place controls on trappers and sport hunters. In California, ALDF successfully stopped the **hunting** of mountain lions and black bears. ALDF is also active internationally bringing legal action against elephant poachers as well as against animal dealers who traffic in **endangered species**.

ALDF's clear goals and swift action have resulted in many court victories. In 1992 alone, the organization won cases involving cruelty to **dolphins**, dogs, horses, birds, and cats. It has also blocked the importation of over 70,000 monkeys from Bangladesh for research purposes, and has filed suit against the National Marine Fisheries Services to stop the illegal gray market in dolphins and other marine mammals. ALDF also publishes a quarterly magazine, *The Animals' Advocate*.

[*Cathy M. Falk*]

RESOURCES
ORGANIZATIONS
Animal Legal Defense Fund, 127 Fourth Street, Petaluma, CA USA 94952 Fax: (707) 769-7771, Toll Free: (707) 769-0785, Email: info@aldf.org, <http://www.aldf.org>

Animal rights

Recent concern about the way humans treat animals has spawned a powerful social and political movement driven by the conviction that humans and certain animals are similar in morally significant ways, and that these similarities oblige humans to extend to those animals serious moral consideration, including rights. Though animal welfare movements, concerned primarily with humane treatment of pets, date back to the 1800s, modern animal rights activism has developed primarily out of concern about the use and treatment of domesticated animals in agriculture and in medical, scientific, and industrial research. The rapid growth in membership of animal rights organizations testifies to the increasing momentum of this movement. The leading animal rights group today, **People for the Ethical Treatment of Animals** (PETA), was founded in 1980 with 100 individuals; today, it has over 300,000 members. The animal rights activist movement has closely followed and used the work of modern philosophers who seek to establish a firm logical foundation for the extension of moral considerability beyond the human community into the animal community.

The nature of animals and appropriate relations between humans and animals have occupied Western thinkers for millennia. Traditional Western views, both religious and philosophical, have tended to deny that humans have any moral obligations to nonhumans. The rise of Christianity and its doctrine of personal immortality, which implies a qualitative gulf between humans and animals, contributed significantly to the dominant Western paradigm. When seventeenth century philosopher René Descartes declared animals mere biological machines, the perceived gap between humans and nonhuman animals reached its widest point. Jeremy Bentham, the father of ethical utilitarianism, challenged this view and fostered a widespread anticruelty movement and exerted powerful force in shaping our legal and moral codes. Its modern legacy, the animal welfare movement, is reformist in that it continues to accept the legitimacy of sacrificing animal interests for human benefit, provided animals are spared any suffering which can conveniently and economically be avoided.

In contrast to the conservatively reformist platform of animal welfare crusaders, a new radical movement began in the late 1970s. This movement, variously referred to as animal liberation or animal rights, seeks to put an end to the routine sacrifice of animal interests for human benefit. In

Animal rights activists dressed as monkeys in prison suits block the entrance to the Department of Health and Human Services in Washington, DC, in protest of the use of animals in laboratory research. (Corbis-Bettmann. Reproduced by permission.)

seeking to redefine the issue as one of rights, some animal protectionists organized around the well-articulated and widely disseminated utilitarian perspective of Australian philosopher **Peter Singer**. In his 1975 classic *Animal Liberation*, Singer argued that because some animals can experience pleasure and pain, they deserve our moral consideration. While not actually a rights position, Singer's work nevertheless uses the language of rights and was among the first to abandon welfarism and to propose a new ethic of moral considerability for all sentient creatures.

To assume that humans are inevitably superior to other **species** simply by virtue of their species membership is an injustice which Singer terms **speciesism**, an injustice parallel to racism and sexism.

Singer does not claim all animal lives to be of equal worth, nor that all sentient beings should be treated identically. In some cases, human interests may outweigh those of nonhumans, and Singer's utilitarian calculus would allow us to engage in practices which require the use of animals

in spite of their pain, where those practices can be shown to produce an overall balance of pleasure over suffering.

Some animal advocates thus reject **utilitarianism** on the grounds that it allows the continuation of morally abhorrent practices. Lawyer Christopher Stone and philosophers Joel Feinberg and **Tom Regan** have focused on developing cogent arguments in support of rights for certain animals. Regan's 1983 book *The Case For Animal Rights* developed an absolutist position which criticized and broke from utilitarianism. It is Regan's arguments, not reformism or the pragmatic principle of utility, which have come to dominate the rhetoric of the animal rights crusade.

The question of which animals possess rights then arises. Regan asserts it is those who, like us, are subjects experiencing their own lives. By "experiencing" Regan means conscious creatures aware of their **environment** and with goals, desires, emotions, and a sense of their own identity. These characteristics give an individual inherent value, and this value entitles the bearer to certain inalienable rights, especially the right to be treated as an end in itself, and never merely as a means to human ends.

The environmental community has not embraced animal rights; in fact, the two groups have often been at odds. A rights approach focused exclusively on animals does not cover all the entities such as ecosystems that many environmentalists feel ought to be considered morally. Yet a rights approach that would satisfy environmentalists by encompassing both living and nonliving entities may render the concept of rights philosophically and practically meaningless. Regan accuses environmentalists of environmental fascism, insofar as they advocate the protection of species and ecosystems at the expense of individual animals. Most animal rightists advocate the protection of ecosystems only as necessary to protect individual animals, and assign no more value to the individual members of a highly **endangered species** than to those of a common or domesticated species. Thus, because of its focus on the individual, animal rights can offer no realistic plan for managing natural systems or for protecting **ecosystem health**, and may at times hinder the efforts of resource managers to effectively address these issues.

For most animal activists, the practical implications of the rights view are clear and uncompromising. The rights view holds that all animal research, factory farming, and commercial or sport **hunting and trapping** should be abolished. This change of moral status necessitates a fundamental change in contemporary Western moral attitudes towards animals, for it requires humans to treat animals as inherently valuable beings with lives and interests independent of human needs and wants. While this change is not likely to occur in the near future, the efforts of animal rights advocates may ensure that wholesale slaughter of these creatures for unnecessary reasons that is no longer routinely the case, and that when such sacrifice is found to be necessary, it is accompanied by moral deliberation.

[*Ann S. Causey*]

RESOURCES
BOOKS

Hargrove, E. C. *The Animal Rights/Environmental Ethics Debate*. New York: SUNY Press, 1992.

Regan, T. *The Case For Animal Rights*. Los Angeles: University of California Press, 1983.

———, and P. Singer. *Animal Rights and Human Obligations*. 2nd ed. Englewood Cliffs, NJ: Prentice-Hall, 1989.

Singer, P. *Animal Liberation*. New York: Avon Books, 1975.

Zimmerman, M. E., et al, eds. *Environmental Philosophy: From Animal Rights To Radical Ecology*. Englewood Cliffs, NJ: Prentice-Hall, 1993.

Animal waste

Animal wastes are commonly considered the excreted materials from live animals. However, under certain production conditions, the waste may also include straw, hay, wood shavings, or other sources of organic debris. It has been estimated that there may be as much as 2 billion tons of animal wastes produced in the United States annually. Application of excreta to **soil** brings benefits such as improved soil **tilth**, increased water-holding capacity, and some plant nutrients. Concentrated forms of excreta or high application rates to soils without proper management may lead to high salt concentrations in the soil and cause serious on- or off-site **pollution**.

Animal Welfare Institute

Founded in 1951, the Animal Welfare Institute (AWI) is a non-profit organization that works to educate the public and to secure needed action to protect animals. AWI is a highly respected, influential, and effective group that works with Congress, the public, the news media, government officials, and the **conservation** community on animal protection programs and projects. Its major goals include improving the treatment of laboratory animals and a reduction in their use; eliminating cruel methods of **trapping wildlife**; saving **species** from **extinction**; preventing painful experiments on animals in schools and encouraging humane science teaching; improving shipping conditions for animals in transit; banning the importation of **parrots** and other exotic wild birds for the pet industry; and improving the conditions under which farm animals are kept, confined, transported, and slaughtered.

In 1971 AWI launched the **Save the Whales** Campaign to help protect **whales**. The organization provides speakers and experts for conferences and meetings around the world, including Congressional hearings and international treaty and commission meetings. Each year, the institute awards its prestigious **Albert Schweitzer** Medal to an individual for outstanding achievement in the advancement of animal welfare. Its publications include *The AWI Quarterly*; books such as *Animals and Their Legal Rights*; *Facts about Furs*; and *The Endangered Species Handbook*; booklets, brochures, and other educational materials, which are distributed to schools, teachers, scientists, government officials, humane societies, libraries, and veterinarians.

AWI works closely with its associate organization, The Society for Animal Protective Legislation (SAPL), a lobbying group based in Washington, D.C. Founded in 1955, SAPL devotes its efforts to supporting legislation to protect animals, often mobilizing its 14,000 "correspondents" in letter-writing campaigns to members of Congress.

SAPL has been responsible for the passage of more animal protection laws than any other organization in the country, and perhaps the world, and it has been instrumental in securing the enactment of 14 federal laws.

Major federal legislation which SAPL has promoted includes the first federal Humane Slaughter Act in 1958 and its strengthening in 1978; the 1959 Wild Horse Act; the 1966 Laboratory Animal Welfare Act and its strengthening in 1970, 1976, 1985, and 1990; the 1969 **Endangered Species Act** and its strengthening in 1973; a 1970 measure banning the crippling or "soring" of Tennessee Walking Horses; measures passed in 1971 prohibiting **hunting** from aircraft, protecting wild horses, and resolutions calling for a moratorium on commercial **whaling**; the 1972 Marine Mammal Protection Act; negotiation of the 1973 Convention on International Trade in **Endangered Species** of **Fauna** and **Flora** (CITES); the 1979 Packwood-Magnuson Amendment protecting whales and other ocean creatures; the 1981 strengthening of the Lacey Act to restrict the importation of illegal wildlife; the 1990 Pet Theft Act; and, in 1992, The Wild Bird Conservation Act, protecting parrots and other exotic wild birds; the International Dolphin Conservation Act, restricting the killing of **dolphins** by tuna fishermen; and the Driftnet Fishery Conservation Act, protecting whales, sea birds, and other ocean life from being caught and killed in huge, 30-mi-long (48-km-long) nets.

Major goals of SAPL include enacting legislation to end the use of cruel steel-jaw leg-hold traps and to secure proper enforcement, funding, administration, and reauthorization of existing animal protection laws. Both AWI and SAPL have long been headed by their chief volunteer, Christine Stevens, a prominent Washington, D.C. humanitarian and community leader.

[*Lewis G. Regenstein*]

RESOURCES
ORGANIZATIONS
Animal Welfare Institute, P.O. Box 3650, Washington, D.C USA 20007 (202) 337-2332, Email: awi@awionline.org, <http://www.awionline.org>
Society for Animal Protective Legislation, P.O. Box 3719, Washington, D.C. USA 20007 (202) 337-2334, Fax: (202) 338-9478, Email: sapl@saplonline.org, <http://www.saplonline.org>

Anion

see **Ion**

Antarctic Treaty (1961)

The Antarctic Treaty, signed in 1961, established an international administrative system for the continent. The impetus for the treaty was the International Geophysical Year, 1957–1958, which had brought scientists from many nations together to study **Antarctica**. The political situation in Antarctica was complex at the time, with seven nations having made sometimes overlapping territorial claims to the continent: Argentina, **Australia**, Chile, France, New Zealand, Norway, and the United Kingdom. Several other nations, most notably the former USSR and the United States, had been active in Antarctic exploration and research and were concerned with how the continent would be administered.

Negotiations on the treaty began in June 1958 with Belgium, Japan, and South Africa joining the original nine countries. The treaty was signed in December 1959 and took effect in June 1961. It begins by "recognizing that it is in the interest of all mankind that Antarctica shall continue forever to be used exclusively for peaceful purposes." The key to the treaty was the nations' agreement to disagree on territorial claims. Signatories of the treaty are not required to renounce existing claims, nations without claims shall have an equal voice as those with claims, and no new claims or claim enlargements can take place while the treaty is in force. This agreement defused the most controversial and complex issue regarding Antarctica, and in an unorthodox way. Among the other major provisions of the treaty are: the continent will be demilitarized; nuclear explosions and the storage of nuclear wastes are prohibited; the right of unilateral inspection of all facilities on the continent to ensure that the provisions of the treaty are being honored is guaranteed; and scientific research can continue throughout the continent.

The treaty runs indefinitely and can be amended, but only by the unanimous consent of the signatory nations. Provisions were also included for other nations to become parties to the treaty. These additional nations can either be "acceding parties," which do not conduct significant research activities but agree to abide by the terms of the treaty, or "consultative parties," which have acceded to the treaty and undertake substantial scientific research on the continent. Twelve nations have joined the original 12 in becoming consultative parties: Brazil, China, Finland, Germany, India, Italy, Peru, Poland, South Korea, Spain, Sweden, and Uruguay.

Under the auspices of the treaty, the Convention on the Conservation of Antarctic Marine Living Resources was adopted in 1982. This regulatory regime is an effort to protect the Antarctic marine **ecosystem** from severe damage due to **overfishing**. Following this convention, negotiations began on an agreement for the management of Antarctic mineral resources. The Convention on the Regulation of Antarctic Mineral Resource Activities was concluded in June 1988, but in 1989 Australia and France rejected the convention, urging that Antarctica be declared an international wilderness closed to mineral development. In 1991 the Protocol on Environmental Protection, which included a 50-year ban on mining, was drafted. At first the United States refused to endorse this protocol, but it eventually joined the other treaty parties in signing the new convention in October 1991.

[*Christopher McGrory Klyza*]

RESOURCES

BOOKS

Shapley, D. *The Seventh Continent: Antarctica in a Resource Age.* Baltimore: Johns Hopkins University Press for Resources for the Future, 1985.

Antarctica

The earth's fifth largest continent, centered asymmetrically around the South Pole. Ninety-eight percent of this land mass, which covers approximately 5.4 million mi^2 (13.8 million km^2), is covered by snow and ice sheets to an average depth of 1.25 mi (2 km). This continent receives very little precipitation, less than 5 in (12 cm) annually, and the world's coldest temperature was recorded here, at -128°F (-89°C). Exposed shorelines and inland mountain tops support life only in the form of **lichens**, two **species** of flowering plants, and several insect species. In sharp contrast, the ocean surrounding the Antarctic continent is one of the world's richest marine habitats. Cold water rich in oxygen and nutrients supports teeming populations of **phytoplankton** and shrimp-like Antarctica **krill**, the food source for the region's legendary numbers of **whales**, **seals**, penguins, and fish. During the nineteenth and early twentieth century, whalers and sealers severely depleted Antarctica's marine mammal populations. In recent decades the whale and seal populations have begun to recover, but interest has grown in new resources, especially oil, minerals, fish, and tourism.

The Antarctic's functional limit is a band of turbulent ocean currents and high winds that circle the continent at about 60 degrees south latitude. This ring is known as the Antarctic convergence zone. Ocean turbulence in this zone creates a barrier marked by sharp differences in **salinity** and water temperature. Antarctic marine habitats, including the limit of krill populations, are bounded by the convergence.

Since 1961 the **Antarctic Treaty** has formed a framework for international cooperation and compromise in the use of Antarctica and its resources. The treaty reserves the Antarctic continent for peaceful scientific research and bans all military activities. Nuclear explosions and **radioactive waste** are also banned, and the treaty neither recognizes nor establishes territorial claims in Antarctica. However, neither does the treaty deny pre-1961 claims, of which seven exist. Furthermore, some signatories to the treaty, including

the United States, reserve the right to make claims at a later date. At present the United States has no territorial claims, but it does have several permanent stations, including one at the South Pole. Questions of territorial control could become significant if oil and mineral resources were to become economically recoverable. The primary resources currently exploited are fin fish and krill fisheries. Interest in oil and mineral resources has risen in recent decades, most notably during the 1973 "oil crisis." The expense and difficulty of extraction and **transportation** has so far made exploitation uneconomical, however.

Human activity has brought an array of environmental dangers to Antarctica. Oil and mineral extraction could seriously threaten marine **habitat** and onshore penguin and seal breeding grounds. A growing and largely uncontrolled fishing industry may be depleting both fish and krill populations in Antarctic waters. The parable of the **Tragedy of the Commons** seems ominously appropriate to Antarctica fisheries, which have already nearly eliminated many whale, seal, and penguin species. **Solid waste** and **oil spills** associated with research stations and with tourism pose an additional threat. Although Antarctica remains free of "permanent settlement," 40 year-round scientific research stations are maintained on the continent. The population of these bases numbers nearly 4,000. In 1989 the Antarctic had its first oil spill when an Argentine supply ship, carrying 81 tourists and 170,000 gal (643,500 l) of diesel fuel, ran aground. Spilled fuel destroyed a nearby breeding colony of Adele penguins (*Pygoscelis adeliae*). With more than 3,000 cruise ships visiting annually, more spills seem inevitable. Tourists themselves present a further threat to penguins and seals. Visitors have been accused of disturbing breeding colonies, thus endangering the survival of young penguins and seals.

[*Mary Ann Cunningham*]

RESOURCES

BOOKS

Child, J. *Antarctica and South American Geopolitics.* New York: Praeger, 1988.

Parsons, A. *Antarctica: The Next Decade.* Cambridge: Cambridge University Press, 1987.

Shapely, D. *The Seventh Continent: Antarctica in a Resource Age.* Baltimore: Johns Hopkins University Press for Resources for the Future, 1985.

Suter, K. D. *World Law and the Last Wilderness.* Sydney: Friends of the Earth, 1980.

Antarctica Project

The Antarctica Project, founded in 1982, is an organization designed to protect **Antarctica** and educate the public, government, and international groups about its current and future status. The group monitors activities that affect the Antarctic region, conducts policy research and analysis in both national and international arenas, and maintains an impressive library of books, articles, and documents about Antarctica. It is also a member of the Antarctic and Southern Ocean Coalition (ASOC), which has 230 member organizations in 49 countries.

In 1988, ASOC received a limited observer status to the Convention on the Conservation of Antarctic Marine Living Resources (CCAMLR). So far, the observer status continues to be renewed, providing ASOC with a way to monitor CCAMLR and to present proposals. In 1989, the Antarctica Project served as an expert adviser to the U.S. Office of Technology Assessment on its study and report of the Minerals Convention. The group prepared a study paper outlining the need for a comprehensive environmental protection convention. Later, a **conservation** strategy on Antarctica was developed with **IUCN—The World Conservation Union**.

Besides continuing the work it has already begun, the Antarctica Project has several goals for the future. One calls for the designation of Antarctica as a world park. Another focuses on developing a bilateral plan to pump out the oil and salvage the *Bahia Parasio*, a ship which sank in early 1989 near the U.S. Palmer Station. Early estimated salvage costs ran at $50 million. One of the more recent projects is the Southern Ocean Fisheries Campaign. This campaign targets the illegal fishing taking place in the Southern Ocean which is depleting the Chilean sea bass population. The catch phrase of this movement is "Take a Pass on Chilean Sea Bass."

Three to four times a year, The Antarctica Project publishes *ECO*, an international publication which covers current political topics concerning the **Antarctic Treaty** System (provided free to members). Other publications include briefing materials, critiques, books, slide shows, videos, and posters for educational and advocacy purposes.

[*Cathy M. Falk*]

RESOURCES

ORGANIZATIONS

The Antarctica Project, 1630 Connecticut Ave., NW, 3rd Floor, Washington, D.C. USA 20009 (202) 234-2480, Email: antarctica@igc.org, <http://www.asoc.org>

Anthracite coal
see **Coal**

Anthrax

Anthrax is a bacterial infection caused by *Bacillus anthracis*. It usually affects cloven-hoofed animals, such as cattle, sheep, and goats, but it can occasionally spread to humans. Anthrax is almost always fatal in animals, but it can be successfully treated in humans if antibiotics are given soon after exposure. In humans, anthrax is usually contracted when spores are inhaled or come in contact with the skin. It is also possible for people to become infected by eating the meat of contaminated animals. Anthrax, a deadly disease in **nature**, gained worldwide attention in 2001 after it was used as a **bioterrorism** agent in the United States. Until the 2001 attack, only 18 cases of anthrax had been reported in the United States in the previous 100 years.

Anthrax occurs naturally. The first reports of the disease date from around 1500 B.C., when it is believed to have been the cause of the fifth Egyptian **plague** described in the Bible. Robert Koch first identified the anthrax bacterium in 1876 and Louis Pasteur developed an anthrax vaccine for sheep and cattle in 1881. Anthrax bacteria are found in nature in South and Central America, southern and eastern Europe, Asia, Africa, the Caribbean, and the Middle East. Anthrax cases in the United States are rare, probably due to widespread vaccination of animals and the standard procedure of disinfecting animal products such as cowhide and wool. Reported cases occur most often in Texas, Louisiana, Mississippi, Oklahoma, and South Dakota.

Anthrax spores can remain dormant (inactive) for years in **soil** and on animal hides, wool, hair, and bones. There are three forms of the disease, each named for its means of transmission: cutaneous (through the skin), inhalation (through the lungs), and intestinal (caused by eating anthrax-contaminated meat). Symptoms appear within several weeks of exposure and vary depending on how the disease was contracted.

Cutaneous anthrax is the mildest form of the disease. Initial symptoms include itchy bumps, similar to insect bites. Within two days, the bumps become inflamed and a blister forms. The centers of the blisters are black due to dying tissue. Other symptoms include shaking, fever, and chills. In most cases, cutaneous anthrax can be treated with antibiotics such as penicillin. Intestinal anthrax symptoms include stomach and intestinal inflammation and pain, nausea, vomiting, loss of appetite, and fever, all becoming progressively more severe. Once the symptoms worsen, antibiotics are less effective, and the disease is usually fatal.

Inhalation anthrax is the form of the disease that occurred during the bioterrorism attacks of October and November 2001 in the eastern United States. Five people died after being exposed to anthrax through contaminated mail. At least 17 other people contracted the disease but survived.

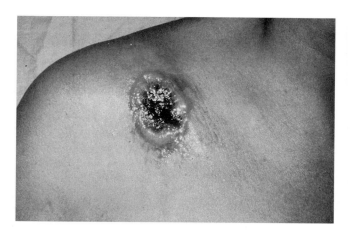

Anthrax lesion on the shoulder of a patient.
(NMSB/Custom Medical Stock Photo. Reproduced by permission.)

One or more terrorists sent media organizations in Florida and New York envelopes containing anthrax. Anthrax-contaminated letters also were sent to the Washington, D.C. offices of two senators. Federal agents were still investigating the incidents as of May 2002 but admitted they had no leads in the case. Initial symptoms of inhalation anthrax are flu-like, but breathing becomes progressively more difficult. Inhalation anthrax can be treated successfully if antibiotics are given before symptoms develop. Once symptoms develop, the disease is usually fatal.

The only natural outbreak of anthrax among people in the United States occurred in Manchester, New Hampshire, in 1957. Nine workers in a textile mill that processed wool and goat hair contracted the disease, five with inhalation anthrax and four with cutaneous anthrax. Four of the five people with inhalation anthrax died. By coincidence, workers at the mill were participating in a study of an experimental anthrax vaccine. No workers who had been vaccinated contracted the disease.

Following this outbreak, the study was stopped, all workers at the mill were vaccinated, and vaccination became a condition of employment. After that, no mill workers contracted anthrax. The mill closed in 1968. However, in 1966 a man who worked across the street from the mill died from inhalation anthrax. He is believed to have contracted it from anthrax spores carried from the mill by the wind. The United States **Food and Drug Administration** approved the anthrax vaccine in 1970. It is used primarily for military personnel and some health care workers. During the 2001 outbreak, thousands of postal workers were offered the vaccine after anthrax spores from contaminated letters were found at several post office buildings.

The largest outbreak worldwide of anthrax in humans occurred in the former Soviet Union in 1979, when anthrax spores released from a military laboratory infected 77 people, 69 of whom died. Anthrax is an attractive weapon to bioterrorists. It is easy to transport and is highly lethal. The World Health Organization (WHO) estimates that 110 lb (50 kg) of anthrax spores released upwind of a large city would kill tens of thousands of people, with thousands of others ill and requiring medical treatment.

The Geneva Convention, which established a code of conduct for war, outlawed the use of anthrax as a weapon in 1925. However, Japan developed anthrax weapons in the 1930s and used them against civilian populations during World War II. During the 1980s, Iraq mass produced anthrax as a weapon.

[*Ken R. Wells*]

RESOURCES
BOOKS

The Parents' Committee for Public Awareness. *Anthrax: A Practical Guide for Citizens.* Cambridge, MA: Harvard Perspectives Press, 2001.

PERIODICALS

Consumers' Research Staff. "What You Need to Know About Anthrax." *Consumers' Research Magazine* (Nov. 2001):10–14.
Belluck, Pam. "Anthrax Outbreak of '57 Felled a Mill but Yielded Answers." *The New York Times* (Oct. 27, 2001).
Bia, Frank, et al. "Anthrax: What You—And Your Patients—Need To Know Now." *Consultant* (Dec. 2001):1797–1804.
Masibay, Kim Y. "Anthrax: Facts, Not Fear." *Science World* (Nov. 26, 2001):4–6.
Spencer, Debbi Ann, et al. "Inhalation Anthrax." *MedSurg Nursing* (Dec. 2001):308ndash;313.

ORGANIZATIONS

Centers for Disease Control and Prevention, 1600 Clifton Road, Atlanta, GA USA 30333 (404)639-3534, Toll Free: (888) 246-2675, Email: cdcresponse@ashastd.org, >http://www.cdc.gov<

Anthropocentrism
see **Environmental ethics**

Anthropogenic

Refers to changes in the natural world due to the activities of people. Such changes may be positive or negative. For example, anthropogenic changes in soils can occur due to plowing, fertilizing, using the **soil** for construction, or long continued manure additions. When manure is added to soils, the change is considered beneficial, but when soils are compacted for use as parking lots, the change is considered negative. Other examples of anthropogenic effects on the **environment** include **oil spills**, **acid rain**, **logging** of old-growth forests, creation of **wetlands**, preservation of **endangered species**, among others.

Antibiotic resistance

Antibiotics are drugs principally derived from naturally occurring **fungi** and **microorganisms** that kill bacteria and can cure patients with bacterial diseases. Before the advent of antibiotics in the 1940s, many common diseases were lethal or incurable. Tuberculosis, pneumonia, scarlet fever, staph and strep infections, typhoid fever, gonorrhea, and syphilis were all dreaded diseases until the development of penicillin and other antibiotics in the middle of the twentieth century. Yet almost as soon as antibiotics came into common use, scientists noticed that some strains of disease-causing bacteria developed **resistance** to the antibiotic used most often against it. People infected with an antibiotic-resistant bacteria must be treated with different antibiotics, often more potent and toxic than the commonly used drug. In some cases, bacteria may be resistant to several antibiotics. Tuberculosis, once the leading killer in the United States at the beginning of the nineteenth century, seemed defeated with the introduction of streptomycin and PAS in the 1940s and early 1950s. But tuberculosis resurged in the United States and worldwide in the 1990s as people came down with antibiotic-resistant strains of the disease. Bacteria that cause salmonella, a food-borne illness, have become increasingly resistant to antibiotics by the early twenty-first century, as have the bacteria that commonly cause early childhood ear infections. Misuse and overuse of antibiotics contribute to the rise of resistant strains.

Bacteria can become resistant to antibiotics relatively quickly. Bacteria multiply rapidly, producing a new generation in as little as a half hour. So evolutionary pressures can produce bacteria with new characteristics in very little time. When a person takes an antibiotic, the drug will typically kill almost all the bacteria it is designed to destroy, plus other beneficial bacteria. Some small percentage of the disease bacteria, maybe as little as 1%, may have a natural ability to resist the antibiotic. So a small number of resistant bacteria may survive drug treatment. When these resistant bacteria are all that are left, they are free to multiply, passing the resistance to their offspring. Physicians warn people to take the full prescribed course of antibiotics even if symptoms of the disease disappear in a day or two. This is to limit the danger of resistant bacteria flourishing. Bacteria can also develop resistance by contact with other **species** of bacteria that are resistant. Neighboring bacteria can pass genetic material back and forth by swapping bits of DNA called plasmids. If bacteria that normally live on the skin and bacteria that live in the intestine should come into contact with each other, they may make a plasmid exchange, and spread antibiotic resistant qualities. Antibiotic-resistant bacteria are often resistant to a whole class of antibiotics, that is, a group of antibiotics that function in a similar way.

People afflicted with a resistant strain of bacteria must be treated with a different class of antibiotics.

Antibiotic resistance was evident in the 1940s, though penicillin had only become available in 1941. By 1946, one London hospital reported that 14% of patients with staph infections had penicillin-resistant strains, and that number rose precipitously over the next decade. In 1943 scientists brought out streptomycin, a new antibiotic that fought tuberculosis (penicillin was found not to work against that disease). But streptomycin-resistant strains of tuberculosis developed rapidly, and other drugs had to be found. In 1959, physicians in Japan found a virulent strain of dysentery that was resistant to four different classes of antibiotic. Some troubling cases of antibiotic resistance have been isolated incidents. But by the 1990s it was clear that antibiotic resistance was a widespread and growing problem. A few cases around the world in 1999 found deadly bacteria resistant to vancomycin, a powerful antibiotic described as a drug of last resort because it is only used when all other antibiotics fail. By this time, scientists in many countries were deeply alarmed about the growing public health threat of antibiotic resistance. A study done by the Mayo clinic and reported in 2001 claimed that deaths from infectious diseases had risen 58% between 1980 and 1992, a rising toll attributed in part to antibiotic resistance. The U.S. **Centers for Disease Control and Prevention** (CDC) claimed in 2001 that antibiotic resistance had spread to "virtually all important human pathogens treatable with antibiotics."

Antibiotic resistance makes treatment of infected patients difficult. The sexually transmitted disease gonorrhea was easily cured with a single dose of penicillin in the middle of the twentieth century. By the 1970s, penicillin-resistant strains of the disease had become prevalent in Asia, and migrated from there to the rest of the world. Penicillin was no longer used to treat gonorrhea in the United States after 1987. Standard treatment was then a dose of either of two classes of antibiotics, fluoroquinolones, or cephalosporins. By the late 1990s, strains of gonorrhea resistant to fluoroquinolones had been detected in Asia. The resistant strains showed up in California in 2001. The California CDC soon recommended not using fluoroquinolones to treat gonorrhea, fearing that use of these drugs would actually strengthen the antibiotic resistance. If patients were only partially cured by fluoroquinolones, yet some infection lingered, they could pass the resistant strain to others. And the resistance could become stronger as only the most resistant bacteria survived exposure to the drug. So public health officials and doctors were left with cephalosporins to treat gonorrhea, more costly drugs with more risk of side effects.

Overuse of antibiotics contributes to antibiotic resistance. The number of antibiotic prescriptions for children rose almost 50% in the United States between 1980 and

1992. Children and the elderly are the most likely to receive antibiotic prescriptions. By 1998 the CDC estimated that approximately half the 100 million prescriptions for antibiotics issued by doctor's offices in the United States annually were unnecessary. Antibiotics work only against bacterial diseases, and are useless against viral infections. Yet physicians frequently prescribe antibiotics for coughs and colds. An article on the problem in *American Family Physician* found that most doctors understood the inappropriateness of their prescriptions, yet feared that patients were unsatisfied with their care unless they received a drug. The CDC launched various state and national initiatives to educate both doctors and their patients about overuse of antibiotics. Other groups took on specific diseases. For example, an association of pediatricians publicized the danger of over prescribing for childhood ear infections in 2001. The common ailment was known to be treatable without antibiotics, but many doctors continued to give antibiotics anyway. By 2001, almost one-third of children in daycare who had ear infections had an antibiotic-resistant form, according to a survey conducted by the National Association of Child Care Professionals (NACCP). The NACCP hoped to convince both pediatricians and parents to use antibiotics only when necessary. The president of the American Medical Association spoke out in 2001 about the number of prescriptions for ciprofloxacin (Cipro) given out in the wake of the mail attacks of inhalation **anthrax**. Tens of thousands of people received prescriptions for Cipro in October 2001, putting them at risk for developing pools of Cipro-resistant bacteria in their bodies. Most bacteria live in the body without causing harm, but can make people ill if they build up to certain levels, or if a person's immune system is weakened. People carrying Cipro-resistant bacteria could potentially come down with a resistant form of pneumonia or some other bacterial illness later in life.

People are also exposed to antibiotics through meat and other food. About half the antibiotics used in the United States go to farm animals, and some are also sprayed on fruits and vegetables. Some farm animals are given antibiotics to cure a specific disease. But other antibiotics are given as preventives, and to promote growth. Animals living in crowded and dirty conditions are more susceptible to disease, and the preventive use of antibiotics keeps such animals healthier than they would otherwise be. The antibiotics prescribed by veterinarians are similar or the same as drugs used in humans. Farm animals in the United States are routinely fed penicillin, amoxicillin, tetracycline, ampicillin, erythromycin, and neomycin, among others, and studies have shown that antibiotic resistance is common in contaminated meat and eggs. One study conducted by the **Food and Drug Administration** (FDA) and the University of Maryland and reported in 2001 found 20% of ground meat samples taken

from several urban supermarkets contained salmonella bacteria. Over 80% of the salmonella bacteria sampled was resistant to at least one antibiotic and over half was resistant to at least three antibiotics. Salmonella bacteria is killed when meat is cooked properly, and most cases of salmonella disease get better without treatment. But for the small percent of cases of more serious infection, multiple antibiotic resistance could make treatment very difficult. A food-borne *E. coli* bacteria was found in 2001 to be causing urinary tract infections which were resistant to the standard antibiotic treatment. Twenty percent of the urinary tract infections studied were resistant to Bactrim, meaning that in most cases physicians would be advised to treat with a stronger antibiotic with more side effects.

Antibiotics in preventive doses or for growth promotion of farm animals were banned by the **European Union** in 1998. Many groups in the United States concerned with antibiotic resistance recommend the United States follow suit. The plan released by the CDC, the Food and Drug Administration (FDA) and the National Institutes of Health (NIH) in 2001 to combat antibiotic resistance called for increased monitoring of antibiotic use in agriculture and in human health. The plan also called for public education on the risks of overuse and improper use of antibiotics, and for more research in combating drug-resistant diseases. The FDA continued to investigate links between agricultural use of antibiotics and human health.

[*Angela Woodward*]

RESOURCES
BOOKS

Moore, Peter. *Killer Germs.* London: Carlton Books, 2001.

PERIODICALS

"Antibiotic Resistance: Appropriate Use, Not "Magic Bullet" Best Bet to Ward Off Dangers." *TB and Outbreaks Week* (June 6, 2000): 24–25.

Barlam, Tamar. "Antibiotics in Jeopardy." *Nutrition Action Health Letter* 29, no. 2 (March 2002): 9.

Brody, Jane E. "Studies Find Resistant Bacteria in Meats." *New York Times* (October 18, 2001): A12.

Colgan, Richard, and Powers, John H. "Appropriate Antimicrobial Prescribing: Approaches that Limit Antibiotic Resistance." *American Family Physician* 64, no. 6 (September 15, 2001): 999.

Cunha, Burke A. "Effective Antibiotic-Resistance Control Strategies." *Lancet* 357, no. 9265 (April 28, 2001): 1307.

"Drug Resistance: Anthrax-Fighting Drugs Apt to Spur Resistance to Common Bugs." *Medical Letter on the CDC and FDA* (November 25, 2001 - December 2, 2001): 18.

"Higher Rates of Repeat Infections Occur Among Children in Group Child Care." *Medical Letter on the CDC and FDA* (December 30, 2001 - January 6, 2002): 15.

Kirby, David. "New Resistant Gonorrhea Migrating to Mainland U.S." *New York Times* (May 7, 2002): D5.

Monroe, Judy. "Antibiotics Vs. the Superbugs." *Current Health 2* 28, no. 2 (October 2001): 24.

ORGANIZATIONS

Alliance for the Prudent Use of Antibiotics, 75 Kneeland Street, Boston, MA USA 02111 (617) 636-0966, Fax: (617) 636-3999, , <http://www.healthsci.tufts.edu/apua>

Ants

see **Fire ants**

ANWR

see **Arctic National Wildlife Refuge**

Apis mellifera scutellata

see **Africanized bees**

AQCR

see **Air Quality Control Region**

Aquaculture

Aquaculture is the husbandry or rearing of aquatic organisms under controlled or semi-controlled conditions. Stated another way, it is the art of cultivating natural plants and animals in water for human consumption or use. It can be considered aquatic agriculture, or as some people wish to call it, underwater agriculture. It is sometimes incorrectly termed aquiculture. Aquaculture involves production in both fresh and salt water. **Mariculture** is aquaculture in saline (**brackish** and marine) water. **Hydroponics** is the raising of aquatic plants in water. Organisms that are grown in aquaculture include fish, shellfish (crustaceans such as crawfish and shrimp, and mollusks such as oysters and clams), algae, and aquatic plants. More people are eating seafood for their added health benefit. Not only can these organisms be raised for human consumption, but they can also be reared for the lucrative baitfish, health food, aquarium, and home garden-pond industries.

Aquaculture dates back more than 3,500 years ago when carp were spawned and reared in China. There have also been records of aquaculture practices being performed in Egypt and Japan nearly that long ago. Mariculture is thought to have been brought to Hawaii about 1,500 years ago. Modern aquaculture finds its roots in the 1960s in the culturing of catfish in the United States and **salmon** in Europe.

There are two general methods used in aquaculture, extensive and intensive. Extensive aquaculture involves the production of low densities of organisms. For example, fingerling fish can be raised in ponds where they feed on the natural foods such as **phytoplankton** and **zooplankton**, and the matured fish are harvested at the end of the growing season. These ponds can also be fertilized to enhance the food chain, thus increasing fish or shellfish production. Typically, several hundred to several thousand pounds of fish are raised per acre annually. Intensive practices utilize much higher densities of organisms. This requires better **water quality**, which necessitates circulation, oxygenation, added commercial foods, and biological **filters** (with bacteria) to remove toxic wastes. These systems can produce more than one million lb of fish per acre (45,000/ha). Intensive aquaculture often utilizes tanks to grow the fish or shellfish, either indoor or outdoor. The water can be recirculated as long as toxic wastes such as ammonia are removed. Some processors run the **effluent** from the fish tanks through greenhouses with plants raised hydroponically to remove these **chemicals**. In this way, added revenue is gained. Some aquaculture operations raise only one type of organism (called **monoculture**) while others grow several **species** together (termed polyculture). An example of polyculture would be growing tilapia and catfish together.

About 86% of the world's aquaculture production (about 22 million tons in 1993) comes from Asia. China is the leading producer nation, with most of their production coming from carp, and secondarily shrimp. India is the second largest producer nation, followed by Japan, Taiwan, and the Philippines. For example, 1991 shrimp production in China reached nearly 250,000 tons. (Note: One English ton equals approximately one metric ton, so they can be thought of as equivalents). Only Taiwan exceeded this with 350,000 tons of shrimp produced. Ecuador produced 100,000 tons. In contrast, total shrimp production for the United States was only 1,600 tons. Salmon is another major fish produced by aquaculture, with most coming from Norway. Nearly 300,000 tons were produced in 1990. Great Britain, Canada, Chile, and Iceland are also major producers of farm-raised salmon. Other important aquaculture species and some of the countries where they are raised include: tilapia (Caribbean countries, Egypt, India, Israel, the Philippines); milkfish (the Philippines); Nile perch (Egypt); sea bass and sea bream (Egypt, Israel); mullet (Egypt, India, Israel); dolphin fish (also known as mahi-mahi; Egypt); grass and silver carp (Israel); halibut and other flatfishes (France, Norway); prawns (Caribbean countries, Israel, the Philippines); mussels and oysters (the Philippines); crabs (India); seaweeds (India, Japan); *Spirulina* (a blue-green algae; India); and many other examples.

Although people in the United States consume less fish and shellfish than people in other parts of the world, each year we spend billions of dollars importing edible fish. Total production from aquaculture in the United States in 1991 was about 540,000 tons, which earned about $750 million. This comprises only about 3% of the world's total aquaculture production. Domestically, catfish is our major aquaculture species ($200 million per year in 1991). Trout (primarily raised in Idaho) is the number two species raised in aquaculture and accounts for a $55 million per year industry. Shrimp, hybrid-striped bass, and tilapia are other species commonly raised in the United States. Of the 16,000 tons of crawfish (more appropriately called crayfish) collected in 1990 in the United States, about 60% came from aquaculture ponds.

In 1986 the U.S. Congress placed aquaculture under the jurisdiction of the **U.S. Department of Agriculture** appropriated $3 million to establish four regional aquaculture centers. These centers are located at the University of Washington, Southeastern Massachusetts University, Mississippi State University, and a Center for Tropical and Subtropical Aquaculture jointly administered by the University of Hawaii and the Oceanic Institute. The following year Congress created a fifth center at Michigan State University. The mandate for these centers was, and still is, to promote aquaculture in their region and to solve some of the problems facing this industry. More recent legislation has created mariculture research centers under the Sea Grant Program within the Department of Commerce.

Aquaculture is an appealing industry because it can help supplement what is caught in natural fisheries and what is produced through agriculture. This has tremendous implications in terms of the growing future dilemma with human over-population, decreased fisheries captured from the oceans, and stabilized agriculture production. Aquaculture is a growing industry in the world, which not only translates into more food, but also more revenue and jobs. This fulfills the old Chinese proverb that you can give a person a fish and feed them for a day, or you can teach a person to fish (or do aquaculture) and feed them for a lifetime.

It is particularly important to look at aquaculture development in third-world countries where protein sources are scarce. Since many of these countries lie in equatorial regions, tilapia (also called St. Peter's fish because they are the likely fish caught by the disciples in the Sea of Galilee) are one of the best prospects to grow because of their tolerance of high temperatures and poor water quality (i.e., high ammonia levels). They are also omnivorous filter feeders and thrive on blue-green algae. One alternative to the **deforestation** of our precious rain forests might be to encourage indigenous people to construct earthen ponds and raise tilapia on the fringes of the forests, thus producing not only a protein source, but also a **cash crop**. An advantage of aquaculture over agriculture is that fish can be grown in ponds overlying **soil** that may be toxic to plants (for example, with high concentrations of sulphur, **aluminum**, or salt). In addition,

aquaculture species can often be held longer if there is a glut on the market, thus allowing more flexible selling times. Despite these positive attributes, there are serious problems involved in aquaculture. These include release of waste into the water (adversely affecting water quality); spread of diseases and **parasites** through concentrated growing conditions; disfavorable taste through chemicals like geosmin in the water being taken up by the fish (termed "off-flavor" in catfish); release of hybrid strains into the natural population which can result in mixed genetics with natural populations; obstruction of the aesthetics along coastlines; and overproduction, which can result in depressed market prices. For example, both the catfish and salmon industries experienced a tremendous drop in their market prices due to overproduction in 1990 and 1991.

Aquaculture seems to have a bright future as long as we address some of the above-mentioned problems. One example of an enterprising solution to one of these problems is that of releasing wrasse (marine parasite-feeding fish) into aquaculture tanks and cages to remove fish lice from salmon. This not only solves the problem of parasites, but also eliminates the need to add chemicals to the water that kill parasites, stress the fish, and pollute the **environment**. The same pitfalls that have plagued agriculture must be avoided in aquaculture. A key concern is that of sustainability—concern for the culture environment and non-crop organisms in measuring the success of the aquaculture industry (discussed in the June 1996 issue of *World Aquaculture*). There is no need for agriculture farmers to feel threatened by aquaculture; on the contrary, it seems that these two areas can actually be brought together, particularly with small local farms in underdeveloped countries supplementing their terrestrial crops with aquatic production. This has already been done quite well in Norway, which has more than 700 fish farms in a country with only 4.2 million people. Aquaculture is evolving from an art, with trial-and-error learning, to more of a science with ongoing research. Countries such as Japan, the United States, Israel, and Norway are developing new technology to bring this field to higher levels. Recent predictions suggest that aquaculture will be able to meet at least 40% of the global demand for fish and shellfish over the next 15 years if current trends prevail. Government, academic institutions, and industry must work hand-in-hand to see that these predictions come to fruition with minimal environmental damage.

[*John Korstad*]

RESOURCES
BOOKS

Bardach, J. "Aquaculture." *BioScience* 37 (1987): 318–19.
Hopkins, J. S. "Aquaculture Sustainability: Avoiding the Pitfalls of the Green Revolution." *World Aquaculture* 27 (1996): 13–5.

Lee, J. S., and M. E. Newman. *Aquaculture Today—An Introduction.* Illinois: Interstate Publishers, 1992.

Parker, N. C. "History, Status, and Future of Aquaculture in the United States." *CRC Critical Reviews in Aquatic Science.* 1 (1989): 97–109.

Stickney, R. R. *Principles of Aquaculture.* New York: Wiley, 1994.

World Aquaculture Society. "World Aquaculture—A Special Section on Aquaculture Research and Development Around the World." *World Aquaculture* 27 (1996): 7–30.

Aquarium trade

International trade in live fish and other marine **species** for international aquarium hobby market. Most fish, corals, and other marine aquarium species traded internationally are collected live from tropical and subtropical coral reefs, especially in the Philippines, Indonesia, and other Asian countries. Nearly half the world aquarium fish market is in the United States. Although the capture and marketing of live reef species provides a significant income source in remote communities, international trade has alarmed conservationists and environmentalists because of harmful collecting methods. A majority of live tropical fish are captured with sodium cyanide (NaCN), a highly toxic and inexpensive powdered poison. Divers use squirt bottles containing a cyanide solution or drop cyanide tablets into reef crevices where fish hide. The poison stuns the fish, making them easy to capture, and the fish usually recover after a few minutes. Fish are transferred to holding tanks on ships and then transported to international wholesalers. From 70–90% of fish captured with cyanide die later, however, due to damage to the liver, stomach, and other tissues. Another problem common in salt water aquaria is "sudden death syndrome," when fish inexplicably die soon after they are introduced to the aquarium. Some researchers attribute the syndrome to cyanide, which is stored in tissues during transport. Fish are often not fed during storage, but when they begin to eat in their new aquarium, or if they undergo even mild shock, the cyanide is released into the blood stream and kills the fish. Although solid figures are impossible to establish, estimates of the proportion of tropical live fish caught with cyanide are 75% or higher worldwide, and 90% in some countries.

The reefs where fish are captured fare as badly as the fish: one study found all cyanide-treated corals dead within three months after treatment. Coral reefs also suffer physical damage as divers break off chunks of coral to retrieve stunned fish. Despite harm to reefs and high fish **mortality** rates, cyanide is considered the most cost effective method to capture popular species such as angel fish and trigger fish.

Peter Rubec, a researcher with the International Marinelife Alliance, reports that an exporter operating from Manila initiated the export of marine fish from the Philippines in 1957. With almost 2,200 fish species, the Philippines has the highest reef fish diversity in the world, including 200 species that are commonly exported for the marine pet fish industry. In 1986 Filipino reefs provided as much as 80% of the world's tropical marine fish, a number that fell only to 70% by 1997. From 1970–1980 exports rose from 1,863,000 lb (845,000 kg) of live fish packed in water to 4.41 million lb (2 million kg) worth $2 billion dollars. Since then export numbers have fallen in the Philippines, as stocks have thinned and the trade has diversified to other countries, but in 1986 the country's marine tropical fish trade was worth $10 million dollars. Worldwide the value of aquarium fish was $100 million dollars, plus corals and other supplies.

Cyanide fishing was first introduced in the Philippines in about 1962. The technique has since spread to Indonesia, New Guinea, and other regions where live fish are caught. Cyanide has since been banned in the Philippines and Indonesia, but enforcement is difficult because fishing vessels are dispersed, often in hard-to-reach sections of remote islands. Despite legal bans, an estimated 375,000 lb (170,000 kg) of cyanide is still used each year in the Philippines. Cyanide fishing boats have even exploited established marine preserves and parks, such as Indonesia's Take Bone Atoll. Furthermore, as enforcement mechanisms develop, cyanide fishing can quickly spread to other countries, where laws are still weak or lacking.

The greatest problems of the live reef fish trade result from its mobility. The frontier of cyanide fishing has moved steadily through Southeast Asia to tropical Pacific islands and even across the Indian Ocean to the Seychelles and Tanzania, as fertile reefs have disappeared and as governments caught on to their fishing practices and began to impose laws restricting the use of cyanide and explosives. Since the market is international and lucrative, it does not matter where the fish are caught. Furthermore, fish exporters succeed best if they continually explore new, unexploited reefs where more unusual and exotic fish can be found. The migratory nature of the industry makes it easy for fishing vessels to move on as soon as a country begins to enact or enforce limits on their activities.

Starting in 1984 international **conservation** groups have worked to introduce net fishing as a less harmful fishing alternative that still allows divers to retain their income from live fish. To catch fish with nets, divers may use a stick to drive fish from their hiding places in the reef and then trap the fish with a fine mesh net. Reports indicate that long-term survival rates of net-caught fish may be as high as 90%, compared to as little as 10% among cyanide-caught fish. In Hawaii and **Australia**, where legal controls are more effective than in Southeast Asia, nets are used routinely. Although the aquarium industry has provided little aid in the effort to increase net-fishing, local communities in the Philippines and Indonesia, with the aid of international conservationist organizations, have worked to encourage net fishing. Increased local control of reef fisheries is also important in helping communities to control cyanide fishing in their nearby reefs. Local communities that have depended on reefs and their fish for generations understand that sustainable use is both possible and essential for their own survival, so they often have a greater incentive to prevent reef damage than either state governments or transient fishing enterprises. In an effort to help coastal villages help themselves, the government of the Philippines has recently granted villagers greater rights to patrol and control nearby fishing areas.

Another step toward the control of cyanide fishing is the development of cyanide detection techniques that allow wholesalers to determine whether fish are tainted with residual cyanide. Simply by sampling the water a fish is carried in, the test can detect if the fish is releasing cyanide from its tissues or metabolizing cyanide. Ideally this test could help control illegal and harmful fish trade, but thus far it is not widely used. Attempts have also begun to establish cyanide-free certification, but this has been slow to take effect because the market is dispersed and there are many different exporters.

Coastal villages also suffer from reef damage. Most coastal communities in **coral reef** regions have traditionally relied on the rich fishery supported by the reef as a principal protein and food source. As reefs suffer, fisheries deteriorate. Some researchers estimate that just 1,300 ft (400 m) of healthy reef the can support 800 people, while the same amount of damaged reef can support only a quarter that many people. Other ecological benefits also disappear as reefs deteriorate, since healthy coral reefs perform critical water clarification functions. Corals, along with sea anemones and other life forms they shelter, filter floating organic matter from the water column, cycling it into the food chain that supports fish, crustaceans, birds, and humans. By breaking ocean waves offshore, healthy and intact corals also control beach **erosion** and reduce storm surges, the unusually large waves and tides associated with storms and high winds.

Divers collecting the fish suffer as well. Exposure to cyanide and inadequate or unsafe breathing hoses are common problems. In addition divers must search deeper waters as more accessible fish are depleted. In 1993, 40 divers in a single small village were reported to have been injured and 10 killed by the bends, which results from rapid changes of pressure as divers rise from deep water.

Unfortunately there has been relatively little breeding of tropical fish in aquaria, at least in part because these fish often have specialized **habitat** needs and life style requirements that are difficult to produce under controlled or domestic conditions. In addition the aquarium trade is specialized and limited in volume, and gearing up to produce fish

can be an expensive undertaking for which a reliable market must be assured. Equally unfortunate is the fact that American and European dealers in aquarium products tend to be elusive about the details of where their fish came from and how they were caught. If pet shops do not mention the source, many aquarium owners are able to ignore the implications of the fish trade they are participating in.

In addition to aquarium fishing, cyanide has now been introduced to the live food-fish market, especially in Asia, where live fish are an expensive delicacy. The live food/fish trade, centered in China and Hong Kong, is estimated to exceed $1 billion per year. The Hong Kong cyanide fleet alone has hundreds of vessels, each employing up to 25 divers to catch fish with cyanide for the Chinese and Hong Kong restaurant market.

Another harmful fishing technique is blast fishing—releasing a small bomb that breaks up coral masses in which fish hide. A single explosive might destroy corals in a circle from 10–33 ft (3–10 m) wide. The fish, briefly stunned by the blast, are easily retrieved from the rubble. Half the countries in the South Pacific have seen coral damage from blasting, including Guam, Indonesia, Malaysia, and Thailand. The technique has also spread to Africa, where it is used in Tanzania and other countries bordering the Indian Ocean.

Some environmentalists stress that simply eliminating the live fish trade is not an adequate solution. Because live fish are so lucrative, much more valuable than the same weight in dead fish, fishermen may be able to produce a better income with fewer fish when they catch live fish. Steering the fish capture methods to a safer alternative, especially the use of nets, could do more to save reef communities than eliminating the trade altogether.

[*Mary Ann Cunningham*]

RESOURCES
BOOKS

Rubec, P.I. "The Effects of Sodium Cyanide on Coral Reefs and Marine Fish in the Philippines". *The First Asian Fisheries Forum*. Manila, Philippines: Asian Fisheries Society, 1986.

PERIODICALS

Ariyoshi, R. "Halting a coral catastrophe." *Nature Conservancy* 47, no.1 (1997): 20–25.
Robinson, S. "A Proposal for the Marine Fish Industry: Converting Philippine Cyanide Users into Netsmen." *Greenfields* 15, no. 9 (1985): 39–45.

Aquatic chemistry

Water can exist in various forms within the **environment**, including: (1) liquid water of oceans, lakes and ponds, rivers and streams, **soil** interstices, and underground aquifers; (2)

solid water of glacial ice and more-ephemeral snow, rime, and frost; and (3) vapor water of cloud, fog, and the general **atmosphere**. More than 97% of the total quantity of water in the hydrosphere occurs in the oceans, while about 2% is glacial ice, and less than 1% is **groundwater**. Only about 0.01% occurs in freshwater lakes, and the quantities in other compartments are even smaller.

Each compartment of water in the hydrosphere has its own characteristic chemistry. Seawater has a relatively large concentration of inorganic solutes (about 3.5%), dominated by the ions chloride (1.94%), sodium (1.08%), sulfate (0.27%), magnesium (0.13%), calcium (0.041%), potassium (0.049%), and bicarbonate (0.014%).

Surface waters such as lakes, ponds, rivers, and streams are highly variable in their chemical composition. Saline and soda lakes of **arid** regions have total salt concentrations that can substantially exceed that of seawater. Lakes such as Great Salt Lake in Utah and the Dead Sea in Israel can have salt concentrations that exceed 25%. The shores of such lakes are caked with a crystalline rime of evaporate minerals, which are sometimes mined for industrial use.

The most chemically dilute surface waters are lakes in watersheds with hard, slowly **weathering** bedrock and soils. Such lakes can have total salt concentrations of less than 0.001%. For example, Beaverskin Lake in Nova Scotia has very clear, dilute water that is chemically dominated by chloride, sodium, and sulfate, in concentrations of two-thirds of the norm for surface water or less, with only traces of calcium, usually most abundant, and no silica. A nearby body of water, Big Red Lake, has a similarly dilute concentration of inorganic ions but, because it receives **drainage** from a bog, its chemistry also includes a large concentration of dissolved organic **carbon**, mainly comprised of humic/fulvic acids that stain the water a dark brown and greatly inhibit the penetration of sunlight.

The water of precipitation is considerably more dilute than that of surface waters, with concentrations of sulfate, calcium, and magnesium of one-fortieth to one-hundredth of surface water levels, but adding small amounts of nitrate and ammonium. Chloride and sodium concentrations depend on proximity to salt water. For example, precipitation at a remote site in Nova Scotia, only 31 mi (50 km) from the Atlantic Ocean, will have six to 10 times as much sodium and chloride as a similarly remote location in northern Ontario.

Acid rain is associated with the presence of relatively large concentrations of sulfate and nitrate in precipitation water. If the negative electrical charges of the sulfate and nitrate anions cannot be counterbalanced by positive charges of the cations sodium, calcium, magnesium, and ammonium, then **hydrogen** ions go into solution, making the water acidic. **Hubbard Brook Experimental Forest**, New Hamp-

shire, within an **airshed** of industrial, **automobile**, and residential emissions from the northeastern United States and eastern Canada, receives a substantially acidic precipitation, with an average **pH** of about 4.1. At Hubbard Brook, sulfate and nitrate together contribute 87% of the anion-equivalents in precipitation. Because cations other than the hydrogen **ion** can only neutralize about 29% of those anion charges, hydrogen ions must go into solution, making the precipitation acidic.

Fogwaters can have much larger chemical concentrations, mostly because the inorganic **chemicals** in fogwater droplets are less diluted by water than in rain and snow. For example, fogwater on Mount Moosilauke, New Hampshire, has average sulfate and nitrate concentrations about nine times more than in rainfall there, with ammonium eight times more, sodium seven times more, and potassium and the hydrogen ion three times more.

The above descriptions deal with chemicals present in relatively large concentrations in water. Often, however, chemicals that are present in much smaller concentrations can be of great environmental importance.

For example, in freshwaters phosphate is the **nutrient** that most frequently limits the productivity of plants, and therefore, of the aquatic **ecosystem**. If the average concentration of phosphate in lake water is less than about 10 μg/l, then the algae productivity will be very small, and the lake is classified as **oligotrophic**. Lakes with phosphate concentrations ranging from about 10–35 μg/l are mesotrophic, those with 35–100 μg/l are eutrophic, and those with more than 100 μ/l are very productive, and very green, hypertrophic waterbodies. In a few exceptional cases, the productivity of freshwater may be limited by **nitrogen**, silica, or carbon, and sometimes by unusual micronutrients. For example, the productivity of **phytoplankton** in Castle Lake, California, has been shown to be limited by the availability of the trace metal, molybdenum.

Sometimes, chemicals present in trace concentrations in water can be toxic to plants and animals, causing substantial ecological changes. An important characteristic of acidic waters is their ability to solubilize **aluminum** from minerals, producing ionic aluminum. In non-acidic waters, ionic aluminum is generally present in minute quantities, but in very acidic waters when pH is less than 2, attainable by **acid mine drainage**, soluble-aluminum concentrations can rise drastically. Although some aquatic biota are physiologically tolerant of these aluminum ions, other **species**, such as fish, suffer toxicity and may disappear from acidified waterbodies. Many aquatic species cannot tolerate even small quantities of ionic aluminum. Many ecologists believe that aluminum ions are responsible for most of the toxicity of acidic waters and also of acidic soils.

Some chemicals can be toxic to aquatic biota even when present in ultra trace concentrations. Many species within the class of chemicals known as **chlorinated hydrocarbons** are insoluble in water but are soluble in biological lipids such as animal fats. These chemicals often remain in the environment because they are not easily metabolized by **microorganisms** or degraded by **ultraviolet radiation** or other inorganic processes. Examples of chlorinated **hydrocarbons** are the insecticides DDT, DDD, dieldrin, and methoxychlor, the class of dielectric fluids known as PCBs, and the chlorinated **dioxin**, TCDD.

These chemicals are so dangerous because they collect in biological tissues, and accumulate progressively as organisms age. They also accumulate into especially large concentrations in organisms at the top of the ecosystem's **food chain/web**. In some cases, older individuals of top predator species have been found to have very large concentrations of chlorinated hydrocarbons in their fatty tissues. The toxicity caused to raptorial birds and other predators as a result of their accumulated doses of DDT, PCBs, and other chlorinated hydrocarbons is a well-recognized environmental problem.

Water pollution can also be caused by the presence of hydrocarbons. Accidental spills of **petroleum** from disabled tankers are the highest profiled causes of oil **pollution**, but smaller spills from tankers disposing of oily bilge waters and chronic discharges from refineries and **urban runoff** are also significant sources of oil pollution. Hydrocarbons can also be present naturally, as a result of the release of chemicals synthesized by algae or during **decomposition** processes in **anaerobic sediment**. In a few places, there are natural seepages from near-surface petroleum reservoirs, as occurs in the vicinity of Santa Barbara, California. In general, the typical, naturally-occurring concentration of hydrocarbons in seawater is quite small. Beneath a surface slick of spilled petroleum, however, the concentration of soluble hydrocarbons can be multiplied several times, sufficient to cause toxicity to some biota. This dissolved fraction does not include the concentration of finely suspended droplets of petroleum, which can become incorporated into an oil-in-water emulsion toxic to organisms that become coated with it. In general, within the very complex mix of hydrocarbons found in petroleum, the smallest molecules are the most soluble in water.

[*Bill Freedman Ph.D.*]

RESOURCES
BOOKS

Bowen, H. J. M. *Environmental Chemistry of the Elements.* San Diego: Academic Press, 1979.

Freedman, B. *Environmental Ecology.* San Diego: Academic Press, 1989.

Aquatic microbiology

Aquatic microbiology is the science that deals with microscopic living organisms in fresh or salt water systems. While aquatic microbiology can encompass all **microorganisms**, including microscopic plants and animals, it more commonly refers to the study of bacteria, viruses, and **fungi** and their relation to other organisms in the aquatic **environment**.

Bacteria are quite diverse in **nature**. The scientific classification of bacteria divides them into 19 major groups based on their shape, cell structure, staining properties (used in the laboratory for identification), and metabolic functions. Bacteria occur in many sizes as well ranging from 0.1 micrometer to greater than 500 micrometers. Some are motile and have flagella, which are tail-like structures used for movement.

Although **soil** is the most common **habitat** of fungi, they are also found in aquatic environments. Aquatic fungi are collectively called *water molds* or *aquatic Phycomycetes*. They are found on the surface of decaying plant and animal matter in ponds and streams. Some fungi are parasitic and prey on algae and protozoa.

Viruses are the smallest group of microorganisms and usually are viewed only with the aid of an electron microscope. They are disease-causing organisms that are very different than bacteria, fungi, and other cellular life-forms. Viruses are infectious **nucleic acid** enclosed within a coat of protein. They penetrate host cells and use the nucleic **acid** of other cells to replicate.

Bacteria, viruses, and fungi are widely distributed throughout aquatic environments. They can be found in fresh water rivers, lakes, and streams, in the surface waters and sediments of the world's oceans, and even in hot springs. They have even been found supporting diverse communities at **hydrothermal vents** in the depths of the oceans.

Microorganisms living in these diverse environments must deal with a wide range of physical conditions, and each has specific adaptations to live in the particular place it calls home. For example, some have adapted to live in fresh waters with very low **salinity**, while others live in the saltiest parts of the ocean. Some must deal with the harsh cold of arctic waters, while those in hot springs are subjected to intense heat. In addition, aquatic microorganisms can be found living in environments where there are extremes in other physical parameters such as pressure, sunlight, organic substances, dissolved gases, and water clarity.

Aquatic microorganisms obtain nutrition in a variety of ways. For example, some bacteria living near the surface of either fresh or marine waters, where there is often abundant sunlight, are able to produce their own food through the process of **photosynthesis**. Bacteria living at hydrothermal vents on the ocean floor where there is no sunlight can produce their own food through a process known as **chemosynthesis**, which depends on preformed organic **carbon** as an energy source. Many other microorganisms are not able to produce their own food. Rather, they obtain necessary nutrition from the breakdown of organic matter such as dead organisms.

Aquatic microorganisms play a vital role in the cycling of nutrients within their environment, and thus are a crucial part of the **food chain/web**. Many microorganisms obtain their nutrition by breaking down organic matter in dead plants and animals. As a result of this process of decay, nutrients are released in a form usable by plants. These aquatic microorganisms are especially important in the cycling of the nutrients **nitrogen**, **phosphorus**, and carbon. Without this **recycling**, plants would have few, if any, organic nutrients to use for growth.

In addition to breaking down organic matter and recycling it into a form of nutrients that plants can use, many of the microorganisms become food themselves. There are many types of animals that graze on bacteria and fungi. For example, some deposit-feeding marine worms ingest sediments and digest numerous bacteria and fungi found there, later expelling the indigestible sediments. Therefore, these microorganisms are intimate members of the food web in at least two ways.

Humans have taken advantage of the role these microorganisms play in **nutrient** cycles. At **sewage treatment** plants, microscopic bacteria are cultured and then used to break down human wastes. However, in addition to the beneficial uses of some aquatic microorganisms, others may cause problems for people because they are pathogens, which can cause serious diseases. For example, viruses such as *Salmonella typhi*, *S. paratyphi*, and the Norwalk **virus** are found in water contaminated by sewage can cause illness. Fecal coliform (*E. coli*) bacteria and Enterococcus bacteria are two types of microorganisms that are used to indicate the presence of disease causing microorganisms in aquatic environments.

[*Marci L. Bortman*]

Aquatic toxicology

Aquatic toxicology is the study of the adverse effects of **toxins** and their activities on aquatic ecosystems. Aquatic toxicologists assess the condition of aquatic systems, monitor trends in conditions over time, diagnose the cause of damaged systems, guide efforts to correct damage, and predict the consequences of proposed human actions so the ecological consequences of those actions can be considered before damage occurs. Aquatic toxicologists study adverse effects at different spatial, temporal, and organizational scales. Because aquatic systems contain thousands of **species**, each of

these species can respond to toxicants in many ways, and interactions between these species can be affected.

Consequently, a virtually unlimited number of responses could be produced by **chemicals**. Scientists study effects as specific as a physiological response of an important fish species or as inclusive as the biological diversity of a large river basin. Generally, attention is first focused on responses that are considered important either socially or biologically. For example, if a societal goal is to have fishable waters, responses to toxicants of game fishes and the organisms they depend on would be of interest.

The two most common tools used in aquatic toxicology are the field survey and the toxicity test. A field survey characterizes the indigenous **biological community** of an aquatic system or **watershed**. Chemistry, geology, and **land use** are also essential components of field surveys. Often, the characteristics of the community are compared to other similar systems that are in good condition. Field surveys provide the best evidence of the existing condition of natural systems. However, when damaged communities and chemical contamination co-occur, it is difficult to establish a cause-and-effect relationship using the field survey alone. Toxicity tests can help to make this connection. The toxicity test excises some replicable piece of the system of interest, perhaps fish or microbial communities of lakes, and exposes it to a chemical in a controlled, randomized, and replicable manner. The most common aquatic toxicity test data provide information about the short-term survival of fish exposed to one chemical. Such tests demonstrate whether a set of chemical conditions can cause a specified response. They also provide information about what concentrations of the chemical are of concern. Toxicity tests can suggest a threshold chemical concentration below which the adverse effect is not expected to occur; they can also present an index of relative toxicity used to rank the toxicity of two chemicals. By using information from the field survey and toxicity tests, along with other tools such as tests on the fate of chemicals released into aquatic systems, aquatic toxicologists can develop models predicting the effects of proposed actions.

The most important challenge to aquatic toxicology will be to develop methods that support sustainable use of the aquatic and other ecosystems. Sustainable use is intended to ensure that those now living do not deprive **future generations** of the essential **natural resources** necessary to maintain a quality lifestyle. Achieving the goal of sustainable use will require aquatic toxicologists to have a much longer temporal perspective than we now have. Additionally, in a more crowded and increasingly affluent world, cumulative effects will become extremely important.

[*John Cairns Jr.*]

Aquatic weed control

A simple definition of an aquatic weed is a plant that grows (usually too densely) in an area such that it hinders the usefulness or enjoyment of that area. Some common examples of aquatic plants that can become weeds are the water milfoils, ribbon weeds, and pondweeds. They may grow in ponds, lakes, streams, rivers, navigation channels, and seashores, and the growth may be due to a variety of factors such as excess nutrients in the water or the introduction of rapidly-growing **exotic species**. The problems caused by aquatic weeds are many, ranging from unsightly growth and nuisance odors to clogging of waterways, damage to shipping and underwater equipment, and impairment of **water quality**.

It is difficult and usually unnecessary to eliminate weeds completely from a lake or stream. Therefore, aquatic weed control programs usually focus on controlling and maintaining the prevalence of the weeds at an acceptable level. The methods used in weed control may include one or a combination of the following: physical removal, mechanical removal, **habitat** manipulation, biological controls, and chemical controls.

Physical removal of weeds involves cutting, pulling, or raking weeds by hand. It is time-consuming and labor-intensive and is most suitable for small areas or for locations that cannot be reached by machinery. Mechanical removal is accomplished by specialized harvesting machinery equipped with toothed blades and cutting bars to cut the vegetation, collect it, and haul it away. It is suitable for off-shore weed removal or to supplement chemical control. Repeated harvesting is usually necessary and often the harvesting blades may be limited in the depth or distance that they can reach. Inadvertent dispersal of plant fragments may also occur and lead to weed establishment in new areas. Operation of the harvesters may disturb fish habitat.

Habitat manipulation involves a variety of innovative techniques to discourage the establishment and growth of aquatic weeds. Bottom liners of plastic sheeting placed on lake bottoms can prevent the establishment of rooted plants. Artificial shading can discourage the growth of shade-intolerant **species**. Drawdown of the water level can be used to eliminate some species by desiccation. **Dredging** to remove accumulated sediments and organic matter can also delay colonization by new plants.

Biological control methods generally involve the introduction of weed-eating fish, insects, competing plant species or weed pathogens into an area of high weed growth. While there are individual success stories (for example, stocking lakes with grass carp), it is difficult to predict the long-term effects of the **introduced species** on the native species and **ecology** and therefore, biological controls should be used with caution.

Chemical control methods consist of the application of herbicides that may be either **systemic** or contact in **nature**. Systemic herbicides are taken up into the plant and cause plant death by disrupting its **metabolism** in various ways. Contact herbicides only kill the directly exposed portions of the plant, such as the leaves. While herbicides are convenient and easy to use, they must be selected and used with care at the appropriate times and in the correct quantities. Sometimes, they may also kill non-target plant species and in some cases, toxic residues from the degrading **herbicide** may be ingested and transferred up the food chain.

[*Usha Vedagiri*]

RESOURCES

BOOKS

Schmidt, J. C. *How to Identify and Control Water Weeds and Algae*. Milwaukee, WI: Applied Biochemists, 1987.

Aquifer

Natural zones below the surface that yield water in sufficient quantities to be economically important for industrial, agricultural, or domestic purposes. Aquifers can occur in a variety of geologic materials, ranging from glacial-deposited outwash to sedimentary beds of limestone and sandstone, and fractured zones in dense igneous rocks. Composition and characteristics are almost infinite in their variety.

Aquifers can be confined or unconfined. Unconfined aquifers are those where direct contact can be made with the **atmosphere**, while confined aquifers are separated from the atmosphere by impermeable materials. Confined aquifers are also artesian aquifers. Though originally artesian was a term applied to water in an aquifer under sufficient pressure to produce flowing **wells**, the term is now generally applied to all confined situations.

Aquifer depletion

An **aquifer** is water-saturated geological layer that easily releases water to **wells** or springs for use as a water supply. Also called ground water reservoirs or water-bearing formations, aquifers are created and replenished when excess precipitation (rain and snowfall) is held in the **soil**. This water is not released through **runoff** nor is removed by the surface flows of rivers or streams. Plants have used what they need (**transpiration**) and little is evaporated from non-living surfaces, such as soil. The remaining excess water slowly percolates downward through the soil and through the air spaces and cracks of the surface **overburden** of rocks into the bedrock. As water collects in this saturated area or **recharge**

zone, it becomes **groundwater**. The uppermost level of the saturated area is called the **water table**.

Groundwater is especially abundant in humid areas where the overburden is relatively thick and the bedrock is porous or fractured, particularly in areas of sedimentary rocks such as sandstone or limestone. Aquifers are extremely valuable **natural resources** in regions where lakes and rivers are not abundant. Groundwater is usually accessed by drilling a well and then pumping the water to the surface.

In less-moist environments, however, the quantity of precipitation available to recharge groundwater is much smaller. Slowly recharging aquifers in **arid** environments are easily depleted if their groundwater is used rapidly by humans. In some cases, groundwater sources may exist in water-bearing geologic areas that make pumping nearly impossible. Moreover, increased **irrigation** use has led to heavy pumping that is draining aquifers and lowering water tables around the world. Aquifer depletion is a growing problem as world populations increase and the need for increased food supplies.

Large, rapidly recharging aquifers underlying humid landscapes can sustain a high rate of pumping of their groundwater. As such, they can be sustainably managed as a renewable resource. Aquifers that recharge very slowly, however, are essentially filled with old, so-called "fossil" water that has accumulated over thousands or more years. This kind of aquifer has little capability of recharging as the groundwater is depleted so rapidly for human use. Therefore, slowly recharging aquifers are essentially non- renewable resources, whose reserves are mined by excessive use.

In 1999, the **Worldwatch Institute** reported that water tables were falling on every continent in the world, mainly because of excessive human consumption. Groundwater in India, in particular, is being pumped at double the rate of the aquifer's ability to recharge from rainfall. The aquifer under the North China Plain is seeing its water table fall at 5 feet (1.5 meters) a year.

In the United Sates, it is similar. The largest aquifer in the world, known as the Ogalalla Aquifer, is located beneath the arid lands of the western United States. The **Ogallala aquifer** is very slowly recharged by underground **seepage** that mostly originates with precipitation falling on a distant recharge zone in mountains located in its extreme western range. Much of the groundwater presently in the Ogalalla is **fossil water** that has accumulated during tens of thousands of years of extremely slow **infiltration**. Although the Ogalalla aquifer is an enormous resource, it is being depleted alarmingly by pumping at more than 150,000 wells. Most of the groundwater being withdrawn by the wells is used in irrigated agriculture, and some for drinking and other household purposes. In recent years, the level of

the Ogalalla aquifer has been decreasing by as much as 3.2 feet (1 meter) per year in intensively utilized zones, while the recharge rate is only of the order of 1 mm/yr. (or a little over 1/32 of an inch). Obviously, the Ogalalla aquifer is being mined on a large scale.

Aquifer depletion brings with it more than the threat of water **scarcity** for human use. Serious environmental consequences can occur when large amounts of water are pumped rapidly from ground water reservoirs. Commonly, the land above an aquifer will subside or sink as the water is drained from the geologic formation and the earth compacts. In 1999, researchers noted that portions of Bangkok, Thailand and **Mexico City, Mexico** were sinking as a result of overexploitation of their aquifers. This can cause foundations of buildings to shift and may even contribute to **earthquake** incidence. Large cities in the United States like Albuquerquer, Phoenix, and Tuscon lie over aquifers that are being rapidly depleted.

Unfortunately, the current solutions to aquifer depletion are to drill wells deeper or abandon irrigated agriculture and import food. Both are costly choices for any country, both in dollars and in economic independence.

[*Bill Freedman Ph.D.*]

RESOURCES
BOOKS

Freeze, R.A. and J.A. Cherry. *Groundwater.*Inglewood Heights, NJ: Prentice Hall, 1979.

Opie, J.*Ogallala: Water for a Dry Land.* Lincoln, NB: University of Nebraska Press, 2000.

Robins, N. (Ed.). *Groundwater Pollution, Aquifer Recharge and Vulnerability.*Special Publication Number 130, London, UK: Geological Society Publishing House, 1998.

ORGANIZATIONS

World Resources Institute, 10 G Street, NE (Suite 800), Washington, DC USA 20002 (202) 729-7600, Fax: (202) 729-7610, Email: front@wri.org, http://www.wri.org/

Worldwatch Institute, 1776 Massachusetts Ave., N.W., Washington, D.C. USA 20036-1904 (202) 452-1999, Fax: (202) 296-7365, Email: worldwatch@worldwatch.org, http://www.worldwatch.org/

Aquifer restoration

Once an **aquifer** is contaminated, the process of restoring the quality of water is generally time-consuming and expensive, and it is often more cost effective to locate a new source of water. For these reasons, the restoration of an aquifer is usually evaluated on the basis of these criteria: 1) the potential for additional contamination; 2) the time period over which the contamination has occurred; 3) the type of contaminant; and 4) the **hydrogeology** of the site. Restoration techniques fall into two major categories, in-situ methods and conventional methods of withdrawal, treatment, and disposal.

Remedies undertaken within the aquifer involve the use of chemical or biological agents which either reduce the toxicity of the contaminants or prevent them from moving any further into the aquifer, or both. One such method requires the introduction of biological cultures or chemical reactants and sealants through a series of injection **wells**. This action will reduce the toxicity, form an impervious layer to prevent the spread of the contaminant, and clean the aquifer by rinsing. However, a major drawback to this approach is the difficulty and expense of installing enough injection wells to assure a uniform distribution throughout the aquifer.

In-situ degradation, another restoration method, can theoretically be accomplished through either biological or chemical methods. Biological methods involve placing **microorganisms** in the aquifer that are capable of utilizing and degrading the hazardous contaminant. A great deal of progress has been made in the development of microorganisms which will degrade both simple and complex organic compounds. It may be necessary to supplement the organisms introduced with additional nutrients and substances to help them degrade certain insoluble organic compounds. Before introduction of these organisms, it is also important to evaluate the intermediate products of the degradation to **carbon dioxide** and water for toxicity. Chemical methods of in-situ degradation fall into three general categories: 1) injection of neutralizing agents for **acid** or caustic compounds; 2) addition of oxidizing agents such as **chlorine** or **ozone** to destroy organic compounds; and 3) introduction of amino acids to reduce PCBs (**polychlorinated biphenyls**).

There are also methods for stabilizing an aquifer and preventing a contaminant **plume** from extending. One stabilizing alternative is the conversion of a contaminant to an insoluble form. This method is limited to use on inorganic salts, and even those compounds can dissolve if the physical or chemical properties of the aquifer change. A change in **pH**, for instance, might allow the contaminant to return to solution. Like other methods of in-situ restoration, conversion requires the contaminant to be contained within a workable area.

The other important stabilizing alternatives are containment methods, which enclose the contaminant in an insoluble material and prevent it from spreading through the rest of the aquifer. There are partial and total containment methods. For partial containment, a clay cap can be applied to keep rainfall from moving additional contaminants into the aquifer. This method has been used quite often where there are sanitary landfills or other types of **hazardous waste** sites. Total containment methods are designed to isolate the area of contamination through the construction

of some kind of physical barrier, but these have limited usefulness. These barriers include **slurry** walls, grout curtains, sheet steel and floor **seals**. Slurry walls are usually made of concrete, and are put in place by digging a trench, which is then filled with the slurry. The depth at which these walls can be use is limited to 80–90 ft (24–27 m). Grout curtains are formed by injecting a cement grout under pressure into the aquifer through a series of injection wells, but it is difficult to know how effective a barrier this is and whether it has uniformly penetrated the aquifer. Depth is again a limiting factor, and this method has a range of 50–60 ft (15–18 m). Steel sheets can be driven to a depth of 100 ft (30 m) but no satisfactory technique for forming impermeable joints between the individual sheets has been developed. Floor seals are grouts installed horizontally, and they are used where the contaminant plume has not penetrated the entire depth of the aquifer. None of these methods offers a permanent solution to the potential problems of contamination, and some type of continued monitoring and maintenance is necessary.

Conventional methods of restoration involve removal of the contaminant followed by withdrawal of the water, and final treatment and disposal. Site characteristics that are important include the **topography** of the land surface, characteristics of the **soil**, depth to the **water table** and how this depth varies across the site, and depth to impermeable layers. The options for collection and withdrawal can be divided into five groups: 1) collection wells; 2) subsurface gravity collection drains; 3) impervious grout curtains (as described above); 4) cut-off trenches; or 5) a combination of these options. Collection wells are usually installed in a line, and are designed to withdraw the contaminant plume and to keep movement of other clean water into the wells at a minimum. Various sorts of drains can be effective in intercepting the plume, but they do not work in deep aquifers or hard rock. Cut-off trenches can be dug if the contamination is not too deep, and water can then be drained to a place where it can be treated before final **discharge** into a lake or stream.

Water taken from a contaminated aquifer requires treatment before final discharge and disposal. To what degree it can be treated depends on the type of contaminant and the effectiveness of the available options. Reverse osmosis uses pressure at a high temperature to force water through a membrane that allows water molecules, but not contaminants, to pass. This process removes most soluble organic **chemicals**, **heavy metals**, and inorganic salts. Ultrafiltration also uses a pressure-driven method with a membrane. It operates at lower temperatures and is not as effective for contaminants with smaller molecules. An **ion exchange** uses a bed or series of tubes filled with a resin to remove selected compounds from the water and replace them with

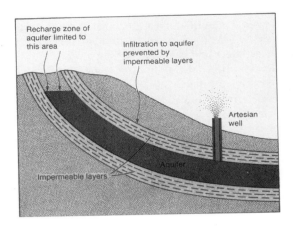

An aquifer contained between two impermeable layers of of rock or clay. (McGraw-Hill Inc. Reproduced by permission.)

harmless ions. The system has the advantage of being transportable, depending on the type of resin, it can be used more than once by flushing the resin with either an acid or salt solution. Both organic and inorganic substances can be removed using this method.

Wet-air oxidation is another important treatment method. It introduces oxygen into the liquid at a high temperature, which effectively treats inorganic and organic contaminants. Combined ozonation/ultraviolet radiation is a chemical process in which the water containing toxic chemicals is brought into contact with ozone and **ultraviolet radiation** to break down the organic contaminants into harmless parts. Chemical treatment is a general name for a variety of processes that can be used to treat water. They often result in the precipitation of the contaminant, and will not remove soluble organic and inorganic substances. **Aerobic** biological treatments are processes that employ microorganisms in the presence of **dissolved oxygen** to convert organic matter to harmless products. Another approach uses activated **carbon** in columns—the water is run over the carbon and the contaminants become attached to it. Contaminants begin to cover the surface area of the carbon over time, and these **filters** must be periodically replaced.

These treatment processes are often used in combination. The methods used depend on the ultimate disposal plan for the end products. The three primary disposal options are: 1) discharge to a **sewage treatment** plant; 2) discharge to a surface water body; and 3) land application. Each option has positive and negative aspects. Any discharge to a municipal treatment plant requires pre-treatment to standards that will allow the plant to accept the waste. Land application requires an evaluation of plant **nutrient** supplying capability and any potentially harmful side-effects on the crops grown.

Discharge to surface water requires that the waste be treated to the standard allowable for that water. For many organics the final disposal method is burning.

[*James L. Anderson*]

RESOURCES
BOOKS

Freeze, R. A., and J. A. Cherry. *Ground Water*. Englewood Cliffs, NJ: Prentice-Hall, 1979.

Pye, V. I., R. Patrick, and J. Quarles. *Ground Water Contamination in the United States*. Philadelphia: University of Pennsylvania Press, 1983.

Arable land

Arable land has **soil** and **topography** suitable for economical and practical cultivation of crops. These range from grains, grasses, and legumes to fruits and vegetables. Permanent pastures and **rangelands** are not considered arable. Forested land is also nonarable, although it can be farmed if the area is cleared and the soil and topography are suitable for cultivation.

Aral Sea

The Aral Sea is a large, shallow, saline lake hidden in the remote deserts of the republics of Uzbekistan and Kazakhstan in the south-central region of the former Soviet Union. Once the world's fourth largest lake in area (smaller only than America's Lake Superior, Siberia's **Lake Baikal**, and East Africa's Lake Victoria), in 1960 the Aral Sea had a surface area of 26,250 mi^2 (68,000 km^2), and a volume of 260 cu mi (1,090 cu km). Its only water sources are two large rivers, the Amu Darya and the Syr Darya. Flowing northward from the Pamir Mountains on the Afghan border, these rivers pick up salts as they cross the Kyzyl Kum and Kara Kum deserts. Evaporation from the landlocked sea's surface (it has no outlet) makes the water even saltier.

The Aral Sea's destruction began in 1918 when plans were made to draw off water to grow cotton, a badly needed **cash crop** for the newly formed Soviet Union. The amount of irrigated cropland in the region was expanded greatly (from 7.2–18.8 million acres; 2.9–7.6 million ha) in the 1950s and 1960s with the completion of the Kara Kum canal. Annual water flows in the Amu Darya and Syr Darya dropped from about 13 cu mi (55 cu km) to less than 1 cu mi (5 cu km). In some years the rivers were completely dry when they reached the lake.

Soviet authorities were warned that the sea would die without replenishment, but sacrificing a remote **desert** lake for the sake of economic development seemed an acceptable tradeoff. Inefficient **irrigation** practices drained away the lifeblood of the lake. Dry years in the early 1970s and mid-1980s accelerated water shortages in the region. Now, in a disaster of unprecedented magnitude and rapidity, the Aral Sea is disappearing as we watch.

Until 1960 the Aral Sea was fairly stable, but by 1990, it had lost 40% of its surface area and two-thirds of its volume. Surface levels dropped 42 ft; 914 m), turning 11,580 mi^2 (30,000 km^2, about the size of the state of Maryland) of former seabed into a salty, dusty desert. Fishing villages that were once at the sea's edge are now 25 mi (40 km) from water. Boats trapped by falling water levels lie abandoned in the sand. **Salinity** of the remaining water has tripled and almost no aquatic life remains. **Commercial fishing** that brought in 48,000 metric tons in 1957 was completely gone in 1990.

Winds whipping across the dried-up seabed pick up salty dust, **poisoning** crops and causing innumerable health problems for residents. An estimated 43 million metric tons of salt are blown onto nearby fields and cities each year. Eye irritations, intestinal diseases, skin infections, **asthma**, **bronchitis**, and a variety of other health problems have risen sharply in the past 20 years, especially among children. Infant **mortality** in the Kara-Kalpak Autonomous Region adjacent to the Aral Sea is 60 per 1,000, twice as high as in other former Soviet Republics.

Among adults, throat cancers have increased five fold in 30 years. Many physicians believe that heavy doses of pesticides used on the cotton fields and transported by **runoff** water to the lake sediments are now becoming airborne in dust storms. Although officially banned, DDT and other persistent pesticides have been widely used in the area and are now found in mothers' milk. More than 35 million people are threatened by this disaster.

A report by the Institute of Geography of the Russian Academy of Sciences predicts that without immediate action, the Aral Sea will vanish by 2010. It is astonishing that such a large body of water could dry up in such a short time. Philip P. Micklin of Western Michigan University, an authority on water issues says this may the world's largest ecological disaster.

What can be done to avert this calamity? Clearly, one solution would be to stop withdrawing water for irrigation, but that would compound the disastrous economic and political conditions in the former Soviet Republics. More efficient irrigation might save as much as half the water now lost without reducing crop yields, but lack of funds and organization in the newly autonomous nations makes new programs and improvements almost impossible. Restoring the river flows to about 5 cu mi (20 cu km) per year would probably stabilize the sea at present levels. It would take perhaps twice as much to return it to 1960 conditions.

Before the dissolution of the Soviet Union, there was talk of grandiose plans to divert part of the northward-flowing Ob and Irtysh rivers in Western Siberia. In the early 1980s a system of **dams**, pumping stations, and a 1,500 mi (2,500 km) canal was proposed to move 25 cu km (6 cu mi) of water from Siberia to Uzbekistan and Kazakhstan. The cost of 100 billion rubles ($150 billion) and the potential adverse environmental effects led President Mikhail Gorbachev to cancel this scheme in 1986.

Perhaps more than technological fixes, what we all need most is a little foresight and humility in dealing with **nature**. The words painted on the rusting hull of an abandoned fishing boat lying in the desert might express it best, "Forgive us Aral. Please come back."

[*William P. Cunningham*]

RESOURCES

PERIODICALS

Ellis, M. S. "A Soviet Sea Lies Dying." *National Geographic* 177 (February 1990): 73–93.

Micklin. P. P. "Dessication of the Aral Sea: A Water Management Disaster in the Soviet Union." *Science* 241 (2 September 1988): 1170–1176.

Arco, Idaho

"Electricity was first generated here from Atomic Energy on December 20, 1951. On December 21, 1951, all of the electrical power in this building was supplied from Atomic Energy."

Those words are written on the wall of a nuclear reactor in Arco, Idaho, a site now designated as a Registered National Historic Landmark by the **U.S. Department of the Interior**. The inscription is signed by 16 scientists and engineers responsible for this event.

The production of electricity from **nuclear power** was truly a momentous occasion. Scientists had known for nearly two decades that such a conversion was possible. Use of nuclear energy as a safe, efficient energy source of power was regarded by many people in the United States and around the world as one of the most exciting prospects for the "world of tomorrow."

Until 1945, scientists' efforts had been devoted to the production of **nuclear weapons**. The conclusion of World War II allowed both scientists and government officials to turn their attention to more productive applications of nuclear energy. In 1949, the United States **Atomic Energy Commission** (AEC) authorized construction of the first nuclear reactor designed for the production of electricity. The reactor was designated Experimental Breeder Reactor No. 1 (EBR-I).

The site chosen for the construction of EBR-I was a small town in southern Idaho named Arco. Founded in the late 1870s, the town had never grown very large. Its population in 1949 was 780. What attracted the AEC to Arco was the 400,000 acres (162,000 ha) of lava-rock-covered wasteland around the town. The area provided the seclusion that seemed appropriate for an experimental nuclear reactor. In addition, AEC scientists considered the possibility that the porous lava around Arco would be an ideal place in which to dump wastes from the reactor.

On December 21, 1951, the Arco reactor went into operation. Energy from a **uranium** core about the size of a football generated enough electricity to light four 200-watt light bulbs. The next day its output was increased to a level where it ran all electrical systems in EBR-I. In July 1953, EBR-I reached another milestone. Measurements showed that breeding was actually taking place within the reactor. The dream of a generation had become a reality.

The success of EBR-I convinced the AEC to expand its breeder experiments. In 1961, a much larger version of the original plant, Experimental Breeder Reactor No. 2 (EBR-II) was also built near Arco on a site now designated as the Idaho National Engineering Laboratory of the **U.S. Department of Energy**. EBR-II produced its first electrical power in August of 1964.

[*David E. Newton*]

RESOURCES
PERIODICALS

"A Village Wakes Up." *Life* (9 May 1949): 98–101.

Crawford, M. "Third Strike for Idaho Reactor." *Science* 251 (18 January 1991): 263.

Elmer-DeWitt, P. "Nuclear Power Plots a Comeback." *Time* 133 (2 January 1989): 41.

Schneider, K. "Idaho Says No." *The New York Times Magazine* 139 (11 March 1990): 50.

U.S. Congress, Office of Technology Assessment. *The Containment of Underground Nuclear Explosions*. OTA-ISC-414. Washington, DC: U.S. Government Printing Office, 1989.

Arctic Council

The Arctic Council is a cooperative environmental protection council composed of eight governments whose goal is to protect the Arctic's fragile **environment** while continuing economic development. Established in 1996, the Council plans to protect the environment by protecting **biodiversity** in the Arctic region and maintaining the sustainable use of its **natural resources**.

The Arctic Council was formed on September 19, 1996, when representatives from Canada, Denmark, Finland, Iceland, Norway, the Russian Federation, Sweden, and the United States signed a *Declaration on the Establishment of the Arctic Council*. The work and programs founded under the former Arctic Environmental Protection Strategy (AEPS) have become integrated into the new Arctic Council, which provides for indigenous representation—including the Inuit Circumpolar Conference, the Saami Council (Scandinavia, Finland, and Russia) for the Nordic areas and the Association of Indigenous Minorities of the North, Siberia, and the Far East of the Russian Federation—at all meetings. The indigenous people cannot vote but may provide knowledge of traditional practices for research and a collective understanding of the Arctic. As members of the Council's board, they can recommended ways to spread the work and benefits of **oil drilling** and mining businesses among indigenous communities while minimizing the negative effects on their land.

The **World Wildlife Fund** oversees the Council's work on protecting the Arctic's **wildlife** and vegetation from various destructive activities, such as diamond mining and **clear-cutting**. There is also evidence that the Arctic area receives a disproportionate amount of **pollution**, as witnessed in the overall land and water degradation.

Former Canadian Environment Minister Sergio Marchi is a supporter of the Council's work, emphasizing "The Arctic is an environmental early warning system for our globe. The Arctic Council will help deliver that warning from pole to pole."

[*Nicole Beatty*]

RESOURCES
ORGANIZATIONS

Arctic Council, Ministry for Foreign Affairs, Unit for Northern Dimension, P.O. Box 176, HelsinkiFinland FIN-00161 +358 9 1605 5562, Fax: +358 9 1605 6120, Email: Petri.Ojanpera@formin.fi, <http://www.arctic-council.org>

Arctic haze

The dry **aerosol** present in arctic regions during much of the year and responsible for substantial loss of **visibility** through the **atmosphere**. The arctic regions are, for the most part, very low in precipitation, qualifying on that basis as deserts. Ice accumulates because even less water evaporates than is deposited. Hence, particles that enter the arctic atmosphere are only very slowly removed by precipitation, a process that removes a significant fraction of particles from the tropical and temperate atmospheres. Thus, relatively small sources can lead to appreciable final atmospheric concentrations.

A **succession** of studies has been conducted on the chemistry of the particles in that **haze**, and those trapped in the snow and ice. Much of the time the mix of trace elements in the particles is very close to that found in the industrial emissions from northern Europe and Siberia and quite different from that in such emissions from northern North America. Concentrations decrease rapidly with depth in the ice layers, indicating that these trace elements began to enter the atmosphere within the past few centuries. It is now generally conceded that most of the haze particles are derived from human activities, primarily—though not exclusively—in northern Eurasia.

Since the haze scatters light, including sunlight, it decreases the **solar energy** received at the ground level in polar regions and may, therefore, have the potential to decrease arctic temperatures. Arctic haze also constitutes a nuisance because it decreases visibility. The trace elements found in arctic haze apparently are not yet sufficiently concentrated in either atmosphere or precipitation to constitute a significant toxic hazard.

[*James P. Lodge Jr.*]

RESOURCES
PERIODICALS

Nriagu, J. O., et al. "Origin of Sulfur in Canadian Arctic Haze From Isotope Measurements." *Nature* 349 (10 January 1991): 142–5.
Soroos, M. S. "The Odyssey of Arctic Haze: Toward a Global Atmospheric Regime." *Environment* 36 (December 1992): 6—11+.

Arctic National Wildlife Refuge

Beyond the jagged Brooks Range in Alaska's far northeastern corner lies one of the world's largest **nature** preserves, the 19.8-million acre (8-million ha) Arctic **National Wildlife Refuge** (ANWR). A narrow strip of treeless coastal plain in the heart of the refuge presents one of nature's grandest spectacles, as well as one of the longest-running environmental battles of the past century. For a few months during the brief arctic summer, the **tundra** teems with **wildlife**. This is the calving ground of the 130,000 caribou of the Porcupine herd, which travels up to 400 mi (640 km) each summer to graze and give birth along the Arctic Ocean shore. It also is important **habitat** for tens of thousands of snow geese, tundra swans, shorebirds and other migratory waterfowl; a denning area for polar bears, arctic foxes and arctic **wolves**; and a year-round home to about 350 shaggy musk ox. In wildlife density and diversity, it rivals Africa's Serengeti.

When Congress established the **Wildlife Refuge** in 1980, a special exemption was made for about 600,000 ha (1.5 million acres) of coastline between the mountains and the Beaufort Sea where geologists think sedimentary strata

may contain billions of barrels of oil and trillions of cubic feet of **natural gas**. Called the 1002 area for the legislative provision that put it inside the wildlife refuge but reserved the right to drill for **fossil fuels**, this narrow strip of tundra may be the last big, on-shore, liquid **petroleum** field in North America. It also is **critical habitat** for one of the richest biological communities in the world. The possibility of extracting the fossil fuels without driving away the wildlife and polluting the pristine landscape is supported by oil industry experts and disputed by biologists and environmentalists.

The amount of oil and gas beneath the tundra is uncertain. Only one seismic survey has been done and a few test **wells** drilled. Industry geologists claim that there may be 16 billion barrels of oil under ANWR, but guessing the size and content of the formation from this limited evidence is as much an art as a science. Furthermore, the amount of oil it's economical to recover depends on market prices and shipping costs. At current wholesale prices of $25 per barrel, the U.S. **Geological Survey** estimates that 7 billion barrels might be pumped profitably. If prices drop below $10 per barrel, as they did in the early 1990s, the economic resource might be only a few hundred million barrels.

Energy companies are extremely interested in ANWR because any oil found there could be pumped out through the existing **Trans-Alaska pipeline**, thus extending the life of their multibillion-dollar investment. The state of Alaska hopes that revenues from ANWR will replenish dwindling coffers as the oil supply from near-by Prudhoe Bay wells dry up. **Automobile** companies, tire manufacturers, filling stations, and others who depend on our continued use of petroleum argue that domestic oil supplies are rapidly being depleted, leaving the United States dependent on foreign countries for more than half of its energy consumption.

Oil drilling proponents point out that prospecting will occur only in winter when the ground is covered with snow and most wildlife is absent or hibernating. Once oil is located, they claim, it will take only four or five small drilling areas, each occupying no more than a few hundred hectares, to extract it. Heavy equipment would be hauled to the sites during the winter on ice roads built with water pumped from nearby lakes and rivers. Each two-meter-thick, gravel drilling pad would hold up to 50 closely spaced wells, which would penetrate the **permafrost** and then spread out horizontally to reach pockets of oil up to 6 mi (10 km) away from the wellhead. A central processing facility would strip water and gas from the oil, which would then be pumped through elevated, insulated pipelines to join oil flowing from Prudhoe Bay.

Opponents of this project argue that the noise, **pollution**, and construction activity accompanying this massive operation will drive away wildlife and leave scars on the landscape that could last for centuries. Pumping the millions

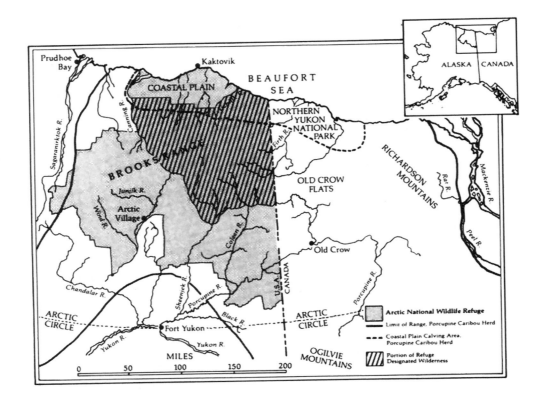

A map of the Arctic National Wildlife Refuge (ANWR) in Alaska. (Map by World Management. Reproduced by permission.)

of gallons of water needed to build ice roads could dry up local ponds on which wildlife depends for summer habitat. Every day six to eight aircraft—some as big as the C-130 Hercules—would fly into ANWR. The smell of up to 700 workers and the noise of numerous trucks and enormous power-generating turbines, each as large and loud as a jumbo aircraft engine, would waft out over the tundra. Pointing to the problems of other Arctic oil drilling operations where drilling crews dumped **garbage**, sewage, and toxic drilling waste into surface pits, environmentalists predict disaster if drilling is allowed in the refuge. Pipeline and drilling spills at Prudhoe Bay have contaminated the tundra and seeped into waterways. And scars from bulldozer tracks made 50 years ago can still be seen clearly today. John Adams, President of the **Natural Resources Defense Council** claims that the once pristine wildlife habitat of Prudhoe Bay has become a toxic, industrial wasteland.

Oil companies planning to drill in ANWR, on the other hand, claim that old, careless ways are no longer permitted in their operations. Wastes are collected and either burned or injected into deep wells. Although some animals do seem to have been displaced at Prudhoe Bay, the Central

Arctic herd with 27,000 caribou is five times larger now than it was in 1978 when drilling operations began there. But in ANWR the Porcupine herd has five times as many animals crowded into one-fifth the area and may be much more sensitive to disturbance than their cousins to the west.

Native people are divided on this topic. The coastal Inupiat people, many of whom work in the oil fields, support opening the refuge to oil exploration. They hope to use their share of the oil revenues to build new schools and better housing. The Gwich'in people, who live south of the refuge, would gain nothing from oil exploitation. They worry that migrating caribou on which they depend might be decline as a result of drilling on critical calving grounds.

Even if ANWR contains seven billion barrels of oil, it will take at least a decade to begin to get it to market and the peak production rate will probably be about one million barrels of oil per day in 2030. Flow from the 1002 area would then meet less than 4% of the U.S. daily oil consumption. Improving the average fuel efficiency of all cars and light trucks in America by just one mile per gallon would save more oil than is ever likely to be recovered from ANWR, and it would do so far faster and cheaper than extracting and transporting crude oil from the arctic. Cutting our fossil fuel consumption also is vital if we are to avoid catastrophic global **climate** change.

In 1995, Congress passed a budget that included a provision to allow drilling in ANWR, but President Bill Clinton vetoed the bill. In 2002, the Republican-controlled House of Representatives once again passed an energy bill that authorized oil and gas exploration in ANWR. President Bush strongly supported this bill and promised to sign it if given an opportunity. The Senate energy bill, however, rejected ANWR drilling by a vote of 54 to 46, and the measure was abandoned by its sponsors.

Undoubtedly, this debate won't disappear anytime soon. With a potential of billions of dollars to be made from oil and gas formations, the industry won't give up easily. For many conservatives, this issue has become a matter of principle. They believe we have both a right and a duty to exploit the resources available to us. Environmental groups feel equally strongly about the value of wildlife, **wilderness**, and one of the "last remaining great places" in the world. Protecting this harsh but beautiful land garnered more donations and public passions than any other environmental issue in the past decade. Clearly, how we decide to manage ANWR will be a landmark in **environmental history**.

[*William P. Cunningham*]

RESOURCES
BOOKS

Lentfer, Hank and Carolyn Servid, eds. *Arctic Refuge: A Circle of Testamony*. Minneapolis, MN: Milkweed Press, 2001.

PERIODICALS

Gibbs, W. Wayt. "The Arctic Oil & Wildlife Refuge." *Scientific American* 284(2001): 62-69.
Miller, Debbie S. "Ground Zero" *The Amicus Journal* 23(2001): 29-34.
McGrath, Susan. "The last great wilderness." *Audubon* 103(2001): 52-65.
Rauber, Paul. "Snake Oil for Fossil Fools." *Sierra* 88(2001): 56-61,86-87.

Arid

Arid lands are dry areas or deserts where a shortage of rainfall prevents permanent rain-fed agriculture. They are bordered by, and interact with, marginal or semi-arid lands where the annual rainfall (still only 10–15 in; 25–38 cm) allows limited agriculture and light grazing. However, in many parts of the world human mismanagement, driven by increasing populations, has degraded these areas into unusable arid lands. For example, clearing natural vegetation and **overgrazing** have led to **soil erosion**, reduced land productivity, and ultimately reduced water availability in the region. This degradation

of semi-arid lands to deserts, which can also occur naturally, is known as **desertification**.

Arid landscaping

Arid landscaping, or xeriscaping (from the Greek word *xeros*, meaning dry), is the integration of practicality and beauty in drought-prone public and private gardens. Xeriscaping is part of a larger trend among environmentalist gardeners to incorporate native rather than imported **species** within local ecosystems.

In drought-prone areas like California, where **water conservation** is imperative and lawns and flower gardens are at risk, gardeners have taken several steps to cope with **drought**. A 20-by-40-ft (6-by-12 m) green lawn requires about 2,500 gal (9,475 l) of water a month, enough for a four-member family for 10 days. Some arid landscapers eliminate lawns altogether and replace them with flagstone or concrete walkways; others aerate the **soil** or cut the grass higher for greater water retention. More popular is replacing imported grasses with native grasses, already adapted to the local **climate**, and letting them grow without being cut.

Some big trees, such as oak, do better in drought than others, such as birches or magnolias. To save big trees that are expensive and important providers of shade, some gardeners place "soaker" hoses, in which water flows out through holes in the length of the hose in concentric circles at the base of trees and water them deeply once or twice a season.

More commonly, arid landscapers replace water-loving shrubs and trees with those best adapted to drought conditions. These include some herbs, such as rosemary, thyme, and lavender, and shrubs like the silvery santolina, which are important in color composition of gardens.

Other plants that grow well in semi-arid and arid conditions are legumes (Leguminosae), poppies (Papaveraceae), daisies (Compositae), black-eyed susans (*Rudbeckia fulgida*), goldenrod (*Solidago*), honeysuckle (*Lonicera*), sunflowers (*Helianthus*), daylilies (*Hemerocallis*), and eucalyptus. Generally, drought-resistant plants have silvery leaves that reflect sunlight (*argentea*); hairy (*tomentosum*) or stiff haired (*hirsuta*) leaves that help retain moisture; long narrow leaves (*angustifolia*), threadlike leaves (*filimentosa*), and aromatic leaves (*aromatica*) that provide a moisture-protecting **haze** in heat.

Drip irrigation, hoses with small regular holes, whose amount is regulated by timers, is the most efficient xeriscape watering technique. Mulching with dead leaves, pine bark, straw, and other organic matter helps retain soil moisture. Soil types are a critical part of arid landscaping: one inch (2.5 cm) of water, for example, will penetrate 6–8 in (15–

20 cm) into loam (a sand, peat, and clay mixture), but only 4–5 in (10–13 cm) into dense clay.

Arid landscapers monitor plant health and soil dryness only when necessary, often in the early morning or in the evening to avoid evaporation by the sun. **Recycling** "gray" water, **runoff** from household sinks and washing machines, is best if the water is used quickly before bacteria collects and if laundry **detergents** are **biodegradable**. Arid landscapers, using native and drought-resistant plant species in conjunction with stone and concrete walkways and **mulch**, create gardens that are not only aesthetically pleasing but ecologically healthy.

[*Stephanie Ocko*]

RESOURCES
BOOKS

Ball, K. *Xeriscape Programs for Water Utilities.* Denver: American Water Works Association, 1990.

PERIODICALS
Ball, K., and G. O. Robinette. "The Water-Saving Garden Landscape." *Country Journal* (September/October 1990): 62–69.

Army Corps of Engineers

The United States Army Corps of Engineers, headquartered in Washington, D.C., is the world's largest engineering organization. The Corps was founded on June 16, 1775, the eve of the Battle of Bunker Hill, as the Continental Army fought cannon bombardments from British ships.

From early in the history of the Corps, the organization has handled both military and civil engineering needs of the United States. In earlier times those needs included coastal fortifications and lighthouses, surveying and exploring the frontier, construction of public buildings, snagging and clearing river channels, and operating early national parks such as Yellowstone. Under Corps' direction, the Panama Canal and **St. Lawrence Seaway** were built, and the Corps administered the Manhattan Project which led to the development of the atomic bomb during World War II.

Today, the Corps of Engineers provides engineering and related services in four areas: water and natural resource management (civil works), military construction and support, engineering research and development, and support to other government agencies.

In its military role, the Corps of Engineers provides support on the battlefield, enhancing movement and operations of American forces while impeding or delaying enemy actions. Corps engineers also plan, design, and supervise construction of military facilities and operate and maintain Army installations worldwide.

In its civil role, the Corps is responsible for the development of the nation's **water resources** and the operation and maintenance of completed water resource projects. Emerging engineering needs include renewal of infrastructure, management and control of **wetlands** development and **ocean dumping**, **waste management**, including hazardous and toxic waste, **solid waste**, and nuclear waste, and disaster response and preparedness.

The Corps undertook the task of "unstraightening" the Kissimmee River in south Florida. The river had been straightened into a canal for flood control between 1961 and 1971, losing half its length and most of its marshes. In the first reversal of a Corps' project, the river is slated to regain its curves. The restoration is due to be completed in 2005, but already there have been an increase in the **flora** and **fauna** around the developed areas.

The Corps employees more than 48,000 military and civilian members, has 13 regional headquarters, 39 district offices, and four major laboratories and research centers throughout the country.

[*Linda Rehkopf*]

RESOURCES
PERIODICALS

Duplaix, N. "Paying the Price." *National Geographic* 178 (July 1990): 96.
Historical Highlights of the United States Army Engineers. Publication EP 360-1-13. Washington, DC: U. S. Government Printing Office, March, 1978.

Svante August Arrhenius (1859 – 1927)
Swedish chemist and physicist

Young Svante gave evidence of his intellectual brilliance at an early age. He taught himself to read by the age of three and learned to do arithmetic by watching his father keep books for the estate of which he was in charge. Arrhenius began school at the age of eight, when he entered the fifth-grade class at the Cathedral School in Uppsala. After graduating in 1876, Arrhenius enrolled at the University of Uppsala.

At Uppsala, Arrhenius concentrated on mathematics, chemistry, and physics and passed the candidate's examination for the bachelor's degree in 1878. He then began a graduate program in physics at Uppsala, but left after three years of study. He was said to be dissatisfied with his physics advisor, Tobias Thalén, and felt no more enthusiasm for the only advisor available in chemistry, Per Theodor Cleve. As a result, he obtained permission to do his doctoral research in absentia with the physicist Eric Edlund at the Physical Institute of the Swedish Academy of Sciences in Stockholm.

The topic Arrhenius selected for his dissertation was the electrical conductivity of solutions. In 1884 Arrhenius submitted his thesis on this topic. In the thesis he hypothesized that when salts are added to water, they break apart into charged particles now known as ions. What was then thought of as a molecule of sodium chloride, for example, would dissociate into a charged sodium atom (a sodium **ion**) and a charged **chlorine** atom (a chloride ion). The doctoral committee that heard Arrhenius's presentation in Uppsala was totally unimpressed by his ideas. Among the objections raised was the question of how electrically charged particles could exist in water. In the end, the committee granted Arrhenius his Ph.D., but with a score so low that he did not qualify for a university teaching position.

Convinced that he was correct, Arrhenius had his thesis printed and sent it to a number of physical chemists on the continent, including Rudolf Clausius, Jacobus van't Hoff, and Wilhelm Ostwald. These men formed the nucleus of a group of researchers working on problems that overlapped chemistry and physics, developing a new discipline that would ultimately be known as physical chemistry. From this group, Arrhenius received a much more encouraging response than he had received from his doctoral committee. In fact, Ostwald came to Uppsala in August 1884 to meet Arrhenius and to offer him a job at Ostwald's Polytechnikum in Riga. Arrhenius was flattered by the offer and made plans to leave for Riga, but eventually declined for two reasons. First, his father was gravely ill (he died in 1885), and second, the University of Uppsala decided at the last moment to offer him a lectureship in physical chemistry.

Arrhenius remained at Uppsala only briefly, however, as he was offered a travel grant from the Swedish Academy of Sciences in 1886. The grant allowed him to spend the next two years visiting major scientific laboratories in Europe, working with Ostwald in Riga, Friedrich Kohlrausch in Würzburg, Ludwig Boltzmann in Graz, and van't Hoff in Amsterdam. After his return to Sweden, Arrhenius rejected an offer from the University of Giessen, Germany, in 1891 in order to take a teaching job at the Technical University in Stockholm. Four years later he was promoted to professor of physics there. In 1903, during his tenure at the Technical University, Arrhenius was awarded the Nobel Prize in chemistry for his work on the dissociation of electrolytes.

Arrhenius remained at the Technical University until 1905 when, declining an offer from the University of Berlin, he became director of the physical chemistry division of the Nobel Institute of the Swedish Academy of Sciences in Stockholm. He continued his association with the Nobel Institute until his death in Stockholm on October 2, 1927.

Although he was always be remembered best for his work on dissociation, Arrhenius was a man of diverse interests. In the first decade of the twentieth century, for example, he became especially interested in the application of physical and chemical laws to biological phenomena. In 1908 Arrhenius published a book entitled *Worlds in the Making* in which he theorized about the transmission of life forms from planet to planet in the universe by means of spores.

Arrhenius's name has also surfaced in recent years because of the work he did in the late 1890s on the **greenhouse effect**. He theorized that **carbon dioxide** in the **atmosphere** has the ability to trap heat radiated from the earth's surface, causing a warming of the atmosphere. Changes over time in the concentration of **carbon** dioxide in the atmosphere would then, he suggested, explain major climatic variations such as the glacial periods. In its broadest outlines, the Arrhenius theory sounds similar to current speculations about **climate** changes resulting from global warming.

Among the honors accorded Arrhenius in addition to the Nobel Prize were the Davy Medal of the Royal Society (1902), the first Willard Gibbs Medal of the Chicago section of the American Chemical Society (1911), and the Faraday Medal of the British Chemical Society (1914).

RESOURCES
BOOKS

Arrhenius, Svante. *Chemistry in Modern Life.* Van Nostrand, 1925.
———. *Theories of Solutions.* Yale University Press, 1912.
———. *Worlds in the Making: The Evolution of the Universe.* Harper, 1908.
Fleck, George. "Svante Arrhenius." In *Nobel Laureates in Chemistry: 1901–1992*, edited by Laylin K. James. American Chemical Society and the Chemical Heritage Foundation, 1993.

Arsenic

Arsenic is an element having an atomic number of 33 and an atomic weight of 74.9216 that is listed by the U.S. **Environmental Protection Agency** (EPA) as a hazardous substance (**Hazardous waste** numbers P010, P012) and as a **carcinogen**. *The Merck Index* states that the symptoms of acute **poisoning** following arsenic ingestion are irritation of the gastrointestinal tract, nausea, vomiting, and diarrhea that can progress to shock and death. According to the 2001 Update Board on Environmental Studies and Toxicology, such toxic presentations generally require weeks to months of exposure to arsenic at high doses, as much as 0.04mg/kg/day (0.02mg/lb/day). Furthermore, long-term, or chronic poisoning can result in skin thickening, exfoliation, and hyperpigmentation. Continuing exposure also has been associated with development of herpes, peripheral neurological manifestations, and degeneration of the liver and kidneys. Of primary importance, however, is the association of chronic arsenic exposure with increased risk of developing high blood pressure, cardiovascular disease, and skin **cancer**, diseases that are increasing public health concerns.

Variations of arsenic were used in everyday life. In the early nineteenth century, it was discovered that a fungus, *Scopulariopsis brevicaulis*, was eating away at the starches found in certain forms of wallpaper. The fungus also altered the arsenate dyes, changing them into trimethylarsine oxide, which is further converted to the extremely toxic trimethylarsine gas. The gas was then released into the room, killing the people who spent long amounts of time there, usually sleeping. This process was discovered in 1897 but the gas itself was not discovered and named until 1945.

Safe drinking water standards are regulated by the EPA in the United States. In February 2002, the EPA revised the former standard for acceptable levels of arsenic in drinking water. The revision of the **Safe Drinking Water Act** reduced the standard from 50 **parts per billion** (ppb) to 10 ppb of arsenic as the acceptable level for drinking water. The regulation requires that all states must comply with the new standard by January 2006. Sources of arsenic contamination of drinking water include **erosion** of natural deposits and run-off of waste from glassmaking and electronics industries. Also, arsenic is used in insecticides and rodenticides, although much less widely than it once was due to new information regarding its toxicity.

Arsenic-treated lumber

Arsenic-treated wood is wood that has been pressure-treated with a **pesticide** containing inorganic **arsenic** (i.e., the arsenic compound does not contain **carbon**) to protect it from dry rot, **fungi**, molds, termites, and other pests. The arsenic can be a part of a CCA (chromated **copper** arsenate) chemical mixture consisting of three pesticidal compounds, copper, chromate, and arsenic; the most commonly used type of CCA contains 34% arsenic as arsenic pentoxide. Less commonly used wood preservatives containing arsenic include the pesticide ACA (ammoniacal copper arsenate), which contains ammonium, copper, and arsenic, and the pesticide ACZA (ammoniacal copper zinc arsenate), which contains ammonia, copper, zinc, and arsenic. In 1996, the United States wood product industry used 30 million pounds of arsenic, or half of all the arsenic produced worldwide.

Inorganic arsenic in CCA has been used since the 1940s. CAA is injected into wood through a process that uses high pressure to saturate wood products with the **chemicals**. Preserved wood products, such as utility poles, highway noise barriers, sign posts, retaining walls, boat bulkheads, dock pilings, and wood decking, are used in the construction, railroad, and utilities industries. Historically CCA has been the principal chemical used to treat wood for outdoor uses

around a home. Residential uses of arsenic-treated woods include play structures, decks, picnic tables, gazebos, landscaping timbers, residential fencing, patios, and walkways/boardwalks. After wood is pressure-treated with arsenic compounds, residues of the preservatives can remain on the surface. The initial residues wash off, but as the wood weathers, new layers of treated wood and pesticides are exposed. Arsenic is also present in paints that are used to cover the cut ends of treated wood. Freshly arsenic-treated wood, if not coated, has a greenish tint, which fades over time.

Arsenic is acutely toxic. Contact with arsenic may cause irritation of the stomach, intestines, eyes, nose, and skin, blood vessel damage, and reduced nerve function. In addition, according to the **National Academy of Sciences** and the **National Research Council**, exposure to arsenic increases the risk of human lung, bladder, and skin **cancer** over a lifetime and is suspected as a cause of kidney, prostate, and nasal passage cancer. The National Academy of Sciences and the Science Advisory Board of the United States **Environmental Protection Agency** has also reported that arsenic may cause high blood pressure, cardiovascular disease, and diabetes.

Arsenic may enter the body through the skin, by ingestion, or by inhalation. Ingestion occurs most frequently when contaminated hands are put in the mouth, or when contaminated hands are used for eating food. Repeated exposure will increase risks of adverse health effects. Splinters of wood piercing the skin may also be a means of entry of arsenic into the body, but the importance of this route has not been well-studied.

The use of most pesticides containing arsenic had been banned by the United States Environmental Protection Agency, but in 1985 CCA was designated as a restricted-use pesticide. However, CCA-treated wood products were not regulated like the pesticides the wood products contained because it was assumed that the pesticides would stay in the wood. Unfortunately adequate information was not available on whether arsenic is fixed in the wood permanently and whether the wood product is safe. It is known that fixation of the chemicals in the wood matrix is enhanced if the treated wood is wrapped in tarps and stored for a sufficient length of time, which varies with the temperature. For example, to achieve fixation, the wood must be stored for 36 days in 50° F (21° C) weather and for 12 days in 70° F (10° C) weather. At 32° F., no fixation occurs.

However, research by the Florida Center for Solid and Hazardous **Waste Management** and the Connecticut Agricultural Experiment Station suggests that **leaching** of arsenic into the soils or into surface water under CCA-treated structures occurs at greater than safe levels. Florida studies showed that arsenic was found in soils underneath the eight pressure treated decks that were investigated. Of

73 samples taken, 61 samples had levels of arsenic higher than the Florida clean up levels for industrial sites. One sample had arsenic levels 300 times higher than the state's mandated clean up level.

Based on the potential for adverse health effects, the United States Protection Agency announced in February 2002 that the use of consumer products made with CCA, including play structures, decks, picnic tables, landscaping timbers, patios, walkways, and board walks, will be phased out voluntarily by the wood treating industry by December 31, 2003. Wood treated prior to December 31, 2003, can still be used, and already-built structures using CCA-treated wood and the **soil** surrounding the structures will not have to be replaced or removed. As of August 2001, Switzerland, Vietnam, and Indonesia had already banned the use of CCA-treated wood, while Germany, Sweden, Denmark, Japan, **Australia**, and New Zealand restricted its use.

The United States Environmental Protection Agency has issued cautionary recommendations to reduce consumer exposure to arsenic-contaminated wood. Treated wood should not be used where the preservative can become a component of water, food or animal feed. Treated wood should not be used where it may come into direct or indirect contact with drinking water, except for uses where there is incidental contact, such as docks and bridges. Treated wood should not be used for cutting boards or counter tops, and food should not be placed directly on treated wood. Children and others should wash their hands after playing or working outdoors, as arsenic may be swallowed from hand-to-mouth activity. Children and pets should be kept out from under-deck areas. Edible plants should not be grown near treated decks or other structures contructed of treated wood. A plastic liner should be placed on the inside of arsenic-treated boards used to frame garden beds. Sawdust from treated wood should not be used as animal bedding, and treated boards should not be used to construct structures for storage of animal feed or human food. Treated wood should not be used in the construction of the parts of beehives that may come into contact with honey.

Only wood that is visibly clean and free of surface residues (e.g., wood that does not show signs of crystallization or resin on its surface) should be used for patios, decks, and walkways. Consumers working with arsenic-treated wood should reduce their exposure by only sawing, sanding, and machining treated wood outdoors and by wearing a dust mask, goggles, and gloves when performing these types of activities. They should also wash all exposed areas of their bodies thoroughly with soap and water before eating, toileting, drinking, or using **tobacco** products. Work clothes should be washed separately from other household clothing before being worn again. Sawdust, scraps, and other construction debris from arsenic-treated wood should not be composted or used as **mulch**, but caught on tarps for disposal off-site. Pressure-treated wood has an exemption from **hazardous waste** regulations and can be disposed of in municipal landfills without a permit. Pressure-treated wood should not be burned in open fires or in stoves, fireplaces, or residential boilers, as both the **smoke** and the ash may contain toxic chemicals. Treated wood from commercial or industrial uses, such as from construction sites, may only be burned in commercial or industrial incinerators or boilers in accordance with state and federal regulations.

The United States Consumer Product Safety Commission has recommended that playground equipment be painted or sealed with a double coat of a penetrating, non-toxic and non-slippery sealant (e.g., oil-based or semi-transparent stains) every one or two years, depending upon wear and **weathering**. However, available data are limited on whether sealants will reduce the **migration** of wood preservative chemicals from CCA-treated wood. Other potential treatments include the use of polyurethane or other hard lacquer, spar varnish, or pain. However, the use of film-forming or non-penetrating stains are not recommended, as subsequent flaking and peeling may later increase exposure to preservatives in the wood. Structures constructed of arsenic-treated wood should be inspected regularly for wood decay and/or structural weakness. If the treated wood cracks and exposes the interior of the wood is still structurally sound, the affected area should be covered with a double coat of a sealant.

Alternatives to the use of arsenic-treated lumber include the use of painted metal, stones or brick, recycled plastic lumber, which may be all plastic or a composite of plastic and wood fiber, lumber that has been treated with an alternative pesticide ACQ (alkaline copper quaternary), which does not contain arsenic, or untreated rot-resistant wood such as cedar and redwood.

Arsenic compounds are referred to as arsenicals.

[*Judith L. Sims*]

RESOURCES

BOOKS

Frankenberger, William T., Jr.*Environmental Chemistry of Arsenic.*???: Marcel Dekker, 2001.

OTHER

*Chromated Copper Arsenate (CCA) and Its Use as a Wood Preservative.*Office of Pesticide Programs, U.S. Environmental Protection Agency, Washington, DC, June 20, 2002. [cited June 25, 2002]. <http://www.epa.gov/pesticides/citizens/1file.htm>.

*Chromated Copper Arsenate (CCA) Wood: WMRC Library Reference Guide.*Waste Management & Research Center, Illinois Department of Natural Resources, Champaign, IL, May 29, 2002. [cited June 25, 2002]. <http://www.wmrc.uiuc.edu/library/libraryinfo/refguides/ccawood.htm>.

Artesian well

A well that discharges water held in a confined **aquifer**. Artesian **wells** are usually thought of as wells whose water is free flowing at the land surface. However, there are many other natural systems that can result in such wells. The classic concept of artesian flow involves a basin with a water-intake area above the level of **groundwater discharge**. These systems can include stabilized sand **dunes**; fractured zones along bedrock faults; horizontally layered rock formations; and the intermixing of **permeable** and impermeable materials along glacial margins.

Asbestos

Asbestos is a fibrous mineral silicate, occurring in numerous forms, of which amosite [$Fe_5Mg_2(Si_8O_{22})(OH)_2$] has been shown to cause mesothelioma, squamous cell carcinoma and adenocarcinoma of the lung after long exposure times. This substance has been listed by the **Environmental Protection Agency** (EPA). Lung **cancer** is most likely to occur in those individuals who are exposed to high air-borne doses of asbestos and who also **smoke**. The pathogenic potential of asbestos appears to be related to its aspect ratio (length-to-diameter ratio), size (particles less than 2 micrometers in length are the most hazardous), and to its surface reactivity.

Asbestos exposure causes thickening of and calcified plaques on the lining of the chest cavity. When inhaled it forms "asbestos bodies" in the lungs, yellowish-brown particles created by reactions between the fibers and lung tissue. This disease was first described by W. E. Cooke in 1921 and given the name **asbestosis**. The **latency** period is generally longer than 20 years—the heavier the exposure, the more likely and the earlier is the onset of the disease. In 1935 an association between asbestos and cancer was noted by Kenneth M. Lynch. However, it was not until 1960 that Christopher Wagner demonstrated a particularly lethal association between cancer of the lining of the lungs and asbestos. By 1973 the **National Institute of Occupational Safety and Health** recommended adoption of an occupational standard of two asbestos fibers per cubic centimeter of air.

During this time, many cases of lung cancer began to surface, especially among asbestos workers who had been employed in shipbuilding during World War II. The company most impacted by lawsuits was the Manville Corporation which had been the supplier to the United States government. Manville Corporation eventually sought Title 11 Federal Bankruptcy protection as a result of these law suits.

More recently, the Reserve Mining Company, a taconite (iron ore) mining operation in **Silver Bay**, Minnesota, was involved in litigation over the dumping of **tailings** (wastes) from their operations into Lake Superior. These tailings contained amositic asbestos particles which appeared to migrate into the Duluth, Minnesota water supply. In an extended law suit, Reserve Mining Company was ordered to shut down their operations. One controversial question raised during the legal action was whether cancer could be caused by from drinking water containing asbestos fibers. In other cancer cases related to asbestos, the asbestos was inhaled rather than ingested. Federal courts held that there is reasonable cause to believe that asbestos in food and drink is dangerous—even in small quantities—and ordered Reserve Mining to stop dumping tailings in the lake.

A significant industry has developed for removing asbestos materials from private and public buildings as a result of the tight standards placed on asbestos concentrations by the **Occupational Safety and Health Administration**. At one time, many steel construction materials, especially horizontal beams, were sprayed with asbestos to enhance their **resistance** to fires. Wherever these materials are now exposed to **ambient air** in buildings, they have the potential to create a hazardous condition. The asbestos must either be covered or removed, and another insulating material substituted as great cost. Removal, however, causes its own problems, releasing high concentrations of fibers into the air. Many experts regard covering asbestos in place with a plastic covering to be the best option in most cases. What was once considered a life-saving material for its flame retardancy, now has become a hazardous substance which must be removed and sequestered in sites specially certified for holding asbestos building materials.

[*Malcolm T. Hepworth*]

RESOURCES
BOOKS

Bartlett, R. V. *The Reserve Mining Controversy: Science, Technology, and Environmental Quality*. Bloomington: Indiana University Press, 1980.
Brodeur, P. *Asbestos and Enzymes*. New York: Ballantine Books, 1972.

Asbestos removal

Asbestos is a naturally occurring mineral, used by humans since ancient times but not extensively until the 1940s. After World War II and for the next 30 years, it was widely used as a construction material in schools and other public buildings. The United States **Environmental Protection Agency** (EPA) estimates that there are asbestos-containing materials in most of the primary and secondary schools as well as in most public and commercial buildings in the nation. It is estimated that 27 million Americans had significant occupational exposure to asbestos between 1940 and 1980. Asbestos has been popular because it is readily available, low in cost, and has very useful properties. It does not burn, conducts heat and electricity poorly, strengthens concrete products into which it is incorporated, and is resistant to chemical corrosion. It has been used in building materials as a thermal and electrical insulator and has been sprayed on steel beams in buildings for protection from heat in fires. Asbestos has also been used as an acoustical plaster. In 1984, an EPA survey found that approximately 66% of those buildings containing asbestos had damaged asbestos-containing materials in them. The EPA distinguishes between two types of asbestos damage. If the material, when dry, can be crumbled by hand pressure, it is called "friable." If not, it is called "non-friable." The friable form is much more likely to release dangerous fibers into the air. Fluffy, spray-applied asbestos fireproofing material is an example of friable asbestos. Vinyl-asbestos floor tile is an example of a non-friable material, although it too can release fibers when sanded, sawed, or aggressively disturbed.

Growing recognition that asbestos dust is a dangerous air pollutant has led to government regulation of its use and provisions for removal of asbestos and asbestos-containing products from dwellings, schools, and the workplace. In the United States, the Federal Government has taken steps to prevent unnecessary exposure to asbestos since the 1970s. Six different agencies have authority to regulate some aspect of asbestos use in the United States. The authorized agencies are the EPA, **Occupational Safety and Health Administration** (OSHA), **Food and Drug Administration**, Consumer Product Safety Commission, Department of **Transportation**, and the Mine Safety and Health Administration.

The EPA has authority to oversee use of asbestos in commerce and has developed standards controlling its handling and use. It administers the Asbestos Hazard Emergency Response Act (AHERA) of 1986 that regulates asbestos in schools. The Act prescribes management practices and abatement standards for both public and private schools, and it follows a 1979 rule that launched a technical assistance program to aid schools in efforts to identify asbestos-containing materials. AHERA requires schools to develop a management plan concerning their asbestos-containing materials. Schools must appoint an asbestos manager to be responsible for a number of activities, which include implementing a plan to manage asbestos-containing building materials and ensuring compliance with federal asbestos regulations. AHERA also requires local officials to carry out inspections to identify asbestos-containing materials in public and private elementary and secondary schools, and to comply with AHERA's record keeping requirements. The emphasis on asbestos abatement in schools arises from the recognition that children are especially vulnerable to a viru-

lent form of lung **cancer** called mesothelioma caused by asbestos.

The EPA also administers the Asbestos School Hazard Abatement Reauthorization Act (ASHARA) of 1990 which requires that accredited personnel be used to work with asbestos in schools and in public and commercial buildings. The Act sets training requirements for asbestos professionals, contractors, and abatement workers. It is estimated that 733,000 public buildings in the United States (about 20% of the total) have some type of defective or deteriorating asbestos-containing material in them. Asbestos is frequently found in pipe and boiler insulation, caulking putties, joint compounds, linoleum, acoustical plaster, ceiling tiles, and other building components. ASHARA does not require owners of public and commercial buildings to conduct inspections for asbestos-containing materials, but it does require owners who opt for inspections to use accredited inspectors. To obtain accreditation, asbestos workers must take a 4-day, 32-hour EPA-approved training course covering topics such as potential health effects of asbestos exposure, the use of personal protective equipment, and up-to-date work practices. Detached single-family homes and residential apartment buildings of fewer than 10 units are not covered by the Act.

The EPA is also responsible for the 1989 Asbestos Ban and Phase out Rule that was partially overturned by a Federal Court in 1991. Originally, the ban applied to a wide variety of asbestos-containing products including roof coatings, floor tile, brake pads and linings, and pipeline wrap. Although these products and many others were removed from the ban, it still applies to all new uses of asbestos. The ban remains in effect for the sprayed-on asbestos fireproofing that has been prohibited since 1978.

Individual home owners may seek advice from local health officials concerning potential asbestos problems in or on their dwelling. State agencies often publish pamphlets or brochures that offer help in identifying and correcting problems. Common places to find asbestos in homes include wall, ceiling, and pipe insulation in structures built between 1930 and 1950, and in sheet vinyl, vinyl tile, and vinyl adhesive used in floor coverings. These materials are considered safe, and need not be removed unless they are damaged or disturbed. Wall and ceiling surfaces may contain troweled-on or sprayed-on surface material containing asbestos. If the material is hard, firmly attached, and does not produce a powder or dust when hand pressure is applied, it is probably not hazardous. On the exterior of many older homes, a cement asbestos board called Transite™ has been used as sheet or lap siding and has sometimes been shaped to mimic wood shingles. This material, too, is considered safe unless it has been damaged or disturbed. Other asbestos-containing objects in the home include older electrical lamp

Asbestos removal from a building. (Photograph by Eugene Smith. Black Star Publishing Company Inc. Reproduced by permission.)

socket collars, switch and receptacle boxes, fuse boxes, and old-fashioned "knob and tube" wiring. Asbestos is found in insulation blankets of older ovens, dishwashers, water heaters, and freezers. These items need not be removed if they are intact and not damaged.

There is no safe and reliable way for an untrained individual to identify asbestos except to have it analyzed by a competent testing laboratory. In taking a sample for laboratory analysis, it is important that one not release fibers into the air or get them into one's clothes. To avoid this, it is often helpful to spray the material with a fine mist of water before sampling. The material should not be disturbed more than necessary, and the sample for testing should be enclosed in a sealed plastic or glass vial. If asbestos-containing materials must be removed, they should not be placed with other household trash. Local health officials should be contacted for instructions on safe disposal.

Most public buildings with asbestos-containing materials do not pose an immediate risk to the occupants. As long as the asbestos is sealed and not releasing fibers, there is no health threat. Since asbestos is easily disturbed during

building alterations, routine maintenance, and even normal use, it is wise to take precautions to control or eliminate exposure. When asbestos-containing objects are identified, removal, encapsulation, and containment may be used to protect occupants. If the objects are damaged or in poor condition, they should be removed. Proper removal techniques must be strictly followed to avoid spreading contamination. Encapsulation is accomplished by sealing objects with an approved chemical sealant. Containment can be accomplished by constructing suitable physical barriers around the objects to contain any fibers that may be released. Air monitoring and inspection of buildings with asbestos-containing material should be performed regularly to be sure that fiber releases have not occurred. Maintenance and operations crews should be notified of an asbestos threat in their building, and they should be taught proper precautions to avoid accidental disturbance.

The Occupational Safety and Health Administration (OSHA), under the jurisdiction of the United States Department of Labor, is charged with protection of worker safety on the job. OSHA estimates that 1.3 million employees in construction and general industry face significant asbestos exposure in the workplace. Greatest exposure occurs in the construction industry, especially while removing asbestos during renovation or demolition. Workers are also apt to be exposed during the manufacture of asbestos products and during **automobile** brake and clutch repair work. OSHA has set standards for maximum exposure limits, protective clothing, exposure monitoring, hygiene facilities and practices, warning signs, labeling, record keeping, medical exams, and other aspects of the workplace. Workplace exposure for workers in general industry and construction must be limited to 0.2 asbestos fibers per 0.061 cu in. of air (0.2 f/cc), averaged over an eight-hour shift, or 1 fiber per 0.3 cu in. of air (1 f/5cc) over a standard 40-hour work week.

[*Douglas C. Pratt*]

RESOURCES
BOOKS

U.S. Environmental Protection Agency. *How to Manage Asbestos in School Buildings: AHERA Designated Person's Self Study Guide.* Washington, D.C., 1996.

U.S. Environmental Protection Agency. *Asbestos in the Home: A Homeowner's Guide.* Washington, D.C., 1992.

U.S. Environmental Protection Agency. *Managing Asbestos in Place: A Building Owners Guide to Operations and Maintenance Programs for Asbestos-Containing Materials.* Washington, D.C., 1990: 207-2003.

U.S. Department of Labor, Occupational Safety and Health Administration. *Better Protection Against Asbestos in the Workplace: Fact Sheet No. OSHA 92-06.* Washington, D.C.: GPO, 1992.

Yang, C. S., and F. W. Piecuch. "Asbestos: A Conscientious Approach." *Enviros* 01, no. 11 (November 1991).

Asbestosis

Asbestos is a fibrous, incombustible form of magnesium and calcium silicate used in making insulating materials. By the late 1970s, over 6 million tons of asbestos were being produced worldwide. About two-thirds of the asbestos used in the United States is used in building materials, brake linings, textiles, and insulation, while the remaining one-third is consumed in such diverse products as paints, **plastics**, caulking compounds, floor tiles, cement, roofing paper, radiator covers, theater curtains, fake fireplace ash, and many other materials.

It has been estimated that of the eight to eleven million current and retired workers exposed to large amounts of asbestos on the job, 30 to 40 percent can expect to die of **cancer**. Several different types of asbestos-related diseases are known, the most significant being asbestosis. Asbestosis is a chronic disease characterized by scarring of the lung tissue. Most commonly seen among workers who have been exposed to very high levels of asbestos dust, asbestosis is

an irreversible, progressively worsening disease. Immediate symptoms include shortness of breath after exertion, which results from decreased lung capacity. In most cases, extended exposure of 20 years or more must occur before symptoms become serious enough to be investigated. By this time the disease is too advanced for treatment. That is why asbestos could be referred to as a silent killer.

Exposure to asbestos not only affects factory workers working with asbestos, but individuals who live in areas surrounding asbestos emissions. In addition to asbestosis, exposure may result in a rare form of cancer called mesothelioma, which affects the lining of the lungs or stomach. Approximately 5–10% of all workers employed in asbestos manufacturing or mining operations die of mesothelioma.

Asbestosis is characterized by dyspnea (labored breathing) on exertion, a nonproductive cough, hypoxemia (insufficient oxygenation of the blood), and decreased lung volume. Progression of the disease may lead to respiratory failure and cardiac complications. Asbestos workers who **smoke** have a marked increase in the risk for developing bronchogenic cancer.

Increased risk is not confined to the individual alone, but there is an extended risk to workers' families, since asbestos dust is carried on clothes and in hair. Consequently, in the fall of 1986, President Reagan signed into law the Asbestos Hazard Emergency Response Act (AHERA), requiring that all primary and secondary schools be inspected for the presence of asbestos; if such materials are found, the school district must file, and carry out an asbestos abatement plan. The **Environmental Protection Agency** (EPA) was charged with the oversight of the project.

[*Brian R. Barthel*]

RESOURCES
BOOKS

Agency for Toxic Substances and Disease Registry. Annual Report. Atlanta, GA: U. S. Department of Health and Human Services, 1989 and 1990.

Nadakavukaren, A. *Man & Environment: A Health Perspective.* Prospect Heights, IL: Waveland, 1990.

PERIODICALS

Mossman, B. T., and J. B. L. Gee. "Asbestos-Related Diseases." *New England Journal of Medicine* 320 (29 June 1989): 1721–30.

Ash, fly

see **Fly ash**

Ashio, Japan

Much could be learned from the Ashio, Japan, mining and smelting operation concerning the effects of **copper poison-**

ing on human beings, rice paddy soils, and the **environment**, including the comparison of long term costs and short term profits.

Copper has been mined in Japan since A.D. 709, and **pollution** has been reported since the sixteenth century. Copper leached from the Ashio Mine pit and **tailings** flowed into the Watarase River killing fish and contaminating rice paddy soils in the mountains of central Honshu. The refining process also released large quantities of sulfur oxide and other waste gases, which killed the vegetation and life in the surrounding streams. In 1790 protests by local farmers forced the mine to close, but it became the property of the government and was reopened to increase the wealth of Japan after Emperor Meiji came to power in 1869. The mine passed into private ownership in 1877, and new technological innovations were introduced to increase the mining and smelting output. A year later, signs of copper pollution were already appearing. Rice yields decreased and people who bathed in the river developed painful sores, but production expanded.

A large vein of copper ore was discovered in 1884, and by 1885 the Ashio Copper mine produced 4,100 tons, about 40% of the total national output per year. **Arsenic** was a byproduct. The piles of slag mounted, and more waste **runoff** polluted the Watarase River and local farmlands. As the crops were damaged and the fish polluted, many people became ill. Consequently, a stream of complaints was heard, and some agreements were made to pay money, not for damages as such but just to "help out" the farmers. Meanwhile, mining and smelting continued as usual. In 1896 a **tailings pond** dam gave way and the deluge of **mine spoil waste** and water contaminated 59,280 acres (24,000 ha) of farm land in six prefectures from Ashio nearly to Tokyo 93 mi (150 km) away. Then the government ordered the Ashio Mining Company to construct facilities to prevent damage by pollutants, but in times of **flooding** these were largely ineffectual. In 1907 the government forced the inhabitants of the Yanaka Village, who had been the most affected by poisoning, to move to Hokkaido, making way for a flood control project.

In 1950, as a result of the Korean War, the Ashio Copper Mine expanded production and upgraded the smelting plant to compete with the high grade ores being processed from other mines. When the Gengorozawa slag pile, the smallest of 14, collapsed and introduced 2,614 cubic yd (2,000 cubic m) of slag into the Watarase River in 1958, it contaminated 14,820 acres (6,000 ha) of rice fields. No remedial action was taken, but in 1967 a maximum average yearly standard of 0.06 mg/l copper in the river water was set. This was meaningless because most of the contamination occurred when large quantities of slag were **leaching** out during the rainy periods and floods. Japanese authorities also

set 125 mg Cu/kg in paddy **soil** as the maximum allowable limit alleged not to damage rice yields, twice the minimum effect level of 56 mg Cu/kg.

In 1972 the government ordered that rice from this area be destroyed, even as the Ashio Mining Company still denied responsibility for its contamination. Testing showed that the soil of the Yanaka Village up to 10 ft (3 m) below the surface still contained 314 mg/kg of copper, 34 mg/kg of **lead**, 168 mg/kg of zinc, 46 mg/kg of arsenic, 0.7 mg/kg of **cadmium**, and 611 mg/kg of manganese. This land drains into the Watarase River, which now provides drinking water for the Tokyo metropolitan area and surrounding prefectures.

That same year the Ashio Mine announced that it was closing due to reduced demand for copper ore and worsening mining conditions; however, smelting continued with imported ores, so the slag piles still accumulated and minerals percolated to the river, especially during spring flooding. In August 1973, the Sabo dam collapsed and released about 2,000 tons of tailings into the river. Later that year the Law Concerning Compensation for Pollution Related Health Damage and Other Measures was passed, and it prompted the Environmental Agency's Pollution Adjustment Committee to begin reviewing the farmers' claims more seriously. For the first time, the company was required to admit being the source of pollution. The farmers' suit was litigated from March 1971 until May 1974, and the plaintiffs were awarded $5 million, much less than they asked for.

As major floods have been impossible to control, some efforts are being made to reforest the mountains. After they were washed bare of soil, the rocks fell and eroded, adding another hazard. So far, large expenditures have produced few results either in flood control or reforestation. The town of Ashio is now trying to attract tourism by billing the denuded mountains as the Japanese Grand Canyon, and the pollution continues.

[*Frank M. D'Itri*]

RESOURCES
BOOKS

Huddle, N., and M. Reich. *Island of Dreams: Environmental Crisis in Japan.* New York: Autumn Press, 1975.

Morishita, T. "The Watarase River Basin: Contamination of the Environment with Copper Discharged from Ashio Mine." In *Heavy Metal Pollution in Soils of Japan*, edited by K. Kitagishi and I. Yamane. Tokyo: Japan Scientific Societies Press. 1981.

Shoji, K., and M. Sugai. "The Ashio Copper Mine Pollution Case: The Origins of Environmental Destruction." In *Industrial Pollution in Japan.* Edited by J. Ui. Tokyo: United Nations University Press. 1992.

Asian longhorn beetle

The Asian longhorn beetle (*Anoplophora glabripennis*) is classified as a **pest** in the United States and their homeland of China. The beetles have the potential to destroy millions of hardwood trees, according to the United States Department of Agriculture (USDA).

Longhorn beetles live for one year. They are 1–1.5 in (2.5–3.8 cm) long, and their backs are black with white spots. The beetles' long antennae are black and white and extend up to 1 in (2.5 cm) beyond the length of their bodies.

Female beetles chew into tree bark and lay from 35 to 90 eggs. After hatching, larvae tunnel into the tree, staying close to the sapwood, and eat tree tissue throughout the fall and winter. After pupating, adults beetles leave the tree through drill-like holes. Beetles feed on tree leaves and young bark for two to three days and then mate. The cycle of infestation continues as more females lay eggs.

Beetle activity can kill trees such as maples, birch, horse chestnut, poplar, willow, elm, ash, and black locust. According to the USDA, beetles came to the United States in wooden packaging material and pallets from China. The first beetles were discovered in 1996 in Brooklyn, New York. Other infestations were found in other areas of the state. In 1998, infestations were discovered in Chicago.

Due to efforts that included quarantines and eradication, infestations have been confined to New York and Chicago. The spread of these pests is always a concern and certain steps are being taken to prevent this, such as the dunnage being heat-treated before leaving China, developing an effective pheromone, biological control agents, and cutting down and burning infected trees. The felled trees are then replaced with ones that are not known to host the beetle. In 1999, 5.5 million dollars was allotted to help finance the detection of infested trees.

[*Liz Swain*]

Asian (Pacific) shore crab

The Asian (Pacific) shore crab (*Hemigrapsus sanguineus*) is a crustacean also known as the Japanese shore crab or Pacific shore crab. It was probably brought from Asia to the United States in ballast water. When a ship's hold is empty, it is filled with ballast water to stabilize the vessel.

The first Asian shore crab was seen in 1988 in Cape May, N.J. By 2001, Pacific shore crabs colonized the East Coast, with populations located from New Hampshire to North Carolina. Crabs live in the sub-tidal zone where low-tide water is several feet deep.

The Asian crab is 2–3 in (5–7.7 cm) wide. Shell color is pink, green, brown, or purple. There are three spines on each side of the shell. The crab has two claws and bands of light and dark color on its six legs.

A female produces 56,000 eggs per clutch. Asian Pacific crabs haves three or four clutches per year. Other crabs produce one or two clutches annually.

At the start of the twenty-first century, there was concern about the possible relationship between the rapidly growing Asian crab population and the decline in native marine populations such as the lobster population in Long Island Sound.

[*Liz Swain*]

Asiatic black bear

The Asiatic black bear or moon bear (*Ursus thibetanus*) ranges through southern and eastern Asia, from Afghanistan and Pakistan through the Himalayas to Indochina, including most of China, Manchuria, Siberia, Korea, Japan, and Taiwan. The usual **habitat** of this bear is angiosperm forests, mixed hardwood-conifer forests, and brushy areas. It occurs in mountainous areas up to the tree-line, which can be as high as 13,000 ft (4,000 m) in parts of their Himalayan range.

The Asiatic black bear has an adult body length of 4.3–6.5 ft (1.3–2.0 m), a tail of 2.5–3.5 ft (75–105 cm), a height at the shoulder of 2.6–3.3 ft (80–100 cm), and a body weight of 110–440 lb (50–200 kg). Male animals are considerably larger than females. Their weight is greatest in late summer and autumn, when the animals are fat in preparation for winter. Their fur is most commonly black, with white patches on the chin and a crescent- or Y-shaped patch on their chest. The base color of some individuals is brownish rather than black.

Female Asiatic black bears, or sows, usually give birth to two small cubs 12–14 oz (350–400 g) in a winter den, although the litter can range from one to three. The gestation period is six to eight months. After leaving their birth den, the cubs follow their mother closely for about 2.5 years, after which they are able to live independently. Asiatic black bears become sexually mature at an age of three to four years, and they can live for as long as 33 years. These bears are highly arboreal, commonly resting in trees, and feeding on fruits by bending branches towards themselves as they sit in a secure place. They may also sleep during the day in dens beneath large logs, in rock crevices, or in other protected places. During the winter Asiatic black bears typically hibernate for several months, although in some parts of their range they only sleep deeply during times of particularly severe weather.

Asiatic black bears are omnivorous, feeding on a wide range of plant and animal foods. They mostly feed on sedges, grasses, tubers, twig buds, conifer seeds, berries and other fleshy fruits, grains, and mast (i.e., acorns and other hard nuts). They also eat insects, especially colonial types such as ants. Because their foods vary greatly in abundance during the year, the bears have a highly seasonal diet. Asiatic black bears are also opportunistic carnivores, and will also scavenge dead animals that they find. In cases where there is inadequate natural habitat available for foraging purposes, these bears will sometimes kill penned livestock, and they will raid bee hives when available. In some parts of their range, Asiatic black bears maintain territories in productive, lowland forests. In other areas they feed at higher elevations during the summer, descending to lower habitats for the winter.

All the body parts of Asiatic black bears are highly prized in traditional Chinese medicine, most particularly the gall bladders. Bear-paw soup is considered to be a delicacy in Chinese cuisine. Most bears are killed for these purposes by shooting or by using leg-hold, dead-fall, or pit traps. Some animals are captured alive, kept in cramped cages, and fitted with devices that continuously drain secretions from their gall bladder, which are used to prepare traditional medicines and tonics.

Populations of Asiatic black bears are declining rapidly over most of their range, earning them a listing of Vulnerable by the IUCN. These damages are being caused by **overhunting** of the animals for their valuable body parts (much of this involves illegal **hunting**, or **poaching**), disturbance of their forest habitats by timber harvesting, and permanent losses of their habitat through its conversion into agricultural land-use. These stresses and ecological changes have caused Asiatic black bears to become endangered over much of their range.

[*Bill Freedman Ph.D.*]

RESOURCES
BOOKS

Grzimek, B (ed.) *Grzimek's Encyclopedia of Mammals.* London: McGraw Hill, 1990.
Nowak, R.M. *Walker's Mammals of the World.* 5th ed. Baltimore: The John Hopkins University Press, 1991.

OTHER

Asiatic Black Bears. [cited May 2002]. <http://www.asiatic-black-bears.com>.

Assimilative capacity

Assimilative capacity refers to the ability of the **environment** or a portion of the environment (such as a stream, lake, air mass, or **soil** layer) to carry waste material without adverse effects on the environment or on users of its resources. **Pollution** occurs only when the assimilative capacity is exceeded. Some environmentalists argue that the concept of assimilative capacity involves a substantial element of value judgement, i.e., pollution **discharge** may alter the **flora** and **fauna** of a body of water, but if it does not effect organisms we value (e.g., fish) it is acceptable and within the assimilative capacity of the body of water.

A classical example of assimilative capacity is the ability of a stream to accept modest amounts of **biodegradable** waste. Bacteria in a stream utilize oxygen to degrade the organic matter (or **biochemical oxygen demand**) present in such a waste, causing the level of **dissolved oxygen** in the stream to fall; but the decrease in dissolved oxygen causes additional oxygen to enter the stream from the **atmosphere**, a process referred to as reaeration. A stream can assimilate a certain amount of waste and still maintain a dissolved oxygen level high enough to support a healthy population of fish and other aquatic organisms. However, if the assimilative capacity is exceeded, the concentration of dissolved oxygen will fall below the level required to protect the organisms in the stream.

Two other concepts are closely related: 1) critical load; and 2) self purification. The term critical load is synonymous with assimilative capacity and is commonly used to refer to the concentration or mass of a substance which, if exceeded, will result in adverse effects, i.e., pollution. Self purification refers to the natural process by which the environment cleanses itself of waste materials discharged into it. Examples include biodegradation of wastes by natural bacterial populations in water or soil, oxidation of organic **chemicals** by photochemical reactions in the atmosphere, and natural **die-off** of disease causing organisms.

Determining assimilative capacity may be quite difficult, since a substance may potentially affect many different organisms in a variety of ways. In some cases, there is simply not enough information to establish a valid assimilative capacity for a pollutant. If the assimilative capacity for a substance can be determined, reasonable standards can be set to protect the environment and the allowable waste load can be allocated among the various dischargers of the waste. If the assimilative capacity is not known with certainty, then more stringent standards can be set, which is analogous to buying insurance (i.e., paying an additional sum of money to protect against potential future losses). Alternatively, if the cost of control appears high relative to the potential benefits to the environment, a society may decide to accept a certain level of risk.

The Federal **Water Pollution** Control Amendments of 1972 established the elimination of discharges of pollution into navigable waters as a national goal. More recently, pollution prevention has been heavily promoted as an appropriate goal for all segments of society. Proper interpretation of these goals requires a basic understanding of the concept of assimilative capacity. The intent of Congress was to prohibit the discharge of substances in amounts that would cause pollution, not to require a concentration of zero. Similarly, Congress voted to ban the discharge of toxic substances in concentrations high enough to cause harm to organisms.

Well meaning individuals and organizations sometimes exert pressure on regulatory agencies and other public and private entities to protect the environment by ignoring the concept of assimilative capacity and reducing waste discharges to zero or as close to zero as possible. Failure to utilize the natural assimilative capacity of the environment not only increases the cost of **pollution control** (the cost to the discharger and the cost to society as a whole); more importantly, it results in the inefficient use of limited resources and, by expending materials and energy for something that **nature** provides free of charge, results in an overall increase in pollution.

[*Stephen J. Randtke*]

St. Francis of Assisi (1181 – 1226)
Italian Saint and religious leader

Born the son of a cloth merchant in the Umbrian region of Italy, Giovanni Francesco Bernardone became St. Francis of Assisi, one of the most inspirational figures in Christian history. As a youth, Francis was entranced by the French troubadours, but then planned a military career. While serving in a war between Assisi and Perugia in 1202, he was captured and imprisoned for a year. He intended to return to combat when he was released, but a series of visions and incidents, such as an encounter with a leper, led him in a different direction.

This direction was toward Jesus. Francis was so taken with the love and suffering of Jesus that he set out to live a life of prayer, preaching, and poverty. Although he was not a priest, he began preaching to the townspeople of Assisi and soon attracted a group of disciples, which Pope Innocent III recognized in 1211, or 1212, as the Franciscan order. The order grew quickly, but Francis never intended to found and control a large and complicated organization. His idea of the Christian life led elsewhere, including a fruitless attempt to end the Crusades peacefully. In 1224 he undertook a 40-day fast at Mount Alverna, from which he emerged bearing stigmata—wounds resembling those Jesus suffered on the cross. He died in 1226 and was canonized in 1228.

Francis's unconventional life has made him attractive to many who have questioned the direction of their own societies. In recent years his rejection of warfare and material goods has brought him the title of "the hippie saint"; Leo-

nardo Boff has seen him as a foreshadowing of liberation theology. Lynn White has proposed him as "a patron saint for ecologists."

There is no doubt that Francis loved **nature**. In his "Canticle of the Creatures," he praises God for the gifts, among others, of "Brother Sun," "Sister Moon," and "Mother Earth, Who nourishes and watches us..." But this is not to say that Francis was a pantheist or nature worshipper. He loved nature not as a whole, but as the assembly of God's creations. As G. K. Chesterton remarked, Francis "did not want to see the wood for the trees. He wanted to see each tree as a separate and almost a sacred thing, being a child of God and therefore a brother or sister of man."

For White, Francis was "the greatest radical in Christian history since Christ" because he departed from the traditional Christian view in which humanity stands over and against the rest of nature—a view, White charges, that is largely responsible for current ecological crises. Against this view, Francis "tried to substitute the idea of the equality of all creatures, including man, for the idea of man's limitless rule of creation."

[*Richard K. Dagger*]

RESOURCES
BOOKS

Boff, L. *St. Francis: A Model for Human Liberation.* Translated by J. W. Diercksmeier. New York: Crossroad, 1982.
Cunningham, L., ed. *Brother Francis.* New York: Harper & Row, 1972.
Chesterton, G. K. *St. Francis of Assisi.* New York: Doubleday Image Books, 1990.

PERIODICALS

White Jr., L. "The Historical Roots of Our Ecological Crisis." *Science* 155 (March 10, 1967): 1203–1207.

Asthma

Asthma is a condition characterized by unpredictable and disabling shortness of breath. It features episodic attacks of bronchospasm (prolonged contractions of the bronchial smooth muscle), and is a complex disorder involving biochemical, autonomic, immunologic, infectious, endocrine, and psychological factors to varying degrees in different individuals.

Asthma occurs in families suggesting that there is a genetic predisposition for the disorder, although the exact mode of genetic transmission remains unclear. The **environment** appears to play an important role in the expression of the disorder. For example, asthma can develop when *predisposed* individuals become infected with viruses or are exposed to allergens or pollutants. On occasion foods or drugs may precipitate an attack. Psychological factors have been investigated but have yet to be identified with any specificity.

The severity of asthma attacks varies among individuals, over time, and with the degree of exposure to the triggering factors. Approximately half of all cases of asthma develop during childhood. Another third develop before the age of 40. There are two basic types of asthma—intrinsic and extrinsic. Extrinsic asthma is triggered by allergens while intrinsic asthma is not. Extrinsic asthma, or allergic asthma, is classified as Type I or Type II, depending on the type of allergic response involved. Type I extrinsic asthma is the classic allergic asthma which is common in children and young adults who are highly sensitive to dust and pollen, and is often seasonal in nature. It is characterized by sudden, brief, intermittent attacks of bronchospasms that readily respond to bronchodilators. Type II extrinsic asthma, or allergic alviolitis, develops in adults under age 35 after long exposure to irritants. Attacks are more prolonged than Type I and are more inflammatory. Fever and infiltrates which are visible on chest x-rays often accompany bronchospasm.

Intrinsic asthma has no known immunologic cause and no known seasonal variation. It usually occurs in adults over the age of 35, many of whom are sensitive to aspirin and have nasal polyps. Attacks are often severe and do not respond well to bronchodilators.

A third type of asthma which occurs in otherwise normal individuals is called exercise induced asthma. Individuals with exercise induced asthma experience mild to severe bronchospasms during or after moderate to severe exertion. They have no other occurrences of bronchospasms when not involved in physical exertion. Although the cause of this type of asthma has not been established it is readily controlled by using a bronchodilator prior to beginning exercise.

[*Brian R. Barthel*]

RESOURCES

BOOKS

Berland, T. *Living With Your Allergies and Asthma*. New York: St. Martin's Press, 1983.

Lane, D. J. *Asthma: The Facts*. New York: Oxford University Press, 1987.

McCance, K. L. *Pathophysiology: The Biological Basis for Disease in Adults and Children*. St. Louis: Mosby, 1990.

Aswan High Dam

A heroic symbol and an **environmental liability**, this dam on the Nile River was built as a central part of modern Egypt's nationalist efforts toward modernization and industrial growth. Begun in 1960 and completed by 1970, the High Dam lies near the town of Aswan, which sits at the Nile's first cataract, or waterfall, 200 river mi (322 km)

from Egypt's southern border. The dam generates urban and industrial power, controls the Nile's annual **flooding**, ensures year-round, reliable **irrigation**, and has boosted the country's economic development as its population climbed from 20 million in 1947 to 58 million in 1990. The Aswan High Dam is one of a generation of huge **dams** built on the world's major rivers between 1930 and 1970 as both functional and symbolic monuments to progress and development. It also represents the hazards of large-scale efforts to control **nature**. Altered flooding, irrigation, and **sediment** deposition patterns have led to the displacement of villagers and farmers, a costly dependence on imported **fertilizer**, **water quality** problems and health hazards, and **erosion** of the Nile Delta.

Aswan attracted international attention in 1956, when planners pointed out that flooding behind the new dam would drown a number of ancient Egyptian tombs and monuments. A worldwide plea went out for assistance in saving the 4000-year old monuments, including the tombs and colossi of Abu Simbel and the temple at Philae. The United Nations Educational and Scientific Organization (**UNESCO**) headed the epic project, and over the next several years the monuments were cut into pieces, moved to higher ground, and reassembled above the water line.

The High Dam, built with international technical assistance and substantial funding from the former Soviet Union, was the second to be built near Aswan. English and Egyptian engineers built the first Aswan dam between 1898 and 1902. Justification for the first dam was much the same as that for the second, larger dam, namely flood control and irrigation. Under natural conditions the Nile experienced annual floods of tremendous volume. Fed by summer rains on the Ethiopian Plateau, the Nile's floods could reach 16 times normal low season flow. These floods carried terrific **silt** loads, which became a rich fertilizer when flood waters overtopped the river's natural banks and sediments settled in the lower surrounding fields. This annual soaking and fertilizing kept Egypt's agriculture prosperous for thousands of years. But annual floods could be wildly inconsistent. Unusually high peaks could drown villages. Lower than usual floods might not provide enough water for crops. The dams at Aswan were designed to eliminate the threat of high water and ensure a gradual release of irrigation water through the year.

Flood control and regulation of irrigation water supplies became especially important with the introduction of commercial cotton production. Cotton was introduced to Egypt by 1835, and within 50 years it became one of the country's primary economic assets. Cotton required dependable water supplies, but with reliable irrigation up to three crops could be raised in a year. Full-year commercial cropping was an important economic innovation, vastly different from traditional seasonal agriculture. By holding back most

An aerial view of the Aswan High Dam in Egypt. (Corbis-Bettmann. Reproduced by permission.)

of the Nile's annual flood, the first dam at Aswan captured 65.4 billion cubic yd (50 billion cubic m) of water each year. Irrigation canals distributed this water gradually, supplying a much greater acreage for a much longer period than did natural flood irrigation and small, village-built water works. But the original Aswan dam allowed 39.2 billion cubic yd (30 billion cubic m) of annual flood waters to escape into the Mediterranean. As Egypt's population, agribusiness, and development needs grew, planners decided this was a loss that the country could not afford.

The High Dam at Aswan was proposed in 1954 to capture escaping floods and to store enough water for long-term **drought**, something Egypt had seen repeatedly in history. Three times as high and nearly twice as long as the original dam, the High Dam increased the reservoir's storage capacity from an original 6.5 billion cubic yd (5 billion cubic m) to 205 billion cubic y (157 billion cubic m). The new dam lies 4.3 mi (7 km) upstream of the previous dam, stretches 2.2 mi (3.6 km) across the Nile and is nearly 0.6 mi (1 km) wide at the base. Because the dam sits on sandstone, gravel, and comparatively soft sediments, an impermeable screen of concrete was injected 590 ft (180 m) into the rock, down

to a buried layer of granite. In addition to increased storage and flood control, the new project incorporates a hydropower generator. The dam's turbines, with a capacity of 8 billion kilowatt hours per year, doubled Egypt's electricity supply when they began operation in 1970.

Lake Nasser, the **reservoir** behind the High Dam, now stretches 311 mi (500 km) south to the Dal cataract in Sudan. Averaging 6.2 mi (10 km) wide, this reservoir holds the Nile's water at 558 ft (170 m) above sea level. Because this reservoir lies in one of the world's hottest and driest regions, planners anticipated evaporation at the rate of 13 cubic yd (10 billion cubic m) per year. Dam engineers also planned for **siltation**, since the dam would trap nearly all the sediments previously deposited on downstream flood plains. Expecting that Lake Nasser would lose about 5% of its volume to siltation in 100 years, designers anticipated a volume loss of 39.2 billion cubic yd (30 billion cubic m) over the course of five centuries.

An ambitious project, the Aswan High Dam has not turned out exactly according to sanguine projections. Actual evaporation rates today stand at approximately 19 billion cubic yd (15 billion cubic m) per year, or half of the water gained by constructing the new dam. Another 1.3–2.6 cubic yd (1–2 billion cubic m) are lost each year through **seepage** from unlined irrigation canals. Siltation is also more severe than expected. With 60–180 million tons of silt deposited in the lake each year, current projections suggest that the reservoir will be completely filled in 300 years. The dam's effectiveness in flood control, water storage, and power generation will decrease much sooner. With the river's silt load trapped behind the dam, Egyptian farmers have had to turn to chemical fertilizer, much of it imported at substantial cost. While this strains commercial **cash crop** producers, a need for fertilizer application seriously troubles local food growers who have less financial backing than agribusiness ventures.

A further unplanned consequence of silt storage is the gradual disappearance of the Nile Delta. The Delta has been a site of urban and agricultural settlement for millennia, and a strong local fishing industry exploited the large schools of sardines that gathered near the river's outlets to feed. Longshore currents sweep across the Delta, but annual sediment deposits counteracted the erosive effect of these currents and gradually extended the delta's area. Now that the Nile's sediment load is negligible, coastal erosion is causing the Delta to shrink. The sardine fishery has collapsed, since river **discharge** and **nutrient** loads have been so severely depleted. Decreased fresh water flow has also cut off water supply to a string of fresh water lakes and underground aquifers near the coast. Salt water **infiltration** and **soil salinization** have become serious threats.

Water quality in the river and in Lake Nasser have suffered as well. The warm, still waters of the reservoir support increasing concentrations of **phytoplankton**, or floating water plants. These plants, most of them microscopic, clog water intakes in the dam and decrease water quality downstream. Salt concentrations in the river are also increasing as a higher percentage of the river's water evaporates from the reservoir.

While the High Dam has improved the quality of life for many urban Egyptians, it has brought hardship to much of Egypt's rural population. Most notably, severe health risks have developed in and around irrigation canal networks. These canals used to flow only during and after flood season; once the floods dissipated the canals would again become dry. Now that they are full year round, irrigation canals have become home to a common tropical snail that carries **schistosomiasis**, a debilitating disease that severely weakens its victims. **Malaria** may also be spreading, since moist mosquito breeding spots have multiplied. Farm fields, no longer washed clean each year, are showing high salt concentrations in the soil. Perhaps most tragic is the displacement of villagers, especially Nubians, who are ethnically distinct from their northern Egyptian neighbors and who lost most of their villages to Lake Nasser. Resettled in apartment blocks and forced to find work in the cities, Nubians are losing their traditional culture.

The Aswan High Dam was built as a symbol of national strength and modernity. By increasing industrial and agricultural output the dam generates foreign exchange for Egypt, raises the national standard of living, and helps ensure the country's high status and profile in international affairs. For all its problems, Lake Nasser now supports a fishing industry that partially replaces jobs lost in the delta fishery, and tourists contribute to the national income when they hire cruise boats on the lake. Most important, the country's expanded population needs a great deal of water. The Egyptian population is currently at 68 million but projected at 90 million by the year 2035, and neither the people nor their necessary industrial activity could survive on the Nile's natural meager water supply in the dry season. The Aswan dam was built during an era when dams of epic scale were appearing on many of the world's major rivers. Such projects were a cornerstone of development theory at the time, and if most rivers were not already dammed today, huge dams might still be central to development theory. Like other countries, including the United States, Egypt experiences serious problems and threats of future problems in its dam, but the Aswan dam is a central part of life and planning in modern Egypt.

[*Mary Ann Cunningham Ph.D.*]

RESOURCES

BOOKS

Säve-Söderbergh, T. *Temples and Tombs of Ancient Nubia: The International Rescue Campaign at Abu Simbel, Philae, and Other Sites.* London: (UN-ESCO) Thames and Hudson, 1987.

Little, T. *High Dam at Aswan—the Subjugation of the Nile.* London: Methuen, 1965.

OTHER

Driver, E. E., and W. O. Wunderlich, eds. *Environmental Effects of Hydraulic Engineering Works.* Proceedings of an International Symposium Held at Knoxville, TN. September 12–14, 1978. Knoxville: Tennessee Valley Authority, 1979.

Atmosphere

The atmosphere is the envelope of gas surrounding the earth, which is for the most part permanently bound to the earth by the gravitational field. It is composed primarily of **nitrogen** (78% by volume) and oxygen (21%). There are also small amounts of argon, **carbon dioxide**, and water vapor, as well as trace amounts of other gases and **particulate** matter.

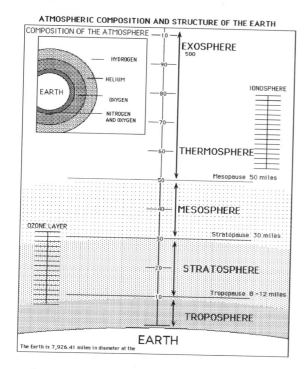

ATMOSPHERIC COMPOSITION AND STRUCTURE OF THE EARTH

(Illustration by Hans & Cassidy.)

Trace components of the atmosphere can be very important in atmospheric functions. **Ozone** accounts on average for two **parts per million** of the atmosphere but is more concentrated in the **stratosphere**. This stratospheric ozone is critical to the existence of terrestrial life on the planet. Particulate matter is another important trace component. **Aerosol loading** of the atmosphere, as well as changes in the tiny **carbon** dioxide component of the atmosphere, can be responsible for significant changes in **climate**.

The composition of the atmosphere changes over time and space. Outside of water vapor (which can vary from 0–4% in the local atmosphere) the concentrations of the major components varies little in time. Above 31 mi (50 km) from sea level, however, the relative proportions of component gases change significantly. As a result, the atmosphere is divided into two compositional components: Below 31 mi (50 km) is the homisphere and above 31 mi (50 km) is the heterosphere.

The atmosphere is also divided according to its thermal behavior. By this criteria, the atmosphere can be divided into several layers. The bottom layer is the *troposphere*; it contains most of the atmosphere and is the domain of weather. Above the **troposphere** is a stable layer called the stratosphere. This layer is important because it contains much of the ozone which **filters** ultraviolet light out of the incident solar radiation. The next layer, is the *mesosphere*,

which is much less stable. Finally, there is the *thermosphere*; this is another very stable zone, but its contents are barely dense enough to cause a visible degree of solar radiation scattering.

[*Robert B. Giorgis Jr.*]

RESOURCES
BOOKS

Anthes, R. A., et al. *The Atmosphere*. 3rd ed. Columbus, OH: Merrill, 1981.
Schaefer, V., and J. Day. *A Field Guide to the Atmosphere*. New York: Houghton Mifflin, 1981.

PERIODICALS

Graedel, T. E., and P. J. Crutzen. "The Changing Atmosphere." *Scientific American* 261 (September 1989): 58–68.

Atmospheric (air) pollutants

Atmospheric pollutants are substances that accumulate in the air to a degree that is harmful to living organisms or to materials exposed to the air. Common air pollutants include **smoke**, **smog**, and gases such as **carbon monoxide, nitrogen** and sulfur oxides, and hydrocarbon fumes. While gaseous pollutants are generally invisible, solid or liquid pollutants in smoke and smog are easily seen. One particularly noxious form of **air pollution** occurs when oxides of sulfur and nitrogen combine with atmospheric moisture to produce sulfuric and nitric **acid**. When the acids are brought to Earth in the form of **acid rain**, damage is inflicted on lakes, rivers, vegetation, buildings, and other objects. Because sulfur and **nitrogen oxides** can be carried for long distances in the **atmosphere** before they are removed in precipitation, damage may occur far from **pollution** sources.

Smoke is an ancient environmental pollutant, but increased use of **fossil fuels** in recent centuries has increased its severity. Smoke can aggravate symptoms of **asthma**, **bronchitis**, and **emphysema**, and long term exposure can lead to lung **cancer**. Smoke toxicity increases when fumes of **sulfur dioxide**, commonly released by **coal combustion**, are inhaled with it. One particularly bad air pollution incident occurred in London in 1952 when approximately 4,000 deaths resulted from high smoke and sulfur dioxide levels that accumulated in the metropolitan area during an **atmospheric inversion**.

Sources of air pollutants are particularly abundant in modem, industrial population centers. Major sources include power and heating plants, industrial manufacturing plants, and **transportation** vehicles. Mexico City is a particularly bad example of a very large metropolitan area that has not adequately controlled harmful emissions into the atmosphere, and as a result many area residents suffer from respiratory ailments. The Mexican government has been forced to shut down a large oil refinery and ban new polluting industries from locating in the city. More difficult for the Mexican government to control are the emissions from over 3.5 million automobiles, trucks, and buses.

Smog (a combination of smoke and fog) is formed from the condensation of moisture (fog) on **particulate** matter (smoke) in the atmosphere. Although smog has been present in urban areas for a long time, **photochemical smog** is an exceptionally harmful and annoying form of air pollution found in large urban areas. It was first recognized as a serious problem in Los Angeles in the late 1940s. Photochemical smog forms by the catalytic action of sunlight on some atmospheric pollutants including unburned **hydrocarbons** evaporated from **automobile** and other fuel tanks. Products of the photochemical reactions include **ozone**, aldehydes, **ketones**, **peroxyacetyl nitrate**, and organic acids. Photochemical smog causes serious eye and lung irritation and other health problems.

Because of the serious damage caused by atmospheric pollution, control has become a high priority in many countries. Several approaches have proven beneficial. An immediate control measure involves a system of air alerts announced when air pollution monitors disclose dangerously high levels of contamination. Industrial sources of pollution are forced to curtail activities until conditions return to normal. These emergency situations are often caused by atmospheric inversions that limit the upward mixing of pollutants. Wider dispersion of pollutants by taller smokestacks can alleviate local pollution. However, dispersion does not lessen overall amounts; instead, it can create problems for communities downwind from the source. Greater **energy efficiency** in vehicles, homes, electrical **power plants**, and industrial plants, lessens pollution from these sources. This reduction in fuel use also reduces pollution created when fuels are produced, and as an added benefit, more energy efficient operation reduces costs. Efficient and convenient **mass transit** also helps to mitigate urban air pollution by cutting the need for personal vehicles.

[*Douglas C. Pratt*]

RESOURCES
BOOKS

Georgii, H. W., ed. *Atmospheric Pollutants in Forest Areas: Their Deposition and Interception*. Boston: Kluwer Academic Publishers, 1986.

Hutchinson, T. C., and K. M. Meema, eds. *Effects of Atmospheric Pollutants on Forests, Wetlands, and Agricultural Ecosystems*. New York: Springer-Verlag, 1987.

Restelli, G., and G. Angeletti, eds. *Physico-Chemical Behaviour of Atmospheric Pollutants*. Boston: Kluwer Academic Publishers, 1990.

PERIODICALS

Graedel, T. E., and P. J Crutzen. "The Changing Atmosphere." *Scientific American* 261, no. 3 (Sept 1989): 58–59.

Atmospheric deposition

Many kinds of particulates and gases are deposited from the **atmosphere** to the surfaces of terrestrial and aquatic ecosystems. Wet deposition refers to deposition occurring while it is raining or snowing, whereas **dry deposition** occurs in the time intervals between precipitation events.

Relatively large particles suspended in the atmosphere, such as dust entrained by strong winds blowing over fields or emitted from industrial smokestacks, may settle gravitationally to nearby surfaces at ground level. Particulates smaller than about 0.5 microns in diameter, however, do not settle in this manner because they behave aerodynamically like gases. Nevertheless, they may be impaction-filtered from the atmosphere when an air mass passes through a physically complex structure. For example, the large mass of foliage of a mature conifer forest provides an extremely dense and complex surface. As such, a conifer canopy is relatively effective at removing particulates of all sizes from the atmosphere, including those smaller than 0.5 microns. Forest canopies dominated by hardwood (or angiosperm) trees are also effective at doing this, but somewhat less-so than conifers.

Dry deposition also includes the removal of certain gases from the atmosphere. For example, **carbon dioxide** gas is absorbed by plants and fixed by **photosynthesis** into simple sugars. Plants are also rather effective at absorbing certain gaseous pollutants such as **sulfur dioxide**, nitric oxide, **nitrogen** dioxide, and **ozone**. Most of this gaseous uptake occurs by **absorption** through the numerous tiny pores in leaves known as stomata. These same gases are also dry-deposited by absorption by moist **soil**, rocks, and water surfaces.

Wet deposition involves substances that are dissolved in rainwater and snow. In general, the most abundant substances dissolved in precipitation water are sulfate, nitrate, calcium, magnesium, and ammonium. At places close to the ocean, sodium and chloride derived from sea-salt aerosols are also abundant in precipitation water. Acidic precipitation occurs whenever the concentrations of the anions (negatively charged ions) sulfate and nitrate occur in much larger concentrations than the cations (positively charged ions) calcium, magnesium, and ammonium. In such cases, the cation "deficit" is made up by **hydrogen** ions going into solution, creating solutions that may have an acidity less than **pH** 4. (Note that the **ion** concentrations in these cases are expressed in units of equivalents, which are molar concentrations multiplied by the number of charges on the ion.)

The deposition of acidified snow or rain can result in the **acidification** of vulnerable surface waters. This is particularly the case of streams, rivers, ponds, and lakes that have low concentrations of alkalinity and consequently little capacity for neutralizing inputs of acidity. Acidification may also be caused by the dry deposition of certain gases, especially sulfur dioxide and oxides of nitrogen. After they are dry-deposited, these gases become oxidized in soil or water into the ions sulfate and nitrate, respectively, a process accompanied by the production of equivalent amounts of hydrogen ions. In some environments the gaseous concentrations of sulfur dioxide and oxides of nitrogen are relatively high, particularly in areas polluted by gaseous emissions from large numbers of automobiles, **power plants**, or other industries. In those circumstances the dry deposition of acidifying substances will be a much more important cause of environmental acidification than the wet deposition of acidic precipitation.

[*Bill Freedman Ph.D.*]

RESOURCES
BOOKS

Freedman, B. *Environmental Ecology*. San Diego, CA: Academic Press, 1995.

Joskow, P.L., A.D. Ellerman, R. Schmalensee, M. Juan-Pablo, and E.M. Bailey, editors. *Markets for Clean Air: The U.S. Acid Rain Program*. Cambridge, UK: Cambridge University Press, 2000.

ORGANIZATIONS

United States Environmental Protection Agency, Clean Air Markets Division, 1200 Pennsylvania Avenue, NW, Washington, D.C. USA 20460 (202) 564-9150, Email: , http://www.epa.gov/airmarkets/acidrain/

Worldwatch Institute, 1776 Massachusetts Ave., N.W., Washington, D.C. USA 20036-1904 (202) 452-1999, Fax: (202) 296-7365, Email: worldwatch@worldwatch.org, http://www.worldwatch.org/

Atmospheric inversion

Atmospheric inversions are horizontal layers of air that increase in temperature with height. Such warm, light air often lies over air that is cooler and heavier. As a result the air has a strong vertical **stability**, especially in the absence of strong winds.

Atmospheric inversions play an important role in **air quality**. They can trap air pollutants below or within them, causing high concentrations in a volume of air that would otherwise be able to dilute air pollutants throughout a large portion of the **troposphere**.

Atmospheric inversions are quite common, and there are several ways in which they are formed. Surface inversions can form during the evening when the radiatively cooling ground becomes a heat sink at the bottom of an air mass immediately above it. As a result heat flows down through

the air which sets up a temperature gradient that coincides with an inversion. Alternatively, relatively warm air may flow over a cold surface with the same results.

Elevated atmospheric inversions can occur when vertical differences in wind direction allow warm air to set up over cold air. However, it is more common in near-subtropical latitudes, especially on the western sides of continents, to get subtropical **subsidence** inversions. Subtropical subsidence inversions often team with mountainous **topography** to trap air both horizontally and vertically. This is the situation on the coast of southern and central California.

Atmospheric pollutants

see **Carbon monoxide; Nitrogen oxides; Particulate; Smog; Sulfur dioxide; Volatile organic compound**

Atomic bomb

see **Nuclear weapons**

Atomic bomb testing

see **Bikini atoll; Nevada Test Site**

Atomic energy

see **Nuclear power**

Atomic Energy Commission

The Atomic Energy Commission (AEC) was established by act of the United States Congress in 1946. It originally had a dual function: to promote the development of **nuclear power** and to regulate the nuclear power industry. These two functions were occasionally in conflict, however. Critics of the AEC claimed that the Commission sometimes failed to apply appropriate safety regulations, for example, because doing so would have proved burdensome to the industry and hindered its growth. In 1975, the AEC was dissolved. Its regulatory role was assigned to the **Nuclear Regulatory Commission** and its research and development role to the **Energy Research and Development Administration**.

Atomic fission

see **Nuclear fission**

Atomic fusion

see **Nuclear fusion**

Atrazine

Atrazine is used as a selective preemergent **herbicide** on crops, including corn, sorghum, and sugar cane, and as a nonselective herbicide along fence lines, right-of-ways, and road sides. It is used in a variety of formulations, both alone and in combination with other herbicides. In the United States it is the single most widely used herbicide, accounting for about 15% of total herbicide application by weight. Its popularity stems from its effectiveness, its selectivity (it inhibits **photosynthesis** in plants lacking a **detoxification** mechanism), and its low mammalian toxicity (similar to that of table salt). Concern over its use stems from the fact that a small percentage of the amount applied can be carried by rainfall into surface waters (stream and lakes), where it may inhibit the growth of plants and algae or contaminate **drinking-water supply**. Because atrazine is classified as a possible human **carcinogen**, the United States **Environmental Protection Agency** (EPA) has limited its concentration in drinking water to 3 μg/L.

Attainment area

An attainment area is a politically or geographically defined region that is in compliance with the National Ambient Air Quality Standards, as set by the **Environmental Protection Agency** (EPA). A region is considered to be an attainment area if it does not exceed the specified thresholds for air pollutants that may be a threat to public health. The standards are set for a variety of ambient air pollutants, including particulates, **ozone**, **carbon monoxide**, sulfur oxides, and **nitrogen oxides**. A region can be an attainment area for one pollutant and a **nonattainment area** for another.

John James Audubon (1785 – 1851)
American naturalist, artist, and ornithologist

John James Audubon, the most renowned artist and naturalist in nineteenth century America, left a legacy of keenly observant writings as well as a portfolio of exquisitely rendered paintings of the birds of North America.

Born April 26, 1785, Audubon was the illegitimate son of a French naval captain and a domestic servant girl from Santo Domingo (now Haiti). Audubon spent his childhood on his father's plantation in Santo Domingo and most of his late teens on the family estate in Mill Grove, Pennsylvania—a move intended to prevent him from being conscripted into the Napoleonic army.

Audubon's early pursuits centered around natural history, and he was continuously collecting and drawing plants, insects, and birds. His habit of keeping meticulous field notes of his observations at a young age. It was at Mill Grove that Audubon, in order to learn more of the movements and habits of birds, tied bits of colored string onto the legs of several Eastern Phoebes and so proved that these birds returned to the same nesting sites the following year. Audubon was the first to use banding to study the movement of birds.

While at Mill Grove, Audubon began courting their neighbor's eldest daughter, Lucy Bakewell, and they were married in 1808. They made their first home in Louisville, Kentucky, where Audubon tried being a storekeeper. He could not stand staying inside, and so he spent most of his time afield to "supply fowl for the table," thus dooming the store to failure.

In 1810, Audubon met by chance Alexander Wilson, who is considered the father of American ornithology. Wilson had finished much of his nine volume *American Ornithology* at this time, and it is believed that his work inspired Audubon to embark on his monumental task of painting the birds of North America.

The task that Audubon undertook was to become *The Birds of America*. Because Audubon decided to depict each **species** of bird life-size, thus rendering each on a 36.5 in x 26.5 in (93 cm x 67 cm) page, this was the largest book ever published up until that time. He was able to draw even larger birds such as the **whooping crane** life-size by depicting them with their heads bent to the ground. Audubon pioneered the use of fresh models instead of stuffed museum skins for his paintings. He would shoot birds and wire them into life-like poses to obtain the most accurate drawings possible. Even though his name is affixed to a modern **conservation** organization, the **National Audubon Society**, it must be remembered that little thought was given to the conservation of birds in the early nineteenth century. It was not uncommon for Audubon to shoot a dozen or more individuals of a species to get what he considered the perfect one for painting.

Audubon solicited subscribers for his *Birds of America* to finance the printing and hand-coloring of the plates. The project took nearly twenty years to complete, but the resulting double elephant folio, as it is known, was truly a work of art, as well as the ornithological triumph of the time. Later in his life, Audubon worked on a book of the mammals of North America with his two sons, but failing health forced him to let them complete the work. He died in 1851, leaving behind a remarkable collection of artwork that depicted the natural world he loved so much.

[*Eugene C. Beckham*]

RESOURCES

BOOKS

Audubon, J. J. *Audubon's Western Journal: 1849–1850*. Irvine: Reprint Services, 1992.

John James Audubon. (Corbis-Bettmann. Reproduced by permission.)

————. *Life of John James Audubon: The Naturalist.* Irvine: Reprint Services, 1993.

Running Press Staff, eds. *Audubon Journal.* Philadelphia: Running Press, 1993.

PERIODICALS

Gopnik, A. "Audubon's Passion." *The New Yorker* 67 (25 February 1991): 96–104.

Audubon Society
see **National Audubon Society**

Australia

The Commonwealth of Australia, a country in the South Pacific Ocean, occupies the smallest continent and covers an area of 2,966,200 mi^2 (7,682,000 km^2). Along the northeast coast of Australia, for a distance of 1,200 mi (2,000 km) in length and offshore as much as 100 mi (160 km) into the Pacific Ocean, lies the **Great Barrier Reef**, the greatest assemblage of coral reefs in the world. The island of Tasmania, which is a part of Australia, lies off the southeast coast of the mainland. Australia consists mainly of plains and plateaus, most of which average 2,000 ft (600 m) above sea level. There are several low mountain ranges in the country.

Only about 16 million people live in Australia, of whom about 150,000 are native people, the Aborigines.

More than a third of Australia receives under 10 in (25 cm) of annual rainfall, while less than a third receives over 20 in (51 cm). Many areas experience prolonged **drought** and frequent heat waves. The Outback is one such area. It is comprised of dry, barren land and vast interior deserts. Soils generally are poor, with fertile land found only in the lowlands and valleys near the east and southeast coast. The largest river is the Murray River in southeastern Australia. In the inland area, few rivers have much water due to the lack of rainfall. In areas with coastal highlands, rivers flow only a short distance to the sea. Lakes frequently dry up and become beds of salt. In contrast, there are lush rain forests in the north and snowy mountains in the southeastern Blue Range.

The most common tree is the eucalyptus, of which there are at least 600 different kinds. Since Australia is the driest of the inhabited continents, there are few forests. The acacia, or wattle, is Australia's national flower. Many of the mammals of Australia (kangaroos, wallabies, and koala bears) are marsupials, unlike those found anywhere else in the world. The young of marsupials are born premature and require continued development outside the womb in pouches on the parent's body. The egg-laying mammal, or monotreme, is also only found in Australia. There are two **species** of monotreme: the duck-billed platypus and the echidna, or spiny anteater. The most famous Australian bird is the kookaburra. There are no hoofed animals in Australia. Two flightless birds, the emu and the cassowary, also are found only in Australia. Snakes are common and include many poisonous species. Two kinds of **crocodiles** are found in the northern part of Australia. All of these unique **fauna** and **flora** species have evolved due to the long separation of the Australian continent from southeast Asia and a general lack of predators.

The goal of environmental protection and management in Australia is integration of the principles of ecologically **sustainable development** into policies and programs encompassing all aspects of environmental issues.

Australia's land resources form the basis of its unique natural heritage and its agricultural industry. However, problems of land degradation are widespread. Land degradation due to dryland **salinity** is a national problem that threatens both **biodiversity** and agricultural productivity. Dryland salinity affects about 5,434,000 acres (2.2 million ha) of once-productive agricultural land and costs an estimated $243 million per year in lost agricultural production. Dryland salinity originates from salt deposited in the landscape over millions of years due to marine **sedimentation**. Land clearing over the last two hundred years and the replacement of deep-rooted perennial native vegetation with shallow-rooted annuals have caused water tables to rise, bringing saline water close to the surface into the root zones of plants. Excessive **irrigation** has also led to increases in levels of water tables. Strategies to address this problem include research for the prediction, prevention, and reversal of dryland salinity. An example of a research effort to ameliorate the effects of salinity is a project that addresses the identification of salt-tolerant plant species and development of appropriate establishment and growth management techniques to ensure their survival. Australia's naturally fire-adapted **ecosystem** has frequent brushfires that now threaten many new developments and even major cities. Other land degradation problems being addressed include the **rehabilitation** of mining sites and **remediation** of sites contaminated by railway operations.

Australia contains about 10% of the world's biological diversity. The Commonwealth is addressing the issues of loss of biodiversity through a program referred to as the National Vegetation Initiative (NVI). About 988 million acres (40 million ha) of land is protected within a terrestrial reserves system, but 1,235 million acres (500 million ha), or more than two-thirds of Australia's land area, are managed by private landholders. Biodiversity outside the reserves has been affected by vegetation clearance and modification. The NVI will attempt to reverse the long-term decline in the quality and extent of Australia's native vegetation by developing programs and incentives to conserve, enhance, and manage remnant native vegetation; to increase revegetation activities; and to encourage the integration of native vegetation into conventional farming systems.

The state of Australia's rivers and other aquatic systems has been declining due to a range of factors. Unsustainable water extractions for agricultural production have resulted in reduced flows in rivers that result in blue-green algae outbreaks, declines in native fish populations, increases in salinity, loss of **wetlands**, and loss of the beneficial aspects of small floods. Research is being conducted to define environmental flow requirements so that water allocations or entitlements will include environmental needs as a legitimate use. Other factors affecting the decline in aquatic systems include poor land and vegetation management, agricultural and urban **pollution**, salinity, destruction of native **habitat**, and spread of exotic pests such as the European carp and the water weed *Salvinia*.

In a 1997 Australian Bureau of **Statistics** survey, the Australian people identified **air pollution** as their greatest environmental concern. The government's primary **air quality** objective is to provide all Australians with equivalent protection from air pollution, which will be achieved by the development of national air quality strategies and standards that minimize the adverse environmental impacts of air pollution. Since over 60% of the Australian population lives in

cities, the Government's first priority is to develop measures to reduce the impact of air pollution on urban areas.

The government of Australia considers **climate** changes as another of its most important environmental issues. A National Greenhouse Response Strategy was adopted in 1992 to respond to the need for greenhouse gas **emission** mitigation. The major thrust of the program is to develop cooperative agreements with industry to abate and reduce greenhouse gas emissions. In addition, large scale vegetation plantings for improvement of land resources will reduce **carbon dioxide**, a major greenhouse gas, in the **atmosphere**. **Methane** emission from intensive agriculture will be reduced through livestock waste treatment. The use of alternative transport fuels such as liquefied **petroleum** gas, compressed **natural gas**, **ethanol**, and other alcohol blends will be encouraged through the use of a fuel excise exemption.

Although **ozone layer depletion** in the atmosphere occurs over most of the Earth, the most dramatic changes are seen when an **ozone** "hole" forms over **Antarctica** each spring. The ozone layer protects life on Earth from harmful **ultraviolet radiation** from the sun, but it can be depleted by the use of some widely used ozone-depleting **chemicals**. The ozone hole sometimes drifts over Australia and exposes residents to dangerous levels of UV radiation. Australia, because of its proximity to Antarctica, has taken a leading role in ozone protection and has been influential in the development and implementation of cost-effective mechanisms to ensure the phase-out of use of ozone-depleting substances by all countries. The government also monitors solar ultraviolet radiation levels and assesses the consequences for public health.

Australia's marine and coastal environments are rich in natural and cultural resources, are adjacent to most of the nation's population (about 86%), and are a focus of much of Australia's economic, social, tourism, and recreational activity. Australia's **Exclusive Economic Zone** (EEZ) is one of the largest in the world, comprising 4.25 million mi^2 (11 million km^2) of marine waters. Uncoordinated and ad hoc development has been identified as a major contributing factor to the decline of coastal **water quality** and marine and estuarine habitats. Programs that include cooperation of the Commonwealth government with state and local governments, industry groups, nongovernmental organizations, and the community have been developed to protect and rehabilitate these environments. These programs include addressing threats to coastal water quality and marine biodiversity from land-based and **marine pollution**; developing integrated management plans for the **conservation** and sustainable use of coastal resources, including fisheries resources; supporting capital works and improvement of technologies to reduce the impacts of sewage and stormwater;

controlling the introduction and spread of exotic marine pests in Australian waters; protecting and restoring fish habitats; and supporting comprehensive and consistent coastal monitoring.

The Great Barrier Reef is facing growing pressures from the reef-based tourism industry, commercial and recreational fishing (which is worth about $1.3 billion per year), expanding coastal urban areas, and downstream effects of **land use** from agricultural activities. Several programs are being developed and implemented to ensure that activities can continue on an ecologically sustainable basis. For example, tourist and recreational impacts are being reduced while providing diverse tourist activities. Fishing catches and efforts are being monitored, major or critical habitats are being identified, and fishing by-catch is being reduced through the development of new methods. **Sediment, nutrient**, and other land-based **runoff** that impacts the health of adjacent marine areas are being controlled in coastal developments. Also, spill contingency planning and other responses to prevent pollution from ships, navigational aids, and ship reporting systems are all being improved. In addition, the state of water quality is being monitored in long-term programs throughout the Reef and threats from pollution are being assessed. Finally, all planning exercises include consultation with Aboriginal and Torres Strait Islander (ATSI) communities to ensure that their interests are considered.

The Commonwealth, through its National and World Heritage programs, is protecting elements of Australia's natural and cultural heritage that are of value for this and **future generations**. The National Estate comprises natural, historic, and indigenous places that have aesthetic, historic, scientific or social significance or other special value. The Register of the National Estate by 1997 had 11,000 places listed. The Register educates and alerts Australians to places of heritage significance. World Heritage properties are areas of outstanding universal cultural or national significance that are included in the **UNESCO** World Heritage List. In 1997, there were 11 Australian properties on the World Heritage List, including the Great Barrier Reef.

[*Judith L. Sims*]

RESOURCES
BOOKS

Hill, R. *Investing in Our Natural Heritage*. Canberra, Australia: Annual Report on the Commonwealth's Environment Expenditure, Minister for the Environment, 1997.

Autecology

A branch of **ecology** emphasizing the interrelationships among individual members of the same **species** and their environment. Autecology includes the study of the life history and/or behavior of a particular species in relation to the environmental conditions that influence its activities and distribution. Autecology also includes studies on the tolerance of a species to critical physical factors (e.g., temperature, **salinity**, oxygen level, light) and biological factors (e.g., predation, **symbiosis**) thought to limit its distribution. Such data are gathered from field measurement or from controlled experiments in the laboratory. Autecology contrasts with *synecology*, the study of interacting groups (i.e., communities) of species.

Automobile

The development of the automobile at the end of the nineteenth century fundamentally changed the structure of society in the developed world and has had wide-ranging effects on the **environment**, the most notable being the increase of **air pollution** in cities. The piston-type internal **combustion** engine is responsible for the peculiar mix of pollutants that it generates. There are a range of other engines suitable for automobiles, but they have yet to displace engines using rather volatile **petroleum** derivatives.

The simplest and most successful way of improving gaseous emissions from automobiles is to find alternative fuels. Diesel fuels have always been popular for larger vehicles, although a few private vehicles in Europe are also diesel-powered. Compressed natural gases have been widely used as fuel in some countries (e.g., New Zealand), while **ethanol** has had a limited success in places such as Brazil, where it can be produced relatively cheaply from sugar cane. There is some enthusiasm for the use of **methanol** in the United States, but it has yet to be seen if this will be widely adopted as a fuel.

Others have suggested that fundamental changes to the engine itself can lower the impact of automobiles on **air quality**. The Wankel rotary engine is a promising power source that offers both low vibration and pollutant emissions from a relatively lightweight engine. Although Wankel engines are found on a number of exotic cars, there are still doubts about long-term engine performance and durability in the urban setting. Steam and gas turbines have many of the advantages of the Wankel engine, but questions of their expense and suitability for automobiles have restricted their use. Electric vehicles have had some impact for special sectors of the market. They have proved ideal for small vehicles within cities where frequent stop-start operation is required (e.g., delivery vans). A few small, one-seat vehicles have been available at times, but they have failed to achieve any enduring popularity. The electric vehicle suffers from low range, low speed and acceleration, and needs heavy batteries. However these vehicles produce none of the conventional combustion-derived pollutants during operation, although the electricity to recharge the batteries requires the use of an electricity generating station. Still, electricity generation can be sited away from the urban center and employ air **pollution** controls. **Fuel cells** are an alternate source of electricity for electric automobiles. It is also possible to power automobiles through the use of flywheels. These are driven up to high speeds by a fixed, probably electric, motor, then the vehicle can be detached and powered using the stored momentum, although range is often limited.

Although **automobile emissions** are of great concern, the automobile has a far wider range of environmental impacts. Large amounts of material are used in their construction, and discarded automobiles can litter the countryside and fill waste dumps. Through much of the twentieth century the vehicles have been made of steel. Increasingly, other materials, such as **plastics** and fiberglass, are used as construction materials. A number of projects, most frequently on the European continent, have tried **recycling** automobile components. Responsible automobile manufacturers are aiming to build vehicles with longer road lives, further aiding **waste reduction**.

The high speeds now possible for automobiles lead to sometimes horrendous accidents, which can involve many vehicles on crowded highways. The number of accidents have been reduced through anti-lock braking systems, thoughtful road design, and imposing harsh penalties for drunk driving. In the future, on-board radar may give warning of impending collisions. Safety features such as seat belts, padding, and collapsible steering columns have helped lower injury during accidents.

The structure of cities has changed with widespread automobile ownership. It has meant that people can live further away from where they work or shop. The need for parking has led to the development of huge parking lots within the inner city. Reaction to crowding in city centers are seen in the construction of huge shopping centers out-of-town, where parking is more convenient (or strip development of large stores along highways). These effects have often caused damage to inner city life and disenfranchised non-car owners, particularly because a high proportion of car ownership often works against the operation of an effective **mass transit** system. For people who live near busy roads, **noise pollution** can be a great nuisance. The countryside has also been transformed by the need for super highways that cope with a large and rapid traffic flow. Such highways have often been built on valuable agricultural land or natural habitats, and once constructed, create both practical and aesthetic nuisances.

[*Peter Brimblecombe*]

RESOURCES
BOOKS

Environmental Effects of Automotive Emissions. Paris: OECD Compass Project, 1986.

Automobile emissions

The **automobile**, powered by piston-type internal **combustion** engine, is so widely used that it has become the dominant source of air pollutants, particularly of **photochemical smog**, in large, urban cities.

Modern internal combustion engines operate through the Otto cycle, which involves rapid batch-burning of **petroleum** vapors. The combustion inside the cylinder is initiated by a spark and proceeds outward through the gas volume until it reaches the cylinder walls where it is cooled. Close to the cylinder wall, where combustion is quenched, a fraction of the fuel remains unburnt. In the next cycle of the engine the hot combusted gases and unburnt fuel vapor are forced out through the exhaust system of the automobile.

Automotive engines generally operate on "fuel rich" mixtures, which means that there is not quite enough oxygen to completely burn the fuel. As a result there is an excess of unburnt **hydrocarbons**, particularly along the cylinder walls, and substantial amounts of **carbon monoxide**. This efficient production of **carbon** monoxide has made automobiles the most important source of this poisonous gas in the urban **atmosphere**.

The high **emission** levels of hydrocarbons and carbon monoxide have caused some engineers to become interested in "lean burn" engines that make more oxygen available during combustion. While such an approach is possible, the process also produces high concentrations of nitric oxide in the exhaust gases.

When the fuel enters the cylinder of an automobile engine in the gaseous form, it generally does not produce **smoke**. However in diesel engines where the fuel is sprayed into the combustion chamber, it will become dispersed as tiny droplets. Sometimes, especially under load, these droplets will not burn completely and are reduced to fine carbon particles, or soot, that are easily visible from diesel exhausts. In many cities this diesel smoke can represent the principle soiling agent in the air. In addition, soot, particularly those from diesel engines, contain small amounts of carcinogenic material.

Many of the carcinogens found in the exhaust from diesel engines are **polycyclic aromatic hydrocarbons** (PAH) and are archetypical carcinogens. Best known of these is benzo-a-pyrene, which was tentatively recognized as a **carcinogen** in the eighteenth century from observations of chimney sweepers who had high incidence of **cancer**.

Some of the PAH can become nitrated during combustion and these may be even more carcinogenic than the unnitrated PAH. Diesel emissions may pose a greater cancer risk than the exhaust gases from **gasoline** engines. There are a number of emissions from gasoline engines that are potentially carcinogenic. **Benzene** represents a large part of the total volatile organic emissions from automobiles. Yet the compound is also recognized by many as imposing a substantial carcinogenic risk to modern society. **Toluene**, although by no means as carcinogenic as benzene, is also emitted in large quantities. Toluene proves a very effective compound at initiating photochemical **smog** and also reacts to form the eye irritant peroxybenzoyl nitrate. The highly dangerous compound **dioxin** can be produced in auto exhausts where **chlorine** is present (anti-knock agents often contain chlorine). Formaldehyde, a suspected carcinogen, is produced in photochemical smog but may also be an enhanced risk from engines burning the otherwise less polluting **methanol** as a fuel.

Many exotic elements that are added to improve the performance of automotive fuels produce their own emissions. The best known is the anti-knock agent **tetraethyl lead**, which was added in such large quantities that it became the dominant source of **lead** particles in the air. A wide range of long-term health effects, such as lowering IQ, have been associated with exposure to lead. Although lead in urban populations are still rather high, the use of unleaded gasoline have decreased the problem somewhat.

Most attention usually focuses on the engine as a source of automobile emissions, but there are other sources. Evaporative loss of volatile materials from the crank case, carburetion system, and fuel tank represent important sources of hydrocarbons for the urban atmosphere. The wear of tires, brake linings, and metal parts contribute to particles being suspended in the air of the near roadside **environment**. The presence of **asbestos** fibers from brake linings in the urban air has often been discussed, although health threats from this source is less serious than from other sources.

There have been relatively few studies of the pollutants inside automobiles, but there has been some concern about the potential hazard of the build-up of automobile emissions from malfunctioning units and from "leaks." These can be from the evaporation of fuel, especially leaded fuels where the volatile tetraethyl lead is present. Carbon monoxide, from the exhaust system, can cause drowsiness and impair judgment. However in many cases the interior of a properly functioning automobile, without additional sources such as smoking, can have somewhat better **air quality** than the air outside. In general, pedestrians, cyclists, and those who work at road sites are likely to experience the worst of automotive

pollutants such as carbon monoxide and potentially carcinogenic hydrocarbons.

Although huge quantities of **fossil fuels** are burnt in power generation and a range of industrial processes, automobiles make a significant and growing contribution to **carbon dioxide** emissions which enhance the **greenhouse effect. Ethanol**, made from sugar cane, is a renewable source and has the advantage of not making as large a contribution to the **greenhouse gases** as gasoline. Automobiles are not large emitters of **sulfur dioxide** and thus do not contribute greatly to the regional **acid rain** problem. Nevertheless, the **nitrogen oxides** emitted by automobiles are ultimately converted to nitric **acid** and these are making an increasing contribution to rainfall acidity. Diesel-powered vehicles use fuel of a higher sulfur content and can contribute to the sulfur compounds in urban air.

Despite the enormous problems created by the automobile, few propose its abolition. The ownership of a car carries with it powerful statements about personal freedom and power. Beyond this, the structure of many modern cities requires the use of a car. Thus while **air pollution** problems might well be cured by a wide range of sociological changes, a technological fix has been favored, such as the use of catalytic converters. Despite this and other devices, cities still face daunting air quality problems. In some areas, most notably the **Los Angeles Basin**, it is clear that there will have to be a wide range of changes if air quality is to improve. Although much attention is being given to lowering emissions of volatile organic compounds, it is likely that non-polluting vehicles will have to be manufactured and a better **mass transit** system created.

[*Peter Brimblecombe*]

RESOURCES

BOOKS

Environmental Effects of Automotive Emissions. Paris: OECD Compass Project, 1986.

Kennedy, D., and R. R. Bates, eds. *Air Pollution, the Automobile, and Public Health.* Washington, DC: National Academy Press, 1988.

PERIODICALS

Renner, M. G. "Car Sick." *World Watch* (November-December 1988): 36–43.

Autotroph

An organism that derives its **carbon** for building body tissues from **carbon dioxide** (CO_2) or carbonates and obtains its energy for bodily functions from radiant sources, such as sunlight, or from the oxidation of certain inorganic substances. The leaves of green plants and the bacteria that oxidize sulfur, iron, ammonium, and nitrite are examples of autotrophs. The oxidation of ammonium to nitrite, and of nitrite to nitrate, a process called **nitrification**, is a critical part of the **nitrogen cycle**. Moreover, the creation of food by photosynthetic organisms is largely an autotrophic process.

Avalanche

A sudden slide of snow and ice, usually in mountainous areas where there is heavy snow accumulation on moderate to steep slopes. Snow avalanches flow at an average speed of 80 mph (130 km/hr), and their length can range from less than 300 ft (100 m) to 2 mi (3.2 km) or more. Generally the term "avalanche" refers to sudden slides of snow and ice, but it can also be used to describe catastrophic debris slides consisting of mud and loose rock. Debris avalanches are especially associated with volcanic activity in which melted snow, earthquakes, and clouds of flowing ash can trigger movement of rock and mud. Snow avalanches generally consist either of loose, fresh snow or of slabs of accumulated snow and ice that move in large blocks. Snow avalanches occur most often where the snow surface has melted under the sun and then refreezes, forming a smooth surface of snow. Later snow falling on this smooth surface tends to adhere poorly, and it may slide off the slick plane of recrystallized snow when it is shaken by any form of vibration—including sound waves, earthquakes, or the movement of skiers.

Several factors contribute to snow avalanches, including snow accumulation, hill slope angle, slope shape (profile), and weather. Avalanches are most common where there is heavy snow accumulation on slopes of 25–65°, and they occur most often on slopes between 30° and 45°. On slopes steeper than 65° snow tends to sluff off rather than accumulate. On shallow slopes avalanches are likely to occur only in wet (melting) conditions, when accumulated snow may be heavy, and when snow melt collecting along a hardened old snow surface within the snowpack can loosen upper layers, allowing them to release easily. Slab avalanches may be more likely to start on convex slopes, where snow masses can be fractured into loose blocks, but they rarely begin on tree-covered slopes. However, loose snow avalanches often start among trees, gathering speed and snow as they cross open slopes. Weather can influence avalanche **probability** by changing the **stability** and cohesiveness of the snow pack. Many avalanches occur during storms when snow accumulates rapidly, or during sustained periods of cold weather when new snow remains loose. Like snow melt, rainfall, can increase chances of avalanche by lubricating the surface of hardened layers within the snow pack. Sustained winds increase snow accumulation on the leeward side of slopes, producing snow masses susceptible to slab movement. When conditions are favorable, an avalanche can be triggered by the weight of a person or by loud noises, earthquakes, or other sources of vibration. Avalanches tend to be most common in mid-winter, when snow accumulation is high, and in late spring, when melting causes instability in the snow pack.

Avalanches play an ecological role by keeping slopes clear of trees, thus maintaining openings vegetated by grasses, forbs, and low brush. They are also a geomorphologic force, since they maintain bare rock surfaces, which are susceptible to **erosion**.

Most research into the dynamics and causes of avalanches has occurred in populous mountain regions such as the Alps, the Cascades, and the Rocky Mountains, where avalanches cause damage and fatalities by crushing buildings and vehicles. Avalanches are very powerful: they can crush buildings, remove full-grown trees from hillsides, and even sweep railroad trains from their tracks. One of the greatest avalanche disasters on record occurred in 1910 in the Cascades near Seattle, Washington, when a passenger train, trapped in a narrow valley in a snow storm for several days, was caught in an avalanche and swept to the bottom of the valley. Ninety-six passengers died as the cars were crushed with snow. Although avalanches are among the more dangerous natural hazards, they have caused fewer than 200 recorded mortalities in North America, and most avalanche victims in North America are caught in slides they triggered themselves by walking or skiing across open slopes with accumulated snow.

[*Mary Ann Cunningham Ph.D.*]

RESOURCES
BOOKS

Brugnot, G., ed. *Snow and Avalanches*. Conference proceedings, Chamonix, Switzerland, 1992. Zurich: Cemagret, 1995.

Salm, B., and H. Gubler. *Avalanche Formation, Movement, and Effects*. Wallingford, Oxfordshire, UK: IAHS Press, 1987.

United States Forest Service. "Snow Avalanche: General rules for Avoiding and Surviving Snow Avalanches." Portland, OR: USDA Forest Service, 1982.

B

Bacillus thuringiensis

Bacillus thuringiensis, or *B.t.*, is a family of bacterial-based, biological insecticides. Specific strains of *B.t.* are used against a wide variety of leaf-eating lepidopteran pests such as European corn borer, tomato hornworms, and **tobacco** moths, and some other susceptible insects such as blackflies and mosquitoes. The active agent in *B.t.* is toxic organic crystals that bind to the gut of an insect and poke holes in cell membranes, literally draining the life from the insect. *B.t.* can be applied using technology similar to that used for chemical insecticides, such as high-potency, low-volume sprays of *B.t.* spores applied by aircraft. The efficacy of *B.t.* is usually more variable and less effective than that of chemical insecticides, but the environmental effects of *B.t.* are considered to be more acceptable because there is little non-target toxicity.

Background radiation

Ionizing radiation has the potential to kill cells or cause somatic or germinal mutations. It has this ability by virtue of its power to penetrate living cells and produce highly reactive charged ions. It is the charged ions which cause cell damage. Radiation accidents and the potential for radiation from nuclear (atomic) bombs create a fear of radiation release due to human activity. Many people are subjected to diagnostic and therapeutic radiation, and older Americans were exposed to **radioactive fallout** from atmospheric testing of **nuclear weapons**. There is some environmental contamination from nuclear fuel used in **power plants**. Accordingly, there is considerable interest in radiation effects on biological systems and the sources of radiation in the **environment**.

Concern for radiation safety is certainly justified and most individuals seek to minimize their exposure to human-generated radiation. However, for most people, exposure levels to radiation from natural sources far exceed exposure to radiation produced by humans. Current estimates of human exposure levels of ionizing radiation suggest that only about 18% is of human origin. The remaining radiation (82%) is from natural sources and is referred to as "background radiation." While radiation doses vary tremendously from person to person, the average human has an annual exposure to ionizing radiation of about 360 millirem. (Millirem or mrem is a measure of radiation absorbed by tissue multiplied by a factor that takes into account the biological effectiveness of a particular type of radiation and other factors such as the competence of radiation repair. One mrem is equal to 10 µSv; µSv is an abbreviation for microSievert, a unit that is used internationally.)

Some radiation has little biological effect. Visible light and infrared radiation do not cause ionization, are not mutagenic and are not carcinogenic. Consequently, background radiation refers to ionizing radiation which is derived from cosmic radiation, terrestrial radiation, and radiation from sources internal to the body. (Background radiation has the potential for producing inaccurate counts from devices such as a Geiger counter. For example, cosmic rays will be recorded when measuring the **radioactive decay** of a sample. This background "noise" must be subtracted from the indicated count level to give a true indication of activity of the sample.)

Cosmic rays are of galactic origin, entering the earth's **atmosphere** from outer space. Solar activity in the form of sunflares and sunspots affects the intensity of cosmic rays. The atmosphere of the earth serves as a protective layer for humans and anything that damages that protective layer will increase the **radiation exposure** of those who live under it. The dose of cosmic rays doubles at 4,920 ft (1,500 m) above sea level. Because of this, citizens of Denver, near the Rocky Mountains, receive more than twice the dose of radiation from cosmic rays as do citizens of coastal cities such as New Orleans. The **aluminum** shell of a jet airplane provides little protection from cosmic rays, and for this reason passengers and crews of high flying jet airplanes receive more radiation than their earth traveling compatriots. Even greater is the cosmic radiation encountered at 60,000 ft

(18,300 m) where supersonic jets fly. The level of cosmic radiation there is 1,000 times that at sea level. While the cosmic ray dose for occasional flyers is minimal, flight and cabin crews of ordinary jet airliners receive an additional exposure of 160 mrem per year, an added radiation burden to professional flyers of more than 40%. Cosmic sources for non-flying citizens at sea level are responsible for about 8% (29–30 mrem) of background radiation exposure per annum.

Another source of background radiation is terrestrial **radioactivity** from naturally occurring minerals, such as **uranium**, thorium, and cesium, in **soil** and rocks. The abundance of these minerals differs greatly from one geographic area to another. Residents of the Colorado plateau receive approximately double the dose of terrestrial radiation as those who live in Iowa or Minnesota. The geographic variations are attributed to the local composition of the earth's crust and the kinds of rock, soil and minerals present. Houses made of stone are more radioactive than houses made of wood. Limestones and sandstones are low in radioactivity when compared with granites and some shales. Naturally occurring **radionuclides** in soil may become incorporated into grains and vegetables and thus gain access to the human body. **Radon** is a radioactive gas produced by the disintegration of radium (which is produced from uranium). Radon escapes from the earth's crust and becomes incorporated into all living matter including humans. It is the largest source of inhaled radioactivity and comprises about 55% of total human radiation exposure (both background and human generated). Energy efficient homes, which do not leak air, may have a higher concentration of radon inside than is found in outside air. This is especially true of basement air. The radon in the home decays into radioactive "daughters" that become attached to **aerosol** particles which, when inhaled, lodge on lung and tracheal surfaces. Obviously, the level of radon in household air varies with construction material and with geographic location. Is radon in household air a hazard? Many people believe it is, since radon exposure (at a much higher level than occurs breathing household air) is responsible for lung **cancer** in non-smoking uranium miners.

Naturally occurring radioactive **carbon** (carbon-14) similarly becomes incorporated into all living material. Thus, external radiation from terrestrial sources often becomes internalized via food, water, and air. Radioactive atoms (radionuclides) of carbon, uranium, thorium, and actinium and radon gas provide much of the terrestrial background radiation. The combined annual exposure to terrestrial sources, including internal radiation and radon, is about 266 mrem and far exceeds other, more feared sources of radiation.

Life on earth evolved in the presence of ionizing radiation. It seems reasonable to assume that mutations can be attributed to this chronic, low level of radiation. Mutations are usually considered to be detrimental, but over the long course of human and other organic **evolution**, many useful mutations occurred, and it is these mutations that have contributed to the evolution of higher forms.

Nevertheless, it is to an organism's advantage to resist the deleterious effects associated with most mutations. The forms of life that inhabit the earth today are descendants of organisms that existed for millions of years on earth. Inasmuch as background ionizing radiation has been on earth longer than life, humans and all other organisms obviously cope with chronic low levels of radiation. Survival of a particular **species** is not due to a lack of genetic damage by background radiation. Rather, organisms survive because of a high degree of redundancy of cells in the body, which enables organ function even after the death of many cells (e.g., kidney and liver function, essential for life, does not fail with the loss of many cells; this statement is true for essentially all organs of the human body). Further, stem cells in many organs replace dead and discarded cells. Naturally occurring antioxidants are thought to protect against free radicals produced by ionizing radiation. Finally, repair mechanisms exist which can, in some cases, identify damage to the double helix and effect DNA repair. Hence, while organisms are vulnerable to background radiation, mechanisms are present which assure survival.

[*Robert G. McKinnell*]

RESOURCES
BOOKS

Benarde, M. A. *Our Precarious Habitat: 15 Years Later.* New York: Wiley, 1989.

Hall, E. J. "Principles of Carcinogenesis: Physical." In *Cancer: Principles and Practice of Oncology,* edited by V. T. DeVita, et al. 4th ed. Philadelphia: Lippincott, 1993.

Knoche, H. W. *Radioisotopic Methods for Biological and Medical Research.* New York: Oxford University Press, 1991.

Sir Francis Bacon (1561 – 1626)
English statesman, author, and philosopher

Sir Francis Bacon, philosopher and Lord Chancellor of England, was one of the key thinkers involved in the development of the procedures and epistemological standards of modern science. Bacon thus has also played a vital role in shaping modern attitudes towards **nature**, human progress, and the **environment**. He inspired many of the great thinkers of the Enlightenment, especially in England and France. Moreover, Bacon laid the intellectual groundwork for the mechanistic view of the universe characteristic of eighteenth and nineteenth century thought and for the explosion of technology in the same period.

Sir Francis Bacon. (Painting by Paul Somer. Corbis-Bettmann. Reproduced by permission.)

In *The Advancement of Learning* (1605) and *Novum Organum* (1620), Bacon attacked all teleological ways of looking at nature and natural processes (i.e., the idea found in Aristotle and in medieval scholasticism that there is an end or purpose which somehow guides or shapes such processes). For Bacon, this way of looking at nature resulted from the tendency of human beings to make themselves the measure of the outer world, and thus to read purely human ends and purposes into physical and biological phenomena. Science, he insisted, must guard against such prejudices and preconceptions if it was to arrive at valid knowledge.

Instead of relying on or assuming imaginary causes, science should proceed empirically and inductively, continuously accumulating and analyzing data through observation and experiment. Empirical observation and the close scrutiny of natural phenomena allow the scientist to make inferences, which can be expressed in the form of hypotheses. Such hypotheses can then be tested through continued observation and experiment, the results of which can generate still more hypotheses. Advancing in this manner, Bacon proposed that science would come to more and more general statements about the laws which govern nature and, eventually, to the secret nature and inner essence of the phenomena it studied.

As Bacon rather famously argued, "Knowledge is power." By knowing the laws of nature and the inner essence of the phenomena studied, human beings can remake things as they desire. All knowledge is for use, and the underlying motivation of science is technical control of nature. Bacon believed that science would ultimately progress to the point that the world itself would be, in effect, merely the raw material for whatever future ideal society human beings decided to create for themselves.

The possible features of this future world are sketched out in Bacon's unfinished utopia, *The New Atlantis* (1627). Here Bacon developed the view that the troubles of his time could be solved through the construction of a community governed by natural scientists and the notion that science and technology indeed could somehow redeem mankind. Empirical science would unlock the secrets of nature thus providing for technological advancement. With technological development would come material abundance and, implicitly, moral and political progress.

Bacon's utopia is ruled by a "Solomon's House"—a academy of scientists with virtually absolute power to decide which inventions, institutions, laws, practices, and so forth will be propitious for society. Society itself is dedicated to advancing the human mastery of nature: "The End of Our Foundation is the Knowledge of Causes and secret motions of things; and the enlarging of the bounds of the human empire, to the effecting of all things possible."

[*Lawrence J. Biskowski*]

RESOURCES
BOOKS

Bacon, F. *The Advancement of Learning.* 1605.
Sibley, M. Q. *Nature and Civilization.* Itasca, IL: Peacock, 1977.
———. *The New Atlantis.* 1627.
———. *Novum Organum.* 1620.

BACT

see **Best available control technology**

Baghouse

An **air pollution control** device normally using a collection of long, cylindrical, fabric **filters** to remove **particulate** matter from an exhaust air stream. The filter arrangement is normally designed to overcome problems of cleaning and handling large exhaust volumes. In most cases, exhaust gas enters long (usually 33–50 ft [10–15 m]), vertical, cylindrical filters on the inside from the bottom. The bags are sealed at the top. As the exhaust air passes through the fabric filter, particles are separated from the air stream by sticking either to the filter fabric or to the cake of particles previously collected on the inside of the filter. The exhaust then passes

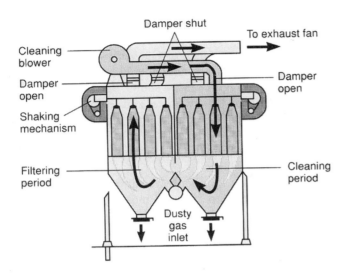

Baghouse. (McGraw-Hill Inc. Reproduced by permission.)

to the **atmosphere** free of most of its original particulate-matter **loading**; collection efficiency usually increases with particle size.

The buildup of particles on the inside of the bags is removed periodically by various methods, such as rapping the bags, pulsing the air flow through the bags, or shaking. The particles fall down the long cylindrical bags and are normally caught in a collection bin, which is unloaded periodically. A baghouse system is usually much cheaper to install and operate than a system using **electrostatic precipitation** to remove particulates.

Balance of nature

The ideal of a balance of **nature** is based on a view of the natural world that is largely an artifact created by the temporal, spatial, and cultural **filters** through which humans respond to the natural world. For a variety of reasons we have interpreted the natural course of events in the world around us to maintain equilibrium, and seek to return it to equilibrium when disturbed.

There are three components to nature's balance: ecological, evolutionary, and population. In an ecological sense,

communities were thought to proceed through successional stages to a steady state climax. When disturbed, the community would return to that climax state. **Stability** was an endpoint, and once reached the community became a partly closed homeostatic system. In an evolutionary sense, the current compliment of **species** is interpreted as the ultimate product of **evolution**, rather than a temporary expression of a continually changing global taxa. In the population sense, concepts like **carrying capacity** and the constant interplay between environmental **resistance** and biotic potential is interpreted as creating a balance of numbers in a population, and between the population and its **environment**. Three ideas are fundamental to the above; that nature undisturbed is constant, when disturbed nature returns to the constant condition, and constancy in nature is the desired endpoint.

This interpretation of nature may be so strongly filtered by our cultural interpretation and idealization of balance, that we tend to produce conclusions not in keeping with our observations of nature. Assumptions of human centrality may be sufficiently strong to bend the usually clear lens supplied by science in this case. Although the theme of balance in nature has been formally criticized in **ecology** for over 65 years (since Frederick Clements and Henry Gleason focused the argument in the 1920s) the core of the science did not change until about 25 years ago. Since that time, a dynamic approach that pays no special attention to equilibrium processes has taken center stage in ecological theorizing.

The primary alternatives are part of the group of ideas termed intermediate disturbance hypotheses. These ideas offer a different view of how communities assemble, suggesting that disturbance is more frequent and/or more influential than performing a routine return to an equilibrium state. Furthermore, disturbance and non-equilibrium situations are responsible for the most diverse communities, tropical rain forests and coral reefs, through the reduction in **competition** caused by disturbance factors.

Few theorists, however, suggest that non-equilibrium settings are the single most powerful explanation, or are mutually exclusive with communities that do have an equilibrium. There are situations that seem to seek equilibrium, and smaller subsystems that appear virtually closed. In local situations, certain levels of resources and disturbance may create long term stability and **niche** differentiation or other mechanisms may be the principal cause of a species diverse situation.

Although theorists have been working to verify, revise, and examine new developments in ecology, very little attention has been given to alternative, more complex theoretical interpretations of nature, in terms of time, space, and dynamism in resource and environmental management. The implications of accepting a non-equilibrium orientation for environmental management are significant. Most of the underpinnings of resource management include steady state carrying capacity, **succession**, predator-prey balance, and community equilibrium as foundations.

There are three major implications of the shift from equilibrium to non-equilibrium approaches to the environment. First, until a more realistic theory is used, the rate of resource extraction from nature will be subject to considerably higher uncertainty than we currently suspect. Since we expect communities to seek equilibrium, we believe we can predict populations and species numbers to a greater degree than may be warranted. Second, we perceive that areas of high **biodiversity** are due to long and short term stability, when in fact the forces may be just the opposite. Therefore, management that attempts to maintain stability is the reverse of what is actually needed. Third, the kinds of disturbance we create in these diverse communities (**deforestation**, introduction of non-native species, or oil **pollution**) may not mimic anything natural and a species may have little defense against them. A characteristic of communities with high species diversity is small population size, thus human disturbance may cause exceptionally high rates of **extinction**. A central facet of the burgeoning practice of ecological restoration should be an ability to accurately mimic disturbance regimes.

Balance in nature has a strong appeal to our sensibilities and has anchored natural resource management practices. Now practice stands well behind theoretical developments and increased integration of more modern ecological science is required to avoid costly resource management mistakes.

[*Dave Duffus*]

RESOURCES
BOOKS

Botkin, D. E. *Discordant Harmonies: A New Ecology for the Twenty First Century*. New York: Oxford University Press, 1990.

Ehrenfeld, D. *The Arrogance of Humanism*. New York: Oxford University Press, 1981.

Pickett, S. T. A., V. T. Parker and P. L. Fiedler. "The New Paradigm in Ecology: Implications for Conservation Biology Above the Species Level." In *Conservation Biology*, edited by P. L. Fiedler and S. K. Jain. New York: Chapman and Hall, 1992.

PERIODICALS

Gleason, H. A. "The Individualistic Concept of Plant Association." *Bulletin of the Torrey Botanical Club* 53 (1926): 7–26.

Bald eagle

The bald eagle (*Haliaeetus leucocephalus*), one of North America's largest birds of prey with a wingspan of up to 7.5 ft (2.3 m), is a member of the family Accipitridae. Adult bald eagles are dark brown to black with a white head and tail; immature birds are dark brown with mottled white wings and are often mistaken for golden eagles (*Aquila chrysaetos*). Bald eagles feed primarily on fish, but also eat rodents, other small mammals and carrion. The bald eagle is the national emblem for the United States, adopted as such in 1782 because of its fierce, independent appearance. This characterization is unfounded, however, as this **species** is usually rather timid.

Formerly occurring over most of North America, the bald eagle's range—particularly in the lower 48 states—had been drastically reduced by a variety of reasons. One being its exposure to DDT and related pesticides, which are magnified in the **food chain/web**. This led to reproductive problems, in particular, thin-shelled eggs that were crushed during incubation. The banning of DDT use in the United States in 1972 may have been a turning point in the recovery of the bald eagle. Eagle populations also were depleted due to **lead poisoning**. Estimates are that for every bird that hunters shot and carried out with them, they left behind about a half pound of **lead shot**, which affects the **wildlife** in that **ecosystem** long after the hunters are gone. Since 1980, more than 60 bald eagles have died from lead poisoning. Other threats facing their populations include **habitat** loss or destruction, human encroachment, collisions with high power lines, and shooting.

In 1982 the population in the lower 48 states had fallen to less than 1,500 pairs, but by 1988 their numbers had risen to about 2,400 pairs. Due to strict **conservation** laws, the numbers have continued to rise and there are now 6,000 pairs. On July 4, 2000, the bald eagle was removed from the Endangered listing and is now listed as Threatened in the lower 48 states. The bald eagle is not endangered in the state of Alaska, since a large, healthy population of about 35,000 birds exists there. During the annual **salmon** run, up to 4,000 bald eagles congregate along the Chilkat River in Alaska to feed on dead and dying salmon.

Bald eagles, which typically mate for life and build huge platform nests in tall trees or cliff ledges, have been aided by several recovery programs, including the construction of artificial nesting platforms. They will **reuse**, add to, or repair the same nest annually, and some pairs have been known to use the same nest for over 35 years. Because the bald eagle is listed as either endangered or threatened throughout most of the United States, the federal government provides some funding for its conservation and recovery projects. In 1989, the federal government spent $44 million on the conservation of threatened and **endangered species**. Of the 554 species listed, $22 million, half of the total allotment, was spent on the top 12 species on a prioritized

Bald eagle (*Haliaeetus leucocephalus*). (Photograph by Robert J. Huffman. Field Mark Publications. Reproduced by permission.)

list. The bald eagle was at the top of that list and received $3 million of those funds.

[*Eugene C. Beckham*]

RESOURCES
BOOKS

Ehrlich, P., D. Dobkin, and D. Wheye. *Birds in Jeopardy*. Stanford, CA: Stanford University Press, 1992.
Temple, S., ed. *Endangered Birds: Management Techniques for Preserving Threatened Species*. Madison: University of Wisconsin Press, 1977.

PERIODICALS

Dunstan, T. C. "Our Bald Eagle: Freedom's Symbol Survives." *National Geographic* 153 (1978): 186–99.

ORGANIZATIONS

American Eagle Foundation, P.O. Box 333, Pigeon Forge, TN USA 37868 (865) 429-0157, Fax: (865) 429-4743, Toll Free: (800) 2EAGLES, Email: EagleMail@Eagles.Org., <http://www.eagles.org>

B(a)P

see **Benzo(a)pyrene**

Barrier island

An elongated island that lies parallel to, but mostly separate from, a coastline. Barrier islands are composed of sediments, mainly sand, deposited by longshore currents, wind, and wave action. Both marine and terrestrial plants and animals find **habitat** on barrier islands or along their sandy beach shorelines. These islands also protect coastal lagoons from ocean currents and waves, providing a warm, quiet **environment** for **species** that cannot tolerate more violent wind and wave conditions. In recent decades these linear, sandy islands, with easy access from the mainland, have proven a popular playground for vacationers. These visitors now pose a significant threat to breeding birds and other coastal species. Building houses, roads, and other disruptive human activities can destabilize **dunes** and expose barrier islands to disastrous storm damage. In some cases, whole islands are swept away, exposing protected lagoons and delicate **wetlands** to further damage. Major barrier island formations in North America include those along the eastern coasts of the mid-Atlantic states, Florida, and the Gulf of Mexico.

Located parallel to coastlines, Barrier islands attract visitors that now pose a significant threat to the birds and other coastal species living there. (JLM Visuals. Reproduced by permission.)

Basel Convention

The Basel Convention on the Control of Transboundary Movements of Hazardous Wastes and their Disposal is a global treaty that was adopted in 1989 and became effective on May 5, 1992. The Basel Convention represents a response by the international community to problems caused by the international shipment of wastes. Of the over 400 million tons of hazardous wastes generated globally every year, an unknown amount is subject to transboundary movement. (United States companies were estimated to export about 140,000 tons of **hazardous waste** in 1993). In the 1980s, several highly publicized "toxic ships" were accused of trying to dump hazardous wastes illegally in developing countries. The uncontrolled movement and disposal of hazardous waste, especially in developing countries which often lacked the know-how and equipment to safely manage and treat hazardous wastes, became a significant problem due to high domestic costs of treating or disposing of wastes.

Through its Secretariat in the **United Nations Environment Programme**, the Convention aims to control the transboundary movement of wastes, monitor and prevent illegal traffic, provide technical assistance and guidelines, and promote cooperation. The Convention imposes obligations on treaty signatories to ensure that wastes are managed and disposed of in an environmentally sound manner. The main principles of the Basel Convention are to (1) reduce transboundary movements of hazardous wastes to a minimum consistent with environmentally sound management; (2) treat and dispose of hazardous wastes as close as possible to their source of generation; and (3) reduce and minimize the generation of hazardous waste. The Convention generally prohibits Parties to the Convention from importing or exporting hazardous wastes or other wastes from or to a non-contracting party. However, a Party to the convention may allow such import or export if the Party has a separate bilateral or multilateral agreement regarding the transboundary movement of hazardous wastes with a non-party and that agreement provides for "environmentally sound management." To date, the Convention has defined environmentally sound management practices for a number of wastes and technologies, including organic solvents, waste oils, pentachloraphenol (PCBs), household wastes, landfills, incinerators, and oil **recycling**.

The Conference of the Parties (COP) is the governing body of the Basel Convention and is composed of all governments that have ratified or acceded to it. The COP has met three times since the Convention entered into force in May 1992 (in Piriapolis, Uruguay, on December 4, 1992; Geneva, Switzerland, on March 25, 1994; and Geneva, Switzerland, September 18–22, 1995). In the most recent meeting, the Parties adopted an amendment to the Convention that will ban the export of hazardous wastes from developed countries to developing ones. The Fourth Meeting of the Conference of the Parties, held in Kuala Lumpur in October 1997, incorporated work on lists of wastes into the system of the Basel Convention. These lists should help to mitigate practical difficulties in determining exactly what is a waste, a problem that has been encountered by a number of Parties.

Currently, the European Community and 108 States are Party to the Basel Convention. Though the United States was among the original signatories to the Basel Convention in 1989, the United States is still not a party to the Convention since Congress has yet to pass implementing legislation. The United States does participate in planning and technical aspects of the Convention.

[*Stuart Batterman*]

Rick Bass. (Photograph by L. L. Griffin. Reproduced by permission.)

Rick Bass (1958 –)

American writer

Bass was born in south Texas and grew up there, absorbing stories and family lore from his grandfather during deer-hunting trips. These early forays into Texas hill country form the basis of the author's first book, *The Deer Pasture*, published when he was 27 years old. At one time or another, Bass has lived in Texas, Mississippi, Vermont, Utah, Arkansas, and Montana.

Bass received a degree in geology from Utah State University in 1979 and went to work as a **petroleum** geologist in Mississippi, prospecting for new oil **wells**. This experience informed one of his better-known nonfiction books, *Oil Notes*. Written in journal form, *Oil Notes* offers meditations on the art and science of finding energy in the ground, as well as reflections on the author's personal life and his outdoor adventures.

Bass is a passionate environmentalist whose nonfiction in particular celebrates efforts to reclaim a wilder America. Books such as *The Lost Grizzlies: A Search for Survivors* demonstrate his conviction that America's larger predators should be allowed to survive and thrive. Bass highlights the plight of the wolf in 1992's *The Ninemile Wolves* and 1998's *The New Wolves*.

Bass features his adopted Montana homeland in some of his publications, including his nonfiction titles *Winter: Notes from Montana*, a 1991 release, and *The Book of Yaak*, which was published in 1996. Bass's essays are often linked by comments about his dog, a German shorthaired pointer named Colter. Bass gave readers a more exclusive look at Colter with his 2000 book *Colter: The True Story of the Best Dog I Ever Had*. The work, however, also stays true to Bass's characteristic nature themes. *Publisher's Weekly* said *Colter* was as much "a book about appreciating nature and life" as it was the story of Bass and his dog.

Although most of Bass's publications are nonfiction, the author has written both short stories and novels. Perhaps not surprisingly, the masculine sports of **hunting**, fishing, and drinking are central to many of Bass's stories. Bass published his first novel in 1998, *Where the Sea Used to Be*. The story involves a young geologist named Wallis who is sent by an oilman named Old Dudley to an isolated part of Montana. While investigating the land, Wallis stays with Mel, Old Dudley's daughter. Wallis comes to love both Mel, who studies **wolves**, and the remote landscape. This book was followed by a collection of short stories in 2002, called *The Hermit's Story*. Bass continued to publish essays and stories in the early 2000s in many popular magazines, including *Sports Afield*, *Audobon*, *National Geographic Traveler*, *Atlantic Monthly*, *Sierra*, and others.

RESOURCES
PERIODICALS

Kinsella, Bridget. "Taking a Title to the Dogs and Beyond" *Publishers Weekly* 247, no. 23 (June 5, 2000): 19.

Bats

Bats, the only mammals that fly, are among nature's least understood and unfairly maligned creatures. They are extremely valuable animals, responsible for consuming huge numbers of insects and pollinating and dispersing the seeds of fruit-bearing plants and trees, especially in the tropics. Yet, superstitions about and fear of these nocturnal creatures have led to their persecution and elimination from many areas, and several **species** of bats are now threatened with **extinction**.

There are over 900 species of bats, representing almost a quarter of all mammal species, and they are found on every continent except **Antarctica**. Most types of bats live in the tropics, and some 40 species are found in the United States and Canada. The largest bats, flying foxes, found on Pacific islands, have wingspreads of five feet (1.5 m). The smallest bats, bamboo bats, are the size of the end of a person's thumb.

Bats commonly feed on mosquitoes and other night-flying insects, especially over ponds and other bodies of water. Some bats consume half of their weight in insects a night, eating up to 5,000 gnat-sized mosquitoes an hour, thus helping to keep insect population under control. Some bats hunt ground-dwelling species, such as spiders, scorpions, large insects, and beetles, and others prey on **frogs**, lizards, small birds, rodents, fish, and even other bats. The infamous vampire bat of Central and South America does actually feed on blood, daily consuming about a tablespoon from cattle and other animals, but it does not generally bother humans.

Bats that live in tropical areas, such as fruit bats (also called flying foxes), often feed on and pollinate plants. Bats are thus extremely important in helping flowers and fruit-bearing plants to reproduce. In tropical rain forests, for example, bats are responsible for pollinating most of the fruit trees and plants.

Bats are usually social animals. Some colonies consist of millions of bats and use the same roost for centuries. Bat manure (**guano**) is often collected from caves and used as **fertilizer**. Most bats come out only at night and spend their days in dark roosts, hanging upside down, sleeping, nursing and tending their young, or grooming their wings and fur. Bats become active an hour or so before dark, and at dusk they leave their roosting areas and fly out to feed, returning home before dawn. Many bats flying at night navigate and locate food, such as flying insects, by echolocation, emitting continuous high frequency sounds that echo or bounce off of nearby objects. Such sounds cannot be heard by humans. Most bats have just one or two young a year, though some have up to four offspring at a time. The newborn must hold onto its mother, sometimes for several weeks, and be nursed for six to eight weeks. Some species of bats live up to 25 years. Most bats in North America migrate or hibernate in caves during the winter, when food is scarce and temperatures reach freezing point. Superstitions about and prejudice against bats have existed for hundreds of years, but most such tales are untrue. Bats do not carry bedbugs or become entangled in women's hair; they are not blind and indeed do not even have poor vision. In fact, except for the occasional rabid bat, these creatures are not dangerous to humans and are quite timid and will try to escape if confronted. In recent years, public education programs and conservationists, such as Dr. Merlin Tuttle, head of Bat Conservation International in Austin, Texas, have helped correct these misconceptions about bats and have increased appreciation for the valuable role these creatures play in destroying pests and pollinating crops. Bracken Cave, located between San Antonio and Austin, is owned by Bat Conservation International and with some 20 million Mexican freetailed bats residing there in the spring and summer, the cave is said to shelter the world's largest bat colony and the largest collection of mammals anywhere on the planet. The pregnant females migrate there in early March from central Mexico to nurse and raise their young, and the colony can consume 250 tons of insects a night.

According to Dr. Tuttle, a colony of just 150 big brown bats can eat almost 40,000 cucumber beetles in a summer, which "means that they've protected local farmers from 18 million root worms, which cost American farmers $1 billion a year," including crop damage and **pesticide** costs. Dr. Tuttle and his organization suggest that people attract the creatures and help provide **habitat** for them by constructing or buying bathouses, which his groups sell. Nevertheless, bats continue to be feared and exterminated throughout the world. Major threats to the survival of bats include intentional killing, loss of habitat (such as old trees, caves, and mines), eviction from barns, attics, and house eaves, pesticide **poisoning**, and vandalism and disturbance of caves where they roost. According to Dr. Tuttle, "Bats are among the most endangered animals in America. Nearly 40 percent of America's 43 species are either endangered or candidates for the list."

Indiana bat (*Myotis sodalis*). (Photograph by Merlin D. Tuttle. Bat Conservation International. Reproduced by permission.)

Over a dozen species of bats worldwide are listed by the **U.S. Department of the Interior** as **endangered species**, including the gray bat (*Myotis grisescens*) of central and southeastern United States; the Hawaiian hoary bat (*Lasiurus cinereus semotus*); the Indiana bat (*Myotis sodalis*) of the eastern and mid-western United States; the Ozark big-eared bat (*Plecotus townsendii ingens*) found in Missouri, Oklahoma, and Arkansas; the Mexican long-nosed bat (*Leptonycteris nivalis*) of New Mexico, Texas, Mexico, and Central America; Sanborn's long-nosed bat (*Leptonycteris sanborni*) of Arizona, New Mexico, Mexico, and Central America; and the Virginia big-eared bat (*Plecotus townsendii virginianus*) found in Kentucky, North Carolina, West Virginia, and Virginia.

[*Lewis G. Regenstein*]

RESOURCES

BOOKS

Neuweiler, Gerhard. *The Biology of Bats.* Translated by Ellen Covey. New York: Oxford University Press, 2000.

Tuttle, Merlin D. *America's Neighborhood Bats.* Austin: University of Texas Press, 1988.

PERIODICALS

Raver, Anne. "Batman Returns (The Real One)." *The New York Times* (May 23, 1993): 18.

OTHER

U.S. Fish & Wildlife Service. December 2001 [cited May 2002]. <http://ecos.fws.gov/webpage>.

Batteries

see **Lead**

Battery recycling

The United States consumes over $10.4 billion in batteries annually, powering everything from children's toys to hearing aids. Because batteries contain certain toxic substances, such as **cadmium**, **lead**, and sulfuric **acid**, introducing them into landfills and other **solid waste** disposal facilities once they are used can be harmful to the **environment** and to public health.

Virtually every type of battery currently in common use—alkaline, lead acid, nickel-cadmium, lithium **ion**, and more—can be recycled to some extent. Even rechargeable batteries, which were designed in part to cut down on the expense and environmental impact of battery consumption, can be recycled after they have lost the ability to hold a charge.

The Universal Waste Rule, an amendment to the **Resource Conservation and Recovery Act** (RCRA) was introduced by the United States **Environmental Protection Agency** (EPA) in 1995 in an effort to lower some of the administrative and financial barriers to collection and **recycling** of batteries and other potentially hazardous household products. It was hoped that making recycling of lead batteries easier and more profitable to recycle would lead to more extensive recycling programs. The rule streamlined the regulatory process for businesses and excluded rechargeable batteries from **hazardous waste** handling requirements. However, individual states had the final determination over whether or not to adopt the amendment.

Further EPA legislation passed in 1996 entitled the Mercury-Containing Rechargeable Battery Management Act (or Battery Act) promotes recycling of rechargeable batteries through a national uniform code that removes obstacles presented by conflicting state recycling laws and regulations. The Battery Act also mandates that manufacturers of portable rechargeable batteries and products use universal recycling labeling, make batteries easy to remove from products, and prohibit the intentional introduction of **mercury**

into batteries (i.e., mercury added beyond the trace amount present naturally).

Some battery manufacturers have created take-back programs, sometimes referred to as product stewardship or extended producer responsibility, to encourage recycling of their products. Manufacturers "take back" used batteries that are returned by consumers, retailers, and community recycling programs. These programs are voluntary in the United States, and are popular in the **European Union** (EU) where they are legally required in many countries. Austria and Germany, for example, require all battery manufacturers to take back batteries at no cost to the consumer.

According to the EPA, over 350 million rechargeable batteries are sold each year in the United States. The industry trade magazine *Electronics Business News* puts worldwide sales of rechargeable batteries at approximately $5.5 billion in 2001. Yet only a small portion of these rechargeable batteries enter the recycling stream, and the batteries have become a significant source of heavy metal contamination at solid waste facilities.

Passage of the Battery Act paved the way for the largest North American take-back battery recycling program, administered by the non-profit Rechargeable Battery Recycling Corporation (RBRC). The RBRC facilitates and promotes awareness of rechargeable battery recycling throughout the United States and Canada. Its activities are funded through licensing fees paid by the by manufacturers of portable rechargeable batteries and products. The RBRC recycles **nickel** cadmium (Ni-Cd), nickel metal hydride (Ni-MH), lithium ion (Li-ion), and small sealed lead (Pb; SSLA) batteries used in cordless phones, power tools, laptop and notebook computers, cell phones, and other portable devices.

Lead-acid batteries, the type of wet-cell battery used in cars, boats, and other vehicles, are one of the most highly recycled products in America. In 1999, 93.3% of lead-acid batteries were recycled, compared to 42% of all paper and 52% of **aluminum** cans. As of early 2002, 42 states had adopted legislation requiring the recycling of lead-acid batteries. Some states charge a deposit fee to consumers who purchase a new lead-acid battery without trading in an old one for recycling. Most states that charge this fee (including Arizona, Minnesota, and New York) refund it if the consumer brings back a used battery for recycling after the purchase. In addition to keeping lead and sulfuric acid out of solid waste facilities, lead-acid battery recycling is a substantial source of recycled polypropylene and generates 2 billion lb (907 million kg) of lead for **reuse** annually.

Most alkaline and carbon-zinc batteries manufactured in the United States since 1993 do not contain added mercury, and therefore are not considered hazardous waste. Because of the perceived reduced risk of these "zero added mercury" batteries to the environment, many communities advise residents to deposit them into their regular trash, where they end up in landfills. This method is sometimes deemed more cost-effective than sorting and transporting these batteries for recycling. However, recycling programs for alkaline batteries are available and growing in popularity.

[*Paula Anne Ford-Martin*]

RESOURCES

BOOKS

Pistoia, J. et al eds. *Used Battery Collection and Recycling.* New York: Elsevier Science, 2001.

PERIODICALS

Sova, Chris and Harve Muellerx. "A Charged-Up Market: The Recycling of Several Types of Batteries Has Created a Number of Established Processes." *Recycling Today* 40, no.3 (March 2002):100.

OTHER

Municipal and Industrial Solid Waste Division, Office of Solid Waste, U.S. Environmental Protection Agency. "Mercury-Containing and Rechargeable Battery Management Act." 42 USC 14301. <http://www.access-.gpo.gov/uscode/title42/chapter137_.html> [May 17, 2002].

ORGANIZATIONS

Rechargeable Battery Recycling Corporation, 1000 Parkwood Circle, Suite 450, Atlanta, GA USA 30339 (678) 419-9990, Fax: (678) 419-9986, Email: corporate@rbrc.com, <http://www.rbrc.org>

Bay of Fundy

The Bay of Fundy is a marine **ecosystem** that lies on the northeastern coast of North America, bordering parts of the Canadian provinces of New Brunswick and Nova Scotia and the U.S. state of Maine. It encompasses about 62.5-thousand mi^2 (180-thousand km^2) of marine coastal-shelf **habitat**, mostly less than about 660 ft (200 m) deep. The Bay is renowned for its exceptionally high tides, which can exceed 53 ft (16 m) in its upper reaches in the Minas Basin. These are higher tides than occur anywhere else in the world. During the peak tidal **flooding** of the Bay the flow of water is about 880-million ft^3/s (25-million m^3/s), equivalent to about 2000 times the average flow of the Saint Lawrence River.

The astonishing tides of the Bay of Fundy occur because its long shape, great size, and increasing up-bay shallowness result in its tidal waters "piling up" to great depths. This effect is amplified by the natural period of tidal oscillation of the Bay of about 13 hours, which further pushes against the natural tidal cycle of 12.4 hours. This rare physical phenomenon is known as a "near-resonant response." The tidal heights are particularly extreme in the upper reaches of the Bay, but even in its lower areas small boats are commonly left high and dry during the twice-daily low tides, and rivers may have a reversing tidal bore (or advancing wave)

The Bay of Fundy with the New Brunswick skyline at its shores. (Communications NB. Reproduced by permission.)

moving upstream with each tide. The "Reversing Falls" near the mouth of the Saint John River is another natural phenomenon associated with the great tides of the Bay of Fundy.

The huge tidal flows of the Bay of Fundy result in great **upwellings** of nutrient-rich bottom waters at some places, allowing high rates of **ecological productivity** to occur. The high productivity of marine **phytoplankton** supports a dense **biomass** of small crustaceans known as **zooplankton**, which are fed upon by great schools of small fishes such as herring. The zooplankton and fishes attract large numbers of such seabirds as gulls, phalaropes, and shearwaters to the Bay during the summer and autumn months, and also abundant fin **whales**, humpback whales, northern right whales (this is the most **endangered species** of large whale), harbor porpoises, and white-sided **dolphins**.

The high productivity of the Bay once also supported large stocks of commercial marine **species**, such as cod, haddock, scallop, and others. Unfortunately, most of these potentially renewable resources have been decimated by overfishing. The Bay still, however, supports large commercial fisheries of lobster and herring. There has also been a huge development of **aquaculture** in the lower Bay, especially in

the Passamaquoddy Bay area of New Brunswick. In the late 1990s the production of Atlantic **salmon** had a value exceeding $100-million (Canadian). The intense management systems associated with salmon aquaculture have caused environmental damages in the vicinity of the holding pens, including the build-up of **organic waste**, contamination with anti-fouling **chemicals** and antibiotics, and genetic "pollution" caused when escaped fish inter-breed with the endangered native salmon of the Bay.

In shallow areas, the extreme tidal ranges of the Bay expose extensive mudflats at low tide. In some parts of the upper Bay these mudflats are utilized by immense numbers of shorebirds during their autumn **migration**. The most abundant of these is the semi-palmated sandpiper, one of the most abundant shorebirds in the world. During its autumn migration, hundreds of thousands of these birds feed on mud shrimp in exposed mudflats at low tide, and then aggregate in dense numbers on shingle beaches at high tide. The sandpipers greatly increase their body weight during the several weeks they spend in the upper Bay of Fundy, and then leave for a non-stop flight to South America, fuelled by the fat laid down in the Bay.

During the early 1970s there was a proposal to develop a huge tidal-power facility at the upper Bay of Fundy, to harvest commercial energy from the immense, twice-daily flows of water. The tidal barrage would have extended across the mouth of Minas Basin, a relatively discrete embayment with a gigantic tidal flow. Partly because of controversy associated with the enormous environmental damages that likely would have been caused by this ambitious development, along with the extraordinary construction costs and untried technology, this tidal-power facility was never built. A much smaller, demonstration project of 20 MW was commissioned in 1984 at Annapolis Royal in the upper Bay, and even this facility has caused significant local damages.

[*Bill Freedman Ph.D.*]

RESOURCES
BOOKS

Thurston, H. *Tidal Life: A Natural History of the Bay of Fundy.* London: Camden House Publishing, 1990.
The 2000 Canadian Encyclopedia. Toronto: McLelland and Stewart.

Beach renourishment

Beach renourishment, also called beach recovery or replenishment, is the act of rebuilding eroded beaches with offshore sand and gravel that is dredged from the sea floor. Renourishment projects are sometimes implemented to widen a beach for more recreational capacity, or to save structures built on an eroding sandy shoreline. The process is one that requires ongoing maintenance; the shoreline created by a renourished beach will eventually erode again. According to the **National Oceanic and Atmospheric Administration** (NOAA), the estimated cost of long-term restoration of a beach is between $3.3 and $17.5 million per mile.

The process itself involves **dredging** sand from an offshore site and pumping it onto the beach. The sand "borrow" or dredging site must also be carefully selected to minimize any negative environmental impact. Dredging can stir up **silt** and bottom **sediment** and cut off oxygen and light to marine **flora** and **fauna**.

Under the Energy and Water Development Appropriations Act of 2002, 65% of beach renourishment costs are paid for with federal funds, and the remaining 35% with state and local monies. Beach renourishment programs are often part of a state's coastal zone management program. In some cases, state and local government work with the U.S. Army Corp of Engineers to evaluate and implement an **erosion** control program such as beach renourishment.

Environmentalists argue that renourishment is for the benefit of commercial and private development, not for the benefit of the beach. Erosion is a natural process governed by weather and sea changes, and critics charge that tampering with it can permanently alter the **ecosystem** of the area in addition to threatening endangered sea turtle and seabird nesting habitats.

In some cases, it is the coastal development that brings about the need for costly renourishment projects. Coastal development can hasten the erosion of beaches, displacing **dunes** and disrupting beach grasses and other natural barriers. Other human constructions, such as sea walls and other armoring, can also alter the shoreline.

Results of a U.S. **Army Corps of Engineers** Study completed in 2001—the Biological Monitoring Program for Beach Nourishment Operations in Northern New Jersey (Manasquan Inlet to Asbury Park Section)—found that although dredging in the borrow area had a negative impact on local marine life, most **species** had fully recovered within 24–30 months. Further studies are needed to determine the full long-range impact of beach renourishment programs on **biodiversity** and local coastal habitats.

[*Paula Anne Ford-Martin*]

RESOURCES
BOOKS

Dean, Cornelia. *Against the Tide: The Battle for America's Beaches* New York: Columbia University Press, 2001.

PERIODICALS

U.S. Army Corp of Engineers, Engineer Research and Development Center. *The New York District's Biological Monitoring Program for the Atlantic Coast of New Jersey, Asbury Park to Manasquan Section Beach Erosion Control Project* Final Report, 2001.
U.S. Department of Commerce, National Oceanic & Atmospheric Administration, National Ocean Service, Office of Ocean & Coastal Resource Management. "State, Territory, and Commonwealth Beach Nourishment Programs: A National Overview." *OCRM Program Policy Series* Technical Document No. 00-01 (March 2000).

OTHER

National Oceanic & Atmospheric Administration, Coastal Services Center. *Beach Renourishment in the Southeast* http://www.csc.noaa.gov/opis/html/beach.htm. Accessed June 5, 2002.

Bear

see **Grizzly bear**

Mollie Beattie (1947 – 1996)

American forester and conservationist

Mollie Hanna Beattie was trained as a forester, worked as a land manager and administrator, and ended her brief career as the first woman to serve as director of the U.S. **Fish and Wildlife Service**. Beattie's bachelor's degree was in philosophy, followed by a master's degree in forestry from the University of Vermont in 1979. Later (in 1991), she used a Bullard Fellowship at Harvard University to add a master's degree in public administration.

Early in her career, Mollie Beattie served in several **conservation** administrative posts and land management positions at the state level. She was commissioner of the Vermont Department of Forests, Parks, and **Recreation** (1985–1989), and deputy secretary of the state's Agency of **Natural Resources** (1989–1990). She also worked for private foundations and institutes: as program director and lands manager (1983–1985), for the Windham Foundation, a private, nonprofit organization concerned with critical issues facing Vermont, and later (1990–1993), as executive director of the Richard A. Snelling Center for Government in Vermont, a public policy institute.

Before becoming director of the Fish and **Wildlife** Service, Beattie was perhaps most widely known (especially in New England) for the book she co-authored on managing woodlots in private ownership, an influential guide reissued in a second edition in 1993. As a reviewer noted about the first edition, for the decade it was in print "thousands of landowners and professional foresters [recommended the book] to others." Especially noteworthy is the book's emphasis on the responsibility of private landowners for effective stewardship of the land. This background was reflected in her thinking as Fish and Wildlife director by making the private lands program—conservation in partnership with private land-owners—central to the agency's conservation efforts.

As director of the Fish and Wildlife Service, Beattie quickly became nationally known as a strong voice for conservation and as an advocate for thinking about land and **wildlife management** in **ecosystem** terms. One of her first actions as director was to announce that "the Service [will] shift to an ecosystem approach to managing our fish and wildlife resources." She emphasized that people use natural ecosystems for many purposes, and "if we do not take an ecosystem approach to conserving **biodiversity**, none of [those uses] will be long lived."

Her philosophy as a forester, conservationist, administrator, and land manager was summarized in her repeated insistence that people must start making better connections—between wildlife and **habitat** health and human health; between their own actions and "the destiny of the ecosystems on which both humans and wildlife depend;" and between the well-being of the **environment** and the well-being of the economy." She stressed "even if not a single job were created, wildlife must be conserved" and that the diversity of natural systems must remain integral and whole because "we humans are linked to those systems and it is in our immediate self interest to care" about them. She reiter-

ated that in any frame greater than the short-term, the economy and the environment are "identical considerations."

Though Beattie's life was relatively short, and her tenure at the Fish and Wildlife Service cut short by illness, her ideas were well received in conservation circles and her influence continues after her death, helping to create what she called "preemptive conservation," anticipating crises before they appear and using that foresight to minimize conflict, to maintain biodiversity and sustainable economies, and to prevent extinctions.

[*Gerald L. Young*]

RESOURCES
BOOKS
Beattie, M., C. Thompson, and L. Levine. *Working With Your Woodland: A Land-Owner's Guide.* Rev. ed. Hanover, N.H.: University Press of New England, 1993.
Beattie, M. "Biodiversity Policy and Ecosystem Management." In *Biodiversity and the Law*, edited by W. J. Snape III. Washington, D.C.: Island Press, 1996.

PERIODICALS
Beattie, M. "A Broader View." *Endangered Species Update* 12, no. 4/5 (April/ May, 1995): 4–5.
———. "An Ecosystem Approach to Fish and Wildlife Conservation." *Ecological Applications* 6, no. 3 (August 1996): 696–699.
———. "The Missing Connection: Remarks to Natural Resources Council of America." *Land and Water Law Review* 29, no. 2 (1994): 407–415.
Cohn, J. P. "Wildlife Warrior." *Government Executive* 26, no. 2 (February 1994): 41–43.

Bees

see **Africanized bees**

Bellwether species

The term bellwether came from the practice of putting a bell on the leader of a flock of sheep. The term is now used to describe an indicator of a complex of ecological changes. Bellwether **species** are also called indicator species and are seen as early warning signs of environmental damage and **ecosystem** change.

Ecologists have identified many bellwether species in various ecosystems, including the Attwater **prairie** chicken(*Tympanuchus cupido attwateri*), the Northern diamondback terrapin(*Malaclemys terrapin terrapin*), the Northern and Southern giant petrels (*Macronectes giganteus* and *M. halli*), and the polar bear(*Ursus maritimus*). There are many more. Problems with these species are extreme indicators of troubles within their ecosystems.

The Attwater prairie chicken is almost extinct. It once numbered in the millions in Louisiana and along the Texas coast. The Northern and Southern giant petrels, large **scavenger** birds of the Antarctic Peninsula, are also heading towards **extinction** as **commercial fishing** fleets head further and further south. In Massachusetts' Outer Cape, conservationists are trying to save the Northern diamondback terrapin. Troubles for the Arctic polar bear are indicating ecosystem change in the Arctic, as humans interact more and more with that ecosystem.

[*Douglas Dupler*]

RESOURCES
PERIODICALS

Helvarg, David. "Elegant Scavengers: Giant Petrels are a Bellwether Species for the Threatened Antarctic Peninsula." *E Magazine*, November/December, 1999. <http://www.emagazine.com>.

Walsh, J. E. "The Arctic as a Bellwether." *Nature*, no. 352 (1999): 19–20.

Below Regulatory Concern

Large populations all over the globe continue to be exposed to low-level radiation. Sources include natural **background radiation**, widespread medical uses of **ionizing radiation**, releases and leakages from **nuclear power** and weapons manufacturing plants and waste storage sites. An added potential hazard to public health in the United States stems from government plans to deregulate **low-level radioactive waste** generated in industry, research, and hospitals and allow it to be mixed with general household trash and industrial waste in unprotected dump sites.

The **Nuclear Regulatory Commission** (NRC) plans to deregulate some of this **radioactive waste** and treat it as if it were not radioactive. The makers of radioactive waste have asked the NRC to treat certain low levels of **radiation exposure** with no regulations. The NRC plan, called Below Regulatory Concern (BRC), will categorize some of this waste as acceptable for regular dumping or **recycling**. It will deregulate radioactive consumer products, manufacturing processes and anything else that their computer models project would, on a statistical average, cause radiation exposures within the acceptable range. The policy sets no limits on the cumulative exposure to radiation from multiple sources or exposures or multiple human use.

In July 1991 the NRC responded to public pressure and declared a moratorium on the implementation of BRC policy. This was a temporary move, leaving the policy intact. The U. S. Department of Energy (DoE) has already dumped radioactive trash on unlicensed incinerators. Although DoE is not regulated by the NRC, this action reflects an adoption of the BRC concept. If the BRC plan is eventually approved, radioactive waste will end up in local landfills, sewage systems, incinerators, recycling centers, consumer products and

building materials, **hazardous waste** facilities and farmland (through **sludge** spreading).

Some scientists maintain that there are acceptable low levels of radiation exposure (in addition to natural background radiation) that do not pose a threat to public health. Other medical studies conclude that there is no safe level of radiation exposure. Critics claim BRC policy is nothing more than linguistic **detoxification** and will, if implemented, inevitably lead to increased radiation exposure levels to the public, and increased risk of **cancer**, **birth defects**, reduced immunity, and other health problems. Implementation of BRC may also mean that efficient cleanup of contaminated **nuclear weapons** plants (Oak Ridge, Savannah River, Fernald), nuclear reactors, and other radioactive facilities might not be completed.

Its critics contend that BRC is basically a financial, not a public health, decision. The nuclear industry has projected that it will save hundreds of millions of dollars if BRC is implemented.

[*Liane Clorfene Casten*]

RESOURCES
BOOKS

Below Regulatory Concern, But Radioactive and Carcinogenic. Washington, DC: U. S. Public Interest Research Group, 1990.

Low-Level Radioactive Waste Regulation—Science, Politics and Fear. Chelsea, MI: Lewis, 1988.

Below Regulatory Concern: A Guide to the Nuclear Regulatory Commission's Policy on the Exemption of Very Low-Level Radioactive Materials, Wastes and Practices. Washington, DC: U. S. Government Printing Office, 1990?.

Hugh Hammond Bennett (1881 – 1960)

American soil conservationist

Dr. Bennett, a noted conservationist, is often called the father of **soil conservation** in the United States. He was born on April 13, 1881, in Anson County, North Carolina, and died July 7, 1960. He is buried in Arlington National Cemetery.

Dr. Bennett graduated from the University of North Carolina in 1903. After completing his university education, he became a **soil** surveyor in the Bureau of Soils of the **U.S. Department of Agriculture**. He recognized early the degradation to the land from soil **erosion** and in 1929 published a bulletin entitled "Soil Erosion, A National Menace." Soon after that, the nation began to heed the admonitions of this young scientist.

In 1933, the U.S. Department of Interior set up a Soil Erosion Service to conduct a nation-wide demonstration program of soil erosion, with Bennett as its head. In 1935, the **Soil Conservation Service** was established as a permanent agency of the U.S. Department of Agriculture, and Bennett became its head, a position he retained until his retirement in April 1951. He was the author of several books and many technical articles on soil **conservation**. He received high awards from many organizations and several foreign countries.

Dr. Bennett traveled widely as a crusader for soil conservation, and was a forceful speaker. Many colorful stories about his speeches exist. Perhaps the most widely quoted concerns the dust storm that hit Washington D.C. in the spring of 1935 at a critical moment in Congressional hearings concerning legislation that, if passed, would establish the Soil Conservation Service. A huge dust cloud that had started in the Great Plains was steadily moving toward Washington. As the storm hit the district, Dr. Bennett, who had been testifying before members of the Senate public lands committee, called the committee members to a window and pointed to the sky, darkened from the dust. It made such an impression that legislation was promptly approved, establishing the Soil Conservation Service.

Dr. Bennett was an outstanding scientist and crusader, but he was also an able administrator. He visualized that if the Soil Conservation Service was to be effective, it must have grass roots support. He laid the foundation for the establishment of the local soil conservation districts with locally elected officials to guide the program and Soil Conservation Service employees to provide the technical support. Currently there are over 3,000 districts (usually by county). The National Association of Conservation Districts is a powerful voice in matters of conservation. Because of Dr. Bennett's leadership in the United States, soil erosion was increasingly recognized worldwide as a serious threat to the long-term welfare of humans. Many other countries then followed the United States' lead in establishing organized soil conservation programs.

[*William E. Larson*]

RESOURCES
BOOKS

Bennett, H. H. *Soil Conservation*. Manchester: Ayer, 1970.

Benzene

A hydrocarbon with chemical formula C_6H_6, benzene contains six **carbon** atoms in a ring structure. A clear volatile liquid with a strong odor, it is one of the most extensively used **hydrocarbons**. Because it is an excellent solvent and a necessary component of many industrial **chemicals**, including **gasoline**, benzene is classified by United States federal agencies as a known human **carcinogen** based on

studies that show an increased incidence of nonlymphocytic **leukemia** from occupational exposure and increased incidence of neoplasia in rats and mice exposed by inhalation and gavage. Because of these cancer-causing properties, benzene has been listed as a hazardous air pollutant under Section 112 of the **Clean Air Act**.

Benzo(a)pyrene

Benzo(a)pyrene [B(a)P] is a polycyclic aromatic hydrocarbon (PAH) having five aromatic rings in a fused, honeycomb-like structure. Its formula and molecular weight are $C_{20}H_{12}$ and 252.30, respectively. It is a naturally occurring and man-made organic compound formed along with other PAH in incomplete **combustion** reactions, including the burning of **fossil fuels**, motor vehicle exhaust, wood products, and cigarettes. It is classified as a known human **carcinogen** by the EPA, and is considered to be one of the primary carcinogens in **tobacco smoke**. Synthesized in 1933, it was the first carcinogen isolated from **coal** tar and often serves as a surrogate compound for **modeling** PAHs.

Wendell Erdman Berry   (1934 –)
American writer, poet, and conservationist

A Kentucky farmer, poet, novelist, essayist and conservationist, Berry has been a persistent critic of large-scale industrial agriculture—which he believes to be a contradiction in terms—and a champion of environmental stewardship and **sustainable agriculture**. Wendell Erdman Berry was born on August 5, 1934, in Henry County, Kentucky, where his family had farmed for four generations. Although he learned the farmer's skills, he did not wish to make his living as a farmer but as a writer and teacher. He earned his B.A. in English at the University of Kentucky in 1956 and an M.A. in 1957. He was awarded a **Wallace Stegner** writing fellow at Stanford University (1958–1959), where he remained to teach in the English Department (1959–1960). He spent the following year in Italy on a Guggenheim fellowship. In 1962 Berry joined the English faculty at New York University.

Dissatisfied with urban life and feeling disconnected from his roots, in 1965 Berry resigned his professorship of English at New York University to return to his native Kentucky to farm, to write, and to teach at the University of Kentucky. His recurring themes—love of the land, of place or region, and the responsibility to care for them—appear in his poems and novels and in his essays. Many modern farming practices, as he argues in *The Unsettling of America* (1977) and *The Gift of Good Land* (1981) and elsewhere, deplete the **soil**, despoil the **environment**, and deny the value of careful husbandry. In relying on industrial scales, techniques and technologies, they fail to appreciate that agriculture is agri-culture—that is, a coherent way of life that is concerned with the care and cultivation of the land—rather than agri-business concerned solely with maximizing yields, efficiency, and short-term profits to the long-term detriment of the land, of family farms, and of local communities. A truly sustainable agriculture, as Berry defines it, "would deplete neither soil, nor people, nor communities."

Berry believes that too few people now live on and farm the land, leaving it to the less-than-tender mercies of corporate "managers" who know more about accounting than about agricultural stewardship. The so-called miracle of modern agriculture has been purchased by selling our birth-right, the God-given gift of good land that Americans have heedlessly traded for the ease, convenience and affluence of urban and suburban living. As a consequence we have become disconnected from our cultural roots and have lost a **sense of place** and purpose and pleasure in work well done. We have also lost a sense of connection with the land and the lore of those who work on and care for it. Instead of food from our own fields, water from rainfall and **wells**, and stories from family and friends, most Americans now get food from the grocery store, water from the faucet, and endless entertainment from television. A wasteful throwaway society produces not only material but cultural junk—throwaway farms, throwaway marriages and children, disposable communities, and a wanton disregard for the natural environment. The transition to a culture of consumption and convenience, Berry believes, does not represent progress so much as it marks a deep and lasting loss.

Although Berry does not believe it possible (or desirable) that all Americans become farmers, he holds that we need think about what we do daily as consumers and citizens and how our choices and activities affect the land. He suggests that the act of planting and tending a garden is a "complete act" in that it enables one to connect consumption with production, and both to a sense of reverence for the fertility and abundance of a world well cared for.

[*Terence Ball*]

RESOURCES
BOOKS

Berry, Wendell. *A Continuous Harmony: Essays Cultural and Agricultural.* New York: Harcourt Brace Jovanovich, 1972.

———. *A Place on Earth.* rev. ed. San Francisco: North Point Press, 1983.

———. *Another Turn of the Crank.* Washington, DC: Counterpoint Press, 1995.

———. *Collected Poems, 1957–1982.* San Francisco: North Point Press, 1985.

———. *The Gift of Good Land.* San Francisco: North Point Press, 1981.

———. *Home Economics.* San Francisco: North Point Press, 1981.

———. *Sex, Economy, Freedom and Community.* New York: Pantheon Books, 1993.

———. *The Unsettling of America.* San Francisco: Sierra Club Books, 1977.

———. *What Are People For?* San Francisco: North Point Press, 1990.

Merchant, Paul, ed. *Wendell Berry.* American Authors Series. Lewiston, Idaho: Confluence Press, 1991.

Best available control technology

Best available control technology (BACT) is a standard used in **air pollution control** in the prevention of significant deterioration (PSD) of **air quality** in the United States. Under the **Clean Air Act**, a major stationary new source of **air pollution**, such as an industrial plant, is required to have a permit that sets **emission** limitations for the facility. As part of the permit, limitations are based on levels achievable by the use of BACT for each pollutant.

To prevent risk to human health and public welfare, each region in the country is placed in one of three PSD areas in compliance with National Ambient Air Quality Standards (NAAQS). Class I areas include national parks and **wilderness** areas, where very little deterioration of air quality is allowed. Class II areas allow moderate increases in ambient concentrations. Those classified as Class III permit industrial development and larger increments of deterioration. Except for national parks, most of the land in the United States is set at Class II.

Under the Clean Air Act, the BACT standards are applicable to all plants built after the effective date, as well as preexisting ones whose modifications might increase emissions of any pollutant. To establish a major new source in any PSD area, an applicant must demonstrate, through monitoring and diffusion models, that emissions will not violate NAAQS or the Clean Air Act. The applicant must also agree to use BACT for all pollutants, whether or not that is necessary to avoid exceeding the levels allowed and despite its cost. The **Environmental Protection Agency** (EPA) requires that a source use the BACT standards, unless it can demonstrate that its use is infeasible based on "substantial and unique local factors." Otherwise BACT, which can include design, equipment, and operational standards, is required for each pollutant emitted above minimal defined levels.

In areas that exceed NAAQS for one or more pollutants, permits are issued only if total allowable emissions of each pollutant are reduced even though a new source is added. The new source must comply with the **lowest achievable emission rate** (LAER), the most stringent limitations possible for a particular plant.

Under the Clean Air Act, if a new stationary source is incapable of BACT standards, it is subject to the **New Source Performance Standard** (NSPS) for a pollutant, as determined by the EPA. NSPSs take into consideration the cost and energy requirements of emission reduction processes, as well as other health and environmental impacts. This contrasts with BACT standards, which are determined without regard to cost. Strict BACT requirements resulted in a more than 20% reduction in particulates and **sulfur dioxide** below NSPS levels.

[*Judith Sims*]

RESOURCES
BOOKS

Findley, R. W., and D. A. Farber. *Environmental Law.* St. Paul, MN: West Publishing Company, 1992.

Plater, Z. J. B., R. H. Abrams, and W. Goldfarb. *Environmental Law and Policy: Nature, Law, and Society.* St. Paul, MN: West Publishing Company, 1992.

Best management practices

Best management practices (BMPs) are methods that have been determined to be the most effective and practical means of preventing or reducing non-point source **pollution** to help achieve **water quality** goals. BMPS include both measures to prevent pollution and measures to mitigate pollution.

BMPS for agriculture focus on reducing non-point sources of pollution from croplands and farm animals. Agricultural **runoff** may contain nutrients, **sediment**, animal wastes, salts, and pesticides. With **conservation tillage**, crop residue, which is plant residue from past harvests, is left on the **soil** surface to reduce runoff and soil **erosion**, conserve soil moisture, and keep nutrients and pesticides on the field. Contour strip farming, where sloping land is farmed across the slopes to impede runoff and soil movement downhill, reduces erosion and sediment production. Managing and accounting for all **nutrient** inputs to a field ensures that there are sufficient nutrients available for crop needs while preventing excessive nutrient **loading**, which may result in **leaching** of the excess nutrients to the ground water. Various BMPs are available for keeping insects, weeds, disease, and other pests below economically harmful levels. **Conservation** buffers, including grassed waterways, **wetlands**, and riparian areas act as an additional barrier of protection by capturing potential pollutants before they move to surface waters. Cows can be kept away from streams by streambank fencing and installation of alternative water sources. Designated stream crossings can provide a controlled crossing or watering access, thus limiting streambank erosion and streambed trampling.

Coastal shorelines can also be protected with BMPs. Shoreline stabilization techniques include headland breaker systems to control shoreline erosion while providing a com-

munity beach. Preservation of shorelines can be accomplished through revegetation, where living plant materials are a primary structural component in controlling erosion caused by land instability.

Stormwater management in urban developed areas also utilize BMPs to remove pollutants from runoff. BMPS include retention ponds, alum treatment systems, constructed wetlands, sand **filters**, baffle boxes, inlet devices, vegetated swales, **buffer** strips, and infiltration/exfiltration trenches. A storm drain stenciling programs is an educational BMP tool to remind persons of the illegality of dumping litter, oil, pesticides, and other toxic substances down **urban runoff drainage** systems.

Logging activities can have adverse impacts on stream water temperatures, stream flows, and water quality. BMPS have been developed that address location of logging roads, skid trails, log landings and stream crossings, riparian management buffer zones, management of litter and fuel and lubricant spills, and reforestation activities.

Successful control of erosion and **sedimentation** from construction and mining activities involves a system of BMPs that targets each stage of the erosion process. The first stage involves minimizing the potential sources of sediment by limiting the extent and duration of land disturbance to the minimum needed, and protecting surfaces once they are exposed. The second stage of the BMP system involves controlling the amount of runoff and its ability to carry sediment by diverting incoming flows and impeding internally generated flows. The third stage involves retaining sediment that is picked up on the project site through the use of sediment-capturing devices. **Acid** drainage from mining activities requires even more complex BMPs to prevent acids and associated toxic pollutants from harming surface waters.

Other pollutant sources for which BMPS have been developed include **atmospheric deposition**, boats and marinas, **habitat** degradation, roads, septic systems, underground storage tanks, and **wastewater** treatment.

[*Judith L. Sims*]

RESOURCES
BOOKS

Urban Water Infrastructure Management Committee. *A Guide for Best Management Practice (BMP) Selection in Urban Developed Areas.* Reston, VA: American Society of Civil Engineers, 2001.

U.S. Environmental Protection Agency. *NPDES Best Management Practices Manual.* Rockville, MD: Government Institutes, ABS Group, Inc., 1995.

OTHER

Logging and Forestry Best Management Practices. Division of Forestry, Indiana Department of Natural Resources. May 30, 2001. [June 15, 2002]. <http://www.state.in.us/dnr/forestry/bmp/logindex.htm>

Best Management Practices (BMPs) for Agricultural Nonpoint Source Pollution Control. North Carolina State University Water Quality Group. June 15,

2002. [June 17, 2002]. <http://h2osparc.wq.ncsu.edu/info/bmps_for_agnps.html>

Best Management Practices (BMPs) for Non-Agricultural Nonpoint Source Pollution Control: Nonpoint Source Pollution Control Measures—Source Categories. North Carolina State University Water Quality Group. June 15, 2002. [June 17, 2002]. <http://h2osparc.wq.ncsu.edu/info/bmps.html>

Best practical technology

Best practical technology (BPT) refers to any of the categories of technology-based **effluent** limitations pursuant to Section 301(b) and Sections 304(b) of the **Clean Water Act** as amended. These categories are the best practicable control technology currently available (BPT); the **best available control technology** (BAT) economically feasible (BAT); and the best **conventional pollutant** control technology (BCT).

Section 301(b) of the Clean Water Act specifies that "in order to carry out the objective of this Act there shall be achieved—(1)(A) not later than July 1, 1977, effluent limitations for point sources, other than publicly owned treatment works (i) which shall require the application of the best practicable control technology currently available as defined by the Administrator pursuant to Section 304(b) of this Act, or (ii) in the case of **discharge** into a publicly owned treatment works which meets the requirements of subparagraph (B) of this paragraph, which shall require compliance with any applicable pretreatment requirements and any requirements under Section 307 of this Act;..."

The BPT identifies the current level of treatment and is the basis of the current level of control for direct discharges. BACT improves on the BPT, and it may include operations or processes not in common use in industry. BCT replaces BACT for the control of conventional pollutants, such as **biochemical oxygen demand** (BOD), total suspended solids (TSS), fecal coliform, and **pH**. Details such as the amount of constituents, and the chemical, physical, and biological characteristics of pollutants, as well as the degree of effluent reduction attainable through the application of the selected technology can be found in the development documents published by the **Environmental Protection Agency** (EPA). These development documents cover different industrial categories such as diary products processing, soap and **detergents** manufacturing, meat products, grain mills, canned and preserved fruits and vegetables processing, and **asbestos** manufacturing.

In accordance with Section 304(b) of the Clean Water Act, the factors to be taken into account in assessing the BPT include the total cost of applying the technology in relation to the effluent reductions to the results achieved from such an application, the age of the equipment and facilities involved, the process employed, the engineering aspects of applying various types of control technologies and process changes, and calculations of environmental impacts other than **water quality** (including energy requirements). As far as evaluating the BCT is concerned, the factors are mostly the same. By they include consideration of the reasonableness of the relationship between the costs of attaining a reduction in effluents and the benefits derived from that reduction, and the comparison of the cost and level of reduction of such pollutants from the discharge from publicly owned treatment works to the cost and level of reduction of such pollutants from a class or category of industrial sources. Control technologies may include in-plant control and preliminary treatment, and end-of-pipe treatment, examples of which are **water conservation** and **reuse**, raw materials substitution, screening, multimedia **filtration**, and activated **carbon absorption**.

[*James W. Patterson*]

Beta particle

An electron emitted by the nucleus of a radioactive atom. The beta particle is produced when a **neutron** within the nucleus decays into a proton and an electron. Beta particles have greater penetrating power than alpha particles but less than x-ray or gamma rays. Although beta particles can penetrate skin, they travel only a short distance in tissue. Beta rays pose relatively little health hazard, therefore, unless they are ingested into the body. Naturally radioactive materials such as potassium-40, carbon-14, and strontium-90 emit beta particles, as do a number of synthetic radioactive materials. *See also* Radioactivity

Beyond Pesticides

Founded in 1981, Beyond Pesticides (originally called the National Coalition Against the Misuse of Pesticides) is a non-profit, grassroots network of groups and individuals concerned with the dangers of pesticides. Members of Beyond Pesticides include individuals, such as "victims" of pesticides, physicians, attorneys, farmers and farmworkers, gardeners, and former chemical company scientists, as well as health, farm, consumer, and church groups. All want to limit **pesticide** use through Beyond Pesticides, which publishes information on pesticide hazards and alternatives, monitors and influences legislation on pesticide issues, and provides seed grants and encouragement to local groups and efforts.

Administered by a 15-member board of directors and a small full-time staff, including a toxicologist and an ecologist, Beyond Pesticides is now the most prominent organization dealing with the pesticide issue. It was established on the premise that much is unknown about the toxic effects of pesticides and the extent of public exposure to them. Because such information is not immediately forthcoming, members of Beyond Pesticides believe the only available way of reducing both known and unknown risks is by limiting or eliminating pesticides. The organization takes a dual-pronged approach to accomplish this. First, Beyond Pesticides draws public attention to the risks of conventional **pest** management; second, it promotes the least-toxic alternatives to current pesticide practices.

An important part of Beyond Pesticides's overall program is the Center for Community Pesticide and Alternatives Information. The Center is a clearinghouse of information, providing a 2,000-volume library about pest control, **chemicals**, and pesticides. To concerned individuals it sells inexpensive brochures and booklets, which cover topics such as alternatives to controlling specific pests and chemicals; the risks of pesticides in schools, to food, and in reproduction; and developments in the **Federal Insecticide, Fungicide and Rodenticide Act** (FIFRA), the national law governing pesticide use and registration in the United States. Through the Center Beyond Pesticides also publishes *Pesticides and You* (*PAY*) five times a year. It is a newsletter sent to approximately 4,500 people, including Beyond Pesticides members, subscribers, and members of Congress. The Center also provides direct assistance to individuals through access to Beyond Pesticides's staff ecologist and toxicologist.

In 1991 Beyond Pesticides also established the Local Environmental Control Project after the Supreme Court decision affirming local communities' rights to regulate pesticide use. Although Beyond Pesticides supported bestowing local control over pesticide use, it needed a new program to counteract the subsequent mobilization of the pesticide industry to reverse the Supreme Court decision. The Local Environmental Control Project campaigns first to preserve the right accorded by the Supreme Court decision and second to encourage communities to take advantage of this right.

Beyond Pesticides marked its tenth anniversary in 1991 with a forum entitled "A Decade of Determination: A Future of Change." It included workshops on **wildlife** and **groundwater** protection, **cancer risk assessment**, and the implications of GATT and free trade agreements. Beyond Pesticides has also established the annual National Pesticide Forum. Through such conferences, its aid to victims and groups, and its many publications, Beyond Pesticides above all encourages local action to limit pesticides and change the methods of controlling pests.

[*Andrea Gacki*]

RESOURCES

ORGANIZATIONS

Beyond Pesticides, 701 E Street, SE, Suite 200, Washington, D.C. USA 20003 (202) 543-5450, Fax: (202) 543-4791, Email: info@beyond pesticides.org, <http://www.beyondpesticides.org>

Bhopal, India

On December 3, 1984, one of the world's worst industrial accidents occurred in Bhopal, India. Along with Three Mile Island and Chernobyl, Bhopal stands as an example of the dangers of industrial development without proper attention to **environmental health** and safety.

A large industrial and urban center in the state of Madhya Pradesh, Bhopal was the location of a plant owned by the American chemical corporation, Union Carbide, Inc. and its Indian subsidiary, Union Carbide India, Ltd. The plant manufactured pesticides, primarily the **pesticide** carbaryl (marketed under the name Sevin), which is one of the most widely used carbamate class pesticides in the United States and throughout the world. Among the intermediate chemical compounds used together to manufacture Sevin is methyl isocyanate (MIC)—a lethal substance that is reactive, toxic, volatile, and flammable. It was the uncontrolled release of MIC from a storage tank in the Bhopal facility that caused the death of 5,000 people, seriously injured another 20,000 and affected an estimated 200,000 people. The actual number of casualties remains unknown; many believe the numbers cited above to be serious underestimates.

MIC (CH3-N=C=O) is highly volatile and has a boiling point of 89°F (39.1°C). In the presence of trace amounts of impurities such as water or metals, MIC reacts to generate heat, and if the heat is not removed, the chemical begins to boil violently. If relief valves, cooling systems and other safety devices fail to operate in a closed storage tank, the pressure and heat generated may be sufficient to cause a release of MIC into the **atmosphere**. Because the vapor is twice as heavy as air, the vapors if released remain close to the ground where they can do the most damage, drifting along prevailing wind patterns. As set by the Occupational Health and Safety Administration (OSHA), the standards for exposure to MIC are set at 0.02 ppm over an eight-hour period. The immediate effects of exposure, inhalation and ingestion of MIC at high concentrations (above 2 ppm) are burning and tearing of the eyes, coughing, vomiting, blindness, massive trauma of the gastrointestinal tract, clogging of the lungs and suffocation of bronchial tubes. When not immediately fatal, the long-term health consequences include permanent blindness, permanently impaired lung functioning, corneal ulcers, skin damage, and potential **birth defects**.

Many explanations for the disaster have been advanced, but the most widely accepted theory is that trace amounts of water entered the MIC storage tank and initiated the hydrolysis reaction, which was followed by MIC's spontaneous reactions. The plant was not well designed for safety, and maintenance was especially poor. Four key safety factors should have contained the reaction, but it was later discovered that they were all inoperative at the time of the accident. The refrigerator that should have slowed the reaction by cooling the chemical was shut off, and, as heat and pressure built up in the tank, the relief valve blew. A vent gas scrubber designed to neutralize escaping gas with caustic soda failed to work. Also, the flare tower that would have burned the gas to harmless by-products was under repair. Yet even if all these features had been operational, subsequent investigations found them to be poorly designed and insufficient for the capacity of the plant. Once the runaway reaction started, it was virtually impossible to contain.

The poisonous cloud of MIC released from the plant was carried by the prevailing winds to the south and east of the city—an area populated by highly congested communities of poorer people, many of whom worked as laborers at the Union Carbide plant and other nearby industrial facilities. Released at night, the silent cloud went undetected by residents who remained asleep in their homes, thus possibly ensuring a maximal degree of exposure. Many hundreds died in their sleep, others choked to death on the streets as they ran out in hopes of escaping the lethal cloud. Thousands more died in the following days and weeks. The Indian government and numerous volunteer agencies organized a massive relief effort in the immediate aftermath of the disaster consisting of emergency medical treatment, hospital facilities, and supplies of food and water. Medical treatment was often ineffective, for doctors had an incomplete knowledge of the toxicity of MIC and the appropriate course of action.

In the weeks following the accident, the financial, legal and political consequences of the disaster unfolded. In the United States Union Carbide's stock dipped 25% in the week immediately following the event. Union Carbide India Ltd. (UCIL) came forward and accepted moral responsibility for the accident, arranging some interim financial compensation for victims and their families. However its parent company, Union Carbide Inc., which owned 50.9% of UCIL, refused to accept any legal responsibility for their subsidiary. The Indian government and hundreds of lawyers on both sides pondered issues of liability and the question of a settlement. While Union Carbide hoped for out-of-court settlements or lawsuits in the Indian courts, the Indian government ultimately decided to pursue class action suits on behalf of the victims in the United States courts in the hope of larger settlements. The United States courts refused to hear the case, and it was transferred to the Indian court system. Warren Anderson, then chairman of Union Carbide, refused to appear in Indian court. The case is still under litigation

and the interim compensation set aside has reached only a fraction of the victims.

The disaster in Bhopal has had far-reaching political consequences in the United States. A number of Congressional hearings were called and the **Environmental Protection Agency** (EPA) and OSHA initiated inspections and investigations. A Union Carbide plant in McLean, Virginia, that uses processes and products similar to those in Bhopal was repeatedly inspected by officials. While no glaring deficiencies in operation or maintenance were found, it was noted that several small leaks and spills had occurred at the plant in previous years that had gone unreported. These added weight to growing national concern about workers' **right-to-know** provisions and emergency response capabilities. In the years following the Bhopal accident, both state and federal environmental regulations were expanded to include mandatory preparedness to handle spills and releases on land, water, or air. These regulations include measures for emergency response such as communication and coordination with local health and law enforcement facilities, as well as community leaders and others. In addition, employers are now required to inform any workers in contact with hazardous materials of the nature and types of hazards to which they are exposed; they are also required to train them in emergency health and safety measures.

The disaster at Bhopal raises a number of critical issues and highlights the wide gulf between developed and developing countries in regard to design and maintenance standards for health and safety. Management decisions allowed the Bhopal plant to operate in an unsafe manner and for a shanty-town to develop around its perimeter without appropriate emergency planning. The Indian government, like many other developing nations in need of foreign investment, appeared to sacrifice worker safety in order to attract and keep Union Carbide and other industries within its borders. While a number of environmental and occupational health and safety standards existed in India before the accident, their inspection and enforcement was cursory or nonexistent. Often understaffed, the responsible Indian regulatory agencies were rife with corruption as well. The Bhopal disaster also raised questions concerning the moral and legal responsibilities of American companies abroad, and the willingness of those corporations to apply stringent United States safety and environmental standards to their operations in the **Third World** despite the relatively free hand given them by local governments.

Although worldwide shock at the Bhopal accident has largely faded, the suffering of many victims continues. While many national and international safeguards on the manufacture and handling of hazardous **chemicals** have been instituted, few expect that lasting improvements will occur in developing countries without a gradual recognition of the

Union Carbide plant in Bhopal, India. (Corbis-Bettmann. Reproduced by permission.)

economic and political values of stringent health and safety standards.

[Usha Vadagiri]

RESOURCES
BOOKS

Diamond, A. *The Bhopal Chemical Leak.* San Diego, CA: Lucent, 1990.
Kurzman, D. *A Killing Wind: Inside the Bhopal Catastrophe.* New York: McGraw-Hill, 1987.

PERIODICALS

"Bhopal Report." *Chemical and Engineering News* (February 11, 1985): 14–65.

Bikini atoll

The primary objective of the Manhattan Project during World War II was the creation of a **nuclear fission** or atomic bomb. Even before that goal was accomplished in 1945, however, some nuclear scientists were thinking about the next step in the development of **nuclear weapons**, a fusion or **hydrogen** bomb.

Progress on a fusion bomb was slow. Questions were raised about the technical possibility of making such a bomb, as well as the moral issues raised by the use of such a destructive weapon. But the detonation of an atomic bomb by the Soviet Union in 1949 placed the fusion bomb in a new perspective. Concerned that the United States was falling behind in its arms race with the Soviet Union, President Harry S. Truman authorized a full-scale program for the development of a fusion weapon.

The first test of a fusion device occurred on October 31, 1952, at Eniwetok Atoll in the Marshall Islands in the Pacific Ocean. This was followed by a series of six more tests, code-named "Operation Castle," at Bikini Atoll in 1954. Two years later, on May 20, 1956, the first **nuclear fusion** bomb was dropped from an airplane over Bikini Atoll.

Bikini Atoll had been selected in late 1945 as the site for a number of tests of fission weapons, to experiment with different designs for the bomb and to test its effects on ships and the natural **environment**.

At that time, 161 people belonging to 11 families lived on Bikini. Since the Bikinians have no written history, little is known about their background. According to their oral tradition, the original home of their ancestors is nearby Wotje Atoll. Until the early 1900s, they had relatively little contact with strangers and were regarded with some disdain even by other Marshall Islanders. After the arrival of missionaries early in the twentieth century, the Bikinians became devout Christians. People lived on coconuts, breadfruits, arrowroot, fish, turtle eggs, and birds, all available in abundance on the atoll. The Bikinians were expert sailors and fishermen. Land ownership was important in the culture, and anyone who had no land was regarded as lacking in dignity.

On January 10, 1946, President Truman signed an order authorizing the transfer of everyone living on Bikini Atoll to the nearly uninhabited Bongerik Atoll. The United States Government asked the Bikinians to give up their native land to allow experiments that would bring benefit to all humankind. Such an action, the Americans argued, would earn for the Bikinians special glory in heaven. The islanders agreed to the request and, along with their homes, church, and community hall, were transported by the United States Navy to Rongerik.

In June and July of 1946, two tests of atomic bombs were conducted at Bikini as part of "Operation Crossroads." More than 90 vessels, including captured German and Japanese ships along with surplus cruisers, destroyers, submarines, and amphibious craft from the United States Navy, were assembled. Following these tests, however, the Navy concluded that Bikini was too small and moved future experiments to Eniwetok Atoll.

The testing of a nuclear fusion device in Operation Castle marked the return of bomb testing to Bikini. The most memorable test of that series took place in 1954 and was code-named "Bravo." Experts expected a yield of six megatons from the hydrogen bomb used in the test, but measured instead a yield of 15 megatons, 250% greater. Bravo turned out to be the largest single explosion in all of human history, producing an explosive force greater than all of the bombs used in all the previous wars in history.

Fallout from Bravo was consequently much larger than had been anticipated. In addition, because of a shift in wind patterns, the fallout spread across an area of about 50,000 mi^2 (11.5 km^2), including three inhabited islands—Rongelap, Itrik, and Rongerik. A number of people living on these islands developed radiation burns and many were evacuated from their homes temporarily. Farther to the east, a Japanese fishing boat which had accidentally sailed into the restricted zone was showered with fallout. By the time the boat returned to Japan, 23 crew members had developed **radiation sickness**. One eventually died of infectious hepa-

titis, probably because of the numerous blood transfusions he received.

The value of Bikini as a test site ended in 1963 when the United States and the Soviet Union signed the Limited Test Ban Treaty which outlawed nuclear weapons testing in the **atmosphere**, the oceans, and outer space. Five years later, the United States government decided that it was safe for the Bikinians to return home. By 1971, some had once again take up residence on their home island. Their return was short-lived. In 1978, tests showed that returnees had ingested quantities of radioactive materials much higher than the levels considered to be safe. The Bikinians were relocated once again, this time to the isolated and desolate island of Kili, 500 mi (804 km) from Bikini.

The primary culprit on Bikini was the radioactive **isotope** cesium-137. It had become so widely distributed in the **soil**, the water, and the crops on the island that no one living there could escape from it. With a half life of 30 years, the isotope is likely to make the island uninhabitable for another century.

Two solutions for this problem have been suggested. The brute-force approach is to scrape off the upper 12 in (30 cm) of soil, transport it to some uninhabited island, and bury it under concrete. A similar burial site, the "Cactus Crater," already exists on Runit Island. It holds radioactive wastes removed from Eniwetok Atoll. The cost of clearing off Bikini's 560 acres (227 ha) and destroying all its vegetation (including 25,000 trees) has been estimated at more then $80 million. A second approach is more subtle and makes use of chemical principles. Since potassium replaces cesium in soil, scientists hope that adding potassium-rich fertilizers to Bikini's soil will leach out the dangerous cesium-137.

At the thirtieth anniversary of Bravo, the Bikinians had still not returned to their home island. Many of the original 116 evacuees had already died. A majority of the 1,300 Bikinians who then lived on Kili no longer wanted to return to their native land. If given the choice, most wanted to make Maui, Hawaii, their new home. But they did not have that choice. The United States continues to insist that they remain somewhere in the Marshall Islands. The only place there they can't go, at least within most of their lifetimes, is Bikini Atoll.

[*David E. Newton*]

RESOURCES

PERIODICALS

Davis, J. "Paradise Regained?" *Mother Jones* 18 (March/April 1993): 17.

Delgado, J. P. "Operation Crossroads." *American History Illustrated* 28 (May/June 1993): 50–59.

Eliot, J. L., and B. Curtsinger. "In Bikini Lagoon Life Thrives in a Nuclear Graveyard." *National Geographic* 181 (June 1992): 70–83.

Lenihan, D. J. "Bikini Beneath the Waves." *American History Illustrated* 28 (May/June 1993): 60–67.

Bioaccumulation

The general term for describing the accumulation of **chemicals** in the tissue of organisms. The chemicals that bioaccumulate are most often organic chemicals that are very soluble in fat and lipids and are slow to degrade. Usually used in reference to aquatic organisms, bioaccumulation occurs from exposure to contaminated water (e.g., gill uptake by fish) or by consuming food that has accumulated the chemical (e.g., **food chain/web** transfer). Bioaccumulation of chemicals in fish has resulted in public health consumption advisories in some areas, and has affected the health of certain fish-eating **wildlife** including eagles, cormorants, terns, and mink.

Bioaerosols

Bioaerosols are airborne particles derived from plants, animals or are living organisms, including viruses, bacteria, **fungi**, and mammal and bird antigens. Bioaerosols can range in size from roughly 0.01 micrometer (**virus**) to 100 micrometer (pollen). These particles can be inhaled and can cause many types of health problems, including allergic reactions (specific activation of the immune system), infectious disease (pathogens that invade human tissues), and toxic effects (due to biologically produced chemical **toxins**). The most common outdoor bioaerosols are pollens from grasses, trees, weeds, and crops. The most common indoor biological pollutants are animal dander (minute scales from hair, feathers, or skin), dust mite and cockroach parts, fungi (molds), infectious agents (bacteria and viruses), and pollen.

Bioassay

Bioassay refers to an evaluation of the toxicity of an **effluent** or other material on living organisms such as fish, rats, insects, bacteria, or other life forms. The bioassay may be used for many purposes, including the determination of: 1) permissible **wastewater discharge** rates; 2) the relative sensitivities of various animals; 3) the effects of physico-chemical parameters on toxicity; 4) the compliance of discharges with effluent guidelines; 5) the suitability of a drug; 6) the safety of an **environment**; and 7) possible synergistic or antagonistic effects.

There are those who wish to reserve the term simply for the evaluation of the potency of substances such as drugs and vitamins, but the term is commonly used as described above. Of course, there are times when it is inappropriate to use bioassay and evaluation of toxicity synonymously, as when the goal of the assay is not to evaluate toxicity.

Bioassays are conducted as static, renewal, or continuous-flow experiments. In static tests, the medium (air or water) about the test organisms is not changed, in renewal tests the medium is changed periodically, and in continuous-flow experiments the medium is renewed continuously. When testing **chemicals** or wastewaters that are unstable, continuous-flow testing is preferable. Examples of instability include the rapid degradation of a chemical, significant losses in **dissolved oxygen**, problems with volatility, and precipitation.

Bioassays are also classified on the basis of duration. The tests may be short-term or acute, intermediate-term, or long-term, also referred to as chronic. In addition, aquatic toxicologists speak of partial- or complete- life-cycle assessments. The experimental design of a bioassay is in part reflected in such labels as range-finding, which is used for preliminary tests to approximate toxicity; screening for tests to determine if toxicity is likely by using one concentration and several replicates; and definitive, for tests to establish a particular end point with several concentrations and replicates).

The results of bioassays are reported in a number of ways. Early aquatic toxicity data were reported in terms of tolerance limits (TL). The term has been superseded by other terms such as effective concentration (EC), inhibiting concentration (IC), and lethal concentration (LC). The results of tests in which an animal is dosed (fed or injected) are reported in terms of effective dosage (ED) or lethal dosage (LD). When the potency of a drug is being studied, a therapeutic index (TI) is sometimes reported, which is the dose needed to cause a certain desirable effect (ED) divided by the lethal dose (LD) or some other ED. Median doses, the amount needed to affect 50% of the test population, are not always used. Needless to say, an evaluation of the response of a control population of organisms not exposed to a test agent or solution during the course of an experiment is very important.

Quality assurance and quality control procedures have become a very important part of bioassay methods. For example, the U.S. **Environmental Protection Agency** (EPA) has very specifically outlined how various aquatic bioassay procedures are to be performed and standardized to enhance reliability, **precision**, and **accuracy**. Other agencies in the United States and around the world have also worked very hard in recent years to standardize and improve quality assurance and control guidelines for the various bioassay techniques.

[*Gregory D. Boardman*]

RESOURCES
BOOKS

Manahan, S. E. *Toxicological Chemistry*. 2nd ed. Ann Arbor, MI: Lewis, 1992.

Rand, G. M., and Petrocelli, S. R. *Fundamentals of Aquatic Toxicology Methods and Applications*. Washington, DC: Hemisphere, 1985.

Bioassessment

According to the official definition of the United States **Environmental Protection Agency** (EPA), bioassessment refers to the process of evaluating the biological condition of a body of water using biological surveys (biosurveys) and other direct measurements of the resident biota—those organisms living in the surface water, including fish, insects, algae, plants and others. Since 1972 when the **Clean Water Act** was passed by Congress in order to clean up America's polluted waterways, the biological integrity of the nation's bodies of water has been the focus of professionals and citizens in ensuring the success of reaching this goal.

The derivation of biocriteria emerges from the bioassessment, and offers a narrative or numeric expression that explains the life surviving in the water. The process of evaluating biocriteria can be in "scores" using a method known as the *Ohio multimetric* approach; or, reported in "clusters" according to the *Maine statistical* approach.

Particularly in the study of **water pollution**, and the examination of ways to correct it, a bioassessment of a body of water measures the composition, diversity, and functional organization of the community living in it. Consequently it lays the groundwork for the biocriteria in the determination of the quality of the desired goal in regard to the condition of the resource, and to what maximum level of management that water can be maintained. The result is not how extensive the pollutant level can be. The result of any bioassessment should be that the unique members of this community of living organisms thrive as they interact with the various elements of their common home. Biocriteria must be maintained separately from any chemical evaluation, whose *whole effluent toxicity* (WET) is a distinct measurement, and plays a different role in water management, along with the physical and toxicity factors also included in a complete **water quality** management program.

Measuring biological integrity is accomplished most successfully when researchers utilize the *Rapid Bioassessment Protocols* (RBPs) established originally in the 1980s by the EPA, and revised and reissued in 1999. The concept at the core of these protocols according to the official EPA information are:

- Cost-effective, scientifically valid procedures for biological surveys

- Provisions for multiple site investigations in a field season
- Quick turn-around of results for management decisions
- Scientific reports easily translated to management and the public

These RBPs resulted from those methods already in use by several state agencies. They are useful in the following areas of study:

- Determining if a stream is supporting a designated aquatic life use specified in state water quality standards
- Characterizing the existence and severity of imppairment to the water resource
- Helping to identify the sources and causes of impairment
- Evaluating the effectiveness of control actions and restoration activities
- Supporting use attainability studies an cumulative impact assessment
- Characterizing regional biotic attributes of reference conditions.

Not only are current living conditions key factors in studying this aquatic life. Physical and biological elements known as stressors are those that exert a negative or adverse effect on the organisms. How those stressors might have affected the organisms over a long period of time is crucial to understanding what actions will be necessary to improve the **ecology** of the waterbody.

Centers and groups for bioassessment are active in each state throughout the United States, and throughout the world. They operate with scientific professionals such as biologists, as well as include trained citizen groups who participate by volunteering their time to monitor local water conditions. In one such program begun in California in 1998, trained citizen groups throughout the state under the training and guidance of a water biologist with the California Department of Fish and Game (CDFG) track the abundance and diversity of macroinvertebrates in order to assess the effects of **pollution**. This method was considered one of the best ways to monitor the health of streams and lakes due to the fact that each **species** is adapted to thrive in a particular habitat—along with an examination of bugs, whose various reactions to different types of pollutions have been monitored and well-established for years.

In an article written for the *EPA Journal*, in 1990, "Measuring Environmental Success," Steve Glomb noted that, "Biological community interactions are probably the most difficult goals to measure, Many scientists, however, think that these interactions are the most important factors to assess." Those interactions are often subtle, such as the example of invertebrates living in the mud on the floor of coastal waters, as Glomb mentioned. "Because they don't move much," he observed, "they can be used as an indication of problems over time." In determining past and current living conditions, bioassessment has proved an invaluable tool in implementing the goals of clean water for all life, including humans.

[*Jane E. Spear*]

RESOURCES
PERIODICALS

California Aquatic Bioassessment Workgroup. *Mission Statement*&lt:http://www.dfg.ca.gov/>

Central Plains Center for BioAssessment. *Home Page.* http://www.cpcb.u-kans.edu/>

Glomb, Steve. "Measuring Environmental Success." *EPA Journal* 16 (Nov/Dec 1990): 57.

Levy, Sharon. "Using Bugs to Bust Polluters." *Bioscience* 48 (May 1998): 342.

U.S. Environmental Protection Agency. *Biocriterial.* http://www.epa.gov/ost/>

Yosemite National Park/U.S. Environmental Protection Agency. *Water.* http://www.yosemite.epa.gov/>

ORGANIZATIONS

California Aquatic Bioassessment Workgroup, Department of Fish and Game, 1416 Ninth Street, Sacramento, CA United States 95814 (916)445-0411, Fax: (916)653-1856, <www.dfg.ca.gov>

Central Plains Center for BioAssessment, 2021 Constant Avenue, Lawrence, KA United States 66047-3729 (785)864-7729, Email: dbaker@ukans.edu, <www.cpcb.ukans.edu>

U. S. Environmental Protection Agency, 1200 Pennsylvania Avenue, NW, Washington, D.C. United States 20460 (202) 260-2090, , <www.epa.gov>

Biocentrism

see **Environmental ethics**

Biochemical oxygen demand

The biochemical oxygen demand (BOD) test is an indirect measure of the **biodegradable** organic matter in an aqueous sample. The test is indirect because oxygen used by **microbes** as they degrade organic matter is measured, rather than the depletion of the organic materials themselves. Generally, the test is performed over a five-day period (BOD$_5$) at 68°F (20°C) in 10 fl oz (300 ml) bottles incubated in the dark to prevent interference from algal growth. **Dissolved oxygen** (DO) levels (in mg/l) are measured on day zero and at the end of the test. The following equation relates how the BOD$_5$ of a sample is calculated:

$$BOD_5, mg/L = \frac{(Do_{initial} - DO_{after\ 5\ days})}{mL\ sample}$$

Note that in the above equation "ml sample" refers to the amount of sample that is placed in a 300 ml BOD bottle. This is critical in doing a test because if too much sample is added to a BOD bottle, microbes will use all the DO

in the water (i.e., DO$_{after\ 5\ days}$=0) and a BOD$_5$ cannot be calculated. Thus, the DO uptake at various dilutions of a given sample are made. One way of making the dilutions is to add different amounts of sample to different bottles. The bottles are then filled with dilution water which is saturated with oxygen, contains various nutrients, is at a neutral **pH**, and is very pure so as not to create any interference. Researchers attempt to identify a dilution which will provide for a DO uptake of greater than or equal to 2 mg/l and a DO residual of greater than or equal to 0.5–1.0 mg/l.

In some cases compounds or ions may be present in a sample that will inhibit microbes from degrading organic materials in the sample, thereby resulting in an artificially low or no BOD. Other times, the right number and/or types of microbes are not present, so the BOD is inaccurate. Samples may be "seeded" with microbes to compensate for these problems. The absence of a required **nutrient** will also make the BOD test invalid.

Thus, a number of factors can influence the BOD, and the test is known to be rather imprecise (i.e., values obtained vary by 10–15% of the actual value in a good test). However, the test continues to be a primary index of how well waters need to be treated or are treated and of the quality of natural waters. If wastes are not treated properly and contain too much BOD, the wastes may create serious oxygen deficits in environmental waters. The saturation concentration of oxygen in water varies somewhat with temperature and pressure but is often only in the area of 8–10 mg/l or less. To serve as a reference, the BOD$_5$ of raw, domestic **wastewater** ranges from about 100–300 mg/l. The demand for oxygen by the sewage is therefore clearly much greater than the amount of oxygen that water can hold in a dissolved form. To determine allowable BOD levels for a wastewater **discharge**, one must consider the relative flows/volumes of the wastewater and receiving (natural) water, temperature, reaeration of the natural water, biodegradation rate, input of other wastes, influence of benthic deposits and algae, DO levels of the wastewater and natural water, and condition/value of the receiving water.

[*Gregory D. Boardman*]

RESOURCES
BOOKS

Corbitt, R. A. *Standard Handbook of Environmental Engineering.* New York: McGraw-Hill, 1990.

Davis, M. L., and D. A. Cornwell. *Introduction to Environmental Engineering.* New York: McGraw-Hill, 1991.

Tchobanoglous, G., and E. D. Schroeder. *Water Quality.* Reading, MA: Addison-Wesley, 1985.

Viessman Jr., W., and M. J. Hammer. *Water Supply and Pollution Control.* 5th ed. New York, Harper Collins, 1993.

Bioconcentration

see **Biomagnification**

Biodegradable

Biodegradable substances are those that can be decomposed quickly by the action of biological organisms, especially **microorganisms**. The term is a process by which materials or compounds are broken down to smaller components, and all living organisms participate to some degree. Foods, for instance, are degraded by living creatures to release energy and chemical constituents for growth. In the sense that the term is usually used, however, it has a more restricted meaning. It refers specifically to the breakdown of undesirable toxic and waste materials or compounds to harmless or tolerable ones. When breakdown results in the destruction of useful objects, it is referred to as biodeterioration.

Although the term has been in common use for only two or three decades, processes of biodegradation have been known and used for centuries. Some of the most familiar are **sewage treatment** of human wastes, **composting** of kitchen, garden and lawn wastes, and spreading of **animal waste** on farm fields. These processes, of course, all mitigate problems with common and ubiquitous byproducts of civilization. The variety of objectionable wastes has greatly increased as human society has become more complex. The **waste stream** now includes items such as plastic bottles, lubricants, and foam packaging. Many of the newer products are virtually non-biodegradable, or they degrade only at a very slow rate. In some instances biodegradability can be greatly enhanced by minor changes in the chemical composition of the product. Biodegradable containers and packaging have been developed that are just as functional for many purposes as their non-degradable counterparts.

Advances in the science of microbiology have greatly expanded the potential for biodegradation and have increased public interest. Examples of new developments include the discovery of hitherto unknown microorganisms capable of degrading crude oil, **petroleum hydrocarbons**, diesel fuel, **gasoline**, industrial solvents, and some forms of synthetic polymers and **plastics**. These discoveries have opened new approaches for cleansing the **environment** of the accumulating **toxins** and debris of human society. Unfortunately, the rate at which the new organisms attack exotic wastes is sometimes quite slow, and dependent on environmental conditions. Presumably, microorganisms have been exposed to common wastes of a very long time and have evolved efficient and rapid ways to attack and use them as food. On the other hand, there has not been sufficient time to develop equally efficient means for degrading the newer wastes.

Research continues on the surprising capabilities of the new **microbes** emphasizing opportunities for genetic control and manipulation of the unique metabolic pathways that make the organisms so valuable. The potential for biodegradation can be improved by increasing the rates at which wastes are attacked, and the range of environmental conditions in which degrading organisms can thrive. The advantages of biological cleanup agents are several. Non-biological techniques are often difficult and expensive. The traditional way of removing petroleum wastes from **soil**, for instance, has been to collect and incinerate it at high temperatures. This is very costly and sometimes impossible. The prospect of accomplishing the same thing by treating the **contaminated soil** with microorganisms without removing it from its location has much appeal. It should have less destructive impact on the contaminated site and be much less expensive.

[Douglas C. Pratt]

RESOURCES
BOOKS

King, R. B., G. M. Long and J. K. Sheldon. *Practical Environmental Bioremediation.* Boca Raton, Florida: CRC Press, Inc., 1992.
Sharpley, J. M. and A. M. Kaplan. *Proceedings of the Third International Biodegradation Symposium.* London: Applied Science Publishers, 1976.

Biodiversity

Biodiversity is an ecological notion that refers to the richness of biological types at a range of hierarchical levels, including: (1) genetic diversity within **species**, (2) the richness of species within communities, and (3) the richness of communities on landscapes. In the context of environmental studies, however, biodiversity usually refers to the richness of species in some geographic area, and how that richness may be endangered by human activities, especially through local or global **extinction**.

Extinction represents an irrevocable and highly regrettable loss of a portion of the biodiversity of Earth. Extinction can be a natural process, caused by: 1) random catastrophic events; 2) biological interactions such as **competition**, disease, and predation; 3) chronic stresses; or 4) frequent disturbance. However, with the recent ascendance of human activities as a dominant force behind environmental changes, there has been a dramatic increase in rates of extinction at local, regional, and even global levels.

The recent wave of **anthropogenic** extinctions includes such well-known cases as the **dodo, passenger pigeon**, great auk, and others. There are many other high-profile species that humans have brought to the brink of extinction, including the plains buffalo, **whooping crane**, eskimo curlew, **ivory-billed woodpecker**, and various marine mammals. Most of these instances were caused by an insatiable over-exploitation of species that were unable to sustain a high rate of **mortality**, often coupled with an intense disturbance of their **habitat**.

Beyond these tragic cases of extinction or endangerment of large, charismatic vertebrates, the earth's biota is experiencing an even more substantial loss of biodiversity caused by the loss of habitat. In part, this loss is due to the conversion of large areas of tropical ecosystems, particularly moist forest, to agricultural or otherwise ecologically degraded habitats. A large fraction of the biodiversity of tropical biomes is comprised of rare, endemic (i.e., with a local distribution) species. Consequently the conversion of **tropical rain forest** to habitats unsuitable for these specialized species inevitably causes the extinction of most of the locally endemic biota. Remarkably, the biodiversity of tropical forests is so large, particularly in insects, that most of it has not yet been identified taxonomically. We are therefore faced with the prospect of a **mass extinction** of perhaps millions of species before they have been recognized by science.

To date, about 1.7 million organisms have been identified and designated with a scientific name. About 6% of identified species live in boreal or polar latitudes, 59% in the temperate zones, and the remaining 35% in the tropics. The knowledge of the global richness of species is very incomplete, particularly in the tropics. If a conservative estimate is made of the number of unidentified tropical species, the fraction of global species that live in the tropics would increase to at least 86%.

Invertebrates comprise the largest number of described species, with insects making up the bulk of that total and beetles (*Coleoptera*) comprising most of the insects. Biologists believe that there still is a tremendous number of undescribed species of insects in the tropics, possibly as many as another 30 million species. This remarkable conclusion has emerged from experiments conducted by the entomologist Terry Erwin, in which scientists "fogged" tropical forest canopies and then collected the "rain" of dead arthropods. This research suggests that: (1) a large fraction of the insect biodiversity of tropical forests is undescribed; (2) most insect species are confined to a single type of forest, or even to particular plant species, both of which are restricted in distribution; and (3) most tropical forest insects have a very limited dispersal ability.

The biodiversity and endemism of other tropical forest biota are better known than that of arthropods. For example, a plot of only 0.2 acres (0.1 ha) in an Ecuadorian forest had 365 species of vascular plants. The richness of woody plants in tropical **rain forest** can approach 300 species per hectare, compared with fewer than 12–15 tree species in a typical temperate forest, and thirty to thirty-five species in the Great

Smokies of the United States, the richest temperate forest in the world.

There have been few systematic studies of all of the biota of particular tropical communities. In one case, D. H. Janzen studied a savanna-like, 67-mi^2 (108-km^2) reserve of dry tropical forest in Costa Rica for several years. He estimated that the site had at least 700 plant species, 400 vertebrate species, and a remarkable 13,000 species of insect, including 3,140 species of moths and butterflies.

Why should one worry about the likelihood of extinction of so many **rare species** of tropical insects, or of many other rare species of plants and animals? There are three classes of reasons why extinctions are regrettable:

(1) There are important concerns in terms of the ethics of extinction. Central questions are whether humans have the "right" to act as the exterminator of unique and irrevocable species of **wildlife** and whether the human existence is somehow impoverished by the tragedy of extinction. These are philosophical issues that cannot be scientifically resolved, but it is certain that few people would applaud the extinction of unique species.

(2) There are utilitarian reasons. Humans must take advantage of other organisms in myriad ways for sustenance, medicine, shelter, and other purposes. If species become extinct, their unique services, be they biological, ecological, or otherwise, are no longer available for exploitation.

(3) The third class of reasons is ecological and involves the roles of species in maintaining the **stability** and integrity of ecosystems, i.e., in terms of preventing **erosion** and controlling **nutrient** cycling, productivity, trophic dynamics, and other aspects of **ecosystem** structure and function. Because we rarely have sufficient knowledge to evaluate the ecological "importance" of particular species, it is likely that an extraordinary number of species will disappear before their ecological roles are understood.

There are many cases where research on previously unexploited species of plants and animals has revealed the existence of products of great utility to humans, such as food or medicinals. One example is the rosy periwinkle (*Catharanthus roseus*), a plant native to **Madagascar**. During a screening of many plants for possible anti-cancer properties, an extract of rosy periwinkle was found to counteract the reproduction of **cancer** cells. Research identified the active ingredients as several alkaloids, which are now used to prepare the important anti-cancer drugs vincristine and vinblastine. This once obscure plant now allows treatment of several previously incurable cancers and is the basis of a multi-million-dollar economy.

Undoubtedly, there is a tremendous, undiscovered wealth of other biological products that are of potential use to humans. Many of these natural products are present in the biodiversity of tropical species that has not yet been "discovered" by taxonomists.

It is well known that extinction can be a natural process. In fact, most of the species that have ever lived on Earth are now extinct, having disappeared "naturally" for some reason or other. Perhaps they could not cope with changes in their inorganic or biotic **environment**, or they may have succumbed to some catastrophic event, such as a meteorite impact.

The rate of extinction has not been uniform over geological time. Long periods characterized by a slow and uniform rate of extinction have been punctuated by about nine catastrophic events of mass extinction. The most intense mass extinction occurred some 250 million years ago, when about 96% of marine species became extinct. Another example occurred 65 million years ago, when there were extinctions of many vertebrate species, including the reptilian orders Dinosauria and Pterosauria, but also of many plants and invertebrates, including about one half of the global **fauna** that existed then.

In modern times, however, humans are the dominant force causing extinction, mostly because of: (1) overharvesting; (2) effects of introduced predators, competitors, and diseases; and (3) habitat destruction. During the last 200 years, a global total of perhaps 100 species of mammals, 160 birds, and many other taxa are known to have become extinct through some human influence, in addition to untold numbers of undescribed, tropical species.

Even pre-industrial human societies caused extinctions. Stone-age humans are believed to have caused the extinctions of large-animal fauna in various places, by the unsustainable and insatiable **hunting** of vulnerable species in newly discovered islands and continents. Such events of mass extinction of large animals, co-incident with human colonization events, have occurred at various times during the last 10–50,000 years in Madagascar, New Zealand, **Australia**, Tasmania, Hawaii, North and South America, and elsewhere.

In more recent times, **overhunting** has caused the extinction of other large, vulnerable species, for example the flightless dodo (*Raphus cucullatus*) of Mauritius. Some North American examples include Labrador duck (*Camptorhynchus labradorium*), passenger pigeon (*Ectopistes migratorius*), Carolina parakeet (*Conuropsis carolinensis*), great auk (*Pinguinus impennis*), and Steller's sea cow (*Hydrodamalis stelleri*). Many other species have been brought to the brink of extinction by overhunting and loss of habitat. Some North American examples include eskimo curlew (*Numenius borealis*), plains **bison** (*Bison bison*), and a variety of marine mammals, including manatee (*Trichechus manatus*), right **whales** (*Euba-*

laena glacialis), bowhead whale (*Balaena mysticetus*), and blue whale (*Balaenoptera musculus*).

Island biotas are especially prone to both natural and anthropogenic extinction. This syndrome can be illustrated by the case of the **Hawaiian Islands**, an ancient volcanic archipelago in the Pacific Ocean, about 994 mi (1,600 km) from the nearest island group and 2,484 mi (4,000 km) from the nearest continental landmass. At the time of colonization by Polynesians, there were at least 68 **endemic species** of Hawaiian birds, out of a total richness of land birds of 86 species. Of the initial 68 endemics, 24 are now extinct and 29 are perilously endangered. Especially hard hit has been an endemic family, the Hawaiian honeycreepers (Drepanididae), of which 13 species are believed extinct, and 12 endangered. More than 50 alien species of birds have been introduced to the Hawaiian Islands, but this gain hardly compensates for the loss and endangerment of specifically evolved endemics. Similarly, the native **flora** of the islands is estimated to have been comprised of 1,765–2,000 taxa of angiosperm plants, of which at least 94% were endemic. During the last two centuries, more than 100 native plants have become extinct, and the survival of at least an additional 500 taxa is threatened or endangered, some now being represented by only single individuals. The most important causes of extinction of Hawaiian biota have been the conversion of natural ecosystems to agricultural and urban landscapes, the introduction of alien predators, competitors, herbivores, and diseases, and to some extent, aboriginal overhunting of some species of bird.

Overhunting has been an important cause of extinction, but in modern times habitat destruction is the most important reason for the event of mass extinction that Earth's biodiversity is now experiencing. As was noted previously, most of the global biodiversity is comprised of millions of as yet undescribed taxa of tropical insects and other organisms. Because of the extreme endemism of most tropical biota, it is likely that many species will become extinct as a result of the clearing of natural tropical habitats, especially forest, and its conversion to other types of habitat.

The amount and rate of **deforestation** in the tropics are increasing rapidly, in contrast to the situation at higher latitudes where forest cover is relatively stable. Between the mid 1960s and the mid 1980s there was little change (less than 2%) in the forest area of North America, but in Central America forest cover decreased by 17%, and in South America by 7% (but by a larger %age in equatorial countries of South America). The global rate of clearing of tropical rain forest in the mid 1980s was equivalent to 6–8% of that **biome** per year, a rate that if projected into the future would predict a biome **half-life** of only nine to 12 years. Some of the cleared forest will regenerate through secondary **succession**, which would ultimately produce another mature forest. Little is known, however, about the rate and biological character of succession in tropical rainforests, or how long it would take to restore a fully biodiverse ecosystem after disturbance.

The present rate of disturbance and conversion of tropical forest predicts grave consequences for global biodiversity. Because of a widespread awareness and concern about this important problem, much research and other activity has recently been directed towards the **conservation** and protection of tropical forests. As of 1985, several thousand sites, comprising more than 640,000 mi² (1 million km), had received some sort of "protection" in low-latitude countries. Of course the operational effectiveness of the protected status varies greatly, depending on the commitment of governments to these issues. Important factors include: (1) political stability; (2) political priorities; (3) finances available to mount effective programs to control **poaching** of animals and lumber and to prevent other disturbances; (4) the support of local peoples and communities for biodiversity programs; (5) the willingness of relatively wealthy nations to provide a measure of debt relief to impoverished tropical countries and thereby reduce their short term need to liquidate **natural resources** in order to raise capital and provide employment; and (6) local **population growth**, which also generates extreme pressures to over exploit natural resources.

The biodiversity crisis is a very real and very important aspect of the global environmental crisis. All nations have a responsibility to maintain biodiversity within their own jurisdictions and to aid nations with less economic and scientific capability to maintain their biodiversity on behalf of the entire planet. The modern biodiversity crisis focuses on species-rich tropical ecosystems, but the developed nations of temperate latitudes also have a large stake in the outcome and will have to substantially subsidize global conservation activities if these are to be successful. Much needs to be done, but an encouraging level of activity in the conservation and protection of biodiversity is beginning in many countries, including an emerging commitment by many nations to the conservation of threatened ecosystems in the tropics.

[*Bill Freedman Ph.D.*]

RESOURCES
BOOKS

Ehrlich, P. R., and A. H. Ehrlich. *Extinction: The Causes and Consequences of the Disappearance of Species.* New York: Ballantyne Books, 1981.

Freedman, B. *Environmental Ecology.* San Diego, CA: Academic Press, 1989.

Peters, R. L., and T. E. Lovejoy. *Global Warming and Biological Diversity.* New Haven, CT: Yale University Press, 1992.

Wilson, E. O. *Biodiversity.* Washington, DC: National Academy Press, 1988.

———. *Biophilia: The Human Bond With Other Species.* Cambridge, MA: Harvard University Press, 1984.

PERIODICALS

Janzen, D. H. "Insect Diversity in a Costa Rican Dry Forest: Why Keep It, and How." *Biological Journal of the Linnaean Society* 30 (1987): 343–356.

Biofilms

Marine microbiology began with the investigations of marine microfouling by L. E. Zobell and his colleagues in the 1930s and 1940s. Their interests focused primarily on the early stages of settlement and growth of **microorganisms**, primarily bacteria on solid substrates immersed in the sea. The interest in the study of marine microfouling was sporadic from that time until the early 1960s when interest in marine bacteriology began to increase. Since 1970 the research on the broad problems of bioadhesion and specifically the early stages of microfouling has expanded tremendously.

The initial step in marine fouling is the establishment of a complex film. This film, which is composed mainly of bacteria and diatoms plus secreted extracellular materials and debris, is most commonly referred to as the "primary film" but may also be called the "bacterial fouling layer," or "slime layer." The latter name is aptly descriptive since the film ultimately becomes thick enough to feel slippery or slimy to touch. In addition to the bacteria and diatoms that comprise most of the biota, the film may also include yeasts, **fungi**, and protozoans.

The settlement sequence in the formation of primary films is dependent upon a number of variables which may include the location of the surface, season of the year, depth, and proximity to previously fouled surfaces and other physiochemical factors.

Many studies have demonstrated the existence of some form of ecological **succession** in the formation of fouling communities, commencing with film forming microorganisms and reaching a climax community of macrofouling organisms such as barnacles, tunicates, mussels, and seaweeds.

Establishment of primary films in marine fouling has two functions: (a) to provide a surface favoring the settlement and adhesion of animal larvae and algal cells, and (b) to provide a **nutrient** source that could sustain or enhance the development of the fouling community.

Formation of a primary film is initiated by a phenomenon known as "molecular fouling" or "surface conditioning." The formation of this molecular film was first demonstrated by Zobell in 1943 and since has been confirmed by many other investigators. The molecular film forms by the **sorption** to solid surfaces of organic matter dissolved or suspended in seawater. The sorption of this dissolved material creates surface changes in the surface of the substrate which are favorable for establishing biological settlement. These dissolved organic materials originate from a variety of sources such as end-products of bacterial decay, excretory products, dissolution from seaweeds, etc., and consist principally of sugars, amino acids, urea, and fatty acids.

This molecular film has been observed to form within minutes after any clean, solid surface is immersed in natural seawater. The role of this film in **biofouling** has been shown to modify the "critical surface tension" or wetability of the immersed surface which than facilitates the strong bonding of the microorganisms through the agency of mucopolysac-charides exuded by film-forming bacteria.

Bacteria have been found securely attached to sub-strates immersed in seawater after just a few hours. Initial colonization is by rod-shaped bacteria followed by stalked forms within 24–72 hours. As many as 40–50 **species** have been isolated from the surface of glass slides immersed in seawater for a few days.

Following the establishment of the initial film of bac-teria and their secreted extracellular polymer on a solid sub-strate, additional bacteria and other microorganisms may attach. Most significant in this population are benthic dia-toms but there are also varieties of filamentous microorgan-isms and protozoans. These organisms, together with debris and other organic particular matter that adhere to the surface create an intensely active biochemical **environment** and form the primary stage in the succession of a typical mac-rofouling community.

Considering the enormous economic consequences of marine fouling it is not at all surprising that there continues to be intense interest in the results of recent research, particu-larly in the conditions and processes of molecular film for-mation.

[*Donald A. Villeneuve*]

RESOURCES
BOOKS

Corpe, W. A. "Primary Bacterial Films and Marine Microfouling." In *Proceedings of the 4th International Congress of Marine Corrosion and Fouling,* edited by V. Romansky. 1977.

PERIODICALS

Zobell, L. E., and E. C. Allen. "The Significance of Marine Bacteria in the Fouling of Submerged Surfaces." *Journal of Bacteriology* 29 (1935): 239–251.

Biofiltration

Biofiltration refers to the removal and oxidation of organic gases (i.e., volatile organic compounds, or VOCs) from con-taminated air by vapor phase biodegradation in beds (biofilt-ers) of compost, **soil,** or other materials such as municipal waste, sand, bark peat, volcanic ash, or diatomaceous earth. As contaminated air (such as air from a soil vapor extraction

process) flows through the biofilter, the VOCs sorb onto surfaces of the pile and are degraded by **microorganisms**. **Nutrient** blends or exogenous microbial cultures can be added to a biofilter to enhance its performance. Moisture needs to be continually supplied to the biofilter to counteract the drying effects of the gas stream. The stationary support media that make up the biofiltration bed should be porous enough to allow gas flow through the biofilter and should provide a large surface area with high wetting and sorptive capacities. This support media should also provide adequate buffering capacity and may also serve as a source of inorganic nutrients. Biofilters, also used to treat odors as well as organic contaminants, have been used in Europe for over twenty years. Compared to **incineration** and **carbon adsorption**, biofilters do not require landfilling of residuals or regenera-tion of spent materials.

Waste gases are moved through the units by induced or forced draft. Biofilters are capable of handling rapid air flow rates (e.g., up to 90,000 cubic ft (2,700 cubic m) per minute (cfm) in **filters** up to 20,000 sq ft (1,800 sq m) in wetted area) and VOC concentrations greater than 1,000 ppm. However, biofiltration removal of more highly haloge-nated compounds such as trichloroethylene (TCE) or carbon tetrachloride, which biodegrade very slowly under **aerobic** conditions, may require very long residence times (i.e., very large biofilters) or treatment of very low flow rates of air containing the contaminants.

The soil-type biofilter is similar in design to a soil compost pile. Fertilizers are preblended into the compost pile to provide nutrients for indigenous microorganisms, which accomplish the biodegradation of the VOCs. In the treatment bed type of biofilter, the waste air stream is humid-ified as it is passed through one or more beds of compost, municipal waste, sand, diatomaceous earth or other materi-als. Another type of biofilter is the disk biofilter, which consists of a series of humidified, compressed disks placed inside a reactor shell. These layered disks contain activated charcoal, nutrients, microbial cultures, and compost material. The waste air stream is passed through the disk system. Collected water condensate from the process is returned to the humidification system for **reuse**.

[*Judith L. Sims*]

RESOURCES
BOOKS

Baker, K. H., and D. S. Herson. *Bioremediation.* New York: McGraw-Hill, 1994.

King, B. R., G. M. Long, and J. K. Sheldon. *Practical Environmental Bioremediation.* Boca Raton, FL: Lewis Publishers, 1992.

Stoner, D. L., ed. "Organic Waste Forms and Treatment Strategies." *Bio-technology for the Treatment of Hazardous Waste.* Boca Raton, FL: Lewis Publishers, 1994.

Biofouling

The term fouling, or more specifically biofouling, is used to describe the growth and accumulation of living organisms on the surfaces of submerged artificial structures as opposed to natural surfaces. Concern over and interest in fouling arises from practical considerations including the enormous costs resulting from fouling of ships, buoys, floats, pipes, cables and other underwater man-made structures.

From its first immersion in the sea, an artificial structure or surface changes through time as a result of a variety of influences including location, season and other physical and biological variables. Fouling communities growing on these structures are biological entities and must be understood developmentally. The development of a fouling community on a bare, artificial surface immersed in the sea displays a form of **succession**, similar to that seen in terrestrial ecosystems, which culminates in a community which may be considered a climax stage. Scientists have identified two distinct stages in fouling community development: 1) the primary or microfouling stage, and 2) the secondary or macrofouling stage.

Microfouling: When a structure is first submerged in seawater, **microorganisms**, primarily bacteria and diatoms, appear on the surface and multiply rapidly. Together with debris and other organic **particulate** matter, these microorganisms form a film on the surface. Although the evidence is not conclusive, it appears that the development of this film is a prerequisite to initiation of the fouling succession.

Macrofouling: The animals and plants that make up the next stages of succession in fouling communities are primarily the attached or sessile forms of animals and plants that occur naturally in shallow waters along the local coast. The development of fouling communities in the sea depends upon the ability of locally-occurring organism to live successfully in the new artificial **habitat**. The first organisms to attach to the microfouled surface are the swimming larvae of **species** present at the time of immersion. The kinds of larvae present vary with the season. Rapidly growing forms that become established first may ultimately be crowded out by others which grow more slowly. A comprehensive list of species making up fouling communities recorded from a wide variety of structures identified 2,000 species of animals and plants. Although the variety of organisms identified seems large it actually represents a very small proportion of the known marine species. Further, only about 50 to 100 species are commonly encountered in fouling, including bivalve mollusks (primarily oysters and mussels), barnacles, aquatic invertebrates in the phylum Bryozoa, tubeworms and other organisms in the class Polychaeta, and green and brown algae.

Control of fouling organisms has long been a formidable challenge resulting in the development and application of a wide variety of toxic paints and greases, or the use of metals which give off toxic ions as they corrode. However, none of the existing methods provide permanent control. Furthermore, the recognition of the potential environmental hazards attendant with the use of materials that leach **toxins** into the marine **environment** has led to the ban of some of the most widely used materials. This has stimulated efforts to develop alternative materials or methods of controlling biofouling that are environmentally safe.

[*Donald A. Villeneuve*]

RESOURCES
BOOKS

Workshop on Preservation of Wood in the Marine Environment. Marine Borers, Fungi and Fouling Organisms of Wood. Paris: Organization for Economic Cooperation and Development, 1971.

Melo, L. F., et al, eds. *Fouling Science and Technology.* Norwell, MA: Kluwer Academic, 1988.

Woods Hole Oceanographic Institution. *Marine Fouling and Its Prevention.* Annapolis, MD: U. S. Naval Institute, 1952.

Biogeochemistry

Biogeochemistry refers to the quantity and cycling of **chemicals** in ecosystems. Biogeochemistry can be studied at various spatial scales, ranging from communities, landscapes (or seascapes), and over Earth as a whole. Biogeochemistry involves the study of chemicals in organisms, and also in non-living components of the **environment**.

An important aspect of biogeochemistry is the fact that elements can occur in various molecular forms that can be transformed among each other, often as a result of biological reactions. Such transformations are an especially important consideration for nutrients, i.e., those chemicals that are required for the healthy functioning of organisms. As a result of biogeochemical cycling, nutrients can be used repeatedly–nutrients contained in dead **biomass** can be recycled through inorganic forms, back into living organisms, and so on. Biogeochemistry is also relevant to the movements and transformations of potentially toxic chemicals in ecosystems, such as metals, pesticides, and certain gases.

Nutrient Cycles

Ecologists have a good understanding of the biogeochemical cycling of the most important nutrients. These include **carbon**, **nitrogen**, **phosphorus**, potassium, calcium, magnesium, and sulfur. Some of these can occur variously as gases in the **atmosphere**, as ions dissolved in water, in minerals in rocks and **soil**, and in a great variety of organic chemicals in the living or dead biomass of organisms. Ecologists study **nutrient** cycles by determining the quantities of the various chemical forms of nutrients in various compartments of ecosystems, and by determining the rates of transformation and cycling among the various compartments.

The **nitrogen cycle** is particularly well understood, and it can be used to illustrate the broader characteristics of nutrient cycling. Nitrogen is an important nutrient, being one of the most abundant elements in the tissues of organisms and a component of many kinds of biochemicals, including amino acids, proteins, and nucleic acids. Nitrogen is also one of the most common limiting factors to **primary productivity**, and the growth rates of plants in many ecosystems will increase markedly if they are fertilized with nitrogen. This is a fairly common characteristic of terrestrial and marine environments, and to a lesser degree of freshwater ones.

Plants assimilate most of their nitrogen from the soil environment, as nitrate (NO_3^-) or ammonium (NH_4^+) dissolved in the water that is taken up by roots. Some may also be taken up as gaseous **nitrogen oxides** (such as NO or NO_2) that are absorbed from the atmosphere. In addition, some plants live in a beneficial **symbiosis** with **microorganisms** that have the ability to fix atmospheric dinitrogen gas (N_2) into ammonia (NH_3), which can be used as a nutrient. In contrast, almost all animals satisfy their nutritional needs by eating plants or other animals and metabolically breaking down the organic forms of nitrogen, using the products (such as amino acids) to synthesize the necessary biochemicals of the animal. When plants and animals die, microorganisms active in the detrital cycle metabolize organic nitrogen in the dead biomass into simpler compounds, ultimately to ammonium.

The nitrogen cycle has always occurred naturally, but in modern times some of its aspects have been greatly modified by human influences. These include the fertilization of agricultural ecosystems, the dumping of nitrogen-containing sewage into lakes and other waterbodies, the **emission** of gaseous forms of nitrogen into the atmosphere, and the cultivation of nitrogen-fixing legumes. In some cases, human effects on nitrogen biogeochemistry result in increased productivity of crops, but in other cases serious ecological damages occur.

Toxic Chemicals in Ecosystems

Some human activities result in the release of toxic chemicals into the environment, which under certain conditions can pose risks to human health and cause serious damages to ecosystems. These damages are called **pollution**, whereas the mere presence of chemicals which cause no damage in the environment is referred to as contamination. Biogeochemistry is concerned with the emissions, transfers, and quantities of these potentially toxic chemicals in the environment and ecosystems.

Certain chemicals have a great ability to accumulate in organisms rather than in the non-living (or inorganic) components of the environment. This tendency is referred to as bioconcentration. Chemicals that strongly bioconcentrate include methylmercury and all of the persistent organochlorine compounds, such as **dichlorodiphenyl-trichloroethane** (DDT), **pentachlorophenol** (PCBs), dioxins, and **furans**. Methylmercury bioconcentrates because it is rather tightly bound in certain body organs of animals. The organochlorines bioconcentrate because they are extremely insoluble in water but highly soluble in fats and lipids, which are abundant in the bodies of organisms but not in non-living parts of the environment.

In addition, persistent organochlorines tend to occur in particularly large concentrations in the fat of top predators, that is, in animals high in the ecological food web, such as marine mammals, predatory birds, and humans. This happens because these chemicals are not easily metabolized into simpler compounds by these animals, so they accumulate in increasingly larger residues as the animals feed and age. This phenomenon is known as food-web magnification (or **biomagnification**). Food-web magnification causes chemicals such as DDT to achieve residues of tens or more **parts per million** (ppm) in the fatty tissues of top predators, even though they occur in the inorganic environment (such as water) in concentrations smaller than one part per billion (ppb). These high body residues can lead to ecotoxicological problems for top predators, some of which have declined in abundance because of their exposure to **chlorinated hydrocarbons**.

Because a few **species** of plants have an affinity for potentially toxic elements, they may bioaccumulate them to extremely high concentrations in their tissues. These plants are genetically adapted ecotypes which are themselves little affected by the residues, although they can cause toxicity to animals that might feed on their biomass. For example, some plants that live in environments in which the soil contains a mineral known as serpentine accumulate **nickel** to concentrations which may exceed thousands of ppm. Similarly, some plants (such as locoweed) growing in semi-arid environments can accumulate thousands of ppm of selenium in their tissues, which can poison animals that feed on the plants.

[*Bill Freedman Ph.D.*]

RESOURCES
BOOKS

Atlas, R.M., and R. Bartha. *Microbial Ecology*. Menlo Park, CA: Benjamin/ Cummings, 1987.

Freedman, B. *Environmental Ecology*, 2nd ed. San Diego, CA: Academic Press, 1995.

Schlesinger, W. H. *Biogeochemistry: An Analysis of Global Change*. San Diego, CA: Academic Press, 1991.

Smith, R. P. *A Primer of Environmental Toxicology*. Philadelphia, PA: Lea & Febiger, 1992.

Biogeography

Biogeography is the study of the spatial distribution of plants and animals, both today and in the past. Developed during the course of nineteenth century efforts to explore, map, and describe the earth, biogeography asks questions about regional variations in the numbers and kinds of **species**: Where do various species occur and why? What physical and biotic factors limit or extend the range of a species? In what ways do species disperse (expand their ranges), and what barriers block their dispersal? How has species distribution changed over centuries or millennia, as shown in the fossil record? What controls the makeup of a **biotic community** (the combination of species that occur together)? Biogeography is an interdisciplinary science: many other fields, including paleontology, geology, botany, oceanography, and climatology, both contribute to biogeography and make use of ideas developed by biogeographers.

Because physical and biotic environments strongly influence species distribution, the study of **ecology** is closely tied to biogeography. Precipitation, temperature ranges, **soil** types, soil or water **salinity**, and insolation (exposure to the sun) are some elements of the physical **environment** that control the distribution of plants and animals. Biotic limits to distribution, constraints imposed by other living things, are equally important. Species interact in three general ways: **competition** with other species (for space, sunlight, water, or food), predation (e.g., an owl species relying on rodents for food), and **mutualism** (e.g., an insect pollenizing a plant while the plant provides nourishment for the insect). The presence or absence of a key plant or animal may function as an important control on another species' spatial distribution. **Community ecology**, the ways in which an assemblage of species coexist, is also important. Biotic communities have a variety of niches, from low to high trophic levels, from generalist roles to specialized ones. The presence or absence of species filling one of these roles influences the presence or survival of a species filling another role.

Two other factors that influence a region's biotic composition or the range of a particular species are dispersal, or spreading, of a species from one place to another;

and barriers, environmental factors that block dispersal. In some cases a species can extend its range by gradually colonizing adjacent, hospitable areas. In other cases a species may cross a barrier, such as a mountain range, an ocean, or a **desert**, and establish a colony beyond that barrier. The cattle egret (*Bubulcus ibis*) exemplifies both types of movement. Late in the nineteenth century these birds crossed the formidable barrier of the Atlantic Ocean, perhaps in a storm, and established a breeding colony in Brazil. During the past one hundred years this small egret has found suitable **habitat** and gradually expanded its range around the coast of South America and into North America, so that by 1970 it had been seen from southern Chile to southern Ontario.

The study of dispersal has special significance in **island biogeography**. The central idea of island biogeography, proposed in 1967 by R. H. MacArthur and **Edward O. Wilson**, is that an island has an equilibrium number of species that increases with the size of the land mass and its proximity to other islands. Thus species diversity should be extensive on a large or nearshore island, with enough complexity to support large carnivores or species with very specific food or habitat requirements. Conversely, a small or distant island may support only small populations of a few species, with little complexity or **niche** specificity in the biotic community.

Principles of island biogeography have proven useful in the study of other "island" ecosystems, such as isolated lakes, small mountain ranges surrounded by deserts, and insular patches of forest left behind by clearcut **logging**. In such threatened areas as the Pacific Northwest and the Amazonian rain forests, foresters are being urged to leave larger stands of trees in closer proximity to each other so that species at high trophic levels and those with specialized food or habitat requirements (e.g., Northern spotted owls and Amazonian monkeys) might survive. In such areas as **Yellowstone National Park**, which national policy designates as an insular unit of habitat, the importance of adjacent habitat has received increased consideration. Recognition that clearcuts and farmland constitute barriers has led some planners to establish forest corridors to aid dispersal, enhance genetic diversity, and maintain biotic complexity in unsettled islands of natural habitat.

[*Mary Ann Cunningham Ph.D.*]

RESOURCES
BOOKS

Brown, J. H., and A. C. Gibson. *Biogeography*. St. Louis: Mosby, 1983.

MacArthur, R. H., and E. O. Wilson. *The Theory of Island Biogeography*. Vol. 1, *Monographs in Population Biology*. Princeton: Princeton University Press, 1967.

Biohydrometallurgy

Biohydrometallurgy is a technique by which **microorganisms** are used to recover certain metals from ores. The technique was first used over 300 years ago to extract **copper** from low-grade ores. In recent years, its use has been extended to the recovery of **uranium** and gold, and scientist believe that it will eventually be applied to the recovery of other metals such as **lead, nickel**, and zinc.

In most cases, biohydrometallurgy is employed when conventional mining procedures are too expensive or ineffective in recovering a metal. For example, dumps of unwanted waste materials are created when copper is mined by traditional methods. These wastes consist primarily of rock, gravel, sand, and other materials that are removed in order to reach the metal ore itself. But the wastes also contain very low concentrations (less than 0.5%) of copper ore.

Until recently, the concentrations of copper ore in a dump were too low to have any economic value. The cost of collecting the ore was much greater than the value of the copper extracted. But, as richer sources of copper ore are used up, low grade reserves (like dumps) become more attractive to mining companies. At this point, biohydrometallurgy can be used to leach out the very small quantities of ore remaining in waste materials.

The extraction of copper by means of biohydrometallurgy involves two types of reactions. In the first, microorganisms operate directly on compounds of copper. In the second, microorganisms operate on metallic compounds other than those of copper. These metallic compounds are then converted into forms which can, in turn, react with copper ores.

The use of biohydrometallurgical techniques on a copper ore waste dump typically begins by spraying the dump with dilute sulfuric **acid**. As the acid seeps into the dump, it creates an **environment** favorable to the growth of acid-loving microorganisms that attack copper ores. As the microorganisms metabolize the ores, they convert copper from an insoluble to a soluble form. Soluble copper is then leached out of the dump with sulfuric acid. It is recovered when the solution is pumped out to a recovery tank.

A second reaction occurs within the dump. Microorganisms also convert ferrous iron (Fe^{2+}) in ores such as pyrite (FeS_2) to ferric iron (Fe^{3+}). The ferric iron, in turn, oxidizes copper in the dump from an insoluble to a soluble form.

The mechanism described here is a highly efficient one. As microorganisms act on copper and iron compounds, they produce sulfuric acid as a by-product, thus enriching the environment in which they live. Ferric iron reduces and oxidizes copper at the same time, making the copper available for attack by microorganisms once again.

A number of microorganisms have been used in biohydrometallurgy. One of the most effective for the **leaching** of copper is *Thiobacillus ferrooxidans*. Research is now being conducted on the development of genetically engineered microorganisms that can be used in the recovery of copper and other metals.

The two other metals for which biohydrometallurgy seems to be most useful are uranium and gold. Waste dumps in South Africa and Canada have been treated to convert insoluble forms of uranium to soluble forms, allowing recovery by a method similar to that used with copper. In the treatment of gold ores, biohydrometallurgy is used in a pretreatment step prior to the conventional conversion of the metal to a cyanide-complex. The first commercial plants for the biohydrometallurgical treatment of gold ores are now in operations in South Africa and Zimbabwe.

[*David E. Newton*]

RESOURCES
BOOKS

McGraw Hill Encyclopedia of Science and Technology. 7th ed. New York: McGraw-Hill, 1992.

Rossi, G. *Biohydrometallurgy.* New York: McGraw-Hill, 1990.

Bioindicator

A bioindicator is a plant or animal **species** that is known to be particularly tolerant or sensitive to **pollution**. Based on the known association of an organism with a particular type or intensity of pollution, the presence of the organism can be used as a tool to indicate polluted conditions relative to unimpacted reference conditions. Sometimes a set of species or the structure and function of an entire **biological community** may function as a bioindicator. In assessing the impacts of pollution, bioindicators are frequently used to evaluate the "health" of an impacted **ecosystem** relative to a reference area or reference conditions. Field-based, site-specific environmental evaluations based on the bioindicator approach generally are complemented with laboratory studies of toxicity testing and **bioassay** experiments.

The use of individual species or a community structure as bioindicators involves the identification, classification and quantification of biota in the affected area. While many species are in use, the most widely used biological communities are the benthic macroinvertebrates. These are the sedentary and crawling worms and insect larvae that reside in the bottom sediments of aquatic systems such as lake and river bottoms. The bottom sediments usually contain most of the pollutants introduced into an aquatic system. Since these macroinvertebrates have limited mobility, they are continually exposed to the highest concentrations of pollutants in the system. Therefore, this benthic community is an ideal bioindicator: stationary, localized and exposed to maximum pollutant concentrations within a specific location.

Often, alterations in community structure due to pollution include a change from a more diverse to a less diverse community with fewer species or taxa. The indicator community may also be composed mostly of species that are tolerant of or adapted to polluted conditions and pollution-sensitive species that are present upstream may be absent in the impacted zones. However, depending on the type of pollutant, the abundance of the pollution-tolerant species may be very high and, therefore, the size of the benthic community may be similar to or exceed the reference community upstream. This is common in cases where pollution from sewage discharges adds organic matter that provides food for some of the tolerant benthic species. In the case of toxic chemical (e.g., **heavy metals**, organic compounds) pollution, the benthic community may show an overall reduction both in diversity and abundance.

Tubificid worms are an example of pollution-tolerant indicator organisms. These worms live in the bottom sediments of streams and lakes and are highly tolerant of the kind of pollution that results from sewage discharges. In a river polluted by **wastewater discharge** from a **sewage treatment** plant, it is common to see a large increase in the number of tubificid worms in the stream sediments immediately downstream of the discharge. Upstream of the discharge, the number of these worms is much lower, reflecting the cleaner conditions. Further downstream, as the discharge is diluted, the number of tubificid worms again decreases to a level similar to the upstream portions of the river. Large populations of these worms dramatically demonstrate that pollution is present, and the location of these populations may also indicate the general area where the pollution enters the **environment**.

Alternatively, pollution-intolerant organisms can also be used to indicate polluted conditions. The larvae of mayflies live in stream sediments and are known to be particularly sensitive to pollution. In a river receiving wastewater discharge, mayflies show a pattern opposite to that of the tubificid worms. The mayfly larvae are normally present in large numbers above the discharge point, decrease or disappear at the discharge point (just where the tubificid worms are most abundant) and reappear further downstream as the effects of the discharge are diluted. In this case, the mayflies are pollution-sensitive indicator organisms and their absence serves as the indication of pollution. Similar examples of indicator organisms can be found among plants, fishes and other biological groups. Giant reedgrass (*Phragmites australis*) is a common marsh plant that is typically indicative of disturbed conditions in **wetlands**. Among fish, disturbed conditions may be indicated by the disappearance of sensitive species like trout which require clear, cold waters to thrive.

The usefulness of indicator organisms is unquestionable but limited. While their presence or absence provides a reliable general picture of polluted conditions, it is often difficult to identify clearly the exact sources of pollution, especially in areas with multiple sources of pollution. In the sediments of New York Harbor, for example, pollution-tolerant insect larvae are overwhelmingly dominant. However, it is impossible to attribute the large larval populations to just one of the numerous possible sources of pollution in this area which include ship traffic, sewage discharge, industrial discharge, and **storm runoff**. As more is learned about the physiology and life-history of an **indicator organism** and its response to different types of pollution, it may be possible to draw more specific conclusions.

Although the two terms are sometimes used interchangeably, indicator organisms should not be confused with monitor organisms (also called biomonitors) which are organisms that bioaccumulate toxic substances present in trace amounts in the environment. For example, when it is difficult to measure directly the low concentrations of a pollutant in water, chemical analysis of shellfish tissues from that location may show much higher, easily detected concentrations of that pollutant. In this case, the shellfish is used to monitor the level of the long-term presence of that pollutant in the area.

In the environmental field, bioindicators are commonly used in field investigations of contaminated sites to document impacts on the biological community and ecosystem. These studies are then followed up with focused laboratory tests to pinpoint the source of toxicity or stress. After clean-up and remedial actions have been implemented at a site, bioindicators are also used to track the effectiveness of the **remediation** activity. In the future, bioindicators may be used more widely as investigative and decision-making tools from the initial pollution and impact assessment stage to the remediation and post-remediation monitoring stages.

[*Usha Vedagiri*]

RESOURCES
BOOKS

Connell, D. W., and G. J. Miller. *Chemistry and Ecotoxicology of Pollution.* New York: Wiley-Interscience, 1984.

Biological community

A biological community is an association or assemblage of populations of organisms living in a localized area or **habitat**. The community is a level of organization incorporating individual organisms, **species**, and populations. A population is an assemblage of one species, and the community is a collage of one or more populations. Communities may be large or small, ranging from the microscopic to the level of **biome** and **biosphere**.

"Community," as contrasted conceptually to **ecosystem**, does not necessarily include consideration of the physical **environment** or the habitat of a particular group of organisms, though it is of course impossible to understand fully the dynamics of a community without reference to the resources on which it exists. The term ecosystem was coined to incorporate study of a community together with its physical environment. Still, communities are adaptive systems, inseparable from and evolving in response to changing environmental conditions. So, the supply and availability of resources in the environment and also time are considerations in the dynamics of community structure and relationships. Individual communities may be relatively stable or in constant flux. Actual equilibrium may never exist but, even in approximation, it must be viewed as a dynamic state—the community in constant, subtle flux and change.

Biological communities are "interactional fields" characterized by a complicated set of interactions among complex assemblages, both within the locale and from without—including trophic relationships, the "who eats who" of energy exchanges. Interactions may be proximal or between locales (close by or quite widely separated), including intensive, extensive, or limited exchanges with the outside world. Organisms in communities interact both functionally and spatially or locationally, both "horizontally" and "vertically," interactions often independent of each other and not reducible to one or the other. Not all organisms necessarily interact with all the others, some may be almost totally uncoupled from some of the others and coexist relatively independently.

Most questions in **community ecology** focus on the "existence, importance, looseness, transience, and contingency of interactions." The degree to which these interactions result in meaningful biological patterns is still an open question—though particular communities are usually identified by some pattern of interactions, if only to set them off from others for the purposes of scientific study.

[*Gerald L. Young*]

RESOURCES
BOOKS

Roughgarden, J. *The Structure and Assembly of Communities*. In *Perspectives in Ecological Theory*, edited by J. Roughgarden, R. M. May, and S. A. Levin. Princeton: Princeton University Press, 1989.

Strong, D. R., Jr., et. al., eds. *Ecological Communities: Conceptual Issues and the Evidence*. Princeton: Princeton University Press, 1984.

Taylor, P. J. "Community." In *Keywords in Evolutionary Biology*, edited by E. Keller and E. Lloyd. Cambridge: Harvard University Press, 1991.

PERIODICALS

Drake, J. A. "The Mechanics of Community Assembly and Succession." *Journal of Theoretical Biology* 147 (1990): 213–233.

Richardson, J. L. "The Organismic Community: Resilience of an Embattled Ecological Concept." *BioScience* 30 (1988): 465–471.

Biological fertility

The number of offspring produced by a female organism. In a population, biological fertility is measured as the general fertility rate (the birth rate multiplied by the number of sexually productive females) or as the total fertility rate (the lifetime average number of offspring per female). General dictionaries list fertility and **fecundity** as synonyms for reproductive fruitfulness. In **population biology**, biological fertility refers to the number of offspring actually produced, while fecundity is merely the biological ability to reproduce. Fecund individuals that fail to mate do not produce offspring (that is, are not biologically fertile), and do not contribute to **population growth**.

Biological integrity
see **Ecological integrity**

Biological magnification
see **Biomagnification**

Biological methylation

The process by which a methyl radical (-CH$_3$) is chemically combined with some other substance through the action of a living organism. One of the most environmentally important examples of this process is the **methylation** of **mercury** in the sediments of lakes, rivers, and other bodies of water. Elementary mercury and many of its inorganic compounds have relatively low toxicity because they are insoluble. However, in sediments, bacteria can convert mercury to an organic form, methylmercury, that is soluble in fat. When ingested by animals, methylmercury accumulates in body fat and exerts highly toxic, sometimes fatal, effects.

Biological oxygen demand
see **Biochemical oxygen demand**

Biological Resources Division

Created to assess, monitor, and research biological resources in United States, the Biological Resources Division (BRD) of the United States **Geological Survey** (USGS) is the non-regulatory biological research component of the United States Department of the Interior. First created in 1994 as the National Biological Survey (NBS), an independent agency within the **U.S. Department of the Interior**, the NBS was merged with the USGS (also part of the Interior Department) in 1996.

The BRD is the principal biological research and monitoring agency of the federal government. It is responsible for gathering, analyzing, and disseminating biological information in order to support sound management and stewardship of nation's biological and **natural resources**. It is also directed to foster understanding of biological systems and their benefits to society, and to make biological information available to the public. Although it was created mainly on the impetus of environmental and scientific organizations, the BRD also supports commercial and economic interests in that it seeks to identify opportunities for sustainable resource use. Agriculture and **biotechnology** are among the industries that stand to benefit from BRD research on new sources of food, fiber, and medicines.

Because it is independent of regulatory agencies—which are responsible for enforcing laws—the BRD has no formal regulatory, management, or enforcement roles. Therefore the BRD does not enforce laws such as the **Endangered Species Act**. Instead the BRD is responsible for gathering data that are scientifically sound and unbiased. The BRD also fosters public-private cooperation. For example, it worked with the International Paper Company in Alabama to develop a management plan for two **species** of pitcher plants found on company land that were candidates for the **Endangered Species** List. If it succeeds, this management plan will both preserve the pitcher plant populations—preventing the legal complications of having it listed as a federally endangered species—and allow continued judicious use of the land and resources.

The Biological Resources Division of the USGS was created in 1994 as an independent agency with the name National Biological Survey. The Survey was established on November 11, 1994, on the recommendations of President Bill Clinton and Interior Secretary Bruce Babbit. Renamed the National Biological Service (NBS) shortly after its creation, the agency was created by combining the biological research, inventory, and monitoring programs of seven agencies within the Department of the Interior. Built on the model of the Geological Survey, the NBS was established to provide accurate baseline data about ecosystems and species in United States territory. The mission of the NBS

was to provide information to support sound stewardship of natural resources on public lands under the administration of the Department of the Interior. Part of its mission was also to foster cooperation among other entities involved in managing, monitoring, and researching natural resources. On October 1, 1996, the NBS was renamed the Biological Resources Division and merged with the United States Geological Survey. The USGS-BRD retains the research and information provision mandates of the NBS. Appointed as the first NBS/BRD director was H. Ronald Pulliam, a professor and research ecologist from of the Institute of Ecology and the University of Georgia in Athens.

The National Biological Service had a predecessor in the Bureau of Biological Survey, which operated as part of the Department of Agriculture from 1885–1939. The Bureau's first director, C. Hart Merriam, revolutionized biological collection techniques and in 15 years nearly quadrupled the number of known American mammal species. In 1939, the Bureau was transferred to the Department of the Interior, where it was the predecessor to the **Fish and Wildlife Service**. After its transfer to the Department of the Interior, the Bureau of Biological Survey's baseline data gathering and survey functions gradually diminished. The Fish and Wildlife Service now has regulatory and management responsibilities as well as research programs.

In recent years the need for a biological survey, and especially for the production of reliable baseline data, has resurfaced. The resurgence of interest in baseline data results in part from an interest in managing ecosystems, with a focus on stewardship and **ecosystem** restoration, in place of previous emphasis on individual species management. Managing and restoring ecosystems requires basic data and information on biological indicators, which are often unavailable. Interest in threatened and endangered species, and public participation in environmental organizations probably also contributed to the widespread support for establishing a biological survey.

The BRD is important because it is the only federal agency whose principal mission is to perform basic scientific, biological, and ecological research. As an explicitly scientific research body, the BRD works to incorporate current ecological theory in its research and monitoring programs. Central to the BRD's mission are ideas such as long-term stewardship of resources, maintaining **biodiversity**, identifying ecological services of biological resources, anticipating **climate** change, and restoring populations and habitats. Because of its emphasis on basic research the BRD can address current scientific themes that pertain to public policy such as fire ecology, **endocrine disruptors** (**chemicals** such as pentachlorophenols [PCBs] that interfere with endocrine and reproductive functions in animals), ecological roles of **wetlands**, and **habitat** restoration. The primary purpose of this research activity is to provide land managers, especially within the Department of the Interior, with sound information to guide their management policies.

The BRD has two broad categories of research activities. One focus is on species for which the Department of Interior has trust responsibilities, including endangered species, marine mammals, migratory birds, and anadromous fish such as **salmon**. Research on these organisms involves studies of physiology, behavior, population dynamics, and **mortality**. The second general focus is on ecosystems. This class of research is directed at using experimentation, **modeling**, and observation to produce practical information concerning the complex interactions and functions of ecosystems, as well as human-induced changes in ecosystems.

In addition to these ecological research questions, the BRD is mandated to perform basic classification, mapping, and description functions. The division is responsible for classifying and mapping species within the United States, as well as prospecting for new species. It is also directed to develop a standard classification system for ecological units, biological indicators for **ecosystem health**, protocols for managing **pollution**, and guidelines for ecosystem restoration.

Part of this basic assessment function is a series of National Status and Trends Reports, to be released by the BRD every two years. These reports are to assess the health of biological resources and to report trends in their decline or improvement. The first two National Status and Trends Reports were released in 1995 and 1997. The 1995 report included more than 200 contributions on monitoring and population trends for plant, invertebrate, and vertebrate species, as well as summaries on the status of several biological communities and ecosystems. The 1997 report, published in two volumes, discusses factors such as natural processes, harvest exploitation, contaminants, **land use**, water use, nonindigenous species, and climate change that affect ecosystems in the United States. This status report also includes a major section discussing marine biological resources.

The USGS-BRD also provides a nationwide coordinated research agenda and a set of priorities for biological research. For example, it has undertaken a set of coordinated regional studies of the wide-ranging biological and ecological impacts of wetland distributions in Colorado, California, Texas, and other regions of the United States. Collectively these studies can provide a picture of nationwide trends and conditions, as well as producing coordinated information on restoration techniques, biodiversity issues, and biological indicators, at the same time as regional wetland problems are being addressed. Similar sets of coordinated studies have begun concerning the effects of endocrine-disrupting chemicals in the **environment** and on the use of prescribed fire as a management technique in a variety of ecosystems.

One of the principal mandates of the BRD is to distribute information. Free sharing of information is intended to maximize the usefulness of research findings. Part of the information sharing program includes substantial use of the Internet to distribute publications. Even the book-size National Status and Trends Reports are being published online to facilitate easy public access to their findings.

In addition to performing its own research the BRD provides coordination and a network of communication between federal agencies, state governments, universities, museums, and private **conservation** organizations. The National Partnership for Biological Survey, coordinated by the BRD, provides a network for sharing information and technology between federal, state, and other agencies. Participants in the Partnership include the United States **Forest Service**, the Natural Resources Conservation Service, the **National Oceanic and Atmospheric Administration**, the **Environmental Protection Agency**, the National Science Foundation, the Department of Defense, and the **Army Corps of Engineers**, as well as the Natural Heritage data centers and programs maintained by many states. The network of Natural Heritage programs is coordinated by **The Nature Conservancy**, a private organization dedicated to preserving habitat and species diversity. Also included in the Partnership are museums and universities nationwide that serve as repositories of information on biological resources and that conduct biological research, a range of non-governmental organizations, international conservation and research groups, Native American groups, private land holders, and resource user groups.

[*Mary Ann Cunningham Ph.D.*]

RESOURCES

OTHER

Biological Resources Division. *Biological Resources Strategic Science Plan.* Washington, D.C.: Government Printing Office, 1996.

National Biological Service. *Our Living Resources.* Washington, DC: Government Printing Office, 1995.

ORGANIZATIONS

US Geological Survey (USGS), Biological Resources Division (BRD), Western Regional Office (WRO), 909 First Ave., Suite #800, Seattle, WA USA 98104 (206) 220-4600, Fax: (206) 220-4624, Email: brd_wro@usgs.gov, <http://biology.usgs.gov>

Biological treatment
see **Bioremediation**

Bioluminescence

Bioluminescence ("living light") is the production of light by living organisms through a biochemical reaction. The general reaction involves a substrate called luciferin and an **enzyme** called luciferase, and requires oxygen. Specifically, luciferin is oxidized by luciferase and the chemical energy produced is transformed into light energy. In **nature**, bioluminescence is fairly widespread among a diverse group of organisms such as bacteria, **fungi**, sponges, jellyfish, mollusks, crustaceans, some worms, fireflies, and fish. It is totally lacking in vertebrate animals. Fireflies are probably the most commonly recognized examples of bioluminescent organisms, using the emitted light for mate recognition.

Biomagnification

The **bioaccumulation** of **chemicals** in organisms beyond the concentration expected if the chemical was in equilibrium between the organism and its surroundings. Biomagnification can occur in both terrestrial and aquatic environments, but it is generally used in relation to aquatic situations. Most often, biomagnification occurs in the higher trophic levels of the **food chain/web**, where exposure to chemicals takes place mostly through food consumption rather than water uptake.

Biomagnification is a specific case of bioaccumulation and is different from bioconcentration. Bioaccumulation describes the accumulation of contaminants in the tissue of organisms. Typical examples of this include the elevated levels of many chlorinated pesticides and **mercury** in fish tissue. Bioconcentration is used to describe the concentration of a chemical in an organism from water uptake alone. This is quantitatively described by the bioconcentration factor, or BCF, which is the chemical concentration in tissue divided by the chemical concentration in water, expressed in equivalent units, at equilibrium. The vast majority of chemicals that bioaccumulate are aromatic organic compounds, particularly those with **chlorine** substituents. For organic compounds, the mechanism of bioaccumulation is thought to be the partitioning or solubilization of chemical into the lipids of the organism. Thus the BCF should be proportional to the lipophilicity of the chemical, which is described by the octanol-water partition coefficient, Kow. The latter is a physical-chemical property of the compound describing its relative solubility in an organic phase and is the ratio of its solubility in octanol to its solubility in water at equilibrium. It is constant at a given temperature. If one assumes that a chemical's solubility in octanol is similar to its solubility in lipid, then we can approximate the lipid-normalized BCF as equal to the Kow. This assumption has been shown to be a reasonable first approximation for most chemicals accumulation in fish tissue.

However, animals are exposed to contaminants by other routes in addition to passive partitioning from water. For instance, fish can take up chemicals from the food they eat. It has been noted in field collections that for certain chemicals, the observed fish-water ratio (BCF) is significantly greater than the theoretical BCF, based on Kow. This indicates that the chemical has accumulated to a greater extent than its equilibrium concentration. This is defined as biomagnification. This condition has been documented in aquatic animals, including fish, shellfish, **seals and sea lions, whales,** and otters, and in birds, mink, rodents, and humans in both laboratory and field studies.

The biomagnification factor, BMF, is usually described as the ratio of the observed lipid-normalized BCF to Kow, which is the theoretical lipid-normalized BCF. This is equivalent to the multiplication factor above the equilibrium concentration. If this ratio is equal to or less than one, then the compound has not biomagnified. If the ratio is greater than one, then the chemicals biomagnified by that factor. For instance, if a chemical's Kow were 100,000, then its lipid normalized BCF should be 100,000 if the chemical were in equilibrium in the organism's lipids. If the fish tissue concentration (normalized to lipids) were 500,000, then the chemical would be said to have biomagnified by a factor of five.

Biomagnification in the aquatic food chain often leads to biomagnification in terrestrial food chains, particularly in the case of bird and **wildlife** populations that feed on fish. Consider the following example that demonstrates the results of biomagnification. The concentrations of the insecticide dieldrin in various trophic levels are determined to be the following: water, 0.1 ng/L; **phytoplankton**, 100 ng/g lipid; **zooplankton**, 200 ng/g lipid; fish, 600 ng/g lipid; terns, 800 ng/g lipid. If the Kow were equal to one million, then the phytoplankton would be in equilibrium with the water, but the zooplankton would have magnified the compound by a factor of 2, the fish by a factor of 6, and the terns by a factor of 8.

The mechanism of biomagnification is not completely understood. To achieve a concentration of a chemical greater than its equilibrium value indicates that the elimination rate is slower than for chemicals that reach equilibrium. Transfer efficiencies of the chemical would affect the relative ratio of uptake and elimination. There are many factors that control the uptake and elimination of a chemical from the consumption of contaminated food, and these include factors specific to the chemical as well as factors specific to the organism. The chemical properties include solubility, Kow, molecular weight and volume, and diffusion rates between organism gut, blood, and lipid pools. The organism properties include the feeding rate, diet preferences, assimilation rate into the gut, rate of chemical's **metabolism**, rate of egestion, and

Biomagnification of the pesticide DDT in the food chain. DDT travels from runoff up through the food chain, accumulating in all exposed species. (McGraw-Hill Inc. Reproduced by permission.)

rate of organism growth. It is thought that the chemical's properties control whether biomagnification will occur, and that it is the transfer rate from lipid to blood that allows the chemical to attain a lipid concentration greater than its equilibrium value. Thus it follows that the chemicals that biomagnify have similar properties. They typically are organic; they have molecular weights between 200 and 600 daltons; they have Kows between 10,000 and 10 million; they are resistant to metabolism by the organism; they are non-ionic, neutral compounds; and they have molecular volumes between 260 and 760 cubic angstroms, a cross sectional width of less than 9.5 angstoms and a molecular surface area between 200 and 460 square angstroms. The latter dimensions allow them to more easily pass through lipid bilayers into cells but perhaps do not allow them to leave the cell easily due to their high lipophilicity. Since this disequilibrium would occur at each **trophic level**, it results in more and more biomagnification at each higher trophic level. Because humans occupy a very high trophic level, we are particularly vulnerable to adverse health effects as a result of exposure to chemicals that biomagnify.

[*Deborah L. Swackhammer*]

RESOURCES
BOOKS

Connell, D. W. *Bioaccumulation of Xenobiotic Compounds.* Boca Raton: CRC Press, 1990.

PERIODICALS

Bierman Jr., V. J. "Equilibrium Partitioning and Biomagnification of Organic Chemicals in Benthic Animals." *Environmental Science and Technology* 24 (September 1990); 1407–12.

Sijm, D., W. Seinen, and A. Opperhuizen. "Life Cycle Biomagnification Study in Fish." *Environmental Science and Technology* 26 (November 1992): 2162–74.

Biomass

Biomass is a measure of the amount of biological substance minus its water content found at a given time and place on the earth's surface. Although sometimes defined strictly as living material, in actual practice the term often refers to living organisms, or parts of living organisms, as well as waste products or non-decomposed remains. It is a distinguishing feature of ecological systems and is usually presented as biomass density in units of dry weight per unit area. The term is somewhat imprecise in that it includes autotrophic plants, referred to as phytomass, heterotrophic **microbes**, and animal material, or zoomass. In most settings, phytomass is by far the most important component. A square meter of the planet's land area has, on average, about 22.05 – 26.46 lb (10 – 12 kg) of phytomass, although values may vary widely depending on the type of **biome**. Tropical rain forests average about 45 kg/m^2 while a **desert** biome may have a value near zero. The global average for heterotrophic biomass is approximately 0.1 kg/m^2, and the average for human biomass has been estimated at 0.5 g/m^2 if permanently glaciated areas are excluded.

The nature of biomass varies widely. Density of fresh material ranges from a low of 0.14 g/cm^3 for floats of aquatic plants to values greater that 1 g/cm^3 for very dense hardwood. The water content of fresh material may be as low as 5% in mature seeds or as high as 95% in fruits and young shoots. Water levels for living plants and animals run from 50 to 80%, depending on the **species**, season, and growing conditions. To insure a uniform basis for comparison, biomass samples are dried at 221°F (105°C) until they reach a constant weight.

Organic compounds typically constitute about 95% by weight of the total biomass, and nonvolatile residue, or ash, about 5%. **Carbon** is the principle element in biomass and usually represents about 45% of the total. An exception occurs in species that incorporate large amounts of inorganic elements such as silicon or calcium, in which case the carbon content may be much lower and nonvolatile residue several times higher. Another exception is found in tissues rich in lipids (oil or fat), where the carbon content may reach values as high as 70%.

Photosynthesis is the principle agent for biomass production. Light energy is used by chlorophyll-containing green plants to remove (or fix) **carbon dioxide** from the **atmosphere** and convert it to energy rich organic compounds or biomass. It has been estimated that on the face of the earth approximately 200 billion tons of carbon dioxide are converted to biomass each year. Carbohydrates are usually the primary constituent of biomass, and cellulose is the single most important component. Starches are also important and predominate in storage organs such as tubers and rhizomes. Sugars reach high levels in fruits and in plants such as sugar cane and sugar beet. Lignin is a very significant non-carbohydrate constituent of woody plant biomass.

[*Douglas C. Pratt*]

RESOURCES
BOOKS

Lieth, H. F. H. *Patterns of Primary Production in the Biosphere*. Stroudsburg, PA: Dowden, Hutchinson, and Ross, distributed by Academic Press, 1978.
Smil, V. *Biomass Energies: Resources, Links, Constraints*. New York: Plenum Press, 1983.

Biomass fuel

A **biomass** fuel is an energy source derived from living organisms. Most commonly it is plant residue, harvested, dried and burned, or further processed into solid, liquid, or gaseous fuels. The most familiar and widely used biomass fuel is wood. Agricultural waste, including materials such as the cereal straw, seed hulls, corn stalks and cobs, is also a significant source. Native shrubs and herbaceous plants are potential sources. **Animal waste**, although much less abundant overall, is a bountiful source in some areas.

Wood accounted for 25% of all energy used in the United States at the beginning of this century. With increased use of **fossil fuels**, its significance rapidly declined. By 1976, only 1–2% of United States energy was supplied by wood, and burning of tree wastes by the forest products industry accounted for most of it. Although the same trend has been evident in all industrialized countries, the decline has not been as dramatic everywhere. Sweden, for instance, still meets 8% of its energy needs with wood, and Finland, 15%.

Globally, it is estimated that biomass supplies about 6 or 7% of total energy, and it continues to be a very important energy source for many developing countries. In the last 15–20 years, interest in biomass has greatly increased even in countries where its use has drastically declined. In the United States rising fuel prices led to a large increase in the use of wood-burning stoves and furnaces for space heating. Impending fossil fuel shortages have greatly increased research on its use in the United States and elsewhere. Because biomass is a potentially renewable resource, it is recognized as a possible replacement of **petroleum** and **natural gas**.

Historically, burning has been the primary mode for using biomass, but because of its large water content it must be dried to burn effectively. In the field, the energy of the sun may be all that is needed to sufficiently lower its water level. When this is not sufficient, another energy source may be needed.

Biomass is not as concentrated an energy source as most fossil fuels even when it is thoroughly dry. Its density may be increased by milling and compressing dried residues. The resulting briquettes or pellets are also easier to handle, store, and transport. Compression has been used with a variety of materials including crop residues, herbaceous native plant material, sawdust, and other forest wastes.

Solid fuels are not as convenient or versatile as liquids or gases, and this is a drawback to the direct use of biomass. Fortunately, a number of techniques are known for converting it to liquid or gaseous forms.

Partial **combustion** is one method. In this procedure, biomass is burned in an **environment** with restricted oxygen. **Carbon monoxide** and **hydrogen** are formed instead of **carbon dioxide** and water. This mixture is called synthetic gas or "syngas." It can serve as fuel although its energy content is lower than natural gas (**methane**). Syngas may also be converted to **methanol**, a one carbon-alcohol that can be used as a **transportation** fuel. Because methanol is a liquid, it is easy to store and transport.

Anaerobic digestion is another method for forming gases from biomass. It uses **microorganisms**, in the absence of oxygen, to convert organic materials to methane. This method is particularly suitable for animal and human waste. Animal **feedlots** faced with disposal problems may install microbial gasifiers to convert waste to gaseous fuel used to heat farm buildings or generate electricity.

For materials rich in starch and sugar, fermentation is an attractive alternative. Through **acid** hydrolysis or enzymatic digestion, starch can be extracted and converted to sugars. Sugars can be fermented to produce **ethanol**, a liquid biofuel with many potential uses.

Cellulose is the single most important component of plant biomass. Like starch, it is made of linked sugar components that may be easily fermented when separated from the cellulose polymer. The complex structure of cellulose makes separation difficult, but enzymatic means are being developed to do so. Perfection of this technology will create a large potential for ethanol production using plant materials that are not human foods.

The efficiency with which biomass may be converted to ethanol or other convenient liquid or gaseous fuels is a major concern. Conversion generally requires appreciable energy. If an excessive amount of expensive fuel is used in the process, costs may be prohibitive. Corn (*Zea mays*) has been a particular focus of efficiency studies. Inputs for the corn system include energy for production and application of **fertilizer** and **pesticide**, tractor fuel, on-farm electricity, etc., as well as those more directly related to fermentation. A recent estimate puts the industry average for energy output at 133% of that needed for production and processing. This net energy gain of 33% includes credit for co-products such as corn oil and protein feed as well as the energy value of ethanol. The most efficient production and conversion systems are estimated to have a net energy gain of 87%. Although it is too soon to make an accurate assessment of the net energy gain for cellulose-based ethanol production, it has been estimated that a net energy gain of 145% is possible.

Biomass-derived gaseous and liquid fuels share many of the same characteristics as their fossil fuel counterparts. Once formed, they can be substituted in whole or in part for petroleum-derived products. **Gasohol**, a mixture of 10% ethanol in **gasoline**, is an example. Ethanol contains about 35% oxygen, much more than gasoline, and a gallon contains only 68% of the energy found in a gallon of gasoline. For this reason, motorists may notice a slight reduction in gas mileage when burning gasohol. However, automobiles burning mixtures of ethanol and gasoline have a lower exhaust temperature. This results in reduced toxic emissions, one reason that clean air advocates often favor gasohol use in urban areas.

Biomass is called as a renewable resource since green plants are essentially solar collectors that capture and store sunlight in the form of chemical energy. Its renewability assumes that source plants are grown under conditions where yields are sustainable over long periods of time. Obviously, this is not always the case, and care must be taken to insure that growing conditions are not degraded during biomass production.

A number of studies have attempted to estimate the global potential of biomass energy. Although the amount of sunlight reaching the earth's surface is substantial, less than a tenth of a percent of the total is actually captured and stored by plants. About half of it is reflected back to space. The rest serves to maintain global temperatures at life-sustaining levels. Other factors that contribute to the small fraction of the sun's energy that plants store include Antarctic and Arctic zones where little **photosynthesis** occurs, cold winters in temperate belts when plant growth is impossible, and lack of adequate water in **arid** regions. The global total net production of biomass energy has been esti-

mated at 100 million megawatts per year per year. Forests and woodlands account for about 40% of the total, and oceans about 35%. Approximately 1% of all biomass is used as food by humans and other animals.

Soil requires some organic content to preserve structure and fertility. The amount required varies widely depending on **climate** and soil type. In tropical rain forests, for instance, most of the nutrients are found in living and decaying vegetation. In the interests of preserving photosynthetic potential, it is probably inadvisable to remove much if any organic matter from the soil. Likewise, in sandy soils, organic matter is needed to maintain fertility and increase water retention. Considering all the constraints on biomass harvesting, it has been estimated that about six million MWyr/yr of biomass are available for energy use. This represents about 60% of human society's total energy use and assumes that the planet is converted into a global garden with a carefully managed "photosphere."

Although biomass fuel potential is limited, it provides a basis for significantly reducing society's dependence on non-renewable reserves. Its potential is seriously diminished by factors that degrade growing conditions either globally or regionally. Thus, the impact of factors like global warming and **acid rain** must be taken into account to assess how well that potential might eventually be realized. It is in this context that one of the most important aspects of biomass fuel should be noted. Growing plants remove **carbon** dioxide from the **atmosphere** that is released back to the atmosphere when biomass fuels are used. Thus the overall concentration of atmospheric carbon dioxide should not change, and global warming should not result. Another environmental advantage arises from the fact that biomass contains much less sulfur than most fossil fuels. As a consequence, biomass fuels should reduce the impact of acid rain.

[*Douglas C. Pratt*]

RESOURCES
BOOKS

Hall, C. W. *Biomass as an Alternative Fuel*. Rockville, Maryland: Government Institutes, Inc., 1981.

Häfele, W. *Energy in a Finite World: A Global Systems Analysis*. Great Britain: Harper & Row Ltd., Inc., 1981.

Lieth, H. F. H. *Patterns of Primary Production in the Biosphere*. Stroudsburg, Pennsylvania: Dowden, Hutchinson and Ross, Inc., 1981.

Morris, D. M., & I. Ahmed, *How Much Energy Does It Take to Make a Gallon of Ethanol?* Washington D.C.: Institute for Local Self-Reliance, 1992.

Smil, V. *Biomass Energies: Resources, Links, Constraints*. New York: Plenum Press, 1983.

Stobaugh, R. & D. Yergin, eds. *Energy Future: Report of the Energy Project at the Harvard Business School*. New York: Random House, Inc., 1979.

Biome

A large terrestrial **ecosystem** characterized by distinctive kinds of plants and animals and maintained by a distinct **climate** and **soil** conditions. To illustrate, the **desert** biome is characterized by low annual rainfall and high rates of evaporation, resulting in dry environmental conditions. Plants and animals that thrive in such conditions include cacti, brush, lizards, insects, and small rodents. Special adaptations, such as waxy plant leaves, allow organisms to survive under low moisture conditions. Other examples of biomes include **tropical rain forest**, arctic **tundra**, **grasslands**, temperate **deciduous forest**, **coniferous forest**, tropical **savanna**, and Mediterranean **chaparral**.

Biomonitoring

see **Bioindicator**

Biophilia

The term biophilia was coined by the American biologist, **Edward O. Wilson** (1929–), to mean "the innate tendency [of human beings] to focus on life and lifelike processes" and "connections that human beings subconsciously seek with the rest of life." Wilson first used the word in 1979 in an article published in the *New York Times Book Review*, and then in 1984 he wrote a short book (157 pages) that explored the notion in more detail.

Clearly, humans have coexisted with other **species** throughout our evolutionary history. In Wilson's view, this relationship has resulted in humans developing an innate, genetically based, or "hard-wired" need to be close to other species and to be empathetic to their needs. The presumed relationship is reciprocal, meaning other animals are also to varying degrees empathetic with the needs of humans. The biophilia hypothesis seems intuitively reasonable, although it is likely impossible that it could ever be proven universally correct. For instance, while many people may feel and express biophilia, some people may not.

There is a considerable body of psychological and medical evidence in support of the notion of biophilia, although most of it is anecdotal. There are, for example, many observations of sick or emotionally distressed people becoming well more quickly when given access to a calming natural **environment**, or when comforted by such pets as dogs and cats. There are also more general observations that many people are more comfortable in natural environments than in artificial ones — it can be much more pleasant to sit in a garden, for instance, than in a windowless room. The aesthetics and comfort of that room can, however, be improved by hanging pictures of animals or landscapes, by watching a nature show on television, or by providing a window that looks out onto a natural scene.

Wilson is a leading environmentalist and a compelling advocate of the need to conserve the threatened **biodiversity** of the world, such as imperiled natural ecosystems and **endangered species**. To a degree, his notion of biophilia provides a philosophical justification for **conservation** actions, by suggesting emotional and even spiritual dimensions to the human relationship with other species and the natural world. If biophilia is a real phenomenon and an integral part of what it is to be human, then our affinity for other species means that willful actions causing endangerment and **extinction** can be judged to be wrong and even immoral. Our innate affinity for other species provides an intrinsic justification for doing them no grievous harm, and in particular, for avoiding their extinction. It provides a justification for enlightened stewardship of the natural world.

Humans also utilize animals in agriculture (we raise them as food), in research, and as companions. Ethical considerations derived from biophilia provides a rationale for treating animals as well as possible in those uses, and may even be used to justify **vegetarianism** and anti-vivisectionism.

[*Bill Freedman Ph.D.*]

RESOURCES
BOOKS

Kellert, S.R. and E.O. Wilson, editors. *The Biophilia Hypothesis. Washington, DC: Island Press, 1993.*
Wilson, E.O. *Biophilia. The Human Bond With Other Species.* Harvard, MA: Harvard University Press, 1984.

OTHER
Arousing Biophilia: A Conversation with E.O. Wilson. EnviroArts: Orion Online. <http://arts.envirolink.org/interviews_and_conversations/EOWilson.html>

Bioregional Project

Based in the Ozark Mountains in Missouri, the Bioregional Project (BP) was founded in 1982 to promote the aims and interests of the bioregional movement in North America. Consisting primarily of a resource center designed to show people how to "come back home to Earth," the Bioregional Project is part of the international campaign to reshape culture and society according to ecological principles: "We work for the honor, protection and healing of the Earth, the Earth's people, and all the Earth's life."

A bioregion is what the group calls a "life region," an area determined by natural rather than historical or political boundaries. It is distinguished by the character of the **flora** and **fauna**, by the landforms, the types of rocks and soils,

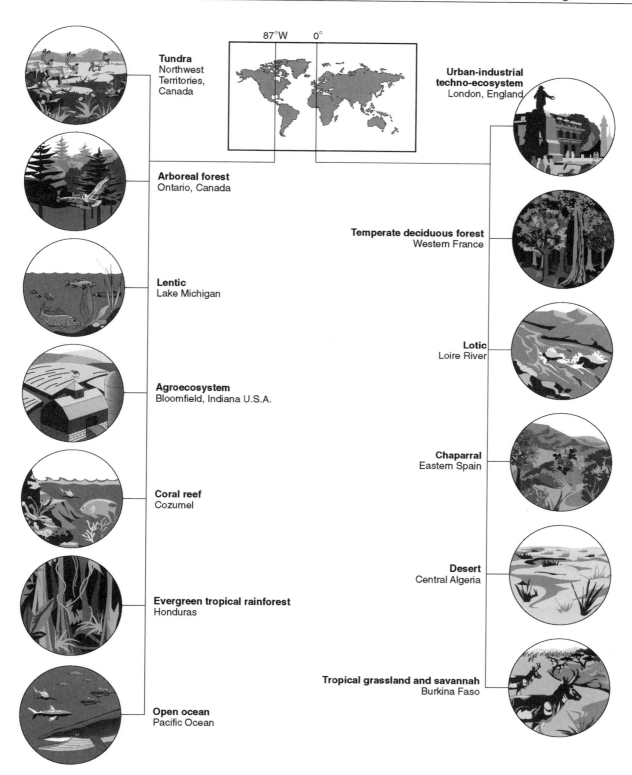

Tundra
Northwest Territories, Canada

Arboreal forest
Ontario, Canada

Lentic
Lake Michigan

Agroecosystem
Bloomfield, Indiana U.S.A.

Coral reef
Cozumel

Evergreen tropical rainforest
Honduras

Open ocean
Pacific Ocean

87°W   0°

Urban-industrial techno-ecosystem
London, England

Temperate deciduous forest
Western France

Lotic
Loire River

Chaparral
Eastern Spain

Desert
Central Algeria

Tropical grassland and savannah
Burkina Faso

Different biomes worldwide. (Illustration by Hans & Cassidy.)

the **climate** in general, and by human habitation as it relates to this **environment**. The Bioregional Project emphasizes the natural logic of these boundaries, and it promotes the development of social and political institutions that take into account the interrelatedness of everything within them. They are working to increase the awareness that bioregions are "living, self-organizing systems," and they value humanity as one **species** among many. The Bioregional Project traces the roots of the movement to native and **indigenous peoples** and the "oldest Earth traditions." The group believes that ecological laws and principles form the basis of society and that the future survival of humanity depends on their ability to cooperate with the environment.

The Bioregional Project and the bioregional movement as a whole have strong ties to the international Green Movement, although bioregionalists consider themselves more "ecologically-centered." The **Greens** are oriented to urban areas, and they work for change in traditional political structures, operating within legislative as opposed to bioregional systems. The chief organizing tool of the bioregional movement is a model known as "the bioregional congress" or "green congress," where participants share information, develop ecological strategies, and draft planning programs and platform statements. The Bioregional Project convened the first bioregional congress in 1980 as the Ozark Community Congress (OACC), and it has since influenced both bioregional and green organizing throughout North America. The Project coordinated the first North American Bioregional Congress in 1984, an international assembly attended by over 200 people representing 130 organizations.

In addition to the assistance it provides for "those organizing bioregionally," the Bioregional Project also publishes books and pamphlets on **bioregionalism** and **ecology**. The organization sponsors lectures and educational presentations on these subjects as well. It supports research and lends technical assistance in a variety of areas from community economic development to **recycling, sustainable agriculture**, and forest protection. It is a subsidiary of the Ozarks Resource Center. Contact: Bioregional Project, Box 3, Brixley, MO 65618. telephone(417) 679 4773

[*Douglas Smith*]

Bioregionalism

Drawing heavily upon the cultures of **indigenous peoples**, bioregionalism is a philosophy of living that stresses harmony with **nature** and the integration of humans as part of the natural **ecosystem**. The keys to bioregionalism involve learning to live off the land, without damaging the **environment** or relying on heavy industrial machines or products.

Bioregionalists believe that if the relationship between nature and humans improves, the society as a whole will benefit.

Environmentalists who practice this philosophy "claim" a bioregion or area. For example, one's place might be a **watershed**, a small mountain range, a particular area of the coast, or a specific **desert**. To develop a connection to the land and a **sense of place**, bioregionalists try to understand the natural history of the area as well as how it supports human life. For example, they study the plants and animals that inhabit the region, the geological features of the land, as well as the cultures of the people who live or have lived in the area.

Bioregionalism also stresses community life where participation, self-determination, and local control play important roles in protecting the environment. Various bioregional groups exist throughout the United States, ranging from the Gulf of Maine to the Ozark Mountains to the San Francisco Bay area. A North American Bioregional Congress loosely coordinates the bioregional movement.

[*Christopher McGrory Klyza*]

RESOURCES
BOOKS

Andruss, V., et al. *Home! A Bioregional Reader*. Philadelphia: New Society Publishers, 1990.
Sale, K. *Dwellers in the Land: The Bioregional Vision*. San Francisco: Sierra Club Books, 1985.
Snyder, G. *The Practice of the Wild*. San Francisco: North Point Press, 1990.

Bioremediation

A number of processes for remediating contaminated soils and **groundwater** based on the use of **microorganisms** to convert contaminants to less hazardous substances. Most commercial bioremediation processes are intended to convert organic substances to **carbon dioxide** and water, although processes for addressing metals are under development. Many bacteria ubiquitously found in soils and groundwater are able to biodegrade a range of organic compounds. Compounds found in **nature**, and ones similar to those, such as **petroleum hydrocarbons**, are most readily biodegraded by these bacteria. Bioremediation of chlorinated solvents, polychlorinated biphenyl (PCB)s, pesticides, and many munitions compounds, while of great interest, is more difficult and has thus been much slower to reach commercialization.

In most bioremediation processes the bacteria use the contaminant as a food and energy source and thus survive and grow in numbers at the expense of the contaminant. In order to grow new cells, bacteria, like other biological **species**, require numerous minerals as well as **carbon** sources. These minerals are typically present in sufficient amounts except for **phosphorus** and **nitrogen**, which are commonly added during bioremediation. If contaminant molecules are to be transformed, species called electron acceptors must also be present. By far the most commonly used electron acceptor is oxygen. Other electron acceptors include nitrate, sulfate, carbon dioxide, and iron. Processes that use oxygen are called **aerobic** biodegradation. Processes that use other electron acceptors are commonly lumped together as **anaerobic** biodegradation.

The vast majority of commercial bioremediation processes use aerobic biodegradation and thus include some method for providing oxygen. The amount of oxygen that must be provided depends not only on the mass of contaminant present but also on the extent of conversion of the contaminants to carbon dioxide and water, other sources of oxygen, and the extent to which the contaminants are physically removed from the soils or groundwater. Typically, designs are based on adding two to three pounds of oxygen for each pound of **biodegradable** contaminant.

The processes generally include the addition of **nutrient** (nitrogen and phosphorus) sources. The amount of nitrogen and phosphorus that must be provided is quite variable and frequently debated. In general, this amount is less than the 100:10:1 ratio of carbon to nitrogen to phosphorus of average cell compositions. It is also important to maintain the **soil** or groundwater **pH** near neutral (pH 6–8.5), moisture levels at or above 50% of **field capacity**, and temperatures between 39°F (4°C) and 95°F (35°C), preferably between 68°F (20°C) and 86°F (30°C).

Bioremediation can be applied in situ and ex situ by several methods. Each of these processes are basically engineering solutions to providing oxygen (or alternate electron acceptors) and possibly, nutrients to the contaminated soils, which already contain the bacteria. The addition of other bacteria is not typically needed or beneficial.

In situ processes have the advantage of causing minimal disruption to the site and can be used to address contamination under existing structures. In situ bioremediation to remediate aquifers contaminated with petroleum hydrocarbons such as **gasoline**, was pioneered in the 1970s and early 1980s by Richard L. Raymond and coworkers. These systems used groundwater recovery **wells** to capture contaminated water which was treated at the surface and reinjected after amendment with nutrients and oxygen. The nutrients consisted of ammonium chloride and phosphate salts and sometimes contained magnesium, manganese, and iron salts. Oxygen was introduced by sparging (bubbling) air into the reinjection water. As the injected water swept through the **aquifer**, oxygen and nutrients were carried to the contaminated soils and groundwater where the indigenous bacteria converted the hydrocarbons to new cell material, carbon dioxide, and water. Variations of this technology include the use of **hydrogen** peroxide as a source of oxygen and direct injection of air into the aquifer.

Bioremediation of soils located between the ground surface and the **water table** is most commonly practiced through **bioventing**. In this method, oxygen is introduced into the contaminated soils by either injecting air or extracting air from wells. The systems used are virtually the same as for vapor extraction. The major difference is in mode of operation and in the fact that nutrients are sometimes added by percolating nutrient amended water through the soil from the ground surface or buried horizontal pipes. Systems designed for bioremediation operate at low air flow rates to replace oxygen consumed during biodegradation and to minimize physical removal of volatile contaminants.

Bioremediation can be applied to excavated soils by landfarming, soil cell techniques, or in soil slurries. The simplest method is landfarming. In this method soils are spread to a depth of 12–18 inches (30–46 cm). Nutrients, usually commercial fertilizers with high nitrogen and low phosphorous content, are added periodically to the soils which are tilled or plowed frequently. In most instances, the treatment area is prepared by grading, laying down an impervious layer (clay or a synthetic liner), and adding a six-inch layer of clean soil or sand. Provisions for treating rainwater **runoff** are typically required. The frequent tilling and plowing breaks up soil clumps and exposes the soils and thus bacteria to air. This method is more suitable for treating silty and clayey soils than are most of the other methods. It is not generally appropriate for soils contaminated with volatile contaminants such as gasoline because vapors can not be controlled unless the process is conducted within a closed structure.

Excavated soils can also be treated in cells or piles. A synthetic liner is placed on a graded area and covered with sand or gravel to permit collection of runoff water. The sands or gravel are covered with a **permeable** fabric and nutrient amended soils are added. Slotted PVC pipe is added as the pile is built. The soils are covered with a synthetic liner and the PVC pipes are connected to a blower. Air is slowly extracted from the soils and, if necessary, treated before being discharged to the **atmosphere**. This method requires less room than landfarming and less maintenance during operations, and can be used to treat volatile contaminants because the vapors can be controlled.

Excavated soils can also be treated in soil/water slurries in either commercial reactors or in impoundments or lagoons. Soils are separated from oversize materials and mixed with water, nutrients are added, and the **slurry** is aerated to provide oxygen. In some cases additional sources of bacteria and/or surfactants are added. These systems are usually capable of attaining more rapid rates of biodegradation than other systems but have limited throughput.

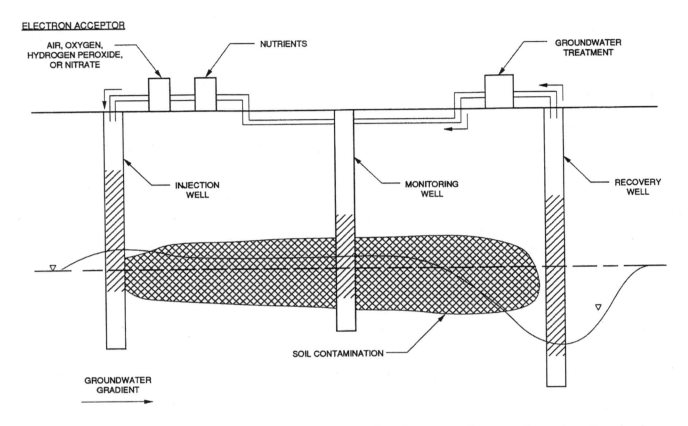

ELECTRON ACCEPTOR

Bioremediation of the saturated zone using hydrogen peroxide in the Raymond process. (Beacon Press. Reproduced by permission.)

Selection and design of a particular bioremediation method requires that the site be carefully investigated to define the lateral and horizontal extent of contamination including the total mass of biodegradable substances. Understanding the soil types and distribution and the site **hydrogeology** is as important as identifying the contaminants and their distribution in both soils and groundwater. Designing bioremediation systems requires the integration of microbiology, chemistry, hydrogeology, and engineering.

Bioremediation is generally viewed favorably by regulatory agencies and is actively supported by the U. S. **Environmental Protection Agency** (EPA). The mostly favorable publicity and the perception of bioremediation as a natural process has led to greater acceptance by the public compared to other technologies, such as **incineration**. It is expected that the use of bioremediation to treat soils and groundwater contaminated with petroleum hydrocarbons and other readily biodegradable compounds will continue to grow. Continued improvements in the design and engineering of bioremediation systems will result from the expanding use of bioremediation in a competitive market. It

is anticipated that processes for treating the more recalcitrant organic compounds and metals will become commercial through greater understanding of microbiology and specific developments in the isolation of special bacteria and **genetic engineering**.

[*Robert D. Norris*]

RESOURCES
BOOKS

Chapelle, F. H. *Ground-Water Microbiology and Geochemistry.* New York: Wiley, 1993.

Hinchee, R. E., and R. F. Olfenbuttel, eds. *In Situ Bioreclamation: Applications and Investigation for Hydrocarbons and Contaminated Site Remediation.* Butterworth-Heinemann, 1991.

Hinchee, R. E., and R. F. Olfenbuttel, eds. *On-Site Bioreclamation: Processes for Xenobiotic and Hydrocarbon Treatment.* Butterworth-Heinemann, 1991.

Matthews, J. E., ed. *In Situ Bioremediation of Groundwater and Geological Materials: A Review of Technologies.* Chelsea, MI: Lewis, 1993.

National Research Council. *In Situ Bioremediation: When Does It Work?* Washington, DC: National Academy Press, 1993.

Biosequence

A sequence of soils that contain distinctly different **soil** horizons because of the influence that vegetation had on the soils during their development. A typical biosequence would be the **prairie** soils in a dry **environment**, oak-savannah soils as a transition zone, and forested soils in a wetter environment. Prairie soils have dark, thick surface horizons while forested soils have a thin, dark surface with a light-colored zone below.

Biosphere

The biosphere is the largest possible earthly organismic community. It is a terrestrial envelope of life, or the total global **biomass** of living matter. The biosphere incorporates every individual organism and **species** on the face of the earth—those that walk on the ground or live in the crevices of rock and down into the **soil**, those that swim in rivers, lakes, and oceans, and those that move in and out of the **atmosphere**.

Bios is the Greek word for life; "sphere" is from the Latin *sphaera*, which means essentially the "circuit or range of action, knowledge or influence," the "place or scene of action or existence," the "natural, normal or proper place." Combined into biosphere, the two ideas define the normal global place of existence for all earthly life-forms and, increasingly, a global area of influence and action for humans. Thinking of the mass of life forms on the earth as the biosphere also provides an impression of circular, cyclic systems and suggests a holistic concept of integration and unity.

A *Scientific American* book on the biosphere described it as "this thin film of air and water and soil and life no deeper than ten miles, or one four-hundredth of the earth's radius [that] is now the setting of the uncertain history of man." G. E. Hutchinson in that same book asked, "What is it that is so special about the biosphere?" He suggested that the answer seems to have three parts: "First, it is a region in which liquid water can exist in substantial quantities. Second, it receives an ample supply of energy from an external source, ultimately from the sun. And third, within it there are interfaces between the liquid, the solid and the gaseous states of matter." Both of these might better describe what LaMont Cole labeled the "ecosphere," the global ecosystem—the biosphere plus its abiotic **environment**. But the significance of Hutchinson's three-part statement is that those three characteristics of the earth's surface make it possible for life to exist. They provide the conditions necessary for the abundant and diverse organisms of the biosphere to live.

Life began in a very different environment than found today: the atmosphere, for example, was mostly **methane**, ammonia, and **carbon dioxide**. As life evolved, it changed the atmosphere (and other abiotic components of the surface of the earth), transforming it into the present oxygen-rich mixture of gases vital to life as it now exists. And those life-forms maintain that critical mixture in a complex, fluctuating system of global cycles.

The diversity and complexity of the biosphere is staggering. The accumulated human knowledge of its workings is prodigious, but even more impressive is the immense ignorance of that complexity. Humans have identified about 1.5 million living members of the biosphere and thus have some knowledge of at least that many. However, conservative estimates of the actual number of species begin at 3 or 3.5–5 million species. Recent and less conservative estimates range up to a possible 100 million. That means humans are totally ignorant of anywhere from 50% to as much as 98.5% of the other members of the earth's **biological community**. Their existence is suspected, but they cannot be identified or their existence documented by even a name.

One of the concerns about large-scale human ignorance of the biosphere is that many species might be extinguished before they are even known. Human activities, especially destruction of **habitat**, are increasing the normal rate of species **extinction**. The diversity of the biosphere may be diminishing rapidly.

Taxonomically, the biosphere is organized into five kingdoms: monera, protista, **fungi**, animalia, and plantae, and a multitude of subsets of these, including the multiple millions of species mentioned above. G. Piel estimates that of the 1,200–1,800 billion tons dry weight of the biosphere, most of it—some 99%—is plant material. All the life-forms in the other four taxons, including animals and obviously the five billion-plus humans alive today, are part of that less than one%.

The biosphere can also be subdivided into biomes: a **biome** incorporates a set of biotic communities within a particular region exposed to similar climatic conditions and which have dominant species with similar life cycles, adaptations, and structures. Deserts, **grasslands**, temperate deciduous forests, coniferous forests, **tundra**, tropical rain forests, tropical seasonal forests, freshwater biomes, estuaries, **wetlands**, and marine biomes, are examples of specific terrestrial or aquatic biomes.

Another indication of the complexity of the biosphere is a measure of the processes that take place within it, especially the essential processes of **photosynthesis** and **respiration**. The sheer size of the biosphere is indicated by the amount of biomass present. Vitousek and his colleagues estimate the net primary production of the earth's biosphere as 224.5 petagrams, one petagram being equivalent to 10^{15} grams.

The biosphere interacts in constant, intricate ways with other global systems: the atmosphere, lithosphere, hydrosphere, and pedosphere. Maintenance of life in the biosphere depends on this complex network of biological-biological, physical-physical, and biological-physical interactions. All the interactions are mediated by an equally complex system of positive and negative feedbacks—and the total makes up the dynamics of the whole system. Since each and all interpenetrate and react on each other constantly, outlining a global **ecology** is a major challenge.

Normally biospheric dynamics are in a rough balance. The **carbon cycle**, for example, is usually balanced between production and **decomposition**, the familiar equation of photosynthesis and respiration. As Piel notes: "The two planetary cycles of photosynthesis and **aerobic metabolism** in the biomass not only secure renewal of the biomass but also secure the steady-state mixture of gases in the atmosphere. Thereby, these life processes mediate the inflow and outflow of **solar energy** through the system; they screen out lethal radiation, and they keep the temperature of the planet in the narrow range compatible with life." But human activities, especially the **combustion** of **fossil fuels**, contribute to increases in **carbon** dioxide, distorting the balance and in the process changing other global relationships such as the **nature** of incoming and out-going radiation and differentials in temperature between poles and tropics.

If humans are to better understand the biosphere, many more studies must be undertaken on many levels. A number of levels of biological integration must be recognized and analyzed, each with different properties and each offering scholars special problems and special insights. The totality of the biosphere can be broken down in many different ways, but life extends from the single cell to the totality of the globe. Though biologists usually define their disciplines within the bounds of one level and though they may study only one level, scholars should recognize context, the full range of levels and the interactions between them.

Humans are, of course, one of the species that make up the living biosphere. *Homo sapiens* fits into the Linnean hierarchy on the primate branch. Using that hierarchy as a connective device, humans may take a first step toward understanding how they relate to the rest of the inhabitants of the biosphere, down to the most remote known species.

Humans are without doubt the dominant species in the biosphere. The transformation of radiant energy into useable biological energy is increasingly being diverted by humans to their own use. A common estimate is that humans are now diverting huge amounts of the net primary production of the globe to their own use: perhaps 40% of terrestrial production and close to 25% of all production is either utilized or wasted through human activity. Net primary production is defined as the amount of energy left after sub-tracting the respiration of primary producers, or plants, from the total amount of energy. It is the total amount of "food" available from the process of photosynthesis—the amount of biomass available to feed organisms, such as humans, that do not acquire food through photosynthesis.

Humans are displacing their neighbors in the biosphere through a multitude of activities: conversion of natural systems to agriculture, direct consumption of plants, consumption of plants by livestock, harvesting and conversion of forests, **desertification**, and many, many others. The biosphere is the source of all good: humans are an integral part of the biosphere and depend on its functioning for their well-being, for their very lives.

[*Gerald L. Young*]

RESOURCES

BOOKS

Bradbury, I. K. *The Biosphere*. London/New York: Belhaven Press, 1991.
Clark, W. C., and R. E. Munn, eds. *Sustainable Development of the Biosphere*. Cambridge: Cambridge University Press, 1986.
Piel, G. "The Biosphere." In *Only One World: Our Own to Make and to Keep*. New York: W. H. Freeman, 1992.

PERIODICALS

Salthe, S. N. "The Evolution of the Biosphere: Towards a New Mythology." *World Futures* 30 (1990): 53–67.
Vitousek, P. M., et al. "Human Appropriation of the Products of Photosynthesis." *BioScience* 36 (1986): 368–373.

Biosphere reserve

A **biosphere** reserve is an area of land recognized and preserved for its ecological significance. Ideally biosphere reserves contain undisturbed, natural environments that represent some of the world's important ecological systems and communities. Biosphere reserves are established in the interest of preserving the genetic diversity of these ecological zones, supporting research and education, and aiding local, **sustainable development**. Official declaration and international recognition of biosphere reserve status is intended to protect ecologically significant areas from development and destruction. Since 1976 an international network of biosphere reserves has developed, with the sanction of the United Nations. Each biosphere reserve is proposed, reviewed, and established by a national biosphere reserve commission in the home country under United Nations guidelines. Communication among members of the international biosphere network helps reserve managers share data and compare management strategies and problems.

The idea of biosphere reserves first gained international recognition in 1973, when the United Nations Educational and Scientific Organization (UNESCO)'s **Man and**

the **Biosphere Program** (MAB) proposed that a worldwide effort be made to preserve islands of the world's living resources from **logging**, mining, urbanization, and other environmentally destructive human activities. The term derives from the ecological word "biosphere," which refers to the zone of air, land, and water at the surface of the earth that is occupied by living organisms. Growing concern over the survival of individual **species** in the 1970s and 1980s led increasingly to the recognition that **endangered species** could not be preserved in isolation. Rather, entire ecosystems, extensive communities of interdependent animals and plants, are needed for threatened species to survive. Another idea supporting the biosphere reserve concept was that of genetic diversity. Generally ecological systems and communities remain healthier and stronger if the diversity of resident species is high. An alarming rise in species extinctions in recent decades, closely linked to rapid **natural resources** consumption, led to an interest in genetic diversity for its own sake. Concern for such ecological principles as these led to UNESCO's proposal that international attention be given to preserving the earth's ecological systems, not just individual species.

The first biosphere reserves were established in 1976. In that year, eight countries designated a total of 59 biosphere reserves representing ecosystems from **tropical rain forest** to temperate sea coast. The following year 22 more countries added another 72 reserves to the United Nations list, and by 2002 there was a network of 408 reserves established in 94 different countries.

Like national parks, **wildlife** refuges, and other **nature** preserves, the first biosphere reserves aimed to protect the natural **environment** from surrounding populations, as well as from urban or international exploitation. To a great extent this idea followed the model of United States national parks, whose resident populations were removed so that parks could approximate pristine, undisturbed natural environments.

But in smaller, poorer, or more crowded countries than the United States, this model of the depopulated reserve made little sense. Around most of the world's nature preserves, well-established populations—often indigenous or tribal groups—have lived with and among the area's **flora** and **fauna** for generations or centuries. In many cases, these groups exploit local resources—gathering nuts, collecting firewood, growing food—without damaging their environment. Sometimes, contrary to initial expectations, the activity of **indigenous peoples** proves essential in maintaining **habitat** and species diversity in preserves. Furthermore, local residents often possess an extensive and rare understanding of plant habitat and animal behavior, and their skills in using resources are both valuable and irreplaceable. At the very least, the cooperation and support of local populations is

essential for the survival of parks in crowded or resource-poor countries. For these reasons, the additional objectives of local cooperation, education, and sustainable economic development were soon added to initial biosphere reserve goals of biological preservation and scientific research. Attention to humanitarian interests and economic development concerns today sets apart the biosphere reserve network from other types of nature preserves, which often garner resentment from local populations who feel excluded and abandoned when national parks are established. United Nations MAB guidelines encourage local participation in management and development of biosphere reserves, as well as in educational programs. Ideally, indigenous groups help administer reserve programs rather than being passive recipients of outside assistance or management.

In an attempt to mesh the diverse objectives of biosphere reserves, the MAB program has outlined a theoretical reserve model consisting of three zones, or concentric rings, with varying degrees of use. The innermost zone, the core, should be natural or minimally disturbed, essentially without human presence or activity. Ideally this is where the most diverse plant and animal communities live and where natural **ecosystem** functions persist without human intrusion. Surrounding the core is a **buffer** zone, mainly undisturbed but containing research sites, monitoring stations, and habitat **rehabilitation** experiments. The outermost ring of the biosphere reserve model is the transition zone. Here there may be sparse settlement, areas of traditional use activities, and tourist facilities.

Many biosphere reserves have been established in previously existing national parks or preserves. This is especially common in large or wealthy countries where well established park systems existed before the biosphere reserve idea was conceived. In 1991 most of the United States' 47 biosphere reserves lay in national parks or wildlife sanctuaries. In countries with few such preserves, nomination for United Nations biosphere reserve status can sometimes attract international assistance and funding. In some instances debt for nature swaps have aided biosphere reserve establishment. In such an exchange, international **conservation** organizations purchase part of a country's national debt for a portion of its face value, and in exchange that country agrees to preserve an ecologically valuable region from destruction. Bolivia's Beni Biosphere Reserve came about this way in 1987 when **Conservation International**, a Washington-based organization, paid $100,000 to Citicorp, an international lending institution. In exchange, Citicorp forgave $650,000 in Bolivian debt, loans the bank seemed unlikely to ever recover, and Bolivia agreed to set aside a valuable tropical mahogany forest. This process has also produced other reserves, including Costa Rica's La Amistad, and Ecuador's Yasuni and Galapagos Biosphere Reserves.

In practice, biosphere reserves function well only if they have adequate funding and strong support from national leaders, legislatures, and institutions. Without legal protection and long-term support from the government and its institutions, reserves have no real defense against development interests.

National parks can provide a convenient institutional niche, defended by national laws and public policing agencies, for biosphere reserves. Pre-existing wildlife preserves and game sanctuaries likewise ensure legal and institutional support. Infrastructure—management facilities, access roads, research stations, and trained wardens—is usually already available when biosphere reserves are established in or adjacent to ready-made preserves.

Funding is also more readily available when an established **national park** or game preserve, with a pre-existing operating budget, provides space for a biosphere reserve. With intense competition from commercial loggers, miners, and developers, money is essential for reserve survival. Especially in poorer countries, international experience increasingly shows that unless there is a reliable budget for management and education, nearby residents do not learn cooperative reserve management, nor do they necessarily support the reserve's presence. Without funding for policing and legal defense, development pressures can easily continue to threaten biosphere reserves. Logging, clearing, and destruction often continue despite an international agreement on paper that resource extraction should cease. Turning parks into biosphere reserves may not always be a good idea. National park administrators in some less wealthy countries fear that biosphere reserve guidelines, with their compromising objectives and strong humanitarian interests, may weaken the mandate of national parks and wildlife sanctuaries set aside to protect endangered species from population pressures and development. In some cases, they argue, there exists a legitimate need to exclude people if **rare species** such as **tigers** or **rhinoceroses** are to survive.

Because of the expense and institutional difficulties of establishing and maintaining biosphere reserves, about two-thirds of the world's reserves exist in the wealthy and highly developed nations of North America and Europe. Poorer countries of Africa, Asia, and South America have some of the most important remaining intact ecosystems, but wealthy countries can more easily afford to allocate the necessary space and money. Developed countries also tend to have more established administrative and protective structures for biosphere reserves and other sanctuaries. An increasing number of developing countries are working to establish biosphere reserves, though. A significant incentive, aside from national pride in indigenous species, is the international recognition given to countries with biosphere reserves. Possession of these reserves grants smaller and

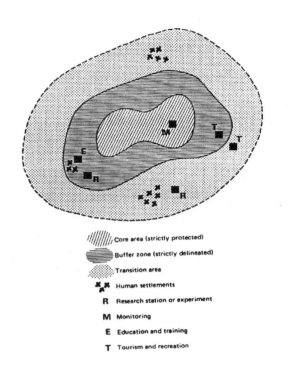

Schematic plan of a biosphere reserve. (Beacon Press. Reproduced by permission.)

Core area (strictly protected)

Buffer zone (strictly delineated)

Transition area

Human settlements

R Research station or experiment

M Monitoring

E Education and training

T Tourism and recreation

less wealthy countries some of the same status as that of more powerful countries such as the United States, Germany, and Russia.

Some difficult issues surround the biosphere reserve movement. One question that arises is whether reserves are chosen for reasons of biological importance or for economic and political convenience. In many cases national biosphere reserve committees overlook critical forests or endangered habitats because logging and mining companies retain strong influence over national policy makers. Another problem is that in and around many reserves, residents are not yet entirely convinced, with some reason, that management goals mesh with local goals. In theory sustainable development methods and education will continue to encourage communication, but cooperation can take a long time to develop. Among reserve managers themselves, great debate continues over just how much human interference is appropriate, acceptable, or perhaps necessary in a place ideally free of human activity. Despite these logistical and theoretical problems, the idea behind biosphere reserves seems a valid one, and the inclusiveness of biosphere planning, both biological and social, is revolutionary.

[*Mary Ann Cunningham Ph.D.*]

RESOURCES

BOOKS

Dogsé, P, and B. Von Droste. *Debt-for-Nature Exchanges and Biosphere Reserves*. Paris: UNESCO, 1990.

OTHER

Biosphere Reserves: Proceedings of the First National Symposium, Udhagaman-dalam, 24-26 Septemeber 1986. New Delhi: Ministry of Environment and Forests.

Biota

see **Biotic community**

Biotechnology

Few developments in science have had the potential for such profound impact on research, technology, and society in general as has biotechnology. Yet authorities do not agree on a single definition of this term. Sometimes, writers have limited the term to techniques used to modify living organisms and, in some instances, the creation of entirely new kinds of organisms.

In most cases, however, a broader, more general definition is used. The Industrial Biotechnology Association, for example, uses the term to refer to any "development of products by a biological process." These products may indeed be organisms or they may be cells, components of cells, or individual and specific **chemicals**. A somewhat more detailed definition is that of the European Federation of Biotechnology, which defines biotechnology as the "integrated use of biochemistry, microbiology, and engineering sciences in order to achieve technological (industrial) application of the capabilities of **microorganisms**, cultured tissue cells, and parts thereof."

By almost any definition, biotechnology has been used by humans for thousands of years, long before modern science existed. Some of the oldest manufacturing processes known to humankind make use of biotechnology. Beer, wine, and breadmaking, for example, all occur because of the process of fermentation. During fermentation, microorganisms such as yeasts, molds, and bacteria are mixed with natural products which they use as food. In the case of wine-making, for example, yeasts live on the sugars found in some type of fruit juice, most commonly, grape juice. They digest those sugars and produce two new products, alcohol and **carbon dioxide**.

The alcoholic beverages produced by this process have been, for better or worse, a mainstay of human civilization for untold centuries. In breadmaking, the products of fermentation are responsible for the wonderful odor (the alcohol) and texture (the **carbon** dioxide) of freshly-baked bread.

Cheese and yogurt are two other products formed when microorganisms act on a natural product, in this case milk—changing its color, odor, texture, and taste.

Biotechnology has long been used in a variety of industrial processes also. As early as the seventeenth century, bacteria were used to remove **copper** from its ores. Around 1910, scientists found that bacteria could be used to decompose organic matter in sewage, thus providing a mechanism for dealing efficiently with such materials in **solid waste**. A few years later, a way was found to use microorganisms to produce glycerol synthetically. That technique soon became very important commercially, since glycerol is used in the manufacture of explosives and World War I was about to begin.

Not all forms of biotechnology depend on microorganisms. Hybridization is an example. Farmers long ago learned that they could control the types of animals bred by carefully selecting the parents. In some cases, they actually created entirely new animal forms that do not occur in **nature**. The mule, a hybrid of horse and donkey, is such an animal.

Hybridization has also been used in plant growing for centuries. Farmers found that they could produce food plants with any number of special qualities by carefully selecting the seeds they plant and by controlling growing conditions. As a result of this kind of process, the 2–3-in (5.1–7.6-cm) vegetable known as maize has evolved over the years into the 12-in (30-cm), robust product called corn. Indeed, there is hardly a fruit or vegetable in our diet today that has not been altered by long decades of hybridization.

Until the late nineteenth century, hybridization was largely a trial-and-error process. Then the work of Gregor Mendel started to become known. Mendel's research on the transmission of hereditary characteristics soon gave agriculturists a solid factual basis on which to conduct future experiments in cross-breeding.

Modern principles of hybridization have made possible a greatly expanded use of biotechnology in agriculture and many other areas. One of the greatest successes of the science has been in the development of new food crops that can be grown in a variety of less-than-optimal conditions. The dramatic increase in harvests made possible by these developments has become known as the **agricultural revolution** or green revolution.

Three decades after the green revolution first changed agriculture in many parts of the world, a number of problems with its techniques have become apparent. The agricultural revolution forced a worldwide shift from subsistence farming to cash farming, and many small farmers in developing countries lack the resources to negotiate this shift. A farmer must make significant financial investments in seed, **agricultural chemicals** (fertilizers and pesticides), and machinery to make use of new farming techniques. In developing countries, peasants do not have and cannot borrow the necessary capital. The seed, chemicals, machinery, and oil to operate the equipment must commonly be imported, adding to already crippling foreign debts. In addition, the new techniques often have harmful effects on the **environment**. In spite of problems such as these, however, the green revolution has clearly made an important contribution to the lessening of world hunger.

Modern methods of hybridization have application in many fields besides agriculture. For example, scientists are now using controlled breeding techniques and other methods from biotechnology to insure the survival of **species** that are threatened or endangered.

The nature of biotechnology has undergone a dramatic change in the last half century. That change has come about with the discovery of the role of **deoxyribosenucleic acid** DNA in living organisms. DNA is a complex molecule that occurs in many different forms. The many forms that DNA can take allow it to store a large amount of information. That information provides cells with the direction they need to carry out all the functions they have to perform in a living organism. It also provides a mechanism by which that information is transmitted efficiently from one generation to the next.

As scientists learned more about the structure of the DNA molecule, they discovered precisely and in chemical terms how genetic information is stored and transmitted. With that knowledge, they have also developed the ability to modify DNA, creating new instructions that direct cells to perform new and unusual functions. The process of DNA modification has come to be known as **genetic engineering**. Since genetic engineering normally involves combining two different DNA molecules, it is also referred to as recombinant DNA research.

There is little doubt that genetic engineering is the best known form of biotechnology today. Indeed, it is easy to confuse the two terms and to speak of one when it is the other that is meant. However, the two terms are different in the respect that genetic engineering is only one type of biotechnology.

In theory, the steps involved in genetic engineering are relatively simple. First, scientists decide what kind of changes they want to make in a specific DNA molecule. They might, in some cases, want to alter a human DNA molecule to correct some error that results in a disease such as diabetes. In other cases, a researcher might want to add instructions to a DNA molecule that it does not normally carry. He or she might, for example, want to include instructions for the manufacture of a chemical such as insulin in the DNA of bacteria that normally lack the ability to make insulin.

Second, scientists find a way to modify existing DNA to correct errors or add new information. Such methods are now well developed. In one approach, enzymes that "recognize" certain specific parts of a DNA molecule are used to cut open the molecule and then insert the new portion.

Third, scientists look for a way to insert the "correct" DNA molecule into the organisms in which it is to function. Once inside the organism, the new DNA molecule may give correct instructions to cells in humans (to avoid genetic disorders), in bacteria (resulting in the production of new chemicals), or in other types of cells for other purposes.

Accomplishing these steps in practice is not always easy. One major problem is to get an altered DNA molecule to express itself in the new host cells. That the molecule is able to enter a cell does not mean that it will begin to operate and function (express itself) as scientists hope and plan. This means that many of the expectations held for genetic engineering may not be realized for many years.

In spite of problems, genetic engineering has already resulted in a number of impressive accomplishments. Dozens of products that were once available only from natural sources and in limited amounts are now manufactured in abundance by genetically engineered microorganisms at relatively low cost. Insulin, human growth hormone, tissue plasminogen activator, and alpha interferon are examples. In addition, the first trials with the alteration of human DNA to cure a genetic disorder were begun in 1991.

The prospects offered by genetic engineering have not been greeted with unanimous enthusiasm by everyone. Many people believe that the hope of curing or avoiding genetic disorders is a positive advance. But they question the wisdom of making genetic changes that are not related to life-threatening disorders. Should such procedures be used for helping short children become taller or for making new kinds of tomatoes? Indeed, there are some critics who oppose *all* forms of genetic engineering, arguing that humans never have the moral right to "play God" with any organism for any reason. As the technology available for genetic engineering continues to improve, debates over the use of these techniques in practical settings are almost certainly going to continue—and to escalate—in the future.

As progress in genetic engineering goes forward, so do other forms of biotechnology. The discovery of monoclonal antibodies is an example. Monoclonal antibodies are cells formed by the combination of tumor cells with animal cells that make one and only one kind of antibody. When these two kinds of cells are fused, they result in a cell that reproduces almost infinitely and that recognizes one and only one kind of antigen. Such cells are extremely valuable in a vast array of medical, biological, and industrial applications, including the diagnosis and treatment of disease, the separation and purification of proteins, and the monitoring of pregnancy.

Biotechnology became a point of contention in 1992 during planning for the **United Nations Earth Summit** in Rio de Janeiro. In draft versions of the treaty on **biodiversity**, developing nations insisted on provisions that would force biotechnology companies in the developed world to pay fees to developing nations for the use of their genetic resources (the plants and animals growing within their boundaries). Currently, companies have free access to most of these raw materials used in the manufacture of new drugs and crop varieties. President George Bush argued that this provision would place an unfair burden on biotechnology companies in the United States, and he refused to sign the biodiversity treaty that contained this clause. For now, it seems that, for a second time, profits to be reaped from biotechnological advances will elude developing countries. (The Clinton administration subsequently endorsed the provisions of the biodiversity treaty and it was signed by Madeleine Albright, U. S. ambassador to the United Nations, on June 4, 1993.)

[*David E. Newton*]

RESOURCES
BOOKS

Fox, M. W. *Superpigs and Wondercorn: The Brave New World of Biotechnology and Where It May All Lead.* New York: Lyons and Buford, 1992.

Mellon, M. *Biotechnology and the Environment.* Washington, D.C.: National Biotechnology Policy Center of the National Wildlife Federation, 1988.

PERIODICALS

Kessler, D. A., et al. "The Safety of Foods Developed by Biotechnology." *Science* (26 June 1992): 1747-1749+.

Bioterrorism

Bioterrorism refers to the use of lethal biological agents to wage terror against a civilian population. It differs from biological warfare in that it also thrives on public fear, which can demoralize a population. An example of bioterrorism is provided by the **anthrax** outbreak which occurred during September-November 2001 in the United States. Anthrax spores intentionally spread in the mail distribution system caused five deaths and a total of 22 infections. The Centers for Disease Control (CDC) classifies bioterror agents into three categories:

- **Category A Diseases/Agents** that can be easily disseminated or transmitted from person to person and that can result in high **mortality** rates while causing public panic and social disruption. Anthrax, botulism, **plague**, small-

pox, tularemis, and viral hemorrhagic fever viruses belong to this category.

- **Category B Diseases/Agents** that are moderately easy to disseminate and that can result in low mortality rates. Brucellosis, food and water safety threats, melioidosis, psittacosis, staphylococcal enterotoxin B, and typhus belong to this category.

- **Category C Diseases/Agents** that include emerging pathogens that could be engineered for mass dissemination in the future because of availability or ease of production and dissemination and that have potential for high mortality rates.

The anthrax attacks of 2001 were very limited in scope compared to the potential damage that could result from large-scale bioterrorism. A large-scale bioterrorism attack on the United States could threaten vital national security interests. Massive civilian casualties, a breakdown in essential services, violation of democratic processes, civil disorder, and a loss of confidence in government could compromise national security, according to a report prepared by four non-profit analytical groups, including the Center for Strategic and International Studies and the John Hopkins Center for Civilian Biodefense Studies.

Probably the first sign of a bioterrorism attack is when people infected during the attack start developing symptoms and showing up in hospital emergency departments, urgent care centers, and doctors' offices. By this time, people infected in the initial attack will have begun spreading it to others.

An added concern is that most physicians have never treated a case of a bioterrorism agent such as smallpox or **Ebola**. This is likely to cause a delay in diagnosis, further promoting the spread of the contagious agent. For example, based on past smallpox history, it is estimated that each person infected during the initial attack will infect another 10–12 persons. In the case of smallpox, only a few **virus** particles are needed to cause infection. One ounce of the smallpox virus could infect 3,000 persons if distributed through an **aerosol** attack, according to William Patrick, senior scientist in the United States biological weapons program before its official termination in 1969, in a 2001 *Washington Post Magazine* interview. Given these numbers, a terrorist with enough smallpox virus to fill a soda can could potentially infect 36,000 people in the initial attack who could then infect another 360,000–432,000. Of these, an estimated 30% or 118,800–140,400, would likely die.

Using disease as a weapon is not a new idea. It goes back at least hundreds of years and possibly much further back. One account of the beginning of the great plague epidemic which occurred in Europe in the fourteenth century and killed a third of the population states that it started with an act of bioterrorism, as reported by A. Daniels in National

Review. The Tartars were attacking a Genoan trading post on the Crimean coast in 1346 when the plague broke out among them. Turning the situation into a weapon, the Tartars catapulted the dead and diseased bodies over the trading post walls. The Genoans soon developed the deadly disease and took it back with them to Genoa, where it soon engulfed all of Europe. Another example from early North American history is provided by the British soldiers who gave smallpox-infected blankets to Native Americans in the 1700s.

The Hague Conventions of 1899 and 1907 included clauses outlawing the deliberate spread of a deadly disease. However, during World War I, German soldiers attempted to infect sheep destined for Russia with anthrax. After the war, 40 members of the League of Nations, the precursor of the United Nations, outlawed biological weapons. But many countries continued biological warfare research. During World War II, the Japanese mass-produced a number of deadly biological agents, including anthrax, typhoid, and plague. They infected water supplies in China with typhoid, killing thousands, including 1,700 Japanese soldiers. Bioterrorism entered popular literature more than a century ago when science fiction writer H. G. Wells wrote "The Stolen Bacillus,", a novel in which a terrorist tries to infect the London water supply with **cholera**, an acute and often deadly disease.

Throughout the Cold War era, several nations, including the United States and Soviet Union, developed sophisticated facilities to produce large amounts of biological agents to be used as weapons. Most nations have renounced the manufacture, possession, or use of biological weapons. However, a few rogue nations, including Iran, Iraq, and North Korea, still have active biological warfare programs according to the United States military. Many experts in the field believe that terrorists could obtain deadly biological agents from these rogue nations, or from other terrorist or criminal groups active in nations of the former Soviet Union.

Among the **Category A Diseases/Agents**, six highly lethal biological agents are most likely to be used by terrorists, according to the CDC. Depending on the biological agent, disease could be spread through the air, or by contaminating the food or water supply.

• Anthrax, caused by *Bacillus anthracis*, is an acute infectious disease that most commonly occurs in hoofed animals but can also infect humans. Initial symptoms are flu-like and can occur up to several weeks after exposure. Treatment with antibiotics after exposure but before symptoms develop is usually successful in preventing infection. There is a anthrax vaccine used by the military but it is not available for civilian use. About 90% of people who are infected die.

• Botulism is a muscle-paralyzing disease caused by a toxin produced by a bacterium called *Clostridium botulinum*. The botulinum toxin is the single most poisonous substance known, according to the Center for Civilian Biodefense Strategies. It is a major bioterrorism threat because of its extreme potency and high rate of death after exposure. It is not contagious and would likely be used by terrorists to contaminate food or water supplies. Flu-like symptoms, along with difficulty speaking, seeing, or swallowing, usually occur 12–72 hours after exposure.

• Plague is a disease caused by *Yersinia pestis*, a bacterium found in rodents and their fleas in many areas around the world. When released into the air, the bacterium can survive for up to an hour. Of the three types of plague (pneumonic, bubonic, and septicemic), pneumonic is the one most likely to be used by terrorists since large stockpiles were developed by the United States and Soviet Union in the 1950s and 1960s. Symptoms include fever, headache, weakness, chest pain, and cough. Early treatment with antibiotics can reduce the risk of death.

• Smallpox is caused by the *variola major* virus and was eliminated from the world in 1977. However, the Soviet Union had large stockpiles of the virus in the 1980s and much of it may still be stored in the former Soviet republics and available to terrorists. Smallpox spreads directly from person to person and can be dispersed in the air. Also, the amount needed to cause infection is very small. Symptoms, including high fever, fatigue, and head and back aches, commonly develop in about 12 days. Flat, red skin lesions follow initial symptoms. Death occurs in about 30% of the cases. There is a vaccine against smallpox but routine vaccinations ended in 1972. The government has an emergency supply of about 15 million doses of the vaccine.

• Tularemia, an infectious disease caused by the bacterium *Francisella tularensis*,, is usually found in animals but can also infect humans. It could be delivered in a terrorist attack through food, water, or air. Symptoms of tularemia include sudden fever, chills, headache, muscle ache, dry cough, weakness, and pneumonia. The disease can be treated with antibiotics if started early. As of May 2002, the U.S. **Food and Drug Administration** (FDA) was reviewing a possible vaccine for the disease.

• Exotic diseases, including viral hemorrhagic fevers, such as Ebola, and arenaviruses, such as the one causing Lassa fever, are also biological agents of interest to terrorists. The Ebola virus is one of the most lethal known, and easily spreads from person to person, with no vaccine or effective treatment presently known.

The possibility that bioterrorists may strike at food and water supplies is of serious concern to health and environmental officials. Such an attack initially could be perceived as unintentional food **poisoning**, which might delay recognition of the outbreak, and complicate identification of the contaminated food. What many consider an act of bioterrorism by domestic terrorists occurred in The Dalles, Oregon, in 1984. Members of a religious cult contaminated restaurant salad bars with *Salmonella typhimurium*, a non-lethal bacterium that nonetheless infected 751 people. The incident was reportedly a trial run for a more extensive attack to disrupt local elections later that year.

"The United States food supply is increasingly characterized by centralized production and wide distribution of products," according to a May 2002 article in *The Lancet*, the "Deliberate contamination of a commercial food product could cause an outbreak of disease, with many illnesses dispersed over wide geographical areas.". The article also stated that the anthrax letter attacks of 2001 have shown that even a small biological attack can produce considerable public nervousness and challenge the health care system.

[*Ken R. Wells*]

RESOURCES
BOOKS

Alexander, Yonah and Stephen Prior. *Terrorism and Medical Responses: U.S. Lessons and Policy Implications.* Ardsley, NY: Transnational Publishers, 2001.

Frist, Bill and William H. Frist. *When Every Moment Counts: What You Need to Know About Bioterrorism from the Senate's Only Doctor.* Totowa, NJ: Rowan & Littlefield, 2002.

Salvucci Jr., Angelo. *Biological Terrorism, Responding to the Threat: A Personal Safety Manual.* Carpinteria, CA: Public Safety Medical, 2001.

U.S. Government CD. *21st Century Complete Guide to Bioterrorism, Biological and Chemical Weapons, Germs and Germ Warfare, Nuclear and Radiation Terrorism - Military Manuals and Federal Documents with Practical Emergency Plans, Protective Measures, Medical Treatment and Survival Information.* New York: Progressive Management, 2001.

PERIODICALS

Daniels, Anthony. "Germs Against Man: Bioterror; A Brief History." *National Review*(Dec. 3, 2001).

Fabian, Nelson. "Post September 11: Some Reflections on the Role of Environmental Health in Terrorism Response." *Journal of Environmental Health*(May 2002): p.77-80.

Guterl, Fred, and Conant, Eve. "In the Germ Labs: The Former Soviet Union Had Huge Stocks of Biological Agents. Assessing the Real Risk." *Newsweek*(Feb. 25, 2002): p. 26.

Maddox, P.J. "Bioterrorism: A Renewed Public Health Threat."*MedSurg Nursing*(Dec. 2001): p. 333-338.

The Lancet(March 9, 2002): p. 874.

Sobel, Jeremy, et al. "Threat of a Biological Terrorist Attack on the U.S. Food Supply: The CDC Perspective."

ORGANIZATIONS

Centers for Disease Control and Prevention - Bioterrorism Preparedness & Response Program (CDC), 1600 Clifton Road, Atlanta, GA USA 30333 (404) 639-3534, Toll Free: (888) 246-2675, Email: cdcresponse@ashastd.org, <http://www.bt.cdc.gov>

Center for the study of bioterrorism and Emergency Infections—Saint Louis University, 3545 Lafayette, Suite 300, St. Louis, MO USA 63104, <http://bioterrorism.slu.edu>

Texas Department of Health Bioterrorism Preparedness Program (TDH), 1100 West 49th Street, Austin, TX USA 78756 (512) 458-7676, Toll Free: (800) 705-8868, <http://www.tdh.state.tx.us/bioterrorism/default.htm>

Biotic community

All the living organisms sharing a common **environment**. The members of a biotic community are usually divided into three major categories: producers, consumers, and **decomposers**, based on the organisms' nutritional habits. Producers (sometimes called autotrophs) include plants and some forms of bacteria that use solar or chemical energy to convert simple compounds into their food. Consumers (sometimes called heterotrophs) obtain the energy they need by eating living plants and animals, or dead plant and animal material (**detritus**). Primary consumers (herbivores) eat plants, while secondary consumers (carnivores) eat other consumers. Consumers that feed on dead plant and animal material are called detrivores. There are two classes of detrivores: detritus feeders and decomposers. Detritus feeders (e.g., crabs, termites, earthworms, vultures) consume dead organisms or organic wastes, while decomposers (**fungi** and bacteria) feed on dead plant material, converting it into simple inorganic compounds such as **carbon dioxide**, water, and ammonia. Decomposers are also an important food sources for other consumers (e.g., worms and insects) living in the **soil** or water.

Biotic impoverishment
see **Biodiversity**

Biotoxins

The term biotoxin refers to naturally occurring, poisonous agents that can cause illness or injury to humans, animals, and marine life. They may come from bacteria, **fungi**, algae, or virii.

Some of the more well-known bacterial biotoxins include *Bacillus anthracis* (**anthrax**), *Brucella melitensis* and *Brucella suis* (brucellosis), *Vibro cholerae* (**cholera**), and *Yersinia pestis* (**plague**).

Diseases spread by viral biotoxins include encephalomyelitis, viral hemorrhagic fever (yellow fever, ebola-marburg, dengue fever), and variola major (smallpox).

Biotoxins may be distributed through wind-borne spores or through contaminated food or water. Some biological agents have also been aerosolized or deliberately introduced into food and water supplies for use as bioweapons.

Marine biotoxins are often responsible for large-scale **fish kills** and can cause severe illness in humans who consume tainted fish or shellfish. Symptoms of shellfish **poisoning** include diarrhea, stomach cramps, headache, nausea, vomiting, and in extreme cases neurotoxic effects including paralysis, seizures, and death. Common marine biotoxins include ciguatera fish poisoning (CFP) toxin and domoic **acid**. Harmful algae blooms (HAB) are sometimes called red tides or brown tides. Some HABs, such as *Pfiesteria piscicida*, can be sources of marine biotoxins that potentially harm marine life and humans.

[*Paula Anne Ford-Martin*]

RESOURCES
BOOKS

Gaffield, William and Anthony Tu, eds. *Natural and Selected Synthetic Toxins: Biological Implications*. Washington, DC: American Chemical Society, 2000.

PERIODICALS

Burrows, W. Dickinson and Sara Renner. "Biological Warefare Agents as Threats to Potable Water." *Environmental Health Perspectives* 107, no.12 (Dec 1999): 975 (10pp).

OTHER

U.S. Environmental Protection Agency, Office of Water. *Pfiesteria piscicida Home Page*. [cited May 15, 2002]. <http://www.epa.gov/owow/estuaries/pfiesteria/>.

ORGANIZATIONS

National Office for Marine Biotoxins and Harmful Algal Blooms; Woods Hole Oceanographic Institution, Biology Dept., MS #32, Woods Hole, MA USA 02543 (508) 289-2252, Fax: (508) 457-2180, Email: jkleindinst@whoi.edu, <http://www.redtide.whoi.edu/hab/>

Bioventing

The process for treating environmental contaminants in the soils located between the ground surface and the **water table** by inducing **aerobic** biodegradation. Air is introduced into the contaminated soils, providing oxygen for native **soil** bacteria to use in the biodegradation of organic compounds. The process is typically accomplished by extracting and/or injecting air from trenches or shallow **wells** which are screened within the unsaturated soils. The systems are similar to those used in vapor extraction. The main difference is that air extraction rates are low to minimize physical removal (stripping) of volatile organic compounds (VOCs) reducing the need for expensive treatment of the off-gases. The process may include the addition of nutrients such as common fertilizers, to provide **nitrogen** and phosphate for the bacteria. Bioventing is particularly attractive around buildings and actively used areas because it is relatively nonintrusive and results in minimal disturbance during installation and operation. The process is most suitable for **petroleum** hydrocarbon blends such as **gasoline**, jet fuel, and diesel oil, for petroleum distillates such as **toluene**, and for nonchlorinated solvents. *See also* Biodegradable; Bioremediation; Vapor recovery system

BirdLife International

"From research...to action. From birds...to people." So reads the cover of the BirdLife International's annual report. This statement perfectly describes the beliefs of BirdLife International, a group founded under the original name International Council for Bird Preservation in 1922 by well-known American and European bird enthusiasts for the **conservation** of birds and their habitats.

Under the leadership of Director and Chief Executive Dr. Michael Rands, the group works to protect endangered birds worldwide and to promote public awareness of their ecological importance. BirdLife International has grown from humble beginnings in England to a federation of over 300 member organizations representing approximately 2.2 million people in 103 countries and territories. This includes developing tropical countries where few, if any, conservation movements existed prior to BirdLife International. There is also a worldwide network of enthusiastic volunteers.

BirdLife International is a key group in international efforts to protect bird **migration** routes, and also works to educate the public about **endangered species** and their ecological importance. The BirdLife International gathers and disseminates information about birds, maintaining a computerized data bank from which it generates reports. It conducts periodic symposiums on bird-related issues, runs the World Bird Club, maintains a Conservation Fund, runs special campaigns when opportunities such as the Migratory Bird Campaign present themselves, and develops and carries out priority projects in their Conservation Program.

The BirdLife International Conservation Program has undertaken many projects on behalf of endangered birds. In 1975, BirdLife International began a captive breeding program for the pink pigeon (*Nesoenas mayeri*), a native of the island of Mauritius in the Indian Ocean, because the total population of this **species** had dwindled to less than 20 birds. As a result of these efforts, well over 100 of the birds were successfully raised in captivity. Later, several pairs were released at Mauritius' Botanic Gardens of Pamplemousses. BirdLife International has focused on other seriously endangered birds as well, such as the imperial parrot (*Amazona imperialis*). In an attempt to protect its threatened **habitat**, BirdLife International has helped buy a forest reserve in Dominica, where only 60 of the **parrots** still survive. With the help of local citizens and educational facilities, BirdLife International hopes that their efforts to save the imperial parrot will be as successful as their work with pink pigeons.

Another important BirdLife International project is the group's work to save the red-tailed parrot (*Amazona brasiliensis*) of southeastern Brazil. This project involves support of an extensive plan to convert an entire nearby island into a refuge for the parrots, which exist in only a very few isolated parts of Brazil. BirdLife International has also focused on islands in other conservation projects. The council purchased Cousin Island (famous for its numerous seabirds), in an effort to save the Seychelles brush warbler (*Acrocephalus sechellensis*). Native only to Cousin Island, this entire brush warbler species numbered only 30 individuals before BirdLife International bought their island. Today, there are more than 300 bush warblers, and BirdLife International continues to be actively involved in helping to breed more.

BirdLife International's publications are many and varied. Quarterly, it issues *Bird Conservation International, U.S. Birdwatch*, and *World Birdwatch* newsletter; and periodically, it issues *Bustard Studies*. It also publishes the well-respected series *Bird Red Data Books*, and such monographs as *Important Bird Areas in Europe* and *Key Forests for Threatened Birds in Africa*. BirdLife International produces numerous technical publications and study reports, and, occasionally, Conservation Red Alert pamphlets on severely threatened birds.

[*Cathy M. Falk*]

RESOURCES

ORGANIZATIONS

BirdLife International, Wellbrook Court, Girton Road, Cambridge, United Kingdom CB3 0NA +44 1 223 277 318, Fax: +44 1 223 277200, Email: birdlife@birdlife.org.uk, <http://www.birdlife.net>

Birth control

see Family planning; Male contraceptives

Birth defects

Birth defects, also known as congenital malformations, are structural or metabolic abnormalities present at birth. While subtle variations from the normal, of no clinical interest, occur in about half of all individuals in the United States, significant congenital defects are found in about 3% of live births. Fortunately, only about half of these require medical attention.

Birth defects may result from genetic causes or environmental insult. Defective genes are not easily repaired and thus are perhaps less interesting than teratogenic substances to environmentalists. It is theoretically possible to limit exposure to teratogens by elimination of the agent in the **environment** or by modification of behavior to prevent contact. It should be noted, however, that the causes of more than half of congenital malformations remain unknown.

Birth defects of genetic origin may be due to aberrant chromosome number or structure, or to a single gene defect.

Normal humans have 46 chromosomes, and variation from this number is referred to as aneuploidy. Down's syndrome, an example of aneuploidy, is usually characterized by an extra chromosome designated number 21. The Down's individual thus has a total of 47 chromosomes, and the presence of the extra chromosome results in multiple defects. These include mental retardation and physical characteristics comprising a small round head, eyes that slant slightly upward, a large and frequently protruding tongue, low set ears, broad hands with short fingers and short stature. People with Down's syndrome are particularly vulnerable to **leukemia**. Children with this condition are rarely born to mothers less than 25 years of age (less than one in 1,500), but the prevalence of Down's syndrome babies increases with mothers older than 45 (about one in 25). Down's syndrome can be detected during pregnancy by chromosome analysis of fetal cells. Fetal chromosomes may be studied by chorionic villus sampling or by amniocentesis.

Other congenital abnormalities with a chromosomal basis include Klinefelter's syndrome, a condition of male infertility associated with an extra X chromosome, and Turner's syndrome, a condition wherein females fail to mature sexually and are characterized by the aneuploid condition of a missing X chromosome. Achondroplasia is a birth defect due to a dominant **mutation** of a Mendelian gene that results in dwarfism. Leg and arm bones are short but, the trunk is normal and the head may be large. Spontaneous mutation accounts for most achondroplasia. The **mortality** rate for affected individuals is so high that the gene would be lost if it were not for mutation. Albinism, a lack of pigment in the skin, eyes, and hair, is another congenital defect caused by a single gene, which in this case is recessive.

The developing fetus is at risk for agents which can pass the placental barrier such as infectious **microbes**, drugs and other **chemicals**, and **ionizing radiation**. Transplacental teratogens exert their effect on incompletely formed embryos or fetuses during the first three months of pregnancy. Organs and tissues in older and full term fetuses appear much as they will throughout life. It is not possible to alter the development of a fully formed structure. However, prior to the appearance of an organ or tissue, or during the development of that structure, teratogenic agents may have a profoundly deleterious effect.

Perhaps the best known **teratogen** is the sedative thalidomide which induces devastating anatomical abnormalities. The limb bones are either shortened or entirely lacking leading to a condition known as phocomelia. Intellectual development of thalidomide babies is unaffected. The experience with this drug, which started in 1959 and ended when it was withdrawn in 1961, emphasizes the fact that medications given to pregnant mothers generally cross the placenta and reach the developing embryo or fetus. Another drug that effects developmental abnormalities is warfarin which is used in anticoagulant therapy. It can cause fetal hemorrhage, mental retardation, and a multiplicity of defects to the eyes and hands when given to pregnant women.

The teratogenic effects of alcohol, or the life style that may accompany alcohol abuse, serve to illustrate that the term environment includes not only air and water but the personal environment as well. Alcoholism during pregnancy can result in "fetal alcohol syndrome" with facial, limb, and heart defects accompanied by growth retardation and reduced intelligence. The effects of alcohol may be magnified by factors associated with alcoholism such as poor diet, altered **metabolism** and inadequate medical care. Because neither the time of vulnerability nor the toxic level of alcohol is known, the best advice is to eschew alcohol as a dietary constituent altogether during pregnancy.

Disease of the mother during pregnancy can present an environmental hazard to the developing fetus. An example of such a hazard is the viral disease German measles, also known as rubella. The disease is characterized by a slight increase in temperature, sore throat, lethargy and a rash of short duration. Greatest hazard to the fetus is during the second and third month. Children born of mothers who had rubella during this period may exhibit cataracts, heart defects, hearing loss and mental retardation. Obviously, the **virus** transverses the placenta to infect the embryo or fetus and that infection may persist in the newborn. Birth defects associated with rubella infection have decreased since the introduction of a rubella vaccine.

The most common viral infection that occurs in human fetuses is that of a herpes virus known as cytomegalovirus. The infection is detected in about 1–2% of all live births. Most newborns, fortunately, do not manifest symptoms of the infection. However, for a very small minority, the effects of congenital cytomegalovirus are cruel and implacable. They include premature birth or growth retardation prior to birth, frequently accompanied by hepatitis, enlarged spleen, and reduction in thrombocytes (blood cells important for clotting). Abnormally small heads, mental retardation, cerebral palsy, heart and cerebral infection, bleeding problems, hearing loss and blindness occur. Exposure of the fetus to the virus occurs during infection of the pregnant woman or possibly from the father, since cytomegalovirus has been isolated from human semen.

Other infections known to provoke congenital defects include herpes simplex virus type II, toxoplasmosis, and syphilis.

Methylmercury is an effective **fungicide** for seed grain. Accidental human consumption of food made from treated seeds has occurred. Industrial **pollution** of sea water with organic **mercury** resulted in the contamination of fish, consumed by humans, from Minamata Bay in Japan. It has been established that organic mercury passes the placental barrier with effects that include mental retardation and a cerebral palsy-like condition due to brain damage. Anatomical birth defects, engendered by organic mercury, include abnormal palates, fingers, eyes and hearts. The toxicity of methylmercury affects both early embryos and developing fetuses. Exclusion of mercury from human food can be effected by not using organic mercury as a fungicide and by ending industrial **discharge** of mercury into the environment.

Of course other chemicals may be hazardous to the offspring of pregnant women. Polychlorinated biphenyl (PCB)s, relatively ubiquitous but low level oily contaminants of the environment, cause peculiar skin pigmentation, low birth weights, abnormal skin and nails, and other defects in offspring when accidentally ingested by pregnant woman.

Uncharacterized mixtures of toxic chemicals which contaminate the environment, are thought to be potential teratogens. Cytogenetic (chromosomal) abnormalities and increased birth defects were detected among the residents of **Love Canal**, New York. **Cigarette smoke** is the most common mixture of **toxic substance** to which fetuses are exposed. **Tobacco smoke** is associated with reduced birth weight but not specific birth anatomical abnormalities.

Much concern has arisen over the damaging effects of ionizing radiation, particularly regarding diagnostic x-rays and **radiation exposure** from nuclear accidents. The latter concern was given international attention following the explosion at the Ukraine's **Chernobyl Nuclear Power Station** in 1986. Fear that birth defects would occur as a result was fueled by reports of defects in Japanese children whose mothers were exposed to radiation at Hiroshima. Scientists believe that radiation to the fetus can result in many defects including various malformations, mental retardation, reduced growth rate and increased risk for leukemia. Fortunately, however, the risk of these effects is exceptionally low. Fetal abnormalities caused by factors other than radiation are thought to be about 10 times greater than those attributed to radiation during early pregnancy. However small the risk, most women choose to limit or avoid exposure to radiation during early pregnancy. This may be part of the reason for the increased popularity of diagnostic ultrasound as opposed to x ray.

While concern is expressed for particular teratogenic agents or procedures, the etiology of most birth defects is unknown. Common defects, with unknown etiology, include hare lip and cleft palate, extra fingers and toes, fused fingers, extra nipples, various defects in the heart and great vessels, cerebral palsy (sometimes as a result of difficult labor and delivery but frequently for no known cause), narrowing of the entrance to the stomach, esophageal abnormalities, spina bifida, clubfoot, hip defects, and many, many others. Since the majority of birth defects are not caused by known effects of disease, drugs, chemicals or radiation, much remains to be learned.

[*Robert G. McKinnell*]

FURTHER READING

Brent, R. L. and J. L. Sever, eds. *Teratogen Update.* New York: Alan R. Liss, 1986.

Moore, K. L. *Essentials of Human Embryology.* Philadelphia: B.C. Decker, Inc., 1988.

Persaud, T. V. N., et al. *Basic Concepts in Teratology.* New York: Alan R. Liss, 1985.

Bison

The American bison (*Bison bison*) or "buffalo" is one of the most famous animals of the American West. Providing food and hides to the early Indians, it was almost completely eliminated by hunters, and now only remnant populations exist though its future survival seems assured.

Scientists do not consider the American bison a true buffalo (like the Asian water buffalo or the African buffalo), since it has a large head and neck, a hump at the shoulder, and 14 pairs of ribs instead of 13. In America, however, the names are used interchangeably. A full-grown American bison bull stand 5.5–6 ft (1.7–1.8 m) at the shoulder, extends 10–12.25 ft (3–3.8 m) in length from nose to tail, and weighs 1,600–3,000 lb (726–1,400 kg). Cows usually weigh about 900 lb (420 kg) or less. Bison are brown-black with long hair which covers their heads, necks, and humps, forming a "beard" at the chin and throat. Their horns can have a spread as large as 35 in (89 cm). Bison can live for 30 or more years, and they are social creatures, living together in herds. Bison bulls are extremely powerful; a charging bull has been known to shatter wooden blanks 2 in (5 cm) thick and 12 in (30 cm) wide.

The American bison is one of the most abundant animals ever to have existed on the North American continent, roaming in huge herds between the Appalachians and the Rockies as far south as Florida. One herd seen in Arkansas in 1870 was described as stretching "from six to 10 mi (9.7 to 16.1 km) in almost every direction." In the far West, the herds were even larger, stretching as far as the eye could see, and in 1871 a cavalry troop rode for six days through a herd of bison.

The arrival of Europeans in America sealed the fate of the American bison. By the 1850s massive slaughters of these creatures had eliminated them from Illinois, Indiana, Kentucky, Ohio, New York, and Tennessee. After the end of the Civil War in 1865, railroads began to bring a massive influx of settlers to the West and bison were killed in enormous numbers. The famous hunter "Buffalo Bill" Cody was able to bag 4,280 bison in just 18 months, and between 1854 and 1856, an Englishman named Sir George Gore killed about 6,000 bison along the lower Yellowstone River. Shooting bison from train windows became a popular **recreation** during the long trip west; there were contests to see who could kill the most animals on a single trip, and on one such excursion a group accompanying Grand Duke Alexis of Russia shot 1,500 bison in just two days. When buffalo tongue became a delicacy sought after by gourmets in the east, even more bison were killed for their tongues and their carcasses left to rot.

In the 1860s and 1870s extermination of the American bison became the official policy of the United States Government in order to deprive the Plains Indians of their major source of food, clothing, and shelter. During the 1870s, two to four million bison were shot each year, and 200,000 hides were sold in St. Louis in a single day. Inevitably, the extermination of the bison helped to eliminate not only the Plains Indians, but also the predatory animals dependent on it for food, such as plains **wolves**. By 1883, according to some reports, only one wild herd of bison remained in the West, consisting of about 10,000 individuals confined to a small part of North Dakota. In September of that year, a group of hunters set off to kill the remaining animals and by November the job was done.

By 1889 or 1890 the entire North American bison population had plummeted to about 500 animals, most of which were in captivity. A group of about 20 wild bison remained in **Yellowstone National Park**, and about 300 wood bison (*Bison bison athabascae*) survived near Great Slave Lake in Canada's Northwest Territories. At that time, naturalist William Temple Hornaday led a campaign to save the **species** from complete **extinction** by the passage of laws and other protective measures. Canada enacted legislation to protect its remnant bison population in 1893 and the United States took similar action the following year.

Today, thousands of bison are found in several national parks, private ranches, and **game preserves** in the United States. About 15,000 are estimated to inhabit Wood Bison **National Park** and other locations in Canada. The few hundred wood bison originally saved around Great Slave Lake also continued to increase in numbers until the population reached around 2,000 in 1922. But in the following years, the introduction of plains bison to the area caused hybridization, and pure specimens of wood bison probably disappeared around Great Slave Lake. Fortunately, a small, previously unknown herd of wood bison was discovered in 1957 on the upper North Yarling River, buffered from hybridization by 75 mi (121 km) of swampland. From this herd (estimated at about 100 animals in 1965) about 24 animals were successfully transplanted to an area near Fort Providence in the Northwest Territories and 45 were relocated to Elk Island National Park in Alberta. Despite these rebuilding programs, the wood bison is still considered endangered, and is listed as such by the U. S. Department of the Interior. It is also listed in Appendix I of the Convention on International Trade in **Endangered Species** of Fauna and Flora (CITES) treaty.

Controversy still surrounds the largest herd of American bison (5,000–6,000 animals in the early 1990s) in Yellowstone National Park. The free-roaming bison often leave the park in search of food in the winter and Montana cattle ranchers along the park borders fear that the bison could infect their herds with brucellosis, a contagious disease that can cause miscarriages and infertility in cows. In an effort to prevent any chance of brucellosis transmission, the **National Park Service** (NPS) and the Montana Department of Fish, **Wildlife** and Parks, along with sport hunters acting in cooperation with these agencies, killed 1,044 bison between 1984 and 1992. Montana held a lottery-type hunt, and 569 bison were killed in the winter of 1988–89, and 271 were killed in the winter of 1991–92. The winter of 1996–97 was exceptionally harsh, and some 850 buffalo of the Park's remaining 3,500 starved or froze to death. In addition, the NPS, the **U.S. Department of Agriculture** (USDA), and the Montana Department of Livestock cooperated in a stepped-up buffalo killing program, in which some 1,080 were shot or shipped off to slaughterhouses. In all, more than half of Yellowstone's bison herd perished that winter.

Wildlife protection groups, such as the **Humane Society of the United States** and the **Fund for Animals**, have protested the **hunting** of these bison—which usually consists of walking up to an animal and shooting it. Animal protection organizations have offered alternatives to the killing of the bison, including fencing certain areas to prevent them from coming into contact with cattle. Conversely, Montana state officials and ranchers, as well as the USDA, have long pressured the National Park Service to eradicate many or all of the Yellowstone bison herd or at least test the animals and eliminate those showing signs of brucellosis. Such an action, however, would mean the eradication of most of the Yellowstone herd, even though no bison have not been known to infect a single local cow.

There is also a species of European bison called the wisent (*Bison bonasus*) which was once found throughout much of Europe. It was nearly exterminated in the early 1900s, but today a herd of about 1,600 animals can be found in a forest on the border between Poland and Russia. The European bison is considered vulnerable by **IUCN—The World Conservation Union**. *See also* Endangered Species Act; Endangered species; Overhunting; Rare species; Wildlife management

[*Lewis G. Regenstein*]

American bison (*Bison bison*). (Photograph by Yoav Levy. Phototake. Reproduced by permission.)

RESOURCES
BOOKS

Grainger, D. *Animals in Peril.* Toronto: Pagurian Press, 1978.
McHugh, T. *Time of the Buffalo.* New York: Knopf, 1972.
Park, E. *The World of the Bison.* New York: Lippincott, 1969.

PERIODICALS
Turbak, G. "When the Buffalo Roam." *National Wildlife* 24 (1986): 30-35.

Bituminous coal
see **Coal**

Black lung disease

Black lung disease, also known as anthracosis or **coal** workers' pneumoconiosis, is a chronic, fibrotic lung disease of coal miners. It is caused by inhaling coal dust which accumulates in the lungs, and forms black bumps or coal macules on the bronchioles. These black bumps in the lungs give the disease its common name. Lung disease among coal miners was first described by German mineralogist Georgius Agricola in the sixteenth century and it is now a widely recognized occupational illness.

Black lung disease occurs most often among miners of anthracite (hard) coal, but it is found among soft coal miners and graphite workers as well. The disease is characterized by gradual onset—the first symptoms usually appear only after 10–20 years of exposure to coal dust. The extent and severity of the disease is clearly related to the length of this exposure. The disease also appears to be aggravated by cigarette smoking. The more advanced forms of black lung disease are frequently associated with **emphysema** or chronic **bronchitis**. There is no real treatment for this disease, but it may be controlled or its development arrested by avoiding exposure to coal dust. Black lung disease is probably the best know occupational illness in the United States. In some regions, more than 50% of coal miners develop the disease after 30 or more years on the job. *See also* Fibrosis; Respiratory diseases

[*Linda Rehkopf*]

RESOURCES
BOOKS

Moeller, D. W. *Environmental Health*. Cambridge: Harvard University Press, 1992.

Black-footed ferret

A member of the Mustelidae (weasel) family, the black-footed ferret (*Mustela nigripes*) is the only ferret native to North America. It has pale yellow fur, an off-white throat and belly, a dark face, black feet, and a black tail. The black-footed ferret usually grows to a length of 18 in (46 cm) and weighs 1.5–3 lb (0.68–1.4 kg), though the males are larger than the females. These ferrets have short legs and slender bodies, and lope along by placing both front feet on the ground followed by both back feet.

Ferrets live in prairie dog burrows and feed primarily upon **prairie dogs**, mice, squirrels, and gophers, as well as small rabbits and carrion. Ferrets are nocturnal animals; activity outside the burrow occurs after sunset until about two hours before sunrise. They do not hibernate and remain active all year long.

Breeding takes place once a year, in March or early April, and during the mating season males and females share common burrows. The gestation period lasts approximately six weeks, and the female may have from one to five kits per litter. The adult male does not participate in raising the young. The kits remain in the burrow where they are protected and nursed by their mother until about four weeks of age, usually sometime in July, when she weans them and begins to take them above ground. She either kills a prairie dog and carries it to her kits or moves them into the burrow with the dead animal. During July and early August, she usually relocates her young to new burrows every three or four days, whimpering to encourage them to follow her or dragging them by the nape of their neck. At about eight weeks old the kits begin to play above ground. In late August and early September the mother positions her young in separate burrows, and by mid-September her offspring have left to establish their own territories.

Black-footed ferret (*Mustela nigripes*). (Visuals Unlimited. Reproduced by permission.)

Black-footed ferrets, like other members of the mustelid family, establish their territories by scent marking. They have well developed lateral and anal scent glands. The ferrets mark their territory by either wiggling back and forth while pressing their pelvic scent glands against the ground, or by rubbing their lateral scent glands against shrubs and rocks. Urination is a third form of scent marking. Males establish large territories that may encompass one or more females of the **species** and exclude all other males. Females establish smaller territories.

Historically, the black-footed ferret was found from Alberta, Canada southward throughout the Great Plains states. The decline of this species began in the 1800s with the settling of the west. Homesteaders moving into the Great Plains converted the prairie into agricultural lands, which led to a decline in the population of prairie dogs. Considering them a nuisance species, ranchers and farmers undertook a campaign to eradicate the prairie dog. The black-footed ferret is dependent upon the prairie dog: it takes 100–150 acres (40–61 ha) of prairie-dog colonies to sustain one adult. Because it takes such a large area to sustain a single adult, one small breeding group of ferrets requires at least 10 mi^2 (26 km^2) of **habitat**. As the prairie dog colonies became scattered, the groups were unable to sustain themselves.

In 1954 the **National Park Service** began capturing black-footed ferrets in an attempt to save them from their endangered status. These animals were released in **wildlife** sanctuaries that had large prairie dog populations. Black-footed ferrets, however, are highly susceptible to canine distemper, and this disease wiped out the animals the park service had relocated.

In September 1981, scientists located the only known wild population of black-footed ferrets near the town of

Meeteetse in northwestern Wyoming. The colony lived in 25 prairie dog towns covering 53 mi² (137 km²). But in 1985 canine distemper decimated the prairie dog towns around Meeteetse and spread among the ferret population, quickly reducing their numbers. Researchers feared that without immediate action the black-footed ferret would become extinct. The only course of action appeared to be removing them from the wild. If an animal had not been exposed to canine distemper, it could be vaccinated and saved. Some animals from the Meeteetse population did survive in captivity.

There is a breeding program and research facility called the National Black-footed Ferret Conservation Center in Wyoming, and in 1987 the Wyoming Fish and Game Department implemented a plan for preserving the black-footed ferret within the state. Researchers identified habitats where animals bred in captivity could be relocated. The program began with the 18 animals from the wild population located at Meeteetse. In 1987 seven kits were born to this group. The following year 13 female black-footed ferrets had litters and 34 of the kits survived. In 1998 about 330 kits survived. Captive propagation efforts have improved the outlook for the black-footed ferret, and captive populations will continue to be used to reestablish ferrets in the wild. Almost 2,000 black-footed ferrets that were bred and raised in captivity have been released into the wild.

[*Debra Glidden*]

RESOURCES
PERIODICALS

"Back Home on the Range." *Environment* 33 (November 1991): 23.

Behler, D. "Baby Black-Footed Ferrets Sighted." *Wildlife Conservation* 95 (November–December 1992): 7.

Cohn, J. "Ferrets Return From Near Extinction." *Bioscience* 41 (March 1991): 132–5.

OTHER

"Black-footed Ferret." *U.S. Fish & Wildlife Service.* October 2001 [cited May 2002]. <http://endangered.fws.gov/i/a07.html>.

Blackout/brownout

A blackout is a total loss of electrical power. A blackout is usually defined as a drop in line voltage below 80 volts (V) (the normal voltage is 120V), since most electrical equipment will not operate below these levels. A blackout may be due to a planned interruption, such as limiting of loads during power shortages by rotating power shutoffs through different areas, or due to an accidental failure caused by human error, a failure of generating or transmission equipment, or a storm. Blackouts can cause losses of industrial production, disturbances to commercial activities, traffic and **transportation**

difficulties, disruption of municipal services, and personal inconveniences. In the summer of 1977, a blackout caused by transmission line losses during a storm affected the New York City area. About nine million people were affected by the blackout, with some areas without power for more than 24 hours. The blackout was accompanied by looting and vandalism. A blackout in northeastern United States and eastern Canada due to a switching relay failure in November of 1965 affected 30 million people and resulted in improved electric utility power system design.

A brownout is a condition (usually temporary, but which may last longer, i.e., from periods ranging from fractions of a second to hours) when the alternating current (AC) electrical utility voltage is lower than normal. If the brownout lasts less than a second, it is called a *sag*. Brownouts may be caused by overloaded circuits, but are sometimes caused intentionally by a utility company in order to reduce the amount of power drawn by users during peak demand periods, or unintentionally when demand for electricity exceeds generating capacity. A sag can also occur when line switching is employed to access power from secondary utility sources. Equipment such as shop tools, compressors, and elevators starting up on a shared power line can cause a sag, which can adversely affect other sensitive electronic equipment such as computers. Generally, electrical utility customers do not notice a brownout except when it does affect sensitive electronic equipment.

Measures to protect against effects of blackouts and brownouts include efficient design of power networks, interconnection of power networks to improve **stability**, monitoring of generating reserve needs during periods of peak demand, and standby power for emergency needs. An individual piece of equipment can be protected from blackouts and brownouts by the use of an uninterruptible power source (UPS). A UPS is a device with internal batteries that is used to guarantee that continuous power is supplied to equipment even if the power supply stops providing power or during line sags. Commonly the UPS will boost voltage if the voltage drops to less than 103V and will switch to battery power at 90V and below. Some UPS devices are capable of shutting down the equipment during extended blackouts.

[*Judith L. Sims*]

RESOURCES
BOOKS

Curvin, R., and B. Porter. *Blackout Looting! New York City, July 13, 1977.* New York: Gardner Press, 1979.

Dugan, R.C. *Electrical Power Systems Quality.* New York, McGraw-Hill, 1996.

Kabisama, H.W. *Electrical Power Engineering.* New York: McGraw-Hill, 1993.

Kazibwe, W.E., and M.H. Sendaula. *Electric Power Quality Control Techniques.* New York: Van Nostrand Reinhold, 1993.

BLM

see **Bureau of Land Management**

Blow-out

A blow-out occurs where the **soil** is left unprotected to the erosive force of the wind. Blow-outs commonly occur as depressional areas, once enough soil has been removed. They most often occur in sandy soils, where vegetation is sparse.

Blue Angel

The best known environmental product certification effort outside the United States, the Blue Angel program was initiated by the German government in 1977. The Blue Angel label features a stylized angel with arms outstretched encircled by a laurel wreath. Since its inception, the program has certified more than 4,000 products, including automobiles, batteries, and deodorants, as environmentally safe. Similar, government-sponsored certification programs also exist in Canada (Environmental Choice) and Japan (Ecomark). *See also* Environmental advertising and marketing; Green Cross; Green Seal

Blue revolution (fish farming)

The blue revolution has been brought about in part by a trend towards more healthy eating which has increased the consumption of fish. Additionally, the supply of wild fish is declining, and some **species**, such as cod, striped sea bass, **salmon**, haddock, and flounder, have already been overfished. **Aquaculture**, or fish farming, appears to be a solution to the problems created by the law of supply and demand. Farm-raised fish currently account for about 15% of the market. There are 93 species of fin fish, seven species of shrimp, and six species of crawfish, along with numerous species of clams, oysters and shellfish that are currently being farm raised worldwide.

There are five central components to all fish farming operations: fish, water supply, nutrition, management and a contained method. Ideally, every aspect of the fishes' **environment** is scientifically controlled by the farmer. The quality of the water should be constantly monitored and adjusted for **pH** and numerous other factors, including oxygen content. Adequate water circulation is also necessary to insure

that waste matter does not accumulate in the cages, for this can lead to outbreaks of disease. The fish are fed formulated diets that contain only enough protein for optimal growth. They are fed regulated amounts that vary according to stage of development, water temperature, and the availability of naturally occurring food in their **habitat**.

Herbicides are used on a regular basis to control any unwanted aquatic vegetation and to prevent fouling of cages. Vaccines are routinely given to the fish to prevent disease, although their effectiveness against most pathogens has yet to be determined. Antibiotics are routinely placed in the food that is fed to farm raised fish, a practice which many have questioned. When given over a prolonged period of time, antibiotics can result in higher incidences of disease because bacterial strains develop **resistance** to them.

Fish that are raised on farms mature in rearing units that are frequently located on shore. These on shore units are typically ponds, large circular tanks or concrete enclosures. Many types of freshwater fin fish are raised in pond systems. Ponds that are easy to harvest, drain and refill are the most economical. Walleye, perch and northern pike are a few of the cool-water species that are raised in pond cultures. Warm-water species such as catfish, carp, and tilapia are also common. A few cold water species, especially trout and salmon, can also be raised in pond systems. Most pond systems are **monoculture** in **nature**, so only one type of fish is raised in each pond.

Silos, raceways, and circular pools are commonly used in fish farming. These are popular because they require a small land base in comparison with most other systems. Trout and salmonoid species are frequently raised in raceways, which are rectangular enclosures usually made of cement, fiberglass or metal, and positioned in a series so that water flows from one into the next. Circular pools are shallow with a center drain. They are easy to clean and maintain, since the growth of aquatic vegetation is usually minimal. Silos are very deep, circular tanks that are similar to silos used for grain storage on traditional farms.

All on shore fish farming operations use large quantities of water, and their operations are either open or closed systems. In open systems the water is used only once, flowing into a pool or through a series of pools before being discharged into a **drainage** ditch, creek, or river. Open systems are used whenever possible because they are relatively inexpensive. In most cases, farmers are not required to treat the water before it is discharged, which poses an environmental hazard because the organic fish wastes, the residues of medications, and the herbicides used in the operation enter the water supply unchecked.

Closed systems are not popular among fish farmers because they are very expensive to build, maintain and operate. In a closed system used water is treated and then reused in the farming operation. The treatment process can include disinfection, removal of organic wastes that have dissolved in the water, and reaeration. The closed system is more environmentally sound than the open systems.

Coastal lakes and estuaries are the most frequently used off-shore sites for rearing units, but it is becoming more common to see units located at sea. There are four basic types of cages used for off-shore fish-farming operations. They are fixed, floating, submersible, and submerged. Fixed cages are made of net or webbing material and supported by posts that are anchored in the river bottom. Floating cages, also known as net pens, are the most common type of cage used. Developed in Norway, they are made of net or rigid mesh and are supported by a buoyant collar. Submersible cages have a frame which enables them to hold their shape made of net or rigid mesh and are supported by a buoyant collar. Submerged cages are usually made of wood and anchored in place.

One major concern about off-shore nets and pens is that they tend to attract marine birds and mammals, which attempt to get at the fish with often fatal results. Fish farmers currently use wire barriers and electrical devices to discourage predators and these devices have killed **sea lions**. A typical four-acre salmon farm holds 75,000 fish, and the amount of **organic waste** produced is equal to that of a town with 20,000 people. Waste matter then settles on the ocean floor, where it disrupts the normal **ecosystem**. Accumulated wastes kill clams, oysters and other shellfish, and also causes a proliferation of algae, **fungi**, and **parasites**, as well as **plankton** bloom. Plankton bloom is dangerous to sea life and to humans. In order to control the algae and plankton farmers treat the fish and the water with numerous **chemicals** such as **copper** sulfate and formalin. These chemicals do not act exclusively on algae, killing many other beneficial forms of aquatic life and disrupting the ecological balance. According to the **Sierra Club** Legal Defense Fund, the pens qualify as point sources of **pollution** and should fall under the **Clean Water Act**. At this point in time, however, the pens do not come under the jurisdiction of the act.

Many fish on farms are raised from eggs imported from other areas. Farm-raised fish can and do escape from their pens, causing havoc with local ecosystems. The interbreeding of imported fish with indigenous species can alter the genetic traits that allow the indigenous species to survive in that particular location. Farm-raised stock that escapes and reproduces may also compete with native species, resulting in the decline of wild fish in that particular area. Off the coast of Norway for example, the offspring of escaped farm-raised salmon outnumber the indigenous species.

There are many questions about health and nutrition that may affect consumers of farm-raised fish. Fish farmers frequently use large quantities of medications to keep the fish healthy and there are concerns over the effects that these medications have on human health. Possible side effects of eating farm-raised fish on a regular basis include allergic reactions, increased incidence of infections by resistant bacterial strains, and suppressed immune system response. If eaten by pregnant women there is evidence of fetal damage, discoloration of infants teeth, and abnormal bone growth. Omega-3 fatty acids are a beneficial part of our diet and they are present in the flesh of wild salmon and in some other fish, but not in their farm-raised counterparts. Recent studies have found that farm-raised salmon and catfish contained twice the amount of fat found in wild species. Other comparative nutritional studies are currently underway.

Fish farming is a relatively new industry that shows a lot of potential, but there are many environmental and health questions that need to be addressed. Monitoring of the industry is virtually non-existent. In the United States, the Joint Subcommittee on Aquaculture (JSA) has made recommendations for additional studies of the industry including the environmental impact of fish farming. JSA has pointed out the need for extensive research into the life cycle of parasites and diseases that **plague** fish. They have recommended drug and chemical testing as well as registration procedures. In the 1983 report issued by JSA every aspect of the farming operation was cited as needing additional studies. *See also* Agricultural pollution; Algal bloom; Aquatic chemistry; Aquatic weed control; Commercial fishing; Feedlot runoff; Marine pollution; Water quality standards

[*Debra Glidden*]

RESOURCES
BOOKS

Beveridge, M. *Cage Aquaculture.* Farham, England: Fishing News Books, 1987.

Brown, E. E. *World Fish Farming: Cultivation and Economics.* Westport, CT: Avi Publishing Company, 1983.

PERIODICALS

Fischetti, M. "A Feast of Gene-Splicing Down on the Fish Farm." *Science* 253 (2 August 1991): 512-3.

Blue-baby syndrome

Blue-baby syndrome (or infant cyanosis) occurs in infants who drink water with a high concentration of nitrate or are fed formula prepared with water containing high nitrate levels. Excess nitrate can result in methemoglobinemia, a condition in which the oxygen-carrying capacity of the blood is impaired by an **oxidizing agent** such as nitrite, which can be reduced from nitrate by bacterial **metabolism** in the human mouth and stomach. Infants in the first three to six months of life, especially those with diarrhea, are particularly susceptible to nitrite-induced methemoglobinemia.

Adults convert about 10% of ingested **nitrates** into **nitrites**, and excess nitrate is excreted by the kidneys. In infants, however, nitrate is transformed to nitrite with almost 100% efficiency. The nitrite and remaining nitrate are absorbed into the body through the intestine. Nitrite in the blood reacts with hemoglobin to form methemoglobin, which does not transport oxygen to the tissues and body organs. The skin of the infant appears blue due to the lack of oxygen in the blood supply, which may lead to asphyxia, or suffocation.

Normal methemoglobin levels in humans range from 1 to 2%; levels greater than 3% are defined as methemoglobinemia. Methemoglobinemia is rarely fatal, readily diagnosed, and rapidly reversible with clinical treatment.

In adults, the major source of nitrate is dietary, with only about 13% of daily intake from drinking water. Nitrates occur naturally in many foods, especially vegetables, and are often added to meat products as preservatives. Only a few cases of methemoglobinemia have been associated with foods high in nitrate or nitrite. Nitrate is also found in air, but the daily respiratory intake of nitrate is small compared with other sources. Nearly all cases of the disease have resulted from ingestion by infants of nitrate in private well water that has been used to prepare infant formula. Levels of nitrate of three times the Maximum Contaminant Levels (MCLs) and above have been found in drinking water **wells** in agricultural areas. Federal MCL standards apply to all public water systems, though they are unenforceable recommendations. Insufficient data are available to determine whether subtle or chronic toxic effects may occur at levels of exposure below those that produce clinically obvious toxicity. If water has or is suspected to have high nitrate concentrations, it should not be used for infant feeding, nor should pregnant women or nursing mothers be allowed to drink it.

Domestic water supply wells may become contaminated with nitrate from mineralization of **soil** organic **nitrogen**, **septic tank** systems, and some agricultural practices, including the use of fertilizers and the disposal of animal wastes. Since there are many potential sources of nitrates in **groundwater**, the prevention of nitrate contamination is complex and often difficult.

Nitrates and nitrites can be removed from drinking water using several types of technologies. The **Environmental Protection Agency** (EPA) has designated reverse osmo-

sis, anion exchange, and electrodialysis as the **Best Available Control Technology** (BAT) for the removal of nitrate, while recommending reverse osmosis and anion exchange as the BAT for nitrite. Other technologies can be used to meet MCLs for nitrate and nitrite if they receive approval from the appropriate state regulatory agency.

[*Judith L. Sims*]

RESOURCES
BOOKS

Clark, R. M. "Water Supply." In *Standard Handbook of Environmental Engineering*, ed. R. A. Corbitt. New York: McGraw-Hill, 1990.
Lappensbusch, W. L. *Contaminated Drinking Water and Your Health*. Alexandria, VA: Lappensbusch Environmental Health, 1986.

PERIODICALS
Pontius, F. W. "New Standards Protect Infants from Blue Baby Syndrome." *Opflow* 19 (1993): 5.

BMP

see **Best management practices**

BOD

see **Biochemical oxygen demand**

Bogs

see **Wetlands**

Bonn Convention

see **Convention on the Conservation of Migratory Species of Wild Animals (1979)**

Murray Bookchin (1921 –)

American social critic, environmentalist, and writer

Born in New York in 1921, Bookchin is a writer, social critic, and founder of "social ecology." He has had a long and abiding interest in the **environment**, and as early as the 1950s he was concerned with the effects of human actions on the environment. In 1951 he published an article entitled "The Problem of Chemicals," which exposed the detrimental effects of **chemicals** on **nature** and on human health. This work predates **Rachel Carson's** famous *Silent Spring* by over 10 years.

Murray Bookchin. (Photograph by Debbie Bookchin. Reproduced by permission.)

In developing his theory of **social ecology**, Bookchin makes the basic point that "you can't have sound ecological practices in nature without having sound social practices in society. Harmony in society means harmony with nature." Bookchin describes himself as an anarchist, contending that "there is a natural relationship between natural **ecology** and anarchy."

Bookchin has long been a critic of modern cities: his widely read and frequently quoted *Crisis in Our Cities* (1965) examines urban life, questioning "the lack of standards in judging the modern metropolis and the society that fosters its growth." His *The Rise of Urbanization and the Decline of Citizenship* (1987) continues the critique by advancing green ideas as a new municipal agenda for the 1990s and the next century. Though he is often mentioned as one of its founding thinkers, Bookchin is an ardent critic of **deep ecology**. He states: "Bluntly speaking, deep ecology, despite all its social rhetoric, has no real sense that our ecological problems have their roots in society and in social problems." Bookchin instead reaffirms a social ecology that is, first of all "social," incorporating people into the calculus needed to solve environmental problems, "avowedly rational," "revolutionary, not merely 'radical,'" and "radically" green.

[*Gerald L. Young*]

RESOURCES
BOOKS

Bookchin, Murray. *Toward an Ecological Society*. Montreal: Black Rose Books, 1980.
———. *The Ecology of Freedom: The Emergence and Dissolution of Hierarchy*. Palo Alto, CA: Cheshire Books, 1982.
———. *Remaking Society: Pathways to a Green Future*. Boston: South End Press, 1990.
Clark, J. "The Social Ecology of Murray Bookchin." In *The Anarchist Moment*. Montreal: Black Rose Books, 1984.

PERIODICALS

Bookchin, Murray. "Social Ecology Versus Deep Ecology." *Socialist Review* 18 (1988): 9–29.

Boreal forest
see **Taiga**

Norman E. Borlaug (1914 –)
American environmental activist

Borlaug, known as the father of the "green revolution" or **agricultural revolution**, was born on March 25, 1914, on a small farm near Cresco, Iowa. He received a B.S. in forestry in 1937 followed by a master's degree in 1940 and a Ph.D. in 1941 in **plant pathology** all from the University of Minnesota.

Agriculture is an activity of humans with profound impact on the **environment**. Traditional cereal grain production methods in some countries have led to recurrent **famine**. Food in these countries can be increased either by expansion of land area under cultivation or by enhancement of crop yield per unit of land. In many developing countries, little if any space for agricultural expansion remains, hence interest is focused on increasing cereal grain yield. This is especially true with regard to wheat, rice, and maize.

Borlaug is associated with the green revolution which was responsible for spectacular increases in grain production. He began working in Mexico in 1943 with the Rockefeller Foundation and the International Maize and Wheat Improvement Center in an effort to increase food crop production. As a result of these efforts, wheat production doubled in the decade after World War II, and nations such as Mexico, India, and Pakistan became exporters of grain rather than importers. Increased yields of wheat came as the result of high genetic yield potential, enhanced disease **resistance** (Mexican wheat was vulnerable to stem rust fungus), responsiveness to fertilizers, the use of pesticides, and the development of dwarf varieties with stiff straw and short blades that resist lodging, i.e., do not grow tall and topple over with the use of **fertilizer**. Further, the new varieties could be used in different parts of the world because they were unaffected

Norman Borlaug. (Corbis-Bettmann. Reproduced by permission.)

by different daylight periods. Mechanized threshing is now replacing the traditional treading out of grain with large animals followed by winnowing because these procedures are slow and leave the grain harvest vulnerable to rain damage. Thus, modern threshers are an essential part of the green revolution.

Borlaug demonstrated that the famine so characteristic of many developing countries could be controlled or eliminated at least with respect to the population of the world at that time. However, respite from famine and poverty is only temporary as the world population relentlessly continues to increase. The sociological and economic conditions which have historically precipitated famine have not been abrogated. Thus, famine will appear again if human population expansion continues unabated despite efforts of the green revolution. Further, critics of Borlaug's agricultural revolution cite the possibility of crop vulnerability because of genetic uniformity—development of crops with high-yield potential have eliminated other varieties, thus limiting **biodiversity**. High-yield crops have not proven themselves hardier in several cases; some varieties are more vulnerable to molds and storage problems. In the meantime, other, hardier varieties are now nonexistent. The environmental effects of fertilizers, pesticides, and energy-dependent mechanized modes of cultivation to sustain the newly developed

crops have also provoked controversy. Such methods are expensive for poorer countries and sometimes create more problems than they solve. For now, however, food from the green revolution saves lives and the benefits currently outweigh liabilities.

At the presentation of the Nobel Peace Prize to Borlaug in 1970, the president of Norway's Lating honored him, saying "more than any other single person of this age, he has helped to provide bread for a hungry world." His alma mater, the University of Minnesota, celebrated his career accomplishments with the award of a Doctor of Science (*honoris causa*) degree in 1984, as have many other academic institutions throughout a grateful world.

In 1992, the Agricultural Council of Experts (ACE) was formed to advise Borlaug and former President Jimmy Carter on information concerning Africa's agricultural and economic issues. They created a model that has since been incorporated into the Global 2000 program, which addresses the African food crisis by experimenting with production test plots (PTPs) in an attempt to extract greater crop yields.

[*Robert G. McKinnell*]

RESOURCES
BOOKS

Wilkes, H. Garrison. "The Green Revolution." In *McGraw-Hill Encyclopedia of Food, Agriculture & Nutrition*, edited by D. N. Lapedes. New York: McGraw-Hill Book Company, 1977.

OTHER

Transcript of Proceedings for the Nobel Prize for Peace, 1970 (Speech by Ms. Aase Lionaes, President of the Lagting, December 11, 1970 and Lecture by Norman E. Borlaug on the Occasion of the Award of the Nobel Peace Prize for 1970, Oslo, Norway, December 11, 1970).

Boston Harbor clean up

Like many harbors near cities along the eastern seaboard, the Boston Harbor in Massachusetts has been used for centuries as a receptacle for raw and partially treated sewage from the city of Boston and surrounding towns. In the late 1800s, Boston designed a sewage and stormwater collection system. This sewage system combined millions of gallons of untreated sewage from homes, schools, hospitals, factories, and other buildings with stormwater collected from streets during periods of moderate to heavy rainfall. The combined sewage and stormwater collected in the sewage system was discharged untreated to Boston Harbor on outgoing tides. Sewage from communities surrounding Boston was also piped to Boston's collection system and discharged to Boston Harbor and the major rivers leading to it: the Charles, Mystic, and Neponset Rivers. Many of the sewage pipes, tunnels, and other infrastructure built in the late 1800s and early 1900s are still in use today.

In the 1950s, the City of Boston was concerned with growing health risks of swimming in and eating shellfish harvested from Boston Harbor as well as odors and aesthetic concerns that resulted from discharging raw sewage into Harbor waters. The city built two advanced primary **sewage treatment** plants during the 1950s and 1960s on two islands located in Boston Harbor. The first sewage treatment plant was built on Nut Island; it treated approximately 110 million gal/per day (416 million l/per day). The second sewage treatment plant was built on Deer Island; it treated approximately 280 million gal/per day (1060 million l/per day). The two sewage treatment plants removed approximately half the total suspended solids and 25% of the biological oxygen demand found in raw sewage. The outgoing tide in Boston Harbor was still used to flush the treated **wastewater** and approximately 50 tons (46 metric tons) per day of sewage **sludge** (also called biosolids), which was produced as a byproduct of the advanced primary treatment process. The sewage sludge forms from solids in the sewage settling in the bottom of tanks.

During the 1960s, the resources devoted to maintaining the City's aging sewage system decreased. As a result, the sewage treatment plants, pipes, pump stations, tunnels, interceptors, and other key components of Boston's sewage infrastructure began to fall into disrepair. Equipment breakdowns, sewer line breaks, and other problems resulted in the **discharge** of raw and partially treated sewage to Boston Harbor and the rivers leading to it. During this time, the Metropolitan District Commission (MDC) was the agency responsible for sewage collection, treatment, and disposal.

In 1972, the United States Congress enacted the **Clean Water Act**. This was landmark legislation to improve the quality of our nation's waters. The Clean Water Act required that sewage discharged to United States waters must meet secondary treatment levels by 1977. Secondary treatment of sewage means that at least 85% of total suspended solids and 85% of biological oxygen demand is removed from sewage. Instead of working towards meeting this federal requirement, the MDC requested a waiver from this new obligation from the United States **Environmental Protection Agency** (EPA). In 1983, the EPA denied the waiver request. The MDC responded by modifying its waiver request to promise that the city would construct a 9.2 mi (14.8 km) outfall to achieve increased dilution of the sewage by discharging to deeper, more flushed waters. As part of its waiver, MDC also promised an end to discharging sewage sludge in the harbor and initiation of a combined sewer overflow abatement project to cease flow into Boston Harbor from 88 overflow pipes. In 1985, EPA denied the second waiver request.

During EPA's consideration of Boston's waiver request, in 1982, the City of Quincy filed a lawsuit against the MDC for violating the Clean Water Act. In 1983, the Conservation Law Foundation filed two lawsuits; one was against MDC for violating the Clean Water Act and the other was against EPA for not fully implementing the Clean Water Act by failing to get Boston to comply with the law. The Massachusetts legislature responded to these pressures by replacing the MDC with the Massachusetts **Water Resources** Authority (MWRA) in 1984. The MWRA was created as an independent agency with the authority to raise water and sewer rates to pay for upgrading and maintaining the collection and treatment of the region's sewage. The following year, the federal court ruled that the MWRA must come into compliance with the Clean Water Act. As a result of this ruling, the MWRA developed a list of sewage improvement projects that were necessary to upgrade the existing sewage treatment system and clean up Boston Harbor.

The Boston Harbor clean up consists of $3.4 billion worth of sewage treatment improvements that include the construction of a 1,270 million-gal/per day (4,800 million-l/per day) primary sewage treatment plant, a 1,080 million-gal/per day (4,080 million-l/per day) secondary sewage treatment plant, a dozen sewage sludge digesters, disinfection basins, a sewage screening facility, an underwater tunnel, a 9.5 mi-long (15.3 km) outfall pipe, 10 pumping stations, a sludge-to-fertilizer facility, and combined sewer overflow treatment facilities.

Today, approximately 370 million gallons/per day (1,400 million l/per day) of sewage **effluent** from over 2.5 million residents and businesses is discharged to Boston Harbor. Almost half the total flow is stormwater **runoff** from streets and **groundwater** infiltrating into cracked sewer pipes. The combined sewage and stormwater is moved through 5,400 mi (8,700 km) of pipes by gravity and with the help of pumps. Five of the 10 pumps have already been replaced. At least two of the pumping stations that were replaced as part of the clean up effort dated back to 1895. The sewage is pumped to Nut Island where more than 10,000 gal/per day (37,800 l/per day) of floatable **pollution** such as grease, oil, and plastic debris are now removed by its new sewage screening facility. The facility also removes any grit, sand, gravel, or large objects. A 4.8-mi (7.7-km) long deep-rock tunnel will be used to transport screened wastewater to Deer Island for further treatment.

One of the most significant changes that has occurred as part of the Boston Harbor clean up project is the reconstruction of Deer Island. Prior to 1989, Deer Island had prison buildings, World War II army bunkers, and an aging sewage treatment plant. All the old buildings and structures have been removed and the island has been reshaped to accommodate 12 massive sewage sludge digesters, 60 acres (24.2 ha) of a new primary sewage treatment plant, and a new secondary treatment facility. The primary sewage treatment plant has reduced the amount of suspended solids discharged to Boston Harbor from 138–57 tons per day (126–52 metric tons).

The secondary sewage treatment plant on Deer Island is still under construction. It will use settling tanks, as are found in primary sewage treatment plants, as well as **microorganisms**, which will consume organic matter in the sewage thereby increasing treatment levels. Secondary treatment of Boston's sewage will result in an increase in removal of total suspended solids from 50–90% and biological oxygen demand from 25–90%. The first phase of the secondary treatment plant construction was completed by the end of 1997.

After the sewage is treated and disinfected to remove any remaining pathogens (disease-causing organisms), the effluent is discharged through a 9.5 mi (15.3 km) long outfall tunnel into the waters of Massachusetts Bay. Massive tunnel boring machines were used to drill the tunnel below the ocean floor. The outfall has 55 diffuser pipes connected at right angles to it along the last 1.25 mi (2.01 km) of the outfall. The diffuser pipes will increase dispersion of the treated sewage in the receiving waters. The tunnel was opened in September of 2000.

The outfall has been a source of controversy for many residents of Cape Cod and for the users of Massachusetts Bay and the Gulf of Maine. There is concern about the long-term impact of contaminants from the treated sewage on the area, which is used for **transportation, recreation,** fishing, and tourism. Some alternatives to the sewage treatment plant and the 9.5-mi (15.3-km) long outfall pipe were developed by civil engineers at the Massachusetts Institute of Technology. The alternatives included modifications to the advanced primary treatment facility and a smaller secondary treatment facility with the effluent discharged to Boston Harbor. Because there was not enough evidence to convince EPA that **water quality standards** would always be met in the harbor, EPA rejected the alternatives.

As part of the Boston Harbor clean up, sewage sludge is no longer discharged to the harbor. Twelve sewage sludge digesters in Deer Island were constructed in 1991. The digesters break down the sewage sludge by using microorganisms such as bacteria. Different types of microorganisms are used in the sewage sludge digestion process than in the secondary treatment process. As the microorganisms consume the sewage sludge, **methane** gas is produced which is used for heat and power. Prior to 1991, all the sewage sludge was discharged to Boston Harbor. Since 1991, the sewage sludge has been shipped to a facility that converts the digested sewage sludge into **fertilizer.**

The sludge-to-fertilizer facility dewaters the sludge and uses rotating, high temperature dryers that produce fertilizer pellets with 60% organic matter, and important nutrients such as **nitrogen, phosphorus,** calcium, sulfur, and iron. The fertilizer is marketed in bulk and also sold as *Bay State Organic,* which is sold locally for use on **golf courses** and landscape.

This massive undertaking to clean up the harbor by upgrading its sewage treatment facilities is one of the world's largest public works projects. Over 80% of the Boston Harbor clean up is completed. These changes have resulted in measurable improvements to Boston Harbor. The harbor sustains a $10 million lobster fishery annually as well as flounder, striped bass, cod, bluefish, and smelt recreational fisheries.

[*Marci L. Bortman Ph.D.*]

RESOURCES
BOOKS

National Research Council. *Managing Wastewater in Coastal Urban Areas.* Washington DC: National Academy Press, 1993.

PERIODICALS

Aubrey, D. G., and M. S. Connor. "Boston Harbor Fallout Over the Outfall." *Oceanus* 36:1, (Spring 1993).
Levy, P. F. "Sewer Infrastructure: An Orphan of Our Times." *Oceanus* 36:1, (Spring 1993).

Botanical garden

A botanical garden is a place where collections of plants are grown, managed, and maintained. Plants are normally labeled and available for scientific study by students and observation by the public. An arboretum is a garden composed primarily of trees, vines and shrubs. Gardens often preserve collections of stored seeds in special facilities referred to as seed banks. Many gardens maintain special collections of preserved plants, known as herbaria, used to identify and classify unknown plants. Laboratories for the scientific study of plants and classrooms are also common.

Although landscape gardens have been known for as long as 4,000 years, gardens intended for scientific study have a more recent origin. Kindled by the need for herbal medicines in the sixteenth century, gardens affiliated with Italian medical schools were founded in Pisa about 1543, and Padua in 1545. The usefulness of these medicinal gardens was soon evident, and similar gardens were established in Copenhagen, Denmark (1600), London, England (1606), Paris, France (1635), Berlin, Germany (1679), and elsewhere. The early European gardens concentrated mainly on **species** with known medical significance. The plant collections were put to use to make and test medicines and to train students in their application.

In the eighteenth and nineteenth centuries, gardens evolved from traditional herbal collections to facilities with broader interests. Some gardens, notably the Royal Botanic Gardens at Kew, near London, played a major role in spreading the cultivation of commercially important plants such as coffee (*Coffea arabica*), **rubber** (*Hevea* spp.), banana (*Musa paradisiaca*), and tea (*Thea sinensis*) from their places of origin to other areas with an appropriate **climate**. Other gardens focused on new varieties of horticultural plants. The Leiden garden in Holland, for instance, was instrumental in stimulating the development of the extensive worldwide Dutch bulb commerce. Many other gardens have had an important place in the scientific study of plant diversity as well as the introduction and assessment of plants for agriculture, horticulture, forestry, and medicine.

The total number of botanical gardens in the world can only be estimated, but not all plant collections qualify for the designation because they are deemed to lack serious scientific purpose. A recent estimate places the number of botanical gardens and arboreta at 1,400. About 300 of those are in the United States. Most existing gardens are located in the North Temperate Zone, but there are important gardens on all continents except **Antarctica**. Although the tropics are home to the vast majority of all plant species, until recently, relatively few gardens were located there. A recognition of the need for further study of the diverse tropical **flora** has led to the establishment of many new gardens. An estimated 230 gardens are now established in the tropics.

In recent years botanical gardens throughout the world have united to address increasing threats to the planet's flora. The problem is particularly acute in the tropics, where as many as 60,000 species, nearly one-fourth of the world's total, risk **extinction** by the year 2050. Botanical gardens have organized to produce, adopt and implement a Botanic Gardens Conservation Strategy to help deal with the dilemma. *See also* Conservation; Critical habitat; Ecosystem; Endangered species; Forest decline; Organic gardening and farming

[*Douglas C. Pratt*]

RESOURCES
BOOKS

Bramwell, D., O. Hamann, V. Heywood, H. Synge. *Botanic Gardens and the World Conservation Strategy*. London: Academic Press, 1987.

Hyams, E. S., and W. MacQuitty. *Great Botanical Gardens of the World*. New York: Macmillan, 1969.

Wyman, D. *The Arboretums and Botanical Gardens of North America*. Jamaica Plain, MA: Arnold Arboretum of Harvard University, 1947.

Kenneth Ewart Boulding (1910 – 1993)

English economist, social scientist, writer, and peace activist

Kenneth Boulding is a highly respected economist, educator, author, and pacifist. In an essay in *Frontiers in Social Thought: Essays in Honor of Kenneth E. Boulding* (1976), Cynthia Earl Kerman described Boulding as "a person who grew up in the poverty-stricken 'inner city' of Liverpool, broke through the class system to achieve an excellent education, had both scientific and literary leanings, became a well-known American economist, then snapped the bonds of economics to extend his thinking into wide-ranging fields—a person who is a religious mystic and a poet as well as a social scientist."

A major recurring theme in Boulding's work is the need—and the quest—for an integrated social science, even a unified science. He does not see the disciplines of human knowledge as distinct entities, but rather a unified whole characterized by "a diversity of methodologies of learning and testing." For example, Boulding is a firm advocate of adopting an ecological approach to economics, asserting that **ecology** and economics are not independent fields of study. He has identified five basic similarities between the two disciplines: 1) both are concerned not only with individuals, but individuals as members of **species**; 2) both have an important concept of dynamic equilibrium; 3) a system of exchange among various individuals and species is essential in both ecological and economic systems; 4) both involve some sort of development—succession in ecology and **population growth** and capital accumulation in economics; 5) both involve distortion of the equilibrium of systems by humans in their own favor.

"If my life philosophy can be summed up in a sentence," Boulding stated, "it is that I believe that there is such a thing as human betterment—a magnificent, multidimensional, complex structure—a cathedral of the mind—and I think human decisions should be judged by the extent to which they promote it. This involves seeing the world as a total system." Boulding's views have been influential in many fields, and he has helped environmentalists reassess and redefine their role in the larger context of science and economics.

[*Gerald L. Young Ph.D.*]

RESOURCES
BOOKS

Boulding, K. E. "Economics As an Ecological Science." In *Economics As a Science*. New York: McGraw-Hill, 1970.

———. *Collected Papers*. Boulder: Colorado Associated University Press, 1971.

Kerman, C. E. *Creative Tension: The Life and Thought of Kenneth Boulding*. Ann Arbor: University of Michigan Press, 1974.

Kenneth Boulding. (Photograph by Ken Abbott. University of Colorado at Boulder. Reproduced by permission.)

Pfaff, M., ed. *Frontiers in Social Thought: Essays in Honor of K. E. Boulding.* Amsterdam and New York: North-Holland, 1976.

Silk, L. "K. E. Boulding: The Economics of Peace and Love." In *The Economists.* New York: Basic Books, 1976.

Wright, R. "Kenneth Boulding." In *Three Scientists and Their Gods: Looking for Meaning in an Age of Information.* New York: Times Books, 1988.

Boundary Waters Canoe Area

The Boundary Waters Canoe Area (BWCA), a federally designated **wilderness** area in northern Minnesota, includes approximately one million acres (410,000 ha) stretching some 200 mi (322 km) along the United States-Canadian border. The BWCA contains more than 1,200 mi (1,932 km) of canoe routes and portages. The second largest expanse in the National Wilderness Preservation system, the BWCA is administered by the United States **Forest Service**. Constituting about one-third of the Superior **National Forest** (established in 1909), the BWCA was set apart as wilderness by an act of Congress in 1958. The 1964 **Wilderness Act** allowed limited **logging** in some parts of the BWCA and the use of motorboats on 60% of the water area. Under pressure from environmental groups—and over objections by developers, logging interests, and many local residents—the Congress finally passed the BWCA Wilderness Act of

1978, which outlawed all logging and limited motorboats to 33% of the water surface area (due to drop to 24% by 1999), and added 45,000 acres (18,450 ha), bringing the total area to 1,075,000 acres (440,750 ha).

Many area residents and resort owners continue to resent and resist efforts to reduce the areas open to motorized watercraft and snowmobile traffic. They have pressed unsuccessfully for federal legislation to that effect. At the urging of Senator Paul Wellstone (D-MN), a mediation panel was convened in 1996 to consider the future of the BWCA. Environmentalists, resort owners, local residents, and representatives of other groups met for several months to try to reconcile competing interests in the area. Unable to reach agreement and arrive at a compromise, the panel disbanded in 1997. The fierce and continuing political quarrels over the future of the BWCA contrast markedly with the silence and serenity of this land of sky-blue waters and green forests.

[*Terence Ball*]

FURTHER READING

Beymer R. *The Boundary Waters Canoe Area*, vol. 1: "The Western Region" and vol. 2: "The Eastern Region." Berkeley: Wilderness Press, 1978.

Box turtle
see **American box turtle**

BPT
see **Best Practical Technology**

Brackish

The **salinity** of brackish water is intermediate between seawater and fresh waters. Brackish water contains too much salt to be drinkable, but not enough salt to be considered seawater. The ocean has an average salinity of 35 parts per thousand (ppt), whereas freshwater contains 0.065–0.30 ppt of salts, primarily chloride, sodium, sulfate, magnesium, calcium, and potassium ions. The salt content of brackish water ranges between approximately 0.50 and 17 ppt. Brackish water occurs where freshwater flows into the ocean, or where salts are dissolved from subsoils and percolate into freshwater basins. The gradient between salt and fresh water in estuaries and deltas varies from sharp distinction to gradual mixing, and different levels of vertical and horizontal mixing depend on the influence of tide, current, and rate of freshwater inflow.

BRDs
see **Bycatch Reduction Devices**

Broad spectrum pesticide
see **Pesticide**

Bromine

Bromine is an element that belongs to Group 17 on the periodic table of elements, the group that contains substances called halogens. Other halogens include fluorine, **chlorine**, and iodine. Halogens are important elements that are used in heavily in industry. Bromine is used in the manufacture of dyes, fumigants, fire-retardant materials, medicines, pesticides, and photographic emulsions. Bromine is also used for water purification. While bromine and bromine-containing products are very useful, some concern surrounds the use of some bromine compounds because of their impact on the **environment**, particularly the **ozone** layer.

The many uses of bromine are due to its chemical properties. The halogens are the most reactive non-metal elements in the periodic table. Halogens characteristically react very rapidly and readily with almost all metals to form salts. The high reactivity of the halogens is due to the presence of seven (rather than the stable noble-gas configuration of eight) electrons in their outer shell. Thus, they have an unpaired electron that is readily available for chemical bonding and reactions with other elements. Unlike chlorine and iodine, bromine is a liquid at room temperature. It is the only liquid nonmetallic element.

Bromine is a fairly heavy inorganic element. First isolated in 1826, the atomic number of bromine is 35, which means it has 35 electrons and 35 protons. Bromine also contains 45 neutrons in its nucleus, making its average atomic mass 79.9 atomic mass units. At room temperature, it is a reddish-brown liquid that emits pungent, noxious fumes that irritate the eyes. Appropriately, the word *bromine* is derived from the Greek word *bromos*, which means "stench." Bromine readily dissolves in water and **carbon** disulfide. It has a boiling point of 138.6°F(59.2°C) and is more reactive than iodine, but less reactive than chlorine. When bromine chemically reacts with other elements or compounds, it has a bleaching effect. It is a fairly strong **oxidizing agent**. Relative to the other halogens, bromine is a stronger oxidizing agent than iodine, but a weaker oxidizer than both chlorine and fluorine. Even so, liquid bro-

mine is very caustic and can seriously damage skin on contact. Bromine, like the other halogens, has a very high affinity for itself, and therefore forms diatomic molecules, which are molecules that contain only two atoms. Molecular bromine exists as two bromine atoms bonded together.

Bromine is obtained from natural salt deposits. Two areas within the United States that have historically been associated with bromine extraction are Arkansas and Michigan. Bromine may also be extracted from seawater. Aqueous bromine may also be produced from the oxidation of bromides during **chlorination** of water, especially when seawater is used as a coolant. Since it contains a very slight concentration of bromine salts (about 85 **parts per million**), seawater is not a major source of industrial bromine. The most common form of mineralized bromine is silver bromide (bromargyrite), found chiefly in Mexico and Chile. Most of the supply of bromine used for industrial purposes, however, comes from the United States and Israel, with smaller amounts produced in France, Japan, and Russia.

Bromine-containing substances have been used for centuries. The first use of a material containing bromine was in ancient Rome. During the Roman Empire, a highly prized purple dye was painstakingly extracted from marine mussels. The dye, a bromine compound, was very expensive because it was so difficult to obtain and only the very wealthy could afford clothing dyed with the hue of this dye, which resulted in the term "royal purple." Nowadays, bromine is used not only the production of dyes, but also in the production of **chemicals** that improve safety, agriculture, and **sanitation**.

Bromine is a very effective agent in controlling the growth of aquatic **microorganisms**. As such, and like chlorine, it is used for **water treatment** because it can kill microorganisms and keep water clear and free from foul odors. For example, bromine salts, like sodium bromide, are used to control the growth of algae and bacteria in recreational spaces like hot tubs and swimming pools. Another common bromine aquatic biocide is bromochlorodimethylhydantoin. Occasionally, bromine is used by municipalities to control disease-causing **microbes** in drinking water because some of these microorganisms are more susceptible to bromine than chlorine.

Bromine compounds are also used as pesticides. Like other halogenated **hydrocarbons** such as DDT, some brominated hydrocarbons are powerful insecticides. A very effective and important bromine-containing hydrocarbon **pesticide** is methyl bromide. This agent, also known as bromomethane, is used to fumigate stored grain and produce to free them from **pest** infestations. Methyl bromide is also used to fumigate **soil** for valuable crops like strawberries, peppers, eggplants, tomatoes, and **tobacco**. Since the 1940s, methyl bromide has been used as a soil pretreatment to kill insect, roundworm, and weed **species** that decrease the productivity of economically important crops. In 1992, approximately 73,000 tons of methyl bromide were used. However it was discovered that methyl bromide, like **chlorofluorocarbons**, also contributes to the depletion of the ozone layer. As a result, the **Environmental Protection Agency** (EPA) made recommendations that its use be phased-out based upon the expert assessments of atmospheric scientists from the World Meteorological Organization and the **National Oceanic and Atmospheric Administration**. Under the **Clean Air Act**, the EPA is enforcing reductions in the use of methyl bromide. A mandatory reduction of 25% was achieved in 1999. By 2001, 50% reduction had been enforced. By the year 2003, a 70% reduction in the use of methyl bromide must be attained with a goal for a complete ban of the compound scheduled for the year 2005.

Despite the ban on methyl bromide, bromine will continue to be a valued chemical. The pharmaceutical industry relies heavily on bromine both in the manufacturing process and as constituent substance in pharmaceutical agents. The general anesthetic, halothane, contains bromine. The production of naproxen sodium, an over-the-counter non-steroidal analgesic, uses bromine in intermediate reactions. Brompheniramine, the widely available antihistamine cold and allergy medication, contains bromine. The synthetic addition of bromine, or other halogens, to medications such as these facilitates their uptake into tissues. Other drugs use bromine to create easily absorbed dosage forms. For example, dextromethorphan hydrobromide, a cough medication, is a soluble bromine salt form of the poorly soluble active drug dextromethorphan, and dissolves easily as a clear solution.

Bromine is also used to create fire-resistant **plastics**. Plastic is made of very flammable hydrocarbon polymers. The addition of bromine compounds creates fire-retardant plastic products. Brominated flame-retardants are used in televisions, stereos, computers, and electrical wiring to reduce fire hazard when these common electronic appliances generate excessive heat. Bromine-containing fire-retardant chemicals are also used in carpeting, draperies, and furniture foam padding. While bromine compounds make products more fire-resistant, they do not make them fire-proof. Rather, they reduce the likelihood that a plastic item will ignite and delay the spread of fire. As bromine-treated plastic products burn, they release brominated hydrocarbons that threaten the ozone layer not unlike chorofluorocarbons (CFCs). For this reason, research is now directed at finding alternatives to bromine flame-retardant chemicals. For example, promising new fire-resistant compounds use silicon.

Aside from its use in dyes, pesticides, water treatment, pharmaceuticals, and fire retardants, bromine compounds are also used in photographic film and print paper emulsions, hydraulic fluids, refrigeration fluids, inks, and hair products.

As useful as bromine is, however, concern for the ozone layer has resulted in heightened vigilance concerning the overuse of bromine-containing chemicals.

[*Terry Watkins*]

RESOURCES
BOOKS

Nilsson, Annika. *Ultraviolet Reflections: Life Under a Thinning Ozone Layer.* New York: John Wiley & Son Ltd, 1996.
Price, D. et al. *Bromine Compounds: Chemistry and Applications.* New York: Elsevier Science Ltd., 1988.

PERIODICALS

Alaee, M. and Wenning, R. J. "The significance of brominated flame retardants in the environment: current understanding, issues and challenges." *Chemosphere.* 46, no. 5 (February, 2000): 579-582.
Anbar, A.D. et al. "Methyl bromide: ocean sources, ocean sinks, and climate sensitivity." *Global Biogeochemistry Cycles* 10, no. 1 (March, 1996): 175-190.
Rhew, R. C. et al. "Natural methyl bromide and methyl chloride emissions from coastal salt marshes." *Nature* 403, no. 6767 (January, 2000): 292-295.

OTHER

Winter, Mark. *Bromine.* WebElements Periodic Table. <http://www.webelements.com/webelements/scholar/index.html>

Bronchial constriction
see **Asthma**

Bronchitis

Chronic bronchitis is a persistent inflammation of the bronchial tubes, airways leading to the lungs. The disease is characterized by a daily cough that produces sputum for at least three months each year for two consecutive years, when no other disease can account for these symptoms. The diagnosis of chronic bronchitis is made by this history, rather than by any abnormalities found on a chest x-ray or through a pulmonary function test.

When a person breathes in, air, **smoke**, germs, allergens, and pollutants pass from the nose and mouth into a large central duct called the trachea. The trachea branches into smaller ducts, the bronchi and bronchioles, which lead to the alveoli. These are the tiny, balloonlike, air sacs, composed of capillaries, supported by connecting tissue, and enclosed in a thin membrane. Bronchitis can permanently damage the alveoli.

Bronchitis is usually caused by **cigarette smoke** or exposure to other irritants or air pollutants. The lungs respond to the irritation in one of two ways. They may become permanently inflamed with fluid, which swells the tissue that lines the airways, narrowing them and making them resist airflow. Or, the mucus cells of the bronchial tree may produce excessive mucus.

The first sign of excessive mucus production is usually a morning cough. As smoking or exposure to air pollutants continues, the irritation increases and is complicated by infection, as excess mucus provides food for bacteria growth. The mucus changes from clear to yellow, and the infection becomes deep enough to cause actual destruction of the bronchial wall. Scar tissue replaces the fine cells, or cilia, lining the bronchial tree, and some bronchioles are completely destroyed. Paralysis of the cilia permits mucus to accumulate in smaller airways, and air can no longer rush out of these airways fast enough to create a powerful cough.

With each pulmonary infection, excess mucus creeps into the alveoli, and on its way, totally blocking portions of the bronchial tree. Little or no gas exchange occurs in the alveoli, and the ventilation-blood flow imbalance significantly reduces oxygen levels in the blood and raises **carbon dioxide** levels. Chronic bronchitis eventually results in airway or air sac damage; the air sacs become permanently hyperinflated because mucus obstructing the bronchioles prevents the air sacs from fully emptying.

Chronic bronchitis usually goes hand-in-hand with the development of **emphysema**, another chronic lung disease. These progressive diseases cannot be cured, but can be treated. Treatment includes avoiding the inhalation of harmful substances such as polluted air or cigarette smoke.

[*Linda Rehkopf*]

RESOURCES
BOOKS

Haas, Francois, et. al. *The Chronic Bronchitis and Emphysema Handbook.* New York: Wiley Science Editions, 1990.

David Ross Brower (1912 – 2000)
American environmentalist and conservationist

David R. Brower, the founder of both **Friends of the Earth** and the **Earth Island Institute**, has long been widely considered to be one of the most radical and effective environmentalists in the United States.

Joining the **Sierra Club** in 1933, Brower became a member of its Board of Directors in 1941 and then its first executive director, serving from 1952 to 1969. In this position, Brower helped transform the group from a regional to a national force, seeing the club's membership expand from 2,000 to 77,000 and playing a key role in the formation of the Sierra Club Foundation. Under Brower's leadership the Sierra Club, among other achievements, successfully opposed the Bureau of Reclamation's plans to build **dams** in

Dinosaur National Monument in Utah and Colorado as well as in Arizona's Grand Canyon, but lost the fight to preserve Utah's Glen Canyon. The loss of Glen Canyon became a kind of turning point for Brower, indicating to him the need to take uncompromising and sometimes militant stands in defense of the natural **environment**. This militancy occasionally caused friction both between the groups he has led and the private corporations and governmental agencies with which they interact and also within the increasingly broad-based groups themselves. In 1969 Brower was asked to resign as executive director of the Sierra Club's Board of Directors, which disagreed with Brower's opposition to a nuclear reactor in California's Diablo Canyon, among other differences. Eventually reelected to the Sierra Club's Board in 1983 and 1986, Brower is now an honorary vice-president of the club and was the recipient, in 1977, of the **John Muir** Award, the organization's highest honor.

After leaving the Sierra Club in 1969, Brower founded Friends of the Earth with the intention of creating an environmental organization that would be more international in scope and concern and more political in its orientation than the Sierra Club. Friends of the Earth, which now is operating in some 50 countries, was intended to pursue a more global vision of **environmentalism** and to take more controversial stands on issues—including opposition to nuclear weapons—than could the larger, generally more conservative organization. But in the early 1980s, Brower again had a falling out with his associates over policy, eventually resigning from Friends of the Earth in 1986 to devote more of his time and energy to the Earth Island Institute, a San Francisco-based organization he founded in 1982 and of which he is presently chairman.

Over the years, Brower played a key role in preserving **wilderness** in the United States, helping to create national park's and national seashores in Kings Canyon, the North Cascades, the **Redwoods**, Cape Cod, Fire Island, and Point Reyes. He also was instrumental in protecting primeval forests in the Olympic **National Park** and wilderness on San Gorgonio Mountain in California and in establishing the National Wilderness Preservation System and the Outdoor **Recreation** Resources Review, which resulted in the Land and **Water Conservation** Fund.

In his youth, Brower was one of this country's foremost rock climbers, leading the historic first ascent of New Mexico's Shiprock in 1939 and making 70 other first ascents in **Yosemite National Park** and the High Sierra as well as joining expeditions to the Himalayas and the Canadian Rockies. A proficient skier and guide as well as a mountaineer, Brower served with the United States Mountain Troops from 1942–45, training soldiers to scale cliffs and navigate in Alpine areas and serving as a combat-intelligence officer in Italy. For his service, Brower was awarded both the Combat

David Ross Brower. (Corbis-Bettmann. Reproduced by permission.)

Infantryman's Badge and the Bronze Star, and rose in rank from private to captain before he left active duty. As a civilian, Brower employed many of the same talents and abilities to show people what he has fought so long and so hard to preserve: he initiated the knapsack, river, and wilderness threshold trips for the Sierra Club's Wilderness Outings Program, and between 1939 and 1956 led some 4,000 people into remote wilderness.

Excluding his military service, Brower was an editor at the University of California Press from 1941 to 1952. Appointed to the *Sierra Club Bulletin*'s Editorial Board in 1935, Brower eventually became the *Bulletin*'s editor, serving in this capacity for eight years. He had been involved with the publication of more than 50 environmentally oriented books each for the Sierra Club and Friends of the Earth, several of which earned him prestigious publishing industry awards. He wrote a two-volume autobiography, *For Earth's Sake* and *Work in Progress*. Brower also made several Sierra Club films, including a documentary of raft trips on the Yampa and Green Rivers designed to show people the stark beauty of Dinosaur National Monument, which at the time was threatened with **flooding** by a proposed dam.

Brower was the recipient of numerous awards and honorary degrees and serves on several boards and councils, including the Foundation on Economic Trends, the Council

on National Strategy, the Council on Economic Priorities, the North Cascades **Conservation** Council, the Fate and Hope of the Earth Conferences, **Zero Population Growth**, the Committee on National Security, and **Earth Day**. He had twice been nominated for the Nobel Peace Prize. Brower has promoted environmental causes around the globe, giving dozens of lectures in 17 different countries and organizing several international conferences. In 1990, Brower's life was the subject of a PBS Video Documentary entitled *For Earth's Sake*. He also was featured in the TV documentary *Green for Life*, which focused on the 1992 Earth Summit in Rio de Janeiro.

Before his death on November 5, 2000, after battling bladder **cancer**, Brower was still actively promoting environmental causes. He devoted much of his time to his duties at the Earth Island Institute, and in promoting the activities of the International Green Circle. In 1990 and 1991, he led Green Circle delegations to Siberia's **Lake Baikal** to aid in its protection and restoration. Brower also lectured to companies and schools throughout the U.S. on Planetary Conservation Preservation and Restoration (CPR). His topics included land conservation, the economics of sustainability and the meaning of wilderness to science. Brower's book, *Let the Mountains Talk, Let the Rivers Run*, includes a credo for the earth, which reflects what Brower had hoped to accomplish with his lectures and publications: "We urge that all people now determine that an untrammeled wilderness shall remain here to testify that this generation had love for the next."

[*Lawrence J. Biskowski*]

RESOURCES
BOOKS

Brower, D. *For Earth's Sake.* Layton, UT: Gibbs Smith, 1990.
————. *Work in Progress.* Layton, UT: Gibbs Smith, 1991.

PERIODICALS

Foster, C. "A Longtime Gadfly Still Stings." *Christian Science Monitor* (8 April 1991): 14.
McKibben, B. "David Brower: Interview." *Rolling Stone* (28 June 1990): 59–62, 87.
Russell, D. "Nicaraguan Journey: The Archdruid at 76." *Amicus Journal* 11 (Summer 1989): 32–37.

Lester R. Brown   (1934 –)

American founder, president and senior researcher, Earth Policy Institute

Lester Brown is a highly respected and influential authority on global environmental issues. He founded the **Worldwatch Institute** in 1974 and served as its president until 2000. In 2001 he launched a new initiative, the Earth

Lester R. Brown. (Corbis-Bettmann. Reproduced by permission.)

Policy Institute Brown is an award-winning author of many books and articles on environmentally sustainable economic development and on environmental, agricultural, and economic problems and trends.

Brown was born in Bridgeton, New Jersey, and during high school and college, he grew tomatoes with his younger brother. At this time he developed his appreciation for nature's ability, if properly treated, to supply us with food on a regular and sustainable basis. After earning a degree in agricultural science from Rutgers University in 1955, he spent six months in rural India studying and working on agricultural projects. In 1959, he joined the U.S. Department of Agriculture's Foreign Agricultural Service as an international agricultural analyst. After receiving an M.S. in agricultural economics from the University of Maryland and a master's degree in public administration from Harvard, he went to work for Orville Freeman, the Secretary of Agriculture, as an advisor on foreign agricultural policy in 1964. In 1969, Brown helped establish the Overseas Development Council and in 1974, with the support of the Rockefeller Fund, he founded the Worldwatch Institute to analyze world conditions and problems such as **famine**, overpopulation, and **scarcity** of **natural resources**.

In 1984, Brown launched Worldwatch's annual *State of the World* report, a comprehensive and authoritative account of worldwide environmental and agricultural trends and problems. Eventually published in over 30 languages, *State of the World* is considered one of the most influential and widely read reports on public policy issues. Other Worldwatch publications initiated and overseen by Brown included *Worldwatch*, a bimonthly magazine, the *Environmental Alert* book series, and the annual *Vital Signs: The Trends That Are Shaping Our Future*. Brown has written or co-authored over a dozen books and some two dozen Worldwatch papers on various economic, agricultural, and environmental topics. Among his many awards, Brown has received a $250,000 "genius award" from the MacArthur Foundation, as well as the United Nation's 1987 environmental prize. *The Washington Post* has described him as "one of the world's most influential thinkers."

Brown has long warned that unless the United States and other nations adopt policies that are ecologically and agriculturally sustainable, the world faces a disaster of unprecedented proportions. In Worldwatch's yearly *State of the World* report, Brown tracked the impact of human activity on the **environment**, listing things like the percentage of bird **species** that were endangered, the number of days China's Yellow River was too depleted to irrigate fields in its lower reaches, the number of females being educated worldwide, and the number of cigarettes smoked per person. His reports made the ecological dimension of global economics clear and concrete. Often these reports were dire. The 1998 report, for example, discussed so-called demographic fatigue. This referred to places where population was falling, not because of a birthrate held in check by **family planning** but because many people were dying through famine, **drought**, and infectious disease. Conventional economics paid little heed to the environmental cost of development, leading to what a reviewer for the *Financial Times* called "a kind of cosy belief that the Earth's resources are unlimited...." Brown struggled to unseat that belief. He also wrote about a way out of the doom his work often foresaw. His 2001 *Eco-Economy: Building an Economy for the Earth* argued that creation of a new, ecologically aware economy, with emphasis on **renewable energy**, tax reform, redesign of cities and **transportation**, better agricultural methods and global cooperation, could alleviate much of the world's ills. Through the Earth Policy Institute's *Earth Policy Alerts*, Brown continued to add to the themes of his book. The mission of Brown's new organization was to reach policy makers and the public with information about building an environmentally sustainable economy.

[*Lewis Regenstein*]

RESOURCES

BOOKS

Brown, Lester. *Eco-Economy: Building an Economy for the Earth.* New York: W.W. Norton, 2001.

PERIODICALS

Hager, Mary. "How 'Demographic Fatigue' Will Defuse the Population Bomb"*Newsweek* (November 2, 1998): 12.

"Lester Brown's Eco-Economy"*Mother Earth News* (February/March 2002): 1.

McWilliam, Fiona. "Sign of the Times"*Geographical Magazine* 70, no. 7. (July 1998): 25.

Tickell, Christian. "The Hidden Costs of Life on Earth"*Financial Times* (February 16, 2002): 4.

ORGANIZATIONS

Earth Policy Institute, 1350 Connecticut Ave. NW, Washington, DC USA 20036 (202) 496-9290, Fax: (202) 496-9325, Email: epi@earth-policy.org, http://www.earth-policy.org

Brown pelican

The brown pelican (*Pelecanus occidentalis*) is a large water bird of the family Pelicanidae that is found along both coasts of the United States, chiefly in saltwater habitats. It weighs up to 8 lb (3.5 kg) and has a wingspan of up to 7 ft (2 m). This pelican has a light brown body and a white head and neck often tinged with yellow. Its distinctive, long, flat bill and large throat pouch are adaptations for catching its primary food, schools of mid-water fishes. The brown pelican hunts while flying a dozen or more feet above the surface of the water, dropping or diving straight down into the water, and using its expandable pouch as a scoop or net to engulf its catch.

Both east and west coast populations, which are considered to be different subspecies, have shown various levels of decline over the later part of the twentieth century. It is estimated that there were 50,000 pairs of nesting brown pelicans along the Gulf coast of Texas and Louisiana in the early part of the twentieth century, but by the early 1960s, most of the Texas and all of the Louisiana populations were depleted. The main reason for the drastic decline was the use of organic pesticides, including DDT and endrin. These pesticides poisoned pelicans directly and also caused thinning of their eggshells. This eggshell thinning led to reproductive failure, because the egg were crushed during incubation. Louisiana has the distinction of being the only state to have its state bird become extinct within its borders. In 1970 the brown pelican was listed as endangered throughout its U.S. range.

During the late 1960s and early 1970s, brown pelicans from Florida were reintroduced to Louisiana, but many of these birds were doomed. Throughout the 1970s these transplanted birds were poisoned at their nesting sites at the

outflow of the Mississippi River by endrin, which was used extensively upriver. In 1972 the use of DDT was banned in the U.S., and the use of endrin was sharply curtailed. Continued reintroduction of the brown pelican from Florida to Louisiana subsequently met with greater success, and the Louisiana population had grown to more than 1,000 pairs by 1989. Although the Texas, Louisiana, and California populations are still listed as endangered, the Alabama and Florida populations of the brown pelican have been removed from the federal list due to recent increases in fledgling success. In 2002, taking the Louisiana populations off of the endangered listing is being considered. There are currently 16,4000 nesting pairs and 35,000 young in that state. The Texas population has only 2,400 pairs.

Other problems that face the brown pelican include **habitat** loss, encroachment by humans, and disturbance by humans. Disturbances have included mass visitation of nesting colonies. This practice has been stopped on federally owned lands and access to nesting colonies is restricted. Other human impacts on brown pelican populations have had a more malicious intent. On the California coast in the 1980s there were cases of pelicans' bills being broken purposefully, so that these birds could not feed and would ultimately starve to death. It is thought that disgruntled commercial fishermen faced with dwindling catches were responsible for at least some of these attacks. The brown pelican was a scapegoat for conditions that were due to weather, **pollution**, or, most likely, **overfishing**.

Recovery for the brown pelican has been slow, but progress is being made on both coasts. The banning of DDT in the early 1970s was probably the turning point for this **species**, and the delisting of the Alabama and Florida populations is a hopeful sign for the future.

[*Eugene C. Beckham*]

RESOURCES

BOOKS

Ehrlich, P., D. Dobkin, and D. Wheye. *Birds in Jeopardy.* Stanford, CA: Stanford University Press, 1992.

PERIODICALS

Schreiber, R. W. "The Brown Pelican: An Endangered Species?" *BioScience* 30 (November 1980): 742–747.

Tompkins, Shannon. "Brown Pelicans Reappearance Proves Lessons can be Learned." *Houston Chronicle* (February 14, 2002) [cited May 2002]. <http://www.chron.com/cs/CDA/story.hts/outdoors/tompkins/1256525>.

Brown tree snake

The brown tree snake *(Boiga irregularis)* has caused major ecological and economic damage in Guam, the largest of the Mariana Islands. The snake is native to New Guinea and northern **Australia**. It has also been introduced to some Pacific islands in addition to Guam.

Brown tree snakes in their natural **habitat** range from 3–6 ft (0.9–1.8 m) in length. Some snakes in Guam are more than 10 ft (3 m) long. The snake's head is bigger than its neck, and its coloring varies with its habitat. In Guam, the snake's brown-and-olive-green pattern blends in with foliage.

The tree snake was accidentally brought to Guam by cargo ships during the years between the end of World War II (1945) through 1952. On Guam, there were no population controls such as predators that eat snakes. As a result, the snake population boomed. During the late 1990s, there were close to 13,000 snakes per square mile in some areas.

The snake's diet includes birds, and the United States Geographical Survey (USGS) said that the brown tree snake "virtually wiped out" 12 of Guam's native forest birds (three of which are now extinct). Furthermore, snakes crawling on electrical lines caused more than 1,200 power outages from 1978 through 1998.

The USGS and other agencies were working to contain snakes on Guam and stop their spread to Hawaii and other islands.

[*Liz Swain*]

Carol Browner (1955 –)

American former administrator of the Environmental Protection Agency

Carol Browner headed the **Environmental Protection Agency** (EPA) under the Clinton administration. She served from 1993 until 2001, making her the longest-serving director the agency had ever had.

Browner was born in Florida on December 17, 1955. Her father taught English and her mother social science at Miami Dade Community College. Browner grew up hiking in the **Everglades**, where her lifelong love for the natural world began. She was educated at the University of Florida in Gainesville, receiving both a BA in English and her law degree there. Her political career began in 1980 as an aide in the Florida House of Representatives. Browner moved to Washington, D.C. a few years later to join the national office of Citizen Action, a grassroots organization that lobbies for a variety of issues, including the **environment**. She left Citizen Action to work with Florida Senator Lawton Chiles, and in 1989 she joined Senator Al Gore's staff as a senior legislative aide. From 1991 to 1993, Browner headed the Department of Environmental Regulation in Florida, the third-largest environmental agency in the country. She

streamlined the process the department used to review permits for expanding manufacturing plants and developing **wetlands**, thus reducing the amount of money and time that process was costing businesses as well as the department. Activists had argued that this kind of streamlining interfered with the ability of government to supervise industries and assess their impact on the environment. But in Florida's business community, Browner built a reputation as a formidable negotiator on behalf of the environment. When the Walt Disney Company filed for state and federal permits to fill in 400 acres of wetlands, she negotiated an agreement which allowed the company to proceed with its development plans in return for a commitment to buy and restore an 8,500-acre ranch near Orlando. She was also the chief negotiator in the settlement of a lawsuit the government had brought against Florida for environmental damage done to Everglades **National Park**. The result was the largest ecological restoration project ever attempted in the United States, a plan to purify and restore the natural flow of water to the Everglades with the cost shared by the state and the federal government, as well as Florida's sugar farmers. Browner was often called "a new type of environmentalist" because of her belief that environmental protection is compatible with economic development and her strong conviction that the stewardship of the environment requires accommodations with industry. "I've found that business leaders don't oppose strong environmental programs," she said in a 1992 interview with the *New York Times*. "What drives them crazy is a lack of certainty."

As director of the EPA, she was determined to protect the environment and public health while not alienating business interests. By many accounts, she was a remarkably successful administrator. She took the job at a time of relatively high environmental fervor, with a Vice President (Al Gore) who had particularly championed the environment, and Democrats in control of both houses of Congress. But a conservative backlash led by Representative Newt Gingrich brought Republican control to the House and Senate in 1994. The political **climate** for environmental reform then became much more embattled. Many conservatives wished to downsize government and cut back the regulatory power of the EPA and other federal agencies. The EPA was shut down twice during Browner's tenure, temporary victim of congressional budgetary squabbles. And appropriations bills for EPA programs were frequently hit with amendments that countermanded the agency's ability to carry out some policies. Despite the hostility of Congress, Browner had several legislative victories. In 1996 Browner led a campaign to have Congress reauthorize the **Safe Drinking Water Act**. That same year she spearheaded the Food Quality Protection Act, which modernized standards that govern **pesticide** use.

Carol Browner. (Corbis-Bettmann. Reproduced by permission.)

This landmark legislation was one of the first environmental laws to specifically protect children's health. The law required scientists to determine what levels of pesticides were safe for children. Browner was also signally successful in getting so-called Superfund sites cleaned up. These were sites that were listed as being particularly polluted. Only 12 Superfund sites had been cleaned up when Browner took over the EPA in 1993. By the time she left the agency, over 700 sites had been cleaned up or were in the process of being cleaned up. After she left the EPA in 2001, Browner continued to work for the environment by becoming a board member of the Audobon Society.

[*Douglas Smith*]

RESOURCES
PERIODICALS

Adler, Jonathan H. "EPA Can't Win for Losing"*Business Journal (Central New York)* (March 24, 2000): 27.

"Babbitt, Browner Take Interior, EPA Posts. *National Parks* 67 (March-April 1993): 9.

"Browner on Board" *Audubon* (July-August 2001): 95.

Grier, Peter, and Sappenfield, Mark. "How a Pesticide Got Banned"*Christian Science Monitor* (August 4, 1999): 1.

Hogue, Cheryl. "Browner Leaves Her Stamp on EPA"*Chemical and Engineering News* (January 15, 2001): 34-37.

Brownfields

By some estimates, as many as 450,000 urban brownfield sites may exist nationwide. These former commercial or industrial properties can be found in nearly every community where, because of real or perceived contamination, the land sits idle or under-used. Brownfield sites can be as small as a gas station with a **leaking underground storage tank** or as large as a 100-acre (40.5-ha) abandoned factory containing dumped waste on the site. The sites are not remarketed or reused because owners or operators fear that, if the properties are contaminated, the regulatory requirements to clean them up could be prohibitively time-consuming and expensive. Furthermore, even if such properties were put on the market, developers, and financial credit sources would shun them in part from concerns that dealing with contamination cleanup could raise the costs and slow the pace of reusing the properties.

But the brownfields situation has been changing as local, state, and federal government officials recognize the dire need to begin cleaning up and returning these contaminated properties to productive use. These officials point out that reusing the brownfield sites not only would reduce the phenomenon of "urban sprawl" but would also furnish communities with the chance to renew the economic vitality of abandoned urban centers. In many metropolitan areas **urban sprawl** has been recognized as a serious problem, with developers turning to rural greenfields for their projects and thereby reducing **habitat** for **wildlife** and contributing to **air pollution** as commuters drive longer distances to outlying greenfield locations. Moreover, environmental justice advocates point to the need for cleaning up contaminated urban sites to reduce exposure to hazardous contaminants faced by residents in these areas. Together, these factors have created momentum for a national brownfields redevelopment initiative. The promise some see in brownfields redevelopment led the Environmental Manager of Portland, Oregon, to state that "Brownfields renewal is one of the most important environmental and economic challenges facing our nation's communities, calling for partnership among our federal and local governments, business and community and environmental leaders."

Although **land use** decisions are made by local authorities, the U.S. **Environmental Protection Agency** (EPA) plays a critical role in the brownfields **reuse** initiative. On the one hand, the EPA's rules governing contaminated site cleanups, based on federal statutes, have played a significant part in deterring brownfields redevelopment; on the other, the Agency has been a major source of funding for brownfields pilots. In particular, the **Comprehensive Environmental Response, Compensation, and Liability Act** (CERCLA)—commonly known as Superfund—has made brownfield sites into expensive liabilities, causing businesses to avoid them because stringent cleanup standards and related costs often exceed the property's value. Under CER-CLA, new property owners can be held liable for contamination created by previous owners. The **Clean Air Act** also drives businesses away from urban brownfield sites because it burdens companies in areas not attaining national **air quality** standards, making it more attractive for them to locate in rural regions where meeting clean air standards may be easier than in the city.

The EPA began its brownfields pilot program in 1993 by supporting pilot projects—each funded up to $200,000 over two years—at eight separate locations. The goal was to learn about brownfields, which the Agency saw as a new and complex territory in which it had little experience. The EPA awarded its first pilot in November 1993 to Cuyahoga County, Ohio, where the county planning commission had formed a broad effort to involve stakeholders in addressing brownfields reuse. The EPA described Cuyahoga County's report on its brownfields initiative as the most complete known to the Agency on the problems urban areas faced in reusing brownfields. Two additional pilots were awarded in 1994 to Bridgeport, Connecticut, and Richmond, Virginia. By October 1996 the EPA had funded 76 brownfields pilots, 39 under its "National Pilots" selected by the EPA Headquarters in Washington, D.C., and 37 under its "Regional Pilots" selected by the EPA's regional offices. An additional 25 National Pilots were scheduled for selection by March 1997.

In addition to funding pilot projects, the EPA took other actions to underscore its commitment to a Brownfields Economic Redevelopment Initiative, intended to help communities revitalize brownfields properties both environmentally and economically. In January 1995, the EPA Administrator **Carol Browner** announced a Brownfields Action Agenda, stating, "We at EPA firmly believe that environmental cleanup is a building block to economic development, not a stumbling block--that identifying and cleaning up contaminated property must go hand-in-hand with bringing life and vitality back to a community." The Brownfields Action Agenda included the Agency's pilots; joint efforts by EPA, states, and local governments to develop guidance documents to clarify burning liability issues; partnership and outreach endeavors with states, cities, and communities; and job training and development activities, including funding towards an **environmental education** program at Cuyahoga Community College in Cleveland, Ohio.

The National Environmental Justice Advisory Council (NEJAC), chartered in September 1993 as a federal advisory committee to the EPA, in a December 1996 report expressed hope that the Agency's brownfields initiative would stem the "ecologically untenable" and "racially divisive" phenome-

non of urban sprawl, in which urban areas were left to stagnate while greenfields were subjected to increasing development. NEJAC also emphasized that the brownfields problem is "inextricably linked to environmental justice" through, for instance, the deterioration of the nation's urban areas. NEJAC reached its hopeful evaluation of the prospects for the brownfields initiative despite misgiving by some NEJAC members that the initiative might be a "smoke screen" for gutting cleanup standards and liability safeguards.

The liability issues addressed under the EPA's Brownfields Action Agenda are among the most important factors affecting these sites. According to the Agency, the fear of inheriting cleanup liabilities for contamination they did not create affects communities, lenders, property owners, municipalities, and others. To help assuage liability concerns, in February 1995 the EPA removed 24,000 of the 40,000 sites listed in its Comprehensive Environmental Response, Compensation and Liability Information System (CERCLIS), a Superfund tracking system. These sites had been screened out of the "Active Investigations" category and designated "No Further Remedial Action Planned," either because they were found to be clean or were being addressed by state cleanup programs. By taking the sites off the CERCLIS list, the EPA hoped to remove any stigma resulting from federal involvement at the sites and thereby to overcome obstacles to their cleanup and reuse.

In May 1995 the EPA issued guidance on "Agreements with Prospective Purchaser of Contaminated Property." The guidance indicated the situations under which the Agency may enter into an agreement to not file a lawsuit against a prospective contaminated property buyer for contamination present at the site before the purchase. Also in May 1995 the EPA issued guidance regarding "Land Use in the CERCLA Remedy Selection Process" to ensure that the Agency considers land use during Superfund cleanups. Land designated for reuse as an industrial site would require less strict cleanup than land designated for a park or school, a recognition that not all contaminated sites should be regarded as equally befouled. In September 1995 the EPA issued guidance that, among other things, released from CERCLA liability governmental units such as municipalities that involuntarily take ownership of property under federal, state, or local law. The EPA hoped this guidance would encourage municipalities to start cleaning up sites without fear of the Superfund liability predicament. These and other guidance documents issued under the Brownfields Action Agenda helped to clarify liabilities facing prospective new owners of private and federally owned contaminated sites.

But the EPA's guidance documents are regarded as only partial solutions to the brownfields problem. A report released February 1997 by the National Association of Local

Government Environmental Professionals (NALGEP), representing 50 cities nationwide, stresses that the EPA's ability to "further clarify and reduce **environmental liability** for brownfields activity is limited," and therefore "legislative solutions" are required to make further progress on the brownfields agenda. Noting that the U.S. House and Senate introduced "numerous brownfields bills" during the 104th Congress and that brownfields legislation is likely in the 105th Congress, NALGEP's report suggests federal law should empower the EPA to delegate authority to the states with cleanup programs that meet minimum requirements to protect health and the **environment**. The EPA would be empowered to authorize qualifying states to limit liability and issue "no action assurances" for less contaminated brownfield sites. Those sites would be clearly differentiated from more seriously contaminated properties that warrant management under the federal Superfund **National Priorities List** of sites.

Going into the 105th Congress, federal lawmakers and President Bill Clinton agreed that brownfields legislation ought to be a priority. Underscoring the close link between brownfields and economic revitalization, Senator Max Baucus (D-Montana), the Ranking Democrat of the Senate Environment and Public Works Committee, said at a press conference in January 1997 that passage of a brownfields bill—apart from Superfund reauthorization legislation—would be a step towards helping create jobs where they are really needed. Welfare reform added to the growing momentum for urban renewal and job opportunities. Representative Michael Oxley (R-Ohio), who proposed Superfund amendments in the 103rd Congress to direct the EPA to help states set up "voluntary cleanup programs" aimed at brownfields sites, compared the federal Superfund liability scheme to the Berlin Wall and said that it should be torn down. The chorus of voices joining Oxley has grown substantially since 1993, and today not only the EPA but also the U.S. Department of Housing and Urban Development, the U.S. Department of Labor, and other federal agencies have joined states and local governments in promoting brownfields redevelopment.

[*David Clarke*]

RESOURCES

BOOKS

National Association of Local Government Environmental Professionals. *Building a Brownfields Partnership from the Ground Up.* Washington, D.C., 1997.

Powers, C. *State Brownfields Policy and Practice.* Boston: Institute for Responsible Management, 1994.

U.S. Environmental Protection Agency. *The Brownfields Action Agenda.* Washington, D.C.: GPO, 1995.

U.S. Environmental Protection Agency. National Environmental Justice Advisory Council. *Environmental Justice, Urban Revitalization, and*

Brownfields: The Search for Authentic Signs of Hope. Washington, D.C.: GPO, 1996.

Gro Harlem Brundtland (1939 –)

Norwegian doctor, former Prime Minister of Norway, Director-General of the World Health Organization

Dr. Gro Harlem Brundtland began serving a five-year term as Director-General of the World Health Organization (WHO) in July 1998. A physician and outspoken politician, Gro Harlem Brundtland had been instrumental in promoting political awareness of the importance of environmental issues. In her view, the world shares one economy and one **environment**. With her appointment to the head of WHO, Brundtland continued to work on global strategies to combat ill health and disease.

Brundtland began her political career as Oslo's parliamentary representative in 1977. She became the leader of the Norwegian Labor Party in 1981, when she first became prime minister. At 42, she was the youngest person ever to lead the country and the first woman to do so. She regained the position in 1986 and held it until 1989; in 1990 she was again elected prime minister. Aside from her involvement in the environmental realm, Brundtland promoted equal rights and a larger role in government for women. In her second cabinet, eight of the 18 positions were filled by women; in her 1990 government, nine of 19 ministers were women.

Brundtland earned a degree in medicine from the University of Oslo in 1963 and a master's degree in public health from Harvard in 1965. She served as a medical officer in the Norwegian Directorate of Health and as medical director of the Oslo Board of Health. In 1974 she was appointed Minister of the Environment, a position she held for four years. This appointment came at a time when environmental issues, especially **pollution**, were becoming increasingly important, not only locally but nationally. She gained international attention and in 1983 was selected to chair the United Nation's World Commission on Environment and Development. The commission published **Our Common Future** in 1987, calling for sustainable development and intergenerational responsibility as guiding principles for economic growth. The report stated that present economic development depletes both nonrenewable and potentially renewable resources that must be conserved for future generations. The commission strongly warned against environmental degradation and urged nations to reverse this trend. The report led to the organization of the so-called Earth Summit in Rio

Gro Harlem Brundtland. (Corbis-Bettmann. Reproduced by permission.)

de Janeiro in 1992, an international meeting led by the United Nation's Conference on Environment and Development.

Bruntland resigned her position as prime minister in October 1996. This was prompted by a variety of factors, including the death of her son and the possibility of appointment to lead the United Nations. Instead she was picked to head the World Health Organization, an assembly of almost 200 member nations concerned with international health issues. Brundtland immediately reorganized WHO's leadership structure, vowed to bring more women into the group, and called for greater financial disclosure from WHO executives. She began dual campaigns called "Rollback Malaria" and "Stop TB." She also launched an unprecedented campaign to combat **tobacco** use worldwide. Brundtland noted that disease due to tobacco was growing enormously. Tobacco-related illnesses already caused more deaths worldwide than **AIDS** and tuberculosis combined. Much of the growth in smoking was in the developing world, particularly China. Thus in 2000, WHO organized the Framework Convention on Tobacco Control to come up with a world-wide treaty governing tobacco sale, advertising, taxation, and labeling. Brundtland broadened the mission of WHO by attempting the treaty. She was credited with making WHO a more active group, and with making health issues a major compo-

nent of global economic strategy for organizations such as the United Nations and the so-called G8 group of developed countries. In 2002 Brundtland announced more new goals for WHO, including renewed work on diet and nutrition.

[*William G. Ambrose and Paul E. Renaud*]

RESOURCES

PERIODICALS

Brundtland, G.H. "Global Change and Our Common Future." *Environment* 31 (1989): 16-34.

———. "The Globalization of Disease." *New Perspectives Quarterly* 16, no. 5 (Fall 1999): 17.

Kapp, Claire. "Brundtland Sets Out Priorities at Annual World Health Assembly." *Lancet* 359, no. 9319, (May 18, 2002): 1758.

McGregor, Alan. "Brundtland Launches New-Look WHO." *Lancet* 352, no. 9124 (July 25, 1998): 300.

"New Director-General Takes Over at WHO." *World Health* 51, no. 4 (July/August 1998): 3.

"The Tobacco War Goes Global." *Economist* (October 14, 2000): 97.

"A Triumph of Experience over Hope." *Economist* (May 26, 2001-June 1, 2001): 79.

Brundtland Report

see Our Common Future (Brundtland Report)

Btu

Abbreviation for "British Thermal Unit," the amount of energy needed to raise the temperature of one pound of water by one degree Fahrenheit. One Btu is equivalent to 1,054 joules or 252 calories. To gain an impression of the size of a Btu, the **combustion** of a barrel of oil yields about 5.6×10^6 joule. A multiple of the Btu, the quad, is commonly used in discussions of national and international energy issues. The term quad is an abbreviation for one quadrillion, or 10^{15}, Btu.

Mikhail I. Budyko (1920 –)

Belarusian geophysicist, climatologist

Professor Mikhail Ivanovich Budyko is regarded as the founder of physical climatology. Born in Gomel in the former Soviet Union, now Belarus, Budyko earned his master of sciences degree in 1942 from the Division of Physics of the Leningrad Polytechnic Institute. As a researcher at the Leningrad Geophysical Observatory, he received his doctorate in physical and mathematical sciences in 1951. Budyko served as deputy director of the Geophysical Observatory until 1954, as director until 1972, and as head of the Division for Physical Climatology at the observatory from 1972 until 1975. In that year he was appointed director of the Division for Climate Change Research at the State Hydrological Institute in St. Petersburg.

During the 1950s, Budyko pioneered studies on global climate. He calculated the energy or heat balance—the amount of the Sun's radiation that is absorbed by the Earth versus the amount reflected back into space—for various regions of the Earth's surface and compared these with observational data. He found that the heat balance influenced various phenomena including the weather. Budyko's groundbreaking book, *Heat Balance of the Earth's Surface*, published in 1956, transformed climatology from a qualitative into a quantitative physical science. These new physical methods based on heat balance were quickly adopted by climatologists around the world. Budyko directed the compilation of an atlas illustrating the components of the Earth's heat balance. Published in 1963, it remains an important reference work for global climate research.

During the 1960s, scientists were puzzled by geological findings indicating that glaciers once covered much of the planet, even the tropics. Budyko examined a phenomenon called planetary **albedo**, a quantifiable term that describes how much a given geological feature reflects sunlight back into space. Snow and ice reflect heat and have a high albedo. Dark seawater, which absorbs heat, has a low albedo. Land formations are intermediate, varying with type and heat-absorbing vegetation. As snow and ice-cover increase with a global temperature drop, more heat is reflected back, ensuring that the planet becomes colder. This phenomenon is called ice-albedo feedback. Budyko found an underlying instability in ice-albedo feedback, called the snowball Earth or white Earth solution: if a global temperature drop caused ice to extend to within 30 degrees of the equator, the feedback would be unstoppable and the Earth would quickly freeze over. Although Budyko did not believe that this had ever happened, he postulated that a loss of atmospheric **carbon dioxide**, for example if severe **weathering** of silicate rocks sucked up the **carbon** dioxide, coupled with a sun that was 6% dimmer than today, could have resulted in widespread **glaciation**.

Budyko became increasingly interested in the relationships between global climate and organisms and human activities. In *Climate and Life*, published in 1971, he argued that mass extinctions were caused by climatic changes, particularly those resulting from volcanic activity or meteorite collisions with the Earth. These would send clouds of particles into the **stratosphere**, blocking sunlight and lowering global temperatures. In the early 1980s Budyko warned that nuclear war could have a similar effect, precipitating a "nuclear winter" and threatening humans with **extinction**.

By studying the composition of the **atmosphere** during various geological eras, Budyko confirmed that increases in atmospheric carbon dioxide, such as those caused by volcanic activity, were major factors in earlier periods of global warming. In 1972, when many scientists were predicting climate cooling, Budyko announced that fossil fuel consumption was raising the concentration of atmospheric carbon dioxide, which, in turn, was raising average global temperatures. He predicted that the average air temperature, which had been rising since the first half of the twentieth century, might rise another 5°F (3°C) over the next 100 years. Budyko has since been examining the potential effects of global warming on rivers, lakes, and ground water, on twenty-first-century food production, on the geographical distribution of vegetation, and on energy consumption.

The author and editor of numerous articles and books, in 1964 Budyko became a corresponding member of the Division of Earth Sciences of the Academy of Sciences of the Union of Soviet Socialist Republics. In 1992 he was appointed academician in the Division of Oceanology, Atmosphere Physics, and Geography of the Russian Academy of Sciences. His many awards include the Lenin National Prize in 1958, the Gold Medal of the World Meteorological Organization in 1987, the A. P. Vinogradov Prize of the Russian Academy of Sciences in 1989, and the A. A. Grigoryev Prize of the Academy of Sciences in 1995. Budyko was awarded the Robert E. Horton Medal of the American Geophysical Union in 1994, for outstanding contribution to geophysical aspects of **hydrology**. In 1998 Dr. Budyko won the Blue Planet Prize of the Asahi Glass Foundation of Japan.

[*Margaret Alic*]

RESOURCES
BOOKS

Andronova, Natalia G. "Budyko, Mikhail Ivanovich." In *Encyclopedia of Global Environmental Change*, edited by Ted Munn, vol. 1. New York: Wiley, 2002.

Budyko, M. I., G. S. Golitsyn, and Y. A. Izrael. *Global Climatic Catastrophes*. New York: Springer-Verlag, 1988.

Budyko, M. I. and Y. A. Izrael., eds. *Anthropogenic Climatic Change*. Tuscon: University of Arizona Press, 1991.

Budyko, M. I., A. B. Ronov, and A. L. Yanshin. *History of the Earth's Atmosphere*. New York: Springer-Verlag, 1987.

OTHER

Budyko, Mikhail I. "Global Climate Warming and its Consequence." *Blue Planet Prize 1998 Commemorative Lectures* . Ecology Symphony. October 30, 1998 [cited May 23, 2002]. <www.ecology.or.jp/special/9902e.html>,

Hoffman, Paul F. and Daniel P. Schrag. "Snowball Earth." *Scientific American* January 2000 [cited May 24, 2002]. <www.sciam.com/2000/0100issue/0100hoffman.html>.

"Dr. Mikhail I. Budyko." *Profiles of the 1998 Blue Planet Prize Recipients*. The Asahi Glass Foundation. 2001 [cited May 23, 2002]. <www.af-info.or.jp/eng/honor/hot/enr-budyko.html>.

ORGANIZATIONS

State Hydrological Institute, 23 Second Line VO, St. Petersburg, Russia 199053.

Buffer

A term in **environmental chemistry** that refers to the capacity of a system to resist chemical change. Most often it is used in reference to the ability to resist change in **pH**. A system that is strongly pH buffered will undergo less change in pH with the addition of an **acid** or a **base** than a less well buffered system. A well buffered lake contains higher concentrations of bicarbonate ions that react with added acid. This kind of lake resists change in pH better than a poorly buffered lake. A highly buffered **soil** contains an abundance of **ion exchange** sites on **clay minerals** and organic matter that react with added acid to inhibit pH reduction. *See also* Acid and base

Bulk density

The mass of **soil** per unit bulk volume. The bulk volume consists of mineral and organic materials, water, and air. Bulk density is affected by external pressures (e.g., weight of tractors and harvesting equipment, mechanical pressures from cultivation machines), and by internal pressures (e.g., swelling and shrinking due to water-content changes, freezing and thawing, and by plant roots). The bulk density of cultivated mineral soils ranges from 1–1.6 mg/m^3.

Burden of proof

The current regulatory system, following traditional legal processes, generally assumes that **chemicals** are innocent, or not harmful, until proven guilty. Thus, the burden to show proof that a chemical is harmful to human and/or **ecosystem health** falls on those who regulate or are affected by these chemicals. As evidence increases that many of the more than 70,000 chemicals in the marketplace today—and the 10,000 more introduced each year—are causing health effects in various **species**, including humans, new regulations are being proposed that would reverse this burden to the manufacturer, importer or user of the chemical and its byproducts. They would then have to prove before its production and distribution that the chemical will not be harmful to human health and the **environment**.

Bureau of Land Management

The Bureau of Land Management (BLM), in the U. S. Department of the Interior, was created by executive reorganization in June 1946. The new agency was a merger of the Grazing Service and the General Land Office (GLO). The GLO was established in the Treasury Department by Congress in 1812 and charged with the administration of the public lands. The agency was transferred to the Department of the Interior when it was established in 1849. Throughout the 1800s and early 1900s, the GLO played the central role in administering the disposal of public lands under a multitude of different laws. But, as the nation began to move from disposal of public lands to retention of them, the services of the GLO became less needed, which helped to pave the way for the creation of the BLM. The Grazing Service was created in 1934 (as the Division of Grazing) to administer the **Taylor Grazing Act**.

In the 1960s, the BLM began to advocate for an organic act that would give it firmer institutional footing, would declare that the federal government planned to retain the BLM lands, and would grant the agency statutory authority to professionally manage these lands (like the **Forest Service**). Each of these goals was achieved with the passage of the **Federal Land Policy and Management Act** (FLPMA) in 1976. The agency was directed to manage these lands and undertake long-term planning for the use of the lands, guided by the principle of multiple use.

The BLM manages 272 million acres (110 million ha) of land, primarily in the western states. This land is of three types: Alaskan lands (92 million acres; 37.2 million ha), which is virtually unmanaged; the Oregon and California lands (2.6 million acres; 1 million ha), prime timber land in western Oregon that reverted back to the government in the early 1900s due to land grant violations; and the remaining land (177 million acres; 71.8 million ha), approximately 70% of which is in grazing districts. As a multiple use agency, the BLM manages these lands for a number of uses: fish and **wildlife**, forage, minerals, **recreation**, and timber. Additionally, FLPMA directed that the BLM review all of its lands for potential **wilderness** designation, a process that is now well underway. (BLM lands were not covered by the **Wilderness Act** of 1964). In addition to these general land management responsibilities, the BLM also issues leases for mineral development on all public lands. FLPMA also directed that all mineral claims under the 1872 Mining Law be recorded with the BLM.

The BLM is headed by a director, appointed by the President, and confirmed by the Senate. The chain of command runs from the Director in Washington to state directors in 12 western states (all but Hawaii and Washington), to district managers, who administer grazing or other districts, to resource area managers, who administer parts of the districts.

The BLM has often been compared unfavorably to the Forest Service. It has received less funding and less staff than its sibling agency, has been less professional, and has been characterized as captured by livestock and mining interests. Recent studies suggest that the administrative capacity of the BLM has improved.

[*Christopher McGrory Klyza*]

RESOURCES

ORGANIZATIONS

U.S. Bureau of Land Management, Office of Public Affairs, 1849 C Street, Room 406-LS, Washington, DC USA 20240 (202) 452-5125, Fax: (202) 452-5124, <http://www.blm.gov>

Bureau of Oceans and International Environmental and Scientific Affairs (OES)

The Bureau of OES was established in 1974 under Section 9 of the Department of State Appropriations Act. It is the lead office in Department of State in the formulation of foreign policy in four areas: 1) International Science and Technology (S&T) Affairs; 2) Environmental, Health, Natural Resource Protection and Global **Climate** Change; 3) Nuclear Energy and Energy Technology Affairs; and 4) Oceans and Fisheries Affairs. Each area is headed by a Deputy Associate Secretary of State. A Bureau coordinator oversees the Office of World Population Affairs. OES provides support and guidance to U. S. Embassy science counselors and attaches in reporting and negotiations. It has responsibility for the International Fisheries Commissions, the Fisherman's Guarantee Fund and the U.S. Secretariat for the **Man and the Biosphere Program**. The bureau prepares an annual Presidential report (the title V Report) on international "Science, Technology and American Foreign Policy."

RESOURCES

ORGANIZATIONS

U.S. Department of State, 2201 C Street NW, Washington, D.C. USA 20520 Toll Free: (202) 647-4000, <http://www.state.gov/g/oes>

Bureau of Reclamation

The U. S. Bureau of **Reclamation** was established in 1902 and is part of the U. S. Department of the Interior. It is primarily responsible for the planning and development of **dams**, **power plants**, and water transfer projects, such as Grand Coulee Dam on the Columbia River, the Central Arizona Project, and Hoover Dam on the **Colorado River**. This latter dam, completed in 1935 between Arizona and Nevada, is the highest arch dam in the Western Hemisphere and is part of the Boulder Canyon Project, the first great multipurpose water development project, providing **irrigation**, electric power, and flood control. It also created Lake Mead, which is supervised by the **National Park Service** to manage boating, swimming, and camping facilities on the 115 mi-long (185-km-long) **reservoir** formed by the dam. The dams on the Colorado River are intended to reduce the impact of the destructive cycle of floods and droughts which makes settlement and farming precarious and to provide electricity and recreational areas; however, the deep canyons and free-flowing rivers with their attendant ecosystems are substantially altered. Along the Columbia River, efforts are made to provide "fish ladders" adjacent to dams to enable **salmon** and other **species** to bypass the dams and spawn up river; however, these efforts have not been as successful as desired and many native species are now endangered.

Problems faced by the Bureau relate to creating a balance between its mandate to provide hydropower, water control for irrigation, and by-product **recreation** areas, and the conflicting need to preserve existing ecosystems. For example, at the **Glen Canyon Dam** on the Colorado River, controls on water releases are being imposed while studies are completed on the best manner of protecting the **environment** downstream in the Grand Canyon **National Park** and the Lake Mead National Recreation Area.

[*Malcolm T. Hepworth*]

RESOURCES

OTHER

The United States Government Manual, 1992/93. Washington, DC: U. S. Government Printing Office, 1992.

ORGANIZATIONS

Bureau of Reclamation, 1849 C Street NW, Washington, D.C. USA 20240-0001, <http://www.usbr.gov>

Buried soil

Buried **soil** is soil that appeared on the surface of the earth and sustained plant life but because of a geologic event has been covered by a layer of **sediment**. Sediments can result from volcanoes, rivers, dust storms, or blowing sand or **silt**.

John Burroughs (left) with naturalist John Muir. (Corbis-Bettmann. Reproduced by permission.)

John Burroughs (1837 – 1921)

American naturalist and writer

A follower of both **Henry David Thoreau** and Ralph Waldo Emerson, Burroughs more clearly defined the **nature** essay as a literary form. His writings provided vivid descriptions of outdoor life and gained popularity among a diverse audience.

Burroughs spent his boyhood exploring the lush countryside surrounding his family's dairy farm in the valleys of the Catskill Mountains, near Roxbury, New York. He left school at age sixteen and taught grammar school in the area until 1863, when he left for Washington, D.C., for a position as a clerk in the U. S. Treasury Department. While in Washington, Burroughs met poet Walt Whitman, through whom he began to develop and refine his writing style. His early essays were featured in the *Atlantic Monthly*. These works, including "With the Birds" and "In the Hemlocks," detailed Borroughs's boyhood recollections as well as recent observations of nature. In 1867 Burroughs published his first book, *Notes on Walt Whitman as a Poet and Person*, to which Whitman himself contributed significantly. Four years later, Burroughs produced *Wake-Robin* independently, followed

by *Winter Sunshine* in 1875, both of which solidified his literary reputation.

By the late 1800s, Burroughs had returned to New York and built a one-room log cabin house he called "Slabsides," where he philosophized with such guests as **John Muir**, **Theodore Roosevelt**, Thomas Edison, and Henry Ford. Burroughs accompanied Roosevelt on many adventurous expeditions and chronicled the events in his book, *Camping and Tramping with Roosevelt* (1907).

Burroughs's later writings took new directions. His essays became less purely naturalistic and more philosophical and deductive. He often sought to reach the levels of spirituality and vision he found in the works of Whitman, Emerson, and Thoreau. Borroughs explored poetry, for example, in *Bird and Bough* (1906), and *The Summit of the Years* (1913), which searched for a less scientific explanation of life, while *Under the Apple Trees* (1916) examined World War I.

Although Burroughs's enthusiastic inquisitiveness prompted travel abroad, he traveled primarily within the United States. He died in 1921 enroute from California to his home in New York. Following his death, the John

Burroughs Association was established through the auspices of the American Museum of Natural History. The association remains active and continues to maintain an exhibit at the museum.

[*Kimberley A. Peterson*]

RESOURCES
BOOKS

Burroughs, J. *In the Catskills: Selections from the Writings of John Burroughs.* Marietta: Cherokee Publishing Co., 1990.

Kanze, E. *The World of John Burroughs.* Bergenfield: Harry N. Abrams, 1993.

McKibben, B., ed. *Birch Browsings: A John Burroughs Reader.* New York: Viking Penguin, 1992.

Renehan Jr., E. J. *John Burroughs: An American Naturalist.* Post Mills: Chelsea Green, 1992.

Bush meat/market

Humans have always hunted wild animals for food, commonly known as bush meat. For many people living in the forests of Africa, South America, and Southeast Asia, **hunting** remains a way of life. However during the 1990s bush meat harvesting, particularly in Africa, was transformed from a local subsistence activity into a profitable commercial enterprise. In 2001, the United Nations Food and Agriculture Organization (FAO) warned that over-hunting in many parts of the world threatened the survival of some animal **species** and the food security of human forest-dwellers.

Although the killing of **endangered species** is illegal, such laws are rarely enforced across much of Africa. Bush meat harvesting is threatening the survival of **chimpanzees** (*Pan troglodytes*), gorillas (*Gorilla gorilla*), and bonobos (pygmy chimps) (*P. paniscus*), as well as some species of smaller monkeys and duikers (small antelope). Colobus monkeys and baboons, including drills and mandrills, are highly-prized as bush meat. In 2000, Miss Waldron's red colobus monkey (*Procolobus badius waldroni*) was declared extinct from over-hunting. Protected **elephants** and wild boar, gazelle, impala, crocodile, lizards, insects, **bats**, birds, snails, porcupine, and squirrel are all killed for bush meat. The chameleons and radiated tortoises of **Madagascar** are threatened with **extinction** from over-hunting.

In Indonesia and Malaysia, particularly on Borneo, orangutans (*Pongo pygmaeus*) are hunted despite their full legal protection. In Southeast Asia, freshwater turtles, harvested for food and medicine, are critically endangered.

Large animals and those that travel in groups are easy targets for hunters whose traditional spears, snares, and nets have been replaced with shotguns and even semi-automatic weapons. Since large animals reproduce slowly, the FAO is encouraging the hunting of smaller animals whose populations are more likely to be able to sustain the harvesting.

With populations of buffalo and other popular African game declining, hunters are beginning to stalk animals such as zebra and hippo that had been protected by local custom and taboos. Hunters also are turning to rodents and reptiles—animals that were previously shunned as food. Local customs against hunting bonobos are collapsing as more humans move into bonobo **habitat**.

Logging in the Congo Basin, the world's second-largest tropical forest, has contributed to the dramatic rise in bush meat harvesting. Logging companies are bringing thousands of workers into sparsely populated areas and feeding them with bush meat. Logging roads open up new areas to hunters and logging trucks carry bush meat to market and into neighboring countries. It has been estimated that about 2,000 bush meat hunters, supported by the logging industry, will kill more than 3,000 gorillas and 4,000 chimpanzees in 2002. Mining and wars also contribute to the increased harvesting of bush meat. Conservationists warn that by 2012 the large mammals of the Congo Basin will be gone.

The expanding market for African bush meat is fueled by human **population growth** and increasing poverty. In rural areas and among the poor, bush meat is often the only source of meat protein, particularly in times of **drought** or **famine**. It also may be the only source of cash income for the rural impoverished. With the urbanization of Africa, the taste for bush meat has moved to the cities. Bush meat is readily available in urban markets, where it is considered superior to domesticated meat. Primate meat is sold fresh, smoked, or dried. Bush meat is on restaurant menus throughout Africa and Europe. Monkey meat is smuggled into the United Kingdom where it is surreptitiously sold in butcher shops. It has been estimated that over one million tons of bush meat is harvested in the Congo Basin every year and that the bush meat trade in central and western Africa is worth more than one billion dollars annually.

Primate meat may contain viruses that are responsible for several emerging human diseases. Scientists believe that humans can contract the deadly **Ebola virus** from infected chimpanzees and gorillas. Furthermore, killing and dressing chimpanzee meat in the wild may contribute to the spread of the human immunodeficiency virus (HIV) that causes **AIDS**.

The Bushmeat Crisis Task Force (BCTF) is working with logging companies in the Congo Basin and West Africa to prevent illegal hunting and the use of company roads for the bush meat trade. The BCTF also works with governments and local communities to develop investment and foreign-aid policies that promote **sustainable development** and **wildlife** protection. Other organizations try to turn hunters into conservationists and educate bush meat consumers. TRAFFIC, which monitors wildlife trade for the **World Wildlife Fund**, and the World Conservation Union are urging that wildlife ownership be transferred from ineffectual African governments to landowners and local communities who have a vested interest in sustaining wildlife populations.

[*Margaret Alic Ph.D.*]

RESOURCES
BOOKS

Robinson, J. G. and E. L. Bennett, eds. *Hunting for Sustainability in Tropical Forests.* New York: Columbia University Press, 2000.

PERIODICALS

McGrath, Susan. "Survival Drill." *Audubon* May–June 2001: 70–77.

McNeil, Donald G. Jr. "The Great Ape Massacre." *New York Times Magazine* May 9, 1999.

OTHER

"Africa's Vanishing Apes." *The Economist*. January 10, 2002 [May 2002]. <http://www.economist.com/world/africa/PrinterFriendly.cfm?Story_ID=930798>.

Lobe, Jim. "Environment-Africa: Increasing Poverty Fuels Bush Meat Market." *World News* Inter Press Service. August 1, 2000 [May 2002]. <www.oneworld.org.ips2/aug00/02_19_004.html>.

Harman, Danna. "Bonobos' Threat: Hungry Humans." *The Christian Science Monitor*. June 7, 2001 [May 2002]. <http://www.csmonitor/com/durable/2001/06/07/p6s1.htm>.

"Bush Meat Utilization—A Critical Issue in East and Southern Africa." *COP11*. TRAFFIC. [May 2002]. <http://www.traffic.org/cop11/newsroom/bushmeat/html>.

Goodall, Jane. "At-Risk Primates" *The Washington Post* April 8, 2000 [May 2002]. <http://www.washingtonpost.com>.

ORGANIZATIONS

Bushmeat Crisis Task Force, 8403 Colesville Road, Suite 710, Silver Spring, MD USA 20910-3314 Fax: (301) 562-0888, Email: info@bushmeat.org, <http://www.bushmeat.org>

The Bushmeat Project, The Biosynergy Institute, P.O. Box 488, Hermosa Beach, CA USA 90254 Email: bushmeat@biosynergy.org, <http://www.bushmeat.net>

The Bushmeat Research Programme, Institute of Zoology, Zoological Society of London, Regents Park, London, UK NW1 4RY 44-207-449-6601, Fax: 44-207- 586-2870, Email: enquiries@ioz.ac.uk, <http://www.zoo.cam.ac.uk/ioz/projects/bushmeat.htm>

TRAFFIC East/Southern Africa - Kenya, Ngong Race Course, PO Box 68200, Ngong Road, Nairobi, Kenya (254) 2 577943, Fax: (254) 2 577943, Email: traffic@iconnect.co.ke, <http://www.traffic.org>

World Wildlife Fund, 1250 24th Street, N.W., P.O. Box 97180, Washington, DC USA 20090-7180 Fax: 202-293-9211, Toll Free: (800)CALL-WWF, <http://www.panda.org>

BWCA
see **Boundary Waters Canoe Area**

Bycatch

The use of certain kinds of **commercial fishing** technologies can result in large bycatches—incidental catches of unwanted fish, **sea turtles**, seabirds, and marine mammals. Because the bycatch animals have little or no economic value, they are usually jettisoned, generally dead, back into the ocean. This non-selectivity of commercial fishing is an especially important problem when trawls, seines, and drift nets are used. The bycatch consists of unwanted **species** of fish and other animals, but it can also include large amounts of undersized, immature individuals of commercially important species of fish.

The global amount of bycatch has been estimated in recent years at about 30 billion tons (27 million tonnes), or more than 1/4 of the overall catch of the world's fisheries. In waters of the United States, the amount of unintentional bycatch of marine life is about 2.2 billion lb per year (1 billion kg per year). However, the bycatch rates vary greatly among fisheries. In the fishery for cod and other groundfish species in the North Sea, the discarded non-market **biomass** averages about 42% of the total catch, and it is 44–72% in the Mediterranean fishery. Discard rates are up to 80% of the catch weight for trawl fisheries for shrimp.

Some fishing practices result in large bycatches of sea turtles, marine mammals, and seabirds. During the fishing year of 1988–1989, for example, the use of pelagic drift nets, each as long as 90 km, may have killed as many as 0.3-1.0 million **dolphins**, porpoises, and other cetaceans. During one 24-day monitoring period, a typical set of a drift-net of 19 km/day in the Caroline Islands of the south Pacific entangled 97 dolphins, 11 larger cetaceans, and 10 sea turtles. It is thought that bycatch-related **mortality** is causing population declines in 13 out of the 44 species of marine mammals that are suffering high death rates from human activities. It is also believed that hundreds of thousands of seabirds have been drowned each year by entanglement in pelagic drift-nets.

In 1991, the United Nations passed a resolution that established a moratorium on the use of drift-nets longer than 1.6 mi (2 km), and most fishing nations have met this guideline. However, there is still some continued use of large-scale drift nets. Moreover, the shorter nets that are still legal are continuing to cause extensive and severe bycatch mortality.

Sea turtles, many of which are federally listed as endangered, appear to be particularly vulnerable to being caught and drowned in the large, funnel-shaped trawl nets used to catch shrimp. Scientists have, however, designed simple, selective, **turtle excluder devices** (TEDs) that can be installed on the nets to allow these animals to escape if caught.

The use of TEDs is required in the United States and many other countries, but not by all of the fishing nations.

Purse seining for tuna has also caused an enormous mortality of certain species of dolphins and porpoises. This method of fishing is thought to have killed more than 200-thousand small cetaceans per year since the 1960s, but perhaps about one-half that number since the early 1990s due to improved methods of deployment used in some regions. Purse seining is thought to have severely depleted some populations of marine mammals.

The use of long-lines also results in enormous by-catches of various species of large fishes, such as tuna, **swordfish**, and **sharks**, and it also kills many seabirds. Long-lines consist of a fishing line up to 80 mi (130 km) long and baited with thousands of hooks. A study in the Southern Ocean reported that more than 44,000 albatrosses of various species are killed annually by long-line fishing for tuna.

In addition, great lengths of fishing nets are lost each year during storms and other accidents. Because the synthetic materials used to manufacture the nets are extremely resistant to degradation, these so-called "ghost nets" continue to catch and kill fish and other marine mammals for many years.

Clearly, fishery bycatches can cause substantial, non-target mortality that is a grave threat to numerous marine species. It is urgent that effective action be taken to curtail this wasteful environmental impact of commercial fishing as soon as possible.

[*Bill Freedman Ph.D.*]

RESOURCES

BOOKS

Alverson, D. L., ed. 1994. *Global Assessment of Fisheries Bycatch and Discards.* Fisheries Technical Paper No. 339. Rome: Food and Agricultural Organization of the United Nations (FAO), 1994.

American Fisheries Society. *Fisheries Bycatch: Consequences and Management.* Fairbanks, AK: University of Alaska Sea Grant, 1998.

Freedman, B. *Environmental Ecology.* San Diego, CA: Academic Press, 1995.

OTHER

The Indiscriminate Slaughter at Sea; Facts About Bycatch and Discards. 1998. World Wildlife Fund. [cited July 9, 2002]. <http://www.panda.org/news/press/archive/news_177f2.htm>.

ORGANIZATIONS

Worldwatch Institute, 1776 Massachusetts Ave., NW, Washington, DC USA 20036-1904 (202) 452-1999, Fax: (202) 296-7365, Email: worldwatch@worldwatch.org, http://www.worldwatch.org/

Bycatch reduction devices

Seabirds, **seals**, **whales**, **sea turtles**, **dolphins**, and non-targeted fish can be unintentionally caught and killed or maimed by modern fish and shrimp catching methods. This phenomenon is called "bycatch" or the unintended capture or **mortality** of living marine resources as a result of fishing. It is managed under such laws as the Migratory Bird Treaty Act, the **Endangered Species Act** of 1973, the **Marine Mammals Protection Act** of 1972 (amended in 1994), and, most recently, the Magnuson-Stevens Fishery Conservation and Management Act of 1996. The 1995 United Nations Code of Conduct for Responsible Fisheries, to which the United States is a signatory, also emphasizes the importance of **bycatch** reduction. Bycatch occurs because most fishing methods are not perfectly "selective," (i.e., they do not catch and retain only the desired size, sex, quality and quantity of **target species**). It also occurs because fishermen often have incentive to catch more fish than they will keep.

Although the exact extent of bycatch-related mortality is uncertain, public awareness of the problem has grown in the 1990s, leading to a deepening public perception that commercial fisheries are wasteful of the world's finite marine resources. According to a 1994 estimate of the United Nations' Food and Agriculture Organization, worldwide **commercial fishing** operations discarded 30 million tons (27 million metric tons) of fish, approximately one-fourth of the world catch, because they were the wrong type, sex, or size. In United States fisheries, a total of 149 **species** groups have been identified as bycatch, 67% of them finfish, crustaceans, or mollusks, and 37% of them protected marine mammals, turtles, or seabirds.

To address this problem, numerous research programs have been established to develop BRDs and other means to reduce bycatch. Much of this research, and the nation's bycatch reduction activities overall, is centered in the Department of Commerce's National Marine Fisheries Service (NMFS), which leads and coordinates the United States' collaborative efforts to reduce bycatch. In March 1997, NMFS proposed a draft long-term strategy, *Managing the Nation's Bycatch*, that seeks to provide structure to the service's diverse bycatch-related research and management programs. These include gear research, technology transfer workshops, and the exploration of new management techniques. NMFS's Bycatch Plan was intended to as a guide for its own programs and for its "cooperators" in bycatch reduction, including eight regional fishery management councils, states, three interstate fisheries commissions, the fishing industry, the conservation community, and other parties. In pursuing its mandate of conserving and managing marine resources, NMFS relies on the direction for bycatch established by the 104th Congress under the new National

Standard 9 of the Magnuson-Stevens Act, which states: "Conservation and management measures shall, to the extent practicable, (A) minimize bycatch and (B) to the extent bycatch cannot be avoided, minimize the mortality of such bycatch."

But, although the national bycatch standard applies across all regions, bycatch issues are not uniform for all fisheries. Indeed, bycatch is not always a problem and can sometimes be beneficial, (e.g., when bycatch species are kept and used as if they had been targeted species). But where bycatch is a problem, the exact nature of the problem and potential solutions will differ depending on the region and fishery. For instance, in the U.S. Gulf of Mexico shrimp fishery, which contributes about 70% of the annual U.S. domestic shrimp production, bycatch of juvenile red snapper (*Lutjanus blackfordi*) by shrimp trawlers reduces red snapper stocks for fishermen who target those fish. According to NMFS, "In the absence of bycatch reduction, red snapper catches will continue to be a fraction of maximum economic or biological yield levels." To address this problem, the Gulf of Mexico Fishery Management Council prepared an amendment to the shrimp fishery management plan for the Gulf of Mexico requiring shrimpers to use BRDs in their nets. But BRD effectiveness differs, and devices often lose significant amounts of shrimp in reducing bycatch. For instance, one device called a "30-mesh fisheye" reduced overall shrimp catches, and revenues, by 3%, an issue that must be addressed in any analysis of the costs and benefits of requiring BRDs, according to NMFS. In Southern New England, the yellowtail flounder (*Pleuronectes ferrugineus*) has been important to the New England groundfish fisheries for several decades. But the stock has been depleted to a record low because, from 1988–1994, most of the catch has been discarded by trawlers. Reasons for treating the catch as bycatch were that most of the fish were either too small for marketing or were smaller than the legal size limit. Among the solutions to this complex problem an increase in the mesh size of nets so smaller fish would not be caught and a redesign of nets to facilitate the escape of undersized yellowtail flounder and other bycatch species.

Although fish bycatch exceeds that of other marine animals, it is by no means the only significant bycatch problem. Sea turtle bycatch has received growing attention in recent years, most notably under the **Endangered Species Act** Amendments of 1988, which mandated a study of sea turtle conservation and the causes and significance of their mortality, including mortality caused by commercial trawlers. That study, conducted by the **National Research Council** (NRC), found that shrimp trawls accounted for more deaths of sea turtle juveniles, subadults, and breeders in coastal waters than all other human activities combined. According to the NRC's 1990 report, some 5,000–50,000

loggerhead turtles (*Caretta caretta*) and 500–5,000 Kemp ridleys (*Lepidochelys kempi*) a year are killed by shrimping operations from Cape Hatteras, North Carolina, to the Mexican border in the Gulf of Mexico. Deaths from drowning increase when trawlers tow their catch for longer than 60 minutes. Winter flounder trawling north of Cape Hatteras, **Chesapeake Bay** passive-gear fishing, and other fisheries were also found to be responsible for some turtle mortality.

To address the turtle bycatch problem, NMFS, numerous Sea Grant programs, and the shrimping industry conducted research that led to the development of several types of net installation devices that were called "turtle excluder devices" (TEDs) or "trawler efficiency devices." In 1983, the only TED approved by NMFS was one developed by the service itself. But in the face of industry concerns about using TEDs, the University of Georgia and NMFS tested devices developed by the shrimping industry, resulting in NMFS certification of new TED designs. Each design was intended to divert turtles out of shrimp nets, excluding the turtles from the catch without reducing the shrimp intake. Over a decade of development, these devices have been made lighter and today at least six kinds of TEDs have NMFS's approval. Early in the development of TEDs, NMFS tried to obtain voluntary use of the devices, but shrimpers considered them an expensive, time-consuming nuisance and feared they could reduce the size of shrimp catches. But NMFS, and environmental groups, countered that the best TEDs reduced turtle bycatch by up to 97% with slight or no loss of shrimp. By 1985, NMFS faced threats of lawsuits to shut down the shrimping industry because trawlers were not using TEDs. In response, NMFS convened mediation meetings that included environmentalists and shrimpers. The meetings led to an agreement to pursue a "negotiated rulemaking" to phase in mandatory TED use, but negotiations fell apart after state and federal legislators, under intense pressure, tried to delay implementation of TED rules. After intense controversy, NMFS published regulations June 28, 1987, on the use of TEDs by shrimp trawlers.

Dolphins are another marine animal that has suffered significant mortality levels as a result of bycatch associated with the eastern tropical Pacific Ocean tuna fishery. Several species of tuna are often found together with the most economically important tuna species, the yellowfin (*Thunnus albacares*). As a result, tuna fishermen have used a fishing technique--called "dolphin fishing," in which they set their nets around herds of dolphins to capture the tuna that are always close by. The spotted dolphin (*Stenella attenuata*) is most frequently associated with tuna. Spinner dolphin (*Stenella longirostris*), and the common dolphin (*Delphinus delphis*) also travel with tuna. According to one estimate, between 1960–1972 the U.S. fleet in the eastern tropical

Pacific Ocean fishery killed more than 100,000 dolphins a year. That number dropped to an estimated 20,000 after 1972, when the Marine Mammals Protection Act was passed. By 1989, the number was estimated at 12,643, largely because the number of boats in the U.S. tuna fishing fleet declined and those that were in operation killed fewer dolphins. U.S. fishing boats are now required to use techniques that allow dolphins to escape from tuna nets before the catch is hauled in. These include having fishermen jump into the ocean to hold the lip of the net below the water surface so the dolphin can jump out.

[*David Clarke*]

RESOURCES
BOOKS

Decline of the Sea Turtles, Causes and Prevention. Washington, D.C.: National Academy Press, 1990.
Dolphins and the Tuna Industry. Washington, D.C.: National Academy Press, 1992.
Managing the Nation's Bycatch: Priorities, Programs and Actions for the National Marine Fisheries Service. Washington, D.C.: National Marine Fisheries Service, National Oceanic and Atmospheric Administration, U.S. Department of Commerce, March 20, 1997.

C

Cadmium

A metallic element that occurs most commonly in **nature** as the sulfide, CdS. Cadmium has many important industrial applications. It is used to electroplate other metals, in the production of paints and **plastics**, and in nickel-cadmium batteries. The metal also escapes into the **environment** during the burning of **coal** and **tobacco**. Cadmium is ubiquitous in the environment, with detectable amounts present in nearly all water, air, and food samples. In high doses, cadmium is toxic. In lower doses, it may cause kidney disease, disorders of the circulatory system, weakening of bones, and, possibly, **cancer**. *See also* Itai-Itai disease

CAFE

see **Corporate Average Fuel Economy standards**

Cairo conference

The 1994 International Conference on Population and Development occurs at a defining moment in the history of international cooperation, the preamble to the Program of Action for the United Nations International Conference on Population and Development (UN ICPD) stated. This historic conference, the fifth such world meeting regarding population issues, was held in Cairo, Egypt September 5th to 13th, 1994, at a time when there was growing, worldwide recognition of the interdependence between global population, development and **environment**. The Cairo Conference considered itself as building upon the foundation already laid in the 1974 World Population Conference in Bucharest and the 1984 International Conference on Population in Mexico City.

Representatives from 180 countries throughout the world gathered to discuss these problems. In order to represent particular world societal blocs, the UN ensured representation from four communities of nations: developing countries, Muslim states, the industrialized West, and Catholic countries. The makeup of attending blocs was intentional, as controlling the world's population involves many hotly debated issues that bring about sharp differences of opinion on religious and cultural beliefs. Conference attendees and those that would follow up on the conference's recommendations in their host nations would have to address issues like marriage, divorce, abortion, contraception and homosexuality.

The call to address **population growth** came at a time when some groups said that the world was already pushing its food supply to natural limits. For example, marine biologists stated that ocean fisheries could not sustain catches upwards of 100 million tons of fish per year. Yet, that level was reached in 1989. Seafood prices have continued to rise since that time but seafood production has continued and so has population growth. In many countries, underground water supplies are reportedly strained and farming has been exploited to its capacity. The world population was estimated that year at 5.6 billion people, and projected to continue to increase at the 1994 level of 86 million people per year. Most frightening of all were the United Nations population projections for the twenty years to follow. These estimates ranged between 7.9 billion to 11.9 billion people inhabiting the earth by 2014.

In his opening address on September 5, UN Secretary-General Boutros-Boutros-Ghali said, **The efficacy of the economic order of the planet on which we live depended in great measure on the conference's outcome..** Leaders of several countries spoke of the many issues surrounding the problem and causing social and ethical dilemmas in setting population policy. Vice-President Albert Gore of the United States spoke of a **holistic** approach that also comprehensively addresses world poverty. Over the course of six days, more than 200 speakers shared their concerns about population and strategies to curb growth. Because the conference was mandated to address the issues and how they impacted population, the debate branched out to a number of economic, reproductive, trade, environmental, family structure, healthcare, and education issues.

The following areas of need were spelled out in the UN ICPD **Programme of Action**:

• Rapid urbanization and movement within and across borders are additional population problems the Cairo conference attempted to address. At that time, more than 93 percent of the 93 million people added to the population each year lived in developing countries. UN data just prior to the conference also showed that 43 percent of the world population resided in urban areas, up from 38 percent in 1975.

• The conference's delegates agreed on common objectives for controlling the world's population and even took on a more ambitious goal of population growth limit than anticipated by some. The conference objectives are targeted for completion by the year 2015. The common strategy limits family size and helps empower women to participate in reproductive decisions.

• The Cairo conference goals also addressed some **mortality** issues, and set an objective of reducing infant mortality worldwide by 45%. In 1994, infant mortality averaged 62 per 1,000 live births. The conference's objective aimed to lower that number to only 12 per 1,000 live births.

• A goal was also outlined to lower maternal mortality to 30 per 100,000 women. These objectives would be furthered by a pledge to offer prenatal care to all pregnant women.

• The conference also included an objective addressing education. It outlined that all children of school age be enabled to complete primary education.

• Contraception availability became another tool to limit population growth. The delegates settled on a goal of making contraception accessible to 70% of the world's population by the year 2015. However, they also added universal access to **family planning** as a final objective.

• In order to meet family planning goals, some leaders in the field suggested a shift in how leaders looked at their policies. Prior to the conference, they tended to focus on organized family planning programs. The Cairo conference objectives shifted the focus to a broadened policy, concerned with issues like gender equality, education, and empowerment of women to better deal with reproductive choices.

• The aging of the populations in more developed countries was also discussed.

Although unprecedented agreement was reached over the few days in Cairo, it didn't happen without heated debate and controversy. Most participants supported the notion that population policies must be based on individual rights and freedom of choice, as well as the rights of women and cul-

tures. Since abortion was considered such a sensitive issue, many delegates agreed that family planning offered women alternative recourse to abortion. The Cairo Conference ended with both hope and concern for the future. If the conference objectives were reached, the delegates projected that world population would rise from 5.7 billion in the mid-1990s to 7.5 billion by the year 2015, and then begin to stabilize. However, if the objectives were not reached, conference attendees concluded that the population would continue to rise to the perilous numbers discussed in the conferences preamble.

Following the Cairo conference, the United States began implementing the conference objectives. The U.S. Agency for International Development (USAID) held a series of meetings between September 1994 and January 1995 to draw upon the expertise and ideas of American leaders in the field and to begin encouraging U.S. participation in population control goals. Among the USAID's own objectives for the United States were new initiatives and expansion of programs in place that emphasized family planning and health, as well as education and empowerment of women in America. For example, USAID suggested expanding access to reproductive health information and services and improved prevention of HIV/AIDS and other sexually transmitted diseases (STDs). It also promoted research on male methods of contraception and new barrier methods to protect from both pregnancy and STDs. The agency focused on improving women's equality and rights beginning with review of existing U.S. programs. A major focus would also include improving women's economic equity in the United States.

Such initiatives cost money. The U.S. Congress appropriated $527 million to family planning and reproductive health in fiscal year 1995 alone. A number of additional programs received substantial funding to support the economic, social, and educational programs needed to complete the United States' objectives related to the Cairo conference. Included in this overall program was establishment of the President's Council on **Sustainable Development**, a combination public/private group charged with making recommendations on U.S. policies and programs that will help balance population growth with consumption of **natural resources**. The United States also worked cooperatively with Japan prior to and following the conference on a common agenda to coordinate population and health assistance goals. The program also was designed to help strengthen relations between Japan and the United States.

Five years after the Cairo conference, a follow-up conference was held in New York City from June 30th to July 2nd, 1999. In preparation for the conference, a forum was held at The Hague, Netherlands. At that forum, delegates of several countries that had expressed concerns about the

Cairo conference objectives five years earlier began trying to renegotiate the program of action. By the time the **Cairo Plus Five** conference convened in 1999, final recommendations had not been completed, as some member delegates, including representatives of the Vatican in Rome, attempted to alter the original Cairo program of action. Money was yet another issue in focus at the 1999 meeting. United Nations Population Fund (UNPFA) Chief Nafis Sadik estimated that $5.7 billion per year was needed to meet the Cairo Conference aims, an amount she described as **peanuts**. The actual average amount available to the **Programme of Action** was far less, $2.2 billion.

The real measure of success lies in whether or not world population stabilizes. There is some evidence to suggest that while still growing, the number of human beings inhabiting this planet is not growing at the nightmare rates projected at the beginning of the Cairo Conference. The United States Bureau of Census projections, revised in 1999, now estimate a world population of 8 billion by 2024, approximately the lowest estimate the Cairo Conference projected for 2014.

[*Joan M. Schonbeck*]

RESOURCES
PERIODICALS

*US Census Bureau*World Population Profile: 1998--Highlights (revised March 18,1999)

United Nations*Programme of Action of the UN ICPD, Preamble*

United Nations Chronicle, On-Line Edition*Population, Progress and Peanuts*Vol.XXXVI, Nov.3, 1999, Dept. of Information

"Focus on Population and Development: Follow-up on Cairo Conference."*US Department of State Dispatch*6, no. 1 (January 2, 1995): 4.

"Cairo Conference Reaches Consensus on Plan to Stabilize World Growth by 2015."*UN Chronicle*31 (December 1994): 63.

ORGANIZATIONS

Population Council, 1 Dag Hammarsrkjold Plz, New York, NY USA 10017 (212) 339-0500, Fax: (212) 755-6052, Email: pubinfo@popinfo.org, <http://www.popcouncil.org>

United Nations, , New York, NY USA 10017 , <http://www.un.org>

Calcareous soil

A calcareous **soil** is soil that has calcium carbonate ($CaCO_3$) in abundance. If a calcareous soil has hydrochloric **acid** added to it, the soil will effervesce and give off **carbon dioxide** and form bubbles because of the chemical reaction. Calcareous soils are most often formed from limestone or in dry environments where low rainfall prevents the soils from being leached of carbonates. Calcareous soils frequently cause **nutrient** deficiencies for many plants.

Dr. Helen Mary Caldicott (1938 –)
Australian physician and activist

Dr. Helen Caldicott is a pediatrician, mother, antinuclear activist, and environmental activist. Born Helen Broinowski in Melbourne, **Australia** on August 7, 1938, she is known as a gifted orator and a tireless public speaker and educator. She traces her activism to age 14 when she read Nevil Shute's *On the Beach*, a chilling novel about nuclear holocaust. In 1961 she graduated from the University of Adelaide Medical School with bachelor of medicine and bachelor of surgery degrees, which are the equivalent of an American M.D. She married Dr. William Caldicott in 1962, and returned to Adelaide, Australia to go into general medical practice. In 1966 she, her husband, and their three children moved to Boston, Massachusetts, where she held a fellowship at Harvard Medical School. Returning to Australia in 1969, she served first as a resident in pediatrics and then as an intern in pediatrics at Queen Elizabeth Hospital. There, she set up a clinic for cystic **fibrosis**, a genetic disease in children.

In the early 1970s, Caldicott led a successful campaign in Australia to ban atmospheric nuclear testing by the French in the South Pacific. Her success in inspiring a popular movement to stop the French testing has been attributed to her willingness to reach out to the Australian people through letters and television and radio appearances, in which she explained the dangers of **radioactive fallout**. Next, she led a successful campaign to ban the exportation of **uranium** by Australia. During that campaign she met strong **resistance** from Australia's government, which had responded to the 1974 international **oil embargo** by offering to sell uranium on the world market. (Uranium is the raw material for nuclear technology.) Caldicott chose to go directly to mine workers, explaining the effects of radiation on their bodies and their genes and talking about the effects of nuclear war on them and their children. As a result, the Australian Council of Trade Unions passed a resolution not to mine, sell, or transport uranium. A ban was instituted from 1975 to 1982, when Australia gave in to international pressure to resume the exportation.

In 1977 Dr. Caldicott and her husband immigrated to the United States, accepting appointments at the Children's Hospital Medical Center and teaching appointments at Harvard Medical School in Boston, Massachusetts. She was a co-founder of Physicians for Social Responsibility (PSR), and she was its president at the time of the March 28, 1979 nuclear accident at the **Three Mile Island Nuclear Reactor** in Pennsylvania. At that time, PSR was a small group of concerned medical specialists. Following the accident, the organization grew rapidly in membership, financial support, and influence. As a result of her television appearances and statements to the media following the Three Mile Island accident, Caldicott became a symbol of the movement to ban all **nuclear power** and oppose **nuclear weapons** in any form. Ironically, she resigned as president of PSR in 1983, when the organization had grown to over 20,000 members. At that time, she began to be viewed as an extreme radical in an organization that had become more moderate as it came to represent a wide, diversified membership.

She also founded Women's Action for Nuclear Disarmament (WAND). WAND has been an effective group lobbying Congress against nuclear weapons.

Throughout her career, Caldicott has considered her family to be her first priority. She has three children and she emphasizes the importance of building and maintaining a strong marriage, believing that good interpersonal relationships are essential before a socially-minded person can work effectively for broad social change.

Caldicott has developed videotapes and films and has written over one hundred articles which have appeared in major newspapers and magazines throughout the world. She has written four books. Her first, *Nuclear Madness: What You Can Do* (1978) is considered important reading in the antinuclear movement. Her second, entitled *Missile Envy*, was published in 1986. In *If You Love This Planet: A Plan to Heal the Earth* (1992), Caldicott discusses the race to save the planet from environmental damage resulting from excess energy consumption, **pollution**, **ozone layer depletion**, and global warming. She urges citizens of the United States to follow the example set by the Australians, who have adopted policies and laws designed to move their society toward greater corporate and institutional responsibility. She urges the various nations of the world to strive for a "new legal world order" by moving toward a sort of transnational control of the world's **natural resources**. One of her books, *A Desperate Passion*, (1996) is an autobiography in which she reflects upon crucial events that have influenced her and talks about people who have inspired her in her life and work. Her latest work was entitled *The New Nuclear Danger: George Bush's Military Industrial Complex* (2002).

Caldicott is the recipient of many awards and prizes including the SANE Peace Prize, the Ghandi Peace Prize, the John-Roger Foundation's Integrity Award (which she shared with Bishop Desmond Tutu), the Norman Cousins Award for Peace-making, the Margaret Mead Award, and many others.

[*Paulette L. Stenzel*]

RESOURCES
BOOKS

Caldicott, H. *If You Love This Planet: A Plan to Heal the Earth.* New York: W. W. Norton, 1992.
————. *Missile Envy.* New York: Bantam Books, 1986.

Dr. Helen Caldicott. (Corbis-Bettmann. Reproduced by permission.)

————. *Nuclear Madness: What You Can Do.* New York: Bantam Books, 1981.
————. *The New Nuclear Danger: George Bush's Military Industrial Complex.* The New Press, 2002.
Caldicott, H. *A Desperate Passion.* New York: Bantam Books, 1996.

PERIODICALS
Nixon, W. "Helen Caldicott: Practicing Global Preventive Medicine." *E Magazine* (September–October 1992): 12–5.

Lynton Keith Caldwell (1913 –)
American scholar and environmentalist

Lynton Caldwell has been a key figure in the development of **environmental policy** in the United States. A longtime advocate for adding an environmental amendment to the Constitution, Caldwell has insisted that the federal government has a duty to protect the **environment** that is akin to the defense of civil rights or freedom of speech.

Caldwell was born November 21, 1913 in Montezuma, Iowa. He received his bachelor of arts from the University of Chicago in 1935 and completed a master of arts at Harvard University in 1938. The same year, Caldwell accepted an assistant professorship in government at Indiana University in Bloomington. In 1943, he attained his doctorate from the University of Chicago and began publishing academic works the following year. The subjects of his early writings were not environmental; he published a study of administrative theory in 1944 and a study of New York state government in 1954. By 1964, however, Caldwell had shifted his emphasis, and he began to receive wide recognition for his work on environmental policy. In that year, he was presented with the William E. Mosher Award from the American Society for Public Administration for his article "Environment: A New Focus for Public Policy."

Caldwell's most important accomplishment was his prominent role in the drafting of the **National Environmental Policy Act** (NEPA) in 1969. As a consultant for the Senate Committee on Interior and Insular Affairs in 1968, he prepared *A Draft Resolution on a National Policy for the Environment*. His special report examined the constitutional basis for a national environmental policy and proposed a statement of intent and purpose for Congress. Many of the concepts first introduced in this draft resolution were later incorporated into the act. As consultant to that committee, Caldwell played a continuing role in the shaping of the NEPA, and he was involved in the development of the **environmental impact statement**.

In past years Caldwell has strongly defended the NEPA, as well as the regulatory agency it created, claiming that they represent "the first comprehensive commitment of any modern state toward the responsible custody of its environment." Although the act has influenced policy decisions at every level of government, the enforcement of its provisions have been limited. Caldwell argues that this is because environmental regulations have no clear grounding in the law. Statutes alone are often unable to withstand the pressure of economic interests. He has proposed an amendment to the Constitution as the best practical solution to this problem and he maintains that without such an amendment, environmental issues will continue to be marginalized in the political arena.

In addition to advising the Senate during the creation of the NEPA, Caldwell has done extensive work on international environmental policy. He has advised the Central Treaty Organization and served on special assignments in countries including Colombia, India, the Philippines, and Thailand. Currently, Caldwell serves as the Arthur F. Bentley Professor of Political Science emeritus and professor of public and environmental affairs at Indiana University in Bloomington, Indiana.

[*Douglas Smith*]

RESOURCES
BOOKS

Metzger, L., ed. "Caldwell, Lynton." In *Contemporary Authors New Revision Series*. Vol. 12. Detroit: Gale Research, 1984.

PERIODICALS

Caldwell, L. K. "20 Years with NEPA Indicate the Need." *Environment* 31 (December 1989): 6–11, 25–28.

California condor

With a wingspan of over 9 ft (about 3 m), the California condor (*Gymnogyps californianus*) has the largest wingspan of any bird in North America. The condor is a **scavenger**. It is a member of the New World vulture family (Cathartidae), and is closely related to the Andean condor found in South America.

The California condor is an **endangered species** barely avoiding **extinction**. While the condor population may have been in decline before the arrival of Europeans in North America, increased settlement of its **habitat, hunting**, and indirect human influences (**pollution**, pesticides, construction hazards such as power lines) have almost wiped out the bird.

The condor was once found in much of North America. Fossil records indicate that it lived from Florida to New York in the East and British Columbia to Mexico in the West. The condor's range became more restricted even before European settlers arrived in North America. By the time Columbus arrived in North America (mid-fifteenth century), the condor had already started its retreat west. At the beginning of the twentieth century it could still be found in southern California and Baja California (northern Mexico), but by 1950 its range had been reduced to a strip of about 150 mi (241 km) at the southern end of the San Joaquin Valley in California.

In the 1940s, the **National Audubon Society** initiated a census of the condor population and recorded approximately 60 birds. In the early 1960s, the population was estimated at 42 birds. A 1966 survey found 51 condors, an increase that may have been due to variability among the sampling systems used. By the end of 1986 there were 24 condors alive, 21 of which were in captivity.

Direct and indirect human stressors are directly responsible for the decline in the condor population during the past three centuries. California condors have been shot by hunters and poisoned by bait set out to kill coyotes. Their food has been contaminated with pesticides such as DDT or **lead** from **lead shot** found in animal carcasses shot and lost by hunters. The condor's rarity has made their eggs a valuable commodity for unscrupulous collectors. Original food sources, such as mammoths and giant camels of the Pleistocene era, have disappeared, but as cattle ranching and sheep farming developed, carcasses of these animals have become suitable substitutes. Habitat destruction and general harassment reduced the condor's range and population to a

California condor (*Gymnogyps californianus*).
(Photograph by Roy Toft. Tom Stack & Associates. Reproduced by permission.)

point where intervention was necessary to halt the bird's rapid decline to extinction.

Most other large birds and mammals in North America that lived during the late Pleistocene period have become extinct. However, there is a small chance that with intervention and restoration projects, the California condor can be saved from extinction. As a result, after much debate, the decision was made to remove all California condors from the wild and initiate a captive breeding and reintroduction program. The United States **Fish and Wildlife Service**, together with the California Fish and Game Commission, captured the last wild condor in April 1987.

The captive breeding program has thus far increased the condor population dramatically. Condors nest only once every two years and typically lay only one egg. By removing the egg from the nest for laboratory incubation, the female can be tricked into laying a replacement egg. This helps accelerate the population increase.

In January 1992, a pair of California condors was released into the wild. In October the male was found dead of kidney failure after apparently drinking from a pool of antifreeze in a parking lot near their sanctuary. Six additional condors were released at the end of 1992. In 1994, the total condor population consisted of nine wild condors and 66 birds in captivity. In 1995 another pair of captive birds was released into the wild, and with the release of more birds in following years, breeding pairs began to establish nest sites. In 2001, an egg that was laid by captive condors in the Los Angeles Zoo was placed in a condor nest in the wild, and the chick hatched successfully, only to die a few days later. The pair of condors then laid their own egg and hatched the first wild chick in April 2002. This was the first chick hatched in the wild in 18 years. As of April 2002,

there were 63 free-living condors in California and Arizona and 18 more were awaiting release from field pens where they were being acclimated to their natural habitat. The captive population totaled 104 birds.

The California Condor Recovery Plan is a consortium of private groups and public agencies. The goal of the recovery plan is to have 150 condors in each of two separate populations with at least 15 breeding pairs in each group. However, there are significant obstacles to the recovery of wild populations of California condors. These include continued pressure from development that causes accidents such as collisions with power lines and **poisoning** by crude oil from drilling operations that the birds mistake for pools of water. Although the capture and release of this **species** is working and the population of captive birds is growing, the survival of the California condor in the wild is still in doubt. Measures must be taken to reduce risks to wild birds from man-made hazards.

[*Eugene C. Beckham and Marie H. Bundy*]

RESOURCES
BOOKS

Darlington, D. *In Condor Country: A Portrait of a Landscape, Its Denizens and Its Defenders*. New York: Henry Holt, 1991.
Snyder, N. F. R., and H. A. Snyder. "Biology and Conservation of the California Condor." In *Current Ornithology*, Vol. 6, edited by D. M. Power. New York: Plenum, 1989.

PERIODICALS
Jurek, R. M. "An Historical Review of California Condor Recovery Programmes." *Vulture News* 23 (1990):3–7.
Kiff, L. "To the Brink and Back: The Battle to Save the California Condor." *Terra* 28 (1990):6–18.
Willwerth, J. "Can They Go Home Again?" *Time* 139 (27 January 1992):56–57.

John Baird Callicot (1941 –)
American environmental philosopher

J. Baird Callicott is a founder and seminal thinker in the modern field of environmental philosophy. He is best known as the leading contemporary exponent of Aldo Leopold's **land ethic**, not only interpreting Leopold's original works but also applying the reasoning of the land ethic to modern resource issues such as **wilderness** designation and **biodiversity** protection.

Callicott's 1987 edited volume, *Companion to A Sand County Almanac*, is the first interpretive and critical discussion of Leopold's classic work. His 1989 collection of essays, *In Defense of the Land Ethic*, explores the intellectual foundations and development of Leopold's ecological and philosophical insights and their ultimate union in his later works.

In 1991 Callicott, with Susan L. Flader, introduced to the public the best of Leopold's remaining unpublished and uncollected literary and philosophical legacy in a collection entitled *The River of the Mother of God and Other Essays* by **Aldo Leopold**.

Since his contribution to the inaugural issue of the journal *Environmental Ethics* in 1979, Callicott's articles and essays have appeared not only in professional philosophical journals and a variety of scientific and technical periodicals, but in a number of lay publications as well. He has contributed chapters to more than twenty books and is internationally known as an author and speaker. Born in Memphis, Tennessee, Callicott completed his Ph.D. in philosophy at Syracuse University in 1971, and has since held visiting professorships at a number of American universities. In 1971, while teaching at the University of Wisconsin-Stevens Point, Callicott designed and taught what is widely acknowledged as the nation's first course in **environmental ethics**. He is currently associate professor of Philosophy and Religion at the University of North Texas. With respect to the major value questions today in environmental ethics, Callicott's position can best be illustrated by his claim that there can be no value without valuers. He thus recognizes the legitimacy of both instrumental and intrinsic valuation on the part of a democracy of valuers, human and nonhuman.

Callicott perceives that we live today on the verge of a profound paradigm shift concerning human interactions with and attitudes toward the natural world. Much of his current work aims at distilling and giving voice to this shift and at articulating an ecologically accurate and philosophically valid concept of sustainability. This concept abandons dualistic (man versus **nature**), ethnocentric (modern Euroamerican), and static (e.g., wilderness frozen in time) elements of current thought in sustainability in favor of those emphasizing dynamic human/nonhuman **mutualism**, both in **ecosystem conservation** and in restoration. Through careful management consistent with the best ecological information and theories, Callicott believes, humans can not only protect but can and should enhance **ecosystem health**. His ideas challenge modern conventional wisdom and could radically alter much current theory in environmental philosophy as well as affect practice in wilderness management and economic development planning.

As one of a handful of scholars who launched the field of environmental ethics as we know it today, and as one of even fewer philosophers who have made their works accessible and pertinent not only to other academicians but to the general public, J. Baird Callicott occupies an important place in the history of modern philosophy. His work will undoubtedly continue to shape thinking on the ethical dimensions of resource management decisions for generations to come.

[*Ann S. Causey*]

RESOURCES
BOOKS

Callicott, J. B. *Companion to A Sand County Almanac.* Madison: University of Wisconsin Press, 1987.

————. *In Defense of the Land Ethic: Essays in Environmental Philosophy.* Albany: State University of New York Press, 1989.

————. *Nature in Asian Tradition of Thought: Essays in Environmental Philosophy.* Albany: State University of New York Press, 1989.

Campephilus principalis
see **Ivory-billed woodpecker**

Canadian Forest Service

The Canadian **Forest Service** (CFS) is a federal government agency with a responsibility to conduct research and analyze information related to the sustainable use of the Canadian forest resource. Because of the importance of the forest industries to the Canadian economy, CFS is a department with its own minister in the federal cabinet.

In Canada, provincial and territorial governments have the responsibility for establishing allowable forest harvests and establishing criteria for acceptable management plans for provincial-crown land, which comprises the bulk of the economic forest resource of the nation. CFS integrates with relevant departments of the provinces and territories through the Canadian Council of Forest Ministers.

CFS operates six forest research centers in the regions of Canada, as well as two research institutes: the Petawawa National Forestry Institute at Chalk River, Ontario, and the Forest **Pest** Management Institute at Sault Ste. Marie, Ontario. Collectively, these institutions and their staff comprise the major force in forestry-related research in Canada.

The Government of Canada has committed to the practice of ecologically **sustainable forestry**, and CFS works with other levels of government, industry, educational institutions, and the public towards this end.

Research programs within CFS are diverse, and include activities related to tree anatomy, physiology, and productivity, techniques of **biomass** and stand inventory, determination of site quality, and research of forest harvesting practices, management practices aimed at increasing productivity and economic value, and forest pest management.

CFS also supports much of the **forest management** and research conducted within each of the provinces and territories, through joint federal-provincial Forest Development Agreements. CFS also supports much of the forestry research conducted at Canadian universities through research contracts and joint-awards programs with other government agencies and industry.

An important initiative, begun in 1992, is the Model Forest Program under Canada's *Green Plan*. The intent of this program is to provide substantial support for innovative forest management and research, conducted by integrated teams of partners involving industry, all levels of government, universities, non-government organizations, woodlot owners, and other groups and citizenry. A major criterion for success in the intense competition for funds towards a model forest is that the consortium of proponents demonstrate a vision of ecologically sustainable forest management, and a likely capability of achieving that result.

CFS's most recent endeavor is titles Forest 2020. This project was established at the 2001 Annual Meeting of Canadian Council of Forest Ministers (CCFM), and revolves around "incorporating wood fibre production through the establishment of plantations of fast growing high-yield tree **species**, and intensified silviculture in previously harvested, or second growth, forest areas."

[*David A. Duffus and Amy Strumolo*]

RESOURCES
BOOKS

The State of Canada's Forests, 1991. Second Report to Parliament. Ottawa: Canadian Forest Service, 1992.

ORGANIZATIONS

Canadian Forest Service, Natural Resources Canada, 580 Booth Street, 8th Floor, Ottawa, OntarioCanada K1A 0E4 (613) 947-7341, Fax: (613) 947-7397, Email: cfs-scf@nrcan.gc.ca, <http://www.nrcan-rncan.gc.ca/cfs-scf>

Canadian Parks Service

The Canadian Parks Service (CPS) is the government agency charged with fulfilling the statutes and policies of Canada's national parks. The first **national park** was established in Banff, Alberta, in 1885, although parks were not institutionalized in Canada until 1911. The original policy and program established national parks to protect characteristic aspects of Canada's natural heritage, and, at the same time, cater to the benefit, enjoyment, and education of its people.

The National Parks Act of 1930 required parliamentary approval of parks, prohibited **hunting**, mining, and exploration, and limited forest harvesting. Other federal statutes gave control over lands and resources to the provinces, which meant land assembly for park development could not proceed without intergovernmental agreement. An amendment to the Parks Act increased the Park Service's ability to enforce regulations and levy meaningful penalties.

In 1971 the park system was expanded to include examples of all 39 terrestrial natural regions of Canada. Currently, 25 regions are represented by 39 parks. The Canadian government has pledged to develop sites in the remaining 14 regions in upcoming years. The CPS hopes that national parks will eventually encompass 12 percent of Canada's land base; they currently occupy only 1.8 percent.

Canada's four marine coasts have been classified into 29 regions, all of which are to be represented in a marine parks program. One site has been formally established, while four others are being considered. The CPS mandate was amended in 1986 to share control of marine parks with two other federal agencies, which manage fisheries and marine **transportation**.

Current park policy emphasizes the maintenance of **ecological integrity** over tourism and **recreation**. Much conflict, however, has centered on development within park boundaries. Townsites, highways, ski resorts, and **golf courses** provide public enjoyment, but conflict with the protective mandate of the parks' charter. Commercial and recreational fishing, **logging**, and **trapping** have also been allowed, despite legal prohibitions. In 1992, the Canadian Parks and **Wilderness Society** and the Sierra Legal Defense Fund successfully sued the Canadian government to stop logging in Wood Buffalo National Park, a protected **habitat** for both the endangered **whooping crane** and wood **bison**.

[*Amy A. Strumolo*]

Canadian Wildlife Service

The Canadian **Wildlife** Service (CWS) was established in 1947. Until 1971 it was governed by the federal government's Parks Branch in the Department of Indian and Northern Development. Since 1971 it has been a branch of **Environment Canada**, most recently under the **Conservation** and Protection division. The Service is served by five regional offices across Canada and is headquartered in Hull, Quebec. There are four branches within the CWS: the Migratory Birds and Conservation Branch; the North American Waterfowl Management Plan Implementation Branch; the Wildlife Toxicology and Surveys Branch/National Wildlife Research Centre; and the Program Analysis and Co-ordination Branch.

Wildlife matters were delegated to the provinces by the Canadian constitution, and responsibility is generally in the hands of provincial administrations. Federal-provincial cooperation was facilitated through meetings of the Federal-Provincial Wildlife Ministers Conferences, ongoing since 1922 and an annual event since 1945. Through those ongoing discussions, federal wildlife policy has developed mainly around areas of transboundary issues and problems deemed in the national interest. Currently, the CWS's primary responsibility is to enforce the Migratory Birds Convention Act, the Canada Wildlife Act, the Convention on International Trade in **Endangered Species** of Fauna and Flora (CITES), and the Ramsar **Convention on Wetlands of International Importance**. Furthermore, it has increased responsibility on federal lands, north of 60°N latitude, although in recent years much responsibility has been delegated to the territorial wildlife services and local co-management agreements with aboriginal peoples.

Within those statutes and agreements CWS is responsible for policy and strategy development, enforcement, research, public relations, education and interpretation, **habitat** classification, and the management of about 98 sanctuaries and 49 wildlife areas. The combination of a national and international mandates and diverse landscapes of Canada make CWS the pivotal **wildlife management** institution in Canada. However, CWS's increasing reliance on cooperative measures with provincial governments and nongovernmental organizations, including **Ducks Unlimited** Canada and **World Wildlife Fund** Canada, has led the organization to less direct management and more coordination activities. Critics of Canada's wildlife management direction have suggested that the once world-renowned repository of research expertise in CWS has suffered in recent years.

[*David A. Duffus*]

RESOURCES

ORGANIZATIONS

Canadian Wildlife Service, Environment Canada, Ottawa, Ontario Canada K1A 0H3 (819) 997-1095, Fax: (819) 997-2756 , Email: cws-scf@ec.gc.ca, <http://www.cws-scf.ec.gc.ca>

Cancer

A malignant tumor, cancer comprises a broad spectrum of malignant neoplasms classified as either carcinomas or sarcomas. Carcinomas originate in the epithelial tissues, while sarcomas originate from connective tissues and structures that have their origin in mesodermal tissue. Cancer is an invasive disease that spreads to various parts of the body. It spreads directly to those tissues immediately surrounding the primary site of the cancer and may spread to remote parts of the body through the lymphatic and circulatory systems.

Cancer occurs in most, if not all, multicellular animals. Evidence from fossil records reveal bone cancer in dinosaurs, and sarcomas have been found in the bones of Egyptian mummies. Hippocrates is credited with coining the term *carcinoma*, the Greek word for crab. Why the word for crab was chosen enjoys much speculation, but may have had to do with the sharp, biting pain and invasive, spreading nature of the disease.

A **carcinogen** is any substance or agent that produces or induces the development of cancer. Carcinogens are known to affect and initiate metabolic processes at the level of DNA (the information-storing molecules in cells). DNA damage (**mutation**) is the development of cancer after exposure to a carcinogen. This kind of mutation is actually reversible; our bodies continually experience DNA damage, which is continually being corrected. It is only when promoter cells intervene during cell proliferation that tumors begin to develop. Although several agents can induce cell division, only promoters induce tumor development.

An example of this process would be what happens in an epidermal cell, when its DNA undergoes rapid, irreversible alteration or mutation after exposure to a carcinogen. The cell undergoes proliferation, producing altered progeny, and it is at this point that the cell may proceed on one of two pathways. The cell may undergo interrupted exposure to promoters and experience early reversible precancerous lesions. Or it may experience continuous exposure to the promoters, thereby causing malignant cell changes. During the late phase of promotion, the primary epidermal cell becomes tumorous and begins to invade normal cells; then it begins to spread. It is at this stage that tumors are identified as malignant.

The spread of tumors throughout the body is believed to be governed by several processes. One possible mechanism is direct invasion of contiguous organs. This mechanism is poorly understood, but it involves multiplication, mechanical pressure, release of lytic enzymes, and increased motility of individual tumor cells. A second process is metastasis. This is the spread of cancer cells from a primary site of origin to a distant site, and it is the life-threatening aspect of malignancy. At present there are many procedures available to surgeons for successfully eradicating primary tumors; however, the real challenge in reducing cancer **mortality** is finding ways to control metastasis.

Clinical manifestations of cancer take on many forms. Usually little or no pain is associated with the early stages of malignant disease, but pain does affect 60–80% of those terminally ill with cancer. General mechanisms causing pain associated with cancer include pressure, obstruction, invasion of a sensitive structure, stretching of visceral surfaces, tissue destruction, and inflammation. Abdominal pain is often caused by severe stretching from the tumor invasion of the hollow viscus, as well as tumors that obstruct and distend the bowel. Tumor compression of nerve endings against a firm surface also creates pain. Brain tumors have very little space to grow without compressing blood vessels and nerve endings between the tumor and the skull. Tissue destruction from infection and necrosis can also cause pain. Frequently infection occurs in the oral area, in which a common cause of pain is ulcerative lesions of the mouth and esophagus.

Cancer treatments involve chemotherapy, radiotherapy, surgery, immunotherapy, and combinations of these

Frequency of Cancer-Related Death	
Cancer Site	**Number of Deaths Per Year**
Lung	160,100
Colon and rectum	56,500
Breast	43,900
Prostate	39,200
Pancreas	28,900
Lymphoma	26,300
Leukemia	21,600
Brain	17,400
Stomach	13,700
Liver	13,000
Esophagus	11,900
Bladder	12,500
Kidney	11,600
Multiple myeloma	11,300

tumor-immune rejection response, and modification of cancer cell susceptibility to the lytic or tumor static effects of the immune system. As with other cancer therapies immunotherapies are not without their own side effects. Most common are flu-like symptoms, skin rashes, and vascular-leak syndrome. At their worst, these symptoms are usually less severe than those of current chemotherapy and radiation treatments. *See also* Hazardous material; Hazardous waste; Leukemia; Radiation sickness

[*Brian R. Barthel*]

RESOURCES
BOOKS

Aldrich, T., and J. Griffith. *Environmental Epidemiology.* New York: Van Nostrad Reinhold, 1993.

Captive propagation and reintroduction

Captive propagation is the deliberate breeding of wild animals in captivity in order to increase their numbers. Reintroduction is the deliberate release of these **species** into their native **habitat**. The Mongolian wild horse, Pere David's deer, and the American **bison** would probably have become extinct without captive propagation. Nearly all cases of captive propagation and reintroduction involve threatened or **endangered species**. Zoos are increasingly involved in captive propagation, sometimes using new technologies. One of these, allows a relatively common species of antelope to act as a surrogate mother and give birth to a **rare species**.

Once suitable sites are selected, a reintroduction can take one of three forms. *Reestablishment reintroductions* take place in areas where the species once occurred but is now entirely absent. Recent examples include the red wolf, the **black-footed ferret**, and the **peregrine falcon** east of the Mississippi River. Biologists use *augmentation reintroduction* to release captive-born wild animals into areas in which the species still occurs but only in low numbers. These new animals can help increase the size of the population and enhance genetic diversity. Examples include a small Brazilian monkey called the golden lion tamarin and the peregrine falcon in the western United States. A third type, *experimental reintroduction*, acts as a test case to acquire essential information for use on larger-scale permanent reintroductions. The red wolf was first released as an experimental reintroduction. A 1982 amendment to the **Endangered Species Act** facilitates experimental reintroductions, offering specific exemptions from the Act's protection, allowing managers greater flexibility should reintroduced animals cause unexpected problems.

modalities. Chemotherapy and its efficacy is related to how the drug enters the cell cycle; the design of the therapy is to destroy enough malignant cells so that the body's own immune system can destroy the remaining cells naturally. Smaller tumors with rapid growth rates seem to be most responsive to chemotherapy. Radiation therapy is commonly used to eradicate tumors without excessive damage to surrounding tissues. Radiation therapy attacks the malignant cell at the DNA level, disrupting its ability to reproduce. Surgery is the treatment of choice when it has been determined that the tumor is intact and has not metastasized beyond the limits of surgical excision. Surgery is also indicated for benign tumors that could progress into malignant tumors. Premalignant and in situ tumors of epithelial tissues, such as skin, mouth, and cervix, can be removed.

Chemotherapy and radiation treatments are the most commonly used therapies for cancer. Unfortunately, both methods produce unpleasant side effects; they often suppress the immune system, making it difficult for the body to destroy the remaining cancer even after the treatment has been successful. In this regard, immunotherapy holds great promise as an alternative treatment, because it makes use of the unique properties of the immune system.

Immunotherapies for the treatment of cancer are generally referred to as biological response modifiers (BRMs). BRMs are defined as mammalian gene products, agents, and clinical protocols that affect biologic responses in host-tumor interactions. Immunotherapies have a direct cytotoxic effect on cancer cells, initiation or augmentation of the host's

The release of baby Kemp's ridley sea turtles on a beach in Mexico. (Photograph by C. Allan Morgan. Peter Arnold Inc. Reproduced by permission.)

Yet captive propagation and reintroduction programs have their drawbacks, the chief one being their high cost. Capture from the wild, food, veterinary care, facility use and maintenance all contribute significant costs to maintaining an animal in captivity. Other costs are incurred locating suitable reintroduction sites, preparing animals for release, and monitoring the results. Some conservationists have argued that the money would be better spent acquiring and protecting habitat in which remnant populations already live.

There are also other risks associated with captive propagation programs such as disease, but perhaps the greatest biological concern is that captive populations of endangered species might lose learned or genetic traits essential to their survival in the wild. Animals fed from birth, for example, might never pick up food-gathering or prey-hunting skills from their parents as they would in the wild. Consequently, when reintroduced such animals may lack the skill to feed themselves effectively. Furthermore, captive breeding of animals over a number of generations could affect their **evolution**. Animals that thrive in captivity might have a selective advantage over their "wilder" cohorts in a **zoo**, but might

be disadvantaged upon reintroduction by the very traits that aided them while in captivity.

Despite these shortcomings, the use of captive propagation and reintroduction will continue to increase in the decades to come. Biologists learned a painful lesson about the fragility of endangered species in 1986 when a sudden outbreak of canine distemper decimated the only known group of black-footed ferrets. The last few ferrets were taken into captivity where they successfully bred. Even as new ferret populations become established through reintroduction, some ferrets will remain as captive breeders for insurance against future catastrophes. Biologists are also steadily improving their methods for successful reintroduction. They have learned how to select the combinations of sexes and ages that offer the best chance of success and have developed systematic ways to choose the best reintroduction sites.

Captive propagation and reintroduction will never become the principal means of restoring threatened and endangered species, but it has been proven effective and will continue to act as insurance against sudden or catastrophic losses in the wild. *See also* Biodiversity; Extinction; Wildlife management; Wildlife rehabilitation

[*James H. Shaw*]

RESOURCES
PERIODICALS

Jones, Suzanne R., ed. "Captive Propagation and Reintroduction: A Strategy for Preserving Endangered Species?" *Endangered Species Update* 8 (1) (1990): 1-88.
Lindburg, Donald G. "Are Wildlife Reintroductions Worth the Cost?" *Zoo Biology* 11 (1992): 1-2.

Carbamates
see **Pesticide**

Carbon

The seventeenth most abundant element on earth, carbon occurs in at least six different allotropic forms, the best known of which are diamond and graphite. It is a major component of all biochemical compounds that occur in living organisms: carbohydrates, proteins, lipids, and nucleic acids. Carbon-rich rocks and minerals such as limestone, gypsum, and marble often are created by accumulated bodies of aquatic organisms. Plants, animals, and **microorganisms** cycle carbon through the **environment**, converting it from simple compounds like **carbon dioxide** and **methane** to more complex compounds like sugars and starches, and then, by the action of **decomposers**, back again to simpler compounds. One of the most important **fossil fuels**, **coal**, is composed chiefly of carbon.

Carbon cycle

Carbon makes up no more than 0.27% of the mass of all elements in the universe and only 0.0018% by weight of the elements in the earth's crust. Yet, its importance to living organisms is far out of proportion to these figures. In contrast to its relative **scarcity** in the **environment**, it makes up 19.4% by weight of the human body. Along with **hydrogen**, carbon is the only element to appear in every organic molecule in every living organism on earth.

The series of chemical, physical, geological, and biological changes by which carbon moves through the earth's air, land, water, and living organisms is called the carbon cycle.

In the **atmosphere**, carbon exists almost entirely as gaseous **carbon dioxide**. The best estimates are that the earth's atmosphere contains 740 billion tons of this gas. Its global concentration is about 350 **parts per million** (ppm), or 0.035% by volume. That makes carbon dioxide the fourth most abundant gas in the atmosphere after **nitrogen**, oxygen and argon. Some carbon is also released as **carbon monoxide** to the atmosphere by natural and human mechanisms. This gas reacts readily with oxygen in the atmosphere, however, converting it to carbon dioxide.

Carbon returns to the hydrosphere when carbon dioxide dissolves in the oceans, as well as in lakes and other bodies of water. The solubility of carbon dioxide in water is not especially high, 88 milliliters of gas in 100 milliliters of water. Still, the earth's oceans are such a vast **reservoir** that experts estimate that approximately 36,000 billion tons of carbon are stored there. They also estimate that about 93 billion tons of carbon flows from the atmosphere into the hydrosphere each year.

Carbon moves out of the oceans in two ways. Some escapes as carbon dioxide from water solutions and returns to the atmosphere. That amount is estimated to be very nearly equal (90 billion tons) to the amount entering the oceans each year. A smaller quantity of carbon dioxide (about 40 billion tons) is incorporated into aquatic plants.

On land, green plants remove carbon dioxide from the air through the process of photosynthesis—a complex series of chemical reactions in which carbon dioxide is eventually converted to starch, cellulose, and other carbohydrates. About 100 billion tons of carbon are transferred to green plants each year, and a total of 560 billion tons of the element is thought to be stored in land plants alone.

The carbon in green plants is eventually converted into a large variety of organic (carbon-containing) compounds. When green plants are eaten by animals, carbohydrates and other organic compounds are used as raw materials for the manufacture of thousands of new organic substances. The total collection of complex organic compounds stored in all kinds of living organisms represents the reservoir of carbon in the earth's **biosphere**.

The cycling of carbon through the biosphere involves three major kinds of organisms. Producers are organisms with the ability to manufacture organic compounds such as sugars and starches from inorganic raw materials such as carbon dioxide and water. Green plants are the primary example of producing organisms. Consumers are organisms that obtain their carbon (that is, their food) from producers: all animals are consumers. Finally, **decomposers** are organisms such as bacteria and **fungi** that feed on the remains of dead plants and animals. They convert carbon compounds in these organisms to carbon dioxide and other products. The carbon dioxide is then returned to the atmosphere to continue its path through the carbon cycle.

Land plants return carbon dioxide to the atmosphere during the process of **respiration**. In addition, animals that eat green plants exhale carbon dioxide, contributing to the 50 billion tons of carbon released to the atmosphere by all forms of living organisms each year. Respiration and **decomposition** both represent, in the most general sense, a reverse of the process of **photosynthesis**. Complex organic compounds are oxidized with the release of carbon dioxide and water—the raw materials from which they were originally produced.

At some point, land and aquatic plants and animals die and decompose. When they do so, some carbon (about 50 billion tons) returns to the atmosphere as carbon dioxide. The rest remains buried in the earth (up to 1,500 billion tons) or on the ocean bottoms (about 3,000 billion tons). Several hundred million years ago, conditions of burial were such that organisms decayed to form products consisting almost entirely of carbon and **hydrocarbons**. Those materials exist today as pockets of the fossil fuels—coal, oil, and **natural gas**. Estimates of the carbon stored in **fossil fuels** range from 5,000 to 10,000 billion tons.

The processes that make up the carbon cycle have been occurring for millions of years, and for most of this time, the systems involved have been in equilibrium. The total amount of carbon dioxide entering the atmosphere from all sources has been approximately equal to the total amount dissolved in the oceans and removed by photosynthesis. However, a hundred years ago changes in human society began to unbalance the carbon cycle. The Industrial Revolution initiated an era in which the burning of fossil fuels became widespread. In a short amount of time, large amounts of carbon previously stored in the earth as **coal**, oil, and natural gas were burned up, releasing vast quantities of carbon dioxide into the atmosphere.

Between 1900 and 1992, measured concentrations of carbon dioxide in the atmosphere increased from about 296 ppm to over 350 ppm. Scientists estimate that fossil fuel **combustion** now released about five billion tons of carbon dioxide into the atmosphere each year. In an equilibrium situation, that additional five billion tons would be absorbed by the oceans or used by green plants in photosynthesis. Yet this appears not to be happening: measurements indicate that about 60% of the carbon dioxide generated by fossil fuel combustion remains in the atmosphere.

The problem is made even more complex because of **deforestation**. As large tracts of forest are cut down and burned, two effects result: carbon dioxide from forest fires is added to that from other sources, and the loss of trees decreases the worldwide rate of photosynthesis. Overall, it appears that these two factors have resulted in an additional one to two billion tons of carbon dioxide in the atmosphere each year.

No one can be certain about the environmental effects of this disruption of equilibria in the carbon cycle. Some authorities believe that the additional carbon dioxide will augment the earth's natural **greenhouse effect**, resulting in long-term global warming and **climate** change. Other argue that we still do not know enough about the way oceans, clouds, and other factors affect climate to allow such predictions.

This controversy involves a difficult choice. Should actions that could potentially cost billions of dollars be taken to reduce the **emission** of carbon dioxide when evidence for climate change is still uncertain? Or should governments wait until that evidence becomes more clear, with the risk that needed actions may then come too late.

[*David E. Newton*]

RESOURCES
BOOKS

McGraw-Hill Encyclopedia of Science & Technology. 7th ed. New York: McGraw-Hill, 1992.

Carbon dating
see **Radiocarbon dating**

Carbon dioxide

The fourth most abundant gas in the earth's **atmosphere**, **carbon** dioxide occurs in an abundance of about 350 **parts per million**. The gas is released by volcanoes and during **respiration**, **combustion**, and decay. Plants convert carbon dioxide into carbohydrates by the process of **photosynthesis**. Carbon dioxide normally poses no health hazard to humans. An important factor in maintaining the earth's **climate**, molecules of carbon dioxide capture heat radiated from the earth's surface, raising the planet's temperature to a level at which life can be sustained, a phenomenon known as the **greenhouse effect**. Some scientists believe that increasing levels of carbon dioxide resulting from human activities are now contributing to a potentially dangerous global warming.

Carbon emissions trading

Carbon emissions trading (CET) is a practice allowing countries—and corporations—to trade their harmful carbon emissions for credit to meet their designated carbon **emission** limits. Most developed countries approved this system in 1992 when the United Nations Framework Convention on Climate Change (UNFCCC) was presented. The document provided for limits on greenhouse gas emissions in an attempt to stem the determination of climate change around the world. The

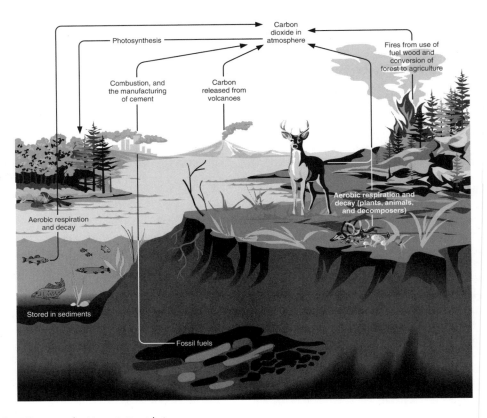

Carbon cycle. (Illustration by Hans & Cassidy.)

United States as of 2001 had backed away from the Kyoto Treaty stating similar restriction but would remain obligated in some regard to the UNFCCC. At least in theory, CET allows one country who might not be maintaining their target to trade credits with others that are well under theirs.

According to information from the Australian Academy of Science, in *Nova: Science in the news,* **carbon dioxide** (CO2) is known as a "greenhouse gas" ones that came out of the Earth's crust before life was known to begin, helping to stabilize the earth's temperatures in order to sustain life. The "greenhouse effect" occurs when the sun's heat energy passes through the **atmosphere** and warms up the earth with no interference. The Earth then radiates that same energy back into space. Other **greenhouse gases** include, water vapor (the primary greenhouse gas), **methane, ozone, carbon monoxide**, and **nitrous oxide**. All of these absorb part of this energy and send it in all directions—including the Earth. By the twenty-first century however, the emissions of greenhouse gases were not in the balance they had been for ages due to human activity, particularly that of burning **fossil fuels** such as **coal**, oil, and **natural gas**. Due to that, and to industrial development as well as residential development

in the developed countries, forests were cleared at such a rate that inclined some scientists to wonder whether these practices were harmful in the significantly increased emissions of greenhouse gases.

The **Environmental Protection Agency** (EPA) launched a method of emissions trading known as, Online Allowance Transfer System (OATS). It is an online system that allows companies engaged in the **sulfur dioxide** and **nitrogen** oxide to record trades directly through the internet without having to file papers with the EPA. No such method yet exists for CET. In view of possible future issues regarding the trades in the United States in February 2002, Senators John McCain (R.-Ariz) and Sam Brownback(R.-Kan.) jointly introduced legislation that propsed establishing a national voluntary registry for companies to register carbon emission reductions. According to *Coal Age* "The propsal is a step toward developing a 'cap-and-trade' carbon emissions control program." Senators McCain and Leiberman (D.-CT) are also working on legislation that supports carbon emissions trading. McCain, noted *Coal Age* said that, "The registry would support current voluntary trading practices in private industry and other nongovernmental organizations."

The real issue that has resulted from CET is the trading market that has emerged. Opinions are varied regarding the advisability, and inherent dangers of CET. The International Carbon Bank & Exchange offered insight on CET through its web site. "Emissions trading reduces costs by allowing a field of players to achieve emissions redutions using market mechanisms. Over time, these mechanisms will drive emissions down and finance the shift to clean energy."

In August 2001, the Joyce Foundation granted $760,100 to fund the design phase of the Chicago Climate Exchange (CCX). The grant was actually directed to the J. L. Kellogg Graduate School of Management at Northwestern University in support of the work being done by Dr. Richard Sandor, an internationally-known trader. With the presidential administration of George W. Bush hesitant to apply a "cap-and-trade" system, the direction of the CCX and open market trading is viewed as providing a desirable alternative—particularly in light of the United States withdrawal from the Kyoto Treaty. According the *Environmental News Network (ENN)* in an article on the new design program, "The CCX's stated goal is to reduce participants' greenhouse gas emissions by five percent below 1999 levels over five years. By comparison, the countries that ratified the Kyoto Protocol must reduce emissions of carbon dioxide to an average of 5.2% below 1990 levels during the five-year period 2008 to 2012." By November 2001, the cities of Chicago and Mexico City announced their participation in CET by joining the CCX, with Chicago becoming the first American city to do so. Chicago's Mayor Richard Daley became honorary chair of the exchange, still in the design phase. He noted that, "For years our financial exchanges have been a vital parto f the local and naitonal economy. This is a good example of the kind of innovation that will help us solve our energy and environmental problems," according to the *Environment News Service.* Other businesses and agencies participating in the design phase of the CCX including, Agrilance, a partnership of agricultural producer-owners, local cooperatives and regional cooperatives; BP; Cinergy; **Ducks Unlimited**; DuPont; the energy company, Exeton; International Paper; the Iowa Farm Bureau Federation; Manitoba (CA) Hydro; National Council of Farmer Cooperatives; PG&E National Energy Group; Suncor Energy; Swiss Re, a reinsurance firm; **The Nature Conservancy**; and, Waste Energy.

On a broader scope of international trading, however, the United States' withdrawal from Kyoto was posing a possible problem. Due to the lack of participation, U.S. companies were seeing a possible short-term advantage in international business competition due to lower costs without the emissions trading—but those same companies, such as DuPont, already cutting emissions, might face the long-

term shortchange depending on which way the international climate change policy could effect trading those emissions. In March 2002, Julie Vorman writing on the CET market for *Reuters* quoted Eileen Claussen, president of the Pew Center on Global Climate Change, noting that, "Despite the United States inaction, it is abundantly clear that we are beginning to see the outlines of a genuine greenhouse gas market." Also, according to Vorman, "The Pew Center report said more than 65 trades of greenhouse gas emissions totaling 55 million to 77 million tons have occurred over the past five years [since the 1997 Kyoto Protocol was introduced] but that those gifures probably underestimate the market activity. The emissions reductions traded for between 60 cents and $3.50 (U.S. dollars) per ton of carbon dioxide equivalent. (The date did not include trades within BP Pic and Royal Dutch Shell, which launched their own internal cap-and-trade programs in 1998 to cut emissions.)

The role of the agricultural industry in CET was being explored along with its options with the CCX. According to the American Farm Bureau, one of the plans under consideration was a plan to pay farmers for agricultural practices reducing carbon emissions into the atmosphere. "Farmers would be compensated for implementing or continuing practices that reduce carbon emissions from the soils. Such practices include reducing tillage, conserving tillage, retiring cropland, fertilizing with livestock manure, decreasing methane and reducing energy use. The compensation could potentially come from the government or companies that are interested in "trading" carbon credits." The Kyoto Protocol does not provide that option; but the Untied States wanted to consider it as a part of the UNFCCC. Jon Doggett, a senior director of governmental relations for the American Farm Bureau Federation (AFBF) added the caveat that such a practice of carbon sequestration in the soil—a practice some Americans support in opposition to European leaders who do not have the amount of land for such a program to succeed as does the United States—could be harmful to farmers in the long-term. Doggett noted that, "The industries that will be required to purchase the carbon credits supply farmers with vital operating materials, including fuel and **fertilizer**. As regulatory costs for companies rise, farmers will pay for that as fuel and fertilizer costs go up. The AFBF also pointed out that such storage might disrupt other environmetally beneficial practices—such as that employed by California producers after harvest when they flood their land and provide a **habitat** for geese and ducks. The issues of trading the benefits of one environmental concern for that of another would remain a matter of debate well into the century, no doubt.

Environmentalists continued to express concern that CET practices, particularly ones that were regulated voluntarily on the open market, would not reduce the greenhouse gases to the extent that some scientists thought crucial to the planet's optimum survival. The ongoing debate on the theory of global warming raged on in the United States particularly. Business and industrial interests questioned the soundness of the theory, and worried that it would create serious obstacles to profits that might be translated into the research necessary for better energy alternatives. Those supporting the theory were concerned with declining standards for **pollution** controls that might enhance the **greenhouse effect**. The voluntary controls industry and countries would place on themselves were nonetheless considered the first step to creating a cleaner **environment**.

[*Jane E. Spear*]

RESOURCES

BOOKS

International Carbon Bank & Exchange. *About Emissions Trading.* "Emission Reduction Credits & Trading." (June 2002). <http://www.icbe.com/>.

OTHER

Australian Academy of Science. *Nova: Science in the News.* "Carbon currency—the credits and debits of carbon emissions trading." (June 2002). <http://www.science.org.au/>

Chicago Climatex. *Emissions Trading.* (June 2002). <http://www.chicagoclimatex.com/>

Coal Age. *Primedia Business Magazines.* "McCain, Brownback Move on Emissions." Feb. 1, 2002. <http://www.industrycli.../>

DMOZ. *Open Directory.* "Emissions Trading." April 16, 2002. <http://www.dmoz.org/Science/Environment/>

Earth Council. *Carbon Trading: A Market Approach to the Environmental Crisis.* July 1997; (June 2002). <http://www.igc.apc.org/globalpolicy/>

Environmental News Network. *Trading for clean air just got easier.* December 5, 2001. <http://www.enn.com/>

Environmental News Network. *First U.S. Carbon Trading Market Enters Design Phase.* August 8, 2001. <http://www.enn.com/>

Environment News Service. *Environment.* "Carbon Trading Market Expands to Chicago, Mexico City." November 13, 2001. <http://www.ens.lycos.com/ens/>

Eye for Energy. *CO2 Trading: The North American Market.* "Opportunities, compliance and profit in domestic and international greenhouse gas emissions trading." June 2002. <http://www.eyeforenergy.com/>

Farm Bureau. *Global Climate Change.* "Agriculture's role discussed in carbon trading." June 22, 2000; (June 2002). <http://www.fb.com/>

Houlder, Vanessa. *FT (Financial Times).* "US opposition to Kyota may sink carbon trading." August 13, 2001. <http://www.ft.com/>

Vorman, Julie. *Reuters.* "Greenhouse trading takes off, U.S. on sidelines." March 20, 2002. <http://www.enn.com/news/wire-stories/>

ORGANIZATIONS

U. S. Environmental Protection Agency, 1200 Pennsylvania Avenue, N.W., Washington, D.C. United States 20460 (202)260-2090, www.epa.gov

Carbon monoxide

A colorless, odorless, tasteless gas that is produced in only very small amounts by natural processes. By far the most important source of the gas is the incomplete **combustion** of **coal**, oil, and **natural gas**. In terms of volume, **carbon monoxide** is the most important single component of **air pollution**. Environmental scientists rank it behind sulfur oxides, **particulate** matter, **nitrogen oxides**, and volatile organic compounds, however, in terms of its relative hazard to human health. In low doses, carbon monoxide causes headaches, nausea, fatigue, and impairment of judgment. In larger amounts, it causes unconsciousness and death.

Carbon offsets (CO$_2$-emission offsets)

Many human activities result in large emissions of **carbon dioxide** (CO$_2$) and other so-called **greenhouse gases** into the **atmosphere**. Especially important in this regard are the use of **fossil fuels** such as **coal**, oil, and **natural gas** to generate electricity, to heat spaces, and as a fuel for vehicles. In addition, the disturbance of forests results in large emissions of CO$_2$ into the atmosphere. This can be caused by the conversion of forests into agricultural or urbanized land uses, and also by the harvesting of timber.

During the past several centuries, human activities have resulted in large emissions of CO$_2$ into the atmosphere, causing substantial increases in the concentrations of that gas. Prior to 1850 the atmospheric concentration of CO$_2$ was about 280 ppm, while in 1997 it was about 360 ppm. Other greenhouse gases (these are also known as radiatively active gases) have also increased in concentration during that same period: **methane** (CH$_4$) from about 0.7 ppm to 1.7 ppm, **nitrous oxide** (N$_2$O) from 0.285 ppm to 0.304 ppm, and **chlorofluorocarbons** (CFCs) from zero to 0.7 ppb.

Many climatologists and environmental scientists believe that the increased concentrations of radiatively active gases are causing an increase in the intensity of Earth's **greenhouse effect**. The resulting climatic warming could result in important stresses for both natural ecosystems and those that humans depend on for food and other purposes (agriculture, forestry, and fisheries). Overall, CO$_2$ is estimated to account for about 60% of the potential enhancement of the greenhouse effect, while CH$_4$ accounts for 15%, CFCs for 12%, **ozone** (O$_3$) for 8%, and N$_2$O for 5%.

Because an intensification of Earth's greenhouse effect is considered to represent a potentially important environmental problem, planning and other actions are being undertaken to reduce the emissions of radiatively active gases. The most important strategy for reducing CO$_2$ emissions is to lessen the use of fossil fuels. This will mostly be accomplished by reducing energy needs through a variety of **conservation** measures, and by switching to non-fossil fuel sources. Another important means of decreasing CO$_2$ emissions is to prevent or slow the rates of **deforestation**, particularly in tropical countries. This strategy would help to maintain organic **carbon** within ecosystems, by avoiding the emissions of CO_2 through burning and **decomposition** that occur when forests are disturbed or converted into other land uses.

Unfortunately, fossil-fuel use and deforestation are economically important activities in the modern world. This circumstance makes it extremely difficult for society to rapidly achieve large reductions in the emissions of CO_2. An additional tactic that can contribute to net reductions of CO_2 emissions involves so-called CO_2 offsets. This involves the management of ecosystems to increase both the rate at which they are fixing CO_2 into plant **biomass** and the total quantity stored as organic carbon. This biological fixation can offset some of the emissions of CO_2 and other greenhouse gases through other activities.

Offsetting CO_2 emissions by planting trees

As plants grow, their rate of uptake of atmospheric CO_2 through **photosynthesis** exceeds their release of that gas by **respiration**. The net effect of these two physiological processes is a reduction of CO_2 in the atmosphere. The biological fixation of atmospheric CO_2 by growing plants can be considered to offset emissions of CO_2 occurring elsewhere—for example, as a result of deforestation or the **combustion** of fossil fuels.

The best way to offset CO_2 emissions in this way is to manage ecosystems to increase the biomass of trees in places where their density and productivity are suboptimal. The carbon-storage benefits would be especially great if a forest is established onto poorer-quality farmlands that are no longer profitable to manage for agriculture (this process is known as afforestation, or conversion into a forest). However, substantial increases in carbon storage can also be obtained whenever the abundance and productivity of trees is increased in "low-carbon ecosystems," including urban and residential areas.

Over the longer term, it is much better to increase the amounts of organic carbon that are stored in terrestrial ecosystems, especially in forests, than to just enhance the rate of CO_2 fixation by plants. The distinction between the amount stored and the rate of fixation is important. Fertile ecosystems, such as marshes and most agroecosystems, achieve high rates of net productivity, but they usually store little biomass and therefore over the longer term cannot sequester much atmospheric CO_2. A less extreme example involves second-growth forests and plantations, which have higher rates of net productivity than do older-growth forests. Averaged over the entire cycle of harvest and regeneration, however, these more-productive forests store smaller quantities of organic carbon than do older-growth forests, particularly in trees and large-dimension woody debris.

Both the greenhouse effect and emissions of CO_2 (and other radiatively active gases) are global in their scale and effects. For this reason, projects to gain CO_2 offsets can potentially be undertaken anywhere on the planet, but tallied as carbon credits for specific utilities or industrial sectors elsewhere. For example, a fossil-fueled electrical utility in the United States might choose to develop an afforestation offset in a less-developed, tropical country. This might allow the utility to realize significant economic advantages, mostly because the costs of labor and land would be less and the trees would grow quickly due to a relatively benign **climate** and long growing season. This strategy is known as a joint-implementation project, and such projects are already underway. These involve United States or European electrical utilities supporting afforestation in tropical countries as a means of gaining carbon credits, along with other environmental benefits associated with planting trees. Although this is a valid approach to obtaining carbon offsets, it can be controversial because some people would prefer to see industries develop forest-carbon offsets within the same country where the CO_2 is being emitted.

Afforestation in rural areas

An estimated 5–8 billion acres (2–3 billion ha) of deforested and degraded agricultural lands may be available world-wide to be afforested. This change in **land use** would allow enormous quantities of organic carbon to be stored, while also achieving other environmental and economic benefits. In North America, millions of acres of former agricultural land have reverted to forest since about the 1930s, particularly in the eastern states and provinces. There are still extensive areas of economically marginal agricultural lands that could be afforested in parts of North America where the climate and soils are suitable for supporting forests.

Agricultural lands typically maintain about one-tenth or less of the plant biomass of forests, while agricultural soils typically contain 60–80% as much organic carbon as forest soils. Because agricultural sites contain a relatively small amount of organic carbon, reforestation of those lands has a great potential for providing CO_2 offsets.

It is also possible to increase the amounts of carbon stored in existing forests. This can be done by allowing forests to develop into an old-growth condition, in which carbon storage is relatively great because the trees are typically big, and there are large amounts of dead biomass present in the surface litter, dead standing trees, and dead logs lying on the forest floor. Once the old-growth condition is reached, however, the **ecosystem** has little capability for accumulating "additional" carbon. Nevertheless, old-growth forests provide an important ecological service by tying up so much carbon in their living and dead biomass. In this sense, maintaining old-growth forests represents a strategy of CO_2-emissions deferral, because if those "high-carbon"

ecosystems were disturbed by timber harvesting or conversion into another kind of land use, a result would be an enormous **emission** of CO_2 into the atmosphere.

Dixon *et al.* (1993) examined carbon-offset projects in various parts of the world, most of which involved afforestation of rural lands. The typical costs of the afforestation projects were $1–10 per ton of carbon fixed. These are only the costs associated with planting and initial tending of the growing trees; there might also be additional expenses for land acquisition and stand management and protection.

Of course, even while rural afforestation provides large CO_2-emission offsets, other important benefits are also provided. In some cases, the forests might be used to provide economic benefits through the harvesting of timber (although the resulting disturbance would lessen the carbon-storage capability). Even if trees are not harvested from the CO_2-offset forests, it would be possible to hunt animals such as deer, and to engage in other sorts of economically valuable outdoor **recreation**. Increasing the area of forests also provides many non-economic benefits, such as providing additional **habitat** for native **species**, and enhancing ecological services related to clean water and air, **erosion** control, and climate moderation.

Urban forests

Urban forests consist of trees growing in the vicinity of homes and other buildings, in areas where the dominant character of land use is urban or suburban. Urban forests may originate as trees that are spared when a forested area is developed for residential land use, or they may develop from saplings that are planted after homes are constructed. Urban forests in older residential neighborhoods generally have a relatively high density and extensive canopy cover of trees. These characters are less well developed in younger neighbourhoods and where land use involves larger buildings used by institutions, business, or industry.

There are about 70 million acres (28-million ha) of urban land in the United States. Its urban forest supports an average density of 20 trees/acre (52 trees/ha), and has a canopy cover of 28%. Nowak *et al.* (1994) estimated that urban areas of the United States contain about 225 million tree-planting opportunities, in which suboptimal tree densities could be subjected to fill-planting.

Urban forests achieve carbon offsets in two major ways. First, as urban trees grow they sequester atmospheric CO_2 into their increasing biomass. The average carbon storage in urban trees in the United States is about 13 tons per acre (33 tonnes per ha). On a national basis that amounts to 0.8 billion tonnes of organic carbon, and an annual rate of uptake of six million tonnes.

In addition, urban trees can offset some of seasonal use of energy for cooling and heating the interior spaces of buildings. Large, well-positioned trees provide a substantial cooling influence through shading. Trees also cool the **ambient air** by evaporating water from their foliage (a process known as **transpiration**). Trees also decrease wind speeds near buildings. This results in decreased heating needs during winter, because less indoor warmth is lost by the **infiltration** of outdoor air into buildings. Over most of North America larger energy offsets associated with urban trees are due to decreased costs of cooling than with decreased heating costs. In both cases, however, much of the energy conserved represents decreased CO_2 emissions through the combustion of fossil fuels.

It is considerably more expensive to obtain CO_2-offset credits using urban trees than with rural trees. This difference is mostly due to urban trees being much larger than rural trees when planted, while also having larger maintenance expenses. In the survey of Dixon *et al.* (1993), the typical costs of rural CO_2-offset projects were $1–10 per ton of carbon fixed, compared with $15–30 per ton for urban trees.

Another study estimated the carbon savings associated with planting 100-million trees in urban areas in the United States (Nowak *et al.*, 1994). In this case, the total CO_2-emission offsets were estimated to be 58.2 kg C per tree per year (for trees at least ten years old). About 90% of the total CO_2 offsets was associated with indirect savings of energy for cooling and heating buildings, and 10% with carbon sequestration into the growing biomass of the trees. The estimated costs of the carbon offsets were $6.6–27.5 per ton of carbon, but these costs would decrease considerably as the trees grew larger. This study estimated that planting trees in urban areas of the United States could potentially offset as much as 2% of this country's emissions of CO_2.

[*Bill Freedman Ph.D.*]

RESOURCES

BOOKS

Nowak, D.J., E.G. McPherson, and R.A. Rowntree, eds. *Chicago's Urban Forest Ecosystem. Results of The Chicago Urban Forest Climate Project.* General Technical Report NE-186, U.S.D.A. Forestry Service, Northeastern Forest Experiment Station, Radnor, PA, 1994.

Trexler, M.C., and C. Haugen. *Keeping It Green: Tropical Forestry Opportunities for Mitigating Climate Change.* Washington, D.C.: World Resources Institute, 1995.

PERIODICALS

Dixon, R.K., et al. "Forest Sector Carbon Offset Projects: Near-term Opportunities To Mitigate Greenhouse Gas Emissions." *Water, Air, & Soil Pollution* 70 (1993): 561-577.

Freedman, B., and T. Keith "Planting Trees For Carbon Credits. A Discussion Of Context, Issues, Feasibility, and Environmental Benefits, With Particular Attention To Canada." *Environmental Review* 4 (1996): 100-111.

Heisler, G.M. "Energy Savings With Trees." *Journal of Arboric* 12 (1986): 113-125.

Kinsman, J.D., and M.C. Trexler. "Terrestrial Carbon Management And Electric Utilities." *Water, Air, Soil Pollution* 70 (1993): 545-560.

McPherson, E.G. "Using Urban Forests For Energy Efficiency And Carbon Storage." *Journal of Forestry* 94 (1994): 36-41.

Sampson, R.N., et al. " Biomass Management and Energy." *Water, Air, Soil Pollution.* 70 (1993): 139-159.

Carbon tax

To limit and control the amount of **carbon dioxide** (CO_2) added to the **atmosphere**, special taxes, called **carbon** taxes, have been proposed and in some cases adopted, on fuels containing carbon. Fuels such as **coal**, **gasoline**, heating oil, and **natural gas**, release energy by combining the carbon they contain with oxygen in the air, to produce carbon dioxide. Increased use of carbon-containing **fossil fuels** in modern times has greatly increased the rate at which carbon dioxide is entering the atmosphere. Measurable increases in atmospheric levels of the gas have been detected. Since carbon dioxide is the principle component of so-called **greenhouse gases**, changes in its concentration in the earth's atmosphere are a concern. Increases in the level of carbon dioxide, and other greenhouse gases such as **methane, nitrous oxide, ozone**, and man-made **chlorofluorocarbons** can be expected to result in warmer temperatures on the surface of the earth, and wide-ranging changes in the global **climate**. Greenhouse gases permit radiation from the Sun to reach the earth's surface but prevent the infrared or heat component of sunlight from re-irradiating into space.

The wide spread concern over the possible climatic effects of increases in heat-trapping gases in the atmosphere was exemplified by a 1994 report of the **Intergovernmental Panel on Climate Change** (IPCC) which concluded that unless greenhouse gas emissions were curtailed, average global temperatures would rise 2.5–8.1° F (1.4–4.5° C) by the year 2100. A 20% reduction in emissions over 20 years was recommended and considered feasible in developed countries. It was recognized that the rapidly expanding economies of many developing countries makes similar reductions less likely.

While limiting the buildup of heat-trapping gases in the atmosphere has been a goal of natural scientists for some time, many economists have now joined the effort. In 1997, 2,000 prominent economists, including six Nobel Laureates, signed the "Economists' Statement on Climate Change" stating that policies to slow global warming are needed and are economically viable. The economists agreed with a review conducted by a distinguished international panel of scientists under the auspices of the IPCC that, "The balance of evidence suggests a discernible human influence on global climate." They further stated that, "As economists, we believe that global climate change carries with it significant environmental, economic, social, and geopolitical risks, and that preventive steps are justified." They state that economic

studies have found that potential policies to reduce greenhouse-gas emissions can be designed so that benefits outweigh costs, and living standards would not be harmed. They claim that United States productivity may actually improve as a result. The statement goes on to claim that, "the most efficient approach to slowing climate change is through market-based policies" rather than limits or regulations. The economists also suggest that, "A cooperative approach among nations is required, such as an international emissions trading agreement." They recommend, "market mechanisms such as carbon taxes or the auction of emissions permits." Their statement goes on to suggest that, "Revenues generated from such policies can effectively be used to reduce budget deficits or lower existing taxes." New taxes are never popular, but are more apt to be accepted when the revenue is used to replace other taxes or to aid the **environment**.

As in any public policy debate, views of even well-qualified experts are often sharply divided. Some industry representatives and scientists are unwilling to accept the premise that greenhouse gas emissions represent a serious threat. Industrial spokesmen continue to claim that efforts to reduce greenhouse gas emissions through taxation or other economic means are not cost effective, and would devastate the economy, cost jobs, and reduce living standards. Individual countries worry about the impacts on their competitiveness if other countries do not adopt similar measures. The potential impact on low-income groups must also be considered. Some are concerned that the taxes have to be high to be effective. The amount of year to year variation in climate and temperature throughout the globe, provide ample room for differences of opinion.

International attention was focused on the need to reduce greenhouse gases at the United Nations sponsored Earth Summit held in Rio de Janeiro in 1992. The conference was held to attempt to reconcile worldwide economic development with the need to preserve the environment. Representatives of 178 nations attended, including 117 heads of state, making it the largest gathering of world leaders in history. Documents and treaties were signed committing most of the world's nations to the economic development in ways that were compatible with a healthy environment. A binding treaty called "The Framework Convention on Climate Change, or Global Warming Convention" was adopted, requiring nations to reduce emissions of greenhouse gases. The treaty failed to set binding targets for **emission** reductions, however. Agreement was hampered by discord between industrialized nations of western Europe, and North America and developing nations in Africa, Latin America, the Middle East and Asia. Developing countries were concerned that environmental restrictions would hamper economic growth unless they received increased aid from developed nations to enable them to grow in an environmentally sound way.

National and regional efforts to limit greenhouse gas has been most evident in western Europe, although disagreements between countries have prevented a unified approach among members of the European Community (EC). Britain has refused to accept a proposed carbon and energy tax, favored by several nations. Spokesmen for Britain expressed doubt that the tax would achieve desired reductions, and claimed that the two-thirds reduction in emissions to which Britain was committed would result from a planned enactment of a value-added tax (VAT) on fuel and energy. Critics of Britain's position argued that their plan was not sufficiently focused because it did not specially target fuels that increase atmospheric carbon dioxide. A 1996 report from the European Environmental Agency (EEA) in Copenhagen concluded that "green taxes" adopted by individual European countries in the last decade have been successful. So-called **green taxes** are a variety of environmentally friendly tax measures intended to reduce **pollution** and toxic wastes and encourage efficient resource use. Carbon taxes, such as those enacted in Norway and Sweden in 1991, are one example. Other examples include taxes on sulfur and **nitrogen** oxide emissions, **water pollution** charges, and charges for **household waste**. The carbon dioxide tax in Norway was reported to have reduced CO_2 emissions by 3–4%. The goal in the Scandinavian countries and the Netherlands has been to move away from taxes on labor or capital to those on **energy and the environment**. The consensus in western Europe seems to be that current environmental policies are weighted too much in the direction of regulation, and that there is a need to provide economic incentives in the form of well-designed environmental taxes.

Although eastern European countries have moved forward more slowly than in the rest of the continent, Poland, Hungary, and Estonia are beginning to adopt environmental charges and taxes. In the far-east, the Japanese Environmental Agency has reported that Japan might be able to keep per capita emissions of carbon dioxide at 1990 levels, which would result in a small increase in total emissions because of **population growth**.

The effectiveness of green taxes, including carbon taxes, is not universally accepted. Some argue that goals need to be defined more carefully with specified amounts or percentages of reduction in targeted pollutants. The EEA report acknowledges that the evaluation of the effectiveness is difficult and that, "Judgments about the performance of green taxes remain at the level of best guesses." Although theoretical evaluation of environmental taxation is well-developed, adequate evaluation of practical experience with such taxes is still relatively rare. Proper evaluation must separate the effects of green taxes from other factors, and

assess what would have happened without the taxes. It is also recognized that it may take as much as a decade for a tax to have the desired effect.

In the United States, the Clinton administration, in 1993, published a 50-point plan to reduce greenhouse gas emissions by 100 million tons through a number of voluntary measures. Among the steps recommended were greater **energy efficiency** in homes and electrical appliances, reduced emissions from **power plants**, greater reliance on hydroelectric power, and increased tree planting. Some observers believe that the prospects for "green taxes" are less favorable in the United States than in Europe because of difficulty of passing legislation through two houses of Congress, which may be controlled by different parties, and obtaining approval by the president. Some individual states, however, have enacted measures that create economic incentives for environmental protection. Louisiana, which has many **petrochemical** companies within its borders, created an "environmental scorecard" in 1991, to keep a tally of companies' toxic discharges, waste volumes, and compliance with environmental laws. Graded exemptions from property taxes are granted to companies with good scores. Oregon's Energy Facility Siting Task Force recommended establishment of a standard to reduce emissions of carbon dioxide, with provisions for offsets for planting trees (which remove CO_2 from the air), and for the use of **renewable energy** sources.

[*Douglas C. Pratt*]

RESOURCES
PERIODICALS

Burke, M. "Environmental Taxes Gaining Ground in Europe." *Environmental Science & Technology* 31 (1997): 84–88.
Environmental Taxes: Implementation and Environmental Effectiveness; *Environmental Issue Series No. 1.* European Environment Agency: Copenhagen, 1996.

Carcinogen

A carcinogen is any substance or agent that produces or induces the development of **cancer**. Carcinogens are known to affect and initiate metabolic processes at the level of cellular DNA.

Cancer accounts for slightly over 20% of all deaths each year. It is estimated that one out of every four Americans will develop cancer eventually and that six out of 10 in this group will die from the disease itself or complications arising from the disease. Half of all cancer deaths occur before the age of 65. Among women between 30 and 40 and children between three and 14, cancer is the leading cause of death next to accidents. It is the most frequent cause of death among Americans under 35 years of age.

The testing of **chemicals** as cancer-producing agents began with the observation of Sir Percival Pott in 1775 that scrotal cancer in young chimney sweeps resulted from the lodgement of soot in the folds of their scrotums. Pott was the first to link an environmental agent, **coal** tar, to cancer growth. In 1918 scientists began to test chemical derivatives for their cancer-causing efficacy. These first experiments looked at polycyclic **hydrocarbons**, specifically benzo(a)-pyrene found in coal tar, and they demonstrated that a certain degree of exposure to coal tar produced cancer in laboratory rats. In 1908, Vilhelm Ellerman and Oluf Bang of Denmark reported that an infectious agent could cause cancer, after they found that a leukemia-like blood disease was transmitted among domestic fowl via a **virus**. In 1911, Peyton Rous established a viral cause for a cancer called sarcoma in domestic fowl, and he was awarded a Nobel Prize for this discovery some 55 years later. In 1932, Lacassagne reported that estrogen injections caused mammary cancer in mice. This opened up investigation into the role hormones played in the development of various types of cancers.

In 1896, Wilhelm Roentgen discovered the x ray, a radioactive **emission** capable of penetrating many solid materials including the human body. x rays quickly found use as a diagnostic tool in medicine; but operators of x ray devices, unaware of their harmful effects, determined the proper intensity of the beams by repeatedly exposing their hands to the rays. Many operators of x ray equipment began to suffer from cancer of the hand, and Roentgen himself died of cancer. The most dramatic environmental link to cancer induced by **radioactivity** was observed after the bombing of **Hiroshima, Japan** when there was a radical increase in **leukemia** type cancers among people exposed to the atomic blast.

Environmental agents such as toxic chemicals and radiation are considered responsible for about 85% of all cancer cases. A great many environmental agents such as synthetic chemicals, sunlight (exposure to UV and UVB rays), air pollutants, **heavy metals**, x rays, high-fat diet, chemical pesticides, and cigarette smoking are known to be carcinogenic. Surveys carried out on the geographic incidence of cancer indicate that certain types of cancer are far more common in heavily industrialized areas. New Jersey, the site of approximately 1,200 chemical plants and related industries, has the highest overall cancer rate in the United States.

Tobacco use, particularly cigarette smoking, is now recognized as the leading contributor to cancer **mortality** in the United States. Currently one third of all cancer deaths are due to lung cancer, and of the 130,000 new lung cancer victims diagnosed each year, 80% are cigarette smokers. Several years ago, the primary cause of cancer among women was breast cancer—but by the late 1980s, lung cancer had surpassed breast cancer as the leading cause of death among

women. Current controversy rages over the role of secondary **smoke** as a contributing cause of cancer among nonsmokers exposed to **cigarette smoke**.

Dietary factors have been extensively investigated, and experiments have implicated everything from coffee to charcoal broiled meat to peanut butter as possible carcinogens. A major concern among meat producers was the use of diethylstilbestrol (DES) as a source for beefing up cattle. DES is a synthetic hormone that increases the rate of growth in cattle. In the 1960s, DES was fed to about three fourths of all the cattle raised in the United States. It was also used to prevent miscarriages in women until 1966, when it was shown to be carcinogenic in mice. DES is now linked to vaginal and cervical cancers in women born between 1950 and 1966 whose mothers took DES during their pregnancies.

In 1971, DES was banned for use in cattle by the **Food and Drug Administration** (FDA), but the federal courts reversed the ban, contending that DES posed no danger since it was not directly added to foods but was administered only to cattle. When the FDA subsequently showed that measurable quantities remained present in slaughtered cattle, the courts reinstated the ban. But the issue of using growth additives in meat production remains unresolved today. Environmentalists are still concerned that known carcinogenic chemicals used to "beef up" cattle are being consumed by humans in various meat products, though no direct links have yet been established. In addition, various **food additives**, such as coal tar dyes used for artificial coloring and food preservatives, have produced cancer in laboratory animals. As yet there is no evidence indicating that human cancer rates are rising because of these substances in food.

Air pollution has been extensively investigated as a possible carcinogen and it is known that people living in cities larger than 50,000 run a 33% higher risk of developing lung cancer than people who live in other areas. The reasons behind this phenomenon, referred to as the "urban factor," have never been conclusively determined. Areas with populations exceeding 50,000 tend to have more industry, and air pollutants can have a profound effect in regions such as New Jersey where they are highly concentrated.

Occupational exposure to carcinogenic substances accounts for an estimated 2–8% of diagnosed cancers in the United States. Until passage of the Toxic Substances Control Act in 1976, which gave the federal government the power to require testing of potentially hazardous substances before they go on the market, hundreds of new chemicals with unknown side effects came into industrial use each year. Substances such as **asbestos** are estimated to cause 30–40% of all deaths among workers who have been exposed to it. **Vinyl chloride**, a basic ingredient in the production of

plastics, was found in 1974 to induce a rare form of liver cancer among exposed workers. Anaesthetic gases used in operating rooms have been traced as the reason nurse anesthetists develop leukemia and lymphoma at three times the normal rate with an associated higher rate of miscarriage and **birth defects** among their children. **Benzene**, an industrial chemical long known as a bone-marrow poison, has been shown to induce leukemia as well. A major step forward in the regulation of these potential cancer causing agents is the implementation by the Occupational Safety and Health Administrations (OSHA) of the Hazard Communication Standard in 1983, intended to provide employees in manufacturing industries access to information concerning hazardous chemicals encountered in the workplace.

With the **erosion** of the **ozone** layer of our **atmosphere**, increased concern about over-exposure to **ultraviolet radiation** and its subsequent effect on the formation of skin cancer has developed. The EPA estimates that a five% ozone depletion in the **stratosphere** would result in a substantial increase in a variety of skin cancers. This would include an average of two million extra cases of basal-cell and squamous-cell skin cancers a year and an additional 30,000 cases of the often fatal melanoma skin cancer, which currently kills 9,000 Americans per year. *See also* Hazardous waste siting; Love Canal, New York; Ozone layer depletion; Radiation exposure; Radiation sickness; Radon; Toxic substance

[*Brian R. Barthel*]

RESOURCES
BOOKS

Aldrich, T., and J. Griffith. *Environmental Epidemiology.* New York: Van Nostrand Reinhold, 1993.

McCance, K. L. *Pathophysiology: the Biological Basis for Disease in Adults and Children.* St. Louis: Mosby, 1990.

National Academy of Sciences. *Ozone Depletion, Greenhouse Gases and Climate Change.* Washington, DC: U. S. Environmental Protection Agency, 1989.

U. S. Environmental Protection Agency. *The Potential Effects of Global Climate Change on the United States.* Washington, DC: U. S. Government Printing Office, 1988.

Carrying capacity

Carrying capacity is a general concept based on the idea that every **ecosystem** has a limit for use that cannot be exceeded without damaging the system. Whatever the specified use of an area might be, whether for grazing, **wildlife habitat**, **recreation**, or economic development, there is a threshold that cannot be breached, except temporarily, without degrading the ability of the **environment** to support that use. Examinations of carrying capacity attempt to determine,

with varying degrees of **accuracy**, where this threshold lies and what the consequences of exceeding it might be.

The concept of carrying capacity was pioneered early this century in studies of range management and **wildlife management**. Range surveys of what was then called "grazing capacity" were carried out on the Kaibab Plateau in Arizona as early as 1911, and this term was used in most of the early bulletins issued by the **U.S. Department of Agriculture** on the subject. In his 1923 classic, *Range and Pasture Management,* Sampson defined grazing capacity as "the number of stock of one or more classes which the area will support in good condition during the time that the forage is palatable and accessible, without decreasing the forage production in subsequent seasons." Sampson was quick to point out that the "grazing capacity equation has not been worked out on any range unit with mathematical precision." In fact, because of the number of variables involved, especially variables stemming from human actions, he did not believe that the "grazing-capacity factor will ever be worked out to a high degree of scientific accuracy." Sampson also pointed out that "grazing the pasture to its very maximum year after year can produce only one result—a sharp decline in its carrying capacity," and he criticized the stocking of lands at their maximum instead of their optimum capacity. Similar discussions of carrying capacity can be found in books about wildlife management from the same period, particularly *Game Management* by **Aldo Leopold**, published in 1933.

Practitioners of applied **ecology** have calculated the number of animal-unit months that any given land area can carry over any given period of time. But there have been some controversies over the application of the concept of carrying capacity. The concept is commonly employed without considering the factor of time, neglecting the fact that carrying capacity refers to **land use** that is sustainable. Another common mistake is to confuse or ignore the implicit distinctions between maximum, minimum, and optimum capacity. In discussions of land use and environmental impact, some officials have drawn graphs with curves showing maximum use of an area and claimed that these figures represent carrying capacity. Such representations are misleading because they assume a perfectly controlled population, one without fluctuation, which is not likely. In addition, the maximum allowable population can almost never be the carrying capacity of an area, because such a number can almost never be sustained under all possible conditions. A population in balance with the environment will usually fluctuate around a mean, higher or lower, depending on seasonal habitat conditions, including factors critical to the support of that particular **species** or community.

The concept of carrying capacity has important ramifications for **human ecology** and **population growth**. Many of the essential systems on which humans depend for sustenance are showing signs of stress, yet demands on these systems are constantly increasing. William R. Catton has formulated an important axiom for carrying capacity: "For any use of any environment there is a use intensity that cannot be exceeded without reducing that environment's suitability for that use." He then defined carrying capacity for humans on the basis of this axiom: "The maximum human population equipped with a given assortment of technologies and a given pattern of organization that a particular environment can support indefinitely."

The concept of carrying capacity is the foundation for recent interest in **sustainable development**, an environmental approach which identifies thresholds for economic growth and increases in human population. Sustainable development calculates the carrying capacity of the environment based on the size of the population, the standard of living desired, the overall quality of life, the quantity and type of artifacts created, and the demand on energy and other resources. With his calculations on sustainable development in Paraguay, Herman Daly has illustrated that it is possible to work out rough estimates of carrying capacity for some human populations in certain areas. He based his figures on the ecological differences between the country's two major regions, as well as on differences among types of settlers, and differences between developed good land and undeveloped marginal lands.

If ecological as well as economic and social factors are taken into consideration, then any given environment has an identifiable tolerance for human use and development, even if that number is not now known. For this reason, many environmentalists argue that carrying capacity should always be the basis for what has been called "demographic accounting."

[*Gerald L. Young and Douglas Smith*]

RESOURCES
BOOKS

Edwards, R. Y., and C. D. Fowle. "The Concept of Carrying Capacity." In *Readings in Wildlife Management*, edited by J. A. Bailey, W. Elder, and T. D. McKinney. Washington, DC: The Wildlife Society, 1974.

PERIODICALS

Budd, W. W. "What Capacity the Land?" *Journal of Soil and Water Conservation* 47 (January-February 1992): 28-31.

Catton, W. R., Jr. "The World's Most Polymorphic Species: Carrying Capacity Transgressed Two Ways." *BioScience* 37 (June 1987): 413-419.

Graefe, A. R., J. V. Vaske, and F. R. Kuss. "Social Carrying Capacity: An Integration & Synthesis of Twenty Years of Research." *Leisure Sciences* 6 (December 1984): 395-431.

Nilsson, S. "The Carrying Capacity Concept." *Interdisciplinary Science Reviews* 9 (June 1984): 137-148.

Rachel Louise Carson (1907 – 1964)
American ecologist, marine biologist, and writer

Rachel Louise Carson was a university-trained biologist, a longtime United States government employee, and a best-selling author of such books as *Edge of the Sea, The Sea Around Us* (a National Book Award winner), and *Silent Spring*.

Her book on the dangers of misusing pesticides, *Silent Spring*, has become a classic of environmental literature and resulted in her recognition as the fountainhead of modern **environmentalism**. *Silent Spring* was reissued in a twenty-fifth anniversary edition in 1987, and remains standard reading for anyone concerned about environmental issues.

Carson grew up in the Pennsylvania countryside and reportedly developed an early interest in **nature** from her mother and from exploring the woods and fields around her home. She was first an English major in college, but a required course in biology rekindled that early interest in the subject and she graduated in 1928 from Pennsylvania College for Women with a degree in zoology and went on to earn a master's degree at Johns Hopkins University. After the publication of *Silent Spring*, she was often criticized for being a "popular science writer" rather than a trained biologist, making it obvious that her critics were unaware of her university work, including a master's thesis entitled "The Development of the Pronephros During the Embryonic and Early Larval Life of the Catfish (*Ictalurus punctatus*)."

Summer work also included biological studies at Woods Hole Marine Biological Laboratory in Massachusetts, where she became more interested in the life of the sea. After doing a stint as a part-time scriptwriter for the Bureau of Fisheries, she was hired full-time as a junior aquatic biologist. When she resigned from the United States **Fish and Wildlife Service** in 1952 to devote her time to her writing, she was biologist and chief editor there. First, as a biologist and writer with the Bureau and then as a free-lance writer and biologist, she successfully combined professionally the two great loves of her life, biology and writing.

Often described as "a book about death which exalts life," *Silent Spring* is the work on which Carson's position as the modern catalyst of a renewed environmental movement rests. The book begins with a shocking fable of one composite town's "silent spring" after pesticides have decimated insects and the birds that feed upon them. The main part of the book is a massive documentation of the effects of organic pesticides on all kinds of life, including birds and humans. The final sections are quite restrained, drawing a hopeful picture of the future, if feasible alternatives to the use of pesticides—such as biological controls—are used in conjunction with and as a partial replacement of chemical sprays.

Carson was quite conservative throughout the book, being careful to limit examples to those that could be verified and defended. In fact, there was very little new in the book; it was all available earlier in a variety of scientific publications. But her science background allowed her to judge the credibility of the facts she uncovered and provided sufficient knowledge to synthesize a large amount of data. Her literary skills made that data accessible to the general public.

Silent Spring was not a polemic against all use of pesticides but a reasoned argument that potential hazards be carefully and sufficiently considered before any such chemical was approved for use. Many people date modern concern with environmental issues from her argument in this book that "future generations are unlikely to condone our lack of prudent concern for the integrity of the natural world that supports all life." It is not an accident that her book is dedicated to **Albert Schweitzer**, because she wrote it from a shared philosophy of reverence for life.

Carson provided an early outline of the potential of using biological controls in place of **chemicals**, or in concert with smaller doses of chemicals, an approach now called **integrated pest management**. She worried that too many specialists were concerned only about the effectiveness of chemicals in destroying pests and "the overall picture" was being lost, in fact not valued or even sought. She pointed out the false safety of assuming that products considered individually were safe, when in concert, or synergistically, they could lead to human health problems.

Her **holistic approach** was one of the real, and unusual, strengths of the book. Prior to the publication of *Silent Spring*, she even refused to appear on a **National Audubon Society** panel on pesticides because such an appearance could provide a forum for only part of the picture and she wanted her material to first appear "as a whole." She did allow it to be partially serialized in *The New Yorker*, but articles in that magazine are long and detailed.

The book was criticized early and often, and often viciously and unfairly. One chemical company, reacting to that pre-publication serialization, tried to get Houghton Mifflin not to publish the book, citing Carson as one of the "sinister influences" trying to reduce the use of **agricultural chemicals** so that United States food supplies would dwindle to the level of a developing nation. The chemical industry apparently united against Carson, distributing critical reviews and threatening to withdraw magazine advertisements from journals deemed friendly to her. Words and phrases used in the attacks included "ignorant," "biased," "sensa-

tional," "unfounded," "distorted," "not written by a scientist," "littered with crass assumptions and gross misinterpretations," to name but a few.

Some balanced reviews were also published, most noteworthy one by Cornell University ecologist LaMont Cole in *Scientific American*. Cole identified errors in her book, but finished by saying "errors of fact are so infrequent, trivial and irrelevant to the main theme that it would be ungallant to dwell on them," and went on to suggest that the book be widely read in the hopes that it "may help us toward a much needed reappraisal of current policies and practices." That was the spirit in which Carson wrote *Silent Spring* and reappraisals and new policies were indeed the result of the myriad of reassessments and studies spawned by its publication. To its credit, it did not take the science community long to recognize her credibility; the President's Science Advisory Committee issued a 1963 report that the journal *Science* suggested "adds up to a fairly thorough-going vindication of Rachel Carson's *Silent Spring* thesis."

While it is important to recognize the importance of *Silent Spring* as a landmark in the environmental movement, one should not neglect the significance of her other work, especially her three books on oceans and marine life and the impact of her writing on people's awareness of one of earth's great natural ecosystems.

Under the Sea Wind (1941) was Carson's attempt "to make the sea and its life as vivid a reality [for her readers] as it has become for me." And readers are given vivid narratives about the shore, including vegetation and birds, on the open sea, especially by tracing the movements of the mackerel, and on the sea bottom, again by focusing on an example, this time the eel. *The Sea Around Us* (1951) continues Carson's treatment of marine biology, adding an account of the history and development of the sea and its physical features such as islands and tides. She also includes human perceptions of and relationships with the sea. *The Edge of the Sea* (1955) was written as a popular guide to beaches and sea shores, but focusing on rocky shores, sand beaches, and coral and mangrove coasts, it complemented the physical descriptions in *The Sea Around Us* with biological data.

Carson was a careful and thorough scientist, an inspiring author, and a pioneering environmentalist. Her groundbreaking book, and the controversy it generated, was the catalyst for much more serious and detailed looks at environmental issues, including increased governmental investigation that led to creation of the **Environmental Protection Agency** (EPA). Her work will remain a hallmark in the increasing awareness modern people are gaining of how humans interact with and impact the **environment** in which they live and on which they depend.

[*Gerald R. Young*]

Rachel Carson. (Corbis-Bettmann. Reproduced by permission.)

RESOURCES

BOOKS

Bonta, M. M. "Rachel Carson, Pioneering Ecologist." In *Women in the Field: America's Pioneering Naturalists.* College Station, TX: Texas A & M University Press, 1991.

Brooks, P. *The House of Life: Rachel Carson at Work.* Boston: Houghton Mifflin, 1972.

Carson, Rachel. *Silent Spring.* Boston: Houghton Mifflin, 1962.

Downs, R. B. "Upsetting the Balance of Nature: Rachel Carson's *Silent Spring.*" In *Books That Changed America.* New York: Macmillan, 1970.

Hynes, H. P. *The Recurring Silent Spring.* New York: Pergamon Press, 1989.

Marco, G. J., R. M. Hollingworth, and W. Durham. *Silent Spring Revisited.* Washington, DC: American Chemical Society, 1987.

PERIODICALS

Graham Jr., F. "Rachel Carson." *EPA Journal* 4 (November–December 1978): 5–7+.

Cash crop

A crop that is produced for the purpose of exporting or selling rather than for consumption by the person who grows it. In many **Third World** countries, cash crops often replace the production of basic food staples such as rice, wheat, or corn in order to generate foreign exchange. For example, in Guatemala, much of the land is devoted to the production of bananas and citrus fruits (97% of the citrus crop is exported), which means that majority of the basic food products needed by the native people are imported from other countries. Often these foods are expensive and difficult for many poor people to obtain. Cash crop agriculture also forces many subsistence and tenant farmers to give up their land in order to make room for industrialized farming.

Catalytic converter

Catalytic converters are devices which employ a catalyst to facilitate a chemical reaction. (A catalyst is a substance that changes the rate of a chemical reaction, but whose own composition is unchanged by that reaction.) For **air pollution control** purposes, such reactions involve the reduction of nitric oxide to molecular oxygen and **nitrogen** or oxidation of **hydrocarbons** and **carbon monoxide** to **carbon dioxide** and water. Using the catalyst, the activation energy of the desired chemical reaction is lowered. Therefore, exothermic chemical conversion will be favored at a lower temperature.

Traditional catalysts have normally been metallic, although nonmetallic materials, such as ceramics, have been coming into use in recent years. Metals used as catalysts may include noble metals, such as platinum, or **base** metals, including **nickel** and **copper**. Some catalysts are more effective in oxidation, others are more effective in reduction. Some metals are effective in both kinds of reactions. The catalyst material is normally coated on a porous, inert support structure of varying design. Examples include honeycomb ceramic structures with long channels and pellet beds. The goal is to channel exhaust over a large surface area of catalyst without an unacceptable pressure drop.

In some cases, reduction and oxidation catalysts are combined to control oxides of nitric oxide, **carbon** monoxide, and hydrocarbon emissions in exhaust from internal **combustion** engines. The reduction and oxidation processes can be conducted sequentially or simultaneously. Dual catalysts are used in sequential reduction-oxidation. In this case, the exhaust gas from a rich-burn engine initially enters the reducing catalyst to reduce nitric oxide. Subsequently, as the exhaust enters an oxidation catalyst, it is diluted with air to provide oxygen for oxidation. Alternatively, three-way catalysts can be used for simultaneous reduction and oxidation. Engines exhausting to such catalysts run slightly rich and require tight regulation of air-fuel ratio.

A catalytic converter at a large petrochemical plant. (Photogtraph by Tom Carroll. Phototake. Reproduced by permission.)

Reducing catalysts can be made more efficient using a reducing agent, such as ammonia. This method of control, referred to as selective catalytic reduction, has been employed successfully on large turbines. In this case, a reducing agent

is introduced upstream of a reducing catalyst, allowing for greater rates of nitric oxide reduction.

[*Robert B. Giorgis Jr.*]

RESOURCES
BOOKS

Silver, R. G., ed. *Catalytic Control of Air Pollution: Mobile and Stationary Sources.* Washington, DC: American Chemical Society, 1992.
Yaverbaum, L. H. *Nitrogen Oxides Control and Removal: Recent Developments.* Park Ridge, NJ: Noyes Data Corp., 1979.

PERIODICALS

Amato, I. "Catalytic Conversion Could Be a Gas." *Science* 259 (15 January 1993): 311.

Cation

see **Ion**

Cation exchange

see **Ion exchange**

Catskill Watershed Protection Plan

New York City has long been proud of its excellent municipal drinking water. Approximately 90% of that water comes from the Catskill/Delaware **Watershed**, which covers about 1,900 square miles (nearly 5,000 square kilometers) of rugged, densely forested land north of the city and west of the **Hudson River**. Stored in six hard-rock reservoirs and transported through enormous underground tunnels, the city water is outstanding for so large an urban area. Yielding 1.2 billion gal (450,000 cubic meters) per day, and serving more than 9 million people, this is the largest surface water storage and supply complex in the world. As the metropolitan **agglomeration** has expended, however, people have moved into the area around the Catskill Forest Preserve, and **water quality** is not as high as it was a century ago.

When the 1986 U.S. **Safe Drinking Water Act** mandated **filtration** of all public surface water systems, the city was faced with building an $8 billion **water treatment** plant that would cost up to $500 million per year to operate. In 1989, however, the **Environmental Protection Agency** (EPA) ruled that the city could avoid filtration if it could meet certain minimum standards for microbial contaminants such as bacteria, viruses, and protozoan **parasites**. In an attempt to limit pathogens and nutrients contaminating surface water and to avoid the enormous cost of filtration, the city proposed land-use regulations for the five counties (Green, Ulster, Sullivan, Schoharie, and Delaware) in the

Catskill/Delaware watershed from which it draws most of its water.

With a population of 50,000 people, the private land within the 200 square mi (520 square km) watershed is mostly devoted to forestry and small dairy farms, neither of which are highly profitable. Among the changes the city called for was elimination of storm water **runoff** from barnyards, **feedlots**, or grazing areas into watersheds. In addition, farmers would be required to reduce **erosion** and surface runoff from crop fields and **logging** operations. Property owners objected strenuously to what they regarded as onerous burdens that would cost enough to put many of them out of business. They also bristled at having the huge megalopolis impose rules on them. It looked like a long and bitter battle would be fought through the courts and the state legislature.

To avoid confrontation, a joint urban/rural task force was set up to see if a compromise could be reached, and to propose alternative solutions to protect both the water supply and the long-term viability of agriculture in the region. The task force agreed that agriculture is the "preferred land use" on private land, and that agriculture has "significant present and future environmental benefits." In addition, the task force proposed a voluntary, locally developed and administered program of "whole farm planning and best management approaches" very similar to ecosystem-based, **adaptive management**.

This grass-roots program, financed mainly by the city, but administered by local farmers themselves, attempts to educate landowners, and provides alternative marketing opportunities that help protect the watershed. Economic incentives are offered to encourage farmers and foresters to protect the water supply. Collecting feedlot and barnyard runoff in **infiltration** ponds together with solid **conservation** practices such as **terracing**, **contour plowing**, strip farming, leaving crop residue on fields, ground cover on waterways, and cultivation of perennial crops such as orchards and sugarbush have significantly improved watershed water quality. As of 1999, about 400 farmers—close to the 85% participation goal—have signed up for the program. The cost, so far, to the city has been about $50 million—or less than 1% of constructing a treatment plant.

In addition to saving billions of dollars, this innovative program has helped create good will between the city and its neighbors. It has shown that upstream cleanup, prevention, and protection are cheaper and more effective than treating water after it's dirty. Farmers have learned they can be part of the solution, not just part of the problem. This experiment serves as an excellent example of how watershed planning through cooperation is effective when local people are given a voice and encouraged to participate.

[*William P. Cunningham Ph.D.*]

RESOURCES

OTHER

About the CWC. July 2002. Catskill Watershed Corporation. [cited July 9, 2002]. <http://www.cwconline.org/>.

New York City Watershed Agreement. New York Public Interest Group. [cited July 9, 2002]. <http://www.nypirg.org/enviro/water/watershed_agreement.html>.

New York City watershed whole farm programme. March 1999. United Nations Sustainable Development Programme. [cited July 9, 2002]. <http://www.un.org/esa/sustdev/success/nyc_wsfp.htm>.

ORGANIZATIONS

Catskill Watershed Corporation, PO Box 569 Main St., Margaretville, NY USA 12455 (845) 586-1400, Fax: (845)586-1401, Toll Free: (877) 928-7433, Email: invest@cwconline.org, http://www.cwconline.org

Center for Environmental Philosophy

The Center for Environmental Philosophy was established in 1989 as an organization dedicated to furthering research, publication, and education in the area of environmental philosophy and ethics. Based at the University of North Texas since 1990, the primary activities of the center are the publication of the journal *Environmental Ethics*, the reprinting of significant books on **environmental ethics** under its own imprint, the sponsorship of various workshops and conferences dedicated to the furthering of research and training in environmental ethics, and the promotion of graduate education, postdoctoral research, and professional development in the field of environmental ethics.

The Center is best known for its journal, which virtually established the field of environmental ethics and remains perhaps the leading forum for serious philosophical work in environmental philosophy. Inspired in part by Aldo Leopold's contention in "A Sand County Almanac" that the roots of most ecological problems were philosophical, Eugene C. Hargrove founded *Environmental Ethics* in 1978. The journal was originally concerned primarily with whether the attribution of rights to animals and to **nature** itself could be coherently defended as a philosophical doctrine. While remaining true to its central preoccupation with ethics, the journal's interests have more recently broadened to include significant essays on such topics as **deep ecology**, **ecofeminism**, **social ecology**, economics, and public policy. Under Hargrove's editorial leadership, the journal has brought environmental ethics to increasing acceptance as a serious field by mainstream academic philosophers. *Environmental Ethics* also is now widely read by researchers concerned with the **environment** in the fields of biological science, economics, and policy science. It also is developing a small but growing following among environmental professionals such as **conservation** biologists.

The Center also helps to sponsor, in conjunction with the Department of Philosophy and Religious Studies at the

University of North Texas, programs in which graduate students may take courses and specialize in the field of environmental ethics.

[*Lawrence J. Biskowski*]

RESOURCES
ORGANIZATIONS

Center for Environmental Philosophy, University of North Texas, 370 EESAT, P.O. Box 310980, Denton, TX USA 76203-0980 (940) 565-2727, Fax: (940) 565-4439, Email: cep@unt.edu, <http://www.cep.unt.edu>

Center for Marine Conservation
see **Ocean Conservatory, The**

Center for Respect of Life and Environment

Formed in 1986, the Center for Respect of Life and **Environment** (CRLE) is a nonprofit group based in Washington, D.C., that works to promote humane and **environmental ethics**, particularly within the academic and religious communities with an emphasis on the links between **ecology**, spirituality, and sustainability.

CRLE describes itself as being committed to "encourage the well-being of life and living systems—plant, animal, and human relationships..." The work of the Center is "to awaken the public's ecological sensibilities, and to transform lifestyles, institutional practices, and social policies to support the community of life..." In order to accomplish these goals, CRLE sponsors conferences of professionals and experts in various fields and puts out a variety of publications, including a quarterly journal, *Earth Ethics*.

Through workshops and conferences, CRLE's Higher Education Project brings together educators from various institutions to discuss greening policies. The Center's Greening of Academia program works with colleges and universities to make the academic curricula, food services, and other campus policies "ecologically sound, socially just, and humane." CRLE also publishes a *Green Guide to Higher Education*, which describes the availability and ecological orientation of courses at various institutions of higher learning.

The Center's Religion in the Ecological Age program focuses on "ecospirituality," stressing that today's religious leaders and institutions must take the responsibility for addressing the issues of environmental justice, the human population explosion, over-consumption of **natural resources**, and other environmental problems that threaten the well-being of the natural environment and of the humans and

wildlife dependent on it. CRLE sponsors conferences and publications on such issues, including a series of international conferences in Assisi, Italy (home of St. Francis of Assisi, the thirteenth century lover of animals and patron saint of **nature**).

Fundamental to the Center's mission was the creation of an **Earth Charter**. The Charter was approved by the Earth Charter Commission in 2000 and hopefully will be backed by the United Nations in 2002. It will "prescribe new norms for state and interstate behavior needed to maintain livelihoods and life on our shared planet." CRLE says that the main purpose of the Earth Charter is to "create a 'soft law' document that sets forth the fundamental principles of this emerging new ethics, principles that include respect for human rights, peace, economic equity, environmental protection and sustainable living... It is hoped that the Charter will become a universal code of conduct for states and people..."

Other activities of the Global Earth Ethic project include sponsoring conferences and publications focusing on ethics as they relate to agriculture, development, the environment, and "the appropriate use of animals in research, education, and agriculture..." This effort to raise consciousness includes a three-year project focused on establishing a new principle for agriculture explained in the "Soul of Agriculture: a Production Ethic for the 21st Century."

CRLE's Sustainable Livelihoods in Sustainable Communities program works with United Nations agencies and other regional and international organizations to promote environmentally sustainable developmental and agricultural practices through publications, conferences and meetings, with particular emphasis on **indigenous peoples** and rural communities.

CRLE's quarterly journal, *Earth Ethics*, offers book reviews and a calendar of upcoming events, and it provides a forum for scholarly and provocative feature articles discussing and debating sustainability and other environmental topics as they affect the fields of religion, agriculture, education, business, and the arts, often "challenging current economic and developmental practices."

The Center is affiliated with and supported by the **Humane Society of the United States**, the nation's largest animal protection organization.

[*Lewis G. Regenstein*]

RESOURCES
BOOKS

Fox, M.W. *The Boundless Circle: Caring for Creatures and Creation.* Wheaton, IL: Quest Books, 1996.

OTHER

Earth Ethics. Washington, D.C.: Center for the Respect for Life and Environment.

ORGANIZATIONS

Center for Respect of Life and Environment, 2100 L Street, NW, Washington, D.C. USA 20037 (202) 778-6133, Fax: (202) 778-6138, Email: info@crle.org, <http://www.crle.org>

Center for Rural Affairs

The Center for Rural Affairs (CRA) is a nonprofit organization dedicated to the social, economic and **environmental health** of rural communities. Founded in 1973, the Center for Rural Affairs includes among its participants farmers, ranchers, business people, and educators concerned with the decline of the family farm.

CRA works to provoke public thought on issues and government policies that affect rural Americans, especially in the Midwest and Plains regions of the country. It sponsors research, education, advocacy, organizing and service projects aimed to improve the life of rural dwellers. CRA's **sustainable agriculture** policy is designed to analyze, propose, and advocate public policies that reward environmental stewardship, promote family farming, and foster responsible technology. CRA assists beginning farmers with design and implement on-site research that helps to make these farms environmentally sound and economically viable. CRA's **conservation** and education programs address the environmental problems caused by agricultural practices in the North Central United States.

Through a rural enterprise assistance program, CRA teaches rural communities to support self-employment, and it provides business assistance and revolving loan funds for the self-employed. It also provides professional farm management and brokerage service to landowners who are willing to rent or sell land to beginning farmers. CRA promotes fair competition in the agriculture marketplace by working to prevent monopolies, encouraging enforcement of laws restricting corporate farming in the United States, and advocating for the role of United States farmers in international markets.

Publications offered by CRA include the *Center for Rural Affairs Newsletter*, a monthly report on policy issues and research findings; the *Rural Enterprise Reporter*, which provides information about developing small local enterprises; and a variety of special reports on topics such as small farm technology and business strategy.

[*Linda Rehkopf*]

RESOURCES

ORGANIZATIONS

Center for Rural Affairs, 101 S Tallman Street, P.O. Box 406, Walthill, NE USA 68067 (402) 846-5428, Fax: (402) 846-5420, Email: info@cfra.org, <http://www.cfra.org>

Center for Science in the Public Interest

The Center for Science in the Public Interest (CSPI) was founded in 1971 by Michael Jacobson, who remains its executive director. It is a consumer advocacy organization principally concerned with nutrition and food safety, and its membership consists of scientists, nutrition educators, journalists, and lawyers.

CSPI has campaigned on a variety of health and nutrition issues, particularly nutritional problems on a national level. It is the purpose of the group to address "deceptive marketing practices, dangerous **food additives** or contaminants, conflicts of interests in the academic community, and flawed science propagated by industries concerned solely with profits." It monitors current research on nutrition and food safety, as well as the federal agencies responsible for these areas. CSPI maintains an office for legal affairs and special projects. It has initiated legal actions to restrict food contaminants and to ban food additives that are either unsafe or poorly tested. The special projects the group has sponsored include: Americans for Safe Food, the Nutrition Project, and the Alcohol Policies Project. The center publishes educational materials on food and nutrition, and it works to influence policy decisions affecting health and the national diet.

CSPI has made a significant impact on food marketing in the past 10 years, and they have successfully contested food labeling practices in many sectors of the industry. They were instrumental in forcing fast-food companies to disclose ingredients, and they have recently pressed the **Food and Drug Administration** to improve regulations for companies which make and distribute fruit juice. Many brands do not reveal the actual percentages of the different juices used to make them, and certain combinations of juices are often misleadingly labeled as cherry juice or kiwi juice, for instance, when they may be little more than a mixture of apple and grape juice. The organization has also taken action against deceptive food advertising, particularly advertising for children's products. It has recently demanded further testing of a sweetener called sucralose, in the wake of studies that have suggested that it could cause shrinkage of the thymus, a gland affecting cellular immune responses.

CSPI is funded mainly by foundation grants and subscriptions to its *Nutrition Action Newsletter*. The newsletter is published ten months out of the year and is intended to increase public understanding of food safety and nutrition issues. It frequently examines the consequences of legislation and regulation at the state and federal level; it has explored the controversy over organic and chemical farming methods, and it has studied how agribusiness has changed the way Americans eat. CSPI also distributes posters, videos, and computer software, and it offers a directory of mail-order sources for organically-grown food. Its brochures and reports include: *Guess What's Coming to Dinner: Contaminants in Our Food* and *Organic Agriculture: What the States Are Doing.* It has a staff of 45, a membership of 800,000, and an annual budget of $10,000,000.

[*Douglas Smith*]

RESOURCES

ORGANIZATIONS

Center for Science in the Public Interest, 1875 Connecticut Ave., NW, Suite 300, Washington, D.C. USA 20009 (202) 332-9110, Fax: (202) 265-4954, Email: cspi@cspinet.org, <http://www.cspinet.org>

Centers for Disease Control and Prevention

The Centers for Disease Control and Prevention (CDC) is the Atlanta, Georgia-based agency of the **Public Health Service** that has led efforts to prevent diseases such as **malaria**, polio, smallpox, tuberculosis, and acquired immuno-deficiency syndrome (**AIDS**). As the nation's prevention agency, the CDC's responsibilities have expanded, and it now addresses contemporary threats to health such as injury, environmental and occupational hazards, behavioral risks, and chronic diseases.

Divisions within the CDC use surveillance, epidemiologic and laboratory studies, and community interventions to investigate and prevent public health threats.

The Center for Chronic Disease Prevention and Health Promotion designs programs to reduce death and disability from chronic diseases—cardiovascular, kidney, liver and lung diseases, and **cancer** and diabetes.

The Center for Environmental Health and Injury Control assists public health officials at the scene of natural or artificial disasters such as **volcano** eruptions, forest fires, hazardous **chemical spills**, and nuclear accidents. Scientists study the effects of **chemicals** and pesticides, reactor accidents, and health threats from **radon**, among others. The National Institute for Occupational Safety and Health helps identify chemical and physical hazards that lead to occupational diseases.

Preventing and controlling infectious diseases has been a goal of the CDC since its inception in 1946. The Center for Infectious Diseases investigates outbreaks of infectious disease locally and internationally. The Center for Prevention Services provides financial and technical assistance to control and prevent diseases. Disease detectives in the **Epidemiology** Program Office investigate outbreaks around the world.

Prevention of **tobacco** use is a critical health issue for CDC because cigarette smoking is the leading preventable cause of death in this country. The Office on Smoking and Health conducts research on the effects of smoking, develops health promotion and education campaigns, and helps health departments with smoking education programs.

CDC researchers have improved technology for lead **poisoning** screening, particularly in children. CDC evidence on environmental lead **pollution** was a key in **gasoline** lead content reduction requirements. The CDC also coordinated and directed health studies of **Love Canal**, New York, residents in the 1980s. The director of the CDC administers the **Agency for Toxic Substances and Disease Registry**, the public health agency created to protect the public from exposure to toxic substances in the **environment**. In 1990, CDC became responsible for energy-related epidemiologic research for the **U.S. Department of Energy** nuclear facilities. This includes studies of people who have been exposed to radiation from materials emitted to the air and water from plant operations.

The CDC today carries out an ever-widening agenda with studies on adolescent health, dental disease prevention, the epidemiology of violence, and categorizing and tracking birth abnormalities and infant **mortality**.

[*Linda Rehkopf*]

RESOURCES

ORGANIZATIONS

Centers for Disease Control and Prevention, Public Inquiries/MASO, Mailstop F07, 1600 Clifton Road, Atlanta, GA USA 30333 Toll Free: (800) 311-3435, , http://www.cdc.gov

CERCLA

see **Comprehensive Environmental Response, Compensation, and Liability Act**

CERES Principles

see **Valdez Principles**

Cesium 137

A radioactive **isotope** of the metallic element cesium. Cesium-137 is one of the major products formed when **uranium** undergoes **nuclear fission**, as in a nuclear reactor or

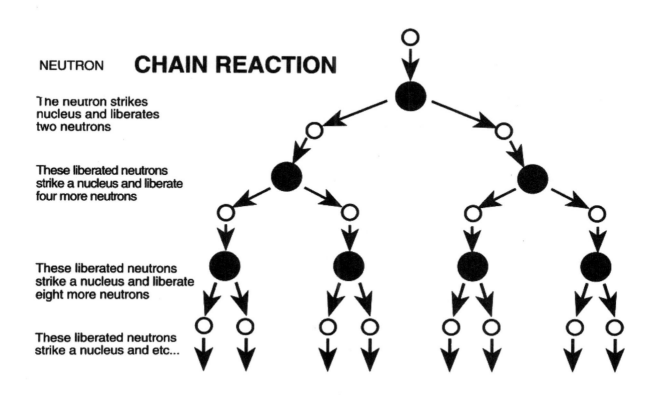

NEUTRON **CHAIN REACTION**

The neutron strikes nucleus and liberates two neutrons

These liberated neutrons strike a nucleus and liberate four more neutrons

These liberated neutrons strike a nucleus and liberate eight more neutrons

These liberated neutrons strike a nucleus and etc...

A chain reaction. (Illustration by Laurence Lawson. (Reproduced by permission.)

nuclear weapons. During atmospheric testing of nuclear weapons after World War II, cesium-137 was a source of major concern to health scientists. Researchers found that the isotope settled to the ground, where it was absorbed by plants and eaten by animals, ultimately affecting humans. Once inside the human body, the isotope releases beta particles, which are carcinogens, teratogens, and mutagens. Large quantities of cesium-137 were released when the Chernobyl nuclear reactor exploded in 1986.

CFCs

see **Chlorofluorocarbons**

CGIAR

see **Consultative Group on International Agricultural Research**

Chain reaction

A situation in which one action causes or initiates a similar action. In a nuclear chain reaction, for example, a **neutron** strikes a uranium-235 nucleus, causing the nucleus to undergo fission, which in turn produces a variety of products. Among these products is one or more neutrons. Thus, the particle needed to initiate this reaction (the neutron) is itself produced as a result of the reaction. Once begun, the reaction continues as long as uranium-235 nuclei are available. Nuclear chain reactions are important sources of fission and fusion energy.

Chaparral

Chaparral is an ecological community consisting of drought-resistant evergreen shrubs and small trees that are adapted to long, hot, dry summers and mild, rainy winters. The chaparral is found in five places on earth where there is a warm land mass and a cool ocean: southern California, the

Cape Town area of South Africa, the western tip of **Australia**, the west coast of South America, and the coastal areas of the Mediterranean in southern Europe. Total annual precipitation ranges between 15 and 40 inches per year, while annual temperatures range from 50–64.4 °F (10–18 °C). Droughts and fires, which are often set by lightning during the summer/autumn dry season, are common in the chaparral. In fact, because the release of minerals occurs as a result of fire, many chaparral plants grow best after a fire.

The chaparral may have many types of terrain, including flat plains, rocky hills, and mountain slopes. The word chaparral comes from the Spanish word *chaparro*, meaning a dry thicket of oak shrubs.

The plants and animals that live in the chaparral are adapted to the characteristic hot and dry climatic conditions. Most of the plants are less than 10 ft tall and have leathery leaves with thick cuticles that hold moisture. Many of the shrub **flora** are aromatic, contain flammable oils, and are adapted to periodic burns. Examples of chaparral plants include poison oak, scrub oak, pine, manzanita, chamise, yucca, and cacti. The animals are mainly grassland and **desert** types, including coyotes, jack rabbits, mule deer, rattlesnakes, mountain lions, kangaroo rats, foxes, bobcats, lizards, horned toads, praying mantis, honey bees, and ladybugs.

[*Judith L. Sims*]

RESOURCES
BOOKS

Collins, Barbara. *Key to Coastal and Chaparral Flowering Plants of Southern California. Third Edition.* Dubuque, IA: Kendall/Hunt Publishing Company, 2000.
Ricciuti, Edward. R. *Chaparral (Biomes of the World.* Tarrytown, NY: Benchmark Books, 1996.

OTHER

Collins, Barbara J. *Wildflowers of Southern California: Photographs of the Chaparral.* California Lutheran University, February 23, 2002. [cited May 27, 2002]. <http://ww1.clunet.edu/wf/>

Chelate

A chemical compound in which one atom is enclosed within a larger cluster of atoms that surrounds it like an envelope. The term comes from the Greek word *chela*, meaning claw. Chelating agents—compounds that can form chelates with other atoms—have a wide variety of environmental applications. For example, the compound ethylenediaminetetraacetic **acid** (EDTA) is used to remove **lead** from the blood. EDTA molecules surround and bind to lead atoms, and the chelate is then excreted in the urine. EDTA can also be used to soften hard water by chelating the calcium and magnesium ions that cause hardness.

Chelyabinsk, Russia

Chelyabinsk is the name of a province and its capital city in west-central Russia. It covers an area of about 34,000 mi^2 (88,060 km^2) and has a population of about 3.6 million. Chelyabinsk city lies on the Miass River on the eastern side of the Ural Mountains. Its population in 1990 was about 1.2 million.

Chelyabinsk is best known today as the home of Mayak, a 77-mi^2 (200-km^2) complex where **nuclear weapons** were built for the former Soviet Union. Because of intentional policy decisions and accidental releases of radioactive materials, Mayak has been called the most polluted spot on Earth.

Virtually nothing was known about Mayak by the outside world, the Russian people, or even the residents of Chelyabinsk themselves until 1991. Then, under the new philosophy of *glasnost*, Soviet president Mikhail Gorbachev released a report on the complex. It listed 937 official cases of chronic **radiation sickness** among Chelyabinsk residents. Medical authorities believe that the actual number is many times larger.

The report also documented the secrecy with which the Soviet government shrouded its environmental problems at Mayak. Physicians were not even allowed to discuss the cause or nature of the radiation sickness. Instead, they had to refer to it as the "ABC disease."

Chelyabinsk's medical problems were apparently the result of three major "incidents" involving the release of radiation at Mayak. The first dated from the late 1940s to the mid-1950s, when **radioactive waste** from nuclear weapons research and development was dumped directly into the nearby Techa River. People downstream from Mayak were exposed to radiation levels 57 times greater than those at the better-known Chernobyl accident in 1986. The Gorbachev report admitted that 28,000 people received radiation doses of "medical consequence." Astonishingly, almost no one was evacuated from the area.

The second incident occurred in 1957, when a nuclear waste dump at Mayak exploded with a force equivalent to a five to 10 kiloton atomic bomb. The site had been constructed in 1953 as an alternative to simply disgorging radioactive wastes into the Techa. When the automatic cooling system failed, materials in the dump were heated to a temperature of 662°F (350°C). In the resulting explosion, 20 million curies of radiation were released, exposing 270,000 people to dangerous levels of **radioactivity**. Neither the Soviet Union nor the United States government, which had detected the accident, revealed the devastation at Mayak.

The third incident happened in 1967. In their search for ways to dispose of radioactive waste, officials at Mayak decided in 1951 to use Lake Karachay as a repository. They realized that dumping into the Techa was not a satisfactory solution, and they hoped that Karachay—which has no natural outlet—would be a better choice.

Unfortunately, radioactive materials began **leaching** into the region's water supply almost immediately. Radiation was eventually detected as far as 2 mi (3 km) away. The 1967 disaster occurred when an unusually dry summer diminished the lake significantly. A layer of radioactive material, deposited on the newly exposed shoreline, was spread by strong winds that blew across the area. This released radiation equivalent to the amount contained in the first atomic bomb explosion over Hiroshima.

[*David E. Newton*]

RESOURCES
PERIODICALS

Cochran, T. B., and R. S. Norris. "A First Look at the Soviet Bomb Complex." *Bulletin of the Atomic Scientists* 47 (May 1991): 25-31.
Hertsgaard, M. "From Here to Chelyabinsk." *Mother Jones* 17 (January-February 1992): 51-55+.
Perea, J. "Soviet Plutonium Plant 'Killed Thousands'." *New Scientist* 134 (20 June 1992): 10.
Wald, M. L. "High Radiation Doses Seen for Soviet Arms Workers." *New York Times* (16 August 1990): A3.

Chemical bond

A chemical bond is any force of attraction between two atoms strong enough to hold the atoms together for some period of time. At least five primary types of chemical bonds are known, ranging from very strong to very weak. They are covalent, ionic, metallic, and **hydrogen** bonds, and London forces.

In all cases, a chemical bond ultimately involves forces of attraction between the positively-charged nucleus of one atom and the negatively-charged electron of a second atom. Understanding the nature of chemical bonds has practical significance since the type of bonding found in a substance explains to a large extent the macroscopic properties of that substance.

An ionic bond is one in which one atom completely loses one or more electrons to a second atom. The first atom becomes a positively charged **ion** and the second, a negatively charged ion. The two ions are attracted to each other because of their opposite electrical charges.

In a covalent bond, two atoms share one or more pairs of electrons. For example, a hydrogen atom and a fluorine atom each donate a single electron to form a shared pair that constitutes a covalent bond between the two atoms. Both electrons in the shared pair orbit the nuclei of both atoms.

In most cases, covalent and ionic bonding occur in such a way as to satisfy the Law of Octaves. Essentially that law states that the most stable configuration for an atom is one in which the outer energy level of the atom contains eight electrons or, in the case of smaller atoms, two electrons.

Ionic and covalent bonds might appear to represent two distinct limits of electron exchange between atoms, one in which electrons are totally gained and lost (ionic bonding) and one in which electrons are shared (covalent bonding). In fact, most chemical bonds fall somewhere between these two extreme cases. In the hydrogen-fluorine example mentioned above, the fluorine nucleus is much larger than the hydrogen nucleus and, therefore, exerts a greater pull on the shared electron pair. The electrons spend more time in the vicinity of the fluorine nucleus and less time in the vicinity of the hydrogen nucleus. For this reason, the fluorine end of the bond is more negative than the hydrogen end, and the bond is said to be a polar covalent bond. A non-polar covalent bond is possible only between two atoms with equal attraction for electrons as, for example, between two atoms of the same element.

Metallic bonds are very different from ionic and covalent bonds in that they involve large numbers of atoms. The outer electrons of these atoms feel very little attraction to any one nucleus and are able, therefore, to move freely throughout the metal.

Hydrogen bonds are very weak forces of attraction between atoms with partial positive and negative charges. Hydrogen bonds are especially important in living organisms since they can be broken and reformed easily during biochemical changes.

London forces are the weakest of chemical bonds. They are forces of attraction between two uncharged molecules. The force appears to arise from the temporary shift of electrical charges within each molecule.

[*David E. Newton*]

RESOURCES
BOOKS

Giddings, J. Calvin. *Chemistry, Man, and Environmental Change: An Integrated Approach.* San Francisco: Canfield Press, 1973.

Chemical oxygen demand

Chemical oxygen demand (COD) is a measure of the ability of chemical reactions to oxidize matter in an aqueous system. The results are expressed in terms of oxygen so that they can be compared directly to the results of **biochemical oxygen demand** (BOD) testing. The test is performed by adding the oxidizing solution to a sample, boiling the mixture on a refluxing apparatus for two hours and then titrating the amount of dichromate remaining after the refluxing period. The titration procedure involves adding ferrous ammonium sulfate (FAS), at a known normality, to reduce the remaining dichromate. The amount of dichromate reduced during the test—the initial amount minus the amount remaining at the end—is then expressed in terms of oxygen. The test has nothing to do with oxygen initially present or used. It is a measure of the demand of a solution or suspension for a strong oxidant. The oxidant will react with most organic materials and certain inorganic materials under the conditions of the test. For example, Fe^{2+} and Mn^{2+} will be oxidized for Fe^{3+} and Mn^{4+}, respectively, during the test.

Generally, the COD is larger than the BOD exerted over a five-day period (BOD_5), but there are exceptions in which **microbes** of the BOD test can oxidize materials that the COD reagents cannot. For a raw, domestic **wastewater**, the COD/BOD_5 ratio is in the area of 1.5–3.0/1.0. Higher ratios would indicate the presence of toxic, non-biodegradable or less readily **biodegradable** materials.

The COD test is commonly used because it is a relatively short-term, precise test with few interferences. However, the spent solutions generated by the test are hazardous. The liquids are acidic, and contain chromium, silver, **mercury**, and perhaps other toxic materials in the sample tested. For this reason laboratories are doing fewer or smaller COD tests in which smaller amounts of the same reagents are used.

[*Gregory D. Boardman*]

RESOURCES

BOOKS

Corbitt, R. A. *Standard Handbook of Environmental Engineering.* New York: McGraw-Hill, 1990.

Davis, M. L., and D. A. Cornwell. *Introduction to Environmental Engineering.* New York: McGraw-Hill, 1991.

Peavy, H. S., D. R. Rowe, and G. Tchobanoglous. *Environmental Engineering.* New York: McGraw-Hill, 1985.

Tchobanoglous, G., and E. D. Schroeder. *Water Quality.* Reading, MA: Addison-Wesley, 1985.

Viessman, W., Jr., and M. J. Hammer. *Water Supply and Pollution Control.* 5th ed. New York: Harper Collins, 1993.

Chemical spills

Chemical spills are any accidental releases of synthetic **chemicals** that pose a risk to the **environment**.

Spills occur at any of the steps between the production of a chemical and its use. A railroad tank car may spring a leak; a pipe in a manufacturing plant may break; or an underground storage tank may corrode allowing its contents to escape into **groundwater**. These spills are often classified into four general categories: the release of a substance into a body of water; the release of a liquid on land; the release of a solid on land; and the release of a gas into the **atmosphere**. The purpose of this method of classification is to provide the basis for a systematic approach to the control of any type of chemical spill.

Some of the most famous chemical spills in history illustrate these general categories. For example, seven cars of a train carrying the **pesticide** metam sodium fell off the tracks near Dunsmuir, California, in August 1991, breaking open and releasing the chemicals into the Sacramento River. Plant and aquatic life for 43 mi (70 km) downriver died as a result of the accident. The pesticide eventually formed a band 225 ft (70 m) wide across Lake Shasta before it could be contained.

In 1983, the **Environmental Protection Agency** (EPA) purchased the whole town of **Times Beach**, Missouri, and relocated more than 2,200 residents because the land was so badly contaminated with highly toxic dioxins. The concentration of these compounds, a by-product of the production of herbicides, was more than a thousand times the maximum recommended level.

In December 1984, a cloud of poisonous gas escaped from a Union Carbide chemical plant in **Bhopal, India**. The plant produced the pesticide Sevin from a number of chemicals, many of which were toxic. The gas that accidentally escaped probably contained a highly toxic mixture of phosgene, methyl isocyanate (MIC), **chlorine**, **carbon monoxide**, and **hydrogen** cyanide, as well as other hazardous gases. The cloud spread over an area of more than 15 mi^2 (40 km^2), exposing more than 200,000 people to its dangers.

Chemists have now developed a sophisticated approach to the treatment of chemical spills, which involves one or more of five major steps: containment, physical treatment, chemical treatment, biological treatment, and disposal or destruction. **Soil** sealants, which can be used to prevent a liquid from sinking into the ground, are an example of containment. One of the most common methods of physical treatment is activated charcoal, because it has the ability to adsorb toxic substances on its surface, thus removing them from the environment. Chemical treatment is possible because many hazardous materials in a spill can be treated by adding some other chemical that will neutralize them, and biological treatment usually involves **microorganisms** that will attack and degrade a toxic chemical. Open burning, **deep-well injection**, and burial in a **landfill** are all methods of ultimate disposal.

[*David E. Newton*]

RESOURCES

BOOKS

Unterberg, W., et al. *How to Respond to Hazardous Chemical Spills.* Park Ridge, N.J.: Noyes Data Corporation, 1988.

Chemicals

The general public often construes the word "chemical" to mean a harmful synthetic substance. In fact, however, the term applies to any element or compound, either natural or synthetic. The thousands of compounds that make up the human body are all chemicals, as are the products of scientific research. A more accurate description, however, can be found in the dictionary. Thus, aspirin is a chemical by this definition, since it is the product of a series of chemical reactions.

The story of chemicals began with the rise of human society. Indeed, early stages of human history, such as the Iron, **Copper**, and Bronze Ages reflect humans' ability to produce important new materials. In the first two eras, people learned how to purify and use pure metals. In the third case, they discovered how to combine two to make an alloy with distinctive properties.

The history of ancient civilizations is filled with examples of men and women adapting **natural resources** for their own uses. Egyptians of the eighteenth dynasty (1700–1500 B.C.), for example, knew how to use cobalt compounds to glaze pottery and glass. They had also developed techniques for making and using a variety of dyes.

Over the next 3,000 years, humans expanded and improved their abilities to manipulate natural chemicals. Then, in the 1850s, a remarkable breakthrough occurred. A discovery by young British scientist William Henry Perkin led to the birth of the synthetic chemicals industry.

Perkin's great discovery came about almost by accident, an occurrence that was to become common in the synthetics industry. As an 18-year-old student at England's Royal College of Chemistry, Perkin was looking for an artificial compound that could be used as a quinine substitute. Quinine, the only drug available for the treatment of **malaria**, was itself in short supply.

Following his teacher's lead, Perkin carried out a number of experiments with compounds extracted from **coal** tar, the black, sticky **sludge** obtained when coal is heated in insufficient air. Eventually, he produced a black powder which, when dissolved in alcohol, created a beautiful purple liquid. Struck by the colorful solution, Perkin tried dyeing clothes with it.

His efforts were eventually successful. He went on to mass produce the synthetic dye—mauve, as it was named—and to create an entirely new industry. The years that followed are sometimes referred to as The Mauve Decade because of the many new synthetic products inspired by Perkin's achievement. Some of the great chemists of that era have been memorialized in the names of the products they developed or the companies they established: Adolf von Baeyer (Bayer aspirin), Leo Baekeland (Baekelite plastic), Eleuthère Irénée du Pont (DuPont Chemical), George East-

man (Eastman 910 adhesive and the Eastman Kodak Company), and Charles Goodyear (Goodyear **Rubber**).

Chemists soon learned that from the gooey, ugly by-products of coal tar, a whole host of new products could be made. Among these products were dyes, medicines, fibers, flavorings, **plastics**, explosives, and **detergents**. They found that the other fossil fuels—petroleum and natural gas—could also produce synthetic chemicals.

Today, synthetic chemicals permeate our lives. They are at least as much a part of the **environment**, if not more, than are natural chemicals. They make life healthier, safer, and more enjoyable. People concerned about the abundance of "chemicals" in our environment should remember that everyone benefits from anti-cancer drugs, pain-killing anesthetics, long-lasting fibers, vivid dyes, sturdy synthetic rubber tires, and dozens of other products. The world would be a much poorer place without them.

Unfortunately, the production, use, and disposal of synthetic chemicals can create problems because they may be persistent and/or hazardous. Persistent means that a substance remains in the environment for a long time: dozens, hundreds, or thousands of years in many cases. Natural products such as wood and paper degrade naturally as they are consumed by **microorganisms**. Synthetic chemicals, however, have not been around long enough for such microorganisms to evolve.

This leads to the familiar problem of **solid waste** disposal. Plastics used for bottles, wrappings, containers, and hundreds of other purposes do not decay. As a result, landfills become crowded and communities need new places to dump their trash.

Persistence is even more of a problem if a chemical is hazardous. Some chemicals are a problem, for example, because they are flammable. More commonly, however, a hazardous chemical will adversely affect the health of a plant or animal. It may be (1) toxic, (2) carcinogenic, (3) teratogenic, or (4) mutagenic.

Toxic chemicals cause people, animals, or plants to become ill, develop a disease, or die. DDT, **chlordane**, heptachlor, and aldrin are familiar, toxic pesticides. Carcinogens cause **cancer**; teratogens produce **birth defects**. Mutagens, perhaps the most sinister of all, inflict genetic damage.

Determining these effects can often be very difficult. Scientists can usually determine if a chemical will harm or kill a person. But how does one determine if a chemical causes cancer twenty years after exposure, is responsible for birth defects, or produces genetic disorders? After all, any number of factors may have been responsible for each of these health problems.

As a result, labeling any specific chemical as carcinogenic, teratogenic, or mutagenic can be difficult. Still, environmental scientists have prepared a list of synthetic chemicals that they believe fall into these categories. Among them are **vinyl chloride**, trichloroethylene, **tetrachloroethylene**, the nitrosamines, and chlordane and heptachlor.

Another class of chemicals are hazardous because they may contribute to the **greenhouse effect** and **ozone layer depletion**. The single most important chemical in determining the earth's annual average temperature is a naturally-occurring compound, **carbon dioxide**. Its increased production is believed to be responsible for a gradual increase in the planet's annual average temperature.

But synthetic compounds may also play a role in global warming. **Chlorofluorocarbons** (CFCs) are widely used in industry because of their many desirable properties, one of which is their chemical **stability**. This very property means, however, that when released into the **atmosphere**, they remain there for many years. Since they capture heat radiated from the earth in much the way **carbon** dioxide does, they are probably important contributors to global warming.

These same chemicals, highly unreactive on earth, decompose easily in the upper atmosphere. When they do so, they react with the **ozone** in the **stratosphere**, converting it to ordinary oxygen. This may have serious consequences, since stratospheric ozone shields the earth from harmful **ultraviolet radiation**.

There are two ways to deal with potentially hazardous chemicals in the environment. One is to take political or legal action to reduce the production, limit the use, and/or control the disposal of such products. A treaty negotiated and signed in Montreal by more than forty nations in 1987, for example, calls for a gradual ban on CFC production. If the treaty is honored, these chemicals will eventually be phased out of use.

A second approach is to solve the problem scientifically. Synthetic chemicals are a product of scientific research, and science can often solve the problems these chemicals produce. For example, scientists are exploring the possibility of replacing CFCs with related compounds called fluorocarbons (FCs) or hydrochlorofluorocarbons (HCFCs). Both are commercially appealing, but they have fewer harmful effects on the environment.

[*David E. Newton*]

RESOURCES
BOOKS

Giddings, J. Calvin. *Chemistry, Man, and Environmental Change: An Integrated Approach.* San Francisco: Canfield Press, 1973.

Joesten, M. D., et al. *World of Chemistry.* Philadelphia: Saunders College Publishing, 1991.

Newton, David E. *The Chemical Elements.* New York: Franklin Watts, 1994.

Chemosynthesis

Chemosynthesis is a metabolic pathway used by some bacteria to synthesize new organic compounds such as carbohydrates by using energy derived from the oxidation of inorganic molecules—hydrogen sulfide (H_2S) or ammonia (NH_3). Chemosynthesis can occur in environments such as the deep ocean around **hydrothermal vents**, where sunlight does not penetrate, but where **chemicals** like **hydrogen** sulfide are available. Chemosynthesis is also a critical part of the **nitrogen cycle**, where bacteria that live in the **soil**, or in special plant structures called heterocysts, utilize ammonia for energy and produce **nitrates and nitrites** which can subsequently be used as nutrients for plants. Some bacteria can also utilize hydrogen gas (H_2) and **carbon dioxide** (CO_2) in a chemosynthetic pathway that results in the production of new organic compounds and **methane** (CH_4).

[*Marie H. Bundy*]

Chernobyl nuclear power station

On April 26, 1986, at precisely 1:24 A.M., the Chernobyl **nuclear power** plant exploded, releasing large amounts of **radioactivity** into the **environment**. The power station is located 9 mi (14.5 km) northwest of the town of Chernobyl, with a population of 12,500, and less than 2 mi (3.2 km) from the town of Pripyat, which contains 45,000 inhabitants. The explosion and its aftermath, including the manner in which the accident was handled, have raised questions about the safety and future of nuclear power.

The Chernobyl accident resulted from several factors: flaws in the engineering design, which were compensated by a strict set of procedures; failure of the plant management to enforce these procedures; and finally the decision of the engineers to conduct a risky experiment. They wanted to test whether the plant's turbine generator—from its rotating inertia—could provide enough power to the reactor in case of a power shutdown. This experiment required disconnecting the reactor's emergency core cooling pump and other safety devices.

The series of critical events, as described by Richard Mould in *Chernobyl, The Real Story*, are as follows: At 1:00 A.M. on April 25, power reduction was started in preparation for the experiment. At 1:40 A.M. the reactors's emergency core cooling system was turned off. At 11:10 P.M. power was further reduced, resulting in a nearly unmanageable situation. At 1:00 A.M. on April 26, power was increased in an attempt to stabilize the reactor; however, cooling pumps were operating well beyond their rated capacity, causing a reduction in steam generation and a fall in stream pressure. By 1:19 A.M., the water in the cooling circuit had approached the boiling point. At 1:23 A.M., the operators tried to control the reaction by manually pushing control rods into the core; however, the rods did not descend their full length into the reactor since destruction of the graphite core was already occurring. In 4.5 seconds, the power level rose two thousandfold. At 1:24 A.M., there was an explosion when the hot reactor fuel elements, lacking enough liquid for cooling, decomposed the water into **hydrogen** and oxygen. The generated pressures blew off the 1,000-ton concrete roof of the reactor, and burning graphite, molten **uranium**, and radioactive ashes spilled out to the **atmosphere**.

The explosion that occurred was not a nuclear explosion such as would occur with an atomic bomb but its effects were just as devastating. In order to put the expulsion of radioactive material from the Chernobyl reactor into perspective, almost 50 tons of fuel went into the atmosphere plus an additional 70 tons of fuel, and 700 tons of radioactive reactor graphite settled in the vicinity of the damaged unit. Some 50 tons of nuclear fuel and 800 tons of reactor graphite remained in the reactor vault, with the graphite burning up completely in the next several days after the accident. The amount of radioactive material that went into the atmosphere was equivalent to 10 Hiroshima bombs.

Officials at first denied that there had been a serious accident at the power plant. The government in Moscow was led to believe for several hours after the explosion and fire at Chernobyl that the reactor core was still intact. This delayed the evacuation for a critical period during which local citizens were exposed to high radiation levels. The evacuation of Chernobyl and local villages was spread out over eight days. A total of 135,000 persons were evacuated from the area, with the major evacuation at Pripyat starting at 2:00 P.M., the day after the explosion. Tests showed that air, water, and **soil** around the plant had significant contamination. Children, in particular, were a matter of concern and were evacuated to the southern Ukraine, the Crimea, and the Black Sea coast.

At the time of the accident, and for several days thereafter, the winds carried the **radioactive waste** to the north. The radioactive cloud split into two lobes, one spreading west and then north through Poland, Germany, Belgium, and Holland, and the other through Sweden and Finland. By the first of May, the wind direction changed and the radioactive fallout—at a diminished rate—went south over the Balkans and then west through Italy. Large areas of Europe were affected, and many farmers destroyed their crops for fear of contamination. Forests have been cleared and large amounts of earth were removed in order to clean up radioactivity. Plastic film has been laid in some areas in an effort to contain radioactive dust.

Officially 31 persons were reported to have been killed at the reactor site by a combination of the explosion and

radiation exposure; another 174 were exposed to high doses of radiation which resulted in **radiation sickness** and long-term illnesses. The maximum permissible dose of radiation for a nuclear power operator is 5 roentgens per year and for the rest of the population, 0.5 roentgens per year. At the Chernobyl plant, the levels of radiation ranged from 1,000 to 20,000 roentgens per hour. One British report estimates that worldwide, the number of persons afflicted with **cancer** which can be attributed to the Chernobyl accident will be about 2,300. Others argued that the number will be much higher. In Minsk, the rate of **leukemia** has more than doubled from 41 per million in 1985 to 93 per million in 1990.

Many heroic deeds were reported during this emergency. Fire fighters exposed themselves to deadly radiation while trying to stop the inferno. Every one eventually died from radiation exposure. Construction workers volunteered to entomb the reactor ruins with a massive concrete sarcophagus. Bus drivers risked further exposure by making repeated trips into contaminated areas in order to evacuate villagers. Over 600,000 people were involved in the decontamination and clean up of Chernobyl. The health effects on them from their exposure are not completely known. The Chernobyl accident focused international attention on the risks associated with operating a nuclear reactor for the generation of power. Public apprehension has forced some governments to review their own safety procedures and to compare the operation of their nuclear reactors with Chernobyl's. In a review of the Chernobyl accident by the Atomic Energy Authority of the United Kingdom, an effort was made to contrast the design of the Chernobyl reactor and management procedures with those in practice in the United States and the United Kingdom.

Three design drawbacks were noted of the Chernobyl nuclear power plant:

- The reactor was intrinsically unstable below 20% power and never should have been operated in that mode. (U.S. and UK reactors do not have this design flaw).

- The shut-down operating system was inadequate and contributed to the accident rather than terminating it. (U.S. and UK control systems differ significantly).

- There were no controls to prevent the staff from operating the reactor in the unstable region or preventing the disabling of existing safeguards.

In addition, the Chernobyl management had no effective watchdog agency to inspect procedures and order closure of the facility. Also in years prior to the accident there was a lack of information given the public of prior nuclear accidents, typical of the press censorship and news management occurring in the period before *glasnost*. The operators were not adequately trained nor were they themselves fully aware of prior nuclear power accidents or near accidents

Chernobyl Nuclear Power Station. Lighter areas of the buidling are part of the original structure, while darker areas are the steel and concrete "sarcophagus" that was added to contain radioactivity leaking from the faulty reactor. (Corbis-Bettmann. Reproduced by permission.)

which would have made them more sensitive to the dangers of a runaway reactor system.

Unfortunately in the former Soviet block nations there are several nuclear reactors that are potentially as hazardous as Chernobyl but which must continue operation to maintain power requirements; however, the operational procedures are under constant review to avoid another accident. Clearly the Western world will have to assist the former Soviet block to bring reactor operating equipment and standards up to a much higher level of safety to avoid a similar and possibly more disastrous accident.

[*Malcolm T. Hepworth*]

RESOURCES
BOOKS

Feshbach, M., and A. Friendly Jr. *Ecocide in the USSR.* New York: Basic Books, 1992.
Fusco, Paul, and Magdalena Caris. *Chernobyl Legacy.* de.MO, 2001.
Medvedev, G. *No Breathing Room: The Aftermath of Chernobyl.* New York: Basic, 1993.

Mould, R. E. *Chernobyl: The Real Story.* New York: Pergamon, 1988.

Chesapeake Bay

The Chesapeake Bay is the largest **estuary** (186 mi [300 km] long) in the United States. The Bay was formed 1500 years ago by the retreat of glaciers and the subsequent sea level rise that inundated the lower Susquehanna River valley. The Bay has a **drainage** basin of 64,076 square miles (166,000 sq km) covering six states and running through Pennsylvania, Maryland, the District of Columbia, and Virginia before entering the Atlantic Ocean. While 150 rivers enter the Bay, a mere eight account for 90% of the freshwater input, with the Susquehanna alone contributing nearly half. Chesapeake Bay is a complex system, composed of numerous habitats and environmental gradients.

Chesapeake Bay's abundant **natural resources** attracted native Americans, first settling on its shores. The first European record of the Bay was in 1572 and the area surrounding Chesapeake Bay was rapidly colonized by Europeans. In many ways, the United States grew up around Chesapeake Bay. The colonists harvested the Bay's resources and used its waterways for **transportation**. Today 10 million people live in the Chesapeake Bay's drainage basin, and many of their activities affect the environmental quality of the Bay as did the activities of their ancestors.

The rivers emptying into the Bay were also used by the colonists to dispose of raw sewage. By the middle 1800s some of the rivers feeding the Bay were polluted: the Potomac was recorded as emitting a lingering stench. The first sewer was constructed in Washington, DC, and it pumped untreated waste into the Bay. It was recognized in 1893 that the diseases suffered by humans consuming shellfish from the Bay were directly related to the **discharge** of raw sewage into the Bay. Despite this recognition, efforts in 1897 by the mayor of Baltimore to oppose the construction of a sewage system that discharged sewage into the Bay in favor of a "land **filtration** technique" failed. Ultimately, a secondary treatment system discharging into the Bay was constructed. In the mid-1970s, a $27 million government-funded study of the Bay's condition concluded that the deteriorating quality of the Chesapeake Bay was a consequence of human impacts. But it was not until the early 1980s that an **Environmental Protection Agency** (EPA) report on the Chesapeake focused interest on saving the Bay, and $500 million was spent on cleanup and construction of **sewage treatment** plants.

While the Chesapeake Bay is used primarily as a transportation corridor, its natural resources rank a close second in importance to humans. The most commercially important fisheries in the Bay are the native American oyster (*Crassostrea virginica*), blue crab (*Callinectes sapidus*), American shad (*Alosa sapidissima*), and striped bass (*Marone saxatilis*). Fisherman first began to notice a decline in fish populations in the 1940s and 1950s, and since then abundances have declined even further. Since the turn of the century, the oyster catch has declined 70%, shad 85%, and striped bass 90%. In the late 1970s, the EPA began to study the declining oyster and striped bass populations and concluded that their decline was due to a combination of over-harvesting and **pollution**.

Work by the EPA and other federal and state agencies has identified six areas of environmental concern for the Bay: (1) excess **nutrient** input from both sewage treatment plants discharging into the Bay and **runoff** from agricultural land; (2) low oxygen levels as a result of increased **biochemical oxygen demand**, which increases dramatically with **loading** of organic material; (3) loss of **submerged aquatic vegetation** due to an increase in turbidity; (4) presence of chemical **toxins**; (5) loss to development of **wetlands** surrounding the Bay that serve as nurseries for juvenile fish

and shellfish and as buffers for runoff of nutrients and toxic **chemicals**; and (6) increasing acidity of water (measured by **pH**) in streams that feed the Bay. These streams are also nursery areas for larval fish that may be adversely affected by decreasing pH.

The increasing growth of phytoplankton—free-floating single-celled plants—in the Bay is generally considered to be a significant contributor to the decline in environmental quality of the Chesapeake Bay. The number of algal blooms has increased dramatically since the 1950s and is attributed to the high levels of the plant nutrients **nitrogen** and **phosphorus** that are discharged into the Bay. In the 1980s and 1990s, it was estimated that discharge from sewage treatment plants and agricultural runoff accounted for 65% of the nitrogen and 22% of the phosphorus found in the Bay. **Acid rain**, formed from discharges from industrial plants in Canada and the northeast United States, contributed 25% of the nitrogen found in the Bay. Excess nutrients encourage **phytoplankton** growth, and as the large number of phytoplankton die and settle to the bottom, their **decomposition** robs the water of oxygen needed by fish and other aquatic organisms. When oxygen levels fall too low, these organisms die or flee from the regions of low oxygen. Decomposition of dead organic matter further reduces the concentration of oxygen. During the late twentieth and early twenty-first centuries, Finfish and shellfish kills became increasingly common in the Bay.

Phytoplankton blooms and the increase in suspended sediments resulting from shoreline development and poor agricultural practices have increased turbidity and led to a decline in submerged aquatic vegetation (SAV) such as **eelgrass**. SAV is extremely important in the prevention of **erosion** of bottom **sediment** and as **critical habitat** for nursery grounds of commercially important fish and shellfish.

Chemicals introduced into the Bay from several sites may have contributed to the decline in the Bay's fish and bird populations. For example, in 1975, the **pesticide Kepone** was leaked or dumped into the James River, **poisoning** fish and shellfish. Harvests of some **species** are still restricted in the area of this spill. **Chlorine** biocides used in **wastewater** treatment plants and **power plants**, which discharge into the Bay, are known human carcinogens and can be toxic to aquatic organisms. **Polycyclic aromatic hydrocarbons** (PAH) have caused dermal lesions in fish populations in the Elizabeth River. PAHs also affect shellfish populations. In the 1990s, public concern focused on **tributyl tin** (TBT) that was used in anti-fouling paint on recreational and commercial boats. TBTs belong to a family of chemicals known as organotins, which are toxic to shellfish and crustaceans. The diversity of chemical pollutants found in the Bay is exemplified by the results of research that identified 100 inorganic and organic contaminants in striped bass caught in the Bay. As of 2002, research efforts were underway to determine the extent of the damage from toxins to natural resources.

Work by private and governmental agencies has started to reverse the declining environmental quality of the Chesapeake Bay. In 1983 Maryland, Virginia, Pennsylvania, the District of Columbia, the Chesapeake Bay Commission, and the EPA signed the Chesapeake Bay Agreement, which outlined procedures to correct many of the Bay's ecological problems, particularly those caused by nutrient enrichment. In 1987 the agreement was significantly expanded and required that the signatories adopt a strategy that would result in at least a 40% reduction in nitrogen and phosphorus entering the Bay by the year 2000. Since 1985, increasing compliance with discharge permits, prohibition of the sale of phosphate-based **detergents**, and the upgrading of wastewater plants has resulted in a 2% reduction in the discharge of nitrogen and a 39% reduction of phosphorous from point sources of pollution. Controls on agriculture and urban development have resulted in approximately 7% reductions in the amount of both nitrogen and phosphorus entering the Bay from nonpoint sources. The amount of toxins entering the Bay have also been reduced. Tributyl tin has been banned for use in anti-fouling paints on non-military vessels, pesticide runoff has been reduced by using alternate strategies for **pest** control, and since 1987, toxic emissions from industrial sources have declined more than 40%. At the same time, some of the Bay's critical habitats are recovering: 22,000 more acres of SAVs are now growing than in 1984, although the total amount of 60,000 acres is still a fraction of the estimated 600,000 acres the historical SAV distribution, man-made oyster reefs are being created to expand suitable **habitat** for oysters, and rivers are being cleared of obstacles such as **dams** and spillways to provide access to spawning areas by migratory fish.

Conflict between commercial and environmental interests have encumbered some of the restoration efforts. Research on the life-history of crabs and oysters shows that limiting the size of crabs that can be sold and the numbers of oysters that can be harvested will help these fisheries rebound, but regulations to limit crab and oyster catches have met with strong **resistance** from watermen who are struggling to survive economically in a declining fishery.

The planned introduction of a non-native Asian oyster (*Crassostrea ariakensis*) by the Virginia Seafood Council, raised hopes that a new oyster fishery could be built around the harvest of this disease-resistant, fast-growing species. In May 2002, the United States **Fish and Wildlife Service** called for a moratorium on the introduction of the Asian oyster, until researchers could determine whether its introduction would endanger native species, or bring new and

deadly disease organisms into Chesapeake Bay. A study by the **National Academy of Sciences** was commenced in 2001 with the goal of making information on these concerns available in summer of 2003.

As of 2002, portions of Chesapeake Bay and its tributaries were still listed as impaired waters under Section 303(d) of the **Clean Water Act**, but progress on arresting the decline of environmental quality of Chesapeake Bay and restoring some of its natural resources continued. In the Chesapeake 2000 Bay Agreement (C2K), Maryland, Pennsylvania, Virginia, and the District of Columbia, together with the Chesapeake Bay Commission and the Federal Government committed to "correct the nutrient- and sediment-related problems in the Chesapeake Bay and its tidal tributaries sufficiently to remove the Bay and the tidal portions of its tributaries from the list of impaired waters under the Clean Water Act." This consortium of stakeholders, jurisdictions, and federal agencies is evidence that citizens, government, and industry can work cooperatively. The Chesapeake Bay program is a national model for efforts to restore other degraded ecosystems.

[*William G Ambrose Jr. and Paul E Renaud and Marie H. Bundy*]

RESOURCES

BOOKS

Brown, L. R. "Maintaining World Fisheries." In *State of the World*, edited by L. Starke. New York: Norton, 1985.

Majumdar, S., et al. *Contaminant Problems and Management of Living Chesapeake Bay Resources.* Easton, NJ: Typehouse of Easton, 1987.

PERIODICALS

D'Elia, C. "Nutrient Enrichment of the Chesapeake Bay." *Environment* 29 (1987): 6-11.

ORGANIZATIONS

Chesapeake Bay Foundation, Philip Merrill Environmental Center, 6 Herndon Avenue, Annapolis, MD 21403 410/268-8816, Email: http://www.cbf.org/about_cbf/contact_us.htm

Child survival revolution

Every year in the developing countries of the world, some 14 million children under the age of five die of common infectious diseases. Most of these children could be saved by simple, inexpensive, preventative medicine. Many public health officials argue that it is as immoral and unethical to allow children to die of easily preventable diseases as it would be to allow them to starve to death or to be murdered. In 1986, the United Nations announced a worldwide campaign to prevent unnecessary child deaths. Called the "child survival revolution," this campaign is based on four principles, designated by the acronym GOBI.

"G" stands for growth monitoring. A healthy child is considered a growing child. Underweight children are much more susceptible to infectious diseases, retardation, and other medical problems than children who are better nourished. Regular growth monitoring is the first step in health maintenance.

"O" stands for oral rehydration therapy (ORT). About one-third of all deaths under five years of age are caused by diarrheal diseases. A simple solution of salts, glucose, or rice powder and boiled water given orally is almost miraculously effective in preventing death from dehydration shock in these diseases. The cost of treatment is only a few cents per child. The British medical journal *Lancet*, called ORT "the most important medical advance of the century."

"B" stands for breast-feeding. Babies who are breast-fed receive natural immunity to diseases from antibodies in their mothers' milk, but infant formula companies have been persuading mothers in many developing countries that bottle-feeding is more modern and healthful than breast-feeding. Unfortunately, these mothers usually do not have access to clean water to combine with the formula and they cannot afford enough expensive synthetic formula to nourish their babies adequately. Consequently, the **mortality** among bottle-fed babies is much higher than among breast-fed babies in developing countries.

"I" is for universal immunization against the six largest, preventable, **communicable diseases** of the world: measles, tetanus, tuberculosis, polio, diphtheria, and whooping cough. In 1975, less than 10% of the developing world's children had been immunized. By 1990, this number had risen to over 50%. Although the goal of full immunization for all children has not yet been reached, many lives are being saved every year. In some countries, yellow fever, typhoid, meningitis, **cholera**, and other diseases also urgently need attention.

Burkina Faso provides an excellent example of how a successful immunization campaign can be carried out. Although this West African nation is one of the poorest in the world (annual gross national product per capita of only $140), and its roads, health care clinics, communication, and educational facilities are either nonexistent or woefully inadequate, a highly successful "vaccination commando" operation was undertaken in 1985. In a single three-week period, one million children were immunized against three major diseases (measles, yellow fever, and meningitis) with only a single injection. This represents 60% of all children under age 14 in the country. The cost was less than $1 per child.

In addition to being an issue of humanity and compassion, reducing child mortality may be one of the best ways to stabilize world **population growth**. There has never been a reduction in birth rates that was not preceded by a reduction

in infant mortality. When parents are confident that their children will survive, they tend to have only the number of children they actually want, rather than "compensating" for likely deaths by extra births. In Bangladesh, where ORT was discovered, a children's health campaign in the slums of Dacca has reduced infant mortality rates 21% since 1983. In that same period, the use of birth control increased 45% and birth rates decreased 21%.

Sri Lanka, China, Costa Rica, Thailand, and the Republic of Korea have reduced child deaths to a level comparable to those in many highly developed countries. This child survival revolution has been followed by low birth rates and stabilizing populations. The United Nations Children's Fund estimates that if all developing countries had been able to achieve similar birth and death rates, there would have been nine million fewer child deaths in 1987, and nearly 22 million fewer births. *See also* Demographic transition

[*William P. Cunningham Ph.D.*]

RESOURCES
BOOKS

UNICEF. *The State of the World's Children*. New York: Oxford University Press, 1987.

Chimpanzees

Common chimpanzees (*Pan troglodytes*) are widespread in the forested parts of West, Central, and East Africa. Pygmy chimpanzees, or bonobos (*P. paniscus*), are restricted to the swampy lowland forests of the Zaire basin. Despite their names, common chimpanzees are no longer common, and pygmy chimpanzees are no smaller than the other **species**.

Chimpanzees are partly arboreal and partly ground-dwelling. They feed in fruit trees by day, nest in other trees at night, and can move rapidly through treetops. On the ground chimpanzees usually walk on all fours (knuckle walking), since their arms are longer than their legs. Their hands have fully opposable thumbs and, although lacking a **precision** grip, can manipulate objects dexterously. Chimpanzees make and use a variety of tools: they shape and strip "fishing sticks" from twigs to poke into termite mounds, and they chew the ends of shoots to fashion fly whisks. They also throw sticks and stones as offensive weapons and hunt and kill young monkeys.

These apes live in small nomadic groups of three to six animals (common chimpanzee) or six to 15 animals (pygmy chimpanzee) which make up a larger community (30–80 individuals) that occupies a territory. Adult males cooperate in defending their territory against predators. Chimpanzee

A chimpanzee (*Pan troglodytes*). (Photograph by Nigel J. Dennis. Photo Researchers Inc. Reproduced by permission.)(Photograph by Nigel J. Dennis. Photo Researchers Inc. Reproduced by permission.)

society consists of fairly promiscuous mixed-sex groups. Female common chimpanzees are sexually receptive for only a brief period in mid-month (estrous), while female pygmy chimpanzees are sexually receptive for most of the month.

Ovulating females capable of fertilization have swollen pink hind quarters and copulate with most of the males in the group. Female chimpanzees give birth to a single infant after a gestation period of about eight months.

Jane Goodall has studied common chimpanzees for almost 30 years in the Gombe Stream **National Park** of Tanzania. She found that chimpanzee personalities are as variable as those of humans, that chimpanzees form alliances, have friendships, have personal dislikes, and run feuds. Chimpanzees also have a cultural tradition, that is, they pass learned behavior and skills from generation to generation. Chimpanzees have been taught complex sign language (the chimpanzee larynx won't allow speech) through which abstract ideas have been conveyed. These studies show that chimpanzees can develop a large vocabulary and that they can manipulate this vocabulary to frame new thoughts.

Humans share 98.4% of their genes with chimpanzees, so only 1.6% of human DNA is responsible for all the differences between the two species. The DNA of gorillas differs 2.3% from chimpanzees, which means that the closest relatives of chimpanzees are humans, not gorillas. Further studies of these close relatives would undoubtedly help to better understand the origins of human social behavior and human **evolution**. Despite this special status, both species of chimpanzees are threatened by the destruction of their forest **habitat** by **hunting** and by capture for research.

[*Neil Cumberlidge Ph.D.*]

RESOURCES
BOOKS

Diamond, J. M. *The Third Chimpanzee: The Evolution and Future of the Human Animal.* New York: Harper Collins, 1992.

Goodall, Jane. *Through a Window: My Thirty Years With the Chimpanzees of Gombe.* Boston: Houghton Mifflin, 1990.

Peterson, Dale, and Jane Goodall. *Visions of Caliban: On Chimpanzees and People.* Boston: Houghton Mifflin, 1993.

PERIODICALS

"Chimp, Human Genes much alike, except for Brain." *USA Today* (February 12, 2002).

Chipko Andolan movement

India has a long history of non-violent, passive **resistance** in social movements rooted in its Hindu concept of *ahimsa*, or "no harm." During the British occupation of India in the early twentieth century, Indian leader Mohandas K. Gandhi began to employ a method of resistance against the British that he called *satyagraha* (meaning "force of truth"). Synthesized from his knowledge of **Henry David Thoreau**, Leo Tolstoy, Christianity and Hinduism, Gandhi's concept of satyagraha involves the absolute refusal to cooperate with a perceived wrong and the use of nonviolent tactics in combination with complete honesty to confront, and ultimately convert, evil.

During the occupation, the rights of peasants to gather products, including forest materials, was severely curtailed. New land ownership systems imposed by the British transformed what had been communal village resources into the private property of newly created landlords. Furthermore, policies that encouraged commercial exploitation of forests were put into place. Trees were felled on a large scale to build ships for the British Royal Navy or to provide ties for the expanding railway network in India, severely depleting forest resources on which traditional cultures had long depended.

In response to British rule with its forest destruction and impoverishment of native people, a series of non-violent movements utilizing satyagraha spread throughout India. The British and local aristocracy suppressed these protests brutally, massacring unarmed villagers by the thousands, and jailing Gandhi a number of times, but Gandhi and his allies remained steadfast in their resistance. The British, forced to comprehend the horror of their actions and unable to scapegoat the nonviolent Indians, at last withdrew from India.

After India gained independence, two of Gandhi's disciples, Mira Behn and Sarala Behn, moved to the foothills of the Himalayas to establish *ashramas* (spiritual retreats) dedicated to raising women's status and rights. Their project was dedicated to four major goals: 1) organizing local women, 2) campaigning against alcohol consumption, 3) fighting for forest protection, and 4) setting up small, local, forest-based industries.

During the 1970s, commercial loggers began large-scale tree felling in the Garhwal region in the state of Uttar Pradesh in northern India. Landslides and floods resulted from stripping the forest cover from the hills. The firewood on which local people depended was destroyed, threatening the way of life of the traditional forest culture.

In April, 1973, village women from the Gopeshwar region who had been educated and empowered by the principles of non-violence devised by the Behns began to confront loggers directly, wrapping their arms around trees to protect them. The outpouring of support sparked by their actions was dubbed the *Chipko Andolan* movement (literally, "movement to hug trees"). This crusade to save the forests eventually prevented **logging** on 4,633 mi² (12,000 km²) of sensitive watersheds in the Alakanada basin. Today, the Chipko Andolan movement has grown to more than four thousand groups working to save India's forests. Their slogan is: "What do the forests bear? **Soil**, water, and pure air."

The successes of this movement, both in empowering local women and in saving the forests on which they depend, are inspiring models for grassroots green movements around the world.

[*William P. Cunningham Ph.D. and Jeffrey Muhr*]

FURTHER READING

Bandyopadhyay, J., and V. Shiva. "Chipko: Rekindling India's Forest Culture." *Ecologist* 17 (1987): 26-34.
————. "Development, Poverty and the Growth of the Green Movement in India." *Ecologist* 19 (1989): 111-17.
Durning, A. B. "Environmentalism South." *Amicus Journal* 12 (Summer 1990): 12-18.

Chisel plow
see **Conservation tillage**

Chisso Chemical Company
see **Minamata disease**

Chlordane

Chlordane and a closely related compound, heptachlor, belong to a group of chlorine-based pesticides known as cyclodienes. They were among the first major **chemicals** to attract national attention and controversy, mainly because of their devastating effects on **wildlife** and domestic animals. By the 1970s, they had become two of the most popular pesticides for home and agricultural uses (especially for termite control), despite links between these chemicals and the **poisoning** of birds and other wildlife, pets and farm animals, as well as links to **leukemia** and other cancers in humans.

In 1975, environmentalists finally persuaded **Environmental Protection Agency** (EPA) to issue an immediate temporary ban on most uses of chlordane and heptachlor based on an "imminent hazard of **cancer** in man." In 1978, when the EPA agreed to phase out most remaining uses of chlordane and heptachlor, the agency stated that "virtually every person in the United States has residues...in his body tissues." Chlordane has now been banned, at least temporarily, for sale or use in the U.S. But potentially dangerous levels of the chemical are still found occasionally in foodstuffs, homes, and the **environment**.

Chlorinated hydrocarbons

Chlorinated **hydrocarbons** are compounds made of **carbon**, **hydrogen**, and **chlorine** atoms. These compounds can be aliphatic, meaning they do not contain **benzene**, or aromatic, meaning they do. The chlorine functional group gives these compounds a certain character; for instance, the aromatic organochlorine compounds are resistant to microbial degradation; the aliphatic chlorinated solvents have certain anesthetic properties (e.g., chloroform); some are known for their antiseptic properties (e.g., hexachloraphene). The presence of chlorine imparts toxicity to many organochlorine compounds (e.g., chlorinated pesticides).

Chlorinated hydrocarbons have many uses, including chlorinated solvents, organochlorine pesticides, and industrial compounds. Common chlorinated solvents are dichloromethane (methylene chloride), chloroform, carbon tetrachloride, trichloroethane, trichloroethylene, tetrachloroethane, **tetrachloroethylene**. These compounds are used in drycleaning solvents, degreasing agents for machinery and vehicles, paint thinners and removers, laboratory solvents, and in manufacturing processes, such as coffee decaffeination. These solvents are hazardous to human health and exposures are regulated in the workplace. Some are being phased out for their toxicity to humans and the **environment**, as molecules have the potential to react with and destroy stratospheric **ozone**.

The organochlorine pesticides include several subgroups, including the cyclodiene insecticides (e.g., **chlordane**, heptachlor, dieldrin), the DDT family of compounds and its analogs, and the hexachlorocyclohexanes (often incorrectly referred to as BHCs, or benzene hexachlorides). These insecticides were introduced and marketed extensively after World War II, but due to their toxicity, persistence, widespread environmental contamination, and adverse ecological impacts, most were banned or restricted for use in the United States in the 1970s and 80s. These insecticides generally have low water solubilities, a high affinity for organic matter, readily bioaccumulate in plants and animals, particularly aquatic organisms, and have long environmental half-lives compared to the currently-used insecticides.

There are many chlorinated industrial products and reagent materials. Examples include **vinyl chloride**, which is used to make PVC (**polyvinyl chloride**) **plastics**; chlorinated benzenes, including hexachlorobenzene; PCB (polychlorinated biphenyl), used extensively in electrical transformers and capacitors; chlorinated phenols, including **pentachlorophenol** (PCP); chlorinated naphthalenes; and chlorinated diphenylethers. They represent a diversity of applications, and are valued for their low reactivity and high insulating properties.

There are also chlorinated byproducts of environmental concern, particularly the polychlorinated dibenzo-p-dioxins (PCDDs) and the polychlorinated dibenzofurans (PCDFs). These families of compounds are products of incomplete **combustion** of organochlorine-containing materials, and are primarily found in the **fly ash** of **municipal solid waste** incinerators. The most toxic component of PCDDs, 2,3,7,8-tetrachlorodibenzo-p-dioxin (2,3,7,8-

TCDD), was a trace contaminant in the production of the **herbicide 2,4,5-T** and is found in trace amounts in 2,4,5-trichlorophenol and technical grade pentachlorophenol. PCDDs and PCDFs can also be formed in the chlorine bleaching process of **pulp and paper mills**, and have been found in their **effluent** and in trace amounts in some paper products.

[*Deborah L. Swackhammer*]

FURTHER READING

Brooks, G. T. *Chlorinated Insecticides.* Cleveland, OH: CRC Press, 1974.

Chau, A. S. Y., and B. K. Afghan. *Analysis of Pesticides in Water*, Vol. II. Boca Raton, FL: CRC Press, 1982.

Fleming, W. J., D. R. Clark, Jr. and C. J. Henny. *Organochlorine Pesticides and PCBs: A Continuing Problem for the 1980s.* Washington, DC: U. S. Fish and Wildlife Service, 1985.

Manahan, S. E. *Environmental Chemistry*, 5th ed. Ann Arbor, MI: Lewis Publishers, 1991.

Chlorination

Chlorination refers to the application of **chlorine** for the purposes of oxidation. The forms of chlorine used for chlorination include: chlorine gas, hypochlorous **acid** (HOCl), hypochlorite **ion** (OCl), and chloramines or combined chlorine (Mono-, di-, and tri-chloramines). The first three forms of chlorine are known as free chlorine.

Chlorine (Cl) has three valences under normal environmental conditions, -1, 0 and +1. Environmental scientists often refer only to the chlorine forms having 0 and +1 valences as chlorine; they refer to the -1 form as chloride. Chlorine with a valence of 0 (Cl_2) and chlorine with a valence of +1 (HOCl) both have the ability to oxidize materials, whereas chlorine at a -1 valence, chloride, is already at its lowest oxidation state and has no oxidizing power.

The functions of chlorination are to disinfect water or **wastewater**, decolorize waters or fabrics, sanitize and clean surfaces, remove iron and manganese, and reduce odors. The fundamental principle of each application is that due to its oxidizing potential, chlorine is able to effect many types of chemical reactions. Chlorine can cause alterations in DNA, cell-membrane porosity, **enzyme** configurations, and other biochemicals; the oxidative process can also lead to the death of a cell or **virus**. Chemical bonds, such as those in certain dyes, can be oxidized, causing a change in the color of a substance. Textile companies sometimes use chlorine to decolorize fabrics or process waters. In some cases, odors can be reduced or eliminated through oxidation. However, the odor of certain compounds, such as some phenolics, is aggravated through a reaction with chlorine. Certain soluble metals can be made insoluble through oxidation by chlorine (soluble Fe^{2+} is oxidized to insoluble Fe^{3+}), making the metal easier to remove through **sedimentation** or **filtration**.

Chlorine is commercially available in three forms; it can also be generated on-site. For treating small quantities of water, calcium hypochlorite ($Ca(OCl)_2$), commonly referred to as high test hypochlorite (HTH) because one mole of HTH provides two OCl^- ions, is sometimes used. For large applications, chlorine gas (Cl_2) is the most wide used source of chlorine. It reacts readily with water to form various chlorine **species** and is generally the least expensive source. There are, however, risks associated with the handling and transport of chlorine gas, and these have convinced some to use sodium hypochlorite ($NaOCl$) instead. Sodium hypochlorite is more expensive than chlorine gas, but less expensive than calcium hypochlorite. Some utilities and industries have generated chlorine on-site for many years, using electrolysis to oxidize chloride ions to chlorine. The process is practical in remote areas where brine, a source of chloride ions, is readily available.

Chlorine has been used in the United States since the early 1900s for disinfection. It is still commonly used to disinfect wastewater and drinking water, but the rules guiding its use are gradually changing. Until recently, chlorine was added to wastewater effluents from treatment plants without great concern over its effects on the **environment**. The environmental impact was thought to be insignificant since chlorine was being used in such low concentrations. However, evidence has accumulated showing serious environmental consequences from the **discharge** of even low levels of various forms of chlorine and chlorine compounds, and many plants now dechlorinate their wastewater after allowing the chlorine to react with the wastewater for 30–60 minutes.

The use of chlorine to disinfect drinking water is undergoing a similar review. Since the 1970s, it has been suspected that chlorine and some by-products of chlorination are carcinogenic. Papers published in 1974 indicated that halogenated methanes are formed during chlorination. During the mid-1970s the **Environmental Protection Agency** (EPA) conducted two surveys of the **drinking-water supply** in the United States, the National Organics Reconnaissance Survey and the National Organics Monitoring Survey, to determine the extent to which **trihalomethanes** (THMs) (chloroform, bromodichloromethane, dibromochloromethane, bromoform) and other halogenated organic compounds were present. The studies indicated that drinking water is the primary route by which humans are exposed to THMs and that THMs are the most commonly detected synthetic organic **chemicals** in United States' drinking water.

Chloroform was the THM found in the highest concentrations during the surveys. The risks associated with drinking water containing high levels of chloroform are not clear. It is known that 0.2 qt (200 ml) of chloroform is usually fatal to humans, but the highest concentrations in the drinking water surveyed fell far below (311 ug/l) this lethal dose. The potential carcinogenic effects of chloroform are more difficult to evaluate. It does not cause *Salmonella typhimurium* in the **Ames test** to mutate, but it does cause mutations in yeast and has been found to cause tumors in rats and mice. However, the ability of chloroform to cause **cancer** in humans is still questionable, and the EPA has classified it and other THMs as probable human carcinogens. Based on these data, the maximum contaminant level for THMs in drinking water is now 100 ug/l. This is an enforceable standard and requires the monitoring and reporting of THM concentrations in drinking water.

There are several ways to test for chlorine, but among the more common methods are iodometric, DPD (N,N-Diethyl-p-phenylenediamine) and amperometric. DPD and amperometric methods are generally used in the water and wastewater treatment industry. DPD is a dye which is oxidized by the presence of chlorine, creating a reddish color. The intensity of the color can then be measured and related to chlorine level; the DPD solution can be titrated with a reducing agent (ferrous ammonium sulfate) until the reddish color dissipates. In the amperometric titration method, an oxidant sets up a current in a solution which is measured by the amperometric titrator. A reducing agent (phenylarsine oxide) is then added slowly until no current can be measured by the titrator. The amount of titrant added is commonly related to the amount of chlorine present.

To minimize the problem of chlorinated by-products, many cities in the United States, including Denver, Portland, St. Louis, Boston, Indianapolis, Minneapolis, and Dallas, use chloramination rather than simple chlorination. Chlorine is still required for chloramination, but ammonia is added before or at the same time to form chloramines. Chloramines do not react with organic precursors to form halogenated by-products including THMs. The problem in using chloramines is that they are not as effective as the free chlorine forms at killing pathogens.

Questions still remain about whether the levels of chlorine currently used are dangerous to human health. The level of chlorine in most water supplies is approximately 1 mg/l, and some scientists believe that the chlorinated by-products formed are not hazardous to humans at these levels. There are some risks, nevertheless, and perhaps the most important question is whether these outweigh the benefits of using chlorine. The final issue concerns the short-term and long-term effects of discharging chlorine into the environment. Dechlorination would be yet another treatment step, requiring the commitment of additional resources. At the present time, the general consensus is that chlorine is more beneficial than harmful. However, it is important to note that a great deal of research is now underway to explore the benefits of using alternative disinfectants such as **ozone**, chlorine dioxide, and ultraviolet light. Each alternative poses some problems of its own, so despite the current availability of a great deal of research data, the selection of an alternative is difficult.

[*Gregory D. Boardman*]

RESOURCES
BOOKS

Tchobanoglous, G., and E. D. Schroeder. *Water Quality.* Reading, MA: Addison-Wesley, 1985.

Chlorine

Chlorine is an element of atomic number 17 and atomic mass of 35.45 atomic mass units. It belongs to Group 17 of the periodic table and is thus a halogen. Halogens are a highly reactive group of elements that, in addition to chlorine, include fluorine, iodine, **bromine** and another element that does not occur in **nature**, astatine, but is produced by bombarding bismuth with alpha particles. Halogens are extremely reactive because they have an unpaired electron in their outermost electron shell. Thus so highly reactive, chlorine is usually not found in a pure form in nature, but is rather typically bound to other elements such as sodium, calcium, or potassium. In its pure form, chlorine exists as a diatomic molecule, meaning a molecule containing two of the same atoms. This form of chlorine is a yellow-green gas at room temperature. Chlorine gas is more dense than air, condenses to form a liquid at -29°F(-34°C), and freezes into a solid at -153°F(-103°C). Because of its reactivity, desirable properties, and abundance, chlorine is an exceptionally useful element. Since it readily combines with other elements and molecules, chlorine is a main component and vital reactant in the manufacture of thousands of useful products. In addition, almost all municipal **water treatment** systems in the United States depend on chlorine **chemicals** to provide clean and safe drinking water. It is estimated that chlorine is used in the production of 85% of all pharmaceuticals, and in 96% of all crop protection chemicals (pesticides and herbicides). Also, chlorine chemicals are powerful bleaching agents used in paper processing and inexpensive but highly effective disinfectants.

Chlorine also has an important impact on the economy. The chlorine industry provides almost 2 million jobs having a combined yearly payroll of over $52 billion. In the United States, 212 industries are direct users of chlorine or related products. The Chlorine Chemistry Council reports that nearly 40% of U.S. jobs and income are directly or indirectly dependent on chlorine. The chlorine-dependent

industries are mostly found in Texas, Michigan, and New Jersey. The **automobile** industry relies heavily on chlorine because cars contain many components that use chlorine in their manufacture. Three chlorine products, PVC, pickled steel, and paint create more than 345,000 jobs in the United States. Polyvinylchloride (or PVC) is a chlorinated hydrocarbon polymer that is used to make dashboards, air bags, wire covers, and sidings. Chlorine is also used in the manufacture of car exterior paints. Automobile coatings typically use titanium dioxide which requires chlorine for synthesis. In addition, chlorine is used in the production of pickled steel for automobile frames and undercarriages. The pickling process provides an impurity-free steel surface for rust **resistance** treatment and painting. Hydrochloric **acid**, a chlorine chemical, is a main component in pickling steel. Roughly 25% of the hydrochloric acid produced in the United States is used in steel pickling for the auto industry. Overall, the auto industry consumes 3.1% of all chlorine produced in the form of hydrochloric acid, 30% of all chlorine produced in the form of PVC, and 4.1% of all chlorine produced involved in the manufacture of titanium dioxide for paints.

Industrially, chlorine is also used in massive quantities for **sanitation** purposes. An important advancement in public health has been the widespread **chlorination** of drinking water. Chlorine-based water purifying chemicals were first introduced in 1908. Chlorine water purification has practically eradicated diseases that were once devastating, such as **cholera** and dysentery. In 1992, the Public Health Advisory Board was created to guide the chlorine industry in issues pertaining to drinking water safety. Because it is so highly reactive, chlorine is one of the most effective germicides available. Chlorine chemicals kill bacteria, algae, **fungi**, and protozoans and inactivate viruses. Simple household bleach, sodium hypochlorite, can even inactivate and sterilize equipment from such deadly viruses as the **Ebola virus**. Chlorine is also used to sanitize pools and spas. Chlorine disinfectants are used to prevent institutional disease transmission in food preparation, day care centers, nursing homes, and hospitals.

Chlorine is a vital chemical in the pharmaceutical and medical industries. Approximately 85% of all drugs either contain chlorine or are manufactured using chlorine. The chemical addition of chlorine to a drug molecule can enhance its **absorption** and delay its elimination from the body, increasing the duration of its action. Also, chlorine can be used to create soluble forms of drugs that dissolve easily into solutions. When hydrochloric acid is reacted with poorly soluble drugs that are weak bases, the result is a chloride salt that is more soluble in water. Greater solubility enhances absorption and allows some drugs to be incorporated into syrups or elixirs. This is very useful since many people prefer oral liquid drug intake. Soluble drugs also have the advantage that they can be inhjected. Finally, about 25% of all medical equipment is made from **chlorinated hydrocarbons**, vinyl or PVC.

Chlorinated **hydrocarbons** are specific hydrocarbon molecules that also have atoms of the element chlorine chemically bonded to them. The number of **carbon** atoms and how they are arranged in three-dimensions determines the chemical and physical properties of these compounds. Because there is a wide variety of possible chlorinated hydrocarbons, this class of useful chemicals has a wide range of applications that are of great economic and practical importance. Chlorinated hydrocarbons include products such as the synthetic rubbers used in car tires and tennis shoes. They are also used in packaging **plastics**, and a variety of products such as fluid pipes, furniture, home siding, credit cards, fences, and toys. Chlorinated hydrocarbons also can be used as precursors in the production of non-stick coatings such as Teflon. Chlorine is also used in the manufacture of solvents such as carbon tetrachloride and trichloroethylene used in **dry cleaning**.

In addition to their use in the manufacture of polymers, rubbers, plastics, solvents, and cleaners, chlorinated hydrocarbons also are powerful pesticides. They rank among the most potent and environmentally persistent insecticides, and when combined with fluorine, they yield the refrigerants called **chlorofluorocarbons**, or CFCs. Because of their wide array of uses, chlorinated hydrocarbons are among the most important industrial organic compounds. Perhaps the best known chlorinated hydrocarbon insecticide is *Dichloro-DiphenylTrichloroethane, or DDT*. DDT was first used as an insecticide in 1939. After its effectiveness and relative safety to humans was established, the use of DDT burgeoned around the world in the war against disease-carrying and agricultural insect pests. However, the very success of DDT in the fight against insect-transmitted diseases (especially **malaria**) subsequently led to its massive overuse. Widespread excessive use led to the emergence of DDT-resistant insects. Additionally, evidence started to show that toxic levels of DDT could accumulate in the fatty tissues of mammals, including humans. Because of such harmful effects, the use of DDT has now been banned in many countries despite its effectiveness. Another important issue with chlorinated hydrocarbons is their environmental persistence. Many, such as DDT, refrigerants, and solvents are not easily broken-down in the environent. Also, because other, less persistent, insecticide alternatives have been developed, the use of chlorinated insecticides has been drastically reduced.

Other toxic chemicals resistant to degradation also contain chlorine such as chlorinated and **polychlorinated biphenyls** (PCBs) and polyaromatic hydrocarbons (PAHs). These substances are known environmental pollutants, with PAHs found for example in diesel exhaust emissions and PCBs in industrial effluents, contaminating rivers as well as the Oceans. They have been shown to accumulate in **whales** and a variety of other marine mammals.

Another environmentally important chlorine compound class are the chemicals known as *ChloroFluoroCarbons*, or CFCs. Chlorofluorocarbons are single carbon atoms chemically bound to both chlorine and fluorine. The compounds trichlorofluoromethane (Freon-11) and dichlorodifluoromethane (Freon-12) are widely used CFCs. They are odorless, nonflammable, very stable compounds that are used as refrigerants in commercial refrigerators and air conditioners. CFCs are also used as **aerosol propellants**. Because CFCs are detrimental to the **ozone** layer, the portion of the **atmosphere** that blocks out the harmful wavelengths of ultraviolet light associated with skin **cancer**, they are being phased out as propellants and refrigerants. An ozone hole detected above **Antarctica** has been attributed by scientists to the release of CFCs into the atmosphere.

The **Environmental Protection Agency** (EPA) is concerned with toxic and persistent chlorine chemicals that are the byproducts of industrial activity. Two such byproducts are dioxins and **furans**, produced when organic (carbon-containing) compounds are heated to high temperatures in the presence of chlorine. Together, dioxins and furans represent a group of over two hundred chemicals. Some dioxins are extremely toxic and are a significant cause for concern to environmental agencies. **Dioxin** compounds are highly persistent in the **environment**. The most toxic dioxin is 2,3,7,8-tetrachlorodibenzo-p-dioxin or TCDD. Dioxins are also by-products of industrial processes involving chlorine such as waste **incineration**, chemical and **pesticide** manufacturing and pulp and paper bleaching. Dioxin is a known **carcinogen** and was the principal toxic component of **Agent Orange**. The EPA reports that incinerators are the largest source of dioxins and furans released into the environment in the United States. Other industrial sources of these toxic compounds include smelters, cement kilns, and fossil fuel burning **power plants**. Because of new standards for incineration, the levels of dioxins and furans have been steadily declining.

[*Terry Watkins*]

Resources

Books

Khanna, N. "Chlorine Dioxide in Food Applications." *In: Proceedings of the Fourth International Symposium, Chlorine dioxide: The state of science, regulatory, environmental issues, and case histories.* Denver: American Water Works Association, 2002.

Stringer, Ruth, and Paul Johnston. *Chlorine and the Environment: An Overview of the Chlorine Industry.* New York: Kluwer Academic Publishers, 2001.

Watts, Susan. *Chlorine (Elements).* Tarrytown: Benchmark Books, 2001.

White, G.C. "Chlorination of Potable Water." *In: The Handbook of Chlorination, 2nd Ed.* New York: Nostrand Reinhold, 1986.

PERIODICALS

Bal'a, M.F.A. et al. "Moderate heat or chlorine destroys Aeromonas hydrophila biofilms on stainless steel." *Dairy, Food, and Environmental Sanitation* 19, (1999): 29-34.

McFarland, M. "Investigations of the environmental acceptability of fluorocarbon alternatives to chlorofluorocarbons." *Proc. Natl. Acad. Sci. U.S.A* 89, no. 3 (February, 1992): 807-811.

OTHER

Winter, Mark. *Chlorine.* WebElements Periodic Table. <http://www.webelements.com/webelements/scholar/index.html>

ORGANIZATIONS

Chlorine Chemistry Council (CCC), 1300 Wilson Boulevard, Arlington, VA USA 22209 703) 741-5000, , http://c3.org/index.html

Euro Chlor: Chlorine Online Information Ressource, Avenue E Van Nieuwenhuyse 4, box 2, Brussels, Belgium B-1160 + 32 2 676 7211, Fax: + 32 2 676 7241, Email: eurochlor@cefic.be , <http://www.eurochlor.org>

Chlorine monoxide

One of two oxides of **chlorine**, either Cl_2O or ClO. When used in **environmental science**, it usually refers to the latter. Chlorine monoxide (ClO) is formed in the **atmosphere** when free chlorine atoms react with **ozone**. In the reaction, ozone is converted to normal, diatomic oxygen (O_2). This process appears to be a major factor in the destruction of ozone in the **stratosphere** observed by scientists in recent years. The most important sources of chlorine by which this reaction is initiated appear to be **chlorofluorocarbons** and other synthetic chlorine compounds released by human activities.

Chlorofluorocarbons

The chlorofluorocarbons (CFCs) are a family of organic compounds containing **carbon**, **hydrogen** (usually), and either **chlorine** or fluorine, or both. The members of this family can be produced by replacing one or more hydrogen atoms in **hydrocarbons** with a chlorine or fluorine atom. In the simplest possible case, treating **methane** (CH_4) with chlorine yields chloromethane, CH_3Cl. Treating this product with fluorine causes the replacement of a second hydrogen atom with a fluorine atom, producing chlorofluoromethane, CH_2ClF.

This process can be continued until all hydrogen atoms have been replaced by chlorine and/or fluorine atoms. By using larger hydrocarbons, an even greater variety of CFCs can be produced. The compound known as CFC-113, for example, is made from ethane (C_2H_6) and has the formula $C_2F_3Cl_3$.

Over the last three decades, the CFCs have become widely popular for a number of commercial applications. These applications fall into four general categories: refrigerants, cleaning fluids, **propellants**, and blowing agents. As refrigerants, CFCs have largely replaced more harmful gases such as ammonia and **sulfur dioxide** in refrigerators, freezers, and air conditioning systems. Their primary application as cleaning fluids has been in the computer manufacturing business where they are used to clean circuit boards. CFCs are used as propellants in hair sprays, deodorants, spray paints, and other types of sprays. As blowing agents, CFCs are used in the manufacture of fast-food take-out boxes and similar containers. By the early 1990s, CFCs had become so popular that their production was a multi-billion dollar business worldwide.

For many years, little concern was expressed about the environmental hazards of CFCs. The very qualities that made them desirable for commercial applications—their **stability**, for example—appeared to make them environmentally benign.

However, by the mid-1970s, the error in that view became apparent. Scientists began to find that CFCs in the **stratosphere** decomposed by sunlight. One product of that **decomposition**, atomic chlorine, reacts with **ozone** (O_3) to form ordinary oxygen (O_2). The apparently harmless CFCs turned out, instead, to be a major factor in the loss of ozone from the stratosphere.

By the time this discovery was made, levels of CFCs in the stratosphere were escalating rapidly. The concentration of these compounds climbed from 0.8 part per billion in 1950 and 1.0 part per billion in 1970 to 3.5 **parts per billion** in 1987.

A turning point in the CFC story came in the mid-1980s when scientists found that a large hole in the ozone layer was opening up over the Antarctic each year. This discovery spurred world leaders to act on the problem of CFC production. In 1987, about 40 nations met in Montreal to draft a treaty that will reduce the production of CFCs worldwide.

This action is encouraging, but it hardly solves the CFC problem. These compounds remain in the **atmosphere** for long periods of time (about 77 years for CFC-11 and 139 years for CFC-12), so they will continue to pose a threat to the ozone layer for many decades to come.

[*David E. Newton*]

RESOURCES
BOOKS

Selinger, B. *Chemistry in the Marketplace.* 4th ed. Sydney: Harcourt Brace Jovanovich Publishers, 1989.

PERIODICALS

O'Sullivan, D. A. "International Gathering Plans Way to Safeguard Atmospheric Ozone." *Chemical & Engineering News* (26 June 1989): 33-36.

Zurer, P. S. "Producers Grapple with Realities of CFC Phaseout." *Chemical & Engineering News* (24 July 1989): 7-13.

Cholera

Cholera is one of the most severe and contagious diseases transmitted by water. It is marked by severe diarrhea, resulting in fluid loss and dehydration, sometimes followed by shock and death. If not treated, **mortality** occurs in over 60% of cases. The Latin American cholera epidemic claimed 4,002 lives in 1991, and resulted in 391,742 reported cases that year, mostly in Peru and Ecuador. Cholera is caused by the bacillus *Vibrio cholerae*, a member of the family Vibrionaceae, which are described as Gram negative, non-sporulating rods that are slightly curved, motile, and have a fermentative **metabolism**.

The natural **habitat** of *V. cholerae* is human feces, but some studies have indicated that natural waters may also be a habitat of the organism. Fecal contamination of water is the most common means by which *V. cholerae* is spread, however, food, insects, soiled clothing, or person-to-person contact may also transmit sufficient numbers of the **pathogen** to cause cholera.

The ability of *V. cholerae* to survive in water is dependent upon the temperature and water type. *V. cholerae* reportedly survive longer at low temperatures, and in seawater, sterilized water, and **nutrient** rich waters. Also, the particular strain of *V. cholerae* affects the survival of the organism in water, since some strains or types are hardier than others. Most methods to isolate *V. cholerae* from water include concentration of the sample, by **filtration**, exposure to high **pH** and selective media. Identification of pathogenic strains of *V. cholerae* is dependent upon agglutination tests. Final confirmation of the strain or type must be done in a specialized laboratory.

Persons infected with *V. cholerae* produce 10^7 to 10^9 organisms per milliliter in the stool at the height of the disease, but the number of excreted organisms drops off quickly as the disease progresses. Asymptomatic carriers of *V. cholerae* excrete 10^2 to 10^5 organisms per gram of feces. The mild form of the illness lasts for five to seven days. Hydration therapy is the treatment of choice for cholera. It is suggested that antibiotics not be used, following the emergence of multiple antibiotic resistant strains in many areas. Vaccines exist to prevent cholera, however, they do not prevent the acquisition of the bacteria in the gastrointestinal tract, do not diminish symptoms in persons already infected, and are effective for well less than a year. Proper **water treatment** should eliminate *V. cholerae* from drinking water, however, the most effective control of this pathogen is dependent upon good sanitary practices.

[*E. K. Black and Gordon R. Finch*]

RESOURCES
BOOKS

Christie, A. B. *Infectious Diseases: Epidemiology and Clinical Practice.* 4th ed. Edinburgh, Scotland: Churchill Livingstone, 1987.
Feachem, R. G., et al. *Sanitation and Disease: Health Aspects of Excreta and Wastewater Management.* New York: Wiley, 1983.
Mitchell, R., ed. *Environmental Microbiology.* New York: Wiley-Liss, 1992.

PERIODICALS

Foliguet, J. M., P. Hartemann, and J. Vial. "Microbial Pathogens Transmitted by Water." *Journal of Environmental Pathology, Toxicology, and Oncology* 7 (1987): 39-114.

Cholinesterase inhibitor

Insecticides kill their target insect **species** in a variety of ways. Two of the most commonly used classes of insecticide are the organophosphates (nerve gases) and the carbamates. These compounds act quickly (in a matter of hours), are lethal at low doses (**parts per billion**), degrade rapidly (in hours to days) and leave few toxic residues in the **environment**. Organophosphates kill insects by inducing loss of control of the peripheral nervous system, leading to uncontrollable spasms followed by paralysis and, ultimately, death. This is often accomplished by a biochemical process called cholinesterase inhibition.

Most animals' nervous systems are composed of individual nerve cells called neurons. Between any two adjacent neurons there is always a gap, called the synaptic cleft; the neurons do not actually touch each other. When an animal senses something—for example, pain—the sensation is transmitted chemically from one neuron to another until the impulse reaches the brain or central nervous system. The first neuron (pre-synaptic neuron) releases a substance, known as a transmitter, into the synaptic cleft. One of the most common chemical transmitters is called acetylcholine. Acetylcholine then diffuses across the gap and binds with receptor sites on the second neuron (post-synaptic neuron). Reactions within the target neuron triggered by occupied receptors result in further transmission of the signal. As soon as the impulse has been transmitted, the acetylcholine in the gap is immediately destroyed by an **enzyme** called cholinesterase; the destruction of the acetylcholine is an absolutely essential part of the nervous process. If the acetylcholine is not destroyed, it continues to stimulate indefinitely the transmission of impulses from one neuron to the next, leading to loss of all control over the peripheral nervous system. When control is lost, the nervous system is first overstimulated and then paralyzed until the animal dies. Thus, **organophosphate** insecticides bind to the cholinesterase enzyme, preventing the cholinesterase from destroying the acetylcholine and inducing the death of the insect.

Some trade names for organophosphate insecticides are malathion and parathion. Carbamates include aminocarb and carbaryl. The carbamates produce the same effect of cholinesterase inhibition as the organophosphates, but the chemical reaction of the carbamates is more easily reversible. The potency or power of these compounds is usually measured in terms of the quantity of the **pesticide** (or inhibitor) that will produce a 50% loss of cholinesterase activity. Since acetylcholine transmission of nervous impulses is common to most vertebrates as well as insects, there is a great potential for harm to non-target species from the use of cholinesterase-inhibiting insecticides. Therefore the use of these insecticides is highly regulated and controlled. Access to treated areas and contact with the compounds is prohibited until the time period necessary for the breakdown of the compounds to non-toxic end products has elapsed.

[*Usha Vedagiri*]

RESOURCES
BOOKS

Connell, D. W., and G. J. Miller. *Chemistry and Ecotoxicology of Pollution.* New York: Wiley, 1984.

Chromatography

Chromatography is the process of separating mixtures of **chemicals** into individual components as a means of identification or purification. It derives from the Greek words *chroma,* meaning color, and *graphy,* meaning writing. The word was coined in 1906 by the Russian chemist Mikhail Tsvett who used a column to separate plant pigments. Currently chromatography is applied to many types of separations far beyond those of just color separations. Common chromatographic applications include gas-liquid chromatography (GC), liquid-solid chromatography (LC), thin layer chromatography (TLC), **ion exchange** chromatography, and gel permeation chromatography (GPC). All of these methods are invaluable in analytical **environmental chemistry**, particularly GC, LC, and GPC.

The basic principle of chromatography is that different compounds have different retentions when passed through a given medium. In a chromatographic system, one has a mobile phase and a stationary phase. The mixture to be separated is introduced in the mobile phase and passed through the stationary phase. The compounds are selectively retained by the stationary phase and move at different rates which allows the compounds to be separated.

In gas chromatography, the mobile phase is a gas and the stationary phase is a liquid fixed to a solid support. Liquid samples are first vaporized in the injection port and carried to the chromatographic column by an inert gas which serves as the mobile phase. The column contains the liquid stationary phase, and the compounds are separated based on their different vapor pressures and their different affinities for the stationary phase. Thus different types of separations can be optimized by choosing different stationary phases, and by altering the temperature of the column. As the compounds elute from the end of the column, they are detected by one of a number of methods that have specificity for different chemical classes.

Liquid chromatography consists of a liquid mobile phase and a solid stationary phase. There are two general types of liquid chromatography: column chromatography and high pressure liquid chromatography (HPLC). In column chromatography, the mixture is eluted through the column containing stationary packing material by passing successive volumes of solvents or solvent mixtures through the column. Separations result as a function of both chemical-solvent interactions as well as chemical-stationary phase interactions. Often this technique is used in a preparative manner to remove interferences from environmental sample extracts. HPLC refers to specific instruments designed to perform liquid chromatography under very high pressures to obtain a much greater degree of resolution. The column outflow is passed through a detector and can be collected for further processing if desired. Detection is typically by ultraviolet light or fluorescence.

A variation of column chromatography is gel permeation chromatography (GPC), which separates chemicals based on size exclusion. The column is packed with porous spheres, which allow certain size chemicals to penetrate the spheres and excludes larger sizes. As the sample mixture traverses the column, larger molecules move more quickly and elute first while smaller molecules require longer elution times. An example of this application in environmental analyses is the removal of lipids (large molecules) from fish tissue extracts being analyzed for pesticides (small molecules).

[*Deborah L. Swackhammer*]

RESOURCES
BOOKS

McNair, H. M., and E. J. Bonelli. *Basic Gas Chromatography.* Palo Alto, CA: Varian Instruments, 1969.

Peters, D. G., J. M. Hayes, and G. M. Hieftje. *Chemical Separations and Measurements.* Philadelphia: Saunders, 1974.

Chronic effects

Chronic effects occur over a long period of time. The length of time termed "long" is dependent upon the life cycle of the organism being tested. For some aquatic **species** a chronic effect might be seen over the course of a month. For animals such as rats and dogs, chronic would refer to a period of several weeks to years.

Chronic effects can be either caused by chronic or acute exposures. Acute exposure to some metals and many carcinogens can result in chronic effects. With certain toxicants, such as cyanide, it is difficult, if not impossible, to cause a chronic effect. However, at a higher dosage, cyanide readily causes **acute effects**. Examples of chronic effects include pulmonary tuberculosis and, in many cases, **lead poisoning**. In each disease the effects are long-term and cause damage to tissues; acute effects generally result in little tissue reaction. Thus, acute and chronic effects are frequently unrelated, and yet it is often necessary to predict chronic toxicity based on acute data. Acute data are more plentiful and easier to obtain. To illustrate the possible differences between acute and chronic effects, consider the examples of halogenated solvents, **arsenic**, and lead.

In halogenated solvents, acute exposure can cause excitability and dizziness, while chronic exposure will result in liver damage. Chronic effects of arsenic poisoning are in blood formation and liver and nerve damage. Acute poisoning affects the gastro-intestinal tract. Lead also effects blood formation in chronic exposure, and damages the gastrointestinal tract in acute exposure. Other chronic effects of exposure to lead include changes in the nervous system and muscles. In some situations, given the proper combination of dose level and frequency, those exposed will experience both acute and chronic effects.

There are **chemicals** that are essential and beneficial for the functions and structure of the body. The chronic effect is therefore better health, although people generally do not refer to chronic effects as being positive. However, vitamin D, fluoride, and sodium chloride are just a few examples of agents that are essential and/or beneficial when administered at the proper dosage. Too much of any of the three or too little, however, could cause acute and/or chronic toxic effects.

In **aquatic toxicology**, chronic toxicity tests are used to estimate the effect and no-effect concentrations of a chemical that is continuously applied over the reproductive life cycle of an organism; for example, the time needed for growth, development, sexual maturity, and reproduction. The range in chemical concentrations used in the chronic tests is determined from acute tests. Criterion for effects might include the number and percent of embryos that de-

velop and hatch, the survival and growth of larvae and juveniles, etc.

The maximum acceptable toxicant concentration (MATC) is defined through chronic testing. The MATC is a hypothetical concentration between the highest concentration of chemical that caused no observed effect (NOEC) and the lowest observed effect concentration (LOEC). Therefore,

$$LOEC > MATC > NOEC$$

Furthermore, the MATC has been used to relate chronic toxicity to acute toxicity through an application factor (AF). AF is defined as follows:

$$AF = \frac{MATC}{LD_{50}}$$

The AF for one aquatic species might then be used to predict the chronic toxicity for another species, given the acute toxicity data for that species.

The major limitations of chronic toxicity testing are the availability of suitable test species and the length of time needed for a test. In animal testing, mice, rats, rabbits, guinea pigs, and/or dogs are generally used; mice and rats being the most common. With respect to aquatic studies, the most commonly used vertebrates are the fathead (fresh water) and sheepshead (saltwater) minnows. The most commonly used invertebrates are freshwater water fleas (*Daphnia*) and the saltwater mysid shrimp (*Mysidopsis*). *See also* Aquatic chemistry; Detoxification; Dose response; Heavy metals and heavy metal poisoning; LD_{50}; Plant pathology; Toxic substance

[*Gregory D. Boardman*]

RESOURCES
BOOKS

Manahan, S. E. *Toxicological Chemistry.* 2nd ed. Ann Arbor, MI: Lewis Publishers, 1992.

Cigarette smoke

Cigarette **smoke** contains more than 4,000 identified compounds. Many are known irritants and carcinogens. Since the first Surgeon General's Report on smoking and health in 1964, evidence linking the use of **tobacco** to illness, injury, and death has continued to mount. Many thousands of studies have documented the adverse health consequences of any type of tobacco, including cigarettes, cigars, and smokeless tobacco.

Specific airborne contaminants from cigarette smoke include respirable particles, nicotine, **polycyclic aromatic hydrocarbons**, **arsenic**, DDT, formaldehyde, **hydrogen**

cyanide, **methane**, **carbon monoxide**, acrolein, and **nitrogen** dioxide. Each one of these compounds impacts some part of the body. Irritating gases like ammonia, hydrogen sulfide and formaldehyde affect the eyes, nose and throat. Others, like nicotine, impact the central nervous system. **Carbon** monoxide reduces the oxygen-carrying capacity of the blood, starving the body of energy. Carcinogenic agents come into prolonged contact with vital organs and with the delicate linings of the nose, mouth, throat, lungs and airways.

Cigarette smoke is one of the six major sources of indoor **air pollution**, along with **combustion** by-products, **microorganisms** and allergens, formaldehyde and other organic compounds, **asbestos** fibers, and **radon** and its airborne decay products. The carbon monoxide concentration in cigarette smoke is more than 600 times the level considered safe in industrial plants, and a smoker's blood typically has 4 to 15 times more carbon monoxide in it than that of a nonsmoker. Airborne particle concentrations in a home with several heavy smokers can exceed **ambient air** quality standards.

Sidestream, or second-hand, smoke actually has higher concentrations of some **toxins** than the mainstream smoke the smoker inhales. Second-hand smoke carries more than 30 known carcinogens. According to a study by the **Centers for Disease Control and Prevention** (CDC) released in 1996, nearly nine out of 10 nonsmoking Americans are exposed to environmental tobacco smoke as measured by the levels of cotinine in their blood. The presence of cotinine, a chemical the body metabolizes from nicotine, is documentation of exposure to cigarette smoke. On the basis of health hazards of second-hand smoke, the **Environmental Protection Agency** has classified second-hand smoke as a Group A **carcinogen**, known to cause **cancer** in humans.

Cigarettes probably represent the single greatest source of **radiation exposure** to smokers in the United States today. Two naturally occurring radioactive materials, lead-210 and polonium-210, are present in tobacco. Both of these long-lived decay products of radon are deposited and retained on the large, sticky leaves of tobacco plants. When the tobacco is made into cigarettes and the smoker lights up, the radon decay products are volatilized and enter the lungs. The resulting dose to small segments of the bronchial epithelium of the lungs of about 50 million smokers in the United States is about 160 mSv per year. (One Sv = 100 rem of radiation.) The dose to the whole body is about 13 mSv, more than 10 times the long-term dose rate limit for members of the public.

The U. S. Department of Health and Human Services reported in 1996 that more than 430,000 Americans die each year from smoking. One in every five deaths in the United States is smoking related, the largest preventable cause of illness and premature death in the United States

About 10 million people in the United States have died from causes attributed to smoking, including heart disease, **emphysema**, and other respiratory disease, since the first Surgeon General's report on smoking and health in 1964. Death is caused primarily by heart disease, lung cancer, heart disease, and chronic obstructive lung diseases such as emphysema or chronic **bronchitis**. In addition, the use of tobacco has been linked to cancers of the larynx, mouth and esophagus, and as a contributory factor in the development of cancers of the bladder, kidney, pancreas, and cervix. Cigarette smoke aggravates **asthma**, triggers allergies, and causes changes in bodily tissues that can leave smokers and non-smokers prone to illness, especially heart disease.

About 180,000 Americans will die prematurely of coronary heart disease every year due to smoking. The risk of a stroke or heart attack is greatly increased by nicotine, which impacts the platelets which enable the blood to clot. Nicotine causes the surface of the platelets to become stickier, thereby increasing the platelets' ability to aggregate. Thus, a blood clot or thrombus forms more easily. A thrombus in an artery of the heart results in a heart attack; in an artery of the brain it results in a stroke.

Epidemiological studies reveal a direct correlation between the extent of maternal smoking and various illnesses in children. Also, studies show significantly lower heights and weights in six to 11-year olds whose mothers smoke. A pregnant woman who smokes faces increased risks of miscarriage, premature birth, stillbirth, infants with low birth weight, and infants with physical and mental impairments. Cigarette smoking also impairs fertility in women and men, contributes to earlier menopause, and increases a woman's risk of osteoporosis.

Cigarette smoke contains **benzene** which, when combined with the radioactive toxins, can cause **leukemia**. Although smoking does not cause the disease, smoking may boost a person's risk of getting leukemia by 30%.

A long-time smoker increases his risk of lung cancer by 1,000 times. In 1986, according to the CDC, about 117,000 people died of lung cancer directly attributed to cigarette smoke. More than 3,000 people each year develop lung cancer from second-hand smoke. Between 1960 and 1990, deaths from lung cancer among women have increased by more than 400%—exceeding breast cancer deaths.

The addiction to nicotine in cigarette smoke, a chemical and behavioral addiction as powerful as that of heroin, is well documented. The immediate effect of smoking a cigarette can range from tachycardia (an abnormally fast heartbeat) to arrhythmia (an irregular heartbeat). Deep inhalations of smoke lower the pressure in a smoker's chest and pulmonary blood vessels, which increases the amount of blood flow to the heart. This increased blood flow is experienced as a relaxed feeling. Seconds later, nicotine enters the liver and causes that organ to release sugar, which leads to a "sugar high." The pancreas then releases insulin to return the blood sugar level to normal, but it makes the smoker irritable and hungry, stimulating a desire to smoke and recover the relaxed, high feeling.

Nicotine also stimulates the nervous system to release adrenaline, which speeds up the heart and respiratory rates, making the smoker feel more tense. Lighting the next cigarette perpetuates the cycle. The greater the number of behaviors linked to the habit, the stronger the habit is and the more difficult to break. Quitting involves combating the physical need and the psychological need, and complete physical withdrawal can take up to two weeks.

From an economic point of view, the Department of Health and Human Services estimates that smoking costs the United States $50 billion in health expenses. That figure is most likely conservative because the medical costs attributable to burn care from smoking-related fires, perinatal care for low birth weight infants of mothers who smoke, and treatment of disease caused by second-hand smoke were not included in the calculation.

[*Linda Rehkopf*]

RESOURCES
BOOKS

Haas, F., and S. Haas. *The Chronic Bronchitis and Emphysema Handbook.* New York: Wiley, 1990.

Moeller, D. W. *Environmental Health.* Cambridge, MA: Harvard University Press, 1992.

PERIODICALS

Baker, S., and S. Carl. "Saving Your Lungs and Your Life." *Health* (June 1991): 64.

CITES

see **Convention on International Trade in Endangered Species of Wild Fauna and Flora (1975)**

Citizen science

Becoming a Citizen Scientist, according to the Cornell University Laboratory of Ornithology, can be as simple as glancing periodically at your backyard bird feeder, or as complicated as getting out in the field, collecting data about the relationship between an environments characteristics and the success of bird-nesting in that milieu. Put simply, citizen science is the practice of involving individual citizens, through their voluntary efforts, in the work of **environmental science**. Such voluntary assistance encompasses a wide variety of environmentally-related issues and projects, in-

cluding the counting of various **species** of bird, monitoring rainfall amounts, or observing and surveying the habits of threatened species of **wildlife**. The goal of all environmental science, whether performed by private citizens or environmental scientists, is the same: to provide for the health of the planet, its **natural resources**, and all living beings.

John Fein, an internationally-renowned science educator, and associate professor at Griffith University (Brisbane, Queensland, **Australia**), described this participatory process as one that can "bridge the gap between science and the community and between scientific research and policy, decision-making and planning." "Bridging these gaps," he went on to note, "involves a process of social learning through sound environmental research, full public participation, the adoption of **adaptive management** practices and the development of the democratic values, skills and institutions for an active civil society."

In the United States, citizen science began in the 1800s, and became more formalized in 1886, when the **National Audubon Society** was created. According to its preamble, the societys aim was to: promote the **conservation** of wildlife and the natural environment, and educate man regarding his relationship with, and place within, the natural environment as an ecological system. Public awareness of environmental concerns increased with the first Earth Day in 1970. The significance of **Earth Day** has served as a reminder to the world community that it is not only the scientists who are responsible for the health of the **environment**. The general public, each individual, has been given the challenge to take measures large or small that might add up to an enormous improvement in the ecological scheme of life.

Today there are so many citizen science projects initiated by various environmental organizations that it would be impossible to list all of them. However, some of the better known include:

• The Christmas Bird Count, the oldest citizen science project in existence, according to the National Audubon Society. It occurs on a daily basis across the United States between December 14th and January 5th each year. Until the recent era of the home computer, it had been the practice for groups to go on birding outings, "into nature" in order to do the count.

• The Great Backyard Bird Count, occurring for three days commencing on Valentine Day involves volunteers counting the birds visiting in their backyards or in nearby parks, and entering the data into their computers. This count is then analyzed and becomes available information to the bird counters" through tables, maps, and in other forms. The director of the Audubon Society in 2002, Frank Gill, worked in coordinated effort with Dr. John W. Fitzpatrick, director of the Cornell Lab of Ornithology, in order to calculate the results.

- Project Feeder Watch, a winter-long survey of birds that visit feeders in backyards, **nature** centers, community areas, and other locales that takes place November through March.
- The Birdhouse Network—involving the installation of bird houses with volunteers monitoring bird activity throughout the breeding season, collecting data on location, **habitat** characteristics, nestings and the number of eggs produced.
- Project Pigeon Watch, counting the number of each different color, recording the colors of courting birds, and helping scientists determine the mystery of why there are so many different colors of pigeons.
- House Finch Disease Survey— the monitoring of backyard feeders, reporting presence or absence of House Finch eye disease.
- Birds in Forested Landscapes— study sites established in forests of varying sizes with volunteers counting the birds during at least two visits (using recordings of vocalizations), and searching for markers that breeding was successful, recording landscape characteristics of the site;
- Golden-winged Warbler Project— the study to survey and conduct point counts at known and potential breeding sites of this bird, using both professionals and volunteers;
- Citizen Science in the Schoolyard— projects in elementary and middle schools that educate children in various aspects of bird-watching and counting;
- BirdSource— an interactive online database operated in conjunction with the Audubon Society collecting information from numerous projects.
- Adirondack Cooperative Loon Program, an ongoing project with volunteers through the Natural History Museum of the Adirondacks, with observation of the common loon as well as the annual count that occurs every July.
- The Mid-Atlantic Integrated Assessment (MAIA). A research, monitoring, and assessment initiative under the auspices of the **Environmental Protection Agency**, in order to provide high-quality scientific information on the condition of natural resources of the Mid-Atlantic region of the east coast, including the Delaware and Chesapeake Bays, Albemarle-Pamlico Sound, and the Delmarva Coastal Bays, utilizing professional researchers and volunteers.
- Smithsonian Neighborhood Nest-watch. Utilizing backyard bird counters during breeding season to assist the Smithsonian Environmental Research Center in gathering scientific data on various birds.
- Citizen Collaborative for **Watershed** Sustainability. A project conducted within the Southeast Minnesota Blufflands region, and coordinating with other farming regions

across the state, in order to conserve the watersheds of southeastern Minnesota.

- Connecticut Department of Environmental Protection. A citizen-based project to monitor streams and rivers, determining the chemical, physical, and biological health of the water.

Citizen science projects such as these have provided information used in decision-making resulting in the purchasing of lands that host certain threatened species during breeding time and the development of bird population management guidelines. Further, data obtained from these surveys is often published in scientific and educational journals.

Citizens respond

The **"explosion"** of popularity and participation in citizen science has been a welcome development. Both the United Nations and individual national governments have become involved, sponsoring events such as the planned 2002 U.N. global summit scheduled for Johannesburg, South Africa, between August 26th and September 4th, 2002 to examine poverty, development, and the environment.

For citizens unaware of environmental issues, it appears that more information and publicity is necessary. In an article for the *Environmental News Network (ENN)* in October 2001, Erica Gies states, "As a person who cares about environmental issues, I often find myself preaching to the choir. But when I have an opportunity to converse with people who are either uninformed or inclined to disagree, it can be difficult to communicate my passion effectively without alienating them, especially if they are people I have a long history with, such as family." To attain that goal of increasing awareness, Gies suggests using art, literature and the media. There are a variety of nonfiction books and essays, poetry; art, and music featuring environmental or nature themes, as well as the films focusing on environmental issues featured each spring at an Environmental Film Festival held in Washington, D.C. But even a comic strip can give an ecological focus. Gies quotes *Dilbert* creator, Scott Adams from his book, *The Dilbert Future* irreverently noting that, *"The children are our future. And that is why, ultimately, we're screwed unless we do something about it. If you haven't noticed, the children who are our future are good looking but they aren't all that bright. As dense as they might be, they will eventually notice that adults have spent all the money, spread disease, and turned the planet into a smoky, filthy ball of death."*

David Suzuki also offered his insights into the challenges facing the environment in an article for *ENN* in June 2002. Enumerating the biggest challenges for the environment in the next century, Suzuki noted, "I'm beginning to think one of the biggest challenges is overcoming the fact that people are tired of all the depressing news about the environment." Citizen science was offering a solution for positive approaches and answers to these issues, giving the public a reason to be hopeful and cherish their involvement in seeing things change. One such person, Robert Boyle, of Cold Spring, NY and author of the 1969 book, *The Hudson River: A Natural and Unnatural History* continued his struggle for 30 years after the publication of his book alerting people to the decay of the **Hudson River**. His book was the waterways equivalent of the *Silent Spring* (Rachel Carson's book that alerted the public to the danger of pesticides and was responsible for much of the environmental movement of the 1960s). Boyles book was instrumental in beginning the process of cleaning up not only the Hudson River, but all other polluted waterways across the United States. In 2002, Boyle continued to fight major industrial polluters and the government in courts in his effort to preserve the natural bounty of the Hudson River.

[*Joan M. Schonbeck*]

RESOURCES
OTHER

Adirondack Cooperative Loon Project. *2002 Annual Census.* [cited June 2002]. <http://www.adkscience.org/>

Citizen Collaborative for Watershed Sustainability. *About the Citizen Collaborative.* [cited June 2002]. <http://www.sustain.org/>

Connecticut Department of Environmental Protection. *A Volunteer opportunity for citizens to monitor wadeable streams and rivers.* 1998 [cited 2002]. <http://www.dep.state.ct.us/>

Cornell Lab of Ornithology. "Citizen Science." *What We Do.* [cited June 2002]. <http://www.birds.cornell.edu/>

Environmental Protection Agency. *Mid-Atlantic Integrated Assessment.* May 15, 2002. <http://www.epa.gov/maia/>

Fein, John; Tim Smith; and, James Whelan. *Cooperative Research Centre for Coastal Zone, Estuary and Waterway Management.* "Citizen Science and Education—Theme 2." January 3, 2002. <http://www.coastal.crc.org.au/>

Gies, Erica. *Environmental News Network.* "Responses to 'Fun literary strategies for environmental debate' An ENN perspective." October 12, 2001. <http://www.enn.com/>

Gorman, James. "Naturalists share their findings online." *New York Times/on the web.* December 13, 2001.

Kansas State University. *Citizen Science.* "Links to help you participate in scientific ornithological research." [cited June 2002]. <http://www.ksu.edu/>

Los Alamos. *Citizen-Based Science: Collaboration between the U.S. Geological Survey (USGS) and the Volunteer Task Force.* [cited June 2002]. <http://www.losalamos.com/mavtf/>

Smithsonian Environmental Research Center. *Avian Ecology.* "Neighborhood Nestwatch." [cited June 2002]. <http://www.serc.si.edu/>

Stoddard, Ed. *Reuters.* "Planet's health source of much debate." April 19, 2002. <http://www.enn.com/news/wire-stories/>

Suzuki, David. *Environmental News Network.* "Time to pull our heads out of the sand." June 13, 2002. <http://www.enn.com/>

University of Massachusetts. *About the Ecological Cities Project.* March 15, 2002. <http://www.umass.edu/ecologicalcities/>

Vermont Institute of Natural Science. *Citizen Science.* [cited June 2002]. <http://www.vinsweb.org/>

Virtanen, Michael. *Associated Press.* "At age 73, advocate refuses to quit fight to keep the Hudson River free of pollution." March 22, 2002. <http://www.enn.com/news/wire-stories/>

ORGANIZATIONS

Cornell Lab of Ornithology, P. O. Box 11, Ithaca, NY USA 14851 Toll Free: (800) 843-2473, <http://www.birds.cornell.edu>

The Natural History Museum of the Adirondacks, P. O. Box 897, Tupper Lake, NY USA 12986 (518) 359-2533, Fax: (518) 523-9841, <http://www.adkscience.org>

Smithsonian Environmental Research Center, 647 Contees Wharf Road, Edgewater, MD USA 21037 (443) 482-2200, Fax: (443) 482-2380, <http://www.serc.si.edu>

Citizen's Clearinghouse for Hazardous Waste

see **Gibbs, Lois**

Citizens for a Better Environment

Citizens for a Better Environment (CBE) is an organization that works to reduce exposure to toxic substances in land, water, and air. Founded in 1971, CBE has 30,000 members and operates with a $1.8 million budget. The organization also maintains regional offices in Minnesota and in Wisconsin.

CBE staff and members focus on research, public information, and advocacy to reduce toxic substances. They also meet with policy-makers on state, regional, and national levels. A staff of scientists, researchers, and policy analysts evaluate specific problems brought to the attention of CBE, testify at legislative and regulatory agency hearings, and file lawsuits in state and federal courts. The organization also conducts public education programs, and provides technical assistance to local residents and organizations that attempt to halt toxic chemical exposures.

The Chicago office won a Supreme Court decision against the construction of an incinerator in a low-income neighborhood in the Chicago area and is researching the issues on volume-based **garbage** for suburban Chicago areas. In Minnesota, the staff has developed a "Good Neighbor" program of agreements between community groups, environmental activists, businesses, and industries along the Mississippi River to reduce **pollution** of the river. In Wisconsin, **transportation** issues under study include selling ride-sharing credits to meet **Clean Air Act** standards. Selling and buying of credits between and among individuals, businesses, and polluting industries has become a lucrative way for polluters to continue their practices. CBE is attempting to close the legislative loopholes that allows this practice to continue.

The Chicago office maintains a library of books, reports, and articles on environmental pollution issues. Publications include CBE's *Environmental Review,* a quarterly journal on the public health effects of pollution; it includes updates of CBE activities and research projects.

[*Linda Rehkopf*]

RESOURCES
ORGANIZATIONS
Citizens for a Better Environment, 1845 N. Farwell Ave., Suite 220, Milwaukee, WI USA 53202 (414) 271-7280, Fax: (866) 256-5988, Toll Free: (414) 271-5904, Email: cbewi@cbemw.org, <http://www.cbemw.org>

Clay minerals

Clay minerals contribute to the physical and chemical properties of most soils and sediments. At high concentrations they cause soils to have a sticky consistency when wet. Individual particles of clay minerals are very small with diameters less than two micrometers. Because they are so finely divided, clay minerals have a very high surface area per unit weight, ranging form 5 to 800 square meters per gram. They are much more reactive than coarser materials in soils and sediments such as **silt** and sand and clay minerals account for much of the reactivity of soils and sediments with respect to **adsorption** and **ion exchange**.

Mineralogists restrict the definition of clay minerals to those aluminosilicates (minerals predominantly composed of **aluminum**, silicon, and oxygen) which in **nature** have particle sizes two micrometers or less in diameter. These minerals have platy structures made up of sheets of silica, composed of silicon and oxygen, and alumina, which is usually composed of aluminum and oxygen, but often has iron and magnesium replacing some or all of the aluminum.

Clay minerals can be classified by the stacking of these sheets. The one to one clay minerals have alternating silica and alumina sheets; these are the least reactive of the clay minerals, and kaolinite is the most common example. The two to one minerals have layers made up of an alumina sheet sandwiched between two silica sheets. These layers have structural defects that result in negative charges, and they are stacked upon each other with interlayer cations between the layers to neutralize the negative layer charges. Common two to one clays are illite and smectite.

In smectites, often called montmorillonite, the interlayer ions can undergo cation exchange. Smectites have the greatest **ion** exchange capacity of the clay minerals and are the most plastic. In illite, the layer charge is higher than for smectite, but the cation exchange capacity is lower because most of the interlayer ions are potassium ions that are trapped between the layers and are not exchangeable.

Some **iron minerals** also can be found in the clay-sized fraction of soils and sediments. These minerals have a low capacity for ion exchange but are very important in some adsorption reactions. Gibbsite, an aluminum hydroxide mineral, is also found in the clay-sized fraction of some soils and sediments. This mineral has a reactivity similar to the iron minerals.

[*Paul R. Bloom*]

Clay-hard pan

A compacted subsurface **soil** layer. Hard pans are frequently found in soils that have undergone significant amounts of **weathering**. Clay will accumulate below the surface and cause the **subsoil** to be dense, making it difficult for roots and water to penetrate. Soils with clay pans are more susceptible to water **erosion**. Clay pans can be broken by cultivation, but over time they will re-form.

Clayoquot Sound

Clayoquot (pronounced CLACK wit) Sound, located about half way up the west coast of British Columbia's Vancouver Island, is a spectacular complex of interconnecting ocean inlets, rocky islands, and narrow fjords cut into densely forested mountains. Home to black bears, cougars, **wolves**, bald eagles, marbled murrelets, five **species** of **salmon**, and lush with ferns, mosses and other water-loving plants, the area contains the largest remaining tract of **temperate rainforest** on Vancouver Island. It also contains the southernmost pristine, coastal rainforest valleys in North America. The forest is dominated by Sitka Spruce, Western Hemlock, Western Red Cedar, and Douglas Fir, many of which can grow to 300 ft (100 m) high and 15 ft (5 m) in diameter, and live over 1,500 years.

Temperate rainforests only occur in areas with a mild, humid **climate** where a cold ocean is close to mountains. They need at least 80 in (190 cm) of rain spread fairly evenly throughout the year to keep the vegetation moist and lush. These conditions occur only along the west coasts of North America, Chile, New Zealand and Tasmania. Today, over half of these magnificent forests are gone and the only remaining large tracts on earth are in southeastern Alaska, coastal British Columbia, and Chile. This thin band of coastal temperate **rain forest** contains some of the oldest and largest trees on Earth and has more **biomass** (volume of organic material) per unit area than any other **ecosystem** in the world.

Countless complex and unique ecological processes take place in ancient forests, most of which scientists don't fully understand. A typical, old growth rain forest contains trees of all ages, but many are between 250 and 1,000 years old. A verdant collection of ferns, mosses, and **lichens** carpets the forest floor and forms thick blankets that smother fallen tree trunks and drape the branches of standing trees. Battered by centuries of wind and rain, the canopy of the old growth forest is ragged and uneven with many broken treetops and snags. A single ancient tree can provide life-sustaining **habitat** for innumerable species of insects, birds, and small mammals that burrow in its bark, nest in its branches, den in hollow cavities in its trunk, and feed on the garden of vegetation that festoons its branches. Many of these animals never descend from the forest canopy and have yet to be studied by scientists.

Dead trees also play important ecological roles in the forest. Standing snags are home to many species of insects and cavity-nesting birds. Fallen tree trunks, decaying slowly on the humid forest floor, provide nutrients and habitat to a flourishing community of **fungi**, insects, and other woodland creatures. Tree seedlings germinate and thrive on the spongy, nutrient-rich wood of fallen "nurse logs." Over 90% of the trees in the temperate rainforest grew initially on these nurse logs. Salmon connect the coastal forest to the sea. Vast schools of salmon ascending the rivers to spawn provide a key food for at least 22 animal species including bears, eagles, and Orca **whales**. When they die, the salmon fertilize streams, providing most of the nutrients that support aquatic life that will feed young salmon until they return to the ocean. Furthermore, salmon carcasses, dragged to the shore by creatures that feed on them, provide up to half the **carbon** and **nitrogen** that nourishes streamside vegetation.

With about one quarter of the world's remaining temperate rain forest, British Columbia has been the site of intense controversy over forest harvest policies and practices, and Clayoquot Sound has been the focal point for much of that debate. For more than 20 years, the area around Clayoquot was the site of confrontations and mass demonstrations both for and against **logging**. In 1993, the area witnessed the largest peaceful civil disobedience action in Canadian history. Clayoquot became an international environmental icon. Protests in support of activists there were held at Canadian Embassies throughout the world, and a boycott of Canadian timber products threatened the country's biggest industry.

Part of the basis for opposing further logging is the **scarcity** of coastal, temperate rain forest. More than 73% of Vancouver Island has either already been logged or is slated for future harvest. While about 13% of the Island's area is set aside in parks or preserves, less than 6% of the original rain forest is protected. Of the 170 watersheds (valleys) on Vancouver Island greater than 10,000 acres (about 4,000 hectares) in size, only 12 remain undeveloped, and half of those are in Clayoquot Sound.

Logging protests began in Clayoquot Sound in the early 1980s, focusing first on Meares Island, where 90% of the forest was designated for harvest. Environmentalists and First Nations people work together to stop this logging. In 1984, the Tla-o-qui-aht and Ahousaht Bands declared the island a Tribal Park, and loggers were turned away. This was the first logging blockade in Canadian history, and it attracted strong local and national support. The two largest logging companies working in Clayoquot Sound at the time were MacMillan Bloedel and the International Forest Products (Interfor). Together, in the 1980s, these two companies were cutting nearly 1 million cubic meters (32,000 large truck loads) per year from Clayoquot Sound forests.

In 1993, the British Columbia government announced a new forestry and land-use decision that aimed to compromise between environmental, economic, and social needs of the area. It would reduce the size of clearcuts, and set aside areas deemed exceptionally valuable or sensitive. A **Sierra Club** mapping project revealed, however, that 74% of the ancient temperate rain forests would be harvested under this plan, and that many of the protected areas were bog and marginal forest. Through the summer of 1993, mass protests and road blockades occurred on a daily basis at Clayoquot Sound. Altogether, more than 12,000 people traveled to this remote location that summer to protest old growth forest logging, and 856 were arrested for blockading roads into logging areas. Rather than simply paying a fine and going home, many of the protesters insisted on going to trial so they could express their love for the forest and their dismay that it was being cut down. A collection of statements made in court was published in *Clayoquot Mass Trials: Defending the Rainforest.*

As protests continued, tempers flared. Radicals on both sides of the controversy engaged in hostile and destructive acts. Property damage included several logging bridges burned, equipment vandalized, and tree spiking. On the other side, angry loggers attacked and injured several protestors. In August, more than 5,000 people from all over British Columbia attended a demonstration in the nearby town of Ucluelet to support logging. A poll conducted about this time found that 52% of British Columbians support the Clayoquot Compromise (in favor of continued logging) while 39% opposed it. Fifty seven percent of the respondents opposed the blockades, with 35% in favor and the remainder undecided. About the same time, the Nuu-Chah-Nulth Tribal Council announced that Native bands would not support any land-use decision in Clayoquot Sound that excludes logging. It said such a decision would affect the area's economic viability and further alienate the land from potential Native use.

Probably the most powerful weapon in the environmentalists arsenal was a worldwide consumer blockade of

Canadian wood products. A highly effective public relations campaign mounted in Europe, **Australia**, and the United States made logging of ancient forests at Clayoquot an international scandal. Several large firms including General Telephone Company, Scott Paper, Walmart and Home Depot announced that they would no longer buy or sell wood or wood products harvested from old-growth forests in an unsustainable manner. To counter the bad publicity, the Provincial Government reduced the harvest at Clayoquot from 959,000 cubic meters in 1988 to 0 in 1998.

In 1999 MacMillan Bloedel signed a Memorandum of Understanding with the First Nations Tribal Council and several national environmental groups. The memorandum set up a joint venture called Iisaak Forest Resources, owned 51% by the the Nuu-chah-nulth Tribal Council, and 49% by MacMillan Bloedel. Iisaak promised it would not log areas larger than 2,470 acres (1,000 hectares) in any of Clayoquot's untouched valleys, and would practice sustainable practices focused on small scale logging, **non-timber forest products** and eco-tourism rather than clearcutting of old-growth trees. However, it said, during a transition period of 50 years or so, some old-growth logging will continue until second-growth timber grows big enough to harvest.

This joint venture presents a dilemma for many environmentalists. On one hand, they want to respect indigenous land rights, and often regard native people as having greater environmental knowledge and sensitivity than those who haven't lived on the land for so long. On the other hand, if the forestry practices of the native corporation turn out to not really be sustainable, it may be difficult for environmentalists to criticize their former allies. The Friends of Clayoquot Sound, for example, refused to sign the Memorandum of Understanding because they don't support any further cutting of old growth trees for any reason.

In 2000, Clayoquot Sound was designated a United Nations Biosphere Reserve. However, this title doesn't give the rain forest any further protection. Perhaps more valuable is the establishment of the Pacific Rim **National Park** along the coast south of the town of Tofino. Nearly 125,000 acres (50,000 hectares), much of it **old-growth forest**, is included in this new national park. MacMillan Bloedel, once the largest logging company in Canada, has been bought by the U.S.-based Weyerhaeuser Corporation. Weyerhaeuser says it intends to continue collaboration with Iisaak joint venture and will honor the memorandum of understanding on management of Clayoquot Sound. Meanwhile, Interfor continues logging ancient forests, and environmentalists fear that the cumulative effect of many small cuts will be the same as if massive clearcuts had continued.

Much of the recent environmental activism in Clayoquot Sound has focused on salmon farming. Protestors claim excess food and feces from caged fish pollutes water in the Sound. Antibiotics and **chemicals** used to prevent diseases in the densely packed fish pens can harm other sea life, and escaped domesticated fish can be a threat to wild populations. Furthermore, fish farm employees often kill marine mammals and sea birds that approach open net-cage salmon pens.

It remains to be seen whether **sustainable forestry** can preserve the ancient forests or new regulations on fish farming can protect **wildlife** and **water quality**. Even in its less than pristine state, however, the Sound is a beautiful place and offers great opportunities for a variety of types of outdoor **recreation**.

[*William P. Cunningham Ph.D.*]

RESOURCES
BOOKS

Berman, Tzeporah. *Clayoquot & Dissent.* Vancouver: Ronsdale Press, 1994.
Breen-Needham, Howard, et al, eds. *Witness to Wilderness: the Clayoquot Sound Anthology.* Vancouver: Arsenal Pulp Press, 1994.
Krawczyk, Betty Shiver. *Clayoquot: the Sound of My Heart.* Custer, WA: Orca Book Publishers, 1996.
MacIsaac, Ronald and Anne Champagne, eds. *Clayoquot Mass Trials: Defending the Rainforest..* Gabriola Island, BC: New Society Publishers, 1994.
Mackenzie, Ian. *Ancient Landscapes of British Columbia: a Photographic Journey Through the Remaining Wilderness of British Columbia.* Edmonton, AB: Lone Pine Publishing, 1995.
McLaren, Jean. *Spirits rising: the story of the Clayoquot Peace Camp, 1993.* Vancouver, BC: Pacific Edge Publishing, 2000.
Streetley, Joanna. *Paddling Through Time: A Sea Kayaking Journey Through Clayoquot Sound.* Raincoast Books, 2000.

OTHER

Ross, Andrew. Chronology of Clayoquot Sound events. October 1996. Department of Political Science, University of Victoria, BC. [cited July 9, 2002]. <http://sitka.dcf.uvic.ca/CLAYOQUOT/chronolo.htm#notes>.

Clean Air Act (1963, 1970, 1990)

The 1970 Clean Air Act and major amendments to the act in 1977 and 1990 serve as the backbone of efforts to control **air pollution** in the United States. This law established one of the most complex regulatory programs in the country. Efforts to control air **pollution** in the United States date back to 1881, when Chicago and Cincinnati passed laws to control industrial **smoke** and soot. Other municipalities followed suit and the momentum continued to build. In 1952, Oregon became the first state to adopt a significant program to control air pollution, and three years later, the federal government became involved for the first time, when the **Air Pollution Control** Act was passed. This law provided funds to assist the states in their air **pollution control** activities.

In 1963, the first Clean Air Act was passed. The act provided permanent federal aid for research, support for the development of state pollution control agencies, and federal involvement in cross-boundary air pollution cases. An amendment to the act in 1965 directed the Department of Health, Education and Welfare (HEW) to establish federal **emission standards** for motor vehicles. (At this time, HEW administered air pollution laws. The **Environmental Protection Agency** (EPA) was not created until 1970.) This represented a significant move by the federal government from a supportive to an active role in setting air pollution policy. The 1967 Air Quality Act provided additional funding to the states, required them to establish Air Quality Control Regions, and directed HEW to obtain and make available information on the health effects of air pollutants and to identify pollution control techniques.

The Clean Air Act of 1970 marked a dramatic change in air pollution policy in the United States. Following enactment of this law, the federal government, not the states, would be the focal point for air pollution policy. This act established the framework that continues to be the foundation for air pollution control policy today. The impetus for this change was the belief that the state-based approach was not working and increased pressure from a developing environmental consciousness across the country. Public sentiment was growing so significantly that environmental issues demanded the attention of high-ranking officials. In fact, the leading policy entrepreneurs on the issue were President Richard Nixon and Senator Edmund Muskie of Maine.

The regulatory framework of The Clean Air Act featured four key components. First, National Ambient Air Quality Standards (NAAQSs) were established for six major pollutants: **carbon monoxide**, **lead** (in 1977), **nitrogen** dioxide, ground-level **ozone** (a key component of **smog**), **particulate** matter, and **sulfur dioxide**. For each of these pollutants, sometimes referred to as criteria pollutants, primary and **secondary standards** were set. The **primary standards** were designed to protect human health; the secondary standards were based on protecting crops, forests, and buildings if the primary standards were not capable of doing so. The Act stipulated that these standards must apply to the entire country and be set by the EPA, based on the best available scientific information. The costs of attaining these standards were not among the factors considered. The EPA was also directed to set standards for less common toxic air pollutants.

Second, New Source Performance Standards (NSPSs) would be set by the EPA. These standards would determine how much air pollution would be allowed by new plants in various industrial sectors. The standards were to be based on the **best available control technology** (BACT) and best available retrofit technology (BART) available for the control of pollutants at sources such as **power plants**, steel factories, and chemical plants.

Third, mobile source **emission** standards were established to control **automobile emissions**. These standards were specified in the statute (rather than left to the EPA), and schedules for meeting them were also written into the law. It was thought that such an approach was crucial to ensure success with the powerful auto industry. The pollutants regulated were **carbon** monoxide, **hydrocarbons**, and **nitrogen oxides**, with goals of reducing the first two pollutants by 90% and nitrogen oxides by 82% by 1975.

The final component of the air protection act dealt with the implementation of the new air quality standards. Each state would be encouraged to devise a state implementation plan (SIP), specifying how the state would meet the national standards. These plans had to be approved by the EPA; if a state did not have an approved SIP, the EPA would administer the Clean Air Act in that state. However, since the federal government was in charge of establishing pollution standards for new mobile and stationary sources, even the states with an SIP had limited flexibility. The main focal point for the states was the control of existing stationary sources, and if necessary, mobile sources. The states had to set limits in their SIPs that allowed them to achieve the NAAQSs by a statutorily determined deadline. One problem with this approach was the construction of tall smokestacks, which helped move pollution out of a particular **airshed** but did not reduce overall pollution levels. The states were also charged with monitoring and enforcing the Clean Air Act.

The 1977 amendments to the Clean Air Act dealt with three main issues: nonattainment, auto emissions, and the prevention of air quality deterioration in areas where the air was already relatively clean. The first two issues were resolved primarily by delaying deadlines and increasing penalties. Largely in response to a court decision in favor of environmentalists (**Sierra Club** v. Ruckelshaus, 1972), the 1977 amendments included a program for the prevention of significant deterioration (PSD) of air that was already clean. This program would prevent polluting the air up to the national levels in areas where the air was cleaner than the standards. In Class I areas, areas with near pristine air quality, no new significant air pollution would be allowed. Class I areas are airsheds over large national parks and **wilderness** areas. In Class II areas, a moderate degree of air quality deterioration would be allowed. And finally, in Class III areas, air deterioration up to the national secondary standards would be allowed. Most of the country that had air cleaner than the NAAQSs was classified as Class II. Related to the prevention of significant deterioration was a provision to protect and enhance **visibility** in national parks and wilderness areas even if the air pollution was not a threat to human health. The impetus for this section of the bill was the growing visibility problem in parks, especially in the Southwest.

Throughout the 1980s, efforts to further amend the Clean Air Act were stymied. President Ronald Reagan was opposed to any strengthening of the Act, which he argued would hurt the economy. In Congress, the controversy over **acid rain** between members from the Midwest and the Northeast further contributed to the stalemate. Gridlock on the issue broke with the election of George Bush, who supported amendments to the Act, and the rise of Senator George Mitchell of Maine to Senate Majority Leader. Over the next two years, the issues were hammered out between environmentalists and industry and between different regions of the country. Major players in Congress were Representatives John Dingell of Michigan and Henry Waxman of California and Senators Robert Byrd of West Virginia and Mitchell.

Major amendments to the Clean Air Act were passed in the fall of 1990. These amendments addressed four major topics: (1) **acid** rain, (2) toxic air pollutants, (3) nonattainment areas, and (4) **ozone layer depletion**. To address acid rain, the amendments mandated a 10 million ton reduction in annual sulfur dioxide emissions (a 40% reduction based on the 1980 levels) and a two million ton annual reduction in nitrogen oxides to be completed in a two-phase program by the year 2010. Most of this reduction will come from old utility power plants. The law also creates marketable pollution allowances, so that a utility that reduces emissions more than required can sell those pollution rights to another source. Economists argue that, to increase efficiency, such an approach should become more widespread for all pollution control.

Due to the failure of the toxic air pollutant provisions of the 1970 Clean Air Act, new, more stringent provisions were adopted requiring regulations for all major sources of 189 varieties of toxic air pollution within ten years. Areas of the country still in nonattainment for criteria pollutants were given from three to twenty years to meet these standards. These areas were also required to impose tighter controls to meet the standards. To help these areas and other parts of the country, the Act required stiffer motor vehicle emissions standards and cleaner **gasoline**. Finally, three chemical families that contribute to the destruction of the stratospheric ozone layer (**chlorofluorocarbons** (CFCs), hydrochlorofluorocarbons (HCFCs), and methyl chloroform) were to be phased out of production and use.

In 1997, the EPA issued revised national ambient air quality standards (NAAQS), setting stricter standards for ozone and particulate matter. The American Trucking Association and other state and industry groups legally challenged the new standards on the grounds that the EPA did not have the authority under the Act to make such changes. On February 27, 2001, the U.S. Supreme Court unanimously upheld the constitutionality of the Clean Air Act as interpreted by the EPA, and all remaining legal challenges to other aspects of the standards change were rejected by a Washington DC District Court ruling in early 2002.

The Clean Air Act has met with mixed success. The national average pollutant levels for the criteria pollutants have decreased. Nevertheless, many localities have not achieved these standards and are in perpetual nonattainment. Not surprisingly, major urban areas are those most frequently in nonattainment. The pollutant for which standards are most often exceeded is ozone, or smog. This is due in part to increases in nitrogen oxides (NOx), which disperse ozone. NOx emissions increased by approximately 20% between 1970 and 2000. As a result, some parts of the country have had worsening ozone levels. According to the EPA, the average ozone levels in 29 national parks increased by over 4% between 1990 and 2000.

The greatest successes of air pollution control have come with lead, which between 1981 and 2000 was reduced by 93% (largely due to the phasing-out of leaded gasoline), and particulates, which were reduced by 47% in the same period. Overall particulate emissions were down 88% since 1970. Carbon monoxide has dropped by 25% and volatile organic compounds and sulfur dioxides have declined by over 40% each between 1970 and 2000. However, air quality analysis is complex, and it is important to note that some changes may be due to shifts in the economy, changes in weather patterns, or other such variables rather than directly attributable to the Clean Air Act.

In February 2002, President Bush introduced the "Clear Skies" legislation, an initiative that, if fully adopted, would make some significant changes to the Clean Air Act. Among them would be a weakening or elimination of new source review regulations and BART rules, and a new "cap and trade" plan that would allow power plants that produced excessive toxic emissions to 'buy credits' from other plants who had levels under the standards. The Bush administration hailed the initiative as a less expensive way to accelerate air pollution clean-up and a more economy-friendly alternative to the Kyoto Protocol for greenhouse gas reductions, while environmental groups and other critics called it a roll-back of the Clean Air Act progress.

[*Christopher McGrory Klyza and Paula Anne Ford-Martin*]

RESOURCES
BOOKS

Belden, Roy S. *The Clean Air Act.* Washington, DC: ABA Publishing, 2001.

Dewey, Scott Hamilton. *Don't Breathe the Air: Air Pollution and U.S. Environmental Politics, 1945–1970.* College Station, TX: Texas A&M University Press, 2000.

Erbes, Russell E. *A Practical Guide to Air Quality Compliance.* New York: John Wiley & Sons, 1996.

U.S. Government Printing Office and R. A. Leiter. *Legislative History of the Clean Air Act Amendments of 1990.* Buffalo, NY: William s Hein & Co, 1998.

PERIODICALS

Adams, Rebecca. " Hill Demands Clean Air Act Rewrite Data." *CQ Weekly.* 60, no.18 (May 4 2002): 1161 (2).

OTHER

Department of Energy, Energy Information Administration. *Emissions of Greenhouse Gases in the United States: 2000.* Report #DOE/EIA-0573(2000): December 2001. Full-text available online at http://www.eia.-doe.gov/oiaf/1605/ggrpt/index.html

Environmental Protection Agency (EPA). *AIRNow* http://www.epa.gov/airnow/index.html.

Environmental Protection Agency (EPA), Office of Air and Radiation. *Latest Findings on National Air Quality: 2000 Status and Trends* [Accessed June 5, 2002.] http://www.epa.gov/oar/aqtrnd00/

Clean coal technology

Coal is rapidly becoming the world's most popular fuel. Today in the United States, more than half of the electricity produced comes from coal-fired **power plants**. The demand for coal is expected to triple by the middle of the next century, making it more widely used than **petroleum** or **natural gas**.

The unpleasant aspect of this trend is that coal is a relatively dirty fuel. When burned, it releases particulates and pollutants such as **carbon monoxide, nitrogen** and sulfur oxides into the **atmosphere**. If the use of coal is to expand continually, something must be done to reduce the hazard its use presents to the **environment**.

Over the past two decades, therefore, there has been an increasing amount of research on clean coal technologies, methods by which the **combustion** of coal releases fewer pollutants to the atmosphere. As early as 1970, the United States Congress acknowledged the need for such technologies in the **Clean Air Act** of that year. One provision of that Act required the installation of **flue gas** desulfurization (FGD) systems ("scrubbers") at all new coal-fired plants.

More than a dozen different technologies are now available on at least an experimental basis for the cleaning of coal. Some of these technologies are used on coal before it is even burned. Chemical, physical, and biological methods have all been developed for pre-combustion cleaning. For example, pyrite (FeS_2) is often found in conjunction with coal when it is mined. When the coal is burned, pyrite is also oxidized, releasing **sulfur dioxide** to the atmosphere. Yet pyrite can be removed from coal by rather simple, straightforward physical means because of differences in the densities of the two substances.

Biological methods for removing sulfur from coal are also being explored. The bacterium *Thiobacillus ferrooxidans* has the ability to change the surface properties of pyrite particles, making it easier to separate them from the coal itself. The bacterium may also be able to extract sulfur that is chemically bound to **carbon** in the coal.

A number of technologies have been designed to modify existing power plants to reduce the release of pollutants produced during the combustion of coal. In an attempt to improve on traditional wet **scrubbers**, researchers are now exploring the use of dry injection as one of these technologies. In this approach, dry compounds of calcium, sodium, or some other element are sprayed directly into the furnace or into the ducts downstream of the furnace. These compounds react with non-metallic oxide pollutants, such as sulfur dioxide and nitrogen dioxide, forming solids that can be removed from the system. A variety of technologies are being developed especially for the release of oxides of nitrogen. Since the amount of this pollutant formed is very much dependent on combustion temperature, methods of burning coal at lower temperatures are also being explored.

Some entirely new technologies are also being developed for installation in power plants to be built in the future. **Fluidized bed combustion**, integrated gasification combined cycle, and improved coal pulverization are three of these. In the first of these processes, coal and limestone are injected into a stream of upward-flowing air, improving the degree of oxidation during combustion. In the second process, coal is converted to a gas that can be burned in a conventional power plant. The third process involves improving on a technique that has long been used in power plants, reducing coal to very fine particles before it is fed into the furnace.

[*David E. Newton*]

RESOURCES
PERIODICALS

Cruver, P. C. "What Will Be the Fate of Clean-Coal Technologies?" *Environmental Science & Technology* (September 1989): 1059-1060.

Shepard, M. "Coal Technologies for a New Age." *EPRI Journal* (January-February 1988): 4-17.

Clean Water Act (1972, 1977, 1987)

Federal involvement in protecting the nation's waters began with the **Water Pollution** Control Act of 1948, the first statute to provide state and local governments with the funding to address water **pollution**. During the 1950s and 1960s, awareness grew that more action was needed and federal funding to state and local governments was increased. In the **Water Quality** Act of 1965 **water quality standards**, to be developed by the newly created Federal Water **Pollution Control** Administration, became an important part of federal water pollution control efforts.

Despite these advances, it was not until the Water Pollution Control Amendments of 1972 that the federal government assumed the dominant role in defining and directing water pollution control programs. This law was the outcome of a battle between Congress and President Richard M. Nixon. In 1970, facing a presidential re-election campaign, Nixon responded to public outcry over pollution problems by resurrecting the Refuse Act of 1899, which authorized the U.S. **Army Corps of Engineers** to issue **discharge** permits. Congress felt its prerogative to set national policy had been challenged. It debated for nearly 18 months to resolve differences between the House and Senate versions of a new law, and on October 18, 1972, Congress overrode a presidential veto and passed the Water Pollution Control Amendments.

Section 101 of the new law set forth its fundamental goals and policies, which continue to this day "to restore and maintain the chemical, physical, and biological integrity of the Nation's waters." This section also set forth the national goal of eliminating discharges of pollution into navigable waters by 1985 and an interim goal of achieving water quality levels to protect fish, shellfish, and **wildlife**. As national policy, the discharge of toxic pollutants in toxic amounts was now prohibited; federal financial assistance was to be given for constructing publicly owned waste treatment works; area-wide pollution control planning was to be instituted in states; research and development programs were to be established for technologies to eliminate pollution; and **nonpoint source** pollution—runoff from urban and rural areas—was to be controlled. Although the federal government set these goals, states were given the main responsibility for meeting them, and the goals were to be pursued through a permitting program in the new **national pollutant discharge elimination system** (NPDES).

Federal grants for constructing publicly owned treatment works (POTWs) had totalled $1.25 billion in 1971, and they were increased dramatically by the new law. The act authorized five billion dollars in fiscal year 1973, six billion for fiscal year 1974, and seven billion for fiscal year 1975, all of which would be automatically available for use without requiring Congressional appropriation action each year. But along with these funds, the act conferred the responsibility to achieve a strict standard of secondary treatment by July 1, 1977. Secondary treatment of sewage consists of a biological process that relies on naturally occurring bacteria and other micro-organisms to break down organic material in sewage. The **Environmental Protection Agency** (EPA) was mandated to publish guidelines on secondary treatment within 60 days after passage of the law. POTWs

also had to meet a July 1, 1983, deadline for a stricter level of treatment described in the legislation as "best practicable **wastewater** treatment." In addition, pretreatment programs were to be established to control industrial discharges that would either harm the treatment system or, having passed through it, pollute receiving waters.

The act also gave polluting industries two new deadlines. By July 1, 1977, they were required to meet limits on the pollution in their discharged **effluent** using Best Practicable Technology (BPT), as defined by EPA. The conventional pollutants to be controlled included **organic waste**, **sediment**, **acid**, bacteria and viruses, nutrients, oil and grease, and heat. Stricter state water quality standards would also have to be met by that date. The second deadline was July 1, 1983, when industrial dischargers had to install **Best Available Control Technology** (BAT), to advance the goal of eliminating all pollutant discharges by 1985. These BPT and BAT requirements were intended to be "technology forcing," as envisioned by the Senate, which wanted the new water law to restore water quality and protect ecological systems.

On top of these requirements for conventional pollutants, the law mandated the EPA to publish a list of toxic pollutants, followed six months later by proposed effluent standards for each substance listed. The EPA could require **zero discharge** if that was deemed necessary. The zero discharge provisions were the focus of great controversy when Congress began oversight hearings to assess implementation of the law. Leaders in the House considered the goal a target and not a legally binding requirement. But Senate leaders argued that the goal was literal and that its purpose was to ensure rivers and streams ceased being regarded as components of the waste treatment process. In some cases, the EPA has relied on the Senate's views in developing effluent limits, but the controversy over what zero discharge means continues to this day.

The law also established provisions authorizing the Army Corps of Engineers to issue permits for discharging dredge or fill material into navigable waters at specified disposal sites. In recent years this program, a key component of federal efforts to protect rapidly diminishing **wetlands**, has become one of the most explosive issues in the Clean Water Act. Farmers and developers are demanding that the federal government cease "taking" their private property through wetlands regulations, and recent sessions of Congress have been besieged with demands for revisions to this section of the law.

In 1977, Congress completed its first major revisions of the Water Pollution Control Amendments (which was renamed the "Clean Water Act"), responding to the fact that by July 1, 1977 only 30 percent of major municipalities were complying with secondary treatment requirements.

Moreover, a National Commission on Water Quality had issued a report which recommended that zero discharge be redefined to stress **conservation** and **reuse** and that the 1983 BAT requirements be postponed for five to ten years. The 1977 Clean Water Act endorsed the goals of the 1972 law, but granted states broader authority to run their construction grants programs. The act also provided deadline extensions and required EPA to expand the lists of pollutants it was to regulate.

In 1981, Congress found it necessary to change the construction grants program; thousands of projects had been started, with $26.6 billion in federal funds, but only 2,223 projects worth $2.8 billion had been completed. The Construction Grant Amendments of 1981 restricted the types of projects that could use grant money and reduced the amount of time it took for an application to go through the grants program.

The Water Quality Act of 1987 phased out the grants program by fiscal year 1990, while phasing in a state revolving loan fund program through fiscal year 1994, and thereafter ending federal assistance for wastewater treatment. The 1987 act also laid greater emphasis on toxic substances; it required, for instance, that the EPA identify and set standards for toxic pollutants in sewage **sludge**, and it phased in requirements for stormwater permits. The 1987 law also established a new toxics program requiring states to identify "toxic hot spots"—waters that would not meet water quality standards even after technology controls have been established—and mandated additional controls for those bodies of water.

These mandates greatly increased the number of NPDES permits that the EPA and state governments issued, stretching budgets of both to the limit. Moreover, states have billions of dollars worth of wastewater treatment needs that remain unfunded, contributing to continued violations of water quality standards. The new permit requirements for stormwater, together with sewer overflow, sludge, and other permit requirements, as well as POTW construction needs, led to a growing demand for more state flexibility in implementing the clean water laws and for continued federal support for wastewater treatment. State and local governments insist that they cannot do everything the law requires; they argue that they must be allowed to assess and prioritize their particular problems.

Yet despite these demands for less prescriptive federal mandates, on May 15, 1991, a bipartisan group of senators introduced the Water Pollution Prevention and Control Act to expand the federal program. The proposal was eventually set aside after intense debate in both the House and Senate over controversial wetlands issues. Amendments to the Clean Water Act proposed in both 1994 and 1995 also failed to make it to a vote.

In 1998, President Clinton introduced a Clean Water Action Plan, which primarily focused on improving compliance, increasing funding, and accelerating completion dates on existing programs authorized under the Clean Water Act. The plan contained over 100 'action items' that involved interagency cooperation of EPA, **U.S. Department of Agriculture** (USDA), Army Corps of Engineers, Department of the Interior, Department of Energy, **Tennessee Valley Authority**, Department of **Transportation**, and the Department of Justice. Key goals were to control nonpoint pollution, provide financial incentives for conservation and stewardship of private lands, restore wetlands, and expand the public's 'right to know' on water pollution issues.

New rules were proposed in 1999 to strengthen the requirements for states to set limits for and monitor the Total Maximum Daily Load (TMDL) of pollution in their waterways. The TMDL rule has been controversial, primarily because states and local authorities lack the funds and resources to carry out this large-scale project. In addition, agricultural and forestry interests, who previously were not regulated under the Clean Water Act, would be affected by TMDL rules. As of May 2002, the Bush administration has delayed the rule for further review.

In May 2002 the EPA announced a new rule changing the definition of "fill material" under the Clean Water Act to allow dirt, rocks, and other displaced material from mountaintop **coal** mining operations to be deposited into rivers as waste under permit from the Army Corps of Engineers, a practice that was previously unlawful under the Clean Water Act. Shortly thereafter, a group of congressional representatives introduced new legislation to overturn this rule, and a federal district court judge in West Virginia ruled that such amendments to the Clean Water Act could only be made by Congress, not by an EPA rule change. Whether or not the fill material rule will remain part of the Act remains to be seen.

As the Federal Water Pollution Control Act marks its thirtieth anniversary, the EPA state and federal regulators are increasingly recognizing the need for a more comprehensive approach to water pollution problems than the current system, which focuses predominantly on POTWs and industrial facilities. Nonpoint source pollution, caused when rain washes pollution from farmlands and urban areas, is the largest remaining source of water quality impairment, yet the problem has not received a fraction of the regulatory attention addressed to industrial and municipal discharges. Despite this fact, the EPA claims that the Clean Water Act is responsible for a one billion ton decrease in annual **soil runoff** from farming, and an associated reduction in **phosphorus** and **nitrogen** levels in water sources. EPA also asserts that wetland loss, while still a problem, has slowed significantly—from 1972 levels of 460,000 acres per year to current losses of 70,000-90,000 annually. However, critics question these figures, citing abandoned mitigation projects and reclaimed wetlands that bear little resemblance to the **habitat** they are supposed to replicate. *See also* Agricultural pollution; Environmental policy; Industrial waste treatment; Sewage treatment; Storm runoff; Urban runoff

[*David Clarke and Paula Anne Ford-Martin*]

RESOURCES
BOOKS

Committee on Mitigating Wetland Losses, Board on Environmental Studies and Toxicology, Water Science and Technology Board, Division on Earth and Life Sciences, National Research Council. *Compensating for Wetland Losses Under the Clean Water Act.* Washington, DC: National Academy Press, 2001. (Full-text available online at http://books.nap.edu/books/0309074320/html/index.html)

PERIODICALS

Kaiser, Jocelyn. "Wetlands Policy is All Wet." *Science Now* (June 26, 2001): 3.

OTHER

Copeland, Claudia. "Water Quality: Implementing the Clean Water Act." *CRS Issue Brief for Congress* IB89102 (August 2001). Online at http://cnie.org/NLE/CRSreports/water/h2o-15.cfm
Sierra Club. *Clean Water* Online [Accessed June 1, 2002]. http://www.sierraclub.org/cleanwater/waterquality/

ORGANIZATIONS

America's Clean Water Foundation, 750 First Street NE, Suite 1030, Washington, DC USA 20002 (202) 898-0908, Fax: (202) 898-0977, Email: , http://yearofcleanwater.org/

Clear-cutting

Webster's Dictionary defines clear-cutting as "removal of all the trees in a stand of timber." This **forest management** technique has been used in a variety of forests around the world. For many years, it was considered the most economical and environmentally sound way of harvesting timber. Since the 1960s and 1970s, the practice has been increasingly called into question as more has been learned about the ecological benefits of old growth timber especially in the Pacific Northwest of the United States and the global ecological value of tropical rain forests.

Foresters and loggers point out that there are practical economic and safety reasons for the practice of clear-cutting. For example, in the Pacific Northwest, these reasons revolve around the volume of wood that is present in the old growth forests. This volume can actually be an obstacle at times to harvesting, so that the most inexpensive way to remove the trees is to cut everything.

During the post-World War II housing boom in the 1950s, clear-cutting overtook selective cutting as the preferred method for harvesting in the Pacific Northwest. Since

that time, worldwide demand for lumber has continued to rise. Practical arguments that the practice should continue despite its ecological implications are tied directly to the volume and character of the forest itself. Wood in temperate climates decays slowly, resulting in a large number of fallen logs—making even walking difficult, much less dragging cut trees to areas where they can be hauled to mills. Tall trees with narrow branch structures produce a lot of stems in a small area, therefore, the total **biomass** (or total input of living matter) in these temperate forests is typically four times that of the densest **tropical rain forest**. Some redwood forest groves in California have been measured with 20 times the biomass of similar sites in the tropics. Those supporting the harvest of these trees and clear-cutting point out that if these trees are not used, it will put increased pressure on other timber producers throughout the world to supply this wood. This could have high environmental costs globally, since it could take 10–30 acres (4–12 ha) of **taiga** forest in northern Canada, Alaska, or Siberia to produce the wood obtainable from one acre in the state of Washington.

Clear-cutting makes harvesting these trees very lucrative. The number of trees and downed logs makes it difficult to cut only some of the trees, and falling trees can damage the survivors; in addition, they can be damaged when dragged across the ground to be loaded onto trucks. This is made more expensive and time consuming by working around the standing trees. In areas that have been selectively logged, the trees left standing are may be knocked down by winds or may drop branches, presenting a considerable safety risk to loggers working in the area.

Old growth forests leave so much woody debris and half-decayed logs on the ground that it can be difficult to walk through a harvested patch to plant seedlings. This is why an accepted part of the clear-cut practice in the past has been to burn the leftover **slash**, after which seedlings are planted. Brush that grows on the site is treated with **herbicide** to allow the seedlings time to grow to the point at which they can compete with other vegetation.

Supporters of clear-cutting contend that it has unique advantages: burning consumes fuel that would otherwise be available to feed forest fires; **logging** roads built to haul the trees out open the forests to recreational uses and provide firebreaks; clear-cutting leaves no snags or trees to interfere with reseeding the forest, and allows use of helicopters and planes to spray herbicides—the brush that grows up naturally in the sunlight and provides browse for big-game **species** such as deer and elk.

Detractors point to a number of very negative environmental and economic consequences that need to be considered: particularly where clear cuts have been done near urban areas, plumes of **smoke** produced by burning slash have polluted cities; animal browse is choked out after a few years

An aerial view of a clear-cut area snaking its way through a South African forest (Photograph by Marco Polo. Phototake. Reproduced by permission.)

by the new seedlings, creating a darkened forest floor that lacks almost any vegetation or **wildlife** for at least three decades; herbicides applied to control vegetation contribute to the degradation of surface **water quality**; mining the forest causes declines in species diversity and loss of habitats (declaration of the **northern spotted owl** as an **endangered species** and the efforts to preserve its **habitat** is an example of the potential loss of diversity); new microclimates appear that promote less desirable species than the trees they replace; so much live and rotting wood is harvested or burned that the **soil** fertility is reduced, affecting the potential for future tree growth.

Critics also point out that **erosion** and **flooding** increases from clear-cut areas have significant economic impact downstream in the watersheds. Studies have shown that since the clear-cut practice increased following World War II, landslides have increased to six times the previous rate. This has resulted in increased **sediment** delivery to rivers and streams where it has a detrimental impact on the stream fishery.

Loss of **critical habitat** for endangered species such as the spotted owl and the impact of sediment on the **salmon** fishery have resulted in government efforts to set aside old growth forest **wilderness** to preserve the unique forest **ecosystem**. The practice of clear-cutting with its pluses and minuses, however, continues not only in the Pacific Northwest, but in other forests in the United States and in other areas of the world.

Clear-cutting and rain forests

Another area where clear-cutting has become the focus of ecological debate is in the tropical rain forests. Many of the same economic pressures that make clear-cutting a lucrative practice in the United States make it equally attractive in the **rain forest**, but there are significant environmental consequences.

Since 1960, the world demand for wood has increased by 90%, and as indicated above, the pressure to supply lumber to meet this demand has also increased. Nearly one-half of the Earth's rain forests have been cut in the last 30 years.

Besides the demand for building material or fuel, these forests are also subject to significant clearing to later be worked to produce food in new agricultural areas and to facilitate the exploration for oil and minerals. If this loss continues at present estimated rates, the rain forests will be totally harvested by the year 2040.

Rain forest activists continually work to remind the world of the importance of the forests. Locally, for instance, the presence or absence of the rain forest can change the **climate** and the local water budget. For example, times of **drought** are more severe and when the rains come, flooding is increased. As well, rain forests can have a major impact on the global climate. When they are cut and the slash burned, significant amounts of **carbon dioxide** are released into the **atmosphere**, which may contribute to the overall **greenhouse effect** and general warming of the earth.

The biological diversity of these forest ecosystems is vast. Rain forests contain about one-half the known plant and animal species in the world and very few of these species have ever been studied by scientists for their potential benefits. (More that 7,000 medicines have already been derived from tropical plants. It is uncertain how many more such uses are yet to be found.)

Most rain forests occur in less-developed countries where it is difficult to meet the expanding needs of rapidly increasing populations for food and shelter. They need the economic benefits that can be derived from the rain forest for their very survival. Any attempt at stopping the clear-cutting practice must provide for their needs to be successful. Until a better practice is developed, clear-cutting will remain an environmental issue on a global scale for the next several decades.

[*James L. Anderson*]

RESOURCES
BOOKS

Dietrich, W. *The Final Forest*. New York: Simon and Schuster, 1992.
Harris, D. *The Last Stand: The War between Wall Street and Main Street*. Times Books, Random House, 1995.

Frederic E. Clements (1874 – 1945)
American ecologist

For Frederick Clements, trained in botany as a plant physiologist, ecology became "the dominant theme in the study of plants, indeed...the central and vital part of botany." He became a leader in the new science, still described as "the leading plant ecologist of the day."

Clements was born in Lincoln, Nebraska, and earned all of his degrees in botany from the University of Nebraska,

attaining his Ph.D. under Charles Bessey in 1989. As a student, he participated in Bessey's famous "Botanical Seminar" and helped carry out an ambitious survey of the vegetation of Nebraska, publishing the results—co-authored with a class-mate—in an internationally recognized volume titled The Phytogeography of Nebraska, out in print the same year (1898) that he received his doctorate. He then accepted a faculty position at the university in Lincoln.

Clements married Edith Schwartz in 1899, described (in Ecology in 1945) as a wife and help-mate, who "unsparingly devoted her unusual ability as an illustrator, linguist, and botanist" to become his life-long field assistant and also a collaborator on research and books on flowers, particularly those of the Rocky Mountains. Clements rose through the ranks as a teacher and researcher at Nebraska and then, in 1907, he was appointed as Professor and Head of the department of botany at the University of Minnesota. He stayed there until 1917, when he moved to the Carnegie Institution in Washington, D.C., where he focused full-time on research for the rest of his career. Retired from Carnegie in 1941, he continued a year-round work-load in research, spending summers on Pikes Peak at an alpine laboratory and winters in Santa Barbara at a coastal laboratory. He died in Santa Barbara on July 26, 1945.

Publication of Research Methods in Ecology in 1905 marked his promotion to full professor at the University of Nebraska, but more importantly it marked his turn from taxonomic and phytogeographical work to ecology. It has been called "the earliest how-to book in ecology," and "a manifesto for the emerging field." More broadly, Arthur Tansley, a leading plant ecologist of the time in Great Britain, though critical of some of Clements' views, described him as "by far the greatest individual creator of the modern science of vegetation." Henry a. Gleason, though an early and severe critic of Clements' ideas, also recognized him as "an original ecologist, one inspired by Europeans but developing his ideas entirely de novo from his own fertile brain."

Clements deplored the "chaotic and unsystematized" state of ecology and assumed the task of remedying it. He did bring rigor, standardization, and an early quantitative approach to research processes in plant ecology, especially through his development of sampling procedures at the turn of the century. Robert McIntosh, writing on the background of ecology, argues that Clements was the "pioneer in developing the quadrat as the basis of quantitative community ecology," was "an earnest cheerleader for quantitative plant ecology, and was the notable American developer and advocate of the quadrat method," though he also declares that "Clements did not get beyond simple counting, and more sophisticated statistical considerations were left to others." However, McIntosh still credits Clements' work as a giant step to quantification in ecology, the crux being that Clements methods "were designed to examine and follow change in vegetation, not only to report the status quo."

Clements is often described as rigid and dogmatic, which seems at odds with his importance in emphasizing change in natural systems; that emphasis on what he called "dynamic ecology" became his research trademark. Clements also stressed the importance of process and function. Clements anticipated ecosystem ecology through his concern for changes through time of plant associations, in correspondence to changes in the physical sites where the communities were found. His rudimentary conception of such systems carried his ecological questions beyond the traditional confines of plant ecology to what McIntosh described as "the larger system transcending plants, or even living organisms" to link to the abiotic environment, especially climate. McIntosh also credits him with a definition of paleoecology that antedated major developments that would studies of vegetation and pollen analysis (or palynology). Clements being Clements, he went into print with a book, Methods and Principles of Paleo-Ecology (1924) appropriate to the new topic. Clements even played a role, if not a major one relative to figures like Forbes, toward understanding what ecologists came to call eutrophic changes in lakes, an area of major current consideration in the environmental science of water bodies.

One of the stimulants for botanical, and then plant ecological, research for Clements and other students of Bessey's at Nebraska was their location in a major agricultural area dependent on plants for crops. Problems of agriculture remained one of Clements' interests for all of his life. During his long career, Clements remained continually involved in various ways of applying ecology to problems of land use such as grazing and soil erosion, even the planning of shelter belts. In large part because of the early work by Bessey and Clements and their colleagues and collaborators, the Midwest became a center for the development of grassland ecology and range management. Worster (in Nature's Economy, 1994) suggests that Clements' "dynamic ecology provided much of the scientific authority for the new ecological conservation movement" which, from the 1930's on, relied heavily on his "climax theory as a yardstick by which man's intrusions into nature could be measured." Though considered misguided by some, pleas for wilderness set-asides still argue today that land-use policy should leave "the climax" undisturbed or preserved areas returned to a perceived climax condition. Clements believed early that homesteaders in Nebraska were wrong to destroy the sod covering the sandhills of Nebraska and that the prairies should be grazed and not tilled. Farmers objected early to the implications of climax theory because they feared threats to their livelihoods from calls for cautious use of marginal lands. Even some scientific attempts to discredit the idea of the climax were based on the desire to undermine its importance to the conservation movement.

Clements even anticipated a century of sporadic connections between biological ecology and human ecology in the social sciences, arguing that sociology "is the ecology of a particular species of animal and has, in consequence, a similar close association with plant ecology." That connection was lost in in professional ecology in the early forties and has still not been reestablished, even at the beginning of the twenty-first century. Though Clements' name is still recognized today as one of ecology's foundation thinkers, and though he gained considerable respect and influence among the ecologists of his day, his work from the beginning was controversial. One reason was that ecologists were skeptical that approaches from plant physiology could be transposed directly to ecology. Another reason, and the idea with which Clements is still associated three-quarters of a century later, was his conviction that the successional process was the same as the development of an organism and that the communities emerging as the end-points of succession were in fact super-organisms. Clements believed that succession was a dynamic process of 'progressive' development of the plant formation, that it was controlled absolutely by climate, developed in an absolutely predictable way (and in the same way in all similar climates) and then was absolutely stable for long periods of time. A cautionary note from MacIntosh: "Clements's ideas are notably resilient and persist, often under different names," even "some of his suspect ideas persist in the new ecology under new rubrics." Synonymizing Gaia with a super organism is one frequently cited example.

Clements' fondness for words, and especially for coining his own nonce terms for about every conceivable nuance of his work also got him in trouble with his colleagues in ecology. As McIntosh terms it, Clements was known for his "labyrinthine logic and proliferation of terminology," characterized by others as "chronic logorrhea." This fondness for coining new words to describe his work, e.g., 'therium' to describe a successional stage caused by animals, added to the fuel for critics who wished to find fault with the substance of his ideas. Unfortunately, Clements also gave further cause to those, then and now, wishing to discount all of his ideas by retaining a belief in Lamarckian evolution, and by doing experiments during retirement at his alpine research station, in which he even claimed he been able to convert "several Linnean species into each other, histologically as well as morphologically."

Kingsland, writing in 1991, in Foundations of Ecology, identified Clements' 1936 article "The Nature and Structure of the Climax" as a classic paper and included its author as among those who defined ecology, but she could still claim that "by the 1950s plant ecologists had abandoned many of the central principles of Clementsian dogma, as well as the more cumbersome features of his classification system, as inappropriate or unproductive." But without the work and thinking of the early ecologists that created a foundation on which to build, ecology might be considerably the lesser today. Missteps certainly were made, but the stones were laid and some of that foundation is based on the dynamics of ecological processes as first conceived by Clements in a considerable body of pioneering work.

[*Gerald L. Young Ph.D.*]

FURTHER READING
BOOKS

Clements, Edith S. *Adventures in Ecology: Half a Million Miles...From Mud to Macadam.* NY: Pageant Press, 1960.
Humphrey, Harry B. "Frederick Edward Clements 1874–1945." In *Makers of North American Botany.* NY: The Ronald Press Company, 1961.

PERIODICALS

Phillips, John. "A Tribute to Frederic E. Clements and His Concepts in Ecology." *Ecology* 35, no. 2 (April 1954): 1114–115.
Pound, Roscoe. "Frederic E. Clements as I Knew Him." *Ecology* 35, no. 2 (April 1954): 112–113.

Climate

Climate is the general, cumulative pattern of regional or global weather patterns. The most apparent aspects of climate are trends in air temperature and humidity, wind, and precipitation. These observable phenomena occur as the **atmosphere** surrounding the earth continually redistributes, via wind and evaporating and condensing water vapor, the energy that the earth receives from the sun.

Although the climate remains fairly stable on the human time scale of decades or centuries, it fluctuates continuously over thousands or millions of years. A great number of variables simultaneously act and react to create **stability** or fluctuation in this very complex system. Some of these variables are atmospheric composition, rates of **solar energy** input, **albedo** (the earth's reflectivity), and terrestrial geography. Extensive research helps explain and predict the behavior of individual climate variables, but the way these variables control and respond to each other remains poorly understood. Climate behavior is often likened to "chaos," changes and movements so complex that patterns cannot be perceived in them, even though patterns may exist. Nevertheless, studies indicate that human activity may be disturbing larger climate trends, notably by causing global warming. This prospect raises serious concern because rapid **anthropogenic** climate change could severely stress ecosystems and **species** around the world.

Solar energy and climate

Solar energy is the driving force in the earth's climate. Incoming radiation from the sun warms the atmosphere and raises air temperatures, warms the earth's surface, and evaporates water, which then becomes humidity, rain, and snow. The earth's surface reflects or re-emits energy back into the atmosphere, further warming the air. Warming air expands and rises, creating convection patterns in the atmosphere that reach over several degrees of latitude. In these convection cells, low pressure zones develop under rising air, and high pressure zones develop where that air returns downward toward the earth's surface. Such differences in atmospheric pressure force air masses to move, from high to low pressure regions. Movement of air masses creates wind on the earth's surface. When these air masses carry evaporated water, they may create precipitation when they move to cooler regions.

The sun's energy comes to the earth in a spectrum of long and short radiation wavelengths. The shortest wavelengths are microwaves and infrared waves. Infrared radiation is felt as heat. A small range of medium wavelength radiation makes up the spectrum of visible light. Longer wavelengths include ultraviolet (UV) radiation and radio waves. These longer wavelengths cannot be sensed, but UV radiation can cause damage as organic tissues (such as skin) absorb them. The difference in wavelengths is important because long and short wavelengths react differently when they encounter the earth and its atmosphere.

Solar energy approaching the earth encounters **filters**, reflectors, and absorbers in the form of atmospheric gases, clouds, and the earth's surface. Atmospheric gases filter incoming energy, selectively blocking some wavelengths and allowing other wavelengths to pass through. Blocked wavelengths are either absorbed and heat the air or scattered and reflected back into space. Clouds, composed of atmospheric water vapor, likewise reflect or absorb energy but allow some wavelengths to pass through. Some energy reaching the earth's surface is reflected; a great deal is absorbed in heating the ground, evaporating water, and conducting **photosynthesis**. Most energy that the earth absorbs is re-emitted in the form of short, infrared wavelengths, which are sensed as heat. Some of this heat energy circulates in the atmosphere for a time, but eventually it all escapes. If this heat did not escape, the earth would overheat and become uninhabitable.

Variables in the climate system

Climate responds to conditions of the earth's energy filters, reflectors, and absorbers. As long as the atmosphere's filtering effect remains constant, the earth's reflective and absorptive capacities do not change, and the amount of incoming energy does not vary, climate conditions should stay constant. Most of the time, though, some or all of these elements fluctuate. The earth's reflectivity changes as the shapes, surface features, and locations of continents change. The atmosphere's composition changes from time to time, so that different wavelengths are reflected or pass through. The amount of energy the earth receives also shifts over time.

During the course of a decade the rate of solar energy input varies by a few watts per square meter. Changes in energy input can be much greater over several millennia. Energy intensity also varies with the shape of the earth's orbit around the sun. In a period of 100 million years the earth's elliptical orbit becomes longer and narrower, bringing the earth closer to the sun at certain times of year, then rounder again, putting the earth at a more uniform distance from the sun. When the earth receives relatively intense energy, heating and evaporation increase. Extreme heating can set up exaggerated convection currents in the atmosphere, with extreme low pressure areas receiving intensified rains and high pressure areas experiencing extreme **drought**.

The earth's albedo depends upon surface conditions. Extensive dark forests absorb a great deal of energy in heating, evaporation of water, and photosynthesis. Light, colored surfaces, such as **desert** or snow, tend to absorb less energy and reflect more. If highly reflective continents are large or are located near the equator, where energy input is great, then they could reflect a great deal of energy back into the atmosphere and contribute to atmospheric heating. However, if those continents are heavily vegetated, their reflective capacity might be lowered.

Other features of terrestrial geography that can influence climate conditions are mountains and glaciers. Both rise and fall over time and can be high enough to interrupt wind and precipitation patterns. For instance, the growth of the Rocky Mountains probably disturbed the path of upper atmospheric winds known as the jet stream. In southern Asia, the Himalayas block humid air masses flowing from the south. Intense precipitation results on the windward side of these mountains, while the downwind side remains one of the driest areas on Erth.

Atmospheric composition is a climate variable that began to receive increased attention during the 1980s. Each type of gas molecule in the atmosphere absorbs a particular range of energy wavelengths. As the mix of gases changes, the range of wavelengths passing through the filter shifts. For instance, the gas **ozone** (O_3) selectively blocks long wave UV radiation. A drop in upper atmospheric ozone levels discovered in the late 1980s has caused alarm because harmful UV rays are no longer being intercepted as effectively before they reach the earth's surface. Water vapor and solid particulates (dust) in the upper atmosphere also block incoming energy. Atmospheric dust associated with ancient meteor impacts is widely thought responsible for climatic cooling that may have killed the earth's dinosaurs 65 million years ago. Climate cooling could occur today if bombs from

a nuclear war threw high levels of dust into the atmosphere. With enough radiation blockage, global temperatures could fall by several degrees, a scenario known as **nuclear winter**.

A human impact on climate that is more likely than nuclear winter is global warming caused by increased levels of **carbon dioxide** (CO_2) in the upper atmosphere. Most solar energy enters the atmospheric system as long wavelengths and is reflected back into space in the form of short wavelength (heat) energy. **Carbon** dioxide blocks these short, warm wavelengths as they leave the earth's surface. Unable to escape, this heat energy remains in the atmosphere and keeps the earth warm enough for life to continue. However, many studies suggest that the burning of **fossil fuels** and **biomass** have raised atmospheric carbon dioxide levels. Rising CO_2 levels could trap excessive amounts of heat and raise global air temperatures to dangerous levels. This scenario is popularly known as the **greenhouse effect**. Extreme amounts of trapped heat could disturb precipitation patterns. Ecosystems could overheat, killing plant and animal species. Polar ice caps could melt, raising global ocean levels and threatening human settlements.

Increased anthropogenic production of other gases such as **methane** (CH_4) also contributes to atmospheric warming, but carbon dioxide has been a focus of concern because it is emitted in much greater volume.

No one knows how seriously human activity may be affecting the large and turbulent patterns of climate. Sometimes a very subtle event can have magnified repercussions in larger wind, precipitation, and pressure systems, disturbing major climate patterns for decades. In many cases the climate appears to have a self-stabilizing capacity—an ability to initiate internal reactions to a destabilizing event that return it to equilibrium. For example, extreme greenhouse heating should cause increased evaporation of water. Resulting clouds could block incoming sunlight, producing an overall cooling effect to counteract heating.

Furthermore, human influences work on climate within a context of continually changing natural conditions and events. On a geologic time scale, temperatures, precipitation, and ocean levels have fluctuated enormously. A long series of ice ages and warmer interglacial periods began 2.5 million years ago and may still be going on. The last glacial maximum, with low sea levels because of extreme ice volumes, ended only 18,000 years ago—an instant in the earth's climate history.

Natural fluctuations occur on a more human time scale, as well. A summer of extreme drought and high temperatures in the United States in 1988 brought threats of global warming to the public's attention, but the drought itself resulted from a temporary aberration in high altitude wind patterns that centered an unusually stable high pressure zone over the Midwest. This temporary departure from normal conditions was simply part of the continual fluctuation within the chaotic climate system. Terrestrial events, such as the 1991 eruption of **Mount Pinatubo** in the Philippines, also cause large, natural disturbances in climate. Dust from its eruption reached the upper atmosphere and was distributed around the globe, blocking enough incoming solar radiation to temporarily cool global temperatures by about 1.8°F (1°C).

No one can yet predict with **precision** how climate variables will respond to human activity. The earth's climate is so complex that human alterations to the atmosphere (such as those caused by carbon dioxide **emission**) amount to an "experiment" having an unknown—and possibly life-threatening—outcome.

[*Mary Ann Cunningham Ph.D.*]

RESOURCES
BOOKS

Henderson-Sellers, A., and P. J. Robinson. *Contemporary Climatology*. London: Longman Scientific and Technical, 1986.

PERIODICALS

Ingersoll, A. P. "The Atmosphere." *Scientific American* 249 (1983): 162-74.
Schneider, S. H. "Climate Modelling." *Scientific American* 256 (1987): 72-80.

Climax (ecological)

Referring to a community of plants and animals that is relatively stable in its **species** composition and **biomass**, ecological climax is the apparent termination of directional succession—the replacement of one community by another. That the termination is only apparent means that the climax may be altered by periodic disturbances such as **drought** or stochastic disturbances such as volcanic eruptions. It may also change extremely slowly owing to the gradual immigration and emigration—at widely differing rates—of individual species, for instance following the retreat of ice sheets during the postglacial period. Often the climax is a shifting mosaic of different stages of **succession** in a more or less steady state overall, as in many climax communities that are subject to frequent fires. Species that occur in climax communities are mostly good competitors and tolerant of the effects (e.g., shade, root **competition**) of the species around them, in contrast to the opportunistic colonists of early successional communities. The latter are often particularly adapted for wide dispersal and abundant reproduction, leading to success in newly opened habitats where competition is not severe.

In a climax community, productivity is in approximate balance with **decomposition**. Biogeochemical cycling of inorganic nutrients is also in balance, so that the stock of **nitrogen, phosphorus,** calcium, etc., is in a more or less steady state.

Frederic E. Clements was the person largely responsible in the early twentieth century for developing the theory of the climax community. Clements regarded **climate** as the predominant determining factor, though he did recognize that other factors—for instance, fire—could prevent the establishment of the theoretical "climatic climax." Later ecologists placed more stress on interactions among several determining factors, including climate, **soil** parent material, **topography,** fire, and the **flora** and **fauna** able to colonize a given site.

[*Eville Gorham Ph.D.*]

RESOURCES
BOOKS

Hagen, J. B. *An Entangled Bank: The Origins of Ecosystem Ecology*. New Brunswick, NJ: Rutgers University Press, 1992.

Clod

A compact, coherent mass of **soil** varying in size from 0.39–9.75 in (10–250 mm). Clods are produced by operations like plowing, cultivation, and digging, especially on soils that are too wet or too dry. They are usually formed by compression, or by breaking off from a larger unit. Tractor attachments like disks, spike-tooth harrows, and rollers are used to break up clods during seedbed preparation.

Cloning

Cloning hit the news headlines in 1997 when scientists in Scotland announced they had successfully cloned a sheep, named Dolly, in 1996. Although several other animal **species** had been cloned in the previous 20 years, it was Dolly that caught the public's attention. Suddenly, the possibility that humans might soon be cloned jumped from the pages of science fiction stories into the mainstream press. Dolly was the first adult mammal ever cloned.

Cloning is the science of using artificial methods to create clones. A clone is a single cell, a group of cells, or an organism produced in a laboratory without sexual reproduction. In effect, the clone is an exact genetic copy of the original source, much like identical twins. There are two types of cloning. Blastomere separation, also called "twinning" after the naturally occurring process that creates identical twins, involves splitting a developing embryo soon after the egg is fertilized by sperm. The result is identical twins with DNA from both parents. The second cloning type, called nuclear transfer, is what scientists used to create Dolly.

In cloning Dolly, scientists transferred genetic material from an adult female sheep to an egg in which the nucleus containing its genetic material had been removed.

Simple methods of cloning plants, such as grafting and stem cutting, have been used for more than 2,000 years. The modern era of laboratory cloning began in 1958 when the English-American plant physiologist Frederick C. Steward cloned carrot plants from mature single cells placed in a **nutrient** culture containing hormones, **chemicals** that play various and significant roles in the body.

The first cloning of animal cells occurred in 1964. In the first step of the experiment, biologist John B. Gurdon destroyed with ultraviolet light the genetic information stored in a group of unfertilized toad eggs. He then removed the nuclei (the part of an animal cell that contains the genes) from intestinal cells of toad tadpoles and injected them into those eggs. When the eggs were incubated (placed in an **environment** that promotes growth and development), Gurdon found that 1–2% of the eggs developed into fertile, adult toads.

The first successful cloning of mammals was achieved nearly 20 years later. Scientists in both Switzerland and the United States successfully cloned mice using a method similar to that of Gurdon. However, the Swiss and American methods required one extra step. After the nuclei were taken from the embryos of one type of mouse, they were transferred into the embryos of another type of mouse. The second type of mouse served as a substitute mother that went through the birthing process to create the cloned mice. The cloning of cattle livestock was achieved in 1988 when embryos from cows were transplanted to unfertilized cow eggs whose own nuclei had been removed.

Since Dolly, the pace and scope of cloning mammals has greatly intensified. In February 2002, scientists at Texas A&M University announced they had cloned a cat, the first cloning of a common domestic pet. Named "CC" (for **carbon** copy or copycat), the cat is an exact genetic duplicate of a two–year–old calico cat. Scientists cloned CC in December 2001 using the nuclear transfer method. In April 2002, a team of French scientists announced they had cloned rabbits using the nuclear transfer process. Out of hundreds of embryos used in the experiment, six rabbits were produced, four that developed normally and two that died. Two of the cloned rabbits mated naturally and produced separate litters of seven and eight babies

The first human embryos were cloned in 1993 using the blastomere technique that placed individual embryonic cells (blastomeres) in a nutrient culture where the cells then divided into 48 new embryos. These experiments were conducted as part of some studies on in vitro (out of the body) fertilization aimed at developing fertilized eggs in test tubes that could then be implanted into the wombs of women having difficulty becoming pregnant. However, these fertilized eggs did not develop to a stage that was suitable for transplantation into a human uterus.

Research into cloning humans also picked up greatly following the success of Dolly. An Italian physician said in April 2002 that a woman was pregnant with what would be the world's first cloned human baby. The doctor, Severino Antinori, operates a fertility clinic near the Vatican in Rome. In March 2002, a Chinese researcher said she had cloned a human embryo to the blastocyst stage, the point at which stem cells can be harvested. Scientists in several other countries also are believed conducting human cloning experiments.

The cloning of cells promises to produce many benefits in farming, medicine, and basic research. In farming, the goal is to clone plants that contain specific traits that make them superior to naturally occurring plants. For example, field tests have been conducted using clones of plants whose genes have been altered in the laboratory by **genetic engineering** to produce **resistance** to insects, viruses, and bacteria. New strains of plants resulting from the cloning of specific traits have led to fruits and vegetables with improved nutritional qualities, longer shelf lives, and new strains of plants that can grow in poor **soil** or even under water.

A cloning technique known as twinning could induce livestock to give birth to twins or even triplets, thus reducing the amount of feed needed to produce meat. Cloning also holds promise for saving certain rare breeds of animals from **extinction**, such as the **giant panda**.

In medicine, gene cloning has been used to produce vaccines and hormones. Cloning techniques have already led to the inexpensive production of the hormone insulin for treating diabetes and of growth hormones for children who do not produce enough hormones for normal growth. The use of monoclonal antibodies in disease treatment and research involves combining two different kinds of cells (such as mouse and human **cancer** cells) to produce large quantities of specific antibodies. These antibodies are produced by the immune system to fight off disease. When injected into the blood stream, the cloned antibodies seek out and attack disease–causing cells anywhere in the body.

Despite the benefits of cloning and its many promising avenues of research, certain moral, religious, and ethical questions concerning the possible abuse of cloning have been raised. At the heart of these questions is the idea of humans tampering with life in a way that could harm society, either morally or in a real physical sense. Some people object to cloning because it allows scientists to "act like God" in manipulating living organisms.

The cloning of Dolly and the fact that some scientists are attempting to clone humans raised the debate over this practice to an entirely new level. A person could choose to make two or 10 or 100 copies of himself or herself by the same techniques used with Dolly. This realization has stirred an active debate about the morality of cloning humans. Some people see benefits from the practice, such as providing a way for parents to produce a new child to replace one dying of a terminal disease. Other people worry about humans taking into their own hands the future of the human race.

Another controversial aspect of cloning deals not with the future but the past. Could Abraham Lincoln or Albert Einstein be recreated using DNA from a bone, hair, or tissue sample? If so, perplexing questions arise about whether this is morally or ethically acceptable? Some scientists say that while it might be possible to do this, the clone might be identical in appearance and in some traits, it would not have the same personality as the original Lincoln. This is because Lincoln, like all people, was greatly shaped from birth by his environment and personal experiences in addition to his genetic coding. Although a duplicate of her mother, CC, the cloned calico cat, has a different color pattern on her fur. This is because environmental factors strongly influence her development in the womb.

Also, since the movie "Jurassic Park" was released in 1993, there has been considerable public discussion about the possibility of cloning dinosaurs and other prehistoric or extinct species. In 1999, the Australian Museum in Sydney, **Australia**, announced scientists were attempting to clone a thylacine (a meat–eating marsupial related to kangaroos and opossums). It has been extinct since 1932 but the museum has the body of a baby thylacine that has been preserved for 136 years. The problem is that today's cloning techniques are possible only with living tissue. Even the head of the project has doubts, saying the chance of cloning a living thylacine is 30% over the next 200 years.

[*Ken R. Wells*]

RESOURCES
BOOKS

Cefrey, Holly. *Cloning and Genetic Engineering (Life in the Future)*. New York: Children's Press, 2002.

Pence, Gregory E. *Who's Afraid of Human Cloning?* Lanham, MD: Rowman & Littlefield Publishers, 1998.

PERIODICALS

Gibbs, Nancy. "Baby, It–s You! And You, And You..." *Time*(Feb. 11, 2001).

Hobson, Katherine. "Pets of the Future." *U.S. News & World Report* (March 11, 2002): p. 46.

Masibay, Kim Y. "Copy Cat." *Science World* (March 25, 2002): p. 6–7.

McGovern, Celeste. "Brave New World." *The Report Newsmagazine* (April 29, 2002).

Pistoi, Sergio. "Father of the Impossible Children." *Scientific American* (April 2002): p. 38–40.

"The Clone Wars.' *Business Week* (March 25, 2002): p. 94.

Dolly, the first genetically reproduced sheep.
(Photograph by Jeff Mitchell. Archive Photos. Reproduced by permission.)

Weidensaul, Scott. "Raising the Dead." *Audubon* (May–June 2002): p. 58–67.

ORGANIZATIONS

The Human Cloning Foundation, <http://www.humancloning.org>

Society for Developmental Biology, 9650 Rockville Pike, Bethesda, MD USA 20814 301–571–0647, Fax: 301–571–5704, Email: ichow@faseb.org, <http://www.sdb.bio.purdue.edu>

Cloud chemistry

One of the exciting new fields of chemical research in the past half century involves chemical changes that take place in the **atmosphere**. Scientists have learned that a number of reactions are taking place in the atmosphere at all times. For example, oxygen (O_2) molecules in the upper **stratosphere** absorb **solar energy** and are converted to **ozone** (O_3). This ozone forms a layer that protects life on Earth by filtering out the harmful **ultraviolet radiation** in sunlight. **Chlorofluorocarbons** and other chlorinated solvents (e.g., **carbon** tetrachloride and methyl chloroform) generated by human activities also trigger chemical reactions in the upper atmosphere including the break up of ozone into the two-atom form of oxygen. This reaction depletes the earth's protective ozone layer.

Clouds are often an important locus for atmospheric chemical reactions. They provide an abundant supply of water molecules that act as the solvent required for many reactions. An example is the reaction between **carbon dioxide** and water, resulting in the formation of carbonic **acid**. The abundance of both carbon dioxide and water in the atmosphere means that natural rain will frequently be somewhat acidic. Although conditions vary from time to time and place to place, the **pH** of natural, unpolluted rain is normally about 5.6. (The pH of pure water is 7.0). Other naturally occurring components of the atmosphere also react with water in clouds. In regions of volcanic activity, for example, **sulfur dioxide** released by outgassing and eruptions is oxidized to sulfur trioxide, which then reacts with water to form sulfuric acid.

The water of which clouds are composed also acts as solvent for a number of other chemical **species** blown into the atmosphere from the earth's surface. Among the most common ions found in solution in clouds are sodium (Na^+), magnesium (Mg^{2+}), chloride (Cl^-), and sulfate (SO_4^{2-}) from sea spray; potassium (K^+), calcium (Ca^{2+}), and carbonate (CO_3^{2-}) from **soil** dust; and ammonium (NH_4^+) from organic decay.

The nature of cloud chemistry is often changed as a result of human activities. Perhaps the best known and most thoroughly studied example of this involves **acid rain**. When **fossil fuels** are burned, sulfur dioxide and **nitrogen oxides** (among other products) are released into the atmosphere. Prevailing winds often carry these products for hundreds or thousands of miles from their original source. Once deposited in the atmosphere, these oxides tend to be absorbed by water molecules and undergo a series of reactions by which they are converted to acids. Once formed in clouds by these reactions, sulfuric and nitric acids remain in solution in water droplets and are carried to earth as fog, rain, snow, or other forms of precipitation.

[*David E. Newton*]

RESOURCES
BOOKS

Harrison, R. M., ed. *Pollution: Causes, Effects, and Control.* Cambridge: Royal Society of Chemistry, 1990.

Club of Rome

In April of 1968, 30 people, including scientists, educators, economists, humanists, industrialists, and government officials, met at the Academia dei Lincei in Rome. The meeting was called by Dr. Aurelio Peccei, an Italian industrialist and economist. The purpose of this meeting was to discuss "the present and future predicament of man." The "Club of Rome" was born from this meeting as an informal organization that has been described as an "invisible college." Its purpose, as described by Donella Meadows, is to foster understanding of the varied but interdependent components—economic, political, natural and social—that make up the global system in which we all live; to bring that new understanding to the attention of policy-makers and the public worldwide; and in this way to promote new policy initiatives and action. The original list of members is listed in the preface to Meadows's book entitled *The Limits to Growth*, in which the basic findings of the group are eloquently explained.

This text is a modern-day equivalent to the hypothesis of Thomas Malthus, who postulated that since increases in food supply cannot keep pace with geometric increases in human population, there would therefore be a time of **famine** with a stabilization of the human population. This eighteenth century prediction has, to a great extent, been delayed by the "green revolution" in which agricultural production has been radically increased by the use of fertilizers and development of special genetic strains of agricultural products. The high cost of **agricultural chemicals** which are generally tied to the price of oil has, however, severely limited the capability of developing nations to purchase them.

The development of the Club of Rome's studies is most potently presented by Meadows in the form of graphs which plot on a time axis the supply of **arable land** needed at several production levels (present, double present, quadruple present, etc.) to feed the world's population based upon growth models.

She states that 7.9 billion acres (3.2 billion ha) of land are potentially suitable for agriculture on the earth; half of that land, the richest and most accessible half, is under cultivation today. She further states that the remaining land will require immense capital inputs to reach, clear, irrigate, or fertilize before it is ready to produce food. One can imagine the impact such conversion will have on the **environment**.

The Club of Rome's studies were not limited to food supply but also considered industrial output per capita, **pollution** per capita, and general resources available per capita. The key issue is that the denominator, per capita, keeps increasing with time, requiring ever more frugal and careful use of the resources; however, no matter how carefully the resources are husbanded, the inevitable result of uncontrolled **population growth** is a catastrophe which can only be delayed. Therefore stabilizing the rate of world population growth must be a continuing priority.

As a follow-up to the Club of Rome's original meeting, a global model for growth was developed by Jay Forrester of the Massachusetts Institute of Technology. This model is capable of update with insertion of information on population, agricultural production, **natural resources**, industrial production, and pollution. Meadows's report *The Limits to Growth* represents a readable summary of the results of this **modeling**.

A new branch of the Club of Rome is the TT30, a group of people around the age of 30 who form a "think tank." This group is primarily concerned with problems of today, future issues, and how to deal with them.

[*Malcolm T. Hepworth*]

RESOURCES

BOOKS

Dror, Yehezkel. *The Capacity to Govern: A Report to the Club of Rome.* Frank Cass & Co, 2001.

Forrester, J. W. *World Dynamics.* Cambridge, MA: Wright-Allen Press, 1971.

Meadows, D. H., et al. *The Limits to Growth: A Report for the Club of Rome's Project on the Predicament of Mankind.* New York: Universe Books, 1974.

ORGANIZATIONS

The Club of Rome, Rissener Landstr 193, Hamburg, Germany 22559 +49 40 81960714, Fax: +49 40 81960715, Email: mail@clubofrome.org, <http://www.clubofrome.org>

C:N ratio

Organic materials are composed of a mixture of carbohydrates, lignins, tannins, fats, oils, waxes, resins, proteins, minerals, and other assorted compounds. With the exception of the mineral fraction, the organic compounds are composed of varying ratios of **carbon** and **nitrogen**. This is commonly abbreviated to the C:N ratio. Carbohydrates are composed of carbon, **hydrogen**, and oxygen and are relatively easily decomposed to **carbon dioxide** and water, plus a small amount of other by-products. Protein-like materials are the prime source of nitrogen compounds as well as sources of carbon, hydrogen, and oxygen and are important to the development of the C:N ratio and the eventual **decomposition** rate of the organic materials.

The **aerobic** heterotrophic bacteria are primarily responsible for the decay of the large amount of organic compounds generated on the earth's surface. These organisms typically have a C:N ratio of about 8:1. When organic residues are attacked by the bacteria under appropriate **habitat** conditions, some of the carbon and nitrogen are assimilated into the new and rapidly increasing microbial population, and copious amounts of carbon dioxide are released to the **atmosphere**. The numbers of bacteria are highly controlled by the C:N ratio of the organic substrate.

As a rule, when organic residues of less than 30:1 ratio are added to a **soil**, there is very little noticeable decrease in the amount of mineral nitrogen available for higher plant

forms. However as the C:N ratio begins to rise to values of greater than 30:1, there may be **competition** for the mineral nitrogen forms. Bacteria are lower in the **food chain/web** and become the immediate beneficiary of available sources of mineral nitrogen, while the higher **species** may suffer a lack of mineral nitrogen. Ultimately, when the carbon source is depleted, the organic nitrogen is released from the decaying **microbes** as mineral nitrogen.

The variation in the carbon content of organic material is reflected in the constituency of the compound. Carbohydrates usually contain less than 45% carbon, while lignin may contain more than 60% carbon. The C:N ratio of plant material may well reflect the kind and stage of growth of the plant. A young plant typically contains more carbohydrates and less lignin, while an older plant of the same species will contain more lignin and less carbohydrate. Ligneous tissue such as found in trees may have a C:N ratio of up to 1000:1.

The relative importance of the C:N ratio addresses two concerns: one, the rate of the organic matter decay to the low C:N ratio of **humus**, (approximately 10:1), and secondly the immediate availability of mineral nitrogen (NH_4^+) to meet the demand of higher plant needs. The addition of mineral nitrogen to organic residues is a common practice to enhance the rate of decay and to reduce the potential for nitrogen deficiency developing in higher plants where copious amounts of organic residue which has a C:N ratio of greater than 30:1 have been added to the soil.

Composting of organic residues permits the breakdown of the residues to occur without competition the of higher plants for the mineral nitrogen and also reduces the C:N ratio of the resulting mass to a C:N value of less than 20:1. When this material is added to a soil, there is little concern about the potential for nitrogen competition between the micro-organisms and the higher plants.

[*Royce Lambert*]

RESOURCES

BOOKS

Brady, N. C. "Soil Organic Matter and Organic Soils." In *The Nature and Properties of Soils.* 10th ed. New York: Macmillan, 1990.

Millar, R. W., and R. L. Donahue. "Organic Matter and Container Media." In *Soils: An Introduction to Plant Growth.* 6th ed. Englewood Cliffs, NJ: Prentice Hall, 1990.

Coagulation
see **Water treatment**

Coal Production From 1950–2000				
Date	**Anthricite**	**Bituminous**	**Lignite**	**Subbituminous**
1950	44.08	516.31	0.00	0.00
1960	18.82	415.51	0.00	0.00
1970	9.73	578.47	8.04	16.42
1980	6.06	628.77	47.16	147.72
1990	3.51	693.21	88.09	244.27
2000	4.51	548.47	88.74	433.78

Coal production from 1950 through 2000. MMst stands for million short tons.

Coal

Consisting of altered remains of plants, coal is a widely used fossil fuel. Generally, the older the coal, the higher the **carbon** content and heating value. Anthracite coal ranks highest in carbon content, then bituminous coal, subbituminous coal, and lignite (as determined by the American Society for Testing Materials). Over 80% of the world's vast reserves occur in the former Soviet Union, the United States, and China. Though globally abundant, it is associated with many environmental problems, including **acid drainage**, degraded land, sulfur oxide emissions, **acid rain**, and heavy **carbon dioxide** emissions. However, clean coal-burning technologies, including liquified or gasified forms, are now available.

Anthracite, or "hard" coal, differs from the less altered bituminous coal by having more than 86% carbon and less than 14% volatile matter. It was formerly the fuel of choice for heat purposes because of high **Btu** (British Thermal Unit) values, minimally 14,500, and low ash content. In the United States, production has dropped from 100 million tons in 1917 to about seven million tons as anthracite has been replaced by oil, **natural gas**, and electric heat. Predomi-

nantly in eastern Pennsylvania's Ridge and Valley Province, anthracite seams have a wavelike pattern, complicating extraction. High water tables and low demand are the main impediments to expansion.

Bituminous coal, or "soft" coal, is much more abundant and easier to mine than anthracite but has lower carbon content and Btu values and higher volatility. Historically dominant, it energized the Industrial Revolution, fueling steam engines in factories, locomotives, and ships. Major coal regions became steel centers because two tons of coal were needed to produce each ton of iron ore. This is the only coal suitable for making coke, needed in iron smelting processes. Major deposits include the Appalachian Mountains and the Central Plains from Indiana through Oklahoma.

Subbituminous coal ranges in Btu values from 10,500 (11,500 if agglomerating) down to 8,300. Huge deposits exist in Wyoming, Montana, and North Dakota with seams 70 ft (21.4 m) thick. Though distant from major utility markets, it is used extensively for electrical power generation and is preferred because of its abundance, low sulfur content, and good grinding qualities. The latter makes it more useful than the higher grade, but harder, bituminous coal because modern plants spray the coal into **combustion** chambers in powder form. Demand for this coal skyrocketed following the 1973 OPEC **oil embargo** and subsequent restrictions on natural gas use in new plants.

Lignite, or "brown" coal, is the most abundant, youngest, and least mature of the coals, with some plant texture still visible. Its Btu values generally range below 8,300. Although over 70% of the deposits are found in North America, mainly in the Rocky Mountain region, there is little production there. It is used extensively in many eastern European countries for heating and steam production. Russian scientists have successfully burned lignite *in situ*, tapping the resultant coal gas for industrial heating. If concerns over global warming are satisfied, future liquefying and gasifying technologies could make lignite a prized resource.

[*Nathan H. Meleen*]

RESOURCES
BOOKS

Hartshorne, T., and J. W. Alexander. *Economic Geography*. 3rd ed. Englewood Cliffs, NJ: Prentice-Hall, 1988.

PERIODICALS

Young, G. "Will Coal Be Tomorrow's Black Gold?" *National Geographic* 148 (August 1975): 234-259.

Coal bed methane

Coal bed **methane** (CBM) is a **natural gas** contained in coal seams. Methane gas is formed during coalification, the transformation of plant material into coal. Also called coal seam gas, CBM is usually not found in the **atmosphere** until it is released during coal mining. The gas released during mining is called coal mine methane (CMM).

According to the United States **Environmental Protection Agency**, 88 billion cubic feet of CMM can provide electricity to more than 1.2 million homes for a year. CBM is another source of methane, which is also used to heat buildings.

At the start of the twenty-first century, CBM drilling was underway in states such as New Mexico, Colorado, Wyoming, and Montana. The gas is often located in aquifers, porous rock that contains water. Gas is obtained by pumping thousands of gallons of water out of a CBM well.

This process could negatively affect the **environment** because **groundwater** frequently contains salt and other minerals. Removed water is reinjected into the ground in New Mexico and Colorado. Reinjection is not always possible, which brings more environmental concerns. In 2002, Mon-

tana residents worried that groundwater released into rivers and streams could hurt crops, **wildlife**, and drinking water.

[*Liz Swain*]

Coal gasification

The term **coal** gasification refers to any process by which coal is converted into some gaseous form that can then be burned as a fuel. Coal gasification technology was relatively well known before World War II, but it fell out of favor after the war because of the low cost of oil and **natural gas**. Beginning in the 1970s, utilities showed renewed interest in coal gasification technologies as a way of meeting more stringent environmental requirements.

Traditionally, the use of **fossil fuels** in **power plants** and industrial processes has been fairly straight-forward. The fuel—coal, oil, or natural gas—is burned in a furnace and the heat produced is used to run a turbine or operate some industrial process. The problem is that such direct use of fuels results in the massive release of oxides of **carbon**, sulfur, and **nitrogen**, of unburned **hydrocarbons**, of **particulate** matter, and of other pollutants. In a more environmentally-conscious world, such reactions are no longer acceptable.

This problem became much more severe with the shift from oil to coal as the fuel of choice in power generating and industrial plants. Coal is "dirtier" than both oil and natural gas and its use, therefore, creates more serious and more extensive environmental problems.

The first response of utilities and industries to new **air pollution** standards was to develop methods of capturing pollutants after **combustion** has occurred. **Flue gas** desulfurization systems, called **scrubbers**, were one approach strongly favored by the United States government. But such systems are very expensive, and utilities and industries rapidly began to explore alternative approaches in which coal is cleansed of material that produce pollutants when burned. One of the most promising of these clean-coal technologies is coal gasification.

A variety of methods are available for achieving coal gasification, but they all have certain features in common. In the first stages, coal is prepared for the reactor by crushing and drying it and then pre-treating it to prevent caking. The pulverized coal is then fed into a boiler where it reacts with a hot stream of air or oxygen and steam. In the boiler, a complex set of chemical reactions occur, some of which are exothermic (heat releasing) and some of which are endothermic (heat-absorbing).

An example of an exothermic reaction is the following.

$$2 C + {}^3/_2 O_2 \rightarrow CO_2 + CO$$

The **carbon monoxide** produced in this reaction may then go on to react with **hydrogen** released from the coal to produce a second exothermic reaction.

$$CO + 3 H_2 \rightarrow CH_4 + H_2O$$

The energy released by one or both of the reactions is then available to initiate a third reaction that is endothermic.

$$C + H_2O \rightarrow > CO + H_2$$

Finally, the mixture of gases resulting from reactions such as these, a mixture consisting most importantly of carbon monoxide, **methane**, and hydrogen, is used as fuel in a boiler that produces steam to run a turbine and a generator.

In practice, the whole series of exothermic and endothermic reactions are allowed to occur within the same vessel, so that coal, air or oxygen, and steam enter through one inlet in the boiler, coal enters at a second inlet, and the gaseous fuel is removed through an outlet pipe.

One of the popular designs for a coal gasification reaction vessel is the Lurgi pressure gasifier. In the Lurgi gasifier, coal enters through the top of a large cylindrical tank. Steam and oxygen are pumped in from the bottom of the tank. Coal is burned in the upper portion of the tank at relatively low temperatures, initiating the exothermic reactions described above. As unburned coal flows downward in the tank, heat released by these exothermic reactions raises the temperature in the tank and brings about the endothermic reaction in which carbon monoxide and hydrogen are produced. These gases are then drawn off from the top of the Lurgi gasifier.

The exact composition of the gases produced is determined by the materials introduced into the tank and the temperature and pressure at which the boiler is maintained. One possible product, chemical synthesis gas, consists of carbon monoxide and hydrogen. It is used primarily by the chemical industry in the production of other **chemicals** such as ammonia and methyl alcohol. A second possible product is medium-Btu gas, made up of hydrogen and carbon monoxide. Medium-Btu gas is used as a general purpose fuel for utilities and industrial plants. A third possible product is substitute natural gas, consisting essentially of methane. Substitute natural gas is generally used as just that, a substitute for natural gas.

Coal gasification makes possible the removal of pollutants before the gaseous products are burned by a utility or industrial plant. Any ash produced during gasification, for example, remains within the boiler, where it settles to the bottom of the tank, is collected, and then removed. **Sulfur dioxide** and **carbon dioxide** are both removed in a much

smaller, less expensive version of the scrubbers used in smokestacks.

Perhaps the most successful test of coal gasification technology has been going on at the Cool Water Integrated Gasification Combined Cycle plant near Barstow, California. The plant has been operating since June 1984 and is now capable of generating 100 megawatts of electricity. The four basic elements of the plant are a gasifier in which combustible gases are produced, a particulate and sulfur removal system, a combustion turbine in which the synthetic gas is burned, and a stem turbine run by heat from the combustion turbine and the gasifier.

The Cool Water plant has been an unqualified environmental success. It has easily met federal and state standards for **effluent** sulfur dioxide, oxides of nitrogen, and particulates, and its solid wastes have been found to be non-hazardous by the California Department of Health.

Waste products from the removal system also have commercial value. Sulfur obtained from the reduction of sulfur dioxide is 99.9% pure and has been selling for about $100 a ton. Studies are also being made to determine the possible use of slag for road construction and other building purposes.

Coal gasification appears to be a promising energy technology for the twenty-first century. One of the intriguing possibilities is to use sewage or hazardous wastes in the primary boiler. In the latter case, hazardous elements could be fixed in the slag drawn off from the bottom of the boiler, preventing their contaminating the final gaseous product.

The major impediment in the introduction of coal gasification technologies on a widespread basis is their cost. At the present time, a plant operated with synthetic gas from a coal gasification unit is about three times as expensive as a comparable plant using natural gas. Further research is obviously needed to make this new technology economically competitive with more traditional technologies.

Another problem is that most coal gasification technologies require very large quantities of water. This can be an especially difficult problem since gasification plants should be built near mines to reduce shipping costs. But most mines are located in Western states, where water supplies are usually very limited.

Finally, coal gasification is an inherently less efficient process than the direct combustion of coal. In most approaches, between 30 and 40% of the heat energy stored in coal is lost during its conversion to synthetic gas. Such conversions would probably be considered totally unacceptable except for the favorable environmental trade-offs they provide.

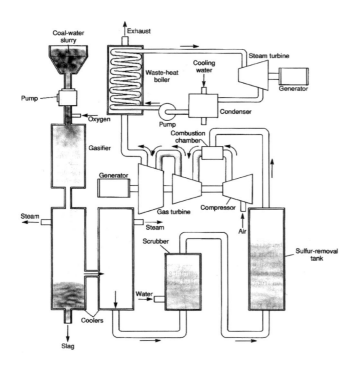

A schematic of coal gasification by heating a coal-and-water slurry in a low-oxygen environment. (McGraw-Hill Inc. Reproduced by permission.)

ogy, **U.S. Department of Energy**, Texaco, Shell, Westinghouse, and Exxon are all studying modifications in the basic coal gasification system to find ways of using a wide range of raw materials, to improve efficiency at various stages in the gasification process, and, in general, to reduce the cost of plant construction. *See also* Alternative fuels; Air pollution control; Air quality; Coal washing; Flue-gas scrubbing; Strip mining; Surface mining

[*David E. Newton*]

RESOURCES
PERIODICALS

Douglas, J. "Quickening the Pace in Clean Coal Technologies." *EPRI Journal* (January-February 1989): 12-15.

Coal mining

see **Black lung disease; Strip mining; Surface mining**

Coal washing

Coal that comes from a mine is a complex mixture of materials with a large variety of physical properties. In addition to the coal itself, pieces of rock, sand, and various minerals are contained in the mixture. Thus, before coal can be sold to consumers, it must be cleaned. The cleaning process consists of a number of steps that results in a product that is specifically suited to the needs of particular consumers. Among the earliest of these steps is crushing and sizing, two processes that reduce the coal to a form required by the consumer.

The next step in coal preparation is a washing or cleaning step. This step is necessary not only to meet consumer requirements, but also to ensure that its **combustion** will conform to environmental standards.

Coal washing is accomplished by one of two major processes, by *density separation* or by *froth flotation*. Both processes depend on the fact that the particles of which a coal sample are made have different densities. When water, for example, is added to the sample, particles sink to various depths depending on their densities. The various components of the sample can thus be separated from each other.

In some cases, a liquid other than water may be used to achieve this separation. In a heavy medium bath, for example, a mineral such as magnetite or feldspar in finely divided form may be mixed with water, forming a liquid medium whose density is significantly greater than that of pure water.

A number of devices and systems have been developed for extracting the various components of coal once they have

One possible solution to the problem described above is to carry out the gasification process directly in underground coal mines. In this process, coal would be loosened by explosives and then burned directly in the mine. The low-grade synthetic gas produced by this method could then be piped out of the ground, upgraded and used as a fuel. Underground gasification is an attractive alternative for many reasons. By some estimates, up to 80% of all coal reserves cannot be recovered through conventional mining techniques. They are either too deep underground or dispersed too thinly in the earth. The development of methods for gasification in coal seams would, therefore, greatly increase the amount of this fossil fuel available for our use.

A great deal of research is now being done to make coal gasification a more efficient process. A promising breakthrough involves the use of potassium hydroxide or potassium carbonate as a catalyst in the primary reactor vessel. The presence of a catalyst reduces the temperature at which gasification occurs and reduces, therefore, the cost of the operation.

Governments in the United States and Europe, energy research institutes, and major energy corporations are actively involved in research on coal gasification technologies. The Electric Power Research Institute, Institute of Gas Technol-

been separated with a water or heavy medium treatment. One of the oldest of these devices is the *jig*. In a jig, the column of water is maintained in a constant up-and-down movement by means of a flow of air. Clean coal particles are carried to the top of the jig by this motion, while heavier refuse particles sink to the bottom.

Another method of extraction, the *cyclone*, consists of a tank in which the working fluid (water or a heavy medium) is kept in a constant circular motion. The tank is constructed so that lighter clean coal particles are thrown out of one side, while heavier refuse particles are ejected through the bottom.

Shaking tables are another extraction method. As the table shakes back and forth, particles are separated by size, producing clean coal at one end and waste products at the other.

In cylindrical separators, a coal mixture is fed into a spinning column of air that throws the heavier waste particles outward. They coat the inner wall of the cylinder and fall to the bottom, where they are drawn off. The clean coal particles remain in the center of the air column and are drawn off at the top of the cylinder.

Froth **flotation** processes depend on the production of tiny air bubbles to which coal particles adhere. The amount of **absorption** onto a bubble depends not only on a particle's density, but also on certain surface characteristics. Separation of clean coal from waste materials can be achieved in froth flotation by varying factors, such as **pH** of the solution, time of treatment, particle size and shape, rate of **aeration**, solution density, and bubble size.

[*David E. Newton*]

RESOURCES
BOOKS

Ward, C. R. *Coal Geology and Coal Technology.* Melbourne, Australia: Blackwell Scientific Publications, 1984.

Coalition for Environmentally Responsible Economies (CERES)
see Valdez Principles

Coase theorem

An economic theorem that is sometimes used in discussions of external costs in environment-related situations. The standard welfare economic view states that in order to make the market efficient, external costs—such as **pollution** produced by a company in making a product—should be internalized by the company in the form of taxes or fees for producing

the pollution. Coase theorem, in contrast, states that the responsibility for the pollution should fall on both the producer and recipient of the pollution. For example, people who are harmed by the pollution can pay companies not to pollute, thereby protecting themselves from any potential harm.

Ronald Coase, the economist who proposed the theorem, further states that government should intervene when the bargaining process or transaction costs between the two parties is high. The government's role, therefore, is not to address external costs which harm bystanders but to help individuals organize for their protection.

Coastal Society, The

The Coastal Society (TCS), founded in 1975, is an international, nonprofit organization which serves as a forum for individuals concerned with problems related to coastal areas. Its members, drawn from university settings, government, and private industry, agree that the **conservation** of coastal resources demands serious attention and high priority.

TCS has four main goals: 1) to foster cooperation and communication among agencies, groups, and private citizens; 2) to promote conservation and intelligent use of coastal resources; 3) to strengthen the education and appreciation of coastal resources; and 4) to help government, industry, and individuals successfully balance development and protection along the world's coastlines. Through these goals, TCS hopes to educate the public and private sectors on the importance of effective coastal management programs and clear policy and law regarding the coasts.

Since its inception, TCS has sponsored numerous conferences and workshops. Individuals from various disciplines are invited to discuss different coastal problems. Past conferences have covered such topics as "Energy Across the Coastal Zone," "Resource Allocation Issues in the Coastal Environment," "The Present and Future of Coasts," and "Gambling with the Shore." Workshops are sponsored in conjunction with government agencies, universities, professional groups, and private organizations. Conference proceedings are subsequently published. TCS also publishes a quarterly magazine, *TCS Bulletin*, which features articles and news covering TCS affairs and the broader spectrum of coastal issues.

TCS representatives present congressional testimony on coastal management, conservation, and **water quality**. Recently the organization drafted a policy statement and it plans to take public positions on proposed policies affecting coastal issues.

[*Cathy M. Falk*]

RESOURCES

ORGANIZATIONS

The Coastal Society, P.O. Box 25408, Alexandria, VA USA 22313-5408 (703) 768-1599 , Fax: (703) 768-1598, Email: info@thecoastalsociety.org, <http://www.thecoastalsociety.org>

Coastal Zone Management Act (1972)

The Coastal Zone Management Act (CZMA) of 1972 established a federal program to help states in planning and managing the development and protection of coastal areas through the creation of a Coastal Zone Management Program (CZMP). The CZMA is primarily a planning act, rather than an environmental protection or regulatory act. Under its provisions, states can receive grants from the federal government to develop and implement coastal zone programs as long as the programs meet with federal approval. State participation in the program is voluntary, and the authority is focused in state governments. In 2002, 99.9% of the national shoreline and coastal waters were managed by state CZMPs.

In the 1960s, public concern began to focus on **dredging** and filling, industrial siting, offshore oil development, and second home developments in the coastal zone. The coastal zone law was developed in the context of increased development of marine and coastal areas, need for more coordinated and consistent governmental efforts, an increase in general environmental consciousness and public **recreation** demands, and a focus on **land-use control** nationally. In 1969, a report by the Commission on Marine Sciences, Engineering, and Resources (the Stratton Commission) recommended a federal grant program to the states to help them deal with coastal zone management. The Commission found that coastal areas were of prime national interest, but development was taking place without proper consideration of environmental and resource values.

During congressional debate over coastal zone legislation, support came primarily from marine scientists and affected state government officials. The opposition emanated from development and real estate interests and industry, who were also concerned with national **land use** bills. The major difference between House and Senate versions of the legislation that passed was which department would administer the program. At the executive level there was no debate: the Office of Coastal Zone Management (now the Office of Ocean and Coastal Resource Management, or OCRM), part of the National Ocean Service of NOAA, was placed in charge of the program. But Congressional opinion varied about the administrative oversight of the Act. The House favored the **U.S. Department of the Interior** (DOI); the Senate, the **National Oceanic and Atmospheric Adminis-**

tration (NOAA), part of the Department of Commerce (DOC). The Senate position was adopted in conference.

The congressional committees and executive branch agencies involved in coastal zone management also greatly varied. The Senate Commerce Committee was selected to have jurisdiction over the legislation. The House originally designated the Merchant Marine and Fisheries Committee as its legislative arm, but it was dissolved in the mid 1990s. In 2002, the current House committee is the Committee on Resources with the Subcommittee on Fisheries Conservation, Wildlife, and Oceans as its working division.

The CZMA declared that "there is a national interest in the effective management, beneficial use, protection, and development of the coastal zone." The purpose of the law is to further the "wise use of land and **water resources** for the coastal zone giving full consideration to ecological, cultural, historic and aesthetic values as well as to needs for economic development." The program is primarily a grant program, and the original 1972 Act authorized the spending of $186 million through 1977.

Under CZMA, the Secretary of Commerce was authorized to make grants to the states with coastal areas, including the **Great Lakes**, to help them develop the coastal zone management programs required by federal standards. The grants would pay for up to two-thirds of a state's program and could be received for no more than three years. In addition to these planning grants, the federal government could also make grants to the states for administering approved coastal zone plans. Again, the grants could not exceed two-thirds of the cost of the state program. With federal approval, the states could forward federal grant money to local governments or regional entities to carry out the act.

The federal government also has oversight responsibilities, to make sure that the states are following the approved plan and administering it properly. The key components of a state plan are to: (1) identify the boundaries of the coastal zone; (2) define the permissible uses in the coastal zone that have a significant effect; (3) inventory and designate areas of particular concern; (4) develop guidelines to prioritize use in particular areas; (5) develop a process for protection of beaches and public access to them; (6) develop a process for energy facility siting; and (7) develop a process to control shoreline **erosion**. The states have discretion in these stages. For instance, some states have opted for coastal zones very close to the water, others have drawn boundaries further inland. The states determine what uses are to be allowed in coastal zones. Developments in the coastal area must demonstrate coastal dependence.

Coastal zone management plans deal primarily with private lands, though the management of federal lands and federal activities within the coastal zone is required by the consistency provision of the CZMA which requires state approval and must be consistent with state legislation. Indeed, this was intended as a major incentive for states to participate in the process. Although federal agencies with management responsibility in coastal zones have input into the plans, this state-federal coordination proved to be a problem in the 1980s and 1990s, especially regarding off-shore oil development, due to differing interpretations of the consistency section of the CZMA.

At first, states were slow to develop plans and have them approved by federal authorities. This was due primarily to the political complexity of the interests involved in the process. The first three states to have their coastal zone management plans approved were California, Oregon, and Washington, which had their final plans approved by 1978. Both California and Washington had passed state legislation on coastal zone management prior to the federal law, California by referendum in 1972 and Washington in 1971. The California program is the most ambitious and comprehensive in the country. Its 1972 act established six regional coastal commissions with permit authority and a state coastal zone agency, which coordinated the program and oversaw the development of a state coastal plan. The California legislature passed a permanent coastal zone act in 1976 based on this plan. The permanent program stemmed from local plans reviewed by the regional commissions and state agency. Any development altering density or intensity of land use requires a permit from the local government, and sensitive coastal resource areas receive additional protection. As of May 2002, of the thirty-five eligible states and territories, thirty-three had approved coastal zone plans, one (Indiana) was in the planning process, and one (Illinois) had chosen not to participate in the program.

Three major issues arose during the state planning processes. Identifying areas of crucial environmental concern was a controversy that pitted environmentalists against developers in many states. In general, developers have proved more successful than environmentalists. A second issue is general development. States that have the most advanced coastal programs, such as California, use a permit system for development within the coastal zone. Environmental concerns and cumulative effects are often considered in these permit decisions. These permit programs often lead developers to alter plans before or during the application process. Such programs have generally served to improve development in coastal zones, and to protect these areas from major abuses.

The final issue, the siting of large scale facilities, especially energy facilities, has proven to be continually controversial as states and localities seek control over siting through their coastal zone plans, while energy companies appeal to the federal government regarding the national need for such facilities. In a number of court cases, the courts ruled that the states did have the power to block energy projects that were not consistent with their approved coastal management plans. This controversy spilled over into offshore oil development in waters in the outer continental shelf (OCS), which were under federal jurisdiction. These waters were often included in state coastal zone plans, many of which sought to prevent offshore oil development. In this case, the courts found in the 1984 ruling (*Secretary of the Interior vs. California*) that such development could proceed over state objections.

Major amendments to the CZMA were passed in 1976, 1980, 1990, and 1996. In 1976, the Coastal Energy Impact Fund was created to sponsor grants and loans to state and local governments for managing the problems of energy development. Other changes included an increase in the federal funding level from two-thirds to 80% of planning and administration, an increase in planning grant eligibility from three to four years, and the addition of planning requirements for energy facilities, shoreline erosion, and beach access.

The 1980 amendments re-authorized the program through 1985 and established new grant programs for revitalizing urban waterfronts and helping coastal cities deal with the effects of energy developments. The amendments also expanded the policies and objectives of the CZMA to include the protection of **natural resources**, the encouragement of states to protect coastal resources of national significance, and the reduction of state-federal conflicts in coastal zone policy.

Amendments to the CZMA in 1990 were included in the budget reconciliation bill. Most importantly, the amendments overturned the 1984 decision of *Secretary of the Interior vs. California*, giving states an increased voice regarding federal actions off their coasts. The law, which was strongly opposed by the Departments of Defense and Interior, gives the states the power to try to block or change federal actions affecting the coastal zones if these actions are inconsistent with adopted plans. The amendments also initiated a **nonpoint source** coastal **water pollution** grant and planning program, repealed the coastal energy impact program, and reauthorized the CZMA through 1995.

In June 2001, a federal district court judge ruled that the Department of the Interior (DOI) must ensure that any oil and gas leases it grants on the outer continental shelf off the coast of California be consistent with the State of California Coastal Management Program (CCMP). The decision requires the Minerals Management Service of the DOI to provide proof that 36 federally-owned oil and gas drilling leases comply with CCMP guidelines. The case is the first to uphold state rights in federal oil leasing activities granted in the 1990 CZMA amendments.

The 1996 reauthorization of the CZMA extended the Act through September 30, 1999. However, several environmental issues—including debate over funding for nonpoint **pollution** programs and lobbying by the oil and gas industry to give the states less control over federal projects such as offshore drilling leases— have delayed its further reauthorization. The continuous **resistance** by industry to curtail the rights of states or the rights of the Federal government to restrict oil and gas exploration and expansion has been in the forefront of the battle to reauthorize the CZMA. New admendments and even new legislation have been proposed that limit states from interfering in industry decisions in certain areas or limit the Federal government in others. It is clear from this contradictory legislation, that industry wants the consistency provision to only work when it is in the industry's interests.

The 106th Congress failed to vote on CZMA amendments entitled the Coastal Management Enhancement Act of 1999, but the legislation was reintroduced as the Coastal Zone Enhancement Reauthorization of 2001 by co-sponsoring Senators Olympia Snowe (R-Maine) and John Kerry (D-Mass.). As of May 2002, it had not yet been put to a Senate vote. *See also* Environmental law; Environmental policy; International Joint Commission; Marine pollution; National lakeshore; Water pollution

[*Christopher McGrory Klyza and Paula A Ford-Martin*]

RESOURCES
BOOKS

Beatley, Timothy, D.J. Brower, and A.K. Schwab.*An Introduction to Coastal Zone Management* 2nd ed. Washington, DC: Island Press, 2001.

PERIODICALS

Durkin, Tish. "Headed for a Beach? Maybe This Yarn's Not For You." *National Journal*(August 18, 2001): 2588.

"States Get More Say in Offshore Activity."*Congressional Quarterly Almanac* 46 (1990): 288–289.

OTHER

U.S. Code Online via GPO Access. "Coastal Zone Management Act of 1972." SEC. 302 [16 U.S.C. 1451].[cited May 27, 2002]. <http://www.access.gpo.gov/uscode/title16/chapter33_.html>

ORGANIZATIONS

Coastal Zone Management Program, OCRM, NOS, NOAA, Coastal Programs Division, N/ORM3; 1305 East-West Highway, SSMC4, Silver Spring, MD USA 20910 (301) 713-3155, Fax: (301) 713-4367, , http://www.ocrm.nos.noaa.gov/czm/

Co-composting

As a form of **waste management**, **composting** is the process whereby **organic waste** matter is microbiologically degraded under **aerobic** conditions to achieve significant volume reduction while also producing a stable, usable end product. Co-composting refers to composting two or more waste types in the same vessel or process, thus providing cost and space savings. The most common type of co-composting practiced by counties and townships in the United States involves mixing sewage **sludge** and **municipal solid waste** to speed the process and increase the usefulness of the end product. The processing and ultimate use or disposal of co-composting end products are regulated by federal and state environmental agencies.

Coevolution

Species are said to "coevolve" when their respective levels of fitness depend not only on their own genetic structure and adaptations but also the development of another species as well. The **gene pool** of one species creates selection pressure for another species. Although the changes are generally reciprocal, they may also be unilateral and still be considered coevolutionary.

The process of coevolution arises from interactions that establish structure in communities. A variety of different types of interactions can occur—symbiotic, where neither member suffers, or parasitic, predatory, and competitive relationships, where one member of a species pair suffers.

Coevolution can result from mutually positive selection pressure. For example, certain plants have in an evolutionary sense created positive situations for insects by providing valuable food sources for them. In return the insects provide a means to distribute pollen that is more efficient than the distribution of pollen by wind. Unfortunately, the plant and the insect species could evolve into a position of total dependency through increased specialization, thus enhancing the risk of **extinction** if either species declines.

Coevolution can also arise from negative pressures. Prey species will continually adapt defensive or evasive systems to avoid predation. Predators respond by developing mechanisms to surmount these defenses. However, these species pairs are "familiar" with one another, and neither of the strategies is perfect. Some prey are always more vulnerable, and some predators are less efficient due to the nature of variability in natural populations. Therefore the likelihood of extinction from this association is limited.

Several factors influence the likelihood and strength of coevolved relationships. Coevolution is more likely to take place in pairs of species where high levels of co-occurrence are present. It is also common in cases where selective pressure is strong, influencing important functions such as reproduction or **mortality**. The type of relationship—be it mutualistic, predator-prey, or competitor—also influences coevolution. Species that have intimate relationships, such as that of a specialist predator or a host-specific parasite, interact actively and thus are more likely to influence each other's selection. Species that do not directly encounter each other but interact through **competition** for resources are less likely candidates to coevolve, but the strength of the competition may influence the situation.

The result of coevolved relationships is structure in communities. Coevolution and **symbiosis** create fairly distinct relationships in communities, relationships that are mirrored in distribution, community energetics, and **resistance** to disturbances in species. Coevolution allows for recurring groupings of plant and animal communities.

[*David A. Duffus*]

RESOURCES
BOOKS

Barth, F. G. *Insects and Flowers: The Biology of a Partnership.* Princeton, NJ: Princeton University Press, 1991.

Erickson, J. *The Living Earth: The Coevolution of the Planet and Life.* Blue Ridge Summit, PA: Tab Books, 1989.

Cogeneration

Cogeneration is the multiple use of energy from a single primary source. In burning **coal**, oil, **natural gas**, or **biomass**, it is possible to produce two usable forms of energy at once, such as heat and electricity. By harnessing heat or exhaust from boilers and furnaces, for example, cogeneration systems can utilize energy that is usually wasted and so operate at a higher efficiency.

The second law of thermodynamics states that in every energy conversion there is a loss of useful energy in the form of heat. It is estimated that nearly half of the energy used in the United States is wasted as heat. Energy conversion efficiencies vary in range but most systems fall below 50%: A **gasoline** internal **combustion** engine is 10–15% efficient, and a steam turbine operates at about 40% efficiency. A simple example of cogeneration would be the heater in an **automobile**, which utilizes the heat of the engine to warm the interior of the car.

Cogeneration is classified into a topping cycle or a bottoming cycle. In the topping cycle, power is generated first, then the spent heat is utilized. In the bottoming cycle, thermal energy is used first, then power is generated from the remaining heat. The basic component of a cogeneration system is the prime mover, such as an internal combustion engine or steam boiler combination, whose function is to convert chemical energy or thermal energy into mechanical energy. The other components are the generator which converts mechanical input into electrical output and a spent

heat recovery system, as well as control and transmission systems. A cogeneration system utilizes the heat which the prime mover component has not converted into mechanical energy, and this can improve the efficiency of a typical gas turbine from approximately 12–30% to an overall rate of 60%.

In the United States, 40% of all electrical power is generated by burning coal, and coal-fired **power plants** lose two-thirds of their energy through the smokestack. Several large **electric utilities** have been using waste heat from their boilers for space heating in their own utility districts. This is known as district heating; zoning regulations and building locations permit this practice in New York City, New York; Detroit, Michigan; and Eugene, Oregon; among other cities. Waste heat from the generation of electricity has long been used in Europe to heat commercial and industrial buildings, and the city of Vestras, Sweden, produces all its electricity as well as all its space heating by utilizing the waste heat from industrial boilers.

In 1978, Congress passed the Public Utilities Regulatory Act, which allowed cogenerators to sell their extra power to utility companies. It has been estimated that if all waste heat generated by industry were used to cogenerate electrical power, there would be no need to construct additional power plants for the next two or three decades. The paper industry in the United States, which must produce steam for its industrial process, often uses cogeneration systems to produce electricity.

Experts maintain that half of the money consumers spend on electric bills pays not for the generation of power but for its distribution and transmission, including losses through transmission. Small, decentralized cogeneration systems that burn biomass, such as **organic waste** in agricultural areas, or **garbage** from large apartment buildings can minimize transmission losses and utilize waste heat for space heating. In Santa Barbara County, California, a hospital which operates a seven-megawatt natural gas turbine generator for its electrical power needs a cogeneration system installed to use thermal energy from the boiler to provide steam for heating and cooling. Extra steam not needed for these purposes is returned to the turbine, and excess electrical power is sold to the local utility.

A new technology for regeneration systems is **coal gasification**. Coal is heated and turned into a gas. The gas is burned to operate two turbines, one fueled by the hot gases and the other by steam generated from the burning gas. **Scrubbers** can remove 95% of the sulfur from the flue gases, and the result is a generating facility with high efficiency and low **pollution**.

[*Muthena Naseri and Douglas Smith*]

RESOURCES
BOOKS

Heating, Ventilating, Air-Conditioning Systems and Equipment. Atlanta: American Society of Heating, Refrigeration, and Air-Conditioning Engineers, 1992.

Kaufman, D. G., and C. M. Franz. *Biosphere 2000.* New York: Harper-Collins, 1993.

Cold fusion

Nuclear fusion has long been thought to have the potential to be a cheap and virtually unlimited source of energy. However, fusion requires extremely high temperatures or pressures. Cold fusion is fusion occurring at moderate temperatures and pressures. Cold fusion could make fusion more practical.

Fusion refers to a process where atomic nuclei collide and fuse together, forming larger particles and releasing energy. The fusion of **hydrogen** nuclei inside the sun is what makes the heat that warms our planet. But outside the enormous pressure in the center of a star like the sun, nuclei are unlikely to fuse. All atomic nuclei are positively charged, and thus repel each other. Some outside force, such as high temperature or pressure, is needed to make the nuclei overcome their natural repulsion and come together. If nuclei could be forced to fuse at low temperatures in a practical device, it would be possible to produce large amounts of energy at a reasonable cost.

One of the fathers of modern physics, Ernest Rutherford, investigated the possibility of cold fusion using deuterium (an **isotope** of hydrogen) atoms in 1934. However, he could find no viable method for achieving fusion at low temperatures.

Meanwhile, experiments with high temperature fusion proceeded. In the 1950s, scientists experimented with magnetic bottles as fusion devices. Theoretically, gas could be heated to such high temperatures inside the magnetic bottle that hydrogen nuclei would fuse, as they do in the sun. A physics laboratory at Princeton University in Princeton, New Jersey, built a magnetic containment device, a massive machine called the Tokamak. It did produce some energy from fusion, but using the Tokamak or anything like it as a power source is not practical, since it requires more energy to make fusion happen than it produce from the fusion reaction. Scientists continued to research various methods of fusion. If energy could be produced from an abundant material like hydrogen or deuterium, the earth would have an almost limitless, new, clean power source, the thinking went. The potential benefits were enormous.

Since cold fusion would operate at low temperatures and pressures, it would not require the expensive and complex equipment needed for high temperature fusion. Cold fusion research was filled with disappointments and dead ends, but scientists continued to experiment. Cold fusion has been called the "Holy Grail" of physics, because it seems virtually unattainable.

In 1989, Stanley Pons and Martin Fleischmann, two electrochemists working at the University of Utah, stunned the world with their announcement that they had performed cold fusion in a plastic dishpan using a small laboratory device. Pons was a prolific researcher in electrochemistry, and Fleischmann was an esteemed member of the British Royal Society and a professor at the University of Southampton in England. The two had collaborated for years, and in the mid-1980s Fleischmann had become a visiting professor at Utah. Their device electrically charged a cathode made of palladium, a metal that contains high concentrations of hydrogen isotopes. Apparently aiming to protect their patent rights to this revolutionary device, Pons and Fleischmann made their first cold fusion claim directly to the popular press, bypassing the scientific community by not submitting their discovery first to a professional journal where it would be reviewed by other scientists.

Pons and Fleischmann were temporarily heroes, appearing before Congress to help the University of Utah secure funding for cold fusion research. Meanwhile, many experts in fusion research and other physicists were less easily convinced. If the palladium cathode was actually fusing hydrogen isotopes, it would have emitted dangerous levels of radiation. The very fact that the clearly healthy scientists had been photographed beaming next to their cold fusion machine alerted knowledgeable people that the device could not be working as Pons and Fleischmann claimed. Within weeks it became clear that Pons and Fleischmann had made some elementary mistakes. It was thought that the energy increase they detected came from pockets of heat in the fluid, caused by the liquid not being stirred enough. Although a number of scientists around the world continued to research it, within ten years of the Utah experiments the discovery of cold fusion seemed to be discredited.

Pons and Fleischmann suffered a serious loss of reputation after the cold fusion debacle, ultimately losing a libel suit they brought against an Italian newspaper that had labeled them frauds. The Japanese government continued to support cold fusion research through much of the 1990s, but finally abandoned all its financial support in 1998. It seemed that only a few hundred scientists worldwide were working on cold fusion in 2001, judging from attendance at a semi-annual conference on the topic. In 2002, a researcher at the Oak Ridge National Laboratory in **Oak Ridge, Tennessee**, claimed to have detected something that might have been cold fusion using a process called acoustic cavitation. This produced high pressure in a liquid by means of sound waves. But with the stark example of Pons and

Fleischmann, any new claim to cold fusion will have to be carefully substantiated by a number of scientists before it can overcome existing skepticism.

[*Angela Woodward*]

RESOURCES
BOOKS

Park, Robert. *Voodoo Science.* Oxford: Oxford University Press, 2000.
Taubes, Gary. *Bad Science: The Short and Weird Times of Cold Fusion.* New York: Random House, 1993.

PERIODICALS

Beals, Gregory. "Pining for a Breakthrough" *Newsweek* 138, no. 16 (October 15, 2001):57.
Goodwin, Irwin. "Washington Dispatches" *Physics Today* 51, no. 7 (July 1998):48.
"Here We Go Again" *Economist* 362, no. 8263 (March 9, 2002):77.
Pollack, Andrew. "Japan, Long a Holdout, Is Ending Its Quest for Cold Fusion" *New York Times* (August 26, 1997):C4.

Coliform bacteria

Coliform bacteria live in the nutrient-rich **environment** of animal intestines. Many **species** fall into this group, but the most common species in mammals is ***Escherichia coli***, usually abbreviated *E. coli.* A typical human can easily have several trillion of these tiny individual bacterial cells inhabiting his or her digestive tract. On a purely numerical basis, a human may have more bacterial than mammalian cells in his or her body. Each person is actually a community or **ecosystem** of diverse species living in a state of cooperation, competition, or coexistence.

The bacterial **flora** of one's gut provides many benefits. They help break down and absorb food, they synthesize and secrete vitamins such as B_{12} and K on which mammals depend, and they displace or help keep under control pathogens that are ingested along with food and liquids. When the pathogens gain control, disagreeable or even potentially lethal diseases can result. A wide variety of diarrheas, dysenteries, and other gastrointestinal diseases afflict people who have inadequate **sanitation**. Many tourists suffer traveler's diseases known by names such as Montezuma's Revenge, La Tourista, or Cairo Crud when they come into contact with improperly sanitized water or food. Some of these diseases, such as **cholera** or food **poisoning** caused by *Salmonella*, *Shigella*, or *Lysteria* species, can be fatal.

Because identifying specific pathogens in water or food is difficult, time-consuming, and expensive, public health officials usually test for coliform organisms in general. The presence of any of these species, whether pathogenic or not, indicates that fecal contamination has occurred and that pathogens are likely present.

Colorado River

One of the major rivers of the western United States, the Colorado River flows for some 1,500 mi (2,415 km) from Colorado to northwestern Mexico. Dropping over 2 mi (3.2 km) in elevation over its course, the Colorado emptied into the Gulf of California until human management reduced its water flow. Over millions of years the swift waters of the Colorado have carved some of the world's deepest and most impressive gorges, including the Grand Canyon.

The Colorado River basin supports an unusual **ecosystem**. Isolated from other **drainage** systems, the Colorado has produced a unique assemblage of fishes. Of the 32 **species** of native fishes found in the Colorado drainage, 21–66%, are endemic species—species that arose in the area and are found nowhere else.

Major projects carried out since the 1920s have profoundly altered the Colorado. When seven western states signed the Colorado River Compact in 1922, the Colorado became the first basin in which "multiple use" of water was initiated. Today the river is used to provide hydroelectric power, **irrigation**, drinking water, and **recreation**; over 20 **dams** have been erected along its length. The river, in fact, no longer drains into the Gulf of Colorado—it simply disappears near the Mexican towns of Tijuana and Mexicali. Hundreds of square miles of land have been submerged by the formation of reservoirs, and the temperature and clarity of the river's water have been profoundly changed by the action of the dams.

Alteration of the Colorado's **habitat** has threatened many of its fragile fishes, and a number are now listed as **endangered species**. The Colorado squawfish serves as an example of how river development can affect native **wildlife**. With the reservoirs formed by the impoundments on the Colorado River also came the introduction of game fishes in the 1940s. One particular species, the Channel catfish, became a prey item for the native squawfish, and many squawfish were found dead, having suffocated due to catfish lodged in their throats with their spines stiffly locked in place. Other portions of the squawfish population have succumbed to diseases introduced by these non-native fishes.

Major projects along the Colorado include the Hoover Dam and its **reservoir**, Lake Mead, as well as the controversial **Glen Canyon Dam** at the Arizona-Utah border, which has a reservoir extending into Utah for over 100 mi (161 km).

[*Eugene C. Beckham and Jeffrey Muhr*]

RESOURCES
BOOKS

Fradkin, P. L. *A River No More: The Colorado River and the West.* New York: Knopf, 1981.

Richardson, J. *The Colorado: A River At Risk*. Englewood, CO: Westcliffe Publishers, 1992.

Combined sewer overflows

In many older coastal cities, especially in the northeastern United States, storm sewers in the street that collect stormwater **runoff** from rainfall are connected to municipal **sewage treatment** plants that process household sewage and industrial **wastewater**. Under normal, relatively dry conditions runoff and municipal waste go to a sewage treatment plant where they are treated. However, when it rains, in some cases less than an inch, the capacity of a sewage treatment plant can be exceeded; the system is overloaded. The mixed urban stormwater runoff and raw municipal sewage is released to nearby creeks, rivers, bays, estuaries or other coastal waters, completely untreated. This is a combined sewer overflow (CSO) event.

Combined sewer overflow events are not rare. In Boston Harbor, for example, there are 88 pipes or outfalls that **discharge** combined stormwater runoff and sewage. It has been estimated that CSO events occur approximately 60 times per year, discharging billions of gallons of untreated runoff and wastewater to Boston Harbor.

Materials released during these CSO events can result in serious **water quality** problems that can be detrimental to both humans and **wildlife**. Toxic **chemicals** from households and industries are released during CSO events. In addition, toxic chemicals found in rainwater runoff, such as oil and antifreeze that have dripped onto roads from cars, will wash into coastal waters during these events.

Harmful bacteria and pathogens in the water are another major problem that can result after a CSO event. Some of these bacteria (coliform), live naturally in the intestinal tracts of humans and other warm blooded animals. After heavy rainfalls, scientists have measured increased levels of **coliform bacteria** in coastal waters near CSO outfalls. These bacteria, which indicate that there are other bacteria and pathogens that can make people sick if they swim in the water or eat contaminated shellfish, come from both animal and human wastes washed in from the streets. The bacteria are not removed or killed because the waters have not been treated in a sewage treatment plant. Because levels of these indicator bacteria are often high after CSO events, many productive shellfish beds are closed to protect human health. This can be a serious economic hardship to the fishing industry.

Combined sewer overflow events also result in increased quantities of trash and **floatable debris** entering coastal waters. When people litter, the trash is washed into storm sewers with rainwater. Since sewage treatment plants cannot handle the volume of water during these rainfall events, this trash is discharged along with the stormwater

and sewage directly into open waters. This floatable debris is unsightly, and can be dangerous to marine animals and birds, which eat it and choke or become entangled within it. This often results in death.

Raw sewage, animal wastes, and runoff from lawns and other fertilized areas contain very high levels of **nitrogen** and **phosphorus**, which are nutrients used by marine and aquatic plants for growth. Therefore, CSO events are major contributors of extra nutrients and organic matter to nearshore waters. These nutrients act as fertilizers for many marine and aquatic algae and plants, promoting extreme growth called blooms. When they eventually die, the bacteria decomposing the plants and algae use up vast quantities of oxygen. This results in a condition known as hypoxia or low **dissolved oxygen** (DO). If DO levels are too low, marine animals will not have enough oxygen to survive and they will either die or move out of the area. Hypoxia has been the cause of some major **fish kills**, and can result in permanent changes in the ecological community if it is persistent.

There are a number of options available to reduce the frequency and impacts of CSO events. Upgrading sewage treatment plants to handle greater flow or constructing new facilities are two of the best, although most costly options. Another possibility is to separate storm sewers and municipal sewage treatment plants. While this would not prevent discharges of stormwater runoff during rainfall events, untreated household and industrial wastewater (i.e., raw sewage) would not be released. In addition, the resulting stormwater could be minimally treated by screening out trash and disinfecting it to kill bacteria. Use of **wetlands** to filter this stormwater has also been considered as an effective alternative and is currently being used in some areas. Another option is to build large storage facilities (often underground) to hold materials that would normally be discharged during CSO events. When dry conditions return, the combined runoff and wastewater are pumped to a nearby sewage treatment plant where they are properly treated. This option is being used in several areas, including some locations in New York City. At a minimum, screening of CSO discharges would reduce the quantity of floatable debris in nearshore waters, even if it did not solve all of the other problems associated with CSOs. Of course **water conservation** is another control that reduces that volume of water treated by sewage treatment plants, and therefore the volume that would be discharged during a CSO event.

[*Max Strieb*]

RESOURCES

BOOKS

National Research Council. "Managing Wastewater In Coastal Urban Areas." Washington, D.C.: National Academy Press, 1993.

Combustion

The process of burning fuels. Traditionally **biomass** was used as fuel, but now **fossil fuels** are the major source of energy for human activities. Combustion is essentially an oxidation process that yields heat and light. Most fuels are **carbon** and **hydrogen** which use oxygen in the air as an oxidant. More exotic fuels are used in some combustion processes, particularly in rockets where metals such as **aluminum** or beryllium or hydrazine (a **nitrogen** containing compound) are well known as effective fuels. As rockets operate beyond the **atmosphere** they carry their own oxidants, which may also be quite exotic.

Combustion involves a mixture of fuel and air, which is thermodynamically unstable. The fuel is then converted to stable products, usually water and **carbon dioxide**, with the release of a large amount of energy as heat. At normal temperatures fuels such as **coal** and oil are quite stable and have to be ignited by raising the temperature. Combustion is said to be spontaneous when the ignition appears to take place without obvious reasons. Large piles of organic material, such as hay, can undergo slow oxidation, perhaps biologically mediated, and increase in temperature. If the amount of material is very large and the heat cannot escape, the whole pile can suddenly burst into flame. Will-o'-the-wisps or jack-o'-lanterns (known scientifically as *ignis fatuus*) are sometimes observed over swamps where **methane** is likely to be produced. The reason these small pockets of gas ignite is not certain, but it has been suggested that small traces of gases such as phosphine that react rapidly with air could ignite the methane.

Typical solid fuels like coal and wood begin to burn with a bright turbulent flame. This forms as volatile materials are driven off and ignited. These vapors burn so rapidly that oxygen can be depleted, creating a smoky flame. After a time the volatile substances in the fuel are depleted. At this point a glowing coal is evident and combustion takes place without a significant flame. Combustion on the surface of the glowing coal is controlled by the diffusion of oxygen towards the hot surface. If the piece of fuel is too small, such as a spark from a fire, it is likely to lose temperature rapidly and combustion will stop. By contrast a bed of coals can maintain combustion because of heat storage and the exchange of radiative heat between the pieces. The most intense combustion takes place between the crevices of a bed of coal. In these regions oxygen may be in limited supply which leads to the production of **carbon monoxide**. This is subsequently oxidized to carbon at the surface of the bed of coals with a faint blue flame. The production of toxic carbon monoxide from indoor fires can occasionally represent a hazard if subsequent oxidation to carbon dioxide is not complete.

Liquid fuels usually need to be evaporated before they burn effectively. This means that it is possible to see liquid combustion and gaseous combustion as similar processes. Combustion can readily be initiated with a flame or spark. Simply heating a fuel-air mixture can cause it to ignite, but temperatures have to be high before reactions occur. A much better way is to initiate combustion with a small number of molecular fragments of radicals. These can initiate chain reactions at much lower temperatures than molecular reactions. In a propane-air flame at about 2000° K, hydrogen and oxygen atoms and hydroxyl radicals account for about 0.3% of a gas mixture. It is these radicals that support combustion. They react with molecules and split them up into more radicals. These radicals can rapidly enter into the exothermic (heat releasing) oxidative processes that lie at the heart of combustion. The reactions also give rise to further radicals that support continued combustion. Under some situations the radicals reaction branch, such that the reaction of each radical produces two new radicals. These can enter further reactions, producing yet further increases in the number of reactions and very soon the system explodes. However the production of radicals can be terminated in a number of ways such as contact with a solid surface. In some systems, such as the internal combustion engine, an explosion is desired, but in others, such as a gas cooker flame, maintaining a stable combustion process is desirable.

In terms of **air pollution** the reaction of oxygen and nitrogen atoms with molecules in air leads to the formation of the pollutant nitric oxide through a set of reactions known as the Zeldovich cycle. It is this process that makes combustion such an important contributor of **nitrogen oxides** to the atmosphere.

[*Peter Brimblecombe*]

RESOURCES
BOOKS

Campbell, I. M. *Energy and the Atmosphere.* New York: Wiley, 1986.

Cometabolism

The partial breakdown of a (usually) synthetic compound by microbiological action. Synthetic **chemicals** are widely used in industry, agriculture, and in the home; many resist complete enzymatic degradation and become persistent environmental pollutants. In cometabolism, the exotic molecule is only partly modified by **decomposers** (bacteria or **fungi**), since they are unable to utilize it either as a source of energy, as a source of **nutrient** elements, or because it is toxic. Cometabolism probably accounts for long-term changes in DDT, dieldrin, and related chlorinated hydrocarbon insecticides in the **soil**. The products of this partial transformation, like the original exotic chemical, usually accumulate in the **environment**.

Commensalism

A type of symbiotic relationship. Many organisms depend on intimate physical relationship with organisms of other **species**, a relationship called **symbiosis**. The larger organism is called the host and the smaller organism, the symbiote. The symbiote always derives some benefit from the relationship. In a commensal relationship, the host organism is neither harmed nor benefitted. The relationship that exists between the clown fish living among the tentacles of sea anemones is one example of commensalism. The host sea anemones can exist without their symbiotes, but the fish cannot exist as successfully without the protective cover of the anemone's stinging tentacles.

Commercial fishing

Because fish have long been considered an important source of food, the fisheries were the first renewable resource to receive public attention in the United States. The National Marine Fisheries Service (NMFS) has existed as such since 1970, but the original Office of Commissioner of Fish and Fisheries was created over 100 years ago, signed into law in 1871 by President Ulysses S. Grant. This office was charged with the study of "the decrease of the food fishes of the seacoasts and lakes of the United States, and to suggest remedial measures." From the beginning, the federal fishery agency has been granted broad powers to study aquatic resources ranging from coastal shallow waters to offshore deepwater habitats.

Worldwide, humans get an average of 16% of their dietary animal protein from fish and shellfish. With human populations ever increasing, the demand for and marketing of seafood has steadily increased, rising over the last half of the twentieth century to a peak in 1994 of about 100 million tons (91 billion kg) per year. The current annual marine fish catch has fallen slightly to around 70 million tons (64 billion kg). The per capita world fish catch has been steadily declining since 1970 as human **population growth** outdistances fish harvests. Scientists have projected that, by 2020, the per capita consumption of ocean fish will be half of what it was in 1988.

To meet the demand for fish, the commercial fishing industry has expanded as well. There are currently about 13 million commercial fishermen in the world. About half of the fish are caught by the vast majority of fishers who use traditional methods. The remainder is harvested by industrial fishing crews, manning about 37,000 vessels that deploy highly innovative methods ranging from enormous nets to sonar and spotting planes. As a result, wild fish populations have been decimated.

In recent decades, the size of the industrial fishing fleet grew at twice the rate of the worldwide catch. The expansion in fishing may be coming to an end, however, as environmental, biological, and economic problems beset the fishing industry. As fish harvests decline, the numbers of jobs also decline. Governments have attempted to prop up the failing fisheries industry: in 1994, fishers worldwide spent $124 billion to catch fish valued at $70 billion, and the shortfall was covered by government subsidies. In recent decades, fishery imports have been one of the top five sources of the United States' trade deficit.

The commercial fisheries industry has contributed to its own problems by **overfishing** certain **species** to the point where those species' populations are too low to reproduce at a rate sufficient to replace the portion of their numbers lost to harvesting. Cod and haddock in the Atlantic Ocean, red snapper in the Gulf of Mexico, and **salmon** and tuna in the Pacific Ocean have all fallen victim to overfishing. The case of the Peruvian anchovy represents a specific example of how several factors may work together to contribute to species decline. Fishing for anchovies began off the coast of Peru in the early 1950s, and, by the late 1960s, as their fishing fleet had grown exponentially, the catch of Peruvian anchovies made up about 20% of the world's annual commercial fish harvest. The Peruvian fishermen were already overfishing the anchovies when meteorological conditions contributed to the problem: in 1972, a strong **El Niño** struck. This phenomenon is a natural but unpredictable warming of the normally cool waters that flow along Peru's coast. The entire food web of the region was altered as a result, and the Peruvian anchovy population plummeted, leading to the demise of Peru's anchovy fishing industry. Peru has made some economic recovery since then by harvesting other species.

Many of the world's major fishing areas have already been fished beyond their natural limits. Different approaches to the problem of overfishing are under consideration to help prevent the collapse of the world's fisheries. **Georges Bank**, once one of the most fertile fishing grounds in the North Atlantic, is now closed and is considered commercially extinct. This area underwent strict controls for scallop fishing in 1996, which proved to be a viable remedy for that species in that locale. The scallop population recovered within five years, reaching levels in excess of the original population, and parts of the bay could be re-opened for scallop fishing. But other species in Georges Bank continue to decline. Rapid and direct replenishment is not possible for slow-growing species that take years to reach maturity. For example, the black sea bass (*Stereolepis gigas*), has a life span comparable to that of humans and adults typically grow to 500 lb (227 kg). The success of a 1982 ban on fishing the black sea bass off the coast of California became evident early this century when significant numbers of these young fish, already weighing as much as 200 lb (91 kg), appeared off the shores of Santa Barbara. Yet, full replenishment of the population remains years away.

Environmental problems also **plague** commercial fishing. Near-shore **pollution** has altered ecosystems, taking a heavy toll on all populations of fish and shellfish, not only those valued commercially. The collective actions of commercial fishermen also create some major environmental problems. The world's commercial fishermen annually catch and then discard about 20 billion lb (9 billion kg) of non-target species of sea life. In addition to fish and shellfish, each year about one million seabirds are caught and killed in fishermen's nets. On average more than 6,000 **seals and sea lions**, about 20,000 **dolphins** and other aquatic mammals, and thousands of **sea turtles** meet the same fate. It is estimated that the amount of fish discarded annually is about 25% of the reported catch, or about 20 million metric tons per year. Ecologically, two major problems arise from this massive disposal of organisms. One is the disruption of predator-prey ratios, and the other is the addition of a tremendous overload of **organic waste** to be dealt with in this **ecosystem**.

In 2001, a $1.6 billion gas pipeline that was proposed to be routed through neighboring waters from Nova Scotia to New Jersey, to be implemented as early as 2005, posed a new environmental threat to the Georges Bank area. Environmentalists, meanwhile, have lobbied the United States government to establish a marine **habitat** protection designation similar to **wilderness** areas and natural parks on land, to provide for the preservation of reefs, marine life, and underwater vegetation. In 2001, less than 1% of **water resources** worldwide had the protection of formal legislation to prevent exploitation.

Habitat destruction is serious environmental concern. Fish and other aquatic **wildlife** rely on the existence of high quality habitat for their survival, and loss of habitat is one of the most pressing environmental threats to shorelines, **wetlands**, and other aquatic habitats. Approaches to the protection of **essential fish habitat** include efforts to strengthen and vigorously enforce the **Clean Water Act** and other protective legislation for aquatic habitats, to develop and implement restoration plans for target regions, to make improved policy decisions based on technical knowledge about shoreline habitats, and to better educate the public on the importance of protecting and restoring habitat. A relatively new approach to habitat recovery is the habitat **conservation** plan (HCP), in which a multi-species ecosystem approach to habitat management is preferred over a

reactive species-by-species plan. Strategies for fish recovery are complex, and, instead of numbers of fish of a given species, the HCP uses quality of habitat to measure the success of restoration and conservation efforts. Long-term situations such as the restoration of black sea bass serve to re-emphasize the importance of resisting the temptation to manage overfishing of single species while failing to address the survival of the ecosystem as a whole.

The Magnuson-Stevens Fishery Conservation and Management Act was passed in 1976 to regulate fisheries resources and fishing activities in Federal waters, those waters extending to the 200-mi (322-km) limit. The act recognizes that commercial fishing contributes to the food supply and is a major source of employment, contributing significantly to the economy of the Nation. However, it also recognizes that overfishing and habitat loss has led to the decline of certain species of fish to the point where their survival is threatened, resulting in a diminished capacity to support existing fishing levels. Further, international fishery agreements have not been effective in ending or preventing overfishing. Fishery resources are limited but renewable and can be conserved and maintained to continue to provide good yields. Also, the act supports the development of underused fisheries, such as bottom-dwelling fish near Alaska.

Another resource to sustain increases in seafood consumption is **aquaculture**, where commercial food-fish species are grown on fish farms. It is estimated that the amount of farm-raised fish has doubled in the past decade and that about 20% of the fish consumed worldwide is raised in captivity.

In the United States, as well as other nations, the commercial fisheries industry faces potential collapse. Severe restrictions and tight controls imposed by the international community may be the only means of salvaging even a portion of this valuable industry. It will be necessary for partnerships to be forged between scientists, fisherman, and the regulatory community to develop and implement measures toward maintaining a sustainable fishery.

[*Eugene C Beckham*]

RESOURCES
BOOKS

Bricklemyer, E., S. Iudicello, and H. Hartmann. "Discarded Catch in U.S. Commercial Marine Fisheries." In *Audubon Wildlife Report 1990-1991*. San Diego: Academic Press, 1990.

Weber, M. "Federal Marine Fisheries Management." In *Audubon Wildlife Report 1986*. San Diego: Academic Press, 1986.

PERIODICALS

Lawren, B. "Net Loss." *National Wildlife* 30 (1992): 46-53.

Associated Press, February 19, 2001.

OTHER

Dudley-Cash, William A. " Aquaculture has been the world's fastest growing food production system." *Factoryfarming.com: Factory Seafood Production*
June 29, 1998 [cited July 9, 2002] < http://www.factoryfarming.com/
fish.html>.

Guinan, John A. and Ralph E. Curtis "A Century Of Conservation."
NMFS National Marine Fisheries Service, April 1971 [cited July 9, 2002]
<http://www.nefsc.nmfs.gov/library/history/century.html>.

Loftas, Tony "Not enough fish in the sea." *Our Planet* 7.6 April 1996
[cited July 9, 2002] <http://www.ourplanet.com/imgversn/76/loftas.html>.

Magnuson-Stevens Fishery Conservation and Management Act, Public Law
94-265 (as amended through October 11, 1996). [cited July 9, 2002] <http://
www.nmfs.noaa.gov/sfa/magact/>.

Vogel, William and Lorin Hicks "Multi-species HCPs:Experiments with
the Ecosystem Approach." *Endangered Species Bulletin* July/August 2000,
Vol. 25, No. 4, pp. 20-22 [cited July 9, 2002] <http://endangered.fws.gov/
esb/2000/07-08/20-22.pdf>.

Commingled recyclables
see Recycling

Commission for Environmental Cooperation

The Commission for Environmental Cooperation (CEC) is
a trilateral international commission established by Canada,
Mexico, and the United States in 1994 to address transboundary environmental concerns in North America. The
original impetus behind the CEC was the perception of
inadequacies in the environmental provisions of the **North
American Free Trade Agreement** (NAFTA). A supplementary treaty, the North American Agreement for Environmental Cooperation (NAAEC) was negotiated to remedy these inadequacies, and it is from the NAAEC that the
CEC derives its formal mandate.

The general goals set forth by the NAAEC are to
protect, conserve, and improve the **environment** for the
benefit of present and **future generations**. More specifically, the three NAFTA signatories agreed to a core set of
actions and principles with regard to environmental concerns
related to trade policy. These actions and principles include
regular reporting on the state of the environment, effective
and consistent enforcement of **environmental law**, facilitation of access to environmental information, the ongoing
improvement of environmental laws and regulations, and
promotion of the use of tax incentives and various other
economic instruments to achieve environmental goals.

The CEC is to function as a forum for the NAFTA
partners to identify and articulate mutual interests and priorities, and to develop strategies for the pursuit or implementation of these interests and priorities. The NAAEC further
specifies the following priorities: identification of appropriate limits for specific pollutants; the protection of endan-

gered and threatened **species**; the protection and **conservation** of wild **flora** and **fauna** and their **habitat**; the
development of new approaches to environmental compliance and enforcement; strategies for addressing environmental issues that have impacts across international borders; the
support of training and education in the environmental field;
and promotion of greater public awareness of North American environmental issues. Central to the CEC's mission is
the facilitation of dialogue among the NAFTA partners in
order to prevent and solve trade and environmental disputes.

The governing body of the CEC is a Council of Ministers consisting of the environment ministers (or equivalent)
from each country. The executive arm of the Commission
is a Secretariat located in Montreal, consisting of a staff of
about 30 members and headed by an Executive Director.
The staff is drawn from all three countries and provides
technical and administrative support to the Council of Ministers and to committees and groups established by the
Council.

Technical and scientific advice is also provided to the
Council of Ministers by a Joint Public Advisory Committee
consisting of five members from each country appointed by
the respective governments. This Committee may, on its
own initiative, advise the Council on any matter within the
scope of the NAAEC, including the annual program and
budget. As a reflection of the CEC's professed commitment
to participation by citizens throughout North America, the
Committee is intended to represent a wide cross-section of
knowledgeable citizens committed to environmental concerns who are willing to volunteer their time in the public
interest. The CEC also accepts direct input from any citizen
or non-governmental organization who believes that a
NAFTA partner is failing to enforce effectively an existing
environmental law.

The NAAEC also contains provisions for dispute resolution in cases in which a NAFTA signatory alleges that
another NAFTA partner has persistently failed to enforce an
existing environmental law, causing specific environmental
damage or trade disadvantages to the claimant. These provisions may be invoked when a lack of effective enforcement
materially affects goods or services being traded between the
NAFTA countries. If the dispute is not resolved through
bilateral consultation, the complaining party may then request a special session of the CEC's Council of Ministers.
If the Council is likewise unable to resolve the dispute,
provisions exist for choosing an Arbitral Panel. Failure to
implement the recommendations of the Arbitral Panel subjects the offending party to a monetary enforcement assessment. Failure to pay this assessment may lead to suspension
of free trade benefits.

The Council is also instructed to develop recommendations on access to courts (and rights and remedies before

courts and administrative agencies) for persons in one country's territory who have suffered or are likely to suffer damage or injury caused by **pollution** originating in the territory of one of the other countries. In 2002, the Council published a five year study which stated that 3.4 million tonnes of **toxins** were produced in North America.

The CEC has been subject to some of the same criticisms leveled at the environmental provisions of NAFTA, particularly that it serves as a kind of environmental window-dressing for a trade agreement that is generally harmful to the environment. The CEC's mandate for conflict resolution is primarily oriented towards consistent enforcement of existing environmental law in the three countries. This law is by no means uniform. By upholding the principles of free trade, and providing penalties for infringements of free trade, NAFTA establishes an environment in which private companies have an economic incentive, other considerations being equal, to locate production where environmental laws are weakest and the costs of compliance are therefore lowest. Countries with stricter environmental regulations face penalties for attempting to protect domestic industries from such comparative disadvantages.

[*Lawrence J. Biskowski*]

RESOURCES

OTHER

Kass, S. L. "First Cases Before New NAFTA Forum Suggest Its Power Will Increase." National Law Journal 18/41 (10 June 1996): C5, C7.

ORGANIZATIONS

Commission for Environmental Cooperation, 393, rue St-Jacques Ouest, Bureau 200, Montréal, QuébecCanada H2Y 1N9 (514) 350-4300, Fax: (514) 350-4314, Email: info@ccemtl.org, <http://www.cec.org>

Barry Commoner (1917 –)

American biologist, environmental scientist, author, and social activist

Born to Russian immigrant parents, Commoner earned a doctorate in biology from Harvard in 1941. As a biologist, he is known for his work with free radicals—chemicals like **chlorofluorocarbons**, which are suspected culprits in **ozone layer depletion**. Commoner led a fairly academic life at first, with research posts at various universities, but rose to some prominence in the late 1950s, when he and others protested atmospheric testing of **nuclear weapons**. He earned a national reputation in the 1960s with books, articles, and speeches on a wide range of environmental concerns, including **pollution, alternative energy sources**, and population. His latest book, *Making Peace with the Planet*, was published in 1990. Commoner's other works include *Science and Survival* (1967), *The Closing Circle*

Barry Commoner speakingto a group of protesters gathered outside a New Jersey hotel where Exxon stockholders met in 1989. (Corbis-Bettmann. Reproduced by permission.)

(1971), *Energy and Human Welfare* (1975), *The Poverty of Power* (1976), and *The Politics of Energy* (1979).

Commoner believes that post-World War II industrial methods, with their reliance on nonrenewable **fossil fuels**

are the root cause of modern environmental pollution. When combined with a myopic view of the bottom line, he states, the devastation is complete: "At present, economic considerations—in particular, the private desire for maximizing short-term profits—govern the choice of productive technology, which in turn determines its environmental impact, generally for the worse." The petrochemicals industry receives the largest share of Commoner's criticism. He refers to "the **petrochemical** industry's toxic invasion of the biosphere" and states flatly that "the petrochemical industry is inherently inimical to environmental quality."

Almost as distressing as environmental pollution is our inability to clean it up. Commoner rejects attempts at environmental regulation as pointless. Far better, he says, to not produce the toxin in the first place. "When a pollutant is attacked at the point of origin—in the production process that generates it—the pollutant can be eliminated; once it is produced, it is too late. This is the simple but powerful lesson of the two decades of intense but largely futile effort to improve the quality of the environment."

Commoner offers radical, sweeping solutions for social and ecological ills. The most urgent of these is a **renewable energy** source, primarily photovoltaic cells powered by **solar energy**. These would not only decentralize **electric utilities** (another target of Commoner's), but would use sunlight to fuel almost any energy need, including smaller, lighter, battery-powered cars. To ease the transition from fossil fuels to solar power, he proposes **methane, cogeneration** (which produces electricity from waste heat), and an organic agriculture system that would "produce enough **ethanol** to replace about 20 percent of the national demand for **gasoline** without reducing the overall supply of food or significantly affecting its price."

Commoner makes few compromises, and his environmental zeal has made him a crusader for social causes as well. Eliminating **Third World** debt, he argues, would improve life in impoverished countries and end the spiral of economic desperation that drives countries to overpopulation. "This [debt forgiveness] should be regarded not as a magnanimous gesture but as partial reparations for the damage inflicted...by the former colonial empires...[T]he cause of poverty is the grossly unequal distribution of the world's wealth...we must redistribute that wealth, among nations and within them."

In 1980, Commoner made a bid for the presidency on the Citizen's Party ticket, a short-lived political attempt to combine environmental and Socialist agendas. Since 1981 he has been the director of the Center for the Biology of Natural Systems at Queens College in New York City.

[*Muthena Naseri and Amy Strumolo*]

RESOURCES
BOOKS

Commoner, B. *The Closing Circle.* New York: Knopf, 1971.
———. *Making Peace With the Planet.* New York: New Press, 1992.

PERIODICALS

Commoner, B. "Ending the War Against Earth." *The Nation* 250 (30 April 1990): 589–90.
———. "The Failure of the Environmental Effort." *Current History* 91 (April 1992): 176–81.
Stone, P. "The Ploughboy Interview." *Mother Earth News* (March–April 1990): 116–26.

Communicable diseases

A communicable disease is any disease that can be transmitted from one organism to another. Agents that cause communicable diseases, called pathogens, are easily spread by direct or indirect contact. These pathogens include viruses, bacteria, **fungi**, and **parasites**. Some pathogens make **toxins** that harm the body's organs. Others actually destroy cells. Some can impair the body's natural immune system, and opportunistic organisms set up secondary infections that cause serious illness or death. Once the pathogens have multiplied inside the body, signs of illness may or may not appear. The human body is adept at destroying most pathogens, but the pathogens may still multiply and spread.

The pathogens responsible for some communicable diseases have been known since the mid-1800s, although they have existed for a much longer period of time. European explorers brought highly contagious diseases such as smallpox, measles, typhus and scarlet fever to the New World, to which Native Americans had never been exposed. These new diseases killed 50–90% of the native population. Native populations in many areas of the Caribbean were totally eliminated.

In some areas of the world, **contaminated soil** or water incubate the pathogens of communicable diseases, and contact with those agents will cause diseases to spread. In the 1800s, when Robert Koch discovered the **anthrax** bacillus (*Bacillus anthracis*), **cholera** bacillus (*Vibrio cholerae*), and tubercle bacillus (*Mycobacterium tuberculosis*), his work ushered in a new era of public **sanitation** by showing how water-borne epidemics, such as cholera and typhoid, could be controlled by water **filtration**.

Malaria, another communicable disease, was responsible for the decline of many ancient civilizations; for centuries, it devitalized vast populations. With the discoveries in the late 1800s of protozoan malarial parasites in human blood, and the discovery of its carrier, the *Anopheles* mosquito, malaria could be combatted by systematic destruction of the mosquitos and their breeding grounds, by the use of barriers between mosquitos and humans such as window screens and mosquito netting, and by drug therapy to kill the parasites in the human host.

Discoveries of the causes of epidemics and transmissible diseases led to the expansion of the fields of sanitation and public health. Draining of marshes, control of the water supply, widespread vaccinations, and quarantine measures improved human health. But, despite the development of advanced detection techniques and control measures to fight pathogens and their spread, communicable diseases still take their toll on human populations. For example, until the 1980s tuberculosis had been declining in the United States due in large part to the availability of effective antibiotic therapy. However, since 1985, the number of tuberculosis cases has risen steadily due to such factors as the emergence of drug-resistant strains of the tubercle bacillus, the increasing incidence of HIV infection which lowers human **resistance** to many diseases, poverty, and immigration.

Epidemiologists track communicable diseases throughout the world, and their work has helped to eradicate smallpox, one of the world's most deadly communicable diseases. A successful global vaccination campaign wiped out smallpox in the 1980s, and today, the **virus** exists only in tightly-controlled laboratories in Moscow and in Atlanta at the **Centers for Disease Control and Prevention**. Scientists are currently debating whether to destroy the viruses or preserve them for study.

Communicable diseases continue to be a major public health problem in developing countries. In the small West African nation of Guinea-Bissau, a cholera epidemic hit in 1990. Epidemiologists traced its outbreak to contaminated shellfish and managed to control the epidemic, but not before it claimed hundreds of lives in that country. Some victims had eaten the contaminated shellfish, others were infected by washing the bodies of cholera victims and then preparing funeral feasts without properly washing their hands. Proper disposal of victims' bodies, coupled with a campaign to encourage proper hygiene, helped stop the epidemic from spreading.

Communicable diseases can be prevented either by eliminating the pathogenic organism from the **environment** (as by killing pathogens or parasites existing in a water supply) or by placing a barrier in the path of its transmission from one organism to another (as by vaccination or by isolating individuals already infected). But identifying and isolating the causal agent and developing weapons to fight it is time consuming, and, as with the **AIDS** virus, thousands of people continue to become infected and many die because educational warnings about ways to avoid infection frequently go unheeded.

AIDS is caused by the human immunodeficiency virus (HIV). Spread by contact with bodily fluids of an HIV-

infected person, the virus weakens and eventually destroys the body's immune system. Researchers think the virus originated in monkeys and was first transmitted to humans about 40 years ago. African **chimpanzees** can be infected with HIV, but don't develop AIDS. This suggests questions that are so far unanswered: Was there a genetic change in the virus, or was there simply more contact between monkeys and people as human populations encroached on their **habitat**?

Some scientists think the AIDS pandemic is just the tip of the iceberg. They worry that new diseases, deadlier than AIDS, will emerge. Virologists point out that viruses, like human populations, constantly change. Rapidly increasing human populations provide fertile breeding grounds for **microbes**, including viruses and bacteria. Pathogens can literally travel the globe in a matter of hours.

For example, the completion in 1990 of a major road through the Amazon **rain forest** in Brazil led to outbreaks of malaria in the region. In 1985, used tires imported to Texas from eastern Asia transported larvae of the Asian tiger mosquito, a dangerous carrier of serious tropical communicable diseases. **Deforestation** and agricultural changes can unleash epidemics of communicable diseases. Outbreaks of Rift Valley fever followed the construction of the **Aswan High Dam**, most likely because breeding grounds were created for mosquitoes which spread the disease. In Brazil, the introduction of cacao farming coincided with epidemics of Oropouche fever, a disease linked to a biting insect that thrives on discarded cacao hulls.

Continued rapid **transportation** of humans around the world is likely to accelerate the movement of communicable diseases. Poverty, lack of adequate sanitation and nutrition, and the crowding of people into megacities in the developing countries of the world only exacerbate the situation. The need for study and control of these disease is likely to grow in the future.

[*Linda Rehkopf*]

RESOURCES
PERIODICALS

Jaret, P. "The Disease Detectives." *National Geographic* 179 (January 1991): 114-40.
Levine, D. "A Killer Returns." *American Health* (April 1992): 9.
Roberts, L. "Disease and Death in the New World." *Science* 246 (8 December 1989): 1245.

Community ecology

In biological **ecology**, the concept of community has been defined in various ways, each definition offering particular advantages and creating its own set of problems. As the authors of one article in a 1987 British Ecological Society symposium on community ecology noted, "Community ecology may be unique amongst the branches of science in lacking a consensus definition of the entity with which it is principally concerned." Daniel Botkin provides a summary of the major alternatives: an ecological community is "either (1) a set of interacting populations of different **species** found in an area, meaning that the community is the living part of an **ecosystem**; or (2) all of the species found in a local area, whether or not they actually interact; or (3) all of the species of the same kind found in a local area, as in a 'plant community' or 'animal community'." Peter Taylor describes an ecological community as consisting "of the populations of different species co-inhabiting a site—a lake, the leaf litter layer in a forest, a dung pat, and so on."

A definition of community is significant because so many different parts of the **environment** are investigated under that rubric in ecology: animal, plant, insect, primate, forest, even herbaceous plant communities. Often what is described is a component of a community rather than a *whole* community made up of many different species of plants and animals. Such a whole assemblage in community ecology has no scale, but it does denote a certain level of organization and is different from components of a community, guilds, or taxons.

Community ecology thus focuses on the living part of ecosystems, mostly on communities of interacting populations, however circumscribed. While a community can include species thrown together by locale that do not necessarily interact, community ecology generally emphasizes the wide diversity of species interactions that exist within the area.

Community ecologists investigate interactions under numerous labels and categories, including on-going studies of traditional topics such as predation, **competition**, trophic exchanges, and others. Borrowing concepts from other disciplines, such as physics, ecologists are also beginning to look closely at the linkages and assemblages that emerge from strong versus weak interactions, or from positive compared to negative interactions. Researchers continue to investigate the relative importance of intra-species and inter-species interactions in terms of their importance to community composition and rates of **succession**. One recent study, for example, analyzed the importance of interspecific interactions in the structuring of the geographical distribution and abundance of a tropical stream fish community.

The importance of trophic relationships and food web dynamics have always been recognized by ecologists, but a recent article in the journal *Ecology* can still suggest that "a step toward understanding the dynamics of communities is to describe the pathways along which feeding interactions occur." Ecologists are beginning to quantify feeding interac-

tions, and are moving far beyond the old, linear food chain studies to recognize the complex, multiple linkages created by the feeding relationships among large assemblages of species. Current research also recognizes the composite effects that trophic interactions have on the whole community.

One example of the significance of feeding relationships is the continuing role of predation in community ecology. A study might conclude that "Alaskan kelp forests are broadly dependent on **sea otter** predation for protection against destructive grazing." Another might look at the timing of predation, and conclude that "temporal variability in predation by whelks can have distinctive effects on prey, create distinctive community compositions, and affect successional paths in [an] intertidal community."

Community ecologists still emphasize species interactions that have been prominent through all the decades of field research. Ecologists still claim, for example, that "assumptions about competition lie at the heart of several models of plant community organization." Other articles in the same journal *Ecology* reinforce this emphasis, stating e.g., "that competition is pervasive in the [Chihuahuan **desert** rodent] community." Another example, in the *Journal of Ecology* indicates that "interspecific competition may limit the growth, branching, survival and berry production by Vaccinium dwarf shrubs in a subarctic plant community."

The degree of pattern or randomness of community structure has long been an issue in community ecology. Natural communities are immensely complex and it is difficult to simplify this complexity down to useable, predictive models. Researchers continue to investigate the extent to which species' interactions can result in a unit organized enough to be considered a coherent community. The mechanics of community assembly depend heavily on invasions, rates of succession, and on changes in the physical environment, as well as a diversity of co-evolutionary patterns.

A debate still continues among community ecologists about the importance of complexity, the role of species diversity and richness in the maintenance of a community. Community structure patterns might be dependent, in part, on factors as seemingly trivial as seed weight. **Accuracy** in estimations of the number of species in a given community is difficult, including the large number of **microorganisms** as yet unidentified and unnamed. Statistical inference can help to dilute this issue somewhat.

Disturbances and perturbations are important factors in the composition and character of communities. One study may focus on the role of fire in tall tussock **grasslands**. Another recent study of old-growth forests in British Columbia concluded that "small-scale natural disturbances involving the death of one to a few trees and creating gaps in the forest canopy are key processes in the...community ecology of many forests." In the recent "landscape view" emerging in ecology, some research indicates that natural communities may persist in patches, as long as they remain interconnected. This underlines again the importance of context in any attempt to understand community ecology.

The most significant source of disturbance and change in natural communities is human activity. Recent studies in community ecology have documented the impact of human removal and then reintroduction of a **keystone species** such as largemouth bass into a lake; have questioned the impact of "importing" an alien seaweed from Japan to expand Brittan's kelp farms; and have indicated that "human fishing pressure has influenced Caribbean **coral reef** community structure by affecting predator-herbivore relationships." Humans can also have positive effects through deliberate attempts to offset destructive impacts. A recent study published in the journal *Nature* described the deliberate modification of **pH** levels in lakes by adding phosphate **fertilizer**, which reversed **acidification** without "drastically altering the community structure." Such manipulations, modifications, alterations, and disruptions by humans could obviously be expanded into a very long list, but the idea should be clear.

One other type of human impact, however, is receiving a lot of attention: evidence is mounting that **climate** change is increasingly impacting natural communities. For example, one recent study suggested that "increased warming could change the dominant vegetation of a widespread meadow **habitat**, changing the competitive relationships between sagebrush and cinquefoil." Another indicated that "regional climatic warming may be altering the species composition of Alaskan arctic tundra" through changes in light, temperature and nutrients that affect community and ecosystem processes.

The information gained in community ecology studies is becoming increasingly important to achieving **conservation** objectives and establishing guidelines to management of the natural systems on which all humans depend. For example, clearer understanding of linkages established in communities through trophic exchanges can help predict the impacts of concentrations of **toxins** and pollutants. Better understanding of organismic interactions in a community context can help in comprehending the processes that lead to **extinction** of species, information critical to attempts to slow the loss of biological diversity. Research into community dynamics can result in better decisions about establishing preserves and refuges, and can create sustainable harvesting strategies.

The professional ecology journals do not publish research on the importance of community for the human species. Yet, few concepts are more significant in human affairs than the role of different kinds of community in establishing ties and linkages among human individuals and between humans and the locales in which they live. Equally important is the lack of community, the looseness of ties and the alienation from locale and from others that stem from failures to establish a sense of community. As in natural communities, interactions among human individuals create community and are in turn shaped by community.

[*Gerald L. Young Ph.D.*]

RESOURCES
BOOKS

Diamond, J., and T. Case, eds. *Community Ecology.* New York: Harper & Row, 1986.
Putman, R. J. *Community Ecology.* New York: Chapman and Hall, 1994.
Taylor, P. "Community," *Keywords in Evolutionary Biology,* ed. by E.F. Keller and E.A. Lloyd. Cambridge, MA: Harvard University Press, 1992.

Community right-to-know
see **Emergency Planning and Community Right-to-Know Act (1986)**

Compaction

Compaction is the mechanical pounding of **soil** and weathered rock into a dense mass with sufficient bearing strength or impermeability to withstand a load. It is primarily used in construction to provide ground suitable for bearing the weight of any given structure. With the advent of huge earth-moving equipment we are now able to literally move mountains. However, such disturbances loosen and expand the soil. Thus soil must be compacted to provide an adequate breathing surface after it has been disturbed. Inadequate compaction during construction results in design failure or reduced service life of a structure. Compaction, however, is detrimental to crop production because it makes root growth and movement difficult, and deprives the soil of access to life-sustaining oxygen.

With proper compaction we can build enduring roadways, airports, **dams**, building foundation pads, or clay liners for secure landfills. Because enormous volumes of ground material are involved, it is far less expensive to use on-site or nearby resources wherever feasible. Proper engineering can overcome material deficiencies in most cases, if rigid quality control is maintained.

Successful compaction requires a combination of proper moisture conditioning, the right placement of material, and sufficient pounding with proper equipment. Moisture is important because dry materials seem very hard, but they may settle or become **permeable** when wet. Because of all the variables involved in the compaction process, standardized laboratory and field testing is essential.

The American Society of State Highway **Transportation** Officials (ASHTO) and the American Society of Testing Materials (ASTM) have developed specific test standards. The laboratory test involves pounding a representative sample with a drop-hammer in a cylindrical mold. Four uniform samples are tested, varying only in moisture content. The sample is trimmed and weighed, and portions oven-dried to determine moisture content.

The results are then graphed. The resultant curve normally has the shape of an open hairpin, with the high point representing the maximum density at the optimum moisture content for the compactive effort used. This curve reflects the fact that dry soils resist compaction, and overly-moistened soils allow the mechanical energy to dissipate. Field densities must normally meet 95% or higher of this lab result.

Because soils are notoriously diverse, several different "curves" may be needed; varied materials require the field engineer to exercise considerable judgment to determine the proper standard. Thus the field engineer and the earth-moving crew must work closely together to establish the procedures for obtaining the required densities.

In the past, density testing has required laboriously digging a hole, and comparing the volume with the weight of the material removed. Nuclear density gages have greatly accelerated this process and allow much more frequent testing if needed or desired. To take a reading, a stake is driven into the ground to form a hole into which a sealed nuclear probe can be lowered.

It is much easier to properly compact materials during construction than to try to make corrections later. A well-compacted, properly tested structure is an investment in the future, well worth the time and effort expended.

[*Nathan H. Meleen*]

RESOURCES
BOOKS

American Society for Testing & Materials. *Compaction of Soils.* ASTM Special Technical Publication, No. 377. Philadelphia: ASTM, 1965.

Daniel, D. P. "Summary Review of Construction Quality Control for Compacted Soil Liners." In *Waste Containment Systems: Construction, Regulation, and Performance,* edited by R. Bonaparte. New York: American Society of Civil Engineers, 1990.

Comparative risk

Initiated by the United States Environmental Protection Agency's (EPA) federal Comparative Risk Project in 1986, comparative risk projects by the end of the twentieth century, there were 46 projects underway. Furthermore, today the priority setting method had gained national attention as many members of Congress, federal professionals, and policy experts agree that environmental protection and public health agencies should, when setting priorities, consider the relative degree of risk their actions will reduce. However, while proponents tout comparative risk as a rational approach to making decisions about priorities, critics among national and grassroots environmental organizations believe that comparative risk ranking can be a "hard" technocratic approach that ignores non-risk factors in decision making, is "undemocratic," and is contrary to **pollution** prevention or the will of the people.

In the past, risk has not been factored into environmental priority setting and, as remarked by William Ruckelshaus, who twice served as EPA administrator, it "was hardly mentioned in the early years of EPA, and it does not have an important place in the Clean Air or Clean Water Acts passed in that period." But, building on Ruckelshaus's groundwork in comparative risk, EPA Administrator Lee Thomas helped thrust the concept onto the national agenda in 1986 when he asked 75 agency professionals to examine 31 environmental problems in the areas of **cancer**, non-cancer, and ecological risks as well as welfare effects (**visibility** impairment, damage to statuary from **acid rain**, etc.). In their 1987 report to Thomas, *Unfinished Business: A Comparative Assessment of Environmental Problems,* the professionals reported that, "Overall, EPA's priorities appear more closely aligned with public opinion than with our estimated risks." For instance, the EPA professionals regarded indoor **air pollution** and global warming as relatively high risks and contaminated **hazardous waste** sites as relatively low risk problems, while opinion polls showed that the public had an opposite ranking. In releasing the report, Thomas noted that EPA has finite resources and therefore must choose its priorities carefully "so that we apply those resources as effectively as possible" in reducing risks. The EPA report became the first significant study suggesting that national environmental priorities were inconsistent with the experts' sense of the most serious environmental problems, and instead the nation's priorities were guided by public perceptions of risk.

Concerns about the growing number of environmental regulations and problems demanding attention, coupled with the decline in resources to spend on these problems, kept up the pressure for EPA, states, and local governments to find a rational approach for deciding where to target finite resources. The publication of a report in 1991, *Environmental Investments: The Cost of a Clean Environment,* reinforced these concerns. That report was a first estimation of the costs that industries and municipalities must incur through compliance with federal **pollution control** programs. In 1991, costs were $115 billion a year—2.1% of the Gross National Product (GNP)—but were projected to rise to $185 billion or 2.8% of the GNP by the year 2000.

In late 1991, Senator Daniel Patrick Moynihan (D-N.Y.) introduced a bill (S.2132) that would have required EPA to seek the advice of experts in ranking relative risks and to use that information in managing its available resources to protect society from the greatest risks. Moynihan's bill, and its successor in the 103rd Congress, became important elements of the debate over whether risk-based "rational" priority setting should be required to ensure more efficient risk reduction.

Even before the 1991 EPA report on the rising costs of environmental protection, EPA Administrator William K. Reilly had taken up comparative risk as a central theme of his administration and had asked the agency's Science Advisory Board to review the report issued by the 75 agency professionals. The board's response, a 1990 report called *Reducing Risk: Setting Priorities And Strategies For Environmental Protection*, also recommended that EPA and other government agencies "assess the range of environmental problems of concern and then target protective efforts at problems that seem to be the most serious." Far and away the greatest expenditures of EPA resources were directed at the agency's construction grants program for publicly owned **wastewater** treatment plants and the cleanup of abandoned Superfund sites—for instance, a full 70% of the agency's fiscal year 1990 budget of $6 billion went to these programs—even though other problems were deemed more serious from a scientific perspective. The science advisors also suggested that EPA should give more attention to the relatively neglected job of protecting ecosystems, which faced high risks from **habitat** alteration, loss of **biodiversity**, global **climate** change, and **ozone layer depletion**.

Reducing Risk, and the general topic of whether "Worst Things First" should be the touchstone of environmental priority setting, was debated at a November 1992 meeting in Annapolis, Maryland, at which critics challenged its "quasi-scientific" claims and offered competing priority setting approaches, including a strategy that would make preventing pollution the fundamental criterion for **environmental policy** decisions founded on the public's decision that pollution is undesirable. In the critics' view, comparative **risk analysis** presumes that some pollution and its risk is acceptable.

Skepticism notwithstanding, EPA chose to foster development of the method. As part of its endorsement of comparative **risk assessment** as an important priority setting tool, EPA promoted a series of comparative risk projects in several of its regions—Region 1, New England; Region 3, the Mid-Atlantic states; and Region 10, the Pacific Northwest—and in three pilot states: Washington, Colorado, and Pennsylvania. Later, Vermont, California, Utah, Michigan, and other states joined in. Sub-state entities also developed projects, including Columbus, Ohio; Atlanta, Georgia; Elizabeth River, Virginia; Houston, Texas; and Wisconsin tribes.

Typically, comparative risk projects follow six basic steps. First, they define and analyze the risks posed by environmental problems facing the jurisdiction, usually working from a list of problems. Second, they rank the problems according to the relative severity of each, using technical information and criteria to grade the negative impacts of individual problems on health, ecosystems, or quality of life. Third, they select priorities for special attention and set goals for reducing risks posed by the problem. Fourth, they propose, analyze, and compare strategies for achieving the goals set in step three. Fifth, they implement strategies having the greatest risk-reduction promise. And, sixth, they monitor results produced by the strategies and adjust jurisdictional policies or budgets based on those results. These steps are carried out through committees composed of state or local officials, industry representatives, environmentalists, and citizens. The projects all seek to broadly incorporate public values—not merely technical scientific information—in assessing and ranking risks.

Among the more widely cited state comparative risk projects is Washington Environment 2010, which, more than any other state project, influenced legislation and state policy. It was initiated in 1988 by Washington's Department of Ecology Director Christine Gregoire and included technical committees with members from 19 state agencies and a steering committee composed of senior managers or heads of those agencies. In addition, it had a public advisory committee composed of 34 prominent legislators, representatives, and important interest groups. Resulting from this project was a ranking of 23 threats to the environment within five priority levels. **Ambient air** pollution, point-source discharges of pollution to water, and polluted **runoff** ranked as the top priorities, and non-ionizing radiation, materials storage in tanks, and litter ranked as the lowest priorities. Based on the project, and a statewide effort to solicit public responses that produced some 300 risk-reduction options, Washington's legislature in 1991 adopted several new environmental laws dealing with clean air, **transportation** demand, **water conservation**, **recycling**, growth management, and the state's **energy policy**. The Department of Ecology redirected $6.8 million of its budget from lower to higher risk priorities. However, in 1993 Gregoire's successor reorganized the department, dismantling an Environment 2010 planning staff that had been established and, with it, the state's institutional memory of the comparative risk project. Washington's experience in this regard, and similar experiences in other states, have made some proponents of comparative risk assessment question how deeply and lastingly its effects will be felt in environmental programs.

Despite such setbacks, the comparative risk concept has continued to find a significant place in governmental discussions of environmental policy. In 1993, the White

House of Science and Technology Policy and the **Office of Management and Budget** invited comparative risk analysis experts to meet with federal government officials to explore how the approach could be used by federal agencies to establish broad priorities more systematically. That effort resulted in a 1996 report, *Comparing Environmental Risks: Tools for Setting Government Priorities*, that proposed a basic framework for using the method in federal agencies. Furthermore, during the 104th Congress, comparative risk analysis was one of numerous risk-related reforms that were included in the Republican Contract With America and its agenda of reducing regulatory burdens through a downsizing of government, the most far-reaching attempt to impose a "rational" risk-based priority setting system on federal agencies.

While the future of comparative risk analysis as a basis for establishing environmental priorities remains uncertain, federal, state, and local interest remains strong. Today, most practitioners and proponents of the approach recognize that comparative risk alone will not suffice to establish priorities. Public values, hard to quantify benefits, and other non-scientific elements that must be weighed in decisions about priorities have a clearly recognized role. That conclusion has been reinforced in all the state and local projects. Some recent studies suggest that the future progress of comparative risk analysis requires more research on how to better inform diverse parts of the affected public throughout the process of setting environmental priorities. Even though federal legislation that would mandate comparative risk analysis as a basis for selecting priorities may never pass, the risk-based approach enunciated in 1987 has emerged today as a central piece of environmental policy and is explicitly discussed in EPA's draft *Environmental Goals for America With Milestones for 2005*, the latest federal effort to clarify and organize environmental priorities.

[*David Clarke*]

RESOURCES
BOOKS

Davies, J. C., ed. *Comparing Environmental Risks: Tools for Setting Government Priorities*. Washington, D.C.: Resources for the Future, 1996.

Finkel, A. M., and D. Golding, eds. *Worst Things First? The Debate over Risk-Based National Environmental Priorities*. Washington, D.C.: Resources for the Future, 1994.

U. S. Environmental Protection Agency. Science Advisory Board. *Reducing Risk: Setting Priorities and Strategies for Environmental Protection*. Washington, D.C.: GOP, 1990.

U. S. Environmental Protection Agency. Office of Policy, Planning, and Evaluation. *Unfinished Business: A Comparative Assessment of Environmental Problems*. Washington, D.C.: GPO, 1987.

Competition

Competition is the interaction between two organisms when both are trying to gain access to the same limited resource. When both organisms are members of the same **species**, such interaction is said to be "intraspecific competition." When the organisms are from different species, the interaction is "interspecific competition."

Intraspecific competition arises because two members of the same species have nearly identical needs for food, water, sunlight, nesting space, and other aspects of the **environment**. As long as these resources are available in abundance, every member of the community can survive without competition. When those resources are in limited supply, however, competition is inevitable. For example, a single nesting pair of bald eagles requires a minimum of 620 acres (250 ha) that they can claim as their own territory. If two pairs of eagles try to survive on 620 acres, competition will develop, and the stronger or more aggressive pair will drive out the other pair.

Intraspecific competition is also a factor in controlling plant growth. When a mature plant drops seeds, the seedlings that develop are in competition with the parent plant for water, sunlight, and other resources. When abundant space is available and the size of the community is small, a relatively large number of seedlings can survive and grow. When population density increases, competition becomes more severe and more seedlings die off.

Competition becomes an important limiting factor, therefore, as the size of a community grows. Those individuals in the community that are better adapted to gain food, water, nesting space, or some other limited resource are more likely to survive and reproduce. Intraspecific competition is thus an important factor in natural selection.

Interspecific competition occurs when members of two different species compete for the same limited resource(s). For example, two species of birds might both prefer the same type of insect as a food source and will be forced to compete for it if it is in limited supply.

Laboratory studies show that interspecific competition can result in the **extinction** of the species less well adapted for a particular resource. However, this result is seldom, if ever, observed in **nature**, at least among animals. The reason is that individuals can adapt to take advantage of slight differences in resource supplies. In the Galapagos Islands, for example, 13 similar species of finches have evolved from a single parent species. Each species has adapted to take advantage of some particular **niche** in the environment. As similar as they are, the finches do not compete with each other to any specific extent.

Interspecific competition among plants is a different matter. Since plants are unable to move on their own, they are less able to take advantage of subtle differences in an environment. Situations in which one species of plant takes over an area, causing the extinction of competitors, are well known.

One mechanism that plants use in this battle with each other is the release of toxic **chemicals**, known as allelochemicals. These chemicals suppress the growth of plants in other—and, sometimes, the same—species. Naturally occurring antibiotics are examples of such allelochemicals.

[*David E. Newton*]

RESOURCES
BOOKS

Moran, J. M., M. D. Morgan, and J. H. Wiersma. *Environmental Science*. Dubuque, IA: William C. Brown, 1993.

Competitive exclusion

Competitive exclusion is the interaction between two or more **species** that compete for a resource that is in limited supply. It is n ecological principle involving competitors with similar requirements for **habitat** or resources; they utilize a similar **niche**. The result of the **competition** is that one or more of the species is ultimately eliminated by the species that is most efficient at utilizing the limiting resource, a driving force of **evolution**. The competitive exclusion principle or "Gause's principle" states that where resources are limiting, two or more species that have the same requirements for the limiting resources cannot co-exist. The co-existing species must therefore adopt strategies that allow resources to be partitioned so that the competing species utilize the resources differently in different parts of the habitat, at different times, or in different parts of the life cycle.

[*Marie H. Bundy*]

Composting

Composting is a fermentation process, the break down of organic material aided by an array of **microorganisms**, earthworms, and other insects in the presence of air and moisture. This process yields compost (residual organic material often referred to as **humus**), ammonia, **carbon dioxide**, sulphur compounds, volatile organic acids, water vapor, and heat. Typically, the amount of compost produced is 40–60% of the volume of the original waste.

For the numerous organisms that contribute to the composting process to grow and function, they must have access to and synthesize components such as **carbon, nitrogen**, oxygen, **hydrogen**, inorganic salts, sulphur, **phospho-**

rus, and trace amounts of micronutrients. The key to initiating and maintaining the composting process is a carbon-to-nitrogen (C:N) ratio between 25:1 and 30:1. When **C:N ratio** is in excess of 30:1, the **decomposition** process is suppressed due to inadequate nitrogen limiting the **evolution** of bacteria essential to break the strong carbon bonds. A C:N ratio of less than 25:1 will produce rapid localized decomposition with excess nitrogen given off as ammonia, which is a source of offensive odors.

Attaining such a balance of ratio and range is possible because all organic material has a fixed C:N ratio in its tissue. For example, **food waste** has a C:N ratio of 15:1, sewage **sludge** has a C:N ratio of 16:1, grass clippings have a C:N ratio of 19:1, leaves have a C:N ratio of 60:1, paper has a C:N ratio of 200:1, and wood has a C:N ratio of 700:1. When these (and other) materials are mixed in the right proportions, they provide optimum C:N ratios for composting. Typically, nitrogen is the limiting component that is encountered in waste materials and, when insufficient nitrogen is present, the composting mixture can be augmented with agricultural fertilizers, such as urea or ammonia nitrate.

In addition to nutrients, the efficiency of the composting process depends on the organic material's size and surface characteristics. Small particles provide multi-faceted surfaces for microbial action. Size also influences porosity (crevices and cracks which can hold water) and permeability (circulation or movement of gases and moisture).

Moisture (water) is an essential element in the biological degradation process. A moisture level of 55–60% by weight is required for optimal microbial, **nutrient**, and air circulation. Below 50% moisture, the nutrients to sustain microbial activity become limited; above 70% moisture, air circulation is inhibited.

Air circulation controls the class of microorganisms that will predominate in the composting process: air-breathing microorganisms are collectively termed **aerobic**, while microorganisms that exist in the absence of air are called **anaerobic**. When anaerobic microorganisms prevail, the composting process is slow, and unpleasant-smelling ammonia or hydrogen sulfide is frequently generated. Aerobic microorganisms will quickly decompose organic material into its principal components of carbon dioxide, heat and water.

The role of acidity and alkalinity in the composting process depends upon the source of organic material and the predominant microorganisms. Anaerobic microorganisms generate acidic conditions which can be neutralized with the addition of lime. However, such adjustments must be done carefully or nitrogen imbalance will occur that can further inhibit biological activity and produce ammonia gas, with its associated unpleasant odor. Organic material with a balanced C:N ratio will initially produce acidic conditions,

6.0 on the **pH** scale. However, at the end of the process cycle, mature compost is alkaline, with a pH reading greater than 7.0 and less than 8.0.

The regulation and measurement of temperature is fundamental to achieving satisfactory processing of organic materials. However, the effect of ambient or surface temperatures on the process is limited to periods of intense cold when biological growth is dormant. Expeditious processing and reduction of herbicides, pathogens, and pesticides is achieved when internal temperatures in the compost pile are maintained at 120–140°F (55–160°C). If the internal temperature is allowed to reach or exceed 150°F (65°C), biological activity is inhibited due to heat stress. As the nutrient content is depleted, the internal temperature decreases to 85°F (30°C) or less—one criteria for defining mature or stabilized compost.

Mature or stabilized compost has physical, chemical, and biological properties which offer a variety of attributes when applied to a host **soil**. For example, adding compost to barren or disturbed soils provides organic and microbial resources. The addition of compost to clay soils enhances the movement of air and moisture. The water retention capacity of sandy soil is enhanced by the addition of compost and **erosion** also is reduced. Soils improved by the addition of compost also display other characteristics such as enhanced retention and exchange of nutrients, improved seed germination, and better plant root penetration. Compost, however, has insufficient nitrogen, phosphorous, and potassium content to qualify as a **fertilizer**. The ultimate *application* or *disposition* of compost depends upon its quality, which is a function of the type of organic material and the method(s) employed to enhance or control processing.

Compost processing can be as simple as a plastic **garbage** bag filled with a mixture of plant waste that has had a couple of ounces of fertilizer, some lime, and sufficient water added to make the material moist. The bag is then sealed and set aside for about 12 months. Faster processing can be achieved with the use of a 55 gal (208 l) drum into which 0.5-in (1.3-cm) holes have been drilled for air circulation. Filled with the same mixture as the garbage bag and rotated at regular intervals, this method will produce compost in two to three months. A multi-compartmentalized container is faster and increases the diversity of materials which can be processed. However, including such items as fruit and vegetable scraps, meat and dairy products, cardboard cartons, and fabrics must be undertaken with caution because they attract and breed vermin.

Such methods are designed for individual use, especially by those who can no longer dispose of garden waste with their **household waste**. Similarly, commercial and government institutions employ natural (low), medium, and advanced technical composting methods, depending on their motivation, i.e., diminishing **landfill** capacity, availability of fiscal resources, and commitment to **recycling**. The simplest composting method currently employed by industry and municipalities entails dumping **organic waste** on a piece of land graded (sloped) to permit precipitation and leachate (excess moisture and organics from composting) to collect in a retention pond. The pond also serves as a source of moisture for the compost process and as a system where **photosynthesis** can oxidize the leachate. The organic material is placed in piles called *windrows* (a ridge pile with a half-cone at each end). The dimensions of a windrow are typically 10–12 ft (3–3.7 m) wide at the base and about 6 ft (1.8 m) high at the top of the ridge. The length is site specific. Windrows are constructed using a front-end loader. A mature compost will be available in 12–24 months, depending on various factors including the care with which the organic material was blended to obtain optimum C:N ratio; the supplementation of the material with additional nutrients; the frequency of **aeration** (mixing and turning); and the moisture content maintained.

Using the same site layout, the next step in mechanization is the use of windrow turners. These turners can be simple aeration machines or machines with the added ability to shred material into smaller particles, while injecting supplemental moisture and/or nutrients. Optimizing the capabilities of such equipment requires close attention to temperature variation within the windrows. Typically, the operator will use a temperature probe to determine when the temperature falls in the range of 100°F (37–38°C). The equipment will then fold the outer surface of the windrow inward, replenishing air and moisture, and mixing in unconsumed or supplemental nutrients. This promotes further decomposition, which is identified by a gradual rise in temperature. Sequential turning and mixing will continue until temperatures are uniformly diminished to levels below 85°F (30°C). This method produces a mature compost in four to eight months.

Two more technologically advanced composting methods are the *in-vessel system* and the *forced-air system*. Both are capable of processing the bulk of all solid and liquid municipal wastes. However, such flexibility imposes substantial capital, technical, and operational requirements. In forced air processing, organic material is placed on top of a series of perforated pipes attached to fans which can either blow air into, or draw air through the pile to control its temperature, oxygen, and carbon dioxide needs. This system is popular for its ability to process materials high in moisture and/or nitrogen content, such as yard wastes. Time to produce a mature compost is measured in days, depending on the class of organic material processed. During in-vessel processing, organic material is continuously fed into an inclined rotating cylinder, where the temperature, moisture,

A man adds moisture to mushroom compost. (Photograph by Holt Confer. Phototake. Reproduced by permission.)

and nutrient air and gas levels are closely controlled to achieve degradation within 24–72 hours. The composted material is then screened to remove foreign or inert materials such as glass, **plastics**, and metals and is allowed to mature for 21 days.

[*George M. Fell*]

RESOURCES
BOOKS

Appelhof, M. *Worms Eat My Garbage.* Kalamazoo, MI: Flower Press, 1982.

The Biocycle Guide to the Art and Science of Composting. Emmaus, PA: JG Press, 1991.

The Biocycle Guide to Yard Waste Composting. Emmaus, PA: JG Press, 1989.

The Rodale Book of Composting. Emmaus, PA: Rodale Press, 1992.

PERIODICALS

Kovacic, D. A., et al. "Compost: Brown Gold or Toxic Trouble?" *Environmental Science and Technology* 26 (January 1992): 38-41.

Lecard, M. "Urban Decay." *Sierra* 76 (September-October 1991): 27-8.

Composting toilets
see **Toilets**

Comprehensive Environmental Response, Compensation, and Liability Act (CERCLA)

In response to **hazardous waste** disasters such as **Love Canal**, New York, in the 1970s, Congress passed the Comprehensive Environmental Response, Compensation, and Liability Act (CERCLA), better known as Superfund, in 1980. The law created a fund of $1.6 billion to be used to clean up hazardous waste sites and hazardous waste spills for a period of five years. The primary source of support for the fund came from a tax on chemical feedstock producers; general revenues supplied the rest of the money needed. CERCLA is different from most environmental laws because it deals with past problems rather than trying to prevent future **pollution**, and because the **Environmental Protec-**

tion Agency (EPA), in addition to acting as a regulatory agency, must clean up sites itself.

Throughout the decade before the creation of Superfund, the public began to focus increasing attention on hazardous wastes. During this period, an increased number of cases involving such wastes contaminating drinking water, streams, rivers, and even homes were reported. Citizens were enraged by dangers posed by leaking landfills, illegal dumping of hazardous wastes along roads or in vacant lots, and explosions and fires at some facilities.

A strong catalyst for hazardous waste regulation was the Love Canal episode near Niagara Falls, New York. In the late 1970s, chemical wastes from an abandoned dump were discovered in the basements of some homes. Studies found significant health effects, including miscarriages and low-weight newborns. Residents worried about increased **cancer** and birth defect rates. In the federal emergency declaration, a school and 200 houses were condemned. The combination of public concern and media coverage, together with EPA interest, brought national attention to this issue.

Debate soon began over proper government response to these problems. Industrial interests argued that a company should not be liable for cleaning up past hazardous waste dumps if it had not violated any law in disposing of the wastes. The industry argued that general taxes, not industry-specific taxes, should be used for the clean-up, and it sought to limit the legal liability of manufacturers in regard to the health effects of their hazardous wastes. Industrial companies also pushed for one national regulatory program, rather than a national program and several state programs to complicate the situation.

When Congress began debating a Superfund program in 1979, EPA officials argued that industry must pay the bulk of the clean-up costs. They based this argument on the philosophy that the polluter should pay, and also the pragmatic reasoning that Congress could not be relied on to continue appropriating the funds needed for such an expensive and lengthy program. The Senate focused on a comprehensive bill that included provisions for liability and **victims' compensation**, but these were dropped in order to secure passage of the program through the House. The act was signed by President Carter in December 1980.

Under the law, the EPA determines the most dangerous hazardous waste sites, based on characteristics like toxicity of wastes and risk of human exposure, and places them on the **National Priorities List** (NPL), which determines priorities for EPA efforts. The EPA has the authority to either force those responsible for the site to clean it up, or clean up the site itself with the Superfund money and then seek to have the fund reimbursed through court action against the responsible parties. If more than one company had dumped wastes at a site, the agency is able to hold one

party responsible for all clean-up costs. It is then up to that party to recover its costs from the other responsible parties. If those identified as responsible by the EPA deny their responsibility in court and lose, they are liable for treble damages. Removal actions, emergency clean-ups, or actions costing less than $2 million and lasting less than a year, can be undertaken by the EPA for any site. For federal hazardous waste sites, the clean-up must be paid for through the appropriation process rather than through Superfund. States are required to contribute 10 to 50 percent of the cost of clean-ups within their boundaries; they are also responsible for all operation and maintenance costs once the job is finished. The EPA can also delegate lead clean-up authority to the states.

Major amendments to CERCLA were passed in 1986. As the scope of the problem grew, Congress re-authorized the Superfund through 1991 and increased its size to $8.5 billion. Plans indicated that the enlarged Superfund would be financed by taxes on **petroleum**, feedstock **chemicals**, corporate income, general revenue, interest from money in the fund, and money recovered from companies responsible for earlier clean-ups. The amendments required several areas of compliance: 1) The clean-ups must meet the applicable state and federal environmental standards; 2) The EPA must begin clean-up on at least 375 sites by 1991; 3) Negotiated settlements for clean-ups are preferred to court litigation; 4) Emergency procedures and community **right-to-know** standards, are required in areas with hazardous waste facilities (largely in response to the **Bhopal, India** toxic gas disaster); and 5) Federal agencies must comply with Superfund Amendments and begin the clean-up of federal facilities and sites. The 1991 Superfund Amendments and Reauthorization Act authorized a four-year extension of the taxes that financed Superfund, but the law was not changed significantly.

As of August 1990, 33,000 sites were listed as being potentially hazardous, 1,082 sites were on the NPL, and over 100 sites were proposed by the EPA to be added to the list. Of the sites that required preliminary EPA investigation, over 90% had been examined, but actual clean-up has been rather slow. In mid-1990, clean-up had been completed at only 54 NPL sites. However, funding had been approved for planning studies at over 1,000 sites, design work at over 400 sites, and remedial work at over 280 sites. Removal actions by the EPA or responsible parties had taken place at over 1,500 sites, most of which were not on the NPL.

Studies of Superfund implementation have been quite critical. Reports by Congressional committees, the General Accounting Office, and the Office of Technology Assessment (OTA) concluded that the EPA relied on temporary rather than permanent treatment methods, took too long to clean up sites because of poor management, too frequently opted to use the Superfund for clean-ups rather than requiring responsible parties to pay, and often lacked the expertise to oversee Superfund clean-up operations. In reality, early implementation efforts of Superfund were hampered by uncommitted EPA officials, lack of financial and staff resources, poor government coordination of policy objective, and by the complexity of identifying and exacting payment from responsible parties.

Another problem had been the amount of expensive and time-consuming litigation involved in the act. In some cases litigation costs to determine responsible parties and recover clean-up costs has exceeded the cost of clean-up itself. Between 1986 and 1988, the EPA only recovered 7% of what it spent on clean-up from private parties.

Implementation of CERCLA has also been marked by charges of corruption and political manipulation. Rita Lavelle, who was in charge of the Superfund program at EPA, resigned in 1983 amid charges that she was giving unduly favorable treatment to industry. She was later convicted on perjury charges. Also in 1983, EPA Administrator Anne Gorsuch Burford resigned, largely in response to the difficulties of Superfund implementation.

Thus far, CERCLA has proved to be a more complicated, costly, and time-consuming process than originally envisioned. A 1988 Congressional report estimated that between $16.7 and $23.8 billion of federal money would be needed to clean up the less than 1,000 sites then on the NPL. A 1989 OTA report estimated the cost of the program to be $500 billion in the long run, with as many as 10,000 sites eventually being placed on the NPL. *See also* Hazardous material; Hazardous Materials Transportation Act; Hazardous Substances Act; Hazardous waste site remediation; Hazardous waste siting; Toxic substances

[*Christopher McGrory Klyza*]

RESOURCES

BOOKS

Davis, C. E. *The Politics of Hazardous Waste.* New York: Prentice Hall, 1993.
Dower, R. C. "Hazardous Wastes." In *Public Policies for Environmental Protection,* edited by P. R. Portney. Washington, DC: Resources for the Future, 1990.
Hays, S. P. *Beauty, Health, and Permanence: Environmental Politics in the United States, 1955-1985.* New York: Cambridge University Press, 1990.
Mazmanian, D., and D. Morell. *Beyond Superfailure: America's Toxics Policy for the 1990s.* Boulder, CO: Westview Press, 1992.

Computer disposal

Because computer technology changes so quickly, the average computer sold in the United States in the early 2000s becomes obsolete in only three years. Consumers were ex-

pected to retire about 50 million computers in 2002. One government survey reported that 75% of all the computers ever sold in the United States were stockpiled by 2001, not disposed of even though their useful life is over. Computers and other electronics account for about 220 million tons of waste annually, according to the United States **Environmental Protection Agency** (EPA). Some older computers find new users when they are passed on to nonprofit groups, schools, or needy families. Some manufacturers arrange to take back their out-of-date products. Approximately 10% of outdated computers are recycled.

Computers contain a variety of materials, some of them toxic. If computers are not recycled but disposed of in landfills, valuable material is wasted, and **toxins**, particularly **lead** and **mercury**, may present a hazard to people and the **environment**. As many consumers are still storing two or three older computers, the number of computers currently in the **waste stream** is only a fraction of what it might be, so computer disposal is a looming problem.

In the United States, the electronics industry, **conservation** groups, and government organizations began working toward a satisfactory system of computer disposal in the early 2000s. The Japanese government enacted a computer **recycling** law in 2001, and the **European Union** passed* similar legislation, that will take effect in the middle of the decade.

A typical computer consists of 30–40% plastic. The plastic may be of several different types. Unlike the plastic in food and beverage containers, which is usually given a number to identify it and make recycling easier, plastic in computers is unlabeled. A computer may contain over four pounds of lead, as well as small amounts of the toxic metals mercury and **cadmium** and traces of other metals including gold, silver, steel, **aluminum**, **copper**, and **nickel**.

Recycling a computer is not a simple process, and is also quite expensive. To recycle a computer it must be broken into parts, and its usable materials separated. Plastic used for computers can be recovered, reprocessed into pellets, and sold for re-use. The metals also can be recovered, although the process itself can generate dangerous waste. For instance, gold can be stripped off computer chips with a wash of hydrochloric **acid**. The acid must then be treated or stored safely, or it can contaminate the environment.

Recycling is not an easy solution to the problem of computer disposal. However, a few companies in the United States have found electronics recycling to be a profitable business. A report compiled by two West Coast environmental groups in 2002 found that up to 80% of computers collected for recycling in California and other western states ended up in **third world** countries, where parts were often salvaged by low-paid workers. Not only were workers often unprotected against toxic materials, but toxic waste was dumped directly into lakes and streams, the report detailed.

Some government and conservation groups have suggested that the computer industry try to reduce toxic waste by redesigning its products. Labeling of **plastics** used would simplify plastic recycling, as would phasing out some more harmful materials and toxic fire-retardant coatings. Making computers with parts that snap together instead of using glue or metal nuts and bolts is another design consideration that can make computers easier to recycle. A number of major manufacturers, including Dell Computer and IBM, began their own recycling programs in the early 2000s. A coalition of industry, environmental, and government groups called the National Electronics Product Stewardship Initiative began meeting in 2001 to come up with national guidelines for computer disposal. The high cost of computer recycling is expected to decline somewhat as the volume of recycled machines rises. Because of the vast backlog of computers in the United States waiting to be thrown out, it is imperative to work out a system for safe disposal quickly.

[*Angela Woodward*]

RESOURCES

PERIODICALS

Chappell, Jeff. "A Growing Problem." *Electronic News* 48, no. 11 (March 11, 2002):1.

"Japan to Mandate Supplier Disposal of Home PCs". *Computergram* (July 2, 2001):N.

Schuessler, Heidi. "All Used Up and Someplace to Go." *New York Times* (November 23, 2001):G1, G9.

Toloken, Steve. "Group Wants Disposal Put on Computer Makers." *Plastics News* 13, no. 41 (December 10, 2001):7.

Truini, Joe. "Electronics Afterlife." *Waste News* 6, no. 32 (January 8, 2001):1

Truini, Joe. "Electronic Waste Spurs California Action." *Waste News* 7, no. 5 (July 9, 2001):1.

Truini, Joe. "Electronic Waste Stream Comes to Fore." *Waste News* 7, no. 17 (December 24, 2001):10.

Wade, Beth. "Life After Death for the Nation's PCs." *American City and County*116, no. 4 (March 2001):22.

ORGANIZATIONS

Electronic Industries Alliance, 2500 Wilson Boulevard, Arlington, VA USA 22201 (703) 907-7500, <http://www.eia.org>

National Electronics Product Stewardship Initiative, <http://eerc.ra. utk.edu/clean/nepsi/index.htm>

Condensation nuclei

When air is cooled below its **dew point**, the water vapor it contains tends to condense as droplets of water or tiny ice crystals. Condensation may not occur, however, in the absence of tiny particles on which the water or ice can form. These particles are known as condensation nuclei. The most common types of condensation nuclei are crystals of salt,

particulate matter formed by the **combustion** of **fossil fuels**, and dust blown up from the earth's surface. In the process of cloud-seeding, scientists add tiny crystals of dry ice or silver iodide as condensation nuclei to the **atmosphere** to promote cloud formation and precipitation.

Condor

see **California condor**

Congenital malformations

see **Birth defects**

Congo River and basin

The Congo River (also known as the Zaire River) is the third longest river in the world, and the second longest in Africa (after the Nile River in northeastern Africa). Its river basin, one of the most humid in Africa, is also the largest on that continent, covering over 12% of the total land area.

History

The equatorial region of Africa has been inhabited since approximately the middle Stone Age. Late Stone Age cultures flourished in the southern savannas after about 10,000 B.C. and remained functional until the arrival of Bantu-speaking peoples during the first millennium B.C. In a series of migrations taking place from about 1,000 B.C. to the mid-first millennium A.D., many Bantu-speakers dispersed from an area west of the Ubangi-Congo River swamp across the forests and savannas of the region known as the modern-day Democratic Republic of the Congo.

In the precolonial era, this region (modern-day Democratic Republic of the Congo) was dominated by three kingdoms: Kongo (late 1300s), the Loango (at its height in the 1600s), and Tio. Portugese navigator Diogo Cam was the first European to sail up the mouth of the Congo in 1482. After meeting with the rulers of the Kingdom of Kongo, Cam negotiated intercontinental trade and commerce agreements—including the slave trade—between Portugal and the region. And a long history of colonialism began.

Over the centuries, the Congo River has inspired both mystery and legend, from the explorations of Henry Morton Stanley and David Livingstone in the 1870s, to Joseph Conrad, whose novel, *Heart of Darkness* transformed the river into an eternal symbol of the "dark continent" of Africa.

Characteristics

The Congo River is approximately 2,720 mi long (4,375 km), and its **drainage** basin consists of about 1.3 million mi^2 (3.6 million km^2). The basin encompasses nearly the entire Democratic Republic of the Congo (capital: Kinshasa), Republic of Congo (capital: Brazzaville), Central African Republic, eastern Zambia, northern Angola, and parts of Cameroon and Tanzania. The river headwaters emerge at the junction of the Lualaba (the Congo's largest tributary) and Luvua rivers. The flow is generally to the northeast first, then west, and finally south to its outlet into the Atlantic Ocean at Banana, Republic of Congo.

The Congo basin comprises one of the most distinct land depressions between the Sahara **desert** to its north, and the Atlantic Ocean to its south and west. The river's tributaries flow down slopes varying from 900 to 1,500 ft (274 to about 457 m) into the central depression forming the basin. This depression extends for more than 1,200 mi (about 1931 km) from the north to the south, from the Congo Lake Chad **watershed**, to the plateaus of Angola. From the east to west of the depression is another 1,200 mi (about 1931 km)—from the Nile-Congo watershed to the Atlantic Ocean. The width of the Congo River ranges from 3.5 mi (about 5.75 km) to 7 mi (about 11.3 km); and its banks contain natural levees formed by **silt** deposits. During floods, however, these levees overflow, widening the river.

With an average annual rainfall of 1,500 mm of rain (about 60 in), about three-quarters returns to the **atmosphere** by **evapotranspiration**; the rest is discharged into the Atlantic. The river is divided into three main regions: the upper Congo, with numerous tributaries, lakes, waterfalls, and rapids; the middle Congo; and, the lower Congo. The middle Congo is characterized by its seven waterfalls, collectively referred to as **Boyoma** (formerly Stanley) Falls. It is below these falls that navigation on the river becomes possible. The river has approximately 10,000 mi (about 16,000 km) of waterways, creating one of the main **transportation** routes in Central Africa.

Economic and environmental impact

Due to its size and other key elements, the Congo River and its basin are crucial to the ecological balance of an entire continent. Although the Congo water **discharge** levels were unstable throughout the second half of the twentieth century—the hydrologic balance of the river has provided some relief from the **drought** that has afflicted the river basin. This relief occurs even with dramatic fluctuations of rainfall throughout the various terrain through which the river passes.

Researchers have suggested that **soil** geology plays a key role in maintaining the river's discharge **stability** despite fluctuations in rainfall. The sandy soils of the Kouyou region, for example, have a stabilizing effect in their ability to store or disperse water.

In 1999, the World Commission on Water for the twenty-first century, based in Paris and supported by the **World Bank** and the United Nations, found that the Congo was one of the world's cleanest rivers—in part due to the lack of industrial development along its shores until that time. However, the situation is changing.

The rapidly increasing human population threatens to compromise the integrity of Congo basin ecosystems. Major threats to the large tropical rainforests and savannas, as well as to **wildlife**, come from the exploitation of **natural resources**. Uncontrolled **hunting** and fishing, **deforestation** (which causes **sedimentation** and **erosion** near **logging** operations) for timber sale or agricultural purposes, unplanned urban expansion (which increases the potential for an increase in untreated sewage and other sources of **pollution** that could harm nearby freshwater systems), and unrestrained extraction of oil and minerals are some of the major economic and environmental issues confronting the region. And these issues are expected to have a global impact as well.

Wildlife

According to the **World Wildlife Fund**, the Congo River and its basin, also known as the "Congo River and Flooded Forests ecoregion," is home to the most diverse and distinctive group of animals adapted to a large-river **environment** in all of tropical Africa.

The Congo river had no outlet to the ocean during the Pliocene Age (5.4–2.4 million years ago) but was instead a large lake. Eventually, the water broke through the rim of the lake, emerging as a river that passed over rocks through a series of rapids, then entered the Atlantic. Except for the beginning and end of its course, the river is uniformly elevated.

With more than 700 fish **species**, 500 of which are endemic to the river, the Congo basin ranks second only to the Amazon in its diversity of species. Nearly 80% of fish species found in the Congo basin exist nowhere else in the world. The various species live both in the river and its attendant habitats—swamps, nearby lakes, and headwater streams. They feed in a variety of ways: scouring the mud at the river's bottom; eating scales off of live fish; and eating smaller fish. Certain fish have even adapted to the river's muddy waters. For example, some have reduced eye size, or no eyes at all, yet easily maneuver through the swift current. The Congo's freshwater fish are a crucial protein source for Central Africa's population; yet the potential for over-fishing near the urban areas along its banks threatens the available supply.

There are also a wide variety of aquatic mammals—such as unusual species of otters, shrews, and monkeys—that are indigenous to the river basin. Rainforests cover over 60% of the Democratic Republic of the Congo, and represents nearly 6% of the world's remaining forested area, and 50% of Africa's remaining forests. Many of the world's **endangered species** live near the river, including gorillas.

River traffic—war, power, and tourism

At the end of May 2001, the United Nations Security Council announced that the Congo River would finally re-open to commercial traffic after a more than two-year blockage due to the war in the Democratic Republic of the Congo. Because of the desperate state of roads throughout the region—or the complete lack of roads—individuals, business, and other agencies have relied primarily on river transportation.

The river had been divided in two at the front line of warring factions—the government forces and their foreign allies on one side, and the rebels backed by Uganda and Rwanda on the other. Massive starvation resulted from the blockage, halting supplies from the United Nations and other humanitarian aid agencies.

With the drought that was running rampant through the early years of the twenty-first century, and those droughts that were anticipated in some areas of southern Africa, distribution of water remains a major challenge. In the fall of 2000, the Southern African Development Community, a group of Congolese business people, began to look to the Congo River for a solution to these water problems.

These developers launched a plan to pump water from the Congo River, by building two long-distance pipelines (known as the Solomon pipelines)— one across the mouth of the Congo River to Walvis Bay in Namibia, 621 mi (about 1,000 km) away; the other, running through civil war zones. They would supply water to the Middle East by way of Port Sudan, a distance of nearly 1,242 mi (about 2,000 km).

The company initiating the plan, Westrac, claimed that the project would create hundreds of jobs, provide for the building of hospitals along the route, and lay fiber-optic communications links as well—thereby boosting the economy of the region and enhancing the lives of the native population. Detractors of the plan countered that the plan would be cost-prohibitive and hazardous to the environment. According to the California-based International Rivers Network, it was "premature to investigate such a complex plan when simpler and cheaper solutions haven't been fully explored," as reported by Radio Netherlands.

With political issues far from resolved as of 2002, and the potential for widespread pollution if industry and the population exploits the resources of the river, future plans could remain unresolved for decades to come. Yet, Salomon Banamuherem, the Democratic Republic of the Congo's minister of tourism, hoped that the project would tap into another great natural resource of the country and its river—tourism. In its entire history, even in times of peace, the country—Africa's third largest country—has attracted no more than100,000 visitors per year.

Despite unresolved political and economic issues, protecting the biodiverse resources and **ecosystem** of the

Congo River and basin is perhaps the most important and challenging task facing this region in the future.

[*Jane E. Spear*]

RESOURCES
BOOKS

Fish, Bruce and Becky-Durost Fish. *Congo: Exploration, Reform, and a Brutal Legacy.* Philadelphia: Chelsea House Pub., 2001.
Tayler, Jeffrey. *Facing the Congo.* St. Paul: Ruminator Books, 2000.
The Congo Basin: Human and Natural Resources. Amsterdam: Netherlands Institute for IUCN, 1998.

OTHER

Congo-pages. *Congo River Basin.* June 3, 2002 [cited July 2, 2002]. <http://www.congo-pages.org/>
Eureka Alert. *Congo River Basin: Geology and soil type influence impact.* January 11, 2002 [cited July 1, 2002]. <http://www.eurekalert.org/>
Johnson, David. *Africana.* "Congo River called one of the world's cleanest." December 3, 1999 [cited June 2002]. <http://www.africana.com/>

ORGANIZATIONS

World Wildlife Fund, 1250 24th St. N.W, P.O. Box 97180., Washington, DC USA 20090-7180 Fax: 202-293-9211, Toll Free: 1-800-CALL-WWF, , http://www.worldwildlife.org

Coniferous forest

Coniferous forests contain trees with cones and generally evergreen needle or scale-shaped leaves. Important genera in the northern hemisphere include pines (*Pinus*), spruces (*Picea*), firs (*Abies*), **redwoods** (*Sequoia*), Douglas firs (*Pseudotsuga*), and larches (*Larix*). Different genera dominate the conifer forests of the southern hemisphere. Conifer forests occupy regions with cool-moist to very cold winters and cool to hot summers. Many conifer forests originated as plantations of **species** from other continents. Among conifer formations in North America are the slow-growing circumpolar **taiga** (boreal), the subalpine-montane, the southern pine, and the Pacific Coast temperate **rain forest**. Softwoods, another name for conifers, are used for lumber, panels, and paper.

Conservation

The philosophy or policy that **natural resources** should be used cautiously and rationally so that they will remain available for **future generations**. Widespread and organized conservation movements, dedicated to preventing uncontrolled and irresponsible exploitation of forests, lands, **wildlife**, and **water resources**, first developed in the United States in the last decades of the nineteenth century. This was a time at which accelerating settlement and resource depletion made conservationist policies appealing both to a large portion of the public and to government leaders. Since then, international conservationist efforts, including work of the United Nations, have been responsible for monitoring natural resource use, setting up **nature** preserves, and controlling environmental destruction on both public and private lands around the world.

The name most often associated with the United States' early conservation movement is that of **Gifford Pinchot**, the first head of the U.S. **Forest Service**. A populist who fervently believed that the best use of nature was to improve the life of the common citizen, Pinchot brought scientific management methods to the Forest Service. He also brought a strongly utilitarian philosophy, which continues to prevail in the Forest Service. Beginning as an advisor to **Theodore Roosevelt**, himself an ardent conservationist, Pinchot had extensive influence in Washington and helped to steer conservation policies from the turn of the century to the 1940s. Pinchot had a number of important predecessors, however, in the development of American conservation. Among these was George Perkins Marsh, a Vermont forester and geographer whose 1864 publication, *Man and Nature*, is widely held as the wellspring of American environmental thought. Also influential was the work of John Wesley Powell, Clarence King, and other explorers and surveyors who, after the Civil War, set out across the continent to assess and catalog the country's physical and biological resources and their potential for development and settlement.

Conservation, as conceived by Pinchot, Powell, and Roosevelt was about using, not setting aside, natural resources. In their emphasis on wise resource use, these early conservationists were philosophically divided from the early preservationists, who argued that parts of the American **wilderness** should be preserved for their aesthetic value and for the survival of wildlife, not simply as a storehouse of useful commodities. Preservationists, led by the eloquent writer and champion of Yosemite Valley, **John Muir**, bitterly opposed the idea that the best vision for the nation's forests was that of an agricultural crop, developed to produce only useful **species** and products. Pinchot, however, insisted that "The object of [conservationist] forest policy is not to preserve the forests because they are beautiful...or because they are refuges for the wild creatures of the wilderness...but the making of prosperous homes...Every other consideration is secondary." Because of its more moderate and politically palatable stance, conservation became, by the turn of the century, the more popular position. By 1905 conservation had become a blanket term for nearly all defense of the **environment**; the earlier distinction was lost until it began to re-emerge in the 1960s as "environmentalists" began once again to object to conservation's anthropocentric (human-centered) emphasis. More recently deep ecologists and bioregionalists have likewise departed from mainstream conserva-

tion, arguing that other species have intrinsic rights to exist outside of human interests.

Several factors led conservationist ideas to develop and spread when they did. By the end of the nineteenth century European settlement had reached across the entire North American continent. The census of 1890 declared the American frontier closed, a blow to the American myth of the virgin continent. Even more important, loggers, miners, settlers, and livestock herders were laying waste to the nation's forests, **grasslands**, and mountains from New York to California. The accelerating, and often highly wasteful, commercial exploitation of natural resources went almost completely unchecked as political corruption and the economic power of timber and lumber barons made regulation impossible. At the same time, the disappearance of American wildlife was starkly obvious. Within a generation the legendary flocks of passenger pigeons disappeared entirely, many of them shot for pig feed while they roosted. Millions of **bison** were slaughtered by market hunters for their skins and tongues or by sportsmen shooting from passing trains. Natural landmarks were equally threatened—Niagara Falls nearly lost its water to hydropower development, and California's Sequoia groves and Yosemite Valley were threatened by **logging** and grazing.

At the same time, post-Civil War scientific surveys were crossing the continent, identifying wildlife and forest resources. As a consequence of this data gathering, evidence became available to document the depletion of the continent's resources, which had long been assumed inexhaustible. Travellers and writers, including John Muir, Theodore Roosevelt, and Gifford Pinchot, had the opportunity to witness the alarming destruction and to raise public awareness and concern. Meanwhile an increasing proportion of the population had come to live in cities. These urbanites worked in occupations not directly dependent upon resource exploitation, and they were sympathetic to the idea of preserving public lands for recreational interests. From the beginning this urban population provided much of the support for the conservation movement.

As a scientific, humanistic, and progressive policy, conservation has led to a great variety of projects. The development of a professionally trained forest service to maintain national forests has limited the uncontrolled "tree mining" practiced by logging and railroad companies of the nineteenth century. Conservation-minded presidents and administrators have set aside millions of acres **public land** for national forests, parks, and other uses for the benefit of the public. A corps of professionally trained game managers and wildlife managers has developed to maintain game birds, fish, and mammals for public **recreation** on federal lands. (For much of its history, federal game conservation has involved extensive predator elimination programs, however several decades of protest have led to more ecological approaches to game management in recent decades.) During the administration of Franklin D. Roosevelt, conservation projects included such economic development projects as the **Tennessee Valley Authority** (TVA), which dammed the Tennessee River for flood control and electricity generation. The Civilian Conservation Corps developed roads, built structures, and worked on **erosion** control projects for the public good. During this time the **Soil Conservation Service** was also set up to advise farmers in maintaining and developing their farmland.

At the same time, voluntary citizen conservation organizations have done extensive work to develop and maintain natural resources. The **Izaak Walton League, Ducks Unlimited**, and scores of local gun clubs and fishing groups have set up game sanctuaries, preserved **wetlands**, campaigned to control **water pollution**, and released young game birds and fish. Other organizations with less directly utilitarian objectives also worked in the name of conservation: the **National Audubon Society**, the **Sierra Club**, the **Wilderness Society**, **The Nature Conservancy**, and many other groups formed between 1895 and 1955 for the purpose of collective work and lobbying in defense of nature and wildlife.

An important aspect of conservation's growth has been the development of professional schools of forestry, game management, and **wildlife management**. When Gifford Pinchot began to study forestry, Yale had only meager resources and he gained the better part of his education at a French school of **forest management** in Nancy, France. Several decades later the Yale School of Forestry (financed largely by the wealthy Pinchot family) was able to produce such well-trained professionals as **Aldo Leopold**, who went on to develop the United States' first professional school of game management at the University of Wisconsin.

From the beginning, American conservation ideas, informed by the science of **ecology** and the practice of resource management on public lands, spread to other countries and regions. It is in recent decades, however, that the rhetoric of conservation has taken a prominent role in international development and affairs. The most visible international conservation organizations today is the United Nations Environment Program (UNEP), the Food and Agriculture Organization of the United Nations (FAO), and the **World Wildlife Fund**. In 1980 the International Union for the Conservation of Nature and Natural Resources (IUCN) published a document entitled the *World Conservation Strategy*, dedicated to helping individual states, and especially developing countries, plan for the maintenance and protection of **soil**, water, forests, and wildlife. A continuation and update of this theme appeared in 1987 with the publication of the UN World Commission on Environment and Develop-

ment's paper, ***Our Common Future***. The idea of sustainable development, a goal of ecologically balanced, conservation-oriented economic development, was introduced in this 1987 paper and has since become a dominant ideal in international development programs of the 1990s.

[*Mary Ann Cunningham Ph.D.*]

RESOURCES
BOOKS

Fox, S. *John Muir and His Legacy: the American Conservation Movement.* Boston: Little, Brown, 1981.

Pinchot, G. *Breaking New Ground.* Washington, DC: Island Press, 1987 (originally 1947).

Marsh, G. P. *Man and Nature.* Cambridge: Harvard University Press, 1965 (originally 1864).

Meine, C. *Aldo Leopold: His Life and Work.* Madison, WI: University of Wisconsin Press, 1988.

Conservation biology

Conservation biology is concerned with the application of ecological and biological science to the conservation and protection of Earth's **biodiversity**. Conservation biology is a relatively recent field of scientific activity, having emerged during the past several decades in response to the accelerating biodiversity crisis. Conservation biology represents an integration of theory, basic and applied research, and broader educational goals. It includes much of **ecology** but extends it with social sciences, policy, and management.

The most important cause of the biodiversity crisis is the disturbance of natural habitats, particularly through the conversion of tropical forests into agricultural habitats. Biodiversity is also greatly threatened by the excessive **hunting** of certain **species**, by commercial forestry, by **climate** change, and by other stressors associated with human activities, such as air and **water pollution**. A central goal of conservation biology is to discover ways of avoiding or repairing the damages that human influences are causing to biodiversity. Important considerations include the development of science-based methods for conserving endangered populations of species on larger landscapes (or seascapes, in marine environments), and of designing systems of protected areas where natural ecosystems and indigenous species can be conserved.

Biodiversity and its importance

Biodiversity can be defined as the total richness of biological variation. The scope of biodiversity ranges from the genetic variation of individuals within and among populations of species, to the richness of species that co-occur in ecological communities. Some ecologists also consider biodiversity to include the spatial and temporal changes of communities on the greater landscape or seascape.

About 1.7 million of Earth's species have been identified and given a scientific name. However, biologists have not yet "discovered" most species, especially those living in tropical habitats. According to some estimates, Earth may support as many as 30–50 million species, of which 90% occur in tropical ecosystems, particularly in old-growth rainforests. Biologists believe that most of the unknown species are invertebrates, especially species of beetles and other insects. Compared with invertebrates, the numbers of species of plants and vertebrate animals are relatively well known.

Biodiversity is valuable for many reasons, but these can be grouped into the following three classes:

(1) *Intrinsic value.* Regardless of its worth in terms of the needs of humans, biodiversity has its own, **intrinsic value**.

(2) *Utilitarian value.* Humans have an undeniable need to harvest wild and domesticated species and their communities as sources of food, materials, and energy. Although harvests of biodiversity can be conducted in ways that foster renewal, these potentially renewable resources are often harvested or managed too intensively, resulting in degradation or **extinction** of the resource.

(3) *Provision of ecological services.* Biodiversity provides numerous ecological services that are directly and indirectly important to human welfare. These services include biological productivity, **nutrient** cycling, cleansing of water and air, control of **erosion**, provision of atmospheric oxygen and removal of **carbon dioxide**, and other functions related to the health and integrity of ecosystems. According to the American biologist Peter Raven: "Biodiversity keeps the planet habitable and ecosystems functional."

Threats to biological diversity

Biodiversity at local, regional, continental, and global scales is critically threatened by human activities. The damages that are being caused to Earth's species and ecosystems are so severe that they are referred to by ecologists as a biodiversity crisis. Many permanent losses of biodiversity have already been caused by human influences, including the extinctions of numerous species and the losses of distinctive, natural communities. Unless there are substantial changes in the ways that humans affect ecosystems, there will be enormously greater losses of biodiversity in the near future.

Earth's natural biodiversity has always been subjected to extinction (that is, the permanent loss of species and other groups) The fossil record shows that species, families, and even entire phyla have appeared and disappeared on Earth. For example, many invertebrate phyla proliferated during an evolutionary radiation at the beginning of the Cambrian era about 570 million years ago, but most of these are now extinct.

Many of the natural extinctions occurred simultaneously, apparently as a result of an unpredictable catastrophe. For instance, about 65 million years ago a **mass extinction** occurred that resulted in the loss of the last of the dinosaurs and as many as 76% of the then-existing species. That catastrophe is believed to have been caused by a meteorite impacting Earth. In other cases, natural extinctions have been caused by more gradual environmental changes, for example in climate or in the intensity of disease or predation.

More recently, however, humans have been responsible for almost all of the extinctions that are occurring. In fact, species are now being lost so quickly that the changes represent a modern mass extinction. Well-known examples of extinctions caused by humans include the **dodo, passenger pigeon**, and great auk. Numerous other species have been taken to the brink of extinction, including the plains **bison, whooping crane, ivory-billed woodpecker**, and right whale. These losses have been caused by over-hunting and the disturbance and conversion of natural habitats.

In addition to these famous cases involving large animals, an even more ruinous damage to Earth's biodiversity is being caused by extensive losses of tropical ecosystems, particularly the conversion of tropical rain forests into agricultural habitats. Because tropical ecosystems are particularly rich in numbers of species, loss of natural tropical **habitat** causes extinctions of numerous species. Many of those species occurred nowhere else but in particular tropical locales.

The mission of conservation biology is to understand the causes and consequences of the modern crisis of extinction and degradation of Earth's biodiversity, and then to apply scientific principles to preventing or repairing the damages. This is largely done by conserving populations and by protecting natural areas.

Conservation at the population level

In some cases, **endangered species** can be enhanced by special programs that increase their breeding success and enhance the survival of their populations. Usually, a variety of actions is undertaken, along with the preservation of appropriate habitat, under a scheme that is known as a population recovery plan. Components of a population recovery plan may include such actions as (1) the careful monitoring of wild populations and the threats that they face; (2) research into the specific habitat needs of the endangered species; (3) the establishment of a captive-breeding program and the release of surplus individuals into the wild; (4) research into genetic variation within the species; and (5) other studies of basic biology and ecology that are considered necessary for preservation of the species, particularly in its natural habitats. Unfortunately, population recovery plans have only been developed for a small fraction of endangered species, and most of these have been prepared for species that occur in relatively wealthy countries.

One example involves the whooping crane (*Grus americana*), an endangered species in North America. Because of excessive hunting and **critical habitat** loss, this species declined in abundance to the point where as few as only 15 individuals were alive in 1941. Since then, however, the wild population of whooping cranes has been vigorously protected in the United States and Canada, and their critical breeding, migratory, and wintering habitats have been preserved. In addition, the basic biology and behaviour of whooping cranes have been studied, and some wild birds have been taken into captivity and used in breeding programs to increase the total population of the species. Some of the captive-bred animals have been released to the wild, and whooping crane eggs have also been introduced into the nests of the closely related sandhill crane (*Grus canadensis*), which serve as foster parents. These applications of conservation biology have allowed the critically endangered population of whooping cranes to increase to more than 150 individuals in the mid-1980s, and to about 300 birds in 1997, of which about half were in captivity. Because of these actions, there is now guarded optimism for the survival of this endangered species.

Protected areas

Protected areas such as parks and ecological reserves are necessary for the conservation of biodiversity in wild, natural ecosystems. Most protected areas are established for the preservation of natural values, particularly the known habitats of endangered species, threatened ecological communities, or representative examples of widespread communities. However, many protected areas (particularly parks) are also used for human activities, as long as they do not severely threaten the ecological values that are being conserved. These uses can include **ecotourism**, and in some cases fishing, hunting, and even timber harvesting. In 1993 there were about 9,000 protected areas globally, with a total area of almost two million acres (792 million ha). Of this total, about 2,500 sites comprising 1.15 million acres (464 million ha) were fully protected, and could be considered to be ecological reserves.

Ideally, a national system of protected areas would provide for the longer-term conservation of all native species and their natural communities, including terrestrial, freshwater, and marine ecosystems. So far, however, no country has implemented a comprehensive system of ecological reserves to fully protect the natural biodiversity of the region. Moreover, many existing reserves are relatively small and are threatened by environmental changes and other disturbances, such as illegal hunting of animals and plants and sometimes intensive tourism.

Ecological knowledge has allowed conservation biologists to make important contributions to the optimized design of networks of protected areas. Important considerations include: (1) the need to protect areas that provide adequate representation of all types of natural ecosystems; (2) the need to preserve all endangered ecosystems and the habitats of threatened species; (3) the requirement of redundancy, so that if one example of an endangered **ecosystem** becomes lost through an unavoidable natural disturbance (such as a **hurricane** or **wildfire**), the type will continue to survive in another protected area; (4) the need to decide whether or not the network of protected areas should be linked by corridors, a matter of some controversy among ecologists.

Conservation biology has also made important contributions towards the spatial design of individual protected areas. Important considerations include (1) the need to make protected areas as large as possible, which will help to allow species and ecosystems to better cope with disturbances and environmental changes; (2) a preference for smaller reserves to have a minimal amount of edge, which helps to avoid damages that can be caused by certain predators and invasive species; (3) the need to take an ecosystem approach which ensues that the reserve and its surrounding area will be managed in an integrated manner.

Although conservation biology is a relatively young field, important progress is being made towards development of the effective ecological and biological tools necessary to preserve biodiversity.

[*Bill Freedman Ph.D.*]

RESOURCES

BOOKS

Freedman, B. *Environmental Ecology*, 2nd edition. Academic Press, San Diego, 1995.

Primack, R.B. *Essentials of Conservation Biology*. Sunderland, MA: Sinauer Associates, 1993.

Wilson, E.O. (ed.). *Biodiversity*. Washington, D.C: National Academy Press, 1988.

Conservation design
see **Urban sprawl**

Conservation easements

A **conservation** easement is a covenant, restriction, or condition in a deed, will, or other legal document that allows the owner to maintain ownership and control of real property, but restricts the use of that property so the land is conserved in its natural state, or, in the case of a historic conservation easement, so that it provides a historic benefit.

The uses allowed by the easement can include **recreation**, agriculture, cultural uses, and establishment of **wildlife habitat**. The federal government allows tax deductions for conservation easements that provide a certified value to the public, such as protecting ecologically valuable natural habitat or, in the case of an easement based on the historical conservation of the property, that contribute to the historic character of the district in which the property is located. Conservation easements are legal instruments enabled by many states as well as non-United States governments, including Canada and most of its provinces.

[*Marie H. Bundy*]

Conservation International

Conservation International (CI) is a non-profit, private organization dedicated to saving the world's endangered rain forests and the plants and animals that rely on these habitats for survival. CI is basically a scientific organization, a fact which distinguishes it from other conservation groups. Its staff includes leading scientists in the fields of botany, ornithology, herpetology, marine biology, entomology, and zoology.

Founded in 1987 when it split off from **the Nature Conservancy**, CI now has over 55,000 members. The group, headed by Peter A. Seligmann, has gathered accolades since its inception. In 1991 *Outside Magazine* gave CI an A- (one of the two highest grades received) in its yearly report card rating 14 leading environmental groups.

The high praise is well founded. CI tends to successfully implement its many projects and goals. Many CI programs focus on building local capacity for conservation in developing countries through financial and technical support of local communities, private organizations, and government agencies. Their "ecosystem conservation" approach balances conservation goals with local economic needs. CI also funds and provides technical support to local communities, private organizations, and government agencies to help build sustainable economies while protecting **rain forest** ecosystems.

Four broad themes underlie all CI projects: 1) a focus on entire ecosystems; 2) integration of economic interests with ecological interests; 3) creation of a base of scientific knowledge necessary to make conservation-minded decisions; and 4) an effort to make it possible for conservation to be understood and implemented at the local level.

CI is involved with projects in 30 countries, including Botswana, Brazil, Canada, Colombia, Indonesia, Mexico, New Guinea, and the Philippines. In 2000, CI expanded into Cambodia and China, among others.

Among CI's many successful projects is the Rapid Assessment Program (RAP), which enlists the world's top field scientists to identify **wilderness** areas in need of urgent conservation attention. RAP teams have completed surveys in Bolivia, Ecuador, Belize, Peru, and Mexico, and CI plans at least 12 more surveys on four continents in the next three to five years. CI has also helped establish important **biosphere** reserves in rain forest countries. These efforts successfully demonstrate CI's **ecosystem** conservation approach, and prove that the economic needs of local communities can be reconciled with conservation needs. No harvesting or **hunting** is allowed in the reserves, but **buffer** zones, which include villages and towns, are located just outside the core areas.

CI strongly supports many educational programs. In 1988, it signed a long-term assistance agreement with Stanford University which involves exchange and training of Costa Rican students and resource managers. In 1989 CI began a program with the University of San Carlos which provides financial and technical support to the research activities in northern Guatemala of the university's Center for Conservation Studies. An educational program of a different kind, the Sorcerer's Apprentice, is designed to record ethnobotanical knowledge, protect useful **species**, and pass this information on to the next generation in indigenous communities. Young men and women in forestry services learn from traditional village healers and midwives.

In 2002, CI has already established several new ways to educate the people and help the **environment**, such as Centers for **Biodiversity** Conservation, the Global Conservation Fund, and a joint venture with Ford Motor Company called the Center for Environmental Leadership in Business. CI also focuses on activities in the major wilderness areas identified as the most endangered. The organization also has expanded its conservation efforts to new ecosystems, including marine, **desert**, and temperate rain forest regions.

[*Cathy M. Falk*]

ORGANIZATIONS
Conservation International, 1919 M Street, NW Suite 600, Washington, D.C. USA 20036 (202) 912-1000, Toll Free: (800) 406-2306, Email: inquiry@conservation.org, <http://www.conservation.org>

Conservation Reserve Program

The Conservation Reserve Program (CRP) is a voluntary program for agricultural landowners, that encourages farmers to plant long-term resource-conserving vegetative ground cover to improve **soil** and water, and create more suitable **habitat** for fish and **wildlife**. Ground cover options include grasses, legumes, shrubs, and tree plantings. The program is authorized by the federal Food Security Act of 1985, as amended, and is implemented through the Commodity

Credit Corporation (CCC). It aims to promote good **land stewardship** and improve rural aesthetics.

The CRP offers annual rental payments, incentive payments, and cost-share assistance to establish approved cover on eligible cropland. The CCC provides assistance of as much as 50% of the landowner's cost in establishing an approved conservation program. Contracts remain in effect for between 10 and 15 years. Annual rental payments are based on the agriculture rental value of the land used in the program. The program provides needed income support for farmers, and helps to curb production of surplus commodities.

Eligibility for participation in CRP extends to individuals, partnerships, associations, Indian tribal ventures corporations, estates, trusts, other business enterprises or legal entities. States, political subdivisions of states, or agencies thereof owning or operating croplands, may also apply.

The CCC via the Farm Service Agency (FSA) manages the CRP. The Natural Resources Conservation Service (NRCS) and the Cooperative State Research and Education Extension Service provide support. State forestry agencies and local soil and **water conservation** districts also provide assistance.

To be eligible for the CRP, cropland should have been planted or considered planted to an agricultural commodity in two of the five most recent crop years. Eligibility encompasses highly **erodible** acreage, cropped **wetlands**, and land surrounding non-cropped wetlands. The cropland must be owned or operated for at least 12 months before the close of the sign-up period. Exceptions can be made for land that was acquired by will or **succession**, or if the FSA determines that ownership was not acquired for the purpose of placing the land in the conservation reserve.

Initially, **erosion** reduction was the sole criterion for acceptance in the CRP, and in 1986–87, 22 million acres (8.9 million ha) were enrolled for this purpose. CRP proved to be effective. According NRCS **statistics**, average erosion on enrolled acres declined by about 90%. It was estimated that the program reduced overall erosion nationwide by more than 22% even though less than 10% of the nation's cropland was enrolled.

An Environmental Benefits Index (EBI) is used to prioritize applications for the CRP. EBI factors include: Wildlife habitat benefits, **water quality** benefits from reduced erosion, **runoff**, and **leaching**, and **air quality** benefits from reduced wind erosion. The NRCS collects data for each of the factors and, based on its analysis, applications are ranked. Selections are made from that ranking.

The CCC bases rental rates on the productivity of soils within a county, and the average rent for the past three years of local agricultural land. The maximum CRP rental rate for each applicant is calculated in advance of enrollment.

Applicants may accept that rate, or may offer a lower rental rate to increase the likelihood that their project will be funded.

The CCC encourages restoration of wetlands by providing a 25% incentive payment of the costs involved to establish approved cover. This is in addition to the normal 50% cost share for non-wetlands. Eligible acreage devoted to special conservation practices, such as riparian buffers, filter strips, grassed waterways, shelter belts, living snow fences, contour grass strips, salt tolerant vegetation, and shallow water areas for wildlife, may be enrolled at any time and is not subject to competitive bidding.

When CRP contracts expire, participants must continue to follow approved conservation plans. They must comply with wetland, **endangered species** and other federal, state, and local environmental laws, and they must respect any continuing conservation easement on the property.

The United States Department of Agriculture (USDA) provides information and technical assistance to CRP participants who wish to return CRP land to row-crop producing status as their contracts expire. This is to ensure that the land is developed in a sound, productive, and sustainable manner, preventing excessive erosion, protecting water quality, and employing other measures that enhance soil moisture-retaining ability.

In states such as North Dakota, the landscape has been changed dramatically since the introduction of the CRP. In the 1970s and early 1980s, fields with steep hills and areas of light soil were often cultivated from fencerow-to-fencerow. The result often was severe erosion and permanent loss of soil fertility. In some areas, winters were dominated by "snirtstorms" when a combination of dirt and snow blew across the landscape depositing a dark coating on countryside downwind. In contrast, today's travelers find these landscapes covered with green vegetation in summer and white snow in winter.

The CRP has particularly benefited migratory birds in states such as North Dakota, South Dakota, and Montana. The perennial vegetation on marginal farmland has provided refuge for migrating birds, and added breeding habitat for others. North Dakota farmers have enrolled about 10% of the state's cropland in the CRP.

The fiscal year 2000 federal agricultural appropriations bill authorized a pilot project of harvesting of **biomass** from CRP land to be used for energy production. Six projects were authorized, no more than one of which could be in any state. Vegetation could not be harvested more often than once every two years, and no commercial use could be made of the harvested biomass other than energy production. Annual CRP rental payments are reduced by twenty five percent during the year the acreage is harvested. Land that is devoted to field windbreaks, waterways, shallow water ways for wildlife, contour grass strips, shelter belts, living snow fences, permanent vegetation to reduce **salinity**, salt tolerant vegetative cover, filter strips, riparian buffers, wetland restoration, and cross-wind trap strips is not eligible for this program. By 2002, contracts had been approved in Iowa, Illinois, Oklahoma, Minnesota, New York, and Pennsylvania.

In June of 2001, the USDA announced a six-state pilot program as part of the CRP to restore up to 500,000 acres (202,000 ha) of farmable wetlands and associated buffers. The Farmable Wetlands Pilot Program is intended to help producers improve the **hydrology** and vegetation of eligible land in Iowa, Minnesota, Montana, Nebraska, North Dakota, and South Dakota. Restoring wetlands in these states should reduce downstream flood damage, improve surface and **groundwater** quality, and recharge groundwater supplies. Essential habitat for migratory birds and many other wildlife **species**, including threatened and endangered species will be created and enhanced. Recreational activities such as hiking and bird watching will also be improved.

In 1985, the year in which CRP originated, spring surveys by the U.S **Fish and Wildlife Service** estimated waterfowl breeding populations at 25.6 million ducks. A fall flight of 54.5 million was predicted. Several species of ducks including mallards, pintails, and blue-winged teal appeared to be fading away. Numbers were at or near their lowest ebb in 30 years. Between 1986 and 1990, farmers enrolled 8.2 million acres (3.3 million hectare) of cropland in CRP within an area known as the **prairie** pothole region. This large glaciated area of the north central United States and southern Canada is where up to 70% of North America's ducks are hatched. Through CRP, nearly 13,000 mi^2 (34,000 km^2) was converted to superior nesting habitat through CRP.

In the early 1990s, increased precipitation filled the prairie potholes and many waterfowl ended their spring **migration** on CRP land, rather than continuing migration to their usual breeding grounds in Canada. Nesting densities increased many fold. Potholes surrounded by CRP grass provided more secure habitat for nests, and hatchlings were no longer easy targets for predators. Nesting success tripled, from 10 to 30%, and waterfowl **mortality** no longer exceeded annual additions. By 1995, ten years after the start of CRP, 36.9 million ducks were included in the annual spring survey numbers, a 40% increase in 10 years.

Waterfowl are not the only bird species to benefit from CRP. In one study, breeding birds were counted in about 400 fields in eastern Montana, North and South Dakota, and western Minnesota. These states have nearly 30% of all land included in the CRP. Fields were planted mostly to mixtures of native and introduced grasses and legumes. For most of the seventy-three different species counted, numbers were far higher in CRP fields than in cropland. Differences were greatest for several grassland species whose numbers had been markedly declining in recent surveys. Two species, lark buntings (*Calamospiza melanocorys*) and grasshopper sparrows (*Ammodramus savannarum*), were 10 and 16 times more common in CRP **environment** than in cropland. The investigators concluded that restoration of suitable habitat in the form of introduced grasses and legumes can have an enormous beneficial effect on populations of grassland birds.

When CRP was due to expire in 1995, restoration of prairie pothole breeding grounds was in jeopardy. United States farm policy was undergoing major change and it appeared that the CRP would not be continued. **Ducks Unlimited** and other wildlife organizations lobbied heavily, and funding of one billion dollars was included in the 1996 Farm Bill to continue the program for another seven years. Moreover, the guidelines for the continuing program were geared more directly to the preservation of wetlands and waterfowl conservation. The prairie pothole region was designated a national conservation priority area, and during the March 1997 sign-up more acres in the prairie pothole region were enrolled in CRP than were due to expire.

In its first 10 years, the CRP cost nearly $2 billion per year. Opponents have argued that this is too expensive, while proponents maintain that the costs are offset by its conservation and environmental benefits. Estimates of the annual value of benefits range from slightly less than $1 billion to more than $1.5 billion. Some analysts claim that the value of benefits approaches or exceeds costs in some sites.

The USDA announced that there would not be general CRP signup for Fiscal Year 2002 although producers could continue to enroll acreage eligible under continuous enrollment provisions. Later, they announced that CRP contracts expiring in 2002 could be extended for another year.

[*Douglas C. Pratt Ph.D.*]

RESOURCES

PERIODICALS

2002 Ducks Unlimited, Inc. May 31, 2002 [June 2002]. <http://www.ducks.org/conservation/crp.asp>.

Conservation Reserve Program Biomass Pilot Projects. U.S. Department of Agriculture, Farm Service Agency Online. Fact Sheet Electronic Edition. November 2000 [May 2002].

Kantrud, Harold A., Rolf R. Koford, Douglas H. Johnson, and Michael D. Schwartz." The Conservation Reserve Program—Good for Birds of Many Feathers." *North Dakota Outdoors* 56, no. 2: 14–17.

Johnson, D. H., and M. D. Schwartz. "The Conservation Reserve Program and Grassland birds." *Conservation Biology* 7: 934–937.

New Conservation Reserve Program The University of Georgia, Cooperative Extension Service Extension, Forest Resources Unit. FOR. 97-003. 1997 [May 2002].

USDA To Help Restore Wetlands Through Six-state Pilot Programs. Farm Service Agency Public Affairs Staff June 4, 2001 [June 2002].

Conservation tillage

Conservation tillage is any **tilth** sequence that reduces loss of **soil** or water in farmland. It is often a form of non-inversion tillage that retains significant amounts of plant residues on the surface. Thirty percent of the soil surface must be covered with plant residues at crop planting time to qualify as conservation tillage under the Conservation Technology Information Center definition. Other forms of conservation tillage include ridge tillage, rough plowing, and tillage that incorporates plant residues in the top few inches of soil.

A number of implements for primary tillage are used to retain all or a part of the residues from the previous crop on the soil surface. These include machines that fracture the soil, such as chisel plows, combination chisel plows, disk harrows, field cultivators, undercutters, and strip tillage machines. In a no-till system, the soil is not disturbed before planting. Most tillage systems that employ the moldboard plow are not considered conservation tillage because the moldboard plow leaves only a small amount of residue on the soil surface (0–10%).

When compared with conventional tillage (moldboard plow with no residue on the surface), various benefits from conservation tillage have been reported. Chief among the benefits are reduced wind and water **erosion** and improved **water conservation**. Erosion reductions from 50–90% as compared with conventional tillage are common. Conservation tillage often relies on herbicides to help control weeds and may require little or no post-planting cultivation for control of weeds in row crops. Depending on the management system used, **herbicide** amounts may or may not be greater than the amounts used on conventionally tilled land. Yields from conservation tillage, particularly corn, may be greater or smaller than from conventional tilled soil. Crop yield problems are most frequent on wet soils in the northern United States. Costs of tillage may be lower, but not always, from conservation tillage as compared with conventional.

[*William E. Larson*]

RESOURCES
BOOKS

Little, C. E. *Green Fields Forever: The Conservation Tillage Revolution in America.* Covelo, CA: Island Press, 1987.

Consultative Group on International Agricultural Research

The Consultative Group on International Agricultural Research (CGIAR) was founded in 1971 to improve food production in developing countries. Research into agricultural productivity and the management of **natural resources** are the two goals of this organization, and it is dedicated to making the scientific advances of industrialized nations available to poorer countries. The CGIAR emphasizes the importance of developing sustainable increases in agricultural yields and creating technologies that can be used by farmers with limited financial resources.

Membership consists of governments, private foundations, and international and regional organizations. The goals of this association are carried out by a network of International Agricultural Research Centers (IARCs). There are currently 18 such centers throughout the world, all but four of them in developing countries, and they are each legally distinct entities, over which the CGIAR has no direct authority. The group has no constitution or by-laws, and decisions are reached by consensus after consultations with its members, either informally or at their semiannual meetings. The function of the CGIAR is to assist and advise the IARCs, and to this end it maintains a Technical Advisory Committee (TAC) of scientists who review ongoing research programs at each center.

Each IARC has its own board of trustees as well as its own management, and they formulate individual research programs. The research centers pursue different goals, addressing problems in a particular sector of agriculture, such as livestock production or agricultural challenges in specific parts of the world, such as crop production in the semi-arid regions of Africa and Asia. Some centers conduct research into integrated plant protection, and others into forestry, while some are more concerned with policy issues, such as food distribution and the international food trade. One of the priorities of the CGIAR is the **conservation** of seed and plant material, known as germplasm, and the development of policies and programs to ensure that these resources are available and fully utilized in developing countries. The International Board for Plant Genetic Resources is devoted exclusively to this goal. Besides research, the basic function of the IARCs is educational, and in the past two decades over 45,000 scientists have been trained in the CGIAR system.

The central challenge facing the CGIAR is world **population growth** and the need to increase agricultural production by nearly 50% in the next 20 years while preserving natural resources. The group was one of the main contributors to the so-called "Green Revolution." It helped develop new high-yielding varieties of cereals and introduced them into countries previously unable to grow the food; some of these countries now have agricultural surpluses. In 2001, CGIAR in South Africa worked on developing two new types of maize, which have a 30–50% larger crop than what is currently being produced by the smaller farmers. The CGIAR is working to increase production even further, narrowing the gap between actual and potential yields, while continuing its efforts to limit **soil erosion, desertification,** and other kinds of **environmental degradation**.

The **World Bank**, the Food and Agriculture Organization (FAO) and the United Nations Development Program (UNDP) are among the original sponsors of the CGIAR, and the organization has its headquarters at the offices of the World Bank, which also funds central staffing positions. Combined funding has grown from $15 million in 1971 to over $340 million in 1999, and the CGIAR has a staff of 12,000 worldwide. The group publishes a newsletter called *CGIAR Highlights*.

[*Douglas Smith*]

RESOURCES
ORGANIZATIONS
CGIAR Secretariat, The World Bank, MSN G6-601, 1818 H Street NW, Washington, D.C. USA 20433 (202) 473-8951, Fax: (202) 473-8110, Email: cgiar@cgiar.org, <http://www.cgiar.org>

Container deposit legislation

Container deposit legislation requires payment of a deposit on the sale of most or all beverage containers and may require that a certain%age of beverage containers be allocated for refillables. The legislation shifts the costs of collecting and processing beverage containers from local governments and taxpayers to manufacturers, retailers and consumers.

While laws vary from state to state, even city to city, container deposit legislation generally provides a monetary incentive for returning beverage cans and bottles for **recycling**. Distributors and bottlers are required to collect a deposit from the retailer on each can and bottle sold. The retailer collects the deposit from consumers, reimbursing the consumer when the container is returned to the store. The retailer then collects the deposit from the distributor or bottler, completing the cycle. Consumers who choose not to return their cans and bottles lose their deposit, which usually becomes the property of the distributors and bottlers, though in some states, unredeemed deposits are collected by the state.

Oregon implemented the first deposit law or "bottle bill" in 1972. In the 1970s and 1980s, industry opponents fought container deposit laws on the grounds that they would result in a loss of jobs, an increase in prices, and a reduction

in sales. Now opponents denounce the legislation as being detrimental to curbside recycling programs. But over the past two decades, container deposit legislation has proven effective not only in controlling litter and conserving **natural resources** but in reducing the **waste stream** as well.

Recovery rates for beverage containers covered under the deposit system depend on the amount of deposit and the size of the container. The overall recovery rate for beverage containers ranges from 75 to 93%. The reduction in container litter after implementation of the deposit law ranges from 42 to 86%, and reduction in total volume ranges from 30 to 60%. Although beverage containers make up just over 5% by weight of all **municipal solid waste** generated in the United States, they account for nearly 10% of all waste recovered, according to the **Environmental Protection Agency** (EPA). While the cans and bottles that are recycled into new containers or new products ease the burden on the **environment**, recycling is a second-best solution.

As recently as 1960, 95% of all soft drinks and 53% of all packaged beer was sold in refillable glass bottles. Those bottles required a deposit and were returned for **reuse** 20 or more times. But the centralization of the beverage industry, the increased mobility of consumers, and the desire for convenience resulted in the virtual disappearance of the reusable beverage container. Today, refillables make up less than 6% by volume of packaged soft drinks and 5% by volume of packaged beer, and these percentages shrink every year, according to the National Soft Drink Association and the Beer Institute.

Reuse is a more environmentally responsible **waste management** option, and is superior to recycling in the **waste reduction** hierarchy established by the EPA. While the container industry has been unwilling to promote refillable bottles, new interest in container deposit legislation may move industries and governments to adopt reuse as part of their waste management practices.

Industry-funded studies have found that a refillable glass bottle used as few as eight times consumes less energy than any other container, including recycled containers. A study conducted for the National Association for Plastic Container Recovery found that the 16-oz (1 pt) refillable bottle produces the least amount of waterborne waste and fewest atmospheric emissions of all container types.

To date, 10 states and one city have enacted beverage container deposit systems, designed to collect and process beverage bottles and cans. A deposit of 5–10 cents per can or bottle is an economic incentive to return the container. The states that have some form of container deposit legislation and accompanying deposit system include Oregon, New York, Connecticut, Maine, Iowa, Vermont, Michigan, Massachusetts, Delaware, and California. Legislation is pending in 25 state legislatures, and on **Earth Day** 1993, a national bottle bill was introduced in the U.S. Congress by sponsors from the House of Representatives.

Despite the fact that opinion polls show the public supports bottle bills by a nearly three-to-one margin, for two decades the beverage and packaging industries have successfully blocked the passage of bottle bills in nearly 40 states and even the most successful container deposit programs have come under attack and are threatened with repeal.

Connecticut has one of the highest percentages of refillable beer bottles in the nation, according to **statistics** from the Beer Institute. Despite the success of the state's five-cents-per-container deposit legislation, Governor Lowell Weicker Jr. has pushed for repeal of the legislation, to be replaced by a five-cents-per-container tax to benefit the state parks system. Opponents to Weicker's plan insist that the repeal of the deposit law would result in greater numbers of bottles and cans left strewn across the state.

Others predict the repeal of the bottle bill would impact the state in other ways: about 1,000 jobs would be lost; small redemption centers would go out of business; and recycling rates for glass, **aluminum** and plastic would drop. In addition, it has been estimated that repeal would cost municipal curbside recycling programs in Connecticut between $5.4 and $12.5 million annually.

The United States has a long way to go to catch up to progressive countries such as Sweden, which does not allow aluminum cans to be manufactured or sold without industry assurances of a 75% recycling rate. Concerned that voluntary recycling would not meet these standards, the beverage, packaging and retail industries in Sweden have devised a deposit-refund system to collect used aluminum cans. Consumers in Sweden return their aluminum cans at a rate that has not been achieved in any other country. The 75% recycling rate was achieved in 1987, and industry experts expect the rate to exceed 86% in 1993. Most North American deposit systems rely on the distributor or bottler, but the deposit in Sweden originates with the can manufacturer or drink importer delivering cans within the country. Also, retail participation is voluntary: retailers collect the deposit but are not required to redeem the containers, though most do.

In 1991, 122 billion containers, weighing about 7.2 million tons, were produced in the United States—a 100% increase in packaging waste since 1960. Containers and packaging are the single largest component of the waste stream and they offer the greatest potential for reduction, reuse, and recycling, according to the EPA. Where individuals, industries, and governments will not voluntarily comply with recycling programs, container deposit legislation has decreased the amount of **recyclables** entering the waste stream.

[*Linda Rehkopf*]

RESOURCES
PERIODICALS

Franklin, P. "Sweden's Aluminum Can Return System." *Resource Recycling* (March 1993): 66.
Langer, G. "Many Happy Returns." *Sierra* 73 (March-April 1988): 19-22.
Williams, T. "The Metamorphosis of Keep America Beautiful." *Audubon* 92 (March 1990): 124-133.

Containment structures
see **Nuclear power**

Contaminated soil

The presence of pollutants in soils at concentrations above background levels that pose a potential health or ecological risk. Soils can be contaminated by many human actions including the **discharge** of solids and liquid pollutants at the **soil** surface; **pesticide** application; subsurface releases from leaks in buried tanks, pipes, and landfills; and deposition of atmospheric contaminants such as dusts and particles containing **lead**. Common contaminants include volatile hydrocarbons—such as **benzene**, **toluene**, ethylene, and **xylene** (BTEX compounds)—found in fuels; heavy paraffins and chlorinated organic compounds such as polychlorinated biphenyl (PCB) and **pentachlorophenol** (PCP); inorganic compounds such as lead, **cadmium**, **arsenic** and **mercury**; and **radionuclides** such as tritium. Often, soil is contaminated with a mixture of contaminants. The nature of soil, the contaminant's chemical and physical characteristics, and environmental factors such as **climate** and **hydrology** interact to determine the accumulation, mobility, toxicity, and overall significance of the contaminant in any specific instance.

Fate of soil contaminants

Contaminants in soils may be present in solid, liquid, and gaseous phases. When liquids are released, they move downward through the soil. Some may fill pore spaces as liquids, some may partition or sorb onto mineral soil surfaces, some may dissolve into water in the soft pores, and some may volatilize. For most **hydrocarbons**, a multiphase system is common. When contaminants reach the **water table** (where the voids between soil particles are completely filled with water), contaminant behavior depends on its density. Light BTEX-type compounds float on the water table while dense chlorinated compounds may sink. While many hydrocarbon compounds are not very soluble, even low levels of dissolved contaminants may produce unsafe or unacceptable **groundwater** quality. Other contaminants such as inorganic salts, such as **nitrates**, may be highly soluble and move rapidly through the **environment**. Metals demonstrate a range of behaviors. Some may be chemically bound and thus quite immobile; some may dissolve and be transported by groundwater and **infiltration**.

Contaminated pore water, called leachate, may be transported in the groundwater flow. Groundwater travels both horizontally and vertically. A portion of the contaminant, called residual, is often left behind as the flow passes, thus contaminating soils after the major contaminant **plume** has passed. If the groundwater velocity is fast, for example, hundreds of feet per year, the zone of contamination may spread quickly, potentially contaminating extraction **wells** or other resources. Over years or decades, especially in sandy or porous soils, groundwater contaminants and leachate may be transported over distances of miles, producing a situation that is very difficult and expensive to remedy. In these cases, immediate action is needed to contain and clean up the contamination. In cases where soils are largely comprised of fine-grained silts and clays, contaminants may spread very slowly. Such examples show the importance of site-specific factors in evaluating the significance of soil contamination as well as the selection of a cleanup strategy.

Superfund and other legislation

Prior to the 1970s, inappropriate land disposal practices, such as dumping untreated liquids in lagoons and landfills, were common and widespread. The presence and potential impacts of soil contamination were brought to public attention after well-publicized incidents at **Love Canal**, New York and the Valley of the Drums. Congress responded by passing the **Comprehensive Environmental Response, Compensation, and Liability Act** in 1980 (commonly known as CERCLA or Superfund) to provide funds with which to cleanup the worst sites. After five years of much litigation but little action, Congress updated this law in 1986 with the Superfund Amendments and Reauthorization Act. At present, about 1,200 sites across the United States have been selected as **National Priorities List** (NPL) sites and are eligible under CERCLA for federal assistance for cleanup. These Superfund sites tend to be the nation's largest and worst sites in terms of the possibility for adverse human and environmental impacts and the most expensive ones to clean up. While not as well recognized, numerous other sites have serious soil contamination problems. The U.S. Office of Technology Assessment and **Environmental Protection Agency** (EPA) estimate that about 20,000 abandoned waste sites and 600,000 other sites of land contamination exist in the United States. These estimates exclude soils contaminated with lead in older city areas, the accumulation of fertilizers, pesticides, and insecticides in agricultural lands,

and other classes of potentially significant soil contamination. Federal and state programs address only some of these sites. Currently, CERCLA is the EPA's largest program, with expenditures exceeding three billion dollars over the last 15 years. However, this is only a fraction of the cost to government and industry that will be needed to cleanup all waste sites. Typical estimates of funds required to mitigate the worst 9,000 waste sites reach at least $500 billion; a 50-year time period is anticipated. The problem of contaminated soil is significant not only in the United States but in all industrialized countries.

U.S. laws such as CERCLA and the **Resource Conservation and Recovery Act** (RCRA) attempt to prohibit practices that have led to extensive soil contamination in the past. These laws restrict disposal practices; mandate financial liability to recover cleanup costs (as well as personal injury and property damage) and criminal liability to discourage willful misconduct or negligence; require record keeping to track waste; and provide incentives to reduce waste generation and improve **waste management**.

Soil cleanups

The cleanup or **remediation** of contaminated soils takes two major approaches: source control and containment, or soil and residual treatment and management. Typical containment approaches include caps and covers over the waste in order to limit infiltration of rain and snow melt and thus decrease the leachate from the contaminated soils. Horizontal transport in near-surface soils and groundwater may be controlled by vertical **slurry** walls. Clay, cement, or synthetic membranes may be used to encapsulate soil contaminants. Contaminated water and leachate may be hydraulically isolated and managed with groundwater pump-and-treat systems, sometimes coupled with the injection of clean water to control the spread of contaminants. Such containment approaches only reduce the mobility of the contaminant, and the barriers used to isolate the waste must be maintained indefinitely.

The second approach treats the soil to reduce the toxicity and volume of contaminants. These approaches may be broken down into extractive and *in situ* methods. Extractive options involve the removal of contaminated soil, generally for treatment and disposal in an appropriate **landfill** or for **incineration** where contaminants are broken down by thermal oxidation. *In situ* processes treat the soil in-place. *In situ* options include thermal, biological, and separation/extraction technologies. Thermal technologies include thermal desorption and in-place vitrification (glassification). Biological treatment includes biodegradation by soil **fungi** and bacteria that ultimately renders contaminants into **carbon dioxide** and water. This process is called mineralization. Biodegradation may produce long-lived toxic intermediate products. Separation technologies include soil vapor extraction for volatile organic compounds that removes a fraction of the contaminants by enhancing volatilization by an induced subsurface air flow; stabilization or chemical fixation that uses additives to bind organics and **heavy metals**, thus eliminating contaminated leachate; soil washing and flushing using dispersants, solvents, or other means to solubilize certain contaminants such as PCBs and enhance their removal; and finally, groundwater pump and treat schemes that purge the contaminated soils with clean water in a flushing action. The contaminated groundwater may then be treated by air stripping, steam stripping, **carbon absorption**, precipitation and flocculation and contaminant removal by **ion exchange**. A number of these approaches—*in situ* soil vitrification and enhanced **bioremediation** using engineered microorganisms—are experimental.

The number of potential remediation options is large and expanding due to an active research program that is driven by the need for low cost and more effective solutions. The selection of an appropriate cleanup strategy for contaminated soils requires a thorough characterization of the site and an analysis of the cost-effectiveness of suitable containment and treatment options. A site-specific analysis is required since the distribution and treatment of wastes may be complicated by variation in geology, hydrology, waste characteristics, and other factors at the site. Often, a demonstration of the effectiveness of an innovative or experimental approach may be required by governmental authorities prior to full scale implementation. In general, large sites use a combination of remediation options. Pump and treat and vapor extraction are the most popular technologies.

Cleanup costs and cleanup standards

The cleanup of contaminated soils can involve significant expense and risk. In general, *in situ* containment is cheaper than soil treatment, at least in the short run. While the Superfund law establishes a preference for permanent remedies, many cleanups that have been funded under this law have used both containment and treatment options. In general, *in situ* treatment methods such as groundwater pump and treat are less expensive than extractive approaches such as soil incineration. *In situ* options, however, may not achieve cleanup goals. Like other processes, costs increase with higher removal levels. Excavation and incineration of contaminated soil can cost $1,500 per ton, leading to total costs of many millions of dollars at large sites. Superfund cleanups have averaged about $26 million (however, a substantial fraction of this is for site investigations). In contrast, small fuel spills at **gasoline** stations may be mitigated using vapor extraction at costs under $50,000.

Unlike air and water, which have specific federal laws and regulations detailing maximum allowable levels of contaminants, no levels have been set for contaminants in soils. Instead, the Environmental Protection Agency and states

A contour farm in southeastern Pennsylvania.
(Photograph by Jerry Irwin. Photo Researchers Inc. Reproduced by permission.)

use several approaches to set specific acceptable contaminant levels. For Superfund sites, cleanup standards must exceed applicable or relevant and appropriate requirements (ARARs) under federal environmental and public health laws. More generally, cleanup standards may be based on achieving background levels, that is, the concentrations found in similar, nearby, and unpolluted soils. Second, soil contaminant levels may be acceptable if the soil does not produce leachate with concentration levels above drinking water standards. Such determinations are often based on a test called the Toxics Characteristic Leaching Procedure which mildly acidifies and agitates the soil. Contaminant levels in the leachate below the maximum contaminant levels (MCLs) in the federal **Safe Drinking Water Act** are acceptable. Third, soil contaminant levels may be set in a determination of health risks based on typical or worst case exposures. Exposures can include inhalation of soils as dust, ingestion of soil (generally by children), and direct skin contact with soil. In part, these various approaches response to the complexity of contaminant mobility and toxicity in soils and the difficulty of pinning down acceptable and safe levels.

[*Stuart Batterman*]

RESOURCES
BOOKS

Jury, W. "Chemical Movement Through Soil." In *Vadose Modeling of Organic Pollutants*, edited by S. C. Hern and S. M. Melancon. Chelsea, MI: Lewis Pub., 1988

PERIODICALS

Chen, C. T. "Understanding the Fate of Petroleum Hydrocarbons in the Subsurface Environment." *Journal of Chemical Education* 5 (1992): 357-59.

Contour plowing

Plowing the **soil** along the contours of the land. For example, rather than plowing up-and-down the hill, the cultivation takes place around the hill. By plowing along the contour, less water runs down the hill, thereby reducing water **erosion**.

Contraceptives

see **Family planning; Male contraceptives**

Convention on International Trade in Endangered Species of Wild Fauna and Flora (1975)

The Convention on International Trade in **Endangered Species** of Wild **Fauna** and Flora (CITES), was signed in 1973 and came into force in 1975. The aim of the treaty is to prevent international trade in listed endangered or threatened animal and plant **species** and products made from them. A full-time, paid secretariat to administer the treaty was initially funded by the United Nations Environ-

mental Program, but has since been funded by the parties to the treaty. By 2002, 158 nations had become party to CITES, including most of the major **wildlife** trading nations, making CITES the most widely accepted wildlife **conservation** agreement in the world. The parties to the treaty meet every two years to evaluate and amend the treaty if necessary.

The species covered by the treaty are listed in three appendices, each of which requires different trade restrictions. Appendix I applies to "all species threatened with extinction," such as African and Asian **elephants** (*Loxodonta africana* and *Elephas maximus*, the hyacinth macaw (*Anodorhynchus hyacinthinus*), and Queen Alexandria's birdwing butterfly (*Ornithoptera alexandrae*). Commercial trade is generally prohibited for the over 900 listed species. Appendix II applies to "all species which although not necessarily now threatened with **extinction** may become so unless trade in specimens of such species is subject to strict regulation," such as the polar bear, giant clams, and Pacific Coast mahogany. Trade in these species requires an export permit from the country of origin. Currently, over 4,000 animals and 22,000 plants (mainly orchids) are listed in Appendix II. Appendix III is designed to help individual nations control the trade of any species. Any species may be listed in Appendix III for any nation. Once listed, any export of this species from the listing country requires an export permit. Usually a species listed in Appendix III is protected within that nation's borders. These trade restrictions apply only to signatory nations, and only to trade between countries, not to practices within countries.

CITES relies on signatory nations to pass domestic laws to carry out the principles included in the treaty. In the United States, the **Endangered Species Act** and the Lacey Act include domestic requirements to implement CITES, and the **Fish and Wildlife Service** is the chief enforcement agency. In other nations, if and when such legislation is passed, effective implementation of these domestic laws is required. This is perhaps the most problematic aspect of CITES, since most signatory nations have poor administrative capacity, even if they have strong desire for enforcement. This is especially a problem in poorer nations of the world, where monies for regulation of trade in endangered species is forced down the agenda by more pressing social and economic issues.

CITES can be regarded as moderately successful. There has been steady progress toward compliance with the treaty, though tremendous enforcement problems still exist. Due to the international scope and size of world wildlife trade, enforcement of CITES is estimated to be only 60–65% effective worldwide. International trade in endangered species is still big business; the profits from such trade can be huge. In 1988, the world market for **exotic species** was estimated to be $5 billion, $1.5 billion of which was estimated to be illegal. Among the most lucrative products traded are reptile skins, fur coats, and ingredients for traditional drugs. If violations are discovered, the goods are confiscated and penalties and fines for the violators are established by each country. Occasionally, sanctions are imposed on a nation if it flagrantly violates the convention.

[*Christopher McGrory Klyza*]

RESOURCES
BOOKS

Fitzgerald, S. *International Wildlife Trade: Whose Business Is It?* Washington, DC: World Wildlife Fund, 1989.

McCormick, J. *Reclaiming Paradise: The Global Environmental Movement.* Bloomington: Indiana University Press, 1989.

Nichol, J. *The Animal Smugglers.* New York: Facts on File, 1987.

ORGANIZATIONS

CITES Secretariat, International Environment House, Chemin des Anémones , Châtelaine, Geneva Switzerland CH-1219 (+4122) 917-8139/40, Fax: (+4122) 797-3417, Email: cites@unep.ch, <http://www.cites.org>

Convention on Long-Range Transboundary Air Pollution (1979)

Held in Geneva in 1979 under the auspices of the United Nations, the goal of the Convention on Long-Range Transboundary **Air Pollution** was to reduce air **pollution** and **acid rain**, particularly in Europe and North America. The accord went into effect in March 1983. It was signed by the United States and Canada, as well as European countries, and the signatories agreed to cooperate in researching and monitoring air pollution and to exchange information on developing technologies for **air pollution control**. This convention established the Cooperative Programme for Monitoring and Evaluating of the Long-Range Transmission of Air Pollutants in Europe, which was first funded in 1984. The countries that signed the treaty also agreed to reduce their sulfur emissions 30% by 1993. All of the countries were able to meet this goal, with many coutries reducing more than 50–60% of their emissions.

[*Douglas Smith*]

RESOURCES
BOOKS

Basic Documents of International Environmental Law. Boston: Graham & Trotman, 1992.

Effectiveness of International Environmental Agreements: A Survey of Existing Legal Instruments. Cambridge, England: Grotius, 1992.

Convention on the Conservation of Migratory Species of Wild Animals (1979)

The first attempt at a global approach to **wildlife management**, the Convention on the **Conservation** of Migratory **Species** of Wild Animals was held in Bonn, Germany in 1979. The purpose of the convention was to reach an agreement on the management of wild animals that migrate "cyclically and predictably" across international boundaries. Egypt, Italy, the United Kingdom, Denmark, and Sweden, among other nations, signed an accord on this issue; the treaty went into effect in 1983, but the United States and Canada, as well as the former Soviet Union, still have not agreed to it.

The challenge facing the convention was how to assist nations that did not have **wildlife** management programs while not disrupting those that had already established them. The United States and Canada did not believe that the agreement reached in Bonn met this challenge. Representatives from both countries argued that the definition of a migratory animal was too broad; it would embrace nearly every game bird or animal in North America, including rabbits, deer, and bear. But both countries were particularly concerned that the agreement did not sufficiently honor national sovereignty, and they believed it threatened the effectiveness of the federal-state and federal-provincial systems that were already in place.

It is widely believed that the principal failure of this convention was its inability to find language dealing with federalist systems that was acceptable to all. The agreement also came into conflict with other laws, particularly laws governing national jurisdictions and territorial boundaries at sea, but it is still considered an important advance in the process of developing international environmental agreements. *See also* Environmental law

[*Douglas Smith*]

RESOURCES
BOOKS

Basic Documents of International Environmental Law. Boston: Graham & Trotman, 1992.

Effectiveness of International Environmental Agreements: A Survey of Existing Legal Instruments. Cambridge, England: Grotius, 1992.

Convention on the Law of the Sea (1982)

During the age of exploration in the seventeenth century, the Dutch lawyer Hugo Grotius formulated the legal principle that the ocean was free (*Mare liberum*), which became the basis of the law of the sea. During the eighteenth century, each coastal nation was granted sovereignty over an offshore margin of three nautical miles to provide for better defense against pirates and other intruders. As undersea exploration advanced in the twentieth century, nations became increasingly interested in the potential resources of the oceans. By the 1970s, many were determined to exploit resources such as **petroleum** on the continental shelf and manganese nodules rich with metals on the deep ocean floor, but uncertainties about the legal status of these resources inhibited investment.

In 1974 the United Nations responded to these concerns and convened a Conference on the Law of the Sea, at which the nations of the world negotiated a consensus dividing the ocean among them. The participants drafted a proposed constitution for the world's oceans, including in it a number of provisions that radically altered their legal status. First, the convention extended the **territorial sea** from three to 22 nautical miles, giving nations the same rights and responsibilities within this zone that they possess over land. Negotiations were necessary to ensure that ships have a right to navigate these coastal zones when going from one ocean to another.

The convention also acknowledged a 200 nautical mile **Exclusive Economic Zone** (EEZ), in which nations could regulate fisheries as well as resource exploration and exploitation. By the late 1970s, over 100 nations had already claimed territorial seas extending from 12 to 200 miles. The United States, for example, claims a 200 mile zone. Today, EEZs cover about one third of the ocean, completely dividing the North Sea, the **Mediterranean Sea**, the Gulf of Mexico, and the Caribbean among coastal states. They also encompass an area that yields more than 90% of the ocean's fish catch, and they are considered a potentially effective means to control **overfishing** because they provide governments clear responsibility for their own fisheries.

The most heated debates during the convention, however, did not concern the EEZs, but involved provisions concerning seabed mining. **Third World** nations insisted that deep-sea metals were the common heritage of mankind, and that their wealth should be shared. After fierce disagreements, a compromise was reached in 1980 that would establish a new International Seabed Authority (ISA). Under the plan, any national or private enterprise could take half the seabed mining sites and the ISA would take the other half. In exchange for mining rights, industrialized nations would underwrite the ISA and sell it the necessary technology. In December 1982, 117 nations signed the Law of the Sea Convention at Montego Bay, Jamaica. President Ronald Reagan refused to sign, citing the seabed-mining provisions as a major reason for United States opposition. Two other

key nations, the United Kingdom and Germany, refused, and 21 nations abstained. In order to become international law, 60 nations must ratify the Law of the Sea Convention, but although 159 nations have now become signatories, only 40 have actually ratified it. *See also* Coastal Zone Management Act; Commercial fishing; Oil drilling

[*David Clarke*]

RESOURCES
BOOKS

Law of the Sea: Protection and Preservation of the Marine Environment. New York: United Nations, 1990.

Porter, G., and J. W. Brown. *Global Environmental Politics.* Boulder, CO: Westview Press, 1991.

Simon, A. W. *Neptune's Revenge, The Ocean of Tomorrow.* New York: Franklin Watts, 1984.

Convention on the Prevention of Marine Pollution by Dumping of Waste and Other Matter (1972)

The 1972 International Convention on the Prevention of **Marine Pollution** by Dumping of Waste and Other Matter, commonly called the London Dumping Convention, entered into force on August 30, 1975. The London Dumping Convention covers "ocean dumping" defined as any deliberate disposal of wastes or other matter from ships, aircraft, platforms, or other human-made structures at sea. The **discharge** of sewage **effluent** and other material through pipes, wastes from other land-based sources, the operational or incidental disposal of material from vessels, aircraft, and platforms (such as fresh or salt water), or wastes from seabed mining are not covered under this convention.

The framework of the London Dumping Convention consists of three annexes. Annex I contains a blacklist of materials that cannot be dumped at sea. Organohalogen compounds, **mercury, cadmium**, oil, high-level radioactive wastes, warfare **chemicals**, and persistent **plastics** and other synthetic materials that may float in such a manner as to interfere with fishing or navigation are examples of substances on the blacklist. The prohibition does not apply to **acid** and alkaline substances that are rapidly rendered harmless by physical, chemical, or biological processes in the sea. Annex I does not apply to wastes such as sewage **sludge** or dredge material that contain blacklist substances in trace amounts.

Annex II of the convention comprises a grey list of materials considered less harmful than the substances on the blacklist. At-sea dumping of grey list wastes requires a special permit from contracting states (countries participating in the convention). These materials include wastes containing

arsenic, **lead**, **copper**, zinc, organosilicon compounds, cyanides, pesticides and radioactive matter not covered in Annex I, and containers, scrap metals, and other bulky debris that may present a serious obstacle to fishing or navigation. Grey list material may be dumped as long as special care is taken with regard to **ocean dumping** sites, monitoring, and methods of dumping to ensure the least detrimental impact on the **environment**. A general permit from the appropriate agencies of the contracting states to the Convention is required for ocean dumping of waste not on either list.

Annex III includes criteria that countries must consider before issuing an ocean dumping permit. These criteria require consideration of the effects dumping activities can have on marine life, amenities, and other uses of the ocean, and they encompass factors related to disposal operations, waste characteristics, attributes of the site, and availability of land-based alternatives.

The International Maritime Organization serves as Secretariat for the London Dumping Convention, undertaking administrative responsibilities and ensuring cooperation among the contracting parties. As of 2001, there were 78 contracting parties to the Convention, including the United States.

The London Dumping Convention covers ocean dumping in all marine waters except internal waters of the contracting states, which are required to regulate ocean dumping consistently with the convention's provisions. However, they are free to impose stricter rules on their own activities than those required by the convention. The London Dumping Convention was developed at the same time as the **Marine Protection, Research and Sanctuaries Act** of 1972 (Public Law 92-532), a law enacted by the United States. The U.S. congress amended this act in 1974 to conform with the London Dumping Convention.

Most nations using the ocean for purposes of dumping waste are developed countries. The United States completely ended its dumping of sewage sludge following the passage of the **Ocean Dumping Ban Act** (1988), which prohibits ocean dumping of all sewage sludge and industrial waste (Public Law 100-688). Britain and the North Sea countries also intend to end ocean dumping of sewage sludge. During the Thirteenth Consultative Meeting of the London Dumping Convention in 1990, the contracting states agreed to terminate all industrial ocean dumping by the end of 1995.

While the volume of sewage sludge and industrial waste dumped at sea is decreasing, ocean dumping of dredged material is increasing. **Incineration** at sea requires a special permit and is regulated according to criteria contained in an addendum to Annex I of the convention.

In 1996, the London Dumping Convention was replaced by the 1996 Protocol to the Convention on the Prevention of Marine **Pollution** by Dumping of Wastes and

Other Matter, 1972. Among the changes is the "reverse list." The Protocol only allows dumping material that is listed in Annex I, it is referred to as the "reverse list" since it completely reverses the original Annex I. Incineration at sea is now completely banned, unless there is an emergency, as is the exportation of waste to other countries for ocean dumping. In 2001, the Protocol had been ratified by 16 states, but needs to be ratified by 26, 15 of which must be Contracting Parties to the Convention, in order to be put into force. *See also* Marine pollution; Seabed disposal

[*Marci L. Bortman*]

RESOURCES
PERIODICALS

Duedall, I. W. "A Brief History of Ocean Disposal." *Oceanus* 33 (Summer 1990): 29-33+.

Kitsos, T. R., and J. M. Bondareff. "Congress and Waste Disposal at Sea." *Oceanus* 33 (Summer 1990): 23-8.

U.S. House of Representatives. House Report No. 100-1090. Ocean Dumping Ban Act Conference Report, 2nd Session, 100th Congress, 1990.

OTHER

National Ocean Servie, International Program Office. *The Convention on the Prevention of Marine Pollution by Dumping of Wastes and Other Matter, London, 1972.* [June 2002]. <http://international.nos.noaa.gov/conv/ldc.html>.

Convention on Wetlands of International Importance (1971)

Also called the Ramsar Convention or **Wetlands** Convention, the Convention on Wetlands of International Importance is an international agreement adopted in 1971 at a conference held in Ramsar, Iran. One of the principal concerns of the agreement was the protection of migratory waterfowl, but it is generally committed, like much wetlands legislation in the United States, to restricting the loss of wetlands in general, because of their ecological functions as well as their economic, scientific, and recreational value. The accord went into effect in 1975, establishing a network of wetlands, primarily across Europe and North Africa.

In 2002, there were 132 Contracting Parties, each of whom was required to set aside at least one wetland reserve. Over 1,178 national wetland sites have been established totaling over 252.3 million acres (102.1 million ha). The convention has secured protection for wetlands around the world, but many environmentalists believe it has the same weakness as many international conventions on the **environment**. There is no effective mechanism for enforcement. **Population growth** continues to increase political and economic pressures to develop wetland areas around the world, and there are no provisions in the agreement strong enough to prevent nations from removing protected status from designated wetlands.

[*Douglas Smith*]

RESOURCES
BOOKS

Basic Documents of International Environmental Law. Boston: Graham & Trotman, 1992.
Effectiveness of International Environmental Agreements: A Survey of Existing Legal Instruments. Cambridge, England: Grotius, 1992.

ORGANIZATIONS
The Ramsar Convention Bureau, Rue Mauverney 28, Gland , Switzerland CH-1196 +41 22 999 0170 , Fax: +41 22 999 0169 , Email: ramsar@ramsar.org, <http://www.ramsar.org/index.html>

Conventional pollutant

Conventional pollutants fall into five categories; the presence of these pollutants is commonly determined by measuring **biochemical oxygen demand**, total suspended solids, **pH** levels, the amount of fecal coliform, and the quantity of oil and grease.

Biochemical oxygen demand (BOD) is the quantity of oxygen required by **microorganisms** to stabilize five-day incubated oxidizable organic matter at 68°F (20°C). Hence, BOD is a measure of the **biodegradable** organic **carbon** and at times, the oxidizable **nitrogen**. BOD is the sum of the oxygen used in organic matter synthesis and in the endogenous **respiration** of microbial cells. Some industrial wastes are difficult to oxidize, and bacterial seed is necessary. In certain cases, an increase in BOD is observed with an increase in dilution. It is hence necessary to determine the detection limits for BOD.

Suspended solids interfere with the transmission of light. Their presence also affects recreational use and aesthetic enjoyment. Suspended solids make fish vulnerable to diseases, reduce their growth rate, prevent successful development of fish eggs and larvae, and reduce the amount of available food. The **Environmental Protection Agency** (EPA) restricts suspended matter to not more than ten percent of the reasonably established amount for aquatic life. This allows sufficient sunlight to penetrate and sustain **photosynthesis**. Suspended solids also cause damage to invertebrates and fill up gravel spawning beds.

The acidity or alkalinity of water is indicated by pH. A pH of seven is neutral. A pH value lower than seven indicates an acidic **environment** and a pH greater than seven indicates an alkaline environment. Most aquatic life is sensitive to changes in pH. The pH of surface waters is specified to protect aquatic life and prevent or control unwanted chemical reactions such as metal **ion** dissolution

in acidic waters. An increase in toxicity of many substances is often observed with changes in pH. For example, an alkaline environment shifts the ammonium ion to a more poisonous form of un-ionized ammonia. EPA criteria for pH are 6.5–9.0 for freshwater life, 6.5–8.5 for marine organisms and 5–9 for domestic consumption.

Fecal **coliform bacteria** are yardsticks for detecting pathogenic or disease causing bacteria. However, this relationship is not absolute because these bacteria can originate from the intestines of humans and other warm blooded animals. Prior knowledge of **river basins** and the possible sources of these bacteria is necessary for a survey to be effective. The strictest EPA criteria for coliforms apply to shellfish, since they are often eaten without being cooked.

Common sources of oil and grease are **petroleum** derivatives and fats from vegetable oil and meat processing. Both surface and domestic waters should be free from floating oil and grease. Limits for oil and grease are based on LC_{50} values. LC_{50} is defined as the concentration at which 50% of an aquatic **species** population perishes. EPA criterion is for a 96-hour exposure, and during this period the concentration of individual petrochemicals should not exceed 0.01 of the LC_{50} median. Oil and grease contaminants vary in physical, chemical, and toxicological properties besides originating from different sources. *See also* Industrial waste treatment; Sewage treatment; Wastewater; Water pollution; Water quality

[*James W. Patterson*]

RESOURCES
BOOKS

Viessman Jr., W., and M. J. Hammer. *Water Supply and Pollution Control.* New York: Harper & Row, 1985.

Copper

A metallic element with an atomic number of 29. It is abbreviated Cu and has an atomic weight of 63.546. Copper is a micronutrient which is needed in many proteins and enzymes. Copper has been and is frequently used in piping systems which convey potable water. Corrosive waters will leach copper from the pipelines, thereby exposing consumers to copper and possibly creating bluish-green stains on household fixtures and clothes. The staining of household fixtures becomes a nuisance when copper levels reach 2–3 mg/l. The drinking water standard for copper is 1 mg/l. It is a secondary standard based on the potential problems of staining and

Copper mining
see **Ashio, Japan; Ducktown, Tennessee;**

Sudbury, Ontario

Coprecipitation

Inorganic contaminants, such as **heavy metals**, which exhibit toxicity effects when present at low levels, can be difficult to treat. The level of toxicity can be well below the metal's solubility concentration, and for this reason precipitation, a common treatment method for the removal of heavy metal, does not provide the needed **effluent** quality. **Cadmium**, for instance, has a reported maximum contaminant level (MCL) of 10 μg/l, with a 5 μg/l proposed level. Based upon theoretical solubility calculations, alkaline precipitation may only reduce the level of cadmium to 140 μg/l. Hence, additional methods of **wastewater** treatment are often required to meet environmental regulations. One such treatment technique is chemical coprecipitation.

Coprecipitation is a process in which a solid is precipitated from a solution containing other ions. These ions are incorporated into the solid by **adsorption** on the surface of the growing particles, physical entrapment in the pore spaces, or substitution in the crystal lattice. Adsorption is one of the principle mechanisms of coprecipitation. It is a process in which the solid **species**, or adsorbent, is added to a solution containing other ions, called adsorbates. In this case, the adsorbates are bound to the solid's surface by physical or chemical interactions between one adsorbate and the adsorbent.

In solution, coprecipitation and adsorption are thus related by the time during which an adsorbate is present. The type of adsorbent present also affects the extent of **ion** uptake from solution. One solid used to perform this treatment is ferric oxide. For example, iron coprecipitation of cadmium has been reported to yield a residual cadmium concentration of about 3 μg/l. Several solid and solution variables must be considered when designing and optimizing coprecipitation as a treatment option. These include the equilibrium concentration of the species in solution, the suspension **pH**, and the presence of other interacting ions. Also, the properties of the solid adsorbent are important. These include type of solid formed, surface area available, age, and surface charge. In short, coprecipitation is controlled by a number of important variables which are specific to a given system. Thus treatability studies must be performed to fully optimize this wastewater treatment process and pro-

duced the necessary effluent quality. *See also* Hazardous waste; Pollution control; Sewage treatment; Toxic substance

[*James W. Patterson*]

RESOURCES
BOOKS

Anderson, M., and A. Rubin. *Adsorption of Inorganics of Solid-Liquid Interfaces*. Ann Arbor, MI: Ann Arbor Science, 1981.
Leckie, J., et al. *Adsorption/Coprecipitation of Trace Elements from Water with Iron Oxyhydroxide*. EPR5 Final Report, 1980.

OTHER

A Review of Solid-Solution Interactions and Implications for the Control of Trace Inorganic Materials in Water Treatment. AWWA Committee Report. Vol. 80, 1988.

Coral bleaching

Coral bleaching is the whitening of coral colonies due to the loss of the symbiotic algae, zooxanthellae, from the tissues of coral polyps. It is mostly caused by stress. The host coral polyp provides the algae with a protected **environment** and a supply of **carbon dioxide** for its photosynthetic processes. The golden-brown algae serve as a major source of nutrition and color for the coral. The loss of the algae exposes the translucent calcium carbonate skeletons of the coral colony, and the corals look "bleached." Corals may recover from short-term bleaching (less than a month), but prolonged bleaching causes irreversible damage and **mortality**, for without the algae, the corals starve and die. However, even a sublethal stress may result in increased susceptibility of corals to infections, with resulting significant mortality. Populations of sea urchins, parrot fish, and worms erode and weaken dead reef skeletons, and the reef can be destroyed by storm surges.

The means by which corals expel the zooxanthellae are not yet known. In laboratory experiments the zooxanthellae are released into the gut of the polyp and then expelled through the mouth; however this method has not been observed in the environment. Another hypothesis is that the stressed corals provide the algae with fewer nutrients, which results in the algae leaving the corals. Another possibility is that the algae may produce oxides under stress, which adversely affect the algae.

Bleaching may be caused by a number of stresses or environmental changes, including disease, excess shade, increased levels of **ultraviolet radiation, sedimentation, pollution, salinity** changes, exposure to air by low tides or low sea level, and increased temperatures. Coral bleaching is most often associated with increased sea surface temperatures, as corals tolerate only a narrow temperature range of between about 77–84°F (25–29°C).

Historically bleaching was observed on a small scale, such as in overheated tide pools. However, in the early to mid 1980s, coral reefs around the world began to experience large scale bleaching, with a bleaching event occurring somewhere in the world almost every year. In 1998, coral reefs around the world suffered the most extensive and severe bleaching and subsequent mortality in recorded history, in the same year that tropical sea surface temperatures were the highest in recorded history. Coral bleaching was reported in at least 60 countries and island nations in the Pacific Ocean, Indian Ocean, Red Sea, Persian Gulf, and the Caribbean. Only the Central Pacific region was spared. About 16% of the world's reefs were lost in a period of only nine months. Previous bleaching events had only affected reefs to a depth of less than 49 ft (15 m), but in 1998, the bleaching extended as deep as 164 ft (50 m). The reason sea temperatures throughout the world were so warm in 1998 remains controversial and uncertain. Three theories have been developed—natural **climate** variability, El Ninño and other climatic variations, and global warming. By the end of 2000, 27% of the world's reefs had been lost, with the largest single cause being the coral bleaching event of 1998. Only half of the reefs lost during 1998 will probably recover, which will add to the 11% already lost to human impacts such as **sediment** and **nutrient** pollution and over-exploitation.

Many of the bleached **coral reef** ecosystems may require decades to recover. Human populations dependent on the reefs will lose fisheries, shoreline protection, and tourism opportunities. Trends of the past century indicate that coral bleaching events may become more frequent and severe if the climate continues to warm.

[*Judith L. Sims*]

RESOURCES

BOOKS

Wilkinson, Clive. ed.*Status of Coral Reefs of the World: 2000.*Queensland, Australia: Australian Institute of Marine Science, 2000.

OTHER

Pomerance, Rafe. "Coral Bleaching, Coral Mortality, and Global Climate Change".*Report to U.S. Coral Reef Task Force, Bureau of Oceans and International Environmental and Scientific Affairs,* U.S. Department of State. March 5, 1999. [cited May 31, 2002]. <http://www.state.gov/www/global/global_issues/coral_reefs/990305_coralreef_rpt.html>.

Coral reef

Coral reefs represent some of the oldest and most complex communities of plants and animals on Earth. About 200–400 million years old, they cover about 231,660 mi^2 (600,000 km^2) worldwide. (The most popular reefs range from 5,000–

10,000 years old.) The primary structure of a coral reef is a calcareous skeleton formed by marine invertebrate organisms known as *cnidarians*, which are relatives of sea anemones. Corals are found in most of the oceans of the world, in deep as well as shallow seas and temperate as well as tropical waters. But corals are most abundant and diverse in relatively shallow tropical waters, where they have adapted to the constant temperatures provided by these waters. The reef-forming corals, or hermatypic corals, have their highest diversity in the Indian and Pacific Oceans, where over 700 **species** are found. By contrast, the Atlantic Ocean provides the **habitat** for less than 40 species. Other physical constraints needed for the success of these invertebrate communities are clear water, a firm substrate, high **salinity**, and sunlight. Clear water and sunlight are required for the symbiotic unicellular plants that live in the surface tissues of the coral polyps. This intimate plant-animal association benefits both participants. Corals obtain oxygen directly from the plants and avoid having to excrete nitrogenous and phosphate waste products because these are absorbed directly as nutrients by the plants. **Respiration** by the coral additionally provides **carbon dioxide** to these plants to be used in the photosynthetic process.

The skeletons of hermatypic coral play a major role in the formation of coral reefs, but contributions to reef structure, in the form of calcium carbonate, come from a variety of other oceanic species. Among these are red algae, green algae, foraminifers, mollusk shells, sea urchins, and the exoskeletons of many other reef-dwelling invertebrates. This limestone infrastructure provides the **stability** needed, not only to support and protect the delicate tissues of the coral polyps themselves, but also to withstand the constant wave action generated in the shallow, near-shore waters of the marine **ecosystem**.

There are essentially three types of coral reefs. These categories are fringing reefs, barrier reefs, and atolls. Fringing reefs form borders along the shoreline. Some of the reefs found in the **Hawaiian Islands** are fringing reefs. Barrier reefs also parallel the shoreline but are found further offshore and are separated from the coast by a **lagoon**. The best example of this type of reef is the **Great Barrier Reef** off the coast of **Australia**. Because the coral colonies form an interwoven network of organisms from one end of the reef to the other, this is the largest individual biological feature on earth. The Great Barrier Reef borders about 1,250 mi (2011 km) of Australia's northeast coast. The second largest continuous barrier reef is located in the Caribbean Sea off the coast of Belize, east of the Yucatan Peninsula. The third type of reef, the atoll, is typically a ring-shaped reef, from which several small, low islands may project above the surface of the ocean. The ring structure is present because it represents the remains of a fringing reef that formed around an oceanic **volcano**. As the volcano eroded or collapsed, the outwardly-growing reef is ultimately all that remains as a circle of coral. Possibly the most infamous atoll is the **Bikini Atoll**, which was the site of the United States' **hydrogen** bomb tests during the 1940s and 1950s.

Besides the physical structure of the coral and the reef itself, the most significant thing about these structures is the tremendous diversity of marine life that exists in, on, and around coral reefs. These highly productive marine ecosystems may contain over 3,000 species of fish, shellfish, and other invertebrates. About 33% of all of the fishes of the world live and depend on coral reefs. This tremendous diversity provides for a huge commercial fishery in countries such as the Philippines and Indonesia. With the advent and availability of SCUBA gear to the general public in this half of this century, the diversity of life exhibited on coral reefs has been a great lure for tourists to these ecosystems throughout the world.

Even with their calcium carbonate skeleton and exquisite beauty, coral reefs are being degraded and destroyed daily, not only by natural events such as constant wave action and storm surges, but, more importantly, by the actions of man. Of the 109 countries that have coral reef formations within their territorial waters, 90 are losing them because of man-induced **environmental degradation**. Most is the result of physical abuse or **pollution** which alters the narrow range of physical and chemical parameters necessary for the coral, or their plant symbionts, to remain viable and thrive. Today, 10% of the world's coral reefs are completely degraded, 30% have reached a critical stage. Scientists have determined that if degradation at this rate continues, 70% of all coral reefs could be gone in 40 years. Reefs can be salvaged however. In 1995, participants from 44 countries representing governments, nongovernmental organizations, international development agencies, and the private sector gathered to launch the International Coral Reef Initiative. In 1997, some 1,400 participants declared the year as the International Year of the Reef, a period they hoped would heighten awareness and further reef salvage activities worldwide.

These reefs, most of which are between 5,000 and 10,000 years old, and some of which have been building on the same site for over a million years, are being degraded and destroyed by a vast array of water pollutants. **Silt**, which washes into the sea from **erosion** of clearcut forests miles inland, cloud the water or smother the coral, thus prohibiting the photosynthetic process from taking place. **Oil spills** and other toxic or hazardous **chemicals** that find their way into the marine ecosystem through man's actions are killing off the coral and/or the organisms associated with the reefs. Mining of coral for building materials takes a massive toll on these communities. Removal of coral to supply the ever

A coral reef. (Phototake. Reproduced by permission.)

increasing demand within the **aquarium trade** is destroying the reefs as well. The tremendous interest in and appeal of marine aquaria has added another problem to this dilemma. In the race to provide the aquarium market with a great variety of beautiful, brilliantly-colored, and often quite rare, marine fishes, unscrupulous collectors, who are selling their catches illegally merely for the short term monetary gain, spray the coral heads with poison solutions (including cyanide) to stun the fishes, causing them to abandon the reefs. This efficient means of collecting reef fishes leaves the coral head enveloped in a cloud of poison, which ultimately kills that entire section of the reef.

An unusual phenomenon has developed within the past decade with regard to coral reefs and pollution. In the Florida Keys in the early 1980s divers began reporting that the coral, sea whips, sea fans, and sponges of the reefs, around which they had been swimming, had turned white. They also reported that the waters felt unusually warm. The same phenomenon occurred in the Virgin Islands in the late 1980s. As much as 50% of the reef was dying due to this bleaching effect. Scientists are still studying these occurrences; however, many feel that it is a manifestation of global warming, and that a mere change in the water temperature around the coral reefs of 2–3°C is inducing the bleaching and death of the coral.

Tourism and **recreation** are inadvertently degrading coral reefs throughout the world as well. Coral is being destroyed by the propellers of recreational boats as well as divers who unintentionally step on coral heads, thus breaking them to pieces, and degrading the very structure of the ecosystem they came to see. Many of the reefs undergoing this degradation are sections that have been set aside for protection. Even with almost a quarter of a million square miles of coral reefs in the world, and about 300 protected regions in 65 countries, ever increasing levels of near-shore pollution, coupled with other acts of man, may be destroying these extremely complex communities of marine organisms at a rate faster than we can control.

[*Eugene C. Beckham*]

RESOURCES
BOOKS

Brown, B., and J. Ogdon. "Coral Bleaching." *Scientific American* 268 (January 1993): 64-70.
Derr, M. "Raiders of the Reef." *Audubon* 94 (February 1992): 48-56.
Falkner, D. *This Living Reef.* New York: The New York Times Book Co., 1974.

Corporate Average Fuel Economy standards

In 1975, as part of the response to the Arab **oil embargo** of 1973–1974, the U.S. Congress and President Richard Nixon passed the **Energy Policy** and **Conservation** Act. One of a number of provisions in this legislation intended to decrease fuel consumption was the establishment of fuel economy standards for passenger cars and light trucks sold in the United States. These Corporate Average Fuel Economy (CAFE) standards require that each manufacturer's entire production of cars or trucks sold in the United States meet a minimum average fuel economy level. Domestic and import cars and trucks are segregated into separate fleets, and each must meet the standards individually. Any manufacturer whose fleet(s) fails to meet the standard is subject to a fine, based on a CAFE shortfall times the total number of vehicles in the particular fleet. Manufacturers are allowed to generate CAFE credits for overachievement in a single year, which may be carried forward or backward up to three years to cover other shortfalls in mileage. This helps to smooth the effects of model introduction cycles or market shifts.

CAFE standards took effect in 1978 for passenger cars and 1979 for light trucks. Truck standards originally covered vehicles under 6,000 lb (2,724 kg) gross vehicle weight (GVW), but in 1980, they were expanded to trucks up to 8,500 lb (3,632 kg). Car standards were set at 18 mpg

for 1978, and increased annually to 27.5 mpg for 1985 and beyond. Manufacturers responded with significant vehicle downsizing, application of new powertrain technology, and reductions in aerodynamic drag and tire friction. These actions helped improve average fuel economy of U.S.-built passenger cars from 13 mpg in 1974 to 25 mpg by 1982. The fall of **gasoline** prices following an embargo in 1980–1981 encouraged consumers to purchase larger vehicles. This mix shift slowed manufacturers' CAFE improvement by the mid-1980s, despite continued improvements in individual model fuel economy, since CAFE is based on all vehicles sold. The Secretary of **Transportation** has the authority to reduce the car CAFE standards from 27.5 mpg to 26 mpg. This authority was used in 1986 when it became clear that market factors would prevent CAFE compliance by a large number of manufacturers.

A separate requirement to encourage new car fuel efficiency was created through the Energy Tax Act of 1978. This act created a "Gas Guzzler" tax on passenger cars whose individual fuel economy value fell below a certain threshold, starting in 1980. The tax is progressively higher at lower fuel economy levels. This act has since been amended both to increase the threshold level and to double the tax. The Gas Guzzler tax is independent of CAFE standards.

Environmental groups have long favored increasing the CAFE standards to reduce oil consumption and **pollution**. At the same time, the **automobile** industry has fought hard to avoid raising mileage standards, claiming increased regulation would reduce competitiveness and cost jobs. After the **Persian Gulf War** in 1991, some members of Congress proposed raising CAFE standards to 45 mpg for cars and 35 mpg for light trucks by 2001. Proponents of the plan claimed it would save 2.8 million barrels of oil per day, or more oil than is imported from Persian Gulf countries (about 2.4 million barrels per day in 2000). Strongly opposed by the automobile industry, the efficiency increase failed in Congress.

Cheap oil prices in the 1990s decreased Americans' concerns about fuel efficiency, and sales of larger cars boomed. Since 1988, nearly all technology gains in automobile efficiency have been offset by increased weight and power in new vehicles. The popular Sport Utility Vehicle (SUV), decried by environmentalists, became the new model of gas guzzler. Due to the "SUV loophole" in the CAFE laws, the SUV was classified as a light truck and was exempted from higher mileage standards and fines for inefficiency, although opponents of the vehicles claim that they are used more like cars and should be classified as such. By law, SUVs can emit up to five times more **nitrogen oxides** than cars, and have a lower fleet mileage standard of 20.7 mpg. By 2001, SUVs and light trucks accounted for nearly half of all new automobiles sold. Because of the high sales of large vehicles, United States automobile efficiency reached its lowest point in 21 years in 2001, despite new technology.

Another battle between environmentalists and the automobile industry occurred in March 2002, when Congress considered raising CAFE standards by 50% by the year 2015. Environmental and other groups claimed that increased efficiency would reduce America's dependence on foreign oil, would reduce pollution and **greenhouse gases** in the **environment**, and eliminate the need to drill for oil in sensitive regions such as the **Arctic National Wildlife Refuge** (ANWR) and offshore areas. The bill was strongly opposed by the automobile industry, and did not pass.

[*Brian Geraghty and Douglas Dupler*]

RESOURCES
PERIODICALS

Gelbspan, Ross. "A Modest Proposal to Stop Global Warming." *Sierra*, May/June 2001, 63.
Hakim, Danny. "M.P.G. or G.P.M.?." *New York Times*, 12 March 2002, WK2.

OTHER
Alliance to Save Energy Factsheet on Transportation and Energy. [2002]. <http://www.ase.org>.

Corrosion and material degradation

Corrosion or degradation involves deterioration of material when exposed to an **environment** resulting in the loss of that material, the most common case being the corrosion of metals and steel by water. The changes brought about by corrosion include weight loss or gain, material loss, or changes in physical and mechanical properties.

Metal corrosion involves oxidation-reduction reactions in which the metal is lost by dissolution at the anode (oxidation). The electrons travel to the cathode where the reduction takes place, while s move through a conducting solution or electrolyte. A positive and a negative pole, called the cathode and the anode respectively, are thereby created with a current flow between them. Thus the process of corrosion is basically electrochemical.

For corrosion to occur, certain conditions must be present. These are: (1) a potential difference between the cathode and the anode to drive the reaction; (2) an anodic reaction; (3) an equivalent cathodic reaction; (4) an electrolyte for the internal circuit; (5) an external circuit where electrons can travel. Sometimes, polarization of the anodic and the cathodic reactions must be taken into consideration. Polarization is a change in equilibrium **electromagnetic field** of a cell due to current flow. It has been reported that polarization may retard corrosion, as in the accumulation of unreacted **hydrogen** on the cathode.

In the corrosion of iron in water, the reactions differ according to whether or not oxygen is present. The common reactions that take place in a deaerated medium are essentially an oxidation reaction releasing ferrous **ion** into solution at the anode and a reduction reaction releasing hydrogen gas at the cathode. In the presence of oxygen, a complementary cathode reaction involves oxygen being reduced to water.

Degradation of concrete, on the other hand, depends on the composition of cement and the aggressive action of the water in contact with it. Some forms of corrosion may be visibly apparent, but some are not. Surface corrosion, corrosion at discrete areas, and anodic attack in a two-metal corrosion may be readily observed. A less identifiable form, erosion-corrosion, is caused by flow patterns that cause abrasion and wear or sweep away protective films and accelerate corrosion. Another form of corrosion which involves the selective removal of an alloy constituent requires another means of examination. Cracking, a form of corrosion which is caused by the simultaneous effects of tensile stress and a specific corrosive medium, could be verified by microscopy.

Some measures adopted to prevent corrosion in metals are cathodic protection, use of inhibitors, coating, and the formation of a passivating film. Protection of concrete, on the other hand, can be achieved by coating, avoiding corrosive **pH** of the water with which the concrete is in contact, avoiding excessive concentrations of ammonia, and avoiding deaeration in pipes. *See also* Hazardous waste sites; Seabed disposal; Waste management

[*James W. Patterson*]

RESOURCES
BOOKS

Dillon, C. P. *Corrosion Control in the Chemical Process Industries.* New York: McGraw-Hill, 1986.
Fontana, M. G., and N. D. Greene. *Corrosion Engineering.* New York: McGraw-Hill, 1967.
Weber, W., Jr. *Physicochemical Processes for Water Quality Control.* New York: Wiley-Interscience, 1972.

Cost-benefit analysis

Environmentalists might believe that total elimination of risk that comes with **pollution** and other forms of **environmental degradation** is possible and even desirable, but economists argue that the benefits of risk elimination have to be balanced against the costs. Measuring risk is itself very complicated. **Risk analysis** in the case of pollution, for instance, involves determining the conditions of exposure, the adverse effects, the levels of exposure, the level of the effects, and the overall contamination. Long **latency** periods, the need to draw implications from laboratory studies

Cost/Benefit Analysis

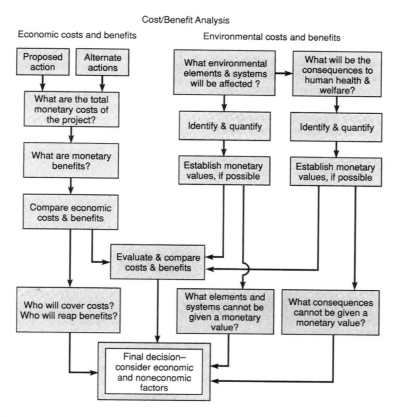

A cost-benefit analysis. (McGraw-Hill Inc. Reproduced by permission.)

of animal **species**, and the impact of background contamination complicate these efforts. Under these conditions, simple cause and effect statements are out of the question.

The most that can be said in health **risk assessment** is that exposure to a particular pollutant *is likely* to cause a particular disease. Risk has to be stated in terms of probabilities, not certainties, and has to be distinguished from safety, which is a societal judgment about how much risk society is willing to bear. When assessing the feasibility of technological systems, different types of risks—from mining, radiation, industrial accidents, or **climate** impacts, for example—have to be compared. This type of comparison further complicates the judgments that have to be made.

Reducing risk involves asking to what extent the proposed methods of reduction are likely to be effective, and how much these proposed methods will cost. In theory, decision making could be left to the individual. Society could provide people with information and each person could then decide whether to purchase a product or service, depending upon the environmental and resource consequences. However, relying upon individual judgments in the market may not adequately reflect society's preference for an amenity

such as **air quality**, if that amenity is a public good with no owner and no price attached to it. Thus, social and political judgments are needed.

However much science reduces uncertainty, gaps in knowledge remain. Scientific limitations open the door for political and bureaucratic biases that may not be rational. In some instances, politicians have framed legislation in ways that seriously hinder if not entirely prohibit the consideration of costs (as in the **Delaney Clause** and the **Clean Air Act**). In other instances, of which the President's **Regulatory Review** Council is a good example, they have explicitly required a consideration of cost factors.

There are different ways that cost factors can be considered. Analysts can carry out cost effectiveness analyses, in which they attempt to figure out how to achieve a given goal with limited resources, or they can carry out more formal risk-benefit and cost-benefit analyses in which they have to quantify both the benefits of risk reduction and the costs.

Economists admit that formal, quantitative approaches to balancing costs and benefits do not eliminate the need for qualitative judgments. Cost-benefit analysis initially was developed for water projects where the issues, while complicated, were not of the same kind as society now faces. For example, how does one assess the value of a magnificent vista obscured by **air pollution**? What is the loss to society if a given genetic strain of grass or animal species becomes extinct? How does one assess the lost opportunity costs of spending vast amounts of money on air pollution that could have been spent on productivity enhancement and global competitiveness?

The most recalcitrant question concerns the value of human life. Cost-benefit analysis requires quantifying the value of a human life in dollars, so that specific health risks can be entered into the calculations against the cost of reducing such risks. Many different methods of arriving at an appropriate figure have been undertaken, all with predictably absurd results. Estimates range from $70,000 to several million dollars a head. Although the question does not admit of a sensible answer, society must nevertheless decide how much it is willing to pay to save a given number of lives (or how much specific polluters should pay for endangering them).

Equity issues (both interpersonal and intergenerational) cannot be ignored when carrying out cost-benefit analysis. The costs of air pollution reduction may have to be borne disproportionately by the poor in the form of higher **gasoline** and **automobile** prices. The costs of **water pollution** reduction, on the other hand, may be borne to a greater extent by the rich because these costs are financed through public spending. Regions dependent on dirty **coal** may find it in their interests to unite with environmentalists in seeking **pollution control** technology. The pollution control technology saves coal mining jobs in West Virginia and the Midwest, where the coal is dirty, but draws away resources from the coal mining industry in the West where large quantities of clean-burning coal are located.

Intergenerational equity also plays a role. **Future generations** have no current representatives in the market system or political process. How their interests are to be taken into account ultimately amounts to a philosophical discussion about altruism: to what extent should current generations hold back on their own consumption for the sake of posterity? Should Jeremy Bentham's utilitarian ideal of "achieving the greatest good for the greatest number" be modified to read "achieving sufficient per capita product for the greatest number over time?" *See also* Environmental economics; Intergenerational justice; Pollution control costs and benefits; Risk analysis

[*Alfred A. Marcus*]

RESOURCES
BOOKS

Buchholz, R., A. Marcus, and J. Post. *Managing Environmental Issues.* Englewood Cliffs, NJ: Prentice-Hall, 1992.

Mann, D., and H. Ingram. "Policy Issues in the Natural Environment." In *Public Policy and the Natural Environment*, edited by H. Ingram and R. K. Goodwin. Greenwich, CT: JAI Press, 1985.

Douglas M. Costle (1939 –)

American former director of Environmental Protection Agency

An educator and an administrator, Douglas M. Costle helped design the **Environmental Protection Agency** (EPA) under Richard Nixon and was appointed to the head of that agency by Jimmy Carter. Costle was born in Long Beach California on July 27, 1939, and spent most of his teenage years in Seattle. He received his B.A. from Harvard in 1961 and his law degree from the University of Chicago in 1964. His career as a trial attorney in the civil rights division of the Justice Department began in Washington in 1965. Later he became a staff attorney for the Economic Development Administration at the Department of Commerce.

In 1969, Costle was appointed to the position of senior staff associate of the President's Advisory Council on Executive Organization, and in this post was instrumental in the formation of the EPA. Although Costle lobbied to be appointed as assistant administrator of the new agency, his strong affiliations with the Democratic party seem to have hindered his bid. Instead he continued as a consultant to the agency for two years and an adviser to the President's **Council on Environmental Quality**.

The road that led Costle back to the EPA took him to Connecticut in 1972, where he became first deputy commissioner and then commissioner of the Department of Environmental Protection in that state. He proved himself an able and efficient administrator there, admired by many for his ability to work with industry on behalf of the **environment**. His most important accomplishment was the development of a structure often called the "Connecticut Plan," where fines for industrial **pollution** were calculated on the basis of the costs that business would have incurred if it had complied with environmental regulations.

Carter appointed Costle head of EPA in 1977 as a compromise candidate during a period of bitter feuding over the direction of the agency (then at the center of a debate over the economic effects of regulation). But many environmentalists believed Costle's record proved he compromised too willingly with business, and they openly questioned whether he had the political strength to support environment protection in the face of fierce political and industrial opposition.

By May of his first year in office he was able to secure funding for 600 additional staff positions in the EPA, and under him much was done to provide a rationale for the regulations he had inherited and base them wherever possible on scientific data. Among other decisions Costle made while head of the EPA, he recommended a delay on the imposition of new auto emissions standards, allowed the construction of the nuclear plant in Seabrook, New Hampshire to continue despite protests, and oversaw the formation of an agreement with U.S. Steel on the reduction of air and **water pollution**.

Throughout his tenure Costle remained a strong proponent of the view that the federal government's responsibility to the environment was not incompatible with the obligations it had to the economy. He often argued that environmental regulation actually assisted economic development. Although conflicts with lobbying groups and hostile litigation, as well as increased controversy over the inflationary effects of environmental regulation, complicated his stewardship of the EPA, Costle continued to believe in what he called a gradual and "quiet victory" for environmental protection.

Costle went on to become chairman of the U.S. Federal Regulatory Council until 1981. He is now a retired dean of Vermont Law School in South Royalton.

[*Douglas Smith*]

RESOURCES
PERIODICALS

Langway, L., and J. Bishop Jr. "The EPA's New Man." *Newsweek* 89 (21 February 1977): 80–82.

Council on Environmental Quality

Until it was abolished by the Clinton Administration in 1993, the President's Council on Environmental Quality (CEQ) was the White House Office that advised the President and coordinated executive branch policy on the **environment**. CEQ was established by the **National Environmental Policy Act** (NEPA) of 1969 to "formulate and recommend national policies" to promote the improvement of the quality of the natural environment.

CEQ had three basic responsibilities: to serve as advisor to the President on **environmental policy** matters; to coordinate the positions of the various departments and agencies of government on environmental issues; and to carry out the provisions of NEPA. The latter responsibility included working with federal agencies on complying with the law and issuing the required regulations for assessing the environmental impacts of federal actions. (NEPA requires that all agencies of the federal government issue "a detailed statement" on "the environmental impact" of "pro-

posals for legislation and other major federal actions significantly affecting the quality of the human environment." This seemingly innocuous provision has been used often by environmental groups to legally challenge federal projects that might damage the environment, on the grounds that the required **Environmental Impact Statement** was inadequate or had not been issued.)

CEQ also prepared and issued the annual Environmental Quality Report; administered the President's Commission on Environmental Quality, an advisory panel involved in voluntary initiatives to protect the environment; and supervised the President's Environment and **Conservation** Challenge Awards, which honored individuals and organizations who achieved significant environmental accomplishments. Under the Nixon and Carter administrations, CEQ had a significant impact on the formulation and implementation of environmental policy. But its role was greatly diminished under the Reagan and Bush administrations, which paid much less attention to environmental considerations.

Perhaps CEQ's best-known and most influential accomplishment was its landmark work, *The Global 2000 Report to the President*, prepared with the U.S. Department of State and other federal agencies, and released in July 1980. This pioneering study was the first report by the U.S. government—or any other—projecting long term environmental, population, and resource trends in an integrated way.

Specifically, the report projected that the world of the future would not be a pleasant place to live for much of humanity, predicting that "if present trends continue, the world in 2000 will be more crowded, more polluted, less stable ecologically and more vulnerable to disruption than the world we live in now. Serious stresses involving population, resources, and environment are clearly visible ahead....the world's people will be poorer in many ways than they are today." CEQ's *Eleventh Annual Report, Environmental Quality—1980*, further warned that "we can no longer assume, as we could in the past, that the earth will heal and renew itself indefinitely. Human numbers and human works are catching up with the earth's ability to recover...The quality of human existence in the future will rest on careful stewardship and husbandry of the earth's resources."

In the following dozen years, CEQ was much more reluctant to speak out about the ecological crisis. Early in his presidency, Bill Clinton abolished CEQ and created the White House Office on Environmental Policy to coordinate the environmental policy and actions of his administration. Clinton said this new body will "have broader influence and a more effective and focused mandate to coordinate policy" than CEQ had.

[*Lewis G. Regenstein*]

RESOURCES
PERIODICALS

"White House Environmentalists." *Buzzworm* 5 (May-June 1993): 72.

OTHER

Environmental Quality: The Eleventh Annual Report of the Council on Environmental Quality. Washington, DC: Council on Environmental Quality, 1980.

The Global 2000 Report to the President. Washington, DC: Council on Environmental Quality, 1980.

ORGANIZATIONS

Council on Environmental Quality, 722 Jackson Place, NW, Washington, D.C. USA 20503 Fax: (202) 395-5750, Toll Free: (202) 456-6546, , <http://www.whitehouse.gov/ceq>

Jacques-Yves Cousteau (1910 – 1997)

French oceanographer, inventor, photographer, explorer, and environmentalist

When most people think of marine biology, the person that immediately comes to mind is Jacques-Yves Cousteau. Whether through invention, research, **conservation**, or education, Cousteau has brought the ocean world closer to scientists and the public, and it is the interest, awareness, and appreciation fostered through Cousteau's work that may ultimately save the marine **environment** from impending destruction.

Born in France in 1910, Cousteau's childhood was full of illnesses that left him anemic. His sickness, however, did not stop him from being an independent thinker, a trait that led to his strong commitment to oceanographic research. He attended the École Navale in Brest, France, the national naval academy, where his interest in marine research was sparked by a cruise around the world on the school ship *Jeanne D'Arc*. Cousteau brought his camera and filmed a rough documentary of the voyage. His fascination with pearl and fish divers, and his ability to use a camera, would revolutionize undersea exploration.

During World War II, Cousteau, his wife, and two friends made masks and snorkels from inner tubes and garden hose. Through experimentation, they discovered that the worst enemy of a diver was the cold, and Cousteau proceeded to work on an effective diving suit. In 1943, Cousteau and Emile Paul Gagnon patented the Aqua Lung, the first self-contained underwater breathing apparatus (SCUBA). It was this invention that led to Cousteau being known as the "father of modern diving." SCUBA has opened a new world to scientific research as it is used extensively not only in marine biology, but also in marine geology, archaeology, and chemical oceanography.

Cousteau's innovations did not end with the Aqua Lung. He combined his interests in diving and photography to develop the first underwater camera housing. While stationed with the Navy in Marseilles, Cousteau continued to develop underwater photographic equipment, including battery packs and lights. In 1943, Cousteau and several friends filmed "Wrecks," a documentary of a sunken ship in the **Mediterranean Sea**. The French navy recognized Cousteau's talent and thought that this technology could be useful in recovering German mines and retrieving lost cargo. He was promoted to commandant and was put in charge of the Undersea Research Group, where he continued to develop diving and photographic techniques.

In 1950, Cousteau realized that the vessel donated to the Undersea Research Group was inadequate and asked the navy to furnish them with another ship. When the French navy refused, Cousteau formed a non-profit organization, Campagnes Oceanographique Françaises, and was able to raise enough money to purchase and refit an old British minesweeper. This would become the most famous and recognized scientific research vessel in the world: *Calypso*.

Cousteau first gained notoriety in the United States in the early 1950s when *Life Magazine* and *National Geographic* introduced Americans to the undersea endeavors of Cousteau and his Undersea Research Group. In 1953 Cousteau was persuaded to translate his journals into English and published them as *The Silent World*. It sold five million copies and was translated into twenty-two languages. It was filmed as a documentary in 1955 and won an Academy Award and the Gold Medal at the Cannes Film Festival that year. Following the commercial success of his film, Cousteau was appointed director of the world's oldest and largest marine research center, the Oceanographic Institute of Monaco. He rebuilt the deteriorating institute, adding aquariums and many live specimens collected during his travels.

While trying unsuccessfully to house **dolphins** in the facility, Cousteau developed a respect for the intelligence of these animals. This prompted him to campaign for stopping the slaughter of dolphins for use in pet foods. This was to be the first of many environmental campaigns for Cousteau, during which he discovered that the public, once educated about the environment, would be willing to try to preserve it. This theme of education has guided **the Cousteau Society**, and many other environmental organizations, ever since.

Exploring depths deeper than possible with SCUBA intrigued Cousteau and, in 1960, he tested the DS-2 diving saucer, in which he dove to over 984 ft (300 m) in the Bay of Ajaccio, near Corsica. After hundreds of successful dives, Cousteau started Continental Shelf Station Number One (Conshelf I) in the Mediterranean. This experiment not only investigated deep undersea habitats, but also determined how divers would respond to life underwater for long periods of time while conducting laboratory experiments. The work

Jacques Cousteau. (Corbis-Bettmann. Reproduced by permission.)

performed by Cousteau and his research group not only laid the groundwork for all subsequent submersible engineering and exploration, but also was one of the first thorough investigations of hyperbaric physiology.

In 1966, Cousteau began that for which he is best known: television documentaries. With the airing of "The World of Jacques Cousteau," millions were introduced to the wonders of the sea. That same year, Cousteau signed a contract and began to film "The Undersea World of Jacques Cousteau," which ran for nearly nine years, giving Americans a glimpse of marine environments and the behaviors of the organisms that live there. The series, now in syndication, continues to fascinate generations of Americans and spawn interest in marine science and conservation.

In the early 1970s, Cousteau became frustrated with increasing **marine pollution** and produced a series of documentaries focusing on the destruction of marine systems. He also expanded his explorations to riverine systems, lakes, **rain forest** destruction and the conflicts between human culture and the environment. The scope of his films demonstrates Cousteau's devotion to preserving all natural systems.

Cousteau's accomplishments are numerous. He had published more than 18 books and contributed many articles to professional and popular journals. His documentary films have won three Academy Awards and his series, "The Un-

dersea World of Jacques Cousteau," has won numerous Emmy Awards. Many organizations have recognized Cousteau for his work in technical fields as well as in conservation. These awards include Gold Medals from the National Geographic Society and the Royal Geographical Society, and the United Nations international environmental award. He has received honorary degrees from the University of California at Berkeley, Brandeis University, Rensselaer Polytechnic Institute, Harvard University, the University of Ghent, and the University of Guadalajara. He was named to the U.S. **National Academy of Sciences** in 1968.

Cousteau's devotion to the natural world has not been with out its costs. His son Philippe was killed in a diving accident and numerous financial difficulties, especially early in his career, set back his work. Whether it is in oceanographic research, marine engineering, the development and manufacture of diving equipment, the production of films and television specials, or **environmental education**, Cousteau continued to inspire and fascinate, motivating scientists and the public alike to work to preserve the ocean world until his death Paris on July 25, 1997.

[*William G. Ambrose Jr. and Paul E. Renaud*]

RESOURCES
BOOKS

Cousteau, J-Y. *The Cousteau Almanac: An Inventory of Life on Our Water Planet.* New York: Doubleday, 1981.
Munson, R. *Cousteau: The Captain and His World.* New York: Morrow, 1989.

Cousteau Society, The

The Cousteau Society is a non-profit organization dedicated to marine research, especially underwater exploration and filmmaking. Created in 1973 by **Jacques-Yves Cousteau** and his son, Jean-Michel Cousteau, the Society provides educational materials, and sponsors or cosponsors scientific, technological and environmental research studies to gauge the health of the world's marine environments.

The Society's educational projects include award-winning television specials that inspire interest in the marine world. Special reports filmed and broadcast have included those on the **Exxon Valdez** oil spills and the natural history of the great white shark (Carcharodon charcharias). Cousteau films are broadcast in more than 100 countries and to schools around the world. Books and filmstrips on marine and environmental issues are produced for colleges, schools, and the public. The Society produces two periodicals that explain scientific and environmental issues, the *Calypso Log* for adults and the *Dolphin Log* for youngsters.

The Society's Project Ocean Search offers field study programs under the supervision of Society educators and scientists. Presently, the Society is developing Cousteau Center sites that will provide visitors with ocean experiences through the use of film and illusion technology. The Society also sponsors research to help the scientific community better understand the nature of a region or phenomenon, as well as studies designed to provide local policy makers with guidelines to protect their **environment**.

Two Cousteau Society research vessels, *Calypso* and *Alcyone*, are circumnavigating the globe to take a fresh look at the planet. Four hours of television documentaries are filmed each year about those expeditions. Scientific teams aboard the two research vessels are measuring the productivity of the oceans, studying the contributions of rivers to ocean vitality, assessing the health of marine and freshwater habitats, and exploring the global connections between major components of the **biosphere** such as tropical forests, rivers, the **atmosphere**, seas, oceans, and humankind.

The Society also provides assistance to the Marine Mammal Stranding Program, a network of scientists and others who study strandings of **whales** and **dolphins**. With the Smithsonian Institution, the Society sponsors the Marine Mammal Events Program, which gathers information on beach stranding reports from the United States and other parts of the world to create a centralized data base. The compilation is expected to lead to a better understanding of the phenomenon of beaching.

Currently, the Society is developing educational computer programs for young people that explore the consequences of various actions on the environment; and it is preparing environmental cartoon books for students in developing countries. It also supports the development of new technologies that will help provide solutions to environmental challenges.

[*Linda Rehkopf*]

RESOURCES
ORGANIZATIONS
The Cousteau Society, 870 Greenbrier Circle, Suite 402, Chesapeake, VA USA 23320 (800) 441-4395, Email: cousteau@cousteausociety.org, <http://www.cousteausociety.org>

Coyote

The name coyote comes from the Aztec name *Coyoti*. The Latin name for the coyote is *Canis latrans* which means barking dog. A member of the family Canidae, the coyote is also called God's Dog, brush wolf, **prairie** wolf, and the Songdog.

Coyotes are especially important to biologists and others studying **wildlife** and the **environment** both because of their ability to control their own population and because of their effect upon other wildlife populations, especially that of the white-tailed deer.

The coyote is roughly the size of a small German Shepherd, weighing from 20–35 lb (9–14 kg) and 4–5 ft (1.2–1.5 m) long, although some can grow as large as 50 lb (23 kg) in the northern and northeastern part of their range. The coat of a coyote can be gray or brownish gray, depending on the season and **habitat**, with a bushy tail that is tipped with black. Coyote fur is long and soft and grows heavier in the winter to protect it from the cold. Its fur also grows lighter in color in the winter and darker in the summer to help it blend in with its natural surroundings.

The early European settlers originally found the coyote only in Central and western North America. However, despite many attempts by humans to get rid of them, coyotes today range from Panama to Northern Alaska, and from California to Newfoundland. They have been seen in every state in the continental United States, but since they are predominately nocturnal, are too elusive to count accurately. Coyotes have even been found in isolated regions such as Cape Cod, as well as in various urban areas such as Chicago, Los Angeles, and New York City.

This ongoing expansion of the coyote's range began in the late 1800s, soon after the United States government began programs of killing them in order to protect domestic livestock, especially the sheep that were **grazing on public lands**. However, unlike the **overhunting** of other **species**, such as the passenger pigeons, grizzly bears, **wolves**, and **whales**, the best efforts to kill off the coyotes have all failed, and coyotes have continued to thrive. Biologists now believe that the coyote population has more than doubled since 1850.

Several factors account for this expansion. Farming and the clearing of the unbroken forests that once covered much of North America have created new and more suitable habitats. The eradication of the wolf in much of North America has removed one of the coyote's major predators and has led to less **competition** for many prey animals. Unlike the wolf, the coyote has more easily adapted to changing habitats, and has adjusted to living close to human populations. In addition, the introduction of sheep and other domestic animals has given the coyote new food sources.

Coyotes are opportunistic carnivores with a variable diet depending on the season or habitat. Although they prefer rodents, rabbits, and hares, they are definitely not picky. They will emulate the larger wolf and hunt deer, elk, and sheep in packs. They will also eat carrion and have been known to eat grasshoppers, beetles, snakes, lizards, **frogs**, rats, domestic cats, porcupines, turtles, and even watermelons and wild blueberries.

Coyotes are one of the most vocal animals of the wild. Through a series of howls, yelps and barks, coyotes communicate to other coyotes in the area, with others in their pack, and with their young.

Studies of the coyote's breeding patterns have shown that they are able to control their own population. To do this the coyote relies upon various breeding strategies. If there are too many coyotes for the food supply, they will continue to mate and have young, but only enough pups will be born and will survive to replace those that have died. However, if they are not being overhunted and if there is plenty of food, there will be more pups born in each litter and the number that will survive into adulthood will increase.

Because of this innate ability to vary their own populations, coyotes have thrived and play a central role in maintaining the overall ecological balance. Extensive studies of the coyote have shown the crucial interaction between them and many other species, both predator and prey. When the jackrabbit population decreases, as it does periodically in the Curlew Valley of Idaho and Utah for example, the coyote population decreases also, usually within the following year. Then, when the jackrabbit population increases again, coyotes become more numerous also.

Coyotes mate for life and the adults breed between January and March. The female maintains the den that she selects from an old badger or woodchuck hole, or a natural cavity in the ground. The female carries her young for over two months, giving birth in April or May. A litter may have from two to 12 pups.

Both parents play an active role in raising the pups. At three weeks old the pups are allowed to leave the den to play and by 12 weeks old they are taught to hunt. The family stays together through the summer but by fall the pups will leave to find their own territories. The survival rate for these pups is low. Between 50 and 70% die before they reach adulthood. Eighty percent of those deaths are due to man.

Coyotes limit the populations of various smaller predators, by preying upon them directly or by competing with them for prey. This in turn indirectly affects various prey communities, especially those of birds. When there are no coyotes to keep them in check, the smaller predators expand and are able to kill off a larger segment of the various bird populations.

The coyote and the deer populations interact in an important way as well. Whenever the coyote population decreases, the deer population, particularly that of the white-tailed deer, increases, leading to unhealthy herds and, according to some biologists, to an increase in Lyme disease-bearing ticks in some regions. Coyotes are thus a major

A coyote. (© U. S. Fish & Wildlife Service. Reproduced by permission.)

factor in maintaining healthy deer populations and in preventing the spread of a disease that is dangerous and potentially fatal to humans.

Coyotes affect humans more directly as well, especially through their impact upon domestic animals. Since Europeans first came to North America, coyotes have preyed upon sheep and other livestock, costing sheep growers, farmers, and ranchers untold millions of dollars. In some areas, livestock may make up 14% of a coyote's diet. This has prompted extensive **hunting** to control the number of coyotes that ranchers and farmers view as pests. Some ranchers keep guard dogs to protect their livestock. New technology, using inaudible ultrasonic sound, also allows ranchers to detect or repel coyotes.

Although there are no documented cases of humans dying from coyote attacks, coyotes will fiercely defend themselves if cornered. However, they are much less dangerous than wild domestic dogs, which have killed more than 300 people in the decade since 1990. Coyotes can coexist with humans. But can people learn to live with them? Humans cause 90% of all adult coyote deaths by hunting, **trapping**, **poisoning**, and **automobile** accidents. Wolves, black bears, mountain lions, and eagles, as well as various **parasites** and diseases, account for the rest.

[*Douglas Dupler*]

RESOURCES

BOOKS

Dobie, J. Frank. *The Voice of the Coyote.* Boston: Little, Brown and Co., 1949.

Leydet, Francois. *The Coyote: Defiant Songdog of the West.* Norman, OK: University of Oklahoma Press, 1988.

PERIODICALS

Finkel, Mike. "Ultimate Survivor." *Aubudon*, May-June 2002.

Foderaro, Lisa W. "Letting Golf Courses Go a Little Wild; Clubs Spray Less and Make Way For Native Plants and Coyotes." *New York Times*, 1 May 2002.

Gompper, Matthew E. "Top Carnivores in the Suburbs? Ecological and Conservation Issues Raised by Colonization of North-Eastern North America by Coyotes." *BioScience*, February 2002, 185–191.

Gregory, Ted. "Hot on the Trail of Nomadic Urban Coyotes." *Chicago Tribune*, 12 May 2002, 1.

Grinder, Martha I. and Paul R. Krausman. "Home Range, Habitat Use, and Nocturnal Activity of Coyotes in an Urban Environment." *Journal of Wildlife Management*, 65, no. 4 (2001): 887–99.

OTHER

Desert Usa. *The Coyote, Canis latrans.* [2002]. <http://www.desertusa.com>.

Senecal, D. "Coyote, Hinterland Who's Who." *Canadian Wildlife Service.* [2002]. <http://www.cws-scf.ec.gc.ca>.

Crane (bird)

see **Whooping crane**

Creutzfeldt jacob disease

see **Mad cow disease**

Criteria pollutant

Criteria pollutants are air pollutants which, at certain levels of exposure, do not threaten human health and meet National Ambient Air Quality Standards. There are two types of standards for such pollutants. National primary ambient **air quality** standards are levels of air quality with a margin of safety adequate to protect public health. National secondary ambient air quality standards are levels of air quality which are necessary to protect the public welfare from any known or anticipated adverse effects of a pollutant. *See also* Air Quality Control Region; Air quality criteria; Primary standards; Secondary standards

Critical habitat

As institutionalized in the U. S. **Endangered Species Act** of 1973, critical **habitat** is considered the area necessary to the survival of a **species**, and, in the case of endangered and threatened species, essential to their recovery. An animal's habitat includes not only the area where it lives, but also its

breeding and feeding grounds, seasonal ranges, and **migration** routes. Critical habitat usually refers to the area that is essential for a minimal viable population to survive and reproduce. The **Endangered Species** Act is intended to conserve "the ecosystems upon which endangered species and threatened species depend." Thus, the Secretary of the Interior is required to identify and designate critical habitats for species that are listed as endangered or threatened under this law. In some cases, areas may be excluded from such designations if the economic, social, or other costs exceed the **conservation** benefits.

The listing of imperiled species and the designation of their critical habitats have become politically sensitive, since these actions can profoundly affect the development and exploitation of areas so designated, and can, under some circumstances, limit such activities as gas and **oil drilling**, timber cutting, dam building, mineral exploration and mining. For this and other reasons, the Department of the Interior often has been reluctant to list certain species, and has excluded species from the protected lists in order not to inconvenience certain commercial interests.

Section 7 of the Endangered Species Act requires all federal agencies and departments to ensure that the activities they carry out, fund, or authorize do not jeopardize the continued existence of listed species or adversely modify or destroy their critical habitat. This provision has proven especially significant, since federal agencies such as the **Forest Service**, **Bureau of Land Management**, and **Fish and Wildlife Service** control vast areas of land that constitute habitat for many listed species and on which a variety of commercial activities, such as **logging** or mining, are undertaken with federal permits.

However, Section 7 of the Endangered Species Act has been implemented in such a way as to generally not affect economic development. The U. S. Fish and Wildlife Service (and, in the case of marine species, the National Marine Fisheries Service) is directed to consult with other federal agencies and review the effects of their actions on listed species. According to a study by the **National Wildlife Federation**, over 99% of the more than 120,000 reviews or consultations conducted between 1979 and 1991 found that no jeopardy to a listed species was involved. In some cases, "economic and technologically feasible" alternatives and modifications, in the words of the act, were suggested that allowed the federal activities to proceed. In only 34 cases were projects cancelled because of threats to listed species. In rare situations, where the conflict between a project and the Endangered Species Act are absolutely irreconcilable, an agency can apply for an exemption from a seven-member Endangered Species Committee.

The earliest major conflict over critical habitat under the Act was the famous 1979 fight over construction of the

$116 million **Tellico Dam** in Tennessee, which would have flooded and destroyed several hundred family farms as well as what was then the only known habitat of a species of minnow, the **snail darter** (*Percina tanasi*). (Since then, snail darters have been found in other areas.) Congress exempted this project from the provisions of the Endangered Species Act, and the dam was built as planned, although many consider it a political boondoggle and a huge waste of taxpayers' money.

More recently, efforts by environmentalists to save the remnants of **old-growth forest** in the Pacific Northwest to preserve habitat for the **northern spotted owl** (*Strix occidentalis caurina*) created tremendous controversy. Thousands of acres of federally-owned forests in Oregon, Washington, and California were placed off-limits to logging, costing jobs in the timber industry in those states. However, conservationists pointed out, if the federal government allowed the last of the ancient forests to be logged, timber jobs would disappear anyway, along with these unique ecosystems and several species dependent upon them. In mid-1993, Interior Secretary announced a compromise decision that allows logging of some ancient forests to continue, but also greatly decreases the areas open to this activity. As natural areas and wildlife habitat continue to be destroyed and degraded, conflicts and controversy over saving critical habitats for listed endangered and threatened species can be expected to continue.

[*Lewis G. Regenstein*]

RESOURCES
BOOKS

Bean, M. J., et al. *Reconciling Conflicts Under the Endangered Species Act: The Habitat Conservation Planning Experience.* Washington, DC: World Wildlife Fund, 1991.

Kohm, K., ed. *Balancing on the Brink of Extinction: The Endangered Species Act and Lessons for the Future.* Covelo, CA: Island Press, 1990.

Crocodiles

The largest of the living reptiles, crocodiles inhabit shallow coastal bodies of water in tropical areas throughout the world, and they are often seen floating log-like in the water with only their eyes and nostrils showing. Crocodiles have long been hunted for their hides, and almost all **species** of crocodilians are now considered to be in danger of **extinction**. Members of the crocodile family, called crocodilians (Crocodylidae), are similar in appearance and include crocodiles, alligators, caimans, and gavials. A crocodile can usually be distinguished from an alligator by its pointed snout (an alligator's is rounded), and by the visible fourth tooth on either side of its snout that protrudes when the jaw is shut.

Crocodiles prey on fish, turtles, birds, crabs, small mammals, and any other animals they can catch, including dogs and occasional humans. They hide at the shore of rivers and water holes and grab an animal as it comes to drink, seizing a leg or muzzle, dragging the prey underwater, and holding it there until it drowns. When seizing larger animals, a crocodile will thrash and spin rapidly in the water and tear its prey to pieces. After eating its fill, a crocodile may crawl ashore to warm itself and digest its food, basking in the sun in its classic "grinning" pose, with its jaws wide open, often allowing a sandpiper or plover to pick and clean its teeth by scavenging meat and **parasites** from between them.

The important role that crocodiles play in the **balance of nature** is not fully known or appreciated, but, like all major predators, their place in the ecological chain is a crucial one. They eat many poisonous water snakes, and during times of **drought**, they dig water holes, thus providing water, food, and **habitat** for fish, birds, and other creatures. When crocodiles were eliminated from lakes and rivers in parts of Africa and **Australia**, many of the food fish also declined or disappeared. It is thought that this may have occurred because crocodiles feed on predatory and scavenging species of fish that are not eaten by local people, and when left unchecked, these fish multiplied out of control and crowded out or consumed many of the food fish.

Crocodiles reproduce by laying eggs and burying them in the sand or hiding them in nests concealed in vegetation. Recent studies of the Nile and American crocodiles show that some of these reptiles can be attentive parents. According to these studies, the mother crocodile carefully watches over the nest until it is time for the eggs to hatch. Then she digs the eggs out and gently removes the young from the shells. After gathering the newborns together, she puts them in her mouth and carries them to the water and releases them, watching over them for some time. American crocodiles are very shy and reclusive, and disturbance during this critical period can disrupt the reproductive process and prevent successful hatchings.

In recent decades, crocodiles and other crocodilians have been intensively hunted for their scaly hides, which are used to make shoes, belts, handbags, wallets, and other fashion products. As a result, they have disappeared or have become rare in most of their former habitats. As of 2001, 12 crocodile species have been designated endangered. These species are found in Africa, the Caribbean, Central and South America, the Middle East, the Philippines, Australia, some Pacific Islands, southeast Asia, the Malay Peninsula, Sri Lanka, and Iran. They are endangered primarily due to overexploitation and habitat loss.

The American crocodile (*Crocodylus acutus*) occurs all along the Caribbean coast, including the shores of Central America, Colombia, Cuba, Hispaniola, Jamaica, Mexico,

An American crocodile. (Photograph by Tom and Pat Leeson. Photo Researchers Inc. Reproduced by permission.)

extreme south Florida, and on the Pacific coast, from Peru north to southern Mexico. The United States population of the American crocodile consists of some 500–1,2000 individuals. This species breeds only in the southern part of **Everglades National Park**, mainly Florida Bay, and perhaps on nearby Key Largo, and at Florida Power and Light Company's Turkey Point plant, located south of Miami. The population is thought to be extremely vulnerable and declining, mainly due to human disturbance, habitat loss (from urbanization, especially real estate development), and direct killing such as on highways and in fishing nets. Predation of hatchlings in Florida Bay mainly by raccoons may also be a factor in the species' decline.

[*Lewis G. Regenstein*]

RESOURCES
BOOKS

Crocodiles: Their Ecology, Management and Conservation. A Special Publication of the Crocodile Specialist Group. Gland, Switzerland: IUCN-The World Conservation Union, 1989.
Ross, C. A., ed. *Crocodiles and Alligators.* New York: Facts on File, 1989.
Thorbjarnarson, J., comp. *Crocodiles. An Action Plan for Their Conservation.* Gland, Switzerland: IUCN-The World Conservation Union, 1992.

William Cronon   (1954 –)

American Frederick Jackson Turner Professor of History, Geography, and Environmental Studies, University of Wisconsin

William Cronon is an environmental historian, and the author of several notable books on **wilderness**, American landscape, and history. Cronon grew up in Madison, Wis-

consin and did his undergraduate work at the University of Wisconsin. He won a Rhodes scholarship and spent two years studying at Oxford University, earning a doctorate. His first book, *Changes in the Land: Indians, Colonists and the Ecology of New England*, was published in 1983. This book was cited by reviewer Jim Miller in *Newsweek* as an "eloquent book" with "the rigor of first-rate history and the power of a tragedy." The book's tragic power lies in its subject-the destruction of the fertile **habitat** of pre-colonial New England by the colonists' farming methods and exploitation of **natural resources**, and the virtual eradication of the native American population by diseases brought from Europe by the colonists. Cronon supports his explication of this devastation with evidence gathered from such sources as original Puritan documents and the literature of **ecology** and anthropology. His next book was *Nature's Metropolis: Chicago and the Great West*, published in 1991. This work extended Cronon's investigation of the way human beings shape landscape, looking at relationships between nineteenth century Chicago and the land around it.

Cronon left a tenured position at Yale in order to take up a post at the University of Wisconsin in 1992, returning to his home town. He put out several more books in the 1990s, including *Under an Open Sky: Rethinking America's Western Past*, and *Uncommon Ground: Rethinking the Human Place in Nature*, and editing a book of John Muir's **nature** writings. Cronon's writings have challenged other environmentalists' conception of wilderness as a pristine place untouched by man. His work focuses on the inextricable relationship between human civilization and nature. Idealizing wilderness as something apart from humankind for Cronon stands in the way of clear thinking about how to preserve the **environment**. For example Cronon wrote about the **Arctic National Wildlife Refuge** in 2001, when the incoming Bush administration sparked renewed calls to drill for oil in the nature preserve. Cronon argued that the word wilderness conjured up for most people a "place remote from human settlements, untouched by human hands." The Arctic **National Wildlife Refuge** was in fact deeply tied to human civilization as a holy place to Alaskan Indians, a breeding ground for their caribou, and a migratory stop for birds found in 49 out of 50 states in the U.S. The Alaskan wilderness was neither as empty nor as distant as it seemed. Pointing out the many complex relationships between human communities and the land surrounding them is the core of Cronon's work.

[*Angela Woodward*]

RESOURCES
PERIODICALS

"An Environmentalist on a Different Path" *New York Times* (April 3, 1999): B7, B9.

Cronon, William. ":Neither Barren Nor Remote" *New York Times* (February 28, 2001): A19.

Cross-Florida Barge Canal

The subject of long and acrimonious debate, this attempt to build a canal across the Florida peninsula began in the 1930s and finally expired in 1990. Although it receives little attention today, the Cross-Florida Barge Canal stands as a landmark because it was one of the early cases in which the **Army Corps of Engineers**, whose primary mission has traditionally been to re-design and alter natural waterways, yielded to environmental pressure. The canal's stated purpose, aside from bringing public works funding to the state, was to shorten the shipping distances from the East Coast to the Gulf of Mexico by bypassing the long water route around the tip of Florida. Rerouting barge traffic would also bring commerce into Florida, directing trade and trans-shipment operations through Floridian hands. An additional supporting argument that persisted into the 1980s was that the existing sea route brought American commerce dangerously close to threatening Cuban naval forces.

Construction on the canal began in 1964, on a route running from the St. Johns River near Jacksonville west to the Gulf of Mexico at Yankeetown, Florida. Canal project plans included three **dams**, five locks, and 110 mi (177 km) of channel 150 ft (46 m) wide and 12 ft (3. 6 m) deep. Twenty-five miles (40 km) of this waterway, along with three locks and three dams, were complete by 1971 when President Richard Nixon, citing economic inefficiency and unacceptable environmental risks, stopped the project by executive order.

From start to finish, the canal's proponents defended the project on economic grounds. The Cross-Florida Canal was proposed as a depression-era job development program. After completion, commerce and recreational fishing would boost the state economy. The Army Corps, well-funded and actively remodelling **nature** in the 1950s and 1960s, took on the project, vastly overestimating economic benefits and essentially dismissing environmental liabilities with the argument that even modest economic gain justified any **habitat** or water loss. After work had begun, further studies concluded that most of the canal's minimal benefits would go to non-Floridian agencies and that environmental dangers were greater than first anticipated. Outcry over environmental costs eventually led to a reappraisal of economic benefits, and the state government rallied behind efforts to halt the canal.

Environmental risks were grave. Although Florida has more **wetlands** than any other state except Alaska, many of the peninsula's natural wetland and riparian habitats had already been lost to development, **drainage**, and channeliza-

tion. Along the canal route these habitats sheltered a rich community of migratory and resident birds, crustaceans, fish, and mammals. Fifteen **endangered species**, including the red-cockaded woodpecker (*Picoides borealis*) and the Florida manatee, stood to loose habitat to channelized rivers and barge traffic. Specialized spring-dwelling mussels and shrimp that depend on reliable and pure water supplies in this porous limestone country were also threatened.

Most serious of all dangers was that to the Floridan **aquifer**, located in northern Florida but delivering water to cities and wetlands far to the south. Like most of Florida, the reach between Jacksonville and Yankeetown consists of extremely porous limestone full of **sinkholes**, springs, and underground channels. The local **water table** is high, often within a few feet of the ground surface, and currents within the aquifer can carry water hundreds of meters or more in a single day. Because the canal route was to cut through 28 mi (45 km) of the Floridan aquifer's **recharge zone**, the area in which water enters the aquifer, any pollutants escaping from barges would disperse through the aquifer with alarming speed. Even a small fuel leak could contaminate millions of gallons of drinking-quality water. In addition, a canal would expose the aquifer to extensive urban and **agricultural pollution** from the surrounding region.

Water loss presented another serious worry. A channel sliced through the aquifer would allow water to drain out into the sea, instead of remaining underground. Evaporation losses from impounded lakes were expected to reach or exceed 40 million gallons of fresh water every day. With water losses at such a rate, water tables would fall, and salt water intrusions into the fresh water aquifer would be highly probable. In 1985, 95% of all Floridians depended on **groundwater** for home and industrial use. The state could ill afford the losses associated with the canal.

Florida water management districts joined environmentalists in opposing the canal. By the mid-1980s the state government, eager to reclaim idle land easements for development, sale, and extension of the Ocala **National Forest**, put its weight against the Corps and a few local development agencies that had been resisting deauthorization for almost 20 years. In 1990 the United States Congress voted to divide and sell the land, effectively eliminating all possibility of completing the canal.

[*Mary Ann Cunningham Ph.D.*]

RESOURCES
PERIODICALS

Hogner, R. H. "Environmentalists Lock Up Canal Development." *Business and Society Review* (Fall 1990): 74-77.

OTHER

Deauthorization Hearings: The Cross-Florida Barge Canal. United States House of Representatives Committee on Public Works 1978. Washington, DC: U. S. Government Printing Office, 1978.

Hearing on the Cross-Florida Barge Canal. United States House of Representatives Committee on Public Works June 10, 1985. Washington, DC: U. S. Government Printing Office, 1985.

CRP

see **Conservation Reserve Program (CRP)**

Paul J. Cruzen (1933 –)

Dutch Meteorologist

Paul Crutzen is one of the world's leading researchers in mapping the chemical mechanisms that affect the **ozone** layer. He has pioneered research on the formation and depletion of the ozone layer and threats placed upon it by industrial society. Crutzen has discovered, for example, that **nitrogen oxides** accelerate the rate of ozone depletion. He has also found that **chemicals** released by bacteria in the **soil** affect the thickness of the ozone layer. For these discoveries he has received the 1995 Nobel Prize in Chemistry, along with Mario Molina and Sherwood Rowland for their separate discoveries related to the ozone and how **chlorofluorocarbons** (CFCs) deplete the ozone layer. According to Royal Swedish Academy of Science, "by explaining the chemical mechanisms that affect the thickness of the ozone layer, the three researchers have contributed to our salvation from a global environmental problem that could have catastrophic consequences."

Paul Josef Crutzen was born December 3, 1933, to Josef C. Crutzen and Anna Gurek in Amsterdam. Despite growing up in a poor family in Nazi-occupied Holland during 1940–1945, he was nominated to attend high school at a time when not all children were accepted into high school. He liked to play soccer in the warm months and ice skate 50–60 mi (80–97 km) a day in the winter. Because he was unable to afford an education at a university, he attended a two-year college in Amsterdam. After graduating with a civil engineering degree in 1954, he designed bridges and homes.

Crutzen met his wife, Tertu Soininen, while on vacation in Switzerland in 1954. They later moved to Sweden where he got a job as a computer programmer for the Institute of **Meteorology** and the University of Stockholm. He started to focus on atmospheric chemistry rather than mathematics because he had lost interest in math and did not want to spend long hours in a lab, especially after the birth of his two daughters, Illona and Sylvia. Despite his busy schedule,

Crutzen obtained his doctoral degree in Meteorology at Stockholm University at the age of 35.

Crutzen's main research focused on the ozone, a bluish, irritating gas with a strong odor. The ozone is a molecule made up of three oxygen atoms (O_3) and is formed naturally in the **atmosphere** by a **photochemical reaction**. The ozone begins approximately 10 mi (16 km) above Earth's surface, reaching between 20–30 miles (32–48 km) in height, and acts as a protective layer that absorbs high-energy **ultraviolet radiation** given off by the sun.

In 1970 Crutzen found that soil **microbes** were excreting **nitrous oxide** gas, which rises to the **stratosphere** and is converted by sunlight to nitric oxide and **nitrogen** dioxide. He determined that these two gases were part of what caused the depletion of the ozone. This discovery revolutionized the study of the ozone and encouraged a surge of research on global biogeochemical cycles.

In 1977, while he was the director of the National Center for Atmospheric Research (NCAR) in Boulder, Colorado, Crutzen studied the effects of burning trees and brush in the fields of Brazil. Every year farmers cleared the forests by burning everything in sight. The theory at the time was that this burning caused more **carbon** compounds or trace gases and **carbon monoxide** to enter the atmosphere. These gases were believed to cause the **greenhouse effect**, or a warming of the atmosphere. Crutzen collected and examined this **smoke** in Brazil and discovered that the complete opposite was occurring. He stated in *Discover* magazine: "Before the industry got started the tropical burning was actually decreasing the amount of **carbon dioxide** in the atmosphere." The study of smoke in Brazil led Crutzen to further examine what effects larger amounts of different kinds of smoke might have on the **environment**, such as smoke from a nuclear war.

The journal *Ambio* commissioned Crutzen and John Birks, his colleague from the University of Colorado, to investigate what effects nuclear war might have on the planet. Crutzen and Birks studied a simulated worldwide nuclear war. They theorized that the black carbon soot from the raging fires would absorb as much as 99% of the sunlight. This lack of sunlight, coined "nuclear inter," would be devastating to all forms of life. For this theory Crutzen was named "Scientist of the Year" by *Discover* magazine in 1984 and awarded the prestigious Tyler Award four years later.

As a result of the discoveries by Crutzen and other environmental scientists, a very crucial international treaty was established in 1987. The Montreal Protocol was negotiated under the auspices of the United Nations and signed by 70 countries to slowly phase out the production of chlorofluorocarbons and other ozone-damaging chemicals by the year 2000. However, the United States had ended the production of CFCs five years earlier, in 1995. According to the

New York Times, "the **National Oceanic and Atmospheric Administration** reported in 1994, while ozone over the South Pole is still decreasing, the depletion appears to be leveling off." Even though the ban has been established, existing CFCs will continue to reach the ozone, so the depletion will continue for some years. The full recovery of the ozone is not expected for at least 100 years.

From 1977–80, Crutzen was director of the Air Quality Division, National Center for Atmospheric Research (NCAR), located in Boulder, Colorado. While at NCAR, he located he taught classes at Colorado State University in the department of Atmospheric Sciences. Since 1980 he has been a member of the Max Planck Society for the Advancement of Science, and he is the director of the Atmospheric Chemistry division at Max Planck Institute for Chemistry. In addition to Crutzen's position at the institute, he is a part-time professor at Scripps Institution of Oceanography at the University of California. In 1995 he was the recipient of the United Nations Environmental Ozone Award for outstanding contribution to the protection of the ozone layer. Crutzen has co-authored and edited several books, as well as having published several hundred articles in specialized publications.

[*Sheila M. Dow*]

RESOURCES
BOOKS

Cruzen, Paul. *Atmosphere, Climate, and Change.* Scientific American Library, 1995.

———. *Atmosphere Change: An Earth System Perspective.* W.H. Freeman, 1993.

———. *Environmental Consequences of Nuclear War 1985.* Schwarzer Himmel, 1986.

Cryptosporidium

Cryptosporidium is a microscopic protozoan parasite that is a significant cause of diarrheal disease (cryptosporidiosis) in humans and animals, including ruminants, swine, cats, and dogs. The word cryptosporidium means hidden spore, referring to the ease with which it can escape detection in environmental samples. Most cryptosporidiosis appears to be caused by the **species** *Cryptosporidium parvum.* The dormant and resistant forms of *Cryptosporidium,* referred to as oocysts, are excreted in the feces of infected humans and animals and can survive under a wide range of environmental conditions. Cryptosporidiosis had long been recognized as a veterinary problem, especially in young farm animals, but was only recognized as a cause of human disease in 1976, when it was diagnosed in animal handlers.

The most common symptom of cryptosporidiosis is watery diarrhea. Other symptoms may include abdominal cramps, nausea, low-grade fever, dehydration, and weight loss. Symptoms usually develop four to six days after infection, but may occur from two to 10 days after infection. In individuals with healthy, normal immune systems, the disease usually only lasts for several days, and rarely more than two weeks. Some infected persons may not even become ill, while others become ill, seem to recover, but become ill again. Infected persons may shed oocysts in their feces for months, even when they are not exhibiting disease symptoms.

Cryptosporidiosis can cause complications in persons with diabetes, alcoholism, or pregnancy. Diarrhea and dehydration can severely affect the very young, the elderly, and the frail. Cryptosporidiosis can be prolonged and life-threatening in immuno-compromised individuals (those who have weak immune systems), such as transplant patients, people infected with the Human Immunodeficiency **Virus** (HIV) (which causes Acquired Immuno-Deficiency Syndrome [AIDS]), **cancer** patients on chemotherapy, and persons who are taking medications that suppress the immune system. By 1983, the number of infections in humans was increasing as the **AIDS** epidemic began; also diagnostic methods were developed to identify the parasite in stool samples, which aided in the identification and documentation of the disease.

In industrialized nations, about 0.4% of the population passes oocysts in the feces at any one time, while about 2–2.5% of patients admitted to hospitals are passing oocysts. About 30–35% of the United States population have antibodies in their blood to *Cryptosporidium.* In developing nations, the prevalence is higher, where up to 60–70% of people have circulating antibodies in their blood to *Cryptosporidium.* In AIDS patients, the numbers of individuals with chronic cryptosporidiosis is about 10% in industrialized nations and up to 40% in some developing nations.

Cryptosporidiosis cannot be diagnosed by symptoms alone, but must be confirmed with a specific diagnostic stool test for the parasite. There is no drug available to cure cryptosporidiosis. Symptoms are treated with anti-diarrheal medicines and fluids to prevent dehydration. People with strong immune systems will recover on their own and may develop limited immunity to recurring infections.

A person can become infected with *Cryptosporidium* by ingesting anything that has been in contact with feces from an infected person or animal. Hands can become contaminated with *Cryptosporidium* through person-to-person contact, such as changing of a child's diaper, caring for someone with diarrhea, or touching a part of the body contaminated with feces. Cryptosporidiosis is easily spread through social groups, such as families, day care centers, and nursing homes. Persons who work with animals or touch soils or objects contaminated with feces can contract cryptosporidiosis.

Infection with *Cryptosporidium* can also occur by drinking water that is contaminated with oocysts. The first report of spread of cryptosporidiosis through a municipal drinking water system was in 1987 in Carrollton, Georgia, where 13,000 people, out of a total population of 65,000, became ill. In 1993, a municipal drinking water system in Milwaukee, Wisconsin, became contaminated with *Cryptosporidium*, and 400,000 people became ill. Four thousand people were hospitalized, at a cost of $54 million, and 100 people died, including some persons with AIDS. Both water systems had met all state and drinking water standards. There have also been more than a dozen outbreaks reported in the United Kingdom. These outbreaks demonstrated the risks of waterborne cryptosporidiosis, for unfortunately *Cryptosporidium* oocysts are resistant to many environmental stresses and chemical disinfectants such as **chlorine** that are used in municipal drinking water systems and swimming pools (swallowing a small amount of water while swimming in a chlorinated pool can cause cryptosporidiosis). The mechanism that protects oocysts from **chlorination** has not yet been positively identified—the oocyst membrane may be protective, or an oocyst may pump **toxins** from its cell before the toxins can cause harm.

Oocysts are present in most surface bodies of water in the United States, many of which are sources of public drinking water. They become more prevalent during periods of **runoff** (generally from March-June during spring rains in North America) or when **wastewater** treatment plants become overloaded or break down. Properly drilled and maintained **groundwater wells**, with intact well casings, proper **seals**, and above-ground caps, are not likely to contain *Cryptosporidium* because of natural **filtration** through **soil** and **aquifer** materials.

The detection of *Cryptosporidium* oocysts is unreliable, for recovery and enumeration of oocysts from water samples is difficult. Concentration techniques for oocysts in water samples are poor, and detection methods often measure algae and other debris in addition to oocysts. The volume of water required to concentrate oocysts for detection can range from 26–264 gal (100–1,000 l). Determination of whether oocysts are infective and viable or a member of the species that causes disease is not easy to accomplish. The development of more accurate, rapid and improved assays for oocysts is required, for present tests are time-consuming, highly subjective, and dependent on the skills of the analyst.

In addition, the number of oocysts (the effective dose) required to cause cryptosporidiosis has not yet been well-defined and requires more investigation. Studies to date have suggested that the 50% infectious dose may be around 132

oocysts, and in some cases, as few as 30 oocysts (infections have also occurred with the ingestion of a single oocyst). Human susceptibility to *Cryptosporidium* likely varies between individuals and between various *Cryptosporidium* strains.

Therefore protection of drinking water supplies from contamination by *Cryptosporidium* requires multiple approaches. Filtration of drinking water supplies is the only reliable conventional treatment method. Water in a treatment plant is mixed with coagulants that aid in the settling of particles in water; removal can be enhanced by using sand filtration. **Ozone** disinfection can kill *Cryptosporidium* but ozone does not leave a residual in the distribution system as chlorine does, which provides protection of treated water to the point of use but does not neessarily kill *Cryptosporidium* anyway.

Watershed protection to prevent contamination from entering water sources is also important in protection of drinking water supplies. Regulation of septic systems and **best management practices** can be used to control runoff of human and animal wastes.

An individual can also take steps to ensure that drinking water is safe. Boiling water (bringing water to a rolling boil for at least one minute) is the best way to kill *Cryptosporidium*. After boiling, the water should be stored in the refrigerator in a clean bottle or pitcher with lid; care should be taken to avoid touching the inside of the bottle or lid to prevent re-contamination. Point-of-use **filters**, either attached to a faucet, or the pour-through type, can also be used to remove *Cryptosporidium* from water. Only filters with an absolute, rather than a nominal, pore size of one micron or smaller should be used to remove oocysts. Reverse osmosis filters are also effective. Lists of filters and reverse osmosis filters that will remove *Cryptosporidium* oocysts can be obtained from NSF International, an independent non-profit testing agency.

The use of bottled water is not necessarily safer than tap water, as water from a surface water source has the same risks of containing oocysts as tap water from that source, unless it has been treated with appropriate treatment technologies, such as distillation, pasteurization, reverse osmosis, or filtration with an absolute one micron rating, before bottling. Bottled water from deep groundwater wells has a low likelihood of being contaminated with oocysts, so the labels on water bottles should be examined before use to determine water source and treatment methods.

Food can also be a source of *Cryptosporidium*. In 1996, in the Northeastern United States, unpasteurized apple cider and juice was associated with *Cryptosporidium* infections. In 1997, in Spokane, Washington, members of a group attending a dinner banquet become ill with cryptosporidiosis. The parasite may be present in uncooked or unwashed fruits and vegetables that are grown in areas where manure was used or animals were grazed or in beverages or ice prepared with contaminated water. Pasteurization of dairy products will kill oocysts. Bottled and canned drinks, such as soda and beer, are usually heated and/or filtered sufficiently to kill or remove *Cryptosporidium* oocysts. Care should be taken to wash hands thoroughly with soap and water before eating, preparing or serving food. Fruits and vegetables that will be eaten raw should be washed or peeled before being eaten. When traveling to areas with poor **sanitation**, extra care should be taken in the selection of food and drink.

[*Judith L. Sims*]

RESOURCES

BOOKS

Fayer, Ron, ed.*Cryptosporidium and Cryptosporidiosis.* Boca Raton, FL: CRC Press, 1997.

Frey, Michelle, Carrie Hancock, and Gary S. Logsdon. *Cryptosporidium: Answers to Questions Commonly Asked by Drinking Water Professionals.* Denver, CO: American Water Works Association, 1998.

Small, Mitchell J., ed. *Protocol for Cryptosporidium Risk Communication.* Denver, CO: American Water Works Association, 2002.

Rose, Joan B., and Huw V. Smith. *Giardia and Cryptosporidium Handbook: A Practical Guide.* Boca Raton, FL: CRC Press, 2000.

PERIODICALS

Conrad, Laura. "The Tedious Hunt for *Cryptosporidium*." *Today's Chemist at Work* 7 (1998): 24–26.

Guerrant, Richard L. "Cryptosporidiosis: An Emerging, Highly Infectious Threat." *Emerging Infectious Diseases* January/March 1997 [June 2002] <http://www.cdc.gov/ncidod/EID/vol3no1/guerrant.htm>.

OTHER

Cryptosporidium: A Waterborne Pathogen. Water Quality Information Center, National Agricultural Library, Agricultural Research Center, U.S. Department of Agriculture. May 1996 [May 2002]. <http://www.nal.usda.gov/wqic/cornell.html>.

"Cryptosporidium parvum." *Bad Bug Book.* Foodborne Pathogenic Microorganisms and Natural Toxins Handbook, Center for Food Safety & Applied Nutrition, U.S. Food and Drug Administration. February 13, 2002 [June 2002]. <http://vm.cfsan.fda.gov/~mow/chap24.html>.

Fact Sheet: Cryptosporidium. Division of Parasitic Diseases, National Center for Infectious Diseases, Centers for Disease Control and Prevention. May 2001 [June 2002]. <http://www.cdc.gov/ncidod/dpd/parasites/cryptosporidiosis/factsht_cryptosporidiosis.htm>.

Fact Sheet: Preventing Cryptosporidiosis—A Guide to Water Filters and Bottled Water. Division of Parasitic Diseases, National Center for Infectious Diseases, Centers for Disease Control and Prevention. August 15, 1999 [June 2002]. <http://www.cdc.gov/ncidod/dpd/parasites/cryptosporidiosis/factsht_crypto_prevent_water.ht>.

CSOs

see **Combined Sewer Overflows**

Cubatao, Brazil

Once called the "valley of death" and the "most polluted place on earth," Cubatao, Brazil, is a symbol both of severe **environmental degradation** and how people can work together to clean up their **environment**. A determined effort to reduce **pollution** and restore the badly contaminated air and water in the past decade has had promising results. While not ideal by any means, Cubatao is no longer among the worst places in the world to live.

Cubatao is located in the state of São Paulo, near the Atlantic coastal city of Santos, just at the base of the high plateau on which São Paulo—Brazil's largest city—sprawls. Thirty years ago, Cubatao was an agreeable, well-situated town. Overlooking Santos Bay with forest-covered mountain slopes rising on three sides around it, Cubatao was well removed from the frantic hustle and bustle of São Paulo on the hills above. Several pleasant little rivers ran through the valley and down to the sea. When the rivers were dammed to generate electricity in the early 1970s, however, the situation changed.

Cheap energy and the good location between São Paulo and the port of Santos attracted industry to Cubatao. An oil refinery, a steel mill, a **fertilizer** plant, and several chemical factories crowded into the valley, while workers and job-seekers scrambled to build huts on the hillsides and the swampy lowlands between the factories. With almost no **pollution control** enforcement, industrial smokestacks belched clouds of dust and toxic effluents into the air while raw sewage and chemical waste poisoned the river. By 1981, the city had 80,000 inhabitants and accounted for 3% of Brazil's industrial output. It was called the most polluted place in the world. More than 1,000 tons of toxic gases were released into the air every day. The steaming rivers seethed with multi-hued chemical slicks, foamy suds, and debris. No birds flew in the air above, and the hills were covered with dying trees and the scars of **erosion** where rains washed dirt down into the valley.

Sulfur dioxide, which damages lungs, eats away building materials, and kills vegetation, was six times higher than World Health Organization guidelines. After a few hours exposure to sunlight and water vapor, sulfur oxides turn into sulfuric **acid**, a powerful and dangerous corrosive agent. Winter air inversions would trap the noxious gases in the valley for days on end. One quarter of all emergency medical calls were related to respiratory ailments. Miscarriages, stillbirths, and deformities rose dramatically. The town was practically uninhabitable.

The situation changed dramatically in the mid-1980s, however. Restoration of democracy allowed citizens to organize to bring about change. Governor Franco Montoro was elected on promises to do something about pollution, and his administration came through on campaign promises. Between 1983 and 1987, the government worked with industry to enforce pollution laws and to share the costs of clean-up. Backed by a **World Bank** loan of $100 million, the state and private industry invested more than $200 million for pollution control. By 1988, 250 out of 320 pollution sources were reduced or eliminated. Ammonia releases were lowered by 97%, **particulate** emissions were reduced 92%, and sulfur dioxide releases were cut 84%. Ozone-producing **hydrocarbons** and volatile organic compounds dropped nearly 80%. The air was breathable again. Vegetation began to return to the hillsides around the valley, and birds were seen once more.

Water quality also improved. Dumping of trash and industrial wastes was cut from some 64 metric tons per day to less than 6 tons. Some 780,000 tons of **sediment** were dredged out of the river bottoms to remove toxic contaminants and to improve water flow. Fish returned to the rivers after a 20-year absence. Reforestation projects are replanting native trees on hillsides where mudslides threatened the town. The government of Brazil now points to Cubatao with pride as an illustration of its concern for environmental protection. This is a heartening example of what can be done to protect the environment, given knowledge, commitment, and cooperation.

[*William P. Cunningham Ph.D.*]

RESOURCES
PERIODICALS

"Cubatao: Brazil's Ecological Success." *Financial Times* (10 June 1988).
"Cubatao: New Life in the Valley of Death." *World Resources 1990–91.* Washington, DC: World Resources Institute, 1991.

Cultivation

see **Agricultural pollution**

Cultural eutrophication

One of the most important types of **water pollution**, cultural eutrophication describes human-generated fertilization of water bodies. Cultural denotes human involvement, and eutrophication means truly nourished, from the Greek word *eutrophic*. Key factors in cultural eutrophication are **nitrates** and **phosphates**, and the main sources are treated sewage and **runoff** from farms and urban areas. The concept of cultural eutrophication is based on anthropocentric values, where clear water with minimal visible organisms is much preferred over water rich in green algae and other **microorganisms**.

Nitrates and phosphates are the most common limiting factors for organism growth, especially in aquatic ecosystems. Most fertilizers are a combination of **nitrogen**, **phosphorus**, and potassium. Nitrates are key components of the amino acids, **peptides**, and proteins needed by all living organisms. Phosphates are crucial in energy transfer reactions within cells. Natural sources of nitrates (and ammonia) are more readily available than phosphates, so the latter is often cited as the crucial limiting factor in plant growth. Nitrates are supplied in limited quantities by decaying plant material and nitrogen-fixing bacteria, but phosphates must come from animal bones, organic matter, or from the breakdown of phosphate-bearing rocks. Consequently, the introduction and widespread use of phosphate **detergents**, combined with excess **fertilizer** in runoff, has produced a near ecological disaster in some waters.

In ecosystems, there is a continuous cycling of matter, with green algae and plants making food from **chemicals** dissolved in water via **photosynthesis**; this provides the food base needed by herbivores and carnivores. Dead plant material and animals are then decomposed by **aerobic** (oxygen using) and **anaerobic decomposers** into the simple elements they came from. Natural water bodies are usually well-suited for handling this matter cycling; however, human impacts often inject large amounts of additional nutrients into the system, changing them from **oligotrophic** (poorly nourished) to eutrophic water bodies. Once present within a relatively closed body of water, such as a lake or **estuary**, these extra nutrients may cycle numerous times before leaving the system.

Green algae and trashy fish may thrive in eutrophic water, but most people are offended by what they perceive as "scum." Nutrient-poor water is usually clear and possesses a rich supply of oxygen, but as the **nutrient** load increases, oxygen levels drop and turbidity rises. For example, when sewage is dumped into a body of water, the sewage fertilizes the algae, and as they multiply and die the aerobic decomposers multiply in turn. The increased demand for oxygen by the decomposers outstrips the system's ability to provide it. As a result, **dissolved oxygen** levels may fall or sag even below 2.0 **parts per million**, the threshold below which even trashy fish cannot a survive (trout need at least 8.0 ppm to survive). Even though the green algae are producing oxygen as a byproduct of photosynthesis, even more oxygen is consumed by decomposers breaking down the dead algae and other organisms.

As water flows downstream the waste is slowly broken down, so there is less for the decomposers to feed on. Biological oxygen demand slowly falls, while the dissolved oxygen levels rise until the river is finally back to normal levels. Most likely, the nutrients recycled by decomposers are either diluted, turned into **biomass** by trees and consumer organ-

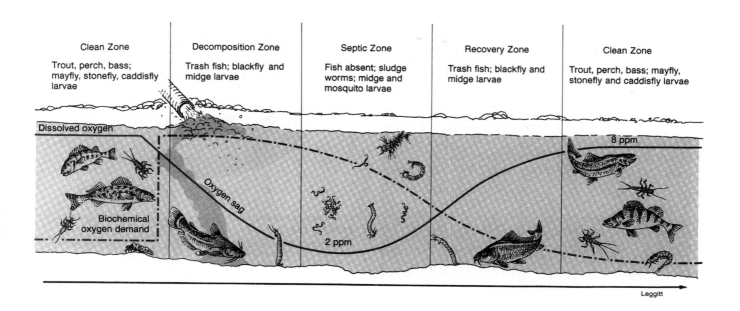

Clean Zone	Decomposition Zone	Septic Zone	Recovery Zone	Clean Zone
Trout, perch, bass; mayfly, stonefly, caddisfly larvae	Trash fish; blackfly and midge larvae	Fish absent; sludge worms; midge and mosquito larvae	Trash fish; blackfly and midge larvae	Trout, perch, bass; mayfly, stonefly and caddisfly larvae

Dissolved oxygen

Oxygen sag

Biochemical oxygen demand

8 ppm

2 ppm

Leggitt

Oxygen sag downstream of a source of organic pollution. (McGraw-Hill Inc. Reproduced by permission.)

isms, or tied up in bottom sediments. Thus a river can naturally cleanse itself of **organic waste** if given sufficient time. Problems arise, however, when discharges are too large and frequent for the river to handle; under extreme conditions it becomes "dead" and is suited only for low-order, often anaerobic, organisms. Municipalities using river water locate their intakes upstream and their **sewage treatment** plants and storm drains downstream. If communities were required to do the reverse, the quality of river water would dramatically improve.

The major sources of nitrates and phosphates in aquatic systems are treated sewage **effluent**; excess fertilizer from farms and urban landscapes; and animal wastes from **feedlots**, pastures, and city streets. In some areas with pristine waters, such as **Lake Tahoe**, tertiary sewage treatment has been added; chemicals are used to reduce nitrate and phosphate levels prior to **discharge**.

Runoff from nonpoint sources is a far more difficult problem because they are harder to control and remediate than runoff from point sources. Point sources can be diverted into treatment plants, but the only feasible way to reduce nonpoint sources is by input reduction or by on-site control.

Less fertilizer more frequently applied, especially on urban lawns, for example, would help reduce runoff. Feedlots are a major concern; runoff needs to be collected and treated in the same manner as human sewage; this may also apply to street runoff. Green cattle ponds are a sure sign of the abundant nutrients supplied by these mobile meat factories.

Phosphate detergents are superior cleaning agents than soap, but the resultant **wastewater** is loaded with this key limiting factor. Phosphate levels in detergents have since been reduced, but its impact is so powerful that abatement may require tertiary treatment.

The battle between Oklahoma and Arkansas over Illinois River **pollution** provides a useful case study of the debate over cultural eutrophication. This river has its headwaters in Arkansas and has become a prime tourist attraction in Oklahoma. Enthusiasts come from all over to canoe the river; summer use is especially heavy. However, economic development within the basin has resulted in a steady decline in river quality.

Arkansas started a legal war with Oklahoma when it sought and obtained **Environmental Protection Agency**

(EPA) approval to dump half of the treated sewage from a new plant in Fayetteville into a tributary of the Illinois. Arkansas argued that its state-of-the-art treatment plant would produce an effluent having little impact on **water quality** by the time it reached the border. Oklahoma countered that it could not risk the potential economic loss if the river became polluted.

This controversy has had a salutary impact on research into the causes of cultural eutrophication within this basin. Scientists from Oklahoma State University and the University of Arkansas are collaborating on a long-term study of river quality and pollution sources. It is highly likely that nonpoint sources within both states will be identified as the key culprits, especially from livestock operations.

There is one major success story in the battle to overcome the effects of cultural eutrophication. The Thames River in England was devoid of aquatic life for centuries. Now a massive cleanup effort is restoring the river to vitality. Many fish have returned, most notably the pollution-sensitive **salmon**, which had not been seen in London for 300 years. However, much work remains, especially in former Warsaw Pact countries and those in the **Third World**.

[*Nathan H. Meleen*]

RESOURCES
BOOKS

Pettyjohn, W. A. *Water Quality in a Stressed Environment*. Minneapolis: Burgess, 1972.

PERIODICALS

Canby, T. "Water: Our Most Precious Resource." *National Geographic* 158 (August 1980): 144-179.
Harleman, D. R. F. "Cutting the Waste in Wastewater Cleanups." *Technology Review* 93 (April 1990): 60-69.
Herber, L. "Cool, Refreshing--and Filthy." In *The Human Habitat: Contemporary Readings*, edited by David L. Wheeler. New York: Van Nostrand Reinhold, 1971.
Maurits la Riviere, J. W. "Threats to the World's Water." *Scientific American* 261 (September 1989): 80-84+.

Cuyahoga River

This 103-mi (166-km) long tributary of **Lake Erie** is a classic industrial river, with, however, one monumental distinction: it caught fire—twice. The first fire, in 1959, burned for eight days; fireboats merely spread the blaze. Typical of the times, a November 1959 article in *Fortune* seemed to glorify the industrial **pollution** here, with words and a portfolio of drawings reminiscent of Charles Dickens. Inspired by feelings of space and excitement, the artist stated: "It is a great expanse, with a smoky cast over everything, smudged with orange dust from the ore—an overall brown color." Cleveland-

ers consoled themselves that the foul water at least symbolized prosperous times.

The second fire occurred on June 22, 1969, as several miles of river along the lowland industrial section called "the Flats" ignited, fed by bunker oil, trash, and tree limbs trapped by a bridge 6 mi (9.7 km) upstream. This fire, along with the Santa Barbara oil well blowout, provided graphic television images which were projected around the world. Jack A. Seamonds's June 18, 1984 article in *U.S. News & World Report* credited the fire with lighting "the fuse that put the bang in the nationwide campaign to clean up the environment." Pamela Brodie described it in the September-October 1983 issue of *Sierra* as "The most infamous episode of **water pollution** in the U.S...It inspired a song...and the Clean Water Act."

Although the fire badly hurt Cleveland's image, steps were already underway to correct the problem. The city government had declared the river a fire hazard and won voter approval in the autumn of 1968 for major funding to correct sewage problems. After the fire, Cleveland's three image-conscious steel companies voluntarily quit dumping cyanide-laced water and two installed cooling towers. As a result of just these actions alone, the river once again began to freeze in the winter. The city of Akron banned phosphate **detergents**, and eventually won the lawsuits brought against it by soap companies. These and other efforts, plus completion in 1983 of the major **sewage treatment** project, brought about far more encouraging reports.

The Cuyahoga was formerly dumping 155 tons of waste per day, and was even devoid of **sludge** worms. But by 1978, *Business Week* could report substantial improvement: **phosphates** were cut in half; **chemicals** were down 20–40%; and **oil spills** went from 300 per year to 25 in 1977. For the first time, local residents saw ducks on the river. By 1984, **point source** pollution had been largely eliminated. The waterfront was rediscovered, with restaurants and trendy stores where once fires had burned

The water is still light brown, churned up by deep-water shipping, and Lake Erie is still polluted. Nonetheless, the Cuyahoga seems to have been largely redeemed, and thus helped revitalize the city of Cleveland. The river became one of 14 American Heritage rivers in 1998. *See also* Great Lakes; Oil Pollution Act (1990); Water quality

[*Nathan H. Meleen*]

RESOURCES
PERIODICALS

Brodie, P. "The Clean Water Act; New Threats From Toxic Waste Demand Stronger Law." *Sierra* 68 (September-October 1983): 39-44.
Lawren, B. "Once Aflame and Filthy, A River Shows Signs of Life." *National Wildlife* 28 (February-March 1990): 24.

Seamonds, J. A. "In Cleveland, Clean Waters Give New Breath of Life." *U. S. News & World Report* 96 (18 June 1984): 68-9.
Wood, W. "Ecological Drums Along the Cuyahoga." *American Education* 9 (January 1973): 15-21.

Cyclodienes

see **Chlorinated hydrocarbons**

Cyclone

see **Tornado and cyclone**

Cyclone collector

A device that removes solids from an **effluent** stream (solid/liquid or solid/gas) using gravitational settling. The input stream enters a vertical tapered cylinder tangentially and spiral fins or grooves cause the stream to swirl. The centrifugal force created by this swirling action causes particles to settle out of the input stream. These solids are concentrated at the walls and move to the bottom of the tapered cone. The solid-free fluid or gas is removed along the center line of the collector via a vortex finder. A **cyclone** collector is a compact device capable of processing large volumes of effluent; however, pressure energy is expended in its operation. It removes particles from the effluent stream less effectively than bag **filters** or **electrostatic precipitation**. *See also* Baghouse

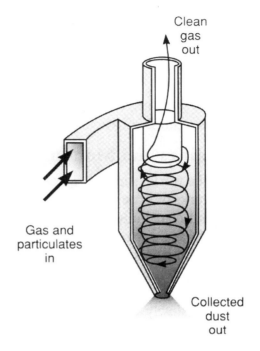

A cyclone collector. (McGraw-Hill Inc. Reproduced by permission.)

D

Dam removal

There are over 77,000 **dams** of significant size (over 6 ft [1.8 m] tall) in the United States, and tens of thousands of additional uncharted smaller dam structures. Constructed for the purposes of harnessing **water resources** for **irrigation**, water supply, and hydroelectric power, dams can be a useful ally for human needs; they also have a major and long-term impact on the entire river **ecosystem**.

Since 1912, over 460 dams have been removed from United States waterways for both safety and environmental reasons. Dams associated with hydroelectric power projects are decommissioned after their useful life and removal is often recommended to return river ecosystems back to their natural state.

Damaged or obsolete dam structures may present a safety hazard, particularly if they are no longer regularly maintained. Upkeep of a dam is expensive, and local taxpayers may bear the brunt of caring for a dam that no longer serves any practical function. Finally, abandoned or unused dams can be aesthetically unpleasant in an otherwise scenic natural area.

Dams and their associated reservoirs can significantly impact river ecosystems and alter their natural course. The structures raise water temperatures, obstruct debris and **nutrient** flow, and prevent **sediment** dispersal. And they have an enormous impact on runs of **salmon**, steelhead, and other migratory fish, often in a relatively short period of time. For example, according to the National **Fish and Wildlife Service** the first dam to be built across the Connecticut River in 1798 resulted in the **extinction** of native Atlantic salmon stock in the river just a few years later. Despite the integration of fish lifts and fish ladders into modern dams, populations of native fish stocks are still greatly depleted by dam structures. Currently, several dams in the northwest are being removed to replenish many of these **species** which are protected under the **Endangered Species Act**.

There can be some short-term negative environmental issues associated with dam removal projects. Contaminated sediment that may collect under a dam can disperse throughout the area, and populations of non-native species that have settled in the dam-altered **habitat** may decline once the dam has been removed. However, in most cases dam removal encourages re-establishment of the native ecosystem and organisms.

Depending on its location and the ownership of the dam, removal may be governed by a variety of federal, state, and local authorities. For federal dam projects, the U.S. **Army Corps of Engineers** and/or the **Bureau of Reclamation** are charged with planning and completion of removal projects. However, only 3% of dams included in the U.S. Army Corps of Engineers national inventory of dams are owned by the federal government, compared with 58% of privately-owned dams.

Even if ownership of a dam is private, the waterways the structure harnesses are public and there is still a significant amount of regulatory oversight on all levels of government. The U.S. **Environmental Protection Agency** (EPA) and some state authorities typically request a full environmental assessment and written report called an "Environmental Impact Statement" (or EIS). The EIS outlines different scenarios for completion of a dam removal project, from no action (letting the dam naturally deteriorate over time) to full dismantling and removal with additional clean-up of dam sediments. It then describes the impact each approach will have on the ecosystem surrounding the dam site.

[*Paula Anne Ford-Martin*]

RESOURCES

BOOKS

McNully, Patrick. *Silenced Rivers: The Ecology and Politics of Large Dams.* 2nd edition. London: Zed Books, 2001.

PERIODICALS

Baish, Sarah K., et al. "The Complex Decision-Making Process for Removing Dams." *Environment* 44, no. 4 (May 2002): 20.

OTHER

American Rivers, Friends of the Earth, & Trout Unlimited. *Dam Removal Success Stories: Restoring Rivers through Selective Removal of Dams that Don't Make Sense.* December 1999.

U.S. Army Corps of Engineers and the Federal Emergency Management Agency. *National Inventory of Dams.* [June 2002]. <http://crunch.tec.army.mil/nid/webpages/nid.cfm>.

ORGANIZATIONS

American Rivers, 1025 Vermont Ave., N.W. Suite 720 , Washington, DC USA 20005 (202)347-7550, Fax: (202)347-9240, Email: amrivers@amrivers.org, <http://www.amrivers.org/damremoval/default.htm>

Dams (environmental effects)

Most dams are built to control flood hazards, to store water for **irrigation** or other uses, or to produce electricity. Along with these benefits come environmental costs including riparian **habitat** loss, water loss through evaporation and **seepage**, **erosion**, and declining **water quality**. Farther-reaching consequences of dams include changes in **groundwater** flow and the displacement of human populations.

Riparian, or stream–side, habitats suffer both above and below dams. Valuable ecological zones that support specialized plants, riparian environments, and nearby shallows provide food and breeding grounds for birds, fish, and many other animals. Upstream of a dam, impounded water drowns riparian communities. Because reservoirs can fill hundreds of miles of river channel, and because many rivers have a long sequence of dams and reservoirs, habitat drowning can destroy a great deal of river **biodiversity**. Downstream, shoreline environments dry up because of water diversions (for irrigation or urban use) or because of evaporation and seepage losses in the **reservoir**. In addition, dams interrupt the annual floods that occur naturally on nearly all rivers. Seasonal **flooding** fertilizes and waters flood plains and clears or redistributes debris in river channels. These beneficial effects of flooding cease once a river is dammed.

Dams and reservoirs alter **sediment** deposition in rivers. Most rivers carry large amounts of suspended **silt** and sand, and dams trap sediments normally deposited downstream. Below the dam, erosion reshapes river channels once sediment deposition ceases. If erosion becomes extreme, bridges, levees, and even river deltas can be threatened. Meanwhile, sediment piling up in the still waters of the reservoir behind the dam decrease water storage capacity. An increasingly shallow reservoir also becomes gradually warmer. Oxygen content decreases as the water temperature rises; fish populations fall, and proliferating algae and aquatic plants can begin to block the dam's water intakes. In **arid** regions a higher percentage of river water evaporates as the reservoir becomes shallower. Evaporating water leaves behind salts, which further decrease water quality in the reservoir and river.

Water losses from evaporation can be extreme: Lakes Powell and Mead on the **Colorado River** lose about 3 billion cu ft (1 billion cu m) of water to evaporation each year; Egypt's Lake Nasser, on the Nile River, loses about 45 billion cu ft (15 billion cu m). Water losses also result from seepage into bedrock. As river water enters groundwater, water tables usually rise around a reservoir. In arid regions increased groundwater can increase local fertility (sometimes endangering delicate dry-land plant **species**), but in moister regions excessive groundwater can cause swamping. Evaporation from exposed groundwater can leave higher salt concentrations in the **soil**. The most catastrophic results of reservoir seepage into groundwater occur when saturated rock loses its strength. In such events valley walls can collapse, causing dam failure and disastrous flooding downstream.

Perhaps the most significant environmental effect of dams results from the displacement of human populations. Because people normally settle along rivers, where water for drinking, irrigation, power, and transport are readily available, reservoir flooding can displace huge populations. The planned **Three Gorges Dam** on China's Chang Jiang (Yangtze River) will displace 1.4 million people and flood some of western China's best agricultural land. A series of dams on India's Narmada river will inundate the homes of 1.5 million people along with 600,000 acres (150,000 ha) of farm land. In both cases, people will need to find new places to live and clear new land to grow food. Such ripple effects carry a dam's influence far beyond its immediate proximity.

Where dams are needed for power, they can have a positive effect in offsetting environmental costs associated with other power sources. Hydropower is cleaner and safer than **nuclear power**. Water turbines are also cleaner than coal-fired generators. Furthermore, both nuclear and **coal** power require extensive mining, with environmental costs far more severe than those of even a large dam.

[*Mary Ann Cunningham Ph.D.*]

RESOURCES
BOOKS

Goldsmith, E., and N. Hidyard, eds. *The Social and Environmental Effects of Large Dams.* (3 Vols.) New York: Wiley, 1986.

PERIODICALS

Esteva, G., and M. S. Prakash. "Grassroots Resistance to Sustainable Development." *The Ecologist* 22 (1992): 45-51.

OTHER

Driver, E. E., and W. O. Wunderlich, eds. *Environmental Effects of Hydraulic Engineering Works.* Proceedings of an International Symposium Held at Knoxville, Tennessee, Sept. 12-14, 1978. Knoxville: Tennessee Valley Authority, 1979.

Danube River
see **Eastern European pollution**

Jay Norwood "Ding" Darling (1876 – 1962)
American environmental cartoonist

Most editorial page cartoonists focus on the tough job of meeting daily deadlines, satisfied that their message will influence public opinion through their drawings. Jay Norwood "Ding" Darling, however, drew Pulitzer prize-winning cartoons, but was also immersed in **conservation** action, emerging as one of the great innovators in the conservation movement of the first half of the twentieth century.

Norwood, Michigan, was Darling's namesake birthplace, but he grew up in Elkhart, Indiana, and attended high school in Sioux City, Iowa, at a time when the area was relatively undeveloped. Wandering the prairies of nineteenth century Iowa, Nebraska, and South Dakota instilled in him a life-long love of the outdoors and **wildlife**.

After an uneven beginning, Darling graduated from Beloit College in Wisconsin with a degree in biology. (At one college, during one semester, biology was the only course he passed). And he was expelled from Beloit for a year for caricaturing individuals in the college faculty and administration. Building on an early interest in sketching (and the cartooning skills developed in college) Darling went on to a half-century of drawing political cartoons, including some of the most memorable conservation/environmental cartoons of the twentieth century. His first job (1900) was as a reporter (and sometimes caricaturist) in Sioux City, but six years later, he was hired by the *Des Moines Register and Leader*, where his primary task was to produce a cartoon each day for the editorial page. He retired from that same paper in 1949. Darling did spend two years as an editorial cartoonist for the *New York Globe*, but returned to Iowa at the first real opportunity.

The only other significant period away from his Des Moines newspaper position illustrates his action mode as a conservationist. Darling was active in political life generally, and this involvement often reflected his love of the outdoors and his dismay at what he felt the United States was doing to destroy its natural resource base. He helped organize the **Izaak Walton League** in Des Moines, was active in the landscaping of local parks, and worked to establish the first cooperative wildlife research unit at Iowa State College in

Ames, a unit that served as a model for the establishment of similar units in other states.

Perhaps his greatest impact as an activist resulted from his appointment, by President Franklin Roosevelt, as head of the U.S. Bureau of Biological Survey, predecessor of the U.S. **Fish and Wildlife Service**. He was only in the position for two years, not being happy as a bureaucrat and not happy away from Iowa. But he was reportedly very effective in the job, "raiding" (in Roosevelt's words) the U.S. Treasury for scarce depression-era funds for waterfowl **habitat** restoration, and initiating the duck stamp program which over the years has funded the acquisition of several million acres added to the **National Wildlife Refuge** system. He also used his drawing skills to design the first duck stamp, as well as the flying goose that has become the signpost and symbol of the national **wildlife refuge** system.

Darling was also one of the founders, and then first President of the **National Wildlife Federation** in 1938. He later criticized the organization, and the proliferation of conservation organizations in general, because he first envisioned the Federation as an umbrella for conservation efforts and thought the emergence of too many groups diluted the focus on solving conservation and environmental problems. Until the end of his life, he tried, and failed, to organize a conservation clearing house that would refocus the conservation effort under one heading.

He put into all his efforts lessons of interdependence learned early from biology classes at Beloit College: "Land, water, and vegetation are [interdependent] with one another. Without these primary elements in natural balance, we can have neither fish nor game, wild flowers nor trees, labor nor capital, nor sustaining habitat for humans."

[*Gerald L. Young Ph.D.*]

RESOURCES
BOOKS

Lendt, David L. *Ding: The Life of Jay Norwood Darling.* Ames, IA: Iowa State University Press, 1979.

PERIODICALS

Dudley, Joseph P. "Jay Norwood 'Ding' Darling: A Retrospective." *Conservation Biology* 7, no. 1 (March 1993): 200–203.
"Jay N. Darling: More Than a Cartoonist." *The World of Comic Art* 1, no. 1 (June 1966): 18–25.

Charles Robert Darwin (1809 – 1882)
English naturalist

Darwin, an English biologist known for his theory of **evolution**, was born at Shrewsbury, England, on February 12, 1809. He was born on the same day and the same year as

Abraham Lincoln, a coincidence that may alert American readers to Darwin's era. He studied, traveled, and published his famous *On the Origin of Species* (1859) just prior to the American Civil War.

Darwin's father was an affluent physician and his mother the daughter of the potter Josiah Wedgwood. Charles married Emma Wedgwood, his first cousin, in 1839. Due to his family's wealth, Darwin was made singularly free to pursue his interest in science.

Darwin entered Edinburgh to study medicine, but, as he described in his autobiography, lectures were "intolerably dull," human anatomy "disgusted" him, and he experienced "nausea" in seeing surgery. He subsequently entered Christ's College, Cambridge, to prepare for Holy Orders in the Church of England. While at Cambridge, Darwin became intensely interested in geology and botany, and because of his knowledge in these sciences he was asked to join the voyage of the HMS *Beagle*. Darwin's experiences during the circumnavigational trek of the Beagle were of seminal importance in his later views on evolution.

Darwin's *On the Origin of Species* is a monumental catalog of evidence that evolution occurs, together with the description of a mechanism that explains such evolution. This "abstract" of his notions on evolution was hurried to publication because of a letter Darwin received from Alfred Russell Wallace expressing similar views. Darwin's evidence for evolution was drawn from comparative anatomy, embryology, distribution of **species**, and the fossil record. He believed that species were not immutable but evolved into other species. But how? His theory of evolution by natural selection is based on the premise that species have a great reproductive capacity. The production of individuals in excess of the number that can survive creates a struggle for survival. Variation between individuals within a species was well documented. The struggle for survival, coupled with variation, led Darwin to postulate that those individuals with favorable variations would have an enhanced survival potential and hence would leave more progeny and this process would lead to new species. This notion is sometimes referred to as "survival of the fittest." While the theory of evolution by natural selection was revolutionary for its day, essentially all biologists in the late twentieth century accept Darwinian evolution as fact.

The first edition of the *Origin* had a printing of 1,250 copies. It sold out the first day. Darwin was an extraordinarily productive author for someone who considered himself to be a slow writer. Among his other books are *Structure and Distribution of Coral Reef* (1842), *Geological Observations on Volcanic Islands* (1844), *On the Various Contrivances by which British and Foreign Orchids are Fertilized by Insects* (1862), *Insectivorous Plants* (1875), and *On the Formation of Vegetable Mould through the Action of Worms* (1881). The last book,

Charles Darwin. (Photo Researchers Inc. Reproduced by permission.)

of interest to ecologists and gardeners, was published only six months prior to Darwin's death.

Darwin died at age 73 and is buried next to Sir Isaac Newton at Westminster Abbey in London.

[*Robert G. McKinnell*]

RESOURCES
BOOKS

Barlow, N. *The Autobiography of Charles Darwin, 1809–1882.* With original omissions restored. Edited with appendix and notes by his granddaughter. New York: Norton, 1958.
Darwin, C. *The Voyage of the Beagle.* New York: Bantam Books, 1958.
Peckham, M., ed. *The Origin of Species by Charles Darwin.* A Variorum Text. Philadelphia: University of Pennsylvania Press, 1959.

DDT

see **Dichlorodiphenyl-trichloroethane**

Dead zones

The term dead zone refers to those areas in aquatic environments where there is a reduction in the amount of **dissolved oxygen** in the water. The condition is more appropriately called hypoxia or hypoxic waters or zones. Hypoxia in marine environments is determined when the dissolved oxygen content is less than 2–3 milligrams/liter. Five to eight milligrams/liter of dissolved oxygen is generally accepted as the normal level for most marine life to survive and reproduce. Dead zones can not only reduce the numbers of marine animals, but they can also change the **nature** of the **ecosystem** within the hypoxic zone.

The main cause of oxygen depletion in aquatic environments is eutrophication. This process is a chain of events that begins with **runoff** rich in **nitrogen** and **phosphorus** that makes its way into rivers that eventually **discharge** into estuaries and river deltas. This nutrient-laden water, combined with sunlight, stimulates plant growth, specifically algae, seaweed, and **phytoplankton**. When these plants die and fall to the ocean floor, they are consumed by bacteria that use large amounts of oxygen, thereby depleting the **environment** of oxygen.

Dead zones can lead to significant shifts in **species** balances throughout an ecosystem. Species of aquatic life that can leave these zones—such as fish and shrimp—do so. Bottom-growing plants, shellfish, and others that cannot leave, die, creating an area devoid of aquatic life. This is the reason the term *dead zone* has been aptly used (especially by the media) to describe these areas.

Eutrophication of estuaries and enclosed coastal seas has often been a natural phenomenon when offshore winds and water currents force deep nutrient-laden waters to rise to the surface, stimulating algae bloom. The timing and duration of these conditions varies from year to year and within seasons. Climatic conditions and catastrophic weather events can influence the rapidity hypoxia can occur. In the past, these natural hypoxic areas were limited, and the marine environment could recover quickly.

Nutrient availability, temperature, energy supply (i.e., soluble **carbon** for most **microorganisms** or light and **carbon dioxide** for plants), and oxygen status all affect the growth and sustainability of aquatic plants and animals. One condition that perpetuates hypoxia is elevated temperatures because the ability of water to hold oxygen (i.e. water solubility) decreases with increasing temperature. Situations that promote consumption of oxygen such as plant and animal **respiration** and **decomposition** can also lead to hypoxic conditions in water. Therefore, situations that stimulate plant (i.e. phytoplankton, benthic algae and macroalgae) growth in water can lead indirectly to hypoxic conditions. Algal growth can be accelerated with elevated levels of carbon dioxide and certain nutrients (especially nitrogen and phosphorus), provided adequate sunlight is available. Low **sediment** loads also support increased algal growth because the water is less turbid allowing more light to penetrate to the bottom.

In the last two decades of the twentieth century, the increased incidence of hypoxia and the expanded size of dead zones have been the result of increased nutrients coming from human sources. Runoff from residential and agricultural activities are loaded with fertilizers, animal wastes, and sewage that have specific nutrients that can stimulate plant growth. In the United States, nutrient runoff has become a major concern for the entire interior watersheds of the Mississippi River Basin which drains into the Gulf of Mexico.

Hypoxic waters occur near the mouths of many large rivers around the world and in coastal estuaries and enclosed coastal seas. In fact, over 40 hypoxic zones have been identified throughout the world. Robert J. Diaz from the Virginia Institute of Marine Science has studied the global patterns of hypoxia and has concluded that the extent of these zones has increased over the past several decades. Hypoxic zones are occurring in the Baltic Sea, Kattegat, Skagerrak Dutch Wadden Sea, **Chesapeake Bay**, Long Island Sound, and northern Adriatic Sea, as well as the extensive one that occurs at the mouth of the Mississippi River in the Gulf of Mexico.

It is interesting to note that one of the largest hypoxic zones documented occurred in conjunction with the increase in ocean temperatures associated with **El Niño**. This weather event occurs periodically off the west coast of North and South America, and influences not only the winter weather in North America, but also affects the anchovy catch off Peru in the Pacific. This in turn, affects the worldwide price for protein meal and has a direct impact on soybean farmers (since this is another major source of protein for this product).

The large hypoxic zone in the northern Gulf of Mexico occurs where the Mississippi and Atchafalya rivers enter the ocean. The zone was first mapped in 1985, and has doubled in size since then. The Gulf of Mexico dead zone fluctuates in size depending on the amount of river flow entering the Gulf, and on the patterns of coastal winds that mix the oxygen-poor bottom waters of the Gulf with Mississippi River water. The zone in 1999 exceeded 8,006 mi^2 (20,720 km^2) in area, making it one of the largest in the world. This zone fluctuates seasonally, as do others. It can form as early as February and last until October. The most widespread and persistent conditions exist from mid-May to mid-September.

Recent research has shown that increased nitrogen concentrations in the river water, which act like **fertilizer**, stimulate massive phytoplankton blooms in the Gulf. Bacteria decomposing the dead phytoplankton consume nearly all of the available oxygen. This, combined with a seasonal layering of the fresh water from the river and salt water in the Gulf, results in the zone of low-dissolved oxygen. Within the zone there are very small fish and shellfish populations.

The effect of periodic events on the transport of nutrients to the Gulf, derived primarily from fertilizer in areas of intensive agriculture in the upper Mississippi River Basin, have been implicated as the most likely cause. The largest amount of nutrients is delivered each year after the spring thaw when streams fill and concentrations of nutrients, such as nitrogen, are highest. In addition, during extreme high-flow events, such as those that occurred during the floods of 1993, very high amounts of nutrients are transported to the Gulf. Levels were 100% higher in that year than in other years.

The nature of the hypoxia problem in the Gulf is complicated by the fact that some nutrient load from the Mississippi River is vital to maintain the productivity of the Gulf fisheries but in levels considerably lower than are now entering the marine system. Approximately 40% of the U.S. fisheries landings comes from this area, including a large amount of the shrimp harvest. In addition, the area also supports a valuable sport fishing industry. The concern is that the hypoxic zone has been increasing in size since the 1960s due to human activities in the Mississippi **Watershed** that have increased nitrogen loads to the Mississippi River. The impact of an expanding Gulf hypoxia include: large algal blooms that affect other aquatic organisms, altered ecosystems with changes in plant and fish populations (i.e., lower **biodiversity**), reduced economic productivity in both commercial and recreational fisheries, and both direct and indirect impact on fisheries such as direct **mortality** and altered **migration** patterns which may lead to declines in populations.

Studies were conducted during the 1990s on the sources of increased nutrient concentrations within the Mississippi River. A significant amount of nutrients delivered to the Gulf come from the Upper Mississippi and Ohio River watersheds. The amount of dissolved nitrogen and phosphorus in the waters of the Mississippi has more than doubled since 1960. The principal areas contributing nutrients are streams draining the corn belt states, particularly Iowa, Illinois, Indiana, Ohio, and southern Minnesota. About 60% of the nitrate transported by the Mississippi River comes from a land area that occupies less than 20% of the basin. These watersheds are predominantly agricultural and contain some of the most productive farmland in the world. This area produces approximately 60% of the nation's corn. The U.S. **Geological Survey** has estimated that 56% of the nitrogen entering the Gulf hypoxic zone originates from fertilizer. Potential agricultural sources within these regions include runoff from cropland, animal grazing areas, **animal waste** facilities, and input from agricultural **drainage** systems. The contributions to nutrient

input from sources such as **atmospheric deposition**, coastal upwelling, and industrial sources within the lower Mississippi Watershed are being evaluated also. It is unclear what effect the damming and channelization of the river for navigation has on nutrient delivery. The dead zone area in the Gulf will continue to be monitored to determine whether it continues to expand.

In the meantime, efforts to reduce nutrient **loading** are being undertaken in agricultural areas across the watershed. Several strategies to reduce nutrient loading have been drafted. They are: a reduction in nitrogen-based fertilizers and runoff from **feedlots**, planting alternative crops that do not require large amounts of fertilizers, removing nitrogen and phosphorus from **wastewater**, and restoring **wetlands** so that they can act as reservoirs and **filters** for nutrients. Depending on whether the zone continues to expand or decreases as nutrient levels diminish will determine whether it remains a significant environmental problem in the Gulf of Mexico. Knowledge gained from the study of changing nutrient loads in the Mississippi River will be useful in addressing similar problems in other parts of the world.

[*James L. Anderson and Marie H. Bundy*]

RESOURCES
PERIODICALS

Rabalais, N. N., R. E. Turner, D. Justic, Q. Dortch, W. J. Wisenman, Jr., and B. K. Sen Gupta. "Nutrient changes in the Mississippi River and septum responses on the adjacent continental shelf."*Estuaries* 19, no. 2B (1996): 386-407.

Turner, R.E., and N.N. Rabalais. "Changes in Mississippi River water quality this century: Implications for coastal food webs."*Bio Science* 41 (1991): 140-147.

Debt for nature swap

Debt for **nature** swaps are designed to relieve developing countries of two devastating problems: spiraling debt burdens and **environmental degradation**. In a debt for nature swap, developing country debt held by a private bank is sold at a substantial discount on the secondary debt market to an environmental **nongovernmental organization** (NGO). The NGO cancels the debt if the debtor country agrees to implement a particular environmental protection or **conservation** project. The arrangement benefits all parties involved in the transaction. The debtor country decreases a debt burden that may cripple its ability to make internal investments and generate economic growth. Debt for nature swaps may also be seen as a good alternative to defaulting on loans, which hurts the country's chances of receiving necessary loans in the future. In addition, the country enjoys the benefits of curbing environmental degradation. The creditor (bank) decreases its holdings of potentially bad debt, which may have to be written off at a loss. The NGO experiences global environmental improvement.

Debt for nature swaps were first suggested by Thomas Lovejoy in 1984. Swaps have taken place between Bolivia, Costa Rica, and Ecuador and NGOs in the United States. The first debt for nature swap was implemented in Bolivia in 1987. **Conservation International**, an American NGO, purchased $650,000 of Bolivia's foreign debt from a private bank in the United States at a discounted price of $100,000. The NGO then swapped the face value of the debt with the Bolivian government for "conservation payments-in-kind," which involved a conservation program in a 3.7 million acre (1.5 million ha) tropical forest region implemented by the government and a local NGO.

Despite the benefits associated with debt for nature swaps, implementation has been minimal so far. Less than two percent of the $38 billion in debt for equity swaps have been debt for nature swaps. A lack of incentives on the part of the debtor or the creditor and the lack of well-developed supporting institutional infrastructure can hinder progress in arranging debt for nature swaps.

If a debtor country is unable to repay foreign debt, it has the option of defaulting on the loans or agreeing to a debt for nature swap. The country has an incentive to agree to a debt for nature swap if defaulting is not a viable option and if the benefits of decreasing debt through a swap outweigh the costs of implementing a particular environmental protection project. The cost of the environmental protection programs can be substantial if the developing country does not have the appropriate institutional infrastructure in place. The program will require the input of professional public administrators and environmental experts. Without institutions to support these individuals, the developing countries may find it impossible to carry out the programs they promise to undertake in exchange for cancellation of the debt. If, in addition, the debtor country is highly capital-constrained, then it might not give high priority to the benefits of an environmental investment.

Whether the creditor has an incentive to sell a debt on the secondary debt market to an NGO depends on the creditor's estimate of the likelihood of receiving payment from the developing country; on the proportion of potentially bad credit the creditor is holding; and on its own financial situation. If the NGOs are willing to pay the price demanded by private banks for developing country debt and swap it for environmental protection projects in the debtor countries, they will have the incentive to pursue debt for nature swaps.

Benefits that may be taken into account by the NGOs are those commonly associated with environmental protection. Many developing countries hold the world's richest tropical rain forests, and the global community will benefit greatly from the preservation of these forests. Tropical forests hold a great deal of **carbon dioxide**, which is released into the **atmosphere** and contributes to the **greenhouse effect** when the forests are destroyed. Another benefit is known as "option value," the value of retaining the option of future use of plant or animal resources that might otherwise become extinct. Although we may not know at present of what use, if any, these **species** might be, there is a value associated with preserving them for unknown future use. Examples of future uses might be pharmaceutical remedies, scientific understanding or **ecotourism**. In addition, NGOs may attach "existence value" to environmental amenities. Existence value refers to the value placed on just knowing that natural environments exist and are being preserved. Many NGOs believe preservation is important so that **future generations** can enjoy the **environment**. This value is known as "bequest value." Finally, the NGO may be interested in decreasing hunger and poverty in developing countries, and both the reduction of external debt and the slowing of the depletion of **natural resources** in developing countries is perceived as a benefit for this purpose.

To make a swap attractive, however, the NGO must be assured that the environmental project will be carried out after the debt has been canceled. Without adequate enforcement and assistance, a country might promise to implement an environmental project without being able or willing to follow through. Again, the solution to this problem lies in the development of institutions that are committed to monitoring and giving assistance in the implementation of the programs. Such institutions might encourage long-term relationships between the debtor and NGO to facilitate a structure by which debt is canceled piecemeal on the condition that the debtor continues to comply with the agreement.

A complicating factor that may affect an NGO's **cost-benefit analysis** of debt for nature swaps in the future is that, as the number of swaps and environmental protection projects increases, the value to be derived from any additional projects will decrease, due to diminishing marginal returns.

As described above, the benefits associated with debt for nature swaps both for the debtor countries and NGOs hinge on the presence of supporting institutions in the developing countries. It is particularly important to promote the establishment of appropriate, professionally managed public agencies with adequate resources to hire and maintain environmental experts and managers. These institutions should be responsible for planning and implementing the programs.

It should be noted that, although large debt burdens and environmental degradation are both serious problems faced by many developing countries, there is no direct linkage between them. Nevertheless, debt for nature swaps are an intriguing remedy that seems to address both problems simultaneously. As the quantity and magnitude of swaps so far have been relatively small, it is impossible to say how successful a remedy it may be on a larger scale. It is clear, though, that the future of debt for nature swaps depends on the development of appropriate incentives to all parties in the swap and on the development of institutions to support the fulfillment of the agreements.

[*Barbara J. Kanninen*]

RESOURCES
PERIODICALS

Hansen, S. "Debt for Nature Swaps: Overview and Discussion of Key Issues." *Ecological Economics* 1 (1989): 77-93.
Lovejoy, T. E. "Aid Debtor Nations' Ecology." *New York Times* (4 October 1984): A31.

Deciduous forest

Deciduous forests are made of trees that lose their leaves seasonally and are leafless for part of each year. The tropical deciduous forest is green during the rainy season and bare during the annual **drought**. The temperate deciduous forest is green during the wet, warm summers and leafless during the cold winters with the leaves turning yellow and red before falling. Temperate deciduous forests once covered large portions of Europe, eastern United States, Japan, and eastern China. **Species** diversity is highest in Asia and lowest in Europe. In the United States deciduous forest, 67 species of trees exist.

Decline spiral

A decline spiral is the destruction of a **species, ecosystem,** or **biosphere** in a continuing downward trend, leading to ecosystem disruption and impoverishment. The term is sometimes used to describe the loss of **biodiversity**, when a catastrophic event has led to a sharp decline in the number of organisms in a **biological community**.

In areas where the **habitat** is highly fragmented either due to human intervention or natural disaster, the loss of species is markedly accelerated. Loss of species diversity often initiates a downward spiral, as the weakening of even one plant or animal in an ecosystem, especially a **keystone species**, can lead to the malfunctioning of the biological community as a whole.

Biodiversity exists at several levels within the same community; it can include ecosystem diversity, species diversity, and genetic diversity. Ecosystem diversity refers to the different types of landscapes that are home to living organisms. Species diversity refers to the different types of species in an ecosystem. Genetic diversity refers to the range of characteristics in the DNA of the plants and animals of a

species. A catastrophic event that affects any aspect of the diversity in an ecosystem can start a decline spiral.

Any major catastrophe that results in a decline in biospheric quality and diversity, known as an ecocatastrophe, may initiate a decline spiral. **Herbicide** and **pesticide** used in agriculture, as well as other forms of **pollution**; increased use of **nuclear power**, and exponential **population growth** are all possible contributing factors. The Lapp reindeer herds were decimated by fallout from the nuclear accident at Chernobyl in 1986. Similarly, the oil spill from the *Exxon Valdez* **in 1989 led to a decline spiral to the Gulf of Alaska ecosystem.**

The force that begins a decline spiral can also be indirect, as when **acid rain**, **air pollution**, **water pollution**, or **climate** change kill off many plants or animals in an ecosystem. Diversity can also be threatened by the introduction of non-native or **exotic species**, especially when these species have no natural predators and are more aggressive than the native species. In these circumstances, native species can enter into a decline spiral that will impact other native species in the ecosystem.

Restoration ecology is a relatively new discipline that attempts to recreate or revive lost or severely damaged ecosystems. It is a hands-on approach by scientists and amateurs alike designed to reverse the damaging trends that can lead toward decline spirals. Habitat rebuilding for **endangered species** is an example of restoration **ecology**. For example, the Illinois chapter of **The Nature Conservancy** has reconstructed an oak-and-grassland **savanna** in Northbrook, Illinois, and a **prairie** in the 100-acre (40-ha) ring formed by the underground Fermi National Accelerator Laboratory at Batavia, Illinois. Since it is easier to reintroduce **flora** than the **fauna** into an ecosystem, practitioners of restoration ecology concentrate on plants first. When the plant mix is right, insects, birds, and small animals return on their own to the ecosystem.

[*Linda Rehkopf*]

RESOURCES
BOOKS

Ehrlich, P., and J. Roughgarden. *The Science of Ecology.* New York: Macmillian, 1987.
May, R. M. *Stability and Complexity in Model Ecosystems.* Princeton, NJ: Princeton University Press, 1973.

Decomposers

Decomposers (also called saprophages, meaning "corpse eating") are the organisms which perform the critical task of **decomposition** in **nature**. They include bacteria, **fungi**, and **detritivores** that break down dead organic matter, releasing nutrients back into the **ecosystem**. Fungi are the dominant decomposers of plant material, and bacteria primarily break down animal matter. Decomposers secrete enzymes into plant and animal material to break down the organic compounds, starting with compounds such as sugars which are easily broken down, and ending with more resistant compounds such as cellulose and lignin. Rates of decomposition are faster at higher values of moisture and temperature. Decomposers thus perform a unique and important function in the **recycling** process in nature.

Decomposition

The chemical and biochemical breakdown of a complex substance into its constituent compounds and elements, releasing energy, and often with the formation of new, simpler substances. Organic decomposition takes place mostly in or on the **soil** under **aerobic** conditions. Dead plant and animal materials are consumed by a myriad of organisms, from mice and moles, to worms and beetles, to **fungi** and bacteria. Enzymes produced by these organisms attack the decaying material, releasing water, **carbon dioxide**, nutrients, **humus**, and heat. New microbial cells are created in the process.

Decomposition is a major process in **nutrient** cycling, including the **carbon** and **nitrogen** cycles. The liberated carbon dioxide can be absorbed by photosynthetic organisms, including green plants, and made into new tissue in the **photosynthesis** process, or it can be used as a carbon source by autotrophic organisms.

Decomposition also acts on inorganic substances in a process called **weathering**. Minerals broken free from rocks by physical disintegration can chemically decompose by reacting with water and other **chemicals** to release elements, including potassium, calcium, magnesium, and iron. These and other elements can be taken up by plants and **microorganisms**, or they can remain in the soil system to react with other constituents, forming clays.

Deep ecology

The term "deep ecology" was coined by the Norwegian environmental philosopher Arne Naess in 1973. Naess drew a distinction between "shallow" and "deep" **ecology**. The former perspective stresses the desirability of conserving **natural resources**, reducing levels of air and **water pollution**, and other policies primarily for promoting the health and welfare of human beings. Deep ecologists maintain that shallow ecology simply accepts, uncritically and without reflection, the homocentric, or human-centered, view that humans are, or ought to be, if not the masters of **nature**, then at least the managers of nature for human ends or purposes.

Defenders of deep ecology, by contrast, claim that shallow **environmentalism** is defective in placing human interests above those of animals and ecosystems. Human beings, like all lower creatures, exist within complex webs of interaction and interdependency. If people insist on conquering, dominating, or merely managing nature for their own benefit or amusement, if people fail to recognize and appreciate the complex webs that hold and sustain them, they will degrade and eventually destroy the natural **environment** that sustains all life.

But, deep ecologists say, if people are to protect the environment for all **species**, now and in the future, they must challenge and change long-held basic beliefs and attitudes about our species place in nature. For example, people must recognize that animals, plants, and the ecosystems that sustain them have intrinsic value—that is, are valuable in and of themselves—quite apart from any use or instrumental value they might have for human beings. The genetic diversity found in insects and plants in tropical rain forests is to be protected not (only or merely) because it might one day yield a drug for curing **cancer**, but also and more importantly because such **biodiversity** is valuable in its own right. Likewise, rivers and lakes should contain clean water not just because humans need uncontaminated water for swimming and drinking, but also because fish do. Like Gandhi, to whom they often refer, deep ecologists teach respect for all forms of life and the conditions that sustain them.

Critics complain that deep ecologists do not sufficiently respect human life and the conditions that promote prosperity and other human interests. Some go so far as to claim that they believe in the moral equivalence of human and all other life-forms. Thus, say the critics, deep ecologists would assign equal value to the life of a disease-bearing mosquito and the child it is about to bite. No human has the right to swat or spray an insect, to kill pests or predators, and so on. But in fact this is a caricature of the stance taken by deep ecology. All creatures, including humans, have the right to protect themselves from harm, even if that means depriving a mosquito of a meal or even eliminating it altogether. **Competition** within and among species is normal, natural, and inevitable. **Bats** eat mosquitoes; bigger fish eat smaller fish; humans eat big fish; and so on. But for one species to dominate or destroy all others is neither natural nor sustainable. Yet human beings have, through technology, an ever-increasing power to destroy entire ecosystems and the life that they sustain. Deep ecologists hold that this power has corrupted human beings and has led them to think—quite mistakenly—that human purposes are paramount and that human interests take precedence over those of lower or lesser species. Human beings cannot exist independently from, but only interdependently with nature's myriad species. Once people recognize the depth and degree of this interdependence, deep ecologists say, they will learn humility and respect. The human species' proper place is not on top, but within nature and with nature's creatures and the conditions that nurture all.

Some cultures and religions have long taught these lessons. Zen Buddhism, Native American religions, and other nature-centered belief systems of belief have counseled humility toward, and respect for, nature and nonhuman creatures. But the dominant Western reaction is to dismiss these teachings as primitive or mystical. Deep ecologists, by contrast, contend that considerable wisdom is to be found in these native and non-Western perspectives.

Deep ecology is at present a philosophical perspective within the environmental movement, and not a movement in itself. This perspective does, however, inform and influence the actions of some radical environmentalists. Organizations such as **Earth First!** and the **Sea Shepherd Conservation Society** are highly critical of moderate shallow environmental groups which are prepared to compromise with loggers, developers, dam builders, strip miners, and oil companies, thus putting the economic interests of some human beings ahead of all others. Such development destroys **habitat**, endangers entire species of animals and plants, and proceeds on the assumption that nature has no **intrinsic value**, but only instrumental value for human beings. It is this assumption, and the actions that proceed from it, that deep ecology is questioning and attempting to change. *See also* Ecosophy; Environmental ethics; Foreman, Dave; Green politics; Greens; Strip mining

[*Terence Ball*]

RESOURCES
BOOKS

Devall, B., and G. Sessions. *Deep Ecology: Living as if Nature Mattered.* Salt Lake City, UT: Gibbs M. Smith, 1985.
Foreman, D. *Confessions of an Eco-Warrior.* New York: Harmony Books, 1991.
Fox, Warwick. *Toward a transpersonal Ecology: Developing New Foundations for Environmentalism.* Albany, NY: State University of New York Press, 1995.
Seed, J., J. Macy, P. Fleming, and A. Naess. *Thinking Like a Mountain.* Philadelphia, PA: New Society Publishers, 1988.

PERIODICALS

Naess, A. "The Shallow and the Deep, Long-Range Ecology Movement," *Inquiry* 16 (1973): 95-100.

Deep-well injection

Injection of liquid wastes into subsurface geologic formations is a technology that has been widely adopted as a waste-disposal practice. The practice entails drilling a well to a

permeable, saline-bearing geologic formation that is confined above and below with impermeable layers known as confining beds. When the injection zones lie below drinking water sources at depths typically between 2,000–5,000 ft (610–1,525 m), they are referred to as Class I disposal **wells**. The liquid **hazardous waste** is injected at a pressure that is sufficient to replace the native fluid and yet not so high that the integrity of the well and confining beds is at risk. Injection pressure is a limiting factor because excessive pressure can cause hydraulic fracturing of the injection zone and confining strata, and the intake rate of most injection wells is less than 400 gal (1,500 l) per minute.

Deep-well injection of liquid waste is one of the least expensive methods of **waste management** because little waste treatment occurs prior to injection. Suspended solids must be removed from **wastewater** prior to injection to prevent them from plugging the pores and reducing permeability of the injection zone. Physical and chemical characteristics of the wastewater must be considered in evaluating its suitability for disposal by injection.

The principal means of monitoring the wastewater injection process is recording the flow rate, the injection and annulus pressures, and the physical and chemical characteristics of the waste. Many consider this inadequate and monitoring is still a controversial subject. The major question which arises concerns the placement of monitoring wells and whether they increase the risk that wastewater will migrate out of the injection zone if they are improperly constructed.

Deep-well injection of wastes began as early as the 1950s, and it was then accepted as a means of alleviating surface **water pollution**. Today, most injection wells are located along the Gulf Coast and near the **Great Lakes**, and their biggest users are the **petrochemical**, pharmaceutical, and steel mill industries.

As with all injection wells, there is a concern that the waste will migrate from the injection zone to the overlying aquifers. *See also* Aquifer restoration; Groundwater monitoring; Groundwater pollution; Hazardous waste siting; Hazardous waste site remediation; Water quality

[*Milovan S. Beljin*]

RESOURCES
BOOKS

Assessing the Geochemical Fate of Deep-Well-Injected Hazardous Wastes: Summaries of Recent Research. Washington, DC: U. S. Environmental Protection Agency, 1990.

OTHER

International Symposium on Subsurface Injection of Liquid Wastes. *Proceedings of the International Symposium on Subsurface Injection of Liquid Waste.* March 3-5, 1986. Dublin, OH: National Water Well Association, 1986.

Defenders of Wildlife

Defenders of Wildlife was founded in 1947 in Washington, D.C. Superseding older groups such as Defenders of Furbearers and the Anti-Steel-Trap League, the organization was established to protect wild animals and the habitats that support them. Today their goals include the preservation of **biodiversity** and the defense of **species** as diverse as gray **wolves** (*Canis lupus*), **Florida panthers** (*Felis concolor coryi*), and grizzly bears (*Ursus arctos*), as well as the western yellow-billed cuckoo (*Coccyzus americanus*), the **desert tortoise** (*Gopherus agassizii*), and Kemp's Ridley sea turtle (*Lepidochelys kempii*).

Defenders of Wildlife employs a wide variety of methods to accomplish their goals, from research and education to lobbying and litigation. They have achieved a ban on livestock grazing on 10,000 acres (4,050 ha) of tortoise **habitat** in Nevada and lobbied for restrictions on the international wildlife trade to protect **endangered species** in other countries. In 1988 they successfully lobbied Congress for funding to expand wildlife refuges throughout the country. Ten million dollars was appropriated for the Lower Rio Grande **National Wildlife Refuge** in Texas, two million dollars to purchase land for a new preserve along the Sacramento River in California, and $1 million for additions to the Rachel Carson National **Wildlife Refuge** in Maine. They are currently seeking passage of the National Biological Diversity Conservation and Research Act and the American Heritage Trust Fund, which would support the continued acquisition of natural habitats.

The organization has been at the forefront of placing preservation on an economic foundation. In Oregon, they have overseen the establishment of a number of areas from which to view wildlife on public and private land, thus improving access to natural habitats and linking the **environment** with the economic benefits of the state's tourism industry. Defenders maintains a speaker's bureau, and they support a number of educational programs for children designed to nourish and expand their interest in wildlife. But the group also participates in more direct action on behalf of the environment. They coordinate grassroots campaigns through their Defenders Activist Network, which has a membership of 9,000. They work with the **Environmental Protection Agency** on a hotline called the "Poison Patrol," which receives calls on the use of pesticides that damages wildlife, and they belong to the *Entanglement Network Coalition*, which works to prevent the loss of animal life through entanglement in nets and plastic refuse.

Restoring wolves to their natural habitats has long been one of the top priorities of Defenders of Wildlife. In 1985, they sponsored an exhibit in **Yellowstone National Park** and at Boise, Idaho, called "Wolves and Humans," which received over 250,000 visitors and won the **Natural Resources** Council of America Award of Achievement for Education. They have helped reintroduce red and gray wolves back into the northern Rockies. In order to assist farmers and the owners of livestock herds that graze in these areas, Defenders has raised funds to compensate them for the loss of land. They are also working to reduce and eventually eliminate the **poisoning** of predators, both by lobbying for stricter legislation and by encouraging Western farmers to participate in their guard dog program for livestock.

In order to conserve land, the Defenders have also launched their own coffee line called "Java Forest." The coffee beans are grown under the forest canopy or on farms which recreate a natural habitat. This reduces the large amount of land that is used for hybrid coffee beans. Twenty-five percent of each purchase is being returned to the Defenders to be used in other programs.

Defenders of Wildlife has 425,000 members and an annual budget of $5.2 million. In addition to wildlife viewing guides for different states, their publications include a bimonthly magazine for members called *Defenders* and *In Defense of Wildlife: Preserving Communities and Corridors*..

[*Douglas Smith*]

RESOURCES

ORGANIZATIONS

Defenders of Wildlife, 1101 14th Street, NW #1400, Washington, D.C. USA 20005 (202) 682-9400, Email: info@defenders.org, <http://www.defenders.org>

Defoliation

Several factors can cause a plant to lose its leaves and become defoliated. Defoliation is a natural and regular occurrence in the case of deciduous trees and shrubs that drop their leaves each year with the approach of winter. This process is aided by an abscission layer that develops at the base of the leaf petiole, weakens the attachment to the plant, and eventually causes the leaf to drop. Severe **drought** may also cause leaves to wilt, dry and drop from a plant. The result of severe dehydration is usually lethal for herbaceous plants, although some woody **species** may survive an episode of drying. Heavy infestation by leaf-eating insects can lead to partial, or complete defoliation. The **gypsy moth** (*Porthetria dispar*) is an important defoliator that attacks many trees, defoliating, and weakening, or killing them. Parasitic wasps that feed on the larvae can help to control gypsy moth outbreaks. Insecticide sprays have also been used to kill the larvae. Spider mites, any of the plant-feeding mites of the family Tetranychidae (subclass Acari), feed on house plants and the foliage and fruit of orchard trees. Heavy infestation can lead to serious or complete defoliation. Spider mites are controlled with pesticides, although growing **resistance** to chemical control agents has made this more difficult, and alternative control measures are under investigation.

Along with natural causes of defoliation, **chemicals** can cause plants to drop their leaves. The best known and most widely used chemical defoliators are 2,4,5 trichlorophenoxyacetic **acid** (**2,4,5-T**) and 2,4 dichlorophenoxyacetic acid (**2,4-D**). Both chemicals are especially toxic to broadleaf plants. The **herbicide** 2,4-D is widely used in lawn care products to rid lawns of dandelions, clover, and other broadleaf plants that interfere with robust turf development. At appropriate application rates, it selectively kills broad-leaf herbaceous plants, and has little effect on narrow-leaf grasses. The uses of 2,4,5-T are similar, although it has been more widely used against woody species.

A mixture of 2,4-D and 2,4,5-T, in a product called **Agent Orange** has been extensively used to control the growth and spread of woody trees and shrubs in sites earmarked for industrial or commercial development. Agent Orange saw extensive use by American forces in Southeast Asia during the Vietnam war, where it was used initially to clear for power lines, roads, railroads, and other lines of communication. Eventually, as the war continued, it was used to spray enemy hiding places, and U.S. military base perimeters to prevent surprise attack. Food crops, especially rice, were also targets for Agent Orange to deprive enemy forces of food. Although Agent Orange and other formulations containing the chlorinated phenoxy acetic acid derivatives are generally lethal to herbaceous plants, woody deciduous plants may survive one or more treatments, depending on the species treated, concentrations used, spacing of applications, and weather. In Vietnam it was found that mangrove forests in the Mekong delta were especially sensitive, and often killed by a single treatment. A member of the **dioxin** family of chemicals, 2,3,7,8-tetrachlorodibenzo-p-dioxin (TCDD) has been found to be an accidental but common contaminant of 2,4,5-T and Agent Orange. Dioxins are very resistant to attack by **microbes** in the **environment**, and are apt to persist in soils for a very long time. Although few disorders have been definitively proven to be caused by dioxins, their effects on laboratory animals have caused some scientists to rank them among the most poisonous substances known. The U.S. government banned some 2,4,5-T containing products in 1979 because of uncertainties regarding its safety, but its use continues in other products.

[*Douglas C. Pratt Ph.D.*]

RESOURCES
BOOKS

Addicott, F. T. *Abscission.* Berkeley: University of California Press, 1952.
Gansner, D. A. *Defoliation potential of gypsy moth.* Radnor, Pa.: U.S. Dept. of Agriculture, Northeastern Forest Experiment Station, 1993.

Teas, H. J. *Herbicide toxicity in mangroves.* U.S. Environmental Protection Agency, Office of Research and Development, Environmental Research Laboratory, Springfield Va.: for sale by the National Technical Information Service, 1976.
Whiteside, T. *The withering rain; Americas herbicidal folly.* New York: Dutton, 1971.

Deforestation

Deforestation is the complete removal of a forest **ecosystem** and conversion of the land to another type of landscape. It differs from **clear-cutting**, which entails complete removal of all standing trees but leaves the **soil** in a condition to regrow a new forest if seeds are available. Humans destroy forests for many reasons. American Indians burned forests to convert them to **grasslands** that supported big game animals. Early settlers cut and burned forest to convert them to croplands. Between 1600 to 1909, European settlement decreased forest cover in the United States by 30%. Since that time, total forest acreage in the United States has actually increased. In Germany about two-thirds of the forest was lost through settlement. Food and Agriculture Organization (FAO) estimated that from 1980 to 1990, 0.9% of remaining tropical forests were deforested annually (65,251 mi^2 [169,000 km^2] per year), an area equivalent to the state of Washington. FAO defines forest as land with more than 10% tree cover, natural understory vegetation, **nature** animals, natural soils, and no agriculture. Analysis of deforestation is difficult because data is unreliable and the definitions for "forest" and "deforestation" keep changing; for example, clear-cuttings which reforest within five years have been considered deforested in some studies but not in others.

The major direct causes of topical deforestation are the expansion of shifting agriculture, livestock production, and fuelwood harvest in drier regions. Forest conversion to permanent cropland, infrastructure, urban areas, and commercial fisheries also occurs. Although not necessarily resulting in deforestation, timber harvest, grazing, and fires can severely degrade the forest. The environmental costs of deforestation can include **species extinction**, **erosion**, **flooding**, reduced land productivity, **desertification**, and **climate** change and increased atmospheric **carbon dioxide**. As more **habitat** is destroyed, more species are facing extinctions. Deforestation of watersheds causes erosion, flooding, and **siltation**. Upstream land loses fertile **topsoil** and downstream crops are flooded, hydroelectric reservoirs are filled with **silt** and fisheries are destroyed. In drier areas, deforestation contributes to desertification.

Deforestation can alter local and regional climates because evaporation of water from leaves makes up as much as two-thirds of the rain that falls in some forest. Without trees to hold back surface **runoff** and block wind, available moisture is quickly drained away and winds dry the soil, sometimes resulting in desert-like conditions. Another potential effect on climate is the large scale release into the **atmosphere** of **carbon** dioxide stored as organic carbon in forests and forest soils. In 1980, tropical deforestation released between 0.4 and 1.6 billion tons of carbon into the atmosphere, an amount equal to 10–40% of that from **fossil fuels**.

As a result of misguided deforestation in the moist and dry tropics, the rural poor are deprived of construction materials, fuel, food, and cash crops harvested from the forest. Species extinctions, siltation, and flooding expand these problems to national and international levels. Despite these human and environmental costs, wasteful deforestation continues. Current actions to halt and reverse deforestation focus on creating economic and social incentives to reduce wasteful land conversion by providing for wiser ways to satisfy human needs. Other efforts are the reforestation of deforested areas and the establishment and maintenance of **biodiversity** preserves.

[*Edward Sucoff*]

RESOURCES
BOOKS

Rowe, R., N. P. Sharma, and J. Browder. "Deforestation: Problems, Causes and Concerns." In *Managing the World's Forests,* edited by N. P. Sharma. Dubuque, IA: Kendall Hunt, 1992.

PERIODICALS

Monastersky, R. "The Deforestation Debate." *Science News* 144 (July 10, 1993): 26-27.

Delaney Clause

The Delaney Clause is a part of the Federal Food, Drug, and Cosmetic Act of 1958, Section 409, and it prohibits the addition to food of any substance that will cause **cancer** in animals or humans. The clause states "no additive will be deemed to be safe if it is found to induce cancer when ingested by man or animal, or if it is found, after tests which are appropriate for the evaluation of the safety of **food additives**, to induce cancer in man or animals..." The clause addresses the safety of food intended for human consumption and few, if any, reasonable individuals would argue with its intent.

There is however, an emerging scientific controversy over its application, and many now question the merits of the clause as it is written. For example, safrole occurs naturally as a constituent in sassafras tea and spices and thus permissibly under the Delaney Clause, but it is illegal and banned as an additive to natural root beer because it has been proven a

carcinogen in animal tests. Coffee is regularly consumed by many individuals, yet more than 70% of the tested **chemicals** that occur naturally in coffee have been shown to be carcinogenic in one or more tests. Naturally occurring carcinogens are found in other foods including lettuce, apples, pears, orange juice, and peanut butter. It is important to note here that the National Cancer Institute recommends the consumption of fruits and vegetables as part of a regimen to reduce cancer risk. This is because it is widely believed that the positive effects of fruits and vegetables far outweigh the potential hazard of trace quantities of naturally occurring carcinogens.

It has been estimated that about 10,000 natural pesticides of plants are consumed in the human diet. These natural pesticides protect the plants from disease and predation by other organisms. Only a few of these natural plant pesticides (less than 60) have been adequately tested for carcinogenic potential and of these about half of them tested positive. Bruce N. Ames and his associates at the University of California estimate that 99.99% of the pesticides ingested by humans are not residues of chemicals applied by humans but are chemicals that occur naturally and therefore legally. It has been argued that such naturally occurring chemicals are less hazardous and thus differ in their cancer-causing potential from synthetic chemicals. But this does not appear to be the case; although the mechanisms for chemical carcinogenesis are poorly understood, there seems to be no fundamental difference in how natural and synthetic carcinogens are metabolized in the body.

The Delaney Clause addresses only the issue of additives to the food supply. It is noteworthy that salt, sugar, corn syrup, citric **acid** and baking soda comprise 98% of the additives listed, while chemical additives, which many fear, constitute only a small fraction. It should also be noted that there are other significant safety issues pertaining to the food supply, including pathogens which cause botulism, hepatitis, and salmonella food **poisoning**. Of similar concern to health are traces of environmental pollutants, such as **mercury** in fish, and cooking-induced production of carcinogens, such as **benzopyrene** in beef cooked over an open flame. Excess fat in the diet is also thought to be a significant health hazard.

Scientists who are rethinking the significance of the "zero risk" requirement of the Delaney Clause do not believe society should be unconcerned about chemicals added to food. They simply believe the clause is no longer consistent with current scientific knowledge, and they argue that chemicals added in trace quantities, for worthwhile reasons, should be considered from a different perspective. *See also* Agricultural chemicals; Agricultural pollution; Drinking-water supply; Food and Drug Administration

[*Robert G. McKinnell*]

RESOURCES
PERIODICALS

Corliss, J. "The Delaney Clause: Too Much of a Good Thing?" *Journal of the National Cancer Institute* 85 (1993): 600-603.
Gold, L. S., et al. "Rodent Carcinogens: Setting Priorities." *Science* 258 (9 October 1992): 261-265.

Demographic transition

Developed by demographer Frank Notestein in 1945, this concept describes the typical pattern of falling death and birth rates in response to better living conditions associated with economic development. This idea is important, for it offers the hope that developing countries will follow the same pathway to population **stability** as have industrialized countries. In response to the Industrial Revolution, for example, Europe experienced a population explosion during the nineteenth century. Emigration helped alleviate overpopulation, but European couples clearly decided on their own to limit family size.

Notestein identified three phases of demographic transition: preindustrial, developing, and modern industrialized societies. Many authors add a fourth phase, postindustrial. In phase one, birth rates and death rates are both high with stable populations. As development provides a better food supply and **sanitation**, death rates begin to plummet, marking the onset of phase two. However, birth rates remain high, as families follow the pattern of preceding generations. The gap between high birth rates and falling death rates produces a population explosion, sometimes doubling in less than 25 years.

After one or two generations of large, surviving families, birth rates begin to taper off, and as the population ages, death rates rise. Finally a new balance is established, phase three, with low birth and death rates. The population is now much larger yet stable. The experience of some European countries, especially in Central Europe and Russia, suggests a fourth phase where populations actually decline. This may be a response to past hardships and oppressive political systems there, however.

Historically, birth rates have always been high. With few exceptions population explosions are linked to declining death rates, not rising birth rates. Infants and young children are especially vulnerable; sanitation and proper food are vital. Infant survival is seen by some as a threat because of the built-in momentum for **population growth**. However, history reveals that there has been no decline in birth rates which has not been preceded by a drop in infant **mortality**. In a burgeoning world this makes infant survival a matter of top priority. To this end, in 1986 the United Nations adopted a program with the acronym GOBI: Growth monitoring, Oral rehydration therapy (to combat killer diarrhea), Breast feeding, and Immunization against major **communicable diseases**. *See also* Child survival revolution; Population Council

[*Nathan H. Meleen*]

RESOURCES
BOOKS

Cunningham, W. P., and B. W. Saigo. *Environmental Science: A Global Concern.* 2nd ed. Dubuque, IA: William C. Brown, 1992.
Maddox, J. *The Doomsday Syndrome.* New York: McGraw-Hill, 1972.

PERIODICALS

Keyfitz, N. "The Growing Human Population." *Scientific American* 261 (September 1989): 7-16.

Dendroica kirtlandii
see **Kirtland's warbler**

Denitrification

A stage in the **nitrogen cycle** in which **nitrates** in the **soil** or in dead organic matter are converted into nitrite, **nitrous oxide**, ammonia, or (primarily) elemental **nitrogen**. The process is made possible by certain types of bacteria, known as denitrifying bacteria. Denitrification is a reduction reaction and occurs, therefore, in the absence of oxygen. For example, flooded soil is likely to experience significant denitrification since it is cut off from atmospheric oxygen. Although denitrification is an important process for the decay of dead organisms, it can also be responsible for the loss of natural and synthetic fertilizers from the soil.

Deoxyribose nucleic acid

Deoxyribose **nucleic acid** (DNA) molecules contain genetic information that is the blueprint for life. DNA is made up of long chains of subunits called nucleotides, which are nitrogenous bases attached to ribose sugar molecules. Two of these chains intertwine in the famous double helix structure discovered in 1953 by James Watson and Francis Crick.

The genetic information contained in DNA molecules is in a code spelled out by the linear sequence of nucleotides in each chain. Each group of three nucleotides makes up a codon, a unit resembling a letter in the alphabet. A string of codons effectively spells a word of the genetic message.

This message is expressed when enzymes (cellular proteins) synthesize new proteins using a copy of a short segment of DNA as a template. Each nucleotide codon specifies which amino **acid** subunit is inserted as the protein is formed, thus determining the structure and function of the proteins. Because the chains are very long, a single DNA strand can contain enough information to direct the synthesis of hundreds of different proteins. Since these proteins make up the cell structure and the machinery (enzymes) by which cells carry out the processes of life, such as synthesizing more molecules including more copies of the DNA itself, DNA can be said to be self-replicating. When cells divide, each of the new cells receives a duplicate set of DNA molecules giving them the necessary information to live and reproduce. *See also* Ribosenucleic acid

Department of Agriculture
see **U.S. Department of Agriculture**

Department of Energy
see **U.S. Department of Energy**

Department of Health and Human Services
see **U.S. Department of Health and Human Services**

Department of the Interior
see **U.S. Department of the Interior**

Desalinization

Desalinization, also known as "desalination," is the process of separating sea water or **brackish** water from their dissolved salts. The average salt content of the ocean water is about 3.4% (normally expressed as 34 parts per thousand). The range of salt content varies from 18 parts per thousand in the North Sea and near the mouths of large rivers to a high of 44 parts per thousand in locked bodies of water such as the Red Sea, where evaporation is very high. The desalination process is accomplished commercially by either distillation or reverse osmosis (RO).

Distillation of sea water is accomplished by boiling water and condensing the vapor. The components of the distillation system consist of a boiler and a condenser with a source of cooling water. Reverse osmosis is accomplished by forcing filtered sea water or brackish water through a reverse osmosis membrane. In a reverse osmosis process, approximately 45% of the pressurized sea water goes through membranes and becomes fresh water. The remaining brine (concentrated salt water) is returned to the sea.

In 1980, the United Nations declared 1981–1990 as the "International Drinking Water Supply and **Sanitation** Decade." The objective was to provide safe drinking water and sanitation to developing nations. Despite some progress in India, Indonesia, and a few other countries, the percentage of the world population with access to safe drinking water has not changed much since that declaration. In the period between 1990 and 2000, the amount of people with access has only increased by 5%.

The World Health Organization (WHO) estimates that only two in five people in the **less developed countries** (LDCs) have access to safe drinking water. The WHO also estimates that at least 25 million people of the LDCs die each year because of polluted water and from water-born diseases such as **cholera**, polio, dysentery, and typhoid. Whether by distillation or by reverse osmosis, desalination of water can transform water that is unusable because of its **salinity** into valuable fresh water. This could be an important water source in many drought-prone areas.

Desalination plants, distribution, and functions

There are approximately 7,500 desalination plants worldwide. Collectively they produce less than 0.1% of the world's fresh water supply. This supply is equal to about 3.5 billion gal per day (13 million l). The cost and the feasibility of producing desalinated water depends upon the cost of energy, labor, and relative costs of desalinated water to that of imported fresh water. It is estimated that in the United States, commercial desalinated water produced from sea water by reverse osmosis costs about $3 per 1,000 gal (3,785 l). This price is four to five times the average price currently paid by urban consumers for drinking water and over 100 times the price paid by farmers for **irrigation** water. The current energy requirement is approximately three kilowatt hours of electricity per one gallon of fresh water extracted from sea water. Currently, using desalinated water for agriculture is cost prohibitive.

About two-thirds of the desalination water is produced in Saudi Arabia, Kuwait, and North Africa. Several small-scale reverse osmosis plants are now operating in the United States, including California (Santa Barbara, Catalina Island, and soon in Ventura and other coastal communities). Generally, desalination plants are used to supplement the existing fresh water supply in areas adjacent to oceans and seas such as southern California, the Persian Gulf region, and other dry coastal areas. Among the advantages of desalinized water are a dependable water supply regardless of rainfall patterns, elimination of **water rights** disputes, and the preservation of the fresh water supply, all of which are essential for existing natural ecosystems.

Reverse osmosis

Reverse osmosis involves forcing water under pressure through a **filtration** membrane that has pores small enough to allow water molecules to pass through but exclude slightly larger dissolved salt molecules. The basic parts of a reverse osmosis system include onshore and offshore components. The onshore components consist of a water pump, an electrical power source, pre-treatment filtration (to remove seaweed and debris), reverse osmosis units connected in series, **solid waste** disposal equipment, and fresh water pumps. The offshore components consist of an underwater intake pipeline, approximately 1,093 yd (1 km) from shore, and a second pipeline for brine **discharge**.

Small reverse osmosis units for home use with a few gallons-per-day capacity are available. These units use a disposable reverse osmosis membrane. Their main drawback is that they waste four to five times the volume of water they purify.

Producing potable water from sea water is an energy intensive, costly process. The high cost of producing desalinized water limits its use to domestic consumption. In areas such as the Persian Gulf and Saudi Arabia where energy is plentiful at a low cost, desalinized water is a viable option for drinking water and very limited greenhouse agriculture. The notion of using desalinized water for wider agricultural purposes is neither practical nor economical at today's energy prices and available technology.

[*Muthena Naseri*]

RESOURCES
BOOKS

Kaufman, D. G., and C. M. Franz. *Biosphere 2000: Protecting Our Global Environment*. New York: Harper-Collins, 1993.

Nebel, B. J., and R. T. Wright. *Environmental Science: The Way the World Works*. 4th Edition. Englewood Cliffs, NJ: Prentice Hall, 1993.

OTHER

Lizarraga, S., and D. Brown. "Fresh Water from Santa Barbara Seas." Reprinted from *Desalination and Water Reuse*, 1992. Santa Barbara: Department of Water Resources.

Desert

Six percent of the world's land surface is desert, a **biome** in which less than 10 in (25 cm) of precipitation occurs per year or any place where evaporation greatly exceeds precipitation, resulting in a lack of available moisture. Sometimes any area lacking the necessary conditions to support life is called a desert. Deserts occur around latitudes 30 degrees north and south where masses of dry circulating air descend to the earth's surface. There are three kinds of deserts—hot (such as Sahara), temperate (such as the Mojave), and cold (such as the Gobi). The area of global desert is increasing yearly, as marginal lands become degraded by human misuse resulting in **desertification**. Deserts are potential sites for the

production of electricity using banks of solar cells or parabolic solar collectors. The lack of water and remoteness of some deserts may make them attractive places to store nuclear and other hazardous wastes.

Desert tortoise

The **desert** tortoise (*Gopherus agassizii*) is a large, herbivorous, terrestrial turtle of the family Testudinidae. It is found in both the southwestern United States and in northwestern Mexico. It is the official reptile in the states of California and Nevada. No other turtle in North America shares the extreme conditions of the habitats occupied by the desert tortoise. It inhabits desert oases, washes, rocky hillsides, and canyon bottoms with sandy or gravelly **soil** under hot, **arid** conditions.

Desert tortoises dig into dry, gravelly soil under bushes in arroyo banks or at the base of cliffs to construct a burrow, which is their home. Climatic conditions dictate daily activity patterns of these tortoises, and they can relieve the problems of high body temperature and evaporative water loss by retreating into their burrows. Since many desert tortoises live in areas devoid of water, except for infrequent rains, they must rely on their food for their water.

The active period for the desert tortoise is from March through September, after which they enter a hibernation period. Nesting and egg laying activities extend from May through July. Desert tortoises lay an average of five moisture-proof eggs, an **adaptation** that helps retain water in its harsh **environment**. These tortoises reach sexual maturity at 15–20 years, and they have a life span of up to 80 years.

The desert tortoise is very sensitive to human disturbances, and this has led to the decimation of many of its populations throughout the desert southwest. The Beaver Dam Slope population of southwestern Utah has been studied over several decades and shows some of the general tendencies of the overall population. In the 1930s and 1940s the desert tortoise population in this area exhibited densities of about 160 adults per square mile. By the 1970s this density had fallen to less than 130 adults per square mile, and more recent studies indicate the level is now about 60 adults per square mile. In southeastern California at least one population reaches densities of 200 adults per square mile, but overall tendencies show that populations are drastically declining. Recent estimates indicate that there are about 100,000 individual desert tortoises existing in the Mojave and Sonoran deserts.

Desert tortoise populations are listed as threatened in Arizona, California, Nevada, and Utah. Numerous factors are contributing to its decline and vulnerability. **Habitat** loss through human encroachment and development, overcol-

A desert tortoise in the Desert Tortoise Natural Area, Mojave Desert, California. (Photograph by Dan Suzio. Reproduced by permission.)

lecting for the **pet trade**, and vandalism—including shooting tortoises for target practice and flipping them over onto their backs, causing them to die from exposure—have decimated populations. Other factors contributing to their decline are grazing livestock, which trample them or their burrows, and mining operations, which also causes respiratory infections among the desert tortoises. Numerous desert tortoises have been killed or maimed by **off-road vehicles**, which also collapse the tortoises' burrows. Concern is mounting as **conservation** efforts seem to be having little effect throughout much of the desert tortoise's range.

[*Eugene C. Beckham*]

RESOURCES
BOOKS

Campbell, F. "The Desert Tortoise." *Audubon Wildlife Report.* San Diego: Academic Press, 1988–89.
Ernst, C., and R. Barbour. *Turtles of the United States.* Lexington: University Press of Kentucky, 1972.
———. *Turtles of the World.* Washington, DC: Smithsonian Institute Press, 1989.

Desertification

About one billion people live in **arid** or semiarid **desert** lands that occupy about one third of the world's land surface. In these drier parts of the world, deserts are increasing rapidly from a combination of natural processes and human activities, a process known as desertification or land degradation. An annual rainfall of less than 10 in (25 cm) will produce a desert anywhere in the world. In the semiarid areas along the desert margins, where the annual rainfall is around 16

in (40 cm), the **ecosystem** is inherently fragile with seasonal rains supporting the temporary growth of plants. Recent changes in the **climate** of these regions have meant that the rains are now unreliable and the lands that were once semiarid are now becoming desert. The process of desertification is precipitated by prolonged droughts, causing the top layers of the **soil** to dry out and blow away. The eroded soils become unstable and compacted and do not readily allow for seeding. This means that desertified areas do not regenerate by themselves but remain bare and continue to erode. Desertification of grazing lands or croplands is accompanied, therefore, by a sharp drop in the productivity of the land.

Natural desertification is greatly accelerated by human activities that leave soils vulnerable to **erosion** by wind and water. The drier **grasslands** with too little rain to support cultivated crops have traditionally been used for grazing livestock. When semiarid land is overgrazed (by keeping too many animals on too little land), plants that could survive moderate grazing are uprooted and destroyed altogether. Since plant roots no longer bind the soil together, the exposed soil dries out and is blown away as dust. The destruction and removal of the **topsoil** means that soil productivity drops drastically. The obvious solution to desertification caused by **overgrazing** is to limit grazing to what the land can sustain, a concept that is easy to espouse but difficult to practice.

In the **Sahel** zone along the southern edge of the Sahara desert, settled agriculture and overgrazing livestock on the fragile scrublands have led to widespread soil erosion. Nomadic pastoralists, who have traditionally followed their herds and flocks in search of new pastures, are now prevented by national borders from reaching their chosen grazing grounds. Instead of migrating, the nomads have been encouraged to settle permanently and this has led to their herds overgrazing available pastures.

Other human factors leading to desertification include over-cultivation, **deforestation**, salting of the soil through **irrigation**, and the plowing of marginal land. These destructive practices are intensified in developing countries by rapid **population growth**, high population density, poverty, and poor land management. The consequences of desertification in some countries mean intensified **drought** and **famine** and lowered standards of living. It is estimated that desertification worldwide has claimed an area the size of Brazil (2 billion acres or 810 million ha) in the past 50 years. Each year new deserts consume an area the size of Belgium (15 million acres or 6 million ha), most of which is in the African Sahel.

In marginal areas throughout the world, traditional farming practices can lead to desertification. Plowing turns the top layer of the soil upside down, burying and killing weeds but exposing bare soil to erosion. In arid areas the

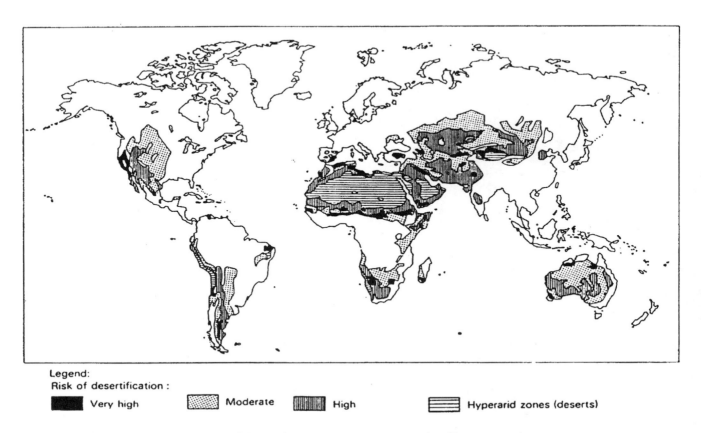

Legend:
Risk of desertification :

| | Very high | | Moderate | | High | | Hyperarid zones (deserts) |

World deserts and areas at risk of desertification. (Beacon Press. Reproduced by permission.)

exposed soil dries out rapidly and is easily lost through wind erosion.

The processes of erosion and soil formation vary with climate and with the composition of the parent material. In balanced ecosystems, soil lost to erosion is replaced by new soil created by natural processes. On average, new soil is formed at a rate of about 5 tons an acre (12.5 tons per ha) per year, which is equivalent to a layer of soil about 0.2 in (0.4 cm) thick. This means that soils can sustain an erosion rate of up to 5 tons per acre per year and still remain in balance. However, much of the world's crop and forest land is not within this balance (with erosion running at two to 10 times the tolerable rate). In the United States about 6 tons of topsoil are lost for every ton of grains produced.

Forests are cut down for many different reasons. Some are cleared for agriculture, some for construction, some for paper products, and some to meet cooking and heating needs. Unfortunately, deforestation results in more than just the loss of trees, for soil is eroded, nutrients are lost to the ecosystem, and the water cycle is disrupted. The roots of the trees serve to bind the soil together and to hold water in the ecosystem, while the leaves of the trees break the

force of the rain and allow it to soak into the topsoil. The result is that surface **runoff** from a forested hillside is half as much as from a grass-covered slope. Additionally, water soaking into the ground (rather than running off) leads to the natural recharge of **groundwater** and other **water resources**. Water and soil runoff from deforested hillsides cause **flooding** and **siltation** of agricultural and aquatic ecosystems in adjacent lowlands. Forests are also more efficient at reabsorbing and **recycling** the nutrients released from decaying **detritus** than are grasslands. Clearing forests therefore exposes the soil to both erosion and **nutrient** loss and alters the recharge of water reserves in the ecosystem.

Industrialized countries experienced a period of intense deforestation during the Industrial Revolution and even today much of the land has low productivity. Fortunately, most of these countries are now reforesting faster than they are deforesting. This is possible because their population growth is low, their agricultural production per acre is high, and their need for fuelwood for cooking and heating is optional, since **fossil fuels** and electricity are widely available. All of these factors release the pressure to further deforest the land.

In developing countries, on the other hand, high population growth rates and widespread poverty put pressure on the forests. Trees are needed for firewood, charcoal, and export, and the land is needed for farmland. In some developing countries deforestation exceeds replanting by five times. In Ethiopia, population and economic pressures pushed people to deforest and cultivate hillsides and marginally dry lands, and more than 1 billion tons of topsoil per year are now lost, resulting in recurrent famines. In parts of India and Africa there is so little wood available that dried animal dung is used to fuel cooking fires, an act that further robs the soil of potential nutrients. In Brazil, soil erosion and desertification have resulted from the conversion of forests to cattle ranches. In China, about one-third of its agricultural land has been lost to erosion, and the story is similar for many other countries.

The answer to erosion from deforestation is reforestation, better forest **conservation**, and better **forest management** to increase productivity. Planting trees on hillsides is particularly effective. Reforesting desertified areas first requires mulching the soil to hold moisture and the protection of the seedlings for several years until natural processes can regenerate the soil. Using these methods, Israel has achieved spectacular success in bringing desertlands (a product of past desertification) back to agriculture.

Desertification and its agents, deforestation and erosion, have been powerful shapers of human history. Agriculture had its roots in the once fertile crescent of the Middle East and in the Mediterranean lands. However, deforestation, overgrazing, and poor agricultural practices have turned once-productive pastureland and farmland into the near deserts of today. It is thought by some that deforestation and desertification may have even contributed to the collapse of the Greek and Roman Empires. Similar fates may have befallen the Harappan civilization in India's Indus Valley, the Mayan civilization in Central America, and the Anasazi civilization of Chaco Canyon, New Mexico, in what is today desertland.

[*Neil Cumberlidge Ph.D.*]

RESOURCES
BOOKS

Dejene, A. *Environment, Famine, and Politics in Ethiopia: A View from the Village.* Boulder, CO: L. Rienner Publishers, 1990.

Grainger, A. *Desertification: How People Make Deserts, How People Can Stop, and Why They Don't.* London: International Institute for Environment and Development, 1982.

McLeish, E. *The Spread of Deserts.* Austin, TX: Steck-Vaughn Library, 1991.

Design for disassembly

For **recycling** to be broadly applicable, many environmentalists believe that products should be cheaper and easier to take apart so that they can be put together again and reused in other products.

Many large companies like Whirlpool, Digital Equipment, 3M, and General Electric have begun to design products in this regard. The BMW car, Z1, is considered to be the first to incorporate the concepts of disassembly. The plastic exterior of the car can be removed from the metal chassis in 20 minutes. The doors, bumpers, and front, rear, and side panels have been manufactured with recyclable plastic, and pop-in/pop-out fasteners replace screws and glues, whenever possible.

Composite materials, on the other hand, which are being used on a more widespread basis are difficult to recycle. Small juice boxes, for instance, which have become very popular are made of many layers of plastic, paper, and **aluminum**, which cannot be separated. *See also* Recyclables; Reuse; Waste management; Waste reduction

Detergents

A group of organic compounds that cause foaming and serve as cleansing agents based on their surface-active properties. A detergent molecule is termed surface-active because a portion of the molecule is hydrophilic and a portion is hydrophobic. It will therefore collect at the interface of water and another medium such as a gas bubble. Bubbles are stabilized by the surface-active molecules so that when the bubbles rise to the top of the bulk liquid, they maintain their integrity and form a foam. Surface-active agents can also agglomerate by virtue of their hydrophilic/hydrophobic **nature** to form *micelles* that can dissolve, trap, and/or envelop **soil** particles, oil, and grease. Surface-active agents are sometimes called *surfactants*. Synthetic detergents are sometimes referred to as *syndets*.

Detoxification

When many toxic substances are introduced into the **environment**, they do not remain in their original form, but are transformed to other products by a variety of biological and non-biological processes. The **chemicals** and their transformation products are degraded, converted progressively to smaller molecules, and eventually utilized in various natural cycles, such as the **carbon cycle**. Toxic metals are not degraded but are interconverted between available and non-available forms. Some organic compounds, such as polychlorinated biphenyl (PCBs), are degraded over a period of many years to less toxic compounds, while compounds such as the **organophosphate** insecticides may break down in only a few hours.

The chemical transformations that occur may either increase (referred to as intoxication, or activation if the parent compound was nontoxic) or decrease (referred to as detoxification) the toxicity of the original compound. For example, elemental **mercury**, which has low toxicity, can be converted to methylmercury, a very hazardous chemical, through **methylation**. Parathion is a fairly nontoxic insecticide until it is converted in a living system or by photochemical reactions to paraoxon, an extremely toxic chemical. However, parathion can also be degraded to less toxic products by the process of hydrolysis.

In microbial degradation, the ultimate fate of the toxic chemicals may be mineralization (that is, the conversion of an organic compound to inorganic products), which results characteristically in detoxification; however, intermediates in the degradation sequence, which may be toxic or have unknown toxicity, may persist for a period of time, or even indefinitely. Likewise, since degradation pathways may contain many steps, detoxification may occur early in the degradation pathway, before mineralization has occurred. Detoxification may also be accomplished biologically through **cometabolism**, the **metabolism** by **microorganisms** of a compound that cannot be used as a **nutrient**; cometabolism does not result in mineralization, and organic transformation products remain. Studies have also shown that the structure and toxicity of many organic compounds can be altered by plants.

In modifying a toxic chemical, detoxifying processes destroy its actual or potential harmful influence on one or more susceptible animal, plant, or microbial **species**. Detoxification may be measured with the use of bioassays. A **bioassay** involves the determination of the relative toxicity of a substance by comparing its effect on a test organism with the conditions of a control. The scale or degree of response may include the rate of growth or decrease of a population, colony, or individual; a behavioral, physiological, or reproductive response; or a response measuring **mortality**. Bioassays can be used for environmental samples through time to determine detoxification of chemicals.

Both acute and chronic bioassays are used to assess detoxification. In an acute bioassay, a severe and rapid response to the toxic chemical is observed within a short period of time (for example, within four days for fish and other aquatic organisms and within 24 hours to two weeks for mammalian species). Detoxification of a chemical may be detected if there is a decrease in the observed toxicity of a test solution over the time of the acute test, indicating removal of the toxic chemical by degradation or other processes.

Similarly, an increase in toxicity could indicate the formation of a more toxic transformation product.

Chronic bioassays are more likely to provide information on the rates of degradation, transformation, and detoxification of toxic compounds. Partial or complete life cycle bioassays may be used, with measurements of growth, reproduction, maturation, spawning, hatching, survival, behavior, and **bioaccumulation**.

Detoxification of chemicals should also be measured by toxicity testing that involves changes in different organisms and interactions among organisms, especially if the chemicals are persistent and stable and may accumulate and magnify in the **food chain/web**. Model ecosystems can be used to simulate processes and assess detoxification in a terrestrial-aquatic **ecosystem**. A typical ecosystem could include **soil** organisms, lake-bottom **fauna**, a plant, an insect, a snail, an alga, a crustacean, and a fish species maintained under controlled conditions for a period of time.

Most major types of reactions that result in transformation and detoxification of toxic chemicals can be accomplished either by biological (enzymatic) or by non-biological (nonenzymatic) mechanisms. Although significant changes in structure and properties of organic compounds may result from non-biological processes, the biological mechanism is the major and often the only mechanism by which organic compounds are converted to inorganic products. Microorganisms are capable of degrading and detoxifying a wide variety of organic compounds; presumably, every organic molecule can be destroyed by one or more types of microorganisms (referred to as the "principle of microbial infallibility.") However, since some organic compounds do accumulate in the environment, there must be factors such as unfavorable environmental conditions that prevent the complete degradation and detoxification of these persistent compounds. There are many examples where certain microorganisms have been identified as capable of detoxifying specific organic compounds. In some cases, these microorganisms can be isolated, cultured, and inoculated into contaminated environments in order to detoxify the compounds of concern.

The major types of transformation reactions include: oxidation, ring scission, photodecomposition, **combustion**, reduction, dehydrohalogenation, hydrolysis, hydration, conjugation, and chelation. Conjugation is the only reaction mediated by enzymes alone, while chelation is strictly nonenzymatic. Primary changes in organic compounds are usually accomplished by oxidative, hydrolytic, or reductive reactions.

Oxidation reactions are reactions in which energy is used in the incorporation of molecular oxygen into the toxic molecule. In most mammalian systems, a monooxygenase system is involved. One atom of molecular oxygen is added to the toxic chemical, which usually results in a decrease in toxicity and an increase in water solubility, as well as provides a reaction group that can be used in further transformation processes such as conjugation. Microorganisms use a dioxygenase system, in which oxidation is accomplished by adding both atoms of molecular oxygen to the double bond present in various aromatic (containing benzene-like rings) **hydrocarbons**.

Ring scission, or opening, of aromatic ring compounds also can occur through oxidation. Though aromatic ring compounds are usually stable in the environment, some microorganisms are able to open aromatic rings by oxidation. After the aromatic rings are opened, the compounds may be further degraded by other organisms or processes. The number, type, and position of substituted molecules on the aromatic ring may protect the ring from enzymatic attack and may retard scission.

Photodecomposition can also result in the detoxification of toxic chemicals in the **atmosphere**, in water, and on the surface of solid materials such as plant leaves and soil particles. The reaction is usually enhanced in the presence of water; photodecomposition is also important in the detoxification of evaporated compounds. The **ultraviolet radiation** in sunlight is responsible for most photodecomposition processes. In photooxidation, for example, photons of light provide the necessary energy to mediate the reactions with oxygen to accomplish oxidation.

Combustion of toxic chemicals involves the oxidation of compounds accompanied by a release of energy. Often combustion does not completely result in the degradation of chemicals, and may result in the production of very toxic combustion products. However, if operating conditions are properly controlled, combustion can result in the detoxification of toxic chemicals.

Under **anaerobic** conditions, toxic compounds may be detoxified enzymatically by reduction. An example of a reductive detoxifying process is the removal of halogens from halogenated compounds. Dehydrohalogenation is another anaerobic process that also results in the removal of halogens from compounds.

Hydrolysis is an important detoxification mechanism in which water is added to the molecular structure of the compound. The reaction can occur either enzymatically or nonenzymatically. Hydration of toxic compounds occurs when water is added enzymatically to the molecular structure of the compound.

Conjugation reactions involve the combination of foreign toxic compounds with endogenous, or internal, compounds to form conjugates that are water soluble and can be eliminated from the biological organism. However, the toxic compound may still be available for uptake by other organisms in the environment. Endogenous compounds used in the conjugation process include sugars, amino **acid** residues, **phosphates**, and sulfur compounds.

Many metals can be detoxified by forming complexes with organic compounds by sharing electrons through the process of chelation. These complexes may be insoluble or nonavailable in the environment; thus the toxicant can not affect the organism. **Sorption** of toxic compounds to solids in the environment, such as soil particles, as well as incorporation into **humus**, may also result in detoxification of the compounds.

Generally, the complete detoxification of a toxic compound is dependent on a number of different chemical reactions, both biological and non-biological, proceeding simultaneously, and involving the original compound as well as the transformation products formed. *See also* Biogeochemical cycles; Biomagnification; Chemical bond; Environmental stress; Incineration; Oxidation reduction reaction; Persistent compound; Water hyacinth

[*Judith Sims*]

RESOURCES
BOOKS

Burnside, O. C. "Prevention and Detoxification of Pesticide Residues in Soils." In *Pesticides in Soil and Water*, edited by W. D. Guenzi. Madison, WI: Soil Science Society of America, 1974.

Dauterman, W. C., and E. Hodgson. "Chemical Transformations and Interactions." In *Introduction to Environmental Toxicology*, edited by F. E. Guthrie, and J. J. Perry. New York: Elsevier, 1980.

Rand, G. M. "Detection: Bioassay." In *Introduction to Environmental Toxicology*, edited by F. E. Guthrie, and J. J. Perry. New York: Elsevier, 1980.

Detritivores

Detritivores are organisms within an **ecosystem** that feed on dead and decaying plant and animal material and waste (called **detritus**); detritivores represent more than half of the living **biomass**. Protozoa, polychaetes, nematodes, Fiddler crabs, and filter-feeders are a few examples of detritivores that live in the salt marsh ecosystem. (Fiddler crabs, for instance, scoop up grains of sand and consume the small particles of decaying organic material between the grains.) While **microbes** would eventually decompose most material, detritivores speed up the process by comminuting, and partly digesting the dead, organic material. This allows the microbes to get at such material more readily. The continuing **decomposition** process is vital to the existence of an ecosystem—essential in the maintenance of **nutrient** cycles, as well as the natural renewal of **soil** fertility.

Detritus

Detritus is dead and decaying matter including the wastes of organisms. It is composed of organic material resulting from the fragmentation and **decomposition** of plants and animals after they die. Detritus is decomposed by bacterial activity, which can help cycle nutrients back into the food chain. In aquatic environments, detritus may make up a substantial percentage of the **particulate** organic **carbon** (POC) that is suspended in the water column. Animals that consume detritus are called "detritivores". Although detritus is available in large quantities in most ecosystems, it is usually not a very high quality food, and may be lacking in essential **nitrogen** or carbon compounds. **Detritivores** generally must expend a larger amount of energy to assimilate carbon and nutrients from detritus than from sources of food based on living plant or animal material. Some detritivores harbor beneficial bacteria or **fungi** in their guts to aid in the digestion of compounds that are difficult to degrade.

[*Marie H. Bundy*]

Development, sustainable
see Sustainable development

Dew point

An expression of humidity defined as the temperature to which air must be cooled to cause condensation of its water vapor content (dew formation) without the adding or subtracting of water vapor or changing its pressure. At this point the air is saturated and relative humidity becomes 100%. When the dew point temperature is below freezing it is also referred to as the frost point. The dewpoint is a conservative expression of humidity because it changes very little across a wide range of temperature and pressure, unlike relative humidity which changes with both. Dew points, however, are affected by water vapor content in the air. High dew points indicate large amounts of water vapor in the air and low dew points indicate small amounts. Scientists measure dew points in several ways: with a dew point hygrometer; from known temperature and relative humidity values; or from the difference between dry and wet bulb temperatures using tables. They use this measurement to predict fog, frost, dew, and overnight minimum temperature.

Diapers
see Disposable diapers

Diazinon

An **organophosphate pesticide**. Malathion and parathion are other well-known organophosphate pesticides. The organophosphates inhibit the action of the **enzyme** cholinesterase, a critical component of the chain by which messages are passed from one nerve cell to the next. They are highly effective against a wide range of pests. However, since they tend to affect the human nervous system in the same way they affect insects, they tend to be dangerous to humans, **wildlife**, and the **environment**. One of the most widely-used lawn **chemicals**, diazinon has been implicated in killing songbirds, pet dogs and cats, and causing near fatal poisonings in humans. The EPA banned its use on **golf courses** and sod farms after numerous reports of its killing ducks, geese, and other birds. *See also* Cholinesterase inhibitor

Dichlorodiphenyl-trichloroethane

Dichlorodiphenyl-trichloroethane (DDT) can be degraded to several stable breakdown products, such as DDE and DDD. Usually DDT refers to the sum of all the DDT-related components.

DDT was first developed for use as an insecticide in Switzerland in 1939, and it was first used on a large scale on the Allied troops in World War II. Commercial, non-military use began in the United States in 1945. The discovery of its insecticidal properties was considered to be one of the great moments in public health disease control, as it was found to be effective on the carriers of many leading causes of death throughout the world including **malaria**, dysentery, dengue fever, yellow fever, filariasis, encephalitis, typhus, **cholera**, and scabies. It could be sprayed to control mosquitoes and flies or applied directly in powder form to control lice and ticks. It was considered the "atomic bomb" of pesticides, as it benefited public health by direct control of more than 50 diseases and enhanced the world's food supply by agricultural **pest** control. It had eliminated mosquito transmission of malaria in the United States by 1953. In the first 10 years of its use, it was estimated to have saved five million lives and prevented 100 million illnesses worldwide.

Use of DDT declined in the mid-1960s due to increased **resistance** of different **species** of mosquitos and flies and other pests, and to increasing concerns regarding the potential harm to ecosystems and human health. Although the potential hazard from dermal **absorption** is small when the compound is in dry or powdered form, if the compound is in oil or an organic solvent it is readily absorbed through the skin and represents a considerable hazard.

Primarily DDT affects the central nervous system, causing dizziness, hyperexcitability, nausea, headaches, tremors, and seizures from acute exposure. Death can result from respiratory failure. It also is a liver toxin, activating microsomal **enzyme** systems and causing liver tumors. DDE is of similar toxicity. It became the focus of much public debate in the United States after the publication of *Silent Spring* by Rachel Carson, who effectively dramatized the harm to birds, **wildlife**, and possibly humans from the widespread use of DDT. Extensive spraying programs to eradicate Dutch elm disease and the **gypsy moth** (*Porthetria dispar*) also caused widespread songbird **mortality**. Its accumulation in the **food chain/web** also led to chronic exposures to certain wildlife populations. Fish-eating birds were subject to reproductive failure, due to egg shell thinning and sterility. DDT is resistant to breakdown and is transported long distances, making it ubiquitous in the world **environment** today. It was banned in the United States in 1972 after extensive government hearings, but is still in use in other parts of the world, mostly in developing countries, where it continues to be useful in the control of carrier-borne diseases. *See also* Acute effects; Chronic effects; Detoxification

[*Deborah L. Swackhammer*]

RESOURCES

BOOKS

Toxicology of Pesticides. Copenhagen, Denmark: World Health Organization, 1982.

OTHER

DDT and Its Derivatives. Environmental Health Criteria 9. Geneva, Switzerland: World Health Organization, 1979.

Dieback

Dieback refers to a rapid decrease in numbers experienced by a population of organisms that has temporarily exceeded, or *overshot*, its **carrying capacity**. Organisms at low trophic levels such as rodents or deer, as well as weed **species** of plants, experience dieback most often. Without pressure from predators or other limiting factors, such "opportunistic" species reproduce rapidly, consume food sources to depletion, and then experience a population crash due chiefly to starvation (though reproductive failure can also play a part in dieback). The presence of predators—for instance, foxes in a meadow inhabited by small rodents—often results in a stabilizing effect on population numbers.

Die-off

Die-offs are the massive, sometimes unexplained but always unexpected, disappearances of plants and animals. The most well-known is probably the die-off of dinosaurs, but die-offs continue today in many regions of the world, including the United States.

Frogs and their kin are mysteriously vanishing in some areas, and scientists suspect that human alteration of ecosystems is partly responsible. On five continents, scientists in 19 countries have reported massive die-offs among amphibians. Frogs are good indicators of environmental change because of the permeability of their skin. They are extremely susceptible to toxic substances on land and in water. Decreasing rainfall in some areas may be a factor in the die-offs, as is **habitat** loss due to **wetlands drainage**. Other scientists are investigating whether increased **ultraviolet radiation** due to **ozone** depletion is killing toads in the Cascade Mountain Range.

Other multiple factors—such as **acid rain**, heavy metal and **pesticide** contamination of ponds and other surface water, and human predation in some areas—may be causing the amphibians to die off. In France, for example, thousands of tons of frog legs imported from Bangladesh and Indonesia are consumed each year.

Scientists also have recorded widespread declines in the numbers of migratory songbirds in the Western Hemisphere and of wild mushrooms in Europe. The decline in these so-called indicator organisms is a sign of declining health of the overall ecosystems.

The decline of songbirds is attributed to the loss or fragmentation of habitat, particularly forests, throughout the songbirds' range from Canada to Central America and northern South America. **Fungi** populations in Europe are dying off, and scientists think it is more than a problem of overharvesting. The health of forests in Europe is closely linked to the fungi populations, which point to the ecological decline of the forests. Some scientists believe that **air pollution** is also playing a role in their decline.

A massive die-off of millions of starfish in the White Sea has been attributed to radioactive military waste in the former Soviet Union. French scientists have recorded growing numbers of dead **dolphins** along the Riviera, probably due to **environmental stress** which left the animals too weak to fight off a **virus**.

[*Linda Rehkopf*]

RESOURCES

BOOKS

Gore, Al. *Earth in the Balance.* New York: Houghton Mifflin, 1992.

PERIODICALS

"Earth Almanac: Why Are Frogs and Toads Knee-Deep in Trouble?" *National Geographic* 183 (April 1993).

Digester

see **Wastewater**

Annie Dillard (1945 –)

American writer

Often compared to the American naturalist **Henry David Thoreau**, Dillard—a novelist, memoir writer, essayist, poet, and author of books about the natural world—is best known for her acute observation of the land, the seasons, the changing weather, and the **wildlife** within her intensely seen **environment**. Though born in Pittsburgh, Pennsylvania on April 30, 1945, Dillard's vision of nature's violence and beauty was most fully developed living in Virginia, where she received her B.A., 1967, and M.A., 1968, from Hollins College. She also lived in the Pacific Northwest from 1975–1979 as scholar-in-residence at the University of Western Washington, in Bellingham, and is adjunct professor of English and writer in residence at Wesleyan University, in Middletown, Connecticut, where she lives with her husband, Bob Richardson, and her daughter, Rosie. Since 1973, she has also been a columnist for *The Living Wilderness*, the magazine of the **Wilderness Society**, the leading organization advocating expansion of the nation's wilderness.

In 1975, Dillard won the Pulitzer Prize for general nonfiction for her first book of prose, *Pilgrim at Tinker Creek* (1974), subtitled "A mystical excursion into the natural world," in which she presented—quoting Henry David Thoreau—"a meteorological journal of the mind" based on her life in the Roanoke Valley, Virginia, where she had lived since 1965. Her vision of "power and beauty, grace tangled in a rapture of violence," of a world in which "carnivorous animals devour their prey alive," is also an intense celebration of the things seen as she wanders the Blue Ridge mountainside, the Roanoke creek banks, observing muskrat, deer, red-winged blackbirds, and the multitude of "free surprises" her environment unexpectedly, and fleetingly, displays. *Seeing* acquires a mystical primacy in Dillard's work.

The urgency of seeing is also conveyed in Dillard's only book of essays, *Teaching A Stone To Talk* (1982), in which she writes: "At a certain point you say to the woods, to the sea, to the mountains, the world, Now I am ready. Now I will stop and be wholly attentive." Dillard suggests that, for the natural phenomena that we do not use or eat, our only task is to witness. But in witnessing, she sees cruelty and suffering, making her question at a religious level what

mystery lies at the heart of the created universe, of which she writes: "The world has signed a pact with the devil...The terms are clear: if you want to live, you have to die."

Unlike some natural historians or writers of environmental books, Dillard is not associated with a specific program for curbing the destructiveness of human civilization. Rather, what has been described as her loving attentiveness to the phenomenal world—"nature seen so clear and hard that the eyes tear," as one reviewer commented—allies her with the broader movement of writers whose works teach some other relationship to **nature** than exploitation. In *Holy the Firm* (1977), her 76-page journal of several days spent in Northern Puget Sound, Dillard records such events as the burning of a seven-year-old girl in a plane crash and a moth's immolation in a candle flame to rehearse her theme of life's harshness, but at the same time, to note, "A hundred times through the fields and along the deep roads I've cried Holy." In the end, Dillard is a sojourner, a pilgrim, wandering the world, ecstatically attentive to nature's bloodiness and its beauty.

The popularity of Dillard's writing during the late 1980s and 1990s can be judged by the frequency with which her work was reprinted during these decades. As well as excerpts included in multi-author collections, the four-volume *Annie Dillard Library* appeared in 1989, followed by *Three by Annie Dillard* (1990), and *The Annie Dillard Reader* (1994). During these years, she also served as the co-editor of two volumes of prose—*The Best American Essays* (1988), with Robert Atwan, and *Modern American Memoirs* (1995), with Cort Conley—and crafted *Mornings Like This: Found Poems* (1995), a collection of excerpts from other writers' prose, which she reformatted into verse.

Though a minor work, *Mornings Like This* could be said to encapsulate all of the qualities that have made Dillard's work consistently popular among readers: clever and playful, it displays her wide learning and eclectic tastes, her interest in the intersection of nature and science with history and art, and her desire to create beauty and unity out of the lost and neglected fragments of human experience.

[*David Clarke*]

RESOURCES

BOOKS

Dillard, Annie. *An American Childhood.* New York: Harper and Row, 1987.

———. *Encounters With Chinese Writers.* Connecticut: Wesleyan University Press, 1984.

———. *Holy the Firm.* New York: Harper & Row Publishers, 1982.

———. *The Living.* New York: Harper Collins, 1992.

———. *Pilgrim at Tinker Creek.* New York: Bantam Books, Inc., 1975.

———. *Teaching A Stone To Talk: Expeditions and Encounters.* New York: Harper & Row Publishers, 1982.

Johnson, Sandra Humble. *The Space Between: Literary Epiphany in the Work of Annie Dillard.* Kent, OH: Kent State University Press, 1992.

Parrish, Nancy C. *Lee Smith, Annie Dillard, and the Hollins Group: A Genesis of Writers.* Baton Rouge: Louisiana State University Press, 1998.

Smith, Linda L. *Annie Dillard.* Boston: Twayne, 1991.

———. *The Annie Dillard Reader.* New York: HarperCollins, 1994.

Dioxin

Dioxin is the chemical byproduct of certain manufacturing processes or products. It develops during the manufacture of two herbicides known as **2,4,5-T** and **2,4-D**, which are the two components found in **Agent Orange**. Dioxin is also manufactured in several other chemical processes, including the chlorinated bleaching of pulp and paper. The issue of dioxin's toxicity is one of the most hotly debated in the scientific community, involving the federal government, industry, the press, and the general public.

While dioxin commonly refers to a particular compound known as tetrachlorodibenzo-p-dioxin, or TCDD, there are actually 75 different dioxins. TCDD has been studied in some manner since the 1940s, and the most recent information indicates that it is capable of interfering with a number of physiological systems. Its toxicity has been compared to **plutonium** and it has proven lethal to a variety of research animals, including guinea pigs, monkeys, rats, and rabbits.

TCDD is also the chemical that some scientists at the **Environmental Protection Agency** (EPA) and a broad spectrum of researchers have called the most carcinogenic ever studied. During the late 1980s and early 1990s, it was linked to an increased risk of rare forms of **cancer** in humans—especially soft tissue sarcoma and non-Hodgkins lymphoma—at very high doses. Some have called dioxin "the most toxic chemical known to man." According to **Barry Commoner**, Director of the Center for the Biology of Natural Systems at Washington University, the chemical is "so potent a killer that just three ounces of it placed in New York City's water supply could wipe out the populace."

Exposure can come from a number of sources. Dioxin can be air-borne and inhaled in drifting incinerator ash, aerial spraying, or the hand spray application of weed killers. Dioxin can be absorbed through the skin as people walk through a recently sprayed area such as a backyard or a golf course. Water **runoff** and **leaching** from agricultural lands treated with pesticides can pollute lakes, rivers, and underground aquifers. Thus, dioxin can be ingested in contaminated water, in fish, and in beef that has grazed on sprayed lands. Residues in plants consumed by animals and humans add to the contaminated **food chain/web**. Research has shown that nursing children now receive trace amounts of dioxin in their mother's milk.

Because it bioaccumulates in the **environment**, TCDD continues to be found in the **soil** and waterways in microscopic quantities over 25 years after its first application. Dioxin is part of a growing family of **chemicals** known as organochlorines—a class of chemicals in which **chlorine** is bonded with **carbon**. These chlorinated substances are created to make a number of products such as **polyvinyl chloride**, solvents, and refrigerants as well as pesticides. Hundreds or thousands of organochlorines are produced as by-products when chlorine is used in the bleaching of pulp and paper or the disinfection of **wastewater** and when chlorinated chemicals are manufactured or incinerated. The by-products of these processes are toxic, persistent, and hormonally active. TCDD is also part of current manufacturing processes, such as the manufacture of the wood preservative, **pentachlorophenol**.

If the exposure to dioxin is intense, there can be an immediate response. Tears and watery nasal **discharge** have been reported, as have intense weakness, giddiness, vomiting, diarrhea, headaches, burning of the skin, and rapid heartbeat. Usually, a weakness persists and a severe skin eruption known as chloracne develops after a period of time. The body excretes very little dioxin, and the chemical can accumulate in the body fat after exposure. Minute quantities may be found in the body years after modest exposure. Since TCDD's **half-life** has been estimated at as much as 10–12 years in the soil, it is possible that some TCDD—suggested to be as much as seven **parts per trillion** (ppt)—is harbored in the bodies of most Americans.

The development of medical problems may appear shortly after exposure, or they may appear 10, 12, or 20 years later. If the exposure is large, the symptoms develop more quickly, but there is a greater **latency** period for smaller exposures. This fact explains why humans exposed to TCDD may appear healthy for years before finally showing what many consider to be typical dioxin-exposure symptoms, such as cancer or immune system dysfunction. There is also a relationship between toxicology and individual susceptibility. Certain people are more susceptible to the effects of dioxin exposure than others. Once a person has become susceptible to the chemical, he or she tends to develop cross reactions to other materials that would not normally trigger any response.

Government publications and research funded by the chemical industry have questioned the relationship between dioxin exposure and many of these symptoms. But a growing number of private physicians treating people exposed to dioxins have become increasingly certain about patterns or clusters of symptoms. They have reported a higher incidence of cancer at sites of industrial accidents, including increases in rates of stomach cancer, lung cancer, soft-tissue sarcomas, and malignant lymphomas. Some reports have indicated that soft-tissue sarcomas in dioxin-exposed workers have increased by a factor of 40, and there have also been indications of psychological and personality changes and an excess of coronary disease.

Many theories about the medical effects of dioxin exposure are based on the case histories of the thousands of American military personnel exposed to Agent Orange during the Vietnam War. Agent Orange, a chemical defoliant, was used despite the fact that certain chemical companies and select members of the military knew about its toxic properties. Thousands of American ground troops were directly sprayed with the chemical. Those in the spraying planes inhaled the chemical directly when some of the herbicides were blown back by the wind into the open doors of their planes. Others were exposed to accidental dumpings from the sky, when planes in trouble had to evacuate their loads during emergency procedures.

Despite what many consider to be the obvious dangers of dioxin, industries continue to produce residues and market products contaminated with the chemical. White bleached paper goods contain quantities of TCDD because no agency has required the paper industry to change its bleaching process. Women use dioxin-tainted, bleached tampons, and infants wear bleached, dioxin-tainted paper diapers. Some scientists have estimated that every person in the United States carries a body burden of dioxin that may already be unacceptable.

Many believe that the EPA has done less to regulate dioxin than it has done for almost any other **toxic substance**. Environmentalists and other activists have argued that any other chemical creating equivalent clusters of problems within specific groups of similarly exposed victims would be considered an epidemic. Industry experts have often downplayed the problems of dioxin. A spokesman for Dow Chemical has stated that "outside of chloracne, no medical evidence exists to link up dioxin exposure to any medical problems." The federal government and federal agencies have also been accused of protecting their own interests. During congressional hearings in 1989 and 1990, the Centers for Disease Control was found to falsify **epidemiology** studies on Vietnam veterans.

In April 1991, the EPA initiated a series of studies intended to revise their estimate of dioxin's toxicity. The agency believed there was new scientific evidence worth considering. Several industries, particularly the paper industry, had also pressured the agency to initiate the studies, in the hope that public fears about dioxin toxicity could be allayed. But the first draft of the revised studies, issued in the summer of 1992, indicated more rather than fewer problems with dioxin. It appears to be the most damaging to animals exposed while still in the uterus. It also seems to affect behavior and learning ability, which suggests that it

may be a **neurotoxin**. These studies have also noted the possibility of extensive effects on the immune system.

Other studies have established that dioxin functions like a steroid hormone. Steroid hormones are powerful chemicals that enter cells, bind to a receptor or protein, form a complex that then attaches to the cell's chromosomes, turning on and off chemical switches that may then affect distant parts of the body. It is not unusual for very small amounts of a steroid hormone to have major effects on the body. Newer studies conducted on **wildlife** around the **Great Lakes** have shown that dioxin has the capacity to feminize male chicks and rats and masculinize female chicks and rats. In male animals, testicle size is reduced as is sperm count.

It is likely that dioxin will remain a subject of considerable controversy both in the public realm and in the scientific community for some time to come. However, even those scientists who question dioxin's long-term toxic effect on humans, agree that the chemical is highly toxic to experimental animals. Dioxin researcher Nancy I. Kerkvliet of Oregon State University in Corvallis characterizes the situation in these terms, "The fact that you can't clearly show the effects in humans in no way lessens the fact that dioxin is an extremely potent chemical in animals—potent in terms of immunotoxicity, potent in terms of promoting cancer." *See also* Bioaccumulation; Hazardous waste; Kepone; Organochloride; Pesticide residue; Pulp and paper mills; Seveso, Italy; Times Beach, Missouri

[*Liane Clorfene Casten*]

RESOURCES

BOOKS

Husar, R. B. *Biological Basis for Risk Assessment of Dioxins and Related Compounds.* Cold Spring Harbor, NY: Cold Spring Harbor Laboratory Press, 1991.

PERIODICALS

"German Dioxin Study Indicates Increased Risk." *BioScience* 42 (February 1992): 151.

Gough, M. "Agent Orange: Exposure and Policy." *American Journal of Public Health* 81 (March 1991): 289-90.

Schmidt, K. F. "Dioxin's Other Face: Portrait of an 'Environmental Hormone'." *Science News* 141 (11 January 1992): 24-7.

Tschirley, F. "Dioxin." *Scientific American* 254 (February 1986): 29-35.

Zumwalt, Admiral Elmo R. Jr. USN (Ret.) "Report to the Secretary of Veterans Affairs, The Hon. Edward J. Derwinski. From the Special Assistant: Agent Orange Issues." First Report, May 5, 1990.

Discharge

A term generally used to describe the release of a gas, liquid, or solid to a treatment facility or the **environment**. For example, **wastewater** may be discharged to a sewer or into a stream, and gas may be discharged into the **atmosphere**.

Disposable diapers

Disposable diapers were introduced by Procter & Gamble in 1961. First used as an occasional convenient substitute for cloth diapers, their popularity has since exploded. By 1990 they were the primary diapering method for 85% of American parents. As a result, 2.7 million tons of disposable diapers are discarded every year, a point decried by environmentalists.

Proponents of reusables argue that this accounts for only two to three% of America's **solid waste**. Although detailed studies have examined the influence of both kinds of diapers on such variables as water consumption, **water pollution**, energy consumption, **air pollution**, and waste generation, there are no indisputable conclusions about which choice is better for the **environment**. Each study was based on different assumptions and came to different conclusions. Most were commissioned by either the disposable-diaper or reusable-diaper industry, and each side put their respective diapers slightly ahead of the other's.

Disposable diapers and their packaging create more solid waste than reusables, and because they are used only once, consume more raw materials—petrochemicals and wood pulp—in their manufacture. And although disposable diapers should be emptied into the toilet before the diapers are thrown away, many people skip this step, which puts feces (that may be contaminated with pathogens) into landfills and incinerators. There is no indication, however, that this practice has resulted in any increase in health problems. But cloth diapers affect the environment as well. They are made of cotton, which is watered with **irrigation** systems and treated with synthetic fertilizers and pesticides. They are laundered and dried up to 78 (commercial) or 180 (home) times, consuming more water and energy than disposables. In fact, home laundering is less energy efficient than commercial because it is done on a smaller scale. Diaper services make deliveries in trucks, which expends another measure of energy and generates more **pollution**. Human waste from cotton diapers is treated in sewer systems. Some disposable diapers are advertised as **biodegradable** and claim to pose less of a solid-waste problem than regular disposables. Their waterproof cover contains a cornstarch derivative that decomposes into water and **carbon dioxide** when exposed to water and air. Unfortunately, modern landfills are airtight and little, if any, degradation occurs. Biodegradable diapers,

therefore, are not significantly different from other disposables.

[*Teresa C. Donkin*]

RESOURCES
PERIODICALS

Poore, P. "Disposable Diapers Are OK." *Garbage* 4 (October-November 1992): 26-8+.
Raloff, J. "Reassessing Costs of Keeping Baby Dry [Cloth vs. Disposable]." *Science News* 138 (1 December 1990): 347.
Rathje, W., and C. Murphy. "Cotton vs. Disposables: What's the Damage." *Garbage* 4 (October-November 1992): 29-30.

OTHER

Lehrburger, C., J. Mullen, and C. V. Jones. *Diapers: Environmental Impacts and Lifecycle Analysis (Summary)*. Report to the National Association of Diaper Services, Philadelphia, PA. January 1991.

Dissolved oxygen

Dissolved oxygen (DO) refers to the amount of oxygen dissolved in water and is particularly important in **limnology** (aquatic **ecology**). Oxygen comprises approximately 21% of the total gas in the **atmosphere**; however, it is much less available in water. The amount of oxygen water can hold depends upon temperature (more oxygen can be dissolved in colder water), pressure (more oxygen can be dissolved in water at greater pressure), and **salinity** (more oxygen can be dissolved in water of lower salinity). Many lakes and ponds have anoxic (oxygen deficient) bottom layers in the summer because of **decomposition** processes depleting the oxygen. The amount of dissolved oxygen often determines the number and types of organisms living in that body of water. For example, fish like trout are sensitive to low DO levels (less than eight **parts per million**) and cannot survive in warm, slow-moving streams or rivers. Decay of organic material in water caused by either chemical processes or microbial action on untreated sewage or dead vegetation can severely reduce dissolved oxygen concentration. This is the most common cause of **fish kills**, especially in summer months when warm water holds less oxygen anyway.

Dissolved solids

Dissolved solids are minerals in solution, typically measured in **parts per million** (ppm) using an electrical conductance meter calibrated to oven-dried samples. In humid regions, dissolved solids are often the dominant form of **sediment** transport. Solution features such as caverns and **sinkholes** are common in limestone regions.

Water that contains excessive amounts of dissolved solids is unfit for drinking. Drinking water standards typically allow a maximum of 250 ppm, the threshold for tasting sodium chloride; by comparison, ocean water ranges from 33,000–37,000 ppm. **Phosphates** and **nitrates** in solution are the major cause of eutrophication (**nutrient** enrichment resulting in excessive growth of algae). Dissolved solids **buffer acid** precipitation; lakes with low levels are especially vulnerable. High levels occur in **runoff** from newly-disturbed landscapes, such as strip mines and road construction.

Diversity

see **Biodiversity**

DNA

see **Deoxyribose nucleic acid**

Dodo

One of the best known extinct **species**, the dodo (*Raphus cucullatus*), a flightless bird native to the Indian Ocean island of Mauritius, disappeared around 1680. A member of the dove or pigeon family, and about the size of a large turkey, the dodo was a grayish white bird with a huge black-and-red beak, short legs, and small wings. The dodo did not have natural enemies until humans discovered the island in the early sixteenth century.

The dodo became extinct due to **hunting** by European sailors who collected the birds for food and to predation of eggs and chicks by introduced dogs, cats, pigs, monkeys, and rats. The Portuguese are credited with discovering Mauritius, where they found a tropical paradise with a unique collection of strange and colorful birds unafraid of humans: **parrots and parakeets**, pink and blue pigeons, owls, swallows, thrushes, hawks, sparrows, crows, and dodos. Unwary of predators, the birds would walk right up to human visitors, making themselves easy prey for sailors hungry for food and sport.

The Dutch followed the Portuguese and made the island a Dutch possession in 1598 after which Mauritius became a regular stopover for ships traversing the Indian Ocean. The dodos were subjected to regular slaughter by sailors, but the species managed to breed and survive on the remote areas of the island.

When the island became a Dutch colony in 1644, the colonists engaged in a seemingly conscious attempt to eradicate the birds, despite the fact that they were not pests or obstructive to human living. But they were easy to kill. The few dodos in inaccessible areas that could not be found by the colonists were eliminated by the animals introduced

The dodo from Mauritius became extinct during the seventeenth century. (Illustration by George Bernard, Science Photo Library. Photo Researchers Inc. Reproduced by permission.)

by the settlers. By 1680, the last remnant survivors of the species were "as dead as a dodo."

Interestingly, while the dodo tree (*Calvaria major*) was once common on Mauritius, the tree seemed to stop reproducing after the dodo disappeared, and the only remaining specimens are about 300 years old. Apparently, a symbiotic relationship existed between the birds and the plants. The fruit of this tree was an important food source for the dodo. When the bird ate the fruit, the hard casing of the seed was crushed, allowing it to germinate when expelled by the dodo.

Three other related species of giant, flightless doves were also wiped out on nearby islands. The white dodo (*Victoriornis imperialis*) inhabited Reunion, 100 mi (161 km) southwest of Mauritius, and seems to have survived up to around 1770. The Reunion solitaire (*Ornithoptera solitarius*) was favored by humans for eating and was hunted to **extinction** by about 1700. The "delightfully beautiful" Rodriguez solitaire (*Pezophaps solitarius*), found on the island of Rodriguez 300 mi (483 km) east of Mauritius, was also widely hunted for food and disappeared by about 1780.

[*Lewis G. Regenstein*]

A playful bottlenosed dolphin. (Photograph by Stephen Frink. Corbis-Bettmann. Reproduced by permission.)

RESOURCES
BOOKS

Day, David. *The Doomsday Book of Animals*. New York: Viking, 1981.
The Dodo. Philadelphia: Wildlife Preservation Trust International, 1985.

Dolphins

There are 32 **species** of dolphins, members of the cetacean family Delphinidae, that are distributed in all of the oceans of the world. These marine mammals are usually found in relatively shallow waters of coastal zones, but some may be found in open ocean. Dolphins are a relatively modern group; they evolved about 10 million years ago during the late Miocene. The Delphinidae represents the most diverse group, as well as the most abundant, of all cetaceans. Among the delphinids are the bottlenose dolphins (*Tursiops truncatus*), best known for their performances in oceanaria; the spinner dolphin (*Stenella longirostris*), which have had their numbers decline due to tuna fishermen's nets; and the orca or the killer whale (*Orcinus orca*), the largest of the dolphins. Dolphins are distinguished from their close relatives, the porpoises, by the presence of a beak.

Dolphins are intelligent, social creatures, and social structure is variously exhibited in dolphins. Inshore species usually form small herds of two to 12 individuals. Dolphins of more open waters have herds comprised of up to 1,000 or more individuals. Dolphins communicate by means of echolocation, ranging from a series of clicks to ultrasonic sounds, which may also be used to stun its prey. By acting cooperatively, dolphins can locate and herd their food using this ability. Aggregations of dolphins also have a negative aspect, however. Mass strandings of dolphins, a behavior in which whole herds beach themselves and die *en mass*, is a well-known phenomenon but little understood by biologists. Theories for this seemingly suicidal behavior include nematode parasite infections of the inner ears, which upsets their balance, orientation, or echolocation abilities; simple disorientation due to unfamiliar waters; or even perhaps magnetic disturbances.

Because of their tendency to congregate in large herds, particularly in feeding areas, dolphins have become vulnerable to large nets of commercial fishermen. **Gill nets**, laid down to catch oceanic **salmon** and capelin, also catch numerous non-target species, including dolphins and inshore species of porpoises. In the eastern Pacific Ocean, especially during the 1960s and 1970s, dolphins have been trapped and drowned in the purse seines of the tuna fishing fleets. This industry was responsible for the deaths of an average of 113,000 dolphins annually and in 1974 alone, killed over half a million dolphins in their nets. Tuna fishermen have recently adopted special nets and different fishing procedures to protect the dolphins. A panel of netting with a finer mesh, the Medina panel, is part of the net furthest from the fishing vessel. Inflatable power boats herd the tuna as the net is pulled under and around the school of fish. As the net is pulled toward the vessel many dolphins are able to escape by jumping over the floats of the Medina panel, but others are assisted by hand from the inflatable boats or by divers. The finer mesh prevents the dolphins from getting tangled in the net, unlike the large mesh which previously snared the dolphins as they sought escape. Consumer pressure and tuna boycotts were major factors behind this shift in techniques on the part of the tuna fishing industry. To advertise this new method of tuna fishing and to try to regain consumer confidence, the tuna fishing industry has begun labeling their products "dolphin safe." This campaign has been successful in that slumping sales from the boycotts have picked up over the last few years.

[*Eugene C. Beckham*]

RESOURCES
BOOKS

Dolphins, Porpoises and Whales of the World. Gland, Switzerland, IUCN—The World Conservation Union, 1991.
Evans, P. *The Natural History of Whales & Dolphins.* New York: Facts on File, 1987.

Dominance

Dominance is an ecological term that refers to the degree that a particular **species** is prevalent within its community, in terms of its relative size, productivity, or cover. Because of their ability to appropriate space, nutrients, and other resources, dominant species have a relatively strong or controlling influence on the structure and functions of their community, and on the fitness and productivity of other species. Ecological communities are often characterized on the basis of their dominant species.

Species may become dominant within their community if they are particularly competitive, that is, if they are relatively successful under conditions in which the availability of resources constrains **ecological productivity** and community structure. As such, competitive species are relatively efficient at obtaining resources. For example, competitive terrestrial plants are relatively efficient at accessing nutrients,

moisture, and space, and they are capable of regenerating beneath their own shade. These highly competitive species can become naturally dominant within their communities.

Many ecological successions are characterized by relatively species-rich communities in the initial stages of recovery after disturbance, followed by the development of less-diverse communities as the most competitive species exert their dominance. In such cases, disturbance can be an important influence that prevents the most competitive species from dominating their community. Disturbance can play this role in two ways: (1) through relatively extensive, or stand-replacing disturbances that set larger areas back to earlier stages of **succession**, allowing species-rich communities to occupy the site for some time, or (2) by relatively local disturbances that create gaps within late-successional stands, allowing species-rich micro-successions to increase the amount of diversity within the larger community.

Some natural examples of species-poor, late-successional communities that are dominated by only a few species include the following: (1) the rocky intertidal of temperate oceans, where mussels (e.g., *Mytilus edulis*) may occupy virtually all available substrate and thereby exclude most other species, (2) eastern temperate forests of North America, where species such as sugar maple (*Acer saccharum*) and eastern hemlock (*Tsuga canadensis*) are capable of forming pure stands that exclude most other species of trees, and (3) a few types of tropical rain forests, such as some areas of Sumatra and Borneo where ironwood (*Eusideroxylon zwageri*) can dominate stands. In the absence of either stand-replacing disturbances or microdisturbances in these sorts of late-successional ecosystems, extensive areas would become covered by relatively simple communities dominated by these highly competitive species.

Humans often manage ecosystems to favor the dominance of one or several species that are economically favored because they are crops grown for food, fibre, or some other purpose. Some examples include: (1) cornfields, in which dominance by *Zea mays* is achieved by plowing, **herbicide** application, insecticide use, and other practices, in order to ensure that few plants other than maize are present, and that the crop species is not unduly threatened by pests or diseases; (2) forestry plantations, in which dominance by the crop species is ensured by planting, herbicide application, and other practices; and (3) grazing lands for domestic cattle, whose dominance may be ensured by fencing lands to prevent wild herbivores from utilizing the available forage, and by shooting native herbivores and predators of cattle. In all of the above cases, continued management of the **ecosystem** by humans is needed to maintain the dominance of the crop species. Otherwise, natural ecological forces would result in diminished dominance or even exclusion of the crop species.

Ecologists have developed a number of quantitative indices to describe the dominance of species within communities. One of these is known as the *community dominance index* (CDI). It is calculated as the percentage of the total abundance of all species in the community that is contributed by the two most abundant species, as in:

$$CDI = 100 \times (y_1 + y_2) / Y$$

where y_1 is the abundance of the most abundant species, y_2 is that of the second-most abundant species, and Y is the total of the abundances of all species in the community. For these purposes, abundance may be estimated by **biomass**, density, cover, productivity, or another suitable measurement. In general, communities with high values of CDI have relatively little community-level diversity. This relationship is being actively explored by ecologists who are interested in the environmental factors that control the levels of **biodiversity** in natural and managed ecosystems.

[*Bill Freedman Ph.D.*]

RESOURCES

BOOKS

Begon, M., Harper, J.L., and Townsend, C.R. *Ecology. Individuals, Populations and Communities.* 3rd ed. London: Blackwell Sci. Pub., 1996.
Ricklefs, R.E. *Ecology.* New York: W.H. Freeman and Co., 1990.

Dose response

Dose response is the relationship between the effect on living things and a stimulus from a physical, chemical, or biological source. Examples of stimuli include therapeutic drugs, pesticides, pathogens, and radiation. A quantal response occurs when the living thing either responds or does not respond to a stimulus of a given dose. Graded responses proportional to the size of the stimulus are also found in environmental applications. Some types of responses have a significant time effect. The Effective Dose for affecting 50% of a population of test subjects, ED_{50}, and the Lethal Dose for killing 50% of a population of test subjects, **LD_{50}**, are commonly used parameters for reporting the toxicity of an environmental pollutant. *See also* Pollution; Radiation exposure; Toxic substance

Double-crested cormorants

The double-crested cormorant (*Phalacrocorax auritus*) is a large (30–36 in; 73–90 cm), blackish water bird that eats fish and crustaceans. The double-crested cormorant population around the **Great Lakes** (but not in other areas) was once in serious decline. It has now has recovered. In fact, some

fishermen feel that there are too many double-crested cormorants. These fishermen believe that an abundance of double-crested cormorants is responsible for a decline in fish populations in the Great Lakes. The **conservation** of this bird entails weighing the interests of the fishing industry against the interests of a now-abundant bird whose range is expanding.

The double-crested cormorant is a large migratory water bird native to North and Central America. The bird is found along the West Coast as far north as southern Alaska, wintering in Southern California and Baja California. On the East Coast it breeds from Maryland to Newfoundland, wintering along the southern coast to Florida, the Bahamas, and Cuba. Inland, double-crested cormorants breed along the upper Mississippi Valley and in the Great Lakes and winter on the Gulf Coast.

The first double-crested cormorant in the great Lakes region were sighted on the western shore of Lake Superior in 1913. By the 1940s, the population around the Great Lakes had increased to about 1,000 nesting pairs. Subsequently, the cormorant population began to decline. By 1973 a survey found only 100 pairs in the region.

In the early 1970s, cormorants were not the only water bird whose population was declining. Around this time, the United States Congress enacted several laws to protect cormorants and other waterfowl. For example, the **pesticide** DDT was banned in 1972. This chemical had entered lakes and rivers in run-off water and was implicated in reducing the birth rate of fish-eating birds, including the **bald eagle**. Congress also amended the Migratory Bird Treaty Act, first passed in 1917, to make it illegal to harm or kill cormorants and other migrating water fowl.

The double-crested cormorant populations around the Great Lakes began to increase in the 1980s, and the birds expanded their range eastward to **Lake Erie** and Lake Ontario. The return of the double-crested cormorant seemed to some conservationists to signal that the Great Lakes, once seriously polluted, were recovering. However, one reason the cormorant population may have increased around the Great Lakes is that **overfishing** during this time seriously depleted the number of large fish in the lakes. This led to an increase in the number of smaller fish such has smelt (*Osmerus mordax*) and alewife (*Alosa pseudoharengus*). Cormorants eat these small fish. An increase in their food supply may have contributed to an increase in the cormorant population.

Another ecological problem, the **zebra mussel**, may also have helped the cormorant. The zebra mussel is an exotic (non-native) invasive mussel that competes with native mussels for food resources. It is a voracious feeder and can clean lakes of green **plankton**, leaving the water particularly clear. Clear water may have helped the cormorant, which hunts fish by sight, find food more easily.

By the 1990s, some local cormorant populations had grown to unprecedented proportions, and at the same time, sport fish populations had declined. As a result, some Great Lakes fishermen, refusing to consider the role of overfishing in the decline of fish populations, blamed double-crested cormorants for steep drops in populations of fish such as smallmouth bass (*Micropterus dolomieui*), rock bass (*Ambloplites rupestris*), and brown bullheads (*Ameiurus nebulosus*). An adult cormorant weighs about four pounds, and eats about one pound of fish a day. Some areas hosted flocks of thousands of cormorants. A 1991 study of cormorants in Lake Ontario estimated that the birds had consumed about five million pounds of fish. In addition, cormorants tend to eat smaller fish, and if enough fish are eaten before they can reproduce, **future generations** of fish are threatened.

Although fishermen have blamed the cormorant for declining fish populations, conservationists have insisted on scientific studies to determine if these birds are indeed causing a decline in fish populations. A 1998 study of cormorants on Galloo Island in Lake Ontario found that the smallmouth bass, a popular sport fish, made up just 1.5% of the cormorant's diet. Yet because the cormorants ate small bass that had not yet grown to reproductive maturity, the bird was considered linked to the smallmouth's decline.

Another 2000 study of the Beaver Islands area in Lake Michigan concluded that it likely the large cormorant population was a factor in the declining numbers of smallmouth bass and other fish. Yet the biologist who led the study was unable to conclude that cormorants were entirely responsible for the decline in the fish population. However, **wildlife** officials took action to manage this cormorant population.

Cormorants also have made themselves unpopular because they nest in large colonies, where thick layers of their droppings can kill off underlying vegetation and leave the area denuded of all but a few trees. Cormorant colonies have endangered some sensitive woodland habitats and contributed to **soil erosion** by killing shoreline plants.

Because federal law protects double-crested cormorants, people cannot legally harass or kill these birds. In 1999, nine men were convicted of slaughtering 2,000 cormorants on Little Galloo Island in Lake Ontario. These men had illegally tried to reduce the cormorant colony by shooting adult birds. Soon after the incident, however, the New York Department of Environmental Conservation enacted a plan to reduce the Galloo colony from 7,500 to 1,500 birds over five years by spraying vegetable oil on cormorant eggs. The oil-coated eggs do not hatch. This method of thinning the flock was considered less disruptive to other wildlife and more humane than killing adult birds.

The New York program generated controversy, as not all scientists who had studied the birds believed that double-crested cormorants were responsible for the decline of the

fish population, and some conservationists feared similar programs would be enacted against other fish-eating birds. Ironically, success in protecting the double-crested cormorant in the 1970s resulted in controversy two decades later. Even though the population of double-crested cormorants substantially increased, the bird is still under federal protection. Any action to manage the cormorant population must be carefully developed, implemented, and evaluated.

[*Angela Woodward*]

RESOURCES
BOOKS

Johnsgard, Paul A. *Cormorants, Darters, and Pelicans of the World.* Washington: Smithsonian Institution Press, 1993.

PERIODICALS

Farquar III, James F. "Balancing Act: Managing Cormorants in Upstate New York." *New York State Conservation* 56, no. 1 (August 2001): 26.

Kloor, Keith. "Killing All Cormorants?" *Audubon* (July/August 1999): 16.

Sharp, Eric. "Controversy Surrounds Cormorants." *Outdoor Life* (August 2000): 119.

"State Studies Show Cormorants' Toll on Fish." *New York Times* (December 20, 1998): 64.

Wallace, Scott. "Who Killed the Cormorants?" *Sports Afield* 222, no. 3 (September 1999): 72.

Marjory Stoneman Douglas (1890 – 1998)
American environmentalist and writer

A newspaper reporter, writer, and environmentalist renowned for her crusade to preserve Florida's **everglades**. Marjory Stoneman Douglas was part of the committee that first advocated formation of the Everglades **National Park** in 1927, and she has been an active advocate of the area's preservation ever since. Born In 1890 in Minneapolis and raised in Taunton, Massachusetts, Marjory Stoneman Douglas graduated from Wellesley College in 1912. After a brief marriage to Kenneth Douglas, a newspaper editor, she moved south to Florida in 1915 to join her father. She soon began to work as a reporter, columnist, and editor for the *Miami Herald*, founded by her father, Judge Frank Bryant Stoneman. During World War I, she left Miami to become the first female enlistee for the Naval Reserves, then joined the Red Cross, for which she worked and traveled in Europe during and after the war. Returning to Coconut Grove in 1920, Douglas remained there for over 75 years.

Douglas was involved very early in efforts to preserve the everglades from agricultural and residential development, as was her father before her. Her book, *The Everglades: River of Grass* was one of the most important statements publicizing the ecological importance and uniqueness of the area, as well as its plight in the face of **drainage**, filling, and water diversion. Her framing of the region as a "river of grass" effectively instilled in the general public an idea of the interconnectedness of the land, water, plants, and animals of the area, helping to raise public awareness of the urgency of preserving the entire **ecosystem**, not just isolated components of it.

Douglas did not start out as a full-time environmental advocate. She wrote *The River of Grass* because she loved the history and the natural history of the area, and she was a long-time supporter of preserving the ecosystem, but she did not become deeply involved in the movement to save the everglades until the 1970s. Friends of Douglas' in the **National Audubon Society**, confronted in 1969 by proposals to build an airport in the everglades, enlisted her aid and she helped organize the Friends of the Everglades, an organization that continues to defend the Park and related south Florida ecosystems.

Douglas proved to be an eloquent and forceful speaker and writer in the cause to save the everglades from development, and since the 1970s the everglades has become her single cause for celebrity. However she had also written poetry, short stories, histories, natural histories and novels, nearly all based in Florida. Initially her writing was popular at least in part because Florida and its history were little known, and rarely written about, when she began her literary career. Among her other publications are *Road to the Sun* (1951), *Hurricane* (1958), *Florida: the Long Frontier* (1967), and *Nine Florida Stories* (1990). She also wrote an autobiography, co-authored by J. Rothchild, *Marjory Stoneman Douglas: Voice of the River.*

In 1990, Douglas was honored on her one hundredth birthday with book signings, interviews, and banquets, and in 1992, she was back in action. That year, Douglas spoke out against President George Bush's proposal to modify the definition of "wetlands," a move that critics pointed out could open the door to future development. President Bill Clinton in 1993 called to wish her a happy birthday as she turned 103, and a few months later, awarded her the Presidential Medal of Freedom. In 1994, Florida state lawmakers passed the Everglades Forever Act, and also in the 1990s the federal government committed hundreds of millions of dollars to restore and protect the area. In 1996, Florida voters passed an amendment to their state's constitution that makes Everglades polluters, particularly sugar farmers, pay for clean-up costs, and more plans to save the **wetlands** were expected. However, voters did not pass a law to tax sugar at a penny a pound to assist with the effort; sugar producers had successfully argued that the ruling would cost many jobs.

Douglas received a bevy of honors in her lifetime, including Floridian of the Year in 1983 and a number of buildings, schools, and parks named after her. The building in Florida's capitol of Tallahassee that is home to the state Department of Natural Resources also bears her name, as does a special **conservation** award. Douglas died in her sleep at home on May 14, 1998.

On October 7, 2000, Douglas was inducted into the National Women's Hall of Fame in Seneca Falls, New York.

[*Mary Ann Cunningham Ph.D.*]

RESOURCES

BOOKS

Douglas, M. S. *The Everglades: River of Grass.* New York: Reinhart, 1947.

PERIODICALS

Chusmir, J. "The Time and Possibilities of Marjory Stoneman Douglas." *Miami Herald*, April 6, 1975.
Soba, D. "Still Fighting the Good Fight." *Audubon* 93 (1991): 30–39.

OTHER

Newsmakers 1998. Issue 4. Farmington Hills, MI: The Gale Group. 2002.

Drainage

The **hydrologic cycle** encompasses all movements of water molecules through the **atmosphere**, hydrosphere, and **groundwater** zones. One key part is the drainage of surface waters and groundwater back to the ocean. Natural drainage systems are well-adjusted to their **climate**, vegetation, and geology. Hydrologists and fluvial geomorphologists have studied these systems in great detail; human disturbances are readily apparent. Three impacts are considered here: wetland drainage and **irrigation**; dam construction and mining alterations; and urbanization.

Robert E. Horton identified many drainage characteristics during the 1930s and 1940s. Later researchers added other useful quantitative parameters. All of these show the predictable patterns found in natural drainage systems. Best known are stream order and drainage density. Although influenced by the detail of the maps used (commonly 1:24,000 or 1:62,500 topographic maps), stream order provides a simple system for ranking tributaries. The smallest ones are labeled first order, second-order streams have at least two first-order tributaries, third-order streams have at least two second-order tributaries, and so forth. The Mississippi River is ranked as a tenth- or twelfth-order stream, depending on the map scale used. Horton also developed a bifurcation ration between stream orders, which is generally around three; that is, on average there are three second-order streams for each third-order one.

Drainage density is a measure of channel length per unit area. This varies widely depending on the nature of the vegetation cover and **soil**. Badlands and deserts have high drainage densities compared to well-vegetated basins with **permeable** soils.

Stream channels are finely adjusted to their drainage requirements, and attain a state of quasi-equilibrium. Human activities disrupt this equilibrium, forcing channel adjustments. This disruption is so pervasive that essentially all channels within settled areas have been impacted.

The earliest human intrusions are wetland drainage and irrigation of **arid** lands for agriculture. Recent litigation has involved decisions by federal officials to deny irrigation water to farmers in order to maintain needed supplies for endangered aquatic organisms. Only recently has the tendency to drain **wetlands** been reversed. With the comeback of the beaver from near **extinction**, wetland proponents now have an efficient ally.

Many large rivers have been dammed, with direct impact on the downstream riparian **environment**. To counteract this, in 1996, an experimental flood was produced on the **Colorado River** through the Grand Canyon, with the desired goal of improving **habitat** for **endangered species** there. Even farm ponds, through their sheer numbers, are major drainage interrupters. Another strong impact has come from mining, through the creation of huge holes, massive **sedimentation**, or increased **runoff**.

Urbanization has had the greatest local impact on drainage. Channels are altered, permeable areas are covered with impervious materials, and storm drains deliver runoff more rapidly to the overwhelmed drainage network. The effect is to deliver much more water in substantial less time, a sure formula for **flooding**.

Tulsa, Oklahoma, is a case study in urban flooding and costly **remediation**. A computerized system now evaluates rain and stream-gage data in real time, and issues flood warning via sirens throughout the city. Construction is banned within the 100-year flood, and valuable land has been set aside for detention ponds so that runoff does not exceed natural conditions up to the level of a 100-year flood. Though costly, these steps have been essential in order to protect lives and reduce property damage.

Changes in the drainage network are prime indicators of human impact. If we are to work with nature rather than fight her, we must understand and reckon with the impacts of our development on this fragile and finely tuned system.

[*Nathan Meleen*]

FURTHER READING

Cooke, R.U., and J.C. Doornkamp. *Geomorphology in Environmental Management.* London: Oxford University Press, 1974.

Leopold, L.B., M.G. Wolman, and J.P. Miller. *Fluvial Processes in Geomorphology.* San Francisco: W.H. Freeman, 1964.

Strahler, A.N., and A.H. Strahler. *Modern Physical Geography* 2nd ed. New York: John Wiley & Sons, 1983.

Dredging

Dredging is a process to remove **sediment**. Dredging sediment to construct new ports and navigational waterways or maintain existing ones is essential for vessels to be able to enter shallow areas. Maintenance dredging is required because sediment suspended in the water eventually settles out, gradually accumulating on the bottom. If dredging were not done, harbors would eventually fill in and marine **transportation** would be severely limited. Dredging is also used to collect sediment (usually sand and gravel) for construction and other commercial uses. Hundreds of millions of cubic feet of sediment are dredged from marine bottoms annually in the United States and throughout the world.

One of the oldest types of dredging is agitation dredging, which uses a combination of mechanical and hydraulic processes and dates back over 2,000 years. An object is dragged along the bottom with the prevailing current; this suspends the sediment and the current carries the suspended material away from the area. Technology currently used to dredge sediment from a harbor, bay, or other marine bottom consists of hydraulic or mechanical devices. Hydraulic dredging involves suspending the sediment, which mixes with water to form a **slurry**, and pumping it to a **discharge** site. Mechanical dredging is typically used to dredge small amounts of material. It lifts sediment from the bottom by metal clamshells or buckets without adding significant amounts of water, and the dredged material is usually transferred to a barge for disposal at a particular site.

Most of the dredging that occurs in the United States is hydraulic dredging. Hopper dredges are vessels that employ hydraulic dredging, and they are often used in the open ocean or in areas where there is vessel traffic. The ship's hull is filled with dredged material and the ship moves the material to a designated disposal site where it is dumped through doors in the hull. Pipeline dredges use hydraulic dredging to remove sediment in nearshore areas, and the dredged material is discharged through a pipeline leading to a beach or diked area. Approximately 550 million wet metric tons of sediment are dredged from the waters of the United States each year, and an estimated one-third is disposed in the marine **environment**, accounting for the greatest amount of waste material dumped in the ocean. Of the dredged material dumped in the marine environment, 66% is disposed in estuaries. Two dozen marine disposal sites in the United States receive approximately 95% of all of the dredged material disposed at sea.

Dredged material is typically composed of **silt**, clay, and sand, and can sometimes include gravel, boulders, organic matter, as well as chemical compounds such as sulfides, hydrous oxides, and metal and organic contaminants. The grain size of the dredged sediment will determine the conditions under which the sediment will be deposited or resuspended if disposed in the marine environment.

The choice of where the dredged material should be placed depends on whether it is uncontaminated or contaminated by pollutants. If contaminated, the level of pollutants in the dredged material can also play a role in the decision of the type and location of disposal. Because many navigational channels and ports are located in industrialized areas, and because sediments are a sink for many pollutants, dredged material may be contaminated with toxic metals, organohalogens, **petrochemical** by-products, or other pollutants. Dredged material can also contain contaminants from agricultural and urban sources.

Dredged material with very little contaminants can be placed in a variety of locations and beneficially reused for beach restoration, construction aggregate, fill material, cover for sanitary landfills, and **soil** supplementation on agricultural land. The primary concerns over the disposal of uncontaminated dredged material in the marine environment are the physical impacts it can have, such as high turbidity in the water column, changes in grain size, and the smothering of bottom dwelling organisms. The ensuing alterations to the bottom **habitat** can lead to changes in the benthic community. Deposited dredged sediment is usually recolonized by different organisms than were present prior to the disposal of the dredged material. For example, disposal of sediment from a dredging project in Narragansett Bay, Rhode Island changed the bottom **topography** and sediment type, and this change in benthic habitat led to a subsequent decline in the clam and finfish fishery at the site and an increase in the lobster fishery. If the dredged material is similar to the sediment on which it is dumped, the area may be recolonized by the same **species** that were present prior to any dumping.

If dredged material is dumped in an area that has less than 197 ft (60 m) of water, most of the material will rapidly descend to the bottom as a high-density mass. A radial gradation of large-to-fine grained sediment usually occurs from the impact area of the deposition outward. Fine-grained material spreads outward from the disposal site, in some cases up to 328 ft (100 m), in the form of a fluid mud. It can range in thickness up to 3.9 in (1 dm). From one to five% of the sediment remains suspended in the water as a **plume**; this sediment plume is transient in **nature** and eventually dissipates by dispersion and gravitational settling. The long-term fate of dredged material dumped in the marine environment depends on the location of the dumping site, its physical characteristics such as bottom topography and currents, and the nature of the sediment. Deep-ocean dumping of dredged material results in wider dispersal of the sediment in the water column. The deposition of the dredged material becomes more widely distributed over the ocean bottom than in nearshore areas.

Dredging contaminated sediment poses a much more severe problem for disposal. Disposing contaminated dredged material in the marine environment can result in long-term degradation to the **ecosystem**. Sublethal effects, **biomagnification** of pollutants, and genetic disorders of organisms are some examples of possible long-term effects from toxic pollutants in contaminated dredged material entering the food chain. However, attributing effects from placement of contaminated dredged material at a marine site to a specific cause can be very difficult if other sources of contaminants are present.

Dredged material must be tested to determine contamination levels and the best method of disposal. These tests include bulk chemical analysis, the elutriate test, selective chemical **leaching**, and bioassays. Bulk chemical analysis involves measurements of volatile solids, **chemical oxygen demand**, oil and grease, **nitrogen**, **mercury**, lead, and zinc. But this chemical analysis does not necessarily provide an adequate assessment of the potential environmental impact on bottom dwelling organisms from disposal of the dredged material. The elutriate test is designed to measure the potential release of chemical contaminants from suspended sediment caused by dredging and disposal activities. However, the test does not take into account some chemical factors governing sediment-water interactions such as complexation, **sorption**, redox, and acid-base reactions.

Selective chemical leaching divides the total concentration of an element in a sediment into identified phases. This test is better than the bulk chemical analysis for providing information that will predict the impact of contaminants on the environment after the disposal of dredged material. **Bioassay** tests commonly use sensitive aquatic organisms to measure directly the effects of contaminants in dredged material as well as other waste materials. Different concentrations of wastes are measured by determining the waste dilution that results in 50% **mortality** of the test organisms. Permissible concentrations of contaminants can be identified using bioassay tests.

If dredged material is considered contaminated, special management and long-term maintenance are required to isolate it from the rest of the environment. Special management techniques can include capping dredged material disposed in water with an uncontaminated layer of sediment, a technique which is recommended in relatively quiescent, shallow water environments. Other management strategies to dispose contaminated dredged material include the use of upland containment areas and containment islands. The use of submarine burrow pits has also been examined as a possible means to contain contaminated dredged material.

There is more than one law in the United States governing dredging and disposal operations. The General Survey Act of 1824 delegates responsibility to the **Army Corps of Engineers** (ACOE) for the improvement and maintenance of harbors and navigation. The ACOE is required to issue permits for any work in navigable waters, according to the Rivers and Harbors Act of 1899. The Marine Protection, Research, and Sanctuaries Act (MPRSA) of 1972 requires the ACOE to evaluate the transportation and **ocean dumping** of dredged material based on criteria developed by the **Environmental Protection Agency** (EPA), and to issue permits for approved non-federal dredging projects. Designating ocean disposal sites for dredged material is the responsibility of EPA. The discharge of dredged material through a pipeline is controlled by the Federal **Water Pollution Control Act**, as amended by the **Clean Water Act** (1977). This act requires the ACOE to regulate ocean discharges of dredged material and evaluate projects based on criteria developed by the EPA in consultation with the ACOE. Other Federal agencies such as the U. S. **Fish and Wildlife Service** and the National Marine Fisheries Service can provide comments and recommendations on any project, but the EPA has the power to veto the use of proposed disposal sites. *See also* Agricultural pollution; Contaminated soil; Hazardous waste; LD50; Runoff; Sedimentation; Synergism; Toxic substance; Urban runoff

[*Marci L. Bortman Ph.D.*]

FURTHER READING

Bokunwiewicz, H. J. "Submarine Borrow Pits as Containment Sites for Dredged Sediment." *Wastes in the Ocean.* Volume 2, *Dredged Material Disposal in the Ocean,* edited by D. R. Kester, et al. New York: Wiley, 1983.

Engler, R. M. "Managing Dredged Materials." *Oceanus* 33 (1990): 63-9.

Kamlet, K. S. "Dredge-Material Ocean Dumping: Perspectives on Legal and Environmental Impacts." *Wastes in the Ocean.* Volume 2, *Dredged Material Disposal in the Ocean,* edited by D. R. Kester, et al. New York: Wiley, 1983.

Kester, D. R., et al. "The Problem of Dredged-Material Disposal." *Wastes in the Ocean.* Volume 2, *Dredged Material Disposal in the Ocean,* edited by D. R. Kester, et al. New York: Wiley, 1983.

Kester, D. R., et al. "Have the Questions Concerning Dredged-Material Disposal Been Answered?" *Wastes in the Ocean.* Volume 2, *Dredged Material Disposal in the Ocean,* edited by D. R. Kester, et al. New York: Wiley, 1983.

Office of Technology Assessment. *Wastes in Marine Environments.* OTA-O-334. Washington, DC: U.S. Government Printing Office, 1987.

Dreissena polymorpha
see **Zebra mussel**

Drift nets

Drift nets are used in large-scale **commercial fishing** operations. Miles-long in length, nets are suspended from floats at various depths and set adrift in open oceans to capture fish or squid. Drift nets are constructed of a light, plastic monofilament that resists rotting. These nets are generally of a type known as gill nets, because fish usually become entangled in them by their bony gill plates. The fishing industry has found these nets to be cost-effective, but their use has become increasingly controversial. They pose a severe threat to many forms of marine life, and they have long been the object of protests and direct action from a range of environmental groups. National and international policies concerning their use have only recently begun to change.

Drift nets are not selective; there is no way to use them to target a particular species of fish. Drift nets can catch almost everything in their path, and there are few protections for species that were never intended to be caught. Although some nets can be quite efficient in capturing only certain species, the **bycatch** from drift nets can include not only non-commercial fish, but sea turtles, seabirds, seals and sea lions, **sharks**, porpoises, **dolphins**, and large whales. Nets that are set adrift from fishing vessels in the open ocean and never recovered pose an even more severe hazard to the marine environment. Lost nets can drift and kill animals for long periods of time, becoming what environmentalists have called "ghost nets."

Drift nets are favored by fishing industries in many countries because of their economic advantages. The equipment itself is relatively inexpensive; it is also less labor intensive than other alternatives, and it supplies larger yields because of its ability to capture fish over such broad areas. Drift-net fisheries can vary considerably, according to the target species and the type of fishing environment. In coastal areas, short nets can be set and recovered in an hour. The nets do not drift very far in this time and the environmental damage can be limited. But in the open ocean, where the target species may be widely dispersed, nets in excess of 31 mi (50 km) in length may be set and allowed to drift for 24 hours before they are recovered and stripped. The primary targets for drift netting include squid in the northern Pacific, salmon in the northeastern Pacific, tuna in the southern Pacific and eastern Atlantic, and swordfish in the Mediterranean.

Because of their cost-effectiveness, the Food and Agricultural Organization (FAO) of the United Nations actively promoted the use of drift nets during the early 1980s. **Earthwatch**, Earth Island Institute, and other environmental groups instituted drift-net monitoring during this period and founded public education programs to pressure drift-

Drift net. (Photograph by Dan Guravich. Photo Researchers Inc. Reproduced by permission.)

net fishing nations. The Sea Shepherd Conservation Society and other direct action groups have actually intervened with drift-net fishing operations on the high seas. Organizations such as these led international awareness about the dangers of drift nets, and their efforts have affected national and international policy. In December of 1991, the United Nations reversed its earlier endorsement of drift-net fishing and adopted General Assembly Regulations 44-225, 45-197, and 46-215 that totally banned high seas drift nets. The regulations applied to all international waters and went into effect on December 31, 1992. An agreement between the United States and the South and Central American countries of Costa Rica, Ecuador, Mexico, Nicaragua, Panama and Venezuela for a moratorium on drift nets was also signed. In 1998, the European Union banned all drift nets in its jurisdiction, with the exception of the Baltic Sea. The ban took effect in 2002. The regulatory body for the Baltic Sea is the International Baltic Sea Fishery Commission, which sets more liberal restrictions on nets and catch limits than the **European Union** (EU).

According to the United States National Marine Fisheries Service, enforcement of the drift net ban in the north

Pacific has been successful, and the U.S. salmon fishery in these waters is no longer impacted by illegal drift nets. Enforcement of this ban in other waters has so far proven difficult, however, as drift net fishing is done in open oceans, far from national jurisdictions. In reaching enforceable international agreements about drift net fishing, the primary problem has been the large investment some nations have in this technology. Japan had 457 fishing vessels using drift nets in 1990, and Taiwan and Korea approximately 140 vessels each. France, Italy, and other nations own smaller fleets.

Japan and many of these nations are primarily concerned with protecting their investment, despite worldwide protest. The United States and Canada have both expressed concern about ecological integrity and the rate of unintended catch in drift nets, particularly the bycatch of North American salmon. In the 1990s, bilateral negotiations were pursued with Korea and Taiwan to control their drift net fleets. The International North Pacific Fisheries Commission has provided a forum for United States, Japan, and Canada to examine and discuss the economic advantages and environmental costs of drift net fishing. A special committee has analyzed bycatch from drift nets used in the northern Pacific. Three avenues are currently being considered to control ecological damage: the use of subsurface nets, research into the construction of biodegradable nets, and alternative gear types for capturing the same species. The Irish Sea Fisheries Board has tested trawl nets as an alternative to drift nets, and found that they are effective in catching tuna. However, environmental groups have raised the concern that trawl nets also capture significant numbers of marine mammals, and have called for restrictions on their use after the Irish Sea Fisheries Board study found that more than 140 whales and dolphins were killed by trawlers during the two year study.

[*Douglas Smith and Marie H. Bundy*]

FURTHER READING
PERIODICALS

Bowden, C. "At Sea With the Shepherd." *Buzzworm* 3 (March-April 1991): 38-47.
McCloskey, W. and C. Wallace. "Casting Drift Nets with the Squidders." *International Wildlife* 21 (March-April 1991): 40-47.
"Net Losses." *Sierra* 76 (March-April 1991): 48-54.
"U.S. Considers Ratification of Driftnet Fishing Treaty." *U. S. Department of State Dispatch* 2 (26 August 1991): 639-40.

Drinking-water supply

The **Safe Drinking Water Act**, passed in 1974, required the **Environmental Protection Agency** (EPA) to develop guidelines for the treatment and monitoring of public water systems. In 1986, amendments to the act accelerated the regulation of contaminants, banning the future use of lead pipe, and requiring surface water from most sources to be filtered and disinfected. The amendments also have provisions for greater **groundwater** protection. Despite the improvement these regulations represent, only public and private systems that serve a minimum of 25 people at least 60 days a year are covered by them. Millions obtain their drinking water from privately owned **wells** that are not covered under the act.

Drinking water comes from two primary sources: surface water and groundwater. Surface water comes from a river or lake, and groundwater, which is pumped from underground sources, generally needs less treatment. Contaminants can originate either from the water source or from the treatment process.

The most common contaminants found in the public water supply are lead, nitrate, and radon–all of which pose substantial health threats. Studies indicate that substances such as **chlorine** and fluoride which are added to water during the treatment process may also have adverse effects on human health. Over 700 different contaminants have been found in water supplies in the United States, yet the EPA has only established maximum containment levels for 30 of them. Drinking water enforcement has been severely limited at both the state and federal levels. In 1990, 38,000 public water systems committed over 100,000 violations of the act, yet state governments only took legal enforcement action against 1,000, and the federal government took action against only 32. In only 6% of these cases were customers informed of the violations.

Chlorinated water was first used in 1908 as a means of reducing diseases in Chicago stockyards. **Chlorination**, which kills some disease-causing **microbes**, is now used to disinfect approximately 75% of the water supply in the United States. Numerous studies conducted over the past 20 years have found that chlorine reacts with organic products such as farm **runoff** or decaying leaves to form by-products that increase the risk of certain kinds of **cancer**. These by-products of chlorination are associated with cancer of the bladder and of the colon, probably because both store concentrated waste products. Research released by the Medical College of Wisconsin suggests that drinking chlorinated water increases the risk of bladder cancer by 20% and the risk of rectal cancer by 38%. Despite the correlation between chlorination by-products and these types of cancer, many still believe that the benefits of chlorine disinfection outweigh the risks. Some hope this study will prompt those in charge of public water systems to investigate other methods of disinfection, such as the use of **ozone** or exposure to ultraviolet light, both of which are currently used in Europe.

The effectiveness of fluoridated water in reducing dental cavities was first noted in communities with a naturally occurring source of fluoride in their drinking water, and controlled studies of communities where fluoride was added to the water confirmed the results. As of 1989, residents in 70% of all cities with populations greater than 100,000 were drinking fluoridated water. The EPA limit for fluoride in water is 4 **parts per million** (ppm), and most cities add only one ppm to their water. Fluoridated water also has adverse effects, and these may include immune system suppression, tooth discoloration, undesirable bone growth, **enzyme** inhibition, and carcinogenesis.

The EPA has set the acceptable level of lead in drinking water at 15 **parts per billion** (ppb), yet according to tests the agency has done, drinking water in almost 20% of cities in the United States exceeds that limit. The EPA has estimated that 25% of a child's lead intake comes from drinking water, and it cautions that the percentage could be much higher if the water contains high levels of lead. Depending on exposure, lead **poisoning** can cause permanent learning disabilities, behavioral and nervous system disorders, as well as severe brain damage and death. Service pipes made of lead and leaded solder used on **copper** plumbing and brass faucets are the main sources of lead in water. Acidic or soft water increases the danger of lead contamination, because it corrodes the plumbing and leeches out the lead. About 80% of homes have water that is moderately to highly acidic. As of January 1, 1993, EPA regulations require all large public water companies to reduce the corrosiveness of water by adding calcium oxide or other hardening agents.

Chlorination and government standards for drinking **water quality** have virtually eliminated the outbreak of the classic water-borne diseases such as **cholera**, typhoid, and **malaria**. According to *The American Journal of Public Health*, however, recent studies have shown that water that meets current drinking water standards can still contain organisms which cause gastrointestinal (GI) disease. In a 15-month study conducted by the University of Quebec in Montreal, researchers equipped 299 homes with reverse-osmosis water **filters**, which remove bacterial and chemical contaminants. Over 600 families participated in the study, about half with the filters and half without them, and they were asked to keep records of all GI illnesses among household members. During this 15 month period, the households equipped with the water filters had 35% fewer incidents of GI illness and diarrhea. In 1992, *The New England Journal of Medicine* published a study showing that drinking water can harbor the bacterium that causes Legionnaire's disease. Some patients diagnosed with Legionnaire's disease were infected with the same type of *Legionella pneumophila* that was found in samples of their drinking water.

Vegetables, drinking water, and meat preservatives are the main sources of **nitrates and nitrites** in our diet. There is a definite link between nitrate and gastric cancer. Nitrate is converted to nitrite by bacteria in the mouth and stomach, and this is in turn converted into N-nitroso compounds, which have been proven highly carcinogenic in laboratory animals. Bottle-fed infants are at additional risk, because once the nitrate is converted to nitrite in the stomach it combines with fetal hemoglobin and converts to methaemoglobin. When 10% of the hemoglobin has been converted, cyanosis or **blue-baby syndrome** occurs; and when 70% of the hemoglobin is in methaemoglobin form, death occurs. According to a recent EPA report, half of the private wells in the United States contain nitrate.

Radioactivity occurs naturally and it can be present in drinking water. Preliminary studies have linked it to increased rates of **leukemia** and cancers of the bladder, breast, and lungs. The EPA has established 5 picocuries per liter (pCi/L) as the safe limit for radium in drinking water. An estimated 100–1,800 deaths per year are attributed to **radon** in tap water. According to EPA estimates over eight million people have excessively high radon levels in their water supply. Unlike most contaminants found in water, radon does not have to be ingested to pose a health hazard; dish washing,

showering, or just running the faucet can agitate the water and release the radon into the air. According to EPA estimates, there are 10,000–40,000 lung-cancer deaths each year from radon inhalation. Radon is most frequently a problem in New England, North Carolina, and Arizona, and it is most likely to be found in well water and small water systems. Most large treatment facilities disperse radon during the treatment process.

Most water-treatment plants in the United States use chemical coagulation to remove impurities and contaminants. **Aluminum** sulfate is often added to the water, causing some contaminants to coagulate with the aluminum and precipitate out. The majority of the aluminum left in the water is removed by subsequent treatment processes, but a residual amount passes through the system to the consumer. Aluminum in drinking water has been linked with neurotoxicity, specifically Alzheimer's disease.

The organic **chemicals** that are found most frequently in drinking water are pesticides, trichloreothylene, and trihalomines. Pesticides usually make their way into drinking water through **seepage** and runoff in agricultural areas, and in high doses they can damage the liver, the kidney, and the nervous system, as well as increase the risk of various cancers. Trichloroethylene are industrial wastes and the populations at highest risk from this chemical have a water supply located near **hazardous waste** sites. The health risks associated with trichloroethylene are nervous system damage and cancer. Chlorination of water that is contaminated with organic matter is responsible for the formation of **trihalomethanes** in water, and preliminary studies suggest that it may increase cancer rates.

The bottled water industry is not sufficiently regulated, and it does not guarantee water purity. Despite the image portrayed by advertising, studies indicate that bottled water is not any safer in most cases than tap water. Home treatment units carry labels which frequently claim they are EPA-approved, but these are not regulated either. Different types of water filters are capable of removing different contaminants, so most experts recommend that anyone planning to install a treatment system have their water tested first.

Though scientific evidence clearly demonstrates that drinking water can be a health hazard, some of the most effective measures are also the easiest to implement. Studies have found that letting tap water run for several minutes reduces the lead content of the water by up to 70%. Companies that supply drinking water can be monitored by requesting copies of their test results and reporting any violations to the EPA. Lobbying for more stringent regulations is widely considered an effective tool for ensuring safe drinking water. Reduction of **pesticide** use and additional measures implemented for the protection of groundwater and surface water would also greatly reduce many of these health risks.

See also Carcinogen; Communicable diseases; Filtration; Groundwater pollution; Hazardous waste siting; Neurotoxin; Toxic substance; Water pollution; Water quality standards; Water resources; Water treatment

[*Debra Glidden*]

RESOURCES

PERIODICALS

Felsenfield, A., and M. A. Roberts. "A Report of Fluorosis in the United States Secondary to Drinking Well Water." *Journal of the American Medical Association* 265 (23/30 January 1991): 486-8.

Stout, J., et al. "Potable Water as a Cause of Sporadic Cases of Community-Acquired Legionnaires' Disease." *New England Journal of Medicine* 326 (16 January 1992): 151-5.

OTHER

Packham, R. F. "Chemical Aspects of Water Quality and Health." *Annual Symposium of the Institution of Water and Environmental Management.* London, England: IWEM, 1990.

Drinking-water treatment
see **Water treatment**

Drip irrigation

Drip **irrigation** maximizes scarce resources by delivering water exactly where it is needed. The **percolation plume** from drip sources follows the bulb shape of the root zone, wasting little water. This also better controls **fertilizer** inputs by applying nutrients with the water application. Though expensive to install, and best-suited to perennials or high-value crops, this system overcomes or minimizes problems from traditional irrigation methods. Spray irrigation maximizes evaporation; while gravity-fed systems contribute to **seepage** losses, **waterlogging**, and **salinization**. Through water budget analyses and weed control, growers can maintain the exact moisture and nutritional levels needed for optimum plant growth.

Drought

Of all natural disasters, drought is the most subtle. Often, farmers cannot tell there is going to be a drought until it is too late. Unlike flash floods, drought is slow to develop. Unlike earthquakes, with destruction to the exterior **environment**, drought does its damage underground long before dust storms rage across the plains.

Technically, drought is measured by the decrease in the amount of **subsoil** moisture that causes crops to die or yield less (agricultural drought) or by a drop in the water level in surface reservoirs and below ground aquifers, causing **wells** to go dry (hydrological drought). Agricultural plus hydrological drought can lead to sociological drought. In this condition, drought effects food and water supplies to the extent that people have to rely on relief donations or are forced to migrate to another area.

Droughts are worldwide, repetitive, and unpredictable. Scientists believe there is a drought somewhere on the earth at any time. Nor are droughts recent developments; analysis of rock cores, glacial ice cores, and tree rings reveal prehistorical and historical droughts, some of which lasted for several decades. Tree rings in California, for example, record a 40-year-drought 300 years ago.

The direct cause of drought is a continued decrease in optimal rainfall. But what causes clouds not to form over an area, or the winds to carry rain-bearing clouds elsewhere, is complex. **Climate** change will alter the location of increased and reduced rainfall, so that some places that have always been well-watered will experience drought.

Some scientists believe that El Niño-La Niña events in the western Pacific Ocean are main drivers in the cause of droughts around the world. **El Niño**, an eastward flow of warm surface waters, creates a high pressure zone over the equator that results in a change in the high and low pressure zones in other parts of the world. This affects the flow of the jet stream and results in a disturbed rainfall pattern, causing, for example, excessive rain in California and drought in southwestern Africa, among other places. The **La Niña**, which usually follows El Niño, is an upwelling of cold deep waters in the western Pacific Ocean. It causes disturbed pressure zones that result in droughts in the Midwest, among other places.

Drought prediction is still in its infancy. Although scientists know that El Niño-La Niña events cause droughts in specific areas, they cannot yet predict when El Niño will occur. Weather satellites can measure subsoil moisture, a good indicator of incipient drought, but other factors also contribute to drought.

Lack of rain, for example, in the **Sahel**, is exacerbated by man-made environmental problems, such as cutting down trees for fuel and not allowing the **soil** to lie fallow, which conserves soil moisture. **Overgrazing** by animals such as cattle, goats, and sheep also contributes to the denuding of **topsoil**, which blows away in the wind, a condition known as **desertification**. Drought then becomes a cycle that feeds on itself: lack of trees reduces the amount of water vapor given off into the **atmosphere**; lack of topsoil reduces water retention. The result is that local rainfall is reduced, and the rain that does fall runs off and is not absorbed. Lack of rain has been the reason for five years of consecutive drought in Texas. The total amount of profit loss from 1998 until 2002 is estimated at $3.7 billion dollars.

Of all the water on the earth, less than 3% is fresh water. A lot of water is lost in evaporation, especially in **arid** climates, not only during rainfall but when it is stored in surface reservoirs. Rainwater or snowmelt that seeps into below-ground **permeable** rock channels, or aquifers, is pumped into wells in many communities. High-tech pumps have contributed to an increased drain on aquifers; if an **aquifer** is pumped too quickly, it collapses, and the ground above sinks. To increase water bank supplies, some communities recharge their aquifers by pumping water into them when they are low.

The only new water introduced into the **hydrologic cycle** is purified ocean water. **Desalinization** plants are expensive to build and maintain and often require burning **fossil fuels** or wood to run. Future plans include perfecting retrieving **solar energy** and **wind energy**.

Currently, farm **irrigation** uses most of the world's fresh water supply, but as city populations grow, they are expected to become the biggest consumers, and urban **conservation** measures will become imperative. Some communities already recycle **wastewater** for small farms and domestic garden use. Drought-causing industrial pollutants that "freeze" the water supply by rendering it toxic are being reduced and resolved under federal law. Reduced or low-flow shower heads and **toilets** are required in new construction in some states.

Distributing water from more to less abundant supplies by laying pipes and installing pumps within a state or a country requires money and management. If water is fed across state or international boundaries, legal and political negotiations are necessary.

During severe drought, sociologists find that people must either adapt, migrate, or die. Death, however, is usually caused by other factors such as war or poverty, as in the Sahel, where relief food supplies have been hijacked and sold at high prices, or where people in remote villages must walk to the distribution centers.

Some migrations have been permanent, as in the **migration** to California during the Midwestern **Dust Bowl** in the 1930s. Others are temporary, as in the Sahel region, where people migrate in search of food and water, crossing country lines.

Most people adapt in drought by making the most of their resources, such as building reservoirs or desalination plants or laying pipes connecting to more abundant water supplies. Farmers often invest in high-tech irrigation techniques or alter their crops to grow low-water plants, such as garbanzo beans.

Drought has also been the inspiration for inventions. The American West at the turn of the twentieth century gave rise to numerous rainmakers who used mysterious **chemicals** or noisemakers to attract rain. Most inventions

failed or were unreliable, but out of the impetus to make rain grew silver iodide cloud-seeding, which now effects a 10–15% increase in local rainfall in some parts of the world.

[*Stephanie Ocko*]

RESOURCES

BOOKS

Glantz, M. H., ed. *Drought and Hunger in Africa: Denying Famine a Future.* New York: Cambridge University Press, 1987.

Dry alkali injection

A method for removing **sulfur dioxide** from **combustion** stack gas. A **slurry** of finely ground alkaline material such as calcium carbonate is sprayed into the **effluent** gases before it enters the smokestack. The material reacts chemically with sulfur dioxide to produce a non-hazardous solid product, such as calcium sulfate, that can then be collected by **filters** or other mechanical means. The technique is called dry injection because the amount of water in the slurry is adjusted so that all moisture evaporates while the chemical reactions are taking place and a dry precipitate results. The use of dry alkali injection can result in a 90% reduction in the **emission** of sulfur dioxide from a stack. It is more expensive than wet alkali injection or simply adding crushed limestone to the fuel, but it is more effective than these techniques and results in a waste product that is relatively easy to dispose of. *See also* Air pollution control

Dry cask storage

Dry cask storage is a method of storing the **radioactive waste** from nuclear reactors. Dry cask storage refers to the containers that hold the waste and the system of storing the waste above ground in containers.

After World War II, nuclear **power plants** began generating electricity. By the close of the twentieth century, reactors at **nuclear power** plants generated 20% of the electricity in the United States. To produce electricity, **uranium** is used as fuel. Each tiny uranium pellet produces almost as much energy as a ton of **coal**. The pellets are contained in long metal rods, and these rods are placed in a fuel assembly that hold from 50 to 300 rods.

The uranium undergoes fission, a process that heats water and converts it to steam. The steam propels the blades of a turbine. This in turn spins the shaft of a generator where electricity is produced.

A large reactor uses 60 assemblies annually, and assemblies must be replaced after several years. The waste called spent fuel is so radioactive that a person standing near an unshielded rod would die within a second. The spent fuel is also extremely hot. The used fuel rods are taken from the reactor core and stored in a concrete pool that is lined with steel. The rods are stored underneath at least 20 ft (6 m) of water. The spent pool set-up serves as a radiation shield while water cools the rods.

Spent fuel pools were regarded as temporary storage facilities. When operators built the first reactors, there were plans to extract and recycle unused uranium and **plutonium** from the fuel. However, the process would consolidate plutonium into a form that could be used in **nuclear weapons**. As a result of that consequence, the process was banned in 1977. By that time the United States had produced enough plutonium to satisfy its own needs for weapons production.

Five years later, Congress passed the Nuclear Waste Policy Act. Among the issues discussed was where to store spent fuel because power plants were starting reach to storage capacity. Congress amended the act in 1987 to designate a permanent waste disposal site in the **Yucca Mountain** area of Nevada.

Yucca Mountain had been proposed as a facility that would open in 1985. However, concerns about a nuclear explosion and other safety issues led to postponement of the opening. The opening was shifted to 1989, 1998, 2003, and 2010.

With no permanent site available, plant operators began to store spent fuel onsite in dry casks. In 1986, the United States **Nuclear Regulatory Commission** (NRC) licensed the first dry storage installation at the Virginia Electric & Power Company Surry Nuclear Plant in Jamestown, Virginia. The utility installed metal casks that were 16 ft (4.9 m) in height. Each dry cask held from 21 to 33 spent fuel assemblies. When filled, each cask weighed 120 tons. Casks were placed vertically on concrete pads that were 3-ft (1 m) thick. Each pad would hold 28 casks.

By 2001, the NRC had approved various dry casks designs. The container is usually steel. After it is filled, the container is either bolted or welded shut. The metal casks are then put inside larger concrete casks to ensure radiation shielding. Some systems involve placing the steel cask vertically in a concrete vault. In other systems, the container is placed horizontally in the concrete vault.

Discussions about dry cask safety in the twenty-first century have centered on the Yucca Mountain proposal. In May of 2002, the United States House of Representatives voted to approve the plan. That vote was an override of Nevada Governor Kenny Guinn's veto of the plan to send waste to Yucca. Opponents of the plan like the Nuclear Information and Resource Service maintained that casks of waste could not be transported safely by train or truck. In 2002, the facility was expected to cost $58 billion. It was scheduled to open in 2010 and hold a maximum of 77,000 tons of waste.

[*Liz Swain*]

RESOURCES
BOOKS

Murray, Raymond, and Judith Powell, ed. *Understanding Radioactive Waste.* Columbus, OH: Battelle Press, 1997.
Saling, James, and Audeen Fentiman. *Radioactive Waste Management.* Philadelphia, PA: Taylor & Francis, Inc., 2001.

ORGANIZATIONS

Nuclear Information and Resource Service., 1424 16th Street NW, #404, Washington, D.C. USA 20036 (202) 328-0002, Fax: (202)462-2183, Email: nirsnet@nirs.org, <http://www.nirs.org>
United States Nuclear Regulatory Commission., One White Flint North, 11555 Rockville Pike , Rockville, MD USA 20852-2738 (301) 415-7000, Toll Free: (800) 368-5642, Email: opa@nrc.gov, <http://www.nrc.gov>

Dry cleaning

Dry cleaning is a process of cleaning clothes and fabrics with solutions that do not contain water. The practice has been traced back to France where around 1825 turpentine was used in the cleaning this process. According to Albert R. Martin and George P. Fulton in *Dry cleaning, Technology and Theory,* published in 1958, the tradition passed down regarding the origins of dry cleaning states that the process was discovered when "a can of 'camphene,' a fuel for oil lamps, was accidentally spilled on a gown and found to clean it, and this discovery led to the first dry cleaning establishment." Because of this, dry cleaning was referred to as "French cleaning" even into the second half of the twentieth century.

By the late 1800s, naphtha, **gasoline**, **benzene**, and benzol—the most common solvent—were being used for dry cleaning. Fire hazards associated with using gasoline for dry cleaning prompted the United States Department of Commerce in March 1928 to issue a standard for dry cleaning specifying that a dry cleaning solvent derived from **petroleum** must have a minimum flash point (the temperature at which it combusts) of 100°F (38°C). This was known as the Stoddard solvent.

The first chlorinated solvent used in dry cleaning was **carbon** tetrachloride. It continued to be used until the 1950s when its toxicity and corrosiveness were determined to be hazardous. By the 1930s, the use of trichloroethylene became common. In the 1990s the chemical was still being used in industrial cleaning plants and on a limited basis in Europe. This chemical's incompatibility with acetate dyes used in the United States brought about the end of its use in the United States. **Tetrachloroethylene** replaced other dry cleaning solvents almost completely by the 1940s and 1950s.

In 1990 about 53% of worldwide demand for tetrachloroethylene was for dry cleaning, and approximately 75% of all dry cleaners used it. However, in Japan petroleum-based solvents continued in use through the 1990s. By the late 1990s, perchloroethylene (perc or PCE) replaced tetrachloroethylene as the predominant cleaning solvent.

When the United States **Environmental Protection Agency** (EPA) issued national regulations to control air emissions of perc from dry cleaners in September 1993, environmental groups and consumers began to pay closer attention to the possible negative impact this chemical could have on human health. In July 2001, the American Council on Science and Health issued a report concluding that perc was not hazardous to humans at the levels most commonly used in dry cleaning. The report noted that, "Perchloroethylene has been the subject of close government and public scrutiny for more than 20 years. But government agencies in the United States and around the world have not agreed about the potential of environmental exposure to PCE to cause adverse health effects, including **cancer**, in humans."

The findings of this report included the following items:

• Inhalation of high levels of PCE and chemically similar solvents can cause neurological effects such as nausea, headache, and dizziness.

• High inhaled doses have been linked to changes in blood chemistry indicating that the liver and kidneys have been affected.

• These effects have been seen almost exclusively in workers, particularly in the dry-cleaning and chemical industries.

• There have been claims that reproductive difficulties are associated with occupational exposure to PCE.

• The claim that PCE is a **carcinogen** (cancer-causing substance) has received the most public and governmental attention. Concern has been expressed that environmental exposures to PCE in outdoor or indoor air and in drinking water can cause cancer in humans.

• Results of some epidemiological studies of dry cleaning and chemical workers exposed to PCE have been interpreted to suggest a relationship between occupational exposure and various types of cancer. Careful examination of the way in which these studies were conducted reveals serious problems including uncertainties about the amount of PCE to which people were exposed, failure to take into account exposure to other **chemicals** at the same time, and failure to take into account known confounders. Due to these deficiencies, these studies do not support a link between PCE and cancer or other adverse effects in humans.

• The differences between humans and rodents in the **metabolism** and mechanisms of action of PCE make it unlikely that the carcinogenic effects seen in mice and rats

administered high levels of PCE will occur in humans exposed at environmentally relevant levels.

The environmental activist association **Greenpeace** also issued a report in July 2001, entitled, *Out of Fashion Moving Beyond Toxic Cleaners.* This report urged the EPA to classify perc as a probable human carcinogen. The report claimed that up to 266 workers' cancer deaths in New York, Chicago, Detroit, and San Francisco were linked to perc.

As of 2002, the dry cleaning industry estimates that approximately 36,000 dry cleaning establishments exist across the United States, with about 200,000 people employed in the industry. Perc is used in at least 85% of dry cleaning shops as the primary solvent. This means that if perc is found to be a cancer causing chemical, many people, including both workers in and people who live near dry cleaning facilities, may be adversely affected.

[*Jane E. Spear*]

RESOURCES
BOOKS

International Fabricare Institute. *Environmental & Health Issues.* 2002. <http://www.ifi.org/industry>

U.S. Environmental Protection Agency. *New Regulation Controlling Emissions from Dry Cleaners.* May 1994; June 2002. <http://www.epa.gov/ttnsbap1>

Occupational Safety and Health Administration. *Dry Cleaning.* 2002. <http://www.osha.gov/SLTC>

National Institute for Occupational Safety and Health. *Drycleaning.* 2002. <http://www.cdc.gov/niosh>

American Council on Science and Health. *The Scientific Facts about the Dry-Cleaning Chemical Perc.* 2001. <http://www.acsh.org/>

American Council on Science and Health. *Science Group States Dry-Cleaning Chemical Poses No Health Threat to Consumers.* July 2001. <http://www.acsh.org/>

Martin, Albert; and George Fulton. *Drycleaning, Technology and Theory.* New York: Textile Book Publishers, Inc., 1958.

Greenpeace USA. *Dry Cleaning Chemical Linked to Hundreds of Deaths, Warrants EPA Listing as Carcinogen.* July 21, 2001. <http://www.greenpeaceuse.org/media/>

ORGANIZATIONS

U.S. Environmental Protection Agency, 1200 Pennsylvania Avenue NW, Washington, D.C. USA 20460 (202) 260-2090, , <www.epa.gov>

International Fabricare Institute, 12251 Tech Road, Silver Spring, MD USA 20904 (301) 622-1900, Fax: (301) 236-9320, Toll Free: (800) 638-2627, Email: techline@ifi.org, <http://www.ifi.org>

Neighborhood Cleaners Association, 252 West 29th Street, New York, NY USA 10001 (212) 967-3002, Fax: (212) 967-2240, Email: sales@nca-i.com, <http://www.nca-i.com>

Dry deposition

A process that removes airborne materials from the **atmosphere** and deposits them on a surface. Dry deposition includes the settling or falling-out of particles due to the influence of gravity. It also includes the deposition of gasphase compounds and particles too small to be affected by gravity. These materials may be deposited on surfaces due to their solubility with the surface or due to other physical and chemical attractions. Airborne contaminants are removed by both wet deposition, such as rainfall scavenging, and by dry deposition. The sum of wet and dry deposition is called total deposition. Deposition processes are the most important way contaminants such as acidic sulfur compounds are removed from the atmosphere; they are also important because deposition processes transfer contaminants to aquatic and terrestrial ecosystems. Cross-media transfers, such as transfers from air to water, can have adverse environmental impacts, and an example of this is how dry deposition of sulfur and **nitrogen** compounds can acidify poorly buffered lakes. *See also* Acid rain; Nitrogen cycle; Sulfur cycle

Dryland farming

Dryland farming is the practice cultivating crops without **irrigation** (rainfed agriculture). In the United States, the term usually refers to crop production in low-rainfall areas without irrigation, using moisture-conserving techniques such as mulches and fallowing. Non-irrigated farming is practiced in the Great Plains, inter-mountain, and Pacific regions of the country, or areas west of the 23.5 in (600 mm) annual precipitation line, where native vegetation was short **prairie** grass. In some parts of the world dryland farming means all rainfed agriculture.

In the western United States, dryland farming has often resulted in severe or moderate wind **erosion**. Alternating seasons of fallow and planting has left the land susceptible to both wind and water erosion. High demand for a crop sometimes resulted in cultivating lands not suitable for long-time farming, degrading the **soil** measurably.

Conservation tillage, leaving all or most of the previous crop residues on the surface, decreases erosion and conserves water. Methods used are stubble **mulch**, mulch, and ecofallow. In the wetter parts of the Great Plains, fallowing land has given over to annual cropping, or three-year rotations with one year of fallow. *See also* Arable land; Desertification; Erosion; Soil; Tilth

[*William E. Larson*]

RESOURCES
BOOKS

Anderson, J. R. *Risk Analysis in Dryland Farming Systems.* Rome: Food and Agriculture Organization of the United Nations, 1992.

René Jules Dubos (1901 – 1982)
French/American microbiologist, ecologist, and writer

Dubos, a French-born microbiologist, spent most of his career as a researcher and teacher at Rockefeller University in New York state. His pioneering work in microbiology, such as isolating the anti-bacterial substance *gramicidin* from a **soil** organism and showing the feasibility of obtaining germ-fighting drugs from **microbes**, led to the development of antibiotics.

Nevertheless, most people know Dubos as a writer. Dubos's books centered on how humans relate to their surroundings, books informed by what he described as "the main intellectual attitude that has governed all aspects of my professional life...to study things, from microbes to man, not *per se* but in their complex relationships." That pervasive intellectual stance, carried throughout his research and writing, reflected what *Saturday Review* called "one of the best-formed and best-integrated minds in contemporary civilization."

A related theme was Dubos's conviction that "the total environment" played a role in human disease. By total **environment**, he meant "the sum of the facts which are not only physical and social conditions but emotional conditions as well." Though not a medical doctor, he became an expert on disease, especially tuberculosis, and headed Rockefeller's clinical department on that disease for several years.

"Despairing optimism" also pervaded Dubos's human-environment writings, his own title for a column he wrote for *The American Scholar*, beginning in 1970. *Time* magazine even labeled him the "prophet of optimism:" "My life philosophy is based upon a faith in the immense resiliency of nature," he once commented.

Dubos held a lifelong belief that a constantly changing environment meant organisms, including humans, had to adapt constantly to keep up, survive, and prosper. But he worried that humans were too good at adapting, resulting in both his optimism and his despair: "Life in the technologized environment seems to prove that [humans] can become adapted to starless skies, treeless avenues, shapeless buildings, tasteless bread, joyless celebrations, spiritless pleasures—to a life without reverence for the past, love for the present, or poetical anticipations of the future." He stated that "the belief that we can manage the earth may be the ultimate expression of human conceit," but insisted that **nature** is not always right and even that humankind often improves on nature. As Thomas Berry suggested, "Dubos sought to reconcile the existing technological order and the planet's survival through the **resilience** of nature and changes in human consciousness."

[*Gerald L. Young Ph.D.*]

René Dubos. (Corbis-Bettmann. Reproduced by permission.)

RESOURCES
BOOKS

Piel, G., and O. Segerberg, eds. *The World of Rene Dubos: A Collection from His Writings.* New York: Henry Holt, 1990.
Ward, B., and R. Dubos. *Only One Earth: The Care and Maintenance of a Small Planet.* New York: Norton, 1972.

PERIODICALS

Culhane, J. "En Garde, Pessimists! Enter Rene Dubos." *New York Times Magazine* 121 (17 October 1971): 44–68.
Kostelanetz, R. "The Five Careers of Rene Dubos." *Michigan Quarterly Review* 19 (Spring 1980): 194–202.

Ducks Unlimited

Ducks Unlimited (DU) is an international (United States, Canada, Mexico, New Zealand, and **Australia**), membership organization founded during the depression years in the United States by a group of sportsmen interested in waterfowl **conservation**. DU was incorporated in early 1937, and DU (Canada) was established later that spring. The organization was established to preserve and maintain waterfowl populations through **habitat** protection and development, primarily to provide game for sport **hunting**. During the **Dust Bowl** of the 1930s, the founding members of DU recognized that most of the continental waterfowl populations were maintained by breeding habitat in the **wetlands** of Canada's southern prairies in Saskatchewan, Manitoba, and Alberta. The organizers established DU Canada and used their resources to protect the Canadian **prairie** breeding grounds. Cross-border funding has since been a fundamental component of DU's operation, although in recent years funds also have been directed to the northern American prairie states. In 1974 Ducks Unlimited de Mexico was established to restore and maintain wetlands south of the U.S.-Mexican border where many waterfowl spend the winter months.

Throughout most of its existence, DU has funded habitat restoration projects and worked with landowners to provide water management benefits on farmlands. But, from its inception DU has been subject to criticism. Early opponents characterized it as an American intrusion into Canada to secure hunting areas. More recently, critics have suggested that DU defines waterfowl habitat too narrowly, excluding upland areas where many ducks and geese nest. The group plans to broaden its focus to encompass preservation of these upland breeding and nesting areas. Since many of these areas are found on private land, DU also plans to expand its cooperative programs with farmers and ranchers. Most commonly, however, DU is criticized for placing the interests of waterfowl hunters above **wildlife management** concerns. The organization does allow duck hunting on its preserves.

Following the fundamental principle of "users pay," duck hunters still provide the majority of DU's funding. For that reason DU has not addressed some issues that have a serious effect on continental waterfowl populations. The combination of illegal hunting and liberal bag limits is blamed by some for the continued decline in waterfowl numbers. DU has not addressed this issue, preferring to leave management issues to government agencies in the United States and Canada, while focusing on habitat preservation and restoration. Critics of DU suggest that the organization will not act on population matters and risk offending the hunters who provide their financial support.

In North America DU has expanded its scope and activities to address ecological and **land use** problems through the work of the North American Waterfowl Management Plan (NAWMP) and the Prairie CARE (Conservation of Agriculture, Resources and **Environment**) program. The wetlands conservation and other habitat projects addressed in these and similar programs, not only benefit game **species**, but other **endangered species** of plants and animals as well. NAWMP (an agreement between the United States and Canada) alone protects over 5.5 million acres (2.2 million ha) of waterfowl habitat. In 2002, the North American Wetlands Conservation Act (NAWCA) granted the DU one million dollars to be put towards a new wetlands in Ohio.

On balance, DU has had a major, positive impact on North American waterfowl habitat and management. Millions of acres of wetlands have been protected, enhanced, and managed in Canada, the United States, and Mexico. However, the continued decline in waterfowl populations may require the organization to redirect some of its efforts to population management and preservation issues.

[*David A. Duffus*]

RESOURCES
ORGANIZATIONS

Ducks Unlimited, Inc., One Waterfowl Way, Memphis, TN USA 38120 (901) 758-3825, Toll Free: (800) 45DUCKS, <http://www.ducks.org>

Ducktown, Tennessee

Tucked in a valley of the Cherokee **National Forest**, on the border of Tennessee, North Carolina, and Georgia, Ducktown once reflected the beauty of the surrounding Appalachian Mountains. Instead, Ducktown and the valley known as the **Copper** Basin now form the only **desert** east of the Mississippi. Mined for its rich copper lode since the 1850s, it had become a vast stretch of lifeless, red-clay hills. It was an early and stark lesson in the devastation that **acid rain** and **soil erosion** can wreak on a landscape, one of the few man-made landmarks visible to the astronauts who landed on the moon.

Prospectors came to the basin during a gold rush in 1843, but the closest thing to gold they discovered was copper, and most went home. But by 1850, entrepreneurs realized the value of the ore, and a new rush began to mine the area. Within five years, 30 companies had dug beneath the **topsoil** and made the basin the country's leading producer of copper.

The only way to separate copper from the zinc, iron, and sulfur present in Copper Basin rock was to roast the ore at extremely high temperatures. Mining companies built giant open pits in the ground for this purpose, some as wide as 600 ft (183 m) and as deep as a 10-story building. Fuel for these fires came from the surrounding forests. The forests must have seemed a limitless resource, but it was not long before every tree, branch, and stump for 50 mi^2 (130 km^2) had been torn up and burned. The fires in the pits emitted great billows of **sulfur dioxide** gas—so thick people could get lost in the clouds even at high noon—and this gas mixed with water and oxygen in the air to form sulfuric **acid**, which is main component in acid rain. Saturated by acidic moisture and choked by the remaining sulfur dioxide gas and dust, the undergrowth died and the soil became poisonous to new plants. **Wildlife** fled the shelterless hillsides. Without root systems, virtually all the soil washed into the Ocoee River,

smothering aquatic life. Open-range grazing of cattle, allowed in Tennessee until 1946, denuded the land of what little greenery remained.

Soon after the turn of the century, Georgia filed suit to stop the **air pollution** which was drifting out of this corner of Tennessee. In 1907, the Supreme Court, in a decision written by Justice Oliver Wendell Holmes, ruled in Georgia's favor, and the sulfur clouds ceased in the Copper Basin. It was one of the first environmental-rights decisions in the United States. That same year, the Tennessee Copper Company designed a way to capture the sulfur fumes, and sulfuric acid, rather than copper, became the area's main product. It remains so today.

Ducktown was the first mining settlement in the area, and residents now take a curious pride not only in the town's history, but in the eerie moonscape of red hills and painted cliffs that surrounds it. Since the 1930s, Tennessee Copper Company, the **Tennessee Valley Authority**, and the **Soil Conservation Service** have worked to restore the land, planting hundreds of loblolly pine and black locust trees. Their efforts have met with little success, but new reforestation techniques such as slow-release **fertilizer** have helped many new plantings survive. Scientists hope to use the techniques practiced here on other deforested areas of the world. Ironically, many of the townspeople want to preserve a piece of the scar, both for its unique beauty and the environmental lesson of what human enterprise can do to **nature**, as well as what it can undo. *See also* Acid waste; Ashio, Japan; Mine spoil waste; Smelter; Sudbury, Ontario; Surface mining; Trail Smelter arbitration

[*L. Carol Ritchie*]

RESOURCES
PERIODICALS

Barnhardt, W. "The Death of Ducktown." *Discover* 8 (October 1987): 34-6+.

Dunes and dune erosion

Dunes are small hills, mounds or ridges of wind-blown **soil** material, usually sand, that are formed in both coastal and inland areas. The formation of coastal or inland dunes requires a source of loose sandy material and dry periods during which the sand can be picked up and transported by the wind. Dunes exist independently of any fixed surface feature and can move or drift from one location to another over time. They are the result of natural **erosion** processes and are natural features of the landscape in many coastal areas and deserts, yet they also can be symptoms of land degradation. Inland dunes are either an expression of aridity or can be indicators of desertification—the result of long-term land degradation in dryland areas.

Coastal dunes are the result of marine erosion in which sand is deposited on the shore by wave action. During low tide, the beach sand dries and is dislodged and transported by the wind, usually over relatively short distances. Depending on the local **topography** and direction of the prevailing winds, a variety of shapes and forms can develop—from sand ridges to parabolic mounds. The upper few centimeters of coastal dunes generally contain chlorides from salt spray and wind-blown salt. As a result, attempts to stabilize coastal dunes with vegetation are often limited to salt-tolerant plants.

The occurrence of beaches and dunes together have important implications for coastal areas. A beach absorbs the energy of waves and acts as a **buffer** between the sea and the dunes behind it. Low lying coastlines are best defended against high tides by consolidated sand dunes. In such cases, maintaining a wide, high beach that is backed by stable dunes is desirable.

Engineering structures along coastal areas and the mouths of rivers can affect the formation and erosion of beaches and coastal dunes. In some instances it is desirable to build and widen beaches to protect coastal areas. This can require the construction of structures that trap littoral drift, rock mounds to check wave action, and sea walls that protect areas behind the beach from heavy wave action. Where serious erosion has occurred, artificial replacement of beach sands may be necessary. Such methods are expensive and require considerable engineering effort and the use of heavy equipment.

The **weathering** of rocks, mainly sandstone, is the origin of material for *inland dunes*. However, whether or not sand dunes form, depends on the vegetative cover condition and use of the land. In contrast to coastal dunes, that are often considered to be beneficial to coastal areas, inland dunes can be indicators of land degradation where the protective cover of vegetation has been removed as a result of inappropriate cultivation, **overgrazing**, construction activities, and so forth. When vegetative cover is absent, soil is highly susceptible to both water and wind erosion. The two work together in drylands to create sources of soil that can be picked up and transported either downwind or downstream. The flow of water moves and exposes sand grains and supplies fresh material that results in deposits of sand in flood plains and ephemeral **drainage** systems. Before dunes can develop in such areas, there must be long dry periods between periodic or episodic sediment-laden flows of water. Wind erosion occurs where such sand deposits from water erosion are exposed to the energy of wind, or in areas that are devoid of vegetative cover.

Where sand is the principle size soil particle and where high wind velocities are common, sand particles are moved by a process called saltation and creep. Sand dunes form under such conditions and are shaped by wind patterns over the landscape. Complex patterns can be formed—the result of interactions of wind, sand, the ground surface topography, and any vegetation or other physical barriers that exist. These patterns can be sword like ridges, called longitudinal dunes, crescentic accumulations or barchans, turret-shaped mounds, shallow sheets of sand, or large seas of transverse dunes. The typical pattern is one of a gradual long slope on the windward side of the dune, dropping off sharply on the leeward side.

Exposed sand dunes can move up to 11 yd (10 m) annually in the direction of the prevailing wind. Such dunes encroach upon areas, covering farmlands, pasture lands, **irrigation** canals, urban areas, railroads and highways. Blowing sand can mechanically injure and kill vegetation in its path and can eventually bury croplands or **rangelands**. If left unchecked, the drifting sand will expand and lead to serious economic and environmental losses.

Worldwide, dryland areas are those most susceptible to wind erosion. For example, 22% of Africa north of the Equator is severely affected by wind erosion as is over 35% of the land area in the Near East. As a result, inland dunes represent a significant landscape component in many **desert** regions. For example, dunes represent 28%, 26%, and 38% of the landscape of the Saharan Desert, Arabian Desert, and **Australia**, respectively (Heathcote 1983). In 1980, Walls estimated that 1.3 billion hectares of land were covered by sand dunes globally. Although dunes can be symptoms of **land use** problems, in some areas they are part of a natural dryland landscape that are considered to be features of beauty and interest. Sand dune have become popular recreational areas in parts of the United States, including the Great Sand Dune National Monument in southern Colorado with its 229-yd (210-m) high dunes that cover a 158-mi^2 (254.4-km^2) area, and the Indiana Dunes State Park along the shore of Lake Michigan.

When dune formation and encroachment represent significant environmental and economic problems, sand dune stabilization and control should be undertaken. Dune stabilization may initially require one or more of the following: applications of water, oil, bitumens emulsions, or chemical stabilizers to improve the cohesiveness of surface sands; the reshaping of the landscape such as construction of foredunes that are upwind of the dunes, and armoring of the surface using techniques such as hydroseeding, jute mats, mulching and asphalt; and constructing fences to reduce wind velocity near the ground surface. Although sand dune stabilization is the necessary first step in controlling this process, the establishment of a vegetative cover is a necessary

The dunes of Nags Head, North Carolina. (Photograph by Jack Dermid, National Audubon Society Collection. Photo Researchers Inc. Reproduced by permission.)

condition to achieve long-term control of sand dune formation and erosion. Furthermore, stabilization and revegetation must be followed with appropriate land management that deals with the causes of dune formation in the first place. Where dune erosion has not progressed to a seriously degraded state, dunes can become reclaimed through natural regeneration simply by protecting the area against livestock grazing, all-terrain vehicles, and foot traffic.

Vegetation stabilizes dunes by decreasing wind speed near the ground and by increasing the cohesiveness of sandy material by the addition of organic colloids and the binding action of roots. Plants trap the finer wind-blown soil particles, which helps improve **soil texture**, and they also improve the **microclimate** of the site, reducing soil surface temperatures. Upwind barriers or windbreak plantings of vegetation, often trees or other woody perennials, can be effective in improving the success of revegetating sand dunes. They reduce wind velocities, help prevent exposure of plant roots from the drifting sand, and protect plantings from the abrasive action of blowing sand. Areas that are susceptible to sand dune encroachment can likewise be protected by using fences or windbreak plantings that reduce wind veloci-

ties near the ground surface. Because of the severity of sand dune environments, it can be difficult to find plant **species** that can be established and survive. In addition, any plantings must be protected against exploitation, for example, from grazing or fuelwood harvesting.

The expansion of sand dunes resulting from **desertification** not only represent environmental problems, but they also represent serious losses of productive land and a financial hardship for farmers and others who depend upon the land for their livelihood. Such problems are particularly acute in many of the poorer dryland countries of the world and deserve the attention of governments, international agencies, and nongovernmental organizations who need to direct their efforts toward the causes of soil erosion and dune formation.

[*Kenneth N. Brooks*]

RESOURCES
BOOKS

Brooks, K.N., P.F. Folliott, H. M. Gregersen and L. F. DeBano. *Hydrology and the Management of Watersheds.* 2nd ed. Ames, Iowa: Iowa State University Press, 1997.

The effects of the Oklahoma dust bowl. (Corbis-Bettmann. Reproduced by permission.)

Folliott, P. F., K. N. Brooks, H.M. Gregersen and A.L. Lundgren. *Dryland Forestry—Planning and Management.* New York: John Wiley & Sons, 1995.
Food and Agricultural Organization (FAO) of the United Nations. *Sand Dune Stabilization, Shelterbelts, and Afforestation in Dry Zones.* Rome: FAO Conservation Guide 10, 1985.

Dust Bowl

"Dust Bowl" is a term coined by a reporter for the *Washington* (D.C.) *Evening Star* to describe the effects of severe wind **erosion** in the Great Plains during the 1930s, caused by severe **drought** and lack of **conservation** practices.

For a time after World War I, agriculture prospered in the Great Plains. Land was rather indiscriminantly plowed and planted with cereals and row crops. In the 1930s, the total cultivated land in the United States increased, reaching 530 million acres (215 million ha), its highest level ever. Cereal crops, especially wheat, were most prevalent in the Great Plains. Summer fallow (cultivating the land, but only planting every other season) was practiced on much of the land. Moisture, stored in the **soil** during the fallow (uncropped) period, was used by the crop the following year.

In a process called dust **mulch**, the soil was frequently clean tilled to leave no crop residues on the surface, control weeds, and, it was thought at the time, preserve moisture from evaporation. Frequent cultivation and lack of crop canopy and residues optimized conditions for wind erosion during the droughts and high winds of the 1930s.

During the process of wind erosion, the finer particles (**silt** and clay) are removed from the **topsoil**, leaving coarser-textured sandy soil. The fine particles carry with them higher concentrations of organic matter and plant nutrients, leaving the remaining soil impoverished and with a lower water storage capacity. Wind erosion of the Dust Bowl reduced the productivity of affected lands, often to the point that they could not be farmed economically.

While damage was particularly severe in Texas, Oklahoma, Colorado, and Kansas, erosion occurred in all of the Great Plains states, from Texas to North Dakota and Montana, even into the Canadian **Prairie** Provinces. The eroding soil not only prevented the growth of plants, it uprooted established ones. **Sediment** filled fence rows, stream channels, road ditches, and farmsteads. Dirt coated the insides of buildings. Airborne dust made travel difficult because of decreased **visibility**; it also impaired breathing and caused **respiratory diseases**.

Dust from the Great Plains was carried high in the air and transported as far east as the Atlantic seaboard. In places, 3–4 in (7–10 cm) of topsoil was blown away, forming **dunes** 15–20 ft (4.6–6.1 m) high where the dust finally came to rest. In a 20-county area covering parts of southwestern Kansas, the Oklahoma strip, the Texas Panhandle, and southeastern Colorado, a soil-erosion survey by the **Soil Conservation Service** showed that 80% of the land was affected by wind erosion, 40% of it to a serious degree.

The droughts and resultant wind erosion of the 1930s created widespread economic and social problems. Large numbers of people migrated out of the Dust Bowl area during the 1930s. The **migration** resulted in the disappearance of many small towns and community services such as churches, schools, and local units of government.

Following the disaster of the Dust Bowl, the 1940s saw dramatically improved economic and social conditions with increased precipitation and improved crop prices. Gradually, changes in farming practices have also taken place. Much of the severely damaged and marginal land has been

returned to grass for livestock grazing. Non-detrimental tillage and management practices, such as **conservation tillage** (stubble mulch, mulch, and residue tillage); use of tree, shrub, and grass windbreaks; maintenance of crop residues on the soil surface; and better machinery have all contributed to improved soil conditions. Annual cropping or a three-year rotation of wheat-sorghum-fallow has replaced the alternate crop-fallow practice in many areas, particularly in the more humid areas of the West.

While the extreme conditions of drought and land mismanagement of the Dust Bowl years have not been repeated since the 1930s, wind erosion is still a serious problem in much of the Great Plains. According to the **Soil Conservation** Service, the states with the most serious erosion per unit area in 1982 were Texas, Colorado, Nevada, and Montana. *See also* Arable land; Desertification; Overgrazing; Soil eluviation; Tilth; Water resources

[*William E. Larson*]

RESOURCES
BOOKS

Hurt, R. D. *The Dust Bowl: An Agricultural and Social History.* Chicago: Nelson-Hall, 1981.

E

Earth Charter

It is the objective of the Earth Charter to set forth an inspiring vision of the fundamental principles of a global partnership for **sustainable development** and environmental **conservation**. The Earth Charter initiative reflects the conviction that a radical change in humanity's attitudes and values is essential to achieve social, economic, and ecological well-being in the twenty-first century. The Earth Charter project is part of an international movement to clarify humanity's shared values and to develop a new global ethics, ensuring effective human cooperation in an interdependent world.

There were repeated efforts to draft the Earth Charter beginning in 1987. Early in 1997 an Earth Charter Commission was formed by the Earth Council and **Green Cross** International. The Commission prepared an Earth Charter which it was circulated as a people's treaty beginning in 1998. The Charter was then submitted to the United Nations General Assembly in the year 2000. On June 29, 2000, the official Earth Charter was established at the Peace Palace in The Hague, Holland.

Historical background, 1945–1992

The role and significance of the Earth Charter are best understood in the context of the United Nations' ongoing efforts to identify the fundamental principles essential to world security. When the U.N. was established in 1945, its agenda for world security emphasized peace, human rights, and equitable socioeconomic development. No mention was made of the **environment** as a common concern, and little attention was given to ecological well-being in the U.N.'s early years. However, since the Stockholm Conference on the Human Environment in 1972, ecological security has emerged as a fourth major concern of the United Nations.

Starting with the Stockholm Declaration, the world's nations have adopted a number of declarations, charters, and treaties that seek to create a global alliance that effectively integrates and balances development and conservation. In addition, a variety of nongovernmental organizations have drafted and circulated their own declarations and people's treaties. These documents reflect a growing awareness that humanity's social, economic, and environmental problems are interconnected and require integrated solutions. The Earth Charter initiative builds on these efforts.

The World Charter for Nature, which was adopted by the U.N. General Assembly in 1982, was a progressive declaration of ecological and ethical principles for its time. It remains a stronger document than any that have followed from the point of view of **environmental ethics**. However, in its 1987 report, *Our Common Future*, the U.N. World Commission on Environment and Development (WCED) issued a call for "a new charter" that would "consolidate and extend relevant legal principles," creating "new norms...needed to maintain livelihoods and life on our shared planet" and "to guide state behavior in the transition to sustainable development." The WCED also recommended that the new charter "be subsequently expanded into a Convention, setting out the sovereign rights and reciprocal responsibilities of all states on environmental protection and sustainable development."

The WCED recommendations, together with deepening environmental and ethical concerns, spurred efforts in the late 1980s to create an Earth Charter. However, before any U.N. action was initiated on the Earth Charter, the Commission on **Environmental Law** of the World Conservation Union (IUCN) drafted the convention proposed in *Our Common Future*. The IUCN Draft International Covenant on Environment and Development presents an integrated legal framework for existing and future international and national environmental and sustainable development law and policy. Even though the IUCN Draft Covenant was presented at the United Nations in 1995 official negotiations have not yet begun on this treaty which many environmentalists believe is urgently needed to clarify, synthesize, and further develop international sustainable development law.

The United Nations Conference on Environment and Development (UNCED), or Earth Summit held in Rio de

Janeiro, Brazil, in 1992 did take up the challenge of drafting the Earth Charter. A number of governments prepared recommendations. Many nongovernmental organizations, including groups representing the major faiths, became actively involved. While the resulting Rio Declaration on Environment and Development is a valuable document, it falls short of the aspirations that many groups have had for the Earth Charter.

The Earth Charter Project, 1994–2000

A new Earth Charter initiative began in 1994 under the leadership of Maurice Strong, the former secretary general of UNCED and chairman of the newly formed Earth Council, and Mikhail Gorbachev, acting in his capacity as chairman of Green Cross International. The Earth Council was created to complete the unfinished business of UNCED and to promote implementation of Agenda 21, the Earth Summit's action plan. Jim MacNeill, former secretary general of the WCED and Prime Minister Ruud Lubbers of The Netherlands were instrumental in facilitating the organization of the new Earth Charter project. Ambassador Mohamed Sahnoun of Algeria served as the executive director of the project during its initial phase, and its first international workshop was held at the Peace Palace in The Hague in May 1995. Representatives from 30 countries and more than 70 different organizations participated in the workshop. Following this event, the secretariat for the Earth Charter project was established at the Earth Council in San José, Costa Rica.

A worldwide Earth Charter consultation process was organized by the Earth Council in connection with the Rio+5 review in 1996 and 1997. The Rio+5 review, which culminated with a special session of the United Nations General Assembly in June 1997, sought to assess progress toward sustainable development since the Rio Earth Summit and to develop new partnerships and plans for implementation of Agenda 21. The Earth Charter consultation process engaged men and women from all sectors of society and all cultures in contributing to the Earth Charter's development. A special program was created to contact and involve the world's religions, interfaith organizations, and leading religious and ethical thinkers. A special indigenous people's network was also organized by the Earth Council.

Early in 1997, an Earth Charter Commission was formed to oversee the project. The 23 members represent the major regions of the world and different sectors of society. The Commission issued a Benchmark Draft Earth Charter in March 1997 at the conclusion of the Rio+5 Forum in Rio de Janeiro. The Forum was organized by the Earth Council as part of its independent Rio+5 review, and it brought together more than 500 representatives from civil society and national councils of sustainable development. The Benchmark Draft reflected the many and diverse contributions received through the consultation process and from the Rio+5 Forum. The Commission extended the Earth Charter consultation until early 1998, and the Benchmark Draft was circulated widely as a document in progress.

The Earth Charter concept

A consensus developed that the Earth Charter should be: a statement of fundamental principles of enduring significance that are widely shared by people of all races, cultures, and religions; a relatively brief and concise document composed in a language that is inspiring, clear, and meaningful in all tongues; the articulation of a spiritual vision that reflects universal spiritual values, including but not limited to ethical values; a call to action that adds significant new dimensions of value to what has been expressed in earlier relevant documents; a people's charter that serves as a universal code of conduct for ordinary citizens, educators, business executives, scientists, religious leaders, nongovernmental organizations, and national councils of sustainable development; and a declaration of principles that can serve as a "soft law" document when adopted by the U.N. General Assembly. The Earth Charter was designed to focus on fundamental principles with the understanding that the IUCN Covenant and other treaties will set forth the more specific practical implications of these principles.

The Earth Charter draws upon a variety of resources, including **ecology** and other contemporary sciences, the world's religious and philosophical traditions, the growing literature on global ethics and the ethics of environment and development, the practical experience of people living sustainably, as well as relevant intergovernmental and nongovernmental declarations and treaties. At the heart of the new global ethics and the Earth Charter is an expanded sense of community and moral responsibility that embraces all people, **future generations**, and the larger community of life on Earth. Among the values affirmed by the Benchmark Draft are: respect for Earth and all life; protection and restoration of the health of Earth's ecosystems; respect for human rights, including the right to an environment adequate for human well-being; eradication of poverty; nonviolent problem solving and peace; the equitable sharing of resources; democratic participation in decision making; accountability and transparency in administration; universal education for sustainable living; and a sense of shared responsibility for the well-being of the Earth community.

[*Steven C. Rockefeller*]

RESOURCES

BOOKS

Earth Ethics, Special Earth Charter Double Issue. Washington, D.C.: Center for Respect of Life and Environment, 7, nos. 3/4 (Spring/Summer, 1996): 1-7 and 8, nos. 2/3 (Winter/Spring, 1997): 3-8.

Rockefeller, S.C. *Principles of Environmental Conservation and Sustainable Development: Summary and Survey.* The Earth Council website, The Earth Charter Consultation page: <http://www.ecouncil.ac.cr>..

ORGANIZATIONS

The Earth Charter Initiative, The Earth Council, P.O. Box 319-6100, San Jose, Costa Rica +506-205-1600 , Fax: +506-249-3500, Email: info@earthcharter.org, <http://www.earthcharter.org>

Earth Day

The first Earth Day, April 22, 1970, attracted over 20 million participants in the United States. It launched the modern environmental movement and spurred the passage of several important environmental laws. It was the largest demonstration in history. People from all walks of life took part in marches, teach-ins, rallies, and speeches across the country. Congress adjourned so that politicians could attend hometown events, and cars were banned from New York's Fifth Avenue.

The event had a major impact on the nation. Following Earth Day, **conservation** organizations saw their memberships double and triple. Within months, the **Environmental Protection Agency** (EPA) was created; Congress also revised the **Clean Air Act**, the **Clean Water Act**, and other environmental laws.

The concept for Earth Day began with Senator Gaylord Nelson, a Wisconsin Democrat, who in 1969 proposed a series of environmental teach-ins on college campuses across the nation. Hoping to satisfy a course requirement at Harvard by organizing a teach-in there, law student Denis Hayes flew to Washington, DC, to interview Nelson. The senator persuaded Hayes to drop out of Harvard and organize the nationwide series of events that were only a few months away. According to Hayes, Wednesday, April 22 was chosen because it was a weekday and would not compete with weekend activities. It also came before students would start "cramming" for finals, but after the winter thaw in the North.

Twenty years later, Earth Day anniversary celebrations attracted even greater participation. An estimated 200 million people in over 140 nations were involved in events ranging from a concert and rally of over a million people in New York's Central Park, to a festival in Los Angeles that attracted 30,000, to a rally of 350,000 at the National Mall in Washington, D.C.

Earth Day 1990 activities included planting trees; cleaning up roads, highways, and beaches; building bird houses; **ecology** teach-ins; and **recycling** cans and bottles. A convoy of **garbage** trucks drove through the streets of Portland, Oregon, to dramatize the lack of **landfill** space. Elsewhere, children wore gas masks to protest **air pollution**, others marched in parades wearing costumes made from recycled materials, and some even released ladybugs into the air to demonstrate alternatives to harmful pesticides. The gas-guzzling car that was buried in San Jose, California, during the first Earth Day was dug up and recycled.

Abroad, Berliners planted 10,000 trees along the East-West border. In Myanmar, there were protests against the killing of **elephants**. Brazilians demonstrated against the destruction of their tropical rain forests. In Japan, there were demonstrations against disposable chopsticks, and 10,000 people attended a concert on an island built on reclaimed land in Tokyo Bay.

The 1990 version was also organized by Denis Hayes, with help from hundreds of volunteers. This time, the event was well organized and funded; it was widely-supported by both environmentalists and the business community. The United Auto Workers Union sent Earth Day booklets to all of its members, the National Education Association sent information to almost every teacher in the country, and the Methodist Church mailed Earth Day sermons to over 30,000 ministers.

The sophisticated advertising and public relations campaign, licensing of its logo, and sale of souvenirs provoked criticism that, Earth Day had become too commercial. Even oil, chemical, and nuclear firms joined in and proclaimed their love for **nature**. But Hayes defended the professional approach as necessary to maximize interest and participation in the event, to broaden its appeal, and to launch a decade of environmental activism that would force world leaders to address the many threats to the planet. He also pointed out that while foundations, corporations, and individuals had donated $3.5 million, organizers turned down over $4 million from companies that were thought to be harming the **environment**.

The 30-year anniversary of the event was also organized by Hayes. Unfortunately, it did not produce the large numbers of the prior anniversary celebration. The movement had reached over 5,000 environmental groups who helped organize local rallies, and hundreds of thousands of people met in Washington to hear political, environmental, and celebrity speakers.

Hayes believes that the long-term success of Earth Day in securing a safe future for the planet depends on getting as many people as possible involved in **environmentalism**. The Earth Day celebrations he helped organize a have been a major step in that direction.

[*Lewis G. Regenstein*]

RESOURCES
PERIODICALS

Borrelli, P. "Can Earth Day Be Every Day?" *Amicus Journal* 12 (Spring 1990): 22-26.

Marchers carry a tree down Fifth Avenue in New York City on the first Earth Day, April 22, 1970. (Corbis-Bettmann. Reproduced by permission.)

Two Earth First! Members hold a sign in front of the statue of Abraham Lincoln at the Lincoln Memorial to protest the destruction of the earth's rain forests.

Hayes, D. "Earth Day, 1990: Threshold of the Green Decade." *Natural History* 99 (April 1990): 55-60.

Stenger, Richard. "Thousands observe Earth Day 2000 in Washington." *CNN.com* (22 April 2000) [June 2002]. <http://www.cnn.com/2000/NATURE/04/22/earth.day/index.html>.

Earth First!

Earth First! is a radical and often controversial environmental group founded in 1979 in response to what **Dave Foreman** and other founders believed to be the increasing co-optation of the environmental movement. For Earth First! members, too much of the environmental movement has become lethargic, compromising, and corporate in its orientation. To avoid a similar fate, Earth First! members have restricted their use of traditional fund-raising techniques and have sought a non-hierarchical organization with neither a professional staff nor formal leadership.

A movement established by and for self-acknowledged environmental hardliners, Earth First!'s general stance is reflected in its slogan, "No compromise in the defense of Mother Earth." Its policy positions are based upon principles of **deep ecology** and in particular on the group's belief in the **intrinsic value** of all natural things. Its goals include preserving all remaining **wilderness**, ending **environmental degradation** of all kinds, eliminating major **dams**, establishing large-scale ecological preserves, slowing and eventually reversing human **population growth**, and reducing excessive and environmentally harmful consumption.

Combining biocentrism with a strong commitment to activism, Earth First! does not restrict itself to lobbying, lawsuits, and letter-writing, but also employs direct action, civil disobedience, "guerrilla theater," and other confrontational tactics, and in fact is probably best known for its various clashes with the **logging** industry, particularly in the Pacific Northwest. Earth First! members and sympathizers have been associated with controversial tactics including the chopping down of billboards and **monkey-wrenching**, which includes pouring sand in bulldozer gas tanks, spiking trees, sabotaging drilling equipment, and so forth. Officially, the organization purports neither to condone nor condemn such tactics.

Earth First! encourages people to respect **species** and wilderness, to refrain from having children, to recycle, to live simpler, less destructive lives, and to engage in civil disobedience to thwart environmental destruction. During the summer of 1990, the group sponsored its most noted event, "Redwood Summer." Activists from around the United States gathered in the Northwest to protest large-scale logging operations, to call attention to environmental concerns, to educate and establish dialogues with loggers and the local public, and to engage in civil disobedience.

Earth First! also sponsors a **Biodiversity** Project for protecting and restoring natural ecosystems. Its Ranching Task Force educates the public about the consequences of **overgrazing** in the American West. The **Grizzly Bear** Task Force focuses on the preservation of the grizzly bear in the Rockies and the reintroduction of the species to its historical range throughout North America. Earth First!'s wider Predator Project seeks the restoration of all native predators to their respective habitats and ecological roles. Carmageddon is an anti-car campaign Earth First! sponsors in the United Kingdom. Other Earth First! projects seek to defend **redwoods** and other native forests, encourage direct action against the fur industry, intervene in government-sponsored wolf-control programs in the U.S. and Canada, and protest government and business decisions which have environmentally destructive consequences for tropical rain forests.

[*Lawrence J. Biskowski*]

Earth Island Institute

The Earth Island Institute (EII) was founded by David Brower in 1982 as a nonprofit organization dedicated to developing innovative projects for the **conservation**, preservation, and restoration of the global **environment**. In its earliest years, the Institute worked primarily with a volunteer staff and concentrated on projects like the first Conference on the Fate of the Earth, publication of *Earth Island Journal*, and the production of films about the plight of **indigenous peoples**. In 1985 and again in 1987, EII expanded its facilities and scope, opening office space and providing support for a number of allied groups and projects. Its membership now numbers approximately 35,000.

EII conducts research on, and develops critical analyses of, a number of contemporary issues. With sponsored projects ranging from saving **sea turtles** to encouraging land restoration in Central America, EII does not restrict its scope to traditionally "environmental" goals but rather pursues what it sees as ecologically-related concerns such as human rights, economic development of the **Third World**, economic conversion from military to peaceful production, and inner city

poverty, among others. But much of its mission is to be an environmental educator and facilitator. In that role EII sponsors or participates in numerous programs designed to provide information, exchange viewpoints and strategies, and coordinate efforts of various groups. EII even produces music videos as part of its **environmental education** efforts.

EII is perhaps best known for its efforts to halt the use of drift nets by tuna boats, a practice which is often fatal to large numbers of **dolphins**. After an EII biologist signed on as a crew member aboard a Latin American tuna boat and managed to document the slaughter of dolphins in drift nets, EII brought a lawsuit to compel more rigorous enforcement of existing laws banning tuna caught on boats using such nets. EII also joined with other environmental groups in urging a consumer boycott of canned tuna. These efforts were successful in persuading the three largest tuna canners to pledge not to purchase tuna caught in drift nets. The monitoring of tuna fishing practices is an ongoing EII project.

EII also sponsors a wide variety of other projects. Its Energy Program promotes energy efficient technology. Its Friends of the Ancient Forest program aims at the protection of old-growth forests on the Pacific Coast. Baikal Watch works for the permanent protection of biologically unique **Lake Baikal**, Russia. The Climate Protection Institute publishes the *Greenhouse Gas-ette* and develops public education material about changes in global climate. EII participates in the International Green Circle, an ecological restoration program which matches volunteers with ongoing projects worldwide. EII's Sea Turtle Restoration Project investigates threats to the world's endangered sea turtles, organizes and educates United States citizens to protect the turtles, and works with Central American sea turtle restoration projects. The **Rain Forest** Health Alliance develops educational materials and programs about the biological diversity of tropical rain forests. The Urban **Habitat** Program develops multicultural environmental leadership and organizes efforts to restore urban neighborhoods.

EII administers a number of funds designed to support creative approaches to environmental conservation, preservation, and restoration to support activists exploring the use of citizen suit provisions of various statutes to enforce environmental laws and to help develop a Green political movement in the United States. EII also sponsors several international conferences, exchange programs, and publication projects in support of various environmental causes.

[*Lawrence J. Biskowski*]

RESOURCES
ORGANIZATIONS
Earth Island Institute, 300 Broadway, Suite 28, San Francisco, CA USA 94133-3312 (415) 788-3666, Fax: (415) 788-7324, <http://www.earthisland.org>

Earth Liberation Front

Earth Liberation Front (ELF) is a grassroots environmental group that the Federal Bureau of Investigation labeled "a serious terrorism threat." Since 1996, ELF and the Animal Liberation Front (ALF) committed more than 600 acts of vandalism that resulted in more than $43 million in damage, an FBI terrorism expert reported to Congress in 2002.

James Jarboe, the FBI section chief for Domestic Terrorism, told Congress that it was hard to track down ELF and AlF members because the two groups have little organized structure. According to the ELF link on the ALF Web site, there is no designated leadership nor formal membership. ELF and ALF claim responsibility for their activities by e-mail, fax, and other communications usually sent to the media.

In 2002, media information was provided by the North American Earth Liberation Front Press Office. Spokesman Leslie James Pickering wrote that he received anonymous communiqués from ELF and further distributed them.

ELF describes itself as an international underground organization dedicated "to stop the continued destruction of the natural environment." Members join anonymous cells that may consist of one person or more. Members of one cell do not know the identity of members in other cells, a structure that prevents activists in one cell from being compromised should members in another cell become disaffected. People act on their own and carry out actions following anonymous ELF postings. ELF actions include sabotage and property damage.

ELF postings include guidelines for taking action. One guideline is to inflict "economic damage on people who profit from the destruction and exploitation of the natural environment." Another guideline is to reveal and educate the public about the "atrocities" committed against the **environment**. The third guideline is to take all needed precautions against harming any animal, human and non-human.

ELF is an outgrowth of **Earth First!**, a group formed during the 1980s to promote environmental causes. Earth First! held protests and civil disobedience events, according to the FBI. In 1984, Earth First! members began a campaign of "tree spiking." to prevent loggers from cutting down trees. Members inserted metal or ceramic spikes into trees. The spikes damaged saws when loggers tried to cut down trees.

When Earth First! members in Brighton, England disagreed with proposals to make their group more mainstream, radical members of the group founded the Earth Liberation Front in 1992. The following year, ELF aligned itself with ALF.

The Animal Liberation Front was started in Great Britain during the mid-1970s. The loosely organized movement had the goal of ending animal abuse and exploitation of animals. An American ALF branch was started in the late 1970s. According to the FBI, people became members by participating in "direct action" activities against companies or people using animals for research or economic gain. ALF activists targeted animal research laboratories, fur companies, mink farms, and restaurants.

ELF and ALF declared mutual solidarity in a 1993 announcement. The following year, the San Francisco branch of Earth First! recommended that the group mainstream itself away from ELF and its unlawful activities. ELF calls its activities "monkeywrenching.", a term that refers to actions such as tree spiking, arson, sabotage of **logging** equipment, and property destruction.

In a 43-page document titled "Year End Report for 2001," ELF and ALF claimed responsibility for 67 illegal acts that year. ELF claimed sole credit for setting a fire that year that caused $5.4 million in damage to a University of Washington horticulture building. The group also took sole credit for a 1998 fire set at a Vail, Colorado ski resort. Damage totaled $12 million for the arson that destroyed four ski lifts, a restaurant, a picnic facility, and a utility building, according to the FBI.

ELF issued a statement after the arson saying that the fire was set to protect the lynx, which was being reintroduced to the Rocky Mountains. "Vail, Inc. is already the largest ski operation in North American and now wants to expand even further...This action is just a warning. We will be back if this greedy corporation continues to trespass into wild and unroaded areas," the statement said.

In 2002, ELF claimed credit for a fire that caused $800,000 in damage at the University of Minnesota Microbial Plant and Genomics Center. ELF targeted the genetic crop laboratory because of its efforts to "control and exploit" **nature**.

"Eco-terrorism" is the term used by the FBI to define illegal activities related to **ecology** and the environment. These activities involve the "use of criminal violence against innocent victims or property by an environmentally-oriented group."

Although ELF members are difficult to track, several arrests have been made. In February 2001, two teen-age boys pleaded guilty to setting fires at a home construction site in Long Island, New York. In December of that year, a man was also charged with spiking 150 trees in Indiana state forests. In his Congress testimony, Jarboe said that cooperation among law enforcement agencies was essential to respond efficiently to eco-terrorism. As of 2002, the FBI has joint terrorism task forces in 44 cities and the bureau plans to have task forces in all 56 of its field offices by the end of 2003.

[*Liz Swain*]

RESOURCES

BOOKS

Alexander, Yonah and Edgar H. Brenner. *U. S. Federal Responses to Terrorism.* Ardsley, NY: Transnational Publishers, 2002.

Newkirk, Ingrid and Chrissie Hynde. *Free the Animals: The Story of the Animal Liberation Front.* New York, NY: Lantern Books, 2000.

PERIODICALS

Society of American Foresters. "Legislator Focuses on Ecoterrorism McInnis Asks Environmental Groups to Denounce Violence." *The Forestry Source*(January, 2002).

ORGANIZATIONS

Federal Bureau of Investigation, 935 Pennsylvania Ave., Washington, DC USA (202) 324-3000, , <http://www.fbi.gov>

North American Earth Liberation Front Press Office, James Leslie Pickering, P.O. Box 14098, Portland, OR USA 97293 (503) 804-4965, Email: elfpress@tao.ca, <http://www.animalliberation.net/library/facts/elf>

Earth Pledge Foundation

Created in 1991 by attorney Theodore W. Kheel, the Earth Pledge Foundation (EPF) is concerned with the impact of technology on society. Recognizing the often delicate balance between economic growth and environmental protection, EPF encourages the implementation of sustainable practices, especially in community development, tourism, cuisine, and architecture.

As a result of the **United Nations Earth Summit** (Rio de Janeiro) the UN pledged its commitment to the principles of sustainable development—to foster "development meeting the needs of the present without compromising the ability of **future generations** to meet their own needs."

Created for the Summit in support of the principles, the Earth Pledge was prominently displayed throughout the event. Heads of state, ambassadors, delegates, and prominent dignitaries from around the world stood in line to sign their names on a large Earth Pledge board. Since the Summit, millions have taken the Earth Pledge: "Recognizing that people's actions towards **nature** and each other are the source of growing damage to the **environment** and to resources needed to meet human needs and ensure survival, I pledge to act to the best of my ability to help make the Earth a secure and hospitable home for present and future generations."

In early 1996, Earth Pledge created the Business Coalition for Sustainable Cities (BCSC) to influence the development of cities as centers of commerce, employment, **recreation**, and settlement. Chaired by William L. Lurie, former president of The Business Roundtable, the BCSC provides business leaders a forum to address issues of major importance to our cities in ways that ensure economic viability

while at the same time promoting respect for the environment. One event sponsored by the BCSC was a seven-course dinner prepared by 12 of the nation's most environmentally-conscious chefs to showing off that restaurants, one of the largest industries and employers, can practice the principles of sustainable cuisine. The BCSC hosted the event with the theme that good food can be well-prepared without adversely impacting health, culture or environment. This theme was elaborated on in 2000 when the Sustainable Cuisine Project was established to develop and teach cooking classes.

The latest development by Earth Pledge is the creation of the web site <farmtotable.org>. This web site highlights local farmers of the New York region and give consumers a direct link to fresh food news.

Earth Pledge has also formed a number of alliances to further their goals with groups such as: The Foundation for Prevention and Resolution of Conflict (PERC—founded by Theodore Kheel); the United Nations Environmental Programme (UNEP); EarthKind International; the New England Aquarium.

As a joint project with the New England Aquarium, EPF sponsors a marine awareness project that educates people on the importance of coastlines and aquatic resources to the **sustainable development** of the world's cities. The project emphasizes that many countries have water shortages due to inefficient use of their water supply, degradation of their water by **pollution** and unsustainable usage of **groundwater** resources.

In late 1995, Earth Pledge co-sponsored the first Caribbean Conference on Sustainable Tourism with the UN Department for Policy Coordination and Sustainable Development, EarthKind International and UNEP. The Conference brought together officials from government and business to discuss strategies for developing a healthy tourist economy, sound infrastructure, environmental protection, and community participation.

The foundation has constructed an environmentally-sensitive building, Foundation House, to display their solutions for improving **air quality** and **energy efficiency**. Sustainable features include heating, cooling, and lighting systems that minimize consumption of **fossil fuels**; increased ventilation and use of natural daylight; an auditorium for conferences with Internet access and computer lab for training. The Foundation House houses exhibits, including one on enhancing efficiency of the workplace for the benefit of the staff.

Earth Pledge continues to develop smaller organizations and promote companies that participate in sustainable practices.

[*Nicole Beatty*]

RESOURCES

ORGANIZATIONS

Earth Pledge Foundation, 122 East 38th Street , New York , NY USA 10016   (212) 725-6611, Fax: (212) 725-6774,   , <http://www.earthpledge.org>

Earth Summit

see **United Nations Earth Summit (1992)**

Earthquake

Earthquakes have been around for as long as the planet and have plagued humans throughout history. With no warning, major earthquakes strike populated areas of the world every year, killing hundreds, injuring thousands, and causing hundreds of millions of dollars in damage. Yet despite millions of dollars and decades of research, seismologists (scientists who study earthquakes) are still unable to predict precisely when and where an earthquake will happen.

An earthquake is a geological event in which rock masses below the surface of the earth suddenly shift, releasing energy and sending out strong vibrations to the surface. Most earthquakes are caused by movement along a fault line, which is a fracture in the earth's crust. Thousands of earthquakes happen each day around the world, but most are too small to be felt.

Earth is covered by a crust of rock that is broken into numerous plates. The plates float on a layer of molten (liquid) rock within the earth called the mantel. This molten rock moves and flows, and this movement is thought to cause the shifting of the plates. When plates move, they either slide past, bump into, overrun, or pull away from each other. The movement of plates is called **plate tectonics**. Boundaries between plates are called faults.

Earthquakes can occur when there are any of the four types of movement along a fault. Earthquakes along the San Andreas and Hayward faults in California occur because of two plates sliding past one another. Earthquakes also occur if one plate overruns another. When this happens one plate is pushed under the other plate, as on the western coast of South America, the northwest coast of North America, and in Japan. If plates collide but neither is pushed downwards, as they do crossing Europe and Asia from Spain to Vietnam, earthquakes result as the plates are pushed into each other and are forced upwards, creating high mountain ranges. Many faults at the floor of the ocean are between two plates moving apart. Many earthquakes with centers at the floor of the ocean are caused by this kind of movement.

The relative size of earthquakes is measured by the Richter Scale, which measures the energy an earthquake releases. Each whole number increase in value on the Richter scale indicates a ten-fold increase in the energy released and a thirty–fold increase in ground motion. An earthquake measuring 8 on the Richter scale is ten times more powerful, therefore, than an earthquake with a Richter magnitude of 7. Another scale, called the Mercalli Scale uses observations of damage (such as fallen chimneys) or people's assessments of effects (such as mild or severe ground shaking) to describe the intensity of a quake. The Richter Scale is open-ended, while the Mercalli scale ranges from 1–12.

Catastrophic earthquakes happened just as often in past human history as they do today. Earthquakes shattered stone-walled cities in the ancient world, sometimes hastening the ends of civilizations. Earthquakes destroyed Knossos, Chattusas, and Mycenae, ancient cities in Europe located in tectonically active mountain ranges. Scribes have documented earthquakes in the chronicles of ancient countries. An earthquake is recorded in the Bible in the Book of Zachariah, and the Apostle Paul wrote that he escaped from jail when the building fell apart around him during an earthquake.

Many faults are located in California because two large plates are sliding past each other there. Of the 15 largest recorded earthquakes ever to hit the continental United States, eight have occurred in California, according to the United States **Geological Survey** (USGS). The San Francisco earthquake of 1906 is perhaps the most famous. It struck on April 4, 1906, killing an estimated 3,000 people, injuring thousands, and causing $524 million in property loss. Many of the casualties and much of the damage resulted from the ensuing fires. This earthquake registered a 7.7 magnitude on the Richter Scale and 11 on the Mercalli Scale. Four other devastating earthquakes have occurred in California in the twentieth century: 1933 in Long Beach, 1971 in the San Fernando Valley, 1989 in the San Francisco Bay area, and 1994 in Los Angeles.

The Long Beach earthquake struck on March 10, 1933, killing 120, injuring hundreds, and causing more than $50 million in property damage. It led to the passage of the state's Field Act, which established strict building code standards designed to make structures better able to withstand strong earthquakes.

Centered about 30 mi (48 km) north of downtown Los Angeles, the San Fernando earthquake killed 65, injured more than 2,000, and caused an estimated $505 million in property damage. The quake hit on February 9, 1971, and registered 6.5 on the Richter Scale and 11 on the Mercalli Scale. Most of the deaths occurred when the Veterans Administration Hospital in San Fernando collapsed.

The Loma Prieta earthquake occurred on October 18, 1989, in the Santa Cruz Mountains about 62 mi (100 km) south of San Francisco. It killed 63, injured 3,757, and caused an estimated $6 billion in property damage, mostly in San Francisco, Oakland, and Santa Cruz. The earthquake was a 6.9 on the Richter Scale and 9 on the Mercalli Scale.

The Northridge earthquake that struck Los Angeles on January 17, 1994, killed 72, injured 11,800, and caused an estimated $40 billion in damage. It registered 6.7 on the Richter Scale and 9 on the Mercalli Scale. It was centered about 30 mi (48 km) northwest of downtown Los Angeles.

In the past 100 years, Alaska has had many more severe earthquakes than California. However, they have occurred in mostly sparsely populated areas, so deaths, injuries, and damage have been light. Of the 15 strongest earthquakes ever recorded in the 50 states, 10 have been in Alaska, with the strongest registering a 9.2 (the second strongest ever recorded in the world) on the Richter Scale and 12 on the Mercalli Scale. It struck the Anchorage area on March 28, 1964, killing 125 (most from a tsunami or tidal wave caused by the earthquake), injuring hundreds, and causing $311 million in property damage.

The strongest earthquake ever recorded in the world registered 9.5 on the Richter scale and 12 on the Mercalli Scale. It occurred on May 22, 1960, and was centered off the coast of Chile. It killed 2,000, injured 3,000, and caused $675 million in property damage. A resulting tsunami caused death, injuries, and significant property damage in Hawaii, Japan, and the West Coast of the United States.

Every major earthquake raises the question of whether scientists will ever be able to predict exactly when and where one will strike. Today, scientists can only make broad predictions. For example, scientists believe there is at least a 50% chance that a devastating earthquake will strike somewhere along the San Andreas fault within the next 100 years. A more precise prediction is not yet possible. However, scientists in the United States and Japan are working on ways they might be able to make predictions more specific.

Ultra sensitive instruments placed across faults at the surface can measure the slow, almost imperceptible movement of fault blocks. This measurement records the great amount of potential energy stored at the fault boundary. In some areas, small earthquakes called foreshocks that precede a larger earthquake may help seismologists predict the larger earthquake. In other areas where seismologists believe earthquakes should be occurring but are not, this discrepancy between what is expected and what is observed may be used to predict an inevitable large-scale earthquake.

Other instruments measure additional fault-zone phenomena that seem to be related to earthquakes. The rate at which **radon** gas issues from rocks near faults has been observed to change before an earthquake. The properties of the rocks themselves (such as their ability to conduct electricity) have been observed to change as the tectonic force exerted on them slowly alters the rocks of the fault zone between earthquakes. Unusual animal behavior has

been reported before many earthquakes, and research into this phenomenon is a legitimate area of scientific inquiry, even though no definite answers have been found.

Techniques of studying earthquakes from space are also being explored. Scientists have found that ground displacements cause waves in the air that travel into the ionosphere and disturb electron densities. By using the network of satellites and ground stations that are part of the Global Positioning System (GPS), and data about the ionosphere that is already being collected, scientists may better understand the energy released from earthquakes. This may help scientists to predict them.

[*Ken R. Wells*]

RESOURCES
BOOKS

Henyey, Tom. *Natural Disasters: Earthquakes: A Reference Handbook.* Santa Barbara, CA: ABC-CLIO, 2002.
Hough, Susan Elizabeth. *Earthshaking Science: What We Know, and Don't Know, About Earthquakes.* Princeton, NJ: Princeton University Press, 2002.
Nicolson, Cynthia Pratt. *Earthquake.* Tonawanda, NY: Kids Can Press, 2002.

PERIODICALS

Chan-Kai, Alex. "Skate Disaster." *Stone Soup* (July 2001):26.
Johnson, Rita. "Whole Lotta Shakin' Goin' On!" *Boys' Life* (Dec. 2001):7–8.
Matty, Jane M. "Recent Quakes." *Rocks & Minerals* (March 2000):90.
Middleton, Nick. "Managing Earthquake Hazards In Los Angeles." *Geography Review* (May 2001):22.
Nur, Amos. "And the Walls Came Tumbling Down." *New Scientist* (July 6, 1991): 45–49.
Thompson, Dick. "Can We Save California? Predicting Earthquakes is One Thing; Preventing Them Would Be Something Else." *Time* (April 10, 2000):104+.

ORGANIZATIONS

National Earthquake Information Center, P.O. Box 25046, DFC, MS 967, Denver, CO USA 80225 (303) 273-8500, Fax: (303) 273-8450, Email: sedas@neis.cr.usgs.gov, <http://www.neic.usgs.gov>

Earthwatch

Earthwatch is a non-profit institution that provides paying volunteers to help scientists around the world conduct field research on environmental and cultural projects. It is one of the world's largest private sponsors of field research expeditions. Its mission is "to improve human understanding of the planet, the diversity of its inhabitants, and the processes which affect the quality of life on earth" by working "to sustain the world's **environment**, monitor global change, conserve endangered habitats and **species**, explore the vast heritage of our peoples, and foster world health and international cooperation."

The group carries out its work by recruiting volunteers to serve in an environmental EarthCorps and to work with research scientists on important environmental issues. The volunteers, who pay from $800 to over $2,500 to join two- or three-week expeditions to the far corners of the globe, gain valuable experience and knowledge on situations that affect the earth and human welfare.

By 2002, Earthwatch has sponsored over 1,180 projects in 50 countries around the world. By the end of the year it expects to have mobilized 4,300 volunteers, ranging in ages from 16 to 85, on 780 research teams. They will address such topics as **tropical rain forest ecology** and **conservation**; marine studies (ocean ecology); geosciences (climatology, geology, oceanography, glaciology, volcanology, paleontology); life sciences (**wildlife management**, biology, botany, ichthyology, herpetology, mammalogy, ornithology, primatology, zoology); social sciences (agriculture, economic anthropology, development studies, nutrition, public health); and art and archaeology (architecture, archaeoastronomy, ethnomusicology, folklore).

Since it was founded in 1971, Earthwatch has organized over 60,000 EarthCorps volunteers, who have contributed over $22 million and more than four million hours on some 1,500 projects in 150 countries and 36 states. No special skills are needed to be part of an expedition, and anyone 16 years or older can apply. Scholarships for students and teachers are also available. Earthwatch's affiliate, The Center for Field Research, receives several hundred grant applications and proposals every year from scientists and scholars who need volunteers to assist them on study expeditions.

Earthwatch publishes *Earthwatch* magazine six times a year, describing its research work in progress and the findings of previous expeditions. The group has offices in Los Angeles, Oxford, Melbourne, Moscow, and Tokyo and is represented in all 50 American states as well as in Germany, Holland, Italy, Spain, and Switzerland by volunteer field representatives.

[*Lewis G. Regenstein*]

RESOURCES

ORGANIZATIONS

Earthwatch, 3 Clock Tower Place, Suite 100 Box 75, Maynard, MA USA 01754 (978) 461-0081, Fax: (978) 461-2332, Toll Free: (800) 776-0188, Email: info@earthwatch.org, <http://www.earthwatch.org>

Eastern European pollution

Between 1987 and 1992 the disintegration of Communist governments of Eastern Europe allowed the people and press of countries from the Baltic to the Black Sea to begin recounting tales of life-threatening **pollution** and disastrous environmental conditions in which they lived. Villages in Czechoslovakia were black and barren because of **acid rain**, **smoke**, and **coal** dust from nearby factories. Drinking water from Estonia to Bulgaria was tainted with toxic **chemicals** and untreated sewage. Polish garden vegetables were inedible because of high **lead** and **cadmium** levels in the soil. Chronic health problems were endemic to much of the region, and none of the region's new governments had the spare cash necessary to alleviate their environmental liabilities.

The air, **soil**, and **water pollution** exposed by new environmental organizations and by a newly vocal press had its roots in Soviet-led efforts to modernize and industrialize Eastern Europe after 1945. (Often the term "Central Europe" is used to refer to Poland, Czech Republic, Slovakia, Hungary, Yugoslavia, and Bulgaria, and "Eastern Europe" to refer to the Baltic states, Belarus, and Ukraine. For the sake of simplicity, this essay uses the latter term for all these states.) Following Stalinist theory that modernization meant industry, especially heavy industries such as coal mining, steel production, and chemical manufacturing, Eastern European leaders invested heavily in industrial buildup. Factories were often built in resource-poor areas, as in traditionally agricultural Hungary and Romania, and they rarely had efficient or clean technology. Production quotas generally took precedence over health and environmental considerations, and billowing smokestacks were considered symbols of national progress. Emission controls on smokestacks and waste **effluent** pipes were, and are, rare. Soft, brown lignite coal, cheap and locally available, was the main fuel source. Lignite contains up to 5% sulfur and produces high levels of **sulfur dioxide**, **nitrogen oxides**, particulates, and other pollutants that contaminate air and soil in population centers, where many factories and **power plants** were built. The region's **water quality** also suffers, with careless disposal of toxic industrial wastes, untreated urban waste, and **runoff** from chemical-intensive agriculture.

By the 1980s the effects of heavy industrialization began to show. Dependence on lignite coal led to sulfur dioxide levels in Czechoslovakia and Poland eight times greater than those of Western Europe. The industrial triangle of Bohemia and Silesia had Europe's highest concentrations of ground-level **ozone**, which harms human health and crops. Acid rain, a result of industrial **air pollution**, had destroyed or damaged half of the forests in the former East Germany and the Czech Republic. Cities were threatened by outdated factory equipment and aging chemical storage containers and pipelines, which leaked **chlorine**, aldehydes, and other noxious gases. People in cities and villages experienced alarming numbers of **birth defects** and short life expectancies. Economic losses, from health care expenses, lost labor, and production inefficiency further handicapped hard-pressed Eastern European governments.

Popular protests against environmental conditions crystallized many of the movements that overturned Eastern and Central European governments. In Latvia, exposés on **petrochemical poisoning** and on environmental consequences of a hydroelectric project on Daugava River sparked the Latvian Popular Front's successful fight for independence. Massive campaigns against a proposed dam on the Danube River helped ignite Hungary's political opposition in 1989. In the same year, Bulgaria's Ecoglasnost group held Sofia's first non-government rally since 1945. The Polish Ecological Club, the first independent environmental organization in Eastern Europe, assisted the Solidarity movement in overturning the Polish government in the mid-1980s.

Citizens of these countries rallied around environmental issues because they had first-hand experience with the consequences of pollution. In Espenhain, of former East Germany, 80% of children developed chronic **bronchitis** or heart ailments before they were eight years old. Studies showed that up to 30% of Latvian children born in 1988 may have suffered from birth defects, and both children and adults showed unusually high rates of **cancer, leukemia,** skin diseases, bronchitis, and asthma. Czech children in industrial regions had acute **respiratory diseases,** weakened immune systems, and retarded bone development, and concentrations of lead and cadmium were found in children's hair. In the industrial regions of Bulgaria skin diseases were seven times more common than in cleaner areas, and cases of rickets and liver diseases were four times as common. Much of the air and soil contamination that produced these symptoms remains today and continues to generate health problems.

Water pollution is at least as threatening as air and soil pollution. Many cities and factories in the region have no facilities for treating **wastewater** and sewage. Existing treatment facilities are usually inadequate or ineffective. Toxic waste dumps containing old and rusting barrels of hazardous materials are often unmonitored or unidentified. Chemical **leaching** from poorly monitored waste sites threatens both surface water and **groundwater**, and water clean enough to drink has become a rare commodity. In Poland untreated sewage, mine **drainage**, and factory effluents make 95% of water unsafe for drinking. At least half of Polish rivers are too polluted, by government assessment, even for industrial use. According to government officials, 70% of all rivers in the industrial Czech region of Bohemia are heavily polluted, 40% of wastewater goes untreated, and nearly a third of the rivers have no fish. In Latvia's port town of Ventspils, heavy oil lies up to 3 ft (1 m) thick on

the river bottom. Phenol levels in the nearby Venta River exceed official limits by 800%.

Few pollution problems are geographically restricted to the country in which they were generated. Shared rivers and aquifers and regional weather patterns carry both airborne and water-borne pollutants from one country to another. The Chernobyl nuclear reactor disaster, which spread radioactive gases and particulates from Belarus across northern Europe and the Baltic Sea to northern Norway and Sweden is one infamous example of trans-border pollution, but other examples are common. The town of Ruse, Bulgaria has long been contaminated by chlorine gas emissions from a Romanian plant just across the Danube. Protests against this poisoning have unsettled Bulgarian and Romanian relations since 1987. Toxic wastes flowing into the Baltic Sea from Poland's Vistula River continue to endanger fisheries and shoreline habitats in Sweden, Germany, and Finland.

The Danube River is a particularly critical case. Accumulating and concentrating urban and industrial waste from Vienna to the Black Sea, this river supports industrial complexes of Austria, Czechia, Hungary, Croatia, Serbia, Bulgaria, and Romania. Before the Danube leaves Budapest, it is considered unsafe for swimming. Like other rivers, the Danube flows through a series of industrial cities and mining regions, each river uniting the pollution problems of several countries. Each city and farm along the way uses the contaminated water and contributes some pollutants of its own. Also like other rivers, the Danube carries its toxic load into the sea, endangering the marine environment.

Western countries from Sweden to the United States have their share of pollution and environmental disasters. The Rhine and the Elbe have disastrous **chemical spills** like those on the Danube and the Vistula. Like recent communist regimes, most western business leaders would prefer to disregard environmental and human health considerations in their pursuit of production quotas. Yet several factors set apart environmental conditions in Eastern Europe. Aside from its aged and outdated equipment and infrastructure, Eastern Europe is handicapped by its compressed geography, intense urbanization near factories, a long-standing lack of information and accurate records on environmental and health conditions, and severe shortages of clean-up funds, especially hard currency.

Eastern Europe's dense settlement crowds all the industrial regions of the Baltic states, Poland, the Czech and Slovak republics, and Hungary into an area considerably smaller than Texas but with a much higher population. This industrial zone lies adjacent to crowded manufacturing regions of Western Europe. In this compact region, people farm the same fields and live on the same mountains that are stripped for mineral extraction. Cities and farms rely on aquifers and rivers that receive factory effluent and **pesticide**

runoff immediately upstream. Furthermore, post-1945 industrialization gathered large labor forces into factory towns more quickly than adequate infrastructure could be built. Expanding urban populations had little protection from the unfiltered pollutants of nearby furnaces. At the same time that many Eastern Europeans were eye witnesses to environmental transgressions, little public discussion about the problem was possible. Official media disliked publicizing health risks or the destruction of forests, rivers, and lakes. Those **statistics** that existed were often unreliable. Air and water quality data were collected and reported by industrial and government officials, who could not afford bad test results.

Now that environmental conditions are being exposed, cleanup efforts remain hampered by a shortage of funding. Poland's long-term environmental restoration may cost $260 billion, or nearly eight times the country's annual GNP in the mid-1980s. Efforts to cut just sulfur dioxide emissions to Western standards would cost Poland about $2.4 billion a year. Hungary, with a mid-1980s GNP of $25 billion, could begin collecting and treating its sewage for about $5 billion. Cleanup in the port of Ventspils, Latvia, is expected to cost 3.6 billion rubles and $1.5 billion in hard currency. East German air, soil, and water **remediation** get a boost from their western neighbors, but the bill is expected to run between $40 and $150 billion.

Ironically, East European leaders see little choice for raising this money aside from expanded industrial production. Meanwhile, business leaders urge production expansion for other capital needs. Some Western investment in cleanup work has begun, especially on the part of such countries as Sweden and Germany, which share rivers and seas with polluting neighbors. Already in 1989 Sweden had begun work on water quality monitoring stations along Poland's Vistula River, which carries pollutants into the Baltic Sea. Capital necessary to purchase mitigation equipment, improve factory conditions, rebuild rusty infrastructure, and train environmental experts will probably be severely limited for decades to come, however.

Meanwhile, western investors are flocking to Eastern and Central Europe in hopes to build or rebuild business ventures for their own gain. The region is seen as one of quick growth and great potential. Manufacturers in heavy and light industries, automobiles, power plants, and home appliances are coming from Western Europe, North America, and Asia. From textile manufacturing to agribusiness, outside investors hope to reshape Eastern economies. Many Western companies are improving and updating equipment and adding **pollution control** devices. In a **climate** of uncertain regulation and rushed economic growth, however, no one knows if the region's new governments will be able or willing to enforce environmental safeguards or if the new investors will take advantage of weak regulations and poor enforcement as did their predecessors.

[*Mary Ann Cunningham Ph.D.*]

RESOURCES
BOOKS

French, H. F. "Restoring the Eastern European and Soviet Environments." In *State of the World 1991*. New York: Norton, 1991.

Feshbach, M., and A. Friendly, Jr. *Ecocide in the USSR*. New York: Basic Books, 1992.

PERIODICALS

Hartsock, J. "Latvia's Toxic Legacy." *Audubon* 94 (1992): 27-8.

Wallich, P. "Dark Days: Eastern Europe Brings to Mind the West's Polluted History." *Scientific American* 263 (1990): 16, 20.

Ebola

Ebola is a highly deadly viral hemorrhagic disease. As the disease progresses, the walls of blood vessels break down and blood gushes from every tissue and organ. The disease is caused by the Ebola **virus**, named after the river in Zaire (now the Democratic Republic of Congo) where the first known outbreak occurred. The disease is extremely contagious and exceptionally lethal. Where a 10% **mortality** rate is considered high for most infectious diseases, Ebola can kill up to 90% of its victims, usually within only a few days after exposure. It seems to take direct contact with contaminated blood or bodily fluids to catch the disease. Health personnel and caregivers are often the most likely to be infected. Even after a patient has died, preparing the body for a funeral can be deadly for families members.

The Ebola virus is one of two members of a family of RNA viruses called the Filoviridae. The other filovirus causes Marburg fever, an equally contagious and lethal hemorrhagic disease, named after a German town where it was first contracted by laboratory workers who handled imported monkeys infected with the virus. Together with members of three other families (arenaviruses, bunyanviruses, and flaviviruses), these viruses cause a group of deadly, episodic diseases including Lassa fever, Rift Valley fever, Bolivian fever, and Hanta or Four-Corners fever (named after the region of the southwestern United States where it was first reported).

The viruses associated with most of these emergent, hemorrhagic fevers are zoonotic. That means a **reservoir** of pathogens naturally resides in an animal host or arthropod vector. We don't know the specific host or vector for Ebola, but monkeys and other primates can contract related diseases. People who initially become infected with Ebola often have been involved in killing, butchering, and eating gorillas, chimps, or other primates. Why the viruses remain peacefully in their hosts for many years without causing much more trouble than a common cold, but then erupt sporadically and unpredictably into terrible human epidemics, is a new and growing question in **environmental health**.

The geographical origin for Ebola is unknown, but all recorded outbreaks have occurred in or around Central Africa, or in animals or people from this area. Ebola appears every few years in Africa. Confirmed cases have occurred in the Democratic Republic of the Congo, Gabon, Sudan, Uganda, and the Ivory Coast. No case of the disease in humans has ever been reported in the United States, but a variant called Ebola-Reston virus killed a number of monkeys in a research facility in Reston, Virginia. A fictionalized account of this outbreak was made into a movie called "Hot Zone." There probably are isolated cases in remote areas that go unnoticed. In fact, the disease may have been occurring in secluded villages deep in the jungle for a long time without outside attention. The most recent Ebola outbreak was in 2002 when about 100 people died in a remote part of Gabon and an adjacent area in Congo.

The worst epidemic of Ebola in humans occurred in 1995, in Kikwit, Zaire (now the Democratic Republic of Congo). Although many more people died in Kikwit that in any other outbreak, in many ways, the medical and social effects of the epidemic there was typical of what happens elsewhere. The first Kikwit victim was a 36-year-old laboratory technician named Kimfumu, who checked into a medical clinic complaining of a severe headache, stomach pains, fever, dizziness, weakness, and exhaustion. Surgeons did an exploratory operation to try to find the cause of his illness. To their horror, they found his entire gastrointestinal tract was necrotic and putrefying. He bled uncontrollably, and within hours was dead. By the next day, the five medical workers who had cared for Kimfumu, including an Italian nun who assisted in the operation, began to show similar symptoms, including high fevers, fatigue, bloody diarrhea, rashes, red and itchy eyes, vomiting, and bleeding from every body orifice. Less than 48 hours later, they, too, were dead, and the disease was spread throughout the city of 600,000.

As panicked residents fled into the bush, government officials responded to calls for help by closing off all travel—including humanitarian aid—into or out of Kikwit, about 250 mi (400 km) from Kinshasa, the national capitol. Fearful neighboring villages felled trees across the roads to seal off the pestilent city. No one dared enter houses where dead corpses rotted in the intense tropical heat. Boats plying the adjacent Kwilu River refused to stop to take on or **discharge** passengers or cargo. Food and clean water became scarce. Hospitals could hardly function as medicines and medical personal became scarce. Within a few weeks, about 400

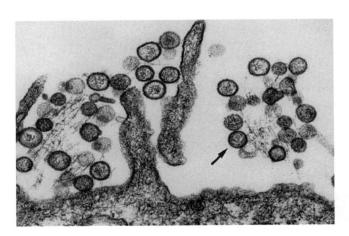

An electron micrograph of the ebola virus and the hanta virus. (Delmar Publishers Inc. Reproduced by permission.)

people in Kikwit had contracted the disease and at least 350 were dead. Eventually, the epidemic dissipated and disappeared. It isn't known why the infection rate dropped or what residents might do to prevent a further reappearance of the terrible disease.

Because health professionals are among the most likely to be exposed to Ebola when an outbreak occurs, it is important for them to have access to rapid antigen or antibody assays and isolation facilities to prevent further spread of the virus. Unfortunately, these advanced medical procedures generally are lacking in the African hospitals where the disease is most likely to occur. There is no standard treatment for Ebola other than supportive therapy. Patients are given replacement fluids and electrolytes, and oxygen levels and blood pressure are stabilized as much as possible. During the Kikwit outbreak, eight patients were given blood of individuals who had been infected with the virus but who had recovered. It was hoped that their blood might have antibodies to fight the infection. Seven of the eight transfusion patients survived, but the number tested is too small to be sure this was statistically significant. There is no vaccine or other antiviral drug available to prevent or halt an infection.

Several factors seem to be contributing to the appearance and spread of highly contagious diseases such as Ebola and Marburg fevers. With 6 billion people now inhabiting the planet, human densities are much higher enabling germs to spread further and faster than ever before. Expanding populations push people into remote areas where they encounter new pathogens and **parasites**. Environmental change is occurring on a larger scale: cutting forests, creating unhealthy urban surroundings, and causing global-climate change, among other things. Elimination of predators and

habitat changes favor disease-carrying organisms such as mice, rats, cockroaches, and mosquitoes.

Another important factor in the spread of many diseases is the speed and frequency of modern travel. Millions of people go every day from one place to another by airplane, boat, train, or **automobile**. Very few places on earth are more than 24 hours by jet plane from any other place. In 2001, a woman flying from the Congo arrived in Canada delirious with a high fever. She didn't, in fact, have Ebola, but Canadian officials were concerned about the potential spread of the disease. Finding ways to cure Ebola and prevent its spread may be more than simply a humanitarian concern for its victims in Central Africa. It might be very much in our own self-interest to make sure that this terrible disease doesn't cross our borders either accidentally or intentionally through actions of a terrorist organization.

[*William P. Cunningham Ph.D.*]

RESOURCES
BOOKS

Close, William T. *Ebola: Through the Eyes of the People.* London: Meadowlark Springs Productions, 2001.
Drexler, Madeline. *Secret Agents: the Menace of Emerging Infections.* Joseph Henry Press, 2002.
Preston, Richard. *The Hot Zone.* Anchor Books, 1995.

OTHER

Disease Information Fact Sheets: Ebola hemorrhagic fever. 2002. Center for Disease Control and Prevention. [cited July 9, 2002]. <http://www.cdc.gov/ncidod/dvrd/spb/mnpages/dispages/ebola.html>.

PERIODICALS

Daszak, P. et al. 2000."Emerging Infectious Diseases of Wildlife-Threats to Biodiversity and Human Health." *Science(US)* 287: 443–449.
Hughes, J.M. "Emerging infectious diseases: a CDC perspective." *Emerging Infectious Diseases.* 7(2001):494-6.
Osterholm, M. T. "Emerging Infections—Another Warning." *The New England Journal of Medicine* 342(17): 4-5.

Eco Mark

The Japanese environmental label known as "Eco Mark" is a relatively new addition to a worldwide effort to designate products that are environmentally friendly. The Eco Mark program was launched in February 1989. The symbol is two arms embracing the world, symbolizing the protection of the earth. The arms create the letter "e" with the earth in the center. Indicating English as the international language, the Japanese use "e" to stand for **environment**, earth, and **ecology**.

The Japanese program is entirely government funded, although a small fee is charged to applicant industries. The annual fee is based on the retail price of a product, not annual product sales as is the case for other national green labeling programs. Products ranging in price from $0–7 are charged an annual fee of $278.00; from $7–70 are charged an annual fee of $417; from $70–700 are charged an annual fee of $556; and products priced over $700 are charged an annual fee of $700. Obviously, those products that are low in price and high in volume sold are most likely to apply for the Eco Mark label.

The Eco Mark program seeks to sanction products with the following four qualities: 1) minimal environmental impact from use; 2) significant potential for improvement of the environment by using the product; 3) minimal environmental impact from disposal after use; and 4) other significant contributions to improve the environment.

In addition, labeled products must comply with the following guidelines: 1) appropriate environmental **pollution control** measures are provided at the stage of production; 2) ease of treatment for disposal of product; 3) energy or resources are conserved with use of product; 4) compliance with laws, standards, and regulations pertaining to quality and safety; 5) price is not extraordinarily higher than comparable products.

The Environment Association, supervised by the Japanese Environment Agency, is in charge of the Eco Mark program. All technical, research, and administrative support is provided by the government. The labeling program is guided by two committees.

The Eco Mark Promotion Committee acts primarily in a supervisory capacity, approving the guidelines for the program's operation and advising on operations, including evaluation of the program categories and criteria. The promotion committee consists of nine members representing industry, marketing groups, local governments, environmental agencies, and the National Institute for Environmental Studies.

In addition to the Promotion Committee there is a committee for approval of products. This committee consists of five members with representation from the science community, the consumer protection community, and, as in the Promotion Committee, a representative each from the Environment Agency and the National Institute for Environmental Studies. The Japanese program is completely voluntary for manufacturers. Once a product is approved by the Approval Committee, a two-year renewable licensing contract for the use of the Eco Mark is signed with the Japan Environment Association.

The Eco Mark program is very goal-oriented and places great emphasis on overall environmental impact. The attention to production impacts, as well as use and disposal impacts, makes the program unique within the family of green labeling programs worldwide. Its primary goals are to encourage innovation by industry and elevate the environmental awareness and consumer behavior of the Japanese people in order to enhance environmental quality.

Japan's Environment Agency claims that responses from consumer and environmental organizations have been positive, while industry has been less than enthusiastic. In fact, the Eco Mark only covered seven products in 1989 and now covers over 5,000. Some scientists have voiced concern over the superficiality of the analysis procedure used to determine Eco Mark products. However, despite criticisms, the Japanese Eco Mark program is a strong national effort to encourage environmentally sound decisions and protect the environment for **future generations** in that country. *See also* Environmental policy; Green packaging; Green products; Precycling; Recycling; Reuse; Waste reduction

[*Cynthia Fridgen*]

FURTHER READING

Salzman, J. *Environmental Labeling in OECD Countries.* Paris, France: OECD Technology and Environmental Program, 1991.

Ecoanarchism

see **Ecoterrorism**

Ecocide

Any substance that enters an ecological system, spreads throughout that system, and kills enough members of the **ecosystem** to disrupt its structure and function. For example, on July 14, 1991, a freight train carrying the **pesticide** metam sodium fell off a bridge near Dunsmuir, California, spilling its contents into the Sacramento River. When mixed with water this pesticide becomes highly poisonous, and all animal life for some distance downstream of the spill site was killed.

Ecofeminism

Coined in 1974 by the French feminist Francoise d'Eaubonne, ecofeminism, or ecological feminism, is a recent movement that asserts that the **environment** is a feminist issue and that feminism is an environmental issue. The term ecofeminism has come to describe two related movements operating at somewhat different levels: (1) the grassroots, women-initiated activism aimed at eliminating the oppression of women and **nature**; and (2) a newly emerging branch of philosophy that takes as its subject matter the foundational questions of meaning and justification in feminism and **environmental ethics**. The latter, more properly termed ecofeminist philosophy, stands in relation to the former as

theory stands to practice. Though closely related, there nevertheless remain important methodological and conceptual distinctions between action- and theory-oriented ecofeminism.

The ecofeminist movement developed from diverse beginnings, nurtured by the ideas and writings of a number of feminist thinkers, including Susan Griffin, Carolyn Merchant, Rosemary Radford Ruether, Ynestra King, Ariel Salleh, and Vandana Shiva. The many varieties of feminism (liberal, marxist, radical, socialist, etc.) have spawned as many varieties of ecofeminism, but they share a common ground. As described by Karren Warren, a leading ecofeminist philosopher, ecofeminists believe that there are important connections—historical, experiential, symbolic, and theoretical—between the domination of women and the domination of nature. In the broadest sense, then, ecofeminism is a distinct social movement that blends theory and practice to reveal and eliminate the causes of the dominations of women and of nature.

While ecofeminism seeks to end all forms of oppression, including racism, classism, and the abuse of nature, its focus is on gender bias, which ecofeminists claim has dominated western culture and led to a patriarchal, masculine value-oriented hierarchy. This framework is a socially constructed mindset that shapes our beliefs, attitudes, values, and assumptions about ourselves and the natural world.

Central to this patriarchal framework is a pattern of thinking that generates normative dualisms. These are created when paired complementary concepts such as male/female, mind/body, culture/nature, and reason/emotion are seen as mutually exclusive and oppositional. As a result of socially-entrenched gender bias, the more "masculine" member of each dualistic pair is identified as the superior one. Thus, a value hierarchy is constructed which ranks the masculine characteristics above the feminine (e.g., culture above nature, man above woman, reason above emotion). When paired with what Warren calls a "logic of domination," this value hierarchy enables people to justify the subordination of certain groups on the grounds that they lack the "superior" or more "valuable" characteristics of the dominant groups. Thus, men dominate women, humans dominate nature, and reason is superior to emotion. Within this patriarchal conceptual framework, subordination is legitimized as the necessary oppression of the inferior. Until we reconceptualize ourselves and our relation to nature in non-patriarchal ways, ecofeminists maintain, the continued dual denigration of women and nature is assured.

Val Plumwood, an Australian ecofeminist philosopher, has traced the roots of the development of the oppression of women and the exploitation of nature to three points, the first two points sharing historical origins, the third having its genesis in human psychology. In the first of these historical women-nature connections, dualism has identified higher and lower "halves." The lower halves, seen as possessing less or no **intrinsic value** relative to their polar opposites, are instrumentalized and subjugated to serve the needs of the members of the "higher" groups. Thus, due to their historical association and supposedly shared traits, women and nature have been systematically devalued and exploited to serve the needs of men and culture.

The second of these historical women-nature connections is said to have originated with the rise of mechanistic science before and during the Enlightenment period. According to some ecofeminists, dualism was not necessarily negative or hierarchical; however, the rise of modern science and technology, reflecting the transition from an organic to a mechanical view of nature, gave credence to a new logic of domination. Rationality and scientific method became the only socially sanctioned path to true knowledge, and individual needs gained primacy over community. On this fertile **soil** were sown the seeds for an ethic of exploitation.

A third representation of the connections between women and nature has its roots in human psychology. According to this account, the features of masculine consciousness which allow men to objectify and dominate are the result of sexually-differentiated personality development. As a result of women's roles in both creating and maintaining/ nurturing life, women develop "softer" ego boundaries than do men, and thus they generally maintain their connectedness to other humans and to nature, a connection which is reaffirmed and recreated generationally. Men, on the other hand, psychologically separate both from their human mothers and from Mother Earth, a process which results in their desire to subdue both women and nature in a quest for individual potency and transcendence. Thus, sex differences in the development of self/other identity in childhood are said to account for women's connectedness with, and men's alienation from, both humanity and nature.

Ecofeminism has attracted criticism on a number of points. One is the implicit assumption in certain ecofeminist writings that there is some connection between women and nature that men either do not possess or cannot experience. And, why female activities such as birth and childcare should be construed as more "natural" than some traditional male activities remains to be demonstrated. This assumption, though, has left some ecofeminists open to charges of having constructed a new value hierarchy to replace the old, rather than having abandoned hierarchical conceptual frameworks altogether. Hints of hierarchical thinking can be found in such ecofeminist practices as goddess worship and in the writings of some radical ecofeminists who advocate the abandonment of reason altogether in the search for an appropriate human-nature relationship. Rather than having destroyed gender bias, some ecofeminists are accused of merely at-

tempting to reverse its polarity, possibly creating new, subtle forms of women's oppression. Additionally, some would argue that ecofeminism runs the risk of oversimplification in suggesting that all struggles between dominator and oppressed are one and the same and thus can be won through unity.

A lively debate is currently underway concerning the compatibility of ecofeminism with other major theories or schools of thought in environmental philosophy. For instance, discussions of the similarities and differences between ecofeminism and **deep ecology** occupy a large portion of the recent theoretical literature on ecofeminism. While deep ecologists are primarily concerned with anthropocentrism as the primary cause of our destruction of nature, ecofeminists point instead to androcentrism as the key problem in this regard. Nevertheless, both groups aim for the expansion of the concept of "self" to include the natural world, for the establishment of a biocentric egalitarianism, and for the creation of connection, wholeness, and empathy with nature.

Given the newness of ecofeminism as a theoretical discipline, it is no surprise that the nature of ecofeminist ethics is still emerging. A number of different feminist-inspired positions are gaining prominence, including feminist **animal rights**, feminist environmental ethics based on caregiving, feminist **social ecology**, and feminist **bioregionalism**. Despite the apparent lack of a unified and overarching environmental philosophy, all forms of ecofeminism do share a commitment to developing ethics which do not sanction or encourage either the domination of any group of humans or the abuse of nature. Already, ecofeminism has shown us that issues in environmental ethics and philosophy cannot be meaningfully or adequately discussed apart from considerations of social domination and control. If ecofeminists are correct, then a fundamental reconstruction of the value and structural relations of our society, as well as a reexamination of the underlying assumptions and attitudes, is necessary.

[*Ann S. Causey*]

RESOURCES
BOOKS

Des Jardins, J. *Environmental Ethics: An Introduction to Environmental Philosophy.* Belmont, CA: Wadsworth, 1993.

Griffin, S. *Woman and Nature: The Roaring Inside Her.* New York: Harper & Row, 1978.

PERIODICALS

Adams, C., and K. Warren. "Feminism and the Environment: A Selected Bibliography." *APA Newsletter on Feminism and Philosophy* (Fall 1991).

Vance, Linda. "Remapping the Terrain: Books on Ecofeminism." *Choice* 30 (June 1993): 1585-93.

Ecojustice

The concept of ecojustice has at least two different usages among environmentalists. The first refers to a general set of attitudes about justice and the **environment** at the center of which is dissatisfaction with traditional theories of justice. With few exceptions (notably a degree of concern about excessive cruelty to animals), anthropocentric and egocentric Western moral and ethical systems have been unconcerned with individual plants and animals, **species**, oceans, **wilderness** areas, and other parts of the **biosphere**, except as they may be used by humans. In general, that which is non-human is viewed mainly as raw material for human uses, largely or completely without moral standing.

Relying upon holistic principles of biocentrism and **deep ecology**, the "ecojustice" alternative suggests that the value of non-human life-forms is independent of the usefulness of the non-human world for human purposes. Antecedents of this view can be found in sources as diverse as Eastern philosophy, Aldo Leopold's "land ethic," Albert Schweitzer's "reverence for life," and Martin Heidegger's injunction to "let beings be." The central idea of ecojustice is that the categories of ethical and moral reflection relevant to justice should be expanded to encompass **nature** itself and its constituent parts, and human beings have an obligation to take the inherent value of other living things into consideration whenever these living things are affected by human actions.

Some advocates of an ecojustice perspective base standards of just treatment on the evident capacity of many life-forms to experience pain. Others assert the equal inherent worth of all individual life-forms. More typically, environmental ethicists assert that all life-forms have at least some inherent worth, and thus deserve moral consideration, although perhaps not the same worth. The practical goals associated with ecojustice include the fostering of **stability** and diversity within and between self-sustaining ecosystems, harmony and balance in nature and within competitive biological systems, and **sustainable development**.

Ecojustice can also refer simply to the linking of environmental concerns with various social justice issues. The advocate of ecojustice typically strives to understand how the logic of a given economic system results in certain groups or classes of people bearing the brunt of **environmental degradation**. This entails, for example, concern with the frequent location of polluting industries and **hazardous waste** dumps near the economically disadvantaged (i.e., those with the least mobility and fewest resources to resist).

In much the same way, ecojustice also involves the fostering of sustainable development in less-developed areas of the globe, so that economic development does not mean the export of polluting industries and other environmental problems to these less-developed areas. An additional point of concern is the allocation of costs and benefits in environmental **reclamation** and preservation—for example, the preservation of Amazonian rain forests affects the global environment and may benefit the whole world, but the costs of this preservation fall disproportionately upon Brazil and the other countries of the region. An advocate of ecojustice would be concerned that the various costs and benefits of development be apportioned fairly. *See also* Biodiversity; Ecological and environmental justice; Environmental ethics; Environmental racism; Environmentalism; Holistic approach

[*Lawrence J. Biskowski*]

RESOURCES
BOOKS

Miller, A. S. *Gaia Connections*. Savage, MD: Rowman and Littlefield, 1991.

Ecological consumers

Organisms that feed either directly or indirectly on producers, plants that convert **solar energy** into complex organic molecules. Primary consumers are animals that eat plants directly. They are also called herbivores. Secondary consumers are animals that eat other animals. They are also called carnivores. Consumers that eat both plants and animals are omnivores. **Parasites** are a type of consumer that lives in or on the plant or animal on which it feeds. Detrivores (**detritus** feeders and **decomposers**) constitute a specialized class of consumers that feed on dead plants and animals. *See also* Biotic community

Ecological integrity

Ecological (or biological) integrity is a measure of how intact or complete an **ecosystem** is. Ecological integrity is a relatively new and somewhat controversial notion, however, which means that it cannot be defined exactly. Human activities cause many changes in environmental conditions, and these can benefit some **species**, communities, and ecological processes, while causing damages to others at the same time. The notion of ecological integrity is used to distinguish between ecological responses that represent improvements, and those that are degradations.

Challenges to ecological integrity

Ecological integrity is affected by changes in the intensity of environmental stressors. Environmental stressors can be defined as physical, chemical, and biological constraints on the productivity of species and the processes of ecosystem development. Many environmental stressors are associated with the activities of humans, but some are also natural factors. Environmental stressors can exert their influence on a local scale, or they may be regional or even global in their effects. Stressors represent environmental challenges to ecological integrity.

Environmental stressors are extremely complex, but they can be categorized in the following ways:

(1) Physical stressors are associated with brief but intense exposures to kinetic energy. Because of its acute, episodic nature, this represents a type of disturbance. Examples include volcanic eruptions, windstorms, and explosions; (2) **Wildfire** is another kind of disturbance, characterized by the **combustion** of much of the **biomass** of an ecosystem, and often the deaths of the dominant plants; (3) **Pollution** occurs when **chemicals** are present in concentrations high enough to affect organisms and thereby cause ecological changes. Toxic pollution may be caused by such gases as **sulfur dioxide** and **ozone**, metals such as **mercury** and **lead**, and pesticides. Nutrients such as phosphate and nitrate can affect ecological processes such as productivity, resulting in a type of pollution known as eutrophication; (4) Thermal stress occurs when releases of heat to the **environment** cause ecological changes, as occurs near natural hot-water vents in the ocean, or where there are industrial discharges of warmed water; (5) Radiation stress is associated with excessive exposures to ionizing energy. This is an important stressor on mountaintops because of intense exposures to **ultraviolet radiation**, and in places where there are uncontrolled exposures to radioactive wastes; (6) Climatic stressors are associated with excessive or insufficient regimes of temperature, moisture, solar radiation, and combinations of these. **Tundra** and deserts are climatically stressed ecosystems, while tropical rain forests occur in places where the climatic regime is relatively benign; (7) Biological stressors are associated with the complex interactions that occur among organisms of the same or different species. Biological stresses result from **competition**, herbivory, predation, parasitism, and disease. The harvesting and management of species and ecosystems by humans can be viewed as a type of biological stress.

All species and ecosystems have a limited capability for tolerating changes in the intensity of environmental stressors. Ecologists refer to this attribute as **resistance**. When the limits of tolerance to **environmental stress** are exceeded, however, substantial ecological changes are caused.

Large changes in the intensity of environmental stress result in various kinds of ecological responses. For example, when an ecosystem is disrupted by an intense disturbance, there will be substantial **mortality** of some species and other damages. This is followed by recovery of the ecosystem through the process of **succession**. In contrast, a longer-term intensification of environmental stress, possibly caused

by chronic pollution or **climate** change, will result in longer lasting ecological adjustments. Relatively vulnerable species become reduced in abundance or are eliminated from sites that are stressed over the longer term, and their modified niches will be assumed by more tolerant species. Other common responses of an intensification of environmental stress include a simplification of species richness, and decreased rates of productivity, **decomposition**, and **nutrient** cycling. These changes represent a longer-term change in the character of the ecosystem.

Components of ecological integrity

Many studies have been made of the ecological responses to both disturbance and to longer-term changes in the intensity of environmental stressors. Such studies have, for instance, examined the ecological effects of air or **water pollution**, of the harvesting of species or ecosystems, and the conversion of natural ecosystems into managed agroecosystems. The commonly observed patterns of change in stressed ecosystems have been used to develop indicators of ecological integrity, which are useful in determining whether this condition is improving or being degraded over time. It has been suggested that greater ecological integrity is displayed by systems that, in a relative sense: (1) *are resilient and resistant to changes in the intensity of environmental stress.* Ecological resistance refers to the capacity of organisms, populations, or communities to tolerate increases in stress without exhibiting significant responses. Once thresholds of tolerance are exceeded, ecological changes occur rapidly. **Resilience** refers to the ability to recover from disturbance; (2) *are biodiverse.* **Biodiversity** is defined as the total richness of biological variation, including genetic variation within populations and species, the numbers of species in communities, and the patterns and dynamics of these over large areas; (3) *are complex in structure and function.* The complexity of the structural and functional attributes of ecosystems is limited by natural environmental stresses associated with climate, **soil**, chemistry, and other factors, and also by stressors associated with human activities. As the overall intensity of stress increases or decreases, structural and functional complexity responds accordingly. Under any particular environmental regime, older ecosystems will generally be more complex than younger ecosystems; (4) *have large species present.* The largest species in any ecosystem appropriate relatively large amounts of resources, occupy a great deal of space, and require large areas to sustain their populations. In addition, large species tend to be long-lived, and consequently they integrate the effects of stressors over an extended time. As a result, ecosystems that are affected by intense environmental stressors can only support a few or no large species. In contrast, mature ecosystems occurring in a relatively benign environmental regime are dominated by large, long-lived species; (5) *have higher-order predators present.* Top predators are sustained by a broad base of **ecological productivity**, and consequently they can only occur in relatively extensive and/or productive ecosystems; (6) *have controlled nutrient cycling.* Ecosystems that have recently been disturbed lose some of their biological capability for controlling the cycling of nutrients, and they may lose large amounts of nutrients dissolved or suspended in stream water. Systems that are not "leaky" of their nutrient capital are considered to have greater ecological integrity; (7) *are efficient in energy use and transfer.* Large increases in environmental stress commonly result in community-level **respiration** exceeding productivity, resulting in a decrease in the standing crop of biomass in the system. Ecosystems that are not losing their capital of biomass are considered to have greater integrity than those in which biomass is decreasing over time; (8) *have an intrinsic capability for maintaining natural ecological values.* Ecosystems that can naturally maintain their species, communities, and other important characteristics, without being managed by humans, have greater ecological integrity. If, for example, a population of a **rare species** can only be maintained by management of its **habitat** by humans, or by a program of captive-breeding and release, then its population, and the ecosystem of which it is a component, are lacking in ecological integrity; (9) *are components of a "natural" community.* Ecosystems that are dominated by non-native, **introduced species** are considered to have less ecological integrity than ecosystems composed of indigenous species.

Indicators (8) and (9) are related to "naturalness" and the roles of humans in ecosystems, both of which are philosophically controversial topics. However, most ecologists would consider that self-organizing, unmanaged ecosystems composed of native species have greater ecological integrity than those that are strongly influenced by humans. Examples of strongly human-dominated systems include agroecosystems, forestry plantations, and urban and suburban areas. None of these ecosystems can maintain their character in the absence of management by humans, including large inputs of energy and nutrients.

Indicators of ecological integrity

Indicators of ecological integrity vary greatly in their intent and complexity. For instance, certain metabolic indicators have been used to monitor the responses by individuals and populations to toxic stressors, as when bioassays are made of **enzyme** systems that respond vigorously to exposures to **dichlorodiphenyl-trichloroethane** (DDT), pentachlorophenols (PCBs), and other **chlorinated hydrocarbons**. Other simple indicators include the populations of **endangered species**; these are relevant to the viability of those species as well as the integrity of the ecosystem of

which they are a component. There are also indicators of ecological integrity at the level of landscape, and even global indicators relevant to climate change, depletion of stratospheric ozone, and **deforestation**.

Relatively simple indicators can sometimes be used to monitor the ecological integrity of extensive and complex ecosystems. For example, the viability of populations of spotted owls (*Strix occidentalis*) is considered to be an indicator of the integrity of the old-growth forests in which this endangered species breeds in the western United States. These forests are commercially valuable, and if plans to harvest and manage them are judged to threaten the viability of a population of spotted owls, this would represent an important challenge to the integrity of the **old-growth forest** ecosystem.

Ecologists are also beginning to develop composite indicators of ecological integrity. These are designed as summations of various indicators, and are analogous to such economic indices such as the Dow-Jones Index of stock markets, the Consumer Price Index, and the gross domestic product of an entire economy. Composite economic indicators of this sort are relatively simple to design, because all of the input data are measured in a common way (for example, in dollars). In **ecology**, however, there is no common currency among the many indicators of ecological integrity. Consequently it is difficult to develop composite indicators that ecologists will agree upon.

Still, some research groups have developed composite indicators of ecological integrity that have been used successfully in a number of places and environmental contexts. For instance, the ecologist James Karr and his co-workers have developed composite indicators of the ecological integrity of aquatic ecosystems, which are being used in modified form in many places in North America.

In spite of all of the difficulties, ecologists are making substantial progress in the development of indicators of ecological integrity. This is an important activity, because our society needs objective information about complex changes that are occurring in environmental quality, including degradations of indigenous species and ecosystems. Without such information, actions may not be taken to prevent or repair unacceptable damages that may be occurring.

Increasingly, it is being recognized that human economies can only be sustained over the longer term by ecosystems with integrity. Ecosystems with integrity are capable of supplying continuous flows of such renewable resources as timber, fish, agricultural products, and clean air and water. Ecosystems with integrity are also needed to sustain populations of native species and their natural ecosystems, which must be sustained even while humans are exploiting the resources of the **biosphere**.

[*Bill Freedman Ph.D.*]

RESOURCES
BOOKS

Freedman, B. *Environmental Ecology,* 2nd ed. San Diego: Academic Press, 1995.
Woodley, S., J. Kay, and G. Francis, eds. *Ecological Integrity and the Management of Ecosystems.* Boca Raton, FL: St. Lucie Press, 1993.

PERIODICALS

Karr, J. "Defining and assessing ecological integrity: Beyond water quality." *Environmental Toxicology and Chemistry* 12 (1993): 1521-1531.

Ecological productivity

One of the most important properties of an **ecosystem** is its productivity, which is a measure of the rate of incorporation of energy by plants per unit area per unit time. In terrestrial ecosystems, ecologists usually estimate plant production as the total annual growth—the increase in plant **biomass** over a year. Since productivity reflects plant growth, it is often used loosely as a measure of the organic fertility of a given area.

The flow of energy through an ecosystem starts with the fixation of sunlight by green plants during **photosynthesis**. Photosynthesis supplies both the energy (in the form of chemical bonds) and the organic molecules (glucose) that plants use to make other products in a process known as biosynthesis. During biosynthesis, glucose molecules are rearranged and joined together to become complex carbohydrates (such as cellulose and starch) and lipids (such as fats and plant oils). These products are also combined with **nitrogen, phosphorus**, sulfur, and magnesium to produce the proteins, nucleic acids, and pigments required by the plant. The many products of biosynthesis are transported to the leaves, flowers, and roots, where they are stored to be used later.

Ecologists measure the results of photosynthesis as increases in plant biomass over a given time. To do this more accurately, ecologists distinguish two measures of assimilated light energy: gross primary production (GPP), which is the total light energy fixed during photosynthesis, and net primary production (NPP), which is the chemical energy that accumulates in the plant over time.

Some of this chemical energy is lost during plant **respiration** (R) when it is used for maintenance, reproduction, and biosynthesis. The proportion of GPP that is left after respiration is counted as net production (NPP). In an ecosystem, it is the energy stored in plants from net production that is passed up the **food chain/web** when the plants are eaten. This energy is available to consumers either directly as plant tissue or indirectly through animal tissue.

One measure of ecological productivity in an ecosystem is the production efficiency. This is the rate of accumula-tion of biomass by plants, and it is calculated as the ratio of net primary production to gross primary production. Production efficiency varies among plant types and among ecosystems. Grassland ecosystems which are dominated by non-woody plants are the most efficient at 60-85%, since grasses and annuals do not maintain a high supporting biomass. On the other end of the efficiency scale are forest ecosystems; they are dominated by trees, and large old trees spend most of their gross production in maintenance. For example, eastern deciduous forests have a production efficiency of about 42%.

Ecological productivity in terrestrial ecosystems is influenced by physical factors such as temperature and rainfall. Productivity is also affected by air and water currents, **nutrient** availability, land forms, light intensity, altitude, and depth. The most productive ecosystems are tropical rain forests, coral reefs, salt marshes and estuaries; the least productive are deserts, **tundra**, and the open sea. *See also* Ecological consumers; Ecology; Habitat; Restoration ecology

[*Neil Cumberlidge Ph.D.*]

Ecological risk assessment

Ecological **risk assessment** is a procedure for evaluating the likelihood that adverse ecological effects are occurring, or may occur, in ecosystems as a result of one or more human activities. These activities may include the alteration and destruction of **wetlands** and other habitats, the introduction of herbicides, pesticides, and other toxic materials into the **environment**, **oil spills**, or the cleanup of contaminated **hazardous waste** sites. Ecological risk assessments consider many aspects of an **ecosystem**, both the biotic plants and animals and the abiotic water, soils, and other elements. Ecosystems can be as small as a pond or stretch of a river or as large as thousands of square miles or lengthy coastlines in which communities exist.

Although closely related to human health risk assessment, ecological risk assessment is not only a newer discipline but also uses different procedures, terminology, and concepts. Both human health and ecological risk assessment provide frameworks for collecting information to define a risk and to help make risk management or regulatory decisions. But human health risk assessment follows four basic steps that were defined in a 1983 by the **National Research Council**: hazard assessment, **dose response** assessment, exposure assessment, and risk characterization. In contrast, ecological risk assessment relies on a *Framework for Ecological Risk Assessment* published by **Environmental Protection Agency** in 1992 as part of a long-term plan to develop ecological risk assessment guidelines. The Framework defines three steps: problem formulation, analysis, and risk characterization.

The problems that human health risk assessments seek to address are clearly defined: **cancer, birth defects, mortality,** and the like. But the problems that ecological risk assessments tries to understand and deal with are less straightforward. For instance, a major challenge ecological **risk assessors** face is distinguishing natural changes in an ecosystem from changes caused by human activities and defining what changes are unacceptable. As a result, the initial problem formulation step of an ecological risk assessment requires extensive discussions between risk assessors and risk managers to define "ecological significance," a key concept in ecological risk assessment. Because it is not immediately clear whether an ecological change is positive or negative—unlike cancer or birth defects, which are known to be adverse—judgments must be made early in the assessment about whether a change is significant and whether it will alter a socially valued ecological condition. For example, **Lake Erie** was declared "dead" in the 1960s as a result of phosphorous loadings from cities and farms. But, in fact, there were more fish in the lake after it was "dead" than before; however, these fish were carp, suckers, catfish, not the walleyed pike, yellow perch, and other fish that had made Lake Erie one of the highest valued freshwater sport fishing lakes in the United States. More recently, with **pollution** inputs greatly reduced, the lake has recovered much of its former productivity. Choosing one ecological condition over the other is a social value of the kind fundamental to ecological risk assessment problem formulation. Once judgments have been made about what values to protect, analysis can proceed to examine the "stressors" that ecosystems are exposed to and a characterization can be made of the "ecological effects" likely from such stressors.

Since 1989, EPA has held workshops on "ecological significance" and other technical issues pertaining to ecological risk assessment and has published the results of its workshops in a series of reports and case studies. In 1996, EPA proposed its first ecological risk assessment guidelines, with final guidelines published in May of 1998. Overall, the direction of ecological protection priorities has been away from earlier concerns with narrow goals (e.g., use of commercially valuable **natural resources**) toward broader interest in protecting natural areas such as National Parks and Scenic Rivers for both present and **future generations** to enjoy.

[*David Clarke*]

RESOURCES
BOOKS

Ecological Risk Assessment Issues Papers. Washington, D.C.: United States Environmental Protection Agency, Risk Assessment Forum, 1994.
Framework for Ecological Risk Assessment. Washington, D.C.: United States Environmental Protection Agency, Risk Assessment Forum, 1992.

Priorities for Ecological Protection: An Initial List and Discussion Document for EPA. Washington, D.C. United States Environmental Protection Agency, 1997.

PERIODICALS
Lackey, R. T. "The Future of Ecological Risk Assessment." *Human and Ecological Risk Assessment, An International Journal* 1, no. 4 (October 1995): 339-343.

Ecological Society of America

The Ecological Society of America (ESA), representing 7,500 ecological researchers in the United States, Canada, Mexico, and 62 other countries, was founded in 1915 as a non-profit, scientific organization and today is the nation's leading professional society of ecologists. Members include ecologists from academia, government agencies, industry, and non-profit organizations. In pursuing its goal of promoting "the responsible application of ecological principles to the solution of environmental problems," the Society publishes reports, membership research, and three scientific journals a year, and it provides expert testimony to Congress. In addition, ESA holds a conference every summer attended by more than 3,000 scientists and students at which members present the latest ecological research. The Society's three journals are: *Ecology* (eight issues per year), *Ecological Monographs* (four issues per year), and *Ecological Applications* (four issues per year). ESA also publishes a bimonthly member newsletter.

A milestone in the Society's development was its 1991 proposal for a **Sustainable Biosphere** Initiative (SBI), which was published in a 1991 issue of *Ecology* as a "call-to-arms for ecologists." Based on research priorities identified in the proposal, ESA chartered the SBI Project Office in the same year to focus on global change, **biodiversity**, and sustainable ecosystems. The SBI marked a commitment by the Society to more actively convey its members' findings to the public and to policy makers, and, as such, included research, education, and environmental decision-making components. The three-pronged SBI proposal grew out of a "period of introspection" during which ESA led its members to examine "the whole realm of ecological activities" in the face of decreasing funds for research, an urgent need to set priorities, and "the need to ameliorate the rapidly deteriorating state of the **environment** and to enhance its capacity to sustain the needs of the world's population."

Since the SBI Project Office was chartered, it has focused on linking the ecological scientific community to other scientists and decision makers through a multi-disciplinary 12-member Steering Committee and five-member staff. For instance, in 1995, the SBI began a series of semi-annual meetings with federal government officials to discuss "overlapping areas of interest and possible collaborative opportunities." In addition, SBI has hosted discussions of key ecological topics, such as a symposium on " The Effects of Fishing Activities on Benthic Habitats" that SBI and the American Fisheries Society, the Ecological Society of America, the **National Oceanic and Atmospheric Administration**, and the US **Geological Survey** organized in 2002 and another SBI-hosted discussion on "Ecosystem Simplification: Why a Patchwork Quilt is More Valuable than a Burlap Sack" at the ESA's 2002 Annual Meeting.

In 1993 ESA chartered a Special Committee on the Scientific Basis of **Ecosystem Management** to establish the scientific grounds for discussing the increasingly prominent ecosystem approach to addressing land and natural resource management problems. The committee published its findings in the August 1996 issue of *Ecological Applications*. Articles discussed the emerging consensus on essential elements of ecosystem management, including its "holistic" nature—incorporating the biological and physical elements of an ecosystem and their interrelationships—and the concept of "sustainability" as the "essential element and precondition" of ecosystem management.

ESA's headquarters in Washington, D.C., consistent with the SBI's goal of broader public education, includes a Public Affairs Office. Its Publications Office is in Ithaca, New York.

[*David Clarke*]

RESOURCES
PERIODICALS

"Forum: Perspectives on Ecosystem Management". *Ecological Applications*, 6, no. 3 (August 1996): 694–747.
"The Sustainable Biosphere Initiative: An Ecological Research Agenda". *Ecology*, 72, no. 2 (1991): 371–412,

ORGANIZATIONS
Ecological Society of America, 1707 H St, NW, Suite 400, Washington, D.C. USA 20006 (202) 833-8773, Fax: (202) 833-8775, Email: esahq@esa.org, <http://www.esa.org>

Ecological economics

Although **ecology** and economics share the common root "eco-" (from Greek *Oikos* or household), these disciplines have tended to be at odds with each other in recent years over issues such as the feasibility of continued economic growth and the value of **natural resources** and environmental services.

Economics deals with resource allocation or trade-offs between competing wants and needs. Economists ask, "what shall we produce, for whom, or for what purpose?" Furthermore, they ask, "when and in what manner should we produce these goods and services?" In mainstream, neoclassical

economics, these questions are usually limited to human concerns: what will it cost to obtain the things we desire and what benefits will we derive from them?

According to classical economists, the costs of goods and services are determined by the interaction of supply and demand in the marketplace. If the supply of a particular commodity or service is high but the demand is low, the price will be low. If the commodity is scarce but everyone wants it, the price will be high. But high prices also encourage invention of new technology and substitutes that can satisfy the same demands. The cyclic relationship of scarce resources and development of new technology or new materials, in this view, allows for unlimited growth. And continued economic growth is seen as the best, perhaps the only, solution to poverty and **environmental degradation**.

Ecologists, however, view the world differently than economists. From their studies of the interactions between organisms and their **environment**, ecologists see our world as a dynamic, but finite system that can support only a limited number of humans with their demands for goods and services. Many ecological processes and the nonrenewable natural resources on which our economy is based have no readily available substitutes. Further, much of the natural world is being degraded or depleted at unsustainable rates. Ecologists criticize the narrow focus of conventional economics and its faith in unceasing growth, market valuation, and endless substitutability. Ecologists warn that unless we change our patterns of production and consumption to ways that protect natural resources and ecological systems, we will soon be in deep trouble.

Ecological economics

Ecological or **environmental economics** is a relatively new field that introduces ecological understanding into our economic discourse. It takes a transdisciplinary, holistic, contextual, value-sensitive approach to economic planning and resource allocation. This view recognizes our dependence on the natural world and the irreplaceable life-support services it renders. Rather than express values solely in market prices, ecological economics pays attention to intangible values, nonmarketed resources, and the needs and rights of **future generations** and other **species**. Issues of equitable distribution of access to resources and the goods and services they provide need to be solved, in this perspective, by means other than incessant growth.

Where neoclassical economics sees our environment as simply a supply of materials, services, and waste sinks, ecological economics regards human activities as embedded in a global system that places limits on what we can and cannot do. Uncertainty and dynamic change are inherent characteristics of this complex natural system. Damage caused by human activities may trigger sudden and irreversible changes. The precautionary principle suggests that we should leave a margin for error in our use of resources and plan for **adaptive management** policies.

Natural capital

Conventional economists see wealth generated by human capital (human knowledge, experience, and enterprise) working with manufactured capital (buildings, machines, and infrastructure) to transform raw materials into useful goods and services. In this view, economic growth and efficiency are best accomplished, by increasing the throughput of raw materials extracted from **nature**. Until they are transformed by human activities, natural resources are regarded as having little value. In contrast, ecological economists see natural resources as a form of capital equally important with human-made capital. In addition to raw materials such as minerals, fuels, fresh water, food, and fibers, nature provides valuable services on which we depend. Natural systems assimilate our wastes and regulate the earth's energy balance, global **climate**, material **recycling**, the chemical composition of the **atmosphere** and oceans, and the maintenance of **biodiversity**. Nature also provides aesthetic, spiritual, cultural, scientific and educational opportunities that are rarely given a monetary value but are, nevertheless, of great significance to many of us.

Ecological economists argue that the value of natural capital should be taken into account rather than treated as a set of unimportant externalities. Our goal, in this view, should be to increase our efficiency in natural resource use and to reduce its throughput. Harvest rates for renewable resources (those like organisms that regrow or those like fresh water that are replenished by natural processes) should not exceed regeneration rates. Waste emissions should not exceed the ability of nature to assimilate or recycle those wastes. **Nonrenewable resources** (such as minerals) may be exploited by humans, but only at rates equal to the creation of renewable substitutes.

Accounting for natural capital

Where neoclassical economics seeks to maximize present value of resources, ecological economics calls for recognition of the real value of those resources in calculating economic progress. A market economist, for example, once argued that the most rational management policy for **whales** was to harvest all the remaining ones immediately and to invest the proceeds in some profitable business. Whales reproduce too slowly, he claimed, and are too dispersed to make much money in the long run by allowing them to remain wild. Ecologists reject this limited view of whales as only economic units of production. They see many other values in these wild, beautiful, sentient creatures. Furthermore whales may play important roles in marine ecology that we don't yet fully understand.

Ecologists are similarly critical of Gross National Product (GNP) as a measure of national progress or well-

being. GNP measures only the monetary value of goods and services produced in a national economy. It doesn't attempt to distinguish between economic activities that are beneficial or harmful. People who develop **cancer** from smoking, for instance, contribute to the GNP by running up large hospital bills. The pain and suffering they experience doesn't appear on the balance sheets. When calculating GNP in conventional economics, a subtraction is made, for capital depreciation in the form of wear and tear on machines, vehicles, and buildings used in production, but no account is made for natural resources used up or ecosystems damaged by that same economic activity.

Robert Repeto of the **World Resources Institute** estimates that **soil erosion** in Indonesia reduces the value of crop production about 40% per year. If natural capital were taken into account, total Indonesian GNP would be reduced by at least 20% annually. Similarly, Costa Rica experienced impressive increases in timber, beef, and banana production between 1970 and 1990. But decreased natural capital during this period represented by soil erosion, forest destruction, biodiversity losses, and accelerated water **runoff** add up to at least $4 billion, or about 25%, of annual GNP. Ecological economists call for a new System of National Accounts that recognizes the contribution of natural capital to economic activity.

Valuation of natural capital

Ecological economics requires new tools and new approaches to represent nature in GNP. Some categories in which natural capital might fit include:

- use values: the price we pay to use or consume a resource
- option value: preserving options for the future
- existence value: those things we like to know still exist even though we may never use or even see them
- aesthetic value: things we appreciate for their beauty
- cultural value: things important for cultural identity
- scientific and educational value: information or experience-rich aspects of nature.

How can we measure this value of natural resources and ecological services not represented in market systems? Ecological economists often have to resort to "shadow pricing" or other indirect valuation methods for natural resources. For instance, what is the worth of a day of canoeing on a **wild river**? We might measure opportunity costs such as how much we pay to get to the river or to rent a canoe. The direct out-of-pocket costs might represent only a small portion, however, of what it is really worth to participants. Another approach is contingent valuation in which potential resource users are asked, "how much would you be willing to pay for this experience?" or "what price would you be willing to accept to sell your access or forego this opportunity?" These approaches are controversial because people may report what they think they ought to pay rather than what they would really pay for these activities.

Carrying capacity and sustainable ddevelopment

Carrying capacity is the maximum number of organisms of a particular species that a given area can sustainably support. Where neoclassical economists believe that technology can overcome any obstacle and that human ingenuity frees us from any constraints on population or economic growth, ecological economists argue that nature places limits on us just as it does on any other species.

One of the ultimate limits we face is energy. Because of the limits of the second law of thermodynamics, whenever work is done, some energy is converted to a lower quality, less useful form and ultimately is emitted as waste heat. This means that we require a constant input of external energy. Many fossil fuel supplies are nearing exhaustion, and continued use of these sources by current technology carries untenable environmental costs. Vast amounts of **solar energy** reach the earth, and this solar energy already drives the generation of all renewable resources and ecological services. By some calculations, humans now control or directly consume about 40% of all the solar energy reaching the earth. How much more can we monopolize for our own purposes without seriously jeopardizing the integrity of natural systems for which there is no substitute? And even if we had an infinite supply of clean, **renewable energy**, how much heat can we get rid of without harming our environment?

Ecological economics urges us to restrain growth of both human populations and the production of goods and services in order to conserve natural resources and to protect remaining natural areas and biodiversity. This does not necessarily mean that the billion people in the world who live in absolute poverty and cannot, on their own, meet the basic needs for food, shelter, clothing, education, and medical care are condemned to remain in that state. Ecological economics calls for more efficient use of resources and more equitable distribution of the benefits among those now living as well as between current generations and future ones.

A mechanism for attaining this goal is **sustainable development**, that is, a real improvement in the overall welfare of all people on a long-term basis. In the words of the World Commission on Economy and Development, sustainable development means "meeting the needs of the present without compromising the ability of future generations to meet their own needs." This requires increased reliance on renewable resources in harmony with ecological systems in ways that do not deplete or degrade natural capital. It doesn't necessarily mean that all growth must cease. There are many human attributes such as knowledge, kindness, compassion, cooperation, and creativity that can expand infinitely without damaging our environment. While ecological economics offers a sensible framework for approaches to

resource use that can be in harmony with ecological systems over the long term, it remains to be seen whether we will be wise enough to adopt this framework before it is too late.

[*William P. Cunningham Ph.D.*]

RESOURCES
BOOKS

Jansson, A.M., et al., eds. *Investing in Natural Capital: the Ecological Economics Approach to Sustainability*. Washington, D.C.: Island Press, 1994.

Krishnan, R., J.M. Harris, and N.R. Goodwin, eds. *A Survey of Ecological Economics*. Washington, D.C.: Island Press, 1995.

Prugh, T. *Natural Capital and Human Economic Survival*. Solomons, MD: International Society for Ecological Economics, 1995.

Turner, R.K., D. Pearce, and I. Bateman. *Environmental Economics: an Elementary Introduction*. Baltimore: The Johns Hopkins University Press, 1993.

Ecological succession

see **Succession**

Ecology

The word ecology was coined in 1870 by the German zoologist Ernst Haeckel from the Greek words *oikos* (house) and *logos* (logic or knowledge) to describe the scientific study of the relationships among organisms and their **environment**. Biologists began referring to themselves as ecologists at the end of the nineteenth century and shortly thereafter the first ecological societies and journals appeared. Since that time ecology has become a major branch of biological science. The contextual, historical understanding of organisms as well as the systems basis of ecology set it apart from the reductionist, experimental approach prevalent in many other areas of science.

This broad ecological view is gaining significance today as modern resource-intensive lifestyles consume much of nature's supplies. Although intuitive ecology has always been a part of some cultures, current environmental crises make a systematic, scientific understanding of ecological principles especially important.

For many ecologists the basic structural units of ecological organization are **species** and populations. A biological species consists of all the organisms potentially able to interbreed under natural conditions and to produce fertile offspring. A population consists of all the members of a single species occupying a common geographical area at the same time. An ecological community is composed of a number of populations that live and interact in a specific region.

This population-community view of ecology is grounded in natural history—the study of where and how organisms live—and the Darwinian theory of natural selection and **evolution**. Proponents of this approach generally view ecological systems primarily as networks of interacting organisms. Abiotic forces such as weather, soils, and **topography** are often regarded as external factors that influence but are apart from the central living core of the system.

In the past three decades the emphasis on species, populations, and communities in ecology has been replaced by a more quantitative, thermodynamic analysis of the processes through which energy flows and the cycling of nutrients and **toxins** are carried out in ecosystems. This process-functional approach is concerned more with the **ecosystem** as a whole than the particular species or populations that make it up. In this perspective, both the living organisms and the abiotic physical components of the environment are equal members of the system.

The feeding relationships among different species in a community are a key to understanding ecosystem function. Who eats whom, where, how, and when determine how energy and materials move through the system. They also influence natural selection, evolution, and species **adaptation** to a particular set of environmental conditions. Ecosystems are open systems, insofar as energy and materials flow through them. Nutrients, however, are often recycled extremely efficiently so that the annual losses to sediments or through surface water **runoff** are relatively small in many mature ecosystems. In undisturbed tropical rain forests, for instance, nearly 100% of leaves and **detritus** are decomposed and recycled within a few days after they fall to the forest floor.

Because of thermodynamic losses every time energy is exchanged between organisms or converted from one form to another, an external energy source is an indispensable component of every ecological system. Green plants capture **solar energy** through **photosynthesis** and convert it into energy-rich organic compounds that are the basis for all other life in the community. This energy capture is referred to as "primary productivity." These green plants form the first trophic (or feeding) level of most communities.

Herbivores (animals that eat plants) make up the next **trophic level**, while carnivores (animals that eat other animals) add to the complexity and diversity of the community. **Detritivores** (such as beetles and earthworms) and **decomposers** (generally bacteria and **fungi**) convert dead organisms or waste products to inorganic **chemicals**. The **nutrient recycling** they perform is essential to the continuation of life. Together, all these interacting organisms form a **food chain/web** through which energy flows and nutrients and toxins are recycled. Due to intrinsic inefficiencies in transferring material and energy between organisms, the energy

content in successive trophic levels is usually represented as a pyramid in which primary producers form the base and the top consumers occupy the apex.

This introduces the problem of persistent contaminants in the food chain. Because they tend not to be broken and metabolized in each step in the food chain in the way that other compounds are, persistent contaminants such as pesticides and **heavy metals** tend to accumulate in top carnivores, often reaching toxic levels many times higher than original environmental concentrations. This **biomagnification** is an important issue in **pollution control** policies. In many lakes and rivers, for instance, game fish have accumulated dangerously high levels of **mercury** and **chlorinated hydrocarbons** that present a health threat to humans and other fish-eating species.

Diversity, in ecological terms, is a measure of the number of different species in a community, while abundance is the total number of individuals. Tropical rain forests, although they occupy only about five% of the earth's land area, are thought to contain somewhere around half of all terrestrial plant and animals species, while coral reefs and estuaries are generally the most productive and diverse aquatic communities. Community complexity refers to the number of species at each trophic level as well as the total number of trophic levels and ecological niches in a community.

Structure describes the patterns of organization, both spatial and functional, in a community. In a **tropical rain forest**, for instance, distinctly different groups of organisms live on the surface, at mid-levels in the trees, and in the canopy, giving the forest vertical structure. A patchy mosaic of tree species, each of which may have a unique community of associated animals and smaller plants living in its branches, gives the forest horizontal structure as well.

For every physical factor in the environment there are both maximum and minimum tolerable limits beyond which a given species cannot survive. The factor closest to the tolerance limit for a particular species at a particular time is the critical factor that will determine the abundance and distribution of that species in that ecosystem. Natural selection is the process by which environmental pressures—including biotic factors such as predation, **competition**, and disease, as well as physical factors such as temperature, moisture, **soil** type, and space—affect survival and reproduction of organisms. Over a very long time, given a large enough number of organisms, natural selection works on the randomly occurring variation in a population to allow evolution of species and adaptation of the population to a particular set of environmental conditions.

Habitat describes the place or set of environmental conditions in which an organism lives; **niche** describes the role an organism plays. A yard and garden, for instance, may provide habitat for a family of cottontail rabbits. Their niche is being primary consumers (eating vegetables and herbs).

Organisms interact within communities in many ways. **Symbiosis** is the intimate living together of two species; **commensalism** describes a relationship in which one species benefits while the other is neither helped nor harmed. **Lichens**, the thin crusty plants often seen on exposed rocks, are an obligate symbiotic association of a fungus and an alga. Neither can survive without the other. Some orchids and bromeliads (air plants), on the other hand, live commensally on the branches of tropical trees. The orchid benefits by having a place to live but the tree is neither helped nor hurt by the presence of the orchid.

Predation—feeding on another organism—can involve pathogens, **parasites**, and herbivores as well as carnivorous predators. Competition is another kind of antagonistic relationship in which organisms vie for space, food, or other resources. Predation, competition, and natural selection often lead to niche specialization and resource partitioning that reduce competition between species. The principle of **competitive exclusion** states that no two species will remain in direct competition for very long in the same habitat because natural selection and adaptation will cause organisms to specialize in when, where, or how they live to minimize conflict over resources. This can contribute to the evolution of a given species into new forms over time.

It is also possible, on the other hand, for species to co-evolve, meaning that each changes gradually in response to the other to form an intimate and often highly dependent relationship either as predator and prey or for mutual aid. Because individuals of a particular species may be widely dispersed in tropical forests, many plants have become dependent on insects, birds, or mammals to carry pollen from one flower to another. Some amazing examples of **coevolution** and mutual dependence have resulted.

Ecological **succession**, the process of ecosystem development, describes the changes through which whole communities progress as different species colonize an area and change its environment. A typical successional series starts with pioneer species such as grasses or fireweed that colonize bare ground after a disturbance. Organic material from these pioneers helps build soil and hold moisture, allowing shrubs and then tree seedlings to become established. Gradual changes in shade, temperature, nutrient availability, wind protection, and living space favor different animal communities as one type of plant replaces its predecessors. Primary succession starts with a previously unoccupied site. Secondary succession occurs on a site that has been disturbed by external forces such as fires, storms, or humans. In many cases, succession proceeds until a mature "climax" community is established. Introduction of new species by natural processes, such as opening of a land bridge, or by human

intervention can upset the natural relationships in a community and cause catastrophic changes for indigenous species.

Biomes consist of broad regional groups of related communities. Their distribution is determined primarily by **climate**, topography, and soils. Often similar niches are occupied by different but similar species (called ecological equivalents) in geographically separated biomes. Some of the major biomes of the world are deserts, **grasslands**, **wetlands**, forests of various types, and **tundra**.

The relationship between diversity and **stability** in ecosystems is a controversial topic in ecology. F. E. Clements, an early biogeographer, championed the concept of climax communities: stable, predictable associations towards which ecological systems tend to progress if allowed to follow natural tendencies. Deciduous, broad-leaved forests are climax communities in moist, temperate regions of the eastern United States according to Clements, while grasslands are characteristic of the dryer western plains. In this view, **homeostasis** (a dynamic steady-state equilibrium), complexity, and stability are endpoints in ecological succession. Ecological processes, if allowed to operate without external interference, tend to create a natural balance between organisms and their environment.

H. A. Gleason, another pioneer biogeographer and contemporary of Clements, argued that ecological systems are much more dynamic and variable than the climax theory proposes. Gleason saw communities as temporary or even accidental combinations of continually changing biota rather than predictable associations. Ecosystems may or may not be stable, balanced, and efficient; change, in this view, is thought to be more characteristic than constancy. Diversity may or may not be associated with stability. Some communities such as salt marshes that have only a few plant species may be highly resilient and stable while species-rich communities such as coral reefs may be highly sensitive to disturbance.

Although many ecologists now tend to agree with the process-functional view of Gleason rather than the population-community view of Clements, some retain a belief in the **balance of nature** and the tendency for undisturbed ecosystems to reach an ideal state if left undisturbed. The efficacy and ethics of human intervention in natural systems may be interpreted very differently in these divergent understandings of ecology. Those who see stability and constancy in **nature** often call for policies that attempt to maintain historic conditions and associations. Those who see greater variability and individuality in communities may favor more activist management and be willing to accept change as inevitable.

In spite of some uncertainty, however, about how to explain ecological processes and the communities they create, we have learned a great deal about the world around us through scientific ecological studies in the past century. This important field of study remains a crucial component in our ability to manage resources sustainably and to avoid or repair environmental damage caused by human actions.

[*William P. Cunningham Ph.D.*]

RESOURCES
BOOKS

Ricklefs, R. E. *Ecology*. 3rd ed. New York: W. H. Freeman, 1990.

Ecology, deep
see **Deep ecology**

Ecology, human
see **Human ecology**

Ecology, restoration
see **Restoration ecology**

Ecology, social
see **Social ecology**

EcoNet

EcoNet is a computer network that focuses on environmental topics and, through the Institute for Global Communications, has links to the international community. Several thousand organizations and individuals have accounts on the network. EcoNet's electronic conferences contain press releases, reports, and electronic discussions on hundreds of topics, ranging from clean air to pesticides. Subscribers can also send e-mail to other users throughout the country and around the world. EcoNet is a branch of ICG Internet, as are PeacNet, WomensNet, and AntiRacismNet.

RESOURCES
ORGANIZATIONS

Institute for Global Communications, P.O. Box 29904, San Francisco, CA USA 94129-0904 Email: support@igc.apc.org, <http://www.igc.org/igc/gateway/enindex.html>

Economic growth and the environment

The issue of economic growth and the **environment** essentially concerns the kinds of pressures that economic growth, at the national and international level, places on the environment over time. The relationship between **ecology** and the economy has become increasingly significant as humans gradually understand the impact that economic decisions have on the sustainability and quality of the planet.

Economic growth is commonly defined as increases in total output from new resources or better use of existing resources; it is measured by increased real incomes per capita. All economic growth involves transforming the natural world, and it can effect environmental quality in one of three ways. Environmental quality can increase with growth. Increased incomes, for example, provide the resources for public services such as **sanitation** and rural electricity. With these services widely available, individuals need to worry less about day-to-day survival and can devote more resources to **conservation**. Second, environmental quality can initially worsen but then improve as the growth rate rises. In the cases of **air pollution**, **water pollution**, and **deforestation** and encroachment there is little incentive for any individual to invest in maintaining the quality of the environment. These problems can only improve when countries deliberately introduce long-range policies to ensure that additional resources are devoted to dealing with them. Third, environmental quality can decrease when the rate of growth increases. In the cases of emissions generated by the disposal of **municipal solid waste**, for example, abatement is relatively expensive and the costs associated with the emissions and wastes are not perceived as high because they are often borne by someone else.

The **World Bank** estimated that, under present productivity trends and given projected population increases, the output of developing countries would be about five times higher by the year 2030 than it is today. The output of industrial countries would rise more slowly, but it would still triple over the same period. If environmental **pollution** were to rise at the same pace, severe environmental hardships would occur. Tens of millions of people would become sick or die from environmental causes, and the planet would be significantly and irreparably harmed.

Yet economic growth and sound environmental management are not incompatible. In fact, many now believe that they require each other. Economic growth will be undermined without adequate environmental safeguards, and environmental protection will fail without economic growth.

The earth's **natural resources** place limits on economic growth. These limits vary with the extent of resource substitution, technical progress, and structural changes. For example, in the late 1960s many feared that the world's supply of useful metals would run out. Yet, today, there is a glut of useful metals and prices have fallen dramatically. The demand for other natural resources such as water, however, often exceeds supply. In **arid** regions such as the Middle

East and in non-arid regions such as northern China, aquifers have been depleted and rivers so extensively drained that not only **irrigation** and agriculture are threatened but the local ecosystems.

Some resources such as water, forests, and clean air are under attack, while others such as metals, minerals, and energy are not threatened. This is because the **scarcity** of metals and similar resources is reflected in market prices. Here, the forces of resource substitution, technical progress, and structural change have a strong influence. But resources such as water are characterized by open access, and there are therefore no incentives to conserve. Many believe that effective policies designed to sustain the environment are most necessary because society must be made to take account of the value of natural resources and governments must create incentives to protect the environment. Economic and political institutions have failed to provide these necessary incentives for four separate yet interrelated reasons: 1) short time horizons; 2) failures in property rights; 3) concentration of economic and political power; and 4) immeasurability and institutional uncertainty.

Although economists and environmentalists disagree on the definition of sustainability, the essence of the idea is that current decisions should not impair the prospects for maintaining or improving future living standards. The economic systems of the world should be managed so that societies live off the dividends of the natural resources, always maintaining and improving the asset base.

Promoting growth, alleviating poverty, and protecting the environment may be mutually supportive objectives in the long run, but they are not always compatible in the short run. Poverty is a major cause of **environmental degradation**, and economic growth is thus necessary to improve the environment. Yet, ill-managed economic growth can also destroy the environment and further jeopardize the lives of the poor. In many poor but still forested countries, timber is a good short-run source of foreign exchange. When demand for Indonesia's traditional commodity export—petroleum—fell and its foreign exchange income slowed, Indonesia began depleting its hardwood forests at non-sustainable rates in order to earn export income.

In developed countries, it is competition that can shorten time horizons. Competitive forces in agricultural markets, for example, induce farmers to take short-term perspectives for financial survival. Farmers must maintain cash flow to satisfy bankers and make a sufficient return on their land investment. They therefore adopt high-yield crops, **monoculture** farming, increased **fertilizer** and **pesticide** use, salinizing irrigation methods, and more intensive tillage practices which cause **erosion**.

"The Tragedy of the Commons" is the classic example of property rights failure. When access to a grazing area, or

commons is unlimited, each herdsman knows that grass not eaten today will not be there tomorrow. As a rational economic being, each herdsman seeks to maximize his gain and adds more animals to his herd. No herdsman has an incentive to prevent his livestock from grazing the area. Degradation follows and the loss of a common resource. In a society without clearly defined property rights, those who pursue their own interests ruin the public good.

In Indonesia, political upheaval can void property rights overnight, and so any individual with a concession to harvest trees is motivated to harvest as many and as quickly as possible. The government-granted timber-cutting concession may belong to someone else tomorrow. The same is true of some developed countries. For example, in Louisiana mineral rights revert to the state when **wetlands** become open water and there has been no mineral development on the property. Thus, the cheapest methods of avoiding loss of mineral revenues has been to hurry the development of oil and gas in areas which might revert to open water, thereby, hastening erosion and saltwater intrusion, or putting up levies around the property to maintain it as private property, thus interfering with normal estuarine processes.

Global or transnational problems such as **ozone layer depletion** or **acid rain** produce a similar problem. Countries have little incentive to reduce damage to the global environment unilaterally when doing so will not reduce the damaging behavior of others or when reduced fossil fuel use would leave that country at a competitive disadvantage. International agreements are thus needed to impose order on the world's nations that would be analogous to property rights.

Concentration of wealth within the industrialized countries allows for the exploitation and destruction of ecosystems in **less developed countries** (LDC) through, for example, timber harvests and mineral extraction. The concentration of wealth inside a less developed country skews public policy toward benefiting the wealthy and politically powerful, often at the expense of the **ecosystem** on which the poor depend. Local sustainability is dependent upon the goals of those who have power—goals which may or may not be in line with a healthy, sustainable ecosystem. Furthermore, when an exploiting party has substitute ecosystems available, it can exploit one and then move to the next. Japanese lumber firms harvest one country and then move on to another. Here the benefits of sustainability are low and exploiters have shorter time horizons than local interests. This is also an example of how the high discount rates in developed countries are imposed on the management of developing countries' assets.

Environmental policy-making is always more complicated than merely measuring the effects that a proposed policy on the environment. But because of scientific uncertainty about biophysical and geological relations and a general inability to measure a policy's effect on the environment, economic rather than ecological effects are more often relied upon to make policy. Policy-makers and institutions are often unable to grasp the direct and indirect effects of policies on ecological sustainability, nor do they know how their actions will affect other areas not under their control.

Many contemporary economists and environmentalists argue that the value of the environment should nonetheless be factored into the economic policy decision-making process. The goal is not necessarily to put monetary values on **environmental resources**; it is rather to determine how much environmental quality is being given up in the name of economic growth, and how much growth is being given up in the name of the environment. A danger always exists that too much income growth may be given up in the future because of a failure to clarify and minimize tradeoffs and to take advantage of policies that are good for both economic growth and the environment. *See also* Energy policy; Environmental economics; Environmental policy; Environmentally responsible investing; Exponential growth; Sustainable agriculture; Sustainable biosphere; Sustainable development

[*Kevin Wolf*]

RESOURCES
BOOKS

Farber, S. "Local and Global Incentives for Sustainability: Failures in Economic Systems." In *Ecological Economics: The Science and Management of Sustainability*, edited by R. Constanza. New York: Columbia University Press, 1991.
World Bank. *World Development Report 1992: Development and the Environment.* New York: Oxford University Press, 1992.

Ecopsychology
see **Roszak, Theodore**

Ecosophy

A philosophical approach to the **environment** which emphasizes the importance of action and individual beliefs. Often referred to as "ecological wisdom," it is associated with other **environmental ethics**, including **deep ecology** and **bioregionalism**.

Ecosophy originated with the Norwegian philosopher Arne Naess. Naess described a structured form of inquiry he called *ecophilosophy*, which examines **nature** and our relationship to it. He defined it as a discipline, like philosophy itself, which is based on analytical thinking, reasoned argument, and carefully examined assumptions. Naess distinguished ecosophy from ecophilosophy; it is not a discipline in the same sense but what he called a "personal philosophy,"

which guides our conduct toward the environment. He defined ecosophy as a set of beliefs about nature and other people which varies from one individual to another. Everyone, in other words, has their own ecosophy, and though our personal philosophies may share important elements, they are based on norms and assumptions that are particular to each of us.

Naess proposed his own ecophilosophy as a model for individual ecosophies, emphasizing the **intrinsic value** of nature and the importance of cultural and natural diversity. Other discussions of ecosophy concentrate on similar issues. Many environmental philosophers argue that all life has a value that is independent of human perspectives and human uses, and that it is not to be tampered with except for the sake of survival. Human **population growth** threatens the integrity of other life systems; they argue that our numbers must be reduced substantially and that radical changes in human values and activities are required to integrate humans more harmoniously into the total system. *See also* Zero population growth

[*Gerald L. Young and Douglas Smith*]

RESOURCES
BOOKS

Naess, A. *Ecology, Community and Lifestyle: Outline of an Ecosophy.* Translated and revised by D. Rothenberg. Cambridge: Cambridge University Press, 1989.

PERIODICALS

Hedgpeth, J. W. "Man and Nature: Controversy and Philosophy." *The Quarterly Review of Biology* 61 (March 1986): 45-67.

Ecosystem

The term ecosystem was coined in 1935 by the Oxford ecologist Arthur Tansley to encompass the interactions among biotic and abiotic components of the **environment** at a given site. It was defined in its presently accepted form by Eugene Odum as follows: "Any unit that includes all of the organisms (i.e, the community) in a given area interacting with the physical environment so that a flow of energy leads to clearly defined trophic structure, biotic diversity, and material cycles (i.e., exchange of materials between living and non-living parts) within the system." Tansley's concept had been expressed earlier in 1913 by the Oxford geographer A. J. Herbertson, who suggested the term "macroorganism" for such a combined biotic and abiotic entity. He was, however, too far in advance of his time and the idea was not taken up by ecologists. On the other hand Tansley's concept—elaborated in terms of the transfer of energy and matter across ecosystem boundaries—was utilized within the next few years by Evelyn Hutchinson, Raymond Lindeman, and the Odum brothers, Eugene and Howard.

The boundaries of an ecosystem can be somewhat arbitrary, reflecting the interest of a particular ecologist in studying a certain portion of the landscape. However, such a choice may often represent a recognizable landscape unit such as a woodlot, a wetland, a stream or lake, or—in the most logical case—a **watershed** within a sealed geological basin, whose exchanges with the **atmosphere** and outputs via stream flow can be measured quite precisely. Inputs and outputs imply an **open system**, which is true of all but the planetary or global ecosystem, open to **energy flow** but effectively closed in terms of materials except in the case of large-scale asteroid impact.

Ecosystems exhibit a great deal of structure, as may be seen in the vertical partitioning of a forest into tree, shrub, herb, and moss layers, underlain by a series of distinctive **soil** horizons. Horizontal structure is often visible as a mosaic of patches, as in forests with gaps where trees have died and herbs and shrubs now flourish, or in bogs with hummocks and hollows supporting different kinds of plants. Often the horizontal structure is distinctly zoned, for instance around the shallow margin of a lake; and sometimes it is beautifully patterned, as in the vast **peatlands** of North America that reflect a very complicated **hydrology**.

Ecosystems exhibit an interesting functional organization in their processing of energy and matter. Green plants, the primary producers of organic matter, are consumed by herbivores, which in turn are eaten by carnivores that may in turn be the prey of other carnivores. Moreover, all these animals may have **parasites** as another set of consumers. Such sequences of producers and successive consumers constitute a food chain, which is always part of a complicated, inter-linked food web along which energy and materials pass. At each step along the food chain some of the energy is egested or passed through the organisms as feces. Much more is used for metabolic processes and—in the case of animals—for seeking food or escaping predators; such energy is released as heat. As a consequence only a small fraction (often of the order of 10%) of the energy captured at a given step in the food chain is passed along to the next step.

There are two main types of food chains. One is made up of plant producers and animal consumers of living organisms, which constitute a grazing food chain. The other consists of organisms that break down and metabolize dead organic matter, such as earthworms, **fungi**, and bacteria. These constitute the **detritus** food chain. Humans rely chiefly on grazing food chains based on **grasslands**, whereas in a forest it is usual for more than 90% of the energy trapped by **photosynthesis** to pass along the detritus food chain.

Whereas energy flows one way through ecosystems and is dispersed finally to the atmosphere as heat, materials are partially and often largely recycled. For example, **nitrogen** in rain and snow may be taken up from the soil by roots, built into leaf protein that falls with the leaves in autumn, there to be broken down by soil **microbes** to ammonia and nitrate and taken up once again by roots. A given molecule of nitrogen may go through this **nutrient** cycle again and again before finally leaving the system in stream outflow. Other nutrients, and **toxins** such as **lead** and **mercury**, follow the same pathway, each with a different **residence time** in the forest ecosystem.

Mature ecosystems exhibit a substantial degree of **stability**, or dynamic equilibrium, as the endpoint of what is often a rather orderly **succession** of **species** determined by the **nature** of the **habitat**. Sometimes this successional process is a result of the differing life spans of the colonizing species, at other times it comes about because the colonizing species alter the habitat in ways that are more favorable to their competitors, as in an **acid** moss bog that succeeds a circumneutral sedge fen that has in its turn colonized a pond as a floating mat. Equilibrium may sometimes be represented on a large scale by a relatively stable mosaic of small-scale patches in various stages of succession, for instance in fire-dominated pine forests. On the millennial time scale, of course, ecosystems are not stable, changing very gradually owing to immigration and emigration of species and to evolutionary changes in the species themselves.

The structure, function, and development of ecosystems are controlled by a series of partially independent environmental factors: **climate**, soil parent material, **topography**, the plants and animals available to colonize a given site, and disturbances such as fire and windthrow. Each factor is, of course, divisible into a variety of components, as in the case of temperature and precipitation under the general heading of climate.

There are many ways to study ecosystems. Evelyn Hutchinson divided them into two main categories, holistic and meristic. The former treats an ecosystem as a "black box" and examines inputs, storages, and outputs, for example in the construction of a lake's heat budget or a watershed's chemical budget. This is the physicist's or engineer's approach to how ecosystems work. The meristic point of view emphasizes analysis of the different parts of the system and how they fit together in their structure and function, for example the various zones of a wetland or a **soil profile**, or the diverse components of food webs. This is the biologist's approach to how ecosystems work.

Ecosystem studies can also be viewed as a series of elements. The first is, necessarily, a description of the system, its location, boundaries, plant and animal communities, environmental characteristics, etc. Description may be followed by any or all of a series of additional elements, including: 1) a study of how a given ecosystem compares with others locally, regionally, or globally; 2) how it functions in terms of hydrology, productivity, and biogeochemical cycling of nutrients and toxins; 3) how it has changed over time; and 4) how various environmental factors have controlled its structure, function, and development. Such studies involve empirical observations about relationships within and among ecosystems, experiments to test the causality of such relationships, and model-building to assist in forecasting what may happen in the future.

The ultimate in ecosystem studies is a consideration of the structure, function, and development of the global or planetary ecosystem, with a view to understanding and mitigating the deleterious impacts upon it of current human activities. *See also* Biotic community

[*Eville Gorham Ph.D.*]

RESOURCES
BOOKS

Hagen, J. B. *An Entangled Bank, the Origins of Ecosystem Ecology.* New Brunswick, NJ: Rutgers University Press, 1992.

PERIODICALS

Herbertson, A. J. "The Higher Units: A Geological Essay." *Scientia* 14 (1913): 199-212.
Tansley, A. G. "The Use and Abuse of Vegetational Concepts and Terms." *Ecology* 16 (1935): 284-307.

Ecosystem health

Ecosystem health is a new concept that ecologists are examining as a tool for use in detecting and monitoring changes in the quality of the **environment**, particularly with regard to ecological conditions.

Ecosystem health (and **ecological integrity**) is an indicator of the well-being and natural condition of ecosystems and their functions. These indicators are influenced by natural changes in environmental conditions, and are related to such factors as **climate** change and disturbances such as **wildfire**, windstorms, and diseases. Increasingly, however, ecosystems are being affected by environmental stressors associated with human activities that cause **pollution** and disturbance, which result in many changes in environmental conditions. Some **species**, communities, and ecological processes benefit from those environmental changes, but others suffer great damages.

The notion of ecosystem health is intended to help distinguish between ecosystem-level changes that represent improvements and those that are considered to be degradations. In the sense meant here, ecosystem-level refers to responses occurring in ecological communities, landscapes, or seascapes. Effects on individual organisms or populations

do not represent an ecosystem-level response to changes in environmental conditions.

The notion of health

The notion of ecosystem health is analogous to that of medical health. In the medical sense, health is a term used to refer to the vitality or well-being of individual organisms. Medical health is a composite attribute, because it is characterized by a diversity of inter-related characteristics and conditions. These include blood pressure and chemistry, wounds and injuries, rational mental function and many other relevant variables. Health is, in effect, a summation of all of these characters related to vitality and well-being. In contrast, a diagnosis of unhealthiness would focus on abnormal values for only one or several variables within the diverse congregation of health-related attributes. For example, an individual might be judged as being unhealthy because they had a broken leg, or a high fever, or unusually high blood pressure, or non-normal behavioral traits, even though they are "normal" with regards to all other traits.

To compare human and ecosystem health is, however, imperfect in some important respects. Health is a relative concept. It depends on what we consider "normal" at a particular stage of development. The aches and pains that are considered normal in a human at age 80 would be a serious concern in a 20-year old. It is much more difficult, however, to say what is to be expected in an ecosystem. Ecosystems don't have a prescribed lifespan and generally don't die but rather change into some other form. Because of these problems, some ecologists prefer the notions of ecological or biological integrity rather than ecosystem health.

It should also be pointed out that many ecologists like none of these notions (that is, ecosystem health, ecological integrity, or biological integrity). The reason is that by their very nature, these concepts are imprecise and difficult to define. For these reasons, scientists have had difficulty in agreeing upon the specific variables that should be included when designing composite indicators of health and integrity in ecological contexts.

Ecosystem health

Ecosystem health is a summation of conditions occurring in communities, watersheds, landscapes, or seascapes. Ecosystem health conditions are higher-level components of ecosystems, in contrast with individual organisms and their populations.

Although ecosystem health cannot be defined precisely, ecologists have identified a number of specific components that are important in this concept. These include the following indicators: (1) an ability of the system to resist changes in environmental conditions without displaying a large response (this is also known as **resistance** or tolerance); (2) an ability to recover when the intensity of **environmental**

stress is decreased (this is known as **resilience**); (3) relatively high degrees of **biodiversity**; (4) complexity in the structure and function of the system; (5) the presence of large species and top predators; (6) controlled **nutrient** cycling and a stable or increasing content of **biomass** in the system; and (7) domination of the system by native species and natural communities that can maintain themselves without management by humans. Higher values for any of these specific elements imply a greater degree of ecosystem health, while decreasing or lower values imply changes that reflect a less healthy condition.

Ecologists are also working to develop composite indicators (or multivariate summations) that would integrate the most important attributes of ecosystem health into a single value. Indicators of this type are similar in structure to composite economic indicators such as the Dow-Jones Stock Market Index and the Consumer Price Index. Because they allow complex situations to be presented in a simple and direct fashion, composite indicators are extremely useful for communicating ecosystem health to the broader public.

[*Bill Freedman Ph.D.*]

RESOURCES
BOOKS

Costanza, R., B.G. Norton, and B.D. Haskell. *Ecosystem Health. New Goals for Environmental Science.* Washington, D.C.: Island Press, 1992.

DiGiulio, R., and E. Monosson. *Interconnections Between Human and Ecosystem Health.* New York: Chapman and Hall, 1996.

Woodley, S., J. Kay, and G. Francis, eds. *Ecological Integrity and the Management of Ecosystems.* Boca Raton, FL: St. Lucie Press, 1993.

Ecosystem management

Ecosystem management (EM) is a concept that has germinated within the past 20 years and continues to increase in popularity across the United States and Canada. It is a concept that eludes one concise definition, however, because it embodies different meanings in different contexts and for different people and organizations. This can be witnessed by the multiple variations on its title, e.g., ecosystem-based management or collaborative ecosystem management. The definitions that have been given for EM, though varied, fall into two distinct groups. One group emphasizes long term ecosystem integrity, while the other group emphasizes an intention to address all concerns equally, be they economic, ecological, political or social, by actively engaging and incorporating the multitude of stakeholders (literally, those who "hold a stake" in the issue) into the decision-making process.

One usable though incomplete definition of EM is provided by R. Edward Grumbine: "Ecosystem management integrates scientific knowledge of ecological relationships within a complex sociopolitical and values framework toward the general goal of protecting native ecosystem integrity over the long term." Ultimately, EM is a new way to make decisions about how we humans should live with each other and with the **environment** that supports us. And, it is best defined not only by articulating an ideal description of its contents, as Grumbine has done, but also through a rigorous analysis of actual EM examples.

EM—a new management Style

Between the years 1992 and 1994, each of the four predominant federal land management agencies in the United States the **National Park Service**, the **Bureau of Land Management**, the **Forest Service** and the **Fish and Wildlife Service** implemented EM as their operative management paradigm. Combined, these four agencies combined control 97% of the 650 million federally-owned acres (267 million ha) in the United States, or roughly 30% of the United States' entire land area. EM has become the primary management style for these agencies because of a fact that became unavoidably apparent in the 1980s and 1990s the traditional resource management style does not work. It is largely ineffective in addressing the loss and fragmentation of wild areas, the increasing number of threatened or **endangered species**, and the increased occurrence of environmental disputes. This ineffectiveness has been attributed to the traditional management style's focus on mainly **species** with economic value, its exclusion of the public from the decision-making process, and its reliance on outdated ecological beliefs. This explicit acknowledgment that the traditional management style is inadequate has coalesced within state and federal agencies, academia, and environmental organizations, and has been bolstered by advances in other relevant fields, such as **ecology** and conflict management.

If we break the traditional management style into individual components, we see that each ineffective attribute has a new or altered counterpart in EM. One of the best ways to describe what EM actually entails is to explicate this juxtaposition between traditional and new management styles.

EM, as its name makes clear, concentrates on managing at the scale of an ecosystem. Alternately, traditional resource management has tended to focus only on one or a handful of species, especially those species that have a utilitarian, or more specifically economic, value. For example, the U.S. Forest Service has traditionally managed the national forests so as to produce a sustained yield of timber. This management style is often harmful to species other than timber and can have negative effects on the entire ecosystem. In EM, all significant biotic and abiotic components of the ecosystem, as well as aspects such as economic factors, are, ideally, reviewed and the important ecological data incorporated into the decision-making process. For example, review of a forest ecosystem may include an analysis of **habitat** for significant song birds, a description of the requirements needed to maintain a healthy black bear (*Ursus americanus*) population, *and* a discussion of acceptable levels of timber production.

A major problem associated with using an ecosystem to define one's management area is that boundaries of jurisdictional authority, or political boundaries, rarely follow ecological ones. This implies that by following political boundaries alone, ecological components may be left out of the management plan one may be forced to manage only part of an ecosystem, that part which is within one's political jurisdiction. For example, the Greater Yellowstone Ecosystem goes far beyond the boundaries of **Yellowstone National Park**. Therefore, a large scale EM project for Yellowstone would require crossing several political boundaries (e.g., **national park** lands and **national forest** lands), which is a difficult task because it entails several political jurisdictions and political entities (e.g., state and federal agencies, and county governments).

EM projects address this obstacle by forming decision teams that include, among others, representatives from all of the relevant jurisdictions. These decision-making bodies can either act as advisory committees without decision-making authority, or they can attempt to become vested with the power to make decisions. This collaborative process involves all of the stakeholders, whether that stakeholder is a **logging** company interested in timber production, or a private citizen concerned with **water quality**. Such collaboration diverges from the traditional resource management method which made most decisions "behind closed doors" asking for and receiving little public input. These agencies traditionally shied away from actively engaging the public because it is easier and faster to make decisions on one's own than to ask for input from many sources, there has often been an antagonistic and distrustful relationship between state and federal agencies and the public, and, there has been little institutional support (i.e., within the structure of the agency itself) encouraging the manager in the field to invite the public into the decision-making process.

EM's more collaborative and inclusive decision-making style ideally fosters a wiser and more effective decision. This happens because as the decision team works toward consensus, personal relationships are established, some trust may form between parties, and, ultimately, people are more likely to support a decision or plan they help create. EM attempts to transcend the traditional antagonistic relationship between agency personnel and the public. Because 70% of the United States is privately owned, many environmental issues arise on private land—land that is partially not affected by federal and state natural resource legislation. EM allows

us to deal with these issues on private lands by establishing a dialogue between private and public decision makers. Finally, even though the EM decision-making style takes longer to conduct, time is saved in the end because the decision achieved is more agreeable to all interested parties. Having all parties agree to a particular management plan decreases the number of potential lawsuits which can arise and delay the plan's implementation.

Nonequilibrium ecology and EM

A change in the dominant theories in ecology has encouraged this switch to EM and an ecosystem-level focus.

The idea that environments achieve a climax state of **homeostasis** has been a significant theory in ecology since the early 1900s. This view, now discredited, was most vigorously articulated by Frederic Clements and holds that all ecosystems have a particular end point to which they each progress, and that ecosystems are closed systems. Disturbances such as fires or floods are considered only temporary setbacks on the ecosystem's ultimate progression to a final state. This theory offers a certain level of predictable **stability** the type of stability desired within traditional resource management. For example, if a forest is in its climax state, that condition can be maintained by eliminating disturbances such as forest fires, and a predictable level of harvestable timber can be extracted (hence, this theory contributed to the creation of "Smoky the Bear" and the national policy of stopping forest fires on **public land**).

This teleological view of **nature** has ebbed and waned in importance, but has lost favor especially within the past two decades. Ecologists, among others, have realized that from certain temporal and spatial points of view ecosystems may seem to be in equilibrium, but in the long term all ecosystems are in a state of nonequilibrium. That is, ecosystems always change. They change because their ecological structure and function is often regulated by dynamic external forces such as storms or droughts and because they comprise of varied habitat types which change and affect one another.

This acknowledgment of a changing ecosystem means that predictable stability does not really exist and that an **adaptive management** style is needed to meet the changing requirements of a dynamic ecosystem. EM is adaptive. After an EM decision team has formulated and implemented a management plan, the particular ecosystem is monitored. The team watches significant biotic and abiotic factors are watched to see if and how they are altered by the management practices. For example, if logging produces changes which effect the fish in one of the ecosystem's streams, the decision team could adapt to the new data and decide to relocate the logging.

The ability to adaptively manage is a crucial aspect of EM, one that is not incorporated in traditional resource management. As stated, the traditional management style emphasizes one or a few species and believes that ecosystems are mostly stable. So, one only needs to view how a particular species fares to determine the necessary management practices little monitoring is conducted and previous management practices are rarely altered. In EM, management practices are constantly reviewed and adjusted as a result of ongoing data gathering and the goals articulated by the decision team.

The future of EM

The future of EM in the United States and Canada looks stable, and other countries such as France are beginning to use EM. There are many impediments, though, to the successful implementation of EM. Institutions, such as the United States' federal land management agencies, are often hesitant to actually change and when they do change it happens very slowly; there are still many legal questions surrounding the legitimacy of implementing EM on federal, state and private lands; and, even though we attempt to review the entire ecosystem in EM examples, we still lack significant understanding of how even the most basic ecosystems operate. Even given these impediments, it looks as if EM will be the primary land management style of the United States and Canada well into the next century.

[*Paul Phifer Ph.D.*]

RESOURCES
BOOKS

Gunderson, L.H., C.S. Holling, and S.S. Light. *Barriers and Bridges to the Renewal of Ecosystems and Institutions.* New York: Columbia University Press, 1995.

Lee, K.N. *Compass and Gyroscope: Integrating Science and Politics for the Environment.* Washington, D.C.: Island Press, 1993.

Yaffee, S.L., et al. *Ecosystem Management in the United States: An Assessment of Current Experience.* Washington, D.C.: Island Press, 1996.

PERIODICALS

Grumbine, R.E. "What is ecosystem management?" *Conservation Biology* 8. (1994):27-38.

Ecotage

see **Ecoterrorism; Monkey-wrenching**

Ecoterrorism

In the wake of the terrorist attacks on the World Trade Center in New York City on September 11, 2001, the line between radical environmental protest (sometimes called ecoanarchism) and terrorism became blurred by strong emotions on all sides. Environmentalists in America have long held passionate beliefs about protecting the **environment** and saving threatened **species** from **extinction**, as well as

treating animals humanely and protesting destructive business practices. One of the first and greatest environmentalists, **Henry David Thoreau** (1817–1862), wrote about the doctrine of "civil disobedience," or using active protest as a political tool. The author Edward Abbey (1927–1989) became a folk hero among environmentalists when he wrote the novel, *The Monkey Wrench Gang*, in 1975. In that book, a group of militant environmentalists practiced **monkey-wrenching**, or sabotaging machinery in desperate attempts to stop **logging** and mining. Monkey-wrenching, in its destruction of private property, goes beyond civil disobedience and is unlawful. The American public has tended to view monkey-wrenchers as idealistic youth fighting for the environment and has not strongly condemned them for their actions. However, the U.S. government views monkey-wrenching as domestic terrorism, and since September 11, 2001, law enforcement activity concerning environmental groups has been significantly increased.

Ecoanarchism is the philosophy of certain environmental or **conservation** groups that pursue their goals through radical political action. The name reflects both their relation to older anarchist revolutionary groups and their distrust of official organizations. Nuclear issues, social responsibility, **animal rights**, and grass-roots democracy are among the concerns of ecoanarchists. Ecoanarchists tend to view mainstream political and environmental organizations as too passive, and those who maintain them as compromising or corrupt. Ecoanarchists may resort to direct confrontation, direct action, civil disobedience, and guerrilla tactics to fight for survival of wild places. Monkey-wrenchers perform sit-ins in front of bulldozers; they disable machinery in various ways including pouring sand in a bulldozer's gas tank; they ram **whaling** ships; and they spike trees by driving metal bars into them to discourage logging. Ecoanarchists may practice ecotage, which is sabotage for environmental ends, often of machines that alter the landscape. In the early 2000s, radical environmentalists committed arson on 35 sport utility vehicles (SUVs) at a car dealership in Eugene, Oregon, to protest the gas-guzzling vehicles, and set fire to buildings at the University of Washington to protest **genetic engineering**, for example.

Ecoanarchists do not necessarily view the destruction of machinery as out-of-bounds, but the U.S. government views it as terrorism. For instance, **Earth First!**, a radical environmental group whose motto is, "No Compromise in Defense of Mother Earth," takes a stand against violence against humans but does approve of monkey-wrenching. The Federal Bureau of Investigation (FBI) defines terrorism as, "the unlawful use, or threatened use, of violence by a group or individual ... committed against persons or property to intimidate or coerce a government, the civilian population or any segment thereof, in furtherance of political or social objectives." The FBI reports that an estimated 600 criminal acts of ecoterrorism have occurred in the United States since 1996, with damages estimated at $43 million. Two groups associated with these acts of sabotage are the militant **Earth Liberation Front** (ELF) and Animal Liberation Front (ALF) groups. In 1998, ELF took credit for arson at Vail Ski Resort, which resulted in damages of $12 million. The act was a protest over the resort's expansion on mountain ecosystems.

Most environmental groups are more peaceful and have voiced concern over radical environmentalists, as well as over law enforcement officials who view all environmental protesters as terrorists. For instance, **Greenpeace** activists have been known to steer boats in the path of whaling ships and throw paint on nuclear vessels as protest, and have been jailed for doing so, although no people were targeted or injured. **People for the Ethical Treatment of Animals** (PETA) members have thrown pies in the faces of business executives whom they found guilty of inhumane treatment of animals, considering the act a form of civil disobedience and not violence.

The issues of ecoterrorism and ecoanarchism become more heated as **environmental degradation** worsens. Environmentalists become more desperate to protect rapidly disappearing endangered areas or species, while industry continually seeks new resources to replace those being used up. Thrown in the middle are law enforcement officials, who must protect against violence and destruction of property, and also uphold citizens' basic rights to protest.

The most famous ecoterrorist has been Theodore Kaczynski, also known as the Unabomber, who was convicted of murder in the mail-bombing of the president of the California Forestry Association. On the other end of the spectrum of environmental protest is Julia Butterfly Hill, whom ecoterrorists and ecoanarchists would do well to emulate. Hill is an activist who lived in a California redwood tree for two years to prevent it from being cut down by the Pacific Lumber Company. Hill practiced a nonviolent form of civil disobedience, and endured what she perceived as violent actions from loggers and timber company actions. Her peaceful protest brought national attention to the issue of logging in ancient growth forests, and a compromise was eventually reached between environmentalists and the timber company. Unfortunately, the tree in which Hill sat, which she named Luna, was damaged by angry loggers.

[*Douglas Dupler*]

RESOURCES
BOOKS

Abbey, Edward. *The Monkey Wrench Gang*. Salt Lake City: Dream Garden Press, 1990.

Hill, Julia Butterfly. *The Legacy of Luna: The Story of a Tree, a Woman, and the Struggle to Save the Redwoods*. San Francisco: Harper, 2000.
Thoreau, Henry David. *Civil Disobedience, Solitude and Life Without Principle*. Amherst, NY: Prometheus Books, 1998.
Whitaker, David J. *The Terrorism Reader*. New York: Routledge, 2001.

PERIODICALS

Chase, Alston. "Harvard and the Making of the Unabomber." *Atlantic Monthly*, June 2000, 41.
Earth First! Journal. P.O. Box 3023, Tucson, AZ 85702. (520) 620-6900.
Richardson, Valerie. "FBI Targets Domestic Terrorists." *Insight on the News*, 22 April 2002, 30.

Ecotone

The boundary between adjacent ecosystems is known as an ecotone. For example, the intermediary zone between a grassland and a forest constitutes an ecotone that has characteristics of both ecosystems. The transition between the two ecosystems may be abrupt or, more commonly, gradual. Because of the overlap between ecosystems, an ecotone usually contains a larger variety of **species** than is to be found in either of the separate ecosystems and often includes species unique to the ecotone. This effect is known as the edge effect. Ecotones may be stable or variable. Over a period of time, for example, a forest may invade a grassland. Changes in precipitation are an important factor in the movement of ecotones.

Ecotourism

Ecotourism is ecology-based tourism, focused primarily on natural or cultural resources such as scenic areas, coral reefs, caves, fossil sites, archeological or historical sites, and **wildlife**, particularly rare and **endangered species**.

The successful marketing of ecotourism depends on destinations which have **biodiversity**, unique geologic features, and interesting cultural histories, as well as an adequate infrastructure. In the United States national parks are perhaps the most popular destinations for ecotourism, particularly **Yellowstone National Park**, the Grand Canyon, the **Great Smoky Mountains** and **Yosemite National Park**. In 1999, there were 300 million recreational visits to the national parks. Some of the leading ecotourist destinations outside the United States include the Galapagos Islands in Ecuador, the wildlife parks of Kenya, Tanzania, and South Africa, the mountains of Nepal, and the national parks and forest reserves of Costa Rica.

Tourism is the second largest industry in the world, producing over $195 billion in domestic and international receipts and accounting for more than 7% of the world's trade in goods and services. There were 693 million international tourists in 2001, creating 74 million tourism jobs. Adventure tourism, which includes ecotourism, accounts for 10% of this market. In developing countries tourism can comprises as much as one-third of trade in goods and services, and much of this is ecotourism. Wildlife-based tourism in Kenya, for example, generates $350 million annually.

Ecotourism is not a new phenomena. In the late 1800s railroads and steamship companies were instrumental in the establishment of the first national parks in the United States, recognizing even then the demand for experiences in **nature** and profiting from transporting tourists to destinations such as Yellowstone and Yosemite. However, ecotourism has recently taken on increased significance worldwide.

There has been a tremendous increase in demand for such experiences, with adventure tourism increasing at a rate of 30% annually. But there is another reason for the increased significance of ecotourism. It is a key strategy in efforts to protect cultural and **natural resources**, especially in developing countries, because resource-based tourism provides an economic incentive to protect resources. For example, rather than converting tropical rain forests to farms which may be short-lived, income can be earned by providing goods and services to tourists visiting the rain forests.

Although ecotourism has the potential to produce a viable economic alternative to exploitation of the **environment**, it can also threaten it. **Water pollution**, litter, disruption of wildlife, trampling of vegetation, and mistreatment of local people are some of the negative impacts of poorly planned and operated ecotourism. To distinguish themselves from destructive tour companies, many reputable tour organizations have adopted environmental codes of ethics which explicitly state policies for avoiding or minimizing environmental impacts. In planning destinations and operating tours, successful firms are also sensitive to the needs and desires of the local people, for without native support efforts in ecotourism often fail.

Ecotourism can provide rewarding experiences and produce economic benefits that encourage **conservation**. The challenge upon which the future of ecotourism depends is the ability to carry out tours which the clients find rewarding, without degrading the natural or cultural resources upon which it is based. *See also* Earthwatch; National Park Service

[*Ted T. Cable*]

RESOURCES
BOOKS

Boo, E. *Ecotourism: The Potentials and Pitfalls.* 2 vols. Washington, DC: World Wildlife Fund, 1990.

Ocko, Stephanie. *Environmental Vacations: Volunteer Projects to Save the Planet.* 2nd ed. Santa Fe, NM: John Muir, 1992.

Whelan, T., ed. *Nature Tourism: Managing for the Environment.* Washington, DC: Island Press, 1991.

A group of tourists visit the Galapagos Islands (Photograph by Anthony Wolff. Phototake. Reproduced by permission.)

Ecotoxicology

Ecotoxicology is a field of science that studies the effects of toxic substances on ecosystems. It analyzes environmental damage from **pollution** and predicts the consequences of proposed human actions in both the short and long term. With more than 100,000 **chemicals** in commercial use and thousands more being introduced each year, the scale of the task is daunting. Ecotoxicologists have a variety of methods with which they measure the impact of harmful substances on people, plants, and animals. Toxicity tests measure the response of biological systems to a substance to determine if it is toxic. A test could study, for example, how well fish live, grow, and reproduce in various concentrations of industrial **effluent**. Another could evaluate the point at which metal contaminants in **soil** damage plants' ability to convert sunlight into food. Still another could measure how various concentrations of pesticides in agricultural **runoff** affect **sediment** and **nutrient absorption** in **wetlands**. Analyses of chemical fate (i.e., where a pollutant goes once it is released into the **environment**) can be combined with toxicity-test information to predict environmental response to pollution. Because toxicity is the interaction between a living system and a substance, only living plants, animals, and systems can be used in these experiments. There is no other way to measure toxicity.

Another tool used in ecotoxicology is the field survey. It describes ecological conditions in both healthy and damaged natural systems, including pollution levels. Surveys often focus on the number and variety of plants and animals supported by the **ecosystem**, but they can also characterize other valued attributes such as crop yield, **commercial fishing**, timber harvest, or aesthetics. Information from a number of field surveys can be combined for an overview of the relationship between pollution levels and the ecological condition.

A logical question might be: why not merely measure the concentration of a **toxic substance** and predict what will happen from that? The answer is because chemical analysis alone cannot predict environmental consequences in most cases. Unfortunately, interactions between toxicants and the components of ecosystems are not clear; in addition, ecotoxicologists have not yet developed simulation models that would allow them to make predictions based on chemical concentration alone. For example:

- An ecosystem's response to toxic materials is greatly influenced by environmental conditions. The concentration of zinc that will kill bluegill sunfish in Virginia's soft waters will not kill them in the much harder waters of Texas. Most of the relationships between environmental conditions and toxicity have not been established.

- Most pollution is a complex mixture of chemicals, not just a single substance. In addition, some chemicals are more harmful when combined with other toxicants.

- Some chemicals are toxic at concentrations and levels too small to be measured.

- An organism's response to toxic materials can be influenced by other organisms in the community. For example, a fish exposed to pollution may be unable to escape from its predators.

Ecotoxicologists use all three kinds of information: field surveys, chemical analyses, and toxicity tests. Field surveys prove that some important characteristic of the ecosystem has been damaged, chemical measurements confirm the presence of a toxicant, and toxicity tests link a particular toxicant to a particular type of damage.

The scope of environmental protection has broadened considerably over the years, and the types of ecotoxicological information required have also changed. Toxicity testing began as an interest in the effects of various substances on human health. Gradually this concern extended to the other organisms that were most obviously important to humans— domestic animals and crop plants—and finally spread to other organisms that are less apparent or universal in their importance. Hunters are interested in deer, fishers in fish,

bird watchers in eagles or pelicans, and beachcombers in loggerhead turtles. Keeping these organisms healthy requires studying the effects of pollution on them. In addition, the toxicant must not eliminate or taint the plants and/or animals upon which they feed nor can it destroy their **habitat**. Indirect effects of toxicants, which can also be devastating, are difficult to predict. A chemical that is not toxic to an organism, but instead destroys the grasses in which it lays eggs or hides from predators, will be indirectly responsible for the death of that organism. Protecting all the **species** that people value, along with their food and habitat, is a small step toward universal protection. An ambitious goal is to prevent the loss of any existing species, regardless of its appeal or known value to human society.

Because each one of the millions of species on this planet cannot be tested before a chemical is used, ecotoxicologists tested a few "representative" species to characterize toxicity. If a pollutant could be found in rivers, testing might be done on an alga, an insect that eats algae, a fish that eats insects, and a fish that eats fish. Other representative species are chosen by their habitat: on the bottom of rivers, midstream, or in the soil. Regardless of the sampling scheme, however, thousands of organisms that will be affected by a pollutant will not be tested.

Some prediction must be made about their response, nonetheless. Statistical techniques can predict the response of organisms in general from information on a few randomly selected organisms. Another approach tests the well-being of higher levels of biological organization. Since natural communities and ecosystems are composed of a large number of interacting species, the response of the whole reflects the responses of its many constituents. The health of a large and complex ecosystem cannot be measured in the same way as the health of a single species, however. Different attributes are important. For example, examining a single species like cattle or trout might require measuring **respiration**, reproduction, behavior, growth, or tissue damage. The condition of an ecosystem, on the other hand, might be determined by measuring production, nutrient spiralling, or colonization. Since people depend on ecosystems for food production, waste processing, and **biodiversity**, keeping them healthy is important.

[*John Cairns, Jr.*]

RESOURCES
BOOKS

Carson, R. *Silent Spring.* Boston: Houghton Mifflin, 1962.

Côté, R. P., and P. G. Wells. *Controlling Chemical Hazards.* Boston: Unwin Hyman, 1991.

Levin, S. A., et al., eds. *Ecotoxicology: Problems and Approaches.* New York: Springer-Verlag, 1989.

Wilson, E. O. *Biodiversity*. Washington, DC: National Academy Press, 1988.

Ecotype

A recognizable geographic variety, population, or ecological race of a widespread **species** that is equivalent to a taxonomic subspecies. Typically, ecotypes are restricted to one **habitat** and are recognized by distinctive characteristics resulting from adaptations to local selective pressures and isolation. For example, a population or ecotype of species found at the foot of a mountain may differ in size, color, or physiology from a different ecotype living at higher altitudes, thus reflecting a sharp change in local selective pressures. Members of an ecotype are capable of interbreeding with other ecotypes within the same species without loss of fertility or vigor.

Ectopistes migratorius
see **Passenger pigeon**

Edaphic

Refers to the concept that soils have influence on living things, particularly plants. For example, soils with a low **pH** will more likely have plants growing on them that are adapted to this level of **soil** acidity. Animals living in the area will most likely eat these acid-loving plants. The more extreme a soil characteristic, the fewer kinds of plants and animals are able to adapt to the soil **environment**.

Edaphology

The ecological study of **soil**, including its role, value, and management as a medium for plant growth and as a **habitat** for animals. This branch of soil science covers physical, chemical, and biological properties, including soil fertility, acidity, water relations, gas and energy exchanges, microbial **ecology**, and organic decay.

Eelgrass

Eelgrass is the common name for a genus of perennial grass-like flowering plants referred to as *Zostera*. *Zostera* is from the Greek word *zoster* meaning belt, which describes the dark green, long, narrow, ribbon shape of the leaves that range in size from 20 to 50 cm in length, but can grow up to 2 m. Eelgrass grows under water in estuaries and in shallow coastal areas. Eelgrass is a member of a group of land plants

that migrated into the sea in relatively recent geologic times, and is not a seaweed.

Eelgrass grows by the spreading of rhizomes and by seed germination. Eelgrass flowers are hidden behind a transparent leaf sheath. Long filamentous pollen is released into the water, where it is spread by waves and currents. Both leaves and rhizomes contain lacunae, which are air spaces that provide buoyancy.

Eelgrass grows rapidly in shallow waters and is highly productive, thus providing habitats and food for many marine organisms in its stems, roots, leaves, and rhizomes. Eelgrass communities (meadows) provide many important ecological functions, including:

- anchoring of sediments with the spreading of rhizomes, which prevents **erosion** and provides stability
- decreasing the impact of waves and currents, resulting in a calm **environment** where organic materials and sediments can be deposited
- providing food, breeding areas, shelter and protective nurseries for marine organisms of commercial, recreational, and ecological importance
- concentrating nutrients from seawater that are then available for use in the food chain
- serving as food for water fowl and other animals such as snails and sea urchins
- as **detritus** (decaying plant matter), providing nutrition to organisms within the eelgrass community, in adjoining marshes, and in offshore sinks at depths up to 30,000 feet.

Eelgrass growth can be adversely impacted by human activities, including **dredging, logging,** shoreline or overwater construction, **power plants, oil spills, pollution,** and **species** invasion. Although transplantation projects designed to restore eelgrass meadows have been initiated on both coasts of the United States, creation of a large-scale meadow with the complex functions and relationships of a natural eelgrass system has not yet been achieved.

[*Judith L. Sims*]

RESOURCES
OTHER

Gussett, Diana. *Eelgrass*. Port Townsend Marine Science Center. [cited May 27, 2002]. <www.ptmsc.org/html/eelgrass.html#lore>.

Effluent

The etymological meaning of this term is "flowing forth out of." For geologists, this word refers to a wide range of situations, from lava flowing from a **volcano** to a river flowing out of a lake. However, effluent is now most commonly used by environmentalists in reference to the **Clean**

Water Act of 1977. In this act, effluent is a **discharge** from a **point source**, and the legislation specifies allowable quantities of pollutants. These discharges are regulated under Section 402 of the act, and these standards must be met before these types of industrial and municipal wastes can be released into surface waters. *See also* Industrial waste treatment; Thermal pollution; Water pollution; Water quality

Effluent tax

Effluent tax refers to the fee paid by a company to **discharge** to a sewer. As originally proposed, the fee would have been paid for the privilege of discharging to the **environment**. However, there is presently no fee structure which allows a company, municipality, or person to contaminate the environment above the levels set by **water quality** criteria and effluent permits, unless fines levied by a regulatory agency could be deemed to be such fees.

Fees are now charged on the bases of simply being connected to a sewer and the types and levels of materials discharged to the sewer. For example, a municipality might charge all sewer customers the same rate for domestic sewage discharges below a certain flowrate. Customers discharging **wastewater** at a higher strength and/or flowrate would be assessed an incremental fee proportional to the increased amount of contaminants and/or flow. This charge is often referred to as a sewer charge, fee or surcharge.

There are cases in which wastewater is collected and tested to ensure that it meets certain criteria (e.g., required level of oil or a toxic metal, etc.) before it is discharged to a sewer. If the criteria are exceeded, the wastewater would require pretreatment. Holding the water before discharge is generally only practical when flowrates are low. In other situations, it may not be possible to discharge to a sewer because the treatment facility is unable to treat the waste; for example, many hazardous materials must be managed in a different manner.

Effluent taxes force dischargers to integrate environmental concerns into their economic plans and operational procedures. The fees cause firms and agencies to re-think **water conservation** policies, waste minimization, processing techniques and additives, and **pollution control** strategies. Effluent taxes, as originally proposed, might be instituted as an alternative to stringent regulation, but the sentiment of the current public and regulatory agencies is to block any significant degradation of the environment. It appears to be too risky to allow **pollution** on a fee basis, even when the fees are high. Thus, in the foreseeable future, effluents to the environment will continue to be controlled by means of effluent limits, water quality criteria, and fines.

However, the taxing of effluents to a sewer is a viable means of challenging industries to enhance pollution control measures and stabilizing the performance of downstream treatment facilities.

[*Gregory D. Boardman*]

RESOURCES
BOOKS

Davis, M. L., and D. A. Cornwell. *Introduction to Environmental Engineering*. New York: McGraw-Hill, 1991.
Peavy, H. S., D. R. Rowe, and G. Tchobanoglous. *Environmental Engineering*. New York: McGraw-Hill, 1985.
Tchobanoglous, G., and E. D. Schroeder. *Water Quality*. Reading, MA: Addison-Wesley, 1985.
Viessman, W., Jr., and M. J. Hammer. *Water Supply and Pollution Control*. 5th ed. New York: Harper Collins, 1993.

Eggshell thinning
see **Dichlorodiphenyl-trichloroethane**

E_H

A measure of the oxidation/reduction status of a natural water, **sediment** or **soil**. It is a relative electrical potential, measured with a potentiometer (e.g., a **pH** meter adjusted to read in volts) using an inert platinum electrode and a reference electrode (calomel or silver/silver chloride). E_H is reported in volts or millivolts, and is referenced to the potential for the oxidation of **hydrogen** gas to hydrogen ions (H^+). This electron transfer reaction is assigned a potential of zero on the relative potential scale. The oxidation and reduction reactions in natural systems are pH dependent and the interpretation of E_H values requires a knowledge of the pH. At pH 7, the E_H in water in equilibrium with the oxygen in air is +0.76v. The lowest possible potential is -0.4v when oxygen and other electron acceptors are depleted and **methane, carbon dioxide**, and hydrogen gas are produced by the decay of organic matter. *See also* Electron acceptor and donor

Paul Ralph Ehrlich (1932 –)
American population biologist and ecologist

Born in Philadelphia, Paul Ehrlich had a typical childhood during which he cultivated an early interest in entomology and zoology by investigating the fields and woods around his home. As he entered his teen years, Ehrlich grew to be an avid reader. He was particularly influenced by ecologist William Vogt's book, *Road to Survival* (1948), in which the author was the outlined the potential global consequences of imbalance between the growing world population and level of food supplies available. This concept is one Ehrlich has discussed and examined throughout his career. After high school, Ehrlich attended the University of Pennsylvania where he earned his undergraduate degree in zoology in 1953. He received his master's degree from University of Kansas two years later and continued at the university to receive his doctorate in 1957. His degrees led to post-graduate work on various aspects of entomological projects, including observing flies on the Bering Sea, the behavioral characteristics of parasitic mites, and (his favorite) the population control of butterfly caterpillars with ants rather than pesticides. Other related field projects have taken him to Africa, Alaska, **Australia**, the South Pacific and South East Asia, Latin America, and **Antarctica**. His travels enabled him to learn first-hand the ordeals endured by those in overpopulated regions.

In 1954, he married Anne Fitzhugh Howland, biological research associate, with whom he wrote the best-selling book, *The Population Bomb* (1968). In the book, the Ehrlichs focus on a variety of factors contributing to overpopulation and, in turn, world hunger. It is evident throughout the book that the words and warnings of *The Survival Game* continued to exert a strong influence on Ehrlich. The authors warned that birth and death rates worldwide need to be "brought into line" before **nature** intervenes and renders (through **ozone layer depletion**, global warming, and **soil** exhaustion, among other **environmental degradation**) the human race extinct. Human reproduction, especially in highly developed countries like the United States, should be discouraged through levying taxes on diapers, baby food, and other related items; compulsory sterilization among the populations of certain countries should be enacted (the authors' feelings on *compulsory* sterilization have relaxed somewhat since 1968). Ehrlich himself underwent a vasectomy after the birth of the couple's first and only child.

In 1968, Ehrlich founded **Zero Population Growth**, Inc., an organization established to create and rally support for balanced population levels and the **environment**. He has been a faculty member at Stanford University (California) since 1959 and currently holds a full professor position there in the Biological Sciences Department. In addition, Ehrlich has been a news correspondent for NBC since 1989. In 1993, he was awarded the Crafoord Prize in **Population Biology** and the **Conservation** of Biological Diversity from the Royal Swedish Academy of Sciences and received the World **Ecology** Medal from the International Center for Tropical Ecology. In 1994, Ehrlich was bestowed with the **United Nations Environment Programme** Sasakawa Environment Prize.

Among Ehrlich's published works are *The Population Bomb* (1968), *The Cold and the Dark: The World After Nuclear War* (1984); *The Population Explosion* (1990); *Healing the Planet* (1991), *Betrayal of Science and Reason* (1996) which was written with his wife, and his most recent work, *Human Natures: Genes, Cultures, and the Human Prospect*, which was published in 2000. In 2001, Ehrlich was the recipient of the Distinguished Scientist Award. He continues his teaching as a professor of Population Studies at Stanford.

[*Kimberley A. Peterson*]

RESOURCES

BOOKS

Ehrlich, P. R. *The Population Bomb*. New York: Ballantine Books, 1968.

————. *Healing the Planet: Strategies for Solving the Environmental Crisis*. Redding: Addison Wesley, 1992.

————. *Human Natures: Genes, Cultures, and the Human Prospect*. Washington, D.C.: Island Press/Shearwater Books, 2000.

————, and A. H. Ehrlich. *The Population Explosion*. New York: Simon & Schuster, 1990.

————, A. H. Ehrlich, and J. Holdren. *Eco-Science: Population, Resources, and Environment*. San Francisco: W. H. Freeman, 1970.

PERIODICALS

Dailey, G. C., and P. R. Ehrlich. "Population Sustainability and Earth's Carrying Capacity." *BioScience* 42 (November 1992): 761–71.

"Distinguished Scientist Award." *BioScience* 51, no. 5 (May 2001): 416.

Eichhornia crassipes
see **Water hyacinth**

EIS
see **Environmental Impact Statement**

El Niño

El Niño is the most powerful weather event on the earth, disrupting weather patterns across half the earth's surface. Its three- to seven- year cycle brings lingering rain to some areas and severe **drought** to others. El Niño develops when currents in the Pacific Ocean shift, bringing warm water eastward from **Australia** toward Peru and Ecuador. Heat rising off warmer water shifts patterns of atmospheric pressure, interrupting the high-altitude wind currents of the jet stream and causing **climate** changes.

El Niño, "Christ child" or "the child" in Spanish, tends to appear in December. The phenomenon was first noted by Peruvian fishermen in the 1700s, who saw a warming of normally cold Peruvian coastal waters and a simultaneous disappearance of anchovy schools that provided their livelihood.

A recent El Niño began to develop in 1989, but significant warming of the Pacific did not begin until late in 1991, reaching its peak in early 1992 and lingering until 1995–the longest runing El Niño on record. Typically, El Niño results in unusual weather and short-term climate changes that cause losses in crops and **commercial fishing**. El Niño contributed to North America's mild 1992 winter, torrential **flooding** in southern California, and severe droughts in southeastern Africa. Wild animals in central and southern Africa died by the thousands, and 20 million people were plagued by **famine**. The dried **prairie** of Alberta, Canada, failed to produce wheat, and Latin America received record flooding. Droughts were felt in the Philippines, Sri Lanka, and Australia, and Turkey experienced heavy snowfall. The South Pacific saw unusual numbers of cyclones during the winter of 1992. El Niño's influence also seems to have suppressed some of the cooling effects of Mount Pinatubo's 1991 explosion.

Scientists mapping the sea floor of the South Pacific near Easter Island found one of the greatest concentration of active volcanoes on Earth. The discovery has intensified debate over whether undersea volcanic activity could change water temperatures enough to affect weather patterns in the Pacific. Some scientists speculate that periods of extreme volcanic activity underwater could trigger El Niño.

El Niño ends when the warm water is diverted toward to the North and South Poles, emptying the moving **reservoir** of stored energy. Before El Niño can develop again, the western Pacific must "refill" with warm water, which takes at least two years. *See also* Atmosphere; Desertification

[*Linda Rehkopf*]

RESOURCES
BOOKS

Diaz, H. R., ed. *El Niño: Historical and Paleoclimatic Aspects of the Southern Oscillation.* New York: Cambridge University Press, 1993.

Glynn, P. W., ed. *Global Ecological Consequences of the El-Niño Southern Oscillation, 1982-1983.* New York: Elsevier Science, 1990.

PERIODICALS

Mathews, N. "The Return of El Niño." *UNESCO Courier* (July-August 1992): 44-46.

Monastersky, R. "Once Bashful El Niño Now Refuses to Go." *Science News* 143 (23 January 1993): 53.

Electric automobiles
see Transportation

Electric utilities

Utilities neither produce energy like oil companies nor consume it like households, but convert it from one form to another. The electricity created is attractive because it is clean and versatile, because it can be moved great distances nearly instantaneously. Demand for electricity has grown even as demand for energy as a whole has contracted, with consumption of electricity increasing from one quarter of total energy consumption in 1973 to about a third.

The major participants in the electric power industry are about 200 investor owned utilities that generate 78% of the power and supply 76% of the customers. The industry is very capital intensive and heavily regulated, and it has a large impact on other industries including **aluminum**, steel, electronics, computers, and robotics. The electrical power industry is the largest consumer of primary energy in the United States: consumes over one-third of the total national energy demand and only supplies one-tenth of that demand, losing from 65–75% of the energy in conversion, transmission, and distribution.

The electrical industry has been subjected to pressures and uncertainties which have had a profound impact on its economic viability, forcing it to reexamine numerous assumptions which previously governed its behavior. In the period after World War II, the main strategy the industry followed was to "grow and build." During this period demand increased at a rate of over 7% per year; new construction was needed to meet the growing demand, and this yielded economies of scale, with greater efficiencies and declining marginal costs. Public utility commissions lowered prices which stimulated additional demand. As long as prices continued to fall, demand continued to rise, and additional construction was necessary. New construction also occurred because the rate of return for the industry was regulated, and the only way for it to increase profits was to expand its rate base by building new plants and equipment.

This period of industry growth came to an end in the 1970s, primarily as a result of the energy crisis. Economic growth slowed and fuel prices escalated, including the weighted average cost of all **fossil fuels** and the spot market price of **uranium** oxide. As fuel prices rose, operating costs went up, and maintenance costs also increased, including the costs of supplies and materials, labor, and administrative expenses. All this led to higher costs per kilowatt hour, and as the price of electricity went up, sales growth declined.

The financial condition of the industry was further affected as capital costs for **nuclear power** and **coal power plants** increased. As the rate of inflation accelerated during this period, interest rates escalated. The rates utilities had to pay on bonds grew, and the costs of construction rose. The average cost of new generating capacity, as well as installed capacity per kilowatt hour, went up. Net earnings and revenue per kilowatt hour went down, as both short-term and long-term debt escalated, and major generating units were cancelled and capital appropriations cut back.

During this decade, many people also came to believe that coal and nuclear power plants were a threat to the **environment**. They argued that new options had to be developed and that **conservation** was important. The federal government implemented new environmental and safety regulations which further increased utility costs. The government affected utility operations in other ways. In 1978 it deregulated interstate power sales, and required utilities to purchase alternative power such as **solar energy** from qualifying facilities at fully avoided costs. But perhaps the greatest transformation took place in the relationship electric power companies had to the public utility commissions. Once friendly, it deteriorated under the many economic and environmental changes that were then taking place. The size and number of requests for rate increases grew but the%age of requests granted actually went down.

By the end of the 1970s, the "grow and build" strategy was no longer tenable for the electric power industry. Since then, the industry has adopted many different strategies, with different segments following different courses based on divergent perceptions of the future. Almost all utilities have tried to negotiate long-term contracts which would lower their fuel-procurement costs, and attempts have also been made to limit the costs of construction, maintenance, and administration. Many utilities redesigned their rate structures to promote use when excess capacity was available and discourage use when it was not. Multiple rate structures for different classes of customers were also implemented for this purpose.

A number of utilities (Commonwealth Edison, Long Island Lighting, Carolina Power and Light, the TVA) have pursued a modified grow and build strategy based on the perception that economic growth would recover and that conservation and **renewable energy** would not be able to handle the increased demand. Some utilities (Consolidated Edison, Duke Power, General Public Utilities, Potomac Electric Power) pursued an option of capital minimization. They were located in areas of the country that were not growing and where the demand for power was decreasing. In areas of rapidly growing energy demand where regulations discouraged nuclear and coal plant construction, utilities such as Southern California Edison and Pacific Gas and Electric have had no option but to rely on their strong internal research and development capabilities and their progressive leadership to explore **alternative energy sources**. They have become energy brokers, buying alternative power from third party producers. Many utilities have also diversified, and the main attraction of diversification has been that

it frees these companies from the profit limitations imposed by the public utility commissions. Outside the utility business (in real-estate, banking, and energy-related services), there was more risk but no limits on making money from profitable ventures. *See also* Alternative fuels; Economic growth and the environment; Energy and the environment; Energy conservation; Energy efficiency; Energy path, hard vs. soft; Energy policy; Geothermal energy; Wind energy

[*Alfred A. Marcus*]

RESOURCES
BOOKS

Anderson, D. *Regulatory Politics and Electric Utilities*. Cambridge, Massachusetts: Auburn House, 1981.

Navarro, P. *The Dimming of America*. Cambridge, Massachusetts: Ballinger, 1985.

Thomas, S. D. *The Realities of Nuclear Power*. New York: Cambridge University Press, 1988.

Three Mile Island: The Most Studied Nuclear Accident in History. Washington, DC: General Accounting Office, 1980.

Zardkoohi, A. "Competition in the Production of Electricity." In *Electric Power*, edited by J. Moorhouse. San Francisco: Pacific Research Institute, 1986.

PERIODICALS

Joskow, D. "The Evolution of Competition in the Electric Power Industry." *Annual Review of Energy* (1988): 215-238.

OTHER

Three Mile Island: A Report to the Commissioners and to the Public, Volumes I and II, Parts 1, 2, and 3. Washington, DC: U.S. Nuclear Regulatory Commission, 1980.

Electromagnetic field

Electromagnetic fields (EMFs) are low-level radiation generated by electrical devices, including power lines, household appliances, and computer terminals. They penetrate walls, buildings, and human bodies, and virtually all Americans are exposed to them, some to relatively high levels. Several dozen studies, conducted mainly over the last 15 years, suggest that exposure to EMFs at certain levels may cause serious health effects, including childhood **leukemia**, brain tumors, and damage to fetuses. But other studies show no such connections.

EMFs are strongest around power stations, high-current electric power lines, subways, movie projectors, handheld radar guns and large radar equipment, and microwave power facilities and relay stations. Common sources of everyday exposure include electric razors, hair dryers, computer video display terminals (VDTs), television screens, electric power tools, electric blankets, cellular telephones, and appliances such as toasters and food blenders.

The electricity used in North American homes, offices, and factories is called alternating current (AC) because it alternates the direction of flow at 60 cycles a second, which is called 60 hertz (Hz) power. Batteries, in contrast, produce direct current (DC). The electric charges of 60 Hz power create two kinds of fields: electric fields, from the strength of the charge, and magnetic fields, from its motion. These fields, taken together, are called electromagnetic fields, and they are present wherever there is electric power. A typical home exposes its residents to electromagnetic fields of 1 or 2 milligauss (a unit of strength of the fields). But under a high voltage power line, the EMF can reach 100–200 milligauss.

The electromagnetic spectrum includes several types of energy or radiation. The strongest and most dangerous are x-rays, gamma rays, and ultraviolet rays, all of which are types of **ionizing radiation**, the kind that contains enough energy to enter cells and atoms and break them apart. Ionizing radiation can cause **cancer** and even instant death at certain levels of exposure. The other forms of radiation are non-ionizing, and do not have enough energy to break up the chemical bonds holding cells together.

Microwave radiation constitutes the middle frequencies of the electromagnetic spectrum and includes radio frequency (RF), radar, and television waves, visible light and heat, and infrared radiation. Microwave radiation is emitted by VDTs, microwave ovens, satellites and earth terminals, radio and television broadcast stations, CB radios, security systems, and sonar, radar, and telephone equipment. Because of the pervasive presence of microwave radiation, virtually the entire American population is routinely exposed to it at some level. It is known that microwave radiation has biological effects on living cells.

The type of radiation generally found in homes is Extremely Low Frequency (ELF), and it has, until recently, not been considered dangerous, since it is non-ionizing and non-thermal. However, numerous studies over the last two decades provide evidence that the ELF range of EMFs can have serious biological effects on humans and can cause cancer and other health problems. Some scientists argue that the evidence is not yet conclusive or even convincing, but others contend that exposure to EMFs may represent a potentially serious public health problem.

After reviewing much of this evidence and data, the U. S. **Environmental Protection Agency** (EPA) released a draft report in December 1990 which determined that significant documentation exists linking EMFs to cancer in humans, and called for additional research on the matter. The study concluded that "...several studies showing leukemia, lymphoma, and cancer of the nervous system in children exposed to magnetic fields from residential 60-Hz electrical power distribution systems, supported by similar findings in adults in several occupational studies also involving electrical power frequency exposures, show a consistent pattern of response which suggests a causal link." The report went on to state that "evidence from a large number of biological test systems shows that ELF electric and magnetic fields induce biological effects that are consistent with several possible mechanisms of carcinogenesis...With our current understanding, we can identify 60-Hz magnetic fields from power lines and perhaps other sources in the home as a possible, but not proven, cause of cancer in humans."

EPA cited nine studies of cancer in children as supporting the strongest evidence of a link between the disease and EMFs stating that "these studies have consistently found modestly elevated risks (some statistically significant) of leukemia, cancer of the nervous system, and...lymphomas," with occupational studies furnishing "additional, but weaker, evidence" of EMFs raising the risk of cancer.

Concerning laboratory studies of the effects of EMFs on cells and the responses of animals to exposure, EPA found that "...there is reason to believe that the findings of carcinogenicity in humans are biologically plausible." EPA scientists further recommended classifying EMFs as a "class B-1 carcinogen," like cigarettes and **asbestos**, meaning that they are a probable source of human cancer.

Several of the studies done on EMFs show that children living near high voltage power lines, and workers occupationally exposed to EMFs from power lines and electrical equipment, are more than twice as likely to contract cancer, especially leukemia, as are children and workers with average exposure. One study found a five-fold increase in childhood cancer among families exposed to strong EMFs, and another study even documented the leukemia rate among children increasing in direct proportion to the strength of the EMFs. There are suspicions that EMFs may also be linked to the apparent dramatic rise in fatal brain tumors over recent years, which sometimes occur in clusters near power substations and other areas where EMFs are high. There is also concern about hand-held cellular telephones, whose antennae emit EMFs very close to the brain. But none of this evidence is considered conclusive.

Fetal damage has been cited as another possible effect of EMFs with higher-than-normal rates of miscarriages reported among pregnant women using electric blankets, waterbeds with electric heaters, and VDTs for a certain number of hours a day. But, other studies have failed to establish a link between EMFs and miscarriages.

In October, 1996, the **National Research Council** of the **National Academy of Sciences** issued an important report on the feared dangers of power lines, examining and analyzing over 500 published studies conducted over the previous 17 years. In announcing the report's findings, the

Council's panel of experts stressed that it could find no discernable hazard to human health from EMFs.

The announcement was hailed by the electric power industry, and widely reported in the news as disputing the alleged link between power lines and cancer. But what the report itself actually said was that there was "no conclusive and consistent evidence that shows that exposure to residential electric magnetic fields produce cancer...," a severe **burden of proof** indeed.

While the study did not find proof that electrical lines pose health hazards to humans, it did conclude that the association between proximity to power lines and leukemia in children was "statistically significant" and "robust." It observed that a dozen or so scientific studies had determined that children living near power lines are 1.5 times more likely to develop childhood leukemia than are other children.

Moreover, the study noted that 17 years of research had not identified any other factors that account for the increased cancer risk among children living near power lines. EMFs thus remain in the minds of many of the most likely albeit unproven cause of these cancers.

There are various theories to account for how EMFs may cause or promote cancer. Some scientists speculate that when cells are exposed to EMFs, normal cell division and DNA function can be disrupted, leading to genetic damage and increased cell growth and, thus, cancer. Indeed, research has shown that ELF fields can speed the growth of cancer cells, and make them more resistant to the body's immune system. The effects of EMFs on the cell membrane, on interaction and communication between groups of cells, and on the biochemistry of the brain are also being studied.

Hormones may also be involved. Weak EMFs are known to lower production of melatonin, a strong hormone secreted by the pineal gland, a tiny organ near the center of the brain. Melatonin strengthens the immune system and depresses other hormones that help tumors grow. This theory might help explain the tremendous increase in female breast cancer in developed nations, where electrical currents are so pervasive, especially the kitchen. The theory may also apply to electric razors, which operate in close proximity to the gland, and whose motors have relatively high EMFs. A 1992 study found that men with leukemia were more than twice as likely to have used an electric razor for over two-and-one-half minutes a day, compared with men who had not. The study also found weaker associations between leukemia and the use of hand-held massagers and hair dryers. Another hypothesis involves the motions of charged particles within EMFs, with calcium ions, for example, accelerating and damaging the structure of cell membranes under such conditions.

With conflicting and inconclusive data being cited by scientists and advocates on both sides of the debate, it is unlikely that the controversy over EMFs wil be resolved any time soon.

In October, 1996, the National Research Council of the National Academy of Sciences issued an important report on the feared dangers of power lines, examining and analyzing over 500 published studies conducted over the previous 17 years. In announcing the report's findings, the Council's panel of experts stressed that it could find no discernable hazard to human health from EMFs.

The announcement was hailed by the electric power industry, and widely reported in the news media as disputing the alleged link between power lines and cancer. But what the report itself actually said was that there was "no conclusive and consistent evidence that shows that exposure to residential electric magnetic fields produce cancer...".

While the study did not find proof that electrical lines pose health hazards to humans, it did conclude that the association between proximity to power lines and leukemia in children was "statistically significant" and "robust." It observed that a dozen or so scientific studies had determined that children living near power lines are 1.5 times more likely to develop childhood leukemia than are other children.

Moreover, the study noted that 17 years of research had not identified any other factors that account for the increased cancer risk among children living near power lines. EMFs thus remained, in the minds of many, the most likely—albeit unproven—cause of these cancers.

And a major five-year study, by the National Cancer Institute, the University of Minnesota, and other childhood leukemia specialists, published in July 1997 in *The New England Journal of Medicine,* found no evidence linking leukemia in children to electric power lines.

[*Lewis G. Regenstein*]

RESOURCES
BOOKS

Brodeur, P. *Currents of Death: Power Lines, Computer Terminals, and the Attempt to Cover Up Their Threat to Your Health.* New York: Simon & Schuster, 1989.

Sugarman, E. *Warning: The Electricity Around You May Be Hazardous to Your Health.* New York: Simon & Schuster, 1992.

U.S. Environmental Protection Agency. *Evaluation of the Potential Carcinogenicity of Electromagnetic Fields.* Review Draft. Washington, DC: U. S. Government Printing Office, 1990.

PERIODICALS
Savitz, D., and J. Chen. "Parental Occupation and Childhood Cancer: Review of Epidemiological Studies." *Environmental Health Perspectives* 88 (1990): 325-337.

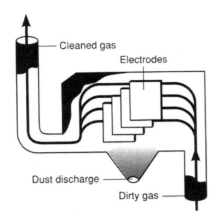

An electrostatic precipitator. (McGraw-Hill Inc. Reproduced by permission.)

Electron acceptor and donor

Electron acceptors are ions or molecules that act as oxidizing agents in chemical reactions. Electron donors are ions or molecules that donate electrons and are reducing agents. In the **combustion** reaction of gaseous **hydrogen** and oxygen to produce water (H_2O), two hydrogen atoms donate their electrons to an oxygen atom. In this reaction, the oxygen is reduced to an oxidation state of -2 and each hydrogen is oxidized to +1. Oxygen is an **oxidizing agent** (electron acceptor) and hydrogen is a reducing agent (electron donor). In **aerobic** (with oxygen) biological **respiration**, oxygen is the electron acceptor accepting electrons from organic **carbon** molecules; and as a result oxygen is reduced to -2 oxidation state in H_2O and organic carbon is oxidized to +4 in CO_2. In flooded soils, after oxygen is used up by aerobic respiration, nitrate, sulfate, as well as iron and manganese oxides can act as electron acceptors for microbial respiration. Other common electron acceptors include peroxide and hypochlorite (household bleach) which are bleaching agents because they can oxidize organic molecules. Other common electron donors include antioxidants like sulfite.

Electrostatic precipitation

A technique for removing **particulate** pollutants from waste gases prior to their exhaustion to a stack. A system of thin wires and parallel metal plates are charged by a high-voltage direct current (DC) with the wires negatively charged and the plates positively charged. As waste gases containing fine particulate pollutants (i.e., **smoke** particles, **fly ash**, etc.) are passed through this system, electrical charges are transferred from the wire to the particulates in the gases. The charged particulates are then attracted to the plates within the device, where they are then shaken off the plates during short intervals when the DC current is interrupted. (Stack gases can be shunted to a second parallel device during this period). They fall to a collection bin below the plates. Under optimum conditions, electrostatic precipitation is 99% efficient in removing particulates from waste gases.

Elemental analysis

Chemists have developed a number of methods by which they can determine the kind of elements present in a material and the amount of each element present. Nuclear magnetic resonance (NMR), flame spectroscopy, and **mass spectrometry** are examples of elemental analysis. These methods have been improved to a point where concentrations of a few **parts per million** of an element or less can be detected with relative ease. Elemental analysis is valuable in environmental work to determine the presence of a contaminant or pollutant. As an example, the amount of **lead** in a paint chip can be determined by means of elemental analysis.

Elephants

The elephant is a large mammal with a long trunk and tusks. The trunk is an elongated nose used for feeding, drinking, bathing, blowing dust, and testing the air. The tusks are upper incisor teeth composed entirely of dentine (ivory) used for defense, levering trees, and scraping for water. Elephants are long-lived (50–70 years) and reach maturity at 12 years. They reproduce slowly (one calf every two to three years) due to a 21-month gestation period and an equally long weaning period. A newborn elephant stands 3 ft (1 m) at the shoulder and weighs 200 lb (90 kg). The Elephantidae includes two living **species** and various extinct relatives.

Asian elephants (*Elephas maximus*) grow to 10 ft (3 m) high and weigh 4 tons. The trunk ends in a single lip, the forehead is high and domed, the back convex, and the ears small. Asian elephants are commonly trained as work animals. They range from India to southeast Asia. There are four subspecies, the most abundant of which is the Indian elephant (*E. m. bengalensis*) with a wild population of about 20,000. The Sri Lankan (*E. m. maximus*), Malayan (*E. m. hirsutus*), and the Sumatran elephants (*E. m. sumatranus*) are all endangered subspecies.

In Africa, adult bush elephants (*Loxodonta africana oxyotis*) are the world's largest land mammals, growing 11 ft (3.3 m) tall and weighing 6 tons. The trunk ends in a double lip, the forehead slopes, the back is hollow, and the ears are large and triangular. African elephants are also endangered and have never been successfully trained to work. The rare round-eared African forest elephant (*L. a. cyclotis*)

Elephants at the Amboseli National Park in Kenya, Africa. (Photograph by Wolfgang Kaehler. Corbis-Bettmann. Reproduced by permission.)

is smaller than the bush elephant and inhabits dense tropical rain forests.

Elephants were once abundant throughout Africa and Asia, but they are now threatened or endangered nearly everywhere because of widespread ivory **poaching**. In 1970 there were about 4.5 million elephants in Africa, by 1990 there were only 600,000. Protection from poachers and the 1990 ban on the international trade in ivory (which caused a drop in the price of ivory) are slowing the slaughter of African bush elephants. However, the relatively untouched forest elephants are now coming under increasing pressure. In West Africa recent **hunting** has reduced forest elephants to less than 3,000.

Elephants are **keystone species** in their ecosystems, and their elimination could have serious consequences for other **wildlife**. For example, wandering elephants disperse fruit seeds in their dung, and the seeds of some plants must pass through elephants to germinate. Elephants are also "bulldozer herbivores," habitually trampling plants and up-rooting small trees. In African forests elephants create open spaces that allow the growth of vegetation favored by gorillas

and forest antelope. In woodland **savanna** elephants convert wooded land into **grasslands**, thus favoring grazing animals. However, large populations of elephants confined to reserves can also destroy most of the vegetation in a region. Culling exploding elephant populations in reserves has been practiced in the past to protect the vegetation for other animals that depend on it.

[*Neil Cumberlidge Ph.D.*]

RESOURCES
BOOKS

———. *Battle for the Elephants.* New York: Viking, 1992.
Martin, C. *The Rainforests of West Africa.* Boston: Birkhauser Verlag, 1991.
Shoshani, J., ed. *Elephants: Majestic Creatures of the Wild.* Emmaus, PA: Rodale Press, 1992.

ELI

see **Environmental Law Institute**

Charles Sutherland Elton (1900–1991)

English ecologist

A factual, accurate, complete history of **ecology** as a discipline has yet to be written. When that history is finally compiled, the British ecologist Charles Sutherland Elton will stand as one of the disciplines leading mentors of the twentieth century.

Charles Elton was born March 29, 1900, in Manchester, England. His interest in what he later called "scientific natural history" was sparked early by his older brother Geoffrey. By the age of 19, Charles Elton was already investigating **species** relationships in ponds, streams, and sand-dune areas around Liverpool.

His formal education in ecology was shaped by an undergraduate education at Oxford University, and by his participation in three scientific expeditions to Spitsbergen in the Arctic (in 1921, 1923, and 1924), the first one as an assistant to Julian Huxley. Even though an undergraduate, he was allowed to begin an ecological survey of animal life in Spitsbergen, a survey completed on the third trip. These experiences and contacts led him to a position as biological consultant to the Hudson's Bay Company, which he used to conduct a long-term study of the fluctuations of fur-bearing mammals, drawing on company records dating back to 1736.

During this period, Elton became a member of the Oxford University faculty (in 1923), and was eventually elected a senior research fellow of Corpus Christie College. His whole academic career was spent at Oxford, from which he retired in 1967.

He applied the skills and insights gained through the Spitsbergen and Hudson Bay studies to work on the fluctuations of mice and voles in Great Britain. To advance and coordinate this work, he started the Bureau of Animal Populations at Oxford. This institution (and Elton's leadership of it) played a vital role in the shaping of research in animal ecology and in the training and education of numerous ecologists in the early twentieth century.

Elton published a number of books, but four proved to be of particular significance in ecology. He published his first book, *Animal Ecology*, in 1929, a volume now considered a classic, its author one of the pioneers in the field of ecology, especially animal ecology. In the preface to a 1966 reissue, Elton suggested that the book "must be read as a pioneering attempt to see...the outlines of the subject at a period when our knowledge [of] terrestrial communities was of the roughest, and considerable effort was required to shake off the conventional thinking of an earlier zoology and enter upon a new mental world of populations, inter-relations, movements and communities—a world [of] direct study of natural processes..." Major topics in that book remain major topics

in ecology today: the centrality of trophic relationships; the significance of **niche** as a functional concept; ecological **succession**; the dynamics of dispersal; and the relationships critical to the fluctuation of animal populations, including interactions with **habitat** and physical **environment**.

His year of work on small mammals in Spitzbergen, for the Hudson's Bay Company, and in British localities accessible to Oxford, culminated in the publication, in 1942, of *Voles, Mice and Lemmings*. This work, still in print almost 60 years later, "brought together...his own work and a collection of observations from all over the world and from ancient history onward." Elton begins the book by establishing a context of "vole and mouse plagues" through history. A second section is on population fluctuations in north-west Europe, voles and mice in Britain, but also lemmings in Scandinavia. The other two sections focus on **wildlife** cycles in northern Labrador, including chapters on fox and marten, voles, foxes, the lemmings again, and caribou herds. In all this work, the emphasis is on the dynamics of change, on the constant interactions and subsequent fluctuations of these various populations and often stringent environments.

Elton's 1958 book, *The Ecology of Invasions by Animals and Plants*, focused on a problem that is of even more concern today—the arrival and impact of **exotic species** introduced from other places, sometimes naturally, increasingly through the actions of humans. As always, Elton is careful to set the historical context by showing how biological "invaders" have been moving around the globe for a long time, but he also emphasizes that "we are living in a period of the world's history when the mingling of thousands of kinds of organisms from different parts of the world is setting up terrific dislocations in nature."

The Pattern of Animal Communities, published in 1966, emerged from years of surveying species, populations, communities and habitats in the Wytham Woods not far from Oxford. In this book, his primary intent was to describe and classify the diverse habitats available to terrestrial animals, and most of the chapters of the book are given to specific habitats for specialized kinds of organisms. Though not generally considered a theoretical ecologist, his early thinking did help to shape the field. In this book, late in his career, he summarized that thinking in a chapter titled "The Whole Pattern," in which he presents a set of fifteen "new concepts of the structure of natural systems," which he stated as a "series of propositions," though some reviewers labeled them "principles" of ecology.

Always the pragmatist, Elton devoted considerable time to the practical, applied aspects of ecology. Nowhere is this better demonstrated than in the work he turned to early in World War II. One of his original purposes in establishing the Bureau of Animal Populations was to better understand the role of disease in the fluctuations of animal numbers. At the beginning of the war, he turned the research focus of the Bureau to the control of rodent pests, especially to help in controlling human disease and to contribute toward the reduction of crop losses to rodents.

Elton was an early conservationist, stating in the preface to his 1927 text that "ecology is a branch of zoology which is perhaps more able to offer immediate practical help to mankind than any of the others [particularly important] in the present rather parous state of civilization." Elton strongly advocated the preservation of biological diversity, and pressed hard for the prevention of extinctions; this is what he emphasizes in his chapter on "The Reasons for Conservation" in the *Invasions* book. But he also expanded his conception of **conservation** to mean looking for some "general basis for understanding what it is best to do" and "looking for some wise principle of co-existence between man and **nature**, even if it has to be a modified kind of man and a modified kind of nature." He even took the unusual step (unusual for professional ecologists in his time and still unusual today) of going into the broadcast booth to popularize the importance of ecology in helping to achieve those goals though environmental management as applied ecology.

Elton's service to ecology as a learned discipline was enormous. In the early twentieth century, ecology was still in its formative years, so Elton's ideas and contributions came at a critical time. He took the infant field of animal ecology to maturity, building it to a status equal that of the more established plant ecology. His research Bureau at Oxford fostered innovative research and survey methods, and provided early intellectual nurture and professional development to ecologists who went on to become major contributors to the field, one example being the American ecologist Eugene Odum. As its first editor, Elton was "in a very real sense the creator" of the *Journal of Animal Ecology*, serving in the position for almost twenty years. He was one of the founders of the British Ecological Society. Elton's books and ideas continue to influence ecologists today.

[*Gerald L. Young Ph.D.*]

RESOURCES

BOOKS

Crowcroft, P. *Elton's Ecologists: A History of the Bureau of Animal Population.* Chicago: University of Chicago Press, 1991.
Elton, C. S. *Animal Ecology.* London: Sidgwick & Jackson, 1927.
———. *The Ecology of Invasions by Animals and Plants.* London: Methuen, 1958.
———. *The Pattern of Animal Communities.* London: Methuen, 1966.
———. *Voles, Mice and Lemmings: Problems in Population Dynamics.* Oxford: The Clarendon Press, 1942.

PERIODICALS

Hardy, A. "Charles Elton's Influence in Ecology." *The Journal of Animal Ecology* 37, no. 1 (February 1968): 1–8.

Emergency Planning and Community Right-to-Know Act (1986)

The Emergency Planning and Community **Right-to-Know** Act (EPCRA), also known as Title III, is a statute enacted by Congress in 1986 as a part of the Superfund Amendments and Reauthorization Act (SARA). It was enacted in response to public concerns raised by the accidental release of poisonous gas from a Union Carbide plant in **Bhopal, India** which killed over 2,000 people.

EPCRA has two distinct yet complementary sets of provisions. First, it requires communities to establish plans for dealing with emergencies created by chemical leaks or spills and defines the general structure these plans must assume. Second, it extends to communities the same kind of right-to-know provisions which were guaranteed to employees earlier in the 1980s. Overall, EPCRA is an important step away from crisis-by-crisis **environmental enforcement** toward a proactive or preventative approach. This proactive approach depends on government monitoring of potential environmental hazards, which is being accomplished by using computerized files of data submitted by businesses.

Under the provisions of EPCRA, the governors of every state were required to establish a State Emergency Response Commission by 1988. Each state commission was required in turn to establish various emergency planning districts and to appoint a local emergency planning committee for each. Each committee was required to prepare plans for potential chemical emergencies in their communities, which includes the identities of facilities, the procedures to be followed in the event of a chemical release, and the identities of community emergency coordinators as well as a facility coordinator from each business subject to EPCRA.

A facility is subject to EPCRA if it has a substance in a quantity equal to or greater than the threshold planning quantity specified on a list of about 400 extremely hazardous substances published by the **Environmental Protection Agency**. Also, after public notice and comment either the state governor or the State Emergency Response Commission may designate facilities to be covered outside of these guidelines. Each covered facility is required to provide facility notification information to the state commission and to designate a facility coordinator to work with the local planning committee.

EPCRA requires these facilities to report immediately any accidental releases of **hazardous material** to the Community Coordinator of its local emergency committee. There are two classifications for such hazardous substances. The substance must be either on the EPA's extremely hazardous substance list, or defined under the Comprehensive Environmental Response Compensation and Liability Act (CER-

CLA). In addition to the initial emergency notice, follow-up notices and information are required.

EPCRA's second major set of provisions is designed to establish and implement a community right-to-know program. Information about the presence of **chemicals** at facilities within the community is collected from businesses and made available to public officials and the general public. Businesses must submit two sets of annual reports: the Hazardous Chemical Inventory and Toxic Chemicals Release Inventories (TRIs), also known as Chemical Release Forms.

For the Hazardous Chemical Inventory, each facility in the community must prepare or obtain a Material Safety Data Sheet for each chemical on its premises meeting the threshold quantity. This information is then submitted to the Local Emergency Planning Committee, the local fire department, and the State Emergency Response Commission. These data sheets are identical to those required under the Occupational Safety and Health Act's worker right-to-know provisions. For each chemical reported in the Hazardous Chemical Inventory, a Chemical Inventory Report must be filed each year.

The second set of annual reports required as a part of the community right-to-know program is the Toxic Release Inventory (TRI), which must be filed annually. Releases reported on this form include even those made legally with permits issued by the EPA and its state counterparts. Releases made by the facility into air, land, and water during the preceding twelve months are summarized in this inventory. The form must be filed by companies having ten or more employees if that company manufactures, stores, imports, or otherwise uses designated toxic chemicals at or above threshold levels.

The information submitted pursuant to both the emergency planning and the right-to-know provisions of EPCRA is available to the general public through the Local Emergency Planning Committees. In addition, health professionals may obtain access to specific chemical identities even if that information is claimed by the business to be a trade secret in order to treat exposed individuals or protect potentially exposed individuals.

During the late 1980s, the EPA and its state counterparts emphasized public awareness and education about the requirements of EPCRA, rather than enforcement. But Congress has provided stiff penalties for noncompliance, and these agencies have now begun to implement their enforcement tools. Civil penalties of up to $25,000 per day for a first violation and up to $75,000 per day for a second may be assessed against a business failing to comply with reporting requirements, and citizens have the right to sue companies that fail to report. Further, enforcement by the government may include criminal prosecution and imprisonment.

In June of 1996, in response to community pressure, the U.S. Congress took up its first major vote on Community Right to Know since 1986. In the 1996 vote, the House of Representatives removed provisions from an EPA budget appropriation which would have made substantial cuts in funds allocated to compiling of Toxics Release Inventories (TRI). In addition, Congress passed an EPA proposal that added seven additional industries to the number of industries which must report under TRI, thus bringing the total number of industries required to report to twenty-seven. Those twenty-seven include more than 31,000 facilities across the United States. These Congressional votes are viewed by environmentalists as victories for right to know. Looking ahead, EPCRA may be further strengthened through provisions included in The Children's Environmental Protection and Right to Know Act of 1997 which was introduced to Congress in May of 1997 by Rep. Henry Waxman (D) and Rep. Jim Saxtan (Republican).

Studies have revealed that EPCRA has had far-reaching effects on companies and that industrial practices and attitudes toward chemical risk management are changing. Some firms have implemented new **waste reduction** programs or adapted previous programs. Others have reduced the potential for accidental releases of hazardous chemicals by developing safety audit procedures, reducing their chemical inventories, and using less hazardous chemicals in their operations.

As information included in reports such as the annual **Toxics Release Inventory** has been disseminated throughout the community, businesses have found they must be concerned with risk communication. Various industry groups throughout the United States have begun making the information required by EPCRA readily available and helping citizens to interpret that information. For example, the Chemical Manufacturers Association has conducted workshops for its members on communicating EPCRA information to the community and on how to communicate about risk in general. Similar seminars are now made available to businesses and their employees through trade associations, universities, and other providers of continuing education. *See also* Chemical spills; Environmental monitoring; Environmental Monitoring and Assessment Program; Hazardous Materials Transportation Act; Toxic substance; Toxic Substances Control Act; Toxics use reduction legislation

[*Paulette L. Stenzel*]

RESOURCES
PERIODICALS

Stenzel, P. L. "Small Business and the Emergency Planning and Community Right-to-Know Act." *Michigan Bar Journal* (February 1990): 181-183.
Stenzel, P. L. "Toxics Use Reduction Legislation: An Important 'Next Step' After Right to Know." *Utah Law Review* 76 (1991): 707-747.

OTHER
Emergency Planning and Community Right-to-Know Act of 1986, 42 U.S.C. Sec. 11001-11050 (1986).

Emergent diseases (human)

Although many diseases such as measles, pneumonia, and pertussis (whooping cough) have probably inflicted humans for millennia, at least 30 new infectious diseases have appeared in the past two decades, In addition, many well-known diseases recently have reappeared in more virulent or drug-resistant forms. An emergent disease is one never known before or one that has been absent for at least 20 years. **Ebola** fever is a good example of an emergent disease. A kind of viral hemorrhagic fever, Ebola is extremely contagious and often kills up to 90% of those who are exposed to it. The disease was unknown until about 20 years ago, but is thought to have been present in monkeys or other primates. Killing and eating chimps, gorillas, and other primates is thought to be the route of infection in humans. **AIDS** is another disease that appears to have suddenly moved from other primates to humans. How pathogens suddenly move across **species** barriers to become highly contagious and terribly lethal is one of the most important questions in **environmental health**.

Some of the most devastating epidemics have occurred when travelers bring new germs to a naïve population lacking immunity. An example was the **plague**, or Black Death, which swept through Europe and Western Asia repeatedly in the fourteenth and fifteenth centuries. During the first- and worst-episode between 1347 and 1355, about half the population of Europe died. In some cities the **mortality** rate was as high as 80%. It's hard to imagine the panic and fear this disease caused. An even worse disaster may have occurred when Europeans brought smallpox, measles, and other infectious diseases to the Americas. By some calculations, up to 90% of the native people perished as diseases swept through their population. One reason European explorers thought the land was an empty **wilderness** was that these diseases spread out ahead of them, killing everyone in their path.

Probably the largest loss of life from an individual disease in a single year was the great influenza pandemic of 1918. Somewhere between 30 and 40 million people succumbed to this **virus** in less than 12 months. This was more than twice the total number killed in all the battles of World War I, which was occurring at the time. Crowded, unsanitary troop ships carrying American soldiers to Europe started the epidemic. War refugees, soldiers from other nations returning home, and a variety of other travelers quickly spread the virus around the globe. Flu is especially contagious, spreading either by direct contact with an infected person or by breathing airborne particles released by coughing or sneezing. Most flu strains are zoonotic (transmitted from an animal host to humans). Pigs, birds, monkeys, and rodents often serve as reservoirs from which viruses can jump to humans. Although new flu strains seem to appear nearly every year, no epidemic has been as deadly as that of 1918.

Malaria, the most deadly of all insect-borne diseases, is an example of the return of a disease that once was thought nearly vanquished. Malaria now claims about 3 million lives every year—90% in Africa and most of them children. With the advent of modern medicines and pesticides, malaria had nearly been wiped out in many places but recently has had a resurgence. The protozoan parasite that causes the disease is now resistant to most antibiotics, while the mosquitoes that transmit it have developed **resistance** to many insecticides. Spraying of DDT in India and Sri Lanka reduced malaria from millions of infections per year to only a few thousand in the 1950s and 1960s. Now South Asia is back to its pre-DDT level of some 2.5 million new cases of malaria every year. Other places that never had malaria or dengue fever now have them because of **climate** change and **habitat** alteration. Gulf-coast states in America, for example, are now home to the *Aedes aegypti* mosquito that carries these diseases.

Why have vectors such as mosquitoes and pathogens such as the malaria parasite become resistant to pesticides and antibiotics? Part of the answer is natural selection and the ability of many organisms to evolve rapidly. Another factor is the human tendency to use control measures carelessly. When we discovered that DDT and other insecticides could control mosquito populations, we spread them indiscriminately without much thought to ecological considerations. In the same way, antimalarial medicines such as chloroquine were given to millions of people, whether they showed symptoms or not. This was a perfect recipe for natural selection. Many organisms were exposed only minimally to control measures. This allowed those with natural resistance to outcompete others and spread their genes through the population. After repeated cycles of exposure and selection, many **microorganisms** and their vectors are insensitive to almost all our weapons against them.

There are many examples of drug resistance in pathogens. Tuberculosis (TB), once the foremost cause of death in the world, had nearly been eliminated—at least from the developed world—by the end of the twentieth century. Drug-resistant varieties of TB are now spreading rapidly, however. One of the places these strains arise is in Russia, where crowded prisons with poor **sanitation**, little medical care, gross overcrowding, and inadequate nutrition serve as a breeding ground for this deadly disease. Inmates who are treated with antibiotics rarely get a complete dose. Those with TB aren't segregated from healthy inmates. Patients with active TB are released from prison and sent home to spread the disease further. And migrants from Russian carry the disease to other countries.

Another development is the appearance of drug-resistant strains of *Staphylococcus aureus*, the most common form of hospital-acquired infections. Staph A has many forms, some of which are extremely toxic—toxic-shock syndrome is one where staphylococcus **toxins** spread through the body and sometimes bring death in a matter of just hours. Another strain of staphylococcus, sometimes called flesh-eating bacteria, causes massive necrosis (cell death) that destroys skin, connective tissue, and muscle. For 40 years vancomycin has been the last recourse against staph infections. Strains resistant to everything else could be controlled by this antibiotic. Now vancomycin-resistant staph strains are being reported in many places.

A number of factors contribute currently to the appearance and spread of these highly contagious diseases. With 6 billion people now inhabiting the planet, human densities are much higher, enabling germs to spread further and faster than ever before. Expanding populations push into remote areas encountering new pathogens and **parasites**. Environmental change on a larger scale, such as cutting forests, creating unhealthy urban surroundings, and causing global-climate change, among other things, eliminates predators and habitat changes favor disease-carrying organisms such as mice, rats, cockroaches, and mosquitoes.

Another important factor in the spread of many diseases is the speed and frequency of modern travel. Millions of people go every day from one place to another by airplane, boat, train, or **automobile**. Very few places on earth are more than 24 hours by jet plane from any other place. Many highly virulent diseases take several days for symptoms to appear.

Humans aren't the only ones to suffer from new and devastating diseases. Domestic animals and **wildlife** also experience sudden and widespread epidemics, sometimes called **emergent ecological diseases**. In 1998, for example, a distemper virus killed half the **seals** in Western Europe. It's thought that toxic pollutants and hormone-disrupting environmental **chemicals** might have made seals and other marine mammals susceptible to infections. In 2002, more dead seals were found in Denmark, raising fears that distemper might be reappearing.

Chronic wasting disease (CWD) is spreading through deer and elk populations in the North America. Caused by a strange protein called a prion, CWD is one of a family of irreversible, degenerative neurological diseases known as transmissible spongiform encephalopathies (TSE) that include **mad cow disease** in cattle, scrapie in sheep, and Creutzfelt-Jacob disease in humans. CWD probably started when elk ranchers fed contaminated animal by-products to their herds. Infected animals were sold to other ranches, and now the disease has spread to wild populations. First recognized in 1967 in Saskatchewan, CWD has been identified in wild deer populations and ranch operations in at least eight American states. No humans are known to have contracted TSE from deer or elk, but there is a concern that we might see something like the mad cow disaster that inflicted Europe in the 1990s. At least 100 people died, and nearly five million European cattle and sheep were slaughtered in an effort to contain that disease.

One of the things all these diseases have in common is that human-caused environmental changes are stressing biological communities and upsetting normal ecological relationships.

[*William P. Cunningham Ph.D.*]

RESOURCES
BOOKS

Diamond, Jared. *Guns, Germs, and Steel: The Fates of Human Societies* New York: W.W. Norton & Company, 1999.

Drexler, Madeline. *Secret Agents: the Menace of Emerging Infections.* Joseph Henry Press, 2002.

Miller, Judith, Stephen Engelberg, and William J. Broad. *Germs: Biological Weapons and America's Secret War.* New York: Simon & Schuster, 2000.

PERIODICALS

Daszak, P. et al. "Emerging Infectious Diseases of Wildlife-Threats to Biodiversity and Human Health." *Science* 287 (2000): 443-449.

Hughes, J. M. "Emerging infectious diseases: a CDC perspective." *Emerging Infectious Diseases.* 7(2001):494-6.

Osterholm, M. T. "Emerging Infections—Another Warning." *The New England Journal of Medicine* 342(2000): 4-5.

Emergent ecological diseases

Emergent ecological diseases are relatively recent phenomena involving extensive damage being caused to natural communities and ecosystems. In some cases, the specific causes of the ecological damage are known, but in others they are not yet understood.

Examples of relatively well-understood ecological diseases mostly involve cases in which introduced, non-native pathogens are causing extensive damage. There are, unfortunately, many examples of this kind of damage caused by invasive organisms. One case involves the introduced chestnut blight fungus (*Endothia parasitica*), which has virtually eliminated the once extremely abundant American chestnut (*Castanea dentata*) from the hardwood forests of eastern North America. A similar ongoing pandemic involves the Dutch elm disease fungus (*Ceratocystis ulmi*), which is removing white elm (*Ulmus americana*) and other native elms from

North America. Some introduced insects are also causing important forest damage, including the effects of the balsam wooly adelgid (*Adelges picea*) on Fraser fir (*Abies fraseri*) in the Appalachian Mountains.

Other cases of ecological diseases involve widespread damages that are well-documented, but the causes of which are not yet understood. One of them affects native forests of Hawaii dominated by the tree ohia (*Metrosideros polymorpha*). For some unknown reason, stands of ohia decline and then die when they reach maturity. This may be caused by the synchronous senescence of a cohort of trees that established following a stand-replacing disturbance, such as a lava flow, and then reached maximum longevity at about the same time. Other causal factors have, however, also been suggested, including **nutrient** dysfunction and pathogens.

Another case is known as birch decline, which occurred over great regions of the northeastern United States and eastern Canada from the 1930s to the 1950s. The disease affected yellow birch (*Betula alleghaniensis*), paper birch (*B. papyrifera*), and grey birch (*B. populifolia*), which suffered **mortality** over a huge area. The specific cause of this extensive forest damage was never determined, but it could have involved the effects of freezing ground conditions during winters with little snow cover.

Rather similar forest declines and diebacks have affected red spruce (*Picea rubens*) and sugar maple (*Acer saccharum*) in the same broad region of eastern North America during the 1970s to 1990s. Although the causes of these forest damages are not yet fully understood, it is thought that **air pollution** or acidifying **atmospheric deposition** may have played a key role. In western Europe, extensive declines of Norway spruce (*Picea abies*) and beech (*Fagus sylvatica*) are also thought to somehow be related to exposure to air **pollution** and **acidification**. In comparison, the damage caused by **ozone** to forests dominated by ponderosa pine (*Pinus ponderosa*) in California is a relatively well-understood kind of emergent ecological disease.

In the marine realm, widespread damage to diverse **species** of corals has been documented in far-flung regions of the world. The phenomenon is known as coral "bleaching," and it involves the corals expelling their symbiotic algae (known as zooxanthellae), often resulting in death of the coral. **Coral bleaching** is thought to possibly be related to **climate** warming, although it can be caused by both unusually high or low water temperatures, changes in **salinity**, and other environmental stresses.

Another unexplained case of an ecological disease appears to be afflicting species of amphibians in many parts of the world. The amphibian declines involve severe population collapses, and have even caused the **extinction** of some species. The specific causes are not yet known, but they likely involve introduced **microbial pathogens**, or possibly

increased exposure to solar **ultraviolet radiation**, climate change, or some other factor.

[*Bill Freedman Ph.D.*]

RESOURCES
BOOKS

Freedman, B. *Environmental Ecology.* San Diego, CA: Academic Press, 1995.

EMF

see **Electromagnetic field**

Emission

Release of material into the **environment** either by natural or human-caused processes. This term is used especially in describing **air pollution** for volatile or suspended contaminants that result from processes such as burning fuel in an engine. Definitions of **pollution** are complicated by the fact that many of the materials that damage or degrade our **atmosphere** have both human and natural origins. Volcanoes emit ash, **acid** mists, **hydrogen** sulfide, and other toxic gases. Natural forest fires release **smoke**, soot, carcinogenic **hydrocarbons**, dioxins, and other toxic **chemicals** as well as large amounts of **carbon dioxide**. Do these emissions constitute pollution when they originate from human sources but not if released by natural processes? Is it reasonable to restrict human emissions if there are already very large natural sources of those same materials in the environment? An important consideration in answering these questions lies in the regenerative capacity of the environment to remove or neutralize contaminants. If we overload that capacity, a marginal additional emission may be important. Similarly, if there are thresholds for response, an incremental addition to ambient levels may be very important.

Emission standards

Federal, state, and local stack and **automobile** exhaust **emission** limits that regulate the quantity, rate, or concentration of emissions. Emission standards can also regulate the opacity of plumes of **smoke** and dust from point and area emission sources. They can also assess the type and quality of fuel and the way the fuel is burned, hence the type of technology used. With the exception of **plume** opacity, such standards are normally applied to the specific type of source for a given pollutant. Federal standards include New Source Performance Standards (NSPS) and **National Emission Standards for Hazardous Air Pollutants** (NESHAPS).

Emission standards may include prohibitory rules that restrict existing and new source emission to specific emission concentration levels, mass emission rates, plume opacity, and emissions relative to process throughput emission rates. They may also require the most practical or best available technology in case of new emission in pristine areas.

New sources and modifications to existing sources can be subject to new source permitting procedures which require technology-forcing standards such as **Best Available Control Technology** and **Lowest Achievable Emission Rate** (LAER). However, these standards are designed to consider the changing technological and economic feasibility of evermore stringent emission controls. As a result, such requirements are not stable and are determined through a process involving discretionary judgements of appropriateness by the governing **air pollution** authority. *See also* Point source

Emissions trading
see **Trade in pollution permits**

Emphysema

Emphysema is an abnormal, permanent enlargement of the airways responsible for gas-exchange in the lungs. Primary emphysema is commonly linked to a genetic deficiency of the **enzyme** &agr;$_1$-antitrypsin which is a major component of &agr;$_1$-globulin, a **plasma** protein. Under normal conditions &agr;$_1$-antitrypsin inhibits the activity of many proteolytic enzymes which breakdown proteins. This results in the increased likelihood of developing emphysema as a result of proteolysis (breakdown) of the lung tissues.

Emphysema begins with destruction of the alveolar septa. This results in "air hunger" characterized by labored or difficult breathing, sometimes accompanied by pain. Although emphysema is genetically linked to deficiency in certain enzymes, the onset and severity of asthmatic symptoms has been definitively linked to irritants and pollutants in the **environment**. A significantly greater proportion of the individuals manifesting emphysemic symptoms is observed in smokers, populations clustered around industrial complexes, and **coal** miners. *See also* Asthma; Cigarette smoke; Respiratory diseases

Endangered species

An "endangered species" under United States law (the **Endangered Species Act** [1973]) is a creature "in danger of **extinction** throughout all or a significant portion of its range." A "threatened" **species** is one that is likely to become endangered in the foreseeable future.

For most people, the endangered species problem involves the plight of such well-known animals as eagles, **tigers**, **whales**, **chimpanzees**, **elephants**, **wolves**, and whooping cranes. However, literally millions of lesser-known or unknown species are endangered or becoming so, and the loss of these life forms could have even more profound effects on humans than that of large mammals with whom we more readily identify and sympathize.

Most experts on species extinction, such as Edward O. Wilson of Harvard and Norman Myers, estimate current and projected *annual* extinctions at anywhere from 15,000 to 50,000 species, or 50 to 150 *per day*, mainly invertebrates such as insects in tropical rain forests. At this rate, 5–10% of the world's species, perhaps more, could be lost in the next decade and a similar percentage in coming decades.

The single most important common threat to **wildlife** worldwide is the loss of **habitat**, particularly the destruction of biologically-rich tropical rain forests. Additional factors have included commercial exploitation, the introduction of non-native species, **pollution**, **hunting**, and **trapping**. Thus, we are rapidly losing a most precious heritage, the diversity of living species that inhabit the earth. Within one generation, we are witnessing the threatened extinction of between one fifth and one half of all species on the planet.

Species of wildlife are becoming extinct at a rate that defies comprehension and threatens our own future. These losses are depriving this generation and future ones of much of the world's beauty and diversity, as well as irreplaceable sources of food, drugs, medicines, and natural processes that are or could prove extremely valuable, or even necessary, to the well-being of our society.

Today's rate of extinction exceeds that of all of the **mass extinction** in geologic history, including the disappearance of the dinosaurs 65 million years ago. It is impossible to know how many species of plants and animals we are actually losing, or even how many species exist, since many have never been "discovered" or identified. What we do know is that we are rapidly extirpating from the face of the earth countless unique life forms that will never again exist.

Most of these species extinctions will occur—and are occurring—in tropical rain forests, which are the richest biological areas on earth and are being cut down at a rate of 1–2 acres (0.4–0.8 ha) a second. Although tropical forests cover only about 5–7% of the world's land surface, they are thought to contain over half of the species on earth.

There are more bird species in one Peruvian preserve than in the entire United States. There are more species of fish in one Brazilian river than in all the rivers of the United States. And a single square mile in lowland Peru or Amazonian Ecuador or Brazil may contain over 1500 species of butterflies, more than twice as many as are found in all of the United States and Canada. Half an acre of Peruvian **rain forest** may contain over 40,000 species of insects.

Eric Eckholm in *Disappearing Species: The Social Challenge* notes that when a plant species is wiped out, some 10–30 dependent species can also be jeopardized, such as insects and even other plants. An example of the complex relationship that has evolved between many tropical species is the 40 different kinds of fig trees native to Central America, each of which has a specific insect pollinator. Other insects, including pollinators for other plants, depend on certain of these fig trees for food.

Thus, the extinction of one species can set off a chain reaction, the ultimate effects of which cannot be foreseen. As Eckholm puts it, "Crushed by the march of civilization, one species can take many others with it, and the ecological repercussion and arrangements that follow may well endanger people." The loss of so many unrecorded, unstudied species will deprive the world not only of beautiful and interesting life forms, but also much-needed sources of medicines, drugs, and food that could be of critical value to humanity. Every day, we could be losing plants that could provide cures for **cancer** or **AIDS** or could become food staples as important as rice, wheat, or corn. We will simply never know the value or importance of the untold thousands of species vanishing each year.

As of spring of 2002, the U.S. Department of the Interior's list of endangered and threatened species included 1,070 animals (mammals, birds, reptiles, amphibians, fish, snails, clams, crustaceans, insects, and arachnids), and 746 plants, for a total of 1816 endangered or threatened species.

Under the Endangered Species Act, the Department of the Interior is given general responsibility for listing and protecting endangered wildlife, except for marine species (such as whales and **seals**), which are the responsibilities of the Commerce Department.

In addition, the United States is subject to the provisions of the Convention on International Trade in Endangered Species of Wild Flora and **Fauna** (CITES), which regulates global commerce in **rare species**. But in many cases, the government has not been enthusiastic about administering and enforcing the laws and regulations protecting endangered wildlife. Conservationists have for years criticized the Interior Department for its slowness and even refusal to list hundreds of endangered species that, without government protection, were becoming extinct. Indeed, the Department admits that some three dozen species have become extinct while undergoing review for listing.

In December 1992, the department settled a lawsuit brought by animal protection groups by agreeing to expedite the listing process for some 1300 species and to take a more comprehensive "multispecies, **ecosystem** approach" to protecting wildlife and their habitat. In October 1992, at

the national conference of the **Humane Society of the United States** held in Boulder, Colorado, the Secretary of the Interior Bruce Babbitt in his keynote address lauded the Endangered Species Act as "an extraordinary achievement," emphasized the importance of preserving endangered species and biological diversity, and noted: "The extinction of a species is a permanent loss for the entire world. It is millions of years of growth and development put out forever." *See also* Biodiversity

[*Lewis G. Regenstein*]

RESOURCES
BOOKS

Mitchell, G. J. *World on Fire: Saving an Endangered Earth*. New York: Charles Scribner's Sons, 1991.
Myers, N. *The Sinking Ark: A New Look at Disappearing Species*. Oxford: Pergamon Press, 1979.
Porritt, J. *Save the Earth*. Atlanta: Turner Publishing, 1991.
Raven, P. H. "Endangered Realm." In *The Emerald Realm*. Washington, DC: National Geographic Society, 1990.
Wilson, E. O., ed. *Biodiversity*. Washington, DC: National Academy Press, 1988.

OTHER

"Endangered and Threatened Wildlife and Plants." U. S. Department of the Interior. *Federal Register* (29 August 1992).

Endangered Species Act (1973)

The **Endangered Species** Act (ESA) is a law designed to save **species** from **extinction**. What began as an informal effort to protect several hundred North American vertebrate species in the 1960s has expanded into a program that could involve hundreds of thousands of plant and animal species throughout the world. As of May 2002, 1,816 species were listed as endangered or threatened, 1,258 in the United States and 558 in other countries. The law has become increasingly controversial as it has been viewed by commercial interests as a major impediment to economic development. This issue recently came to a head in the Pacific Northwest, where the **northern spotted owl** has been listed as threatened. This action has had significant affects on the regional forest products industry. The ESA was due to be re-authorized in 1992, but this was postponed due to that year's election. Although it expired on October 1, 1992, Congress has allotted enough funds to keep the ESA active.

Government action to protect endangered species began in 1964, with the formation of the Committee on Rare and Endangered Wildlife Species within the Bureau of Sport Fisheries and Wildlife (now the **Fish and Wildlife Service** [FWS]) in the **U.S. Department of the Interior**. In 1966, this committee issued a list of 83 native species (all verte-

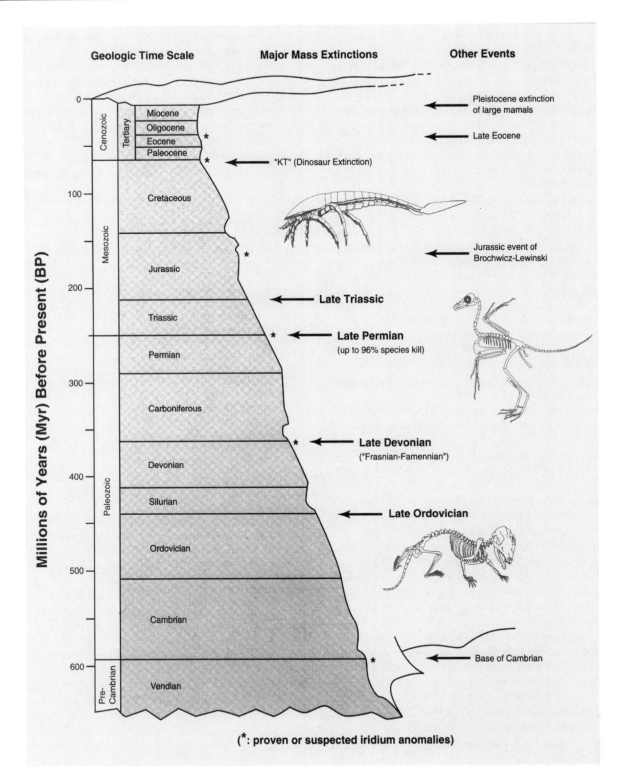

Geologic Time Scale

Major Mass Extinctions

Other Events

Millions of Years (Myr) Before Present (BP)

Cenozoic
Tertiary
Miocene
Oligocene
Eocene
Paleocene

Mesozoic
Cretaceous
Jurassic
Triassic

Paleozoic
Permian
Carboniferous
Devonian
Silurian
Ordovician
Cambrian

Pre-Cambrian
Vendian

Pleistocene extinction of large mamals

Late Eocene

"KT" (Dinosaur Extinction)

Jurassic event of Brochwicz-Lewinski

Late Triassic

Late Permian
(up to 96% species kill)

Late Devonian
("Frasnian-Famennian")

Late Ordovician

Base of Cambrian

(*: proven or suspected iridium anomalies)

A chart of the estimated annual rate of species loss from 1700–2000. (Beacon Press. Reproduced by permission.)

brates) that it considered endangered. Two years later, the first act designed to protect species in danger of extinction, the Endangered Species Preservation Act of 1966, was passed. The Secretary of the Interior was to publish a list, after consulting the states, of native vertebrates that were endangered. This law directed federal agencies to protect endangered species when it was "practicable and consistent with the primary purposes" of these agencies. The taking of listed endangered species was prohibited only within the **national wildlife refuge** system; that is, species could be killed almost anywhere in the United States. Finally, the law authorized the acquisition of **critical habitat** for these endangered species.

In 1969, the Endangered Species Conservation Act was passed, which included several significant amendments to the 1966 Act. Species could now be listed if they were threatened with worldwide extinction. This substantially broadened the scope of species to be covered, but it also limited the listing of specific populations that might be endangered in some parts of the United States but not in danger elsewhere (e.g., grizzly bears, bald eagles, timber **wolves**, all of which flourish in Canada and Alaska). The 1969 law stated that mollusks and crustaceans could now be included on the list, further broadening of the scope of the law. Finally, trade in illegally taken endangered species was prohibited. This substantially increased the protection offered such species, compared to the 1966 law.

The Endangered Species Act of 1973 built upon and strengthened the previous laws. The impetus for the law was a call by President Nixon in his state of the union message for further protection of endangered species and the concern in Congress that the previous acts were not working well enough. The goal of the ESA was to protect all endangered species through the use of "all methods and procedures necessary to bring any endangered or threatened species to the point at which the measures provided pursuant to [the] Act are no longer necessary." In other words, the goal was to bring endangered species to full recovery. This goal, like others included in environmental legislation at the time, was unrealistic. The ESA also expanded the number of species that could be considered for listing to all animals (except those considered pests) and plants. It stipulated that the listing of such species should be based on the best scientific data available. Additionally, it included a provision that allowed groups or individuals to petition the government to list or de-list a species. If the petition contained reasonable support, the agency had to respond to it.

The law created two levels of concern: endangered and threatened. An endangered species was "in danger of extinction throughout all or a significant portion of its range." A threatened species was "likely to become an endangered species within the foreseeable future throughout all

or a significant portion of its range." Also, the species did not have to face worldwide extinction before it could be listed. No taking of any kind was allowed for endangered species; limited taking could be allowed for threatened species. Thus, the distinction between "endangered" and "threatened" species allowed for some flexibility in the program.

The 1973 Act divided jurisdiction of the program between the FWS and the National Marine Fisheries Service (NMFS), an agency of the **National Oceanic and Atmospheric Administration** in the Department of Commerce. The NMFS would have responsibility for species that were primarily marine; responsibility for marine mammals (**whales, dolphins**, etc.) was shared by the two agencies. The law also provided for the establishment of cooperative agreements between the federal government and the states on endangered species protection. This has not proved very successful due to a lack of funds to entice the states to participate and due to frequent conflict between the states (favoring development and **hunting**) and the FWS.

The most controversial aspect of the Act was Section 7, which required that no action by a federal agency, such as the destruction of critical **habitat**, jeopardize any endangered species. So, before undertaking, funding, or granting a permit for a project, federal agencies had to consult with the FWS as to the effect the action might have on endangered species. This provision proved to have enormous consequences, as many federal developments could be halted due to their affect on endangered species. The most famous and controversial application of this provision involved the **snail darter** and **Tellico Dam** in Tennessee. The **Tennessee Valley Authority** (TVA) was building the nearly completed dam when the snail darter was listed as endangered. Its only known habitat would be destroyed if the dam was completed. The TVA challenged the FWS evidence, but the TVA was itself soon challenged in the courts by environmentalists. In a case that was appealed through the Supreme Court, *TVA v. Hill*, the courts ruled that the ESA was clear: no federal action could take place that would jeopardize an endangered species. The dam could not be completed.

In response to the conflicts that followed the passage of the ESA, especially the swelling estimates of the number of endangered species and the snail darter-Tellico Dam issue, the 1978 re-authorization of the ESA included heated debate and a few significant changes in the law. Perhaps the most important change was the creation of the Endangered Species Committee, sometimes referred to as the "God Committee." Created in response to the snail darter controversy and the *TVA v. Hill* decision, this committee could approve federal projects that were blocked due to their harmful effects on endangered species. If an agency's actions were blocked due to an endangered species, they could appeal to this committee for an exemption from the ESA. The committee, which consists of three cabinet secretaries, the administrators of the EPA and NOAA, the chair of the Council of Economic Advisors, and the governor from the state in which the project is located, weigh the advantages and disadvantages of the project and then make a decision as to the appeal. Ironically, the committee heard the appeal for Tellico Dam and rejected it. The committee has only been used three times, and only once, regarding the northern spotted owl and 1,700 acres (689 ha) of land in Oregon, has it approved an appeal. Nonetheless, the creation of the "God Committee" demonstrated that values beyond species survival had to be weighed into the endangered species equation.

Additionally, the 1978 amendments mandated: increased public participation and hearings when species were proposed for listing; a five-year review of all species on the list (to determine if any had improved to the point that they could be removed from the list); a requirement that the critical habitat of a species must be specified at the time the species is listed and that an economic assessment of the critical habitat designation must be done; the mandatory development of a recovery plan for each listed species; and a time limit between when a species was proposed for listing and when the final rule listing the species must be issued. These amendments were designed to do two things: to provide a loophole for projects that might be halted by the ESA and to speed up the listing process. Despite this latter goal, the many new requirements included in the amendments led to a further slowing of the listing process. It should also be noted that these amendments passed with overwhelming majorities in both the House and the Senate; there was still strong support for protecting endangered species, at least in the abstract, in Congress.

The 1982 re-authorization of the ESA did not lead to significant changes in the Act. The law was to have been re-authorized again in 1985, but opposition in the Senate prevented re-authorization until 1988. This demonstrated the growing uneasiness in Congress to the economic repercussions of the ESA. In addition to re-authorizing spending to implement the ESA through 1992, the 1988 amendments also increased the procedural requirements for recovery plans. This further underscored the tension between the desire for public participation at all stages of ESA implementation and the need for the government to move quickly to protect endangered species. Overall, the implementation of the ESA has not been successful. The law has suffered from two main problems: poor administrative capacity and fragmentation. The FWS has suffered from its lack of stature within the bureaucracy, its conflicting institutional mission, a severe lack of funds and personnel, and limited public support. Species assigned to the NMFS have fared even worse, as the agency has shown little interest in the ESA.

Fragmentation is demonstrated with the division of responsibilities between the FWS and NMFS, the conflict with other federal agencies due to Section 7, and the federal-state conflicts over jurisdictional responsibility.

[*Christopher McGrory Klyza*]

RESOURCES
BOOKS

Harrington, W., and A. C. Fisher. "Endangered Species." In *Current Issues in Natural Resource Policy*, edited by P. R. Portney. Baltimore: Johns Hopkins University Press, 1982.
Tobin, R. *The Expendable Future: U. S. Politics and the Protection of Biological Diversity.* Durham, NC: Duke University Press, 1990.

PERIODICALS

Egan, T. "Strongest U. S. Environmental Law May Become Endangered Species." *New York Times* (26 May 1992): A1.

Endemic species

Endemic **species** are plants and animals that exist only in one geographic region. Species can be endemic to large or small areas of the earth: some are endemic to a particular continent, some to part of a continent, and others to a single island. Usually an area that contains endemic species is isolated in some way, so that species have difficulty spreading to other areas, or it has unusual environmental characteristics to which endemic species are uniquely adapted. Endemism, or the occurrence of endemic animals and plants, is more common in some regions than in others. In isolated environments such as the **Hawaiian Islands**, **Australia**, and the southern tip of Africa, as many of 90% of naturally occurring species are endemic. In less isolated regions, including Europe and much of North America, the%age of endemic species can be very small.

Biologists who study endemism do not only consider species, the narrowest classification of living things; they also look at higher level classifications of genus, family, and order. These hierarchical classifications are nested so that, in most cases, an order of plants or animals contains a number of families, each of these families includes several genera (plural of "genus"), and each genus has a number of species. These levels of classification are known as "taxonomic" levels.

Species is the narrowest taxonomic classification, with each species closely adapted to its particular **environment**. Therefore species are often endemic to small areas and local environmental conditions. Genera, a broader class, are usually endemic to larger regions. Families and orders more often spread across continents. As an example, the order Rodentia, or rodents, occurs throughout the world. Within this order, the family Heteromyidae occurs only in western North America and the northern edge of South America.

One member of this family, the genus *Dipodomys*, or kangaroo rats, is restricted to several western states and part of Mexico. Finally, the species *Dipodomys ingens*, occurs only in a small portion of the California coast. Most often endemism is considered on the lowest taxonomic levels of genus and species.

Animals and plants can become endemic in two general ways. Some evolve in a particular place, adapting to the local environment and continuing to live within the confines of that environment. This type of endemism is known as "autochthonous," or native to the place where it is found. An "allochthonous" endemic species, by contrast, originated somewhere else but has lost most of its earlier geographic range. A familiar autochthonous endemic species is the Australian koala, which evolved in its current environment and continues to occur nowhere else. A well-known example of allochthonous endemism is the California coast redwood (*Sequoia sempervirens*), which millions of years ago ranged across North America and Eurasia, but today exists only in isolated patches near the coast of northern California. Another simpler term for allochthonous endemics is "relict," meaning something that is left behind.

In addition to geographic relicts, plants or animals that have greatly restricted ranges today, there are what is known as "taxonomic relicts." These are species or genera that are sole survivors of once-diverse families or orders. **Elephants** are taxonomic relicts: millions of years ago the family Elephantidae had 25 different species (including woolly mammoths) in five genera. Today only two species remain, one living in Africa (*Loxodonta africana*) and the other in Asia (*Elephas maximus*). Horses are another familiar species whose family once had many more branches. Ten million years ago North America alone had at least 10 genera of horses. Today only a few Eurasian and African species remain, including the zebra and the ass. Common horses, all members of the species *Equus caballus*, returned to the New World only with the arrival of Spanish conquistadors.

Taxonomic relicts are often simultaneously geographic relicts. The ginkgo tree, for example was one of many related species that ranged across Asia 100 million years ago. Today the family Ginkgoales contains only one genus, *Ginkgo*, with a single species, *Ginkgo biloba*, that occurs naturally in only a small portion of eastern China. Similarly the coelacanth, a rare fish found only in deep waters of the Indian Ocean near **Madagascar**, is the sole remnant of a large and widespread group that flourished hundreds of millions of years ago.

Where living things become relict endemics, some sort of environmental change is usually involved. The redwood, the elephant, the ginkgo, and the coelacanth all originated in the Mesozoic era, 245–65 million years ago, when the earth was much warmer and wetter than it is today. All of these species managed to survive catastrophic environmental

change that occurred at the end of the Cretaceous period, changes that eliminated dinosaurs and many other terrestrial and aquatic animals and plants. The end of the Cretaceous was only one of many periods of dramatic change; more recently two million years of cold ice ages and warmer interglacial periods in the Pleistocene substantially altered the distribution of the world's plants and animals. Species that survive such events to become relicts do so by adapting to new conditions or by retreating to isolated refuges where habitable environmental conditions remain.

When endemics evolve in place, isolation is a contributing factor. A species or genus that finds itself on a remote island can evolve to take advantage of local food sources or environmental conditions, or its characteristics may simply drift away from those of related species because of a lack of contact and interbreeding. Darwin's Galapagos finches, for instance, are isolated on small islands, and on each island a unique species of finch has evolved. Each finch is now endemic to the island on which it evolved. Expanses of water isolated these evolving finch species, but other sharp environmental gradients can contribute to endemism, as well. The humid southern tip of Africa, an area known as the Cape region, has one of the richest plant communities in the world. A full 90% of the Cape's 18,500 plant species occur nowhere else. Separated from similar **habitat** for millions of years by an expanse of dry **grasslands** and **desert**, local families and genera have divided and specialized to exploit unique local niches. Endemic speciation, or the **evolution** of locally unique species, has also been important in Australia, where 32% of genera and 75% of species are endemic. Because of its long isolation, Australia even has family-level endemism, with 40 families and sub-families found only on Australia and a few nearby islands.

Especially high rates of endemism are found on long-isolated islands, such as St. Helena, New Caledonia, and the Hawaiian chain. St. Helena, a volcanic island near the middle of the Atlantic, has only 60 native plant species, but 50 of these exist nowhere else. Because of the island's distance from any other landmass, few plants have managed to reach or colonize St. Helena. Speciation among those that have reached the remote island has since increased the number of local species. Similarly Hawaii and its neighboring volcanic islands, colonized millions of years ago by a relatively small number of plants and animals, now has a wealth of locally-evolved species, genera, and sub-families. Today's 1,200–1,300 native Hawaiian plants derive from about 270 successful colonists; 300–400 arthropods that survived the journey to these remote islands have produced over 6,000 descendent species today. Ninety-five percent of the archipelago's native species are endemic, including all ground birds. New Caledonia, an island midway between Australia and Fiji, consists partly of continental rock, suggesting that at one time the island was attached to a larger landmass and its resident species had contact with those of the mainland. Nevertheless, because of long isolation 95% of native animals and plants are endemic to New Caledonia.

Ancient, deep lakes are like islands because they can retain a stable and isolated habitat for millions of years. Siberia's **Lake Baikal** and East Africa's Lake Tanganyika are two notable examples. Lake Tanganyika occupies a portion of the African Rift Valley, 0.9 mi (1.5 km) deep and perhaps 6 million years old. Fifty percent% of the lake's snail species are endemic, and most of its fish are only distantly related to the fish of nearby Lake Nyasa. Siberia's Lake Baikal, another rift valley lake, is 25 million years old and 1 mi (1.6 km) deep. Eighty-four percent of the lake's 2,700 plants and animals are endemic, including the nerpa, the world's only freshwater seal.

Because endemic animals and plants by definition have limited geographic ranges, they can be especially vulnerable to human invasion and habitat destruction. Island species are especially vulnerable because islands commonly lack large predators, and many island endemics evolved without defenses against predation. Cats, dogs, and other carnivores introduced by sailors have decimated many island endemics. The **flora** and **fauna** of Hawaii, exceptionally rich before Polynesians arrived with pigs, rats, and agriculture, were severely depleted because their range was limited and they had nowhere to retreat as human settlement advanced. Tropical rain forests, with extraordinary species diversity and high rates of endemism, are also vulnerable to human invasion. Many of the species eliminated daily in Amazonian rain forests are locally endemic, so that their entire range can be eliminated in a short time.

[*Mary Ann Cunningham Ph.D.*]

RESOURCES
BOOKS

Berry, E. W. "The Ancestors of the Sequoias." *Natural History* 20 (1920): 153-55.
Brown, J. H., and A. C. Gibson. *Biogeography*. St. Louis: Mosby, 1983.
Cox, G. W. *Conservation Biology*. Dubuque, IA: William C. Brown, 1993.
Kirch, P. "The Impact of the Prehistoric Polynesians on the Hawaiian Ecosystem." *Pacific Science* 36 (1982): 1-14.
Nitecki, M. W., ed. *Extinctions*. Chicago: University of Chicago Press, 1984.

Endocrine disruptors

In recent years, scientists have proposed that **chemicals** released into the **environment** may be disrupting the endocrine system of humans and **wildlife**. The endocrine system is a network of glands and hormones that regulates many of the body's functions, such as growth, development, behavior,

and maturation. The endocrine glands include the pituitary, thyroid, adrenal, thymus, pancreas, and the male and female gonads (testes and ovaries). These glands secrete regulated amounts of hormones into the bloodstream, where they act as chemical messengers as they are carried throughout the body to control and regulate many the body's functions. The hormones bind to specific cell sites called receptors. By binding to the receptors, the hormones trigger various responses in the tissues that contain the receptors.

An endocrine disruptor is an external agent that interferes in some way with the role of the hormones in the body. The agent might disrupt the endocrine system by affecting any of the stages of hormone production and activity, such as preventing the synthesis of a hormone, directly binding to hormone receptors, or interfering with the breakdown of a natural hormone. Disruption in endocrine function during highly sensitive prenatal periods is especially critical, as small changes in endocrine functions may have delayed consequences that may become evident later in adult life or in a subsequent generation. Adverse effects that might be a result of endocrine disruption include the development of cancers, reproductive and developmental effects, neurological effects (effects on behavior, learning and memory, sensory function, and psychomotor development), and immunological effects (immunosuppression, with resulting disease susceptibility).

Exposure to suspected endocrine disruptors may occur through direct contact with the chemicals or through ingestion of contaminated water, food, or air. Suspected endocrine disruptors can enter air or water from chemical and manufacturing processes and through **incineration** of products. Industrial workers may be exposed in work settings. Documented examples of health effects of humans exposed to endocrine disrupting chemicals include shortened penises in the sons of women exposed to dioxin-contaminated rice oil in China and reduced sperm count in workers exposed to high doses of **Kepone** in a Virginia **pesticide** factory. Diethylstilbestrol (DES), a synthetic estrogen, was used in the 1950s and 1960s by pregnant women to prevent miscarriages. Unfortunately it did not prevent miscarriages, but the teenage daughters of women who had taken DES suffered high rates of vaginal cancers, **birth defects** of the uterus and ovaries, and immune system suppression. These health effects were traced to their mothers' use of DES.

A variety of chemicals, including some pesticides, have been shown to result in endocrine disruption in animal laboratory studies. However, except for the incidences of endocrine disruption due to chemical exposures in the workplace and to the use of DES, causal relationships between exposure to specific environmental agents and adverse health effects in humans due to endocrine disruption have not yet been firmly established.

There is more evidence that the endocrine systems of fish and wildlife have been affected by chemical contamination in their habitats. Groups of animals that have been affected by endocrine disruption include snails, oysters, fish, alligators and other reptiles, and birds, including gulls and eagles. Whether effects on individuals of a particular **species** impact populations of that organism is difficult to prove. Whether endocrine disruption is confined to specific areas or is more widespread is also not known. In addition, proving that a specific chemical causes a particular endocrine effect is difficult, as animals are exposed to a variety of chemicals and non-chemical stressors. However, some persistent organic chemicals such as DDT (dichlorodiphenyltrichloroethane), PCBs (**polychlorinated biphenyls**), **dioxin**, and some pesticides have been shown to act as endocrine disruptors in the environment. Adverse effects seen that may be caused by endocrine disrupting mechanisms include abnormal thyroid function and development in fish and birds, decreased fertility in shellfish, fish, birds, and mammals, decreased hatching success in fish, birds, and reptiles, demasculinization and feminization of fish, birds, reptiles, and mammals, defeminization and masculinization of **gastropods**, fish, and birds, and alteration of immune and behavioral function in birds and mammals. Many potential endocrine disrupting chemicals are persistent and bioaccumulate in fatty tissues of organisms and increase in concentration as they move up through the food web. Because of this persistence and mobility, they can accumulate and harm organisms far from their original source.

More information is needed to define the ecological and human health risks of endocrine disrupting chemicals. Epidemiological investigations, exposure assessments, and laboratory testing studies for a wide variety of both naturally occurring and synthetic chemicals are tools that are being used to determine whether these chemicals as environmental contaminants have the potential to disrupt hormonally mediated processes in humans and animals.

[*Judith L. Sims*]

RESOURCES
BOOKS

Colburn, Theo, Dianne Dumanoski, and John Peterson Myers.*Our Stolen Future: Are We Threatening Our Fertility, Intelligence, and Survival? A Scientific Detective Story.*New York, NY: Penguin Books USA, 1997.

Gillette, Louis J., and D. Andrew Crain. ed.*Environmental Endocrine Disruptors.*London, England: Taylor & Francis Group, 2000.

Krimsky, Sheldon, and Lynn Goldman.*Hormonal Chaos: The Scientific and Social Origins of the Environmental Endocrine Hypothesis.*Baltimore, MD: Johns Hopkins Press, 1999.

Weyer, Peter, and David Riley.*Endocrine Disruptors and Pharmaceuticals in Drinking Water.*Denver, CO: American Water Works Association, 2001.

PERIODICALS

Kavlock, Robert J. et al. "Research Needs for the Risk Assessment of Health and Environmental Effects of Endocrine Disruptors: A Report of the U.S. EPA-Sponsored Workshop." *Environmental Health Perspectives* 104 (1996): 715–740.

OTHER

*Committee on Environment and Natural Resources, National Science and Technology Council. Endocrine Disruptors Research Initiative.*U.S. Environmental Protection Agency, Washington, DC, June 29, 1999. [cited June 1, 2002]. <http://www.epa.gov/endocrine/>.

Endocrine Disruptors. Natural Resources Defense Council, November 25, 1998. [cited June 1, 2002]. <http://nrdc.org/health/effects/qendoc.asp>.

Endocrine Disruptors. World Wildlife Fund, [cited June 1, 2002], <http://www.wildlife.org/toxics/progareas/ed/>.

Technical Panel, Office of Research and Development, Office of Prevention, Pesticides, and Toxic Substances. Special Report on Environmental Endocrine Disruption: An Effects Assessment and Analysis. EPA/630/R-96/012, U.S. Environmental Protection Agency, Washington, DC, 1997.

Energy and the environment

Energy is a prime factor in environmental quality. Extraction, processing, shipping, and **combustion** of **coal**, oil, and **natural gas** are the largest sources of air pollutants, thermal and chemical **pollution** of surface waters, accumulation of mine **tailings** and toxic ash, and land degradation caused by **surface mining** in the United States.

On the other hand, a cheap, inexhaustible source of energy would allow us people to eliminate or repair much of the environmental damage done already and to improve the quality of the **environment** in many ways. Often, the main barrier to reclaiming degraded land, cleaning up polluted water, destroying wastes, restoring damaged ecosystems, or remedying most other environmental problems is that solutions are expensive—and much of that expense is energy costs. Given a clean, sustainable, environmentally benign energy source, people could create a true utopia and extend its benefits to everyone.

Our ability to use external energy to do useful work is one of the main characteristics that distinguishes humans from other animals. Clearly, technological advances based on this ability have made our lives much more comfortable and convenient than that of our early ancestors. They have also allowed us to make bigger mistakes, faster than ever before. A large part of our current environmental crisis is that our ability to modify our environment has outpaced our capacity to use energy and technology wisely.

In the United States, **fossil fuels** supply about 85% of the commercial energy. This situation cannot continue for very long because the supplies of these fuels are limited and their environmental effects are unacceptable. Americans now get more than half of their oil from foreign sources at great economic and political costs. At current rates of use, known, economically extractable world supplies of oil and natural gas will probably last only a century or so. Reserves of coal are much larger, but coal is the dirtiest of all fuels. Its contribution of **greenhouse gases** that cause global warming are reason enough to curtail our coal use. In addition, coal burning is the largest single source in the United States of **sulfur dioxide** and **nitrogen oxides** (which cause respiratory health problems, **ecosystem** damage, and **acid** precipitation). Paradoxically, coal-burning **power plants** also release **radioactivity**, since radioactive minerals such as **uranium** and thorium are often present in low concentrations in coal deposits.

Nuclear power was once thought to be an attractive alternative to fossil fuels. Billed as "the clean energy alternative" and as an energy source "too cheap to bother metering," nuclear power was promoted in the 1960s as the energy source for the future. The disastrous consequences of accidents in nuclear plants, such as the explosion and fire at Chernobyl in the Ukraine in 1986, problems with releases of radioactive materials in mining and processing of fuels, and the inability to find a safe, acceptable permanent storage of nuclear waste have made nuclear power seem much less attractive in recent years. Between seventy and ninety percent of the citizens of most European and North American countries now regard nuclear power as unacceptable.

The United States Government once projected that 1,500 nuclear plants would be built. In 2002, only 105 plants were in operation and no new construction has been undertaken since 1975. Many of these aging plants are now reaching the end of their useful life. There will be enormous costs and technical difficulties in dismantling them and disposing of the radioactive debris. Some reactor designs are inherently safer than those now in operation, but public confidence in nuclear power technology is at such a low level that it seems unlikely that it will never supply much energy. Damming rivers to create hydroelectric power from spinning water turbines has the attraction of providing a low-cost, renewable, air pollution-free energy source. Only a few locations remain in the United States, however, where large hydroelectric projects are feasible. Many more sites are available in Canada, Brazil, India, and other countries, but the social and ecological effects of building large **dams, flooding** valuable river valleys, and eliminating free-flowing rivers are such that opposition is mounting to this energy source.

An example of the ecological and human damage done by large hydroelectric projects is seen in the James Bay region of Eastern Quebec. A series of huge dams and artificial lakes have flooded thousands of square miles of forest. **Migration** routes of caribou are disrupted, the **habitat** for game on which indigenous people depended is destroyed, and decaying vegetation has acidified waters, releasing **mercury** from the bedrock and raising mercury concentrations in fish

to toxic levels. The **hunting** and gathering way of life of local Cree and Inuit people has probably been destroyed forever. This kind of tragedy has been repeated many times around the world by ill-conceived hydro projects.

There are several sustainable, environmentally benign energy sources that should be developed. Among these are wind power, **biomass** (burning **renewable energy** crops such as fast-growing trees or shrubs), small-scale hydropower (low head or run-of-the- river turbines), passive-solar space heating, active-solar water heaters, photovoltaic energy (direct conversion of sunlight to electricity), and ocean tidal or **wave power**. There may be unwanted environmental consequences of some of these sources as well, but they seem much better in aggregate than current energy sources. A big disadvantage is that most of these **alternative energy sources** are diffuse and not always available when or where we want to use energy.

We need ways to store and ship energy generated from these sources. There have been many suggestions that a breakthrough in battery technology could be on the **horizon**. Other possibilities include converting biomass into **methane** or **methanol** fuels or using electricity to generate **hydrogen** gas through electrolysis of water. These fuels would be easily storable, transportable, and used with current technology without great alterations of existing systems. It is estimated that some combination of these sustainable energy sources could supply all of American energy needs by utilizing only a small fraction (perhaps less than one percent) of United States land area. If means are available to move this energy efficiently, these energy farms could be in remote locations with little other value.

Clearly, the best way to protect the environment from damage associated with energy production is to use energy more efficiently. Many experts estimate that people could enjoy the same comfort and convenience but use only half as much energy if they practiced **energy conservation** using currently available technology. This would not require great sacrifices economically or in one's lifestyle. *See also* Acid rain; Air pollution; Greenhouse effect; Photovoltaic cell; Solar energy; Thermal pollution; Wind energy

[*William P. Cunningham Ph.D.*]

RESOURCES
BOOKS

Davis, G. R. *Energy for Planet Earth*. New York: W. H. Freeman, 1991.

PERIODICALS

Weinberg, C. J., and R. H. Williams. "Energy from the Sun." *Scientific American* 263 (September 1990): 146-55.

Energy conservation

Energy **conservation** was a concept largely unfamiliar to America—and to much of the rest of the world—prior to 1973. Certainly some thinkers prior to that date thought about, wrote about, and advocated a more judicious use of the world's energy supplies. But in a practical sense, it seemed that the world's supply of **coal**, oil, and **natural gas** was virtually unlimited.

In 1973, however, the **Organization of Petroleum Exporting Countries** (OPEC) placed an arbitrary limit on the amount of **petroleum** that non-producing nations could buy from them. Although the OPEC embargo lasted only a short time, the nations of the world were suddenly forced to consider the possibility that they might have to survive on a reduced and ultimately finite supply of the **fossil fuels**.

In the United States, the OPEC embargo set off a flurry of administrative and legislative activity, designed to ensure a dependable supply of energy for the nation's further needs. Out of this activity came acts such as the **Energy Policy** and Conservation Act of 1976, the Energy Conservation and Production Act of 1976, and the National Energy Act of 1978.

An important feature of the nation's (and the world's) new outlook on energy was the realization of how much energy is wasted in **transportation**, residential and commercial buildings, and industry. When energy supplies appeared to be without limit, waste was a matter of small concern. However, when energy shortages began to be a possibility, conservation of energy sources assumed a high priority.

Energy conservation is certainly one of the most attainable goals the federal government can set for the United States. Almost every way we use energy results in enormous waste. Only about 20% of the energy content of **gasoline**, for example, is actually put to productive work in an **automobile**. Each time we make use of electricity, we produce waste when coal is burned to heat water to drive a turbine to operate a generator to make electricity. No wonder more than ninety% of the energy generated in the electrical process is wasted.

Fortunately, a vast array of conservation techniques are available in each of the major categories of energy use: transportation, residential and commercial buildings, and industry. In the area of transportation, conservation efforts focus on the nation's use of the private automobile for most personal travel. Certainly, the private automobile is an enormously wasteful method for moving people from one place to another. It is hardly surprising, therefore, that conservationists have long argued for the development of alternative means of transportation: bicycles, motorcycles, mopeds, carpools and van-pools, dial-a-rides, and various forms of **mass transit**. The amount of energy needed to move a single individual on the average is about one-third on a bus what it is in a private car. One need only compare the relative energy cost per passenger for eight people travelling in a commuter van-pool to the cost for a single individual in her private automobile to see the advantages of some form of mass transit.

For a number of reasons, however, mass transit systems in the United States are not very popular. While the number of new cars sold continues to rise year after year, investment in and use of heavy and light rail systems, trolley systems, subways and various types of pools remain modest.

Many authorities believe that the best hope for energy conservation in the field of transportation is to make private automobiles more efficient or to increase the tax on their use. Some experts argue that technology already exists for the construction of 100-mi-per-gal (42.5 km/l) automobiles if industry will make use of that technology. They also argue for additional research on electric cars as an energy-saving and pollution-reducing alternative to internal **combustion** vehicles.

Increasing the cost of using private automobiles has also been explored. One approach is to raise the tax on gasoline to a point where commuters begin to consider mass transit as an economical alternative. Increased parking fees and more aggressive traffic enforcement have also been tried. Such approaches often fail—or are never attempted—because public officials are reluctant to anger voters.

Other methods that have been suggested for increasing **energy efficiency** in automobiles include the design and construction of smaller, lighter cars, extending the useful life of a car, improving the design of cars and tires, and encouraging the design of more efficient cars through federal grants or tax credits.

A number of techniques are well known and could be used, however, to conserve energy in large buildings or even small two-room cottages. A thorough insulation of floors, walls, and ceilings, for example, can save up to 80% of the cost of heating and cooling a building.

In addition, buildings can be designed and constructed to take advantage of natural heating and cooling factors in the **environment**. A home in Canada, for example, should be oriented with windows tilting toward the south so as to take advantage of the sun's heating rays. A home in Mexico might have quite a different orientation.

One of the most extreme examples of environmental-friendly buildings are those that have been constructed at least partially underground. The earthen walls of these buildings provide a natural cooling effect in the summer and provide excellent insulation during the winter.

The kind, number, and placement of trees around a building can also contribute to energy efficiency. Trees that lose their leaves in the winter will allow sunlight to heat a building during the coldest months, but will shield the building from the sun during the hot summer months.

Energy can also be conserved by modifying appliances used within a building. Prior to 1973, consumers became enamored with all kinds of electrical devices, from electric toothbrushes to electric shoe-shine machines to trash compactors. As convenient as these appliances may be, they are energy wasteful and do not always meet a basic human need.

Even items as simple as light bulbs can become a factor in energy conservation programs. Fluorescent light bulbs use at least 75% less energy than do incandescent bulbs, and they often last 20 times longer. Although many commercial buildings now use fluorescent lighting exclusively, it still tends to be relatively less popular in private homes.

As the largest single user of energy in American society, industry is a prime candidate for conservation measures. Always sensitive to possible money-saving changes, industry has begun to develop and implement energy savings devices and procedures. One such idea is **cogeneration**, the use of waste heat from an industrial process for use in the generation of electricity.

Another approach is the expanded use of **recycling** by industry. In many cases, re-using a material requires less energy than producing it from raw materials. Finally, researchers are continually testing new designs for equipment that will allow that equipment to operate on less energy.

Governments and utilities have two primary methods by which they can encourage energy conservation. One approach is to penalize individuals and companies that use too much energy. For example, an industry that uses large amounts of electricity might be charged at a higher rate per kilowatt hour than one that uses less electricity, a policy just the opposite of that now in practice in most places.

A more positive approach is to encourage energy conservation by techniques such as tax credits. Those who insulate their homes might, for example, be given cash bonuses by the local utility or a tax deduction by state or federal government.

In recent years, another side of energy conservation has come to the fore, its environmental advantages. Obviously, the less coal, oil, and natural gas that humans use, the fewer pollutants are released into the environment. Thus, a practice that is energy-wise, conservation, can also provide environmental benefits. Those concerned with global warming and **climate** change have been especially active in this area. They point out that reducing our use of fossil fuels will both reduce our consumption of energy and our release of **carbon dioxide** to the **atmosphere**. We can take a step toward heading off climate change, they point out, by taking the wise step of wasting less energy.

Energy conservation does not yet appear to have won the heart of most Americans. The general public concern about energy waste engendered by the 1973 OPEC **oil embargo** eventually dissolved into complacency. To be sure, some of the sensitivity to energy conservation created by that event has not been lost. Many people have switched to more energy-efficient forms of transportation, think more carefully about leaving house lights on all night, and take energy efficiency into consideration when buying major appliances.

But some of the more aggressive efforts to conserve energy have become stalled. Higher taxes on gasoline, for example, still are certain to raise an enormous uproar among the populace. And energy-saving construction steps that might well be mandated by law still remain optional, and frequently ignored.

In an era of apparently renewed confidence in an endless supply of fossil fuels, many people are no longer convinced that energy conservation is very important or have the will to act on their suspicion that it is. And governments, reflecting the will of the people, do not take leadership action to change that trend.

[*David E. Newton*]

RESOURCES
BOOKS

Fardo, S. *Energy Conservation Guidebook.* Englewood Cliffs, NJ: Prentice Hall, 1993.

PERIODICALS

Reisner, M. "The Rise and Fall and Rise of Energy Conservation." *Amicus Journal* 9 (Spring 1987): 22-31.

Energy crops
see **Biomass**

Energy efficiency

The utilization of energy for human purposes is a defining characteristic of industrial society. The conversion of energy from one form to another and the efficient production of mechanical work for heat energy has been studied and improved for centuries. The science of thermodynamics deals with the relationship between heat and work and is based on two fundamental laws of **nature**, the first and second **laws of thermodynamics**. The utilization of energy and the **conservation** of critical, nonrenewable energy resources are controlled by these laws and the technological improvements in the design of energy systems.

The First Law of Thermodynamics states the principle of conservation of energy: energy can be neither created nor destroyed by ordinary chemical or physical means, but it can be converted from one form to another. Stated another way, *in a closed system, the total amount of energy is constant.* An interesting example of energy conversion is the incandescent light bulb. In the incandescent light bulb, electrical energy is used to heat a wire (the bulb filament) until it is hot enough to glow. The bulb works satisfactorily except that the great majority (95%) of the electrical energy supplied to the bulb is converted to heat rather than light. The incandescent bulb is not very efficient as a source of light. In contrast, a fluorescent bulb uses electrical energy to excite atoms in a gas, causing them to give off light in the process at least four times more efficiently than the incandescent bulb. Both light sources, however, conform to the First Law in that no energy is lost and the total amount of heat and light energy produced is equal to the amount of electrical energy flowing to the bulb.

The Second Law of Thermodynamics states that whenever heat is used to do work, some heat is lost to the surrounding **environment**. The complete conversion of heat into work is not possible. This is not the result of inefficient engineering design or implementation but, rather, a fundamental and theoretical thermodynamic limitation. The maximum, theoretically possible efficiency for converting heat into work depends solely on the operating temperatures of the heat engine and is given by the equations: $E = 1 - T_2/T_1$. T_1 is the absolute temperature at which heat energy is supplied and T_2 is the absolute temperature at which heat energy is exhausted.

The maximum possible thermodynamic efficiency of a four-cycle internal **combustion** engine is about 54%; for a diesel engine, the limit is about 56%; and for a steam engine, the limit is about 32%. The actual efficiency of real engines, which suffer from mechanical inefficiencies and parasitic losses (eg. friction, drag, etc.) is significantly lower than these levels. Although thermodynamic principles limit maximum efficiency, substantial improvements in energy utilization can be obtained through further development of existing equipment such as **power plants**, refrigerators, and automobiles and the development of new energy sources such as solar and geothermal.

Experts have estimated the efficiency of other common energy systems. The most efficient of these appear to be electric power generating plants (33% efficient) and steel plants (23% efficient). Among the least efficient systems are those for heating water (1.5–3%), for heating homes and buildings (2.5–9%), and refrigeration and air-conditioning systems (4–5%). It has been estimated that about 85% of the energy available in the United States is lost due to inefficiency.

The predominance of low efficiency systems reflects the fact that such systems were invented and developed when energy costs were low and there was little customer demand for energy efficiency. It made more sense then to build appliances that were inexpensive rather than efficient because the cost to operate them was so low. Since the 1973 **oil embargo** by the **Organization of Petroleum Exporting Countries** (OPEC), that philosophy has been carefully re-examined. Experts began to point out that more expensive appliances could be designed and built if they were also more efficient. The additional cost to the manufacturer, industry and homeowner could usually be recovered within a few years because of the savings in fuel costs.

The concept of energy efficiency suggests a new way of looking at energy systems and that is the examination of the total lifetime energy use and cost of the system. Consider the common light bulb. The total cost of using a light bulb includes both its initial price and the cost of operating it throughout its lifetime. When energy was cheap, this second factor was small. There was little motivation to make a bulb that was more efficient when the life-cycle savings for its operation was minimal.

But as the cost of energy rises, that argument no longer holds true. An inefficient light bulb costs more and more to operate as the cost of electricity rises. Eventually, it makes sense to invent and produce more efficient light bulbs. Even if these bulbs cost more to buy, they pay back that cost in long-term operating savings.

Thus, consumers might balk at spending $25 for a fluorescent light bulb unless they knew that the bulb would last ten times as long as an incandescent bulb that costs $3.75. Similar arguments can and have been used to justify the higher initial cost of energy-saving refrigerators, solar-heating systems, household insulation, improved internal combustion engines and other energy-efficient systems and appliances.

Governmental agencies, utilities, and industries are gradually beginning to appreciate the importance of increasing energy efficiency. The 1990 amendments to the **Clean Air Act** encourage industries and utilities to adopt more efficient equipment and procedures. Certain leaders in the energy field, such as Pacific Gas and Electric and Southern California Edison have already implemented significant energy efficiency programs.

[*David E. Newton and Richard A. Jeryan*]

RESOURCES
BOOKS

Miller, G. T., Jr. *Energy and Environment: The Four Energy Crises.* 2nd edition. Belmont, CA: Wadsworth Publishing Company, 1980.

Sears, F. W. and M. W. Zemansky. *University Physics.* 2nd edition. Reading, MA: Addison-Wesley Publishing, 1957.

OTHER

Council on Environmental Quality. *Environmental Quality*, 21st annual report. Washington, DC: U. S. Government Printing Office, 1990.

Energy flow

Understanding energy flow is vital to many environmental issues. One can describe the way ecosystems function by saying that matter cycles and energy flows. This is based on the laws of **conservation** of matter and energy and the second law of thermodynamics, or the law of energy degradation.

Energy flow is strictly one way, such as from higher to lower or from hotter to colder. Objects cool only by loss of heat. All cooling units, such as refrigerators and air conditioners, are based on this principle: they are essentially heat pumps, absorbing heat in one place and expelling it to another.

This heat flow is explained by the laws of radiation, as seen in fire and the color wheel. All objects emit radiation, or heat loss, but the hotter the object the greater the amount of radiation, and the shorter and more energetic the wavelength. As energy intensities rise and wavelengths shorten, the radiation changes from infrared to red, then orange, yellow, green, blue, violet, and ultraviolet. A blue flame, for example, is desired for gas appliances. A well-developed wood fire is normally yellow, but as the fire dies out and cools, the color gradually changes to orange, then red, then black. Black coals may still be very hot, giving off invisible, infrared radiation. These varying wavelengths are the main differences seen in the electromagnetic spectrum.

All chemical reactions and **radioactivity** emit heat as a by-product. Because this heat radiates out from the source, the basis of the second law of thermodynamics, one can never achieve 100% **energy efficiency**. There will always be a heat-loss tax. One can slow down the rate of heat loss through insulating devices, but never stop it. As the insulators absorb heat, their temperatures rise and they in turn lose heat.

There are three main applications of energy flow to environmental concerns. First, only 10% of the food passed on up the **food chain/web** is retained as body mass; 90% flows to the **atmosphere** as heat. In terms of caloric efficiency, more calories are obtained by eating plant food than meat. Since fats are more likely to be part of the 10% retained as body mass, pesticides dissolved in fat are subject to **bioaccumulation** and **biomagnification**. This explains the high levels of DDT in birds of prey like the **peregrine falcon** (*Falco peregrinus*) and the **brown pelican** (*Pelecanus occidentalis*).

Second, the%age of waste heat is an indicator of energy efficiency. In light bulbs, 5% produces light and 95% heat, just the opposite of the highly efficient fire fly. Electrical generation from **fossil fuels** or **nuclear power** produces vast amounts of waste heat.

Third, control of heat flow is a key to comfortable indoor air and solving global warming. Well-insulated buildings retard heat flow, reducing energy use. Atmospheric **greenhouse gases**, such as **anthropogenic carbon dioxide** and **methane**, retard heat flow to space, which theoretically should cause global temperatures to rise. Policies that reduce these greenhouse gases allow a more natural flow of heat back to space. *See also* Greenhouse effect

[*Nathan H. Meleen*]

RESOURCES
PERIODICALS

"Energy." *National Geographic* 159 (February 1981): 2-23.

Energy Information Admistration
see U.S. Department of Energy

Energy path, hard vs. soft

What will energy use patterns in the year 2100 look like? Such long-term predictions are difficult, risky, and perhaps impossible. Could an American citizen in 1860 have predicted what the pattern of today's energy use would be like?

Yet, there are reasons to believe that some dramatic changes in the ways we use energy may be in store over the next century. Most importantly, the world's supplies of non-renewable energy—especially, **coal**, oil, and natural gas—continue to decrease. Critics have been warning for a least two decades that time was running out for the **fossil fuels** and that we could not count on using them as prolifically as we had in the past.

For at least two decades, experts have debated the best way to structure our energy use patterns in the future. The two most common themes have been described (originally by physicist Amory Lovins) as the "hard path" and the "soft path."

Proponents of the hard path argue essentially that we should continue to operate in the future as we have in the past, except more efficiently. They point out that predictions from the 1960s and 1970s that our oil supplies would be depleted by the end of the century have been proved wrong. If anything, our reserves of fossil fuels may actually have increased as economic incentives have encouraged further exploration.

Our energy future, the hard-pathers say, should focus on further incentives to develop conventional energy sources such as fossil fuels and **nuclear power**. Such incentives might include tax breaks and subsidies for coal, **uranium** and **petroleum** companies. When our supplies of fossil fuels do begin to be depleted, our emphasis should shift to a greater reliance on nuclear power.

An important feature of the hard energy path is the development of huge, centralized coal-fired and nuclear-powered plants for the generation of electricity. One characteristic of most hard energy proposals, in fact, is the emphasis on very large, expensive, centralized systems. For example, one would normally think of **solar energy** as a part of the soft energy path. But one proposal developed by the National Aeronautics and Space Administration (NASA) calls for a gigantic solar power station to be orbited around the earth. The station could then transmit power via microwaves to centrally-located transmission stations at various points in the earth's surface.

Those who favor a soft energy path have a completely different scenario in mind. Fossil fuels and nuclear power must diminish as sources of energy as soon as possible, they say. In their place, alternative sources of power such as hydropower, **geothermal energy**, **wind energy**, and photovoltaic cells must be developed.

In addition, the soft-pathers say, we should encourage **conservation** to extend coal, oil, and **natural gas** supplies as long as possible. Also since electricity is one of the most wasteful of all forms of energy, its use should be curtailed.

Most importantly, soft-path proponents maintain energy systems of the future should be designed for small-scale use. The development of more efficient solar cells, for example, would make it possible for individual facilities to generate a significant portion of the energy they need.

Underlying the debate between hard- and soft-pathers is a fundamental question as to how society should operate. On the one hand are those who favor the control of resources in the hands of a relatively small number of large corporations. On the other hand are those who prefer to have that control decentralized to individual communities, neighborhoods, and families. The choice made between these two competing philosophies will probably determine which energy path the United States and the world will ultimately follow. *See also* Alternative energy sources; Alternative fuels; Energy and the environment

[*David E. Newton*]

RESOURCES
BOOKS

Lovins, A. *Soft Energy Paths.* San Francisco: Friends of the Earth, 1977.

Energy policy

Energy policies are the actions governments take to affect the demand for energy as well as the supply of it. These actions include the ways in which governments cope with energy supply disruptions and their efforts to influence energy consumption and economic growth.

The energy policies of the United States government have often worked at cross purposes, both stimulating and suppressing demand. Taxes are perhaps the most important kind of energy policy, and **energy taxes** are much lower in the U. S. than in other countries. This is partially responsible for the fact that energy consumption per capita is higher than elsewhere, and there is less incentive to invest in **conservation** or alternative technologies. Following the 1973 Arab **oil embargo**, the federal government instituted price controls which kept energy prices lower than they would otherwise have been, thereby stimulating consumption. Yet the government also instituted policies at the same time, such as fuel-economy standards for automobiles, which were designed to increase conservation and lower energy use. Thus, policies in the period after the embargo were contradictory: what one set of policies encouraged, the other discouraged.

The United States government has a long history of different types of interference in energy markets. The **Natural Gas** Act of 1938 gave the **Federal Power Commission** the right to control prices and limit new pipelines from entering the market. In 1954 The Supreme Court extended price controls to field production. Before 1970, the Texas Railroad Commission effectively controlled oil output (in the United States) through prorationing regulations that provided multiple owners with the rights to underground pools. The federal government provided tax breaks in the form of intangible drilling expenses and gave the oil companies a depletion allowance. A program was also in place from 1959 to 1973 which limited oil imports and protected domestic producers from cheap foreign oil. The ostensible purpose of this policy was maintaining national security, but it contributed to the depletion of national reserves.

After the oil embargo, Congress passed the Emergency **Petroleum** Allocation Act giving the federal government the right to allocate fuel in a time of shortage. In 1974 President Gerald Ford announced Project Independence which was designed to eliminate dependence on foreign imports. Congress passed the Federal Non-Nuclear Research and Development Act in 1974 to focus government efforts on non-nuclear research. Finally, in 1977 Congress approved the cabinet-level creation of the U. S. Department of Energy (DOE) which had a series of direct and indirect policy approaches at its disposal, designed to encourage and coerce both the energy industry as well as the commercial and residential sectors of the country to make changes. After

Ronald Reagan became president, many DOE programs were abolished, though DOE continued to exist, and the net impact has probably been to increase economic uncertainty.

Energy policy issues have always been very political in nature. Different segments of the energy industry have often been differently affected by policy changes, and various groups have long proposed divergent solutions. The energy crisis, however, intensified these conflicts. Advocates of strong government action called for policies which would alter consumption habits, reducing dependence on foreign oil and the nation's vulnerability to an oil embargo. They have been opposed by proponents of free markets, some of whom considered the government itself responsible for the crisis. Few issues were subject to such intensive scrutiny and fundamental conflicts over values as energy policies were during this period. Interest groups representing causes from **energy conservation** to **nuclear power** mobilized. Business interests also expanded their lobbying efforts.

An influential advocate of the period was Amory Lovins, who helped create the **renewable energy** movement. His book, *Soft Energy Paths: Toward A Durable Peace* (1977), argued that energy problems existed because large corporations and government bureaucracies had imposed expensive centralized technologies like nuclear power on society. Lovins argued that the solution was in small scale, dispersed, technologies. He believed that the "hard path" imposed by corporations and the government led to an authoritarian, militaristic society while the "soft path" of small-scale dispersed technologies would result in a diverse, peaceful, self-reliant society.

Because **coal** was so abundant, many in the 1970s considered it a solution to American dependence on foreign oil, but this expectation has proved to be mistaken. During the 1960s, the industry had been controlled by an alliance between management and the union, but this alliance disintegrated by the time of the energy crisis, and wildcat strikes hurt productivity. Productivity also declined because of the need to address safety problems following passage of the 1969 Coal Mine Health and Safety Act. Environmental issues also hurt the industry following passage of the **National Environmental Policy Act** of 1969, the **Clean Air Act** of 1970, the **Clean Water Act** of 1972, and the 1977 **Surface Mining Control and Reclamation Act**. Worker productivity in the mines dropped sharply from 19 tons per worker day to 14 tons, and this decreased the advantage coal had over other fuels. The 1974 Energy Supply and Environmental Coordination Act and the 1978 Fuel Use Act, which required utilities to switch to coal, had little effect on how coal was used because so few new plants were being built.

Other energy-consuming nations responded to the energy crises of 1973–74 and 1979–80 with policies that were

different from the United States. Japan and France, although via different routes, made substantial progress in decreasing their dependence on Mideast oil. Great Britain was the only major industrialized nation to become completely self-sufficient in energy production, but this fact did not greatly aid its ailing economy. When energy prices declined and then stabilized in the 1980s, many consuming nations eliminated the conservation incentives they had put in place.

Japan is the most heavily petroleum-dependent industrialized nation. To pay for a high level of energy and raw material imports, Japan must export the goods which it produces. When energy prices increased after 1973, it was forced to expand exports. The rate of economic growth in Japan began to decline. Annual growth in GNP averaged nearly 10% from 1963–1973, and from 1973–1983 it was just under 4%, although the association between economic growth and energy consumption has weakened.

The Energy Rationalization Law of 1979 was the basis for Japan's energy conservation efforts, providing for the financing of conservation projects and a system of tax incentives. It has been estimated that over 5% of total Japanese national investment in 1980 was for energy-saving equipment. In the cement, steel, and chemical industries over 60% of total investment was for energy conservation, Japanese society shifted from petroleum to a reliance on other forms of energy including nuclear power and liquefied natural gas.

In France, energy resources at the time of the oil embargo were extremely limited. It possessed some natural gas, coal, and hydropower, but together these sources constituted only 0.7% of the world's total energy production. By 1973, French dependence on foreign energy had grown to 76.2%: oil made up 67% of the total energy used in France, up from 25% in 1960.

France had long been aware of its dependence on foreign energy and had taken steps to overcome it. Political instability in the Mideast and North Africa had led the government to take a leading role in the development of civilian nuclear power after World War II. In 1945 Charles de Gaulle set up the French **Atomic Energy Commission** to develop military and peaceful uses for nuclear power. The nuclear program proceeded at a very slow pace until the 1973 embargo, after which there was rapid growth in France's reliance on nuclear power. By 1990, more than 50 reactors had been constructed and over 70% of France's energy came from nuclear power. France now exports electricity to nearly all its neighbors, and its rates are about the lowest in Europe. Starting in 1976 the French government also subsidized 3,100 conservation projects at a cost of more than 8.4 billion francs, and these subsidies were particularly effective in encouraging energy conservation.

Concerned about oil supplies during World War I, the British government had taken a majority interest in British Petroleum and tried to play a leading role in the search for new oil. After the World War II, the government nationalized the coal, gas, and electricity industries, creating, for ideological reasons as well as for postwar reconstruction, the National Coal Board, British Gas Corporation, and Central Electricity Generating Board. After the discovery of oil reserves in the 1970s in the North Sea, the government established the British National Oil Company. This government corporation produced about 7% of North Sea oil and ultimately handled about 60% of the oil produced there.

All the energy sectors in the United Kingdom were thus either partially or completely nationalized. Government relations with the nationalized industries often were difficult, because the two sides had different interests. The government intervened to pursue macroeconomic objectives such as price restraint, and it attempted to stimulate investment at times of unemployment. The electric and gas industries had substantial operating profits and they could finance their capital requirements from their revenues, but profits in the coal industry were poor, the work force was unionized, and opposition to the closure of uneconomic mines was great. Decision-making was highly politicized in this nationalized industry, and the government had difficulty addressing the problems there. It was estimated that 90% of mining losses came from 30 of the 190 pits in Great Britain, but only since 1984–85 has there been rapid mine closure and enhanced productivity. New power-plant construction was also poorly managed, and comparable coal-fired power stations cost twice as much in Great Britain as in France or Italy.

The Conservative Party proposed that the nationalized energy industries be privatized. However, with the exception of coal, these energy industries had natural monopoly characteristics: economies of scale and the need to prevent duplicate investment in fixed infrastructure. The Conservative Party called for regulation after privatization to deal with the natural monopoly characteristics of these industries, and it took many steps toward privatization. In only one area, however, did it carry its program to completion, abolishing the British National Oil Company and transferring its assets to private companies. *See also* Alternative energy sources; Corporate Average Fuel Efficiency Standards; Economic growth and the environment; Electric utilities; Energy and the environment; Energy efficiency; Energy path, hard vs. soft

[*Alfred A. Marcus*]

RESOURCES
BOOKS

Marcus, A. A. *Controversial Issues in Energy Policy*. Phoenix, AZ: Sage Press, 1992.

Energy recovery

A fundamental fact about energy use in modern society is that huge quantities are lost or wasted in almost every field and application. For example, the series of processes by which nuclear energy is used to heat a home with electricity results in a loss of about 85 percent of all the energy originally stored in the **uranium** used in the nuclear reactor. Industry, utilities, and individuals could use energy far more efficiently if they could find ways to recover and **reuse** the energy that is being lost or wasted.

One such approach is **cogeneration**, the use of waste heat for some useful purpose. For example, a factory might be redesigned so that the steam from its operations could be used to run a turbine and generate electricity. The electricity could then be used elsewhere in the factory or sold to power companies. Cogeneration in industry can result in savings of between 10 and 40 percent of energy that would otherwise be wasted.

Cogeneration can work in the opposite direction also. Hot water produced in a utility plant can be sold to industries that can use it for various processes. Proposals have been made to use the wasted heat from electricity plants to grow flowers and vegetables in greenhouses, to heat water for commercial fish and shell-fish farms, and to maintain warehouses at constant temperatures. The total **energy efficiency** resulting from this sharing is much greater than it would be if the utility's water was simply discarded.

Another possible method of recovering energy is by generating or capturing **natural gas** from **biomass**. For example, as organic materials decay naturally in a **landfill**, one of the products released is **methane**, the primary component of natural gas. Collecting methane from a landfill is a relatively simple procedure. Vertical holes are drilled into the landfill and porous pipes are sunk into the holes. Methane diffuses into the pipes and is drawn off by pumps. The recovery system at the Fresh Kills landfill on Staten Island, New York, for example, produces enough methane to heat 10,000 homes.

Biomass can also be treated in a variety of ways to produce methane and other combustible materials. Sewage, for example, can be subjected to **anaerobic digestion**, the primary product of which is methane. Pyrolysis is a process in which organic wastes are heated to high temperatures in the absence of oxygen. The products of this reaction are solid, liquid, and gaseous **hydrocarbons** whose composition is similar to those of **petroleum** and natural gas. Perhaps the most known example of this approach is the manufacture of **methanol** from biomass. When mixed with **gasoline**, a new fuel, **gasohol**, is obtained.

Energy can also be recovered from biomass simply by **combustion**. The waste materials left after sugar is extracted from sugar cane, known as *bagasse*, have long been used as a fuel for the boilers in which the sugar extraction occurs. The burning of **garbage** has also been used as an energy source in a wide variety of applications such as the heating of homes in Sweden, the generation of electricity to run streetcars and subways in Milan, Italy, and the operation of a desalination plant in Hempstead, Long Island.

The recovery of energy that would otherwise be lost or wasted has a secondary benefit. In many cases, that wasted energy might cause **pollution** of the **environment**. For example, the wasted heat from an electric power plant may result in **thermal pollution** of a nearby waterway. Or the escape of methane into the **atmosphere** from a landfill could contribute to **air pollution**. Capture and recovery of the waste energy not only increases the efficiency with which energy is used, but may also reduce some pollution problems.

[*David E. Newton*]

RESOURCES
BOOKS

Franke, R. G., and D. N. Franke. *Man and the Changing Environment.* New York: Holt, Rinehart and Winston, 1975.
Moran, J. M., M. D. Morgan, and J. H. Wiersma. *Introduction to Environmental Science.* 2nd ed. New York: W. H. Freeman, 1986.

Energy Reorganization Act (1973)

Passed in 1974 during the Ford Administration, this act created the **Energy Research and Development Administration** (ERDA) and **Nuclear Regulatory Commission** (NRC). The purpose of the act was to begin an extensive non-nuclear federal research program, separating the regulation of the **nuclear power** from research functions. Regulation was carried out by the NRC, while nuclear power research was carried out by the ERDA. The passage of the 1974 Energy Reorganization Act ended the existence of the **Atomic Energy Commission**, which had been the main instrument to implement nuclear policy. In 1977 ERDA incorporated into the newly created **U.S. Department of Energy**. *See also* Alternative energy sources; Energy policy

Energy Research and Development Administration

This agency was created in 1974 from the non-regulatory parts of the **Atomic Energy Commission** (AEC), and it existed until 1977, when it was incorporated into the **U.S. Department of Energy**. In its short life span, the Energy Research and Development Administration (ERDA) started to diversify U.S. energy research outside of **nuclear power**. Large-scale demonstration projects were begun in numerous

areas. These included projects to convert **coal** and solid wastes into liquid and gaseous fuels; experiments on methods to extract and process **oil shale** and **tar sands**, as well as an effort to develop a viable breeder reactor that would ensure a virtually inexhaustible source of **uranium** for electricity. The agency also supported research on **solar energy** for space heating, industrial process heat, and electricity. In the short time available, basic problems could not be solved, and the achievements of many of the demonstration projects were disappointing. Nevertheless, many important advances were made in commercializing cost-effective technologies for **energy conservation**, such as energy-efficient lighting systems, improved heat pumps, and better heating systems. The agency also conducted successful research in environmental, safety, and health areas.

Energy taxes

The main energy tax levied in the United States is the one on **petroleum**, though the United States tax is half of the amount levied in other major industrialized nations. As a result, **gasoline** prices in the United States are much lower than elsewhere, and both environmentalists and others have argued that this encourages energy consumption and **environmental degradation** and causes national and international security problems.

In 1993, the House passed a **Btu** tax while the Senate passed a more modest tax on **transportation** fuels. A Btu tax would restrict the burning of **coal** and other **fossil fuels** and proponents maintain that this would be both environmentally and economically beneficial. Every barrel of oil and every ton of coal that is burned adds **greenhouse gases** to the **atmosphere**, increasing the likelihood that **future generations** will face a global climatic calamity. United States dependence on foreign oil, much of it from potentially unstable nations like Iraq, now approaches 50%. A Btu tax would create incentives for **energy conservation**, and it would help stimulate the search for alternatives to oil. It would also help reduce the burgeoning trade deficit, of which foreign petroleum and petroleum-based products now constitute nearly 40%.

President Bill Clinton urged Americans to support higher energy taxes because of the considerable effect they could have on the federal budget deficit. For instance, if the government immediately raised gasoline prices to levels commonly found in other industrial nations (about $4.15 a gallon), the budget deficit would almost be eliminated. It is estimated that every penny increase in gasoline taxes, yields a billion dollars in revenue for the federal treasury, and in June 2002 the budget deficit was estimated to be about $6 trillion.

Of course, to raise gasoline taxes immediately to these levels is utterly impractical, as the effects on the economy would be catastrophic. It would devastate the economics of rural and western states. Inflation across the country would soar and job losses would skyrocket. Supporters of increasing energy taxes agree that the increases must be gradual and predictable, so people can adjust. Many believe they should take place over a 15 year period, after which energy prices in the United States would be roughly equivalent to those in other industrial nations.

Many economists emphasize that the positive effects of higher energy taxes will be felt only if there are no increases in government spending. It is, they believe, ultimately a question of how Americans want to be taxed. Do they want wages, profits, and savings to be taxed, as they are now, or their use of energy? In the former case, the government is taxing a desirable activity which should be encouraged for the sake of job creation and economic expansion. In the latter case, it is taxing undesirable activity which should be discouraged for the sake of protecting the **environment** and preserving national security. *See also* Energy policy; Environmental economics

[*Alfred A. Marcus*]

RESOURCES
BOOKS

Marcus, A. A. *Controversial Issues in Energy Policy.* Phoenix, AZ: Sage Press, 1992.

Enhydra lutris
see **Sea otter**

Eniwetok Atoll
see **Bikini atoll**

Enteric bacteria

Enteric bacteria are defined as bacteria which reside in the intestines of animals. Members of the Enterobacteriaceae family, enteric bacteria are important because some of them symbiotically aid the digestion of their hosts, while other pathogenic **species** cause disease or death in their host organism. The pathogenic members of this family include species from the genera *Escherichia, Salmonella, Shigella, Klebsiella,* and *Yersinia.* All of these pathogens are closely associated with fecal contamination of foods and water. In North America, the reported incidence of salmonellosis outweighs the occurrence of all of the other reportable diseases by other enteric bacteria combined.

Most infections from enteric pathogens require large numbers of organisms to be ingested by immunocompetent adults, with the exception of *Shigella.* Symptoms include gastrointestinal distress and diarrhea. The enteric bacteria related to ***Escherichia coli*** are known as the **coliform** bacteria. Coliform bacteria are used as indicators of pathogenic enteric bacteria in drinking and recreational waters. *See also* Sewage treatment

Entrainment
see **Air pollution**

Environment

When people say "I am concerned about the environment," what do they mean? What does the use of the definite article mean in such a statement? Is there such a thing as "the" environment?

Environment is derived from the French words *environ* or *environner,* meaning "around," which in turn originated from the Old French *virer* and *viron* (together with the prefix *en*), which mean "a circle, around, the country around, or circuit." Etymologists frequently conclude that, in English usage at least, *environment* is the total of the things or circumstances around an organism—including humans—though *environs* is limited to the "surrounding neighborhood of a specific place, the neighborhood or vicinity."

Even a brief etymological encounter with the word environment provokes two persuasive suggestions for possible structuring of a contemporary definition. First, the word environment is identified with a totality, the everything that encompasses each and all of us, and this association is established enough to be not lightly dismissed. The very notion of "environment," as Anatol Rapoport indicated, suggest the partitioning of a "portion of the world into regions, an inside and an outside." The environment is the outside. Second, the word's origin in the phrase "to environ" indicates a process derivative, one that alludes to some sort of action or interaction, at the very least inferring that the encompassing is active, in some sense reciprocal, that the environment, whatever its **nature**, is not simply an inert phenomenon to be impacted without response or without affecting the organism in return. Environment must be a relative word, because it always refers to something "environed" or enclosed.

Ecology as a discipline is focused on studying the interactions between an organism of some kind and its environment. So ecologists must be concerned with what H. L. Mason and J. H. Langenheim described as a "key concept in the structure of ecological knowledge," but a concept with which ecologists continue to have problems of confusion between ideas and reality—the concept of environment. Mason and Langenheim's article "Language Analysis and the Concept Environment" continues to be the definitive statement on the use of the word environment in experimental ecology.

The results of Mason and Langenheim's analysis were essentially four-fold: 1) they limited environmental phenomena "in the universal sense" to only those phenomena that have an operational relation with any organism: other phenomena present that do not enter a reaction system are excluded, or dismissed as not "environmental phenomena"; 2) they restricted the word environment itself to mean "the class composed of the sum of those phenomena that enter a reaction system of the organism or otherwise directly impinge upon it" so that physical exchange or impingement becomes the clue to a new and limited definition; 3) they specifically note that their definition does not allude to the larger meaning implicit in the etymology of the word; and 4) they designate their limited concept as *operational environment* but state that when the word environment is used with qualification, then it still refers to the operational construct, establishing that "'environment' per se is synonymous with 'operational environment'."

This definition does allow a prescribed and limited conception of environment and might work for experimental ecology but is much too limited for general usage. Environmental phenomena of relevance to the aforementioned concern for "the" environment must incorporate a multitude of things other than those that physically impinge on each human being. And it is much more interactive and overlapping than a restricted definition would have people believe. To better understand contemporary human interrelationships with the world around them, environment must be an incorporative, holistic term and concept.

Thinking about the environment in the comprehensive sense—with the implication that *everything* is the environment with each entity connected to each of a multitude of others—makes environment what David Currie in a book of case studies and material on **pollution** described "as not a modest concept." But, such scope and complexity, difficult as they are to resolve, intensify rather than eliminate the very real need for a kind of transcendence. The assumption seems valid that human consciousness regarding environment needs to be raised, not restricted. Humans need increasingly to comprehend and care about what happens in far away places and to people they do not know but that do affect them, that do impact even their localized environments, that do impinge on their individual well-being. And they need to incorporate the reciprocal idea that their actions impact people and environments outside the immediate in place and time: in the world today, environmental impacts

transcend the local. Thus it is necessary that human awareness of those impacts also be transcendent.

It is uncertain that confining the definition of environment to operationally narrow physical impingement could advance this goal. One suspects instead that it would significantly retard it, a retardation that contemporary human societies can ill afford. Internalization of a larger environment, including an understanding of common usages of the word, might on the other hand aid people in caring about, and assuming responsibility for, what happens to that environment and to the organisms in it.

An operational definition can help people find the mechanisms to deal with problems immediate and local, but can, if they are not careful, limit them to an unacceptable mechanistic and unfeeling approach to problems in the environment-at-large.

Acceptance of either end of the spectrum—a limited operational definition or an incorporative holistic definition—as the only definition creates more confusion than clarification. Both are needed. Outside the laboratory, however, in study of the interactional, interdependent world of contemporary humankind, the holistic definition must have a place. A sense of the comprehensive "out there," of the totality of world and people as a functionally significant, interacting unit should be seeping into the consciousness of every person.

Carefully chosen qualifiers can help deal with the complexity: "natural" or "built" or "perceptual" all specify aspects of human surroundings more descriptive and less incorporative than "environment" used alone, without adjectives. Other noun can also pick up some of the meanings of environment, though none are direct synonyms: **habitat**, milieu, mis en scence, ecumene all designate specified and limited aspects of the human environment, but none except "environment" are incorporative of the whole complexity of human surroundings.

An understanding of environment must not be limited to an abstract concept that relates to daily life only in terms of whether to recycle cans or walk to work. The environment is the base for all life, the source of all goods. Poor people in underdeveloped nations know this; their day-to-day survival depends on what happens in their local environments. Whether it rains or does not, whether commercial seiners move into local fishing grounds or leave them alone, and whether local forest products are lost to the cause of world timber production affect these people more directly. What they, like so many other humans around the world, may not also recognize, is that "environment" now extends far beyond the bounds of the local: environment is the intimate enclosure of the individual or a local human population *and* the global domain of the human **species**.

The Brundtland report ***Our Common Future*** recognized this with a healthy, modern definition: "The environment does not exist as a sphere separate from human actions, ambitions, and needs, and attempts to defend it in isolation from human concerns have given the word 'environment' a connotation of naivety in some political circles." The report goes on to note that "the 'environment' is where we all live...and 'development' is what we all do in attempting to improve our lot within that abode. The two are inseparable."

Each human being lives in a different environment than any other human because every single one screens their surroundings through their own individual experience and perceptions. Yet all human beings live in the same environment, an external reality that all share, draw sustenance from, and excrete into. So understanding environment becomes a dialectic, a resolution and synthesis of individual characteristics and shared conditions. Solving environmental problems depends on the intelligence exhibited in that resolution.

[*Gerald L. Young Ph.D.*]

RESOURCES
BOOKS

Bates, M. *The Human Environment*. Berkeley: University of California, School of Forestry, 1962.

Dubos, R. "Environment." *Dictionary of the HistoryIn of Ideas*, edited by P. P. Wiener. New York: Charles Scribner's Sons, 1973.

PERIODICALS

Mason, H. L., and J. H. Langenheim. "Language Analysis and the Concept Environment." *Ecology* 38 (April 1957): 325-340.

Patten, B. C. "Systems Approach to the Concept of Environment." *Ohio Journal of Science* 78 (July 1978): 206-222.

Young, G. L. "Environment: Term and Concept in the Social Sciences." *Social Science Information* 25 (March 1986): 83-124.

Environment Canada

Environment Canada is the agency with overall responsibility for the development and implementation of policies related to environmental protection, monitoring, and research within the government of Canada. Parts of this mandate are shared with other federal agencies, including those responsible for agriculture, forestry, fisheries, and **nonrenewable resources** such as minerals. Environment Canada also works with the environment-related agencies of Canada's 10 provincial and two territorial governments through such groups as the Canadian Council of Ministers of the Environment.

The head of Environment Canada is a minister of the federal cabinet, who is "responsible for policies and actions to preserve and enhance the quality of the environment for the benefit of present and **future generations** of Canadians."

In 1990, following a lengthy and extensive consultation process organized by Environment Canada, the Government of Canada released its *Canada's Green Plan for a Healthy Environment*, which details the broader goals, as well as many specific objectives, to be pursued towards achieving a state of ecologically sustainable economic development in Canada. The first and most general of the national objectives under the *Green Plan* is to "secure for current and future generations a safe and healthy environment, and a sound and prosperous economy."

The *Green Plan* is intended to set a broad environmental framework for all government activities and objectives, including the development of policies. The government of Canada has specifically committed to working toward the following priority objectives: (1) clean air, water, and land; (2) **sustainable development** of renewable resources; (3) protection of special places and **species**; (4) preserving the integrity of northern Canada; (5) global environmental security; (6) environmentally responsible decision making at all levels of society; and (7) minimizing the effects of environmental emergencies. Environment Canada will play the lead role in implementing the vision of the *Green Plan* and in coordinating the activities of the various agencies of the government of Canada.

In order to be able to integrate the dual challenges of a new environmental agenda (set by the expectations of Canadians in general and the federal government in particular) and the need to continue to deliver traditional programs, Environment Canada is moving from a three-program to a one-program administrative structure. Under that single program, six activities are coordinated: (1) the Atmospheric Environment Service activity, through which information is provided and research conducted on weather, **climate**, oceanic conditions, and **air quality**; (2) the Conservation and Protection Service activity, which focuses on special species and places, global environmental integrity, the integrity of Canadian ecosystems, environmental emergencies, and ecological and economic interdependence; (3) the **Canadian Parks Service** activity, concentrating on the ecological and cultural integrity of special places, as well as on environmental and cultural citizenship; (4) the Corporate Environmental Affairs activity, dealing with environmentally responsible decision making and ecosystem-science leadership; (5) the State of the Environment Reporting activity, through which credible and comprehensive environmental information, linked with socio-economic considerations, is provided to Canadians; and (6) the Administration activity, covering corporate management and services. *See also* Environment; Future generations

[*Bill Freedman Ph.D.*]

RESOURCES
BOOKS

Canada's Green Plan for a Healthy Environment. Ottawa: Government of Canada, 1990.
Environment Canada. *Annual Report, 1988-1990*. Ottawa: Government of Canada, 1990.

Environmental accounting

A system of national or business accounting where such environmental assets as air, water, and land are not considered to be free and abundant resources but instead are considered to be scarce economic assets. Any environmental damage caused by the production process must be treated as an economic expense and entered on the balance sheet accordingly. It is important to include in this framework the full environmental cost occurring over the full life cycle of a product, including not only the environmental costs incurred in the production process, but also the environmental costs resulting from use, **recycling** and disposal of products. This is also known as the cradle-to-grave approach. *See also* Ecological economics

Environmental aesthetics

In his journal, **Henry David Thoreau** asked in 1859 "In what book is this world and its beauty described? Who has plotted the steps toward the discovery of beauty?" Almost a 100 years later, in his book *The Sand County Almanac*, ecologist **Aldo Leopold** addressed Thoreau's question by advocating what he called a "conservation esthetic" as the door to appreciating the richness of the natural world, and a resulting **conservation** ethic.

Leopold suggested that increased ecological awareness would more finely tune people's perception of the world around them. The word aesthetics is, after all, derived from the Greek word *aisthesis*, literally "perception by the senses." Leopold claimed that perception, "like all real treasures of the mind, can be split into infinitely small fractions without losing its quality," a necessity if we are to revive our appreciation of the richness and diversity of the world that surrounds us. Instead, he thought that most recreationists are like "motorized ants who swarm the continents before learning to see [their] own back yard," so that recreational development becomes "a job not of building roads into lovely country, but of building receptivity into the still unlovely human mind."

Despite the fact that Thoreau and Leopold are both widely read by environmentalists and others, aesthetics has remained largely the domain of philosophy, and environmental aesthetics until recently a neglected domain in gen-

eral. In philosophy, aesthetics has developed quite narrowly as an esoteric subdiscipline focused on theories of the arts, but a philosophy exclusive of **nature** and **environment**. An environmental philosophy has emerged in recent years, but this has developed mainly from the ethics tradition in philosophy and has, for some reason, neglected the aesthetic tradition (though Sepänmaa, and some other philosophers, have tried to encourage an environmental aesthetics at the intersection of traditions: on the environment as an aesthetic object and on environmental aesthetics as the philosophy of environmental criticism). While extensive literature has emerged on **environmental ethics**, the gap left by philosophy for environmental aesthetics has been filled by writers in a number of disciplines, notably psychology and geography, especially through empirical research on perception. Yi Fu Tuan's pioneering works (e.g., *Topophilia*, in 1974) broadened the study of aesthetics, transcended his own discipline of geography, and continue to shape the debate today about the content and concepts appropriate to an environmental aesthetic.

In classical aesthetics, sight and hearing are considered the primary senses since they are most immediately involved in the contemplation and appreciation of art and music. Arguably, the other three senses--touch, smell, and taste-- must also be brought to bear for a true sensing of all the arts. Certainly, all five of the basic senses are central to an environmental aesthetic, an aesthetic sense, as Berleant notes, of all "the continuities that join integrated human persons with their natural and cultural condition." To achieve this, Berleant suggests, requires what he calls "an integrated sensorium," a "recognition of synaesthesia," i.e., a "fusion of the sense modalities." He claims that "perception is not passive but an active, reciprocal engagement with environment"; environmental aesthetics is what Berleant calls "an aesthetics of engagement."

Aesthetic engagement with environment involves not only the five senses, but the individual, behavioral and personal history of each person, as well as the cultural context and continuity in which that person has developed. An environmental aesthetic is "always contextual, mediated by the [long and complex] variety of conditions and influences that shape all [human] experience." An environmental aesthetic is also contingent on place. A **sense of place**, or a developed, knowledgeable understanding of the place in which one lives, is central to a well-developed environmental aesthetic. This can lead to greater engagement with place, and more involvement with local planning and land-use decisions, among other issues.

An understanding of environmental aesthetics cannot be attained without some notion of how the word and concept 'environment' itself is defined. Before recent publications on environmental aesthetics, most writers on the subject focused on the beauty of nature, for example as experienced in the national parks of the United States. As a more formal environmental aesthetics has emerged, the definition of environment (and environmental beauty) has enlarged. Sepänmaa, for example, defines "environmental aesthetics [as] the aesthetics of the real world" and includes in this "all of the observer's external world: the natural environment, the cultural environment, and the constructed environment."

Berleant also claims that "the idea of an aesthetic environment is a new concept that enlarges the meaning of environment." But he considers environment not only as everything "out there," but as "everything that there is; it is all-inclusive, a total, integrated, continuous process." This conception is hard to grasp but "nonetheless soberly realistic, for it recognizes that ultimately everything affects everything else, that humans along with all other things inhabit a single intraconnected realm." Berleant's conception of environment "does not differentiate between the human and the natural and...interprets everything as part of a single, continuous whole." He advocates "the largest idea of environment [which] is the natural process as people live it, however they live it. Environment is nature experienced, nature lived." An environmental aesthetic then, depends on education in **human ecology**, on more complete understanding of how people connect with other people, and of how they interact with a wide range of environments, local to global.

Better understanding is both a result of, and a cause of, more engagement with place and surroundings. The resulting connections and commitment have ramifications for the design and planning of the built environment, and for exploiting and managing the natural environment. Aesthetics, in this ecological, integrated sense, can contribute to cost-benefit decisions about how we use the environments in which we live, including city and regional planning, and in design at all levels, from individual artifact to our grasp of the global realities of the contemporary world.

Berleant argues that environmental perception becomes an aesthetic only through a complete understanding of the totality. Though that is not possible for any one person, an appreciation of the richness and diversity of the earth's natural and human domain is necessary to achieve such an aesthetic. An environmental aesthetic then is necessarily complex, but it can also provide simple commandments for everyday life: as Thoreau suggested, "to affect the quality of the day, that is the highest of arts."

[*Gerald L. Young Ph.D.*]

RESOURCES
BOOKS

Berleant, A. *The Aesthetics of Environment.* Philadelphia, PA: Temple University Press, 1992.

Nasar, J. L., ed. *Environmental Aesthetics: Theory, Research, and Applications.* New York: Cambridge University Press, 1988.

Environmental auditing

The environmental auditing movement gained momentum in the early 1980s as companies beset by new liabilities associated with Superfund and old **hazardous waste** sites wanted to insure that their operations were adhering to federal and local policies and company procedures. Most audits were initiated to avoid legal conflicts, and many companies brought in outside consultants to do the audits. The audits served many useful functions, including increasing management and employee awareness of environmental issues and initiating data collection and central monitoring of matters previously not watched as carefully. In many companies environmental auditing played a useful role in organizing information about the **environment**. It paved the way for the **pollution** prevention movement which had a great impact on company environmental management in the late 1980s when some companies started to view all of their pollution problems comprehensively and not in isolation from one another.

Environmental chemistry

Environmental chemistry refers to the occurrence, movements, and transformations of **chemicals** in the **environment**. Environmental chemistry deals with naturally occurring chemicals such as metals, other elements, organic chemicals, and biochemicals that are the products of biological **metabolism**. Environmental chemistry also deals with synthetic chemicals that have been manufactured by humans and dispersed into the environment, such as pesticides, **polychlorinated biphenyls** (PCBs), dioxins, **furans**, and many others.

The occurrence of chemicals refers to their presence and quantities in various compartments of the environment and ecosystems. For example, in a terrestrial **ecosystem** such as a forest, the most important compartments to consider are the mineral **soil**, water and air present in spaces within the soil, the above-ground **atmosphere**, dead **biomass** within the soil and lying on the ground as logs and other organic debris, and living organisms, the most abundant of which are trees. Each of these components of the forest ecosystem contains a wide variety of chemicals in some concentration, and in some amount. Chemicals move between all of these compartments, as fluxes that represent elements of **nutrient** and mineral cycles.

The movements of chemicals within and among compartments often involve a complex of transformations among

potential molecular states. There may also be changes in physical states, such as evaporation of liquids, or crystallization of dissolved substances. The transformations of chemicals among molecular states can be illustrated by reference to the environmental cycling of sulfur. Sulfur (S) is commonly emitted to the atmosphere as the gases **sulfur dioxide** (SO_2) or **hydrogen** sulfide (H_2S), which are transformed by photochemical reactions into the negatively-charged **ion**, sulfate (SO_4^{-2}). The sulfate may eventually be deposited with precipitation to a terrestrial ecosystem, where it may be absorbed along with soil water by tree roots, and later used to synthesize biochemicals such as proteins and amino acids. Eventually, the plant may die and its biomass deposited to the soil surface as litter. **Microorganisms** can then metabolize the organic matter as a source of energy and nutrients, eventually releasing simple inorganic compounds of sulfur such as sulfate or hydrogen sulfide into the environment. Alternatively, the plant biomass may be harvested by humans and used as a fuel, with the organic sulfur being oxidized during **combustion** and emitted to the atmosphere as sulfur dioxide. Organic and mineral forms of sulfur also occur in **fossil fuels** such as **petroleum** and **coal**, and the combustion of those materials also results in an **emission** of sulfur dioxide to the atmosphere.

Contamination and pollution

Contamination and **pollution** both refer to the presence of chemicals in the environment, but it is useful to distinguish between these two conditions. Contamination refers to the presence of one or more chemicals in concentrations higher than normally occurs in the ambient environment, but not high enough to cause biological or ecological damages. In contrast, pollution occurs when chemicals occur in the environment in concentrations high enough to cause damages to organisms. Pollution results in toxicity and ecological changes, but contamination does not cause those damages.

Chemicals that are commonly involved in pollution include the gases sulfur dioxide and **ozone**, diverse kinds of pesticides, elements such as **arsenic**, **copper**, **mercury**, **nickel**, and selenium, and some naturally occurring biochemicals. In addition, large concentrations of nutrients such as phosphate and nitrate can cause eutrophication, a type of pollution associated with excessive **ecological productivity**. Although any of these chemicals can cause pollution in certain situations, they most commonly occur in concentrations too small to cause toxicity or other ecological damages.

Modern analytical chemistry has become extremely sophisticated, and this allows trace contamination of potentially toxic chemicals to be measured at levels that are much smaller than what is required to cause demonstrable physiological or ecological damages.

Environmental chemistry of the Aamosphere

Nitrogen gas (N_2) comprises about 79% of the mass of Earth's atmosphere, while 20% is oxygen (O_2), 0.9% argon (Ar), 0.035% **carbon dioxide** (CO_2), and the remainder composed of a variety of trace gases. The atmosphere also contains variable concentrations of water vapor, which can range from 0.01% in frigid arctic air to 5% in humid tropical air.

The atmosphere also can contain high concentrations of gases, vapors, or particulates that are potentially harmful to people, other animals, or vegetation, or that cause damages to buildings, art, or other materials. The most important gaseous air pollutants (listed alphabetically) are ammonia (NH_3), **carbon monoxide** (CO), fluoride (F, usually occurring HF), nitric oxide and nitrogen dioxide (NO and NO_2, together known as oxides of nitrogen, or NO_x), ozone (O_3), **peroxyacetyl nitrate** (PAN), and sulfur dioxide (SO_2).

Vapors of elemental mercury and **hydrocarbons** can also be air pollutants. Particulates with tiny diameters (less than $1\mu m$) can also be important, including dusts containing such toxic elements as arsenic, copper, **lead**, nickel, and vanadium, organic aerosols that are emitted as **smoke** during combustions (including **toxins** known as **polycyclic aromatic hydrocarbons**), and non-reactive minerals such as silicates.

Some so-called "trace toxics" also occur in the atmosphere in extremely small concentrations. The trace toxics include persistent organochlorine chemicals such as the pesticides DDT and dieldrin, polychlorinated biphenyls (PCBs), and the **dioxin**, TCDD. Other, less persistent pesticides may also be air pollutants close to places where they are used.

Environmental chemistry of water

Earth's surface waters vary enormously in their concentrations of dissolved and suspended chemicals. Other than the water, the chemistry of oceanic water is dominated by sodium chloride (NaCl), which has a typical concentration of about 3.5% or 35 g/l. Also important are sulfate (2.7 g/l), magnesium (1.3 g/l), and potassium and calcium (both 0.4 g/l). Some saline lakes can have much larger concentrations of dissolved ions, such as Great Salt Lake in Utah, which contains more than 20% salts.

Fresh waters are much more dilute in ions, although the concentrations are variable among waterbodies. The most important cations in typical fresh waters are calcium (Ca^{2+}), magnesium (Mg^{2+}), sodium (Na^+), ammonium (NH_4^+), and hydrogen ion (H^+; this is only present in acidic waters, otherwise hydroxy ion or OH^- occurs). The most important anions are bicarbonate (HCO_3^-) sulfate (SO_4^{2+}), chloride (Cl^-), and nitrate (NO_3^-). Some fresh waters have high concentrations of dissolved organic compounds, known as humic substances, which can stain the water a tea-like color. Typical concentrations of major ions in fresh water are: calcium 15 mg/l, sulfate 11 mg/l, chloride 7 mg/l, silica 7 mg/l, sodium 6 mg/l, magnesium 4 mg/l, and potassium 3 mg/l.

The water of clean precipitation is considerably more dilute than that of surface waters such as lakes. For example, precipitation at a remote place in Nova Scotia contained 1.6 mg/l of sulfate, 1.3 mg/l chloride, 0.8 mg/l sodium, 0.7 mg/l nitrate, 0.13 mg/l calcium, 0.08 mg/l ammonium, 0.08 mg/l magnesium, and 0.08 mg/l potassium. Because that site is about 31 mi (50 km) from the Atlantic Ocean, its precipitation is influenced by sodium and chloride originating with sea spray. In comparison, a more central location in North America had a sodium concentration of 0.09 mg/l and chloride 0.15 mg/l.

Pollution of surface waters is most often associated with the dumping of human or industrial sewage, nutrient inputs from agriculture, **acidification** caused by acidic precipitation or by acid-mine **drainage**, and industrial inputs of toxic chemicals. Eutrophication is caused when nutrient inputs cause large increases in aquatic productivity, especially in fresh waters and shallow marine waters into which sewage is dumped or that receive **runoff** containing agricultural fertilizers. In general, marine ecosystems become eutrophic when they are fertilized with nitrate, and freshwater systems with phosphate. Only 35–100 µg/l or more of phosphate is enough to significantly increase the productivity of most shallow lakes, compared with the background concentration of about 10 µg/l or less.

Freshwater ecosystems can become acidified by receiving drainage from bogs, by the deposition of acidifying substances from the atmosphere (such as acidic rain), and by acid-mine drainage. Atmospheric depositions have caused a widespread acidification of surface waters in eastern North America, Scandinavia, and other places. Surface waters acidified by atmospheric depositions commonly develop pHs of about 4.5–5.5. Tens of thousands of lake and running-water ecosystems have been damaged in this way. Acidification has many biological consequences, including toxicity caused to many **species** of plants and animals, including fish.

Some industries emit metals to the environment, and these may pollute fresh and marine waters. For instance, lakes near large smelters at **Sudbury, Ontario**, have been polluted by sulfuric **acid**, copper, nickel, and other metals, which in some cases occur in concentrations large enough to cause toxicity to aquatic plants and animals.

Mercury contamination of fish is also a significant problem in many aquatic environments. This phenomenon is significant in almost all large fish and **sharks**, which accumulate mercury progressively during their lives and commonly have residues in their flesh that exceed 0.5 ppm (this

is the criterion set by the World Health Organization for the maximum concentration of mercury in fish intended for human consumption). It is likely, however, that the oceanic mercury is natural in origin, and not associated with human activities. Many fresh-water fish also develop high concentrations of mercury in their flesh, also commonly exceeding the 0.5 ppm criterion. This phenomenon has been demonstrated in many remote lakes. The source of mercury may be mostly natural, or it may originate with industrial sources whose emissions are transported over a long distance in the atmosphere before they are deposited to the surface. Severe mercury pollution has also occurred near certain factories, such as chlor-alkali plants and pulp mills. The most famous example occurred at Minamata, Japan, where industrial discharges led to the pollution of marine organisms, and then resulted in the **poisoning** of fish-eating animals and people.

Environmental chemistry of soil and rocks

The most abundant elements in typical soils and rocks are oxygen (47%), silicon (28%), **aluminum** (8%), and iron (3–4%). Virtually all of the other stable elements are also present in soil and rocks, and all of these can occur in a great variety of molecular forms and minerals. Under certain circumstances, some of these chemicals can occur in relatively high concentrations, sometimes causing ecological damages.

This can occur naturally, as in the case of soils influenced by so-called serpentine minerals, which can contain hundreds to thousands of ppm of nickel. In addition, industrial emissions of metals from smelters have caused severe pollution. Soils near Sudbury, for example, can contain nickel and copper concentrations up to 5,000 ppm each. Even urban environments can be severely contaminated by certain metals. Soils collected near urban factories for **recycling** old **automobile** batteries can contain lead in concentrations in the percent range, while the edges of roads can contain thousands of ppm of lead emitted through the use of leaded **gasoline**.

Trace toxics

Some chemicals occur in minute concentrations in water and other components of the environment, yet still manage to cause significant damages. These chemicals are sometimes referred to as trace toxics. The best examples are the numerous compounds known as halogenated hydrocarbons, particularly **chlorinated hydrocarbons** such as the insecticides DDT, DDD, and dieldrin, the dielectric fluids PCBs, and the chlorinated dioxin, TCDD. These chemicals are not easily degraded by either **ultraviolet radiation** or by metabolic reactions, so they are persistent in the environment. In addition, chlorinated hydrocarbons are virtually insoluble in water, but are highly soluble in lipids such as fats and oils. Because most lipids in ecosystems occur within the bodies of organisms, chlorinated hydrocarbons have a marked tendency to bioaccumulate (i.e., to occur preferentially in organisms rather than in the non-living environment). This, coupled with the persistence of these chemicals, results in their strong tendency to food-chain/web accumulate or biomagnify (i.e., to occur in their largest concentrations in top predators).

Fish-eating birds are examples of top predators that have been poisoned by exposure to chlorinated hydrocarbons in the environment. Some examples of species that have been affected by this type of ecotoxicity include the **peregrine falcon** (*Falco peregrinus*), **bald eagle** (*Haliaeetus leucocephalus*), osprey (*Pandion haliaetus*), **brown pelican** (*Pelecanus occidentalis*), **double-crested cormorant** (*Phalacrocorax auritus*), and western grebe (*Aechmophorus occidentalis*). Concentrations of chlorinated hydrocarbons in the water of aquatic habitats of these birds is generally less than 1 μg/l (part per billion, or ppb), and less than 1 ng/l (part per trillion, or ppt) in the case of TCDD. However, some of the chlorinated hydrocarbons can biomagnify to tens to hundreds of mg/kg (ppm) in the fatty tissues of fish-eating birds. This can cause severe toxicity, characterized by reproductive failures, and even the deaths of adult birds, both of which can cause populations to collapse.

Other trace toxics also cause ecological damages. For example, although it is only moderately persistent in aquatic environments, the insecticide carbofuran can accumulate in acidic standing water in recently treated fields. If geese, ducks, or other birds or mammals utilize those temporary aquatic habitats, they can be killed by the carbofuran residues. Large numbers of **wildlife** have been killed this way in North America.

Petroleum

Water pollution can also result from the occurrence of hydrocarbons in large concentrations, especially after spills of crude oil or its refined products. Oil pollution can result from accidental spills of petroleum from wrecked tankers, offshore drilling platforms, broken pipelines, and from spills during warfare, as occurred during the Gulf War of 1991. Other important sources of oil pollution include operational discharges from tankers disposing oily bilge waters, and chronic releases from oil refineries and **urban runoff**.

The concentration of natural hydrocarbons in seawater is about 1 ppb, mostly due to releases from **phytoplankton** and bacteria. Beneath a slick of petroleum spilled at sea, however, the concentration of dissolved hydrocarbons can exceed several ppm, enough to cause toxicity to some organisms. There are also finely suspended droplets of petroleum in water beneath slicks, as a result of wave action on the floating oil. The slick and the sub-surface emulsion of oil-in-water are highly damaging to organisms that become coated with these substances.

[*Bill Freedman Ph.D.*]

RESOURCES

BOOKS

Freedman, B. *Environmental Ecology*, 2nd ed. San Diego: Academic Press, 1995.

Hemond, H.F., and E.J. Fechner. *Chemical Fate and Transport in the Environment*. San Diego: Academic Press, 1994.

Manahan, S. *Environmental Chemistry*, 6th ed. Boca Raton, FL: Lewis Publishers, 1994.

Environmental Defense

Environmental Defense is a **public interest group** founded in 1967 and concerned primarily with the protection of the **environment** and the concomitant improvement of public health. Originally called Environmental Defense Fund, the name was shortened to end confusion related to the use of the word fund. In the beginning a group of Long Island scientists organized to oppose local spraying of the **pesticide** DDT, and Environmental Defense is still staffed by scientists as well as lawyers and economists. Over time the group has expanded its interests to include **air quality**, energy, **solid waste**, **water resources**, agriculture, **wildlife**, habitats, and international environmental issues. Environmental Defense presently has a membership of approximately 300,000, an annual budget of $39.1 million, and a staff of 200 working out of eight regional offices.

Environmental Defense seeks to protect the environment by initiating legal action in environment-related matters and also by conducting public service and educational campaigns. It publishes a newsletter detailing the organization's activities, as well as occasional books, reports, and monographs. Environmental Defense also conducts and encourages research relevant to environmental issues and promotes administrative, legislative, and corporate actions and policies in defense of the environment.

Environmental Defense's strategies and orientation have changed somewhat over the years from the early days when the group's motto was "Sue the bastards!" At about the time that Frederic D. Krupp became its executive director in 1984, Environmental Defense began to view environmental problems more in view of economic needs. As Krupp put it, the practical effectiveness of the environmental movement in the future would depend on its realization that behind environmental problems "there are nearly always legitimate social needs–and that long-term solutions lie in finding alternative ways to meet those underlying needs."

With this in mind, Krupp proposed a "third stage of environmentalism" which combined direct opposition to environmentally harmful practices with proposals for realistic, economically-viable alternatives. This strategy was first applied successfully to large-scale power production in California, where utilities were planning a massive expansion of generating capacity. Environmental Defense demonstrated that this expansion was largely unnecessary (thereby saving the utilities and their customers a considerable amount of money while also protecting the environment) by showing that the use of existing and well-established technology could greatly reduce the need for new capacity without affecting the utility's customers. Environmental Defense also showed that it was economically effective to buy power generated from **renewable energy** resources, including **wind energy**.

The Environmental Defense worked with the Mc-Donalds Corporation in 1991 on a task force to reduce the fast food giant's estimated two million lb-per-day (907,184 kg) effusion of waste. One of the most widely publicized results of these efforts was that McDonald's was convinced to stop packaging its hamburgers in **polystyrene** containers. Combined with other strategies, McDonald's estimates that the task force's recommendations will eventually reduce its waste flow by 75%. More recently in 2000, the Environmental Defense joined with eight companies to reduce the amount of **greenhouse gases** in the environment. In 2001, the Action Network (which has 750,000 members) and For My World projects were set up to inform the public on environmental and health issues. Scorecard.org rates area pollutants based on the most recent findings by the **Environmental Protection Agency** (EPA).

Environmental Defense continues to search for ways to harness economic forces in ways that destroy incentives to degrade the environment. This approach has made Environmental Defense one of the most respected and heeded environmental groups among United States corporations. Environmental Defense virtually ghost-wrote the Bush Sr. Administration's **Acid Rain** Bill, which makes considerable use of market-oriented strategies such as the issuing of tradable emissions permits. These permits allow for a set amount of **pollution** per firm based on ceilings set for entire industries. Companies can buy, sell, and trade these permits, thus providing them with a profit motive to reduce harmful emissions.

[*Lawrence J. Biskowski*]

RESOURCES
ORGANIZATIONS
Environmental Defense, 257 Park Avenue South, New York, NY USA 10010 (212) 505-2100, Fax: (212) 505-2375, Email: webmaster@ environmentaldefense.org, <http://www.environmentaldefense.org>

Environmental Defense Fund
see **Environmental Defense**

Environmental degradation

Degradation is the act or process of reducing something in value or worth. Environmental degradation, therefore, is the de-valuing of and damage to the **environment** by natural or **anthropogenic** causes. The loss of **biodiversity, habitat** destruction, depletion of energy or mineral sources, and exhaustion of **groundwater** aquifers are all examples of environmental degradation.

Presently there are four major areas of global concern due to environmental degradation: marine environment, **ozone** layer, **smog** and **air pollution**, and the vanishing **rain forest. Pollution**, at some level, is found throughout the world's oceans, which cover two-thirds of the planet's surface. Marine debris, farm **runoff**, industrial waste, sewage, dredge material, stormwater runoff, and **atmospheric deposition** all contribute to **marine pollution**. The level of degradation varies from region to region, but its effects are seen in such remote places as **Antarctica** and the Bering Sea.

Issues of **waste management** and disposal have had a large degrading impact on these areas. There are major national and international efforts to control pollution from shipping (including **oil spills** and general pollution due to ship ballast) and direct ocean or **estuary** discharges. Clean up has been started in some areas with some initial success.

Another major problem facing the world is the depletion of the ozone layer, which is linked to the use of a group of **chemicals** called **chlorofluorocarbons** (CFCs). These chemicals are widely used by industry as refrigerants and in **polystyrene** products. Once released into the air they rise to the **stratosphere** and eat away at the ozone layer. This layer is important because it protects us from harmful **ultraviolet radiation**, which is the chief cause of skin **cancer**.

Smog in urban areas and **air quality** in general have become crucial issues in the last few decades. **Acid rain**, which occurs when **sulfur dioxide** and **nitrogen** oxide—as emissions from **power plants** and industries—change in the **atmosphere** to form harmful compounds that fall to earth in rain, fog, and snow. **Acid** rain damages lakes and streams as well as buildings and monuments. Air **visibility** is curtailed and the health of humans as well as plants and trees can be affected.

Automobile emissions of **nitrogen oxides, carbon monoxide**, and volatile organic compounds—although much reduced in the United States—also contribute to smog and acid rain. **Carbon** monoxide continues to be a problem in cities such as Los Angeles where there is heavy **automobile** congestion.

The vanishing rain forest is also of major global concern. The degradation of the rain forest—with its extensive **logging, deforestation**, and massive destruction of habitat—has threatened the survival of many **species** of plants and animals as well as disrupting **climate** and weather patterns locally and globally. Although tropical rain forests cover only about five to seven percent of the world's land surface, they contain about one-half to two-thirds of all species of plants and animals, some which have never been studied for their medicinal or food properties.

The problem of environmental degradation has been addressed by various environmental organizations throughout the world. Environmentalists are no longer solely concerned with the local region and efforts to stop or at least slow down environmental degradation has taken on a global significance.

[*James L. Anderson*]

RESOURCES
BOOKS

The Global Ecology Handbook: What You Can Do About the Environmental Crisis. Boston: Beacon Press, 1990.
Our Common Future. World Commission on Environmental Development. New York: Oxford University Press, 1987.
Preserving Our Future Today. U.S. Environmental Protection Agency. Washington, DC: U.S. Government Printing Office, 1991.
Silver, C. S., and R. S. Defries. *One Earth, One Future: Our Changing Global Environment.* Washington, DC: National Academy Press, 1990.
Triedjell, S. T. "Soil and Vegetative Systems." In *Contemporary Problems in Geography.* Oxford: Clarendon Press, 1988.

Environmental design

Environmental design is a new approach in planning consumer products and industrial processes that are ecologically intelligent, sustainable, and healthy for both humans and our **environment**. Based on the work of innovative thinkers such as architect Bill McDonough, chemist Michael Braungart, physicist Amory Lovins, Swedish physician Dr. Karl-Henrik Robert, and business executive Paul Hawken, this movement is an effort to rethink our whole industrial economy. During the first Industrial Revolution 200 years ago, raw materials such as lumber, minerals, and clean water seemed inexhaustible, while **nature** was regarded as a hostile force to be tamed and civilized. We use materials to make the things we wanted, then discard them when they no longer are useful. "Dilution is the solution to pollution," suggests that if we just spread our wastes out in the environment widely enough, no one will notice.

This approach has given us an abundance of material things, but also has produced massive **pollution** and **environmental degradation**. It also is incredibly wasteful. On average, for every truckload of products delivered in the United States, 32 truckloads of waste are produced along the way. The **automobile** is a typical example. Industrial ecologist, Amory Lovins, calculates that for every 100 gallons

(380 l) of **gasoline** burned in your car engine, only 1% (0 gal or 3.8 l) actually moves the passengers inside. All the rest is used to move the vehicles itself. The wastes produced—carbon dioxide, **nitrogen oxides**, unburned hydrocarbon, **rubber** dust, heat—are spread through the environment where they pollute air, water, and **soil**. And when the vehicle wears out after only a few years of service, thousands of pounds of metal, rubber, plastic, and glass become part of our rapidly growing waste steam.

This isn't the way things work in nature, environmental designers point out. In living systems, almost nothing is discarded or unused. The wastes from one organism become the food of another. Industrial processes, to be sustainable over the long term, should be designed on similar principles, designers argue. Rather than following current linear patterns in which we try to maximize the throughput of materials and minimize labor, products and processes should be designed to be energy efficient and use renewable materials. They should create products that are durable and reusable or easily dismantled for repair and remanufacture, and are non-polluting throughout their entire life cycle. We should base our economy on renewable **solar energy** rather than **fossil fuels**. Rather than measure our economic progress by how much material we use, we should evaluate productivity by how many people are gainfully and meaningfully employed. We should judge how well we're doing by how many factories have no smokestacks or dangerous effluents. We ought to produce nothing that will require constant vigilance from **future generations**.

Inspired by how ecological systems work, Bill McDonough proposes three simple principles for designing processes and products:

• Waste equals food. This principle encourages elimination of the concept of waste in industrial design. Every process should be designed so that the products themselves, as well as leftover **chemicals**, materials, and effluents, can become "food" for other processes.

• Rely on current solar income. This principle has two benefits: First, it diminishes, and may eventually eliminate, our reliance on hydrocarbon fuels. Second, it means designing systems that sip energy rather than gulping it down.

• Respect diversity. Evaluate every design for its impact on plant, animal, and human life. What effects do products and processes have on identity, independence, and integrity of humans and natural systems? Every project should respect the regional, cultural, and material uniqueness of its particular place.

According to McDonough, our first question about a product is whether it is really needed. Could we obtain the same satisfaction, comfort, or utility in another way that would have less environmental and social impacts? Can the things we design be restorative and regenerative: that is, can they help reduce the damage done by earlier, wasteful approaches, and can they help nature heal rather than simply adding to existing problems? McDonough invites us to reinvent our businesses and institutions to work with nature, and redefine ourselves as consumers, producers, and citizens to promote a new sustainable relationship with the Earth. In an eco-efficient economy, he says, products might be divided into three categories:

• Consumables are products like food, natural fabrics, or paper that are produced from renewable materials and can go back to the soil as compost.

• Service products are durables such as cars, televisions, and refrigerators. These products should be leased to the customer to provide their intended service, but would always belong to the manufacturer. Eventually, they would be returned to the maker, who would be responsible for **recycling** or remanufacturing.

• Unmarketables are materials like radioactive isotopes, persistent **toxins**, and bioacumulative chemicals. Ideally, no one would make or use these products. But because eliminating their use will take time, McDonough suggests that for now, these materials should belong to the manufacturer and be molecularly tagged with the maker's mark. If they are discovered to be discarded illegally, the manufacturer would be liable.

Following these principles McDonough Braungart Design Chemistry has created nontoxic, easily recyclable, healthy materials for buildings and for consumer goods. Rather than design products for a "cradle to grave" life cycle, MBDC aims for a fundamental conceptual shift to a Cradle to Cradle® processes whose materials perpetually circulate in closed systems that create value and are inherently healthy and safe. Among some important examples are carpets designed to be recycled at the end of their useful life, paints and adhesives that are non-toxic and non-allergenic, and clothing that is both healthy for the wearer and that has minimal environmental impact in its production.

In his architecture firm, McDonough + partners, these new design models and environmentally friendly materials have been used in a number of innovative building projects. A few notable examples include: The Gap Inc. offices in California and the Environmental Studies building at Oberlin College in Ohio.

Built in 1994, The Gap building in San Bruno, California, is designed to maintain the unique natural features of the site while providing comfortable, healthy, and flexible office spaces. Intended to promote employee well-being and productivity as well as eco-efficiency, The Gap building has high ceilings, open, airy spaces, a natural ventilation system including operable windows, a full-service fitness center (including a pool), and a landscaped atrium for each office bay that brings the outside in. Skylights in the roof deliver daylight to interior offices and vent warm, stale air. Warm interior tones and natural woods (all wood used in the building was harvested by certified sustainable methods) give a friendly feel. Paints, adhesives, and floor coverings are low toxicity to maintain a healthy indoor environment. A pleasant place to work, the offices help recruit top employees and improve both effectiveness and retention.

The roof of The Gap Building is planted with native grasses and wildflowers that absorb rainwater and help improve ambient environmental quality. The grass roof also is beautiful and provides thermal and acoustic insulation. At night, cool outdoor air is flushed through the building to provide natural cooling. By providing abundant daylight, high-efficiency fluorescent lamps, fresh air ventilation, and other energy-saving measures, this pioneering building is more than 30% more energy efficient than required by California law. Operating savings within the first four to eight years of occupancy are expected to repay the initial costs of these design innovations.

An even more environmentally friendly building was built at Oberlin College in 2001 to house its Environmental Studies Program. Under the leadership of Dr. David Orr, the Adam Joseph Lewis Center is planned around the concept of ecological stewardship, and is intended to be both "restorative" and "regenerative" rather than merely non-damaging to the environment. The building is designed to be a net energy exporter, generating more power from renewable sources than it consumes annually. More than 3,700 sq ft (roughly 350 sq m) of photovoltaic panels on the roof are expected to generate 75,000 kilowatt hours of energy per year. The building also draws on geothermal **wells** for heating and cooling, and features use of natural daylight and ventilation to maintain interior comfort levels and a healthy interior environment. High efficiency insulation in walls and windows are expected to make energy consumption nearly 80% lower than standard academic buildings in the area.

The Lewis Center also incorporates an innovative "living machine" for internal waste **water treatment**, a constructed wetland for storm water management, and a landscape that provides social spaces, learning opportunities with live plants, and **habitat** restoration. It is expected that all water used in the building will be returned to the environment in as good quality as when it entered. The water produced by natural cleaning processes should be of high enough quality for drinking, although doing so isn't planned at present.

Taken together, these restorative and regenerative environmental design approaches could bring about a new industrial revolution. The features of environment design are incorporated in McDonough's "Hanover Principles" prepared for the 2000 World Fair in Hanover, Germany. This manifesto for green design urges us to recognize how humans interact with and depends on the natural world. According to McDonough, we need to recognize even distant effects and consider all aspects of human settlement, including community, dwelling, industry and trade, in terms of existing and evolving connections between spiritual and material consciousness. We should accept responsibility for the consequences of design decisions upon human well being, the viability of natural systems. We have to understand the limitations of design. No human creation lasts forever, and design doesn't solve all problems. Those who create and plan should practice humility in the face of nature. We should treat nature as a model and mentor, not an inconvenience to be evaded and controlled. If we can incorporate these ecologically intelligent principles in our practice, we may be able to link long-term, sustainable considerations with ethical responsibility, and to reestablish the integral relationship between natural processes and human activity.

[*William P. Cunningham Ph.D.*]

RESOURCES
BOOKS

Hawken, Paul, Amory Lovins, and L. Hunter Lovins. *Natural Capitalism: Creating the Next Industrial Revolution.* Back Bay Books, 2000.

Hawken, Paul. *The Ecology of Commerce: A Declaration of Sustainability.* New York: Harperbusiness, 1994.

Hutchison, Colin. *Building to Last: The Challenge for Business Leaders.* London: Earthscan, 1997.

McDonough, William *The Hanover Principles.* 2000. [cited July 9, 2002].<http://www.mcdonoughpartners.com/principles.pdf>.

McDonough, William and Michael Braungart. *Cradle to Cradle: A Blueprint for the Next Industrial Revolution.* San Francisco, CA: North Point Press, 2002.

Environmental dispute resolution

Environmental Dispute Resolution (EDR) or Alternative Dispute Resolution (ADR), as it is more generally known, is an out-of-court alternative to litigation to resolve disputes between parties. Although ADR can be used with virtually any legal dispute, it is often used to resolve environmental disputes. There are several types of ADR, ranging from the least formal to the most formal process: (a) negotiation, (b) mediation, (c) adjudication, (d) arbitration, (e) minitrial, and (f) summary jury trial.

Negotiation: Negotiation is the simplest and most often practiced form of ADR. The parties do not enter the judicial system, but rather settlements are reached in an informal setting and then reduced to written terms.

Mediation: Mediation is an extension of the direct negotiation process. The term is loosely used and is often confused with arbitration or informal processes in general. Mediation is a process in which a neutral third-party inter-venes to help disputants reach a voluntary settlement. The mediator has no authority to force the parties to reach an agreement.

Mediation is often the most appropriate technique for environmental disputes because the parties often have no prior negotiating relationship and, because there are often many technical and scientific uncertainties, the assistance of a qualified professional is helpful.

Mediation is also used with varying success in environmental policy-making, standard setting, determination of development choices, and the enforcement of environmental standards. Many states explicitly recognize mediation as the primary method for initially dealing with environmental disputes, and mediation procedures are written into federal **environmental policy**, specifically in the regulations dealing with the **Comprehensive Environmental Response, Compensation and Liability Act** (CERCLA) and the **Resource Conservation and Recovery Act** (RCRA). Mediation is not appropriate, however, with all environmental disputes because some environmental laws were designed to encourage a slower examination of issues that impact society.

Adjudication: Adjudication is sometimes referred to as "private judging." It is an ADR process in which the parties give their evidence and arguments to a neutral third-party who then renders an objective, binding decision. It is a voluntary procedure and private unless one party seeks judicial enforcement or review after the decision is made. The parties must agree on the adjudication and procedural rules for the process and each side is contractually bound for the length of the proceeding.

The advantage of adjudication is that a law- and/or environment-trained third party renders an objective decision based on the presented facts and legal arguments. The parties set their own rules so an adjudicator is not bound to legal principles of any particular jurisdiction. Private organizations provide adjudication services for fees but they can be expensive.

Arbitration: Arbitration is a process whereby a private judge, or arbitrator, hears the arguments of the parties and renders a judgment. The process works much like a court except that the parties choose the arbitrator and the substantive law he or she should apply. The arbitrator also has much more latitude in creating remedies which are fair to both parties. People often confuse the responsibilities of arbitrators and mediators. Arbitrators are passive functionaries who determine right or wrong; mediators are active functionaries who attempt to move the parties to reconciliation and agreement, regardless of who is right or wrong.

Parties cannot be forced into arbitration unless the contract in question includes an arbitration clause or the parties consented to enter into arbitration after the dispute developed. Since arbitration is a contractual remedy, the arbitrator can consider only those disputes and remedies which the parties agreed to submit to arbitration.

Minitrial: A minitrial is a private process in which parties agree to voluntarily reach a negotiated settlement. They present their cases in summary form before a panel of designated representatives of each party. The panel offers non-binding conclusions on the probable outcome of the case, were it to be litigated. The parties may then use the results to assist with negotiation and settlement.

Summary Jury Trial: A summary jury trial is similar to a minitrial except that the evidence is presented to a non-expert, impartial jury, rather than a panel chosen by the parties, which subsequently prepares non-binding conclusions on each of the issues in dispute. Parties may then use the assessment of the jury's "verdict" to help with negotiation and settlement.

[*Kevin Wolf*]

RESOURCES
BOOKS

Loew, W. R., and A. M. Ramirex. "Resolving Environmental Disputes with ADR." *The Practical Real Estate Lawyer* 8 (May 1992): 15-23.

PERIODICALS

Kubasek, N., and G. Silverman, "Environmental Mediation." *American Business Law Journal* 26 (Fall 1988): 533-555.

Environmental economics

Environmental economics is a relatively new field, but its roots go back to the end of the nineteenth century when economists first discussed the problem of **externality**. Economic transactions have external effects which are not captured by the price system. Prime examples of these externalities are **air pollution** and **water pollution**. The absence of a price for nature's capacity to absorb wastes has an obvious solution in economic theory. Economists advocate the use of surrogate prices in the form of **pollution** taxes and **discharge** fees. The non-priced aspect of the transaction then has a price, which sends a signal to the producers to economize on the use of the resource.

In addition to the theory of externalities, economists have recognized that certain goods, such as those provided by **nature**, are common property. Lacking a discrete owner, they are likely to be over-utilized. Ultimately, they will be depleted. Few will be left for **future generations**, unless common property goods like the air and water are protected.

Besides pollution taxes and discharge fees, economists have explored the use of marketable **emission** permits as a means of rectifying the market imperfection caused by pollution. Rather than establishing a unit charge for pollu-

tion, government would issue permits equivalent to an agreed-upon environmental standard. Holders of the permits would have the right to sell them to the highest bidder. The advantage of this system, wherein a market for pollution rights has been established, is that it achieves environmental quality standards. Under a charge system, trial-and-error tinkering would be necessary to achieve the standards.

Besides discharge fees and markets for pollution rights, economists have advocated the use of **cost-benefit analysis** in environmental decision making. Since control costs are much easier to measure than pollution benefits, economists have concentrated on how best to estimate the benefits of a clean **environment**. They have relied on two primary means of doing so. First, they have inferred from the actual decisions people make in the marketplace what value they place on a clean and healthy environment. Second, they have directly asked people to make trade-off choices. The inference method might rely on residential property values, decomposing the price of a house into individual attributes including **air quality**, or it might rely on the wage premium risky jobs enjoy. Despite many advances, the problem of valuing environmental benefits continues to be controversial with special difficulties surrounding the issues of quantifying the value of a human life, recreational benefits, and ecological benefits including **species** and **habitat** survival.

For instance, the question of how much a life is worth is repellent and absurd since human worth cannot be truly captured in monetary terms. Nonetheless it is important to determine the benefits for cost-benefit purposes. The costs of reducing pollution often are immediate and apparent, while the benefits are far-off and hard to determine. So, it is important to try to gauge what these benefits might be worth.

Economists call for a more rational ordering of risks. The funds for risk reduction are not limitless, and the costs keep mounting. Risks should be viewed in a detached and analytical way. Polls suggest that Americans worry most about such dangers as **oil spills**, **acid rain**, pesticides, **nuclear power**, and hazardous wastes, but scientific risk assessments show that these are only low or medium-level dangers. The greater hazards come from **radon, lead**, indoor air pollution, and fumes from **chemicals** such as **benzene** and formaldehyde. Radon, the odorless gas that naturally seeps up from the ground and is found in people's homes, causes as many as 20,000 lung **cancer** deaths per year, while **hazardous waste** dumps cause at most 500 cancer deaths. Yet the **Environmental Protection Agency** (EPA) spends over $6 billion a year to clean up hazardous waste sites while its spends only $100 million a year for radon protection. To test a home for radon costs about $25, and to clean it up if it is found contaminated costs $1,000. To make the entire national housing stock free from radon would cost a few billion dollars. In contrast, projected spending for cleaning up hazardous waste sites is likely to exceed $500 billion despite the fact that only about 11 percent of such sites pose a measurable risk to human health.

Greater rationality would mean that less attention would be paid to some risks and more attention to others. For instance, scientific **risk assessment** suggests that sizable new investments will be needed to address the dangers of **ozone layer depletion** and greenhouse warming. **Ozone** depletion is likely to result in 100,000 more cases of skin cancer by the year 2050. Global warming has the potential to cause massive catastrophe.

For businesses, risk assessment provides a way to allocate costs efficiently. They are increasingly using it as a management tool. To avoid another accident like **Bhopal, India**, Union Carbide has set up a system by which it rates its plants "safe," "made safer," or "shut down." Environmentalists, on the other hand, generally see risk assessment as a tactic of powerful interests used to prevent regulation of known dangers or permit building of facilities where there will be known fatalities. Even if the chances of someone contracting cancer and dying is only one in a million, still someone will perish, which the studies by **risk assessors** indeed document. Among particularly vulnerable groups of the population (allergy sufferers exposed to benzene for example) the risks are likely to be much greater, perhaps as great as one fatality for every 100 persons. Environmentalists conclude that the way economists present their findings is too conservative. By treating everyone alike, they overlook the real danger to particularly vulnerable people. Risk assessment should not be used as an excuse for inaction.

Environmentalists have also criticized environmental economics for its emphasis on economic growth without considering the unintended side-effects. Economists need to supplement estimates of the economic costs and benefits of growth with estimates of the effects of that growth that cannot be measured in economic terms. Many environmentalists also believe that the **burden of proof** should rest with new technologies, in that they should not be allowed simply because they advance material progress. In affluent societies especially, economic expansion is not necessary.

Growth is promoted for many reasons to restore the balance of payments, to make the nation more competitive, to create jobs, to reduce the deficit, to provide for the old and sick, and to lessen poverty. The public is encouraged to focus on **statistics** on productivity, balance of payments, and growth, while ignoring the obvious costs. Environmental groups, on the other hand, have argued for a **steady-state economy** in which population and per capita resource consumption stabilize. It is an economy with a constant number of people and goods, maintained at the lowest feasible flows of matter and energy. Human services would play a large

role in a steady-state economy because they do not require much energy or material throughput and yet contribute to economic growth. Environmental clean-up and **energy conservation** also would contribute, since they add to economic growth while also having a positive effect on the environment.

Growth can continue, according to environmentalists, but only if the forms of growth are carefully chosen. Free time, in addition, would have to be a larger component of an environmentally-acceptable future economy. Free time removes people from potentially harmful production. It also provides them with the time needed to implement alternative production processes and techniques, including **organic gardening**, **recycling**, public **transportation**, and home and appliance maintenance for the purposes of energy **conservation**.

Another requirement of an environmentally acceptable economy is that people accept a *new frugality*, a concept that also has been labeled *joyous austerity*, *voluntary simplicity*, and *conspicuous frugality*.

Economists represent the environment's interaction with the economy as a materials balance model. The production sector, which consists of mines and factories, extracts materials from nature and processes them into goods and services. Transportation and distribution networks move and store the finished products before they reach the point of consumption. The environment provides the material inputs needed to sustain economic activity and carries away the wastes generated by it. People have long recognized that nature is a source of material inputs to the economy, but they have been less aware that the environment plays an essential role as a receptacle for society's unwanted by-products. Some wastes are recovered by recycling, but most are absorbed by the environment. They are dumped in landfills, treated in incinerators, and disposed of as ash. They end up in the air, water, or **soil**.

The ultimate limits to economic growth do not come only from the availability of raw materials from nature. Nature's limited capacities to absorb wastes also set a limit on the economy's ability to produce. Energy plays a role in this process. It helps make food, forest products, chemicals, **petroleum** products, metals, and structural materials such as stone, steel, and cement. It supports materials processing by providing electricity, heating, and cooling services. It aids in transportation and distribution. According to the law of the conservation of energy, the material inputs and energy that enter the economy cannot be destroyed. Rather they change form, finding their way back to nature in a disorganized state as unwanted and perhaps dangerous by-products.

Environmentalists use the laws of physics (the notion of entropy) to show how society systematically dissipates low entropy, highly concentrated forms of energy by converting it to high entropy, little concentrated waste that cannot be used again except at very high cost. They project current resource use and **environmental degradation** into the future to demonstrate that civilization is running out of critical resources. The earth cannot tolerate additional contaminants. Human intervention in the form of technological innovation and capital investment complemented by substantial human ingenuity and creativity is insufficient to prevent this outcome unless drastic steps are taken soon. Nearly every economic benefit has an environmental cost, and the sum total of the costs in an affluent society often exceed the benefits. The notion of **carrying capacity** is used to show that the earth has a limited ability to tolerate the disposal of contaminants and the depletion of resources.

Economists counter these claims by arguing that limits to growth can be overcome by human ingenuity, that benefits afforded by environmental protection have a cost, and that government programs to clean up the environment are as likely to fail as the market forces that produce pollution. The traditional economic view is that production is a function of labor and capital and, in theory, that resources are not necessary since labor and/or capital are infinitely substitutable for resources. Impending resource **scarcity** results in price increases which lead to technological substitution of capital, labor, or other resources for those that are in scarce supply. Price increases also create pressures for efficiency-in-use, leading to reduced consumption. Thus, resource scarcity is reflected in the price of a given commodity. As resources become scarce, their prices rise accordingly. Increases in price induce substitution and technological innovation.

People turn to less scarce resources that fulfill the same basic technological and economic needs provided by the resources no longer available in large quantities. To a large extent, the energy crises of the 1970s (the 1973 price shock induced by the Arab **oil embargo** and 1979 price shock following the Iranian Revolution) were alleviated by these very processes: higher prices leading to the discovery of additional supply and to conservation. By 1985, energy prices in real terms were lower than they were in 1973.

Humans respond to signals about scarcity and degradation. Extrapolating past consumption patterns into the future without considering the human response is likely to be a futile exercise, economists argue. As far back as the end of the eighteenth century, thinkers such as Thomas Malthus have made predictions about the limits to growth, but the lesson of modern history is one of technological innovation and substitution in response to price and other societal signals, not one of calamity brought about by resource exhaustion. In general, the prices of **natural resources** have been declining despite increased production and demand. Prices have fallen because of discoveries of new resources and because of innovations in the extraction and refinement pro-

cess. *See also* Greenhouse effect; Trade in pollution permits; Tragedy of the Commons

[*Alfred A. Marcus*]

RESOURCES
BOOKS

Ekins, P., M. Hillman, and R. Hutchinson. *The Gaia Atlas of Green Economics*. New York: Doubleday, 1992.

Kneese, A., R. Ayres, and R. D'Arge. *Economics and the Environment: A Materials Balance Approach*. Washington, DC: Resources for the Future, 1970.

Marcus, A. A. *Business and Society: Ethics, Government, and the World Economy*. Homewood, IL: Irwin Publishing, 1993.

PERIODICALS

Cropper, M. L., and W. E. Oates. "Environmental Economics." *Journal of Economic Literature* (June 1992): 675-740.

Environmental education

Environmental education is fast emerging as one of the most important disciplines in the United States and in the world. Merging the ideas and philosophy of **environmentalism** with the structure of formal education systems, it strives to increase awareness of environmental problems as well as to foster the skills and strategies for solving those problems. Environmental issues have traditionally fallen to the state, federal, and international policymakers, scientists, academics, and legal scholars. Environmental education (often referred to simply as "EE") shifts the focus to the general population. In other words, it seeks to empower *individuals* with an understanding of environmental problems and the skills to solve them.

Background

The first seeds of environmental education were planted roughly a century ago and are found in the works of such writers as George Perkins Marsh, **John Muir**, **Henry David Thoreau**, and **Aldo Leopold**. Their writings served to bring the country's attention to the depletion of **natural resources** and the often detrimental impact of humans on the **environment**. In the early 1900s, three related fields of study arose that eventually merged to form the present-day environmental education.

Nature education expanded the teaching of biology, botany, and other natural sciences out into the natural world, where students learned through direct observation. **Conservation** education took root in the 1930s, as the importance of long-range, "wise use" management of resources intensified. Numerous state and federal agencies were created to tend public lands, and citizen organizations began forming in earnest to protect a favored animal, park, river, or other resource. Both governmental and citizen entities included an educational component to spread their message to the general public. Many states required their schools to adopt conservation education as part of their curriculum. Teacher training programs were developed to meet the increasing demand. The Conservation Education Association formed to consolidate these efforts and help solidify citizen support for natural resource management goals. The third pillar of modern EE is outdoor education, which refers more to the method of teaching than to the subject taught. The idea is to hold classrooms outdoors; the topics are not restricted to environmental issues but includes art, music, and other subjects.

With the burgeoning of industrial output and natural resource depletion following World War II, people began to glimpse the potential environmental disasters looming ahead. The environmental movement exploded upon the public agenda in the late 1960s and early 1970s, and the public reacted emotionally and vigorously to isolated environmental crises and events. Yet it soon became clear that the solution would involve nothing short of fundamental changes in values, lifestyles, and individual behavior, and that would mean a comprehensive educational approach.

In August 1970, the newly-created **Council on Environmental Quality** called for a thorough discussion of the role of education with respect to the environment. Two months later, Congress passed the Environmental Education Act, which called for EE programs to be incorporated in all public school curricula. Although the act received little funding in the following years, it energized EE proponents and prompted many states to adopt EE plans for their schools. In 1971, the National Association for Environmental Education formed, as did myriad of state and regional groups.

Definition

What EE means depends on one's perspective. Some see it as a teaching method or philosophy to be applied to all subjects, woven into the teaching of political science, history, economics, and so forth. Others see it as a distinct discipline, something to be taught on its own. As defined by federal statute, it is the "education process dealing with people's relationships and their natural and manmade surroundings, and includes the relation of population, **pollution**, resource allocation and depletion, conservation, **transportation**, technology and urban and rural planning to the total human environment."

One of the early leaders of the movement is William Stapp, a former professor at the University of Michigan's School of Natural Resources and the Environment. His three-pronged definition has formed the basis for much subsequent thought: "Environmental education is aimed at producing a citizenry that is *knowledgeable* concerning the biophysical environment and its associated problems, *aware*

of how to help solve these problems, and *motivated* to work toward their solution."

Many environmental educators believe that programs covering kindergarten through twelfth grade are necessary to successfully instill an environmental ethic in students and a comprehensive understanding of environmental issues so that they are prepared to deal with environmental problems in the real world. Further, an emphasis is placed on problem-solving, action, and informed behavioral changes. In its broadest sense, EE is not confined to public schools but includes efforts by governments, interest groups, universities, and news media to raise awareness. Each citizen should understand the environmental issues of his or her own community: land-use planning, traffic congestion, economic development plans, **pesticide** use, **water pollution** and **air pollution**, and so on.

International level

Concurrently with the emergence of EE in this country, other nations began pushing for a comprehensive approach to environmental problems within their own borders and on a global scale. In 1972, at the **United Nations Conference on the Human Environment** in Stockholm, the need for an international EE effort was clearly recognized and emphasized. Three years later, an International Environmental Education Workshop was held in Belgrade, from which emerged an eloquent, urgent mandate for the drastic reordering of national and international development policies. The "Belgrade Charter" called for an end to the military arms race and a new global ethic in which "no nation should grow or develop at the expense of another nation." It called for the eradication of poverty, hunger, illiteracy, pollution, exploitation, and domination. Central to this impassioned plea for a better world was the need for environmental education of the world's youth. That same year, the UN approved a $2 million budget to facilitate the research, coordination, and development of an international EE program among dozens of nations.

Effectiveness

There has been criticism over the last 15 years that EE too often fails to educate students and makes little difference in their behavior concerning the environment. Researchers and environmental educators have formulated a basic framework for how to improve EE: 1) Reinforce individuals for positive environmental behavior over an extended period of time. 2) Provide students with positive, informal experiences outdoors to enhance their "environmental sensitivity." 3) Focus instruction on the concepts of "ownership" and "empowerment." The first concept means that the learner has some personal interest or investment in the environmental issues being discussed. Perhaps the student can relate more readily to concepts of **solid waste** disposal if there is a **landfill** in the neighborhood. Empowerment gives learners the sense that they can make changes and help resolve environmental problems. 4) Design an exercise in which students thoroughly investigate an environmental issue and then develop a plan for citizen action to address the issue, complete with an analysis of the social, cultural, and ecological consequences of the action.

Despite the efforts of environmental educators, the movement has a long way to go. The scope and number of critical environmental problems facing the world today far outweigh the successes of EE. Further, most countries still do not have a comprehensive EE program that prepares them, as future citizens, to make ecologically sound choices and to participate in cleaning up and caring for the environment. Lastly, educators, including the media, are largely focused on explaining the problems but fall short on explaining or offering possible solutions. The notion of "empowerment" is often absent.

Recent developments and successes in the United States

Project WILD, based in Boulder, Colorado, is a K–12 supplementary conservation and environmental education program emphasizing **wildlife** protection, sponsored by fish and wildlife agencies and environmental educators. The project sets up workshops in which teachers learn about wildlife issues. They in turn teach children and help students understand how they can act responsibly on behalf of wildlife and the environment. The program, begun in 1983, has grown tremendously in terms of the number of educators reached and the monetary support from states, which, combined, are spending about $3.6 million annually.

The Global Rivers Environmental Education Network (GREEN), begun at the University of Michigan under the guidance of William Stapp, has likewise been enormously successful, perhaps more so. Teachers all over the world take their students down to their local river and show them how to monitor **water quality**, analyze **watershed** usage, and identify socioeconomic sources of river degradation. Lastly, and most importantly, the students then present their findings and recommendations to the local officials. These students also exchange information with other GREEN students around the world via computers.

Another promising development is the National Consortium for Environmental Education and Training (NCEET), also based at the University of Michigan. The partnership of academic institutions, non-profit organizations, and corporations, NCEET was established in 1992 with a three-year, $4.8 million grant from the **Environmental Protection Agency** (EPA). Its main purpose is to dramatically improve the effectiveness of environmental education in the United States. The program has attacked its mission from several angles: to function as a national clearinghouse for K–12 teachers, to make available top-quality

EE materials for teachers, to conduct research on effective approaches to EE, to survey and assess the EE needs of all 50 states, to establish a computer network for teachers needing access to information and resources, and to develop a teacher training manual for conducting EE workshops around the country.

[*Cathryn McCue*]

RESOURCES
BOOKS

Gerston, R. *Just Open the Door: A Complete Guide to Experiencing Environmental Education.* Danville, IL: Interstate Printers and Publishers, 1983.
Swan, M. "Forerunners of Environmental Education." In *What Makes Education Environmental?*, edited by N. McInnis and D. Albrect. Louisville, KY: Data Courier and Environmental Educators, 1975.

PERIODICALS

Hungerford, H. R., and T. L. Volk. "Changing Learner Behavior Through Environmental Education." *Journal of Environmental Education* 21 (Spring 1990): 8-21.
"The Belgrade Charter." *Connect: Unesco-UNEP Environmental Education Newsletter* 1 (January 1976).

Environmental enforcement

Environmental enforcement is the set of actions that a government takes to achieve full implementation of environmental requirements (compliance) within the regulated community and to correct or halt situations or activities that endanger the **environment** or public health. Experience with environmental programs has shown that enforcement is essential to compliance because many people and institutions will not comply with a law unless there are clear consequences for noncompliance. Enforcement by the government usually includes inspections to determine the compliance status of the regulated community and to detect violations; negotiations with individuals or facility managers who are out of compliance to develop mutually agreeable schedules and approaches for achievement of compliance; legal action when necessary to compel compliance and to impose some consequences for violation of the law or for posing a threat to public health and the environment; and compliance promotion, such as educational programs, technical assistance, and subsidies, to encourage voluntary compliance.

Nongovernmental groups may become involved in enforcement by detecting noncompliance, negotiating with violators, and commenting on governmental enforcement actions. They may also, if the law allows, take legal actions against a violator for noncompliance or against the government for not enforcing environmental requirements. The banking and insurance industries may be indirectly involved with enforcement by requiring assurance of compliance with environmental requirements before they issue a loan or an insurance policy to a facility. Strong social sanctions for noncompliance with environmental requirements can also be effective to ensure compliance. For example, the public may choose to boycott a product if they believe the manufacturer is harming the environment.

Environmental enforcement is based on environmental laws. An **environmental law** provides the vision, scope, and authority for environmental protection and restoration. Some environmental laws contain requirements while others specify a structure and criteria for establishing requirements, which are then developed separately. Requirements may be general, in which they apply to a group of facilities, or facility-specific.

Examples of environmental enforcement programs include those that govern the ambient environment, performance, technology, work practices, dissemination of information and product or use bans.

Ambient standards (media quality standard) are goals for the quality of the ambient environment (that is, air and **water quality**). Ambient standards are usually written in units of concentration, and they are used to plan the levels of emissions that can be accommodated from individual sources while still meeting an area-wide goal. Ambient standards can also be used as triggers, i.e., when a standard is exceeded, monitoring or enforcement efforts are increased. Enforcement of these standards involves relating an ambient measurement to emissions or activities at a specific facility, which can be difficult.

Performance standards, widely used for regulations, permits, and monitoring requirements, limit the amount or rate of particular **chemicals** or discharges that a facility can release into the environment in a given period of time. These standards allow sources to choose which technologies they will use to meet the standards. Performance standards are often based on output that can be achieved by using the **best available control technology**. Some standards allow a source with multiple emissions to vary its emissions from each stack as long as the total sum of emissions does not exceed the permitted total. Compliance with **emission standards** is accomplished by sampling and monitoring, which in some cases may be difficult and/or expensive.

Technology standards require the regulated community to use a particular type of technology (i.e., "best available technology") to control and/or monitor emissions. Technology standards are effective if the equipment specified is known to perform well under the range of conditions experienced by the source. Compliance is measured by whether the equipment is installed and operating properly. However, proper operation over a long period of time is more difficult to monitor. The use of technology standards can inhibit technological innovation.

Practice standards require or prohibit work activities that may have environmental impacts (e.g., prohibition of carrying hazardous liquids in uncovered containers). Regulators can easily inspect for compliance and take action against noncomplying sources, but ongoing compliance is not easy to ensure.

Dissemination of information and product or use bans are also governed by environmental enforcement programs. Information standards require a source of potential **pollution** (e.g., a manufacturer or facility involved in generating, transporting, storing, treating, and disposing of hazardous wastes) to develop and submit information to the government. For example, a source generating pollution may be required to monitor, maintain records, and report on the level of pollution generated and whether or not the source exceeds performance standards. Information requirements are also used when a potential pollution source is a product such as a new chemical or **pesticide**. The manufacturer may be required to test and report on the potential of the product to cause harm if released into the environment. Finally, product or use bans are used to prohibit a product (i.e., ban the manufacture, sale, and/or use of a product) or they may prohibit particular uses of a product.

An effective environmental law should include the authority or power necessary for its own enforcement. An effective authority should govern implementation of environmental requirements, inspection, and monitoring of facilities, and legal sanctions for noncompliance. One type of authority that is used is guidance for the implementation of environmental laws by issuance of regulations, permits, licenses, and/or guidance policies. Regulations establish, in greater detail than is specified by law, general requirements that must be met by the regulated community. Some regulations are directly enforced while others provide criteria and procedures for developing facility-specific requirements utilizing permits and licenses to provide the basis of enforcement. Permits are used to control activities related to construction or operation of facilities that generate pollutants. Requirements in permits are based on specific criteria established in laws, regulations, and/or guidance. General permits specify what a class of facilities is required to do, while a facility-specific permit specifies requirements for a particular facility, often taking into account the conditions there. Licenses are permits to manufacture, test, sell, and/or distribute a product that may pose an environmental or public health risk if improperly used. Licenses may be general or facility-specific. Written guidance and policies, which are prepared by the regulator, are used to interpret and implement requirements to ensure consistency and fairness. Guidance may be necessary because not all applications of requirements can be anticipated, or when regulation is achieved by the use of facility-specific permits or licenses.

Authority is also required to provide for inspection and monitoring of facilities, with legal sanctions for noncompliance. Requirements may either be waived or prepared for facility-specific conditions. The authority will inspect regulated facilities and gain access to their records and equipment to determine if they are in compliance.

Authority is necessary to ensure that the regulated community monitors its own compliance, maintains records of its compliance activities and status, reports this information periodically to the enforcement program, and provides information during inspections.

An effective law should also include the authority to take legal action against noncomplying facilities, imposing a range of monetary penalties and other sanctions on facilities that violate the law, as well as criminal sanctions on those facilities or individuals who deliberately violate the law (e.g., facilities that knowingly falsify data). Also, power should be granted to correct situations that pose an immediate and substantial threat to public health and/or the environment.

The range and types of environmental enforcement response mechanisms available depend on the number and types of authorities provided to the enforcement program by environmental and related laws. Enforcement mechanisms may be designed to return violators to compliance, impose a sanction, or remove the economic benefits of noncompliance. Enforcement may require that specific actions be taken to test, monitor, or provide information. Enforcement may also correct environmental damages and modify internal company management problems.

Enforcement response mechanisms include informal responses such as phone calls, site visits and inspections, warning letters, and notices of violations, which are more formal than warning letters. They provide the facility manager with a description of the violation, what should be done to correct it, and by what date. Informal responses do not penalize but can lead to more severe responses if ignored. The more formal enforcement mechanisms are backed by law and are accompanied by procedural requirements to protect the rights of the individual. Authority to use formal enforcement mechanisms for a specific situation must be provided in the applicable environmental law. Civil administrative orders are legal, independently enforceable orders issued directly by enforcement program officials that define the violation, provide evidence of the violation, and require the recipient to correct the violation within a specified time period. If the recipient violates the order, program managers can take further legal action using additional orders or the court system to force compliance with the order.

Further legal action includes the use of field citations, which are administrative orders issued by inspectors in the field. They require the violator to correct a clear-cut violation and pay a small monetary fine. Field citations are used to handle more routine types of violations that do not pose a major threat to the environment. Legal action may also lead to civil judicial enforcement actions, which are formal lawsuits before the courts. These actions are used to require action to reduce immediate threats to public health or the environment, to enforce administrative orders that have been violated, and to make final decisions regarding orders that have been appealed. Finally, a criminal judicial response is used when a person or facility has knowingly and willfully violated the law or has committed a violation for which society has chosen to impose the most serious legal sanctions available. This response involves criminal sanction, which may include monetary penalties and imprisonment. The criminal response is the most difficult type of enforcement, requiring intensive investigation and case development, but it can also create a significant deterrence.

Environmental enforcement must include processes to balance the rights of individuals with the government's need to act quickly. A notice of violation should be issued before any action is taken so that the finding of violation can be contested, or so that the violation can be corrected before further government action. Appeals should be allowed at several stages in the enforcement process so that the finding of violation, the required remedial action, or the severity of the proposed sanction can be reviewed. There should also be dispute resolution processes for negotiations between program officials and the violator, which may include face-to-face discussions, presentations before a judge or hearing examiner, or use of third party mediators, arbitrators, or facilitators.

[*Judith Sims*]

RESOURCES
BOOKS

Wasserman, C. E. *Principles of Environmental Enforcement.* Washington, D.C.: U.S. Environmental Protection Agency, 1992.

Environmental engineering

The development of environmental engineering as a discipline is a reflection of the modern need to maintain public health by providing safe drinking water and **sanitation**, and by treating and disposing of sewage, **municipal solid waste**, and **pollution**. Originally, sanitary engineering, a limited subdiscipline of civil engineering, performed some of these functions. But with the growth of concern for protecting the **environment** and the passage of laws regulating disposal of wastes, environmental engineering has grown into a discrete discipline encompassing a wide range of activities including: "proper disposal or **recycling** of **wastewater** and solid wastes, adequate **drainage** of urban and rural areas for proper sanitation, control of water, **soil** and atmospheric pollution and the social and environmental impact of these solutions." Education for environmental engineers requires that they be "well informed concerning engineering problems in the field of public health, such as control of insect-borne diseases, the elimination of industrial health hazards, and the provision of adequate sanitation in urban, rural and recreational areas, and the effect of technological advance on the environment." More broadly environmental engineering is defined by W. E. Gilbertson as "that branch of engineering which is concerned with the application of scientific principles to (1) the protection of human populations from the effects of adverse environmental factors, (2) the protection of environments, both local and global, from the potentially deleterious effects of human activities, and (3) the improvement of environmental quality of man's health and well-being."

The American Academy of Environmental Engineers (AAEE) has defined environmental engineering as "the application of engineering principles to the management of the environment for the protection of human health; for the protection of nature's beneficial ecosystems and for environment-related enhancement of the quality of human life."

Degree-granting institutions in the United States do not necessarily consider environmental engineering as a separate discipline. A report by the U. S. Engineering Manpower Commission found that only 192 baccalaureate environmental engineering degrees were granted in 1988; however C. Robert Baillod estimates that at least 10% of the 8,800 annual graduates from baccalaureate civil engineering programs are educated to function as environmental engineers. If similar estimates are made for chemical, mechanical, geological and other engineers who function as environmental engineers, 1,000–2,000 graduates are entering the profession each year. Data collected by Baillod indicate that the supply of environmental engineers will satisfy half the demand for 2,000–5,000 new environmental engineering graduates per year for the next decade.

From 1970 to 1985, an increasing number of environmental statutes passed at the federal and state level in the United States while parallel legislation was being established internationally to regulate and control environmental pollution. In the United States, the establishment of the **Comprehensive Environmental Response, Compensation and Liability Act** (CERCLA)—known as *Superfund* for short—has provided the impetus for significant activity in **remediation** as well as providing industries and municipalities with incentives (such as liability for environmental damage) to clean up and avoid pollution. In order to comply with environmental laws and also to maintain good business practice, corporations are including impacts on the environment in planning for their process engineering. A serious potential

for fines or costly liability suits exists if the design of processes is not carefully conducted with environmental safeguards.

In addition to employment by the private sector, state and federal governments also employ environmental engineers. The primary function of environmental engineers in government is research and development for implementation of regulations and their enforcement. At the federal level, agencies such as the **Environmental Protection Agency** (EPA), as well as the Departments of Commerce, Energy, Interior and Agriculture, employ environmental engineers.

Environmental engineering is proving crucial in addressing an array of environmental needs. Techniques recently conceived by environmental engineers include:

- The development of an oil-absorbing, floating sponge-like material as a re-usable first response material for remediating **oil spills** on open bodies of water.
- Design and operation of a plant to process electrolytic plating wastes collected from a large urban area to **reuse** the metals and detoxify the cyanide as a means of avoiding the discharges of these wastes into the sewer system.
- Development of a process for reusing the **lead**, zinc, and **cadmium** which would otherwise be lost as a fume in the remelting of automotive scrap to form steel products.
- Use of naturally-occurring bacterial agents in cleanup of underground aquifers contaminated by prior discharges of creosote, a substance used to preserve wood products.
- Development of processes for removal of and destruction of PCBs and other hazardous organic agents from spills into soil.
- Development of sensing techniques which enable tracing of pollution to point sources and the determination of the degree of pollution which has occurred for application of legal remedies.
- Development of process design and control instrumentation in nuclear reactors to prevent, contain and avoid nuclear releases.
- Design and development of **feedlots** for animals in which the waste products are made into reusable agricultural products.
- Development of sterilization, **incineration** and gas cleanup systems for treatment of hospital wastes.
- Certification of properties to verify the absence of factors which would make new owners liable to environmental litigation (for example, absence of **asbestos**, absence of underground storage tanks for hazardous material)s.
- Redesign of existing chemical plants to recycle or eliminate waste streams.
- Development of processes for recycling wastes (for example, processes for de-inking and reuse of newsprint or reuse of plastics).

These wide-ranging examples are typical of the solutions which are being developed by a new generation of technically-trained individuals. In an increasingly populated and industrialized world, environmental engineers will continue to play a pivotal role in devising technologies needed to minimize the impact of humans on the earth's resources.

[*Malcolm T. Hepworth*]

RESOURCES
BOOKS

American Academy of Environmental Engineers. AAEE Bylaws. Annapolis, Maryland: 1990.

Cartledge, B., ed. *Monitoring the Environment.* New York: Oxford University Press, 1992.

Corbitt, Robert A.: *Standard Handbook of Environmental Engineering.* New York: McGraw-Hill, 1989.

Jacobsen, J., ed. *Human Impact on the Environment: Ancient Roots, Current Challenges.* Boulder, CO: Westview Press, 1992.

PERIODICALS

Crucil, C. "Environmentally Sound Buildings Now Within Reach." *Alternatives* 19 (January-February 1993): 9-10.

OTHER

Gilbertson, W. E. "Environmental Quality Goals and Challenges." *Proceedings of the Third National Environmental Engineering Education Conference* edited by P. W. Purdon. American Academy of Environmental Engineers and the Association of Environmental Engineering Professors, Drexel University, 1973.

Environmental estrogens

The United States **Environmental Protection Agency (EPA)** defines an environmental endocrine disruptor—the term the Agency uses for environmental estrogens—as "an exogenous agent that interferes with the synthesis, secretion, transport, binding, action, or elimination of natural hormones in the body that are responsible for the maintenance of **homeostasis**, reproduction, development, and/or behavior." Dr. Theo Colborn, a zoologist and senior scientist with the **World Wildlife Fund**, and the person most credited with raising national awareness of the issue, describes these **chemicals** as "hand-me-down poisons" that are passed from mothers to offspring and may be linked to a wide range of adverse effects, including low sperm counts, infertility, genital deformities, breast and prostate **cancer**, neurological disorders in children such as hyperactivity and attention deficits, and developmental and reproductive disorders in **wildlife**. Colborn discusses these effects in her 1996 book, *Our Stolen Future*—co-authored with Dianne Dumanoski and John Peterson Myers—which asks: "Are we threatening our fertility, intelligence, and survival?" Some other names used for the same class of chemicals are hormone disruptors, estrogen

mimics, endocrine disrupting chemicals, and endocrine modulators.

While EPA takes the position that it is "aware of and concerned" about data indicating that exposure to environmental **endocrine disruptors** may cause adverse impacts on human health and the **environment**, the Agency at present does not consider endocrine disruption to be "an adverse endpoint *per se.*" Rather, it is "a mode or mechanism of action potentially leading to other outcomes"—such as the health effects Colborn described drawing from extensive research of numerous scientists—but, in EPA's view, the link to human health effects remains an unproven hypothesis. For Colborn and a significant number of other scientists, however, enough is known to support prompt and far-reaching action to reduce exposures from these chemicals and myriad products that are manufactured using them. Foods, plastic packaging, and pesticides are among the sources of exposure Colborn raises concerns about in her book.

Ultimately, the environmental estrogens issue is about whether these chemicals are present in the environment at high enough levels to disrupt the normal functioning of wildlife and human endocrine systems and thereby cause harmful effects. The endocrine system is one of at least three important regulatory systems in humans and other animals (the nervous and immune systems are the other two) and includes such endocrine glands as the pituitary, thyroid, pancreas, adrenal, and the male and female gonads, or testes and ovaries. These glands secrete hormones into the bloodstream where they travel in very small concentrations and bind to specific sites called "cell receptors" in target tissues and organs. The hormones affect development, reproduction, and other bodily functions. The term "endocrine disruptors" includes not only estrogens but also antiandrogens and other agents that act on the endocrine system.

The question of whether environmental endocrine disruptors may be causing effects in humans has arisen over the past decade based on a growing body of evidence about effects in wildlife exposed to dichlorodiphenyl-trichlorethane (DDT), **polychlorinated biphenyls** (PCBs), and other chemicals. For instance, field studies have proven that tributyltin (TBT), which is used as an antifouling paint on ships, can cause "imposex" in female snails, which are now commonly found with male genitalia, including a penis and vas deferens, the sperm-transporting tube. TBT has also been shown to cause decreased egg production by the periwinkle (*Littorina littorea*). As early as 1985, concerns arose among scientists and the public in the United Kingdom over the effects of synthetic estrogens from birth control pills entering rivers, a concern that was heightened when anglers reported catching fish with both male and female characteristics. Other studies have found **Great Lakes salmon** to invariably have thyroids that were abnormal in appearance, even when there were no overt goiters. Herring gulls (*Larus argentatus*) throughout the Great Lakes have also been found with enlarged thyroids. In the case of the salmon and gulls, no agent has been determined to be causing these effects. But other studies have linked DDT exposure in the Great Lakes to eggshell thinning and breakage among bald eagles and other birds. In Lake Apopka, Florida, male alligators (*Alligator mississippiensis*) exposed to a mixture of dicofol, DDT, and dichlorodiphenyldichloroethylene (DDE) have been "demasculinized," with phalluses one-half to one-fourth the normal size. Red-eared turtles (*Trachemys scripta*) in the lake have also been demasculinized. One 1988 study reported that four of 15 female black bears (*Ursus americanus*) and one of four female brown bears (*Ursus arctos*) had, to varying degrees, male sex organs. These and nearly 300 other peer-reviewed studies have led EPA—in conjunction with the multi-agency White House Committee on Environment and Natural Resources—to develop a "framework for planning" and an extensive research agenda to answer questions about the effects of endocrine disruptors. The goal is to better understand the potential effects of such chemicals on human beings before implementing regulatory actions.

The federal research agenda has been evolving through a series of workshops. As early as 1979, the **National Institute of Environmental Health Sciences** (NIEHS), based in Research Triangle Park, North Carolina, held an "Estrogens in the Environment" conference to evaluate the chemical properties and diverse structures among environmental estrogens. NIEHS held a second conference in 1985 that addressed numerous potential toxicological and biological effects from exposure to these chemicals. NIEHS's third conference, held in 1994, focused on detrimental effects in wildlife. At an April 1995 EPA-sponsored workshop on "Research Needs for the **Risk Assessment** of Health and Environmental Effects of Endocrine Disruptors," a number of critical research questions were discussed: What do we know about the carcinogenic effects of endocrine-disrupting agents in humans and wildlife? What are the research needs in this area, including the highest priority research needs? Similar questions were discussed for reproductive effects, neurological effects, immunological effects, and a variety of risk assessment issues. Drawing on the preceding conferences and workshops, in February 1997 EPA issued a *Special Report on Environmental Endocrine Disruption: An Effects Assessment and Analysis* that recommended key research needs to better understand how environmental endocrine disruptors may be causing the variety of specific effects in human beings and wildlife hypothesized by some scientists. For instance, male reproductive research should include tests that evaluate both the quantity and quality of sperm produced. Furthermore, when testing the endocrine-disrupting potential of chemicals, it is important to test for both estrogenic and

antiandrogenic activity because new data suggest that it is possible the latter—antiandrogenic activity—not estrogenic activity, is causing male reproductive effects. In the area of ecological research, EPA's special report highlighted the need for research on such issues as what chemicals or class of chemicals can be considered genuine endocrine disruptors and what dose is needed to cause an effect.

Even before environmental estrogens received a place on the federal environmental agenda as a priority concern, Colborn and other scientists first met in July 1991 in Racine, Wisconsin, to discuss their misgivings about the prevalence of estrogenic chemicals in the environment. From that meeting came the landmark "Wingspread Consensus Statement" of 21 leading researchers. The statement asserted that the scientists were certain that a large number of human-made chemicals that have been released into the environment, as well as a few natural ones, "have the potential to disrupt the endocrine system of animals, including humans," and that many wildlife populations are already affected by these chemicals. Furthermore, the scientists expressed certainty that the effects may be entirely different in the embryo, fetus, or perinatal organisms than in the adult; that effects are more often manifested in offspring than in exposed parents; that the timing of exposure in the developing organism is crucial; and that, while embryonic development is the critical exposure period, "obvious manifestations may not occur until maturity." Besides these and other "certain" conclusions, the scientists estimated with confidence that "some of the developmental impairments reported in humans today are seen in adult offspring of parents exposed to synthetic hormone disruptors (agonists and antagonists) released in the environment" and that "unless the environmental load of synthetic hormone disruptors is abated and controlled, large scale dysfunction at the population level is possible." The Wingspread Statement included numerous other consensus views on what models predict and the judgment of the group on the need for much greater research and a comprehensive inventory of these chemicals.

The Food Quality Protection Act of 1996 (FQPA) and the **Safe Drinking Water Act** Amendments of 1996 require EPA to develop a screening program to determine whether pesticides or other substances cause effects in humans similar to effects produced by naturally occurring estrogens and other endocrine effects. The FQPA requires **pesticide** registrants to test their products for such effects and submit reports, and it requires that registrations be suspended if registrants fail to comply. Besides the EPA screening program, the **United Nations Environment Programme** is pursuing a multinational effort to manage "persistent organic pollutants," including DDT and PCBs, which, though banned in the United States, are still used elsewhere and can persist in the environment and be transported long-distance. In February 1997, Illinois became the first state to issue a strategy for endocrine disruptors that requires every Illinois EPA program to assess its current activities affecting these chemicals and to begin monitoring a list of known, probable, and suspected chemicals in case further action is needed in the future.

[*David Clarke*]

RESOURCES
BOOKS

Colborn, T., D. Dumanoski, and J. P. Myers. *Our Stolen Future*. New York: Penguin Books, 1996.
"Estrogens in the Environment." *Environmental Health Perspectives Supplements* 3, supplement 7. North Carolina: Research Triangle Park, 1995.
National Science and Technology Council. Committee on Environment and Natural Resources. *The Health and Ecological Effects of Endocrine Disrupting Chemicals, A Framework for Planning*. Washington, D.C., 1996.
U.S. Environmental Protection Agency. *Special Report on Environmental Endocrine Disruption: An Effects Assessment and Analysis*. (EPA/630/R-96/012). Washington, D.C.: GPO, 1997.

Environmental ethics

Ethics is a branch of philosophy that deals with morals and values. Environmental ethics refers to the moral relationships between humans and the natural world. It addresses such questions as, do humans have obligations or responsibilities toward the natural world, and if so, how are those responsibilities balanced against human needs and interests? Are some interests more important than others?

Efforts to answer such ethical questions have led to the development of a number of schools of ethical thought. One of these is **utilitarianism**, a philosophy associated with the English eccentric Jeremy Bentham and later modified by his godson John Stuart Mill. In its most basic terms, utilitarianism holds that an action is morally right if it produces the greatest good for the greatest number of people. The early environmentalist **Gifford Pinchot** was inspired by utilitarian principles and applied them to **conservation**. Pinchot proposed that the purpose of conservation is to protect **natural resources** to produce "the greatest good for the greatest number for the longest time." Although utilitarianism is a simple, practical approach to human moral dilemmas, it can also be used to justify reprehensible actions. For example, in the nineteenth century many white Americans believed that the extermination of native peoples and the appropriation of their land was the right thing to do. However, most would now conclude that the good derived by white Americans from these actions does not justify the genocide and displacement of native peoples.

The tenets of utilitarian philosophy are presented in terms of human values and benefits, a clearly anthropocentric

world view. Many philosophers argue that only humans are capable of acting morally and of accepting responsibility for their actions. Not all humans, however, have this capacity to be moral agents. Children, the mentally ill, and others are not regarded as moral agents, but, rather, as moral subjects. However, they still have rights of their own—rights that moral agents have an obligation to respect. In this context, moral agents have **intrinsic value** independent of the beliefs or interests of others.

Although humans have long recognized the value of non-living objects, such as machines, minerals, or rivers, the value of these objects is seen in terms of money, aesthetics, cultural significance, etc. The important distinction is that these objects are useful or inspiring to some person—they are not ends in themselves but are means to some other end. Philosophers term this instrumental value, since these objects are the instruments for the satisfaction of some other moral agent. This philosophy has also been applied to living things, such as domestic animals. These animals have often been treated as simply the means to some humanly-desired end without any inherent rights or value of their own.

Aldo Leopold, in his famous essay on environmental ethics, pointed out that not all humans have been considered to have inherent worth and intrinsic rights. As examples he points to children, women, foreigners, and indigenous peoples—all of whom were once regarded as less than full persons; as objects or the property of an owner who could do with them whatever he wished. Most civilized societies now recognize that all humans have intrinsic rights, and, in fact, these intrinsic rights have also been extended to include such entities as corporations, municipalities, and nations.

Many environmental philosophers argue that we must also extend recognition of inherent worth to all other components of the natural world, both living and non-living. In their opinion, our anthropocentric view, which considers components of the natural world to be valuable only as the means to some human end, is the primary cause of **environmental degradation**. As an alternative, they propose a biocentric view which gives inherent value to all the natural world regardless of its potential for human use.

Paul Taylor outlines four basic tenets of biocentrism in his book, *Respect for Nature*. These are: 1) Humans are members of earth's living community in the same way and on the same terms as all other living things; 2) Humans and other **species** are interdependent; 3) Each organism is a unique individual pursuing its own good in its own way; 4) Humans are not inherently superior to other living things. These tenets underlie the philosophy developed by Norwegian Arne Naess known as **deep ecology**.

From this biocentric philosophy Paul Taylor developed three principles of ethical conduct: 1) Do not harm any natural entity that has a good of its own; 2) Do not try to manipulate, control, modify, manage or interfere with the normal functioning of natural ecosystems, biotic communities, or individual wild organisms; 3) Do not deceive or mislead any animal capable of being deceived or misled. These principles led Professor Taylor to call for an end to **hunting**, fishing and **trapping**, to espouse **vegetarianism**, and to seek the exclusion of human activities from **wilderness** areas. However, Professor Taylor did not extend intrinsic rights to non-living natural objects, and he assigned only limited rights to plants and domestic animals. Others argue that all natural objects, living or not, have rights.

Regardless of the appeal that certain environmental philosophies may have in the abstract, it is clear that humans must make use of the natural world if they are to survive. They must eat other organisms and compete with them for all the essentials of life. Humans seek to control or eliminate harmful plants or animals. How is this intervention in the natural world justified? Stewardship is a principle that philosophers use the justify such interference. Stewardship holds that humans have a unique responsibility to care for domestic plants and animals and all other components of the natural world. In this view, humans, their knowledge, and the products of their intellect are an essential part of the natural world, neither external to it nor superfluous. Stewardship calls for humans to respect and cooperate with **nature** to achieve the greatest good. Because of their superior intellect, humans can improve the world and make it a better place, but only if they see themselves as an integral part of it.

Ethical dilemmas arise when two different courses of action each have valid ethical underpinnings. A classic ethical dilemma occurs when any course of action taken will cause harm, either to oneself or to others. Another sort of dilemma arises when two parties have equally valid, but incompatible, ethical interests. To resolve such competing ethical claims Paul Taylor suggests five guidelines: 1) it is usually permissible for moral agents to defend themselves; 2) basic interests, those interests necessary for survival, take precedence over other interests; 3) when basic interests are in conflict, the least amount of harm should be done to all parties involved; 4) whenever possible, the disadvantages resulting from competing claims should be borne equally by all parties; 5) the greater the harm done to a moral agent, the great is the compensation required.

Ecofeminists do not find that utilitarianism, biocentrism or stewardship provide adequate direction to solve environmental problems or to guide moral actions. In their view, these philosophies come out of a patriarchal system based on domination—of women, children, minorities and nature. As an alternative, ecofeminists suggest a pluralistic, relationship-oriented approach to human interactions with the **environment. Ecofeminism** is concerned with nurturing, reciprocity, and connectedness, rather than with rights, re-sponsibilities, and ownership. It challenges humans to see themselves as related to others and to nature. Out of these connections, then, will flow ethical interactions among individuals and with the natural world. *See also* Animal rights; Bioregionalism; Callicott, J. Baird; Ecojustice; Environmental racism; Environmentalism; Future generations; Humanism; Intergenerational justice; Land stewardship; Rolston, Holmes; Speciesism

[*Christine B. Jeryan*]

RESOURCES
BOOKS

Devall, B., and G. Sessions. *Deep Ecology.* Layton, UT: Gibbs M. Smith, 1985.
Odell, R. *Environmental Awakening: The New Revolution to Protect the Earth.* Cambridge, MA: Ballinger, 1980.
Olson, S. *Reflections From the North Country.* New York: Knopf, 1980.
Plant, J. *Healing the Wounds: The Promise of Ecofeminism.* Santa Cruz, CA: New Society Publishers, 1989.
Rolston, H. *Environmental Ethics.* Philadelphia, Temple University Press, 1988.
Taylor, P. *Respect for Nature.* Princeton, NJ: Princeton University Press, 1986.

Environmental health

Environmental health is concerned with the medical effects of **chemicals**, pathogenic (disease-causing) organisms, or physical factors in our **environment**. Because our environment affects nearly every aspect of our lives in some way or other, environmental health is related to virtually every branch of medical science. The special focus of this discipline, however, tends to be health effects of polluted air and water, contaminated food, and toxic or hazardous materials in our environment. Concerns about these issues makes environmental health one of the most compelling reasons to be interested in **environmental science**.

For a majority of humans, the most immediate environmental health threat has always been pathogenic organisms. Improved **sanitation**, nutrition, and modern medicine in the industrialized countries have reduced or eliminated many of the **communicable diseases** that once threatened us. But for people in the **less developed countries** where nearly 80% of the world's population lives, bacteria, viruses, **fungi, parasites**, worms, flukes, and other infectious agents remain major causes of illness and death. Hundreds of millions of people suffer from major diseases such as **malaria**, gastrointestinal infections (diarrhea, dysentery, **cholera**), tuberculosis, influenza, and pneumonia spread through the air, water, or food. Many of these terrible diseases could be eliminated or greatly reduced by a cleaner environment, inexpensive dietary supplements, and better medical care.

For the billion or so richest people in the world—including most of the population of the United States and Canada—diseases related to lifestyle or longevity tend to be much greater threats than more conventional environmental concerns such as dirty water or polluted air. Heart attacks, strokes, **cancer**, depression and hypertension, traffic accidents, trauma, and **AIDS** lead as causes of sickness and death in wealthy countries. These diseases are becoming increasingly common in the developing world as people live longer, exercise less, eat a richer diet, and use more drugs, **tobacco**, and alcohol. Epidemiologists predict that by the middle of the next century, these diseases of affluence will be leading causes of sickness and death everywhere.

Although a relatively minor cause of illness compared to the factors above, toxic or hazardous synthetic chemicals in the environment are becoming an increasing source of concern as industry uses more and more exotic materials to manufacture the goods we all purchase. There are many of these compounds to worry about. Somewhere around five million different chemical substances are known, about 100,000 are used in commercial quantities, and about 10,000 new ones are discovered or invented each year. Few of these materials have been thoroughly tested for toxicity. Furthermore, the process of predicting what our chances of exposure and potential harm might be from those re-leased into the environment remains highly controversial. **Toxins** are poisonous, which means that they react specifically with cellular components or interfere with unique physiological processes. A particular chemical may be toxic to one organism but not another, or dangerous in one type of exposure but not others. Because of this specificity, they may be harmful even in very dilute concentrations. Ricin, for instance, is a protein found in castor beans and one of the most toxic materials known. Three hundred picograms (trillionths of a gram) injected intravenously is enough to kill an average mouse. A single molecule can kill an individual cell. If humans were as sensitive as mice, a few teaspoons of this compound, divided evenly and distributed uniformly could kill everyone in the world. By the way, this points out that not all toxins are produced by industry. Many natural products are highly toxic.

Toxins that have chronic (long-lasting) or irreversible effects are of special concern. Among some important examples are neurotoxins (attack nerve cells), mutagens (cause genetic damage), teratogens (result in **birth defects**), and carcinogens (cause cancer). Many pesticides and metals such as **mercury**, **lead**, and chromium are neurotoxins. Loss of even a few critical neurons can be highly noticeable or may even be lethal making this category of great importance. Chemicals or physical factors such as radiation that damage genetic material can harm not only cells produced in the exposed individual, but also the offspring of those individuals as well.

Among the most dread characteristics of all these chronic environmental health threats are that the initial exposure may be so small or have results so unnoticeable that the victim doesn't even know that anything has happened until years later. Furthermore the results may be catastrophic and irreversible once they do appear. These are among our worst fears and are powerful reasons that we are so apprehensive about environmental contaminants. There may be no exposure—no matter how small—of some chemicals that is absolutely safe. Because of these fears, we often demand absolute protection from some of the most dread contaminants. Unfortunately, this may not be possible. There may be no way to insure that we are never exposed to any amount of some hazards. It may be that our only recourse is to ask how we can reduce our exposure or mitigate the consequences of that exposure.

In spite of the foregoing discussion of the dangers of **chronic effects** from minute exposures to certain materials or factors, not all pollutants are equally dangerous nor is every exposure an unacceptable risk. Our fear of unknown and unfamiliar industrial chemicals can lead to hysterical demands for zero exposure to risks. The fact is that life is risky. Furthermore, some materials are extremely toxic while others are only moderately or even slightly so.

This is expressed in the adage of the German physician, Paracelsus, who said in 1540 that "The dose makes the poison." It has become a basic principle of toxicology that nearly everything is toxic at some concentration but most materials have some lower level at which they present an insignificant risk. Sodium chloride (table salt), for instance, is essential for human life in small doses. If you were forced to eat a kilogram all at once, however, it would make you very sick. A similar amount injected all at once into your blood stream would be lethal.

How a material is delivered—at what rate, through which route of entry, in what form—is often as important as what the material is. The movement, distribution, and fate of materials in the environment are important aspects of environmental health. Solubility is one of the most important characteristics in deter- mining how, when, and where a material will travel through the environment and into our bodies. Chemicals that are water soluble move more rapidly and extensively but also are easier to wash off, excrete, or eliminate. Oil or fat soluble chemicals may not move through the environment as easily as water-soluble materials but may penetrate very efficiently through the skin and into tissues and organs. They also may be more likely to be concentrated and stored permanently in fat deposits in the body.

The most common route of entry into the body for many materials is through ingestion and **absorption** in the gastrointestinal (GI) tract. The GI tract, along with the urinary system are the main routes of excretion of dangerous materials. Not surprisingly, those cells and tissues most intimately and continuously in contact with dangerous materials are among the ones most likely to be damaged. Ulcers, infections, lesions, or tumors of the mouth, esophagus, stomach, intestine, colon, kidney, bladder, and associated glands are among the most common manifestations of environmental toxins. Other common routes of entry for toxins are through the respiratory system and the skin. These also are important routes for excreting or discharging unwanted materials.

Some of our most convincing evidence about the toxicity of particular chemicals on humans has come from experiments in which volunteers (students, convicts, or others) were deliberately given measured levels under controlled conditions. Because it is now considered unethical to experiment on living humans, we are forced to depend on proxy experiments using computer models, tissue cultures, or laboratory animals. These proxy tests are difficult to interpret. We can't be sure that experimental methods can be extrapolated to how real living humans would react. The most commonly used laboratory animals in toxicity tests are rodents like rats and mice. However, different **species** can react very differently to the same compound. Of some 200 chemicals shown to be carcinogenic in either rats or mice, for instance, about half caused cancer in one species but not the other. How should we interpret these results? Should we assume that we are as sensitive as the most susceptible animal, as resistant as the least sensitive, or somewhere in between?

It is especially difficult to determine responses to very low levels of particular chemicals, especially when they are not highly toxic. The effects of random events, chance, and unknown complicating factors become troublesome, often resulting in a high level of uncertainty in predicting risk. The case of the sweetener saccharin is a good example of the complexities and uncertainties in **risk assessment**. Studies in the 1970s suggested a link between saccharin and bladder cancer in male rats. Critics pointed out that humans would have to drink 800 cans of soft drink per day to get a dose equivalent to that given to the rats. Furthermore, they argued, most people are not merely large rats.

The **Food and Drug Administration** uses a range of estimates of the probable toxicity of saccharine in humans. At current rates of consumption, the lower estimate predicts that only one person in the United states will get cancer every 1,000 years from saccharine. That is clearly inconsequential considering the advantages of reduced weight, fewer cases of diabetes, and other benefits from this sugar substitute. The upper estimate, however, suggests that 3,640 people

will die each year from this same exposure. That is most certainly a risk worth worrying about.

An emerging environmental health concern with a similarly high level of uncertainty but potentially dire consequences is the disruption of endocrine hormone functions by synthetic chemicals. About ten years ago, **wildlife** biologists began to report puzzling evidence of reproductive failures and abnormal development in certain wild animal populations. Alligators in a lake in central Florida, for instance, were reported to have a 90% decline in egg hatching and juvenile survival along with feminization of adult males including abnormally small penises and lack of sperm production. Similar reproductive problems and developmental defects were reported for trout in the **Great Lakes**, seagulls in California, panthers in Florida, and a number of other species. Even humans may be effected if reports of global reduction of sperm counts and increases of hormone-dependent cancers prove to be true.

Both laboratory and field studies point to a possible role of synthetic chemicals in these problems. More than 50 chemicals, if present in high enough concentrations, are now known to mimic or disrupt the signals conveyed by naturally occurring endocrine hormones that control almost every aspect of development, behavior, immune functions, and **metabolism**. Among these chemicals are **dioxin**, polychlorinated biphenyl, and several persistent pesticides. This new field of research promises to be of great concern in the next few years because it combines dread factors of great emotional power such as undetectable exposure, threat to **future generations**, unknown or delayed consequences, and involuntary or inequitable distribution of risk.

In spite of the seriousness of the concerns expressed above, the **Environmental Protection Agency** warns that we need to take a balanced view of environmental health. The risks associated with allowable levels of certain organic solvents in drinking water or some pesticides in food are thought to carry a risk of less than one cancer in a million people in a lifetime. Many people are outraged about being exposed to this risk, yet they cheerfully accept risks thousands of times as higher from activities they enjoy such as smoking, driving a car, or eating an unhealthy diet. According to the EPA, the most important things we as individuals can do to improve our health are to reduce smoking, drive safely, eat a balanced diet, exercise reasonably, lower stress in our lives, avoid dangerous jobs, lower indoor pollutants, practice safe sex, avoid sun exposure, and prevent household accidents. Many of these factors over which we have control are much more risky than the unknown, uncontrollable, environmental hazards we fear so much.

[*William P. Cunningham Ph.D.*]

RESOURCES

BOOKS

Foster, H. D. *Health, Disease and the Environment.* Boca Raton: CRC Press, 1992.

Moeller, D. W. *Environmental Health.* Cambridge: Harvard University Press, 1992.

Morgan, M. T. *Environmental Health.* Madison: Brown & Benchmark, 1993.

PERIODICALS

Hall, J. V., et al. "Valuing the Health Benefits of Clean Air." *Science* 255 (February 14, 1991): 812–17.

Environmental history

Much of human history has been a struggle for food, shelter, and survival in the face of nature's harshness. Three major events or turning points have been the use of fire, the development of agriculture, and the invention of tools and machines. Each of these advances has brought benefits to humans but often at the cost of **environmental degradation**. Agriculture, for instance, increased food supplies but also caused **soil erosion**, population explosions, and support of sedentary living and urban life. It was the Industrial Revolution that gave humankind the greater power to conquer and devastate our **environment**. Jacob Bronowski called it an energy revolution, with power as the prime goal. As he noted, it is an ongoing revolution, with the fate of literally billions of people hanging on the outcome.

The Industrial Revolution, with its initial dependence on the steam engine, iron works, and heavy use of **coal**, made possible our modern lifestyle with its high consumption of energy and material resources. With it, however, has come devastating levels of air, water, land, and chemical **pollution**. In essence, environmental history is the story of the growing recognition of our negative impact upon **nature** and the corresponding public interest in correcting these abuses. Cunningham and Saigo describe four stages of **conservation** history and environmental activism: 1) pragmatic resource conservation; 2) moral and aesthetic resource preservation; 3) growing concern over the impact of pollution on health and ecosystems; and 4) global environmental citizenship.

Environmental history, like all history, is very much a study of key individuals and events. Included here are Thomas Robert Malthus, George Perkins Marsh, **Theodore Roosevelt**, and Rachel Carson. Writing at the end of the eighteenth century, Malthus was the first to develop a coherent theory of population, arguing that growth in food supply could not keep up with the much larger growth in population. Of cruel necessity, **population growth** would inevitably be limited by **famine**, pestilence, disease, or war. Modern supporters are labeled "neoMalthusians" and include notable spokespersons Paul Erlich and Lester Brown.

In his 1864 book *Man in Nature*, George Perkins Marsh was the first to attack the American myth of super-abundance and inexhaustible resources. Citing many examples from Mediterranean lands and the United States, he described the devastating impact of land abuse through **deforestation** and soil erosion. Lewis Mumford called this book "the fountainhead of the conservation movement," and Stewart Udall described it as the beginning of land wisdom in this country. Marsh's work led to forest preservation and influenced President Theodore Roosevelt and his chief forester, **Gifford Pinchot**.

Effective forest and **wildlife** protection began during Theodore Roosevelt's presidency whose term of office (1901–1909) has been called "The Golden Age of Conservation." His administration established the first wildlife refuges and national forests.

At this time key differences emerged between proponents of conservation and preservation. Pinchot's policies were utilitarian, emphasizing the wise use of resources. By contrast, preservationists led by **John Muir** argued for leaving nature untouched. A key battle was fought over the **Hetch Hetchy Reservoir** in **Yosemite National Park**, a proposed water supply for San Francisco, California. Although Muir lost, the **Sierra Club** (founded in 1882) gained national prominence. Similar battles are now being waged over **petroleum** extraction in Alaska's **Arctic National Wildlife Refuge** and mining permits on federal lands, including **wilderness** areas.

Rachel Carson gained widespread fame through her battle against the indiscriminate use of pesticides. Her 1962 book *Silent Spring* has been hailed as the "fountainhead of the modern environmental movement." It has been translated into over 20 languages and is still a best seller. She argued against **pesticide** abuse and for the right of common citizens to be safe from pesticides in their own homes. Though vigorously opposed by the chemical industry, her views were vindicated by her overpowering reliance on scientific evidence, some given surreptitiously by government scientists. In effect, *Silent Spring* was the opening salvo in the battle of ecologists against chemists. Much of the current mistrust of **chemicals** stems from her work.

Several historical events are relevant to environmental history. The closing of the American frontier at the end of the nineteenth century gave political strength to the Theodore Roosevelt presidency. The 1908 White House Conference on Conservation, organized and chaired by Gifford Pinchot, is perhaps the most prestigious and influential meeting ever held in the United States.

During the 1930s, the **drought** in the American **Dust Bowl** awakened the country to the soil erosion concerns first voiced by Marsh. The establishment of the **Soil Conservation Service** in 1935 was a direct response to this national tragedy.

In 1955, an international symposium entitled "Man's Role in Changing the Face of the Earth" was held at Princeton University. An impressive assemblage of scholars led by geographer Carl Ortwin Sauer, zoologist Marston O. Bates, and urban planner Lewis Mumford documented the history of human impact on the earth, the processes of human alteration of the environment, and the prospects for future habitability.

The *Apollo* moon voyages, especially *Apollo 8* in December 1968 and the dramatic photos taken of Earth from space, awakened the world to the concept of "spaceship earth." It was as though the entire human community gave one collective gasp at the small size and fragile beauty of this one planet we call home.

The two energy price shocks of the 1970s spawned by the 1973 Arab-Israeli War and the 1979 Iranian Islamic Revolution fundamentally altered American energy habits. Our salvation came in part from our horrific waste. Simple **energy conservation** measures and more fuel-efficient automobiles, predominantly of foreign manufacture, produced enough savings to cause a mid-1980s crash in petroleum prices. Nonetheless, many efficiencies begun during the 1970s remain.

The Montreal Accord of 1988 is notable for being the first international agreement to phase out a damaging chemical, **chlorofluorocarbons** (CFCs). This was a direct response to evidence from satellite data over **Antarctica** that chlorine-based compounds were destroying the stratospheric **ozone** shield, which provides vital protection against damaging **ultraviolet radiation**.

Key accidents and corresponding media coverage of environmental concerns have powerfully influenced public opinion. Common themes during the 1960s and 1970s included the "death of Lake Erie," the ravages of **strip mining** for coal, the confirmation of **automobile** exhaust as a key source of **photochemical smog**, destruction of the ozone layer, and the threat of global warming. The ten-hour Annenberg CPB project, *Race to Save the Planet*, is now common fare in **environmental science** telecourses.

Media coverage of specific accidents or sites has had some of the greatest impact on public awareness of environmental problems. **Oil spills** have provided especially vivid and troubling images. The wreck of the *Torrey Canyon* in 1967 off southern England was the first involving a supertanker, and a harbinger of even larger disasters to come, such as the *Exxon Valdez* spill in Alaska. The blowout of an oil well in California's Santa Barbara Channel played nightly on network television news programs. Scenes of oil-covered birds and muddy shorelines were powerful images in the battle for environmental awareness and commitment.

The **Cuyahoga River** in Cleveland was so polluted with petroleum waste that it actually caught fire twice. **Love Canal**, a forgotten chemical waste dump in Niagara Falls, New York, became the inspiration for passage of the Superfund Act, a tax on chemical companies to pay for cleanup of abandoned **hazardous waste** sites. It was also personal vindication for **Lois Marie Gibbs**, leader of the Love Canal Homeowners Association.

One **air pollution** episode in New York City was blamed for the deaths of about 300 people. In response to another in Birmingham, Alabama, a federal judge ordered the temporary shutdown of local steel mills.

The loudest alarms raised against the growing use of **nuclear power** in the U.S. were sounded by the combination of the 1979 near-meltdown at **Three Mile Island Nuclear Reactor** in Pennsylvania and the subsequent (1986) **Chernobyl Nuclear Power Station** disaster in the Ukraine.

Media coverage and growing public awareness of pollution problems, has led to widespread support for corrective legislation. A large body of such legislation was passed between 1968 and 1980. Especially notable were the **National Environmental Policy Act**, which created the **Environmental Protection Agency**, clean air and water acts, Superfund, and the **Surface Mining Control and Reclamation Act**. The latter required miners to reshape the land to near its original contour and to replace **topsoil**, both essential keys to the long-term recovery of the land.

Earlier noteworthy legislation includes the establishment in 1872 of Yellowstone as the world's first **national park**; establishment of national wildlife refuges, forests, and parks, and the agencies to oversee them; and the creation of the **Soil Conservation** Service. The **Wilderness Act** of 1964 sought to set aside government land for nondestructive uses only.

Much has been accomplished, and environmental issues now command widespread public attention. A list of books, journals, environmental organizations, and relevant government agencies now fills six pages of small print in one popular environmental textbook. Nonetheless, important challenges lie ahead in the pursuit of a quality environment that will tax environmental organizations, government policymakers, and voters.

Some key issues for the future can be grouped into the following four categories. 1) Rapidly increasing costs as control standards reach higher and higher levels. The inexpensive and easy solutions have mostly been tried. Solving the air pollution problems within the **Los Angeles basin** is a prime example of this challenge. 2) Control of **phosphates** and **nitrates** in our waterways will require increasing commitment to tertiary (or chemical) **sewage**

treatment plants. We may also find it necessary to reroute all **urban runoff** through such plants. 3) Solutions to the global warming problem, if supported by ongoing scientific research, will require alternative energy strategies, especially as large, newly emerging economies of China, India, Brazil, and other countries seek a growing share of the total energy pie. 4) There is a growing conservative trend and related hostility to environmental concerns among our younger population. Consequently, the need for meaningful **environmental education** and dialogue will only continue to increase.

[*Nathan H. Meleen*]

RESOURCES
BOOKS

The American Experience. "Rachel Carson's *Silent Spring*." Boston: Public Broadcasting System, 1993.
Cunningham, W. P., and B. W. Saigo. *Environmental Science: A Global Concern.* 4th ed. Dubuque, IA: Wm. C. Brown, 1997.
Marsh, G. P. *Man and Nature, or Physical Geography as Modified by Human Action.* 1864. Reprint, Cambridge, MA: Harvard University Press, 1965.
Miller Jr., G. T. *Living in the Environment.* 9th ed. Belmont, CA: Wadsworth, 1996.
Thomas Jr., W. L., ed. *Man's Role in Changing the Face of the Earth.* Chicago: University of Chicago Press, 1956.

Environmental impact assessment

A written analysis or process that describes and details the probable and possible effects of planned industrial or civil project activities on the **ecosystem**, resources, and **environment**. The **National Environmental Policy Act** (NEPA) first promulgated guidelines for environmental impact assessments with the intention that the environment receive proper emphasis among social, economic, and political priorities in governmental decision-making. This act required environmental impact assessments for major federal actions affecting the environment. Many states now have similar requirements for state and private activities. Such written assessments are called Environmental Impact Statements or EISs.

EISs range from brief statements to extremely detailed multi-volume reports that require many years of data collection and analysis. In general, the environmental impact assessment process requires consideration and evaluation of the proposed project, its impacts, alternatives to the project, and mitigating strategies designed to reduce the severity of adverse effects. The assessments are completed by multi-disciplinary teams in government agencies and consulting firms. The experience of the United States **Army Corps of Engineers** in detailing the impacts of projects such as **dams** and waterways is particularly noteworthy, as the Corps has

developed comprehensive methodologies to assess impacts of such major and complex projects. These include evaluation of direct environmental impacts as well as social and economic ramifications.

The content of the assessments generally follows guidelines in the National Environmental Policy Act. Assessments usually include the following sections:

• Background information describing the affected population and the environmental setting, including archaeological and historical features, public utilities, cultural and social values, **topography**, **hydrology**, geology and **soil**, climatology, **natural resources**, and terrestrial and aquatic communities;

• Description of the proposed action detailing its purpose, location, time frame, and relationship to other projects;

• The environmental impacts of proposed action on natural resources, ecological systems, population density, distribution and growth rate, **land use**, and human health. These impacts should be described in detail and include primary and secondary impacts, beneficial and adverse impacts, short and long term effects, the rate of recovery, and importantly, measures to reduce or eliminate adverse effects;

• Adverse impacts that cannot be avoided are described in detail, including a description of their magnitude and implications;

• Alternatives to the project are described and evaluated. These must include the "no action" alternative. A comparative analysis of alternatives permits the assessment of environmental benefits, risks, financial benefits and costs, and overall effectiveness;

• The reason for selecting the proposed action is justified as a balance between risks, impacts, costs, and other factors relevant to the project;

• The relationship between short and long term uses and maintenance is described, with the intent of detailing short and long term gains and losses;

• Reversible and irreversible impacts;

• Public participation in the process is described;

• Finally, the EIS includes a discussion of problems and issues raised by interested parties, such as specific federal, state, or local agencies, citizens, and activists.

The environmental impact assessment process provides a wealth of detailed technical information. It has been effective in stopping, altering, or improving some projects. However, serious questions have been raised about the adequacy and fairness of the process. For example, assessments may be too narrow or may not have sufficient depth. The alternatives considered may reflect the judgment of decision-makers who specify objectives, the study design, and the alternatives considered. Difficult and important questions exist regarding the balance of environmental, economic, and other interests. Finally, these issues often take place in a politicized and highly-charged **atmosphere** that may not be amenable to negotiation. Despite these and other limitations, environmental impact assessments help to provide a systematic approach to sharing information that can improve public decision-making. *See also* Risk assessment

[*Stuart Batterman*]

RESOURCES
BOOKS

Rau, J., and D. G. Wooten, eds. *Environmental Impact Analysis Handbook.* New York, McGraw-Hill, 1980.

OTHER

The National Environmental Policy Act of 1969, as Amended, P.L. 91-140 (1 January 1970), amended P.L. 94-83 (9 August 1975).

Environmental Impact Statement

The **National Environmental Policy Act** (1969) made all federal agencies responsible for analyzing any activity of theirs "significantly affecting the quality of the human environment." Environmental Impact Statements (EIS) are the assessments stipulated by this act, and these reports are required for all large projects initiated, financed, or permitted by the federal government. In addition to examining the damage a particular project might have on the **environment**, federal agencies are also expected to review ways of minimizing or alleviating these adverse effects a review which can include consideration of the environmental benefits of abandoning the project altogether. The agency compiling an EIS is required to hold public hearings; it is also required to submit a draft to public review, and it is forbidden from proceeding until it releases a final version of the statement.

The NEPA has been called "the first comprehensive commitment of any modern state toward the responsible custody of its environment," and the EIS is considered one of the most important mechanisms for its enforcement. It is often difficult to identify environmental damages with remedies that can be pursued in court, but the filing of an EIS and the standards the document must meet are clear and definite requirements for which federal agencies can be held accountable. These requirements have allowed environmental groups to focus legal challenges on the adequacy of the report, contesting the way an EIS was prepared or identifying environmental effects that were not taken into account. The expense and the delays involved in defending against these challenges have often given these groups powerful leverage for convincing a company or an agency to change or omit particular elements of a project. Many environmental organizations have taken advantage of these op-

portunities; between 1974 and 1983, over 100 such suits were filed every year.

Although litigation over impact statements can have a decisive influence on a wide range of decisions in government and business, the legal status of these reports and the legal force of the NEPA itself are not as strong as many environmentalists believe they should be. The act does not require agencies to limit or prevent the potential environmental damage identified in an EIS. The Supreme Court upheld this interpretation in 1989, deciding that agencies are "not constrained by NEPA from deciding that other values outweigh the environmental costs." The government, in other words, is required only to identify and evaluate the adverse impacts of proposed projects; it is not required, at least by NEPA, to do anything about them. Environmentalists have long argued that environmental protection needs a stronger legal grounding than this act provides; some such as Lynton Caldwell, who was originally involved in the drafting of the NEPA, maintain that only a constitutional amendment will serve this purpose.

In addition to the controversies over what should be included in these reports and what should be done about the information, there have also been a number of debates over who is required to file them. Environmental groups have filed suit in the Pacific Northwest alleging that the government should require **logging** companies to file impact statements. And many people have observed that an EIS is not actually required of all government agencies; the U. S. Department of Agriculture, for instance, is not required to file such reports on its commodity support programs.

Impact statements have been opposed by business and industrial groups since they were first introduced. An EIS can be extremely costly to compile, and the process of filing and defending them can take years. Businesses can be left in limbo over projects in which they have already invested large amounts of money, and the uncertainties of the process itself have often stopped development before it has begun. In the debate about these statements, many advocates for business interests have pointed out that environmental regulation accounts for 23% of the 400 billion dollars the federal government spends on regulation each year. They argue that impact statements restrict the ability of the United States to compete in international markets by forcing American businesses to spend money on compliance that could be invested in research or capital improvements. Many people believe that impact statements seriously delay many aspects of economic growth, and business leaders have questioned the priorities of many environmental groups, who seem to value **conservation** over social benefits such as high-levels of employment.

In July of 1993, a judge in a federal district court ruled that the **North American Free Trade Agreement**

(NAFTA) could not be submitted to Congress for approval until the Clinton Administration had filed an EIS on the treaty. The controversy over whether an EIS should be required for NAFTA is a good example of the battle between those who want to extend the range of the EIS and those who want to limit it, as well as the practical problems with positions held by both sides.

Environmentalists fear the consequences of free trade in North America, particularly free trade with Mexico. They believe that most industries would not take any precautions about the environment unless they were forced to observe them. Environmental protection in Mexico, when it exists, is corrupt and inefficient; if NAFTA is approved by Congress, many believe that businesses in the United States will move south of the border to escape environmental regulations. This could have devastating consequences for the environment in Mexico, as well as an adverse impact on the United States economy. It is also possible that an extensive economic downturn, if perceived to be the result of such relocations, could affect the future of environmental regulation in this country as the United States begins to compete with Mexico over the incentives it can offer industry.

Opponents of the decision to require an EIS for NAFTA insist that such a document would be almost impossible to compile. The statement would have to be enormously complex; it would have to consider a range of economic as well as environmental factors, projecting the course of economic development in Mexico before predicting the impact on the environment. Extending the range of impact statements and the NEPA would cause the same expensive delays for this treaty that these statutes have caused for projects within the United States, and critics have focused mainly on the effect such an extension would have on our international competitiveness. They argue that this decision, if upheld, could have broad ramifications for American foreign policy. An EIS could be required for every treaty the government signs with another country, including negotiations over fishing rights, arms control treaties, and other trade agreements. Foreign policy decisions could then be subject to extensive litigation over the adequacy of the EIS filed by the appropriate agency. Many environmentalists would view this as a positive development, but others believe it could prevent us from assuming a leadership role in international affairs.

Carol Browner, the former chief administrator of the **Environmental Protection Agency** (EPA), announced that the agency was determined to reduce some of the difficulties of complying with environmental regulations. She was especially concerned with increasing efficiency and limiting the delays and uncertainties for business. But whatever changes she was able to make, the process of compiling an EIS will never seem cost effective to business, at least in the short

term, and the controversy over these statements continues. *See also* Economic growth and the environment; Environmental auditing; Environmental economics; Environmental impact assessment; Environmental Monitoring and Assessment Program; Environmental policy; Life-cycle assessment; Risk analysis; Sustainable development

[*Douglas Smith*]

RESOURCES
PERIODICALS

Burck, C. "Surprise Judgement on NAFTA." *Fortune* 128 (July 26, 1993): 12.
Davies, J. "Suit Threatens Washington State Industry." *Journal of Commerce* 390 (October 21, 1991): 9A.
Dentzer, S. "Hasta la Vista in Court, Baby." *U.S. News and World Report* 115 (12 July 1993): 47.
Ember, L. "EPA's Browner to Take Holistic Approach to Environmental Protection." *Chemical and Engineering News* 71 (1 March 1993): 19.
Gregory, R., R. Keeney, and D. von Wintervelt. "Adapting the Environmental Impact Statement Process to Inform Decisionmakers." *Journal of Policy Analysis and Management* (Winter 1992): 58.

Environmental labeling
see **Blue Angel; Eco Mark; Green Cross; Green Seal**

Environmental law

Environmental law has been defined as the law of planetary housekeeping. It is concerned with protecting the planet and its people from activities that upset the earth and its life-sustaining capabilities, and it is aimed at controlling or regulating human activity toward that end.

Until the 1960s, most environmental legal issues in the United States involved efforts to protect and conserve **natural resources**, such as forests and water. Public debate focused on who had the right to develop and manage those resources. In the succeeding decades, lawyers, legislators and environmental activists increasingly turned their attention to the growing and pervasive problem of **pollution**. In both instances, environmental law—a term not coined until 1969—evolved mostly from a grassroots movement that forced Congress to pass sweeping legislation, much of which contained provisions for citizen suits. As a result, the courts were thrust into a new era of judicial review of the administrative processes and of scientific uncertainty.

Initially, environmental law formed around the principles of common law, which is law created by courts and judges that rests upon a foundation of judicial precedents. However, environmental law soon moved into the arena of administrative and legislative law, which encompasses most of today's environmental law. The following discussion looks at both areas of law, reviews some of the basic issues involved in environmental law, and outlines some landmark cases.

Generally speaking, common law is based on the notion that one party has done harm to another, in legal terms called a *tort*. There are three broad types of torts, all of which have been applied in environmental law with varying degrees of success. Trespass is the physical invasion of one's property, which has been interpreted to include situations such as **air pollution, runoff** of liquid wastes, or contamination of **groundwater**.

Closely associated with trespass are the torts of private and public nuisance. *Private nuisance* is interference with the use of one's property. Environmental examples include **noise pollution**, odors and other air pollution, and **water pollution**. The operation of a **hazardous waste** site fits the bill for private nuisance, where the threat of personal discomfort or disease interferes with the enjoyment of one's home. A *public nuisance* adversely affects the safety or health of the public or causes substantial annoyance or inconvenience to the public. In these situations, the courts tend to balance the plaintiff's interest against the social and economic need for the defendant's activity.

Lastly, *negligence* involves the defendant's conduct. To prove negligence it must be shown that the defendant was obligated to exercise *due care*, that the defendant breached that duty, that the plaintiff suffered actual loss or damages, and that there is a reasonable connection between the defendant's conduct and the plaintiff's injury.

These common law remedies have not been very effective in protecting the overall quality of our **environment**. The lawsuits and resulting decisions were fragmented and site specific as opposed to issue oriented. Further, they rely heavily on a level of hard scientific evidence that is elusive in environmental issues. For instance, a trespass action must be based on a somehow visible or tangible invasion, which is difficult if not impossible to prove in pollution cases. Common law presents other barriers to action. Plaintiffs must prove actual physical injury (so-called "aesthetic injuries" don't count) and a causal relationship to the plaintiff's activity, which again, is a difficult task in environmental issues.

In the early 1970s, environmental groups, aided by the media, focused public attention on the broad scope of the environmental crisis, and Congress reacted. It passed a host of comprehensive laws, including amendments to the **Clean Air Act** (CAA), the **Endangered Species Act** (ESA), the **National Environmental Policy Act** (NEPA), the **Resource Conservation and Recovery Act** (RCRA), the Toxic Substances Control Act (TSCA), and others. These laws, or statutes, are implemented by federal agencies, who gain their authority through "organic acts" passed by Congress or by executive order.

As environmental problems grew more complicated, legislators and judges increasingly deferred to the agencies' expertise on issues such as the health risk from airborne **lead**, the threshold at which a **species** should be considered endangered, or the engineering aspects of a hazardous waste incinerator. Environmental and legal activists then shifted their focus toward administrative law—challenging agency discretion and procedure as opposed to specific regulations—in order to be heard. Hence, most environmental law today falls into the administrative category.

Most environmental statutes provide for administrative appeals by which interest groups may challenge agency decisions through the agency hierarchy. If no solution is reached, the federal Administrative Procedures Act provides that any person aggrieved by an agency decision is entitled to judicial review.

The court must first grant the plaintiff "standing," the right to be a party to legal action against an agency. Under this doctrine, plaintiffs must show they have been injured or harmed in some way. The court must then decide the level of judicial review based on one of three issues—interpretation of applicable statutes, factual basis of agency action, and agency procedure—and apply a different level of scrutiny in each instance.

Generally, courts are faced with five basic questions when reviewing agency action: Is the action or decision constitutional? Did the agency exceed its statutory authority or jurisdiction? Did if follow legal procedure? Is the decision supported by substantial evidence in the record? Is the decision arbitrary or capricious? Depending on the answers, the court may uphold the decision, modify it, remand or send it back to the agency to redo or reverse it.

By far the most important statute that cracked open the administrative process to judicial review is NEPA. Passed in 1969, the law requires all agencies to prepare an **Environmental Impact Statement** (EIS) for all major federal actions, including construction projects and issuing permits. Environmental groups have used this law repeatedly to force agencies to consider the environmental consequences of their actions, attacking various procedural aspects of EIS preparation. For example, they often claim that a given agency failed to consider alternative actions to the proposed one, which might reduce environmental impact.

In filing a lawsuit, plaintiffs might seek an injunction against a certain action, say, to stop an industry from dumping toxic waste into a river, or stop work on a public project such as a dam or a timber sale that they claim causes environmental damage. They might seek compensatory damages to make up for a loss of property or for health costs, for instance, and punitive damages, money awards above and beyond repayment of actual losses.

Boomer v. Atlantic Cement Co. (1970) is a classic common law nuisance case. The neighbors of a large cement plant claimed they had incurred property damage from dirt, **smoke** and vibrations. They sued for compensatory damages and to enjoin or stop the polluting activities, which would have meant shutting down the plant, a mainstay of the local economy. The New York court rejected a long-standing practice and denied the injunction. Further, in an unusual move, the court ordered the company to pay the plaintiffs for present *and* future economic loss to their properties. A dissenting judge said the rule was a virtual license for the company to continue the nuisance so long as it paid for it.

Sierra Club v. Morton (1972) opened the way for environmental groups to act on behalf of the public interest, and of **nature**, in the courtroom. The **Sierra Club** challenged the U.S. Forest Service's approval of Walt Disney Enterprises' plan to build a $35 million complex of motels, restaurants, swimming pools and ski facilities that would accommodate up to 14,000 visitors daily in Mineral King Valley, a remote, relatively undeveloped national game refuge in the Sierra Nevada Mountains of California. The case posed the now-famous question: Do trees have standing? The Supreme Court held that the Sierra Club was not "injured in fact" by the development and therefore did not have standing. The Sierra Club reworded its petition, gained standing and stopped the development.

Citizens to Preserve Overton Park v. Volpe (1971) established the so-called "hard look" test to which agencies must adhere even during informal rule making. It opened the way for more intense judicial review of the administrative record to determine if an agency had made a "clear error of judgment." The plaintiffs, local residents and conservationists, sued to stop the U.S. Department of **Transportation** from approving a six-lane interstate through a public park in Memphis, Tennessee. The court found that Secretary Volpe had not carefully reviewed the facts on record before making his decision and had not examined possible alternative routes around the park. The case was sent back to the agency, and the road was never built.

Tennessee Valley Authority v. Hill (1978) was the first major test of the **Endangered Species** Act and gained the tiny **snail darter** fish fame throughout the land. The Supreme Court authorized an injunction against completion of a multi-million dollar dam in Tennessee because it threatened the snail darter, an endangered species. The court balanced the act against the money that had already been spent and ruled that Congress's intent in protecting endangered species was paramount.

Just v. Marinette County (1972) involved **wetlands**, the **public trust** doctrine and private property rights. The plaintiffs claimed that the county's ordinance against filling in wetlands on their land was unconstitutional, and that the restrictions amounted to taking their property without compensation. The county argued it was exercising its normal police powers to protect the health, safety and welfare of citizens by protecting its **water resources** through zoning measures. The Wisconsin appellate court ruled in favor of the defendant, holding that the highest and best use of land does not always equate to monetary value, but includes the natural value. The opinion reads, "...we think it is not an unreasonable exercise of that [police] power to prevent harm to public rights by limiting the use of private property to its natural uses."

Although some progress was made in curbing **environmental degradation** through environmental law in the 1970s, environmental legislation was significantly weakened by the Supreme Court in the 1980s.

[*Cathryn McCue*]

RESOURCES
BOOKS

Anderson, F., D. R. Mandelker, and A. D. Tarlock. *Environmental Protection: Law and Policy.* New York: Little, Brown, 1984.

Findley, R., and D. Farber. *Environmental Law in a Nutshell.* St. Paul, MN: West Publishing Co., 1988.

Plater, Z., R. Abrams, and W. Goldfarb. *Environmental Law and Policy: Nature, Law and Society.* St. Paul, MN: West Publishing Co., 1992.

Environmental Law Institute

Environmental Law Institute (ELI) is an independent research and education center involved in developing environmental laws and policies at both national and international levels. The institute was founded in 1969 by the Public Law Education Institute and the Conservation Foundation to conduct and promote research on environmental law. In the ensuing years it has maintained a strong and effective presence in forums ranging from college courses to law conferences. For example, ELI has organized instructional courses at universities for both federal and non-governmental agencies. In addition, it has sponsored conferences in conjunction with such bodies as the American Bar Association, the American Law Institute, and the Smithsonian Institute.

Within the field of environmental law, ELI provides a range of educational programs and services. In 1991, for instance, the institute helped develop an environmental law course for practicing judges in the New England area. Through funding and endowments, the institute has since managed to expand this particular judicial education program into other regions. A similar program enables ELI to offer training courses to federal judges currently serving in district, circuit, and even bankruptcy courts.

ELI also offers various workshops to the general public. In New Jersey, the institute provided a course designed to guide citizens through the state's environmental laws and thus enable them to better develop pollution-prevention programs in their communities. Broader **right-to-know** guidance has since been provided—in collaboration with the **World Wildlife Fund**—at the international level.

ELI's endeavors at the federal level include various interactions with the **Environmental Protection Agency** (EPA). The two groups worked together to develop the National **Wetlands** Protection Hotline, which answers public inquiries on wetlands protection and regulation, and to assess the dangers of exposure to various pollutants.

Since its inception, ELI has evolved into a formidable force in the field of environmental law. In 1991, it drafted a statute to address the continuing problem of **lead poisoning** in children. The institute has also worked—in conjunction with federal and private groups, including scientists, bankers, and even realtors—to address health problems attributable to **radon** gas.

ELI has compiled and produced several publications. Among the leading ELI books are *Law of Environmental Protection*, a two-volume handbook (updated annually) on **pollution control** law, and *Practical Guide to Environmental Management*, a resource book on worker health and safety. In addition, ELI has worked with the EPA in producing *Environmental Investments: The Cost of a Clean Environment*. The institute's principal periodical is *Environmental Law Reporter*, which provides analysis and coverage of topics ranging from courtroom decisions to regulation developments. ELI also publishes *Environmental Forum*—a policy journal intended primarily for individuals in environmental law, policy, and management—and *National Wetlands Newsletter*, which reports on ensuing developments—legal, scientific, regulatory—related to wetlands management.

[*Les Stone*]

RESOURCES

ORGANIZATIONS

Environmental Law Institute, 1616 P St., NW, Suite 200, Washington, D.C. USA 20036 (202) 939-3800, Fax: (202) 939-3868, Email: law@eli.org, <http://www.eli.org>

Environmental liability

Environmental liability refers primarily to the civil and criminal responsibility in the environmental issues of hazardous substances that threaten to endanger public health. Compliance with the standards issued by the U.S. **Environmental**

Protection Agency (EPA) became a major issue following the December 11, 1980 enactment by Congress of the original **Comprehensive Environmental Response, Compensation, and Liability Act** (CERCLA). In 1986, the **Superfund Amendments and Reauthorization Act** (SARA) provided an amendment to CERCLA. The initial legislation created a tax on chemical and **petroleum** companies, and gave the federal government authority to handle the releases or threatened releases of the **hazardous waste**. That tax created $1.6 billion over the first five years of the act, put into a trust fund that would cover the costs of cleaning up abandoned or uncontrolled hazardous waste sites. When "Superfund" was created, changes to the original legislation, as well as additions, were made that reflected the experience gained from the first years of administering the program. It also raised the trust fund to $8.5 billion.

The complex issue of liability was made more complex following 1986 when regulations increased state involvement, and encouraged greater citizen participation in the decisions of site cleanup. In addition to the civil liability of claims due to federal, state, and local governments, a possibility of criminal liability emerged as a matter of particular concern. As those who might be responsible, CERCLA defines four categories of individuals and corporations against whom judgment could be rendered, referred to as potentially responsible parties (PRPs):

- current owners or operators of a specific piece of real estate
- past owners if they owned or operated the property at the time of the hazardous contamination
- generators and possessors of hazardous substances who arranged for disposal, treatment, or transport
- certain transporters of hazardous substances

In acting under EPA-expanded powers, some states have provided exemptions from liability in certain cases. For example, the state of Wisconsin has provided that the person responsible for the **discharge** of a hazardous substance is the one who is required to report it, investigate, it and clean up the contamination. According to Wisconsin Department of **Natural Resources** information, the state defines the responsible person as the one who "causes, possesses, or controls" the contamination—the one who owns the property with a contaminant discharge or owns a container that has ruptured. Other Wisconsin exemptions include limiting liability for parties who voluntarily remediate contaminated property; limiting liability for lenders and representatives, such as banks, credit unions, mortgage bankers, and similar financial institutions, or insurance companies, pension funds or government agencies engaged in secured lending; limiting liability for local government units; and, limiting liability for property affected by off-site discharge.

Courts have also acted in finding persons responsible not specifically listed as PRPs:

- lessees of contaminated property
- lessors of the contaminated property for contamination caused by their lessees
- Landlords and lessees for the contamination cause by their sub-lessees
- corporate officers in their personal capacity
- shareholders
- parent corporations liable for their subsidiaries
- trustees and personal representatives personally liable for contaminated property owned by a trust or estate
- successor corporation
- donees
- lenders who foreclose on and subsequently manage contaminated property

In an environmentally aware, health-conscious society, both in America and throughout the world, environmental liability continues into the early twenty-first century to be a matter of grave concern not only with land, but also in maritime issues.

[*Jane E. Spear*]

RESOURCES
PERIODICALS

American Insurance Association. *Asbestos and Environmental Liability.* [cited June 2002]. <http://www.aiadc.org/IndustryIssues/Asbestos/>.

Amos, Bruce. "A Free Market for Environmental Liability." *Pollution Engineering Online* 1998 [June 2002]. <http://www.pollutionengineering.com>.

Battelle. *Managing Corporate Environmental Liability.* 2001 [June 2002]. <http://www.battelle.org/Environment>.

Environmental Liability. 1993 [cited July 2002]. <http://www.law-text.com>.

Goldstein, Michael R., and Howard D. Rosen. "Environmental Liability in the 90s." *Asset Protection News* 2, no. 3 (April/May 1993).

Wisconsin Department of Natural Resources. *Environmental Liability Exemptions.* April 12, 2002 [June 2002]. <http://www.dnr.state.wi.us/org>.

ORGANIZATIONS

U.S. Environmental Protection Agency, 1200Pennsylvania Avenue, NW, Washington, DC USA 20460 202-260-2090, , <http://www.epa.gov>

Environmental literacy and ecocriticism

Environmental literacy and ecocriticism refer to the work of educators, scholars, and writers to foster a critical understanding about environmental issues. Environmental literacy includes educational materials and programs designed to provide lay citizens and students with a broad understanding of the relationship between humans and the natural world, borrowing from the fields of science, politics, economics, and the arts. Environmental literacy also seeks to develop the knowledge and skills citizens and students may need to

identify and resolve environmental crises, individually or as a group. Ecocriticism is a branch of literary studies that offers insights into the underlying philosophies in literature that address the theme of **nature** and have been catalysts for change in public consciousness concerning the **environment**.

Americans have long turned to literature and popular culture to develop, discuss, and communicate various ideals about the natural world and their relationship to how Americans see themselves and function together. This literature has also made people think about the idea of progress: what constitutes advancement in culture, what are the goals of a healthy society, and how nature would be considered and treated by such a society. In contemporary times, the power and **visibility** of modern media in influencing these debates is also widely recognized. Given this trend, understanding how these forms of communication work and developing them further to broaden public participation, which is a task of environmental literacy and ecocriticism, is vital to the environmental movement.

Educators and ecocritics take diverse approaches to the task of raising consciousness about environmental issues, but they share a collective concern for the global environmental crisis and begin with the understanding that nature and human needs require rebalancing. In that, they become emissaries, as writer Barry Lopez suggests in *Orion* magazine, who have to "reestablish good relations with all the biological components humanity has excluded from its moral universe." For Lopez, as with many generations of nature writers, including **Henry David Thoreau**, **John Muir**, Edward Abbey, and Terry Tempest Williams, and Annie Dilliard, the lessons to be imparted are learned from long experience with and observation of nature. Lopez suggests another pervasive theme, that observing the ever-changing natural world can be a humbling experience, when he writes of "a **horizon** rather than a boundary for knowing, toward which we are always walking."

The career of Henry David Thoreau was one of the most influential and early models for being a student of the natural world and for the development of an environmental awareness through attentive participation within nature. Thoreau also made a fundamental contribution to American's identification with the ideals of individualism and self-sufficiency. His most important work, *Walden*, was a book developed from his journal written during a two-and-a-half-year experiment of living alone and self-sufficiently in the woods near Concord, Massachusetts. Thoreau describes his process of education as an awakening to a deep sense of his interrelatedness to the natural world and to the sacred power of such awareness. This is contrasted to the human society from which he isolated himself, of whose **utilitarianism**, materialism, and consumerism he was extremely critical.

Thoreau famously writes in *Walden*: "I went to the woods because I wished to live deliberately, to front only the essential facts of life, and see if I could not learn what it had to teach, and not, when I came to die, discover that I had not lived." For Thoreau, living with awareness of the greater natural world became a matter of life and death.

Many educators have also been influenced by two founding policy documents, created by commissions of the United Nations, in the field of environmental literacy. The Belgrade Charter (UNESCO-UNEP, 1976) and the Tbilisi Declaration (**UNESCO**, 1978) share the goal "to develop a world population that is aware of, and concerned about, the environment and its associated problems." Later governmental bodies such as the Brundtland Commission (Brundtland, 1987), the United Nations Conference on Environment and Development in Rio (UNCED, 1992), and the Thessaloniki Declaration (UNESCO, 1997) have built on these ideas.

One of the main goals of environmental literacy is to provide learners with knowledge and experience to assess the health of an ecological system and to develop solutions to problems. Models for environmental literacy include curriculums that address key ecological concepts, provide hands-on opportunities, foster collaborative learning, and establish an **atmosphere** that strengthens a learner's belief in responsible living. Environmental literacy in such programs is seen as more than the ability to read or write. As in nature writing, it is also about a sensibility that views the natural world with a sense of wonder and experiences nature through all the senses. The element of direct experience of the natural world is seen as crucial in developing this sensibility. The Edible Schoolyard program in the Berkeley, California, school district, for example, integrates an organic garden project into the curriculum and lunch program, where students become involved in the entire process of farming, while learning to grow and prepare their own food. The program aims to promote participation and awareness to the workings of the natural world, and also to awaken all the senses to enrich the process of an individual's development.

Public interest in **environmental education** came to the forefront in the 1970s. Much of the impetus as well as the funding for integrating environmental education into school curriculums comes from non-profit foundations and educators' associations such as the Association for Environmental and Outdoor Education, the Center for Ecoliteracy, and The Institute for Earth Education. In 1990, the United States Congress created the National Environmental Education and Training Foundation (NEETF) whose efforts include expanding environmental literacy among adults and providing funding opportunities for school districts to advance their environmental curriculums. The National Environmental Education Act of 1990 directed the **Environmental Protection Agency** (EPA) to provide national leadership

in the environmental literacy arena. To that end, the EPA established several initiatives including the Environmental Education Center as a resource for educators, and the Office of Environmental Education, which provides grants, training, fellowships, and youth awards.

The Public Broadcasting System also plays an active role in the promotion of environmental literacy as evidenced by the partnership of the Annenberg Foundation and the Corporation for Public Broadcasting to create and disseminate educational videos for students and teachers, and grant programs such as that sponsored by New York's Channel 13/WNET Challenge Grants.

A common thread woven through these organizations is a definition of environmental learning that goes beyond simple learning to an appreciation of nature. However, appreciation is measured differently by each organization and segments of the American population differ on which aspects of the environment should be preserved. At the end of the 1990s, the George C. Marshall Institute directed an independent commission to study whether the goals of environmental education were being met. The Commission's 1997 report found that curricula and texts vary widely on many environmental concepts, including what constitutes **conservation**. Although thirty-one states have academic standards for environmental education, a national cohesiveness is lacking.

Thus, the main challenges to environmental literacy are the lack of unifying programs that would bring together the many approaches to environmental education, and the fact that there is inconsistent support for these programs from the government and public school system. Observers of environmental literacy movements suggest that the new perspectives that learners gain may often be at odds with the concerns and ethics of mainstream society, issues that writers such as Thoreau grappled with. For instance, consumerism and conservationism may be at opposite ends of the spectrum of how people interact with the natural world and its resources. To be effective, literacy initiatives must address these dilemmas and provide tools to solve them. Environmental literacy is thus about providing new ways of seeing the world, about providing language tools to address these new perceptions, and to provide ethical frameworks through which people can make informed choices on how to act.

Ecocriticism develops the tools of literary criticism to understand how the relationship of humans to nature is addressed in literature, as a subject, character, or as a component of the setting. Ecocritics also highlight the ways in which literature is a vehicle to create environmental consciousness. For critic William Rueckert, the scholar who coined the term *ecocriticism* in 1978, poetry and literature are the "verbal equivalent of fossil fuel, only renewable," through which abundant energy is transferred between nature and the reader.

Ecocritics highlight aspects of nature described in literature, whether frontiers, rivers, regional ecosystems, cities, or **garbage**, and ask what the purposes of these descriptions are. Their interests have included understanding how historical movements such as the Industrial Revolution have changed the relationship between human society and nature, giving people the false illusion that they can completely control nature, for instance. Ecocriticism also brings together perspectives from various academic disciplines and draws attention to their shared purposes. Studies in **ecology** and cellular biology, for example, echo the theme of interconnectedness of the individual and the natural world seen in poetry, by demonstrating how the life of all organisms is dependent upon their on-going interactions with the environment around them.

Although nature writers have expressed their philosophies of nature and reflected on their modes of communication since the nineteenth century, as a self-conscious practice, ecocriticism's history did not began until the late 1970s. By the 1990s, it had gained wide currency. In his 1997 article "Wild Things," published in *Utne Reader*, Gregory McNamee notes that courses in environmental literature are available at colleges across the nation and that "'ecocritcism' has become something of an academic growth industry." In 1992, the Association for the Study of Literature and Environment (ASLE) was founded with the mission "to promote the exchange of ideas and information about literature and other cultural representations that consider that human relationships with the natural world".

Nature's role in theatre and film are also popular ecocriticism topics for academic study in the form of seminars on, for example, The Nature of Shakespeare, and suggested lists of commercial films for class discussion that include *Chinatown, Deliverance, The China Syndrome, Silkwood, A Civil Action* and *Jurassic Park*.

Ecocritics Carl Herndl and Stuart Brown suggest that there are three underlying philosophies in evaluating nature in modern society. The language used by institutions that make government policies usually regards nature as a resource to be managed for greater social welfare. This is described as an ethnocentric perspective, which begins with the idea that one opinion or way of looking at the world is superior to others. Thus, the benefits of environmental issues are always measured against various political and social interests, and not seen as important simply in themselves.

Another viewpoint is the anthropocentric perspective, wherein human perspectives are central in the world and are the ultimate source of meaning. The specialized language of the sciences, which treats nature as an object of study, is an example of this. The researcher is seen as existing outside of or above nature, and science is grounded on the faith that humans can come to know all of nature's secrets.

In contrast, poetry often describes nature in terms of its beauty and emotional and spiritual power. This language sees man as part of the natural world and seeks to harmonize human values and actions with a respect for nature. This is the ecocentric perspective, which means putting nature and ecology as the central viewpoint when considering the various interactions in the world, including human ones. That is, this perspective acknowledges that humans are a part and parcel of nature and ultimately depend upon the ecology's living and complex interactions for survival.

Scholars make the distinction between environmental writing and other kinds of literature that use images of nature in some fashion or another. Environmental writing explores at length ecocentric perspectives. They include discussions about human ethical responsibility towards the natural world, such as in Aldo Leopold's *A Sand County Almanac*, considered one of the best explorations of **environmental ethics**. Many ecocritics also share a concern for the environment, and one aim of eco-criticism is to raise awareness within the literary world about the environmental movement and nature-centered perspectives in understanding human relationships and cultural practices.

In *Silent Spring*, a major text in the field of environmental literacy and ecocriticism, Rachel Carson writes that society faces two choices: to travel as we now do on a super-highway at high speed but ending in disaster, or to walk the less traveled "other road" which offers the chance to preserve the earth. The challenge of ecocriticism is to spread the word of the "other road," and to simultaneously offer constructive criticism to the environmental movement from within.

[*Douglas Dupler*]

RESOURCES
BOOKS

Carson, Rachel. *Silent Spring*. New York: Houghton Mifflin, 1994.

Finch, Robert, and John Elder, eds. *The Norton Book of Nature Writing*. New York: W.W. Norton & Co., 1990.

Herndl, Carl, and Stuart Brown, eds. *Green Culture: Environmental Rhetoric in Contemporary America*. Madison, WI: University of Wisconsin Press, 1996.

Leopold, Aldo. *A Sand County Almanac*. New York: Oxford University Press, 1966.

Rueckert, William. "Literature and Ecology: An Experiment in Ecocriticism." *The Ecocriticism Reader*, edited by Cheryll Glotfelty and Harold Fromm. Athens, GA: University of Georgia Press, 1996.

Snyder, Gary. *Practice of the Wild*. San Francisco: North Point Press, 1990.

Thoreau, Henry David. *Walden*. 1854; Reprint, Boston: Beacon Press, 1997.

Williams, Terry Tempest. *Refuge: An Unnatural History of Family and Place*. New York: Pantheon, 1991.

PERIODICALS

Lopez, Barry. "The Naturalist." *Orion* Autumn 2001 [cited June 2002]. <http://www.oriononline.org>.

McNamee, Gregory. "Wild Things." *Utne Reader* November-December 1997 [cited July 2002]. <http://www.asle.umn.edu/archive/intro/utne.html>.

OTHER

Association for the Study of Literature and Environment. [cited July 2002]. <http://www.asle.umn.edu>.

Center for Ecoliteracy. [cited June 2002]. <http://www.ecoliteracy.org>.

Environmental Education Page U.S. Environmental Protection Agency. [cited July 2002]. <http://www.governmentguide.com/officials_and _agencies/u.s./indepdendent/govsite.adp>.

Institute for Earth Education. [cited July 2002]. <http://www.eartheducation.org>.

National Environmental Education and Training Foundation. [cited July 2002]. <http://www.neetf.org>.

North American Association for Environmental Education. [cited June 2002]. <http://www.naaee.org>.

Environmental mediation and arbitration

see **Environmental dispute resolution**

Environmental monitoring

Environmental monitoring detects changes in the health of an **ecosystem** and indicates whether conditions are improving, stable, or deteriorating. This quality, too large to gauge as a whole, is assessed by measuring indicators, which represent more complex characteristics. The concentration of **sulfur dioxide**, for example, is an indicator that reflects the presence of other air pollutants. The abundance of a predator indicates the health of the larger **environment**. Other indicators include **metabolism**, population, community, and landscape. All changes are compared to an ideal, pristine ecosystem. The SER (stressor-exposure-response) model, a simple but widely used tool in environmental monitoring, classifies indicators as one of three related types:

- Stressors, which are agents of change associated with physical, chemical, or biological constraints on environmental processes and integrity. Many stressors are caused by humans, such as **air pollution**, the use of pesticides and other toxic substances, or **habitat** change caused by forest clearing. Stressors can also be natural processes, such as **wildfire**, hurricanes, volcanoes, and **climate** change.

- Exposure indicators, which link a stressor's intensity at any point in time to the cumulative dose received. Concentrations or accumulations of toxic substances are exposure indicators; so are **clear-cutting** and urbanization.

- Response indicators, which shows how organisms, communities, processes, or ecosystems react when exposed to a stressor. These include changes in physiology, productivity, or **mortality**, as well as changes in **species** diversity within communities and in rates of **nutrient** cycling.

The SER model is useful because it links ecological change with exposure to **environmental stress**. Its effectiveness is limited, however. The model is a simple one, so it cannot be used for complex environmental situations. Even with smaller-scale problems, the connections between stressor, exposure, and response are not understood in many cases, and additional research is required.

Environmental monitoring programs are usually one of two types, extensive or intensive. Extensive monitoring occurs at permanent, widely spaced locations, sometimes using remote-sensing techniques. It provides an overview of changes in the ecological character of the landscape, often detecting regional trends. It measures the effects of human activities like farming, forestry, mining, and urbanization. Information from extensive monitoring is often collected by the government to determine such variables as water and **air quality**, to calculate allowable forest harvests, set bag limits for **hunting** and fishing, and establish the production of agricultural commodities.

Extensive monitoring usually measures stressors (such as emissions) or exposure indicators (concentration of pollutants in the air). Response indicators, if measured at all in these programs, almost always have some economic importance (damage to forest or agricultural crops). Distinct species or ecological processes do not have economic value and are not usually assessed in extensive-monitoring programs, even though these are the most relevant indicators of **ecological integrity**.

Intensive monitoring is used for detailed studies of structural and functional **ecology**. Unlike extensive monitoring, a relatively small number of sites provide information on stressors such as climate change and **acid rain**. Intensive monitoring is also used to conduct experiments in which stressors are manipulated and the responses studied, for example by acidifying or fertilizing lakes, or by conducting forestry over an entire **watershed**. This research, aimed at understanding the dynamics of ecosystems, helps develop ecological models that distinguish between natural and **anthropogenic** change.

Support for ecological monitoring of either kind has been weak, although more countries are beginning programs and establishing networks between monitoring sites. The United States has founded the Long-Term Ecological Research (LTER) network to study extensive ecosystem function, but little effort is directed toward understanding environmental change. The **Environmental Monitoring and Assessment Program** (EMAP) of the **Environmental Protection Agency** (EPA) studies intensive environmental change, but its activities are not integrated with LTER. In comparison, an ecological-monitoring network being designed by the government of Canada to study changes in the environment will integrate both extensive and intensive monitoring.

Communication between the two types of monitoring is important. Intensive information provides a deeper understanding of the meaning of extensive-monitoring indicators. For example, it is much easier to measure decreases in surface-water **pH** and alkalinity caused by **acid** rain than to monitor resulting changes in fish or other biological variables. These criteria can, however, be measured at intensive-monitoring sites, and their relationships to pH and alkalinity used to predict effects on fish and other **fauna** at extensive sites where only pH and alkalinity are monitored.

The ultimate goal of environmental monitoring is to measure, anticipate, and prevent the deterioration of ecological integrity. Healthy ecosystems are necessary for healthy societies and sustainable economic systems. Environmental monitoring programs can accomplish these goals, but they are expensive and require a substantial commitment by government. Much has yet to be accomplished.

[*Bill Freedman and Cynthia Staicer*]

RESOURCES
BOOKS

Freedman, B., C. Staicer, and N. Shackell. *A Framework for a National Ecological-Monitoring Program for Canada.* Ottawa: Environment Canada, 1992.

PERIODICALS

Franklin, J. F., C. S. Bledsoe, and J. T. Callahan. "Contributions of the Long-term Ecological Research Program." *Bioscience* 40 (1990): 509–524.
Odum, E. P. "Trends Expected in Stressed Ecosystems." *Bioscience* 35 (1985): 419–422.
Schindler, D. W. "Experimental Perturbations of Whole Lakes as Tests of Hypotheses Concerning Ecosystem Structure and Function." *Oikos* 57 (1990): 25–41.

Environmental Monitoring and Assessment Program

The **Environmental Monitoring** and Assessment Program (EMAP), established in 1990 by the **Environmental Protection Agency** (EPA), is a federal project designed to create a continually updated survey of ecological resources in the United States. This comprehensive list monitors and links resource data from several U.S. agencies, including the **National Oceanic and Atmospheric Administration**, the **Fish and Wildlife Service**, and the **U.S. Department of Agriculture**.

Research from the program is intended to illustrate changes in specific ecosystems in the U.S. and to determine

if those changes could have resulted from "human-induced stress."

RESOURCES
ORGANIZATIONS

Environmental Monitoring and Assessment Program, , Email: emap@epa.gov, <http://www.epa.gov/emap>

Environmental policy

Strictly, an environmental policy can be defined as a government's chosen course of action or plan to address issues such as **pollution**, **wildlife** protection, **land use**, energy production and use, waste generation, and waste disposal. In reality, the way a particular government handles environmental problems is most often not a result of a conscious choice from a set of alternatives. More broadly, then, a government's environmental policy may be characterized by examining the overall orientation of its responses to environmental challenges as they occur, or by defining its policy as the sum of plans for, and reactions to, environmental issues made by any number of different arms of government.

A society's environmental policy will be shaped by the actions of its leaders in relation to the five following questions:

• Should government intervene in the regulation of the **environment** or leave resolution of environmental problems to the legal system or the market?

• If government intervention is desirable, at what level should that intervention take place? In the United States, for example, how should responsibility for resolution of environmental problems be divided between and among federal, state and local governments and who should have *primary* responsibility?

• If government intervenes at some level, how much protection should it give? How safe should the people be and what are the economic trade-offs necessary to ensure that level of safety?

• Once environmental standards have been set, what are the methods to attain them? How does the system control the sources of environmental destruction so that the environmental goals are met?

• Finally, how does the system monitor the environment for compliance to standards and how does it punish those who violate them?

Policy in the United States

The United States has no single, overarching environmental policy and its response to environmental issues—subject to conflicting political, corporate and public influence, economic limitation and scientific uncertainty—is rarely monolithic. American environmental policies are an amalgamation of Congressional, state and local laws, regulations and rules formulated by agencies to implement those laws, judicial decisions rendered when those rules are challenged in court, programs undertaken by private businesses and industry, as well as trends in public concerns.

In Congress, many environmental policies were originally formed by what are commonly known as "iron triangles." These involve three groups of actors who form a powerful coalition: the Congressional committee with jurisdiction over the issue; the relevant federal agency handling the problem; and the interest group representing the particular regulated industry. For example, the key actors in forming policy on **clear-cutting** in the national forests are the House subcommittee on Forests, Family Farms and Energy, the U.S. **Forest Service** (USFS), and the National Forest Products Association, which represents many industries dependent on timber.

For more than a century, **conservation** and environmental groups worked at the fringes of the traditional "iron triangle." Increasingly, however, these public interest groups—which derived their financial support and sense of mission from an increasing number of citizen members—began gaining more influence. Scientists, whose studies and research today play a pivotal role in decision-making, also began to emerge as major players.

The Watershed years

Catalyzed by vocal, energetic activists and organizations, the emergence of an "environmental movement" in the late 1960s prompted the government to grant environmental protection a greater priority and **visibility**. 1970, the year of the first celebration of **Earth Day**, saw the federal government's landmark passage of the **Clean Air Act** and the **National Environmental Policy Act**, as well as Richard Nixon's creation of an **Environmental Protection Agency** (EPA) which was given the control of many environmental policies previously administered by other agencies. In addition, some of the most serious problems such as DDT and **mercury** contamination began to be addressed between 1969 and 1972. Yet, environmental policies in the 1970s developed largely in an adversarial setting pitting environmental groups on one side and the traditional iron triangles on the other.

The first policies that came out of this era were designed to clean up visible pollution—clouds of industrial soot and dust, detergent-filled streams and so forth—and employed "end-of-pipe" solutions to target point sources, such as **wastewater discharge** pipes, smokestacks, and other easily identifiable emitters.

An initial optimism generated by improvements in air and **water quality** was dashed by a series of frightening environmental episodes at **Times Beach**, Missouri, Three Mile Island, **Love Canal**, New York and other locations.

Such incidents (as well as memory of the devastation caused by the recently-banned DDT) shifted the focus of public concern to specific toxic agents. By the early 1980s, a fearful public led by environmentalists had steered governmental policy toward tight regulation of individual, invisible toxic substances—dioxin, PCBs and others—by backing measures limiting emissions to within a few **parts per million**. Without an overall governmental framework for action, the result has been a multitude of regulations and laws that address specific problems in specific regions that sometimes conflict and often fail to protect the environment in a comprehensive manner. "It's been reactionary, and so we've lost the integration of thought and disciplines that is essential in environmental policy making," says **Carol Browner**, administrator of the U.S. EPA.

One example of policy-making gone awry is the 1980 **Comprehensive Environmental Response, Compensation and Liability Act** (CERCLA), or *Superfund* toxic waste program. The law grew as much out of the public's perception and fear of toxic waste as it did from crude scientific knowledge of actual health risks. Roughly $2 billion dollars a year has been spent cleaning up a handful of the nation's worst toxic sites to near pristine condition. EPA officials now believe the money could have been better spent cleaning up *more* sites, although to a somewhat lesser degree.

Current trends in environmental policy

Today, governmental bodies and public interest groups are drawing back from "micro management" of individual **chemicals**, individual **species** and individual industries to focus more on the interconnections of environmental systems and problems. This new orientation has been shaped by several (sometimes conflicting) forces, including:

- industrial and public **resistance** to tight regulations fostered by fears that such laws impact employment and economic prosperity; (2) financial limitations that prevent government from carrying out tasks related to specific contaminants, such as cleaning up waste sites or closely monitoring toxic discharges; (3) a perception that large-scale, global problems such as the **greenhouse effect**, **ozone layer depletion**, **habitat** destruction and the like should receive priority; (4) the emergence of a "preventative" orientation on the part of citizen groups that attempts to link economic prosperity with environmental goals. This approach emphasizes **recycling**, efficiency, and environmental technology and stresses the prevention of problems rather than their **remediation** after they reach a critical stage. This strategy also marks an attempt by some citizen organizations to a more conciliatory stance with industry and government.

This new era of environmental policy is underscored by the election of Bill Clinton and Albert Gore, who made the environment a cornerstone of their campaign. In all likelihood, the Clinton administration will transform the

EPA into the cabinet-level position of Department of the Environment, giving the agency more stature and power. The EPA, the USFS and other federal environmental agencies have announced a new "ecosystem" approach to resource management and **pollution control**. In a bold first move, Congressional Democratic leaders are simultaneously reviewing four major environmental statutes (the **Resource Conservation and Recovery Act** [RCRA], **Clean Water Act** [CWA], **Endangered Species Act** [ESA] and Superfund) in the hopes of integrating the policies into a comprehensive program. *See also* Pollution Prevention Act

[*Cathryn McCue and Kevin Wolf and Jeffrey Muhr*]

RESOURCES

BOOKS

Lave, Lester B. *The Strategy of Social Regulation.* Washington, DC: Brookings Institution, 1981.

Logan, Robert, Wendy Gibbons, and Stacy Kingsbury. *Environmental Issues for the '90s: A Handbook for Journalists.* Washington DC: The Media Institute, 1992.

Portney, Paul R., ed. *Public Policies for Environmental Protection.* Washington, DC: Resources for the Future, 1991.

Wolf Jr., Charles. *Markets or Government.* Cambridge, Massachusetts: MIT Press, 1988.

World Resources Institute. *1992 Environmental Almanac.* Boston: Houghton Mifflin Co., 1992.

PERIODICALS

Schneider, Keith. "What Price Clean Up?" *New York Times*, March 21–26, 1993.

Smith, Fred. "A Fresh Look at Environmental Policy." *SEJ Journal* 3 (Winter 1993).

OTHER

Browner, Carol. Administrator of U.S. Environmental Protection Agency, comments during a press conference in Ann Arbor, MI. March 23, 1993.

Environmental and Energy Study Institute. *Special Report.* October 14, 1992.

Environmental Protection Agency (EPA)

The Environmental Protection Agency (EPA) was established in July of 1970, a landmark year for environmental concerns, having been preceded by the passing of the **National Environmental Policy Act** in January and the first **Earth Day** celebrations in April. President Richard Nixon and Congress, working together in response to the growing public demand for cleaner air, land, and water, sought to create a new agency of the federal government structured to make a coordinated attack on the pollutants that endanger human health and degrade the **environment**. The EPA was charged with repairing the damage already done to the

environment and with instituting new policies designed to maintain a clean environment.

The EPA's mission is "to protect human health and to safeguard the natural environment." At the time the EPA was formed, at least fifteen programs in five different agencies and cabinet-level departments were handling **environmental policy** issues. For the EPA to work effectively, it was necessary to consolidate the environmental activities of the federal government into one agency. **Air pollution control**, solid **waste management**, radiation control, and the drinking water program were transferred from the U.S. Department of Health, Education and Welfare (currently known as the **U.S. Department of Health and Human Services**). The **water pollution** control and pesticides research programs were acquired from the **U.S. Department of the Interior**. Registration and regulation of pesticides was transferred from the **U.S. Department of Agriculture**, and the responsibility for setting tolerance levels for pesticides in food was acquired from the **Food and Drug Administration**. The EPA also took over from the **Atomic Energy Commission** the responsibility for setting some environmental radiation protection standards and assumed some of the duties of the Federal Radiation Council. For some environmental programs, the EPA works with other agencies: for example, the United States Coast Guard and the EPA work together on flood control, shoreline protection, and **dredging** and filling activities. And, since most state governments in the United States have their own environmental protection departments, the EPA delegates the implementation and enforcement of many federal programs to the states.

The EPA's headquarters is in Washington DC, and there are ten EPA regional offices and field laboratories. The main office develops national environmental policy and programs, oversees the regional offices and laboratories, requests an annual budget from Congress, and conducts research. The regional offices implement national policies, oversee the environmental programs that have been delegated to the states, and review Environmental Impact Statements for federal actions. The field laboratories conduct research, the data from which are used to develop policies and provide analytical support for monitoring and enforcement of EPA regulations, and for the administration of permit programs.

The administrator of the EPA is appointed by the President, subject to approval by the Senate. The same procedure is used to appoint a deputy administrator, who assists the administrator, and nine assistant administrators, who oversee programs and support functions. Other posts include the chief financial officer, who manages the EPA' budget and funding operations, the inspector general, who is responsible for investigating environmental crimes, and a general counsel, who provides legal support.

In addition to the administrative offices, the EPA is organized into the following program offices: The Office of Air and Radiation, the **American Indian Environmental Office**, the Office of Enforcement and Compliance Assurance, the Office of Environmental Justice, the Office of Environmental Information, the History Office, the Office of International Affairs, the Office of Prevention, Pesticides and Toxic Substances, the Office of Research and Development, the Science Policy Council, the office of **Solid Waste** and Emergency Response, and the Office of Water.

The current EPA Administrator, elected by President George W. Bush, is Christie Whitman, formerly Governor of New Jersey, who was sworn in on January 31, 2001. Whitman's official administrative philosophy is that environmental goals are compatible with and are connected to economic goals, and that relationships between citizens, policy makers, and private sector must be strengthened. The nomination of Linda J. Fisher to the post of EPA Deputy Administrator was confirmed by the U.S. Senate on May 24, 2001. Fisher, who formerly practiced law in Washington DC and has served as Vice President and Corporate Officer at the Monsanto Co. in St. Louis MO, is Whitman' top managerial and policy assistant.

One of the major activities of the EPA is the management of Superfund sites. For many years, uncontrolled dumping of hazardous chemical and industrial wastes in abandoned warehouses and landfills continued without concern for the potential impact on public health and the environment. Concern over the extent of the hazardous-waste-site problem led Congress to establish the Superfund Program in 1980 to locate, investigate, and clean up the worst such sites. The EPA' Office of Emergency and Remedial Response (OERR) oversees management of the program in cooperation with individual states. When a hazardous-waste site is discovered, the EPA is notified. The EPA makes a preliminary assessment of the site and gives a numerical score according to **Hazard Ranking System** (HRS), which determines whether the site is placed on the **National Priorities List** (NPL). As of May 29, 2002, 1221 sites were listed on the final NPL, with 74 new sites proposed, 812 sites were reported to have completed construction, and 258 sites had been deleted from the list. The final NPL lists Superfund sites in which the clean-up plan is under construction or ongoing. NPL proposed sites include sites for which the HRS indicates that placement on the final NPL is appropriate. Among currently proposed NPL sites are Air Force Plant 85 near Columbus OH (**coal** deposits **leaching** sulfuric **acid**, ammonia, and **heavy metals**), Blackbird Mine in Lemhi ID (high levels of **arsenic, copper**, cobalt, and **nickel** in surface water and sediments of Meadow and Blackbird

Creeks downstream from mining tunnels and waste-rock piles), the Omaha **Lead** site in Omaha, NE (lead **contaminated soil** near populated areas and **wetlands**) and the Libby **Asbestos** site in Libby, MT (heavy asbestos exposures and chromium, copper, and nickel deposits near wetlands and fisheries). A final NPL site is deleted from the list when it is determined that no further clean up is needed to protect human health or the environment Finally, under Superfund's redevelopment program, former hazardous-waste sites have been remade into office buildings, parking lots, or even **golf courses** to be re-integrated as productive parts of the community.

The offices and programs of the EPA recognize a set of main objectives, or "core functions." These core functions help define the agency's mission and provide a common focus for all agency activities. The core functions are:

- **Pollution** Prevention—taking measures to prevent pollution from being created rather than only cleaning up what has already been released, also known as source reduction
- Risk Assessment and Risk Reduction—identifying problems that pose the greatest risk to human health and the environment and taking measures to reduce those risks
- Science, Research and Technology—conducting research that will help in developing environmental policies and promoting innovative technologies to solve environmental problems
- Regulatory Development—developing requirements such as operating procedures for facilities and standards for emissions of pollutants
- Enforcement—assuring compliance with established regulations
- Environmental Education—developing educational materials, serving as an information clearinghouse, and providing grant assistance to local educational institutions

Many EPA programs are established by legislation enacted by Congress. For example, many of the activities carried out by the Office of Solid Waste and Emergency Response originated in the **Resource Conservation and Recovery Act** (RCRA). Among other laws that form the legal basis for the programs of the EPA are the National Environmental Policy Act (NEPA) of 1969, which represents the basic national charter of the EPA, the **Clean Air Act** (CAA) of 1970, the **Occupational Safety and Health Act** (OSHA) of 1970, the **Endangered Species Act** (ESA) of 1973, the **Safe Drinking Water Act** (SDWA) of 1974, the Toxic Substances Control Act (TSCA) of 1976, the **Clean Water Act** (CWA) of 1977, the Superfund Amendments and Reauthorization Act (SARA) of 1986, and the **Pollution Prevention Act** (PPA) of 1990. It is through such legislation that the EPA obtains authority to develop and enforce regulations. Environmental regulations drafted by

Environmental Protection Agency

Administrative
  Law Judges
Civil Rights
Small and
  Disadvantaged
  Business Utilization
Science Advisory
  Board
Cooperative Environ-
  mental Management
Executive Support
  Office
Executive Secretariat

ADMINISTRATOR

- - - - - - - -

DEPUTY ADMINISTRATOR

Regional Operations and
State/Local Relations

Communications and
Public Affairs

Congressional and
Legislative Affairs

International
Activities

Administration and
Resource Management

Enforcement

General Counsel

Policy Planning,
and Evaluation

Inspector General

Air and
Radiation

Water

Pesticides and
Toxic Substances

Research and
Development

Solid Waste and
Emergency Response

Regional Offices

EPA organization chart. (Photograph by Tom Pantages. Reproduced by permission.)

the agency are subjected to intense review before being finalized. This process includes approval by the President's **Office of Management and Budget** and input from the private sector and from other government agencies.

Public concern over the environmental changes with time, and the EPA alters its policy priorities in response. For example, in answer to growing concern regarding environmental impact to children's health, the Office of Children's Health Protection (OCHP) was created in May 1997. On February 14, 2002, President George W. Bush announced the Clear Skies Initiative, which contain the farthest reaching legislative changes to the Clean Air Act since 1990.

Growing public concern about water pollution led to a landmark piece of legislation, the Federal Water **Pollution Control** Act of 1972, amended in 1977 and commonly known as the Clean Water Act. The Clean Water Act gives the EPA the authority to administer pollution control programs and to set **water quality standards** for contaminants of surface waters. October 18, 2002, marks the thirtieth Anniversary of the enactment of the Clean Water Act. In continuing support of the goals of this Act, Congress has proclaimed 2002 as the "Year of Clean Water."

As part of its mission as an information clearing house, the EPA maintains an excellent web site at <http://www.epa.gov> with countless links to supporting information of all kinds. The web site also includes Spanish translations of many documents, and links to children's activities.

[*Teresa C. Donkin*]

RESOURCES
BOOKS

Environmental Management. Washington, DC: U.S. Environmental Protection Agency, October 1991.

Keating, B., and D. Russell. "Inside the EPA: Yesterday and Today...Still Hazy After All These Years." *E Magazine* 3 (August 1992): 30–37.

OTHER

U.S. Environmental Protection Agency Laws and Regulations. *Clean Water Act.* March 26, 2002 [cited July 10, 2002]. <http://www.epa.gov/region5/water/cwa.htm>.

U.S. Environmental Protection Web Site. [cited July 10, 2002]. <http://www.epa.gov>.

Environmental racism

The term environmental racism was coined in a 1987 study conducted by the United Church of Christ that examined the location of **hazardous waste** dumps and found an "insidious form of racism." Concern had surfaced five years before, when opposition to a polychlorinated biphenyl (PCB) **landfill** prompted Congress to examine the location of hazardous waste sites in the Southeast, the **Environmental Protection Agency** (EPA)'s Region IV. They found that three of the four facilities in the area were in communities primarily inhabited by people of color. Subsequent studies, such as Ben Goldman's *The Truth about Where You Live*, have contended that exposure to environmental risks is significantly greater for racial and ethnic minorities than for nonminority populations. However, an EPA study contends that there is not enough data to draw such broad conclusions.

The *National Law Journal* found that official response to environmental problems may be racially biased. According to their study, penalties for environmental crimes were higher in white communities. They also found that the EPA takes 20% longer to place a hazardous waste site in a minority community on the Superfund's **National Priorities List** (NPL). And, once assigned to the NPL, these clean ups are more likely to be delayed.

Advocates also contend that environmentalists and regulators have tried to solve environmental problems without regard for the social impact of the solutions. For example, the Los Angeles Air Quality Management District wanted to require businesses to set up programs that would discourage their employees from driving to work. As initially conceived, employers could have simply charged fees for parking spaces without helping workers set up car pools. The Labor-Community Strategy Center, a local activist group, pointed out that this would have disproportionately affected people who could not afford to pay for parking spaces. As a compromise, regulators will now review employers' plans and only approve those that mitigate the any unequal effects on poor and minority populations.

In response to the concern that traditional **environmentalism** does not recognize the social and economic components of environmental problems and solutions, a national movement for "environmental and economic justice" has spread across the country. Groups like the Southwest Network for Environmental and Economic Justice attempt to frame the **environment** as part of the fight against racism and other inequalities.

In addition, the federal government has begun to address the debate over environmental racism. In 1992, the EPA established an Environmental Equity office. In addition, several bills that advocate environmental justice have been introduced. *See also* Comprehensive Environmental Response, Compensation, and Liability Act (CERCLA); Environmental economics; Environmental law; Hazardous waste siting; South

[*Alair MacLean*]

RESOURCES
BOOKS

Alston, D., ed. *We Speak for Ourselves: Social Justice, Race and Environment.* Washington, DC: The Panos Institute, 1990.

Goldman, B. A. *The Truth About Where You Live: An Atlas for Action on Toxins and Mortality.* New York: Times Books, 1991.

Lee, C. *Toxic Wastes ad Race in the United States: A National Report on the Racial and Socio-Economic Characteristics of Communities with Hazardous Waste Sites.* New York: United Church of Christ, Commission for Racial Justice, 1987.

U.S. Environmental Protection Agency. *Environmental Equity: Reducing Risk for All Communities: Report to the Administrator from the EPA Environmental Equity Workgroup.* Washington, DC: U.S. Government Printing Office, 1992.

U.S. General Accounting Office. *Siting of Hazardous Waste Landfills and Their Correlation with Racial and Economic Status of Surrounding Communities.* Washington, DC: U.S. Government Printing Office, 1983.

PERIODICALS

Russell, D. "Environmental Racism." *Amicus Journal* 11 (Spring 1989): 22–32.

Environmental refugees

The term environmental refugee was coined in the late 1980s by the **United Nations Environment Programme** and refers to people who are forced to leave their community of origin because the land can no longer support them. Environmental factors such as **soil erosion**, **drought**, or floods, which are often coupled with poor socioeconomic conditions, are the cause of this loss of **stability** and security. Many environmental influences may cause such a displacement and include the deterioration of agricultural land, natural or "unnatural" disasters, **climate** change, the destruction resulting from war, and environmental **scarcity**.

Environmental scarcity can be supply induced, demand induced, or structural. Supply induced scarcity refers to the depletion of agricultural resources, as in the erosion of cropland or **overgrazing**. Demand induced scarcity occurs when consumption of a resource increases or when **population growth** occurs, as is occurring in countries such as Philippines, Kenya, and Costa Rica. Structural scarcity results from the unequal social distribution of a resource within a community. The causes of environmental scarcity can occur simultaneously and in combination with each other, as seen in South Africa during the years of apartheid. Approximately 60% of the cropland in South Africa is marked by low organic content, and half of the country receives less than 19.5 in (500 mm) of annual precipitation. When these factors were coupled with the rapid soil erosion, overcrowding, and unequal social distribution of resources experienced at this time, environmental scarcity resulted. Other environmental influences such as climate change and natural disasters can greatly compound the problems related to scarcity. Those countries which are especially vulnerable to these other influences are those which are already experiencing the precursors to scarcity, for example, highly populated countries such as Egypt and Bangladesh. On Haiti, per capita grain production is half of what it was only 40 years ago and residents only get about 80% of their minimum nutritional needs. Environmental problems place an added burden on a situation that is already under pressure. When this combination occurs in societies without strong social ties or political and economic stability, many times the people within the population have no choice but to relocate.

Environmental refugees tend to come from rural areas and developing countries--those most vulnerable to the influences of scarcity, climate change, and natural disasters. According to the **Centers for Disease Control and Prevention**, since the early 1960s most emergencies involving refugees have taken place in these **less developed countries** where resources are inadequate to support the population during times of need. In 1995, CDC directed 45 relief missions to developing countries such as Angola, Bosnia, Haiti, and Sierra Leone.

The number of displaced people is rising worldwide. Of these, the number forced to migrate because of economic and environmental conditions is growing more rapidly than refugees from political strife. According to Dr. Norman Myers at the University of Oxford, there are 25 million environmental refugees today, compared with 20 million officially recognized refugees migrating due to political, religious, or ethical problems. It has been predicted that by the year 2010, this number could rise to 50 million. The number of migrants seeking environmental refuge is grossly underestimated because many do not actually cross borders but are forced to wander within their own country. As global warming causes major climate changes, these numbers could increase even more. Climate change alone may displace 150 million more people by the middle of the next century. Not only would a global climate change increase the number of refugees, it could have a negative impact on agricultural production which would seriously limit the amount of food surpluses available to help displaced people.

Although approximately three out of every five refugees are fleeing from environmental hardships, this group of refugees is not legally recognized. According to the 1951 Convention on the Status of Refugees as modified by the 1967 Protocol, a legal refugee is a person who escapes a country and cannot re-enter due to fear of persecution for reasons of race, religion, nationality, social affiliation, or political opinion. This definition requires both the element of persecution as well as cross-border **migration**. Because of these two requirements, states are not legally compelled to recognize environmental refugees; as mentioned above, many of these refugees never leave their own country, and it is unreasonable to expect them to prove fear of persecution. Environmental refugees are often forced to enter a country illegally since they cannot be granted protection or asylum. Many of the Mexican immigrants who enter the United States are escaping the sterile, unproductive land they have been living on. Over 60% of the land in Mexico is degraded, with soil erosion contributing to over 494,000 acres (200,000 ha) more land being rendered unproductive every year. Many of the people considered to be economic migrants are actually environmental refugees. Those that are recognized as political refugees often must live in overcrowded refugee camps which are no more prepared to sustain the population than the land they are escaping from. Two thousand Somali refugees forced to live in such camps on the border of Kenya were displaced once more when **flooding** in 1994 ended a long drought but destroyed relief food. Many environmental refugees never resettle and must live the rest of their lives migrating from place to place, looking for land that can sustain them.

Researchers are currently working on ways to predict where the next large migrations will come from and how to prevent them from occurring. Predictive models are extremely difficult to produce because of the interaction between the socioeconomic status of the people and the environmental influences on the land. Stuart Liederman, an environmental scientist at the University of New Hampshire, is developing a model which will predict which areas are at risk of producing environmental refugees. This model is a mathematical formula which could be used with any population. One side of the equation combines the rate of environmental decay, the amount of time over which this deterioration will take place, and the susceptibility of the people and the **environment**. The other side of the equation combines the restoration rate, the time it would take to reestablish the environment, and the potential for recovery of the people and the land. These predictions will allow for preventive measures to be taken, for preparations to be made for future refugees, and for restoring the devastated homelands of past migrants. Creation of a working model may also help convince policymakers that those escaping unlivable environmental conditions need to be legally recognized as refugees.

Until environmental refugees are granted legal status, there will be no protection or compensation granted these individuals. The most effective way to deal with the growing number of displaced persons is to concentrate on the reasons they are being forced to leave their homelands. The increasing number of people forced to leave their homeland due to ecological strife is an indicator of environmental quality. Environmental protection is necessary to prevent the situation in many countries from getting worse. In the meantime, measures must be taken to accommodate the needs of these refugees, and the environment must be recognized as a legitimate source of conflict for individuals seeking protection in other lands.

[*Jennifer L. McGrath*]

RESOURCES

BOOKS

Jacobson, J. L. *Environmental Refugees: A Yardstick of Habitability.* Worldwatch Institute, 1988.

Tickell, C. "Environmental Refugees: The Human Impact of Global Environmental Change." In *Greenhouse Glasnost: The Crisis of Global Warming,* edited by T. Minger. New York: Ecco Press, 1990.

Lindahl-Kiessling, K., and H. Landberg, eds. *Population, Economic Development, and the Environment.* New York: Oxford University Press, 1994.

PERIODICALS

Fell, N. "Outcasts from Eden." *New Scientist Magazine* 151, no. 2045 (August 1996): 24–7.

Homer-Dixon, T. F. "Environmental Change and Economic Decline in Developing Countries." *International Studies Notes* 16, no. 1 (Winter 1991): 18.

Myers, N. "Environmental Refugees in a Globally Warmed World." *BioScience* 43, no. 11 (December 1993): 752–61.

OTHER

Centers for Disease Control. *Famine Affected, Refugee, and Displaced Populations: Recommendations For Public Health Issues.* MMWR 1992; 41 (No RR-13).

Environmental resources

An environmental resource is any material, service, or information from the **environment** that is valuable to society. This can refer to anything that people find useful in their environs, or surroundings. Food from plants and animals, wood for cooking, heating, and building, metals, **coal**, and oil are all environmental resources. Clean land, air, and water are environmental resources, as are the abilities of land, air, and water to absorb society's waste products. Heat from the sun, **transportation** and **recreation** in lakes, rivers, and oceans, a beautiful view, or the discovery of a new **species** are all environmental resources.

The environment provides a vast array of materials and services that people use to live. Often these resources have competing uses and values. A piece of land, for instance, could be used as a farm, a park, a parking lot, or a housing development. It could be mined or used as a **garbage** dump. The topic of environmental resources, then, raises the question, what do people find valuable in their environment,

and how do people choose to use the resources that their environment provides?

Some resources are renewable, or infinite, and some are non-renewable, or finite. Renewable resources like energy from the sun are plentiful and will be available for a long time. Finite resources, like oil and coal, are non-renewable because once they are extracted from the earth and burned they cannot be used again. These resources are in limited supply and need to be used carefully. Many resources are becoming more and more limited, especially as population and industrial growth place increasing pressure on the environment. Before the Industrial Revolution, for example, people relied on their own strength and their animals for work and transportation. The invention of the steam engine in the 1850s radically altered peoples' ability to do work and to consume energy. Today we have transformed our environment with machines, cars, and **power plants** and in the process we have burnt extraordinary amounts of coal, oil, and **natural gas**. Some predict that world coal deposits will last another 200 years, while oil and natural gas reserves will last another one hundred years at current rates of consumption. This rate of use is clearly not sustainable. The terms finite and infinite are important because they indicate how much of a given resource is available, and how fast people can use that resource without limiting future supplies.

Some resources that were once taken for granted are now becoming more valuable. One of these resources is the environment's ability to absorb the waste that people produce. In Jakarta, Indonesia, people living in very close quarters in small shanties along numerous tidal canals use their only water supply for bathing, washing clothes, drinking water, fishing, and as a toilet. It is common to see people bathing just down stream from other people who are defecating directly into the river. This scene illustrates a central problem in environmental resource management. These people have only one water source and many needs in order to live. The demands that they place on these resources seriously affect the health and quality of life for all the people, but all of the needs must be met in some way. Thoughtful management of these environmental resources, like building latrines, could alleviate some of the strain on the river and improve other uses of the same resource. People all over the world have taken for granted the valuable resources of air, land, and **water quality** so that many rivers are undrinkable and unswimable because they contain raw sewage, chemical fertilizers, and industrial wastes. As people make decisions about what they will take from their environment, they also must be conscious of what they intend to put back into that environment.

Resource economics was established during a time in human history when environmental resources were thought to be limitless and without value until they were harvested and brought to market. From this viewpoint, the world is big enough that when one resource is exhausted another resource can be found to take its place. Land is valuable according to what can be taken from it in order to make a profit. This kind of management leads to enormous short term gains and is responsible for the speed and efficiency of economic growth throughout the world. One the other hand, this view overlooks longer term profits and the reality that the world is an increasingly small, interconnected, and fragile system. People can no longer assume that they can find fresh new supplies when they use up what they have. Very few places on earth remain untouched and unexploited.

The world's remaining forests, if managed with care, could supply all of society's needs for timber and still remain relatively healthy and intact. Forest resources can be renewable, since forests grow quickly enough to replace themselves if used in moderation. Unfortunately, in many places forests are being destroyed at an alarming rate. In Costa Rica, Central America, 25% of the remaining forest land has disappeared since 1970. These forests have been cleared to harvest tropical hardwoods, to create farmland and pasture for animals, and to forage wood for cooking and heating. In a country struggling for economic growth, these are all important needs, but they do not always make long term economic sense. Farmers who graze cattle in tropical rain forests or who clear trees off of steep hillsides destroy their land in a matter of years with the idea that this is the fastest way to make money. In the same way, loggers harvest trees for immediate sale, even though many of these trees take hundreds of years to replenish themselves. In fact, the price for tropical hardwoods has gone up four-fold since 1970. The trees cut and sold in 1970 represent a huge economic loss to the Costa Rican economy, since they were sold for a fraction of their present value. Often, the **soil** on this land quickly erodes downhill into streams and rivers, clogging the rivers with **sediment** and killing fish and other **wildlife**. This has the added drawback of damaging hydroelectric and **irrigation dams** and hurting the fishing industry.

Despite these tragic losses, Costa Rica is a model in Central America and in the world for finding alternative uses for its **natural resources**. Costa Rica has set aside one fifth of its total land area for **nature** preserves and **national park** lands. These beautiful and varied parks are valuable for several reasons. First, they help to protect and preserve a huge diversity of tropical species, many undiscovered and unstudied. Second, they protect a great deal of vegetation that is important in producing oxygen, stabilizing atmospheric chemistry, and preventing global **climate** change. Third, the natural beauty of these parks attracts many international tourists. Tourism is one of Costa Rica's major industries, providing much needed economic development. People from around the world appreciate the beauty and the wonder—the intangi-

ble values—of these resources. Local people who would have been hired one time to cut down a forest can now be hired for a lifetime to work as park rangers and guides. Some would also argue that these nature preserves have value in themselves without reference to human needs, simply because they are filled with beautiful living birds, insects, plants, and animals.

Much of the dialogue in environmental resource management is about the need to balance the needs for economic growth and prosperity with needs for sustainable resource use. In a limited, finite world, there is a need to close the gap between the rates of consumption and rates of supply. The debate over how to assign value to different environmental resources is a lively one because the way that people think about their environment directly affects how they interact with the world.

[*John Cunningham*]

RESOURCES
BOOKS

Ahmad, Y., et al. *Environmental Accounting and Sustainable Development: A UNEP World Bank Symposium.* Washington, DC: World Bank, 1989.

PERIODICALS

Repetto, R. "Accounting for Environmental Assets." *Scientific American* 266 (June 1992): 94–8+.

Environmental restoration
see **Restoration ecology**

Environmental risk analysis
see **Risk analysis**

Environmental science

Environmental science is often confused with other fields of related interest, especially **ecology**, environmental studies, **environmental education**, and **environmental engineering**. Renewed interest in environmental issues in the late 1960s and early 1970s, gave rise to numerous programs at many universities in the United States and other countries, most under two rubrics: environmental science or environmental studies. The former focused, as might be expected, on scientific questions and issues of environmental interest; the latter were often courses, with the emphasis on questions of **environmental ethics**, aesthetics, literature, etc.

These new academic units marked the first formal appearance of environmental science on most campuses, at least by that label. But environmental science is essentially the application of scientific methods and principles to the study of environmental questions, so it has probably been around in some form as long as science itself. Air and **water quality** research, for example, have been carried on in many universities for many decades: that research is environmental science.

By whatever label and in whatever unit, environmental science is not constrained within any one discipline; it is a comprehensive field. A considerable amount of environmental research is accomplished in specific departments such as chemistry, physics, civil engineering, or the various biology disciplines. Much of this work is confined to a single field, with no interdisciplinary perspective. These programs graduate scientists who build on their specific training to continue work on environmental problems, sometimes in a specific department, sometimes in an interdisciplinary environmental science program.

Many new academic units are interdisciplinary, their members and graduates specifically designated as environmental scientists. Most have been trained in a specific discipline, but they may have degrees from almost any scientific background. In these units, the degrees granted—from B.S. to Ph.D.—are in Environmental Science, not in a specific discipline.

Environmental science is not ecology, though that discipline may be included. Ecologists are interested in the interactions between some kind of organism and its surroundings. Most ecological research and training does not focus on environmental problems except as those problems impact the organism of interest. Environmental scientists may or may not include organisms in their field of view: they mostly focus on the environmental problem, which may be purely physical in **nature**. For example, **acid deposition** can be studied as a problem of emissions and characteristics of the **atmosphere** without necessarily examining its impact on organisms. An alternate focus might be on the **acidification** of lakes and the resulting implications for resident fish. Both studies require expertise from more than one traditional discipline; they are studies in environmental science. *See also* Air quality; Environment; Environmental ethics; Nature; Water quality

[*Gerald L. Young Ph.D.*]

RESOURCES
BOOKS

Cunningham, W. P. *Environmental Science: A Global Concern.* Dubuque, IA: William C. Brown, 1992.

Henry, J. G., and G. W. Heinke. *Environmental Science and Engineering.* Englewood Cliffs, NJ: Prentice-Hall, 1989.

Jorgensen, S. E., and I. Johnson. *Principles of Environmental Science and Technology.* 2nd ed. Amsterdam, NY: Elsevier, 1989.

Environmental stress

In the ecological context, environmental stress can be considered any environmental influence that causes a discernible ecological change, especially in terms of a constraint on **ecosystem** development. Stressing agents (or stressors) can be exogenous to the ecosystem, as in the cases of long-range transported acidifying substances, toxic gases, or pesticides. Stress can also cause change as a result of an accentuation of some pre-existing site factor beyond a threshold for biological tolerance, for example thermal **loading**, **nutrient** availability, wind, or temperature extremes.

Often implicit within the notion of environmental stress, particularly from the perspective of ecosystem managers, is a judgement about the quality of the ecological change. That is, from the human perspective, whether the effect is "good" or "bad."

Environmental stressors can be divided into several, not necessarily exclusive, classes of causal agencies:

• "Physical stress" refers to episodic events (or disturbance) associated with intense but usually brief loadings of kinetic energy, perhaps caused by a windstorm, volcanic eruption, tidal wave, or an explosion.

• Wildfire is another episodic stress, usually causing a mass **mortality** of ecosystem dominants such as trees or shrubs and a rapid **combustion** of much of the **biomass** of the ecosystem.

• Pollution occurs when certain **chemicals** are bio-available in a sufficiently large amount to cause toxicity. Toxic stressors include gaseous air pollutants such as **sulfur dioxide** and **ozone**, metals such as **lead** and **mercury**, residues of pesticides, and even nutrients that may be beneficial at small rates of supply but damaging at higher rates of loading.

• Nutrient impoverishment implies an inadequate availability of physiologically essential chemicals, which imposes an **oligotrophic** constraint upon ecosystem development.

• Thermal stress occurs when heat energy is released into an ecosystem, perhaps by aquatic discharges of low-grade heat from **power plants** and other industrial sources.

• Exploitative stress refers to the selective removal of particular **species** or size classes. Exploitation by humans includes the harvesting of forests or wild animals, but it can also involve natural herbivory and predation, as with infestations of defoliating insects such as locusts, spruce budworm, or **gypsy moth**, or irruptions of predators such as crown-of-thorns starfish.

• *Climatic stress* is associated with an insufficient or excessive regime of moisture, solar radiation, or temperature. These can act over the shorter term as weather, or over the longer term as **climate**.

Within most of these contexts, stress can be exerted either chronically or episodically. For example, the toxic gas sulfur dioxide can be present in a chronically elevated concentration in an urbanized region with a large number of point sources of **emission**. Alternatively, where the emission of sulfur dioxide is dominated by a single, large **point source** such as a **smelter** or power plant, the toxic stress associated with this gas occurs as relatively short-term events of **fumigation**.

Environmental stress can be caused by natural agencies as well as resulting directly or indirectly from the activities of humans. For example, sulfur dioxide can be emitted from smelters, power plants, and homes, but it can also be emitted in large quantities by volcanoes. Similarly, climate change has always occurred naturally, but it may also be forced by human activities that result in emissions of **carbon dioxide**, **methane**, and **nitrous oxide** into the **atmosphere**.

Over most of Earth's history, natural stressors have been the dominant constraints on ecological development. Increasingly, however, the direct and indirect consequences of human activities are becoming dominant environmental stressors. This is caused by both the increasing human population and by the progressively increasing intensification of the per-capita effect of humans on the **environment**.

[*Bill Freedman Ph.D.*]

RESOURCES
BOOKS

Freedman, B. "Environmental Stress and the Management of Ecological Reserves." In *Science and the Management of Protected Areas*, edited by J. H. M. Willison, et al. Amsterdam: Elsevier, 1992.
Grime, J. P. *Plant Strategies and Vegetation Processes.* New York: Wiley, 1979.

Environmental Stress Index

Environmental Stress Index (ZPG) is a survey to determine the quality of life in American cities. Zero Population Growth, Inc. based in Washington D.C., conducted this "Urban Stress Test" in the late 1980s. One hundred and ninety-two cities were selected throughout the United States. The population-linked survey was based on 11 criteria: Population change; Population density; Education; Violent crime; Community economics; Individual economics (percent below federal poverty level and per capita income); Births (percent of teenage births and infant **mortality**); **Air quality** (meeting **Environmental Protection Agency** (EPA) standards); **Hazardous waste** (number of EPA-designated hazardous waste sites); Water (quality and supply); Sewage (model cities provide better than secondary treatment of their **wastewater**).

Cites were ranked one to six with number one being best. The cities with the lower scores were called model cities. The cities with the higher scores were called the stressed cities. Among the model cities were Abilene, Texas, with an index of 1.6; Roanoke, Virginia, 1.6; Berkeley, California, 2.0; Colorado Springs, Colorado, 2.0; and Peoria, Illinois, 2.0.

Among America's worst cities were Phoenix, Arizona, 5; Houston, Texas, 4.5; Los Angeles, 4.3; Honolulu, Hawaii, 4.3; and Baltimore, Maryland, 4.3.

Environmental Working Group

The Environmental Working Group is a public interest research group that monitors public agencies and public policies on topics relating to environmental and social justice. EWG publicizes its findings in research reports that emphasize both national and local implications of federal laws and activities. These research reports are based on analysis of public databases, often obtained through the Freedom of Information Act. In operation since 1993, EWG is a nonprofit organization funded by grants from private foundations. EWG is based in Washington, D.C. and Oakland, California, and is associated with the Tides Center of San Francisco. The organization performs its research both independently and in collaboration with other public interest research groups such as the **Sierra Club** and the Surface **Transportation** Policy Project (a nonprofit coalition focusing on social and environmental quality in transportation policy).

EWG specializes in analyzing large computer databases maintained by government agencies, such as the Toxic Release Inventory database maintained by the **Environmental Protection Agency** to record spills of toxic **chemicals** into the air or water, or the Regulatory Analysis Management System, maintained by the **Army Corps of Engineers** for internal tracking of permits granted for filling and draining **wetlands**. Because many of these data sources are capable of exposing actions or policies embarrassing to public agencies, EWG has often obtained data by using the Freedom of Information Act, which legally enforces public release of data belonging to the public domain. Many of the databases researched by EWG have never been thoroughly analyzed before, even by the agency collecting the data. EWG is unusual in working with primary data—going directly to the original database—rather than basing its research on secondary sources, anecdotal information, or interviews.

Research findings are published both in print and electronically on the Internet. Electronic publishing allows immediate and inexpensive distribution of reports that concern issues of current interest. EWG is a prolific source of information, producing extensive, detailed reports, often at a rate of more than one a month. Among the environmental topics on which the EWG has reported are drinking **water quality**, wetland protection and destruction, and the impacts of agricultural pesticides on both farm workers and consumers. Social justice and policy issues that EWG has researched include campaign finance reform, inequalities and inefficiency in farm subsidy programs, and threats to public health and the **environment** from **medical waste**. For each general topic, EWG usually publishes a series of articles ranging from the nature and impacts of a federal law to what individuals can do about the current problem. Also included with research reports are state-by-state and county summaries of **statistics**, which provide details of local implications of the general policy issues.

[*Mary Ann Cunningham Ph.D.*]

RESOURCES
BOOKS

Cook, K. A., and A. Art. *City Slickers: Farm Subsidy Recipients in America's Big Cities.* Washington, DC: Environmental Working Group, 1995.
Environmental Working Group. *Dishonorable Discharge: Toxic Pollution of America's Waters.* Washington, DC: Environmental Working Group, 1996.

ORGANIZATIONS

Environmental Working Group, 1718 Connecticut Ave., NW, Suite 600, Washington, D.C. USA 20009 (202) 667-6982, Fax: (202) 232-2592, Email: info@ewg.org, <http://www.ewg.org>

Environmentalism

Environmentalism is the ethical and political perspective that places the health, harmony, and integrity of the natural **environment** at the center of human attention and concern. From this perspective human beings are viewed as *part of nature* rather than as overseers. Therefore to care for the environment is to care about human beings since we cannot live without the survival of the natural **habitat**.

Although there are many different views within the very broad and inclusive environmentalist perspective, several common features can be discerned. The first is environmentalism's emphasis on the interdependence of life and the conditions that make life possible. Human beings, like other animals, need clean air to breathe, clean water to drink, and nutritious food to eat. Without these necessities, life would be impossible. Environmentalism views these conditions as being both basic and interconnected. For example, fish contaminated with polychlorinated biphenyl (PCB), **mercury**, and other toxic substances are not only hazardous to humans but to bears, eagles, gulls, and other predators. Likewise, mighty **whales** depend on tiny **plankton**, cows on corn,

koala bears on eucalyptus leaves, bees on flowers, and flowers on bees and birds, and so on through all **species** and ecosystems. All animals, human and nonhuman alike, are interdependent participants in the cycle of birth, life, death, decay, and rebirth.

A second emphasis of environmentalism is on the sanctity of life—not only human life but all life, from the tiniest microorganism to the largest whale. Since the fate of our species is inextricably tied with theirs and since life requires certain conditions to sustain it, environmentalists contend that we have an obligation to respect and care for the conditions that nurture and sustain life in its many forms.

While environmentalists agree on some issues, there are also a number of disagreements about the purposes of environmentalism and about how to best achieve those ends. Some environmentalists emphasize the desirability of conserving **natural resources** for **recreation**, sightseeing, **hunting**, and other human activities, both for present and **future generations**. Such a utilitarian view has been sharply criticized by Arne Naess and other proponents of **deep ecology** who claim that the natural environment has its own **intrinsic value** apart from any aesthetic, recreational, or other value assigned to it by human beings. Bears, for example, have their own intrinsic value or worth, quite apart from that assigned to their existence via **shadow pricing** or other mechanism by bear-watchers, hunters, or other human beings.

Environmentalists also differ on how best to conserve, reserve, and protect the natural environment. Some groups, such as the **Sierra Club** and **the Nature Conservancy**, favor gradual, low-key legislative and educational efforts to inform and influence policy makers and the general public about environmental issues. Other more radical environmental groups, such as the **Sea Shepherd Conservation Society** and **Earth First!**, favor carrying out direct action by employing the tactics of ecotage (ecological sabotage), or **monkey-wrenching**, to stop **strip mining**, **logging**, drift net fishing, and other activities that they deem dangerous to animals and ecosystems. Within this environmental spectrum are many other groups, including the **World Wildlife Fund**, **Greenpeace**, **Earth Island Institute**, Clean Water Action, and other organizations which use various techniques to inform, educate, and influence public opinion regarding environmental issues and to lobby policy makers.

Despite these and other differences over means and ends, environmentalists agree that the natural environment, whether valued instrumentally or intrinsically, is valuable and worth preserving for present and future generations.

[*Terence Ball*]

RESOURCES

BOOKS

Chase, S., ed. *Defending the Earth: A Dialogue Between Murray Bookchin and Dave Foreman.* Boston: South End Press, 1991.

Devall, B., and G. Sessions. *Deep Ecology: Living as if Nature Mattered.* Layton, UT: Gibbs M. Smith, 1985.

Eckersley, R. *Environmentalism and Political Theory.* Albany, NY: State University of New York Press, 1992.

Naess, A. *Ecology, Community and Lifestyle.* New York: Cambridge University Press, 1989.

Worster, D. *Nature's Economy: A History of Ecological Ideas.* New York: Cambridge University Press, 1977.

Environmentally preferable purchasing

Environmentally preferable purchasing (EPP) invokes the practice of buying products with environmentally-sound qualities—reduced packaging, reusability, **energy efficiency**, recycled content and rebuilt or re-manufactured products. It was first addressed officially with Executive Order (EO) 12873 in October 1993, "Federal Acquisition, **Recycling** and Waste Prevention," but was further enhanced in September 14, 1998, in by EO 13101 also signed by President Clinton. Entitled, "Greening the Government through Waste Prevention, Recycling and Federal Acquisition," it superseded EO 12873, but retained similar directives for purchasing. The "Final Guidance" of directives was issued through the **Environmental Protection Agency** in 1995.

What the federal government would adopt as a guideline for its purchases also would mark the beginning of environmentally preferable purchasing for the private sector, and create an entirely new direction for individuals and businesses as well as governments. At the federal level, the EPA's "Final Guidance" was issued to apply to all acquisitions, from supplies and services to buildings and systems. It developed five "guiding principles" for incorporating the plan into the federal government setting.

The five guiding principles are listed as follows:

- Environment + Price + Performance = Environmentally Preferable Purchasing
- **Pollution** prevention
- Life cycle perspective/multiple attributes
- Comparison of environmental impacts
- Environmental performance information

Through its web site, in an entire section devoted to environmentally preferable purchasing, product and service information is provided that includes alternative fuels, buildings, cleaners, conferences, electronics, food serviceware, carpets, and copiers.

In the private world of business, environmentally preferable purchasing has promised to save money, in addition to meeting EPA regulations and improving employee safety and health. In an age of **environmental liability**, EPP can make the difference when a question of **environmental ethics**, or damage arises.

For the private consumer, purchasing "green" in the late 1960s and 1970s tended to mean something as simple as recycled paper used in Christmas cards, or unbleached natural fibers for clothing. By 2002, the average American home is affected in countless additional ways—energy efficient kitchen appliances and personal computers; environmentally-sound household cleaning products; and, neighborhood recycling centers. To be certified as "green" things such as recyclability, biodegradability, organic ingredients, and no **ozone** depleting **chemicals** are tested.

Of those everyday uses, the concern over cleaning products for home, industrial, and commercial use has been the focus of particular attention. Massachusetts has been one of the state's that has taken a lead in providing leadership on the issue of toxic chemicals with its **Massachusetts Toxic Use Reduction Act.** With a focus on products that have known carcinogens and ozone-depleting substances, excessive phosphate concentrations, and volatile organic compounds, testing has been continued to provide alternative products that are more environmentally acceptable—and safer for humans and all forms of life, as well. By 2002, the state had awarded contracts to six firms selling environmentally preferred cleaning agents.

The products approved for purchasing must follow the following mandated criteria:

- contain no ingredients from the Massachusetts Toxic Use Reduction Act list of chemicals
- contain no carcinogens appearing on lists established by the International Agency for Research on **Cancer**, the National Toxicology Program, or the **Occupational Safety and Health Administration**; and not contain any chemicals defined as Class A, B. or C carcinogens by the EPA
- contain no ozone-depleting ingredients
- must be compliant with the phosphate content levels stipulated in Massachusetts law
- must be compliant with the **Volatile Organic Compound** (VOC) content levels stipulated in Massachusetts law

The National Association of Counties offers an extensive list of EPP resources links through its web site. In addition to offices and agencies of the federal government, the states of Minnesota and Massachusetts, and the local governments of King County, Washington and Santa Monica, California, the list includes such organizations as, Buy Green, Earth Systems' Virtual Shopping Center for the Environment (an online database of recycling industry products and services), the **Environmental Health** Coalition, **Green Seal**, the National Institute of Government Purchasing, Inc., and the National Pollution Prevention Roundtable. Businesses and business-related companies mentioned include the **Chlorine** Free Products Association, **Pesticide Action Network** of North America, Chlorine-Free Paper Consortium, and the Smart Office Resource Center.

[*Jane E. Spear*]

RESOURCES

OTHER

Argonne National Laboratory, (U.S. Department of Energy). *Green Purchasing Links.* [cited June 2002]. <http://www.anl.gov/P2>.

Commonwealth of Massachusetts. *Environmentally Preferable Products Procurement Program.* [cited June 2002]. <http://www.state.ma.us/osd>.

Environmental Protection Agency. *Environmentally Preferable Purchasing.* 1998 [cited April 2002]. <http://www.epa.gov/opptintr/epp/finalguidance.htm>.

National Association of Counties. *Environmentally Preferred Purchasing Resources.* [cited July 2002]. <http://www.naco.org/links/env_pur.cfm>.

National Safety Council/Environmental Health Center. *Environmentally Preferable Purchasing.* May 8, 2001 [June 2002]. <http://www.nsc.org/ehc>.

NYCWasteLe$$ Government. *Environmentally Preferable Purchasing.* October 2001 [June 2002]. <http://www.nycwasteless.org/gov-bus/citysense/epp.htm>.

Ohio Environmental Protection Agency. *Environmentally Preferable Purchasing.* [cited June 2002]. <http://www.epa.state.oh.us/opp/epp-main.html>.

ORGANIZATIONS

Earth Systems, 508 Dale Avenue, Charlottesville, Virginia USA 22903 434-293-2022, <http://www.earthsystems.org>

National Association of Counties, 440 First Street, NW, Washington, D.C. USA 20001 202-393-6226, Fax: 202-393-2630, <http://www.naco.org>

U.S. Environmental Protection Agency, 1200 Pennsylvania Avenue, NW, Washington, D.C. USA 20460 202-260-2090, <http://www.epa.gov>

Environmentally responsible investing

Environmentally responsible investing is one component of a larger phenomenon known as *socially responsible investing.* The idea is that investors should use their money to support industries whose operations accord with the investors' personal ethics. This concept is not a new one. In the early part of the century, Methodists, Presbyterians, and Baptists shunned companies that promoted sinful activities such as smoking, drinking, and gambling. More recently, many investors chose to protest apartheid by divesting from companies with operations in South Africa. Investors today might arrange their investment portfolios to reflect companies' commitment to affirmative action, human rights, **animal rights**, the **environment**, or any other issues the investors believe to be important.

The purpose of environmentally responsible investing is to encourage companies to improve their environmental records. The recent emergence and growth of mutual funds identifying themselves as environmentally oriented funds indicates that environmentally responsible investing is a popular investment area for the 1990s. In 1990, around $1 billion were invested in environmentally oriented mutual funds. The naming of these funds can be misleading, however. Some funds have been developed for the purpose of being environmentally responsible; others have been developed for the purpose of reaping the profits anticipated to occur in the environmental services sector as environmentalists in the marketplace and environmental regulations encourage the purchasing of **green products** and technology. These funds are not necessarily environmentally responsible; some companies in the environmental clean-up industry, for example, have less than perfect environmental records.

As the idea of environmentally responsible investing is still new, a generally accepted set of criteria for identifying environmentally responsible companies has not yet emerged. The fact is that everyone pollutes to some extent. The question is where to draw the line between acceptable and unacceptable behavior toward the environment.

When grading a company in terms of its behavior toward the environment, one could use an absolute standard. For example, one could exclude all companies that have violated any **Environmental Protection Agency** (EPA) standards. The problem with such a standard is that some companies that have very good overall environmental records have sometimes failed to meet certain EPA standards. Alternatively, a company could be graded on its efforts to solve environmental problems. Some investors prefer to divest of all companies in heavily polluting industries, such as oil and chemical companies; others might prefer to use a relative approach and examine the environmental records of companies within industry groups. By directly comparing oil companies with other oil companies, for example, one can identify the particular companies committed to improving the environment.

For consistency, some investors might choose to divest from all companies that supply or buy from an environmentally irresponsible company. It then becomes an arbitrary decision as to where this process stops. If taken to an extreme, the approach rejects holding United States treasury securities, since public funds are used to support the military, one of the world's largest polluters and a heavy user of nonrenewable energy.

A potential new indicator for identifying environmentally responsible companies has been developed by the Coalition for Environmentally Responsible Economies (CERES); it is a code called the **Valdez Principles**. The principles are the environmental equivalent of the Sullivan Principles, a code of conduct for American companies operating in South Africa. The Valdez Principles commit companies to strive to achieve sustainable use of **natural resources** and the reduction and safe disposal of waste. By signing the principles, companies commit themselves to continually improving their behavior toward the environment over time. So far, however, few companies have signed the code, possibly because it requires companies to appoint environmentalists to their boards of directors.

As there is no generally accepted set of criteria for identifying environmentally responsible companies, investors interested in such an investment strategy must be careful about accepting "environmentally responsible" labels. Investors must determine their own set of screening criteria based on their own personal beliefs about what is appropriate behavior with respect to the environment.

[*Barbara J. Kanninen*]

RESOURCES
BOOKS

Brill, J. A., and A. Reder. *Investing From the Heart.* New York: Crown, 1992.
Harrington, J. C. *Investing With Your Conscience.* New York: Wiley, 1992.

PERIODICALS

McMurdy, D. "Green Is the Color of Money [Environmental Investing in Canada]." *Maclean's,* December 1991, 49–50.
Rauber, P. "The Stockbroker's Smile [Environmental Sector Funds]." *Sierra* 75 (July-August 1990): 18–21.

Enzyme

Enzymes are catalysts, compounds (a protein) that speed up the rate at which chemical reactions occur within living organisms without undergoing any permanent change themselves. They are crucial to life since, without them, the vast majority of biochemical reactions would occur too slowly for organisms to survive.

In general, enzymes catalyze two quite different kinds of reactions. The first type of reaction includes those by which simple compounds are combined with each other to make new tissue from which plants and animals are made. For example, the most common enzyme in **nature** is probably carboxydismutase, the enzyme in green plants that couples **carbon dioxide** with an acceptor molecule in one step of the **photosynthesis** process by which carbohydrates are produced.

Enzymes also catalyze reactions by which more complex compounds are broken down to provide the energy needed by organisms. The principal digestive enzyme in the human mouth, for example, is ptyalin (also known as a -amylase), which begins the digestion of starch.

Enzymes have both beneficial and harmful effects in the **environment**. On the one hand, environmental hazards such as **heavy metals**, pesticides, and radiation often exert their effects on an organism by disabling one or more of its critical enzymes. As an example, **arsenic** is poisonous to animals because it forms a compound with the enzyme glutathione. The enzyme is disabled and prevented from carrying out its normal function, the maintenance of healthy red blood cells.

On the other hand, uses are now being found for enzymes in cleaning up the environment. For example, the Novo Nordisk company has discovered that adding an enzyme known as Pulpzyme® can vastly reduce the amount of **chlorine** needed to bleach wood pulp in the manufacture of paper. Since chlorine is a serious environmental contaminant, this technique may represent a significant improvement on present pulp and paper manufacturing techniques.

[*David E. Newton*]

EPA

see **Environmental Protection Agency**

Ephemeral species

Ephemeral **species** are plants and animals whose lifespan lasts only a few weeks or months. The most common types of ephemeral species are **desert** annuals, plants whose seeds remain dormant for months or years but which quickly germinate, grow, and flower when rain does fall. In such cases the amount and frequency of rainfall determine entirely how frequently ephemerals appear and how long they last. Tiny, usually microscopic, insects and other invertebrate animals often appear with these desert annuals, feeding on briefly available plants, quickly reproducing, and dying in a few weeks or less. Ephemeral ponds, short-duration desert rain pools, are especially noted for supporting ephemeral species. Here small insects and even amphibians have ephemeral lives. The spadefoot toad (*Scaphiopus multiplicatus*), for example, matures and breeds in as little as eight days after a rain, feeding on short-lived brine shrimp, which in turn consume algae and plants that live as long as water or **soil** moisture lasts. Eggs, or sometimes the larvae of these animals, then remain in the soil until the next moisture event.

Ephemerals play an important role in many plant communities. In some very dry deserts, as in North Africa, ephemeral annuals comprise the majority of living species—although this rich **flora** can remain hidden for years at a time. Often widespread and abundant after a rain, these plants provide an essential food source for desert animals, including domestic livestock. Because water is usually unavailable in such environments, many desert perennials also behave like ephemeral plants, lying dormant and looking dead for months or years but suddenly growing and setting seed after a rare rain fall.

The frequency of desert ephemeral recurrence depends upon moisture availability. In the Sonoran Desert of California and Arizona, annual precipitation allows ephemeral plants to reappear almost every year. In the drier deserts of Egypt, where rain may not fall for a decade or more, dormant seeds must survive for a much longer time before germination. In addition, seeds have highly sensitive germination triggers. Some annuals that require at least one inch (two to three cm) of precipitation in order to complete their life cycle will not germinate when only one centimeter has fallen. In such a case seed coatings may be sensitive to soil **salinity**, which decreases as more rainfall seeps into the ground. Annually-recurring ephemerals often respond to temperature, as well. In the Sonoran Desert some rain falls in both summer and winter. Completely different summer and winter floral communities appear in response. Such **adaptation** to different temporal niches probably helps decrease **competition** for space and moisture and increase each species' odds of success.

Although they are less conspicuous, ephemeral species also occur outside of desert environments. Short-duration food supplies or habitable conditions in some marine environments lead to ephemeral species growth. Ephemerals successfully exploit such unstable environments as volcanoes and steep slopes prone to slippage. More common are spring ephemerals in temperate deciduous forests. For a few weeks between snow melt and closure of the overstory canopy, quick-growing ground plants, including small lilies and violets, sprout and take advantage of available sunshine. Flowering and setting seed before they are shaded out by larger vegetation, these ephemerals disappear by mid-summer. Some persist in the form of underground root systems, but others are true ephemerals, with only seeds remaining until the next spring. *See also* Adaptation; Food chain/web; Opportunistic organism

[*Mary Ann Cunningham Ph.D.*]

RESOURCES
BOOKS

Whitford, W. G. *Pattern and Process in Desert Ecosystems.* Albuquerque: University of New Mexico Press, 1986.

Zahran, M. A., and A. J. Willis. *The Vegetation of Egypt.* London: Chapman and Hall, 1992.

PERIODICALS

Hughes, J. "Effects of Removal of Co-Occurring Species on Distribution and Abundance of *Erythronium americanum* (Liliaceae), a Spring Ephemeral." *American Journal of Botany* 79 (1990): 1329–39.

Went, F. W. "The Ecology of Desert Plants." *Scientific American* 192 (1955): 68–75.

Epidemiology

Epidemiology, the study of epidemics, is sometimes called the medical aspect of **ecology** because it is the study of diseases in animal populations, including humans. The epidemiologist is concerned with the interactions of organisms and their environments as related to the presence of disease. Environmental factors of disease include geographical features, **climate**, and concentration of pathogens in **soil** and water. Epidemiology determines the numbers of individuals affected by a disease, the environmental circumstances under which the disease may occur, the causative agents, and the transmission of disease.

Epidemiology is commonly thought to be limited to the study of infectious diseases, but that is only one aspect of the medical specialty. The epidemiology of the **environment** and lifestyles has been studied since Hippocrates's time. More recently, scientists have broadened the worldwide scope of epidemiology to studies of violence, of heart disease due to lifestyle choices, and to the spread of disease because of **environmental degradation**.

Epidemiologists at the Epidemic Intelligence Service (EIS) of the **Centers for Disease Control and Prevention** have played important roles in landmark epidemiologic investigations. Those include the identification in 1955 of a lot of poliovirus vaccine, supposedly dead, that was contaminated with live polio **virus**; an investigation of the definitive epidemic of Legionnaires' disease in 1976; identification of tampons as a risk factor for toxic-shock syndrome; and investigation of the first cluster of cases that came to be called acquired immunodeficiency syndrome (**AIDS**). EIS officers are increasingly involved in the investigation of noninfectious disease problems, including the risk of injury associated with all-terrain vehicles and cluster deaths related to flour contaminated with parathion.

The epidemiological classification of disease deals with the incidence, distribution, and control of disorders of a population. Using the example of typhoid, a disease spread through contaminated food and water, scientists first must establish that the disease observed is truly caused by *Salmonella typhosa*, the typhoid organism. Investigators then must know the number of cases, whether the cases were scattered over the course of a year or occurred within a short period, and the geographic distribution. It is critical that the precise locations of the diseased patients be established. In a hypothetical case, two widely separated locations within a city might be found to have clusters of cases of typhoid arising simultaneously. It might be found that each of these clusters revolved around a family unit, suggesting that personal relationships might be important. Further investigation might disclose that all of the infected persons had dined at one time or at short intervals in a specific home, and that the person who had prepared the meal had visited a rural area, suffered a mild attack of the disease, and now was spreading it to family and friends by unknowing contamination of food.

One very real epidemic of **cholera** in the West African nation of Guinea-Bissau was tracked by CDC researchers using maps, interviews, and old-fashioned footwork door-to-door through the country. An investigator eventually tracked the source of the cholera outbreak to contaminated shellfish.

Epidemic diseases result from an ecological imbalance of some kind. Ecological imbalance, and hence, epidemic disease may be either naturally caused or induced by man. A breakdown in **sanitation** in a city, for example, offers conditions favorable for an increase in the rodent population, with the possibility that diseases may be introduced into and spread among the human population. In this case, an epidemic would result as much from an alteration in the environment as from the presence of a causative agent. For example, an increase in the number of epidemics of viral encephalitis, a brain disease, in man has resulted from the ecological imbalance of mosquitoes and wild birds caused by man's exploitation of lowland for farming. Driven from their natural **habitat** of reeds and rushes, the wild birds, important natural hosts for the virus that causes the disease, are forced to feed near farms; mosquitoes transmit the virus from birds to cattle to man.

Lyme disease, which was tracked by epidemiologists from man to deer to the ticks which infest deer, is directly related to environmental changes. The lyme disease spirochete probably has been infecting ticks for a long time; museum specimens of ticks collected on Long Island in the 1940s were found to be infected. Since then, tick populations in the Northeast have increased dramatically, triggering the epidemic.

There are more ticks because many of the forests that had been felled in the Northeast have returned to forestland. Deer populations in those areas have exploded, close to concentrated human populations, as have the numbers of *Ixodes dammini* ticks which feed on deer. The deer do not become ill, but when a tick bite infects a human host, the result can be a devastating disease, including crippling arthritis and memory loss.

Disease detectives, as epidemiologists are called, are taking on new illnesses like heart disease and **cancer**, diseases that develop over a lifetime. In 1948, epidemiologists enrolled 5,000 people in Framingham, Massachusetts, for a study on heart disease. Every two years the subjects have undergone physicals and answered survey questions. Epidemiologists began to understand what factors put people at risk, such as high blood pressure, elevated cholesterol levels, smoking, and lack of exercise.

CDC epidemiologists are now tracking the pattern of violence, traditionally a matter for police. If a pattern is found, then young people who are at risk can be taught to stop arguments before they escalate to violence, or public health workers can recognize behaviors that lead to spouse abuse, or the warning signs of teenage suicide, for example.

In the 1980s, classic epidemiology discovered that a puzzling array of illnesses was linked, and it came to be known as AIDS. Epidemiologists traced the disease to sexual contact, then to contaminated blood supplies, then proved the AIDS virus could cross the placental barrier, infecting babies born to HIV-infected mothers.

The AIDS virus, called human immunodeficiency virus, may have existed for centuries in African monkeys and apes. Perhaps 40 years ago, this virus crossed from monkey to man, although researchers do not know how or why. African **chimpanzees** can be infected with HIV, but they don't develop the disease, suggesting that chimps have developed protective immunity. Eventually AIDS, over centuries, probably will develop into a less deadly disease in humans. But before then, researchers fear that new, more deadly, diseases will evolve.

As human communities change and create new ways for diseases to spread, viruses and bacteria constantly evolve as well. Rapidly increasing human populations prove a fertile breeding ground for **microbes**, and as the planet becomes more crowded, the distances that separate communities become smaller.

Epidemiology has become one of the important sciences in the study of nutritional and biotic diseases around the world. The United Nations supports, in part, a World Health Organization investigation of nutritional diseases.

Epidemiologists have also been called upon in times of natural emergencies. When **Mount St. Helens** erupted on May 18, 1980, CDC epidemiologists were asked to assist in an epidemiologic evaluation. The agency funded and assisted in a series of studies on the health effects of dust exposure, occupational exposure, and mental health effects of the volcanic eruption.

In 1990, CDC epidemiologists began research for the Department of Energy to study people who have been exposed to radiation. A major task of the study is to quantify exposures based on historical reconstructions of emissions from nuclear plant operations. Epidemiologists have undertaken a major thyroid disease study for those people exposed to radioactive iodine as a result of living near the **Hanford Nuclear Reservation** in Richland, Washington, during the 1940s and 1950s.

[*Linda Rehkopf*]

RESOURCES
BOOKS

Friedman, G. D. *Primer of Epidemiology.* 3rd ed. New York: McGraw-Hill, 1987.

Goldsmith, J. R., ed. *Environmental Epidemiology: Epidemiological Investigations of Community Environmental Problems.* St. Louis: CRC Press, 1986.

Kopfler, F. C., and G. Craun, eds. *Environmental Epidemiology.* Chelsea, MI: Lewis, 1986.

Erodible

Susceptible to **erosion** or the movement of **soil** or earth particles due to the primary forces of wind, moving water, ice and gravity. Tillage implements may also move soil particles, but this transport is usually not considered erosion.

Erosion

Erosion is the wearing away of the land surface by running water, wind, ice, or other geologic agents, including such processes as gravitational creep.

The term geologic erosion refers to the normal, natural erosion caused by geological processes acting over long periods of time, undisturbed by humans. Accelerated erosion is a more rapid erosion process influenced by human, or sometimes animal, activities. Accelerated erosion in North America has only been recorded for the past few centuries, and in research studies, postsettlement erosion rates were found to be eight to 350 times higher than presettlement erosion rates.

Soil erosion has been both accelerated and controlled by humans since recorded history. In Asia, the Pacific, Africa, and South America, complex **terracing** and other erosion control systems on **arable land** go back thousands of years. Soil erosion and the resultant decreased food supply have been linked to the decline of historic, particularly Mediterranean, civilizations, though the exact relationship with the decline of governments such as the Roman Empire is not clear.

A number of terms have been used to describe different types of erosion, including gully erosion, rill erosion, interrill erosion, sheet erosion, splash erosion, saltation, surface creep, suspension, and **siltation**. In gully erosion, water accumulates in narrow channels and, over short periods, removes the soil from this narrow area to considerable depths, ranging from 1.5 ft (0.5 m) to as much as 82–98 ft (25–30 m).

Rill erosion refers to a process in which numerous small channels of only a few inches in depth are formed, usually occurring on recently cultivated soils. Interrill erosion is the removal of a fairly uniform layer of soil on a multitude of relatively small areas by rainfall splash and film flow.

Usually interpreted to include rill and interril erosion, sheet erosion is the removal of soil from the land surface by rainfall and surface **runoff**. Splash erosion, the detachment and airborne movement of small soil particles, is caused by the impact of raindrops on the soil.

Saltation is the bouncing or jumping action of soil and mineral particles caused by wind, water, or gravity. Saltation occurs when soil particles 0.1–0.5 mm in diameter are blown to a height of less than 6 in (15 cm) above the soil surface for relatively short distances. The process includes gravel or stones effected by the energy of flowing water, as well as any soil or mineral particle movement downslope due to gravity.

Surface creep, which usually requires extended observation to be perceptible, is the rolling of dislodged particles 0.5–1.0 mm in diameter by wind along the soil surface. Suspension occurs when soil particles less than 0.1 mm diameter are blown through the air for relatively long distances, usually at a height of less than 6 in (15 cm) above the soil surface. In siltation, decreased water speed causes deposits water-borne sediments, or **silt**, to build up in stream channels, lakes, reservoirs, or flood plains.

In the water erosion process, the eroded **sediment** is often higher (enriched) in organic matter, **nitrogen**, **phosphorus**, and potassium than in the bulk soil from which it came. The amount of enrichment may be related to the soil, amount of erosion, the time of sampling within a storm, and other factors. Likewise, during a wind erosion event, the eroded particles are often higher in clay, organic matter, and plant nutrients. Frequently, in the Great Plains, the surface soil becomes increasingly more sandy over time as wind erosion continues.

Erosion estimates using the Universal Soil Loss Equation (USLE) and the Wind Erosion Equation (WEE) estimate erosion on a point basis expressed in mass per unit area. If aggregated for a large area (e.g., state or nation), very large numbers are generated and have been used to give misleading conclusions. The estimates of USLE and WEE indicate only the soil moved from a point. They do not indicate how far the sediment moved or where it was deposited. In cultivated fields, the sediment may be deposited in other parts of the field with different crop cover or in areas where the land slope is less. It may also be deposited in **riparian land** along stream channels or in flood plains.

Only a small fraction of the water-eroded sediment leaves the immediate area. For example, in a study of five river watersheds in Minnesota, it was estimated that from less than 1–27% of the eroded material entered stream channels, depending on the soil and topographic conditions. The deposition of wind-eroded sediment is not well quantified, but much of the sediment is probably deposited in nearby

Soil erosion on a trail in the Adirondack Mountains (Photograph by Yoav Levy. Phototake. Reproduced by permission.)

areas more protected from the wind by vegetative cover, stream valleys, road ditches, woodlands, or farmsteads.

While a number of national and regional erosion estimates for the United States have been made since the 1920s, the methodologies of estimation and interpretations have been different, making accurate time comparisons impossible. The most extensive surveys have been made since the Soil, Water and Related Resources Act was passed in 1977. In these surveys a large number of points were randomly selected, data assembled for the points, and the Universal Soil Loss Equation (USLE) or the Wind Erosion Equation (WEE) used to estimate erosion amounts. While these equations were the best available at the time, their results are only estimations, and subject to interpretation. Considerable research on improved methods of estimation is underway by the **U.S. Department of Agriculture**.

In the cornbelt of the United States, water erosion may cause a 1.7–7.8% drop in soil productivity over the next one hundred years, as compared to current levels, depending on the **topography** and soils of the area. The U.S.D.A. results, based on estimated erosion amounts for 1977, only included sheet erosion, not losses of plant nutrients. Though the figures may be low for this reason, other surveys have produced similar estimates.

In addition to depleting farmlands, eroded sediment causes off-site damages that, according to one study, may exceed on-site loss. The sediment may end up in a domestic water supply, clog stream channels, even degrade **wetlands**, **wildlife** habitats, and entire ecosystems. *See also* Environmental degradation; Gillied land; Soil eluviation; Soil organic matter; Soil texture

[*William E. Larson*]

RESOURCES

BOOKS

Paddock, J. N., and C. Bly. *Soil and Survival: Land Stewardship and the Future of American Agriculture.* San Francisco: Sierra Club Books, 1987.
Resource Conservation Glossary. 3rd ed. Ankeny, IA: Soil Conservation Society of America, 1982.

PERIODICALS

Steinhart, P. "The Edge Gets Thinner." *Audubon* 85 (November 1983): 94–106+.

OTHER

Brown, L. R., and E. Wolf. "Soil Erosion: Quiet Crisis in the World Economy." *Worldwatch Paper #60.* Washington DC: Worldwatch Institute, 1984.

Escherichia coli

Escherichia coli, or *E. coli* is a bacterium in the family *Enterobacteriaceae* that is found in the intestines of warm-blooded animals, including humans. *E. coli* represent about 0.1% of the total bacteria of an adult's intestines (on a Western diet). As part of the normal **flora** of the human intestinal tract, *E. coli* aids in food digestion by producing vitamin K and B-complex vitamins from undigested materials in the large intestine and suppresses the growth of harmful bacterial **species**. However, *E. coli* has also been linked to diseases in about every part of the body. Pathogenic strains of *E. coli* have been shown to cause pneumonia, urinary tract infections, wound and blood infections, and meningitis.

Toxin-producing strains of *E. coli* can cause severe gastroenteritis (hemorrhagic colitis), which can include abdominal pain, vomiting, and bloody diarrhea. In most people, the vomiting and diarrhea stop within two to three days. However, about 5–10% of the those affected will develop hemolytic-uremic syndrome (HUS), which is a rare condition that affects mostly children under the age of 10, but also may affect the elderly as well as persons with other illnesses. About 75% of HUS cases in the United States are caused by an enterohemorrhagic (intestinally-related organism that causes hemorrhaging) strain of *E. coli* referred to as *E. coli* O157:H7, while the remaining cases are caused by non-O157 strains. *E. coli.* O157:H7 is found in the intestinal tract of cattle. In the United States, the **Centers for Disease**

Control and Prevention estimates that there are about 10,000–20,000 infections and 500 deaths annually that are caused by *E. coli* O157:H7.

E. coli O157:H7, first identified in 1982, and isolated with increasing frequency since then, is found in contaminated foods such as meat, dairy products, and juices. Symptoms of an *E. coli* O157:H7 infection start about seven days after infection with the bacteria. The first symptom is sudden onset of severe abdominal cramps. After a few hours, watery diarrhea begins, causing lose of fluids and electrolytes (dehydration), which causes the person to feel tired and ill. The watery diarrhea lasts for about a day, and then changes to bright red bloody stools, as the infection causes sores to form in the intestines. The bloody diarrhea lasts for two to five days, with as many as 10 bowel movements a day. Additional symptoms may include nausea and vomiting, without a fever, or with only a mild fever. After about five to 10 days, HUS can develop. HUS is characterized by destruction of red blood cells, damage to the lining of blood vessel walls, reduced urine production, and in severe cases, kidney failure. **Toxins** produced by the bacteria enter the blood stream, where they destroy red blood cells and platelets, which contribute to the clotting of blood. The damaged red blood cells and platelets clog tiny blood vessels in the kidneys, or cause lesions to form in the kidneys, making it difficult for the kidneys to remove wastes and extra fluid from the body, resulting in hypertension, fluid accumulation, and reduced production of urine. The diagnosis of an *E. coli* infection is made through a stool culture.

Treatment of HUS is supportive, with particular attention to management of fluids and electrolytes. Some studies have shown that the use of antibiotics and antimotility agents during an *E. coli* infection may worsen the course of the infection and should be avoided. Ninety percent of children with HUS who receive careful supportive care survive the initial acute stages of the condition, with most having no long-term effects. In about 50% of the cases, short term replacement of kidney function is required in the form of dialysis. However, between 10 and 30% of the survivors will have kidney damage that will lead to kidney failure immediately or within several years. These children with kidney failure require on-going dialysis to remove wastes and extra fluids from their bodies, or may require a kidney transplant.

The most common way an *E. coli* O157:H7 infection is contracted is through the consumption of undercooked ground beef (e.g., eating hamburgers that are still pink inside). Healthy cattle carry *E. coli* within their intestines. During the slaughtering process, the meat can become contaminated with the *E. coli* from the intestines. When contaminated beef is ground up, the *E. coli* bacteria are spread throughout the meat. Additional ways to contract an *E. coli*

infection include drinking contaminated water and unpasteurized milk and juices, eating contaminated fruits and vegetables, and working with cattle. The infection is also easily transmitted from an infected person to others in settings such as day care centers and nursing homes when improper sanitary practices are used.

Prevention of HUS caused by ingestion of foods contaminated with *E. coli* O157:H7 and other toxin-producing bacteria is accomplished through practicing hygienic food preparation techniques, including adequate hand washing, cooking of meat thoroughly, defrosting meats safely, vigorous washing of fruits and vegetables, and handling leftovers properly. Irradiation of meat has been approved by the United States **Food and Drug Administration** and the United States Department of Agriculture in order to decrease bacterial contamination of consumer meat supplies.

The presence of *E. coli* in surface waters indicates that there has been fecal contamination of the water body from agricultural and/or urban and residential areas. However, the contribution from human vs. agricultural sources is difficult to determine. Since the concentration of *E. coli* in a surface water body is dependent on **runoff** from various sources of contamination, it is related to the **land use** and **hydrology** of the surrounding **watershed**. *E. coli* concentrations at a specific location in a water body will vary depending on the bacteria levels already in the water, inputs from various sources, dilution with precipitation and runoff, and **die-off** or multiplication of the organism within the water body. Sediments can act as a **reservoir** for *E. coli*, as the sediments protect the organisms from bacteriophages and microbial toxicants. The *E. coli* can persist in the sediments and contribute to concentrations in the overlying waters for months after the initial contamination.

Routine monitoring for enteropathogens, which cause gastrointestinal diseases and are a result of fecal contamination, is necessary to maintain water that is safe for drinking and swimming. Many of these organisms are hard to detect, so monitoring of an **indicator organism** is used to determine fecal contamination. To provide safe drinking water, the water is treated with **chlorine**, ultra-violet light, and/or **ozone**. Traditionally fecal **coliform bacteria** have been used as the indicator organisms for monitoring, but the test for these bacteria also detects thermotolerant non-fecal coliform bacteria.

Therefore, the U.S. **Environmental Protection Agency** (EPA) is recommending that *E. coli* as well as enterococci be used as indicators of fecal contamination of a water body instead of fecal coliform bacteria. The test for *E. coli* does not include non-fecal thermotolerant coliforms. An epidemiological study has shown that even though the strains of *E. coli* present in a water body may not be pathogenic, these organisms are the best predictor of swimming-associated gastrointestinal illness.

The U.S. EPA recreational **water quality** standard is based on a threshold concentration of *E. coli* above which the health risk from waterborne disease is unacceptably high. The recommended standard corresponds to approximately 8 gastrointestinal illnesses per 1000 swimmers. The standard is based on two criteria: 1) a geometric mean of 126 organisms per 100 ml, based on several samples collected during dry weather conditions, or 2) 235 organisms/100 ml sample for any single water sample. During 2002, the U.S. EPA finalized guidance on the use of *E. coli* as the basis for bacterial water quality criteria to protect recreational freshwater bodies.

[*Judith L. Sims*]

RESOURCES

BOOKS

Bell, Chris, and Alec Kyriakides. *E. coli.* Boca Raton: Chapman & Hall, 1998.

Burke, Brenda Lee. *Don't Drink the Water: The Waterton Tragedy.* Victoria, BC: Trafford Publishing, 2001.

Parry, Sharon, and S. Palmer. *E. coli: Environmental Health Issues of Vtec 0157.* London, UK:Spon Press, 2001.

Sussman, Max, ed. *Escherichia coli: Mechanisms of Virulence.* Cambridge, UK: Cambridge University Press, 1997.

U.S. Environmental Protection Agency. *Implementation Guidance for Ambient Water Quality Criteria for Bacteria.* Draft, EPA-823-B-02-003. Washington, DC: U.S. Environmental Protection Agency, 2002.

PERIODICALS

Koutkia, Polyxeni, Eleftherios Mylonakis, and Timothy Flanigan. "Enterohemorrhagic *Escherichia coli*: An Emerging Pathogen." *American Family Physician*, 56, no. 3 (September 1, 1997): 853–858.

OTHER

"*Escherichia coli* and Recreational Water Quality in Vermont." *Bacterial Water Quality.* February 7, 2000 [June 2002]. <http://snr.uvm.edu/www/pc/sal/ecoli/index.htm>.

U.S. Food and Drug Administration. "*Escherichia coli.*" *Bad Bug Book.* February 13, 2002 [cited May 25, 2002]. <http://vm.cfsan.fda.gov/~mow/chap15.html>.

Essential fish habitat

Essential Fish Habitat (EFH) is a federal provision to conserve and sustain the habitats that fish need to go through their life cycles. The United States Congress in 1996 added the EFH provision to the Magnuson Fishery **Conservation** and Management Act of 1976. Renamed the Magnuson-Stevens Conservation and Management Act in 1996, the act is the federal law that governs marine (sea) fishery management in the United States.

The amended act required that fishery management plans include designations and descriptions of essential fish habitats. The plan is a document describing the strategy to reach management goals in a fishery, an area where fish breed and people catch them. The Magnuson-Stevens Act covers plans for waters located within the United States' **exclusive economic zone**. The zone extends offshore from the coastland for three to 200 miles.

The designation of EFH was necessary because the continuing loss of aquatic **habitat** posed a major longterm threat to the viability of commercial and recreational fisheries, Congress said in 1996. Lawmakers defined EFH as "those waters and substrate necessary to the fish for spawning, breeding, feeding, or growth to maturity". Substrate consists of **sediment** and structures below the water.

The Magnuson-Stevens Act called for identification of EFH by eight regional fishery management councils and the Highly Migratory Species Division of the National Marine Fisheries Service (NMFS), an agency of the Commerce Department's **National Oceanic and Atmospheric Administration** (NOAA). Under the Magnuson-Stevens Act, NOAA manages more than 700 **species**. These species range from tiny reef fish to large tuna.

NOAA Fisheries and the councils are required by the act to minimize "to the extent practicable" the adverse effects of fishing on EFH. The act also directed the councils and NOAA to devise plans to conserve and enhance EFH. Those plans are included in the management plans. Also in the plan are "habitat areas of particular concern." These areas within an EFH include rare habitat or habitat that is ecologically important.

Furthermore, the act required federal agencies to work with NMFS when the agencies plan to authorize, finance, or carry out activities that could adversely affect EFH. This process called an EFH consultation is required if the agency plans an activity like **dredging** near an essential fishing habitat. While NMFS does not have veto power over the project, NOAA Fisheries will provide conservation recommendations.

The eight regional fishery management councils were established by the 1976 Magnuson Fishery and Conservation Management Act. That legislation also established the exclusive economic zone and staked the United States' claim to it. The 1976 act also addressed issues such as foreign fishing and how to connect the fishing community to the management process, according to an NOAA report. The councils manage living marine resources in their regions and address issues such as EFH.

The New England Fishery Management Council manages fisheries in federal waters off the coasts of Maine, New Hampshire, Massachusetts, Rhode Island, and Connecticut. New England fish species include Atlantic cod, Atlantic halibut, and white hake.

The Mid-Atlantic Fishery Management Council manages fisheries in federal waters off the mid-Atlantic coast. Council members represent the states of New York,

New Jersey, Pennsylvania, Delaware, Maryland, and Virginia. North Carolina is represented on this council and the South Atlantic Council. Fish species found within this region include ocean quahog, Atlantic mackerel, and butterfish.

The South Atlantic Fishery Management Council is responsible for the management of fisheries in the federal waters within a 200-mile area off the coasts of North Carolina, South Carolina, Georgia, and east Florida to Key West. Marine species in this area include cobia, golden crab, and Spanish mackerel.

The Gulf of Mexico Fishery Management Council draws its membership from the Gulf Coast states of Florida, Alabama, Mississippi, Louisiana, and Texas. Marine species in this area include shrimp, red drum, and stone crab.

The Caribbean Fishery Management Council manages fisheries in federal waters off the Commonwealth of Puerto Rico and the U.S. Virgin Islands. The management plan covers coral reefs and species including queen triggerfish and spiny lobster.

The North Pacific Fishery Management Council includes representatives from Alaska and Washington state. Species within this area include **salmon**, scallops, and king crab.

The Pacific Fishery Management Council draws its members from Washington, Oregon, and California. Species in this region include salmon, northern anchovy, and Pacific bonito.

The Western Pacific Fishery Management Council is concerned with the United States exclusive economic zone that surrounds Hawaii, American Samoa, Guam, the Northern Mariana Islands, and other U.S. possessions in the Pacific. Fishery management encompasses coral and species such as **swordfish** and striped marlin.

[*Liz Swain*]

RESOURCES

BOOKS

Dobbs, David. *The Great Gulf: Fishermen, Scientists, and the Struggle to Revive the World's Greatest Fishery.* Washington, DC: Island Press, 2000.

Hanna, Susan. *Fishing Grounds: Defining a New Era for American Fishing Management.* Washington, DC: Island Press, 2000.

ORGANIZATIONS

National Marine Fisheries Service Office of Habitat Conservation, 1315 E. West Highway, 15th Floor, Silver Springs, MD 20910 (301) 713-2325, Fax: (301) 713-1043, Email: cyber.fish@noaa.gov, http://www.nmfs.noaa.gov

National Oceanic and Atmospheric Administration, 14th Street & Constitution Avenue, NW, Room 6013, Washington, D.C. 20230 (202) 482-6090, Fax: (202) 482-3154, Email: answers@noaa.gov, http://www.noaa.gov

Estuary

Estuaries represent one of the most biologically productive aquatic ecosystems on Earth. An estuary is a coastal body of water where chemical and physical conditions modulate in an intermediate range between the freshwater rivers that feed into them and the salt water of the ocean beyond them. It is the point of mixture for these two very different aquatic ecosystems. The freshwater of the rivers mix with the salt water pushed by the incoming tides to provide a **brackish** water **habitat** ideally suited to a tremendous diversity of coastal marine life.

Estuaries are nursery grounds for the developing young of commercially important fish and shellfish. The young of any **species** are less tolerant of physical extremes in their **environment** than adults. Many species of marine life cannot tolerate the concentrations of salt in ocean water as they develop from egg to subadult, and by providing a mixture of fresh and salt water, estuaries give these larval life forms a more moderate environment in which to grow. Because of this, the adults often move directly into estuaries to spawn.

Estuaries are extremely rich in nutrients, and this is another reason for the great diversity of organisms in these ecosystems. The flow of freshwater and the periodic **flooding** of surrounding marshlands provides an influx of nutrients, as does the daily surges of tidal fluctuations. Constant physical movement in this environment keeps valuable **nutrient** resources available to all levels within the **food chain/web**.

Coastal estuaries also provide a major **filtration** system for waterborne pollutants. This natural **water treatment** facility helps maintain and protect **water quality**, and studies have shown that one acre of tidal estuary can be the equivalent of a $75,000 waste treatment plant. When its value for **recreation** and seafood production are included in this estimate, a single acre of estuary has been valued at $83,000. An acre of farmland in America's Corn Belt, for comparison, has a top value of $1,200 and an annual income through crop production of $600.

Throughout the ages man has settled near bodies of water and utilized the bounty they provide, and the economic value of estuaries, as well as the fact that they are a coastal habitat, has made them vulnerable to exploitation. **Chesapeake Bay** on the Atlantic coast is the largest estuary in the United States, draining six states and the District of Columbia. It is the largest producer of oysters in the country; it is the single largest producer of blue crabs in the world, and 90 percent of the striped bass found on the East Coast hatch there. It one of the most productive estuaries in the world, yet its productivity has declined in recent decades due to a huge increase in the number of people in the region. Between the 1940s and the 1990s, the population jumped from about three and a half million to over 15 million, bringing with it an increase in **pollution** and **overfishing** of the bay. **Sewage treatment** plants contribute large amounts of **phosphates**, while agricultural, urban, and suburban discharges deposit **nitrates**, which in turn contribute to algal blooms and oxygen depletion. Pesticides and industrial toxics also contribute to the bay's problems. Since the early 1980s concerted efforts to clean up the Chesapeake Bay and restore its seafood productivity have been undertaken by state and federal agencies. Progress has been made but there is still much to be done, both to restore this vital **ecosystem** and to insure prolonged cooperation between government agencies, industry, and the general populace. *See also* Agricultural pollution; Aquatic chemistry; Commercial fishing; Dissolved oxygen; Nitrates and nitrites; Nitrogen cycle; Restoration ecology

[*Eugene C. Beckham*]

RESOURCES
BOOKS
Baretta, J. W., and P. Ruardij, eds. *Tidal Flat Estuaries.* New York: Springer-Verlag, 1988.
McLusky, D. *The Estuarine Ecosystem.* New York: Chapman & Hall, 1989.

PERIODICALS
Horton, T. "Chesapeake Bay: Hanging in the Balance." *National Geographic* 183 (1993): 2–35.

Ethanol

Ethanol is an organic compound with the chemical formula C_2H_5OH. Its common names include ethyl alcohol and grain alcohol. The latter term reflects one method by which the compound can be produced: the distillation of corn, sugar cane, wheat and other grains. Ethanol is the primary component in many alcoholic drinks such as beer, wine, vodka, gin, and whiskey. Many scientists believe that ethanol can and should be more widely used in automotive fuels. When mixed in a one to nine ratio with **gasoline**, it is sold as **gasohol**. The reduced costs of producing gasohol have only recently made it a viable economic alternative to other automotive fuels. *See also* Alternative fuels

Ethnobotany

The field of ethnobotany is concerned with the relationship between indigenous cultures and plants. Plants play a major and complex role in the lives of **indigenous peoples**, providing nourishment, shelter, and medicine. Some plants have had such a major effect on traditional cultures that religious ceremonies and cultural beliefs were developed around their use. Ethnobotanists study and document these relationships.

The discovery of many plant-derived foods and medicines first used by indigenous cultures has changed the modern world. On the economic side, the field of ethnobotany determines the traditional uses of plants in order to find other potential applications for food, medicine, and industry. As an academic discipline, ethnobotany studies the interaction between peoples and plant life to learn more about human culture, history, and development. Ethnobotany draws upon many academic areas including anthropology, archeology, biology, **ecology**, chemistry, geography, history, medicine, religious studies, and sociology to help understand the complex interaction between traditional human cultures and the plants around them.

Early explorers who observed how native peoples used plants then carried those useful plants back to their own cultures might be considered the first ethnobotanists, although that is not a word they would have used to describe themselves. The plant discoveries these explorers made caused the expansion of trade between many parts of the globe. For example, civilizations were changed by the discovery and subsequent trade of sugar, tea, coffee, and spices including cinnamon and black pepper.

During his 1492 voyage to the Caribbean, Christopher Columbus discovered **tobacco** in Cuba and took it back with him to Europe, along with later discoveries of corn, cotton, and allspice. Other Europeans traveling to the Americas discovered tomatoes, potatoes, cocoa, bananas, pineapples and other useful plants and medicines. Latex **rubber** was discovered in South America when European explorers observed native peoples dipping their feet in the rubber compound before walking across hot coals.

The study of plants and their place in culture has existed for centuries. In India the *vedas*, which are long religious poems that have preserved ancient beliefs for thousands of years, contain descriptions of the use and value of certain plants. Descriptions of how certain plants can be used have been found in the ancient Egyptian scrolls. More than 500 years ago the Chinese made records of medicinal plants. In ancient Greece, Aristotle wrote about the uses of plants in early Greek culture, and theorized that each plant had a unique spirit. The Greek surgeon Dioscorides in A.D. 77 recorded the medicinal and folk use of nearly 600 plants in the Mediterranean region. During the Middle Ages, there existed many accounts of the folk and medicinal use of plants in Europe in books called herbals.

As the study of plants became more scientific, ethnobotany evolved as a field. One of the most important contributors to the field was Carl Linnaeus, a Swedish botanist who developed the system of naming organisms that is still used today. This system of binomial classification gives each organism a Latin genus and **species** name. It was the first system that enabled scientists speaking different languages to organize and accurately record new plant discoveries. In addition to naming the 5,900 plants known to European botanists, Linnaeus sent students around the world looking for plants that could be useful to European civilization. This was an early form of economic botany, an offshoot of ethnobotany that is interested primarily in practical developments of new uses for plants. Pure ethnobotany is more sociology-based, and is primarily interested in the cultural relationship of plants to the indigenous peoples who use them.

In the nineteenth century, bioprospecting, (the active search for valuable plants in other cultures) multiplied as exploration expanded across the globe. The most famous ethnobotanist of the 1800s was Richard Spruce, an Englishman who spent 17 years in the Amazon living among the native people and observing and recording their use of plants. Observing a native treatment for fever, Spruce collected cinchona bark from which the drug quinine was developed. Quinine has saved the lives of millions of people infected with **malaria**. Spruce also documented the native use of hallucinogenic plants used in religious rituals and the accompanying belief systems associated with these hallucinogenic plants.

Many native cultures on all continents believed that certain psychoactive plants that caused visions or hallucinations gave them access to spiritual and healing powers. Spruce discovered that for some cultures, plants were central to the way human life and the world was perceived. Thus, the field of ethnobotany began expanding from the study of the uses of plants to how plants and cultures interacted on more sociological or philosophical/religious grounds.

An American botanist named John Harshberger first coined the term "ethnobotany" in 1895, and the field evolved into an academic discipline in the twentieth century. The University of New Mexico offered the first master's degree in ethnobotany. Their program concentrated on the ethnobotany of the Native Americans of the southwestern United States.

In the twentieth century one of the most famous ethnobotanists and plant explorers was Harvard professor Richard Evans Schultes. Inspired by the story of Richard Spruce, Schultes lived for 12 years with several indigenous tribes in South America while he observed and documented their use of plants as medicine, poison, and for religious purposes. He discovered many plants that have been investigated for pharmaceutical use. His work also provided further insight into the use of hallucinogenic plants by native peoples and the cultural significance of that practice.

Interest in ethnobotany increased substantially in the 1990s. One study showed that the number of papers in the field doubled in the first half of the 1990s, when compared to the last half of the 1980s. By the early 2000s, several universities around the world offered graduate programs in ethnobotany, and undergraduate programs were becoming more common. By the start of twenty-first century, ethnobotany was an accepted academic field. In addition, the pharmaceutical industry continued to research plants used by indigenous cultures in a quest for new or more effective drugs.

Many countries conduct ethnobotanical studies of their native cultures. A research survey showed that nearly 40% of ethnobotanical studies are being conducted in North America, but other countries including Brazil, Columbia, Germany, France, England, India, China, and Kenya, are also active in the field. For instance, ethnobotanists in China and India are recording traditional medicine systems that use native plants. Researchers are also studying traditional **sustainable agriculture** methods in Africa in the hope of applying them to new drought-management techniques.

Interest in the field of ethnobotany has increased as scientists have learned more about indigenous cultures. "Primitive" people were once viewed as backward and lacking in knowledge compared to the "civilized" world, but ethnobotanists and other scientists have shown that native peoples often have developed sophisticated knowledge about preparing and using plants for food and medicine.

The ways in which people and plants interact are broad and complex, and the modern field of ethnobotany, in its many areas of inquiry, reflects this complexity. Areas of interest to ethnobotanists include traditional peoples' knowledge and interaction with plant life, traditional agricultural practices, indigenous peoples' conception of the world around them and the role of plants in their religious belief systems and rituals, how plants are used to make products and art objects, the knowledge and use of plants for medicine (a sub-field called ethnopharmacology), and the historical relationship between plants and past peoples (a sub-field called paleoethnobotany).

At the beginning of the twenty-first century, ethnobotanists are fighting a war against time, because the world's native cultures and stores of plant life are disappearing at a rapid pace. Language researchers have estimated that the world lost over half of its spoken languages in the twentieth century, and of the languages that remain, only 20% were being taught to young children in the early 2000s. The loss of traditional languages indicates that indigenous cultures are disappearing quickly as well. Historically, up to 75% of all useful plant-derived compounds have been discovered through the observation of folk and traditional plant use. Ethnobotanists are concerned that modern people will lose a storehouse of valuable plant knowledge, as well as rare cultural and historical information, as these indigenous cultures disappear.

Ethnobotanists, like many scientists, are concerned about environmental devastation from development pressure and slash-and-burn agriculture in less developed regions that are rich and diverse in plant life, such as the Amazon rainforest. Harvard biologist E. O. Wilson has estimated that the world is losing as many as 74 species per day. To put ethnobotanists' concerns about forest loss in perspective, a single square kilometer of rainforest may contain more plant species than an area in North America equaling size of Vermont. And, of all the plant species in the rainforest, fewer than 1% have been scientifically analyzed for potential uses. Considering that 25% of all prescription drugs in the United States contain active ingredients derived from plants, the continuing loss of the world's plant communities could be an incalculable loss.

In the face of disappearing indigenous peoples and the loss of botanically rich **habitat**, ethnobotanists must work quickly. Active areas in the field of ethnobotany in the early 2000s include the identification of new plant-derived foods, fibers and pharmaceuticals, investigation of traditional agricultural methods and vegetation management, using traditional technologies for **sustainable development**, and the deeper understanding of ecology and ways to preserve the **biodiversity** of ecosystems.

[*Douglas Dupler*]

RESOURCES
BOOKS

Cotton, C. M. *Ethnobotany: Principles and Applications.* New York: John Wiley & Sons, 1996.

Davis, Wade. *Light at the Edge of the World: A Journey Through the Realm of Vanishing Cultures.* Washington, DC: National Geographic Society, 2001.

Moerman, Daniel E. *Native American Ethnobotany.* Portland, OR: Timber Press, 1998.

Plotkin, Mark J. *Tales of a Shaman's Apprentice: An Ethnobotanist Searches for New Medicines in the Amazon Rainforest.* New York: Viking, 1993.

Schultes, Richard Evans, and Albert Hoffman. *Plants of the Gods.* Rochester, VT: Healing Arts Press, 1992.

Schultes, Richard Evans, and Siri von Reis, eds. *Ethnobotany: Evolution of a Discipline.* Portland, OR: Dioscorides Press, 1995.

Wilson, E. O. *Biophilia.* Cambridge: Harvard University Press, 1984.

PERIODICALS

Cox, Paul Alan. "Carl Linnaeus: The Roots of Ethnobotany." *Odyssey* (February 2001): 6.

Magee, Bernice E. "Hunting for Poisons." *Appleseeds*, April 2001, 20.

OTHER

Society for Economic Botany Home Page. [cited July 2002]. <http://www.econbot.org>.

EU

see **European Union**

Eurasian milfoil

Eurasian milfoil (*Myriophyllum spicatum*) is a feathery-looking underwater plant that has become a major nuisance in waterways in the United States. Introduced into the country over 70 years ago as an ornamental tank plant by the aquarium industry, it has since spread to the East Coast from Vermont to Florida and grows as far west as Wisconsin and Texas. It is also found in California. It is particularly abundant in the **Chesapeake Bay**, the Potomac River, and in some Tennessee Valley reservoirs.

Eurasian milfoil is able to tolerate a wide range of **salinity** in water and grows well in inland fresh waters as well as in **brackish** coastal waters. The branched stems grow from 1–10 ft (0.3–3 m) in length and are upright when young, becoming horizontal as they get older. Although most of the plant is underwater, the tips of the stems often project out at the surface. Where the water is hard, the stems may get stiffened by calcification. The leaves are arranged in whorls of four around the branches with 12–16 pairs of leaflets per leaf. As with other milfoils, the red flowers of Eurasian milfoil are small and inconspicuous. As fragments of the plant break off, they disperse quickly through an area, with each fragment capable of putting out roots and developing into a new plant. Due to its extremely dense growth, Eurasian milfoil has become a nuisance that often impedes the passage of boats and vessels along navigable waterways. It can cause considerable damage to underwater equipment and lead to losses of time and money in the maintenance and operation of shipping lanes. It also interferes with recreational water uses such as fishing, swimming, and diving. In the **prairie** lakes of North Dakota, the woody seeds of this plant are eaten by some waterfowl such as ducks. However, this plant appears to have little **wildlife** value, although it may provide some cover to fishlife.

Aquatic weed control experts recommend that vessels that have passed through growths of Eurasian milfoil be examined and plant fragments removed. They also advocate the physical removal of plant growths from shorelines and other areas with a high potential for dispersal. Additionally, chemical agents are also often used to control the weed. Some of the commonly used herbicides include **2,4-D** and Diquat. These are usually diluted with water and applied at the rate of one to two gallons per acre. The application of herbicides to control aquatic weed growth is regulated by environmental agencies due to the potential of harm to nontarget organisms. Application times are restricted by season and duration of exposure and depend on the kind of **herbi-**

cide used, its mode of action, and its impacts on the other biota in the aquatic system. *See also* Aquatic chemistry; Commercial fishing

[*Usha Vedagiri*]

RESOURCES
BOOKS

Gunner, H. B. *Microbiological Control of Eurasian Watermilfoil.* Washington, DC: U.S. Army Engineer Waterways Experiment Station, 1983.

Hotchkiss, N. *Common Marsh Underwater and Floating-Leaved Plants of the United States and Canada.* Mineola, NY: Dover Publications, 1972.

Schmidt, J. C. *How to Identify and Control Water Weeds and Algae.* Milwaukee, WI: Applied Biochemists, Inc., 1987.

European Economic Community (EEC)
see European Union

European Greens
see Green politics

European Union

The European Union (EU) is a political and monetary organization of European nations. Its members as of 2002 were Austria, Belgium, Denmark, Finland, France, Germany, Greece, Ireland, Italy, Luxembourg, the Netherlands, Portugal, Spain, Sweden, and the United Kingdom. While member states retain their national governments, the EU promulgates transnational laws and treaties, has a unified agricultural policy, and has removed trade barriers between members. The EU uses a common currency, the euro. The EU represents about a quarter of the global economy, roughly equal to that of the United States.

The European Union developed gradually in the years after World War II. It began in 1957 as the European Economic Community, or EEC, which originally included France, West Germany, Italy, Belgium, the Netherlands, and Luxembourg. The EEC was superceded by the European Community, or EC, in 1967. The principal aim of the EC was to foster free trade among member countries, eliminating tariffs and customs barriers. The EC expanded its role in the 1970s and 1980s. As more states joined, the EC began to strengthen political cooperation among members, holding meetings to coordinate foreign policy and international law. By the mid-1980s, plans were underway to transform the EC into a single European market. This single market would have a central banking system, a transnational currency, and a unified foreign policy. These goals were articulated in the Single European Act of 1987, which

also for the first time set out environmental protection as an important goal of European economic development. A treaty written in 1991, called the Maastricht Treaty or the Treaty on European Union, spelled out the future shape of the European Union. All the member states had ratified the treaty by November, 1993, when the European Union officially superceded the EC. The EU went on to launch a new currency, the euro, in 1999. In January 2002 the euro replaced existing EU member currencies.

The EU had incorporated **environmental policy** into its economic plans beginning in 1987. New members were required to conform to certain environmental standards. And the EU was able to make regional environmental policy, which often made more sense than national laws for problems like air and **water pollution** that extended across national borders. Environmental regulation in the EU was generally stricter than in the United States, though member states did not always comply with the law. The EU was often quicker than the United States on implementing environmental strategy, such as rules for computer and cell phone disposal. The EU approved the 1997 Kyoto Protocol, an international agreement to reduce the production of so-called **greenhouse gases** in order to arrest global warming. As details of the implementation of the plan were being worked out over the next five years, EU negotiators remained committed to curtailing European gas emissions even after it became clear that the United States would not sign the Kyoto treaty. Sweden took over the rotating presidency of the EU in 2002, and planned to make environmental policy a high priority. "The **environment** is more global than anything else except foreign policy," Sweden's environmental minister and EU spokesman told *Harper's Bazaar* (March 2001). With the EU representing 25% of the global economy, the organization's stance on environmental issues was always significant.

[*Angela Woodward*]

RESOURCES
PERIODICALS

Andrews, Edmund L. "Bush Angers Europe by Eroding Pact on Warming"- *New York Times*, April 1, 2001, 3.

Milmo, Sean. "EU Lacks Industry Support for EU-Wide Emissions Trading System." *Chemical Market Reporter* 260, no. 5 (July 30, 2001): 6.

Sains, Ariane. "EU News: Sweden's EU Agenda." *Harper's Bazaar* 134, no. 3472 (March 2001): 287.

Scott, Alex. "EU Mulls Criminal Sanctions for Eco-Crimes." *Chemical Week* 164, no. 16 (April 17, 2002): 15.

"What Next, Then?" *Economist* 360, no. 8232 (July 28, 2001): 69.

Eutectic

Refers to a type of solar heating system which makes use of phase changes in a chemical storage medium. At its melting point, any chemical compound must absorb a quantity of heat in order to change phase from solid to liquid, and at its boiling point, it must absorb an additional quantity of heat to change to a gas. Conversely, a compound *releases* a quantity of heat when condensing or freezing. Therefore, a chemical warmed by **solar energy** through a phase change releases far more energy when it cools than, for example, water heated from a cool liquid to a warm liquid state. For this reason, solar heating systems that employ phase-changing **chemicals** can store more energy in a compact space than a water-based system.

Eutrophication
see **Cultural eutrophication**

Evapotranspiration

Evapotranspiration is a key part of the **hydrologic cycle**. Some water evaporates directly from soils and water bodies, but much is returned to the **atmosphere** by **transpiration** (a word combining transport and evaporation) from plants via openings in the leaves called stomata. Within the same climates, forests and lakes yield about the same amount of water vapor. The amount of evapotranspiration is dependent on energy inputs of heat, wind, humidity, and the amount of stored **soil** water. In **climate** studies this term is used to indicate levels of surplus or deficit in water budgets. Aridity may be defined as an excess of potential evapotranspiration over actual precipitation, while in humid regions the amount of **runoff** correlates well with the surplus of precipitation over evapotranspiration.

Everglades

A swampy region in southern Florida, the Everglades are described as a vast, shallow sawgrass (*Cladium effusum*) marsh with tree islands, wet prairies, and aquatic sloughs, the Everglades historically covered most of southeastern Florida, prior to massive **drainage** and **reclamation** projects launched at the turn of the century. The glades constitute the southern end of the Kissimmee Lake Okeechobee Everglades system, which encompasses most of south and central Florida below Orlando. Originally, the Everglades covered an area approximately 40 mi (64 km) wide and 100 mi (161 km) long, or 2.5 million acres, but large segments have been isolated by canals and levees. Today, intensive agriculture

in the north and rapid urban development in the east are among the Everglades' various land uses.

Two general **habitat** regions can be demarcated in the Everglades. The first includes three **water conservation** areas, basins created to preserve portions of the glades and provide multiple uses, such as water supply. This region is located in the northern Everglades and contains most of the intact natural marsh. The second is the southern habitat, which includes the Everglades **National Park** and the southern third of the three water **conservation** areas. The park has been designated a World Heritage Site of international ecological significance, and the Everglades as a whole are one of the outstanding freshwater ecosystems in the United States.

Topographically flat, elevations in the Everglades are generally less than 20 ft (6.1 m). The ground slopes north to south at an average gradient of 0.15 ft per mile, with the highest elevations in the north and the lowest in the south. The **climate** is generally subtropical, with long, hot, humid, and wet summers from May to October followed by mild, dry winters from November to April. During the wet season severe storms can result in lengthy periods of **flooding**, while during the dry season cool, sometimes freezing temperatures can be accompanied by thunderstorms, tornadoes, and heavy rainfall.

Before the Everglades were drained, large areas of the system were inundated each year as Lake Okeechobee overflowed its southern rim. The "River of Grass" flowed south and was in constant flux through **evapotranspiration**, rainfall, and water movement into and out of the Everglades' **aquifer**. The water discharged into the tidewaters of south Biscayne Bay, Florida Bay, and Ten Thousand Islands.

In the early 1880s, Philadelphia industrialist Hamilton Disston began draining the Everglades under a contract with Florida trustees. Disston, whose work ceased in 1889, built a substantial number of canals, mainly in the upper waters of the Kissimmee River, and constructed a canal between Lake Okeechobee and the Calooshatchee River to provide an outlet to the Gulf of Mexico. The Miami River was channelized beginning in 1903, and other canals—the Snapper Creek Canal, the Cutler Canal, and the Coral Gables Waterway—were opened to help drain the Everglades. Water tabless in south Florida fell 5–6 ft (1.5–1.8 m) below 1900 levels, causing stress to **wetlands** systems and losses of peat up to 6 ft (1.8 m) in depth.

Full-scale drainage and reclamation occurred under Governor W.S. Jennings (1901–1905) and Governor Napoleon Bonaparte Broward (1905–1909). In 1907, the Everglades Drainage District was created and built six major canals over 400 miles long before it suffered financial collapse in 1928. These canals enabled agriculture to flourish within the region. In the late 1920s, when settlers realized better water control and flood protection were needed, low muck levees were built along Lake Okeechobee's southwest shore, eliminating the lake's overflow south to the Everglades. But hurricanes in 1926 and 1928 breached the levees, destroying property and killing 2,100 people. As a result, the Lake Okeechobee Flood Control District was established in 1929, and over the following fifteen years the United States **Army Corps of Engineers** constructed and enlarged flood control canals.

It was only in the mid-1950s, with the development and implementation of the Central and Southern Florida Project for Flood Control & Other Purposes (C&SF Project), that water control took priority over uncontrolled drainage of the Everglades. The project, completed by 1962, was to provide flood protection, water supply, and environmental benefits over a 16,000 square-mile (41,440 sq-km) area. It consists of 1,500 miles (2,415 km) of canals and levees, 125 major water control structures, 18 major pumping stations, 13 boat locks, and several hundred smaller structures. Interspersed throughout the Everglades is a series of habitats, each dominated by a few or in some cases a single plant **species**. Seasonal wetlands and upland pine forests, which once dominated the historic border of the system, have come under the heaviest pressure from urban and agricultural development. In the system's southern part, freshwater wetlands are superseded by muhly grass (*Muhlenbergia filipes*), prairies, upland pine and tropical hardwood forests, and mangrove forests that are influenced by the tides.

Attached algae, also known as periphyton, are an important component of the Everglades food web, providing both organic food matter and habitat for various grazing invertebrates and forage fish that are eaten by wading birds, reptiles, and sport fish. These algae include calcareous and filamentous algae (*Scytonema hoffmani*, *Schizothrix calcicola*) and diatoms (*Mastogloia smithii v. lacustris*).Sawgrass (*Cladium jamaicense*) constitutes one of the main plants occurring throughout the Everglades, being found in 65–70% of the remaining freshwater marsh. In the north, the sawgrass grows in deep **peat soils** and is both dense and tall, reaching up to 10 ft (3 m) in height. In the south, it grows in low-**nutrient** marl soils and is less dense and shorter, averaging 2.5–5 ft (0.75–1.5 m). Sawgrass is adapted to survive both flooding and burning. Stands of pure sawgrass as well as mixed communities are found in the Everglades. The mixed communities can include maidencane (*Panicum hemitomon*) arrowhead (*Sagittaria lancifolia*), water hyssop (*Bacopa caroliniana*), and spikerush (*Eleocharis cellulosa*).

Wet prairies, which together with aquatic sloughs provide habitat during the rainy season for a wide variety of aquatic invertebrates and forage fish, are another important habitat of the Everglades system. They are seasonally inundated wetland communities that require certain standing water for six to ten months. Once common, today more than 1,500 square miles (3,885 sq km) of these prairies have been drained or destroyed. The lowest elevations of the Everglades are ponds and sloughs, which have deeper water and longer inundation periods. They occur throughout the system, and in some cases can be formed by alligators in peat soils. Among the types of emergent vegetation commonly found in these areas are white water lily (*Nymphaea odorata*), floating heart (*Nymphoides aquatica*), and spatterdock (*Nuphar luteum*). Common submerged species include bladderwort (*Utricularia*) and the periphyton mat community. Ponds and sloughs serve as important feeding areas and habitat for Everglades **wildlife**.

At the highest elevations are found communities of isolated trees surrounded by marsh called tree islands. These provide nesting and roosting sites for colonial birds and habitat for deer and other terrestrial animals during high-water periods. Typical dominant species constituting tree islands are red bay (*Persa borbonia*), swamp bay (*Magnolia virginiana*), dahoon holly (*Ilex cassine*), pond apple (*Annona glabra*), and wax myrtle (*Myrica cerifera*). Beneath the canopy grows a dense shrub layer of cocoplum (*Chrysobalanus icacao*), buttonbush (*Cephalanthus accidentalis*), leather leaf fern (*Acrostichum danaeifolium*), royal fern (*Osmunda regalis*), cinnamon fern (*O. cinnamonea*), chain fern (*Anchistea virginica*), bracken fern (*Pteridium aquilinium*), and lizards tail (*Saururus cernuus*).

In addition to the indigenous plants of the Everglades, numerous exotic and nuisance species have been brought into Florida and have now spread in the wild. Some threaten to invade and displace indigenous species. Brazilian pepper (*Schinus terebinthifolius*), Australian pine (*Casuarina equisetifolia*), and melaleuca (*Melaleuca quinquenervia*) are three of the most serious **exotic species** that have gained a foothold and are displacing native plants.

The Florida Game and Fresh Water Fish Commission has identified 25 threatened or endangered species with the Everglades. Mammals include the **Florida panther** (*Felis concolor coryi*), mangrove fox squirrel (*Sciurus niger avicennia*), and black bear (*Ursus americanus floridanus*). Birds include the wood stork (*Mycteria americana*), snail kite (*Rostrhamus sociabilis*), and the red-cockaded (*Picoides borealis*). Endangered or threatened reptiles and amphibians include the gopher tortoise (*Gopherus polyphemus* the eastern indigo snake (*Drymarchon corais couperi*), and the loggerhead **sea turtle** (*Caretta caretta*).

The alligator (*Alligator mississippiensis*) was once endangered due to excessive alligator hide **hunting**. In 1972, the state made alligator product sales illegal. Protection allowed the species to recover, and it is now widely distributed in wetlands throughout the state. It is still listed as threatened

A swampy area in the Everglades. (Photograph by Gerald Davis. Phototake. Reproduced by permission.)

by the federal government, but in 1988 Florida instituted an annual alligator harvest.

Faced with pressures on Everglades habitats and the species within them, as well as the need for water management within the rapidly developing state, in 1987 the Florida legislature passed the Surface Water Improvement and Management Act (1987). The law requires the state's five water management districts to identify areas needing preservation or restoration. The Everglades Protection Area was identified as a priority for preservation and improvement planning. Within the state's protection plan, excess nutrients, in large part from agriculture, have been targeted as a major problem that causes natural periphyton to be replaced by species more tolerant of **pollution**. In turn, sawgrass and wet **prairie** communities are overrun by other species, impairing the Everglades' ability to serve as habitat and forage for higher **trophic level** species.

A federal lawsuit was filed against the South Florida Management District in 1988 for **phosphorus** pollution, and in 1989, President George Bush authorized the addition of more than 100,000 acres to the Everglades National Park. The law that authorized this addition was Public Law 101-229 or the Everglades National Park Protection and Expansion Act of 1989. Included in this legislation was the stipulation that the Army Corps of Engineers improve water flow

to the Park. In 1994, the Everglades Forever Act was passed by the Florida State Legislature. The Act called for construction of experimental marshes called Stormwater Treatment Areas that were designed to remove phosphorus from water entering the Everglades. In 1997, six more Stormwater Treatment Areas were constructed and **phosphorus removal** was estimated to be as much as 50%, due in part to better management practices that were mandated by the Everglades Forever Act. In 2000, President Clinton authorized the spending of billions of federal dollars to restore the Everglades, while Florida Governor Jeb Bush agreed to a state commitment of 50% of the cost, in a bill called the Florida Investment Act. In 2001 and 2002, the state of Florida, under Governor Jeb Bush, and the federal government, under President George W. Bush, committed $7.8 billion dollars to implement the Comprehensive Everglades Restoration Plan (CERP).

[*David Clarke and Marie H. Bundy*]

RESOURCES

BOOKS

Douglas, M. S. *The Everglades: River of Grass.* New York: H. Wolff, 1947.

PERIODICALS

Johnson, R. "New Life for the 'River of Grass.'" *American Forests* 98 (July-August 1992): 38–43.

Stover, D. "Engineering the Everglades." *Popular Science* 241 (July 1992): 46–51.

Evolution

Evolution is an all-pervading concept in modern science applied to physical as well as biological processes. The phrase *theory of evolution*, however, is most commonly associated with organic evolution, the origin and evolution of the enormous diversity of life on this planet.

The idea of organic evolution arose from attempts to explain the immense diversity of living organisms and dates from the dawn of culture. Attempts to formulate a naturalistic explanation of the phenomena of evolution in one form or another had been proposed, and by the end of the eighteenth century a number of naturalists from Carolus Linnaeus to Jean-Baptiste Lamarck had questioned the prevailing doctrine of fixity of **species**. Lamarck came the closest to proposing a complete theory of evolution, but for several reasons his theory fell short and was not widely accepted.

Two English naturalists conceived the first truly complete theory of evolution. In a most remarkable coincidence, working quite independently, Charles Darwin and Alfred Russell Wallace arrived at essentially the same thesis, and their ideas were initially publicized jointly in 1858. However, it is to Charles Darwin that the recognition of the founder of modern evolution is attributed. His continued efforts toward a detailed development of the evidence and the explication of a complete, convincing theory of organic evolution resulted in the publication in 1859 of his book *On the Origin of Species by Natural Selection*. With the publication of this volume, widely regarded as one of the most influential books ever written, regardless of subject matter, the science of biology would never be the same.

Darwin was the first evolutionist whose theories carried conviction to a majority of his contemporaries. He set forth a formidable mass of supporting evidence in the form of direct observations from **nature**, coupled with a comprehensive and convincing synthesis of current knowledge. Most significantly, he proposed a rational, plausible instrumentality for evolutionary change: natural selection.

The theory Darwin presented is remarkably simple and rational. In brief, the theoretical framework of evolution presented by Darwin contained three basic elements. First is the existence of variation in natural populations. In nature there are no individuals who are exactly alike, therefore natural populations always consist of members who all differ from one another to some degree. Many of these differences are heritable and are passed on from generation to generation. Second, some of these varieties are better adapted to their **environment** than others. Third, the reproductive potential of populations is unlimited. All populations have the capacity to overproduce, but populations in nature are limited by high **mortality**. Thus, if all offspring cannot survive each generation, the better-adapted members have a greater **probability** of surviving each generation than those less adapted.

Thomas Henry Huxley, when first presented with Darwin's views, was supposed to have exclaimed: "How extremely stupid not to have thought of that." This apocryphal statement expresses succinctly the immediate intelligibility of this momentous discovery.

Darwin's theory as presented was incomplete, as he was keenly aware, but the essential parts were adequate at the time to demonstrate that the material causes of evolution are possible and can be investigated scientifically.

One of the most critical difficulties with Darwin's theory was the inability to explain the source of hereditary variations observed in populations and, in particular, the source of hereditary innovations which, through the agency of natural selection, could eventually bring about the transformation of a species. The lack of existing knowledge of the mechanism of heredity in the mid-1800s precluded a scientifically sound solution to this problem.

The re-discovery in 1900 of Gregor Mendel's experiments demonstrating the underlying process of inheritance did not at first provide a solution to the problem facing

Darwinian evolutionary theory. In fact, in one of the most extraordinary paradoxes in the history of science, during the first decades of Mendelian genetics, a temporary decline occurred in the support of Darwinian selectionism. This was primarily because of the direction and level of research as much as the lack of knowledge in further details of the underlying mechanisms of inheritance.

It was not until the publication of Theodosius Dobzhansky's *Genetics and the Origin of Species* (1937) that the modern synthetic theory of evolution began to take form, renewing confidence in selectionism and the Darwinian theory. In his book, Dobzhansky, for the first time, integrated the newly developed mathematical models of population genetics with the fats of chromosomal theory and **mutation** together with observations of variation in natural populations.

Dobzhansky also stimulated a series of books by several major evolutionists which provided the impetus and direction of research over the next decade, resulting in a reformulation of Darwin's theory. This became known as the *synthetic theory of evolution*. Synthetic is used here to indicate that there has been a convergence and synthesis of knowledge from many disciplines of biology and chemistry.

The birth of molecular genetics in 1950 and the enormous expansion of knowledge in this area have affected the further development and refinement of the synthetic theory markedly. But it has not altered the fundamental nature of the theory. Darwin's initial concept in this expanded, modernized form has taken on such stature as to be generally considered as the most encompassing theory in biology. To the majority of biologists, the synthetic theory of evolution is a grand theory that forms the core of modern biology. It also provides the theoretical foundation upon which most all other theories find support and comprehension. Or, as the distinguished evolutionary biologist Dobzhansky so succinctly put it: "Nothing in life makes sense except in the light of evolution."

If the three main elements of the Darwinian theory were to be restated in terms of the modern synthetic theory, they would be as follows. First, genetic systems governing heredity in each organism are composed of genes and chromosomes which are discrete but tightly interacting units of different levels of complexity. The genes, their organized associations in chromosomes, and whole sets of chromosomes, have a large degree of **stability** as units. But these units are shuffled and combined in various ways by sexual processes of reproduction in most organisms. The result of this process, called *recombination*, maintains a considerable amount of variation in any population in nature without introducing any new hereditary factors. Genetic experimentation, is, in fact, occurring all the time in natural populations. Each generation, as a result of recombination, is made up of members who carry different hereditary combinations drawn from the common pool of the population. Each member generation in turn interacts within the same environmental matrix which, in effect, "tests" the fitness of these hereditary combinations. The sole source of hereditary innovation from within the population, other than recombination, would come from mutation affecting individual genes, chromosomes or sets of chromosomes, which are then fed into the process of recombination.

Second, populations of similar animals—members of the same species—usually interbreed among themselves making up a common pool of genes. The gene pools of established populations in nature tend to remain at equilibrium from generation to generation. An abundance of evidence has established the fact, as Darwin and others before him had observed, that populations have greater reproductive capacity than they can sustain in nature.

Third, change, over time, in the makeup of the **gene pool**, may occur by one of four known processes: mutation, fluctuation in gene frequencies—known as "sampling errors," or "genetic drift," which is generally effective only in small, isolated populations—introduction of genes from other populations, and differential reproduction.

The first three of these processes do not produce adaptive trends. These are essentially random in their effects, usually nonadaptive and only rarely and coincidentally produce **adaptation**. Of the four processes, differential reproduction results in adaptive trends. Differential reproduction describes the consistent production of more offspring, on an average, by individuals with certain genetic characteristics than those without those characteristics. This is the modern understanding of natural selection, with emphasis upon success in reproduction. The Darwinian and Neo-Darwinian concepts, on the other hand, placed greater emphasis upon "survival of the fittest" in terms of mortality and survival. Natural selection, in the Darwinian sense, was non-random and produced trends that were adaptive. Differential reproduction produces the same result. In what may appear to be a paradox, differential selection also acts in stabilizing the gene pool by reducing the reproductive success of those members whose hereditary variation result in non-adaptive traits.

As Darwin first recognized, the link between variability or constancy of the environment and evolutionary change or stability is provided by natural selection. The interaction between organisms and their environment will inevitably affect the genetic composition of each new generation. As a result of differential selection, the most fit individuals, in terms of reproductive capacity, contribute the largest proportion of genes or alleles to the next generation. The direction in which natural selection guides the population depends upon both the nature of environmental change or stability and the content of the gene pool of the population.

From the very beginning, evolutionists have recognized the significance of the environment/evolution linkage. This linkage always involves the interactions of a great many individuals both within populations of a single species (intraspecific) and among members of different species (interspecific linkage). These are inextricably linked with changes in the physical environment.

Dobzhansky estimated that the world may contain four to five million different kinds of organisms, each exploiting in various ways a large number of habitats. The complexity of ecosystems is based upon thousands of different ways of exploiting the same habitats in which each species is adapted to its specifically defined **niche** within the total **habitat**. The diversity of these population-environment interactions is responsible for the fact that natural selection can be conservative—(normalizing selection)—and can promote constancy in the gene pool. It directs continuous change (directional selection), or promotes diversification into new species (diversifying selection).

Recent trends in research have been of special significance in understanding the dynamics of these processes. They have greatly increased the body of information related to the ecological dimension of **population biology**. These studies have developed methods of analysis and sophisticated models which have made major advances in the study and understanding of the nature of the selective forces that direct evolutionary change. It is increasingly clear that a reciprocal imperative exists between ecological and environmental studies to further our understanding of the role of evolutionary and counterrevolutionary processes affecting stability in biological systems over time. The urgency of the need for an expanded synthesis of ecological and evolutionary studies is driven by the accelerating and expanding counterrevolutionary changes in the world environment resulting from human actions.

Evolutionary processes are tremendously more complex in detail than this brief outline suggests. This spare outline, however, should be sufficient to show that here is a mechanism, involving materials and processes known beyond doubt to occur in nature, capable of shaping and changing species in response to the continuously dynamic universe in which they exist.

[*Donald A. Villeneuve*]

RESOURCES
BOOKS

Dobzhansky, T., et al. *Evolution.* San Francisco: W. H. Freeman, 1977.
Simpson, G. G. *The Meaning of Evolution.* New Haven, CT: Yale University Press, 1967.

Exclusive economic zone

The Exclusive Economic Zone (EEZ) is a 200-mile boundary off the coasts of a nation that provides the exclusive right to fish, mine, and otherwise utilize the **natural resources** located within the zone. After World War II, many nations became concerned about the intrusion of foreign fishing vessels into the rich and productive waters off of their coasts. The 1958 Convention on Fishing and Conservation of Living Resources of the High Seas, established rules that allowed a coastal nation impose terms on the uses of natural resources of their coastal oceans. This effort at regulation was unsuccessful, and the rules were never put in place. The policy that ultimately established exclusive economic zones around sovereign nations resulted from provisions of the Third United Nations Conference on the Law of the Sea, held in 1973. As a result of the actions taken in 1958, virtually all of the world's fisheries are controlled by one nation or another. However, the controlling nation must provide access to the resources that it cannot harvest, and is responsible for preventing **pollution** and depletion of the resources in the waters that it controls. In the United States, the Fishery Conservation and Management Act of 1976 established the fishery conservation zone out to 200 miles, however, it wasn't until 1983 that the EEZ was created by presidential proclamation.

[*Marie H. Bundy*]

Existence value
see Debt for nature swap

Exotic species

Exotic **species** are organisms that are introduced to a region or **ecosystem**, often unintentionally, through human **migration** or trade. Some exotic species are useful to man, such as horses, goats, pigs, and edible plants including wheat and oats. These are examples of species that were brought to the Americas intentionally by European colonists. Other exotic species that were introduced accidentally such as the **Mediterranean fruit fly** (*Ceratitis capitata*), Japanese beetle (*Popillia japonica*), **Africanized bees** (*Apis mellifera scutellata*)(sometimes called killer bees), and Norway rat (*Rattus norvegicus*) have become pests. Many exotic species, including most tropical fish, birds, and houseplants brought to colder climates, can survive only under continuous care. A few prove extremely adaptable and thrive in their new **environment**, sometime becoming invasive and out competing native species.

The federal government's Office of Technology Assessment has estimated that more than 2,000 plant species introduced from around the world currently live and thrive in the United States, and that 15 of these have caused more than $500 million worth of damage. Economic costs associated with exotic species include agricultural losses, damage to infrastructure, as when aquatic plants clog water intakes, and the costs of attempts to restore native species whose survival is endangered by **introduced species**.

Exotic species are most are most destructive when they adapt readily to their new environment and compete successfully with native species. Unfortunately there is no sure way to know which introduced species will become invasive. Often these plant and animals turn out to be better competitors because, unlike native species, they have no natural pests, **parasites**, diseases, or predators in their new home.

Purple loosestrife (*Lythrum salicaria*) is an example of a plant that is kept in check in its native **habitat**, but is invasive in North America. This showy wetland plant with tall purple flowers may have arrived accidentally or been intentionally imported to North America as a garden ornamental. In its native northern Europe, resident beetles feed on its roots and leaves, keeping the loosestrife in check, so that it only appears occasionally and temporarily in disturbed sites. When loosestrife arrived in North America, the beetles, along with other pests and diseases, were left behind. Loosestrife has proven to be extremely adaptable and has become an aggressive weed across much of the American East and Midwest, often taking over an entire wetland and choking out other plants, eliminating much of the wetland **biodiversity**.

In addition to the competitive advantage of having few predators, invasive exotic plants and animals may have ecological characteristics that make them especially competitive. They can be hardy, adaptable to diverse habitat conditions, and able to thrive on a variety of food sources. They may reproduce rapidly, generating large numbers of seeds or young that spread quickly. If they are aggressive colonizers adapted to living in marginal habitat, introduced species can drive resident natives from their established sites and food sources, especially around the disturbed environments of human settlement.

This competitiveness has been a problem, for example, with house sparrows (*Passer domesticus*). These birds were intentionally introduced from Europe to North America in 1850 to control insect pests. Their aggressive foraging and breeding habits often drive native sparrows, martins, and bluebirds from their nests, and today they are one of the most common birds in North America. Exotic plants can also become nuisance species when they crowd, shade, or out-propagate their native competitors. They can be extraordinarily effective colonists, spreading quickly and eliminating **competition** as they become established.

The list of species introduced to the Americas from Europe, Asia, and Africa is immense, as is the list of species that have made the reverse trip from the Americas to Europe, Asia, and Africa. Some notable examples are **kudzu** (*Pueraria lobata*), the **zebra mussel** (*Dreissena polymorpha*), Africanized bees *Apis mellifera scutellata*), and **Eurasian milfoil** (*Myriophyllum spicatum*).

Kudzu is a cultivated legume in Japan. It was intentionally brought to the southern United States for ground cover and **erosion** control. Fast growing and tenacious, kudzu quickly overwhelms houses, tangles in electric lines, and chokes out native vegetation.

Africanized "killer" bees were accidentally released in Brazil by a beekeeper in 1957. These aggressive insects have no more venom than standard honey bees (also an Old World import), but they attack more quickly and in great numbers. Breeding with resident bees and sometimes traveling with cargo shipments, Africanized bees have spread north from Brazil at a rate of up to 200 miles (322 km) each year and now threaten to invade commercially valuable fruit orchards and domestic bee hives in Texas and California.

The zebra mussel, accidentally introduced to the **Great Lakes** around 1985 presumably in ballast water dumped by ships arriving from Europe, colonizes any hard surface, including docks, industrial water intake pipes, and the shells of native bivalves. Each female zebra mussel can produce 50,000 eggs a year. Growing in masses with up to 70,000 individuals per square foot, these mussels clog pipes, suffocate native clams, and destroy breeding grounds for other aquatic animals. They are also voracious feeders, competing with fish and native mollusks for **plankton** and microscopic plants. The economy and environment of the Great Lakes now pay the price of zebra mussel infestations. Area industries spend hundreds of millions of dollars annually unclogging pipes and equipment, and commercial fishermen complain of decreased catches.

Eurasian milfoil is a common aquarium plant that can propagate from seeds or cuttings. A tiny section of stem and leaves accidentally introduced into a lake by a boat or boat trailer can grow into a huge mat covering an entire lake. When these mats have consumed all available nutrients in the lake, they die and rot. The rotting process robs fish and other aquatic animals of oxygen, causing them to die.

Exotic species have brought ecological disasters to every continent, but some of the most extreme cases have occurred on isolated islands where resident species have lost their defensive strategies. For example, rats, cats, dogs, and mongooses introduced by eighteenth century sailors have devastated populations of ground-breeding birds on Pacific islands. Rare flowers in Hawaii suffer from grazing goats

and rooting pigs, both of which were brought to the island for food, but have escaped and established wild populations. Grazing sheep threaten delicate plants on ecologically fragile North Atlantic islands, while rats, cats, and dogs endanger northern seabird breeding colonies. Rabbits introduced into **Australia** overran parts of the island and wiped out hundreds of acres of grassland.

Humans have always carried plants and animals as they migrated from one region to another with little regard to the effects of these introductions might have on their new habitat. Many introduced species seem benign, useful, or pleasing to have around, making it difficult to predict which imports will become nuisance species. When an exotic plant or animal threatens human livelihoods or economic activity, as do kudzu, zebra mussels, and "killer" bees, people begin to seek ways to control these invaders.

Control efforts include using pesticides and herbicides, and introducing natural predators and parasites from the home range of the exotic plant or animal. For example, beetles that naturally prey on purple loosestrife have been experimentally introduced in American loosestrife populations. This deliberate introduction requires a great deal of care, research, and monitoring, however, to ensure that an even worse problem does not result, as happened with the house sparrow. Such solutions, and the time and money to develop them, are usually elusive and politically controversial, so in many cases effective control methods remain unavailable.

In 1999, President Bill Clinton signed the Executive Order on Invasive Species. This order established the Invasive Species Council to coordinate the activities of federal agencies, such as the Aquatic Nuisance Species Task Force, the Federal Interagency Committee for the Management of Noxious and Exotic Weeds, and the Committee on Environment and **Natural Resources**. The Invasive Species Council is responsible for the development of a National Invasive Species Management Plan. This plan is intended to be updated very two years to provide guidance and recommendations about the identification of pathways by which invasive species are introduced, and measures that can be taken for their control.

Non-profit environmental organizations across the globe are leading the effort for control of exotic species. For example, **The Nature Conservancy** has established Landscape Conservation Networks to address issues of land conservation that include invasive species management. These networks bring in outside experts and land conservation partners to develop innovative and cost effective means of controlling exotic species. The Great Lakes information Network, managed by the Great Lakes Commission based in Ann Arbor, Michigan, provides online access about environmental issues, including exotics species, in the Great

Lakes region. The Florida Exotic **Pest** Plant Council, founded in 1984, provides funding to organizations that educate the public about the impacts of exotic invasive plants in the State of Florida.

[*Mary Ann Cunningham and Marie H. Bundy*]

RESOURCES

BOOKS

Cunningham, W. *Understanding Our Environment: An Introduction.* Dubuque, IA: William C. Brown, 1993.

PERIODICALS

Barrett, S. C. H. "Waterweed Invasions." *Scientific American* 261 (October 1989): 90–7.

Rendall, Jay. " Invasive Species". *Imprint* 7, no. 4 (1990): 1–8, 1990.

Walker, T. "Dreissena Disaster—Scientists Battle an Invasion of Zebra Mussels." *Science News* 139 (May 4, 1991): 282–84.

Experimental Lakes Area

The Experimental Lakes Area (ELA) in northwestern Ontario is in a remote landscape characterized by Precambrian bedrock, northern mixed-species forests, and **oligotrophic** lakes, bodies of water deficient in plant nutrients. The Canadian Department of Fisheries and Oceans began developing a field-research facility at ELA in the 1960s, and the area has become the focus of a large number of investigations by D. W. Schindler and others into chemical and biological conditions in these lakes.

Of the limnological investigations conducted at ELA, the best known is a series of whole-lake experiments designed to investigate the ecological effects of perturbation by a variety of **environmental stress** factors, including eutrophication, **acidification**, metals, **radionuclides**, and **flooding** during the development of reservoirs.

The integrated, whole-lake projects at ELA were initially designed to study the causes and ecological consequences of eutrophication. In one long-term experiment, Lake 227 was fertilized with phosphate and nitrate. This experiment was designed to test whether **carbon** could limit algal growth during eutrophication, so none was added. Lake 227 responded with a large increase in **primary productivity** by drawing on the **atmosphere** for carbon, but it was not possible to determine which of the two added nutrients, phosphate or nitrate, had acted as the primary limiting factor.

Observations from experiments at other lakes in ELA, however, clearly indicated that phosphate is the primary limiting **nutrient** in these oligotrophic water bodies. Lake 304 was fertilized for two years with **phosphorus**, **nitrogen**, and carbon, and it became eutrophic. It recovered its oligotrophic condition again when the phosphorus fertilization was stopped, even though nitrogen and carbon fertilization

were continued. Lake 226, an hourglass-shaped lake, was partitioned with a vinyl curtain into two basins, one of which was fertilized with carbon and nitrogen, and the other with phosphorus, carbon, and nitrogen. Only the latter treatment caused an **algal bloom**. Lake 302 received an injection of all three nutrients directly into its hypolimnion during the summer. Because the lake was thermally stratified at that time, the hypolimnetic nutrients were not available to fertilize plant growth in the epilimnetic euphotic zone, and no algal bloom resulted. Nitrogen additions to Lake 227 were reduced in 1975 and eliminated in 1990. The lake continued with high levels of productivity by fixing nitrogen from the atmosphere.

Research of this sort was instrumental in confirming conclusively the identification of phosphorus as the most generally limiting nutrient to eutrophication of freshwaters. This knowledge allowed the development of **waste management** systems which reduced eutrophication as an environmental problem by reducing the phosphorus concentration in **detergents**, removing phosphorus from sewage, and diverting sewage from lakes.

Another well known ELA project was important in gaining a deeper understanding of the ecological consequences of the acidification of lakes. Sulfuric **acid** was added to Lake 223, and its acidity was increased progressively, from an initial **pH** near 6.5 to pH 5.0–5.1 after six years. Sulfate and **hydrogen** ions were also added to the lake in increasing concentrations during this time. Other chemical changes were caused indirectly by acidification: manganese increased by 980%, zinc by 550%, and **aluminum** by 155%.

As the acidity of Lake 223 increased, the **phytoplankton** shifted from a community dominated by golden-brown algae to one dominated by chlorophytes and dinoflagellates. **Species** diversity declined somewhat, but productivity was not adversely affected. A mat of the green alga *Mougeotia* sp. developed near the shore after the pH dropped below 5.6. Because of reduced predation, the density of cladoceran **zooplankton** was larger by 66% at pH 5.4 than at pH 6.6, and copepods were 93% more abundant. The nocturnal zooplankton predator *Mysis relicta*, however, was an important **extinction**. The crayfish *Orconectes virilis* declined because of reproductive failure, inhibition of carapace hardening, and effects of a parasite. The most acid-sensitive fish was the fathead minnow (*Pimephales promelas*), which declined precipitously when the lake pH reached 5.6.

The first of many year-class failures of lake trout (*Salvelinus namaycush*) occurred at pH 5.4, and failure of white sucker (*Catastomus commersoni*) occurred at pH 5.1. One minnow, the pearl dace (*Semotilus margarita*), increased markedly in abundance but then declined when pH reached 5.1. Adult lake trout and white sucker were still abundant, though emaciated, at pH 5.0–5.5, but in the absence of successful reproduction they would have become extinct. Overall, the Lake 223 experiment indicated a general sensitivity of many organisms to the acidification of lake water. However, within the limits of physiological tolerance, the tests showed that there can be a replacement of acid-sensitive species by relatively tolerant ones.

In a similarly designed experiment in Lake 302, nitric acid was shown to be nearly as effective as sulfuric acid in acidifying lakes, thereby alerting the international community to the need to control atmospheric emissions of gaseous nitrogen compounds. *See also* Acid rain; Algicide; Aquatic chemistry; C:N ratio; Cultural eutrophication; Water pollution

[*Bill Freedman Ph.D.*]

RESOURCES
BOOKS

Freedman, B. *Environmental Ecology*. San Diego: Academic Press, 1995.
Schindler, D. W. "The Coupling of Elemental Cycles by Organisms: Evidence from Whole-Lake Chemical Perturbations." In *Chemical Processes in Lakes*, edited by W. Stumm. New York: Wiley, 1985.

PERIODICALS

Schindler, D. W., et al. "Long-Term Ecosystem Stress: The Effects of Years of Experimental Acidification of a Small Lake." *Science* 228 (June 21, 1985): 1395–1401.

Exponential growth

The distinction between arithmetic and exponential growth is crucial to an understanding of the nature of growth. Arithmetic growth takes place when a constant amount is being added, as when a child puts a dollar a week in a piggy-bank. Although the total amount increases, the amount being added remains the same. Exponential growth, on the other hand, is characterized by a constant or even accelerating rate of growth.

At a constant rate of increase, measured in percentages, the amounts added grow themselves. Growth is then usually measured in doubling times because these remain constant while the amounts added increase. When the annual rate of increase is 1%, the doubling time will be 70 years. From this fact, a simple formula to calculate doubling times given a rate of increase can be derived: dividing 70 by the percentage rate will yield the number of years it takes to double the original amount.

A savings account with, say, a fixed annual interest rate of 5% furnishes a convenient example. If the original deposit is $1,000, then the growth over the first year is $50. Over the second year, growth will be $52.50. In 14 years, there will be $2,000 in the account (70 divided by 5 equals 14). In the first period of 14 years, then, total growth will

be $1,000, but in the second period of 14 years total growth will be $2,000, and so on. During the tenth 14-year period, $512,000 is added, and at the end of that period the total amount in the account will be $1,024,000. As this example illustrates, growth will be relatively slow initially, but it will start speeding up dramatically over time. When growth takes place at an accelerating rate of increase, doubling times of course will become shorter and shorter.

The notion of exponential growth is of particular interest in **population biology** because all populations of organisms have the capacity to undergo exponential growth. The biotic potential or maximum rate of reproduction for all living organisms is very high, that is to say that all **species** theoretically have the capacity to reproduce themselves many, many times over during their lifetimes. In actuality, only a few of the offspring of most species survive, due to reproductive failure, limited availability of space and food, diseases, predation, and other mishaps. A few species, such as the lemming, go through cycles of exponential **population growth** resulting in severe overpopulation. A catastrophic **dieback** follows, during which the population is reduced enormously, readying it for the next cycle of growth and dieback. Interacting species will experience related fluctuations in population levels. By and large, however, populations are held stable by environmental **resistance**, unless an environmental disturbance takes place.

Climatological changes and other natural phenomena may cause such **habitat** disturbances, but more usually they result from human activity. **Pollution, predator control**, and the introduction of foreign species into habitats that lack competitor or predator species are a few examples among many of human activities that may cause declines in some populations and exponential growth in others.

An altogether different case of exponential population growth is that of humans themselves. The human population has grown at an accelerating rate, starting at a low average rate of 0.002% per year early in its history and reaching a record level of 2.06% in 1970. Since then the rate of increase has dropped below 2%, but human population growth is still alarming and many scientists predict that humans are headed for a catastrophic dieback.

[*Marijke Rijsberman*]

RESOURCES
BOOKS

Cunningham, W., and B. W. Saigo. "Dynamics of Population Growth." In *Environmental Science: A Global Concern*. Dubuque, IA: Wm. C. Brown Publishers, 1990.

Miller, G. T. "Human Population Growth." In *Living in the Environment*. Belmont, CA: Wadsworth Publishing, 1990.

External costs

see **Internalizing costs**

Externality

Most economists argue that markets ordinarily are the superior means for fulfilling human wants. In a market, deals are ideally struck between consenting adults only when the parties feel they are likely to benefit. Society as a whole is thought to gain from the aggregation of individual deals that take place. The wealth of a society grows by means of what is called the hidden hand of free market mechanisms, which offers spontaneous coordination with a minimum of coercion and explicit central direction. However, the market system is complicated by so-called externalities, which are effects of private market activity not captured in the price system.

Economics distinguishes between positive and negative externalities. A positive externality exists when producers cannot appropriate all the benefits of their activities. An example would be research and development, which yields benefits to society that the producer cannot capture, such as employment in subsidiary industries. **Environmental degradation**, on the other hand, is a negative externality, or an imposition on society as a whole of costs arising from specific market activities. Historically, the United States have encouraged individuals and corporate entities to make use of **natural resources** on public lands, such as water, timber, and even the land itself, in order to speed development of the country. Many undesirable by-products of the manufacturing process, in the form of exhaust gases or toxic waste, for instance, were simply released into the **environment** at no cost to the manufacturer. The **agricultural revolution** brought new farming techniques that relied heavily on fertilizers, pesticides, and **irrigation**, all of which affect the environment. **Automobile** owners did not pay for the **air pollution** caused by their cars. Virtually all human activity has associated externalities in the environmental arena, which do not necessarily present themselves as costs to participants in these activities. Over time, however, the consequences have become unmistakable in the form of a serious depletion of renewable resources and in **pollution** of the air, water, and **soil**. All citizens suffer from such environmental degradation, though not all have benefited to the same degree from the activities that caused them.

In economic analysis, externalities are closely associated with common property and the notion of **free riders**. Many natural resources have no discrete owner and are therefore particularly vulnerable to abuse. The phenomenon of degradation of common property is known as the **Tragedy of the Commons**. The costs to society are understood as costs to nonconsenting third parties, whose interests in the environment have been violated by a particular market activity. The consenting parties inflict damage without compensating third parties because without clear property rights there is no entity that stands up for the rights of a violated environment and its collective owners.

Nature's owners are a collectivity which is hard to organize. They are a large and diverse group that cannot easily pursue remedies in the legal system. In attempting to gain compensation for damage and force polluters to pay for their actions in the future the collectivity suffers from the free rider problem. Although everyone has a stake in ensuring, for example, good **air quality**, individuals will tend to leave it to others to incur the cost of pursuing legal redress. It is not sufficiently in the interest of most members of the group to sue because each has only a small amount to gain. Thus, government intervention is called for to protect the interests of the collectivity, which otherwise would be harmed.

The government has several options in dealing with externalities such as pollution. It may opt for regulation and set standards of what are considered acceptable levels of pollution. It may require reduced **lead** levels in **gasoline** and require automakers to manufacture cars with greater fuel economy and reduced emissions, for instance. If manufacturers or social entities such as cities exceed the standards set for them, they will be penalized. With this approach, many polluters have a direct incentive to limit their most harmful activities and develop less environmentally costly technologies. So far, this system has not proved to be very effective. In practice, it has been difficult (or not politically expedient) to enforce the standards and to collect the fines. Supreme Court decisions since the early 1980s have reinterpreted some of the laws to make standards much less stringent. Many companies have found it cheaper to pay the fines than to invest in reducing pollution. Or they evade fines by declaring bankruptcy and reorganizing as a new company.

Economists tend to favor pollution taxes and **discharge** fees. Since external costs do not enter the calculations a producer makes, the producer manufactures more of the good than is socially beneficial. When polluters have to absorb the costs themselves, to internalize them, they have an incentive to reduce production to acceptable levels or to develop alternative technologies. A relatively new idea has been to give out marketable pollution permits. Under this system, the government sets the maximum levels of pollution it will tolerate and leaves it to the market system to decide who will use the permits. The costs of past pollution (in the form of permanent environmental damage or costly cleanups) will still be borne disproportionately by society as a whole. The government generally tries to make responsible parties pay for clean-ups, but in many cases it is impossible to determine who the culprit was and in others the parties responsible for the pollution no longer exist.

A special case is posed by externalities that make themselves felt across national boundaries, as is the case with **acid rain**, **ozone layer depletion**, and the pollution of rivers that run through more than one country. Countries that suffer from environmental degradation caused in other countries receive none of the benefits and often do not have the leverage to modify the polluting behavior. International **conservation** efforts must rely on agreements specific countries may or may not follow and on the mediation of the United Nations. *See also* Internalizing cost; Trade in pollution permits

[*Alfred A. Marcus and Marijke Rijsberman*]

RESOURCES

BOOKS

Mann, D., and H. Ingram. "Policy Issues in the Natural Environment." In *Public Policy and the Natural Environment*, edited by H. Ingram and R. K. Goodwin. Greenwich, CT: JAI Press, 1985.

Marcus, A. A. *Business and Society: Ethics, Government, and the World Economy*. Homewood, IL:. Irwin Press, 1993.

Extinction

Extinction is the complete disappearance of a **species**, when all of its members have died or been killed. As a part of natural selection, the extinction of species has been ongoing throughout the earth's history. However, with modern human strains on the **environment**, plants, animals, and invertebrates are becoming extinct at an unprecedented rate of thousands of species per year, especially in tropical rain forests. Many thousands more are threatened and endangered.

Scientists have determined that mass extinctions have occurred periodically in prehistory, coming about every 50 million years or so. The greatest of these came at the end of the Permian period, some 250 million years ago, when up to 96% of all species on the earth may have died off. Dinosaurs and many ocean species disappeared during a well-documented **mass extinction** at the end of the Cretaceous period (about 65 million years ago). It is estimated that of the billions of species that have lived on the earth during the last 3.5 billion years, 99.9% are now extinct.

It is thought that most prehistoric extinctions occurred because of climatological changes, loss of food sources, destruction of **habitat**, massive volcanic eruptions, or asteroids or meteors striking the earth. Extinctions, however, have never been as rapid and massive as they have been in the modern era. During the last two centuries, more than 75

species of mammals and over 50 species of birds have been lost, along with countless other species that had not yet been identified. James Fisher has estimated that since 1600, including species and subspecies, the world has lost at least 100 types of mammals and 258 kinds of birds.

The first extinction in recorded history was the European lion, which disappeared around A.D. 80. In 1534, seamen first began slaughtering the great auk, a large, flightless bird once found on rocky North Atlantic islands, for food and oil. The last two known auks were killed in 1844 by an Icelandic fisherman motivated by rewards offered by scientists and museum collectors for specimens. Humans have also caused the extinction of many species of marine mammals. Steller's sea cow, once found on the Aleutian Islands off Alaska, disappeared by 1768. The sea mink, once abundant along the coast and islands of Maine, was hunted for its fur until about 1880, when none could be found. The Caribbean monk seal, hunted by sailors and fishermen, has not been found since 1962.

The early European settlers of America succeeded in wiping out several species, including the Carolina parakeet and the **passenger pigeon**. The pigeon was one of most plentiful birds in the world's history, and accounts from the early 1800s describe flocks of the birds blackening the sky for days at a time as they passed overhead. By the 1860s and 1870s tens of millions of them were being killed every year. As a result of this **overhunting**, the last passenger pigeon, Martha, died in the Cincinnati Zoo in 1914. The pioneers who settled the West were equally destructive, causing the disappearance of 16 separate types of **grizzly bear**, six of **wolves**, one type of fox, and one cougar. Since the Pilgrims arrived in North America in 1620, over 500 types of native American animals and plants have disappeared.

In the last decade of the twentieth century, the rate of species loss was unprecedented and accelerating. Up to 50 million species could be extinct by 2050, with a rate of three per day. Most of these species extinctions will occur—and are occurring—in tropical rain forests, the richest biological areas on the earth. Rain forests are being cut down at a rate of one to two acres per second.

In 1988, Harvard professor and biologist **Edward O. Wilson** estimated the current annual rate of extinction at up to 17,500 species, including many unknown **rain forest** plants and animals that have never been studied or even seen, by humans. Botanist Peter Raven, director of the Missouri **Botanical Garden**, calculated that a total of one-quarter the world's species could be gone by 2010. A study by the **World Resources Institute** pointed out that humans have accelerated the extinction rate to 100 to 1,000 times its natural level.

While it is impossible to predict the magnitude of these losses or the impact they will have on the earth and its **future generations**, it is clear that the results will be profound, possibly catastrophic. In his book, *Disappearing Species: The Social Challenge*, Eric Eckholm of the **Worldwatch Institute** observed that humans, in their ignorance, have changed the natural course of **evolution** with current mass-extinction rates. "Should this biological massacre take place, evolution will no doubt continue, but in a grossly distorted manner. Such a multitude of species losses would constitute a basic and irreversible alteration in the nature of the **biosphere** even before we understand its workings..."

Eckholm further notes that when a plant species is wiped out, some 10 to 30 dependent species, such as insects and even other plants, can also be jeopardized. An example of the complex relationship that has evolved between many tropical species is the 40 different kinds of fig trees native to Central America, each of which has a specific insect pollinator. Other insects, including pollinators of other plants, depend on these trees for food. Thus, the extinction of one species can set off a **chain reaction**, the ultimate effects of which cannot be foreseen.

Although scientists know that human life will be harmed by these losses, the weight of the impact is unclear. As the **Council on Environmental Quality** states in its book *The Global Environment and Basic Human Needs*, over the next decade or two, "unique ecosystems populated by thousands of unrecorded plant and animal species face rapid destruction—irreversible genetic losses that will profoundly alter the course of evolution." This report also cautions that species extinction entails the loss of many useful products. Perhaps the greatest industrial, agricultural and medical costs of species reduction will stem from future opportunities unknowingly lost. Only about 5% of the world's plant species have yet been screened for pharmacologically active ingredients. Ninety percent of the food that humans eat comes from just 12 crops, but scores of thousands of plants are edible, and some will undoubtedly prove useful in meeting human food needs. *See also* Biodiversity; Climate; Dodo; Endangered species

[*Lewis G. Regenstein*]

RESOURCES
BOOKS

Etheredge, N. *The Miner's Canary: A Paleontologist Unravels the Mysteries of Extinction.* Englewood Cliffs, NJ: Prentice-Hall, 1991.

Raup, D. M. *Extinction: Bad Genes or Bad Luck?* New York: Norton, 1991.

Tudge, C. *Last Animals at the Zoo: How Mass Extinction Can Be Stopped.* Washington, DC: Island Press, 1992.

Exxon Valdez

On March 24, 1989, the 987-foot super tanker *Exxon Valdez* outbound from Port Valdez, Alaska, with a full load of oil from Alaska's Prudhoe Bay passed on the wrong side of a lighted channel marker guarding a shallow stretch of **Prince William Sound**. The momentum of the large ship carried it onto Bligh Reef and opened a 6 x 20 ft hole in the ship's hull. Through this hole poured 257,000 barrels (11 million gallons) of crude oil, approximately 21% of the ship's 1.26 million barrel (53 million gallon) cargo, making it the largest oil spill in the history of the United States.

The oil spill resulting from the *Exxon Valdez* accident spread 38,0000 metric tonnes of oil along 1,500 miles of pristine shoreline on Prince William Sound and the Kenai Peninsula, covering an area of 460 miles. Oil would eventually reach shores southwest of the spill up to 600 miles away.

The *Exxon Valdez* Oil Spill Trustee Council estimates that 250,000 seabirds, 2,800 sea otters, 300 harbor **seals**, 250 bald eagles, and 22 killer **whales**, were killed. These figures may be an underestimate of the animals killed by the oil because many of the carcasses likely sank or washed out to sea before they could be collected. Most of the birds died from hypothermia due to the loss of insulation caused by oil-soaked feathers. Many predatory birds, such as bald eagles, died as a result of ingesting contaminated fish and birds. Hypothermia affected sea otters as well, and many of the dead mammals suffered lung damage due to oil fumes. Billions of **salmon** eggs were also lost to the spill. While a record 43 million pink salmon were caught in Prince William Sound in 1990, by 1993 the harvest had declined to a record low of three million.

Response to the oil spill was slow and generally ineffective. The Alyeska Oil Spill Team responsible for cleaning up **oil spills** in the region took more than 24 hours to respond, despite previous assurances that they could mount a response in three hours. Much of the oil containment equipment was missing, broken, or barely operable. By the time oil containment and recovery equipment were in place, 42 million liters of oil had already spread over a large area. Ultimately, less than 10% of this oil was recovered, the remainder dispersing into the air, water, and **sediment** of Prince William Sound and adjacent sounds and fjords. Exxon reports spending a total of $2.2 billion to clean up the oil. Much of this money employed 10,000 people to clean up oil-fouled beaches; yet after the first year, only 3% of the soiled beaches had been cleaned.

In response to public concern about the poor response time and uncoordinated initial cleanup efforts following the Valdez spill, the Oil **Pollution** Act (OPA; part of the **Clean Water Act**) was signed into law in August 1990. The Act

After hitting Bligh Reef on March 24, 1989, the *Exxon Valdez* sits near Naked Island, Alaska, swirls of crude oil surrounding the ship. Containment booms can be seen attached to the ship. (AP/Wide World Photos. Reproduced by permission.)

established a Federal trust fund to finance clean-up efforts for up to $1 billion per spill incident.

Ultimately, **nature** was the most effective surface cleaner of beaches; winter storms removed the majority of oil and by the winter of 1990, less than 6 miles (10 km) of shoreline was considered seriously fouled. Cleanup efforts were declared complete by the U.S. Coast Guard and the State of Alaska in 1992.

On October 9, 1991, a settlement between Exxon and the State of Alaska and the United States government was approved by the U.S. District Court. Under the terms of the agreement, Exxon agreed to pay $900 million in civil penalties over a 10-year period. The civil settlement also provides for a window of time for new claims to be made should unforeseen environmental issues arise. That window is from September 1, 2002 to September 1, 2006.

Exxon was also fined $150 million in a criminal plea agreement, of which $125 million was forgiven in return for the company's cooperation in cleanup and various private settlements. Exxon also paid $100 million in criminal restitution for the environmental damage caused by the spill.

A flood of private suits against Exxon have also deluged the courts in the years since the spill. In 1994, a district court ordered Exxon to pay $287 million in compensatory damages to a group of commercial fishermen and other Alaskan natives who were negatively impacted by the spill. The jury who heard the case also awarded the plaintiffs $5 billion in punitive damages. However, in November 2001 a federal appeals judge overturned the $5 billion punitive award, deeming it excessive and ordering the district court to reevaluate the settlement. As of May 2002, the final punitive settlement had not been determined.

The Captain of the *Exxon Valdez*, Joseph Hazelwood, had admitted to drinking alcohol the night the accident occurred, and had a known history of alcohol abuse. Nevertheless, he was found not guilty of charges that he operated a shipping vessel under the influence of alcohol. He was found guilty of negligent **discharge** of oil, fined $50,000, and sentenced to 1,000 hours of community service work.

Despite official cleanup efforts by Exxon having ended, the environmental legacy of the *Valdez* spill lives on. The Auke Bay Laboratory of the Alaskan Fisheries Science Center conducted a beach study of Prince William Sound in the summer of 2001. Researchers found that approximately 20 acres of Sound shoreline is still contaminated with oil, the majority of which has collected below the surface of the beaches where it continues to pose a danger to **wildlife**. Of the 30 **species** of wildlife affected by the spill, only two—the American **bald eagle** and the river otter— were considered recovered in 1999. Preliminary 2002 reports from the Exxon Valdez Oil Spill Trustee Council reflect progress is being made in the area of wildlife recovery, however, black oystercatchers, common murres, killer whales, subtidal communities, sockeye salmon, and pink salmon are all classified as recovered in the April draft of the organization's "Update on Injured Resources and Services," bringing the total recovered species to eight.

[*William G Ambrose and Paul E Renaud and Paula A Ford-Martin*]

RESOURCES
BOOKS

Keeble, J. *Out of the Channel: The Exxon Valdez Oil Spill in Prince William Sound, 10th Anniversary Edition.* Spokane, WA: Eastern Washington University Press, 1999.

Picou, J. S., et al. *The Exxon Valdez Disaster: Readings on a Modern Social Problem, 2nd ed.* Dubuque, IA: Kendall/Hunt Publishing, 1999.

PERIODICALS

Berg, Catherine. "The Exxon Valdez Spill: 10 Years Later." *Endangered Species Bulletin* 26, no.2 (March/April 1999): 18–9.

OTHER

Alaska Fisheries Science Center, National Marine Fisheries Service, NOAA. "The Exxon Valdez Oil Spill: How Much Oil Remains?" *AFSC Quarterly*(July-September 2001). http://www.afsc.noaa.gov/Quarterly/jas2001/feature_jas01.htm. Accessed May 28, 2002.

U.S. Environmental Protection Agency Oil Spill Program. [cited May 28, 2002]. <http://www.epa.gov/oilspill/index.htm>.

ORGANIZATIONS

The Exxon Valdez Oil Spill Trustee Council, 441 West Fifth Avenue, Suite 500, Anchorage, AL USA 99501 (907) 278-8012 , Fax: (907) 276-7178 , Toll Free: (800) 478-7745 (within Alaska), Toll Free: (800) 283-7745 (outside Alaska), Email: restoration@oilspill.state.ak.us, http://www.oilspill.state.ak.us/

F

Falco peregrinus
see **Peregrine falcon**

Falcon
see **Peregrine falcon**

Fallout
see **Radioactive fallout**

Family planning

The exaxt population of the world is unknown but believed at about 6.2 billion; it continues in its unrelenting growth especially in developing countries. Worldwide **famine** has been postponed thanks to modern agricultural procedures, known as the Green Revolution, which have greatly increased grain production. Nevertheless, with limited land and resources that can be devoted to food production—and increasing numbers of humans who need both space and food—there appears to be a significant risk of catastrophe by overpopulation. Because of this, there is increased interest in family planning. Family planning in this context means birth control to limit family size.

The subject of "family planning" is not limited to birth control but includes procedures designed to overcome difficulties in becoming pregnant. About 15% of couples are unable to conceive children after a year of sexual activity without using birth control. Many couples feel an intense desire and need to conceive children. Aid to these couples is thus a reasonable part of family planning. However, for most discussions of family planning, the emphasis is on limitation of family size, not augmentation.

Birth control procedures have evolved rapidly in this century. Further, utilization of existing procedures is chang-ing with different age groups and different populations. Thus any account of birth control is likely to become rapidly obsolete. An example of the changing technology includes oral contraception with pills containing hormones. Birth control pills have been marketed in the United States since the 1960s. Since that time there have been many formulations with significant reductions in dosage. There was much greater acceptance of pills by American women under the age of 30. The intrauterine device (IUD) was much more popular in Sweden than in the United States whereas sterilization was more common in the United States than in Sweden.

A very common form of birth control is the condom, which is a thin **rubber** sheath worn by men during sexual intercourse. They are generally readily accessible, cheap, and convenient for those individuals who may not have sexual relations regularly. Sperm cannot penetrate the thin (0.3—0.8 mm thick) latex. Neither the human immunodeficiency **virus** (HIV) associated with **AIDS** nor the other pathogenic agents of sexually transmitted diseases (STDs) are able to penetrate the latex barrier. Some individuals are opposed to treating healthy bodies with drugs (hormones) for birth control, and for these individuals, condoms have a special appeal. Natural "skin" (lamb's intestine) condoms are still available for individuals who may be allergic to latex, but this product provides less protection to HIV and other STDs.

The reported failure rate of condoms is high and is most likely due to improper use. Yet during the Great Depression in the 1930s—when pills and other contemporary birth control procedures were not available—it is thought that the proper use of condoms caused the birth rate in the United States to plummet.

Spermicides—surface active agents which inactivate sperm and STD pathogens—can be placed in the vagina in jellies, foam, and suppositories. Condoms used in conjunction with spermicides have a failure rate lower than either method used alone and may provide added protection against some infectious agents.

Types of Contraceptives

Effectiveness	Predicted (%)	Actual (%)
Birth control pills	99.9	97
Condoms	98	88
Depo Provera	99.7	99.7
Diaphragm	94	82
IUDS	99.2	97
Norplant	99.7	99.7
Tubal sterilization	99.8	99.6
Spermicides	97	79
Vasectomy	99.9	99.9

The vaginal diaphragm, like the condom, is another form of barrier. The diaphragm was in use in World War I and was still used by about one third of couples by the time of World War II. However, because of the efficacy and ease of use of oral contraceptives, and perhaps because of the protection against disease by condoms, the use of vaginal diaphragms is down. The diaphragm, which must be fitted by a physician, is designed to prevent sperm access to the cervix and upper reproductive tract. It is used in conjunction with spermicides. Other similar barriers include the cervical cap and the contraceptive sponge. The cervical cap is smaller than the diaphragm and fits only around that portion of the uterus that protrudes into the vagina. The contraceptive sponge, which contains a spermicide, is inserted into the vagina prior to sexual intercourse and retained for several hours afterwards to insure that no living sperm remain.

Intrauterine devices (IUDs) were popular during the 1960s and 1970s in the United States, but their use today has dwindled. However in China, a nation which is rapidly attending to its population problems, about 60 million women use IUDs. The failure rate of IUDs in **less developed countries** is reported to be less than that with the pill. The devices may be plastic, **copper**, or stainless steel. The plastic versions may be impregnated with barium sulfate to permit visualization by x ray and also may slowly release hormones such as progesterone. Ovulation continues with IUD use. Efficacy probably results from a changed uterine **environment** which kills sperm.

Oral contraception is by means of the "pill." Pills contain an estrogen and a progestational agent, and current dosage is very low compared with several decades ago. The combination of these two agents is taken daily for three weeks followed by one week with neither hormone. Fre-

quently a drug-free pill is taken for the last week to maintain the pill-taking habit and thus enhance the efficacy of the regimen. The estrogenic component prevents follicle maturation, and the progestational component prevents ovulation. Pill-taking women who have multiple sexual partners may wish to consider the addition of a barrier method to minimize risk for STDs. The reliability of the pill reduces the need for abortion or surgical sterilization. There may be other salutary health effects which include less endometrial and ovarian **cancer** as well as fewer uterine fibroids. Use of oral contraceptives in women over the age of 35 who also **smoke** is thought to increase the risk of heart and vascular disease.

Contraceptive hormones can be administered by routes other than oral. Subdermal implants of progestin-containing tubules have been available since 1990 in the United States. In this device familiarly known as Norplant, six tubules are surgically placed on the inside of the upper arm, and the hormone diffuses through the wall of the tubules to provide long term contraceptive activity. Another form of progestin-only contraception is by intramuscular injection which must be repeated every three months.

Fears engendered by IUD litigation are thought to have increased the reliance of many American women on surgical sterilization (tubal occlusion). Whatever the reason, more American women rely on the procedure than do their European counterparts. Tubal occlusion involves the mechanical disruption of the oviduct, the tube that leads from the ovary to the uterus, and prevents sperm from reaching the egg. Inasmuch as the fatality rate for the procedure is lower than that of childbirth, surgical sterilization is now the safest method of birth control. Tubal occlusion is far more common now that it was in the 1960s because of the lower cost and reduced surgical stress. Use of the laparoscope and very small incisions into the abdomen have allowed the procedure to be completed during an office visit.

Male sterilization, another method, involves severing the vas deferens, the tube that carries sperm from the testes to the penis. Sperm comprise only a small portion of the ejaculate volume, and thus ejaculation is little changed after vasectomy. The male hormone is produced by the testes and production of that hormone continues as does erection and orgasm.

Most abortions would be unnecessary if proper birth control measures were followed. That of course is not always the case. Legal abortion has become one of the leading surgical procedures in the United States. Morbidity and **mortality** associated with pregnancy have been reduced more with legal abortion than with any other event since the introduction of antibiotics to fight puerperal fever.

Other methods of birth control are used by individuals who do not wish to use mechanical barriers, devices, or drugs (hormones). One of the oldest of these methods is withdrawal (*coitus interruptus*), in which the penis is removed from the vagina just before ejaculation. Withdrawal must be exquisitely timed, is probably frustrating to both partners, is not thought to be reliable, and provides no protection against HIV and other STD infections. Another barrier-and-drug-free procedure is natural family planning (also known as the rhythm method). Abstinence of sexual intercourse is scheduled for a period of time before and after ovulation. Ovulation is calculated by temperature change, careful record keeping of menstruation (the calendar method), or by vaginal mucous inspection. Natural family planning has appeal for individuals who wish to limit their exposure to drugs, but it provides no protection against HIV and other STDs.

The population of the world increases by about 140 million every year, while the world is unable to sustain its new residents adequately. That increase signals the need for family planning education and the continued development of ever more efficient birth control methods. *See also* Population Council; Population growth; Population Institute

[*Robert G. McKinnell*]

RESOURCES
BOOKS

Sitruk-Ware, R., and C. W. Bardin. *Contraception*. New York: Marcel Dekker, 1992.
Speroff, L., and P. D. Darney. *A Clinical Guide for Contraception*. Baltimore: Williams & Wilkins, 1992.

Famine

Famine is widespread hunger and starvation. A region struck with famine experiences acute shortages of food, massive loss of lives, social disruption, and economic chaos. Images of starving mothers and children with emaciated eyes and swollen bellies during recent crises in Ethiopia and Somalia have brought international attention to the problem of famine. Other well-known famines include the great Irish potato famines of the 1850s that drove millions of immigrants to America, and a Russian famine during Stalin's **agricultural revolution** that killed 20 million people in the 1930s. The worst recorded famine in recent history occurred in China between 1958 and 1961, when 23–30 million people died as a result of the failed agricultural program, "the great leap forward."

Even though we may think of these tragedies as single, isolated events, famine, and chronic hunger continue to be serious problems. Between 18 and 20 million people, three-quarters of them children, die each year of starvation or diseases caused by malnourishment. How can this be? Environmental problems like overpopulation, scarce resources,

and natural disasters affect people's ability to produce food. Political and economic problems like unequal distribution of wealth, delayed or insufficient action by local governments, and imbalanced trade relationships between countries affect people's ability to buy food when they cannot produce it.

Perhaps the most common explanation for famine is overpopulation. The world's population, now more than 6.2 billion people, grows by 250,000 people every day. It seems impossible that the natural world could support such rapid growth. Indeed, the pressures of rapid growth have had a devastating impact on the **environment** in many places. Land that once fed one family must now feed 10 families and resulting over-use harms the quality of the land. The world's deserts are rapidly expanding as people destroy fragile **topsoil** by poor farming techniques, clearing vegetation, and **overgrazing**.

Although the demands of **population growth** and industrialization are straining our environment, we have yet to exceed the limits of growth. Since the 1800s some have predicted that humans, like rabbits living without predators, would foolishly reproduce far beyond the **carrying capacity** of their environment and then die in masses from lack of food. This argument assumes that the supply of food will remain the same as populations grow, but as populations have grown, people have learned to grow more food. World food production increased two-and-a-half times between 1950 and 1980. Alter World War II, agriculture specialists caused a "green revolution," developing new crops and farming techniques that radically increased food production per acre. Farmers began to use special hybrid crop strains, chemical fertilizers, pesticides, and advanced **irrigation** systems. Today, there is more than enough food available to feed everyone. In fact, the United States government spends billions of dollars every year to store excess grain and to keep farmers from farming portions of their land.

Many famines occur in the aftermath of natural disasters like floods and droughts. In times of **drought**, crops cannot grow because they do not have enough water. In times of flood, excess water washes out fields, destroying crops and disrupting farm activity. These disasters have several effects. First, damaged crops cause food shortages, making nutrients difficult to find and making any food available too expensive for many people. Second, reduced food production means less work for those who rely on temporary farm work for their income. Famines usually affect only the poorest five to ten percent of a country's population. They are most vulnerable because during a crisis wages for the poorest workers go down as food prices go up.

Famine is a problem of distribution as well as production. Environmental, economic, and political factors together determine the supply and distribution of food in a country. Starvation occurs when people lose their ability to obtain food by growing it or by buying it. Often, poor decisions and organizations aggravate environmental factors to cause human suffering. In Bangladesh, floods during the summer of 1974 interfered with rice transplantation, the planting of small rice seedlings in their rice patties. Although the crop was only partly damaged, speculators hoarded rice, and fears of a shortage drove prices beyond the reach of the poorest in Bangladesh. At the same time, disruption of the planting meant lost work for the same people. Even though there was plenty of rice from the previous year's harvest, deaths from starvation rose as the price of rice went up. In December of 1974, when the damaged rice crop was harvested, the country found that its crop had been only partly ruined. Starvation resulted not from a shortage of rice, but from price speculation. The famine could have been avoided completely if the government had responded more quickly, acting to stabilize the rice market and to provide relief for famine victims.

In other cases, governments have acted to avoid famine. The Indian state of Maharahtra offset affects of a severe drought in 1972 by hiring the poorest people to work on public projects like roads and **wells**. This provided a service for the country and at the same time diverted a catastrophe by providing an income for the most vulnerable citizens to compete with the rest of the population for a limited food supply. At the same time, the countries of Mali, Niger, Chad, and Senegal experienced severe famine, even though the average amount of food per person in these countries was the same as in Maharahtra. The difference, it would seem, lies in the actions and intentions of the governments. The Indian government provides a powerful example. Although India lags behind many countries in economic development, education, and health care, the Indians have managed to avert serious famine since 1943, four years before they gained independence from the British.

Responsibility for hunger and famine rests also with the international community. Countries and peoples of the world are increasingly interconnected, continuously exchanging goods and services. We are more and more dependent on one another for success and for survival. The world economic and political order dramatically favors the wealthiest industrialized countries, in Europe and North America. Following patterns established during colonial expansion, **Third World** nations often produce raw materials, unfinished goods, and basic commodities like bananas and coffee that they sell to the **First World** at low prices. The First World nations then manufacture and refine these products and sell back information and technology, like machinery and computers for a very high price. As a result, the wealthiest nations amass capital and resources, and enjoy very high standards of living, while the poorest nations retain huge

national debts and struggle to remain stable economically and politically. The word's poorest countries, then, are left vulnerable to all of the conditions which cause famine, economic hardship, political instability, overpopulation, and over-taxed resources.

Furthermore, large colonial powers often left behind unjust political and social hierarchies that are very good at extracting resources and sending them north, but not as good at promoting social justice and human welfare. Many Third World countries are dominated by a small ruling class who own most of the land, control industry, and run the government. Since the poorest people, who suffer most in famines, have little power to influence government policies and manage the countries economy, their needs are often unheard and unmet. A government that rules without democratic support of its people has less incentive to protect those who would suffer in times of famine. In addition, the poorest often do not benefit from the industry and agriculture that does exist in a developing country. Large corporate farms often force small subsistence farmers off of their land. These farmers must then work for day wages, producing food for export, while local people go without adequate nutrition.

Economic and social arrangements, as well as environmental conditions, are central to the problems of hunger and starvation. Famine is much less likely to occur in countries that are concerned with issues of social justice. In the same way, famine is much less likely to occur in a world that is concerned with issues of social justice. Environmental pressures of population growth and human use of **natural resources** will continue to be issues of great concern. Natural disasters like droughts and floods will continue to occur. The best response to the problem of famine lies in working to better manage **environmental resources** and crisis situations and to change political and economic structures that cause people to go without food.

[*John Cunningham*]

RESOURCES
BOOKS

Lappe, F. M. *World Hunger: Twelve Myths* New York: Grove Press, 1986.

PERIODICALS

Sen, A. "Economics of Life and Death." *Scientific American* 268 (May 1993): 40–47.

Farming

see **Agricultural revolution; Conservation tillage; Dryland farming; Feedlots; Organic gardening and farming; Shifting cultivation; Slash and burn agriculture; Strip-farming; Sustainable agriculture**

Fauna

All animal life that lives in a particular geographic area during a particular time in history. The type of fauna to be found in any particular region is determined by factors such as plant life, physical **environment**, topographic barriers, and evolutionary history. Zoologists sometimes divide the earth into six regions inhabited by distinct faunas: Ethiopian (Africa south of the Sahara, **Madagascar**, Arabia), Neotropical (South and Central America, part of Mexico, the West Indies), Australian (**Australia**, New Zealand, New Guinea), Oriental (Asia south of the Himalaya Mountains, India, Sri Lanka, Malay Peninsula, southern China, Borneo, Sumatra, Java, the Philippines), Palearctic (Europe, Asia north of the Himalaya Mountains, Afghanistan, Iran, North Africa), and Nearctic (North America as far south as southern Mexico).

Fecundity

Fecundity comes from the Latin word *fecundus*, meaning fruitful, rich, or abundant. It is the rate at which individual organisms in the population produce offspring. Although the term can apply to plants, it is typically restricted to animals.

There are two aspects of reproduction: 1) fertility, referring to the physiological ability to breed, and 2) fecundity, referring to the ecological ability to produce offspring. Thus, higher fecundity is dependent on advantageous conditions in the **environment** that favor reproduction (e.g., abundant food, space, water and mates; limited predation, parasitism, and **competition**). The intrinsic rate of increase (denoted as "r") equals the birth rate minus the death rate. It is a population characteristic that takes into account that not all individuals have equal birth rates and death rates. It therefore refers to the reproductive capacity in the population made up of individual organisms. Fecundity, on the other hand, is an individual characteristic. It can be further subdivided into potential and realized fecundity. For example, deer can potentially produce four or more fawns per year, but they typically give birth to only one or two per year. In good years with ample food, they often have only two fawns.

Animals in **nature** are limited by environmental conditions that control their life history characteristics such as birth, **survivorship**, and death. A graph of the number of offspring per female per age class (e.g., year) is a fecundity curve. This can then be used to interpret the individuals of a certain age class who contribute more to the **population growth** than others. In other words, certain age classes have a greater reproductive output than others. **Wildlife** managers often use this type of information in deciding which individ-

uals in a population can be hunted verses those that should be protected so they can reproduce.

As the number of animals increase, competition for food may become more intense and, therefore, growth and reproduction may decrease. The result is an example of density-dependent fecundity. Fecundity in predators typically increases with an increase in the prey population. Conversely, fecundity in prey **species** typically increases when predation pressure is low.

Some scientists have found that fecundity is inversely related to the amount of parental care given to the young. In other words, small organisms such as insects and fish which typically invest less time and energy into caring for the young usually have higher fecundity. Larger organisms such as birds and mammals which expend a lot of energy on caring for the young through building of nests, feeding, protecting, and caring have lower fecundity rates.

[John Korstad]

RESOURCES
BOOKS

Colinvaux, P. A. *Ecology.* New York: Wiley, 1986.

Smith, R. E. *Ecology and Field Biology.* 4th ed. New York: Harper and Row, 1990.

Ricklefs, R. E. *Ecology.* 3rd ed. New York: W. H. Freeman, 1990.

Krebs, C. J. *Ecology: The Experimental Analysis of Distribution and Abundance.* 3rd ed. New York: Harper and Row, 1985.

Federal Energy Regulatory Commission

The Federal Energy Regulatory Commission (FERC) is an independent, five-member commission within the **U.S. Department of Energy**. The Commission was created in October 1977 as a part of the Federal government's mammoth effort to restructure and reorganize its energy program. The Commission was assigned many of the functions earlier assigned to the **Federal Power Commission** (FPC).

The Federal Power Commission had existed since 1920 when it was created to license and regulate hydroelectric **power plants** situated on interstate streams and rivers. Over the next half century, the Power Commission was assigned more and more responsibility for the management of United States energy reserves. In the Public Utilities Holding Company Act of 1935, for example, Congress gave to the Commission responsibility for setting rates for the wholesale pricing of electricity shipped across state lines. FPC's mission was expanded even further in the 1935 Natural Gas Act. That act gave the Commission the task of regulating the nation's natural gas pipelines and setting rates for the sale of **natural gas**.

Regulating energy prices in the pre-1970s era was a very different problem than it is in the 1990s. That era was one of abundant, inexpensive energy. Producers, consumers and regulators consistently dealt with a surplus of energy. There was more energy of almost any kind than could be consumed. That situation began to change in the early 1970s, especially after the **oil embargo** instituted by the **Organization of Petroleum Exporting Countries** in 1973. The resulting energy crisis caused the United States government to re-think carefully its policies and practices regarding energy production and use.

One of the studies that came out of that re-analysis was the 1974 Federal Energy Regulation Study Team report. The team found a number of problems in the way energy was managed in the United States. They reported that large gaps existed in some areas of regulation, with no agency responsible, while other areas were characterized by overlaps, with two, three, or more agencies all having some responsibility for a single area. The team also found that regulatory agencies were more oriented to the past than to current problems or future prospects, worked with incomplete or inaccurate date, and employed procedures that were too lengthy and drawn out.

As one part of the Department of Energy Organization Act of 1977, then, the FPC was abolished and replaced by the Federal Energy Regulatory Commission. The Commission is now responsible for setting rates and charges for the **transportation** and sale of natural gas and for the transmission and sale of electricity. It continues the FPC's old responsibility for the licensing of hydroelectric plants.

In addition, the Commission has also been assigned responsibility for establishing the rates for the transportation of oil by pipelines as well as determining the value of the pipe lines themselves. Overall, the Commission now controls the pricing of 60% of all the natural gas and 30% of all the electricity in the United States.

Ordinary citizens sometimes do not realize the power and influence of independent commissions like the FERC. But they can have significant impact on federal policy and practice. As one writer has said, "While Energy Secretaries come and go, and Congress can do little more than hold hearings, the five-member FERC is making national **energy policy** by itself."

[*David E. Newton*]

RESOURCES

ORGANIZATIONS

Federal Energy Regulatory Commission, 888 First Street, NE, Washington, D.C. USA 20009 , <http://www.ferc.gov>

Federal Insecticide, Fungicide and Rodenticide Act (1972)

The **Environmental Protection Agency** (EPA) is the primary regulatory agency of pesticides. The EPA's authority on pesticides is given in the Congressionally-enacted Federal Insecticide, **Fungicide** and Rodenticide Act (FIFRA)—a comprehensive regulatory program for pesticides and herbicides enacted in 1972 and amended nearly 50 times over the years. The goal of FIFRA is to regulate the use of pesticides through registration.

Section 3 of FIFRA mandates that the EPA must first determine that the product "will perform its intended function without unreasonable adverse effects on the environment" before it is registered. The Act defines adverse effects as "any unreasonable effects on man or the **environment**, taking into account the economic, social and environmental costs and benefits of using any pesticide." To further this objective, Congress placed a number of regulatory tools at the disposal of the EPA. Congress also made clear that the public was not to bear the risk of uncertainty concerning the safety of a **pesticide**. To grant registration, the EPA must conclude that the food production benefits of a pesticide outweigh any risks.

To make the cost-benefit determination required to register a pesticide as usable, manufacturers must submit to EPA detailed tests on the chemical's health and environmental effects. The burden rests on the manufacturer to provide the data needed to support registration for use on a particular crop. The pesticide manufacturer is required to submit certain health and safety data to establish that the use of the pesticide will not generally cause unreasonable adverse effects. Data required include disclosure of the substance's chemical and toxicological properties, likely distribution in the environment, and possible effects on **wildlife**, plants, and other elements in the environment.

FIFRA is a licensing law. Pesticides may enter commerce only after they are approved or "registered following an evaluation against statutory risk/benefit standards." The Administrator may take action to terminate any approval whenever it appears, on the basis of new information or a re-evaluation of information, that the pesticide no longer meets the statutory standard. These decisions are made on a use-by-use basis since the risks and benefits of a pesticide vary from one use to another.

FIFRA is also a control law. Special precautions and instructions may be imposed. For example, pesticide applicators may be required to wear protective clothing, or the use of certain pesticides may be restricted to trained and certified applicators. Instructions, warnings, and prohibitions are incorporated into product labels, and these labels may not be altered or removed.

FIFRA embodies the philosophy that those who would benefit by government approval of a pesticide product should bear the **burden of proof** that their product will not pose unreasonable risks. This burden of proof applies both when initial marketing approval is sought and in any proceeding initiated by the Administrator to interrupt or terminate registration through suspension or cancellation. Of course, while putting the burden on industry, the assumption is that industry will be honest in its research and reports.

Licensing decisions are usually based on tests furnished by an applicant for registration. The tests are performed by the petitioning company in accordance with testing guidelines prescribed by the EPA. Requirements for the testing of pesticides for major use can be met only through the expenditure of several millions of dollars and up to four years of laboratory and field testing.

However, major changes in test standards, advances in testing methodology, and the heightened awareness of the potential chronic health effects of long-term, low-level exposure to **chemicals** which have come into the marketplace within the past several decades have brought the need to update EPA mandates. Thus, Congress directed that the EPA reevaluate its licensing decisions through a process of re-registration. That means if the government once approved a certain product for domestic use, it does not mean the EPA can be confident today that its use can continue.

The EPA has the power to suspend or cancel products. Cancellation means the product in question is no longer considered safe. It must be taken off the commercial market and is no longer available for use. Suspension means the product in question is not to be sold or used under certain conditions or in certain places. This may be a temporary decision, usually dependent on further studies.

There may be certain products that, in the opinion of the administrator, may be harmful. But no action is taken if and when the action is incompatible with administration priorities. That is because the EPA administrator is responsible to the directives and priorities of the President and the executive branch. Thus, some regulatory decisions are political.

There is a statutory way to avoid the re-registration process. It is called Section 18. Section 18 under FIFRA allows for the use of unregistered pesticides in certain emergency situations. It provides that "the Administrator may, at his discretion, exempt any Federal or State agency from any provisions of this subchapter if he determines that emergency conditions exist which requires such exemption. The Administrator, in determining whether or not such emergency conditions exist, shall consult with the Secretary of Agriculture and the Governor of any state concerned if they request such determination."

From 1978 to 1983, the General Accounting Office (GAO) and the House Subcommittee on Department Operations, Research and Foreign Agriculture of the Committee on Agriculture (DORFA Subcommittee) thoroughly examined the EPA's implementation of Section 18. Under the auspices of Chair George E. Brown, Jr., (D-CA) the DORFA Subcommittee held a series of hearings which revealed numerous abuses in EPA's administration of Section 18. A report was issued in 1982 and reprinted as part of the Committee's 1983 Hearings.

The Subcommittee found that "the rapid increase in the number and volume of pesticides applied under Section 18 was clearly the most pronounced trend in the EPA's pesticide regulatory program." According to the Subcommittee report, "a primary cause of the increase in the number of Section 18 exemptions derived from the difficulty the Agency had in registering chemicals under Section 3 of FIFRA in a timely manner." The DORFA committee stated: "Regulatory actions involving suspect human carcinogens which meet or exceed the statute's "unreasonable adverse effects" criterion for chronic toxicity often become stalled in the Section 3 review process for several years. The **risk assessment** procedures required by States requesting Section 18 actions, and by EPA in approving them, are generally less strict. For example, a relatively new insecticide, first widely used in 1977, was granted some 140 Section 18 emergency exemptions and over 300 (Section 24c) Special Local Needs registrations in the next four years while the Agency debated the significance to man of positive evidence of oncogenicity in laboratory animals."

The EPA's practices changed little over the next eight years. In the spring of 1990, the Subcommittee on Environment of the House Science, Space and Technology Committee (Chair: James H. Scheuer, D-NY) investigated the EPA's procedures under Section 18 of FIFRA and found if anything, the problem had gotten worse. The report states: "Since 1973, more than 4,000 emergency exemptions have been granted for the use of pesticides on crops for which there is no registration." A large number of these emergency exemptions have been repeatedly granted for the same uses for anywhere from fourteen, to ten, to eight, to five years. The House Subcommittee also found that the EPA required less stringent testing procedures for pesticides under the exemption, which put companies that follow the normal procedure at a disadvantage. The Subcommittee concluded that the large numbers of emergency exemptions arose from "the EPA's failure to implement its own regulations."

The Subcommittee identified "emergencies" as "routine predicted outbreaks and foreign competition" and "a company's need to gain market access for use of a pesticide on a new crop, although the company often never intends to submit adequate data to register the chemical for use."

As for the re-registration requirement, the Subcommittee observed: "The EPA's reliance on Section 18 may be related to the Agency's difficulty in re-registering older chemical substances. Often, Section 18 requests are made for the use of older chemicals on crops for which they are not registered. These older chemicals receive repetitive exemptions for use despite the fact that many of these substances may have difficulty obtaining re-registration since they have been identified as potentially carcinogenic. Thus, by liberally and repetitively granting exemptions to potentially carcinogenic substances, little incentive is provided to encourage companies to invest in the development of newer, safer pesticides or alternative agricultural practices."

The report concluded, "...Allowing these exemptions year after year in predictable situations provides 'back-door' pre-registration market access to potentially dangerous chemicals." *See also* Environmental law; Integrated pest management; National Coalition Against the Misuse of Pesticides; Pesticide Action Network; Risk analysis

[*Liane Clorfene Casten*]

RESOURCES
BOOKS

Rodgers Jr., W. H. *Environmental Law: Pesticides and Toxic Substances.* 3 vols. St. Paul, MN: West, 1988.

U.S. Environmental Protection Agency. *Federal Insecticide, Fungicide, and Rodenticide Act: Compliance-Enforcement Guidance Manual.* Rockville, MD: Government Institutes, 1984.

PERIODICALS

"EPA Data Is Flawed, Says GAO." *Chemical Marketing Reporter* 243 (January 11, 1993): 7+.

"Controlling the Risk in Biotech." *Technology Review* 92 (July 1989): 62–9.

Federal Land Policy and Management Act (1976)

The Federal Land Policy and Management Act (FLPMA), passed in 1976, is the statutory grounding for the **Bureau of Land Management** (BLM), giving the agency authority and direction for the management of its lands. The initiative leading to the passage of FLPMA can be traced to the BLM itself. The agency was concerned about its insecure status—it was formed by executive reorganization rather than by a congressional act, it lacked a clear mandate for land management, and it was uncertain of the federal government's plans to retain the lands it managed. This final point can be traced to the **Taylor Grazing Act**, which included a clause that these public lands would be managed for grazing "pending final disposal." The BLM wanted a law that would address each of these issues, so that the agency could undertake

long-range, multiple use planning like their colleagues in the **Forest Service**.

Agency officials drafted the first "organic act" in 1961, but two laws passed in 1964 served to really get the legislative process moving. The **Public Land** Law Review Commission (PLLRC) Act established a commission to examine the body of public land laws and make recommendations as to how to proceed in this policy area. The Classification and Multiple Use Act instructed the BLM to inventory its lands and classify them for disposal or retention. This would be the first inventory of these lands and resources, and suggested that at least some of these lands would be retained in federal ownership.

The PLLRC issued its report in 1970. In the following years, Congress began to consider three general types of bills in response to the PLLRC report. The administration and the BLM supported a BLM organic act without additional major reforms of other public land laws. The second approach provided the BLM with an organic act, but also made significant revisions in the Mining Law of 1872 and included environmental safeguards for BLM activities. This variety of bill was supported by environmentalists. The final type of bill provided a general framework for more detailed legislation in the future. This general framework tended to support commodity production, and was favored by livestock, mining, and timber interests.

In 1973, a bill of the second variety, introduced by Henry Jackson of Washington, passed the Senate. A similar bill died in the House, though, when it was denied a rule, and hence a trip to the floor, by the Rules Committee. Jackson re-introduced a bill that was nearly identical to the bill previously passed, and the Senate passed this bill in February 1976. In the House, things did not move as quickly. The main House bill, drafted by four western members of the Interior and Insular Affairs Committee, included significant provisions dealing with grazing—most importantly, a provision to adopt a statutory grazing fee formula based upon beef prices and private forage cost. This bill had the support of commodity interests, but was opposed by the administration and environmental groups. The bill passed the full House by fourteen votes in July 1976.

The major differences that needed to be addressed in the conference committee included law enforcement, the grazing provisions, mining law provisions, wild horses and burros, unintentional trespass, and the California **Desert Conservation** Area. By late September, four main differences remained, three involving grazing, and one dealing with mining. For a period it appeared that the bill might die in committee, but final compromises on the grazing and mining issues were made and a bill emerged out of conference. The bill was signed into law in October 1976 by President Gerald Ford.

As passed, FLPMA dealt with four general issue areas: 1) the organic act sections, giving the BLM authority and direction for managing the lands under its control; 2) grazing policy; 3) preservation policy; and 4) mining policy. The act begins by stating that these lands will remain in public ownership: "The Congress declares that it is the policy of the United States that...the public lands be retained in public ownership." This represented the true, final closing of the public domain; the federal government would retain the vast majority of these lands. To underscore this point, FLPMA repealed hundreds of laws dealing with the public lands that were no longer relevant. The BLM, under the authority of the Secretary of the Interior, was authorized to manage these lands for multiple use and sustained yield and was required to develop **land use** plans and resource inventories for the lands based on long-range planning. A director of the BLM was to be appointed by the President, subject to confirmation by the Senate.

FLPMA limited the withdrawal authority of the Secretary, often used to close lands to mineral development or to protect them for other environmental reasons, by repealing many of the sources of this authority and limiting its uses in other cases. The act allowed for the sale of public lands under a set of guidelines. In a section of the law that received much attention, the BLM was authorized to enforce the law on the lands it managed. The agency was directed to cooperate with local law enforcement agencies as much as possible in this task. It was these agencies, and citizens who lived near BLM lands, who were skeptical of this new BLM enforcement power. Other important provisions of the law allowed for the capture, removal, and relocation of wild horses and burros from BLM lands and authorized the Secretary of the Interior to grant rights-of-way across these lands for most pipelines and electrical **transmission lines**.

The controversial grazing fee formula in the House bill, favored by the livestock industry, was dropped in the conference committee. In its place, FLPMA froze grazing fees at the 1976 level for one year and directed the Secretaries of Agriculture and the Interior to undertake a comprehensive study of the grazing fee issue so that an equitable fee could be determined. This report was completed in 1977, and Congress established a statutory fee formula in 1978. That formula was only binding until 1985, though, and since that time Congress has debated the grazing fee issue numerous times, but the issue remains unsettled.

FLPMA also provided that grazing permits be for ten year periods, and that at least two year notice be given before permits were cancelled (except in an emergency). At the end of the ten year lease, if the lands are to remain in grazing, the current permittee has the first priority on renewing the lease to those lands. This virtually guarantees a rancher the use of certain public lands as long as they are to be used for grazing. The permittee is also to receive compensation for private improvements on public lands if the permit is cancelled. These provisions, advocated by livestock interests, further demonstrated their belief, and the belief of their supporters in Congress, that these grazing permits were a type of property right. Grazing advisory boards, originally started after the Taylor Grazing Act but terminated in the early 1970s, were resurrected. These boards consist of local grazing permittees in the area, and advise the BLM on the use of range improvement funds and on allotment management plans.

Important provisions regarding the preservation of BLM lands were also included in FLPMA. BLM lands were not covered in the **Wilderness Act** of 1964, and FLPMA dealt with this omission by directing that these lands be reviewed for potential **wilderness** designation, and that recommendations be made by the agency of which lands should be designated as wilderness. These designations would then be acted upon by Congress. This process is well underway. As has been the case with additions to the National Wilderness Preservation System on **national forest** lands since RARE II, BLM wilderness designation is being considered on a state-by-state basis. Thus far, a comprehensive wilderness designation law has only been passed for Arizona and California. Recent controversy has centered over the designation of wilderness in Utah.

FLPMA established a special California Desert Conservation Area, and directed the BLM to study this area and develop a long-range plan for its management. In 1994, after eight years of consideration, Congress passed the California Desert Protection Act. Senator Dianne Feinstein of California played the major role in guiding the legislation to passage, including overcoming an opposition-led filibuster against the act in October. The act, which included a number of compromises with desert users, established two new national parks and a new national preserve as well as designating approximately 7.5 million acres (3 million ha) of California desert as wilderness (in the two parks, the preserve, and nearly 70 new wilderness areas). The new national parks were created by enlarging and upgrading the existing Death Valley and Joshua Tree National Monuments. The Mojave National Preserve was originally to be a third **national park** in the desert, but its status was reduced to a national preserve to allow continued **hunting**, a compromise that helped gain further support for the bill. This law protected more wilderness than any law since the 1980 Alaska Lands Act. The following year, however, there was a move to alter these provisions. As part of the 1996 fiscal year Interior Appropriations bill, Congress directed that the BLM—not the National Park Service—manage the new Mojave National Preserve. According to Republican supporters, the BLM would allow for more use of the land. President Clinton vetoed this appropriations bill in December 1995, in part due to this change in California Desert management. When the final Interior Appropriations Act was passed in April 1996, it included a provision requiring the Park Service to manage the Mojave under the less restrictive BLM standards, but it also allowed the President to waive this provision. Clinton signed the bill, and then immediately waived the provision, so the Mojave is being managed by the **National Park Service** under its own standards.

FLPMA required that all mining claims, based on the 1872 Mining Law, be recorded with the BLM within three years. Claims not recorded were presumed abandoned. In the past, such claims only had to be recorded at the county courthouse in the county in which the claim was located. This allowed for increased knowledge about the number and location of such claims. The law also included amendments to the **Mineral Leasing Act** of 1920, increasing the share of the revenues from such leases that went to the states, allowing the states to spend these funds on any public facilities needed (rather than just roads and schools), and reducing the amount of revenues going to the fund to reclaim these mineral funds.

The implementation of FLPMA has been problematic. One consequence of the act, and the planning and management that it has required, was the stimulation of western hostility to the BLM and the existence of so much federal lands. According to a number of analysts, FLPMA was largely responsible for starting the **Sagebrush Rebellion**, the movement to have federal lands transferred to the states. The foremost implementation problems have been due to the poor bureaucratic capacity of the BLM: the lack of adequate funding, the lack of an adequate number of employees, poor standing within the **U.S. Department of the Interior** and presidential administrations, and its history of subservience to grazing and mining interests.

[*Christopher McGrory Klyza*]

RESOURCES

BOOKS

Dana, S. T., and S. K. Fairfax. *Forest and Range Policy.* 2nd ed. New York: McGraw-Hill, 1980.

PERIODICALS

"Fragile California Desert Bill Blooms Late in Session." *Congressional Quarterly Almanac* 50 (1994): 227–231.

"Public Land Management." *Congressional Quarterly Almanac* 32 (1976): 182–188.

Senzel, I. "Genesis of a Law, Part 1." *American Forests* (January 1978): 30–32+.

———. "Genesis of a Law, Part 2." *American Forests* (February 1978): 32–39.

Federal Power Administration

see **U.S. Department of Energy**

Federal Power Commission

The Federal Power Commission was established June 23, 1930, under the authority of the Federal Water Power Act, which was passed on March 3, 1921. The commission was terminated on August 4, 1977, and its functions were transferred to the **Federal Energy Regulatory Commission** under the umbrella of the **U.S. Department of Energy**.

The most important function of the commission during its 57-year existence was the licensing of water-power projects. It also reviewed plans for water-development programs submitted by major federal construction agencies for conformance with the interests of public good. In addition, the commission retained responsibility for interstate regulation of **electric utilities** and the siting of hydroelectric **power plants** as well as their operation. It also set rates and charges for the **transportation** and sale of **natural gas** and electricity. The five members of the commission were appointed by the president with approval of the Senate; three of the members were the Secretaries of the Interior, Agriculture, and War (later designated as U.S. Department of the Army). The commission retained its status as an independent regulatory agency for decision making, which is considered necessary for national security purposes.

Feedlot runoff

Feedlots are containment areas used to raise large numbers of animals to an optimum weight within the shortest time span possible. Most feedlots are open air, and are thereby subject to variable weather conditions. A substantial portion of the feed is not converted into meat, and is excreted, thus degrading the air, ground, and surface **water quality**. The issues of odor and **water pollution** from such facilities center on the traditional attitudes of producers that farming has always produced odors, and manure is a **fertilizer**, not a waste from a commercial undertaking.

Animal excrement is indeed rich in nutrients, particularly **nitrogen**, **phosphorus**, and potassium. A single 1,300-lb (590 kg) steer will excrete about 150 lb (68 kg) of nitrogen; 50 lb (23 kg) of phosphorus; and 100 lb (45 kg) of potassium in the course of a year. That is almost as much **nutrient** as would be required to grow one acre of corn, which needs 185 lb (84 kg) of nitrogen; 80 lb (36 kg) of phosphorus; and 215 lb (98 lb) of potassium. Unfortunately, manure is costly to transport, difficult to apply, and its nutrient quality is inconsistent. Artificial fertilizers, on the other hand, offer ease of application and storage and proven quality and plant growth.

Legislative and regulatory action have increased with encroachment of urban population and centers of high sensitivity, such as shopping malls and **recreation** facilities. Since odor is difficult to measure, control of these facilities is being achieved on the grounds that they must not pose a "nuisance," a principle that is being sustained by the courts.

Odor is influenced by feed, number and **species** of animal, lot surface and manure removal frequency, wind, humidity, and moisture. These factors, individually and collectively, influence the type of **decomposition** that will occur. Typically, it is an **anaerobic** process which produces a sharp pungent odor of ammonia, the nauseating odor of rotten eggs from **hydrogen** sulfide, and the smell of decaying cabbage or onions from methyl mercaptan.

Odorous compounds seldom reach concentrations that are dangerous to the public. However, levels can become dangerously elevated with reduced ventilation in winter months or during pit cleaning. It is this latter activity, in conjunction with disposal onto the surface of the land, that is most frequently the cause of complaints. Members of the public respond to feedlot odors depending on their individual sensitivity, previous experience, and disposition. It can curtail outdoor activities and require windows to be closed, which means the additional use of air purifiers or air-conditioning systems.

Surface water contamination is the problem most frequently attributed to open feedlot and manure spreading activities. It is due to the dissolving, eroding action of rain striking the manured-covered surface. Duration and intensity of rainfall dictates the concentration of contaminants that will flow into surface waters. Their dilution or retention in ponds, rivers, and streams depends on area **hydrology** (dry or wet conditions) and **topography** (rolling or steeply graded landscape). Such factors also influence conditions in those parts of the continent where precipitation is mainly in the form of snow. Large snow drifts form around wind breaks, and in the early spring, substantial volumes of snowmelt are generated.

Odor and water **pollution control** techniques include simple operational changes, such as increasing the frequency of removing manure, scarifying the surface to promote **aerobic** conditions, and applying disinfectants and feed-digestion supplements. Other control measures require construction of additional structures or the installation of equipment at feedlots. These measures include installing water spargelines, adding impervious surfaces, drains, pits and roofs, and installing extraction fans. *See also* Animal waste; Odor control

[*George M. Fell*]

RESOURCES
BOOKS

Larson, R. E. *Feedlot and Ranch Equipment for Beef Cattle.* Washington, DC: U.S. Government Printing Office, 1976.
Peters, J. A. *Source Assessment: Beef Cattle Feedlots.* Research Triangle Park, NC: U.S. Environmental Protection Agency, 1977.

Feedlots

A feedlot is an open space where animals are fattened before slaughter. Beef cattle usually arrive at the feedlot directly from the ranch or farm where they were raised, while poultry and pigs often remain in an automated feedlot from birth until death. Feed (often grains, alfalfa, and molasses) is provided to the animals so they do not have to forage for their food. This feeding regimen promotes the production of higher quality meat more rapidly. There are no standard parameters for the number of animals per acre in a feedlot, but the density of animals is usually very high. Some feedlots can contain 100,000 cows and steers. **Animal rights** groups actively campaign against confining animals in feedlots, a practice they consider inhumane, wasteful, and highly polluting.

Feedlots were first introduced in California in the 1940s, but many are now found in the Midwest, closer to grain supplies. Feedlot operations are highly mechanized and large numbers of animals can be handled with relatively low labor input. About half of the beef produced in the United States is feedlot-raised.

Feedlots are a significant **nonpoint source** of the **pollution** flowing into surface waters and **groundwater** in the United States. At least half a billion tons of **animal waste** are produced in feedlots each year. Since this waste is concentrated in the feedlot rather than scattered over grazing lands, it overwhelms the soil's ability to absorb and **buffer** it and creates nitrate-rich, bacteria-laden **runoff** to pollute streams, rivers, and lakes. Dissolved pollutants can also migrate down through the **soil** into aquifers, leading to **groundwater pollution** over wide areas. To protect surface waters, most states require that **feedlot runoff** be collected. However, protection of groundwater has proved to be a more difficult problem, and successful regulatory and technological controls have not yet been developed.

[*Christine B. Jeryan*]

RESOURCES
BOOKS

Kerr, R. S. *Livestock Feedlot Runoff Control By Vegetative Filters.* Ada, OK: U.S. Environmental Protection Agency, 1979.
Larson, R. E. *Feedlot and Ranch Equipment for Beef Cattle.* Washington, DC: U.S. Government Printing Office, 1976.

Peters, J. A. *Source Assessment: Beef Cattle Feedlots.* Research Triangle Park, NC: U.S. Environmental Protection Agency, 1977.

Felis concolor coryi

see **Florida panther**

Fens

see **Wetlands**

Ferret

see **Black-footed ferret**

Fertility

see **Biological fertility**

Fertilizer

Any substance that is applied to land to encourage plant growth and produce higher crop yield. Fertilizers may be made from organic material—such as recycled waste, animal manure, compost, etc.—or chemically manufactured. Most fertilizers contain varying amounts of **nitrogen**, **phosphorus**, and potassium, inorganic nutrients that plants need to grow.

Since the 1950s crop production worldwide has increased dramatically because of the use of fertilizers. In combination with the use of pesticides and insecticides, fertilizers have vastly improved the quality and yield of such crops as corn, rice, wheat, and cotton. However overuse and improper use of fertilizers have also damaged the **environment** and affected the health of humans, animals, and plants.

In the United States, it is estimated that as much as 25% of fertilizer is carried away as **runoff**. Fertilizer runoff has contaminated **groundwater** and polluted bodies of water near and around farmlands. High and unsafe nitrate concentrations in drinking water have been reported in countries that practice intense farming, including the United States. Accumulation of nitrogen and phosphorus in waterways from chemical fertilizers has also contributed to the eutrophication of lakes and ponds. Ammonia, released from the decay of fertilizers, causes minor irritation to the respiratory system.

While very few advocate the complete eradication of chemical fertilizers, many environmentalists and scientists urge more efficient ways of using them. For example, some farmers use up to 40% more fertilizer than they need. Frugal applications—in small doses and on an as-needed-basis on specific crops—helps reduce fertilizer waste and runoff. The use of organic fertilizers, including **animal waste**, crop residues, or grass clippings, is also encouraged as an alternative to chemical fertilizers. *See also* Cultural eutrophication; Recycling; Sustainable agriculture; Trace element/micronutrient

Fibrosis

A medical term that refers to the excessive growth of fibrous tissue in some part of the body. Many types of fibroses are known, including a number that affect the respiratory system. A number of these respiratory fibroses, including such conditions as **black lung disease**, silicosis, **asbestosis**, berylliosis, and byssinosis, are caused by environmental factors. A fibrosis develops when a person inhales very tiny solid particles or liquid droplets over many years or decades. Part of the body's reaction to these foreign particles is to enmesh them in fibrous tissue. The disease name usually suggests the agent that causes the disease. Silicosis, for example, is caused by the inhalation of silica, tiny sand-like particles. Occupational sources of silicosis include rock mining, quarrying, stone cutting, and sandblasting. Berylliosis is caused by the inhalation of beryllium particles over a period of time, and byssinosis (from byssos, the Greek word for flax)is found among textile workers who inhale flax, cotton or hemp fibers.

Field capacity

Field capacity refers to the amount of water that can be held in the **soil** after all the gravitational water has drained away. Sandy soils will have less water held at field capacity than clay soils. The more water a soil can hold at field capacity the more water is available for plants.

Filters

Primarily devices for removing particles from aerosols. Filters utilize a variety of microscopic forms and a variety of mechanisms to accomplish this. Most common are fibrous filters, in which the fibers are of cellulose (paper filters), but almost any fibrous material, including glass fiber, wool, **asbestos**, and finely spun polymers, has been used. Microscopically, these fibers collect fine particles because fine particles vibrate around their average position due to collision with air molecules (Brownian motion). These vibrations are likely to cause them to collide with the fibers as they pass through the filter. Larger particles are removed because, as the air stream carrying them passes through the filter, some of the particles are intercepted as they pass close to the fibers and touch them. Other particles are in air streams that would cause them to miss the fibers, but when the air stream bends to go around the fibers the momentum, of the particles is too much to let them remain with the stream, so that they are "centrifuged out" onto the fibers (impaction). By electrophoresis, still other particles may be attracted to the fibers by electric charges of opposite sign on the particles and on the fibers. Finally, particles may simply be larger than the space between fibers, and be sifted out of the air in a process called sieving.

Filters are also formed by a process in which polymers such as cellulose esters are made into a film out of a solution in an organic solvent containing water. As the solvent evaporates, a point is reached at which the water separates out as microscopic droplets, in which the polymer is not soluble. The final result is a film of polymer full of microscopic holes where the water droplets once were. Such filters can have pore sizes from a small fraction of a micrometer to a few micrometers. (One micrometer equals 0.00004 in) These are called membrane filters.

Another form of membrane filter is formed from the polymer called polycarbonate. A thin film of this material is fastened to a surface of **uranium** metal and placed in a nuclear reactor for a time. In the reactor, the uranium undergoes **nuclear fission**, and gives off particles called fission fragments, atoms of the elements formed when the uranium atoms split. Every place that an atom from the fissioning uranium passes through the film is disturbed on a molecular scale. After removal from the reactor, if the polymer sheet is soaked in alkali, the disturbed material is dissolved. The amount of material dissolved is controlled by the temperature of the solution and the amount of time the film is treated. Since the fission fragments are very energetic, they travel in straight lines, and so the holes left after the alkali treatment are very straight and round. Again, pore sizes can be from a small fraction of a micrometer to a few micrometers. These filters are known by their trade name, *Nuclepore*.

In both types of membrane filters, the small pore size increases the role of sieving in particle removal. Because of their very simple structure, Nuclepore filters have been much studied to understand **filtration** mechanisms, since they are far easier to represent mathematically than a random arrangement of fibers.

It was mentioned above that small particles are collected because of their Brownian motion, while larger particles are removed by interception, impaction, and sieving. Under many conditions, a particle of intermediate size may pass through, too large for Brownian diffusion, and too small for impaction, interception, or sieving. Hence many filters may show a penetration maximum for particles of a few tenths of a micrometer. For this reason, standard methods of filter testing specify that the **aerosol** test for determining the efficiency of filters should contain particles in that size

range. This phenomenon has also been used to select relatively uniform particles of that size out of mixtures of many sizes.

In circumstances where filter strength is of paramount importance, such as in industrial filters where a large air flow must pass through a relatively small filter area, filters of woven cloth are used, made of materials ranging from cotton to glass fiber and asbestos, these last for use when very hot gases must be filtered. The woven fabric itself is not a particularly good filter, but it retains enough particles to form a particle cake on the surface, and that soon becomes the filter. When the cake becomes thick enough to slow airflow to an unacceptable degree, the air flow is interrupted briefly, and the filters are shaken to dislodge the filter cake, which falls into bins at the bottom of the filters. Then filtration is resumed, allowing the cloth filters to be used for months before being replaced. A familiar domestic example is the bag of a home vacuum cleaner. Cement plants and some electric **power plants** use dozens of cloth bags up to several feet in diameter and more than ten feet (three meters) in length to remove particles from their waste gases.

Otherwise poor filters can be made efficient by making them thick. A glass tube can be partially plugged with a wad of cotton or glass fiber, then nearly filled with crystals of sugar or naphthalene and used as a filter; this is advantageous since sugar can be dissolved in water, or naphthalene will sublime away if gently heated, leaving behind the collected particles. *See also* Baghouse; Electrostatic precipitation; Odor control; Particulate

[*James P. Lodge Jr.*]

Filtration

A common technique for separating substances in two physical states. For example, a mixture of solid and liquid can be separated into its components by passing the mixture through a filter paper. Filtration has many environmental applications. In water purification systems, impure water is often passed through a charcoal filter to remove the solid and gaseous contaminants that give water a disagreeable odor, color, or taste. Trickling **filters** are used to remove solid wastes in plants. Solid and liquid contaminants in waste industrial gases can be removed by passing them through a filter prior to **discharge** in a smokestack.

Fire

see **Prescribed burning; Wildfire**

Fire ants

Two distinct **species** of fire ants (genus *Solenopsis*) from South America have been introduced into the United States this century. The South American black fire ant (*S. richteri*) was first introduced into the United States in 1918. Its close relative, the red fire ant (*S. wagneri*), was introduced in 1940, probably escaping from a South American freighter docked in Mobile, Alabama. Both species became established in the southeastern United States, spreading into nine states from Texas across to Florida and up into the Carolinas. It is estimated that they have infested over 320 million acres (130 million ha) covering 13 states as well as Puerto Rico.

Successful **introduced species** are often more aggressive than their native counterparts, and this is definitely true of fire ants. They are very small, averaging 0.2 in (5 mm) in length, but their aggressive, swarming behavior makes them a threat to livestock and pets as well as humans. These industrious, social insects build their nests in the ground—the location is easily detected by the elevated earthen mounds created from their excavations. The mounds are 18–36 in (46–91 cm) in diameter and may be up to 36 in (91 cm) high, although mounds are generally 6–10 in (15–25 cm) high. Each nest contains as many as 25,000 workers, and there may be over 100 nests on an acre of land.

If the nest is disturbed, fire ants swarm out of the mound by the thousands and attack with swift ferocity. As with other aspects of ant behavior, a chemical alarm pheromone is released that triggers the sudden onslaught. Each ant in the swarm uses its powerful jaws to bite and latch onto whatever disturbed the nest, while using the stinger on the tip of its abdomen to sting the victim repeatedly. The intruder may receive thousands of stings within a few seconds.

The toxin produced by the fire ant is extremely potent, and it immediately causes an intense burning pain that may continue for several minutes. After the pain subsides, the site of each sting develops a small bump which expands and becomes a tiny, fluid-filled blister. Each blister flattens out several hours later and fills with pus. These swollen pustules may persist for several days before they are absorbed and replaced by scar tissue. Fire ants obviously pose a problem for humans. Some people may become sensitized to fire ant venom, have a generalized systematic reaction, and go into anaphylactic shock. Fire-ant induced deaths have been reported. Because these species prefer open, grassy yards or fields, pets and livestock may fall prey to fire ant attacks as well.

Attempts to eradicate this **pest** involved the use of several different generalized pesticides, as well as the widespread use of **gasoline** either to burn the nest and its inhabitants or to kill the ants with strong toxic vapors. Another approach involved the use of specialized crystalline pesticides which were spread on or around the nest mound. The workers collected them and took them deep into the nest, where they were fed to the queen and other members of the colony, killing the inhabitants from within. A more recent method involves the release of a natural predator of the fire ant, the "phorid" fly. The fly releases an egg into the fire ant. The larva then eats the ant's brain while releasing an **enzyme**. The enzyme systematically destroys the joints causing the ant's head to fall off. The flies were released in 11 states as of 2001 and seem to be slowly inhibiting the growth of the fire ant population. As effective as some of these methods are, fire ants are probably too numerous and well established to be completely eradicated in North America.

[*Eugene C. Beckham*]

RESOURCES
BOOKS

Holldobler, B. *The Ants*. Cambridge: Harvard University Press, 1990.
Taber, Stephen Welton. *Fire Ants*. Texas A&M University Press, 2000.

PERIODICALS
Vergano, Dan. "Decapitator Flies will Fight Fire Ants." *USA Today*, November 20, 2000.

OTHER
"Imported Fire Ant." *NAPIS Page for BioControl Host*. April 2001 [cited May 2002]. <http://www.ceris.purdue.edu/napis/pests/ifa/ifabc.html>.
"Invasive Species and Pest Management: Imported Fire Ant." *Animal and Plant Health Inspection Service*. May 2002 [cited May 2002]. <http://www.aphis.usda.gov>.

First World

The world's more wealthy, politically powerful, and industrially developed countries are unofficially, but commonly, designated as the First World. The term differentiates the powerful, capitalist states of Western Europe and North America and Japan from the (formerly) communist states (**Second World**) and from the nonaligned, developing countries (**Third World**) in world systems theory. In common usage, First World refers mainly to a level of economic strength. The level of industrial development of the First World, characterized by an extensive infrastructure, mechanized production, efficient and fast transport networks, and pervasive use of high technology, consumes huge amounts of **natural resources** and requires an educated and skilled work force. However, such a system is usually highly profitable. Often depending upon raw materials imported from poorer countries (wood, metal ores, **petroleum**, food, and so on), First World countries efficiently produce goods that **less developed countries** desire but cannot produce themselves, including computers, airplanes, optical equipment, and military hardware. Generally, high domestic and international demand for such specialized goods keeps First World countries wealthy, allowing them to maintain a high standard of material consumption, education, and health care for their citizens.

Fish and Wildlife Service

The United States Fish & Wildlife Service based in Washington, D.C., is charged with conserving, protecting, and enhancing fish, **wildlife**, and their habitats for the benefit of the American people. As a division of the **U.S. Department of the Interior**, the Service's primary responsibilities are for the protection of migratory birds, **endangered species**, freshwater and anadromous (saltwater **species** that spawn in freshwater rivers and streams) fisheries, and certain marine mammals.

In addition to its Washington, D.C., headquarters, the Service maintains seven regional offices and a number of field units. Those include national wildlife refuges, national fish hatcheries, research laboratories, and a nationwide network of law enforcement agents.

The Service manages 530 refuges that provide habitats for migratory birds, endangered species, and other wildlife. It sets migratory bird **hunting** regulations, and leads an effort to protect and restore endangered and threatened animals and plants in the United States and other countries.

Service scientists assess the effects of contaminants on wildlife and habitats. Its geographers and cartographers work with other scientists to map **wetlands** and carry out programs to slow wetland loss, or preserve and enhance these habitats. Restoring fisheries that have been depleted by **overfishing**, **pollution**, or other **habitat** damage is a major program of the Service. Efforts are underway to help four important species: lake trout in the upper **Great Lakes**; striped bass in both the **Chesapeake Bay** and Gulf Coast; Atlantic **salmon** in New England; and salmonid species of the Pacific Northwest.

Fish and Wildlife biologists working with scientists from other federal and state agencies, universities, and private organizations develop recovery plans for endangered and threatened species. Among its successes are the **American alligator**, no longer considered endangered in some areas, and a steadily increasing **bald eagle** population.

Internationally, the Service cooperates with 40 wildlife research and **wildlife management** programs, and provides technical assistance to many other countries. Its 200 special agents and inspectors help enforce wildlife laws and treaty obligations. They investigate cases ranging from individual migratory bird hunting violations to large-scale **poaching** and commercial trade in protected wildlife.

It its "Vision for the Future" statement, the Fish and Wildlife Service states its mission to "provide leadership to achieving a national net gain of fish and wildlife and the natural systems which support them." Into the twenty-first century, this vision statement calls for new **conservation** compacts with all citizens to increase the value of the United States wildlife holdings in number and **biodiversity**, and to provide increased opportunities for the public to use, associate with, learn about and enjoy America's wildlife wealth.

[*Linda Rehkopf*]

RESOURCES

OTHER

U.S. Fish and Wildlife Service. *Vision for the Future.* Washington, DC: U.S. Government Printing Office, 1991.

ORGANIZATIONS

U.S. Fish and Wildlife Service, Email: contact@fws.gov, <http://www.fws.gov>

Fish farming
see **Blue revolution (fish farming)**

Fish kills

Fishing has long been a major provider of food and livelihood to people throughout the world. In the United States, 50 million people enjoy fishing as an outdoor recreation—38 million in fresh water and 12 million in salt water. Combined, they spend over $315 million annually on this sport. It is no surprise, then, that public attitude towards factors that influence fishing is strong.

The **Environmental Protection Agency** (EPA) is charged with overseeing the quality of the nation's waterways. In 1977 they received information on 503 separate incidents in which 16.5 million fish were killed. In 1974, a record 47 million fish were killed in the Black River near Essex, Maryland, by a **discharge** from a sewage plant.

Fish kills can result from natural as well as human causes. Natural causes include sudden changes in temperature, oxygen depletion, toxic gases, epidemics of viruses and bacteria, infestations of **parasites**, toxic algal blooms, lightning, **fungi**, and other similar factors. Human influences that lead to fish kills include **acid rain**, sewage **effluent**, and toxic spills.

In a 10-year study of the causes of 409 documented fish kills totaling 3.6 million fish in the state of Missouri,

S. M. Czarnezki determined the percentage contributions as: 26% municipal-related (sewage effluent), 17% from agricultural activities, 11% from industrial operations, 8% by **transportation** accidents, 7% each by oxygen depletions, nonindustrial operations, and mining, 4% by disease, 3% by "other" factors, and 10% as undetermined.

Fish kills may occur quite rapidly, even within minutes of a major toxic spill. Usually, however, the process takes days or even months, especially in natural causes. Experienced fishery biologists usually need a wide variety of physical, chemical, and biological tests of the **habitat** and fish to determine the exact causative agent or agents. The investigative procedure is often complex and may require a lot of time.

Species of fish vary in their susceptibility to the different factors that contribute to die-offs. Some species are sensitive to almost any disturbance, while other fish are tolerant of changes. As discussed below, predatory fish at the top of the **food chain/web** are typically the first fish affected by toxic substances that accumulate slowly in the water.

The most common contributor to fish kills by natural causes is oxygen depletion, which occurs when the amount of oxygen utilized by **respiration**, **decomposition**, and other processes exceeds oxygen input from the **atmosphere** and **photosynthesis**. Oxygen is more soluble in cold than warm water. Summer fish kills occur when lakes are thermally stratified. If the lake is eutrophic (highly productive), dead plant and animal matter that settles to the bottom undergoes decomposition, utilizing oxygen. Under windless conditions, more oxygen will be used than is gained, and animals like fish and **zooplankton** often die from suffocation.

Winter fish kills can also occur. Algae can photosynthesize even when the lake is covered with ice because sunlight can penetrate through the ice. However, if heavy snowfall accumulates on top of the ice, light may not reach the underlying water, and the **phytoplankton** die and sink to the bottom. **Decomposers** and respiring organisms again use up the remaining oxygen and the animals eventually die. When the ice melts in the spring, dead fish are found floating on the surface. This is a fairly common occurrence in many lakes in Michigan, Wisconsin, Minnesota, and surrounding states. For example, dead alewives (*Alosa pseudoharengus*) often wash up on the southwestern shore of Lake Michigan near Chicago during spring thaws following harsh winters.

In summer and winter, artificial **aeration** can help prevent fish kills. The addition of oxygen through aeration and mixing is one of the easiest and cheapest methods of dealing with low oxygen levels. In intensive **aquaculture** ponds, massive fish deaths from oxygen depletion are a constant threat. Oxygen sensors are often installed to detect low oxygen levels and trigger the release of pure oxygen gas from nearby cylinders.

Natural fish kills can also result from the release of toxic gases. In 1986, 1,700 villagers living on the shore of Lake Nyos, Cameroon, mysteriously died. A group of scientists sent to investigate determined that they died of asphyxiation. Evidently a **landslide** caused the trapped **carbon** dioxide-rich bottom waters to rapidly rise to the surface much like a popped champagne bottle. The poisonous gas killed everyone in its downwind path. Fish in the upper oxygenated waters of the lake were also killed as the **carbon dioxide** passed through.

Hydrogen sulfide (H_2S), a foul-smelling gas naturally produced in the oxygen-deficient sediments of eutrophic lakes, can also cause fish deaths. Even in oxygenated waters, high H_2S levels can cause a condition in fish called "brown blood." The brown color of the blood is caused by the formation of sulfhemoglobin, which inhibits the blood's oxygen-carrying capacity. Some fish survive, but sensitive fish such as trout usually die.

Fish kills can also result from toxic algal blooms. Some bluegreen algae in lakes and dinoflagellates in the ocean release **toxins** that can kill fish and other vertebrates, including humans. For example, dense blooms of bluegreen algae such as *Anabaena*, *Aphanizomenon*, and *Microcystis* have caused fish kills in many farm ponds during the summer. Fish die not only from the toxins but also from asphyxiation resulting from decomposition of the mass of algae that also die due to lack of sunlight in the densely-populated lake water. In marine waters, toxic dinoflagellate blooms called red tides are notorious for causing massive fish kills. For example, blooms of *Gymnodinium* or *Gonyaulax* periodically kill fish along the East and Gulf Coasts of the United States. Die-offs of **salmon** in aquaculture pens along the southwestern shoreline of Norway have been blamed on these organisms. Millions of dollars can be lost if the fish are not moved to clear waters. Saxitoxin, the toxic chemical produced by *Gonyaulax*, is 50 times more lethal than strychnine or curare.

Pathogens and parasites can also contribute to fish kills. Usually the effect is more secondary than direct. Fish weakened by parasites or infections of bacteria or viruses usually are unable to adapt to and survive changes in water temperature and chemistry. Under stressful conditions of over-crowding and malnourishment, gizzard shad often die from minor infestations of the normally harmless bacterium *Aeromonas hydrophila*. In the same way, fungal infections such as *Ichthyophonus hoferia* can contribute to fish kills. Most fresh water aquarium keepers are familiar with the threat of "ick" for their fish. The telltale white spots under the epithelium of the fins, body, and gills are caused by the protozoan parasite *Ichthyophthirius multifiliis*.

Changes in **pH** of lakes resulting from **acid** rain are a modern example of how humans can cause fish kills. **Atmospheric pollutants** such as **nitrogen** dioxide and **sulfur dioxide** released from automobiles and industries mix with water vapor and cause the rainwater to be more acid than normal (>pH 6.5). Nonprotected lakes downwind that receive this rainfall increase in acidity, and sensitive fish eventually die. Most of the once-productive trout streams and lakes in the southern half of Norway are now devoid of these prized fish. Sweden has combatted this problem by adding enormous quantities of lime to their affected lakes in the hope of neutralizing the acid's effects.

Sewage treatment plants add varying amounts of treated effluent to streams and lakes. Sometimes during heavy rainfall raw sewage escapes the treatment process and pollutes the aquatic **environment**. The greater the organic matter that comprises the effluent, the more decomposition occurs, resulting in oxygen usage. Scientists call this the biological or **biochemical oxygen demand** (BOD), the quantity of oxygen required by bacteria to oxidize the **organic waste** aerobically to carbon dioxide and water. It is measured by placing a sample of the **wastewater** in a glass-stoppered bottle for five days at 71 degrees Fahrenheit (20 degrees Celsius) and determining the amount of oxygen consumed during this time. Domestic sewage typically has a BOD of about 200 milligrams per liter, or 200 **parts per million** (ppm); rates for industrial waste may reach several thousand milligrams per liter. Reports of fish kills in industrialized countries have greatly increased in recent years. Sewage effluent not only kills fish; it can also create a barrier to fish migrating upstream because of the low oxygen levels. For example, coho salmon will not pass through water with oxygen levels below 5 ppm. Oxygen depletion is often more detrimental to fish than thermal shock.

Toxic **chemical spills**, whether via sewage treatment plants or other sources, are the major cause of fish kills. Sudden discharges of large quantities of highly toxic substances usually cause massive death of most aquatic life. If they enter the **ecosystem** at sublethal levels over a long time, the effects are more subtle. Large predatory or omnivorous fish are typically the first ones affected. This is because toxic **chemicals** like methyl **mercury**, DDT, PCBs, and other organic pollutants have an affinity for fatty tissue and progressively accumulate in organisms up the food chain. This is called the principle of **biomagnification**. Unfortunately for human consumers, these fish do not usually die right away, so people who eat a lot of tainted fish become sick and possibly die. Such is the case for **Minamata disease**, named for the first documented connection between the death of fishermen and methyl mercury contamination.

[*John Korstad*]

RESOURCES
BOOKS

Czarnezki, J. M. *A Summary of Fish Kill Investigations in Missouri, 1970–1979.* Columbia, MO: Missouri Dept. of Conservation, 1983.

Ehrlich, P. R., A. H. Ehrlich, and J. P. Holdren. *Ecoscience: Population, Resources, Environment.* San Francisco: W. H. Freeman, 1977.

Goldman, C. R., and A. J. Horne. *Limnology.* New York: McGraw-Hill, 1983.

Hill, D. M. "Fish Kill Investigation Procedures." In *Fisheries Techniques*, edited by L. A. Nielson and D. L. Johnson. Bethesda, MD: American Fisheries Society, 1983.

Meyer, F. P., and L. A. Barclay, eds. *Field Manual for the Investigation of Fish Kills.* Washington, DC: U.S. Fish and Wildlife Service, 1990.

Moyle, P. B., and J. J. Cech Jr. *Fishes: An Introduction to Ichthyology.* 2nd ed. New York: Prentice-Hall, 1988.

PERIODICALS

Keup, L. E. "How to 'Read' A Fish Kill." *Water and Sewage Works* 12 (1974): 48–51.

Fish nets

see **Drift nets; Gill nets**

Fisheries and Oceans Canada

The Department of Fisheries and Oceans (DFO) in Canada was created by the Department of Fisheries and Oceans Act on April 2, 1979. This act formed a separate government department from the Fisheries and Marine Service of the former Department of Fisheries and the Environment. The new department was needed, in part, because of increased interest in the management of Canada's oceanic resources, and also because of the mandate resulting from the unilateral declaration of the 200-nautical-mi Exclusive Economic Zones in 1977.

At its inception, the DFO assumed responsibility for seacoast and inland fisheries, fishing and recreational vessel harbors, hydrography and ocean science, and the coordination of policy and programs for Canada's oceans. Four main organizational units were created: Atlantic Fisheries, Pacific and Freshwater Fisheries, Economic Development and Marketing, and Ocean and Aquatic Science. Among the activities included in the department's original mandate were: comprehensive husbandry of fish stocks and protection of **habitat**; "best use" of fish stocks for optimal socioeconomic benefits; adequate hydrographic surveys; the acquisition of sufficient knowledge for defense, **transportation**, energy development and fisheries, with provision of such information to users; and continued development and maintenance of a national system of harbors.

Since its inception, the department's mandate has changed in minor ways, to include new terminology such as "sustainability" and to include Canada's "ecological interests." Recently, attention has been given to support those who make their living or benefit from the sea. This constituency includes the public first, but the DFO also directs its efforts toward commercial fishers, fish plant workers, importers, aquaculturists, recreational fishers, native fishers, and the ocean manufacturing and service sectors. There are now six DFO divisions: Science, Atlantic Fisheries, Pacific Fisheries, Inspection Services, International, and Corporate Policy and Support administered through six regional offices.

A primary focus of DFO's current work is the failing cod and groundfish stocks in the Atlantic; the department has commissioned two major inquiries in recent years to investigate those problems. In addition, the DFO has increased regulation of foreign fleets, and works to manage straddling stocks in the Atlantic **Exclusive Economic Zone** through the North Atlantic Fisheries Organization, the Pacific drift nets fisheries, recreational fishing and **aquaculture** development. In 1992, management problems in the major fisheries arose on both the Pacific and Atlantic coasts. American fisheries managers reneged on quotas established through the Pacific **Salmon** Treaty, northern cod stocks in Newfoundland virtually failed, and the Aboriginal Fishing Strategy was adopted as part of a land claim settlement on the Pacific coast.

There are several major problems associated with ocean resource and environment management in Canada—problems that the DFO has neither the resources, the legislative infrastructure, nor the political will to address. One result of this has been the steady decline of commercial fish stocks, highlighted by the virtual collapse of Atlantic cod (*Gadus callarias*), which is Canada's, and perhaps, the Atlantic's most historically significant fishery. A second result has been an increased need to secure international agreements with Canada's ocean neighbors. A third result of social significance is the perception that fisheries have been used in a political sense in cases of regional economic incentives and land claims settlements. *See also* Commercial fishing; Department of Fisheries and Oceans (DFO), Canada

[*David A. Duffus*]

Fishing

see **Commercial fishing; Drift nets; Gill nets**

Fission

see **Nuclear fission**

Floatable debris

Floatable debris is buoyant **solid waste** that pollutes waterways. Sources include boats and shipping vessels, storm water **discharge**, sewer systems, industrial activities, offshore drilling, recreational beaches, and landfills. Even waste dumped far from a water source can end up as floatable debris when **flooding**, high winds, or other weather conditions transport it into rivers and streams.

According to the U.S. **Environmental Protection Agency** (EPA), floatable debris is responsible for the death of over 100,000 marine mammals and one million seabirds annually. **Seals, sea lions, manatees, sea turtles,** and other marine creatures often mistake debris for food, eating objects that block their intestinal tract or cause internal injury. They can also become entangled in lost fishing nets and line, six-pack rings, or other objects.

Fishing nets lost at sea catch tons of fish that simply decompose, a phenomenon known as "ghost fishing." Often, seabirds are ensnared in these nets when they try to eat the fish.

Lost nets and other entrapping debris are also a danger for humans who swim, snorkel, or scuba dive. And biomedical waste and sewage can spread disease in recreational waters. Floatable debris takes a significant financial toll as well. It damages boats, deters tourism, and negatively impacts the fishing industry.

[*Paula Anne Ford-Martin*]

RESOURCES
BOOKS

Coe, James, and Donald Rogers, eds. *Marine Debris: Sources, Impacts, and Solutions* New York: Springer-Verlag, 1996.

PERIODICALS

Miller, John. "Solving the Mysteries of Ocean-borne Trash." *U.S. News & World Report* 126, no.14 (April 1999): 48.

OTHER

U.S. Environmental Protection Agency, Office of Water, Oceans and Coastal Protection Division. *Assessing and Monitoring Floatable Debris—Draft.* [cited May 11, 2002]. <http://www.epa.gov/owow/oceans/debris/floatingdebris>.

U.S. Environmental Protection Agency, Office of Water, Oceans and Coastal Protection Division. *Turning the Tide on Trash: A Marine Debris Curriculum* [cited May 2002]. <http://www.epa.gov/owow/OCPD/Marine/contents.htm>.

ORGANIZATIONS

The Center for Marine Conservation, 1725 DeSales Street, N.W., Suite 600, Washington, DC USA 20036 (202) 429-5609, Fax: (202) 872-0619, Email: cmc@dccmc.org, http://www.cmc-ocean.org

Flooding in Texas caused by Hurricane Beulah in 1967. (National Oceanic and Atmospheric Administration.)

Flooding

Technically, flooding occurs when the water level in any stream, river, bay, or lake rises above bank full. Bays may flood as the result of a tsunami or tidal wave induced by an **earthquake** or volcanic eruption; or as a result of a tidal storm surge caused by a **hurricane** or tropical storm moving inland. Streams, rivers and lakes may be flooded by high amounts of surface **runoff** resulting from widespread precipitation or rapid snow melt. On a smaller scale, flash floods due to extremely heavy precipitation occurring over a short period of time can flood streams, creeks, and low lying areas in a matter of a few hours. Thus, there are various temporal and spatial scales of flooding. Historical evidence suggests that flooding causes greater loss of life and property than any other natural disaster. The magnitude, seasonality, frequency, velocity, and load are all properties of flooding which are studied by meteorologists, climatologists, and hydrologists.

Spring and winter floods occur with some frequency primarily in the mid-latitude regions of the earth, and particularly where continental **climate** is the norm. Five climatic

features contribute to the spring and winter flooding potential of any individual year or region: 1) heavy winter snow cover; 2) saturated soils or soils at least near their **field capacity** for storing water; 3) rapid melting of the winter's snow pack; 4) frozen **soil** conditions which limit **infiltration**; and 5) somewhat heavy rains, usually from large scale cyclonic storms. Any combination of three of these five climatic features usually leads to some type of flooding. This type of flooding can cause hundreds of millions of dollars in property damage, but it can usually be predicted well in advance, allowing for evacuation and other protective action to be taken (sandbagging, for instance). In some situations flood control measures such as stream or channel diversions, **dams**, and levees can greatly reduce the risk of flooding. This is more often done in **floodplain** areas with histories of very damaging floods. In addition, **land use** regulations, encroachment statutes and building codes are often intended to protect the public from the risk of flooding.

Flash flooding is generally caused by violent weather, such as severe thunderstorms and hurricanes. This type of flooding more frequently occurs during the warm season when convective thunderstorms develop more frequently.

Rainfall intensity is so great that the **carrying capacity** of streams and channels is rapidly exceeded, usually within hours, resulting in sometimes life-threatening flooding. It is estimated that the average death toll in the United States exceeds 200 per year as a result of flash flooding. Many government weather services provide the public with flash flood watches and warnings to prevent loss of life. Many flash floods occur as the result of afternoon and evening thundershowers which produce rainfall intensities ranging from a few tenths of an inch per hour to several inches per hour. In some highly developed urban areas, the risk of flash flooding has increased over time as the native vegetation and soils have been replaced by buildings and pavement which produce much higher amounts of surface runoff. In addition, the increased usage of parks and recreational facilities which lie along stream and river channels has exposed the public to greater risk. *See also* Urban runoff

[*Mark W. Seeley*]

RESOURCES
BOOKS

Battan, L. J. *Weather In Your Life*. San Francisco: W. H. Freeman, 1983.
Critchfield, H. J. *General Climatology*. 4th ed. Englewood Cliffs, NJ: Prentice-Hall, 1983.

Floodplain

An area that has been built up by stream deposition, generally represented by the main **drainage** channel of a **watershed**, is called a floodplain. This area, usually relatively flat with respect to the surrounding landscape, is subject to periodic **flooding**, with

return periods ranging from one year to 100 years. Floodplains vary widely in size, depending on the area of the drainage basin with which they are associated. The soils in floodplains are often dark and fertile, representing material lost the to erosive forces of heavy precipitation and **runoff**. These soils are often farmed, though subject to the risk of periodic crop losses due to flooding. In some areas, floodplains are protected by flood control measures such as reservoirs and levees and are used for farming or residential development. In other areas, land-use regulations, encroachment statutes and local building codes often prevent development on floodplains.

Flora

All forms of plant life that live in a particular geographic region at a particular time in history. A number of factors determine the flora in any particular area, including temperature, sunlight, **soil**, water, and evolutionary history. The flora in any given area is a major factor in determining the type of **fauna** found in the area. Scientists have divided the earth's surface into a number of regions inhabited by distinct flora. Among these regions are the African-Indian **desert**, western African **rain forest**, Pacific North American region, Arctic and Sub-arctic region, and the Amazon.

Florida panther

The Florida panther (*Felis concolor coryi*), a subspecies of the mountain lion, is a member of the cat family, Felidae, and is severely threatened with **extinction**. Listed as endangered, the Florida panther population currently numbers between 30 and 50 individuals. Its former range probably extended from western Louisiana and Arkansas eastward through Mississippi, Alabama, Georgia, and southwestern South Carolina to the southern tip of Florida. Today the Florida panther's range consists of the Everglades-Big Cypress Swamp area. The preferred **habitat** for this large cat is subtropical forests comprised of dense stands of trees, vines, and shrubs, typically in low, swampy areas.

Several factors have contributed to the decline of the Florida panther. Historically the most significant factors have been habitat loss and persecution by humans. **Land use** patterns have altered the **environment** throughout the former range of the Florida panther.

With shifts to cattle ranching and agriculture, lands were drained and developed, and with the altered vegetation patterns came a change in the prey base for this top carnivore. The main prey item of the Florida panther is white-tailed deer (*Odocoileus virginianus*). Formerly, the spring and summer rains kept the area wet, and then, as it dried out, fires would renew the grassy meadows at the forest edges, creating an ideal habitat for the deer. With development and increased deer **hunting** by humans, the panther's prey base declined and so did the number of panthers. Prior to the 1950s, Florida had a bounty on Florida panthers because the animal was considered a "threat" to humans and livestock. During the 1950s, state law protected the dwindling population of panthers. In 1967 the Florida panther was listed by the U. S. **Fish and Wildlife Service** as an **endangered species**.

Land development is still moving southward in Florida. With the annual influx of new residents, fruit orchards being moved south due to recent freezes, and continued draining and clearing of land, panther habitat continues to be destroyed. The Florida panther is forced into areas that are not good habitat for white-tailed deer, and the panthers are catching armadillos and raccoons for food. The panthers then become underweight and anemic due to poor nutrition.

A Florida panther (*Felis concolor coryi*). (Photograph by Tom and Pat Leeson. Photo Researchers Inc. Reproduced by permission.)

Development contributes to the Florida panther's decline in other ways, too. Its range is currently split in half by the east-west highway known as Alligator Alley. During peak seasons, over 30,000 vehicles traverse this stretch of highway daily, and, since 1972, 44 panthers have been killed by cars, the largest single cause of death for these cats in recent decades.

Biology is also working against the Florida panther. Because of the extremely small population size, **inbreeding** of panthers has yielded increased reproductive failures, due to deformed or infertile sperm. The spread of feline distemper **virus** also is a concern to **wildlife** biologists. All these factors have led officials to develop a recovery plan that includes a captive breeding program using a small number of injured animals, as well as a mark and recapture program, using radio collars, to **inoculate** against disease and track young panthers with hopes of saving this valuable part of the biota of south Florida's **Everglades ecosystem**.

[*Eugene C. Beckham*]

RESOURCES
BOOKS

Belden, R. "The Florida Panther." *Audubon Wildlife Report 1988/1989*. San Diego: Academic Press, 1988.

Fergus, Charles. *Swamp Screamer: At Large with the Florida Panther*. New York: North Point Press, 1996.

Miller, S. D., and D. D. Everett, eds. *Cats of the World: Biology, Conservation, and Management*. Washington, DC: National Wildlife Federation, 1986.

OTHER

Florida Panther Net. [cited May 2002]. <http://www.panther.state.fl.us>.

Florida Panther Society. [cited May 2002]. <http://www.atlantic.net/~oldfla/panther/panther.html>.

Flotation

An operation in which submerged materials are floated, by means of air bubbles, to the surface of a water and removed. Bubbles are generated through a system called dissolved air flotation (DAF), which is capable of producing clouds of very fine, very small bubbles. A large number of small-sized bubbles is generally most efficient for removing material from water.

This process is commonly used in **wastewater** treatment and by industries, but not in **water treatment**. For example, the mining industry uses flotation to concentrate fine ore particles, and flotation has been used to concentrate **uranium** from sea water. It is commonly used to thicken the sludges and to remove grease and oil at wastewater treatment plants. The textile industry often uses flotation to treat process waters resulting from dyeing operations. Flotation might also be used to remove surfactants. Materials that are denser than water or that dissolve well in water are poor candidates for flotation. Flotation should not be confused with foam separation, a process in which surfactants are added to create a foam that affects the removal or concentration of some other material.

Flu pandemic

The influenza outbreak of 1918–1919 carried off between 20 to 40 million people worldwide. The Spanish flu outbreak differed significantly from other influenza (flu) epidemics. It was much more lethal, and it killed a high proportion of otherwise healthy adults. Most flu outbreaks kill only the very young, the elderly, and people with weakened immune systems. Scientists and public health officials have been trying to learn more about Spanish flu in the hopes of preventing a similar outbreak.

The Spanish flu **virus** caused one of the worst pandemic of an infectious disease ever recorded. And while the threat of many infectious diseases, including tuberculosis and smallpox, have been contained by antibiotics and vaccination programs, influenza remains a difficult disease. There are worldwide outbreaks of influenza every year, and the flu typically reaches pandemic proportions (lethally afflicting an unusually high portion of the population) every 10–40 years. The last influenza pandemic was the Hong Kong flu of 1968–69, which caused 700,000 deaths worldwide, and killed 33,000 Americans. The influenza virus is highly mutable, so each year's flu outbreak presents the human body with a slightly different virus. Because of this, people do not build an immunity to influenza. Vaccines are successful in protecting people against influenza, but vaccine manufacturers must prepare a new batch each year, based on their best supposition of which particular virus will spread. Most influenza viruses originate in China, and doctors, scientists, and public health officials closely monitor flu cases there in order to make the appropriate vaccine. The two main organizations tracking influenza are the Centers for Disease Control (CDC) and the World Health Organization (WHO). The CDC and other government agencies have been preparing for a flu pandemic on the level of Spanish flu since the early 1990s.

Spanish flu did not originate in Spain, but presumably in Kansas, where the first case was recorded in March, 1918, at the army base Camp Funston. It quickly spread across the United States, and then to Europe with American soldiers who were fighting in the last months of World War I. Infected ships brought the outbreak to India, New Zealand, and Alaska. Spanish flu killed quickly. People often died within 48 hours of first feeling symptoms. The disease afflicted the lungs, and caused the tiny air sacs, called alveoli, to fill with fluid. Victims were soon starved of oxygen, and sometimes effectively drowned on the fluid clogging their lungs. Children and old people recovered from the Spanish flu at a much higher rate than young adults. In the United States, the death rate from Spanish flu was several times higher for men aged 25–29 than for men in their seventies.

Social conditions at the time probably contributed to the remarkable power of the disease. The flu struck just at the end of World War I, when thousands of soldiers were moving from America to Europe and across that continent. In a peaceful time, sick people may have gone home to bed, and thus passed the disease only to their immediate family. But in 1918, men with the virus were packed in already crowded hospitals and troop ships. The unrest and devastation left by the war probably hastened the spread of Spanish flu. So it is possible that if a similarly virulent virus were to arise again soon, it would not be quite as destructive.

Researchers are concerned about a return of Spanish flu because little is known about what made it so virulent. The flu virus was not isolated until 1933, and since then, there have been several efforts to collect and study the 1918 virus by exhuming graves in Alaska and Norway, where bodies were preserved in permanently frozen ground. In 1997, a Canadian researcher, Kirsty Duncan, was able to extract tissue samples from the corpses of seven miners who

had died of Spanish flu in October 1918 and were buried in frozen ground on a tiny island off Norway. Duncan's work allowed scientists at several laboratories around the world to do genetic work on the Spanish flu virus. But by 2002, there was still no conclusive agreement on what was so different about the 1918 virus.

The influenza virus is believed to originate in migratory water fowl, particularly ducks. Ducks carry influenza viruses without becoming ill. They excrete the virus in their feces. When their feces collect in water, other animals can become infected. Domestic turkeys and chickens can easily become infected with influenza virus borne by wild ducks. But most avian (bird-borne) influenza does not pass to humans, or if it does, is not particularly virulent. But other mammals too can pick up influenza from either wild birds or domestic fowl. **Whales, seals,** ferrets, horses, and pigs are all susceptible to bird-borne viruses. When the virus moves between **species,** it may mutate. Human influenza viruses most likely pass from ducks to pigs to humans. The 1918 virus may have been a particularly unusual combination of avian and swine virus, to which humans were unusually vulnerable.

Enacting controls on pig and poultry farms may be an important way to prevent the rise of a new influenza pandemic. Some influenza researchers recommend that pigs and domestic ducks and chickens not be raised together. Separating pigs and fowl at live markets may also be a sensible precaution. With the concentration of poultry and pigs at huge "factory" farms, it is important for farmers, veterinarians, and public health officials to monitor for influenza. A flu outbreak among chickens in Hong Kong in 1997 eventually killed six people, but the epidemic was stopped by the quick slaughter of millions of chickens in the area. Any action to control flu of course must be an international effort, since the virus moves rapidly without respect to national borders.

[*Angela Woodward*]

RESOURCES
PERIODICALS

Gladwell, Malcolm. "The Dead Zone." *New Yorker* (September 29, 1997): 52–65.

Henderson, C. W. "Spanish Flu Victims Hold Clues to Fight Virus." *Vaccine Weekly* (November 29, 1999/December 6, 1999): 10.

Koehler, Christopher S. W. "Zeroing in on Zoonoses." *Modern Drug Discovery* 8, no. 4 (August 2001): 44–50.

Lauteret, Ronald L. "A Short History of a Tragedy" *Alaska* (November 1999): 21–23.

Pickrell, John. "Killer Flu with a Human-Pig Pedigree?" *Science* 292 (May 11, 2001): 1041.

Shalala, Donna E. "Collaboration in the Fight Against Infectious Diseases." *Emerging Infectious Diseases* 4, no. 3 (July/September 1998): 354.

Webster, Robert G. "Influenza: An Emerging Disease." *Emerging Infectious Diseases* 4, no. 3 (July-September 1998).

Westrup, Hugh. "Debugging a Killer Virus." *Current Science* 84, no. 9 (January 8, 1999): 4.

Flue gas

The exhaust gas vented from **combustion**, a chemical reaction, or other physical process, which passes through a duct into the **atmosphere**. Exhaust air is usually captured by an enclosure and brought into the exhaust duct through induced or forced ventilation. Induced ventilation is created by lowering the pressure in the duct using fans at the end of the duct. Forced ventilation occurs when exhaust air is forced into the duct using high pressure inlet air. Flues are valuable because they not only direct polluted air to a **pollution control** device, but also keep the air pollutant concentrations high. High concentrations can be important if the air pollutant removal process is concentration dependent. *See also* Air pollution control

Flue-gas scrubbing

Flue-gas scrubbing is a process for removing oxides of sulfur and **nitrogen** from the waste gases emitted by various industrial processes. Since the oxides of sulfur and nitrogen have been implicated in a number of health and environmental problems, controlling them is an important issue. The basic principle of scrubbing is that flue gases are forced through a system of baffles within a smokestack. The baffles contain some chemical or **chemicals** that remove pollutants from these gases.

A number of scrubbing processes are available, all of which depend on the reaction between the oxide and some other chemical to produce a harmless compound that can then be removed from the smokestack. For example, currently the most common scrubbing reaction involves the reaction between **sulfur dioxide** and lime. In the first step of this process, limestone is heated to produce lime. The lime then reacts with sulfur dioxide in flue gases to form calcium sulfite, which can be removed with **electrostatic precipitation**.

Many other scrubbing reactions have been investigated. For example, magnesium oxide can be used in place of calcium oxide in the scrubber. The advantage of this reaction is that the magnesium sulfite that is formed decomposes readily when heated. The magnesium oxide that is regenerated can then be reused in the scrubber while the sulfur dioxide can be used to make sulfuric **acid**. In yet another process, a mixture of sodium citrate and citric acid is used in the scrubber. When sulfur dioxide is absorbed by the mixture, a reaction occurs in which elemental sulfur is precipitated out.

Although the limestone/lime process is by far the most popular scrubbing reaction, it has one serious disadvantage. The end product, calcium sulfite, is a solid that must be disposed of in some way. **Solid waste** disposal is already a serious problem in many areas, so adding to that problem is not desirable. For that reason, reactions such as those involving magnesium oxide, sodium citrate and citric acid have been carefully studied. The products of these reactions, sulfuric acid and elemental sulfur, are valuable raw materials that can be sold and used. In spite of that fact, the limestone/lime scrubbing process, or some variation of it, remains the most popular method of extracting sulfur dioxide from flue gases today.

Scrubbing to remove **nitrogen oxides** is much less effective. In principle, reactions like those used with sulfur dioxide are possible. For example, experiments have been conducted in which ammonia or **ozone** is used in the scrubber to react with and remove oxides of nitrogen. But such methods have had relatively little success and are rarely used by industry.

Flue gas scrubbing has long met with **resistance** from utilities and industries. For one thing, they are not convinced that oxides of sulfur and nitrogen are as dangerous as environmentalists sometimes claim. In addition, they argue that the cost of installing **scrubbers** is often too great to justify their use. *See also* Air pollution control; Dry alkali injection; Stack emissions

[*David E. Newton*]

RESOURCES

BOOKS

American Chemical Society. *Cleaning Our Environment: A Chemical Perspective.* 2nd edition. Washington, DC: American Chemical Society, 1978.

PERIODICALS

Bretz, E. A. "Efficient Scrubbing Begins With Proper Lime Prep, Handling." *Electrical World* 205 (March 1991): 21–22.

"New Choices in FGD Systems Offer More Than Technology." *Electrical World* 204 (November 1990): 46–47.

"Scrubbers, Low-Sulfur Coal, of Plant Retirements?" *Electrical World* 204 (June 1990): 18.

Fluidized bed combustion

Of all **fossil fuels, coal** exists in the largest amount. In fact, the world's coal resources appear to be sufficient to meet our energy needs for many hundreds of years. One aim of energy technology, therefore, is to find more efficient ways to make use of these coal reserves. One efficiency procedure that has been under investigation for at least two decades is known as fluidized bed **combustion**.

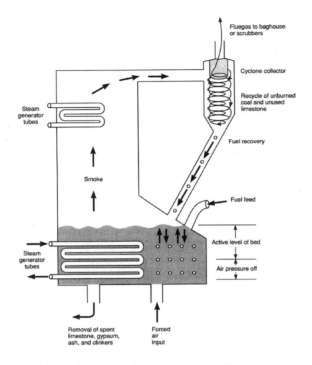

Fluidized bed combustion. Fuel is lifted by a stream of air from underneath the bed. Fuel efficiency is good and sulfur dioxide and nitrogen oxide emissions are lower than with conventional boilers. (McGraw-Hill Inc. Reproduced by permission.)

In a fluidized bed boiler, granulated coal and limestone are fed simultaneously onto a moving grate. A stream of air from below the grate lifts coal particles so that they are actually suspended in air when they begin to burn. The flow of air, small size of the coal particles, and exposure of the particles on all sides contribute to an increased rate of combustion. Heat produced by burning the coal is then used to boil water, run a turbine, and drive a generator, as in a conventional power plant.

The fluidized bed process has much to recommend it from an environmental standpoint. Sulfur and **nitrogen oxides** react with limestone added to the boiler along with the coal. The product of this reaction, primarily calcium sulfite and calcium sulfate, can be removed from the bottom of the boiler. However, disposing of large quantities of this waste product represents one of the drawbacks of the fluidized bed system.

Fly ash is also reduced in the fluidized bed process. As coal burns in the boiler, particles of fly ash tend to adhere to each other, forming larger particles that eventually settle out at the bottom of the boiler. Conventional methods of removal in the stack, such as **electrostatic precipitation**, can further increase efficiency with which particles are removed.

Incomplete combustion of coal is common in the fluidized bed process. However, **carbon monoxide** and **hydrogen** sulfide formed in this way are further oxidized in the space above the moving grate. The products of this further oxidation are then removed by the limestone (which reacts with **sulfur dioxide**) or allowed to escape harmlessly in to the air (in the case of the **carbon dioxide**). After being used as a **scavenger** in the process, the limestone, calcium sulfite, and calcium sulfate can be treated to release sulfur dioxide and regenerate the original limestone. The limestone can then be re-used and the sulfur dioxide employed to make sulfuric **acid**.

A further advantage of the fluidized bed process is that it operates at a lower temperature than does a conventional power plant. Thus, the temperature of cooling water ejected from the plant is lower, and the amount of **thermal pollution** of nearby waterways correspondingly lessened.

Writers in the 1970s expressed high hopes for the future of fluidized bed combustion systems, but the cost of such systems is still at least double that of a conventional plant. They are also only marginally more efficient than a conventional plant. Still, their environmental assets are obvious. They should reduce the amount of sulfur dioxide emitted by up to 90% and the amount of **nitrogen** oxides by more than 60%. *See also* Air pollution control; Stack emissions

[*David E. Newton*]

RESOURCES
BOOKS

Electric Power Research Institute. *Atmospheric Fluidized Bed Combustion Development.* Palo Alto, CA: EPRI, 1982.

Government Institutes, Inc. Staff, eds. *Evaluating the Fluidized Bed Combustion Option, 1988.* Rockville, MD: Government Institutes, 1988.

Marshall, A. R. *Reduced NOx Emissions and Other Phenomena in Fluidized Bed Combustion.* Lanham: UNIPUB, 1992.

PERIODICALS

Balzhiser, R. E., and K. E. Yeager. "Fluidized Bed Combustion." *Scientific American* 257 (September 1987): 100–107.

Fluoridation

Fluoridation is the precise adjustment of the concentration of the essential trace element fluoride in the public water supply to protect teeth and bones. Advocates of fluoridation such as the American Dental Association (ADA) and the National Center for Chronic Disease Prevention and Health Promotion (CDC) state that fluoridation is a safe and effective method of preventing tooth decay. Opponents of fluoridation however, such as Citizens for Health and the Fluoride Action Metwork, maintain that the role fluoridation in the decline of tooth decay is in serious doubt and that more research is required before placing a compound in water reservoirs that could cause **cancer**, brittle bones, and neurological problems.

Fluoride is any compound that contains fluorine, a corrosive, greenish-yellow element. Tooth enamel contains small amounts of fluoride. In addition, fluoride is found in varying amounts in water and in all food and beverages, according to the ADA.

A Colorado dentist discovered the effects of fluoride on teeth in the 1900s. When Frederick McKay began practicing in Colorado Springs, he established a connection between a substance in the water and the condition of residents' teeth. People did not have cavities, but their teeth were stained brown. Dental research started on the substance that was identified as fluoride during the 1930s. Researchers concluded that a concentration of fluoride in drinking water at a ratio of 1 part per million (ppm) prevented tooth decay without staining teeth. In 2000, the ADA stated that a fluoride concentration ranging from 0.7 ppm to 1.2 ppm was sufficient to fight tooth decay.

The first community to try fluoridation was Grand Rapids, Michigan. The city fluoridated the community water supply in 1945. Ten years later, Grand Rapids reported that incidents of tooth decay had declined by 60% in the children raised on fluoridated water. During the 1950s, Chicago, Philadelphia, and San Francisco also started to fluoridate their water supply. Cities including New York and Detroit opted for fluoridation during the 1960s. However, not all Americans advocated fluoridation. During the 1950s and 1960s, members of the John Birch Society maintained that fluoridation was a form of mass medication by the government. Some members charged that fluoridation was part of a Communist plot to take over the country. In the decades that followed, fluoridation was no longer associated with conspiracy theories. However, opinion about fluoridation was divided at the close of the twentieth century. By 2000, public water systems served 246.1 million Americans, according to the federal CDC. Of that amount, 65.8% of Americans used fluoridated water.

In Washington D.C., 100% of the water is fluoridated, according to a CDC report on the percentage of state populations with fluoridated public water systems in 2000. The top 10 on the list were: Minnesota (98.2%), Kentucky (96.1%), North Dakota (95.4%), Indiana (95.3%), Tennessee (94.5%), Illinois (93.4%), Virginia (93.4%), Georgia (92.9%), Iowa (91.3%), and South Carolina (91.2%). At the other end of the spectrum in terms of fluoridated public water usage were: Louisiana (53.2%), Mississippi (46%), Idaho (45.4%), New Hampshire (43%), Wyoming (30.3%), California (28.7%),

Oregon (22.7%), Montana (22.2%), New Jersey (15.5%), Hawaii (9%), and Utah (2%). The CDC estimated the cost of fluoridation at 50 cents per year in communities of more than 20,000 residents. The annual cost was estimated at $1 in communities of 10,000 to 20,000 residents. In communities numbering less than 5,000 people, the yearly cost was estimated at $3.

The CDC reported in 2000 that extensive research during the previous 50 years proved that fluoridation was safe. Fluoridation was also endorsed by groups including the American Medical Association, the American Academy of Pediatrics, the National PTA, and the American Cancer Society. Advocates of fluoridation state that it especially benefits people who may not be able to afford dental care. Opponents however, counter that toothpaste with fluoride is available for people who believe that fluoride fights tooth decay. Furthermore, opponents point out that toothpaste with fluoride contains a warning label advising users to "seek professional assistance or contact a poison control center" if they accidentally swallow more than the amount used for brushing teeth. Lastly, their question the research methodology used to conclude that fluoridation is responsible for decreased tooth decay. Fluoridation critics include consumer advocates **Ralph Nader** and Jim Turner. Turner chairs the board of Citizens for Health, a grassroots organization that is asking Congress to hold hearings and review fluoridation policy. Citizens for Health belongs to the groups that believe more research is required to determine the risks and benefits of fluoridation.

[*Liz Swain*]

RESOURCES
BOOKS

American Water Works Association. *Water Fluoridation Principles and Practices.* Denver: AWWA, 1996.

Health Research Staff. *Facts You Should Know About Fluoridation.* Pomeroy, WA: Health Research Books, 1996.

Martin, Brian. *Scientific Knowledge in Controversy: The Social Dynamics of the Fluoridation Debate (Science, Technology and Society).* Albany, NY: SUNY Press, 1991.

ORGANIZATIONS

American Dental Association, 211 E. Chicago Avenue, Chicago, IL USA 60611 (312) 440-2500, Fax: (312) 440-2800, Email: publicinfo@ada.org, <http://www.ada.org>

Centers for Disease Control and Prevention, 1600 Clifton Road, Atlanta, GA USA 30333 (404) 639-3534, Toll Free: (800) 311-3435, , <http://www.cdc.gov>

Citizens for Health, 5 Thomas Circle, NW, Suite 500, Washington, D.C. USA 20005 (202) 483-1652, Fax: (202) 483-7369, Email: cfh@citizens.org, <http://www.citizens.org>

National Center for Fluroridation Policy and Research (NCFPR) at the University of Buffalo, 315 Squire Hall, Buffalo, NY USA 14214 (716) 829-2056 , Fax: (716) 833-3517, Email: mweasley@buffalo.edu, <http://fluoride.oralhealth.org>

Fly ash

The fine ash from **combustion** processes that becomes dispersed in the air. To the casual observer it might appear as **smoke**, and indeed it is often found mixed with smoke. Fly ash arises because fuels contain a small fraction of incombustible matter. In a fuel like **coal**, ash has a rock-like siliceous composition, but the high temperatures at which it is formed often means that metals such as iron are incorporated into the ash particles, which take on the appearance of small, colored, glassy spheres. **Petroleum** produces less ash, but it is often associated with a range of oxides such as vanadium (in the case of fuel oils) and, more noticeably, hollow spheres of **carbon**. In traditional furnaces, much ash remained on the grate, but modern furnaces produce such fine ash that it is carried away in the hot exhaust gas.

Early power stations dispersed so much fine fly ash throughout the nearby **environment** that they were soon forced to adopt early **pollution** abatement techniques. They adopted "cyclones" in which centrifugal force removes the particles by causing the waste gas stream to flow on a curved or vortical course. This technique is effective down to particle sizes of about 10 μm, but smaller particles are not removed well by **cyclone** collectors and here **electrostatic precipitation** process often proves more successful, coping with a size range 30—0.1 μm. In electrostatic precipitators the particles are given a negative charge and then attracted to a positive electrode where they are collected and removed. Cloth or paper **filters** and spraying water through the exhaust gases can be useful in removing fly ash.

Fly ash is a nuisance at high concentrations because it accumulates as grit on the surfaces of buildings, clothes, cars, and outdoor furnishings. It is a highly visible and very annoying aspect of industrial **air pollution**. The deposition of fly ash increases cleaning costs incurred by people who live near poorly controlled combustion sources. Fly ash also has health impacts because the finer particles can penetrate into the human lung. If the deposits are especially heavy, fly ash can also inhibit plant growth.

Each year millions of tons of fly ash are produced from coal-powered furnaces, most of which are dumped in waste tips. Care needs to be taken that toxic metals and alkalis are not leached from these disposal sites into watercourses. Fly ash may be used as a low grade cement in road building because it contains a large amount of calcium oxide but generally, the demand is rather low.

[*Peter Brimblecombe Ph.D.*]

RESOURCES
BOOKS

Sellers, B. H. *Pollution of Our Atmosphere*. Bristol: Adam Hilger, 1984.

Flyway

The route taken by migratory birds and waterfowl when they travel between their breeding grounds and their winter homes. Flyways often follow geographic features such as mountain ranges, rivers, or other bodies of water. Protecting flyways is one of the many responsibilities of **wildlife** managers. Draining of **wetlands**, residential and commercial development, and **overhunting** are some of the factors that threaten flyway sites visited by birds for food and rest during **migration**. In most cases, international agreements are needed to guarantee protection along the entire length of a flyway. In the United States, flyway protection is financed to a large extent by funds produced through the Migratory Bird **Hunting** Stamp Act passed by the U.S. Congress in 1934.

Food additives

Food additives are substances added to food as flavorants, nutrients, preservatives, emulsifiers, or colorants. In addition, foods may contain residues of **chemicals** used during the production of plant or animal crops, including pesticides, antibiotics, and growth hormones. The use of most food additives is clearly beneficial because it results in improved public health and prevention of spoilage, which enhances the food supply. Nevertheless, there is controversy about the use of many common additives and over the presence of contaminants in food. This is partly because some people are hypersensitive and suffer allergic reactions if they are exposed to certain of these chemicals. In addition, some people believe that low levels of chronic toxicity and diseases may be caused in the larger population by exposure to some of these substances. Although there is no compelling scientific evidence that this is indeed the case, the possibility of chronic damage caused by food additives and chemical residues is an important social and scientific issue.

The use of food additives in the United States is closely regulated by the government agencies responsible for health, consumer safety, and agriculture. This is also the case of other developed countries in Europe, Canada, and elsewhere. Chemicals cannot be used as additives in those countries unless regulators are convinced that they have been demonstrated to be toxicologically safe, with a wide margin of security. In addition, chemicals added to commercially prepared foods must be listed on the packaging so that consumers can know what is present in the foodstuffs that they choose to eat. Because of the intrinsic nature of low-level, toxicological risks, especially those associated with diseases that may take a long time to develop, scientists are never able to demonstrate that trace exposures to any chemical are absolutely safe—there is always a level of risk, however small.

UNHEALTHY FOOD ADDITIVES		
Name	**Description**	**Example products**
Aspartame	An artificial sweetener associated with rashes, headaches, dizziness, depression, etc.	Diet sodas, sugar substitutes, etc.
Brominated vegetable oil (BVO)	Used as an emulsifier and clouding agent. Its main ingredient, bromate, is a poison.	Sodas, etc.
Butylated hydroxyanisole (BHA)/ butylated hydroxytoluene (BHT)	Prevents rancidity in foods and is added to food packagings. It slows the transfer of nerve impulses and affects sleep, aggressiveness, and weight in test animals.	Cereal and cheese packaging
Citrus red dye #2	Used to color oranges, it is a probable carcinogen. The FDA has recommended it be banned.	Oranges
Monosodium gltamate (MSG)	A flavor enhancer that can cause headaches, heart palpitations, and nausea.	Fast food, processed and packaged food
Nitrites	Used as preservatives, nitrites form cancer-causing compounds in the gastrointestinal tract and have been associated with cancer and birth defects.	Cured meats and wine
Saccharin	An artificial sweetener that may be carcinogenic.	Diet sodas and sugar substitutes
Sulfites	Used as a food preservative, sulfites have been linked to at least four deaths reported to the FDA in the United States.	Dried fruits, shrimp, and frozen potatoes
Tertiary butyhydroquinone (TBHQ)	It is extremely toxic in low doses and has been linked to childhood behavioral problems.	Candy bars, baking sprays, and fast foods
Yellow dye #6	Increases the number of kidney and adrenal gland tumors in lab rats. It has been banned in Norway and Sweden.	Candy and sodas

Because some people object to these potential, low-level, often involuntary risks, a certain degree of controversy will always be associated with the use of food additives. This is also true of the closely related topic of residues of pesticides, antibiotics, and growth hormones in foods.

Flavorants

Certain chemicals are added to foods to enhance their flavor. This is particularly true of commercially processed or prepared foods, such as canned vegetables and frozen foods and meals. One of the most commonly added flavorants is table salt (or sodium chloride), a critical **nutrient** for humans and other animals. In large amounts, however, sodium chloride can predispose people to developing high blood pressure, a factor that is important in strokes and other circulatory and heart diseases.

Table sugar (or sucrose), manufactured from sugar cane or sugar beets, and fructose, or fruit sugar, are commonly used to sweeten prepared foods. Such foods include sugar candies, chocolate products, artificial drinks, sweetened fruit juices, peanut butter, jams, ketchup, and most commercial breads. Sugars are easily assimilated from foods and are a useful form of metabolic energy. In large amounts, however, sugars can lead to weight gain, tooth decay, and hypoglycemia and diabetes in genetically predisposed people. Artificial sweeteners such as saccharine, aspartame, and sodium cyclamate avoid the nutritional problems associated with eating too much sugar. These nonsugar sweeteners may have their own problems, however, and some people consider them to be a low-level health hazard.

Monosodium glutamate (or MSG) is commonly used as a flavor enhancer, particularly in processed meats, prepared soups, and oriental foods. Some people are relatively sensitive to this chemical, developing headaches and other symptoms that are sometimes referred to as "Chinese food syndrome." Other flavorants used in processed foods include many kinds of spices, herbs, vanilla, mustard, nuts, peanuts, and wine. Some people are extremely allergic to even minute exposures to peanuts or nuts in food and can rapidly develop a condition known as anaphylactic shock, which is life-threatening unless quickly treated with medicine. This is one of the reasons why any foods containing peanuts or nuts as a flavoring ingredient must be clearly labeled as such.

Many flavorants are natural in origin. Increasingly, however, synthetic flavorants are being discovered and used. For example, vanilla used to be extracted from a particular **species** of tropical orchid and was therefore a rather expensive flavorant. However, a synthetic vanilla flavorant can now be manufactured from wood-pulp lignins, and this has made this pleasant flavor much more readily available than it used to be.

Nutrients

Many foods are fortified with minerals, vitamins, and other micronutrients. One such example is table salt, which has iodine added (as potassium iodide) to help prevent **goiter** in the general population. Goiter used to be relatively common but is now rare, in part because of the widespread use of iodized salt.

Other foods that are commonly fortified with minerals and vitamins include milk and margarine (with vitamins A and D), flour (with thiamine, riboflavin, niacin, iron), and some commercial breads and breakfast cereals (with various vitamins and minerals, particularly in some commercial cereal preparations). Micronutrient additives in these and other commercial foods are carefully formulated to help contribute to a balanced diet in their consumers. Nevertheless, some people believe that it is somehow un-natural and un-healthy to consume foods that have been adulterated in this manner, and they prefer to eat "natural" foods that do not have any vitamins or minerals added to them.

Preservatives

Preservatives are substances added to foods to prevent spoilage caused by bacteria, **fungi**, yeasts, insects, or other biological agents. Spoilage can lead to a decrease in the nutritional quality of foods, to the growth of food-poisoning **microorganisms** such as the botulism bacterium, or to the production of deadly chemicals, such as **aflatoxin**, that can be produced in stored grains and seeds (such as peanuts) by species of fungi.

Salt has long been used to preserve meat and fish, either added directly to the surface or by immersing the food in a briny solution. **Nitrates and nitrites** (such as sodium nitrate, or saltpetre) are also used to preserve meats, especially cured foods such as sausages, salamis, and hams. These chemicals are especially useful in inhibiting the growth of *Clostridium botulinum*, the bacterium that causes deadly botulism. Vinegar and wood **smoke** are used for similar purposes. **Sulfur dioxide**, sodium sulfite, and benzoic **acid** are often used as preservatives in fruit products, such beverages as wine and beer, and in ketchup, pickles, and spice preparations.

Anti-oxidants are chemicals added to certain foods to prevent a deterioration in their quality or flavor, occurring due to the exposure of fats and oils to atmospheric oxygen. Examples of commonly used antioxidants are ascorbic acid (or vitamin C), butylated hydroxyanisole (or BHA), butylated hydroxytoluene (or BHT), gallates, and ethoxyquin.

Stabilizers and emulsifiers

Stabilizers and emulsifiers are added to prepared foods to maintain suspensions of fats or oils in water matrices (or vice versa), or to prevent the caking of ingredients during storage or preparation. One example of an emulsifying additive is glyceryl monostearate, often added to stored starch products to maintain their texture. Alginates are compounds added to commercial ice cream, salad dressing, and other foods to stabilize emulsions of oil- or fat-in-water during storage.

Colorants

Some prepared foods have colors added to "improve" their aesthetic qualities and thereby to make them more attractive to consumers. This practice is especially common in the preparation of confectionaries such as candies, chocolate bars, ice creams, and similar products, and in fancy cakes and pastries. Similarly, products such as ketchup and strawberry preserves have red dyes added to enhance their color, relishes and tinned peas have green colors added, and dark breads may contain brown colorants. Most margarines have yellow colors added, to make them appear more similar to butter. Artificial drinks and drink-mixes contain food colorants appropriate to their flavor--cherry and raspberry contain red dyes, and so forth.

Various chemicals are used as food colorants, some of them being extracted from plants (for example, yellow and orange carotenes), while many others are synthetic chemicals derived from **coal** tars and other organic substances. The acute toxicity (i.e., short-term **poisoning**) and chronic toxicity (i.e., longer-term damage associated with diseases, cancers, and developmental abnormalities) of these colorants are stringently tested on animals in the laboratory, and the substances must be demonstrated to be safe before they are allowed to be used as food additives. Still, some people object to having these chemicals in their food, and choose to consume products that are not adulterated with colorants.

Residues of pesticides, antibiotics, and growth hormones

Insecticides, fungicides, herbicides, and other pesticides are routinely used in modern, industrial agriculture. Some of these chemicals are persistent, because they do not quickly break down in the **environment** to simpler substances, and/or they do not readily wash off produce. The chemicals in such cases are called residues, and it is not unusual for them to be present on or in foodstuffs in low concentrations. The permissible residue levels allowed in foodstuffs intended for human consumption are closely regulated by government. However, not all foods can be properly inspected, so it is common for people to be routinely exposed to small concentrations of these chemicals in their diet.

In addition, most animals cultivated in intensive agricultural systems, such as **feedlots** and factory farms, are routinely treated with antibiotics in their feed. This is done to prevent outbreaks of **communicable diseases** under densely crowded conditions. Antibiotic use is especially common during the raising of chickens, turkeys, pigs, and cows. Small residues of these chemicals remain in the meat, eggs, milk, or other products of these animals, and are ingested

by human consumers. Also, growth hormones are given to beef and dairy cows to increase their productivity. Small residues of these chemicals also occur in products eaten by consumers.

Strictly speaking, residues of pesticides, antibiotics, and growth hormones are not additives because they are not added directly to foodstuffs. Nevertheless, these chemicals are present in foods eaten by people, and many consumers find this to be objectionable. So-called organic foods are cultivated without the use of synthetic pesticides, antibiotics, or growth hormones, and many people prefer to eat these foods instead of the much more abundantly available foodstuffs that are typically sold in commercial outlets. (Note that the term "organic foods" is somewhat of a misnomer, because all foods are organic in nature. The phrase "organic" in this sense is used to refer to foods containing additives and/or residues, etc.)

Irradiation of food

Irradiation is a new technology that can be used to prevent spoilage of foods by sterilizing most or all of the microorganisms and insects that they may contain. This process utilizes gamma radiation, and it is not known to cause any chemical or physical changes in foodstuffs, other than the intended benefit of killing organisms that can cause spoilage. Although this process displaces some of the uses of preservative chemicals as food additives, irradiation itself is somewhat controversial. Even though there is no scientific evidence that **food irradiation** poses tangible risks to consumers, some people object to the use of this technology and prefer not to consume foodstuffs processed in this manner.

[*Bill Freedman Ph.D.*]

RESOURCES

BOOKS

British Nutrition Foundation. *Why Additives? The Safety of Foods*. London: Forbes, 1977.

Freed, D. L. J. *Health Hazards of Milk*. London: Bailliere Tindall, 1984.

Marcus, A. I. *Cancer from Beef: DES, Federal Food Regulation, and Consumer Confidence*. Baltimore: Johns Hopkins University Press, 1994.

Miller, M. *Danger! Additives at Work: A Report on Food Additives, Their Use and Control*. London: London Food Commission, 1985.

Safety and Nutritional Adequacy of Irradiated Food. Geneva, Switzerland: World Health Organization, 1994.

PERIODICALS

Etherton, T. D. "The Impact Of Biotechnology On Animal Agriculture And The Consumer." *Nutrition Today* 29, no. 4 (1994): 12–18.

Food and Drug Administration

Founded in 1927, the Food and Drug Administration (FDA) is an agency of the Untied States **Public Health**

Service. One of the nation's oldest consumer protection agencies, the FDA is charged with enforcing the Federal Food, Drug, and Cosmetics Act, and other related public health laws. The agency assesses risks to the public posed by foods, drugs, and cosmetics, as well as medical devices, blood, and medications such as insulin which are made from living organisms. It also tests food samples for contaminants, sets labeling standards, and monitors the public health effects of drugs given to animals raised for food.

To carry out its mandate of consumer protection, the FDA employs over 9,000 investigators, inspectors, and scientists who collect domestic and imported product samples for examination by FDA scientists. The FDA has the power to remove from the market those foods, drugs, **chemicals**, or medical devices it finds unsafe. The FDA often seeks voluntary recall of the product by manufacturers, but the agency can also stop sales and destroy products through court action. About 3,000 products a year are found to be unfit for consumers and are withdrawn from the marketplace based on FDA action. Also, about 30,000 import shipments each year are detained at the port of entry on FDA orders.

FDA scientists analyze samples of products to detect contamination, or review test results submitted by companies seeking agency approval for drugs, vaccines, **food additives**, dyes, and medical devices. The FDA also operates the National Center for Toxicological Research at Jefferson, Arkansas, which conducts research to investigate the biological effects of widely used chemicals. The Agency's Engineering and Analytical Center at Winchester, Massachusetts, tests medical devices, radiation-emitting products, and radioactive drugs. The Bureau of Radiological Health was formed in 1971 to protect against unnecessary human exposure to radiation from electronic products such as microwave ovens.

The FDA is one of several federal organizations that oversees the safety of **biotechnology**, such as the industrial use of **microorganisms** to processes waste and water products.

In 1996, when the FDA declared that cigarettes and smokeless **tobacco** are nicotine-delivery devices, it took responsibility for regulating those products under the authority of the Federal Food, Drug, and Cosmetics Act. With regard to these products, the FDA has issued federal mandates concerning sales to minors, sales from vending machines, and advertising campaigns.

[*Linda Rehkopf*]

RESOURCES
PERIODICALS

Gibbons, A. "Can David Kessler Revive the FDA?" *Science* 252 (April 12, 1991): 200–3.

Iglehart, J. K. "The Food and Drug Administration and Its Problems." *New England Journal of Medicine* 325 (July 18, 1991): 217–20.

ORGANIZATIONS
U.S. Food and Drug Administration, 5600 Fishers Lane, Rockville, MD USA 20857-0001 Toll Free: (888) INFO-FDA, <http://www.fda.gov>

Food chain/web

Food chains and food webs are methods of describing an **ecosystem** by describing how energy flows from one **species** to another.

First proposed by the English zoologist Charles Elton in 1927, food chains and food webs describe the successive transfer of energy from plants to the animals that eat them, and to the animals that eat those animals, and so on. A food chain is a model for this process which assumes that the transfer of energy within the community is relatively simple. A food chain in a grassland ecosystem, for example, might be: Insects eat grass, and mice eat insects, and fox eat mice. But such an outline is not exactly accurate, and many more species of plants and animals are actually involved in the transfer of energy. Rodents often feed on both plants and insects, and some animals, such as predatory birds, feed on several kinds of rodents. This more complex description of the way energy flows through an ecosystem is called a food web. Food webs can be thought of as interconnected or intersecting food chains.

The components of food chains and food webs are producers, consumers, and **decomposers**. Plants and chemosynthetic bacteria are producers. They are also called primary producers or autotrophs ("self-nourishing") because they produce organic compounds from inorganic **chemicals** and outside sources of energy. The groups that eat these plants are called primary consumers or herbivores. They have adaptations that allow them to live on a purely vegetative diet which is high in cellulose. They usually have teeth modified for chewing and grinding; ruminants such as deer and cattle have well-developed stomachs, and lagomorphs such as rabbits have caeca which aid their digestion. Animals that eat herbivores are called secondary consumers or primary carnivores, and predators that eat these animals are called tertiary consumers. Decomposers are the final link in the **energy flow**. They feed on dead organic matter, releasing nutrients back into the ecosystem. Animals that eat dead plant and animal matter are called scavengers, and plants that do the same are known as saprophytes.

The components of food chains and food webs exist at different stages in the transfer of energy through an ecosystem. The position of every group of organisms obtaining their food in the same manner is known as a **trophic level**. The term comes from a Greek word meaning "nursing," and the implication is that each stage nourishes the next. The

In an ecosystem, food chains become interconnected to form food webs. (Illustration by Hans & Cassidy.)

first tropic level consists of autotrophs, the second herbivores, the third primary carnivores. At the final trophic level exists what is often called the "top predator." Organisms in the same trophic level are not necessarily connected taxonomically; they are connected ecologically, by the fact they obtain their energy in the same way. Their trophic level is determined by how many steps it is above the primary producer level. Most organisms occupy only one trophic level; however some may occupy two. Insectivorous plants like the venus flytrap are both primary producers and carnivores. Horseflies are another example: the females bite and draw blood, while the males are strictly herbivores.

In 1942, Raymond Lindeman published a paper entitled "The Tropic-Dynamic Aspect of Ecology." Although a young man and only recently graduated from Yale University, he revolutionized ecological thinking by describing ecosystems in the terminology of energy transformation. He used data from his studies of Cedar Bog Lake in Minnesota to construct the first energy budget for an entire ecosystem. He measured harvestable net production at three trophic levels, primary producer, herbivore, and carnivore. He did this by measuring gross production minus growth, reproduc-

tion, **respiration**, and excretion. He was able to calculate the assimilation efficiency at each tropic level, and the efficiency of energy transfers between each level. Lindeman's calculations are still widely regarded today, and his conclusions are usually generalized by saying that the ecological efficiency of energy transfers between trophic levels averages about 10%.

Lindeman's calculations and some basic laws about physics reveal important truths about food chains, food webs, and ecosystems in general. The First Law of Thermodynamics states that energy cannot be created or destroyed; energy input must equal energy output. The Second Law of Thermodynamics states that all physical processes proceed in such a way that the availability of the energy involved decreases. In other words, no transfer of energy is completely efficient. Using the generalized 10% figure from Lindeman's study, a hypothetical ecosystem with 1,000 kcal of energy available (net production) at the primary-producer level would mean that only 100 kcal would be available to the herbivores at the second trophic level, 10 kcal to the primary carnivores at the third level, and 1 kcal to the secondary carnivores at the fourth level. Thus, no matter how much energy is assimilated by the autotrophs at the first level of an ecosystem, the eventual number of trophic levels is limited by the laws which govern the transfer of energy. The number of links in most natural food chains is four.

The relationships between trophic levels has sometimes been compared to a pyramid, with a broad base which narrows to an apex. Trophic levels represent successively narrowing sections of the pyramid. These pyramids can be described in terms of the number of organisms at each trophic level. This was first proposed by Charles Elton, who observed that the number of plants usually exceeded the number of herbivores, which in turn exceeded the number of primary carnivores, and so on. Pyramids of number can be inverted, particularly at the base; an example of this would be the thousands of insects which might feed on a single tree. The pyramid-like relationship between trophic levels can also be expressed in terms of the accumulated weight of all living matter, known as **biomass**. Although upper-level consumers tend to be large, the population of organisms at lower trophic levels are usually much higher, resulting in a larger combined biomass. Pyramids of biomass are not normally inverted, though they can be under certain conditions. In aquatic ecosystems, the biomass of the primary producers may be less than that of the primary consumers because of the rate at which they are being consumed; **phytoplankton** can be eaten so rapidly that the biomass of **zooplankton** and other herbivores are greater at any particular time. The relationship between trophic levels can also be described in terms of energy, but pyramids of energy cannot be inverted. There will always be more energy at the bottom than there is at the top.

Humans are the top consumer in many ecosystems, and they exert strong and sometimes damaging pressures on food chains. For example, **overfishing** or **overhunting** can cause a large drop in the number of animals, resulting in changes in the food-web interrelationships. On the other hand, overprotection of some animals like deer or moose can be just as damaging. Another harmful influence is that of **biomagnification**. Toxic chemicals such as **mercury** and DDT released into the **environment** tend to become more concentrated as they travel up the food chain. Some ecologists have proposed that the **stability** of ecosystems is associated with the complexity of the internal structure of the food web and that ecosystems with a greater number of interconnections are more stable. Although more studies must be done to test this hypothesis, we do know that food chains in constant environments tend to have a greater number of species and more trophic links, whereas food chains in unstable environments have fewer species and trophic links.

[*John Korstad and Douglas Smith*]

RESOURCES
BOOKS

Krebs, C. J. *Ecology: The Experimental Analysis of Distribution and Abundance.* 3rd ed. New York: Harper & Row, 1985.

PERIODICALS

Hairston, N. G., F. E. Smith, and L. B. Slobodkin. "Community Structure, Population Control, and Competition." *American Naturalist* 94 (1960): 421–25.

Lindeman, R. L. "The Trophic-Dynamic Aspect of Ecology." *Ecology* 23 (1942): 399–418.

Food irradiation

The treatment of food with **ionizing radiation** has been in practice for nearly a century since the first irradiation process patents were filed in 1905. Regular use of the technology in food processing started in 1963 when the U.S. **Food and Drug Administration** (FDA) approved the sale of irradiated wheat and wheat flour. Today irradiation treatment is used on a wide variety of food products and is regulated in the United States by the FDA under a Department of Health and Human Services regulation.

Irradiation of food has three main applications: extension of shelf life, elimination of insects, and the destruction of bacteria and other pathogens that cause foodborne illness. This final goal may have the most far-reaching implications for Americans; the U.S. Centers for Disease Control (CDC) estimate that 76 million Americans get sick, and 5,000 die each year from illnesses caused by foodborne **microorganisms**, such as *E. coli, Salmonella,* the botulism toxin, and other pathogens responsible for food **poisoning**.

Irradiation technology involves exposing food to ionizing radiation. The radiation is generated from gamma rays emitted by cobalt-60 or cesium-137, or from x rays or electron beams. The amount of radiation absorbed during irradiation processing is measured in units called RADs (radiant energy absorbed). One hundred RADs is equivalent to one Gray (Gy). Depending on the food product being irradiated, treatment can range from 0.05 to 30 kGy. A dosimeter, or film badge, verifies the kGy dose. The ionizing radiation displaces electrons in the food, which slows cell division and kills bacteria and pests.

The irradiation process itself is relatively simple. Food is packed in totes or containers, which are typically placed on a conveyer belt. Beef and other foods that require refrigeration are loaded into insulated containers prior to treatment. The belt transports the food bins through a lead-lined irradiation cell or chamber, where they are exposed to the ionizing radiation that kills the microorganisms. Several trips through the chamber may be required for full irradiation. The length of the treatment depends upon the food being processed and the technology used, but each rotation takes only a few minutes.

The FDA has approved the use of irradiation for wheat and wheat powder, spices, **enzyme** preparations, vegetables, pork, fruits, poultry, beef, lamb, and goat meat. In 2000, the FDA also approved the use of irradiation to control salmonella in fresh eggs.

Labeling guidelines introduced by the Codex Alimentarius Commission, an international food standards organization sponsored jointly by the United Nations Food and Agricultural Organization (FAO) and the World Health Organization (WHO), requires that all irradiated food products and ingredients be clearly labeled as such for consumers. Codex also created the radura, a voluntary international symbol that represents irradiation. In the United States, the food irradiation process is regulated jointly by FDA and the **U.S. Department of Agriculture** (USDA). Facilities using radioactive sources such as cobalt-60 are also regulated by the **Nuclear Regulatory Commission** (NRC. The FDA regulates irradiation sources, levels, food types and packaging, as well as required recordkeeping and labeling. Records must be maintained and made available to FDA for one year beyond the shelf-life of the irradiated food to a maximum of three years. They must describe all aspects of the treatment and foods that have been irradiated must be denoted with the radura symbol and by the statement "Treated with radiation" or "Treated by irradiation". As of 2002, food irradiation is allowed in some 50 countries and is endorsed by the World Health Organization (WHO), and many other organizations.

New legislation entitled the "Farm Security and Rural Investment Act of 2002" (the Farm Bill) passed in May 2002 may soften the food irradiation standards. The Farm Bill calls for the Secretary of Health and Human Services and FDA to implement a new regulatory program for irradiated foods. The program will allow the food industry to instead label irradiated food as "pasteurized" as long as they meet appropriate food safety standards. As of the writing of this entry, these new guidelines had not been implemented.

Food that has been treated with ionizing energy typically looks and tastes the same as non-irradiated food. Just like a suitcase going through an airport x-ray machine, irradiated food does not come into direct contact with a radiation source and is not radioactive. However, depending on the strength and duration of the irradiation process, some slight changes in appearance and taste have been reported in some foods after treatment. Some of the flavor changes may be attributed to the generation of substances known as radiolytic products in irradiated foods.

When food products are irradiated the energy displaces electrons in the food and forms compounds called free radicals. The free radicals react with other molecules to form new stable compounds, termed radiolytic products. **Benzene**,

formaldehyde, and **hydrogen** peroxide are just a few of the radiolytic products that may form during the irradiation process. These substances are only present in minute amounts, however, and the FDA reports that 90% of all radiolytic products from irradiation are also found naturally in food.

The chemical change that creates radiolytic products also occurs in other food processing methods, such as canning or cooking. However, about 10% of the radiolytic products found in irradiated food are unique to the irradiation process and little is known about the effects that they may have on human health. It should be noted, however, that the World Health Organization, the American Medical Association, the American Dietetic Association, and a host of other professional healthcare organizations endorse the use of irradiation as a food safety measure.

Treating fruit and vegetables with irradiation can also eliminate the need for chemical **fumigation** after harvesting. Produce shelf life is extended by the reduction and elimination of organisms that cause spoilage. It also slows cell division, thus delaying the ripening process, and in some types of produce irradiation extends the shelf life for up to a week. Advocates of irradiation claim that it is a safe alternative to the use of fumigants, several of which have been banned in the United States.

Nevertheless, irradiation removes some of the nutrients from foods, particularly vitamins A, C, E, and the B-Complex vitamins. Whether the extent of this **nutrient** loss is significant enough to be harmful is debatable. Advocates of irradiation say the loss is insignificant, and standard methods of cooking can destroy these same vitamins. However, research suggests that cooking an irradiated food may further increase the loss of nutrients.

Critics of irradiation also question the long-term safety of consumption of irradiated food and their associated radiolytic products. They charge that the technology does nothing to address the unsanitary food processing practices and inadequate inspection programs that breed foodborne pathogens.

Even if irradiation is 100% safe and beneficial there are numerous environmental concerns. Many opponents of irradiation cite the proliferation of radioactive material and the environmental hazards. The mining and on-site processing of radioactive materials are devastating to regional ecosystems. There are also safety hazards associated with the **transportation** of radioactive material, production of isotopes, and disposal.

[*Paula Ford-Martin and Debra Glidden*]

RESOURCES
BOOKS

Molins, Ricardo A. *Food Irradiation: Principles and Applications.* New York: John Wiley and Sons, 2001.

Stuart, Thorne, ed. *Food Irradiation (Elsevier Applied Food Science Series)*. New York: Elsevier, 1992.

PERIODICALS

Ennen, Steve. "Irradiation Plant Gets USDA Approval." *Food Processing* (March 2001): 90–2.

Louria, Donald. "Food Irradiation: Unresolved Issues." *Clinical Infectious Diseases* 33 (August 2001): 378–80.

Steele, J. H. "Food Irradiation: A Public Health Challenge for the 21st Century." *Clinical Infectious Diseases* 33 (August 2001): 376–7.

U.S. General Accounting Office. "Food Irradiation: Available Research Indicates That Benefits Outweigh Risks." *GAO Report to Congressional Requesters.* GAO/RCED-00-217 (August 2000).

OTHER

U.S. Department of Agriculture, Food Safety and Inspection Service. *Irradiation of Meat and Poultry Products.* [cited July 2002]. <http://www.fsis.usda.gov/OA/topics/irrmenu.htm>.

U.S. Department of Agriculture, Food Safety and Inspection Service. "Irradiation of Meat Food Products; Final Rule." *Federal Register* 64, no. 246 (December 23, 1999): 72150–66.

ORGANIZATIONS

The Food Irradiation Website, <http://www.food-irradiation.com>

National Food Processors Association, 1350 I Street, NW Suite 300, Washington, DC USA 20005 (202) 639-5900, Fax: (202) 639-5932, Email: nfpa@nfpa-food.org, <http://www.nfpa-food.org>

Public Citizen, Critical Mass Energy & Environmental Program, 1600 20th St. NW, Washington, DC USA 20009 (202) 588-1000, Email: CMEP@citizen.org, <http://www.citizen.org/cmep/foodsafety/food_irrad>

Food policy

Through a variety of agricultural, economic, and regulatory programs which support or direct policies related to the production and distribution of food, the United States government has a large influence on how agriculture business is conducted. The government's major impact on agriculture is the setting of prices and mandates regarding how land can be used by farmers that participate in government programs. These policies can also have a large impact on the adoption or use of alternative practices and technologies that may be more efficient and sustainable in the world marketplace.

In the past, farm commodity programs have had a large influence in the past on the kinds and amounts of crops grown as well as on the choice of management practices used to grow them. Prices under government commodity programs have often been above world market prices which meant that many farmers felt compelled to preserve or build their farm commodity program base acres, since acreage determines program eligibility and future income. These programs strongly influenced land-use decisions on about two-thirds of the harvested cropland in the United States.

Price and income support programs for major commodities also influence growers not in the programs. For example, pork producers are not a part of a government program, and in the past they have paid higher feed prices because of high price supports on production of feed grains. At other times, particularly after the Food Security Act of 1985, they benefited from policies resulting in lower food costs. So as the government changes policy in one area, there can be widespread indirect impacts in other areas. For example, the federal dairy termination program, which ran from 1985–1987 was designed to reduce overproduction of milk. Those farmers who sold their milk cows and decided to produce hay for local cash markets caused a steep decline in the prices received by other established hay producers.

Federal policy evolved as a patchwork of individual programs, each created to address individual problems. There was not a coherent strategy to direct programs toward a common set of goals. Many programs such as **soil conservation** and export programs have had conflicting objectives, but attempts have now been made in the most current farm legislation to address some of these problems.

Government food policy has produced a wide variety of results. The policy has not only affected commodity prices and the level of output, but it has also shaped technological change, encouraged uneconomical capital investments in machinery and facilities, inflated the value of land, subsidized crop production practices that have led to resource degradation (such as **soil erosion** and surface and **groundwater pollution**), expanded the interstate highway system, financed **irrigation** projects, and promoted farm commodity exports. Together with other economic forces, government policy has had a far-reaching structural influence on agriculture, much of it unintended and unanticipated.

Federal commodity programs were put into place beginning in the 1930s, primarily with the Agriculture Adjustment Act of 1938. The purpose of these programs born out of the Depression was primarily to protect crop prices and farmer income which has been done by a number of programs over the years. A variety of methods have been used including setting prices, making direct payments to farmers, and subsidizing exports. However, by the mid-1980s and early 1990s, an increasing number of people felt that these programs impeded movement toward alternative types of agriculture, to the detriment of family farms and society in general.

Two components in particular were highlighted as being problems: base acre requirements and cross-compliance. All crop price and income support programs relied on the concept of an acreage base planted with a given commodity that would produce a predictable yield. Most of this acreage was planted to maximize benefits and was based on a five-year average. A farmer knew that if he/she reduced their acres for a given crop, they would not only lose the current year's benefits, but would also lose future benefits.

Cross-compliance was instituted in the Food Security Act of 1985. It was designed to control government payments and production by attaching financial penalties to the expansion of program crop base acres. It served as an effective financial barrier to diversification of crops by stipulating that to receive any benefits from an established crop acreage base, the farmer must not exceed his or her acreage base for any other program crop. This had a profound impact on farmers and crop growers since about 70% of the United States' cropland acres were enrolled in the programs.

In addition to citing these problem areas, critics of food policy programs argued that many farmers faced economic penalties for adopting beneficial practices such as crop rotation or strip cropping, practices that in general reduce soil erosion and improve environmental quality. The economic incentives built into commodity programs, for example, encouraged heavier use of **fertilizer**, pesticides, and irrigation. These programs also encouraged surplus production, subsidized inefficient use of inputs, and they resulted in increased government expenditures. Critics argued that the rules associated with these programs discouraged farmers from pursuing alternative practices or crops, or technologies that might have proved more effective in the long term or that were more environmentally friendly.

Critics also contend that much of the research conducted over the past 40 years has responded to the needs of farmers operating under a set of economic and policy incentives that encouraged high yields without regard to the long-term environmental impacts. During the late 1980s and early 1990s, several **U.S. Department of Agriculture** research and education programs were instituted to determine whether current levels of production can be maintained with reduced levels of fertilizers and pesticides, to examine more intensive management practices, to increase understanding of biological principles, and to improve profitability per unit of production with less government support. As the impacts of the alternative production systems on the **environment** are evaluated, it will be important to have policies in place that will allow the farmer to easily adopt those practices that increase efficiency and reduce impacts. In a farm bill passed in 1996 there are provisions that change these commodity programs. Labeled the "right to farm provisions," these allow farmers to make decisions on what they grow and establish a phased seven-year reduction in price supports.

Food quality and safety are major concerns addressed as a part of federal policy. Programs addressing these concerns are primarily designed to prevent health risks and acute illnesses from chemical and microbial contaminants in food. Supporters say that this has provided the country with the safest food supply in the world. However, critics contend that a number of regulations do not enhance quality or safety and put farmers that use or would adopt alternative agricultural practices at a disadvantage. Several examples can be cited. Critics of government food policy point out that until recently, meat grading standards awarded producers of fatty beef which has been linked to the increased likelihood of heart disease.

The use of pesticides provides another example. The **Environmental Protection Agency** establishes **pesticide** residual tolerance levels in food which are monitored for compliance. For some types of risk, **cancer** in particular, there is a great deal of uncertainty, and critics point out which cosmetic standards that increase prices for fruits and vegetables may encourage higher risks of disease among consumers. Also, certain poultry slaughter practices can result in microbiological contamination. Of particular concern is salmonella food **poisoning** which has become widespread.

Government food policy heavily influences on-farm decision making. In some cases one part of the policy has negative or unintended consequences for another policy or segment of farmers. As we look to the future, the struggle will be to continue to provide a coherent, coordinated policy. The recent changes in policy will need to be evaluated from the standpoint of sustainability and environmental impacts.

[*James L. Anderson*]

RESOURCES
BOOKS

Agriculture and the Environment. The 1991 Yearbook of Agriculture. U.S. Government Printing Office. Washington, D.C.
Alternative Agriculture. Board on Agriculture, National Research Council., Washington, DC: National Academy Press, 1989.

Food waste

Waste food from residences, grocery stores, and food services accounts for nearly seven percent of the **municipal solid waste** stream. The per capita amount of food waste in municipal **solid waste** has been declining since 1960 due to increased use of **garbage** disposals and increased consumption of processed foods. Food waste ground in garbage disposals goes into sewer systems and thus ends up in **wastewater**. Waste generated by the food processing industry is considered to be industrial waste, and is not included in municipal solid waste estimates.

Waste from the food processing industry includes: vegetables and fruits unsuitable for canning or freezing; vegetable, fruit, and meat trimmings; and pomace from juice manufacturing. Vegetable and fruit processing waste is sometimes used as animal feed and waste from meat and seafood processing can be composted. Liquid waste from juice manufacturing can be applied to cropland as a **soil** amendment. Much of the waste generated by all types of food processing is wastewater due to such processes as washing, peeling, blanching, and cooling. Some food industries recycle wastewaters back into their processes, but there is potential for more of this wastewater to be reused.

Grocery stores generate food waste in the form of lettuce trimmings, excess foliage, unmarketable produce, and meat trimmings. Waste from grocery stores located in rural areas is often used as hog or cattle feed, whereas grocery waste in urban areas is usually ground in garbage disposals. There is potential for more urban grocery store waste to be either used on farms or composted, but lack of storage space, odor, and **pest** problems prevent most of this waste from being recycled.

Restaurants and institutional cafeterias are the major sources of food service industry waste. In addition to food preparation wastes, they also generate large amounts of cooking oil and grease waste, post-consumer waste (uneaten food), and surplus waste. In some areas small amounts of surplus waste is utilized by feeding programs, but most waste generated by food services goes to landfills or into garbage disposals.

Most food waste is generated by sources other than households. However, a greater percentage of household food waste is disposed of because there is a higher rate of **recycling** of industrial and commercial food waste. Only a very small segment of households compost or otherwise recycle their food wastes.

[*Teresa C. Donkin*]

RESOURCES
BOOKS

U.S. Environmental Protection Agency. Solid Waste and Emergency Response. *Characterization of Municipal Solid Waste in the United States: 1990 Update.* Washington, DC: U.S. Government Printing Office, June 1990.
U.S. Environmental Protection Agency. Solid Waste and Emergency Response. *Characterization of Municipal Solid Waste in the United States: 1992 Update (Executive Summary).* Washington, DC: U.S. Government Printing Office, July 1992.

PERIODICALS

Youde, J., and B. Prenguber. "Classifying the Food Waste Stream." *Biocycle* 32 (October 1991): 70–71.

Food-borne diseases

Food-borne diseases are illnesses caused when people consume contaminated food or beverages. Contamination is frequently caused by disease-causing **microbes** called pathogens. Other causes of food-borne diseases are poisonous **chemicals** or harmful substances in food and beverages. There are so many different pathogens that more than 250 food-borne illnesses have been described, according to the United States Centers for Disease Control and Prevention(CDC). CDC estimated that food-borne pathogens

cause approximately 76 million illnesses, 5,000 deaths, 325,000 hospitalizations in the United States each year.

Most food-borne illnesses are infections caused by bacteria, viruses, and **parasites** such as **Cryptosporidium**. Harmful toxins cause food poisonings. Since there are so many different types of food-borne illnesses, symptoms will vary. However, some early symptoms are similar because the microbe or toxin travels through the gastrointestinal tract. The initial symptoms of food-borne diseases are nausea, vomiting, abdominal cramps, and diarrhea.

According to CDC, the most common food-borne viruses are caused by three bacteria and a group of viruses. *Campylobacter* is a bacterium that lives in the intestines of healthy birds. It is also found in raw poultry meat. An infection is caused by eating undercooked chicken or food contaminated by juices from raw chicken. The bacterial **pathogen** causes fever, diarrhea, and abdominal cramps. *Campylobacter* is the primary cause of bacteria-related diarrhea illness throughout the world.

Salmonella is a bacterium that is prevalent in the intestines of birds, mammals, and reptiles. The bacterium spreads to humans through various foods. *Salmonella* causes the illness *salmonellosis*. Symptoms include fever, diarrhea, and abdominal cramps. This illness can result in a life-threatening infection for a person who is in poor health or has a weakened immune system.

E. coli O157:H7 is a bacterial pathogen that has a **reservoir** in cattle and similar animals. *E. coli* causes a serious illness. People become ill after eating food or drinking water that was contaminated with microscopic amounts of cow feces, according to CDC. A person often experiences severe and bloody diarrhea and painful abdominal cramps. Hemolytic urine syndrome (HUS) occurs in 3–5% of *E. coli* cases. This complication may occur several weeks after the first symptoms. HUS symptoms include temporary **anemia**, profuse bleeding, and kidney failure.

Food-borne illnesses are also caused by *Calicivirus*, which is also known as the Norwalk-like **virus**. This group of viruses is believed to spread from one person to another. An infected health service worker preparing a salad or sandwich could contaminate the food. According to CDC, infected fishermen contaminated oysters that they handled. These viruses are characterized by severe gastrointestinal illness. There is more vomiting than diarrhea, and the person usually recovers in two days.

The types and causes of food-borne illnesses have changed through the years. The pasteurization of milk, improved **water quality**, and safer canning techniques led to a reduction in the number of cases of common food-borne illness like typhoid fever, tuberculosis, and **cholera**. Cause of contemporary food-borne illnesses range from parasites living in imported food to food-processing techniques. An outbreak of a diarrheal illness in 1996 and 1997 was attributed to *Cyclsopora*, a parasite that contaminated raspberries grown in Guatemala. That led to 2,500 confirmed infection cases in 21 states.

Food may be contaminated during processing. For example, the meat contained in one hamburger may come from hundreds of animals, according to CDC. An *E. coli* outbreaks during the 1990s were linked to hamburgers purchased at fast food restaurants.

Technology in the form of **food irradiation** may eliminate the pathogens that cause food-borne disease. Advocates say that radiating food with gamma rays is effective and can be done when the food is packaged. Opponents say the process is dangerous and could produce the free radicals that cause **cancer**.

CDC ranks "raw foods of animal origin" as the foods most likely to be contaminated. This category includes meat, poultry, raw eggs, raw shellfish, and unpasteurized milk. Furthermore, raw fruit and vegetables could also pose a health risk. Vegetables fertilized by manure can also be contaminated by the **fertilizer**. CDC said that some outbreaks of food-borne illness were traced to unsanitary processing procedure. Water quality was crucial when washing vegetables, as was chilling the produce after it was harvested.

Of particular concern are alfalfa sprouts and other raw sprouts. These vegetables sprout in conditions that are favorable to microbes. Sprouts are eaten raw, and the microbes can grow into a large number of pathogens.

CDC advises consumers to thoroughly cook meat, poultry, and eggs. Produce should be washed. Leftovers should be chilled promptly. CDC is part of the United States **Public Health Service**. It researches and monitors health issues. Federal regulation of food safety is the responsibility of agencies such as the **Food and Drug Administration**, the United States Department of Agriculture, and the National Marine Fisheries Service.

[*Liz Swain*]

RESOURCES
BOOKS

Fox, Nichols. *It Was Probably Something You Ate*. New York: Penguin Books, 1999.

Hubbert, William, et al. *Food Safety and Quality Assurance*. Ames, IA: Iowa State University Press, 1996.

Satin, Morton. *Food Alert! The Ultimate Sourcebook for Food Safety*. New York: Checkmark Books, 1999.

ORGANIZATIONS
Centers for Disease Control and Prevention, 1600 Clifton Road, Atlanta, GA USA 30333 (404) 639-3311, Toll Free: (800) 311-3435, <http://www.cdc.gov>

Foot and mouth disease

Foot-and-mouth disease (FMD), also called hoof-and-mouth disease, is a highly contagious and economically devastating viral disease of cattle, swine, and other cloven-hoofed (split-toed) ruminants, including sheep, goats, and deer. The disease is highly contagious, for nearly 100% of exposed animals become infected, and it spreads rapidly through susceptible populations. Although there is no cure for FMD, it is seldom fatal, but it can kill very young animals.

The initial symptoms of the disease include fever and blister-like lesions (vesicles). The vesicles rupture into erosions on the tongue, in the mouth, on the teats, and between the hooves. Vesicles that rupture **discharge** clear or cloudy fluid and leave raw, eroded areas with ragged fragments of loose tissue. Erosions in the mouth result in excessive production of sticky, foamy, stringy saliva, which is a characteristic of FMD. Another characteristic symptom is lameness with reluctance to move. Other possible symptoms and effects of FMD include elevated temperatures in the early stages of the disease for two to three days, abortion, low conception rates, rapid weight loss, and drop in milk production. FMD lasts for two to three weeks, with most animals recovering within six months. However, it can leave some animals debilitated, thus causing severe losses in the production of meat and milk. Even cows that have recovered seldom produce milk at their original rates. Animals grown for meat do not usually regain lost weight for many months. FMD can also lead to myocarditis, which is an inflammation of the muscular walls of the heart, and death, especially in newborn animals. Infected animals can spread the disease throughout their lives so the only way to stop an outbreak is to destroy the animals.

The **virus** that causes the disease survives in lymph nodes and bone marrow at neutral **pH**. There are at least seven types and many subtypes of the FMD virus. The virus persists in contaminated fodder and in the **environment** for up to one month, depending on the temperature and pH. FMD thrives in dark, damp places, like barns, and can be destroyed with heat, sunlight, and disinfectants.

The disease is not likely to affect humans, either directly or indirectly through eating meat from an infected animal, but humans can spread the virus to animals. FMD can remain in human nasal passages for up to 28 hours. FMD viruses can be spread by other animals and materials to susceptible animals. The viruses can also be carried for several miles on the wind if environmental conditions are appropriate for virus survival. Specifically, an outbreak can occur when:

• people wearing contaminated clothes or footwear or using contaminated equipment pass the virus to susceptible animals

- animals carrying the virus are introduced into susceptible herds
- contaminated facilities are used to hold susceptible animals
- contaminated vehicles are used to move susceptible animals
- raw or improperly cooked **garbage** containing infected meat or animal products is fed to susceptible animals
- susceptible animals are exposed to contaminated hay, feedstuffs, or hides
- susceptible animals drink contaminated water
- a susceptible cow is inseminated by semen from an infected bull

Widespread throughout the world, FMD has been identified in Africa, South America, Middle East, Asia, and parts of Europe. North America, Central America, **Australia**, New Zealand, Chile, and some European countries are considered to be free of FMD. The United States has been free of FMD since 1929, when the last of nine outbreaks that occurred during the nineteenth and early twentieth centuries was eradicated.

In 2001, an FMD outbreak was confirmed in the United Kingdom, France, the Netherlands, the Republic of Ireland, Argentina, and Uruguay. Officials in the United Kingdom detected 2,030 cases of FMD, slaughtered almost four million animals, and took seven months to control the outbreak. The economic losses were estimated to be in the billions of pounds, and tourism in the affected countries was adversely affected. The outbreak was detected on February 20, 2001; no new cases were reported after September 30, 2001; on January 15, 2002, the British government declared the FMD outbreak to be over. The outbreak appeared to have started in a pig finishing unit in Northumberland, which was licensed to feed processed waste food. The disease appeared to have spread through two routes: through infected pigs who were sent to a slaughterhouse and through windborne spread to sheep on a nearby farm. These sheep entered the marketing chain and were sold in markets and through dealers, where they infected other sheep, people, and vehicles, spreading the FMD virus throughout England, Wales, southern Scotland. As the outbreak continued, cases were detected in other European countries.

The Animal and Plant Health Inspection Service (APHIS) of the United States Department of Agriculture (USDA) has developed a on-going comprehensive prevention program to protect American agriculture from FMD. APHIS continuously monitors for FMD cases worldwide. When FMD outbreaks are identified, APHIS initiates regulations that prohibit importation of live ruminants and swine and many animal products from the affected countries. In the 2001 outbreak in some **European Union** member countries, APHIS temporarily restricted importation of live ruminants and swine and their products from all European Union mem-

ber states. APHIS officials are on duty at all United States land and maritime ports-of-entry to ensure that passengers, luggage, cargo, and mail are checked for prohibited agricultural products or other materials that could carry FMD. The USDA Beagle Brigade, dogs trained to sniff out prohibited meat products and other contraband items, are also on duty at airports to check incoming flights and passengers.

The cooperation of private citizens is a crucial component of the protection program. APHIS prevents travelers entering the United States from bringing any agricultural products that could spread FMD and other harmful agricultural pests and diseases. Therefore passengers must declare all food items and other materials of plant or animal origin that they are carrying. Prohibited agricultural products that are found are confiscated and destroyed. Passengers must also report any visits to farms or livestock facilities. Failure to declare any items may result in delays and fines up to $1,000. Individuals traveling from countries that have been designated as FMD-affected must have shoes disinfected if they have visited farms, ranches, or other high risk areas, such as zoos, circuses, fairs, and other facilities and events where livestock and animals are exhibited.

APHIS recommends that travelers should shower and shampoo prior to and again after returning to the United States from an FMD-affected country. They should also launder or dry clean clothes before returning to the United States. Full-strength vinegar can be used by passengers to disinfect glasses, jewelry, watches, belts, hats, cell phones, hearing aids, camera bags, backpacks, and purses. If travelers had visited a farm or had any contact with livestock on their trip, they should avoid contact with livestock, **zoo** animals, or **wildlife** for five days after their return. Although dogs and cats cannot become infected with FMD, their feet, fur, and bedding should be cleaned of excessive dirt or mud. Pet bedding should not contain straw, hay, or other plant materials. The pet should be bathed as soon as it reaches its final destination and be kept away from all livestock for at least five days after entering the United States.

In the United States, animal producers and private veterinarians also monitor domestic livestock for symptoms of FMD. Their surveillance activities are supplemented with the work of 450 specially trained animal disease diagnosticians from federal, state, and military agencies. These diagnosticians are responsible for collecting and analyzing samples from animals suspected of FMD infection. If an outbreak were confirmed, APHIS would quickly try to identify infected and exposed animals, establish and maintain quarantines, destroy all infected and exposed animals using humane euthanization procedures as quickly as possible, and dispose of the carcasses by an approved method such as **incineration** or burial. After cleaning and disaffection of facilities where the infected animals were housed, the facility would be left vacant for several weeks. After this period, a few susceptible animals would be placed in the facility and observed for signs of FMD. A large area around the facility would be quarantined, where animal and human movement would be restricted or prohibited. In some cases, all susceptible animals within a two-mile radius would be also euthanized and disposed of properly by incineration or burial. In addition, APHIS has developed plans to pay affected producers the fair market value of their animals.

APHIS would consider vaccinating animals against FMD to enhance other eradication activities as well as to prevent spread to disease-free areas. However, vaccinated animals may still become infected and serve as a **reservoir** for the disease, even though they do not develop the disease symptoms themselves. Also for continued protective immunity, the vaccines must be given every few months.

APHIS is working with the U.S. Armed Forces to ensure that military vehicles and equipment are cleaned and disinfected before returning to the United States.

Preventing FMD from infecting and becoming established in an FMD-free area requires constant vigilance and a well-developed, thorough plan to control and eradicate any cases that might occur.

[*Judith L. Sims*]

RESOURCES
BOOKS

Haynes, N. Bruce, and Robert F. Kahrs. *Keeping Livestock Healthy.* North Adams, MA: Storey Communications, Inc., 2001.

OTHER

Animal and Plant Health Inspection Service, U.S. Department of Agriculture. *Foot-and-Mouth Disease.* [cited June 2002]. <http://www.aphis.usda.gov/oa/fmd>.

Department for Environment, Food and Rural Affairs, United Kingdom." *Origin of the UK Foot and Mouth Disease Epidemic in 2001.* June 2002 [cited June 2002].

"History of the World's Worst Ever Foot and Mouth Epidemic: United Kingdom 2001." [cited June 2002]. <http://www.pighealth.com/fmdoutbreaks.htm>.

Stephen Alfred Forbes (1844 – 1930)
American entomologist and naturalist

Stephen Alfred Forbes, an entomologist and naturalist, was the son of Isaac Sawyer and Agnes (Van Hoesen) Forbes. Forbes was born at Silver Creek, Ill. His father was a farmer, and died when Stephen was 10 years old. An older brother, Henry, then 21 years old, had been independent since he was 14, working his way toward a college education, but on his father's death he abandoned his career, took the burden of his father's family on his shoulders, and supported and

educated the children. He taught Stephen to read French, sent him to Beloit to prepare for college; and when the Civil War came he sold the farm and gave the proceeds (after the mortgage was paid) to his mother and sister for their support. Both brothers then joined the 7th Illinois Cavalry, Henry having retained enough money to buy horses for both. Stephen, enlisting at 17, was rapidly promoted, and at 20 became a captain in the regiment of which his brother ultimately became colonel. In 1862, while carrying dispatches, he was captured and held in a Confederate prison for four months. After liberation and three months in the hospital recuperating, he rejoined his regiment and served until the end of the war. He had learned to read Italian and Spanish in addition to French, before the war, and studied Greek while in prison.

He was a born naturalist. His farm life as a boy and his open-air life in the army intensified his interest in **nature**. After the close of the war, he began at once the study of medicine, entering the Rush Medical College where he nearly completed the course. His biographers have not as yet given the reason for the radical change in his plans which caused him to abandon medicine at this late stage in his education; but the writer has been told by his son, that it was "because of a series of incidents having to do mainly with operations without the use of anesthetics which convinced him that he was not temperamentally adapted to medical practice." His scientific interests, however, had been thoroughly aroused, and for several years while he taught school in southern Illinois, he carried on studies in natural history. In 1872 through the interest and influence of Dr. George Vasey, the well-known botanist, he was made curator of the Museum of State Natural History at Normal, Ill., and three years later was made instructor in zoology at the normal school. In 1877 the Illinois State Museum was established at Springfield; and the museum at Normal, becoming the property of the state, was made the Illinois State Laboratory of Natural History. Forbes was made its director. During these years he had been publishing the results of his researches rather extensively, and had gone into a most interesting and important line of investigation, namely the food of birds and fishes. He studied intensively the food of the different **species** of fish inhabiting Illinois waters and the food of the different birds. This study, of course, kept him close to entomology, and in 1884 he was appointed professor of zoology and entomology in the University of Illinois. The State Laboratory of Natural History was transferred to the university and in 1917 was renamed the Illinois Natural History Survey. He retained his position as chief, and held it up to the time of his death. He was appointed state entomologist in 1882 and served until 1917, when the position was merged in the survey. He retired from his teaching position as an emeritus professor in 1921. He served as dean of the College of Science of the university from 1888 to 1905.

All through his career Forbes had been publishing his writings actively. As early as 1895, Samuel Henshaw, in his *Bibliography of the more Important Contributions to American Economic Entomology*, listed 101 titles. It is said that his bibliography runs to more than 500 titles. And the range of these titles is extraordinary; they include papers on entomology, ornithology, **limnology**, ichthyology, **ecology**, and other phases of biology. All of his work was characterized by remarkable originality and depth of thought. Forbes was the first writer and teacher in America to stress the study of ecology, and thus began a movement which has gained great headway. He published 18 annual entomological reports, all of which have been models. He was the first and leading worker in America on hydrobiology. He studied the fresh-water organisms of the inland waters and was the first scientist to write on the **fauna** of the **Great Lakes**. His work on the food of fishes was pioneer work and has been of very great practical value. Forbes was a charter member of the American Association of Economic Entomologists and served twice as its president. He was also a charter member of the Illinois Academy of Science; a member of the **National Academy of Sciences** and of the American Philosophical Society; and in 1928 was made an honorary member of the Fourth International Congress of Entomology. Indiana University gave him the degree of Ph.D., in 1884, on examination and presentation of a thesis. He married, on December 25, 1873, Clara Shaw Gaston, whose death preceded his by only six months. A son, Dr. Ernest B. Forbes of State College, Pa., and three daughters survived him.

[*Leland Ossian Howard*]

RESOURCES
BOOKS

Croker, Robert A. *Stephen Forbes and the Rise of American Ecology.* Washington, DC: Smithsonian Institution Press, 2001.

PERIODICALS

Dictionary of American Biography. Base set. American Council of Learned Societies, 1928–1936. Reproduced in Biography Resource Center. Farmington Hills, MI: The Gale Group. 2002.
Schneider, Daniel W. "Local Knowledge, Environmental Politics, and the Founding of Ecology in the United States: Stephen Forbes and 'The Lake as a Microcosm' 1887." *Isis* 91, no. 4 (December 2000): 681–705.

Francois-Alphonse Forel (1841 – 1912)

Swiss professor of medicine

Francois-Alphonse Forel created a legacy by spending a lifetime studying a lake near which he lived. As the recog-

nized founder of the science of the study of lakes, **limnology** Forel meticulously observed Lake Geneva. His observations not only contributed directly to the modern-day ecological and environmental direction. He helped to uncover the mysteries of the *seiche*—movements caused by wind or air pressure and that occur in lakes.

Forel was born in Morges, Switzerland on February 2, 1841. As a professor of anatomy and physiology at the University of Geneva, (also known as Lake Le man) he spent most of his time investigating the life and movement of the lake near his home, world-famous Lake Geneva. To that purpose he created the word "limnology" referring to the study of lakes. On April 2, 1869, he found a nematode worm 40 meters down in the lake. That discovery was the beginning of his lifelong work. Before publishing his first major work, Forel recruited help in describing the types of algae and invertebrate animals he had found. A preliminary study he published in 1874, *La Faune Profonde des Lacs Suisses*, (The Bottom Fauna of Swiss Lakes), provided the basis for his later work.

When his book on the lake, *Le Le man*, was published Forel himself noted there that, "This book is called *Limnological Monography*. I have to explain this new word and apologize, if it would be necessary. The aim of my description is a part of the Earth, that's geography. But the geography of the ocean is called oceanography. But a lake, as big it is, is by not means an ocean; its limited area give it a special characteristic, which is very different of the one of the endless ocean. I had to find a more modest word to describe my work, like the word *limnography*. But, because a limnograph is a tool to measure the water level of lakes, I had to fabricate the new word **limnology**. The limnology is in fact the *oceanography of the lakes.*" A second book, *Handbuch der Seenkunde* (Handbook of Lake Studies) also became a standard work in the study of lakes. According to Ayer Company Publishers—the publisher that reprinted the work in 1978— Forel realized that his original work was an encyclopedia which might not hold the interest of students. He published his handbook to be used as the first textbook for the study of limnology.

Forel died on August 7, 1912. In 1970 his work was honored when the F.-A. Forel Institute was founded in his honor by Professor Jean-Pierr Vernet who directed the laboratory until 1995. The institute became a part of the Earth Sciences department of the University of Geneva in 1980, and plays a vital role in the International Commission for the Protection of Lake Geneva's water (CIPEL). The active research that occurs through the institute includes limnology, environmental geochemistry, **ecotoxicology**, and quaternary geology. It is also the home of the secretariat of the Center of Earth and Environmental Sciences Studies

(CESNE) and the Diplome d'etudes en sciences naturelles de l'environnement (DESNE).

One of the other ways in which Forel has been honored is the use of his name added to a device known as the **Forel-Ule Color Scale.** According to the Marine Educatin and Research Program at Occidental College, the scale is ued in marine studies, and consists of a series of numbered tubes (from 1 to 22) that contain various shades of colored fluids that range from blue (lowest numbers) through green, yellow, brown, and red (highest numbers).

[*Jane Spear*]

RESOURCES
BOOKS

Forel, Francois-Alphonse. *Le Leman.* (Reprint) Geneva: Slatkine Reprints, 1969.
Forel, Francois-Alphonse. *La Faune Profonde des Lacs Suisse.* (Reprint) Manchester, NH: Ayer Company Publishers.
Forel, Francois-Alphonse. *Handbuch der Seenkunde.* (Reprint) Manchester, NH: Ayer Company Publishers, 1978.

PERIODICALS

Korgen, Ben. *The Seiche Newsletter.* ldquo;Bonanza for Lake Superior,rdquo; February 2000.

OTHER

F/-A. Forel Institute. "Dr. F.-A. Forel." *History.* [cited July 2002]. <http://www.unige.ch/forel>.
Marine Education and Research Program, Occidental College. "Water Cooler." *Equipment and Methodology.* 2002 [cited July 2002]. <http://tethys.oxy.edu/MERP>.

ORGANIZATIONS

F.-A. Forel Institute, 10 route de Suise, Geneva, Switzerland +4122-950-92-10, Fax: +4122-755-13-82, Email: wyss@terre.unige.ch, www.unige.ch/forel

Dave Foreman (1946 –)
American radical environmental activist

Dave Foreman is a self-described radical environmentalist, Co-founder of **Earth First!**, and a leading defender of "monkey-wrenching" as a direct-action tactic to slow or stop **strip mining**, clear-cut **logging** of old-growth forests, the damming of wild rivers, and other environmentally destructive practices.

The son of a United States Air Force employee, Foreman traveled widely while growing up. In college he chaired the conservative Young Americans for Freedom and worked in the 1964 presidential election campaign of Senator Barry Goldwater. In the 1970s Foreman was a conservative Republican and moderate environmentalist who worked for the **Wilderness Society** in Washington, D.C. He came to believe that the **petrochemical**, logging, and mining interests

were "extremists" in their pursuit of profit, and that government agencies—the **Forest Service**, the **Bureau of Land Management**, the **U.S. Department of Agriculture**, and others—were "gutless" and unwilling or unable to stand up to wealthy and powerful interests intent upon profiting from the destruction of American **wilderness**. Well-meaning moderate organizations like the **Sierra Club**, **Friends of the Earth**, and the Wilderness Society were, with few exceptions, powerless to prevent the continuing destruction. What was needed, Foreman reasoned, was an immoderate and unrespectable band of radical environmentalists like those depicted in Edward Abbey's novel *The Monkey Wrench Gang* (1975) to take direct action against anyone who would destroy the wilderness in the name of "development." With several like-minded friends, Foreman founded Earth First!, whose motto is "No compromise in defense of Mother Earth."

From the beginning, Earth First! was unlike any other radical group. It did not issue manifestoes or publish position papers; it had "no officers, no bylaws or constitution, no incorporation, no tax status; just a collection of women and men committed to the Earth." Earth First!, Foreman wrote, "would be big enough to contain street poets and cowboy bar bouncers, agnostics and pagans, vegetarians and raw steak eaters, pacifists and those who think that turning the other cheek is a good way to get a sore face." Its weapons would include "monkey-wrenching," civil disobedience, music, "media stunts [to hold] the villains up to ridicule," and self-deprecating humor: "Radicals frequently verge toward a righteous seriousness. But we felt that if we couldn't laugh at ourselves we would be merely another bunch of dangerous fanatics who should be locked up (like the oil companies). Not only does humor preserve individual and group sanity, it retards hubris, a major cause of environmental rape, and it is also an effective weapon." But besides humor, Foreman called for "fire, passion, courage, and emotionalism...We [environmentalists] have been too reasonable, too calm, too understanding. It's time to get angry, to cry, to let rage flow at what the human **cancer** is doing to Mother Earth."

In 1987 Foreman published *Ecodefense: A Field Guide to Monkeywrenching*, in which he described in detail the tools and techniques of environmental sabotage or **monkey-wrenching**. These techniques included "spiking" old-growth **redwoods** and Douglas firs to prevent loggers from felling them; "munching" logging roads with nails; sabotaging bulldozers and other earth-moving equipment; pulling up surveyors' stakes; and toppling high-voltage power lines. These tactics, Foreman said, were aimed at property, not at people. But critics quickly charged that loggers' lives and jobs were endangered by tree-spiking and other techniques that could turn deadly. Moderate or mainstream environmental organizations joined in condemning the confrontational tactics favored by Foreman and Earth First!

In his autobiography *Confessions of an Eco-Warrior* (1991) Foreman defends monkey-wrenching as an unfortunate tactical necessity that has achieved its primary purpose of attracting the attention of the American people and the media to the destruction of the nation's remaining wilderness. It also attracted the attention of the FBI, whose agents arrested Foreman at his home in 1989 for allegedly financing and encouraging *ecoteurs* (ecological saboteurs) to topple high-voltage power poles. Foreman was put on trial to face felony charges, which he denied. The charges were questioned when it was disclosed that an FBI informant had infiltrated Earth First! with the intention of framing Foreman and discrediting the organization. In a plea bargain, Foreman pleaded guilty to a lesser charge and received a suspended sentence.

Foreman left Earth First! to found and direct The Wildlands Project in Tucson, Arizona. He continues to lecture and write about the protection of the wilderness.

[*Terence Ball*]

RESOURCES
BOOKS

Foreman, D. *Confessions of an Eco-Warrior.* New York: Harmony Books, 1991.

————. "Earth First!" In *Ideals and Ideologies: A Reader*, edited by T. Ball and R. Dagger. New York: Harper-Collins, 1991.

————, and B. Haywood. *Ecodefense: A Guide to Monkeywrenching.* Tucson: Ned Ludd Books, 1985.

List, P. C., ed. *Radical Environmentalism: Philosophy and Tactics.* Belmont, CA: Wadsworth, 1993.

Manes, C. *Green Rage: Radical Environmentalism and the Unmaking of Civilization.* Boston: Little, Brown, 1990.

Scarce, R. *Eco-Warriors: Understanding the Radical Environmental Movement.* Chicago: Noble Press, 1990.

Forest and Rangeland Renewable Resources Planning Act (1974)

The Forest and Rangeland Renewable Resources Planning Act (RPA) was passed in response to the growing tension between the timber industry and environmentalists in the late 1960s and the early 1970s. These tensions can be traced to increased controversy over and restrictions on timber harvesting on the national forests, due especially to **wilderness** designations and study areas and **clear-cutting**. These environmental restrictions, coupled with a dramatic increase in the price of timber in 1969, made Congress receptive to timber industry demands for a steadier supply of timber. Numerous bills addressing timber supply were introduced

and debated in Congress, but none passed due to strong environmental pressure. A task force appointed by President Richard Nixon, the President's Panel on Timber and the Environment, delivered its recommendations in 1973, but these were geared toward dramatically increased harvests from the national forests, and hence were also unacceptable to environmentalists.

One aspect of the various proposals that proved to be acceptable to all interested parties—the timber industry, environmentalists, and the Forest Service—was increased long-range resource planning. Senator Hubert Humphrey of Minnesota drafted a bill creating such a program and helped guide it to passage in Congress. RPA planning is based on a two-stage process, with a document accompanying each stage. The first stage is called the *Assessment*, which is an inventory of the nation's forest and range resources (public and private). The second stage, which is based on the *Assessment*, is referred to as the *Program*. Based on the completed inventory, the **Forest Service** provides a plan for the use and development of the available resources. The *Assessment* is to be done every 10 years. A *Program* based on the *Assessment* will be completed every five years. This planning was to be done by interdisciplinary teams and to incorporate widespread public involvement.

The RPA was quite popular with the Forest Service since the plans generated through the process gave the agency a solid foundation to base its budget requests on, increasing the likelihood of increased funding. This has proved to be successful, as the Forest Service budget increased dramatically in 1977, and the agency fared much better than other resource agencies in the late twentieth century.

The RPA was amended by the **National Forest Management Act** of 1976. Based on this law, in addition to the broad national planning mandated in the 1974 law, an *Assessment* and *Program* was required for each unit of the **national forest** system. This has allowed the Forest Service to use the plans to help shield itself from criticism. Since these plans address all uses of the forests, and make budget recommendations for these uses, if Congress does not fund these recommendations, the Forest Service can point to Congress as the culprit. However, the plans have also been more visible targets for interest group criticism.

Overall, the RPA has met with mixed results. The Forest Service has received increased funds and the planning process has been expanded to each national forest unit, but planning at such a scale is a difficult task. The act has also led to increased controversy and to increased bureaucracy. Perhaps most importantly, planning cannot solve a problem based on conflicting values, commodity use versus forest preservation, which is at the heart of **forest management** policy. *See also* Old-growth forest

[*Christopher McGrory Klyza*]

RESOURCES
BOOKS

Clary, D. A. *Timber and the Forest Service*. Lawrence: University Press of Kansas, 1986.

Dana, S. T., and S. K. Fairfax. *Forest and Range Policy*. 2nd ed. New York: McGraw-Hill, 1980.

Stairs, G. R., and T. E. Hamilton, eds. *The RPA Process: Moving Along the Learning Curve*. Durham, NC: Center for Resource and Environmental Policy Research, Duke University.

Forest decline

In recent decades there have been observations of widespread declines in vigor and **dieback** of mature forests in many parts of the world. In many cases, **pollution** may be a factor contributing to forest decline, for example in regions where **air quality** is poor because of acidic deposition or contamination with **ozone**, **sulfur dioxide**, **nitrogen** compounds, or metals. However, forest decline also occurs in some places where the air is not polluted, and in these cases it has been suggested that the phenomenon is natural.

Forest decline is characterized by a progressive, often rapid deterioration in the vigor of trees of one or several **species**, sometimes resulting in mass **mortality** (or dieback) within stands over a large area. Decline often selectively affects mature individuals, and is thought to be triggered by a particular stress or a combination of stressors, such as severe weather, **nutrient** deficiency, toxic substances in **soil**, and **air pollution**. According to this scenario, excessively stressed trees suffer a large decline in vigor. In this weakened condition, trees are relatively vulnerable to lethal attack by insects and **microbial pathogens**. Such secondary agents may not be so harmful to vigorous individuals, but they can cause the death of severely stressed trees.

The preceding is only a hypothetical etiology of forest dieback. It is important to realize that although the occurrence and characteristics of forest decline can be well documented, the primary environmental variable(s) that triggers the decline disease are not usually known. As a result, the etiology of the decline syndrome is often attributed to a vague but unsubstantiated combination of biotic and abiotic factors.

The symptoms of decline differ among tree species. Frequently observed effects include: (1) decreased productivity; (2) chlorosis, abnormal size or shape, and premature abscission of foliage; (3) a progressive death of branches that begins at the extremities and often causes a "stag-headed" appearance; (4) root dieback; (5) an increased frequency of secondary attack by fungal pathogens and defoliating or wood-boring insects; and (6) ultimately mortality, often as a stand-level dieback.

One of the best-known cases of an apparently natural forest decline, unrelated to human activities, is the widespread dieback of birches that occurred throughout the northeastern United States and eastern Canada from the 1930s to the 1950s. The most susceptible species were yellow (*Betula alleghaniensis*) and paper birch (*B. papyrifera*), which were affected over a vast area, often with extensive mortality. For example, in 1951 at the time of peak dieback in Maine, an estimated 67% of the birch trees had been killed. In spite of considerable research effort, a single primary cause has not been determined for birch dieback. It is known that a heavy mortality of fine roots usually preceded deterioration of the above-ground tree, but the environmental cause(s) of this effect are unknown, although deeply frozen soils caused by a sparse winter snow cover are suspected as being important. No biological agent was identified as a primary predisposing factor, although fungal pathogens and insects were observed to secondarily attack weakened trees and cause their death.

Another apparently natural forest decline is that of ohia (*Metrosideros polymorpha*), an **endemic species** of tree usually occurring in monospecific stands, that dominates the native forest of **Hawaiian Islands**. There are anecdotal accounts of events of widespread mortality of ohia extending back at least a century, but the phenomenon is probably more ancient than this. The most recent widespread decline began in the late 1960s and resulted in about 200 mi^2 (518 km^2) of forest with symptoms of ohia decline in a 1982 survey of 308 mi^2 (798 km^2). In most declining stands only the canopy individuals were affected. Understory saplings and seedlings were not in decline, and in fact were released from competitive stresses by dieback of the overstory.

An hypothesis to explain the cause of ohia decline has been advanced by D. Mueller-Dombois and co-workers, who believe that the stand-level dieback is caused by the phenomenon of "cohort senescence." This is a stage of the life history of ohia characterized by a simultaneously decreasing vigor in many individuals, occurring in old-growth stands. The development of senescence in individuals is governed by genetic factors, but the timing of its onset can be influenced by environmental stresses. The decline-susceptible, over mature, life history stage follows a more vigorous, younger, mature stage in an even-aged stand of individuals of the same generation (i.e., a cohort) that had initially established following a severe disturbance. In Hawaii, lava flows, events of deposition of volcanic ash, and hurricanes are natural disturbances that initiate **succession**. Sites disturbed in this way are colonized by a cohort of ohia individuals, which produce an even-aged stand. If there is no intervening catastrophic disturbance, the stand matures, then becomes senescent and enters a decline and dieback phase. The original stand is then replaced by another ohia forest comprised of an advance regeneration of individuals released from the understory. Therefore, according to the cohort senescence theory, the ohia dieback should be considered to be a characteristic of the natural population dynamics of the species.

Other forest declines are occurring in areas where the air is contaminated by various potentially toxic **chemicals**, and these cases might be triggered by air pollution. In North America, prominent declines have occurred in ponderosa pine (*Pinus ponderosa*), red spruce (*Picea rubens*) and sugar maple (*Acer saccharum*). In western Europe, Norway spruce (*Picea abies*) and beech (*Fagus sylvatica*) have been severely affected.

The primary cause of the decline of ponderosa pine in stands along the western slopes of the mountains of southern California is believed to be the toxic effects of ozone. Ponderosa pine is susceptible to the effects of this gas at the concentrations that are commonly encountered in the declining stands, and the symptomalogy of damage is fairly clear.

In the other cases of decline noted above that are putatively related to air pollution, the evidence so far is less convincing. The recent forest damage in Europe has been described as a "new" decline syndrome that may in some way be triggered by stresses associated with air pollution. Although the symptoms appear to be similar, the "new" decline is believed to be different from diebacks that are known to have occurred historically and are believed to have been natural. The modern decline syndrome was first noted in fir (*Abies alba*) in Germany in the early 1970s. In the early 1980s a larger-scale decline was apparent in Norway spruce, the most commercially-important species of tree in the region, and in the mid 1980s decline became apparent in beech and oak (*Quercus* spp.).

Decline of this type has been observed in countries throughout Europe, extending at least to western Russia. The decline has been most intensively studied in Germany, which has many severely damaged stands, although a widespread dieback has not yet occurred. Decline symptoms are variable in the German stands, but in general: (1) mature stands older than about 60 years tend to be more severely affected; (2) dominant individuals are relatively vulnerable; and (3) individuals located at or near the edge of the stand are more-severely affected, suggesting that a shielding effect may protect trees in the interior. Interestingly, epiphytic **lichens** often flourish in badly damaged stands, probably because of a greater availability of light and other resources caused by the diminished cover of tree foliage. In some respects this is a paradoxical observation, since lichens are usually hypersensitive to air pollution, especially toxic gases.

From the information that is available, it appears that the "new" forest decline in Europe is triggered by a variable combination of environmental stresses. The weakened trees then decline rapidly, and may die as a result of attack by secondary agents such as fungal disease or insect attack. Suggestions of the primary inducing factor include gaseous air pollutants, **acidification**, toxic metals in soil, nutrient imbalance, and a natural climatic effect, in particular **drought**. However, there is not yet a consensus as to which of these interacting factors is the primary trigger that induces forest decline in Europe, and it is possible that no single stress will prove to be the primary cause. In fact, there may be several "different" declines occurring simultaneously in different areas.

The declines of red spruce and sugar maple in eastern North America involve species that are long-lived and shade-tolerant, but shallow-rooted and susceptible to drought. The modern epidemic of decline in sugar maple began in the late 1970s and early 1980s, and has been most prominent in Quebec, Ontario, New York, and parts of New England. During the late 1980s and early 1990s, the decline appeared to reverse, and most stands became more healthy. The symptoms are similar to those described for an earlier dieback, and include abnormal coloration, size, shape, and premature abscission of foliage, death of branches from the top of the tree downward, reduced productivity, and death of trees. There is a frequent association with the pathogenic fungus *Armillaria mellea*, but this is believed to be a secondary agent that only attacks weakened trees. Many declining stands had recently been severely defoliated by the forest tent caterpillar (*Malacosoma disstria*), and many stands were tapped each spring for sap to produce maple sugar. Because the declining maple stands are located in a region subject to a high rate of **atmospheric deposition** of acidifying substances, this has been suggested as a possible predisposing factor, along with soil acidification and mobilization of available **aluminum**. Natural causes associated with **climate**, especially drought, have also been suggested. However, little is known about the modern sugar maple decline, apart from the fact that it occurred extensively; no conclusive statements can yet be made about its causal factor(s).

The stand-level dieback of red spruce has been most frequent in high-elevation sites of the northeastern United States, especially in upstate New York, New England, and the mid- and southern-Appalachian states. These sites are variously subject to acidic precipitation (mean annual **pH** about 4.0–4.1), to very acidic fog water (pH as low as 3.2–3.5), to large depositions of sulfur and nitrogen from the **atmosphere**, and to stresses from metal toxicity in acidic soil.

Declines of red spruce are anecdotally known from the 1870s and 1880s in the same general area where the modern decline is occurring. Up to one-half of the mature red spruce in the Adirondacks of New York was lost during that early episode of dieback, and there was also extensive damage in New England. As with the European forest decline, the "old" and "new" episodes appear to have similar symptoms, and it is possible that both occurrences are examples of the same kind of disease.

The hypotheses suggested to explain the initiation of the modern decline of red spruce are similar to those proposed for European forest decline. They include acidic deposition, soil acidification, aluminum toxicity, drought, winter injury exacerbated by insufficient hardiness due to nitrogen fertilization, **heavy metals** in soil, nutrient imbalance, and gaseous air pollution. Climate change, in particular a long term warming that has occurred subsequent to the end of the Little **Ice Age** in the early 1800s, may also be important.

At present, not enough is known about the etiology of the forest declines in Europe and eastern North America to allow an understanding of possible role(s) of air pollution and of natural environmental factors. This does not necessarily mean that air pollution is not involved. Rather, it suggests that more information is required before any conclusive statements can be made regarding the causes and effects of the phenomenon of forest decline. *See also* Forest management

[*Bill Freedman Ph.D.*]

RESOURCES
BOOKS

Barnard, J. E. "Changes in Forest Health and Productivity in the United States and Canada." In *Acidic Deposition: State of Science and Technology.* Vol. 3, *Terrestrial, Materials, Health, and Visibility Effects.* Washington DC: U. S. Government Printing Office, 1990.
Freedman, B. *Environmental Ecology.* Second Edition. San Diego: Academic Press, 1995.

PERIODICALS

Mueller-Dombois, D. "Natural Dieback in Forests." *Bioscience* 37 (1987): 575–583.

Forest management

The question of how forest resources should be used goes beyond the science of growing and harvesting trees; forest management must solve the problems of balancing economic, aesthetic, and biological value to entire ecosystems. The earliest forest managers in North America were native peoples, who harvested trees for building and burned forests to make room for grazing animals. But many native populations were wiped out by European diseases soon after Europeans arrived. By the mid-nineteenth century, it became apparent to many Americans that overharvesting of timber along with wasteful practices, such as uncontrolled burning of **logging** waste, was denuding forests and threatening future ecological and economic **stability**. The **Forest Service** (established in 1905) began studying ways to preserve

forest resources for their economic as well as aesthetic, recreational, and **wilderness** value.

From the 1600s to 1820s, 370 million acres (150 million ha) of forests—about 34% of the United States' total—were cleared, leaving about 730 million acres (296 million ha) today. Only 10–15% of the forests have never been cut. Many previously harvested areas, however, have been replanted and the annual growth now exceeds harvest overall. But the nature of the forests has been altered, many believe for the worse. If logging of old-growth forests were to continue at the rate maintained during the 1980s, all remaining unprotected stands would be gone by 2015. Some 33,000 timber-related jobs could also be lost during that time, not just from environmental protection but also from over-harvesting, increased mechanization, and increasing reliance on foreign processing of whole logs cut from private lands. Recent federal and court decisions, most notably to protect the **northern spotted owl** in the United States, have slowed the pace of old-growth harvesting and for now has put more old forests under protection. But the questions of how to use forest resources is still under fierce debate.

For decades, **clear-cutting** of tracts has been the standard forestry management practice. Favored by timber companies, clear-cutting takes virtually all material from a tract. But clear-cutting has come under increasing criticism from environmentalists, who point out that the practice replaces mixed-age, biologically diverse forests with single-age, single or few **species** plantings. Clear-cutting also relies heavily on roads to haul out timber, causing root damage, **topsoil erosion**, and **siltation** of streams. Industry standards such as "best management practices (BMPs)" prevent most erosion and siltation by keeping roads away from stream beds. But BMPs only address **water quality**. Clear-cutting also removes small trees, snags, boles, and woody debris that are important to invertebrates and **fungi**.

Rather than focusing on what is removed from a forest, sustainable forest management focuses on what is left behind. In **sustainable forestry**, tracts are never clear-cut: instead, individual trees are selected and removed to maintain diversity and health of the remaining **ecosystem**. Such methods avoid artificial replanting, herbicides, insecticides, and fertilizers. However, much debate remains on which trees and how many are chosen for harvesting under sustainable forestry.

In a new management style known most commonly as new forestry, 85–90% of trees on a site are harvested, and the land is left alone for decades to recover. Proponents say this method would cut down on erosion and increase diversity left behind on a tract, especially where one or two species dominate. The Forest Service and some Northwest states are studying new forestry, but environmentalists say too little is known about its effects on old-growth stands to use the practice. Timber companies say more and larger tracts would have to be harvested under new forestry to meet demand.

Those who make their living from America's forests, and those who place value on the biological ecosystems they support, must resolve the debate on how to best preserve our forests. One-third of forest resources now come from Forest Service lands, and the debate is an increasingly public one, involving interests ranging from the **Sierra Club** and sporting clubs to the **Environmental Protection Agency** (EPA), the Agriculture Department (and its largest agency, the Forest Service), and timber companies and their employees. The future of our forests depends on balancing real short-term needs with the high price of long-term forest health.

[*L. Carol Ritchie*]

RESOURCES
BOOKS

Lansky, M. *Beyond the Beauty Strip: Saving What's Left of Our Forests.* Gardiner, ME: Tilbury House Publishers, 1992.

PERIODICALS

Franklin, K. "Timber! The Forest Disservice." *The New Republic* 200 (January 2, 1989): 12–14.
Gillis, A. M. "The New Forestry: An Ecosystem Approach to Land Management." *BioScience* 40 (September 1990): 558–62.
McLean, H. E. "Paying the Price for Old Growth." *American Forests* 97 (September-October 1991): 22–25.
Steen, H. K. "Americans and Their Forests: A Love-Hate Story." *American Forests* 98 (September-October 1992): 18–20.

Forest Service

The **national forest** system in the United States must be considered one of the great success stories of the **conservation** movement. This remains true despite the continual controversies that seem to accompany administration of national forest lands by the United States Forest Service.

The roots of the Forest Service began with the appointment in 1876 of Franklin B. Hough as a "forestry agent" in the **U.S. Department of Agriculture** to gather information about the nation's forests. Ten years later, Bernhard E. Fernow was appointed chief of a fledgling Division of Forestry. Part way through Fernow's tenure, Congress passed the Forest Reserve Act of 1891, which authorized the president to withdraw lands from the public domain to establish federal forest reserves. The public lands were to be administered, however, by the General Land Office in the **U.S. Department of the Interior**.

Gifford Pinchot succeeded Fernow in 1899 and was Chief Forester when President **Theodore Roosevelt** approved the transfer of 63 million acres (25 million ha) of

forest reserves into the Department of Agriculture in 1905. That same year, the name of the Bureau of Forestry was changed to the United States Forest Service. Two years later, the reserves were redesignated national forests.

The Forest Service today is organized into four administrative levels: the office of the Chief Forester in Washington, D.C.; nine regional offices; 155 national forests; and 637 ranger districts. The Forest Service also administers twenty national **grasslands**. In addition, a research function is served by a forest products laboratory in Madison, Wisconsin, and eight other field research stations.

These lands are used for a wide variety of purposes and given official statutory status with the passage of the **Multiple Use-Sustained Yield Act** of 1960. That act officially listed five uses--timber, water, range, **wildlife**, and recreation--to be administered on national forest lands. **Wilderness** was later included. Forest Service now administers more than 34 million acres (14 million ha) in 387 units of the wilderness preservation system.

Despite a professionally trained staff and a sustained spirit of public service, Forest Service administration and management of national forest lands has been controversial from the beginning. The agency has experienced repeated attempts to transfer it to the Department of the Interior (or once to a new Department of Natural Resources); its authority to regulate grazing and timber use, including attempts to transfer national forest lands into private hands, has been frequently challenged and some of the Service's management policies have been the center of conflict. These policies have included **clear-cutting** and "subsidized" **logging**, various **recreation** uses, preservation of the **northern spotted owl**, and the cutting of old-growth forests in the Pacific Northwest.

[*Gerald L. Young Ph.D.*]

RESOURCES
BOOKS

Clary, D. A. *Timber and the Forest Service.* Lawrence: University Press of Kansas, 1986.

O'Toole, R. *Reforming the Forest Service.* Washington, DC: Island Press, 1988.

Steen, H. K. *The United States Forest Service: A History.* Seattle: University of Washington Press, 1976.

———. *The Origins of the National Forests.* Durham, NC: Forest History Society, 1992.

ORGANIZATIONS

Forest Service, U. S. Department of Agriculture, Sidney R. Yates Federal Building 201 14th Street, SW at Independence Ave., SW, Washington, D.C. USA 20250 Email: wo_fs-contact@fs.fed.us, <http://www.fs.fed.us>

Forestry Canada

see **Canadian Forest Service**

Forests

see **Coniferous forest; Deciduous forest; Hubbard Brook Experimental Forest; National forest; Old-growth forest; Rain forest; Taiga; Temperate rain forest; Tropical rain forest**

Dr. Dian Fossey (1932 – 1985)

American naturalist and primatologist

Dian Fossey is remembered by her fellow scientists as the world's foremost authority on mountain gorillas. But to the millions of **wildlife** conservationists who came to know Fossey through her articles and book, she will always be remembered as a martyr. Throughout the nearly 20 years she spent studying mountain gorillas in central Africa, the American primatologist tenaciously fought the poachers and bounty hunters who threatened to wipe out the endangered primates. She was brutally murdered at her research center in 1985 by what many believe was a vengeful poacher.

Fossey's dream of living in the wilds of Africa dates back to her lonely childhood in San Francisco. She was born in 1932, the only child of George, an insurance agent, and Kitty, a fashion model, (Kidd) Fossey. The Fosseys divorced when Dian was six years old. A year later, Kitty married a wealthy building contractor named Richard Price. Price was a strict disciplinarian who showed little affection for his stepdaughter. Although Fossey loved animals, she was allowed to have only a goldfish. When it died, she cried for a week.

Fossey began her college education at the University of California at Davis in the preveterinary medicine program. She excelled in writing and botany, she failed chemistry and physics. After two years, she transferred to San Jose State University, where she earned a bachelor of arts degree in occupational therapy in 1954. While in college, Fossey became a prize-winning equestrian. Her love of horses in 1955 drew her from California to Kentucky, where she directed the occupational therapy department at the Kosair Crippled Children's Hospital in Louisville.

Fossey's interest in Africa's gorillas was aroused through primatologist George Schaller's 1963 book, *The Mountain Gorilla: Ecology and Behavior.* Through Schaller's book, Fossey became acquainted with the largest and rarest of three subspecies of gorillas, *Gorilla gorilla beringei.* She learned that these giant apes make their home in the mountainous forests of Rwanda, Zaire, and Uganda. Males grow

up to 6 ft (1.8 m) tall and weigh 400 lb (182 kg) or more. Their arms span up to 8 ft (2.4 m). The smaller females weigh about 200 lb (91 kg).

Schaller's book inspired Fossey to travel to Africa to see the mountain gorillas in their homeland. Against her family's advice, she took out a three-year bank loan for $8,000 to finance the seven-week safari. While in Africa, Fossey met the celebrated paleoanthropologist Louis Leakey, who had encouraged **Jane Goodall** in her research of **chimpanzees** in Tanzania. Leakey was impressed by Fossey's plans to visit the mountain gorillas.

Those plans were nearly destroyed when she shattered her ankle on a fossil dig with Leakey. But just two weeks later, she hobbled on a walking stick up a mountain in the Congo (now Democratic Republic of the Congo) to her first encounter with the great apes. The sight of six gorillas set the course for her future. "I left Kabara (gorilla site) with reluctance but with never a doubt that I would, somehow, return to learn more about the gorillas of the misted mountains," Fossey wrote in her book, *Gorillas in the Mist.*

Her opportunity came three years later, when Leakey was visiting Louisville on a lecture tour. Fossey urged him to hire her to study the mountain gorillas. He agreed, if she would first undergo a preemptive appendectomy. Six weeks later, he told her the operation was unnecessary; he had only been testing her resolve. But it was too late. Fossey had already had her appendix removed.

The L.S.B. Leakey and the Wilkie Brothers foundations funded her research, along with the National Geographic Society. Fossey began her career in Africa with a brief visit to Jane Goodall in Tanzania to learn the best methods for studying primates and collecting data.

Fossey set up camp early in 1967 at the Kabara meadow in Zaire's Parc National des Virungas, where Schaller had conducted his pioneering research on mountain gorillas a few years earlier. The site was ideal for Fossey's research. Because Zaire's park system protected them against human intrusion, the gorillas showed little fear of Fossey's presence. Unfortunately, civil war in Zaire forced Fossey to abandon the site six months after she arrived.

She established her permanent research site September 24, 1967, on the slopes of the Virunga Mountains in the tiny country of Rwanda. She called it the Karisoke Research Centre, named after the neighboring Karisimbi and Visoke mountains in the Parc National des Volcans. Although Karisoke was just five miles from the first site, Fossey found a marked difference in Rwanda's gorillas. They had been harassed so often by poachers and cattle grazers that they initially rejected all her attempts to make contact.

Theoretically, the great apes were protected from such intrusion within the park. But the government of the impov-

**Dian Fossey photographing mountain gorillas
in Zaire.** (AP/Wide World Photos. Reproduced by permission.)

erished, densely populated country failed to enforce the park
rules. Native Batusi herdsmen used the park to trap antelope
and buffalo, sometimes inadvertently snaring a gorilla. Most
trapped gorillas escaped, but not without seriously mutilated
limbs that sometimes led to gangrene and death. Poachers
who caught gorillas could earn up to $200,000 for one by
selling the skeleton to a university and the hands to tourists.
From the start, Fossey's mission was to protect the endangered gorillas from extinction—indirectly, by researching
and writing about them, and directly, by destroying traps
and chastising poachers.

Fossey focused her studies on some 51 gorillas in four
family groups. Each group was dominated by a sexually
mature silverback, named for the characteristic gray hair on
its back. Younger, bachelor males served as guards for the
silverback's harem and their juvenile offspring.

When Fossey began observing the reclusive gorillas,
she followed the advice of earlier scientists by concealing
herself and watching from a distance. But she soon realized
that the only way she would be able to observe their behavior
as closely as she wanted was by "habituating" the gorillas to
her presence. She did so by mimicking their sounds and
behavior. She learned to imitate their belches that signal
contentment, their barks of curiosity, and a dozen other

sounds. To convince them she was their kind, Fossey pre-
tended to munch on the foliage that made up their diet.
Her tactics worked. One day early in 1970, Fossey made
history when a gorilla she called Peanuts reached out and
touched her hand. Fossey called it her most rewarding moment with the gorillas.

She endeared laymen to Peanuts and the other gorillas
she studied through her articles in National Geographic
magazine. The apes became almost human through her descriptions of their nurturing and playing. Her early articles
dispelled the myth that gorillas are vicious. In her 1971
National Geographic article she described the giant beasts as
ranking among "the gentlest animals, and the shiest." In
later articles, Fossey acknowledged a dark side to the gorillas.
Six of 38 infants born during a 13-year-period were victims
of infanticide. She speculated the practice was a silverback's
means of perpetuating his own lineage by killing another
male's offspring so he could mate with the victim's mother.

Three years into her study, Fossey realized she would
need a doctoral degree to continue receiving support for
Karisoke. She temporarily left Africa to enroll at Cambridge
University, where she earned her Ph.D. in zoology in 1974.
In 1977, Fossey suffered a tragedy that would permanently
alter her mission at Karisoke. Digit, a young male she had
grown to love, was slaughtered by poachers. Walter Cronkite
focused national attention on the gorillas' plight when he
reported Digit's death on the CBS Evening News. Interest
in gorilla **conservation** surged. Fossey took advantage of
that interest by establishing the Digit Fund, a non-profit
organization to raise money for anti-poaching patrols and
equipment.

Unfortunetly, the money wasn't enough to save the
gorillas from poachers. Six months later, a silverback and
his mate from one of Fossey's study groups were shot and
killed defending their three-year-old son, who had been shot
in the shoulder. The juvenile later died from his wounds. It
was rumored that the gorilla deaths caused Fossey to suffer
a nervous breakdown, although she denied it. What is clear
is that the deaths prompted her to step up her fight against
the Batusi poachers by terrorizing them and raiding their
villages. "She did everything short of murdering those poachers," Mary Smith, senior assistant editor at National Geographic, told contributor Cynthia Washam in an interview.
A serious calcium deficiency that causes bones to snap and
teeth to rot forced Fossey to leave Africa in 1980. She spent
her three-year sojourn as a visiting associate professor at
Cornell University. Fossey completed her book, *Gorillas in
the Mist,* during her stint at Cornell. It was published in
1983. Although some scientists criticized the book for its
abundance of anecdotes and lack of scientific discussion, lay
readers and reviewers received it warmly.

When Fossey returned to Karisoke in 1983, her scientific research was virtually abandoned. Funding had run dry. She was operating Karisoke with her own savings. "In the end, she became more of an animal activist than a scientist," Smith said. "Science kind of went out the window."

On Dec. 27, 1985, Fossey, 54, was found murdered in her bedroom at Karisoke, her skull split diagonally from her forehead to the corner of her mouth. Her murder remains a mystery that has prompted much speculation. Rwandan authorities jointly charged American research assistant Wayne McGuire, who discovered Fossey's body, and Emmanuel Rwelekana, a Rwandan tracker Fossey had fired several months earlier. McGuire maintains his innocence. At the urging of U.S. authorities, he left Rwanda before the charges against him were made public. He was convicted in absentia and sentenced to die before a firing squad if he ever returns to Rwanda.

Farley Mowat, the Canadian author of Fossey's biography, *Woman in the Mists*, believes McGuire was a scapegoat. He had no motive for killing her, Mowat wrote, and the evidence against him appeared contrived. Rwelekana's story will never be known. He was found dead after apparently hanging himself a few weeks after he was charged with the murder. Smith, and others, believe Fossey's death came at the hands of a vengeful poacher. "I feel she was killed by a poacher," Smith said. "It definitely wasn't any mysterious plot."

Fossey's final resting place is at Karisoke, surrounded by the remains of Digit and more than a dozen other gorillas she had buried. Her legacy lives on in the Virungas, as her followers have taken up her battle to protect the endangered mountain gorillas. The Dian Fossey Gorilla Fund, formerly the Digit Fund, finances scientific research at Karisoke and employs camp staff, trackers and anti-poaching patrols.

The Rwanda government, which for years had ignored Fossey's pleas to protect its mountain gorillas, on September 27, 1990, recognized her scientific achievement with the Ordre (sic) National des Grandes Lacs, the highest award it has ever given a foreigner. Gorillas in Rwanda are still threatened by cattle ranchers and hunters squeezing in on their **habitat**. According to the Colorado-based Dian Fossey Gorilla Fund, by the early 1990s, fewer than 650 mountain gorillas remained in Rwanda, Zaire, and Uganda. The Virunga Mountains is home to about 320 of them. Smith is among those convinced that the number would be much smaller if not for Fossey's 18 years of dedication to save the great apes. "Her conservation efforts stand above everything else (she accomplished at Karisoke)," Smith said. "She single-handedly saved the mountain gorillas."

[*Cynthia Washam*]

RESOURCES
BOOKS

Brower, Montgomery. "The Strange Death of Dian Fossey." *People*, February 17, 1986, 46–54.
Fossey, D. *Gorillas in the Mist.* Houghton Mifflin Company, 1983.
Hayes, H. T. P. *The Dark Romance of Dian Fossey.* Simon and Schuster, 1990.
Mowat, Farley. *Woman in the Mists.* Warner Books, 1987.
Schoumatoff, Alex. *African Madness.* Alfred A. Knopf, 1988.

Fossil fuels

In early societies, wood or other biological fuels were the main energy source. Today in many non-industrial societies, they continue to be used widely. Biological fuels may be seen as part of a solar economy where energy is extracted from the sun in a way that makes them renewable. However industrialization requires energy sources at much higher density and these have generally been met through the use of fossil fuels such as **coal**, gas, or oil. In the twentieth century a number of other options such as nuclear or higher density **renewable energy** sources (wind power, hydro-electric power, etc.) have also been available. Nevertheless fossil fuels represent the principal source of energy for most of the industrialized world.

Fossil fuels are types of sedimentary organic materials, often loosely called bitumens, with asphalt, a solid, and **petroleum**, the liquid form. More correctly bitumens are sedimentary organic materials that are soluble in **carbon** disulfide. It is this that distinguishes asphalt from coal, which is an organic material largely insoluble in carbon disulfide.

Petroleum can probably be produced from any kind of organism, but the fact that these sedimentary deposits are more frequent in marine sediments has suggested that oils arise from the fats and proteins in material deposited on the sea floor. These fats would be stable enough to survive the initial decay and burial but sufficiently reactive to undergo conversion to petroleum **hydrocarbons** at low temperature. Petroleum consists largely of paraffins or simple alkanes, with smaller amounts of napthenes. There are traces of aromatic compounds such as **benzene** present at the percent level in most crude oils. **Natural gas** is an abundant fossil fuel that consists largely of **methane** and ethane, although traces of higher alkanes are present. In the past, natural gas was regarded very much as a waste product of the petroleum industry and was simply burnt or flared off. Increasingly it is being seen as the favored fuel.

Coal, unlike petroleum, contains only a little **hydrogen**. Fossil evidence shows that coal is mostly derived from the burial of terrestrial vegetation with its high proportion of lignin and cellulose.

Most sediments contain some organic matter, and this can rise to many percent in shales. Here the organic matter can consist of both coals and bitumens. This organic material, often called sapropel, can be distilled to yield petroleum. Oil shales containing the sapropel kerogen are very good sources of petroleum. Shales are considered to have formed where organic matter was deposited along with fine grain sediments, perhaps in fjords, where restricted circulation keeps the oxygen concentrations low enough to prevent decay of the organic material.

Fossil fuels are mined or pumped from geological reservoirs where they have been stored for long periods of time. The more viscous fuels, such as heavy oils, can be quite difficult to extract and refine, which has meant that the latter half of the twentieth century has seen lighter oils being favored. However, in recent decades natural gas has been popular because it is easy to pipe and has a somewhat less damaging impact on the **environment**. These fossil fuel reserves, although large, are limited and non-renewable. The total recoverable light to medium oil reserves are estimated at about 1.6 trillion barrels, of which about a third has already been used. Natural gas reserves are estimated at the equivalent of 1.9 trillion barrels of oil and about a sixth has already been used. Heavy oil and bitumen amount to about 0.6 and 0.34 trillion barrels, most of which has remained unutilized. The gas and lighter oil reserves lie predominantly in the eastern hemisphere, which accounts for the enormous petroleum industries of the Middle East. The heavier oil and bitumen reserves lie mostly in the western hemisphere. These are more costly to use and have been for the most part untapped. Of the 7.6-trillion ton coal reserve, only 2.5% has been used. Almost two thirds of the available coal is shared between China, the former Soviet Union, and the United States.

Petroleum is not burnt in its crude form but must be refined, which is essentially a distillation process that splits the fuel into batches of different volatility. The lighter fuels are used in automobiles, with heavier fuels used as diesel and fuel oils. Modern refining can use chemical techniques in addition to distillation to help make up for changing demands in terms of fuel type and volatility.

The **combustion** of fuels represents an important source of air pollutants. Although the combustion process itself can lead to the production of pollutants such as carbon or **nitrogen oxides**, it has often been the trace impurities in fossil fuels that have been the greatest source of **air pollution**. Natural gas is a much favored fuel because it has only traces of impurities such as hydrogen sulfide. Many of these impurities are removed from the gas by relatively simple scrubbing techniques, before it is distributed. Oil is refined, so although it contains more impurities than natural gas, these become redistributed in the refining process. Sulfur compounds tend

to be found only in trace amounts in the light automotive fuels. Thus automobiles are only a minor source of **sulfur dioxide** in the **atmosphere**. Diesel oil can have as much as a percent of sulfur, and heavier fuel oils can have even more, so in some situations these can represent important sources of sulfur dioxide in the atmosphere. Oils also dissolve metals from the rocks in which they are present. Some of the organic compounds in oil have a high affinity for metals, most notably **nickel** and vanadium. Such metals can reach high concentration in oils, and refining will mean that most become concentrated in the heavier fuel oils. Combustion of fuel oil will yield ashes that contain substantial fractions of the trace metals present in the original oil. This means that an element like vanadium is a useful marker of fuel oil combustion.

Coal is often seen as the most polluting fuel because low grade coals can contain large quantities of ash, sulfur, and **chlorine**. However, it should be emphasized that the quantity of impurities in coal can vary widely, depending on where it is mined. The sulfur present in coal is found both as iron pyrites (inorganic) and bound up with organic matter. The **nitrogen** in coal is almost all organic nitrogen. Coal users are often careful to choose a fuel that meets their requirements in terms of the amount of ash, **smoke** or **pollution** risk it imposes. High rank coals such as anthracite have a high carbon content. They are mined in locations such as Pennsylvania and South Wales and contain little volatile matter and burn almost smokelessly. Much of the world's coal reserve is bituminous, which means that it contains about 20–25% volatile matter.

The fuel industry is often seen as responsible for pollutants and environmental risks that go beyond those produced by the combustion of its products. Mining and extraction processes result in **spoil** heaps, huge holes in open cast mining, and the potential for slumping of land (conventional mining). Petroleum refineries are large sources of hydrocarbons, although not usually the largest **anthropogenic** source of volatile organic compounds in the atmosphere. Refineries also release sulfur, carbon, and nitrogen oxides from the fuel that they burn. Liquid natural gas and **oil spills** are experienced both in the refining and transport of petroleum.

Being a solid, coal presents somewhat less risk when being transported, although wind-blown coal dust can cause localized problems. Coal is sometimes converted to coke or other refined products such as Coalite, a smokeless coal. These derivatives are less polluting, although much concern has been expressed about the pollution damage that occurs near the factories that manufacture them. Despite this, the conversion of coal to less polluting synthetic solid, and liquid and gaseous fuels would appear to offer much opportunity for the future.

One of the principal concerns about the current reliance on fossil fuels relates not so much to their limited supply, but more to the fact that combustion releases such large amounts of **carbon dioxide**. Our use of fossil fuels over the last century has increased the concentration of carbon dioxide in the atmosphere. Already there is mounting evidence that this has increased the temperature of the earth through an enhanced **greenhouse effect**.

[*Peter Brimblecombe*]

RESOURCES
BOOKS

Campbell, I. M. *Energy and the Atmosphere.* New York: Wiley, 1986.

Fossil water

Water that occurs in an **aquifer** or **zone of saturation** protected or isolated from the current **hydrologic cycle**. This water, because it is old, does not have the levels of **chemicals** or contaminants used in of our industrialized society, and its unblemished **nature** often makes it prized drinking water. In other cases, however, these aquifers are so isolated and the original condition of the water is so high in inorganic salts that scientists have suggested using them for waste disposal or containment areas. *See also* Drinking-water supply; Groundwater; Hazardous waste siting; Water quality

Four Corners

The Hopi believe that the Four Corners—where Colorado, Utah, New Mexico and Arizona meet—is the center of the universe and holds all life on Earth in balance. It also has some of the largest deposits of **coal, uranium,** and **oil shale** in the world. According to the **National Academy of Sciences** the Four Corners is a "national sacrifice area." This ancestral home of the Hopi and Dineh (Navajo) people is the center of the most intense energy development in the United States. Traditional grazing and farm land are being swallowed up by uranium mines, coal mines, and **power plants**.

The Four Corners, sometimes referred to as the "joint-use area," is comprised of 1.8 million acres (729,000 ha) of high **desert** plateau where Navajo sheep herders have grazed their flocks on idle Hopi land for generations. In 1972 Congress passed Public Law (PL) 93-531, which established the Navajo/Hopi Relocation Commission who had the power to enforce livestock reduction and the removal of over 10,000 traditional Navajo and Hopi, the largest forced relocation within the United States since the Japanese intern-

ment during World War II. Elders of both Nations issued a joint statement that officially opposed the relocation: "The traditional Hopi and Dineh (Navajo) realize that the so-called dispute is used as a disguise to remove both people from the JUA (joint use area), and for non-Indians to develop the land and mineral resources...Both the Hopi and Dineh agree that their ancestors lived in harmony, sharing land and prayers for more than four hundred years...and cooperation between us will remain unchanged."

The traditional Navajo and Hopi leaders have been replaced by Bureau of Indian Affairs (BIA) tribal councils. These councils, in association with the **U.S. Department of the Interior**, Peabody Coal, the Mormon Church, attorneys and public relation firms, created what is commonly known as the "Hopi-Navajo land dispute" to divide the joint-use area, so that the area could be opened up for energy development.

In 1964, 223 Southwest utility companies formed a consortium known as the Western Energy and Supply Transmission Associates (WEST) which includes water and power authorities on the West Coast as well as Four Corners area utility companies. WEST drafted plans for massive coal **surface mining** operations and six coal-fired, electricity-generating plants on Navajo and Hopi land. By 1966 John S. Boyden, attorney for the Bureau of Indian Affairs Hopi Tribal Council, secured lease arrangements with Peabody Coal to surface mine 58,000 acres (23,490 ha) of Hopi land and contracted WEST to build the power plants. This was done despite objections by the traditional Hopi leaders and the self-sufficient Navajo shepherds. Later that same year Kennecott **Copper**, owned in part by the Mormon Church, bought Peabody Coal. Peabody supplies the Four Corners' power plant with coal. The plant burned 5 million tons of coal a year which is the equivalent of ten tons per minute. It emits over 300 tons of **fly ash** and other particles into the San Juan River Valley every day. Since 1968 the coal mining operations and the power plant have extracted over 60 million gal (227 million l) of water a year from the Black Mesa **water table**, which has caused extreme **desertification** of the area, causing the ground in some areas to sink by up to 12 ft (3.6 m).

The worst nuclear accident in American history occurred at Church Rock, New Mexico, on July 26, 1979, when a Kerr-McGee uranium **tailings pond** spilled over into the Rio Puerco. The spill contaminated drinking water from Church Rock to the **Colorado River**, over 200 mi (322 km) to the west. The mill **tailings** dam broke—two months prior to the break cracks in the dam structure were detected yet repairs were never made—and discharged over 100 million gal (379 million l) of highly radioactive water directly into the Rio Puerco River. The main source of potable water for over 1,700 Navahoes was contaminated.

When Kerr-McGee abandoned the Shiprock site in 1980 they left behind 71 acres (29 ha) of "raw" uranium tailings, which retained 85% of the original **radioactivity** of the ore at the mining site. The tailings were at the edge of the San Juan River and have since contaminated communities located downstream.

What is the future of the Four Corners area, with its 100 plus uranium mines, uranium mills, five power plants, depleted **watershed** and radioactive contamination? One "solution" offered by the United States government is to zone the land into uranium mining and milling districts so as to forbid human habitation.

[*Debra Glidden*]

RESOURCES
BOOKS

Garrity, M. "The U.S. Colonial Empire Is As Close As the Nearest Reservation." In *Trilateralism: The Trilateral Commission and Elite Planning For World Management.* Boston: South End Press, 1980.

Kammer, J. *The Second Long Walk: The Navajo-Hopi Land Dispute.* Albuquerque: University of New Mexico Press, 1980.

Moskowitz, M. *Everybody's Business.* New York: Harper and Row, 1980.

Scudder, T., et al. *Expected Impacts of Compulsory Relocation on Navajos, with Special Emphasis on Relocation from the Former Joint Use Area Required by Public Law 93-531.* Binghamton, NY: Institute for Development of Anthropology, 1979.

PERIODICALS

Tso, H., and L. Shields. "Navajo Mining Operations: Early Hazards and Recent Interventions." *New Mexico Journal of Science* 20 (June 1980): 13.

Fox hunting

Fox **hunting** is the sport of mounted riders chasing a wild fox with a pack of hounds. The sport is also known as riding to the hounds, because the fox is pursued by horseback riders following the hounds that chase the fox. The specially trained hounds pursue the fox by following its scent. The riders are called the "field" and their leader is called the "master of the foxhounds.rdquo; A huntsman manages the pack of hounds.

Foxhunting originated in England, and dates back to the Middle Ages. People hunted foxes because they were predators that killed farm animals such as chickens and sheep. Rules were established reserving the hunt to royalty, the aristocracy (people given titles by royalty), and landowners. As the British Empire expanded, the English brought fox hunting to the lands they colonized. The first fox hunt in the United States was held in Maryland in 1650, according to the Masters of Foxhounds Association (MFHA), the organization that controls foxhunting in the United States.

Although the objective of most fox hunts is to kill the fox, some hunts do not involve any killing. In a drag hunt, hounds chase the scent of a fox on a trail prepared before the hunt. In the United States, a hunt ends successfully when the fox goes into a hole in the ground called the "earth."

In Great Britain, a campaign to outlaw fox hunting started in the late twentieth century. The subject drew strong debate on an issue that had not been resolved by March of 2002.

Organized hunt supporters like the Countryside Alliance said a hunting ban would result in the loss of 14,000 jobs. In addition, advocates said that hunts help to eliminate a rural threat by controlling the fox population. Hunt supporters described foxes as vermin, a category of destructive, disease-bearing animals. Hunting was seen as less cruel than other methods of eliminating foxes.

Opponents called foxhunting a "blood sport" that was cruel to animals. According to the International Fund for Animal Welfare (IAFW), more than 15,000 foxes are killed during the 200 hunts held each year. The IAFW reported that young dogs are trained to hunt by pitting them against fox cubs. The group wants the British Parliament to outlaw fox hunting. Drag hunting was suggested as a more humane alternative.

Another opposition group, the Nottingham Hunt Saboteurs Association, attempts to disrupt "blood sports" through "non-violent direct action." Saboteurs' methods include trying to distract hounds by laying false scent trails, shouting, and blowing horns.

Both supporters and opponents of fox hunting claim public support for their positions. In 1997, the issue of a ban was brought to Parliament, the British national legislative body consisting of the House of Lords and the House of Commons. That year, the Labour Party won the general election and proposed a bill to ban hunting with hounds. The following year, the bill passed through some legislative readings. However, time ran out before a final vote was taken.

In July 1999, Prime Minister Tony Blair promised to make fox hunting illegal. That November, Home Secretary Jack Straw called for an inquiry to study the effect of a ban on the rural economy. The Burns Inquiry concluded in June 2000 that a hunting ban would result in the loss of between 6,000 and 8,000 jobs. The inquiry did not find evidence that being chased was painful for foxes. However, the inquiry stated that foxes did not die immediately. This conclusion about a slower, painful death echoed opponents' charges that hunting was a cruel practice.

Two months before the inquiry was released, Straw proposed that lawmakers should have several options to vote on. One choice was a ban on fox hunting; another was to make no changes. A third option was to tighten fox hunting regulations.

In March 2002, Parliament cast non-binding opinion votes on the options. The House of Commons voted for a ban. The House of Lord voted for licensed hunting. After the vote, Rural Affairs Minister Alun Michael said that the government would try to find a common ground before trying to legislate fox hunting. The process of trying to reach agreement was expected to take six months at most.

Fox hunting in other countries

The Scottish Parliament banned hunting in February of 2002. The ban was to take effect on Aug, 1, 2002. The Countryside Alliance announced plans to legally challenge that ruling.

Fox hunting is legal in the following countries: Ireland, Belgium, Portugal, Italy, and Spain. Hunting with hounds is banned in Switzerland. In the United States, the MFHA was established in 1907. In March of 2002, the MFHA reported that there were 171 organized hunt clubs in North America.

[*Liz Swain*]

RESOURCES
BOOKS

Pool, Daniel. *What Jane Austen Ate and Charles Dickens Knew: From Fox Hunting to Whist—The Facts of Daily Life in Nineteenth-Century Enland.* Carmichael, CA: Touchstone Books, 1994.

Robards, Hugh J. *Foxhunting in England, Ireland, and North America.* Lanham, MD: Derrydale Press, 2000.

Thomas, Joseph B., and Mason Houghland. *Hounds and Hunting Through the Ages.* Lanham, MD: Derrydale Press, 2001.

ORGANIZATIONS

British Government Committee of Inquiry into Hunting with Dogs in England and Wales, , England Email: huntingwithdogs@defra.gsi.gov.uk, <http://www.huntinginquiry.gov.uk/mainsections/huntingframe.htm>

Countryside Alliance, The OldTown Hall, 367 Kensington Road, London SE11 4PT, England (011) 44-020-7840-9200, Fax: (011) 44-020-7793-8899, Email: info@countryside-alliance.org, <http://www.countryside-alliance.org>

International Fund for Animal Welfare, 411 Main Street, P.O. Box 193, Yarmouth Port, MA USA 02675 (508) 744-2000, Fax: (508) 744-2009, Toll Free: (800) 932-4329, Email: info@iafw.org, <http://www.iafw.org>

Masters of Foxhounds Association, Morven Park, P.O. Box 2420, Leesburg, VA USA 20177 Email: office@mfha.com, <http://www.mfha.com>

Nottingham Hunt Saboteurs, The Sumac Centre, 245 Gladstone Street, Nottingham NG7 6HX, England

Free riders

A free rider, in the broad sense of the term, is anyone who enjoys a benefit provided, probably unwittingly, by others. In the narrow sense, a free rider is someone who receives the benefits of a cooperative venture without contributing to the provision of those benefits. A person who does not participate in a cooperative effort to reduce **air pollution** by driving less, for instance, will still breathe cleaner air—and thus be a free rider—if the effort succeeds.

In this sense, free riders are a major concern of the theory of collective action. As developed by economists and social theorists, this theory rests on a distinction between private and public (or collective) goods. A public good differs from a private good because it is indivisible and nonrival. A public good, such as clean air or national defense, is indivisible because it cannot be divided among people the way food or money can. It is nonrival because one person's enjoyment of the good does not diminish anyone else's enjoyment of it. Smith and Jones may be rivals in their desire to win a prize, but they cannot be rivals in their desire to breathe clean air, for Smith's breathing clean air will not deprive Jones of an equal chance to do the same.

Problems arise when a public good requires the cooperation of many people, as in a campaign to reduce **pollution** or conserve resources. In such cases, individuals have little reason to cooperate, especially when cooperation is burdensome. After all, one person's contribution—using less **gasoline** or electricity, for example—will make no real difference to the success or failure of the campaign, but it will be a hardship for that person. So the rational course of action is to try to be a free rider who enjoys the benefits of the cooperative effort without bearing its burdens. If everybody tries to be a free rider, however, no one will cooperate and the public good will not be provided. If people are to prevent this from happening, some way of providing selective or individual incentives must be found, either by rewarding people for cooperating or punishing them for failing to cooperate.

The free rider problem posed by public goods helps to illuminate many social and political difficulties, not the least of which are environmental concerns. It may explain why voluntary campaigns to reduce driving and to cut energy use so often fail, for example. As formulated in Garrett Hardin's **Tragedy of the Commons**, moreover, collective action theory accounts for the tendency to use common resources—grazing land, fishing banks, perhaps the earth itself—beyond their **carrying capacity**. The solution, as Hardin puts it, is "mutual coercion, mutually agreed upon" to prevent the overuse and destruction of vital resources. Without such action, the desire to ride free may lead to irreparable ecological damage.

[*Richard K. Dagger*]

RESOURCES
BOOKS

Hardin, R. *Collective Action.* Baltimore: Johns Hopkins University Press, 1982.

Olson, M. *The Logic of Collective Action.* New York: Schocken Books, 1971.

PERIODICALS

Hardin, G. "The Tragedy of the Commons." *Science* 162 (December 13, 1968): 1243–48.

Freon

The generic name for several **chlorofluorocarbons** (CFCs) widely used in refrigerators and air conditioners, including the systems in houses and cars. Freon—comprised of **chlorine**, fluorine, and **carbon** atoms—is a non-toxic gas at room temperature. It is environmentally significant because it is extremely long-lived in the **atmosphere**, with a typical **residence time** of 70 years. This long life-span permits CFCs to disperse, ultimately reaching the **stratosphere** 19 mi (30 km) above the earth's surface. Here, high energy photons in sunlight break down freon, and chlorine atoms liberated during this process participate in other chemical reactions that consume **ozone**. The final result is to decrease the stratospheric ozone layer that shields the earth from damaging **ultraviolet radiation**. Under the 1987 Montreal Protocol, 31 industrialized countries agreed to phase out CFC freon production. Freon substitutes use **bromine** atoms to replace the chlorine atoms, providing a substitute refrigeration compound that appears less damaging, although considerably more expensive and less energy efficient.

Fresh water ecology

The study of fresh water habitats is called **limnology**, coming from the Greek word *limnos*, meaning "pool, lake, or swamp." Fresh water habitats are normally divided into two groups: the study of standing bodies of water such as lakes and ponds (called lentic ecosystems) and the study of rivers, streams, and other moving sources of water (called lotic ecosystems). Another important area that should be included is fresh water **wetlands**.

The historical roots of limnology go back to F. A. Forel, who studied Lake Geneva, Switzerland, in the late 1800s and E. A. Birge and C. Juday, who studied lakes in Wisconsin in the early 1900s. More recently, the modern "father" of limnology can arguably be attributed to G. Evelyn Hutchinson, who died in 1991 after teaching at Yale University for more than 40 years. Among his prolific writings are four treatises on limnology which offer the most detailed descriptions of lakes that have been published.

Fresh water **ecology** is an intriguing field because of the great diversity of aquatic habitats. For example, lakes can be formed in different ways: volcanic origin such as Crater Lake, Oregon; tectonic (earth movement) origin like **Lake Tahoe** in California/Nevada and **Lake Baikal** in Siberia; glacially-derived lakes like the **Great Lakes** or smaller kettle hole or cirque lakes; oxbow lakes which form as rivers change their meandering paths; and human-created reservoirs. Aquatic habitats are strongly influenced by the surrounding **watershed**, and lakes in the same geographic area tend to be of the same origin and have similar **water quality**.

Lakes are characteristically non-homogeneous. Physical, chemical, and biological factors contribute to both horizontal and vertical zonations. For example, light penetration creates an upper photic (lighted) zone and a deeper aphotic (unlit) zone. **Phytoplankton** (microscopic algae such as diatoms, desmids, and filamentous algae) inhabit the **photic zone** and produce oxygen through **photosynthesis**. This creates an upper productive area called the trophogenic (productive) zone. The deeper area where **respiration** prevails is called the tropholytic (unproductive) zone. **Zooplankton** (microscopic invertebrates such as cladocerans, copepods, and rotifers) and **nekton** (free-swimming animals such as fish) inhabit both of these zones. The boundary area where oxygen produced from photosynthesis equals that consumed through respiration is called the compensation depth. Near-shore areas where light penetrates to the bottom and aquatic macrophytes such as cattails and bulrushes grow, are called the **littoral zone**. This is typically the most productive area, and it is more pronounced in ponds than in lakes. Open water areas are called the limnetic zone, where most **plankton** inhabit. Some **species** of zooplankton are found more concentrated in deeper waters during the day and in greater numbers in the upper waters during the night. One explanation for this vertical **migration** is that these zooplankton, often large and sometimes pigmented, are avoiding visually-feeding planktivorous fish. These zooplankton are thus able to feed on phytoplankton in the trophogenic zone during periods of darkness, and then swim to deeper waters during daylight hours. Phytoplankton are also adapted for existence in the limnetic zone. Some species are quite small in size (less than 20 microns in diameter and called nannoplankton), allowing them to be competitive at **nutrient** uptake due to their high surface-to-volume ratio. Other groups form large colonies, often with spines, lessening the negative impacts of herbivory and sinking. Blue-green algae produce oils that help them float on or near the water's surface. Some are able to fix atmospheric **nitrogen** (called **nitrogen fixation**), giving them a competitive advantage in low-nitrogen conditions. Other species of blue-greens produce toxic **chemicals**, making them inedible to herbivores. There have even been reports of cattle deaths following ingestion of water with dense growths of these algae.

Lakes can be isothermal (uniform temperature from top to bottom) during some times of the year, but during the summer months they are typically thermally stratified with an upper, warm layer called the epilimnion (upper lake), and a colder, deeper layer called the hypolimnion (lower lake). These zones are separated by the metalimnion, which is determined by the depths with a temperature change of more than 1°C per meter depth, called the **thermocline**. The summer temperature **stratification** creates a density gradient that effectively prevents mixing between zones.

Wind is another physical factor that influences aquatic habitats, particularly in lakes with broad surface areas exposed to the main direction of the wind, called fetch. Strong winds can produce internal standing waves called seiches that create a rocking motion in the water once the wind dies down. Other types of wind-generated water movements include Ekman spirals and Langmuir cells. Deep or chemically-stratified lakes that never completely mix are called meromictic. Lakes in tropical regions that mix several times a year are called polymictic. In regions with severe winters resulting in ice covering the surface of the lake, mixing normally occurs only during the spring and fall when the water is isothermal. These lakes are called dimictic, and the mixing process is called overturn. People living downwind of these lakes often notice a rotten egg smell caused by the release of **hydrogen** sulfide. This gas is a product of benthic (bottom-dwelling) bacteria that inhabit the **anaerobic** muds of productive lakes (discussed below in more detail).

Lakes that receive a low-nutrient input remain fairly unproductive, and are called **oligotrophic** (low nourished). These lakes typically have low concentrations of phytoplankton, with diatoms being the main representative. Moderately productive lakes are called mesotrophic. Eutrophic (well nourished) lakes receive more nutrient input and are therefore more productive. They are typically shallower than oligotrophic lakes and have more accumulated bottom sediments that often experience summer anoxia. These lakes have an abundance of algae, particularly blue-greens (Cynaobacteria) which are often considered nuisance algae because they float on the surface and out-compete the other algae for nutrients and light.

Most lakes naturally age and become more productive over time; however, large, deep lakes may remain oligotrophic. The maturing process is called eutrophication and it is regulated by the input of nutrients which are needed by the algae for growth. Definitive limnological studies done in the 1970s concluded that **phosphorus** is the key limiting nutrient in most lakes. Thus, the accelerated input of this chemical into streams, rivers, and eventually lakes by excess fertilization, sewage input (both human and animal), and **erosion** is called **cultural eutrophication**. Much debate and research has been spent on how to slow down or control this process. One interesting management tool is called biomanipulation, in which piscivorous (fish-eating) fish are added to lakes to consume planktivorous (plankton-eating) fish. Because the planktivores are visual feeders on the largest prey, this allows higher numbers of large zooplankton to

thrive in the water, which consume more phytoplankton, particularly non-toxic blue-green algae. A more practical approach to controlling cultural eutrophication is by limiting the nutrient **loading** into our bodies of water. Although this isn't easy, we must consider ways of limiting excessive uses of fertilizers, both at home and on farms, as well as more effectively regulating the release of treated sewage into rivers and lakes, particularly those which are vulnerable to eutrophication. Another lake management tool is to aerate the bottom (hypolimnetic) water so that it remains oxygenated. This keeps iron in the oxidized state (Fe^{+3}), which chemically binds with phosphate (PO_4) and prevents it from being available for algal uptake. Lakes that have anoxic bottom water keep iron in the reduced state (Fe^{+2}), and phosphate is released from the **sediment** into the water. Fall and spring turnover then returns this limiting nutrient to the photic zone, promoting high algal growth. When these organisms eventually die and sink to the bottom, **decomposers** use up more oxygen, and we get a "snow ball" effect. Thus, productive lakes can become more eutrophic with time, and may eventually develop into hypertrophic (overly productive) systems.

Lotic ecosystems differ from lakes and ponds in that currents are more of a factor and primary production inputs are generally external (allochthonous) instead of internal (autochthonous). Thus, a river or stream is considered a heterotrophic **ecosystem** along most of its length. Gradients in physical and chemical parameters also tend to be more horizontal than vertical in running water habitats and organisms living in lotic ecosystems are specially adapted for surviving in these conditions. For example, trout require higher amounts of **dissolved oxygen**, and are primarily found in relatively cold, fast-moving water with low nutrient input. Carp are able to tolerate warmer, slower, more productive bodies of running water. Darters are fish that quickly dart back and forth behind rocks in the bottom of fast-moving streams and rivers as they feed on aquatic insects. Many of these insects are shredders and **detritivores** on the organic material like leaves that enter the water. Other groups specialize by scraping algae and bacteria off rocks in the water.

Recently, ecologists have begun to take a greater interest in studying fresh water wetlands. These areas are defined as being inundated or saturated by surface or ground water for most of the year, and therefore having characteristic wetland vegetation. Although some people consider these areas a nuisance and prefer them being drained, ecologists realize that they are valuable habitats for migrating water fowl. They also serve as major "adsorptive" areas for nutrients, which are particularly useful around sewage lagoons. We must therefore take greater care in preserving these habitats.

[*John Korstad*]

RESOURCES
BOOKS

Horne, A. J., and C. R. Goldman. *Limnology*. New York: McGraw Hill, 1994.
Hutchinson, G.E. *A Treatise on Limnology*. vol. 1: *Geography, Physics, Chemistry*; vol. 2: *Introduction to Lake Biology and Limnoplankton*; vol. 3: *Limnological Botany*; vol. 4: *The Zoobenthos*. New York: Wiley, 1993.
Smith, R. L. *Ecology and Field Biology*. 4th ed. New York: Harper Collins, 1996.
Wetzel, R. *Limnology*. 2nd ed. Philadelphia: Saunders, 1983.

Friends of the Earth

Friends of the Earth (FOE) is a public interest environmental group committed to the **conservation**, restoration, and rational use of the **environment**. Founded by David Brower and other militant environmentalists in San Francisco in 1969, FOE works on the local, national, and international levels to prevent and reverse **environmental degradation**, and to promote the wise use of **natural resources**.

FOE has an international membership of one million. Its particular areas of interest include **ozone layer depletion, greenhouse effect**, toxic chemical safety, **coal** mining, coastal and ocean **pollution**, the destruction of tropical forests, **groundwater** contamination, corporate accountability, and **nuclear weapons** production. In addition to its efforts to influence policy and increase public awareness of environmental issues, FOE's ongoing activities include the operation of the Take Back the Coast Project and the administration of the Oceanic Society. Over the years, FOE has published numerous books and reports on various topics of concern to environmentalists.

FOE was originally organized to operate internationally and now has national organizations in some 63 countries. In several of these, most notably the United Kingdom, FOE is considered to be the best-known and most effective **public interest group** concerned with environmental issues.

The organization has changed its strategies considerably over the years, and not without considerable controversy within its own ranks. Under Brower's leadership, FOE's tactics were media-oriented and often confrontational, sometimes taking the form of direct political protests, boycotts, sit-ins, marches, and demonstrations. Taking a **holistic approach** to the environment, the group argued that fundamental social change was required for lasting solutions to many environmental problems.

FOE eventually moved away from confrontational tactics and towards a new emphasis on lobbying and legislation, which helped provoke the departure of Brower and some of the group's more radical members. FOE began downplaying several of its more controversial stances (for example, on the control of nuclear weapons) and moved its headquarters from San Francisco to Washington, D.C. More recent controversies have concerned FOE's endorsement of so-called **green products** and its acceptance of corporate financial contributions.

FOE remains committed, however, to most of its original goals, even if it has foresworn its earlier illegal and disruptive tactics. Relying more on the technical competence of its staff and the technical rationality of its arguments than on idealism, FOE has been highly successful in influencing legislation and in creating networks of environmental, consumer, and human rights organizations worldwide. Its publications and educational campaigns have been quite effective in raising public consciousness of many of the issues with which FOE is concerned.

[*Lawrence J. Biskowski*]

RESOURCES
ORGANIZATIONS
Friends of the Earth, 1025 Vermont Ave. NW, Washington, D.C. USA 20005 (202) 783-7400, Fax: (202) 783-0444, Email: foe@foe.org, <http://www.foe.org>

Frogs

Frogs are amphibians belonging to the order anura. The anuran group has nearly 2,700 **species** throughout the world and includes both frogs and toads. The word *anura* means "without a tail," and the term applies to most adult frogs. The anura are distinguished from tailed amphibians (urodeles) such as salamanders because the latter retain a tail as an adult.

One of the most studied and best understood frogs is the northern leopard frog, *Rana pipiens*. This species is well-known to most children, to people who love the outdoors, and to scientists. Leopard frogs live throughout much of the United States as well as Canada and northern Mexico. Inhabiting a diverse array of environmental conditions, the order anura exhibits an impressive display of anatomical and behavioral variations among its members. Despite such diversity, the leopard frog is often used as a model that represents all members of the group.

Leopard frogs mate in the early spring. The frogs deposit their eggs in jelly-like masses. These soft formless clumps may be seen in temporary ponds where frogs are common. Early embryonic development occurs within the jelly mass after which the eggs hatch releasing small swimming tadpoles. The tadpoles feed on algal periphyton, fine organic **detritus**, and yolk reserves through much of the spring and early summer. Next, metamorphosis begins a few weeks or up to two years after the eggs is hatched depending of the species. Metamorphosis is the process whereby amphibious tadpoles lose their gills and tails and develop arms and legs. The process of metamorphosis is complex and involves not only the loss of the tail and the development of limbs, but also a fundamental reorganization of the gut. For example, in the leopard frog, the relatively long intestine of the vegetarian tadpole is reorganized to form the short intestine of the carnivorous adult frog because nutrients are more difficult to extract from plant sources. Additionally, metamorphosis profoundly changes the method of **respiration** in the leopard frog. As the animal loses it gills, air-breathing lungs are formed and become functional. A significant portion of respiration and gas exchange will occur through the skin of the adult frog. When metamorphosis of a tadpole is complete, the result is a terrestrial, insect-eating, air-breathing frog.

Frogs are important tolls for biological learning and research. Many students first encounter vertebrate anatomy with the dissection of an adult leopard frog. Consequently physiologists have used frogs for the study of muscle contraction and the circulation of blood which is easily seen in the webbing between the toes of a frog. Embryologists have used frogs for study because they lay an abundance of eggs. A mature *R. pipiens* female may release 3,000 or more eggs during a single spawning. Frog eggs are relatively large and abundant which simplifies manipulation and experimentation. Another anuran, the South African clawed frog, *Xenopus laevis*, is useful in research because it can be cultivated to metamorphosis easily and can be bred any time of year. For these reasons, frogs emerge as extremely useful laboratory test animals that provide valuable information about human biology.

Frogs and biomedical research

Many biological discoveries have been made or enhanced by using frogs. For example, the role of sperm in development was studied in the late 1700s in frogs. Amphibians in general, including frogs, have significantly more deoxyribonucleic **acid** (DNA) per cell than do other chordates. Thus, their chromosomes are large and easy to see with a microscope. In addition, induced ovulation in frogs by pituitary injection was developed during the early 1930s. Early endocrinologists studied the role of the hormone thyroxine (also found in human beings) in vertebrate development in 1912. The role of viruses in animal and human **cancer** is receiving renewed interest—the first herpes **virus** known to cause a cancer was the frog cancer herpes virus. Furthermore, mammalian and human **cloning**, a controversial topic, has its foundations in the cloning of frogs. The first vertebrate ever cloned was *Rana pipiens*, in an experiment published by Thomas King and Robert Briggs in 1952. As such, experimentation with frogs has contributed greatly to biomedical research.

Unfortunately, the future of amphibians, like the northern leopard frog, appears to be jeopardized. Most amphibians in the world are frogs and toads. Since the 1980s, scientists have noted a distinct decline in amphibian populations. Worldwide, over 200 species of amphibians have experienced recent population declines. At least 32 documented species extinctions have occurred. Of the 242 native North American amphibian species, the U.S. Nature Conservancy has identified three species that are presumed to be extinct, another three classified as possibly extinct, with an additional 38% categorized as vulnerable to **extinction**. According to the United States **Fish and Wildlife Service**, there are four frog species listed as endangered and three listed as threatened.

The actual cause of the reductions in frog and amphibian populations remains unclear, but many well-supported hypotheses exist. One speculation is the run-off of **chemicals** that poison ponds producing defective frogs that cannot survive well. A second possibility is an increase in amphibian disease caused by virulent pathogens. Viral and fungal infection of frogs has led to recent declines in many populations. A third explanation involves parasitic infections of frogs with flatworms, causing decreased survival. Another theory blames atmospheric **ozone** depletion for frog decline. Yet another implicates a toxic combination of intensified agriculture, **drainage** of **habitat**, and predator changes as the cause for the drastic declines in frog populations. While each argument has merits of its own, it is unlikely that any single cause can adequately explain all amphibian decline.

The poisoned pond hypothesis brought the issue of frog population decline to the forefront. Several years ago, students on a field trip at a pond near the Minnesota community of Henderson discovered a number of frogs with extremely abnormal limbs. Some frogs had three or more hind legs, some had fused appendages, and some had no legs at all. Concerned with what they had found, their teacher, Cindy Reinitz, and her students contacted the Minnesota **Pollution Control** Agency. Soon, state agency biologists and officials from the University of Minnesota confirmed the presence of many abnormal frogs at the site. The word spread, and by late 1996, many sites in Minnesota were known to have abnormal frogs.

Frog developmental abnormalities are not new to science. What made the Minnesota case of frog malformations different, however, was the extraordinary concentration of abnormal frogs in an unusually large number of locations. Concern became more urgent when abnormal animals also were reported in several other states, then Canada, and finally in Japan. Many of the abnormal frogs were found in agricultural areas where large amounts of fertilizers and pesticides were used.

One reason for concern is that frogs and humans metabolize toxic **xenobiotic** chemicals, such as fertilizers or pesticides, in similar ways. Also, human beings and frogs have very similar early developmental stages, which is why frogs are used as model animals in embryology. Because of this, worry exists regarding the potential for human developmental abnormalities from such chemicals. Some of the chemicals implicated in the frog malformations were retinoids. Retinoids are derivatives of vitamin A that are potent and crucial hormonal regulators of vertebrate embryological development. Numerous laboratory studies using retinoic acid (a form of retinoid) have reproduced the abnormalities seen in the Minnesota frogs, and the role of retinoids in the regulation of genes involved in limb development is well-characterized. Frog anatomical defects are regarded as a warning sign for potential problems in humans exposed to the same chemicals, since retinoic acid regulates human limb development as well.

Disease is a growing problem for frogs. Recently, two important frog pathogens have been identified and are suspected to play a role in global frog decline. Iridiovirus, a virus that commonly infects fish and insects, has now been found to infect frogs. An aquatic fungus, chytrid, is also implicated in frog decline. Infection with the fungus is called chytridiomycosis. The chytrid fungus, *Batrachochytrium dendrobatidis*, which normally resides in decaying plant material, causes the degeneration of tadpole mouthparts. This results in the death of post-metamorphic frogs, and is reportedly involved in amphibian decline in **Australia**, Europe, and recently in the Sierra Nevada region of California. In Australia the depletion of the frog population may have been caused by a disease introduced through infected fish. The Australian cane toad, *Bufo marinus*, may be responsible for spreading a virus to other frogs outside its native range.

Another potential cause for the striking decline of frog populations involves parasitism, the act of being a parasite. Parasitism is a type of **symbiosis** in which a parasitic organism gains benefit from living within or upon a host organism. The host, in turn, gains no benefit, and in fact may be harmed by the symbiosis. Scientists have discovered that a parasitic trematode or flatworm is threatening many frog populations. Trematode larvae erupt from snails inhabiting ponds, which then infect frog tadpoles. Once inside the tadpoles, the flatworm larvae multiply and physically scramble developing limb bud cells in the hind quarters of the tadpole. If the limb bud cells are not in their proper places, malformations are the consequence. The result of such parasitism by flatworms is adult frogs with multiple legs or fused legs. It is believed that the flatworms derive benefit from the relationship because frogs with defective legs are easier for birds to prey upon. Birds that eat infected frogs in-turn become infected. The cycle is complete when bird droppings littered with flatworm larvae are dropped into pond water.

A fourth possible explanation for the decline of frogs involves ozone depletion. Some scientists believe that atmospheric ozone loss has led to an increase in ultraviolet light penetration to the surface of the earth. Frogs are exquisitely sensitive to ultraviolet light. It is believed that due to increased UV-B penetration, mutations in frogs has been increased, resulting in limb malformations. Evidence for this hypothesis exists in laboratory experiments that have been able to reliably replicate the limb malformations observed in the Minnesota ponds using UV-B radiation on experimental frogs.

Many scientists believe, however, that multiple factors are to blame for frog decline. They believe that synergy, or the combination of many factors is the most plausible explanation for the decrease in the number and diversity of frogs. **Climate** change, urbanization, prolonged **drought**, global warming, secondary **succession**, drainage of habitat for housing developments, **habitat fragmentation**, introduced predators, loss of territory, and the aforementioned infectious and **poisoning** reasons may all simultaneously contribute to the decline of frog populations worldwide. Humans perpetuate the decline by **hunting** frogs for food. Because the decrease in numbers and diversity of frogs is so striking, conservationists are concerned that it is an early indicator of the consequences of human progress and overpopulation.

[*Robert G McKinnell*]

RESOURCES

BOOKS

Behler, J. L., and F. W. King. *National Audubon Society Field Guide to North American Reptiles & Amphibians*. New York: Alfred A. Knopf, Inc., 1997.

DiBerardino, M. A. *Genomic Potential of Differentiated Cells*. New York: Columbia University Press, 1997.

Duellman, W. E., and L. Trueb. *Biology of Amphibians*. New York: McGraw-Hill Book Company, 1986.

Gilbert, S. F. *Development Biology*. 5th ed. Sunderland, MA: Sinauer Associates, Inc., 1997.

Stebbins, R. C., and N. W. Cohen. *A Natural History of Amphibians*. Princeton: Princeton University Press, 1995.

PERIODICALS

Carlson, D. L., L. A. Rollins-Smith, and R. G. McKinnell. "The Lucké Herpesvirus Genome: Its Presence in Neoplastic and Normal Kidney Tissue," *Journal of Comparative Pathology* 110 (1995): 349–355.

Cheh, A. M., et al. "A Comparison of the Ability of Frog and Rat S-9 to Activate Promutagens in the Ames Test," *Environmental Mutagenesis* 2 (1980): 487–508.

OTHER

CSIRO Australian Animal Health Laboratory. December 1995 [June 2002]. <http://www2.open.ac.uk/biology/froglog/FROGLOG-15-4.html>.

Tennessee Wildlife Resources Agency. March 22, 2002 [cited June 2002]. <http://www.state.tn.us/twra/lifecyc.html>.

United State Fish and Wildlife Services, Division of Endangered Species. December 7, 2001 [cited June 2002]. <http://ecos.fws.gov/servlet/TESSSpeciesReport/generate>.

Frontier economy

An economy similar to that which was prevalent at the "frontier" of European settlement in North America in the eighteenth and nineteenth centuries. A frontier economy is characterized by relative scarcities (and high prices) of capital equipment and skilled labor, and by a relative abundance (and low prices) of **natural resources**. Because of these factors, producers will look to utilize natural resources instead of capital and skilled labor whenever possible. For example, a sawmill might use a blade that creates large amounts of wood waste since the cost of extra logs is less than the cost of a better blade. The long-term environmental effects of high natural resource use and **pollution** from wastes are ignored since they seem insignificant compared to the vastness of the natural resource base.

A frontier economy is sometimes contrasted with a spaceship economy, in which resources are seen as strictly limited and need to be endlessly recycled from waste products.

Frost heaving

The lifting of earth by **soil** water as it freezes. Freezing water expands by approximately nine percent and exerts a pressure of about fifteen tons per square inch. Although this pressure and accompanying expansion are exerted equally in all directions, movement takes place in the direction of least **resistance**, namely upward. As a result, buried rocks, varying from pebbles to boulders, can be raised to the ground surface; small mounds and layers of soil can be heaved up; young plants can be ripped from the earth or torn apart below ground; and pavement and foundations can be cracked and lifted. Newly planted tree seedlings, grass, and agricultural crops are particularly vulnerable to being lifted by acicular ice crystals during early fall and late spring frosts. Extreme cases in cold climates at high latitudes or high elevation at mid-latitudes result in characteristic patterned ground.

Fuel cells

Fuel cells produce energy through electrochemical reactions rather than through the process of **combustion**. They convert **hydrogen** and oxygen into electricity and heat. Fuel cells are sometimes compared to batteries because, like batteries, they have two electrodes, an anode and a cathode, through which an electrical current flows into and out of the cell. But fuel cells are fundamentally different electrical devices from batteries since the latter simply store electrical energy, while the former are a source of electrical energy.

In a fuel cell, chemical energy is converted directly into electrical energy by means of an oxidation-reduction reaction. The British physicist Sir William Grove developed the fuel cell concept in 1839. A practical, working model of the concept was not constructed until a century later, however. The earliest fuel cells carried out this energy conversion by means of the reaction between hydrogen gas and oxygen gas, and formed water as a by-product. In this chemical reaction, each hydrogen atom loses one electron to an oxygen atom. The exchange of electrons, from hydrogen to oxygen, is what characterizes an oxidation-reduction reaction. A fuel cell thus creates a pathway through which electrons lost by hydrogen atoms must flow before they reach oxygen atoms. The cell typically consists of four basic parts: an anode, a cathode, an electrolyte, and an external circuit.

In the simplest fuel cell, hydrogen gas is pumped into one side of the fuel cell where it passes into a hollow, porous anode. At the anode, hydrogen atoms lose an electron to hydroxide ions present in the electrolyte. The source hydroxide ions is a solution of potassium hydroxide in water. The electrons released in this reaction travel up the anode, out of the fuel cell, and into the external circuit, which carries the flow of electrons (an electric current) to some device such as a light bulb where it can be used. Meanwhile, a second reaction is taking place at the opposite pole of the fuel cell. Oxygen gas is pumped into this side of the fuel cell where it passes into the hollow, porous cathode. Oxygen atoms pick up electrons from the cathode and react with water in the electrolyte to regenerate hydroxide ions.

As a result of the two chemical reactions taking place at the two poles of the fuel cell, electrons are removed from hydroxide ions at the anode, passed through the external circuit where they can be used to do work, returned to the cell through the cathode, and then returned to water molecules in the electrolyte. Meanwhile, oxygen and hydrogen are used up in the production of water. A fuel cell such as the one described here should have a voltage of 1.23 volts and a theoretical efficiency (based on the heat of combustion of water) of 83%. The actual voltage of a typical hydrogen/oxygen fuel cell normally ranges between 0.6 to 1.1 volts depending on operating conditions.

Fuel cells have many advantages as energy sources. They are significantly more efficient than energy sources such as nuclear or fossil-fueled **power plants**. In addition, a fuel cell is technically simple and lightweight. Also, the product of the fuel cell reaction —water— is of course harmless to humans and the rest of the **environment**. Finally, both hydrogen and oxygen, the raw materials used in a fuel cell, are abundant in **nature**. They can both be obtained from water, the most common single compound on the planet.

Until recently, electricity produced from fuel cells was more expensive than that obtained from other sources, and they have been used in only specialized applications. One of these is in spacecrafts, where their light weight represents an important advantage. For example, the fuel cell used on an 11-day Apollo moon flight, weighed 500 lb (227 kg), while a conventional generator would have weighed several tons. In addition, the water produced in the cell was purified and then used for drinking.

A great many variations on the simple hydrogen/oxygen fuel cell have been investigated. In theory, any fuel that contains hydrogen can be used at the anode, while any **oxidizing agent** can be used at the cathode. Elemental hydrogen and oxygen are only the simplest, most fundamental examples of each.

Among the hydrogen-containing compounds explored as possible fuels are hydrazine, **methanol**, ammonia, and a variety of **hydrocarbons**. The order in which these fuels are listed here corresponds to the efficiency with which they react in a fuel cell, with hydrazine being most reactive (after hydrogen itself) and the hydrocarbons being least reactive. Each of these potential alternatives has serious disadvantages. Hydrazine, for example, is expensive to manufacture and dangerous to work with.

In addition to oxygen, liquid oxidants such as hydrogen peroxide and nitric **acid** have also been investigated as possible cathode reactants. Again, neither compound works as efficiently as oxygen itself, and each presents problems of its own as a working fluid in a fuel cell.

The details of fuel cell construction often differ depending on the specific use of the cell. Cells used in spacecraft, for example, use cathodes made of **nickel** oxide or gold and anodes made of a platinum alloy. The hydrogen and oxygen used in such cells are supplied in liquid form that must be maintained at high pressure and very low temperature.

Fuel cells can also operate with an acidic electrolyte such as phosphoric or a fluorinated sulfuric acid. The chemical reactions that occur in such a cell are different from those described for the alkaline (potassium hydroxide) cell described above. Fuel cells of the acidic type are more commonly used in industrial applications. They operate at a higher temperature than alkaline cells, with slightly less efficiency.

Another type of fuel cell makes use of a molten salt instead of a water solution as the electrolyte. In a typical cell of this kind, hydrogen is supplied at the anode of the cell, **carbon dioxide** is supplied at the cathode, and molten potassium carbonate is used as the electrolyte. In such a cell, the fundamental process is the same as in an aqueous electrolyte cell. Electrons are released at the anode. The electrons then travel through the external circuit where they can do work. They return to the cell through the cathode where they make the cathodic reaction possible.

Yet a third type of fuel cell is now being explored, one that makes use of state-of-the-art and sometimes exotic solid-state technology. This is the high- temperature solid ionic cell. In one design of this cell, the anode is made of nickel metal combined with zirconium oxide while the cathode is composed of a lanthanum-manganese alloy doped with strontium. The electrolyte in the cell is a mixed oxide of yttrium and zirconium. The fuel provided to the cell is **carbon monoxide** or hydrogen, either of which is oxidized by the oxygen in ordinary air. A series of these cells are connected to each other by connections made of lanthanum chromite doped with magnesium metal.

The solid ionic cell is particularly attractive to **electric utilities** and other industries because it contains no liquid, as the other fuel cell designs do. The presence of such liquids creates problems in the handling and maintenance of conventional fuel cells that are eliminated with the all-solid cell.

The Proton Exchange Membrane (PEM) fuel cell and Direct Methanol Fuel cells (DMFC) are similar in that they both use a polymer membrane as the electrolyte, but a DMFC has an anode catalyst that utilizes hydrogen from liquid methanol, rather than from hydrogen gas. This improvement avoids the storage problems that are inherent with the use of hydrogen gas. Other possibilities are made up of stacks of fuel cells that use PEM-type technology. In 2001, the **U.S. Department of Energy** started a new program designed to implement the development of efficient and low-cost versions of these "planar solid-oxide fuel cells", or SOFCs.

Some experts are enthusiastic about the future role of fuel cells in the world's energy equation. If costs for all these emerging technologies can be reduced, their high efficiency should make them attractive alternatives to fossil fuel- and nuclear-generated electricity. The major concerns about fuel cell technologies include the inability of fuel cells to store power and the associated requirement that energy be generated on an "as needed" basis. Proposals have been made to use pumped water, compressed air, and batteries to store energy that is generated when times of demand are low.

For that reason, still more variations in the fuel cell are being explored. One of the most intriguing of these future possibilities is a cell no more than 0.04–0.07 in (1–2 mm) in diameter. These cells have the advantage of a greater electrode surface area on which oxidation and reduction occur than do conventional cells. The distance that electrons have to travel in such cells is also much shorter, resulting in their having a greater power output per unit volume than that of conventional cells. The technology for constructing and maintaining such small cells is not, however, fully developed. Because of more stringent clean air standards, the **transportation** industry is especially interested in increasing **energy efficiency** of the **automobile** industry. The utility of **hybrid vehicles** that use conventional fuels when accelerating and fuel cells for highway driving is being investigated. These cars would use **gasoline** or methanol as a fuel source for both the conventional engine and as a source of hydrogen for the fuel cells. Fuel cells are also being researched for the aviation industry. An innovative approach is to couple photovoltaic or solar cell technology with fuel cell technology to provide a source of energy for the plane that could be used day or night.

[*David E Newton and Marie H. Bundy*]

RESOURCES

BOOKS

Hoffmann, Peter, and Tom Harkin. *Tomorrow's Energy: Hydrogen, Fuel Cells, and the Prospects for a Cleaner Planet.* Cambridge, MA: MIT Press, 2001.

Hoogers, G., ed. *Fuel Cell Technology Handbook.* Boca Raton: CRC Press, 2002.

Joesten, M. D., et al. *World of Chemistry.* Philadelphia: Saunders College Publishing, 1991.

Larminie, James, and Andrew Dicks. *Fuel Cell Systems Explained.* New York: John Wiley & Sons, 2000.

ORGANIZATIONS

Argonne National Laboratory, 9700 S. Cass Avenue, Argonne, IL USA 60439 (630) 252-2000, <http://www.anl.gov>

DOE Office of Energy Efficiency and Renewable Energy (EERE), Department of Energy, Mail Stop EE-1, Washington, DC USA 20585 (202) 586-9220, <http://www.eren.doe.gov>

The Online Fuel Cell Information Center, <http://www.fuelcells.org>

Fuel switching

Fuel switching is the substitution of one energy source for another in order to meet requirements for heat, power, and/or electrical generation. Generally, this term refers to the practices of some industries that can substitute among **natural gas**, electricity, **coal**, and LPG within 30 days without modifying their fuel-consuming equipment and that can resume the same level of production following the change. The **U.S. Department of Energy** estimates that among manufacturers in 1991, 2.8 quadrillion Btus of energy consumption could be switched from a total consumption of about 20.3 quadrillion Btus, representing about 14%. Price is the primary reason for fuel switching; however, additional factors may include environmental regulations, agreements with energy or fuel suppliers, and equipment capabilities.

Fugitive emissions

Contaminants that enter the air without going through a smokestack and, thus, are often not subject to control by conventional **emission** control equipment or techniques. Most fugitive emissions are caused by activities involving the production of dust, such as **soil erosion** and **strip mining**, or building demolition, or the use of volatile compounds. In a steel-making complex, for example, there are several identifiable smokestacks from which emissions come, but there are also numerous sources of fugitive emissions, which escape into the air as a result of processes such as producing coke, for which there is no identifiable smokestack. The control of fugitive emissions is generally much more complicated and costly than the control of smokestack emissions for which known add-on technologies to the smokestack have been developed. Baghouses and other costly mechanisms typically are needed to control fugitive emissions.

Fumigation

Most commonly, fumigation refers to the process of disinfecting a material or an area by using some type of toxic material in gaseous form. The term has a more specialized meaning in **environmental science**, where it refers to the process by which pollutants are mixed in the **atmosphere**. Under certain conditions, emitted pollutants rise above a stable layer of air near the ground. These pollutants remain aloft until convective currents develop, often in the morning, at which time the cooler pollutants "trade places" with air at ground level as it is warmed by the sun and rises. The resulting damage to ecosystems from the pollutants is most obvious around metal smelters.

Fund for Animals

Founded in 1967 by author and humorist **Cleveland Amory**, the Fund is one of the most activist of the national animal protection groups. Formed "to speak for those who can't," it has led sometimes militant campaigns against sport **hunting**, **trapping**, and wearing furs, as well as the killing of **whales**, **seals**, bears, and other creatures. Amory, in particular, has campaigned tirelessly against these activities on television and radio, and in lectures, articles, and books.

In the early 1970s, the Fund worked effectively to rally public opinion in favor of passage of the Marine Mammal Protection Act, which was signed into law in October 1972. This act provides strong protection to whales, **seals and sea lions**, **dolphins**, sea otters, polar bears, and other ocean mammals. In 1978, the Fund bought a British trawler and renamed it *Sea Shepherd*. Under the direction of its captain,

Paul Watson, they used the ship to interfere with the baby seal kill on the ice floes off Canada. Activists sprayed some 1,000 baby harp seals with a harmless dye that destroyed the commercial value of their white coats as fur, and the ensuing publicity helped generate worldwide to the seal kill and a ban on imports into Europe. In 1979, *Sea Shepherd* hunted down and rammed *Sierra*, an outlaw **whaling** vessel that was illegally killing protected and **endangered species** of whales. After *Sea Shepherd* was seized by Portuguese authorities, Watson and his crew scuttled the ship to prevent it from being given to the owners of *Sierra* for use as a whaler.

Also in 1979, the Fund used helicopters to airlift from the Grand Canyon almost 600 wild burros that were scheduled to be shot by the **National Park Service**. The airlift was so successful, and generated so much favorable publicity, that it led to similar rescues of feral animals on public lands that the government wanted removed to prevent damage to vegetation. Burros were also airlifted by the Fund from Death Valley National Monument, as were some 3,000 wild goats on San Clemente Island, off the coast of California, scheduled to be shot by the United States Navy.

Many of the wild horses, burros, goats, and other animals rescued by the Fund end up, at least temporarily before adoption, at Black Beauty Ranch, a 1430-acre (578-ha) sanctuary near Arlington, Texas. The ranch has provided a home for abused race and show horses, a non-performing elephant, and Nim, the famous signing chimpanzee who was saved from a medical laboratory.

Legal action initiated by the Fund has resulted in the addition of almost 200 **species** to the U.S. Department of the Interior's list of threatened and endangered species, including the **grizzly bear**, the Mexican wolf, the Asian elephant, and several species of kangaroos. The Fund is also active on the grassroots level, working on measures to restrict **hunting and trapping**. A recent example is the passage of an initiative in Colorado in November 1992 banning the use of dogs and bait to hunt bears, and halting the spring bear hunt, when mothers are still nursing their cubs.

[*Lewis G. Regenstein*]

RESOURCES
ORGANIZATIONS
The Fund for Animals, 200 West 57th Street, New York, NY USA 10019 (212) 246-2096, Fax: (212) 246-2633, Email: hdquarters@fund.org, <http://fund.org>

Fungi

Fungi are one of the five Phyla of organisms. Fungi are broadly characterized by cells that possess nuclei and rigid cell walls but lack chlorophyll. Fungal spores germinate and grow slender tube-like structures called *hyphae*, separated by cell walls called *septae*. The vegetative **biomass** of most fungi in **nature** consists of masses of hyphae, or *mycelia*. Most **species** of fungi inhabit **soil**, where they are active in the **decomposition** of organic matter. The biologically most complex fungi periodically form spore-producing fruiting structures, known as mushrooms. Some fungi occur in close associations, known as *mycorrhizae*, with the roots of many species of vascular plants. The plant benefits mostly through an enhancement of **nutrient** uptake, while the fungus benefits through access to metabolites. Certain fungi are also partners in the symbioses with algae known as **lichens**. *See also* Fungicide

Fungicide

A fungus is a tiny plant-like organism that obtains its nourishment from dead or living organic matter. Some examples of **fungi** include mushrooms, toadstools, smuts, molds, rusts, and mildew.

Fungi have long been recognized as a serious threat to natural plants and human crops. They attack food both while it is growing and also after it has been harvested and placed into storage. One of the great agricultural disasters of the second half of the twentieth century was caused by a fungus. In 1970, the fungus that causes southern corn-leaf blight swept through the southern and midwestern United States and destroyed about 15% of the nation's corn crop. Potato blight, wheat rust, wheat smut, and grape mildew are other important disasters caused by fungi.

Chestnut blight is another example of the devastation that can be caused by fungi. Until 1900, chestnut trees were common in many parts of the United States. In 1904, however, chestnut trees from Asia were imported and planted in parts of New York. The imported trees carried with them a fungus that attacked and killed the native chestnut trees. Over a period of five decades, the native trees were all but totally eliminated from the eastern part of the country.

It is hardly surprising that humans began looking for fungicides—substances that will kill or control the growth of fungi—early on in history. The first of these fungicides was a naturally occurring substance, sulfur. One of the most effective of all fungicides, Bordeaux mixture was invented in 1885. Bordeaux mixture is a combination of two inorganic compounds, **copper** sulfate and lime.

With the growth of the chemical industry during the twentieth century, a number of synthetic fungicides have been invented; these include ferbam, ziram, captan, naban, dithiocarbonate, quinone, and 8-hydroxyquinoline.

For a period of time, compounds of **mercury** and **cadmium** were very popular as fungicides. Until quite recently, for example, the compound methylmercury was widely used by farmers in the United States who used it to protect growing plants and to treat stored grains.

During the 1970s, however, evidence began to accumulate about a number of adverse effects of mercury- and cadmium-based fungicides. The most serious effects were observed among birds and small animals who were exposed to sprays and dusting or who ate treated grain. A few dramatic incidents of methylmercury **poisoning** among humans, however, were also recorded. The best known of these was the 1953 disaster at Minamata Bay, Japan.

At first, scientists were mystified by an epidemic that spread through the Minamata Bay area between 1953 and 1961. Some unknown factor caused serious nervous disorders among residents of the region. Some sufferers lost the ability to walk, others developed mental disorders, and still others were permanently disabled. Eventually researchers traced the cause of these problems to methylmercury in fish eaten by residents of the area. For the first time, the terrible effects of the compound had been confirmed.

As a result of the problems with mercury and cadmium compounds, scientists have tried to develop less toxic substitutes for the more dangerous fungicides. Dinocap, binapacryl, and benomyl are three examples of such compounds.

Another approach has been to use **integrated pest management** and to develop plants that are resistant to fungi. The latter approach was used with great success during the corn blight disaster of 1970. Researchers worked quickly to develop strains of corn that were resistant to the corn-leaf blight fungus and by 1971 had provided farmers with seeds of the new strain. *See also* Minamata disease

[*David E. Newton*]

RESOURCES
BOOKS

Chemistry and the Food System. A Study by the Committee on Chemistry and Public Affairs. Washington, DC: American Chemical Society, 1980.

Fletcher, W. W. *The Pest War.* New York: Wiley, 1974.

Selinger, B. *Chemistry in the Marketplace.* 4th ed. Sydney: Harcourt Brace Jovanovich, 1989.

Furans

Furans are by-products of natural and industrial processes and are considered environmental pollutants. They are chemical substances found in small amounts in the **environment**, including air, water and **soil**. They are also present in some foods. Although the amounts are small, they are persistent and remain in the environment for long periods of time, also accumulating in the food chain. The U.S. Environmental Protection Agency's (EPA) Persistent Bioaccumulative and Toxic (PBT) Chemical Program classifies furans as priority PBTs.

Furans belong to a class of organic compounds known as heterocyclic aromatic **hydrocarbons**. The basic furan structure is a five-membered ring consisting of four atoms of **carbon** and one oxygen. Various types of furans have additional atoms and rings attached to the basic furan structure. Some furans are used as solvents or as raw materials for synthesizing **chemicals**.

Polychlorinated dibenzofurans (PCDFs) are of particular concern as environmental pollutants. These are three-ringed structures, with two rings of six carbon atoms each (**benzene** rings) attached to the furan. Between one and eight **chlorine** atoms are attached to the rings. There are 135 types of PCDFs, whose properties are determined by the number and position of the chlorine atoms. PCDFs are closely related to polychlorinated dioxins (PCDDs) and **polychlorinated biphenyls** (PCBs). These three types of toxic compounds often occur together and PCDFs are major contaminants of manufactured PCBs. In fact, the term **dioxin** commonly refers to a subset of these compounds that have similar chemical structures and toxic mechanisms. This subset includes 10 of the PCDFs, as well as seven of the PCDDs and 12 of the PCBs. Less frequently, the term dioxin is used to refer to all 210 structurally-related PCDFs and PCDDs, regardless of their toxicities.

Furans are present as impurities in various industrial chemicals. PCDFs are trace byproducts of most types of **combustion**, including the **incineration** of chemical, industrial, medical, and municipal waste, the burning of wood, **coal**, and peat, and **automobile emissions**. Thus most PCDFs are released into the environment through smokestacks. However the backyard burning of common household trash in barrels has been identified as potentially one of the largest sources of dioxin and furan emissions in the United States. Because of the lower temperatures and inefficient combustion in burn barrels, they release more PCDFs than municipal incinerators. Some industrial chemical processes, including chlorine bleaching in **pulp and paper mills**, also produce PCDFs.

PCDFs that are released into the air can be carried by currents to all parts of the globe. Eventually they fall to earth and are deposited in soil, sediments, and surface water. Although furans are slow to volatilize and have a low solubility in water, they can wash from soils into bodies of water, evaporate, and be re-deposited elsewhere. Furans have been detected in soils, surface waters, sediments, plants, and animals throughout the world, even in arctic organisms. They are very resistant to both chemical breakdown and biological degradation by **microorganisms**.

Most people have low but detectable levels of PCDDs and PCDFs in their tissues. Furans enter the food chain from soil, water, and plants. They accumulate at the higher levels of the food chain, particularly in fish and animal fat. The concentrations of PCDDs and PCDFs may be hundreds or thousands of times higher in aquatic organisms than in the surrounding waters. Most humans are exposed to furans through animal fat, milk, eggs, and fish. Some of the highest levels of furans are found in human breast milk. The presence of dioxins and furans in breast milk can lead to the development of soft, discolored molars in young children. Industrial workers can be exposed to furans while handling chemicals or during industrial accidents.

Furans bind to aromatic hydrocarbon receptors in cells throughout the body, causing a wide range of deleterious effects, including developmental defects in fetuses, infants, and children. Furans also may adversely affect the reproductive and immune systems. At high exposure levels, furans can cause chloracne, a serious acne-like skin condition. Furan itself, as well as PCDFs, are potential cancer-causing agents.

The switch from leaded to unleaded **gasoline**, the halting of PCB production in 1977, changes in paper manufacturing processes, and new air and **water pollution** controls, have reduced the emissions of furans. In 1999 the United States, Canada, and Mexico agreed to cooperate to further reduce the release of dioxins and furans.

On May 23, 2001, EPA Administrator Christie Whitman, along with representatives from more than 90 other countries, signed the global treaty on **Persistent Organic Pollutants**. The treaty phases out the manufacture and use of 12 toxic chemicals, the so-called "dirty dozen," that includes furans. The United States opposed a complete ban on furans and dioxins; thus, unlike eight of the other chemicals that were banned outright, the treaty calls for the use of dioxins and furans to be minimized and eliminated where feasible.

[*Margaret Alic Ph.D.*]

RESOURCES

BOOKS

Lippmann, Morton, ed. *Environmental Toxicants: Human Exposures and their Health Effects*. New York: Wiley-Interscience, 2000.

Paddock, Tod. *Dioxins and Furans: Questions and Answers*. Philadelphia: Academy of Natural Sciences, 1989.

Wittich, Rolf-Michael, ed. *Biodegradation of Dioxins and Furans*. Austin: R. G. Landes Co., 1998.

OTHER

"Dioxins, PCBs, Furans, and Mercury." *Fox River Watch*. [cited May 15, 2002]. <http://www.foxriverwatch.com/dioxins_pcb_pcbs_1.html?source= overture>.

"Polychlorinated Dibenzo-p-dioxins and Related Compounds Update: Impact on Fish Advisories." *EPA Fact Sheet*. United States Environmental Protection Agency. September 1999. [cited May 8, 2002]. <http://www.ep-a.gov/ost/fish/dioxin.pdf>.

U.S. Environmental Protection Agency. "Priority PBTs: Dioxins and Furans." *Persistent Bioaccumulative and Toxic (PBT) Chemical Program*. March 9, 2001 [cited May 13,2002]. <http://www.epa.gov/pbt/dioxins.htm>.

ORGANIZATIONS

Clean Water Action Council, 1270 Main Street, Suite 120, Green Bay, WI USA 54302 (920) 437-7304, Fax: (920) 437-7326, Email: CleanWater@cwac.net, <http://www.cwac.net/index.html>

United Nations Environment Programme: Chemicals, 11-13, chemin des Anémones, 1219 Châtelaine, Geneva, Switzerland, Email: opereira@ unep.ch, <http://www.chem.unep.ch>

United States Environmental Protection Agency, 1200 Pennsylvania Avenue, NW, Washington , DC USA 20460 Toll Free: (800) 490-9198, Email: public-access@epa.gov, <http://www.epa.gov>

Fusion
see **Nuclear fusion**

Future generations

According to demographers, a generation is an age-cohort of people born, living, and dying within a few years of each other. Human generations are roughly defined categories, and the demarcations are not as distinct as they are in many other **species**. As the Scottish philosopher David Hume noted in the eighteenth century, generations of human beings are not like generations of butterflies, who come into existence, lay their eggs, and die at about the same time, with the next generation hatching thereafter. But distinctions can still be made, and future generations are all age-cohorts of human beings who have not yet been born.

The concept of future generations is central to **environmental ethics** and **environmental policy**, because the health and well-being—indeed the very existence—of human beings depends on how people living today care for the natural **environment**.

Proper stewardship of the environment affects not only the health and well-being of people in the future but their character and identity. In *The Economy of the Earth*, Mark Sagoff compares environmental damage to the loss of our rich cultural heritage. The loss of all our art and literature would deprive future generations of the benefits we have enjoyed and render them nearly illiterate. By the same token, if we destroyed all our wildernesses and dammed all our rivers, allowing **environmental degradation** to proceed at the same pace, we would do more than deprive people of the pleasures we have known. We would make them into what Sagoff calls "environmental illiterates," or "yahoos" who would neither know nor wish to experience the beauties and pleasures of the natural world. "A pack of yahoos," says Sagoff, "will like a junkyard environment" because they will have known nothing better.

The concept of future generations emphasizes both our ethical and aesthetic obligations to our environment. In relations between existing and future generations, however, the present generation holds all the power. While we can affect them, they can do nothing to affect us. Though, as some environmental philosophies have argued, our moral code is in large degree based on reciprocity, the relationship between generations cannot be reciprocal. Adages such as "like for like," and "an eye for an eye," can apply only among contemporaries. Since an adequate environmental ethic would require that moral consideration be extended to include future people, views of justice based on the norm of reciprocity may be inadequate.

A good deal of discussion has gone into what an alternative environmental ethic might look like and on what it might be based. But perhaps the important point to note is that the treatment of generations yet unborn has now become a lively topic of philosophical discussion and political debate. *See also* Environmental education; Environmentalism; Intergenerational justice

[*Terence Ball*]

RESOURCES
BOOKS

Barry, B., and R. I. Sikora, eds. *Obligations to Future Generations.* Philadelphia: Temple University Press, 1978.

Fishkin, J., and P. Laslett, eds. *Justice Between Age Groups and Generations.* New Haven, CT: Yale University Press, 1992.

Partridge, E., ed. *Responsibilities to Future Generations.* Buffalo, NY: Prometheus Books, 1981.

Sagoff, M. *The Economy of the Earth: Philosophy, Law and the Environment.* Cambridge and New York: Cambridge University Press, 1988.

G

Gaia hypothesis

The Gaia hypothesis was developed by British biochemist James Lovelock, and it incorporates two older ideas. First, the idea implicit in the ancient Greek term *Gaia*, that the earth is the mother of all life, the source of sustenance for all living beings, including humans. Second, the idea that life on earth and many of earth's physical characteristics have coevolved, changing each other reciprocally as the generations and centuries pass.

Lovelock's theory contradicts conventional wisdom, which holds "that life adapted to the planetary conditions as it and they evolved their separate ways." The Gaia hypothesis is a startling break with tradition for many, although ecologists have been teaching the **coevolution** of organisms and **habitat** for at least several decades, albeit more often on a local than a global scale.

The hypothesis also states that Gaia will persevere no matter what humans do. This is undoubtedly true, but the question remains: in what form, and with how much diversity? If humans don't change the nature and scale of some of their activities, the earth could change in ways that people may find undesirable—loss of **biodiversity**, more "weed" **species**, increased **desertification**, etc.

Many people, including Lovelock, take the Gaia hypothesis a step further and call the earth itself a living being, a long-discredited organismic analogy. Recently a respected **environmental science** textbook defined the Gaia hypothesis as a "proposal that Earth is alive and can be considered a system that operates and changes by feedbacks of information between its living and nonliving components." Similar sentences can be found quite commonly, even in the scholarly literature, but upon closer examination they are not persuasive. A furnace operates via a positive and negative feedback system—does that imply it is alive? Of course not. The important message in Lovelock's hypothesis is that the health of the earth and the health of its inhabitants are inextricably intertwined. *See also* Balance of nature; Biological community; Biotic community; Ecology; Ecosystem; Environment; Environmentalism; Evolution; Nature; Sustainable biosphere

[*Gerald L. Young*]

RESOURCES
BOOKS

Schneider, S. H., and P. J. Boston, eds. *Scientists on Gaia*. Cambridge: MIT Press, 1991.

Joseph, L. E. *Gaia: The Growth of an Idea*. New York: St. Martin's Press, 1990.

Lovelock, J. E. *Gaia: A New Look at Life on Earth*. Oxford: Oxford University Press, 1979.

———. *The Ages of Gaia: A Biography of Our Living Earth*. New York: Norton, 1988.

PERIODICALS

Lyman, F. "What Gaia Hath Wrought: The Story of a Scientific Controversy." *Technology Review* 92 (July 1989): 54–61.

Galápagos Islands

Within the theory of **evolution**, the concept of adaptive radiation (evolutionary development of several **species** from a single parental stock) has had as its prime example, a group of birds known as Darwin's finches. Charles Darwin discovered and collected specimens of these birds from the Galápagos Islands in 1835 on his five-year voyage around the world aboard the HMS *Beagle*. His cumulative experiences, copious notes, and vast collections ultimately led to the publication of his monumental work, *On the Origin of Species*, in 1859. The Galápagos Islands and their unique assemblage of plants and animals were an instrumental part of the development of Darwin's evolutionary theory.

The Galápagos Islands are located at 90° W longitude and 0° latitude (the equator), about 600 mi (965 km) west Ecuador. These islands are volcanic in origin and are about 10 million years old. The original colonization of the Galápagos Islands occurred by chance transport over the ocean as indicated by the gaps in the **flora** and **fauna** of this archipelago compared to the mainland. Of the hundreds of species of birds along the northwestern South American coast, only seven species colonized the Galápagos Islands. These evolved into 57 resident species, 26 of which are endemic to the islands, through adaptive radiation. The only native land mammals are a rat and a bat. The land reptiles include iguanas, a single species each of snake, lizard, and gecko, and the Galápagos giant tortoise (*Geochelone elephantopus*). No amphibians and few insects or mollusks are found in the Galápagos. The flora has large gaps as well—no conifers or palms have colonized these islands. Many of the open niches have been filled by the colonizing groups. The tortoises and iguanas are large and have filled niches normally occupied by mammalian herbivores. Several plants, such as the prickly pear cactus, have attained large size and occupy the ecological position of tall trees.

The most widely known and often used example of adaptive radiation is Darwin's finches, a group of 14 species of birds that arose from a single ancestor in the Galápagos Islands. These birds have specialized on different islands or into niches normally filled by other groups of birds. Some are strictly seed eaters, while others have evolved more warbler-like bills and eat insects, still others eat flowers, fruit, and/or nectar, and others find insects for their diet by digging under the bark of trees, having filled the **niche** of the woodpecker. Darwin's finches are named in honor of their discoverer, but they are not referred to as Galápagos finches because there is one of their numbers that has colonized Cocos Island, located 425 mi (684 km) north-northeast of the Galápagos.

Because of the Galápagos Islands' unique **ecology**, scenic beauty and tropical **climate**, they have become a mecca for tourists and some settlement. These human activities have introduced a host of environmental problems, including **introduced species** of goats, pigs, rats, dogs, and cats, many of which become feral and damage or destroy nesting bird colonies by preying on the adults, young, or eggs. Several races of giant tortoise have been extirpated or are severely threatened with **extinction**, primarily due to exploitation for food by humans, destruction of their food resources by goats, or predation of their hatchlings by feral animals. Most of the 13 recognized races of tortoise have populations numbering only in the hundreds. Three races are tenuously maintaining populations in the thousands, one race has not been seen since 1906, but it is thought to have disappeared due to natural causes, another race has a population of about 25 individuals, and the Abingdon Island tortoise is represented today by only one individual, "Lonesome George," a captive male at the Charles Darwin Biological Station. For most of these tortoises to survive, an active capture or extermination program of the feral animals will have to continue. One other potential threat to the Galápa-

Coastline in the Galápagos Islands. (Photograph by Anthony Wolff. Phototake NYC. Reproduced by permission.)

gos Islands is tourism. Thousands of tourists visit these islands each year and their numbers can exceed the limit deemed sustainable by the Ecuadoran government. These tourists have had, and will continue to have, an impact on the fragile habitats of the Galápagos. *See also* Endemic species; Ecotourism

[*Eugene C. Beckham*]

RESOURCES
BOOKS

Harris, M. *A Field Guide to the Birds of the Galápagos.* London: Collins, 1982.
Root, P., and M. McCormick. *Galápagos Islands.* New York: Macmillan, 1989.
Steadman, D. W., and S. Zousmer. *Galápagos: Discovery on Darwin's Islands.* Washington, DC: Smithsonian Institution Press, 1988.

Birute Marija Filomena Galdikas (1948 –)
Lithuanian/Canadian primatologist

The world's leading expert on orangutans, Birute Galdikas has dedicated much of her life to studying the orangutans of Indonesia's Borneo and Sumatra islands. Her work, which has complemented that of such other scientists as **Dian Fossey** and **Jane Goodall**, has led to a much greater understanding of the primate world and more effective efforts to protect orangutans from the effects of human infringement. Galdikas has also been credited with providing valuable insights into human culture through her decades of work with primates. She discusses this aspect of her work in her 1995 autobiography, Reflections of Eden: My Years with the Orangutans of Borneo.

Galdikas was born on May 10, 1948 in Wiesbaden in what was then West Germany, while her family was en route from their native Lithuania. She was the first of four children. The family moved to Toronto, Canada when she was two, and she grew up in that city. As a child, Galdikas was already enamored of the natural world, and spent much of her time in local parks and reading books on jungles and their inhabitants. She was already especially interested in orangutans. The Galdikas family eventually moved to Los Angeles, where Birute attended the local campus of the University of California. She earned a BA there in 1966 and immediately began work on a master's degree in anthropology. Galdikas had already decided to start a long-term study of orangutans in the rain forests of Indonesia, where most of the world's last remaining wild individuals live.

Galdikas began to realize her dream in 1969 when she approached famed paleoanthropologist Louis Leakey after he gave a lecture at UCLA. Leakey had helped launch the research efforts of Fossey and Goodall, and she asked him to do the same for her. He agreed, and by 1971 had helped her raise enough money to get started. With her first husband, Galdikas traveled to southern Borneo's Tanjung Puting **National Park** in East Kalimantan to start setting up her

research station. Such challenges as huge leeches, extremely toxic plants, perpetual dampness, swarms of insects, and aggressive viruses slowed Galdikas down, but did not ruin her enthusiasm for her new project.

After finally locating the area's elusive **orangutan** population, Galdikas faced the difficulty of getting the shy animals accustomed enough to her presence that they would permit her to watch them even from a distance.

Once Galdikas accomplished this, she was able to begin documenting some of the traits and habits of the little-studied orangutans. She compiled a detailed list of staples in the animals' diets, discovered that they occasionally eat meat, and recorded their complex behavioral interactions.

Eventually, the animals came to accept Galdikas and her husband so thoroughly that their camp was often overrun by them. Galdikas recalled in a 1980 National Geographic article that she sometimes felt as though she were "surrounded by wild, unruly children in orange suits who had not yet learned their manners." Meanwhile, she applied her findings to her UCLA education, earning both her master's degree and doctorate in 1978.

During her first decade on Borneo, Galdikas founded the Orangutan Project, which has since been funded by such organizations as the National Geographic Society, the **World Wildlife Fund**, and **Earthwatch**. The Project not only carries out primate research, but also rehabilitates hundreds of former captive orangutans. She also founded the Los Angeles-based nonprofit Orangutan Foundation International in 1987.

From 1996 to 1998, Galdikas served as a senior adviser to the Indonesian Forestry Ministry on orangutan issues as that government attempted to rectify the mistreatment of the animals and the mismanagement of their dwindling **rain forest habitat**. As part of these efforts, the Jakarta government also helped Galdikas establish the Orangutan Care Center and Quarantine near Pangkalan Bun, which opened in 1999. This center has since cared for many of the primates injured or displaced by the devastating fires in the Borneo rain forest in 1997–1998.

Divorced in 1979, Galdikas married a native Indonesian man of the Dayak tribe in 1981. She has one son with her first husband and two children with her second. Galdikas and her second husband currently live in a Dayak village, but Galdikas travels to Canada once a year to visit her first son. She became a visiting professor at Simon Fraser University in 1981, but since then has been appointed full professor. Besides her work, Galdikas reportedly most enjoys playing with her children, reading, walking, and listening to native Indonesian music. She has been featured on such popular television shows as "Good Morning, America" and "Eye to Eye," using such exposure to increase public aware-

Birute Galdikas being embraced by two orang-utans in Borneo. (The Liaison Network. ©Liaison Agency. Reproduced by permission.)

ness of her programs and the danger faced by the world's remaining orangutan population.

RESOURCES
BOOKS

Galdikas, Birute. *Reflections of Eden: My Years with the Orangutans of Borneo.* Little, Brown: New York, 1995.
Montgomery, Sy. *Walking with the Great Apes: Jane Goodall, Dian Fossey, Birute Galdikas.* 1991.
Notable Scientists: From 1900 to the Present. Farmington Hills, MI: Gale Group, 2001.

Game animal

Birds and mammals commonly hunted for sport. The major groups include upland game birds (quail, pheasant, and partridge), waterfowl (ducks and geese), and big game (deer, antelope, and bears). Game animals are protected to varying degrees throughout most of the world, and **hunting** levels are regulated through the licensing of hunters as well as by seasons and bag limits. In the United States, state **wildlife** agencies assume primary responsibility for enforcing hunting regulations, particularly for resident or non-migratory **species**. The **Fish and Wildlife Service** shares responsibility with state agencies for regulating harvests of migratory game animals, principally waterfowl.

Game preserves

Game preserves (also known as game reserves, or **wildlife** refuges) are a type of protected area in which **hunting** of certain **species** of animals is not allowed, although other kinds of resource harvesting may be permitted. Game preserves are usually established to conserve populations of larger game species of mammals or waterfowl. The protection from hunting allows the hunted species to maintain relatively large populations within the sanctuary. However, animals may be legally hunted when they move outside of the reserve during their seasonal migrations or when searching for additional **habitat**.

Game preserves help to ensure that populations of hunted species do not become depleted through excessive harvesting throughout their range. This **conservation** allows the species to be exploited in a sustainable fashion over the larger landscape.

Although hunting is not allowed, other types of resource extraction may be permitted in game reserves, such as timber harvesting, livestock grazing, some types of cultivated agriculture, mining, and exploration and extraction of **fossil fuels**. However, these land-uses are managed to ensure that the habitat of game species is not excessively damaged. Some game preserves are managed as true ecological reserves, where no extraction of **natural resources** is allowed. However, low-intensity types of land-use may be permitted in these more comprehensively protected areas, particularly non-consumptive **recreation** such as hiking and wildlife viewing.

Game preserves as a tool in conservation

The term "conservation" refers to the wise (i.e., sustainable) use of natural resources. Conservation is particularly relevant to the use of renewable resources, which are capable of regenerating after a portion has been harvested. Hunted species of animals are one type of renewable resource, as are timber, flowing water, and the ability of land to support the growth of agricultural crops. These renewable resources have the potential to be harvested forever, as long as the rate of exploitation is equal to or less than the rate of regeneration. However, potentially renewable resources can also be harvested at a rate exceeding their regeneration. This is known as over-exploitation, a practice that causes the stock of the resource to decline and may even result in its irretrievable collapse.

Wildlife managers can use game preserves to help conserve populations of hunted species. Other methods of conservation of game animals include: (1) regulation of the time of year when hunting can occur; (2) setting of "bag limits" that restrict the maximum number of animals that any hunter can harvest; (3) limiting the total number of animals that can be harvested in a particular area, and; (4) restricting the hunt to certain elements of the population. Wildlife managers can also manipulate the habitat of game species so that larger, more productive populations can be sustained, for example by increasing the availability of food, water, shelter, or other necessary elements of habitat. In addition, wildlife managers may cull the populations of natural predators to increase the numbers of game animals available to be hunted by people.

Some or all of these practices, including the establishment of game preserves, may be used as components of an integrated game management system. Such systems may be designed and implemented by government agencies that are responsible for managing game populations over relatively large areas such as counties, states, provinces, or entire countries.

Conservation is intended to benefit humans in their interactions with other species and ecosystems, which are utilized as valuable natural resources. When defined in this way, conservation is very different from the preservation of indigenous species and ecosystems for their ecocentric and biocentric values, which are considered important regardless of any usefulness to humans or their economic activities.

Examples of game preserves

The first **national wildlife refuge** in the United States was established by President **Theodore Roosevelt** in 1903. This was a breeding site for brown pelicans (*Pelecanus occidentalis*) and other birds in Florida. The U.S. national system of wildlife refuges now totals some 437 sites covering 91.4 million acres (37 million ha); an additional 79 million acres (32 million ha) of habitat are protected in national parks and monuments. The largest single wildlife reserve is the Alaska Maritime Wildlife Refuge, which covers 3.5 million acres (1.4 million ha); in fact, about 85% of the national area of **wildlife refuges** is in Alaska. Most of the national wildlife refuges protect migratory, breeding, and wintering habitats for waterfowl, but others are important for large mammals and other species. Some wildlife refuges have been established to protect **critical habitat** of **endangered species**, such as the Aransas Wildlife Refuge in coastal Texas, which is the primary wintering grounds of the **whooping crane** (*Grus americana*). Since 1934, sales of *Migratory Bird Hunting Stamps*, or "duck stamps," have been critical to providing funds for the acquisition and management of federal wildlife refuges in the United States.

Although hunting is not permitted in many National wildlife refuges, in 1988, closely regulated hunting was permitted in 60% of the refuges, and fishing was allowed in 50%. In addition, some other resource-related activities are allowed in some refuges. Depending on the site, it may be possible to harvest timber, graze livestock, engage in other kinds of agriculture, or explore for or mine metals or fossil fuels. The various components of the multiple-use plans of particular national wildlife refuges are determined by the Secretary of the Interior. Any of these economically important activities may cause damage to wildlife habitats, and this has resulted in intense controversy between economic interests and some environmental groups. Environmental organizations such as the **Sierra Club**, **Ducks Unlimited**, and the Audubon Society have lobbied federal legislators to further restrict exploration and extraction in national wildlife refuges, but business interests demand greater access to valuable resources within national wildlife refuges.

Many states and provinces also establish game preserves as a component of wildlife-management programs on their lands. For example, many jurisdictions in eastern North America have set up game preserves for management of populations of white-tailed deer (*Odocoileus virginianus*), a widely hunted species. Game preserves are also used to conserve populations of mule deer (*Odocoileus hemionus*) and elk (*Cervus canadensis*) in more western regions of North America.

Some other types of protected areas, such as state, provincial, and national parks are also effective as wildlife preserves. These protected areas are not primarily established for the conservation of natural resources—rather, they are intended to preserve natural ecosystems and wild places for their **intrinsic value**. Nevertheless, relatively large and productive populations of hunted species often build up within parks and other large ecological reserves, and the surplus animals are commonly hunted in the surrounding areas. In addition, many protected areas are established by nongovernmental organizations, such as **The Nature Conservancy**, which has preserved more than 16 million acres (6.5 million ha) of natural habitat throughout the United States.

Yellowstone National Park is one of the most famous protected areas in North America. Hunting is not allowed in Yellowstone, and this has allowed the build-up of relatively large populations of various species of big-game mammals, such as white-tailed deer, elk, **bison** (*Bison bison*), and **grizzly bear** (*Ursus arctos*). Because of the large populations in the park, the overall abundance of game species on the greater landscape is also larger. This means that a relatively high intensity of hunting can be supported. This is considered important because it provides local people with meat and subsistence as well as economic opportunities through guiding and the marketing of equipment, accommodations, food, fuel, and other necessities to non-local hunters.

By providing a game-preserve function for the larger landscape, wildlife refuges and other kinds of protected areas help to ensure that hunting can be managed to allow populations of exploited species to be sustained, while providing opportunities for people to engage in subsistence and economic activities.

[*Bill Freedman Ph.D.*]

RESOURCES
BOOKS

Freedman, B. *Environmental Ecology.* 2nd ed. San Diego: Academic Press, 1995.

Miller, G. T. *Resource Conservation and Management.* Belmont, CA: Wadsworth Publishing Co., 1990.

Owen, O. S., and D. D. Chiras. *Natural Resource Conservation. Management for a Sustainable Future.* Englewood Cliffs, NJ: Prentice Hall, 1995.

Robinson, W.L., and E. G. Bolen. *Wildlife Ecology and Management.* 3rd ed. New York: MacMillan Publishing Co., 1995.

Gamma ray

High energy forms of electromagnetic radiation with very short wavelengths. Gamma rays are emitted by cosmic sources or by **radioactive decay** of atomic nuclei which occurs during nuclear reactions or the detonation of **nuclear weapons**. Gamma rays are the most penetrating of all forms of nuclear radiation. They travel about 100 times deeper into human tissue than beta particles and 10,000 times deeper than alpha particles. Gamma rays cause chemical changes in cells through which they pass. These changes can result in the cells' death or the loss of their ability to function properly. Organisms exposed to gamma rays may suffer illness, genetic damage, or death. Cosmic gamma rays do not usually pose a danger to life because they are absorbed as they travel through the **atmosphere**. *See also* Ionizing radiation; Radiation exposure; Radioactive fallout

Mohandas Karamchand Gandhi (1869 – 1948)
Indian Religious leader

Mohandas Karamchand Gandhi led the movement that freed India from colonial occupation by the British. His leadership was based not only on his political vision but also on his moral, economic, and personal philosophies. Gandhi's beliefs have influenced many political movements throughout the world, including the civil rights movement in the United States, but their relevance to the modern environmental movement has not been widely recognized or understood until recently.

In developing the principles that would enable the Indian people to form a united independence movement, one of Gandhi's chief concerns was preparing the groundwork for an economy that would allow India to be both self-sustaining and egalitarian. He did not believe that an independent economy in India could be based on the Western model; he considered a consumer economy of unlimited growth impossible in his country because of the huge population base and the high level of poverty. He argued instead for the development of an economy based on the careful use of indigenous **natural resources**. His was a philosophy of **conservation**, and he advocated a lifestyle based on limited consumption, **sustainable agriculture**, and the utilization of labor resources instead of imported technological development.

Gandhi's plans for India's future were firmly rooted both in moral principles and in a practical recognition of its economic strengths and weaknesses. He believed that the key to an independent national economy and a national sense of identity was not only indigenous resources but indigenous products and industries. Gandhi made a point of wearing only homespun, undyed cotton clothing that had been handwoven on cottage looms. He anticipated that the practice of wearing homespun cotton cloth would create an industry for a product that had a ready market, for cotton was a resource that was both indigenous and renewable. He recognized that India's major economic strength was its vast labor pool, and the low level of technology needed for this product would encourage the development of an industry that was highly decentralized. It could provide employment without encouraging mass **migration** from rural to urban areas, thus stabilizing rural economies and national demography. The use of cotton textiles would also prevent dependence on expensive synthetic fabrics that had to be imported from Western nations, consuming scarce foreign exchange. He also believed that synthetic textiles were not suited to India's **climate**, and that they created an undesirable distinction between the upper classes that could afford them and the vast majority that could not.

The essence of his economic planning was a philosophical commitment to living a simple lifestyle based on need. He believed it was immoral to kill animals for food and advocated **vegetarianism**; advocated walking and other simple forms of **transportation** contending that India could not afford a car for every individual; and advocated the integration of ethical, political, and economic principles into individual lifestyles. Although many of his political tactics, particularly his strategy of civil disobedience, have been widely embraced in many countries, his economic philosophies have had a diminishing influence in a modern, independent India, which has been pursuing sophisticated technologies and a place in the global economy. But to some,

Mohandas Karamchand Gandhi. (AP/Wide World Photos. Reproduced by permission.)

his work seems increasingly relevant to a world with limited resources and a rapidly growing population.

[*Usha Vedagiri and Douglas Smith*]

RESOURCES
BOOKS

Chadha, Yogesh. *Gandhi: A Life*. Wiley, 1998.
Fischer, L. *The Life of Mahatma Gandhi*. New York: Harper, 1950.
Mehta, V. *Mahatma Gandhi and His Apostles*. New York: Viking, 1977.

Garbage

In 1999, the United States generated 230 million tons of **municipal solid waste**, compared with 195 million tons in 1990, according to **Environmental Protection Agency** (EPA) estimates. On average, each person generated 4.6 lb (2.1 kg) of such waste per day in 1999, and the EPA expects that amount to continue increase. That waste includes cans, bottles, newspapers, paper and plastic packages, uneaten food, broken furniture and appliances, old tires, lawn clip-

pings, and other refuse. This waste can be placed in landfills, incinerated, recycled, or in some cases composted.

Landfilling—waste disposed of on land in a series of layers that are compacted and covered, usually with soil—is the main method of **waste management** in this country, accounting for about 57% of the waste. But old landfills are being closed and new ones are hard to site because of community opposition. Landfills once were open dumps, causing unsanitary conditions, **methane** explosions, and releases of hazardous **chemicals** into **groundwater** and air. Old dumps make up 22% of the sites on the Superfund **National Priorities List**. Today, landfills must have liners, gas collection systems, and other controls mandated under Subtitle D of the **Resource Conservation and Recovery Act** (RCRA).

Incineration has been popular among **solid waste** managers because it helps to destroy bacteria and toxic chemicals and to reduce the volume of waste. But public opposition, based on fears that toxic metals and other chemical emissions will be released from incinerators, has made the siting of new facilities extremely difficult. In the past, garbage burning was done in open fields, in dumps, or in backyard drums, but the **Clean Air Act** (1970) banned open burning, leading to new types of incinerators, most of which are designed to generate energy.

Recycling, which consists of collecting materials from waste streams, preparing them for market, and using those materials to manufacture new products, is catching national attention as a desirable waste management method. As of 1999, all states and the District of Columbia had some type of statewide recycling law aimed at promoting greater recycling of glass, paper, metals, **plastics**, and other materials. Used oil, household batteries, and lead-acid automotive batteries are recyclable waste items of particular concern because of their toxic constituents.

Composting is a waste management approach that relies on heat and microorganisms—mostly bacteria and fungi—to decompose yard wastes and food scraps, turning them into a nutrient-rich mix called **humus** or compost. This mix can be used as **fertilizer**. However, as with landfills and incinerators, composting facilities have been difficult to site because of community opposition, in part because of the disagreeable smell generated by some composting practices.

Recently, waste managers have shown interest in source reduction, reducing either the amount of garbage generated in the first place or the toxic ingredients of garbage. Reusable blankets instead of throw-away cardboard packaging for protecting furniture is one example of source reduction. Businesses are regarded as a prime target for source reduction, such as implementing double-sided photocopying to save paper, because the approach offers potentially large cost savings to companies.

[*David Clarke*]

RESOURCES
BOOKS

Blumberg, L., and R. Gottlieb. *War on Waste: Can America Win Its Battle With Garbage?* Washington, DC: Island Press, 1989.
Underwood, J., A. Hershkowitz, and M. de Kadt. *Garbage—Practices, Problems, Remedies.* New York: INFORM, 1988.
U.S. Office of Technology Assessment. *Facing America's Trash: What Next For Municipal Solid Waste?* Washington, DC: U.S. Government Printing Office, 1989.

Garbage Project

The **Garbage** Project was founded in 1973, shortly after the first **Earth Day**, by William Rathje, professor of anthropology, and fellow archaeologists at the University of Arizona. The objective was to apply the techniques and tools of their science to the study of modern civilization by analyzing its garbage.

Using sample analysis and assessing biodegradation, they also hoped to increase their understanding of resource depletion and environmental and landfill-related problems. Because it requires sunlight, moisture, and oxygen, as well as organic material and bacteria, little biodegradation actually takes place in landfills, resulting in perfectly preserved heads of lettuce, 40-year-old hot dogs, and completely legible 50-year-old newspapers.

In *Rubbish: The Archaeology of Garbage*, published in 1992, Rathje and *Atlantic Monthly* managing editor Cullen Murphy discuss some of the data gleaned from the project. For example, the accumulation of refuse has raised the City of New York 6–30 ft (1.8–9 m) since its founding. Also, the largest proportion—40%—of landfilled garbage is paper, followed by the leftovers from building construction and demolition. In fact, newspapers alone make up about 13% of the total volume of trash.

Just as interesting as what they found was what they did not find. Contrary to much of public opinion, fast-food packaging made up only one-third of 1% of the total volume of trash landfilled between 1980 and 1989, while expanded **polystyrene** foam accounted for no more than 1%. Even **disposable diapers** averaged out at only 1% by weight of the total **solid waste** contents (1.4% by volume). Of all the garbage examined, **plastics** constituted from 20–24%. Surveys of several national landfills revealed that organic materials made up 40–52% of the total volume of waste.

The Garbage Project also debunked the idea that the United States is running out of space for landfills. While it is true that many landfills have been shut down, it is also true that many of those were quite small to begin with and that they now pose fewer environmental hazards. It is estimated that one **landfill** 120 ft (35.4 m) deep and measuring 44 mi^2 (71 km^2) would adequately handle the needs of the entire nation for the next 100 years (assuming current levels of waste production).

In "A Perverse Law of Garbage," Rathje extrapolated from "Parkinson's Law" to define his Parkinson's Law of Garbage: "Garbage expands so as to fill the receptacles available for its containment." As evidence he cites a Garbage Project study of the recent mechanization of garbage pick-up in some larger cities and the ensuing effects. As users were provided with increasingly larger receptacles (in order to accommodate the mechanized trucks), they continued to fill them up. Rathje attributes this to the newfound convenience of disposing of that which previously had been consigned to the basement or secondhand store, and concludes that the move to automation may be counterproductive to any attempt to reduce garbage and increase **recycling**.

[*Ellen Link*]

RESOURCES
BOOKS

Rathje, W. L., and C. Murphy. *Rubbish!: The Archaeology of Garbage.* New York: Harper Collins, 1992.

PERIODICALS

Lilienfeld, R. M. "Six Enviro-Myths: Recycling is the Key." *New York Times* (January 21, 1995): 23(L).
Rathje, W. L. "A Perverse Law of Garbage." *Garbage* 4, no. 6 (December-January 1993): 22.

Garbology

The study of **garbage**, through either archaeological excavation of landfills or analysis of fresh garbage, to determine what the composition of **municipal solid waste** says about the society that generated it. The term is associated with the **Garbage Project** of the University of Arizona (Tucson), co-directed by William Rathje and Wilson Hughes, which began studying trash in 1973 and excavating landfills in 1987. They found that little degradation occurs in landfills; New York newspapers from 1949 and an ear of corn discarded in 1971 were found intact. Unexpectedly, the project also found that **plastics** make up less than one percent of the total volume of landfills.

Gardens

see **Botanical garden; Organic gardening and farming**

Gasohol

Gasohol is a term used for the mixture of 10% ethyl alcohol (also called **ethanol** or grain alcohol) with **gasoline**. Ethanol raises the **octane rating** of lead-free **automobile** fuel and significantly decreases the **carbon monoxide** released from tailpipes. It has also been promoted as a means of reducing corn surpluses. By 2001, 2.2 billion gal (8.3 billion l) were being produced a year and this number is expected to rise to 4.6 billion gal (17.4 billion l). However, ethanol also raises the vapor pressure of gasoline, and it has been reported to increase the release of "evaporative" volatile **hydrocarbons** from the fuel system and oxides of **nitrogen** from the exhaust. These substances are components of urban **smog**, and thus the role of ethanol in reducing **pollution** is controversial.

Gasoline

Crude oil in its natural state has very few practical uses. However, when it is separated into its component parts by the process of fractionation, or refining, those parts have an almost unlimited number of applications.

In the first 60 years after the process of **petroleum** refining was invented, the most important fraction produced was kerosene, widely used as a home heating product. The petroleum fraction slightly lighter than kerosene — gasoline — was regarded as a waste product and discarded. Not until the 1920s, when the **automobile** became popular in the United States, did manufacturers find any significant use for gasoline. From then on, however, the importance of gasoline has increased with automobile use.

The term gasoline refers to a complex mixture of liquid **hydrocarbons** that condense in a fractionating tower at temperatures between 100° and 400°F (40° and 205°C). The hydrocarbons in this mixture are primarily single- and double-bonded compounds containing five to 12 **carbon** atoms.

Gasoline that comes directly from a refining tower, known as naphtha or "straight-run" gasoline, was an adequate fuel for the earliest motor vehicles. But as improvements in internal **combustion** engines were made, problems began to arise. The most serious problem was "knocking."

If a fuel burns too rapidly in an internal combustion engine, it generates a shock wave that makes a "knocking" or "pinging" sound. The shock wave will, over time, also cause damage to the engine. The hydrocarbons that make up straight-run gasolines proved to burn too rapidly for automotive engines developed after 1920.

Early in the development of automotive fuels, engineers adopted a standard for the amount of knocking caused by a fuel and, hence, for the fuel's efficiency. That standard

was known as "octane number." To establish a fuel's octane number, it is compared with a very poor fuel (n-heptane), assigned an octane number of zero, and a very good fuel (isooctane), assigned an octane number of 100. The octane number of straight-run gasoline is anywhere from 50 to 70.

As engineers made more improvements in automotive engines after the 1920s, chemists tried to keep pace by developing better fuels. One approach they used was to subject straight-run gasoline (as well as other crude oil fractions) to various treatments that changed the shape of hydrocarbon molecules in the gasoline mixture. One such method, called cracking, involves the heating of straight-run gasoline or another petroleum fraction to high temperatures. The process results in a better fuel from newly-formed hydrocarbon molecules.

Another method for improving the quality of gasoline is catalytic reforming. In this case, the cracking reaction takes place over a catalyst such as **copper**, platinum, rhodium, or other "noble" metal, or a form of clay known as zeolite. Again, hydrocarbon molecules formed in the fraction are better fuels than straight-run gasoline. Gasoline produced by catalytic cracking or reforming has an octane number of at least eighty.

A very different approach to improving gasoline quality is the use of additives, **chemicals** added to gasoline to improve the fuel's efficiency. Automotive engineers learned more than 50 years ago that adding as little as two grams of **tetraethyl lead**, the best-known additive, to one gallon of gasoline raises its octane number by as much as ten points.

Until the 1970s, most gasolines contained tetraethyl **lead**. Then, concerns began to grow about the release of lead to the **environment** during the combustion of gasoline. Lead concentrations in urban air had reached a level five to 10 times that of rural air. Residents of countries with few automobiles, such as Nepal, had only one-fifth the lead in their bodies as did residents of nations such as the United States, with many automotive vehicles.

The toxic effects of lead on the human body have been known for centuries, and risks posed by leaded gasoline became a major concern. In addition, leaded gasoline became a problem because it damaged a car's **catalytic converter**, which reduced air pollutants in exhaust.

Finally, in 1973, the **Environmental Protection Agency** (EPA) acted on the problem and set a time-scale for the gradual elimination of leaded fuels. According to this schedule, the amount of lead was to be reduced from 2 to 3 grams per gallon (the 1973 average) to 0.5 g/gal by 1979. Ultimately, the additive was to be totally eliminated from all gasoline.

The elimination of leaded fuels has been made possible by the invention of new and safer additives. One of the most popular is methyl-t-butyl ether (MTBE). By 1988 MTBE had become so popular that it was among the 40 most widely produced chemicals in the United States. In 2001, MTBE was asked to be phased out of production in California by 2003.

Yet another approach to improving fuel efficiency is the mixing of gasoline and ethyl or **methanol**. This product, known as **gasohol**, has the advantage of high **octane rating**, lower cost, and reduced **emission** of pollutants, compared to normal gasoline. *See also* Air pollution; Alternative fuels

[*David E. Newton*]

RESOURCES
BOOKS

Joesten, M. D., et al. *World of Chemistry*. Philadelphia: Saunders, 1991.
Lapedes, D. N., ed. *McGraw-Hill Encyclopedia of Energy*. New York: McGraw-Hill, 1976.

PERIODICALS

"MBTE Growth Limited Despite Lead Phasedown in Gasoline." *Chemical & Engineering News* (July 15, 1985): 12.
Williams, R. "On the Octane Trail." *Technology Illustrated* (May 1983): 52–53.

Gasoline tax

Gasoline taxes include federal, state, county, and municipal taxes imposed on gasoline motor vehicle fuel. In the United States, most of the federal tax is used to fund maintenance and improvements in such **transportation** infrastructures as interstate highways. As of mid-2002, the federal excise tax for gasoline stood at 18.4 cents per gallon, and state excise taxes ranged from 7.5 cents in Georgia to 29 cents in Rhode Island (making the weighted national average state tax 19.97 cents per gallon). In total, the U.S. national average gasoline tax (combining federal and state) was 38 cents per gallon.

Oregon became the first state to institute a tax on gasoline in 1919. By the time the federal government established its own 1 cent gas tax in 1932, every state had a gas tax. After several small increases in the 1930s and 1940s, the gas tax was raised to 3 cents to finance the Highway Trust Fund in 1956. The Trust Fund was earmarked to pay for federal interstate construction and other road work. In 1982, the federal gasoline tax was increased to 9 cents to fund road maintenance and **mass transit**. The tax was hiked again in 1990 to 14.1 cents, and to 18.4 cents in 1993—where it remained as of July 2002.

Over this time, gasoline prices increased from about 20 cents per gallon in 1938 to a U.S. national average of 139.2 cents in May of 2002. The average national gasoline tax (both federal and a weighted average of state taxes) accounts for 30.2% of the retail price of a gallon of gas.

In some countries, diesel fuels are taxed and priced less than gasoline. Commercial vehicles are major consumers of diesel, and lower taxes avoided undue impacts on trucking and commerce. In the United States, diesel is taxed at a higher rate than gasoline—an average of 44 cents per gallon (including 24.3 cents federal tax and the weight average of state taxes). In contrast, **gasohol**, an alternative fuel consisting of 90% gasoline and 10% **ethanol**, has a lower rate of taxation (with a 13.1 cent federal excise tax).

Although federal gasoline taxes are a manufacturer's excise tax, meaning that the government collects the tax directly from the manufacturer, rate hikes are often passed on to consumers at the pump. In this light, gasoline taxes have been criticized as regressive and thus inequitable—i.e., lower income individuals pay a greater share of their income as tax than higher income individuals. Also, the tax as a share of the pump price has been increasing. However, it should be noted that the general trend of the real price of gasoline (adjusted for inflation and including taxes) has been declining for many decades. In addition, **automobile** efficiency (mileage) has been improving and thus the fuel requirements and cost-per-mile-traveled have declined.

These factors can be viewed as largely or completely offsetting the impact of gasoline taxes. For example, Congress attempted in May 1996 to rollback the 4.3 cents tax increase of 1993. The impact of this repeal for a family of four who drive 12,000 miles a year at 20 miles per gallon is a savings of $26, which in the House debates was compared to the cost of a family dinner at McDonald's. On the other hand, a rollback could have bigger consequences for the future upkeep of the country's highways and interstates; a 2000 Congressional report estimated that a repeal of the federal gasoline tax would mean a $5.2 billion annual loss in revenues for the Highway Trust Fund.

Outside of the United States, both gasoline prices and gas tax rates are typically far higher (e.g., gasoline taxes were 307 cents per gallon in the United Kingdom as of March 2002). In addition to funding governments, high gasoline taxes form part of a strategy to encourage the use of public transportation, reduce **pollution**, conserve energy, and improve national security (since most gasoline is imported).

[*Stuart Batterman and Paula Anne Ford-Martin*]

RESOURCES
PERIODICALS

Talley, Louis Allen. Congressional Research Service. "The Federal Excise Tax on Gasoline and the Highway Trust Fund: A Short History" *CRS Report for Congress*. Washington, DC: Congressional Research Service, March 2000.

OTHER

American Petroleum Institute, Policy Analysis and Statistics Department. *How Much We Pay for Gasoline: April 2001 Review*. Washington, DC: API, 2001.

U.S. Department of Energy, Energy Information Administration. *Gasoline and Diesel Fuel Update*. [cited July 9, 2002]. <http://tonto.eia.doe.gov/oog/info/gdu/gasdiesel.asp>.

U.S. Department of Transportation, Federal Highway Administration. *Our Nation's Highways: Selected Facts and Figures 2000*. Publication No. FHWA-PL-01-1012. [cited July 6, 2002]. <http://www.fhwa.dot.gov/ohim/onh00/onh.htm>.

Gastropods

Gastropods are invertebrate animals that make up the largest class in the phylum Mollusca. Examples of common gastropods include all varieties of snails, abalone, limpets, and land and sea slugs. There are over 35,000 existing **species**, as well an additional 15,000 separate fossil species. Gastropods first appeared in the fossil record during the early Cambrian period, approximately 550 million years ago.

This diverse group of animals is characterized by a soft body, made up of three main parts: the head, foot, and visceral mass. The head contains a mouth and often sensing tentacles. The lower portion of the body makes up the foot, which allows slow creeping along rocks and other solid surfaces. The visceral mass is the main part of the body, containing most of the internal organs. In addition to these body parts, gastropods possess a mantle, or fold which secretes a hard, calcium carbonate shell. The single, asymmetrical shell of a gastropod is most often spiral shaped, however, it can be flattened or cone-like. This shell is an important source of protection. Predators have a difficult time accessing the soft flesh inside, especially if there are sharp points on the outside, as there are on the shells of some of the more ornate gastropods. There are also some gastropods, such as slugs and sea hares, that do not have shells or have greatly reduced shells. Some of the shelless types that live in the ocean (i.e., nudibranchs or sea slugs) are able to use stinging cells from prey that they have consumed as a means of protection.

In addition to a spiraling of their shells, the soft bodies of most gastropods undergo 180 degrees of twisting, or torsion, during early development, when one side of the visceral mass grows faster than the other. This characteristic distinguishes gastropods from other molluscs. Torsion results in a U-shaped digestive tract, with the anal opening slightly behind the head. The torsion of the soft body and the spiraling of the shell are thought to be unrelated evolutionary events.

Gastropods have evolved to live in a wide variety of habitats. The great majority are marine, living in the world's oceans. Numerous species live in fresh water, while others live entirely on land. Of those that live in water, most are found on the bottom, attached to rocks or other surfaces. There are even a few species of gastropods without shells, including sea butterflies, that are capable of swimming. Living in different habitats has resulted in a wide variety of structural adaptations within the class Gastropoda. For example, those gastropods that live in water use gills to obtain the oxygen necessary for **respiration**, while their terrestrial relatives have evolved lungs to breathe.

Gastropods are important links in food webs in the habitats in which they live, employing a wide variety of feeding strategies. For example, most gastropods move a rasping row of teeth on a tongue like organ called a radula back and forth to scrape microscopic algae off rocks or the surface of plants. Because the teeth on the radula gradually wear away, new teeth are continuously secreted. Other gastropods have evolved a specialized radula for drilling through shells of animals to get at their soft flesh. For example, the oyster drill, a small east coast gastropod, bores a small hole in the shell of neighboring molluscs such as oysters and clams so that it can consume the soft flesh. In addition, some terrestrial gastropods such as snails use their radula to cut through pieces of leaves for food.

Gastropods are eaten by numerous animals, including various types of fish, birds, and mammals. They are also eaten by humans throughout the world. Abalone, muscular shelled gastropods that cling to rocks, are consumed on the west coast of the United States and in Asia. Fritters and chowder are made from the large, snail-like queen conch on many Caribbean islands. Escargot (snails in a garlicky butter sauce) are a European delicacy.

[*Max Strieb*]

RESOURCES
BOOKS

Barnes, R. D. *Invertebrate Zoology*. 5th ed. Philadelphia: Saunders College Publishing, 1987.

Gene bank

The term gene bank refers to any system by which the genetic composition of some population is identified and stored. Many different kinds of gene banks have been established for many different purposes. Perhaps the most numerous gene banks are those that consist of plant seeds, known as germ banks.

The primary purpose for establishing a gene bank is to preserve examples of threatened or **endangered species**. Each year, untold numbers of plant and animal **species** become extinct because of natural processes and more commonly, as the result of human activities. Once those species become extinct, their gene pools are lost forever.

Scientists want to retain those gene pools for a number of reasons. For example, agriculture has been undergoing a dramatic revolution in many parts of the world over the past half century. Scientists have been making available to farmers plants that grow larger, yield more fruit, are more disease-resistant, and have other desirable characteristics. These plants have been produced by agricultural research in the United States and other nations. Such plants are very attractive to farmers, and they are also important to governments as a way of meeting the food needs of growing populations, especially in **Third World** countries.

When farmers switch to these new plants, however, they often abandon older, more traditional crops that may then become extinct. Although the traditional plants may be less productive, they have other desirable characteristics. They may, for example, be able to survive droughts or other extreme environmental conditions that new plants cannot.

Placing seeds from traditional plants in a gene bank allows them to be preserved. At some later time, scientists may want to study these plants further and perhaps identify the genes that are responsible for various desirable properties of the plants. The **U.S. Department of Agriculture** (USDA) has long maintained a **seed bank** of plants native to the United States. About 200,000 varieties of seeds are stored at the USDA's Station at Fort Collins, Colorado, and another 100,000 varieties are kept at other locations around the country.

Efforts are now underway to establish gene banks for animals, too. Such banks consist of small colonies of the animals themselves. Animal gene banks are desirable as a way of maintaining species whose natural population is very low. Sometimes the purpose of the bank is simply to maintain the species to prevent its becoming extinct. In other cases, species are being preserved because they were once used as farm animals although they have since been replaced by more productive modern hybrid species. The Fayoumi chicken native to Egypt, for example, has now been abandoned by farmers in favor of imported species. The Fayoumi, without some form of protection, is likely to become extinct. Nonetheless, it may well have some characteristics (genes) that are worth preserving.

In recent years, another type of gene bank has become possible. In this kind of gene bank, the actual base sequence of important genes in the human body will be determined, collected, and catalogued. This effort, begun in 1990, is a part of the Human Genome Project effort to map all human genes. *See also* Agricultural revolution; Extinction; Genetic engineering; Population growth

[*David E. Newton*]

RESOURCES
PERIODICALS

Anderson, C. "Genetic Resources: A Gene Library That Goes 'Moo'." *Nature* 355 (January 30, 1992): 382.

Crawford, M. "USDA Bows to Rifkin Call for Review of Seed Bank." *Science* 230 (December 6, 1985): 1146–1147.

Roberts, L. "DOE to Map Expressed Genes." *Science* 250 (November 16, 1990): 913.

Gene pool

The term gene pool refers to the sum total of all the genetic information stored within any given population. A gene is a specific portion of a DNA (**deoxyribose nucleic acid**) molecule, so a gene pool is the sum total of all of the DNA contained within a population of individuals.

The concept of gene pool is important in ecological studies because it reveals changes that may or may not be taking place within a population. In a population living in an ideal **environment** for its needs, the gene pool is likely to undergo little or no change. If individuals are able to obtain all the food, water, energy, and other resources they need, they experience relatively little stress and there is no pressure to select one or another characteristic.

Changes do occur in gene frequency because of natural factors in the environment. For example, natural **radiation exposure** causes changes in DNA molecules that are revealed as genetic changes. These natural mutations are one of the factors that make possible continuous changes in the genetic constitution of a population that, in turn, allows for **evolution** to occur.

Natural populations seldom live in ideal situations, however, and so they experience various kinds of stress that lead to changes in the gene pool. A classical example of this kind of change was reported by J. B. S. Haldane in 1937. Haldane found that a population of moths gradually became darker in color over time as the trees on which they lived also became darker because of **pollution** from factories. Moths in the population who carried genes for darker color were better able to survive and reproduce than were their lighter-colored cousins, so the composition of the gene pool changed to relieve stress.

Humans have the ability to make conscious changes in gene pools that no other **species** has. Sometimes we make those changes in the gene pools of plants or animals to serve our own needs for food or other resource. Hybridization of plants to produce populations that have some desirable quality such as **resistance** to disease, shorter growing season, or better-tasting fruit. The modern science of **genetic engineering** is perhaps the most specific and deliberate way of changing in gene pools today.

Humans can also change the gene pool of their own species. For example, individuals with various genetic disorders were at one time doomed to death. Our inability to treat diabetes, sickle-cell **anemia**, phenylketonuria, and other hereditary conditions meant that the frequency of the genes causing those disorders in the human gene pool was kept under control by natural forces.

Today, many of those same disorders can be treated by medical or genetic techniques. That results in positive benefit for the individuals who are cured, but raises questions about the quality of the human gene pool overall. Instead of having many of those deleterious genes being lost naturally by an individual's death, they are now retained as part of the gene pool. This fact has at times raised questions about the best way in which medical science should deal with genetic disorders. *See also* Agricultural revolution; Birth defects; Extinction; Gene bank; Population growth

[*David E. Newton*]

RESOURCES

BOOKS

Patt, D. I., and G. R. Patt. *An Introduction to Modern Genetics.* Reading, MA: Addison-Wesley, 1976.

Genetic engineering

Genetic engineering is the manipulation of the hereditary material of organisms at the molecular level. The hereditary material of most cells is found in the chromosomes, and it is made of **deoxyribose nucleic acid** (DNA). The total DNA of an organism is referred to as its genome. In the 1950s, scientists first discovered how the structure of DNA molecules worked and how they stored and transmitted genetic information.

Genetic engineering relies on recombinant DNA technology to manipulate genes. Methods are now available for rapidly sequencing the nucleotides of pieces of DNA, as well as for identifying particular genes of interest, and for isolating individual genes from complex genomes. This allows genetic engineers to alter genetic materials to produce new substances or create new functions.

The biochemical tools used by genetic engineers or molecular biologists include a series of enzymes that can "cut and paste" genes. Enzymes are used to cut a piece of DNA, insert into it a new piece of DNA from another organism, and then seal the joint.

One important group of these are restriction enzymes, of which well over 500 are known. Most restriction enzymes

are endonucleases—enzymes that break the double helix of DNA within the molecule, rather than attacking the ends of the helix. Every restriction **enzyme** is given a specific name to identify it uniquely. The first three letters, in italics, indicate the biological source of the enzyme, the first letter being the initial of the genus, the second and third letters being the first two letters of the **species** name. Thus restriction enzymes from **Escherichia coli** are called *Eco*, those from *Haemophilus influenzae* are *Hin*, from *Diplococcus pneumoniae* comes *Dpn*, and so on.

The genetic engineer can use a restriction enzyme to locate and cut almost any sequence of bases. Cuts can be made anywhere along the DNA, dividing it into many small fragments or a few longer ones. The results are repeatable: cuts made by the same enzyme on a given sort of DNA will always be the same. Some enzymes recognize sequences as long as six or seven bases; these are used for opening a circular strand of DNA at just one point. Other enzymes have a smaller recognition site, three or four bases long; these produce small fragments that can then be used to determine the sequence of bases along the DNA.

The cut that each enzyme makes varies from enzyme to enzyme. Some, like *Hin* dII, make a clean cut straight across the double helix, leaving DNA fragments with ends that are flush. Other enzymes (*Eco* RI) make a staggered break, leaving single strands with protruding cohesive ends ("sticky ends") that are complementary in base sequence. Following breakage, and under the right conditions, the complementary bases from different sources can be rejoined to form recombinant DNA.

Another important biochemical tool used by genetic engineers is DNA polymerase, an enzyme that normally catalyses the growth of a **nucleic acid** chain. DNA polymerase is used by genetic engineers to seal the gaps between the two sets of fragments in newly joined chimera molecules of recombinant DNA. DNA polymerase is also used to label DNA fragments, for DNA polymerase labels practically every base, allowing minute quantities of DNA to be studied in detail. If a piece of **ribonucleic acid** (RNA) of the target gene is the starting point, then the enzyme reverse transcriptase is used to produce a strand of complementary DNA (cDNA).

Genetic engineers usually need large numbers of genetically identical copies of the DNA fragment of interest. One way of doing this is to insert the gene into a suitable gene carrier, called a **cloning** vector. Common cloning vectors are bacterial plasmids or viruses such as the bacteriophage lambda, which are small circles of DNA found in bacterial cells independently of the main DNA molecule. When the cloning vectors divide, they replicate both themselves and the foreign DNA segment linked to it.

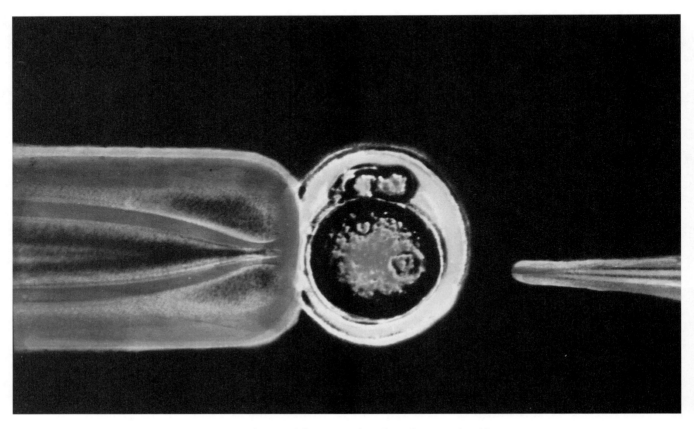

DNA injected into a mouse embryo. (Photograph by Jon Gordon. Phototake. Reproduced by permission.)

In the plasmid insertion method, restriction enzymes are used to cleave the plasmid double helix so that a stretch of DNA (previously cleaved with the same enzyme) can be inserted into the plasmid. As a result, the "sticky ends" of the plasmid DNA and the foreign DNA are complementary and base-pair when mixed together. The fragments held together by base pairing are permanently joined by DNA ligase. The host bacterium, with its 20- to 30-minute reproductive cycle, is like a manufacturing plant. With repeated doublings of its offspring on a controlled culture medium, millions of clones of the purified DNA fragments can be produced overnight.

Similarly, if viruses (bacteriophages) are used as cloning vectors, the gene of interest is inserted into the phage DNA, and the **virus** is allowed to enter the host bacterial cell where it multiplies. A single parental lambda phage particle containing recombinant DNA can multiply to several hundred progeny particles inside the bacterial cell (*E. coli*) within roughly 20 minutes.

Cosmids are another type of viral cloning vehicle that attaches foreign DNA to the packaging sites of a virus and thus introduces the foreign DNA into an infective viral particle. Cosmids allow researchers to insert very long stretches of DNA into host cells where cell multiplication amplifies the amount of DNA available. Large artificial chromosomes of yeast (called megaYACs) are also used as cloning vehicles, since they can store even larger pieces of DNA, 35 times more than can be stored conventionally in bacteria.

The polymerase **chain reaction** (PCR) technique is an important new development in the field of genetic engineering, since it allows the mass production of short segments of DNA directly, and offers the advantage of bypassing the several steps involved in using bacterial and viruses as cloning vectors.

DNA fragments can be introduced into mammalian cells, but a different method must be used. Here, genes packed in solid calcium phosphate are placed next to a cell membrane that surrounds the fragment and transfers it to the cytoplasm. The gene is delivered to the nucleus during mitosis (when the nuclear membrane has disappeared) and the DNA fragments are incorporated into daughter nuclei, then into daughter cells. A mouse containing human **cancer** genes (the onchomouse) was patented in 1988.

The potential benefits of recombinant DNA research are enormous. In recent years, scientists have identified the genetic basis of a number of medical disorders. Genetic engineering helps scientists replace a particular missing or defective gene with correct copies of that gene. If that gene then begins functioning in an individual, a genetic disorder may be cured. Researchers have already met with great success in finding the single genes that are responsible for common diseases like cystic **fibrosis** and hemochromatosis. In the early twenty-first century, scientists were applying knowledge from the human genome project to try to map multiple genes responsible for diseases like diabetes, hypertension, and schizophrenia.

Genetic engineering has also led to a great deal of controversy, however. Since scientists can duplicate genes, they can also duplicate, or "clone" animals and humans. In December 2001, scientists announced the birth of the first cloned female cat, although the 1990s saw big headlines for the first big mammal cloning of sheep. The announcement was made in April 2002 that the first human baby produced by a human cloning program would be born in November, sparking considerable medical and ethical controversy. In August 2001, President George W. Bush struggled with the debate over allowing federal funding for embryonic stem cell research. He allowed funding for only existing lines of cells, leaving further research in the hands of those who could seek private funding.

Genetic engineering also presents positive and negative controversy in its application to agriculture and the **environment**. Recombinant DNA techniques help scientists produce plants that offer medicinal value too. For example, calcium-fortified orange juice or vitamin-enriched milk boost the nutritional value of these foods. However, scientists can now further genetically modify foods, offering benefits and risks to the environment. The benefits mean possible cheaper and easier production of certain medicines, while the risks include unnatural introduction of plants and animals produced in a laboratory environment. At the University of Arizona, scientists have modified tomatoes to produce vaccines for diarrhea and hepatitis B, and that these vaccines will likely be much cheaper to produce than current drugs. Some critics worry that once the new crops move out of the safety of locked greenhouses and into crop fields, them may cross-pollinate with conventional tomato crops and contaminate them with modified genes.

A poll reported on in 2002 showed that Americans were fairly evenly split on their feelings about the risks and benefits of genetically modified foods and **biotechnology**. Most like the idea that scientists can create plants that will help clean up toxic soils, reduce **soil erosion** and reduce **fertilizer** run-off into streams and lakes. They also favor production of genetically engineered methods to reduce the amount of water used to grow crops and development of disease-resistant tree varieties to replace those that might be threatened or endangered. Americans also favor use of genetic engineering to reduce the need to log in native forests and to reduce the amount of chemical pesticides used by farmers. On the other hand, Americans express concern over possible environmental effects of genetically modified plants or fish contaminating ordinary plants and fish, reducing genetic diversity, increasing the number of insects that might become resistant to pesticides, or changing the **ecosystem**. *See also* Gene bank; Gene pool

[*Neil Cumberlidge Ph.D.*]

RESOURCES

BOOKS

Cherfas, J. *Man Made Life: An Overview of the Science and Technology and Commerce of Genetic Engineering.* New York: Pantheon Books, 1982.
Watson, J. D, J. Tooze, and D. T. Kurtz. *Recombinant DNA: A Short Course.* San Francisco: W. H. Freeman, 1983.
Wheale, P. R., and R. M. McNally. *Genetic Engineering: Catastrophe or Utopia?* New York: St. Martin's Press, 1988.

PERIODICALS

"Americans Evenly Divided Over Environmental Risks, Benefits of Genetically Modified Food and Biotech." *Health and Medicine Week*, March 4, 2002, 18.
Coghlan, A. "Engineering the Therapies of Tomorrow." *New Scientist* 137 (April 24, 1993): 26–31.
Kahn, P. "Genome on the Production Line." *New Scientist* 137 (April 24, 1993): 32–36.
Miller, S. K. "To Catch a Killer Gene." *New Scientist* 137 (April 24, 1993): 37–40.
Primedia Intertec, Inc. "Cloning Controversy." *Better Nutrition* 64 (July 2002): 18–21.
Verma, I. M. "Gene Therapy." *Scientific American* 263 (November 1990): 68–84.

OTHER

"Genetic Engineers Blurring Lines Between Kitchen Pantry and Medicine Cabinet." *Pew Initiative on Food and Biotechnology Newsroom.* [cited July 9, 2002]. <http://pewagbiotech.org/newsroom/summaries/display.-php3?NewsID=170>.

ORGANIZATIONS

Pew Initiative on Food and Biotechnology, 1331 H Street, Suite 900, Washington, DC USA 20005 (202) 347-9044, Fax: (202) 347-9047, Email: inquiries@pewagbiotech.org, http://pewagbiotech.org

Genetic resistance (or genetic tolerance)

Genetic **resistance** (or genetic tolerance) refers to the ability of certain organisms to endure environmental conditions that are extremely stressful or lethal to non-adapted individuals of the same **species**. Such tolerance has a genetic basis, and it evolves at the population level in response to intense selection pressures.

Genetic resistance occurs when genetically variable populations contain some individuals that are relatively tolerant of an exposure to some environmental factor, such as the presence of a high concentration of a specific chemical. If the tolerance is genetically based (i.e., due to specific information embodied in the DNA of the organism's chromosomes), some or all of the offspring of these individuals will also be tolerant. Under conditions in which the chemical occurs in concentrations high enough to cause toxicity to non-tolerant individuals, the resistant ones will be relatively successful. As time passes their offspring will become increasingly more prominent in the population. Acquiring genetic resistance is an evolutionary process, involving increased tolerance within a population, for which there is a genetic basis, and occurring in response to selection for resistance to the effects of a toxic chemical. Some of the best examples of genetic resistance involve the tolerance of certain bacteria to antibiotics and of certain pests to pesticides.

Resistance to antibiotics

Antibiotics are **chemicals** used to treat bacterial infections of humans and domestic animals. Examples of commonly used antibiotics include various kinds of penicillins, streptomycins, and tetracyclines, all of which are metabolic byproducts created by certain **microorganisms**, especially **fungi**. There are also many synthetic antibiotics.

Antibiotics are extremely toxic to non-resistant strains of bacteria, and this has been very beneficial in the control of bacterial infections and diseases. However, if even a tiny fraction of a bacterial population has a genetically based tolerance to a specific antibiotic, **evolution** will quickly result in the development of a population that is resistant to that chemical. Bacterial resistance to antibiotics was first demonstrated for penicillin, but the phenomenon is now quite widespread. This is an important medical problem because some serious pathogens are now resistant to virtually all of the available antibiotics, which means that infections by these bacteria can be extremely difficult to control. Bacterial resistance has recently become the cause of infections by some virulent strains of *Staphylococcus* and other potentially deadly bacteria. Some biologists believe that this problem has been made worse by the failure of many people to finish their course of prescribed antibiotic treatments, which can allow tolerant bacteria to survive and flourish. Also possibly important has been the routine use of antibiotics to prevent diseases in livestock kept under crowded conditions in industrial farming. The small residues of antibiotics in meat, eggs, and milk may be resulting in low-level selection for resistant bacteria in exposed populations of humans and domestic animals.

Resistance to Pesticides

The insecticide **dichlorodiphenyl-trichloroethane** (DDT) was the first **pesticide** to which insect pests developed resistance. This occurred because the exposure of insect populations to toxic DDT results in intense selection for resistant genotypes. Tolerant populations can evolve because genetically resistant individuals are not killed by the pesticide and therefore survive to reproduce. Almost 500 species of insects and mites have populations that are known to be resistant to at least one insecticide. There are also more than 100 examples of fungicide-resistant plant pathogens and about 50 herbicide-resistant weeds. Insecticide resistance is most frequent among species of flies and their relatives (order Diptera), including more than 50 resistant species of malaria-carrying *Anopheles* mosquitoes. In fact, the progressive evolution of insecticide resistance by *Anopheles* has been an important factor in the recent resurgence of **malaria** in countries with warm climates. In addition, the protozoan *Plasmodium*, which actually causes malaria, has become resistant to some of the drugs that used to effectively control it.

Crop geneticists have recently managed to breed varieties of some plant species that are resistant to glyphosate, a commonly used agricultural **herbicide** that is effective against a wide range of weeds, including both monocots and dicots. The development of glyphosate-tolerant varieties of such crops as rapeseed means that this effective herbicide can be used to control difficult weeds in planted fields without causing damage to the crop.

[*Bill Freedman Ph.D.*]

RESOURCES
BOOKS

Freedman, B. *Environmental Ecology*. 2nd ed. San Diego: Academic Press, 1995.

Hayes, W. C., and E. R. Laws, eds. *Handbook of Pesticide Toxicology*. San Diego: Academic Press, 1991.

National Research Council (NRC). *Pesticide Resistance*. Washington, DC: National Academy Press, 1986.

Raven, P. H., and G. B. Johnson. *Biology*. 3rd ed. St. Louis: Mosby Year Book, 1992.

Genetically engineered organism

The modern science of genetics began in the mid-nineteenth century with the work of Gregor Mendel, but the nature of the gene itself was not understood until James Watson and Francis Crick announced their findings in 1953. According to the Watson and Crick model, genetic information is stored in molecules of DNA (**deoxyribose nucleic acid**) by means of certain patterns of **nitrogen** base that occur in such molecules. Each set of three such nitrogen bases were codes, they said, for some particular amino **acid**, and a long series of nitrogen bases were codes for a long series of amino acids or a protein.

Deciphering the genetic code and discovering how it is used in cells has taken many years of work since that of Watson and Crick. The basic features of that process, however, are now well understood. The first step involves the construction of a RNA (**ribonucleic acid**) molecule in the nucleus of a cell, using the code stored in DNA as a template. The RNA molecule then migrates out of the nucleus to a ribosome in the cell cytoplasm. At the ribosome, the sequence of nitrogen bases stored in RNA act as a map that determines the sequence of amino acids to be used in constructing a new protein.

This knowledge is of critical importance to biologists because of the primary role played by proteins in an organism. In addition to acting as the major building materials of which cells are made, proteins have a number of other crucial functions. All hormones and enzymes, for example, are proteins, and therefore nearly all of the chemical reactions that occur within an organisms are mediated by one protein or another.

Our current understanding of the structure and function of DNA makes it at least theoretically possible to alter the biological characteristics of an organism. By changing the kind of nitrogen bases in a DNA molecule, their sequence, or both, a scientist can change the genetic instructions stored in a cell and thus change the kind of protein produced by the cell.

One of the most obvious applications of this knowledge is in the treatment of genetic disorders. A large majority of genetic disorders occur because an organism is unable to correctly manufacture a particular protein molecule. An example is Lesch-Nyhan syndrome. It is a condition characterized by self-mutilation, mental retardation, and cerebral palsy which arises because a person's body is unable to manufacture an **enzyme** known as hypoxanthine guanine phosphoribosyl transferase (HPRT).

The general principles of the techniques required to make such changes are now well understood. The technique is referred to as **genetic engineering** or genetic surgery because it involves changes in an organism's gene structure. When used to treat a particular disorder in humans, the procedure is also called human gene therapy. Developing specific experimental techniques for carrying out genetic engineering has proved to be an imposing challenge, yet impressive strides have been made. A common procedure is known as recombinant DNA (rDNA) technology.

The first step in an rDNA procedure is to collect a piece of DNA that carries a desired set of instructions. For a genetic surgery procedure for a person with Lesch-Nyhan syndrome, a researcher would need a piece of DNA that

codes for the production of HPRT. That DNA could be removed from the healthy DNA of a person who does not have Lesch-Nyhan syndrome, or the researcher might be able to manufacture it by chemical means in the laboratory.

One of the fundamental tools used in rDNA technology is a closed circular piece of DNA found in bacteria called a plasmid. Plasmids are the vehicle or vector that scientists use for transferring new pieces of DNA into cells. The next step in an rDNA procedure, then, would be to insert the correct DNA into the plasmid vector. Cutting open the plasmid can be accomplished using certain types of enzymes that recognize specific base sequences in a DNA molecule. When these enzymes, called restriction enzymes, encounter the recognized sequence in a DNA molecule, they cleave the molecule. After the plasmid DNA has been cleaved and the correct DNA mixed with it, a second type of enzyme is added. This kind of enzyme inserts the correct DNA into the plasmid and closes it up. The process is known as gene splicing.

In the final step, the altered plasmid vector is introduced into the cell where it is expected to function. In the case of a Lesch-Nyhan patient, the plasmid would be introduced into the cells where it would start producing HPRT from instructions in the correct DNA. Many technical problems remain with rDNA technology, and this last step has caused some of the greatest obstacles. It has proven very difficult to make introduced DNA function. Even when the plasmid vector with its new DNA gets into a cell, it may never actually begin to function.

Any organism whose cells contain DNA altered by this or some other technique is called a genetically engineered organism. The first human patient with a genetic disorder who is treated by human gene therapy will be a genetically engineered organism. The use of genetic engineering on human subjects has gone forward very slowly for a number of reasons. One reason is that humans are very complex organisms. Another reason is that changing the genetic make-up of a human involves more ethical questions and more difficult questions than does the genetic engineering of bacteria, mice, or cows.

Most of the existing examples of genetically engineered organisms, therefore, involve plants, non-human animals, or **microorganisms**. One of the earliest success stories in genetic engineering involved the altering of DNA in microorganisms to make them capable of producing **chemicals** they do not normally produce. Recombinant DNA technology can be used, for instance, to insert the DNA segment or gene that codes for insulin production into bacteria. When these bacteria are allowed to grow and reproduce in large fermentation tanks, they produce insulin. The list of chemicals produced by this mechanism now includes somatostatin, alpha interferon, tissue plasminogen activator (tPA), Factor VIII, erythroprotein, and human growth hormone, and this list continues to grow each year.

[*David E. Newton*]

RESOURCES
PERIODICALS

Hoffman, C. A. "Ecological Risks of Genetic Engineering of Crop Plants." *BioScience* 40 (June 1990): 434–437.

Kessler, D. A., et al. "The Safety of Foods Developed by Biotechnology." *Science* 256 (June 1992): 1747–1749+.

Kieffer, G. H. *Biotechnology, Genetic Engineering, and Society*. Reston, VA: National Association of Biology Teachers, 1987.

Mellon, M. *Biotechnology and the Environment*. Washington, DC: National Biotechnology Policy Center of the National Wildlife Federation, 1988.

Pimentel, D., et al. "Benefits and Risks of Genetic Engineering in Agriculture." *BioScience* 39 (October 1989): 606–614.

Weintraub, P. "The Coming of the High-Tech Harvest." *Audubon* 94 (July–August 1992): 92–4+.

Wheale, P. R., and R. M. McNally. *Genetic Engineering: Catastrophe or Utopia?* New York: St. Martin's Press, 1988.

Genetically modified organism

A genetically modified organism, or GMO, is an organism whose genetic structure has been altered by incorporating a single gene or multiple genes—from another organism or species—that adds, removes, or modifies a trait in the organism by a technique called gene splicing. An organism that has been genetically modified—or engineered—to contain a gene from another **species** is also called a transgenic organism (because the gene has been transferred) or a living modified organism (LMO). Most often the transferred gene allows an organism—such as a bacterium, fungus, **virus**, plant, insect, fish, or mammal—to express a trait that enhances its desirability to producers or consumers of the final product.

Overview

Plants and livestock have been bred for desired qualities (selective breeding) for thousands of years—long before people knew anything about the science of genetics. As technology advanced, however, so did the means by which people could select desired traits. Modern **biotechnology** represents a significant step in the history of genetic modification.

Until the final decades of the twentieth century, breeding techniques were limited to the transfer of desired traits within the same or closely related species. Genetically modified organisms, however, may contain traits transferred from completely dissimilar species. For example, before modern biotechnology, apple breeders could only cross-breed apples with apples or other closely related species. So, if a breeder wanted to make a certain tasty apple variety that was more

tolerant to the cold, a cold-tolerant apple variety had to be hybridized with a tasty variety. This process usually involved significant trial and error because there was little assurance the cold-tolerance ability would be transferred in any individual attempt to hybridize the two varieties.

Development of modern biotechnology

The characteristics of all organisms are determined by genes—the basic units of heredity. A gene—a segment of deoxyribonucleic **acid** (DNA)—is capable of replication and **mutation**, occupying a fixed position on a chromosome (a group of several thousand genes; humans are defined by 22 pairs of chromosomes plus X and Y), and is passed on from parents to offspring during reproduction. A gene determines the structure of a protein or a **ribonucleic acid** (RNA) molecule. Found in all cells, DNA carries the genetic instructions for creating proteins; RNA decodes those instructions. Proteins perform diverse biological functions in the body, from helping muscles contract, to enabling blood to clot, to allowing biochemical reactions to proceed quickly enough to sustain life. By modifying a protein, particular phenotypic (physical) or physiologic changes—such as the color of a rose or the ability to bioluminesce (glow like a firefly)—are created.

A fundamental aspect of modern biotechnology is the belief that the essential genetic elements of all life are the same. Since the 1950s, molecular biologists have known that the DNA in every organism is made up of pairs of four nitrogen-containing bases, or building blocks: adenine (A), thymine (T), cytosine (C), and guanine (G).

In 1953, scientists James Watson and Francis Crick discovered that DNA is constructed in a double helix pattern—sort of a twisted ladder. Although A always pairs with T, and G with C, there is significant variety in how the pairs stack. The variable sequence of these DNA base pairs constitutes, in effect, the variety of life. So even all organisms are made from the same basic building blocks, their differences are a result of varying DNA sequences. A principle of modern molecular biology and biotechnology is that because these genetic building blocks are the same for all species, DNA can be extracted and inserted across species.

Genetic engineering techniques

The tools of modern biotechnology allow the transfer of specific genes, hence specific traits, to occur with more **precision** than ever before. The individual gene conveying a trait can be identified, isolated, replicated, and the inserted into another organism. This process is called **genetic engineering**, or recombinant DNA (rDNA) technology—that is, recombining DNA from multiple organisms.

Through such engineering, the apple breeder who wants the tasty apple variety to be cold-tolerant as well has the potential to find a gene that conveys cold tolerance and insert that gene directly into the tasty variety. Although there is still some trial and error in this process, overall, there is greater precision in the ability to move genes from one organism to another. The gene that conveys a cold-tolerance ability does not need to come from another apple variety; it may come from any other organism. A cold-tolerant fish, for example, might have a suitable gene.

There are multiple methods by which genetic material may be transferred from one organism to another. Before a gene is moved between organisms, however, it is necessary to identify the individual gene that confers the desired trait. This stage is often quite time-consuming and difficult because often it is not clear what gene is needed or where to find it. Finding the gene entails using or creating a library of the species' DNA, specifying the amino acid sequence of the desired protein, and then devising a probe (any biochemical agent labeled or tagged in some way so that it can be used to identify or isolate a gene, RNA, or protein) for that sequence. As of 2002, isolating the desired gene is one of the most limiting aspects to creating a GMO.

Once the desired gene has been identified it must be extracted from the organism, which is usually done with a restriction endonuclease (an **enzyme**). Restriction endonucleases recognize particular base sequences in a DNA molecule and cut and isolate these sequences in a predictable and consistent manner. Once isolated, the gene must be replicated to generate sufficient usable material, as more than one copy of the gene is needed for the next steps in the engineering process.

One common method of gene replication is called polymerase **chain reaction** (PCR). Through the PCR method, the strands of the DNA are broken apart—in effect, the ladder is divided down the middle—and then exact copies of the opposite sides of the ladder are produced, creating thousands or millions of copies of the complete gene.

After replication, the gene must be inserted into the new organism via a vector (an agent that transfers material—typically DNA—from one host to another). A common vector is *Agrobacterium tumefaciens*, a bacterium that normally inserts itself into plants, causing a tumor. By genetically engineering *A. tumefaciens*, however, the bacterium can be used to insert the desired gene into a plant, replacing its tumor-causing genetic material.

Another vector, called ballistics, involves coating a microprojectile—usually a heavy metal such as tungsten—with the desired gene, then literally shooting the microprojectile into material from the new host organism. The microprojectile breaks apart the DNA of the host organism. Then when the DNA reassembles, some of the new, desired genetic material is inserted. Once the DNA has been inserted, the host organism can be grown, raised, or produced normally and tests can be performed to observer whether the desired trait manifests.

Varieties of GMOs

In 2000, of the total land planted with GMO crops worldwide, the United States occupied 68%, Argentina 23%, and China 1%. The United States produces many types of GMOs for commercial and research purposes.

Agricultural crops

A common type of GMO is the modified agricultural seed. Corn, soybeans, and cotton are a few examples of staple agricultural GMO products grown in the United States. Genetic modifications to these products may, for example, alter a crop's nutritional content, storage ability, or taste.

With the advent of genetic engineering, for example, common hybrid crops such as the tomato were launched into a new era. The first food produced from gene splicing and evaluated by the **Food and Drug Administration** (FDA) was the Flavr Savr Tomato in 1994. Tomatoes usually get softer as they ripen because of a protein in the tomato that breaks down the cell walls of the tomato. Because it is difficult to ship a quality ripe tomato across the country before the tomato spoils, tomatoes are usually shipped unripened. Engineers of the Flavr Savr Tomato spliced a gene into its deoxyribonucleic acid (DNA) to prevent the breakdown of the tomato's cell walls. The result of adding the new gene was a firm ripe tomato that was more desirable to consumers than the tasteless variety typically found on store shelves, particularly in the winter months.

Genetic modifications may also confer to a species an ability to produce its own **pesticide** biologically, thereby potentially reducing or even eliminating the need to apply external pesticides. For example, ***Bacillus thuringiensis*** (Bt) is a soil bacterium that produces toxins against insects (mainly in the genera *Lepidoptera*, *Diptera*, and *Coleoptera*).-When they are genetically modified to carry genetic material from the Bt bacterium, plants such as soybeans will able to produce their own Bt toxin and be resistant to insects such as the cornstalk borer and velvetbean catepillar. Researchers at the Monsanto chemical company estimate that Bt soybeans will be commercially available by about 2006.

By means of genetic engineering, a gene from a soil bacterium called *Agrobacterium* sp confers glyphosate **resistance** to a plant. Glyphosate (brand names include Roundup, Rodeo, and Accord) is a broad-spectrum **herbicide** (kills all green plants). As of 2002, Monsanto was the sole developer of all glyphosate resistant crops on the market. These crops (often called "Roundup Ready") include corn, soybeans, cotton, and canola.

Pharmaceuticals

Genetically engineered pharmaceuticals are also useful GMO products. One of the first GMOs was a bacterium with a human gene inserted into its genetic code to produce a very high quality human insulin for diabetes. Vaccines against diseases such as meningitis or hepatitis B, for example, are produced by genetically engineered yeast or bacteria. Other pharmaceuticals produced by using GMO **microbes** include interferon for **cancer**, erythropoetin for **anemia**, growth hormone for the treatment of dwarfism, tissue plasminogen activator for heart attack victims. Through genetic engineering, transgenic plants are likely to become a commercially viable source of pharmaceuticals.

Benefits and risks

Although the technological advances that allowed the creation of GMOs are impressive, as with any new technology, the benefits must be weighed against the risks. Useful pharmaceutical products have been created as a result of genetic engineering. Also, GMOs have produced and new and improved agricultural products that are resistant to crop pests, thus improving production and reducing chemical pesticide usage. These developments have had a major impact on food quality and nutrition.

The possibility that biotech crops could make a substantial contribution to providing sufficient food for an expanding world is also a reason given for engaging in the research that underlies their development. However, the debate over GMOs continues among scientists and between consumers and modern agricultural producers throughout the world regarding issues such as regulation, labelling, human health risk, and environmental impact.

In the United States, for example, there has been much controversy over the health risks of canola oil. Canola oil is genetically engineered rapeseed oil or "LEAR" oil (low erucic acid rape), a semi-drying industrial oil used as a lubricant, fuel, soap, and synthetic **rubber** base, and as an illuminant to give color pages in magazines their slick look. It was first developed in Canada (thus the name "canola oil").

Canola oil is derived from the mustard family and is considered a toxic and poisonous weed. In 1998, the EPA classified canola oil as a biopesticide with "low chronic toxicities," yet placed it on the "Generally Considered Safe" list of foods. Proponents, who tout the oil's health benefits, claim that due to genetic engineering and irradiation, it is completely safe, pointing to its unsaturated structure and digestibility.

Widely used as a cooking oil, and because it is so inexpensive, it is used in thousands of processed foods in the United States and North America. Thus, millions of people have been exposed—most of them unknowingly—to genetically engineered foods. However, there has been little research on the potential adverse effects of these products on humans.

When processed, canola oil becomes rancid very easily. It has been shown to cause health problems such as lung cancer and has been associated with loss of vision, disruption of the central nervous system, respiratory illness, anemia,

constipation, increased incidence of heart disease and cancer, low birth weight in infants, and irritability. It has a tendency to inhibit proper **metabolism** of foods and curbs normal enzyme function. Generally, rapeseed has a cumulative effect, often taking nearly ten years for symptoms to manifest.

The dangers of introducing genes that may cause undesirable effects in the **environment** is also a concern among many. For example, when farmers spray an herbicide to remove weeds growing among crops, the sprayed chemical often damages the crop plants. If the crop is genetically engineered to be resistant to the chemical, the weeds are killed, but the crop plants remain undamaged.

Although this situation appears to be beneficial, it is likely to lead to greater use of the particular herbicide, which would have several negative effects: the crop is likely to contain greater herbicide residues; and the increased spraying contaminates the rest of the environment. Although not all herbicides are dangerous, the safer choice seems to minimize rather than maximize their use. Also, if genes added to produce pesticide resistance in crop plants jumped to a weed species, then weeds would thrive and be difficult to control.

Thus, as debate over GMOs rages, it is important to note that a wide variety of ecological and human health concerns exist side by side with the new advances made possible by genetic engineering.

[*Paul R. Phifer Ph.D.*]

RESOURCES
BOOKS

Anderson, Luke. *Genetic Engineering, Food, and Our Environment.* White River Junction, VT: Chelsea Green Publishing Co., 1999.

Ho, Mae-Wan. *Genetic Engineering Dream or Nightmare: Turning the Tide on the Brave New World of Bad Science and Big Business.* New York: Continuum Publishing Group, 2000.

Kreuzer, H., and A. Massey. *Recombinant DNA and Biotechnology: A Guide for Students, 2nd Edition.* Washington, DC: ASM Press, 2001.

PERIODICALS

Brown, K. "Seeds of Concern." *Scientific American* (April 2001): 52–57.

Nemecek, S. "Does the World Need GM Foods?" *Scientific American* (April 2001): 48–51.

Wolfenbarger, L., and P. Phifer. "The Ecological Risks and Benefits of Genetically Engineered Plants." *Science* 290 (2000): 2088–2093.

ORGANIZATIONS

Environmental Protection Agency (EPA), 1200 Pennsylvania Avenue, NW, Washington, DC USA 20460 (202) 260-2090, Email: public-access@epa.gov, <http://www.epa.gov>

Human Genome Project Information, Oak Ridge National Laboratory, 1060 Commerce Park MS 6480, Oak Ridge, TN USA 37830 (865) 576-6669, Fax: (865) 574-9888, Email: genome@science.doe.gov, <http://www.ornl.gov/hgmis>

Geodegradable

The term geodegradable refers to a material that could degrade in the **environment** over a geologic time period. While **biodegradable** generally refers to items that may degrade within our lifetime, geodegradable material does not decompose readily and may take hundreds or thousands of years. **Radioactive waste**, for example, is degraded only over thousands of years. The glass formed as an end result of a **hazardous waste** treatment technology known as "in situ vitrification," is considered geodegradable only after a million years. *See also* Half-life; Hazardous waste site remediation; Hazardous waste siting; Waste management

Geographic information systems

A Geographic Information System (GIS) is a computer system capable of assembling, storing, mapping, **modeling**, manipulating, querying, analyzing, and displaying geographically referenced information (i.e., data that are identified according to their locations). Some practitioners expand the definition of GIS to include the personnel involved, the data that are entered into the system, and the uses, decisions, and interpretations that are made possible by the system. A GIS can be used for scientific investigations, resource management, and planning. The development of GIS was made possible from innovations in many different disciplines, including geography, cartography, remote sensing, surveying, civil engineering, **statistics**, computer science, operations research, artificial intelligence, and demography.

Even early man used GIS-type systems. Thirty-five thousand years ago Cro-Magnon hunters in Lascaux, France, drew pictures of the animals they hunted. Along with the animal drawings were track lines that are thought to show **migration** routes. These early records included two essential elements of modern GIS: a graphic file linked to an attribute data base.

In a GIS, maps and other data from many different sources and in many different forms are stored or filed as layers of information. A GIS makes it possible to link and integrate information that is difficult to associate through other means. A GIS can combine mapped variables to build and analyze new variables. For example, by knowing and entering data for water use for a specific residence, predictions can be made on the amount of **wastewater** contaminants generated and released to the **environment** from that residence.

The primary requirement for the source data is that the locations for the variables are known. Any variable that can be located spatially can be entered into a GIS. Location may be determined by x, y, and z coordinates of longitude, latitude, and elevation, or by other systems such as ZIP codes or highway mile markers. A GIS can convert existing digital information, which may not be in map form, into forms it can recognize and utilize. For example, census data can be converted into map form, such that different types of information can be presented in layers.

If data are not in digital form, (i.e., not in a form that the computer can utilize), various techniques are available to capture the information. Maps can be hand traced with a computer mouse to collect the coordinates of features. Electronic scanning devices can be used to convert map lines and points to digits. Data capture, i.e., entering the information into the system, is the most time-consuming component of a GIS. Identities of the objects on the map and their spatial relationships must be specified. Editing of information that is captured can be difficult. For example, an electronic scanner will record blemishes on a map just as it will record actual map features. A fleck of dirt that is scanned may connect two lines that should not be connected. Such extraneous information must be identified and removed from the digital data file.

Different maps may be at different scales, so map information in a GIS must be manipulated so that it fits with information collected from other maps. Data may also have to undergo projection conversion before being integrated into a GIS. Projection, a fundamental component of map making, is a mathematical means of transferring information from the earth's three-dimensional curved surface to a two-dimensional medium (i.e., paper or a computer screen). Different projections are used for different types of maps based on the type of projection appropriate for a specific use. Much of the information in a GIS comes from existing maps, so the computing power of the GIS can integrate digital information from different sources with different projections into a common projection.

Digital data are collected and stored in various ways, and so different data sources may not be compatible. A GIS must be able to convert data from one structure to another. For example, image data from a satellite that has been interpreted by a computer to produce a **land use** map can be entered into the GIS in raster format. Raster format is like a spreadsheet, with rows and columns into which numbers are placed. These rows and columns are linked to x,y coordinates, with the intersection of each row and column forming a cell that corresponds to a specific point in the world. These cells contain numbers that can represent such features as elevation, soils, archeological sites, etc. Maps in a raster GIS can be handled like numbers (e.g., added, subtracted, etc.) to form new maps. Raster data files can be quickly manipulated by computer but are less detailed and may be less visually appealing than vector data files, which appear more like traditional hand-drafted maps. Vector digital data are captured as points, lines (a series of point coordinates), or areas (shapes bounded by lines). A Vector GIS output looks more like a traditional paper map.

A GIS can be used to depict two- and three-dimensional characteristics of the earth's surface, subsurface, and **atmosphere** from information points. Each thematic map (i.e., a map displaying information about one characteristic of a region) is referred to as a layer, coverage, or level. A specific thematic map can be overlain and analyzed with any other thematic map covering the same area. Not all analyses may require using all of the map layers at the same time. A researcher may use information selectively to consider relationships among specific layers. Information from two or more layers may be combined and transformed into a new layer for use in subsequent analyses. For example, with maps of **wetlands**, slopes, streams, land use, and soils, a GIS can produce a new overlay that ranks the wetlands according to the relative sensitivity to damage from nearby factories or homes. This process of combining and transforming information from different layers is referred to as map "algebra," as it involves adding and subtracting information.

Recorded information from off-screen files can be retrieved from the maps by pointing at a location, object, or area on the screen. Conditions of adjacency (what is next to what) containment (what is enclosed by what), and proximity (how close something is to something else) can be determined using a GIS.

A GIS can also be used for "what if" scenarios by simulating the route of materials along a linear network. For example, an evaluation could be made on how long it would take for **chemicals** accidentally released into a river from a factory to move into a wetland area. Direction and speed can be assigned to the digital stream, and the contaminants can be traced through the stream system.

An important component of a GIS is the ability to produce graphics on the computer screen or to output to paper (e.g., wall maps) to inform resource decision makers about the results of analyses. The viewers of such materials can visualize and thus better understand the results of analyses or simulations of potential events. A GIS can be used to produce not only maps, but also drawings and animations that allow different ways of viewing information. These types of images are especially helpful in conveying technical concepts to non-scientists.

GIS technology is an improvement in the efficiency and analytical power of cartographic science. Traditional maps are abstractions of the real world, with important elements portrayed on a sheet of paper with symbols to represent physical objects. For example, topographic maps show the shape of the land surface with contour lines, but the actual shape of the land can only be imagined. Graphic display techniques in a GIS illustrate more clearly relationships among the map elements, increasing the ability to extract and analyze information.

Many commercial GIS software systems are available, some of which are specialized for use in specific types of decision-making situations. Other programs are more general and can be used for a wide number of applications or can be customized to meet individual requirements.

GIS applications are useful for a wide range of disciplines, including urban planning, environmental and natural resource management, facilities management, **habitat** studies, archaeological analyses, hazards management, emergency planning, marketing and demographic analyses, and **transportation** planning. The ability to separate information in layers and then combine the layers with other layers of information is the key to making GIS technology a valuable tool for the analysis and display of large volumes of data, thus allowing better management and understanding of information and increased scientific productivity.

[*Judith L. Sims*]

RESOURCES
BOOKS

Heywood, D. Ian, Ian Heywood, Sarah Cornelius, and Steve Carver. *An Introduction to Geographical Information Systems.*: Prentice Hall, 2000.

OTHER

"GIS WWW Resource List." *University of Edinburgh Web Page.* 1996 [cited June 1, 2002]. <http://www.geo.ed.ac.uk/home/giswww.html>.

"Geographic Information Systems." *United States Geological Survey Web Page.* April 24, 2001 [cited June 1, 2002]. <http://www.usgs.gov/research/gis/title/html>.

*Geographic Information Systems as an Integrating Technology: Context, Concepts, and Definitions.*Kenneth E. Foote and Margaret Lynch, The Geographer's Craft Project, Department of Geography, University of Colorado at Boulder, october 12, 1997. [cited June 1, 2002], <http://www.Colorado.EDU/geography/gcraft/notes/intro/intro_f.html>.

*Geographic Information Systems: Internet Web Sites.*Federal Facilities Restoration and Reuse Office, U.S. Environmental Protection Agency, Washington, DC, November 5, 2000 [cited June 23, 2002], <http://www.epa.gov/swerffrr/compend/intrnt_b.htm>.

Geological Survey

The United States Geological Survey (USGS) is the federal agency responsible for surveying and publishing maps of **topography** (giving landscape relief and elevation), geology, and natural resources—including minerals, fuels, and water. The USGS, part of the U. S. Department of the Interior, was formed in 1879 as the United States began systematically to explore its newly expanded western territories. Today it has an annual budget of about $700 million, which is devoted to primary research, resource assessment and monitoring, map production, and providing information to the public and to other government agencies.

The United States Geological Survey, now based in Reston, Virginia, originated in a series of survey expeditions sent to explore and map western territories and rivers after the Civil War. Four principal surveys were authorized between 1867 and 1872: Clarence King's exploration of the fortieth parallel, Ferdinand Hayden's survey of the Rocky Mountain territories, John Wesley Powell's journey down the **Colorado River** and through the Rocky Mountains, and George Wheeler's survey of the 100th meridian. Twelve years later, in 1879, these four ongoing survey projects were combined to create a single agency, the United States Geological Survey. The USGS' first director was Clarence King. In 1881 his post was taken by John Wesley Powell, whose name is most strongly associated with the early Survey. It was Powell who initiated the USGS topographic mapping program, a project that today continues to produce the most comprehensive map series available of the United States and associated territories.

In addition to topographic mapping, the USGS began detailed surveys and mapping of mineral resources in the 1880s. Mineral exploration led to mapping geologic formations and structures and a gradual reconstruction of geologic history in the United States. Research and mapping of glacial history and fossil records naturally followed from mineral explorations, so that the USGS became the primary body in the United States involved in geologic field research and laboratory research in experimental geophysics and geochemistry. During World Wars I and II, the USGS' role in identifying and mapping tactical and strategic resources increased. Water and fuel resources (**coal**, oil, **natural gas**, and finally **uranium**) were now as important as **copper**, gold, and mineral ores, so the Survey took on responsibility for assessing these resources as well as topographic and geologic mapping.

Today the USGS is one of the world's largest earth science research agencies and the United States' most important map publisher. The Survey conducts and sponsors extensive laboratory and field research in geology, **hydrology**, oceanography, and cartography. The agency's three divisions, **Water Resources**, Geology, and National Mapping, are responsible for basic research. They also publish, in the form of maps and periodic written reports, information on the nation's topography, geology, fuel and mineral resources, and other aspects of earth sciences and **natural resources**. Most of the United States' hydrologic records and research, including streamflow rates, **aquifer** volumes, and **water quality**, are produced by the USGS. In addition, the USGS publishes information on natural hazards, including earthquakes, volcanoes, landslides, floods, and droughts. The Survey is the primary body responsible for providing basic earth

science information to other government agencies, as well as to the public. In addition, the USGS undertakes or assists research and mapping in other countries whose geologic survey systems are not yet well developed.

[*Mary Ann Cunningham Ph.D.*]

RESOURCES
BOOKS

U.S. Geological Survey. *Maps for America.* Reston, VA: U.S. Government Printing Office, 1981.
USGS Yearbook: Fiscal Year 1985. Washington, DC: U.S. Government Printing Office, 1985.

Georges Bank (collapse of the ground fishery)

Until the 1990s Georges Bank, off the coasts of New England and Nova Scotia, was one of the world's most valuable fisheries. A bank is a plateau found under the surface of shallow ocean water. Georges Bank is the southernmost and the most productive of the banks that form the continental shelf. A majority of the $800 million northeastern fishery industry comes from Georges Bank. The oval shaped bank is 149 mi long and 74.5 mi wide (240 km by 120 km). Georges Bank covers an area larger than the state of Massachusetts. The ocean bottom of Georges Bank formed ideal **habitat** for favorable quantities of groundfish, *demersal finfishes*, fish which feed off or near the ocean's floor mdash; cod, *Gadidae*, clam, *Pelecypoda*, haddock, *Melanogrammus aeglefinus*, hake, *Merlucciidae*, herring, *Clupea harengus*, lobster, *Homarus*, pollock, *Pollachaius*, flounder, and scallops, *Pectinidae*. But by 1994 the Georges Bank ground fishery had collapsed.

Cod were by far the most numerous and valuable of the Georges Bank's fish. Atlantic cod form distinct stocks and the Georges Bank stock grows faster than those of the colder waters further north. They are the world's largest and thickest cod. In 1938 a cod weighing 180 lb (82 kg) was caught off the bank. The cod move in schools from feeding to spawning grounds, in dense aggregates of hundreds of millions of fish, making them easy prey for fishing nets.

During the second half of the twentieth century, gigantic trawlers towing enormous nets could haul in 200 tons (181.4 metric tons) of fish an hour off Georges Bank. At its peak in 1968, 810,000 tons (734,827 metric tons) of cod were harvested. By the 1970s, fleets of Soviet, European, and Japanese factory ships were trawling the cod-spawning grounds, scooping up the fish before they could reproduce. If the catch was of mixed **species** or the wrong size, the nets were dumped, leaving the ocean surface teaming with dead fish. After the catch was sorted, many species of dead

bycatch, including young cod, flounder, and crabs, were discarded. For every three tons of processed fish, at least a ton of bycatch died. These ships also trawled for herring, capelin, mackerel, and other small fish that the cod and other groundfish depend on for food.

The Fisheries Conservation and Management Act of 1976 extended exclusive American fishing rights from 12–200 mi (19–322 km) offshore. Since much of Georges Bank is within the 200-mi (322 km) limit of Nova Scotia, conflict erupted between American and Canadian fishermen. International arbitration eventually gave Canada the northeast corner of the bank. The legislation also established the New England Fishery Management Council to regulate fishing. Although the goal was to conserve fisheries as well as to create exclusive American fishing grounds, the council was controlled by commercial interests. The result was the development of financial incentives and boat-building subsidies to modernize the fishing fleet.

Soon the New England fleet surpassed the fishing capacities of the foreign fleets it replaced and every square foot of Georges Bank had been scraped with the heavy chains that hold down the trawling nets and stir up fish. This destroyed the rocky bottom structure of the bank and the vegetation and marine invertebrates that created habitat for the groundfish. Cod, pollack, and haddock were replaced by dogfish and skates, so-called "trash fish."

During the 1990s tiny hydroids, similar to jellyfish, began appearing off Georges Bank, in concentrations as high as 100 per gal of water. Although they were drifting in the water, they were in their sedentary life-stage form, indicating that they may have been ripped from their attachments by storms or commercial trawlers. These hydroids ate most of the small crustaceans that the groundfish larvae depend on. They also directly killed cod larvae.

In 1994 the National Marine Fisheries Service found that the Georges Bank cod stock had declined by 40% since 1990, the largest decline ever recorded. Furthermore, the yellowtail flounder stock had collapsed. In a given year, only eight out of 100 flounder survived and the breeding population had fallen 94% in three years. The last successful flounder spawning was in 1987; but 60% of the catch from that year's group were too small to sell and were discarded. In response the Fisheries Service closed large areas of Georges Bank, but fishing continued in the Canadian sector and western portions of the American sector. With the goal of annually harvesting only 15% of the remaining stock, each vessel was restricted to 139 days of ground fishing. Nevertheless by 1996, 55% of the remaining Georges Bank cod stock — the only surviving North Atlantic population — had been caught. Fishing was restricted to 88 days. A satellite-based vessel monitoring system is used to detect fishing boats that enter closed areas of Georges Bank.

At the time of the cod moratorium, it was argued that the population would recover in five years; however there were few signs of recovery as of 2002. Not only is the cod stock near an all-time low, but so are populations of other commercial fish and many other species. The average size of the bottom-dwelling fish of Georges Bank is a fraction of what it was twenty years ago.

Georges Bank is just one example of a eastern coastal area negatively affected by excessive trawling. Even though there is $800 million worth of fish extracted from Georges Bank and the surrounding area, there is an overall decline in groundfish stock along the entire boreal and sub-arctic coast of eastern North America. American and Canadian moratoriums on gas and oil exploration and extraction from Georges Bank—activities that could further disrupt the fishery—are in effect until at least 2012.

[*Margaret Alic Ph.D.*]

RESOURCES

BOOKS

Dobbs, David. *The Great Gulf: Fishermen, Scientists, and the Struggle to Revive the World's Greatest Fishery.* Washington, DC: Island Press, 2000.
Kurlansky, Mark. *Cod: A Biography of the Fish that Changed the World.* New York: Walker and Company, 1997.

PERIODICALS

Hattam, Jennifer. "Victory at Sea." *Sierra* 85, no. 3 (May/June 2000): 91.
Molyneaux, Paul. "Vessel Monitor Convicts New Bedford Scalloper." *National Fisherman* 82, no. 11 (March 2002): 50.

OTHER

American Museum of Natural History. *Georges Bank—The Sorry Story of Georges Bank.* [cited June 2002]. <http://www.sciencebulletins.amnh.org/biobulletin/biobulletin/story1209.html>.
Public Broadcasting System. *Empty Oceans, Empty Nets.* 2002 [cited May 2002]. <http://www.pbs.org/empty oceans>.
Status of the Fishery Resources off the Northeastern United States. Resource Evaluation and Assessment Division, Northeast Fisheries Science Center. June 2001 [cited May 2002]. <www.nefsc.nmfs.gov/sos/index.html>.
United States Geological Survey. *Geology and the Fishery of Georges Bank.* January 3, 2001 [cite June 2002]. <http://www.marine.usgs.gov/fact-sheets/georges-bank/title.html>.

ORGANIZATIONS

Cape Cod Commercial Hook Fishermen's Association, 210 Orleans Road, North Chatham, MA USA 02650 (508) 945-2432, Fax: (508) 945-0981, Email: enichols@ccchfa.org, <http//www.ccchfa.org>
Coastal Waters Project/Task Force Atlantis, 418 Main Street, Rockland, ME USA 04841 (207) 594-5717, Email: coastwatch@acadia.net, <http://www.atlantisforce.org>
Northeast Fisheries Science Center, 166 Water Street, Woods Hole, MA USA 02543-1026 (508) 495-2000, Fax: (508) 495-2258, , <http://www.nefsc.nmfs.gov>
U.S. GLOBEC Georges Bank Program, Woods Hole Oceanographic Institution, Woods Hole, MA USA 02543-1127 (508) 289-2409, Fax: (508) 457-2169, Email: rgroman@whoi.edu, <http://globec.whoi.edu>

Geosphere

The solid portion of the earth. It is also known as the lithosphere. From a technical standpoint, the geosphere includes inner parts of the earth virtually inaccessible to human study, the inner and outer core and mantle, as well as the outermost crust. For the most part, however, environmental scientists are primarily interested in the relatively thin outer layer of the crust on which plants and animals live, in the ores and minerals that occur within the crust, and in the changes that take place in the crust as a result of **erosion** and mountain-building.

Geothermal energy

Geothermal energy is obtained from hot rocks beneath the earth's surface. The planet's core, which may generate temperatures as high as 8,000°F (4,500°C), heats its interior, whose temperature increases, on an average, by about 1°C (2°F) for every 60 ft (18 m) nearer the core. Some heat is also generated in the mantle and crust as a result of the **radioactive decay** of **uranium** and other elements.

In some parts of the earth, rocks in excess of 212°F (100°C) are found only a few miles beneath the surface. Water that comes into contact with the rock will be heated above its boiling point. Under some conditions, the water becomes super-heated, that is, is prevented from boiling even though its temperature is greater than 212°F (100°C). Regions of this kind are known as wet steam fields. In other situations the water is able to boil normally, producing steam. These regions are known as dry steam fields.

Humans have long been aware of geothermal energy. Geysers and fumaroles are obvious indications of water heated by underground rock. The Maoris of New Zealand, for example, have traditionally used hot water from geysers to cook their food. Natural hot spring baths and spas are a common feature of many cultures where geothermal energy is readily available.

The first geothermal well was apparently opened accidentally by a drilling crew in Hungary in 1867. Eventually, hot water from such **wells** was used to heat homes in some parts of Budapest. Geothermal heat is still an important energy source in some parts of the world. More than 99% of the buildings in Reykjavik, the capital of Iceland, are heated with geothermal energy.

The most important application of geothermal energy today is in the generation of electricity. In general, hot steam or super-heated water is pumped to the planet surface where it is used to drive a turbine. Cool water leaving the generator is then pumped back underground. Some water is lost by evaporation during this process, so the energy that comes

from geothermal wells is actually non-renewable. However, most zones of heated water and steam are large enough to allow a geothermal mine to operate for a few hundred years.

A dry steam well is the easiest and least expensive geothermal well to drill. A pipe carries steam directly from the heated underground rock to a turbine. As steam drives the turbine, the turbine drives an electrical generator. The spent steam is then passed through a condenser where much of it is converted to water and returned to the earth.

Dry steam fields are relatively uncommon. One, near Larderello, Italy, has been used to produce electricity since 1904. The geysers and fumaroles in the region are said to have inspired Dante's *Inferno.* The Larderello plant is a major source of electricity for Italy's electric railway system. Other major dry steam fields are located near Matsukawa, Japan, and at Geysers, California. The first electrical generating plant at the Geysers was installed in 1960. It and companion plants now provide about 5% of all the electricity produced in California.

Wet steam fields are more common, but the cost of using them as sources of geothermal energy is greater. The temperature of the water in a wet steam field may be anywhere from 360–660°F (180–250°C). When a pipe is sunk into such a reserve, some water immediately begins to boil, changing into very hot steam. The remaining water is carried out of the reserve with the steam.

At the surface, a separator is used to remove the steam from the hot water. The steam is used to drive a turbine and a generator, as in a dry steam well, before being condensed to a liquid. The water is then mixed with the hot water (now also cooled) before being returned to the earth.

The largest existing geothermal well using wet steam is in Wairakei, New Zealand. Other plants have been built in Russia, Japan, and Mexico. In the United States, pilot plants have been constructed in California and New Mexico. The technology used in these plants is not yet adequate, however, to allow them to compete economically with fossil-fueled **power plants**.

Hot water (in contrast to steam) from underground reserves can also be used to generate electricity. Plants of this type make use of a binary (two-step) process. Hot water is piped from underground into a heat exchanger at the surface. The heat exchanger contains some low-boiling point liquid (the "working fluid"), such as a **freon** or isobutane. Heat from the hot water causes the working fluid to evaporate. The vapor then produced is used to drive the turbine and generator. The hot water is further cooled and then returned to the rock **reservoir** from which it came.

In addition to dry and wet steam fields, a third kind of geothermal reserve exists: pressurized hot water fields located deep under the ocean floors. These reserves contain **natural gas** mixed with very hot water. Some experts believed that these geopressurized zones are potentially rich energy sources although no technology currently exists for tapping them.

Another technique for the capture of geothermal energy makes use of a process known as hydrofracturing. In hydrofracturing, water is pumped from the surface into a layer of heated dry rock at pressures of about 7,000 lb/in^2 (500 kg/cm^2). The pressurized water creates cracks over a large area in the rock layer. Then, some material such as sand or plastic beads is also injected into the cracked rock. This material is used to help keep the cracks open.

Subsequently, additional cold water can be pumped into the layer of hot rock, where it is heated just as natural **groundwater** is heated in a wet or dry steam field. The heated water is then pumped back out of the earth and into a turbine-generator system. After cooling, the water can be re-injected into the ground for another cycle. Since water is continually re-used in this process and the earth's heat is essentially infinite, the hydrofracturing system can be regarded as a renewable source of energy.

Considerable enthusiasm was expressed for the hydrofracturing approach during the 1970s and a few experimental plants were constructed. But, as oil prices dropped and interest in **alternative energy sources** decreased in the 1980s, these experiments were terminated.

Geothermal energy clearly has some important advantages as a power source. The raw material—heated water and steam—is free and readily available, albeit in only certain limited areas. The technology for extracting hot water and steam is well developed from petroleum-drilling experiences, and its cost is relatively modest. Geothermal mining, in addition, produces almost no **air pollution** and seems to have little effect on the land where it occurs.

On the other hand, geothermal mining does have its disadvantages. One is that it can be achieved in only limited parts of the world. Another is that it results in the release of gases, such as **hydrogen** sulfide, **sulfur dioxide**, and ammonia, that have offensive odors and are mildly irritating. Some environmentalists also object that geothermal mining is visually offensive, especially in some areas that are otherwise aesthetically attractive. **Pollution** of water by **runoff** from a geothermal well and the large volume of cooling water needed in such plant are also cited as disadvantages.

At their most optimistic, proponents of geothermal energy claim that up to 15% of the United States' power needs can be met from this source. Lagging interest and research in this area over the past decade have made this goal unreachable. Today, no more than 0.1% of the nation's electricity comes from geothermal sources. Only in California is geothermal energy a significant power source. As an example, GeoProducts Corporation, of Moraga, California,

has constructed a $60 million geothermal plant near Lassen **National Park** that generates 30 megawatts of power.

Until the government and the general public becomes more concerned about the potential of various types of alternative energy sources, however, geothermal is likely to remain a minor energy source in the country as a whole. *See also* Alternative fuels; Fossil fuels; Renewable resources; Water pollution

[*David E. Newton*]

RESOURCES
BOOKS

Moran, J. M., M. D. Morgan, and J. H. Wiersma. *Environmental Science.* Dubuque, IA: W. C. Brown, 1993.

National Academy of Sciences. *Geothermal Energy Technology.* Washington, DC: National Academy Press, 1988.

Rickard, G. *Geothermal Energy.* Milwaukee, WI: Gareth Stevens, 1991.

U.S. Department of Energy. *Geothermal Energy and Our Environment.* Washington, DC: U.S. Government Printing Office, 1980.

PERIODICALS

Fishman, D. J. "Hot Rocks." *Discover* 12 (July 1991): 22–23.

Giant panda

Today, the giant panda (*Ailuropoda melanoleuca*) is one of the best known and most popular large mammals among the general public. Although its existence was known long ago, having been mentioned in a 2,500-year-old Chinese geography text, Europeans did not learn of its existence until its discovery by a French missionary in 1869. The first living giant panda did not reach the Western Hemisphere until 1937. The giant panda, variously classified with the true bears or, often, in a family of its own, once ranged throughout much of China and Burma, but is now restricted to a series of 13 **wildlife** reserves totaling just over 2,200 mi^2 (5,700 km^2) in three central and western Chinese provinces. The giant panda population has been decimated over the past 2,000 years by **hunting** and **habitat** destruction. In the years since 1987, they have lost more than 30% of their habitat. Giant pandas are one of the rarest mammals in the world, with current estimates of their population size at about 1,000 individuals (150 in captivity). Today human pressure on giant panda populations has diminished, although **poaching** continues. Giant pandas are protected by tradition and sentiment, as well as by law in the Chinese mountain forest reserves. Despite this progress, however, IUCN—The World Conservation Union—and the U. S. **Fish and Wildlife Service** consider the giant panda to be endangered. Some of this species' unique requirements and habits do seem to put them in jeopardy.

The anatomy of the giant panda indicates that it is a carnivore, however, its diet consists almost entirely of bamboo, whose cellulose cannot be digested by the panda. Since the giant panda obtains so little **nutrient** value from the bamboo, it must eat enormous quantities of the plant each day, about 35 lb (16 kg) of leaves and stems, in order to satisfy its energy requirements. Whenever possible, it feeds solely on the young succulent shoots of bamboo, which, being mostly water, requires it to eat almost 90 lb (41 kg) per day. This translates into 10–12 hours per day that pandas spend eating. Giant pandas have been known to supplement their diet with other plants such as horsetail and pine bark, and they will even eat small animals, such as rodents, if they can catch them, but well over 95% of their diet consists of the bamboo plant.

Bamboo normally grows by sprouting new shoots from underground rootstocks. At intervals from 40 to 100 years, the bamboo plants blossom, produce seeds, then die. New bamboo then grows from the seed. In some regions it may take up to six years for new plants to grow from seed and produce enough food for the giant panda. Undoubtedly this has produced large shifts in panda population size over the centuries. Within the last quarter century, two bamboo flowerings have caused the starvation of nearly 200 giant pandas, a significant portion of the current population. Although the wildlife reserves contain sufficient bamboo, much of the vast bamboo forests of the past have been destroyed for agriculture, leaving no alternative areas to move to should bamboo blossoming occur in their current range.

Low **fecundity** and limited success in captive breeding programs in zoos does not bode well for replenishing any significant losses in the wild population. Although there are 150 pandas in captivity, only about 28% are breeding. In 1999, the first giant panda to live more than a few days was born in captivity. For the time being, the giant panda population appears stable, a positive sign for one of the world's scarcest and most popular animals.

[*Eugene C. Beckham*]

RESOURCES

BOOKS

Nowak, R. M., ed. *Walker's Mammals of the World.* 5th ed. Baltimore: Johns Hopkins University Press, 1991.

PERIODICALS

"China goes High Tech to Help Panda Population." *USA Today*, August 13, 2001.

Drew, L. "Are We Loving the Panda to Death?" *National Wildlife* 27 (1989): 14–17.

"Pandas Still under Threat of Extinction." *USA Today*, February 16, 2001.

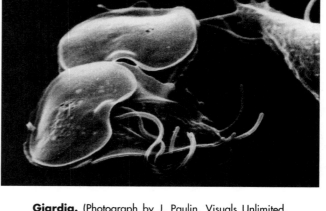

Giardia. (Photograph by J. Paulin. Visuals Unlimited. Reproduced by permission.)

Giardia

Giardia is the genus (and common) name of a protozoan parasite in the phylum Sarcomastigophora. It was first described in 1681 by Antoni van Leeuwenhoek (called "The Father of Microbiology"), who discovered it in his own stool. The most common **species** is *Giardia intestinalis* (also called *lamblia*), which is a fairly common parasite found in humans. The disease it causes is called giardiasis.

The trophozoite (feeding) stage is easily recognized by its pear-shaped, bilaterally-symmetrical form with two internal nuclei and four pairs of external flagella; the thin-walled cyst (infective) stage is oval. Both stages are found in the upper part of the small intestine in the mucosal lining. The anterior region of the ventral surface of the troph stage is modified into a sucking disc used to attach to the host's abdominal epithelial tissue. Each troph attaches to one epithelial cell. In extreme cases, nearly every cell will be covered, causing severe symptoms. Infection usually occurs through drinking contaminated water. Symptoms include diarrhea, flatulence (gas), abdominal cramps, fatigue, weight loss, anorexia, and/or nausea and may last for more than five days. Diagnosis is usually done by detecting cysts or trophs of this parasite in fecal specimens.

Giardia has a worldwide distribution. It is more common in warm, tropical regions than in cold regions. Hosts include **frogs**, cats, dogs, beaver, muskrat, horses, and humans. Children as well as adults can be affected, although it is more common in children. It is highly contagious. Normal infection rate in the United States ranges from 1.5 to 20%. In one case involving scuba divers from the New York City police and fire fighters, 22–55% were found to be infected, presumably after they accidentally drank contaminated water in the local rivers while diving. In another case, an epidemic of giardiasis occurred in Aspen, Colorado, in 1965 during the popular ski season and 120 people were infected. Higher infection rates are common in some areas of the world, including Iran and countries in Sub-Saharan Africa.

Giardia can typically withstand sophisticated forms of **sewage treatment**, including **filtration** and **chlorination**. It is therefore hard to eradicate and may potentially increase in polluted lakes and rivers. For this reason, health officials should make concerted efforts to prevent contaminated feces from infected animals (including humans) from entering lakes used for drinking water.

The most effective treatment for giardiasis is the drug Atabrine (quinacrine hydrochloride). Adult dosage is 0.1 g taken after meals three times each day. Side effects are rare and minimal. *See also* Cholera; Coliform bacteria

[*John Korstad*]

RESOURCES
BOOKS

Markell, E. K., M. Voge, and D. T. John. *Medical Parasitology.* 7th ed. Philadelphia: W. B. Saunders, 1992.

Schmidt, G. D., and L. S. Roberts. *Foundations of Parasitology.* 4th ed. St. Louis: Times Mirror/Mosby, 1989.

U.S. Department of Health and Human Services. *Health Information for International Travel.* Washington, DC: U.S. Government Printing Office, 1991.

Gibbons

Gibbons (genus *Hylobates*, meaning "dweller in the trees") are the smallest members of the ape family which also includes gorillas, **chimpanzees**, and orangutans. They spend most of their lives at the tops of trees in the jungle, eating leaves and fruit. They are extremely agile, swinging at speeds of 35 mph (56 km/h) with their long arms on branches to move from tree to tree. The trees can even be 50 ft (15 m) apart. They have no tails and are often seen walking upright on tree branches. Gibbons are known for their loud calls and songs, which they use to announce their territory and warn away others. They are devoted parents, raising usually one or two offspring at a time and showing extraordinary affection in caring for them. Conservationists and animal protectionists who have worked with gibbons describe them as extremely intelligent, sensitive, and affectionate.

Gibbons have long been hunted for food, for medical research, and for sale as pets and **zoo** specimens. A common method of collecting them is to shoot the mother and capture the nursing or clinging infant, if it is still alive. The **mortality** rate in collecting and transporting gibbons to areas where they can be sold is extremely high. This, coupled with the fact that their jungle **habitat** is being destroyed at a rate of

A gibbon. (©Breck P. Kent/JLM Visuals. Reproduced by permission.)

32 acres (13 ha) per minute, has resulted in severe depletion of their numbers.

Gibbons are found in southeast Asia, China, and India, and nine **species** are recognized. All nine species are considered endangered by the **U.S. Department of the Interior** and are listed in the most endangered category of the

Convention on International Trade in Endangered Species of Wild Fauna and Flora (CITES). **IUCN—The World Conservation Union** considers three species of gibbon to be endangered and two species to be vulnerable. Despite the ban on international trade in gibbons conferred by listing in Appendix I of CITES, illegal trade in gibbons, particularly babies, continues on a wide scale in markets throughout Asia.

[*Lewis G. Regenstein*]

RESOURCES
BOOKS

Benirschke, K. *Primates: The Road to Self-sustaining Populations.* New York: Springer-Verlag, 1986.
Preuschoft, H., et al. *The Lesser Apes: Evolutionary and Behavioral Biology.* Edinburgh: Edinburgh University Press, 1984.

OTHER

International Center for Gibbon Studies. [cited May 2002]. <http://www.gibboncenter.org>.

Lois Marie Gibbs (1951 –)
American environmentalist and community organizer

An activist dedicated to protecting communities from hazardous wastes, Lois Gibbs began her political career as a housewife and homeowner near the **Love Canal**, New York. She was born in Buffalo on June 25, 1951, the daughter of a bricklayer and a full-time homemaker. Gibbs was 21 and a mother when she and her husband bought their house near a buried dump containing hazardous materials from industry and the military, including wastes from the research and manufacture of chemical weapons.

From the time the first articles about Love Canal began appearing in newspapers in 1978, Gibbs has petitioned for state and federal assistance. She began when she discovered the school her son was attending had been built directly on top of the buried canal. Her son had developed epilepsy and there were many similar, unexplained disorders among other children at the school, yet the superintendent was refusing to transfer anyone. The New York State Health Department then held a series of public meetings in which officials appeared more committed to minimizing the community perception of the problem than to solving the problem itself. The governor made promises he was unable to keep, and Gibbs herself was flown to Washington to appear at the White House for what she later decided was little more than political grandstanding. In the book she wrote about her experience, *Love Canal: My Story*, Gibbs describes her frustration and her increasing disillusionment with government, as the threats to the health of both adults and children in the community became more obvious and as it became clearer that no one would be able to move because no one could sell their homes.

While state and federal agencies delayed, the media took an increasing interest in their plight, and Gibbs became more involved in political action. To force federal action, Gibbs and a crowd of supporters took two officers from the **Environmental Protection Agency** (EPA) hostage. A group of heavily armed FBI agents occupied the building across the street and gave her seven minutes before they stormed the offices of the Homeowners' Association, where the men were being held. With less than two minutes left in the countdown, Gibbs appeared outside and released the hostages in front of a national television audience. By the middle of the next week, the EPA had announced that the Federal Disaster Assistance Administration would fund immediate evacuation for everyone in the area.

But the families who left the Love Canal area still could not sell their homes, and Gibbs fought to force the federal government to purchase them and underwrite low-interest loans. After she accused President Jimmy Carter of inaction on a national talk show, in the midst of an approaching election, he agreed to purchase the homes. But he refused to meet with her to discuss the loans. Carter signed the appropriations bill in a televised ceremony at the Democratic National Convention in New York City, and Gibbs simply walked onstage in the middle of it and repeated her request for mortgage assistance. The president could do nothing but promise his political support, and the assistance she had been asking for was soon provided.

Gibbs was divorced soon after her family left the Love Canal area. She moved to Washington D.C. with her two children and founded the Citizen's Clearinghouse for Hazardous Wastes in 1981 (later renamed the Center for Health, Environment and Justice in 1997). Its purpose is to assist communities in fighting toxic waste problems, particularly plans for toxic waste dumping sites, and the organization has worked with over 7,000 neighborhood and community groups. Gibbs has also published *Dying from Dioxin, A Citizen's Guide to Reclaiming Our Health and Rebuilding Democracy*. She has appeared on many television and radio shows and has been featured in hundreds of newspaper and magazine articles. Gibbs has also been the subject of several documentaries and television movies. She often speaks at conferences and seminars and has been honored with numerous awards, including the prestigious Goldman Environmental Prize in 1991. Because of Gibbs' activist work, no commercial sites for hazardous wastes have been opened in the United States since 1978.

[*Lewis G. Regenstein and Douglas Smith*]

Lois Gibbs at her desk during her fight to win permanent relocation for the families living at Love Canal. (Corbis-Bettmann. Reproduced by permission.)

RESOURCES
BOOKS

Gibbs, L. *Love Canal: My Story*. Albany: State University of New York Press, 1982.
Wallace, A. *Eco-Heroes*. San Francisco: Mercury House, 1993.

Gill nets

Gill nets are panels of diamond-shaped mesh netting used for catching fish. When fish attempt to swim through the net their gill covers get caught and they cannot back out. Depending on the **target species**, different mesh sizes are available for use. The top line of the net has a series of floats attached for buoyancy, and the bottom line has **lead** weights to hold the net vertically in the water column.

Gill nets have been in use for many years. They became popular in commercial fisheries in the nineteenth century, evolving from cotton twine netting to the more modern nylon twine netting and monofilament nylon netting. As with many other aspects of **commercial fishing**, the use of gill nets has developed from minor utilization to a major environmental issue. Coupled with **overfishing**, the use of gill nets has caused serious concern throughout the world.

Because gill nets are so efficient at catching fish, they are just as efficient at catching many non-target **species**, including other fishes, **sea turtles**, sea mammals, and sea birds. Gill nets have been used extensively in the commercial fishery for **salmon** and capelin (*Mallotus villosus*). **Dolphins**, **seals**, and sea otters (*Enhydra lutris*) get tangled in the nets, as do diving sea birds such as murres, guillemots, auklets, and puffins that rely on capelin as a mainstay in their diet. Sea turtles are also entangled and drown.

The problem has gotten worse over the last decade with the introduction and extensive use, primarily by foreign fishing fleets, of drift nets. Described as "the most indiscriminate killing device used at sea," drift nets are monofilament gill nets up to 40 mi (64 km) in length. Left at sea for several days and then hauled on board a fishing vessel, these drift nets contain vast numbers of dead marine life, besides the target species, that are simply discarded over the side of the boat. The outrage expressed regarding these "curtains of death" led to a United Nations resolution banning their use in commercial fisheries after the end of 1992. Commercial fishermen who use other types of nets for catching fish, such as the purse seines used in the tuna fishing industry and the bag trawls used in the shrimping industry, have modified their nets and fishing techniques to attempt to eliminate the killing of dolphins and sea turtles, respectively. Unfortunately, such modifications of gill nets are nearly impossible due to the nets' design and the way these nets are used. *See also* Turtle excluder device

[*Eugene C. Beckham*]

RESOURCES
PERIODICALS

Norris, K. "Dolphins in Crisis." *National Geographic* 182 (1992): 2–35.

GIS

see **Geographic information systems**

Glaciation

The covering of the earth's surface with glacial ice. The term also includes the alteration of the surface of the earth by glacial **erosion** or deposition. Due to the passage of time, ice erosion can be almost unidentifiable; the **weathering** of hard rock surfaces often eliminates minor scratches and other evidence of such glacial activities as the carving of deep valleys. The evidence of deposition, known as depositional imprints, can vary. It may consist of specialized features a few meters above the surrounding terrain, or it may consist of ground materials several meters in thickness covering wide areas of the landscape.

Only 10% of the earth's surface is currently covered with glacial ice, but it is estimated that 30% had been covered with glacial ice at some time. During the last major glacial period, most of Europe and more than half of the North American continent were covered with ice. The glacial ice of modern day is much thinner than it was in the **ice age**, and the majority of it (85%) is found in **Antarctica**. About 11% of the remaining glacial ice is in Greenland, and the rest is scattered in high altitudes throughout the world.

Moisture and cold temperatures are the two main factors for the formation of glacial ice. Glacial ice in Antarctica is the result of relatively small quantities of snow deposition and low loss of ice because of the cold **climate**. In the middle and low latitudes where the loss of ice, known as ablation, is higher, snowfall tends to be much higher and the glaciers are able overcome ablation by generating large amounts of ice. These types of systems tend to be more active than the glaciers in Antarctica, and the most active of these are often located at high altitudes and in the path of prevailing winds carrying marine moisture.

The **topography** of the earth has been shaped by glaciation. Hills have been reduced in height and valleys created or filled in the movement of glacial ice. *Moraine* is a French term used to describe the ridges and earthen dikes formed near the edges of regional glaciers. Ground moraine is material that accumulates beneath a glacier and has low-relief characteristics, and end moraine is material that builds up along the extremities of a glacier in a ridge-like appearance.

In England, early researchers found stones that were not common to the local ground rock and decided they must have "drifted" there, carried by icebergs on water. Though geology has changed since that time, the term remains and all deposits made by glacial ice are usually identified as drift. These glacial deposits, also known as till, are highly varied in composition. They can be a fine grained deposit, or very coarse with rather large stones present, or a combination of both.

Rock and other soil-like debris are often crushed and ground into very small particles, and they are commonly found as **sediment** in waters flowing from a glacial mass. This material is called glacial flour, and it is carried downstream to form another kind of glacial deposit. During certain cold, dry periods of the year winds can pick up portions of this deposit and scatter it for miles. Many of the different soils in the American "Corn Belt" originated in this way, and they have become some of the more important agricultural soils in the world.

[*Royce Lambert*]

RESOURCES
BOOKS

Flint, R. F. *Glacial and Pleistocene Geology*. New York: Wiley, 1972.

Henry A. Gleason (1882 – 1975)
American ecologist

Henry A. Gleason was a half generation after that small group of midwesterners who founded **ecology** as a discipline in the United States. He was a student of **Stephen Forbes** and his early work in ecology was influenced strongly by Cowles and **Frederic E. Clements**. He did later, however, in 1935, claim standing—bowing only to Cowles and Clements and for some reason not including Forbes—as "the only other original ecologist in the country." And he was original. His work built on that of the founders, but he quickly and actively questioned their ideas and concepts, especially those of Clements, in the process creating controversy and polarization in the ecological community. Gleason called himself an "ecological outlaw" and probably over-emphasized his early lack of acceptance in ecology, but in a resolution of respect from the Ecological Society of America after his death, he was described as a revolutionary and a heretic for his skepticism toward 'established' ideas in ecology. Stanley Cain said that Gleason was never "impressed nor fooled by the philosophical creations of other ecologists, for he has always tested their ideas concerning the association, **succession**, the climax, environmental controls, and **biogeography** against what he knew in **nature**." Gleason was a critical rather than negative thinker and never fully rejected the utility of the idea of community, but the early marginalization of his ideas by mainstream ecologists, and the controversy they created, may have played a role in his later concentration on taxonomy over ecology.

He claimed that if plant associations did exist, they were individualistic and different from area to area, even where most of the same species were present. Gleason did clearly reject, however, Clements' idea of a monoclimax, proclaiming that "the Clementsian concept of succession, as an irreversible trend leading to the climax, was untenable." His field observations also led him to repudiate Clements' organismic concept of the plant community, asking "are we not justified in coming to the general conclusion, far removed from the prevailing opinion, that an association is not an organism?" He went on to say that it is "scarcely even a vegetation unit." He also pointed out the errors in 'Raunkiaer's Law,' on frequency distribution, which as Robert McIntosh noted, was "widely interpreted [in early ecology] as being a fundamental community characteristic indicating homogeneity," and questioned Jaccard's comparison of two communities through a coefficient of similarity that Gleason believed unduly gave as much weight to rare as to common species.

Gleason's own approach to the study of vegetation emerged from his skills as a floristic botanist, an approach rejected as "old botany" by the founders of ecology. As Nicolson suggests, "a floristic approach entailed giving primacy to the study of the individual plants and their species. This was the essence of [Gleason's] individualistic concept." In hindsight, somewhat ironically then, Gleason used old botany to create a new alternative to what had quickly become dogma in ecology, the centrality of the idea that units of vegetation were real, that the plant association was indispensable to an ecological approach.

Clements was more accepted in the early part of the twentieth century than Gleason, though many ecologists at the time considered both too extreme, just in opposite ways. Today, Clements' theories remain out of favor and some of Gleason's have been revived, though not all of them. Contrary to his own observations, he was persuaded that plants are distributed randomly, at least over small areas, which is seldom if ever the case, though he later backed away from this assertion. He could not accept the theory of continental drift, stating that "the theory requires a shifting of the location of the poles in a way which does considerable violence to botanical and geological facts," and therefore should have few adherents among botanists.

Despite Gleason's skepticism about some of Clements' major ideas, the older botanist was a major influence, especially early in Gleason's career. Especially influential was Clements' rudimentary development of the quadrat method of sampling vegetation, which shaped Gleason's approach to field work; Gleason took the method much further than Clements, and though not trained in mathematics, was the first ecologist to employ a number of quantitative approaches and methods. As McIntosh demonstrated, Gleason, following Forbes lead in aquatic ecology "was clearly one of the earliest and most insightful proponents of the use of quantitative methods in terrestrial ecology."

Gleason was born in the heart of the area where ecology first flourished in the United States. His interest in vegetation and his contributions to ecology were both stimulated by growing up in and doing research on the dynamics of the prairie-forest border. He won bachelor's and master's degrees from the University of Illinois and a Ph.D. from Columbia University. He returned to the University of Illinois as an instructor in botany (1901–1910), where he worked with Stephen Forbes at one of the major American centers of ecological research at the time. In 1910, he moved to the University of Michigan (1910) and while in Ann Arbor, married Eleanor Mattei. Then, in 1919, he moved to the New York Botanical Garden, where he spent the rest of his career, sometimes (reluctantly) as an administrator,

always as a research taxonomist. He retired from the Garden in 1951.

Moving out of the Midwest, Gleason also moved out of ecology. Most of his work at the Botanic Garden was taxonomic. He did some ecological work, such as a three-month ecological survey of Puerto Rico in 1926, and a restatement of his "individualistic concept of the plant association, (also in 1926 and also in the *Bulletin of the Torrey Botanical Club*), in which he posed what Nicolson described as "a radical challenge" to the basis of contemporary ecological practice. Gleason's challenge to his colleagues and critics in ecology was to "demolish our whole system of arrangement and classification and start anew with better hope of success." His reasoning was that ecologists had "attempted to arrange all our facts in accordance with older ideas, and have come as a result into a tangle of conflicting ideas and theories." He anticipated twenty-first century thinking that identification on the ground of community and ecosystem as ecological units is arbitrary, noting that vegetation was too continuously varied to identify recurrent associations. He claimed, for example, that "no ecologist would refer the alluvial forests of the upper and lower Mississippi to the same association, yet there is no place along their whole range where one can logically mark a boundary between them. As Mcintosh suggests, "one of Gleason's major contributions to ecology was that he strove to keep the conceptual mold from hardening prematurely."

In his work as a taxonomist for the Garden, Gleason traveled as a plant collector, becoming what he described as "hooked" on tropical American botany, specializing in the large family of melastomes, tropical plants ranging from black mouth fruits to handsome cultivated flowers, a group which engaged him for the rest of his career. His field work, on this family but especially many others, was reinforced by extensive study and identification on material collected by others and made available to him at the Garden.

A major assignment during his New York years, emblematic of his work as a taxonomist, was a revision of the Britton and Brown *Illustrated Flora of the Northeastern United States* (1952) which Maguire describes as a "heavy duty [that] intervened and essentially brought to a close Gleason's excellent studies of the South American floras and [his] detailed inquiry into the Melastomataceae...this great work...occupied some ten years of concentrated, self-disciplined attention." He did publish a few brief pieces on the melastomes after the Britton and Brown, and also two books with Arthur Cronquist, *Manual of Vascular Plants of Northeastern United States and Adjacent Canada* (1963) and the more general *The Natural Geographyof Plants* (1964). The latter, though co-authored, was an overt attempt by Gleason to summarize a life's work and make it accessible to a wider public.

Gleason's early ecological work on species-area relations, the problem of rare species, and his extensive taxonomic work all laid an initial base for contemporary concern among biologists (especially) about the threat to the earth's bio-diversity. Gleason wrote that analysis of "the various species in a single association would certainly show that their optimum environments are not precisely identical," a foreshadowing of later work on niche separation.

McIntosh claimed in 1975 that Gleason's individualistic concept "must be seen not simply as one of historical interest but very likely as one of the key concepts of modern and, perhaps, future ecological thought." A revival of Gleason's emphasis on the individual at mid-twentieth century became one of the foundations for what some scientists in the second half of the twentieth century called a "new ecology," one that rejects imposed order and system and emphasizes the chaos, the randomness, the uncertainty and the unpredictability of natural systems. A call for 'adaptive' resource and environmental management policies flexible enough to respond to unpredictable change in individually variant natural systems is one outgrowth of such changes in thinking in ecology and the **environmental sciences**.

[*Gerald L. Young*]

FURTHER READING
BOOKS

Gleason, H. A. "Twenty-Five Years of Ecology, 1910–1935." Vol. 4, *Memoirs, Brooklyn Botanic Garden*. Brooklyn: Brooklyn Botanic Garden, 1936.

PERIODICALS

Cain, Stanley A. "Henry Allan Gleason: Eminent Ecologist 1959." *Bulletin of the Ecological Society of America* 40, no. 4 (December 1959): 105–110.

Gleason, H. A. "Delving Into the History of American Ecology—Reprint of 1952 Letter to C. H. Muller." *The Bulletin of the Ecological Society of America* 56, no. 4 (December 1975): 7–10.

Maguire, Bassett. "Henry Allan Gleason—1881–1975." *Bulletin of the Torrey Botanical Club* 102, no. 5 (September/October 1975): 274–282.

McIntosh, Robert P. "H.A. Gleason—"Individualistic Ecologist" 1882–1975: His Contributions to Ecological Theory." *Bulletin of the Torrey Botanical Club* 102, no. 5 (September/October 1975): 253–273.

Nicolson, Malcolm. "Henry Allan Gleason and the Individualistic Hypothesis: The Structure of a Botanist's Career." *The Botanical Review* 56, no. 2 (April/June 1990): 91–161.

Glen Canyon Dam

Until 1963, Glen Canyon was one of the most beautiful stretches of natural scenery in the American West. The canyon had been cut over thousands of years as the **Colorado River** flowed over sandstone that once formed the floor of an ancient sea. The colorful walls of Glen Canyon were often compared to those of the Grand Canyon, only about 50 mi (80 km) downstream.

Humans have long seen more than beauty in the canyon, however. They have envisioned the potential value of a water **reservoir** that could be created by damming the Colorado. In a region where water can be as valuable as gold, plans for the construction of a giant **irrigation** project with water from a Glen Canyon dam go back to at least 1850.

Flood control was a second argument for the construction of such a dam. Like most western rivers, the Colorado is wild and unpredictable. When fed by melting snows and rain in the spring, its natural flow can exceed 300,000 ft^4 (8,400 m^4) per second. At the end of a hot dry summer, flow can fall to less than 1% of that value. The river's water temperature can also fluctuate widely, by more than 36°F (20°C) in a year. A dam in Glen Canyon held the promise of moderating this variability.

By the early 1900s, yet a third argument for building the dam was proposed—the generation of hydroelectric power. Both the technology and the demand were reaching the point that power generated at the dam could be supplied to Phoenix, Los Angeles, San Diego, and other growing urban areas in the Far West.

Some objections were raised in the 1950s when construction of a Glen Canyon Dam was proposed, and environmentalists fought to protect this unique natural area. The 1950s and early 1960s were not, however, an era of high environmental sensitivity, and plans for the dam eventually were approved by the U. S. Congress. Construction of the dam, just south of the Utah-Arizona border, was completed in 1963 and the new lake it created, Lake Powell, began to develop. Seventeen years later, the lake was full holding a maximum of 27 million acre-feet of water.

The environmental changes brought about by the dam are remarkable. The river itself has changed from a muddy brown color to a clear crystal blue as the sediments it carries are deposited behind the dam in Lake Powell. **Erosion** of river banks downstream from the dam has lessened considerably as spring floods are brought under control. Natural beaches and sandbars, once built up by deposited **sediment**, are washed away. River temperatures have stabilized at an annual average of about 50°F (10°C). These physical changes have brought about changes in **flora** and **fauna** also. Four **species** of fish native to the Colorado have become extinct, but at least 10 species of birds are now thriving where they barely survived before. The **biotic community** below the dam is significantly different from what it was before construction.

During the 1980s, questions about the dam's operation began to grow. A number of observers were especially concerned about the fluctuations in flow through the dam, a pattern determined by electrical needs in distant cities. During peak periods of electrical demand, operators increase the flow of water though the dam to a maximum of 30,000 ft^4

(840 m⁴) per second. At periods of low demand, that flow may be reduced to 1,000 ft⁴ (28 m⁴) per second. As a result of these variations, the river below the dam can change by as much as 13 ft (4 m) in height in a single 24-hour period. This variation can severely damage riverbanks and can have unsettling effects on **wildlife** in the area as, for example, fish are stranded on the shore or swept away from spawning grounds. River-rafting is also severely affected by changing river levels as rafters can never be sure from day to day what water conditions they may encounter.

Operation of the Glen Canyon Dam is made more complex by the fact that control is divided up among at least three different agencies in the U. S. Department of the Interior, the **Bureau of Reclamation**, the **Fish and Wildlife Service**, and the **National Park Service**, all with somewhat different missions. In 1982, a comprehensive re-analysis of the Glen Canyon area was initiated. A series of environmental studies called the Glen Canyon Environmental Studies were designed and carried out over much of the following decade. In addition, Interior Secretary Manuel Lujan announced in 1989 that an **environmental impact statement** on the downstream effects of the dam would be conducted.

The purpose of the environmental impact statement was to find out if other options were available for operating the dam that would minimize harmful effects on the **environment**, recreational opportunities, and Native American activities while still allowing the dam to produce sufficient levels of hydroelectric power. The effects studied included water, sediment, fish, vegetation, wildlife and **habitat**, endangered and other special-status species, cultural resources, **air quality**, **recreation**, hydropower, and non-use value (i.e., general appreciation of **natural resources**).

Nine different operating options for the dam were considered. These options fell into three general categories: unrestricted fluctuating flows (two alternative modes); restricted fluctuating flows (four modes); and steady flows (three modes).

The final choice made was one that involves "periodic high, steady releases of short duration" that reduce the dam's performance significantly below its previous operating level. The criterion for this decision was the protection and enhancement of downstream resources while continuing to permit a certain level of flexibility in the dam's operation.

A later series of experiments was designed to see what could be done to restore certain downstream resources that have been destroyed or damaged by the dam's operation. Between March 26 and April 2, 1996, the U.S. Bureau of **Reclamation** released unusually large amounts of water from the dam. The intent was to reproduce the large scale **flooding** on the Colorado River that had normally occurred every spring before the dam was built.

The primary focus of this project was to see if downstream sandbars could be restored by the flooding. The sandbars have traditionally been used as campsites and have been a major mechanism for the removal of **silt** from backwater channels used by native fish. Depending on the final results of this study, the Bureau will determine what changes, if any, should be taken in adjusting flow patterns over the dam to provide for maximum environmental benefit downstream along with power output. *See also* Alternative energy sources; Riparian land; Wild river

[*David E. Newton*]

RESOURCES
PERIODICALS

Elfring, C. "Conflict in the Grand Canyon." *BioScience* (November 1990): 709–711.
Udall, J. R. "A Wild, Swinging River." *Sierra* (May 1990): 22–26.

Global Environment Monitoring System

A data-gathering project administered by the **United Nations Environment Programme**. The Global Environment Monitoring System (GEMS) is one aspect of the modern understanding that environmental problems ranging from the **greenhouse effect** and **ozone layer depletion** to the preservation of **biodiversity** are international in scope. The system was inaugurated in 1975, and it monitors weather and **climate** changes around the world, as well as variations in soils, the health of plant and animal **species**, and the environmental impact of human activities.

GEMS was not intended to replace any existing systems; it was designed to coordinate the collection of data on the environment, encouraging other systems to supply information it believed was being omitted. In addition to coordinating the gathering of this information, the system also publishes it in an uniform and accessible fashion, where it can be used and evaluated by environmentalists and policy makers.

GEMS operates 25 information networks in over 142 countries. These networks monitor **air pollution**, including the release of **greenhouse gases** and changes in the **ozone** layer, and **air quality** in various urban center; they also gather information on **water quality** and food contamination in cooperation with the World Health Organization and the Food and Agriculture Organization of the United Nations.

Global Forum

The Global Forum of Spiritual and Parliamentary Leaders on Human Survival is a worldwide organization of scientists, leaders of world religions, and parliamentarians who are attempting to change environmental and developmental values in their countries. Members include local, national, and international leaders in the arts, business, community action, education, faith, government, media, and youth sectors.

Historically, lawmakers and spiritual leaders have differed in their views toward stewardship of the earth. A conference held in Oxford, England, in 1988, and attended by 200 spiritual and legislative leaders brought these groups together with scientists to discuss solutions to worldwide environmental problems. Speakers included the Dalai Lama, Mother Teresa, and the Archbishop of Canterbury, who conferred with experts such as Carl Sagan, Kenyan environmentalist Wangari Maathai, and **Gaia hypothesis** scientist James Lovelock. As a result of the Oxford conference, the Soviet Union invited the Global Forum to convene an international meeting on critical survival issues. The Moscow conference, called the Global Forum on Environment and Development, took place in January 1990. Over 1,000 spiritual and parliamentary leaders, scientists, artists, journalists, businessmen, and young people from eighty-three countries attended the Moscow Forum. One initiative of the Moscow Forum was a joint commitment by scientists and religious leaders to preserve and cherish the earth.

The Global Forum tries not to duplicate the activities of other environmental groups but works to relate global issues to local environments. For example, participants at the first U.S.-based Global Forum conference in Atlanta in May 1992, learned about the local effects of global problems such as **tropical rain forest** destruction, global warming, and **waste management**. The Global Forum has initiated seminars worldwide on ethical implications of the environmental crisis. Artists learn about the role of the arts in communicating global survival issues. Business leaders promote **sustainable development** at the highest levels of business and industry. Young people petition their schools to include curriculum on environmental issues as required subjects.

[*Linda Rehkopf*]

RESOURCES

ORGANIZATIONS

Global Forum, East 45th St., 4th Floor , New York , NY USA 10017

Global Releaf

Global Releaf, an international citizen action and education program, was initiated in 1988 by the 115-year-old American Forestry Association in response to the worldwide concern over global warming and the **greenhouse effect**. Campaigning under the slogan "Plant a tree, cool the globe," its over 112,000 members began the effort to reforest the earth one tree at a time.

In 1990, Global Releaf began Global Releaf Forest, an effort to restore damaged **habitat** on public lands through tree plantings Global Releaf Fund is its urban counterpart. Using each one-dollar donation to plant one tree resulted in the planting of more than four million trees on 70 sites in 33 states. By involving local citizens and resource experts in each project, the program ensures that the right **species** are planted in the right place at the right time. Results include the protection of endangered and threatened animals, restoration of native species, and improvement of recreational opportunities.

Funding for the program has come largely from government agencies, corporations, and non-profit organizations. Chevrolet-Geo celebrated the planting of its millionth tree in October 1996. The Texaco/Global Releaf Urban Tree Initiative, utilizing more than 6,000 Texaco volunteers, has helped local groups plant more than 18,000 large trees and invested over $2.5 million in projects in twelve cities. Outfitter Eddie Bauer began an "Add a Dollar, Plant a Tree" program to fund eight Global Releaf Forest sites in the United States and Canada, planting close to 350,000 trees.

The Global Releaf Fund also helps finance urban and rural reforestation on foreign **soil** in projects undertaken with its international partners. Engine manufacturer, Briggs & Stratton, for example, has made possible tree plantings both in the United States and in Ecuador, England, Germany, Poland, Romania, Slovakia, South Africa, and Ukraine, while Costa Rica, Gambia, and the Philippines have benefitted from picture-frame manufacturer Larsen-Juhl.

Unfortunately, not enough funding exists to grant all the requests; in 1996, only 40% of the proposed projects received financial backing. Forced to pick and choose, the review board favors those projects which aim to protect endangered and threatened species. Burned forests and natural disaster areas—like the Francis Marion **National Forest** in South Carolina, devastated by 1989's **Hurricane** Hugo—are also high on the priority list, as are streamside woodlands and landfills.

Looking to the future, Global Releaf 2000 was launched in 1996 with the aim of encouraging the planting of 20 million trees, increasing the canopy in select cities by 20%, and expanding the program to include private lands and sanitary landfills. A 20-city survey done in 1985 by

American Forests showed that four trees die for every one planted in United States cities and that the average city tree lives only 32 years (just seven years, downtown). With these facts in mind, Global Releaf asks that communities plant twice as many trees as are lost in the next decade. In August of 2001, more than 19 million trees had been planted.

[*Ellen Link*]

RESOURCES
BOOKS

Sobel K. L., S. Orrick, and R. Honig. *Environmental Profiles: A Global Guide to Projects and People.* New York: Garland, 1993.

PERIODICALS

"Global Releaf 2000." *American Forests* 103, no. 4 (Autumn 1996): 30.
"A Helping Hand for Damaged Land." *American Forests* 102, no. 3 (Summer 1996): 33–35.
"Planting One for the Millennium." *American Forests* 102, no. 3 (Summer 1996): 13–15.

ORGANIZATIONS

American Forests, P.O. Box 2000, Washington , D.C. USA 20013 (202) 955-4500, Fax: (202) 955-4588, Email: info@amfor.org, <http://www.americanforests.org>

Global 2000 Report
see *The Global 2000 Report*

Global warming
see **Greenhouse effect**

GOBO
see **Child survival revolution**

Goiter

Generally refers to any abnormal enlargement of the thyroid gland. The most common type of goiter, the simple goiter, is caused by a deficiency of iodine in the diet. In an attempt to compensate for this deficiency, the thyroid gland enlarges and may become the size of a large softball in the neck. The general availability of table salt to which potassium iodide has been added ("iodized" salt) has greatly reduced the incidence of simple goiter in many parts of the world. A more serious form of goiter, toxic goiter, is associated with hyperthyroidism. The etiology of this condition is not well understood. A third form of goiter occurs primarily in women and is believed to be caused by changes in hormone production.

Golf courses

The game of golf appears to be derived from ancient stick-and-ball games long played in western Europe. However,

the first documented rules of golf were established in 1744, in Edinburgh, Scotland. Golf was first played in the United States in the 1770s, in Charleston, South Carolina. It was not until the 1880s, however, that the game began to become widely popular, and it has increasingly flourished since then. In 2002, there were about 16,000 golf courses in the United States, and thousands more in much of the rest of the world.

Golf is an excellent form of outdoor **recreation**. There are many health benefits of the game, associated with the relatively mild form of exercise and extensive walking that can be involved. However, the development and management of golf courses also results in environmental damage of various kinds. The damage associated with golf courses can engender intense local controversy, both for existing facilities and when new ones are proposed for development.

The most obvious environmental affect of golf courses is associated with the large amounts of land that they appropriate from other uses. Depending on its design, a typical 18-hole golf course may occupy an area of about 100-200 acres. If the previous use of the land was agricultural, then conversion to a golf course results in a loss of food production. Alternatively, if the land previously supported forest or some other kind of natural **ecosystem**, then the conversion results in a large, direct loss of **habitat** for native **species** of plants and animals.

In fact, some particular golf courses have been extremely controversial because their development caused the destruction of the habitat of **endangered species** or rare kinds of natural ecosystems. For instance, the Pebble Beach Golf Links course, one of the most famous in the world, was developed in 1919 on the Monterey Peninsula of central California, in natural coastal and forest habitats that harbor numerous rare and endangered species of plants and animals. Several additional gold courses and associated tourist facilities were subsequently developed nearby, all of them also displacing natural ecosystems and destroying the habitat of **rare species**. Most of those recreational facilities were developed at a time when not much attention was paid to the needs of endangered species. Today, however, the **conservation** of **biodiversity** is considered an important issue. It is quite likely that if similar developments were now proposed in such critical habitats, citizen groups would mount intense protests and government regulators would not allow the golf courses to be built.

The most intensively modified areas on golf courses are the fairways, putting greens, aesthetic lawns and gardens, and other highly managed areas. Because these kinds of areas are intrinsic to the design of golf courses, a certain amount of loss of natural habitat is inevitable. To some degree, however, the net amount of habitat loss can be decreased by attempting, to the degree possible, to retain natural community types within the golf course. This can be done particularly effectively in the brushy and forested areas between the holes and their approaches. The habitat quality in these less-intensively managed areas can also be enhanced by providing nesting boxes and brush piles for use by birds and small mammals, and by other management practices known to favor **wildlife**. Habitat quality is also improved by planting native species of plants wherever it is feasible to do so.

In addition to land appropriation, some of the management practices used on golf courses carry the risk of causing local environmental damage. This is particularly the case of putting greens, which are intensively managed to maintain an extremely even and consistent lawn surface.

For example, to maintain a **monoculture** of desired species of grasses on putting greens and lawns, intensive management practices must be used. These include frequent mowing, **fertilizer** application, and the use of a variety of pesticidal **chemicals** to deal with various pests affecting the turfgrass. This may involve the application of such herbicides as Roundup (glyphosate), **2,4-D**, MCPP, or Dicamba to deal with undesirable weeds. **Herbicide** application is particularly necessary when putting greens and lawns are being first established. Afterward their use can be greatly reduced by only using spot-applications directly onto turf-grass weeds. Similarly, fungicides might be used to combat infestations of turf-grass disease **fungi**, such as the fusarium blight (*Fusarium culmorum*), take-all patch (*Gaeumannomyces graminis*), and rhizoctonia blight (*Rhizoctonia solani*).

Infestations by turf-damaging insects may also be a problem, which may be dealt with by one or more insecticide applications. Some important insect pests of golf-course turf-grasses include the Japanese beetle (*Popillia japonica*), chafer beetles (*Cyclocephala* spp.), June beetles (*Phyllophaga* spp.), and armyworm beetle (*Pseudaletia unipuncta*). Similarly, rodenticides may be needed to get rid of moles (*Scalopus aquaticus*) and their burrows.

Golf courses can also be a major user of water, mostly for the purposes of **irrigation** in dry climates or during droughty periods. This can be an important problem in semi-arid regions, such as much of the southwestern U.S., where water is a scare and valuable commodity with many competing users. To some degree, water use can be decreased by ensuring that irrigation is only practiced when necessary, and only in specific places where it is needed, rather than according to a fixed schedule and in a broadcast manner. In some climatic areas, nature-scaping and other low-maintenance practices can be used over extensive areas of golf courses. This can result in intensive irrigation only being practiced in key areas, such as putting greens, and to a lesser degree fairways and horticultural lawns.

Many golf courses have ponds and lakes embedded in their spatial design. If not carefully managed, these waterbodies can become severely polluted by nutrients, pesticides, and eroded materials. However, if care is taken with golf-course management practices, their ponds and lakes can sustain healthy ecosystems and provide refuge habitat for local native plants and animals.

Increasingly, golf-course managers and industry associations are attempting to find ways to support their sport while not causing an unacceptable amount of environmental damage. One of the most important initiatives of this kind is the Audubon Cooperative Sanctuary Program for Golf Courses, run by the Audubon International, a private conservation organization. Since 1991, this program has been providing **environmental education** and conservation advice to golf-course managers and designers. By 2002, membership in this Audubon program had grown to more than 2,300 courses in North America and elsewhere in the world.

The Audubon Cooperative Sanctuary Program for Golf Courses provides advice to help planners and managers with: (a) environmental planning; (b) wildlife and habitat management; (c) chemical use reduction and safety; (d) **water conservation**; and (e) outreach and education about environmentally appropriate management practices. If a golf course completes recommended projects in all of the components of the program, it receives recognition as a Certified Audubon Cooperative Sanctuary. This allows the golf course to claim that it is conducting its affairs in a certifiably "green" manner. This results in tangible environmental benefits of various kinds, while being a source of pride of accomplishment for employees and managers, and providing a potential marketing benefit to a clientele of well-informed consumers.

There are many specific examples of environmental benefits that have resulted from golf courses engaged in the Audubon Cooperative Sanctuary Program. For instance, seven golf courses in Arizona and Washington have allowed the installation of 150 artificial nesting burrows for burrowing owls (*Athene cunicularia*), an endangered species, on suitable habitat on their land. In 2000, Audubon International conducted a survey of cooperating golf courses, and the results were rather impressive. About 78% of the respondents reported that they had decreased the total amount of turf-grass area on their property; 73% had taken steps to increase the amount of wildlife habitat; 45% were engaged in an ecosystem restoration project; 90% were attempting to use native plants in their horticulture; and 85% had decreased their use of pesticides and 91% had switched to lower-toxicity chemicals. Just as important, about half of the respondents believed that there had been an improvement in the playing quality of their golf course and in the satisfaction of both employees and their client golfers. Moreover, none of the respondents believed that any of these values had been degraded as a result of adopting the management practices advised by the Audubon International program.

These are all highly positive indicators. They suggest that the growing and extremely popular sport of golf can, within limits, potentially be practiced in ways that do not cause unacceptable levels of environmental and ecological damage.

[*Bill Freedman Ph.D.*]

RESOURCES
BOOKS

Balogh, J. C., and W. J. Walker, eds. *Golf Course Management and Construction: Environmental Issues.* Leeds, UK: Lewis Publishers, 1992.

Gillihan, S. W. *Bird Conservation on Golf Courses: A Design and Management Manual.* Ann Arbor, MI: Ann Arbor Press, 1992.

Sachs, P. D., and R.T. Luff. *Ecological Golf Course Management.* Ann Arbor, MI: Ann Arbor Press, 2002.

OTHER

"Audubon Cooperative Sanctuary Program for Golf." *Audubon International.* 2002 [cited July 2002]. –http://www.audubonintl.org/programs/acss/golf.htm>.

United States Golf Association. [cited July 2002]. <http://www.usga.org>.

ORGANIZATIONS

United States Golf Association, P.O. Box 708, Far Hills, N.J. USA 07931-0708, Fax: 908-781-1735, Email: usga.org, http://www.usga.org/

Good wood

Good wood, or smart wood, is a term certifying that the wood is harvested from a forest operating under environmentally sound and sustainable practices. A "certified wood" label indicates to consumers that the wood they purchase comes from a forest operating within specific guidelines designed to ensure future use of the forest. A well-managed forestry operation takes into account the overall health of the forest and its ecosystems, the use of the forest by indigenous people and cultures, and the economic influences the forest has on local communities. Certification of wood allows the wood to be traced from harvest through processing to the final product (i.e., raw wood or an item made from wood) in an attempt to reduce uncontrollable **deforestation**, while meeting the demand for wood and wood products by consumers around the world.

Public concern regarding the disappearance of tropical forests initially spurred efforts to reduce the destruction of vast acres of rainforests by identifying environmentally responsible forestry operations and encouraging such practices by paying foresters higher prices. Certification, however, is not limited to tropical forests. All forest types—tropical, temperate, and boreal (those located in northern climes)—from all countries may apply for certification. Plantations (stands of timber that have been planted for the purpose of **logging** or that have been altered so that they no longer

support the ecosystems of a natural forest) may also apply for certification.

Certification of forests and forest owners and managers is not required. Rather, the process is entirely voluntary. Several organizations currently assess forests and **forest management** operations to determine whether they meet the established guidelines of a well-managed, sustainable forest. The Forest Stewardship Council (FSC), founded in 1993, is an organization of international members with environmental, forestry, and socioeconomic backgrounds that monitors these organizations and verifies that the certification they issue is legitimate.

A set of 10 guiding principles known as Principles and Criteria (P&C) were established by the FSC for certifying organizations to utilize when evaluating forest management operations. The P&C address a wide range of issues, including compliance with local, national, and international laws and treaties; review of the forest operation's management plans; the religious or cultural significance of the forest to the indigenous inhabitants; maintenance of the rights of the indigenous people to use the land; provision of jobs for nearby communities; the presence of threatened or **endangered species**; control of excessive **erosion** when building roads into the forest; reduction of the potential for lost **soil** fertility as a result of harvesting; protection against the invasion of non-native **species**; **pest** management that limits the use of certain chemical types and of genetically altered organisms; and protection of forests when deemed necessary (for example, a forest that protects a **watershed** or that contains threatened and/or endangered species).

Guarding against illegal harvesting is a major hurdle for those forest managers working to operate within the established regulations for certification. Forest devastation occurs not only from harvesting timber for wood sales but when forests are clear cut to make way for cattle crazing or farming, or to provide a fuel source for local inhabitants. Illegal harvesting often occurs in developing countries where enforcement against such activities is limited (for example, the majority of the trees harvested in Indonesia are done so illegally).

Critics argue against the worthiness of managing forests, suggesting that the logging of select trees from a forest should be allowed and that once completed, the remaining forest should be placed off limits to future logging. Nevertheless, certified wood products are in the market place; large wood and wood product suppliers are offering certified wood and wood products to their consumers. In 2001 the Forest Leadership Forum (a group of environmentalists, forest industry representatives, and retailers) met to identify how wood retailers can promote sustainable forests. It is hoped that consumer demand for good wood will drive up the number of forests participating in the certification program, thereby reducing the rate of irresponsible deforestation of the world's forests.

[*Monica Anderson*]

RESOURCES
BOOKS

Bass, Stephen, et al. *Certification's Impact on Forests, Stakeholders and Supply Chains.* London: IIED, 2001.

ORGANIZATIONS

Forest Stewardship Council United States, 1155 30th Street, NW, Suite 300, Washington, DC USA 20007 (202) 342 0413, Fax: (202) 342 6589, Email: info@foreststewardship.org, <http://www.fscus.org<

Jane Goodall (1934 –)
English primatologist and ethnologist

Jane Goodall is known worldwide for her studies of the **chimpanzees** of the Gombe Stream Reserve in Tanzania, Africa. She is well respected within the scientific community for her ground-breaking field studies and is credited with the first recorded observation of chimps eating meat and using and making tools. Because of Goodall's discoveries, scientists have been forced to redefine the characteristics once considered as solely human traits. Goodall is now leading efforts to ensure that animals are treated humanely both in their wild habitats and in captivity.

Goodall was born in London, England, on April 3, 1934, to Mortimer Herbert Goodall, a businessperson and motor-racing enthusiast, and the former Margaret Myfanwe Joseph, who wrote novels under the name Vanne Morris Goodall. Along with her sister, Judy, Goodall was reared in London and Bournemouth, England. Her fascination with animal behavior began in early childhood. In her leisure time, she observed native birds and animals, making extensive notes and sketches, and read widely in the literature of zoology and ethnology. From an early age, she dreamed of traveling to Africa to observe exotic animals in their natural habitats.

Goodall attended the Uplands private school, receiving her school certificate in 1950 and a higher certificate in 1952. At age eighteen she left school and found employment as a secretary at Oxford University. In her spare time, she worked at a London-based documentary film company to finance a long-anticipated trip to Africa. At the invitation of a childhood friend, she visited South Kinangop, Kenya. Through other friends, she soon met the famed anthropologist Louis Leakey, then curator of the Coryndon Museum in Nairobi. Leakey hired her as a secretary and invited her to participate in an anthropological dig at the now famous Olduvai Gorge, a site rich in fossilized prehistoric remains of early ancestors of humans. In addition, Goodall was sent to study the vervet monkey, which lives on an island in Lake Victoria.

Leakey believed that a long-term study of the behavior of higher primates would yield important evolutionary information. He had a particular interest in the chimpanzee, the second most intelligent primate. Few studies of chimpanzees had been successful; either the size of the safari frightened the chimps, producing unnatural behaviors, or the observers spent too little time in the field to gain comprehensive knowledge. Leakey believed that Goodall had the proper temperament to endure long-term isolation in the wild. At his prompting, she agreed to attempt such a study. Many experts objected to Leakey's selection of Goodall because she had no formal scientific education and lacked even a general college degree.

While Leakey searched for financial support for the proposed Gombe Reserve project, Goodall returned to England to work on an animal documentary for Granada Television. On July 16, 1960, accompanied by her mother and an African cook, she returned to Africa and established a camp on the shore of Lake Tanganyika in the Gombe Stream Reserve. Her first attempts to observe closely a group of chimpanzees failed; she could get no nearer than 500 yd (457 m) before the chimps fled. After finding another suitable group of chimpanzees to follow, she established a non-threatening pattern of observation, appearing at the same time every morning on the high ground near a feeding area along the Kakaombe Stream valley. The chimpanzees soon tolerated her presence and, within a year, allowed her to move as close as 30 ft (9 m) to their feeding area. After two years of seeing her every day, they showed no fear and often came to her in search of bananas.

Goodall used her newfound acceptance to establish what she termed the "banana club," a daily systematic feeding method she used to gain trust and to obtain a more thorough understanding of everyday chimpanzee behavior. Using this method, she became closely acquainted with more than half of the reserve's one hundred or more chimpanzees. She imitated their behaviors, spent time in the trees, and ate their foods. By remaining in almost constant contact with the chimps, she discovered a number of previously unobserved behaviors. She noted that chimps have a complex social system, complete with ritualized behaviors and primitive but discernible communication methods, including a primitive "language" system containing more than twenty individual sounds. She is credited with making the first recorded observations of chimpanzees eating meat and using and making tools. Tool making was previously thought to be an exclusively human trait, used, until her discovery, to distinguish man from animal. She also noted that chimpanzees throw stones as weapons, use touch and embraces to comfort one another, and develop long-term familial bonds. The male plays no active role in family life but is part of the group's social **stratification**. The chimpanzee "caste" system places the dominant males at the top. The lower castes often act obsequiously in their presence, trying to ingratiate themselves to avoid possible harm. The male's rank is often related to the intensity of his entrance performance at feedings and other gatherings.

Ethologists had long believed that chimps were exclusively vegetarian. Goodall witnessed chimps stalking, killing, and eating large insects, birds, and some bigger animals, including baby baboons and bushbacks (small antelopes). On one occasion, she recorded acts of cannibalism. In another instance, she observed chimps inserting blades of grass or leaves into termite hills to lure worker or soldier termites onto the blade. Sometimes, in true toolmaker fashion, they modified the grass to achieve a better fit. Then they used the grass as a long-handled spoon to eat the termites.

In 1962 Baron Hugo van Lawick, a Dutch **wildlife** photographer, was sent to Africa by the National Geographic Society to film Goodall at work. The assignment ran longer than anticipated; Goodall and van Lawick were married on March 28, 1964. Their European honeymoon marked one of the rare occasions on which Goodall was absent from Gombe Stream. Her other trips abroad were necessary to fulfill residency requirements at Cambridge University, where she received a Ph.D. in ethnology in 1965, becoming only the eighth person in the university's long history who was allowed to pursue a Ph.D. without first earning a baccalaureate degree. Her doctoral thesis, "Behavior of the Free-Ranging Chimpanzee," detailed her first five years of study at the Gombe Reserve.

Van Lawick's film, *Miss Goodall and the Wild Chimpanzees,* was first broadcast on American television on December 22, 1965. The film introduced the shy, attractive, unimposing yet determined Goodall to a wide audience. Goodall, van Lawick (along with their son, Hugo, born in 1967), and the chimpanzees soon became a staple of American and British public television. Through these programs, Goodall challenged scientists to redefine the long-held "differences" between humans and other primates.

Goodall's fieldwork led to the publication of numerous articles and five major books. She was known and respected first in scientific circles and, through the media, became a minor celebrity. *In the Shadow of Man,* her first major text, appeared in 1971. The book, essentially a field study of chimpanzees, effectively bridged the gap between scientific treatise and popular entertainment. Her vivid prose brought the chimps to life, although her tendency to attribute human behaviors and names to chimpanzees struck some critics being as manipulative. Her writings reveal an animal world of social drama, comedy, and tragedy where distinct and varied personalities interact and sometimes clash.

From 1970 to 1975 Goodall held a visiting professorship in psychiatry at Stanford University. In 1973 she was appointed honorary visiting professor of Zoology at the University of Dar es Salaam in Tanzania, a position she still holds. Her marriage to van Lawick over, she wed Derek Bryceson, a former member of Parliament, in 1973. He has since died. Until recently, Goodall's life has revolved around Gombe Stream. But after attending a 1986 conference in Chicago that focused on the ethical treatment of chimpanzees, she began directing her energies more toward educating the public about the wild chimpanzee's endangered **habitat** and about the unethical treatment of chimpanzees that are used for scientific research.

To preserve the wild chimpanzee's **environment**, Goodall encourages African nations to develop nature-friendly tourism programs, a measure that makes wildlife into a profitable resource. She actively works with business and local governments to promote ecological responsibility. Her efforts on behalf of captive chimpanzees have taken her around the world on a number of lecture tours. She outlined her position strongly in her 1990 book *Through a Window:* "The more we learn of the true nature of non-human animals, especially those with complex brains and corresponding complex social behaviour, the more ethical concerns are raised regarding their use in the service of man-whether this be in entertainment, as 'pets,' for food, in research laboratories or any of the other uses to which we subject them. This concern is sharpened when the usage in question leads to intense physical or mental suffering-as is so often true with regard to vivisection."

Goodall's stance is that scientists must try harder to find alternatives to the use of animals in research. She has openly declared her opposition to militant **animal rights** groups who engage in violent or destructive demonstrations. Extremists on both sides of the issue, she believes, polarize thinking and make constructive dialogue nearly impossible. While she is reluctantly resigned to the continuation of animal research, she feels that young scientists must be educated to treat animals more compassionately. "By and large," she has written, "students are taught that it is ethically acceptable to perpetrate, in the name of science, what, from the point of view of animals, would certainly qualify as torture."

Goodall's efforts to educate people about the ethical treatment of animals extends to young children as well. Her 1989 book, *The Chimpanzee Family Book,* was written specifically for children, to convey a new, more humane view of wildlife. The book received the 1989 Unicef/Unesco Children's Book-of-the-Year award, and Goodall used the prize money to have the text translated into Swahili. It has been distributed throughout Tanzania, Uganda, and Bu-

————. *Through a Window: My Thirty Years with the Chimpanzees of Gombe.* Houghton, 1990.

Montgomery, S. *Walking with the Great Apes: Jane Goodall, Dian Fossey, Birute Galdikas.* Houghton Mifflin, 1991.

Smith, W. "The Wildlife of Jane Goodall." *USAir*, February 1991, 42–47.

Gopherus agassizii
see **Desert tortoise**

Albert Gore Jr. (1948 –)

American former U.S. representative, senator and vice president of the United States

Albert Gore, Jr. was born and raised in Washington, D.C., where his father was a well-known and widely respected representative and later senator from Tennessee. Gore attended St. Alban's Episcopal School for Boys, where he excelled both academically and athletically. He later went to Harvard, earning a bachelor's degree in government. After graduation he enlisted in the army, serving as a reporter in Vietnam in a war he was opposed to. After completing his tour of duty, in 1974 Gore entered the law school at Vanderbilt University. Following in his father's footsteps, Gore ran for Congress, was elected, and served five terms before running for and winning a Senate seat in 1984. He served in the Senate until 1992, when then-governor and Democratic presidential candidate Bill Clinton selected him as his vice presidential running-mate. After winning the 1992 election Vice President Gore became the Clinton administration's chief environmental advisor. He was also largely responsible for President Clinton's selection of **Carol Browner** as head of the **Environmental Protection Agency** (EPA) and Bruce Babbit as Secretary of the Interior.

A self-described "raging moderate," Gore for more than two decades championed environmental causes and drafted and sponsored environmental legislation in the Senate. He was one of two U.S. senators to attend and take an active part in the 1992 U.N.-sponsored Rio Summit on the **environment**, and after he became vice president in 1993, he took a leading role in shaping the Clinton administration's environmental agenda. As vice president, Gore's ability to shape that agenda was perhaps somewhat limited. At the close of Clinton's first term, most environmental organizations rated his administration's record on environmental matters as mixed, at best. Yet clearly environmental groups had more access to the White House than they had ever had before. And in his second term, Gore was thought to be responsible for salvaging the 1997 Kyoto agreement on **climate** change when he flew to Japan as negotiations were falling apart and personally represented a new American position.

Jane Goodall. (The Library of Congress.)

rundi to educate children who live in or near areas populated by chimpanzees. A French version has also been distributed in Burundi and Congo.

In recognition of her achievements, Goodall has received numerous honors and awards, including the Gold Medal of **Conservation** from the San Diego Zoological Society in 1974, the J. Paul Getty Wildlife Conservation Prize in 1984, the Schweitzer Medal of the **Animal Welfare Institute** in 1987, the National Geographic Society Centennial Award in 1988, and the Kyoto Prize in Basic Sciences in 1990. In 1995, Goodall was presented with a CBE (Commander of the British Empire) from Queen Elizabeth II. Many of Goodall's endeavors are conducted under the auspices of the Jane Goodall Institute for Wildlife Research, Education, and Conservation, a nonprofit organization located in Ridgefield, Connecticut. In April of 2002, it was revealed that Goodall was chosen to be a United Nations Messenger of Peace.

[*Tom Crawford*]

RESOURCES
BOOKS

Green, T. *The Restless Spirit: Profiles in Adventure.* Walker, 1970.
Goodall, J. *Jane Goodall's Animal World: Chimpanzees.* Macmillan, 1989.

Gore's most notable contribution to the environmental movement might be as an author. With a wealth of statistical and scientific evidence, Gore's 1992 book *Earth in the Balance* makes the case that careless development and growth have damaged the natural environment; but better policies and regulations will supply incentives for more environmentally responsible actions by individuals and corporations. For example, a so-called **carbon tax** could provide financial incentives for developing new, nonpolluting energy sources such as solar and wind power. Corporations could also be given tax credits for using these new sources. By structuring a system of incentives that favors the protection and restoration of the natural environment, government at the local, national, and (through the United Nations) international levels can restore the balance between satisfying human needs and protecting the earth's environment. But restoring this balance, Gore maintained in his book, requires more than public policy and legislation; it requires changes in basic beliefs and attitudes toward **nature** and all living creatures. More specifically, environmental protection and restoration requires a willingness on the part of individuals to accept responsibility for their actions (or inaction). At the individual level, environmental protection means living, working, eating, and recreating responsibly, with an eye to one's effects on the natural and social environment, now and in the future in which our children and their children will survive.

Despite the strong stances Gore articulated in *Earth in the Balance*, his achievements as vice president under Clinton were more moderate. His former staffer Carol Browner headed the EPA, and she prevailed in promoting several major pieces of environmental legislation. But she had to withstand concerted attacks on her and her office by the Republican majority in the House and Senate after 1994. Gore had managed to save the Kyoto treaty on climate change by his personal efforts in 1997, but the Clinton administration never committed to specific actions to cut carbon-dioxide emissions. In August 1998 Gore met with the leaders of several environmental groups and explained to them that he had no political backing for advocating controversial actions like limiting **pollution** from coal-burning **power plants**. When he ran for president in 2000, the environment was not a strong feature in his campaign, though this was an area where he had sharp differences with his opponent, George W. Bush. In the last months of the campaign, Gore found himself defending his record on the environment against Green Party candidate **Ralph Nader**. Though Gore had the backing of the **Sierra Club** and other major environmental groups, Nader accused Gore of "eight years of principles betrayed and promises broken." In the end Gore lost the election to Bush, who quickly reversed Clinton administration environmental policies such as new

Albert Gore, Jr. (Corbis-Bettmann. Reproduced by permission.)

standards on drinking water safety, and refused to endorse the Kyoto climate change treaty.

After losing the election, Gore began lecturing at Fisk University in Nashville, at UCLA, and at the School of Journalism at Columbia University. Out of office, he remained for the most part out of the public eye. But he did speak at Vanderbilt University on **Earth Day** in 2002, roundly criticizing George W. Bush for his **environmental policy**. Gore noted that some of his more extreme environmental stances, such as calling for new technology to replace the internal **combustion** engine, were now being considered even by Republican lawmakers. In mid-2002, Gore would not commit to another run for the presidency. But it did seem that he stood by his earlier ideas, and would continue to identify himself with the environmental movement.

[*Terence Ball*]

RESOURCES
BOOKS

Gore Jr., Albert. *Earth in the Balance*. Boston: Houghton Mifflin, 1992.

PERIODICALS
Barro, Robert J. "Gore's 'Reckless and Offensive' Passion for the Environment." *Business Week*, November 6, 2000, 32.

Berke, Richard L. "Lieberman Has One Eye on '04 Run, the Other, Quite Expectably, on Gore." *New York Times*, May 2, 2002, A25.

Branegan, Jay, and Dick Thompson. "Is Al Gore a Hero or a Traitor?" *Time* , April 26, 1999, 66.

"How Green Is Al Gore?" *Economist* (April 22, 2000): 30.

Jehl, Douglas. "On a Favorite Issue, Gore Finds Himself on a 2-Front Defense." *New York Times*, November 3, 2000, A28.

Seelye, Katherine Q. "Gore, on Earth Day, Says Bush Policies Help Polluters." *New York Times*, April 23, 2002, A16.

Gorillas

Gorillas (*Gorilla gorilla*) inhabit the forests of Central Africa and are the largest and most powerful of all primates. Adult males stand 6 ft (1.8 m) upright (an unnatural position for a gorilla) and weigh up to 450 lb (200 kg), while females are much smaller. Gorillas live to about 44 years and mature males (those usually over 13 years), or silverbacks, are marked by a band of silver-gray hair on their backs.

Gorillas live in small family groups of several females and their young, led by a dominant silverback male. The females comprise a harem for the silverback, who holds the sole mating rights in the troop. Like humans, female gorillas produce one infant after a gestation period of nine months. The large size and great strength of the silverback are advantages in competing with other males for leadership of the group and in defending the group against outside threats.

During the day these ground-living apes move slowly through the forest, selecting **species** of leaves, fruit, and stems from the surrounding vegetation. Their home range is about 9–14 mi^2 (25–40 km^2). At night the family group sleeps in trees, resting on platform nests that they make from branches; silverbacks usually sleep at the foot of the tree.

Gorillas belong to the family Pongidae (which includes **chimpanzees** *Pan*], orangutans [*Pongo pygmaeus*], and **gibbons** [genus *Hylobates*). Together with chimpanzees, gorillas are the animal species most closely related to man. Like most megavertebrates, gorilla numbers are declining rapidly and only about 40,000 remain in the wild. There are three subspecies, the western lowland gorilla (*G. g. gorilla*), the eastern lowland gorilla (*G. g. graueri*), and the mountain gorilla (*G. g. beringei*). The rusty-gray western lowland gorillas are found in Nigeria, Cameroon, Equatorial Guinea, Gabon, Congo, Angola, Central African Republic, and Zaire (now Democratic Republic of the Congo [DRC]). The black-haired eastern lowland gorillas are found in eastern DRC. **Deforestation** and **hunting** now threaten lowland gorillas throughout their range.

The mountain gorilla has been intensely studied in the field, notably by George Schaller and Dian Fossey, upon whose life the film *Gorillas in the Mist* is based. This endangered subspecies is found in the misty mountains of eastern

A silverback male gorilla. (Photograph by Jerry L. Ferrara. Photo Researchers Inc. Reproduced by permission.)

Zaire, Rwanda, and Uganda at altitudes of up to 9,000 ft (3,000 m) and in the Impenetrable Forest in southwest Uganda. Field research has shown these powerful primates to be intelligent, peaceful and shy, and of little danger to humans.

Other than humans, gorillas have no real predators, although leopards will occasionally take young apes. Hunting, **poaching** (a mountain gorilla is worth $150,000), and **habitat** loss are causing gorilla populations to decline. The shrinking forest refuge of these great apes is being progressively felled in order to accommodate the ever-expanding human population. Mountain gorillas are somewhat safeguarded in the Virunga Volcanoes **National Park** in Rwanda. Their protection is funded by strictly controlled small-group gorilla-viewing tourist experiences that exist alongside long-term field research programs. Recent population estimates are 10,000–35,000 (with 550 in captivity) western lowland gorillas, 4,000 (with 24 in captivity) eastern lowlands gorillas, and 620 mountain gorillas.

[*Neil Cumberlidge Ph.D.*]

RESOURCES
BOOKS

Fossey, D. *Gorillas in the Mist*. Boston: Houghton Mifflin, 1983.

Schaller, G. B. *The Mountain Gorilla: Ecology and Behavior.* Chicago: University of Chicago Press, 1988.

———. *The Year of the Gorilla.* Chicago: University of Chicago Press, 1988.

OTHER

The Dian Fossey Gorilla Fund International. [cited May 2002]. <http://www.gorillafund.org/000_core_frmset.html>.

Gorilla Aid. [cited May 2002]. <http://www.gorillaaid.org>.

Grand Canyon
see **Colorado River; Glen Canyon Dam**

Grand Staircase-Escalante National Monument

The Grand Staircase-Escalante National Monument, encompassing 1.7 million acres (700,000 ha) of public lands on the Colorado Plateau in south-central Utah, was created on September 18, 1996, by presidential proclamation under authority of the Antiquities Act of 1906 (34 Stat. 225, 16 U.S.C. 431). The **U.S. Department of the Interior** had first recommended the creation of the Escalante National Monument along the Colorado and Green Rivers in 1936. In 1937, Capitol Reef National Monument was established in the area northeast of the Escalante Canyons along the upper portion of Waterpocket Fold. In 1941, the **National Park Service** studied the basin in conjunction with a comprehensive study of **water resources** in the **Colorado River** Basin. The study, published in 1946, identified the Aquarius Plateau/Escalante River Basin as "a little known, but potentially important **recreation** area." The area was recognized as a strategic link between the national parks in southwestern Utah and the canyon country of southeastern Utah.

A national monument is the designation given to a particular area to protect "historic landmarks, historic and prehistoric structures, and objects of historic or scientific interest that are situated upon the lands owned or controlled by the government of the United States." President **Theodore Roosevelt** exercised this authority to ensure protection for the Grand Canyon. More than 100 national monuments have been established by Presidents over the past 90 years, including Zion, Bryce Canyon, Glacier Bay, Death Valley, and Grand Teton. The Grand Staircase-Escalante National Monument was dedicated from the Grand Canyon **National Park** in Arizona, and no elected official from Utah was at the ceremony because of the controversy in Utah about the monuments designation.

The Grand Staircase-Escalante National Monument was created to preserve geological, paleontological, archaeological, biological, and historical features of the area. Geological features include clearly exposed stratigraphy and structures. The sedimentary rock layers are relatively undeformed

and unobscured by vegetation, providing a view to understanding of the processes of the formation of the earth. A wide variety of geological formations in colors such as red, pink, orange, and purple have been exposed by millennia of **erosion**. The monument contains significant portions of a vast geological stairway, which was named the Grand Staircase by the geologist Clarence Dutton. This stairway rises 5,500 ft (1,678 m) to the rim of Bryce Canyon in an unbroken sequence of cliffs and plateaus. The monument also includes the canyon country of the upper Paria Canyon system, major components of the White and Vermillion Cliffs and associated benches, and the Kaiparowits Plateau. The Kaiparowits Plateau includes about 1,600 mi^2 (2,574 km^2) of sedimentary rock and consists of south-to-north ascending plateaus or benches, deeply cut by steep-walled canyons. Naturally burning underground **coal** seams have changed the tops of the Burning Hills to brick-red. A major landmark, the East Kaibab Monocline, or Cockscomb, is aligned with the Paunsaugant, Sevier, and **Hurricane** Faults, which may indicate that it may also be a fault at depth. The Circle Cliffs, which features intensively colored red, orange, and purple mounds and ledges at the base of the Wingate Sandstone Cliffs, are one of the most distinctive landscapes of the Colorado Plateau. Inclusion of part of the Waterpocket Fold completes the protection of this geologic feature, which was begun with the establishment of the Capitol Reef National Monument in 1936. There are many arches and natural bridges within the monument boundaries, including the 130-ft (39.4-m) high Escalante Natural Bridge, with a 100-ft (30.3-m) span and the Grosvenor Arch, a double arch. The upper Escalante Canyons, in the northeastern part of the monument, include several major arches and bridges and geological features in narrow, serpentine canyons, where erosion has exposed sandstone and shale deposits in colors of red, maroon, brown, tan, gray, and white.

Paleontological features include petrified wood, such as large unbroken logs more than 30 ft (9 m) in length. The stratigraphy of the Kaiparowits Plateau provides one of the best and most continuous records of the paleontology of the late Cretaceous Era. Fossils of marine and **brackish** water mollusks, turtles, **crocodiles**, lizards, dinosaurs, fishes, and mammals (including a marsupial primitive mammal) have been recovered from the Dakota, Tropic Shale, and Wahweap Formations and the Tibbett Canyon, Smoky Hollow, and John Henry members of the Straight Cliffs Formation.

Archeological inventories show extensive use of places within the monument by Native American cultures. Recorded sites include rock art panels, occupation sites, rock shelters, campsites, and granaries.

Historical evidence indicates that the monument was occupied by both Kayenta and Fremont agricultural cultures for a period of several hundred years centered around A.D.

1100. The area has been used by modern tribal groups, including the Southern Paiute and the Navajo. In 1872, an expedition of John Wesley Powell did initial mapping and scientific field work in the area. The expedition discovered the Escalante River, naming it in honor of the Friar Silvester Valez de Escalante expedition of 1776. The Escalante River Canyons have been a major barrier to east-west travel in the region in historic times. The river is presently bridged only at its upper end. Early Mormon pioneers left many historic objects, including trails, inscriptions, ghost towns such as the Old Paria townsite (built in 1874 and abandoned in 1890), rock houses, cowboy camps, and they built the Hole-in-the Rock Trail in 1879–1880 as part of their colonization activities. Sixty miles (96.6 km) of the Hole-in-the-Rock Trail are within the monument, as well as Dance Hall Rock, used by Mormon pioneers for meetings and dances, and now a National Historic Site.

As a biological resource, the Grand Staircase-Escalante National Monument spans five life zones, from low-lying **desert** to **coniferous forest**. Remoteness, limited travel corridors, and low visitation have helped to preserve the ecological features, such as areas of relict vegetation, many of which have existed since the Pleistocene. Pinon-juniper communities containing trees up to 1,400 years old and relict sagebrush-grass park vegetation can be found on No Man's Mesa, Little No Man's Mesa, and Four Mile Bench Old Tree Area. These relict areas can be used to establish a baseline against which to measure changes in community dynamics and biogeochemical cycles in areas impacted by human activity. The monument contains an abundance of unique isolated communities such as hanging gardens and canyon bottom communities, with riparian plants and their pollinators; tinajas, which contain tadpoles, fairy and clam shrimp, amphibians, and snails; saline seeps, with plants and animals adapted to highly saline conditions; dunal pockets, with **species** adapted to shifting sands; rock crevice communities, consisting of slow-growing species that can thrive in extremely infertile sites; and cryptobiotic crusts, which stabilize the highly **erodible** desert soils and provide nutrients for plants. The **wildlife** of the monument is characterized by a diversity of species, where both northern and southern **habitat** species intermingle. Mountain lions, bears, and desert bighorn sheep, as well as over 200 species of birds, including bald eagles and peregrine falcons, can be found within the monument. The wildlife concentrates around the Paria and Escalante Rivers and other riparian areas.

The Secretary of the Interior, through the **Bureau of Land Management** (BLM), which in the past has been responsible for the public lands included within the monument area, will manage the monument. This is the first national monument that will be managed by the BLM. The BLM will develop a management plan to address measures necessary to protect the scientific and historic features within the monument by three years after the date of establishment of the monument. The BLM will consult with state and local governments, other federal agencies, and tribal governments to prepare the **land use** plan.

The boundaries of the Grand Staircase-Escalante National Monument were drawn to exclude as much private land as possible, as well as the towns of Escalante, Boulder, Kanab, and Tropic. The national monument designation applies only to Federal land and not to the approximately 9,000 acres (3,644 ha) of private land remaining within the boundaries of the monument. Private landowners continue to have existing rights of access to their property. The landowners may participate in land exchanges with the BLM to trade land within the monument for land of equal value outside the area. The State of Utah owns about 180,000 acres (73,800 ha) of isolated, 640-acre (262 ha) sections of school lands within the boundaries of monument. The State will be allowed to exchange these isolated school lands for federal lands of equal value outside the monument boundaries. All federal lands were withdrawn from sale or leasing under the **public land** laws upon designation as a national monument. The designation also prohibited the issuance of any new mineral leases in the area, including new claims made under the Mining Law of 1872. Existing uses under federal or state laws, such as **hunting**, camping, travel, hiking, backpacking, and other recreational activities, as well as grazing permits, continue under current policies and rules. The proclamation did not reserve water or make any federal **water rights** claims. As part of the management plan, the BLM will evaluate the extent to which water is necessary for the care and management of objects of the monument and the extent to which further action may be necessary under federal or state law to ensure availability of water.

[*Judith Sims*]

RESOURCES
BOOKS

Bureau of Land Management. *List of Historic and Scientific Objects of Interest: Grand Staircase-Escalante National Monument.* Salt Lake City: Bureau of Land Management, 1997.

———. *Questions and Answers on the Grand Staircase-Escalante National Monument.* Salt Lake City: Bureau of Land Management, 1997.

Clinton, W. J. *Establishment of the Grand Staircase-Escalante National Monument: A Proclamation.* Washington, DC: The White House, 1996.

Grasslands

Grasslands are environments in which herbaceous **species**, especially grasses, make up the dominant vegetation. Natural grasslands, commonly called **prairie**, pampas, shrub steppe, palouse, and many other regional names, occur in regions where rainfall is sufficient for grasses and forbs but too sparse or too seasonal to support tree growth. Such conditions occur at both temperate and tropical latitudes around the world. In addition, thousands of years of human activity—clearing pastures and fields, burning, or harvesting trees for materials or fuel—have extended and maintained large expanses of the world's grasslands beyond the natural limits dictated by **climate**. Precipitation in temperate grasslands (those lying between about 25 and 65 degrees latitude) usually ranges from approximately 10–30 in (25–75 cm) per year. At tropical and subtropical latitudes, annual grassland precipitation is generally between 24 and 59 in (60 and 150 cm). Besides its relatively low volume, precipitation on natural grasslands is usually seasonal and often unreliable. Grasslands in **monsoon** regions of Asia can receive 90% of their annual rainfall in a few weeks; the remainder of the year is dry. North American prairies receive most of their moisture in spring, from snow melt and early rains that are followed by dry, intensely hot summer months. Frequently windy conditions further evaporate available moisture.

Grasses (family Gramineae) can make up 90% of grassland **biomass**. Long-lived root masses of perennial bunch grasses and sod-forming grasses can both endure **drought** and allow asexual reproduction when conditions make reproduction by seed difficult. These characteristics make grasses especially well suited to the dry and variable conditions typical of grasslands. However, a wide variety of grass-like plants (especially sedges, Cyperaceae) and leafy, flowering forbs contribute to species richness in grassland **flora**. Small shrubs are also scattered in most grasslands, and **fungi**, mosses, and **lichens** are common in and near the **soil**. The height of grasses and forbs varies greatly, with grasses of more humid regions standing 7 ft (2 m) or more, while **arid** land grasses may be less than one-half meter tall. Wetter grasslands may also contain scattered trees, especially in low spots or along stream channels. As a rule, however, trees do not thrive in grasslands because the soil is moist only at intervals and only near the surface. Deeper tree roots have little access to water, unless they grow deep enough to reach **groundwater**.

Like the plant community, grassland animal communities are very diverse. Most visible are large herbivores—from American **bison** and elk to Asian camels and horses to African kudus and wildebeests. Carnivores, especially **wolves**, large cats, and bears, historically preyed on herds of these herbivores. Because these carnivores also threatened domestic herbivores that accompany people onto grasslands, they have been hunted, trapped, and poisoned. Now most wolves, bears, and large cats have disappeared from the world's grasslands. Smaller species compose the great wealth

of grassland **fauna**. A rich variety of birds breed in and around ponds and streams. Rodents perform essential roles in spreading seeds and turning over soil. Reptiles, amphibians, insects, snails, worms, and many other less visible animals occupy important niches in grassland ecosystems.

Grassland soils develop over centuries or millennia along with regional vegetation and according to local climate conditions. Tropical grassland soils, like tropical forest soils, are highly leached by heavy rainfall and have moderate to poor **nutrient** and contents. In temperate grasslands, however, generally light precipitation lets nutrients accumulate in thick, organic upper layers of the soil. Lacking the acidic leaf or pine needle litter of forests, these soils tend to be basic and fertile. Such conditions historically supported the rich growth of grasses on which grassland herbivores fed. They can likewise support rich grazing and crop lands for agricultural communities. Either through crops or domestic herbivores, humans have long relied on grasslands and their fertile, loamy soils for the majority of their food.

Along a moisture gradient, the margins of grasslands gradually merge with moister savannas and woodlands or with drier, **desert** conditions. As grasslands reach into higher latitudes or altitudes and the climate becomes to cold for grasses to flourish, grasslands grade into **tundra**, which is dominated by mosses, sedges, willows, and other cold-tolerant plants.

[*Mary Ann Cunningham Ph.D.*]

RESOURCES
BOOKS

Coupland, R. T., ed. *Grassland Ecosystems of the World: Analysis of Grasslands and Their Uses.* London: Cambridge University Press, 1979.
Cushman, R. C., and S. R. Jones. *The Shortgrass Prairie.* Boulder: Pruett Publishing Co., 1988.

Grazing on public lands

Grazing on public lands is the practice of raising livestock on land that is not privately owned. Livestock such as cattle and sheep eat forage (grass and other herbage) on the **public land**. Through the twentieth century and into the twenty-first, ranchers grazed livestock on federal and state public land in the western states.

The western livestock industry developed during the decades after the Civil War, according to a **Bureau of Land Management** (BLM) report. People headed west where the land was open. A prospective rancher just needed a headquarters, some horses and cowboy employees. Some ranches consisted of a dugout shelter for the people and a horse corral.

Livestock grazed on open land called the range. When livestock ate all the forage in one area, the rancher moved the herd and headquarters to another area. By 1870, there were 4.1 million beef cattle and 4.8 million sheep in 17 states, according to BLM. Thirty years later, there were 19.6 million cattle and 25.1 million sheep. As the number of livestock increased, the range became crowded and there was less forage.

The twentieth century opened with discussion in Congress about how to regulate use of public lands. After years of debate, Congress approved the **Taylor Grazing Act** in 1934. The act established a permit requirement and "significantly reduced" the number of ranchers and livestock, according to BLM. A portion of fees was allocated to the grazing district for improvements like fencing.

The act only addressed grazing as a **land use**. That policy changed with the approval of the 1976 **Federal Land Policy and Management Act**. FLPMA stated that federal public lands and their resources would be managed for multiple uses that "best meet the present and future needs for the American people."

Passage of the act reflected concerns about environmental protection and **conservation**. At the same time, the increase in population brought more demand for recreational uses of public land. In the years since the passage of FLPMA and environmental protection laws, the land-use debate intensified.

Federal public lands include the 264 million acres managed by BLM in 12 western states and the 191 million acres of United States **Forest Service** land in 44 states. Federal grazing permits were issued by those jurisdictions, the **National Park Service**, and the **Fish and Wildlife Service**. BLM manages about half of the 270 million acres where grazing is allowed. Ranchers pay a fee called an animal unit month (AUM). This is based on the monthly amount of forage needed to sustain one cow and her calf, one horse, or five sheep or goats. The AUM in 2002 was $1.43. Opponents called the fee a government subsidy. They estimated an AUM of $11 for grazing on private land.

BLM records for fiscal year 2001 showed 15,643 operators (ranchers) of cattle, yearling, and buffalo; 1,232 operators of horses and burros; and 1,225 operators of sheep and goats. Those figure were not a total count since some operators may have raised more than one type of livestock. The BLM figures were from the administrative jurisdiction of Arizona, California, Colorado, Idaho, Montana, Nevada, New Mexico, Oregon, Utah, and Wyoming. California BLM administers some land in Nevada, and Montana administers all public land in North and South Dakota.

Within those states, active AUMs totaled 11.3 million as of November 29, 2001, according to BLM. Livestock cause more damage to public land than the chainsaw or the bulldozer, according to anti-grazing activists. Damage includes trampled vegetation, **soil** damage, **water pollution**, and the spread of invasive weeds, according to grazing opponents.

The beef and sheep industries are represented by the **Public Lands Council** (PLC), an organization that said grazing helped to preserve open space. PLC said that opponents incorrectly blamed ranchers for **environmental degradation**. According to PLC, research indicated that conservation goals on public land were easier to achieve when grazing was properly managed. Livestock managers could make improvements that would not be done if grazing was prohibited.

Grazing opponents disagreed and used various methods to limit grazing. Lawsuits claimed that grazing violated environmental regulations. Successful cases include the 1998 removal of livestock from the Gila River Basin in Arizona and New Mexico. The ruling was based on the **Endangered Species Act** and the protection of the spotted owl and other **species**.

Furthermore, activists tried to buy leases. They urged ranchers to end leases by retiring their permits. Another solution was proposed in 2002 by the National Public Lands Grazing Campaign. An umbrella group of various organizations, NPLGC wanted federal law changed to allow the voluntary retirement of grazing permits. The federal government would pay $175 per AUM to buy out leases. NPLGC said the plan would cost an average of $13.45 per acre.

The livestock industry rejected the plan. PLC said in April of 2002 that the plan could cost taxpayers more than $3.2 billion. Plan supporters countered that grazing "subsidies" cost taxpayers up to $460 million annually. The federal government had not taken action on the proposal as of May of 2002.

[*Liz Swain*]

RESOURCES
BOOKS

Holechek, Jerry, Rex Pieper, and Carlton Herbel. *Range Management: Principles and Practices.* Upper Saddle River, NJ: Prentice Hall, 1998.

Wuerthner, George, Mollie Matheson, eds. *Welfare Ranching: The Subsidized Destruction of the West.* Washington, DC: Island Press, 2002.

ORGANIZATIONS
Public Policy Center, National Cattlemen's Beef Association (Public Lands Council)., 1301 Pennsylvania Avenue, NW, Suite 300, Washington, D.C. 20004-1701 (202) 347-0228, Fax: (202) 638-0607, Email: jcampbell@beef.org, htttp://hill.beef.org/files/fedlnds.htm

United States Bureau of Land Management., 1849 C Street, Washington, D.C. 20240 (202) 452-5125, Fax: (202) 452-5124, , http://www.blm.gov

Western Watersheds Project (National Public Lands Grazing Campaign)., P.O. Box 1770, Hailey, ID 83333 Email: andykerr@andykerr.net, http://www.publiclandsranching.org

Great Barrier Reef

The Great Barrier Reef, located in the Coral Sea off the eastern coast of Queensland in northeastern **Australia**, consists of more than 2,800 reefs that range in size from 1 hectare to over 10,000 hectares in area. As the largest reef in the world, the Great Barrier Reef is 1,250 mi (2,011 km) in length, extending from a point near McKay, Queensland in the **south** to the Torres Strait in the north, which lies between Australia and New Guinea. The Reef is 45 mi (72 km) across at its widest point, with a total area of more than 300,000 square kilometers. Approximately 20% of the reefs are submerged reefs or shoals, while about 26% are fringing reefs around continental islands or along the mainland coast. The remaining reefs are shelf reef platforms. Water depths range from 65.6–98 ft (20–30 m) in lagoonal areas to 131–197 ft (40–60 m) between the reefs on the outer shelf. Starting in the south near the Tropic of Capricorn, the Great Barrier Reef is a wide scattering of reefs about 186 mi (300 km) out from the coast; moving north the Reef becomes more continuous and is within 10–12 mi (16–19 km) of the coast, with a few individual reefs in the inner **lagoon** (the area between the reef and the coast). North of Cairns the reef is almost a continuous barrier between the coast and the Coral Sea. Drilling has indicated that in places the reef is over 1,640.5 ft (500 m) thick.

At the end of the last **ice age**, about 20,000 years ago, the sea level was about 394 ft (120 m) lower than it is today, and dry land extended from the present day coast to the location of today's outer Barrier Reef. As the ice melted, the sea rose, and by 13,000 years ago the coastal plain had become a submerged continental shelf. Corals and reefs grow best where there is water movement, so the reefs tended to form on submerged hills in the **flooding** coastal plain. By 6000 to 7000 years ago, the sea had reached its present level, and the reef began to assume its present shape. The Spanish mariner Luis Vaáez de Torres was the first European known to have sailed the northern reef, but the Spanish kept the route a secret to protect their route between Europe and the Orient. The Reef was explored and charted by Captain James Cook, whose ship in 1770 ran aground on the reef that now is called the Endeavor, after the name of his ship.

The Great Barrier Reef provides homes and shelter for a wide diversity of life. The Great Barrier Reef is home to corals, which form the reefs, **dolphins** and **whales**, six **species** of **sea turtles**, more than 1,500 species of fish, 4,000 types of mollusks, 500 species of seaweed, and more than 200 species of birds. The corals that make up the Reef consist of individual living coral polyps, which as they divide in a process called budding, form colonies in fan, antler, brain, and plate shapes. Each polyp, which is a tiny jelly-like, sack-like animal with a mouth surrounded by tentacles, lives inside a shell of aragonite, a type of calcium carbonate

that is the hard shell typically recognized as coral. The polyps are joined together to form a colony. Coral polyps obtain food by catching **plankton** in their tentacles as well as deriving nutrients from symbiotic algae, the zooxanthellae, that live within their tissues. These zooxanthellae produce nutrients through **photosynthesis**, which are then available for use by the coral. Association with the algae allow corals to build skeletons three times faster in the light than in the dark, and faster than storms and waves can break it down. **Coral bleaching** occurs when stressed corals expel the zooxanthellae and turn white, or bleached. If the zooxanthellae do not return to the coral, the coral will die.

Every year about one-third of the 350 species of coral reproduce sexually during a mass spawning event. Spawning occurs in most of the inner reefs around November and in the outer reefs in December. The spawning occurs at night, up to six days after the full moon. Eggs and sperm are released into the water, where they combine to form a free-swimming larval stage.

The Great Barrier Reef is completely within the tropics. The **climate** of the Reef is a typical **monsoon** weather pattern, which consists of strong south-easterly winds dominating during the dry winter months and weaker variable winds occurring during the summer wet season. Both air and sea temperatures exhibit seasonal variations. Mean sea temperatures in inshore areas exhibit a range of temperatures of 69.8°F (21°C) in July and August to 86°F (30°C) in January and February. Temperatures on offshore reefs exhibit less seasonal difference, varying from 73.4°F (23°C) in winter to 82°F (28°C) in summer.

There are two types of islands along the Great Barrier Reef. The larger islands, such as those of the Whitsunday group, are the tops of submerged mountains that at one time were high points of a range running along the coast. These islands have vegetation similar to the mainland. Some of these islands that have fertile soils and are affected by heavy monsoonal rains are covered with rain forests. Other islands are isolated low-lying coral cays. Cays are formed when coral grow to levels that are higher than sea level, even at low tide. Coral can only survive a few hours out of water. The dead coral on the cays is worn down and broken off by waves and storm action. Eventually the coral debris is ground into sand. Through time the sand stabilizes, seabirds start to nest, and hardy vegetation begins to grow. Decomposing plant materials and bird droppings change the sand into a more developed **soil**, providing an **environment** for a wider diversity of plant life. Cays are found more frequently on inner reefs in the Great Barrier Reef, where waves and currents are less strong than on reefs further away from the coastline. However, vegetated cays are not sufficiently stable to withstand severe weather, such as storms and cyclones.

Comprehensive protection of the Great Barrier Reef was accomplished in 1975 by the formation of the Great

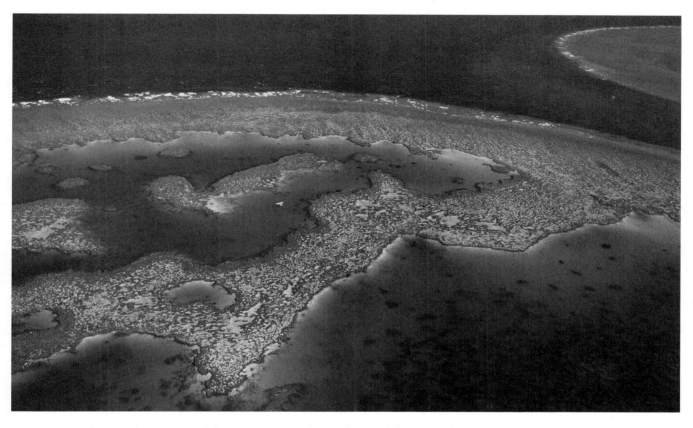

An aerial view of a portion of the Great Barrier Reef. (Photograph by Yann Arthus-Bertrand. Corbis-Bettmann. Reproduced by permission.)

Barrier Reef Marine Park, administered by the Great Barrier Reef Marine Park Authority (GBRMPA). Over 98% of the area, with the exception of a few coastal areas, is now included within the Marine Park. Drilling and mining for minerals within all areas of the Marine Park is forbidden. In 1981, in recognition of its outstanding natural heritage value, most of the Great Barrier Reef area was added to the World Heritage List by the United Nations Education, Scientific, & Cultural Organization (**UNESCO**). The Great Barrier Reef World Heritage Area is the largest of the world's over 550 World Heritage Areas. In addition to the **coral reef** area, the Great Barrier Reef World Heritage Area also contains the continental slope, the inter-reefal areas, and the Great Barrier Reef lagoon.

The Great Barrier Reef Marine Park has been classified into 70 bioregions, which are areas of differing marine **biodiversity** with plant and animal communities that, together with physical features, are significantly different than the surrounding areas and the rest of the Marine Park. Based on these bioregions, the GBRMPA has divided the Marine Park into six zones: (1) the Preservation Zone, which in-

cludes areas that are intended to be kept completely untouched; (2) the Scientific Research Zone, which includes areas set aside exclusively for scientific research; (3) the Marine **National Park** A Zone, which includes areas where recreational use (i.e., fishing with one line and one hook) is permitted, but **commercial fishing** is not permitted; (4) the General Use B Zone, where reasonable recreational and commercial uses are permitted, but trawling and shipping are prohibited; and (5) the General Use A Zone, where all reasonable uses are permitted, including trawling and shipping, but mining, **oil drilling**, commercial spear fishing and fishing with scuba equipment are not permitted.

More than two million people a year visit the reef, generating more than $AU1 billion in tourism dollars. Tourists visit the reef using private charter boats, regular scheduled daily cruises, seaplanes, air charter, and helicopter flights. They view the reefs through reef walks, glass bottom or semi-submersible boats, snorkeling, or scuba diving.

The Great Barrier Reef is sensitive to climate changes and to changes in patterns of water movement. In 1998 scientists announced that about 60% of the reef had been affected by coral bleaching due to increasing sea temperatures and fresh water flooding.

The Reef is also susceptible to physical damage. Human damage to the Great Barrier Reef has been occurring due to carelessness, where people walk on the reefs, drop anchors onto them, drag diving gear over them, break them into pieces to take as souvenirs, and knock into and ground boats onto them.

Since the 1960s, during periodic outbreaks, the Crown of Thorns starfish has been destroying corals that make up the reef. The Crown of Thorns is a large, thorny, brown-colored starfish. The starfish eats coral by turning its stomach out through its mouth, in a process called stomach eversion, wraps the stomach around a coral, and secretes an **enzyme** to digest the living coral polyps. When the stomach is drawn back in, all that is left is the limestone coral skeleton. This skeleton becomes covered with green algae and other types of encrusting plants and animals, which give the coral skeleton a dull, gray appearance. Some types of corals become infested with boring organisms. During periods of rough weather (e.g., during a **cyclone**), the dead colony and the newly attached organisms may collapse. Some scientists feel that outbreaks of the Crown of Thorns starfish may be a natural part of the lifecycle of the reef. By feeding on fast-growing species of corals, the starfish may act to preserve coral diversity, for fast-growing populations, if left unchecked, could dominate the reef.

Water quality has been identified by the Great Barrier Reef Marine Park Authority as a critical management issue, as **pollution** by sediments, toxic metal and organic contaminants, **oil spills**, and pesticides, can potentially harm the Reef. Sewage **discharge** is regulated in the Marine Park area, with tertiary treatment (i.e., **nutrient** reduction) required before marine discharge; alternatively secondary or tertiary treated **wastewater** can be land applied with minimal marine discharge. However, the primary water quality concern is nutrient **loading** from inshore ecosystems. Programs are being developed to reduce nutrient inputs through regulation of discharges, integrated catchment management, improved land management practices, and wetland protection.

[*Judith L. Sims*]

RESOURCES
BOOKS

Cates, Alison. *The Great Barrier Reef: Australia's Tropical Paradise.* Sydney: New Holland/Struik, 1998.
Doubilet, David. *Great Barrier Reef.* Washington, DC: National Geographic Society, 2002.
Lawrence, David, et al. *The Great Barrier Reef: Managing an Ecosystem.* Melbourne: Melbourne University Press, 2002.

Smith, Fay. *Australia's Great Barrier Reef.* Seven Hills, Australia: Cimino Publishing Group, 1998.

Great Lakes

The advance and retreat of glaciers over millions of years scraped and scoured the Great Lakes basins until they attained their present form about 10,000 years ago. Forming the largest system of inland lakes in the world, the Great Lakes have a surface area of 94,200 mi^2 (244,000 km^2) and a volume of more than 28 trillion yd^4 (22,000 km^3) of water, 20 percent of the world's surface freshwater.

Lake Superior, with more than 31,660 mi^2 (82,000 km^2) of water, has the largest surface area of freshwater on earth. Lake Huron, the world's fifth largest lake, is at the same elevation and about the same size as Lake Michigan, the world's sixth largest lake. The two are joined by the narrow, deep Straits of Mackinac. Their accumulated waters empty into the St. Clair River which flows into the 460 mi^2 (1,190 km^2) Lake St. Clair. The water continues its flow into the Detroit River before entering **Lake Erie**, the eleventh largest lake in the world. It is the oldest, shallowest, busiest, and most eutrophic of the Great Lakes. The waterway continues on into the Niagara River, then to the famous Niagara Falls, where the water descends a total of 325 ft (99 m) before it empties into the last Great Lake, Ontario. The fourteenth largest lake on earth, Lake Ontario is the smallest in surface area but the second deepest of the Great Lakes. It discharges into the St. Lawrence River, which flows into the Atlantic Ocean at the Gulf of St. Lawrence.

The first European explorers discovered a great variety of native fish. Approximately 153 **species** were eventually identified before human interference disrupted the **ecosystem**, first by **overfishing**, and then by lumbering and industrial development. As many species of fish have disappeared, about twenty new species have been introduced. Some, such as the Pacific salmonids, carp, and smelt, were introduced intentionally. Others, such as the sea lamprey, alewife, and **zebra mussel**, gained access through the Erie and Welland Canals or by release with the ballast water of vessels transporting other cargo.

Today, lake trout, burbot, and whitefish are the principal catches of a once extraordinarily rich fisheries enterprise. Despite the decline in the quality and numbers of suitable fish, sport and **commercial fishing** are still vital Great Lakes industries. The sport fishery consists primarily of coho, chinook **salmon**, steelhead trout, walleye, and perch. They now attract about five million anglers annually with a regional economic benefit of about $2 billion.

Besides directly water-related activities, presently, one-fifth of the industry and commerce of the United States is located in the Great Lakes catchment basin because of the availability of abundant cheap and clean freshwater and accessible, efficient water **transportation** among the lakes and to the oceans. As a consequence, **pollution** has taken some obvious as well as more subtle forms. Using the lakes as a cheap sewage disposal site for shoreline city populations began in the early seventeenth and eighteenth centuries and continued until the early 1970s. To improve the quality of the Great Lakes, the first efforts concentrated on preventing or removing conventional pollutants such as **phosphates**, suspended solids, and **nitrogen**.

More deadly toxic contaminants often are not visible and so initially attracted less attention. Over the past fifty years municipal and industrial wastes so polluted the waters, especially the lower Great Lakes, that, beginning in the middle 1960s organochlorides were identified as serious contaminants. Fish were collecting, through **bioaccumulation**, relatively large concentrations of agricultural pesticides such as DDT and dieldrin as well as the industrial chemical polychlorinated biphenyl (PCB) in their tissues. These were passed into the human **food chain/web**. By 1980 more than 400 organic and **heavy metals** contaminants had been found in fish, and fishermen were warned to limit their consumption. The effects of pollutants are seen primarily at the tops of food chains and are usually discovered through changes in population levels of predator species. Organochlorines and methylated **mercury**, for example, bioaccumulate to levels that may cause reproductive failures in fish-eating birds and animals such as cormorants, eagles, and mink.

Between 1969 and 1972 legislation was enacted in several states bordering the Great Lakes basin to restrict or ban the use of dieldrin, DDT, PCBs, mercury and other toxic **chemicals**. After **point source** discharges were regulated, lake trout and chub, especially in Lake Michigan, showed dramatic declines in these contaminants. By 1978–79, however, the fish contaminant declines were only slight; or the levels remained relatively constant, reflecting airborne inputs as well as the remobilization of contaminants from the **sediment**.

This problem is likely to continue because the turnover rates of the Great Lakes are very slow; and mercury, PCBs and the pesticides DDT, dieldrin and **chlordane** are very resistant to degradation in the **environment**. Also, these compounds continue to enter the Great Lakes ecosystem from highly diffuse nonpoint sources such as airborne deposition, agricultural and **urban runoff**, remobilization from the sediments, **leaching** from municipal and industrial landfills, municipal and industrial discharges, and illegal dumping. *See also* Agricultural chemicals; Agricultural pollution; Great Lakes Water Quality Agreement; Heavy metals and heavy metal poisoning; Industrial waste treatment; Methylation; Water pollution

[*Frank M. D'Itri*]

RESOURCES
BOOKS

Ashworth, W. *The Late, Great Lakes: An Environmental History.* New York: Knopf, 1986.

Great Lakes, Great Legacy. Washington, DC: Conservation Foundation, 1989.

Hough, J. L. *Geology of the Great Lakes.* Urbana, IL: University of Illinois Press, 1958.

Sixth Biennial Report on Great Lakes Water Quality. Windsor, Ont.: International Joint Commission, 1992.

Weller, P. *Fresh Water Seas: Saving the Great Lakes, Between the Lines.* Toronto: Publishers, 1990.

Great Lakes Water Quality Agreement (1978)

The **Great Lakes Water Quality** Agreement of 1978 amended and strengthened the International Great Lakes Water Quality Agreement between Canada and the United States, which was signed in 1972. The original agreement established a framework for research, clean-up, and **pollution control** based on goals determined by the two nations. The existing **International Joint Commission**, in cooperation with the newly-created International Great Lakes Water Quality Board, was to oversee the implementation of the agreement. Chief goals of the agreement were to reduce the amount of **phosphorus** being dumped into the lakes by 50 percent, to require all municipal **sewage treatment** plants to be at the secondary level (removing nutrients such as phosphorous and **nitrogen**), and to control toxic **water pollution**.

The impetus for action in 1972, and again in 1978, was the decreasing water quality in the Great Lakes. Two incidents best symbolized this to the nation. First, the massive algal blooms in **Lake Erie**, due to eutrophication, led many commentators to declare that the lake was dead. And second, when the **Cuyahoga River** in Cleveland caught fire in 1969, it provided a clear visual message to the rest of the country about the state of **pollution** in the Great Lakes. Perhaps not as visible, but just as serious, the Great Lakes were also suffering from toxic chemical pollution, which was a problem for water use, and also for sports and recreational fishing. Limits, and in some cases, bans, were placed on how much fish from the lakes could be consumed.

The 1978 Agreement substantially strengthened the 1972 accord. The new agreement focused especially on toxic pollutants and phosphorous, the chief **nutrient** responsible for eutrophication in the Great Lakes. A stricter definition of a **toxic substance** was included, as well as specific water quality objectives for certain **heavy metals**, polychlorinated biphenyl (PCB), and pesticides, and a list of hundreds of hazardous and potentially hazardous materials. Phosphorous entering the lakes would be reduced further in certain lakes, primarily by improved **municipal solid waste** plants. Additional sections of the agreement dealt with pollution resulting from airborne **toxins**, **dredging**, shipping, and **nonpoint source** pollution. A contingency plan between the two nations to respond to spills or other severe pollution episodes was also to be created. The Agreement is an Executive Agreement with no force of law in either Canada or the United States. Rather, each nation must depend on existing laws to meet these joint goals.

The Great Lakes Critical Programs Act, passed in 1990, was designed to increase the efforts of the **Environmental Protection Agency** (EPA) and the states in cleaning up the Great Lakes. Evaluations of the 1978 Agreement indicated that neither the EPA nor the states had been putting sufficient effort into the implementation of the necessary programs.

By the middle 1990's, significant progress had been made on reducing phosphorous pollution and improving municipal sewage treatment by building new plants and improving existing plants. Phosphorous loadings were below 1978 target loads for Superior, Huron, and Michigan, and at or near target loads for Erie and Ontario. Hence, this aspect of the water quality agreement has largely been achieved. There has also been significant improvement on **point source** pollution of toxic **chemicals** such as PCBs. Toxic substances in the lakes remain a major problem due to **air pollution**, **runoff** (non-point source pollution), and releases from already contaminated sediments. For instance, the source of over 75% of the PCBs entering Lake Superior and 95% of the lead entering the Great Lakes is air pollution. In 1995, Canada and the United States agreed to develop a coordinated binational strategy to virtually eliminate persistent toxic substances from the Great Lakes Basin. That same year, the EPA launched the Great Lakes Water Quality Initiative as a framework for achieving that goal. *See also* Agricultural runoff; Chemical spills; Heavy metals and heavy metal poisoning; Oil spills

[*Christopher McGory Klyza*]

RESOURCES
BOOKS

Ashworth, W. *The Late, Great Lakes*. New York: Knopf, 1986.
Council on Environmental Quality. *Environmental Quality: 21st Annual Report*. Washington DC: U.S. Government Printing Office, 1990.

PERIODICALS
"U. S., Canada Reach Pollution Pact." *New York Times*, June 1, 1978, B6.

Great Smoky Mountains

The Great Smoky Mountains (GSM) **National Park** covers approximately 543,400 acres (220,000 ha) of western North Carolina and southeastern Tennessee, making the park the single largest **wilderness** area in the eastern United States. The mountains received their name from the Cherokee for the smoke-like clouds which frequently shroud their peaks and the grey-green **haze** of the wooded slopes. In early fall, this grey-green haze becomes a blaze of autumnal colors. The GSM are known for their large stands of some of the world's best remaining examples of temperate **deciduous forest**, and for their "balds" or treeless mountain tops. The park's size, scenic beauty, and geographical proximity to large population centers are features that contribute to attracting more visitors annually than any other national park (including the somewhat better-known Yellowstone, Yosemite, and Grand Canyon National Parks).

The GSM are situated in what had been one of the least accessible areas of the eastern United States. Natural barriers contributed to relatively low levels of development and exploitation through the eighteenth and nineteenth centuries. GSM National Park was authorized in 1926. Land purchase (funded in part by a five million dollar donation from foundations associated with the Rockefeller family) began in 1930 and was established through an act of Congress in June of 1934. The Park was formally dedicated by President Franklin D. Roosevelt in 1940. In 1976, the Park was recognized as a **biosphere reserve**, and in 1983 the GSM were selected as a World Heritage Site.

The GSM contain several of the highest mountain peaks in the United States east of the Mississippi River. The Park's Mt. Guyot, Mt. Chapman, and Mt. Leconte all tower over 6,000 ft (1,830 m). At 6,643 ft (2,025 m), Clingman's Dome, on the Tennessee side, is the highest point in the Park, as well as the crest of the Appalachian Trail. This peak was named after General Thomas L. Clingman, a soldier and scientist who, while exploring the area with Arnold Guyot in the late nineteenth century, was the first to measure the summit's elevation.

Among the oldest mountains in the world, the GSM also contain some of the oldest rock strata in the eastern United States. Sheetlike layers of slate (a fine-grained rock originally formed from mud **sediment**) create jagged outcrops of spectacularly-angled rock, glinting in areas where mica deposits (a sparkling mineral formed when slate is heated) show through to the surface. The slate that makes up much of the jagged central ridge line tells the story of how the mountains were formed by repeated elevations rather than by one single geological event. The bedrock ranges in age from about 600 million–1 billion years and exhibits a range of different characteristics owing to metamorphoses caused by the heat and violence of geological shifts.

More than 148,200 acres (60,000 ha) of GSM National Park are pristine or primitive. Early in the summer, the characteristic "balds" of the GSM (clear, treeless patches on the summits) are brightened by rhododendron blooms around their peripheries. Several varieties of spruce, buckeye, birch, hemlock, dogwood are also common. The abundant sunshine and frequent rain of the region help support over 1,500 **species** of flowering plants, including eight species of trillium, 32 species of violets, and 29 species of orchids. During the last **ice age**, the highlands were a refuge for northern plant species displaced by the continental ice sheet. Remnants of some of these species, including red spruce and Fraser fir, can still be found at the higher elevations but virtually nowhere else at this latitude. The Park contains the largest virgin forests of red spruce and balsam fir, and the finest virgin hardwood forests left in the United States. Indeed, the slopes of the GSM may support more virgin timber than the rest of the eastern United States combined. In the fall, when the green chlorophyll in deciduous leaves begins to break down, these forests put on a spectacular show of color. Fall colors generally peak between October 15–25, when up and down the mountains may be seen the brilliant reds of maples, golden yellow of beech, and the deeper hues of oaks and more southerly species. But Park visitation also peaks at this time, and those visitors who use the main roads through the Park frequently find themselves in bumper-to-bumper traffic jams with other nature-lovers.

Despite the huge annual invasion of tourists and hikers, GSM National Park nevertheless still supports abundant **wildlife**. Wild hogs, white-tailed deer, turkey, groundhogs, and ruffed grouse are among the species that can be seen and/or heard in the back country. Black bear have done particularly well under the protection and management of the **National Park Service**, to the extent that they have become habituated to the presence of humans and frequent rest-stops, campsites, and picnic areas. Campers, hikers, and even day trippers in vehicles need to be knowledgeable about bear behavior and about food storage and **garbage** disposal procedures which help to minimize unwanted attention from the bears.

Several species of trout are found in abundance in the cool and clear mountain streams. Estimating conservatively, about 250 species of birds may be seen in the Park at various times during the year, 165 of these species being resident. The avian population includes twenty varieties of warbler, thirty of finch, and twenty of geese and ducks. The higher elevations provide **habitat** for ravens and hawks, as well as occasional peregrine falcons and eagles.

GSM National Park also contains approximately 72 mi (116 km) of the Appalachian Trail, which runs along the crest of the Appalachians all the way from Mt. Katahdin, a granite monolith in Maine, south to Springer Mountain in Georgia. In the Park, the Appalachian Trail runs from the Pigeon River to the Little Tennessee River, and here can be found the highest, roughest, and some of the most scenic segments of the entire route. Over 900 mi (1,450 km) of other trails provide hikers and horseback riders with access to dozens of waterfalls, mountain meadows, and scenic, grassy balds.

[*Lawrence J. Biskowski*]

Green advertising and marketing

In the last decade growing consumer interest in environmental issues has significantly impacted how advertisers market their products and companies. The evidence regarding this greater concern for the environmental impact of commercial goods has been documented by several marketing groups. A 1989 survey by Michael Peters consultants found that 53% of the Americans asked had refused to buy a product in that year because of the effect of the product or package on the **environment**; 75% indicated that they would purchase a product with **biodegradable** or recyclable packaging even if it meant spending more money. In 1990 an Abt Associates study of American consumers showed that 90% of those interviewed were willing to pay more for environmentally-friendly products. For many years, German, Scandinavian, and Dutch consumers had shown a willingness to buy phosphate-free **detergents** and other so-called environmentally-friendly products. Indeed, a German business was saved from bankruptcy by offering a washing machine that consumed less water, detergent, and energy than its rivals. In England, *The Green Consumer Guide*, by John Elkington and Julia Hailes, was a best-seller for four weeks after its publication. Generally, the most environmentally-concerned consumers were well-to-do with the most discretionary income and the highest educational level. In short, they were trend-setters that advertising and marketing people could not ignore.

Marketers began to commonly use the terms "environmentally friendly," "safe for the environment," "recycled," "degradable," "biodegradable," "compostable," and "recyclable." Cause-related marketing also became popular as companies promised to support moderate environmental organizations such as **World Wildlife Fund**. While the advertising practices of many companies went uncontested by environmental groups, concerns arose regarding the claims of certain companies. For example, the Mobil Oil Corporation was sued for misleading advertising after claiming that its plastic Hefty **garbage** bags were recyclable. After suffering much embarrassment, British **Petroleum** was forced to withdraw its claim that its new brand of unleaded **gasoline** caused no **pollution**. Reacting to these and similar findings, 10 state Attorney Generals issued a report in 1990 calling for greater accountability in "green" marketing. The **Environmental Protection Agency** (EPA) and the Federal Trade

Commission also devised standards to evaluate the claims made by advertisers.

Often, the issue has been whether one product is really better for the environment than another. For instance, phosphate-free detergents created a controversy when they were introduced in France. Some companies claimed that they were no more environmentally benign than detergents that had **phosphates**. Rhone-Poulenc, the French producer of the detergents with phosphates, ran ads of dead fish apparently killed by the substances in the detergents which did not have phosphates. Proctor & Gamble launched a campaign which claimed that **disposable diapers** actually had less negative environmental impacts than reusable diapers. They pointed to the detergents, hot water, and energy used in washing cloth diapers, the energy needed to bring them to consumers, and the pesticides that were in the cotton out of which they were made.

Life-cycle assessments came into vogue as companies argued about the relative environmental merits of various products. Assessments exam the total environmental impact of using the product and how it rates—environmentally—to other similar products. Migros, the large Swiss retailer, has developed an "eco-balance" or life-cycle program to analyze the impact of its packaging in terms of the resources used and how they are disposed of.

Green labeling programs exist in Germany (**Blue Angel**), Canada (Environmental Choice), and Japan (Eco-mark). They are run by the governments of these countries, but the United States government has not been willing to give this kind of endorsement to commercial products. Instead, various environmental groups have **seals** of approval which they have applied to selected goods that pass their tests of environmental acceptability. *See also* Environmentalism; Green packaging; Green products; Recycling

[*Alfred A. Marcus*]

RESOURCES
BOOKS
Cairncross, F. *Costing the Earth*. Boston: Harvard Business School Press, 1992.
Elkington, J., and J. Hailes. *The Green Consumer Guide*. London: Gollancz, 1988.

Green belt/greenway

A green belt is an area of land that usually surrounds a town or city, and is kept open by government restrictions on further development. Often it comprises both public and private land.

Green belts provide both recreational areas and landscape, but their main purpose is to contain cities, diverting future growth, and preventing cities and towns from merging.

It grew out of Sir Ebenezer Howard's "Garden City" approach to town planning, which aimed to provide natural areas for residents of cities. Garden cities were surrounded by countryside; city expansion was only to take place by developing new garden cities on the other side of the green belt. The result would be clusters of cities grouped around a central city. Several garden cities (e.g., Welwyn in England) were built in Britain.

In the 1920s, London's government called for study of an "agricultural belt," although the farming function was considered incidental. Its key purpose was to act as a barrier to growth. Ultimately, these were to serve as open spaces. Sir Raymond Unwin proposed a "green girdle" of spaces easily accessible to towns and cities. This plan led to the 1938 Green Belt Act, and about 38,000 acres (15,580 ha) were purchased. After World War II, green belts were again proposed as a means to limit development. A green belt about 5 mi deep (8 km) separated the inner city and the suburbs from the outer countryside. Land was both purchased and controlled through zoning compensation and by purchasing development rights. Typically, farming was allowed, but housing was not. By 1959, 840 mi^2 (2,175 km^2) had been secured. The green belt was not considered completely successful, as it did not halt London's growth, and residents' use of the lands was not high, as they were not easily accessible.

Other difficulties, according to William H. Whyte, author of *The Last Landscape*, were the time and expense of controlling land and maintaining it, arbitrary placement of the green belt areas (not following natural boundaries) and lack of function attached to the land.

Green belts today emphasize the landscape and recreational values of open areas. Containment is negative and it does not work, a conclusion the Japanese came to in 1965 when they abandoned a proposed London-type green belt to contain Tokyo.

The green belt concept came to the United States much later, and many would say that it grew into the *greenways* concept, which have the specific purpose of providing open space and recreational areas. In the United States, the term is often used interchangeably with the term greenway, although the meaning is somewhat different.

Greenways, according to Jennifer Howard, author of *Greenways: A Citizen's Guide*, are linear corridors of land and water and the natural, cultural, and recreational resources they link together. They help conserve a variety of resources and create recreational opportunities.

Greenways were inspired both by green belts and by the work of landscape architect Frederick Law Olmstead, who, in the late 1800s, designed a number of urban park systems he called "parkways." One of his most notable is the Emerald Necklace, a string of park areas around Boston.

They may take the form of a riverfront walkway, a bicycle path, an urban walking trail connecting historic sites and neighborhood parks, a **wildlife migration** corridor, among others. Greenways are typically categorized as river greenways, paths and trails, cultural and historic greenways, wildlife corridors. They are typically comprising both public and private lands, and can include trails, riverways, **habitat** and resource **conservation** areas, notable natural features, scenic roads, historic structures, vacant urban lands, forestland, and farm fields—basically any resource that is significant to a community. Greenways help make these resources accessible to residents without need for a car. Four categories of land and **water resources** can be incorporated into a greenway system: resource-conservation areas, parks and open spaces, cultural and historic resources, and the corridors of land and water which connect these other elements together.

In the last 20 years, the greenway concept has increased in recognition and popularity. Today, hundreds are underway. In 1987, The President's Commission on Americans Outdoors focused national attention on greenways, and recommended their establishment "to link together the rural and urban spaces in the American landscape."

In Massachusetts, several greenways exist. The Connecticut River Greenway State Park in Massachusetts comprises about 3,500 acres (14,350 ha) of land owned or controlled by the Department of Environmental Management along a 70-mi (113-km) stretch of the Connecticut River. Much of the land is in private ownership and includes farms and woodlots.

The Massachusetts Bay Circuit encircles the entire Boston Metropolitan area; within Boston, the Emerald Necklace, a string of parkways, was planned by Frederick Law Olmsted. Other notable green belts exist in Portland, Oregon; Vienna; and Tokyo.

[*Carol Steinfeld*]

RESOURCES
BOOKS

Howard, J., *Creating Greenways: A Citizen's Guide.* The Department of Environmental Management, 1997.
Whyte, W. H. *The Last Landscape.* New York: Doubleday & Company, 1968.

Green consumerism
see **Green advertising and marketing; Green Cross; Green products**

Green Cross

Long used by several organizations, the Green Cross has become a symbol of environmental awareness and responsibility.

A religiously oriented group officially called the American Association of the Green Cross is based in Colorado Springs, Colorado, and describes itself as "a new Christian environmental organization whose purpose is to address the ethical and moral issues underlying ecological issues, and to mobilize volunteers in service to Creation." The group's motto is "serving and keeping Creation," and it encourages such activities as "the development of every church as a Creation-awareness center, education about Christian responsibility for the earth, local action to address ecological issues." The group also promotes tree planting, urban gardening, **habitat** restoration, resource **conservation**, and **waste reduction**. It is associated with the North American Conference on Christianity and Ecology, and it plans to establish a network of chapters in churches, schools, youth groups, and colleges.

The International Green Cross and Green Crescent is a new environmental group formed in April 1993 and headed by former Soviet President Mikhail S. Gorbachev. Hoping to do for the **environment** what the Red Cross and Red Crescent have done for disaster relief, the International Green Cross will work to coordinate environmental efforts on a global scale. Mr. Gorbachev has said that he accepted leadership of the group because "I am convinced that saving the environment is the number one priority for all countries." Sponsors of the organization include an array of distinguished world political and spiritual leaders, including India's Mother Teresa and Javier Perez de Cuellar, the former Secretary General of the United Nations.

The Green Cross Certification Company is the former name of a non-profit group that awarded certifications to manufacturers whose products met certain limited environmental standards. The certification program is now administered by Scientific Certification Systems, Inc. (SCS), a private, for-profit laboratory that charges manufacturers a fee to research products and to verify their performance and claims. SCS says that it is "committed to developing programs that motivate private industry to work toward an environmentally sustainable future" by "conducting independent, unbiased evaluations of products and product claims, and recognizing products achieving exceptional environmental performance goals." The SCS Environmental Report Card summarizes the environmental performance of a product, including the amount of "environmental burden" associated with the product and its packaging.

SCS emphasizes that it does not approve products as "green" or environmentally acceptable, but rather verifies the environmental claims that companies make for their products and analyses their environmental impact. SCS tries to evaluate a product's life cycle program, the impact it has from manufacture to disposal. Factors considered usually include the toxic waste generated and the energy used in production, the recycled content of the product, and its recyclability or biodegradability upon disposal. Different product categories have varying standards of acceptability depending on the state of technology for the above factors for the particular product or industry. *See also* Environmental consumers; Environmental ethics; Green advertising and marketing; Green products; Nongovernmental organization

[*Lewis G. Regenstein*]

Green packaging

Packaging is the largest form of domestic **garbage**. In 1999 it amounted to 33.1% of **solid waste** as measured by weight in the United States. Significant waste prevention implies reductions in packaging. There simply is not enough room in landfills or incinerators for all the excess packaging the industry produces.

People value products not only for their content but also for the packaging. It gives products a better feel and a more attractive appearance and suggests less of a risk of contamination. It prolongs the life of the product and allows people to make fewer trips to the supermarket.

For purely economic reasons most packages are becoming lighter. **Aluminum** cans, for instance, are 45% lighter today than they once were. Shrink wrap film and a plastic base are increasingly taking the place of corrugated boxes. Some companies are trying to eliminate packaging entirely. Outer boxes were once thought to be absolutely essential for the sale of toothpaste, but the giant Swiss retailer, Migros, discovered that consumers ultimately became accustomed to unboxed tubes and that sales did not suffer as a result. McDonald's and other fast food restaurants in the United States have stopped using **polystyrene** boxes to package their sandwiches, turning instead to paper wraps. Other innovations in packaging are also occurring. For instance, Procter & Gamble is no longer using metal-based inks for printing on packages. *See also* Container deposit legislation; Green advertising and marketing; Waste reduction

[*Alfred A. Marcus*]

RESOURCES
BOOKS

Cairncross, F. *Costing the Earth*. Boston: Harvard Business School Press, 1992.

Green plans

Green plans are comprehensive environmental strategies that are intended to improve environmental quality and make rapid progress towards sustainability. (In its use here, the word "green" is non-political and non-ideological, and merely refers to a context of environmental protection and sustainable development.) Green plans are characterized by a longer-term view, while being thorough in their consideration and integration of environmental issues. Green plans also take account of economic realities, while consistently ensuring an appropriate degree of protection of environmental quality and natural ecological values (such as the needs of **endangered species** and rare ecosystems).

Green plans represent an extremely important tool for the longer-term protection of environmental quality, **conservation** of **natural resources** and ecological values, and achievement of a sustainable economic system. Green plans do this by proposing sustainable alternatives to the many kinds of modern activities that are causing damages to the **environment** and **biosphere**. Green plans attempt to integrate the economic, scientific, and political interests of society to develop a strategy that can achieve a sustainable prosperity for present and **future generations** of humans and their economic systems, while also supporting other **species** and natural ecosystems.

Green plans are designed to replace more conventional methods for protecting the environment. These conventional methods include the following:

(1) *Sectoral structures of government and administration.* This type of structure can be a problem because there is often a lack of integration among sectors, even though there may be important environmental linkages. For example, environmental management by government typically involves separate agencies responsible for **air pollution, water pollution**, forestry, agriculture, fisheries, metal resources, fossil-fuel resources, industrial development, human health, **biodiversity** (i.e., the conservation of indigenous species and ecosystems), and other environment-related mandates. Because the responsibilities of these agencies are not well integrated, they often work at cross-purposes. For instance, agencies responsible for managing harvests of timber may not take adequate consideration of the interests of agencies concerned with **pollution**, endangered species, or rare ecosystems (such as old-growth forests). This commonly results in environmental and ecological damages being caused by timber harvesting, with attendant controversies.

(2) *Single-issue policies.* Single-issue policies may result from the actions of special-interest groups, which are seeking to advance their particular environmental, ideological, or socio-economic agendas. This can, however, result in poor integration among issues, and divisive political and social

controversies that can impair the development of policies that would achieve balanced levels of environmental protection. Single-issue policies often result in regulation through short-term objectives and standards that focus on individual issues, such as the concentrations of particular **chemicals**, or the abundances of certain species. This can result in actions that focus on compliance with narrow regulatory criteria, rather than the more comprehensive, longer-term environmental goals that are pursued by green plans.

Green plans can be implemented by agencies of government at all levels (that is, federal, state or provincial, county, and city or town), and by companies of any size. Green plans can be designed by any of these organizations, but this is done in close consultation with the public, non-governmental organizations, and environmental specialists. In fact, multi-sectoral and multi-organizational discussions are one of the most important aspects of green plans. This process allows a broad degree of understanding to be reached among the spectrum of interest groups, allowing green plans to bridge political and ideological differences concerning environmental issues.

Green plans have already been implemented by the national governments of Canada, the Netherlands, New Zealand, and Singapore. Green plans are also being developed or seriously considered by the national governments of Austria, Denmark, Germany, Norway, Sweden, and the United Kingdom, and by the European Community as a whole. In addition, many state and provincial governments, municipalities, and companies in some of those countries have implemented green plans. Non-governmental environmental organizations are also advocating that green plans be designed by federal and state governments and companies in the United States. Unfortunately, significant actions in this regard have not yet been undertaken in that country. It is likely, however, that green plans will also be developed in the United States, once their benefits become more broadly recognized.

[*Bill Freedman Ph.D.*]

RESOURCES
BOOKS

Johnson, H. D. *Green Plans. Greenprint for Sustainability.* Lincoln, NB: University of Nebraska Press, 1995.

OTHER
"A Green Plan Primer." *Green Plan Center.* May 1997 [cited July 2002]. <http://www.rri.org/gparchive/gp-primer.html>.

Green politics

Green Politics is a relatively recent political movement that places a concern for **nature** and its myriad **species** at the top of its agenda. New social and political movements arise in response to crises that are perceived to be both long-term and **systemic**. The crisis out of which the broadly based Green movement has emerged is the environmental crisis, which is actually a series of interconnected crises caused by **population growth**, air and **water pollution**, the destruction of the tropical and temperate rain forests, the rapid **extinction** of entire species of plants and animals, the **greenhouse effect**, **acid rain**, **ozone layer depletion**, and other now familiar instances of **environmental degradation**. Many are by-products of technological innovations, such as the internal **combustion** engine. But the causes of these environmental crises are not only technological but are also broadly cultural and political. They stem from beliefs and attitudes that place human beings above or apart from nature. Despite their differences, the major mainstream political perspectives–liberalism, socialism, and conservatism—are alike in viewing nature as either a hostile force to be conquered or a resource base to be exploited for human purposes. All, in short, share an anthropocentric, or human-centered, bias.

Against these views, the modern environmental or Green movement counterpoises its own perspective. Many **Greens** prefer not to call their perspective a political ideology, but an environmental ethic. Earlier ecological thinkers, such as **Aldo Leopold**, spoke of a "land ethic." Others, such as Christopher stone, speak of an ethic with the earth itself at its center, while still others, such as Hans Jonas, speak of an emerging "planetary ethic." Despite differences of accent and emphasis, however, all are alike in several important respects. An environmental or Ecological ethics, they say, would include several key features. First, such an ethic would emphasize the web of interconnections and mutual dependence within which we and other species live. From the this recognition of interconnectedness a second feature follows: a respect for all life, however humble humans may believe it to be, because the fate of our species is tied in with theirs. And since life requires certain conditions to sustain it, the third feature follows: we have an obligation to respect and care for the conditions that nurture and sustain life in its myriad forms. Since nature nourishes her creatures within a complex web of interconnected conditions, to damage one part of this life-sustaining web is to damage the others, and to endanger the existence of the creatures that depend upon it.

Green thinkers hold that the enormous power that humans have over nature imposes on our species a special responsibility for restraining our reach and using our power wisely and well. Greens point out that the fate of the earth and all its creatures now depends, to an unprecedented degree, on human decisions and actions. For not only do we depend on nature, but nature depends on our care and restraint and forbearance. Humans have the nuclear means to destroy in mere minutes the earth's inhabitants and the ecosystems that sustain them. From this emerges a fourth feature of a "green" political perspective: Greens must oppose militarism and work for peace.

But the earth is in danger not only from global thermonuclear war but from the slower destruction of the natural **environment**. Such destruction is a consequence not only of large-scale policies but of small-scale, everyday acts. And each of us bears full responsibility for our actions and also, since we live in a democracy, some share of responsibility for cumulative effects and collective outcomes. Each of us has, or can have, a hand in making the laws under which we live. It is for this reason that Greens give equal emphasis to our collective and individual responsibility for protecting the environment that protects us. The fifth feature of the Green political perspective, then, is to emphasize the importance of informed and active democratic citizenship at the grass-roots level. Hence the Green adage, "Think globally and act locally."

On this much most Greens agree. But there are also a number of unresolved differences of approach, emphasis, and political strategy. The internal ideological spectrum ranges from "light green" conservationists to "dark green" radicals, and includes assorted anarchist beliefs, **deep ecology**, **ecofeminism**, **social ecology**, **bioregionalism**, New Age Gaia worship, and other groupings, each differing in various ways form the others. Among these are differences regarding the basic beliefs underlying and motivating the green movement. Some New Age Greens envision an environmental ethic grounded in spiritual or religious values. We should, they say, look upon the earth as a benevolent and kindly deity—the goddess *Gaia* (from the Greek word for "earth")—to be worshiped in reverence and awe. In this way we can liberate ourselves from the restrictive rationalism that characterizes modern science. Other Greens disapprove of such a spiritual or religious orientation, contending that such beliefs are politically pernicious and inimical to the rational scientific thinking required to diagnose and solve environmental problems.

Other differences have to do with the political strategies and tactics to be employed by the environmental movement. Some say that Greens should take an active part in electoral politics, perhaps even following the lead of Greens in Germany and organizing a Green Party. Aware of the formidable obstacles facing minority third parties, most have favored other strategies, such as working within existing mainstream parties (especially the Democratic Party in the United States), or hiring lobbyists to influence legislation. Still other Greens favor working outside of traditional interest group politics, believing the earth and its inhabitants hardly constitute a special interest. Others, such as social

ecologists, tend to favor local, grass-roots campaigns which involve neighbors, friends, and fellow citizens in efforts to protect the environment. Some social ecologists are anarchists who see the state and its pro-business and pro-growth policies as the problem, rather than the solution, and seek its eventual replacement by a decentralized system of communes and cooperatives. Greens of the "bioregionalist" persuasion add that such social and political organization ought to be based on biological or natural, rather than artificial or political, boundaries and regions.

Although all Greens agree on the importance of informing and educating the public, they disagree as to how this might best be done. Some groups, such as **Greenpeace**, favor dramatic direct action calculated to make headlines and capture public attention. Even more militant groups, such as the Sea Shepherd Society and **Earth First!**, have advocated **monkey-wrenching** as a morally justifiable means of publicizing and protesting practices destructive of the natural environment.

Such militant tactics are decried by moderate or mainstream groups, which tend to favor subtle, low-key efforts to influence legislation and inform the public on environmental matters. The **Sierra Club**, for example, lobbies Congress and state legislatures to pass environmental legislation. It also publishes books and produces films and videos about a wide variety of environmental issues. Similar strategies are followed by other groups, such as the **Environmental Defense** Fund. Another group, **The Nature Conservancy**, solicits funds to buy land for nature preserves.

Differences over strategy and tactics are, however, differences about means and not necessarily about basic assumptions and ends. Despite their political differences, Greens are alike in assuming that all things are connected— ecology is, after all, the study of interconnections—and they agree that complex ecosystems and the myriad life-forms they sustain are valuable and worthy of protection by political and other means. *See also* Abbey, Edward; Bioregional Project; Bookchin, Murray; Brower, David Ross; Environmental attitudes/values; Environmental Defense Fund; Foreman, Dave; Green advertising and marketing; Green products; Sea Shepherd Society; Sierra Club

[*Terence Ball*]

RESOURCES
BOOKS

Bahro, R. *Building the Green Movement.* London: G.R.P., 1978.

Biehl, J. *Rethinking Ecofeminist Politics.* Boston: South End Press, 1991.

Bookchin, M. *The Modern Crisis.* Philadelphia: New Society Publishers, 1986.

Capra, F., and C. Spritnak. *Green Politics.* New York: Dutton, 1984.

Foreman, D. *Confessions of an Eco-Warrior.* New York: Harmony Books, 1991.

Jonas, H. *The Imperative of Responsibility.* Chicago: University of Chicago Press, 1984.

Leopold, A. *The Sand County Almanac.* New York: Oxford University Press, 1948.

Manes, C. *Green Rage: Radical Environmentalism and the Unmaking of Civilization.* Boston: Little Brown, 1990.

Milbrath, L. W. *Envisioning a Sustainable Society.* Albany: State University of New York Press, 1989.

Paehlke, R. *Environmentalism and the Future of Progressive Politics.* New Haven, CT: Yale University Press, 1989.

Porritt, J. *Seeing Green: The Politics of Ecology Explained.* Oxford: Basil Blackwell, 1984.

Seed, J., et al. *Thinking Like a Mountain.* Philadelphia: New Society Publishers, 1988.

Worster, D. *Nature's Economy: A History of Ecological Ideas.* Cambridge: Cambridge University Press, 1977.

Green products

Some companies have thrived by marketing product lines as environmentally correct or "green." A prime example is Body Shop, a cosmetics company that is strongly and explicitly pro-environment with regard to its products. It strives, for instance, to develop products made with substances derived from threatened tropical rain forests so that they can be preserved.

The American ice cream manufacturer, Ben and Jerry's, has adopted a similar approach to using **rain forest** products in what it sells. Mercury- and cadmium-free batteries have been marketed by Varta, a German company. Ecover, a small Belgian company, made major sales gains when it began to market a line of phosphate-free **detergents**. Wal-Mart is another company that provides its customers with green products. Loblaw, a Canadian grocery chain, has introduced a "green-line" of environmentally-friendly products and has sold more than twice the amount than it had initially projected. Seventh Generation, a mail-order company based in California, has successfully marketed its own line of recycled toilet paper, **biodegradable** soaps and cleansers, and phosphate-free laundry and dishwashing detergent.

Many factors comprise a green product. The product has to be made with the fewest raw materials and produced with the least amount of contaminants released into the **environment** and with the smallest effect on human health.

Consideration must also be given to how consumers will use the product and how they will dispose of it when they are finished. To reduce its waste potential, a product must often last a significant amount of time or be reusable or recyclable.

As consumers become more aware of environmental issues, they will likely look to producers and governments to provide more products that will permit them to maintain a life-style that is less harmful to the environment. Therefore, the very nature of products will have to change. They will have to be lighter, smaller, and more durable so that they can consume fewer resources in their production and use and take up less space when they are disposed of.

Ultimately, a real revolution in the use of green products would mean replacing or substantially modifying virtually the capital stock of society—appliances, automobiles, housing, highways, etc.—with a different type of product. In contrast to old smokestack industries, new technologies and emerging industries—such as telecommunications, computers, and information—should be able to offer products that are less environmentally harmful. They should be able to produce many new types of green products and modify existing products so that they are less damaging to the environment.

[*Alfred A. Marcus*]

RESOURCES
BOOKS

Buchholz, R., A. Marcus, and J. Post. *Managing Environmental Issues.* Englewood Cliffs, NJ: Prentice-Hall, 1992.

Cairncross, F. *Costing the Earth.* Boston: Harvard Business School Press, 1992.

Green revolution

see **Agricultural revolution; Borlaug, Norman E.; Consultative Group on International Agricultural Research**

Green Seal

An independent, non-profit group that encourages the production and sale of consumer products that are environmentally responsible, Green Seal allows the use of its certification mark on products that meet its strict environmental standards. The mark has a green check over a blue globe.

A growing number of people are becoming aware that consumer demand for certain products causes great harm to the **environment** and provides an economic incentive for activities that damage the planet. Some examples are products that contain **chemicals** such as **chlorofluorocarbons** (CFCs), which deplete the earth's protective **ozone** layer; mahogany and other kinds of wood from rapidly-disappearing tropical rain forests; fur coats made from rare and **endangered species**; tuna caught using techniques that kill **dolphins**; and products that waste energy or water, are overpackaged, cannot be recycled, or are harmful when disposed of.

Green Seal symbol. (Photograph by Tom Pantages. Reproduced by permission.)

By avoiding these products and buying those that do not cause harm to **wildlife** or degrade the environment, consumers can encourage corporations to make and sell goods that are environmentally responsible. With the public's growing commitment to protecting the environment, store shelves are now full of products that claim to be earth friendly, environmentally friendly, recycled, **biodegradable**, natural, organic, or are labeled in such a way as to take advantage of **green advertising and marketing**.

The Green Seal certification mark helps consumers choose those products that actually are less harmful to the planet and are not simply marketed in a clever way. Green Seal uses the highly respected Underwriters Laboratory (UL) for most of its product testing and certification. Through its certification process and its educational activities, Green Seal encourages people to think about how they can help protect the environment in their everyday activities and their daily lives.

Green Seal points out that its research shows that four out of five consumers are more likely to buy a product with its certification mark when choosing between similar products. A Gallup survey found that the Green Seal certification would have more impact on consumers than would government guidelines. Thus, consumers, guided by Green Seal, have the opportunity to influence the actions of major corporations and their impact on the environment through their purchasing decisions.

Green Seal is headed by Arthur B. Weissman, who serves as its President and Chief Executive Officer. Its Chairman of the Board is Denis Hayes, the well-known environmentalist and **solar energy** advocate who organized the 1970 and 1990 **Earth Day** celebrations.

[*Lewis G. Regenstein*]

RESOURCES
ORGANIZATIONS
Green Seal, 1001 Connecticut Avenue, NW, Suite 827, Washington, D.C. USA 20036-5525 (202) 872-6400, Fax: (202) 872-4324, Email: greenseal@greenseal.org, <http://www.greenseal.org>

Green taxes

The search for alternatives to legislation and enforcement of **environmental policy** led a 1988 bipartisan Congressional study group (Project 88) to call for the use of market forces, including taxes, to protect the **environment**. Project 88's advocacy of these "green" taxes and other economic incentives for reducing **pollution** is actually an old idea. Charles Schultze, chairman of the Council of Economic Advisers under President Jimmy Carter, maintained in the 1976 Godkin Lectures delivered at Harvard University (later published as a book titled *The Public Use of the Private Interest*) that detailed laws and bureaucratic requirements were a costly and ineffective way to control pollution. Instead, reliance should be placed on taxes and subsidies that would make private interests more congruent with public goals.

The economists' argument, posed by Project 88, Schultze, and others, is that the harm pollution causes to health, property, and aesthetics is not paid for by business. Industries have no reason to consider this harm in their production decisions. By taxing pollution, the government would make polluters pay for the damage they inflict. External production costs would be incorporated into ordinary production decisions. This would correct a market defect and the market would become more efficient. Green taxes would not lower environmental standards; they would provide more protection at the same level of expenditure, or the same protection with less money.

The present regulatory system is expensive and inefficient. The amount of litigation is excessive, and relations between business and government suffer. The uniform standards often do not make sense, because different companies have different removal costs depending on their production process and other factors. For example, a study in the St. Louis area found that removing a ton of **particulate** matter from a paper factory cost $4, while removing the same material from a brewery cost $600.

Industries that can easily reduce pollution should be encouraged to go beyond the standard and not stop at mere compliance. Businesses for whom pollution reduction is a great burden should be able to pay a fine equivalent to the damage caused. To impose the same requirements on all businesses regardless of cost is unfair. Moreover, regulations do not permit pollution-reducing experiments. Companies should be allowed to choose the lowest-cost method whether it means treating wastes, modifying production processes, substituting less-polluting raw materials, or other innovations. By shifting away from uniform standards, pollution-control costs can be cut drastically. According to one study, an equivalent **air quality** can be achieved at 10 percent of existing costs.

Why have governments been so slow to implement green taxes? The primary reason is because taxes have never been popular with legislators or their constituents. Nor do interest groups support them. Businesses prefer court delays and lobbyists, to the certainty of taxes. Environmentalists argue that pollution must be eliminated and that companies should not be given the right to pollute for a fee. Bureaucrats, moreover, have been comfortable with the existing system. Pollution-tax proposals proposed by Presidents Lyndon Johnson and Richard Nixon were almost immediately dismissed.

During the Clinton administration there was new interest and support for pollution tax proposals. However, the necessary Congressional support did not materialize. Moreover, industry spokespersons and lobbyists promoted the idea that definitive scientific proof for global warming had yet to be established.

However, in 2002, during the Bush administration, which had previously stated that there was not enough evidence to link industrial emissions to global warming, the **Environmental Protection Agency** released a report that endorsed what many scientists had argued, i.e., that oil refinery, **power plants**, and auto emissions were important causes of global warming. In 2001, President Bush had caused international outrage when he said that he would not join other nations in ratifying the Kyoto Protocol, a United Nations plan to cut emissions of green house gases, because he claimed it would be too costly to the American economy.

Yet by 2002 at least nine European nations had implemented environmental taxes as a means of reducing air pollutants, the growing shortage of **landfill** space, and to promote the **conservation** of water and electricity. But in spite of the growing concern over global warming and pollution, efforts to implement environmental taxes have not found the needed support in this country due in large part to a general anti-tax sentiment. At this point there still remains a majority interest in preserving the status quo as opposed to addressing needed environmental concerns through such measures as environmental tax reform. With the growing body of evidence supporting the need for action, and pressure from other nations, that sentiment may change in the future. However, the issue at that time will more likely be whether or not it is too little reform, too late. *See also* Corporate Average Fuel Efficiency Standards; Environmental economics; Externality; Internalizing costs; Pollution control; Pollution control costs and benefits

[*Bill Asenjo Ph.D.*]

RESOURCES
PERIODICALS

Franz, D. "The Environmental Tax Shift." *E/The Environmental Magazine*, May 10, 2002.
———, and B. W. Whitehead. "Dealing With Pollution: Market-Based Incentives for Environmental Protection." *Environment* 34 (September 1992): 7–42.

OTHER

Stavins, R. N., ed. *Project 88—Harnessing Market Forces to Protect the Environment*. A Public Policy Study Sponsored by Senator Timothy E. Wirth and Senator John Heinz. Washington, DC: U.S. Government Printing Office, 1988.

Greenhouse effect

The greenhouse effect is a natural phenomenon that traps radiation within the earth's **atmosphere**. Natural **greenhouse gases** include water vapor, **carbon dioxide**, **nitrous oxide**, **methane**, and **ozone**, all essential to support life. The enhanced greenhouse effect, the direct result of human activities, increases concentrations of these gases in the atmosphere, and leads to **pollution** of the lower atmosphere and contributes to global warming. These gases let in sunlight but tend to insulate Earth against the loss of heat, as do the glass walls of a greenhouse. A higher concentration of the greenhouse gases means a warmer **climate**. For example, the twentieth century was been 1° warmer on worldwide average than the nineteenth century—warming at a rate 20 times faster than average.

Carbon dioxide (CO_2) is considered the predominant greenhouse gas and has the greatest impact on global heat. From April 1958, when monthly measurements of CO_2 from atop the Mauna Loa **volcano** began, through June 1991, the CO_2 concentration in **parts per million** went from 316 ppm to almost 360 ppm. The peak concentration is due to the destruction of tropical rain forests and the burning of **fossil fuels**, which accounts for half of the greenhouse gases added to the atmosphere. CO_2 is dumped into the atmosphere at a much faster rate than it can be withdrawn or absorbed by the oceans or living things in the **biosphere**; since 1765, its presence in the atmosphere has increased by over 27%. CO_2 buildup in the next few decades to centuries could be one of the principal controlling factors of the near-future climate.

Methane, another greenhouse gas, is produced when oxygen is not freely available and bacteria have access to organic matter, such as in swamps, bogs, rice paddies and moist soils. Methane also is produced in the guts of termites and cows, in **garbage** dumps, landfills, emissions from **coal** mining, **natural gas** production and distribution, and changing **land use**. Methane concentrations have increased over 100% since 1765.

Nitrous oxide concentrations have increased in recent years due to **fertilizer** use and chemical production, such as in the manufacture of nylon. Nitrous oxide is also dispersed during fossil fuel **combustion, biomass** burning and changing land use.

CFCs (**chlorofluorocarbons**), also implicated in **ozone layer depletion**, act as greenhouse gases. While useful and widely used as refrigerants, their total effect is significant because compared to a molecule of carbon dioxide, each molecule of CFC absorbs much more radiation, thereby **trapping** heat in the atmosphere.

Other greenhouse gases are ground-level ozone (sunlight reacting with **automobile emissions**). and water vapor. Water vapor represents about two% of total atmospheric composition, and is the most abundant greenhouse gas. With methane and carbon dioxide, it plays an important role in regulating the temperature of the planet through the production of clouds.

Rain forest destruction also contributes to global warming. When the canopy of leaves is removed through **clear-cutting** or burning, the sudden warming of the forest floor releases methane and CO_2, in a kind of biochemical burning. The massive increase in the number of dead tree trunks and branches leads to a population explosion of termites, which themselves produce methane. Dead trees can no longer store CO_2 or convert it to oxygen.

Two factors which appear to mitigate the effect of enhanced greenhouse gases are aerosols and dust. Aerosols, minute solid particles, are finely dispersed in the atmosphere and have become an influence on the greenhouse effect. Aerosols are produced by combustion, but they also come from natural sources, primarily volcanoes. By blocking light, aerosols and dust can offset warming from greenhouse gasses. For example, a significant cooling trend in the spring and summer of 1992 seemed to correlate with the eruption of **Mount Pinatubo** in the Philippines. The fall and winter of 1992 were fairly mild on worldwide average. As all the **particulate** matter from the Mt. Pinatubo eruption settled

out of the atmosphere, the surface cooling effects abated and the global warming trend resumed.

Anthropogenic (human-caused) greenhouse gases now appear responsible for increasing the global average temperature. According to current projections, global temperatures may rise as much as 35.6–37.4° F (2–3° C) above the pre-industrial temperatures by the year 2100. To place this change in perspective, the temperature rise that brought the planet out of the most recent **ice age** was only about 37.4–39.2° F (3–4° C).

The top 10 warmest years of average global recorded temperatures were in the last 15 years of the twentieth centry and saw devastating fires in **Yellowstone National Park**, **flooding** in Bangladesh, record number of hurricanes and tornados, and a deadly heat wave and **drought** in the southeastern United States. It is probable, based on computer models, that a resumption in warming will accompany changes in regional weather. A 40-year trend of increased precipitation in Europe and decreased precipitation in the African **Sahel** (Ethiopia, the Sudan, Somalia) may be an early consequence of global warming due to the greenhouse effect. Longer and more frequent heat waves would result in public health threats as well as inconveniences such as road buckling, electrical brownouts, or blackouts.

Precipitation is likely to increase regionally because as the temperature increases, more evaporation takes place, leading to more precipitation. The average precipitation event is likely to be heavier: wetter monsoons in coastal subtropics; more frequent and heavier winter snows at high altitudes and high latitudes; an earlier snowmelt, and a wetter spring.

Increases in rain- or snowfall are not expected to offset the effects of higher temperatures on **soil**, however. Higher temperatures are expected to dry the soil in North America and southern Europe, among other places, by boosting the rates of evaporation and **transpiration** through plants. More favorable agricultural conditions in high latitudes could move the center of agriculture farther north into Canada and Siberia and out of the United States.

Other consequences of global warming from the enhanced greenhouse effect include the reduction of sea ice, coastal sea level rises of several feet per century, more frequent and powerful hurricanes, and more frequent and severe forest fires. In the United States, the frequency of tornadoes is near or above record levels for the years 1990–1994.

The rise of sea level is the most easily predicted consequence. The one-degree increase in temperature over the past century contributed to a 4–-in (10–20-cm) rise in mean sea level. This could lead to severe and frequent storm damage, flooding and disappearance of **wetlands** and lowlands, coastal **erosion**, loss of beaches and low islands, **wildlife** extinctions, and increased **salinity** of rivers, bays and aqui-

The greenhouse effect. (Illustration by Hans & Cassidy.)

fers. However, because the global atmosphere operates as a complex system, it is difficult, even with today's sophisticated computer models, to predict the exact nature of the changes we are likely to cause with increased greenhouse gases. Scientists have predicted that low-lying areas and islands, including the Seychelles, the Maldives, the Marshall Islands, and large areas of Bangladesh, Egypt, Florida, Louisiana, and North Carolina will disappear over the next few decades.

The earth's natural atmospheric cleanser — rain — may wash excess greenhouse gases out of the atmosphere. But until rates of greenhouse gases slow their rapid increases or actually begin to decrease, the planet will get warmer. In response to climate projections, the United Nations Framework Convention on Climate Change (UNFCC), adopted and signed by 162 countries in 1992 at the Rio Earth Summit, sets country-by-country standards to reduce the emissions of greenhouse gases, particularly carbon dioxide.

Policy-makers in the United States, including Vice President Albert Gore Jr., propose stricter requirements for more fuel-efficient cars, "environment taxes" that penalize heavy polluters and help pay for cleansing the atmosphere, and trading technological advances for rain forest protection in **Third World** countries. However, because global warming often is made a political ping-pong ball, changes in political administrations worldwide can extend to policy makers and climate researchers, who depend on government assistance for research.

The greatest controversy over slowing the rate of greenhouse gases injected into the atmosphere seems to be how to do it. Some scientists advocate increased use of **nuclear power** to reduce dependence on fossil fuels, but that carries its own controversies. Nuclear **power plants** are so energy-intensive just to build, the trade-off is negligible. **Conservation** and a switch from a dependence on fossil fuels to dependence on renewable resources such as wind and **solar energy**, slows the rate of increase of carbon dioxide emissions into the atmosphere. *See also* Environmental economics; Environmental policy; Global Tomorrow Coalition; *The Global 2000 Report*

[*Linda Rehkopf*]

RESOURCES
BOOKS

Bates, A. X. *Climate in Crisis.* Summertown, TN: The Book Publishing Co., 1990.

Houghton, J. T., and L. G. Meira Filho, ed. *Climate Change 1995: The Science of Climate Change. Contribution of Working Group I to the Second Assessment Report of the Intergovernmental Panel on Climate Change.* CambridgeUniversityPress,1996.

PERIODICALS

"Indices of Climate for the United States." *Bulletin of the American Meteorological Society* 77, no. 2 (February 1996): 279–292.

OTHER
Changing by Begrees: Steps to Reduce Greenhouse Gases. U.S. Congress, Office of Technology Asessment, 1991.

Greenhouse gases

Greenhouse gases are gases in the **atmosphere** that absorb and re-emit energy from the sun. They are believed to cause the global climatic changes known as the **greenhouse effect**.

The earth's **climate** depends on a wide variety of gases, vapors, and aerosols, and many of these contribute to global warming. **Carbon dioxide** is the most abundant; the atmosphere contains about 700 billion tons of this gas, and the oceans contain about 50 times this amount. Water vapor also contributes to global warming, and other important greenhouse gases include **ozone**, **methane**, **nitrous oxide**, and **chlorofluorocarbons**. Halogenated gases and a variety of volatile organic **hydrocarbons** are also important trace gases. Volatile compounds can absorb solar and infrared radiation directly; they can also affect the photochemistry of ozone, increasing the transmission of heat and thus indirectly affecting the climate. While not gases, small long-lived ambient particles, such as particulates, agricultural fields, **wetlands**, and oceans. Industrial emissions of greenhouses gases consist largely of **carbon** dioxide; these arise from burning **fossil fuels** such as **coal**, oil, and **natural gas**.

Increases in the concentrations of carbon dioxide and methane in the atmosphere during this century have been attributed in part to rapid increases in the utilization of fossil fuels. Efforts to reduce greenhouse gases have focused on limiting and controlling the burning of these fuels. There have been programs to encourage the utilization of other sources of energy such as **nuclear power**, or **alternative energy sources** such as **solar energy** or hydropower. Technologies for controlling fossil fuel emissions and sequestering ambient carbon dioxide have also been developed, and researchers have emphasized the importance of improving **energy efficiency** and **energy conservation**. *See also* Air pollution; Air pollution control; Flue gas; Pollution control

[*Stuart Batterman and Douglas Smith*]

RESOURCES
PERIODICALS

Dickinson, R. E., R. J. Cicerone. "Future Global Warming from Atmospheric Trace Gases." *Nature* 319 (1986): 109–115.

Hansen, J., A. Lacis, and M. Prather. "Greenhouse Effect of Chlorofluorocarbons and Other Trace Gases." *Journal of Geophysical Research* 94 (1989): 16417–21.

The Greenpeace ship, *Rainbow Warrior*, sailing up the St. George's Channel offthe west coast of England. Photograph by Noble. Greenpeace. Reproduced by permission.)

Greenpeace

Founded in 1971, Greenpeace is an international environmental organization dedicated to protecting the global **environment** through non-violent direct action, public education, and legislative lobbying. With a worldwide membership of over 2.5 million (approximately 250,000 in Greenpeace USA), Greenpeace operates offices in some 30 countries and maintains a scientific base in **Antarctica**.

Having mounted successful campaigns on a wide variety of environmental issues, Greenpeace is perhaps best known for its direct and often confrontational crusades against nuclear testing and commercial **whaling**. The group has also garnered wide publicity for protesting various environmental abuses by hanging enormous banners from smokestacks, buildings, bridges, and the scaffolding used in the renovation of the Statue of Liberty.

Greenpeace is presently active in four broadly defined environmental issue areas—Atmosphere and Energy; Ocean Ecology and Forests; **Toxins**; and Disarmament. In the area of **Atmosphere** and Energy, Greenpeace works to eliminate widespread dependence upon **fossil fuels** and lobbies for laws and policies encouraging **energy efficiency** and **renewable energy** sources. The group is also working to halt the spread of **nuclear power** and the dumping of **radioactive waste** as well as to ban ozone-depleting **chemicals** such as **chlorofluorocarbons** (CFCs).

With regard to Ocean Ecology and Forests, Greenpeace seeks to protect both habitats and threatened **species**, including **whales**, harp **seals**, **dolphins**, **sea turtles**, **elephants**, and birds of prey. It works to discourage **overfishing** and other wasteful fishing practices, particularly the killing of dolphins in tuna nets. Greenpeace was instrumental in protecting Antarctica by persuading 23 nations to sign an accord banning all mining in Antarctica for at least 50 years. Supporting the principle of **biodiversity**, the group also works to protect tropical and temperate forests around the world. In 2002, the ships *MV Esperanza* and *Rainbow Warrior* stopped illegally logged timber from Africa and the Amazon from being imported.

In the area of Toxins, Greenpeace is especially concerned with stopping the use of unneeded **chlorine** in the bleaching of paper and with preventing the dumping of **hazardous waste** in **Third World** nations. Particularly concerned in recent years with **dioxin**, polychlorinated biphenyl (PCB), CFCs, and pesticides, the group regularly investigates, publicizes, and lobbies against chemical **pollution**. Greenpeace also conducts research on the effects of toxic substances on human beings and the environment and encourages **recycling** as a means of reducing pollution. In 1998, Greenpeace activists prevented a PVC plant from opening in Convent, Louisiana.

Also concerned with Disarmament issues, Greenpeace conducts research into the effects of warfare on human beings and the environment and advocates the global elimination of **nuclear weapons**. More immediately, the group also urges the cessation of all nuclear and chemical weapons testing and is trying to persuade the major powers to agree to a global ban on naval nuclear propulsion.

In an effort to avoid compromising its goals and activities, Greenpeace does not seek corporate or government funding. Nor does it become directly involved in the electoral process in any of the nations in which it is active. Greenpeace's frequently confrontational tactics have on occasion provoked angry responses from various governmental authorities, including the bombing and sinking of Greenpeace's flagship vessel *Rainbow Warrior* by agents of the French government in 1985. The *Rainbow Warrior* had been in New Zealand preparing to protest French nuclear testing in the South Pacific when it was sabotaged. In October 1992 one of Greenpeace's ships was seized by the Russian coast guard while investigating Russian nuclear waste dumping in Arctic waters.

[*Lawrence J. Biskowski*]

RESOURCES
ORGANIZATIONS
Greenpeace, 702 H Street NW, Washington, D.C. USA 20001 Toll Free: (800) 326-0959, , <http://www.greenpeaceusa.org>

Greens

The name given to those who engage in **green politics**. The term originated in Germany, where members of the environmentally-oriented Green Party were quickly dubbed *die Grünen*, or "the Greens." In the United States, greens refers not to a particular political party, but to any individual or group making environmental issues the central focus and main political concern. Thus the term covers a wide array of political perspectives and organizations, ranging from moderate or mainstream "light green" groups such as the **Sierra Club** and **Greenpeace** to more militant or "dark green" movements and direct-action organizations such as the **Sea Shepherd Conservation Society** and **Earth First!**, as well as ecofeminists, bioregionalists, social ecologists, and deep ecologists. Greens in the United States are divided over many issues. Some, for example, are in favor of organizing as interest groups to lobby for environmental legislation, while others reject politics in favor of a more spiritual orientation. Some greens (for example, social ecologists and ecofeminists) see their cause as connected to questions of social justice—the elimination of exploitation, militarism, racism, sexism, and so on—while others (deep ecologists, for instance) seek to separate their cause from such humanistic concerns, favoring a biocentric instead of an anthropocentric orientation. Despite such differences, however, all greens agree that the preservation and protection of the natural **environment** is a top priority and a precondition for every other human endeavor. *See also* Environmental ethics; Environmentalism

Jacques Grinevald
French university professor

Jacques Grinevald is recognized as a key expert on **biospheres** His studies and publications regarding the concept's initiator, **Vladimir Vernadsky** who published his work on the subject first in 1926, have comprised the substance of Grinevald's work as a scientific historian and philosopher.

Grinevald has been on the faculty of the University of Geneva (Switzerland) since 1987, serving as a part-time lecturer. He received a science degree in policies from the University Institute of High International Studies in Geneva in 1970; and his doctorate of 3rd cycle of philosophy, Paris X-Nanterre, in 1979. His early career included positions at the University of Geneva as an assistant in charge of research and teaching to the faculty of law, in addition to duties as the person in charge of press information; part-time lecturer position at the federal polytechnic school of Lausanne, serving as program man technique **environment** beginning in 1981; and, serving as an invited professor at the Federal Universidade of Rio de Janeiro (Brazil) in 1980 and 1984. His active schedule has taken him all over the world for conferences and seminars. Grinevald retains membership in several professional societies, including International Society for Ecological Economics; European Association for Bioeconomic Studies; and, World Council for the **Biosphere**.

Grinevald has been published extensively. His writings—among them chapters and articles in various books and journals—include, *The Greening of Europe* in 1990; "The Revolution Carnotienne: thermodynamics, economy and ideology," from the *European Review of Social Sciences*, 1976; and, "There is holistic total concept for deep and **ecology**: the Biosphere," for *Fundamenta Scientiae*, 1987. He has lectured and written on subjects that include the biosphere, the **greenhouse effect**, and famous scientists such as Stephen H. Schneider, a native New Yorker, whose research has focused on the greenhouse effect on civilization. He has been a regular contributor to a journal established by a group at the University of Geneva in 1990, *Strategies Energetiques Biosphere et Society* (Energy Strategies, Biosphere and Company) (SEBES). What began as a special volume became a publication devoted to the biosphere.

Writing for the publication *Etat De La Planete* (State of the Planet), Grinevald discussed the key issues of the biosphere. "This concept underlines the fac that the Life exceeds the individuals and is an ecological phenomenon of solidarity on various scales, microbial communities on a planetary scale of the **Biosphere**. It is the observer which decides scale of observation, so much [more] at the geographical level than at the temporal level. It is our world civilization which discovers the Biosphere as a phenomenon characteristic of the face of the Earth in cosmos. That implies a certain responsibility. The interdisciplinary and holistic concept of Biosphere associates astronomy, geophysics, **meteorology**, **biogeography**, evolutionary biology, geology, the geochemistry and, in fact, all science of the ground and the living."

[*Jane Spear*]

RESOURCES
BOOKS

Freeman, William. *The Biosphere*. San Francisco: Scientific American, 1970.

PERIODICALS

Grinevald, Jacques. "Biodiversity and Biosphere." *Etat De La Planete*. No. 1.
Grinevald, Jacques. "On Holistic Aconcept for Deep and Global Ecology: The Biosphere." *Fundamenta Scientiae* (1987): 197–226.

OTHER

University of Geneva. "The Effect Greenhouse of the Biosphere Thermo-industrial Revolution with Total Ecology." 1990 [cited July 2002]. &lt:http://unige.ch/sebes>.
University of Geneva. "Stephen H. Schneider." *SEBES*. 1995 [cited July 2002]. &lt:http://unige.ch/sebes>
University of Geneva Faculty Web Page. [cited June 2002]. &lt:http://w3.unige.ch>.

ORGANIZATIONS

University of Geneva, 24 street Rothschild, Geneva, Switzerland Email: Jacques.Grinevald@iued.unige.ch, www.unige.ch

Grizzly bear

The grizzly bear (*Ursus arctos*), a member of the family Ursidae, is the most widely distributed of all bear **species**. Although reduced from prehistoric times, its range today extends from Scandinavia to eastern Siberia, Syria to the Himalayan Mountains, and, in North America, from Alaska and northwest Canada into the northwestern portion of the lower 48 states. Even though the Russian, Alaskan, and Canadian populations remain fairly large, the grizzly bear population in the northwestern continental United States represents only about 1% of its former size of less than 200 years ago. Grizzly bears occupy a variety of habitats, but in North America they seem to prefer open areas including **tundra**, meadows, and coastlines. Before the arrival of Europeans on the continent, grizzlies were common on the Great Plains. Now they are found primarily in **wilderness** forests with open areas of moist meadows or **grasslands**.

Female grizzly bears vary in size from 200–450 lb (91–204 kg), whereas the much larger males can weigh up to 800 lb (363 kg). The largest individuals—from the coast of southern Alaska—weigh up to 1,720 lb (780 kg). Grizzly bears measure from 6.5–9 ft (2–2.75 m) tall when standing erect. To maintain these tremendous body sizes, grizzly bears must eat large amounts of food daily. They are omnivorous and are highly selective feeders. During the six or seven months spent outside their den, grizzly bears will consume up to 35 lb (16 kg) of food, chiefly vegetation, per day. They are particularly fond of tender, succulent vegetation, tubers, and berries, but also supplement their diet with insect grubs, small rodents, carrion, **salmon**, trout, young deer, and livestock, when the opportunity presents itself. In Alaska, along the McNeil River in particular, when the salmon are migrating upstream to spawn in July and August, it is not unusual to see congregations of dozens of grizzly bears, along the riverbank or in the river, catching and eating these large fish.

Grizzly bears breed during May or June, but implantation of the fertilized egg is delayed until late fall when the female retreats to her den in a self-made or natural cave, or a hollow tree. Two or three young are born in January,

A grizzly bear (*Ursus arctos*). (Reproduced by permission.)

February, or March, and are small (less than 1 lb/0.45 kg) and helpless. They remain in the den for three or four months before emerging, and stay with their mother for one and a half to four years. The age at which a female first reproduces, litter size, and years between litters are determined by nutrition, which induces females to establish foraging territories which exclude other females. These territories range from 10–75 mi² (26–194 km²). Males tend to have larger ranges extending up to 400 mi² (1,036 km²) and incorporate the territories of several females. Young females, however, often stay within the range of their mother for some time after leaving her care, and one case was reported of three generations of female grizzly bears living within the same range.

Grizzly bear populations have been decimated over much of their original range. **Habitat** destruction and **hunting** are the primary factors involved in their decline. The North American population, particularly in the lower 48 states, has been extremely hard hit. Grizzly bears numbered near 100,000 in the lower 48 states as little as 180 years ago, but, today, fewer than 1,000 remain on less than 2% of their original range. This population has been further fragmented into seven small, isolated populations in Washington, Idaho,

Montana, Wyoming, and Colorado. This decline and fragmentation makes their potential for survival tenuous. The U. S. **Fish and Wildlife Service** considers the grizzly bear to be threatened in the lower 48 states.

Little has been done to protect this declining species in the lower 48 states. In 1999 it was agreed to begin slowly reintroducing grizzlies into the 1.2 million acre (49,000 ha) area of the Selway-Bitterroot Wilderness on the border of Idaho and Montana. Unfortunately, the project was put on hold as of 2001 due to unfounded fear of the animal. Habitat loss due to timbering, road building, and development in this region is still a major problem and will continue to impact these threatened populations of bears.

[*Eugene C. Beckham*]

RESOURCES
BOOKS

Nowak, R. M., ed. *Walker's Mammals of the World* 5th ed. Baltimore: Johns Hopkins University Press, 1991.

PERIODICALS

"Interior Department Caves in to Grizzly Bear Scare." *USA Today*, August 6, 2001.

OTHER

Craighead Environmental Research Institute. [cited May 2002]. <http://www.grizzlybear.org>.

"New Grizzly Bear Recovery Plan: Bad News for Bears." *Wild Forever*, 1993.

Groundwater

Groundwater occupies the void space in a geological strata. It is one element in the continuous process of moisture circulation on Earth, termed the **hydrologic cycle**.

Almost all groundwater originates as surface water. Some portion of rain hitting the earth runs off into streams and lakes, and another portion soaks into the **soil**, where it is available for use by plants and subject to evaporation back into the **atmosphere**. The third portion soaks below the root zone and continues moving downward until it enters the groundwater. Precipitation is the major source of groundwater. Other sources include the movement of water from lakes or streams and contributions from such activities as excess **irrigation** and **seepage** from canals. Water has also been purposely applied to increase the available supply of groundwater. Water-bearing formations called aquifers act as reservoirs for storage and conduits for transmission back to the surface.

The occurrence of groundwater is usually discussed by distinguishing between a **zone of saturation** and a zone of **aeration**. In the zone of saturation the pores are entirely filled with water, while the zone of aeration has pores that are at least partially filled by air. Suspended water does occur in this zone. This water is called vadose, and the zone of aeration is also known as the **vadose zone**. In the zone of aeration, water moves downward due to gravity, but in the zone of saturation it moves in a direction determined by the relative heights of water at different locations.

Water that occurs in the zone of saturation is termed groundwater. This zone can be thought of as a natural storage area of **reservoir** whose capacity is the total volume of the pores of openings in rocks.

An important exception to the distinction between these zones is the presence of ancient sea water in some sedimentary formations. The pore spaces of materials that have accumulated on an ocean floor, which has then been raised through later geological processes, can sometimes contain salt water. This is called connate water.

Formations or strata within the saturated zone from which water can be obtained are called aquifers. Aquifers must yield water through **wells** or springs at a rate that can serve as a practical source of water supply. To be considered an **aquifer** the geological formation must contain pores or open spaces filled with water, and the openings must be large enough to permit water to move through them at a measurable rate. Both the size of pores and the total pore volume depends on the type of material. Individual pores in fine-grained materials such as clay, for example, can be extremely small, but the total volume is large. Conversely, in coarse material such as sand, individual pores may be quite large but total volume is less. The rate of movement from fine-grained materials, such as clay, will be slow due to the small pore size, and it may not yield sufficient water to wells to be considered an aquifer. However, the sand is considered an aquifer even though they yield a smaller volume of water because, they will yield water to a well.

The **water table** is not stationary but moves up or down depending on surface condition such as excess precipitation, **drought**, or heavy use. Formations where the top of the saturated zone or water table define the upper limit of the aquifer are called unconfined aquifers. The hydraulic pressure at any level with an aquifer is equal to the depth from the water table, and there is a type known as a water-table aquifer, where a well drilled produces a static water level which stands at the same level as the water table.

A local zone of saturation occurring in an aerated zone separated from the main water table is called a perched water table. These most often occur when there is an impervious strata or significant particle-size change in the zone of aeration which causes the water to accumulate. A confined aquifer is found between impermeable layers. Because of the confining upper layer, the water in the aquifer exists within the pores at pressures greater than the atmosphere. This is termed an artesian condition and gives rise to an **artesian well**.

Groundwater has always been an important resource, and it will become more so in the future as the need for good quality water increases due to urbanization and agricultural production. It has recently been estimated that 50% of the drinking water in the United States comes from groundwater; 75% of the nation's cities obtain all or part of their supplies from groundwater, and rural areas are 95% dependent upon it. For these reasons, it is widely believed that every precaution should be taken to protect groundwater purity. Once contaminated, groundwater is difficult, expensive, and sometimes impossible to clean up. The most prevalent sources of contamination are waste disposal, the storage, **transportation** and handling of commercial materials, mining operations, and nonpoint sources such as agricultural activities. *See also* Agricultural pollution; Aquifer restoration; Contaminated soil; Drinking water supply; Safe Drinking Water Act; Water quality; Water table draw-down

[*James L. Anderson*]

RESOURCES
BOOKS

Collins, A. G., and A. I. Johnson, eds. *Ground-Water Contamination: Field Methods.* Philadelphia: American Society for Testing and Materials, 1988.

Davis, S. N., and R. J. M. DeWiest. *Hydrogeology.* New York: Wiley, 1966.

Fairchild, D. M. *Ground Water Quality and Agricultural Practices.* Chelsea, MI: Lewis, 1988.

Freeze, R. A., and J. A. Cherry *Ground Water.* Englewood Cliffs, NJ: Prentice-Hall, 1979.

Ground Water and Wells. St. Paul: Edward E. Johnson, 1966.

Groundwater monitoring

Monitoring **groundwater** quality and **aquifer** conditions can detect contamination before it becomes a problem. The appropriate type of monitoring and the design of the system depends upon **hydrology, pollution** sources, and the population density and **climate** of the region. There are four basic types of groundwater monitoring systems: ambient monitoring, source monitoring, enforcement monitoring, and research monitoring.

Ambient monitoring involves collection of background **water quality** data for specific aquifers as a way to detect and evaluate changes in water quality. Source monitoring is performed in an area surrounding a specific, actual, or potential source of contamination such as a **landfill** or spill site. Enforcement monitoring systems are installed at the direction of regulatory agencies to determine or confirm the origin and concentration gradients of contaminants relative to regulatory compliance. Research monitoring **wells** are installed for detection and assessment of cause and effect relationships between groundwater quality and specific **land use** activities. *See also* Aquifer restoration; Contaminated soil; Drinking-water supply; Hazardous waste siting; Leaching; Water quality standards

Groundwater pollution

When contaminants in **groundwater** exceed the levels deemed safe for the use of a specific **aquifer** use the groundwater is considered polluted. There are three major sources of groundwater **pollution**. These include natural sources, waste disposal activities, and spills, leaks, and **nonpoint source** activities such as agricultural management practices.

All groundwater naturally contains some dissolved salts or minerals. These salts and minerals may be leached from the **soil** and from the aquifer materials themselves and can result in water that poses problems for human consumption, is considered polluted, or does not meet the **secondary standards** for **water quality**. Natural minerals or salts that may result in polluted ground water include chloride, nitrate, fluoride, iron and sulfate.

There are currently no feasible methods for the large-scale disposal of waste that do not have the potential for serious pollution of the **environment**, and there are a number of waste-disposal practices the specifically threaten groundwater. These include activities which range from separate **sewage treatment** systems for individual residences, used by 30% of the population in the United States, to the storage and disposal of industrial wastes. Many of the problems posed by industrial waste arise from the use of surface storage facilities that rely on evaporation for disposal. These facilities are also known as **discharge** ponds, and there are other types in which waste is treated to standards suitable for discharge to surface water. But in the use of both facilities the potential exists for the movement of contaminants into groundwater. Many of the numerous sanitary landfills in the country are in the same situation. Water moving down and away from these sites into groundwater aquifers carries with it a variety of **chemicals** leached from the material deposited in the landfills. The liquid that moves out of landfills is called leachate.

Agricultural practices also contribute to groundwater pollution, and there have been increases in nitrate concentrations and low-level concentrations of pesticides. For control of groundwater pollution, one of the most important agricultural practices is the management of **nitrogen** from all sources--fertilizer, nitrogen fixing plants, and **organic waste**. Once nitrogen is in the nitrate form it is subject to **leaching**, so it is important that the amount applied not exceed the crops' ability to use it. At the same time crops need adequate nitrogen to obtain high yields, and a good balance must be maintained. Low-level **pesticide** contamination occurs in areas where aquifers are sensitive to surface activity, particularly areas of shallow aquifers beneath rapidly **permeable** soils, and regions of "karst" **topography** where deep and wide range pollution can occur due to fractures in the bedrock.

Except in cases of **deep-well injection** waste or substances contained in sanitary landfills, most contaminants move from the land surface to aquifers. The water generally moves through an unsaturated zone, in which biological and chemical processes may act to degrade or change the contaminant. Plant uptake can also act to reduce some of the pollution. Once in the aquifer, how the contaminant moves with the water will depend the solubility of the compound, and the speed of contamination will depend on how fast water moves through the aquifer. Chemical and biological degradation of contaminants can occur in the aquifer, but usually at a slower rate than it does on the surface due to lower temperatures, less available oxygen, and reduced biological activity. In addition aquifer contaminants exist in

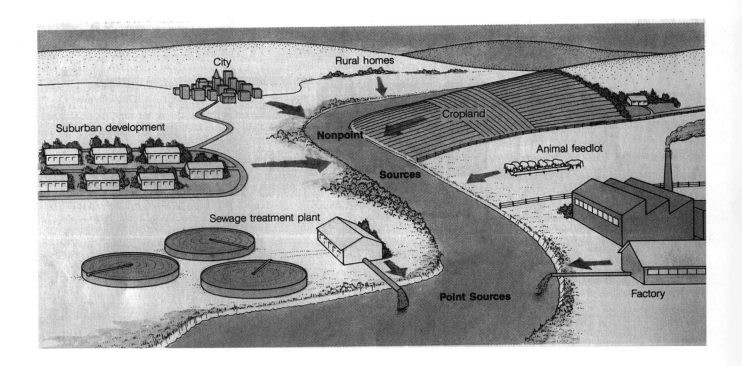

Sources of groundwater pollution. (McGraw-Hill Inc. Reproduced by permission.)

lower concentrations, diluted by the large water volume. Most pollution remains relatively localized in aquifers, since movement of the contaminants usually occurs in plumes that have definite boundaries and do not mix with the rest of the water. This does provide an advantage for isolation and treatment.

The types of chemicals that pollute groundwater are as varied as their sources. They range from such simple inorganic materials as nitrate from fertilizers, septic tanks, and **feedlots**, chloride from high salt, and **heavy metals** such as chromium from metal plating processes, to very complex organic chemicals used in manufacturing and household cleaners.

One of the main criteria in judging risk is public health. **Acute effects** from immediate exposure to a concentrated product is often well documented, but little written evidence exists to link physiological effects with long-term chronic exposure. About all that is available are epidemiological data that suggest possible effects, but are by no means conclusive. Environmental effects are even less well understood, but some have suggested that the best way to determine the potential danger is to look at how long a substance persists.

Those that remain the longest are most likely to pose long-term risks.

The most efficient way to protect groundwater is to limit activities in recharge areas. For confined aquifers it may be possible to control activities that can result in pollution, but this is extremely difficult for unconfined aquifers which are essentially open systems and subject to effects from any land activity. In areas of potential salt-water intrusion excess pumping can be regulated, and this can also be done where water is being used for **irrigation** faster than the recharge rate, so that the water is becoming saline. Another important activity for the protection of groundwater is the proper sealing of all **wells** that are not currently being used.

Classification of aquifers according to their predominant use is another management tool now employed in a number of states. This establishes water-quality goals and standards for each aquifer, and means that aquifers can be regulated according to their major use. This protects the most valuable aquifers, but leaves the problem of predicting future needs. Once an aquifer is contaminated, it is very expensive if not impossible to restore, and this management tool may have serious drawbacks in the future.

In rural areas of the United States, 95% of the population draws their drinking water from the groundwater supply. With a growing population, continued industrialization, and increasing agricultural reliance on the use of chemicals, many believe it is now more important than ever to protect groundwater. Contamination problems have been encountered in every state, but prevention is far more efficient and effective than restoration after damage has been done. Prevention can be achieved through regional planning and enforcement of state and federal regulations. *See also* Agricultural pollution; Aquifer restoration; Contaminated soil; Drinking-water supply; Feedlot runoff; Groundwater monitoring; Hazardous waste site remediation; Hazardous waste siting; Heavy metals and heavy metal poisoning; Waste management; Water quality; Water quality standards; Water treatment

[*James L. Anderson*]

RESOURCES
BOOKS

Freeze, R. A., and J. A. Cherry. *Ground Water.* Englewood Cliffs, NJ: Prentice-Hall, 1979.

Pye, V. I., R. Patrick, and J. Quarles. *Ground Water Contamination in the United States.* Philadelphia: University of Pennsylvania Press, 1983.

PERIODICALS

Hallberg, G. R. "From Hoes to Herbicides: Agriculture and Groundwater Quality." *Journal of Soil and Water Conservation* (November-December 1986): 358–59.

Growth curve

A graph in which the number of organisms in a population is plotted against time. Such curves are amazingly similar for populations of almost all organisms from bacteria to human beings and are considered characteristic of populations.

Growth curves typically have a sigmoid or S-shaped curve. When a few individuals enter a previously unoccupied area, growth is at first slow during the positive acceleration phase. The growth then becomes rapid and increases exponentially, called the logarithmic phase. The growth rate eventually slows down as environmental **resistance** gradually increases; this phase is called the negative acceleration phase. It finally reaches an equilibrium or saturation level. The final stage of the growth curve is termed the **carrying capacity** of the **environment**.

A good example of a species' growth curve is demonstrated by the sheep population in Tasmania. Sheep were introduced into Tasmania in 1800. Careful records of their numbers were kept, and by 1850 the sheep population had

reached 1.7 million. The population remained more or less constant at this carrying capacity for nearly a century.

The figures used to plot a growth curve—time and the total number in the population—vary from one **species** to another, but the shape of the growth curve is similar for all populations. Once a population has become established in a certain region and has reached the equilibrium level, the numbers of individuals will vary from year to year depending on various environmental factors. Comparing these variations for different species living in the same region is helpful to scientists who manage **wildlife** areas or who track factors that affect populations.

For example, a study of the population variations of the snowshoe hare and the lynx (*Lynx canadensis*) in Canada is a classic example of species interaction and interdependence. The peak of the hare population comes about a year before the peak of the lynx population. Since the lynx feeds on the hare, it is obvious that the lynx cycle is related to the hare cycle. This leads to a decline in the population of hares and secondarily to a decline in the lynx population. This permits the plants to recover from the overharvesting by the hares, and the cycle can begin again.

Growth curves are just one of the characteristics of populations. Other characteristics that are a function of the whole group and not of the individual members include population density, birth rate, death rate, age distribution, biotic potential, and rate of dispersion. *See also* Population growth

[*Linda Rehkopf*]

Growth, exponential
see Exponential growth

Growth limiting factors

There are a number of essential conditions which all organisms, both plants and animals, require to grow. These are known as growth factors. Plants, for example, require sunlight, water, and **carbon dioxide** in order to perform **photosynthesis**. They require nutrients such as **nitrogen, phosphorus**, and various trace elements in order to form tissues. The **environment** in which the plant is growing does not contain a unlimited supply of these growth factors. When one or more of them is present in levels or concentrations low enough to constrain the growth of the plant, it is known as a growth limiting factor. The rate or magnitude of the growth of any organism is controlled by the growth factor that is available in the lowest quantities. This concept is analogous to the saying that a chain is only as strong as its weakest link.

These factors limit **population growth**. If they did not exist, a population could increase exponentially, limited only by its own intrinsic lifespan. Growth limiting factors are essential to the traditional concept of **carrying capacity**, which rests on the assumption that the available resources limit the population that can be sustained in that area. Advances in technology have enabled people to increase the carrying capacity in certain areas by manipulating the growth limiting factors. Perhaps the best example of this is the use of fertilizers on farmland.

In the field of population **ecology**, identifying growth limiting factors is part of establishing the constraints and pressures on populations and predicting growth in various conditions. Algal growth in New York Harbor provides an example of the importance of identifying growth limiting factors. In New York Harbor, several billion gallons of untreated **wastewater** are released daily, bringing enormous quantities of nutrients and suspended solids into the water. Algae in the harbor take advantage of the **nutrient** loads and grow more than they would under nutrient-poor conditions. At the same time, however, the suspended solids and silts brought into harbor cause the water to become very turbid, limiting the amount of sunlight that penetrates it. Sunlight is rarely a growth limiting factor for algae; nutrients are usually what limits their growth, but in this case nutrients are in excess supply. This means that if **pollution control** in the harbor ever results in control of the turbidity in the water, there will probably be a sharp increase in the growth of algae.

Consideration of growth limiting factors is also very important in the field of **conservation biology** and **habitat** protection. If the goal is to protect a bird such as the heron, which may feed on fish from a lake and nest in upland trees nearby, limiting factors must be taken into account not only for the growth of the individual but also for the population. **Conservation** efforts must not be directed only toward ensuring there are enough fish in the lake. Enough trees must also be left uncut and undisturbed for nesting in order to address all of the growth requirements for the population. Regardless of how abundant the fish are, the number of herons will only grow to the extent allowed by the number of available nesting sites.

Environmentalists use growth limiting factors to distinguish between undisturbed ecosystems and unstable or stressed systems. In an **ecosystem** that has been distressed or disturbed, the nature of growth limiting factors changes, and these changes are often human-induced, as they are in New York Harbor. Though the change in circumstances may not always appear negative in impact, it still represents a shift away from the original balance, and it may have effects on other **species** or lead to subtle long-term changes in the system. Any cleanup or management strategy must

use these new growth limiting factors to identify the nature of the imbalance that has occurred and develop a procedure to restore the system to its original condition.

Growth limiting factors are extensively used in the field of **bioremediation**, in which **microbes** are used to clean up environmental contaminants by breakdown and **decomposition**. **Oil spills** are a good example. Bacteria that can break down and degrade oils are naturally present in small quantities in **soil**, but under normal conditions their growth is limited by both the availability of essential nutrients and the availability of oil. In the event of an oil spill on land, the only growth limiting factor for these bacteria is nutrients. Bioremediation scientists can add nitrogen and phosphorus to the soil in these circumstances to stimulate growth, which increases degradation of the oil. Techniques such as these, which use naturally occurring bacterial populations to control contamination, are still in development; they are most useful when the contaminants are present in high concentrations and confined to a limited area. *See also* Algal bloom; Decline spiral; Ecological productivity; Exponential growth; Food chain/web; Restoration ecology

[Usha Vedagiri and Douglas Smith]

RESOURCES
BOOKS

Mayer, G. *Ecological Stress and the New York Bight: Science and Management.* Columbia, SC: Estuarine Research Federation, 1982.
Smith, R. L. *Ecology and Field Biology.* New York: Harper and Row Publishers, 1980.

Growth, logistic
see **Logistic growth**

Growth, population
see **Population growth**

Grus americana
see **Whooping crane**

Guano

Manure created by flying animals that is deposited in a central location because of nesting habits. Guano can occur in caves from **bats** or in nesting grounds where large populations of birds congregate. Guano was frequently used as a source of **nitrogen fertilizer** prior to the time when nitrogen fertilizer was commercially manufactured from **natural gas**. *See also* Animal waste

Guinea worm eradication

In 1986, the world health community began a campaign to eliminate the guinea worm (*Dracunculus medinensis*) from the entire world. If successful, this will be only the second global disease ever completely eradicated (smallpox, which was abolished in 1977 was first), and the only time that a human parasite will have been totally exterminated worldwide. Known as the fiery serpent, the guinea worm has been a terrible scourge in many tropical countries. Dracunculiasis (pronounced dra-KUNK-you-LIE-uh-sis) or guinea worm disease starts when people drink stagnant water contaminated with tiny copepod water fleas (called cyclops) containing guinea worm larvae. Inside the human body, the worms grow to as long as 3 ft (1 m). After a year of migrating through the body, a threadlike adult worm emerges slowly through a painful skin blister. Most worms come out of the legs or feet but they can appear anywhere on the body. The eight to 12 weeks of continuous emergence are accompanied by burning pain, fever, nausea, and vomiting. Many victims bathe in a local pond or stream to soothe their fever and pain. When the female worm senses water, she releases tens of thousands of larvae, starting the cycle once again. Once the worms become established in local ponds, infections among people living nearby are at high risk for further infections.

As the worm emerges from the wound, it can be rolled around small stick and pulled out a few centimeters each day. Sometimes the entire worm can be extracted in a few days, but the process usually takes weeks. Unfortunately, if you pull too fast and the worm breaks off, the part left in your body can die and fester, leading to serious secondary infections. If the worm comes out through a joint, permanent crippling can occur. There is no cure for guinea worm disease once the larvae are ingested. There is no vaccine, and having been infected once doesn't give you immunity. Many people in affected villages suffer the disease repeatedly year after year. The only way to break the cycle is through behavioral changes. Community health education, providing clean water from **wells** or by filtering or boiling drinking water, eliminating water fleas by chemical treatment, and teaching infected victims to stay out of drinking supplies are the only solutions to this dreadful problem.

Although people rarely die as a direct effect of the parasite, the social and economic burden at both the individual and community level is great. During the weeks that worms are emerging, victims usually are unable to work or carry out family duties. This debilitation often continues for several months after worms are no longer visible. In severe cases, arthritis-like conditions can develop in infected joints, and the patient may be permanently crippled.

When the eradication campaign was started in 1986, guinea worms were endemic to 16 countries in sub-Saharan Africa as well as Yemen, India, and Pakistan. Every year about 3.5 million people were stricken and at least 100 million people were at risk. With the leadership of former U.S. President Jimmy Carter, a consortium of agencies, institutions, and organizations—the World Health Organization (WHO), UNICEF, the United Nations Development Program (UNDP), the **World Bank**, bilateral aid agencies, and the governments of many developed countries—banded together to fight this disease. Although complete success has not yet occurred, encouraging progress has been made. Already the guinea worm infections are down more than 96%. Pakistan was the first formerly infested country to be declared completely free of these **parasites**. Infection rates in Kenya, Senegal, Cameroon, Chad, India, and Yemen are down below 100 cases per year. More than 80% of all remaining cases occur in Sudan, where civil war, poverty, **drought**, and governmental **resistance** to outside aid have made treatment difficult.

An encouraging outcome of this crusade is the demonstration that public health education and community organization can be effective, even in some of the poorest and most remote areas. Village-based health workers and volunteers conduct disease surveillance and education programs, allowing funds and supplies to be distributed in an efficient manner. Once people understand how the disease spreads and what they need to do to protect themselves and their families, they do change their behavior. A great advantage of this community health approach is educating villagers about the importance of proper **sanitation** and clean drinking water is effective not only against dracunculiasis, but also can help eliminate **malaria**, shistosomiasis, and many other water-borne diseases.

[*William P. Cunningham Ph.D.*]

RESOURCES

OTHER

The Carter Center. *Eradicating Guinea Worm Disease.* 2002 [cited July 9, 2002]. <http://www.cartercenter.org/healthprograms>.

Center for Disease Control. *Fact Sheets: Dracunculiasis (Guinea Worm Disease).* January 2001 [cited July 9, 2002]. <http://www.cdc.gov/ncidod/dvrd/spb/mnpages/dispages/dracuculiasis.html>.

UNICEF. *Guinea Worm in Retreat.* March/April 1995 [cited July 9, 2002]. <http://www.unicef.org/pon95/heal0013.html>.

ORGANIZATIONS

The Carter Center, One Copenhill 453 Freedom Parkway, Atlanta, GA USA 30307 Email: carterweb@emory.edu, http://www.cartercenter.org

Gulf War syndrome

Approximately 697,000 U.S. service members were deployed to the Persian Gulf from January to March 1991 as part of a multinational effort to stop Iraq's attack against Kuwait. And while the war itself was short, a long battle has been taking place ever since by veterans, the government and scientists to determine what has caused "Gulf War Syndrome," a mysterious collection of symptoms reported by as many as 70,000 U.S. men and women who served in the war. They are joined by British veterans in their health complaints, and in smaller numbers by Canadians, Czechs, and Slovaks.

Gulf War Syndrome is a complex array of symptoms, including chronic fatigue, rashes, headaches, diarrhea, sleep disorders, joint and muscle pain, digestive problems, memory loss, difficulty concentrating, and depression. A small percentage of veterans have had babies born with twisted limbs, congestive heart failure, and missing organs. The veterans blamed these abnormalities on their service in the Gulf. The U.S. **Environmental Protection Agency** (EPA) has also found high rates for brain and nervous system cancers among these veterans, up to seven to 14 times higher than among the general population, depending on the age group. Considering that most soldiers and veterans are younger and in better physical shape than the general population, researchers find such figures more than surprising.

Collectively, these ailments suggest that neurological processes may have been altered, or immune systems damaged. While no single cause has been identified, various analyses of the Gulf War experience point to low-level exposure of chemical weapons, combined with other environmental and medical factors, as key contributors to the health problems triggered years after exposure.

The war was unique in the levels of physical and emotional stresses created for those who served, as well as for their families. A significant portion of troops were from the reserves, rather than active enlistees. Deployment occurred at unprecedented speed. Most troops were given multiple vaccinations that singularly do not have adverse effects; their combined effects were not tested before distribution. Detectors often signaled the presence of chemical weapons during the conflict, but were mostly ignored as inaccurate. The soldiers worked long hours in extreme temperatures, lived in crowded and unsanitary conditions where pesticides were used indiscriminately to rid areas of flies, snakes, spiders, and scorpions, and breathed and had dermal exposures to **chemicals** from the continuous oil fires—burning trash, feces, fuels, and solvents. Blazing sun, blowing sand and biting sandflies further increased the discomfort and stress of military life in the **desert**. Exposures to the various fumes often exceeded federal standards and World Health Organi-

zation health guidelines; these alone could have caused "permanent impairment," according to a 1994 National Institutes of Health report.

The U.S. military now admits it was inadequately prepared for chemical and biological warfare, which it knew Iraq had previously used. Three of four reserve units, for example, didn't have protective gear. The drug pyridostigmine bromide (PB, 3-dimethylaminocarbonyloxy-N-methylpyridium bromide) was given to almost 400,000 troops before and during the Gulf War to combat the effects of nerve gas, even though it is approved by the **Food and Drug Administration** (FDA) only for treatment of the neurological disorder myasthenia gravis. The FDA agreed on the condition that commanders inform troops what they were taking and what the potential side effects were. One survey, however, found that 63 of 73 veterans who had taken the drug did not receive such information. Records were not kept on who took which drugs or vaccines, as required by FDA and Defense Department guidelines.

While the Defense Department and other government agencies have spent more than $80 million to try to identify the cause of veterans' ailments, a privately funded team of toxicologists and epidemiologists may have discovered an explanation for at least some problems experienced by Gulf War veterans. Researchers treated chickens in 1996 with nonlethal doses of three chemicals veterans were exposed to: DEET (N,N-diethyl-m-toluamide) and chlorpyrifos (O,O-diethyl O-3,5,6-trichloropyridinyl phosphorothioate), used topically or sprayed on uniforms as insecticides, and the anti-nerve gas drug pyridostigmine bromide. They found that simultaneous exposure to two or more of the insecticides and drugs damaged the chickens' nervous system, even though none of the chemicals caused problems by itself. The range of symptoms the chickens developed is similar to those the veterans describe. A similar study by the Defense Department found that the chemicals were more toxic to rats when given together than individually. Follow up studies are underway to determine if this also holds for humans.

The researchers hypothesize that multiple chemicals overwhelmed the animals' ability to neutralize them. The **enzyme** butyrylcholinesterase, which circulates in the blood, breaks down a variety of nitrogen-containing organic compounds, including the three substances tested. But the anti-nerve gas drug, in particular, can monopolize the enzyme, preventing it from dealing with the insecticides. Those chemicals could then sneak into the brain, and cause damage they would not produce on their own.

Many veterans believe that, while the drugs and pesticides may have played a role in their ailments, so have chemical weapons. Troops could have been subjected to much more low-level exposure of chemical weapons than previously believed, either directly or via air plumes, because

75% of Iraq's chemical weapons production capability, along with 21 chemical weapons storage sites, were destroyed by allied air raids.

In addition, U.S. battalions blew up an Iraqi arms dump soon after the war was over, before many troops had left the Gulf. Khamisiyah, an enormous ammunition storage site, covered 20 mi^2 (50 km^2) with 100 ammunition bunkers and other storage facilities. Two large explosions were set off, one on March 4 and a second on March 10, 1991. Smaller demolition operations continued in the area through most of April 1991.

While the site was not believed to have contained chemical weapons at the time, the Defense Department admitted in June 1996 that the complex had included nerve and mustard gases. The Central Intelligence Agency also admitted in April 1997 that it knew in 1986 that thousands of mustard gas weapons had been stored at the Khamisiyah depot, but the agency failed to include it on a list of suspected sites provided to the Defense Department before the 1991 war, which led troops to assume it was safe to blow it up.

Weather data shows that upper-level winds in the gulf were blowing in a southerly direction during and after the bombing. Thus, vapors carried by these winds could have contaminated troops hundreds of miles away. A 1974 report, *Delayed Toxic Effects of Chemical Warfare Agents*, found that chemicals weapons plant workers suffer as many chronic symptoms as those now suffered by Gulf War veterans, including neurological, gastrointestinal and heart problems, loss of memory and a greater risk of **cancer**; exposure to these chemicals may also create birth effects in children. A 1995 study by a British medical researcher found many of the same symptoms in **Third World** people exposed to **organophosphate** insecticides, like DEET used in the Gulf War, which are diluted versions of chemical weapons.

While British, Canadian and Slovak veterans have reported similar ailments, albeit in smaller numbers, no French veterans have complained of such illnesses, despite extensive publicity. This is also providing valuable clues to the U.S. veterans' maladies, in several ways. For example, the French did not use many of the vaccines that the British and Americans used, including pyridostigmine bromide. French camps were not sprayed with insecticides as a preventive measure, rather only when needed to control **pest** populations. When they did spray, they did not use organophosphates. Finally, the French were nowhere near the Khamisiyah munitions depot when the destruction occurred.

In February 1997, a series of study results established the most definitive links between Gulf War syndrome and chemicals to date. The research identifies six "syndromes," or clusters of like symptoms in discrete groups of veterans, and associates each with distinct events during the war. Troops who reported exposure to chemical weapons, for example, are likely to suffer from confusion, balance problems, impotence and depression. Other sets of symptoms correspond to the use of insect repellants and anti-nerve gas drugs. While not conclusive, these findings will likely spur further research into the effects of low-level exposure to certain chemicals.

Such research was advocated by the presidential advisory committee, a 12-member panel of veterans, scientists, and health care and policy experts established in 1995. The committee held 18 public meetings between August 1995 and November 1996 to investigate the nature of Gulf War veterans' illnesses, health effects of Gulf War risk factors, and the government's response to Gulf War illnesses. While the committee's final report in January 1997 concluded that no single, clinically recognizable disease can be attributed to Gulf War service, it recommended additional research on the long-term health effects of low-level exposures to chemical weapons, on the synergistic effects of pyridostigmine bromide with other Gulf War risk factors, and on the body's physical response to stress.

While the debate continues, the Veterans Affairs and the Defense Departments are providing free medical help to any veteran who believes he or she is suffering from Gulf War Syndrome. In January 1997, President Clinton proposed new regulations that would extend the time available to veterans to prove their disabilities are related to Gulf War service from two to 10 years. He also initiated a presidential review to ensure that in any future deployments the health of service men and women and their families is better protected.

Definitive answers as to the causes and treatments for veterans' ailments may be years away. What is clear is that the complex biological, chemical, physical and psychological stresses of the **Persian Gulf War** appear to have produced a variety of complex adverse health effects. No single disease or syndrome is apparent, but rather multiple illnesses with overlapping symptoms and causes. If what had been considered acceptable trace levels of chemical agents in the war **environment** are found to be harmful, the U.S. military will have to revamp the way it protects its forces against even those tiny amounts. Tragically, that would mean that not only did "friendly fire" account for nearly 25% of the 146 U.S. deaths, but also that allied actions were responsible for the war's most persistent and haunting pain.

[*Sally Cole-Misch*]

RESOURCES
PERIODICALS

Abou-Donia, M.B., et. al. "Increased Neurotoxicity Following Concurrent Exposure to Pyridostigmine Bromide, DEET and Chlorpyrifos." *Fundamental and Applied Toxicology* 34, no. 190 (1996): 201–.222.

Gully erosion in Australia. (Photograph by A. B. Joyce. Photo Researchers Inc. Reproduced by permission.)

Barry, J., and R. Watson. "Scent of War." *Newsweek*, September 30, 1996, 38–39.

Cary, P., and M. Tharp. "The Gulf War's Grave Aura." *U.S. News & World Report*, 121, no. 2, July 8, 1996, 38–39.

Cowley, G., and M. Hager. "Poisoned in the Gulf?" *Newsweek* 127, no. 18, April 29, 1996, 74.

———. "A Gulf War Cover-up?" *Newsweek*, November 11, 1996, 48–49.

"Darkness at Noon." *The Economist* 342, no. 7999 (January 1, 1997): 71–74.

National Institutes of Health Technology Assessment Workshop Panel. "The Persian Gulf Experience and Health." *Journal of the American Medial Association* 272, no. 5 (August 3, 1994): 391–395.

Pennisi, E. "Chemicals Behind Gulf War Syndrome." *Science* 272, no. 5261 (April 26, 1996): 479–480.

Shenon, P. "CIA Says It Knew of Iraq Chemicals." *Detroit Free Press*, April 10, 1997, 1–3A.

Thompson, M. "The Silent Treatment." *Time*, 148, no. 28, December 23, 1996, 33–37.

Waldman, A. "Credibility Gulf." *The Washington Monthly*, 28, no. 12, December 1996, 28–35.

Gullied land

Areas where all diagnostic **soil** horizons have been removed by flowing water, resulting in a network of V-shaped or U-shaped channels. Generally, gullies are so deep that extensive reshaping is necessary for most uses. They cannot be crossed with normal farm machinery. While gullied land can occur on any land, they are often most prevalent on loess, sandy, or other soils with low cohesion. *See also* Erosion; Soil profile; Soil texture

Gymnogyps californianus
see **California condor**

Gypsy moth

The gypsy moth (*Portheria dispar*), a native of Europe and parts of Asia, has been causing both ecological and economic damage in the eastern United States and Canada since its introduction in New England in the 1860s.

In 1869, french entomologist Leopold Trouvelet brought live specimens of the insect to Medford, Massachusetts for experimentation with silk production. Several individual specimens escaped and became an established popula-

Gypsy moth females laying eggs. (Photograph by Michael Gadumski. Photo Researchers Inc. Reproduced by permission.)

tion over the next 20 years. The destructive abilities of the gypsy moth became readily apparent to area residents, who watched large sections of forest be destroyed by the larvae. From the initial infestation in Massachusetts, the gypsy moth spread throughout the northeastern United States and southeastern Canada. As of 2001, the following states were quarantined: Connecticut, Delaware, Maryland, Michigan, New Hampshire, New Jersey, New York, Pennsylvania, Rhode Island, and Vermont. More that two-thirds of Virginia is quarantined, as well as sections of Ohio, Indiana, Maine, North Carolina, Wisconsin, and Wyoming. The U. S. Department of Agriculture reports that virtually all areas that have experienced an invasion of gypsy moths continue to have extremely high levels of infestation.

Gypsy moths have a voracious appetite for leaves, and the primary environmental problem caused by them is the destruction of huge areas of forest. Gypsy moth caterpillars defoliate a number of **species** of broadleaf trees including birches, larch, and aspen, but prefer the leaves of several species of oaks, though they have also been found to eat some evergreen needles. One caterpillar can consume up to one square foot of leaves per day. In 2001, 84.9 million acres (34.4 million ha) had been defoliated by these insects. The sheer number of gypsy moth caterpillars produced in one generation can create other problems as well. Some areas become so heavily infested that the insects have covered houses and yards, causing psychological difficulties as well as physical.

There are few natural predators of the gypsy moth in North America, and none that can keep its population under control. Attempts have been made since the 1940s to control the insect with pesticides, including DDT, but these efforts have usually resulted in further contaminating of the **environment** without controlling the moths or their caterpillars. Numerous attempts have also been made to introduce species from outside the region to combat it, and almost 100 different natural enemies of the gypsy moth have been introduced into the northeast United States. Most of these have met, at best, with limited success. Recent progress has been made in experiments with a Japanese fungus that attacks and kills the gypsy moth. The fungus enters the body of the caterpillar through its pores and begins to destroy the insect from the inside out. It is apparently non-lethal to all other species in the infested areas, and its use has met with limited success in parts of Rhode Island and upstate New York. It remains unknown however, whether it will control the gypsy moth, or at least stem the dramatic population increases and severe infestations.

[*Eugene C. Beckham*]

RESOURCES
PERIODICALS

Pond, D., and W. Boyd. "Gypsy on the Move." *American Forests* 98 (March–April 1992): 21–25.

OTHER

Gypsy Moth in North America. September 15, 1998 [cited May 2002]. <http://www.fs.fed.us/ne/morgantown/4557/gmoth>.

Gypsy Moth in Virginia. April 10, 2001 [cited May 2002]. <http://www.gypsymoth.ento.vt.edu/vagm/index.html>.

U.S. Department of Agriculture. "Plant Protection and Quarantine: Gypsy Moth Quarantine." *Animal and Plant Health Inspection Service.* June 15, 2001 [cited May 2002]. <http://www.aphis.usda.gov/ppq/maps/gypmoth.pdf>.

H

Arie Jan Haagen-Smit (1900 – 1977)
Dutch atmospheric chemist

The discoverer of the causes of **photochemical smog**, Haagen-Smit was one of the founders of atmospheric chemistry, but first made significant contributions to the chemistry of essential oils. Haagen-Smit was born in Utrecht, Holland, in 1900, and graduated from the University of Utrecht. He became head assistant in organic chemistry there and later, served as a lecturer until 1936. He came to the United States as a lecturer in biological chemistry at Harvard, then became associate professor at the California Institute of Technology in Pasadena. He retired as Professor of Biochemistry in 1971, having served as executive officer of the department of biochemistry and director of the plant **environment** laboratory. Haagen-Smit also contributed to the development of techniques for decreasing **nitrogen** oxide formation during **combustion** in electric **power plants** and autos, and to studies of damage to plants by **air pollution**.

While early workers on the **haze** and eye irritation that developed in Los Angeles, California, tried to treat it as identical with the **smog** (**smoke** + fog) then prevalent in London, England, Haagen-Smit knew at once it was different. In his reading, Haagen-Smit had encountered a 1930s Swiss patent on a process for introducing random oxygen functions into **hydrocarbons** by mixing the hydrocarbons with nitrogen dioxide and exposing the mixture to ultraviolet light. He thought this mixture would smell much more like a smoggy day in Los Angeles than would some sort of mixture containing **sulfur dioxide**, a major component of London smog. He followed the procedure and found his supposition was correct. Simple analysis showed the mixture now contained **ozone**, organic peroxides, and several other compounds. These findings showed that the sources of the problems of Los Angeles were **petroleum** refineries, **petrochemical** industries, and ubiquitous **automobile** exhaust.

Haagen-Smit was immediately attacked by critics, who set up laboratories and developed instruments to prove him wrong. Instead the research proved him right, except in minor details.

Haagen-Smit had a long and distinguished career. In addition to his work on the chemistry of essential flower oils and famous findings on smog, he also contributed to the chemistry of plant hormones and plant alkaloids and the chemistry of **microorganisms**. He was a founding editor of the *International Journal of Air Pollution*, now known as *Atmospheric Environment*, one of the leading air **pollution** research journals. Though he found the work uncongenial, he stayed with it for the first year, then retired to the editorial board, where he served until 1976.

Once it was obvious that he had correctly identified the cause of the Los Angeles smog, he was showered with honors. These included membership in the National Academy of Science, receipt of the Los Angeles County Clean Air Award, the Chambers Award of the **Air Pollution Control** Association (now the **Air and Waste Management Association**), the Hodgkins Medal of the Smithsonian Institution, and the National Medal of Science. In his native Netherlands he was made a Laureate of Labor by the Netherlands Chemical Society, and Knight of the Order of Orange Nassau.

[*James P. Lodge Jr.*]

RESOURCES
PERIODICALS

Lodge, J. P. "Obituary: A. J. Haagen-Smit." *Nature* 267 (1977): 565–566.

Habitat

Refers to the type of **environment** in which an organism or **species** occurs, as defined by its physical properties (e.g., rainfall, temperature, topographic position, **soil texture**, **soil** moisture) and its chemical properties (e.g., soil acidity, concentrations of nutrients and **toxins**, oxidation reduction status). Some authors include broad biological characteristics in their definitions, for instance forest versus **prairie** habitats, referring to the different types of environments occupied by trees and grasses. Within a given habitat there may be different micro-habitats, for example the hummocks and hollows on bogs or the different soil horizons in forests.

Habitat conservation plans

Protection of the earth's **flora** and **fauna**, the myriad of plants and animals that inhabit our planet, is totally dependent upon preserving their habitats. This is because a **habitat**, or natural **environment** for a specific variety of plant or animal, provides everything necessary for sustaining life for that **species**. Habitat **conservation** is part of a larger picture involving interdependency between all living things. Human life has been sustained since its dawning through the utilization of both plants and animals for food, clothing, shelter and medicines. It follows then that the destruction of any specie's environment, which will result in the eventual destruction of the species itself, adversely affects us all.

Habitat destruction occurs for a number of reasons. Human industry has usually resulted in **pollution** that has often destroyed the balance of natural elements in **soil**, water and air that are necessary to life. The need for forest products such as lumber has threatened woodlands in more ways than one. A lumbering practice called **clear-cutting**, in addition to over-harvesting a forest, leaves behind barren ground that results in **erosion** and threatens species dependent on the vegetation that grows on the forest floor. This wearing away of soil often results in negative changes to nearby streams and thus the water supply to multitudes of living things. The introduction of non-native species into an area can also threaten a habitat, as plants and animals with no natural enemies may thrive abnormally. This in turn will throw off the delicate balance and natural biological controls on **population growth** that each environment provides for its inhabitants.

As the understanding of these facts became more widespread, the demand for habitat conservation throughout the world increased. In the United States, in 1973, Congress passed the **Endangered Species Act** to protect both at risk species and their environments. In order to control activities by private and non-federal government landowners that might disturb habitats, the Act included a section outlining Habitat Conservation Plans (HCPs).

These HCPs were not implemented without a good deal of controversy. Some environmentalists were critical of the plans, believing that they failed to actually preserve the habitats and species they were designed to protect. Both conservationists and landowners and industrialists argued that these HCPs' were based upon faulty science and invalid and insufficient data. Impartial reviews did indicate that there were flaws. One cited weakness was that a species could theoretically be added to the endangered list but have no modification made in the plan to protect that species' living-space.

It is clearly a measure of this controversy that in the law's first twenty years, only fourteen such plans were developed and approved nationwide. During the administration of President Bill Clinton, the federal government revised policies to encourage more numerous and more effective HCP applications. A more responsive to change *"No surprises"* policy helped encourage participation in habitat conservation plans.

This policy recognizes that **natural resources** change and environments require continual monitoring and alterations in plans, that those trying to balance environments can do so with **adaptive management**. Such alterations encouraged participation in HCPs, and by 1997, more than 200 plans covering nine million acres of land had been approved. As of April of 2002, the United States **Fish and Wildlife Service** notes that nearly double that number, 379 plans covering nearly 30 million acres and protecting more than 200 endangered and threatened species have now been approved.

One positive example of the possibilities created by such plans is the work of the Plum Creek Timber Company, a nationwide timberland company whose corporate offices are located in Seattle, Washington. Self-described as "the second largest private timberland owner in the United States, with 7.8 million acres located in the northwestern, southern and northeastern regions of the country", the company lists many initiatives it has undertaken to preserve the environments under its charge. In the early 1990's, Plum Creek Timber Company developed a plan to provide a *ladder-like* framework in Coquille River tributaries in Oregon to aid fish in reaching upper areas blocked for many years by a culvert. Spawning surveys completed the next year showed that coho **salmon** and steelhead trout, for the first time in forty years, were present above the culvert.

HCPs were created to focus attention on the problem of declining **wildlife** on land not owned and protected by the federal government and to attempt to maintain the **biodiversity** so necessary for all life. This goal would ideally assure that land is developed in such a way that it serves both the needs of the landowner and of threatened wildlife. However, in reality, it is not always possible to achieve such a goal. Often the more appropriate aim, if habitat conservation is to be successful, will be total protection of a wildlife environment even at the price of banning all development.

[*Joan M. Schonbeck*]

RESOURCES

BOOKS

Gilbert Oliver L., and Penny Anderson. *Habitat Creation and Repair.* New York: Oxford University Press, 1998.

Walters, C. *Adaptive Management of Renewable Resources.* New York: Mac-Millan, 1986.

PERIODICALS

Mann, C., and M. Plummer. "Qualified Thumbs Up for Habitat Plan Science." *Science* 278, no. 5346 (December 19, 1997): 2052.

ORGANIZATIONS

Fish and Wildlife Reference Service, 54300 Grosvenor Lane, Suite 110, Bethesda, MD USA 20814 (301) 492-6403, Toll Free: (800) 582-3421, Email: fw9fareferenceservice@fws.gov, <http://www.lib.iastate.edu/collections/db/usfwrs.html>

National Wildlife Federation, 111000 Wildlife Center Drive, Reston, VA USA 20190 (703)438-6000, Toll Free: (800) 822-9919, Email: info@nwf.org, <http://www.nwf.org>

Habitat fragmentation

The **habitat** of a living organism, plant, animal, or microbe, is a place, or a set of environmental conditions, where the organism lives. Net loss of habitat obviously has serious implications for the survival and well-being of dependent organisms, but the nature of remaining habitat is also very important. One factor affecting the quality of surviving habitat is the size of remaining pieces. Larger areas tend to be more desirable for most **species**. Various influences, often a result of human activity, cause habitat to be divided into smaller and smaller, widely separated pieces. This process of habitat fragmentation has profound implications for species living there.

Each patch created when larger habitat areas are fragmented results in more edge area where patches interface with the surrounding **environment**. These smaller patches with a relatively large ratio of edge to interior area have some unique characteristics. They are often distinguished by increased predation when predators are able to hunt or forage along this edge more easily. The decline of songbirds throughout the United States is due in part to the increase of the brown-headed cowbird competing with other birds along habitat edges. The cowbird acts as a parasite by laying its eggs in other birds' nests and leaving them for other birds to hatch and raise. After hatching, the young cowbirds compete with the smaller birds of the nest, almost always killing them.

In the smaller patches formed from fragmentation, habitat is less protected from adverse environmental events, and a single storm may destroy the entire area. A disease outbreak may eliminate the entire population. When the number of breeding adults becomes very low, some species can no longer reproduce successfully.

Some songbirds found in the United States are declining in number as their habitat shrinks or disappears. When they migrate south in the winter they find that habitat is more scarce and fragmented. When they return from the tropics in the spring, they discover that the nesting territory that they used the previous year has disappeared.

Species dispersal is decreased as organisms must travel farther to go from one habitat area to another, increasing their exposure to predation and possibly harmful environmental conditions. Populations become increasingly insular as they become separated from related populations, losing the genetic benefits of a larger interbreeding population.

Road building often divides habitat areas, seriously disrupting **migration** of some mammals and herptiles (**frogs**, snakes, and turtles). Large swaths of land used by modern freeways are particularly effective in this regard. In earlier times, railroads built across the Great Plains to connect the west coast of the United States with states east of the Mississippi, divided **bison** habitat and hampered their migration from one grazing area to another. This was one of the factors that led to their near **extinction**.

Habitat fragmentation, usually a result of human activity, is found in all major habitat types around the world. Rain forests, **wetlands**, **grasslands**, and hardwood and conifer forests are all subject to various degrees of fragmentation. Globally, rain forests are currently by far the most seriously impacted **ecosystem**. Because they contain 50% or more of the world's species, the resulting number of species extinctions is particularly disturbing. It is estimated that 25% of the world's rainforests disappeared during the twentieth century, and another 25% were seriously fragmented and degraded. In the last two centuries, nearly all of the **prairie** grassland once found in the United States has disappeared. Remaining remnants occur in small, scattered, isolated patches. This has resulted in the extinction or near extinction of many plant and animal species.

Ability to survive habitat fragmentation and other environmental changes varies greatly among species. Most find the stress overwhelming and simply disappear. A few of the common species that have been very successful in adapting to changing conditions include animals such as the opossum, raccoon, gray squirrel, and European starling. Plant examples include dandelions, crab grass, creeping Charlie, and many other weed species. The wetland invader, **purple loosestrife**, originally imported from Europe to the United States, is rapidly spreading into disrupted habitat previously occupied by native emergent aquatic vegetation such as reeds and cattails. Animal species that once found a comfortable home in cattail stands must move on.

[Douglas C. Pratt Ph.D.]

RESOURCES
BOOKS

Cunningham, W. P., and Cunningham, M. A. *Principals of Environmental Science: Inquiry and Applications.* McGraw-Hill, 2002.

Ernst Heinrich Philipp August Haeckel (1834 – 1919)

German naturalist, scientist, biologist, philosopher, and professor

Ernst Haeckel was born in Potsdam, Germany. As a young boy he was interested in **nature**, particularly botany, and kept a private herbarium, where he noticed that plants varied more than the conventional teachings of his day advocated. Despite these natural interests, he studied medicine—at his father's insistence—at Würzburg, Vienna, and Berlin between 1852 and 1858. After receiving his license he practiced medicine for a few years, but his desire to study *pure science* won over, and he enrolled at the University of Jena to study zoology. Following completion of his dissertation, he served as professor of zoology at the university from 1862 to 1909. The remainder of his adult life was devoted to science.

Haeckel was considered a liberal non-conformist of his day. He was a staunch supporter of Charles Darwin, one of his contemporaries. Haeckel was a prolific researcher and writer. He was the first scientist to draw a "family tree" of animal life, depicting the proposed relationships between various animal groups. Many of his original drawings are still used in current textbooks. One of his books, The Riddle of the Universe (1899), exposited many of his theories on **evolution**. Prominent among these was his theory of recapitulation, which explained his views on evolutionary vestiges in related animals. This theory, known as the *Biogenic Law*, stated that "ontogeny recapitulates phylogeny"—the development of the individual (ontogeny) repeats the history of the race (phylogeny). In other words, he argued that when an embryo develops, it passes through the various evolutionary stages that reflect its evolutionary ancestry. Although this theory was widely prevalent in biology for many years, scientists today consider it inaccurate or only partially correct. Some even argue that Haeckel falsified his diagrams to prove his theory.

In **environmental science**, Haeckel is perhaps best known for coining the term **ecology** in 1869, which he defined as "the body of knowledge concerning the economy of nature—the investigation of the total relationship of the animal both to its organic and its inorganic **environment** including, above all, its friendly and inimical relations with those animals and plants with which it directly or indirectly comes into contact—in a word, ecology is the study of all

Ernst Haeckel. (Corbis-Bettmann. Reproduced by permission.)

those complex interrelations referred to by Darwin as the conditions for the struggle for existence."

[*John Korstad*]

RESOURCES
BOOKS

Gasman, D. *The Scientific Origins of National Socialism: Social Darwinism in Ernst Haeckel and the German Monist League.* New York: American Elsevier, 1971.
Hitching, F. *The Neck of the Giraffe: Darwin, Evolution, and the New Biology.* Chula Vista, CA: Mentor, 1982.
Smith, R. E. *Ecology and Field Biology.* 4th ed. New York: Harper and Row, 1990.

Half-life

A term primarily used to describe the physical half-life, how **radioactive decay** processes cause unstable atoms to be transformed into another element, but can also refer to the biological half-life of substances that are not radioactive.

Specifically, the physical half-life is the time required for half of a given initial quantity to disappear (or be converted into something else). This description is useful because radioactive decay proceeds in such a way that a fixed percentage of the atoms present are transmuted during a given period of time (a second, a minute, a day, a year). This means that many atoms are removed from the population when the total number present is high, but the number removed per unit of time decreases quickly as the total number falls.

For all practical purposes, the number of radioactive atoms in a population never reaches zero because decay affects only a fraction of the number present. Some infinitesimal number will still be present even after an infinite number of half-lives because with each time period half of what was present before decays and is lost from the sample. Therefore, determining the point at which half of the original number disappears (the half-life) is usually the most accurate way of describing this process. *See also* Radioactivity

Haliaeetus leucocephalus
see **Bald eagle**

Halons

Halons are **chemicals** that contain **carbon**, fluorine, and **bromine**. They are used in fire extinguishers and other firefighting equipment. Because of their bromine content, halons can destroy **ozone** molecules(O_3) very effectively, thereby contributing to the depletion of ozone and the creation of holes in the ozone layer of the **stratosphere**. The ozone layer is located 10–28 mi (16–47 km) above the surface of the earth and it protects humans and the **environment** from the Sun's ultraviolet-B radiation. Halons account for approximately 20% of the ozone depletion.

Halons have been used since the 1940s, when they were discovered by U.S. Army researchers looking for a fire-extinguishing agent to replace carbon tetrachloride. Halons are very effective against most types of fires, are nonconductive, and dissipate without leaving a residue. They are also economical, very stable, and safe for human use.

Halons consist of carbon atom chains with attached **hydrogen** atoms that are replaced by the halogens fluorine (F) and bromine (Br). Some also contain **chlorine** (Cl). Halon-1211 ((bromochlorodifluoromethane, CF_2ClBr), halon-1301 (bromotrifluoromethane, CF_3Br), and halon-2402 (dibromotetrafluoroethane, $C_2F_4Br_2$) are the major fire-suppressing halons. Halon-1211 is discharged as a liquid and vaporizes into a cloud within a few feet. Halon-1301 is stored as a liquid but discharges as a gas. Halons suppress fires because they bond with the free radicals and intermediates of the decomposing fuel molecules that propagate fire, thus rendering the fuel inert. They also lower the temperature the fire.

Halons may take up to seven years to drift up and distribute themselves throughout the stratosphere, with the highest concentrations over the poles. High-energy **ultraviolet radiation** breaks their bromine and chlorine bonds thus releasing these very reactive halogen molecules, which in turn break down the ozone molecules and react with free oxygen to interfere with ozone creation. Although chlorine is more abundant, bromine is more than 100 times more damaging to ozone.

Halons are categorized as class I ozone-depleting substances, along with **chlorofluorocarbons** (CFCs) and other substances with ozone-depleting potentials (ODPs) of 0.2 or greater. The Ozone Depletion Potential(ODP)is a number that refers to the amount of ozone depletion caused by a substance. Halon-1211 has an ODP of 3.0 and an atmospheric lifetime of 11 years. It also has a global warming potential (GWP) of 1300. Halon-1301 has an ODP of 10.0, an average lifetime of 65 years, and a GWP of 5600 to 6900. Halon-2402 has an ODP of 6.0. In contrast, CFC-11, a common refrigerant, has an ODP of 1. Halon-1211 and halon-1301 are the most common halons in the United States. Halon-2402 is widely used in Russia and the developing world. Although total halon production between 1986 and 1991 accounted for only about 2% of the total production of class I substances, it accounted for about 23% of the ozone depletion caused by class I substances.

Halon production in the United States ended on December 31, 1993 because they contribute to ozone depletion. Under the Montreal Protocol on Ozone Depleting Substances, first negotiated in 1987 and now including more than 172 countries, halons became the first ozone-depleting substances to be phased out in industrialized nations, with production stopped in 1994. Under the **Clean Air Act**, the United States banned the production and importation of halons as of January 1, 1994. The use of existing halons in fire protection systems continues and recycled halons can be purchased to recharge such systems. It is estimated that about 50% of all halons ever produced currently exist in portable fire extinguishers and firefighting equipment. The United States holds 40% of the world's supply of halon-1301. In 1997, approximately 1,080 tons (977 metric tons) of halon-1211 and 790 tons (717 metric tons) of halon-1301 were released in the United States. In 1998, the U.S. **Environmental Protection Agency** (EPA) prohibited the venting of halons during training, testing, repair, or disposal of equipment, and banned the blending of halons, to prevent the accumulation of nonrecyclable stocks.

The **European Union** has gone beyond the Montreal Protocol, banning the sale and non-critical use of halons after December 31, 2002, and requiring the decommissioning of non-critical halon systems by December 31, 2003.

Halon production and consumption continues in developing countries, especially China, the Republic of Korea, India, and Russia. Under the Montreal Protocol, developing countries were to freeze halon consumption by January 2002. A 50% reduction in halon consumption is required by January 2005 and the complete halt of halon production and use is slated for January 2010. However halon-1211 emissions increased by about 25% between 1988 and 1999. Since most of the increased manufacture and release of halon-1211 occurs in China, the **United Nations Environment Programme** is helping China to phase-out production by 2006.

Atmospheric levels of halon-1202, which is not covered by the Montreal Protocol, increased fivefold between the late 1970s and 1999, and by 17% annually in the late 1990s. It is not known whether this increase is a byproduct of the inefficient production of other halons in developing countries or whether some countries are manufacturing it for military applications. Halon-1202's ODP is about one-half that of the common CFCs.

Alternatives are now available for most halon applications. Existing halon supplies from fire suppression systems are being recycled for critical uses where no alternative exists.

[*Margaret Alic Ph.D.*]

RESOURCES

BOOKS

Peterson, Eric. *Standards and Codes of Practice to Eliminate Dependency on Halons: Handbook of Good Practices in the Halon Sector.* Paris: United Nations Environment Programme, 2001.

PERIODICALS

Chang, Lisa. "Regulating Halon Emissions." *NFPA Journal* 93, no. 4 (July/August 1999): 90.

Poynter, Ronald J. "Halon Replacements: Chemistry and Applications." *Professional Safety* 44, no. 3 (March 1999): 46–50.

Zurer, Pamela. "Slow Road to Ozone Recovery." *Chemical and Engineering News* 77, no. 17 (April 26, 1999): 8–9.

OTHER

U.S. Environmental Protection Agency. *Ozone Depletion.* May 15, 2002 [cited May 19, 2002]. <http://www.epa.gov/ozone/index.html>.

ORGANIZATIONS

Halon Alternatives Research Corporation, Halon Recycling Corporation, 2111 Wilson Boulevard, Eighth Floor, Arlington, VA USA 22201 (703) 524-6636, Fax: (703) 243-2874, Toll Free: (800) 258-1283, Email: harc@harc.org, <http://www.harc.org>

Stratospheric Ozone Information Hotline, United States Environmental Protection Agency, 1200 Pennsylvania Avenue, NW, Washington, D.C. USA 20460 (202) 775-6677, Toll Free: (800) 296-1996, Email: public-access@epa.gov, <http://www.epa.gov/ozone>

United Nations Environment Programme, Division of Technology, Industry and Economics, Energy and OzonAction Programme, Tour Mirabeau, 39-43 quai André Citroën, 73759 Paris Cedex 15, France (33-1) 44 37 14 50, Fax: (33-1) 44 37 14 74, Email: ozonaction@unep.fr, <http://www.uneptie.org/ozonaction<

Hanford Nuclear Reservation

The Hanford Engineering Works was conceived in June 1942 under the direction of Major General Leslie R. Groves, head of the famous Manhattan Project, to produce **plutonium** and other materials for use in the development of **nuclear weapons**. By December 1942, a decision was reached to proceed with the construction of three plants—two to be located at the Clinton Engineering Works in Tennessee and a third at the Hanford Engineering Works in Washington.

Hanford was established in the southeastern portion of Washington state between the Yakima Range and the Columbia River, about 15 mi (24 km) northwest of Pasco, Washington. The site occupies approximately 586 mi^2 (1,517 km^2) of **desert** with the Columbia River flowing through its northern region. Once a linchpin of United States nuclear weapons production during the Cold War era, Hanford has now become the world's largest environmental cleanup project.

Hanford Engineering Works, known as HEW or "site W" in classified terms, was originally under the control of the Manhattan District of the **Army Corps of Engineers** (MED) until the **Atomic Energy Commission** (now the Department of Energy, DOE) took over in 1947. The actual operation of the site has been managed by a series of contractors since its inception. The first organization granted a contract to run site operations at Hanford was E.I. DuPont de Nemours and Company. In 1946, General Electric took over, and with the aid of several subcontractors ran construction and operation of the site through 1965. A series of contractors have directed operations at both the main DOE-Richland Operations Office and the DOE-Office of River Protection (ORP), the agency responsible for overseeing **hazardous waste** tank farm clean up along the Columbia River, since then.

In 1965, Battelle Memorial Institute, a non-profit organization, assumed management of the federal government's DOE research laboratories on the Hanford Site. The newly formed Pacific Northwest Laboratory (which became Pacific Northwest National Laboratory, or PNNL, in 1995) supports the Hanford site cleanup through the development and testing of new technologies. Battelle still runs the PNNL today.

A number of contractors have directed operations at HEW throughout its history. As of early 2002, the prime contractors at the DOE-Richland Operations Office included Battelle Memorial Institute (BMI); Bechtel Hanford, Inc.(BHI); Fluor Hanford, Inc. (FHI); and the Hanford **Environmental Health** Foundation (HEHF). The DOE-Office of River Protection (ORP), responsible for overseeing hazardous waste tank farm clean up along the Columbia

River, is managed by prime contractors CH2M Hill Hanford Group, Inc. (CHG) and Bechtel National, Inc. (BNI).

Over the period from 1943 to 1963, a total of nine plutonium-production reactors and five processing centers were built at Hanford, with the last of the reactors ceasing operations in 1987 (permanent shut-down of the Fast Flux Test Facility [FFTF], a sodium-cooled breeding reactor that produced isotopes for medical and industrial use, was ordered to close in 2001). Plutonium produced at Hanford was used in the world's first atomic explosion at Alamogordo, New Mexico in 1945. The Hanford Site processed an estimated 74 tons of plutonium during its years of active operation, accounting for approximately two-thirds of all plutonium produced for United States military use.

The plutonium fuel was processed on site at the Plutonium and **Uranium** Extraction Plant (PUREX). In the processing of plutonium, a substantial quantity of **radioactive waste** is produced; at Hanford, it amounts to 40% of the nation's one billion Curies of **high-level radioactive waste** from weapons production. PUREX ceased regular production in 1988 and was officially closed in 1992. Deactivation of the plant was completed in 1997.

During the HEWs operational lifetime, some high-level radioactive wastes were diverted from the relatively safe underground storage to surface trenches. Lower-level contaminated water was also released into ditches and the nearby Columbia River. The DOE reports that over 450 billion gallons of waste liquid from the Hanford plants was improperly discharged into the **soil** column during their operational lifetime. These wastes contained cesium-137, technetium-99, plutonium-239 and -240, strontium-90, and cobalt-60. So much low-level waste was dumped that the **groundwater** under the reservation was observed to rise by as much as 75 ft (23 m).

Today more than 50 million gal (189.3 million l) of high-level radioactive waste is contained in approximately 177 underground storage tanks, 67 of which have known leaks. One million gallons (3.8 million l) of this tank waste has already leaked into the surrounding ground and groundwater. An estimated 25 million ft^4 of radioactive solid wastes remains buried in trenches which are similar to septic-tank **drainage** fields. Spent nuclear fuel basins have leaked over 15 million gal (56.8 million l) of radioactive waste. In total, over 1,900 waste sites and 500 contaminated facilities have been identified for clean up at Hanford. The contaminated zone encompasses a variety of ecosystems and the nearby Columbia River.

In the early 1990s, Congress passed a bill requiring the DOE to create a watchdog list of those Hanford tanks at risk for explosion or other potential release of high-risk radioactive waste. By 1994, the list had 56 tanks listed as a high-risk potential. By late 2001, the final 24 tanks had been removed from the congressional watch list, which is sometimes referred to as the "Wyden Watch List" (after sponsoring Senator Ron Wyden).

In 1989, the DOE, U.S. **Environmental Protection Agency** (EPA), and the State of Washington Department of Ecology signed the Hanford Federal Facility Agreement and Consent Order ("Tri-Party Agreement"). The Tri-Party Agreement (TPA) outlines cleanup efforts required to achieve compliance with the Comprehensive Environmental Response Compensation and Liability Act (CERCLA; or Superfund) and the **Resource Conservation and Recovery Act** (RCRA).

Compliance with federal and state authorities under the TPA has not always progressed without problems for Hanford officials. The EPA fined DOE for poor **waste management** practices in 1999, levying $367,078 in civil penalties that were later reduced to $25,000 and a promise to spend $90,000 on additional clean-up activities. The following year EPA issued an additional $55,000 fine against the DOE for non-compliance with the Tri-Party agreement. However, the five-year review of the project completed by EPA in 2001, while specifying 18 "action items" for DOE compliance, also states that cleanup of soil waste sites and burial grounds "are proceeding in a protective and effective manner." In March 2002, DOE was fined $305,000 by the Washington Department of Ecology for failing to start construction of a waste treatment facility as outlined in the Tri-Party agreement. As of May 2002, the State had agreed to forgive the fine if the DOE began construction on the facility by the end of the year.

Even though the Hanford site poses significant environmental concerns, the facility has generated a number of innovative technological advances in the clean up and immobilization of radioactive wastes. For instance, one process developed in Hanford's Pacific Northwest Laboratory is *in situ vitrification*, which uses electricity to treat waste and surrounding **contaminated soil**, melting it into a glass material that is more easily disposed. The William R. Wiley Environmental and Molecular Sciences Laboratory (EMSL), which opened in 1997, houses a wide range of experimental programs aimed at solving nuclear waste treatment issues.

And despite the extent of its contamination, the isolation and security of the Hanford site has made the area home to a large and diverse population of **flora** and **fauna**. A 1997 Nature Conservancy study at Hanford found dozens of previously undiscovered and rare plant and animal **species** in various site habitats. The **biodiversity** study led to the establishment of the federally-protected 195,000-acre (79,000-ha) Hanford Reach National Monument in June 2000.

In May 2002 Department of Energy officials introduced a "Performance Management Plan for the Accelerated Cleanup of the Hanford Site." The new plan sets significantly revised goals for the completion of the site cleanup, setting some projects ahead 35 years or more. While the plan was still in its draft stage at the writing of this entry, DOE officials anticipate its approval by the end of 2002. Strategic initiatives outlined in the plan include restoration of the Columbia River Corridor by 2012 and completion of tank decontamination and closure by 2035.

[*Paula A Ford-Martin*]

RESOURCES

BOOKS

Gerber, Michele S. *On the Home Front: The Cold War Legacy of the Hanford Nuclear Site, 2nd edition.* Lincoln, NE: University of Nebraska Press, 2002.

PERIODICALS

"Three-Way Agreement Reduces Length of Required Cleanup Time." *Hazardous Waste Superfund Week* 24, no. 12 (March 25, 2002).

"Washington State Freezes Fine Against Hanford Nuclear-Cleanup Site." *Tri-City Herald*, March 12, 2002 [cited May 2002]. <http://www.tri-cityherald.com/news/2002/0312/story4.html>.

OTHER

Washington State Department of Ecology Nuclear Waste Program. [cited May 2002]. <http://www.ecy.wa.gov/programs/nwp>.

United States Department of Energy, Richland Operations Office and Office of River Protection. *Performance Management Plan for the Accelerated Cleanup of the Hanford Site.* [cited May 2002]. <http://www.hanford.gov/docs/hpmp/hpmp.pdf>.

ORGANIZATIONS

U.S. Department of Energy, Hanford Site, Richland Operations Office, Freedom of Information Office, 825 Jadwin Avenue, P.O. Box 550, Richland, WA USA 99352 (509) 376-6288 , Fax: (509)376-9704 , Email: FOIA@rl.gov, <http://www.hanford.gov/FOIA/index.cfm>

Dr. Garrett Hardin (1915 –)

American environmentalist and writer

Trained as a biologist (University of Chicago undergraduate degree; Ph.D. from Stanford in 1942 in microbial **ecology**) Garrett Hardin spent most of his career at the University of California at Santa Barbara, where his title was Professor of **Human Ecology**. He was born in Dallas, Texas, and grew up in various places in the Midwest, spending summers on his grandparents' farm in Missouri.

Very few biologists, short of Charles Darwin, have generated the levels of controversy that Hardin's thinking and writing have. The controversies, which continue today, center on two metaphors of human-ecological relationships; "the tragedy of the commons" and "the lifeboat ethic."

Hardin is widely credited with inventing the idea of the **tragedy of the commons**, but his work was long pre-

ceded by an ancient rhyme about the tragedy that results from stealing the commons from the goose. He did popularize the idea, though, and it made him a force to reckon with in population studies. Seldom is an academic author so identified with one article (though his thoughts on lifeboat ethics have since become almost equal in identification and impact). The idea is very simple: resources held in common will be exploited by individuals for personal gain in disregard of public impacts; individual profit belongs to individual exploiters while they bear the brunt of only part of the impacts. Much of the controversy centers on Hardin's solutions to the tragedy: first, that private property owners "recognize their responsibility to care for" the land, thus lending at least implicit support to privatization efforts and second, the paradoxical idea that since exercise of individual freedom leads to ruin, such freedom cannot be tolerated, thus we must turn to "mutual coercion mutually agreed upon." As Hardin noted, "if everyone would restrain himself, all would be well; but it takes only one less than everyone to ruin a system of voluntary restraint. In a crowded world of less than perfect human beings, mutual ruin is inevitable if there are no controls. This is the tragedy of the commons." Hardin has since expanded on his thesis, answering his critics by incorporating the differences between an open access system and an closed one. But the debate continues.

His other widely debated metaphor was of a lifeboat (standing for a nation's land and resources) occupied by rich people in an ocean of poor people (who have fallen out of their own, inadequate lifeboats). If the rich boat is close to its margin of safety, what should its occupants do about the poor people in the ocean? Or in the other boats? If the occupants let in even a few more people, the boat may be swamped, thus creating an ethical dilemma for the rich occupants. How do these occupants of the already full lifeboat justify taking in additional people if it guarantees the collapse for all? Ever since, Hardin has been accused of social Darwinism, of racism, of ignoring the possibility that "the poverty of the poor may be caused in part by the affluence of the rich," and of isolationism (since he argued that "for the foreseeable future survival demands that we govern our actions by the ethics of a [sovereign] lifeboat" rather than "space-ship" ethics that [arguably] try to care for all equitably).

What Hardin was trying to do in both of these cases was to look unemotionally at **population growth** and resource use, employing the rationale and language of a scientist viewing human populations in an objective, evolutionary perspective. Biologists know that "natural" populations that overuse their resources, that exceed the **carrying capacity** of their range, are then adjusted—in numbers or in resource use levels—also naturally and often brutally (from the perspective of many people). The problems remain that the division between the rich and the poor in the world is widening, and that many resources and ecosystems on earth are being stressed, though how close to the breaking point no one really knows. Also, nation-states are arbitrary divisions, and perhaps—in a corollary to the space-ship metaphor—the ultimate life-boat is the earth and all the inhabitants must exist in it together.

Hardin retired as a professor emeritus of human ecology, a rare title that reflects his attempts to fold the human **species** into the evolutionary ecology perspective developed in biology. Ultimately what he tried to get across was what he called an *ecolate* view: *ecolacy* asks the question "and then what?" The basic insight of the ecolate citizen is that the world is a complex of systems so intricately interconnected that we can seldom be very confident that a proposed intervention in this system of systems will produce the consequences we want." Hardin's views are rich and varied—the themes presented here are only two of many—but he still summarizes the human condition by what he labels "the ecolate predicament": that all human interventions are doomed to failure if a population exceeds its carrying capacity, whether of a region or of the world. Doing so "will bring everyone down to a level of poverty."

[*Gerald L. Young Ph.D.*]

RESOURCES
BOOKS

Hardin, G. "Living on a Lifeboat." *BioScience* 24, no. 10, (October 1974): 561–568.

———. *Living Within Limits: Ecology, Economics, and Population Taboos.* New York: Oxford University Press, 1993.

———. *The Ostrich Factor: Our Population Myopia.* New York: Oxford University Press, 1999.

PERIODICALS

Bajema, C. J. "Garrett James Hardin: Ecologist, Educator, Ethicist, and Environmentalist." *Population and Environment* 12, no. 3, (Spring 1991): 193–212.

———. "The Tragedy of the Commons." *Science* 162, (13 December 1968): 1243–1248.

Hawaiian Islands

The Hawaiian Islands are made up of a chain of ancient volcanic islands that have formed at irregular intervals over the last ten million years. There are over 100 islands in the chain, eight of which are considered major. Of the eight major islands, Kauai is the oldest, and the island of Hawaii is the youngest, having formed within the past million years. This vast discrepancy in age has contributed to the tremendous **biodiversity** that has evolved there. The other factor

responsible for Hawaii's great **species** diversity is the island chain's remoteness. The nearest continent is 2,400 mi (3,862 km) away, thus limiting the total colonization that could, or has, taken place. The niches available to these colonizing species are very diverse due to geophysical events. For example, on the island of Kauai, the average annual rainfall on the windward side of Mount Waialeale is 460 in (1,169 cm), whereas on its leeward side, it is only 19 in (48 cm). The temperature on the islands ranges from 75–90°F (24–32°C) for 300 days each year. The lowest temperature ever recorded in Hawaii was 54°F (12°C). Thus with its tropical **climate** and unique biotic communities, it is easy to understand why Hawaii has been considered a paradise. But now this paradise is threatened by serious environmental problems caused by humans.

The impact of humans has been felt in the Hawaiian Islands since their first arrival, but perhaps never more so than today. In the last quarter century tourism has replaced sugar cane and pineapple as the islands' main revenue source. Over six million tourists visit Hawaii each year spend $11 million a day. With tourism comes development that often destroys natural habitats and strains existing **natural resources**. Hawaii has more than 60 **golf courses**, with plans to develop at least 100 more. This would destroy thousands of acres of natural vegetation, require the use of millions of gallons of freshwater for **irrigation**, and necessitate the use of tons of chemical fertilizers and pesticides in their maintenance. The increased number of tourists, many of whom visit the islands to enjoy its natural beauty, are often destroying the very thing they are there to see. Along the shore of Hanauma Bay, just south of Honolulu, over 90% of the **coral reef** is dead, primarily because people have trampled it in their desire to see and experience this unique piece of **nature**.

Much of Hawaii's **fauna** and **flora** are unique. Over 10,000 species are native to the Hawaiian Islands, having evolved and filled specialized niches there through the process of adaptive radiation from the relatively few original colonizing species. Examples are found in virtually every group of plants and animals in Hawaii. The avian adaptive radiation in the Hawaiian Islands surpasses even the best-known example, Darwin's finches of the Galapagos Islands. From as few as 15 colonizing species evolved 90 native species of birds in the Hawaiian archipelago. Included in this number are the Hawaiian honeycreepers of the family Drepanididae. This endemic family of birds arose from a single ancestral species, which gave rise to 23 species, including 24 subspecies, of honeycreepers spread throughout the main islands. **Niche** availability and reproductive isolation contributed greatly to the spectacular diversity of forms that evolved. These birds developed adaptations such as finch-like bills for crushing seeds, parrot-like bills for foraging on larvae in wood, long decurved bills for taking nectar and insects from specialized flowers, and small forcep-like bills for capturing insects.

Introductions of vast numbers of alien species is taking its toll on native Hawaiian species, a process that began when the first humans settled the islands over 1,500 years ago. Many native species of birds had experienced drastic population declines by the time Europeans first encountered the islands over 200 years ago. Since then at least 23 species of Hawaii's native birdlife have become extinct, and currently over 30 additional avian species are threatened with **extinction**. Through the process of primary ecological **succession**, one new species became established in the Hawaiian Islands every 70,000 years. Introductions of alien species are taking place a million times faster, and are thus eliminating native species at an unprecedented rate. Recent estimates indicate that there are over 8,000 **introduced species** of plants and animals throughout the Hawaiian Island chain. The original Polynesian settlers brought with them pigs, dogs, chickens, and rats, along with a variety of plants they had cultivated for fiber, food, and medicine. With the Europeans came cattle, horses, sheep, cats, and additional rodents. The mongoose was introduced purposefully to control the rat populations; however, they presented more of a threat to ground-nesting birds. The introduction of rabbits caused the loss of vast quantities of vegetation and ultimately the extinction of three bird species on Laysan Island. Over 150 species of birds, including escaped cage birds, have been introduced to the islands. However, most of these have not established breeding populations.

Although Hawaii represents only 0.2% of the United States's land base, almost 75% of the total extinctions of birds and plants of the nation have occurred in this state. Hawaii's endemic flora are disappearing rapidly. Introduced mammalian browsers are decimating the native plantlife, which never needed to evolve defense mechanisms against such predators, and other introduced animals are destroying populations of native pollinators. **Habitat** destruction and opportunistic non-native vegetation are also working against the endemic Hawaiian plant species. Organizations such as the Hawaii Plant Conservation Center are working to preserve the state's floral diversity by collecting and propagating as many of the rare and **endangered species** as possible. Their greenhouses now contain over 2,000 plants representing almost two thirds of Hawaii's native species, and their goal is to propagate 400 of the state's most endangered plants.

Introduced plant and animal species and their assault on native species are by no means the only environmental problems facing the Hawaiian Islands. Because Hawaii's population is growing at a higher rate than the national average, and in part due to its isolation, it is facing many environmental problems on a grander scale and at a more rapid rate than its sister states on the mainland. Energy is one of the primary problems. Much of Hawaii's electricity is produced by burning imported oil, which is extremely expensive. Because of limited reserve capabilities with regard to electrical generation, Hawaii faces the potential for blackouts.

Planners have been, over the past several decades, looking at the feasibility of tapping into Hawaii's seemingly vast **geothermal energy** resources. Hawaii has the most active **volcano** in the world, Kilauea, whose underground network of geothermal reserves is the largest in the state. The proposed Hawaii Geothermal Deep Water Cable project would supply the energy for a 500-megawatt power plant, the electrical power of which would be transmitted from the island of Hawaii to Oahu by three undersea cables. This would, of, course, be of economic and environmental benefit by reducing dependence on oil reserves. Opponents of this geothermal power plant point to several problems. They are concerned with the potential for the release of toxic substances, such as **hydrogen** sulfide, **lead**, **mercury**, and chromium, into the **environment** from well-heads. They also have voiced negative opinions concerning construction of a geothermal plant so near the lava flows and fissures of Hawaii's two active volcanoes. An alternately proposed site would have the facility located in Hawaii's last major inholding of lowland **tropical rain forest**. To avoid economic disaster from escalating prices for imported oil and the problem of frequent blackouts, Hawaii must reach some compromise on geothermal energy and/or research the potential for getting its electricity from one or more of wind, wave, or **solar energy**.

Hawaii is also faced with environmental problems of another sort—natural disasters. Volcanoes not only provide the potential for geothermal energy, they also have the potential for massive destruction. Over the past 200 years, Hawaii's two active volcanoes, Kilauea and Mauna Loa, have covered nearly 200,000 acres (81,000 ha) of land with lava, and geologists expect them to remain active for centuries to come. Severe hurricanes and tidal waves also hold the potential for vast destruction, not only of human property and lives, but of natural areas and the **wildlife** it holds. Many of the **endemic species** of Hawaii are threatened or endangered, and their populations are often so low that a single storm could wipe out most or all of its numbers.

There are efforts to reverse the environmental destruction in the Hawaiian Islands. Several organizations comprised of native Hawaiians are working to stem the destruction and loss of the **wilderness** paradise discovered by their ancestors.

[*Eugene C. Beckham*]

RESOURCES

BOOKS

Pratt, D., P. Brunner, and D. Berret. *The Birds of Hawaii and the Tropical Pacific.* Princeton, NJ: Princeton University Press, 1987.

Scott, J., and J. Sincock. "Hawaiian Birds." *Audubon Wildlife Report 1985.* New York: National Audubon Society, 1985.

Shallenberger, R., ed. *Hawaii's Birds.* 3rd ed. Honolulu: Hawaii Audubon Society, 1981.

Stewart, F., ed. *A World Between Waves: Writings on Hawaii's Rich Natural History.* Washington, DC: Island Press, 1992.

PERIODICALS

White, D. "Plants in a Precarious State." *National Wildlife* 31 (May 1993): 30–35.

Denis Allen Hayes (1944 –)

American environmental activist and Earth Day organizer

As executive director of the first **Earth Day** in April 1970, Hayes helped launch the modern movement of **environmentalism**, and has promoted the use of **solar energy** and other renewable resources. A native of Camas, Washington, Hayes acquired his appreciation of **nature** exploring and enjoying the mountains, lakes, and beaches of the Pacific Northwest. At age 19, he dropped out of Clark College in Vancouver, Washington, and spent the next three years traveling the world. He installed church pews in Honolulu, taught swimming and modeled in Tokyo, and hitchhiked through Africa.

After returning to the United States, Hayes enrolled at Stanford as a history major, was elected student body president, and became active in the anti–Viet Nam War movement, occupying laboratories that researched military projects. After graduating from Stanford, he went to Harvard Law School, but dropped out in 1970 to help organize the first Earth Day. During the Carter administration, Hayes headed the federal Solar Energy Research Institute. He left after the agency's $120 million budget was cut by the Reagan administration. From 1983 to 1992, after completing his law degree at Stanford, he served there as an adjunct professor of engineering.

In 1990, as the international chairman of Earth Day on its twentieth anniversary, Hayes helped to organize participation by over 200 million people in 141 countries. This event generated extraordinary publicity for and concern about global environmental problems. In 1992, Hayes was named president of the Seattle-based Bullitt Foundation, which works to protect the **environment** of the Pacific Northwest and to help disadvantaged children. He also serves as chairman of the board of **Green Seal**, a group that endorses consumer products meeting strict environmental standards and co-chairs the group promoting the "Valdez Principles" of corporate responsibility.

Denis Hayes. (Corbis-Bettmann. Reproduced by permission.)

Hayes has received awards and honors from many groups, including the **Humane Society of the United States**, the Interfaith Center for Corporate responsibility, the **National Wildlife Federation**, and the **Sierra Club**. In 1990, *Life* magazine named him one of the 18 Americans most likely to have an impact on the twenty-first century.

Hayes has written over 100 papers and articles, and his book on solar energy, *Rays of Hope: The Transition to a Post-Petroleum World*, has been published in six languages. He has long advocated increased development of solar and other **renewable energy** sources, which he believes could provide most of the nation's energy supply within a few years.

In 1990, Hayes wrote that "Time is running out. We have at most ten years to embark on some undertakings if we are to avoid crossing some dire environmental thresholds. Individually, each of us can do only a little. Together, we can save the world."

[*Lewis G. Regenstein*]

RESOURCES

BOOKS

Hayes, D. *Rays of Hope: The Transition to a Post-Petroleum World.* New York: Norton, 1977.

PERIODICALS

Hayes, D. "Earth Day 1990: Threshold of the Green Decade." *Natural History* 99 (April 1990): 55–60.
———. "The Green Decade." *Amicus Journal* 12 (Spring 1990): 10–21.
Reed, S. "Twenty Years After He Mobilized Earth Day, Denis Hayes Is Still Racing to Save Our Planet." *People Weekly*, April 2, 1990, 96–99.
Ridenour, J. M., et al. "Global Prescription: Leading Conservationists Look to the Future and Speak Their Minds." *National Parks* 64 (March–April 1990): 16–18.

Hazard ranking system

The Hazard ranking system (HRS) is a numerical scoring procedure that the federal **Environmental Protection Agency** (EPA) uses to place and prioritize waste sites on the **National Priorities List**. Only these priority sites can be cleaned up through the Superfund Trust Fund program.

The HRS score is based on an evaluation of threats related to the release or potential release of hazardous substances. The HRS assessment of a site ranks public health factors such as threats to drinking water, the food chain, and populations exposed through occupational and ambient environments. Also evaluated are environmental threats like the effect of substances on **air quality**, resources, and sensitive ecosystems.

Federal investigators score a site by evaluating four pathways that could be affected by hazardous releases. The pathways are ground water **migration**, surface water migration, air migration, and **soil** exposure.

The pathway scores are combined using a root-mean-square equation. This calculation produces the overall score for a site. A high HRS score does not guarantee immediate action because clean-up work may be going on at other sites. The decision to take action on a site is based on additional research that includes a remedial investigation of what corrective action is needed.

[*Liz Swain*]

Hazardous materials, solidification of
see Solidification of hazardous materials

Hazardous chemicals
see Hazardous material; Hazardous waste

Hazardous material

Any agent that presents a risk to life-forms or the **environment** can be considered a hazardous material. This is a very broad term which encompasses pure compounds and mixtures, raw materials, and other naturally occurring substances, as well as industrial products and wastes. Depending on the nature and the length of exposure, virtually all substances can have toxic effects, ranging from headaches and dizziness to **cancer**. The challenge facing any legislation is not only to devise regulations for the safe handling of hazardous materials but also to define the term itself.

The legislation that offers the most detailed and comprehensive definition of hazardous materials is the **Resource Conservation and Recovery Act** (RCRA), enacted in 1976. RCRA classifies a waste mixture or compound as hazardous if it fails what is called a *characteristic test* or appears on one of a few lists. The lists of hazardous wastes include those from specific and nonspecific sources and those which are acutely hazardous and generally hazardous. There are four characteristic tests: ignitability, reactivity, corrosivity, and extraction-procedure toxicity.

A waste fails the ignitable test if it is a liquid with flash point below 140°F (60°C); or a solid that, under standard temperature and pressure, causes fire through friction, absorbing moisture, or spontaneous changes and burns vigorously and persistently; or a compressed gas defined by the Department of **Transportation** (DOT) as an oxidizer or as being ignitable. Spent solvents, paint removers, epoxy resins, and waste inks are often classified as hazardous under this definition.

A waste fails the corrosivity test if it is aqueous and has a **pH** of either 2 or less or 12.5 or more, or if it is a liquid that corrodes steel at a rate equal to or more than 0.25 in (6.35 mm) per year. Examples of corrosive wastes include various acids and bases such as nitric **acid**, ammonium hydroxide, perchloric acid, sulfuric acid, and sodium hydroxide, though a waste will not be classified as hazardous in this test if these acids and bases are neutralized or present at low levels.

Reactivity is related to one of the following criteria: If the material is unstable, with the potential for violent reactions with water, or generates toxic fumes; if it has cyanide and sulfide content; if it can be easily detonated; or is defined by DOT as a Class A or B explosive. Compounds commonly causing wastes to fall into this category are chromic acid, hypochlorites, picric acid, nitroglycerin, dinitrophenol, and organic peroxides.

Extraction-Procedure toxicity is determined through the extraction of **solid waste**, in a procedure referred to as the *Toxicity Characteristic Leaching Procedure* (TCLP). Liquid waste can also fail this test, although a liquid containing less than 0.5% solids after **filtration** does not need to undergo it. Such a liquid can be directly analyzed for contaminants. In the TCLP, solids are extracted in a mildly acidic medium (pH 5.0) for 18 hours using a tumbling apparatus, and the

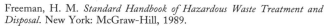
resulting liquid extract is then analyzed for contaminants. The waste is deemed to be hazardous if the level of a contaminant detected in this extraction procedure exceeds a given level. If liquid extracts contain **arsenic** at a concentration greater than 5.0 mg/l, for example, the waste is classified as hazardous. Different methods are recommended by the **Environmental Protection Agency** (EPA) for measuring the contaminants in each class.

The Resource Conservation and Recovery Act (1976) and the Hazardous and Solid Waste Amendments added to it in 1984 were intended to protect **groundwater**, surface water, and land from improper management of solid wastes. The act and the amendments defined the responsibilities of industries and others who generate and transport **hazardous waste**. They also set standards for land disposal facilities and underground storage tanks, as well as standards for proper management of hazardous materials from "cradle to grave."

The Toxic Substances Control Act (1976) is another important piece of legislation concerning hazardous materials. It was intended to regulate the introduction and use of new, potentially hazardous substances. The bill requires industry to test extensively **chemicals** that may be harmful, and it requires industries to provide the EPA with information about the production, use, and health effects of any new substances or mixtures before they are manufactured.

The Comprehensive Environmental Response, Compensation and Liabilities Act (CERCLA) (1980) and the Superfund Amendments and Reauthorization Act (SARA) (1986) set policy for situations in which hazardous materials have been mismanaged in the past. These legislations have established a system for ranking sites that need **remediation**, called the **Hazard Ranking System**, as well as a procedure for raising funds to support these efforts. The bills also impose schedules for site investigations, feasibility studies, and remedial action.

The definition of hazardous materials remains a difficult and complex issue. There are changes in public perceptions about certain materials, such as **asbestos** or **lead**, and there are scientific contributions which result in the addition or deletion of various chemicals or compounds from lists of hazardous materials. The definition is therefore dynamic, but changes are made within a regulatory framework designed to protect both life and the environment.

[*Gregory D. Boardman*]

RESOURCES
BOOKS

Corbitt, R. A. *Standard Handbook of Environmental Engineering*. New York: McGraw-Hill, 1990.

Freeman, H. M. *Standard Handbook of Hazardous Waste Treatment and Disposal*. New York: McGraw-Hill, 1989.

Martin, E. J., and J. H. Johnson. *Hazardous Waste Management Engineering*. New York: Van Nostrand Reinhold, 1987.

Sax, N. I. *Dangerous Properties of Industrial Materials*. 6th ed. New York: Van Nostrand Reinhold, 1984.

Wagner, T. P. *Hazardous Waste Identification and Classification Manual*. New York: Van Nostrand Reinhold, 1990.

Wentz, C. A. *Hazardous Waste Management*. New York: McGraw-Hill, 1989.

Hazardous materials, storage and transport

see **Storage and transport of hazardous material**

Hazardous Materials Transportation Act (1975)

The Hazardous Materials **Transportation** Act, enacted in 1975 as part of a law dealing with transportation safety, strengthened the 1970 Hazardous Materials Transportation Control Act. The impetus for this act was increased illegal, or midnight, dumping; increasing spills; and poor enforcement. Illegal dumping increased in the 1970s as many landfills began to refuse to take **hazardous waste**, thus dramatically increasing the costs of disposal. The illegal dumping took place in vacant lots, along highways, or actually on the highways. In the congressional debate on the act, the U.S. Department of Transportation (DOT), which administers the law, estimated that 75% of all hazardous waste shipments violated the existing regulations. This poor enforcement was due to a lack of inspection personnel, fragmented jurisdiction and lack of coordination among the Coast Guard, the Federal Aviation Administration, the Federal Highway Administration, and the Federal Railroad Administration.

The law establishes minimum standards of regulation for the transport of hazardous materials by air, ship, rail, and motor vehicle. The DOT regulates the packing, labeling, handling, vehicle routing, and manufacture of packing and transport containers for hazardous materials transportation. The hazardous materials and wastes covered by the law, based on DOT regulations, are those on the **Resource Conservation and Recovery Act** (RCRA) list and certain substances designated by the **Environmental Protection Agency** under the authority of Superfund. All hazardous waste transporters must register as such with the proper state and federal agencies; they must use the RCRA uniform manifest system to track the pick-up and delivery of all shipments; they must only deliver to permitted hazardous waste facilities; they must notify the proper agencies of any accidents; they must clean up any discharges that occur during the transportation process. The law also provides a significant role for the states, though there are provisions in the act to prevent overly strict state and local regulations of hazardous waste transport. The Hazardous Materials Transportation Act includes numerous information requirements, also designed to increase public safety. Each vehicle carrying hazardous materials must display a sign identifying the hazard class of the cargo, and emergency response information has been required since 1990. Each shipment must also be accompanied by its RCRA hazardous waste manifest.

The manifest system is part of the RCRA "cradle-to-grave" approach to regulating hazardous materials. The system is supposed to prevent illegal dumping, since hazardous waste transporters could not accept hazardous waste without a manifest, and, similarly, hazardous waste treatment and disposal facilities could not accept waste from transporters without a manifest. Since all hazardous materials could be traced and accounted for through such manifests, illegal dumping should stop. Nevertheless, it is unclear how much effect the manifest system has had on illegal dumping since such dumping is still less costly than proper disposal.

In 1990, the Hazardous Materials Transportation Uniform Safety Act was passed, the first major amendments to the 1975 Act. Poor enforcement of the existing law was the stimulus for action. The law focused on better enforcement by increasing the number of inspectors, increasing the civil and criminal penalties for violation of the regulations, and helping states better respond to accidents involving hazardous materials. *See also* Chemical spills; Comprehensive Environmental Response, Compensation and Liability Act (CERLA) (1980); Hazardous waste siting; Solidification of hazardous material; Storage and transport of hazardous materials

[*Christopher McGrory Klyza*]

RESOURCES

BOOKS

Dower, R. C. "Hazardous Wastes." *Public Policies for Environmental Protection*, edited by P. R. Portney. Washington, DC: Resources for the Future, 1990.

Mazmanian, D., and D. Morell. *Beyond Superfailure: America's Toxics Policy for the 1990s.* Denver: Westview Press, 1992.

PERIODICALS

"Hazardous Materials Law Strengthened." *Congressional Quarterly Almanac* 46 (1990): 380–82.

Hazardous Substances Act (1960)

This law was one of Congress's first forays into consumer protection, and it helped to pave the way for the explosion in consumer protection legislation that began in the mid-

1960s. The Hazardous Substances Labeling Act was passed in 1960 (the word "Labeling" was deleted by the 1966 amendments the act). The law authorized the Secretary of the Department of Health, Education, and Welfare (HEW) to require warning labels for household substances that were deemed hazardous. These substances were categorized as: toxic, corrosive, irritant, strong sensitizer, flammable or combustible, pressure generating, or radioactive. The law does not cover pesticides (which are regulated by the Federal Insecticide, **Fungicide**, and Rodenticide Act); food, drugs, or cosmetics (which are covered by the Federal Food, Drug, and Cosmetics Act); radioactive materials related to **nuclear power**; fuels for cooking, heating, or refrigeration; or **tobacco** products.

A product is defined to be hazardous if it might lead to personal injury or substantial illness, especially if there is a reasonable danger that a child might ingest the substance. When the HEW Secretary declares a substance to be hazardous, a label is required. The label must include a description of the chief hazard and first aid instructions, along with handling and storage instructions. If its chief hazard is flammable, corrosive, or toxic, it must say DANGER on the label; other hazardous substances require either CAUTION or WARNING. In addition, all labels must include the statement "Keep out of the reach of children."

Major amendments to the act, the Child Protection Act, were passed in 1966. These amendments, largely in response to the message on consumer issues by President Lyndon Johnson, expanded federal control over hazardous substances. The **Food and Drug Administration** (FDA), which administered the law, could now ban substances (after formal hearings) that were deemed too hazardous, even if they had a warning label, if "the degree or **nature** of the hazard involved in the presence or use of such substance in households is such that the objective of the protection the public health and safety can be adequately served" only by such a ban. The amendments also extended the scope of the law to pay greater attention to toys and children's articles. This meant the government could require a warning on all household items, rather than just packaged items.

The Child Protection and Toy Safety Act of 1969 further amended the Hazardous Substances Act. Toys could be declared hazardous if they presented electrical, mechanical, or thermal dangers. Also, substances that were hazardous to children, including toys, could be banned automatically. As the titles of these amendments suggest, the Hazardous Substances Act became the primary vehicle to protect children from dangerous substances and toys. Administration of the act has been shifted to the Consumer Product Safety Commission. If the Commission finds a "substantial risk of injury," children's clothes, furniture, and toys can be pulled from the market immediately. *See also* Environmental law; Hazardous material; Hazardous Materials Transportation Act; Hazardous waste; Hazardous waste siting; Radioactive waste; Radioactivity

[*Christopher McGrory Klyza*]

RESOURCES

PERIODICALS

"Child Protection." *Congressional Quarterly Almanac* 22 (1966): 325–7.
"Toy Safety," *Congressional Quarterly Almanac* 25 (1969): 248–50.

Hazardous waste

Of the thousands of millions of tons of waste generated in the United States annually, approximately 60 million tons are classified as hazardous. Hazardous waste is legally defined by the **Resource Conservation and Recovery Act** (RCRA) of 1976. The RCRA defines hazardous waste as any waste or combination of wastes, which because of its quantity, concentration, or physical, chemical, or infectious characteristics may: A) cause, or significantly contribute to, an increase in **mortality** or an increase in serious irreversible or incapacitating illness; or, B) pose a substantial present or potential hazard to human health or the **environment** when improperly treated, stored, transported, disposed of, or otherwise managed.

In the Code of Federal Regulations, the **Environmental Protection Agency** (EPA) specifies that a **solid waste** is hazardous if it meets any of four conditions: 1) It exhibits ignitability corrosivity, reactivity, or EP toxicity; 2) has been listed as a hazardous waste; 3) is a mixture containing a listed hazardous waste and a nonhazardous waste, unless the mixture is specifically excluded or no longer exhibits any of the four characteristics of hazardous waste; 4) is not specifically excluded from regulation as a hazardous waste.

The EPA established two criteria for selecting the characteristics given above. The first criterion is that the characteristic is capable of being defined in terms of physical, chemical, or other properties. The second criterion is that the properties defining the characteristic must be measurable by standardized and available test procedures. For example under the term ignitability (Hazard code label "I"), any one of four criteria can be met: 1) A liquid with a flash point less than 60°F (16°C); 2) If not a liquid, then it is capable under standard temperature and pressure of causing fire through friction, **absorption** of moisture, or spontaneous chemical changes, and when ignited, burns so vigorously and persistently that it creates a hazard; 3) It may be an ignitable compressed gas; 4) It is an oxidizer.

Similarly under the characteristics of corrosivity, reactivity, and toxicity, there are specifically defined requirements which are spelled out in the Code of Federal Register (CFR). Further examples are given below:

Corrosivity (Hazard code "C") has either of the following properties: an aqueous waste with a **pH** equal to or less than 2.0 or greater than 12.5; or a liquid which will corrode **carbon** steel at a rate greater than 0.25 in (0.64 cm) per year.

Reactivity (Hazard code "R") has at least one of the following properties: a substance which is normally unstable and undergoes violent physical and/or chemical change without being detonated; a substance which reacts violently with water (for example, sodium metal); a substance which forms a potentially explosive mixture when mixed with water; a substance which can generate harmful gases, vapors, or fumes when mixed with water; a cyanide- or sulfide-bearing waste which can generate harmful gases, vapors, or fumes when exposed to pH conditions between 2 and 12.5; a waste which, when subjected to a strong initiating source or when heated in confinement, will detonate and/or generate an explosive reaction; a substance which is readily capable of detonation at standard temperature and pressure.

Toxicity (Hazard code "E") has the properties such that an aqueous extract contains contamination in excess of that allowed (e.g., **arsenic** >5 mg/l; barium 0.100 mg/l; **cadmium** >1 mg/l; chromium >5 mg/l; **lead** >5 mg/l). Additional codes under toxicity include an "acute hazardous waste" with code "H": a substance which has been found to be fatal to humans in low doses or has been found to be fatal in corresponding human concentrations in laboratory animals. Toxic waste (hazard code "T") designates wastes which have been found through laboratory studies to be a **carcinogen, mutagen,** or **teratogen** for humans or other life forms.

Certain wastes are specifically excluded from classification as hazardous wastes under RCRA, including domestic sewage, **irrigation** return flows, **household waste**, and nuclear waste. The latter is controlled via other legislation. The impetus for this effort at legislation and classification comes from several notable cases such as **Love Canal**, New York; **Bhopal, India; Stringfellow Acid Pits** (Glen Avon, California); and **Seveso, Italy**; which have brought media and public attention to the need for identification and classification of dangerous substances, their effects on health and the environment, and the importance of having knowledge about the potential risk associated with various wastes.

A notable feature of the legislation is its attempt at defining terms so that professionals in the field and government officials will share the same vocabulary. For example, the difference between "toxic" and "hazardous" has been established; the former denotes the capacity of a substance to produce injury and the latter denotes the *probability* that injury will result from the use of (or contact with) a substance.

The RCRA legislation on hazardous waste is targeted toward larger generators of hazardous waste rather than small operations. The small generator is one who generates less than 2,205 lb (1,000 kg) per month; accumulates less than

2,205 lb (1,000 kg); produces wastes which contain no more than 2.2 lb (1 kg) of acutely hazardous waste; has containers no larger than 5.3 gal (20 l) or contained in liners less than 22 lb (10 kg) of weight of acutely hazardous waste; has no greater than 220 lb (100 kg) of residue or **soil** contaminated from a spill, etc. The purpose of this exclusion is to enable the system of regulations to concentrate on the most egregious and sizeable of the entities that contribute to hazardous waste and thus provide the public with the maximum protection within the resources of the regulatory and legal systems.

[*Malcolm T. Hepworth*]

RESOURCES
BOOKS

Dawson, G. W., and B. W. Mercer. *Hazardous Waste Management.* New York: Wiley, 1986.
Dominguez, G. S., and K. G. Bartlett. *Hazardous Waste Management.* Vol. 1, *The Law of Toxics and Toxic Substances.* Boca Raton: CRC Press, 1986.
U.S. Environmental Protection Agency. *Hazardous Waste Management: A Guide to the Regulations.* Washington, DC: U.S. Government Printing Office, 1980.
Wentz, C. A. *Hazardous Waste Management.* New York: McGraw-Hill, 1989.

Hazardous waste site remediation

The overall objective in remediating **hazardous waste** sites is the protection of human health and the **environment** by reducing risk. There are three primary approaches which can be used in site **remediation** to achieve acceptable levels of risk:

• the hazardous waste at a site can be contained to preclude additional **migration** and exposure
• the hazardous constituents can be removed from the site to make them more amenable to subsequent ex situ treatment, whether in the form of **detoxification** or destruction
• the hazardous waste can be treated in situ (in place) to destroy or otherwise detoxify the hazardous constituents

Each of these approaches has positive and negative ramifications. Combinations of the three principal approaches may be used to address the various problems at a site. There is a growing menu of technologies available to implement each of these remedial approaches. Given the complexity of many of the sites, it is not uncommon to have treatment trains with a sequential implementation of various in situ and/or ex situ technologies to remediate a site.

Hazardous waste site remediation usually addresses soils and **groundwater**. However, it can also include wastes, surface water, **sediment**, sludges, bedrock, buildings, and other man-made items. The hazardous constituents may be organic, inorganic and, occasionally, radioactive. They may be elemental ionic, dissolved, sorbed, liquid, gaseous, vaporous, solid, or any combination of these.

Hazardous waste sites may be identified, evaluated, and if necessary, remediated by their owners on a voluntary basis to reduce environmental and health effects or to limit prospective liability. However, in the United States, there are two far-reaching federal laws which may mandate entry into the remediation process: the **Comprehensive Environmental Response, Compensation, and Liability Act** (CERCLA, also called the Superfund law), and the **Resource Conservation and Recovery Act** (RCRA). In addition, many of the states have their own programs concerning abandoned and uncontrolled sites, and there are other laws that involve hazardous site remediation, such as the clean-up of polychlorinated biphenyl (PCB) under the auspices of the federal Toxic Substances Control Act (TSCA).

Potential sites may be identified by their owners, by regulatory agencies, or by the public in some cases. Site evaluation is usually a complicated and lengthy process. In the federal Superfund program, sites at which there has been a release of one or more hazardous substances that might result in a present or potential future threat to human health and/or the environment are first evaluated by a Preliminary Assessment/Site Inspection (PA/SI). The data collected at this stage is evaluated, and a recommendation for further action may be formulated. The **Hazard Ranking System** (HRS) of the U.S. **Environmental Protection Agency** (EPA) may be employed to score the site with respect to the potential hazards it may pose and to see if it is worthy of inclusion on the **National Priorities List** (NPL) of sites most deserving the attention of resources.

Regardless of the HRS score or NPL status, the EPA may require the parties responsible for the release (including the present property owner) to conduct a further assessment, in the form of the two-phase Remedial Investigation/Feasibility Study (RI/FS). The objective of the RI is to determine the nature and extent of contamination at and near the site. The RI data is next considered in a baseline **risk assessment**. The risk assessment evaluates the potential threats to human health and the environment in the absence of any remedial action, considering both present and future conditions. Both exposure and toxicity are considered at this stage.

The baseline risk assessment may support a decision of no action at a site. If remedial actions are warranted, the second phase of the RI/FS, an engineering Feasibility Study, is performed to allow for educated selection of an appropriate remedy. The final alternatives are evaluated on the basis of nine criteria in the federal Superfund program. The EPA selects the remedial action it deems to be most appropriate and describes it and the process which led to its selection in the **Record of Decision** (ROD). Public comments are solicited on the proposed clean-up plan before the ROD is issued. There are also other public comment opportunities during the RI/FS process. Once the ROD is issued, the project moves to the Remedial Design/Remedial Action (RD/RA) phase unless there is a decision of no action. Upon design approval, construction commences. Then, after construction is complete, long-term operation, maintenance, and monitoring activities begin.

For Superfund sites, the EPA may allow one or more of the responsible parties to conduct the RI/FS and RD/RA under its oversight. If possibly responsible parties are not willing to participate or are unable to be involved for technical, legal, or financial reasons, the EPA may choose to conduct the project with government funding and then later seek to recover costs in lawsuits against the parties. Other types of site remediation programs often replicate or approximate the approaches described above. Some states, such as Massachusetts, have very definite programs, while others are less structured.

Containment is one of the available treatment options. There are several reasons for using containment techniques. A primary reason is difficulty in excavating the waste or treating the hazardous constituents in place. This may be caused by construction and other man-made objects located over and in the site. Excavation could also result in uncontrollable releases at concentrations potentially detrimental to the surrounding area. At many sites, the low levels of risks posed, in conjunction with the relative costs of treatment technologies, may result in the selection of a containment remedy.

One means of site containment is the use of an impermeable cap to reduce rainfall **infiltration** and to prevent exposure of the waste through **erosion**. Another means of containment is the use of cut-off walls to restrict or direct the movement of groundwater. In situ solidification can also be used to limit the mobility of contaminants. Selection among alternatives is very site specific and reflects such things as the site **hydrogeology**, the chemical and physical nature of the contamination, proposed **land use**, and so on. Of course, the resultant risk must be acceptable.

As with any in situ approach, there is less control and knowledge of the performance and behavior of the technology than is possible with off-site treatment. Since the use of containment techniques leaves the waste in place, it usually results in long-term monitoring programs to determine if the remediation remains effective. If a containment remedy were to fail, the site could require implementation of another type of technology.

The ex situ treatment of hazardous waste provides the most control over the process and permits the most detailed assessments of its efficacy. Ex situ treatment technologies offer the biggest selection of options, but include an additional risk factor during transport. Examples of treatment options include **incineration**; innovative thermal destruc-

tion, such as infrared incineration; **bioremediation**; stabilization/solidification; **soil** washing; chemical extraction; chemical destruction; and thermal desorption. Another approach to categorizing the technologies available for hazardous waste site remediation is based upon their respective levels of demonstration. There are existing technologies, which are fully demonstrated and in routine commercial use. Performance and cost information is available. Examples of existing technologies include **slurry** walls, caps, incineration, and conventional solidification/stabilization.

The next level of technology is innovative and has grown rapidly as the number of sites requiring remediation grew. Innovative technologies are characterized by limited availability of cost and performance data. More site-specific testing is required before an innovative technology can be considered ready for use at a site. Examples of innovative technologies are vacuum extraction, bioremediation, soil washing/flushing, chemical extraction, chemical destruction, and thermal desorption. Vapor extraction and in situ bioremediation are expected to be the next innovative technologies to reach "existing" status as a result of the growing base of cost and performance information generated by their use at many hazardous waste sites.

The last category is that of emerging technologies. These technologies are at a very early stage of development and therefore require additional laboratory and pilot scale testing to demonstrate their technical viability. No cost or performance information is available. An example of an emerging technology is electrokinetic treatment of soils for metals removal.

Groundwater contaminated by hazardous materials is a widespread concern. Most hazardous waste site remediations use a pump and treat approach as a first step. Once the groundwater has been brought to the surface, various treatment alternatives exist, depending upon the constituents present. In situ air sparging of the groundwater using pipes, **wells**, or curtains is also being developed for removal of volatile constituents. The vapor is either treated above ground with technologies for off-gas emissions, or biologically in the unsaturated or **vadose zone** above the **aquifer**. While this approach eliminates the costs and difficulties in treating the relatively large volumes of water (with relatively low contaminant concentrations) generated during pump-and-treat, it does not necessarily speed up remediation.

Contaminated bedrock frequently serves as a source of groundwater or soil recontamination. Constituents with densities greater than water enter the bedrock at fractures, joints or bedding planes. From these locations, the contamination tends to diffuse in all directions. After many years of accumulation, bedrock contamination may account for the majority of the contamination at a site. Currently, little can be done to remediate contaminated bedrock. Specially designed vapor stripping applications have been proposed when the constituents of concern are volatile. Efforts are on-going in developing means to enhance the fractures of the bedrock and thereby promote removal. In all cases, the ultimate remediation will be driven by the diffusion of contaminants back out of the rock, a very slow process.

The remediation of buildings contaminated with hazardous waste offers several alternatives. Given the cost of disposal of hazardous wastes, the limited disposal space available, and the volume of demolition debris, it is beneficial to determine the extent of contamination of construction materials. This contamination can then be removed through traditional engineering approaches, such as scraping or sand blasting. It is then only this reduced volume of material that requires treatment or disposal as hazardous waste. The remaining building can be reoccupied or disposed of as nonhazardous waste. *See also* Hazardous material; Hazardous waste siting; Solidification of hazardous materials; Vapor recovery system

[*Ann N. Clarke and Jeffrey L. Pintenich*]

RESOURCES
BOOKS

U.S. Environmental Protection Agency. Office of Emergency and Remedial Response. *Guidance for Conducting Remedial Investigations and Feasibility Studies Under CERCLA.* Washington, DC: U.S. Government Printing Office, 1988.

U.S. Environmental Protection Agency. Office of Emergency and Remedial Response. *Handbook: Remedial Action at Waste Disposal Sites.* Washington, DC: U.S. Government Printing Office, 1985.

U.S. Environmental Protection Agency. Office of Emergency and Remedial Response. *Guidance on Remedial Actions for Contaminated Groundwater at Superfund Sites.* Washington, DC: U.S. Government Printing Office, 1988.

U.S. Environmental Protection Agency. Office of Environmental Engineering. *Guide for Treatment Technologies for Hazardous Waste at Superfund Sites.* Washington, DC: U.S. Government Printing Office, 1989.

U.S. Environmental Protection Agency. Office of Solid Waste and Emergency Response. *Innovative Treatment Technologies.* Washington, DC: U.S. Government Printing Office, 1991.

U.S. Environmental Protection Agency. Risk Reduction Engineering Laboratory. *Handbook on In Situ Treatment of Hazardous Waste: Contaminated Soils.* Washington, DC: U.S. Government Printing Office, 1990.

U.S. Environmental Protection Agency. *Technology Screening Guide for Treatment of CERCLA Soils and Sludges.* Washington, DC: U.S. Government Printing Office, 1988.

Hazardous waste siting

Regardless of the specific technologies to be employed, there are many technical and nontechnical considerations to be addressed before **hazardous waste** can be treated or disposed of at a given location. The specific nature and relative importance of these considerations to the successful siting reflect the chemodynamic behavior (i.e., transport and fate of the waste and/or treated residuals in the **environment**

after **emission**) as well as the specifics of the location and associated, proximate areas. Examples of these considerations are: the nature of the **soil** and hydrogeological features such as depth to and quality of **groundwater**; quality, use, and proximity of surface waters; and **ambient air** quality and meteorological conditions; and nearby critical environmental areas (**wetlands**, preserves, etc.), if any. Other considerations include surrounding **land use**; proximity of residences and other potentially sensitive receptors such as schools, hospitals, parks, etc.; availability of utilities; and the capacity and quality of the roadway system. It is also critical to develop and obtain the timely approval of all appropriate local, state, and federal permits. Associated with these permits is the required documentation of financial viability as established by escrowed closure funds, site insurance, etc. Site-specific standard operating procedures as well as contingency plans for use in emergencies are also required. Additionally, there needs to be baseline and ongoing monitoring plans developed and implemented to determine if there are any releases to or general degradation of the environment. One should also anticipate public hearings before permits are granted. Several states in the United States have specific regulations which restrict the siting of hazardous **waste management** facilities.

Haze

An **aerosol** in the **atmosphere** of sufficient concentration and extent to decrease **visibility** significantly when the relative humidity is below saturation is known as haze. Haze may contain dry particles or droplets or a mixture of both, depending on the precise value of the humidity. In the use of the word, there is a connotation of some degree of permanence. For example, a dust storm is not a haze, but the coarse particles may settle rapidly and leave a haze behind once the velocity drops.

Human activity is responsible for many hazes. Enhanced **emission** of **sulfur dioxide** results in the formation of aerosols of sulfuric **acid**. In the presence of ammonia, which is excreted by most higher animals including humans, such emissions result in aerosols of ammonium sulfate and bisulfate. Organic hazes are part of **photochemical smog**, such as the **smog** often associated with Los Angeles, and they consist primarily of polyfunctional, highly oxygenated compounds with at least five **carbon** atoms. Such hazes can also form if air with an enhanced **nitrogen** oxide content meets air containing the natural terpenes emitted by vegetation.

All hazes, however, are not products of human activity. Natural hazes can result from forest fires, dust storms, and the natural processes that convert gaseous contaminants into particles for subsequent removal by precipitation or deposition to the surface or to vegetation. Still other hazes are of mixed origin, as noted above, and an event such as a dust storm can be enhanced by human-caused devegetation of **soil**.

Though it may contain particles injurious to health, haze is not of itself a health hazard. It can have a significant economic impact, however, when tourists cannot see scenic views, or if it becomes sufficiently dense to inhibit aircraft operations. *See also* Air pollution; Air quality; Air quality criteria; Los Angeles Basin; Mexico City, Mexico

[*James P. Lodge Jr.*]

RESOURCES
BOOKS

Husar, R. B. *Trends in Seasonal Haziness and Sulfur Emissions Over the Eastern United States.* Research Triangle Park, NC: U. S. Environmental Protection Agency, 1989.

PERIODICALS

Husar, R. B., and W. E. Wilson. "Haze and Sulfur Emission Trends in the Eastern United States." *Environmental Science and Technology* 27 (January 1993): 12–16.
Malm, W. C. "Characteristics and Origins of Haze in the Continental United States." *Earth-Science Reviews* 33 (August 1992): 1–36.
Raloff, J. "Haze May Confound Effects of Ozone Loss." *Science News* 141 (4 January 1992): 5.

Heat (stress) index

The heat index (HI) or heat stress index—sometimes called the apparent temperature or comfort index—is a temperature measure that takes into account the relative humidity. Based on human physiology and on clothing science, it measures how a given air temperature feels to the average person at a given relative humidity. The HI temperature is measured in the shade and assumes a wind speed of 5.6 mph (9 kph) and normal barometric pressure.

At low relative humidity, the HI is less than or equal to the air temperature. At higher relative humidity, the HI exceeds the air temperature. For example, according to the National Weather Service's (NWS) HI chart, if the air temperature is 70°F (21°C), the HI is 64°F (18°C) at 0% relative humidity and 72°F (22°C) at 100% relative humidity. At 95°F (35°C) and 55% relative humidity, the HI is 110°F (43°C). In very hot weather, humidity can raise the HI to extreme levels: at 115°F (46°C) and 40% relative humidity, the HI is 151°F (66°C). This is because humidity affects the body's ability to regulate internal heat through perspiration. The body feels warmer when it is humid because perspiration evaporates more slowly; thus the HI is higher.

The HI is used to predict the risk of physiological heat stress for an average individual. Caution is advised at an HI of 80–90°F (27–32°C): fatigue may result with prolonged exposure and physical activity. An HI of 90–105°F (32–41°C) calls for extreme caution, since sunstroke, muscle cramps, and heat exhaustion are possible. Danger warnings are issued at HIs of 105–130°F (41–54°C), when sunstroke and heat exhaustion are likely and there is a potential for heat stroke. Category IV, extreme danger, occurs at HIs above 130°F (54°C), when heatstroke and sunstroke are imminent.

Individual physiology influences how people are affected by high HIs. Children and older people are more vulnerable. Acclimatization (being used to the **climate**) can alleviate some of the danger. However sunburn can increase the effective HI by slowing the skin's ability to shed excess heat from blood vessels and through perspiration. Exposure to full sunlight can increase HI values by as much as 15°F (8°C). Winds, especially hot dry winds, also can increase the HI. In general, the NWS issues excessive heat alerts when the daytime HI reaches 105°F (41°C) and the nighttime HI stays above 80°F (27°C) for two consecutive days; however these values depend somewhat on the region or metropolitan area. In cities, high HIs often mean increased **air pollution**. The concentration of **ozone**, the major component of **smog**, tends to rise at ground level as the HI increases, causing respiratory problems for many people.

The National Center for Health **Statistics** estimates that heat exposure results in an average of 371 deaths annually in the United States. About 1,700 Americans died in the heat waves of 1980. In Chicago in 1995, more than 700 people died during a five-day heat wave when the nighttime HI stayed above 89°F (32°C).

At higher temperatures, the air can hold more water vapor; thus humidity and HI values increase as the **atmosphere** warms. Since the late nineteenth century, the mean annual surface temperature of the earth has risen between 0.5 and 1.0°F (0.3 and 0.6°C). According to the National Aeronautics and Space Administration, the five-year-mean temperature increased about 0.9°F (0.5°C) between 1975 and 1999, the fastest rate of recorded increase. In 1998 global surface temperatures were the warmest since the advent of reliable measurements and the 1990s accounted for seven of the 10 warmest years on record. Nighttime temperatures have been increasing twice as fast as daytime temperatures.

Greenhouse gases, including **carbon dioxide**, **methane**, **nitrous oxide**, and **chlorofluorocarbons**, increase the heat-trapping capabilities of the atmosphere. Evaporation from the ocean surfaces increased during the twentieth century, resulting in higher humidity that enhanced the **greenhouse effect**. It is projected that during the twenty-first century greenhouse gas concentrations will double or even quadruple from pre-industrial levels. Increased urbanization also contributes to global warming, as buildings and roads hold in the heat. Climate simulations predict an average surface air temperature increase of 4.5–7°F (2.5–4°C) by 2100. This will increase the number of extremely hot days and, in temperate climates, double the number of very hot days, for an average increase in summer temperatures of 4–5°F (2–3°C). More heat-related illnesses and deaths will result.

The **National Oceanic and Atmospheric Administration** projects that the HI could rise substantially in humid regions of the tropics and sub-tropics. Warm, humid regions of the southeastern United States are expected to experience substantial increases in the summer HI due to increased humidity, even though temperature increases may be smaller than in the continental interior. Predictions for the increase in the summer HI for the Southeast United States over the next century range from 8–20°F (4–11°C).

[*Margaret Alic Ph.D.*]

RESOURCES
BOOKS

Wigley, T. M. L. *The Science of Climate Change: Global and U.S. Perspectives.* Arlington, VA: Pew Center on Global Climate Change, 1999.

PERIODICALS

Delworth, T. L., J. D. Mahlman, and T. R. Knutson. "Changes in Heat Index Associated with CO₂-Induced Global Warming." *Climatic Change* 43 (1999): 369–86.

OTHER

Darling, Allan. "Heat Wave." *Internet Weather Source.* June 27, 2000 [cited May 2002]. <http://weather.noaa.gov/weather/hwave.html>.

Davies, Kert. "Heat Waves and Hot Nights." *Ozone Action and Physicians for Social Responsibility.* July 26, 2000 [cited May 2002]. <http://www.psr.org/heatreport.pdf>.

National Assessment Synthesis Team. *Climate Change Impacts on the United States: The Potential Consequences of Climate Variability and Change. Overview: Southeast.* U. S. Global Change Research Program. March 5, 2002 [May 2002]. <http://www.usgcrp.gov/usgcrp/Library/nationalassessment/overviewsoutheast.htm>.

Union of Concerned Scientists. *Early Warning Signs of Global Warming: Heat Waves and Periods of Unusually Warm Weather.* 2000 [cited May 2002]. <http://www.ucsusa.org/balance/gw_heat.html>.

Trenberth, Kevin E. "The IPCC Assessment of Global Warming 2001." *FailSafe.* Spring 2001 [cited May 2002]. <http://www.felsef.org/spring01.htm#3>.

ORGANIZATIONS

National Weather Service, National Oceanic and Atmospheric Administration, U. S. Department of Commerce, 1325 East West Highway, Silver Spring, USA 20910 , <http://www.nws.noaa.gov>

Physicians for Social Responsibility, 1875 Connecticut Avenue, NW, Suite 1012, Washington , DC USA 20009 (202) 667-4260, Fax: (202) 667-4201, Email: psrnatl@psr.org, <http://www.psr.org>

Union of Concerned Scientists, 2 Brattle Square, Cambridge, MA USA 02238 (617) 547-5552, Email: ucs@ucsusa.org, <http://www.ucsusa.org>

Heavy metals and heavy metal poisoning

Heavy metals are generally defined as environmentally stable elements of high specific gravity and atomic weight. They have such characteristics as luster, ductility, malleability, and high electric and thermal conductivity. Whether based on their physical or chemical properties, the distinction between heavy metals and non-metals is not sharp. For example, **arsenic**, germanium, selenium, tellurium, and antimony possess chemical properties of both metals and non-metals. Defined as metalloids, they are often loosely classified as heavy metals. The category "heavy metal" is, therefore, somewhat arbitrary and highly non-specific because it can refer to approximately 80 of the 103 elements in the periodic table. The term "trace element" is commonly used to describe substances which cannot be precisely defined but most frequently occur in the **environment** in concentrations of a few **parts per million** (ppm) or less. Only a relatively small number of heavy metals such as **cadmium, copper,** iron, cobalt, zinc, **mercury,** vanadium, **lead, nickel,** chromium, manganese, molybdenum, silver, and tin as well as the metalloids arsenic and selenium are associated with environmental, plant, animal, or human health problems.

While the chemical forms of heavy metals can be changed, they are not subject to chemical/biological destruction. Therefore, after release into the environment they are persistent contaminants. Natural processes such as bedrock and **soil weathering,** wind and water **erosion,** volcanic activity, sea salt spray, and forest fires release heavy metals into the environment. While the origins of **anthropogenic** releases of heavy metals are lost in antiquity, they probably began as our prehistoric ancestors learned to recover metals such as gold, silver, copper, and tin from their ores and to produce bronze. The modern age of heavy metal **pollution** has its beginning with the Industrial Revolution. The rapid development of industry, intensive agriculture, **transportation,** and urbanization over the past 150 years, however, has been the precursor of today's environmental contamination problems. Anthropogenic utilization has also increased heavy metal distribution by removing the substances from localized ore deposits and transporting them to other parts of the environment. Heavy metal by-products result from many activities including: ore extraction and smelting, fossil fuel **combustion,** dumping and landfilling of industrial wastes, exhausts from leaded gasolines, steel, iron, cement and **fertilizer** production, refuse and wood combustion. Heavy metal cycling has also increased through activities such as farming, **deforestation,** construction, **dredging** of harbors, and the disposal of municipal sludges and industrial wastes on land.

Thus, anthropogenic processes, especially combustion, have substantially supplemented the natural atmospheric emissions of selected heavy metals/metalloids such as selenium, mercury, arsenic, and antimony. They can be transported as gases or adsorbed on particles. Other metals such as cadmium, lead, and zinc are transported atmospherically only as particles. In either state heavy metals may travel long distances before being deposited on land or water.

The heavy metal contamination of soils is a far more serious problem than either air or **water pollution** because heavy metals are usually tightly bound by the organic components in the surface layers of the soil and may, depending on conditions, persist for centuries or millennia. Consequently, the soil is an important geochemical sink which accumulates heavy metals rapidly and usually depletes them very slowly by **leaching** into **groundwater** aquifers or bioaccumulating into plants. However, heavy metals can also be very rapidly translocated through the environment by erosion of the soil particles to which they are adsorbed or bound and redeposited elsewhere on the land or washed into rivers, lakes or oceans to the **sediment**.

The cycling, bioavailability, toxicity, transport, and fate of heavy metals are markedly influenced by their physico-chemical forms in water, sediments, and soils. Whenever a heavy metal containing **ion** or compound is introduced into an aquatic environment, it is subjected to a wide variety of physical, chemical, and biological processes. These include: hydrolysis, chelation, complexation, redox, biomethylation, precipitation and **adsorption** reactions. Often heavy metals experience a change in the chemical form or speciation as a result of these processes and so their distribution, bioavailability, and other interactions in the environment are also affected.

The interactions of heavy metals in aquatic systems are complicated because of the possible changes due to many dissolved and **particulate** components and non-equilibrium conditions. For example, the speciation of heavy metals is controlled not only by their chemical properties but also by environmental variables such as: 1) **pH**; 2) redox potential; 3) **dissolved oxygen**; 4) ionic strength; 5) temperature; 6) **salinity**; 7) alkalinity; 8) hardness; 9) concentration and **nature** of inorganic ligands such as carbonate, bicarbonate, sulfate, sulfides, chlorides; 10) concentration and nature of dissolved organic chelating agents such as organic acids, humic materials, **peptides**, and polyamino-carboxylates; 11) the concentration and nature of particulate matter with surface sites available for heavy metal binding; and 12) biological activity.

In addition, various **species** of bacteria can oxidize arsenate or reduce arsenate to arsenite, or oxidize ferrous iron to ferric iron, or convert mercuric ion to elemental mercury or the reverse. Various **enzyme** systems in living organisms can biomethylate a number of heavy metals. While it had been known for at least 60 years that arsenic

and selenium could be biomethylated, **microorganisms** capable of converting inorganic mercury into monomethyl and dimethylmercury in lake sediments were not discovered until 1967. Since then, numerous heavy metals such as lead, tin, cobalt, antimony, platinum, gold, tellurium, thallium, and palladium have been shown to be biomethylated by bacteria and **fungi** in the environment.

As environmental factors change the chemical reactivities and speciation of heavy metals, they influence not only the mobilization, transport, and bioavailability, but also the toxicity of heavy metal ions toward biota in both freshwater and marine ecosystems. The factors affecting the toxicity and **bioaccumulation** of heavy metals by aquatic organisms include: 1) the chemical characteristics of the ion; 2) solution conditions which affect the chemical form (speciation) of the ion; 3) the nature of the response such as acute toxicity, bioaccumulation, various types of **chronic effects**, etc.; 4) the nature and condition of the aquatic animal such as age or life stage, species, or **trophic level** in the food chain. The extent to which most of the methylated metals are bioaccumulated and/or biomagnified is limited by the chemical and biological conditions and how readily the methylated metal is metabolized by an organism. At present, only methylmercury seems to be sufficiently stable to bioaccumulate to levels that can cause adverse effects in aquatic organisms. All other methylated metal ions are produced in very small concentrations and are degraded naturally faster than they are bioaccumulated. Therefore, they do not biomagnify in the food chain.

The largest proportion of heavy metals in water is associated with suspended particles, which are ultimately deposited in the bottom sediments where concentrations are orders of magnitude higher than those in the overlying or interstitial waters. The heavy metals associated with suspended particulates or bottom sediments are complex mixtures of: 1) weathering and erosion residues such as iron and **aluminum** oxyhydroxides, clays and other aluminosilicates; 2) methylated and non-methylated forms in organic matter such as living organisms, bacteria and algae, **detritus** and **humus**; 3) inorganic hydrous oxides and hydroxides, **phosphates** and silicates; and 4) diagenetically produced iron and manganese oxyhydroxides in the upper layer of sediments and sulfides in the deeper, anoxic layers.

In anoxic waters the precipitation of sulfides may control the heavy metal concentrations in sediments while in oxic waters adsorption, **absorption**, surface precipitation and **coprecipitation** are usually the mechanisms by which heavy metals are removed from the water column. Moreover, physical, chemical and microbiological processes in the sediments often increase the concentrations of heavy metals in the pore waters which are released to overlying waters by diffusion or as the result of consolidation and bioturbation.

Transport by living organisms does not represent a significant mechanism for local movement of heavy metals. However, accumulation by aquatic plants and animals can lead to important biological responses. Even low environmental levels of some heavy metals may produce subtle and chronic effects in animal populations.

Despite these adverse effects, at very low levels, some metals have essential physiological roles as micronutrients. Heavy metals such as chromium, manganese, iron, cobalt, molybdenum, nickel, vanadium, copper, and selenium are required in small amounts to perform important biochemical functions in plant and animal systems. In higher concentrations they can be toxic, but usually some biological regulatory mechanism is available by means of which animals can speed up their excretion or retard their uptake of excessive quantities.

In contrast, non-essential heavy metals are primarily of concern in terrestrial and aquatic systems because they are toxic and persist in living systems. Metal ions commonly bond with sulfhydryl and carboxylic **acid** groups in amino acids, which are components of proteins (enzymes) or polypeptides. This increases their bioaccumulation and inhibits excretion. For example, heavy metals such as lead, cadmium, and mercury bind strongly with -SH and -SCH$_3$ groups in cysteine and methionine and so inhibit the **metabolism** of the bound enzymes. In addition, other heavy metals may replace an essential element, decreasing its availability and causing symptoms of deficiency.

Uptake, translocation, and accumulation of potentially toxic heavy metals in plants differ widely depending on soil type, pH, redox potential, moisture, and organic content. Public health officials closely regulate the quantities and effects of heavy metals that move through the agricultural food chain to be consumed by human beings. While heavy metals such as zinc, copper, nickel, lead, arsenic, and cadmium are translocated from the soil to plants and then into the animal food chain, the concentrations in plants are usually very low and generally not considered to be an environmental problem. However, plants grown on soils either naturally enriched or highly contaminated with some heavy metals can bioaccumulate levels high enough to cause toxic effects in the animals or human beings that consume them.

Contamination of soils due to land disposal of sewage and industrial effluents and sludges may pose the most significant long term problem. While cadmium and lead are the greatest hazard, other elements such as copper, molybdenum, nickel, and zinc can also accumulate in plants grown on sludge-treated land. High concentrations can, under certain conditions, cause adverse effects in animals and human beings that consume the plants. For example, when soil contains high concentrations of molybdenum and selenium,

they can be translocated into edible plant tissue in sufficient quantities to produce toxic effects in ruminant animals. Consequently, the U. S. **Environmental Protection Agency** has issued regulations which prohibit and/or tightly regulate the disposal of contaminated municipal and industrial sludges on land to prevent heavy metals, especially cadmium, from entering the food supply in toxic amounts. However, presently, the most serious known human toxicity is not through bioaccumulation from crops but from mercury in fish, lead in **gasoline**, paints and water pipes, and other metals derived from occupational or accidental exposure. *See also* Aquatic chemistry; Ashio, Japan; Atmospheric pollutants; Biomagnification; Biological methylation; Contaminated soil; Ducktown, Tennessee; Hazardous material; Heavy metals precipitation; Itai-Itai disease; Methylmercury seed dressings; Minamata disease; Smelters; Sudbury, Ontario; Xenobiotic

[*Frank M. D'Itri*]

RESOURCES
BOOKS

Craig, P. J. "Metal Cycles and Biological Methylation." *The Handbook of Environmental Chemistry.* Vol. 1, Part A, edited by O. H. Hutzinger. Berlin: Springer Verlag, 1980.
Förstner, U., and G. T. W. Wittmann. *Metal Pollution in the Aquatic Environment.* 2nd ed. Berlin: Springer Verlag, 1981.
Kramer, J. R., and H. E. Allen, eds. *Metal Speciation: Theory, Analysis and Application.* Chelsea, MI: Lewis, 1988.

Heavy metals precipitation

The principle technology to remove metals pollutants from **wastewater** is by chemical precipitation. Chemical precipitation includes two secondary removal mechanisms, **coprecipitation** and **adsorption**. Precipitation processes are characterized by the solubility of the metal to be removed. They are generally designed to precipitate trace metals to their solubility limits and obtain additional removal by coprecipitation and adsorption during the precipitation reaction.

There are many different treatment variables that affect these processes. They include the optimum **pH**, the type of chemical treatments used, and the number of treatment stages, as well as the temperature and volume of wastewater, and the chemical specifications of the pollutants to be removed. Each of these variables directly influences treatment objectives and costs. Treatability studies must be performed to optimize the relevant variables, so that goals are met and costs minimized.

In theory, the precipitation process has two steps, nucleation followed by particle growth. Nucleation is represented by the appearance of very small particle seeds which are generally composed of 10–100 molecules. Particle growth involves the addition of more atoms or molecules into this particle structure. The rate and extent of this process is dependent upon the temperature and chemical characteristics of the wastewater, such as the concentration of metal initially present and other ionic **species** present, which can compete with or form soluble complexes with the target metal species.

Heavy metals are present in many industrial wastewaters. Examples of such metals are **cadmium, copper, lead, mercury, nickel,** and zinc. In general, these metals can be complexed to insoluble species by adding sulfide, hydroxide, and carbonate ions to a solution. For example, the precipitation of copper (Cu) hydroxide is accomplished by adjusting the pH of the water to above 8, using precipitant **chemicals** such as lime ($Ca(OH)_2$) or sodium hydroxide (N_aOH). Precipitation of metallic carbonate and sulfide species can be accomplished by the addition of calcium carbonate or sodium sulfide. The removal of coprecipitive metals during precipitation of the soluble metals is aided by the presence of solid ferric oxide, which acts as an adsorbent during the precipitation reaction. For example, hydroxide precipitation of ferric chloride can be used as the source of ferric oxide for coprecipitation and adsorption reactions. Precipitation, coprecipitation, and adsorption reactions generate suspended solids which must be separated from the wastewater. Flocculation and clarification are again employed to assist in solids separation. The treatment is an important variable which must be optimized to effect the maximum metal removal possible.

Determining the optimal pH range to facilitate the maximum precipitation of metal is a difficult task. It is typically accomplished by laboratory studies, such as by-jar tests rather than theoretical calculations. Often the actual wastestream behaves differently, and the theoretical metal solubilities and corresponding optimal pH ranges can vary considerably from theoretical values. *See also* Heavy metals and heavy metal poisoning; Industrial waste treatment; Itai-itai disease; Minamata disease; Sludge; Waste management

[*James W. Patterson*]

RESOURCES
BOOKS

Nemerow, N. L., and A. Dasgupta. *Industrial and Hazardous Waste Treatment.* New York: Van Nostrand Reinhold, 1991.

Robert Louis Heilbroner (1919 –)

American economist and author

An economist by profession, Robert Heilbroner is the author of a number of books and articles that put economic theories

and developments into historical perspective and relate them to contemporary social and political problems. He is especially noteworthy for his gloomy speculations on the future of a world confronted by the environmental limits to economic growth.

Born in New York City in 1919, Heilbroner received a bachelor's degree from Harvard University in 1940 and a Bronze Star for his service in World War II. In 1963 he earned a Ph.D. in economics from the New School for Social Research in New York, and in 1972, became the Norman Thomas Professor of Economics there. His books include *The Worldly Philosophers* (1955), *The Making of Economic Society* (1962), *Marxism: For and Against* (1980), and *The Nature and Logic of Capitalism* (1985). He has also served on the editorial board of the socialist journal *Dissent*.

In 1974, Heilbroner published *An Inquiry into the Human Prospect*, in which he argues that three "external challenges" confront humanity: the population explosion, the threat of war, and "the danger...of encroaching on the **environment** beyond its ability to support the demands made on it." Each of these problems, he maintains, arises from the development of scientific technology, which has increased human life span, multiplied weapons of destruction, and encouraged industrial production that consumes **natural resources** and pollutes the environment. Heilbroner believes that these challenges confront all economies, and that meeting them will require more than adjustments in economic systems. Societies will have to muster the will to make sacrifices.

Heilbroner goes on to argue that persuading people to make these sacrifices may not be possible. Those living in one part of the world are not likely to give up what they have for the sake of those in another part, and people living now are not likely to make sacrifices for **future generations**. His reluctant conclusion is that coercion is likely to take the place of persuasion. Authoritarian governments may well supplant democracies because "the passage through the gantlet ahead may be possible only under governments capable of rallying obedience far more effectively than would be possible in a democratic setting. If the issue for mankind is survival, such governments may be unavoidable, even necessary."

Heilbroner wrote *An Inquiry into the Human Prospect* in 1972 and 1973, but his position had not changed by the end of the decade. In a revised edition written in 1979, he continued to insist upon the environmental limits to economic growth: "the industrialized capitalist and socialist worlds can probably continue along their present growth paths" for about 25 years, at which point "we must expect...a general recognition that the possibilities for expansion are limited, and that social and economic life must be maintained within fixed...material boundaries." Heilbroner has published a number of books, including *21st Century Capitalism*

Robert L. Heilbroner. (Photograph by Jose Pelaez. W. W. Norton. Reproduced by permission.)

and *Visions of the Future*. He also received the New York Council for the Humanities Scholar of the Year award in 1994. Heilbroner currerently holds the position of Norman Thomas Professor of Economics, Emeritus, at the New School for Social Research, in New York City.

[*Richard K. Dagger*]

RESOURCES
BOOKS

Heilbroner, Robert L. *An Inquiry into the Human Prospect*. Rev. ed. New York: Norton, 1980.
————. *The Making of an Economic Society*. 6th ed. Englewood Cliffs, NJ: Prentice-Hall, 1980.
————. *The Nature and Logic of Capitalism*. New York: Norton, 1985.
————. *Twenty-First Century Capitalism*. Don Mills, Ont.: Anansi, 1992.
————. *The Worldly Philosophers: The Lives, Times and Ideas of the Great Economic Thinkers*. 6th ed. New York: Simon & Schuster, 1986.
Straub, D., ed. *Contemporary Authors: New Revision Series*. Vol. 21. Detroit, MI: Gale Research, 1987.

Hells Canyon

Hells Canyon is a stretch of canyon on the Snake River between Idaho and Oregon. This canyon, deeper than the Grand Canyon and formed in ancient basalt flows, contains some of the United States' wildest rapids and has provided extensive recreational and scenic boating since the 1920s. The narrow canyon has also provided outstanding dam sites. Hells Canyon became the subject of nationwide controversy between 1967 and 1975, when environmentalists challenged hydroelectric developers over the last stretch of free-flowing water in the Snake River from the border of Wyoming to the Pacific.

Historically Hells Canyon, over 100 mi (161 km) long, filled with rapids, and averaging 6,500 ft (1,983 m) deep, presented a major obstacle to travelers and explorers crossing the mountains and deserts of southern Idaho and eastern Oregon. Nez Percé, Paiute, Cayuse, and other Native American groups of the region had long used the area as a mild wintering ground with good grazing land for their horses. European settlers came for the modest timber and with cattle and sheep to graze. As early as the 1920s travelers were arriving in this scenic area for recreational purposes, with the first river runners navigating the canyon's rapids in 1928. By the end of the Depression the **Federal Power Commission** was urging regional utility companies to tap the river's hydroelectric potential, and in 1958 the first dam was built in the canyon.

Falling from the mountains in southern **Yellowstone National Park** through Idaho, and into the Columbia River, the Snake River drops over 7,000 vertical ft (2,135 m) in 1,000 mi (1,609 km) of river. This drop and the narrow gorges the river has carved presented excellent dam opportunities, and by the end of the 1960s there were 18 major **dams** along the river's course. By that time the river was also attracting great numbers of whitewater rafters and kayakers, as well as hikers and campers in the adjacent national forests. When a proposal was developed to dam the last free-running section of the canyon, protesters brought a suit to the United States Supreme Court. In 1967, Justice William O. Douglas led the majority in a decision directing the utilities to consider alternatives to the proposed dam.

Hells Canyon became a national environmental issue. Several members of Congress flew to Oregon to raft the river. The **Sierra Club** and other groups lobbied vigorously. Finally, in 1975 President Gerald Ford signed a bill declaring the remaining stretch of the canyon a National Scenic Waterway, creating a 650,000-acre (260,000-ha) Hells Canyon National **Recreation** Area, and adding 193,000 acres (77,200 ha) of the area to the National **Wilderness** Preservation System. *See also* Wild and Scenic Rivers Act; Wild river

[*Mary Ann Cunningham Ph.D.*]

RESOURCES
BOOKS

Collins, R. O., and R. Nash. *The Big Drops*. San Francisco: Sierra Club Books, 1978.

OTHER

Hells Canyon Recreation Area. "Hells Canyon." Washington, DC: U.S. Government Printing Office, 1988.

Hazel Henderson (1933 –)

English/American environmental activist and writer

Hazel Henderson is an environmental activist and futurist who has called for an end to current "unsustainable industrial modes" and urges redress for the "unequal access to resources which is now so dangerous, both ecologically and socially."

Born in Clevedon, England, Henderson immigrated to the United States after finishing high school; she became a naturalized citizen in 1962. After working for several years as a free-lance journalist, she married Carter F. Henderson, former London bureau chief of the *Wall Street Journal* in 1957. Her activism began when she became concerned about **air quality** in New York City, where she was living. To raise public awareness, she convinced the FCC and television networks to broadcast the **air pollution index** with the weather report. She persuaded an advertising agency to donate their services to her cause and teamed up with a New York City councilman to co-found Citizens for Clean Air. Her endeavors were rewarded in 1967, when she was commended as Citizen of the Year by the New York Medical Society.

Henderson's career as an advocate for social and environmental reform took flight from there. She argued passionately against the spread of industrialism, which she called "pathological" and decried the use of an economic yardstick to measure quality of life. Indeed, she termed economics "merely politics in disguise" and even "a form of brain damage." Henderson believed that society should be measured by less tangible means, such as political participation, literacy, education, and health. "Per-capita income," she felt, is "a very weak indicator of human well-being."

She became convinced that traditional industrial development wrought little but "ecological devastation, social unrest, and downright hunger...I think of development, instead,...as investing in ecosystems, their restoration and management."

Even the fundamental idea of labor should, Henderson argued, "be replaced by the concept of 'Good Work'—which challenges individuals to grow and develop their faculties; to overcome their ego-centeredness by joining with others in common tasks; to bring forth those goods and services needed for a becoming existence; and to do all this with an ethical concern for the interdependence of all life forms..."

To advance her theories, Henderson has published several books, *Creative Alternative Futures: The End of Economics* (1978), *The Politics of the Solar Age: Alternatives to Economics* (1981), *Building a Win-Win World* (1996), *Toward Sustainable Communities: Resources for Citizens and Their Governments* (1998), and *Beyond Globalization: Shaping a Sustainable Global Economy* (1999). She has also contributed to several periodicals, and lectured at colleges and universities. In 1972 she co-founded the Princeton Center for Alternative Futures, of which she is still a director. She is a member of the board of directors for **Worldwatch Institute** and the Council for Economic Priorities, among other organizations. In 1982 she was appointed a Horace Allbright Professor at the University of California at Berkeley. In 1996, Henderson was awarded the Global Citizen Award.

[*Amy Strumolo*]

RESOURCES

BOOKS

Henderson, H. *Beyond Globalization: Shaping a Sustainable Global Economy.* 1999.

———. *Building a Win-Win World.* 1996.

———. *Creative Alternative Futures: The End of Economics.* 1978.

———. *The Politics of the Solar Age: Alternatives to Economics.* 1981.

———. *Toward Sustainable Communities: Resources for Citizens and Their Governments.* 1998.

Telephone Interview with Hazel Henderson. *Whole Earth Review* (Winter 1988): 58–59.

PERIODICALS

Henderson, H. "The Legacy of E. F. Schumacher." *Environment* 20 (May 1978): 30–36.

Holden, C. "Hazel Henderson: Nudging Society Off Its Macho Trip." *Science* 190 (November 28, 1975): 863–64.

Herbicide

Herbicides are chemical pesticides that are used to manage vegetation. Usually, herbicides are used to reduce the abundance of weedy plants, so as to release desired crop plants from **competition**. This is the context of most herbicide use in agriculture, forestry, and for lawn management. Sometimes herbicides are not used to protect crops, but to reduce the quantity or height of vegetation, for example along highways and transmission corridors. The reliance on herbicides to achieve these ends has increased greatly in recent decades, and the practice of chemical weed control appears to have become an entrenched component of the modern technological culture of humans, especially in agroecosystems.

The total use of pesticides in the United States in the mid-1980s was 957 million lb per year (434 million kg/year), used over 592,000 mi^2 (1.5 million km^2). Herbicides were most widely used, accounting for 68% of the total quantity [646 million lb per year [293 million kg/year]), and applied to 82% of the treated land [484,000 square miles per year (121 million hectares/year)]. Note that especially in agriculture, the same land area can be treated numerous times each year with various pesticides.

A wide range of **chemicals** is used as herbicides, including:

- chlorophenoxy acids, especially **2,4-D** and **2,4,5-T**, which have an auxin-like growth-regulating property and are selective against broadleaved angiosperm plants;
- triazines such as **atrazine**, simazine, and hexazinone;
- chloroaliphatics such as dalapon and trichloroacetate;
- the phosphonoalkyl chemical, glyphosate, and
- inorganics such as various arsenicals, cyanates, and chlorates.

A "weed" is usually considered to be any plant that interferes with the productivity of a desired crop plant or some other human purpose, even though in other contexts weed **species** may have positive ecological and economic values. Weeds exert this effect by competing with the crop for light, water, and nutrients. Studies in Illinois demonstrated an average reduction of yield of corn or maize (*Zea mays*) of 81% in unweeded plots, while a 51% reduction was reported in Minnesota. Weeds also reduce the yield of small grains, such as wheat (*Triticum aestivum*) and barley (*Hordeum vulgare*), by 25–50%.

Because there are several herbicides that are toxic to dicotyledonous weeds but not grasses, herbicides are used most intensively used in grain crops of the Gramineae. For example, in North America almost all of the area of maize cultivation is treated with herbicides. In part this is due to the widespread use of no-tillage cultivation, a system that reduces **erosion** and saves fuel. Since an important purpose of plowing is to reduce the abundance of weeds, the no-tillage system would be impracticable if not accompanied by herbicide use. The most important herbicides used in maize cultivation are atrazine, propachlor, alachlor, 2,4-D, and butylate. Most of the area planted to other agricultural grasses such as wheat, rice (*Oryza sativa*), and barley is also treated with herbicide, mostly with the phenoxy herbicides 2,4-D or MCPA.

The intended ecological effect of any **pesticide** application is to control a **pest** species, usually by reducing its abundance to below some economically acceptable threshold. In a few situations, this objective can be attained without important nontarget damage. For example, a judicious spot-application of a herbicide can allow a selective kill of large lawn weeds in a way that minimizes exposure to nontarget plants and animals.

Of course, most situations where herbicides are used are more complex and less well-controlled than this. Whenever a herbicide is broadcast-sprayed over a field or forest, a wide variety of on-site, nontarget organisms is affected, and sprayed herbicide also drifts from the target area. These cause ecotoxicological effects directly, through toxicity to nontarget organisms and ecosystems, and indirectly, by changing **habitat** or the abundance of food species of **wildlife**. These effects can be illustrated by the use of herbicides in forestry, with glyphosate used as an example.

The most frequent use of herbicides in forestry is for the release of small coniferous plants from the effects of competition with economically undesirable weeds. Usually the silvicultural use of herbicides occurs within the context of an intensive harvesting-and-management system, which may include **clear-cutting**, scarification, planting seedlings of a single desired species, spacing, and other practices.

Glyphosate is a commonly used herbicide in forestry and agriculture. The typical spray rate in silviculture is about 2.2–4.9 mi (1–2.2 kg) active ingredient/ha, and the typical projection is for one to two treatments per forest rotation of 40–100 years.

Immediately after an aerial application in forestry, glyphosate residues are about six times higher than litter on the forest floor, which is physically shielded from spray by overtopping foliage. The persistence of glyphosate residues is relatively short, with typical half-lives of two to four weeks in foliage and the forest floor, and up to eight weeks in **soil**. The disappearance of residues from foliage is mostly due to translocation and wash-off, but in the forest floor and soil glyphosate is immobile (and unavailable for root uptake or **leaching**) because of binding to organic matter and clay, and residue disappearance is due to microbial oxidation. Residues in oversprayed waterbodies tend to be small and short-lived. For example, two hours after a deliberate overspray on Vancouver Island, Canada, residues of glyphosate in stream water rose to high levels, then rapidly dissipated through flushing to only trace amounts 94 hours later.

Because glyphosate is soluble in water, there is no propensity for **bioaccumulation** in organisms in preference to the inorganic **environment**, or to occur in larger concentrations at higher levels of the **food chain/web**. This is in marked contrast to some other pesticides such as DDT, which is soluble in organic solvents but not in water, so it has a strong tendency to bioaccumulate into the fatty tissues of organisms.

As a plant poison, glyphosate acts by inhibiting the pathway by which four essential amino acids are synthesized. Only plants and some **microorganisms** have this metabolic pathway; animals obtain these amino acids from food. Consequently, glyphosate has a relatively small acute toxicity to animals, and there are large margins of toxicological safety in comparison with environmental exposures that are realistically expected during operational silvicultural sprays.

Acute toxicity of chemicals to mammals is often indexed by the oral dose required to kill 50% of a test population, usually of rats (i.e., rat LD_{50}). The LD_{50} value for pure glyphosate is 5,600 mg/kg, and its silvicultural formulation has a value of 5,400 mg/kg. Compare these to LD_{50}s for some chemicals which many humans ingest voluntarily: nicotine 50 mg/kg, caffeine 366, acetylsalicylic **acid** (ASA) 1,700, sodium chloride 3,750, and **ethanol** 13,000. The documented risks of longer-term, chronic exposures of mammals to glyphosate are also small, especially considering the doses that might be received during an operational treatment in forestry.

Considering the relatively small acute and chronic toxicities of glyphosate to animals, it is unlikely that wildlife inhabiting sprayed clearcuts would be directly affected by a silvicultural application. However, glyphosate causes large habitat changes through species-specific effects on plant productivity, and by changing habitat structure. Therefore, wildlife such as birds and mammals could be secondarily affected through changes in vegetation and the abundance of their arthropod foods. These indirect effects of herbicide spraying are within the context of **ecotoxicology**. Indirect effects can affect the abundance and reproductive success of terrestrial and aquatic wild life on a sprayed site, irrespective of a lack of direct, toxic effects.

Studies of the effects of habitat changes caused by glyphosate spraying have found relatively small effects on the abundance and species composition of wildlife. Much larger effects on wildlife are associated with other forestry practices, such as clear-cutting and the broadcast spraying of insecticides. For example, in a study of clearcuts sprayed with glyphosate in Nova Scotia, Canada, only small changes in avian abundance and species composition could be attributed to the herbicide treatment. However, such studies of bird abundance are conducted by enumerating territories, and the results cannot be interpreted in terms of reproductive success. Regrettably, there are not yet any studies of the reproductive success of birds breeding on clearcuts recently treated with a herbicide. This is an important deficiency in terms of understanding the ecological effects of herbicide spraying in forestry.

An important controversy related to herbicides focused on the military use of herbicides during the Viet Nam war. During this conflict, the United States Air Force broadcast-sprayed herbicides to deprive their enemy of food production and forest cover. More than 5,600 mi^2 (14,503 km^2) were sprayed at least once, about 1/7 the area of South Viet Nam. More than 55 million lb (25 million kg) of 2,4-D, 43 million lb (21 million kg) of 2,4,5-T, and 3.3 million lb (1.5 million kg) of picloram were used in this military program. The most frequently used herbicide was a 50:50 formulation of 2,4,5-T and 2,4-D known as **Agent Orange**. The rate of application was relatively large, averaging about 10 times the application rate for silvicultural purposes. About 86% of

spray missions were targeted against forests, and the remainder against cropland.

As was the military intention, these spray missions caused great ecological damage. Opponents of the practice labelled it "ecocide," i.e., the intentional use of anti-environmental actions as a military tactic. The broader ecological effects included severe damage to mangrove and tropical forests, and a great loss of wildlife habitat.

In addition, the Agent Orange used in Viet Nam was contaminated by the **dioxin** isomer known as TCDD, an incidental by-product of the manufacturing process of 2,4,5-T. Using post-Vietnam manufacturing technology, the contamination by TCDD in 2,4,5-T solutions can be kept to a concentration well below the maximum of 0.1 **parts per million** (ppm) set by the United States **Environmental Protection Agency** (EPA). However, the 2,4,5-T used in Viet Nam was grossly contaminated with TCDD, with a concentration as large as 45 ppm occurring in Agent Orange, and an average of about 2.0 ppm. Perhaps 243–375 lb (110–170 kg) of TCDD was sprayed with herbicides onto Vietnam. TCDD is well known as being extremely toxic, and it can cause **birth defects** and miscarriage in laboratory mammals, although as is often the case, toxicity to humans is less well understood. There has been great controversy about the effects on soldiers and civilians exposed to TCDD in Vietnam, but epidemiological studies have been equivocal about the damages. It seems likely that the effects of TCDD added little to human **mortality** or to the direct ecological effects of the herbicides that were sprayed in Vietnam.

A preferable approach to pesticide use is **integrated pest management** (IPM). In the context of IPM, pest control is achieved by employing an array of complementary approaches, including:

- use of natural predators, **parasites**, and other biological controls;
- use of pest-resistant varieties of crops;
- environmental modifications to reduce optimality of pest habitat;
- careful monitoring of pest abundance; and
- a judicious use of pesticides, when necessary as a component of the IPM strategy.

A successful IPM program can greatly reduce, but not necessarily eliminate, the reliance on pesticides.

With specific relevance to herbicides, more research into organic systems and into procedures that are pest-specific are required for the development of IPM systems. Examples of pest-specific practices are the biological control of certain introduced weeds, for example:

- St. John's wort (*Hypericum perforatum*) is a serious weed of pastures of the United States Southwest because it is toxic to cattle, but it was controlled by the introduction in 1943 of two herbivorous leaf beetles;

• the prickly pear cactus (*Opuntia* spp.) became a serious weed of Australian **rangelands** after it was introduced as an ornamental plant, but it has been controlled by release of the moth *Cactoblastis cactorum*, whose larvae feed on the cactus.

Unfortunately, effective IPM systems have not yet been developed for most weed problems for which herbicides are now used. Until there are alternative, pest-specific methods to achieve an economically acceptable degree of control of weeds in agriculture and forestry, herbicides will continue to be used for that purpose. *See also* Agricultural chemicals

[*Bill Freedman Ph.D.*]

RESOURCES
BOOKS

Freedman, B. *Environmental Ecology.* 2nd Edition. San Diego, CA: Academic Press, 1995.

McEwen, F. L., and G. R. Stephenson. *The Use and Significance of Pesticides in the Environment.* New York: Wiley, 1979.

PERIODICALS

Pimentel, D., et al. "Environmental and Economic Effects of Reducing Pesticide Use." *Bioscience* 41 (1991): 402–409.

———. "Controversy Over the Use of Herbicides in Forestry, With Particular Reference to Glyphosate Usage." *Environmental Carcinogenesis Reviews* C8 (1991): 277–286.

Heritage Conservation and Recreation Service

The Heritage Conservation and **Recreation** Service (HCRS) was created in 1978 as an agency of the U. S. Department of the Interior (Secretarial Executive Order 3017) to administer the National Heritage Program initiative of President Carter. The new agency was an outgrowth of and successor to the former Bureau of Outdoor Recreation. The HCRS resulted from the consolidation of some 30 laws, executive orders and interagency agreements that provided federal funds to states, cities and local community organizations to acquire, maintain, and develop historic, natural and recreation sites. HCRS focused on the identification and protection of the nation's significant natural, cultural and recreational resources. It classified and established registers for heritage resources, formulated policies and programs for their preservation, and coordinated federal, state and local resource and recreation policies and actions. In February

1981 HCRS was abolished as an agency and its responsibilities were transferred to the **National Park Service**.

Hetch Hetchy Reservoir

The Hetch Hetchy Reservoir, located on the Tuolumne River in **Yosemite National Park**, was built to provide water and hydroelectric power to San Francisco. Its creation in the early 1900s led to one of the first conflicts between preservationists and those favoring utilitarian use of **natural resources**. The controversy spanned the presidencies of Roosevelt, Taft, and Wilson.

A prolonged conflict between San Francisco and its only water utility, Spring Valley Water Company, drove the city to search for an independent water supply. After surveying several possibilities, the city decided to build a dam and reservoir in the Hetch Hetchy Valley because the river there could supply the most abundant and purest water. This option was also the least expensive, since the city planned to use the dam to generate hydroelectric power. It would also provide an abundant supply of **irrigation** water for area farmers and the **recreation** potential of a new lake.

The city applied to the U. S. Department of the Interior in 1901 for permission to construct the dam, but the request was not approved until 1908. The department then turned the issue over to Congress to work out an exchange of land between the federal government and the city. Congressional debate spanned several years and produced a number of bills. Part of the controversy involved the Right of Way Act of 1901, which gave Congress power to grant rights of way through government lands; some claimed this was designed specifically for the Hetch Hetchy project.

Opponents of the project likened the valley to Yosemite on a smaller scale. They wanted to preserve its high cliff walls, waterfalls, and diverse plant **species**. One of the most well-known opponents, **John Muir**, described the Hetch Hetchy Valley as "a grand landscape garden, one of Nature's rarest and most precious mountain temples." Campers and mountain climbers fought to save the campgrounds and trails that would be flooded.

As the argument ensued, often played out in newspapers and other public forums, overwhelming national opinion appeared to favor the preservation of the valley. Despite this public support, a close vote in Congress led to the passage of the Raker Act, allowing the O'Shaughnessy Dam and Hetch Hetchy Reservoir to be constructed. President Woodrow Wilson signed the bill into law on December 19, 1913.

The Hetch Hetchy Reservoir was completed in 1923 and still supplies water and electric power to San Francisco. In 1987, Secretary of the Interior Donald Hodel created a brief controversy when he suggested tearing down O'Shaughnessy Dam. *See also* Economic growth and the environment; Environmental law; Environmental policy

[*Teresa C. Donkin*]

RESOURCES
BOOKS

Jones, Holway R. *John Muir and the Sierra Club: The Battle for Yosemite.* San Francisco: Sierra Club, 1965.
Nash, Roderick. "Conservation as Anxiety." In *The American Environment: Readings in the History of Conservation.* 2nd ed. Reading, Mass: Addison-Wesley Publishing Company, 1976.

Heterotroph

A heterotroph is an organism that derives its nutritional **carbon** and energy by oxidizing (i.e., decomposing) organic materials. The higher animals, **fungi**, actinomycetes, and most bacteria are heterotrophs. These are the biological consumers that eventually decompose most of the organic matter on the earth. The **decomposition** products then are available for chemical or biological **recycling**. *See also* Biogeochemical cycles; Oxidation reduction reactions

High-grading (mining, forestry)

The practice of high-grading can be traced back to the early days of the California gold rush, when miners would sneak into claims belonging to others and steal the most valuable pieces of ore. The practice of high-grading remains essentially unchanged today. An individual or corporation will enter an area and selectively mine or harvest only the most valuable specimens, before moving on to a new area. High-grading is most prevalent in the mining and timber industries. It is not uncommon to walk into a forest, particularly an **old-growth forest**, and find the oldest and finest specimens marked for harvesting. *See also* Forest management; Strip mining

High-level radioactive waste

High-level **radioactive waste** consists primarily of the byproducts of nuclear **power plants** and defense activities. Such wastes are highly radioactive and often decay very slowly. They may release dangerous levels of radiation for hundreds or thousands of years. Most high-level radioactive wastes have to be handled by remote control by workers who are protected by heavy shielding. They present, therefore, a serious health and environmental hazard. No entirely satisfactory method for disposing of high-level wastes has as yet

been devised. Currently, the best approach seems to involve immobilizing the wastes in a glass-like material and then burying them deep underground. *See also* Low-level radioactive waste; Radioactive decay; Radioactive pollution; Radioactive waste management; Radioactivity

High-solids reactor

Solid waste disposal is a serious problem in the United States and other developed countries. Solid waste can constitute valuable raw materials for commercial and industrial operations, however, and one of the challenges facing scientists is to develop an economically efficient method for utilizing it.

Although the concept of bacterial waste conversion is simple, achieving an efficient method for putting the technique into practice is difficult. The main problem is that efficiency of conversion requires increasing the ratio of solids to water in the mixture, and this makes mixing more difficult mechanically. The high-solids reactor was designed by scientists at the **Solar Energy** Research Institute (SERI) to solve this problem. It consists of a cylindrical tube on a horizontal axis, and an agitator shaft running through the middle of it, which contains a number of Teflon-coated paddles oriented at 90 degrees to each other. The pilot reactors operated by SERI had a capacity of 2.6 gal (10 l).

SERI scientists modeled the high solids reactor after similar devices used in the **plastics** industry to mix highly viscous materials. With the reactor, they have been able to process materials with 30–35% solids content, while existing reactors normally handle wastes with five to eight% solid content. With higher solid content, SERI reactors have achieved a yield of **methane** five to eight times greater than that obtained from conventional mixers. Researchers hope to be able to process wastes with solid content ranging anywhere from zero to 100%. They believe that they can eventually achieve 80% efficiency in converting **biomass** to methane.

The most obvious application of the high-solids reactor is the processing of municipal solid wastes. Initial tests were carried out with **sludge** obtained from **sewage treatment** plants in Denver, Los Angeles, and Chicago. In all cases, conversion of solids in the sludge to methane was successful, and other applications of the reactor are also being considered. For example, it can be used to leach out **uranium** from mine wastes: **anaerobic** bacteria in the reactor will reduce uranium in the wastes and the uranium will then be absorbed on the bacteria or on **ion exchange** resins. The use of the reactor to clean **contaminated soil** is also being considered in the hope is that this will provide a desirable alternative to current processes for cleaning **soil**, which create large volumes of contaminated water. *See also* Biomass fuel; Solid waste incineration; Solid waste recycling and recovery; Solid waste volume reduction; Waste management

[*David E. Newton*]

RESOURCES
PERIODICALS

"High Solids Reactor May Reduce Capital Costs." *Bioprocessing Technology* (June 1990).
"SERI Looking for Partners for Solar-Powered High Solids Reactor." *Waste Treatment Technology News* (October 1990).

High-voltage power lines
see **Electromagnetic field**

High-yield crops
see **Borlaug, Norman E.; Consultative Group on International Agricultural Research**

Hiroshima, Japan

Hiroshima is a beautiful modern city located near the southwestern tip of the main Japanese island of Honshu. It had been a military center with the headquarters of the Japanese southern army and a military depot prior to the end of World War II. The city is now a manufacturing center with a major university and medical school. It is most profoundly remembered because it was the first city to be exposed to the devastation of an atomic bomb.

At 8:15 A.M. on the morning of August 6, 1945, a single B29 bomber flying in from Tinian Island in the Marinas released the bomb at 2,000 ft (606.6 m) above the city. The target was a "T"-shaped bridge near the city center. The only surviving building in the city center after the atomic bomb blast was a domed cement building at ground zero, just a few yards from the bridge. An experimental bomb developed by the Manhattan Project had been exploded at Alamagardo, New Mexico, only a few weeks earlier. The Alamagardo bomb had the explosive force of 15,000 tons of TNT. The Hiroshima uranium-235 bomb, with the explosive power of 20,000 tons of TNT, was even more powerful than the New Mexico bomb. The immediate effect of the bomb was to destroy by blast, winds, and fire an area of 4.4 mi² (7 km²). Two-thirds of the city was destroyed. A portion of Hiroshima was protected from the blast by hills, and this is all that remains of the old city. Destruction of human lives was caused immediately by the blast force

of the bomb or by burns or **radiation sickness** later. Seventy-five thousand people were killed or were fatally wounded; there was an equal number of wounded survivors. Nagasaki, to the south and west of Hiroshima, was bombed on August 9, 1945, with much loss of life. The bombing of these two cities brought World War II to a close. The lessons that Hiroshima (and Nagasaki) teach are the horrors of war with its random killing of civilian men, women, and children. That is the major lesson—war is horrible with its destruction of innocent lives.

The why of Hiroshima should be taken in the context of the battle for Okinawa which occurred only weeks before. America forces suffered 12,000 dead with 36,000 wounded in the battle for that small island 350 mi (563.5 km) from the mainland of Japan. The Japanese were reported to have lost 100,000 men. The determination of the Japanese to defend their homeland was well known, and it was estimated that the invasion of Japan would cost no less than 500,000 American lives. Japanese casualties were expected to be larger. It was the military judgment of the American President that a swift termination of the war would save more lives than it would cost; both American and Japanese. Whether this rationale for the atomic bombing of Hiroshima was correct, i.e., whether more people would have died if Japan was invaded, will never be known. However, it certainly is a fact that the war came to a swift end after the bombing of the two cities.

The second lesson to be learned from Hiroshima is that **radiation exposure** is hazardous to human health and radiation damage results in radiation sickness and increased **cancer** risk. It had been known since the development of x rays at the turn of the century that radiation has the potential to cause cancer. However, the thousands of survivors at Hiroshima and Nagasaki were to become the largest group ever studied for radiation damage. The Atomic Bomb Casualty Commission, now referred to as the Radiation Effects Research Foundation (RERF), was established to monitor the health effects of radiation exposure and has studied these survivors since the end of World War II. The RERF has reported a 10–15 times excess of all types of **leukemia** among the survivors compared with populations not exposed to the bomb. The leukemia excess peaked four to seven years after exposure but still persists among the survivors. All forms of cancer tended to develop more frequently in heavily irradiated individuals, especially children under the age of 10 at the time of exposure. Thyroid cancer was also increased in these children survivors of the bomb.

War is destructive to human life. The particular kind of destruction at Hiroshima, due to an atomic bomb, continues to be relentlessly destructive. The city of Hiroshima is known as a Peace City.

[*Robert G. McKinnell*]

A painting by Yasuko Yamagata depicting the Japanese city of Hiroshima after the atomic bomb was dropped on it. (Reproduced by permission.)

Holistic approach

First formulated by Jan Smuts, holism has been traditionally defined as a philosophical theory that states that the determining factors in **nature** are wholes which are irreducible to the sum of their parts and that the **evolution** of the universe is the record of the activity and making of such wholes. More generally, it is the concept that wholes cannot be analyzed into parts or reduced to discrete elements without unexplainable residuals. Holism may also be defined by what it is not: it is not synonymous with organicism; holism does not require an entity to be alive or even a part of living processes. And neither is holism confined to spiritual mysticism, unaccessible to scientific methods or study.

The holistic approach in **ecology** and **environmental science** derives from the idea proposed by Harrison Brown that "a precondition for solving [complex] problems is a realization that all of them are interlocked, with the result that they cannot be solved piecemeal." For some scholars holism is the rationale for the very existence of ecology. As

David Gates notes, "the very definition of the discipline of ecology implies a holistic study."

The holistic approach has been successfully applied to environmental management. The United States **Forest Service**, for example, has implemented a multi-level approach to management that takes into account the complexity of forest ecosystems, rather than the traditional focus on isolated incidents or problems.

Some people believe that a holistic approach to nature and the world will counter the effects of "reductionism"—excessive individualism, atomization, mechanistic worldview, objectivism, materialism, and anthropocentrism. Advocates of holism claim that its emphasis on connectivity, community, processes, networks, participation, synthesis, systems, and emergent properties will undo the "ills" of reductionism. Others warn that a balance between reductionism and holism is necessary. American ecologist Eugene Odum mandated that "ecology must combine holism with reductionism if applications are to benefit society." Parts and wholes, at the macro- and micro-level, must be understood.

The basic lesson of a combined and complementary parts-whole approach is that every entity is both part *and* whole—an idea reenforced by Arthur Koestler's concept of a *holon*. A holon is any entity that is both a part of a larger system and itself a system made up of parts. It is essential to recognize that holism can include the study of *any* whole, the entirety of any individual in all its ramifications, without implying any organic analogy other than organisms themselves. A holistic approach alone, especially in its extreme form, is unrealistic, condemning scholars to an unproductive wallowing in an unmanageable complexity. Holism and reductionism are both needed for accessing and understanding an increasingly complex world. *See also* Environmental ethics

[*Gerald L. Young Ph.D.*]

RESOURCES
BOOKS

Bowen, W. "Reductions and Holism." In *Thinking About Nature: An Investigation of Nature, Value and Ecology.* Athens: University of Georgia Press, 1988.

Johnson, L. E. "Holism." In *A Morally Deep World: An Essay on Moral Significance and Environmental Ethics.* Cambridge: Cambridge University Press, 1991.

Savory, A. *Holistic Resource Management.* Covelo, CA: Island Press, 1988.

PERIODICALS

Krippner, S. "The Holistic Paradigm." *World Futures* 30 (1991): 133–40.

Marietta Jr., D. E. "Environmental Holism and Individuals." *Environmental Ethics* 10 (Fall 1988): 251–58.

McCarty, D. C. "The Philosophy of Logical Wholism." *Synthese* 87 (April 1991): 51–123.

Van Steenbergen, B. "Potential Influence of the Holistic Paradigm on the Social Sciences." *Futures* 22 (December 1990): 1071–83.

Homeostasis

Humans, all other organisms, and even ecological systems, live in an **environment** of constant change. The persistently shifting, modulating, and changing milieu would not permit survival, if it were not for the capacity of biological systems to respond to this constant flux by maintaining a relatively stable internal environment. An example taken from mammalian biology is temperature which appears to be *"fixed"* at approximately 98.6°F (37°C). While humans can be exposed to extreme summer heat, and arctic mammals survive intense cold, body temperature remains constant within vary narrow limits. Homeostasis is the sum total of all the biological responses that provide internal equilibrium and assure the maintenance of conditions for survival.

The human **species** has a greater variety of living conditions than any other organism. The ability of humans to live and reproduce in such diverse circumstances is due

to a combination of homeostatic mechanisms coupled with cultural (behavioral) responses.

The scientific concept of homeostasis emerged from two scientists: Claude Bernard, a French physiologist, and Walter Bradford Cannon, an American physician. Bernard contrasted the external environment which surrounds an organism and the internal environment of that organism. He was, of course, aware that the external environment fluctuated considerably in contrast to the internal environment which remained remarkably constant. He is credited with the enunciation of the constancy of the internal environment ("*La fixité du milieu intérieur...*") in 1859. Bernard believed that the survival of an organism depended upon this constancy, and he observed it not only in temperature control but in the regulation of all of the systems that he studied. The concept of the stable "*milieu intérieur*" has been accepted and extended to the many organ systems of all higher vertebrates. This precise control of the internal environment is effected through hormones, the autonomic nervous system, endocrines, etc.

The term "homeostasis," derived from the Greek *homoios* meaning similar and *stasis* meaning to stand, suggests an internal environment which remains relatively similar or the same through time. The term was devised by Cannon in 1929 and used many times subsequently. Cannon noted that, in addition to temperature, there were complex controls involving many organ systems that maintained the internal **stability** within narrow limits. When those limited are exceeded, there is a reaction in the opposite direction that brings the condition back to normal, and the reactions returning the system to normal is referred to as negative feedback. Both Bernard and Cannon were concerned with human physiology. Nevertheless, the concept of homeostasis is applied to all levels of biological organization from the molecular level to ecological systems, including the entire **biosphere**. Engineers design self-controlling machines known as servomechanisms with feedback control by means of a sensing device, an amplifier which controls a servomotor which in turn runs the operation of the device. Examples of such devices are the thermostats which control furnace heat in a home or the more complicated automatic pilots of aircraft. While the human-made servomechanisms have similarities to biological homeostasis, they are not considered here.

As indicated above, temperature is closely regulated in humans and other homeotherms (birds and mammals). The human skin has thermal receptors sensitive to heat or cold. If cold is encountered, the receptors notify an area of the brain known as the hypothalamus via a nerve impulse. The hypothalamus has both a heat-promoting center and a heat-losing center, and, with cold, it is the former which is stimulated. Thyroid-releasing hormone, produced in the hypothalamus, causes the anterior pituitary to release thyroid stimulating hormone which, in turn, causes the thyroid gland to increase production of thyroxine which results in increased **metabolism** and therefore heat. Sympathetic nerves from the hypothalamus stimulate the adrenal medulla to secrete epinephrine and norepinephrine into the blood which also increases body metabolism and heat. Increased muscle activity will generate heat and that activity can be either voluntary (stamping the feet for instance) or involuntary (shivering). Since heat is dissipated via body surface blood vessels, the nervous system causes surface vasoconstriction to decrease that heat loss. Further, the small quantity of blood that does reach the surface of the body, where it is chilled, is reheated by countercurrent heat exchange resulting from blood vessels containing cold blood from the limbs running adjacent to blood vessels from the body core which contain warm blood. The chilled blood is prewarmed prior to returning to the body core. A little noted response to chilling is the voluntary reaching for a jacket or coat to minimize heat loss.

The body responds with opposite results when excessive heat is encountered. The individual tends to shed unnecessary clothing, and activity is reduced to minimize metabolism. Vasodilation of superficial blood vessels allows for radiation of heat. Sweat is produced, which by evaporation reduces body heat. It is clear that the maintenance of body temperature is closely controlled by a complex of homeostasis mechanisms.

Each step in temperature regulation is controlled by negative feedback. As indicated above, with exposure to cold the hypothalamus, through a series of steps, induces the synthesis and release of thyroxine by the thyroid gland. What was not indicated above was the fact that elevated levels of thyroxine control the level of activity of the thyroid by negative feedback inhibition of thyroid stimulating hormone. An appropriate level of thyroid hormone is thus maintained. In contrast, with inadequate thyroxine, more thyroid stimulating hormone is produced. Negative feedback controls assure that any particular step in homeostasis does not deviate too much from the normal.

Historically, biologists have been particularly impressed with mammalian and human homeostasis. Lower vertebrates have received less attention. However, while internal physiology may vary more in a frog than in a human, there are mechanisms which assure the survival of **frogs**. For instance, when the ambient temperature drops significantly in the autumn in northern latitudes, leopard frogs move into lakes or rivers which do not freeze. Moving into lakes and rivers is a behavioral response to a change in the external environment which results in internal temperature stability. The metabolism and structure of the frog is inadequate to protect the frog from freezing, but the specific heat of the water is such that freezing does not occur except at the surface of the overwintering lake or river. Even though life at the bottom of a lake with an ice cover moves at a slower pace than during the warm summer months, a functioning circulatory system is essential for survival. In general, frog blood (not unlike crankcase oil prior to the era of multiviscosity oil) increases in viscosity with as temperature decreases. Frog blood, however, decreases in viscosity with the prolonged autumnal and winter cold temperatures, thus assuring adequate circulation during the long nights under an ice cover. This is another control mechanism that assures the survival of frogs by maintaining a relatively stable internal environment during the harsh winter. With a return of a warm external environment, northern leopard frogs leave cold water to warm up under the spring sun. Warm temperature causes frog blood viscosity to increase to summer levels. It may be that the behavioral and physiological changes do not prevent oscillations that would be unsuitable for warm blooded animals but, in the frog, the fluctuations do not interfere with survival, and in biology, that is all that is essential.

There is homeostasis in ecological systems. Populations of animals in complex systems fluctuate in numbers, but the variations in numbers are generally between limits. For example, predators survive in adequate numbers as long as prey are available. If predators become too great in number, the population of prey will diminish. With fewer prey, the numbers of predators plummet through negative feedback thus permitting recovery of the preyed upon species. The situation becomes much more complex when other food sources are available to the predator.

Many organisms encounter a negative feedback on growth rate with crowding. This density dependent population control has been studied in larval frogs, as well as many other organisms, where excretory products seem to specifically inhibit the crowded species but not other organisms in the same environment. Even with adequate food, high density culture of laboratory mice results in negative feedback on reproductive potential with abnormal gonad development and delayed sexual maturity. Density independent factors affecting populations are important in population control but would not be considered homeostasis. **Drought** is such a factor, and its effects can be contrasted with crowding. Populations of tadpoles will drop catastrophically when breeding ponds dry. Instead of fluctuating between limits (with controls), all individuals are affected the same (i.e., they die). The area must be repopulated with immigrants at a subsequent time, and the **migration** can be considered a population homeostatic control. The inward migration results in maintenance of population within the geographic area and aids in the survival of the species.

[*Robert G. McKinnell*]

RESOURCES
BOOKS

Hardy, R. N. *Homeostasis.* London: Edward Arnold, Ltd., 1976.
Langley, L. L. *Homeostasis.* New York: Reinhold Publishing Co., 1965.
Tortora, G. J., and N. P. Anagnostakos. *Principles of Anatomy and Physiology.*
5th ed. New York: Harper and Row, 1987.

Homestead Act (1862)

The Homestead Act was signed into law in 1862. It was a legislative offer on a vast scale of free homesteads on unappropriated public lands. Any citizen (or alien who filed a declaration of intent to become a citizen), who had reached the age of 21, and was the head of a family could acquire title to a stretch of **public land** of up to 160 acres (65 ha) after living on it and farming it for five years. The only payment required was administrative fees. The settler could also obtain the land without the requirement of residence and cultivation for five years, against payment of $1.25 per acre. With the advent of machinery to mechanize farm labor, 160-acre (65 ha) tracts soon became uneconomical to operate, and Congress modified the original act to allow acquisition of larger tracts. The Homestead Act is still in effect, but good unappropriated land is scarce. Only Alaska still offers opportunities for homesteaders.

The Homestead Act was designed to speed development of the United States and to achieve an equitable distribution of wealth. Poor settlers, who lacked the capital to buy land, were now able to start their own farms. Indeed, the act contributed greatly to the growth and development of the country, particularly in the period between the Civil War and World War I, and it did much to speed settlement west of the Mississippi River. In all, well over a quarter of a billion acres of land has been distributed under the Homestead Act and its amendments. However, only a small percentage of land granted under the act between 1862 and 1900 was in fact acquired by homesteaders. According to estimates, only at most 1 of every 6 acres (0.4 of every 2.4 ha) and possibly only 1 in 9 acres (0.4 in 3.6 ha) passed into the hands of family farmers.

The railroad companies and land speculators obtained the bulk of the land, sometimes through gross fraud using dummy entrants. Moreover, the railroads often managed to get the best land while the homesteaders, ignorant of farming conditions on the Plains, often ended up with tracts least suitable to farming. Speculators frequently encouraged settlement on land that was too dry or had no sources of water for domestic use. When the homesteads failed, many settlers sold the land to speculators.

The environmental consequences of the Homestead Act were many and serious. The act facilitated railroad development, often in excess of **transportation** needs. In many instances, competing companies built lines to connect the same cities. Railroad development contributed significantly to the destruction of **bison** herds, which in turn led to the destruction of the way of life of the Plains Indians. Cultivation of the Plains caused wholesale destruction of the vast prairies, so that whole ecological systems virtually disappeared. Overfarming of semi-arid lands led to another environmental disaster, whose consequences were fully experienced only in the 1930s. The great **Dust Bowl**, with its terrifying dust storms, made huge areas of the country unlivable.

The Homestead Act was based on the notion that land held no value unless it was cultivated. It has now become clear that reckless cultivation can be self-destructive. In many cases, unfortunately, the damage can no longer be undone.

[*William E. Larson and Marijke Rijsberman*]

RESOURCES
PERIODICALS

Shimkin, M. N. "Homesteading on the Republican River." *Journal of the West* 26 (October 1987): 58–66.

Horizon

Layers in the **soil** develop because of the additions, losses, translocations, and transformations that take place as the soil ages. The soil layers occur as a result of water percolating through the soil and **leaching** substances downward. The layers are parallel to the soil surface and are called horizons. Horizons will vary from the surface to the **subsoil** and from one soil to the next because of the different intensities of the above processes. Soils are classified into different groups based on the characteristics of the horizons.

Horseshoe crabs

The horseshoe crab (*Limulus polyphemus*) is the American **species** of a marine animal that is only a distant relation of crustaceans like crabs and lobsters. Horseshoe crabs are more closely related to spiders and scorpions. The crabs have been called "living fossils" because the genus dates back millions of years, and *Limulus* evolved very little over the years.

Fossils found in British Columbia indicate that the ancestors of horseshoe crabs were in North America about 520 million years ago. During the late twentieth century, the declining horseshoe crab population concerned environmentalists. Horseshoe crabs are a vital food source for dozens of species of birds that migrate from South America to the Arctic Circle. Furthermore, crabs are collected for medical research. After blood is taken from the crabs, they are returned to the ocean.

American horseshoe crabs live along the Atlantic Ocean coastline. Crab **habitat** extends south from Maine to the Yucatán in the Gulf of Mexico. Several other crab species are found in Southeast Asia and Japan.

The American crab is named for its helmet-like shell that is shaped like a horseshoe. *Limulus* has a sharp tail shaped like a spike. The tail helps the crab move through the sand. If the crab tips over, the tail serves as a rudder so the crab can get back on its feet. The horseshoe crab is unique; its blood is blue and contains **copper**. The blood of other animals is red and contains iron.

Mature female crabs measure up to 24 inches in length. Males are about two-thirds smaller. Horseshoe crabs can live for 19 years, and they reach sexual maturity in 10 years. The crabs come to shore to spawn in late May and early June. They spawn the during the phases of the full and new moon. The female digs nests in the sand and deposits from 200 to 300 eggs in each pit. The male crab fertilizes the eggs with sperm, and the egg clutch is covered with sand.

During the spawning season, a female crab could deposit as many as 90,000 eggs. This spawning process coincides with the **migration** of shorebirds. Flocks of birds like the red knot and the sandpiper eat their fill of crab eggs before continuing their northbound migration.

Through the years, people found a variety of uses for horseshoe crabs. During the sixteenth century, Native Americans in South Carolina attached the tails to the spears that they used to catch fish. In the nineteenth century, people ground the crabs up for use as **fertilizer** or food for chickens and hogs.

During the twentieth century, researchers learned much about the human eye by studying the horseshoe crab's compound eye. Furthermore, researchers discovered that the crab's blood contained a special clotting agent that could be used to test the purity of new drugs and intravenous solutions. The agent called *Limulus Amoebocyte Lysate* is obtained by collecting horseshoe crabs during the spawning season. Crabs are bled and then returned to the beach.

Horseshoe crabs are also used as bait. The harvesting of crabs increased sharply during the 1990s when people in the fishing industry used crabs as bait to catch eels and conch. The annual numbers of crabs harvested jumped from the thousands to the millions during the 1990s, according to environmental groups and organizations like the **National Audubon Society**.

The declining horseshoe crab population could affect millions of migrating birds. The Audubon Society reported seeing fewer birds at the Atlantic beaches where horseshoe

A horseshoe crab (*Limulus polyphemus*). (©John M. Burnley, National Audubon Society Collection. Photo Researchers Inc. Reproduced by permission.)

crabs spawn. In the spring of 2000, scientists said that birds appeared undernourished. Observers doubted that they would complete their journey to the Arctic Circle.

The Audubon Society and environmental groups have campaigned for state and federal regulations to protect horseshoe crabs. By 2002, coastal states and the Atlantic States Marine Fisheries Commission had set limits on the amount of crabs that could be harvested. The state of Virginia made bait bags mandatory when fishing with horseshoe crab bait. The mesh bag made of hard plastic holds the crab. That made it more difficult for predators to eat the crab so fewer *Limulus* crabs were needed as bait.

Furthermore, the federal government created a 1,500-square-mile refuge for horseshoe crabs in Delaware Bay. The refuge extends from Ocean City, New Jersey to north of Ocean City, Maryland. As of March of 2002, harvesting was banned in the refuge. People who took crabs from the area faced a fine of up to $100,000, according to the National Marine Fisheries Service.

As measures like those were enacted, marine biologists said that it could be several decades before the crab population increased. One reason for slow **population growth** was that it takes crabs 10 years to reach maturity.

[*Liz Swain*]

RESOURCES
BOOKS

Fortey, Richard. *Trilobite!: Eyewitness to Evolution.* New York: Alfred A. Knopf, 2000.

Tanacredi, John, Ed. *Limulus in the Limelight: 350 Million Years in the Making and in Peril?* New York: Kluwer Academic Publishers, 2001.

ORGANIZATIONS

Atlantic States Marine Fisheries Commission., 1444 Eye Street, NW, Sixth Floor, Washington, D.C. 20005 (202) 289-6400, Fax: (202) 289-6051, Email: comments@asmfc.org, http://www.asmfc.org

National Audubon Society Horseshoe Crab Campaign., 1901 Pennsylvania Avenue NW, Suite 1100, Washington, D.C. 20006 (202) 861-2242, Fax: (202) 861-4290, Email: pplumart@audubon.org, http://www.audubon.org/campaign/horseshoe/contacts.htm

National Marine Fisheries Service., 1315 East West Highway, SSMC3, Silver Spring, MD 20910 (301) 713-2334, Fax: (301) 713-0596, Email: cyber.fish@noaa.gov, hhttp://www.nmfs.noaa.gov

Hospital wastes

see **Medical waste**

Household waste

Household waste is commonly referred to as **garbage** or trash. As the population of the world expands, so does the amount of waste produced. Generally, the more automated and industrialized human societies become, the more waste they produce. For example, the industrial revolution introduced new manufactured products and new manufacturing processes that added to household **solid waste** and industrial waste. Modern consumerism and the excess packaging of many products also contribute significantly to the increasing amount of solid waste.

Much of the trash Americans produce (about 40%) is paper and paper products. Paper accounts for more than 71 million tons of garbage. Yard wastes are the next most common waste, contributing more than 31 million tons of solid waste. Metals account for more than 8% of all household waste, and **plastics** are close behind with another 8% or 14 million tons. America's trash also contains about 7% glass and nearly 20 million tons of other materials like **rubber**, textiles, leather, wood, and inorganic wastes. Much of the waste comes from packaging materials. Other types of waste produced by consumers are durable goods such as tires, appliances, and furniture, while other household solid waste is made up of non-durable goods such as paper, disposable products, and clothing. Many of these items could be recycled and reused, so they also can be considered a non-utilized resource.

In less industrialized times and even today in many developing countries, households and industries disposed of unwanted materials in bodies of water or in land dumps.

However, this practice creates undesirable effects such as health hazards and foul odors. Open dumps serve as breeding grounds for disease-carrying organisms such as rats and insects. As the **first world** became more alert to environmental hazards, methods for waste disposal were studied and improved. Today, however, governments, policymakers, and individuals still wrestle with the problem of how to improve methods of waste disposal, storage, and **recycling**.

In 1976, the United States Congress passed the **Resource Conservation and Recovery Act** (RCRA) in an effort to protect human health and the **environment** from hazards associated with waste disposal. In addition, the act aims to conserve energy and **natural resources** and to reduce the amount of waste Americans generate. Further, the RCRA promotes methods to manage waste in an environmentally sound manner. The act covers regulation of solid waste, **hazardous waste**, and underground storage tanks that hold **petroleum** products and certain **chemicals**.

Most household solid waste is removed from homes through community garbage collection and then taken to landfills. The garbage in landfills is buried, but it can still produce noxious odors. In addition, rainwater can seep through **landfill** sites and leach out pollutants from the landfill trash. These are then carried into nearby bodies of water. Pollutants can also contaminate **groundwater**, which in turn leads to contamination of drinking water.

In order to fight this problem, sanitary landfills were developed. Clay or plastic liners are placed in the ground before garbage is buried. This helps prevent water from seeping out of the landfill and into the surrounding environment. In sanitary landfills, each time a certain amount of waste is added to the landfill, it is covered by a layer of **soil**. At a predetermined height the site is capped and covered with dirt. Grass and trees can be planted on top of the capped landfill to help prevent **erosion** and to improve the look of the site. Sanitary landfills are more expensive than open pit dumps, and many communities do not want the stigma of having a landfill near them. These factors make it politically difficult to open new landfills. Landfills are regulated by state and local governments and must meet minimum requirements set by the United States **Environmental Protection Agency** (EPA). Some household hazardous wastes such as paint, used motor oil, or insecticides can not be accepted at landfills and must be handled separately.

Incineration (burning) of solid waste offers an alternative to disposal in landfills. Incineration converts large amounts of solid waste to smaller amounts of ash. The ash must still be disposed of, however, and it can contain toxic materials. Incineration released **smoke** and other possible pollutants into the air. However, modern incinerators are equipped with smokestack **scrubbers** that are quite effective in trapping toxic emissions. Many incinerators have the added benefit of generating electricity from the trash they burn.

Composting is a viable alternative to landfills and incineration for some **biodegradable** solid waste. Vegetable trimmings, leaves, grass clippings, straw, horse manure, wood chippings, and similar plant materials are all biodegradable and can be composted. Compost helps the environment because it reduces the amount of waste going into landfills. Correct composting also breaks down biodegradable material into a nutrient-rich soil additive that can be used in gardens or for landscaping. In this way, nutrients vital to plants are returned to the environment. To successfully compost biodegradable wastes, the process must generate high enough temperatures to kill seeds or organisms in the composted material. If done incorrectly, compost piles can give off foul odors.

Families and communities can help reduce household waste by making some simple lifestyle changes. They can reduce solid waste by recycling, repairing rather than replacing durable goods, buying products with minimal packaging, and choosing packaging made from recycled materials. Reducing packaging material is an example of source reduction. Much of the responsibility for source reduction with manufacturers. Businesses need to be encouraged to find smart and cost effective ways to manufacture and package goods in order to minimize waste and reduce the toxicity of the waste created. Consumers can help by encouraging companies to create more environmentally responsible packaging through their choice of products. For example, consumers successfully pressured McDonalds to change from serving their sandwiches in non-biodegradable Styrofoam boxes to wrapping them in biodegradable paper.

For individual households can reduce the amount of waste they send to landfills by recycling. Paper, **aluminum**, glass, and plastic containers are the most commonly recycled household materials. Strategies for household recycling vary from community to community. In some areas materials must separated by type before collection. In others, the separation occurs after collection.

Recycling preserves natural resources by providing an alternative supply of raw materials to industries. It also saves energy and eliminates the emissions of many toxic gases and water pollutants. In addition, recycling helps create jobs, stimulates development of more environmentally sound technologies, and conserves resources for **future generations**. For recycling to be successful, there must be an end market for goods made from recycled materials. Consumers can support recycling by buying "green" products made of recycled materials.

Battery recycling is also becoming increasingly common in the United States and is required by law in many European countries. In 2001, a nonprofit organization called Rechargeable Battery Recycling Corporation (RBRC) began offering American communities cost-free recycling of portable rechargeable batteries such as those used in cell phones, camcorders, and laptop computers. These batteries contain **cadmium**, which is recycled back into other batteries or used in certain coatings or color pigments.

Household waste disposal is an international problem that is being attacked in many ways in many countries. In Tripoli, Libya, a plant exploits household waste, converting it to organic **fertilizer**. The plant recycles 500 tons of household waste, producing 212 tons of fertilizer a day. In France, a country with less available space for landfills than the United States, incineration is proving a desirable alternative. The French are turning household waste into energy through **combustion** and are developing technologies to control the residues that occur from incineration.

In the United States, education of consumers is a key to reducing the volume and toxicity of household waste. The EPA promotes four basic principles for reducing solid waste: reduce the amount of trash discarded, **reuse** products and containers, recycle and compost, and reconsider activities that produce waste.

[*Teresa G. Norris*]

RESOURCES
OTHER

"Communities Invited to Recycle Rechargables Cost Free." *Environmental News Network.* October 11, 2001 [cited July 2002]. <http://www.enn.com/news>.

ORGANIZATIONS

U.S. Environmental Protection Agency Office of Solid Waste, 1200 Pennsylvania Avenue NW, Washington, DC USA 20460 (703) 412-9810, Toll Free: (800) 424-9346, , <http://www.epa.gov>

HRS

see **Hazard Ranking System**

Hubbard Brook Experimental Forest

The Hubbard Brook Experimental Forest is located in West Thornton, New Hampshire. It is an experimental area established in 1955 within the White Mountains **National Forest** in New Hamphire's central plateau and is administered by the U.S. **Forest Service**. Hubbard Brook was the site of many important ecological studies beginning in the 1960s which established the extent of **nutrient** losses when all the trees in a **watershed** are cut.

Hubbard Brook is a north temperate watershed covered with a mature forest, and it is still accumulating **bio-**

mass. In one early study, vegetation cut in a section of Hubbard Brook was left to decay while nutrient losses were monitored in the **runoff**. Total **nitrogen** losses in the first year were twice the amount cycled in the system during a normal year. With the rise of nitrate in the runoff, concentrations of calcium, magnesium, sodium, and potassium rose. These increases caused eutrophication and **pollution** of the streams fed by this watershed. Once the higher plants had been destroyed, the **soil** was unable to retain nutrients.

Early evidence from the studies indicated that total losses in the **ecosystem** due to the **clear-cutting** were a large number of the total inventory of **species**. The site's ability to support complex living systems was reduced. The lost nutrients could accumulate again, but **erosion** of primary minerals would limit the number of plants and animals sustained in the area.

Another study at the Hubbard Brook site investigated the effects of forest cutting and **herbicide** treatment on nutrients in the forest. All of the vegetation in one of Hubbard Brook's seven watersheds was cut and then the area was treated with the herbicides. At the time the conclusions were startling: **deforestation** resulted in much larger runoffs into the streams. The **pH** of the **drainage** stream went from 5.1 to 4.3, along with a change in temperature and electrical conductivity of the stream water. A combination of higher nutrient concentration, higher water temperature, and greater solar radiation due to the loss of forest cover produced an **algal bloom**, the first sign of eutrophication. This signaled that a change in the ecosystem of the watershed had occurred. It was ultimately demonstrated at Hubbard Brook that the use of herbicides on a cut area resulted in their transfer to the outgoing water.

Hubbard Brook Experimental Forest continues to be an active research facility for foresters and biologists. Most current research focuses on **water quality** and nutrient exchange. The Forest Service also maintains an **acid rain** monitoring station, and conducts research on old-growth forests. The results from various studies done at Hubbard Brook have shown that mature forest ecosystems have a greater ability to trap and store nutrients for **recycling** within the ecosystem. In addition, mature forests offer a higher degrees of **biodiversity** than do forests that are clear-cut. *See also* Aquatic chemistry; Cultural eutrophication; Decline spiral; Experimental Lakes Area; Nitrogen cyle

[*Linda Rehkopf*]

RESOURCES
BOOKS

Bormann, F. H. *Pattern and Process in a Forested Ecosystem: Disturbance Development and the Steady State Based on the Hubbard Brook Ecosystem Study.* New York: Springer-Verlag, 1991.

Botkin, D. B. *Forest Dynamics: An Ecological Model.* New York: Oxford University Press, 1993.

PERIODICALS

Miller, G. "Window Into a Water Shed." *American Forests* 95 (May-June 1989): 58–61.

Hudson River

Starting at Lake Tear of the Clouds, a two-acre (0.8-ha) pond in New York's **Adirondack Mountains**, the Hudson River runs 315 miles (507 km) to the Battery on Manhattan Island's southern tip, where it meets the Atlantic Ocean. Although polluted and extensively dammed for hydroelectric power, the river still contains a wealth of aquatic **species**, including massive sea sturgeon (*Acipenser oxyrhynchus*) and short-nosed sturgeon (*A. brevirostrum*). The upper Hudson is fast-flowing trout stream, but below the Adirondack Forest Preserve, **pollution** from municipal sources, paper companies, and industries degrades the water. Stretches of the upper Hudson contain so-called warm water fish, including northern pike (*Esox lucius*), chain pickerel (*E. niger*), smallmouth bass (*Micropterus dolomieui*), and largemouth bass (*M. salmoides*). These latter two fish swam into the Hudson through the **Lake Erie** and Lake Champlain canals, which were completed in the early nineteenth century.

The Catskill Mountains dominate the mid-Hudson region, which is rich in fish and **wildlife**, though dairy farming, a source of **runoff** pollution, is strong in the region. American shad (*Alosa sapidissima*), historically the Hudson's most important commercial fish, spawn on the river flats between Kingston and Coxsackie. Marshes in this region support snapping turtles (*Chelydra serpentina*) and, in the winter, muskrat (*Ondatra zibethicus*) and mink (*Mustela vison*). Water chestnuts (*Trapa natans*) grow luxuriantly in this section of the river.

Deep and partly bordered by mountains, the lower Hudson resembles a fiord. The unusually deep lower river makes it suitable for navigation by ocean-going vessels for 150 miles (241 km) upriver to Albany. Because the river's surface elevation does not drop between Albany and Manhattan, the tidal effects of the ocean are felt all the way upriver to the Federal Lock and Dam above Albany. These powerful tides make long stretches of the lower Hudson saline or **brackish**, with saltwater penetrating as high as 60 miles (97 km) upstream from the Battery.

The Hudson contains a great variety of botanical species. Over a dozen oaks thrive along its banks, including red oaks (*Quercus rubra*), black oaks (*Q. velutina*), pin oaks (*Q. palustris*), and rock chestnut (*Q. prinus*). Numerous other trees also abound, from mountain laurel (*Kalmia latifolia*) and red pine (*Pinus resinosa*) to flowering dogwood (*Cornus*

florida), together with a wide variety of small herbaceous plants.

The Hudson River is comparatively short. More than 80 American rivers are longer than it, but it plays a major role in New York's economy and **ecology**. Pollution threats to the river have been caused by the **discharge** of industrial and municipal waste, as well as pesticides washed off the land by rain. From 1930 to 1975, one chemical company on the river manufactured approximately 1.4 billion pounds of **polychlorinated biphenyls** (PCBs), and an estimated 10 million pounds a year entered the **environment**. In all, a total of 1.3 million pounds of PCB contamination allegedly occurred during the years prior to the ban, with the pollution originating from plants at Ford Edward and Hudson Falls. A ban was put in place for a time prohibiting the possession, removal, and eating of fish from the waters of the upper Hudson River. A proposed cleanup was designated, to proceed by means of a 40-mile **dredging** and sifting of 2.65 million cubic yards of **sediment** north of Albany, with an anticipated yield of 75 tons of PCBs.

In February of 2001 the U.S. **Environmental Protection Agency** (EPA), having invoked the Superfund law, required the chemical company to begin planning the cleanup. The company was given several weeks to present a viable plan of attack, or else face a potential $1.5 billion fine for ignoring the directive in lieu of the cost of cleanup. The cleanup cost, estimated at $500 million was presented as the preferred alternative. The engineering phase of the cleanup project was expected to take three years of planning and was to be scheduled after the offending company filed a response to the EPA. The company responded within the allotted time frame in order to placate the EPA, although the specifics of a drafted work plan remained undetermined, and the company refused to withdraw a lawsuit filed in November of 2000, which challenged the constitutionality of the so-called Superfund law that authorized the EPA to take action. The river meanwhile was ranked by one environmental watchdog group as the fourth most endangered in the United States, specifically because of the PCB contamination. Environmental groups demanded also that attention be paid to the issues of **urban sprawl**, noise, and other pollution, while proposals for potentially polluting projects were endorsed by industrialists as a means of spurring the area's economy. Among these industrial projects, the construction of a cement plant in Catskill where there is easy access to a limestone quarry, and the development of a power plant along the river in Athens generated controversy, stemming from the industrial asset afforded by development along the river versus the advantages of a less fouled environment. Additionally, the power plant, which threatened to add four new smokestacks to the skyline and to aggravate pollution, was seen as potentially detrimental to

tourism in that area. Also in recent decades, **chlorinated hydrocarbons**, dieldrin, endrin, DDT, and other pollutants have been linked to the decline in populations of the once common Jefferson salamander (*Ambystoma jeffersonianum*), fish hawk (*Pandion haliaetus*), and **bald eagle** (*Haliaeetus leucocephalus*).

Concerns over the condition of the lower river spread anew following a severe September 11 terrorist attack on New York City in 2001. In this coastal tri-state urban area where anti-dumping laws were put in place in the mid twentieth century to protect the river from deterioration due to pollution, new threats of pollution surfaced regarding the potential for assorted types of leakage into the river caused when the integrity of some land-based structures including seawalls and underwater tunnels was compromised by the impact of exploding commercial jetliners involved in the attack. *See also* Agricultural pollution; Dams; Estuary; Feedlot runoff; Fertilizer runoff; Industrial waste treatment; Sewage treatment; Wastewater

[*David Clarke*]

RESOURCES
BOOKS

Boyle, R. H. *The Hudson River, A Natural and Unnatural History.* New York: Norton, 1979.
Peirce, N. R., and J. Hagstrom. *The Book of America, Inside the Fifty States Today.* New York: Norton, 1983.

PERIODICALS
The Scientist, March 19, 2001.

Human ecology

Human **ecology** may be defined as the branch of knowledge concerned with relationships between human beings and their environments. Among the disciplines contributing seminal work in this field are sociology, anthropology, geography, economics, psychology, political science, philosophy, and the arts. Applied human ecology emerges in engineering, planning, architecture, landscape architecture, **conservation**, and public health. Human ecology, then, is an interdisciplinary study which applies the principles and concepts of ecology to human problems and the human condition. The notion of interaction—between human beings and the **environment** and between human beings—is fundamental to human ecology, as it is to biological ecology.

Human ecology as an academic inquiry has disciplinary roots extending back as far as the 1920s. However, much work in the decades prior to the 1970s was narrowly drawn and was often carried out by a few individuals whose intellectual legacy remained isolated from the mainstream of their disciplines. The work done in sociology offers an exception to the latter (but not the former) rule; sociological human ecology is traced to the Chicago school and the intellectual lineage of Robert Ezra Park, his student Roderick D. Mackenzie, and Mackenzie's student Amos Hawley. Through the influence of these men and their school, human ecology, for a time, was narrowly identified with a sociological analysis of spatial patterns in urban settings (although broader questions were sometimes contemplated).

Comprehensive treatment of human ecology is first found in the work of Gerald L. Young, who pioneered the study of human ecology as an interdisciplinary field and as a conceptual framework. Young's definitive framework is founded upon four central themes. The first of these is interaction, and the other three are developed from it: levels of organization, functionalism (part-whole relationships), and holism. These four basic concepts form the foundation for a series of field derivatives (**niche**, community, and **ecosystem**) and consequent notions (institutions, proxemics, alienation, ethics, world community, and stress/capacitance). Young's emphasis on linkages and process set his approach apart from other synthetic attempts in human ecology, which were largely cumbersome classificatory schemata. These were subject to harsh criticism because they tended to embrace virtually all knowledge, resolve themselves into superficial lists and mnemonic "building blocks," and had little applicability to real-world problems.

Generally, comprehensive treatment of human ecology is more advanced in Europe than it is in the United States. A comprehensive approach to human ecology as an interdisciplinary field and conceptual framework gathered momentum in several independent centers during the 1970s and 1980s. Among these have been several college and university programs and research centers, including those at the University of Göteborg, Sweden, and, in the United States, at Rutgers University and the University of California at Davis. Interdisciplinary programs at the undergraduate level were first offered in 1972 by the College of the Atlantic (Maine) and The Evergreen State College (Washington). The Commonwealth Human Ecology Council in the United Kingdom, the International Union of Anthropological and Ethnological Sciences' Commission on Human Ecology, the Centre for Human Ecology at the University of Edinburgh, the Institute for Human Ecology in California, and professional societies and organizations in Europe and the United States have been other centers of development for the field.

Dr. Thomas Dietz, President of the Society for Human Ecology, defined some of the priority research problems which human ecology addresses in recent testimony before the U.S. House of Representatives Subcommittee on Environment and the **National Academy of Sciences** Committee on Environmental Research. Among these, Dietz listed

global change, values, post-hoc evaluation, and science and conflict in **environmental policy**. Other human ecologists would include in the list such items as commons problems, **carrying capacity**, **sustainable development**, human health, ecological economics, problems of resource use and distribution, and family systems. Problems of epistemology or cognition such as environmental perception, consciousness, or paradigm change also receive attention.

Our Common Future, the report of the United Nation's World Commission on Environment and Development of 1987, has stimulated a new phase in the development of human ecology. A host of new programs, plans, conferences and agendas have been put forth, primarily to address phenomena of global change and the challenge of sustainable development. These include the *Sustainable Biosphere Initiative* published by the **Ecological Society of America** in 1991 and extended internationally; the United Nations Conference on Environment and Development; the proposed new United States National Institutes for the Environment; the Man and the **Biosphere** Program's Human-Dominated Systems Program; the report of the **National Research Council** Committee on Human Dimensions of Global Change and the associated National Science Foundation's Human Dimensions of Global Change Program; and **green plans** published by the governments of Canada, Norway, the Netherlands, the United Kingdom, and Austria. All of these programs call for an integrated, interdisciplinary approach to complex problems of human-environmental relationships. The next challenge for human ecology will be to digest and steer these new efforts and to identify the perspectives and tools they supply.

[*Jeremy Pratt*]

RESOURCES
BOOKS

Jungen, B. "Integration of Knowledge in Human Ecology." In *Human Ecology: A Gathering of Perspectives*, edited by R. J. Borden, et al., Selected papers from the First International Conference of the Society for Human Ecology, 1986.
———. *Origins of Human Ecology*. Stroudsberg, PA: Hutchinson & Ross, 1983.

PERIODICALS

Young, G. L. "Human Ecology As An Interdisciplinary Concept: A Critical Inquiry." *Advances in Ecological Research* 8 (1974): 1–105.
———. "Conceptual Framework For An Interdisciplinary Human Ecology." *Acta Oecologiae Hominis* 1 (1989): 1–136.

Humane Society of the United States

The largest animal protection organization in the United States, Humane Society of the United States (HSUS) works to preserve **wildlife** and **wilderness**, save **endangered species**, and promote humane treatment of all animals. Formed in 1954, HSUS specializes in education, cruelty investigations and prosecutions, wildlife and **nature** preservation, environmental protection, federal and state legislative activities, and other actions designed to protect animal welfare and the **environment**.

Major projects undertaken by HSUS in recent years have included campaigns to stop the killing of **whales**, **dolphins, elephants**, bears, and **wolves**; to help reduce the number of animals used in medical research and to improve the conditions under which they are used; to oppose the use of fur by the fashion industry; and to address the problem of pet overpopulation.

The group has worked extensively to ban the use of tuna caught in a way that kills dolphins, largely eliminating the sale of such products in the United States and western Europe. It has tried to stop international airlines from transporting exotic birds into the United States. Other high priority projects have included banning the international trade in elephant ivory, especially imports into the United States, and securing and maintaining a general worldwide moratorium on commercial **whaling**.

HSUS companion animals section works on a variety of issues affecting dogs, cats, birds, horses, and other animals commonly kept as pets, striving to promote responsible pet ownership, particularly the spaying and neutering of dogs and cats to reduce the tremendous overpopulation of these animals. HSUS works closely with local shelters and humane societies across the country, providing information, training, evaluation, and consultation.

Several national and international environmental and animal protection groups are affiliated and work closely with HSUS. Humane Society International works abroad to fulfill HSUS's mission and to institute reform and educational programs that will benefit animals. EarthKind, a global environmental protection group that emphasizes wildlife protection and humane treatment of animals, has been active in Russia, India, Thailand, Sri Lanka, the United Kingdom, Romania, and elsewhere, working to preserve forests, **wetlands**, wild rivers, natural ecosystems, and endangered wildlife.

The National Association for Humane and **Environmental Education** is the youth education division of HSUS, developing and producing periodicals and teaching materials designed to instill humane values in students and young people, including *KIND (Kids in Nature's Defense) News*, a newspaper for elementary school children, and *KIND TEACHER*, an 80-page annual full of worksheets and activities for use by teachers.

The Center for Respect of Life and the Environment works with academic institutions, scholars, religious leaders and organizations, arts groups, and others to foster an ethic of respect and compassion towards all creatures and the natural environment. Its quarterly publication, *Earth Ethics*, examines such issues as earth education, sustainable communities, ecological economics, and other values affecting our relationship with the natural world. The Interfaith Council for the Protection of Animals and Nature promotes **conservation** and education mainly within the religious community, attempting to make religious leaders, groups, and individuals more aware of our moral and spiritual obligations to preserve the planet and its myriad life forms.

HSUS has been quite active, hard-hitting, and effective in promoting its animal protection programs, such as leading the fight against the fur industry. It accomplishes its goals through education, lobbying, grassroots organizing, and other traditional, legal means of influencing public opinion and government policies.

With over 3.5 million members or "constituents" and an annual budget of over $35 million, HSUS is considered the largest and one of the most influential animal protection groups in the United States and, perhaps, the world.

[*Lewis G. Regenstein*]

RESOURCES

ORGANIZATIONS

The Humane Society of the United States, 2100 L Street, NW, Washington, D.C. USA 20037 (202) 452-1100, <http://www.hsus.org>

Humanism

A perspective or doctrine that focuses primarily on the interests, capacities, and achievements of human beings. This focus on human concerns has led some to conclude that human beings have rightful dominion over the earth and that their interests and well-being are paramount and take precedence over all other considerations. *Religious humanism*, for instance, generally holds that God made human beings in His own image and put them in charge of His creation. *Secular humanism* views human beings as the source of all value or worth. Some environmentally-minded critics, such as Lynn White Jr., and David Ehrenfeld claim that much environmental destruction can be traced to "the arrogance of humanism."

Human-powered vehicles

Finding easy modes of **transportation** seems to be a basic human need, but finding easy *and clean* modes is becoming imperative. Traffic congestion, overconsumption of **fossil fuels** and **air pollution** are all direct results of automotive

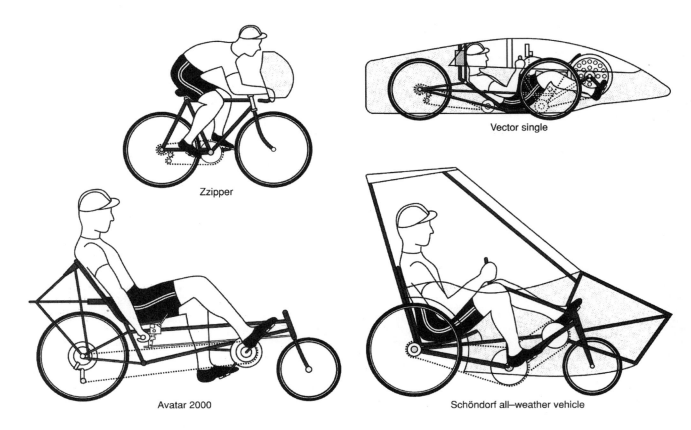

Zzipper

Vector single

Avatar 2000

Schöndorf all–weather vehicle

Innovative bicycle designs. (McGraw-Hill Inc. Reproduced by permission.)

lifestyles around the world. The logical alternative is human-powered vehicles (HPVs), perhaps best exemplified in the bicycle, the most basic HPV. New high-tech developments in HPVs are not yet ready for mass production, nor are they able to compete with cars. Pedal-propelled HPVs in the air, on land, or under the sea are still in the expensive, design-and-race-for-a-prize category. But the challenge of human-powered transport has inspired a lot of inventive thinking, both amateur and professional.

Bicycles and rickshaws comprise the most basic HPVs. Of these two vehicles, bicycles are clearly the most popular, and production of these HPVs has surpassed production of automobiles in recent years. The number of bicycles in use throughout the world is roughly double that of cars; China alone contains 270 million bicycles, or one third of the total bicycles worldwide. Indeed the bicycle has overtaken the **automobile** as the preferred mode of transportation in many nations. There are many reasons for the popularity of the bike: it fulfills both recreational and functional needs, it is an economical alternative to automobiles, and it does not contribute to the problems facing the **environment**.

Although the bicycle provides a healthy and scenic form of **recreation**, people also find it useful in basic transportation. In the Netherlands, bicycle transportation accounts for 30% of work trips and 60% of school trips. One-third of commuting to work in Denmark is by bicycle. In China, the vast majority of all trips there are made via bicycle.

A surge in bicycle production occurred in 1973, when in conjunction with rising oil costs, production doubled to 52 million per year. Soaring fuel prices in the 1970s inspired people to find inexpensive, economical alternatives to cars, and many turned to bicycles. Besides being efficient transportation, bikes are simply cheaper to purchase and to maintain than cars. There is no need to pay for parking or tolls, no expensive upkeep, and no high fuel costs.

The lack of fuel costs associated with bicycles leads to another benefit: bicycles do not harm the environment. Cars consume fossil fuels and in so doing release more than two-thirds of the United States' smog-producing **chemicals**. They are furthermore considered responsible for many other environmental ailments: depletion of the **ozone** layer through release of **chlorofluorocarbons** from automobile air conditioning units; cause of **cancer** through toxic emissions; and consumption of the world's limited fuel resources. With human energy as their only requirement, bicycles have none of these liabilities.

Nevertheless, in many cases—such as long trips or travelling in inclement weather—cars are the preferred form of transportation. Bicycles are not the optimal choice in many situations. Thus engineers and designers seek to improve on the bicycle and make machines suitable for transport under many different conditions. They are striving to produce new human-powered vehicles—HPVs that maximize air and sea currents, that have reasonable interior ergonomics, and that can be inexpensively produced. Several machines designed to fit this criteria exist.

As for developments in human-powered aircraft, success is judged on distance and speed, which depend on the strength of the pedaller and the lightness of the craft. The current world record holder is Greek Olympic cyclist Kanellos Kanellopoulos who flew *Daedalus 88*. *Daedalus 88* was created by engineer John Langford and a team of MIT engineers and funded by American corporations. Kanellopoulos flew *Daedalus 88* for 3 hours and 54 minutes across the Aegean Sea between Crete and Santorini, a distance of 74 mi (119 km), in April 1988. The craft averaged 18.5 mph (29 kph) and flew 15 ft (4.6 m) above the water. Upon arrival at Santorini, however, the sun began to heat up the black sands and generate erratic shore winds and *Daedalus 88* plunged into the sea. It was a few yards short of its goal, and the tailboom of the 70-lb (32-kg) vehicle was snapped by the wind. But to cheering crowds on the beach, Kanellopoulos rose from the sea with a victory sign and strode to shore.

In the creation of a human-powered helicopter, students at California Polytechnic State University had been working on perfecting one since 1981. In 1989 they achieved liftoff with Greg McNeil, a member of the United States National Cycling Team, pedalling an astounding 1.0 hp. The graphite epoxy, wood, and Mylar craft, *Da Vinci III*, rose 7 in (17.7 cm) for 6.8 seconds. But rules for the $10,000 Sikorsky prize, sponsored by the American Helicopter Society, stipulate that the winning craft must rise nearly 10 ft, or 3 m, and stay aloft one minute.

On land, recumbent vehicles, or recumbents, are wheeled vehicles in which the driver pedals in a semi-recumbent position, contained within a windowed enclosure. The world record was set in 1989 by American Fred Markham at the Michigan International Speedway in an HPV named *Goldrush*. Markham pedalled more than 44 mph (72 kph).

Unfortunately, the realities of road travel cast a long shadow over recumbent HPVs. Crews discovered that they tended to be unstable in crosswinds, distracted other drivers and pedestrians, and lacked the speed to correct course safely in the face of oncoming cars and trucks.

In the sea, being able to maneuver at your own pace and be in control of your vehicle—as well as being able to beat a fast retreat undersea—are the problems faced by HPV submersible engineers. Human-powered subs are not a new idea. The Revolutionary War created a need for a bubble sub that was to plant an explosive in the belly of a British ship in New York Harbor. (The naval officer, breathing one-half hour's worth of air, failed in his night mission, but survived).

The special design problems of modern two-person HP-subs involve controlling buoyancy and ballast, pitch and yaw (nose up/down/sideways), reducing drag, increasing thrust, and positioning the pedaller and the propulsor in the flooded cockpit (called "wet") in ways that maximize air intake from scuba tanks and muscle power from arms and legs.

Depending on the design, the humans in HP-subs lie prone, foot to head or side by side, or sit, using their feet to pedal and their hands to control the rudder through the underwater currents. Studies by the United States Navy Experimental Dive Unit indicate that a well-trained athlete can sustain 0.5 hp for 10 minutes underwater.

On the surface of the water, fin-propelled watercraft—lightweight inflatables that are powered by humans kicking with fins—are ideal for fishermen whom maneuverability, not speed, is the goal. Paddling with the legs, which does not disturb fish, leaves the hands free to cast. In most designs, the fisherman sits on a platform between tubes, his feet in the water. Controllability is another matter, however: in open windy water, the craft is at the mercy of the elements in its current design state. Top speed is about 50 yd (46 m) in three minutes.

Finally, over the surface of the water, the first human-powered hydrofoil, *Flying Fish*, with national track sprinter Bobby Livingston, broke a world record in September 1989 when it traveled 100 m over Lake Adrian, Michigan, at 16.1 knots (18.5 mph). A vehicle that pedalled like a bicycle, resembled a model airplane with a two-blade propeller and a 6-ft (1.8-m) **carbon** graphite wing, *Flying Fish* sped across the surface of the lake on two pontoons.

[*Stephanie Ocko and Andrea Gacki*]

RESOURCES

BOOKS

Lowe, M. "Bicycle Production Outpaces Autos." In *Vital Signs 1992: The Trends That Are Shaping Our Future*, edited by L. R. Brown, C, Flavin, and H. Hane. New York: Norton, 1992.

PERIODICALS

Banks, R. "Sub Story." *National Geographic World* (July 1992): 8–11.

Blumenthal, T. "Outer Limits." *Bicycling*, December 1989, 36.

Britton, P. "Muscle Subs." *Popular Science*, June 1989, 126–129.

———. "Technology Race Beneath the Waves." *Popular Science*, June 1991, 48–54.

Horgan, J. "Heli-Hopper: Human-powered Helicopter Gets Off the Ground." *Scientific American* 262 (March 1990): 34.

Kyle, C. R. "Limits of Leg Power." *Bicycling*, October 1990, 100–101.

Langley, J. "Those Flying Men and Their Magnificent Machines." *Bicycling*, April 1992, 74–76.

"Man-Powered Helicopter Makes First Flight." *Aviation Week and Space Technology*, December 1989, 115.

Martin, S. "Cycle City 2000." *Bicycling*, March 1992, 130–131.

Humus

Humus is essentially decomposed organic matter in **soil**. Humus can vary in color but is often dark brown. Besides containing valuable nutrients, there are many other benefits of humus: it stabilizes soil mineral particles into aggregates, improves pore space relationships and aids in air and water movement, aids in water holding capacity, and influences the **absorption** of **hydrogen** ions as a **pH** regulator.

Hunting and trapping

Wild animals are a potentially renewable natural resource. This means that they can be harvested in a sustainable fashion, as long as then birth rate is greater than the rate of exploitation by humans. In the sense meant here, "harvesting" refers to the killing of wild animals as a source of meat, fur, antlers, or other useful products, or as an outdoor sport. The harvesting can involve trapping, or hunting using guns, bows-and-arrows, or other weapons. (Fishing is also a kind of hunting, but it is not dealt with here). From the ecological perspective, it is critical that the exploitation is undertaken in a sustainable fashion; otherwise, serious damages are caused to the resource and to ecosystems more generally.

Unfortunately, there have been numerous examples in which wild animals have been harvested at grossly unsustainable rates, which caused their populations to decline severely. In a few cases this caused **species** to become extinct—they no longer occur anywhere on Earth. For example, commercial hunting in North America resulted in the extinctions of the great auk (*Pinguinnis impennis*), **passenger pigeon** (*Ectopistes migratorius*), and Steller's sea cow (*Hydrodamalis stelleri*). Unsustainable commercial hunting also brought other species to the brink of **extinction**, including the Eskimo curlew (*Numenius borealis*), northern right whale (*Eubalaena glacialis*), northern fur seal (*Callorhinus ursinus*), grey whale (*Eschrichtius robustus*), and American **bison** or buffalo (*Bison bison*).

Fortunately, these and many other examples of over-exploitation of wild animals by humans are regrettable cases from the past. Today, the exploitation of wild animals in North America is undertaken with a view to the longer-term **conservation** of their stocks, that is, an attempt is made to manage the harvesting in a sustainable fashion. This means that trapping and hunting are much more closely regulated than they used to be.

If harvests of wild animals are to be undertaken in a sustainable manner, it is critical that harvest levels are determined using the best available understanding of population-level productivity and stock sizes. It is also essential that harvest quotas are respected by trappers and hunters and that illegal exploitation (or **poaching**) does not compromise what might otherwise be a sustainable activity. The challenge of modern **wildlife management** is to ensure that good conservation science is sensibly integrated with effective monitoring and management of the rates of exploitation.

Ethics of trapping and hunting

From the strictly ecological perspective, sustainable trapping and hunting of wild animals is no more objectionable than the prudent harvesting of timber or agricultural crops. However, people have widely divergent attitudes about the killing of wild (or domestic) animals for meat, sport, or profit. At one end of the ethical spectrum are people who see no problem with the killing wild animals as a source of meat or cash. At the other extreme are individuals with a profound respect for the rights of all animals, and who believe that killing any sentient creature is ethically wrong. Many of these latter people are animal-rights activists, and some of them are involved in organizations that undertake high-profile protests and other forms of advocacy to prevent or restrict trapping and hunting. In essence, these people object to the lethal exploitation of wild animals, even under closely regulated conditions that would not deplete their populations. Most people, of course, have attitudes that are intermediate to those just described.

Trapping

The fur trade was one a very important commercial activity during the initial phase of the colonization of North America by Europeans. During those times, as now, furs were a valuable commodity that could be obtained from **nature** and could be sold at a great profit in urban markets. In fact, the quest for furs was the most important reason for much of the early exploration of the interior of North America, as fur traders penetrated all of the continent's great rivers seeking new sources of pelts and profit. Most fur-bearing animals are harvested by a form of hunting known as trapping.

Until recently, most trapping involved leg-hold traps, a relatively crude method that results in many animals enduring cruel and lingering deaths. Fortunately, other, more humane alternatives now exist in which most trapped animals are killed quickly and do not suffer unnecessarily. In large part, the movement towards more merciful trapping methods has occurred in response to effective, high-profile lobbying by organizations that oppose trapping, and the trapping industry has responded by developing and using more humane methods of killing wild furbearers.

Various species of furbearers are trapped in North America, particularly in relatively remote, wild areas, such as the northern and montane forests of the continental United States, Alaska, and Canada. Among the most valuable fur-bearing species are beaver (*Castor canadensis*), muskrat (*Ondatra zibethicus*), mink (*Mustela vison*), river otter (*Enhydra lutris*), bobcat (*Lynx rufus*), lynx (*Lynx canadensis*), red fox (*Vulpes vulpes*), wolf (*Canis lupus*), and **coyote** (*Canis latrans*). The hides of other species are also valuable, such as black bear (*Ursus americanus*), white-tailed deer (*Odocoileus virginianus*), and moose (*Alces alces*), but these species are not hunted primarily for their pelage.

Some species of **seals** are hunted for their fur, although this is largely done by shooting, clubbing, or netting, rather than by trapping. The best examples of this are the harp seal (*Phoca groenlandica*) of the northwestern Atlantic Ocean and the northern fur seal (*Callorhinus ursinus*) of the Bering Sea. Many seal pups are killed by commercial hunters in the spring when their coats are still white and soft. This harvest has been highly controversial and is the subject of intense opposition from **animal rights** groups.

Game Mammals

Hunting is a popular sport in North America, enjoyed by about 16 million people each year, most of them men. In 1991, hunting contributed more than $12 billion to the United States economy, about half of which was spent by big-game hunters.

Various species of terrestrial animals are hunted in large numbers. This is mostly done by stalking the animals and shooting them with rifles, although shotguns and bow-and-arrow are sometimes used. Some hunting is done for subsistence purposes, that is, the meat of the animals is used to feed the family or friends of the hunters. Subsistence hunting is especially important in remote areas and for aboriginal hunters. Commercial or market hunts also used to be common, but these are no longer legal in North America (except under exceptional circumstances) because they have generally proven to be unsustainable. However, illegal, semi-commercial hunting (or poaching) still takes place in many remote areas where game animals are relatively abundant and where there are local markets for wild meat.

In addition, many people hunt as a sport, that is, for the excitement and accomplishment of tracking and killing wild animals. In such cases, using the meat of the hunted animals may be only a secondary consideration, and in fact the hunter may only seek to retain the head, antlers, or horns of the prey as a trophy (although the meat may be kept by the hunter's guide). Big-game hunting is an economically important activity in North America, with large amounts of money being spent on the equipment, guides, and **transportation** necessary to undertake this sport.

The most commonly hunted big-game mammal in North America is the white-tailed deer. Other frequently hunted ungulates include the mule deer (*Odocoileus hemionus*), moose, elk or wapiti (*Cervus canadensis*), caribou (*Rangifer tarandus*), and pronghorn antelope (*Antilocapra americana*). Black bear and **grizzly bear** (*Ursus arctos*) are also hunted, as are bighorn sheep (*Ovis canadensis*) and mountain goat (*Oreamnos americanus*). Commonly hunted small-game species include various species of rabbits and hares, such as the cottontail rabbit (*Sylvilagus floridanus*), snowshoe hare (*Lepus americanus*), and jackrabbit (*Lepus californicus*), as well as the grey or black squirrel (*Sciurus carolinensis*) and woodchuck (*Marmota monax*). Wild boar (*Sus scrofa*) are also hunted in some regions—these are feral animals descended from escaped domestic pigs.

Game birds

Various larger species of birds are hunted in North America for their meat and for sport. So-called upland game birds are hunted in terrestrial habitats and include ruffed grouse (*Bonasa umbellus*), willow ptarmigan (*Lagopus lagopus*), bobwhite quail (*Colinus virginianus*), wild turkey (*Meleagris gallopava*), mourning dove (*Zanaidura macroura*), and woodcock (*Philohela minor*). Several **introduced species** of upland gamebirds are also commonly hunted, particularly ring-necked pheasant (*Phasianus colchicus*) and Hungarian or grey partridge (*Perdix perdix*).

Much larger numbers of waterfowl are hunted in North America, including millions of ducks and geese. The most commonly harvested species of waterfowl are mallard (*Anas platyrhynchos*), wood duck (*Aix sponsa*), Canada goose (*Branta canadensis*), and snow and blue goose (*Chen hyperborea*), but another 35 or so species in the duck family are also hunted. Other hunted waterfowl include coots (*Fulica americana*) and moorhens (*Gallinula chloropus*).

[*Bill Freedman Ph.D.*]

RESOURCES
BOOKS

Halls, L. K., ed. *White-tailed Deer: Ecology and Management.* Harrisburg: Stackpole Books, 1984.

Novak, M., et al. *Wild Furbearer Management and Conservation in North America*. North Bay: Ontario Trappers Association, 1987.

Phillips, P. C. *The Fur Trade* (2 vols.). Norman: University of Oklahoma, 1961.

Robinson, W. L., and E. G. Bolen. *Wildlife Ecology and Management*. 3rd ed. New York: Macmillan, 1996.

PERIODICALS

Freedman, B. *Environmental Ecology*. 2nd ed. San Diego: Academic Press, 1995.

Hurricane

Hurricanes, called typhoons or tropical cyclones in the Far East, are intense cyclonic storms which form over warm tropical waters, and generally remain active and strong only while over the oceans. Their intensity is marked by a distinct spiraling pattern of clouds, very low atmospheric pressure at the center, and extremely strong winds blowing at speeds greater than 74 mph (120 kph) within the inner rings of clouds. Typically when hurricanes strike land and move inland, they immediately start to disintegrate, though before they do they bring widespread destruction of property and loss of life. The radius of such a storm can be 100 mi (160 km) or greater. Thunderstorms, hail, and tornados frequently are imbedded in hurricanes.

Hurricanes occur in every tropical ocean except the South Atlantic, and with greater frequency from August through October than any other time of year. The center of a hurricane is called the eye. It is an area of relative calm, few clouds and higher temperatures, and represents the center of the low pressure pattern. Hurricanes usually move from east to west near the tropics, but when they migrate poleward to the mid-latitudes they can get caught up in the general west to east flow pattern found in that region of the earth. *See also* Tornado and cyclone

George Evelyn Hutchinson (1903 – 1991)

American ecologist

Born January 30, 1903, in Cambridge, England, Hutchinson was the son of Arthur Hutchinson, a professor of mineralogy at Cambridge University, and Evaline Demeny Shipley Hutchinson, an ardent feminist. He demonstrated an early interest in **flora** and **fauna** and a basic understanding of the scientific method. In 1918, at the age of 15, he wrote a letter to the *Entomological Record and Journal of Variation* about a grasshopper he had seen swimming in a pond. He described an experiment he performed on the insect and included it for taxonomic identification.

In 1924, Hutchinson earned his bachelor's degree in zoology from Emmanuel College at Cambridge University, where he was a founding member of the Biological Tea Club. He then served as an international education fellow at the Stazione Zoologica in Naples from 1925 until 1926, when he was hired as a senior lecturer at the University of Witwatersrand in Johannesburg, South Africa. He was apparently fired from this position two years later by administrators who never imagined that in 1977 the university would honor the ecologist by establishing a research laboratory in his name.

Hutchinson earned his master's degree from Emmanuel College *in absentia* in 1928 and applied to Yale University for a fellowship so he could pursue a doctoral degree. He was instead appointed to the faculty as a zoology instructor. He was promoted to assistant professor in 1931 and became an associate professor in 1941, the year he obtained his United States citizenship. He was made a full professor of zoology in 1945, and between 1947 and 1965 he served as director of graduate studies in zoology. Hutchinson never did receive his doctoral degree, though he amassed an impressive collection of honorary degrees during his lifetime.

Hutchinson was best known for his interest in **limnology**, the science of freshwater lakes and ponds. He spent most of his life writing the four-volume *Treatise on Limnology*, which he completed just months before his death. The research that led to the first volume—covering geography, physics, and chemistry—earned him a Guggenheim Fellowship in 1957. The second volume, published in 1967, covered biology and **plankton**. The third volume, on water plants, was published in 1975, and the fourth volume, about invertebrates, appeared posthumously in 1993.

The *Treatise on Limnology* was among the nine books, nearly 150 research papers, and many opinion columns which Hutchinson penned. He was an influential writer whose scientific papers inspired many students to specialize in **ecology**. Hutchinson's greatest contribution to the science of ecology was his broad approach, which became known as the "Hutchinson school." His work encompassed disciplines as varied as biochemistry, geology, zoology, and botany. He pioneered the concept of **biogeochemistry**, which examines the exchange of **chemicals** between organisms and the **environment**. His studies in biogeochemistry focused on how **phosphates** and **nitrates** move from the earth to plants, then animals, and then back to the earth in a continuous cycle. His **holistic approach** influenced later environmentalists when they began to consider the global scope of environmental problems.

In 1957, Hutchinson published an article entitled "Concluding Remarks," considered his most inspiring and intriguing work, as part of the Cold Spring Harbor Symposia on Quantitative Biology. Here, he introduced and described the ecological **niche**, a concept which has been the source of much research and debate ever since. The article was one of only three in the field of ecology chosen for the 1991 collection *Classics in Theoretical Biology*.

Hutchinson won numerous major awards for his work in ecology. In 1950, he was elected to the National Academy of Science. Five years later, he earned the Leidy Medal from the Philadelphia Academy of Natural Sciences. He was awarded the Naumann Medal from the International Association of Theoretical and Applied Limnology in 1959. This is a global award, granted only once every three years, which Hutchinson earned for his contributions to the study of lakes in the first volume of his treatise. In 1962, the **Ecological Society of America** chose him for its Eminent Ecologist Award.

Hutchinson's research often took him out of the country. In 1932, he joined a Yale expedition to Tibet, where he amassed a vast collection of organisms from high-altitude lakes. He wrote many scientific articles about his work in North India, and the trip also inspired his 1936 travel book, *The Clear Mirror*. Other research projects drew Hutchinson to Italy, where, in the **sediment** of Lago di Monterosi, a lake north of Rome, he found evidence of the first case of artificial eutrophication, dating from around 180 B.C.

Hutchinson was devoted to the arts and humanities, and he counted several musicians, artists, and writers among his friends. The most prominent of his artistic friends was English author Rebecca West. He served as her literary executor, compiling a bibliography of her work which was published in 1957. He was also the curator of a collection of her papers at Yale's Beinecke Library. Hutchinson's writing reflected his diverse interests. Along with his scientific works and his travel book, he also wrote an autobiography and three books of essays, *The Itinerant Ivory Tower* (1953), *The Enchanted Voyage and Other Studies* (1962), and *The Ecological Theatre and the Evolutionary Play* (1965). For 12 years, beginning in 1943, Hutchinson wrote a regular column titled "Marginalia" for the *American Scientist*. His thoughtful columns examined the impact on society of scientific issues of the day.

Hutchinson's skill at writing, as well as his literary interests, was recognized by Yale's literary society, the Elizabethan Club, which twice elected him president. He was also a member of the Connecticut Academy of Arts and Sciences and served as its president in 1946.

While Hutchinson built his reputation on his research and writing, he also was considered an excellent teacher. His teaching career began with a wide range of courses including beginning biology, entomology, and vertebrate embryology. He later added limnology and other graduate courses to his areas of expertise. He was personable as well as innovative, giving his students illustrated note sheets, for example, so they could concentrate on his lectures without worrying about taking their own notes. Leading oceanographer Linsley Pond was among the students whose careers were changed by Hutchinson's teaching. Pond enrolled in Yale's doctoral program with the intention of becoming an experimental embryologist. But after one week in Hutchinson's limnology class, he had decided to do his dissertation research on a pond.

Hutchinson loved Yale. He particularly cherished his fellowship in the residential Saybrook College. He was also very active in several professional associations, including the American Academy of Arts and Sciences, the American Philosophical Society, and the **National Academy of Sciences**. He served as president of the American Society of Limnology and Oceanography in 1947, the American Society of Naturalists in 1958, and the International Association for Theoretical and Applied Limnology from 1962 until 1968.

Hutchinson retired from Yale as professor emeritus in 1971, but continued his writing and research for 20 more years, until just months before his death. He produced several books during this time, including the third volume of his treatise, as well as a textbook titled *An Introduction to Population Ecology* (1978), and memoirs of his early years, *The Kindly Fruits of the Earth* (1979).

He also occasionally returned to his musings on science and society, writing about several topical issues in 1983 for the *American Scientist*. Here, he examined the question of nuclear disarmament, speculating that "it may well be that total nuclear disarmament would remove a significant deterrent to all war." In the same article, he also philosophized on differences in behavior between the sexes: "On the whole, it would seem that, in our present state of **evolution**, the less aggressive, more feminine traits are likely to be of greater value to us, though always endangered by more aggressive, less useful tendencies. Any such sexual difference, small as it may be, is something on which perhaps we can build."

Several of Hutchinson's most prestigious honors, including the Tyler Award, came during his retirement. Hutchinson earned the $50,000 award, often called the Nobel Prize for **conservation**, in 1974. That same year, the National Academy of Sciences gave him the Frederick Garner Cottrell Award for Environmental Quality. He was awarded the Franklin Medal from the Franklin Institute in 1979, the Daniel Giraud Elliot Medal from the National Academy of Sciences in 1984, and the Kyoto Prize in Basic Science from Japan in 1986. Having once rejected a National Medal of Science because it would have been bestowed on him by President Richard Nixon, he was awarded the medal posthumously by President George Bush in 1991.

Hutchinson's first marriage, to Grace Evelyn Pickford, ended with a divorce in 1933. During the six weeks residence the state of Nevada then required to grant divorces, he studied the lakes near Reno and wrote a major paper on freshwater ecology in **arid** climates. Later that year, Hutchinson married Margaret Seal, who died in 1983 from Alzheimer's disease. Hutchinson cared for her at home during her illness. In 1985, he married Anne Twitty Goldsby, whose care enabled him to travel extensively and continue working in spite of his failing health. When she died unexpectedly in December 1990, the ailing widower returned to his British homeland. He died in London on May 17, 1991, and was buried in Cambridge.

[*Cynthia Washam*]

RESOURCES

BOOKS

Hutchinson, George A. *A Preliminary List of the Writings of Rebecca West.* Yale University Library, 1957.

———. *A Treatise on Limnology.* Wiley, Vol. 1, 1957; Vol. 2, 1967; Vol. 3, 1979; Vol. 4, 1993.

———. *An Introduction to Population Ecology.* Yale University Press, 1978.

———. *The Clear Mirror.* Cambridge University Press, 1937.

———. *The Ecological Theater and the Evolutionary Play.* Yale University Press, 1965.

———. *The Enchanted Voyage and Other Studies.* Yale University Press, 1962.

———. *The Itinerant Ivory Tower.* Yale University Press, 1952.

———. *The Kindly Fruits of the Earth.* Yale University Press, 1979.

PERIODICALS

———. "Marginalia." *American Scientist* (November–December 1983): 639–644.

Edmondson, Y. H., ed. "G. Evelyn Hutchinson Celebratory Issue." *Limnology and Oceanography* 16 (1971): 167–477.

Edmondson, W. T. "Resolution of Respect." *Bulletin of the Ecological Society of America* 72 (1991): 212–216.

Hutchinson, George E. "A Swimming Grasshopper." *Entomological Record and Journal of Variation* 30 (1918): 138.

———. "Marginalia." *American Scientist* 31 (1943): 270.

———. "Concluding Remarks." *Bulletin of Mathematical Biology* 53 (1991): 193–213.

———. "Lanula: An Account of the History and Development of the Lago di Monterosi, Latlum, Italy." *Transactions of the American Philosophical Society* 64 (1970): part 4.

Hybrid vehicles

The roughly 200 million automobiles and light trucks currently in use in the United States travel approximately 2.4 trillion miles every year, and consume almost two-thirds of the U.S. oil supply. They also produce about two-thirds of the **carbon monoxide**, one-third of the **lead** and **nitrogen oxides**, and a quarter of all volatile organic compounds (VOCs). More efficient **transportation** energy use could have dramatic effects on environmental quality as well as

saving billions of dollars every year in our payments to foreign governments. In response to high **gasoline** prices in the 1970s and early 1980s, **automobile** gas-mileage averages in the United States more than doubled from 13 mpg in 1975 to 28.8 mpg in 1988.

Unfortunately, cheap fuel prices and the popularity of sport utility vehicles (SUVs) and light trucks in the 1990s caused fuel efficiency to slide back below where it was 25 years earlier. By 2002, the average mileage for U.S. cars and trucks was only 27.6 mpg. Amory B. Lovins of the **Rocky Mountain Institute** in Colorado estimated that raising the average fuel efficiency of the United States car and light truck fleet by one mile per gallon would cut oil consumption about 295,000 barrels per day. In one year, this would equal the total amount the Interior Department hopes to extract from the **Arctic National Wildlife Refuge** (ANWR) in Alaska.

It isn't inevitable that we consume and pollute so much. A number of alternative transportation options already are available. Of course the lowest possible fossil fuel consumption option is to walk, skate, ride a bicycle, or other forms of human-powered movement. Many people, however, want or need the comfort and speed of a motor vehicle. Several models of battery-powered electric automobiles have been built, but the batteries are heavy, expensive, and require more frequent recharging than most customers will accept. Even though 90% of all daily commutes are less than 50 mi (80 km), most people want the capability to take a long road trip of several hundred miles without needing to stop for fuel or recharging.

An alternative that appears to have much more customer appeal is the hybrid gas-electric vehicle. The first hybrid to be marketed in the United States was the two-seat Honda Insight. A 3-cylinder, 1.0 liter gas engine is the main power source for this sleek, lightweight vehicle. A 7-hp (horsepower) electric motor helps during acceleration and hill climbing. When the small battery pack begins to run down, it is recharged by the gas engine, so that the vehicle never needs to be plugged in. More electricity is captured during "regenerative" braking further increasing efficiency. With a streamlined lightweight plastic and **aluminum** body, the Insight gets about 75 mpg (33.7 km/l) in highway driving and has low-enough emissions to qualify as a "super low **emission** vehicle." It meets the most strict **air quality** standards anywhere in the United States. Quick acceleration and nimble handling make the Insight fun to drive. Current cost is about $20,000.

Perhaps the biggest drawback to the Insight is its limited passenger and cargo capacity. Although the vast majority of all motor vehicle trips made in the United States involve only a single driver, most people want the ability to have more than one passenger or several suitcases at least occasionally. To meet this need, Honda introduced a hybrid-engine version of its popular Civic line in 2002. With four doors and ample space for four adults plus a reasonable amount of luggage. The 5-speed manual version of the Civic hybrid gets 48 mpg in both city and highway driving. With a history of durability and consumer satisfaction in other Honda models, and a 10-year warranty on its battery and drive train, the hybrid Civic appears to offer the security that consumers will want in adopting this new technology.

Toyota also has introduced a hybrid vehicle called the Prius. Similar in size to the Honda Civic, the Prius comes in a four-door model with enough room for the average American family. During most city driving, it depends only on its quiet, emission-free, electric motor. The batteries needed to drive the 40-hp are stacked up behind the back seat providing a surprisingly large trunk for luggage. The 70-hp, 1.5 liter gas engine kicks in to help accelerate or when the batteries need recharging. Getting about 52 mpg (22 km/l) in city driving, the Prius is one of the most efficient cars on the road and can travel more than 625 mi (1,000 km) without refueling. Some drivers are unnerved by the noiseless electric motor. Sitting at a stoplight, it makes no sound at all. You might think it was dead, but when the light changes, you glide off silently and smoothly.

Introduced in Japan in 1997, the Prius sells in the United States for about the same price as the Honda hybrids. The **Sierra Club** estimates that in 100,000 mi (160,000 km), a Prius will generate 27 tons of CO_2, a Ford Taurus will generate 64 tons, while the Ford Excursion SUV will produce 134 tons. In 1999, the Sierra Club awarded both the Insight and the Prius an "excellence in engineering" award, the first time this organization has ever endorsed commercial products.

Both Ford and General Motors (GM) have announced intentions to build hybrid engines for their popular sport utility vehicles and light trucks. This program may be more for public relations, however, than to save fuel or reduce **pollution**. The electrical generators coupled to engines of these vehicles will produce only 12 volts of power. This is far less than the 42 volts needed to provide drive the wheels. Instead, the electricity generated by the gasoline-burning engine will only be used to power accessories such as video recorders, computers, on-board refrigerators, and the like. Having this electrical power available will probably actually increase fuel consumption rather than reduce it. For uncritical consumers, however, it provides a justification for continuing to drive huge, inefficient vehicles.

In 2002, President G. W. Bush announced he was abandoning the $1.5 billion government-subsidized project to develop high-mileage gasoline-fueled vehicles started with great fanfare eight years earlier by the Clinton/Gore administration. Instead, Bush was throwing his support behind a plan to develop hydrogen-based **fuel cells** to power the automobiles of the future. Fuel cells use a semi-permeable film or electrolyte that allows the passage of charged atoms, called ions, but is impermeable to electrons to generate an electrical current between an anode and cathode.

A fuel cell run on pure oxygen and **hydrogen** produces no waste products except drinkable water and radiant heat. **Fossil fuels** can be used as the source for the hydrogen, but some pollutants are released (most commonly **carbon dioxide**) in the process of hydrogen generation. Currently, the fuel cells available need to be quite large to provide enough energy for a vehicle. Fuel cell-powered buses and vans that have space for a large power system are currently being tested, but a practical, family vehicle appears to be years away.

While they agree that fuel cells offer a wonderful option for cars of the future, many environmentalists regard putting all our efforts into this one project to be misguided at best. It probably will be at least a decade before a fuel-cell vehicle is commercially available.

[*William P. Cunningham Ph.D.*]

RESOURCES

BOOKS

Hodkinson, Ron, and John Fenton. *Lightweight Electric/Hybrid Vehicle Design.* Warrendale, PA: Society of Automotive Engineers, 2001.

Jurgen., Ronald K., ed. *Electric and Hybrid-electric Vehicles.* Warrendale, PA: Society of Automotive Engineers, 2002.

Koppel, Tom. *Powering the Future: The Ballard Fuel Cell and the Race to Change the World.* New York: John Wiley & Sons, 1999.

PERIODICALS

"Dark Days For Detroit—The Big Three's Gravy Train in Recent Years—Fat Profits from Trucks—is Being Derailed by a New Breed of Hybrid Vehicles from Europe and Japan." *Business Week*, January 28, 2002, 61.

Ehsani, M., K. M. Rahman, and H. A. Toliyat. "Propulsion System Design of Electric and Hybrid Vehicles." *IEEE Transactions on Industrial Electronics* 44 (1997): 19.

Hermance, David, and Shoichi Sasaki. "Special Report on Electric Vehicles—Hybrid Electric Vehicles take to the Streets." *IEEE Spectrum* 35 (1998): 48.

Jones, M. "Hybrid Vehicles—The Best of Both Worlds?" *Chemistry and Industry* 15 (1995): 589.

Maggetto, G. and J. Van Mierlo. "Fuel cells: Systems and applications—Electric vehicles, hybrid vehicles and fuel cell electric vehicles: State of the art and perspectives." *Annales de chimie—science des matériaux.* 26 (2001): 9.

Hydrocarbons

Any compound composed of elemental **carbon** and **hydrogen**, hydrocarbons may also contain **chlorine**, oxygen, **nitrogen**, and other atoms. Hydrocarbons are classified according to the arrangement of carbon atoms and the types of chemical bonds. The major classes include *aromatic* or carbon ring compounds, *alkanes* (also called aliphatic or paraffin) compounds with straight or branched chains and single bonds, and *alkenes* and *alkynes* with double and triple bonds, respectively. Most hydrocarbon fuels are a mixture of many compounds. **Gasoline**, for example, includes several hundred hydrocarbon compounds, including paraffins, olefins, and aromatic compounds, and consequently exhibits a host of possible environmental effects. All of the **fossil fuels**, including crude oils and **petroleum**, as well as many other compounds important to industries, are hydrocarbons. Hydrocarbons are environmentally important for several reasons. First, hydrocarbons give off **greenhouse gases**, especially **carbon dioxide**, when burned and are important contributors to **smog**. In addition, many aromatic hydrocarbons and hydrocarbons containing halogens are toxic or carcinogenic.

CHF_3	HFC-23
$CHCl_2CF_3$	HCFC-123
CH_2FCClF_2	HCFC-133b
CH_3CHClF	HCFC-151a

Hydrochlorofluorocarbons

The term hydrochlorofluorocarbon (HCFC) refers to halogenated **hydrocarbons** that contain **chlorine** and/or fluorine in place of some **hydrogen** atoms in the molecule. They are chemical cousins of the **chlorofluorocarbons** (CFCs), but differ from them in that they have less chlorine. A special subgroup of the HCFCs is the hydrofluorocarbons (HFCs), which contain no chlorine at all.

A total of 53 HCFCs and HFCs are possible.

The HCFCs and HFCs have become commercially and environmentally important since the 1980s. Their growing significance has resulted from increasing concerns about the damage being done to stratospheric **ozone** by CFCs.

Significant production of the CFCs began in the late 1930s. At first, they were used almost exclusively as refrigerants. Gradually other applications—especially as **propellants** and blowing agents—were developed. By 1970, the production of CFCs was growing by more than 10% per year, with a worldwide production of well over 662 million lb (300 million kg) of one family member alone, CFC-11.

Environmental studies began to show, however, that CFCs decompose in the upper **atmosphere**. Chlorine atoms produced in this reaction attack ozone molecules (O_3), converting them to normal oxygen (O_2). Since stratospheric ozone provides protection for humans against solar **ultraviolet radiation**, this finding was a source of great concern. By 1987, 31 nations had signed the Montreal Protocol, agreeing to cut back significantly on their production of CFCs.

The question became how nations were to find substitutes for the CFCs. The problem was especially severe in developing nations where CFCs are widely used in refrigeration and air-conditioning systems. Countries like China and India refused to take part in the CFC-reduction plan unless developed nations helped them switch over to an equally satisfactory substitute.

Scientists soon learned that HCFCs were a more benign alternative to the CFCs. They discovered that compounds with less chlorine than the amount present in traditional CFCs were less stable and often decomposed before they reached the **stratosphere**. By mid 1992, the United States **Environmental Protection Agency** (EPA) had selected 11 **chemicals** that they considered to be possible replacements for CFCs. Nine of those compounds are HFCs and two are HCFCs.

The HCFC-HFC solution is not totally satisfactory, however. Computer models have shown that nearly all of the proposed substitutes will have at least some slight effect on the ozone layer and the **greenhouse effect**. In fact, the British government considered banning one possible substitute for CFCs, HCFC-22, almost as soon as the compound was developed. In addition, one of the most promising candidates, HCFC-123, was found to be carcinogenic in rats.

Finally, the cost of replacing CFCs with HCFCs and HFCs is expected to be high. One consulting firm, Metroeconomica, has estimated that CFC substitutes may be six to 15 times as expensive as CFCs themselves. *See also* Aerosol; Air pollution; Air pollution control; Air quality; Carcinogen; Ozone layer depletion; Pollution; Pollution control

[*David E. Newton*]

RESOURCES

PERIODICALS

Johnson, J. "CFC Substitutes Will Still Add to Global Warming." *New Scientist* 126 (April 14, 1990): 20.

MacKenzie, D. "Cheaper Alternatives for CFCs." *New Scientist* 126 (June 30, 1990): 39–40.

Pool, R. "Red Flag on CFC Substitute." *Nature* 352 (July 11, 1991): 352.

Stone, R. "Ozone Depletion: Warm Reception for Substitute Coolant." *Science* 256 (April 3, 1992): 22.

Hydrogen

The lightest of all chemical elements, hydrogen has a density about one-fourteenth that of air. It has a number of special chemical and physical properties. For example, hydrogen has the second lowest boiling and freezing points of all elements. The **combustion** of hydrogen produces large quantities of heat, with water as the only waste product. From an environmental standpoint, this fact makes hydrogen a highly desirable fuel. Many scientists foresee the day when hydrogen will replace **fossil fuels** as our most important source of energy.

Hydrogeology

Sometimes called **groundwater hydrology** or geohydrology, this branch of hydrology is concerned with the relationship of subsurface water and geologic materials. Of primary interest is the saturated zone of subsurface water, called groundwater, which occurs in rock formations and in unconsolidated materials such as sands and gravels. Groundwater is studied in terms of its occurrence, amount, flow, and quality. Historically, much of the work in hydrogeology centered on finding sources of groundwater to supply water for drinking, **irrigation**, and municipal uses. More recently, groundwater contamination by pesticides, chemical fertilizers, toxic wastes, and **petroleum** and **chemical spills** have become new areas of concern for hydrogeologists.

Hydrologic cycle

The natural circulation of water on the earth is called the hydrologic cycle. Water cycles from bodies of water, via evaporation to the **atmosphere**, and eventually returns to the oceans as precipitation, **runoff** from streams and rivers, and **groundwater** flow. Water molecules are transformed from liquid to vapor and back to liquid within this cycle. On land, water evaporates from the **soil** or is taken up by plant roots and eventually transpired into the atmosphere through plant leaves; the sum of evaporation and **transpiration** is called **evapotranspiration**.

Water is recycled continuously. The molecules of water in a glass used to quench your thirst today, at some point in time may have dissolved minerals deep in the earth as groundwater flow, fallen as rain in a tropical typhoon, been transpired by a tropical plant, been temporarily stored in a mountain glacier, or quenched the thirst of people thousands of years ago.

The hydrologic cycle has no real beginning or end but is a circulation of water that is sustained by **solar energy** and influenced by the force of gravity. Because the supply of water on the earth is fixed, there is no net gain or loss of water over time. On an average annual basis, global evaporation must equal global precipitation. Likewise, for any

body of land or water, changes in storage must equal the total inflow minus the total outflow of water. This is the hydrologic or water balance.

At any point in time, water on the earth is either in active circulation or in storage. Water is stored in icecaps, soil, groundwater, the oceans, and other bodies of water. Much of this water is only temporarily stored. The **residence time** of water storage in the atmosphere is several days and is only about 0.04% of the total freshwater on the earth. For rivers and streams, residence time is weeks; for lakes and reservoirs, several years; for groundwater, hundreds to thousands of years; for oceans, thousands of years; and for icecaps, tens of thousands of years. As the driving force of the hydrologic cycle, solar radiation provides the energy necessary to evaporate water from the earth's surface, almost three-quarters of which is covered by water. Nearly 86% of global precipitation originates from ocean evaporation. Energy consumed by the conversion of liquid water to vapor cools the temperature of the evaporating surface. This same energy, the latent heat of vaporization, is released when water vapor changes back to liquid. In this way, the hydrologic cycle globally redistributes heat energy as well as water.

Once in the atmosphere, water moves in response to weather circulation patterns and is transported often great distances from where it was evaporated. In this way, the hydrologic cycle governs the distribution of precipitation and hence, the availability of fresh water over the earth's surface. About 10% of atmospheric water falls as precipitation each day and is simultaneously replaced by evaporation. This 10% is unevenly distributed over the earth's surface and, to a large extent, determines the types of ecosystems that exist at any location on the earth and likewise governs much of the human activity that occurs on the land.

The earliest civilizations on the earth settled in close proximity to fresh water. Subsequently, and for centuries, humans have been striving to correct, or cope with, this uneven distribution of water. Historically, we have extracted stored water or developed new storages in areas of excess, or during periods of excess precipitation, so that water could be available where and when it is most needed.

Understanding processes of the hydrologic cycle can help us develop solutions to water problems. For example, we know that precipitation occurs unevenly over the earth's surface because of many complex factors that trigger precipitation. For precipitation to occur, moisture must be available and the atmosphere must become cooled to the **dew point**, the temperature at which air becomes saturated with water vapor. This cooling of the atmosphere occurs along storm fronts or in areas where moist air masses move into mountain ranges and are pushed up into colder air. However, atmospheric particles must be present for the moisture to condense upon, and water droplets must coalesce until they are large enough to fall to the earth under the influence of gravity.

Recognizing the factors that cause precipitation has resulted in efforts to create conditions favorable for precipitation over land surfaces via cloud seeding. Limited success has been achieved by seeding clouds with particles, thus promoting the condensation-coalescence process. Precipitation has not always increased with cloud seeding and questions of whether cloud seeding limits precipitation in other downwind areas is of both economic and environmental concern.

Parts of the world have abundant moisture in the atmosphere, but it occurs as fog because the mechanisms needed to transform this moisture into precipitation do not exist. In dry coastal areas, for example, some areas have no measurable precipitation for years, but fog is prevalent. By placing huge sheets of plastic mesh along coastal areas, fog is intercepted, condenses on the sheets, and provides sufficient drinking water to supply small villages.

Total rainfall alone does not necessarily indicate water abundance or **scarcity**. The magnitude of evapotranspiration compared to precipitation determines to some extent whether water is abundant or in short supply. On a continent basis, evapotranspiration represents from 56 to 80% of annual precipitation. For individual watersheds within continents, these%ages are more extreme and point to the importance of evapotranspiration in the hydrologic cycle.

Weather circulation patterns responsible for water shortages in some parts of the world are also responsible for excessive precipitation, floods, and related catastrophes in other parts of the world. Precipitation that falls on land, but that is not stored, evaporated or transpired, becomes excess water. This excess water eventually reaches groundwater, streams, lakes, or the ocean by surface and subsurface flow. If the soil surface is impervious or compacted, water flows over the land surface and reaches stream channels quickly. When surface flow exceeds a channel's capacity, flash **flooding** is the result. Excessive precipitation can saturate soils and cause flooding no matter what the pathway of flow. For example, in 1988 catastrophic flooding and mudslides in Thailand caused over 500 fatalities or missing persons, nearly 700 people were injured, 4,952 homes were lost, and 221 roads and 69 bridges were destroyed. A three-day rainfall of over nearly 40 in (1,000 mm) caused hillslopes to become saturated. The effects of heavy rainfall were exacerbated by the removal of natural forest cover and conversion to **rubber** plantations and agricultural crops.

Although floods and mudslides occur naturally, many of the pathways of water flow that contribute to such occurrences can be influenced by human activity. Any time vegeta-

tive cover is severely reduced and soil exposed to direct rainfall, surface water flow and soil **erosion** can degrade **watershed** systems and their aquatic ecosystems.

The implications of global warming or greenhouse effects on the hydrologic cycle raise several questions. The possible changes in frequency and occurrence of droughts and floods are of major concern, particularly given projections of **population growth**. Global warming can result in some areas becoming drier while others may experience higher precipitation. Globally, increased temperature will increase evaporation from oceans and ultimately result in more precipitation. The pattern of precipitation changes over the earth's surface, however, cannot be predicted at the present time.

The hydrologic cycle influences **nutrient** cycling of ecosystems, processes of soil erosion and transport of **sediment**, and the transport of pollutants. Water is an excellent liquid solvent; minerals, salts, and nutrients become dissolved and transported by water flow. The hydrologic cycle is an important driving mechanism of nutrient cycling. As a transporting agent, water moves minerals and nutrients to plant roots. As plants die and decay, water leaches out nutrients and carries them downstream. The physical action of rainfall on soil surfaces and the forces of running water can seriously erode soils and transport sediments downstream. Any minerals, nutrients, and pollutants within the soil are likewise transported by water flow into groundwater, streams, lakes, or estuaries.

Atmospheric moisture transports and deposits **atmospheric pollutants**, including those responsible for **acid rain**. Sulfur and **nitrogen oxides** are added to the atmosphere by the burning of **fossil fuels**. Being an excellent solvent, water in the atmosphere forms acidic compounds that become transported via the atmosphere and deposited great distances from their original site. Atmospheric pollutants and **acid** rain have damaged freshwater lakes in the Scandinavian countries and terrestrial vegetation in eastern Europe. In 1983, such **pollution** caused an estimated $1.2 billion loss of forests in the former West Germany alone. Once pollutants enter the atmosphere and become subject to the hydrologic cycle, problems of acid rain have little chance for resolution. However, programs that reduce atmospheric emissions in the first place provide some hope.

An improved understanding of the hydrologic cycle is needed to better manage **water resources** and our **environment**. Opportunities exist to improve our global environment, but better knowledge of human impacts on the hydrologic cycle is needed to avoid unwanted environmental effects. *See also* Estuary; Leaching

[*Kenneth N. Brooks*]

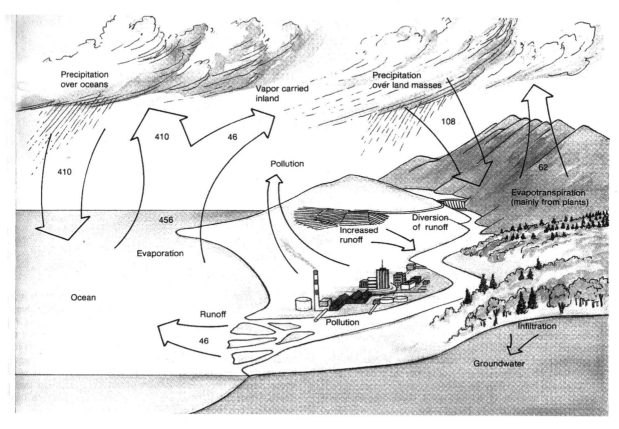

The hydrologic or water cycle. (McGraw-Hill Inc. Reproduced by permission.)

RESOURCES

BOOKS

Committee on Opportunities in the Hydrologic Sciences, Water Sciences Technology Board. *Opportunities in the Hydrologic Sciences.* National Research Council. Washington, DC: National Academy Press, 1991.

Lee, R. *Forest Hydrology.* New York: Columbia University Press, 1980.

Postel, S. "Air Pollution, Acid Rain, and the Future of Forests." *Worldwatch Paper 58.* Washington, DC: Worldwatch Institute, 1984.

Van der Leeden, F., F. L. Troise, and D. K. Todd. *The Water Encyclopedia.* 2nd ed. Chelsea, MI: Lewis Publishers, 1990.

PERIODICALS

Nash, N. C. "Chilean Engineers Find Water for Desert by Harvesting Fog in Nets." *New York Times*, July 14, 1992, B5.

OTHER

Rao, Y. S. "Flash Floods in Southern Thailand." *Tiger Paper* 15 (1988): 1–2. Regional Office for Asia and the Pacific (RAPA), Food and Agricultural Organization of the United Nations. Bangkok.

Hydrology

The science and study of water, including its physical and chemical properties and its occurrence on earth. Most commonly, hydrology encompasses the study of the amount, distribution, circulation, timing, and quality of water. It includes the study of rainfall, snow accumulation and melt, water movement over and through the **soil**, the flow of water in saturated, underground geologic materials (**groundwater**), the flow of water in channels (called streamflow), evaporation and **transpiration**, and the physical, chemical and biological characteristics of water. Solving problems concerned with water excesses, **flooding**, water shortages, and **water pollution** are in the domain of hydrologists. With increasing concern about water **pollution** and its effects on humans and on aquatic ecosystems, the practice of hydrology has expanded into the study and management of chemical and biological characteristics of water.

Hydroponics

Hydroponics is the practice of growing plants in water as opposed to **soil**. It comes from the Greek *hydro* ("water") and *ponos* ("labor"), implying "water working." The essential macro- and micro- (trace) nutrients needed by the plants are supplied in the water.

Hydroponic methods have been used for more than 2,000 years, dating back to the Hanging Gardens of Babylon. More recently, it has been used by plant physiologists to discover which nutrients are essential for plant growth. Unlike soil, where **nutrient** levels are unknown and variable, precise amounts and kinds of minerals can be added to deionized water, and removed individually, to find out their role in plant growth and development. During World War II hydroponics was used to grow vegetable crops by U.S. troops stationed on some Pacific islands.

Today, hydroponics is becoming a more popular alternative to conventional agriculture in locations with low or inaccessible sources of water or where land available for farming is scarce. For example, islands and **desert** areas like the American Southwest and the Middle East are prime regions for hydroponics. Plants are typically grown in greenhouses to prevent water loss. Even in temperate areas where fresh water is readily available, hydroponics can be used to grow crops in greenhouses during the winter months.

Two methods are traditionally used in hydroponics. The original technique is the water method, where plants are supported from a wire mesh or similar framework so that the roots hang into troughs which receive continuous supplies of nutrients. A recent modification is a nutrient-film technique (NFT), also called the nutrient-flow method, where the trough is lined with plastic. Water flows continuously over the roots, decreasing the stagnant boundary layer surrounding each root, and thus enhances nutrient uptake. This provides a versatile, lightweight, and inexpensive system. In the second method, plants are supported in a growing medium such as sterile sand, gravel, crushed volcanic rock, vermiculite, perlite, sawdust, peatmoss, or rice hulls. The nutrient solution is supplied from overhead or underneath holding tanks either continuously or semi-continuously using a drip method. The nutrient solution is usually not reused.

On some Caribbean Islands like St. Croix, hydroponics is being used in conjunction with intensive fish farms (e.g., tilapia) which use recirculated water (a practice is more recently known as aquaponics). This is a "win-win" situation because the nitrogenous wastes, which are toxic to the fish, are passed through large greenhouses with hydroponically-grown plants like lettuce. The plants remove the nutrients and the water is returned to the fish tanks. There is a sensitive balance between stocking density of fish and lettuce production. Too high a ratio of lettuce plants to fish results in lower lettuce production due to nutrient limitation. Too low a ratio also results in low vegetable production, but this time as a result of the buildup of toxic **chemicals**. The optimum yield came from a ratio of 1.9 lettuce plants to 1 fish. One pound (0.45 kg) of feed per day was appropriate to feed 33

lb (15 kg) of tilapia fingerlings, which sustained 189 lettuce plants and produced nearly 3,300 heads of lettuce annually. When integrated systems (fish-hydroponic recirculating units) are compared to separate production systems, the results clearly favor the former. The combined costs and chemical requirements of the separate production systems was nearly two to three times greater than that of the recirculating system to produce the same amount of lettuce and fish. However, there are some drawbacks that must be considered—disease outbreaks in plants and/or fish; the need to critically maintain proper nutrient (especially trace element), plant, and fish levels; uncertainties in fish and market prices; and the need for highly-skilled labor. The integrated method can be adapted to grow other types of vegetables like strawberries, ornamental plants like roses, and other types of animals such as shellfish. Some teachers have even incorporated this technique into their classrooms to illustrate ecological as well as botanical and culture principles.

Some proponents of hydroponic gardening make fairly optimistic claims and state that a sophisticated unit is no more expensive than an equivalent parcel of farmed land. They also argue that hydroponic units (commonly called "hydroponicums") require less attention than terrestrial agriculture. Some examples of different types of "successful" hydroponicums are: a person in the desert area of southern California has used the NFT system for over 18 years and grows his plants void of substate in water contained in open cement troughs that cover 3 acres (7.5 ha); a hydroponicum in Orlando, Florida, utilizes the Japanese system of planting seedlings on styrofoam boards that float on the surface of a nutrient bath which is constantly aerated; an outfit in Queens, New York, uses the Israeli Ein-Gedi system which allows plant roots to hang free inside a tube which is sprayed regularly with a nutrient solution, yielding 150,000 lbs (68,000 kg) of tomatoes, 100,000 lb (45,500 kg) of cucumbers, and one million heads of lettuce per acre (0.4 ha) each year; and finally, a farmer in Blooming **Prairie**, Minnesota, uses the NFT system in a greenhouse to grow Bibb and leafy lettuce year-round so he can sell his produce to area hospitals, some supermarkets, and a few produce warehouses.

Most people involved in hydroponics agree that the main disadvantage is the high cost for labor, lighting, water, and energy. Root fungal infections can also be easily spread. Advantages include the ability to grow crops in **arid** regions or where land is at a premium; more controlled conditions, such as the ability to grow plants indoors, and thus minimize pests and weeds; greater planting densities; and constant supply of nutrients. Hydroponic gardening is becoming more popular for home gardeners. It may also be a viable option to growing crops in some developing countries. Overall, the future looks bright for hydroponics.

[John Korstad]

RESOURCES

BOOKS

Resh, H. M. *Hydroponic Food Production: A Definitive Guidebook for the Advanced Home Gardener and Commercial Hydroponci Grower*, 5th ed. Santa Barbara: Woodbridge Press, 1995.

Saffell, H. L. *How to Start on a Shoestring and Make a Profit with Hydroponics.* Franklin, TN: Mayhill Press, 1994.

PERIODICALS

Nicol, E. "Hydroponics and Aquaculture in the High School Classroom." *The American Biology Teacher* 52 (1990): 182–4.

Rakocy, J. E. "Hydroponic Lettuce Production in a Recirculating Fish Culture System." *Island Perspectives* 3 (1988–89): 5–10.

Hydropower

see Aswan High Dam; Dams (environmental effects); Glen Canyon Dam; James Bay hydropower project; Low-head hydropower; Tellico Dam; Tennessee Valley Authority; Three Gorges Dam

Hydrothermal vents

Hydrothermal vents are hot springs located on the ocean floor. The vents spew out water heated by magma, molten rock from below the earth's crust. Water temperatures of higher than 660°F. have been recorded at some vents.

Water flowing from vents contains minerals such as iron, **copper**, and zinc. The minerals fall like rain and settle on the ocean floor. Over time, the mineral deposits build up and form a chimney around the vent.

The first hydrothermal vents were discovered in 1977 by scientists aboard the submersible *Alvin*. The scientists found the vents near the **Galápagos Islands** in the eastern Pacific Ocean. Other vents were discovered in the Pacific, Atlantic, and Indian oceans.

In 2000, scientists discovered a field of hydrothermal vents in the Atlantic Ocean. The area called the "Lost City" contained 180-feet tall chimneys. These were the largest known chimneys.

Hydrothermal vents are located at ocean depths of 8,200 to 10,000 feet. The area near a hydrothermal vent is home to unique animals. They exist without sunlight and live in mineral-levels that would poison animals living on land. These unique animals include 10-foot-long tube worms, 1-foot-long clams, and shrimp.

[*Liz Swain*]

Hypolimnion: Lakes
see **Great Lakes**

I

IAEA

see **International Atomic Energy Agency**

Ice age

Ice age usually refers to the Pleistocene epoch, the most recent occurrence of continental **glaciation**. Beginning several million years ago in **Antarctica**, it is marked by at least four major advances and retreats (excluding Antarctica). Ice ages occur during times when more snow falls during the winter than is lost by melting, evaporation, and loss of ice chunks in water during the summer. Alternating glacial and interglacial stages are best explained by a combination of Earth's orbital cycles and changes in **carbon dioxide** levels. These cycles operate on time scales of tens of millennia. By contrast, global warming projections involve decades, a far more imminent concern for humankind.

Ice age refugia

The series of ice ages that occurred between 2.4 million and 10,000 years ago had a dramatic effect on the **climate** and the life forms in the tropics. During each glacial period the tropics became both cooler and drier, turning some areas of **tropical rain forest** into dry seasonal forest or **savanna**. For reasons associated with local **topography**, geography, and climate, some areas of forest escaped the dry periods, and acted as refuges (refugia) for forest biota. During subsequent interglacials, when humid conditions returned to the tropics, the forests expanded and were repopulated by plants and animals from the species-rich refugia.

Ice age refugia today correspond to present day areas of tropical forest that typically receive a high rainfall and often contain unusually large numbers of **species**, including a high proportion of **endemic species**. These species-rich refugia are surrounded by relatively species-poor areas of forest. Refugia are also centers of distribution for obligate forest species (such as the gorilla [*Gorilla gorilla*]) with a present day narrow and disjunct distribution best explained by invoking past episodes of **deforestation** and reforestation. The location and extent of the forest refugia have been mapped in both Africa and South America. In the African rain forests there are three main centers of species richness and endemism recognized for mammals, birds, reptiles, amphibians, butterflies, freshwater crabs, and flowering plants. These centers are in Upper Guinea, Cameroon, and Gabon, and the eastern rim of the Zaire basin. In the **Amazon Basin** more than 20 refugia have been identified for different groups of animals and plants in Peru, Columbia, Venezuela, and Brazil.

The precise effect of the ice ages on **biodiversity** in tropical rain forests is currently a matter of debate. Some have argued that the repeated fluctuations between humid and **arid** phases created opportunities for the rapid **evolution** of certain forest organisms. Others have argued the opposite — that the climatic fluctuations resulted in a net loss of species diversity through an increase in the **extinction** rate. It has also been suggested that refugia owe their species richness not to past climate changes but to other underlying causes such as a favorable local climate, or **soil**.

The discovery of centers of high biodiversity and endemism within the tropical **rain forest biome** has profound implications for **conservation biology**. A "refuge rationale" has been proposed by conservationists, whereby ice age refugia are given high priority for preservation, since this would save the largest number of species, (including many unnamed, threatened, and **endangered species**), from extinction.

Since refugia survived the past dry-climate phases, they have traditionally supplied the plants and animals for the restocking of the new-growth forests when wet conditions returned. Modern deforestation patterns, however, do not take into account forest history or biodiversity, and both forest refugia and more recent forests are being destroyed equally. For the first time in millions of years, future tropical forests which survive the present mass deforestation episode could have no species-rich centers from which they can be restocked. *See also* Biotic community; Deciduous forest; Desertification; Ecosystem; Environment; Mass extinction

[*Neil Cumberlidge Ph.D.*]

RESOURCES
BOOKS

Collins, Mark, ed. *The Last Rain Forests*. London: Mitchell Beazley Publishers, 1990.

Kingdon, Jonathan. *Island Africa: The Evolution of Africa's Rare Animals and Plants*. Princeton: Princeton University Press, 1989.

Sayer, Jeffrey A., et al., eds. *The Conservation Atlas of Tropical Forests*. New York: Simon and Schuster, 1992.

Whitmore, T. C. *An Introduction to Tropical Rain Forests*. Oxford, England: Clarenden Press, 1990.

Wilson, E. O., ed. *Biodiversity*. Washington DC: National Academy Press, 1988.

OTHER

"Biological Diversification in the Tropics." *Proceedings of the Fifth International Symposium of the Association for Tropical Biology, at Caracas, Venezuela, February 8-13, 1979*, edited by Ghillean T. Prance. New York: Columbia University Press, 1982.

Impervious material

As used in **hydrology**, this term refers to rock and **soil** material that occurs at the earth's surface or within the subsurface which does not permit water to enter or move through them in any perceptible amounts. These materials normally have small-sized pores or have pores that have become clogged (sealed) which severely restrict water entry and movement. At the ground surface, rock outcrops, road surfaces, or soil surfaces that have been severely compacted would be considered impervious. These areas shed rainfall easily, causing overland flow or surface **runoff** which pick up and transport soil particles and cause excessive soil **erosion**. Soils or geologic strata beneath the earth's surface are considered impervious, or impermeable, if the size of the pores is small and/or if the pores are not connected.

Improvement cutting

Removal of crooked, forked, or diseased trees from a forest in which tree diameters are 5 in (13 cm) or larger. In forests where trees are smaller, the same process is called cleaning or weeding. Both have the objective of improving **species** composition, stem quality and/or growth rate of the forest. Straight, healthy, vigorous trees of the desired species are favored. By discriminating against certain tree species and eliminating trees with cavities or insect problems, improvement cuts can reduce the variety of habitats and thereby diminish **biodiversity**. An improvement cut is the initial step to prepare a neglected or unmanaged stand for future harvest. *See also* Clear-cutting; Forest management; Selection cutting

In situ mining
see **Bureau of Mines**

Inbreeding

Inbreeding occurs when closely related individuals mate with one another. Inbreeding may happen in a small population or due to other isolating factors; the consequence is that little new genetic information is added to the **gene pool**. Thus recessive, deleterious alleles become more plentiful and evident in the population. Manifestations of inbreeding are known as *inbreeding depression*. A general loss of fitness often results and may cause high infant **mortality**, and lower birth weights, **fecundity**, and longevity. Inbreeding depression is a major concern when attempting to protect small populations from **extinction**.

Incidental catch
see **Bycatch**

Incineration

As a method of **waste management**, incineration refers to the burning of waste. It helps reduce the volume of **landfill** material and can render toxic substances non-hazardous, provided certain strict guidelines are followed. There are two basic types of incineration: municipal and **hazardous waste** incineration.

Municipal waste incineration

The process of incineration involves the combination of organic compounds in solid wastes with oxygen at high temperature to convert them to ash and gaseous products. A municipal incinerator consists of a series of unit operations which include a loading area under slightly negative pressure to avoid the escape of odors, a refuse bin which is loaded by a grappling bucket, a charging hopper leading to an inclined feeder and a furnace of varying type—usually of a horizontal burning grate type—a **combustion** chamber equipped with a bottom ash and clinker **discharge**, followed by a gas flue system to an expansion chamber. If byproduct stream is to be produced either for heating or power generation purposes, then the downstream flue system includes heat exchanger tubing as well. After the heat has been exchanged, the **flue gas** proceeds to a series of gas cleanup

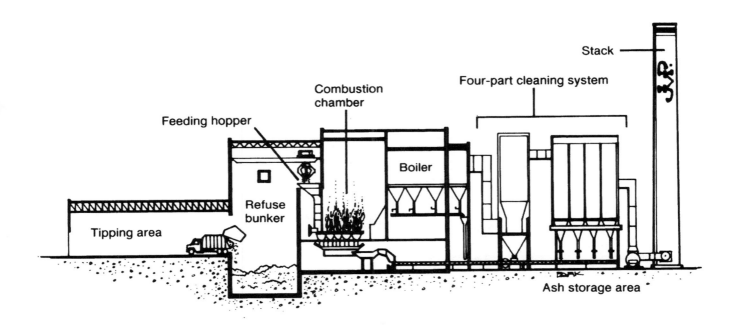

Diagram of a municipal incinerator. (McGraw-Hill Inc. Reproduced by permission.)

systems which neutralizes the **acid** gases (**sulfur dioxide** and hydrochloric acid, the latter resulting from burning chlorinated plastic products), followed by gas **scrubbers** and then solid/gas separation systems such as baghouses before dischargement to **tall stacks**. The stack system contains a variety of sensing and control devices to enable the furnace to operate at maximum efficiency consistent with minimal **particulate** emissions. A continuous log of monitoring systems is also required for compliance with county and state environmental quality regulations.

There are several products from a municipal incinerator system: items which are removed before combustion such as large metal pieces; grate or bottom ash (which is usually water-sprayed after removal from the furnace for safe storage); fly (or top ash) which is removed from the flue system generally mixed with products from the acid neutralization process; and finally the flue gases which are expelled to the **environment**. If the system is operating optimally, the flue gases will meet **emission** requirements, and the **heavy metals** from the wastes will be concentrated in the **fly ash**. (Typically these heavy metals, which originate from volatile metallic constituents, are **lead** and arsenic.) The fly ash typically is then stored

in a suitable landfill to avoid future problems of **leaching** of heavy metals. Some municipal systems blend the bottom ash with the top ash in the plant in order to reduce the level of heavy metals by dilution. This practice is undesirable from an ultimate environmental viewpoint.

There are many advantages and disadvantages to municipal waste incineration. Some of the advantages are as follows: 1) The waste volume is reduced to a small fraction of the original. 2) Reduction is rapid and does not require semi-infinite residence times in a landfill. 3) For a large metropolitan area, waste can be incinerated on site, minimizing **transportation** costs. 4) The ash residue is generally sterile, although it may require special disposal methods. 5) By use of gas clean-up equipment, discharges of flue gases to the environment can meet stringent requirements and be readily monitored. 6) Incinerators are much more compact than landfills and can have minimal odor and vermin problems if properly designed. 7) Some of the costs of operation can be reduced by heat-recovery techniques such as the sale of steam to municipalities or electrical energy generation.

There are disadvantages to municipal waste incineration as well. For example: 1) Generally the capital cost is high and is escalating as **emission standards** change. 2) Permitting requirements are becoming increasingly more difficult to obtain. 3) Supplemental fuel may be required to burn municipal wastes, especially if **yard waste** is not removed prior to collection. 4) Certain items such as mercury-containing batteries can produce emissions of **mercury** which the gas cleanup system may not be designed to remove. 5) Continuous skilled operation and close maintenance of process control is required, especially since stack monitoring equipment reports any failure of the equipment which could result in mandated shut down. 6) Certain materials are not burnable and must be removed at the source. 7) Traffic to and from the incinerator can be a problem unless timing and routing are carefully managed. 8) The incinerator, like a landfill, also has a limited life, although its lifetime can be increased by capital expenditures. 9) Incinerators also require landfills for the ash. The ash usually contains heavy metals and must be placed in a specially-designed landfill to avoid leaching.

Hazardous waste incineration

For the incineration of hazardous waste, a greater degree of control, higher temperatures, and a more rigorous monitoring system are required. An incinerator burning hazardous waste must be designed, constructed, and maintained to meet **Resource Conservation and Recovery Act** (RCRA) standards. An incinerator burning hazardous waste must achieve a destruction and removal efficiency of at least 99.99 percent for each principal organic hazardous constituent. For certain listed constituents such as polychlorinated biphenyl (PCB), mass air emissions from an incinerator are required to be greater than 99.9999%. The Toxic Substances Control Act requires certain standards for the incineration of PCBs. For example, the flow of PCB to the incinerator must stop automatically whenever the combustion temperature drops below the specified value; there must be continuous monitoring of the stack for a list of emissions; scrubbers must be used for hydrochloric acid control; among others.

Recently medical wastes have been treated by steam sterilization, followed by incineration with treatment of the flue gases with activated **carbon** for maximum **absorption** of organic constituents. The latter system is being installed at the Mayo Clinic in Rochester, Minnesota, as a model medical disposal system. *See also* Fugitive emissions; Solid waste incineration; Solid waste volume reduction; Stack emissions

[*Malcolm T. Hepworth*]

RESOURCES
BOOKS

Brunner, C. R. *Handbook of Incineration Systems*. New York: McGraw-Hill, 1991.

Edwards, B. H., et al. *Emerging Technologies for the Control of Hazardous Wastes*. Park Ridge, NJ: Noyes Data Corporation, 1983.
Hickman Jr., H. L., et al. *Thermal Conversion Systems for Municipal Solid Waste*. Park Ridge, NJ: Noyes Publications, 1984.
Vesilind, R. A., and A. E. Rimer. *Unit Operations in Resource Recovery Engineering*. Englewood Cliffs, NJ: Prentice-Hall, 1981.
Wentz, C. A. *Hazardous Waste Management*. New York: McGraw-Hill, 1989.

Incineration, solid waste
see **Solid waste incineration**

Indicator organism

Indicator organisms, sometimes called bioindicators, are plant or animal **species** known to be either particularly tolerant or particularly sensitive to **pollution**. The health of an organism can often be associated with a specific type or intensity of pollution, and its presence can then be used to indicate polluted conditions relative to unimpacted conditions.

Tubificid worms are an example of organisms that can indicate pollution. Tubificid worms live in the bottom sediments of streams and lakes, and they are highly tolerant of sewage. In a river polluted by **wastewater discharge** from a **sewage treatment** plant, it is common to see a large increase in the numbers of tubificid worms in stream sediments immediately downstream. Upstream of the discharge, the numbers of tubificid worms are often much lower or almost absent, reflecting cleaner conditions. The number of tubificid worms also decreases downstream, as the discharge is diluted.

Pollution-intolerant organisms can also be used to indicate polluted conditions. The larvae of mayflies live in stream sediments and are known to be particularly sensitive to pollution. In a river receiving wastewater discharge, mayflies will show the opposite pattern of tubificid worms. The mayfly larvae are normally present in large numbers above the discharge point; they decrease or disappear at the discharge point and reappear further downstream as the effects of the discharge are diluted.

Similar examples of indicator organisms can be found among plants, fish, and other biological groups. Giant reedgrass (*Phragmites australis*) is a common marsh plant that is typically indicative of disturbed conditions in **wetlands**. Among fish, disturbed conditions may be indicated by the disappearance of sensitive species like trout which require clear, cold waters to thrive.

The usefulness of indicator organisms is limited. While their presence or absence provides a reliable general picture of polluted conditions, they are often little help in identifying the exact sources of pollution. In the sediments of New York Harbor, for example, pollution-tolerant insect larvae are overwhelmingly dominant. However, it is impossible to attribute the large larval populations to just one of the sources of pollution there, which include ship traffic, sewage and industrial discharge, and **storm runoff**.

The U.S. **Environmental Protection Agency** (EPA) is working diligently to find reliable predictors of aquatic **ecosystem health** using indicator species. Recently, the EPA has developed standards for the usefulness of species as ecological indicator organisms. A potential indicator species for use in evaluating **watershed** health must successfully pass four phases of evaluation. First, a potential indicator organism should provide information that is relevant to societal concerns about the **environment**, not simply academically interesting information. Second, use of a potential indicator organism should be feasible. Logistics, sampling costs, and timeframe for information gathering are legitimate considerations in deciding whether an organism is a potential indicator species or not. Thirdly, enough must be known about a potential species before it may be effectively used as an indicator organism. Sufficient knowledge regardin! g the natural variations to environmental flux should exist before incorporating a species as a true watershed indicator species. Lastly, the EPA has set a fourth criterion for evaluation of indicator species. A useful indicator should provide information that is easily interpreted by policy makers and the public, in addition to scientists.

Additionally, in an effort to make indicator species information more reliable, the creation of indicator species indices are being investigated. An index is a formula or ratio of one amount to another that is used to measure relative change. The major advantage of developing an indicator organism index that is somewhat universal to all aquatic environments is that it can be tested using **statistics**. Using mathematical statistical methods, it may be determined whether a significant change in an index value has occurred. Furthermore, statistical methods allow for a certain level of confidence that the measured values repres! ent what is actually happening in **nature**. For example, a study was conducted to evaluate the utility of diatoms (a kind of microscopic aquatic algae) as an index of aquatic system health. Diatoms meet all four criteria mentioned above, and various species are found in both fresh and salt water. An index was created that was calculated using various measurable characteristics of diatoms that could then be evaluated statistically over time and among varying sites. It was determined that the diatom index was sensitive enough to reliably reflect three categories of the health of an aquatic **ecosystem**. The diatom index showed that values obtained from areas impacted by human activities had greater variability over time than diatom indices obtained from less disturbed loca-

tions. Many such indices are being developed using different species, and multiple species in an effort to create reliable information from indicator organisms. As more is learned about the physiology and life history of indicator organisms and their individual responses to different types of pollution, it may be possible to draw more specific conclusions. *See also* Algal bloom; Nitrogen cycle; Water pollution

[*Terry Watkins*]

RESOURCES
BOOKS

Browder, J. A., ed. *Aquatic Organisms As Indicators of Environmental Pollution.* Bethesda, MD: American Water Resources Association, 1988.
Connell, D. W., and G. J. Miller. *Chemistry and Ecotoxicology of Pollution.* New York: Wiley-Interscience, 1984.

Indigenous peoples

Cultural or ethnic groups living in an area where their culture developed or where their people have existed for many generations. Most of the world's indigenous peoples live in remote forests, mountains, deserts, or arctic **tundra**, where modern technology, trade, and cultural influence are slow to penetrate. Many had much larger territories historically but have retreated to, or been forced into, small, remote areas by the advance of more powerful groups. Indigenous groups, also sometimes known as native or tribal peoples, are usually recognized in comparison to a country's dominant cultural group. In the United States the dominant, non-indigenous cultural groups speak English, has historic roots in Europe, and maintain strong economic, technological, and communication ties with Europe, Asia, and other parts of the world. Indigenous groups in the United States, on the other hand, include scores of groups, from the southern Seminole and Cherokee to the Inuit and Yupik peoples of the Arctic coast. These groups speak hundreds of different languages or dialects, some of which have been on this continent for thousands of years. Their traditional economies were based mainly on small-scale subsistence gathering, **hunting**, fishing, and farming. Many indigenous peoples around the world continue to engage in these ancient economic practices.

It is often difficult to distinguish who is and who is not indigenous. European-Americans and Asian-Americans are usually not considered indigenous even if they have been here for many generations. This is because their cultural roots connect to other regions. On the other hand, a German residing in Germany is also not usually spoken of as indigenous, even though by any strict definition she or he *is* indigenous. This is because the term is customarily reserved to denote economic or political minorities—groups that are relatively powerless within the countries where they live.

Historically, indigenous peoples have suffered great losses in both population and territory to the spread of larger, more technologically advanced groups, especially (but not only) Europeans. Hundreds of indigenous cultures have disappeared entirely just in the past century. In recent decades, however, indigenous groups have begun to receive greater international recognition, and they have begun to learn effective means to defend their lands and interests—including attracting international media attention and suing their own governments in court. The main reason for this increased attention and success may be that scientists and economic development organizations have recently become interested in biological diversity and in the loss of world rain forests. The survival of indigenous peoples, of the world's forests, and of the world's gene pools are now understood to be deeply interdependent. Indigenous peoples, who know and depend on some of the world's most endangered and biologically diverse ecosystems, are increasingly looked on as a unique source of information, and their subsistence economies are beginning to look like admirable alternatives to large-scale **logging**, mining, and conversion of jungles to monocrop agriculture.

There are probably between 4,000 and 5,000 different indigenous groups in the world; they can be found on every continent (except **Antarctica**) and in nearly every country. The total population of indigenous peoples amounts to between 200 million and 600 million (depending upon how groups are identified and their populations counted) out of a world population just over 6.2 billion. Some groups number in the millions; others comprise only a few dozen people. Despite their world-wide distribution, indigenous groups are especially concentrated in a number of "cultural diversity hot spots," including Indonesia, India, Papua New Guinea, **Australia**, Mexico, Brazil, Zaire, Cameroon, and Nigeria. Each of these countries has scores, or even hundreds, of different language groups. Neighboring valleys in Papua New Guinea often contain distinct cultural groups with unrelated languages and religions. These regions are also recognized for their unusual biological diversity. Both indigenous cultures and **rare species** survive best in areas where modern technology does not easily penetrate. Advanced technological economies involved in international trade consume tremendous amounts of land, wood, water, and minerals. Indigenous groups tend to rely on intact ecosystems and on a tremendous variety of plant and animal **species**. Because their numbers are relatively small and their technology simple, they usually do little long-lasting damage to their **environment** despite their dependence on the resources around them. The remote areas where indigenous peoples and their natural environment survive, however, are also the richest remaining reserves of **natural resources** in most countries. Frequently state governments claim all timber, mineral,

water, and land rights in areas traditionally occupied by tribal groups. In Indonesia, Malaysia, Burma (Myanmar), China, Brazil, Zaire, Cameroon, and many other important cultural diversity regions, timber and mining concessions are frequently sold to large or international companies that can quickly and efficiently destroy an ecological area and its people. Usually native peoples, because they lack political and economic clout, have no recourse to losing their homes. Generally they are relocated, attempts are made to integrate them into mainstream culture, and they join laboring classes in the general economy.

Indigenous rights have begun to strengthen in recent years. As long as international media attention continues to give them the attention they need—especially in the form of international economic and political pressure on state governments—and as long as indigenous leaders are able to continue developing their own defense strategies and legal tactics, the survival rate of indigenous peoples and their environments may improve significantly.

[*Mary Ann Cunningham Ph.D.*]

RESOURCES

BOOKS

Redford, K. H., and C. Padoch. *Conservation of Neotropical Forests: Working from Traditional Resources Use.* New York: Columbia University Press, 1992.

OTHER

Durning, A. T. "Guardians of the Land: Indigenous Peoples and the Health of the Earth." *Worldwatch Paper* 112. Washington, DC: Worldwatch Institute, 1992.

Indonesian forest fires

For several months in 1997 and 1998, a thick pall of **smoke** covered much of Southeast Asia. Thousands of forest fires burning simultaneously on the Indonesian islands of Kalimantan (Borneo) and Sumatra, are thought to have destroyed about 8,000 mi^2 (20,000 km^2) of primary forest, or an area about the size of New Jersey. The smoke generated by these fires spread over eight countries and 75 million people, covering an area larger than Europe. Hazy skies and the smell of burning forests could be detected in Hong Kong, nearly 2,000 mi (3,200 km) away. The **air quality** in Singapore and the city of Kuala Lumpur, Malaysia, just across the Strait of Malacca from Indonesia, was worse than any industrial region in the world. In towns such as Palembang, Sumatra, and Banjarmasin, Kalimantan, in the heart of the fires, the **air pollution index** frequently passed 800, twice the level classified in the United States as an air quality emergency, hazardous to human health. Automobiles had to drive with their headlights on, even at noon. People groped along smoke-darkened streets unable to see or breathe normally.

At least 20 million people in Indonesia and Malaysia were treated for illnesses such as **bronchitis**, eye irritation, **asthma**, **emphysema**, and cardiovascular diseases. It's thought that three times that many who couldn't afford medical care went uncounted. The number of extra deaths from this months-long episode is unknown, but it seems likely to have been hundreds of thousands, mostly elderly or very young children. Unable to see through the thick **haze**, several boats collided in the busy Straits of Malacca, and a plane crashed on Sumatra, killing 234 passengers. Cancelled airline flights, aborted tourist plans, lost workdays, medical bills, and ruined crops are estimated to have cost countries in the afflicted area several billion dollars. **Wildlife** suffered as well. In addition to the loss of **habitat** destroyed by fires, breathing the noxious smoke was as hard on wild **species** as it was on people. At the Pangkalanbuun Conservation Reserve, weak and disoriented orangutans were found suffering from **respiratory diseases** much like those of humans.

Geographical isolation on the 16,000 islands of the Indonesian archipelago has allowed **evolution** of the world's richest collection of **biodiversity**. Indonesia has the second largest expanse of tropical forest and the highest number of **endemic species** anywhere. This makes destruction of Indonesian plants, animals, and their habitat of special concern. The dry season in tropical Southeast Asia has probably always been a time of burning vegetation and smoky skies. Farmers practicing traditional **slash and burn agriculture** start fires each year to prepare for the next growing season. Because they generally burn only a hectare or two at a time, however, these shifting cultivators often help preserve plant and animal species by opening up space for early successional forest stages. Globalization and the advent of large, commercial plantations, however, have changed agricultural dynamics. There is now economic incentive for clearing huge tracts of forestland to plant oil palms, export foods such as pineapples and sugar cane, and fast-growing eucalyptus trees. Fire is viewed as the only practical way remove **biomass** and convert wild forest to into domesticated land. While it can cost the equivalent of $200 to clear a hectare of forest with chainsaws and bulldozers, dropping a lighted match into dry underbrush is essentially free.

In 1997 to 1998, the Indonesian forest was unusually dry. A powerful El Niño/Southern Oscillation weather pattern caused the most severe droughts in 50 years. Forests that ordinarily stay green and moist even during the rainless season became tinder dry. Lightning strikes are thought to have started many forest fires, but many people took advantage of the **drought** for their own purposes. Although the government blamed traditional farmers for setting most of

the fires, environmental groups claimed that the biggest fires were caused by large agribusiness conglomerates with close ties to the government and military. Some of these fires were set to cover up evidence of illegal **logging** operations. Others were started to make way for huge oil-palm plantations and fast-growing pulpwood trees,

Neil Byron of the Center for International Forestry Research was quoted as saying that "fire crews would go into an area and put out the fire, then come back four days later and find it burning again, and a guy standing there with a petrol can." According to the World Wide Fund for Nature, 37 plantations in Sumatra and Kalimantan were responsible for a vast majority of the forest burned on those islands. The plantation owners were politically connected to the ruling elite, however, and none of them was ever punished for violation of **national forest** protection laws. Indonesia has some of the strongest land-use management laws of any country in the world, but these laws are rarely enforced. In theory, more than 80% of its land is in some form of protected status, either set aside as national parks or classified as selective logging reserves where only a few trees per hectare can be cut. The government claims to have an ambitious reforestation program that replants nearly 1.6 million acres (1 million hectares) of harvested forest annually, but when four times that amount is burned in a single year, there's not much to be done but turn it over to plantation owners for use as agricultural land.

Aquatic life, also, is damaged by these forest fires. Indonesia, Malaysia, and the Philippines have the richest **coral reef** complexes in the world. More than 150 species of coral live in this area, compared with only about 30 species in the Caribbean. The clear water and fantastic biodiversity of Indonesia's reefs have made it an ultimate destination for scuba divers and snorkelers from around the world. Unfortunately, **soil** eroded from burned forests clouds coastal waters and smothers reefs.

Perhaps one of the worst effects of large tropical forest fires is that they may tend to be self-reinforcing. Moist tropical forests store huge amounts of **carbon** in their standing biomass. When this carbon is converted into CO_2 by fire and released to the **atmosphere**, it acts as a greenhouse gas to trap heat and cause global warming. All the effects of human-caused global **climate** change are still unknown, but we stronger climatic events such as severe droughts may make further fires even more likely. Alarmed by the magnitude of the Southeast Asia fires and the potential they represent for biodiversity losses and global climate change, world leaders have proposed plans for international intervention to prevent A recurrence. Fears about imposing on national sovereignty, however, have made it difficult to come up with a plan for how to cope with this growing threat.

[*William P. Cunningham Ph.D.*]

RESOURCES

BOOKS

Glover, David, and Timothy Jessup, eds. *Indonesia's Fires and Haze: The Cost of Catastrophe*. Singapore: International Development Research Centre, 2002.

PERIODICALS

Aditama, Tjandra Yoga. "Impact of Haze from Forest Fire to Respiratory Health: Indonesian Experience." *Respirology* (2000): 169–174.

Chan, C. Y., et al. "Effects of 1997 Indonesian forest fires on tropospheric ozone enhancement, radiative forcing, and temperature change over the Hong Kong region" *Journal of Geophysical Research-Atmospheres* 106 (2001):14875-14885.

Davies, S. J. and L. Unam. "Smoke-haze from the 1997 Indonesian forest fires: effects on pollution levels, local climate, atmospheric CO2 concentrations, and tree photosynthesis." *Forest Ecology & Management* 124(1999):137-144.

Murty, T. S., D. Scott, and W. Baird. "The 1997 El Niño, Indonesian Forest Fires and the Malaysian Smoke Problem: A Deadly Combination of Natural and Man-Made Hazard" *Natural Hazards* 21 (2000): 131–144.

Tay, Simon. "Southeast Asian Fires: The Challenge Over Sustainable Environmental Law and Sustainable Development." *Peace Research Abstracts* 38 (2001): 603–751.

Indoor air quality

An assessment of **air quality** in buildings and homes based on physical and chemical monitoring of contaminants, physiological measurements, and/or psychosocial perceptions. Factors contributing to the quality of indoor air include lighting, ergonomics, thermal comfort, **tobacco smoke**, noise, ventilation, and psychosocial or work-organizational factors such as employee stress and satisfaction. "Sick building syndrome" (SBS) and "building-related illness" (BRI) are responses to indoor **air pollution** commonly described by office workers. Most symptoms are nonspecific; they progressively worsen during the week, occur more frequently in the afternoon, and disappear on the weekend.

Poor indoor air quality (IAQ) in industrial settings such as factories, **coal** mines, and foundries has long been recognized as a health risk to workers and has been regulated by the U.S. **Occupational Safety and Health Administration** (OSHA). The contaminant levels in industrial settings can be hundreds or thousands of times higher than the levels found in homes and offices. Nonetheless, indoor air quality in homes and offices has become an environmental priority in many countries, and federal IAQ legislation has been introduced in the U.S. Congress for the past several years. However, none has yet passed, and currently the U.S. **Environmental Protection Agency** (EPA) has no enforcement authority in this area.

Importance of IAQ

The prominence of IAQ issues has risen in part due to well-publicized incidents involving outbreaks of Legionnaires' disease, Pontiac fever, **sick building syndrome**, mul-

tiple **chemical sensitivity**, and **asbestos** mitigation in public buildings such as schools. Legionnaire's disease, for example, caused twenty-nine deaths in 1976 in a Philadelphia hotel due to infestation of the building's air conditioning system by a bacterium called *Legionella pneumophila*. This microbe affects the gastrointestinal tract, kidneys, and central nervous system. It also causes the non-fatal Pontiac fever.

IAQ is important to the general public for several reasons. First, individuals typically spend the vast majority of their time—80–90%—indoors. Second, an emphasis on **energy conservation** measures, such as reducing air exchange rates in ventilation systems and using more energy efficient but synthetic materials, has increased levels of air contaminants in offices and homes. New "tight" buildings have few cracks and openings so minimal fresh air enters such buildings. Low ventilation and exchange rates can increase indoor levels of **carbon monoxide**, **nitrogen oxides**, **ozone**, volatile organic compounds, **bioaerosols**, and pesticides and maintain high levels of second-hand tobacco smoke generated inside the building. Thus, many contaminants are found indoors at levels that greatly exceed outdoor levels. Third, an increasing number of synthetic chemicals—found in building materials, furnishing, cleaning and hygiene products—are used indoors. Fourth, studies show that exposure to indoor contaminants such as **radon**, asbestos, and tobacco smoke pose significant health risks. Fifth, poor IAQ is thought to adversely affect children's development and lower productivity in the adult population. Demands for indoor air quality investigations of "sick" and problem buildings have increased rapidly in recent years, and a large fraction of buildings are known or suspected to have IAQ problems.

Indoor contaminants

Indoor air contains many contaminants at varying but generally low concentration levels. Common contaminants include radon and radon progeny from the entry of **soil** gas and **groundwater** and from concrete and other mineral-based building materials; tobacco smoke from cigarette and pipe smoking; formaldehyde from polyurethane foam insulation and building materials; volatile organic compounds (VOCs) emitted from binders and resins in carpets, furniture, or building materials, as well as VOCs used in **dry cleaning** processes and as **propellants** and constituents of personal use and cleaning products, like hair sprays and polishes; pesticides and insecticides; **carbon** monoxide, **nitrogen** oxides, and other **combustion** productions from gas stoves, appliances, and vehicles; asbestos from high temperature insulation; and biological contaminants including viruses, bacteria, molds, pollen, dust mites, and indoor and outdoor biota. Many or most of these contaminants are present at low levels in all indoor environments.

Some major indoor air pollutants.

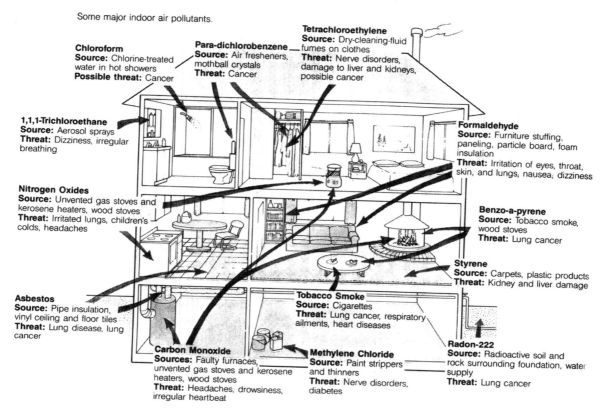

Chloroform
Source: Chlorine-treated water in hot showers
Possible threat: Cancer

Para-dichlorobenzene
Source: Air fresheners, mothball crystals
Threat: Cancer

Tetrachloroethylene
Source: Dry-cleaning-fluid fumes on clothes
Threat: Nerve disorders, damage to liver and kidneys, possible cancer

1,1,1-Trichloroethane
Source: Aerosol sprays
Threat: Dizziness, irregular breathing

Formaldehyde
Source: Furniture stuffing, paneling, particle board, foam insulation
Threat: Irritation of eyes, throat, skin, and lungs; nausea; dizziness

Nitrogen Oxides
Source: Unvented gas stoves and kerosene heaters, wood stoves
Threat: Irritated lungs, children's colds, headaches

Benzo-a-pyrene
Source: Tobacco smoke, wood stoves
Threat: Lung cancer

Styrene
Source: Carpets, plastic products
Threat: Kidney and liver damage

Asbestos
Source: Pipe insulation, vinyl ceiling and floor tiles
Threat: Lung disease, lung cancer

Tobacco Smoke
Source: Cigarettes
Threat: Lung cancer, respiratory ailments, heart diseases

Carbon Monoxide
Sources: Faulty furnaces, unvented gas stoves and kerosene heaters, wood stoves
Threat: Headaches, drowsiness, irregular heartbeat

Methylene Chloride
Source: Paint strippers and thinners
Threat: Nerve disorders, diabetes

Radon-222
Source: Radioactive soil and rock surrounding foundation, water supply
Threat: Lung cancer

Some major indoor air pollutants. (Wadsworth Inc. Reproduced by permission.)

The quality of indoor air can change rapidly in time and from room to room. There are many diverse sources that emit various physical and chemical forms of contaminants. Some releases are slow and continuous, such as outgassing associated with building and furniture materials, while others are nearly instantaneous, like the use of cleaners and aerosols. Many building surfaces demonstrate significant interactions with contaminants in the form of sorption-desorption processes. Building-specific variation in air exchange rates, mixing, **filtration**, building and furniture surfaces, and other factors alter dispersion mechanisms and contaminant lifetimes. Most buildings employ **filters** that can remove particles and aerosols. Filtration systems do not effectively remove very small particles and have no effect on gases, vapors, and odors. Ventilation and air exchange units designed into the heating and cooling systems of buildings are designed to diminish levels of these contaminants by dilution. In most buildings, however, ventilation systems are turned off at night after working hours, leading to an increase in contaminants through the night. Though operation and maintenance issues are estimated to cause the bulk of indoor air quality problems, deficiencies in the design of the heating,

ventilating and air conditioning (HVAC) system can cause problems as well. For example, locating a building's fresh air intake near a truck loading dock will bring diesel fumes and other noxious contaminants into the building.

Health impacts

Exposures to indoor contaminants can cause a variety of health problems. Depending on the pollutant and exposure, health problems related to indoor air quality may include non-malignant respiratory effects, including mucous membrane irritation, allergic reactions, and **asthma**; cardiovascular effects; infectious diseases such as Legionnaires' disease; immunologic diseases such as hypersensitivity pneumonitis; skin irritations; malignancies; neuropsychiatric effects; and other non-specific **systemic** effects such as lethargy, headache, and nausea. In addition indoor air contaminants such as radon, formaldehyde, asbestos, and other **chemicals** are suspected or known carcinogens. There is also growing concern over the possible effects of low level exposures on suppressing reproductive and growth capabilities and impacting the immune, endocrine, and nervous systems.

Solving IAQ problems

Acute indoor air quality problems can be greatly eliminated by identifying, evaluating, and controlling the sources of contaminants. IAQ control strategies include the use of higher ventilation and air exchange rates, the use of lower **emission** and more benign constituents in building and consumer products (including product use restriction regulations), air cleaning and filtering, and improved building practices in new construction. Radon may be reduced by inexpensive subslab ventilation systems. New buildings could implement a day of "bake-out," which heats the building to temperatures over 90°F (32°C) to drive out volatile organic compounds. Filters to remove ozone, organic compounds, and sulfur gases may be used to condition incoming and recirculated air. Copy machines and other emission sources should have special ventilation systems. Building designers, operators, contractors, maintenance personnel, and occupants are recognizing that healthy buildings result from combined and continued efforts to control emission sources, provide adequate ventilation and air cleaning, and good maintenance of building systems. Efforts toward this direction will greatly enhance indoor air quality.

[*Stuart Batterman*]

RESOURCES
BOOKS

Godish, T. *Indoor Air Pollution Control.* Chelsea, MI: Lewis, 1989.
Kay, J. G., et al. *Indoor Air Pollution: Radon, Bioaerosols and VOCs.* Chelsea, MI: Lewis, 1991.
Samet, J. M., and J. D. Spengler. *Indoor Air Pollution: A Health Perspective.* Baltimore: Johns Hopkins University Press, 1991.

PERIODICALS

Kreiss, K. "The Epidemiology of Building-Related Complaints and Illnesses." *Occupational Medicine: State of the Art Reviews* 4 (1989): 575–92.

Industrial waste treatment

Many different types of solid, liquid, and gaseous wastes are discharged by industries. Most industrial waste is recycled, treated and discharged, or placed in a **landfill**. There is no one means of managing industrial wastes because the nature of the wastes varies widely from one industry to another. One company might generate a waste that can be treated readily and discharged to the **environment** (direct **discharge**) or to a sewer in which case final treatment might be accomplished at a publicly owned treatment works (POTW). Treatment at the company before discharge to a sewer is referred to as pretreatment. Another company might generate a waste which is regarded as hazardous and therefore requires special management procedures related to storage, **transportation** and final disposal.

The pertinent legislation governing to what extent wastewaters need to be treated before discharge is the 1972 **Clean Water Act** (CWA). Major amendments to the CWA were passed in 1977 and 1987. The **Environmental Protection Agency** (EPA) was also charged with the responsibility of regulating the priority pollutants under the CWA. The CWA specifies that toxic and nonconventional pollutants are to be treated with the Best Available Technology (BAT). Gaseous pollutants are regulated under the **Clean Air Act** (CAA), promulgated in 1970 and amended in 1977 and 1990. An important part of the CAA consists of measures to attain and maintain National Ambient Air Quality Standards (NAAQS). Hazardous air pollutant (HAP) emissions are to be controlled through Maximum Achievable Control Technology (MACT) which can include process changes, material substitutions and/or **air pollution control** equipment. The "cradle to grave" management of hazardous wastes is to be performed in accordance with the **Resource Conservation and Recovery Act** (RCRA) of 1976 and the Hazardous and **Solid Waste** Amendments (HSWA) of 1984.

In 1990, the United States, through the **Pollution Prevention Act**, adopted a program designed to reduce the volume and toxicity of waste discharges. **Pollution** prevention (P2) strategies might involve changing process equipment or chemistry, developing new processes, eliminating products, minimizing wastes, **recycling** water or **chemicals**, trading wastes with another company, etc. In 1991, the EPA instituted the 33/50 program which was to result in an overall 33% reduction of 17 high priority pollutants by 1992 and a 50% reduction of the pollutants by 1995. Both goals were surpassed. Not only has this program been successful, but it sets an important precedence because the participating companies volunteered. Additionally, P2 efforts have led industries to rigorously think through product life cycles. A Life Cycle Analysis (LCA) starts with consideration for acquiring raw materials, moves through the stages related to processing, assembly, service and **reuse**, and ends with retirement/disposal. The LCA therefore reveals to industry the costs and problems versus the benefits for every stage in the life of a product.

In designing a **waste management** program for an industry, one must think first in terms of P2 opportunities, identify and characterize the various solid, liquid and gaseous waste streams, consider relevant legislation, and then design an appropriate waste management system. Treatment systems that rely on physical (e.g., settling, floatation, screening, **sorption**, membrane technologies, air stripping) and chemical (e.g., coagulation, precipitation, chemical oxidation and reduction, **pH** adjustment) operations are referred to as physicochemical, whereas systems in which **microbes** are cultured to metabolize waste constituents are known as biological processes (e.g., **activated sludge**, trickling **filters**, biotowers, aerated lagoons, **anaerobic digestion**, **aerobic** digestion, **composting**). Oftentimes, both physicochemical and biological systems are used to treat solid and liquid waste streams. Biological systems might be used to treat certain gas streams, but most waste gas streams are treated physicochemically (e.g., cyclones, electrostatic precipitators, **scrubbers**, bag filters, thermal methods). Solids and the sludges or residuals that result from treating the liquid and gaseous waste streams are also treated by means of physical, chemical, and biological methods.

In many cases, the systems used to treat wastes from domestic sources are also used to treat industrial wastes. For example, municipal wastewaters often consist of both domestic and industrial waste. The local POTW therefore may be treating both types of wastes. To avoid potential problems caused by the input of industrial wastes, municipalities commonly have pretreatment programs which require that industrial wastes discharged to the sewer meet certain standards. The standards generally include limits for various toxic agents such as metals, organic matter measured in terms of **biochemical oxygen demand** (bod) or **chemical oxygen demand**, nutrients such as **nitrogen** and **phosphorus**, pH and other contaminants that are recognized as having the potential to impact on the performance of the POTW. At the other end of the spectrum, there are wastes that need to be segregated and managed separately in special systems. For example, an industry might generate a **hazardous waste** that needs to be placed in barrels and transported to an EPA approved treatment, storage or disposal facility (TSDF).

Thus, it is not possible to simply use one train of treatment operations for all industrial waste streams, but an effective, generic strategy has been developed in recent years for considering the waste management options available to an industry. The basis for the strategy is to look for P2 opportunities and to consider the life cycle of a product. An awareness of **waste stream** characteristics and the potential benefits of stream segregation is then melded with the knowledge of regulatory compliance issues and treatment system capabilities/performance to minimize environmental risks and costs.

[*Gregory D. Boardman*]

RESOURCES
BOOKS

Freeman, H. M. *Industrial Pollution Prevention Handbook.* New York: McGraw-Hill, Inc., 1995.
Haas, C.N., and R.J. Vamos. *Hazardous and Industrial Waste Treatment.* Englewood Cliffs: Prentice Hall, Inc., 1995.
LaGrega, M. D., P. L. Buckingham, and J. C. Evans. *Hazardous Waste Management.* New York: McGraw-Hill, Inc., 1994.

Metcalf and Eddy, Inc. *Wastewater Engineering Treatment, Disposal and Reuse.* Revised by G. Tchobanoglous and F. Burton. New York: McGraw-Hill, Inc., 1991.
Nemerow, N. L., and Dasgupta, A. *Industrial and Hazardous Waste Treatment.* New York: Van Nostrand Reinhold, 1991.
Peavy, H. S., D. R. Rowe, and G. Tchobanoglous. *Environmental Engineering.* New York: McGraw-Hill Book, 1995.
Tchobanoglous, G., et al. *Integrated Solid Waste Management Engineering Principles and Management Issues.* New York: McGraw-Hill, Inc., 1993.

PERIODICALS

Romanow, S., and T. E. Higgins. "Treatment of Contaminated Groundwater from Hazardous Waste Sites—Three Case Studies." Presented at the *60th Water Pollution Control Federation Conference*, Philadelphia (October 5-8, 1987).

Inertia

see **Resistance (inertia)**

Infiltration

In **hydrology**, infiltration refers to the maximum rate at which a **soil** can absorb precipitation. This is based on the initial moisture content of the soil or on the portion of precipitation that enters the soil. In soil science, the term refers to the process by which water enters the soil, generally by downward flow through all or part of the soil surface. The rate of entry relative to the amount of water being supplied by precipitation or other sources determines how much water enters the root zone and how much runs off the surface. *See also* Groundwater; Soil profile; Water table

INFORM

INFORM was founded in 1973 by environmental research specialist Joanna Underwood and two colleagues. Seriously concerned about **air pollution**, the three scientists decided to establish an organization that would identify practical ways to protect the **environment** and public health. Since then, their concerns have widened to include **hazardous waste**, solid **waste management**, **water pollution**, and land, energy, and **water conservation**. The group's primary purpose is "to examine business practices which harm our air, water, and land resources" and pinpoint "specific ways in which practices can be improved."

INFORM's research is recognized throughout the United States as instrumental in shaping environmental policies and programs. Legislators, **conservation** groups and business leaders use INFORM's authority as an acknowledged basis for research and conferences. Source reduction has become one of INFORM's most important projects. A decrease in the amount and/or toxicity of waste entering the **waste stream**, source reduction includes any activity by an individual, business, or government that lessens the amount of solid waste—or garbage—that would otherwise have to be recycled or incinerated. Source reduction does not include **recycling**, **municipal solid waste composting**, household hazardous waste collection, or beverage container deposit and return systems.

The first priority in source reduction strategies is elimination; the second, **reuse**. Public education is a crucial part of INFORM's program. To this end INFORM has published *Making Less Garbage: A Planning Guide for Communities*. This book details ways to achieve source reduction including buying reusable, as opposed to disposable, items; buying in bulk; and maintaining and repairing products to extend their lives.

INFORM's outreach program goes well beyond its source reduction project. The staff of over 25 full-time scientists and researchers and 12 volunteers and interns makes presentations at national and international conferences and local workshops. INFORM representatives have also given briefings and testimony at Congressional hearings and produced television and radio advertisements to increase public awareness on environmental issues. The organization also publishes a quarterly newsletter, *INFORM Reports*.

[*Cathy M. Falk*]

RESOURCES

ORGANIZATIONS

INFORM, Inc., 120 Wall Street, New York, NY USA 10005 (212) 361-2400, Fax: (212) 361-2412, Email: brown@informinc.org, <http://www.informinc.org>

INFOTERRA (U.N. Environment Program)

INFOTERRA has its international headquarters in Kenya and is a global information network operated by the **Earthwatch** program of the United Nations Environment Program (UNEP).

Under INFOTERRA, participating nations designate institutions to be national focal points, such as the **Environmental Protection Agency** (EPA) in the United States. Each national institution chosen as a focal point, prepares a list of its national environmental experts and selects what it considers the best sources for inclusion in INFOTERRA's international directory of experts.

INFOTERRA initially used its directory only to refer questioners to the nearest appropriate experts, but the organization has evolved into a central information agency. It consults sources, answers public queries for information, and analyzes the replies. INFOTERRA is used by governments, industries, and researchers in 177 countries.

[*Linda Rehkopf*]

RESOURCES

ORGANIZATIONS

UNEP-Infoterra/USA, MC 3404 Ariel Rios Building, 1200 Pennsylvania Avenue, Washington, D.C. USA 20460 Fax: (202) 260-3923, Email: library-infoterra@epa.gov

Injection well

Injection **wells** are used to dispose waste into the subsurface zone. These wastes can include brine from oil and gas wells, liquid hazardous wastes, agricultural and **urban runoff**, municipal sewage, and return water from air-conditioning. Recharge wells can also be used for injecting fluids to enhance oil recovery, injecting treated water for artificial **aquifer** recharge, or enhancing a pump-and-treat system. If the wells are poorly designed or constructed, or if the local geology is not sufficiently studied, injected liquids can enter an aquifer and cause **groundwater** contamination. Injection wells are regulated under the Underground Injection Control Program of the **Safe Drinking Water Act**. *See also* Aquifer restoration; Deep-well injection; Drinking-water supply; Groundwater monitoring; Groundwater pollution; Water table

Inoculate

To inoculate involves the introduction of **microorganisms** into a new **environment**. Originally the term referred to the insertion of a bud or shoot of one plant into the stem or trunk of another to develop new strains or hybrids. These hybrid plants would be resistant to botanic disease or they would allow greater harvests or range of climates. With the advent of vaccines to prevent human and animal disease, the term inoculate has come to represent injection of a serum to prevent, cure, or make immune from disease.

Inoculation is of prime importance in that the introduction of specific microorganism **species** into specific macroorganisms may establish a symbiotic relationships where each organism benefits. For example, the introduction of **mycorrhiza** fungus to plants improves the plants' ability to absorb nutrients from the **soil**. *See also* Symbiosis

Insecticide

see Pesticide

Integrated pest management

Integrated **pest** management (IPM) is a newer science that aims to give the best possible pest control while minimizing damage to human health or the **environment**. IPM means either using fewer **chemicals** more effectively or finding ways, both new and old, that substitute for **pesticide** use.

Technically, IPM is the selection, integration and implementation of pest control based on predicted economic, ecological and sociological consequences. IPM seeks maximum use of naturally occurring pest controls, including weather, disease agents, predators and **parasites**. In addition, IPM utilizes various biological, physical, and chemical control and **habitat** modification techniques. Artificial controls are imposed only as required to keep a pest from surpassing intolerable population levels which are predetermined from assessments of the pest damage potential and the ecological, sociological, and economic costs of the control measures. Farmers have come to understand that the presence of a pest **species** does not necessarily justify action for its control. In fact, tolerable infestations may be actually desirable, providing food for important beneficial insects. Why this change in farming practices?

The introduction of synthetic organic pesticides such as the insecticide DDT, and the **herbicide 2,4-D** (half the formula in **Agent Orange**) after World War II began a new era in pest control. These products were followed by hundreds of synthetic organic fungicides, nematicides, rodenticides and other chemical controls. These chemical materials were initially very effective and very cheap. Synthetic chemicals eventually became the primary means of pest control in productive agricultural regions, providing season-long crop protection against insects and weeds. They were used in addition to fertilizers and other treatments.

The success of modern pesticides led to widespread acceptance and reliance upon them, particularly in this country. Of all the chemical pesticides applied worldwide in agriculture, forests, industry and households, one-third to one-half were used in the United States. Herbicides have been used increasingly to replace hand labor and machine cultivation for control of weeds in crops, in forests, on the rights-of-way of highways, utility lines, railroads and in cities. Agriculture consumes perhaps 65% of the total quantity of synthetic organic pesticides used in the United States each year. In addition, chemical companies export an increasingly larger amount to **Third World** countries. Pesticides banned in the United States such as DDT, EDB and **chlordane**, are exported to countries where they are applied to crops imported by the United States for consumption.

For more than a decade, problems with pesticides have become increasingly apparent. Significant groups of pests have evolved with **genetic resistance** to pesticides. The increase in **resistance** among insect pests has been exponential, following extensive use of chemicals in the last forty years. Ticks, insects and spider mites (nearly 400 species) are now especially resistant, and the creation of new insecticides to combat the problem is not keeping pace with the emergence of new strains of resistant insect pests. Despite the advances in modern chemical control and the dramatic increase in chemical pesticides used on U.S. cropland, annual crop losses from all pests appear to have remained constant or to have increased. Losses caused by weeds have declined slightly, but those caused by insects have nearly doubled. The price of synthetic organic pesticides has increased significantly in recent years, placing a heavy financial burden on those who use large quantities of the materials. As farmers and growers across the United States realize the limitations and human health consequences of using artificial chemical pesticides, interest in the alternative approach of integrated pest management grows.

Integrated pest management aims at management rather than eradication of pest species. Since potentially harmful species will continue to exist at tolerable levels of abundance, the philosophy now is to manage rather than eradicate the pests. The **ecosystem** is the management unit. (Every crop is in itself a complex ecological system.) Spraying pesticides too often, at the wrong time, or on the wrong part of the crop may destroy the pests' natural enemies ordinarily present in the ecosystem. Knowledge of the actions, reactions, and interactions of the components of the ecosystems is requisite to effective IPM programs. With this knowledge, the ecosystem is manipulated in order to hold pests at tolerable levels while avoiding disruptions of the system.

The use of natural controls is maximized. IPM emphasizes the fullest practical utilization of the existing regulating and limiting factors in the form of parasites, predators, and weather, which check the pests' **population growth**. IPM users understand that control procedures may produce unexpected and undesirable consequences, however. It takes time to change over and determination to keep up the commitment until the desired results are achieved.

An interdisciplinary systems approach is essential. Effective IPM is an integral part of the overall management of a farm, a business or a forest. For example, timing plays an important role. Certain pests are most prevalent at particular times of the year. By altering the date on which a crop is planted, serious pest damage can be avoided. Some farmers simultaneously plant and harvest, since the procedure prevents the pests from migrating to neighboring fields after the harvest. Others may plant several different crops in the same field, thereby reducing the number of pests. The variety of crops harbor greater numbers of natural enemies and make it more difficult for the pests to locate and colonize their host plants. In Thailand and China, farmers flood their fields for several weeks before planting to destroy pests. Other farmers turn the **soil**, so that pests are brought to the surface and die in the sun's heat.

The development of specific IPM program depends on the pest complex, resources to be protected, economic values, and availability of personnel. It also depends upon adequate funding for research and to train farmers. Some of the techniques are complex, and expert advice is needed. However, while it is difficult to establish absolute guidelines, there are general guidelines that can apply to the management of any pest group.

Growers must analyze the "pest" status of each of the reputedly injurious organisms and establish economic thresholds for the "real" pests. The economic threshold is, in fact, the population level, and is defined as the density of a pest population below which the cost of applying control measures exceeds the losses caused by the pest. Economic threshold values are based on assessments of the pest damage potential and the ecological, sociological, and economic costs associated with control measures. A given crop, forest area, backyard, building, or recreational area may be infested with dozens of potentially harmful species at any one time. For each situation, however, there are rarely more than a few pest species whose populations expand to intolerable levels at regular and fairly predictable intervals. Key pests recur regularly at population densities exceeding economic threshold levels and are the focal point for IPM programs.

Farmers must also devise schemes for lowering equilibrium positions of key pests. A key pest will vary in severity from year to year, but its average density, known as the equilibrium position, usually exceeds its economic threshold. IPM efforts manipulate the environment in order to reduce a pest's equilibrium position to a permanent level below the economic threshold. This reduction can be achieved by deliberate introduction and establishment of natural enemies (parasites, predators, and diseases) in areas where they did not previously occur. Natural enemies may already occur in the crop in small numbers or can be introduced from elsewhere. Certain **microorganisms**, when eaten by a pest, will kill it.

Newer chemicals show promise as alternatives to synthetic chemical pesticides. These include insect attractant chemicals, weed and insect disease agents and insect growth regulators or hormones. A **pathogen** such as **Bacillus thuringiensis** (BT), has proven commercially successful. Since certain crops have an inbuilt resistance to pests, pest-resistant or pest-free varieties of seed, crop plants, ornamental plants, orchard trees, and forest trees can be used. Growers can also modify the pest environment to increase the effectiveness of the pest's biological control agents, to destroy its breeding, feeding, or shelter habitat or otherwise render it harmless.

This includes crop rotation, destruction of crop harvest residues and soil tillage, and selective burning or mechanical removal of undesirable plant species and pruning, especially for forest pests.

While nearly permanent control of key insect and plant disease pests of agricultural crops has been achieved, emergencies will occur, and all IPM advocates acknowledge this. During those times, measures should be applied that create the least ecological destruction. Growers are urged to utilize the best combination of the three basic IPM components: natural enemies, resistant varieties and environmental modification. However, there may be a time when pesticides may be the only recourse. In that case, it is important to coordinate the proper pesticide, the dosage and the timing in order to minimize the hazards to nontarget organisms and the surrounding ecosystems.

Pest management techniques have been known for many years and were used widely before World War II. They were deemphasized by insect and weed control scientists and by corporate pressures as the synthetic chemicals became commercially available after the war. Now there is a renewed interest in the early control techniques and in new chemistry.

Reports detailing the success of IPM are emerging at a rapid rate as thousands of farmers yearly join the ranks of those who choose to eliminate chemical pesticides. Sustainable agricultural practice increases the richness of the soil by replenishing the soil's reserves of fertility. IPM does not produce secondary problems such as pest resistance or resurgence. It also diminishes soil **erosion**, increases crop yields and saves money over the long haul. Organic foods are reported to have better cooking quality, better flavor and greater longevity in the storage bins. And with less **pesticide residue**, our food is clearly more healthy to eat. *See also* Sustainable agriculture

[*Liane Clorfene Casten*]

RESOURCES
BOOKS

Baker, R. R., and P. Dunn. *New Directions in Biological Control: Alternatives for Supressing Agricultural Pests and Diseases.* New York: Wiley, 1990.

Burn, A. J., et al. *Integrated Pest Management.* New York: Academic Press, 1988.

DeBach, P., and D. Rosen. *Biological Control By Natural Enemies.* 2nd ed. Cambridge, MA: Cambridge University Press, 1991.

Pimentel, D. *The Pesticide Question: Environment, Economics and Ethics.* New York: Chapman & Hall, 1992.

PERIODICALS

Bottrell, D. G., and R. F. Smith. "Integrated Pest Management." *Environmental Science & Technology* 16 (May 1982): 282A–288A.

Intergenerational justice

One of the key features of an environmental ethic or perspective is its concern for the health and well-being of **future generations**. Questions about the rights of future people and the responsibilities of those presently living are central to environmental theory and practice and are often asked and analyzed under the term *intergenerational justice*. Most traditional accounts or theories of justice have focused on relations between contemporaries: What distribution of scarce goods is fairest or optimally just? Should such goods be distributed on the basis of merit or need? These and other questions have been asked by thinkers from Aristotle through John Rawls. Recently, however, some philosophers have begun to ask about just distributions over time and across generations.

The subject of intergenerational justice is a key concern for environmentally-minded thinkers for at least two reasons. First, human beings now living have the power to permanently alter or destroy the planet (or portions thereof) in ways that will affect the health, happiness, and well-being of people living long after we are all dead. One need only think, for example, of the radioactive wastes generated by nuclear **power plants** which will be intensely "hot" and dangerous for many thousands of years. No one yet knows how to safely store such material for a hundred, much less many thousands, of years. Considered from an intergenerational perspective then, it would be unfair—that is, unjust—for the present generation to enjoy the benefits of **nuclear power**, passing on to distant posterity the burdens and dangers caused by our (in)action.

Second, we not only have the power to affect future generations, but we *know* that we have it. And with such knowledge comes the moral responsibility to act in ways that will prevent harm to future people. For example, since we know about the health effects of radiation on human beings, our having that knowledge imposes upon us a moral obligation not to needlessly expose anyone—now or in the indefinite future—to the harms or hazards of radioactive wastes. Many other examples of intergenerational harm or hazard exist: global warming, **topsoil erosion**, disappearing tropical rain forests, depletion and/or **pollution** of aquifers, among others. But whatever the example, the point of the intergenerational view is the same: the moral duty to treat people justly or fairly applies not only to people now living, but to those who will live long after we are gone. To the extent that our actions produce consequences that may prove harmful to people who have not harmed (and in the nature of the case cannot harm) us is, by any standard, unjust. And yet it seems quite clear that we in the present generation are in many respects acting unjustly toward distant posterity. This is true not only for harms or hazards bequeathed to future people, but the point applies also to deprivations of various kinds.

Consider, for example, the present generation's profligate use of **fossil fuels**. Reserves of oil and **natural gas** are both finite and nonreplaceable; once burned (or turned into plastic or some other petroleum-based material), a gallon of oil is gone forever; every drop or barrel used now is therefore unavailable for future people. As Wendell Berry observed, the claim that fossil fuel energy is *cheap* rests on a simplistic and morally doubtful assumption about the *rights* of the present generation: "We were able to consider [fossil fuel energy] "cheap" only by a kind of moral simplicity: the assumption that we had a "right" to as much of it as we could use. This was a "right" made solely by might. Because fossil fuels, however abundant they once were, were nevertheless limited in quantity and not renewable, they obviously did not "belong" to one generation more than another. We ignored the claims of posterity simply because we could, the living being stronger than the unborn, and so worked the "miracle" of industrial progress by the theft of energy from (among others) our children."

And that, Berry adds, "is the real foundation of our progress and our affluence. The reason that we are a rich nation is not that we have earned so much wealth — you cannot, by any honest means, earn or deserve so much. The reason is simply that we have learned, and become willing, to market and use up in our own time the birthright and livelihood of posterity."

These and other considerations have led some environmentally-minded philosophers to argue for limits on present-day consumption, so as to save a fair share of scarce resources for future generations. John Rawls, for instance, constructs a *just savings principle* according to which members of each generation may consume no more than their fair share of scarce resources. The main difficulty in arriving at and applying any such principle lies in determining what counts as fair share. As the number of generations taken into account increases, the share available to any single generation then becomes smaller; and as the number of generations approaches infinity, any one generation's share approaches zero.

Other objections have been raised against the idea of intergenerational justice. These objections can be divided into two groups, which we can call conceptual and technological. One conceptual criticism is that the very idea of intergenerational justice is itself incoherent. The idea of justice is tied with that of reciprocity or exchange; but relations of reciprocity can exist only between contemporaries; therefore the concept of justice is inapplicable to relations between existing people and distant posterity. Future people are in no position to reciprocate; therefore people now living cannot be morally obligated to do anything for them. Another conceptual objection to the idea of intergenerational justice is concerned with rights. Briefly, the objection runs as follows: future people do not (yet) exist; only actually existing people have rights, including the right to be treated justly; therefore future people do not have rights which we in the present have a moral obligation to respect and protect. Critics of this view counter that it not only rests on a too-restrictive conception of rights and justice, but that it also paves the way for grievous intergenerational injustices.

Several arguments can be constructed to counter the claim that justice rests on reciprocity (and therefore applies only to relations between contemporaries) and the claim that future people do not have rights, including the right to be treated justly by their predecessors. Regarding reciprocity: since we acknowledge in ethics and recognize in law that it is possible to treat an infant or a mentally disabled or severely retarded person justly or unjustly, even though they are in no position to reciprocate, it follows that the idea of justice is not necessarily connected with reciprocity. Regarding the claim that future people cannot be said to have rights that require our recognition and respect: one of the more ingenious arguments against this view consists of modifying John Rawls's imaginary *veil of ignorance*. Rawls argues that principles of justice must not be partisan or favor particular people but must be blind and impartial. To ensure impartiality in arriving at principles of justice, Rawls invites us to imagine an original position in which rational people are placed behind a veil of ignorance wherein they are unaware of their age, race, sex, social class, economic status, etc. Unaware of their own particular position in society, rational people would arrive at and agree upon impartial and universal principles of justice. To ensure that such impartiality extends across generations, one need only *thicken* the veil by adding the proviso that the choosers be unaware of the generation to which they belong. Rational people would not accept or agree to principles under which predecessors could harm or disadvantage successors.

Some critics of intergenerational justice argue in technological terms. They contend that existing people need not restrict their consumption of scarce or **nonrenewable resources** in order to save some portion for future generations. For, they argue, substitutes for these resources will be discovered or devised through technological innovations and inventions. For example, as fossil fuels become scarcer and more expensive, new fuels—gasohol or fusion-derived nuclear fuel—will replace them. Thus we need never worry about depleting any particular resource because every resource can be replaced by a substitute that is as cheap, clean, and accessible as the resource it replaces. Likewise, we need not worry about generating nuclear wastes that we do not yet know how to store safely. Some solution is bound to be devised sometime in the future.

Environmentally-minded critics of this technological line of argument claim that it amounts to little more than wishful thinking. Like Charles Dickens's fictional character Mr. Micawber, those who place their faith in technological solutions to all environmental problems optimistically expect that "something will turn up." Just as Mr. Micawber's faith was misplaced, so too, these critics contend, is the optimism of those who expect technology to solve all problems, present and future. Of course such solutions may be found, but that is a gamble and not a guarantee. To wager with the health and well-being of future people is, environmentalists argue, immoral.

There are of course many other issues and concerns raised in connection with intergenerational justice. Discussions among and disagreements between philosophers, economists, environmentalists, and others are by no means purely abstract and academic. How these matters are resolved will have a profound effect on the fate of future generations.

[*Terence Ball*]

RESOURCES
BOOKS

Auerbach, B. E. *Unto the Thousandth Generation: Conceptualizing Intergenerational Justice.* New York: Peter Lang, 1995.

Ball, T. *Transforming Political Discourse.* Oxford, England: Blackwell, 1988.

Barry, B., and R. I. Sikora, eds. *Obligations to Future Generations.* Philadelphia: Temple University Press, 1978.

Barry, B. *Theories of Justice.* Berkeley: University of California Press, 1988.

Berry, W. *The Gift of Good Land.* San Francisco: North Point Press, 1981.

De-Shalit, A. *Why Posterity Matters: Environmental Policies and Future Generations.* London and New York: Routledge, 1995.

Fishkin, J., and P. Laslett, eds. *Justice Between Age Groups and Generations.* New Haven, CT: Yale University Press, 1991.

MacLean, D., and P. G. Brown, eds. *Energy and the Future.* Totawa, NJ: Rowman & Littlefield, 1983.

Partridge, E., ed. *Responsibilities to Future Generations.* Buffalo, NY: Prometheus Books, 1981.

Rawls, J. *A Theory of Justice.* Cambridge: Harvard University Press, 1971.

Wenz, P. S. *Environmental Justice.* Albany: State University of New York Press, 1988.

Intergovernmental Panel on Climate Change (IPCC)

The Intergovernmental Panel on Climate Change (IPCC) was established in 1988 as a joint project of the **United Nations Environment Programme** (UNEP) and the World Meteorological Organization (WMO). The primary mission of the IPCC is to bring together the world's leading experts on the earth's climate to gather, assess, and disseminate scientific information about climate change, with a view to informing international and national policy makers. The

IPCC has become the highest-profile and best-regarded international agency concerned with the climatic consequences of "greenhouse gases," such as **carbon dioxide** and **methane**, that are a byproduct of the **combustion** of **fossil fuels**. The IPCC is an organization that has been and continues to be at the center of a great deal of controversy.

The IPCC was established partly in response to Nobel Laureate Mario Molina's 1985 documentation of chemical processes which occur when human-made **chemicals** deplete the earth's atmospheric **ozone** shield. Ozone depletion is likely to result in increased levels of **ultraviolet radiation** reaching the earth's surface, producing a host of health, agricultural, and environmental problems. Molina's work helped to persuade most of the industrialized nations to ban **chlorofluorocarbons** and several other ozone-depleting chemicals. It also established a context in which national and international authorities began to pay serious attention to the global environmental consequences of atmospheric changes resulting from industrialization and reliance on fossil fuels.

Continuing to operate under the auspices of the United Nations and headquartered in Geneva, Switzerland, the IPCC is organized into three working groups and a task force, and meets about once a year. The first group gathers scientific data and analyzes the functioning of the climate system with special attention to the detection of potential changes resulting from human activity. The second group's assignment is to assess the potential socioeconomic impacts and vulnerabilities associated with climate change. It is also charged with exploring options for humans to adapt to potential climate change. The third group focuses on ways to reduce greenhouse gas emissions and to stop or reduce climate change. The task force is charged with maintaining inventories of greenhouse emissions for all countries.

The IPCC has published its major findings in "Full Assessment" Reports, first issued in 1990 and 1995. The Tenth Session of the IPCC (Nairobi, 1994) directed that future full assessments should be prepared approximately every five years. The Third Assessment Report was entitled "Climate Change 2001". Special reports and technical papers are also published as the panel identifies issues.

The IPCC has drawn a great deal of criticism virtually from its inception. Massive amounts of money are at stake in policy decisions which might seek to limit greenhouse gas emissions, and much of the criticism directed at the IPCC tends to come from lobbying and research groups mostly funded by industries that either produce or use large quantities of fossil fuels. Thus, a lobbying group sponsored by energy, **transportation**, and manufacturing interests called the Global Climate Coalition attacked parts of the 1995 report as unscientific. At the core of the controversy was Chapter Eight of the report, "Detection of Climate

Change and Attribution of Causes". Although the IPCC was careful to hedge its conclusions in various ways, acknowledging difficulties in measurement, disagreements over methodologies for interpreting data, and general uncertainty about the conclusions of its findings, it nevertheless suggested a connection between greenhouse gas emissions and global warming. Not satisfied with such caveats, the Global Climate Coalition charged that the IPCC's conclusions had been presented as far less debatable than they actually were. This cast a cloud of uncertainty over the report, at least for some United States policymakers. However, other leaders took the report more seriously. The Second Assessment Report provided important input to the negotiations that led to the development of the Kyoto Protocol in 1997, a treaty aimed at reducing the global output of **greenhouse gases**.

In the summer of 1996, results of new studies of the upper **atmosphere** were published which provided a great deal of indirect support for the IPCC's conclusions. Investigators found significant evidence of cooling in the upper atmosphere and warming in the lower atmosphere, with this effect being especially pronounced in the southern hemisphere. These findings confirmed the predictions of global warming models such as those employed by the IPCC.

Perhaps emboldened by this confirmation, but still facing a great deal of political opposition, the IPCC released an unequivocal statement about global warming and its causes in November 1996. The IPCC declared that "the balance of evidence suggests that there is a discernible human influence on global climate". The statement made clear that a preponderance of evidence and a majority of scientific experts indicated that observable climate change was a result of human activity. The IPCC urged that all nations limit their use of fossil fuels and develop more energy-efficient technologies.

These conclusions and recommendations provoked considerable criticism from less-developed countries. Leaders of the less-industrialized areas of the world tend to view potential restrictions on the use of fossil fuels as unfair hindrance of their efforts to catch up with the United States and Western Europe in industry, transportation, economic infrastructure, and standards of living. The industrialized nations, they point out, were allowed to develop without any such restrictions and now account for the vast majority of the world's energy consumption and greenhouse gas emissions. These industrialized nations therefore should bear the brunt of any efforts to protect the global climate, substantially exempting the developing world from restrictions on the use of fossil fuels.

The IPCC's conclusions and recommendations have also drawn strong opposition from industry groups in the United States, such as the American **Petroleum** Institute, and conservative Republican politicians. These critics charge that the IPCC's new evidence is only fashionable but warmed-over theory, and that no one has yet proven conclusively that climate change is indeed related to human influence. In view of the likely massive economic impact of any aggressive program aimed at the reduction of emissions, there is no warrant for following the IPCC's dangerous and ill-considered advice. Under Republican leadership, Congress slashed funds for **Environmental Protection Agency** and Department of Energy programs concerned with global warming and its causes, as well as funds for researching alternative and cleaner sources of energy.

These funding cuts and the signals they sent created foreign relations problems for the Clinton Administration. The United States was unable to honor former President Bush's 1992 pledge (at the Rio de Janeiro Earth Summit) to reduce the country's **emission** of **carbon** dioxide and methane to 1990 levels by the year 2000. Indeed, owing in part to low oil prices and a strong domestic economy, the United States was consuming more energy and emitting more greenhouse gases than ever before by 2000.

In the summer of 2001, the IPCC released its strongest statement to date on the problem of global warming, in its Third Assessment Report. The report, "Climate Change 2001", provides further evidence for global warming and its cause—the widescale burning of fossil fuels by humans. The report projects that global mean surface temperatures on earth will increase by 2.5–10.4°F (1.5–5.9°C) by the year 2100, unless greenhouse gas emissions are reduced well below current levels. The report also notes that this warming trend will represent the fasting warming of the earth in 10,000 years, with possible dire consequences to human society and the **environment**.

In the early 2000s, the administration of President George W. Bush, a former oilman, was resistant to the ideas of global warming and reducing greenhouse gas emissions. The administration strongly opposed the Kyoto Treaty and domestic **pollution** reduction laws, claiming such measures would cost jobs and reduce the standard of living, and that the scientific evidence was inconclusive. In June 2001, a National Academy of Science (NAS) panel reported to President Bush that the IPCC's studies on global warming were scientifically valid. In April 2002, with pressure from the oil industry, the Bush administration forced the removal of IPCC Chairman Robert Watson, an American atmospheric scientist who had been outspoken over the issue of climate change and the need for greenhouse gas reduction in industrialized countries.

The IPCC elected Dr. Rajendra K. Pachauri as its next Chairman at its nineteenth session in Geneva. Dr. Pachauri, a citizen of India, is a well-known world-class

expert in economics and technology, with a strong commitment to the IPCC process and to scientific integrity.

[*Lawrence J. Biskowski and Douglas Dupler*]

RESOURCES
BOOKS

McKibbin, Warwick J., and Peter Wilcoxen. *Climate Change Policy After Kyoto: Blueprint for a Realistic Approach.* Washington DC: The Brookings Institution Press, 2002.

PERIODICALS

McKibben, Bill. "Climate Change 2001: Third Assessment Report." *New York Review of Books,* July 5, 2001, 35.
Trenberth, Kevin E. "Stronger Evidence of Human Influences on Climate: The 2001 IPCC Assessment." *Environment,* May 2001, 8.

OTHER

Intergovernmental Panel on Climate Change Home Page. [cited July 2002]. <http://www.ipcc.ch>.
Union of Concerned Scientists Global Warming Web Page. [cited July 2002]. <http://www.ucsusa/warming>.
World Meteorological Organization Home Page. [cited July 2002]. <http://www.wmo.ch>.

ORGANIZATIONS

IPCC Secretariat, C/O World Meteorological Organization, 7bis Avenue de la Paix, C.P. 2300, CH- 1211, Geneva, Switzerland 41-22-730-8208, Fax: 41-22-730-8025, Email: ipcc.sec@gateway.wmo.ch, <http://www.ipcc.ch>

Internal costs
see **Internalizing costs**

Internalizing costs

Private market activities create so-called externalities. An example of a negative **externality** is **air pollution**. It occurs when a producer does not bear all the costs of an activity in which he or she engages. Since external costs do not enter into the calculations producers make, they will make few attempts to limit or eliminate **pollution** and other forms of **environmental degradation**.

Negative externalities are a type of market defect all economists believe is appropriate to try to correct. Milton Friedman refers to such externalities as "neighborhood effects," (although it must be kept in mind that some forms of pollution have an all but local effect). The classic neighborhood effect is pollution. The premise of a free market is that when two people voluntarily make a deal, they both benefit. If society gives everyone the right to make deals, society as a whole will benefit. It becomes richer from the aggregation of the many mutually beneficial deals that are made. However, what happens if in making mutually beneficial deals there is a waste product that the parties release into the **environment** and that society must either suffer from or clean up? The two parties to the deal are better off, but society as a whole has to pay the costs. Friedman points out that individual members of a society cannot appropriately charge the responsible parties for external costs or find other means of redress.

Friedman's answer to this dilemma is simple: society, through government, must charge the responsible parties the costs of the clean-up. Whatever damage they generate must be internalized in the price of the transaction. Polluters can be forced to internalize environmental costs through pollution taxes and **discharge** fees, a method generally favored by economists. When such taxes are imposed, the market defect (the price of pollution which is not counted in the transaction) is corrected. The market price then reflects the true social costs of the deal, and the parties have to adjust accordingly. They will have an incentive to decrease harmful activities and develop less environmentally damaging technology. The drawback of such a system is that society will not have direct control over pollution levels, although it will receive monetary compensation for any losses it sustains. However, if the government imposed a tax or charge on the polluting parties, it would have to place a monetary value on the damage. In practice, this is difficult to do. How much for a human life lost to pollution? How much for a vista destroyed? How much for a plant or animal **species** brought to **extinction**? Finally, the idea that pollution is all right as long as the polluter pays for it is unacceptable to many people.

In fact, the government has tried to control activities with associated externalities through regulation, rather than by supplementing the price system. It has set standards for specific industries and other social entities. The standards are designed to limit environmental degradation to acceptable levels and are enforced through the **Environmental Protection Agency** (EPA). They prohibit some harmful activities, limit others, and prescribe alternative behaviors. When market actors do not adhere to these standards they are subject to penalties. In theory, potential polluters are given incentives to reduce and treat their waste, manufacture less harmful products, develop alternative technologies, and so on. In practice, the system has not worked as well as it was hoped in the 1960s and 1970s, when much of the environmental legislation presently in force was enacted. Enforcement has been fraught with political and legal difficulties. Extensions on deadlines are given to cities for not meeting clean air standards and to the **automobile** industry for not meeting standards on fuel economy of new cars, for instance. It has been difficult to collect fines from industries found to have been in violation. Many cases are tied up in the courts through a lengthy appeals process. Some companies simply declare bankruptcy to evade fines. Others continue

polluting because they find it cheaper to pay fines than to develop alternative production processes.

Alternative strategies presently under debate include setting up a **trade in pollution permits**. The government would not levy a tax on pollution but would issue a number of permits that altogether set a maximum acceptable pollution level. Buyers of permits can either use them to cover their own polluting activities or resell them to the highest bidder. Polluters will be forced to internalize the environmental costs of their activities so that they will have an incentive to reduce pollution. The price of pollution will then be determined by the market. The disadvantage of this system is that the government will have no control over where pollution takes place. It is thinkable that certain regions will have high concentrations of industries using the permits, which may result in local pollution levels that are unacceptably high. Whether marketable pollution permits address present pollution problems more satisfactorily than does regulation alone has yet to be seen. *See also* Environmental economics

[*Alfred A. Marcus and Marijke Rijsberman*]

RESOURCES
BOOKS

Friedman, M. *Capitalism and Freedom*. Chicago: University of Chicago Press, 1962.
Marcus, A. *Business and Society: Ethics, Government, and the World Economy*. Homewood, IL: Irwin Press, 1993.

International Atomic Energy Agency

The first decade of research on **nuclear weapons** and nuclear reactors was characterized by extreme secrecy, and the few nations that had the technology carefully guarded their information. In 1954, however, that philosophy changed, and the United States, in particular, became eager to help other nations use nuclear energy for peaceful purposes. A program called "Atoms for Peace" brought foreign students to the United States for the study of nuclear sciences and provided enriched **uranium** to countries wanting to build their own reactors, encouraging interest in nuclear energy throughout much of the world.

But this program created a problem. It increased the potential diversion of nuclear information and nuclear materials for the construction of weapons, and the threat of nuclear proliferation grew. The United Nations created the International Atomic Energy Agency (IAEA) in 1957 to address this problem. The agency had two primary objectives: to encourage and assist with the development of peaceful applications of **nuclear power** throughout the world and to prevent the diversion of nuclear materials to weapons research and development.

The first decade of IAEA's existence was not marked by much success. In fact, the United States was so dissatisfied with the agency's work that it began signing bilateral non-proliferation treaties with a number of countries. Finally, the 1970 nuclear non-proliferation treaty more clearly designated the IAEA's responsibilities for the monitoring of nuclear material.

Today the agency is an active division of the United Nations Educational, Scientific, and Cultural Organization (UNESCO), and its headquarters are in Vienna. The IAEA operates with a staff of more than 800 professional workers, about 1,200 general service workers, and a budget of about $150 million. To accomplish its goal of extending and improving the peaceful use of nuclear energy, IAEA conducts regional and national workshops, seminars, training courses, and committee meetings. It publishes guidebooks and manuals on related topics and maintains the International Nuclear Information System, a bibliographic database on nuclear literature that includes more than 1.2 million records. The database is made available on magnetic tape to its 42-member states.

The IAEA also carries out a rigorous program of inspection. In 1987, for example, it made 2,133 inspections at 631 nuclear installations in 52 non-nuclear weapon nations and four nuclear weapon nations. In a typical year, IAEA activities include conducting safety reviews in a number of different countries, assisting in dealing with accidents at nuclear **power plants**, providing advice to nations interested in building their own nuclear facilities, advising countries on methods for dealing with radioactive wastes, teaching nations how to use radiation to preserve foods, helping universities introduce nuclear science into their curricula, and sponsoring research on the broader applications of nuclear science.

[*David E. Newton*]

RESOURCES
ORGANIZATIONS
International Atomic Energy Agency, P.O. Box 100, Wagramer Strasse 5, Vienna, Austria A-1400 (413) 2600-0, Fax: (413) 2600-7, Email: official.mail@iaea.org

International Cleaner Production Cooperative

The International Cleaner Production Cooperative is an Internet resource (<http://es.epa.gov/cooperative/international>) that was implemented to provide access to globally relevant information about cleaner production and **pollution** prevention to the international community. The site is hosted by the U.S. **Environmental Protection Agency** and gives access to a consortium of World Wide Web sites that provides information to businesses, professional, and local, regional, national, and international agencies that are striving for cleaner production. The cooperative provides links to people and businesses involved with cleaner production and pollution prevention, and to sources of technical assistance and information on international policy. The **United Nations Environment Programme** (UNEP) is one of the primary members of the cooperative.

[*Marie H. Bundy*]

International Convention for the Regulation of Whaling (1946)

The International **Whaling** Commission (IWC) was established in 1949 following the inaugural International Convention for the Regulation of Whaling, which took place in Washington, D.C., in 1946. Many nations have membership in the IWC, which primarily sets quotas for **whales**. The purpose of these quotas is twofold: they are intended to protect the whale **species** from **extinction** while allowing a limited whaling industry. In recent times, however, the IWC has come under attack. The vast majority of nations in the Commission have come to oppose whaling of any kind and object to the IWC's practice of establishing quotas. Furthermore, some nations—principally Iceland, Japan, and Norway—wish to protect their traditional whaling industries and are against the quotas set by the IWC. With two such divergent factions opposing the IWC, its future is as doubtful as that of the whales.

Since its inception, the Commission has had difficulty implementing its regulations and gaining approval for its recommendations. In the meantime whale populations have continued to dwindle. In its original design, the IWC consisted of two sub-committees, one scientific and the other technical. Any recommendation that the scientific committee put forth was subject to the politicized technical committee before final approval. The technical committee evaluated the recommendation and changed it if it was not politically or economically viable; essentially, the scientific committee's recommendations have often been rendered powerless. Furthermore, any nation that has decided an IWC recommendation was not in its best interest could have dismissed it by simply registering an objection. In the 1970s this gridlock and inaction attracted public scrutiny; people objected to the IWC's failure to protect the world's whales. Thus in 1972 the **United Nations Conference on the Human Environment** voted overwhelmingly to stop commercial whaling.

Nevertheless, the IWC retained some control over the whaling industry. In 1974 the Commission attempted to bring scientific research to management strategies in its

"New Management Procedure." The IWC assessed whale populations with finer resolution, scrutinizing each species to see if it could be hunted and not die out. It classified whales as either "initial management stocks" (harvestable), "sustained management stocks" (harvestable), or "protection stocks" (unharvestable). While these classifications were necessary for effective management, much was unknown about whale population **ecology**, and quota estimates contained high levels of uncertainty.

Since the 1970s, public pressure has caused many nations in the IWC to oppose whale **hunting** of any kind. At first, one or two nations proposed a whaling moratorium each year. Both pro- and anti-whaling countries began to encourage new IWC members to vote for their respective positions, thus dividing the Commission. In 1982, the IWC enacted a limited moratorium on commercial whaling, to be in effect from 1986 until 1992. During that time it would thoroughly assess whale stocks and afterward allow whaling to resume for selected species and areas. Norway and Japan, however, attained special permits for whaling for scientific research: they continued to catch approximately 400 whales per year, and the meat was sold to restaurants. Then in 1992—the year when whaling was supposed to have resumed—many nations voted to extend the moratorium. Iceland, Norway, and Japan objected strongly to what they saw as an infringement on their traditional industries and eating customs. Iceland subsequently left the IWC, and Japan and Norway have threatened to follow. These countries intend to resume their whaling programs. Members of the IWC are torn between accommodating these nations in some way and protecting the whales, and amid such controversy it is unlikely that the Commission can continue in its present mission.

Although the IWC has not been able to marshall its scientific advances or enforce its own regulations in managing whaling, it is broadening its original mission. The Commission may begin to govern the hunting of small cetaceans such as **dolphins** and porpoises, which are believed to suffer from **overhunting**.

[*David A. Duffus and Andrea Gacki*]

RESOURCES
BOOKS

Burton, R. *The Life and Death of Whales*. London: Andre Deutsch Ltd., 1980.

Kellog, R. *The International Whaling Commission*. International Technical Conference on Conservation of Living Resources of the Sea. New York: United Nations Publications, 1955.

PERIODICALS

Holt, S. J. "Let's All Go Whaling." *The Ecologist* 15 (1985): 113–124.

Pollack, A. "Commission to Save Whales Endangered, Too." *The New York Times*, May 18, 1993, B8.

International Council for Bird Preservation
see **BirdLife International**

International Geosphere-Biosphere Programme (U.N. Environmental Programme)

Research scientists from all countries have always interacted with each other closely. But in recent decades, a new type of internationalism has begun to evolve, in which scientists from all over the world work together on very large projects concerning the planet.

An example is research on global change. A number of scientists have come to believe that human activities, such as the use of **fossil fuels** and **deforestation** of tropical rain forests, may be altering the earth's **climate**. To test that hypothesis, a huge amount of meteorological data must be collected from around the world, and no single institution can possibly obtain and analyze it all.

A major effort to organize research on important, worldwide scientific questions such as climate change was begun in the early 1980s. Largely through the efforts of scientists from two United States organizations, the National Aeronautics and Space Administration (NASA) and the **National Research Council**, a proposal was developed for the creation of an International Geosphere-Biosphere Programme (IGBP). The purpose of the IGBP was to help scientists from around the world focus on major issues about which there was still too little information. Activity funding comes from national governments, scientific societies, and private organizations.

IGBP was not designed to be a new organization, with new staff, new researchers, and new funding problems. Instead, it was conceived of as a coordinating program that would call on existing organizations to attack certain problems. The proposal was submitted in September 1986 to the General Assembly of the International Council of Scientific Unions (ICSU), where it received enthusiastic support.

Within two years, more than 20 nations agreed to cooperate with IGBP, forming national committees to work with the international office. A small office, administered by Harvard oceanographer James McCarthy, was installed at the Royal Swedish Academy of Sciences in Stockholm.

IGBP has moved forward rapidly. It identified existing programs that fit the Programme's goals and developed new research efforts. Because many global processes are gradual, a number of IGBP projects are designed with time frames of ten to twenty years.

By the early 1990s, IGBP had defined a number of projects, including the Joint Global Ocean Flux Study, the Land-Ocean Interactions in the Coastal Zone study, the Biospheric Aspects of the Hydrological Cycle research, Past Global Changes, Global Analysis, Interpretation and **Modeling**, and Global Change System for Analysis, Research and Training.

[*David E. Newton*]

RESOURCES

BOOKS

Kupchella, C. E. *Environmental Science: Living within the System of Nature.* Boston: Allyn and Bacon, Inc., 1986.

PERIODICALS

Edelson, E. "Laying the Foundation." *Mosaic* (Fall/Winter 1988): 4–11.

Perry, J. S. "International Institutions for the Global Environment." *MTS Journal* (Fall 1991): 27–8.

OTHER

International Geosphere-Biosphere Programme. [cited June 2002]. <http://www.igbp.kva.se>.

International Institute for Sustainable Development

The International Institute for **Sustainable Development** (IISD) is a nonprofit organization that serves as an information and resources clearinghouse for policy makers promoting sustainable development. IISD aims to promote sustainable development in decision making worldwide by assisting with policy analysis, providing information about practices, measuring sustainability, and building partnerships to further sustainability goals. It serves businesses, governments, communities, and individuals in both developing and industrialized nations. IISD's stated aim is to "create networks designed to move sustainable development from concept to practice." Founded in 1990 and based in Winnipeg, Canada, IISD is funded by foundations, governmental organizations, private sector sources, and revenue from publications and products.

IISD works in seven program areas. The Business Strategies program focuses on improving competitiveness, creating jobs, and protecting the **environment** through sustainability. Projects include several publications and the EarthEnterprise program, which offers entrepreneurial and employment strategies. IISD's Trade and Sustainable Development program works on building positive relationships between trade, the environment, and development. It examines how to make international accords, such as those made by the **World Trade Organization**, compatible with the goals of sustainable development. The Community **Adapta-**

tion and Sustainable Livelihoods program identifies adaptive strategies for drylands in Africa and India, and it examines the influences of policies and new technology on local ways of life. The Great Plains program works with community, farm, government, and industry groups to assist communities in the Great Plains region of North America with sustainable development. It focuses on government policies in agriculture, such as the Western Grain **Transportation** Act, as well as loss of transportation subsidies, the **North American Free Trade Agreement** (NAFTA), **soil** salination and loss of **wetlands**, job loss, and technological advances. Measurement and Indicators aims to set measurable goals and progress indicators for sustainable development. As part of this, IISD offers information about the successful uses of taxes and subsidies to encourage sustainability worldwide. Common Security focuses on initiatives of peace and consensus-building.

IISD's Information and Communications program offers several publications and Internet sites featuring information on sustainable development issues, terms, events, and media coverage. This includes *Earth Negotiations Bulletin*, which provides on-line coverage of major environmental and development negotiations (especially United Nations conferences), and IISDnet, with information about sustainable development worldwide. IISD also publishes more than 50 books, monographs, and discussion papers, including *Sourcebook on Sustainable Development*, which lists organizations, databases, conferences, and other resources. IISD produces five journals that include *Developing Ideas*, published bimonthly both in print and electronically, and featuring articles on sustainable development terms, issues, resources, and recent media coverage. *Earth Negotiations Bulletin* reports on conferences and negotiations meetings, especially United Nations conferences. IISD's Internet journal, */linkages/journal/*, is a bimonthly electronic multi-media subscription magazine focusing on global negotiations. Its reporting service, Sustainable Developments, reports on environmental and development negotiations for meetings and symposia via the Internet. IISD also operates IISDnet (http://iisd1.iisd.ca/), an Internet information site featuring research, new trends, global activities, contacts, information on IISD's activities and projects, including United Nations negotiations on environment and development, corporate environmental reporting, and information on trade issues.

IISD is the umbrella organization for Earth Council, an international **nongovernmental organization** (NGO) created in 1992 as a result of the **United Nations Earth Summit**. The organization creates measurements for achieving sustainable development and assesses practices and economic measures for their effects on sustainable development. Earth Council coordinated the Rio+5 Forum in Rio de Janeiro in March 1997, which assessed progress towards sustainable development since the Earth Summit in 1992. IISD has produced two publications, *Trade and Sustainable Development* and *Guidelines for the Practical Assessment of Progress Toward Sustainable Development.*

[*Carol Steinfeld*]

RESOURCES
ORGANIZATIONS
International Institute for Sustainable Development, 161 Portage Avenue East, 6th Floor, Winnipeg, ManitobaCanada R3B 0Y4 (204) 958-7700, Fax: (204) 958-7710, Email: info@iisd.ca, <http://www.iisd.org>

International Joint Commission

The International Joint Commission (IJC) is a permanent, independent organization of the United States and Canada formed to resolve trans-boundary ecological concerns. Founded in 1912 as a result of provisions under the Boundary Waters Treaty of 1909, the IJC was patterned after an earlier organization, the Joint Commission, which was formed by the United States and Britain.

The IJC consists of six commissioners, with three appointed by the President of the United States, and three by the Governor-in-Council of Canada, plus support personnel. The commissioners and their organizations generally operate freed from direct influence or instruction from their national governments. The IJC is frequently cited as an excellent model for international dispute resolution because of its history of successfully and objectively dealing with **natural resources** and environmental disputes between friendly countries.

The major activities of the IJC have dealt with apportioning, developing, conserving, and protecting the binational **water resources** of the United States and Canada. Some other issues, including transboundary **air pollution**, have also been addressed by the Commission.

The power of the IJC comes from its authority to initiate scientific and socio-economic investigations, conduct quasi-judicial inquiries, and arbitrate disputes.

Of special concern to the IJC have been issues related to the **Great Lakes**. Since the early 1970s, IJC activities have been substantially guided by provisions under the 1972 and 1978 **Great Lakes Water Quality Agreement** plus updated protocols. For example, it is widely acknowledged, and well documented, that environmental quality and **ecosystem health** have been substantially degraded in the Great Lakes. In 1985, the **Water Quality** Board of the IJC recommended that states and provinces with Great Lakes boundaries make a collective commitment to address this communal problem, especially with respect to **pollution**. These governments agreed to develop and implement remedial ac-

tion plans (RAPs) towards the restoration of **environmental health** within their political jurisdictions. Forty-three areas of concern have been identified on the basis of environmental pollution, and each of these will be the focus of a remedial action plan.

An important aspect of the design and intent of the overall program, and of the individual RAPs, will be developing a process of integrated **ecosystem management**. **Ecosystem** management involves systematic, comprehensive approaches toward the restoration and protection of environmental quality. The ecosystem approach involves consideration of interrelationships among land, air, and water, as well as those between the inorganic **environment** and the biota, including humans. The ecosystem approach would replace the separate, more linear approaches that have traditionally been used to manage environmental problems. These conventional attempts have included directed programs to deal with particular resources such as fisheries, migratory birds, **land use**, or point sources and area sources of toxic emissions. Although these non-integrated methods have been useful, they have been limited because they have failed to account for important inter-relationships among environmental management programs, and among components of the ecosystem.

[*Bill Freedman Ph.D.*]

RESOURCES
ORGANIZATIONS
International Joint Commission, 1250 23rd Street, NW, Suite 100, Washington, D.C. USA 20440 (202) 736-9000, Fax: (202) 735-9015, , <http://www.ijc.org>

International Primate Protection League

Founded in 1974 by Shirley McGreal, International Primate Protection League (IPPL) is a global **conservation** organization that works to protect nonhuman primates, especially monkeys and apes (**chimpanzees**, orangutans, **gibbons**, and gorillas).

IPPL has 30,000 members, branches in the United Kingdom, Germany, and **Australia**, and field representatives in 31 countries. Its advisory board consists of scientists, conservationists, and experts on primates, including the world-renowned primatologist **Jane Goodall**, whose famous studies and books are considered the authoritative texts on chimpanzees. Her studies have also heightened public interest and sympathy for chimpanzees and other nonhuman primates.

IPPL runs a sanctuary and **rehabilitation** center at its Summerville, South Carolina headquarters, which houses two dozen gibbons and other abandoned, injured, or traumatized primates who are refugees from medical laboratories or abusive pet owners. IPPL concentrates on investigating and fighting the multi-million dollar commercial trafficking in primates for medical laboratories, the **pet trade**, and zoos, much of which is illegal trade and smuggling of **endangered species** protected by international law. IPPL is considered the most active and effective group working to stem the cruel and often lethal trade in primates.

IPPL's work has helped to save the lives of literally tens of thousands of monkeys and apes, many of which are threatened or endangered **species**. For example, the group was instrumental in persuading the governments of India and Thailand to ban or restrict the export of monkeys, which were being shipped by the thousands to research laboratories and pet stores across the world.

The trade in primates is especially cruel and wasteful, since a common way of capturing them is by shooting the mother, which then enables poachers to capture the infant. Many captured monkeys and apes die enroute to their destinations, often being transported in sacks, crates, or hidden in other devices.

IPPL often undertakes actions and projects that are dangerous and require a good deal of skill. In 1992, its investigations have led to the conviction of a Miami, Florida, animal dealer for conspiring to help smuggle six baby orangutans captured in the jungles of Borneo. The endangered **orangutan** is protected by the Convention on International Trade in Endangered Species of **Fauna** and Flora (CITES), as well as by the United States **Endangered Species Act**. In retaliation, the dealer unsuccessfully sued McGreal, as did a multi-national corporation she once criticized for its plan to capture chimpanzees and use them for hepatitis research in Sierra Leone.

A more recent victory for IPPL occurred in April 2002. In 1997, Chicago O'Hare airport received two shipments from Indonesia, each of which contained more than 250 illegally imported monkeys. Included in the shipments were dozens of unweaned baby monkeys. After several years of pursuing the issue, the U.S. **Fish and Wildlife Service** and the U.S. Federal prosecutors charged the LABS Company (a breeder of monkeys for research based in the United States) and several of its employees, including its former president, on eight felonies and four misdemeanors.

IPPL publishes *IPPL News* several times a year and sends out periodic letters alerting members of events and issues that affect primates.

[*Lewis G. Regenstein*]

RESOURCES
ORGANIZATIONS
International Primate Protection League, P.O. Box 766, Summerville, SC USA 29484 (843) 871-2280, Fax: (843) 871-7988, Email: ippl@awod.com, <http://ippl.org>

International Register of Potentially Toxic Chemicals (U. N. Environment Programme)

The International Register of Potentially Toxic **Chemicals** is published by the **United Nations Environment Programme** (UNEP). Part of UNEP's three-pronged **Earthwatch** program, the register is an international inventory of chemicals that threaten the **environment**. Along with the **Global Environment Monitoring System** and **INFOTERRA**, the register monitors and measures environmental problems worldwide. Information from the register is routinely shared with agencies in developing countries. **Third World** countries have long been the toxic dumping grounds for the world, and they still use many chemicals that have been banned elsewhere. Environmental groups regularly send information from the register to toxic chemical users in developing countries as part of their effort to stop the export of toxic **pollution**.

RESOURCES
ORGANIZATIONS
International Register of Potentially Toxic Chemicals, Chemin des Anémones 15, Genève, Switzerland CH-1219 +41-22-979 91 11, Fax: +41-22-979 91 70, Email: chemicals@unep.ch

International Society for Environmental Ethics

The International Society for **Environmental Ethics** (ISEE) is an organization that seeks to educate people about the environmental ethics and philosophy concerning **nature**. An environmental ethic is the philosophy that humans have a moral duty to sustain the natural **environment** and attempts to answer how humans should treat other **species** (plant and animal), use Earth's **natural resources**, and place value on the aesthetic experiences of nature.

The society is an auxiliary organization of the American Philosophical Association, with about 700 members in over 20 countries. Many of ISEE's current members are philosophers, teachers, or environmentalists. The ISEE officers include president Mark Sagoff (Institute for Philosophy and Public Policy, University of Maryland) and John Baird Callicott, vice president, (professor of philosophy at the University of North Texas). Two other key members are editors of the ISEE newsletter, Jack Weir and **Holmes Rolston**, III (Professor of Philosophy, Colorado State University). All have contributed to the ongoing ISEE Master Environmental Ethics Bibliography.

ISEE publishes a quarterly newsletter available to members in print form and maintains an Internet site of back issues. Of special note is the ISEE Bibliography, an ongoing project that contains over 5,000 records from journals such as *Environmental Ethics*, *Environmental Values*, and the *Journal of Agricultural and Environmental Ethics*.

Another work in progress, the ISEE Syllabus Project, continues to be developed by Callicott and Robert Hood, doctoral candidate at Bowling Green State University. They maintain a database of course offerings in environmental philosophy and ethics, based on information from two-year community colleges and four-year state universities, private institutions, and master's- and doctorate-granting universities. ISEE supports the enviroethics program which has spurred many Internet discussion groups and is constantly expanding into new areas of communication.

[*Nicole Beatty*]

RESOURCES

ORGANIZATIONS

International Society for Environmental Ethics, Environmental Philosophy Inc., Department of Philosophy, University of North Texas, P.O. Box 310980, Denton, TX USA 76203-0980, <http://www.cep.unt.edu/ISEE.html>

International trade in toxic waste

Just as VCRs, cars, and laundry soap are traded across borders, so too is the waste that accompanies their production. In the United States alone, industrial production accounts for at least 500 million lb (230 million kg) of **hazardous waste** a year. The industries of other developed nations also produce waste. While some of it is disposed within national borders, a portion is sent to other countries where costs are cheaper and regulations less stringent than in the waste's country of origin.

Unlike consumer products, internationally traded hazardous waste has begun to meet local opposition. In some recent high-profile cases, barges filled with waste have traveled the world looking for final resting places. In at least one case, a ship may have dumped about ten tons of toxic municipal incinerator ash in the ocean after being turned away from dozens of ports. In recent years national and international bodies have begun to voice official opposition to this dangerous trade through bans and regulations.

The international trade in toxic wastes is, at bottom, waste disposal with a foreign-relations twist. Typically a manufacturing facility generates waste during the production process. The facility manager pays a waste-hauling firm to dispose of the waste. If the landfills in the country of origin cost too much, or if there are no landfills that will take the waste, the disposal firm will find a cheaper option, perhaps a **landfill** in another country. In the United States, the shipper must then notify the **Environmental Protection Agency** (EPA), which then notifies the State Department. After ascertaining that the destination country will indeed accept the waste, American regulators approve the sale.

Disposing of the waste overseas in a landfill is only the most obvious example of this international trade. Waste haulers also sell their cargo as raw materials for **recycling**. For example, used lead-acid batteries discarded by American consumers are sent to Brazil where factory workers extract and resmelt the **lead**. Though the lead-acid alone would classify as hazardous, whole batteries do not. Waste haulers can ship these batteries overseas without notification to Mexico, Japan, and Canada, among other countries. In other cases, waste haulers sell products, like DDT, that have been banned in one country to buyers in another country that has no ban. Whatever the strategy for disposal, waste haulers are most commonly small, independent operators who provide a service to waste producers in industrialized countries.

These haulers bring waste to other countries to take advantage of cheaper disposal options and less stringent regulatory climates. Some countries forbid the disposal of the certain kinds of waste. Countries without such prohibitions will import more waste. Cheap landfills depend on cheap labor and land. Countries with an abundance of both can become attractive destinations. Entrepreneurs or government officials in countries, like Haiti, or regions within countries, such as Wales, that lack a strong manufacturing base, view waste disposal as a viable, inexpensive business. Inhabitants may view it as the best way to make money and create jobs. Simply by storing hazardous waste, the country of Guinea-Bissau could have made $120 million, more money than its annual budget.

Though the **less developed countries** (LDC) predictably receive large amounts of toxic waste, the bulk of the international trade occurs between industrialized nations. Canada and the United Kingdom in particular import large volumes of toxic waste. Canada imports almost 85% of the waste sent abroad by American firms, approximately 150,000 lb (70,000 kg) per year. The bulk of the waste ends up at an incinerator in Ontario or a landfill in Quebec. Because Canada's disposal regulations are less strict than United States laws, the operators of the landfill and incinerator can charge lower fees than similar disposal sites in the United States.

A waste hauler's life becomes complicated when the receiving country's government or local activists discover that the waste may endanger health and the **environment**. Local regulators may step in and forbid the sale. This happened many times in the case of the *Khian Sea*, a ship that had contracted to dispose of Philadelphia's incinerator ash. The ship was turned away from Haiti, from Guinea-Bissau, from Panama, and from Sri Lanka. For two years, beginning in 1986, the ship carried the toxic ash from port to port looking for a home for its cargo before finally mysteriously losing the ash somewhere in the Indian Ocean.

This early **resistance** to toxic-waste dumping has since led to the negotiation of international treaties forbidding or regulating the trade in toxic waste. In 1989, the African, Caribbean, and Pacific countries (ACP) and the countries belonging to the European Economic Community (EEC) negotiated the Lome IV Convention, which bans shipments of nuclear and hazardous waste from the EEC to the ACP countries. ACP countries further agreed not to import such waste from non-EEC countries. Environmentalists have encouraged the EEC to broaden its commitment to limiting waste trade.

In the same year, under the auspices of the **United Nations Environment Programme** (UNEP), the **Basel Convention** on the Control of Transboundary Movements of Hazardous Wastes and Their Disposal was negotiated. This requires shippers to obtain government permission from the destination country before sending waste to foreign landfills or incinerators. Critics contend that Basel merely formalizes the trade.

In 1991, the nations of the Organization of African Unity negotiated another treaty restricting the international waste trade. The Bamako Convention on the Ban of the Import into Africa and the Control of Transboundary Movement and Management of Hazardous Wastes within Africa criminalized the import of all hazardous waste. Bamako further forbade waste traders from importing to Africa materials that had been banned in one country to a country that has no such ban. Bamako also radically redefined the assessment of what constitutes a health hazard. Under the treaty, all **chemicals** are considered hazardous until proven otherwise.

These international strategies find their echoes in national law. Less developed countries have tended to follow the Lome and Bamako examples. At least eighty-three African, Latin-Caribbean, and Asian-Pacific countries have banned hazardous waste imports. And the United States, in a policy similar to the Basel Convention, requires hazardous waste shipments to be authorized by the importing country's government.

The efforts to restrict toxic waste trade reflect, in part, a desire to curb environmental inequity. When waste flows from a richer country to a poorer country or region, the inhabitants living near the incinerator, landfill, or recycling facility are exposed to the dangers of toxic compounds. For example, tests of workers in the Brazilian lead resmelting operation found blood-lead levels several times the United States standard. Lead was also found in the water supply of a nearby farm after five cows died. The loose regulations that keep prices low and attract waste haulers mean that there are fewer safeguards for local health and the environment. For example, leachate from unlined landfills can contaminate local **groundwater**. Jobs in the disposal industry tend to be lower paying than jobs in manufacturing. The inhabitants of the receiving country receive the wastes of industrialization without the benefits.

Stopping the waste trade is a way to force manufacturers to change production processes. As long as cheap disposal options exist, there is little incentive to change. A waste-trade ban makes hazardous waste expensive to discard, and will force business to search for ways to reduce this cost.

Companies that want to reduce their hazardous waste may opt for source reduction, which limits the hazardous components in the production process. This can both reduce production costs and increase output. A Monsanto facility in Ohio saved more than $3 million dollars a year while eliminating more than 17 million lb (8 million kg) of waste. According to officials at the plant, average yield increased by 8%. Measures forced by a lack of disposal options can therefore benefit the corporate bottom line, while reducing risks to health and the environment. *See also* Environmental law; Environmental policy; Groundwater pollution; Hazardous waste siting; Incineration; Industrial waste treatment; Leaching; Ocean dumping; Radioactive waste; Radioactive waste management; Smelter; Solid waste; Solid waste incineration; Solid waste recycling and recovery; Solid waste volume reduction; Storage and transport of hazardous materials; Toxic substance; Waste management; Waste reduction

[*Alair MacLean*]

RESOURCES

BOOKS

Dorfman, M., W. Muir, and C. Miller. *Environmental Dividends: Cutting More Chemical Waste.* New York: INFORM, 1992.

Moyers, B. D. *Global Dumping Ground: The International Traffic in Hazardous Waste.* Cabin John, MD: Seven Locks Press, 1990.

Vallette, J., and H. Spalding. *The International Trade in Wastes: A Greenpeace Inventory.* Washington, DC: Greenpeace, 1990.

PERIODICALS

Chepesiuk, R. "From Ash to Cash: The International Trade in Toxic Waste." *E Magazine* 2 (July-August 1991): 30–37.

International Union for the Conservation of Nature and Natural Resources

see **IUCN—The World Conservation Union**

International Voluntary Standards

International Voluntary Standards are industry guidelines or agreements that provide technical specifications so that products, processes, and services can be used worldwide. The need for development of a set of international standards to be followed and used consistently for environmental management systems was recognized in response to an increased desire by the global community to improve environmental management practices. In the early 1990s, the International Organization for Standardization or ISO, which is located in Geneva, Switzerland, began development of a strategic plan to promote a common international approach to environmental management. ISO 14000 is the title of a series of voluntary international environmental standards that is under development by ISO and is 142 member nations, including the United States. Some of the standards developed by ISO include standardized sampling, testing and analytical methods for use in the monitoring of environmental variables such as the quality of air, water and **soil**.

[*Marie H. Bundy*]

International Whaling Commission

see **International Convention for the Regulation of Whaling (1946)**

International Wildlife Coalition

The International **Wildlife** Coalition (IWC) was established by a small group of individuals who came from a variety of environmental and **animal rights** organizations in 1984. Like many NGOs (nongovernmental organizations) that arose in the 1970s and 1980s their initial work involved the protection of **whales**. The IWC raised money for whale **conservation** programs on endangered Atlantic humpback whale populations. This was one of the first **species** where researchers identified individual animals through tail photographs. Using this technique the IWC developed what is now a common tool, a whale adoption program based on individual animals with human names.

From that basis, the fledgling group established itself in an advocacy role with three principles in their mandate: to prevent cruelty to wildlife, to prevent killing of wildlife, and to prevent destruction of wildlife **habitat**. In light of those principles, the IWC can be characterized as an extended animal rights organization. They maintain the "prevention of cruelty" aspect common to humane societies, perhaps the oldest progenitor of animal rights groups. In standing by an ethic of preventing killing, they stand with animal rights groups, but by protecting habitat they take a more significant step by acting in a broad way to achieve their initial two principles.

The program thus works at both ends of the spectrum, undertaking **wildlife rehabilitation** and other programs dealing with the individual animals, as well as lobbying and promoting letter writing campaigns to improve wildlife legislation. For example, they have used their Brazilian office to create pressure to combat the international trade in exotic pets, and their Canadian office to oppose the harp seal hunt and the deterioration of Canada's impending **endangered species** legislation. Their United States-based operation has built a reputation in the research field working with government agencies to ensure that whale-watching on the eastern seaboard does not harm the whales. Offices in the United Kingdom are a focus for the IWC concern over the European Community policies, such as lifting their ban on importing fur from animals killed in leg hold traps.

It has become evident that the diversity within the varied groups that constitute the environmental community is a positive force, however, most conservation NGOs do not cross the gulf between animal rights and habitat conservation. A clear distinction exists between single animal approaches and broader conservation ideals, as they appeal to different protection strategies, and potentially different donors. Although the emotional appeal of releasing porpoises from fishing nets alive outranks backroom lobbying for changes in fishing regulations, the lobbying effort protect more porpoises. The IWC may be deemed more successful by exploiting a range of targets, or less so than a dedicated advocacy group applying all its focus to one issue. They can point to a growth from a modest 3,000 supporters in the beginning, to over 100,000 people supporting the International Wildlife Coalition today.

[*David Duffus*]

RESOURCES

ORGANIZATIONS

International Wildlife Coalition, 70 East Falmouth Highway, East Falmouth, MA USA 02536 (508) 548-8328, Fax: (508) 548-8542, Email: iwchq@iwc.org, <http://www.iwc.org>

Intrinsic value

Saying that an object has intrinsic value means that, even though it has no specific use, market, or monetary value, it nevertheless can be valuable in and of itself and for its own sake. The **Northern spotted owl** (*Strix occidentalis caurina*) for example, has no instrumental or market value; it is not a means to any human end, nor is it sold or traded in any market. But, environmentalists argue, utility and price are not the only measures of worth. Indeed, they say, some of the things humans value most—truth, love, respect—are not for sale at any price, and to try to put a price on them would only tend to cheapen them. Such things have "intrinsic value."

Similarly, environmentalists say, the natural **environment** and its myriad life-forms are valuable in their own right. **Wilderness**, for instance, has intrinsic value and is worthy of protecting for its own sake. To say that something has intrinsic value is not necessarily to deny that it may also have instrumental value for humans and non-human animals alike. Deer, for example, have intrinsic value; but they also have instrumental value as a food source for **wolves** and other predator **species**. *See also* Shadow pricing

Introduced species

Introduced **species** (also called invasive species) are those that have been released by humans into an area to which they are not native. These releases can occur accidently, from places such as the cargo holds of ships. They can also occur intentionally, and species have been introduced for a range of ornamental and recreational uses, as well as for agricultural, medicinal, and **pest** control purposes.

Introduced species can have dramatically unpredictable effects on the **environment** and native species. Such effects can include overabundance of the introduced species, competitive displacement, and disease-caused **mortality** of the native species. Numerous examples of adverse consequences associated with the accidental release of species or the long term effects of deliberately introduced species exist in the United States and around the world. Introduced species can be beneficial as long as they are carefully regulated. Almost all the major varieties of grain and vegetables used in the United States originated in other parts of the world. This includes corn, rice, wheat, tomatoes, and potatoes.

The **kudzu** vine, which is native to Japan, was deliberately introduced into the southern United States for **erosion** control and to shade and feed livestock. It is, however, an extremely aggressive and fast-growing species, and it can form continuous blankets of foliage that cover forested hillsides, resulting in malformed and dead trees. Other species introduced as ornamentals have spread into the wild, displacing or outcompeting native species. Several varieties of cultivated roses, such as the multiflora rose, are serious pests and nuisance shrubs in field and pastures. The **purple loos-**estrife, with its beautiful purple flowers, was originally brought from Europe as a garden ornamental. It has spread rapidly in freshwater **wetlands** in the northern United States, displacing other plants such as cattails. This is viewed with concern by ecologists and **wildlife** biologists since the food value of loosestrife is minimal, while the roots and starchy tubes of cattails are an important food source to muskrats. Common ragweed was accidently introduced to North America, and it is now a major health irritant for many people.

Introduced species are sometimes so successful because human activity has changed the conditions of a particular environment. The Pine Barrens of southern New Jersey form an **ecosystem** that is naturally acidic and low in nutrients. Bogs in this area support a number of slow-growing plant species that are adapted to these conditions, including peat moss, sundews, and pitcher plants. But **urban runoff**, which contain fertilizers, and **wastewater effluent**, which is high in both **nitrogen** and **phosphorus**, have enriched the bogs; the waters there have become less acidic and shown a gradual elevation in the concentration of nutrients. These changes in **aquatic chemistry** have resulted in changes in plant species, and the acidophilus mosses and herbs are being replaced by fast-growing plants that are not native to the Pine Barrens.

Zebra mussels were transported by accident from Europe to the United States, and they are causing severe problems in the **Great Lakes**. They proliferate at a prodigious rate, crowding out native species and clogging industrial and municipal water-intake pipes. Many ecologists fear that shipping traffic will transport the **zebra mussel** to harbors all over the country. Scattered observations of this tiny crustacean have already been made in the lower **Hudson River** in New York.

Although introduced species are usually regarded with concern, they can occasionally be used to some benefit. The **water hyacinth** is an aquatic plant of tropical origin that has become a serious clogging nuisance in lakes, streams, and waterways in the southern United States. Numerous methods of physical and chemical removal have been attempted to eradicate or control it, but research has also established that the plant can improve **water quality**. The water hyacinth has proved useful in the withdrawal of nutrients from sewage and other wastewater. Many constructed wetlands, polishing ponds, and waste lagoons in waste treatment plants now take advantage of this fact by routing wastewater through floating beds of water hyacinth.

The reintroduction of native species is extremely difficult, and it is an endeavor that has had low rates of success. Efforts by the **Fish and Wildlife Service** to reintroduce the endangered **whooping crane** into native **habitat** in the southwestern United States were initially unsuccessful because of the fragility of the eggs, as well as the poor parenting skills of birds raised in captivity. The service then devised a strategy of allowing the more common sandhill crane to incubate the eggs of captive whooping cranes in **wilderness** nests, and the fledglings were then taught survival skills by their surrogate parents. Such projects, however, are extremely time and labor intensive; they are also costly and difficult to implement for large numbers of most species.

Due to the difficulties and expense required to protect native species and to eradicate introduced species, there are not many international laws and policies that seek to prevent these problems before they begin. Thus customs agents at ports and airports routinely check luggage and cargo for live plant and animal materials to prevent the accidental or deliberate transport of non-native species. Quarantine policies are also designed to reduce the **probability** of spreading introduced species, particularly diseases, from one country to another.

There are similar concerns about genetically engineered organisms, and many have argued that their creation and release could have the same devastating environmental consequences as some introduced species. For this reason, the use of bioengineered organisms is highly regulated; both the **Food and Drug Administration** and the **Environmental Protection Agency** (EPA) impose strict controls on the field testing of bioengineered products, as well as on their cultivation and use.

Conservation policies for the protection of native species are now focused on habitats and ecosystems rather than single species. It is easier to prevent the encroachment of introduced species by protecting an entire ecosystem from disturbance, and this is increasingly well recognized both inside and outside the conservation community. *See also* Bioremediation; Endangered species; Fire ants; Gypsy moth; Rabbits in Australia; Wildlife management

[*Usha Vedagiri and Douglas Smith*]

RESOURCES
BOOKS

Common Weeds of the United States. United States Department of Agriculture. New York: Dover Publications, 1971.

Forman, R. T. T., ed. *Pine Barrens: Ecology and Landscape.* New York: Academic Press, 1979.

Inversion

see **Atmospheric inversion**

Iodine 131

A radioactive **isotope** of the element iodine. During the 1950s and early 1960s, iodine-131 was considered a major health hazard to humans. Along with cesium-137 and strontium-90, it was one of the three most abundant isotopes found in the fallout from the atmospheric testing of **nuclear weapons**. These three isotopes settled to the earth's surface and were ingested by cows, ultimately affecting humans by way of dairy products. In the human body, iodine-131, like all forms of that element, tends to concentrate in the thyroid, where it may cause **cancer** and other health disorders. The Chernobyl nuclear reactor explosion is known to have released large quantities of iodine-131 into the **atmosphere**. *See also* Radioactivity

Ion

Forms of ordinary chemical elements that have gained or lost electrons from their orbit around the atomic nucleus and, thus, have become electrically charged. Positive ions (those that have lost electrons) are called *cations* because when charged electrodes are placed in a solution containing ions the positive ions migrate to the cathode (negative electrode). Negative ions (those that have gained extra electrons) are called *anions* because they migrate toward the anode (positive electrode). Environmentally important cations include the **hydrogen** ion (H^+) and dissolved metals. Important anions include the hydroxyl ion (OH^-) as well as many of the dissolved ions of nonmetallic elements. *See also* Ion exchange; Ionizing radiation

Ion exchange

The process of replacing one **ion** that is attached to a charged surface with another. A very important type of ion exchange is the exchange of cations bound to **soil** particles. Soil **clay minerals** and organic matter both have negative surface charges that bind cations. In a fertile soil the predominant exchangeable cations are Ca^{2+}, Mg^{2+} and K^+. In **acid** soils Al^{3+} and H^+ are also important exchangeable ions. When materials containing cations are added to soil, cations **leaching** through the soil are retarded by cation exchange.

Ionizing radiation

High-energy radiation with penetrating competence such as x rays and gamma rays which induces ionization in living material. Molecules are bound together with covalent bonds, and generally an even number of electrons binds the atoms together. However, high-energy penetrating radiation can fragment molecules resulting in atoms with unpaired electrons known as "free radicals." The ionized "free radicals" are exceptionally reactive, and their interaction with the macromolecules (DNA, RNA, and proteins) of living cells can, with high dosage, lead to cell death. Cell damage (or death) is a function of penetration ability, the kind of cell exposed, the length of exposure, and the total dose of ionizing radiation. Cells that are mitotically active and have a high oxygen content are most vulnerable to ionizing radiation. *See also* Radiation exposure; Radiation sickness; Radioactivity

Iron minerals

The oxides and hydroxides of ferric iron (Fe(III)) are very important minerals in many soils, and are important suspended solids in some fresh water systems. Important oxides and hydroxides of iron include goethite, hematite, lepidocrocite, and ferrihydrite.

These minerals tend to be very finely divided and can be found in the clay-sized fraction of soils, and like other clay-sized minerals, are important adsorbers of ions. At high **pH** they adsorb hydroxide (OH⁻) ions creating negatively charged surfaces that contribute to cation exchange surfaces. At low pH they adsorb **hydrogen** (H⁺) ions, creating anion exchange surfaces. In the pH range between 8 and 9 the surfaces have little or no charge. Iron hydroxide and oxide surfaces strongly adsorb some environmentally important anions, such as phosphate, arsenate and selanite, and cations like **copper**, **lead**, manganese and chromium. These ions are not exchangeable, and in environments where iron oxides and hydroxides are abundant, surface **adsorption** can control the mobility of these strongly adsorbed ions.

The hydroxides and oxides of iron are found in the greatest abundance in older highly weathered landscapes. These minerals are very insoluble and during **soil weathering** they form from the iron that is released from the structure of the soil-forming minerals. Thus, iron oxide and hydroxide minerals tend to be most abundant in old landscapes that have not been affected by **glaciation**, and in landscapes where the rainfall is high and the rate of soil mineral weathering is high. These minerals give the characteristic red (hematite or ferrihydrite) or yellow-brown (goethite) colors to soils that are common in the tropics and subtropics. *See also* Arsenic; Erosion; Ion exchange; Phosphorus; Soil profile; Soil texture

Irradiation of food
see **Food irradiation**

Irrigation

Irrigation is the method of supplying water to land to support plant growth. This technology has had a powerful role in the history of civilization. In **arid** regions sunshine is plentiful and **soil** is usually fertile, so irrigation supplies the critical factor needed for plant growth. Yields have been high, but not without costs. Historic problems include **salinization** and water **logging**; contemporary difficulties include immense costs, spread of water-borne diseases, and degraded aquatic environments.

One geographer described California's Sierra Nevada as the "mother nurse of the San Joaquin Valley." Its heavy winter snowpack provides abundant and extended **runoff** for the rich valley soils below. Numerous irrigation districts, formed to build diversion and storage **dams**, supply water through gravity-fed canals. The snow melt is low in nutrients, so salinization problems are minimal. Wealth from the lush fruit orchards has enriched the state.

By contrast, the **Colorado River**, like the Nile, flows mainly through arid lands. Deeply incised in places, the river is also limited for irrigation by the high salt content of **desert** tributaries. Still, demand for water exceeds supply. Water crossing the border into Mexico is so saline that the federal government has built a **desalinization** plant at Yuma, Arizona. Colorado River water is imperative to the Imperial Valley, which specializes in winter produce in the rich, delta soils. To reduce salinization problems, one-fifth of the water used must be drained off into the growing Salton Sea.

Salinization and water logging have long plagued the Tigris, Euphrates, and Indus River flood plains. Once fertile areas of Iraq and Pakistan are covered with salt crystals. Half of the irrigated land in our western states is threatened by salt buildup.

Some of the worst problems are degraded aquatic environments. The **Aswan High Dam** in Egypt has greatly amplified surface evaporation, reduced nutrients to the land and to fisheries in the delta, and has contributed to the spread of **schistosomiasis** via water snails in irrigation ditches. Diversion of **drainage** away from the **Aral Sea** for cotton irrigation has severely lowered the shoreline, and threatens this water body with ecological disaster.

Spray irrigation in the High Plains is lowering the Ogallala Aquifer's **water table**, raising pumping costs. Kesterson Marsh in the San Joaquin Valley has become a hazard to **wildlife** because of selenium **poisoning** from irrigation drainage. The federal **Bureau of Reclamation** has invested huge sums in dams and reservoirs in western states. Some question the wisdom of such investments, given the past century of farm surpluses, and argue that water users are not paying the true cost.

A farm irrigation system. (U. S. Geological Survey
Reproduced by permission.)

Irrigation still offers great potential, but only if used
with wisdom and understanding. New technologies may yet
contribute to the world's ever-increasing need for food. *See
also* Climate; Commercial fishing; Reclamation

[*Nathan H. Meleen*]

RESOURCES

BOOKS

Huffman, R. E. *Irrigation Development and Public Water Policy*. New York:
Ronald Press, 1953.
Powell, J. W. "The Reclamation Idea." In *American Environmentalism:
Readings in Conservation History*. 3rd ed., edited by R. F. Nash. New York:
McGraw-Hill, 1990.
Wittfogel, K. A. "The Hydraulic Civilizations." In *Man's Role in Changing
the Face of the Earth*, edited by W. L. Thomas Jr. Chicago: University of
Chicago Press, 1956.
Zimmerman, J. D. *Irrigation*. New York: Wiley, 1966.

OTHER

U.S. Department of Agriculture. *Water: 1955 Yearbook of Agriculture*. Washington, DC: U.S. Government Printing Office, 1955.

Island biogeography

Island **biogeography** is the study of past and present animal
and plant distribution patterns on islands and the processes

that created those distribution patterns. Historically, island
biogeographers mainly studied geographic islands—continental islands close to shore in shallow water and oceanic
islands of the deep sea. In the last several decades, however,
the study and principles of island biogeography have been
extended to ecological islands such as forests and **prairie**
fragments isolated by human development. Biogeographic
"islands" may also include ecosystems isolated on mountaintops and landlocked bodies of water such as Lake Malawi
in the African Rift Valley. Geographic islands, however,
remain the main laboratories for developing and testing the
theories and methods of island biogeography.

Equilibrium theory

Until the 1960s, biogeographers thought of islands as
living museums—relict (persistent remnant of an otherwise
extinct **species** of plant or animal) scraps of mainland ecosystems in which little changed—or closed systems mainly
driven by **evolution**. That view began to radically change
in 1967 when Robert H. MacArthur and Edward O. Wilson
published *The Theory of Island Biogeography*.

In their book, MacArthur and Wilson detail the equilibrium theory of island biogeography—a theory that became
the new paradigm of the field. The authors proposed that
island ecosystems exist in dynamic equilibrium, with a steady
turnover of species. Larger islands—as well as islands closest
to a source of immigrants—accommodate the most species
in the equilibrium condition, according to their theory. Mac-
Arthur and Wilson also worked out mathematical models
to demonstrate and predict how island area and isolation
dictate the number of species that exist in equilibrium.

Dispersion

The driving force behind species distribution is dispersion—the means by which plants and animals actively leave
or are passively transported from their source area. An island
ecosystem can have more than one source of colonization,
but nearer sources dominate. How readily plants or animals
disperse is one of the main reasons equilibrium will vary
from species to species.

Birds and **bats** are obvious candidates for anemochory
(dispersal by air), but some species normally not associated
with flight are also thought to reach islands during storms
or even normal wind currents. Orchids, for example, have
hollow seeds that remain airborne for hundreds of kilometers. Some small spiders, along with other insects like bark
lice, aphids, and ants (collectively knows as aerial **plankton**)
often are among the first pioneers of newly formed islands.

Whether actively swimming or passively floating on
logs or other debris, dispersal by sea is called thallasochory.
Crocodiles have been found on Pacific islands 600 miles
(950 km) from their source areas, but most amphibians,
larger terrestrial reptiles, and, in particular, mammals, have
difficulty crossing even narrow bodies of water. Thus, thalla-

sochory is the medium of dispersal primarily for fish, plants, and insects. Only small vertebrates such as lizards and snakes are thought to arrive at islands by sea on a regular basis.

Zoochory is transport either on or inside an animal. This method is primarily a means of plant dispersal, mostly by birds. Seeds ride along either stuck to feathers or survive passage through a bird's digestive tract and are deposited in new territory.

Anthropochory is dispersal by human beings. Although humans intentionally introduce domestic animals to islands, they also bring unintended invaders, such as rats.

Getting to islands is just the first step, however. Plants and animals often arrive to find harsh and alien conditions. They may not find suitable habitats. Food chains they depend on might be missing. Even if they manage to gain a foothold, their limited numbers make them more susceptible to **extinction**. Chances of success are better for highly adaptable species and those that are widely distributed beyond the island. Wide distribution increases the likelihood a species on the verge of extinction may be saved by the rescue effect, the replenishing of a declining population by another wave of immigration.

Challenging established theories

Many biogeographers point out that isolated ecosystems are more than just collections of species that can make it to islands and survive the conditions they encounter there. Several other contemporary theories of island biogeography build on MacArthur and Wilson's theory; other theories contradict it.

Equilibrium theory suggests that species turnover is constant and regular. Evidence collected so far indicates MacArthur and Wilson's model works well in describing communities of rapid dispersers who have a regular turnover, such as insects, birds, and fish. However, this model may not apply to species who disperse more slowly.

Proponents of historical legacy models argue that communities of larger animals and plants (forest trees, for example) take so long to colonize islands that changes in their populations probably reflect sudden climactic or geological upheaval rather than a steady turnover. Other theories suggest that equilibrium may not be dynamic, that there is little or no turnover. Through **competition**, established species keep out new colonists; the newcomers might occupy the same ecological niches as their predecessors. Established species may also evolve and adapt to close off those niches. Island resources and habitats may also be distinct enough to limit immigration to only a few well-adapted species.

Thus, in these later models, dispersal and colonization are not nearly as random as in MacArthur and Wilson's model. These less random, more deterministic theories of island ecosystems conform to specific assembly rules—a complex list of factors accounting for the species present in the source areas, the niches available on islands, and competition between species.

Some biogeographers suggest that every island—and perhaps every **habitat** on an island—may require its own unique model. Human disruption of island ecosystems further clouds the theoretical picture. Not only are habitats permanently altered or lost by human intrusion, but anthropochory also reduces an island's isolation. Thus, finding relatively undisturbed islands to test different theories can be difficult.

Since the time of naturalists Charles Darwin and his colleague, Alfred Wallace, islands have been ideal "natural laboratories" for studying evolution. Patterns of evolution stand out on islands for two reasons: island ecosystems tend to be simpler than other geographical regions, and they contain greater numbers of **endemic species**, plant, and animal species occurring only in a particular location.

Many island endemics are the result of adaptive radiation—the evolution of new species from a single lineage for the purpose of filling unoccupied ecological niches. Many species from mainland source areas simply never make it to islands, so species that can immigrate find empty ecological niches where once they faced competition. For example, monitor lizards immigrating to several small islands in Indonesia found the **niche** for large predators empty. Monitors on these islands evolved into Komodo Dragons, filling the niche.

Conservation of biodiversity

Theories of island biogeography also have potential applications in the field of **conservation**. Many conservationists argue that as human activity such as **logging** and ranching encroach on wild lands, remaining parks and reserves begin to resemble small, isolated islands. According to equilibrium theory, as those patches of wild land grow smaller, they support fewer species of plants and animals. Some conservationists fear that plant and animal populations in those parks and reserves will sink below minimum viable population levels—the smallest number of individuals necessary to allow the species to continue reproducing. These conservationists suggest that one way to bolster populations is to set aside larger areas and to limit species isolation by connecting parks and preserves with **wildlife** corridors.

Islands with greatest variety of habitats support the most species; diverse habitats promotes successful dispersal, survival, and reproduction. Thus, in attempting to preserve island **biodiversity**, conservationists focus on several factors: the size (the larger the island, the more habitats it contains), **climate**, geology (**soil** that promotes or restricts habitats), and age of the island (sparse or rich habitats). All of these factors must be addressed to ensure island biodiversity.

[Darrin Gunkel]

RESOURCES

BOOKS

Harris, Larry D. *The Fragmented Forest: Island Biogeography Theory and the Preservation of Biotic Diversity.* Chicago: University of Chicago Press, 1984.

Mac Arthur, Robert H., and Edward O. Wilson. *The Theory of Island Biogeography.* Princeton: Princeton University Press, 1967.

Quaman, David. *Song of the Dodo: Island Biogeography in an Age of Extinction.* New York: Scribner, 1996.

Whittaker, Robert J. *Island Biogeography: Ecology, Evolution and Conservation.* London: Oxford University Press, 1999.

PERIODICALS

Grant, P. R. "Competition Exposed by Knight?" *Nature* 396: 216–217.

OTHER

"Island Biogeography." *University of Oxford School of Geography and the Environment.* August 7, 2000 [cited June 26, 2002]. <http://www.geog.ox.-ac.uk/research/bie/islandbio/index.html>.

ORGANIZATIONS

Environmental Protection Agency (EPA), 1200 Pennsylvania Avenue, NW, Washington, DC USA 20460 (202) 260-2090, Email: public-access@epa.gov, <http://www.epa.gov>

ISO 14000: International Environmental Management Standards

ISO 14000 refers to a series of environmental management standards that were adopted by the International Organization for Standardization (ISO) in 1996 and are beginning to be implemented by businesses across the world. Any person or organization interested in the **environment** and in the goal of improving environmental performance of businesses should be interested in ISO 14000. Such interested parties include businesses themselves, their legal representatives, environmental organizations and their members, government officials, and others.

What is the ISO and what are ISO standards?

The International Organization for Standardization (ISO) is a private (nongovernmental) worldwide organization whose purpose is to promote uniform standards in international trade. Its members are elected representatives from national standards organizations in 111 countries. The ISO covers all fields involving promoting goods, services, or products and where a Member Body suggests that standardization is desirable, with the exception of electrical and electronic engineering, which are covered by a different organization called the International Electrotechnical Commission (IEC). However, the ISO and the IEC work closely together.

Since the ISO began operations in 1947, its Central Secretariat has been located in Geneva, Switzerland. Between 1951 (when it published its first standard) and 1997, the ISO issued over 8,800 standards. Standards are docu-

ments containing technical specifications, rules, guidelines, and definitions to ensure equipment, products, and services perform as specified.

Among the best known standards published by the ISO are those that comprise the ISO 9000 series, which was developed between 1979–1986, and published in 1987. Because ISO 9000 is a forerunner to ISO 14000, it is important to understand the basic structure and function of ISO 9000. The ISO 9000 series is a set of standards for quality management and quality assurance. The standards apply to processes and systems that produce products; they do not apply to the products themselves. Further, the standards provide a general framework for any industry; they are not industry-specific. A company that has become registered under ISO 9000 has demonstrated that it has a documented system for quality that is in place and consistently applied. ISO 9000 standards apply to all kinds of companies whether large or small, in services or manufacturing.

The latest major set of standards published by the ISO is the ISO 14000 series. The impetus for that series came from the United Nations Conference on the Environment and Development (UNCED), which was held in Rio De Janeiro in 1992 and attended by representatives of over one hundred nations. One of the documents resulting from that conference was the *Global Environmental Initiative*, which prompted the ISO to develop its ISO 14000 series of international environmental standards. The ISO's goal is to insure that businesses adopt common internal procedures for environmental controls including, but not limited to, audits. It is important to note that the standards are process standards, not performance standards. The goal is to ensure that businesses are in compliance with their own national and local applicable environmental laws and regulations. The initial standards in the series include numbers 14001, 14004, and 14010-14012; all of them adopted by the ISO in 1996.

Provisions of ISO 14000 Series standards

ISO 14000 sets up criteria pursuant to which a company may become registered or certified as to its environmental management practices.

Central to the process of registration pursuant to ISO 14000 is a company's Environmental Management System (EMS). The EMS is a set of procedures for assessing compliance with environmental laws and company procedures for environmental protection, identifying and resolving with problems, and engaging the company's workforce in a commitment to improved environmental performance by the company.

ISO 14001 series can be divided into two groups: guidance documents and specification documents. The series sets out standards against which a company's EMS will be evaluated. For example, it must include an accurate summary of the legal standards with which the company must comply, such as permit stipulations, and relevant provisions of statutes and regulations, and even provisions of administrative or court-certified consent judgments. To become certified, the EMS must: (1) include an **environmental policy**; (2) establish plans to meet environmental goals and comply with legal requirements; (3) provide for implementation of the policy and operation under it including training for personnel, communication, and document control; (4) set up monitoring and measurement devices and an audit procedure to insure continuing improvement; and (5) provide for management review. The EMS must be certified by a registrar who has been qualified under ISO 13012, a standard that predates the ISO 14000 series.

The ISO 14004 series is a guidance document that gives advice that may be followed but is not required. It includes five principles each of which corresponds to one of the five areas of ISO 14001 listed above.

ISO 14010, 14011, and 14012 are auditing standards. For example, 14010 covers general principles of **environmental auditing**, and 14011 provides guidelines for auditing of an Environmental Management System (EMS). ISO 14012 provides guidelines for establishing qualifications for environmental auditors, whether those auditors are internal or external.

Plans for additional standards within the ISO 14000 Series standards

The ISO is considering proposals for standards on training and certifying independent auditors (called registrars) who will certify that ISO 14000-certified business have established and adhere to stringent internal systems to monitor and improve their own environmental protection actions. Later the ISO may also establish standards for assessing a company's environmental performance. Standards may be adopted to for use of eco-labeling and **life cycle assessment** of goods involved in international trade.

Benefits and consequences of ISO 14000 Series standards

A company contemplating obtaining ISO 14000 registration must evaluate its potential advantages as well as its costs to the company.

ISO 14000 certification may bring various rewards to companies. For example, many firms are hoping that, in return for obtaining ISO 14000 certification (and the actions required to do so), regulatory agencies such as the U.S. **Environmental Protection Agency** (EPA) will give them more favorable treatment. For example, leniency might be shown in less stringent filing or monitoring requirements or even less severe sanctions for past or present violations of environmental statutes and regulations.

Further, compliance may be merely for good public relations, leading consumers to view the certified company

as a good corporate citizen that works to protect the environment. There is public pressure on companies to demonstrate their environmental stewardship and accountability; obtaining ISO 14000 certification is one way to do so.

The costs to the company will depend on the scope of the Environmental Management System (EMS). For example, the EMS might be international, national, or limited to individual plants operated by the company. That decision will affect the costs of the environmental audit considerably. National and international systems may prove to be costly.

On the other hand, a company may realize cost savings. For example, an insurance company may give reduced rates on insurance to cover accidental **pollution** releases to a company that has a proven environmental management system in place. Internally, by implementing an EMS, a company may realize cost savings as a result of **waste reduction**, use of less fewer toxic **chemicals**, less energy use, and **recycling**.

A major legal concern raised by lawyers studying the ISO 14000 standards relates to the question of confidentiality. There are serious questions as to whether a governmental regulatory agency can require disclosure of information discovered during a self-audit by a company. The use of a third-party auditor during preparation of the EMS process may weaken a company's argument that information discovered is privileged.

ISO 14000 has potential consequences with respect to international law as well as international trade. ISO 14000 is intended to promote a series of universally accepted EMS practices and lead to consistency in environmental standards between and among trading partners. Some developing countries such as Mexico are reviewing ISO 14000 standards and considering incorporating their provisions within their own environmental laws and regulations. On the other hand, some developing countries have suggested that environmental standards created by ISO 14000 may constitute non-tariff barriers to trade in that costs of ISO 14000 registration may be prohibitively high for small- to medium-size companies.

Companies that have implemented ISO 9000 have learned to view their companies' operations through a "quality of management" lens and implementation of ISO 14000 may lead to use of "environmental quality" lens. ISO 14000 has the potential to lead to two kinds of cultural changes. First, within the corporation, it has the potential to lead to consideration of environmental issues throughout the company and its business decisions ranging from hiring of employees to marketing. Second, ISO 14000 has the potential to become part of a global culture as the public comes to view ISO 14000 certification as a benchmark connoting good environmental stewardship by a company.

[*Paulette L. Stenzel*]

RESOURCES

BOOKS

Tibor, T., and I. Feldman. *ISO 14000: A Guide to the New Environmental Management Standards.* Irwin Publishing Company, 1996.

von Zharen, W. M. *ISO 14000: Understanding the Environmental Standards.* Government Institutes, 1996.

PERIODICALS

Kass, S. L. "The Lawyer's Role in Implementing ISO 14000." *Natural Resources & Environment* 3, no. 5 (Spring 1997).

Isotope

Different forms of atoms of the same element. Atoms consist of a nucleus, containing positively-charged particles (protons) and neutral particles (neutrons), surrounded by negatively-charged particles (electrons). Isotopes of an element differ only in the number of neutrons in the nucleus and hence in atomic weight. The nuclei of some isotopes are unstable and undergo **radioactive decay**. An element can have several stable and radioactive isotopes, but most elements have only two or three isotopes that are of any importance. Also, for most elements the radioactive isotopes are only of concern in material exposed to certain types of radiation sources. **Carbon** has three important isotopes with atomic weights of 12, 13, and 14. C-12 is stable and represents 98.9% of natural carbon. C-13 is also stable and represents 1.1% of natural carbon. C-14 represents an insignificant fraction of naturally-occurring carbon, but it is radioactive and important because its radioactive decay is valuable in the dating of fossils and ancient artifacts. It is also useful in tracing the reactions of carbon compounds in research. *See also* Nuclear fission; Nuclear power; Radioactivity; Radiocarbon dating

Itai-itai disease

The symptoms of Itai-Itai disease were first observed in 1913 and characterized between 1947 and 1955; it was 1968, however, before the Japanese Ministry of Health and Welfare officially declared that the disease was caused by chronic **cadmium poisoning** in conjunction with other factors such as the stresses of pregnancy and lactation, aging, and dietary deficiencies of vitamin D and calcium. The name arose from the cries of pain, "itai-itai" (ouch-ouch) by the most seriously stricken victims, older Japanese farm women. Although men, young women, and children were also exposed, 95% of the victims were post-menopausal women over 50 years of age. They usually had given birth to several children and had lived more than 30 years within 2 mi (3 km) of the lower stream of the Jinzu River near Toyama.

The disease started with symptoms similar to rheumatism, neuralgia, or neuritis. Then came bone lesions, osteomalacia, and osteoporosis, along with renal disfunction and proteinuria. As it escalated, pain in the pelvic region caused the victims to walk with a duck-like gait. Next, they were incapable of rising from their beds because even a slight strain caused bone fractures. The suffering could last many years before it finally ended with death. Overall, an estimated 199 victims have been identified, of which 162 had died by December 1992.

The number of victims increased during and after World War II as production expanded at the Kamioka Mine owned by the Mitsui Mining and Smelting Company. As 3,000 tons of zinc-lead ore per day were mined and smelted, cadmium was discharged in the **wastewater**. Downstream, farmers withdrew the fine particles of **flotation tailings** in the Jinzu River along with water for drinking and crop **irrigation**. As rice plants were damaged near the irrigation inlets, farmers dug small **sedimentation** pools that were ineffective against the nearly invisible poison.

Both the numbers of Itai-Itai disease patients and the damage to the rice crops rapidly decreased after the mining company built a large settling basin to purify the wastewater in 1955. However, even after the **discharge** into the Jinzu River was halted, the cadmium already in the rice paddy soils was augmented by airborne exhausts. Mining operations in several other Japanese prefectures also produced cadmium-contaminating rice, but afflicted individuals were not certified as Itai-Itai patients. That designation was applied only to those who lived in the Jinzu River area.

In 1972 the survivors and their families became the first **pollution** victims in Japan to win a lawsuit against a major company. They won because in 1939 Article 109 of the Mining Act had imposed strict liability upon mining facilities for damages caused by their activities. The plaintiffs had only to prove that cadmium discharged from the mine caused their disease, not that the company was negligent. As epidemiological proof of causation sufficed as legal proof in this case, it set a precedent for other pollution litigation as well.

Despite legal success and compensation, the problem of contaminated rice continues. In 1969 the government initially set a maximum allowable standard of 0.4 **parts per million** (ppm) cadmium in unpolished rice. However, because much of the contaminated farmland produced grain in excess of that level, in 1970 under the Foodstuffs Hygiene Law this was raised to 1 ppm cadmium for unpolished rice and 0.9 ppm cadmium for polished rice. To restore contaminated farmland, Japanese authorities instituted a program in which, each year, the most highly contaminated soils in a small area are exchanged for uncontaminated soils. Less contaminated soils are rehabilitated through the addition of lime, phosphate, and a cadmium sequestering agent, EDTA.

By 1990 about 10,720 acres (4,340 ha), or 66.7% of the approximately 16,080 acres (6,510 ha) of the most highly cadmium contaminated farmland had been restored. In the remaining contaminated areas where farm families continue to eat homegrown rice, the symptoms are alleviated by treatment with massive doses of vitamins B1, B12, D, calcium, and various hormones. New methods have also been devised to cause the cadmium to be excreted more rapidly. In addition, the high costs of compensation and restoration are leading to the conclusion that prevention is not only better but cheaper. This is perhaps the most encouraging factor of all. *See also* Bioaccumulation; Environmental law; Heavy metals and heavy metal poisoning; Mine spoil waste; Smelter; Water pollution

[*Frank M. D'Itri*]

RESOURCES
BOOKS

Kobayashi, J. "Pollution by Cadmium and the Itai-Itai Disease in Japan." In *Toxicity of Heavy Metals in the Environment, Part 1*, edited by F. W. Oehme. New York: Marcel Dekker, 1978.
Kogawa, K. "Itai-Itai Disease and Follow-Up Studies." In *Cadmium in the Environment, Part II*, edited by J. O. Nriagu. New York: Wiley, 1981.
Tsuchiya, K., ed. *Cadmium Studies in Japan: A Review*. Tokyo, Japan, and Amsterdam, Netherlands: Kodansha and Elsevier/North-Holland Biomedical Press, 1978.

IUCN—The World Conservation Union

Founded in 1948 as the International Union for the Conservation of Nature and Natural Resources (IUCN), IUCN works with governments, conservation organizations, and industry groups to conserve **wildlife** and approach the world's environmental problems using "sound scientific insight and the best available information." Its membership, currently over 980, comes from 140 countries and includes 56 sovereign states, as well as government agencies and nongovernmental organizations. IUCN exists to serve its members, representing their views and providing them with the support necessary to achieve their goals. Above all, IUCN works with its members "to achieve development that is sustainable and that provides a lasting improvement in the quality of life for people all over the world." IUCN's three basic conservation objectives are: (1) to secure the conservation of nature, and especially of biological diversity, as an essential foundation for the future; (2) to ensure that where the earth's natural resources are used this is done in a wise, equitable, and sustainable way; (3) to guide the development of human communities toward ways of life that are both of

good quality and in enduring harmony with other components of the **biosphere**.

IUCN is one of the few organizations to include both governmental agencies and nongovernmental organizations. It is in a unique position to provide a neutral forum where these organizations can meet, exchange ideas, and build partnerships to carry out conservation projects. IUCN is also unusual in that it both develops environmental policies and then implements them through the projects it sponsors. Because the IUCN works closely with, and its membership includes, many government scientists and officials, the organization often takes a conservative, pro-management, as opposed to a "preservationist," approach to wildlife issues. It may encourage or endorse limited **hunting** and commercial exploitation of wildlife if it believes this can be carried out on a sustainable basis.

IUCN maintains a global network of over 5,000 scientists and wildlife professionals who are organized into six standing commissions that deal with various aspects of the union's work. There are commissions on Ecology, Education, Environmental Planning, **Environmental Law**, National Parks and Protected Areas, and **Species** Survival. These commissions create action plans, develop policies, advise on projects and programs, and contribute to IUCN publications, all on an unpaid, voluntary basis.

IUCN publishes an authoritative series of "Red Data Books," describing the status of rare and endangered wildlife. Each volume provides information on the population, distribution, **habitat** and ecology, threats, and protective measures in effect for listed species. The "Red Data Books" concept was originated in the mid-1960s by the famous British conservationist Sir Peter Scott, and the series now includes a variety of publication on regions and species. Other titles in the series of "Red Data Books" include *Dolphins, Porpoises, and Whales of the World*; *Lemurs of Madagascar and the Comoros*; *Threatened Primates of Africa*; *Threatened Swallowtail Butterflies of the World*; *Threatened Birds of the Americas*; and books on plants and other species of wildlife, including a series of conservation action plans for threatened species.

Other notable IUCN works include *World Conservation Strategy: Living Resources Conservation for Sustainable Development* and its successor document *Caring for the Earth—A Strategy for Sustainable Living*; and the *United Nations List of Parks and Protected Areas*. IUCN also publishes books and papers on regional conservation, habitat preservation, environmental law and policy, ocean ecology and management, and conservation and development strategies.

[*Lewis G. Regenstein*]

RESOURCES

ORGANIZATIONS

IUCN—The World Conservation Union Headquarters, Rue Mauverney 28, Gland, Switzerland CH-1196 ++41 (22) 999-0000, Fax: ++41 (22) 999-0002, Email: mail@hq.iucn.org, <http://www.iucn.org>

Ivory-billed woodpecker

The ivory-billed woodpecker (*Campephilus principalis*) is one of the rarest birds in the world and is considered by most authorities to be extinct in the United States. The last confirmed sighting of ivory-bills was in Cuba in 1987 or 1988. Though never common, the ivory-billed woodpecker was rarely seen in the United States after the first years of the twentieth century. Some were seen in Louisiana in 1942, and since then, occasional sightings have been unverified. Interest in the bird rekindled in 1999, when a student at Louisiana State University claimed to have seen a pair of ivory-billed woodpeckers in a **wilderness** preserve. Teams of scientists searched the area for two years. No ivory-billed woodpecker was sighted, though some evidence made it plausible the bird was in the vicinity. By mid-2002, the ivory-billed woodpecker's return from the brink of **extinction** remained a tantalizing possibility, but not an established fact.

The ivory-billed woodpecker was a huge bird, averaging 19–20 in (48–50 cm) long, with a wingspan of over 30 in (76 cm). The ivory-colored bills of these birds were prized as decorations by native Americans. The naturalist John James Audubon found ivory-billed woodpecker in swampy forest edges in Texas in the 1830s. But by the end of the nineteenth century, the majority of the bird's prime **habitat** had been destroyed by **logging**. Ivory-billed woodpeckers required large tracts of land in the bottomland cypress, oak, and black gum forests of the Southeast, where they fed off insect larva in mature trees. This **species** was the largest woodpecker in North America, and they preferred the largest of these trees, the same ones targeted by timber companies as the most profitable to harvest. The territory for breeding pairs of ivory-billed woodpeckers consists of about three square miles of undisturbed, swampy forest, and there was little prime habitat left for them after 1900, for most of these areas had been heavily logged. By the 1930s, one of the only virgin cypress swamps left was the Singer Tract in Louisiana, an 80,000-acre (32,375-ha) swathe of land owned by the Singer Sewing Machine Company. In 1935 a team of ornithologists descended on it to locate, study, and record some of the last ivory-billed woodpeckers in existence. They found the birds and were able to film and photograph them, as well as make the only sound recordings of them in existence. The Audubon Society, the state of Louisiana, and the U.S. **Fish and Wildlife Service** tried to buy the land from Singer to make it a refuge for the rare birds. But Singer

had already sold timber rights to the land. During World War II, when demand for lumber was particularly high, the Singer Tract was leveled. One of the giant cypress trees that was felled contained the nest and eggs of an ivory-billed woodpecker. Land that had been virgin forest then became soybean fields.

Few sightings of these woodpeckers were made in the 1940s, and none exist for the 1950s. But in the early 1960s ivory-billed woodpeckers were reported seen in South Carolina, Texas, and Louisiana. Intense searches, however, left scientists with little hope by the end of that decade, as only six birds were reported to exist. Subsequent decades yielded a few individual sightings in the United States, but none were confirmed.

In 1985 and 1986, there was a search for the Cuban subspecies of the ivory-billed woodpecker. The first expedition yielded no birds, but trees were found that had apparently been worked by the birds. The second expedition found at least one pair of ivory-billed woodpeckers. Most of the land formerly occupied by the Cuban subspecies was cut over for sugar cane plantations by the 1920s, and surveys in 1956 indicated that this population had declined to about a dozen birds. The last reported sightings of the species occurred in the Sierra de Moa area of Cuba. They are still considered to exist there, but the health of any remaining individuals must be in question, given the **inbreeding** that must occur with such a low population level and the fact that so little suitable habitat remains.

In 1999, a student at Louisiana State University (LSU) claimed to have seen a pair of ivory-billed woodpeckers while he was **hunting** for turkey in the Pearl River **Wildlife Management** Area near the Louisiana-Mississippi border. The student, David Kulivan, was a credible witness, and he soon convinced ornithologists at LSU to search for the birds. News of the sighting attracted thousands of amateur and professional birders over the next two years. Scientists from LSU, Cornell University, and the Louisiana Department of **Wildlife** and Fisheries organized an expedition that included posting of high-tech listening devices. Over more than two years, no one else saw the birds, though scientists found trunks stripped of bark, characteristic of the way the ivory-billed woodpecker feeds, and two groups heard the distinct double rapping sound the ivory-billed woodpecker makes when it knocks a trunk. No one heard the call of the ivory-billed woodpecker, though this sound would have been considered definitive evidence of the ivory-billed woodpecker's existence. Hope for confirmation of Kulivan's sighting rested on deciphering the tapes made by a dozen recording devices. This was being done at Cornell University, and was expected to take years.

By mid-2002, the search for the ivory-billed woodpecker in Louisiana had wound down, disappointingly in-

Ivory-billed woodpecker (*Campephilus*). (Photograph by James Tanner. Photo Researchers Inc. Reproduced by permission.)

conclusive. While some scientists remained skeptical about the sighting, others believed that the forest in the area may have regrown enough to support an ivory-billed woodpecker population. *See also* Deforestation; Endangered species; Extinction; International Council for Bird Preservation; Wildlife management

[*Eugene C Beckham*]

RESOURCES
BOOKS
Collar, N. J., et al. *Threatened Birds of the Americas: The ICBP/IUCN Red Data Book.* Washington, DC: Smithsonian Institution Press, 1992.
Ehrlich, P. R., D. S. Dobkin, and D. Wheye. *The Birder's Handbook.* New York: Simon & Schuster, 1988.
Ehrlich, P. R., D. S. Dobkin, and D. Wheye. *Birds in Jeopardy: The Imperiled and Extinct Birds of the United States and Canada, Including Hawaii and Puerto Rico.* Stanford: Stanford University Press, 1992.

PERIODICALS
Gorman, James. "Listening for the Call of a Vanished Bird" *New York Times*, March 5, 2002, F1.
Graham, Frank Jr. "Is the Ivorybill Back?" *Audubon* (May/June 2000): 14.

Pianin, Eric. "Scientists Give Up Search for Woodpecker; Some Signs Noted of Ivory-Billed Bird Not Seen Since '40s" *Washington Post*, February 21, 2002, A2.
Tomkins, Shannon. "Dead or Alive?" *Houston Chronicle*, April 14, 2002, 8.

Izaak Walton League

In 1922, 54 sportsmen and sportswomen—all concerned with the apparent destruction of American fishing waterways—established the Izaak Walton League of America (IWLA). They looked upon Izaak Walton, a seventeenth-century English fisherman and author of *The Compleat Angler*, as inspiration in protecting the waters of America. The Izaak Walton League has since widened its focus: as a major force in the American **conservation** movement, IWLA now pledges in its slogan "to defend the nation's **soil**, air, woods, water, and wildlife."

When sportsmen and sportswomen formed IWLA approximately 70 years ago, they worried that American industry would ruin fishing streams. Raw sewage, soil **erosion**, and rampant **pollution** threatened water and **wildlife**.

Initially the League concentrated on preserving lakes, streams, and rivers. In 1927, at the request of President Calvin Coolidge, IWLA organized the first national **water pollution** inventory. Izaak Walton League members (called "Ikes") subsequently helped pass the first national water **pollution control** act in the 1940s. In 1969 IWLA instituted the Save Our Streams program, and this group mobilized forces to pass the groundbreaking **Clean Water Act** of 1972. The League did not only concentrate on the preservation of American waters, however. From its 1926 campaign to protect the black bass, to the purchase of a helicopter in 1987 to help game law officers protect waterfowl from poachers in the Gulf of Mexico, IWLA has also been instrumental in the preservation of wildlife. In addition, the League has fought to protect public lands such as the National Elk Refuge in Wyoming, the **Everglades National Park**, and the Isle Royale National Park.

IWLA currently sponsors several environmental programs designed to conserve **natural resources** and educate the public. The aforementioned Save Our Streams (SOS) program is a grassroots organization designed to monitor **water quality** in streams and rivers. Through 200 chapters nationwide, SOS promotes "stream rehabilitation" through stream adoption kits and water pollution law training. Another program, **Wetlands** Watch, allows local groups to purchase, adopt, and protect nearby wetlands. Similarly, the Izaak Walton League Endowment buys land to save it from unwanted development. IWLA's Uncle Ike Youth Education program aims to educate children and convince them of the necessity of preserving the **environment**. A last major program from the League is its internationally acclaimed Outdoor Ethics program. Outdoor Ethics works to stop **poaching** and other illegal and unsportsmanlike outdoor activities by educating hunters, anglers, and others.

The League also sponsors and operates regional conservation efforts. Its Midwest Office, based in Minnesota, concentrates on preservation of the Upper Mississippi River region. The **Chesapeake Bay** Program is a major regional focus. Almost 25% of the "Ikes" live in the region of this **estuary**, and public education, awards, and local conservation projects help protect Chesapeake Bay. In addition the **Soil Conservation** Program focuses on combating soil erosion and **groundwater pollution**, and the Public Lands Restoration Task Force works out of its headquarters in Portland, Oregon, to strike a balance between forests and the desire for their natural resources in the West.

IWLA makes its causes known through a variety of publications. *Splash*, a product of SOS, enlightens the public as to how to protect streams in America. *Outdoor Ethics*, a newsletter from the program of the same name, educates recreationists to responsible practices of **hunting**, boating, and other outdoor activities. The League also publishes a membership magazine, *Outdoor America*, and the *League Leader*, a vehicle of information for IWLA's 2,000 chapter and division officers. IWLA has also produced the longest-running weekly environmental program on television. Entitled *Make Peace with Nature*, the program has aired on PBS for almost 20 years and presents stories of environmental interest.

Having expanded its scope from water to the general environment, IWLA has become a vital force in the national conservation movement. Through its many and varied programs, the League continues to promote constructive and active involvement in environmental problems.

[*Andrea Gacki*]

RESOURCES

ORGANIZATIONS

Izaak Walton League of America, 707 Conservation Lane, Gaithersburg, MD USA 20878 (301) 548-0150, Fax: (301) 548-0146, Toll Free: (800) IKE-LINE, Email: general@iwla.org, <http://www.iwla.org>

J

Wes Jackson (1936 –)
American environmentalist and writer

Wes Jackson is a plant geneticist, writer, and co-founder, with his wife Dana Jackson, of the **Land Institute** in Salina, Kansas. He is one of the leading critics of conventional agricultural practices, which in his view are depleting **topsoil**, reducing genetic diversity, and destroying small family farms and rural communities. Jackson is also critical of the culture that provides the pretext and the context within which this destruction occurs and is justified as "necessary," "efficient," and "economical." He contrasts a culture or mind-set that emphasizes humanity's mastery or dominion over **nature** with an alternative vision that takes "nature as the measure" of human activity. The former viewpoint can produce temporary triumphs but not long-lasting or sustainable livelihood; only the latter holds out the hope that humans can live with nature, on nature's terms.

Jackson was born in 1936 in Topeka, Kansas, the son of a farmer. Young and restless, Jackson held various jobs—welder, farm hand, ranch hand, teacher—before devoting his time to the study of agricultural practices in the United States and abroad. He attended Kansas Wesleyan University, the University of Kansas, and North Carolina State University, where he earned his doctorate in plant genetics in 1967.

According to Jackson, agriculture as we know it is unnatural, artificial, and, by geological time-scales, of relatively recent origin. It requires plowing, which leads to loss of topsoil, which in turn reduces and finally destroys fertility. Large-scale "industrial" agriculture also requires large investments, complex and expensive machinery, fertilizers, pesticides, and herbicides, and leads to a loss of genetic diversity, to **soil erosion** and **compaction**, and other negative consequences. It is also predicated on the planting and harvesting of annual crops—corn, wheat, soybeans—that leaves fields uncovered for long periods and thus leaves precious topsoil unprotected and vulnerable to erosion by wind and water. For every bushel of corn harvested, a bushel of topsoil is lost. Jackson estimates that America has lost between one-third and one-half of its topsoil since the arrival of the first European settlers.

At the Land Institute, Jackson and his associates are attempting to re-think and revise agricultural practices so as to "make nature the measure" and enable farmers to "meet the expectations of the land," rather than the other way around. In particular, they are returning to, and attempting to learn from, the native **prairie** plants and the ecosystems that sustain them. They are also exploring the feasibility of alternative farming methods that might minimize or even eliminate entirely the planting and harvesting of annual crops, favoring instead the use of perennials that protect and bind topsoil.

Jackson's emphasis is not exclusively scientific or technical. Like his long-time friend Wendell Berry, Jackson emphasizes the *culture* in agriculture. Why humans grow food is not at all mysterious or problematic: we must eat in order to live. But how we choose to plant, grow, harvest, distribute, and consume food is clearly a cultural and moral matter having to do with our attitudes and beliefs. Our contemporary consumer culture is out of kilter, Jackson contends, in various ways. For one, the economic emphasis on minimizing costs and maximizing yields ignores longer-term environmental costs that come with the depletion of topsoil, the diminution of genetic diversity, and the depopulation of rural communities. For another, most Americans have lost (and many have never had) a sense of connectedness with the land and the natural **environment**; Jackson contends that they are unaware of the mysteries and wonder of birth, death and rebirth, and of cycles and seasons, that are mainstays of a meaningful human life. To restore this sense of mystery and meaning requires what Jackson calls *homecoming* and "the resettlement of America," and "becoming native to this place." More Americans need to return to the land, to repopulate rural communities, and to re-learn the wealth of skills that we have lost or forgotten or never acquired. Such skills are more than matters of method or technique, they also have to do with ways of relating to nature and to each other. Jackson has received several awards, such as the Pew

Conservation Scholars award (1990) and a MacArthur Fellowship (1992).

Wes Jackson has been called, by critics and admirers alike, a radical and a visionary. Both labels appear to apply. For Jackson's vision is indeed radical, in the original sense of the term (from the Latin *radix*, or root). It is a vision not only of "new roots for agriculture" but of new and deeper roots for human relationships and communities that, like protected prairie topsoil, will not easily erode.

[*Terence Ball*]

RESOURCES
BOOKS

Berry, W. "New Roots for Agricultural Research." In *The Gift of Good Land*. San Francisco: North Point Press, 1981.
Eisenberg, E. *New Roots for Agriculture*. San Francisco: Friends of the Earth, 1980.
Jackson, Wes. *Altars of Unhewn Stone*. San Francisco: North Point Press, 1987.
————. *Becoming Native to this Place*. Lexington, KY: University Press of Kentucky, 1994.
————, W. Berry, and B. Coleman, eds. *Meeting the Expectations of the Land*. San Francisco: North Point Press, 1984.

PERIODICALS
Eisenberg, E. "Back to Eden." *The Atlantic* (October 1989): 57–89.

James Bay hydropower project

James Bay forms the southern tip of the much larger Hudson Bay in Quebec, Canada. To the east lies the Quebec-Labrador peninsula, an undeveloped area with vast expanses of pristine **wilderness**. The region is similar to Siberia, covered in **tundra** and sparse forests of black spruce and other evergreens. It is home to roughly 100 **species** of birds, twenty species of fish and dozens of mammals, including muskrat, lynx, black bear, red fox, and the world's largest herd of caribou. The area has also been home to the Cree and other Native Indian tribes for centuries. Seven rivers drain the wet, rocky region, the largest being the La Grande.

In the 1970s, the government-owned Hydro-Quebec electric utility began to divert these rivers, **flooding** 3,861 square miles (10,000 km²) of land. They built a series of reservoirs, **dams** and dikes on La Grande that generated 10,300 megawatts of power for homes and businesses in Quebec, New York, and New England. With its $16 billion price tag, the project is one of the world's largest energy projects. The complex generates a total of 15,000 megawatts. A second phase of the project added two more hydroelectric complexes, supplying another 12,000 megawatts of power--

the equivalent of more than thirty-five nuclear **power plants**.

But the project has had many opponents. The Cree and other Inuit tribes joined forces with American environmentalists to protest the project. Its environmental impact has had scant analysis; in fact, damage has been severe. Ten thousand caribou drowned in 1984, while crossing one of the newly-dammed rivers on their **migration** route. When the utility flooded land, it destroyed **habitat** for countless plants and animals. The graves of Cree Indians, who for millennia, had hunted, traveled, and lived along the rivers, were inundated. The project also altered the **ecology** of the James and Hudson bays, disrupting spawning cycles, **nutrient** systems, and other important maritime resources. Naturally-occurring **mercury** in rocks and **soil** is released as the land is flooded and accumulates as it passes through the food chain from microscopic organisms, to fish, to humans. A majority of the native people in villages where fish are a main part of the diet show symptoms of mercury **poisoning**.

Despite these problems, Hydro-Quebec pursued the project, partly because of Quebec's long-standing struggle for independence from Canada. The power is sold to corporate customers, providing income for the province and attracting industry to Quebec.

The Cree and environmentalists, joined by New York congressmen, took their fight to court. On **Earth Day** 1993, they filed suit against New York Power Authority in United States District Court in New York, challenging the legality of the agreement, which was to go into effect in 1999. Their claim was based on the United States Constitution and the 1916 Migratory Bird Treaty with Canada.

In February 2002, nearly 70 percent of Quebec's James Bay Cree Indians endorsed a 2.25 billion dollar deal with the Quebec government for hydropower development on their land. Approval for the deal ranged from a low of 50 percent, to a high of 83 percent, among the nine communities involved. Some Cree spokespersons considered the agreement a vindication of the long campaign, waged since 1975, to have Cree rights respected.

Under the deal, the James Bay Cree would receive $16 million in 2002, $30.7 million in 2003, then $46.5 million a year for 48 years. In return, the Cree would drop environmental lawsuits totaling $2.4 billion. The Cree also agreed to hydroelectric plants along the Eastman River and Rupert River, subject to environmental approval. The deal guarantees the Cree jobs with the hydroelectric authority and gives them more control over **logging** and other areas of their economy. *See also* Environmental law; Hetch Hetchy Reservoir; Nuclear energy

[*Bill Asenjo Ph.D.*]

RESOURCES
BOOKS

McCutcheon, S. *Electric Rivers: The Story of the James Bay Project.* Montreal: Black Rose Books, 1991.

PERIODICALS

Associated Press. "James Bay Cree Approve Deal with Quebec on Hydropower Development." February 05, 2002.
Picard, A. "James Bay II." *Amicus Journal* 12 (Fall 1990): 10–16.

Japanese logging

In recent decades the timber industry has intensified efforts to harvest logs from tropical, temperate, and boreal forests worldwide to meet an increasing demand for wood and wood products. Japanese companies have been particularly active in **logging** and importing timber from around the world. Because of wasteful and destructive logging practices that result from efforts to maximize corporate financial gains, those interested in reducing **deforestation** have raised many concerns about Japan's logging industry.

The world's forests, especially tropical rain forests, are rich in **species**, including plants, insects, birds, reptiles, and mammals. Many of these species exist only in very limited areas where conditions are suitable for their existence. These endangered forest dwellers provide unique and irreplaceable genetic material that can contribute to the betterment of domestic plants and animals. The forests are a valuable resource for medically useful drugs. Healthy forests stabilize watersheds by absorbing rainfall and retarding **runoff**. Mat roots help control **soil erosion**, preventing the silting of waterways and damage to reefs, fisheries, and spawning grounds.

The United Nations Food and Agriculture Organization reports tropical deforestation rates of 42 million acres (over 17 million ha) per year. More than half of the Earth's primary tropical forest area has vanished, and more than half of the remaining forest has been degraded. While Brazil contains about a third of the world's remaining **tropical rain forest**, southeast Asia is now a major supplier of tropical woods. Burma, Thailand, Laos, Vietnam, Kampuchea, Malaysia, Indonesia, Borneo, New Guinea, and the Philippines contain 20% of the world's remaining tropical forests. With current rates of deforestation, it is estimated that almost all of Southeast Asia's primary forests will be gone by the year 2010. While a number of countries make use of **rain forest** wood, Japan is the number one importer of tropical timber. Japan's imports account for about 30% of the world trade in tropical lumber.

Japan also imports large amounts of timber from temperate and boreal forests in Canada, Russia, and the United States. These three countries contain most of the remaining boreal forests, and they supply more than half of the world's industrial wood. As demand for timber continues to climb, previously undisturbed virgin forests are increasingly being used. To speed harvesting, logging roads are built to provide access, and heavy equipment is brought in to hasten work. In the process, soil is compacted, making plant re-growth difficult or impossible. Although these practices are not limited to one country, Japanese firms have been cited by environmentalists as particularly insensitive to the environmental impact of logging.

The globe's forests are sometimes referred to as the 'lungs of the planet,' exchanging **carbon dioxide** for oxygen. Critics claim that the wood harvesting industry is destroying this vital natural resource, and in the process this industry is endangering the planet's ability to nurture and sustain life. Widespread destruction of the world's forests is a growing concern. Large Japanese companies, and companies affiliated with Japanese firms, have logged old growth forests in several parts of the globe to supply timber for the Japanese forest products industry. **Clear-cutting** of trees over large areas in tropical rain forests has made preservation of the original **flora** and **fauna** impossible. Many species are becoming extinct. Replanting may in time restore the trees, but it will not restore the array of organisms that were present in the original forest.

Large scale logging activities have had a largely negative impact on the local economies in exporting regions because whole logs are shipped to Japan for further processing. Developed countries such as the United States and Canada, which in the past harvested timber and processed it into lumber and other products, have lost jobs to Japan. Indigenous cultures that have thrived in harmony with their forest homelands for eons are displaced and destroyed. Provision has not been made for the survival of local flora and fauna, and provision for forest re-establishment has thus far proven inadequate. As resentment has grown in impacted areas, and among environmentalists, efforts have emerged to limit or stop large-scale timber harvesting and exporting.

Although concern has been voiced over all large-scale logging operations, special concern has been raised over harvesting of tropical timber from previously undisturbed primary forest areas. Tropical rain forests are especially unique and valuable **natural resources** for many reasons, including the density and variety of species within their borders. The exploitation of these unique ecosystems will result in the **extinction** of many potentially valuable species of plants and animals that exist nowhere else. Many of these forms of life have not yet been named or scientifically studied. In addition, over-harvesting of tropical rainforests has a negative effect on weather patterns, especially by reducing rainfall.

Japan is a major importer of tropical timber from Malaysia, New Guinea, and the Solomon Islands. Although the number of imported logs has declined in recent years, this has been matched by an increase in imported tropical plywood manufactured in Indonesia and Malaysia. As a result, the total amount of timber removed has remained fairly constant. An environmentalist group called the **Rainforest Action Network** (RAN) has issued an alarm concerning recent expansion of logging activity by firms affiliated with Japanese importers. The RAN alleges that: "After laying waste to the rain forests of Asia and the Pacific islands, giant Malaysian logging companies are setting their sights on the Amazon. This past year, some of Southeast Asia's biggest forestry conglomerates have moved into Brazil, and are buying controlling interests in area logging companies, and purchasing rights to cut down vast rain forest territories for as little as $3 U.S. dollars per acre. In the last few months of 1996 these companies quadrupled their South American interests, and now threaten 15% of the Amazon with immediate logging. According to *The Wall Street Journal*, up to 30 million acres (12.3 million ha) are at stake. Major players include the WTK Group, Samling, Mingo, and Rimbunan Hijau."

The RAN claims that "the same timber companies in Sarawak, Malaysia, worked with such rapacious speed that they devastated the region's forest within a decade, displacing traditional peoples and leaving the landscape marred with silted rivers and eroded soil."

One large Japanese firm, the Mitsubishi Corporation, has been targeted for criticism and boycott by the RAN, as one of the world's largest importers of timber. The boycott is an effort to encourage environmentally-conscious consumers to stop buying products marketed by companies affiliated with the huge conglomerate, including automobiles, cameras, beer, cell phones, and consumer electronics equipment. Through its subsidiaries, Mitsubishi has logged or imported timber from the Philippines, Malaysia, Papua New Guinea, Bolivia, Indonesia, Brazil, Chile, Canada (British Columbia and Alberta), Siberia, and the United States (Alaska, Oregon, Washington, and Texas). The RAN charges that "Mitsubishi Corporation is one of the most voracious destroyers of the world's rain forests. Its timber purchases have laid waste to forests in the Philippines, Malaysia, Papua New Guinea, Indonesia, Brazil, Bolivia, **Australia**, New Zealand, Siberia, Canada, and even the United States." The Mitsubishi Corporation itself does not sell consumer products, but it consists of 190 interlinked companies and hundreds of associated firms that do market to consumers. This conglomerate forms one of the world's largest industrial and financial powers. The Mitsubishi umbrella includes Mitsubishi Bank, Mitsubishi Heavy Industries, Mitsubishi Electronics, Mitsubishi Motors, and other major components. To force Mitsubishi and other corporations involved with timber harvesting to operate in a more environmentally responsible way and to end "their destructive logging and trading practices," an international boycott was organized in 1990 by the World Rainforest Movement (tropical forests) and the **Taiga** Rescue Network (boreal forests).

The Mitsubishi Corporation has countered criticism by launching a program "to promote the regeneration of rain forests...in Malaysia that plants seedlings and monitors their development." In 1990, the corporation formed an Environmental Affairs Department, one of the first of its kind in Japan, to draft environmental guidelines, and coordinate corporate environmental activities. In the words of the Mitsubishi Corporation Chairman, "A business cannot continue to exist without the trust and respect of society for its environmental performance." Mitsubishi Corporation reports that they have launched a program to support experimental reforestation projects in Malaysia, Brazil, and Chile. In Malaysia, the company is working with a local agricultural university, under the guidance of a professor from Japan. About 300,000 seedlings were planted on a barren site in 1991. Within five years, the trees were over 33 feet (10 m) in height and the corporation claimed that they were "well on the way to establishing techniques for regenerating tropical forest on burnt or barren land using indigenous species." Similar projects are underway in Brazil and Chile. The company is also conducting research on sustainable management of the Amazon rain forests. In Canada, Mitsubishi Corporation has participated in a pulp project called Al-Pac to start a mill "which will supply customers in North America, Europe, and Asia," meeting "the strictest environmental standards by employing advanced, environmentally safe technology. Al-Pac harvests around 0.25% of its total area annually and all harvested areas will be reforested."

[*Bill Asenjo Ph.D.*]

RESOURCES

BOOKS

Marx, M. J. *The Mitsubishi Campaign: First Year Report*. Rainforest Action Network, San Francisco, 1993.

Mitsubishi Corporation Annual Report 1996. Mitsubishi Corporation, Tokyo, 1996.

Wakker, E. "Mitsubishi's Unsustainable Timber Trade: Sarawak." *In Restoration of Tropical Forest Ecosystems*. L. and M. Lohmann, eds. Netherlands: Kluwer Academic Publishers, 1993.

PERIODICALS

Marshall, G. "The Political Economy of the Logging: The Barnett Inquiry into Corruption in the Papua New Guinea Timber Industry," *The Ecologist* 20, no. 5 (1990).

Neff, R., and W. J. Holstein. "Mitsubishi is on the Move," *Business Week*, September 24, 1990.

World Rainforest Report XII, no. 4 (October-December 1995). San Francisco: Rainforest Action Network.

K

Kapirowitz Plateau

The Kapirowitz Plateau, a **wildlife refuge** on the northern rim of the Grand Canyon, has come to symbolize **wildlife management** gone awry, a classic case of misguided human intervention intended to help **wildlife** that ended up damaging the animals and the **environment**. The Kapirowitz is located on the **Colorado River** in northwestern Arizona, and is bounded by steep cliffs dropping down to the Kanab Canyon to the west, and the Grand and Marble canyons to the south and southeast. Because of its inaccessibility, according to naturalist James B. Trefethen, the Plateau was considered a "biological island," and its deer population "evolved in almost complete genetic isolation."

The lush grass meadows of the Kapirowitz Plateau supported a resident population of 3,000 mule deer (*Odocoileus hemionus*), which were known and renowned for their massive size and the huge antlers of the old bucks. Before the advent of Europeans, Paiute and Navajo Indians hunted on the Kapirowitz in the fall, stocking up on meat and skins for the winter. In the early 1900s, in an effort to protect and enhance the magnificent deer population of the Kapirowitz, the federal government prohibited all killing of deer, and even eliminated the predator population in the area. As a result, the deer population exploded, causing massive overbrowsing, starvation, and a drastic decline in the health and population of the herd.

In 1893, when the Kapirowitz and surrounding lands were designated the Grand Canyon **National Forest** Reserve, hundreds of thousands of sheep, cattle, and horses were grazing on the Plateau, resulting in **overgrazing, erosion**, and large-scale damage to the land. On November 28, 1906, President **Theodore Roosevelt** established the one million-acre (400,000-ha) Grand Canyon National Game Preserve, which provided complete protection of the Kapirowitz's deer population. By then, however, overgrazing by livestock had destroyed much of the native vegetation and changed the Kapirowitz considerably for the worse. Contin-

ued pasturing of over 16,000 horses and cattle degraded the Kapirowitz even further.

The **Forest Service** carried out President Roosevelt's directive to emphasize "the propagation and breeding" of the mule deer by not only banning **hunting**, but also natural predators as well. From 1906 to 1931, federal agents poisoned, shot, or trapped 4,889 coyotes (*Canis latrans*), 781 mountain lions (*Puma concolor*, 554 bobcats (*Felis rufus*), and 20 **wolves** (*Canis lupus*). Without predators to remove the old, the sick, the unwary, and other biologically unfit animals, and keep the size of the herd in check, the deer herd began to grow out of control, and to lose those qualities that made its members such unique and magnificent animals. After 1906, the deer population doubled within 10 breeding seasons, and by 1918 (two years later), it doubled again. By 1923, the herd had mushroomed to at least 30,000 deer, and perhaps as many as 100,000 according to some estimates.

Unable to support the overpopulation of deer, range grasses and land greatly deteriorated, and by 1925, 10,000–15,000 deer were reported to have died from starvation and malnutrition. Finally, after relocation efforts mostly failed to move a significant number of deer off of the Kapirowitz, hunting was reinstated, and livestock grazing was strictly controlled. By 1931, hunting, disease, and starvation had reduced the herd to under 20,000. The range grasses and other vegetation returned, and the Kapirowitz began to recover. In 1975 James Trefethen wrote, "the Kapirowitz today again produces some of the largest and heaviest antlered mule deer in North America."

In the fields of wildlife management and biology, the lessons of the Kapirowitz Plateau are often cited (as in the writings of naturalist **Aldo Leopold**) to demonstrate the valuable role of predators in maintaining the **balance of nature** (such as between herbivores and the plants they consume) and survival of the fittest. The experience of the Kapirowitz shows that in the absence of natural predators, prey populations (especially ungulates) tend to increase beyond the **carrying capacity** of the land, and eventually the results are overpopulation and malnutrition. *See also* Predator control; Predator-prey interactions

[*Lewis G. Regenstein*]

RESOURCES
BOOKS

Leopold, A. *A Sand County Almanac.* New York: Oxford University Press, 1949.

Trefethen, J. B. *An American Crusade for Wildlife.* New York: Winchester Press, 1975.

PERIODICALS

Rasmussen, D. I. "Biotic Communities of the Kapirowitz Plateau," *Ecological Monographs* 3 (1941): 229–275.

Robert Francis Kennedy Jr. (1954 –)

American environmental lawyer

Robert "Bobby" Kennedy Jr. had a very controversial youth. Kennedy entered a drug **rehabilitation** program, at the age of 28, after being found guilty of drug possession following the South Dakota incident. He was sentenced to two years probation and community service.

Clearly the incident was a turning point in Bobby's life. "Let's just say, I had a tumultuous adolescence that lasted until I was 29," he told a reporter for New York magazine, which ran a long profile of Kennedy in 1995, entitled "Nature Boy." The title refers to the passion which has enabled Kennedy to emerge from his bleak years as a strong and vital participant in environmental causes.

A Harvard graduate and published author, Kennedy serves as chief prosecuting attorney for a group called the Hudson Riverkeeper (named after the famed New York river) and senior attorney for the **Natural Resources Defense Council**. Kennedy, who earlier in his career served as assistant district attorney in New York City after passing the bar, is also a clinical professor and supervising attorney at the Environmental Litigation Clinic at Pace University School of Law in New York.

While Kennedy appeared to be following in the family's political footsteps, working, for example, on several political campaigns and serving as a state coordinator for his Uncle Ted's 1980 presidential campaign, it is in environmental issues that Bobby Jr. has found himself. He has worked on environmental issues across the Americas and has assisted several indigenous tribes in Latin America and Canada in successfully negotiating treaties protecting traditional homelands. He is also credited with leading the fight to protect New York City's water supply, a battle which resulted in the New York City **Watershed** Agreement, regarded as an

international model for combining development and environmental concerns.

Opportunity was always around the corner for a young, confident and intelligent Kennedy. After Harvard, Bobby Jr. earned a law degree at the University of Virginia. In 1978, the subject of Kennedy's Harvard thesis—a prominent Alabama judge—was named head of the FBI. A publisher offered Bobby money to expand his previous research into a book, published in 1978, called Judge Frank M. Johnson, Jr.: A Biography.

Bobby did a publicity tour which included TV appearances, but the reviews were mixed. In 1982, Bobby married Emily Black, a Protestant who later converted to Catholicism. Two children followed: Robert Francis Kennedy III and Kathleen Alexandra, named for Bobby's Aunt Kathleen, who died in a plane crash in 1948.

The marriage, however, coincided with Bobby's fallout through drug addiction. In 1992, Bobby and Emily separated, and a divorce was obtained in the Dominican Republic. In 1994, Bobby married Mary Richardson, an architect, with whom he would have two more children.

During this time Bobby emerged as a leading environmental activist and litigator. Kennedy is quoted as saying: "To me...this is a struggle of good and evil between short-term greed and a long-term vision of building communities that are dignified and enriching and that meet the obligations of **future generations**. There are two visions of America. One is that this is just a place where you make a pile for yourself and keep moving. And the other is that you put down roots and build communities that are examples to the rest of humanity." Kennedy goes on: "The **environment** cannot be separated from the economy, housing, civil rights. How we distribute the goods of the earth is the best measure of our democracy. It's not about advocating for fishes and birds. It's about human rights."

RESOURCES
BOOKS

Young Kennedys: The New Generation. Avon, 1998.

OTHER

Biography Resource Center Online. Biography Resource Center. Farmington Hills, MI: The Gale Group. 2002.

ORGANIZATIONS

Riverkeeper, Inc., 25 Wing & Wing, Garrison, NY USA 10524-0130 (845) 424-4149, Fax: (845) 424-4150, Email: info@riverkeeper.org, <http://www.riverkeeper.org>

Kepone

Kepone ($C_{10}Cl_{10}O$) is an organochlorine **pesticide** that was manufactured by the Allied Chemical Corporation in Virginia from the late 1940s to the 1970s. Kepone was responsible for human health problems and extensive contamination of the James River and its **estuary** in the **Chesapeake Bay**. It is a milestone in the development of a public environmental consciousness, and its history is considered by many to be a classic example of negligent corporate behavior and inadequate oversight by state and federal agencies.

Kepone is an insecticide and **fungicide** that is closely related to other chlorinated pesticides such as DDT and aldrin. As with all such pesticides, Kepone causes lethal damage to the nervous systems of its target organisms. A poorly water-soluble substance, it can be absorbed through the skin, and it bioaccumulates in fatty tissues from which it is later released into the bloodstream. It is also a contact poison; when inhaled, absorbed, or ingested by humans, it can damage the central nervous system as well as the liver and kidneys. It can also lead to neurological symptoms such as tremors, muscle spasms, sterility, and **cancer**. Although the manufacture and use of Kepone is now banned by the **Environmental Protection Agency** (EPA), organochlorines have long half-lives, and these compounds, along with their residues and degradation products, can persist in the **environment** over many decades.

Allied Chemical first opened a plant to manufacture nitrogen-based fertilizers in 1928 in the town of Hopewell, on the banks of the James River in Virginia. This plant began producing Kepone in 1949. Commercial production was subsequently begun, although a battery of toxicity tests indicated that Kepone was both toxic and carcinogenic and that it caused damage to the functioning of the nervous, muscular, and reproductive systems in fish, birds, and mammals. It was patented by Allied in 1952 and registered with federal agencies in 1957. The demand for the pesticide grew after 1958, and Allied expanded production by entering into a variety of subcontracting agreements with a number of smaller companies, including the Life Science Products Company.

In 1970, a series of new environmental regulations came into effect which should have changed the way wastes from the manufacture of Kepone were discharged. The Refuse Act Permit Program and the National Pollutant **Discharge** Elimination Program (NPDES) of the **Clean Water Act** required all dischargers of effluents into United States waters to register their discharges and obtain permits from federal agencies. At the time these regulations went into effect, Allied Chemical had three pipes discharging Kepone and plastic wastes into the Gravelly Run, a tributary of the James River, about 75 mi (120 km) north of Chesapeake Bay.

A regional **sewage treatment** plant which would accept industrial wastes was then under construction but not scheduled for completion until 1975. Rather than installing expensive **pollution control** equipment for the interim pe-

riod, Allied chose to delay. They adopted a strategy of misinformation, reporting the releases as temporary and unmonitored discharges, and they did not disclose the presence of untreated Kepone and other process wastes in the effluents. The Life Science Products Company also avoided the new federal permit requirements by discharging their wastes directly into the local Hopewell sewer system. These discharges caused problems with the functioning of the biological treatment systems at the sewage plant; the company was required to reduce concentrations of Kepone in sewage, but it continued its discharges at high concentrations, violating these standards with the apparent knowledge of plant treatment officials.

During this same period, an employee of Life Science Products visited a local Hopewell physician, complaining of tremors, weight loss, and general aches and pains. The physician discovered impaired liver and nervous functions, and a blood test revealed an astronomically high level of Kepone—7.5 **parts per million**. Federal and state officials were contacted, and the epidemiologist for the state of Virginia toured the manufacturing facility at Life Science Products. This official reported that "Kepone was everywhere in the plant," and that workers wore no protective equipment and were "virtually swimming in the stuff." Another investigation discovered 75 cases of Kepone **poisoning** among the workers; some members of their families were also found to have elevated concentrations of the chemical in their blood. Further investigations revealed that the environment around the plant was also heavily contaminated. The **soil** contained 10,000–20,000 ppm of Kepone. Sediments in the James River, as well as local landfills and trenches around the Allied facilities were just as badly contaminated. Government agencies were forced to close 100 mi (161 km) of the James River and its tributaries to commercial and recreational fishing and shellfishing.

In the middle of 1975, Life Science Products finally closed its manufacturing facility. It has been estimated that since 1966, it and Allied together produced 3.2 million lb (1.5 million kg) of Kepone and were responsible for releasing 100,000–200,000 lb (45,360–90,700 kg) into the environment. In 1976, the Northern District of Virginia filed criminal charges against Allied, Life Science Products, the city of Hopewell, and six individuals on 1,097 counts relating to the production and disposal of Kepone. The indictments were based on violations of the permit regulations, unlawful discharge into the sewer systems, and conspiracy related to that discharge.

The case went to trial without a jury. The corporations and the individuals named in the charges negotiated lighter fines and sentences by entering pleas of "no contest." Allied ultimately paid a fine of 13.3 million dollars, although their annual sales reach three billion dollars. Life Science Products was fined four million dollars, which it could not pay due to lack of assets. Company officers were fined 25,000 dollars each, and the town of Hopewell was fined 10,000 dollars. No one was sentenced to a jail term. Civil suits brought against Allied and the other defendants resulted in a settlement of 5.25 million dollars to pay for cleanup expenses and to repair the damage that had been done to the sewage treatment plant. Allied paid another three million dollars to settle civil suits brought by workers for damage to their health.

Environmentalists and many others considered the results of legal action against the manufacturers of Kepone unsatisfactory. Some have argued that these results are typical of environmental litigation. It is difficult to establish criminal intent beyond a reasonable doubt in such cases, and even when guilt is determined, sentencing is relatively light. Corporations are rarely fined in amounts that affect their financial strength, and individual officers are almost never sent to jail. Corporate fines are generally passed along as costs to the consumer, and public bodies are treated even more lightly, since it is recognized that the fines levied on public agencies are paid by taxpayers.

Today, the James River has been reopened to fishing for those **species** that are not prone to the **bioaccumulation** of Kepone. Nevertheless, sediments in the river and its estuary contain large amounts of deposited Kepone which is released during periods of turbulence. Scientists have published studies which document that Kepone is still moving through the **food chain/web** and the **ecosystem** in this area, and Kepone toxicity has been demonstrated in a variety of invertebrate test species. There are still deposits of Kepone in the local sewer pipes in Hopewell; these continue to release the chemical, endangering treatment plant operations and polluting receiving waters.

[*Usha Vedagiri and Douglas Smith*]

RESOURCES
BOOKS

Goldfarb, W. *Kepone: A Case Study.* New Brunswick, NJ: Rutgers University, 1977.

Sax, N. I. *Dangerous Properties of Industrial Materials.* 6th ed. New York: Van Nostrand Reinhold, 1984.

Kesterson National Wildlife Refuge

One of a dwindling number of freshwater marshes in California's San Joaquin Valley, Kesterson **National Wildlife Refuge** achieved national notoriety in 1983 when refuge managers discovered that agricultural **runoff** was **poisoning** the area's birds. Among other elements and **agricultural chemicals** reaching toxic concentrations in the **wetlands**,

Breeding populations of American coots were affected by selenium poisoning at Kesterson National Wildlife Refuge. (Photograph by Leonard Lee Rue III. Visuals Unlimited. Reproduced by permission.)

the naturally occurring element selenium was identified as the cause of falling fertility and severe **birth defects** in the refuge's breeding populations of stilts, grebes, shovellers, coots, and other aquatic birds. Selenium, **lead**, boron, chromium, molybdenum, and numerous other contaminants were accumulating in refuge waters because the refuge had become an evaporation pond for tainted water draining from the region's fields.

The soils of the **arid** San Joaquin valley are the source of Kesterson's problems. The flat valley floor is composed of ancient sea bed sediments that contain high levels of trace elements, **heavy metals**, and salts. But with generous applications of water, this sun-baked **soil** provides an excellent medium for food production. Perforated pipes buried in the fields drain away excess water—and with it dissolved salts and trace elements—after flood **irrigation**. An extensive system of underground piping, known as tile **drainage**, carries **wastewater** into a network of canals that lead to Kesterson Refuge, an artificial basin constructed by the **Bureau of Reclamation** to store irrigation runoff from central California's heavily-watered agriculture. Originally a final drainage canal from Kesterson to San Francisco Bay was planned, but because an outfall point was never agreed upon, contaminated drainage water remained trapped in Kesterson's 12 shallow ponds. In small doses, selenium and other trace elements are not harmful and can even be dietary necessities. But steady evaporation in the refuge gradually concentrated these contaminants to dangerous levels.

Wetlands in California's San Joaquin valley were once numerous, supporting huge populations of breeding and migrating birds. In the past half century drainage and the development of agricultural fields have nearly depleted the area's marshes. The new ponds and cattail marshes at Kesterson presented a rare opportunity to extend breeding **habitat**, and the area was declared a national **wildlife refuge** in 1972, one year after the basins were constructed. Eleven years later, in the spring of 1983, observers discovered that a shocking 60% of Kesterson's nestlings were grotesquely deformed. High concentrations of selenium were found in their tissues, an inheritance from parent birds who ate algae, plants, and insects—all tainted with selenium—in the marsh.

Following extensive public outcry the local water management district agreed to try to protect the birds. Alternate drainage routes were established, and by 1987 much of the most contaminated drainage had been diverted from the **wildlife** refuge. The California Water Resource Control Board ordered the Bureau of **Reclamation** to drain the ponds and clean out contaminated sediments, at a cost of well over $50 million. However, these contaminants, especially in such large volumes and high concentrations, are difficult to contain, and similar problems could quickly emerge again. Furthermore, these problems are widespread. Selenium poisoning from irrigation runoff has been discovered in least nine other national wildlife refuges, all in the arid west, since it appeared at Kesterson. Researchers continue to work on affordable and effective responses to such contamination in wetlands, an increasingly rare habitat in this country.

[*Mary Ann Cunningham Ph.D.*]

RESOURCES
BOOKS

Harris, T. *Death in the Marsh*. Washington, DC: Island Press, 1991.

PERIODICALS

Claus, K. E. "Kesterson: An Unsolvable Problem?" *Environment* 89 (1987): 4–5.
Harris, T. "The Kesterson Syndrome." *Amicus Journal* 11 (Fall 1989): 4–9.
Marshal, E. "Selenium in Western Wildlife Refuges." *Science* 231 (1986): 111–12.
Tanji, K., A. Läuchli, and J. Meyer. "Selenium in the San Joaquin Valley." *Environment* 88 (1986): 6–11.

ORGANIZATIONS
Kesterson National Wildlife Refuge, c/o San Luis NWR Complex, 340 I Street, P.O. Box 2176, Los Banos, CA USA 93635 (209) 826-3508,

Ketones

Ketones belong to a class of organic compounds known as carbonyls. They contain a **carbon** atom linked to an oxygen atom with a double bond ($C=O$). **Acetone** (dimethyl ketone) is a ketone commonly used in industrial applications. Other ketones include methyl ethyl ketone (MEK), methyl isobutyl ketone (MIBK), methyl amyl ketone (MAK), isophorone, and diacetone alcohol.

As solvents, ketones have the ability to dissolve other materials or substances, particularly polymers and adhesives. They are ingredients in lacquers, epoxies, polyurethane, nail polish remover, degreasers, and cleaning solvents. Ketones are also used in industry for the manufacture of **plastics** and composites and in pharmaceutical and photographic film manufacturing. Because they have high evaporation rates and dry quickly, they are sometimes employed in drying applications.

Some types of ketones used in industry, such as methyl isobutyl ketone and methyl ethyl ketone, are considered both hazardous air pollutants (HAP) and volatile organic compounds (VOC) by the EPA. As such, the **Clean Air Act** regulates their use.

In addition to these industrial sources, ketones are released into the **atmosphere** in **cigarette smoke** and car and truck exhaust. More "natural" environmental sources such as forest fires and volcanoes also emit ketones. Acetone, in particular is readily produced in the atmosphere during the oxidation of organic pollutants or natural emissions. Ketones (in the form of acetone, beta-hydroxybutyric **acid**, and acetoacetic acid) also occur in the human body as a byproduct of the **metabolism**, or break down, of fat.

[*Paula Anne Ford-Martin*]

RESOURCES
PERIODICALS

Wood, Andrew. "Cleaner Ketone Oxidation." *Chemical Week* (Aug 1, 2001).

OTHER
U.S. National Library of Medicine. *Hazardous Substances Data Bank*. [cited May 2002]. <http://toxnet.nlm.nih.gov/cgi-bin/sis/htmlgen?HSDB>.

ORGANIZATIONS
American Chemical Society, 1155 Sixteenth St. NW, Washington, D.C. USA 20036 (202) 872-4600, Fax: (202) 872-4615, Toll Free: (800) 227-5558, Email: help@acs.org, <http://www.chemistry.org>

Keystone species

Keystone **species** have a major influence on the structure of their ecological community. The profound influence of keystone species occurs because of their position and activity within the **food chain/web**. In the sense meant here, a "major influence" means that removal of a keystone species would result in a large change in the abundance, and even the local extirpation, of one or more species in the community. This would fundamentally change the structure of the overall community in terms of species composition, productivity, and other characteristics. Such changes would have substantial effects on all of the species that are present, and could allow new species to invade the community.

The original use of the word "keystone" was in architecture. An architectural keystone is a wedge-shaped stone that is strategically located at the summit of an arch. The keystone serves to lock all other elements of the arch together, and it thereby gives the entire structure mechanical integrity. Keystone species play an analogous role in giving structure to the "architecture" of their ecological community.

The concept of keystone species was first applied to the role of certain predators (i.e., keystone predators) in their community. More recently, however, the term has been extended to refer to other so-called "strong interactors." This has been particularly true of keystone herbivores that have a relatively great influence on the species composition and relative abundance of plants in their community.

Keystone species directly exert their influence on the populations of species that they feed upon, but they also have indirect effects on species lower in the food web. Consider, for example, a hypothetical case of a keystone predator that regulates the population of a herbivore. This effect will also, of course, indirectly influence the abundance of plant species that the herbivore feeds upon. Moreover, by affecting the competitive relationships among the various species of plants in the community, the abundance of plants that the herbivore does not eat will also be indirectly affected by the keystone predator. Although keystone species exert their greatest influence on species with which they are most closely linked through feeding relationships, their influences can ramify throughout the food web.

Ecologists have documented the presence of keystone species in many types of communities. The phenomenon does not, however, appear to be universal, in that keystone species have not been identified in many ecosystems.

Predators as keystone species

The term "keystone species" was originally used by the American ecologist Robert Paine to refer to the critical influence of certain predators. His original usage of the concept was in reference to rocky intertidal communities of western North America, in which the predatory starfish *Pisaster ochraceous* prevents the mussel *Mytilus californianus* from monopolizing the available space on rocky habitats and thereby eliminating other, less-competitive herbivores and even seaweeds from the community. By feeding on mussels, which are the dominant competitor among the herbivores

in the community, the starfish prevents these shellfish from achieving the **dominance** that would otherwise be possible. This permits the development of a community that is much richer in species than would occur in the absence of the predatory starfish. Paine demonstrated the keystone role of the starfish by conducting experiments in which the predator was excluded from small areas using cages. When this was done, the mussels quickly became strongly dominant in the community and eliminated virtually all other species of herbivores. Paine also showed that once mussels reached a certain size they were safe from predation by the starfish. This prevented the predator from eliminating the mussel from the community.

Sea otters (*Enhydra lutris*) of the west coast of North America are another example of a keystone predator. This species feeds heavily on sea urchins when these invertebrates are available. By greatly reducing the abundance of sea urchins, the sea otters prevent these herbivores from **overgrazing** kelps and other seaweeds in subtidal habitats. Therefore, when sea otters are abundant, urchins are not, and this allows luxurious kelp "forests" to develop. In the absence of otters, the high urchin populations can keep the kelp populations low, and the **habitat** then may develop as a rocky "barren ground." Because sea otters were trapped very intensively for their fur during the eighteenth and nineteenth centuries, they were extirpated over much of their natural range. In fact, the species had been considered extinct until the 1930s, when small populations were "discovered" off the coast of California and in the Aleutian Islands of Alaska. Thanks to effective protection from **trapping**, and deliberate reintroductions to some areas, populations of sea otters have now recovered over much of their original range. This has resulted in a natural depletion of urchin populations, and a widespread increase in the area of kelp forests.

Herbivores as Keystone Species

Some herbivorous animals have also been demonstrated to have a strong influence on the structure and productivity of their ecological community. One such example is the spruce budworm (*Choristoneura fumiferana*), a moth that occasionally irrupts in abundance and becomes an important **pest** of conifer forests in the northeastern United States and eastern Canada. The habitat of spruce budworm is mature forests dominated by balsam fir (*Abies balsamea*), white spruce (*Picea glauca*), and red spruce (*P. rubens*). This native species of moth is always present in at least small populations, but it sometimes reaches very high populations, which are known as irruptions. When budworm populations are high, many species of forest birds and small mammals occur in relatively large populations that subsist by feeding heavily on larvae of the moth. However, during irruptions of budworm most of the fir and spruce foliage is eaten by the abundant larvae, and after this happens for several years many of the trees die. Because of damages caused to mature trees in the forest the budworm epidemic collapses, and then a successional recovery begins. The plant communities of early **succession** contain many species of plants that are uncommon in mature conifer forests. Eventually, however, another matures, conifer forest redevelops, and the cycle is primed for the occurrence of another irruption of the budworm. Clearly, spruce budworm is a good example of a keystone herbivore, because it has such a great influence on the populations of plant species in its habitat, and also on the many animal species that are predators of the budworm.

Another example of a keystone herbivore concerns snow geese (*Chen caerulescens*) in salt marshes of western Hudson Bay. In the absence of grazing by flocks of snow geese this **ecosystem** would become extensively dominated by several competitively superior species, such as the salt-marsh grass *Puccinellia phryganodes* and the sedge *Carex subspathacea*. However, vigorous feeding by the geese creates bare patches of up to several square meters in area, which can then be colonized by other species of plants. The patchy disturbance regime associated with goose grazing results in the development of a relatively complex community, which supports more species of plants than would otherwise be possible. In addition, by manuring the community with their droppings, the geese help to maintain higher rates of plant productivity than might otherwise occur. In recent years, however, large populations of snow goose have caused severe damages to the salt-marsh habitat by over-grazing. This has resulted in the development of salt-marsh "barrens" in some places, which may take years to recover.

Plants as keystone species

Some ecologists have also extended the idea of keystone species to refer to plant species that are extremely influential in their community. For example, sugar maple (*Acer saccharum*) is a competitively superior species that often strongly dominates stands of forest in eastern North America. Under these conditions most of the community-level productivity is contributed by sugar-maple trees. In addition, most of the seedlings and saplings are of sugar maple. This is because few seedlings of other species of trees are able to tolerate the stressful conditions beneath a closed sugar-maple canopy.

Other ecologists prefer to not use the idea of keystone species to refer to plants that, because of their competitive abilities, are strongly dominant in their community. Instead, these are sometimes referred to as "foundationstone species." This term reflects the facts that strongly dominant plants contribute the great bulk of the **biomass** and productivity of their community, and that they support almost all herbivores, predators, and **detritivores** that are present.

[*Bill Freedman Ph.D.*]

RESOURCES

BOOKS

Begon, M., J. L. Harper, and C. R. Townsend. *Ecology. Individuals, Populations and Communities.* 3rd ed. London: Blackwell Sci. Pub., 1996.

Krebs, C. J. *Ecology. The Experimental Analysis of Distribution and Abundance.* San Francisco: Harper and Row, 1985.

Ricklefs, R. E. *Ecology.* New York: W. H. Freeman and Co., 1990.

PERIODICALS

Paine, R. T. "Intertidal Community Structure: Experimental Studies of the Relationship Between A Dominant Competitor and Its Principal Predator." *Oecologia* 15 (1974): 93–120.

Killer bees

see **Africanized bees**

Kirtland's warbler

Kirtland's warbler (*Dendroica kirtlandii*) is an **endangered species** and one of the rarest members of the North American wood warbler family. Its entire breeding range is limited to a seven-county area of north-central Michigan. The restricted distribution of the Kirtland's warbler and its specific **niche** requirements have probably contributed to low population levels throughout its existence, but human activity has had a large impact on their numbers over the past hundred years.

The first specimen of Kirtland's warbler was taken by Samuel Cabot in October 1841, and brought on ship in the West Indies during an expedition to the Yucatan. But this specimen went unnoticed until 1865, long after the **species** had been formally described. Charles Pease is credited with discovering Kirtland's warbler. He collected a specimen on May 13, 1851 near Cleveland, Ohio, and gave it to his father-in-law, Dr. Jared P. Kirtland, a renowned naturalist. Kirtland sent the specimen to his friend, ornithologist Spencer Fullerton Baird, who described the new species the following year and named it in honor of the naturalist.

The wintering grounds of Kirtland's warbler is the Bahamas, a fact which was well established by the turn of the century, but its nesting grounds went undiscovered until 1903, when Norman Wood found the first nest in Oscoda County, Michigan. Every nest found since then has been within a 60 mile (95 km) radius of this spot.

In 1951 the first exhaustive census of singing males was undertaken in an effort to establish the range of Kirtland's warblers as well as its population level. Assuming that numbers of males and females are approximately equal and that a singing male is defending an active nesting site, the total of 432 in this census indicated a population of 864 birds. Ten years later another census counted 502 singing males, indicating the population was over 1,000 birds. In 1971, annual counts began, but for the next 20 years these counts revealed that the population had dropped significantly, reaching lows of 167 singing males in 1974 and 1987. In the early 1990s, **conservation** efforts on behalf of the species began to bear fruit and the population began to recover. By 2001 the annual census counted 1,085 singing males or a total population of over 2,000 birds.

The first problem facing this endangered species centers on its specialized nesting and **habitat** requirements. The Kirtland's warbler nests on the ground, and its reproductive success is tied closely to its selection of young jack pine trees as nesting sites. When the jack pines are 5–20 ft (1.5–6 m) tall, at an age of 8–20 years, their lower branches are at ground level and provide the cover this warbler needs. The life cycle of the pine, however, is dependent on forest fires, as the intense heat is needed to open the cones for seed release. The advent of fire protection in **forest management** reduced the production of the number of young trees the warblers needed and the population suffered. Once this relationship was fully understood, jack pine stands were managed for Kirtland's warbler, as well as commercial harvest, by instituting controlled burns on a 50 year rotational basis.

The second problem is the population pressures brought to bear by a nest parasite, the brown-headed cowbird (*Molothrus ater*), which lays its eggs in the nests of other songbirds. Originally a bird of open plains, it did not threaten Kirtland's warbler until Michigan was heavily deforested, thus providing it with appropriate habitat. Once established in the warbler's range, it has increasingly pressured the Kirtland's population. Cowbird chicks hatch earlier than other birds and they compete successfully with the other nestlings for nourishment. Efforts to trap and destroy this nest parasite in the warbler's range have resulted in improved reproductive success for Kirtland's warbler. *See also* Deforestation; Endangered Species Act; International Council on Bird Preservation; Rare species; Wildlife management

[*Eugene C. Beckham*]

RESOURCES

BOOKS

Ehrlich, P. R., D. S. Dobkin, and D. Wheye. *Birds in Jeopardy: The Imperiled and Extinct Birds of the United States and Canada, Including Hawaii and Puerto Rico.* Stanford: Stanford University Press, 1992.

PERIODICALS

Weinrich, J. A. "Status of Kirtland's Warbler, 1988." *Jack-Pine Warbler* 67 (1989): 69–72.

OTHER

U.S. Fish and Wildlife Service. *Kirtland"s Warbler (Dendroica kirtlandii).* April 2, 2002 [cited June 19, 2002]. <http://endangered.fws.gov/i/b0d.html>.

Krakatoa

The explosion of this triad of volcanoes on August 27, 1883, the culmination of a three-month eruptive phase, astonished the world because of its global impact. Perhaps one of the most influential factors, however, was its timing. It happened during a time of major growth in science, technology, and communications, and the world received current news accompanied by the correspondents' personal observations. The explosion was heard some 3,000 mi (4,828 km) away, on the Island of Rodriguez in the Indian Ocean. The glow of sunsets was so vivid three months later that fire engines were called out in New York City and nearby towns.

Krakatoa (or Krakatau), located in the Sunda Strait between Java and Sumatra, is part of the Indonesian volcanic system, which was formed by the subduction of the Indian Ocean plate under the Asian plate. A similar explosion occurred in A.D. 416, and another major eruption was recorded in 1680. Now a new **volcano** is growing out of the caldera, likely building toward some future cataclysm.

This immense natural event, perhaps twice as powerful as the largest **hydrogen** bomb, had an extraordinary impact on the solid earth, the oceans, and the **atmosphere**, and demonstrated their interdependence. It also made possible the creation of a **wildlife refuge** and **tropical rain forest** preserve on the Ujung Kulon Peninsula of southwestern Java.

Studies revealed that this caldera, like Crater Lake, Oregon, resulted from Krakatoa's collapse into the now empty magma chamber. The explosion produced a 131-ft (40-m) high tsunami, or tidal wave, which carried a steamship nearly 2 mi (3.2 km) inland, and caused most of the fatalities resulting from the eruption. Tidal gauges as far away as San Francisco Bay and the English Channel recorded fluctuations.

The explosion provided substantial benefits to the young science of **meteorology**. Every barometer on Earth recorded the blast wave as it raced towards its antipodal position in Columbia, and then reverberated back and forth in six more recorded waves. The distribution of ash in the **stratosphere** gave the first solid evidence of rapidly flowing westerly winds, as debris encircled the equator over the next 13 days. Global temperatures were lowered about 0.9°F (0.5°C), and did not return to normal until five years later.

An ironic development is that the Ujung Kulon Peninsula was never resettled after the tsunami killed most of the people. Without Krakatoa's explosion, the population would have most likely grown significantly and much of the **habitat** there would likely have been altered by agriculture. Instead, the area is now a **national park** that supports a variety of **species**, including the Javan rhino (*Rhinoceros sondaicus*), one of Earth's rarest and most **endangered species**. This park has provided a laboratory for scientists to study nature's

Krill. (McGraw-Hill Inc. Reproduced by permission.)

healing process after such devastation. *See also* Mount Pinatubo, Philippines; Mount Saint Helens, Washington; Volcano

[*Nathan H. Meleen*]

RESOURCES
BOOKS

Nardo, D. *Krakatoa*. World Disasters Series. San Diego: Lucent Books, 1990.
Simkin, T., and R. Fiske. *Krakatau 1883: The Volcanic Eruption and Its Effects*. Washington, DC: Smithsonian Books, 1983.

PERIODICALS

Ball, R. "The Explosion of Krakatoa," *National Geographic* 13 (June 1902): 200–203.
Plage, D., and M. Plage. "Return of Java's Wildlife," *National Geographic* 167 (June 1985): 750–71.

Krill

Marine crustaceans in the order Euphausiacea. Krill are **zooplankton**, and most feed on microalgae by filtering them from the water. In high latitudes, krill may account for a large proportion of the total zooplankton. Krill often occur in large swarms and in a few **species** these swarms may reach several hundred square meters in size with densities over 60,000 individuals per square meter. This swarming behavior makes them valuable food sources for many species of **whales** and seabirds. Humans have also begun to harvest krill for use as a dietary protein supplement.

Joseph Wood Krutch (1893 – 1970)
American literary critic and naturalist

Through much of his career, Krutch was a teacher of criticism at Columbia University and a drama critic for *The Nation*. But then respiratory problems led him to early retirement in the **desert** near Tucson, Arizona. He loved the desert and there turned to biology and geology, which he applied to maintain a consistent, major theme found in all of his writings, that of the relation of humans and the universe. Krutch subsequently became an accomplished naturalist.

Readers can find the theme of man and universe in Krutch's early work, for example *The Modern Temper* (1929), and in his later writings on human-human and human-nature relationships, including natural history—what Rene Jules Dubos described as "the social philosopher protesting against the follies committed in the name of technological progress, and the humanist searching for permanent values in man's relationship to nature." Assuming a pessimistic stance in his early writings, Krutch despaired about lost connections, arguing that for humans to reconnect, they must conceive of themselves, **nature**, "and the universe in a significant reciprocal relationship."

Krutch's later writings repudiated much of his earlier despair. He argued against the dehumanizing and alienating forces of modern society and advocated systematically reassembling—by reconnecting to nature—"a world man can live in." In *The Voice of the Desert* (1954), for instance, he claimed that "we must be part not only of the human community, but of the whole community." In such books as *The Twelve Seasons* (1949) and *The Great Chain of Life* (1956), he demonstrated that humans "are a part of Nature...whatever we discover about her we are discovering also about ourselves." This view was based on a solid anti-deterministic approach that opposed mechanistic and behavioristic theories of **evolution** and biology.

His view of modern technology as out of control was epitomized in the **automobile**. Driving fast prevented people from reflecting or thinking or doing anything except controlling the monster: "I'm afraid this is the metaphor of our society as a whole," he commented. Krutch also disliked the proliferation of suburbs, which he labeled "affluent slums." He argued in *Human Nature and the Human Condition* (1959) that "modern man should be concerned with achieving the good life, not with raising the [material] standard of living."

An editorial ran in *The New York Times* a week after Krutch's death: today's younger generation, it read, "unfamiliar with Joseph Wood Krutch but concerned about the **environment** and contemptuous of materialism," should "turn to a reading of his books with delight to themselves and profit to the world."

[*Gerald L. Young Ph.D.*]

RESOURCES

BOOKS

Krutch, J. W. *The Desert Year.* New York: Viking, 1951.

Margolis, J. D. *Joseph Wood Krutch: A Writer's Life.* Knoxville: The University of Tennessee Press, 1980.

Pavich, P. N. *Joseph Wood Krutch.* Western Writers Series, no. 89. Boise: Boise State University, 1989.

PERIODICALS

Gorman, J. "Joseph Wood Krutch: A Cactus Walden." *MELUS: The Journal of the Society for the Study of the Multi-Ethnic Literature of the United States* 11 (Winter 1984): 93–101.

Holtz, W. "Homage to Joseph Wood Krutch: Tragedy and the Ecological Imperative." *The American Scholar* 43 (Spring 1974): 267–279.

Lehman, A. L. "Joseph Wood Krutch: A Selected Bibliography of Primary Sources." *Bulletin of Bibliography* 41 (June 1984): 74–80.

Kudzu

Pueraria lobata or kudzu, also jokingly referred to as "foot-a-night" and "the vine that ate the South," is a highly aggressive and persistent semi-woody vine introduced to the United States in the late nineteenth century. It has since become a symbol of the problems possible for native ecosystems caused by the introduction of **exotic species**. Kudzu's best known characteristic is its extraordinary capacity for rapid growth, managing as much as 12 in (30.5 cm) a day and 60–100 ft (18–30 m) a season under ideal conditions. When young, kudzu has thin, flexible, downy stems that grow outward as well as upward, eventually covering virtually everything in its path with a thick mat of leaves and tendrils. This lateral growth creates the dramatic effect, common in southeastern states such as Georgia, of telephone poles, buildings, neglected vehicles, and whole areas of woodland being enshrouded in blankets of kudzu. Kudzu's tendency towards aggressive and overwhelming colonization has many detrimental effects, killing stands of trees by robbing them of sunlight and pulling down or shorting out utility cables. Where stem nodes touch the ground, new roots develop which can extend 10 ft (3 m) or more underground and eventually weigh several hundred pounds. In the nearly ideal **climate** of the Southeast, the prolific vine easily overwhelms virtually all native competitors and also infests cropland and yards.

A member of the pea family, kudzu is itself native to China and Japan. Introduced to the United States at the Japanese garden pavilion during the 1876 Philadelphia Centennial Exhibition, kudzu's broad leaves and richly fragrant reddish-purple blooms made it seem highly desirable as an ornamental plant in American gardens. It now ranges along the eastern seaboard from Florida to Pennsylvania, and westward to Texas. Although hardy, kudzu does not tolerate cold weather and prefers acidic, well-drained soils and bright sunlight. It rarely flowers or sets seed in the northern part of its range and loses its leaves at first frost.

For centuries, the Japanese have cultivated kudzu for its edible roots, medicinal qualities, and fibrous leaves and stems, which are suitable for paper production. After its initial introduction as an ornamental, kudzu also was touted as a forage crop and as a cure for **erosion** in the United States. Kudzu is nutritionally comparable to alfalfa, and its tremendous durability and speed of growth were thought to outweigh the disadvantages caused for cutting and baling by its rope-like vines. But its effectiveness as a ground cover, particularly on steeply-sloped terrain, is responsible for kudzu's spectacular spread. By the 1930s, the United States **Soil Conservation Service** was enthusiastically advocating kudzu as a remedy for erosion, subsidizing farmers as well as highway departments and railroads, with as much as $8 an acre to use kudzu for **soil** retention. The Depression-era Civilian Conservation Corps also facilitated the spread of kudzu, planting millions of seedlings as part of an extensive erosion control project.

Kudzu also has had its unofficial champions, the best known of whom is Channing Cope of Covington, Georgia. As a journalist for Atlanta newspapers and popular radio broadcaster, Cope frequently extolled the virtues of kudzu, dubbing it the "miracle vine" and declaring that it had replaced cotton as "King" of the South. The spread of the vine was precipitous. In the early 1950s, the federal government began to question the wisdom of its support for kudzu. By 1953, the Department of Agriculture stopped recommending the use of kudzu for either fodder or ground cover. In 1982, kudzu was officially declared a weed.

Funding is now directed more at finding ways to eradicate kudzu or at least to contain its spread. Continuous overgrazing by livestock will eventually eradicate a field of kudzu, as will repeated applications of defoliant herbicides. Even so, stubborn patches may take five or more years to be completely removed. Controlled burning is usually ineffective and attempting to dig up the massive root system is generally an exercise in futility, but kudzu can be kept off lawns and fences (as an ongoing project) by repeated mowing and enthusiastic pruning.

A variety of new uses are being found for kudzu, and some very old uses are being rediscovered. Kudzu root can be processed into flour and baked into breads and cakes; as a starchy sweetener, it also may be used to flavor soft drinks. Medical researchers investigating the scientific bases of traditional herbal remedies have suggested that isoflavones found in kudzu root may significantly reduce craving for alcohol in alcoholics. Eventually, derivatives of kudzu may also prove to be useful for treatment of high blood pressure. **Methane** and **gasohol** have been successfully produced from kudzu, and kudzu's stems may prove to be an economically viable source of fiber for paper production and other purposes. The prolific vine has also become something of a humorous cultural icon, with regional picture postcards throughout the south portraying spectacular and only somewhat exaggerated images of kudzu's explosive growth. Fairs, festivals, restaurants, bars, rock groups, and road races have all borrowed their name and drawn some measure of inspiration from kudzu, poems have been written about it, and kudzu cookbooks and guides to kudzu crafts are readily available in bookstores.

[*Lawrence J. Biskowski*]

RESOURCES

PERIODICALS

Dolby, V. "Kudzu Grows Beyond Erosion Control to Help Control Alcoholism." *Better Nutrition* 58, no. 11 (November 1996): 32.

Hipps, C. "Kudzu." *Horticulture* 72, no. 6 (June 1994): 36–9.

Tenenbaum, D. "Weeds from Hell," *Technology Review* 99, no. 6 (August 1996): 32–40.

Kwashiorkor

One of many severe protein energy malnutrition disorders that are a widespread problem among children in developing countries. The word's origin is in Ghana, where it means a deposed child, or a child that is no longer suckled. The disease usually affects infants between one and four years of age who have been weaned from breast milk to a high starch, low protein diet. The disease is characterized by lethargy, apathy, or irritability. Over time the individual will experience retarded growth processes both physically and mentally. Approximately 25% of children suffer from recurrent relapses of kwashiorkor, interfering with their normal growth.

Kwashiorkor results in amino **acid** deficiencies which inhibit protein synthesis in all tissues. The lack of sufficient **plasma** proteins, specifically albumin, results in **systemic** pressure changes, ultimately causing generalized edema. The liver swells with stored fat because there are no hepatic proteins being produced for digestion of fats. Kwashiorkor additionally results in reduced bone density and impaired renal function. If treated early on in its development the disease can be reversed with proper dietary therapy and treatment of associated infections. If the condition is not reversed in its early stages, prognosis is poor and physical and mental growth will be severely retarded. *See also* Sahel; Third World

Kyoto Protocol/Treaty

In the mid-1980s, a growing body of scientific evidence linked man-made greenhouse gas emissions to global warming. In 1990, the United Nations General Assembly issued a report that confirmed this link. The Rio Accord of 1992 resulted from this report. Formally called the United Nations Framework Convention on Climate Change (UNFCC), the accord was signed by various nations in Rio de Janeiro, Brazil and committed industrialized nations to stabilizing their emissions at 1990 levels by 2000.

In December 1997, representatives of 160 nations met in Kyoto, Japan, in an attempt to produce a new and improved treaty on climate change. Major differences occurred between industrialized and still developing countries with the United States perceived, particularly by representatives of the **European Union** (EU), as not doing its share to reduce emissions, especially those of **carbon dioxide**.

The outcome of this meeting, the Kyoto Protocol to the United Nations Framework Convention on Climate Change (UNFCCC), required industrialized nations to reduce their emissions of **carbon** dioxide, **methane**, **nitrous oxide**, hydrofluorocarbons, sulfur dioxides, and perfluorocarbons below 1990 levels by 2012. The requirements would be different for each country and would have to begin by 2008 and be met by 2012. There would be no requirements for the developing nations. Whether or not to sign and ratify the treaty was left up to the discretion of each individual country.

Global warming

The organization that provided the research for the Kyoto Protocol was the **Intergovernmental Panel on Climate Change** (IPCC), set up in 1988 as a joint project of the **United Nations Environment Programme** (UNEP) and the World Meteorological Organization (WMO). In 2001 the IPCC released a report, "Climate Change, 2001". Using the latest climatic and atmospheric scientific research available, the report predicted that global mean surface temperatures on earth would increase by 2.5–10.4°F (1.5–5.9 °C) by the year 2100, unless greenhouse gas emissions were reduced well below current levels. This warming trend was seen as rapidly accelerating, with possible dire consequences to human society and the **environment**. These accelerating temperature changes were expected to lead to rising sea levels, melting glaciers and polar ice packs, heat waves, droughts and wildfires, and a profound and deleterious effect on human health and well-being.

Some of the effects of these temperature changes may already be occurring. Most of the United States has already experienced increases in mean annual temperature of up to 4°F (2.3°C). Sea ice is melting in both the **Antarctica** and the Arctic. Ninety-eight percent of the world's glaciers are shrinking. The sea level is rising at three times its historic rate. Florida is already feeling the early effects of global warming with shorelines suffering from **erosion**, with dying coral reefs, with saltwater polluting the fresh water sources, with an increase in wildfires, and with higher air and water temperatures. In Canada, forest fires have more than doubled since 1970, water **wells** are going dry, lake levels are down, and there is less rainfall.

Controversy

Since its inception, the Kyoto Protocol has generated a great deal of controversy. Richer nations have argued that the poorer, less developed nations are getting off easy. The developing nations, on the other hand, have argued that they will never be able to catch up with the richer nations unless they are allowed to develop with the same degree of **pollution** as that which let the industrial nations become rich in the first place.

Another controversy rages between environmentalists and big business. Environmentalists have argued that the Kyoto Protocol doesn't go far enough, while **petroleum** and industry spokespersons have argued that it would be impossible to implement without economic disaster.

In the United States, the controversy has waxed especially high. The Kyoto Protocol was signed under the administration of President Bill Clinton, but was never ratified by the Republican-dominated U.S. Senate. Then in 2001, President George W. Bush, a former Texas oilman, backed out of the treaty, saying it would cost the U.S. economy 400 billion dollars and 4.9 million jobs. Bush unveiled an alternative proposal to the Kyoto accord that he said would reduce **greenhouse gases**, curb pollution and promote **energy efficiency**. But critics of his plan have argued that by the year 2012 it would actually increase the 1990 levels of greenhouse gas emissions by more than 30%.

Soon after the Kyoto Protocol was rejected by the Bush administration, the European Union criticized the action. In particular, Germany was unable to understand why the Kyoto restrictions would adversely effect the American economy, noting that Germany had been able to reduce their emissions without serious economic problems. The Germans also suggested that President Bush's program to induce voluntary reductions was politically motivated and was designed to prevent a drop in the unreasonably high level of consumption of greenhouse gases in the United States, a drop that would be politically damaging for the Bush administration.

In rejecting the Kyoto Protocol, President Bush claimed that it would place an unfair burden on the United States. He argued that it was unfair that developing countries such as India and China should be exempt. But China had already taken major steps to affect climate change. According to a June report by the **World Resources Institute**, a Washington, D.C.-based environmental think tank, China voluntarily cut its carbon dioxide emissions by 19% between 1997 and 1999. Contrary to Bush's fears that cutting carbon dioxide output would inevitably damage the United States economy, China's economy grew by 15% during this same two-year period.

Politics has always been at the forefront of this debate. The IPCC has provided assessments of climate change that have helped shape international treaties, including the Kyoto Protocol. However, the Bush administration, acting at the request of ExxonMobil, the world's largest oil company, and attempting to cast doubts upon the scientific integrity of the IPCC, was behind the ouster in 2002 of IPCC chairperson Robert Watson, an atmospheric scientist who supported implementing actions against global warming.

The ability of trees and plants to fix carbon through the process of **photosynthesis**, a process called *carbon or C sequestration*, results in a large amount of carbon stored in **biomass** around the world. In the framework of the Kyoto Protocol, C sequestration to mitigate the **greenhouse effect** in the terrestrial **ecosystem** has been an important topic of discussion in numerous recent international meetings and reports. To increase C sequestration in soils in the dryland and tropical areas, as a contribution to global reductions of atmospheric CO_2, the United States has promoted new strategies and new practices in agriculture, pasture use and forestry, including **conservation** agriculture and **agroforestry**. Such practices should be facilitated particularly by the application of article 3.4 of the Kyoto Protocol covering the additional activities in agriculture and forestry in the developing countries and by appropriate policies.

Into the future

In June 2002, the 15 member nations of the European Union formally signed the Kyoto Protocol. The ratification by the 15 EU countries was a major step toward making the 1997 treaty effective. Soon after, Japan signed the treaty, and Russia was expected to follow suit.

To take effect, the Kyoto Protocol must be ratified by 55 countries, but these ratifications have to include industrialized nations responsible for at least 55% of the 1990 levels of greenhouse gases. As of 2002, over 70 countries had already signed, exceeding the minimum number of countries needed. If Russia signs the treaty, nations responsible for over 55% of the 1990 levels of greenhouse gas pollution will have signed, and the Kyoto Protocol will take effect.

Before the EU ratified the protocol, the vast majority of countries that had ratified were developing countries. With the withdrawal of the United States, responsible for 36.1% of greenhouse gas emissions in 1990, ratification by industrialized nations was crucial. For example, environmentalists hope that Canada will ratify the treaty as it has already committed compliance.

Although the Bush administration opposed the Kyoto Protocol, saying that its own plan of voluntary restrictions would work as well without the loss of billions of dollars and without driving millions of Americans out of work, the EPA, under its administrator Christine Todd Whitman, in 2002 sent a climate report to the United Nations detailing specific, far-reaching, and disastrous effects of global warming upon the American environment and its people. The EPA report also admitted that global warming is occurring because of man-made carbon dioxide and other greenhouse gases. However, it offered no major changes in administration policies, instead recommending adapting to the inevitable and catastrophic changes.

Although the United States was still resisting the Kyoto Protocol in mid-2002, and the treaty's implications for radical and effective action, various states and communities decided to go it alone. Massachusetts and New Hampshire enacted legislation to cut carbon emissions. California was considering legislation limiting emissions from cars and small trucks. Over 100 U.S. cities had already opted to cut carbon emissions. Even the U.S business community, because of their many overseas operations, was beginning to voluntarily cut back on their greenhouse emissions.

[*Douglas Dupler*]

RESOURCES
BOOKS

Brown, Paige. *Climate, Biodiversity and Forests: Issues and Opportunities Emerging from the Kyoto Protocol.* Washington, DC: World Resources Institute, 1998.

Gelbspan, Ross. *The Heat is On: The Climate Crisis, the Cover-up, the Prescription.* New York: Perseus Book, 1998.

McKibbin, Warwick J., and Peter Wilcoxen. *Climate Change Policy After Kyoto: Blueprint for a Realistic Approach.* Washington DC: The Brookings Institution Press, 2002.

Victor, David G. *Collapse of the Kyoto Protocol and the Struggle to Slow Global Warming.* Boston: Princeton University Press, 2002.

PERIODICALS

Benedick, Richard E. "Striking a New Deal on Climate Change." *Issues in Science and Technology,* Fall 2001, 71.

Gelbspan, Ross. "A Modest Proposal to Stop Global Warming." *Sierra,* May/June 2001, 63.

McKibben, Bill. "Climate Change 2001: Third Assessment Report." *New York Review of Books,* July 5, 2001, 35.

Rennie, John. "The Skeptical Environmentalist Replies." *Scientific American,* May 2002, 14.

OTHER

"Guide to the Kyoto Protocol." *Greenpeace International Web Site.* 1998 [cited July 2002]. <http://www.greenpeace.org>.

Intergovernmental Panel on Climate Change Web Site. [cited July 2002]. <http://www.ipcc.ch>.

"Kyoto Protocol." *United Nations Convention of Climate Change.* 1997 [cited July 1997]. <http://unfccc.int>.

Union of Concerned Scientists Global Warming Web Site. [cited July 2002]. <http://www.ucsusa/warming>.

ORGANIZATIONS

IPCC Secretariat, C/O World Meteorological Organization, 7bis Avenue de la Paix, C.P. 2300, CH- 1211, Geneva, Switzerland 41-22-730-8208, Fax: 41-22-730-8025, Email: ipcc.sec@gateway.wmo.ch, <http://www.ipcc.ch>

UNIDO—Climate Change/Kyoto Protocol Activities, UNIDO New York Office, New York, NY USA 10017 (212) 963-6890, Email: office.newyork@unido.org, <http://www.unido.org/doc/310797.htmls>

L

La Niña

La Niña, Spanish for "the little girl," is also called a cold episode, "El Viejo" (The Old Man), or anti-El Niño. It is one of two major changes in the Pacific Ocean surface temperature that affect global weather patterns. La Niña and **El Niño** ("the little boy") are the extreme phases of the El Niño/Southern Oscillation, a **climate** cycle that occurs naturally in the eastern tropical Pacific Ocean. The effects of both phases are usually strongest from December to March. In some ways, La Niña is the opposite of El Niño. For example, La Niña usually brings more rain to **Australia** and Indonesia, areas that are susceptible to **drought** during El Niño.

La Niña is characterized by unusually cold ocean surface temperatures in the equatorial region. Ordinarily, the sea surface temperature off the western coast of South America ranges from 60–70°F (15–21°C). According to the **National Oceanic and Atmospheric Administration** (NOAA), the temperature dropped by up to 7°F (4°C) below normal during the 1988–1989 La Niña. In the United States, La Niña usually brings cooler and wetter than normal conditions in the Pacific Northwest and warmer and drier conditions in the Southeast. In contrast, during El Niño, surface water temperatures in the tropical Pacific are unusually warm. Because water temperatures increase around Christmas, people in South America called the condition "El Niño" to honor the Christ Child.

The two weather phenomena are caused by the interaction of the ocean surface and the **atmosphere** in the tropical Pacific. Changes in the ocean affect the atmosphere and climate patterns around the world, with changes in the atmosphere in turn affecting the ocean temperature and currents. Before the onset of La Niña, there is usually a build-up of cooler than normal subsurface water in the tropical Pacific. The cold water is brought to the surface by atmospheric waves and ocean waves. Winds and currents push warm water towards Asia. In addition, the system can drive the polar jet stream (a stream of winds at high altitude) to the north; this affects weather in the United States.

The effects of La Niña and El Niño are generally seen in the United States during the winter. The two conditions usually occur every three to five years. However, the period between episodes may be from two to seven years. The conditions generally last from nine to 12 months, but episodes could last as long as two years.

Since 1975, El Niños have occurred twice as frequently as La Niñas. While both conditions are cyclical, a La Niña episode does not always follow an El Niño episode. La Niñas in the twentieth century occurred in 1904, 1908, 1910, 1916, 1924, 1928, 1938, 1950, 1955, 1964, 1970, 1973, 1975, 1988, 1995, and 1998. Effects of the 1998 La Niña included **flooding** in Mozambique in 2000 and a record warm winter in the United States. Nationwide temperatures averaged 38.4°F (3.5°C) from December 1999 through February 2000, according to the NOAA. In addition, that three-month period was the sixteenth driest winter in the 105 years that records have been kept by National Climatic Data Center. The 1998 La Niña was diminishing by November 2000; another La Niña had not been projected as of May 2002.

Scientists from the NOAA and other agencies use various tools to monitor La Niña and El Niño. These tools include satellites and data buoys that are used to monitor sea surface temperatures. Tracking the two weather phenomena can help nations to prepare for potential disasters such as floods. In addition, knowledge of the systems can help businesses plan for the future. In a March 1999 article in *Nation's Restaurant News*, writer John T. Barone described the impact that La Niña could have on food and beverage prices. Barone projected that drought conditions in Brazil could bring an increase in the price of coffee.

[*Liz Swain*]

RESOURCES

BOOKS

Caviedes, Cesar. *El Niño in History: Storming Through the Ages.* Gainesville, FL: University Press of Florida, 2001.

Glantz, Michael. *Currents of Change: Impacts of El Niño and La Niña on Climate and Society.* New York: Cambridge University Press, 2001.

PERIODICALS

Barone, John T. "La Niña to Put a Chill in Prices this Winter." *Nation's Restaurant News* (December 6, 1999): 78.
Le Comte, Douglas. "Weather Around the World." *Weatherwise* (March 2001): 23.

ORGANIZATIONS

National Oceanic and Atmospheric Administration, 14th Street & Constitution Avenue, NW, Room 6013, Washington, DC USA 20230 (202) 482-6090, Fax: (202) 482-3154, Email: answers@noaa.gov, <http://www.noaa.gov>

La Paz Agreement

The 1983 La Paz Agreement between the United States and Mexico is a pact to protect, conserve, and improve the **environment** of the border region of both countries. The agreement defined the region as the 62 mi (100 km) to the north and south of the international border. This area includes maritime (sea) boundaries and land in four American states and six Mexican border states.

Representatives from the two countries signed the agreement on Aug. 14, 1983, in La Paz, Mexico. The agreement took effect on Feb. 16, 1984. It established six workgroups, with each group concentrating on an environmental concern. Representatives from both countries serve on the workgroups that focus on water, air, hazardous and **solid waste**, **pollution** prevention, contingency planning and emergency response, and cooperative enforcement and compliance.

In February of 1992, environmental officials from the two countries released the Integrated Environmental Plan for the Mexican-U.S. Border Area. The Border XXI Program created nine additional workgroups. These groups focus on environmental information resources, **natural resources**, and **environmental health**.

Border XXI involves federal, state, and local governments on both sides of the border. Residents also participate through activities such as public hearings.

[*Liz Swain*]

Lagoon

A lagoon is a shallow body of water separated from a larger, open body of water. It is typically associated with the ocean, such as coastal lagoons and **coral reef** lagoons. Lagoon also can be used to describe shallow areas of liquid waste material as in sewage lagoons. Oceanic lagoons can be formed in several ways. Coastal lagoons are typically found along coastlines where there are sand bars or barrier islands that separate the open ocean from the near shore body of water. Coral reef lagoons can form in two ways. The first type is found in barrier reefs such as those in **Australia** and Belize, where there is a body of water (lagoon) which is separated from the open ocean by the reef formed many miles off shore. Another type of lagoon is that formed in the center of atolls, which are circular or horse-shoe shaped bodies of water in the middle of partially sunken volcanic islands with coral reefs growing around their periphery. Some of these atoll lagoons are more than 30 mi (50 km) across and have breathtaking **visibility**, thus providing superb sites for SCUBA diving.

Lake Baikal

The **Great Lakes** are a prominent feature of the North American landscape, but Russia holds the distinction of having the "World's Great Lake." Called the "Pearl of Siberia" or the "Sacred Sea" by locals, Lake Baikal is the world's deepest and largest lake by volume. It has a surface area of 12,162 sq miles (31,500 sq km), a maximum depth of 5,370 ft (1,637 m), or slightly more than 1 mile, an average depth of 2,428 ft (740 m), and a volume of 30,061 cu yd (23,000 cu m). It thus contains more water than the combined volume of all of the North American Great Lakes—20 percent of the world's fresh water (and 80 percent of the fresh water of the former Soviet Union).

Lake Baikal is located in Russia in south-central Siberia near the northern border of Mongolia. Scientists estimate that the lake was formed 25 million years ago by tectonic (**earthquake**) displacement, creating a crescent-shaped, steep-walled basin 395 miles (635 km) long by 50 miles (80 km) wide and nearly 5.6 miles (9 km) deep. In contrast, the Great Lakes were formed by glacial scouring a mere 10,000 years ago.

Although **sedimentation** has filled in 80 percent of the basin over the years, the lake is believed to be widening and deepening ever so slightly with time because of recurring crustal movements. The area surrounding Lake Baikal is underridden by at least three crustal plates, causing frequent earthquakes. Fortunately, most are too weak to feel.

Like similarly ancient Lake Tanganyika in Africa, the waters of Lake Baikal host a great number of unique **species**. Of the 1,200 known animal and 600 known plant species, more than 80 percent are endemic to this lake. These include many species of fish, shrimp, and the world's only fresh water sponges and **seals**. Called *nerpa* or *nerpy* by the natives, these seals (*Phoca sibirica*) are silvery-gray in color and can grow to 5 ft (1.5 m) long and weigh up to 286 lb (130 kg). Their diet consists almost exclusively of a strange-looking relict fish called *golomyanka* (*Comephorus baicalensis*), ren-

A view of the Strait of Olkhon, on the west coast of Lake Baikal, Russia. (Photograph by Press Agency, Science Photo Library. Photo Researchers Inc. Reproduced by permission.)

dered translucent by its fat-filled body. Unlike other fish, they lack scales and swim bladders and give birth to live larvae rather than eggs. The seal population is estimated at 60,000. Commercial hunters are permitted to kill 6,000 each year.

Although the waters of Lake Baikal are pristine by the standards of other large lakes, increased **pollution** threatens its future. Towns along its shores and along the stretches of the Selenga River, the major tributary flowing into Baikal, add human and industrial wastes, some of which is nonbiodegradable and some highly toxic. A hydroelectric dam on the Angara River, the lake's only outlet, raised the water level and placed spawning areas of some fish below the optimum depth. Most controversial to the people who depend on this lake for their livelihood and pleasure, however, was the construction of a large cellulose plant at the southern end near the city of Baikalsk in 1957. Built originally to manufacture high-quality aircraft tires (ironically, synthetic tires proved superior), today it produces clothing from bleached cellulose and employs 3,500 people. Uncharacteristic public outcry over the years has resulted in the addition

of advanced **sewage treatment** facilities to the plant. Although some people would like to see it shut down, the local (and national) economy has taken precedence.

In 1987 the Soviet government passed legislation protecting Lake Baikal from further destruction. **Logging** was prohibited anywhere close to the shoreline and **nature** reserves and national parks were designated. However, with the recent political turmoil and crippling financial situation in the former Soviet Union, these changes have not been enforced and the lake continues to receive pollutants. Much more needs to be done to assure the future of this magnificent lake. *See also* Endemic species

[*John Korstad*]

RESOURCES
BOOKS

Feshbach, M., and A. Friendly, Jr. *Ecocide in the USSR.* New York: Basic Books, 1992.

Matthiessen, P. *Baikal: Sacred Sea of Siberia.* San Francisco: Sierra Club Books, 1992.

PERIODICALS

Belt, D. "Russia's Lake Baikal, the World's Great Lake." *National Geographic* 181 (June 1992): 2–39.

Lake Erie

Lake Erie is the most productive of the **Great Lakes**. Located along the southern fringe of the Precambrian Shield of North America, Lake Erie has been ecologically degraded by a variety of **anthropogenic** stressors including **nutrient loading**; extensive **deforestation** of its **watershed** that caused severe **siltation** and other effects; vigorous **commercial fishing**; and **pollution** by toxic **chemicals**.

The watershed of Lake Erie is much more agricultural and urban in character than are those of the other Great Lakes. Consequently, the dominant sources of **phosphorus** (the most important nutrient causing eutrophication) to Lake Erie are agricultural **runoff** and municipal point sources. The total input of phosphorus to Lake Erie (standardized to watershed area) is about 1.3 times larger than to Lake Ontario and more than five times larger than to the other Great Lakes.

Because of its large loading rates and concentrations of nutrients, Lake Erie is more productive and has a larger standing crop of **phytoplankton**, fish, and other biota than the other Great Lakes. During the late 1960s and early 1970s, the eutrophic western basin of Lake Erie had a summer-chlorophyll concentration averaging twice as large as in Lake Ontario and 11 times larger than in **oligotrophic** Lake Superior. However, since that time the eutrophication of Lake Erie has been alleviated somewhat, in direct response to decreased phosphorus inputs with sewage and **detergents**. A consequence of the eutrophic state of Lake Erie was the development of anoxia (lack of oxygen) in its deeper waters during summer **stratification**. In the summer of 1953, this condition caused a collapse of the population of benthic mayfly larvae (*Hexagenia* spp.), a phenomenon that was interpreted in the popular press as the "death" of Lake Erie.

Large changes have also taken place in the fish community of Lake Erie, mostly because of its fishery, the damming of streams required for spawning by anadromous fishes (fish that ascend rivers or streams to spawn), and **sedimentation** of shallow-water **habitat** by **silt** eroded from deforested parts of the watershed. Lake Erie has always had the most productive fishery on the Great Lakes, with fish landings that typically exceed the combined totals of all the other Great Lakes. The peak years of commercial fishery in Lake Erie were in 1935 and 1956 (62 million lb/28 million kg), while the minima were in 1929 and 1941 (24 million lb/11 million kg). Overall, the total catch by the commercial fishery has been remarkably stable over time, despite large changes in **species**, effort, eutrophication, toxic pollution, and other changes in habitat.

The historical pattern of development of the Lake Erie fishery was characterized by an initial exploitation of the most desirable and valuable species. As the populations of these species collapsed because of unsustainable fishing pressure, coupled with habitat deterioration, the fishery diverted to a progression of less-desirable species. The initial fishery focused on lake white fish (*Coregonus clupeaformis*), lake trout (*Salvelinus namaycush*), and lake herring (*Leucichthys artedi*), all of which rapidly declined to **scarcity** or **extinction**. The next target was "second-choice" species, such as blue pike (*Stizostedion vitreum glaucum*) and walleye (*S. v. vitreum*), which are now extinct or rare. Today's fishery is dominated by species of much smaller economic value, such as yellow perch (*Perca flavescens*), rainbow smelt (*Osmerus mordax*), and carp (*Cyprinus carpio*).

In 1989 an invasive species—the **zebra mussel** (*Dreissena polymorpha*—reached Lake Erie and began to have a significant ecological impact on the lake. Zebra mussels are filter feeders, and each adult mussel can filter a liter of water per day, removing every microscopic plant (phytoplankton or algae) and animal (**zooplankton**) in the process. Zebra mussel densities in Lake Erie have reached such a level that the entire volume of the lake's western basin is filtered each week. This has increased water clarity up to 600 percent and reduced some forms of phytoplankton in the lake's food web by as much as 80 percent. In addition, the increased clarity of the water allows light to penetrate deeper into the water, thus facilitating the growth of rooted aquatic plants and increasing populations of some bottom-dwelling algae and tiny animals. Zebra mussels also concentrate 10 times more **toxins** than do native mussels, and these contaminants are passed up the food chain to the fish and birds that eat zebra mussels. Since **bioaccumulation** of toxins has already led to advisories against eating some species of Great Lakes fish, the contribution of zebra mussels to contaminant cycling in lake species is a serious concern. *See also* Cultural eutrophication; Water pollution

[*Bill Freedman Ph.D.*]

RESOURCES

BOOKS

Ashworth, W. *The Late, Great Lakes: An Environmental History.* New York: Knopf, 1986.

Freedman, B. *Environmental Ecology.* 2nd edition San Diego: Academic Press, 1995.

PERIODICALS

Regier, H. A., and W. L. Hartman. "Lake Erie's Fish Community: 150 Years of Cultural Stresses." *Science* 180 (1973): 1248–55.

OTHER

"Zebra Mussels and Other Nonindigenous Species." *Sea Grant Great Lakes Network.* August 15, 2001 [June 19, 2002]. <http://www.seagrant.wisc.edu/greatlakes/glnetwork/exotics.html>.

Lake Tahoe

A beautiful lake 6,200 ft (1,891 m) high in the Sierra Nevada, straddling the California-Nevada state line, Lake Tahoe is a jewel to both nature-lovers and developers. It is the tenth deepest lake in the world, with a maximum depth of 1,600 ft (488 m) and a total volume of 37 trillion gallons. At the south end of the lake sits a dam that supplies up to six feet of Lake Tahoe's water flow into the outlet of the Truckee River. The U.S. **Bureau of Reclamation** controls water diversion into the Truckee, which is used for **irrigation**, power, and recreational purposes throughout Nevada.

Tahoe and Crater Lake are the only two large alpine lakes remaining in the United States. Visitors have expressed their awe of the lake's beauty since it was discovered by General John Frémont in 1844. Mark Twain wrote that it was "the fairest sight the whole Earth affords."

The arrival of Europeans in the Tahoe area was quickly followed by environmental devastation. Between 1870 and 1900, forests around the lake were heavily logged to provide timber for the mine shafts of the Comstock Lode. While this **logging** dramatically altered the area's appearance for years, the natural **environment** eventually recovered and no long-term logging-related damage to the lake can now be detected.

The same can not be said for a later assault on the lake's environment. Shortly after World War II, people began moving into the area to take advantage of the region's natural wonders—the lake itself and superb snow skiing—as well as the young casino business on the Nevada side of the lake. The 1960 Winter Olympics, held at Squaw Valley, placed Tahoe's recreational assets in the international spotlight. Lakeside population grew from about 20,000 in 1960 to more than 65,000 today, with an estimated tourist population of 22 million annually.

The impact of this rapid **population growth** soon became apparent in the lake itself. Early records showed that the lake was once clear enough to allow **visibility** to a depth of about 130 ft (40 m). By the late 1960s, that figure had dropped to about 100 ft (30 m).

Tahoe is now undergoing eutrophication at a fairly rapid rate. Algal growth is being encouraged by sewage and **fertilizer** produced by human activities. Much of the area's natural **pollution** controls, such as trees and plants, have been removed to make room for residential and commercial development. The lack of significant flow into and out of the lake also contributes to a favorable environment for algal growth.

Efforts to protect the pristine beauty of Lake Tahoe go back at least to 1912. Three efforts were made during that decade to have the lake declared a **national park**, but all failed. By 1958, concerned conservationists had formed the Lake Tahoe Area Council to "promote the preservation and long-range development of the Lake Tahoe basin." The Council was followed by other organizations with similar objectives, the League to Save Lake Tahoe among them.

An important step in resolving the conflict between preservationists and developers occurred in 1969 with the creation of the Tahoe Regional Planning Agency (TRPA). The agency was the first and only **land use** commission with authority in more than one state. It consisted of fourteen members, seven appointed by each of the governors of the two states involved, California and Nevada. For more than a decade, the agency attempted to write a land-use plan that would be acceptable to both sides of the dispute. The conflict became more complex when the California Attorney General, John Van de Kamp, filed suit in 1985 to prevent TRPA from granting any further permits for development. Developers were outraged but lost all of their court appeals.

By 2000, the strain of tourism, development, and non-point **automobile** pollution was having a visible impact on Lake Tahoe's legendary deep blue surface. A study released by the University of California—Davis and the University of Nevada—Reno reported that visibility in the lake had decreased to 70 ft (21 m), an average decline of a foot a year since the 1960s. As part of a renewed effort to reverse Tahoe's environmental decline, President Clinton signed the Lake Tahoe Restoration Act into law in late 2000, authorizing $300 million towards restoration of **water quality** in Lake Tahoe over a period of 10 years. *See also* Algal bloom; Cultural eutrophication; Environmental degradation; Fish kills; Sierra Club; Water pollution

[*David E. Newton and Paula Anne Ford-Martin*]

RESOURCES
BOOKS

Strong, Douglas. *Tahoe: From Timber Barons to Ecologists.* Lincoln, NE: Bison Books, 1999.

OTHER

United States Department of Agriculture (USDA) Forest Service. *Lake Tahoe Basin Management Unit.* [cited July 8, 2002]. <http://www.r5.fs.fed.us/ltbmu>.

University of California-Davis. *Tahoe Research Group.* [cited July 8, 2002]. <http://trg.ucdavis.edu/default.html>.

United States Geological Survey (USGS) Lake Tahoe Data Clearinghouse. *Lake Tahoe Data Clearinghouse.* [cited July 8, 2002]. <http://tahoe.usgs.gov/intro.html>.

Lake Washington

One of the great messages to come out of the environmental movement of the 1960s and 1970s is that, while humans can cause **pollution**, they can also clean it up. Few success stories illustrate this point as clearly as that of Lake Washington. Lake Washington lies along the state of Washington's west coastline, near the city of Seattle. It is 24 miles (39 km) from north to south and its width varies from 2–4 miles (3–6 km).

For the first half of this century, Lake Washington was clear and pristine, a beautiful example of the Northwest's spectacular natural scenery. Its shores were occupied by extensive wooded areas and a few small towns with populations of no more than 10,000. The lake's purity was not threatened by Seattle, which dumped most of its wastes into Elliot Bay, an arm of Puget Sound. This situation changed rapidly during and after World War II. In 1940, the spectacular Lake Washington Bridge was built across the lake, joining its two facing shores with each other and with Seattle. Population along the lake began to boom, reaching more than 50,000 by 1950.

The consequence of these changes for the lake are easy to imagine. Many of the growing communities dumped their raw sewage directly into the lake or, at best, passed their wastes though only preliminary treatment stages. By one estimate, 20 million gallons (76 million liters) of wastes were being dumped into the lake each day. On average these wastes still contained about half of their pollutants when they reached the lake. In less than a decade, the effect of these practices on lake **water quality** were easy to observe. Water clarity was reduced from at least 15 ft (4.6 m) to 2.5 ft (0.8 m) and levels of **dissolved oxygen** were so low that some **species** of fish disappeared. In 1956, W. T. Edmonson, a zoologist and pollution authority, and two colleagues reported their studies of the lake. They found that eutrophication of the lake was taking place very rapidly as a result of the dumping of domestic wastes into its water.

Solving this problem was especially difficult because **water pollution** is a regional issue over which each individual community had relatively little control. The solution appeared to be the creation of a new governmental body that would encompass all of the Lake Washington communities, including Seattle. In 1958, a ballot measure establishing such an agency, known as Metro, was passed in Seattle but defeated in its suburbs. Six months later, the Metro concept was redefined to include the issue of sewage disposal only. This time it passed in all communities.

Metro's approach to the Lake Washington problem was to construct a network of sewer lines and **sewage treatment** plants that directed all sewage away from the lake and delivered it instead to Puget Sound. The lake's

pollution problems were solved within a few years. By 1975 the lake was back to normal, water clarity returned to 15 ft and levels of potassium and **nitrogen** in the lake decreased by more than 60 percent. Lake Washington's biological oxygen demand (BOD), a critical measure of water purity, decreased by 90 percent and fish species that had disappeared were once again found in the lake. *See also* Aquatic chemistry; Cultural eutrophication; Waste management; Water quality standards

[*David E. Newton*]

RESOURCES
BOOKS

Edmonson, W. T. "Lake Washington." In *Environmental Quality and Water Development,* edited by C. R. Goodman, et al. San Francisco: W. H. Freeman, 1973.

———. *The Uses of Ecology: Lake Washington and Beyond.* Seattle: University of Washington Press, 1991.

OTHER

Li, Kevin. "The Lake Washington Story." *King County Web Site.* May 2, 2001 [June 19,2002]. <http://dnr.metrokc.gov/wlr/waterres/lakes/biolake.htm>.

Lakes

see **Experimental Lakes Area; Lake Baikal; Lake Erie; Lake Tahoe; Lake Washington; Mono Lake; National lakeshore**

Land degradation

see **Desertification**

Land ethic

Land ethic refers to an approach to issues of **land use** that emphasizes **conservation** and respect for our natural **environment**. Rejecting the belief that all **natural resources** should be available for unchecked human exploitation, a land ethic advocates land use without undue disturbances of the complex, delicately balanced ecological systems of which humans are a part. Land ethic, **environmental ethics**, and ecological ethics are sometimes used interchangeably.

Discussions of land ethic, especially in the United States, usually begin with a reference of some kind to **Aldo Leopold**. Many participants in the debate over land and resource use admire Leopold's prescient and pioneering quest and date the beginnings of a land ethic to his *A Sand County Almanac,* published in 1949. However, Leopold's earliest formulation of his position may be found in "A Conservation Ethic," a benchmark essay on ethics published in 1933.

Even recognizing Leopold's remarkable early contribution, it is still necessary to place his pioneer work in a larger context. Land ethic is not a radically new invention of the twentieth century but has many ancient and modern antecedents in the Western philosophical tradition. The Greek philosopher Plato, for example, wrote that morality is "the effective harmony of the whole"—not a bad statement of an ecological ethic. Reckless exploitation has at times been justified as enjoying divine sanction in the Judeo-Christian tradition (man was made master of the creation, authorized to do with it as he saw fit). However, most Christian thought through the ages has interpreted the proper human role as one of careful husbandry of resources that do not, in fact, belong to humans. In the nineteenth century, the Huxleys, Thomas and Julian, worked on relating **evolution** and ethics. The mathematician and philosopher Bertrand Russell wrote that "man is not a solitary animal, and so long as social life survives, self-realization cannot be the supreme principle of ethics." **Albert Schweitzer** became famous—at about the same time that Leopold formulated a land ethic—for teaching reverence for life, and not just human life. Many non-western traditions also emphasize harmony and a respect for all living things. Such a context implies that a land ethic cannot easily be separated from age-old thinking on ethics in general. *See also* Land stewardship

[*Gerald L. Young and Marijke Rijsberman*]

RESOURCES

BOOKS

Bormann, F. H., and S. R. Kellert, eds. *Ecology, Economics, Ethics: The Broken Circle*. New Haven, CT: Yale University Press, 1991.

Kealey, D. A. *Revisioning Environmental Ethics*. Albany: State University of New York Press, 1989.

Leopold, A. *A Sand County Almanac*. New York: Oxford University Press, 1949.

Nash, R. F. *The Rights of Nature: A History of Environmental Ethics*. Madison: University of Wisconsin Press, 1989.

Rolston, H. *Environmental Ethics*. Philadelphia: Temple University Press, 1988.

Turner, F. "A New Ecological Ethics." In *Rebirth of Value*. Albany: State University of New York Press, 1991.

OTHER

Callicott, J. Baird. "The Land Ethic: Key Philosophical and Scientific Challenges." October 15, 1998 [June 19, 2002]. <http://www.orst.edu/dept/philosophy/ideas/leopold/presentations/callicott/pres-03.html>.

Land Institute

Founded in 1976 by Wes and Dana Jackson, the Land Institute is both an independent agricultural research station and a school devoted to exploring and developing alternative agricultural practices. Located on the Smoky Hill River near Salina, Kansas, the Institute attempts—in Wes Jackson's words—to "make **nature** the measure" of human activities so that humans "meet the expectations of the land," rather than abusing the land for human needs. This requires a radical rethinking of traditional and modern farming methods. The aim of the Land Institute is to find "new roots for agriculture" by reexamining its traditional assumptions.

In traditional tillage farming, furrows are dug into the **topsoil** and seeds planted. This leaves precious topsoil exposed to **erosion** by wind and water. Topsoil loss can be minimized but not eliminated by **contour plowing**, the use of windbreaks, and other means. Although critical of traditional tillage agriculture, Jackson is even more critical of the methods and machinery of modern industrial agriculture, which in effect trades topsoil for high crop yields (roughly one bushel of topsoil is lost for every bushel of corn harvested). It also relies on plant monocultures—genetically uniform strains of corn, wheat, soybeans, and other crops. These crops are especially susceptible to disease and insect infestations and require extensive use of pesticides and herbicides which, in turn, kill useful creatures (for example, worms and birds), pollute streams and **groundwater**, and produce other destructive side effects. Although spectacularly successful in the short run, such an agriculture is both non-sustainable and self-defeating. Its supposed strengths—its productivity, its efficiency, its economies of scale—are also its weaknesses. Short-term gains in production do not, Jackson argues, justify the longer term depletion of topsoil, the diminution of genetic diversity, and such social side-effects as the disappearance of small family farms and the abandonment of rural communities.

If these trends are to be questioned—much less slowed or reversed—a practical, productive, and feasible alternative agriculture must be developed. To develop such a workable alternative is the aim of the Land Institute. The Jacksons and their associates are attempting to devise an alternative vision of agricultural possibilities. This begins with the important but oft-neglected truism that agriculture is not self-contained but is intertwined with and dependent on nature. The Institute explores the feasibility of alternative farming methods that might minimize or even eliminate the planting and harvesting of annual crops, turning instead to "herbaceous perennial seed-producing polycultures" that protect and bind topsoil. Food grains would be grown in pasture-like fields and intermingled with other plants that would replenish lost **nitrogen** and other nutrients, without relying on chemical fertilizers. Covered by a rooted living net of diverse plant life, the **soil** would at no time be exposed to erosion and would be aerated and rejuvenated by natural means. And the farmer, in symbiotic partnership, would take nature as the measure of his methods and results.

The experiments at the Land Institute are intended to make this vision into a workable reality. It is as yet too early to tell exactly what these continuing experiments might yield. But the re-visioning of agriculture has already begun and continues at the Land Institute.

[*Terence Ball*]

RESOURCES

ORGANIZATIONS

The Land Institute, 2440 E. Water Well Road, Salina, KS USA 67401 (785) 823-5376, Fax: (785) 823-8728, Email: thelandweb@landinstitute.org, <http://www.landinstitute.org>

Land reform

Land reform is a social and political restructuring of the agricultural systems through redistribution of land. Successful land reform policies take into account the political, social, and economic structure of the area.

In agrarian societies, large landowners typically control the wealth and the distribution of food. Land reform policies in such societies allocate land to small landowners, to farm workers who own no land, to collective farm operations, or to state farm organizations. The exact nature of the allocation depends on the motivation of those initiating the changes. In areas where absentee ownership of farmland is common, land reform has become a popular method for returning the land to local ownership. Land reforms generally favor the family-farm concept, rather than absentee landholding.

Land reform is often undertaken as a means of achieving greater social equality, but it can also increase agricultural productivity and benefit the **environment**. A tenant farmer may have a more emotional and protective relation to the land he works, and he may be more likely to make agricultural decisions that benefit the **ecosystem**. Such a farmer might, for instance, opt for natural **pest** control. An absentee owner often does not have the same interest in **land stewardship**.

Land reform does have negative connotations and is often associated with the state collective farms under communism. Most proponents of land reform, however, do not consider these collective farms good examples, and they argue that successful land reform balances the factors of production so that the full agricultural capabilities of the land can be realized. Reforms should always be designed to increase the efficiency and economic viability of farming.

Land reform is usually more successful if it is enacted with agrarian reforms, which may include the use of agricultural extension agents, agricultural cooperatives, favorable labor legislation, and increased public services for farmers,

such as health care and education. Without these measures land reform usually falls short of redistributing wealth and power, or fails to maintain or increase production. *See also* Agricultural pollution; Sustainable agriculture; Sustainable development

[*Linda Rehkopf*]

RESOURCES

BOOKS

Mengisteab, K. *Ethiopia: Failure of Land Reform and Agricultural Crisis.* Westport, CT: Greenwood Publishing Group, 1990.

PERIODICALS

Perney, L. "Unquiet on the Brazilian Front." *Audubon* 94 (January-February 1992): 26–9.

Land stewardship

Little has been written explicitly on the subject of land stewardship. Much of the literature that does exist is limited to a biblical or theological treatment of stewardship. However, literature on the related ideas of sustainability and the **land ethic** has expanded dramatically in recent years, and these concepts are at the heart of land stewardship.

Webster's and the *Oxford English Dictionary* both define a "steward" as an official in charge of a household, church, estate, or governmental unit, or one who makes social arrangements for various kinds of events; a manager or administrator. Similarly, stewardship is defined as doing the job of a steward or, in ecclesiastical terms, as "the responsible use of resources," meaning especially money, time and talents, "in the service of God."

Intrinsic in those restricted definitions is the idea of responsible caretakers, of persons who take good care of the resources in their charge, including **natural resources**. "Caretaking" universally includes caring for the material resources on which people depend, and by extension, the land or **environment** from which those resources are extracted. Any concept of steward or stewardship must include the notion of ensuring the essentials of life, all of which derive from the land.

While there are few works written specifically on land stewardship, the concept is embedded implicitly and explicitly in the writings of many articulate environmentalists. For example, Wendell Berry, a poet and essayist, is one of the foremost contemporary spokespersons for stewardship of the land. In his books, *Farming: A Handbook* (1970), *The Unsettling of America* (1977), *The Gift of Good Land* (1981), and *Home Economics* (1987), Berry shares his wisdom on caring for the land and the necessity of stewardship. He finds a mandate for good stewardship in religious traditions, including Judaism and Christianity: "The divine mandate to use the world justly and charitably, then, defines every person's moral predicament as that of a steward." Berry, however, does not leave stewardship to divine intervention. He describes stewardship as "hopeless and meaningless unless it involves long-term courage, perseverance, devotion, and skill" on the part of individuals, and not just farmers. He suggests that when we lost the skill to use the land properly, we lost stewardship.

However, Berry does not limit his notion of stewardship to a biblical or religious one. He lays down seven rules of land stewardship—rules of "living right." These are:

• using the land will lead to ruin of the land unless it "is properly cared for;"

• if people do not know the land intimately, they cannot care for it properly;

• motivation to care for the land cannot be provided by "general principles or by incentives that are merely economic;"

• motivation to care for the land, to live with it, stems from an interest in that land that "is direct, dependable, and permanent;"

• motivation to care for the land stems from an expectation that people will spend their entire lives on the land, and even more so if they expect their children and grandchildren to also spend their entire lives on that same land;

• the ability to live carefully on the land is limited; owning too much acreage, for example, decreases the quality of attention needed to care for the land;

• a nation will destroy its land and therefore itself if it does not foster rural households and communities that maintain people on the land as outlined in the first six rules.

Stewardship implies at the very least then, an attempt to reconnect to a piece of land. Reconnecting means getting to know that land as intimately as possible. This does not necessarily imply ownership, although enlightened ownership is at the heart of land stewardship. People who own land have some control of it, and effective stewardship requires control, if only in the sense of enough power to prevent abuse. But, ownership obviously does not guarantee stewardship—great and widespread abuses of land are perpetrated by owners. Absentee ownership, for example, often means a lack of connection, a lack of knowledge, and a lack of caring. And public ownership too often means non-ownership, leading to the "Tragedy of the Commons." Land ownership patterns are critical to stewardship, but no one type of ownership guarantees good stewardship.

Berry argues that true land stewardship usually begins with one small piece of land, used or controlled or owned by an individual who lives on that land. Stewardship, however, extends beyond any one particular piece of land. It implies

knowledge, and caring for, the entire system of which that land is a part, a knowledge of a land's context as well as its content. It also requires understanding the connections between landowners or land users and the larger communities of which they are a part. This means that stewardship depends on interconnected systems of **ecology** and economics, of politics and science, of sociology and planning. The web of life that exists interdependent with a piece of land mandates attention to a complex matrix of connections. Stewardship means keeping the web intact and functional, or at least doing so on enough land over a long-enough period of time to sustain the populations dependent on that land.

Berry and many other critics of contemporary land-use patterns and policies claim that little attention is being paid to maintaining the complex communities on which sustenance, human and otherwise, depends. Until holistic, ecological knowledge becomes more of a basis for economic and political decision-making, they assert, stewardship of the critical land-base will not become the norm. *See also* Environmental ethics; Holistic approach; Land use; Sustainable agriculture; Sustainable biosphere; Sustainable development

[*Gerald L. Young Ph.D.*]

RESOURCES
BOOKS

Byron, W. J. *Toward Stewardship: An Interim Ethic of Poverty, Power and Pollution*. New York: Paulist Press, 1975.

de Jouvenel, B. "The Stewardship of the Earth." In *The Fitness of Man's Environment*. New York: Harper & Row, 1968.

Knight, Richard L., and Peter B. Landres, eds. *Stewardship Across Boundaries*. Washington, DC: Island Press, 1998.

Paddock, J., N. Paddock, and C. Bly. *Soil and Survival: Land Stewardship and the Future of American Agriculture*. San Francisco: Sierra Club Books, 1986.

Land Stewardship Project

The **Land Stewardship** Project (LSP) is a nonprofit organization based in Minnesota and committed to promoting an ethic of environmental and agricultural stewardship. The group believes that the natural **environment** is not an exploitable resource but a gift given to each generation for safekeeping. To preserve and pass on this gift to **future generations**, for the LSP, is both a moral and a practical imperative.

Founded in 1982, the LSP is an alliance of farmers and city-dwellers dedicated both to preserving the small family farm and practicing **sustainable agriculture**. Like Wendell Berry and **Wes Jackson** (with whom they are affiliated), the LSP is critical of conventional agricultural practices that emphasize plant monocultures, large acreage, intensive tillage, extensive use of herbicides and pesticides, and the economies of scale that these practices make possible. The group believes that agriculture conducted on such an industrial scale is bound to be destructive not only of the natural environment but of family farms and rural communities as well. The LSP accordingly advocates the sort of smaller scale agriculture that, in Berry's words, "depletes neither **soil**, nor people, nor communities."

The LSP sponsors legislative initiatives to save farmland and **wildlife habitat**, to limit **urban sprawl** and protect family farms, and to promote sustainable agricultural practices. It supports educational and outreach programs to inform farmers, consumers, and citizens about agricultural and environmental issues. The LSP also publishes a quarterly *Land Stewardship Letter* and distributes video tapes about sustainable agriculture and other environmental concerns.

[*Terence Ball*]

RESOURCES

ORGANIZATIONS

The Land Stewardship Project, 2200 4th Street, White Bear Lake, MN USA 55110 (651) 653-0618, Fax: (651) 653-0589, Email: lspwbl@landstewardshipproject.org, <http://www.landstewardshipproject.org>

Land trusts

A land trust is a private, legally incorporated, nonprofit organization that works with property owners to protect open land through direct, voluntary land transactions. Land trusts come in many varieties, but their intent is consistent. Land trusts are developed for the purpose of holding land against a development plan until the public interest can be ascertained and served. Some land trusts hold land open until public entities can purchase it. Some land trusts purchase land and manage it for the common good. In some cases land trusts buy development rights to preserve the land area for **future generations** while leaving the current use in the hands of private interests with written documentation as to how the land can be used. This same technique can be used to adjust **land use** so that some part of a parcel is preserved while another part of the same parcel can be developed, all based on land sensitivity.

There is a hierarchy of land trusts. Some trusts protect areas as small as neighborhoods, forming to address one land use issue after which they disband. More often, land trusts are local in nature but have a global perspective with regard to their goals for future land protection. The big national trusts are names that we all recognize such as the Conservation Fund, **The Nature Conservancy**, the **Ameri-**

can **Farmland Trust**, and the Trust for **Public Land**. The Land Trust Alliance coordinates the activities of many land trusts. Currently, there are over 1,200 local and regional land trusts in the United States. Some of these trusts form as a direct response to citizen concerns about the loss of open space. Most land trusts evolve out of citizen's concerns over the future of their state, town, and neighborhood. Many are preceded by failures on the part of local governments to respond to stewardship mandates by the voters.

Land trusts work because they are built by concerned citizens and funded by private donations with the express purpose of securing the sustainability of an acceptable quality of life. Also, land trusts are effective because they purchase land (or development rights) from local people for local needs. Transactions are often carried out over a kitchen table with neighbors discussing priorities. In some cases the trust's board of directors might be engaged in helping a citizen to draw up a will leaving farmland or potential **recreation** land to the community. This home rule concept is the backbone of the land trust movement. Additionally, land trusts gain strength from public/private partnerships that emerge as a result of shared objectives with governmental agencies. If the work of the land trust is successful, part of the outcome is an enhanced ability to cooperate with local government agencies. Agencies learn to trust the land trust staff and begin to rely on the special expertise that grows within a land trust organization. In some cases the land trust gains both opportunities and resources as a result of its partnership with governmental agencies. This public/private partnership benefits citizens as projects come together and land use options are retained for current and future generations.

Flexibility is an important and essential quality of a land trust that enables it to be creative. Land trusts can have revolving accounts, or lines of credit, from banks that allow them to move quickly to acquire land. Compensation to landowners who agree to work with the trust may come in the form of extended land use for the ex-owner, land trades, tax compensation, and other compensation packages. Often some mix of protection and compensation packages will be created that a governmental agency simply does not have the ability to implement. A land trust's flexibility is its most important attribute. Where a land trust can negotiate land acquisition based on a discussion among the board members, a governmental agency would go through months or even years of red tape before an offer to buy land for the public domain could be made. This quality of land trusts is one reason why many governmental agencies have built relationships with land trusts in order to protect land that the agency deems sensitive and important.

There are some limiting factors constraining what land trusts can do. For the more localized trusts, limited volunteer staff and extremely limited budgets cause fund raising to become a time-consuming activity. Staff turnover can be frequent so that a knowledge base is difficult to maintain. In some circumstances influential volunteers can capture a land trust organization and follow their own agenda rather than letting the agenda be set by affected stakeholders.

Training is needed for those committed to working within the legal structure of land trusts. The national Trust for Public Lands has established training opportunities to better prepare local land trust staff for the complex negotiations that are needed to protect public lands. Staff that work with local citizenry to protect local needs must be aware of the costs and benefits of land preservation mechanisms. Lease purchase agreements, limited partnerships, and fee simple transactions all require knowledge of real estate law. Operating within enterprise zones and working with economic development corporations requires knowledge of state and federal programs that provide money for projects on the urban fringe. In some cases urban renewal work reveals open space within the urban core that can be preserved for community gardens or parks if that land can be secured using HUD funds or other government financing mechanisms. A relatively new source of funding for land acquisition is mitigation funds. These funds are usually generated as a result of settlements with industry or governmental agencies as compensation for negative land impacts. Distinguishing among financing mechanisms requires specialized knowledge that land trust staff need to have available within their ranks in order to move quickly to preserve open space and enhance the quality of life for urban dwellers. On the other hand some land trusts in rural areas are interested in conserving farmlands using preserves that allow farmers to continue to farm while protecting the rural character of the countryside. Like their urban counterparts, these farmland preserve programs are complex, and if they are to be effective the trust needs to employ its solid knowledge of economic trends and resources.

The work that land trusts do is varied. In some cases a land trust incorporates as a result of a local threat, such as a pipeline or railway coming through an area. In some cases a trust forms to counter an undesirable land use such as a **landfill** or a **low-level radioactive waste** storage facility. In other instances, a land trust comes together to take advantage of a unique opportunity, such as a family wanting to sell some pristine forest close to town or an industry deciding to relocate leaving a lovely waterfront location with promise as a riverfront recreation area. It is rare that a land trust forms without a focused need. However, after the initial project is completed, its success breeds self-confidence in those who worked on the project and new opportunities or challenges may sustain the goals of the fledgling organization.

There are many examples of land trusts and the few highlighted here may help to enhance understanding of the value of land trust activities and to offer guidance to local groups wanting to preserve land. One outstanding example of land trust activity is the Rails to Trails program in Michigan. Under this program, abandoned railroad right of ways are preserved to create green belts for recreation use through agreements with the railroad companies. The Trust for Public Lands (TPL) has assisted many local land trusts to implement a wide variety of land acquisition projects. One such complex agreement took place in Tucson, Arizona. In this case, the Tucson city government wanted to acquire seven parcels of land that totaled 40 acres (164 ha). For financial reasons the city was not able to acquire the land. At that point the Trust for Public Land was asked to become a private nonprofit partner and to work with the city to acquire the land. TPL used its creative expertise to help each of the landowners make mutually beneficial arrangements with the city so that a large urban park could become a reality. In some cases the TPL offered a life tenancy to the current owners in exchange for a reduced land price. In another case they offered a five-year tenancy and a job as caretaker, in exchange for a reduced purchase price. As the community worked on the future of the park, another landowner who owned a contiguous parcel stepped forward with an offer to sell. Each of these transactions was successful because the land trust was flexible, considerate of the land owners and up front about the goals of their work, and responsive to their public partner, the city government.

Our current land trust effort in the United States has affected the way we protect our sensitive lands, reclaim damaged lands, and respond to local needs. Land trusts conserve land, guide future planning, educate local citizens and government agencies to a new way of doing business, and do it all with a minimum amount of confrontation and legal interaction. These private, non-profit organizations have stepped in and filled a **niche** in the environmental conservation movement started in the 1970s and have gotten results through a system of cooperative and well-informed action.

[*Cynthia Fridgen*]

RESOURCES

BOOKS

Diamond, H. L., and P. F. Noonan. *Land Use in America: The Report of the Sustainable Use of Land Project.* Lincoln Institute of Land Policy, Washington, DC: Island Press, 1996.

Endicott, E., ed. *Land Conservation Through Public/Private Partnerships.* Lincoln Institute of Land Policy, Washington DC: Island Press, 1993.

Platt, R. H. *Land Use and Society: Geography, Law, and Public Policy.* Washington DC: Island Press, 1996.

OTHER

Land Trust Alliance. 2002 [June 20, 2002]. <http://www.lta.org>.

Trust for Public Land. 2002 [June 20, 2002]. <http://www.tpl.org>.

Land use

Land is any part of the earth's surface that can be owned as property. Land comprises a particular segment of the earth's crust and can be defined in specific terms. The location of the land is extremely important in determining land use and land value.

Land is limited in supply, and, as our population increases, we have less land to support each person. Land nurtures the plants and animals that provide our food and shelter. It is the **watershed** or **reservoir** for our water supply. Land provides the minerals we utilize, the space on which we build our homes, and the site of many recreational activities. Land is also the depository for much of the waste created by modern society. The growth of human population only provides a partial explanation for the increased pressure on land resources. Economic development and a rise in the standard of living have brought about more demands for the products of the land. This demand now threatens to erode the land resource.

We are terrestrial in our activities and as our needs have diversified, so has land use. Conflicts among the competing land uses have created the need for land-use planning. Previous generations have used and misused the land as though the supply was inexhaustible. Today, goals and decisions about land use must take into account and link information from the physical and biological sciences with the current social values and political realities.

Land characteristics and ownership provide a basis for the many uses of land. Some land uses are classified as irreversible, for example, when the application of a particular land use changes the original character of the land to such a extent that reversal to its former use is impracticable. Reversible land uses do not change the **soil** cover or landform, and the land manager has many options when overseeing reversible land uses.

A framework for land-use planning requires the recognition that plans, policies, and programs must consider physical and biological, economical, and institutional factors. The physical framework of land focuses on the inanimate resources of soil, rocks and geological features, water, air, sunlight, and **climate**. The biological framework involves living things such as plants and animals. A key feature of the physical and biological framework is the need to maintain healthy ecological relationships. The land can support many human activities, but there are limits. Once the resources are brought to these limits, they can be destroyed and replacing them will be difficult.

The economic framework for land use requires that operators of land be provided sufficient returns to cover the cost of production. Surpluses of returns above costs must be realized by those who make the production decisions and

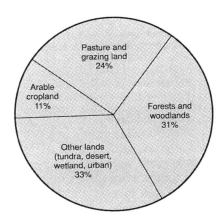

World land use. (McGraw-Hill Inc. Reproduced by permission.)

by those who bear the production costs. The economic framework provides the incentive to use the land in a way that is economically feasible. The institutional framework requires that programs and plans be acceptable within the working rules of society. Plans must also have the support of current governments. A basic concept of land use is the *right of land*—who has the right to decide the use of a given tract of land. Legal decisions have provided the framework for land resource protection.

Attitudes play an important role in influencing land use decisions, and changes in attitudes will often bring changes in our institutional framework. Recent trends in land use in the United States show that substantial areas have shifted to urban and **transportation** uses, state and national parks, and **wildlife** refuges since 1950. The use of land has become one of our most serious environmental concerns. Today's land use decisions will determine the quality of our future life styles and **environment**. The land use planning process is one of the most complex and least understood domestic concerns facing the nation. Additional changes in the institutional framework governing land use are necessary to allow society to protect the most limited resource on the planet—the land we live on.

[*Terence H. Cooper*]

RESOURCES
BOOKS

Beatty, M. T. *Planning the Uses and Management of Land.* Series in Agronomy, no. 21. Madison, WI: American Standards Association, 1979.

Davis, K. P. *Land Use.* New York: McGraw-Hill, 1976.

Fabos, J. G. *Land-Use Planning: From Global to Local Challenge.* New York: Chapman and Hall, 1985.

Lyle, John T., and Joan Woodward. *Design for Human Ecosystems: Landscape, Land Use and Natural Resources.* Washington, DC: Island Press, 1999.

McHarg, I. L. *Design With Nature.* New York: John Wiley and Sons, 1995.
Silber, Jane, and Chris Maser. *Land-Use Planning for Sustainable Development.* Boca Raton: CRC Press, 2000.

Landfill

Surface water, oceans and landfills are traditionally the main repositories for society's solid and **hazardous waste**. Landfills are located in excavated areas such as sand and gravel pits or in valleys that are near waste generators. They have been cited as sources of surface and **groundwater** contamination and are believed to pose a significant health risk to humans, domestic animals, and **wildlife**. Despite these adverse effects and the attendant publicity, landfills are likely to remain a major waste disposal option for the immediate future.

Among the reasons that landfills remain a popular alternative are their simplicity and versatility. For example, they are not sensitive to the shape, size, or weight of a particular waste material. Since they are constructed of **soil**, they are rarely affected by the chemical composition of a particular waste component or by any collective incompatibility of co-mingled wastes. By comparison, **composting** and **incineration** require uniformity in the form and chemical properties of the waste for efficient operation. Landfills also have been a relatively inexpensive disposal option, but this situation is rapidly changing. Shipping costs, rising land prices, and new landfill construction and maintenance requirements contribute to increasing costs.

About 57% of the **solid waste** generated in the United States still is dumped in landfills. In a sanitary landfill, refuse is compacted each day and covered with a layer of dirt. This procedure minimizes odor and litter, and discourages insect and rodent populations that may spread disease. Although this method does help control some of the **pollution** generated by the landfill, the fill dirt also occupies up to 20 percent of the landfill space, reducing its waste-holding capacity. Sanitary landfills traditionally have not been enclosed in a waterproof lining to prevent **leaching** of **chemicals** into groundwater, and many cases of **groundwater pollution** have been traced to landfills.

Historically landfills were placed in a particular location more for convenience of access than for any environmental or geological reason. Now more care is taken in the siting of new landfills. For example, sites located on faulted or highly **permeable** rock are passed over in favor of sites with a less-permeable foundation. Rivers, lakes, floodplains, and groundwater recharge zones are also avoided. It is believed that the care taken in the initial siting of a landfill will reduce the necessity for future clean-up and site **rehabilitation**. Due to these and other factors, it is becoming increasingly difficult to find suitable locations for new landfills.

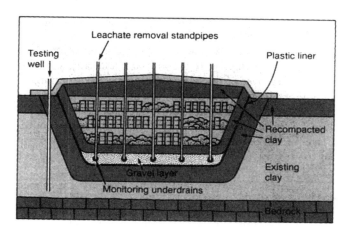

A secure landfill. (McGraw-Hill Inc. Reproduced by permission.)

Easily accessible open space is becoming scarce and many communities are unwilling to accept the siting of a landfill within their boundaries. Many major cities have already exhausted their landfill capacity and must export their trash, at significant expense, to other communities or even to other states and countries.

Although a number of significant environmental issues are associated with the disposal of solid waste in landfills, the disposal of hazardous waste in landfills raises even greater environmental concerns. A number of urban areas contain hazardous waste landfills. **Love Canal** is, perhaps, the most notorious example of the hazards associated with these landfills. This Niagara Falls, New York neighborhood was built over a dump containing 20,000 metric tons of toxic chemical waste. Increased levels of **cancer**, miscarriages, and **birth defects** among those living in Love Canal led to the eventual evacuation of many residents. The events at Love Canal were also a major impetus behind the passage of the Comprehensive Environmental Response, Compensation and Liability Act in 1980, designed to clean up such sites. The U. S. **Environmental Protection Agency** estimates that there may be as many as 2,000 hazardous waste disposal sites in this country that pose a significant threat to human health or the **environment**.

Love Canal is only one example of the environmental consequences that can result from disposing of hazardous waste in landfills. However, techniques now exist to create secure landfills that are an acceptable disposal option for hazardous waste in many cases. The bottom and sides of a secure landfill contain a cushion of recompacted clay that is flexible and resistant to cracking if the ground shifts. This clay layer is impermeable to groundwater and safely contains the waste. A layer of gravel containing a grid of perforated drain pipes is laid over the clay. These pipes collect any

seepage that escapes from the waste stored in the landfill. Over the gravel bed a thick polyethylene liner is positioned. A layer of soil or sand covers and cushions this plastic liner, and the wastes, packed in drums, are placed on top of this layer.

When the secure landfill reaches capacity it is capped by a cover of clay, plastic and soil, much like the bottom layers. Vegetation in planted to stabilize the surface and make the site more attractive. Sump pumps collect any fluids that filter through the landfill either from rainwater or from waste leakage. This liquid is purified before it is released. Monitoring **wells** around the site ensure that the groundwater does not become contaminated. In some areas where the **water table** is particularly high, above-ground storage may be constructed using similar techniques. Although such facilities are more conspicuous, they have the advantage of being easier to monitor for leakage.

Although technical solutions have been found to many of the problems associated with secure landfills, several non-technical issues remain. One of these issues concerns the **transportation** of hazardous waste to the site. Some states do not allow hazardous waste to be shipped across their territory because they are worried about the possibility of accidental spills. If hazardous waste disposal is concentrated in only a few sites, then a few major transportation routes will carry large volumes of this material. Citizen opposition to hazardous waste landfills is another issue. Given the past record of corporate and governmental irresponsibility in dealing with hazardous waste, it is not surprising that community residents greet proposals for new landfills with the NIMBY (**Not In My BackYard**) response. However, the waste must go somewhere. These and other issues must be resolved if secure landfills are to be a viable long-term solution to hazardous waste disposal. *See also* Groundwater monitoring; International trade in toxic waste; Storage and transportation of hazardous materials

[*George M. Fell and Christine B. Jeryan*]

Resources

Books

Bagchi, A. *Design, Construction and Monitoring of Landfills.* 2nd ed. New York: Wiley, 1994.

Neal, H. A. *Solid Waste Management and the Environment: The Mounting Garbage and Trash Crisis.* Englewood Cliffs, NJ: Prentice-Hall, 1987.

Noble, G. *Siting Landfills and Other LULUs.* Lancaster, PA: Technomic Publishing, 1992.

Requirements for Hazardous Waste Landfill Design, Construction and Closure. Cincinnati: U.S. Environmental Protection Agency, 1989.

Periodicals

"Experimental Landfills Offer Safe Disposal Options." *Journal of Environmental Health* 51 (March-April 1989): 217–18.

Loupe, D. E. "To Rot or Not; Landfill Designers Argue the Benefits of Burying Garbage Wet vs. Dry." *Science News* 138 (October 6, 1990): 218–19+.

Wingerter, E. J., et al. "Are Landfills and Incinerators Part of the Answer? Three Viewpoints." *EPA Journal* 15 (March-April 1989): 22–26.

Landscape ecology

Landscape **ecology** is an interdisciplinary field that emerged from several intellectual traditions in Europe and North America. An identifiable landscape ecology started in central Europe in the 1960s and in North America in the late 1970s and early 1980s. It became more visible with the establishment, in 1982, of the International Association of Landscape Ecology, with the publication of a major text in the field, *Landscape Ecology*, by Richard Forman and Michel Godron in 1984, and with the publication of the first issue of the association's journal, *Landscape Ecology* in 1987.

The phrase 'landscape ecology' was first used in 1939 by the German geographer, Carl Troll. He suggested that the "concept of landscape ecology is born from a marriage of two scientific outlooks, the one geographical (landscape), the other biological (ecology)." Troll coined the term landscape ecology to denote "the analysis of a physico-biological complex of interrelations, which govern the different area units of a region." He believed that "landscape ecology...must not be confined to the large scale analysis of natural regions. Ecological factors are also involved in problems of population, society, rural settlement, **land use**, transport, etc."

Landscape has long been a unit of analysis and a conceptual centerpiece of geography, with scholars such as Carl Sauer and J. B. Jackson adept at "reading the landscape," including both the natural landscape of landforms and vegetation, and the cultural landscape as marked by human actions and as perceived by human minds.

Zev Naveh has been working on his own version of landscape ecology in Israel since the early 1970s. Like Troll, Naveh includes humans in his conception, in fact enlarges landscape ecology to a global human **ecosystem** science, sort of a "bio-cybernetic systems approach to the landscape and the study of its use by [humans]." He sees landscape ecology first as a **holistic approach** to biosystems theory, the centerpiece being "recognition of the total human ecosystem as the highest level of integration," and, second, as playing a central role in cultural **evolution** and as a "basis for interdisciplinary, task-oriented, environmental education."

Landscape architecture is also to some extent landscape ecology, since landscape architects design complete vistas, from their beginnings and at various scales. This concern with designing and creating complete landscapes from bare ground can certainly be considered ecological, as it includes creating or adapting local land forms, planting appropriate vegetation, and designing and building various kinds of 'furniture' and other artifacts on site. The British Landscape Institute and the British Ecological Society held a joint meeting in 1983, recognizing "that the time for ecology to be harnessed for the service of landscape design has arrived." The meeting produced the twenty-fourth symposium of the British Ecological Society titled *Ecology and Design in Landscape.*

Landscape planning can also to some degree be considered landscape ecology, especially in the ecological approach to landscape planning developed by Ian McHarg and his students and colleagues, and the LANDEP, or Landscape Ecological Planning, approach designed by Ladislav Miklos and Milan Ruzicka. Both of these ecological planning approaches are complex syntheses of spatial patterns, ecological processes, and human needs and wants.

Building on all of these traditions, yet slowly finding its own identity, landscape ecology is considered by some as a sub-domain of biological ecology and by others as a discipline in its own right. In Europe, landscape ecology continues to be an extension of the geographical tradition that is preoccupied with human-landscape interactions. In North America, landscape ecology has emerged as a branch of biological ecology, more concerned with landscapes as clusters of interrelated natural ecosystems. The European form of landscape ecology is applied to land and resource **conservation**, while in North America it focuses on fundamental questions of spatial pattern and exchange. Both traditions can address major environmental problems, especially the **extinction** of **species** and the maintenance of biological diversity.

The term landscape, despite the varied traditions and emerging disciplines described above, remains somewhat indeterminate, depending on the criteria set by individual researchers to establish boundaries. Some consensus exists on its general definition in the new landscape ecology, as described in the composite form attempted here: a terrestrial landscape is miles- or kilometers-wide in area; it contains a cluster of interacting ecosystems repeated in somewhat similar form; and it is a heterogeneous mosaic of interconnected land forms, vegetation types, and land uses. As Risser and his colleagues emphasize, this interdisciplinary area focuses explicitly on spatial patterns: "Specifically, landscape ecology considers the development and dynamics of spatial hetereogeneity, spatial and temporal interactions and exchanges across heterogeneous landscapes, influences of spatial heterogeneity on biotic and abiotic processes, and management of spatial heterogeneity." Instead of trying to identify homogeneous ecosystems, landscape ecology focuses particularly on the heterogeneous patches and mosaics created by human disruption of natural systems, by the intermixing of cultural and natural landscape patterns. The real rationale for a land-

scape ecology perhaps should be this acknowledgment of the heterogeneity of contemporary landscape patterns, and the need to deal with the patchwork mosaics and intricate matrices that result from long-term human disturbance, modification, and utilization of natural systems.

Typical questions asked by landscape ecologists include these formulated by Risser and his colleagues: "What formative processes, both historical and present, are responsible for the existing pattern in a landscape?" "How are fluxes of organisms, of material, and of energy related to landscape heterogeneity?" "How does landscape heterogeneity affect the spread of disturbances?" While the first question is similar to ones long asked in geography, the other two are questions traditional to ecology, but distinguished here by the focus on heterogeneity.

Richard Forman, a prominent figure in the evolving field of landscape ecology, thinks the field has matured enough for general principles to have emerged; not ecological laws as such, but principles backed by enough evidence and examples to be true for 95 percent of landscape analyses. His 12 principles are organized by four categories: landscapes and regions; patches and corridors; mosaics; and applications. The principles outline expected or desirable spatial patterns and relationships, and how those patterns and relationships affect system functions and flows, organismic movements and extinctions, resource protection, and optimal environmental conditions. Forman claims the principles "should be applicable for any environmental or societal land-use objective," and that they are useful in more effectively "growing wood, protecting species, locating houses, protecting **soil**, enhancing game, protecting **water resources**, providing **recreation**, locating roads, and creating sustainable environments."

Perhaps Andre Corboz provided the best description when he wrote of "the land as palimpsest:" landscape ecology recognizes that humans have written large on the land, and that behind the current writing visible to the eye, there is earlier writing as well, which also tells us about the patterns we see. Landscape ecology also deals with gaps in the text and tries to write a more complete accounting of the landscapes in which we live and on which we all depend.

[*Gerald L. Young Ph.D.*]

RESOURCES
BOOKS

Farina, Almo. *Landscape Ecology in Action.* New York: Kluwer, 2000.

Forman, R. T. T., and M. G. *Landscape Ecology.* New York: Wiley, 1986.

Risser, P. G., J. R. Karr, and R. T. T. Forman. *Landscape Ecology: Directions and Approaches.* Champaign: Illinois Natural History Survey, 1983.

Tjallingii, S. P., and A. A. de Veer, eds. *Perspectives in Landscape Ecology: Contributions to Research, Planning and Management of Our Environment.*

Troll, C. *Landscape Ecology.* Delft, The Netherlands: The ITC-UNESCO Centre for Integrated Surveys, 1966.

Turner, Monica, R. H. Gardner, and R. V. O'Neill. *Landscape Ecology in Theory and Practice: Patterns and Processes.* New York: Springer Verlag, 2001.

Wageningen, The Netherlands: Pudoc, 1982. (Proceedings of the International Congress Organized by the Netherlands Society for Landscape Ecology, Veldhoven, The Netherlands, 6-11 April, 1981).

Zonneveld, I. S., and R. T. T. Forman, eds. *Changing Landscapes: An Ecological Perspective.* New York: Springer-Verlag, 1990.

PERIODICALS

Forman, R. T. T. "Some General Principles of Landscape and Regional Ecology." *Landscape Ecology* (June 1995): 133–142.

Golley, F.B. "Introducing Landscape Ecology." *Landscape Ecology* 1, no. 1 (1987): 1–3.

Naveh, Z. "Landscape Ecology as an Emerging Branch of Human Ecosystem Science." *Advances in Ecological Research* 12 (1982): 189–237.

Landslide

A general term for the discrete downslope movement of rock and **soil** masses under gravitational influence along a failure zone. The term "landslide" can refer to the resulting land form, as well as to the process of movement. Many types of landslides occur, and they are classified by several schemes, according to a variety of criteria. Landslides are categorized most commonly on basis of geometric form, but also by size, shape, rate of movement, and water content or fluidity. Translational, or planar, failures, such as debris avalanches and earth flows, slide along a fairly straight failure surface which runs approximately parallel to the ground surface. Rotational failures, such as rotational slumps, slide along a spoon shaped failure surface, leaving a hummocky appearance on the landscape. Rotational slumps commonly transform into earthflows as they continue down slope. Landslides are usually triggered by heavy rain or melting snow, but major earthquakes can also cause landslides.

Land-use control

Land-use control is a relatively new concept. For most of human history, it was assumed that people could do whatever they wished with their own property. However, societies have usually recognized that the way an individual uses private property can sometimes have harmful affects on neighbors.

Land-use planning has reached a new level of sophistication in developed countries over the last century. One of the first restrictions on **land use** in the United States, for example, was a 1916 New York City law limiting the size of skyscrapers because of the shadows they might cast on adjacent property. Within a decade, the federal government began to act aggressively on land control measures. It passed the **Mineral Leasing Act** of 1920 in an attempt to control the exploitation of oil, **natural gas**, phosphate, and potash.

It adopted the Standard State Zoning Act of 1922 and the Standard City Planning Enabling Act of 1928 to promote the concept of zoning at state and local levels. Since the 1920s, every state and most cities have adopted zoning laws modeled on these two federal acts.

Often detailed, exhaustive, and complex zoning regulations now control the way land is used in nearly every governmental unit. They specify, for example, whether land can be used for single-dwelling construction, multiple-dwelling construction, farming, industrial (heavy or light) development, commercial use, **recreation** or some other purpose. Requests to use land for purposes other than that for which it is zoned requires a variance or conditional use permit, a process that is often long, tedious, and confrontational.

Many types of land require special types of zoning. For example, coastal areas are environmentally vulnerable to storms, high tides, **flooding**, and strong winds. The federal government passed laws in 1972 and 1980, the National Coastal Zone Management Acts, to help states deal with the special problem of protecting coastal areas. Although initially slow to make use of these laws, states are becoming more aggressive about restricting the kinds of construction permitted along seashore areas.

Areas with special scenic, historic, or recreational value have long been protected in the United States. The nation's first **national park**, **Yellowstone National Park**, was created in 1872. Not until 44 years later, however, was the **National Park Service** created to administer Yellowstone and other parks established since 1872. Today, the National Park Service and other governmental agencies are responsible for a wide variety of national areas such as forests, wild and scenic rivers, historic monuments, trails, battlefields, memorials, seashores and lakeshores, parkways, recreational areas, and other areas of special value.

Land-use control does not necessarily restrict usage. Individuals and organizations can be encouraged to use land in certain desirable ways. An enterprise zone, for example, is a specifically designated area in which certain types of business activities are encouraged. The tax rate might be reduced for businesses locating in the area or the government might relax certain regulations there.

Successful land-use control can result in new towns or planned communities, designed and built from the ground up to meet certain pre-determined land-use objectives. One of the most famous examples of a planned community is Brasilia, the capital of Brazil. The site for a new capital—an undeveloped region of the country—was selected and a totally new city was built in the 1950s. The federal government moved to the new city in 1960, and it now has a population of more than 1.5 million. *See also* Bureau of Land Management; Riparian rights

[David E. Newton]

RESOURCES

BOOKS

Becker, Barbara, Eric D. Kelly, and Frank So. *Community Planning: An Introduction to the Comprehensive Plan.* Washington, DC: Milldale Press, 2000.

Newton, D. E. *Land Use, A–Z.* Hillside, NJ: Enslow Press, 1991.

Platt, Rutherford H. *Land Use and Society: Geography, Law, and Public Policy.* Washington, DC: Island Press, 1996.

Latency

Latency refers to the period of time it takes for a disease to manifest itself within the human body. It is the state of seeming inactivity that occurs between the instant of stimulation or initiating event and the beginning of response. The latency period differs dramatically for each stimulation and as a result, each disease has its unique time period before symptoms occur.

When pathogens gain entry into a potential host, the body may fail to maintain adequate immunity and thus permits progressive viral or bacterial multiplication. This time lapse is also known as the incubation period. Each disease has definite, characteristic limits for a given host. During the incubation period, dissemination of the **pathogen** takes place and leads to the inoculation of a preferred or target organ. Proliferation of the pathogen, either in a target organ or throughout the body, then creates an infectious disease.

Botulism, tetanus, gonorrhea, diphtheria, staphylococcal and streptococcal disease, pneumonia, and tuberculosis are among the diseases that take varied periods of time before the symptoms are evident. In the case of the childhood diseases—measles, mumps, and chicken pox—the incubation period is 14–21 days.

In the case of **cancer**, the latency period for a small group of transformed cells to result in a tumor large enough to be detected is usually 10–20 years. One theory postulates that every cancer begins with a single cell or small group of cells. The cells are transformed and begin to divide. Twenty years of cell division ultimately results in a detectible tumor. It is theorized that very low doses of a **carcinogen** could be sufficient to transform one cell into a cancerous tumor.

In the case of **AIDS**, an eight- to eleven-year latency period passes before the symptoms appear in adults. The length of this latency period depends upon the strength of the person's immune system. If a person suspects he or she has been infected, early blood tests showing HIV antibodies or antigens can indicate the infection within three months of the stimulation. The three-month period before the appearance of HIV antibodies or antigens is called the "window period."

In many cases, doctors may fail to diagnose the disease at first, since AIDS symptoms are so general they may be confused with the symptoms of other, similar diseases. Childhood AIDS symptoms appear more quickly since young children have immune systems that are less fully developed.

[Liane Clorfene Casten]

Lawn treatment

Lawn treatment in the form of pesticides and inorganic fertilizers poses a substantial threat to the **environment**. Homeowners in the United States use approximately three times more pesticides per acre than the average farmer, adding up to some 136 million pounds (61.7 kg) annually. Home lawns occupy more acreage in the United States than any agricultural crop, and a majority of the **wildlife pesticide** poisonings tracked by the **Environmental Protection Agency** (EPA) annually are attributed to **chemicals** used in lawn care. The use of grass **fertilizer** is also problematic when it runs off into nearby waterways. Lawn grass in almost all climates in the United States requires watering in the summer, accounting for some 40 to 60 percent of the average homeowner's water use annually. Much of the water sprinkled on lawns is lost as **runoff**. When this runoff carries fertilizer, it can cause excess growth of algae in downstream waterways, clogging the surface of the water and depleting the water of oxygen for other plants and animals. Herbicides and pesticides are also carried into downstream water, and some of these are toxic to fish, birds, and other wildlife.

Turf grass lawns are ubiquitous in all parts of the United States, regardless of the local **climate**. From Alaska to Arizona to Maine, homeowners surround their houses with grassy lawns, ideally clipped short, brilliantly green, and free of weeds. In almost all cases, the grass used is a hybrid of several **species** of grass from Northern Europe. These grasses thrive in cool, moist summers. In general, the United States experiences hotter, dryer summers than Northern Europe. Moving from east to west across the country, the climate becomes less and less like that the common turf grass evolved in. The ideal American lawn is based primarily on English landscaping principals, and it does not look like an English lawn unless it is heavily supported with water.

The prevalence of lawns is a relatively recent phenomenon in the United States, dating to the late nineteenth century. When European settlers first came to this country, they found indigenous grasses that were not as nutritious for livestock and died under the trampling feet of sheep and cows. Settlers replaced native grasses with English and European grasses as fodder for grazing animals. In the late eighteenth century, American landowners began surrounding their estates with lawn grass, a style made popular earlier in England. The English lawn fad was fueled by eighteenth century landscaper Lancelot "Capability" Brown, who removed whole villages and stands of mature trees and used sunken fences to achieve uninterrupted sweeps of green parkland. Both in England and the United States, such lawns and parks were mowed by hand, requiring many laborers, or they were kept cropped by sheep or even deer. Small landowners meanwhile used the land in front of their houses differently. The yard might be of stamped earth, which could be kept neatly swept, or it may have been devoted to a small garden, usually enclosed behind a fence. The trend for houses set back from the street behind a stretch of unfenced lawn took hold in the mid-nineteenth century with the growth of suburbs. Frederick Law Olmsted, the designer of New York City's Central Park, was a notable suburban planner, and he fueled the vision of the English manor for the suburban home. The unfenced lawns were supposed to flow from house to house, creating a common park for the suburb's residents. These lawns became easier to maintain with the invention of the lawn mower. This machine debuted in England as early as 1830, but became popular in the United States after the Civil War. The first patent for a lawn sprinkler was granted in the United States in 1871. These developments made it possible for middle class home owners to maintain lush lawns themselves.

Chemicals for lawn treatment came into common use after World War II. Herbicides such as **2,4-D** were used against broadleaf weeds. The now-banned DDT was used against insect pests. Homeowners had previously fertilized their lawns with commercially available organic formulations like dried manure, but after World War II inorganic, chemical-based fertilizers became popular for both agriculture and lawns and gardens. Lawn care companies such as Chemlawn and Lawn Doctor originated in the 1960s, an era when homeowners were confronted with a bewildering array of chemicals deemed essential to a healthy lawn. Rachel Carson's 1962 book *Silent Spring* raised an alarm about the prevalence of lawn chemicals and their environmental costs. Carson explained how the insecticide DDT builds up in the food chain, passing from insects and worms to fish and small birds that feed on them, ultimately endangering large predators like the eagle. DDT was banned in 1972, and some lawn care chemicals were restricted. Nevertheless, the lawn care industry continued to prosper, offering services such as combined seeding, **herbicide**, and fertilizer at several intervals throughout the growing season. Lawn care had grown to a $25 billion industry in the United States by the 1990s. Even as the perils of particular lawn chemicals became clearer, it was difficult for homeowners to give them up. **Statistics** from the United States National **Cancer** Institute show that the incidence of childhood **leukemia** is 6.5% greater in familes that use lawn pesticides than in those who

do not. In addition, 32 of the 34 most widely used lawn care pesticides have not been tested for health and environmental issues. Because some species of lawn grasses grow poorly in some areas of the United States, it does not thrive without extra water and fertilizer. It is vulnerable to insect pests, which can be controlled with pesticides, and if a weed-free lawn is the aim, herbicides are less labor-intensive than digging out dandelions one by one.

Some common pesticides used on lawns are acephate, bendiocarb, and **diazinon**. Acephate is an **organophosphate** insecticide which works by damaging the insect's nervous system. Bendiocarb is called a carbamate insecticide, sold under several brand names, which works in the same way. Both were first developed in the 1940s. These will kill many insects, not only pests such as leafminers, thrips, and cinch bugs, but also beneficial insects, such as bees. Bendiocarb is also toxic to earthworms, a major food source for some birds. Birds too can die from direct exposure to bendiocarb, as can fish. Both these chemicals can persist in the **soil** for weeks. Diazinon is another common pesticide used by homeowners on lawns and gardens. It is toxic to humans, birds, and other wildlife, and it has been banned for use on **golf courses** and turf farms. Nevertheless, homeowners may use it to kill **pest** insects such as **fire ants**. Harmful levels of diazinon and were found in metropolitan storm water systems in California in the early 1990s, leached there from orchard run-off. Diazinon is responsible for about half of all reported wildlife poisonings involving lawn and garden chemicals.

Common lawn and garden herbicides appear to be much less toxic to humans and animals than pesticides. The herbicide 2,4-D, one of the earliest herbicides used in this country, can cause skin and eye irritation to people who apply it, and it is somewhat toxic to birds. It can be toxic to fish in some formulations. Although contamination with 2,4-D has been found in some urban waterways, it has only been in trace amounts not thought to be harmful to humans. Glyphosate is another common herbicide, sold under several brand names, including the well-known Roundup. It is considered non-toxic to humans and other animals. Unike 2,4-D, which kills broadleaf plants, glyphosate is a broad spectrum herbicide used to control control a great variety of annual, biennial, and perennial grasses, sedges, broad leafed weeds and woody shrubs.

Common lawn and garden fertilizers are generally not toxic unless ingested in sufficient doses, yet they can have serious environmental effects. Run-off from lawns can carry fertilizer into nearby waterways. The **nitrogen** and **phosphorus** in the fertilizer stimulates plant growth, principally algae and microscopic plants. These tiny plants bloom, die, and decay. Bacteria that feed off plant decay then also undergo a surge in population. The overabundant bacteria consume oxygen, leading to oxygen-depleted water. This condition is called hypoxia. In some areas, fertilized run-off from lawns is as big a problem as run-off from agricultural fields. Lawn fertilizer is thought to be a major culprit in **pollution** of the **Everglades** in Florida. In 2001 the Minnesota legislature debated a bill to limit homeowners' use of phosphorus in fertilizers because of problems with algae blooms on the state's lakes.

There are several viable alternatives to the use of chemicals for lawn care. Lawn care companies often recommend multiple applications of pesticides, herbicides, and fertilizers, but an individual lawn may need such treatment on a reduced schedule. Some insects such as thrips and mites are suceptible to insecticidal soaps and oils, which are not long-lasting in the environment. These could be used in place of diazinon, acephate and other pesticides. Weeds can be pulled by hand, or left alone. Homeowners can have their lawn evaluated and their soil tested to determine how much fertilizer is needed. Slow-release fertilizers or organic fertilizers such as compost or seaweed emulsion do not give off such a large concentration of nutrients at once, so these are gentler on the environment. Another way to cut back on the excess water and chemicals used on lawns is to reduce the size of the lawn. The lawn can be bordered with shrubbery and perennial plants, leaving just enough open grass as needed for **recreation**. Another alternative is to replace non-native turf grass with a native grass. Some native grasses stay green all summer, can be mown short, and look very much like a typical lawn. Native buffalo grass (Buchloe dactyloides) has been used successfully for lawns in the South and Southwest. Other native grass species are adapted to other regions. Another example is blue grama grass (Bouteloua gracilis), native to the Great Plains. This grass is tolerant of extreme temperatures and very little rainfall. Some native grasses are best left unmowed, and in some regions homeowners have replaced their lawns with native grass prairies or meadows. In some cases, homeowners have done away with their lawns altogether, using stone or bark **mulch** in its place, or planting a groundcover plant like ivy or wild ginger. These plants might grow between trees, shrubs and perennials, creating a very different look than the traditional green carpet.

For areas with water shortages, or for those who are concerned about conserving **natural resources**, xeriscape landscaping should be considered. Xeriscape comes from the Greek word xeros, meaning dry. Xeriscaping takes advantage of using plants, such as cacti and grasses, such as Mexican feather grass and blue oat grass that thrive in **desert** conditions. Xeriscaping can also include rock gardening as part of the overall landscape plan.

[*Angela Woodward*]

RESOURCES

BOOKS

Bormann, F. Herbert, Diana Balmori, and Gordon T. Geballe. *Redesigning the American Lawn*. New Haven and London: Yale University Press, 1993.
Jenkins, Virginia Scott. *The Lawn: A History of an American Obsession.* Washington and London: Smithsonian Institution Press: 1994.
Stein, Sara. *Planting Noah's Garden: Further Adventures in Backyard Ecology*. Boston: Houghton Mifflin Co., 1997.
Wasowski, Andy, and Sally Wasowski. *The Landscaping Revolution*. Chicago: Contemporary Books, 2000.

PERIODICALS

Bourne, Joel. "The Killer in Your Yard." *Audubon* (May-June 2000): 108.
"Easy Lawns." *Brooklyn Botanic Garden Handbook* 160 (Fall 1999).
Simpson, Sarah. "Shrinking the Dead Zone." *Scientific American* (July 2001): 18.
Stewart, Doug. "Our Love Affair with Lawns." *Smithsonian* (April 1999): 94.
Xeriscaping Tips Page. 2002 [cited June 18, 2002]. <http://ct.essortment.com/xeroscaping_rksh.htm>.

LDC

see **Less developed countries**

LD₅₀

LD₅₀ is the dose of a chemical that is lethal to 50 percent of a test population. It is therefore a measure of a particular median response which, in this case, is death. The term is most frequently used to characterize the response of animals such as rats and mice in acute toxicity tests. The term is generally not used in connection with aquatic or inhalation toxicity tests. It is difficult, if not impossible, to determine the dosage of an animal in such tests; results are most commonly represented in terms of lethal concentrations (LC), which refer to the concentration of the substance in the air or water surrounding an animal.

In LD testing, dosages are generally administered by means of injection, food, water, or forced feeding. Injections are used when an animal is to receive only one or a few dosages. Greater numbers of injections would disturb the animal and perhaps generate some false-positive types of responses. Food or water may serve as a good medium for administering a chemical, but the amount of food or water wasted must be carefully noted. Developing a healthy diet for an animal which is compatible with the chemical to be tested can be as much art as science. The chemical may interact with the foods and become more or less toxic, or it may be objectionable to the animal due to taste or odor. Rats are often used in toxicity tests because they do not have the ability to vomit. The investigator therefore has the option of gavage, a way to force-feed rats with a stomach tube or other device when a chemical smells or tastes bad.

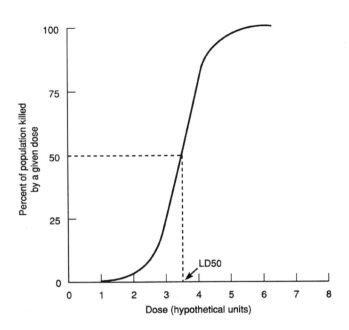

A chart showing the increased dose of LD$_{50}$.
(McGraw-Hill Inc. Reproduced by permission.)

Toxicity and LD$_{50}$ are inversely proportional, which means that high toxicity is indicated by a low LD$_{50}$ and vice versa. LD$_{50}$ is a particular type of effective dose (ED) for 50 percent of a population (ED$_{50}$). The midpoint (or effect on half of the population) is generally used because some individuals in a population may be highly resistant to a particular toxicant, making the dosage at which all individuals respond a misleading data point. Effects other than death, such as headaches or dizziness, might be examined in some tests, so EDs would be reported instead of LDs. One might also wish to report the response of some other percent of the test population, such as the 20 percent response (LD$_{20}$ or ED$_{20}$) or 80 percent response (LD$_{80}$ or ED$_{80}$).

The LD is expressed in terms of the mass of test chemical per unit mass of the test animals. In this way, dose is normalized so that the results of tests can be analyzed consistently and perhaps extrapolated to predict the response of animals that are heavier or lighter. Extrapolation of such data is always questionable, especially when extrapolating from animal response to human response, but the system appears to be serving us well. However, it is important to note that sometimes better dose-response relations and extrapolations can be derived through normalizing dosages based on surface area or the weight of target organs. *See also*

Bioassay; Dose response; Ecotoxicology; Hazardous material; Toxic substance

[*Gregory D. Boardman*]

RESOURCES
BOOKS

Casarett, L. J., J. Doull, and C. D. Klaassen, eds. *Casarett and Doull's Toxicology: The Basic Science of Poisons.* 6th ed. New York: McGraw Hill, 2001.
Hodgson, E., R. B. Mailman, and J. E. Chambers. *Dictionary of Toxicology.* 2nd ed. New York: John Wiley and Sons, 1997.
Lu, F. C. *Basic Toxicology: Fundamentals, Target Organs, and Risk Assessment.* 3rd ed. Hebron, KY: Taylor & Francis, 1996.
Rand, G. M., ed. *Fundamentals of Aquatic Toxicology: Effects, Environmental Fate, and Risk Assessment.* 2nd ed. Hebron, KY: Taylor & Francis, 1995.

Leachate

see **Contaminated soil; Landfill**

Leaching

The process by which soluble substances are dissolved out of a material. When rain falls on farmlands, for example, it dissolves weatherable minerals, pesticides, and fertilizers as it soaks into the ground. If enough water is added to the **soil** to fill all the pores, then water carrying these dissolved materials moves to the groundwater—the soil becomes leached. In soil chemistry, leaching refers to the process by which nutrients in the upper layers of soil are dissolved out and carried into lower layers, where they can be a valuable **nutrient** for plant roots. Leaching also has a number of environmental applications. For example, toxic **chemicals** and radioactive materials stored in sealed containers underground may leach out if the containers break open over time. *See also* Landfill; Leaking underground storage tank

Lead

One of the oldest metals known to humans, lead compounds were used by Egyptians to glaze pottery as far back as 7000 B.C. The toxic effects of lead also have been known for many centuries. In fact, the Romans limited the amount of time slaves could work in lead mines because of the element's harmful effects. Some consequences of lead **poisoning** are **anemia**, headaches, convulsions, and damage to the kidneys and central nervous system. The widespread use of lead in plumbing, **gasoline**, and lead-acid batteries, for example, has made it a serious **environmental health** problem. Bans on the use of lead in motor fuels and paints attempt to deal with this problem. *See also* Heavy metals and heavy metal poisoning; Lead shot

Lead management

Lead, a naturally occurring bluish gray metal, is extensively used throughout the world in the manufacture of storage batteries, **chemicals** including paint and **gasoline**, and various metal products including sheet lead, solder, pipes, and ammunition. Due to its widespread use, large amounts of lead exist in the **environment**, and substantial quantities of lead continue to be deposited into air, land, and water. Lead is a poison that has many adverse effects, and children are especially susceptible. At present, the production, use, and disposal of lead are regulated with demonstrably effective results. However, because of its previous widespread use and persistence in the environment, lead exposure is a pervasive problem that affects many populations. Effective management of lead requires an understanding of its effects, blood action levels, sources of exposure, and policy responses, topics reviewed in that order.

Effects of Lead

Lead is a strong toxicant that adversely affects many systems in the body. Severe lead exposures can cause brain and kidney damage to adults and children, coma, convulsions, and death. Lower levels, e.g., lead concentrations in blood (PbB) below 50 μg/dL, may impair hemoglobin synthesis, alter the central and peripheral nervous systems, cause hypertension, affect male and female reproductive systems, and damage the developing fetus. These effects depend on the level and duration of exposure and on the distribution and kinetics of lead in the body. Most lead is deposited in bone, and some of this stored lead may be released long after exposure due to a serious illness, pregnancy, or other physiological event. Lead has not been shown to cause **cancer** in humans, however, tumors have developed in rats and mice given large doses of lead and thus several United States agencies consider lead acetate and lead phosphate as human carcinogens.

Children are particularly susceptible to lead **poisoning**. PbB levels as low as 10 μg/dL are associated with decreased intelligence and slowed neurological development. Low PbB levels also have been associated with deficits in growth, vitamin **metabolism**, and effects on hearing. The neurological effects of lead on children are profound and are likely persistent. Unfortunately, childhood exposures to chronic but low lead levels may not produce clinical symptoms, and many cases go undiagnosed and untreated. In recent years, the number of children with elevated blood lead levels has declined substantially. For example, the average PbB level has decreased from over 15 μg/dL in the

1970s to about 5 μg/dL in the 1990s. As described later, these decreases can be attributed to the reduction or elimination of lead in gasoline, food can and plumbing solder, and residential paint. Still, childhood lead poisoning remains the most widespread and preventable childhood health problem associated with environmental exposures, and childhood lead exposure remains a public health concern since blood levels approach or exceed levels believed to cause effects. Though widely perceived as a problem of inner city minority children, lead poisoning affects children from all areas and from all socioeconomic groups.

The definition of a PbB level that defines a level of concern for lead in children continues to be an important issue in the United States. The childhood PbB concentration of concern has been steadily lowered by the Centers for Disease Control (CDC) from 40 μg/dL in 1970 to 10 μg/dL in 1991. The **Environmental Protection Agency** lowered the level of concern to 10 μg/dL ("10-15 and possibly lower") in 1986, and the **Agency for Toxic Substances and Disease Registry** (ATSDR) also identified 10 μg/dL in its 1988 Report to Congress on childhood lead poisoning.

In the workplace, the medical removal PbB concentration is 50 μg/dL for three consecutive checks and 60 μg/dL for any single check. Blood level monitoring is triggered by an air lead concentration above 30 μg/m^3. A worker is permitted to return to work when his blood lead level falls below 40 μg/dL. In 1991, the National Institute for Occupational Safety and Health (NIOSH) set a goal of eliminating occupational exposures that result in workers having PbB levels greater than 25 μg/dL.

Exposure and Sources

Lead is a persistent and ubiquitous pollutant. Since it is an elemental pollutant, it does not dissipate, biodegrade, or decay. Thus, the total amount of lead pollutants resulting from human activity increases over time, no matter how little additional lead is added to the environment. Lead is a multi-media pollutant, i.e., many sources contribute to the overall problem, and exposures from air, water, **soil**, dust, and food pathways may be important.

For children, an important source of lead exposure is from swallowing nonfood items, an activity known as pica (an abnormal eating habit e.g., chips of lead-containing paint), most prevalent in 2 and 3 year-olds. Children who put toys or other items in their mouths may also swallow lead if lead-containing dust and dirt are on these items. Touching dust and dirt containing lead is commonplace, but relatively little lead passes through the skin. The most important source of high-level lead exposure in the United States is household dust derived from deteriorated lead-based paint. Numerous homes contain lead-based paint and continue to be occupied by families with small children, including 21 million pre-1940 homes and rental units which, over time, are rented to different families. Thus, a single house with deteriorated lead-based paint can be the source of exposure for many children.

In addition to lead-based paint in houses, other important sources of lead exposure include (1) **contaminated soil** and dust from deteriorated paints originally applied to buildings, bridges, and water tanks; (2) drinking water into which lead has leached from lead, bronze, or brass pipes and fixtures (including lead-soldered pipe joints) in houses, schools, and public buildings; (3) occupational exposures in smelting and refining industries, steel welding and cutting operations, battery manufacturing plants, gasoline stations, and radiator repair shops; (4) airborne lead from smelters and other point sources of **air pollution**, including vehicles burning leaded fuels; (5) **hazardous waste** sites which contaminate soil and water; (6) food cans made with lead-containing solder and pottery made with lead-containing glaze; and (7) food consumption if crops are grown using fertilizers that contain sewage **sludge** or if much lead-containing dust is deposited onto crops. In the **atmosphere**, the use of leaded gasoline has been the single largest source of lead (90%) since the 1920s, although the use of leaded fuel has been greatly curtailed and gasoline contributions are now greatly reduced (35%). As discussed below, leaded fuel and many other sources have been greatly reduced in the United States, although drinking water and other sources remain important in some areas. A number of other countries, however, continue to use leaded fuel and other lead-containing products.

Government Responses

Many agencies are concerned with lead management. Lead agencies in the United States include the Environmental Protection Agency, the Centers for Disease Control, the **U.S. Department of Health and Human Services**, the Department of Housing and Urban Development, the **Food and Drug Administration**, the Consumer Product Safety Commission, the National Institute for Occupational Safety and Health, and the **Occupational Safety and Health Administration**. These agencies have taken many actions to reduce lead exposures, several of which have been very successful. General types of actions include: (1) restrictions or bans on the use of many products containing lead where risks from these products are high and where substitute products are available, e.g., interior paints, gasoline fuels, and solder; (2) **recycling** and safer ultimate disposal strategies for products where risks are lower, or for which technically and economically feasible substitutes are not available, e.g., lead-acid automotive batteries, lead-containing wastes, pigments and used oil; (3) **emission** controls for lead smelters, primary metal industries, and other industrial point sources, including the use of the best practicable control technology (BPCT) for new lead smelting and processing facilities and

reasonable available control technologies (RACT) for existing facilities, and; (4) education and abatement programs where exposure is based on past uses of lead.

The current goals of the Environmental Protection Agency (EPA) strategy are to reduce lead exposures to the fullest extent practicable, to significantly reduce the incidence of PbB levels above 10 μg/dL in children, and to reduce lead exposures that are anticipated to pose risks to children, the general public, or the environment. Several specific actions of this and other agencies are discussed below.

The Residential Lead-based Paint Hazard Reduction Act of 1992 (Title X) provides the framework to reduce hazards from lead-based paint exposure, primarily in housing. It establishes a national infrastructure of trained workers, training programs and proficient laboratories, and a public education program to reduce hazards from lead exposure in paint in the nation's housing stock. Earlier, to help protect small children who might swallow chips of paint, the Consumer Product Safety Commission (CPSC) restricted the amount of lead in most paints to 0.06 percent by weight. CDC further suggests that inside and outside paint used in buildings where people live be tested for lead. If the level of lead is high, the paint should be removed and replaced with a paint that contains an allowable level of lead. CPSC published a consumer safety alert/brochure on lead paint in the home in 1990, and has evaluated lead test kits for safety, efficacy, and consumer-friendliness. These kits are potential screening devices that may be used by the consumer to detect lead in paint and other materials. Title X also requires EPA to promulgate regulations that ensure personnel engaged in abatement activities are trained, to certify training programs, to establish standards for abatement activities, to promulgate model state programs, to establish a laboratory accreditation program, to establish a information clearinghouse, and to disclose lead hazards at property transfer.

The Department of Housing and Urban Development (HUD) has begun activities that include updating regulations dealing with lead-based paint in HUD programs and federal property; providing support for local screening programs; increasing public education; supporting research to reduce the cost and improve the reliability of testing and abatement; increasing state and local support; and providing more money to support abatement in low and moderate income households. HUD estimated that the total cost of testing and abatement in high-priority hazard homes will be $8 to 10 billion annually over 10 years, although costs could be substantially lowered by integrating abatement with other renovation activities.

CPSC, EPA, and states are required by the Lead Contamination Control Act of 1988 to test drinking water in schools for lead and to remove lead if levels are too high.

Drinking water coolers must also be lead-free and any that contain lead must be removed. EPA regulations limit lead in drinking water to 0.015 mg/L.

To manage environmental exposures resulting from inhalation, EPA regulations limit lead to 0.1 and 0.05 g/gal (0.38 and 0.19 g/L) in leaded and unleaded gasoline, respectively. Also, the National Ambient Air Quality Standards set a maximum lead concentrations of 1.5 $\mu g/m^3$ using a three month average, although typical levels are far lower, 0.1 or 0.2 $\mu g/m^3$.

To identify and mitigate sources of lead in the diet, the Food and Drug Administration (FDA) has undertaken efforts that include voluntary discontinuation of lead solder in food cans by the domestic food industry, and elimination of lead in glazing on ceramic ware. Regulatory measures are being introduced for wine, dietary supplements, crystal ware, **food additives**, and bottled water.

For workers in lead-using industries, the Occupational Safety and Health Administration (OSHA) has established environmental and biological standards that include maximum air and blood levels. This monitoring must be conducted by the employer, and elevated PbB levels may require the removal of an individual from the work place (levels discussed previously). The Permissible Exposure Level (PEL) limits air concentrations of lead to 50 $\mu g/m^3$, and, if 30 $\mu g/m^3$ is exceeded, employers must implement a program that includes medical surveillance, exposure monitoring, training, regulated areas, respiratory protection, protective work clothing and equipment, housekeeping, hygiene facilities and practices, signs and labels, and record keeping. In the construction industry, the PEL is 200 $\mu g/m^3$. The National Institute for Occupational Safety and Health (NIOSH) recommends that workers not be exposed to levels of more than 100 $\mu g/m^3$ for up to 10 hours, and NIOSH has issued a health alert to construction workers regarding possible adverse health effects from long-term and low-level exposure. NIOSH has also published alerts and recommendations for preventing lead poisoning during blasting, sanding, cutting, burning, or welding of bridges and other steel structures coated with lead paint.

Finally, lead screening for children has recently increased. The CDC recommends that screening (testing) for lead poisoning be included in health care programs for children under 72 months of age, especially those under 36 months of age. For a community with a significant number of children having PbB levels between 10-14 $\mu g/dL$, community-wide lead poisoning prevention activities should be initiated. For individual children with PbB levels between 15-19 $\mu g/dL$, nutritional and educational interventions are recommended. PbB levels exceeding 20 $\mu g/dL$ should trigger investigations of the affected individual's environment and medical evaluations. The highest levels, above 45 $\mu g/$dL, require both medical and environmental interventions, including chelation therapy. CDC also conducts studies to determine the impact of interventions on children's blood lead levels.

These regulatory activities have resulted in significant reductions in average levels of lead exposure. Nevertheless, lead management remains an important public health problem.

[*Stuart Batterman*]

RESOURCES

BOOKS

Breen, J. J., and C. R. Stroup, eds. *Lead Poisoning: Exposure, Abatement, Regulation.* Lewis Publishers, 1995.

Kessel, I., J. T. O'Connor, and J. W. Graef. *Getting the Lead Out: The Complete Resource for Preventing and Coping with Lead Poisoning.* Rev. ed. Cambridge, MA: Fisher Books, 2001.

Pueschel, S. M., J. G. Linakis, and A. C. Anderson. *Lead Poisoning in Childhood.* Baltimore: Paul H. Brookes Publishing Co., 1996.

OTHER

Farley, Dixie. "Dangers of Lead Still Linger." *FDA Consumer* January-February 1998 [cited July 2002]. <http://www.cfsan.fda.gov/~dms/fda-lead.html>.

Lead shot

Lead shot refers to the small pellets that are fired by shotguns while **hunting** waterfowl or upland fowl, or while skeet shooting. Most lead shots miss their target and are dissipated into the **environment**. Because the shot is within the particle-size range that is favored by medium-sized birds as grit, it is often ingested and retained in the gizzard to aid in the mechanical abrasion of plant seeds, the first step in avian digestion. However, the shot also abrades during this process, releasing toxic lead that can poison the bird. It has been estimated that as much as 2–3 percent of the North American waterfowl population, or several million birds, may die from shot-caused lead **poisoning** each year. This problem will decrease in intensity, however, because lead shot is now being substantially replaced by steel shot in North America. *See also* Heavy metals and heavy metal poisoning

Leafy spurge

Leafy spurge (*Euphorbia esula* L.), a perennial plant from Europe and Asia, was introduced to North America through imported grain products by 1827. It is 12 in (30.5 cm) to 3 ft (1 m) in height. Stems, leaves, and roots contain milky white latex which contains toxic cardiac glycosides that is distasteful to cattle, who will not eat it. Considered a noxious, or destructive, weed in southern Canada and the northern

Great Plains of the United States, it crowds out native range-land grasses, reducing the number of cattle that can graze the land. It is responsible for losses of approximately 35–45 million dollars per year to the United States cattle and hay industries. Its aggressive root system makes controlling spread difficult. Roots spread vertically to 15 ft (5 m) with up to 300 root buds, and horizontally to nearly 30 ft (9 m). It regenerates from small portions of root. Tilling, burning, and **herbicide** use are ineffective control methods as roots are not damaged and may prevent immediate regrowth of the desired **species**. The introduction of specific herbivores of the leafy spurge from its native range, including certain species of beetles and moths, may be an effective means of control, as may be certain pathogenic **fungi**. Studies also indicate sheep and Angora goats will eat it. To control this plant's rampant spread in North America, a combination of methods seems most effective.

[*Monica Anderson*]

League of Conservation Voters

In 1970 Marion Edey, a House committee staffer, founded the League of Conservation Voters (LCV) as the non-partisan political action arm of the United States' environmental movement. LCV works to establish a pro-environment—or "green"—majority in Congress and to elect environmentally conscious candidates throughout the country. Through campaign donations, volunteers, and endorsements, pro-environment advertisements, and annual publications such as the *National Environmental Scorecard*, the League raises voter awareness of the environmental positions of candidates and elected officials.

Technically it has no formal membership, but the League's supporters—who make donations and purchase its publications—number 100,000. The board of directors is comprised of 24 important environmentalists associated with such organizations as the **Sierra Club**, the **Environmental Defense** Fund, and **Friends of the Earth**. Because these organizations would endanger their charitable tax status if they participated directly in the electoral process, environmentalists developed the League. Since 1970 LCV has influenced many elections.

From its first effort in 1970—wherein LCV successfully prevented Rep. Wayne Aspinall of Colorado from obtaining a democratic nomination—the League has grown to be a significant force in American politics. In the 1989–90 elections LCV supported 120 pro-environment candidates and spent approximately $250,000 on their campaigns. In 1990 the League developed new endorsement tactics. First

it invented the term "greenscam" to identify candidates who only appear green. Next LCV produced two generic television advertisements for candidates. One advertisement, entitled "Greenscam," attacked the aforementioned candidates; the other, entitled "Decisions," was an award-winning, positive advertisement in support of pro-environment candidates. By the 2000 campaign the League had attained an unprecedented degree of influence in the electoral process. That year LCV raised and donated 4.1 million dollars in support of both Democratic and Republican candidates in a variety of ways.

In endorsing a candidate the League no longer simply contributes money to a campaign. It provides "in-kind" assistance—for example, it places a trained field organizer on a staff, creates radio and television advertisements, or develops grassroots outreach programs and campaign literature. In addition to supporting specific candidates, LCV holds all elected officials accountable for their track records on environmental issues. The League's annual publication *National Environmental Scorecard* lists the voting records of House and Senate members on environmental legislation. Likewise, the *Presidential Scorecard* identifies the positions that presidential candidates have taken. Through these publications and direct endorsement strategies, the League continues to apply pressure in the political process and elicit support for the **environment**

[*Andrea Gacki*]

RESOURCES

ORGANIZATIONS

League of Conservation Voters, 1920 L Street, NW, Suite 800, Washington, D.C. USA 20036 (202) 785-8683, Fax: (202) 835-0491, <http://www.lcv.org>

Louis Seymour Bazett Leakey (1903 – 1972)

African-born English paleontologist and anthropologist

Louis Seymour Bazett Leakey was born on August 7, 1903, in Kabete, Kenya. His parents, Mary Bazett (d. 1948) and Harry Leakey (1868–1940) were Church of England missionaries at the Church Missionary Society, Kabete, Kenya. Louis spent his childhood in the mission, where he learned the Kikuyu language and customs (he later compiled a Kikuyu grammar book). As a child, while pursuing his interest in ornithology—the study of birds—he often found stone tools washed out of the **soil** by the heavy rains, which Leakey believed were of prehistoric origin. Stone tools were primary evidence of the presence of humans at a particular site, as toolmaking was believed at the time to be practiced only by humans and was, along with an erect posture, one of the chief characteristics used to differentiate humans from non-humans. Scientists at the time, however, did not consider East Africa a likely site for finding evidence of early humans; the discovery of *Pithecanthropus* in Java in 1894 (the so-called Java Man, now considered to be an example of *Homo erectus*) had led scientists to assume that Asia was the continent from which human forms had spread.

Shortly after the end of World War I, Leakey was sent to a public school in Weymouth, England, and later attended St. John's College, Cambridge. Suffering from severe headaches resulting from a sports injury, he took a year off from his studies and joined a fossil-hunting expedition to Tanganyika (now Tanzania). This experience, combined with his studies in anthropology at Cambridge (culminating in a degree in 1926), led Leakey to devote his time to the search for the origins of humanity, which he believed would be found in Africa. Anatomist and anthropologist Raymond A. Dart's discovery of early human remains in South Africa was the first concrete evidence that this view was correct. Leakey's next expedition was to northwest Kenya, near Lakes Nakuru and Naivasha, where he uncovered materials from the Late Stone Age; at Kariandusi he discovered a 200,000-year-old hand ax.

In 1928 Leakey married Henrietta Wilfrida Avern, with whom he had two children: Priscilla, born in 1930, and Colin, born in 1933; the couple was divorced in the mid-1930s. In 1931 Leakey made his first trip to Olduvai Gorge—a 350-mi (564-km) ravine in Tanzania—the site that was to be his richest source of human remains. He had been discouraged from excavating at Olduvai by Hans Reck, a German paleontologist who had fruitlessly sought evidence of prehistoric humans there. Leakey's first discoveries at that site consisted of both animal fossils, important in the attempts to date the particular stratum (or layer of earth) in which they were found, and, significantly, flint tools. These tools, dated to approximately one million years ago, were conclusive evidence of the presence of hominids—a family of erect primate mammals that use only two feet for locomotion—in Africa at that early date; it was not until 1959, however, that the first fossilized hominid remains were found there.

In 1932, near Lake Victoria, Leakey found remains of *Homo sapiens* (modern man), the so-called Kanjera skulls (dated to 100,000 years ago) and Kanam jaw (dated to 500,000 years ago); Leakey's claims for the antiquity of this jaw made it a controversial find among other paleontologists, and Leakey hoped he would find other, independent, evidence for the existence of *Homo sapiens* from an even earlier period—the Lower Pleistocene.

In the mid-1930s, a short time after his divorce from Wilfrida, Leakey married his second wife, Mary Douglas Nicol; she was to make some of the most significant discoveries of Leakey's team's research. The couple eventually had three children: Philip, Jonathan, and Richard E. Leakey. During the 1930s, Leakey also became interested in the study of the Paleolithic period in Britain, both regarding human remains and geology, and he and Mary Leakey carried out excavations at Clacton in southeast England.

Until the end of the 1930s, Leakey concentrated on the discovery of stone tools as evidence of human habitation; after this period he devoted more time to the unearthing of human and prehuman fossils. His expeditions to Rusinga Island, at the mouth of the Kavirondo Gulf in Kenya, during the 1930s and early 1940s produced a large number of finds, especially of remains of Miocene apes. One of these apes, which Leakey named *Proconsul africanus,* had a jaw lacking in the so-called simian shelf that normally characterized the jaws of apes; this was evidence that *Proconsul* represented a stage in the progression from ancient apes to humans. In 1948 Mary Leakey found a nearly complete *Proconsul* skull, the first fossil ape skull ever unearthed; this was followed by the unearthing of several more *Proconsul* remains.

Louis Leakey began his first regular excavations at Olduvai Gorge in 1952; however, the Mau Mau (an anti-white secret society) uprising in Kenya in the early 1950s disrupted his paleontological work and induced him to write *Mau Mau and the Kikuyu,* in an effort to explain the rebellion from the perspective of a European with an insider's knowledge of the Kikuyu. A second work, *Defeating Mau Mau,* followed in 1954.

During the late 1950s, the Leakeys continued their work at Olduvai. In 1959, while Louis was recuperating from an illness, Mary Leakey found substantial fragments of a hominid skull that resembled the robust australopithecines—African hominids possessing small brains and near-human dentition—found in South Africa earlier in the century. Louis Leakey, who quickly reported the find to the journal *Nature,* suggested that this represented a new genus, which he named *Zinjanthropus boisei,* the genus name meaning "East African man," and the **species** name commemorating Charles Boise, one of Leakey's benefactors. This species, now called *Australopithecus boisei,* was later believed by Leakey to have been an evolutionary dead end, existing contemporaneously with *Homo* rather than representing an earlier developmental stage.

In 1961, at Fort Ternan, Leakey's team located fragments of a jaw that Leakey believed were from a hitherto unknown genus and species of ape, one he designated as *Kenyapithecus wickeri,* and which he believed was a link between ancient apes and humans, dating from 14 million years ago; it therefore represented the earliest hominid. In 1967, however, an older skull, one that had been found two decades earlier on Rusinga Island and which Leakey had

Louis Leakey. (The Library of Congress.)

originally given the name *Ramapithecus africanus*, was found to have hominid-like lower dentition; he renamed it *Kenyapithecus africanus*, and Leakey believed it was an even earlier hominid than *Kenyapithecus wickeri*. Leakey's theories about the place of these Lower Miocene fossil apes in human **evolution** have been among his most widely disputed.

During the early 1960s, a member of Leakey's team found fragments of the hand, foot, and leg bones of two individuals, in a site near where *Zinjanthropus* had been found, but in a slightly lower and, apparently, slightly older layer. These bones appeared to be of a creature more like modern humans than *Zinjanthropus*, possibly a species of *Homo* that lived at approximately the same time, with a larger brain and the ability to walk fully upright. As a result of the newly developed potassium-argon dating method, it was discovered that the bed from which these bones had come was 1.75 million years old. The bones were, apparently, the evidence for which Leakey had been searching for years: skeletal remains of *Homo* from the Lower Pleistocene. Leakey designated the creature whose remains these were as *Homo habilis* ("man with ability"), a creature who walked upright and had dentition resembling that of modern humans, hands capable of toolmaking, and a large cranial capacity. Leakey saw this hominid as a direct ancestor of *Homo erectus* and modern humans. Not unexpectedly, Leakey was attacked by other scholars, as this identification of the frag-

ments moved the origins of the genus *Homo* back substantially further in time. Some scholars felt that the new remains were those of australopithecines, if relatively advanced ones, rather than very early examples of *Homo*.

Health problems during the 1960s curtailed Leakey's field work; it was at this time that his Centre for Prehistory and Paleontology in Nairobi became the springboard for the careers of such paleontologists as **Jane Goodall** and Dian Fossey in the study of nonhuman primates. A request came in 1964 from the Israeli government for assistance with the technical as well as the fundraising aspects involved in the excavation of an early Pleistocene site at Ubeidiya. This produced evidence of human habitation dating back 700,000 years, the earliest such find outside Africa.

During the 1960s, others, including Mary Leakey and the Leakeys' son Richard, made significant finds in East Africa; Leakey turned his attention to the investigation of a problem that had intrigued him since his college days: the determination of when humans had reached the North American continent. Concentrating his investigation in the Calico Hills in the Mojave **Desert**, California, he sought evidence in the form of stone tools of the presence of early humans, as he had done in East Africa. The discovery of some pieces of chalcedony (translucent quartz) that resembled manufactured tools in **sediment** dated from 50,000 to 100,000 years old stirred an immediate controversy; at that time, scientists believed that humans had settled in North America approximately 20,000 years ago. Many archaeologists, including Mary Leakey, criticized Leakey's California methodology—and his interpretations of the finds—as scientifically unsound, but Leakey, still charismatic and persuasive, was successful in obtaining funding from the National Geographic Society and, later, several other sources. Human remains were not found in conjunction with the supposed stone tools, and many scientists have not accepted these "artifacts" as anything other than rocks.

Shortly before Louis Leakey's death, Richard Leakey showed his father a skull he had recently found near Lake Rudolf (now Lake Turkana) in Kenya. This skull, removed from a deposit dated to 2.9 million years ago, had a cranial capacity of approximately 800 cubic centimeters, putting it within the range of *Homo* and apparently vindicating Leakey's long-held belief in the extreme antiquity of that genus; it also appeared to substantiate Leakey's interpretation of the Kanam jaw. Leakey died of a heart attack in early October, 1972, in London.

Some scientists have questioned Leakey's interpretations of his discoveries. Other scholars have pointed out that two of the most important finds associated with him were actually made by Mary Leakey, but became widely known when they were interpreted and publicized by him; Leakey had even encouraged criticism through his tendency to publicize his somewhat sensationalistic theories before they had been sufficiently tested. Critics have cited both his tendency toward hyperbole and his penchant for claiming that his finds were the "oldest," the "first," the "most significant"; in a 1965 *National Geographic* article, for example, Melvin M. Payne pointed out that Leakey, at a Washington, D.C., press conference, claimed that his discovery of *Homo habilis* had made all previous scholarship on early humans obsolete. Leakey has also been criticized for his eagerness to create new genera and species for new finds, rather than trying to fit them into existing categories. Leakey, however, recognized the value of publicity for the fundraising efforts necessary for his expeditions. He was known as an ambitious man, with a penchant for stubbornly adhering to his interpretations, and he used the force of his personality to communicate his various finds and the subsequent theories he devised to scholars and the general public.

Leakey's response to criticism was that scientists have trouble divesting themselves of their own theories in the light of new evidence. "Theories on prehistory and early man constantly change as new evidence comes to light," Leakey remarked, as quoted by Payne in *National Geographic*. "A single find such as *Homo habilis* can upset long-held—and reluctantly discarded—concepts. A paucity of human fossil material and the necessity for filling in blank spaces extending through hundreds of thousands of years all contribute to a divergence of interpretations. But this is all we have to work with; we must make the best of it within the limited range of our present knowledge and experience." Much of the controversy derives from the lack of consensus among scientists about what defines "human"; to what extent are toolmaking, dentition, cranial capacity, and an upright posture defining characteristics, as Leakey asserted?

Louis Leakey's significance revolves around the ways in which he changed views of early human development. He pushed back the date when the first humans appeared to a time earlier than had been believed on the basis of previous research. He showed that human evolution began in Africa rather than Asia, as had been maintained. In addition, he created research facilities in Africa and stimulated explorations in related fields, such as primatology (the study of primates). His work is notable as well for the sheer number of finds—not only of the remains of apes and humans, but also of the plant and animal species that comprised the ecosystems in which they lived. These finds of Leakey and his team filled numerous gaps in scientific knowledge of the evolution of human forms. They provided clues to the links between prehuman, apelike primates, and early humans, and demonstrated that human evolution may have followed more than one parallel path, one of which led to modern humans, rather than a single line, as earlier scientists had maintained.

[*Michael Sims*]

RESOURCES

BOOKS

Cole, S. *Leakey's Luck: The Life of Louis Seymour Bazett Leakey, 1903–1972.* Harcourt, 1975.

Isaac, G., and E. R. McCown, eds., *Human Origins: Louis Leakey and the East African Evidence.* Benjamin-Cummings, 1976.

Johanson, D. C., and M. A. Edey. *Lucy: The Beginnings of Humankind.* Simon & Schuster, 1981.

Leakey, M. *Disclosing the Past.* Doubleday, 1984.

Leakey, R. *One Life: An Autobiography.* Salem House, 1984.

Malatesta, A., and R. Friedland, *The White Kikuyu: Louis S. B. Leakey.* McGraw-Hill, 1978.

Mary Douglas Nicol Leakey (1913 – 1996)

English paleontologist and anthropologist

For many years Mary Leakey lived in the shadow of her husband, Louis Leakey, whose reputation, coupled with the prejudices of the time, led him to be credited with some of his wife's discoveries in the field of early human archaeology. Yet she has established a substantial reputation in her own right and has come to be recognized as one of the most important paleoanthropologists of the twentieth century. It was Mary Leakey who was responsible for some of the most important discoveries made by Louis Leakey's team. Although her close association with Louis Leakey's work on Paleolithic sites at Olduvai Gorge—a 350-mi (564-km) ravine in Tanzania—has led to her being considered a specialist in that particular area and period, she has in fact worked on excavations dating from as early as the Miocene Age (an era dating to approximately 18 million years ago) to those as recent as the Iron Age of a few thousand years ago.

Mary Leakey was born Mary Douglas Nicol on February 6, 1913, in London. Her mother was Cecilia Frere, the great-granddaughter of John Frere, who had discovered prehistoric stone tools at Hoxne, Suffolk, England, in 1797. Her father was Erskine Nicol, a painter who himself was the son of an artist, and who had a deep interest in Egyptian archaeology. When Mary was a child, her family made frequent trips to southwestern France, where her father took her to see the Upper Paleolithic cave paintings. She and her father became friends with Elie Peyrony, the curator of the local museum, and there she was exposed to the vast collection of flint tools dating from that period of human prehistory. She was also allowed to accompany Peyrony on his excavations, though the archaeological work was not conducted in what would now be considered a scientific way—artifacts were removed from the site without careful study of the place in the earth where each had been found, obscuring valuable data that could be used in dating the artifact and analyzing its context. On a later trip, in 1925, she was taken to Paleolithic caves by the Abbe Lemozi of France, parish priest of Cabrerets, who had written papers on cave art. After her father's death in 1926, Mary Nicol was taken to Stonehenge and Avebury in England, where she began to learn about the archaeological activity in that country and, after meeting the archaeologist Dorothy Liddell, to realize the possibility of archaeology as a career for a woman.

By 1930 Mary Nicol had undertaken coursework in geology and archaeology at the University of London and had participated in a few excavations in order to obtain field experience. One of her lecturers, R. E. M. Wheeler, offered her the opportunity to join his party excavating St. Albans, England, the ancient Roman site of Verulamium; although she only remained at that site for a few days, finding the work there poorly organized, she began her career in earnest shortly thereafter, excavating Neolithic (early Stone Age) sites in Henbury, Devon, where she worked between 1930 and 1934. Her main area of expertise was stone tools, and she was exceptionally skilled at making drawings of them. During the 1930s Mary met Louis Leakey, who was to become her husband. Leakey was by this time well known because of his finds of early human remains in East Africa; it was at Mary and Louis's first meeting that he asked her to help him with the illustrations for his 1934 book, *Adam's Ancestors: An Up-to-Date Outline of What Is Known about the Origin of Man.*

In 1934 Mary Nicol and Louis Leakey worked at an excavation in Clacton, England, where the skull of a hominid—a family of erect primate mammals that use only two feet for locomotion—had recently been found and where Louis was investigating Paleolithic geology as well as **fauna** and human remains. The excavation led to Mary Leakey's first publication, a 1937 report in the *Proceedings of the Prehistoric Society.*

By this time, Louis Leakey had decided that Mary should join him on his next expedition to Olduvai Gorge in Tanganyika (now Tanzania), which he believed to be the most promising site for discovering early Paleolithic human remains. On the journey to Olduvai, Mary stopped briefly in South Africa, where she spent a few weeks with an archaeological team and learned more about the scientific approach to excavation, studying each find *in situ*—paying close attention to the details of the geological and faunal material surrounding each artifact. This knowledge was to assist her in her later work at Olduvai and elsewhere.

At Olduvai, among her earliest discoveries were fragments of a human skull; these were some of the first such remains found at the site, and it would be twenty years before any others would be found there. Mary Nicol and Louis Leakey returned to England. Leakey's divorce from his first wife was made final in the mid-1930s, and he and Mary Nicol were then married; the couple returned to Kenya

in January of 1937. Over the next few years, the Leakeys excavated Neolithic and Iron Age sites at Hyrax Hill, Njoro River Cave, and the Naivasha Railway Rock Shelter, which yielded a large number of human remains and artifacts.

During World War II, the Leakeys began to excavate at Olorgasailie, southwest of Nairobi, but because of the complicated geology of that site, the dating of material found there was difficult. It did prove to be a rich source of material, however; in 1942 Mary Leakey uncovered hundreds, possibly thousands, of hand axes there. Her first major discovery in the field of prehuman fossils was that of most of the skull of a *Proconsul africanus* on Rusinga Island, in Lake Victoria, Kenya, in 1948. *Proconsul* was believed by some paleontologists to be a common ancestor of apes and humans, an animal whose descendants developed into two branches on the evolutionary tree: the *Pongidae* (great apes) and the *Hominidae* (who eventually evolved into true humans). *Proconsul* lived during the Miocene Age, approximately 18 million years ago. This was the first time a fossil ape skull had ever been found—only a small number have been found since—and the Leakeys hoped that this would be the ancestral hominid that paleontologists had sought for decades. The absence of a "simian shelf," a reinforcement of the jaw found in modern apes, is one of the features of *Proconsul* that led the Leakeys to infer that this was a direct ancestor of modern humans. *Proconsul* is now generally believed to be a **species** of *Dryopithecus*, closer to apes than to humans.

Many of the finds at Olduvai were primitive stone hand axes, evidence of human habitation; it was not known, however, who had made them. Mary's concentration had been on the discovery of such tools, while Louis's goal had been to learn who had made them, in the hope that the date for the appearance of toolmaking hominids could be moved back to an earlier point. In 1959 Mary unearthed part of the jaw of an early hominid she designated *Zinjanthropus* (meaning "East African Man") and whom she referred to as "Dear Boy"; the early hominid is now considered to be a species of *Australopithecus*—apparently related to the two kinds of australopithecine found in South Africa, *Australopithecus africanus* and *Australopithecus robustus*— and given the species designation *boisei* in honor of Louis Leakey's sponsor Charles Boise. By means of potassium-argon dating, recently developed, it was determined that the fragment was 1.75 million years old, and this realization pushed back the date for the appearance of hominids in Africa. Despite the importance of this find, however, Louis Leakey was slightly disappointed, as he had hoped that the excavations would unearth not another australopithecine, but an example of *Homo* living at that early date. He was seeking evidence for his theory that more than one hominid form lived at Olduvai at the same time; these forms were the australopithecines, who eventually died out, and some early form of *Homo*, which survived—owing to toolmaking ability and larger cranial capacity—to evolve into *Homo erectus* and, eventually, the modern human. Leakey hoped that Mary Leakey's find would prove that *Homo* existed at that early level of Olduvai. The discovery he awaited did not come until the early 1960s, with the identification of a skull found by their son Jonathan Leakey that Louis designated as *Homohabilis* ("man with ability"). He believed this to be the true early human responsible for making the tools found at the site.

In her autobiography, *Disclosing the Past*, released in 1984, Mary Leakey reveals that her professional and personal relationship with Louis Leakey had begun to deteriorate by 1968. As she increasingly began to lead the Olduvai research on her own, and as she developed a reputation in her own right through her numerous publications of research results, she believes that her husband began to feel threatened. Louis Leakey had been spending a vast amount of his time in fundraising and administrative matters, while Mary was able to concentrate on field work. As Louis began to seek recognition in new areas, most notably in excavations seeking evidence of early humans in California, Mary stepped up her work at Olduvai, and the breach between them widened. She became critical of his interpretations of his California finds, viewing them as evidence of a decline in his scientific rigor. During these years at Olduvai, Mary made numerous new discoveries, including the first *Homo erectus* pelvis to be found. Mary Leakey continued her work after Louis Leakey's death in 1972. From 1975 she concentrated on Laetoli, Tanzania, which was a site earlier than the oldest beds at Olduvai. She knew that the lava above the Laetoli beds was dated to 2.4 million years ago, and the beds themselves were therefore even older; in contrast, the oldest beds at Olduvai were two million years old. Potassium-argon dating has since shown the upper beds at Laetoli to be approximately 3.5 million years old. In 1978 members of her team found two trails of hominid footprints in volcanic ash dated to approximately 3.5 million years ago; the form of the footprints gave evidence that these hominids walked upright, thus moving the date for the development of an upright posture back significantly earlier than previously believed. Mary Leakey considers these footprints to be among the most significant finds with which she has been associated.

In the late 1960s Mary Leakey received an honorary doctorate from the University of the Witwatersrand in South Africa, an honor she accepted only after university officials had spoken out against apartheid. Among her other honorary degrees are a D.S.Sc. from Yale University and a D.Sc. from the University of Chicago. She received an honorary D.Litt. from Oxford University in 1981. She has also received the Gold Medal of the Society of Women Geographers.

Louis Leakey was sometimes faulted for being too quick to interpret the finds of his team and for his propensity for developing sensationalistic, publicity-attracting theories. In recent years Mary Leakey had been critical of the conclusions reached by her husband—as well as by some others—but she did not add her own interpretations to the mix. Instead, she has always been more concerned with the act of discovery itself; she wrote that it is more important for her to continue the task of uncovering early human remains to provide the pieces of the puzzle than it is to speculate and develop her own interpretations. Her legacy lies in the vast amount of material she and her team have unearthed; she leaves it to future scholars to deduce its meaning.

[*Michael Sims*]

RESOURCES
BOOKS

Isaac, G., and E. R. McCown, eds. *Human Origins: Louis Leakey and the East African Evidence*. Benjamin-Cummings, 1976.

Reader, J. *Missing Links*. Little, Brown, 1981.

Moore, R. E., *Man, Time, and Fossils: The Story of Evolution*. Knopf, 1961.

Malatesta, A., and R. Friedland, *The White Kikuyu: Louis S. B. Leakey*. McGraw-Hill, 1978.

Leakey, R. *One Life: An Autobiography*. Salem House, 1984.

Johanson, D. C., and M. A. Edey, *Lucy: The Beginnings of Humankind*. Simon & Schuster, 1981.

Cole, S. *Leakey's Luck: The Life of Louis Seymour Bazett Leakey, 1903–1972*. Harcourt, 1975.

Leakey, L. *By the Evidence: Memoirs, 1932–1951*. Harcourt, 1974.

Richard Erskine Frere Leakey (1944 –)
African-born English paleontologist and anthropologist

Richard Erskine Frere Leakey was born on December 19, 1944, in Nairobi, Kenya. Continuing the work of his parents, Leakey has pushed the date for the appearance of the first true humans back even further than they had, to nearly three million years ago. This represents nearly a doubling of the previous estimates. Leakey also has found more evidence to support his father's still controversial theory that there were at least two parallel branches of human **evolution**, of which only one was successful. The abundance of human fossils uncovered by Richard Leakey's team has provided an enormous number of clues as to how the various fossil remains fit into the puzzle of human evolution. The team's finds have also helped to answer, if only speculatively, some basic questions: When did modern human's ancestors split off from the ancient apes? On what continent did this take place? At what point did they develop the characteristics now considered as defining human attributes? What is the relationship among and the chronology of the various genera and **species** of the fossil remains that have been found?

While accompanying his parents on an excavation at Kanjera near Lake Victoria at the age of six, Richard Leakey made his first discovery of fossilized animal remains, part of an extinct variety of giant pig. Richard Leakey, however, was determined not to "ride upon his parents' shoulders," as Mary Leakey wrote in her autobiography, *Disclosing the Past*. Several years later, as a young teenager in the early 1960s, Richard demonstrated a talent for **trapping wildlife**, which prompted him to drop out of high school to lead photographic safaris in Kenya. His paleontological career began in 1963, when he led a team of paleontologists to a fossil-bearing area near Lake Natron in Tanganyika (now Tanzania), a site that was later dated to approximately 1.4 million years ago. A member of the team discovered the jaw of an early hominid—a member of the family of erect primate mammals that use only two feet for locomotion—called an *Australopithecus boisei* (then named *Zinjanthropus*).) This was the first discovery of a complete *Australopithecus* lower jaw and the only *Australopithecus* skull fragment found since Mary Leakey's landmark discovery in 1959. Jaws provide essential clues about the nature of a hominid, both in terms of its structural similarity to other species and, if teeth are present, its diet. Richard Leakey spent the next few years occupied with more excavations, the most important result of which was the discovery of a nearly complete fossil elephant.

In 1964 Richard married Margaret Cropper, who had been a member of his father's team at Olduvai the year before. It was at this time that he became associated with his father's Centre for Prehistory and Paleontology in Nairobi. In 1968, at the age of 23, he became administrative director of the National Museum of Kenya.

While his parents had mined with great success the fossil-rich Olduvai Gorge, Richard Leakey concentrated his efforts in northern Kenya and southern Ethiopia. In 1967 he served as the leader of an expedition to the Omo Delta area of southern Ethiopia, a trip financed by the National Geographic Society. In a site dated to approximately 150,000 years ago, members of his team located portions of two fossilized human skulls believed to be from examples of *Homo sapiens*, or modern humans. While the prevailing view at the time was that *Homo sapiens* emerged around 60,000 years ago, these skulls were dated 130,000 years old.

While on an airplane trip, Richard Leakey flew over the eastern portion of Lake Rudolf (now Lake Turkana) on the Ethiopia-Kenya border, and he noticed from the air what appeared to be ancient lake sediments, a kind of terrain that he felt looked promising as an excavation site. He used his next National Geographic Society grant to explore this area. The region was Koobi Fora, a site that was to become Richard Leakey's most important area for excavation. At Koobi Fora his team uncovered more than four hundred hominid fossils and an abundance of stone tools, such tools being a primary indication of the presence of early humans. Subsequent excavations near the Omo River in Kenya, from 1968, unearthed more examples of early humans, the first found being another *Australopithecus* lower jaw fragment. At the area of Koobi Fora known as the KBS tuff (tuff being volcanic ash; KBS standing for the Kay Behrensmeyer Site, after a member of the team) stone tools were found. Preliminary dating of the site placed the area at 2.6 million years ago; subsequent tests over the following few years determined the now generally accepted age of 1.89 million years.

In July of 1969, Richard Leakey came across a virtually complete *Australopithecus boisei* skull—lacking only the teeth and lower jaw—lying in a river bed. A few days later a member of the team located another hominid skull nearby, comprising the back and base of the cranium. The following year brought the discovery of many more fossil hominid remains, at the rate of nearly two per week. Among the most important finds was the first hominid femur to be found in Kenya, which was soon followed by several more. It was at about this time that Leakey obtained a divorce from his first wife, and in October of 1970, he married Meave Gillian Epps, who had been on the 1969 expedition.

In 1972, Richard Leakey's team uncovered a skull that appeared to be similar to the one identified by his father and called *Homo habilis* ("man with ability"). This was the early human that Louis Leakey maintained had achieved the toolmaking skills that precipitated the development of a larger brain capacity and led to the development of the modern human—*Homo sapiens*. This skull was more complete and apparently somewhat older than the one Louis Leakey had found and was thus the earliest example of the species *Homo* yet discovered. They labeled the new skull, which was found below the KBS tuff, "Skull 1470," and this proved to among Richard Leakey's most significant discoveries. The fragments consisted of small pieces of all sides of the cranium, and, unusually, the facial bones, enough to permit a reasonably complete reconstruction. Larger than the skulls found in 1969 and 1970, this example had approximately twice the cranial capacity of *Australopithecus* and more than half that of a modern human—nearly 800 cubic centimeters. At the time, Leakey believed the fragments to be 2.9 million years old (although a more recent dating of the site would place them at less than 2 million years old). Basing his theory in part on these data, Leakey developed the view that these early hominids may have lived as early as 2.5 or even 3.5 million years ago and gave evidence to the theory that *Homo habilis* was not a descendant of the australopithecines, but a contemporary.

By the late 1960s, relations between Richard Leakey and his father had become strained, partly because of real or imagined competition within the administrative structure of the Centre for Prehistory, and partly because of some divergences in methodology and interpretation. Shortly before Louis Leakey's death, however, the discovery of Skull 1470 by Richard Leakey's team allowed Richard to present his father with apparent corroboration of one of his central theories.

Richard Leakey did not make his theories of human evolution public until 1974. At this time, scientists were still grappling with Louis Leakey's interpretation of his findings that there had been at least two parallel lines of human evolution, only one of which led to modern humans. After Louis Leakey's death, Richard Leakey reported that, based on new finds, he believed that hominids diversified between 3 and 3.5 million years ago. Various lines of australopithecines and *Homo* coexisted, with only one line, *Homo,* surviving. The australopithecines and *Homo* shared a common ancestor; *Australopithecus* was not ancestral to *Homo.* As did his father, Leakey believes that *Homo* developed in Africa, and it was *Homo erectus* who, approximately 1.5 million years ago, developed the technological capacity to begin the spread of humans beyond their African origins. In Richard Leakey's scheme, *Homo habilis* developed into *Homo erectus,* who in turn developed into *Homo sapiens,* the present-day human.

As new finds are made, new questions arise. Are newly discovered variants proof of a plurality of species, or do they give evidence of greater variety within the species that have already been identified? To what extent is sexual dimorphism responsible for the apparent differences in the fossils? In some scientific circles, the discovery of fossil remains at Hadar in Ethiopia by archaeologist Donald Carl Johanson and others, along with the more recent revised dating of Skull 1470, cast some doubt on Leakey's theory in general and on his interpretation of *Homo habilis* in particular. Johanson believed that the fossils he found at Hadar and the fossils Mary Leakey found at Laetoli in Tanzania, and which she classified as *Homo habilis,* were actually all australopithecines; he termed them *Australopithecus afarensis* and claimed that this species is the common ancestor of both the later australopithecines and *Homo.* Richard Leakey has rejected this argument, contending that the australopithecines were not ancestral to *Homo* and that an earlier common ancestor would be found, possibly among the fossils found by Mary Leakey at Laetoli.

The year 1975 brought another significant find by Leakey's team at Koobi Fora: the team found what was apparently the skull of a *Homo erectus,* according to Louis Leakey's theory a descendent of *Homo habilis* and probably dating to 1.5 million years ago. This skull, labeled "3733," represents the earliest known evidence for *Homo erectus* in Africa.

Richard Leakey began to suffer from health problems during the 1970s, and in 1979 he was diagnosed with a serious kidney malfunction. Later that year he underwent a

kidney transplant operation, his younger brother Philip being the donor. During his recuperation Richard completed his autobiography, *One Life,* which was released in 1984, and following his recovery, he renewed his search for the origins of the human species. The summer of 1984 brought another major discovery: the so-called Turkana boy, a nearly complete skeleton of a *Homo erectus,* missing little but the hands and feet, and offering, for the first time, the opportunity to view many bones of this species. It was shortly after the unearthing of Turkana boy—whose skeletal remains indicate that he was a twelve-year-old youngster who stood approximately five-and-a-half feet tall—that the puzzle of human evolution became even more complicated. The discovery of a new skull, called the Black Skull, with an *Australopithecus boisei* face but a cranium that was quite apelike, introduced yet another complication, possibly a fourth branch in the evolutionary tree. Leakey became the Director of the Wildlife **Conservation** and Management Department for Kenya (Kenya Wildlife Service) in 1989 and in 1999 became head of the Kenyan civil service.

[*Michael Sims*]

RESOURCES
BOOKS

Leakey, M. *Disclosing the Past: An Autobiography.* Doubleday, 1984.
Leakey, R., and R. Lewin. *Origins Reconsidered: In Search of What Makes Us Human.* Doubleday, 1992.
Reader, J. *Missing Links.* Little, Brown, 1981.

Leaking underground storage tank

Leaking underground storage tanks (LUST) that hold toxic substances have come under new regulatory scrutiny in the United States because of the health and environmental hazards posed by the materials that can leak from them. These storage tanks typically hold **petroleum** products and other toxic **chemicals** beneath gas stations and other petroleum facilities. An estimated 63,000 of the nation's underground storage tanks have been shown to leak contaminants into the **environment** or are considered to have the potential to leak at any time. One reason for the instability of underground storage tanks is their construction. Only five percent of underground storage tanks are made of corrosion-protected steel, while 84 percent are made of bare steel, which corrodes easily. Another 11 percent of underground storage tanks are made of fiberglass.

Hazardous materials seeping from some of the nation's six million LUSTs can contaminate aquifers, the water-bearing rock units that supply much of the earth's drinking water. An **aquifer**, once contaminated, can be ruined as a source of fresh water. In particular, **benzene** has been found to be a contaminant of **groundwater** as a result of leaks from underground **gasoline** storage tanks. Benzene and other volatile organic compounds have been detected in bottled water despite manufacturers' claims of purity.

According to the **Environmental Protection Agency** (EPA), more than 30 states reported groundwater contamination from petroleum products leaking from underground storage tanks. States also reported water contamination from **radioactive waste leaching** from storage containment facilities. Other reported **pollution** problems include leaking hazardous substances that are corrosive, explosive, readily flammable, or chemically reactive. While **water pollution** may be the most visible consequence of leaks from underground storage tanks, fires and explosions are dangerous and sometimes real possibilities in some areas.

The EPA is charged with exploring, developing, and disseminating technologies and funding mechanisms for cleanup. The primary job itself, however, is left to state and local governments. Actual cleanup is sometimes funded by the Leaking Underground Storage Tank trust fund established by Congress in 1986. Under the Superfund Amendment and Reauthorization Act, owners and operators of underground storage tanks are required to take corrective action to prevent leakage. *See also* Comprehensive Environmental Response, Compensation and Liability Act; Groundwater monitoring; Groundwater pollution; Storage and transport of hazardous materials; Toxic Substances Control Act

[*Linda Rehkopf*]

RESOURCES
BOOKS

Epstein, L., and K. Stein. *Leaking Underground Storage Tanks—Citizen Action: An Ounce of Prevention.* New York: Environmental Information Exchange (Environmental Defense Fund), 1990.

PERIODICALS

Breen, B. "A Mountain and a Mission." *Garbage* 4 (May-June 1992): 52–57.
Hoffman, R. D. R. "Stopping the Peril of Leaking Tanks." *Popular Science* 238 (March 1991): 77–80.

OTHER

U.S. Environmental Protection Agency. *Office of Underground Storage Tanks (OUST).* June 13, 2002 [June 21, 2002]. <http://www.epa.gov/swerust1>.

Aldo Leopold (1886 – 1978)

American conservationist, ecologist, and writer

Leopold was a noted forester, game manager, conservationist, college professor, and ecologist. Yet he is known worldwide for *A Sand County Almanac*, a little book considered an important, influential work to **conservation** movement of the twentieth century. In it, Leopold established the **land ethic**, guidelines for respecting the land and preserving its integrity. Leopold grew up in Iowa, in a house overlooking the Mississippi River, where he learned **hunting** from his father and an appreciation of **nature** from his mother. He received a master's degree in forestry from Yale and spent his formative professional years working for the United States **Forest Service** in the American Southwest.

In the Southwest, Leopold began slowly to consider preservation as a supplement to Gifford Pinchot's "conservation as wise use—greatest good for the greatest number" land management philosophy that he learned at Yale and in the Forest Service. He began to formulate arguments for the preservation of **wilderness** and the **sustainable development** of wild game. Formerly a hunter who encouraged the elimination of predators to save the "good" animals for hunters, Leopold became a conservationist who remembered with sadness the "dying fire" in the eyes of a wolf he had killed. In the *Journal of Forestry*, he began to speculate that perhaps Pinchot's principle of highest use itself demanded "that representative portions of some forests be preserved as wilderness."

Leopold must be recognized as one of a handful of originators of the wilderness idea in American conservation history. He was instrumental in the founding of the **Wilderness Society** in 1935, which he described in the first issue of *Living Wilderness* as "one of the focal points of a new attitude—an intelligent humility toward man's place in nature." In a 1941 issue of that same journal, he asserted that wilderness also has critical practical uses "as a base-datum of normality, a picture of how healthy land maintains itself," and that wilderness was needed as a living "land laboratory."

This thinking led to the first large area designated as wilderness in the United States. In 1924, some 574,000 acres (232,000 ha) of the Gila **National Forest** in New Mexico was officially named a wilderness area. Four years before, the much smaller Trappers Lake valley in Colorado was the first area designated "to be kept roadless and undeveloped."

Aldo Leopold is also widely acknowledged as the founder of **wildlife management** in the United States. His classic text on the subject, *Game Management* (1933), is still in print and widely read. Leopold tried to write a general management framework, drawing upon and synthesizing **species** monographs and local manuals. "Details apply to game alone, but the principles are of general import to all fields of conservation," he wrote. He wanted to coordinate "science and use" in his book and felt strongly that land managers could either try to apply such principles, or be reduced to "hunting rabbits." Here can be found early uses of concepts still central to conservation and management, such as limiting factor, **niche**, saturation point, and **carrying capacity**. Leopold later became the first professor of game management in the United States at the University of Wisconsin.

Leopold's *A Sand County Almanac*, published in 1949, a year after his death, is often described as "the bible of the environmental movement" of the second half of the twentieth century. The *Almanac* is a beautifully written source of solid ecological concepts such as trophic linkages and **biological community**. The book extends basic ecological concepts, forming radical ideas to reformulate human thinking and behavior. It exhibits an ecological conscience, a conservation aesthetic, and a land ethic. He advocated his concept of ecological conscience to fill in a perceived gap in conservation education: "Obligations have no meaning without conscience, and the problem we face is the extension of the social conscience from people to land." Lesser known is his attention to the aesthetics of land: according to the *Almanac*, an acceptable land aesthetic emerges only from learned and sensitive perception of the connections and needs of natural communities. The last words in the *Almanac* are that a true conservation aesthetic is developed "not of building roads into lovely country, but of building receptivity into the still unlovely human mind."

Leopold derived his now famous land ethic from an ecological conception of community. All ethics, he maintained, "rest upon a single premise: that the individual is a member of a community of interdependent parts." He argued that "the land ethic simply enlarges the boundaries of the community to include soils, waters, plants, and animals, or collectively: the land." Perhaps the most widely quoted statement from the book argues that "a thing is right when it tends to preserve the integrity, **stability**, and beauty of the **biotic community**. It is wrong when it tends otherwise."

Leopold's land ethic was first proposed in the *Journal of Forestry* article in 1933 and later expanded in the *Almanac*. It is a plea to care for land and its biological complex, instead of considering it a commodity. As Wallace Tegner noted, Leopold's ideas were heretical in 1949, and to some people still are. "They smack of socialism and the public good," he wrote. "They impose limits and restraints. They are anti-Progress. They dampen American initiative. They fly in the face of the faith that land is a commodity, the very foundation stone of American opportunity." As a result, Stegner and others do not think Leopold's ethic had much influence on public thought, though the book has been widely read. Leopold recognized this. "The case for a land ethic would appear hopeless but for the minority which is in obvious revolt against these 'modern' trends," he commented. Nevertheless, the land ethic is alive and still flourishing, in an ever-growing minority. Even Stegner argued that "Leopold's land ethic is not a fact but [an on-going] task." Leopold did not shrink from that task, being actively involved in many

Aldo Leopold examining a gray partridge. (Photograph by Robert Oetking. University of Wisconsin-Madison Archives. Reproduced by permission.)

conservation associations, teaching management principles and the land ethic to his classes, bringing up all five of his children to become conservationists, and applying his beliefs directly to his own land, a parcel of "logged, fire-swept, overgrazed, barren" land in Sauk County, Wisconsin. As his work has become more recognized and more influential, many labels have been applied to Leopold by contemporary writers. He is a "prophet" and "intellectual touchstone" to Roderick Nash a "founding genius" to J. Baird Callicott "an American Isaiah" to Stegner the "Moses of the new conservation impulse" to Donald Fleming. In a sense, he may have been all of these, but more than anything else, Leopold was an applied ecologist who tried to put into practice the principles he learned from the land.

[*Gerald R. Young Ph.D.*]

RESOURCES
BOOKS

Callicott, J. B., ed. *Companion to A Sand County Almanac: Interpretive and Critical Essays.* Madison: University of Wisconsin Press, 1987.

Flader, S. L., and J. B. Callicott, eds. *The River of the Mother of God and Other Essays by Aldo Leopold.* Madison: University of Wisconsin Press, 1991.

Fritzell, P. A. "A Sand County Almanac and The Conflicts of Ecological Conscience." In *Nature Writing and America: Essays Upon a Cultural Type.* Ames: Iowa State University Press, 1990.

Leopold, A. *Game Management.* New York: Charles Scribner's Sons, 1933.

———. *A Sand County Almanac.* New York: Oxford University Press, 1949.

Meine, C. *Aldo Leopold: His Life and Work.* Madison: University of Wisconsin Press, 1988.

Oelschlaeger, M. "Aldo Leopold and the Age of Ecology." In *The Idea of Wilderness: From Prehistory to the Age of Ecology.* New Haven, CT: Yale University Press, 1991.

Strong, D. H. "Aldo Leopold." In *Dreamers and Defenders: American Conservationists.* Lincoln: University of Nebraska Press, 1988.

Less developed countries

Less developed countries (LDCs) have lower levels of economic prosperity, health care, and education than most other countries. Development or improvement in economic and social conditions encompasses various aspects of general welfare, including infant survival, expected life span, nutrition, literacy rates, employment, and access to material goods. Less developed countries (LDCs) are identified by their relatively poor ratings in these categories. In addition, most LDCs are marked by high **population growth**, rapidly expanding cities, low levels of technological development, and weak economies dominated by agriculture and the export of **natural resources**. Because of their limited economic and technological development, LDCs tend to have relatively little international political power compared to more developed countries (MDC) such as Japan, the United States, and Germany.

A variety of standard measures, or development indices, are used to assess development stages. These indices are generalized statistical measures of quality of life for individuals in a society. Multiple indices are usually considered more accurate than a single number such as Gross National Product, because such figures tend to give imprecise and simplistic impressions of conditions in a country. One of the most important of the multiple indices is the infant **mortality** rate. Because children under five years old are highly susceptible to common diseases, especially when they are malnourished, infant mortality is a key to assessing both nutrition and access to health care. Expected life span, the average age adults are able to reach, is used as a measure of adult health. Daily calorie and protein intake per person are collective measures that reflect the ability of individuals to grow and function effectively. Literacy rates, especially among women, who are normally the last to receive an education, indicate access to schools and preparation for technologically advanced employment.

Fertility rates are a measure of the number of children produced per family or per woman in a population and are regarded as an important measure of the confidence parents have in their childrens' survival. High birth rates are associated with unstable social conditions because a country with a rapidly growing population often cannot provide its citizens with food, water, **sanitation**, housing space, jobs, and other basic needs. Rapidly growing populations also tend to undergo rapid urbanization. People move to cities in search of jobs and educational opportunities, but in poor countries the cost of providing basic infrastructure in an expanding city can be debilitating. As most countries develop, they pass from a stage of high birth rates to one of low birth rates, as child survival becomes more certain and a family's investment in educating and providing for each child increases.

Most LDCs were colonies under foreign control during the past 200 years. Colonial powers tended to undermine social organization, local economies, and natural resource bases, and many recently independent states are still recovering from this legacy. Thus, much of Africa, which provided a wealth of natural resources to Europe between the seventeenth and twentieth centuries, now lacks the effective and equitable social organization necessary for continuing development. Similarly, much of Central America (colonized by Spain in the fifteenth century) and portions of South and Southeast Asia (colonized by England, France, the Netherlands, and others) remain less developed despite their wealth of natural resources. The development processes necessary to improve standards of living in LDCs may involve more natural resource extraction, but usually the most important steps involve carefully choosing the goods to be produced, decreasing corruption among government and business leaders, and easing the social unrest and conflicts that prevent development from proceeding. All of these are extraordinarily difficult to do, but they are essential for countries trying to escape from poverty. *See also* Child survival revolution; Debt for nature swap; Economic growth and the environment; Indigenous peoples; Shanty towns; South; Sustainable development; Third World; Third World pollution; Tropical rain forest; World Bank

[*Muthena Naseri*]

RESOURCES

BOOKS

Gill, S., and D. Law. *The Global Political Economy.* Baltimore: Johns Hopkins University Press, 1991.

World Bank. *World Development Report: Development and the Environment.* Oxford, England: Oxford University Press, 1992.

World Bank. *World Development Report 2000/2001: Attacking Poverty.* Oxford, England: Oxford University Press, 2000.

Leukemia

Leukemia is a disease of the blood-forming organs. Primary tumors are found in the bone marrow and lymphoid tissues, specifically the liver, spleen, and lymph nodes. The characteristic common to all types of leukemia is the uncontrolled proliferation of leukocytes (white blood cells) in the blood stream. This results in a lack of normal bone marrow growth, and bone marrow is replaced by immature and undifferentiated leukocytes or "blast cells." These immature and undifferentiated cells then migrate to various organs in the body, resulting in the pathogenesis of normal organ development and processing.

Leukemia occurs with varying frequencies at different ages, but it is most frequent among the elderly who experience 27,000 cases a year in the United States to 2,200 cases a year for younger people. Acute lymphoblastic leukemia, most common in children, is responsible for two-thirds of all cases. Acute nonlymphoblastic leukemia and chronic lymphocytic leukemia are most common among adults—they are responsible for 8,000 and 9,600 cases a year respectively. The geographical sites of highest concentration are the United States, Canada, Sweden, and New Zealand. While there is clear evidence that some leukemias are linked to genetic traits, the origins of this disease in most cases is mysterious. It seems clear, however, that environmental exposure to radiation, toxic substances, and other risk factors plays an important role in many leukemias. *See also* Cancer; Carcinogen; Radiation exposure; Radiation sickness

Lichens

Lichens are composed of **fungi** and algae. Varying in color from pale whitish green to brilliant red and orange, lichens usually grow attached to rocks and tree trunks and appear as thin, crusty coatings, as networks of small, branched strands, or as flattened, leaf-like forms. Some common lichens are reindeer moss and the red "British soldiers." There are approximately 20,000 known lichen **species**. Because they often grow under cold, dry, inhospitable conditions, they are usually the first plants to colonize barren rock surfaces.

The fungus and the alga form a symbiotic relationship within the lichen. The fungus forms the body of the lichen, called the thallus. The thallus attaches itself to the surface of a rock or tree trunk, and the fungal cells take up water and nutrients from the **environment**. The algal cells grow inside the fungal cells and perform **photosynthesis**, as do other plant cells, to form carbohydrates.

Lichens are essential in providing food for other organisms, breaking down rocks, and initiating **soil** building. They are also important indicators and monitors of **air pollution** effects. Since lichens grow attached to rock and tree surfaces, they are fully exposed to airborne pollutants, and chemical analysis of lichen tissues can be used to measure the quantity

of pollutants in a particular area. For example, **sulfur dioxide**, a common **emission** from **power plants**, is a major air pollutant. Many studies show that as the concentrations of sulfur dioxide in the air increase, the number of lichen species decreases. The disappearance of lichens from an area may be indicative of other, widespread biological impacts.

Sometimes, lichens are the first organisms to transfer contaminants to the food chain. Lichens are abundant through vast regions of the arctic **tundra** and form the main food source for caribou (*Rangifer tarandus*) in winter. The caribou are hunted and eaten by northern Alaskan Eskimos in spring and early summer. When the effects of **radioactive fallout** from weapons-testing in the arctic tundra were studied, it was discovered that lichens absorbed virtually all of the **radionuclides** that were deposited on them. Strontium-90 and cesium-137 were two of the major radionuclide contaminants. As caribou grazed on the lichens, these radionuclides were absorbed into the caribous' tissues. At the end of the winter, caribou flesh contained three to six times as much cesium-137 as it did in the fall. When the caribou flesh was consumed by the Eskimos, the radionuclides were transferred to them as well. *See also* Indicator organism; Symbiosis

[*Usha Vedagiri*]

RESOURCES
BOOKS

Connell, D. W., and G. J. Miller. *Chemistry and Ecotoxicology of Pollution.* New York: Wiley, 1984.

Smith, R. L., and T. M. Smith *Ecology and Field Biology.* 6th ed. Upper Saddle River, NJ: Prentice Hall, 2002.

Weier, T. E., et al. *Botany: An Introduction to Plant Biology.* New York: Wiley, 1982.

Life cycle assessment

Life cycle assessment (or LCA) refers to a process in industrial **ecology** by which the products, processes, and facilities used to manufacture specific products are each examined for their environmental impacts. A balance sheet is prepared for each product that considers: the use of materials; the consumption of energy; the **recycling**, re-use, and/or disposal of non-used materials and energy (in a less-enlightened context, these are referred to as "wastes"); and the recycling or re-use of products after their commercial life has passed. By taking a comprehensive, integrated look at all of these aspects of the manufacturing and use of products, life cycle assessment finds ways to increase efficiency, to re-use, reduce, and recycle materials, and to lessen the overall environmental impacts of the process.

Lignite

see **Coal**

Limits to Growth (1972) and *Beyond the Limits* (1992)

Published at the height of the oil crisis in the 1970s, the *Limits to Growth* study is credited with lifting environmental concerns to an international and global level. Its fundamental conclusion is that if rapid growth continues unabated in the five key areas of population, food production, industrialization, **pollution**, and consumption of nonrenewable **natural resources**, the planet will reach the limits of growth within 100 years. The most probable result will be a "rather sudden and uncontrollable decline in both population and industrial capacity."

The study grew out of an April 1968 meeting of 30 scientists, educators, economists, humanists, industrialists, and national and international civil servants who had been brought together by Dr. Aurelio Peccei, an Italian industrial manager and economist. Peccei and the others met at the Accademia dei Lincei in Rome to discuss the "present and future predicament of man," and from their meeting came the **Club of Rome**. Early meetings of the club resulted in a decision to initiate the Project on the Predicament of Mankind, intended to examine the array of problems facing all nations. Those problems ranged from poverty amidst plenty and **environmental degradation** to the rejection of traditional values and various economic disturbances.

In the summer of 1970, Phase One of the project took shape during a series of meetings in Bern, Switzerland and Cambridge, Massachusetts. At a two-week meeting in Cambridge, Professor Jay Forrester of the Massachusetts Institute of Technology (MIT) presented a global model for analyzing the interacting components of world problems. Professor Dennis Meadows led an international team in examining the five basic components, mentioned above, that determine growth on this planet and its ultimate limits. The team's research culminated in the 1972 publication of the study, which touched off intense controversy and further research.

Underlying the study's dramatic conclusions is the central concept of **exponential growth**, which occurs when a quantity increases by a constant percentage of the whole in a constant time period. "For instance, a colony of yeast cells in which each cell divides into two cells every ten minutes is growing exponentially," the study explains. The model used to capture the dynamic quality of exponential growth is a System Dynamics model, developed over a 30-year period at MIT, which recognizes that the structure of any

system determines its behavior as much as any individual parts of the system. The components of a system are described as "circular, interlocking, sometimes time-delayed." Using this model (called World3), the study ran scenarios—what-if analyses—to reach its view of how the world will evolve if present trends persist.

"Dynamic **modeling** theory indicates that any exponentially growing quantity is somehow involved with a positive feedback loop," the study points out. "In a positive feedback loop a chain of cause-and-effect relationships closes on itself, so that increasing any one element in the loop will start a sequence of changes that will result in the originally changed element being increased even more."

In the case of world **population growth**, the births per year act as a positive feedback loop. For instance, in 1650, world population was half a billion and was growing at a rate of 0.3 percent a year. In 1970, world population was 3.6 billion and was growing at a rate of 2.1 percent a year. Both the population and the rate of population growth have been increasing exponentially. But in addition to births per year, the dynamic system of population growth includes a negative feedback loop: deaths per year. Positive feedback loops create runaway growth, while negative feedback loops regulate growth and hold a system in a stable state. For instance, a thermostat will regulate temperature when a room reaches a certain temperature, the thermostat shuts off the system until the temperature decreases enough to restart the system. With population growth, both the birth and death rates were relatively high and irregular before the Industrial Revolution. But with the spread of medicines and longer life expectancies, the death rate has slowed while the birth rate has risen. Given these trends, the study predicted a worldwide jump in population of seven billion over 30 years.

This same dynamic of positive and negative feedback loops applies to the other components of the world system. The growth in world industrial capital, with the positive input of investment, creates rising industrial output, such as houses, automobiles, textiles, consumer goods, and other products. On the negative feedback side, depreciation, or the capital discarded each year, draws down the level of industrial capital. This feedback is "exactly analogous to the death rate loop in the population system," the study notes. And, as with world population, the positive feedback loop is "strongly dominant," creating steady growth in worldwide industrial capital and the use of raw materials needed to create products.

This system in which exponential growth is occurring, with positive feedback loops outstripping negative ones, will push the world to the limits of exponential growth. The study asks what will be needed to sustain world economic and population growth until and beyond the year 2000 and concludes that two main categories of ingredients can be defined. First, there are physical necessities that support all physiological and industrial activity: food, raw materials, fossil and nuclear fuels, and the ecological systems of the planet that absorb waste and recycle important chemical substances. **Arable land**, fresh water, metals, forests, and oceans are needed to obtain those necessities. Second, there are social necessities needed to sustain growth, including peace, social **stability**, education, employment, and steady technological progress.

Even assuming that the best possible social conditions exist for the promotion of growth, the earth is finite and therefore continued exponential growth will reach the limits of each physical necessity. For instance, about 1 acre (0.4 ha) of arable land is needed to grow enough food per person. With that need for arable land, even if all the world's arable land were cultivated, current population growth rates will still create a "desperate land shortage before the year 2000," the study concludes. The availability of fresh water is another crucial limiting factor, the study points out. "There is an upper limit to the fresh water **runoff** from the land areas of the earth each year, and there is also an exponentially increasing demand for that water."

This same analysis is applied to **nonrenewable resources**, such as metals, **coal**, iron, and other necessities for industrial growth. World demand is rising steadily and at some point demand for each nonrenewable resource will exceed supply, even with **recycling** of these materials. For instance, the study predicts that even if 100 percent recycling of chromium from 1970 onward were possible, demand would exceed supply in 235 years. Similarly, while it is not known how much pollution the world can take before vital natural processes are disrupted, the study cautions that the danger of reaching those limits is especially great because there is usually a long delay between the time a pollutant is released and the time it begins to negatively affect the **environment**.

While the study foretells worldwide collapse if exponential growth trends continue, it also argues that the necessary steps to avert disaster are known and are well within human capabilities. Current knowledge and resources could guide the world to a sustainable equilibrium society provided that a realistic, long-term goal and the will to achieve that goal are pursued.

The sequel to the 1972 study, *Beyond the Limits*, was not sponsored by the Club of Rome, but it is written by three of the original authors. While the basic analytical framework remains the same in the later work—drawing upon the concepts of exponential growth and feedback loops to describe the world system—its conclusions are more severe. No longer does the world only face a potential of "overshooting" its limits. "Human use of many essential resources and generation of many kinds of pollutants have

already surpassed rates that are physically sustainable," according to the 1992 study. "Without significant reductions in material and energy flows, there will be in the coming decades an uncontrolled decline in per capita food output, energy use, and industrial output."

However, like its predecessor, the later study sounds a note of hope, arguing that decline is not inevitable. To avoid disaster requires comprehensive reforms in policies and practices that perpetuate growth in material consumption and population. It also requires a rapid, drastic jump in the efficiency with which we use materials and energy.

Both the earlier and the later study were received with great controversy. For instance, economists and industrialists charged that the earlier study ignored the fact that technological innovation could stretch the limits to growth through greater efficiency and diminishing pollution levels. When the sequel was published, some critics charged that the World3 model could have been refined to include more realistic distinctions between nations and regions, rather than looking at all trends on a world scale. For instance, different continents, rich and poor nations, North, South, and East, various regions—all are different, but those differences are ignored in the model, thereby making it unrealistic even though modeling techniques have evolved significantly since World3 was first developed. *See also* Sustainable development

[*David Clarke*]

RESOURCES
BOOKS

Meadows, D., et al. *The Limits to Growth: A Report for The Club of Rome's Project on the Predicament of Mankind.* New York: Universe Books, 1972.

Meadows, D., D. L. Meadows, and J. Randers. *Beyond the Limits: Confronting Global Collapse, Envisioning a Sustainable Future.* Post Mills, VT: Chelsea Green, 1992.

Limnology

Derived from the Greek word *limne*, meaning marsh or pond, the term limnology was first used in reference to lakes by F. A. Forel (1841–1912) in 1892 in a paper titled "Le Léman: Monographie Limnology," a study of what we now call Lake Geneva in Switzerland. Limnology, also known as aquatic **ecology**, refers to the study of fresh water communities within continental boundaries. It can be subdivided into the study of lentic (standing water habitats such as lakes, ponds, bogs, swamps, and marshes) and lotic (running water habitats such as rivers, streams, and brooks) environments. Collectively, limnologists study the morphological, physical, chemical, and biological aspects of these habitats.

Raymond L. Lindeman (1915 – 1942)
American ecologist

Few scholars or scientists, even those much published and long-lived, leave singular, indelible imprints on their disciplines,. Yet, Raymond Lindeman, in 26 short years, who published just six articles, was described shortly after his death by G. E. Hutchinson as "one of the most creative and generous minds yet to devote itself to ecological science," and the last of those six papers, "The Trophic-Dynamic Aspect of Ecology,"—published posthumously in 1942—continues to be considered one of the foundational papers in **ecology**, an article "path-breaking in its general analysis of ecological **succession** in terms of **energy flow** through the ecosystem," an article based on an idea that Edward Kormondy has called "the most significant formulation in the development of modern ecology."

Immediately after completing his doctorate at the University of Minnesota, Lindeman accepted a one-year Sterling fellowship at Yale University to work with G. Evelyn Hutchinson, the Dean of American limnologists. He had published chapters of his thesis one by one and at Yale worked to revise the final chapter, refining it with ideas drawn from Hutchinson's lecture notes and from their discussions about the ecology of lakes. Lindeman submitted the manuscript to Ecology with Hutchinson's blessings, but it was rejected based on reviewers' claims that it was speculation far beyond the data presented from research on three lakes, including Lindeman's own doctoral research on Cedar Bog Lake. After input from several well-known ecologists, further revisions, and with further urging from Hutchinson, the editor finally overrode the reviewers' comments and accepted the manuscript; it was published in the October, 1942 issue of *Ecology*, a few months after Lindeman died in June of that year.

The important advances made by Lindeman's seminal article included his use of the **ecosystem** concept, which he was convinced was "of fundamental importance in interpreting the data of dynamic ecology," and his explication of the idea that "all function, and indeed all life" within ecosystems depends on the movement of energy through such systems by way of trophic relationships. His use of ecosystem went beyond the little attention paid to it by Hutchinson and beyond Tansley's labeling of the unit seven years earlier to open up "new directions for the analysis of the functioning of ecosystems." More than half a century after Lindeman's article, and despite recent revelations on the uncertainty and unpredictability of natural systems, a majority of ecologists probably still accept ecosystem as the basic unit in ecology and, in those systems, energy exchange as the basic process.

Lindeman was able to effectively demonstrate a way to bring together or synthesize two quite separate traditions in ecology, **autecology**, dependent on physiological studies of individual organisms and **species**, and synecology focused on studies of communities, aggregates of individuals. He believed, and demonstrated, that ecological research would benefit from a synthesis of these organism-based approaches and focus on the energy relationships that tied organism and **environment** into one unit—the ecosystem—suggesting as a result that biotic and abiotic could not realistically be disengaged, especially in ecology.

Half a decade or so of work on cedar bog lakes, and half a dozen articles would seem a thin stem on which to base a legacy. But it really boils down to Lindeman's synthesis of that work in that one singular, seminal paper, in which he created one of the significant stepping stones from a mostly descriptive discipline toward a more sophisticated and modern theoretical ecology.

[*Gerald J. Young Ph.D.*]

RESOURCES
PERIODICALS

Cook, Robert E. "Raymond Lindeman and the Trophic-Dynamic Concept in Ecology." *Science* 198, no. 4312 (October 1977): 22–26.

Lindsey, Alton A. "The Ecological Way." *Naturalist-Journal of the Natural History Society of Minnesota* 31 (Spring 1980): 1–6.

Reif, Charles B. "Memories of Raymond Laurel Lindeman." *The Bulletin of the Ecological Society of America* 67, no. 1 (March 1986): 20–25.

Liquid metal fast breeder reactor

The liquid metal fast breeder reactor (LMFBR) is a nuclear reactor that has been modified to increase the efficiency at which non-fissionable uranium-238 is converted to fissionable plutonium-239, which can be used as fuel in the production of **nuclear power**. The reactor uses "fast" rather than "slow" neutrons to strike a uranium-238 nucleus, resulting in the formation of plutonium-239. In a second modification, it uses a liquid metal, usually sodium, rather than neutron-absorbing water as a more efficient coolant. Since the reactor produces new fuel as it operates, it is called a breeder reactor.

The main appeal of breeder reactors is that they provide an alternative way of obtaining fissionable materials. The supply of natural **uranium** in the earth's crust is fairly large, but it will not last forever. Plutonium-239 from breeder reactors might become the major fuel used in reactors built a few hundred or thousand years from now.

However, the potential of LMFBRs has not as yet been realized. One serious problem involves the use of liquid sodium as coolant. Sodium is a highly corrosive metal and in an LMFBR it is converted into a radioactive form, sodium-24. Accidental release of the coolant from such a plant could, therefore, constitute a serious environmental hazard.

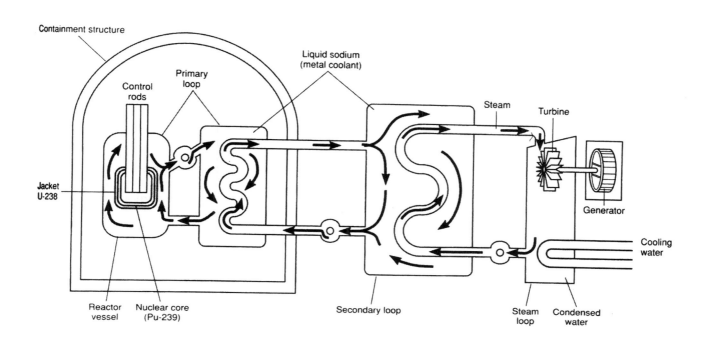

Liquid metal fast breeder reactor. (McGraw-Hill Inc. Reproduced by permission.)

In addition, **plutonium** itself is difficult to work with. It is one of the most toxic substances known to humans, and its **half-life** of 24,000 years means that its release presents long-term environmental problems.

Small-scale pilot LMFBR reactors have been tested in the United States, Saudi Arabia, Great Britain, and Germany since 1966, and all have turned out to be far more expensive than had been anticipated. The major United States research program based at Clinch, Tennessee, began in 1970. By 1983, the U. S. Congress refused to continue funding the project due to its slow and unsatisfactory progress. *See also* Nuclear fission; Nuclear Regulatory Commission; Radioactivity; Radioactive waste management

[*David E. Newton*]

RESOURCES
BOOKS

Cochran, Thomas B. *The Liquid Metal Fast Breeder Reactor: An Environmental and Economic Critique.* Baltimore: Johns Hopkins University Press, 1974.

Mitchell III, W., and S. E. Turner. *Breeder Reactors.* Washington, DC: U.S. Atomic Energy Commission, 1971.

OTHER

International Nuclear Information System. *Links to Fast Reactor Related Sites.* June 7, 2002 [June 21, 2002]. <http://www.iaea.or.at/inis/ws/fnss/fr.html>.

Liquified natural gas

Natural gas is a highly desirable fuel in many respects. It burns with the release of a large amount of energy, producing almost entirely **carbon dioxide** and water as waste products. Except for possible greenhouse effects of **carbon** dioxide, these compounds produce virtually no environmental hazard. Transporting natural gas through transcontinental pipelines is inexpensive and efficient where **topography** allows the laying of pipes. Oceanic shipping is difficult, however, because of the flammability of the gas and the high volumes involved. The most common way of dealing with these problems is to condense the gas first and then transport it in the form of liquified natural gas (LNG). But LNG must be maintained at temperatures of about -260°F (-160°C) and protected from leaks and flames during **loading** and unloading. *See also* Fossil fuels

Lithology

Lithology is the study of rocks, emphasizing their macroscopic physical characteristics, including grain size, mineral composition, and color. Lithology and its related field, petrography (the description and systematic classification of rocks), are subdisciplines of petrology, which also considers microscopic and chemical properties of minerals and rocks as well as their origin and decay.

Littoral zone

In marine systems, littoral zone is synonymous with intertidal zone and refers to the area on marine shores that is periodically exposed to air during low tide. The freshwater littoral zone is that area near the shore characterized by submerged, floating, or emergent vegetation. The width of a particular littoral zone may vary from several miles to a few feet. These areas typically support an abundance of organisms and are important feeding and nursery areas for fishes, crustaceans, and birds. The distribution and abundance of individual **species** in the littoral zone is dependent on predation and **competition** as well as tolerance of physical factors. *See also* Neritic zone; Pelagic zone

Loading

The term loading has a wide variety of specialized meanings in various fields of science. In general, all refer to the addition of something to a system, just as loading a truck means filling it with objects. In the science of acoustics, for example, loading refers to the process of adding materials to a speaker in order to improve its acoustical qualities. In **environmental science**, loading is used to describe the contribution made to any system by some component. One might analyze, for example, how an increase in chlorofluorocarbon (CFC) loading in the **stratosphere** might affect the concentration of **ozone** there.

Logging

Logging is the systematic process of cutting down trees for lumber and wood products. The method of logging called clearcutting, in which entire areas of forests are cleared, is the most prevalent practice used by lumber companies. Clearcutting is the cheapest and most efficient way to harvest a forest's available resources. This practice drastically alters the forest **ecosystem**, and many plants and animals are displaced or destroyed by it. After clearcutting is performed on forests, forestry management techniques may be intro-

duced in order to manage the growth of new trees on the cleared land. Selective logging is an alternative to clearcutting. In selective logging, only certain trees in a forest are chosen to be logged, usually on the basis of their size or **species**. By taking a smaller percentage of trees, the forest is protected from destruction and fragile plants and animals in the forest ecosystem are more likely to survive. New, innovative techniques offer alternatives for preserving the forest. For example, the Shelterwood Silvicultural System harvests mature trees in phases. First, part of the original stand is removed to promote growth of the remaining trees. After this occurs, regeneration naturally follows using seeds provided by the remaining trees. Once regeneration has occurred, the remaining mature trees are harvested.

Early logging equipment included long, two-man straight saws and teams of animals to drag trees away. After World War II, technological advances made logging easier. The bulldozer and the helicopter allowed loggers to enter into new and previously untouched areas. The chainsaw allowed loggers to cut down many more trees each day. Today, enormous machines known as feller-bunchers take the place of human loggers. These machines use a hydraulic clamp that grasps the individual tree and huge shears that cut through it in one swift motion.

High demands for lumber and forest products have caused prolific and widespread commercial logging. Certain methods of timber harvesting allow for subsequent regeneration, while others cause **deforestation**, or the irreversible creation of a non-forest condition. Deforestation significantly changed the landscape of the United States. Some observers remarked as early as the mid-1700s upon the rapid changes made to the forests from the East Coast to the Ohio River Valley. Often, the lumber in forests was burned away so that the early settlers could build farms upon the rich **soil** that had been created by the forest ecosystem. The immediate results of deforestation are major changes to Earth's landscapes and diminishing **wildlife** habitats. Longer-range results of deforestation, including unrestrained commercial logging, may include damage to Earth's **atmosphere** and the unbalancing of living ecosystems. Forests help to remove **carbon dioxide** from the air. Through the process of **photosynthesis**, forests release oxygen into the air. A single acre of temperate forest releases more than six tons of oxygen into the atmosphere every year. In the last 150 years, deforestation, together with the burning of **fossil fuels**, has raised the amount of **carbon** dioxide in the atmosphere by more than 25%. It has been theorized that this has contributed to global warming, which is the accumulation of gasses leading to a gradual increase in Earth's surface temperature. Human beings are still learning how to measure their need for wood against their need for a viable **environment** for themselves and other life forms.

Although human activity, especially logging, has decimated many of the world's forests and the life within them, some untouched forests still remain. These forests are known as old-growth or ancient-growth forests. Old-growth forests are at the center of a heated debate between environmentalists, who wish to preserve them, and the logging industry, which continually seeks new and profitable sources of lumber and other forest products.

Very little of the original uncut North American forest still remains. It has been estimated that the United States has lost over 96% of its old-growth forests. This loss continues as logging companies become more attracted to ancient-growth forests, which contain larger, more profitable trees. A majority of old-growth forests in the United States are in Alaska and Pacific Northwest. On the global level, barely 20% of the old-growth forests still remain, and the South American rainforests account for a significant portion of these. About 1% of the Amazon rainforest is deforested each year. At the present rate of logging around the world, old-growth forests could be gone within the first few decades of the twenty-first century unless effective **conservation** programs are instituted.

As technological advancements of the twentieth century dramatically increased the efficiency of logging, there was also a growth in understanding about the contribution of the forest to the overall health of the environment, including the effect of logging upon that health. Ecologists, who are scientists that study the complex relationships within natural systems, have determined that logging can affect the health of air, soil, water, plant life, and animals. For instance, clearcutting was at one time considered a healthy forestry practice, as proponents claimed that clearing a forest enabled the growth of new plant life, sped the process of regeneration, and prevented fires. The American Forest Institute, an industry group, ran an ad in the 1970s that stated, "I'm clear-cutting to save the forest." Ecologists have come to understand that clearcutting old-growth forests has a devastating effect on plant and animal life, and affects the health of the forest ecosystem from its rivers to its soil. Old-growth trees, for example, provide an ecologically diverse **habitat** including woody debris and **fungi** that contribute to nutrient-rich soil. Furthermore, many species of plants and wildlife, some still undiscovered, are dependent upon old-growth forests for survival. The huge canopies created by old-growth trees protect the ground from water **erosion** when it rains, and their roots help to hold the soil together. This in turn maintains the health of rivers and streams, upon which fish and other aquatic life depend. In the Pacific Northwest, for example, ecologists have connected the health of the **salmon** population with the health of the forests and the logging practices therein. Ecologists now understand that clear-cutting and the planting of new trees, no matter how scientif-

ically managed, cannot replace the wealth of **biodiversity** maintained by old-growth forests.

The pace of logging is dictated by the consumer demand for lumber and wood products. In the United States, for instance, the average size of new homes doubled between 1970 and 2000, and the forests ultimately bear the burden of the increasing consumption of lumber. In the face of widespread logging, environmentalists have become more desperate to protect ancient forests. There is a history of controversy between the timber industry and environmentalists regarding the relationship between logging and the care of forests. On the one hand, the logging industry has seen forests as a source of wealth, economic growth, and jobs. On the other hand, environmentalists have viewed these same forests as a source of **recreation**, spiritual renewal, and as living systems that maintain the overall **environmental health**. In the 1980s, a controversy raged between environmentalists and the logging industry over the protection of the **northern spotted owl**, a threatened species of bird whose habitat is the **old-growth forest** of the Pacific Northwest. Environmentalists appealed to the **Endangered Species Act** of 1973 to protect some of these old-growth forests. In other logging controversies, some environmentalists chained themselves to old-growth trees to prevent their destruction, and one activist, Julia Butterfly Hill, lived in an old-growth California redwood tree for two years in the 1990s to prevent it from being cut down. The clash between environmentalists and the logging industry may become more intense as the demand for wood increases and supplies decrease. However, in recent years these opposing views have been tempered by discussion of concepts such as responsible **forest management** to create sustainable growth, in combination with preservation of protected areas.

Most of the logging in the United States occurs in the national forests. From the point of view of the U.S. **Forest Service**, logging provides jobs, helps manage the forest in some respects, prevents logging in other parts of the world, and helps eliminate the danger of forest fires. To meet the demands of the logging industry, the national forests have been developed with a labyrinth of logging roads and contain vast areas that have been devastated by clearcutting. In the United States there are enough logging roads in the National Forests to circle the earth 15 times, roads that speed up soil erosion that then washes away fertile **topsoil** and pollutes streams and rivers.

The Roadless Initiative was established in 2001 to protect 60 million acres (24 million ha) of national forests. The initiative was designed by the Clinton administration to discourage logging and taxpayer-supported road building on public lands. The goal was to establish total and permanent protection for designated roadless areas. Advocates of the initiative contended that roadless areas encompassed some of the best wildlife habitats in the nation, while forest service officials argued that banning road building would significantly reduce logging in these areas. Under the initiative, more than half of the 192 million acres (78 million ha) of **national forest** would still remain available for logging and other activities. This initiative was considered one of the most important environmental protection measures of the Clinton administration.

Illegal logging has become a problem with the growing worldwide demand for lumber. For example, the **World Bank** predicted that if Indonesia does not halt all current logging, it would lose its entire forest within the next 10 to 15 years. Estimates indicate that up to 70% of the wood harvested in Indonesia comes from illegal logging practices. Much of the timber being taken is sent to the United States. **Indigenous peoples** of Indonesia are being displaced from their traditional territories. Wildlife, including endangered **tigers**, **elephants**, rhinos, and orangutans are also being displaced and may be threatened with **extinction**. In 2002 Indonesia placed a temporary moratorium on logging in an effort to stop illegal logging.

Other countries around the world were addressing logging issues in the early twenty-first century. In China, 160 million acres (65 million ha) out of 618 million acres (250 million ha) were put under state protection. Loggers turned in their tools to become forest rangers, working for the government in order to safeguard trees from illegal logging. China has set aside millions of acres of forests for protection, particularly those forests that are crucial sources of fresh water. China also announced that it was planning to further reduce its timber output in order to restore and enhance the life-sustaining abilities of its forests.

[*Douglas Dupler*]

RESOURCES
BOOKS

Dietrich, William. *The Final Forest: The Battle for the Last Great Trees of the Pacific Northwest.* New York: Simon & Schuster, 1992.

Durbin, Kathie. *Tree Huggers: Victory, Defeat and Renewal in the Northwest Ancient Forest Campaign.* Seattle, WA: Mountaineers, 1996.

Hill, Julia Butterfly. *The Legacy of Luna: The Story of a Tree, a Woman, and the Struggle to Save the Redwoods.* San Francisco: Harper, 2000.

Luoma, Jon R. *The Hidden Forest.* New York: Henry Holt and Co., 1999.

Nelson, Sharlene P., and Ted Nelson. *Bull Whackers to Whistle Punks: Logging in the Old West.* New York: Watts, 1996.

PERIODICALS

Alcock, James. "Amazon Forest Could Disappear, Soon." *Science News,* July 14, 2001.

De Jong, Mike. "Optimism Over Lumber." *Maclean's,* November 29, 2001, 16.

Kerasote, Ted. "The Future of our Forests." *Audubon,* January/February 2001, 44.

Murphy, Dan. "The Rise of Robber Barons Speeds Forest Decline." *Christian Science Monitor,* August 14, 2001, 8.

OTHER

American Lands Home Page. [cited July 2002]. <http://www.americanlands.org>.

Global Forest Watch Home Page. [cited July 2002]. <http://www.globalforest-watch.org>.

SmartWood Program of the Rainforest Alliance. [cited July 2002]. <http://www.smartwood.org>.

Logistic growth

Assuming the rate of immigration is the same as emigration, population size increases when births exceed deaths. As population size increases, population density increases, and the supply of limited available resources per organism decreases. There is thus less food and less space available for each individual. As food, water, and space decline, fewer births or more deaths may occur, and this imbalance continues until the number of births are equal to the number of deaths at a population size that can be sustained by the available resources. This equilibrium level is called the **carrying capacity** for that **environment**.

A temporary and rapid increase in population may be due to a period of optimum growth conditions including physical and biological factors. Such an increase may push a population beyond the environmental carrying capacity. This sudden burst will be followed by a decline, and the population will maintain a steady fluctuation around the carrying capacity. Other population controls, such as predators and weather extremes (**drought**, frost, and floods), keep populations below the carrying capacity. Some environmentalists believe that the human population has exceeded the earth's carrying capacity.

Logistic growth, then, refers to growth rates that are regulated by internal and external factors that establish an equilibrium with **environmental resources**. The sigmoid (idealized S-shaped) curve illustrates this logistic growth where environmental factors limit **population growth**. In this model, a low-density population begins to grow slowly, then goes through an exponential or geometric phase, and then levels off at the environmental carrying capacity. *See also* Exponential growth; Growth limiting factors; Sustainable development; Zero population growth

[*Muthena Naseri*]

Dr. Bjørn Lomborg (1965 –)
Danish political scientist

In 2001, Cambridge University Press published *The Skeptical Environmentalist: Measuring the Real State of the World* by the Danish statistician Bjørn Lomborg. The book triggered a firestorm of criticism, with many well-known scientists denouncing it as an effort to "confuse legislators and regulators, and poison the well of public environmental information." In January 2002, *Scientific American* published a series of articles by five distinguished environmental scientists contesting Lomborg's claims. To some observers, the ferocity of the attack was surprising. Why so much furor over a book that claims to have good news about our environmental condition?

Lomborg portrays himself as an "left-wing, vegetarian, **Greenpeace** member," but says he worries about the unrelenting "doom and gloom" of mainstream **environmentalism**. He describes what he regards as an all-pervasive ideology that says, among other things, "Our resources are running out. The population is ever growing, leaving less and less to eat. The air and water are becoming ever more polluted. The planet's **species** are becoming extinct in vast numbers. The forests are disappearing, fish stocks are collapsing, and coral reefs are dying." This ideology has pervaded the environmental debate so long, Lomborg says, "that blatantly false claims can be made again and again, without any references, and yet still be believed."

In fact, Lomborg tells us, these allegations of the collapse of ecosystems are "simply not in keeping with reality. We are not running out of energy or **natural resources**. There will be more and more food per head of the world's population. Fewer and fewer people are starving. In 1900 we lived for an average of 30 years; today we live 67. According to the UN we have reduced poverty more in the last 50 years than in the preceding 500, and it has been reduced in practically every country." He goes on to challenge conventional scientific assessment of global warming, forest losses, fresh water **scarcity**, energy shortages, and a host of other environmental problems. Is Lomborg being deliberately (and some would say, hypocritically) optimistic, or are others being unreasonably pessimistic? Is this simply a case of regarding the glass as half full versus half empty?

The inspiration to look at environmental **statistics**, Lomborg says, was a 1997 interview with the controversial economist Dr. Julian L. Simon in *Wired* magazine. Simon, who died in 1998, spent a good share of his career arguing that the "litany" of the Green movement—human overpopulation leading to starvation and resource shortages—was premeditated hyperbole and fear mongering. The truth, Simon, claimed is that the quality of human life is improving, not declining.

Lomborg felt sure that Simon's allegations were "simple American right-wing propaganda." It should be a simple matter, he thought, to gather evidence to show how wrong Simon was. Back at his university in Denmark, Lomborg set out with 10 of his sharpest students to study Simon's

claims. To their surprise, the group found that while not everything Simon said was correct, his basic conclusions seemed sound. When Lomborg began to publish these findings in a series of newspaper articles in the *London Guardian* in 1998, he stirred up a hornet's nest. Some of his colleagues at the University of Aarhus set up a website to denounce the work. When the whole book came out, their fury only escalated. Altogether, between 1998 and 2002, more than 400 articles appeared in newspapers and popular magazines either attacking or defending Lomborg and his conclusions.

In general, the debate divides between mostly conservative supporters on one side and progressive, environmental activists and scientists on the other. The *Wall Street Journal* described the *Skeptical Environmentalist* as "superbly documented and readable." *The Economist* called it "a triumph." A review in the *Daily Telegraph* (London) declared it "the most important book on the **environment** ever written." A review in the *Washington Post* said it is a "richly informative, lucid book, a magnificent achievement." And, *The Economist*, which started the debate by publishing his first articles, announced that, "this is one of the most valuable books on public policy—not merely on environmental policy—to have been written in the past ten years."

Among most environmentalists and scientists, on the other hand, Lomborg has become an anathema. A widely circulated list of "Ten things you should know about the Skeptical Environmentalist" charged that the book is full of pseudo-scholarship, statistical fallacies, distorted quotations, inaccurate or misleading citations, misuse of data, interpretations that contradict well-established scientific work, and many other serious errors. This list accuses Lomborg of having no professional credentials or training—and having done no professional research—in **ecology**, **climate** science, resource economic, **environmental policy**, or other fields covered by his book. In essence, they complain, "Who is this guy, and how dare he say all this terrible stuff?"

Harvard University Professor **E. O. Wilson**, one of the world's most distinguished biologists, deplores what he calls "the Lomborg scam," and says that he and his kind "are the parasite load on scholars who earn success through the slow process of peer review and approval." It often seems that more scorn and hatred is focused on those, like Lomborg, who are viewed as a turncoats and heretics, than for those who are actually out despoiling the environment and squandering resources.

Perhaps the most withering criticism of Lomborg comes from his reporting of statistics and research results. Stephen Schneider, a distinguished climate scientist from Stanford University, for instance, writes in *Scientific American* "most of [Lomborg's] nearly 3,000 citations are to secondary literature and media articles. Moreover, even when cited, the peer-reviewed articles come elliptically from those studies that support his rosy view that only the low end of the uncertainty ranges [of climate change] will be plausible. IPCC authors, in contrast, were subjected to three rounds of review by hundreds of outside experts. They didn't have the luxury of reporting primarily from the part of the community that agrees with their individual views."

Lomborg also criticizes **extinction** rate estimates as much too large, citing evidence from places like Brazil's Atlantic Forest, where about 90% of the forest has been cleared without large numbers of recorded extinctions. Thomas Lovejoy, chief **biodiversity** adviser to the **World Bank**, responds, "First, this is a region with very few field biologists to record either species or their extinction. Second, there is abundant evidence that if the Atlantic forest remains as reduced and fragmented as it is, will lose a sizable fraction of the species that at the moment are able to hang on."

Part of the problem is that Lomborg is unabashedly anthropocentric. He dismisses the value of biodiversity, for example. As long as there are plants and animals to supply human needs, what does it matter if a few non-essential species go extinct? In Lomborg's opinion, poverty, hunger, and human health problems are much more important problems than **endangered species** or possible climate change. He isn't opposed to reducing greenhouse gas emissions, for instance, but argues that rather than spend billions of dollars per year to try to meet Kyoto standards, we could provide a healthy diet, clean water, and basic medical services to everyone in the world, thereby saving far more lives than we might do by reducing global climate change. Furthermore, Lomborg believes, **solar energy** will probably replace **fossil fuels** within 50 years anyway, making worries about increasing CO_2 concentrations moot.

Lomborg infuriates many environmentalists by being intentionally optimistic, cheerfully predicting that progress in population control, use of **renewable energy**, and unlimited water supplies from desalination technology will spread to the whole world, thus avoiding crises in resource supplies and human impacts on our environment. Others, particularly Lester Brown of the **Worldwatch Institute** and Professor Paul Ehrlich of Stanford University, according to Lomborg, seem to deliberately adopt worst-case scenarios.

Protagonists on both sides of this debate use statistics selectively and engage in deliberate exaggeration to make their points. As Stephen Schneider, one of the most prominent anti-Lomborgians, said in an interview in *Discover* in 1989, "[We] are not just scientists but human beings as well. And like most people we'd like to see the world a better place. To do that we need to get some broad-based support, to capture the public's imagination. That, of course, entails getting loads of media coverage. So we have to offer up scary scenarios, make simplified, dramatic statements, and make little mention of any doubts we might have. Each of us has

to decide what the right balance is between being effective and being honest."

As is often the case in complex social issues, there are both truth and error on both sides in this debate. It takes good critical thinking skills to make sense out of the flurry of charges and counter charges. In the end, what you believe depends on your perspective and your values. Future events will show us whether Bjørn Lomborg or his critics are correct in their interpretations and predictions. In the meantime, it's probably healthy to have the vigorous debate engendered by strongly held beliefs and articulate partisans from many different perspectives.

In November 2001, Lomborg was selected Global Leader for Tomorrow by the World Economic Forum, and in February 2002, he was named director of Denmark's national Environmental Assessment Institute. In addition to the use of statistics in environmental issues, his professional interests are simulation of strategies in collective action dilemmas, simulation of party behavior in proportional voting systems, and use of surveys in public administration.

[*William Cunningham Ph.D.*]

RESOURCES

BOOKS

Lomborg, Bjorn. *The Skeptical Environmentalist: Measuring the Real State of the World.* Cambridge University Press, 2001.

PERIODICALS

Bell, Richard C. "Media Sheep: How did The Skeptical Environmentalist Pull the Wool over the Eyes of so Many Editors?" *Worldwatch* 15, no. 2 (2002): 11–13.

Dutton, Denis. "Greener than you think." *Washington Post*, October 21, 2001.

Schneider, Stephen. "Global Warming: Neglecting the Complexities." *Scientific American* 286 (2002): 62–65.

Wade, Nicholas. "Bjørn Lomborg: A Chipper Environmentalist." *The New York Times*, August 7, 2001.

OTHER

Anti-Lomborgian Web Site. December 2001 [cited July 9, 2002]. <http://www.anti-lomborg.com>.

Bjørn Lomborg Home Page. 2002 [cited July 9, 2002]. <http://www.lomborg.com>.

Regis, Ed. "The Doomslayer: The environment is going to hell, and human life is doomed to only get worse, right? Wrong. Conventional wisdom, meet Julian Simon, the Doomslayer." February 1997. *Wired.* [cited July 9, 2002]. <http://www.wired.com>.

Wilson, E. O. "Vanishing Point: On Bjørn Lomborg and Extinction." *Grist* December 12, 2001 [cited July 9, 2002]. <http://www.gristmagazine.com/books/wilson121201.asp>.

World Resources Institute and World Wildlife Fund. *Ten Things Environmental Educators Should Know About The Skeptical Environmentalist.* January 2002 [cited July 9, 2002]. <http://www.wri.org/press/mk_lomborg_10_things.html>.

London Dumping Convention

see **Convention on the Prevention of Marine Pollution by Dumping of Waste and Other Matter (1972)**

Barry Holstun Lopez (1945 –)

American environmental writer

Barry Lopez has often called his own nonfiction writing natural history, and he is often categorized as a **nature** writer. This partially describes his work, but limiting him to that category is misleading, partly because his work transcends the kinds of subjects implicit in that classification. He could as well, for example, be called a travel writer, though that label also does not completely describe his work. He has in addition published a number of unusual works of fiction, and even has one children's book (on *Crow and Weasel*) to his credit.

Barry Lopez was born in Port Chester, New York, but he spent several early childhood years in rural southern California and, at the age of 10, returned to the East to grow up in New York City. He earned a BA degree from the University of Notre Dame, followed by an MAT from the University of Oregon in 1968. His initial goal was to teach, but in the late 1960s he set out to become a professional writer and since 1970 has earned his living writing (as well as by lecturing and giving readings).

Lopez's nonfiction writing transcends the category of natural history because his real topic, as Rueckert suggests, is the search for human relationships with nature, relationships that are "dignified and honorable." Natural history as a category of literature implies a focus on primeval nature undisturbed by human activity, or at least on nature as it exists in its own right rather than as humans relate to it. He is a practitioner of what some have called "the new naturalism—a search for the human as mirrored in nature." Lopez's focus then is **human ecology**, the interactions of human beings with the world around them, especially the natural world. Even his most "natural" book of natural history, *Of Wolves and Men*, is not just about the natural history of **wolves** but about how that species' existence or "being" in the wild is affected by human perceptions and actions, and it is as well about the image of the wolf in human minds.

His fiction works can be called unusual, partly because they are often presented as brief notes or sketches, and partly because they are frequently blended into and combined with legends, factual observations of nature, and personal meditations. Everything Lopez writes, however, is in the form of a story, whether fiction, natural history, folklore, or travel writing. His story-telling makes all of his writing enjoyable to read, easy to access and ingest, and often memorable.

Lopez as a story teller, occupies the spaces between truth-teller and mystic, between natural scientist and folklorist. He has written of wolves and humans, and then of people with blue skins who could not speak and, apparently, subsisted only on clean air. He writes of the land in reality and the land in our imaginations, frequently in the same text. His writings on natural history provide the reader with great detail about the places in the world he describes, but his fiction can force the shock of recognition of places in the mind. In 1998, Lopez was a National Magazine Award in Fiction finalist for *The Letters of Heaven* and in 1999 he received the Lannan residency fellowship.

Barry Lopez is in part a naturalist, in the best sense of that word. He is also something of an anthropologist. He is a student of folklore and mythology. He travels widely, but he studies his own home place and local **environment** intently. And, of course, he is a writer.

[*Gerald L. Young Ph.D.*]

RESOURCES
BOOKS

Lopez, Barry. *About this Life: Journeys on the Threshold of Memory.* Random, 1998.

———. *Arctic Dreams: Imagination and Desire in a Northern Landscape.* New York: Scribner, 1986.

———. *Crossing Open Ground.* New York: Scribner, 1988.

———. *Desert Notes; Reflections in the Eye of a Raven.* Kansas City: Sheed, Andrews & McMeel.

———. *Lessons from the Wolverine.* Illustrated by Tom Pohrt. University of Georgia Press, 1997.

———. *Light Action in the Caribbean.* Knopf, 2000.

Rueckert, W. H. "Barry Lopez and the Search for a Dignified and Honorable Relationship With Nature." In *Earthly Words: Essays on Contemporary American Nature and Environmental Writers.* Ed. J. Cooley. Ann Arbor: University of Michigan Press, 1994.

PERIODICALS

Paul, S. "Barry Lopez." *Hewing to Experience: Essays and Reviews on Recent American Poetry and Poetics, Nature and Culture.* Iowa City: University of Iowa Press, 1989.

Los Angeles Basin

The second most populous city in the United States, Los Angeles has perhaps the most fascinating **environmental history** of any urban area in the country. The Los Angeles Basin, into which more than 80 communities of Los Angeles County are crowded, is a trough-shaped region bounded on three sides by the Santa Monica, Santa Susana, San Gabriel, San Bernadino, and Santa Ana Mountains. On its fourth side, the county looks out over the Pacific Ocean.

The earliest settlers arrived in the Basin in 1769 when Spaniard Gaspar de Portolá and his expedition set up camp along what is now known as the Los Angeles River. The site was eventually given the name El Pueblo de la Reyna de Los Angeles (the Town of the Queen of the Angels).

For the first century of its history, Los Angeles grew very slowly. Its population in 1835 was only 1,250. By the end of the century, however, the first signs of a new trend appeared. In response to the promises of sunshine, warm weather, and "easy living," immigrants from the East Coast began to arrive in the Basin. Its population more than quadrupled between 1880 and 1890, from 11,183 to 50,395.

The rush was on, and it has scarcely abated today. The metropolitan population grew from 102,000 in 1900 to 1,238,000 in 1930 to 3,997,000 in 1950 to 9,838,861 in 2000.

The **pollution** facing Los Angeles today results from a complex mix of natural factors and intense **population growth**. The first reports of Los Angeles's famous **photochemical smog** go back to 1542. The "many smokes" described by Juan Cabrillo in that year were not the same as today's **smog**, but they occurred because of geographic and climatic conditions that are responsible for modern environmental problems.

The Los Angeles Basin has one of the highest probabilities of experiencing thermal inversions of any area in the United States. An inversion is an atmospheric condition in which a layer of cold air becomes trapped beneath a layer of warm air. That situation is just the reverse of the most normal atmospheric condition in which a warm layer near the ground is covered by a cooler layer above it. The warm air has a tendency to rise, and the cool air has a tendency to sink. As a result, natural mixing occurs. In contrast, when a thermal inversion occurs, the denser cool air remains near the ground while the less dense air above it tends to stay there.

Smoke and other pollutants released into a thermal inversion are unable to rise upward and tend to be trapped in the cool lower layer. Furthermore, horizontal movements of air, which might clear out pollution in other areas, are blocked by the mountains surrounding LA county. The lingering **haze** of the "many smokes" described by Cabrillo could have been nothing more than the smoke from campfires trapped by inversions that must have existed even in 1542.

As population and industrial growth occurred in Los Angeles during the second half of the twentieth century, the amount of pollutants trapped in thermal inversions also grew. By the 1960s, Los Angeles had become a classic example of how modern cities were being choked by their own wastes.

The geographic location of the Los Angeles Basin contributes another factor to Los Angeles's special environmental problems. Sunlight warms the Basin for most of the

year and attracts visitors and new residents. **Solar energy** fuels reactions between components of Los Angeles's polluted air, producing **chemicals** even more toxic than those from which they came. The complex mixture of noxious compounds produced in Los Angeles has been given the name *smog*, reflecting the combination of human (*smoke*) and natural factors (*fog*) that make it possible. Smog, also called ground level **ozone**, can cause a myriad of health problems including breathing difficulties, coughing, chest pains, and congestion. It may also exacerbate **asthma**, heart disease, and **emphysema**.

As Los Angeles grew in area and population, conditions which guaranteed a continuation of smog increased. The city and surrounding environs eventually grew to cover 400 square miles (1,036 square kilometers), a widespread community held together by freeways and cars. A major oil company bought the city's public transit system, then closed it down, ensuring the wide use of **automobile transportation**. Thus, gases produced by the **combustion** of **gasoline** added to the city's increasing pollution levels.

Los Angeles and the State of California have been battling **air pollution** for over 20 years. California now has some of the strictest **emission standards** of any state in the nation, and LA has begun to develop **mass transit** systems once again. For an area that has long depended on the automobile, however, the transition to public transportation has not been an easy one. But some measurable progress has been made in controlling ground level ozone. In 1976, smog was detectable at levels above the state standard acceptable average of 0.09 ppm a staggering 237 days out of the year. By 2001, the number had dropped to 121 days. Still, much work remains to be done; in 2000, 2001, and 2002 Los Angeles topped the American Lung Association's annual list of most ozone polluted cities and counties.

Another of Los Angeles's population-induced problems is its enormous demand for water. As early as 1900, it was apparent that the Basin's meager **water resources** would be inadequate to meet the needs of the growing urban area. The city turned its sights on the Owens Valley, 200 mi (322 km) to the northeast in the Sierra Nevada. After a lengthy dispute, the city won the right to tap the water resources of this distant valley. A 200-mile water diversion public works project, the Los Angeles Aqueduct, was completed in 1913.

This development did not satisfy the area's growing need for water, however, and in the 1930s, a second canal was built. This canal, the **Colorado River** Aqueduct, carries water from the Colorado River to Los Angeles over a distance of 444 mi (714 km). Even this proved to be inadequate, however, and the search for additional water sources has gone on almost without stop. In fact, one of the great ongoing debates in California is between legislators from

Northern California, where the state's major water resources are located, and their counterparts from Southern California, where the majority of the state's people live. Since the latter contingent is larger in number, it has won many of the battles so far over distribution of the state's water resources.

Of course, Los Angeles has also experienced many of the same problems as urban areas in other parts of the world, regardless of its special geographical character. For example, the Basin was at one time a lush agricultural area, with some of the best **soil** and growing conditions found anywhere. From 1910 to 1950, Los Angeles County was the wealthiest agricultural region in the nation. But as urbanization progressed, more and more farmland was sacrificed for commercial and residential development. During the 1950s, an average of 3,000 acres (1,215 hectares) of farmland per day was taken out of production and converted to residential, commercial, industrial, or transportation use.

One of the mixed blessings faced by residents of the Los Angeles Basin is the existence of large oil reserves in the area. On the one hand, the oil and **natural gas** contained in these reserves is a valuable natural resource. On the other hand, the presence of working oil **wells** in the middle of a modern metropolitan area creates certain problems. One is aesthetic, as busy pumps in the midst of barren or scraggly land contrasts with sleek new glass and steel buildings.

Another petroleum-related difficulty is that of land **subsidence**. As oil and gas are removed from underground, land above it begins to sink. This phenomenon was first observed as early as 1937. Over the next two decades, subsidence had reached 16 ft (5 m) at the center of the Wilmington oil fields. Horizontal shifting of up to 9 ft (2.74 m) was also recorded.

Estimates of subsidence of up to 45 ft (14 m) spurred the county to begin remedial measures in the 1950s. These measures included the construction of levees to prevent seawater from flowing into the subsided area and the repressurizing of oil zones with water injection. These measures have been largely successful, at least to the present time, in halting the subsidence of the oil field.

Los Angeles' annual ritual of pumping and storing water into underground aquifers in anticipation of the long, dry summer season has also been responsible for elevation shifts in the region. Researchers with the United States **Geological Survey** (USGS) observed that the ground surface of a 20 by 40 km area of Los Angeles rises and falls approximately 10–11 centimeters annually in conjunction with the water storage activities.

As if population growth itself were not enough, the Basin poses its own set of natural challenges to the community. For example, the area has a typical Mediterranean **climate** with long hot summers, and short winters with little rain. Summers are also the occasion of Santa Ana winds,

severe windstorms in which hot air sweeps down out of the mountains and across the Basin. Urban and forest fires that originate during a Santa Ana wind not uncommonly go out of control causing enormous devastation to both human communities and the natural **environment**.

The Los Angeles Basin also sits within a short distance of one of the most famous fault systems in the world, the San Andreas Fault. Other minor faults spread out around Los Angeles on every side. Earthquakes are common in the Basin, and the most powerful **earthquake** in Southern California history struck Los Angeles in 1857 (8.25 magnitude). Sixty miles (97 kilometers) from the quake's epicenter, the tiny community of Los Angeles lost the military base at Fort Tejon although only two lives were lost in the disaster. Like San Franciscans, residents of the Los Angeles Basin live not wondering if another earthquake will occur, but only when "The Big One" will hit. *See also* Air quality; Atmospheric inversion; Environmental Protection Agency (EPA); Mass transit; Oil drilling

[*David E. Newton and Paula Anne Ford-Martin*]

RESOURCES
BOOKS

Davis, Mike. *Ecology of Fear: Los Angeles and the Imagination of Disaster.* New York: Vintage Books, 1999.
Gumprecht, Blake. *The Los Angeles River: Its Life, Death, and Possible Rebirth* Baltimore: John Hopkins University Press, 2001.

PERIODICALS
Hecht, Jeff. "Finding Fault" *New Scientist* 171 (August 2001): 8.

OTHER
American Lung Association. *State of the Air 2002 Report.* [cited July 9, 2002]. <http://www.stateoftheair.org>.
South Coast Air Quality Management District. *Smog Levels.* [cited July 9, 2002]. <http://ozone.aqmd.gov/smog>.
United States Geological Survey (USGS). *Earthquake Hazards Program: Northern California.* [cited July 9, 2002]. <http://quake.usgs.gov>.

Love Canal

Probably the most infamous of the nation's **hazardous waste** sites, the Love Canal neighborhood of Niagara Falls, New York, was largely evacuated of its residents in 1980 after testing revealed high levels of toxic **chemicals** and genetic damage.

Between 1942 and 1953, the Olin Corporation and the Hooker Chemical Corporation buried over 20,000 tons of deadly chemical waste in the canal, much of which is known to be capable of causing **cancer**, **birth defects**, miscarriages, and other health disorders. In 1953, Hooker deeded the land to the local board of education but did not clearly warn of the deadly nature of the chemicals buried

Weeds grow around boarded up homes in Love Canal, New York in 1980. (Corbis-Bettmann. Reproduced by permission.)

there, even when homes and playgrounds were built in the area.

The seriousness of the situation became apparent in 1976, when years of unusually heavy rains raised the **water table** and flooded basements. As a result, houses began to reek of chemicals, and children and pets experienced chemical burns on their feet. Plants, trees, gardens, and even some pets died.

Soon neighborhood residents began to experience an extraordinarily high number of illnesses, including cancer, miscarriages, and deformities in infants. Alarmed by the situation, and frustrated by inaction on the part of local, state, and federal governments, a 27-year-old housewife named Lois Gibbs began to organize her neighbors. In 1978 they formed the Love Canal Homeowners Association and began a two-year fight to have the government relocate them into another area.

In August 1978 the New York State Health Commissioner recommended that pregnant women and young children be evacuated from the area, and subsequent studies documented the extraordinarily high rate of birth defects, miscarriages, genetic damage and other health affects. In

1979, for example, of 17 pregnant women in the neighborhood, only two gave birth to normal children. Four had miscarriages, two suffered stillbirths, and nine had babies with defects.

Eventually, the state of New York declared the area "a grave and imminent peril" to human health. Several hundred families were moved out of the area, and the others were advised to leave. The school was closed and barbed wire placed around it. In October 1980 President Jimmy Carter declared Love Canal a national disaster area.

In the end, some 60 families decided to remain in their homes, rejecting the government's offer to buy their properties. The cost for the cleanup of the area has been estimated at $250 million. Ironically, twelve years after the neighborhood was abandoned, the state of New York approved plans to allow families to move back to the area, and homes were allowed to be sold.

Love Canal is not the only hazardous waste site in the country that has become a threat to humans--only the best known. Indeed, the United States **Environmental Protection Agency** has estimated that up to 2,000 hazardous waste disposal sites in the United States may pose "significant risks to human health or the environment," and has called the toxic waste problem "one of the most serious problems the nation has ever faced." *See also* Contaminated soil; Hazardous waste site remediation; Hazardous waste siting; Leaching; Storage and transport of hazardous material; Stringfellow Acid Pits; Toxic substance

[*Lewis G. Regenstein*]

RESOURCES
BOOKS

Gibbs, Lois. *Love Canal: My Story.* Albany: State University of New York Press, 1982.

Regenstein, L. G. *How to Survive in America the Poisoned.* Washington, DC: Acropolis Books, 1982.

PERIODICALS

Brown, M. H. "Love Canal Revisited." *Amicus Journal* 10 (Summer 1988): 37–44.

Kadlecek, M. "Love Canal—10 Years Later." *Conservationist* 43 (November-December 1988): 40–43.

———. "A Toxic Ghost Town: Ten Years Later, Scientists Are Still Assessing the Damage From Love Canal." *The Atlantic* 263 (July 1989): 23–26.

Sir James Ephraim Lovelock (1919 –)
English chemist

Sir James Lovelock is the framer of the **Gaia hypothesis** and developer of, among many other devices, the electron capture gas chromatographic detector. The highly selective nature and great sensitivity of this detector made possible not only the identification of **chlorofluorocarbons** in the **atmosphere**, but led to the measurement of many pesticides, thus providing the raw data that underlie Rachel Carson's *Silent Spring*.

Sir James Lovelock was born in Letchworth Garden City, earned his degree in chemistry from Manchester University, and took a Ph.D. in medicine from the London School of Hygiene and Tropical Medicine. His early studies in medical topics included work at Harvard University Medical School and Yale University. He spent three years as a professor of chemistry at Baylor University College of Medicine in Houston, Texas. It was from that position that his work with the Jet Propulsion Laboratory for NASA began.

The Gaia hypothesis, Sir James Lovelock's most significant contribution to date, grew out of the work of Sir James Lovelock and his colleagues at the lab. While attempting to design experiments for life detection on Mars, Sir James Lovelock, Dian Hitchcock, and later Lynn Margulis, posed the question, "If I were a Martian, how would

Sir James E. Lovelock at his home in Cornwall with thedevice he invented to measure chlorofluorocarbons in the atmosphere. (Photograph by Anthony Howarth. Photo Researchers Inc. Reproduced by permission.)

I go about detecting life on Earth?" Looking in this way, the team soon realized that our atmosphere is a clear sign of life and it is totally impossible as a product of strictly chemical equilibria. One consequence of viewing life on this or another world as a single homeostatic organism is that energy will be found concentrated in certain locations rather than spread evenly, frequently, or even predominantly, as chemical energy. Thus, against all **probability**, the earth has an atmosphere containing about 21% free oxygen and has had about this much for millions of years. Sir James Lovelock bestowed on this superorganism comprising the whole of the **biosphere** the name "Gaia," one spelling of the name of the Greek earth-goddess, at the suggestion of a neighbor, William Golding, author of *Lord of the Flies*.

Sir James Lovelock's hypothesis was initially attacked as requiring the whole of life on earth to have purpose, and hence in some sense, common intelligence. Sir James Lovelock then developed a computer model called "Daisyworld" in which the presence of black and white daisies controlled the global temperature of the planet to a nearly constant value despite a major increase in the heat output of its sun. The concept that the biosphere keeps the **environment** constant has been attacked as sanctioning

environmental degradation, and accusers took a cynical view of Sir James Lovelock's service to the British **petrochemical** industry. However, the hypothesis has served the environmental community well in suggesting many ideas for further studies, virtually all of which have given results predicted by the hypothesis.

Since 1964, Sir James Lovelock has operated a private consulting practice, first out of his home in Bowerchalke, near Salisbury, England, and later from a home near Launceston, Cornwall. He has authored over 200 scientific papers, covering research that ranged from techniques for freezing and successfully reviving hamsters to global systems science, which he has proposed to call geophysiology. Sir James Lovelock has been honored by his peers worldwide with numerous awards and honorary degrees, including Fellow of the Royal Society. He was also named a Commander of the British Empire by Queen Elizabeth in 1990.

[*James P. Lodge Jr.*]

RESOURCES
BOOKS

Joseph, L. E. *Gaia: The Growth of an Idea*. New York: St. Martin's Press, 1990.
Lovelock, J. *The Ages of Gaia, A Biography of Our Living Earth*. New York: Norton, 1988.
———. *Gaia, a New Look at Life on Earth*. Oxford: Oxford University Press, 1979.
———, and M. Allaby. *The Greening of Mars*. New York: St. Martin's Press, 1984.

Amory Bloch Lovins (1947 –)

American physicist and energy conservationist

Amory Lovins is a physicist specializing in environmentally safe and sustainable energy sources. Born in 1947 in Washington, D.C., Lovins attended Harvard and Oxford universities. He has had a distinguished career as an educator and scientist. After resigning his academic post at Oxford in 1971, Lovins became the British representative of **Friends of the Earth**. He has been Regents' Lecturer at the University of California at Berkeley and has served as a consultant to the United Nations and other international and environmental organizations. Lovins is a leading critic of hard energy paths and an outspoken proponent of soft alternatives.

According to Lovins, an energy path is "hard" if the route from source to use is complex and circuitous requires extensive, expensive, and highly complex technological means and centralized power to produce, transmit, and store the energy produces toxic wastes or other unwanted side effects has hazardous social uses or implications and tends over time to harden even more, as other options are fore-

closed or precluded as the populace becomes ever more dependent on energy from a particular source. A hard energy path can be seen in the case of fossil fuel energy. Once readily abundant and cheap, **petroleum** fueled the internal **combustion** engines and other machines on which humans have come to depend, but as that energy source becomes scarcer, oil companies must go farther afield to find it, potentially causing more environmental damage. As oil supplies run low and become more expensive, the temptation is to sustain the level of energy use by turning to another, and even harder, energy path—nuclear energy.

With its complex technology, its hazards, its long-lived and highly toxic wastes, its myriad military uses, and the possibility of its falling into the hands of dictators or terrorists, **nuclear power** is perhaps the hardest energy path. No less important are the social and political implications of this hard path: radioactive wastes will have to be stored somewhere nuclear **power plants** and **plutonium** transport routes must be guarded we must make trade-offs between the ease, convenience, and affluence of people presently living and the health and well-being of **future generations** and so on. A hard energy path is also one that, once taken, forecloses other options because, among other considerations, the initial investment and costs of entry are so high as to render the decision, once made, nearly irreversible. The longer term economic and social costs of taking the hard path, Lovins argues, are astronomically high and incalculable.

Soft energy paths, by contrast, are shorter, more direct, less complex, cheaper (at least over the long run), are inexhaustible and renewable, have few if any unwanted side-effects, have minimal military uses, and are compatible with decentralized local forms of community control and decision-making. The old windmill on the family farm offers an early example of such a soft energy source newer versions of the windmill, adapted to the generation of electricity, supply a more modern example. Other soft technologies include **solar energy**, **biomass** furnaces burning peat, dung or wood chips, and **methane** from the rotting of vegetable matter, manure, and other cheap, plentiful, and readily available organic material.

Much of Lovins's work has dealt with the technical and economic aspects, as well as the very different social impacts and implications, of these two competing energy paths. A resource consultant agency, The **Rocky Mountain Institute**, was founded by Lovins and his wife, Hunter in 1982. In 1989, Lovins and his wife won the Onassis Foundation's first Delphi Prize for their "essential contribution towards finding alternative solutions to energy problems."

[*Terence Ball*]

Amory Lovins. (Reproduced by permission of the Rocky Mountain Institute.)

RESOURCES
BOOKS

Nash, H., ed. *The Energy Controversy: Amory B. Lovins and His Critics.* San Francisco: Friends of the Earth, 1979.
Lovins, A. B. *Soft Energy Paths.* San Francisco: Friends of the Earth, 1977.
———, and L. H. Lovins. *Energy Unbound: Your Invitation to Energy Abundance.* San Francisco: Sierra Club Books, 1986.

PERIODICALS

Louma, J. R. "Generate 'Nega-Watts' Says Fossil Fuel Foe." *New York Times* (April 2, 1993): B5, B8.

Lowest Achievable Emission Rate

Governments have explored a number of mechanisms for reducing the amount of pollutants released to the air by factories, **power plants**, and other stationary sources. One mechanism is to require that a new or modified installation releases no more pollutants than determined by some law or regulation determining the lowest level of pollutants that can be maintained by existing technological means. These limits are known as the Lowest Achievable Emission Rate (LAER). The **Clean Air Act** of 1970 required, for example, that any new source in an area where minimum **air pollution** standards were not being met had to conform to the LAER

standard. *See also* Air quality; Best Available Control Technology (BAT); Emission standards

Low-head hydropower

The term hydropower often suggests giant **dams** capable of transmitting tens of thousands of cubic feet of water per minute. Such dams are responsible for only about six percent of all the electricity produced in the United States today.

Hydropower facilities do not have to be massive buildings. At one time in the United States--and still, in many places around the world--electrical power is generated at low-head facilities, dams where the vertical drop through which water passes is a relatively short distance and/or where water flow is relatively modest. Indeed, the first commercial hydroelectric facility in the world consisted of a waterwheel on the Fox River in Appleton, Wisconsin. The facility, opened in 1882, generated enough electricity to operate lighting systems at two paper mills and one private residence.

Electrical demand grew rapidly in the United States during the early twentieth century, and hydropower supplied much of that demand. By the 1930s, nearly 40 percent of the electricity used in this country was produced by hydroelectric facilities. In some Northeastern states, hydropower accounted for 55-85 percent of the electricity produced.

A number of social, economic, political, and technical changes soon began to alter that pattern. Perhaps most important was the vastly increased efficiency of **power plants** operated by **fossil fuels**. The fraction of electrical power from such plants rose to more than 80 percent by the 1970s.

In addition, the United States began to move from a decentralized energy system in which many local energy companies met the needs of local communities, to large, centralized utilities that served many counties or states. In the 1920s, more than 6,500 electric power companies existed in the nation. As the government recognized power companies as monopolies, that number began to drop rapidly. Companies that owned a handful of low-head dams on one or more rivers could no longer compete with their giant cousins that operated huge plants powered by oil, **natural gas**, or **coal**.

As a result, hundreds of small hydroelectric plants around the nation were closed down. According to one study, over 770 low-head hydroelectric plants were abandoned between 1940 and 1980. In some states, the loss of low-head generating capacity was especially striking. Between 1950 and 1973, Consumers Power Company, one of Michigan's two **electric utilities**, sold off 44 hydroelectric plants.

Some experts believe that low-head hydropower should receive more attention today. Social and technical factors still prevent low-head power from seriously competing with other forms of energy on a national scale. But it may meet the needs of local communities in special circumstances. For example, a project has been undertaken to rehabilitate four low-head dams on the Boardman River in northwestern Michigan. The new facility is expected to increase the electrical energy available to nearby Traverse City and adjoining areas by about 20 percent.

Low-head hydropower appears to have a more promising future in less-developed parts of the world. For example, China has more than 76,000 low-head dams that generate a total of 9,500 megawatts of power. An estimated 50 percent of rural townships depend on such plants to meet their electrical needs. Low-head hydropower is also of increasing importance in nations with fossil-fueled plants and growing electricity needs. Among the fastest growing of these are Peru, India, the Philippines, Costa Rica, Thailand, and Guatemala. *See also* Alternative energy sources; Electric utilities; Wave power

[*David E. Newton*]

RESOURCES
BOOKS

Lapedes, D. N., ed. *McGraw-Hill Encyclopedia of Energy*. New York: McGraw-Hill, 1976.

PERIODICALS

Kakela, P., G, Chilson, and W. Patric. "Low-Head Hydropower for Local Use." *Environment* (January-February 1984): 31–38.

Low-inut agriculture
see **Sustainable agriculture**

Low-level radioactive waste

Low-level **radioactive waste** consists of materials used in a variety of medical, industrial, commercial, and research applications. They tend to release a low level of radiation that dissipates in a relatively short period of time. Although care must be taken in handling such materials, they pose little health or environmental risk. Among the most common low level radioactive materials are rags, papers, protective clothing, and **filters**. Such materials are often stored temporarily in sealed containers at their use site. They are then disposed of by burial at one of three federal sites: Barnwell, South Carolina, Beatty, Nevada, or Hanford, Washington. *See also* Hanford Nuclear Reservation; High-level radioactive waste; Radioactive waste; Radioactivity

LUST
see **Leaking underground storage tank**

Sir Charles Lyell (1797 – 1875)
Scottish geologist

Lyell was born in Kinnordy, Scotland, the son of well-to-do parents. When Lyell was less than a year old, his father moved his family to the south of England where he leased a house near the New Forest in Hampshire. Lyell spent his boyhood there, surrounded by his father's collection of rare plants. At the age of seven, Lyell became ill with pleurisy and while recovering began to collect and study insects. As a young man he entered Oxford to study law, but he also became interested in mineralogy after attending lectures by the noted geologist William Buckland. Buckland advocated the theories of Abraham Gottlob Werner, a neptunist who postulated that a vast ocean once covered the earth and that the various rocks resulted from chemical and mechanical deposition underwater, over a long period of time. This outlook was more in keeping with the Biblical story of Genesis than that of the vulcanists or plutonists who suscribed to the idea that volcanism, along with **erosion** and deposition, were the major forces sculpting the Earth. While on holidays with his family, Lyell made the first of many observations in hopes of confirming the views of Buckland and Werner. However, he continued to study law and was eventually called to the bar in 1822. Lyell practiced law until 1827 while still devoting time to geology.

Lyell traveled to France and Italy where he collected extensive data which caused him to reject the neptunist philosophy. He instead drew the conclusion that volcanic activity and erosion by wind and weather were primarily responsible for the different strata rather than the deposition of sediments from a "world ocean." He also rejected the catastrophism of Georges Cuvier, who believed that global catastrophes, such as the biblical Great Flood, periodically destroyed life on Earth, thus accounting for the different fossils found in each rock layer. Lyell believed change was a gradual process that occurred over a long period of time at a constant rate. This theory, known as uniformitarianism, had been postulated 50 years earlier by Scottish geologist James Hutton. It was Lyell, though, who popularized uniformitarianism in his work *The Principles of Geology*, which is now considered a classic text in this field. By 1850 his views and those of Hutton had become the standard among geologists. However, unlike many of his colleagues, Lyell adhered so strongly to uniformitarianism that he rejected the possibility of even limited catastrophe. Today most scientists accept that catastrophes such as meteor impacts played an important, albeit supplemental, role in the earth's **evolution**.

In addition to his championing of uniformitarianism, Lyell named several divisions of geologic time such as the Eocene, Miocene, and Pliocene Epochs. He also estimated the age of some of the oldest fossil-bearing rocks known at that time, assigning them the then startling figure of 240 million years. Even though Lyell came closer than his contemporaries to guessing the correct age, it is still less than half the currently accepted figure used by geologists today. While working on *The Principles of Geology*, Lyell formed a close friendship with Charles Darwin who had outlined his evolutionary theory in *The Origin of Species*. Both scientists quickly accepted the work of the other (Lyell was one of two scientists who presented Darwin's work to the influential Linnaean Society). Lyell even extended evolutionary theory to include humans at a time when Darwin was unwilling to do so. In his *The Antiquity of Man* (1863), Lyell argued that humans were much more ancient than creationists (those who interpreted Book of Genesis literally) and catastrophists believed, basing his ideas on archaeological artifacts such as ancient ax heads. Lyell was knighted for his work in 1848 and created a baronet in 1864. He also served as president of the Geological Society and set up the Lyell Medal and the Lyell Fund. He died in 1875 while working on the twelfth edition of his *Principles of Geology*.

RESOURCES
BOOKS

Adams, Alexander B. "Reading the Earth's Story: Charles Lyell—1979–1875." In *Eternal Quest: The Story of the Great Naturalists*. NY: G. P. Putnam's Sons, 1969.

Wilson, Leonard G. *Charles Lyell—The Years to 1841: The Revolution in Geology*. New Haven, CN: Yale University Press, 1972.

PERIODICALS

Camardi, Giovanni. "Charles Lyell and the Uniformity Principle." *Biology and Philosophy* 14, no. 4 (October 1999): 537–560.

Kennedy, Barbara A. "Charles Lyell and 'Modern Changes of the Earth': the Milledgeville Gully." *Geomorphology* 40 (2001): 91–98.

Lysimeter

A device for 1) measuring **percolation** and **leaching** losses from a column of **soil** under controlled conditions, or 2) for measuring gains (precipitation, **irrigation**, and condensation) and losses (**evapotranspiration**) by a column of soil. Many kinds of lysimeters exist: weighing lysimeters record the weight changes of a block of soil; non-weighing lysimeters enclose a block of soil so that losses or gains in the soil must occur through the surface suction lysimeters are devises for removing water and dissolved **chemicals** from locations within the soil.

Lythrum salicaria
see **Purple loosestrife**

M

Robert Helmer MacArthur (1930 – 1972)

Canadian biologist and ecologist

Few scientists have combined the skills of mathematics and biology to open new fields of knowledge the way Robert H. MacArthur did in his pioneering work in evolutionary **ecology**. Guided by a wide-ranging curiosity for all things natural, MacArthur had a special interest in birds and much of his work dealt primarily with bird populations. His conclusions, however, were not specific to ornithology but transformed both **population biology** and **biogeography** in general.

Robert Helmer MacArthur was born in Toronto, Ontario, Canada, on April 7, 1930, the youngest son of John Wood and Olive (Turner) MacArthur. While Robert spent his first seventeen years attending public schools in Toronto, his father shuttled between the University of Toronto and Marlboro College in Marlboro, Vermont, as a professor of genetics. Robert MacArthur graduated from high school in 1947 and immediately immigrated to the United States to attend Marlboro College. He received his undergraduate degree from Marlboro in 1951 and a master's degree in mathematics from Brown University in 1953. Upon receiving his doctorate in 1957 from Yale University under the direction of G. Evelyn Hutchinson, MacArthur headed for England to spend the following year studying ornithology with David Lack at Oxford University. When he returned to the United States in 1958, he was appointed Assistant Professor of Biology at the University of Pennsylvania.

As a doctoral student at Yale, MacArthur had already proposed an ecological theory that encompassed both his background as a mathematician and his growing knowledge as a naturalist. While at Pennsylvania, MacArthur developed a new approach to the frequency distribution of **species**. One of the problems confronting ecologists is measuring the numbers of a specific species within a geographic area—one cannot just assume that three crows in a 10-acre corn field means that in a 1000-acre field there will be 300 crows.

Much depends on the number of species occupying a **habitat**, species **competition** within the habitat, food supply, and other factors. MacArthur developed several ideas relating to the measurement of species within a known habitat, showing how large masses of empirical data relating to numbers of species could be processed in a single model by employing the principles of information theory. By taking the sum of the product of the frequencies of occurrences of a species and the logarithms of the frequencies, complex data could be addressed more easily.

The most well-known theory of frequency distribution MacArthur proposed in the late 1950s is the so-called broken stick model. This model had been suggested by MacArthur as one of three competing models of frequency distribution. He proposed that competing species divide up available habitat in a random fashion and without overlap, like the segments of a broken stick. In the 1960s, MacArthur noted that the theory was obsolete. The procedure of using competing explanations and theories simultaneously and comparing results, rather than relying on a single hypothesis, was also characteristic of MacArthur's later work.

In 1958, MacArthur initiated a detailed study of warblers in which he analyzed their **niche** division, or the way in which the different species will to be best suited for a narrow ecological role in their common habitat. His work in this field earned him the Mercer Award of the **Ecological Society of America**. In the 1960s, he studied the so-called "species-packing problem." Different kinds of habitat support widely different numbers of species. A **tropical rain forest** habitat, for instance, supports a great many species, while arctic **tundra** supports relatively few. MacArthur proposed that the number of species crowding a given habitat correlates to niche breadth. The book *The Theory of Island Biogeography*, written with **biodiversity** expert Edward O. Wilson and published in 1967, applied these and other ideas to isolated habitats such as islands. The authors explained the species-packing problem in an evolutionary light, as an equilibrium between the rates at which new species arrive or develop and the **extinction** rates of species already present.

These rates vary with the size of the habitat and its distance from other habitats.

In 1965 MacArthur left the University of Pennsylvania to accept a position at Princeton University. Three years later, he was named **Henry Fairfield Osborn** Professor of Biology, a chair he held until his death. In 1971, MacArthur discovered that he suffered from a fatal disease and had only a few years to live. He decided to concentrate his efforts on encapsulating his many ideas in a single work. The result, *Geographic Ecology: Patterns in the Distribution of Species*, was published shortly before his death the following year. Besides a summation of work already done, *Geographic Ecology* was a prospectus of work still to be carried out in the field.

MacArthur was a Fellow of the American Academy of Arts and Science. He was also an Associate of the Smithsonian Tropical Research Institute, and a member of both the Ecological Society and the National Academy of Science. He married Elizabeth Bayles Whittemore in 1952; they had four children: Duncan, Alan, Donald, and Elizabeth. Robert MacArthur died of renal **cancer** in Princeton, New Jersey, on November 1, 1972, at the age of 42.

RESOURCES
BOOKS

Carey, C. W. "MacArthur, Robert Helmer." Vol. 14, *American National Biography* edited by J. A. Garraty and M. C. Carnes. NY: Oxford University Press, 1999.

Gillispie, Charles Coulson, ed. *Dictionary of Scientific Biography.* Vol. 17–18: Scribner, 1990.

MacArthur, Robert. *Geographic Ecology: Patterns in the Distribution of Species.* Harper, 1972.

———. *The Biology of Populations.* Wiley, 1966.

———. *The Theory of Island Biogeography.* Princeton University Press, 1967.

Notable Scientists: From 1900 to the Present. Farmington Hills, MI: Gale Group, 2002.

Mad cow disease

Mad cow disease, a relatively newly discovered malady, was first identified in Britain in 1986, when farmers noticed that their cows' behavior had changed. The cows began to shake and fall, became unable to walk or even stand, and eventually died or had to be killed. It was later determined that a variation of this fatal neurological disease, formally known as Bovine Spongiform Encephalopathy (BSE), could be passed on to humans.

It is still not known to what extent the population of Britain and perhaps other countries is at risk from consumption of contaminated meat and animal by-products. The significance of the BSE problem lies in its as yet unquantifiable potential to not only damage Britain's $7.5 billion beef industry, but also to endanger millions of people with the threat of a fatal brain disease.

A factor that stood out in the autopsies of infected animals was the presence of holes and lesions in the brains, which were described as resembling a sponge or Swiss cheese. This was the first clue that BSE was a subtype of untreatable, fatal brain diseases called transmissible spongiform encephalopathies (TSEs). These include a very rare human malady known as Creutzfeldt-Jakob Disease (CJD), which normally strikes just one person in a million, usually elderly or middle-aged. In contrast to previous cases of CJD, the new bovine-related CJD in humans is reported to affect younger people, and manifests with unusual psychiatric and sensory abnormalities that differentiate it from the endemic CJD. The BSE-related CJD has a delayed onset that includes shaky limb movements, sudden muscle spasms, and dementia.

As the epidemic of BSE progressed, the number of British cows diagnosed began doubling almost yearly, growing from some 7,000 cases in 1989, to 14,000 in 1990, to over 25,000 in 1991. The incidence of CJD in Britain was simultaneously increasing, almost doubling between 1990 and 1994 and reaching 55 cases by 1994. In response to the problem and growing public concern, the government's main strategy was to issue reassurances. However, it did undertake two significant measures to try to safeguard public health. In July 1988, it ostensibly banned meat and bone meal from cow feed, but failed to strictly enforce the action. In November 1989, a law that intended to remove those bovine body parts considered to be the most highly infective (brain, spinal cord, spleen, tonsils, intestines, and thymus) from the public food supply was passed. A 1995 government report revealed that half of the time, the law was not being adhered to by slaughterhouses. Thus, livestock—and the public—continued to be potentially exposed to BSE.

As the disease continued to spread, so did public fears that it might be transmissible to humans, and could represent a serious threat to human health. But the British government, particularly the Ministry of Agriculture, Fisheries, and Food (MAFF), anxious to protect the multibillion dollar cattle industry, insisted that there was no danger to humans. However, on March 20, 1996, in an embarrassing reversal, the government officially admitted that there could be a link between BSE and the unusual incidence of CJD among young people. At the time, 15 people had been newly diagnosed with CJD. Shocking the nation and making headlines around the world, the Minister of Health Stephen Dorrell announced to the House of Commons that consumption of contaminated beef was "the most likely explanation" for the outbreak of a new variant CJD in 10 people under the age of 42, including several teenagers. Four dairy farmers, including some with infected herds, had also contracted CJD, as had a Frenchman who died in January 1996.

British authorities estimated that some 163,000 British cows had contracted BSE. But other researchers, using the same database, put the figure at over 900,000, with 729,000 of them having been consumed by humans. In addition, an unknown number had been exported to Europe, traditionally a large market for British cattle and beef. Many non-meat products may also have been contaminated. Gelatin, made from ligaments, bones, skin, and hooves, is found in ice cream, lipstick, candy, and mayonnaise; keratin, made from hooves, horns, nails, and hair, is contained in shampoo; fat and tallow are used in candles, cosmetics, deodorants, soap, margarine, detergent, lubricants, and pesticides; and protein meal is made into medical and pharmaceutical products, **fertilizer**, and **food additives**. Bone meal from dead cows is used as fertilizer on roses and other plants, and is handled and often inhaled by gardeners.

In reaction to the government announcement, sales of beef dropped by 70%, cattle markets were deserted, and even hamburger chains stopped serving British beef. Prime Minister Major called the temporary reaction "hysteria" and blamed the press and opposition politicians for fanning it.

On March 25, 1996, the **European Union** banned the import of British beef, which had since 1990 been excluded from the United States and 14 other countries. Shortly afterwards, in an attempt to have the European ban lifted, Britain announced that it would slaughter all of its 1.2 million cows over the age of 30 months (an age before which cows do not show symptoms of BSE), and began the arduous task of killing and incinerating 22,000 cows a week. The government later agreed to slaughter an additional 100,000 cows considered most at risk from BSE.

A prime suspect in causing BSE is a by-product derived from the rendering process, in which the unusable parts of slaughtered animals are boiled down or "cooked" at high temperatures to make animal feed and other products. One such product, called meat and bone meal (MBM), is made from the ground-up, cooked remains of slaughtered livestock—cows, sheep, chicken, and hogs—and made into nuggets of animal feed. Some of the cows and sheep used in this process were infected with fatal brain disease. (Although MBM was ostensibly banned as cattle feed in 1988, spinal cords continued to be used.)

It is theorized that sheep could have played a major role in initially infecting cows with BSE. For over 200 years, British sheep have been contracting scrapie, another TSE that results in progressive degeneration of the brain. Scrapie causes the sheep to tremble and itch, and to "scrape" or rub up against fences, walls, and trees to relieve the sensation. The disease, first diagnosed in British sheep in 1732, may have recently jumped the **species** barrier when cows ate animal feed that contained brain and spinal cord tissue from diseased sheep. In 1999 the World Health Organization

(WHO) implored high-risk countries to assess outbreaks of BSE-like manifestations in sheep and goat stocks. In August 2002, sheep farms in the United Kingdom demonstrated to the WHO that no increase in illnesses potentially linked to BSE occurred in non-cattle livestock. However, that same year, the European Union Scientific Steering Committee (SSC) on the risk of BSE identified the United Kingdom and Portugal as hotspots for BSE infection of domestic cattle relative to other European nations.

Scrapie and perhaps these other spongiform brain diseases are believed to be caused not by a **virus** (as originally thought) but rather by a form of infectious protein-like particles called prions, which are extremely tenacious, surviving long periods of high intensity cooking and heating. They are, in effect, a new form of contagion. The first real insights into the origins of these diseases were gathered in the 1950s by Dr. D. Carleton Gajdusek, who was awarded the 1976 Nobel Prize in Medicine for his work. His research on the fatal degenerative disease "kuru" among the cannibals of Papua, New Guinea, which resulted in the now-familiar brain lesions and cavities, revealed that the malady was caused by consuming or handling the brains of relatives who had just died.

In the United States, Department of Agriculture officials say that the risk of BSE and other related diseases is believed to be small, but cannot be ruled out. No BSE has been detected in the United States, and no cattle or processed beef is known to have been imported from Britain since 1989. However, several hundred mink in Idaho and Wisconsin have died from an ailment similar to BSE, and many of them ate meat from diseased "downer" cows, those that fall and cannot get up. Some experts believe that BSE can occur spontaneously, without apparent exposure to the disease, in one or two cows out of a million every year. This would amount to an estimated 150–250 cases annually among the United States cow population of some 150 million. Moreover, American feed processors render the carcasses of some 100,000 downer cows every year, thus utilizing for animal feed cows that are possibly seriously and neurologically diseased.

In June 1997, the **Food and Drug Administration** (FDA) announced a partial ban on using in cattle feed remains from dead sheep, cattle, and other animals that chew their cud. But the ruling exempts from the ban some animal protein, as well as feed for poultry, pigs, and pets. In March of that year, a coalition of consumer groups, veterinarians, and federal meat inspectors had urged the FDA to include pork in the animal feed ban, citing evidence that pigs can develop a form of TSE, and that some may already have done so. The coalition had recommended that the United States adopt a ban similar to Britain's, where protein from

French farmers protest against an allowance they must pay to bring dead animals to the knackery—a service that was free of charge prior to mad cow disease. (AP/Wide World Photos. Reproduced by permission.)

The United States Centers for Disease Control (CDC) has reclassified CJD that is associated with the interspecies transmission of BSE disease-causing factor. The current categorization of CJD is termed new variant CJD (nvCJD) to distinguish it from the extremely rare form of CJD that is not associated with BSE contagion. According to the CDC, there have been 79 nvCJD deaths reported worldwide. By April 2002, the global incidence of nvCJD increased to 125 documented reports. Of these, most (117) were from the United Kingdom. Other countries reporting nvCJD included France, Ireland, and Italy. The CDC stresses that nvCJD should not be confused with the endemic form of CJD. In the United States, CJD seldom occurs in adults under 30 years old, having a median age of death of 68 years. In contrast, nvCJD, associated with BSE, tends to affect a much younger segment of society. In the United Kingdom, the median age of death from nvCJD is 28 years. As of April 2002, no cases of nvCJD have been reported in the United States, and all known worldwide cases of nvCJD have been associated with countries where BSE is known to exist. The first possible infection of a U.S. resident was documented and reported by the CDC in 2002. A 22-year-old citizen of the United Kingdom living in Florida was preliminarily diagnosed with nvCJD during a visit abroad. Unfortunately, the only way to verify a diagnosis with nvCJD is via brain biopsy or autopsy. If confirmed, the CDC and Florida Department of Health claim that this case would be the first reported in the United States.

The outlook for BSE is uncertain. Since tens of millions of people in Britain may have been exposed to the infectious agent that causes BSE, plus an unknown number in other countries, some observers fear that a latent epidemic of serious proportions could be in the offing. (There is also concern that some of the four million Americans alone now diagnosed with Alzheimer's disease may actually be suffering from CJD.) There are others who feel that a general removal of most infected cows and animal brain tissue from the food supply has prevented a human health disaster. But since the incubation period for CJD is thought to be 7–40 years, it will be some time before it is known how many people are already infected and the extent of the problem becomes apparent.

[*Lewis G. Regenstein*]

RESOURCES

BOOKS

Rhodes, R. *Deadly Feasts*. New York: Simon & Schuster, 1997.

PERIODICALS

Blakeslee, S. "Fear of Disease Prompts New Look at Rendering." *The New York Times*, March 11, 1997.

all mammals is excluded from animal feed, and some criticized the FDA's action as "totally inadequate in protecting consumers and public health."

Lanchester, J. "A New Kind of Contagion." *The New Yorker*, December 2, 1996.

Madagascar

Described as a crown jewel among earth's ecosystems, this 1,000-mi long (1,610-km) island-continent is a microcosm of **Third World** ecological problems. It abounds with unique **species** which are being threatened by the exploding human population. Many scientists consider Madagascar the world's foremost **conservation** priority. Since 1984 united efforts have sought to slow the island's deterioration, hopefully providing a model for treating other problem areas.

Madagascar is the world's fourth largest island, with a **rain forest climate** in the east, **deciduous forest** in the west, and thorn scrub in the south. Its Malagasy people are descended from African and Indonesian seafarers who arrived about 1,500 years ago. Most farm the land using ecologically devastating **slash and burn agriculture** which has turned Madagascar into the most severely eroded land on earth. It has been described as an island with the shape, color, and fertility of a brick; second growth forest does not do well.

Having been separated from Africa for 160 million years, this unique land was sufficiently isolated during the last 40 million years to become a laboratory of **evolution**. There are 160,000 unique species, mostly in the rapidly disappearing eastern rain forests. These include 65 percent of its plants, half of its birds, and all of its reptiles and mammals. Sixty percent of the earth's chameleons live here. Lemurs, displaced elsewhere by monkeys, have evolved into 26 species. Whereas Africa has only one species of baobab tree, Madagascar has six, and one is termite resistant. The thorn scrub abounds with potentially useful poisons evolved for plant defense. One species of periwinkle provides a substance effective in the treatment of childhood (lymphocytic) **leukemia**.

Humans have been responsible for the loss of 93 percent of tropical forest and two-thirds of rain forest. Four-fifths of the land is now barren as the result of **habitat** destruction set in motion by the exploding human population (3.2 percent growth per year). Although **nature** reserves date from 1927, few Malagasy have ever experienced their island's biological wonders; urbanites disdain the bush, and peasants are driven by hunger. If they can see Madagascar's rich ecosystems first hand, it may engender respect which, in turn, may encourage understanding and protection.

The people are awakening to their loss and the impact this may have on all Madagascar's inhabitants. Pride in their island's unique **biodiversity** is growing. The **World Bank** has provided $90 million to develop and implement a 15-year Environmental Action Plan. One private preserve in the south is doing well and many other possibilities exist for the development of **ecotourism**. If **population growth** can be controlled, and high yield farming replaces **slash** and burn agriculture, there is yet hope for preserving the diversity and uniqueness of Madagascar. *See also* Deforestation; Erosion; Tropical rain forest

[*Nathan H. Meleen*]

RESOURCES
BOOKS

Attenborough, D. *Bridge to the Past: Animals and People of Madagascar*. New York: Harper, 1962.

Harcourt, C., and J. Thornback. *Lemurs of Madagascar and the Comoros: The IUCN Red Data Book*. Gland, Switzerland: IUCN, 1990.

Jenkins, M. D. *Madagascar: An Environmental Profile*. Gland, Switzerland: IUCN, 1987.

PERIODICALS

Jolly, A. "Madagascar: A World Apart." *National Geographic* 171 (February 1987): 148–83.

Magnetic separation

An on-going problem of environmental significance is **solid waste** disposal. As the land needed to simply throw out solid wastes becomes less available, **recycling** becomes a greater priority in **waste management** programs. One step in recycling is the magnetic separation of ferrous (iron-containing) materials. In a typical recycling process, wastes are first shredded into small pieces and then separated into organic and inorganic fractions. The inorganic fraction is then passed through a magnetic separator where ferrous materials are extracted. These materials can then be purified and reused as scrap iron. *See also* Iron minerals; Resource recovery

Malaria

Malaria is a disease that affects hundreds of millions of people worldwide. In the developing world malaria contributes to a high infant **mortality** rate and a heavy loss of work time. Malaria is caused by the single-celled protozoan parasite, *Plasmodium*. The disease follows two main courses: tertian (three day) malaria and quartan (four day) malaria. *Plasmodium vivax* causes benign tertian malaria with a low mortality (5%), while *Plasmodium falciparum* causes malignant tertian malaria with a high mortality (25%) due to interference with the blood supply to the brain (cerebral malaria). Quartan malaria is rarely fatal.

Plasmodium is transmitted from one human host to another by female mosquitoes of the genus *Anopheles*. Thousands of **parasites** in the salivary glands of the mosquito are injected into the human host when the mosquito takes blood. The parasites (in the sporozoite stage) are carried to the host's liver where they undergo massive multiplication into the next stage (cryptozoites). The parasites are then released into the blood stream, where they invade red blood cells and undergo additional division. This division ruptures the red blood cells and releases the next stage (the merozoites), which invade and destroy other red blood cells. This red blood cell destruction phase is intense but short-lived. The merozoites finally develop into the next stage (gametocytes) which are ingested by the biting mosquito.

The pattern of chills and fever characteristic of malaria is caused by the massive destruction of the red blood cells by the merozoites and the accompanying release of parasitic waste products. The attacks subside as the immune response of the human host slows the further development of the parasites in the blood. People who are repeatedly infected gradually develop a limited immunity. Relapses of malaria long after the original infection can occur from parasites that have remained in the liver, since treatment with drugs kills only the parasites in the blood cells and not in the liver. Malaria can be prevented or cured by a wide variety of drugs (quinine, chloroquine, paludrine, proguanil, or pyrimethamine). However, resistant strains of the common **species** of *Plasmodium* mean that some prophylactic drugs (chloroquine and pyrimethamine) are no longer totally effective.

Malaria is controlled either by preventing contact between humans and mosquitoes or by eliminating the mosquito vector. Outdoors, individuals may protect themselves from mosquito bites by wearing protective clothing, applying mosquito repellents to the skin, or by burning mosquito coils that produce **smoke** containing insecticidal pyrethrins. Inside houses, mosquito-proof screens and nets keep the vectors out, while insecticides (DDT) applied inside the house kill those that enter. The aquatic stages of the mosquito can be destroyed by eliminating temporary breeding pools, by spraying ponds with synthetic insecticides, or by applying a layer of oil to the surface waters. Biological control includes introducing fish (*Gambusia*) that feed on mosquito larvae into small ponds. Organized campaigns to eradicate malaria are usually successful, but the disease is sure to return unless the measures are vigilantly maintained. *See also* Epidemiology; Pesticide

[*Neil Cumberlidge Ph.D.*]

RESOURCES
BOOKS

Bullock, W. L. *People, Parasites, and Pestilence: An Introduction to the Natural History of Infectious Disease.* Minneapolis: Burgess Publishing Company, 1982.

Knell, A. J., ed. *Malaria: A Publication of the Tropical Programme of the Wellcome Trust.* New York: Oxford University Press, 1991.

Markell, E. K., M. Voge, and D. T. John. *Medical Parasitology.* 7th ed. Philadelphia: Saunders, 1992.

Phillips, R. S. *Malaria.* Institute of Biology's Studies in Biology, No. 152. London: E. Arnold, 1983.

Male contraceptives

Current research into male contraceptives will potentially increase the equitability of **family planning** between males and females. This shift will also have the potential to address issues of **population growth** and its related detrimental effects on the **environment**.

While prophylactic condoms provide good barrier protection from unwanted pregnancies, they are not as effective as oral contraceptives for women. Likewise, vasectomies are very effective, but few men are willing to undergo the surgery. There are three general categories of male contraceptives that are being explored. The first category functionally mimics a vasectomy by physically blocking the vas deferens, the channel that carries sperm from the seminiferous tubules to the ejaculatory duct. The second uses heat to induce temporary sterility. The third involves medications to halt sperm production. In essence, this third category concerns the development of "The Pill" for men.

Despite its near 100% effectiveness, there are two major disadvantages to vasectomy that make it unattractive to many men as an option for contraception. The first is the psychological component relating to surgery. Although vasectomies are relatively non-invasive, when compared to taking a pill the procedure seems drastic. Second, although vasectomies are reversible, the rate of return to normal fertility is only about 40%. Therefore, newer "vas occlusive" methods offer alternatives to vasectomy with completely reversible effects. Vas occlusive devices block the flow of or render dysfunctional the sperm in the vas deferens. The most recent form of vas occlusive male contraception, called Reversible Inhibition of Sperm Under Guidance (RISUG), involves the use of a **styrene** that is combined with the chemical DMSO (dimethyl sulfoxide). The complex is injected into the vas deferens. The complex then partially occludes passage of sperm and also causes disruption of sperm cell membranes. As sperm cells contact the RISUG complex, they rupture. It is believed that a single injection of RISUG may provide contraception for up to 10 years. Large safety and efficacy trials examining RISUG are being conducted in India.

Two additional vas occlusive methods of male contraception involve the injection of polymers into the vas deferens. Both methods involve injection of a liquid form of polymer, microcellular polyurethane (MPU) or medical-grade silicon **rubber** (MSR), into the vas deferens where it hardens within 20 minutes. The resulting plug provides a barrier to sperm. The technique was developed in China, and since 1983 some 300,000 men have reportedly undergone this method of contraception. Reversal of MPU and MSR plugs requires surgical removal of the polymers. Another method involving silicon plugs (called the Shug for short) offers an alternative to injectable plugs. This double-plug design offers a back-up plug should sperm make their way past the first.

Human sperm is optimally produced at a temperature that is a few degrees below body temperature. Infertility is induced if the temperature of the testes is elevated. For this reason, men trying to conceive are often encouraged to avoid wearing snugly-fitting undergarments. The thermal suspensory method of male contraception utilizes specially designed suspensory briefs to use natural body heat or externally applied heat to suppress spermatogenesis. Such briefs hold the testes close to the body during the day, ideally near the inguinal canal where local body heat is greatest. Sometimes this method is also called artificial cryptorchidism since is simulates the infertility seen in men with undescended testicles. When worn all day, suspensory briefs lead to a gradual decline in sperm production. The safety of briefs that contain heating elements to warm the testes is being evaluated. Externally applied heat in such briefs would provide results in a fraction of the time required using body heat. Other forms of thermal suppression of sperm production utilize simple hot water heated to about 116°F(46.7°C). Immersion of the testicles in the warm water for 45 minutes daily for three weeks is said to result in six months of sterility followed by a return to normal fertility. A newer, but essentially identical, method of thermal male contraception uses ultrasound. This simple, painless, and convenient method using ultrasonic waves to heat water results in six-month, reversible sterility within only 10 minutes.

Drug therapy is also being evaluated as a potential form of male contraception. Many drugs have been investigated in male contraception. An intriguing possibility is the observation that a particular class of blood pressure medications, called calcium channel blockers, induces reversible sterility in many men. One such drug, nifedipine, is thought to induce sterility by blocking calcium channels of sperm cell membranes. This reportedly results in cholesterol deposition and membrane instability of the sperm, rendering them incapable of fertilization. Herbal preparations have also been used as male contraceptives. Gossypol, a constituent of cottonseed oil, was found to be an effective and reliable male contraceptive in very large-scale experiments conducted in China. Unfortunately, an unacceptable number of men experienced persistent sterility when gossypol therapy was discontinued. Additionally, up to 10% of men treated with gossypol experienced kidney problems in the studies conducted in China. Because of the potential toxicity of gossypol, the World Health Organization concluded that research on this form of male contraception should be abandoned. Most recently, a form of sugar that sperm interact with in the fertilization process has been isolated from the outer coating of human eggs. An **enzyme** in sperm, called N-acetyl-beta-D-hexosaminidase (HEX-B) cuts through the protective outer sugar layer of the egg during fertilization. A decoy sugar molecule that mimics the natural egg coating is being investigated. The synthetic sugar would bind specifically to sperm HEX-B enzyme, curtailing the sperm's ability to penetrate the egg's outer coating. Related experiments in male rats have shown effective and reversible contraceptive properties.

Perhaps one of the most researched methods of male contraception using drugs involves the use of hormones. Like female contraceptive pills, Male Hormone Contraceptives (MHCs) seek to stop the production of sperm by stopping the production of hormones that direct the development of sperm. Many hormones in the human body work by feedback mechanisms. When levels of one hormone are low, another hormone is released that results in an increase in the first. The goal of MHCs is to artificially raise the levels of hormone that would result in suppression of hormone release required for sperm production. The best MHC produced only provides about 90% sperm suppression, which is not enough to reliably prevent conception. Also, for poorly understood reasons, some men do not respond to the MHC preparations under investigation. Despite initial promise, more research is needed to make MHCs competitive with female contraception. Response failure rates for current MHC drugs range from 5–20%.

[*Terry Watkins*]

RESOURCES

ORGANIZATIONS

Contraceptive Research and Development Program (CONRAD), Eastern Virginia Medical School, 1611 North Kent Street, Suite 806, Arlington, VA USA 22209 (703) 524-4744, Fax: (703) 524-4770, Email: info@conrad.org, <http://www.conrad.org>

Malignant tumors

see **Cancer**

Man and the Biosphere Program

The Man and the **Biosphere** (MAB) program is a global system of biosphere reserves begun in 1986 and organized by the United Nations Educational, Social, and Cultural Organization (**UNESCO**). MAB reserves are designed to conserve natural ecosystems and **biodiversity** and to incorporate the sustainable use of natural ecosystems by humans in their operation. The intention is that local human needs will be met in ways compatible with resource **conservation**. Furthermore, if local people benefit from tourism and the harvesting of surplus **wildlife**, they will be more supportive of programs to preserve **wilderness** and protect wildlife.

MAB reserves differ from traditional reserves in a number of ways. Instead of a single boundary separating **nature** inside from people outside, MAB reserves are zoned into concentric rings consisting of a core area, a **buffer** zone, and a transition zone. The core area is strictly managed for wildlife and all human activities are prohibited, except for restricted scientific activity such as **ecosystem** monitoring. Surrounding the core area is the buffer zone, where non-destructive forms of research, education, and tourism are permitted, as well as some human settlements. Sustainable light resource extraction such as **rubber** tapping, collection of nuts, or selective **logging** is permitted in this area. Pre-existing settlements of **indigenous peoples** are also allowed. The transition zone is the outermost area, and here increased human settlements, traditional **land use** by native peoples, experimental research involving ecosystem manipulations, major restoration efforts, and tourism are allowed.

The MAB reserves have been chosen to represent the world's major types of regional ecosystems. Ecologists have identified some 14 types of biomes and 193 types of ecosystems around the world and about two-thirds of these ecosystem types are represented so far in the 276 biosphere reserves now established in 72 countries. MAB reserves are not necessarily pristine wilderness. Many include ecosystems that have been modified or exploited by humans, such as **rangelands**, subsistence farmlands, or areas used for **hunting** and fishing. The concept of biosphere reserves has also been extended to include coastal and marine ecosystems, although in this case the use of core, buffer, and transition areas is inappropriate.

The establishment of a global network of biosphere reserves still faces a number of problems. Many of the MAB reserves are located in debt-burdened developing nations, because many of these countries lie in the biologically rich tropical regions. Such countries often cannot afford to set aside large tracts of land, and they desperately need the short-term cash promised by the immediate exploitation of their lands. One response to this problem is the debt for nature swaps in which a conservation organization buys the debt of a nation at a discount rate from banks in exchange for that nation's commitment to establish and protect a nature reserve.

Many reserves are effectively small, isolated islands of natural ecosystems surrounded entirely by developed land. The protected organisms in such islands are liable to suffer genetic **erosion**, and many have argued that a single large

reserve would suffer less genetic erosion than several smaller reserves which cumulatively protect the same amount of land. It has also been suggested that reserves sited as close to each other as possible, and corridors that allow movement between them, would increase the **habitat** and **gene pool** available to most **species**.

[*Neil Cumberlidge Ph.D.*]

RESOURCES
BOOKS
Gregg, W. P., and S. L. Krugman, eds. *Proceedings of the Symposium on Biosphere Reserves.* Atlanta, GA: U.S. National Park Service, 1989.
Office of Technology Assessment. *Technologies to Maintain Biological Diversity.* Philadelphia: Lippincott, 1988.

PERIODICALS
Batisse, M. "Developing and Focusing the Biosphere Reserve Concept. *Nature and Resources* 22 (1986): 1–10.

Manatees

A relative of the elephant, manatees are totally aquatic, herbivorous mammals of the family Trichechidae. This group arose 15–20 million years ago during the Miocene period, a time which also favored the development of a tremendous diversity of aquatic plants along the coast of South America. Manatees are adapted to both marine and freshwater habitats and are divided into three distinct **species**: the Amazonian manatee (*Trichechus inunguis*), restricted to the freshwaters of the Amazon River; the West African manatee (*Trichechus senegalensis*), found in the coastal waters from Senegal to Angola; and the West Indian manatee (*Trichechus manatus*), ranging from the northern South American coast through the Caribbean to the southeastern coastal waters of the United States. Two other species, the dugong (*Dugong dugon*) and Steller's sea cow (*Hydrodamalis gigas*), along with the manatees, make up the order Sirenia. Steller's sea cow is now extinct, having been exterminated by man in the mid-1700s for food.

These animals can weigh 1,000–1,500 lb (454–680 kg) and grow to be more than 12 ft (3.7 m) long. Manatees are unique among aquatic mammals because of their herbivorous diet. They are non-ruminants, therefore, unlike cows and sheep, they do not have a chambered stomach. They do have, however, extremely long intestines (up to 150 ft/46 m) that contain a paired blind sac where bacterial digestion of cellulose takes place. Other unique traits of the manatee include horizontal replacement of molar teeth and the presence of only six cervical, or neck, vertebrae, instead of seven as in all other mammals. The intestinal sac and tooth replacement are adaptations designed to counteract the defenses evolved by the plants that the manatees eat. Several plant

Manatee with a researcher, Homosassa Springs, Florida. (Photograph by Faulkner. Photo Researchers Inc. Reproduced by permission.)

species contain tannins, oxalates, and **nitrates**, which are toxic, but which may be detoxified in the manatee's intestine. Other plant species contain silica spicules, which, due to their abrasiveness, wear down the manatee's teeth, necessitating the need for tooth replacement. The life span of manatees is long, greater than 30 years, but their reproductive rate is low, with gestation being 13 months and females giving birth to one calf every two years. Because of this the potential for increasing the population is low, thus leaving the population vulnerable to environmental problems.

Competition for food is not a problem. In contrast to terrestrial herbivores, which have a complex division of food resources and competition for the high-energy level land plants, manatees have limited competition from **sea turtles**. This is minimized by different feeding strategies employed within the two groups. Sea turtles eat blades of seagrasses at greater depths than manatees feed, and manatees tend to eat not only the blades, but also the rhizomes of these plants, which contain more energy for the warm-blooded mammals.

Because manatees are docile creatures and a source of food, they have been exploited by man to the point of **extinction**. There are currently between 1,500 and 3,000 in the U.S. Also because manatees are slow moving, a more recent threat is taking its toll on these shallow-swimming animals. Power boat propellers have struck hundreds of manatees in recent years, causing 90% of the man-related manatee deaths. This has also resulted in permanent injury or scarring to others. **Conservation** efforts, such as the **Marine Mammals Protection Act** of 1972 and the **Endangered Species Act** of 1973, have helped reduce some of these problems but much more will have to be done to prevent the extirpation of the manatees.

[*Eugene C. Beckham*]

RESOURCES
BOOKS

Ridgway, S. H., and R. Harrison, eds. *Handbook of Marine Mammals.* Vol. 3, *The Sirenians and Baleen Whales.* London: Academic Press, 1985.

OTHER

Manatees of Florida. [cited May 2002]. <http://www.xtalwind.net/~cfa>.
Save the Manatees Club. [cited May 2002]. <http://www.savethemanatee.org>.

Mangrove swamp

Mangrove swamps or forests are the tropical equivalent of temperate salt marshes. They grow in protected coastal embayments in tropical and subtropical areas around the world, and some scientists estimate that 60-75 percent of all tropical shores are populated by mangroves.

The term "mangrove" refers to individual trees or shrubs that are angiosperms (flowering plants) and belong to more than 80 **species** within 12 genera and five families. Though unrelated taxonomically, they share some common characteristics. Mangroves only grow in areas with minimal wave action, high **salinity**, and low **soil** oxygen. All of the trees have shallow roots, form pure stands, and have adapted to the harsh **environment** in which they grow. The mangrove swamp or forest community as a whole is called a mangal.

Mangroves typically grow in a sequence of zones from seaward to landward. This zonation is most highly pronounced in the Indo-Pacific regions, where 30-40 species of mangroves grow. Starting from the shore-line and moving inland, the sequence of genera there is *Avicennia* followed by *Rhizophora*, *Bruguiera*, and finally *Ceriops*. In the Caribbean, including Florida, only three species of trees normally grow: red mangroves (*Rhizophora mangle*) represent the pioneer species growing on the water's edge, black mangroves (*Avicennia germinans*) are next, and white mangroves (*Laguncularia racemosa*) grow mostly inland. In addition, buttonwood (*Conocarpus erectus*) often grows between the white mangroves and the terrestrial vegetation.

Mangrove trees have made special adaptations to live in this environment. Red mangroves form stilt-like prop roots that allow them to grow at the shoreline in water up to several feet deep. Like cacti, they have thick succulent leaves which store water and help prevent loss of moisture. They also produce seeds which germinate directly on the tree, then drop into the water, growing into a long, thin seedling known as a "sea pencil." These seedlings are denser at one end and thus float with the heavier hydrophilic (water-loving) end down. When the seedlings reach shore, they take root and grow. One acre of red mangroves can produce three tons of seeds per year, and the seeds can survive floating on the ocean for more than 12 months. Black mangroves produce straw-like roots called pneumatophores which protrude out of the **sediment**, thus enabling them to take oxygen out of the air instead of the **anaerobic** sediments. Both white and black mangroves have salt glands at the base of their leaves which help in the regulation of osmotic pressure.

Mangrove swamps are important to humans for several reasons. They provide water-resistant wood used in construction, charcoal, medicines, and dyes. The mass of prop roots at the shoreline also provides an important **habitat** for a rich assortment of organisms, such as snails, barnacles,

Mangrove creek in the Everglades National Park. (Photograph by Max & Bea Hunn. Visuals Unlimited. Reproduced by permission.)

oysters, crabs, periwinkles, jellyfish, tunicates, and many species of fish. One group of these fish, called mud skippers (*Periophthalmus*), have large bulging eyes, seem to skip over the mud, and crawl up on the prop roots to catch insects and crabs. Birds such as egrets and herons feed in these productive waters and nest in the tree branches. Prop roots tend to trap sediment and can thus form new land with young mangroves. Scientists reported a growth rate of 656 feet (200 m) per year in one area near Java. These coastal forests can be helpful **buffer** zones to strong storms.

Despite their importance, mangrove swamps are fragile ecosystems whose ecological importance is commonly unrecognized. They are being adversely affected worldwide by increased **pollution**, use of herbicides, filling, **dredging**, channelizing, and **logging**. *See also* Marine pollution; Wetlands

[*John Korstad*]

RESOURCES
BOOKS

Castro, P., and M. E. Huber. *Marine Biology*. St. Louis: Mosby, 1992.

Nybakken, J. W. *Marine Biology: An Ecological Approach*. 2d ed. New York: Harper & Row, 1988.

Tomlinson, P. B. *The Botany of Mangroves*. Cambridge: Cambridge University Press, 1986.

Smith, R. E. *Ecology and Field Biology.* 4th ed. New York: Harper & Row, 1990.

PERIODICALS

Lugo, A. E., and S. C. Snedaker. "The Ecology of Mangroves." *Annual Review of Ecology and Systematics* 5 (1974): 39–64.
Rützler, K., and C. Feller. "Mangrove Swamp Communities." *Oceanus* 30 (1988): 16–24.

Manure
see **Animal waste**

Manville Corporation
see **Asbestos**

Marasmus

A severe deficiency of all nutrients, categorized along with other protein energy malnutrition disorders. Marasmus, which means "to waste" can occur at any age but is most commonly found in neonates (children under one year old). Starvation resulting from marasmus is a result of protein and carbohydrate deficiencies. In developing countries and impoverished populations, early weaning from breast feeding and over dilution of commercial formulas places neonates at high risk for getting marasmus.

Because of the deficiency in intake of all dietary nutrients, metabolic processes--especially liver functions--are preserved, while growth is severely retarded. Caloric intake is too low to support metabolic activity such as protein synthesis or storage of fat. If the condition is prolonged, muscle tissue wasting will result. Fat wasting and **anemia** are common and severe. Severe vitamin A deficiency commonly results in blindness, although if caught early, this process can be reversed. Death will occur in 40% of children left untreated.

Mariculture

Mariculture is the cultivation and harvest of marine **flora** and **fauna** in a controlled saltwater **environment**. Sometimes called marine fish farming, marine **aquaculture**, or aquatic farming, mariculture involves some degree of human intervention to enhance the quality and/or quantity of a marine harvest. This may be achieved by feeding practices, protection from predators, breeding programs, or other means.

Fish, crustaceans, salt-water plants, and shellfish may be farm raised for bait, fishmeal and fish oil production, scientific research, **biotechnology** development, and repopulating threatened or **endangered species**. Ornamental

fish are also sometimes raised by fish farms for commercial sales. The most widespread use of aquaculture, however, is the production of marine life for human food consumption. With seafood consumption steadily rising and **overfishing** of the seas a growing global problem, mariculture has been hailed as a low-cost, high-yield source of animal-derived protein.

According to the Fisheries Department of the United Nations' Food and Agriculture Organization (FAO), over 33 million metric tons of fish and shellfish encompassing 220 different **species** are cultured (or farmed) worldwide, representing an estimated \$49 billion in 1999. Pound for pound, China leads the world in aquaculture production with 32.5% of world output. In comparison, the United States is only responsible for 4.4% of global aquaculture output by weight. Just 7% of the total U.S. aquatic production (both farmed and captured resources) is attributable to aquaculture (compared to 62% of China's total aquatic output).

In the United States, Atlantic **salmon** and channel catfish represent the largest segments of aquaculture production (34% and 40%, respectively, in 1997). Though most farmed seafood is consumed domestically, the United States imports over half of its total edible seafood annually, representing a \$7 billion annual trade deficit in 2001. The Department of Commerce launched an aquaculture expansion program in 1999 with the goal of increasing domestic seafood supply derived from aquaculture production to \$5 billion annually by the year 2025. According to the U.S. Joint Subcommittee on Aquaculture, U.S. aquaculture interests harvested 842 million pounds of product at an estimated value of \$987 million in 1999.

To encourage further growth of the U.S. aquaculture industry, the National Aquaculture Act was passed in 1980 (with amendments in 1985 and 1996). The Act established funding and mandated the development of a national aquaculture plan that would encourage "aquaculture activities and programs in both the public and private sectors of the economy; that will result in increased aquacultural production, the coordination of domestic aquaculture efforts, the **conservation** and enhancement of aquatic resources, the creation of new industries and job opportunities, and other national benefits."

In the United States, aquaculture is regulated by the **U.S. Department of Agriculture** (USDA) and the Department of Commerce through the National Marine Fisheries Service (NMFS), **National Oceanic and Atmospheric Administration** (NOAA). State and local authorities may also have some input into the location and practices of mariculture facilities if they are located within an area governed by a Coastal Zone Management Plan (CZMP). Coastal Zone Management Plans, as authorized by the **Coastal Zone**

Management Act (CZMA) of 1972, allow individual states to determine the appropriate use and development of their respective coastal zones. Subsequent amendments to the CZMA have also made provision for states to be eligible for federal funding for the creation of state plans, procedures, and regulations for mariculture activities in the coastal zone.

Tilapia, catfish, striped bass, yellow perch, walleye, salmon, and trout are just a few of the fresh and salt-water finned fish species farmed in the United States. Crawfish, shrimp, and shellfish are also cultured in the U.S. Some shellfish, such as oysters, mussels, and clams, are "planted" as juveniles and farmed to maturity, when they are harvested and sold. Shellfish farmers buy the juvenile shellfish (known as "spat") from shellfish hatcheries and nurseries. Oysters and mussels are attached to lines or nets and put in a controlled ocean environment, while clams are buried in the beach or in sandy substrate below low tide. All three of these shellfish species feed on **plankton** from salt water.

But just as overfishing takes a toll on terrestrial **natural resources**, aquaculture without appropriate environmental management can damage native ecosystems. Farmed fish are raised in open-flow pens and nets. Because large populations of farmed fish are often raised in confined areas, disease spreads easily and rapidly among them. And farmed fish often transmit sea lice and other **parasites** and diseases to wild fish, wiping out or crippling native stock. Organic **pollution** from **effluent**, the waste products from farmed fish, can build up and suffocate marine life on the sea floor below farming pens. This waste includes fish feces, which contributes to **nutrient loading**, and **chemicals** and drugs used to keep the fish disease free and promote growth. It also contains excess fish food, which often contains dyes to make farmed fish flesh more aesthetically analogous to its wild counterparts.

Farmed fish that escape from their pens interbreed with wild fish and weaken the genetic line of the native stock. If escaped fish are diseased, they can trigger outbreaks among indigenous marine life. Infectious Salmon **Anemia**, a viral disease that has plagued fish farms in New Brunswick and Scotland in the 1990s, was eventually found in native salmon. In 2001, the disease first appeared at U.S. Atlantic salmon farms off the coast of Maine.

The use of drugs in farmed fish and shellfish intended for human consumption is regulated by the U.S. **Food and Drug Administration** (FDA). In recent years, **antibiotic resistance** has been a growing issue in aquaculture, as fish have been treated with a wide array of human and veterinary antibiotic drugs to prevent disease.

The commercial development of genetically-engineered, or transgenic, farmed fish is also regulated by FDA. As of May 2002, no transgenic fish had been cleared by FDA for human consumption. The impact transgenic fish may have on the survival and reproduction of native species will have to be closely followed if and when commercial farming begins.

As mandated by the 1938 Mitchell Act, the NMFS funds 25 salmon hatcheries in the Columbia River Basin of the Pacific Northwest, the largest federal marine fishery program in the United States. These aquaculture facilities were originally introduced to assist in repopulation of salmon stocks that had been lost to or severely hampered by hydroelectric dam projects. However, some environmentalists charge that the salmon hatcheries may actually be endangering wild salmon further, by competing for local **habitat** and weakening the genetic line of native species.

Without careful resource management, aquaculture may eventually take an irreversible toll on other non-farmed marine species. Small pelagic fish, such as herring, anchovy, and chub, are captured and processed into fish food compounds for high-density carnivorous fish farms. According to the FAO, at its current rate, fish farming is consuming twice as many wild fish in feeding their domestic counterparts as aquaculture is producing in fish harvests—an average of 1.9 kg of wild fish required for every kilogram of fish farmed.

No matter how economically sound mariculture has been, it has also led to serious habitat modification and destruction, especially in mangrove forests. In the past twenty years, the area of mangrove forests have dwindled by 35% worldwide. Though some of that loss is due to active **herbicide** control of mangroves, their conversion to salt flats, and the industrial harvesting of forest products (wood chips and lumber), mariculture is responsible for 52% of the world's mangrove losses.

Mangrove forests are important to the environment because these ecosystems are **buffer** zones between saltwater areas and freshwater/land areas. Mangroves act as **filters** for agricultural nutrients and pollutants, **trapping** these contaminants before they reach the deeper waters of the ocean. They also prevent coastal **erosion**, provide spawning and nursery areas for fish and shellfish, host a variety of migratory **wildlife** (birds, fish, and mammals), and support habitats for a number of endangered species.

Shrimp farming, in particular, has played a major role in mangrove forest reduction. Increasing from 3% to 30% in less than 15 years, commercial shrimp farming has impacted coastal mangroves profoundly by cutting down mangrove trees to create shrimp and prawn ponds. In the Philippines, 50% of the mangrove environments were converted to ponds and between 50% and 80% of those in Southeast Asia were lost to pond culture as well.

Shrimp mariculture places high demands on resources. It requires large supplies of juvenile shrimp, which can seriously deplete natural shrimp stocks, and large quantities of shrimp meal to feed them. There also is considerable waste derived from shrimp production. This can pump organic matter and nutrients into the ponds, causing eutrophication, which causes algae bloom and oxygen depletion in the ponds themselves or even downstream. Many shrimp farmers need to pump pesticides, fungicides, parasiticides, and algicides into the ponds between harvests to sterilize them and mitigate the effects of nutrient loading. Shrimp ponds also have extremely short life spans, usually about 5–10 years, forcing their abandonment and the cutting of more mangrove forests to create new ponds.

Mariculture also limits other marine activities along coastal waters. Some aquaculture facilities can occupy large expanses of ocean along beaches which become commercial and recreational no-fish zones. These nursery areas are also sensitive to disturbances by recreational activities like boating or swimming and the introduction of pathogens by domestic or farm animals.

[*Paula Anne Ford-Martin*]

RESOURCES
BOOKS

Food and Agriculture Organization of the United Nations. *The State of World Fisheries and Aquaculture* Rome, Italy: FAO, 2000. [cited June 5, 2002]. <http://www.fao.org/DOCREP/003/X8002E/X8002E00.htm>.

Jahncke, Michael L. et al.*Public, Animal, and Environmental Aquaculture Health Issues.* New York: John Wiley & Sons, 2002.

Olin, Paul. "Current Status of Aquaculture in North America." From *Aquaculture in the Third Millennium: Technical Proceedings of the Conference on Aquaculture in the Third Millennium, Bangkok, Thailand. 20-25 February 2000.*Rome, Italy: FAO, 2000.

PERIODICALS

Barcott, Bruce, and Natalie Fobes. "Aquaculture's Troubled Harvest." *Mother Jones* 26, no.6 (November –December 2001): 38 (8).

Naylor, Rosamond, et al. "Effect of Aquaculture on World Fish Supplies."-*Nature*(June 29, 2000).

OTHER

"National Aquaculture Policy, Planning, and Development." 16 USC 2801. [cited June 4, 2002]. <http://www.access.gpo.gov/uscode/title16/chapter48_.html>.

National Marine Fisheries Service, National Oceanic and Atmospheric Administration. *Aquaculture.* [cited July 2002]. <http://www.nmfs.noaa.-gov/aquaculture.htm>.

ORGANIZATIONS

The Northeast Regional Aquaculture Center, University of Massachusetts Dartmouth, Violette Building, Room 201 285 Old Westport Road, Dartmouth, MA USA 02747-2300 (508) 999-8157, Fax: (508) 999-8590, Toll Free: (866) 472-NRAC (6722), Email: nrac@umassd.edu, http://www.umassd.edu/specialprograms/NRAC/

Marine ecology and biodiversity

Understanding the nature of ocean life and the patterns of its diversity represents a difficult challenge. Not only are there technical difficulties involved with studying life under water (high pressure, need for oxygen tanks, lack of light), there is an urgency to develop a greater understanding of marine life as links between ocean processes and the larger patterns of terrestrial life become more well known. Our current understanding of oceanic life is based on three principal concepts: size, complexity, and spatial distribution.

Our knowledge about the size of the ocean's domain is grounded in three great discoveries of the past few centuries. When Magellan first circumnavigated the earth he inadvertently found that the oceans were a continuous water body, rather than a series of discrete bodies of water. Some time later, the encompassing nature of the world oceans was further clarified by the discovery that the oceans were a chemically uniform aqueous system. All of the principal ions (sodium, chloride, and sulfate) exist in the same concentrations. The third discovery, still underway, is that the ocean is composed of comparatively immense ecological systems. Thus in most ways the oceans are a unified system which is the first defining characteristic of the marine **environment**.

There is, however, a dichotomy between the integral nature of the ocean and the external forces played upon it. Mechanical, thermodynamic, chemical, and biological forces create variation through such things as differential heating, Coriolis force, wind, dissolved gases, **salinity** differences, and evaporation. The actions in turn set controls in motion which move toward physical equilibrium through feedback mechanisms. Those physical changes then interact with biological systems in nonlinear ways, that is, out of synchronization with the external stimuli and become quite difficult to predict. Thus, we have the second broad characteristic of the oceans, complexity.

The third major aspect of ocean life is that life itself is sparse in terms of the overall volume of the oceans, but locally productive systems can create immense populations and/or sites with exceptionally high **species** diversity. Life is arranged in active layers dictated by nutrients and light in the horizontal planes, and by vertical current (downwelling and upwelling) in the vertical planes. Life decreases through the depth zones from the epipelagic zone in the initial 328 ft (100 m) of the water column to the bottom layers of water, and then jumps again at the benthic layer at the water-substrate interface. Life also decreases from the littoral zones along the world's shorelines to the open ocean, interrupted by certain areas with special life supporting systems, like floating sargasso weed beds.

In the past twenty years the focus of **conservation** has shifted to include not only individual species or habitats, but to a phenomenon called biological diversity, or **biodiversity** for short. Biological diversity encompasses from three to four levels. Genetic diversity is the level of genotypic differences within all the individuals that constitute a population of organisms; species diversity refers to the number of species in an area; and community diversity to the number of different community types in a landscape. The highest level, landscape diversity has not frequently been applied in aqueous environments and will not be discussed here.

Commonly, species diversity is interpreted as biological diversity, and since few marine groups, except marine mammals, have had very much genetic work done, and community functions are only well known from a few systems, it is the taxonomic interpretation of diversity that is most commonly discussed (e.g., species or higher taxonomic levels such as families and classes, orders and phyla). Of all the species that we know, roughly 16% are from the seas. General diversity patterns in the sea are similar to those on land, there are more smaller than larger species, and there are more tropical species than temperate or polar species. There are centers of diversity for specific taxa, and the structure of communities and ecosystems is based on particular patterns of energy availability. For example, **estuary** systems are productive due to importation of **nitrogen** from the land, coral reefs are also productive, but use scarce nutrients efficiently by specially adapted filter feeding mechanisms. Abyssal communities, on the other hand, depend on their entire energy supply from **detritus** fall from upper levels in the ocean. Perhaps the most specifically adapted of all life forms are the hydrothermal vent communities that employ **chemosynthesis** rather than **photosynthesis** for primary production. Water temperature, salinity, and pressure create differing ecosystems in ways that are distinctly different from terrestrial systems. In addition, the boundaries between systems may be dynamic, and are certainly more difficult to detect than on land.

Most marine biodiversity occurs at higher taxonomic levels, while the land holds more species, most of them are arthropods. Most of the phyla (32 of 33) that we now know are either marine, or both marine and non-marine, while only one is exclusively non-marine. Thus most of the major life plans exist in the sea.

We are now learning that the number of species in the ocean is probably underestimated as we discover more cryptic species, very similar organisms that are actually distinct, many of which have been discovered on the deep ocean floor. This is one of the important diversity issues in the marine environment. Originally, the depths were cast as biological deserts, however, that view may have been promoted by a lack of samples, the small size of many benthic invertebrates, and the low density of benthic populations in the deep sea beds.

Improved sampling since the 1960s changed that view to one of the ocean floor as a highly species diverse environment. The deep sea is characterized by a few individuals in each of many species; rarity dominates. Whereas, in shallow water benthic environments, there are large, dense populations dominated by a few species. At least three theoretical explanations for this pattern have been made. The stability-time hypothesis suggests that ocean bottoms have been very stable environments over long periods of time. This condition causes very finely tuned adaptations to narrow niches, and results in many closely related species. The disturbance or "cropper" hypothesis suggests that intense predation of limited food sources prevents populations from reaching high levels and because the food source is dominated by detrital rain, generalist feeders abound and compete for the same food, which results in only small differences between species. A third hypothesis is that the area of the deep sea bed is so large it supports many species, following from generalizations made by the species area relationship concept used in **island biogeography** theory. The picture of species number and relative rarity is still not clearly understood.

In general, some aspects of marine biology are well studied. Rocky intertidal life has been the subject of major ecological research and yielded important theoretical advances. Similarly, coral reefs are the subject of many studies of life history **adaptation** and evolutionary biology and **ecology**. Physiology and morphology research has used many marine animals as examples of organisms' functions under extreme conditions. This new found knowledge is timely. Up until now we have considered the oceans as an inexhaustible source of food and a sink for our wastes, yet we now realize they are neither. Relative to the land, the sea is in good ecological condition, but to prevent major ecological problems in the marine environment we need to increase human knowledge rapidly and manage our behavior toward the oceans very conservatively, which is a difficult task under the conditions where the ocean is treated as a common resource.

[*David Duffus*]

Marine Mammals Protection Act (1972)

The Marine Mammals Protection Act (MMPA) was initially passed by Congress in 1972 and is the most comprehensive federal law aimed at the protection of marine mammals. The MMPA prohibits the taking (i.e., harassing, **hunting**, capturing, or killing, or attempting to harass, hunt, capture, or kill) on the high seas of any marine mammal by persons or vessels subject to the jurisdiction of the United States. It also prohibits the taking of marine mammals in

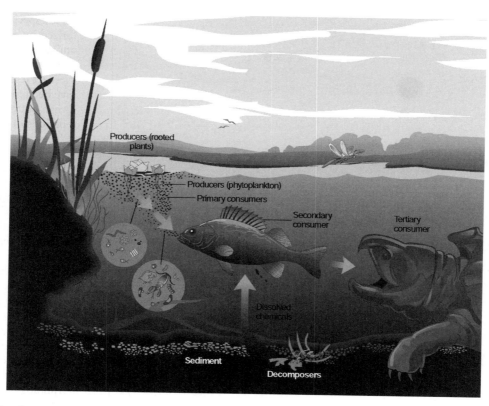

An example of a marine ecosystem. (Illustration by Hans & Cassidy.)

waters or on land subject to United States jurisdiction and the importation into the United States of marine mammals, parts thereof, or products made from such animals. The MMPA provides that civil and criminal penalties apply to illegal **takings**.

The MMPA specifically charges the National Marine Fisheries Service (NMFS) with responsibility for the protection and **conservation** of marine mammals. The NMFS is given statutory authority to grant or deny permits to take **whales**, **dolphins**, and other mammals from the oceans.

The original legislation established "a moratorium on the taking of marine mammals and marine mammal products...during which time no permit may be issued for the taking of any marine mammal and no marine mammal may be imported into the United States." Four types of exceptions allowed for limited numbers of marine mammals to be taken: (1) animals taken for scientific review and public display, after a specified review process; (2) marine mammals taken incidentally to **commercial fishing** operations prior to October 21, 1974; (3) animals taken by Native Americans and Inuit Eskimos for subsistence or for the production of traditional crafts or tools; and (4) animals taken under a temporary

exemption granted to persons who could demonstrate economic hardship as a result of MMPA (this exemption was to last for no more than a year and was to be eliminated in 1974). MMPA also sought specifically to reduce the number of marine mammals killed in purse-seine or drift net operations by the commercial tuna industry.

The language used in the legislation is particularly notable in that it makes clear that the MMPA is intended to protect marine mammals and their supporting **ecosystem**, rather than to maintain or increase commercial harvests: "[T]he primary objective in management should be to maintain the health and **stability** of the marine ecosystem. Whenever consistent with this primary objective, it should be the goal to obtain an optimum sustainable population keeping in mind the optimum **carrying capacity** of the habitat."

All regulations governing the taking of marine mammals must take these considerations into account. Permits require a full public hearing process with the opportunity for judicial review for both the applicant and any person opposed to the permit. No permits may be issued for the taking or importation of a pregnant or nursing female, for taking in an inhumane manner, or for taking animals on the **endangered species** list.

Subsidiary legislation and several court decisions have modified, upheld, and extended the original MMPA: *Globe Fur Dyeing Corporation v. United States* upheld the constitutionality of the statutory prohibition of the killing of marine mammals less than eight months of age or while still nursing.

In *Committee for Humane Legislation v. Richardson*, the District of Columbia Court of Appeals ruled that the NMFS had violated MMPA by permitting tuna fishermen to use the purse-seine or drift net method for catching yellowfin tuna, which resulted in the drowning of hundreds of thousands of porpoises.

Under the influence of the Reagan Administration, MMPA was amended in 1981 specifically to allow this type of fishing, provided that the fishermen employed "the best marine mammal safety techniques and equipment that are economically and technologically practicable." The Secretaries of Commerce and the Interior were empowered to authorize the taking of small numbers of marine mammals, provided that the **species** or population stocks of the animals involved were not already depleted and that either Secretary found that the total of such taking would have a negligible impact.

The 1984 reauthorization of MMPA continued the tuna industry's general permit to kill incidentally up to 20,500 porpoises per year, but provided special protection for two threatened species. The new legislation also required that yellowfin tuna could only be imported from countries that have rules at least as protective of porpoises as those of the United States.

In *Jones v. Gordon* (1985), a federal district court in Alaska ruled in effect that the **National Environmental Policy Act** provided regulations which were applicable to the MMPA permitting procedure. Significantly, this decision made an **environmental impact statement** mandatory prior to the granting of a permit.

Presumably owing to the educational, organizing, and lobbying efforts of environmental groups and the resulting public outcry, the MMPA was amended in 1988 to provide a three-year suspension of the "incidental take" permits, so that more ecologically responsible standards could be developed. Subsequently, Congress decided to prohibit the drift netting method as of the 1990 season.

In 1994, the MMPA was amended to include more concrete definitions of harassment levels and grouped them as level A harassment (potential to hurt a wild marine mammal) and level B harassment (potential to disrupt their **environment** or biology). The 1994 amendments also constructed new restrictions of photography permits and states

under harassment level B that scientific research must limit its influence on the marine mammals being studied.

[*Lawrence J. Biskowski*]

RESOURCES

BOOKS

Dolgin, E. L., and T. G. P. Guilbert, eds. *Federal Environmental Law.* St. Paul, MN: West Publishing Co., 1974.

Freedman, W. *Federal Statutes on Environmental Protection.* New York: Quorum Books, 1987.

PERIODICALS

Hofman, J. "The Marine Mammals Protection Act: A First of Its Kind Anywhere." *Oceanus* 32 (Spring 1989): 7–16.

OTHER

National Marine Fisheries Services. [cited June 2002]. <http://www.nmfs.noaa.gov>.

Marine pollution

Marine **pollution** is a major threat to any organism living in or depending upon the ocean. Human impact on coastal and open ocean habitats comes in many forms: **nutrient loading** from agricultural **runoff** and sewage discharges, toxic chemical inputs from industry and agriculture, **petroleum** spills, and inert solid wastes. While there has been some recognition of the destruction of marine systems from pollution, regulations are often weak or are not enforced. Recent efforts have lead to slow recovery of some coastal areas, but many of the detrimental practices continue and some systems may never recover.

Nutrient loading is perhaps the most well-studied form of pollution, and its biological consequences have been observed and documented. Algal blooms, including red tides, have been attributed to elevated nutrient levels in coastal systems. These blooms, through their **respiration** and **decomposition**, can deplete the levels of **dissolved oxygen** in the waters to almost zero, killing **zooplankton**, fish, and shellfish. These nutrients often come from runoff of agricultural fertilizers, the use of which has increased sevenfold since 1950. It is estimated that approximately 25 percent of the 46.8 million metric tons of **fertilizer** used in the United States annually enters rivers and coastal waters.

Another source of nutrients is the **discharge** of sewage. The United States and England are the only countries that dump **sludge** into the oceans, averaging 17 million metric tons annually between them. Throughout the world, however, raw sewage is released into rivers and coastal habitats by many countries, leading to algal blooms and increased **biochemical oxygen demand**. BOD, the rate at which oxygen disappears from a sample of water, increases dramatically with loading of organic material such as sewage, and results in the lowering of dissolved oxygen levels. The wastes of domesticated animals may also have a major impact in some systems. A single cow produces approximately 31 lb (14 kg) of waste per day, the equivalent of ten people. When discharged into rivers or coastal waters, wastes produced by large herds or **feedlots** may have substantial effects.

Boston Harbor is considered to be one of the most polluted harbors in the United States. Since Europeans settled in the Boston area, domestic wastes have been discharged directly into the harbor. As Boston's population grew, so did the dumping, a problem exacerbated by the growth of industry in the late nineteenth and early twentieth centuries. Fish and shellfish in the harbor contain toxic levels of polychlorinated biphenyl (PCB) and **heavy metals**, BOD levels are astronomical, and dissolved oxygen levels are low. In 1984, the Massachusetts **Water Resources** Authority (MWRA) was formed and has managed to improve **water quality** of the harbor slightly. Recommendations of the MWRA have been hard to enact, however. **Sewage treatment** modifications have been met with **Environmental Protection Agency** (EPA) and State objections, shelving the project while courts decide the fate of environmental quality in Boston Harbor.

Toxic substances are introduced into the marine **environment** from various sources, some of which may be hundreds of kilometers away. Heavy metals, pesticides, and **acid rain** threaten not only coastal and estuarine systems, but also life in the open ocean. Heavy metals occur in many forms, some of which are soluble in seawater. These soluble compounds may not be the forms in which they were originally released into the environment, often making their sources difficult to determine. Many heavy metals are released in industrial effluents, especially from chemical plants, smelters, and mining runoff. These compounds may affect humans directly through contact, or indirectly, from the consumption of fish and shellfish, where metals often accumulate in tissues.

Between the late 1930s and the mid-1950s, a Japanese chemical company manufacturing acetaldehyde discharged **mercury** into Minamata Bay, where it formed a soluble compound that accumulated in fish. It was not until 50 people died of **Minamata disease** and hundreds were left with debilitating nervous disorders from eating poisoned fish that environmental studies were initiated. Finally, in 1969, the plant was closed. Other metals, such as zinc, **cadmium**, **copper**, and silver, are commonly discharged into marine systems by industry. Not only are these metals toxic by themselves, but synergistic effects compound their toxicity.

Acid precipitation is rain, snow, or fog that has a lower **pH** than normal and is caused by inputs of nitric and sulfuric acid into the **atmosphere** from manufacturing and the burning of **fossil fuels**, as in automobiles. These causes may be far from the area of impact since prevailing winds can carry pollutants considerable distances. The ocean has a high **buffer** capacity, that is, the ability to neutralize many of the acid inputs. Therefore, most of the severe effects of acid rain are observed in freshwater lakes and rivers. Some estuaries, however, may also be seriously impacted, as has been observed in the upper **Chesapeake Bay**. The pH of a river feeding the Bay dropped from 6.3 to 5.8 between 1972 and 1978. Juvenile and spawning striped bass may not be able to tolerate such high acidity. Precipitation of nitrogen-based acid also increases nutrient loading in aquatic systems.

Toxic organic compounds, especially pesticides and a family of **chemicals** known as PCBs, have been shown to have serious effects on marine systems. Runoff has introduced considerable amounts of PCBs, DDT, and many other synthetic organic compounds into coastal areas. These compounds may persist for many years. In 1976, Congress banned the manufacture of PCBs, but they are still found today as coolants in older transformers and buried in sediments. It is estimated that 1 percent of PCBs used have reached the ocean. In 1987–1988, 700 bottlenose **dolphins** washed up on the U.S. Atlantic coast and some were found to have elevated levels of PCBs and DDE (a form of DDT). Biologists claim that these compounds inhibited the dolphins' immune systems, making them vulnerable to infections. In 1975, workers at a chemical plant on the Chesapeake were poisoned when the insecticide **Kepone** was leaked or dumped into the bay. The Kepone spread downstream where it poisoned fish and shellfish.

One of the most publicized sources of marine pollution is that caused by petroleum products. While large **oil spills** can devastate a local area, equally important is the discharge of crude oil while cleaning bilges and emptying tanks at sea. Over the past decade, an average of approximately 32 million gallons (120 million l) of oil have been spilled annually. The Gulf of Mexico has recorded the most spills, while spills in the Persian Gulf average 2 million gallons (7 million l) per year. Since tanker ports and refineries are, by necessity, located on the coast, these sensitive areas receive considerable damage from the spills. The damage to marine life is staggering. Seabirds are killed by the hundreds of thousands annually, their oil-matted plumage making flight impossible and exposing them to hypothermia. Oil-soaked fur of marine mammals loses its water repellency, also leading to death by hypothermia. Ingestion of oil by fishes, birds, and mammals also may result in death.

Hundreds of tons of inert solid wastes are dumped into the oceans from ships annually. Of these, **plastics** and **polystyrene** (styrofoam) are deadly to marine life. Often floating for hundreds of miles and lasting for many years, plastics are frequently mistaken for food by fishes, turtles, and mammals, and proceed to either interfere with subsequent feeding or strangle the consumer. It has been estimated that plastics and discarded fishing gear, such as monofilament line and discarded nets, kill one million seabirds and 150,000 marine mammals each year.

Beginning with the Refuse Act of 1899 and the **Water Pollution** Act of 1948, there have been efforts to remedy the polluted oceans. Not until 1972, however, was legislation drafted that was powerful enough, and sufficient monies appropriated, to effect any change. With the Federal Water Pollution Act, or **Clean Water Act**, improvement of water quality in the United States began. By 1985, five of the goals of the 1972 Act were reached, and since then, the Clean Water Act has been rewritten to reflect increasing national concerns.

Pollution of the world's oceans had become so pervasive that an international convention was convened in 1973 to establish laws governing the discharge of all wastes into the ocean. The International Convention for the Prevention of Pollution from Ships (1978), commonly known as MARPOL, and subsequent annexes to the convention, covers all **garbage** discharged by ships at sea beyond three miles from shore. While it is too soon to tell, adherence to the MARPOL agreement should significantly reduce the amount of solid wastes polluting the world's oceans.

Boston Harbor, the Chesapeake Bay, and the **Mediterranean Sea** are the most publicized examples of marine pollution. However, the problem is widespread. Many estuaries in Southeast Asia, many coastal areas in Japan, and the waters off Rio de Janeiro are also areas where historical and current use of the waters as dumping grounds has left them in disastrous conditions. The Golden Horn estuary around Istanbul, Turkey, has been declared dead, containing no living organisms and posing a threat to surrounding waters at the eastern end of the Mediterranean.

Since humans have been living along the coasts and using the seas as waste dumping grounds for thousands of years, these areas are the most heavily impacted. What makes matters worse is that coastal waters are vital spawning and nursery grounds for most of the commercially harvested fish and shellfish in marine systems. Destruction of these areas has significant nutritional and economic repercussions. Examples of the world-wide destruction of marine habitats due to pollution are numerous, but, with increased international efforts to limit inputs and rehabilitate damaged areas, improvements in the health of marine systems may be seen. *See also* Bioaccumulation; Estuary; Food chain/web; Heavy metals and heavy metal poisoning

[*William G. Ambrose Jr. and Paul E. Renaud*]

RESOURCES
BOOKS

Clark, R. B. *Marine Pollution.* New York: Oxford University Press, 1989.
Nybakken, J. W. *Marine Biology: An Ecological Approach.* 2nd ed. New York: Harper and Row, 1988.

PERIODICALS
Dolin, E. J. "Boston's Murky Political Waters." *Environment* 34 (1992): 6–33.

Marine protected areas

Marine protected areas (MPAs) are marine environments that enjoy certain federal, state, local, or tribal regulation and **conservation** management programs by virtue of their unique ecosystems and/or cultural resources.

Known variously as marine reserves, marine sanctuaries, and/or no-take areas, marine protected areas may include national marine sanctuaries, marine features of national parks and monuments, marine **wildlife** refuges, fishery reserves, estuarine reserves, or state conservation areas and reserves. They may be regulated as no-take areas, meaning that marine resources and/or marine life cannot be removed from these sanctuaries. Or, they may allow fishing and other commercial or recreational activities within certain guidelines and limitations. Management of an MPA is determined on a site-by-site basis by one or more federal, state, territorial, local, and/or tribal regulatory authorities.

Marine protected areas are not just a reaction to the effects of industrialization. They have existed in some form among traditional fishing cultures for centuries. Tabu or kapu areas, controlled by clans or chiefs in Oceania, created no-take zones. When westernization occurred, many of these areas disappeared. In recent years, some of these tabu areas have been revived.

Some countries created protected areas along their coastlines through governmental decrees or legislation. The Royal **National Park** in New South Wales, **Australia**, is perhaps the oldest protected area in the world. Designated in 1879, the park prohibits the use of explosives, net-fishing, and the commercial harvesting of oysters. Certain areas restrict all **commercial fishing**. Other Australian reserves followed. Today, another Australian MPA, the **Great Barrier Reef** Marine Park, is considered one of the most successful multiple-use MPAs in the world.

Other governments soon instituted marine protected areas on their shores. New Zealand has led the world with the establishment of 291 MPAs. Africa has established 92 reserves, and Europe is rapidly catching up, with MPAs popping up in France, Spain, Greece, Albania, Bosnia, and Croatia.

Though the United States has 9,3208 mi.(150,000 km) of coastline and had some initial legislation in place, it waited until serious damage was occurring before developing a viable plan to coordinate the nation's MPA-related initiatives. Unfortunately, much of the focus of the US MPAs will first be on remediaton of many of its fragile coastal environments before concentrating on developing an inclusive plan for its extensive coastline.

In May 2000, President Clinton signed Executive Order 13158 on Marine Protected Areas, which was designed to ensure the long-term security of critical ocean habitats, marine life, and cultural resources through an integrated national network of science-based MPA selection and management. The order designated the Department of Interior and Department of Commerce as jointly responsible for the MPA mission to "(a) strengthen the management, protection, and conservation of existing marine protected areas and establish new or expanded MPAs; (b) develop a scientifically based, comprehensive national system of MPAs representing diverse U.S. marine ecosystems, and the Nation's natural and cultural resources; and (c) avoid causing harm to MPAs through federally conducted, approved, or funded activities."

Executive Order 13158 mandated the creation of a national database of MPAs. This register of nationally-recognized MPAs would promote integrated conservation efforts between federal, state, and local authorities and stakeholders. Order 13158 defined an MPA as "any area of the marine **environment** that has been reserved by federal, state, territorial, tribal or local laws or regulations to provide lasting protection for part or all of the natural and cultural resources therein." To further hone this relatively broad-based definition, the **National Oceanic and Atmospheric Administration** (NOAA) has established five criteria for inclusion in the MPA inventory:

- The site must have defined geographical boundaries and must be in federal, U.S. territorial, state, or tribal marine areas.

- The site must be marine in **nature**, meaning that it includes ocean or coastal waters, including intertidal areas, bays, and estuaries. It may also be a part of the **Great Lakes** or their connecting waters, and can include land as an integral component of the site.

- Must be established as a reserved area by Federal, state, territorial, local, or tribal law or regulation.

- Must be lasting, as in a permanent and year-round protected area.

- Must have existing laws or regulations that are designed and applied to afford the site with increased protection for part or all of the natural and cultural resources therein.

The MPA list had not yet been developed as of May 2002.

The establishment of a MPA list or inventory would bring further clarity and regulatory integration to a legacy of environmental legislation affecting ocean habitats that began with the National Marine Sanctuaries Act (NMSA) in 1972 (and its amendments and reauthorizations in 1980, 1984, 1988, 1992, 1996, and 2000). NMSA, Title III of the **Marine Protection, Research and Sanctuaries Act**, was signed by President Nixon on the 100-year anniversary of the **National Park Service**. The Act set up the criteria for national sanctuaries that today only include 13 marine areas spanning approximately 18,000 mi^2 (28,967.4 km^2). As of May 2002, U.S. nationally- designated sanctuaries include:

- *Channel Islands.* Designated as a marine sanctuary in 1980, the Channel Islands are located off the coast of Santa Barbara, California. The unique **ecosystem** of the sanctuary is home to giant kelp forests and 27 **species** of **whales** and dolphins.

- *Cordell Bank.* Cordell Bank is a submerged island on the edge of the continental shelf off the coast of San Francisco. The sanctuary, created in 1989, covers 526 mi^2 (846.5 km^2) that are home to a rich and diverse array of marine life.

- *Fagatele Bay.* A tropical **coral reef** formed by a now-extinct **volcano**, Fagatele Bay is located off of Tutuila, an island of American Somoa. The sanctuary was designated in 1986.

- *Florida Keys.* The Florida Keys National Marine Sanctuary encompasses 28,000 nautical mi^2 (51,856 km^2) surrounding the Florida Keys, and includes such unique features as coral reefs and mangroves. It was established in 1990.

- *Flower Garden Banks.* A series of three sites off the coasts of Texas and Louisiana, Flower Garden Banks was established in 1992 and 1996 as a national marine sanctuary (two sites in 1992 and the third in 1996). The area's unique coral reefs and twin salt domes host an abundance of tropical marine life.

- *Gerry E. Studds-Stellwagen Bank National Marine Sanctuary.* This nutrient-rich expanse of Massachusetts Bay is home to the endangered right and humpback whales. It was designated in 1992.

- *Gray's Reef.* Given sanctuary status in 1981 by President Carter, Gray's Reef is a sandstone reef with a "live-bottom habitat"—attached corals, sponges, and other invertebrates. The sanctuary encompasses 17 mi^2 (27.4 km^2) off the coast of Georgia.

- *Gulf of the Farallones.* The 1,255 mi^2 (2,019.7 km^2) Gulf of the Farallones National Marine Sanctuary was created in 1981. Located 30 mi. (48 km) west of San Francisco Bay's Golden Gate Bridge, the Farallones sanctuary is home to a variety of **seals**, **sea lions**, and seabirds.

- *Hawaiian Islands Humpback Whale.* An estimated two-thirds of the entire Pacific humpback whale population return to the **Hawaiian Islands** in the winter to breed, calve, and raise their young. The sanctuary was designated in 1992 to protect this important **habitat**.

• *Monitor.* The first of only two archaeological sites of the 13 U.S. marine sanctuaries, the Monitor National Marine Sanctuary is the location of the sunken remains of the U.S.S. Monitor, the first ironclad Civil War battleship. The Monitor site was the first U.S. marine sanctuary, designated in 1975 in the coastal waters off North Carolina.

• *Monterey Bay.* Another California coastal sanctuary, Monterey Bay is the largest of the marine sanctuaries at over 5,200 mi^2 (8,368.4 km^2). It was established in 1992.

• *Olympic Coast.* The Olympic Coast National Marine Sanctuary was designated in 1994 and covers 3,300 mi^2 (5,310.7 km^2) off the Washington State coast and Olympic Peninsula. It is home to a rich and complex intertidal ecosystem and a variety of seabirds and marine mammals.

• *Thunder Bay.* Thunder Bay National Marine Sanctuary and Underwater Preserve is the only designated sanctuary located in the Great Lakes. The 448 mi^2 (721 km^2) Lake Huron location, a historical shipping alley, is the site of at least 116 nineteenth and twentieth- century shipwrecks. It became a national sanctuary in 2000.

As of May 2002, NOAA had begun the designation process for a fourteenth national marine sanctuary, the Hawaiian Islands Coral Reef Ecosystem Reserve, home to an estimated 7,000 marine species.

The NMSA calls for marine use management that balances the public's right for appropriate recreational and commercial use of protected areas with habitat conservation and ecosystem needs. Some areas may allow recreational use with permits, or during specific times of the year. Others may restrict certain uses completely. Because human use of marine resources frequently conflicts with **biodiversity** and conservation goals, this is a major challenge of the ongoing management of marine protected areas.

[*Paula Anne Ford-Martin*]

RESOURCES
BOOKS

Committee on the Evaluation, Design, and Monitoring of Marine Reserves and Protected Areas in the United States, Ocean Studies Board, National Research Council. *Marine Protected Areas: Tools for Sustaining Ocean Ecosystems.* Washington, D.C.: National Academy Press, 2001.

PERIODICALS

"Presidential Documents. Executive Order 13158 of May 26, 2000." *Federal Register* Washington, DC: U.S. Government Printing Office, 2000.

OTHER

"National Marine Sanctuaries Act (NMSA) of 1972." *Code of Federal Regulations* 16 USC 1431 et seq.

U.S. Department of Commerce and U.S. Department of the Interior. *Marine Protected Areas of the United States.* [cited June 2, 2002]. <http://www.mpa.gov>.

ORGANIZATIONS

National Marine Sanctuary System, National Ocean Service, National Oceanic and Atmospheric Administration, 1305 East-West Highway, 11th Floor, Silver Spring, MD USA 20910 (301) 713-3125, Fax: (301) 713-0404, Email: nmscomments@noaa.gov, http://www.sanctuaries.nos.noaa.gov/

Marine Protection, Research and Sanctuaries Act (1972)

The Marine Protection, Research and Sanctuaries Act of 1972 (also known as the National Marine Sanctuaries Act) is a comprehensive law designed to deal with ocean resources. The law has three main sections: first, it regulates **ocean dumping**; second, it authorizes **marine pollution** research; third, it establishes the marine sanctuary program. These sanctuaries can be established to protect areas of significant **conservation**, cultural, ecological, educational, esthetic, historical, or recreational values. A fourth component to the law, added in 1990, establishes regional marine research programs.

The law established a permit process administered by the **Environmental Protection Agency** (EPA) to regulate all ocean dumping, with the exception of **dredging** materials, which require a permit from the **Army Corps of Engineers**. The act mandated that ocean dumping of sewage **sludge** and industrial waste end by 1981, but this deadline was missed. The **Ocean Dumping Ban Act** of 1988 amended the 1972 law, establishing a deadline of December 31, 1991, for the end of ocean dumping of sewage sludge. A system of escalating fees and fines was incorporated into the act to help reach the deadline. The law also immediately prohibits the dumping of **medical waste** in the ocean

The marine sanctuaries can be designated in coastal waters or in the **Great Lakes**. Thus far, 13 National Marine Sanctuaries have been designated: Channel Islands (off the coast of California), Cordell Bank (California), Fagatele Bay (American Samoa), Florida Keys (Florida), Flower Garden Banks (Louisiana/Texas), Gray's Reef (Georgia), Gulf of the Farallones (California), Humpback Whale (Hawaii), Key Largo (Florida), Looe Key (Florida), Monitor (North Carolina), Monterey Bay (California), and Sellwagen Bank (Massachusetts). Numerous additional sites are being studied for designation. The sanctuaries range in size from the 0.25 mi^2 (0.65 km^2). Fagatele Bay to the 5,312 mi^2 (13,758 km^2) at Monterey Bay (larger than Connecticut).

Designation is usually made by the Secretary of Commerce, after consultation with other agencies and the affected states' governments. State waters can be included in a sanctuary if the state agrees.

In the 1980s Congress and environmental groups were concerned with the slow pace of sanctuary designation. Congress acted to expedite the designation process, and in 1990 by-passed the usual process completely to designate the Florida Keys Marine Sanctuary, largely in response to a number of ship groundings in the area.

The Sanctuaries and Reserves Division of the **National Oceanic and Atmospheric Administration** (NOAA), in the Department of Commerce, is responsible for the administration of these sanctuaries. Management of the sanctuaries is based on a multiple use approach: the significant resources within the sanctuaries are to be protected, but uses such as diving and sport and **commercial fishing** may also continue in most cases if such uses do not harm the significant resources. Regulations as to what is allowed and not allowed are established for each particular sanctuary, depending on what resources are present and being protected. Offshore gas and **oil drilling**, for example, is not allowed in sanctuaries, but shipping (including oil tankers) often is. In addition to protection and use, the sanctuaries are also sites for research and marine education.

The act has had several amendments added, with the a complete overhaul happening in 1992. Two of those changes were that the management plans of the sanctuaries must be reviewed every five years by the Secretary and the violation fine was raised to $100,000. The most recent amendments were made in 2000. The most significant of those was the clarification of illegal acts on the sanctuaries. *See also* Environmental education; Environmental science; Recreation

[*Christopher McGrory Klyza*]

RESOURCES

BOOKS

National Oceanic and Atmospheric Administration. *National Marine Sanctuary Program.* Washington, DC: U.S. Government Printing Office, 1990.

PERIODICALS

James, A. "Watery Keep." *Outside* (January 1993): 19–20.
"Ocean Pollution Controlled." *Congressional Quarterly Almanac* 44 (1988): 160–61.

OTHER

National Marine Sanctuaries. [cited July 2002]. <http://www.sanctuaries.-nos.noaa.gov>.

Marine provinces

Marine provinces are specific geographic areas of the ocean delineated by common landform (i.e., coastal geology and position of the continental shelf) and water conditions (i.e., temperature, currents, wind patterns, and **salinity**). In addition, marine provinces may be categorized by the unique **flora** and **fauna** (or biota) they support.

With less than five percent of the ocean mapped as of early 2002, the exact number of marine provinces worldwide is unknown. The United States **Geological Survey** (USGS) specifies ten coastal marine and estuarine provinces bordering on the United States—the Acadian, Arctic, Californian, Carolinian, Columbian, Fjord, Louisianian, Pacific Insular, Virginian, and West Indian provinces. The U.S. **Environmental Protection Agency** (EPA) also uses these classifications for its **Environmental Monitoring and Assessment Program** (EMAP), but further divides the Arctic marine province surrounding Alaska into the Arctic, Bering, Aleutian, and Alaskan Provinces.

Marine provinces may be subdivided into marine biogeographical regions or ecoregions. In their classification system for marine ecosystems and habitats, the **National Oceanic and Atmospheric Administration** (NOAA) defines marine provinces as benthic (bottom floor) or pelagic (water column) zones of the ocean. These zones are then categorized by depth (i.e., intertidal, sublittoral, bathyal, abyssal, hadal, neritic, and oceanic) and ocean floor topographical features (e.g., reef, sandbar, crevice, etc.).

[*Paula Anne Ford-Martin*]

RESOURCES
BOOKS

Gibson, G.R., M.L. Bowman, J. Gerritsen, and B.D. Snyder. *Estuarine and Coastal Marine Waters: Bioassessment and biocriteria technical guidance-* Washington, DC: U.S. Environmental Protection Agency, Office of Water, EPA 822-B-00-024, 2000. Available online at http://www.epa.gov/ost/biocriteria/States/estuaries/estbiogd.html. Accessed May 15, 2002.

Sullivan Sealey, K. and G. Bustamante. *Setting Geographic Priorities for Marine Conservation in Latin America and the Caribbean.* Arlington, VA: The Nature Conservancy, 1999. Available online at http://www.bsponline.org/bsp/publications/lac/marine/Titlepage.htm Accessed May 15, 2002.

OTHER

Allee, Rebecca, et al. National Ocean and Atmospheric Administration (NOAA). *Marine and Estuarine Ecosystem and Habitat Classification.* Technical Memorandum, July 2000, NMFS-F/SPO-43.

ORGANIZATIONS

UNEP World Conservation Monitoring Centre, 219 Huntingdon Road, Cambridge, UK CB3 0DL +44 (0)1223 277314, Fax: +44 (0)1223 277136, Email: info@unep-wcmc.org, http://www.unep-wcmc.org

Marsh

see **Wetlands**

George Perkins Marshall (1801 – 1882)

American diplomat, philologist, conservationist, politician, and lawyer

A long-time diplomat, Marsh served 21 years as ambassador to Italy and a shorter term in Turkey. He was a skilled lawyer, a Congressman from Vermont, a many-times-failed businessman, a learned scholar, and master of numerous languages. He was also author of *Man and Nature: Physical Geography as Modified by Human Action*, a book that **Gifford Pinchot** called "epoch-making" and that Lewis Mumford, in *The Brown Decades*, described as "the fountainhead of the **conservation** movement." Rene Jules Dubos described Marsh himself as "the first American prophet of ecology" and not a few have ascribed to him the actual founding of the science.

Marsh was born and grew up in Woodstock, Vermont, or as he put it, he was "born in the woods." As a young person, poor eyesight turned him from avid book-worm to student of **nature**, and created his lifelong attitude: "the power most important to cultivate and, at the same time, hardest to acquire, is that of seeing what is before [you]." What he saw, especially in the over-grazed, deforested lands of Italy and Turkey, is what inspired the writing of *Man and Nature*.

The emphasis in Marsh's book is on human beings as agents of change, too often change that is detrimental to nature. "Man is everywhere a disturbing agent. Wherever he plants his foot, the harmonies of nature are turned to discords," he wrote. He went on to emphasize that "not a sod has been turned, not a mattock struck into the ground, without leaving its enduring record of the human toils and aspirations that accompany the act." The consequence of this change, Marsh wrote, is that "the earth is fast becoming an unfit home" for its human inhabitants.

Nevertheless, Marsh was also an early humanist, believing that human dominion over nature could be used constructively: he saw the purpose of his book as not only tracing human ravages of nature, but as also suggesting "the possibility and the importance of the restoration of disturbed harmonies and the material improvements of waste and exhausted regions." His writings and activities spurred others into action.

His reaction to **deforestation** directly influenced the United States Congress to establish a federal forest commission and, eventually, to the reserves now part of the vast system of national forests in the United States. While ambassador to Italy, he wrote a report on **irrigation** for the U.S. Commissioner of Agriculture that led to the formation of the **Bureau of Reclamation** in 1902. His claim that it is "desirable that some large and easily accessible region of American **soil** should remain as far as possible in its primitive

condition" influenced **wilderness** advocates such as **John Muir** and **Aldo Leopold** and was an early clarion call for the setting aside of pristine wildlands.

Though not a trained scientist nor a particularly skilled writer, Marsh sensed early the destructive capability and impact of human activities. He assessed those impacts with great clarity and did much to establish a base for intelligent restorative actions.

[*Gerald L. Young Ph.D.*]

RESOURCES
BOOKS

Curtis, J., W. Curtis, and F. Lieberman. *The World of George Perkins Marsh: America's First Conservationist and Environmentalist.* Woodstock, VT: The Countryman Press, 1982.
Strong, D. H. "The Forerunners: Thoreau, Olmsted, Marsh." In *Dreamers and Defenders: American Conservationists.* Lincoln: University of Nebraska Press, 1988.

PERIODICALS
Gade, D. W. "The Growing Recognition of George Perkins Marsh." *The Geographical Review* 73 (July 1983): 341–344.

Marshall Islands

see **Bikini atoll**

Robert Marshall (1901 – 1939)

American wilderness advocate

Bob Marshall was the son of a prominent New York lawyer who helped insert the phrase "forever wild" into the state constitution, a key to preserving the forests of the Adirondack region. His experiences in the **Adirondack Mountains** as a youth, at his family's summer place, and the early habit of taking long walks, gave early shape to his preoccupation with the importance of **wilderness**. Marshall worked diligently for the protection of wilderness in the United States for aesthetic reasons and for the psychological health of the American people by counteracting the "drabness" and "horrible banality" of modern civilization.

Marshall earned a forestry degree from Syracuse University, a Masters of Forestry from Harvard, and ultimately a doctorate in plant physiology from Johns Hopkins. He first worked for the United States **Forest Service** at an experimental station in Missoula, Montana, before leaving to continue his education in graduate school. His experience in Montana exposed him to "the working man's" view of forestry, views that conditioned his later opinions on social reform.

The death of his father left him financially independent, but Marshall returned to the Forest Service in 1932 and 1933 to work on the National Plan for American Forestry. He later left to become director of forestry for the Office of Indian Affairs and then returned as chief of the Division of **Recreation** and Lands. In his Indian Affairs position, he worked for a greater role by Native Americans in the management of the forests on their reservations. As head of the Forest Service Division of Recreation and Lands, he became one of the twentieth century's best known voices in support of protection of wilderness areas in national forests. His last achievement may have been the adoption in 1939 of the Regulation U1, governing wilderness areas and Regulation U2 governing wild areas (the two known together as the U Regulations). The Regulations were signed by Agriculture Secretary, Henry Wallace, as a Forest Service policy to endorse the recommendations developed by Marshall (and others on a committee) in an effort to protect these areas from development, including—and especially—road building (but also, **logging**, building, etc.), and to give such areas more permanence by requiring public notice and a waiting period before any alterations.

Bob Marshall's travels to Alaska in the late twenties and thirties resulted in detailed mapping of parts of the Brooks Range. This led in part to their preservation as wilderness and to a better understanding of tree growth at **timberline** in the Arctic.

Two books were written based on his travels to Alaska. The first, *Arctic Village* (1933), based on a year's residence in the Arctic village of Wiseman, earned him wide recognition, both critically and popularly, and was almost an ethnographic study of life among both the Eskimo and the white settlers in the wilderness back country of northern Alaska. The second, *Arctic Wilderness*, published in 1956 (well after his death), was more in line with his well-deserved reputation as a scholar and activist with vast knowledge of wilderness conditions.

Bob Marshall contributed also as one of the founders and financial supporters of the **Wilderness Society**. The Wilderness Society was a culmination of his call for "an organization of spirited people who will fight for the freedom of the wilderness." The Society helped to implement his dream of wilderness preservation through its decisive role in the eventual passage of the **Wilderness Act**, though that did not take place until more than two decades after Marshall's death. The Bob Marshall Wilderness system in Montana was later named in his honor.

[*Gerald L. Young Ph.D.*]

RESOURCES

BOOKS

Glover, J. M. *A Wilderness Original: The Life of Bob Marshall*. Seattle: The Mountaineers, 1986.

Marshall, R. *The People's Forests*. New York: H. Smith and R. Haas, 1933.

PERIODICALS

Jackson, D. D. "Just Plain Bob Was the Best Friend Wilderness Ever Had." *Smithsonian* 25, no. 5 (August 1994): 92–100.

Marshall, R. "The Problem of the Wilderness." *Scientific Monthly* 30, (February 1930): 141–148.

Marshall, R. "The Universe of the Wilderness is Vanishing." *The Living Wilderness* 35, no. 114 (Summer 1971): 8–14.

Nash, R. "The Strenuous Life of Bob Marshall." *Forest History* 10, no. 3 (October 1966): 18–25.

Mass burn

Mass burn refers to the **incineration** of unsorted municipal waste in a Municipal Waste Combustor (MWC) or other incinerator designated to burn only waste from municipalities. This **waste management** method avoids the expensive and unpleasant task of sorting through the **garbage** for unburnable materials. All waste received at the facility is shredded into small pieces and fed into the incinerator. Steam produced in the incinerator's boiler can be used to generate electricity or to heat nearby buildings. The residual ash and unburnable materials, representing about 10-20 percent of the original volume of waste, are taken to a **landfill** for disposal. Mass burn incineration also has several drawbacks. Since the waste is unsorted, it often generates more polluting emissions than sorted waste, and it is more likely to corrode burner grates and chimneys. The residual ash and unburned materials may be toxic and require special treatment.

The **Environmental Protection Agency** (EPA) establishes standards for the burning of municipal waste. A MWC is designed so that it cannot burn continuously 24 hours a day. Waste cannot be fed to the unit nor can ash be removed while **combustion** is occurring. As a result, burning occurs in batches or spurts. A large MWC plant has a capacity greater than 250 tons per day. A very large MWC plant has a capacity of about 1,100 tons per day.

Municipal solid waste includes household, commercial or retail, and institutional waste. **Household waste** includes material discarded by single and multiple residential dwellings, hotels, motels, and other similar permanent or temporary housing establishments. Commercial or retail waste includes material discarded by stores, offices, restaurants, warehouses, and nonmanufacturing activities at industrial facilities. Institutional waste includes material discarded by schools, hospitals, and nonmanufacturing activities at prisons, government facilities, and other similar establishments.

Household, commercial or retail, and institutional wastes do not include sewage, wood pallets, construction and demolition wastes, industrial process or manufacturing wastes, or motor vehicles (including motor vehicle parts or vehicle fluff). Municipal **solid waste** also does not include solely segregated medical wastes. However, any mixture of segregated medical wastes and other wastes which contains more than 30 percent **medical waste** discards is considered to be acceptable as municipal solid waste.

The *Code of Federal Regulations* (CFR) specifically defines a Municipal Waste Combustor (MWC) unit as any device that combusts solid, liquid or gasified municipal solid waste including but not limited to field-erected incinerators (with or without heat recovery), modular incinerators (starved air or excess air), boilers (steam generating units), furnaces (whether suspension-fired, grate-fired, mass-fired or fluidized bed-fired), and gasification/combustion units. This does not include combustion units, engines or other devices that combust landfill gases collected by landfill gas collection systems.

According to the CFR, terms connected with a MWC must be clearly defined. MWC **acid** gases are defined as all acid gases emitted in the exhaust gases from MWC units including **sulfur dioxide** and **hydrogen** chloride gases. MWC organics are defined as organic compounds emitted in the exhaust gases from MWC units including total tetra-(4) through octa- (8) chlorinated di-benzo-p-dioxins and dibenzofurans. This class of chlorinated **chemicals** (where **chlorine** gas binds with certain organic matter, especially **carbon** atoms) includes **dioxin** or TCDD.

EPA also identifies some of following as potential discharges from a MWC: hydrogen chloride, sulfur dioxide, refuse-derived fuel (RDF), **nitrogen oxides**, and **carbon monoxide**. The amounts of these emissions discharged are closely monitored. EPA sets allowable standards and compliance schedules for each **emission**. These standards are legally enforceable regulations.

EPA also sets standards for facility operations. These standards include guidelines for monitoring, for handling ash, for reporting and recordkeeping, for operating the MWC unit within acceptable standards, for maintaining proper combustion air supply levels, and for start-up, shutdown and malfunction of the unit. Compliance and performance testing guidelines are also included in the CFR. These standards and guidelines apply to MWCs constructed, modified, or reconstructed after December 20, 1989. The Code pertaining to MWC is updated periodically. The latest update was issued in February 1991.

However, there are many older MWC units that still burn municipal wastes. These are not as technologically sophisticated or as rigorously monitored as are the newer incinerators. Emissions tests are not as stringent or comprehensive for these older units. They are listed as Sub Part E under the 1971 Code (amended in 1974). These MWC units were constructed, modified, or reconstructed before December 20, 1989, and burn up to 50 tons per day. *See also* Fly ash; Hazardous waste; Scrubbers; Solid waste; Solid waste incineration

[*Liane Clorfene Casten*]

RESOURCES

BOOKS

Clarke, M. J. *Burning Garbage in the U.S.: Practice vs. State of the Art*. New York: INFORM, 1991.

Denison, R. *Recycling and Incineration: Evaluating the Choices*. Covelo, CA: Island Press, 1990.

PERIODICALS

Griffin, R. D. "Garbage Crisis." *CQ Researcher* 2 (March 20, 1992): 243–248.

"So Much Smoke? Incineration." *The Economist* 326 (February 13, 1993): A26–27.

Mass extinction

Periodically, every fifty to one hundred million years or so, the earth has experienced mass extinctions, the relatively rapid, large-scale disappearance of many if not most living creatures. The world may be experiencing this phenomenon now at a far more rapid pace than ever before.

Of the five largest mass extinctions, the oldest occurred during the lower Ordovician period some 435 million years ago, when approximately one-quarter of all ocean families, and half of ocean genera, disappeared. (Creatures living in the sea are often used to measure the severity of mass extinctions, because the fossil record of marine sediments is more complete than that for terrestrial animals.) Among the creatures lost during this time were many types of trilobites, cephalopods, and crinoids, all of which are thought to have died out in shallow tropical waters because of sudden changes in ocean levels.

During the late Devonian period, about 357 million years ago, over a fifth of marine families and more than half of marine genera gradually died out over what may have been a ten-million-year interval. **Climate** and sea-level changes apparently doomed many types of corals, trilobites, fish, and brachiopods.

The most severe mass **extinction** of all took place at the end of the Permian period 250 million years ago. This destroyed as much as 96% of all plant and animal **species**, probably over an interval of at least a million years. Over half of all ocean families were wiped out, as were up to 80% of the marine genera. This was fatal for the remaining trilobites and most land species. But the "great dying," as the event is called, also created ecological niches for future life forms, including the dinosaurs.

In the Permian extinctions the suspected but unproven cause is a massive and extremely violent volcanic eruption in Siberia, which catapulted huge quantities of sunlight-blocking dust and **aerosol** droplets into the **atmosphere**. This, in turn, cooled the climate abruptly, expanding the polar icecaps, shrinking the oceans, and causing a new **ice age**. The rapid fluctuations in sea levels decimated the marine creatures of this period. To make matters worse, volcanic explosions could also have turned sulfate minerals into sulfuric **acid** and **sulfur dioxide** gas, which would have produced a ruinous rain of acidic precipitation. Over a 600,000 year period, these eruptions spewed a flood of molten basalt that created rock formations 870 miles (1400 km) in diameter next to **Lake Baikal**, an area called the Siberian Traps.

Another mass extinction occurred about 198 million years ago, during the late Triassic period, when a quarter of marine families and half or more of marine genera disappeared. Cephalopods, **gastropods**, bivalves, brachiopods, and reptiles, were destroyed, as were the conondonts, the fish from which vertebrates may have descended.

The most recent and best known mass extinction occurred a mere 65 million years ago, at the end of the Cretaceous period. It brought about the end of many creatures, including the dinosaurs and ammonites (shellfish). Sixteen percent of marine families and up to 46% of marine genera were lost at this time.

Scientists are still debating the causes of this catastrophic event. It could have coincided with (and may have been caused or accelerated by) a massive volcanic eruption and lava flood in what is now India, where it created a rock formation called the Deccan Traps. Some scientists suggest that a catastrophic climate change and a drastic cooling of the atmosphere made survival impossible for the dinosaurs, since they had no fur, feathers, or hibernating dens to shield them from the cold, nor could they migrate to a warmer climate. Other experts believe that the explosion of a nearby star cooled the atmosphere and emitted deadly radiation for thousands of years. Or perhaps herbivorous dinosaurs starved because they could not adapt to the new types of vegetation that developed. Their extinction would have killed off the carnivorous dinosaurs that preyed on the plant-eaters. Maybe the dinosaurs could not compete for food with the newly-emerging mammals, which ultimately replaced them as the dominant creatures on the planet.

In truth, much is unknown about the causes of the past mass extinctions. A theory with the weight of some evidence behind it raises the possibility that one or more comets or asteroids may have hit the earth, flinging billions of tons of dust and other debris—or ice crystals, if the object impacted in the oceans—into the atmosphere. By blocking out the sunlight for several months, and plunging the earth into darkness, the collision could have lowered temperatures below freezing, killing off the dinosaurs and the plants on which they fed. This hypothesis is based in part on the

1979 discovery, by Dr. Walter Alvarez of the University of California at Berkeley, of the unusual presence of the precious metal iridium in sediments from the time of the dinosaurs' extinction. Iridium, which belongs to the platinum family of elements, is much more plentiful in meteorites than planetary rocks, but it is also frequently found in strata from volcanic eruptions. Other scientists contend that virtually all mass extinctions could have been caused by volcanic eruptions and the huge floods of basalt resulting from such explosions. Evidence supports both theories. University of Chicago paleobiologist Dr. David M. Raup may have put it best when he wrote, "The disturbing reality is that for none of the thousands of well-documented extinctions in the geologic past do we have a solid explanation of why the extinction occurred."

Mass extinctions are not just a phenomenon of ancient geologic history; the last few thousand years have witnessed several such events, albeit on a far smaller geographic scale than those previously discussed. For example, the extinction of the giant animals of North America appears to have coincided with the arrival of humans on the continent, with the changing climate at the end of the last ice age also a possible factor. Crossing the Bering land bridge linking Siberia and Alaska about 12,000 years ago, primitive tribespeople, possibly from Siberia, found a population of large mammals, including giant beavers, **bison**, camels, mammoths, mastodons, and even lions in North America. These prehistoric hunters, moving south from Alaska towards South America, may have wiped out most of these species within 1,000 years.

Fossil remains of these Pleistocene mammals, some with spear points imbedded in them, reveal such creatures as the elephant-like mastodon; the huge-tusked, shaggy red-haired mammoth; a giant beaver weighing over 400 pounds (181 kg); the fabled saber toothed "tiger"; a 7 foot (2.1 m) wide, long-necked camel; 20-foot (6 m) long ground sloths; a bison 7 feet (2.1 m) high at the hump, with a 6 foot (1.8 m) spread between its horns; and huge **wolves**, lions, bears, and horses.

As the twenty-first century approaches, the earth faces another threat of mass extinctions which may occur over a few years or decades, instead of thousands or millions of years. Particularly endangered are those species that are less well known, and in many cases not yet identified by science. Many of these are plants, animals, and invertebrates found in tropical rain forests, which are rapidly being cut and destroyed.

For example, the 1980 *Global 2000 Report* to the President, compiled by the President's **Council on Environmental Quality** and the State Department, with the help of other federal agencies, projected that "...between half a million and two million species—15–20 percent of all species on earth—could be extinguished by the year 2000." And a report issued by the **World Resources Institute** in 1989 predicted that "between 1990 and 2020, species extinctions caused primarily by tropical **deforestation** may eliminate somewhere between 5 and 15% of the world's species...This would amount to a potential loss of 15,000 to 50,000 species per year, or 50 to 150 species per day." In 1988, Harvard biologist **Edward O. Wilson** put the annual rate of species extinction at up to 17,500.

Many of these lost species may be potential sources of food or medicines, and may perform essential ecological functions critical to sustaining life on the planet. It is impossible to predict precisely what the effects of these modern mass extinctions will be, but the ecological disruptions that will inevitably occur could have drastic consequences for human welfare and survival. *See also* Acid rain; Biodiversity; Endangered species; Food chain/web; Greenhouse effect; Ice age refugia; Radioactivity; Radiocarbon dating; Rare species; Volcano

[*Lewis G. Regenstein*]

RESOURCES

BOOKS

Donovan, S. K., ed. *Mass Extinctions: Processes and Evidence.* New York: Columbia University Press, 1991.

Tudge, C. *Last Animals at the Zoo: How Mass Extinction Can Be Stopped.* Covelo, CA: Island Press, 1992.

PERIODICALS

Hecht, J. "Global Catastrophes and Mass Extinctions." *Analog: Science Fiction-Science Fact* 111 (May 1991): 78–89.

Mass spectrometry

A technique of **elemental analysis** first developed by Sir Francis Aston in the early twentieth century. In a mass spectrometer, a sample is first vaporized and then converted to positively charged ions. These ions are accelerated to a high speed and then passed through a magnetic field. Since ions of different weight are bent by different amounts in the magnetic field, elements can be identified on the basis of how far they are bent in the field. Mass spectrometry is a very sensitive analytical technique that permits the detection of trace amounts of a substance, such as the amount of **ozone** in a sample of air. *See also* Measurement and sensing

Mass transit

Mass transit systems transport large numbers of people simultaneously in single vehicles. Examples of mass transit systems include buses, ferries, rapid rail, light rail, commuter rail and intercity rail systems. Although popular before the age of the **automobile**, mass transit systems have become marginal **transportation** modes in many cities in the United States. Recently however, as the negative impacts of automobile use have become of greater public concern, a renewed interest in encouraging mass transit has emerged. The Intermodal Surface Transportation Efficiency Act (ISTEA) of 1991 explicitly allocated funds toward improving mass transit systems.

Transportation congestion is worsening in the United States. In addition to the opportunity costs of time associated with traffic congestion, there are environmental costs associated with automobile use: **emission** of **greenhouse gases**, **air pollution**, **noise pollution**, and increased suburban sprawl and **land use**. Other social costs include increased **probability** of traffic accidents, increased occurrence of stress-related disorders, and increased social severance. Yet in most cities, people seem to be unwilling to use alternative modes of transportation, such as public transit, carpooling, or bicycling, which might impose lower environmental and time costs on society. Nationally, the automobile occupancy rate during commute hours was 1.1 in 1990. Trends indicate the situation will worsen over time.

The social costs associated with automobile use, including environmental **pollution**, congestion, accidents and public health effects, are known in the economics literature as "externalities." These are costs associated with automobile use that are not borne directly by the individuals making the decisions to use their automobiles. Although everyone experiences the negative effects of air pollution, and all drivers experience the negative effects of congestion, each individual driver only experiences a fraction of the total social cost produced when he or she drives. When a driver makes the decision about whether or not to drive an automobile, the full costs of driving are not taken into account. The result is that too many commuters choose to drive, and the social costs far outweigh the private benefits associated with the existing level of driving.

The solution from an economic standpoint is to force drivers to "internalize the externalities" or bear the full cost of their driving. This can be done through policy tools such as congestion pricing or **gasoline** taxes, both of which discourage driving. An alternative solution to the inefficiency is to encourage more use of mass transit by making mass transit less costly to commuters. Many people believe driving is more safe, comfortable, and convenient than mass transit. In order to compete with automobile use, mass transit must either be less expensive than automobiles or more attractive in amenities.

A commonly-held misconception about mass transit is that it must pay for itself. Mass transit is a public good that benefits all commuters, including those who use mass transit and those driving who enjoy less-congested roads and less-polluted communities. Because benefits are received by both users and non-users, it makes sense for the public sector to be subsidizing mass transit. In fact, because the external benefits of mass transit are difficult to quantify, it is unclear how much subsidization of mass transit is optimal. More subsidization creates lower transit fares and a more convenient system which encourages more riders; both riders and non-riders benefit from this.

In order to be an effective solution to the transportation problem, mass transit should satisfy several criteria. It should be cost-effective and less polluting per passenger mile than automobile use. Its public benefits should outweigh the public costs of operating the system. Investment in mass transit is only worthwhile if it will cause a significant number of people to switch to transit from using automobiles. National **statistics** show the number of vehicle occupants moving closer to one over time, demonstrating the high value the general public seems to place on driving their cars. Alternative modes such as mass transit must be competitive with automobiles in order to cause a decrease in automobile usage and a subsequent improvement in the **environment**.

Mass transit systems have significant environmental benefits, including substantial reductions in greenhouse gases, energy consumption and traffic congestion, provided enough commuters switch to using it. A fully occupied train can remove 100 cars from the road during rush hour, a bus forty, and a vanpool thirteen. Associated energy savings can be around 40–60%, and cleaner fuels can reduce emissions even further. Mass transit systems in different ways. Park and ride systems for example, are much less environmentally friendly than transit-only options. During the first few miles of an automobile trip, emissions and fuel use are very high, due to the cold start. In a 10-mile trip for example, 90% of emissions are produced in the first few miles. Although they offer fewer environmental benefits than full transit systems, they still decrease traffic congestion and accidents.

Not every city in the United States is suited for mass transit systems. Large cities such as New York City and Washington, D.C. have the density of people and businesses necessary to make the systems viable, and mass transit systems are heavily used. Many cities have grown in a sprawling manner, due mainly to the dominance of the automobile, so that the density of the population makes good mass transit infeasible. On the other hand, if a good mass transit system were in place, one could envision the development of business and residential hubs around the transit system which would eventually increase population density and ridership. An addition to the above-mentioned benefits of mass transit, the revitalization of downtowns, many of which have deteriorated with the growth of the suburbs, might occur as development adapts to improved mass transit systems.

Future technological developments might improve the attractiveness of mass transit relative to automobile use. Intelligent Vehicle Highway Systems (IVHS) refer to the application of advanced technologies, especially communications technologies, to improve the efficiency of the nation's surface transportation system. Potential outcomes of this endeavor include the creation and implementation of "smart cars," "smart transit," and "smart streets."

A great deal of IVHS is devoted to improving the efficiency of automobile transportation. IVHS might offer traffic signal timing and smart car technology which would reduce congestion and traffic accidents. These technologies will make automobile use more convenient and safe than before, encouraging more automobile travel. On the other hand, IVHS offers the possibility of implementing a congestion pricing system where automobile users pay a fee when using certain roads at congested times. This cost would make transit more competitive relative to automobiles during heavy commute hours.

Certain components of IVHS might improve convenience to make mass transit more competitive with automobiles. A component of IVHS is the development of communications systems controlling traffic patterns and flows on streets and highways. These systems can be used in conjunction with HOV (high-occupancy vehicle) lanes to make transit systems more competitive in travel time. Also, computerized fare-collection devices speed transit boarding, a time-consuming aspect of present transit travel. These technologies make transit and ridesharing more convenient and attractive to commuters, increasing transit use relative to individual vehicle use.

IVHS could encourage or discourage mass transit use. For planning purposes, it is important to consider the implications of the different components of IVHS to the future of mass transit systems. The benefits associated with increased mass transit use are generally public as opposed to private. Yet, the individual decision about which transportation mode to use is a private decision. Increasing mass transit use requires the development of attractive, competitive mass transit systems that are a convenient, safe, and comfortable option to automobile travel. Improved mass transit in these ways should pay off in terms of an improved environment and quality of life. *See also* Environmental economics; Externality; Urban sprawl

[*Barbara J. Kanninen*]

RESOURCES
BOOKS

Brown, Lester R., ed. *The World Watch Reader on Global Environmental Issues.* New York: W.W. Norton & Co., 1991.
Gordon, Deborah. *Steering a New Course: Transportation, Energy, and the Environment.* Washington, DC: Island Press, 1991.

Small, Kenneth A. *Urban Transportation Economics*. Chur, Switzerland: Harwood Academic Publishers, 1992.

Material Safety Data Sheets

The use of a Material Safety Data Sheet (MSDS), as mandated by the Occupational Health and Safety Administration (OSHA), is an integral part of hazard communication and "right-to-know" in the work place. An MSDS for a chemical substance is designed to provide both workers and emergency personnel with procedures for handling or working with that substance. An MSDS provides information concerning safe handling procedures, first aid measures, and procedures to be taken when the substance is accidentally spilled or released. MSDSs are provided by the manufacturer and supplier of **chemicals** and should be available to all employees at all times. MSDSs for many substances are available on the Internet. Many MSDSs are complete, accurate, and informative, while others, especially for those chemicals a manufacturer considers to be proprietary, may be missing key pieces of information. An MSDS must contain (written in English) the information described below.

An MSDS contains information on ingredients and their physical and chemical characteristics. Except for trade secrets, the MSDS gives the specific chemical name, chemical formula, and common names for the ingredients that have been determined by OSHA to be hazardous chemicals (i.e., any chemical that is defined as a physical or health hazard). If the substance is a mixture that has been tested as a whole to determine its hazardous properties, the chemical and common names of the ingredients that contribute to those known hazards and the common names for the mixture are listed.

If the substance is a mixture that has not been tested as a whole, the chemical and common names are listed for all ingredients that are determined to be health hazards and that comprise 1% or more of the mixture; identified as carcinogens and present at 0.1% or more; or determined to present a physical hazard when present in the mixture.

The MSDS also gives information about physical and chemical characteristics that indicate what a material or mixture is like and how it behaves, including boiling point, vapor pressure, vapor density, water solubility, specific gravity, evaporation rate, and reactivity with water and other substances.

An MSDS also contains information describing hazards, precautions, control measures, and emergency procedures. It describes physical hazards such as combustibility, flammability, and explosiveness, and it identifies health hazards, which include acute and chronic hazards that result from exposure, as well as whether the substance is listed as a confirmed or potential **carcinogen**. In addition, the MSDS gives the applicable precautions for safe handling and use, including steps to be followed if material is spilled or released; waste disposal methods for spilled substances, which must follow federal, state, and local regulations; precautions to be taken in handling and storage; personal protective equipment that are required; signs and symptoms of exposure; and description of routes of entry into the human body during normal usage or a foreseeable emergency. The MSDS describes the applicable control measures such as special fire fighting procedures and extinguishing media and emergency first aid procedures.

Finally, MSDSs describe exposure limits and give information about responsible parties. The document lists the OSHA Permissible Exposure Limit (PEL), the American Conference of Governmental Industrial Hygienists (AC-GIH) Threshold Limit Value (TLV), and any other recommended exposure limits that might apply. The MSDS includes the date of preparation or latest revision, with the name of the person responsible for preparing the MSDS being optional. It also lists the name, address, and telephone number of the manufacturer, importer or other responsible party.

[*Judith L. Sims*]

Materials balance approach

A materials (or mass) balance approach for contaminants of public health and/or environmental concern is used to determine the presence, fate and transport of contaminants in the **environment**. The materials balance approach, a fundamental principle of science, engineering, and industrial research and **risk analysis**, is familiar to professionals trained in the physical or life sciences and engineering. The use of a materials balance approach provides a technique to describe the environment as it is today and as it might be under conditions resulting from remedial actions or from changes in the way society produces, uses, and disposes of **chemicals** of environmental concern. It also provides a rational and fundamental basis for asking specific questions and for obtaining specific information, which is necessary for determining fate and transport of contaminants, selecting and evaluating remedial treatment options, and monitoring treatment effectiveness.

The determination or construction of a materials balance is dependent on **conservation** of materials. Material is not created nor is it destroyed by ordinary processes, but it is transformed. When applied to chemicals of environmental concern, conservation of materials requires that a chemical entering a specific environment must be transformed, held in, or transported out of the environment. The amount of a chemical that leaves any process, environmental compartment, or area must be exactly balanced by the amount that enters minus net accumulation within the process or compartment boundaries. This materials balance can be stated as a simple equation:

Change in mass in a volume = Mass entering a volume - Mass leaving a volume

The control volume (compartment) for analysis, the shape of the volume, and the identification of input and output flows as well as the processes acting within the control volume all must be chosen carefully in order for the materials balance to provide useful information for the assessment of environmental impacts and the control of environmental effects.

The steps involved in determining a materials balance of contamination in an environment include:

- Where is the contamination, in what form(s) does it exist and in what concentrations is it present?
- Where is the contamination going under the influence of natural geochemical and geobiological processes (that is, what are key pathways of transport and fate; what are the processes that affect mobility and degradation of contaminants)?
- Based on the answers to questions 1 and 2, how can the contamination be contained, destroyed, or immobilized in the specific phases in which it is found? Through time, as natural and remedial processes act upon specific contaminants, additional determination of the materials balance is required to assess the fate and transport of the contaminants or their transformation products and to develop further treatment or containment methodologies.
- Based on the answers to questions 1 and 2, what environmental phases should be monitored through time to assess the fate and transport of the contamination and effectiveness of treatment or containment under both natural and remedial processes?

Data to answer questions 1 and 2 may be obtained from three categories:

- direct data from administrative records regarding the type and quantity of contamination, as well as where it entered the environment;
- direct data from chemical measurements;
- indirect data from modelling and simulations of natural processes.

Types of direct data may include an assessment of the presence of contaminants in all environmental compartments present in a specific environment, which may include the air phase, aqueous and non-aqueous liquid phases, and solid phases. Contaminants may be transformed partially or completely in the **respiration** process to obtain energy to synthesize new microbial cells, so an estimate of mineralization (for example, by measuring oxygen utilization or **carbon**

dioxide production in **aerobic systems**) may also provide important information.

A determination of the **accuracy** of a materials balance assessment is indicated by a good agreement between direct and indirect measurements, suggesting that the control volume was well-bounded and the processes acting within the volume were well-defined. The study of the materials balance approach involves both art and science, but if well-implemented, can provide an understanding of the environment of concern. *See also* Biogeochemical cycles; Risk assessment; Waste management

[*Judith Sims*]

RESOURCES
PERIODICALS

Sims, R. C. "Soil Remediation Techniques at Uncontrolled Hazardous Waste Sites: A Critical Review." *Journal of the Air & Waste Management Association* 40 (1990): 703–732.

Maximum permissible concentration

In radiology, the maximum permissible concentration refers to the recommended upper limit for the dose which may be safely received during a specific period by a person exposed to **ionizing radiation**. It is also sometimes called permissible dose. *See also* LD$_{50}$; Threshold dose

Maximum social welfare

see **Pareto optimality (Maximum social welfare)**

Ian Lennox McHarg (1920 – 2001)
Scottish design ecologist and writer

Clydebank, Scotland, where Ian McHarg was born and raised (and where he received education through high school) produced one of America's best known design ecologists. Before his first move to the United States, McHarg had attended two colleges in Scotland and spent eight years in the British army, including active battle service in World War II. He entered the army as an enlisted man, after four years attended Officer Training School, and was demobilized in 1946 at the rank of major.

That same year, McHarg entered Harvard University to study landscape architecture, from which he graduated with a bachelor's degree in landscape architecture 1949 and a masters in the same subject, one year later. He returned to Scotland for a short while and then earned another Harvard degree, a master's, in city planning (1951). For a short time, he returned to Scotland, serving as a planning officer in the country's Department of Health and teaching a course in landscape architecture at Edinburgh College of Art and then at Glasgow College of Art. His career in the United States began in 1954 when he accepted an appointment as an assistant professor of landscape architecture and city planning at the University of Pennsylvania, where a year later he was instrumental in creating the Department of Landscape Architecture.

While many practitioners still consider landscape architecture a traditional design field that creates back yards for the well-to-do, McHarg is widely credited with a revolution in the field. At the university, he taught landscape architecture and planning from an interdisciplinary, ecological perspective, bringing in ecologists, geologists, anthropologists, and even lawyers to participate in team-taught courses. From the university, his students have emerged to foment McHarg's ecological revolution in university design and planning departments, in planning agencies, and in design firms across the country.

As an active designer and planner outside the university McHarg put his ecological ideas into practice in such places as Hazleton, Pennsylvania, Medford, New Jersey, Amelia Island, Florida, Teheran, Iran, and Taroko, Taiwan. What one person can plan and build in a lifetime is limited, so it is McHarg's ideas (and the students, readers, and practitioners instilled with those ideas) that will remain significant far into the future.

The best known distillation and presentation of those ideas is in his *Design With Nature*, published in 1969 and reissued in a twenty-fifth anniversary edition in 1992. Lewis Mumford, in his preface to the first edition, best describes McHarg's accomplishment:

"McHarg's emphasis is not on either design or nature by itself, but upon the preposition *with*, which implies human cooperation and biological partnership. He seeks, not arbitrarily to impose design, but to use to the fullest the potentialities—and with them, necessarily, the restrictive conditions—that nature offers. So, too, in embracing nature, he knows that man's own mind, which is part of nature, has something precious to add that is not to be found at such a high point of development in raw nature untouched by man." Later, Mumford exclaims "here are the foundations for a civilization."

In *Design With Nature*, McHarg tried, and to a large degree succeeded, to direct planning away from its socioeconomic preoccupation toward an in-depth consideration of **environment**; to help integrate the physical and life sciences into an applied **environmental science**; to explore the role of values in planning, and to present a theory and method of planning as a way to understand and channel human adaptations to the environment.

McHarg's ideas changed both landscape architecture and planning, but perhaps the latter more so. In planning, the human ecological philosophy he developed and advocated—in *Design With Nature* (and in numerous articles)—and the very practical method of comprehensive data collection and utilization he worked out in many years of classes and design projects, are today employed worldwide, including integrated applications in new GIS systems. Known by many as the father of ecological planning, McHarg moved from a limited environmental determinism to a contemporary interdisciplinary approach that defines planning as interactive, as applied **human ecology**.

[*Gerald L. Young Ph.D.*]

RESOURCES
BOOKS

Landecker, H. "Ian McHarg: In Search of an Arbiter." *Profiles in Landscape Architecture*. Washington, DC: American Society of Landscape Architects, 1992.

McHarg, Ian. *A Quest for Life: An Autobiography*. John Wiley, 1996.

———. *Design With Nature, 25th Anniversary Edition*. New York: John Wiley & Sons, 1992.

———. *To Heal the Earth: Selected Writings of Ian L. McHarg*. Island Press, 1998.

Miller, E. L., and S. Pardal. *Classic McHarg: An Interview*. Lisbon, Portugal: CESUR, Technical University of Lisbon, 1992.

PERIODICALS

McHarg, Ian. "Human Ecological Planning at Pennsylvania." *Landscape Planning* 8 (1981): 109–120.

Bill Ernest McKibben (1960 –)
American environmentalist and writer

William E. "Bill" McKibben is an American **nature** writer who was born in Palo Alto, California. He graduated from Harvard University in 1982 and was a staff writer and editor at *The New Yorker* until 1987 when he began his freelance career. He now lives with his wife in the **Adirondack Mountains** in New York, and this region figures prominently in his writings.

In *The End of Nature*, McKibben's first book, he argues that the **greenhouse effect** is not only part of humanity's destruction of the **environment**, but a symptom of Man's alienation from the natural world. He calls for an end to practices that contribute to the greenhouse effect, such as burning **fossil fuels**. Such practices, he writes, "will lead us, if not straight to hell, then straight to a place with a similar temperature."

Some critics dismiss McKibben's arguments as "too absolute," and "more a slogan and hyperbole than scientific insight." Environmentalists, however, praise the book, calling it the "*Silent Spring* of the '90s"—a reference to the book by Rachel Carson which in 1962 awakened the nation to the indiscriminate use of pesticides. However, McKibben's lament over the "end of nature" is a personal account of how human activity has changed not just the industrialized centers of the world, but his own backyard.

McKibbon's second book, *The Age of Missing Information* (1993), is a study of the seduction of television. On one day he videotaped every show on every cable channel available in his area, more than 1,000 hours of television, and then watched it all. To contrast that experience, he spent a day camping in the woods near his Adirondack home. "This book is about the results of that experiment—about the information that each day imparted," he says. Talking about this book, McKibben says, "Television...is a very private experience, with an almost constant message of 'you are the most important, this Bud's for you.' And if you are at the center of the world, it's hard to live environmentally aware."

McKibben has also edited *Birch Browsings: A John Burroughs Reader* (1992). **John Burroughs**, a nature writer whose career spanned 60 years beginning in 1865, is credited with establishing the nature essay as a literary genre. Many environmentalists are returning to the writings of such early nature writers as Burroughs, **John Muir**, and **Aldo Leopold** to study their ideas about **land stewardship**.

[*Linda Rehkopf*]

RESOURCES
BOOKS

McKibben, Bill. *Birch Browsings: A John Burroughs Reader*. New York: Penguin Books, 1992.

———. *The Age of Missing Information*. New York: Plume, 1993.

———. *The End of Nature*. New York: Random House, 1989.

———. *Long Distance: A Year of Living Strenuously*. Simon & Schuster, 2000.

———. *Maybe One: A Case for Smaller Families*. Simon & Schuster, 1998.

———. *The Return of the Wolf: Reflections on the Future of Wolves in the Northeast*. Middlebury: University Press of New England, 2000.

———. *Wolves in the Northeast*. University Press of New England, 2000.

Magill, Frank N., ed. *Magill's Literary Annual*. Vol. l. Pasadena: Salem Press, 1990.

Trosky, Susan M., ed. *Contemporary Authors*. Vol. 130. Detroit: Gale Research, 1990.

PERIODICALS

Walljasper, Jay. "Checking in with Bill McKibben." *Conscious Choice* (July 1999).

MDC

see **More developed country**

Measurement and sensing

A fundamental premise of all scientific research is that scientists can make measurements, expressed in precise mathematical terms, of conditions and events. Indeed, some argue that a field of study can only be called a science to the extent that the data being collected can be mathematically measured. These criteria apply to **environmental science** as they do to other physical and biological sciences.

Measurements serve a number of functions in environmental studies. One is to provide baseline information about certain aspects of the **environment**. A source of concern to scientists and non-scientists alike is the change that appears to be taking place in the earth's **ozone** layer. Mounting evidence appears to suggest that, as a result of human activities, the concentration of ozone in the **stratosphere** seems to be decreasing. That finding is significant because the ozone layer acts as a shield against potentially harmful **ultraviolet radiation** from the sun.

But how do scientists know that changes are taking place in the ozone layer? Such a conclusion can be drawn only if information exists regarding the "normal" concentrations of ozone in the stratosphere. One purpose of measurements, therefore, is to accumulate a huge volume of information regarding the "normal" condition of the environment.

A second purpose of measuring is to determine how some aspect of the environment may be changing, often as a result of human activities. Our concern about the ozone layer, for example, arises out of readings taken by high-flying aircraft over a period of more than a decade. Those readings show consistent changes (decreases) from the level that is regarded as natural or normal. These levels are not fluctuating, that is, decreasing and regaining normalcy. Rather, there is a trend toward decreased ozone levels.

Dozens of techniques are available for measuring environmental characteristics. Many of these techniques are not unique to environmental science. For example, the use of thermometers to record temperatures is vitally important in many environmental studies. But temperature-taking is not a specialized procedure used by environmental scientists. The measurement of **radioactivity** is another example. Geiger counters are used to determine the level of radiation in environmental studies just as they are in scientific research, medical applications, industrial processes and other situations.

Over the years, a number of techniques have been developed for measuring specific environmental characteristics. Many of these techniques can be classified according to their use in the measurement of **air quality** or **water quality**.

An example of the former is the high-volume sampler (HVS) used in measuring particulates in the **atmosphere**.

An HVS is essentially a modified vacuum cleaner that sucks air into a hose and forces it through a filter paper. The difference in the weight of the paper before and after it has trapped the **particulate** matter is used to determine the weight of the particulates. Since the total volume of air passing through the HVS can be easily measured, the systems provides of measure of particulates in mass per volume.

Techniques for measuring various components of a gas are also becoming more sophisticated. At one time, for example, the determination of ozone levels was made simply by suspending a piece of natural **rubber** of known size and weight in the air. Since ozone causes rubber to crack, the speed at which cracking occurred was taken as a measure of ozone levels in the air.

As with most other gases, ozone is now measured by chemical means. A sample of gas is passed through some device, often a "bubbler," that contains a compound that will react with the gas being tested. The concentration of that gas can be measured, then, by determining the amount of chemical change that takes place in the measuring device.

For example, a reaction with which many high school students are familiar can be used to determine the level of **sulfur dioxide** in the air. Sulfur dioxide reacts with **lead** dioxide (PbO_2) to form a characteristic black precipitate of lead sulfate.

$$SO_2 + PbO_2 \rightarrow PbSO_4$$

If lead dioxide is added to a bubbler through which a sample of gas is passed, the amount of darkening in the device (due to the formation of lead sulfate) is a measure of the concentration of sulfur dioxide in the gas.

Many environmental measurements now make use of sophisticated chemical techniques. One example is infrared spectrometry. The term *spectrometry* refers to the measurement of energy absorbed or emitted by various compounds. All molecules are held together by electrons that vibrate with characteristic frequencies. If energy is added to those molecules, they will absorb frequencies of energy that match their characteristic frequencies, but no others. Each kind of molecule can be identified, therefore, by a "map" of the frequencies that it does and does not absorb.

Infrared spectrometry is the most common form of the technique used because most molecules vibrate with frequencies in the infrared region. Techniques such as infrared spectrometry are valuable because they can detect concentrations of a material at much lower levels than can most chemical systems.

The increasing sophistication of measuring techniques does not mean that all simple procedures have been abandoned. For example, one method for measuring the level of **air pollution** requires no more than good eyesight and a reference card. The reference card contains the Ringelmann scale, a set of six squares that range from pure white to pure black. Each square contains an amount of hatching that corresponds to completely pure air (pure white, no hatching) to badly polluted air (totally black). The four intermediary squares contain increasingly more hatching, constituting equivalents of 20 percent, 40 percent, 60 percent and 80 percent "blackness."

For comparison, opacity (darkening) of air can also be determined by electronic means more precisely. A photometer, for example, is a device that records the amount of light transmitted by a sample of air. The light is converted to an electric current that can be read on a meter. In some industrial and **power plants**, a photometer is attached to the smokestack to obtain a continuous record of the opacity of the gases being emitted.

The two most common types of air quality measurements are those for **ambient air** quality and emissions. Ambient air refers to the outdoor air that surrounds us. Ambient air quality measurements provide information on the possible accumulation of harmful compounds such as **carbon monoxide**, sulfur dioxide, **nitrogen oxides**, and **hydrocarbons**.

Emission measurements reveal the level of such compounds being released from a power plant, a factory, or some other source. In many cases, emissions can be studied by simply drilling a hole in a smokestack, extracting a small sample of exhaust gases, and studying their composition by methods described above.

Many biological, chemical, and physical methods are available for measuring water quality. The presence of pathogens in a sample of water can be determined, for example, by standard bacteriological techniques in which a water sample is allowed to incubate for some period of time and the number of bacteria produced counted visually or electronically.

At one time, most water tests were fairly straightforward chemical tests. The concentration of **nitrogen** in a water sample, for example, can be determined by a standard procedure known as the Kjeldahl test and the amount of **chlorination** by precipitation with a silver salt. Today, most water tests can be conducted by more sophisticated instrumental techniques. A photometer can be used to compare an unknown water sample with a known standard to determine the concentration of nitrogen, **phosphorus**, **chlorine**, or some other component.

One of the most basic measurements of water quality is that of oxygen demand. The more polluted a water sample is, the more organic matter it is likely to contain. The more organic matter, the greater the amount of oxygen the sample will require to decompose the organic material.

Traditionally, the method for determining this characteristic was **biochemical oxygen demand** (BOD), a test in which a sample of water is studied over a five-day period. To overcome the lengthy time required to conduct this test, modifications such as total organic **carbon** (TOC) have been developed.

Specialized measuring techniques are sometimes required for particular types of environmental studies. Determining the amount of **noise pollution** in an area, for example, requires the use of a sound level meter, a device consisting of a microphone, amplifier, frequency-measuring circuit, and read-out screen.

One consequence of the improved technology now available for making measurements is that smaller and smaller concentrations of a substance can be detected. Chemical means can routinely detect the presence of a substance to the level of one part in a thousand or one part per million (ppm). The most advanced technologies today have stretched that sensitivity to levels of one part per billion (ppb) and even one part per trillion (ppt).

The efficiency of these measuring devices poses some new issues for environmentalists. What does it mean, for example, to learn that a **toxic substance** exists in the **soil** at a level of one ppb if we know the material is harmful only in much higher concentrations? Does any level of exposure to the substance pose a hazard, or can we safely ignore such a minuscule quantity of the substance?

A field of measurement of increasing importance in environmental studies is remote sensing. The term refers to any method by which an object or an area is studied at some great distance. Most commonly today, the term is used to describe surveys of the earth's surface by satellites orbiting around it.

Remote sensing procedures depend on the fact that various types of materials absorb and reflect **solar energy** in different ways. Instruments in satellites that can measure these differences can, therefore, detect variations in land and ocean surfaces.

Remote sensing uses three parts of the electromagnetic spectrum: the visible, infrared, and microwave regions. Perhaps the most common form of remote sensing is that which uses photography. Photographic equipment has now been developed to the point where objects of no more than a few yards apart can be distinguished from outer space.

Since some objects and features emit radiation in electromagnetic regions other than the visual, infrared, and microwave techniques are also used. The images obtained from any one of these methods can be further improved by computer enhancement of the original photographs.

Remote sensing has now been used for a number of environmental applications. Some examples include the lo-

cating of possible mineral reserves, the tracing of water **drainage** patterns, the determination of soil moisture, the tracing of plant diseases, the calculation of snow and ice masses, and the measurement of biological productivity in the oceans. *See also* Drinking-water supply; Emission standards; Greenhouse gases; National Ambient Air Quality Standards; Ozone layer depletion; Radiocarbon dating; Water pollution

[*David E. Newton*]

RESOURCES
BOOKS

McGraw-Hill Encyclopedia of Science & Technology. 7th ed. New York: McGraw-Hill, 1992.
Vesilind, P. A., J. J. Peirce, and R. Weiner. *Environmental Engineering.* 2nd ed. Boston: Butterworths, 1988.

Medical waste

Medical waste is a subcategory of **hazardous waste** that is attracting increasing concern. The **Environmental Protection Agency** (EPA) lists the following categories of medical waste: cultures and stocks; pathological wastes which includes body parts; blood and blood products; used "sharps" such as needles and scalpels; **animal waste** or animal corpses which have been inoculated with infectious substances in medical research; isolation wastes, which come from people with highly contagious diseases; and unused, discarded "sharps."

Ten to fifteen percent of medical waste is considered infectious, although guidelines on just what is infectious medical waste vary from state to state. As a result, state and federal guidelines on disposal of these wastes are a hodgepodge of confusing laws and regulations. And the guidelines that do exist usually exempt generators of 50 pounds (23 kg) or less per month from any regulatory action.

The need to address medical waste began soon after such items as syringes, IV bags, and scalpels were observed washing up on ocean beaches in the summer of 1988. Congress directed the EPA to gather data on their sources, associated health hazards, and current procedures and regulations for management and disposal and to evaluate the health hazards associated with transporting them, incinerating them, burying them in a **landfill**, and disposing of them in a sanitary sewer system.

A two-year, voluntary program enacted by Congress, the Medical Waste Tracking Act of 1988, was instituted in response to public concern over the treatment of medical waste. The Act created a "cradle-to-grave" tracking system based on detailed shipping records, similar to the program in place for hazardous waste. The pilot program has now expired, the EPA data has been sent to Washington administrators, and any action on the findings of the EPA's study is in limbo. It is doubtful that the EPA report on medical waste will ever be submitted to Congress or find its way into the *Federal Register* because many of the states that participated in the program went on to pass strict medical-waste management guidelines of their own. All the states participating at least revised their **municipal solid waste** guidelines to include the category of medical waste, whether or not management practices were included.

Another complicating factor in the regulation of medical waste is that at least four different federal agencies are involved with medical-waste issues: the EPA, the **Occupational Safety and Health Administration** (OSHA), the Centers for Disease Control (CDC), and the **Agency for Toxic Substances and Disease Registry**.

The EPA's study reported that 3.2 million tons of medical waste was produced in the United States each year. Not surprisingly, hospitals, long-term health care facilities, and physician's offices are the major producers of medical wastes, which account for about 0.3% by weight of all municipal **solid waste**. The EPA found out that current practices of management and disposal range from handling the waste as nonhazardous municipal solid waste to strict segregation, packaging, labeling, and tracking from the generator to the disposal site, the so-called "cradle-to-grave" management.

Health care workers are required by federal law to segregate medical waste in special containers; if the waste is transported for treatment, it must be labeled with the generator's name and carry a biological hazard symbol. But facilities that produce less than 50 lb (23 kg) of medical waste per month are exempt from most requirements. Home and small generator medical wastes fall through the cracks. Insulin-dependent diabetics, for instance, typically dispose of used syringes in their household trash or flush them down the toilet. In some northeastern cities, New York City for example, antiquated sewage systems pour material into rivers and oceans during heavy rainfalls. Depending on tides and currents, syringes can end up on beaches, which is what New York and New Jersey coastlines experienced in 1988 and 1989.

A growing source of medical waste that is not regulated is that generated by home health-care providers. As a result of new patient-care strategies and rising medical costs, more long-term illnesses are being treated at home. That means that medical waste is being disposed of in ordinary household trash. Medical clinics and intravenous drug users are also suspected of contributing to unsafe dumping of medical wastes. Even U.S. Navy vessels contributed to beach medical waste.

The need to address the problem evolved long before wastes washed up on northeastern beaches, however. Over the last 10 years, and partly in response to the **AIDS** and hepatitis epidemics, the use of disposable health care products has contributed to the increased volume of medical waste. Cost-containment measures have increased the use of plastic disposables in health care settings. From syringes to bedpans, health care aids are increasingly thrown away.

Common treatment techniques of medical waste include steam sterilization and **incineration**, although some waste is discarded into sewage systems. During autoclaving, the waste is exposed to steam at a temperature of 250°F (121°C) for at least 45 minutes. Autoclaving fails to reduce the volume of waste that must be landfilled, and is only preferred in areas where there is no appropriate incineration equipment. Proper incineration efficiently destroys all categories of infectious wastes, effectively kills live and dormant forms of pathogenic organisms, and alters the waste volume.

About 60 percent of medical waste is treated on site at hospital facilities. After treatment, ash residues or sterilized and disinfected materials must be transported to commercial treatment facilities or taken directly to landfills. Other options for medical waste treatment include **compaction**, microwaving, and mechanical or chemical disinfection, which could be less costly and alleviate concerns about emissions from incinerators.

According to the congressional Office of Technology Assessment (OTA), air emissions of **dioxin** and **heavy metals** from hospital incinerators average 10 to 100 times more per gram of waste burned than emissions from well-controlled municipal waste incinerators. And while waste generators are generally required to track with manifests the route their wastes take to reach disposal, strict requirements do not exist for on-site treatment facilities.

Some industry watchers, environmentalists, and policy makers are advocating a medical waste disposal program similar to some in Europe. Virtually all medical wastes in Switzerland and Germany, for example, go to regional incineration facilities. Stringent air-quality regulations in these countries make it impractical for hospitals to do on-site incineration, but the regional facilities made advanced **air pollution control** devices cost-effective. In Munich, operators dump the resulting ash in specially lined landfills. Hospitals in Canada are turning away from disposable supplies in favor of products that can be cleaned and reused.

For a regional system to work, however, medical waste must be safely transported. In the United States, there are no federal statutes governing the transport of medical refuse. Refrigerated trucks legally can and do carry food after transporting medical wastes.

A proposal to manage home users of syringes and medicines that end up in municipal solid waste landfills would include a pharmacy "swap" system. In Switzerland and Germany, for example, pharmacies accept old medicines for appropriate disposal, then send the medicine to regional incineration outlets. A deposit-and-return system on syringes has been advocated in this country to insure the safe disposal of syringes.

[*Linda Rehkopf*]

RESOURCES
BOOKS

Moeller, D. W. *Environmental Health.* Cambridge: Harvard University Press, 1992.

PERIODICALS

Carlile, J. "Finding Disposal Options for Medical Waste." *American City and County* (November 1989): 66.
Groves, L. "Hospitals Re-Think Disposables." *Alternatives* (January-February 1993): 13.
Hershkowitz, A. "Without a Trace: Handling Medical Waste Safely." *Technology Review* (August-September 1990): 35.
"Managing Medical Waste." *The Futurist* (September-October 1991): 49.
"Tracking Seaside Medical Wastes." *Science News* (September 16, 1989): 191.

Mediterranean fruit fly

The Mediterranean fruit fly (referred to as Med fly, or Moscamed in Spanish), *Ceratitis capitata*, is one of the most destructive fruit pests in the world. The United State has suffered infestations of Med fly in California, Texas, and Florida, but with aggressive detection and eradication programs, the Med fly has not yet become established in the United States. However, eradication efforts are difficult and expensive. There is a loss of crop yield associated with the infestation, control measures are costly, and both fresh and processed fruit and vegetables must be sorted. Some countries maintain quarantines against the Med fly, which eliminates potential markets when Med fly becomes established in an area.

The Med fly, which originated in Africa, is widely distributed throughout the world, as it has a wide range of hosts and is able to tolerate colder climates better than most other **species** of fruit flies. It has caused infestations in over 85 countries in tropical and subtropical regions. Since 1929, the United States has been involved in several outbreak and eradication programs. The Med fly became established in Hawaii in 1910. Hawaii remains infested, and there is no eradication program currently under way.

Med fly larvae can develop and feed on most deciduous, subtropical, and tropical fruits and some vegetables. It has been shown to attack more than 260 different fruits, flowers, vegetables, and nuts. The Med fly prefers thin-skinned, ripe succulent fruits such as deciduous fruits (for example, pear, peach, and apple) more than citrus fruits.

The larvae feed on the pulp of fruits, tunneling through it, and reducing the fruit to an inedible juicy mass. In the Mediterranean countries, often only the earlier maturing fruits are grown, because later maturing fruits would be too heavily infested to be marketable. Harvesting before complete maturity is also practiced in Mediterranean areas with Med fly infestations.

The adult Med fly is slightly smaller than a common housefly. It is colorful, with dark blue eyes, shiny black thorax (back), and a yellowish abdomen with silvery cross bands. Its droopy wings are covered with yellow, brown, and black blotchy spots and bands.

The life cycle of the Med fly has fives phases: (1) the adult female pierces the skin of a fruit or vegetable with a needle-like ovipositor and deposits one to 10 eggs (other Med flies may deposit in the same puncture); (2) eggs hatch into maggots (worm-like, legless larvae); (3) larvae feed upon the pulp of the fruit or vegetable before dropping to the ground; (4) the larvae transform into pupae into the **soil**; and (5) the pupae mature into adults and emerge from the soil. The life cycle is completed in 21–30 days under tropical summer weather conditions.

An eradication program consists of three areas of action: survey, regulation, and control.

The Animal and Plant Health Inspection Service (APHIS) of the **U.S. Department of Agriculture** in cooperation with state departments of agriculture survey high-risk areas of states susceptible to Med fly infestations by conducting **trapping** programs. If one or more Med flies is collected in a trap, APHIS and state officials implement a survey to determine if there is an infestation and if so, to define the limits of the infested area. Using the original detection site as the focal point, additional traps are placed and monitored.

If an infestation is identified, then federal and state regulations dictate that a quarantine must be imposed to prevent the spread of the Med fly. Federal quarantine laws prevent the interstate movement of any article that might harbor the fly. State regulations control the movement of these articles going to uninfested areas within the same state. The regulated articles include all of the fruits and vegetables within the area that could serve as a Med fly host. Open-air vegetable and fruit stands are required to cover produce to prevent infestation. All commercial and home-grown produce may not be transported without inspection and treatment.

To eradicate a Med fly infestation, three kinds of treatment are used alone or in combination. An aerial and ground bait spray can be used that contains a protein/sugar bait to attract the fly and an insecticide in minimal amounts to kill the flies.

In the Sterile Insect Technique (SIT), Med flies are raised in large numbers, sterilized with a non-lethal dose of irradiation, and released into infested areas, where they mate with the wild Med flies. As these matings do not produce offspring, the wild population is eliminated through attrition. This technique is most effective in areas with low Med fly populations where a high proportion of sterile to wild Med flies can be achieved. Before the use of SIT, bait spray can be applied to decrease local populations down to lower densities.

The third method is to apply insecticide to the soil under host trees, where the insecticide kills some larvae as the enter the soil to pupate but most of the adults as they emerge from the ground. Application of insecticide to the soil is used only when larvae are detected. The preferred method of eradication control is an integrated approach, using all three treatments, with the use of SIT emphasized.

To prevent Med flies from being brought into the United States, APHIS administers agricultural quarantine laws to keep foreign plant pests and diseases from being brought into the country. Travelers coming to the continental United States from Hawaii or a foreign country are not allowed to bring into the country fresh fruits, meats, plants, birds, and plant and animal products. In fiscal year 1998, when 400,000 aircraft were cleared to bring travelers and cargo into the United States, APHIS officers intercepted more than 1.8 million illegal plants, animals, or plant and animal byproducts. More than 52,000 plant pests and diseases known to be dangerous to the U.S. agricultural industry were also intercepted.

If Med fly were to become established in the United States, prices of fruits and vegetables would increase, and produce would become less available. Both commercial production areas and backyard gardens would require the application of more pesticides on a regular basis. In 1993, APHIS estimated that annual losses due to Med fly infestations would be about $1.5 billion annually, if Med fly were to become established in the United States. These losses would be due to export sanctions, lost markets, treatment costs, reduced crop yields, and premature fruit drop.

[*Judith L. Sims*]

Mediterranean Sea

For centuries, the Mediterranean Sea has been the focal point of western civilization. It is an area rich in history and has played critical roles in the development of shipping and trade, as a resource for feeding growing populations, and as an aid to the spread and mingling of races and cultures.

The Mediterranean began to form about 250 million years ago when the Eurasian and African continental plates began moving toward each other, pinching off the Tethys Sea, an extensive shallow sea that separated Europe and much of Asia from Africa and India. It now has only two outlets, the Straits of Gibraltar and the Bosporus, a narrow strait between the Mediterranean and Black Seas. While the central basin of the Mediterranean reaches depths of several thousand yards, there is a sill under the Straits of Gibraltar that is only 1,970 feet (600 m) below the surface. Through this passageway flows surface water from the Atlantic Ocean.

Since the Mediterranean is situated in one of the world's **arid** belts, the inputs from precipitation and rivers is far less than the water lost through evaporation. If the straight at Gibraltar were to close due to further plate movements, the Mediterranean would dry up. In fact, data from the Deep Sea Drilling Project, seismic surveys, and fossil analysis have found evidence of salt deposits, ancient river valleys, and fresh water animals, all suggesting that this has occurred at least once. Since the African and Eurasian plates are moving together, this will probably happen again.

Humans can do nothing about this impending geological disaster. There are, however, events that people can influence. Domestic sewage, industrial **discharge**, agricultural **runoff**, and **oil spills** are seriously threatening the Mediterranean, fouling its once clear waters, altering its chemical cycling, and killing its organisms. Along its northern coastline are some of the most heavily industrialized nations in the world, whose industries are destroying nearshore nursery habitats, damaging fisheries. **Dams** on inflowing rivers reduce the **sediment** inputs, making coastal **erosion** a major problem. Shipping, once the hallmark of Mediterranean civilization, releases every manner of waste into the Sea, including oil. Annually, 6 million barrels of oil end up in the Mediterranean. The limited water circulation patterns of the Mediterranean compound this problem as pollutants accumulate.

Today seafood contamination and eye, skin, and intestinal diseases are frequently experienced by coastal residents. Marine mammal and sea turtle populations are threatened by **habitat** loss and nondegradable pollutants dumped into the waters. Sea grass (*Posidonia oceanica*), which provides food and habitat for some 400 **species** of algae and thousands of species of fish and invertebrates, is disappearing. **Nutrient** enrichment of the Mediterranean results in large **plankton** blooms which, combined with destructive fishing practices, contribute to the demise of the sea grass beds.

These problems have been recognized, and efforts are being made to reverse the declining health of the Mediterranean. Early efforts included the 1910 construction of one of the first institutions for study of the seas, the Musée Oceanographique by Prince Albert I of Monaco. Since then, the conflicts between the political and religious ideologies of the 18 nations surrounding the Mediterranean have been major hurdles in completing cleanup plans. In 1976, the *Mediterranean Action Plan* was signed by 13 of the nations. A major component of this agreement was the Blue Plan, a study of future effects of increasing coastal populations. Other efforts include the Genoa Declaration in 1985 and the Nicosia Charter in 1990. The latter commits resources of the community, the **World Bank**, the European Investment Bank, and the **United Nations Environment Programme** to achieve a Mediterranean **environment** compatible with **sustainable development** by 2025. Hopefully, these efforts can reverse the decline of this natural wonder. *See also* Algal bloom; Biofouling; Commercial fishing; Environmental degradation; Ocean dumping; Water pollution

[*William G. Ambrose and Paul E. Renaud*]

RESOURCES
BOOKS

Heezen, B. C., and C. D. Hollister. *Faces of the Deep.* London: Oxford University Press, 1971.
Thurman, H. V. *Essentials of Oceanography.* Columbus: Merrill, 1983.

PERIODICALS

Batisse, M. "Probing the Future of the Mediterranean Basin." *Environment* 32 (1990): 4–15.

Megawatt (MW)

A megawatt is a unit of power equivalent to one million watts (10^6 watts), or one thousand kilowatts. As a unit of power, a megawatt expresses the rate at which energy is produced. A megawatt is equivalent to one million joules per second. A megawatt is a fairly large unit of power and is used, therefore, when discussing the size of a power plant, a nation's total energy-generating capacity, or some other such large statistic. For example, the total electrical energy generating capacity for the state of California in 1988 was 45,900 megawatts. *See also* Energy and the environment; Energy efficiency

Chico Mendes (1944 – 1988)
Brazilian union leader, conservationist, and activist

Francisco Alvo Mendes Filho, known as Chico Mendes, was a defender of the tropical rain forests and a champion of the concept of sustainable harvest as a means of saving and protecting that threatened **ecosystem**. As president of the local Rural Workers Union, representing **rubber** tappers in his native Brazil, Mendes became too powerful and politically influential for ranchers who wanted to turn the **rain forest** into grazing land for their cattle. The struggle between them ended in 1988, when Mendes was assassinated.

Chico Mendes. (Corbis-Bettmann. Reproduced by permission.)

Chico Mendes was born in 1944 in Acre Province of Brazil, along the upper reaches of the Amazon River not far from the border with Peru and Bolivia. Following his father, he was a *seringueiro*, a rubber tapper. He farmed a small clearing, but relied on the sale of rubber from several hundred native rubber trees in the rain forest itself to provide income for him and his family. Mendes inherited the land and the trees from his father who had begun tapping them in the 1930s. Two long v-shaped cuts made with care in the bark of each rubber tree would yield one or two cups of the milky latex sap each week, which could then be dried to make natural rubber.

Mendes would also collect other natural forest products, such as fruits and Brazil nuts, to supplement his income. There are approximately 100,000 other rubber tappers living throughout the rain forest, and this is what they do as well. It is sustainable harvest which does not destroy the forest and provides a substantial income. On average, **logging** yields a one-time profit of $1,290 per acre; if the land is converted to cattle pasture, it yields an income of about $61 per acre per year. The sustainable yield of forest products, on the other hand, provides an income of $2,762 per acre per year.

Land speculators and large cattle ranching concerns are more interested in the short-term profits they can realize by cutting down the forest, selling the timber, and converting the land for cattle grazing. Ranching requires huge tracts of land, and satellite images indicate that in 1988 alone about 30 million acres of forest was destroyed for this industry. Low land prices, low tax rates, and direct subsidies to ranchers further encourage this practice. The Brazilian government had further aided ranching by building and maintaining roads into the forest, which are then used to ship cattle to market.

Chico Mendes fought to end this destruction of the **tropical rain forest**. He made many political inroads, gaining influence with the Interior Ministry as well as the public. His main adversary was Darli Alves da Silva, a cattle rancher who had begun acquiring forest land in Acre through strong-arm tactics and he vowed that Mendes would not live out 1988. Mendes had helped establish several forest reserves that year, thus all but ending forest clearing in Acre.

Mendes, his wife, and two policemen assigned to guard him were playing cards at his home on the night of December 22, 1988. Mendes stepped outside for a moment and was killed by a shotgun blast to the chest from a waiting assassin. The local police claimed no clues or suspects in the case, but local and international protests forced the Brazilian government to enter the investigation. Evidence led them to the ranch of Darli da Silva. In the summer of 1989 indictments for murder were handed down to Darli da Silva, his son Darci Pereia da Silva, and Jerdeir Pereia, one of da Silva's ranch hands. Testimony indicated that Darli ordered the murder and that Darci supervised as Jerdeir carried out the plot. There is evidence that other prominent ranchers may have been involved in the plot, and they are currently under investigation.

[*Eugene C. Beckham*]

RESOURCES
BOOKS

Dwyer, A. *Into the Amazon: The Struggle for the Rain Forest.* San Francisco: Sierra Club Books, 1990.
Revkin, A. *The Burning Season: The Murder of Chico Mendes and the Fight for the Amazon Rain Forest.* New York: Houghton Mifflin, 1991.

PERIODICALS
Willrich, M. "Murder in Acre." *Amicus Journal* 11 (Spring 1989): 10–13.

Mercury

Mercury is a naturally occurring element in minerals, rocks, **soil**, water, air, plants, and animals. The predominant forms in the **atmosphere**, water, and **aerobic** soils and sediments are elemental and mercuric mercury; while cinnabar is commonly found in mineralized ore deposits and **anaerobic** soils and sediments. Mercury is present throughout the atmosphere because of its relatively high vapor pressure. It vaporizes from the earth's surface and is transported in a global cycle, sometimes for hundreds of kilometers, before being deposited again with particulates, rain, or snow. The background concentrations in rocks and soils typically range between 20 and 100 µg Hg/kg with a worldwide average of about 50 µg Hg/kg. Natural background concentrations in the uncontaminated atmosphere are in the order of between 1 and 10 ng/m^3 increasing to between 50 and 1,000,000 ng/m^3 or more over mineralized areas. Mercury is transported to aquatic ecosystems via surface **runoff** and **atmospheric deposition**. Airborne concentrations associated with **anthropogenic** activities such as **coal** burning, smelting, industry, and **incineration** range between 100 and 100,000 ng/m^3.

The annual worldwide production from cinnabar was about 11,500 metric tons in 1990. The element can be divided into two major categories, organic and inorganic. Inorganic mercury includes the elemental (Hg^0) silvery liquid metal (mp, 38°C; bp, 357°C) as well as mercurous **ion** (Hg$^+$), mercuric ion (Hg^{++}), and their compounds. Organic mercury includes chemical compounds which contain **carbon** atoms that are covalently bound to a mercury atom, such as methylmercury (CH$_3$-Hg$^+$).

During the latter half of the twentieth century, inorganic mercury was used extensively to produce caustic soda and **chlorine** as well as to manufacture batteries, switches, street lamps, and fluorescent lamps. Gold mining, dental amalgams, pharmaceuticals, and other consumer items also consume inorganic mercury. Organic mercury applications have mostly been eliminated in agricultural fungicides, slimicides in paper pulp production, bacteriostats in water based paints, and industrial catalysts.

Over the centuries the symptoms of inorganic mercury **poisoning** were well documented by the exposure of miners and industrial workers as mercury accumulated in their brains, kidneys, and livers. Loose teeth, tremors, and psychopathological symptoms were common at low exposure, but removal from the source would often enable the victims to recover. However, the effects of organic alkyl mercurials, such as methylmercury, were more severe. With a **half-life** in the human body of about seventy days, continued exposure elevates the levels. It also crosses the blood/brain and placental barriers, attacking the central nervous system and inducing teratogenic changes in the fetus. The neurological symptoms include: loss of coordination in walking; slurred speech; constriction of the field of vision; loss of sensation, especially in the fingers, toes, and lips; and loss of hearing. Severe poisoning can cause coma, blindness, and death.

The concentrations of mercury in the ocean and uncontaminated freshwater are generally believed to be less than 300 and 200 ng/l respectively. However, new ultra clean analytical techniques indicate that the actual concentrations may be three to five fold lower. In contaminated aquatic systems concentrations as high as 5 μg Hg/l have been reported. In the water column, mercury readily adsorbs onto organic particulates, metal oxides, and clays. Then they settle into the sediments. Historically, depending on their location, the natural background concentrations of mercury in sediments have ranged between 10 and 200 μg/kg. However, most aquatic systems have received some mercury contamination, and the rate has increased during the past century. Among sites that have been measured, the total concentrations have usually been from five to ten times greater than background and ranged from less than 0.5 mg Hg/kg (dry weight) in remote areas to 2010 mg Hg/kg (dry weight) in Minamata Bay, Japan.

In the aquatic **ecosystem** inorganic mercury is converted to methylmercury by both biotic and abiotic processes. It is then released, and aquatic organisms bioaccumulate it easily and metabolize and excrete it very poorly. The biological half-life in fish may be as long as one to three years. Exposed organisms at each level of the food chain bioconcentrate methylmercury and pass it on to animals at the higher trophic levels.

Depending on the **species** of fish and the type and amount of mercury being released from the sediments, it may be magnified biologically from 1,000 and 100,000 times or more. While background levels of total mercury in freshwater and marine fishes from unpolluted waters typically range from less than 0.1 to about 0.2 mg Hg/kg, higher concentrations are found in some pelagic top predator ocean fishes such as tuna and shark, sometimes exceeding 1.5 mg/kg. Conversely, fish from contaminated waters typically contain levels between 0.5 and 5.0 mg Hg/kg and up to 35 to 50 mg Hg/kg in highly contaminated areas.

Several standards have been developed to protect the public's health from the threat of mercury poisoning. The **maximum permissible concentration** allowed by the United States **Environmental Protection Agency** (EPA) under its drinking water standards is 2 μg Hg/l. The United States **Food and Drug Administration** guideline for mercury in seafood is 1 mg Hg/kg freshweight; however, some states, such as Michigan, adhere to a more restrictive guideline of 0.5 mg Hg/kg freshweight. The Food and Agriculture Organization of the United Nations (FAO), on the other hand, recommends a provisional tolerable intake (PTI) of 0.3 mg mercury per week for a person weighing 154 lb (70 kg), of which no more than 0.2 should be in the methylated form. *See also* Biological methylation; Birth defects; Food chain/web; Minamata disease; Teratogen; Water pollution; Xenobiotic

[*Frank M. D'Itri*]

RESOURCES

BOOKS

D'Itri, F. M., et al. *An Assessment of Mercury in the Environment*. Washington, DC: National Academy of Sciences, 1978.

PERIODICALS

D'Itri, F. M. "Mercury Contamination: What We Have Learned Since Minamata." *Environmental Monitoring and Assessment* 19 (1991): 165–82.

Metabolism

The sum total of biochemical reactions occurring in living organisms by which energy is made available to the organism. Metabolism consists of *catabolism*, the chemical breakdown of large molecules into their smaller molecular components (e.g., from proteins to amino acids), and *anabolism*, the reconstruction or chemical synthesis of large molecules. In physiology, metabolism also describes regulated sequences of chemical reactions (physiological pathways), such as protein metabolism or urea metabolism. In **ecology**, the metabolism of lakes or ponds is the sum of the chemical reactions taking place between the inhabitants and the **environment**.

Metals, as contaminants

Metals, objects made from metal, and chemical compounds containing metal are pervasive in the home, workplace, and **environment**. Although metals play an essential role in modern human society, there are many instances where they occur as unneeded and harmful contaminants. **Heavy metals** such as **mercury**, **cadmium**, **lead**, **uranium**, chromium, manganese, **nickel**, thallium, and bismuth are all industrially useful, but they are also potentially harmful at high concentrations in the environment or when found in food or drinking water. Environmental contamination with heavy metals can result from many activities including mining, large-scale **combustion** of **coal** in heating or **power plants**, mineral and metal processing, and use of fungicides and pesticides.

The metal mercury (Hg), a useful component of many products including thermometers, pesticides, drugs, paints, and batteries, is unusual in that it is a liquid at ordinary temperatures. It forms a variety of useful salts and organic compounds. Misuse of products containing mercury can lead to serious health problems. Disorders caused by mercury depend on the form of mercury ingested and the method of exposure. Ingestion of mercury salts can lead to kidney damage and death. Exposure to mercury vapor can lead to inhalation or **absorption** through the skin. Brain function is affected, and loss of memory, depression, anxiety, and other personality changes may result. Ingestion of organic mercury compounds used in pesticides can result in permanent neurological damage. One very serious mercury **pollu-**

tion incident occurred in the 1950s at Minamata Bay, Japan, where releases of mercury in **effluent** from a manufacturing plant were ingested by fish. People eating the fish developed serious, and in some cases fatal, neurological maladies. Infants and children were particularly vulnerable to this "Minimata disease."

Lead (Pb) has been used for many purposes for thousands of years. However, toxicity has been largely unnoticed or ignored until recent decades. Lead uptake by the body is quite slow, but its rate of excretion is even slower. Thus in cases of long term exposure, lead levels in the body gradually increase. There are many sources of contamination. Lead water pipes were once used to carry domestic water. Until recently, lead-containing solder was use to join **copper** piping in homes. Lead contamination of drinking water resulted, and symptoms of lead toxicity were noted in those who used the water. Until recently, white lead pigment was widely used in house paint. Deteriorating and chipping paint poisoned occupants, particularly children, who accidentally ingested or inhaled dust and fragments. Symptoms of lead toxicity can be both chronic and acute. Weakness, loss of appetite, **anemia**, vomiting, and convulsions have all been reported. Lead causes lesions in the central nervous system and serious long-term damage.

Cadmium (Cd) is useful both as an element and in compounds. The element is a lustrous silvery metal that is usually associated in **nature** with zinc. It is used to plate steel, copper, brass, and other alloys and serves to retard corrosion. Cadmium oxide is used in nickel-cadmium storage batteries and as a pigment in paints and inks. Although metallic cadmium is not dangerous under ordinary conditions, cadmium ions, produced when the metal is attacked by mineral acids, are very toxic. Toxicity symptoms resemble those of mercury.

Uranium (U) with an atomic number of 92, is one of the heaviest of all elements. Once thought to be rare, it is now known to be widely distributed in the Earth's crust. Its concentration in sea water exceeds those of mercury and cadmium and is at about the same level as copper and lead. Interest in uranium and its use expanded dramatically with the development of atomic power and **nuclear weapons**. Environmental and health concerns have centered on the radioactive nature of uranium, but chemical toxicity is also a potential problem. Uranium contamination surrounding mining and mill **tailings** in southwestern United States has been a major concern for Native Americans living nearby.

Each of the heavy metals has its own specific toxic effects on human health and the environment. Before this fact was generally realized, mining, smelting and manufacturing activities often resulted in appalling instances of environmental contamination. While greater concern has resulted in improvement, serious problems remain in instances where proper control has not been practiced.

[*Douglas C. Pratt Ph.D.*]

RESOURCES
BOOKS

Baselt, R. C. *Disposition of Toxic Drugs and Chemicals in Man.* 3rd ed. Chicago: Year Book Medical Publishers, 1989.

Eichstaedt, P. H. *If You Poison Us: Uranium and Native Americans.* Santa Fe.: Red Crane Books, 1994.

Friberg, L., et al., eds. *Handbook on the Toxicology of Metals.* 2nd ed. Amsterdam, New York: Elsevier, 1986.

Meteorology

Meteorology is derived from the Greek words *meteora* meaning things in the air or things above, and *logy* meaning science or discourse. It is a branch of physics concerned with the study and theory of atmospheric phenomena and is frequently equated to atmospheric science. One of the earliest references to this branch of physics is Aristotle's *Meteorologica* written around 340 B.C.

In the modern context meteorology is founded upon the basic physical principles and laws governing the energy and mass exchanges within the earth's **atmosphere** and involves the study of short term variations of atmospheric properties (temperature, moisture, wind) and interactions with the earth's surface. The ability to predict and explain short term changes in the atmosphere from observations and numerical models (using the laws of physics) is an important dimension of meteorology as well. Thus the words meteorologist and forecaster are often used interchangeably to describe someone who can predict the weather.

Meteorologists are trained in observations, instrumentation, data processing, and **modeling** techniques for the purpose of analyzing and predicting trajectories of major weather systems, including their associated temperature, precipitation, wind, and sky conditions. Modern methods include the use of automated surface observation systems, radar, satellites, radiosondes, wind profilers, and high resolution computer models (sometimes called global circulation models) to estimate temporal and spatial variability. *See also* Acid rain; Climate; Cloud chemistry; Hydrologic cycle; Photochemical smog

Methane

An organic compound with the chemical formula CH_4, methane occurs naturally in air at a concentration of about 0.0002 percent. It is produced in processes such as the **anaerobic** decay of organic matter, the growth of certain types of plants, and the belching of cattle. Methane is the major component of **natural gas**, making up about 85 percent of that fuel. Environmental scientists are increasingly concerned about methane as a possible greenhouse gas. Like **carbon dioxide**, methane traps heat reflected from the earth and, therefore, may contribute to global warming. Increases in agricultural and dairying activities have resulted in an increase in methane production, possibly contributing to **climate** change. *See also* Greenhouse effect; Greenhouse gases

Methane digester

Methane digesters are systems that use anaerobes to produce methane through fermentation. Methane is a main constituent of **natural gas** and can be readily substituted for that nonrenewable resource.

The anaerobes used in methane digesters are methanogenes, bacteria belonging to the genera *Methanobacterium*; *Methanosarcina*; and *Methanoccus*. They can be found in the gastrointestinal tracts of animals such as cows and other ruminants, as well as in **soil**, water, and sewage. In septic tanks, bacteria liquefy some of the organic matter; which releases energy for the bacteria and by products such as methane and **carbon dioxide**.

Methane digesters are also known as biogass digesters and organic digesters. The central portion is an airtight drum, called a digester unit, which contains the methanogenes. Raw material is place into the drum, and the unit is kept at a constant temperature of about 95°F (35°C). Some of the methane produced by the digester heats the water. Outlets on the digester unit take away the various products of the system. Liquid and solid fertilizers are collected to be used for crops and other plants. Methane is stored in a tank, from which it can be drawn off for fuel for a variety of purposes. **Carbon** dioxide and **hydrogen** sulfide can be filtered out of the methane and put under pressure for use in turning turbines.

Common household organic wastes can be put into a digester, and the methane produced can be used to make electricity. It can also be used for cooking, illumination, heating, and **automobile** fuel. One system at the University of Maine produced over $8,000 worth of power a year in the early 1990s; the digester also produced a **sludge** that could be used as a nutritious and relatively odorless plant **fertilizer**. Some sludges, however can contain high level of metals if the original material is unsorted municipal waste.

Methane is a clean and nontoxic automobile fuel, and it produces no pollutants when burned. It has an octane number of 130. Italy has used it as a motor fuel for over 40 years, and Modesto, California, has a small fleet of methane-powered cars. Because of its cleanliness, it extends the life of engines as well as making starting easier.

The organic matter this system converts into other uses would otherwise have decomposed in a **landfill, leaching** into the surrounding **environment** and contaminating **groundwater** supplies. In a digester, it becomes a useful resource. *See also* Alternative energy sources; Alternative fuels; Anaerobic digestion; Bioremediation; Groundwater pollution

[*Nikola Vrtis*]

RESOURCES
PERIODICALS

Trans, W. B. "Just Plug It Into That Cherry Tomato Over There." *Sierra* 75 (May-June 1990): 20–21.

Methanol

Methanol is an organic compound with the chemical form CH_3OH. It is also known as methyl alcohol or wood alcohol. Like most alcohols, methanol is very toxic. Its ingestion can cause severe nerve damage leading to blindness, insanity, and death. Methanol can be prepared through destructive distillation of **coal**, wood and wood products, **garbage**, sewage **sludge**, and other forms of **biomass**. It is an excellent automotive fuel and has long been used to power racing cars. Existing methods of production are still too expensive, however, to make it an economically viable alternative to **gasoline** in the general market. *See also* Alternative fuels; Fuel-switching

Methyl tertiary butyl ether

Methyl tertiary butyl ether (MTBE) is a flammable, volatile, and colorless liquid fuel additive that is manufactured by the chemical reaction of **methanol** and isobutylene. It is one of a group of **chemicals** referred to as oxygenates, because they raise the level of oxygen in **gasoline**. Oxygen helps gasoline burn more completely, thus reducing emissions.

MTBE has a strong odor similar to a general anesthetic. Humans can detect it by smell at very low concentrations: 53 **parts per billion** (ppb) in air and 20 to 40 ppb in water. MTBE is very soluble in water and more soluble than other gasoline constituents. It is also persistent and resistant to degradation.

As a gasoline additive, it is used to increase the octane level (replacing the use of **lead** additives) and to reduce vehicular emissions of **carbon monoxide** (CO) and ozone-forming pollutants. It was first used in the United States in the 1970s, with its use increasing during the 1990s, when the **Clean Air Act** Amendments of 1990 created the Oxygenated Fuel Program. Areas of the country with severe **air pollution** problems are required to add oxygenates to gasoline during winters to reduce vehicular emissions of CO. Although regulatory requirements do not specify the types of oxygenates required, MTBE and **ethanol** are the main oxygenates used. Ethanol is the primary oxygenate used in the Winter Oxyfuel Program. Areas with problems meeting **ozone** standards require the use of reformulated gasoline (RFG) year round to reduce ozone-forming pollutants. RFG is oxygenated gasoline (with a minimum of 2% oxygen by weight) that is specially blended to have fewer polluting compounds than conventional gasoline. As the addition of ethanol increases the vapor pressure of gasoline such that it is difficult for ethanol-containing gasoline to comply with summertime volatility requirements, MTBE has been the preferred oxygenate for use during the summer. In 1970, MTBE was the 39th-highest produced organic chemical in the United States; by 1998, it was the 4th-highest produced organic chemical. In 1999 over 200,000 barrels of MTBE per day were produced.

The use of RFG has improved the quality of air in the United States. The U.S. **Environmental Protection Agency** estimates that smog-forming pollutants are being reduced annually bey at least 105 million tons and toxic chemicals by at least 24 million tons.

However, the success of the **air quality** program that utilizes MTBE to reduce emissions has been offset by contamination of ground and surface waters with MTBE. MTBE has been detected in **soil**, surface water, and ground water throughout the United States. Point sources of MTBE include releases from underground and above-ground storage tanks and pipelines due to leaks, overfilling, and faulty construction. Non-point sources of MTBE include **urban runoff**, precipitation, vehicle accidents, and motorized water craft. MTBE may also be released to the **environment** through gasoline fumes during vehicle refueling. A study by the U.S. **Geological Survey** in 1993–1994 showed that MTBE was the second most frequently detected volatile organic chemical of 60 chemicals in samples collected from shallow ground water in eight urban areas. Another report in 1999 by the U.S. Environmental Protection Agency's Blue Ribbon Panel on Oxygenates in Gasoline indicated that between five and ten percent of drinking water **wells** in areas with high use of MTBE had detectable levels of MTBE, while one percent had levels higher than 20 micrograms per liter. During the summer of 1996, the City of Santa Monica in California stopped pumping ground water from two of its well fields because of persistent and increasing levels of MTBE. These two wells provided about 50% of the city's drinking water supply.

MTBE has been shown to have the potential to produce adverse effects associated with central nervous system depression, including headaches, dizziness, nausea, and disorientation. These effects are reversible if exposure is discontinued. Evidence from animal studies suggests that MTBE is a potential human **carcinogen**. When it enters the human body through inhalation or **absorption** through the skin, it may metabolize into two compounds, tertiary butyl alcohol and formaldehyde, that are carcinogenic in animals and are also classified as potential human carcinogens.

In 1997 the U.S. Environmental Protection Agency issued a Drinking Water Advisory for MTBE. The Advisory recommended that the levels of MTBE in drinking water by limited to 20 to 40 micrograms per liter. This limit was set to assure consumer acceptance with regards to taste and odor and to provide an adequate margin of safety from toxic effects. Based on this Advisory, several states have set their own drinking water standards for MTBE. A nation-wide standard for MTBE under the **Safe Drinking Water Act** is not expected until after February 2005, due to lack of information regarding health effects and occurrence data. In March of 2000, the U.S. Environmental Protection Agency formally began regulatory action to eliminate or phase down MTBE, issuing an Advance Notice of Proposed Rulemaking under Section 6 of the Toxic Substances Control Act, which gives the U.S. Environmental Protection Agency authority to ban, phase out, limit or control the manufacture of any chemical substance deemed to pose an unreasonable risk to the public or the environment. The State of California is considering a ban on the use of MTBE by the end of 2003. As of 2002, bans on MTBE are also under consideration in at least 16 other states. Research is being conducted to develop alternative oxygenates so if MTBE is banned, air quality benefits of the use of oxygenated fuels will be preserved.

Because of its physical properties, during a release MTBE migrates rapidly through the soil column. Upon reaching the ground water, MTBE moves at the same velocity as water and faster and farther than other gasoline constituents. As the gasoline contaminant **plume** degrades over time, **benzene, toluene**, ethyl benzene, and **xylene** (BTEX) decrease in concentration more than MTBE, which is more resistant to degradation. Eventually MTBE may be the only contaminant remaining from the release.

The extent of MTBE contamination in the subsurface has led to the investigation of treatment technologies. If MTBE in soil is not dissolved in water, it can be removed by such treatment methods as soil vapor extraction. However, the high solubility of MTBE in water and its **resistance** to degradation makes it difficult and time consuming to remove from ground and surface waters using many common remedial technologies. Promising technologies for removal of MTBE that are being studied include chemical oxidation using ultraviolet light and **hydrogen** peroxide, **phytoremediation** using deep-rooted trees, and cometabolic degradation processes.

[*Judith L. Sims*]

RESOURCES

BOOKS

Jacobs, James J., Jacques Guertin, and Christy Herron. *MTBE: Effects on Soil and Groundwater Resources.* Boca Raton: Lewis Publishers, 2000.

OTHER

Report on Methyl Tertiary Butyl Ether (MTBE). Arizona Department of Environmental Quality, October 1, 1999.

U.S. Environmental Protection Agency. *Methyl Tertiary Butyl Ether (MTBE).* June 22, 2001 [cited June 23, 2002]. <http://www.epa.gov/mtbe>.

Methylation

A chemical reaction in which the methyl radical ($-CH_3$) is added to some substance. The most common mechanism by which methylation occurs in the **environment** is **biological methylation**, which involves the action of living organisms. Bacteria in oxygen-poor soils, for example, can convert metallic **mercury** to an organic compound, methyl mercury. Similar reactions occur with other metals, including **arsenic**, selenium, tin, and **lead**. These reactions are significant because they convert non-soluble metals of low toxicity into soluble forms with high toxicity. Light can also induce methylation. Photomethylation has occurred in the laboratory, but its relative importance in environmental systems is not yet well understood.

Methylmercury seed dressings

Seed dressings were devised to prevent diseases caused by a wide variety of seed-borne plant-pathogenic **fungi**, to protect the germinating seeds against secondary infections, and to increase crop yields. Various **chemicals**, including several **heavy metals**, have been used as fungicides to treat seeds since the end of the nineteenth century. The effectiveness of these fungicides was greatly increased when aryl organomercurials were introduced around 1914. They had a wider spectrum of fungicidal activity than nonmercurial formulations and were used extensively until the mid-1980s to control fungus diseases through their application as seed dressings on many grains, vegetables, and nuts, as well as to protect fruit trees, rice, turf grasses, and **golf courses**. However, with the introduction of the more effective alkyl **mercury** compounds in the 1930s, especially methyl mercury

and ethyl mercury, severe **poisoning** incidents followed. In developing countries hundreds of people died or became incapacitated due to either the consumption of grains treated with alkylmercury compounds or meat from animals that had eaten such treated seeds.

Poisonings from eating alkylmercury-treated grains occurred on several occasions in various parts of Iraq. Destitute, illiterate rural families either did not understand the words or poison symbols on the bags of grain or did not believe government warnings that it was unsafe to eat. In some instances the families fed some of the grain to chickens and swine first. When they did not observe poisoning symptoms in the livestock and poultry after a few days, the farmers became convinced that the warnings were false and the grain was safe to eat. However, depending on the amount of methyl mercury consumed, there is a **latency** period of weeks or months between exposure and the development of poisoning symptoms.

When the seed grain was ground into flour, baked into bread, and consumed by the rural victims, both sexes and all ages were affected. The ingested quantities of methylmercury ranged from small amounts that produced no overt effects to lethal doses. Fetuses suffered the most damage. Among the rest of the population, the severity of the neurological and psychiatric symptoms was almost directly proportional to the amount of bread consumed. In cities where bread was produced from government-inspected flour mills, not a single case of poisoning was reported.

The first documented incident occurred in Iraq in 1956. Of the two hundred persons afflicted, seventy died. In 1960 an estimated 1000 persons were affected in a similar incident and over 200 died. During the 1960s other smaller but similar episodes of alkylmercury poisoning occurred on a more limited scale in West Pakistan, Guatemala, Ghana, and in other countries such as Mexico and the United States. The total death toll in these countries was 42 with approximately 197 individuals less seriously affected.

But the most serious outbreak of poisoning from eating methyl mercury-poisoned bread occurred in Iraq early in 1972. In October and November of 1971 a total of about 73,000 tons of high-yield Mexipac wheat seed grain and 22,000 tons of barley seed grain, all treated with alkyl mercury **fungicide**, had been distributed by cooperatives to farmers for planting throughout the country. Some of this grain, as before, was used to prepare homemade bread. The Iraqi government estimated that the treated grain was distributed to no more than five percent of the rural population, about 200,000 people. Most of the fatalities occurred within three months after the end of the exposure, although a few long-term illnesses resulted in fatalities as well.

By March 1973, up to 40,000 persons, residents of every province, were unofficially estimated to have been poisoned. The total number of casualties will never be known precisely because hospitals were quickly overloaded, and many victims did not have access to them. In addition, most of the poisonings occurred in rural areas, and many were not reported to authorities. The government officially acknowledged that 6,530 persons were hospitalized and 459 died. These figures were not confirmed because news reporters were denied entry to the country and the movements of foreigners were restricted. However, tourists reported that large numbers of Iraqis suffered brain damage, blindness, and paralysis. Since then, with the exception of a few follow-up scientific reports published between 1985 and 1989, hardly any new information relative to the long term health effects on the thousands of victims has been published or released by the Iraqi government although this was the largest such tragedy of this kind. Because of the highly toxic nature of the alkyl mercurials, as well as the severity of the accidents caused by misuse of the treated seeds, they were banned in 1970 in the United States and many other countries. *See also* Agricultural chemicals; Birth defects; Heavy metals and heavy metal poisoning; Minamata disease; Plant pathology; Teratogen; Xenobiotic

[*Frank M. D'Itri*]

RESOURCES
BOOKS

D'Itri, P. A., and F. M. D'Itri. *Mercury Contamination: A Human Tragedy.* New York: Wiley-Interscience, 1977.

PERIODICALS

Bakir, F., et al. "Methylmercury Poisoning in Iraq." *Science* 181 (1973): 230–241.

"Conference on Intoxication Due to Alkyl Mercury-Treated Seed." *World Health Organization* 53 (Suppl.) (1976): 138.

Greenwood, M. R. "Methylmercury Poisoning in Iraq: An Epidemiological Study of the 1971-72 Outbreak." *Journal of Applied Toxicology* 5 (1985): 148–159.

Mexico City, Mexico

Founded in the fourteenth century, Mexico City has been a center for three great civilizations: the Aztecs, the Spanish, and the modern-day Mexicans. But in addition to an imposing political background, its geographical location has assured the city a fascinating ecological and **environmental history**. Mexico City lies in a basin 7,350 feet (2,240 m) high. It is surrounded by mountain ranges on all sides, and the presence of the extinct volcanoes Ixtacihuatl and Popocatepetl to the east are a reminder that the city lies on an active **earthquake** fault.

Most of Mexico City's environmental problems are caused by a combination of its geographical location and growing population. In 1900, the population was estimated at 350,000; the rest of the century has seen nothing but rapid growth. Population has leapt from 1,029,000 in 1930 to 4,871,000 by 1960, and then to 12,000,000 in the mid-1970s, and then 15,000,000 by 1981. According to some estimates, Mexico City will have more than 32 million inhabitants by the year 2000, making it the most populous urban area in the world.

The amount of **pollution** produced by a city of this size would be difficult to control in even the most favorable geographic circumstances. But the basin in which Mexico City is built traps the **ozone**, **nitrogen oxides**, sulfur, and particulates that are released each day. Soft **coal**, wood, and low-grade **gasoline** and oil are burned widely throughout the city, contributing greatly to this problem. In addition, prevailing winds from the northeast carry dust particles into the city from rural areas, further degrading **air quality**. The city has gone from having one of the most perfect natural settings and ideal climates in the world to being the most heavily polluted. Pollution levels can rise so high that schools are occasionally forced to close so students can remain indoors and not breathe the polluted air. Entrepreneurs have even set up booths on city streets where people can pay to breath oxygen from tanks.

Lying near the boundary of the Pacific and North American geological plates, Mexico City has long been at risk for major earthquakes, a risk which has been greatly increased by the city's history. When the Aztecs first settled the area, it was largely covered by an enormous lake, Lake Tenochititlán, which either was filled in or dried out as the city began to grow. Today, Mexico City sits on a soft **subsoil** that is highly unstable. Some parts it are actually sinking into the old lake bed, while the whole area rides out each earthquake like a boat on an unsettled ocean.

The geographical instability of the area has been worsened by the fact that residents traditionally obtain their water from **wells**. As the population grows, more water is removed and **subsidence** increases. In some parts of the old city, buildings have actually sunk more than 6 ft (2 m) below street level. Since measurements were first made in 1891, subsidence in some areas has exceeded 26 ft (8 m), and it measures at least 13 ft (4 m) in nearly all parts of the city. *See also* Air pollution control; Air quality criteria; Los Angeles Basin; Nitrogen cycle; Population growth; Sulfur cycle; Sulfur dioxide

[*David E. Newton*]

RESOURCES
BOOKS

Poland, J. F., and G. H. Davis, "Land Subsidence Due to Withdrawal of Fluids." In *Man's Impact on the Environment*, edited by T. R. Detwyler. New York: McGraw-Hill, 1971.

Polluted air over Mexico City, Mexico. (Photograph by A. J. Copley. Visuals Unlimited. Reproduced by permission.)

PERIODICALS

"Line Up to Breathe." *BioScience* (September 1991): 591.

"School Days and Lethal Haze." *Environment* 30 (March 1988): 23.

Microbes (microorganisms)

Microbes, also known as microorganisms, are defined solely on the basis of their size—they are too small to be seen by the naked eye. Microorganisms can only be viewed if they are magnified in size, using an optical or electron microscope.

Apart from their size, the major groups of microorganisms have little affinity in terms of their evolutionary history and systematics. Included among the microorganisms are viruses, bacteria, blue-green bacteria, some algae, some **fungi**, yeasts, and protozoans.

Viruses

Although viruses are commonly considered to be microorganisms, they are actually "pseudo-organisms" because they do not display all of the characteristics of life. Viruses consist only of bits of **nucleic acid** (either deoxyribonucleic **acid** (DNA), or **ribonucleic acid** (RNA)) surrounded by a protein capsule (called a capsid). Viruses are not capable of independent reproduction, and they cannot perform other important metabolic functions. To reproduce and grow viruses must invade and parasitize the living cells of other organisms, and appropriate the metabolic capabilities of their host.

It is probable that all living cells are infected by viruses of various sorts, and in some cases serious diseases are caused. Viral diseases of humans include colds, flu, and more deadly ailments such as smallpox, yellow fever, rabies, herpes, polio, and human immunodeficiency syndrome (HIV, also known as **AIDS**). Some viral diseases can be controlled by vaccination, a practice that involves infecting hosts with killed but intact **virus** bodies. These are non-virulent but nevertheless cause the host to develop **resistance** to pathogenic forms of the strain. Vaccination made it possible to achieve the eradication of smallpox, a disease that had long been a scourge of humans.

Bacteria and Blue-green Bacteria

Bacteria and blue-green bacteria are in the kingdom of life known as Monera. Most **species** in this group are bacteria; these have rigid or semi-rigid cell walls, propagate by binary division of the cell, and do not display mitosis during cell division. Blue-green bacteria or cyanobacteria (sometimes incorrectly referred to as blue-green algae) utilize chlorophyll dispersed within their cytoplasm as a pigment for capturing light energy during **photosynthesis**. The genetic material of monerans is organized as a single strand of DNA, and it is not contained within a membrane-bounded organelle called a nucleus, so these microorganisms are referred to as being prokaryotic. (All of the other kingdoms have a nucleus within their cells, and are known as eukaryotic.) In addition, prokaryotes do not display meiosis or mitosis, their reproduction is by asexual cellular division, and they do not have organelles such as chloroplasts, mitochondria, or flagella. Prokaryotes were the first organisms to evolve, about 3.5 million years ago. It was not until 2 million years later that the first eukaryotes evolved.

About 4,800 species of bacteria have been named, but microbiologists believe that there are many additional species that have not yet been discovered. Bacteria are capable of exploiting an astonishing range of ecological and metabolic opportunities. Some species can only function in the presence of oxygen, and others only under **anaerobic** conditions, although some species are able to opportunistically switch between these metabolic types. Some bacteria can tolerate very extreme environments, surviving in hot springs at temperatures as hot as 172°F (78°C), while other species have been found active in the frigid conditions that occur as deep as 436 yd (400 m) in glaciers.

Most free-living bacteria are heterotrophic, and among the diversity of bacterial species are some that are capable of metabolizing virtually any organic substances as a source of nutrition. Other species of bacteria are photosynthetic (including blue-green bacteria), capable of capturing sunlight and using it to reduce **carbon dioxide** and water into simple sugars, which are used as a source of energy in more complex biochemical syntheses. Other bacteria are chemosynthetic, coupling their biosynthetic abilities to energy released during the oxidation of certain inorganic compounds, as when pyritic sulfur or sulfide are oxidized to sulfate.

Many species of bacteria are not free-living, and instead live in a mutualistic **symbiosis** with more complex organisms, such as plants or invertebrate or vertebrate animals. For example, numerous species of bacteria live as a microbial community within the rumen of cows and sheep, while others live in the gut of humans and other primates. These gut bacteria help with the digestion of complex organic foods, and they also synthesize vitamins and micronutrients that are useful to their host. Other bacteria in the genus *Rhizobium* live in a **mutualism** with the roots of peas, clovers, and other leguminous plants, fixing atmospheric dinitrogen gas (N_2) into ammonia (NH_3), which after con-

version to ammonium (NH_4), it becomes a source of **nitrogen** that plants can utilize as a **nutrient**.

Many bacteria are **parasites** of other organisms, and some cause important diseases. Significant diseases of humans caused by bacteria include various kinds of infections, bacterial pneumonia, **cholera**, diphtheria, gastric ulcers, gonorrhea, Legionnaire's disease, leprosy, scarlet fever, syphilis, tetanus, tooth decay, tuberculosis, whooping cough, most types of food **poisoning**, and the "flesh-eating disease," which is caused by a virulent strain of *Streptococcus*. Bacteria also cause diseases of other species, and this is sometimes used to the advantage of humans. For example, **Bacillus thuringiensis** is a pathogen of many species of moths, butterflies, and blackflies, and strains of this bacterium have been used as a biological insecticide against certain insect pests in agriculture and forestry.

Microscopic Protists

The kingdom Protista consists of a wide range of simple eukaryotic organisms, including numerous unicellular and multicellular species. Microbial protists include protozoans, foraminifera, slime molds, single-celled algae, and multicellular algae. (Some other multicellular algae are not microscopic—the largest seaweeds, known as kelps, can grow fronds longer than 11 yd [10 m]!) Eukaryotic organisms have their genetic material organized within a membrane-bounded nucleus, containing paired chromosomes of DNA. Protists have flagellated spores, and mitochondria and plastids are often, but not always, present.

Recent systematic treatments have divided the kingdom Protista into about 14 phyla, consisting of about 40,000 named species. Many of these phyla, however, differ enormously from the others in basic elements of their biology. It is likely that additional systematic research of this extremely diverse group will result in the Protista being divided into several kingdoms.

Several phyla of protists are photosynthetic, and these are collectively known as algae. Most species in the following groups are microscopic: diatoms (phylum Bacillariophyta), green algae (Chlorophyta), dinoflagellates (Dinoflagellata), euglenoids (Euglenophyta), and red algae (Rhodophyta). The brown algae and kelps (Phaeophyta) are macroscopic, as are some colonial species in several of the other groups. Algae are important primary producers in marine and freshwater ecosystems. Uncommon phenomena known as "red tides" are natural blooms of certain species of marine dinoflagellates that produce toxic biochemicals.

Other phyla of microbial protists have a heterotrophic nutrition. Protozoans are single-celled microorganisms, generally considered to be microscopic animals. Protozoans reproduce by binary fission, and are often motile, usually using cilia or flagella for propulsion. Some protozoans are colonial. Protozoans are abundant in most aquatic environments.

Other heterotrophic protists include the ciliates (Ciliophora), forams (Foraminifera), amoebae (Rhizopoda), and unicellular flagellates (Zoomastigina). Forams are unicellular microorganisms that form a "shell" of calcium carbonate. Chalk is a mineral that is formed from the remains of forams that have accumulated over geologically long periods of time. (The white cliffs of Dover, England, consist of the fossil remains of innumerable forams).

Some species of protists cause diseases of humans, such as *Plasmodium*, a unicellular blood parasite that causes **malaria** and is spread to people and other vertebrate animals by mosquitoes. Some species of amoebae are parasites of animals, causing amoebic dysentery in humans. Hiker's diarrhea (or beaver fever) is a water-borne disease caused by the ciliate *Giardia*, which is the reason why even the cleanest-looking natural waters should be boiled or otherwise disinfected before drinking.

Microscopic Fungi

The kingdom Fungi consists of single-celled microorganisms known as yeasts, plus the fungi, which are multi-celled, filamentous in their growth form, and reproduce by budding or by cellular fission. All yeasts and fungi are heterotrophic, excreting enzymes that digest complex organic materials, allowing simple organic compounds to be ingested.

A few species of yeasts are economically important because they have the ability to ferment sugars under anaerobic (O_2-deficient) conditions, yielding gaseous CO_2 and ethyl alcohol (or **ethanol**). The CO_2 produced during fermentation can be utilized to raise bread dough prior to baking, while the alcohol can be used by brewers to produce beer, wine, and other intoxicating beverages.

Microorganisms are characterized only by their microscopic size. Otherwise they comprise a wide range of diverse but not necessarily related groups of organisms. As a larger group, microorganisms are extremely important as primary producers in ecosystems, as agents of decay of dead organisms and **recycling** of the nutrients contained in their **biomass**, and as parasites and diseases of humans and other species.

[*Bill Freedman Ph.D.*]

RESOURCES
BOOKS

Atlas, R. M., and R. Bartha. *Microbial Ecology*. Menlo Park, CA: Benjamin/Cummings, 1987

Raven, P. H,. and G.B. Johnson. *Biology*. 3rd ed, St. Louis: Mosby Year Book, 1992.

Starr, C., and R. Taggart. *Biology. The Unity and Diversity of Life*. Belmont, CA: Wadsworth Pub. Co., 1992.

Microbial pathogens

Microbial pathogens are **microorganisms** that are capable of producing disease. Virtually all groups of bacteria have some members that are pathogens. One notable exception is the Kingdom Archaea, where there are no known pathogenic members. Other disease-causing microbial agents are viruses and parasitic protozoa. Earlier methods of detecting and identifying microbial pathogens involved culturing and isolating bacterial colonies in growth media in the lab. With the advent of polymerase **chain reaction** (PCR) assays, identification of microorganisms that were difficult or impossible to culture became possible. Many microbial pathogens can be controlled with antimicrobial drugs called antibiotics. However, these drugs are not effective against viruses or **parasites**, and indiscriminant use may cause resistant strains of pathogens to evolve. **Antibiotic resistance** has resulted in the reemergence of several disease, and as of 2002, most of the major bacterial diseases that infect humans are becoming resistant to antibiotics.

[*Marie H. Bundy*]

Microclimate

In general, **climate** conditions near the ground are called microclimates. More specifically, microclimate refers to the climate characteristics of highly localized areas, ranging from the area around an individual plant to a field of crops or a small forested area. The horizontal area considered may be less than one square meter or up to several thousands of square meters. The vertical extent may range from a few centimeters involving the still layers of air within a plant canopy, for instance, to 100 meters or more, when the **atmosphere** surrounding a forested area is studied.

Microclimates are governed to a large extent by the interactions of surface features with the overlying atmosphere, and their characteristics may differ markedly from those of the surrounding large-scale climate. Microclimates exhibit great ranges in environmental conditions depending on the moisture and radiation properties of the surface. They typically show large diurnal temperature ranges and are highly influenced by slope, aspect, and elevation. Most plants and animals are adapted to highly specific microclimatic conditions.

Micronutrient
see **Trace element/micronutrient**

Migration

Although some scientists define animal migration as any animal movement, this definition becomes cumbersome because it does not distinguish between small-scale daily movements, annual migrations, and irrupting dispersions. Mobile animals tend to move frequently, and migration should be distinct from emigration (directional one-way movement) and dispersal (non-directional one-way movement), and it should refer primarily to regular round trip movement that happens at least once in the life span of the organism. Migration is a spatial behavior pattern that allows animals to locate themselves in the most favorable portions of their **habitat** for as long as necessary. Such favorable conditions may vary according to season or life history, but in both cases it is related to adaptive fitness. This allows the organism to take in nutrients in excess of energy expenditures and to successfully reproduce.

Generally animal migration can be divided into two areas of study: the behavioral aspects, which concentrates on "how" migration happens; and the ecological aspects, which addresses "why" migrations takes place. The ecological questions also concern **evolution**, for spatial behavior is an evolved compromise between differing requirements of an organism's life. Most migratory behavior depends on food abundance. In most habitats productivity varies with the seasons, and thus energy availability also varies at all upper levels of food chains. Migration must often accommodate several energy and reproductive requirements. As a result migration patterns tend to be complicated with subsections of migrants taking slightly differing paths at various times to serve different needs.

Many **species** of North American waterfowl use several breeding areas, from the northern prairies to the Arctic coast, and travel along several major flyways to wintering areas that range from the southern prairies to the estuaries of northern South America. Within that framework, males **desert** the hens during nesting season and make shorter migrations to molting areas, afterward meeting with the females and newly fledged young during the fall migration. In either of those sites, requirements for habitat differ in terms of water-cover ratios, water permanence, and food preferences.

The altitudinal migration of mountain sheep (*Ovis canadensis*) in the Rocky Mountains brings them into more productive habitats during the summer in the alpine meadows and into the mountain forest during the winter. Thus, food intake requirements are accommodated and, simultaneously, the sheep take advantage of the **microclimate** of the forested area during the winter to supplement energy losses from body temperature maintenance.

Migration can also be a response to particular breeding site requirements, mate location, and a combination of several forces acting together. The longest mammal migration, that of the California gray whale (*Eschrichtius robustus*), places the breeding-ready adults and newly impregnated females in the productive shallows of the Bering and Chukchi Seas, the non-breeding animals in the more patchy feeding grounds of the northeast Pacific Coast, and the calving females in the warm shallow lagoons of the Baja Peninsula.

[*David A. Duffus*]

RESOURCES
PERIODICALS

Clark, C. W. "Moving With the Heard (Hydrophonic Monitoring of Migratory Bowhead Whales)." *Natural History* (March 1991): 38–42.

Dybas, C. L. "Secret Creatures of the Nigh; When the Moon is New and Darkness Falls, American Eels Begin Their Eerie Autumn Migration." *National Wildlife* 28 (October-November 1990): 18–23.

Hansson, L. "The Lemming Phenomenon: Or Why the Legendary Mass Migrations of Rodents Are Restricted to the Extreme North." *Natural History* (December 1989): 38–43.

Pallace, D. R. "Avian Nations: The Patterns and Problems of Migrating Birds." *Wilderness* 54 (Fall 1990): 42–9+.

"Recent Developments in the Study of Animal Migration." [Symposium on Recent Developments in the Study of Animal Migration.] *American Zoologist* 31 (1991): 151–276.

Milankovitch weather cycles

According to the theory of the Milankovitch weather cycles, ice ages are cyclical, caused by changes in the earth's orbit. The theory was developed by Serbian geophysicist Milutin Milankovitch (1879–1958) in the 1930s, and it postulates that the amount of available sunlight in the northern hemisphere is affected by the earth's orientation in space. Because the earth is a globe in motion, sunlight strikes the earth differently depending on the following factors: The eccentricity of its orbit, which returns to the same point every 100,000 years; the tilt of the axis of its rotation, a 41,000-year cycle; and the precession of the equinoxes, a 23,000-year cycle. Milankovitch proposed that decreased sunlight prevents ice and snow from melting in the summer in the northern latitudes. This, in turn, cools the **atmosphere**, because ice reflects 90 percent of solar radiation back into space, and over a long period of time, the ice accumulates and moves south.

The theory has gone through its own cycles of acceptance and rejection. The first problem for proponents of the theory was to prove the occurrence of glacial-interglacial patterns during the Pleistocene era, the epoch preceding ours. It was also necessary to date the onset of each **ice age** in order to calculate the date of the next one. During the last fifty years, data collection improvements in astronomy and geology, especially in satellites and deep-sea **sediment**

coring, have challenged and refined the theory. Supported first by astronomical and geological data, the theory encountered opposition in the 1950s when scientists used the newly-developed Carbon-14 method to date warm-era fossils in supposed ice-age deposits. But examination of fossilized microscopic sea creatures on the ocean bed, known as foraminifera, revealed two types, the larger of which flourished when the ocean was warm. In 1955, geologist Cesare Emiliani refined dating techniques of the foraminifera retrieved by deep-sea coring. He isolated oxygen isotopes. Oxygen 16, a lighter **isotope**, evaporates more quickly and becomes trapped in ice, while oxygen 18, which thrives in warm waters, was absorbed by the shells of the foraminifera. From the evidence of theses oxygen isotopes, Emiliani was able to reconstruct glacial-interglacial periods for 300,000 years. His date agreed with Milankovitch's orbital dates.

In 1971 John Imbrie of Brown University created Project CLIMAP, which coordinated all of the data relevant to Milankovitch's cycles. From this data, scientists were able to develop a "rosetta stone" for Pleistocene glacial-interglacial dates. In the 1980s, Project SPECMAP brought together all of the deep-sea core data. It was from this data that scientists predicted the next ice age was imminent.

But critics have argued that the Milankovitch orbital theory does not take into account the concept of chaos as well as the climatic influence of the heat transfer between the atmosphere and the ocean. In 1992, United States **Geological Survey** hydrogeologist Isaac Winograd challenged the Milankovitch theory when a scuba team with a submersible drill retrieved a 14-in (36-cm) core of calcite in a water-filled fault called Devils Hole in Nevada. The ice age dates for this core of calcite differed from the Milankovitch dates, and this suggested to Winograd that ice ages were tied less to orbital cycles than to an interaction of heat and moisture between the atmosphere, the ocean, and ice sheets. Imbrie and many others defended Milankovitch against these findings. They stood on their mountain of ocean-sediment core data, and they argued that the calcite core from Devils Hole reflected local rather than global changes.

Another problem for the Milankovitch theory is identifying the effects of human activity on the **environment**. The current interglacial period began 10,000 years ago, and according to Imbrie, it might become a "super interglacial" period due to the burning of **fossil fuels** and the subsequent **greenhouse effect**. Imbrie has argued that the natural cooling cycle which began 7,000 years ago will be postponed until the excess **carbon dioxide** is exhausted in 2,000 years. Then, after another 1,000 years, serious cooling will set in, expected to last for 23,000 years. *See also* Climate; Glaciation; Ice age refugia; Ozone layer depletion; Radiocarbon dating

[*Stephanie Ocko*]

RESOURCES
BOOKS
Imbrie, J., and K. P. Imbrie. *Ice Ages: Solving the Mystery.* Short Hills, NJ: Enslow, 1979.

PERIODICALS
Kerr, R. A. "Milankovitch Climate Cycles Through the Ages." *Science* 235 (February 27, 1987): 973–4.

Winograd, I. J., et al. "Continuous 500,000-Year Climate Record from Vein Calcite in Devils Hole, Nevada." *Science* 258 (October 9, 1992): 255–260.

OTHER
Berger, A., et al. *Milankovitch and Climate.* Proceedings of the NATO Advanced Research Workshop on Milankovitch and Climate. Dordrecht: D. Reidel Publishing, 1982.

Milfoil

see **Eurasian milfoil**

Minamata disease

The town of Minamata, near the southern tip of Japan's Kyushu Island, gave its name to one of the most notorious examples of environmental contamination known. Minamata disease is really alkylmercury **poisoning**, caused by eating food such as fish or grain contaminated with **mercury** or its derivatives. At Minamata people were poisoned when they ate large quantities of methylmercury-contaminated fish. Often the victims were the poorest members of society, who could not afford to stop eating the cheap fish known to be affected by **effluent** discharged from the local chemical company. The first victims were reported in 1956; before that, cats were seen moving strangely, sometimes flinging themselves into the sea, and birds were observed flying awkwardly or even falling out of the sky.

The Chisso Company, a leading chemical manufacturer, produced acetaldehyde by passing acetylene gas across an inorganic mercury catalyst, leaving methylmercury as a by-product. The company was a major local employer and the only significant major source of industrial waste discharged into Minamata Bay. There, methylmercury in the water biomagnified to high levels in fish and shellfish consumed by the local inhabitants. Methylmercury concentrates in specific regions of the central nervous system and readily crosses the blood/brain barrier as well as the placental barrier. It is a human **teratogen** which causes brain damage during prenatal exposures, resulting in congenital or fetal Minamata disease. The compound has a **half-life** of about 70 days in the human body, and the damage is generally irreversible. Although company officials knew that similar symptoms could be induced in cats by feeding them fish taken from the bay, Chisso was not blamed by the Japanese government

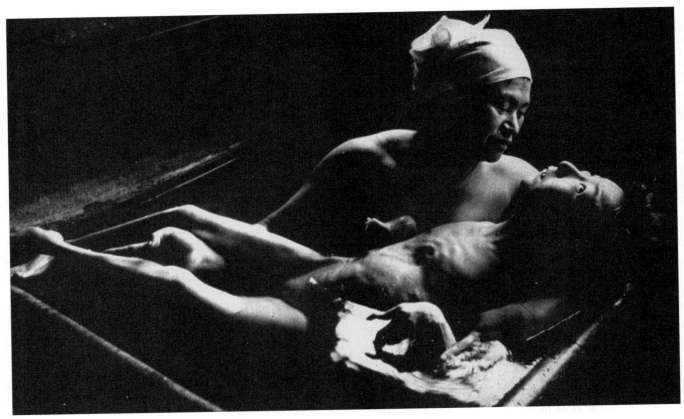

A mother from Minamata, Japan, bathes her daughter who suffers from Minamata disease. The daughter's brain damage and birth defects were caused by alkylmercury poisoning from mercury-contaminated seafood that her mother ate while pregnant. (Photograph by Eugene Smith. Black Star Publishing Company Inc. Reproduced by permission.)

for the disaster until 1968. Despite the evidence, it was difficult to be certified as a Minamata disease victim. Those affected were not compensated until after another epidemic occurred at Niigata, Japan, in 1965. That one was caused by methylmercury waste discharged from the Showa Denko Corporation Kanose factory into the Agano River. The company also manufactured acetaldehyde using the same process as Chisso.

Nearly four decades after Minamata disease was identified, the full range of its neurological symptoms is still not known, nor have the total number of sufferers been determined. While some children were born with contorted bodies and severe retardation, in milder cases the victims may be moderately retarded or exhibit only sensory disturbances. Even as the symptoms advance in adult patients, it is often difficult to determine whether they are the result of long-term poisoning, delayed effects of residual methylmercury, aging, or other complications. Therefore, even thirty years later, as of December 31, 1992, only 2,945 individuals were officially certified as Minamata Disease victims and 1,343 of

them had died. Another 13,761 had been denied certification and the fate of 2,430 was still pending.

Lawsuits dragged on for more than twenty-five years, and victims seeking compensation tried confrontation tactics like an encampment that lasted a year and a half in front of Chisso's main headquarters in Tokyo. **Reclamation** of the **environment** has been even slower. The Chisso Company installed waste treatment equipment in 1966 and stopped making acetaldehyde in 1968. By then 400–600 tons of mercury and 60,000 tons of **sludge** had been dumped into the shallow Minamata Bay. Mercury levels in fish remained elevated. **Dredging** scheduled to begin in 1975 did not get underway until 1982. Since then, mercury levels in ten **species** of fish have dropped; but as of March 1993, they were still two or three times higher than what the government deems acceptable. Consequently, fishermen have lost their livelihood and health and gained a social stigma. The Chisso company's fortunes also declined and were further strained by lawsuits that mandated restitution. *See also* Agricultural chemicals; Biomagnification; Birth defects; Environmental law; Food chain/web; Heavy metals and heavy metal poisoning; Marine pollution; Methylmercury seed dressings; Water pollution; Xenobiotic

[*Frank M. D'Itri*]

RESOURCES
BOOKS

Harada, M. "Methyl Mercury Poisoning Due to Environmental Contamination (Minamata Disease)." In *Toxicity of Heavy Metals in the Environment, Part 1*, edited by F. W. Oehme. New York: Marcel Dekker, 1978.

Ui, J. "Minamata Disease." In *Industrial Pollution in Japan*. Tokyo, Japan: United Nations University Press, 1992.

PERIODICALS

D'Itri, F. M. "Mercury Contamination—What We Have Learned Since Minamata." *Environmental Monitoring and Assessment* 19 (1991): 165–182.

Mine drainage
see **Acid mine drainage**

Mine spoil waste

Most human extractions of earth materials, such as clay for pottery or **coal** for power generation, produce some waste. The raw materials are rarely pure, so unwanted **detritus** is discarded, usually close to the extraction site. Over the last two centuries, exponential industrial growth has resulted in huge increases in the production of mine **spoil** waste. The management of this waste has become an increasingly important issue.

New technologies allow the mining of ever lower grades of ore, with mounting waste as a byproduct. Early mining actually wasted ore, since only the richest veins were extracted. However, less waste was generated by this high grade ore. Now operations tend to remove varying grades of ore *en mass*, yielding a higher return, but multiplying the waste produced. Where concentrations are high, it has even been profitable to rework older **tailings**.

Surface mining accounts for most mining waste, but underground work also contributes. Metallic ores, sand, gravel, and building stone, including aggregate for concrete, are usually extracted from open pit mines. **Strip mining** is effective where resources lie in sedimentary layers, and is particularly used for coal, phosphate, and gypsum. Much of the waste from strip mining comes from **overburden** removal. **Dredging** is used to extract sand and heavy placer deposits such as gold and tin; it reinjects large amounts of fluvial **sediment** into the flowing water. Hydraulic mining with high pressure hoses is common in gold fields; devastating effects are still visible in the foothills of California's Sierra Nevada Mountains, more than a century later.

Environmental impacts of mining include the creation of new landforms, severe **ecosystem** disruption, and the formation of dangerous **chemicals**. The scale of operation varies enormously, but some projects are immense. The holes and mountains of overburden created may become useful as chat for railroad beds, sub-base material for roads, or recreational lakes. But, severe environmental problems are a more common result. Ecosystem disruption stems mainly from loss of **topsoil**, rich in organisms and nutrients; sterile landscapes with high sediment **runoff** are a common outcome.

The most serious problem resulting from mine wastes is **acid drainage** and **leaching** of hazardous substances. When these wastes are moved from a reducing **environment** (oxygen deficient) to an oxidizing environment, sulfuric acids are formed by the oxidation of the sulfides in metallic ores or the sulfur that commonly accompanies coal deposits. These acids may flow into surface waters or may leach hazardous metals from the waste. The best solution is to minimize exposure to oxygen, usually by burial.

Mining presents the dilemma of short term gains versus long term losses, especially of land suitable for agriculture or forestry. The goal of sustainability makes **reclamation** of mine spoil wastes imperative. *See also* Acid mine drainage; Erosion; Hazardous material; Sustainable development; Waste management

[*Nathan H. Meleen*]

RESOURCES
BOOKS

Caudill, H. M. *Night Comes to the Cumberlands*. Boston: Atlantic-Little Brown, 1963.

Meleen, N. H. "Mining Wastes and Reclamation." In *Magill's Survey of Science: Earth Science Series*, edited by Frank N. Magill. Pasadena, CA: Salem Press, 1990.

OTHER

U.S. Department of the Interior. *Surface Mining and Our Environment*. Washington, DC: U.S. Government Printing Office, 1967.

Mineral Leasing Act (1920)

The Mineral Leasing Act of 1920 regulates the exploitation of fuel and **fertilizer** minerals on the public lands. The act resulted from the perceived failure of existing federal laws dealing with **coal** and oil resources. Coal lands had been managed under an 1873 law, allowing the coal to be mined for either $10 or $20 per ton on tracts of 160 or 320 acres (64.8 or 129.6 ha). These acreage limitations led to abuse of the law, and in 1906 over 65 million acres (26.3 million ha) of land were withdrawn from coverage under of the coal lands law. These lands were then reclassified according to whether they contained coal or not, and the price for lands containing coal was increased. The withdrawal and reclassification process slowed development and led many to argue for a new approach. The fate of these coal lands was soon tied to the fate of oil lands.

Oil resources were being managed based on the Mining Law of 1870 and the Oil Placer Act of 1897. Neither law, however, sufficiently recognized the difference between **petroleum** and other minerals this in turn resulted in major problems in the exploitation of petroleum. These problems included overproduction, market instability, claim jumping, and national security concerns. In 1909, President William Howard Taft withdrew over three million acres of **public land** from oil development in California and Wyoming, thereby initiating a policy debate over how to manage petroleum resources on the public lands.

The first leasing bill, supported by the Taft administration, was introduced in Congress in 1913, but because it was so controversial in the western states, the Mineral Leasing Act was not passed until 1920. The law, which applies to deposits of coal, oil, gas, **oil shale**, phosphate, potash, sodium, and sulfur on the public lands, has two main features: federal regulatory authority and conditional access to the public lands. Controversy centered on the oil provisions of the Act. These gave the Secretary of the Interior the authority to issue permits for prospecting on land that was not known to have any oil. If oil were discovered, the prospector acquired lease rights for 20 years and paid a royalty fee of five percent. For proven oil-producing lands, tracts were offered under a system of competitive bidding based on royalty payments. The minimum area covered was 640 acres (259.2 ha), and the minimum royalty accepted was 12.5%. These royalties were to be divided among the **Reclamation** Fund (for western water projects), the states in which the land is located (for education and roads), and the federal government.

Amendments to the act have done away with prospecting permits, increased the size of tracts that can be leased, and changed the bidding procedures for leases. Most lands that are not known to contain oil are leased through a lottery conducted by the U. S. Department of the Interior. For known oil lands, a competitive bidding procedure is used. In addition to the bid fee, a 12.5% royalty and an annual rental fee of $2.00 per acre is also required.

The Mineral Leasing Act was a significant departure from past mining policy, based on the Mining Law of 1872, which granted free access to the public lands, the potential for inexpensive purchase of mineral lands, and included no royalty payments to the government.

[*Christopher McGrory Klyza*]

RESOURCES

BOOKS

Hays, S. P. *Conservation and the Gospel of Efficiency: The Progressive Conservation Movement, 1890–1920.* New York: Atheneum, 1975.

Mayer, C. J., and G. A. Riley. *Public Domain, Private Dominion: A History of Public Mineral Policy in America.* San Francisco: Sierra Club Books, 1985.

Swenson, R. W. "Legal Aspects of Mineral Resources Exploitation." In *History of Public Land Law Development,* by Paul W. Gates. Washington, DC: U.S. Government Printing Office, 1968.

Minerals, strategic

see **Strategic minerals**

Minimum-tillage agriculture

see **Conservation tillage**

Mining

see **Acid mine drainage; Ashio, Japan; Ducktown, Tennessee; Itai-itai disease; Mine spoil waste; Placer mining; Silver Bay; Strip mining; Sudbury, Ontario; Surface mining; Surface Mining Control and Reclamation Act (1977)**

Mining, undersea

Practically all of the mineral and energy resources found on land are present under the sea. Development, however, is limited by extraction costs that increase with depth of water, by the relative abundance of resources on land, and by political questions involving ownership of deep ocean resources.

Worth $80 billion, the most valuable undersea commodity is oil and **natural gas**, representing 90% of the resources obtained from the seafloor. Continental-margin deposits comprise about one-third of the world's estimated oil and gas reserves. Large deposits have been found on the continental shelves in the Gulf of Mexico, the Persian Gulf, the North Sea, and off the coasts of northern **Australia**, southern California, and the Arctic Ocean. Other sites are promising, and many other continental shelves remain relatively unexplored. The harsh conditions found in the North Sea have fostered huge investments in colossal offshore rigs. Extraction costs on the continental shelves are several times higher than on land, and projected costs in the Arctic Ocean are several times higher again. Such fields are profitable only because of their high flow rates and reserves. Oil exploration continues to push the limits of profitable extraction into deeper and deeper waters. One commercially promising well in the Gulf of Mexico was drilled in waters well over 1 mi (1.6 km).

Nearshore resources are much more likely to be mined than distant ones. Sand and gravel predominate in volume. Though in great demand for use in concrete and as fill material and beach sand, their low value restricts usage to local areas.

Delta sediments are often mined for valuable deposits washed from the land, including gold, platinum, tin, diamonds, iron, and **uranium**. Japan, an island country with limited land resources, is especially eager to exploit such deposits; **coal** beds are also being mined.

Phosphorite is the source rock for phosphate, a key **fertilizer**. Phosphorite deposits occur where upwelling brings dissolved phosphate to the surface, where, in turn, rising temperature and **pH** cause them to precipitate out. Their value comes from their high grade, since they are not diluted by land-derived sediments.

Manganese nodules were first discovered on the deep-ocean floor during the 1873–76 voyage of the British ship, *Challenger*. Although rich in valuable ferroalloys used in steelmaking, these potato-like ores have yet to be mined commercially. The deposits are immense, but the technology for profitable mining does not currently exist.

Another tantalizing resource is the sulfide ores precipitated out of ocean water around **hydrothermal vents** in deep-ocean rift systems. Commercial extraction is being attempted in the Red Sea, and experimental vehicles have been built for use in the deep ocean, but technological and economic limitations hinder commercial development.

Economic considerations are one of the greatest limitations to the growth of undersea mining, along with uncertain environmental impact. Only highly profitable ventures, as in oil and gas extraction, or readily accessible, near-shore, bulk resources, such as sand and gravel, are currently feasible. Undersea mining appears to hold great potential **resources for the future**, when supplies and extraction costs of land-based resources tip the scales in its favor.

Political considerations, especially questions of ownership and the sharing of profits, are also key factors here, with developing countries clamoring for their share. During the 1970s, the United Nations strove for a treaty to regulate deep-ocean exploitation. The 1982 Law of the Sea Treaty claimed manganese nodules as the heritage of all humankind, to be regulated by the United Nations' seabed authority and with profits to be shared. The United States chose not to sign the treaty, and has joined with other developed countries in a provisional understanding to award exploitation licenses.

These political and environmental issues are insignificant, so long as the economic realities favor resources on land. It will be beneficial if the time spent waiting for devel-opment of undersea mining creates the opportunity for more productive international discussions and agreements.

[*Nathan Meleen*]

RESOURCES
BOOKS

Duxbury, A. C., and A. B. Duxbury. *An Introduction to the World's Oceans.* 3rd ed. Dubuque, IA: Wm. C. Brown Publishers, 1991.

Ingmanson, D. E., and W. J. Wallace. *Oceanograph.* 5th ed. Belmont, CA: Wadsworth Publishing, 1995.

Mirex

An organic compound that was manufactured for use as an insecticide against imported **fire ants** and, secondarily, as a fire retardant for **plastics**, **rubber**, paint, paper, and electrical products. It has a molecular weight of 545.59 and consists of twelve **chlorine** atoms attached to a ten **carbon** cage. Its full name is dodecachloro-octahydro-1,3,4-metheno-1H-cyclobuta[c,d]pentalene. The United States **Environmental Protection Agency** (EPA) has classified it as a **carcinogen**. Mirex can be degraded to the toxic **pesticide Kepone** in the **environment**. Due to discharges from manufacturing facilities in the state of New York, it is found in the water, sediments, and biota of Lake Ontario at levels of concern. *See also* Cancer; Great Lakes; Toxic substance

Mission to Planet Earth (NASA)

The Mission to Planet Earth (MTPE), now officially called the Earth Science Enterprise (ESE), is a program of the National Aeronautics and Space Administration (NASA) of the U.S. government. The mission is to use spacecraft and space technology to provide a comprehensive scientific study of the earth's living systems as viewed from space. Using satellites with precise measuring equipment, the mission will provide information about weather patterns, **climate**, oceans, coastlines, surface activity, atmospheric conditions, natural disasters, **pollution**, and hundreds of other measurements. The goals of the program, according to NASA, are to find out how the living system of the earth is changing, and to determine the causes and consequences of this change. NASA also has the goal of using the program's results to develop a "predictive capability" for climate, weather, and natural hazards. Finally, the program has an end-to-end strategy of performing studies and gathering data as well as making the findings readily available to the public and scientific community. Other countries are also taking part in the mission, including Canada, Japan, Russia, and many major European countries, providing funding and scientific studies.

MTPE was initiated by President George Bush in 1989, on the recommendation of the **National Research Council**. At first, NASA and the U.S. government were ambitious about the plan to study the earth from space, and proposed a budget of $35 billion for its first 15 years. The original concept of the program was to build a huge observatory that would orbit and monitor the earth. The program was scaled down during budget battles in the early 1990s and received about $8 billion for its first decade of operation. The scientific concept of the program was changed as well due to the revised budget; MTPE will utilize up to 18 small satellites that will monitor the earth from space, called the Earth Observing System, and collect data that will be coordinated by computers. In 1998 NASA changed the name of MTPE to the Earth Science Enterprise.

By 2002, the program had several satellites and measuring instruments in space. The Earth Resources Observation System (EROS) has mapped cities and recorded urban growth and **habitat** destruction. The Geostationary Environmental Satellite (GOES) has tracked hurricanes and weather patterns. The Upper **Atmosphere** Research Satellite was launched in 1991 to study the **ozone** layer. In 1996 a Japanese satellite carried NASA instruments to study global wind patterns, and in 1997 a joint United States/Japanese satellite was launched to study how tropical rainfall affects the world's climate. In April 1999, NASA launched the Landsat 7 satellite to map and study environmental changes on the earth's surface, including mapping the world's forest canopy. In November 2000 a new generation of satellite was launched, the EO-1, which was only one-seventh the size of the Landsat satellites. The new satellite had sophisticated instruments, some of which can measure parameters such as the earth's gravity field. This measurement may help scientists better understand ocean currents and heat movement between the poles. Other governmental departments besides NASA are also contributing to the project. The National Oceanic and Atmospheric Administration's (NOAA) Polar-orbiting Operational Environmental Satellite is monitoring factors such as global vegetation, ocean currents, ice warming, weather patterns, the ozone layer, and solar storms. The MTPE was projected to consist of 25 measuring instruments on 10 satellites by 2002, depending on launch schedules and other factors.

NASA plans to make all of the data from MTPE satellites available to the public on computers. The information will be stored in NASA's Earth Observing System Data and Information System (EOSDIS). NASA estimates that MTPE satellites will send enough data back to Earth during its first 15 years of operation to fill nearly 6.5 million books

per day. The management of this information will be as demanding a project as sending the satellites to space, and users will be able access the data from the Internet, CD-ROMs, tapes, microfiche, and photographs. By 2002 MTPE data was available via the Internet, including live broadcasts of the earth from satellites. Eventual users of MTPE data may include atmospheric scientists studying the ozone layer, meteorologists predicting weather changes, ecologists monitoring tropical rain forests, scholars performing environmental research, and many others interested in environmental patterns and problems.

[*Douglas Dupler*]

RESOURCES

BOOKS

Burrows, William E. *This New Ocean: The Story of the First Space Age*. New York: Random House, 1998.

Walter, William. *Space Age*. New York: Random House, 1992.

PERIODICALS

Lawler, Andrew. "Climate Warms a Bit for NASA Mission" *Science*, March 28, 1997, 1870.

Stevens, William K. "NASA Readies a 'Mission to Planet Earth.'" *New York Times*, February 17, 1998, B9.

OTHER

Earth Observing System Project Office. [cited July 2002]. <http://www.eos.nasa.gov>.

Earth Science Enterprise Home Page. [cited July 2002]. <http://www.earth.nasa.gov>.

Mitsui Mining and Smelting Company
see **Itai-itai disease**

Mixing zones

Most **wastewater** treatment plants and industrial facilities **discharge** their **effluent** into the nearest available body of surface water such as a stream or river. The mixing zone is the localized area in the receiving stream within which the mixing, dispersal, or dissipation of the effluent can be detected. For example, cooling water discharges from **power plants** typically create a mixing zone in the receiving stream where the temperature is higher than the ambient background temperature of the stream waters. Environmental regulations usually require that the mixing zone be limited in size and not create a nuisance or hazardous conditions.

MMPA
see **Marine Mammals Protection Act (1972)**

Modeling (computer applications)

Computer modeling is a mathematical tool used to analyze complicated systems or to predict events such as floods, **climate** changes, or population changes. Computer models are applied in many disciplines, from engineering (predicting the strength of a bridge or dam) to economics (predicting inflation rates) to **ecology** (describing a **food chain/web** or projecting the survival chances of an **endangered species**). A model may consist of a few simple calculations on a spreadsheet, or it may involve millions of calculations on hundreds of thousands of input parameters, as in the case of general circulation models used to describe and predict climate conditions. Both simple and complex models involve quantifying input variables and the relationships between them, in order to produce numeric output that can describe or predict some real world phenomenon or condition. Models are often criticized because they can generalize real-world phenomena unrealistically, or because any errors in initial assumptions are expanded as assumptions are added and multiplied together. However, modeling remains a common tool because it is often the only available method to find answers to important research questions.

Models let researchers "experiment" with systems that do not allow manipulative experimentation. You cannot actually build a power plant just to test its effect on a population, and you cannot deliberately change the temperature of the **atmosphere** in order to see what happens to plant populations and sea levels. Instead a model can be used to predict probable outcomes of known conditions. Different variables can also be easily introduced in a model, for example, on the computer a programmer can easily add **pollution control** equipment to the power plant and observe its effect on lung disease rates in the surrounding population, or a modeler can easily reduce worldwide **carbon dioxide** emissions and calculate the impact on atmospheric temperatures.

Computer models have many applications. Some aid in assessing or visualizing a phenomenon. Three-dimensional geological models are used to help visualize or approximate the volume and dimensions of an **aquifer**, an oil reserve, or a magma chamber beneath a **volcano**. A simple **ecosystem** model can help identify general relationships between populations and resources, by letting the modeler adjust variables and see which arrangements most closely approximate observed data. Other models are used to predict the outcome of known conditions: a demographic model representing known rates of **population growth**, reproduction, and death can reasonably predict a country's population a century from now. Still other models are used to test "what if" questions. A **pollution** dispersion model showing how much pollution could spread from a power plant to a town can be used to test different scenarios: What if there were a north-west wind for an entire week? What if the power plant needed to double its energy production? What if the town expanded in the direction of the power plant? Models used to describe or visualize a system are known as descriptive models. Models that project or predict the results of input conditions are known as predictive models.

Anything that approximates or simulates a real-world phenomenon could be considered a model. A plaster replica of a landform, a diagram drawn on paper to describe links in a food web, or a verbal description of a storm could be considered models, because each of these describes or simulates a real phenomenon. Computer models approximate phenomena in the real world by performing calculations on input variables. These calculations produce numeric output that predicts the outcome of the variables in the model. For example, a wolf population model might multiply the number of breeding adults (A) by the average number of pups per litter (P) to produce the expected number of young **wolves** in a year (N): A x P= N.

This is a very simple, and not very realistic, mathematical model. A computer model can produce a more realistic representation of a wolf population, by incorporating more variables, by using randomized variables, by performing more complicated mathematical functions, and by repeating calculations many times in order to produce a statistically significant number of "experiments."

Models can be composed of variables, parameters, and constants. A variable is any factor that is changeable. The number of breeding adults in a wolf population might change from year to year, as might the number of pups per adult. Parameters are sometimes distinguished from variables as input factors that are unlikely to change: the number of litters in a wolf pack is unlikely to be more than one per year, or the month in which pups are born might reasonably be expected not to change from one year to the next. Constants are input factors that are assumed to be unchanging. In a model of a dam's structure and strength, the force of gravity on the water behind the dam would be a constant. In addition, randomized (stochastic), values can also be used to introduce an element of chance into the model. For example, in a simulation of population change in wolves a realistic birth rate may be somewhere between 4-9 pups per breeding female per year. A computer model can randomly select a number between 4-9, for this variable each time the model is run. In addition there might be a 70% chance that each pup would survive its first year. This stochastic variable could be added to increase the **precision** of the model's prediction of population at the end of the year. As the computer processes the model it arbitrarily picks a value for the stochastic variable. Stochastic variables can improve a model by making it better represent real events that are somewhat random, or at least unpredictable.

Deterministic and Stochastic Models

A deterministic model produces a single answer for each set of variables; that is, the input variables determine the outcome. Different results can be produced only by entering different values for the input variables. Deterministic models are useful where the values of input variables are reliably known. For example, if an engineer knows with reasonable certainty the strength of a concrete dam and the force applied by a full **reservoir**, a deterministic model should be used to calculate the water level that would cause the dam to break. A stochastic model, in contrast, includes an element of randomness, so that there will probably be a different result each time the model is run. For example, a population model might include a slightly randomized birth rate, or a reservoir model might include a randomized variable for rainfall, which in fact varies unpredictably from year to year. An advantage of using a stochastic (random) variable is that it helps the model approximate the fluctuations that can occur in a real system: there is no way to predict the actual number of pups a wolf pack will have each year, but it is possible to predict with some precision the minimum and maximum number that is likely. Running a stochastic model repeatedly can produce a statistically significant sample of experimental results. Statistical analysis can be used to assess the **probability** of various outcomes, such as a 25% probability that a wolf population will rise over the next 100 years.

Development of Modeling

Although mathematical models and analog models (physical structures such as clay models of landforms or **river basins**) have been used in environmental problem solving for a long time, the growth of modern computer modeling began in the 1960s with the development of computers. One of the first publicized computer models was a climate model that calculated climate conditions based on three input parameters: temperature, air pressure, and wind speed. Since then models have been used to predict weather, stream **discharge**, **soil erosion**, population growth, pollution impacts, and many other occurrences. The ability to use a computer greatly increased the complexity of models that could be developed, and the widespread availability of computers increased the number of people developing models for a growing number of purposes. Models are used in **environmental monitoring** and management, physics, engineering, **hydrology**, economics, demographics, and many other fields. One reason modeling developed more or less simultaneously in many disciplines is that the techniques of programming a computer to simulate real-world variables are highly portable: the structure of a model used to interpret water flow might be modified and applied to magma movement in the earth or to **nutrient** flows through a landscape. Modeling is now a widespread technique that is used to help explain

how natural systems function and to inform public policy concerning the **environment**, the economy, and many other issues.

Sensitivity Analysis, Calibration, and Validation

Once a model is developed it is usually tested against observed events or data to indicate how reliable it is or where its weaknesses lie. One of the first tests usually run is sensitivity analysis. A model's outcome may be more sensitive to changes in one variable or process than another: in a beach erosion model, for example, the predicted erosion rate is likely to be influenced more by wave energy than by sand particle size, even though both factors need to be considered. It is essential to know to which factors the model is most sensitive, because error in estimating those factors could introduce significant error into the results. Sensitivity analysis involves adjusting variables or processes and then running the model to see which factors cause the greatest variability in outcomes.

Once the input variables and relationships in a model are established, the model can be calibrated, or tuned, to make it better represent reality. Calibration usually entails adjusting different parameters in order to produce a result similar to some observed results in the real world. Calibrating a model for forest growth, for example, might involve running the model repeatedly with different growth rates until the model approximates observed historic growth rates and densities.

Validation is like calibration, in that it involves comparing a model's results to observed data. The objective of validation, though, is to demonstrate that the model is reliable. The assumption of model validation is that if the model predicts the correct results in a known situation, then it will probably predict correctly in an unknown, experimental situation.

General Circulation Models

A type of model that has received much attention since the mid-1980s is the general circulation model, or GCM. This is a class of highly complex models designed to predict with reasonable **accuracy** the behavior of the earth's atmosphere (climate) or oceans. GCMs have been used to predict worldwide global warming as a result of increased concentrations of **carbon** dioxide in the atmosphere. GCMs are complex because they simulate the behavior of the atmosphere and the oceans, which have complex three-dimensional movements or flows of air masses or water. Predicting how changes in heat input at one location impacts temperatures and flow rates at another location requires thousands of calculations performed on many interrelated variables at thousands of points in space. Because GCMs must keep track of so many variables and so many points in space, they are usually run on supercomputers that can perform (billions) of calculations per second.

The appropriate complexity of a model depends on its intended use. A model designed to describe ocean circulation over space and time to produce realistic results on which to base public policy might incorporate many variables. A model built to test very general relationships between variables, in order to stimulate a researcher's insight into a small aspect of ecosystem functions, may more appropriately be quite simple, with only a few input parameters and equations.

[*Mary Ann Cunningham Ph.D.*]

RESOURCES
BOOKS

Gordon, S. I. *Computer Models in Environmental Planning.* New York: Van Nostrand Reinhold, 1985.

Hardisty, J., D. M. Taylor, and S. E. Metcalfe. *Computerised Environmental Modelling.* New York: Wiley, 1993.

Starfield, A. M., K. A. Smith, and A. L. Bleloch. *Model It: Problem Solving for the Computer Age.* New York: McGraw-Hill, 1990.

Dr. Mario Jose Molina (1943 –)
Mexican chemist

Mario Molina was born on March 19, 1943, in Mexico. He received his bachelor's degree from the National Autonomous University of Mexico in 1965 and his Ph.D. in physical chemistry from the University of California at Berkeley in 1972. After teaching for a year at the National Autonomous University, he returned to Berkeley as a research associate for one year.

In 1973, Molina joined the research laboratory of F. Sherwood Rowland at the University of California at Irvine. Molina was looking for a topic on which he could do his post-doctoral research with Rowland, and Rowland was ready with a suggestion because he had just come from a scientific meeting where he became interested in the possible effects of an important commercial chemical, trichlorofluoromethane, also known as chlorofluorocarbon-11, or CFC-11. The compound was a member of a widely-successful group of **chemicals**, called freons, produced by Dow Chemical Company.

In particular, the question that interested Rowland was what effects, if any, this compound would have on atmospheric gases. CFC-11 was rapidly becoming very popular as a propellant in hair sprays, spray paints, and other **aerosol** products. By 1974, more than $2 billion of CFC-11 and related **chlorofluorocarbons** were being used each year.

Rowland and Molina developed a theory about the fate of CFC molecules released in the **troposphere**, the layer of the **atmosphere** in which we live. They predicted that those molecules would rise into the **stratosphere**, the layer of air above the troposphere. There, they said, **solar energy** would cause CFC molecules to decompose, releasing free **chlorine** atoms.

If that were to happen, they hypothesized, the free chlorine would be likely to attack **ozone** molecules, converting them to ordinary oxygen. The chlorine oxide formed in that reaction might then react with single oxygen atoms, to form more oxygen and regenerate the original chlorine.

Two important conclusions can be drawn from this series of reactions. First, a single atom of chlorine would be capable of destroying many (Rowland and Molina predicted about 100,000) molecules of ozone. Second, since ozone in the stratosphere absorbs **ultraviolet radiation**, this process would result in more ultraviolet radiation reaching the earth's surface and causing an increase of skin **cancer** among humans.

When Rowland and Molina first proposed this theory, no measurements had ever been made of chlorine in the stratosphere. By 1979, they had carried out the first of those measurements and obtained results that closely matched their predictions. An important new environmental problem, **ozone layer depletion**, had been identified.

Molina held positions as a research associate, assistant professor, and associate professor at the University of California at Irvine. In 1983, he left Irvine to become Senior Research Scientist at the Jet Propulsion Laboratory at the California Institute of Technology in Pasadena. Molina was awarded the Society of Hispanic Engineers Award in 1983. He is also the recipient of more than a dozen awards including the 1987 American Chemical Society Esselen Award, the 1988 American Association for the Advancement of Science Newcomb-Cleveland Prize, the 1989 NASA Medal for Exceptional Scientific Advancement, and the 1989 United Nations Environmental Programme Global 500 Award. In 1995, he received the Nobel Prize in Chemistry along with F. S. Rowland and P. Crutzen.

[*David E. Newton*]

RESOURCES
PERIODICALS

Molina, M. J., et al. "Experimental Study of Intermediates from OH-initiated Reactions of Toluene." *Journal American Chemical Society* 121 (1999): 10225—10226.

———, and M. J. Molina. "Chlorofluoromethanes in the Environment." *Review of Geophysical Space Research* (January 1975): 1–35.

Rowland, F. S. "Atmospheric Chemistry: Causes and Effects." *MTS Journal* (Fall 1991): 12–18.

OTHER

Mario Molina Web Page. [cited July 2002]. <http://eaps.mit.edu/molina>.

Molluscicide
see Pesticide

Monarch butterfly

Like all butterflies, moths, and other insects with a lifecycle that involves complete metamorphosis, individual monarch butterflies (*Danaus plexippus*) go through four stages from eggs to larvae (caterpillars) to pupae to adults. Unlike other insects, as a **species** monarchs also undergo an annual cycle that involves several generations and a **migration** that covers thousands of miles.

Monarch butterflies are native to North and South America, but have were spread by humans throughout much of the world in the 1800s. They first appeared in Hawaii in the 1840s, then spread throughout the rest of the South Pacific in the 1850s and 1860s. In the early 1870s, the first monarchs were reported in **Australia** and New Zealand. These different populations have adapted to their new habitats with an amazing range of behaviors, but the annual migration undergone by the North American monarchs makes them unique among insects.

It is thought that there are three distinct populations of monarchs in the United States. One breeds east of the Rocky Mountains and overwinters in the mountains of central Mexico. Another breeds west of the Rocky Mountains and overwinters on the California coast. These populations constitute separate breeding pools with little genetic exchange. A third population is found in southern Florida. This population is genetically less isolated and probably receives significant influx from the eastern population in the fall and possibly the spring.

The monarch life cycle

An individual monarch's life cycle begins when a female lays an egg on a plant in the milkweed family (Asclepiadacae). Dozens of milkweed species are found throughout North America, and most of these are suitable hosts for monarchs. Common milkweed (*Asclepias syriaca*) in the northern United States and Canada, is probably host to more monarchs than any other species. It is difficult to tell just how many eggs each butterfly lays during her life, but the average female in the wild is thought to lay 300–500 eggs. In captive females average about 700 eggs over two to five weeks of egg laying. Eggs usually hatch three to seven days after they are laid.

The newly hatched larva's first food is its egg-casing, but it soon begins eating the plant on which it was hatched. Only after consuming milkweed does the larva develop the characteristic yellow, white, and black striped pattern recognized by children and adults throughout the world. Monarch larvae undergo five stages called instars, shedding their old skin between stages to allow for growth. The main activity of the larva is eating, and during the larval period of about eight to 14 days, they increase their mass about 2,000-fold.

The third stage in the monarch life cycle is the pupa, or chrysalis. When the monarch is ready to pupate, it seeks a protected location and spins a white silk pad using a spinneret located just under its mouth. It attaches itself to this pad and dangles head down. About a day later it sheds its skin for the last time and the gold-spotted green chrysalis, or pupa, is complete. This stage lasts eight to 15 days. The day before the adult butterfly emerges, its folded wings are visible through the pupa case, and it is possible to see the black, orange and white color pattern that is the monarch's trademark. This pattern is due to tiny scales that cover the monarch's wings and body. On the day that the butterfly emerges, the colors are very distinct and the pupa no longer has any green coloring.

Adults live for two to six weeks, spending their time gathering nectar from flowers, mating, and laying eggs. Four to five generations repeat this cycle throughout the spring and summer.

The monarch's annual cycle

Butterflies that emerge in late August and September put reproduction on hold. Instead of eating, mating, and laying eggs, the adults fly south. Individuals in the eastern North American population, live up to nine times as long as their spring and summer counterparts. They travel to old growth oyamel fir (*Abies religiosa*) forests high in the transvolcanic mountains of Michoacan, Mexico. West coast monarchs also begin a migration that will end in the eucalyptus and Monterey pine (*Pinus radiata*)groves of Pacific Grove, Santa Cruz, and Fremont on the Central California coast.

Starting in early November, the monarchs form compact roosts thickly covering the trunks and branches of hundreds of trees. They remain in a nearly dormant state until the temperatures warm in February, when they become more active, mating, and seeking water and nectar. In mid-March, they fly north to look for the milkweed plants in order to produce new generations of monarchs that will continue the migration cycle. Recolonization of the summer breeding grounds is a two-step process. Monarchs that have overwintered make part of the return trip. Then they lay eggs and die. Their offspring continue the journey north until the breeding range is reoccupied by late May or early June.

Because of their incredible yearly migration cycle, monarchs depend on resources in widely dispersed locations, making them especially vulnerable to disruption of their **habitat**. In their breeding range throughout the United States and southern Canada, they need fields, roadsides, gardens, and prairies filled with milkweed for larvae and nectar sources for adults. During their fall migration monarchs need safe flyways and nectar sources to fuel their long journey. Finally, the migratory generation requires an intact habitat in their wintering one that will protect them from temperature and humidity extremes during their winter rest. Just as with migratory song birds, preserving the monarch's migratory life cycle requires international interest, knowledge, and cooperation.

Monarch population size

There are no reliable estimates of monarch population size during the breeding stage of the annual cycle. However, estimates of absolute numbers during the overwintering period exist. During the winters of 1985 to 2002, researchers estimated that overwintering monarch densities are approximately 10 million butterflies per hectare, with 6–12 hectares of land used for roosting. However, estimates of monarch **mortality** during a major winter storm in 2002 suggested that this estimate is too low, and that densities may be more like 50 or 60 million monarchs per hectare of occupied forest.

Sources of monarch mortality

About 90–95% of monarch eggs never reach adulthood, and predators, milkweed defenses, and environmental conditions can all cause mortality during the immature stages. Predators and environmental conditions can also kill adults. Habitat destruction limits the number of monarchs reaching maturity during the summer breeding period, and also affects the number that survive the overwintering period.

Invertebrate predators such as insects and spiders, and diseases caused by bacteria, viruses, **fungi**, and other organisms kill many monarchs in natural populations. Invertebrate predators of monarchs include both native and introduced flies and wasps that lay eggs in living larvae or on the leaves that larvae eat. The wasp or fly larvae that emerge from these eggs ultimately kill the monarch larvae.

Monarchs accumulate **chemicals** called cardenolides (also called cardiac glycosides) that are present in the milkweed. These cardenolides are poisonous to most vertebrates, and monarchs face little predation from **frogs**, lizards, mice, or birds during the adult breeding stage. However, vertebrate predation is a major source of mortality for overwintering monarchs. Both birds and mice consume significant numbers of overwintering adults.

The milky latex present in milkweed leaves is contained under pressure in a system of vesicles. When the plant is punctured, the latex flows out of these vesicles and coagulates on contact with air. The latex serves as a defense to the plant, since it is similar to glue after coagulation. It can kill small larvae by sealing their mandibles or gluing their entire body to the leaf. While monarch larvae can cut off latex flow by trenching the leaves, about 25% of first instar larvae die after becoming mired in the latex of some milkweed species.

Monarch eggs do not hatch in very dry conditions. Dry weather can also reduce the population of monarch by killing milkweed and reducing the amount of nectar in flowers. Very hot weather also causes mortality. Studies have shown that prolonged temperatures above approximately 95°F (35°C) can be lethal to all stages. Likewise, temperatures below freezing can kill monarchs. All stages can survive if temperatures are only a few degrees below freezing for short periods, but extended periods of severe cold and wet kill them. In the overwintering colonies, the highest mortality occurs when adult monarchs become wet from rain or snowfall, then are subjected to freezing temperatures. Since temperatures extremes are likely to be greater in forests that have been thinned, **logging** can exacerbate the effects of winter storms.

The most important source of human-caused mortality in the breeding range of monarch is habitat loss, especially the destruction of milkweed and nectar sources. Milkweed is considered a noxious weed by some people, and is often destroyed. Herbicides used in agriculture and on roadsides and lawns kill both nectar plants needed by adults and milkweed needed by larvae. Monarchs are also exposed to insecticides used to control insects such as mosquitoes. Research conducted in 2001 found that most monarchs in the upper Midwest probably originate in agricultural fields, especially corn and soybean fields. This means that any agricultural practices that affect milkweed abundance or monarch survival in these fields could have large impacts on monarch populations.

There is also concern that genetically corn containing Bt toxin is a hazard to monarchs. Pollen produced by this corn contains the toxin. Pollen may be blown onto milkweed plants growing in and near cornfields. Since many monarchs that migrate to Mexico breed in the corn belt of the central United States, and milkweed growing within cornfields constitutes an important food source for these monarchs, exposure to this risk is likely to be high. The most commonly used Bt strains express levels of toxin in their pollen that are unlikely to kill monarch larvae outright. Sublethal impacts have not been closely studied.

[*Karen S. Oberhauser Ph.D.*]

RESOURCES

BOOKS

Pyle, R. M. *Chasing Monarchs: Migrating with the Butterflies of Passage.* Boston, MA: Houghton Mifflin, 1999.

PERIODICALS

Anderson, J. B., and L. P. Brower. "Freeze-protection of Overwintering Monarch Butterflies in Mexico: Critical Role of the Forest as a Blanket and an Umbrella." *Ecological Entomology* 21 (1996): 107–116.

Borkin, S. S. "Notes on Shifting Distribution Patterns and Survival of Immature *Danaus plexippus* Lepidoptera: Danaidae) on the Food Plant *Asclepias syriaca.*" *Great Lakes Entomologist* (East Lansing) 15 (1982):199–206

Brower, L. P. "Canary in the Cornfield: The Monarch and the Bt Corn Controversy." *Orion* 20 (2001): 32–14.

Brower, L. P. "Monarch Butterfly Orientation: Missing Pieces of a Magnificent Puzzle." *Journal of Experimental Biology* 199 (1996): 93–103.

Heat Stress on Monarch Butterfly (Lepidoptera: Danaidae) Development."

Dussourd, D. E. "The Vein Drain, or How Insects Outsmart Plants." *Natural History* 90 (1990): 44–49.

Oberhauser, K. S. "Effects of Spermatophores on Male and Female Monarch Butterfly Reproductive Success." *Behavioral Ecology and Sociobiolgy* 25 (1989): 237–246.

Oberhauser, K. S., M. D. Prysby, H. R. Mattila, et al. "Temporal and Spatial Overlap Between Monarch Larvae and Corn Pollen." *Proceedings of the National Academy of Science* 98 (2001):11913–11918.

Prysby, M. D., and K. S. Oberhauser. 1999. Large-scale Monitoring of Larval Monarch Populations and Milkweed Habitat in North America. In *Proceedings of the 1997 North American Conference on the Monarch Butterfly.* edited by J. L. Hoth, et al., Commission for Environmental Cooperation: Montreal, Canada. 1997: 379–383.

Wassenaar, L. I., and K. A. Hobson. "Natal Origins of Migratory Monarch Butterflies at Wintering Colonies in Mexico: New Isotopic Evidence." *Proceedings of the National Academy of Sciences* 95 (1998):15436–15439.

York, H., and K. S. Oberhauser. "Effects of Duration and Timing of *Journal of the Kansas Entomological Society.*

OTHER

Journey North: Monarch Migration. "Comparative Migration Maps." 2002 [cited 26 April 2002]. <http://www.learner.org/jnorth/tm/monarch/MigrationMaps.html>.

Monarch Larva Monitoring Project 2002. April 2002 [cited April 2002]. <http://www.mlmp.org>.

Prysby, M. and K. Oberhauser. *Temporal and Geographical Variation in Monarch Densities: Citizen Scientists Document Monarch Population Patterns.*

Monkey-wrenching

Also called *ecotage* (ecological sabotage), monkey-wrenching refers to techniques used by some radical environmentalists to stop or slow the machinery used in **logging, strip mining**, and other sorts of environmentally destructive activities. The term was popularized by Edward Abbey's novel, *The Monkey Wrench Gang* (1975) and the concept was developed by **Dave Foreman** in *Ecodefense: A Field Guide to Monkeywrenching* (1987). The techniques of monkey-wrenching include "spiking" old-growth trees to prevent loggers from cutting them down, "munching" logging roads with nails to puncture the tires of logging vehicles, pulling up surveyors' stakes, putting sand or grinding compound in the gas tanks or oil intakes (or oatmeal or Minute Rice in the radiators) of bulldozers and logging trucks, and other forms of disruption or destruction.

Mono Lake

Clear, cold water tumbles from snowcapped peaks and alpine fields at 13,000 ft (4,000 m) down the precipitous eastern escarpment of California's Sierra Nevada. The water feeds semi-arid, sagebrush–covered Mono Basin and, at the basin's heart, majestic Mono Lake. This is a salt lake with an area of 60 square miles (155 km^2) at an elevation of 6,400 ft (2,000 m), and mountain waters have flowed to it for at least the past half million years, making it one of the oldest lakes in North America.

Mono Basin lies in the **rain shadow** of the Sierra Nevada; shielded from moist Pacific air masses to the west, it has a semi-arid **climate**. Being surrounded by higher ground, it also has no natural outlet. Over the millennia, the major water loss from Mono Lake has been by evaporation, a process that removes pure water and leaves dissolved salts behind, and the water in the lake has become alkaline as a result and two and a half times saltier than sea water.

The salty, alkaline lake water excludes fish, but it does provide ideal conditions for several life forms that normally are uncommon to inland waters. Algae blooms in abundance during winter, sometimes turning the lake water pea–soup green, and it provides summer sustenance to a profusion of brine flies and tiny brine shrimp. Brine flies and brine shrimp make Mono Lake a summer haven for hundreds of thousands of nesting and migratory birds. The lake's islands provide safe nesting for tens of thousands of California gulls and snowy plovers. Nearly 90% of the state's population of California gulls, which live mainly along the Pacific coast, are hatched on these islands. More than 70 **species** of migratory birds use Mono Lake—most notably, several hundred thousand eared grebes which stop over during their fall **migration**, and more than 100,000 phalaropes which come from South America for the summer.

Mono Lake's **aquatic chemistry**, combined with its unusual geologic and topographic **environment**, has given rise to the lake's signature feature: tufa towers. Tufa is a type of limestone which consists of calcium carbonate that forms around fresh water springs emanating from the lake bottom. Calcium carried by the spring water combines with carbonate in the alkaline lake to form soft rock masses that slowly grow into underwater pinnacles. These towers were almost entirely covered by water until recently; they were first exposed when the lake level started dropping in the early 1940s, because the growing city of Los Angeles had begun intercepting fresh water from streams feeding the lake and diverting it south to the city in a 250–mi (400 km) long aqueduct.

By the mid-1980s, Mono Basin supplied nearly 20% of the water used by Los Angeles, and the lake level had dropped more than 40 ft (12 m). As the water level dropped, the lake's **salinity** increased and shoreline **habitat** for brine flies decreased. Tufa formations became more exposed, dust storms grew more frequent and severe, and predators occasionally found access to island nesting sites. Meanwhile, California created the Mono Lake Tufa State Reserve, and the fish habitat of streams feeding the lake. Lawsuits have been fought in state and federal courts, and in 1989 California's State Supreme Court ordered the Los Angeles Department of Water and Power (LADWP) to reduce the amount of water it was diverting from the lake. In 1993, the State **Water Resources** Control Board recommended that the diversion be cut again, this time by half. LADWP has disagreed, arguing that Los Angeles needs the water and that the reduction is neither ecologically necessary nor economically wise.

The war over eastern Sierra water began at the turn of the twentieth century, when Los Angeles acquired rights to water previously used by farmers and ranchers in the Owens Valley, just south of Mono Lake. By mid–century the battleground had spread north to Mono Basin, and the war promises to continue well into the next century. *See also* Drinking-water supply; Hydrologic cycle; Los Angeles Basin; National forest; Water allocation; Water resources; Water rights

[*Ronald D. Taskey*]

RESOURCES
BOOKS

Lane, P. H., and A. Rossmann. "Owens Valley Groundwater Conflict." In *Deepest Valley*, edited by Genny Smith. Los Altos, CA: William Kaufmann, 1978.
Patten, D. T. *The Mono Basin Ecosystem*. Washington, DC: National Academy Press, 1987.

Monoculture

The agricultural practice of planting only one or two crops over large areas. In the United States, corn and soybean are the only crops grown on most farms in the central Midwest, while on the Great Plains wheat is almost exclusively grown. Although it minimizes farmers' investments in large, expensive implements, the practice exposes crops to the risk of being wiped out by a single predator. This happened with the Irish potato blight of the 1840s and the corn leaf blight of 1970 in the United States, which destroyed millions of acres of corn. Ecologists warn against monoculture's oversimplification of the **food chain/web**, arguing that complex webs are more stable.

Monsoon

Monsoon (from Arabic, *mausim*, season) technically means a reversal of winds, that point between the dry and the wet seasons in tropical and subtropical India, Southeast Asia, and parts of Africa and **Australia**, when seasonal winds change their direction. When the land heats up, the hot air

A view of frost covered rocks in Mono Lake, California. (Photograph by R. Rowan. Photo Researchers Inc. Reproduced by permission.)

the United States Congress established the Mono Basin **National Forest** Scenic Area.

Ultimately, the demand for water by Los Angeles clashed with the demands of environmental groups, who sought to maintain Mono Lake's **ecological integrity** and

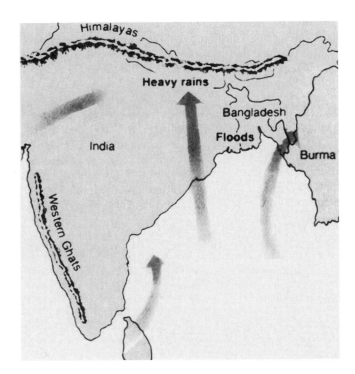

Each summer, warming air rises over the plains of central India, creating a low-pressure cell that draws in warm, moisture-laden air from the ocean. Rising over the Western Ghats or the Himalayas, the air cools causing heavy monsoon rains. (McGraw-Hill Inc. Reproduced by permission.)

rises, causing a low pressure zone that sucks in moisture-filled cooler ocean air, creating clouds and producing rain. In winter, the opposite happens: warm air over the ocean rises and makes a low pressure zone that draws the cooler air off the land.

Although monsoon winds have always been watched by traders and sailors in the Eastern Hemisphere, their arrival is critical to millions of people who depend on agriculture. Cultural and religious customs, especially in India and southeast Asia, are tied to the monsoon rains that bring a season of fertility after a long hot and sterile dry period.

Coastal radar and satellites aid in weather prediction, but the climatological components of monsoons are complex. Tied to the heat and moisture exchange between land and oceans, their effect can be altered by changes in the circulation of hemispheric winds at the equator, as well as by precessional changes in the orbit of the earth.

Environmental changes such as **deforestation** or **soil erosion** can invite severe **flooding**, as in Bangladesh during the 1980s. Scientists believe a rise in sea surface temperature in the Atlantic Ocean, possibly related to the **greenhouse**

effect, prevented the monsoon rain from reaching the African **Sahel** and contributed to recent droughts. This ocean temperature rise may also be tied to the **El Niño** event in the Pacific Ocean.

Any fluctuations in monsoon rain patterns can cause disease and death, along with millions of dollars in damage. If the rains are delayed, or never come, or fall too heavily in the beginning or at the end of the growing season, disastrous results often follow. *See also* Climate; Cloud chemistry; Meteorology

Montreal Protocol on Substances That Deplete the Ozone Layer (1987)

A historical agreement made in 1987 by members of the United Nations to phase out substances that are harmful to the earth's **ozone** layer. The ozone layer protects life on earth by blocking out the sun's harmful **ultraviolet radiation**. Since the 1970s scientists have documented the depletion of the ozone by **chlorofluorocarbons** (CFCs), commonly used for refrigeration and as solvents and **aerosol propellants**. Alarmed by this growing global trend, scientists and policymakers urged a decrease in the use and production of CFCs as well as other ozone-damaging **chemicals**. Ratifying the 1987 Montreal Protocol was a difficult process, however, with the European Community, the former Soviet Union, and Japan reluctant to pose strict controls on chemicals reduction. United States, Canada, Norway, and Sweden, among others, favored stronger control and negotiated with these nations to cut back and eventually phase out completely ozone-depleting substances.

An amendment of the Montreal Protocol was made in 1990 by 93 nations, including China and India, who had not previously participated, to eliminate the use of CFCs, **carbon** tetrachloride, and halon gases by the year 2000 and eliminate the production of methyl chloroform by 2005. Some countries, like the United States, accelerated the schedule to 1995. This 1990 amendment also established the "Montreal Protocol Multilateral Fund" to help developing countries become less dependent on ozone-depleting chemicals.

In November 1992 delegates from all over the world met again in Copenhagen, Denmark, to further revise the Montreal Protocol and accelerate the phase-out of ozone-damaging substances and regulate three additional chemicals. Some of those provisions were as follows: phase out production of CFCs and carbon tetrachloride by 1996; ban **halons** by 1994 (the production of halogen was ended in 1994 in most industrialized nations and is expected to be halted in China, Korea, India, and the former Soviet Union by 2010); end production of methyl chloroform by 1996;

control the use of hydrochloroflurocarbons (HCFCs) and eliminate them by 2030; and increase funding for the Multilateral Fund (between $340 and $500 million by 1996).

Since the Copenhagen Amendments there have been other amendments, such as the Montreal Amendment of 1997, which according to the *Journal of Environmental Law & Policy* "adjusted the timetable for phaseout of some substances and modified trade restrictions, including the creation of a licensing system to attempt to decrease the black market in ozone depleting substances;" and the Beijing Amendment in 2002, which closely monitors bromochloromethane and the trade of hydrochloroflurocarbons.

As of July 2002, 175 nations have ratified the Montreal Protocol. However, while countries have volunteered to control ozone-damaging chemicals, individual companies can still produce the banned chemicals for "essential uses and for servicing certain existing equipment." The Alliance for Responsible CFC Policy in Arlington, Vermont, praised the concession for balancing environmental and economic concerns. Others, such as members of the **Friends of the Earth**, decry the provision as a "big loophole" that undermines the initiative of the Montreal Protocol. *See also* Ozone layer depletion

[*Kyung-Sun Lim*]

RESOURCES
PERIODICALS

Benedick, R.E. "Ozone Diplomacy." *Issues in Science and Technology* 6 (Fall 1989): 43–50.

DeSombre, Elizabeth R. "The Experience of the Montreal Protocol: Particularly Remarkable, and Remarkably Particular." *UCLA Journal of Environmental Law & Policy* 19, no. 1 (Summer 2001): 49.

"EU/UN: Change to Montreal Protocol Outlawing HCFCS due to Enter into Force." *European Report* (January 9, 2002): 515.

"Ozone-Protection Treaty Strengthened." *Science News* 142 (December 12, 1992): 415.

More developed country

The terms more developed countries (MDCs) and **less developed countries** (LDC) were coined by economists to classify the world's 183 countries on the basis of economic development (average annual per capita income and gross national product). The 33 countries (including the United States, Canada, Japan, **Australia**, New Zealand and all the western European countries) in the MDC group are wealthy and industrially-developed. They tend to have temperate climates and fertile soils. About 23 percent of the world's population live in MDCs, but they consume about 80 percent of its mineral and energy resources. In contrast the LDCs are poorer and less industrially-developed. They tend to be located in the Southern Hemisphere where the **climate** is

less favorable and soils are generally less fertile. Though the boundaries are purposely vague, this dichotomy is useful for contrasting the economic and social welfare of the richer and poorer countries and in critical environmental categories involving mainly demographic, economic, and social **statistics**. *See also* Environmental economics; First World; Third World

[*Nathan H. Meleen*]

RESOURCES
PERIODICALS

Ehrlich, P. R., and A. H. Ehrlich. "Growing, Growing, Gone (Rich Nations Must Recognize Their Responsibility to Aid Overpopulated Third World)." *Sierra* 75 (March-April 1990): 36–40.

Preston, S. H. "Population Growth and Economic Development." *Environment* (March 1986): 6–9+.

Mortality

A measure of the death rate in a biological population, usually presented in terms of the number of deaths per hundred or per thousand. If there are 100 mice at the beginning of the year and fifteen of them die by the end of the year, the group's mortality rate is fifteen per 100 individuals (the initial population), or 15 percent. In ecological and demographic studies of populations mortality is an important measurement, along with birth rates (natality), immigration, and emigration, used to assess changes in population size over time. In human populations mortality rates are often figured for specific age and gender groups, or for other population categories including race, income level, occupation, and so on. This way group mortality rates can be compared and risks for each subgroup can be evaluated. *See also* Evolution; Extinction; Population growth; Zero population growth

Mount Pinatubo

Mount Pinatubo in the Philippines erupted on June 15, 1991. When the 5,770-ft (1760 m) mountain shot **sulfur dioxide** 25 mi (40 km) into the **atmosphere**, the cloud mixed with water vapor and circled the globe in 21 days, temporarily offsetting the effects of global warming. Satellite images taken of the area after the eruption showed a dustlike smudge in the **stratosphere**. The sulfur dioxide cloud deflected 2% of the earth's incoming sunlight and lowered temperatures on worldwide average. Although the effects on global temperatures were significant, they are thought to be temporary. These light sulfur dioxides are expected to remain in the stratosphere for years and contribute to damage to the **ozone** layer.

The Philippine islands originated as volcanoes built up from the ocean floor. Most volcanoes erupt along plate edges where ocean floors plunge under continents and melting rock rises to the surface as magma. The earth's crust pulls apart, creating gaps where the magma can rise. The island of Luzon, where Mount Pinatubo is located, has thirteen active volcanoes. The pattern of volcanoes around the rim of the Pacific Ocean is called the Ring of Fire.

Mount Pinatubo sits in the center of a 3-mi (5-km) wide *caldera*, a depression from an earlier eruption that made the **volcano** collapse in on itself. A new cone formed over time, and geothermal vents gave a clue that the volcano was active. Before the 1991 eruption, Mount Pinatubo last erupted 600 years ago. In April 1991 steam eruptions, earthquakes, increasing sulfur dioxide emissions, and rapid growth of a lava dome all indicated a powerful impending blast. Minor explosions began on the mountain on June 12, 1991.

This major eruption occurred in a country with an already shaky economy, and the human effects are likely to be significant for a long period of time. A total of 42,000 homes and 100,000 acres (40,500 ha) of crops were destroyed. Nine hundred people died and 200,000 were relocated, with 20,000 people remaining in tent cities. More than 500 of those have died from disease and exposure. The country suffered over $1 billion in economic losses. Nevertheless, many lives were saved as a consequence of scientists' predictions of the eruption.

The destruction from Mount Pinatubo came mostly from lahars, rushs of cementlike mud, formed when heavy rains loosened the tons of ash dumped on the mountain's sides. These lahars can bury towns and roads and virtually anything else in the way. Also deadly were the pyroclastic flows, killer clouds of hot gases, pumice, and ash that traveled up to 80 mi (130 km) per hour across the countryside, up to 11 mi (18 km) away from the volcano. *See also* Geothermal energy; Greenhouse effect; Ozone layer depletion; Plate tectonics

[*Linda Rehkopf*]

RESOURCES
PERIODICALS

Berreby, D. "Acid-Flecked Candy-Colored Sunscreen." *Discover* 13 (January 1992): 44–46.

Brasseur, G. "Mount Pinatubo Aerosols, Chlorofluorocarbons, and Ozone Depletion." *Science* 257 (28 August 1992): 1239–42.

Grove, N. "Volcanoes: Crucibles of Creation." *National Geographic* 182 (December 1992): 5–41.

Kerr, R. A. "Pinatubo Global Cooling on Target." *Science* 259 (January 29, 1993): 594.

Monastersky, R. "Pinatubo Deepens the Antarctic Ozone Hole." *Science News* 142 (October 24, 1992): 278–79.

Mount St. Helens

On May 18, 1980, Mount St. Helens exploded with a force comparable to 500 Hiroshima-sized atom bombs. David Johnston, a United States **Geological Survey** (USGS) geologist based at a monitoring station six miles (9.7 km) away announced the eruption with his final words, "Vancouver, Vancouver, this is it." Dramatic photograph's provided the public with an awesome display of nature's power.

Mount St. Helens, in southwestern Washington near Portland, Oregon, is part of the Cascade Range, a chain of subduction volcanoes running from northern California through Washington. The Mount St. Helens eruption was instrumental in the expansion of the USGS **Volcano** Hazards Program. Research at the new Cascades Volcano Observatory in Vancouver, Washington, has strengthened basic understanding of volcanic processes and the ability to predict eruptions. Highly relevant ecological studies have corrected previous errors and misconceptions, leading to a new theory about nature's ability to recover after such events.

Research has heightened public awareness of the inherent instability of high, snow-covered volcanoes, where even small eruptions can almost instantaneously melt large volumes of snow. A relatively small 1985 eruption at Nevado del Ruiz in central Colombia killed more than 23,000 people. The Mount St. Helens blast and subsequent collapse generated a 0.7 cubic mile (2.8 km³) mud flow which raced 22 miles (35 km) at speeds as high as 157 mph (253 kph). This caused massive problems, even halting traffic on the Columbia River. These flows may also create unstable **dams**, which may burst years after the intial eruption.

In addition to its awesome power and destructive force, the Mount St. Helens eruption has provided rich material for research. Conductivity studies have located a large rotating block under Mounts Rainier, Adams, and St. Helens, the friction from which is a likely source of eruptions. Geologic mapping and historical research, coupled with field studies of current volcanism, have corrected misconceptions and given clues to hazard frequency. Studies of nature's recovery efforts have produced surprises, notably the early arrival in the eruption zone of predatory insects; elk grazing in open, reforested areas; and the explosive growth of uncommon, dangerous bacteria due to the high temperatures generated by the eruption. *Biological legacy* has emerged as the unifying theory describing nature's recovery capabilities, an idea with direct applications to forestry practices and **reclamation** of human-disturbed land. Nature's mess provides valuable nutrients and nurseries; furthermore, old growth areas within managed ecosystems nurture the recovery of **biodiversity**.

Mount St. Helens has provided a unique laboratory for study of volcano hazards and nature's ability to recover

Mount St. Helens erupting with Mt. Hood in the background. (UPI/Corbis-Bettmann Newsphotos. Reproduced by permission.)

from the devastation caused by volcanic eruptions. *See also* Mount Pinatubo, Philippines; Reclamation; Topography

[*Nathan H. Meleen*]

RESOURCES
BOOKS

Bilderback, D. E., ed. *Mount St. Helens, 1980: Botanical Consequences of the Explosive Eruptions.* Berkeley: University of California Press, 1987.

PERIODICALS

Decker, R., and B. Decker. "Eruption of Mount St. Helens." *Scientific American* 244 (March 1981): 68–80.

OTHER

Tilling, R. I., L. J. Topinka, and D. A. Swanson. *Eruptions of Mount St. Helens: Past, Present, and Future.* USGS General Interest Publication. Washington, DC: U. S. Government Printing Office, 1990.

Wright, T. L., and T. C. Pierson. *Living With Volcanoes: The USGS Volcano Hazards Program.* USGS Circular 1073. Washington, DC: U. S. Government Printing Office, 1992.

MTBE

see **Methyl tertiary butyl ether**

John Muir (1838 – 1914)

American naturalist and writer

John Muir is considered one of the towering giants of the conservation/environmental movement in the United States. Anyone seriously interested in natural history, **conservation**, **wilderness** preservation, or the national parks in this country should be aware of John Muir's work. He was a spirited, joyous naturalist, a serious student of glaciers, an influential advocate of wilderness preservation, and the acknowledged founder of the **national park** idea. Born in Dunbar, Scotland, Muir emigrated with his family to the United States in 1849 when he was 11 years old. He spent his youth working on a farm in the Wisconsin wilderness, trying to please his father, who was a deeply religious man. The wilderness, his religious background, and the hard labor influenced his thinking the rest of his life.

Muir's father believed the Bible to be the only book necessary for a young person, but Muir managed to educate himself and to spend several years at the University of Wisconsin (where he chose his own curriculum and so left without a degree). After school, he worked at various jobs, generally quite successfully, until a factory accident temporarily blinded him. He vowed that if his sight returned, he would leave the factory and see as much of the world as possible. After about a month, his sight did return and he left for various jaunts in the wilderness, including a famous 1,000-mi (1,609 km) walk through the country to the Gulf of Mexico, an account recorded in *A Thousand Mile Walk to the Gulf* (1916).

His eventual goal was to reach South America and wander through the Amazonian tropical rain forests. He reached Cuba, but a bout with fever (carried over from the humid lowlands of Florida) turned him instead toward the drier West, especially California and the Yosemite Valley, about which he had seen a brochure and which he determined to see for himself. "Seeing for himself" also became a life-time habit, and he eventually traveled over much of the world. As he had planned, he did make it up the Amazon, in 1911, at the age of 73.

Arriving in California in 1868, he made his way to Yosemite and spent several years studying its landforms, **wildlife**, and waterways, earning his living by herding sheep, working in a sawmill, and other odd jobs. As Edward Hoagland noted, Muir "lived to hike," a mode of **transportation** that involved him intimately in the landscape. He traveled light, and alone, often with little more than some dry bread in a sack, tea in a pocket, a few matches and a tin cup, and perhaps a plant press.

Through his travels in Yosemite, he became convinced that the spectacular land forms of Yosemite had been carved by glaciers or, as he put it, "nature chose for a tool...the tender snow-flowers noiselessly falling through unnumbered centuries." His belief in glacial origins placed him in conflict with the established scientific ideas of the time, especially those held by the California state geologist. But Muir eventually prevailed, his ideas vindicated when he found the first known glacier in the Sierra range. The results of his years of intense glacial investigations are available in *Studies in the Sierra* (1950). Current views have verified his theories, changing only the number of glacial events and emphasizing the role of water in cutting the canyons. Muir also made five trips to Alaska to study glaciers there, one of which is named for him.

His glacial studies were the principle contributions Muir made as an original scientist, most of his life being devoted to travel, writing, and conservation activism. Even as early as 1868, Muir was concerned with the effects sheepherding had on plant life and **soil erosion**.

Muir's travels were interrupted for a time when, in early 1880, he married Louise Strentzel, the daughter of a fruit rancher in the Alhambra valley. Muir helped run the ranch, first rented and then bought some of the acreage, applying his inventiveness and hard work to fruit growing.

Reportedly, he was a good businessman, prospering after only a decade to ensure a measure of financial independence. He then sold part of the ranch and leased the rest, which allowed him time with his daughters, to return to his beloved wilderness, and to write and actively promote his wilderness ideas. Muir's intimate acquaintance with the Yosemite area and the Sierra Nevada exposed him not only to the depredations of sheep but also to the rapid felling of giant old Sequoias, cut up for shingles and grape stakes. Muir's response: "As well sell the rain clouds, and the snow, and the river, to be cut up and carried away...." In 1889, he escorted the editor of *Century* magazine to Yosemite and showed him the negative impacts of sheep, which he called "hoofed locusts." A series of articles in that magazine alerted the public to the destruction of the land, and they eventually pressured Congress to establish the Yosemite area a national park in 1890.

An earlier attempt to rally interest in the plight of the western forests—a suggestion for a government commission to survey the forests and recommend conservation measures—was also realized with the appointment of such a commission in 1896. Charles Sargent, the chair of the commission, invited Muir to participate and, on the basis of the Sargent Commission's recommendation, President Grover Cleveland created thirteen forest preserves, setting aside 21 million acres. Negative reaction from commercial interests, however, nullified most of these gains. Muir responded by writing two articles on forest reserves and parks in *Harper's Weekly* and *Atlantic Monthly* in 1897. These articles helped to rally public support and in 1898, the annulments were reversed by Congress.

Muir influenced the public and extended his influence by friendships and correspondence with some of the most powerful people of his time. A number of the successes of the early conservation movement, for example, can be attributed to his influence on such figures as **Theodore Roosevelt**. After a three-day camping trip with Muir under the Big Trees in 1903, Roosevelt added many millions more acres to the **national forest** system, as well as national monuments and national parks and created what became the **national wildlife refuge** system.

Known for his many successes, Muir was much saddened by his one big loss: the damming of Hetch Hetchy valley in **Yosemite National Park** as a **reservoir** to supply water to San Francisco. Muir's public image was damaged by the excessive vehemence of his attacks upon the citizens of San Francisco, whom he denounced as "satanic," and following the Hetch Hetchy incident, Muir retired to his ranch to edit his journals for publication.

Muir never considered himself much of a writer and begrudged the time it took away from his beloved mountains and forests. Most of his books were published late in life,

John Muir and his dog. (Corbis-Bettmann. Reproduced by permission.)

after the turn of the century. His writings are still widely read today by students, scholars, activists, and philosophers. In the opinion of most observers, the primary importance of Muir's writings lies not in their literary quality but in the fact that they persuaded a large number of Americans to regard scenic wilderness areas as irreplaceable **natural resources** which must be protected and preserved.

[*Gerald L. Young Ph.D.*]

RESOURCES

BOOKS

Browning, P., ed. *John Muir in His Own Words: A Book of Quotations.* Lafayette, CA: Great West Books, 1988.

Cohen, M. P. *The Pathless Way: John Muir and the American Wilderness.* Madison: University of Wisconsin Press, 1984.

Fox, S. *John Muir and His Legacy: The American Conservation Movement.* Madison: University of Wisconsin Press, 1985.

Turner, F. *Rediscovering America: John Muir in His Time and Ours.* New York: Viking Penguin, 1985.

PERIODICALS

Hoagland, E. "In Praise of John Muir." *Anteus* no. 52 (Spring 1984): 170–83.

Wadden, K. A. "John Muir and the Community of Nature." *The Pacific Historian* 19 (Summer–Fall 1985): 94–102.

Mulch

Material applied to the surface of a **soil** to protect the soil or to improve the **environment** of the soil's surface. Mulch can be made from many different kinds of organic or inorganic materials like stones, bark, compost, leaves, wood chips, and manure. The benefits of using mulch include the following: protection of soil from **erosion**, evaporation reduction, increased water **infiltration**, reduction in weed seed germination, increased seed germination, and reduction of **compaction** of soil. *See also* Animal waste; Composting; Fertilizer; Soil organic matter; Soil texture; Topsoil

Multiple chemical sensitivity

Regarded by skeptics as a manifestation of "technophobia" and "chemophobia," multiple chemical sensitivity is a highly controversial disorder associated with low levels of environmental **chemicals** in general and volatile organic chemicals in particular. Sufferers experience fatigue, malaise, headache, dizziness, lack of concentration, and loss of memory, and are often so disabled that they cannot live or work except in completely chemical-free environments. Critics argue that this condition should not receive clinical recognition as a disease, insisting that the condition has no uniform cause or consistent, measurable features. Advocates for sufferers say it is a chronic condition marked by heightened sensitivity to even slight chemical exposures; and that multiple symptoms occur in multiple organ systems, usually caused by large "triggering" exposures such as to new carpeting that emits chemicals.

Multiple Use-Sustained Yield Act (1960)

On June 12, 1960, Congress passed the Multiple Use-Sustained Yield Act, designed to prevent the obliteration of national forests by **logging** and **water reclamation** projects. This law officially mandated the management of national forests to "best meet the needs of the American people." The forests were to be used not primarily for economic gain, but for a balanced combination of "outdoor **recreation**, range, timber, **watershed**, and **wildlife** and fish purposes."

The Multiple Use Act emerged from a strong history of diverse uses of the federal reserves. Early settlers assumed access to and free use of public lands. The text of the Sundry Civil Act of 1897 mandated that no public forest reservation was to be established except to improve and protect forests and water flow. The act also provided for free use of timber and stone and of all reservation waters by miners and residents. The Act of February 28, 1899, strengthened use policy by providing for recreational use of the reserves.

When the reserves were transferred from the General Land Office to the Bureau of Forestry, the Secretary of Agriculture signed a letter (actually written by **Gifford Pinchot**) dictating formal forest policy: "all the resources of the forest reserves are for use... you will see to it that the water, wood, and forage of the reserves are conserved and wisely used." The first and subsequent editions of the *Use Book*—an extensive guide to management and use of **national forest** lands—stated concisely that the aim of **Forest Service** policy was that "the timber, water, pasture, mineral and other resources of the forest reserves are for the use of the people."

The first use of the term "multiple use" appears to be in two Forest Service reports of 1933. That year's chief's report reaffirmed that the "principle to govern the use of land...is multiple-purpose use." The "National Plan for American Forestry" (the Copeland Report) emphasized that "the peculiar and highly important multiple use characteristics of forest land [involve] five major uses--timber production, watershed protection, recreation, production of forage, and **conservation** of wildlife."

Multiple use has always been controversial. Some critics argued that multiple uses meant that the Forest Service was losing sight of its original protective function: a 1927 article in the *Outlook* claimed that "the Forest Service will have to be called from its enthusiasm for entertaining visitors to the original but more somber work of forestry." Many similar statements can be identified. A writer in the *Journal of Forestry* in 1946 went so far as to propose separating all the lands in each use class, consolidating each type into a separate bureau under one cabinet officer. The **Sierra Club** proposed a vast land exchange in 1959, hoping to move scenic areas out of the Forest Service and into the **U.S. Department of the Interior**.

Critics have argued, and feel that subsequent policy and events on the ground bear them out, that the Multiple Use Act did not reduce confusion because it did not eliminate water and timber as priority uses. Proponents of the act feel that it did give statutory authority to all uses as equal to timber and did give the other four uses more stature and **visibility**. *See also* Bureau of Land Management; Commercial fishing; Public Lands Council

[*Gerald L. Young Ph.D.*]

RESOURCES
BOOKS

Bowes, M. D., and J. V. Krutilla. *Multiple-Use Management: The Economics of Public Forestlands.* Washington, DC: Resources for the Future, 1989.
Frederick, K. D., and R. A. Sedjo, eds. *America's Renewable Resources: Historical Trends and Current Challenges.* Washington, DC: Resources for the Future, 1991.

Steen, H. K. "Multiple Use: The Greatest Good of the Greatest Number" and "Multiple Use Tested: An Environmental Epilogue." In *The United States Forest Service: A History.* Seattle, WA: University of Washington Press, 1976.

Multi-species management

Multi-species management (MSM) specifies the development of a particular ecologically balanced assessment and operation in protecting fish and **wildlife**. Such protection of **species** and the **environment** that supports and sustains them relies on understanding the species' interaction with each other, and within a particular environment—whether a body of water, a marshland, or other natural surroundings. The **Environmental Protection Agency** (EPA) and the U.S. **Fish and Wildlife Service** (FWS) play key roles in overseeing this management due to the imperative for restoring and maintaining an effective **ecosystem**. In concert with the Endangerd Species Act (ESA), the focus in MSM is on lessening the possibility of certain species being added to the *Endangered Species* list in addition to the management of those species currently thriving.

In the fall of 2002, the College of Foresty at Oregon State University was hosting a symposium, *Innovations in Species Conservation* for the purpose of studying the issues of strategies for conserving species, ecological risks and uncertainties associated with certain strategies, and examining social and legal contexts of **conservation** strategies. The announcement for the symposium pointed out that, "In the past, efforts to conserve species have focused on providing appropriate **habitat** for, and population management of, individual high-profile species protected by laws and regulations. Some regional plans have been designed to conserve a broad array of species and biological diversity by specifying protection of rare and uncommon species." But according to the research indicated, there also remain some questions regarding the multi-species direction. The discussion goes on to say that, "Such approaches have proven to be complex and expensive, and have placed constraints on the ability to meet other important management objectives. Multiple-species or ecosystem approaches addressing species assemblages at regional scales may be more efficient and lessen management constraints, but the degree to which they protect individual species rests more on hypothesis than on systematic testing. Such multi-system approaches may also be more susceptible to legal challenges due to a lower level of certainty regarding the outcomes for a particular species."

Still, by 2002, evidence continues to emerge from extensive research over the last few decades, that MSM has produced some results that indicate success. The theory is that if one species, or aspect of an ecosystem is out of balance, therefore the entire system is out of balance. In a 1999 article written by Dick Monroe, Vice President of Environmental Affairs for Darden Restaurants, he noted that global management of **whales** and **seals** directly affected the seafood and restaurant industry. It was his concern—in opposition to the philosophy of many **animal rights** and whale watch groups—that whales were consuming between three and six times the total annual catch of all commercial fisheries. He noted that an MSM approach must be supported because when the focus is simply on one species, another is sure to lose out and the supply depleted.

The Kyoto Treaty—the United States backed out of support of the treaty under the George W. Bush administration in 2001—also specifies MSM as one of its components. Throughout the United States, under the sponsorship of various government and private agencies, MSM had become an important research approach to environmental concerns. It represents a worldwide effort as well as one for the United States—especially in countries that rely heavily on the fishing industry as a support of their economic system. In Norway, for instance, in order to manage certain fishing stocks in the Barents Sea, it was crucial for researchers to determine how much of that species waaas eaten by another fish. The Institute of Marine Research in Bergen (Norway) developed a multi-species model to address this issue, among others. In managing the stock of capelin, it was crucial to know how much of it was consumed by the cod, and also in marine mammals. Beginning in 1987, Norwegian and Russian scientists cooperated in collecting this information. By 1993, the groups had collected samples from the stomach content of over 50,000 cod. This example of the mechanism by which MSM operates is typical of similar programs throughout the United States, and around the world.

[*Jane E. Spear*]

RESOURCES

PERIODICALS

Gutting, Richard. "Action Needed to Avert Supply/Demand Gap." *Seafood Industry* (August 1996): 43.

Russell, Dick. "Hitting Bottom.&rdquo *Amicus Journal* February 3, 1997 [cited July 2002].

Valles, Colleen. "Panel Passes Restrictions on West Coast Fishing to Protect Depleted Species." *Associated Press/Environmental News Network* [cited June 21, 2002]. <http://www.enn.com/news>.

OTHER

Cardin, Ben. *Testimony in Support of the Reauthorization of the Chesapeake Bay Office of the National Oceanic and Atmospheric Administration HR 4789.* September 21, 2000 [cited July 2002]. <http://www.house.gov/ca>.

Chesapeake Bay Program. *Chesapeake 2000 Agreement.* [cited June 2002]. <http://www.chesapeake.org>.

Chesapeake Bay Program. *Scientific and Technical Advisory Committee.* [cited June 2002]. <http://www.chesapeake.org/stac>.

Food and Agriculture Organization. *Sustainable Contribution of Fisheries to Food Security.* [cited June 2002]. <http://www.fao.org/fi>.

IWMC (formerly International Wildlife Management Consortium). *Environment; What Do Whales have to do with Your Menu?* September 25, 1999 [cited July 2002]. <http://www.iwmc.org>.

Norwegian-scenery. *Norwegian Management of Marine Resources.* [cited June 2002]. <http://www.norwegian-sceneery.com>.

Oregon State University, College of Forestry. *Innovations in Species Conservation.* [cited June 2002]. <http://www.outreach.cof.orst.edu>.

Stefansson, Gunnar. "Management in the Multi-species Context." *Hafrannsoknastofnunin (Marine Research Institute).* February 28, 2002 [July 2002]. <http://www.hafro.is>.

U.S. Fish and Wildlife Service. *National Fish Hatchery System.* [cited June 2002]. <http://www.fisheries.fws.gov>.

White, Ph.D., Michael D. *The Lower Colorado River Multi-Species Conservation Program.* [cited June 2002]. <http://www.sci.sdsu.edu/salton>.

World Wildlife Fund. *The Food-Web Effect.* [cited June 2002]. <http://www.panda.org/resources>.

ORGANIZATIONS

U.S. Environmental Protection Agency, 1200 Pennsylvania Avenue, N.W., Washington, D.C. USA 20460 202-260-2090, www.epa.gov

U.S. Fish and Wildlife Service, Washington, D.C. USA, www.fws.gov

Municipal solid waste

Americans generate about 160 million tons of municipal **solid waste** (MSW) per year not counting construction debris. That's enough **garbage** to fill a convoy of trash trucks reaching half way from the earth to the moon. That much garbage equals about 1,300 pounds (590 kg) of waste per year for every person in the United States, or about 25 pounds (11.5 kg) per person per week. The **Environmental Protection Agency** (EPA) tells us that every year Americans throw away 60 billion cans, 28 million bottles, 4 million tons of plastic, 40 million tons of paper, 100 million tires, and 3 million cars. If growth in disposal rates continue, Americans may generate nearly 2,000 million tons of MSW by the year 2000.

Municipal solid waste is waste generated by households, commercial businesses, institutions, light industry and agricultural enterprises and nontoxic wastes from hospitals and laboratories. Municipal waste is composed of (in order of volume contribution) paper/packaging, **yard waste**, **food waste**, magazines/newspapers, **plastics**, glass, wood/fabric, **disposable diapers**, and other contributions such as tires, appliances, and nontoxic home maintenance supplies. In addition to this relatively benign list, there is other municipal waste that ends up in a **landfill** inappropriately—e.g., leftover paint, crankcase oil, batteries, and parts of some appliances such as capacitors in refrigerators and air conditioners. In some cases sewage **sludge** from the local **wastewater** treatment facility ends up in the landfill with other municipal solid waste. It is these inappropriate wastes that cause concern in the minds of local officials as they ponder the siting of landfills.

Since municipal waste is a non-homogeneous stream of materials, its impact is unpredictable. Although only about two percent of the municipal **waste stream** is toxic, even that small amount has the potential of contaminating large amounts of **groundwater** if the leachate from landfills reach the **aquifer**. In addition, landfills situated near a lake or a stream can leak laterally and contaminate surface water. In spite of the fact that the engineering of landfills has become much more sophisticated in the last 10 years, there is still sincere concern about the impact of municipal waste on ground and surface water.

Several methods for managing municipal solid waste have been designed and implemented over the last 20 years. In the United States, 80% or more of the municipal solid waste ends up in landfills, about 10% is incinerated, and about 11% is recycled. Waste-to-energy **incineration** is an option aimed at reducing the volume of municipal waste and producing an ongoing supply of energy for nearby markets. The complication of this treatment method is the ash residue and its disposal as well as **air pollution**. Probably the most desirable management technique for MSW is **recycling**.

Recycling captures the embodied energy in the products to be discarded. The conversion of the waste stream into a second or third generation of usable products is far superior to simple volume destruction or land burial. Unfortunately, recycling requires behavioral change on the part of the waste generator and behavioral and infrastructure change on the part of industry in order to create markets for post consumer materials. Individuals must learn to clean and separate household garbage and make it ready for recycling. Industry must make the investment to change manufacturing processes to accommodate a heterogeneous "raw" material.

Even more desirable than recycling is source reduction and **reuse** of household products and industrial and business supplies. Buying durable and re-usable products reduces the volume of the MSW stream. Packaging that can be reused after the product has been used is another way of shrinking municipal solid waste volume. In some European countries like Germany, manufacturers are required to take back all their packaging materials.

The cost of MSW disposal has increased dramatically over the last 10 years. Part of this cost is due to the pressure being put on fewer and fewer landfills as older landfills close. Many metropolitan areas of the United States have exhausted their landfill capacity and are transporting waste into rural areas. Unfortunately, in rural areas the volume of MSW is not sufficient for landfill companies to make the investment and build state-of-the-art landfills. Furthermore there is increasing concern that many towns and cities are dumping their MSW in areas where people have less political power or need the revenue from tipping fees. Consequently the

management of waste has become a social issue as well as an environmental concern.

Currently there are approximately 3,500 licensed landfills operating in the United States. Tipping fees have increased as have transport fees. States are transporting MSW hundreds of miles to other states because they can't build landfills due to citizen opposition. The NIMBY (**Not In My Backyard**) syndrome makes it difficult to create more landfills. **Waste management** companies are offering extensive incentive packages to local communities in order to get approval for landfill siting. Yet, almost without exception, local citizens attempt to block the construction of landfills.

Municipal solid waste is a renewable resource. We will always have new supplies of it. Even though many citizens are beginning to turn to nondisposable items and there is a real effort to launch recycling programs nationwide, we are still faced with an increasing volume of municipal waste. It is a complex problem that will require behavioral change on the part of individuals, a commitment on the part of local city planners, and an investment on the part of industry in order to recycle the waste stream. *See also* Medical waste

[*Cynthia Fridgen*]

RESOURCES
BOOKS

Blumberg, L., and R. Gottlieb. *War on Waste: Can America Win Its Battle with Garbage?* Washington, DC: Island Press, 1989.

Rathje, W., and C. Cullen. *Rubbish! The Archaelogy of Garbage.* New York: Harper-Collins Publishers, 1992.

The Solid Waste Dilemma: An Agenda for Action. Washington, DC: Environmental Protection Agency, 1989.

Municipal solid waste composting

Municipal Solid Waste (MSW) composting is a rapidly growing method of solid **waste management** in the United States. MSW includes the residential, commercial, and institutional **solid waste** generated within a community. MSW **composting** is the process by which the organic, **biodegradable** portion of MSW is microbiologically degraded under **aerobic** conditions.

During the process of degradation, bacteria are used to decompose and break down the organic matter into water and **carbon dioxide**, which produces large amounts of heat and water vapor in the process. Given sufficient oxygen and optimum temperatures, the composting process achieves a high degree of volume reduction and also generates a stable end product called compost that can be used for mulching, **soil** amendment, and soil enhancement. As a form of solid waste management, MSW composting reduces the amount of waste that would otherwise end up in landfills.

Although composting has been practiced by humans for centuries, the concept of composting mixed solid waste as a form of large-scale solid waste management is still in an early stage of development in the United States. MSW generally consists of a mixture of organic compostable materials such as **food waste** and paper and inert, nonbiodegradable materials such as **plastics** and glass. The introduction of non-compostable materials may pose problems in materials handling during the composting process and also hinder the formation of a uniform, homogeneous compost. Therefore, in order to practice MSW composting as a form of waste management, the composting system must be designed to remove the non-compostable materials either by presorting and screening or by sifting and removal at the end of the process.

The three most common methods of MSW composting are closed in-vessel, windrow, and static aerated pile composting. The method of choice depends on the volume of waste to be composted and the availability of space for composting. In the closed in-vessel method, the MSW is physically contained within large drums or cylinders and all necessary **aeration** and agitation is supplied to the vessel. In windrows, the MSW is heaped in long rows of material approximately four to seven feet high. Air and ventilation are supplied by physically turning over the piles with mechanical windrow turners. In static aerated compost piles, the MSW piles are not physically agitated, rather air is supplied and excess heat is removed by a system of sensors and pipes within the pile. In all cases, the goal is to ensure a steady, optimum rate of composting by providing adequate oxygen and ventilation to remove excess heat and water so that microbiological action is not impaired. When there is insufficient air supply, microbial action is unable to fully decompose the waste and the piles become **anaerobic** and unpleasant odors and putrefaction may result. Strong neighborhood complaints against odors is the single most common reason for the failure and shutdown of composting plants.

In the United States, numerous, small-scale, pilot projects have demonstrated the feasibility of MSW composting for townships and municipalities. However, there are relatively few operations that successfully carry out MSW composting on a large, commercial scale. The operational large-scale facilities are located mainly in Florida and Minnesota, two states that have traditionally shown interest in innovative waste management options. *See also* Aerobic sludge digestion; Solid waste incineration; Solid waste recycling and recovery; Waste reduction

[*Usha Vedagiri*]

RESOURCES
PERIODICALS

Goldstein, N., and R. Steuteville. "Solid Waste Composting in the United States." *Biocycle* 33 (1992): 44–52.

Mustela nigripes
see **Black-footed ferret**

Mutagen

Any agent, chemical or physical, that has the potential for inducing permanent change to the genetic material of an organism by altering its DNA. The alteration may be either a point **mutation** (nucleotide substitution, insertion, or deletion) or a chromosome aberration (translocation, inversion, or altered chromosome complement). There are long lists of chemical mutagens which include such diverse agents as formaldehyde, mustard gas, triethylenemelamine, **vinyl chloride**, **aflatoxin** B, benzo(a)pyrene, and acridine orange. Chemical mutagens may be direct acting, or they may have to be converted by metabolic activity to the ultimate mutagen. Physical mutagens include (but are not limited to) x rays and **ultraviolet radiation**. *See also* Agent Orange; Birth defects; Chemicals; Gene pool; Genetic engineering; Love Canal, New York

Mutation

A mutation is a change in the DNA of an organism, which is genetically transmitted, and may give rise to a heritable variation. Mutagens, substances that have the competence to produce a mutation, may be subject to chromosomal changes such as deletions, translocations, or inversions. Mutations may also be more subtle, resulting in changes of only one or a few nucleotides in the sequence of DNA. These more subtle mutations are called as "point mutations," and it is these that most people refer to when discussing mutation.

Ordinarily, mutation is thought of as a genetic change that results in alterations in a subsequent generation. Germinal tissue, which gives rise to spermatozoa and ova, is the tissue in which such mutations occur. However, mutations can arise in many cell types in addition to the germ line, and these changes to non-germinal DNA are referred to as somatic mutations. Although somatic mutations are not passed to subsequent generations, they are not less important than germinal mutations. **Ionizing radiation** and certain **chemicals** can have mutagenic effects on somatic cells which often result in **cancer**.

A six-legged green frog— the result of a genetic mutation. (JLM Visuals. Reproduced by permission.)

Mutations may be either spontaneous or induced. The term spontaneous is perhaps misleading; even these mutations have a physical basis in cosmic rays, natural **background radiation**, or simply kinetic effects of molecular motion. Induced mutations ensue from known exposure to a diversity of chemicals and certain ionizing radiations.

DNA is composed of a linear array of nucleotides, which are translocated through RNA into protein. The genetic code consists of consecutive nucleotide triplets which correspond to particular amino acids, and it may be altered by substitution, insertion, or deletion of individual nucleotides. Substitution of a nucleotide in some cases can be inconsequential, since the substituted nucleotide may specify an amino **acid** which does not affect the function of the protein. On the other hand, substitution of one nucleotide for another one can result in a protein gene product with a changed amino acid sequence that has a major biological effect. An example of such a single nucleotide substitution is sickle cell hemoglobin, which differs from normal hemoglobin by a single amino acid. The substitution of a valine for a glutamic acid in the mutant hemoglobin molecule results in sickle cell disease, which is characterized by chronic hemolytic **anemia**.

The insertion or deletion of a single nucleotide pair results in what is known as a frameshift mutation; this is because the mutation changes the sequence of molecules beyond the point at which it occurs, causing them to be read in different groups of three. Consequently, there is a miscoding of the nucleotides into their gene product. Such proteins are usually shortened in length and no longer functional.

[*Robert G. McKinnell*]

RESOURCES
BOOKS

Mutagenic Effects of Environmental Contaminants. New York: Academic Press, 1972.

Mutualism

A mutualism is a **symbiosis** where two or more **species** gain mutual benefit from their interactions, and suffer negative impacts when the mutualistic interactions are prevented from occurring. Mutualism is a form of symbiosis where the interactions are frequently obligatory, with neither species being capable of surviving without the other. A well-known example of mutualism is the relationship between certain species of algae or blue-green bacteria and **fungi** that results in organisms called **lichens**. The fungal member of the relationship provides a spatial **habitat** for the algae, which in turn provide energy from **photosynthesis** to the fungus. Mutualistic interactions are thought to be the origin of the many cell organelles like mitochondria and chloroplasts, which may have resulted from the acquisition of free-living **phytoplankton** and other single-celled organisms by host species. Both the incorporated cell and the host soon evolved so that neither could exist without the other.

[*Marie H. Bundy*]

MW

see **Megawatt**

Mycorrhiza

Refers to a close, symbiotic relationship between a fungus and the roots of a higher plant. Mycorrhiza (from the Greek *myketos* meaning fungus and *rhiza* meaning root) are common among trees in temperate and tropical forests. There are generally two forms—ectomycorrhiza, where the fungus forms a sheath around the plant roots, and endomycorrhiza, where the fungus penetrates into the cells of the plant roots. In both cases, the fungus acts as extended roots for the plant and therefore increase its total surface area. This allows for greater **adsorption** of water and nutrients vital to growth.

Mycorrhiza even allow plants to utilize nutrients bound up in silicate minerals and phosphate-containing rocks that are normally unavailable to plant roots. They also can stimulate the plants to produce **chemicals** that hinder invading pathogens in the **soil**. In addition to the physical support, the mycorrhiza obtain carbohydrates from the higher, photosynthetic plant. This obligate relationship between **fungi** and plant roots is especially important in nutrient-impoverished soils. In fact, many trees will not grow without mycorrhiza. *See also* Symbiosis; Temperate rain forest; Tropical rain forest

Mycotoxin

Mycotoxins are toxic biochemical substances produced by **fungi**. They are produced on grains, fishmeal, peanuts, and many other substances, including all kinds of decaying vegetation. Mycotoxins are produced by several **species** of

fungi—especially *Aspergillus*, *Penicillium*, and *Fusarium*—under appropriate environmental conditions of temperature, moisture, and oxygen on crops in the field or in storage bins. In recent years, research on this subject has indicated considerable specialization in mycotoxin production by fungi. For example, aflatoxins B_1, B_2, G_1, and G_2 are relatively similar mycotoxins produced by the fungus *Aspergillus flavus* under conditions of temperatures ranging from 80-100°F (27-38°C) and 18-20% moisture in the grain. Aflatoxins are among the most potent carcinogens among naturally occurring products. Head scab on wheat in the field is produced by the fungus *Fusarium graminearum* which produces a mycotoxin known as DON (*deoxynivalenol*), also known as Vomitoxin.

Myriophyllum spicatum
see **Eurasian milfoil**